PHYLOGENY OF THE LIVING WORLD
BACTERIA

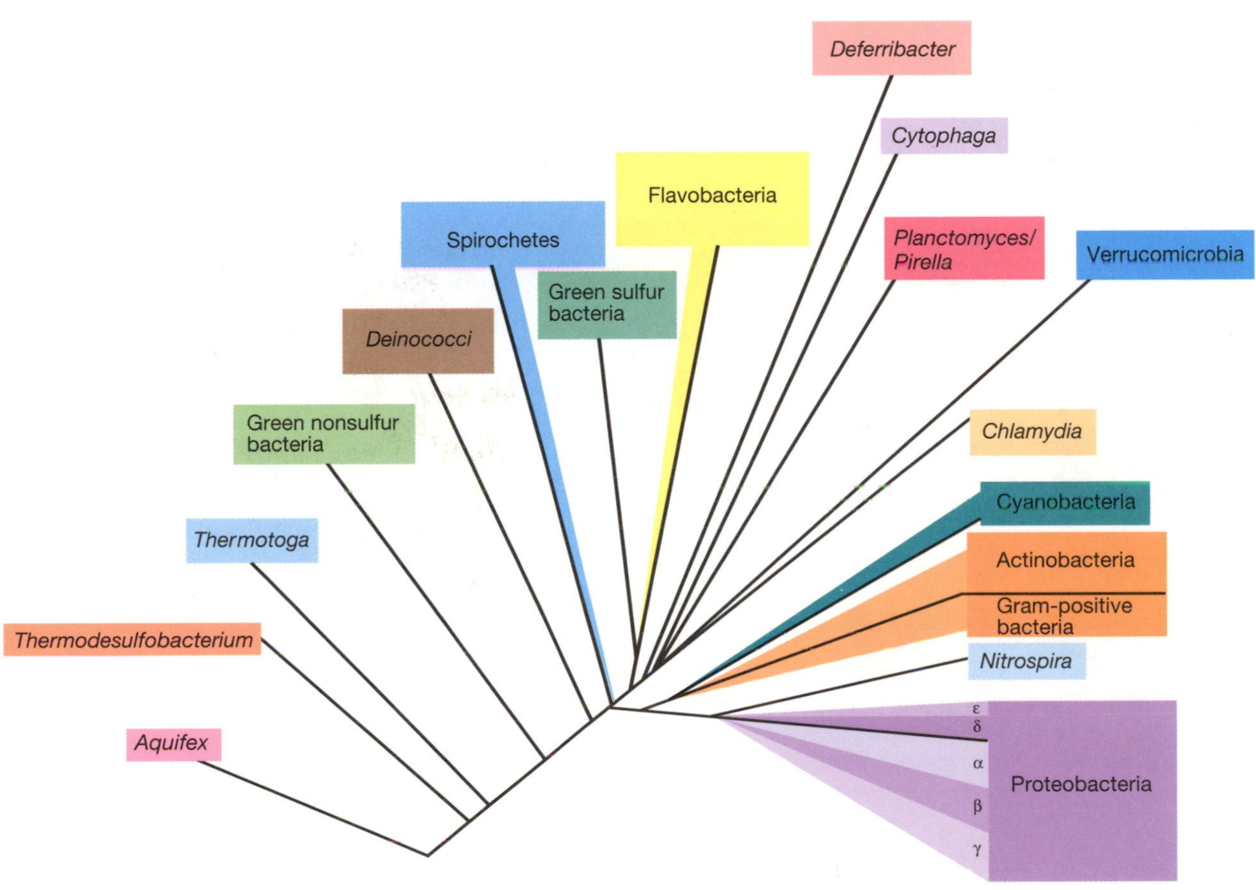

Phylogenetic Tree of *Bacteria* This tree is derived from 16S ribosomal RNA sequences. At least 17 major groups of *Bacteria* can be defined as indicated. Many other groups are known from environmental DNA but do not yet contain cultured representatives. See Sections 14.5–14.9 for further information on ribosomal RNA-based phylogenies. *Data for the tree obtained from the Ribosomal Database project* **http://rdp.cme.msu.edu**

A WEALTH OF ONLINE RESOURCES AND TOOLS

The Microbiology Place Website

This companion website for ***Brock Biology of Microorganisms*, Twelfth Edition** includes chapter guides, chapter quizzes, chapter practice tests, online tutorials, animations with quizzes, videos, study tools (interactive flashcards and a glossary), and an e-book. This greatly expanded website also includes a password-protected Instructor's Resource Section with a gradebook and other time-saving features.
www.microbiologyplace.com

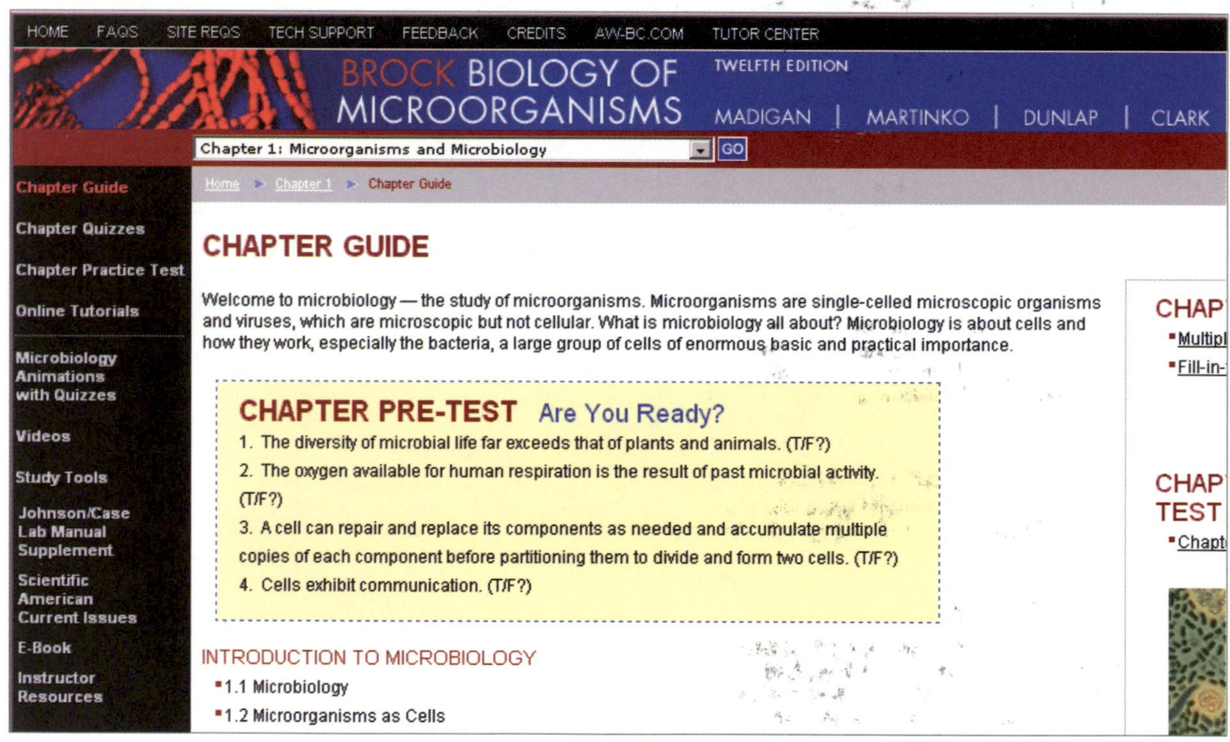

The CourseCompass™ Course Management System

CourseCompass™ combines the strength of the content of The Microbiology Place companion website with state-of-the-art eLearning tools. CourseCompass™ is a nationally hosted, dynamic, interactive online course management system powered by Blackboard, the leading platform for web-based learning tools. This easy-to-use and customizable program enables professors to tailor content and functionality to meet individual course needs. The course includes all of the content found on The Microbiology Place website together with the complete Test Bank and also course management functionality (such as discussion boards).
www.coursecompass.com

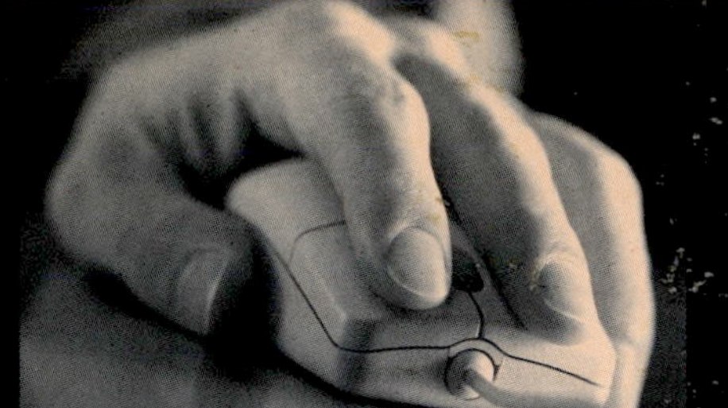

Log on.

Tune in.

Succeed.

Your steps to success.

STEP 1: Register

All you need to get started is a valid email address and the access code below. To register, simply:

1. Go to www.microbiologyplace.com
2. Click the appropriate book cover.
   Cover must match the textbook edition being used for your class.
3. Click **"Register"** under **"First-Time User?"**
4. Leave **"No, I Am a New User"** selected.
5. Using a coin, scratch off the silver coating below to reveal your access code.
   Do not use a knife or other sharp object, which can damage the code.
6. Enter your access code in lowercase or uppercase, without the dashes.
7. Follow the on-screen instructions to complete registration.
   During registration, you will establish a personal login name and password to use for logging into the website. You will also be sent a registration confirmation email that contains your login name and password.

Your Access Code is:

Note: If there is no silver foil covering the access code, it may already have been redeemed, and therefore may no longer be valid. In that case, you can purchase access online using a major credit card. To do so, go to www.microbiologyplace.com, click the cover of your textbook, click "**Buy Now**", and follow the on-screen instructions.

STEP 2: Log in

1. Go to www.microbiologyplace.com and click the appropriate book cover.
2. Under **"Established User?"** enter the login name and password that you created during registration. *If unsure of this information, refer to your registration confirmation email.*
3. Click **"Log In"**.

STEP 3: (Optional) Join a class

Instructors have the option of creating an online class for you to use with this website. If your instructor decides to do this, you'll need to complete the following steps using the Class ID your instructor provides you. By "joining a class," you enable your instructor to view the scored results of your work on the website in his or her online gradebook.

To join a class:

1. Log into the website. For instructions, see "STEP 2: Log in."
2. Click **"Join a Class"** near the top right.
3. Enter your instructor's **"Class ID"** and then click **"Next"**.
4. At the Confirm Class page you will see your instructor's name and class information. If this information is correct, click **"Next"**.
5. Click **"Enter Class Now"** from the Class Confirmation page.

- *To confirm your enrollment in the class, check for your instructor and class name at the top right of the page. You will be sent a class enrollment confirmation email.*
- *As you complete activities on the website from now through the class end date, your results will post to your instructor's gradebook, in addition to appearing in your personal view of the Results Reporter.*

To log into the class later, follow the instructions under "STEP 2: Log in."

Got technical questions?

Customer Technical Support: To obtain support, please visit us online anytime at http://247.aw.com where you can search our knowledgebase for common solutions, view product alerts, and review all options for additional assistance.

SITE REQUIREMENTS

For the latest updates on Site Requirements, go to www.microbiologyplace.com, choose your text cover, and click Site Reqs.

WINDOWS
OS: Windows 2000, XP
Resolution: 1024 x 768
Plugins: Latest version of Flash/QuickTime/Shockwave (as needed)
Browsers: Internet Explorer 6.0; Firefox 1.0
Internet Connection: 56k minimum

MACINTOSH
OS: 10.2.4, 10.3.2
Plugins: Latest version of Flash/QuickTime/Shockwave (as needed)
Browsers: Firefox 1.0; Safari 1.3
Internet Connection: 56k minimum

Register and log in

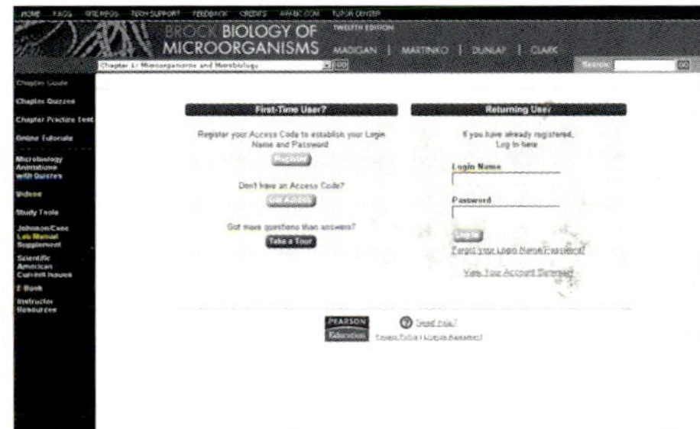

Join a Class

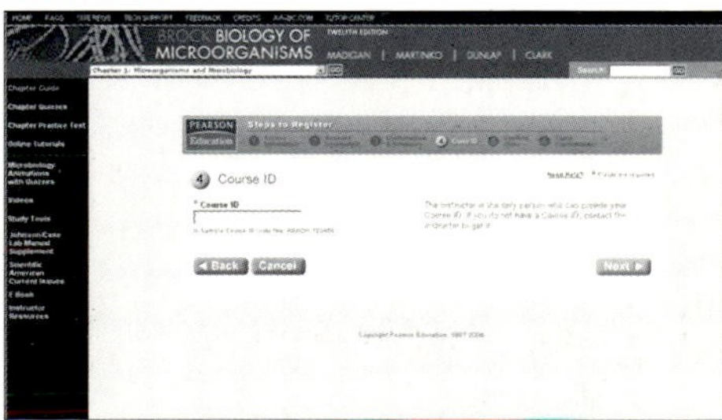

Biology of Microorganisms

TWELFTH EDITION

Michael T. Madigan
Southern Illinois University Carbondale

John M. Martinko
Southern Illinois University Carbondale

Paul V. Dunlap
University of Michigan, Ann Arbor

David P. Clark
Southern Illinois University Carbondale

San Francisco Boston New York
Cape Town Hong Kong London Madrid Mexico City
Montreal Munich Paris Singapore Sydney Tokyo Toronto

Executive Editors: Leslie Berriman and Gary Carlson
Development Editor: Elmarie Hutchinson
Editorial Assistant: Kelly Reed
Managing Editor: Wendy Earl
Production Supervisor: Karen Gulliver
Director, Media Development: Lauren Fogel
Media Producer: Sarah Young-Dualan
Associate Media Producer: Suzanne Rasmussen
Copy Editor: Jane Loftus
Art: Imagineering Media Services, Inc.
Art Coordinator: Jean Lake
Director, Image Resource Center: Melinda Patelli
Image Rights and Permissions Manager: Zina Arabia
Photo Researcher: Elaine Soares
Text Design: tani hasegawa
Manufacturing Buyer: Evelyn Beaton
Marketing Manager: Gordon Lee
Compositor: Aptara
Cover Design: Riezebos Holzbaur Design Group
Cover Image: Michael Wagner, University of Vienna (Austria)

Credits for selected images can be found on page P-1.

Library of Congress Cataloging-in-Publication Data

Brock biology of microorganisms / Michael T. Madigan . . . [et al.]—12th ed.
p. ; cm.
Includes index.
Rev. ed. of: Brock biology of microorganisms / Michael T. Madigan. 11th ed. 2006.
ISBN 0-132-32460-1 (hardcover)
1. Microbiology. I. Madigan, Michael T., 1949– II. Madigan, Michael T., 1949–. Biology of microorganisms. III. Brock, Thomas D. Biology of microorganisms. IV. Title: Biology of microorganisms.
[DNLM: 1. Microbiology. QW 4 B864 2009]
QR41.2.B77 2009
579—dc22

2007048239

ISBN 0-132-32460-1 (student edition)
ISBN 978-0-132-32460-1 (student edition)
ISBN 0-132-55077-3 (professional copy)
ISBN 978-0-132–55077-4 (professional copy)

1 2 3 4 5 6 7 8 9 10—VHP—12 11 10 09 08

www.pearsonhighered.com

About the Authors

Michael T. Madigan received a bachelor's degree in biology and education from Wisconsin State University at Stevens Point in 1971 and M.S. and Ph.D. degrees in 1974 and 1976, respectively, from the University of Wisconsin, Madison, Department of Bacteriology. His graduate work centered on hot spring phototrophic bacteria under the direction of Thomas D. Brock.

Following three years of postdoctoral training in the Department of Microbiology, Indiana University, where he worked on phototrophic bacteria with Howard Gest, Dr. Madigan moved to Southern Illinois University Carbondale, where he has been a professor of microbiology for nearly 30 years. He has coauthored *Biology of Microorganisms* since the fourth edition (1984) and teaches courses in introductory microbiology, bacterial diversity, and diagnostic and applied microbiology. In 1988 he was selected as the outstanding teacher in the SIU College of Science and in 1993 its outstanding researcher. In 2001 he received the university's Outstanding Scholar Award. In 2003 he received the Carski Award for Distinguished Undergraduate Teaching from the American Society for Microbiology.

Dr. Madigan's research has primarily dealt with anoxygenic phototrophic bacteria, especially species that inhabit extreme environments, and he has graduated over 20 masters and doctoral students. He has published over 110 research papers, has coedited a major treatise on phototrophic bacteria, and has served as chief editor of the journal *Archives of Microbiology*. He currently serves on the editorial board of the journal *Environmental Microbiology*.

Mike's nonscientific interests include tree planting and caring for his dogs and horses. He lives beside a quiet lake about five miles from the SIUC campus with his wife, Nancy, four shelter dogs (Gaino, Snuffy, Pepto, and Merry), and three horses (Springer, Feivel, and Festus).

John M. Martinko received his B.S. in biology from The Cleveland State University. As an undergraduate student he participated in a cooperative education program, gaining experience in several microbiology and immunology laboratories. He worked for two years at Case Western Reserve University, conducting research on the structure, serology, and epidemiology of *Streptococcus pyogenes*. He did his graduate work at the State University of New York at Buffalo, investigating antibody specificity and antibody idiotypes for his M.A. and Ph.D. in microbiology. As a postdoctoral fellow, he worked at Albert Einstein College of Medicine in New York on the structure of major histocompatibility complex proteins. Since 1981, he has been in the Department of Microbiology at Southern Illinois University Carbondale where he is an associate professor and director of the Molecular Biology, Microbiology, and Biochemistry Graduate Program.

Dr. Martinko's current research involves manipulating immune reactions by inducing structural mutations in single-chain peptide–major histocompatibility protein complexes. He teaches undergraduate and graduate courses in immunology, and he also teaches immunology, host defense, and infectious disease topics in a general microbiology course as well as to medical students.

John has been active in educational outreach programs for pre-university students and teachers. For his educational efforts, he won the 2007 Southern Illinois University Outstanding Teaching Award. He is also an avid golfer and cyclist. John lives in Carbondale with his wife, Judy, a high school science teacher.

continued

Paul V. Dunlap received his B.S. degree in microbiology from Oregon State University in 1975. As an undergraduate student, he participated in research in marine microbiology in the laboratory of R.Y. Morita and served in his senior year as a teaching assistant for courses in microbiology, gaining experience in laboratory and field research and in teaching. He then taught English in Japan until 1978, when he returned to the United States for graduate studies in biology with J.G. Morin at UCLA. Research for his Ph.D. degree, awarded in 1984, addressed the ecology and physiology of bioluminescent symbiosis. He then moved to Cornell University in Ithaca, New York, for postdoctoral studies with E.P. Greenberg on the genetic regulation of bacterial luminescence.

In 1986 Dr. Dunlap joined the faculty at New Mexico State University and in 1989 moved to the Biology Department at the Woods Hole Oceanographic Institution, where he worked for several years on quorum sensing and symbiosis in luminous bacteria before moving in 1996 to the University of Maryland's Center of Marine Biotechnology in Baltimore. In 2001, he joined the faculty of the University of Michigan in Ann Arbor, where he is an associate professor in the Department of Ecology and Evolutionary Biology.

Dr. Dunlap's research focuses on the systematics of luminous bacteria, microbial evolution, bioluminescent symbiosis, and quorum sensing. He teaches an undergraduate majors course in introductory microbiology and a senior/graduate level course in microbial diversity.

Paul's nonscientific interests include family history research and the practice of aikido, a Japanese martial art. He lives in Ann Arbor with his wife, daughter, and their Australian terrier.

David P. Clark grew up in Croydon, a London suburb. He won a scholarship to Christ's College, Cambridge, where he received his B.A. degree in natural sciences in 1973. In 1977 he received his Ph.D. from Bristol University, Department of Bacteriology, for work on the effect of cell envelope composition on the entry of antibiotics into *Escherichia coli*. He then left England to become a postdoctoral researcher studying the genetics of lipid metabolism in the laboratory of John Cronan at Yale University. A year later he moved with the same laboratory to the University of Illinois at Urbana-Champaign. He joined the faculty of Southern Illinois University Carbondale in 1981.

Dr. Clark's research has focused on the growth of bacteria by fermentation under anaerobic conditions. He has published over 70 research articles and graduated over 20 masters and doctoral students. In 1989 he won the College of Science Outstanding Researcher Award. In 1991 he was the Royal Society Guest Research Fellow at the Department of Molecular Biology and Biotechnology, Sheffield University, England. He is the author of two books: *Molecular Biology, Made Simple and Fun,* now in its third edition, and *Molecular Biology, Understanding the Genetic Revolution*.

David is unmarried and lives with two cats, Little George, who is orange and very nosey, and Mr. Ralph, who is mostly black and eats cardboard.

Dedications

Michael T. Madigan dedicates this book to his wife Nancy—the apple of his eye and the love of his life.

John M. Martinko dedicates this book to his wife, Judy. Even the tough stuff is easy with you to share the load!

Paul V. Dunlap dedicates this book to R.Y. Morita with gratitude and appreciation for his mentoring and guidance in marine microbiological research.

David P. Clark dedicates this book to his father, Leslie, who set him the example of reading as many books as possible.

Preface

The authors are proud to present the 12th edition of *Brock Biology of Microorganisms* (*BBOM 12/e*). The roots of this book go back nearly 40 years, but its main objective has never wavered: to present the basic principles of microbiology in a clear and exciting way. We hope that the 12th edition of what is now a textbook classic continues in this vein and inspires a new generation of microbiologists to choose careers in this fascinating field.

Microbiology today places unusual demands on students and instructors alike. It also places unusual demands on textbook authors. The knowledge base in microbiology is simply enormous, and as authors, our job is to describe the principles along with a reasonable amount of supporting material. Thus, to maintain the authority that users of this book have become accustomed to, and also to keep the book within bounds, two new coauthors have joined the *BBOM* author team. David Clark, an expert in bacterial genetics and molecular biology and the author of two popular textbooks in molecular biology, has authored the chapters in *BBOM* that deal with these topics. Paul Dunlap, an expert in bacterial systematics and microbial evolution, has authored the microbial diversity and evolution chapters. Madigan and Martinko have authored the remaining chapters. Along with our publisher, Benjamin Cummings, who took over this project midstream from Prentice Hall, we feel we have produced a book that both students and instructors will find enticing and exciting.

WHAT'S NEW IN THE 12TH EDITION?

Instructors who have taught from *BBOM* in the past will be quite comfortable with this new edition. As usual, the chapters are organized into modules by numbered head, which allows instructors to build their courses as they see fit. In addition, study aids and review tools are an integral part of the text. Our new MiniReviews feature, which debuts with the 12th edition, is intended to refresh students' understanding and challenge their mastery of the principles as they work their way through a chapter. Also new to this edition is the location of the newly-named Review of Key Terms feature, which now summarizes a chapter's key terms and complements the challenging review questions that form the heart of the end-of-chapter review materials. A comprehensive glossary and index wrap up the learning package.

All of these learning aids are displayed in a new, more open design that gives the figures and other elements in the book some breathing room and has allowed us to present the material in a more visually appealing way. Supporting the narrative is a spectacular illustration program, with every piece of art revised as needed for content, clarity, or color contrast. The distinctive illustrations are accurate and highly instructive. The art complements (and in many cases integrates) the hundreds of photos in this text, many of which are new to this edition.

Our Microbial Sidebars in *BBOM 12/e* are fun reads of enrichment material related to chapter themes. Several new Microbial Sidebars were written for this edition, including "Microbial Growth in the Real World: Biofilms" (Chapter 6); "Did Viruses Invent DNA?" (Chapter 10); "Mimivirus and Virus Evolution" (Chapter 19); "Synthetic Biology and Bacterial Photography" (Chapter 26); "Probiotics" (Chapter 28); "SARS as an Example of Epidemiological Success" (Chapter 33); "Special Pathogens and Viral Hemorrhagic Fevers" (Chapter 35); and "Spinach and *Escherichia coli* O157:H7" (Chapter 37).

Although the 12th edition is about the same length as the 11th edition, careful editing by the author team has allowed several new chapters to debut in this edition. Chapter 8 (Archaeal and Eukaryotic Molecular Biology) focuses on the molecular biology of *Archaea* and their relationships to eukaryotic cells. Chapter 12 (Genetic Engineering) describes the tools that underlie the revolution in molecular biology. Chapter 14 (Microbial Evolution and Systematics) is a fresh new approach to issues surrounding the origin of life and microbial evolution. Chapter 26 (Biotechnology) reviews the basic science behind the biotech revolution and presents several spectacular examples of the benefits of biotechnology to human medicine, and plant and animal agriculture. Chapter 30 (Immunology in Host Defense and Disease) pulls together many of the applied aspects of immunology and shows how the immune response keeps us healthy. Coupled with these new chapters are many heavily reworked chapters. Readers will especially appreciate the more phylogenetic approach to the systematics of both prokaryotes and eukaryotes in this new edition as well as the expanded ecology material that includes some exciting and experimentally tractable microbial symbioses. Material on microbial genomics has also been greatly updated to reflect the pace and excitement of research in this area today.

CHAPTER-BY-CHAPTER REVISIONS

BBOM 12/e is *the* textbook of microbiology for both the beginning student and the seasoned researcher; it displays the perfect mix of principles and details. Major topics include general principles (structure/function, metabolism and growth), molecular biology/genetics, genomics, evolution and microbial diversity, metabolic diversity, microbial ecology, biotechnology and industrial microbiology, control of microbial growth, basic and applied immunology, and complete coverage of infectious diseases organized by mode of transmission. In addition, *BBOM 12/e* maintains a theme of evolution and ecology from beginning to end and is the only book that recognizes the importance and unique biology of *Archaea*. See highlights as follows.

Chapter 1

- New coverage on the antiquity and extent of microbial life and on careers in microbiology
- Expanded coverage of the contributions of Louis Pasteur

Chapter 2

- Broad coverage of all forms of microscopy
- An updated snapshot of microbial diversity

Chapter 3

- The essentials of cell chemistry with an emphasis on what students really need to know to understand how cells work

Chapter 4

- A streamlined chapter focused on the principles of prokaryotic cell structure and function

Chapter 5

- New discussion on the chemical elements of life
- Major coverage of catabolic principles plus an overview of essential anabolic reactions
- A new box entitled "The Products of Fermentation and the Pasteur Effect" that ties together the concepts of fermentation and respiration

Chapter 6

- Expanded coverage of cell division processes supported by spectacular new art and color photos
- A new box entitled "Microbial Growth in the Real World: Biofilms"

Chapter 7

- A substantially revised treatment that brings together the essential principles of molecular biology in *Escherichia coli,* the model species of *Bacteria*

Chapter 8

- A brand-new chapter that compares/contrasts the molecular biology of *Archaea* with that of *Eukarya* and *Bacteria*
- A new box entitled "Inteins and Protein Splicing"
- Coverage of RNA interference (RNAi), research that garnered a 2006 Nobel Prize

Chapter 9

- Major updates of the regulation of gene expression, one of the hottest areas in microbiology today
- New material on negative and positive control, global control, quorum sensing, gene regulation in *Archaea,* regulation of sporulation in *Bacillus,* the *Caulobacter* life cycle, and RNA-based regulation

Chapter 10

- The essential principles of virology plus a snapshot of viral diversity
- New coverage of subviral entities, including prions and viroids
- An intriguing new box entitled "Did Viruses Invent DNA?"

Chapter 11

- A streamlined chapter that now deals exclusively with bacterial genetics: chromosomes and plasmids, mutation, and genetic exchange
- New coverage of genetic exchange in *Archaea*

Chapter 12

- A discussion of modern *in vitro* molecular methods, molecular cloning, and genetic manipulations, all brought together in one chapter as a prelude to genomics

Chapter 13

- Microbial genomics as we know it today. Coverage of genome function and regulation, the evolution of genomes, and the exciting field of metagenomics

Chapter 14

- New detailed coverage of the origin of life and endosymbiosis
- Expanded coverage of theoretical and analytical aspects of molecular evolution and microbial phylogeny

Chapters 15–17

- Prokaryotic diversity presented in a phylogenetic context. Chapter 15 covers all five subdivisions of *Proteobacteria*, and Chapter 16 covers other well-studied lineages of *Bacteria*. Chapter 17 updates the biology of *Archaea*.
- Spectacular photo- and electron micrographs that guide the reader through the diversity of *Bacteria* and *Archaea*
- All bacterial names updated to the latest in nomenclatural standards

Chapter 18

- A revised treatment of the cell biology and diversity of microbial eukaryotes
- A truly phylogenetic rather than taxonomic treatment of eukaryotic microbial diversity based on multiple gene and protein analyses

Chapter 19

- Viral diversity from bacteriophages through animal and plant viruses organized by replication mechanism
- An exciting new box focused on giant viruses entitled "Mimivirus and Virus Evolution"

Chapters 20–21

- Coverage of metabolic diversity in two reorganized chapters that group related processes: Chapter 20, phototrophy, chemolithotrophy, and important biosyntheses; Chapter 21, catabolism of organic compounds
- Coverage of new metabolic processes, such as proton reduction as an anaerobic respiration and the mechanism of anoxic hydrocarbon oxidation

Chapter 22

- All the latest methods in microbial ecology, including new coverage of cutting-edge methods such as phylochips, T-RFLP, and stable isotope probing

Chapter 23

- Principles of microbial ecology and descriptions of major microbial habitats
- Expanded coverage of biofilms, including new photos of the development of *Pseudomonas aeruginosa* biofilms

Chapter 24

- Microbial ecology with a focus on nutrient cycles, bioremediation, and microbial animal and plant symbioses
- New, richly illustrated coverage of the *Aliivibrio*–squid symbiosis

Chapter 25

- The principles behind major commercial microbial fermentations
- Expanded coverage of wine production

Chapter 26

- Biotechnology with the emphasis on products rather than methods
- Expanded coverage of transgenic plants and animals
- A fascinating new box entitled "Synthetic Biology and Bacterial Photography" that ties basic research into a unique microbial application

Chapter 27

- An introduction to microbial growth control, in particular, antibiotics and their mode of action
- Expanded coverage of antibiotic resistance and infection control

Chapter 28

- Revised treatment of the human normal flora based on molecular microbial community analyses
- Expanded discussion of virulence factors using *Salmonella* as an example and microbial toxins and toxin mechanisms
- A new box entitled "Probiotics" that answers the question: Do live bacterial supplements really have any benefit?

Chapters 29–31

- The world of immunology: Chapter 29 (the essentials), Chapter 30 (host defenses and applications), and Chapter 31 (immune mechanisms at the molecular level). All chapters are modular so that instructors can pick and choose topics that work best in their courses.
- Major revisions to art in all of these chapters in order to greatly improve consistency

Chapter 32

- A streamlined treatment of modern immunological and molecular methods used in clinical diagnostics

Chapter 33

- Updates of the HIV–AIDS pandemic and other important emerging infectious diseases
- A snapshot of the rapidly changing picture in healthcare-associated disease and infection control
- A new box entitled "SARS as an Example of Epidemiological Success" that connects SARS surveillance with the control of other rapidly emerging diseases

Chapter 34

- Major human diseases organized by mode of transmission
- Updated and expanded coverage of streptococcal and staphylococcal diseases and increased antibiotic resistance
- Expanded coverage of influenza and scenarios for a pandemic
- A new section on human papillomavirus infection as a preventable and treatable sexually transmitted disease that can lead to cancer

Chapter 35

- An updated chronicle of the spread of West Nile virus across the United States and the implications of this avian disease for human health
- An exciting new box entitled "Special Pathogens and Viral Hemorrhagic Fevers" that describes an important branch of the U.S. Centers for Disease Control (CDC) that deals with extremely dangerous pathogens

Chapter 36

- Expanded coverage of water quality assessment methods, including current U.S. Environmental Protection Agency standards

Chapter 37

- Discussion of "aseptic processing," a new preservation method in the food industry
- An update on *Salmonella* as a food pathogen
- A timely new box entitled "Spinach and *Escherichia coli* O157:H7" that explores the transmission of this food pathogen by plant as well as animal food products

Three New Chapters on Rapidly Developing Fields

Three new chapters focus on the rapidly developing fields of archaeal and eukaryotic molecular biology, biotechnology, and immunology in host defense and disease.

CHAPTER 8: ARCHAEAL AND EUKARYOTIC MOLECULAR BIOLOGY

Molecular biology has revealed the three domains of life (*Bacteria*, *Archaea*, and *Eukarya*) and given us an unprecedented snapshot of microbial evolution. From a molecular perspective, cells of *Archaea* and *Eukarya* share much in common. The all-new Chapter 8 is the first of its kind in a microbiology textbook and compares and contrasts the molecular biology of these two domains with that of *Bacteria* (Chapter 7). Together, Chapter 7 and the new Chapter 8 give students the molecular background they need to understand microbiology today and to appreciate the science behind photos such as the one here of telomeres on eukaryotic chromosomes.

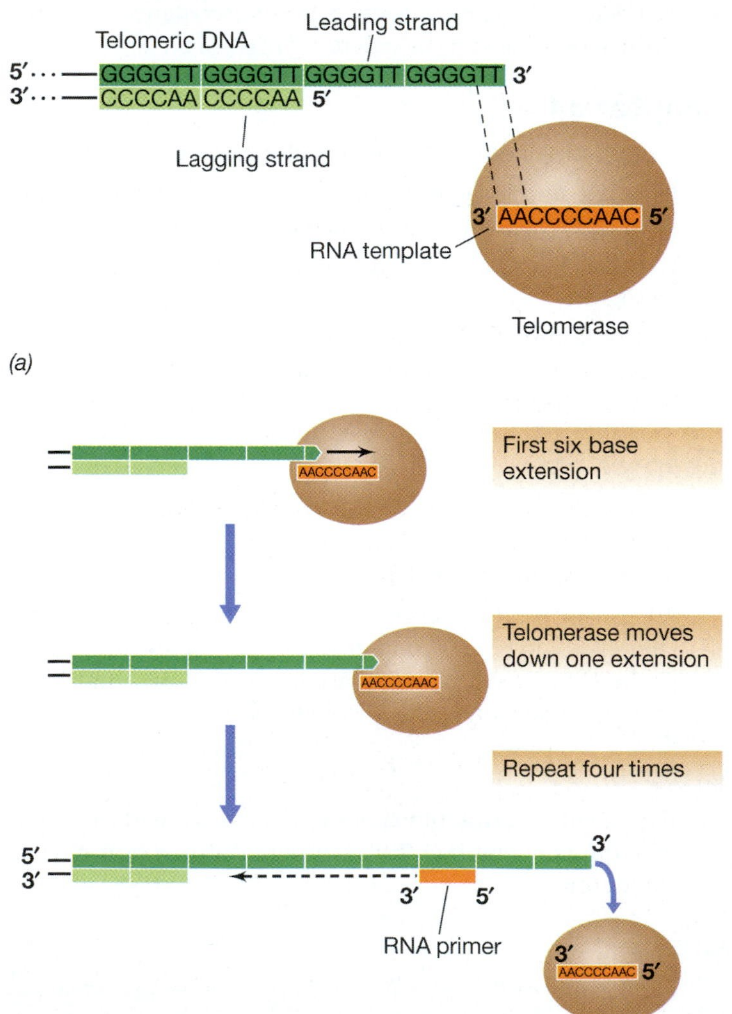

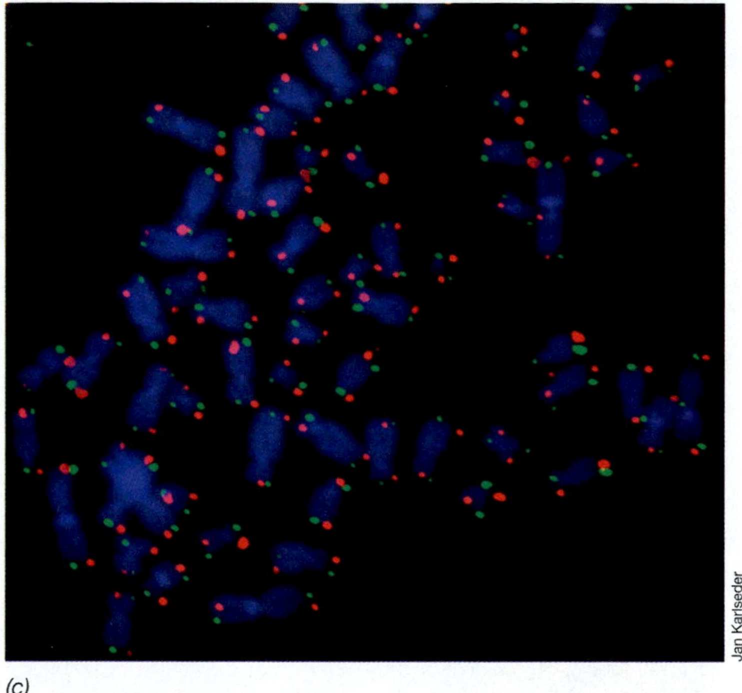

(c)

Figure 8.9 Model for the activity of telomerase at one end of a eukaryotic chromosome. *(a)* A diagram of the sequence of the end of the DNA in a telomere, with four of the guanine-rich repeats and the enzyme telomerase, which contains a short RNA template. *(b)* Steps in elongation of the guanine-rich strand catalyzed by telomerase. After telomerase finishes, the lagging strand can be primed with an RNA primer by primase followed by completion of the lagging strand by DNA polymerase and ligase. *(c)* A preparation of HeLa cell chromosomes stained with fluorescent dyes. The red dots are leading strand telomeres and the green dots are lagging strand telomeres.

CHAPTER 26: BIOTECHNOLOGY

The new Chapter 26 examines the advances in biotechnology that have generated a host of new tools for molecular biology and genomics as well as several important products for human health. Biotechnology is yielding new perspectives on genes and gene expression (see the fluorescent pig) and is playing a greater role in the food we eat (see the transgenic salmon).

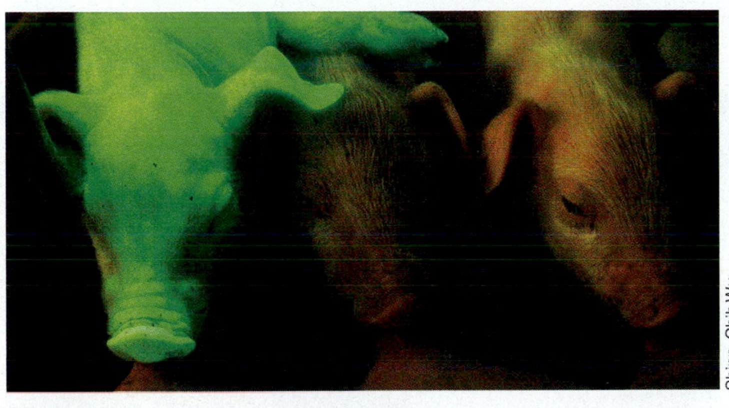

(a)

(b)

Figure 26.8 Transgenic animals. *(a)* A piglet (left) that has been genetically engineered to express the green fluorescent protein and thus fluoresces green under blue light. Control piglets are shown in the center and right. *(b)* Fast-growing salmon. The *AquAdvantage*™ Salmon" (top) was engineered by Aqua Bounty Technologies (St. Johns, Newfoundland, Canada). Both the transgenic and the control fish are 18 months old and weigh 4.5 kg and 1.2 kg, respectively.

CHAPTER 30: IMMUNOLOGY IN HOST DEFENSE AND DISEASE

Immunology is a major sub-discipline of microbiology, but many instructors struggle to present the key concepts. The new Chapter 30 is designed as an overview of immunology with a focus on the applied aspects that all students need to know: different forms of the immune response, principles of immunization, and autoimmunity. Instructors will find Chapter 30 to be the "one-stop shop" they need to teach immunology.

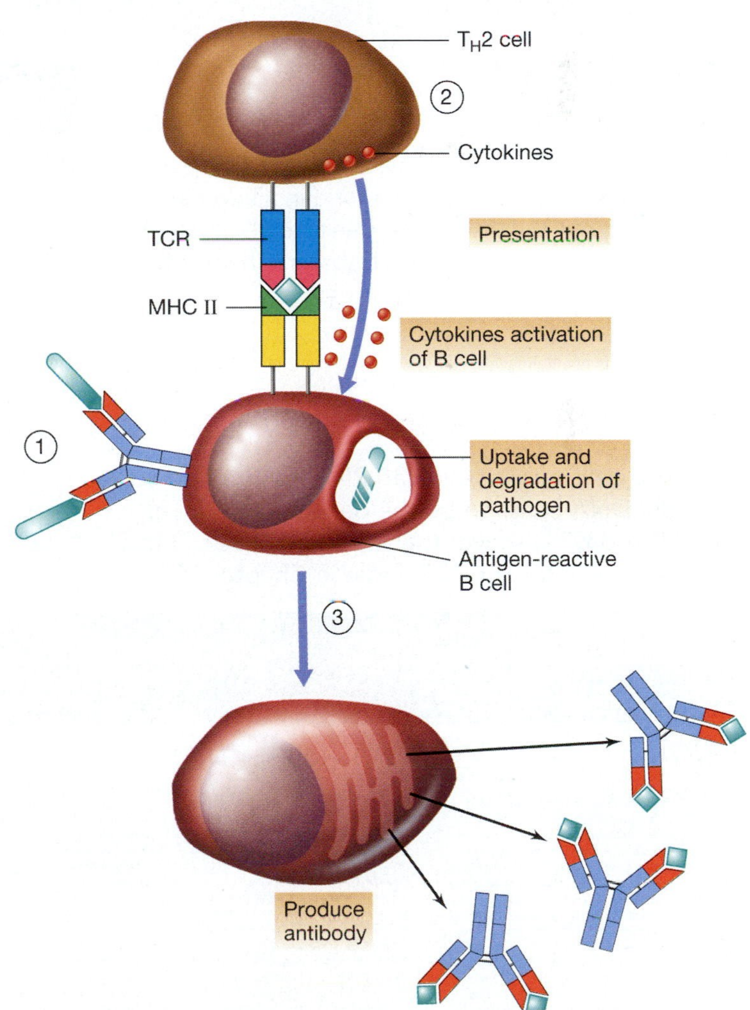

Figure 30.3 Antibody-mediated immunity. ① Antibody on B cells binds to a pathogen. The B cell ingests, degrades, and processes the pathogen. ② The B cell presents pathogen antigen to a T_{H2} cell, activating it to produce cytokines that in turn influence the B cell to develop into a plasma cell. ③ The plasma cell produces antibodies.

Hot Topics Integrated Throughout the Chapters

Hot topics draw students in to the chapters and introduce them to current issues in microbiology. The number of hot topics has been increased in this edition.

Microbial Sidebars present enrichment material related to a chapter's central theme.

BIOFILMS

Biofilms develop when bacterial cells attach and grow on surfaces. They are associated with some human diseases, such as cystic fibrosis. Exciting new coverage of biofilms is included in Chapters 6 and 23 and is illustrated with several spectacular photos.

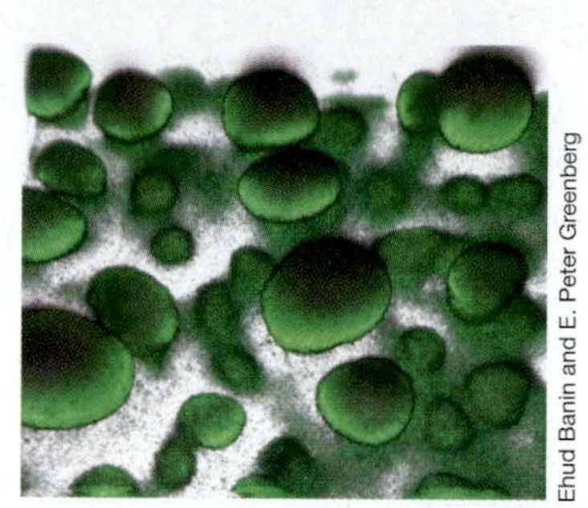

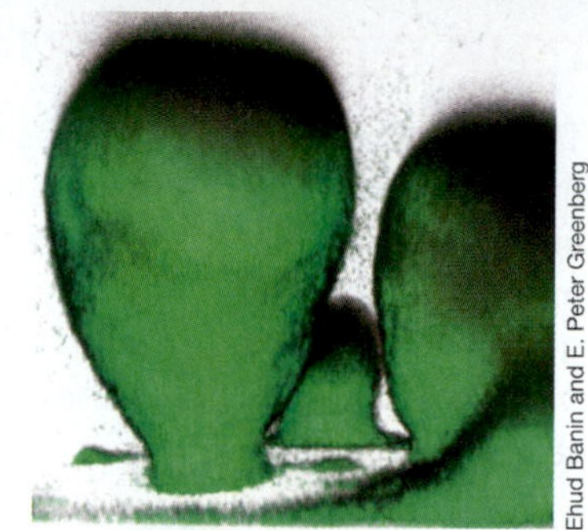

Figure 23.7 Biofilms of *Pseudomonas aeruginosa.*

INTEINS AND PROTEIN SPLICING

Self-splicing RNAs (ribozymes) have a counterpart in self-splicing proteins (inteins), found in all domains of life. Find out the "what, why, and how" of inteins in an exciting new Microbial Sidebar in Chapter 8.

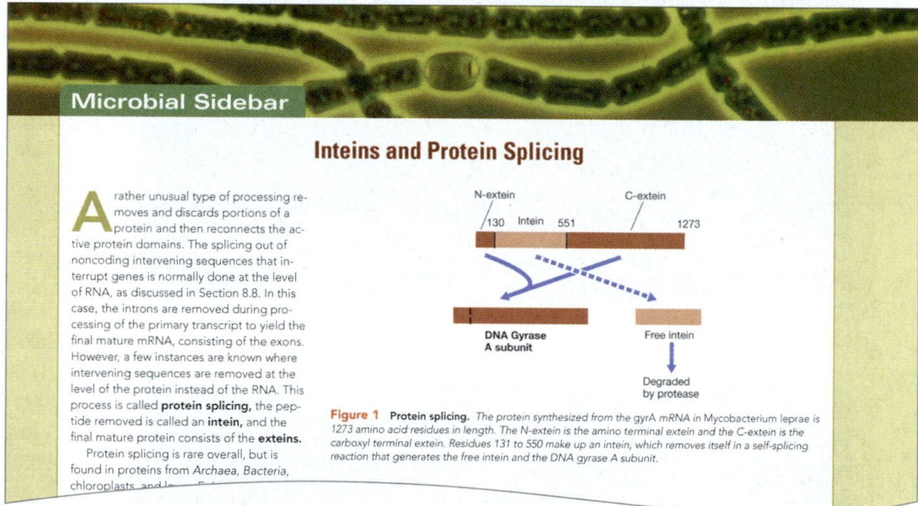

Microbial Sidebar

Inteins and Protein Splicing

A rather unusual type of processing removes and discards portions of a protein and then reconnects the active protein domains. The splicing out of noncoding intervening sequences that interrupt genes is normally done at the level of RNA, as discussed in Section 8.8. In this case, the introns are removed during processing of the primary transcript to yield the final mature mRNA, consisting of the exons. However, a few instances are known where intervening sequences are removed at the level of the protein instead of the RNA. This process is called **protein splicing,** the peptide removed is called an **intein,** and the final mature protein consists of the **exteins.**

Protein splicing is rare overall, but is found in proteins from *Archaea, Bacteria,* chloroplasts

Figure 1 Protein splicing. *The protein synthesized from the gyrA mRNA in* Mycobacterium leprae *is 1273 amino acid residues in length. The N-extein is the amino terminal extein and the C-extein is the carboxyl terminal extein. Residues 131 to 550 make up an intein, which removes itself in a self-splicing reaction that generates the free intein and the DNA gyrase A subunit.*

DID VIRUSES INVENT DNA?

Primitive cells might have had RNA genomes. If so, how did DNA come to be the genetic material of all cells today? Discover the virus connection in this intriguing new Microbial Sidebar in Chapter 10.

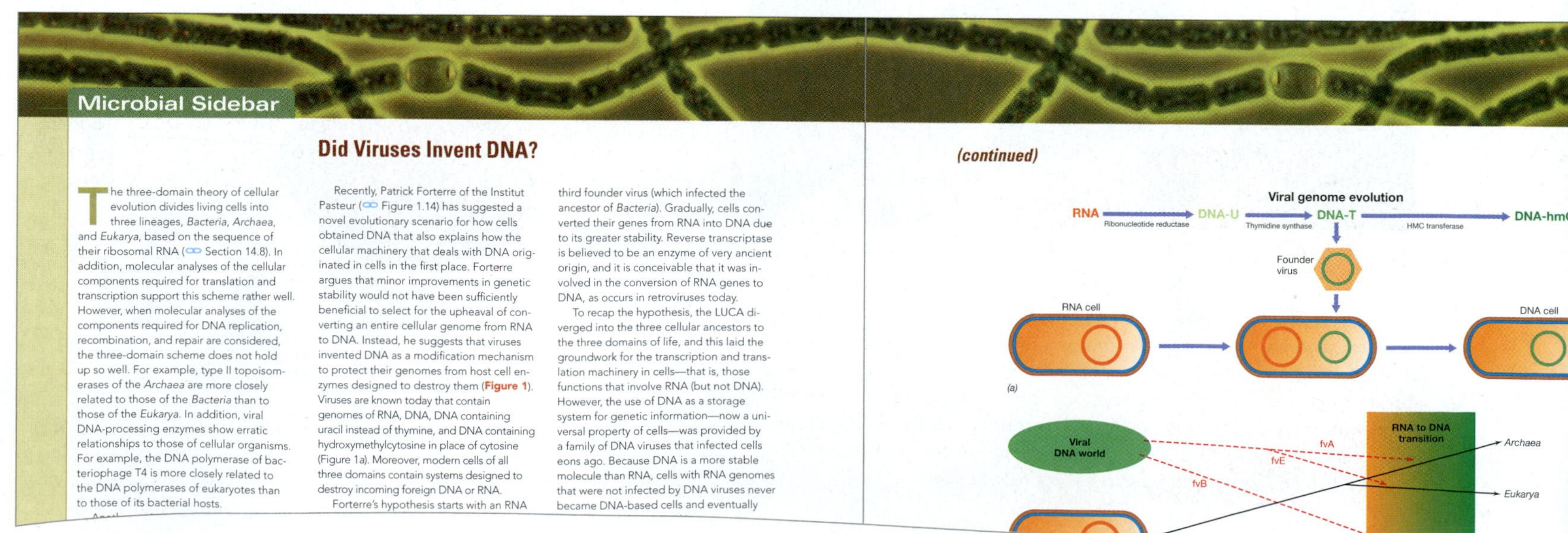

Microbial Sidebar

Did Viruses Invent DNA?

The three-domain theory of cellular evolution divides living cells into three lineages, *Bacteria, Archaea,* and *Eukarya,* based on the sequence of their ribosomal RNA (Section 14.8). In addition, molecular analyses of the cellular components required for translation and transcription support this scheme rather well. However, when molecular analyses of the components required for DNA replication, recombination, and repair are considered, the three-domain scheme does not hold up so well. For example, type II topoisomerases of the *Archaea* are more closely related to those of the *Bacteria* than to those of the *Eukarya.* In addition, viral DNA-processing enzymes show erratic relationships to those of cellular organisms. For example, the DNA polymerase of bacteriophage T4 is more closely related to the DNA polymerases of eukaryotes than to those of its bacterial hosts.

Recently, Patrick Forterre of the Institut Pasteur (Figure 1.14) has suggested a novel evolutionary scenario for how cells obtained DNA that also explains how the cellular machinery that deals with DNA originated in cells in the first place. Forterre argues that minor improvements in genetic stability would not have been sufficiently beneficial to select for the upheaval of converting an entire cellular genome from RNA to DNA. Instead, he suggests that viruses invented DNA as a modification mechanism to protect their genomes from host cell enzymes designed to destroy them (**Figure 1**). Viruses are known today that contain genomes of RNA, DNA, DNA containing uracil instead of thymine, and DNA containing hydroxymethylcytosine in place of cytosine (Figure 1*a*). Moreover, modern cells of all three domains contain systems designed to destroy incoming foreign DNA or RNA.

Forterre's hypothesis starts with an RNA third founder virus (which infected the ancestor of *Bacteria*). Gradually, cells converted their genes from RNA into DNA due to its greater stability. Reverse transcriptase is believed to be an enzyme of very ancient origin, and it is conceivable that it was involved in the conversion of RNA genes to DNA, as occurs in retroviruses today.

To recap the hypothesis, the LUCA diverged into the three cellular ancestors to the three domains of life, and this laid the groundwork for the transcription and translation machinery in cells—that is, those functions that involve RNA (but not DNA). However, the use of DNA as a storage system for genetic information—now a universal property of cells—was provided by a family of DNA viruses that infected cells eons ago. Because DNA is a more stable molecule than RNA, cells with RNA genomes that were not infected by DNA viruses never became DNA-based cells and eventually

(continued)

Exceptionally Clear Illustrations and Photos

Unparalleled illustration and photo program gives students a clear and fascinating view into the microbial world.

ILLUSTRATIONS

The cytoplasmic membrane with its dense array of proteins comes alive with spectacular art.

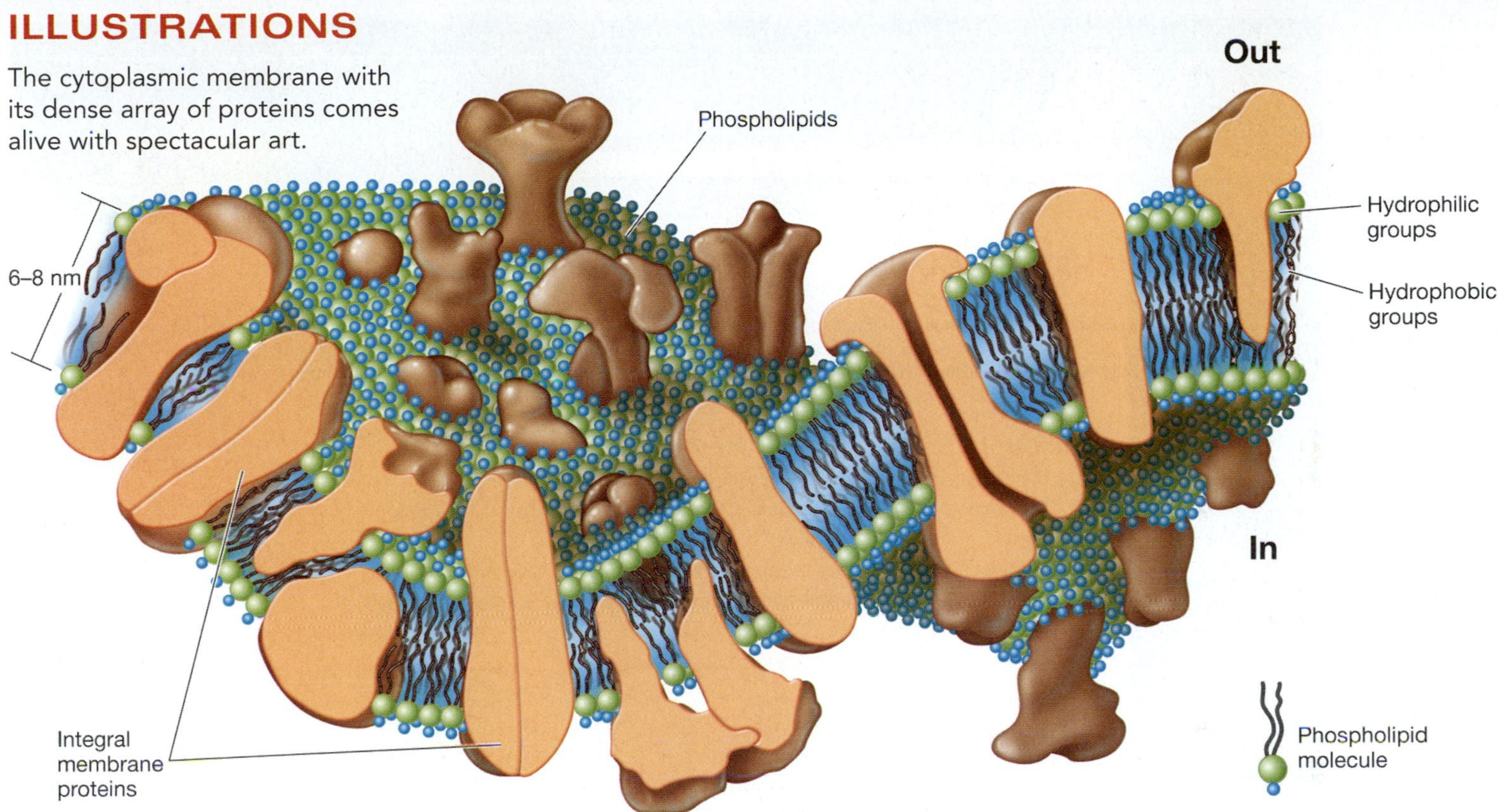

Figure 4.5 Structure of the cytoplasmic membrane.

PHOTOMICROGRAPHS

The best photomicrographs of any microbiology textbook are exemplified here by a cell of the protist *Tetrahymena* stained to reveal special cellular structures.

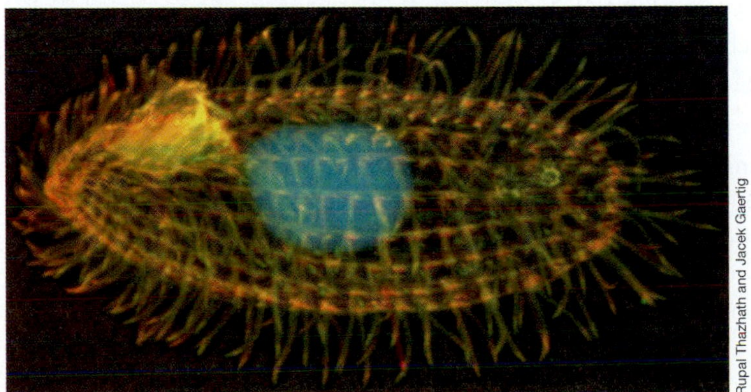

Figure 18.10 Tubulin of *Tetrahymena thermophila.*

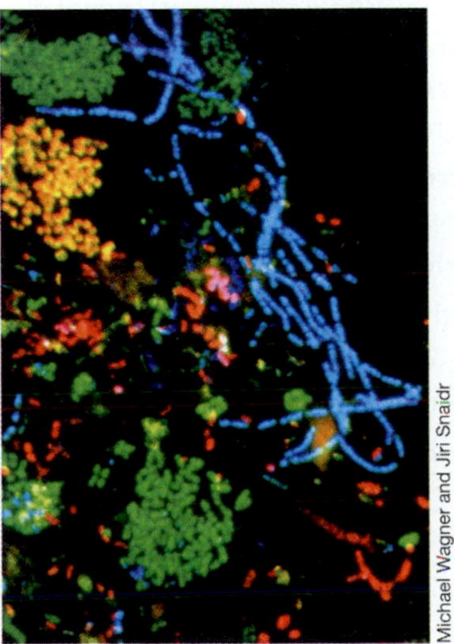

Figure 22.11 FISH analysis of sewage sludge.

Phylogenetic stains, used here on a sewage sample, give microbial ecologists the power to both quantify microorganisms in natural samples and identify them phylogenetically.

The Microbiology Place Website

The Microbiology Place Website is rich with media assets to ensure student success.

www.microbiologyplace.com

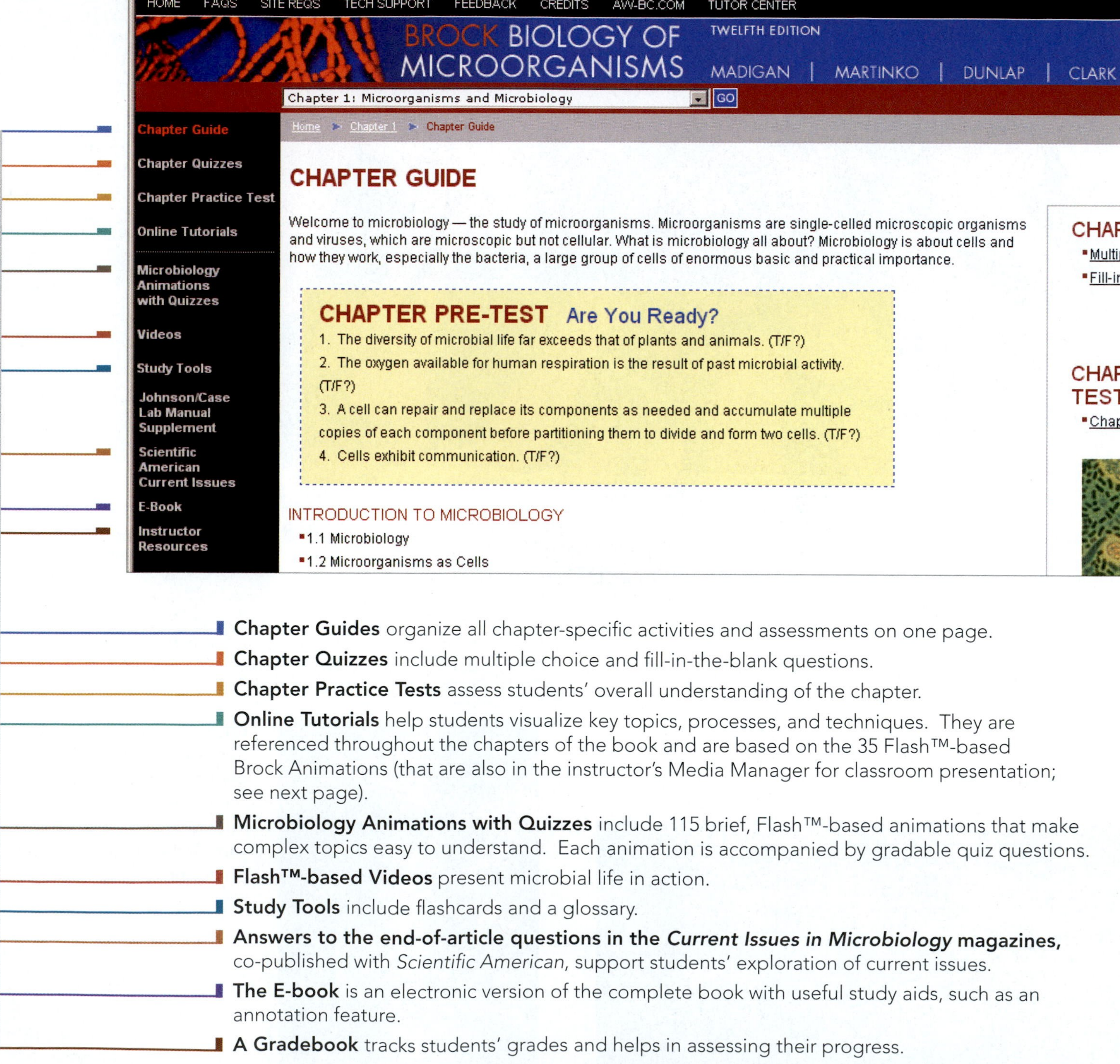

- **Chapter Guides** organize all chapter-specific activities and assessments on one page.
- **Chapter Quizzes** include multiple choice and fill-in-the-blank questions.
- **Chapter Practice Tests** assess students' overall understanding of the chapter.
- **Online Tutorials** help students visualize key topics, processes, and techniques. They are referenced throughout the chapters of the book and are based on the 35 Flash™-based Brock Animations (that are also in the instructor's Media Manager for classroom presentation; see next page).
- **Microbiology Animations with Quizzes** include 115 brief, Flash™-based animations that make complex topics easy to understand. Each animation is accompanied by gradable quiz questions.
- **Flash™-based Videos** present microbial life in action.
- **Study Tools** include flashcards and a glossary.
- **Answers to the end-of-article questions in the *Current Issues in Microbiology* magazines,** co-published with *Scientific American*, support students' exploration of current issues.
- **The E-book** is an electronic version of the complete book with useful study aids, such as an annotation feature.
- **A Gradebook** tracks students' grades and helps in assessing their progress.

Media Manager 2.1

Media Manager 2.1, in DVD/CD format, organizes all instructor media resources by chapter in one convenient and easy-to-use package. It includes the following assets and features.

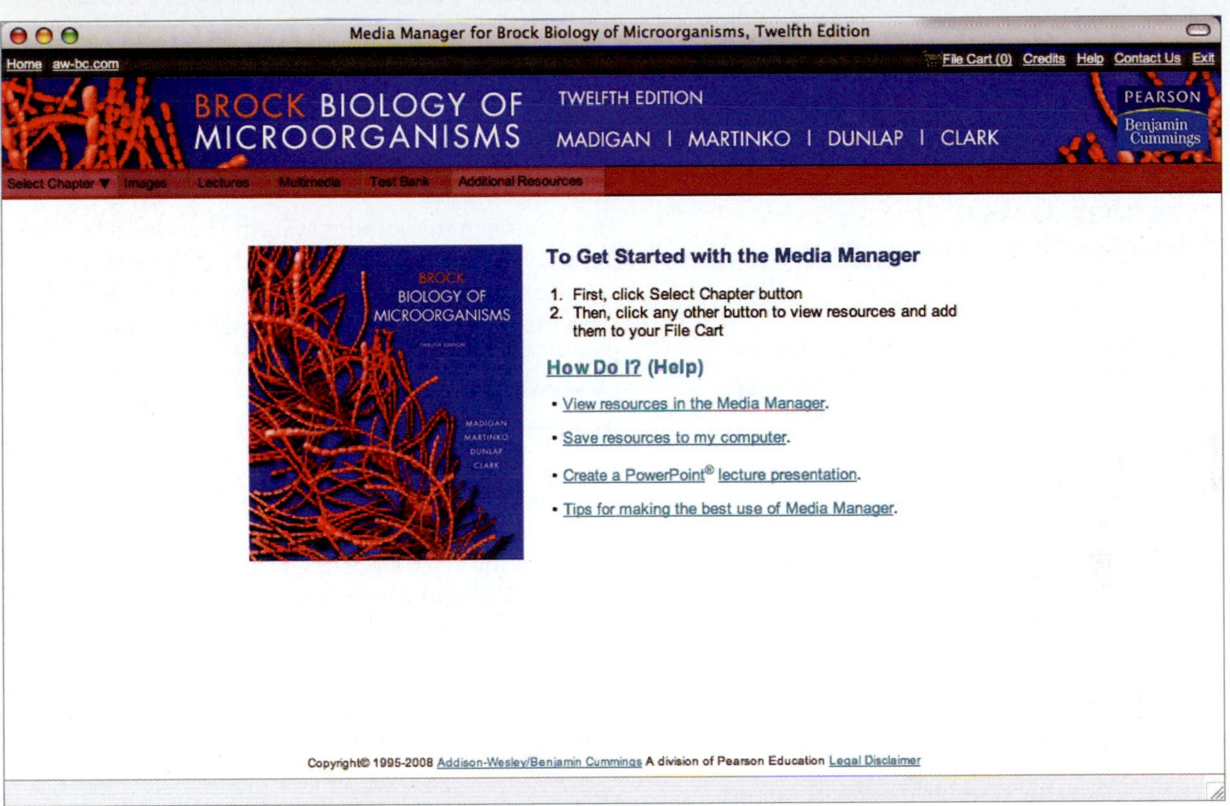

- **All illustrations and photos from the book** in both JPEG and PowerPoint® formats
- **A "Label Edit" feature** for customizing labels and leader lines on all illustrations and photos so that instructors can control what they present in the classroom (and what they use for student assessment)
- **All tables from the book**
- **35 Flash™-based Brock Animations** that are also the basis of the Online Tutorials for students on The Microbiology Place website; see previous page
- **115 Flash™-based Microbiology Animations** on all major topics (with Quizzes in PRS-enabled Clicker Question format)
- **25 Flash™-based videos** of microorganisms in action
- **Customizable PowerPoint® Lecture Outlines** for all chapters, combining lecture notes, illustrations, photos, tables, and links to animations
- **Active Lecture Questions** (in PRS-enabled Clicker Question format), which stimulate effective classroom discussions and check student comprehension
- **Quiz Show chapter reviews** (in PRS-enabled Clicker Question format), which encourage student interaction
- **The Computerized Test Bank**
- **"Shopping Cart" functionality,** allowing instructors to quickly search, select, and download any item
- **A printed Quick Reference Guide** that shows all media assets for each chapter at a glance

Supplements

FOR THE INSTRUCTOR

Media Manager
0-132-32498-9 / 978-0-132-32498-4
See page xiii for a description. Note that the Media Manager also includes the Computerized Test Bank.

Instructor's Manual/Test Bank
0-132-32496-2 / 978-0-132-32496-0
by Thomas M. Wahlund and Elizabeth F. McPherson
The Instructor's Manual provides chapter summaries that help with class preparation as well as the answers to the end-of-chapter Review Questions and Application Questions. The Test Bank contains 3,000 questions for use in quizzes, tests, and exams.

Computerized Test Bank
by Elizabeth F. McPherson
This cross-platform, easy-to-use testing program (on CD-ROM) allows instructors to view and edit electronic questions from the Test Bank, create tests, and print them in a variety of formats. The Computerized Test Bank is included in the Media Manager package.

Full-Color Transparency Acetates
0-132-32500-4 / 978-0-132-32500-4
All illustrations, photos, and tables from the textbook are included in this transparency package. Label sizes have been increased for easy viewing from any location in the classroom.

FOR THE STUDENT

Current Issues in Microbiology
Volume 1: 978-0-8053-4623-7 / 0-8053-4623-6
Volume 2: 978-0-321-53816-1 / 0-321-53816-1
This four-color magazine includes articles from *Scientific American* magazine selected especially for students of microbiology. End-of-article questions help students check their knowledge and connect science to society. Answers to the questions appear on The Microbiology Place website.

The Microbe Files: Cases in Microbiology for the Undergraduate
***With Answers:* 978-0-8053-4927-6 / 0-8053-4927-8**
***Without Answers:* 978-0-8053-4928-3 / 0-8053-4928-6**
by Marjorie K. Cowan
The first of its kind, this brief book provides students with a fascinating series of short cases that help them apply what they have learned by placing them in real-life situations that health professionals face every day.

WEBSITES AND COURSE MANAGEMENT PROGRAMS

The Microbiology Place
See page xii for a description.
www.microbiologyplace.com

CourseCompass
This course management system includes all of the assets from The Microbiology Place website as well as all of the test questions from the Computerized Test Bank. It also features class management tools such as discussion boards and email functionality.
www.aw-bc.com/coursecompass

WebCT
This open-access course management system contains preloaded content of the Chapter Quizzes, Chapter Practice Tests, Online Tutorials, flashcards, and glossary from The Microbiology Place website as well as all of the test questions from the Computerized Test Bank.
www.aw-bc.com/webct

Blackboard
This open-access course management system contains preloaded content of the Chapter Quizzes, Chapter Practice Tests, Online Tutorials, flashcards, and glossary from The Microbiology Place website as well as all of the test questions from the Computerized Test Bank.
www.aw-bc.com/blackboard

Acknowledgments

This book is the collective effort of many people. These include a number of folks at Prentice Hall/Benjamin Cummings, but especially executive editors Leslie Berriman and Gary Carlson and their assistants, Kelly Reed, Jessica Young, and Lisa Tarabokjia. The wisdom and enthusiasm of Leslie and Gary were the keys to making this project go, and they are gratefully acknowledged. Our excellent production editor, Karen Gulliver, carefully guided the transformation of a raw manuscript into a polished book, and our art coordinator, Jean Lake, maintained art flow and handled all of the behind-the-scenes interactions with the art studio. In addition, our art liaison, Jack Haley, did an absolutely superb job of converting the author's rough sketches into a form that artists in the studio could convert into eye-appealing and pedagogically strong art. The authors also give hearty thanks to our production and design team, headed by Wendy Earl of Wendy Earl Productions, and to the first-rate efforts of Jane Loftus for copyediting the manuscript. Gordon Lee in Benjamin Cummings Marketing headed up our superb marketing campaign.

The authors especially wish to thank the extremely helpful input early on in this project of our developmental editor, Elmarie Hutchinson, and the expert assistance with electronic photographs provided by Deborah O. Jung. In addition, the input of Elizabeth McPherson as an accuracy checker during the production of *BBOM 12/e* was very helpful and is heartily acknowledged. All authors also wish to thank their graduate students, colleagues, and departmental staff for their assistance, patience, and cooperation during the busy time of preparing a book of this scope. Finally, MTM wishes to acknowledge his coauthors, John, Paul, and David; it has been both a pleasure and a privilege to have worked with you guys on this edition.

No textbook in microbiology could be published without thorough reviewing of the manuscript and the contribution of photos from experts in the field. We are therefore extremely grateful for the kind help of the many individuals who provided general or technical reviews or who supplied new photos for the 12th edition. They are listed below. We give special thanks to Professor Dr. Michael Wagner, University of Vienna, Austria, for the spectacular image used on the cover of *BBOM 12/e* and also for contributing several other color photos to the book.

Morad Abou-Sabe, *Rutgers University*
Richard Adler, *University of Michigan–Dearborn*
Gladys Alexandre-Jouline, *University of Tennessee*
Stephen Aley, *University of Texas at El Paso*
Thomas Alton, *Western Illinois University*
Linda Amaral Zettler, *The Josephine Bay Paul Center for Comparative Molecular Biology and Evolution*
Rodney Anderson, *Ohio Northern University*
Esther Angert, *Cornell University*
Marie Asao, *Southern Illinois University*
Jennifer Ast, *University of Michigan*
Susan Bagley, *Michigan Technological University*
Tamar Barkay, *Rutgers University*
Dale Barnard, *Utah State University*
Carl Bauer, *Indiana University*
Dennis A. Bazylinski, *University of Nevada, Las Vegas*
Michael Benedik, *Texas A&M University*
Nicholas H. Bergman, *University of Michigan*
Robert Blankenship, *Washington University*
Paul Blum, *University of Nebraska–Lincoln*
Katherine Boettcher, *University of Maine*
Susan Borrstein-Forst, *Marian College*
David Bradley, *Iowa State University*
Derrick Brazill, *Hunter College*
Forrest Brem, *Memphis, Tennessee*
Robert Britton, *Michigan State University*
Mary Burke, *Oregon State University*
Peter D. Burrows, *University of Alabama at Birmingham*
Donald Canfield, *University of Southern Denmark*
Richard Castenholz, *University of Oregon*
Centers for Disease Control and Prevention Public Health Image Library, *Atlanta, Georgia*
Delphi Chatterjee, *Colorado State University*
Aaron Chevalier, *University of Texas*
Sallie Chisholm, *Massachusetts Institute of Technology*
Rhonda Clark, *University of Calgary*
Morris Cooper, *Southern Illinois University School of Medicine*
Diana Cundell, *Philadelphia University*
Philip Cunningham, *Wayne State University*
Mike Dalbey, *University of California–Santa Cruz*
Dennis Dean, *Virginia Polytechnic Institute and State University*
Edward F. DeLong, *Massachusetts Institute of Technology*
David DeRosier, *Brandeis University*
Don Deters, *Bowling Green State*
Thomas DiChristina, *Georgia Institute of Technology*
Ulrich Dobrindt, *University of Wuerzburg (Germany)*
William Ebomoyi, *Chicago State University*
Tassos Economou, *Institute of Molecular Biology and Biotechnology, Iraklio-Crete (Greece)*
Edward P. Ewing, Jr., *Centers for Disease Control, Atlanta*
Mekomen Fekadu, *Ethiopian Health and Nutrition Institute, Addis Ababa*
Leanne Field, *University of Texas*
Douglas Fix, *Southern Illinois University*
Alex Formstone, *University of Oxford (England)*
Larry Forney, *University of Idaho*
Christopher Francis, *Stanford University*
Chris Frazee, *University of Wisconsin*
S. Marvin Friedman, *Hunter College*

Jed Fuhrman, *University of Southern California*
Clay Fuqua, *Indiana University*
Michelle Furlong, *Clayton State University*
Jacek Gaertig, *University of Georgia*
Daniel Gage, *University of Connecticut*
Bruce Geller, *Oregon State University*
Howard Gest, *Indiana University*
Stephen Giovannoni, *Oregon State University*
Donald Glassman, *Des Moines Area Community College*
Bonnie Glatz, *Iowa State University*
Steven Glenn, *Centers for Disease Control, Atlanta*
Peter Greenberg, *University of Washington*
Ricardo Guerrero, *University of Barcelona (Spain)*
John Haddock, *Southern Illinois University*
Lidija Halda-Alija, *University of Mississippi*
Lydia Halden, *HA Micro Gene Technology International, Inc.*
Robin Harris, *University of Mainz (Germany)*
John Hayes, *Woods Hole Oceanographic Institution*
Jim Holden, *University of Massachusetts*
Margaret Hollingsworth, *University of Buffalo*
Jarmo Holopainen, *University of Kuopio (Finland)*
Scott M. Holt, *Western Illinois University*
Michael Ibba, *Ohio State University*
Johannes Imhoff, *University of Kiel (Germany)*
James Imlay, *University of Illinois*
Nicholas J. Jacobs, *Dartmouth Medical School*
Christine Jacobs-Wagner, *Yale University*
Ken Jarrell, *Queen's University (Canada)*
John Jones, *Saint Xavier University*
Deborah O. Jung, *Southern Illinois University*
Judith Kandel, *California State University, Fullerton*
Jan Karlseder, *The Salk Institute for Biological Sciences*
Kazem Kashefi, *Michigan State University*
Parjit Kaur, *Georgia State University*
David Kehoe, *Indiana University*
Nemat O. Keyhani, *University of Florida*
Joan Kiely, *SUNY, Stony Brook*
David Kirchman, *University of Delaware*
Gregory Kleinheinz, *University of Wisconsin–Oshkosh*
Allan Konopka, *Pacific Northwest Laboratories*
Susan F. Koval, *University of Western Ontario*
Robert Krasner, *Providence College*
Bassam Lahoud, *Lebanese American University*
John Lammert, *Gustavus Adolphus College*
Brian Lanoil, *University of California, Riverside*
Jennifer Leavey, *Georgia Institute of Technology*
John E. Lennox, *Pennsylvania State University Altoona*
Matt Levy, *University of Texas*
Karen Lips, *Southern Illinois University*
Carol Litchfield, *George Mason University*
Alexander Loy, *University of Vienna (Austria)*
Jean Lu, *Kennesaw State University*
Bonnie Lustigman, *Montclair State University*
Ann Matthysse, *University of North Carolina*
Edward M. Marcotte, *University of Texas*
William Margolin, *University of Texas Health Sciences Center*
Cortez McBerry, *University of Cincinnati*
Mark McBride, *University of Wisconsin–Milwaukee*
William McCleary, *Brigham Young University*
Colleen McDermott, *University of Wisconsin–Oshkosh*
John McEncroe, *Colorado School of Mines*
Margaret McFall-Ngai, *University of Wisconsin*
Joseph McGonigle, *Aqua Bounty Technologies Waltham, Massachusetts*
Elizabeth McPherson, *University of Tennessee*
Michael McInerney, *University of Oklahoma*
Jay Mellies, *Reed College*
William Metcalf, *University of Illinois*
Eric S. Miller, *North Carolina State University*
Christine Moe, *Emory University*
William W. Mohn, *University Of British Columbia*
Søren Molin, *Danish Technical University (Denmark)*
Melanie Moser, *Centers for Disease Control, Atlanta*
Scott Mulrooney, *Michigan State University*
Frederick Murphy, *Centers for Disease Control, Atlanta*
Colin Murrell, *University of Warwick (England)*
Marc Mussman, *University of Vienna (Austria)*
Moon H. Nahm, *University of Alabama at Birmingham*
Aurora Nedelcu, *University of New Brunswick (Canada)*
Daniela Nicastro, *Brandeis University*
Daniel Nickrent, *Southern Illinois University*
Andrew Ogram, *University of Florida*
Aharon Oren, *Hebrew University, Jerusalem*
Mike Pentella, *University of Iowa*
Mary Perille Collins, *University of Wisconsin–Milwaukee*
David Prangishvili, *Institut Pasteur, Paris*
John C. Priscu, *Montana State University*
Vasu Punj, *University of Illinois College of Medicine*
Reinhard Rachel, *Universität Regensburg (Germany)*
Peter Radstrom, *Lund University, Sweden*
Didier Raoult, *CNRS, Marseille (France)*
Robert Reeves, *Florida State University*
Christopher Rensing, *University of Arizona*
Todd B. Reynolds, *University of Tennessee*
Anna-Louise Reysenbach, *Portland State University*
Charles Rice, *Kansas State University*
Gary P. Roberts, *University of Wisconsin*
Amy Rosenzweig, *Northwestern University*
Frank Rosenzweig, *University of Montana–Missoula*
Silvia Rossbach, *Western Michigan University*
Kathleen Ryan, *University of California, Berkeley*
Frances Sailer, *University of North Dakota*
Vladimir Samarkin, *University of Georgia*
Mahfuzar R. Sarker, *Oregon State University*
Matt Sattley, *Washington University*
Simon Scheuring, *Institut Curie, Paris*
Bernhard Schink, *Universität Konstanz (Germany)*
Tom Schmidt, *Michigan State University*
Thomas Sellers, *Centers for Disease Control, Atlanta*
Hua Shen, *Virginia State University*
Jolynn Smith, *Southern Illinois University*
Dieter Soll, *Yale University*
Nancy Spear, *Murphysboro, Illinois*

Charlotte Spencer, *University of Alberta, Canada*
James Staley, *University of Washington*
Kenneth Stedman, *Portland State University*
John G. Steiert, *Missouri State University*
John F. Stolz, *Duquesne University*
Michael Sulzinski, *University of Scranton*
Jeff Tabor, *University of California, San Francisco*
Ralph Tanner, *University of Oklahoma*
Louise Teel, *Uniformed Services University of the Health Sciences*
Rupal Thazhath, *University of Georgia*
Beth Traxler, *University of Washington*
Amy Treonis, *University of Richmond*
Henryk Urbanczyk, *University of Michigan*
Ed W. J. van Niel, *Lund University, Sweden*
Manuel Varela, *Eastern New Mexico University*
Michael Wagner, *University of Vienna (Austria)*
David Ward, *Montana State University*
Thomas Wahlund, *University of California, San Marcos*
James Walker, *University of Texas at Austin*
Valerie Watson, *West Virginia University*
Maribeth Watwood, *Northern Arizona University*
Allen T. Weber, *University of Nebraska at Omaha*
David Wessner, *Davidson College*
Kristi Westover, *Winthrop University*
Elizabeth H. White, *Centers for Disease Control, Atlanta*
William Whitman, *University of Georgia*
Carl Woese, *University of Illinois*
Shinn-Chih Wu, *National Taiwan University*
Vladimir Yurkov, *University of Manitoba (Canada)*
John Zamora, *Middle Tennessee State University*
Jill Zeilstra-Ryalls, *Oakland University*
Stephen Zinder, *Cornell University*
David A. Zuberer, *Texas A&M University*
Patty Zwollo, *College of William and Mary*

As hard as authors may try otherwise, no published textbook is completely error free. Any errors in this book, either ones of commission or omission, are solely the responsibility of the authors. In past editions, users have been kind enough to contact us when they found an error. Users should feel free to continue to do so and to contact the authors directly about errors, concerns, or any questions they may have about the book.

Michael T. Madigan (madigan@micro.siu.edu)
John M. Martinko (martinko@micro.siu.edu)
Paul V. Dunlap (pvdunlap@umich.edu)
David P. Clark (clark@micro.siu.edu)

Brief Contents

UNIT 1 Principles of Microbiology

UNIT 2 Molecular Biology of Microorganisms

UNIT 3 Microbial Diversity

UNIT 4 Metabolic Diversity and Microbial Ecology

UNIT 5 Putting Microorganisms to Work

UNIT 6 Antimicrobial Agents and Pathogenicity

UNIT 7 Immunology

UNIT 8 Diagnosing and Tracking Infectious Diseases

UNIT 9 Microbial Diseases

Contents

▶

1

Microorganisms and Microbiology

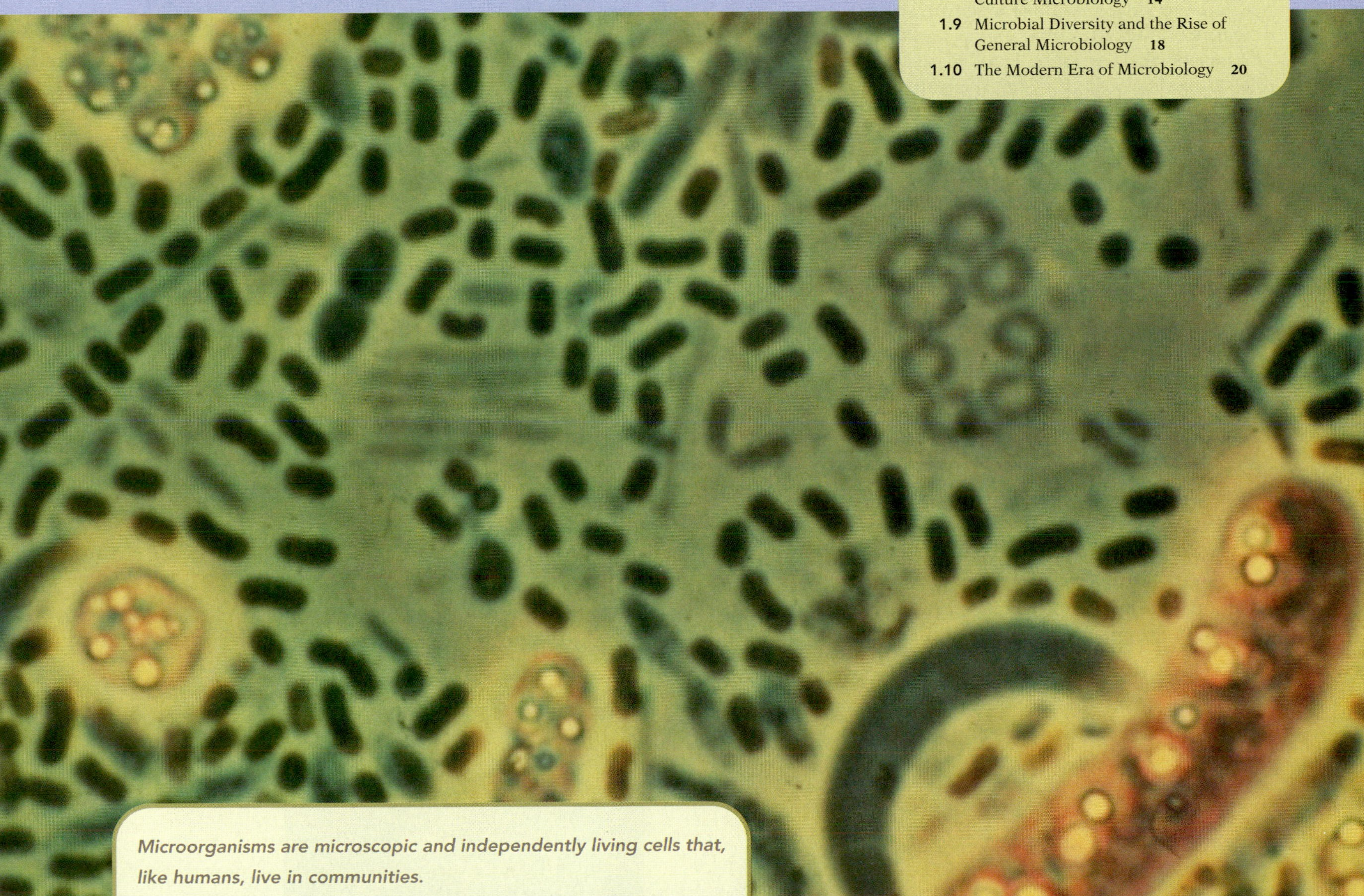

Microorganisms are microscopic and independently living cells that, like humans, live in communities.

Welcome to microbiology—the study of microorganisms. **Microorganisms** are single-celled microscopic organisms and viruses, which are microscopic but not cellular.

What is microbiology all about? Microbiology is about cells and how they work, especially the bacteria, a large group of cells of enormous basic and practical importance (**Figure 1.1**). Microbiology is about diversity and evolution, about how different kinds of microorganisms arose and why. It is about what microorganisms do in the world at large, in soils and waters, in the human body, and in animals and plants. One way or another, microorganisms affect all other life forms on Earth (Figure 1.1*b*), and thus we may think of microbiology as the foundation of the biological sciences.

Microorganisms differ from the cells of macroorganisms. The cells of macroorganisms such as plants and animals are unable to live alone in nature and exist only as parts of multicellular structures, such as the organ systems of animals or the leaves of leafy plants. By contrast, most microorganisms can carry out their life processes of growth, energy generation, and reproduction independently of other cells.

This chapter begins our journey into the microbial world. Here we discover what microorganisms are and their impact on life. We set the stage for consideration of the structure and evolution of microorganisms that will unfold in the next chapter. We also place microbiology in historical perspective, as a process of scientific discovery. From the landmark contributions of both early microbiologists and scientists practicing today, we can see microbiology at work in medicine, agriculture, the environment, and other everyday aspects of our lives.

I INTRODUCTION TO MICROBIOLOGY

In the first five sections of this chapter we introduce the field of microbiology, look at microorganisms as cells, examine where and how microorganisms live in nature, review the evolutionary history of microbial life, and examine the impact that microorganisms have had and continue to have on human affairs.

1.1 Microbiology

The science of microbiology revolves around two themes: (1) understanding basic life processes, and (2) applying our understanding of microbiology for the benefit of humankind.

As a basic biological science, microbiology uses and develops tools for probing the fundamental processes of life. Scientists have been able to gain a sophisticated understanding

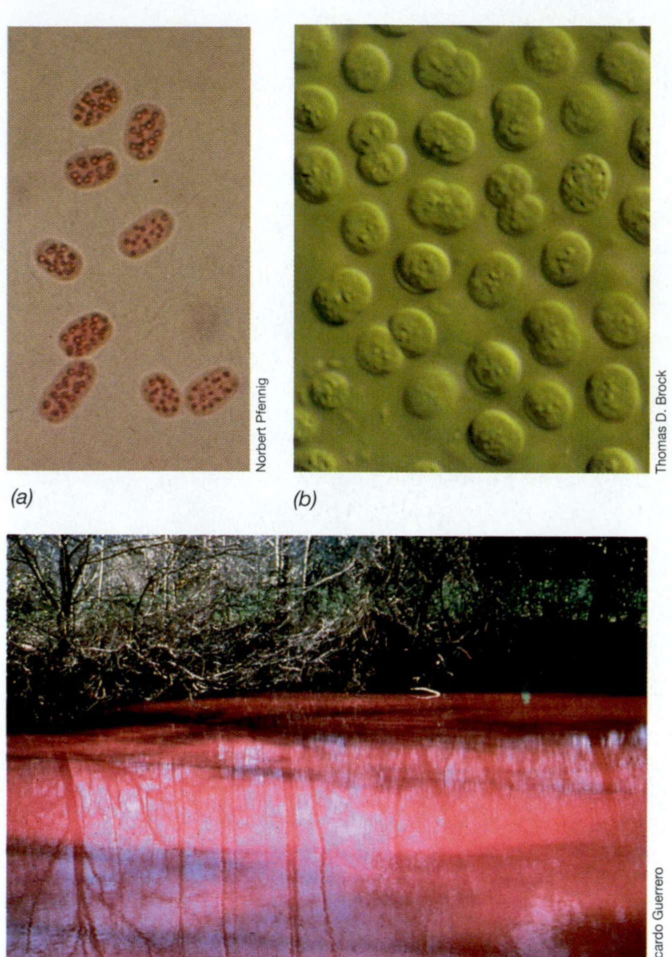

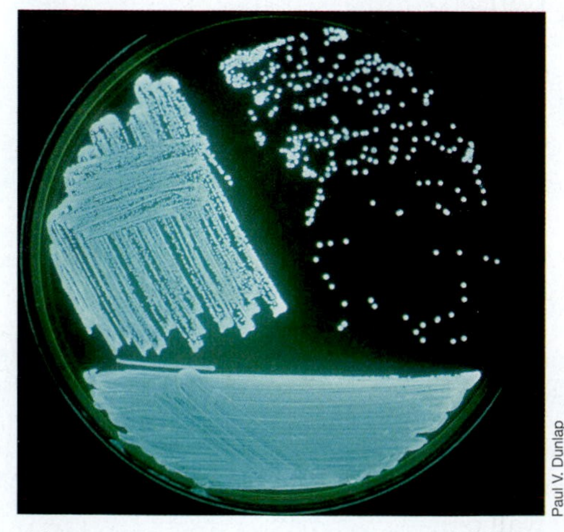

Figure 1.1 **Microorganisms.** *(a, b)* A single microbial cell can have an independent existence. Shown are photomicrographs of phototrophic (photosynthetic) microorganisms called (*a*) purple bacteria and (*b*) cyanobacteria. Purple bacteria were among the first phototrophs on Earth; cyanobacteria were the first oxygen-evolving phototrophs. Cyanobacteria oxygenated the atmosphere, opening the way for the evolution of other life forms. *(c, d)* In nature or in the laboratory, bacterial cells can grow to form extremely large populations. Shown are (*c*) a bloom of purple bacteria (compare with Figures 1.1*a* and 1.18) in a small lake in Spain, Lake Cisó, and (*d*) bioluminescent (light-emitting) cells of the bacterium *Photobacterium leiognathi* grown in laboratory culture. One milliliter of water from the lake (*c*) or one colony from the plate (*d*) contains more than 1 billion (10^9) individual cells.

of the chemical and physical basis of life from studies of microorganisms because microbial cells share many characteristics with cells of multicellular organisms; indeed, *all* cells have much in common. Moreover, microbial cells can grow to extremely high densities in laboratory culture (Figure 1.1*d*), making them readily amenable to biochemical and genetic study. These features make microorganisms excellent models for understanding cellular processes in multicellular organisms, including humans.

As an applied biological science, microbiology deals with many important practical problems in medicine, agriculture, and industry. For example, most animal and plant diseases are caused by microorganisms. Microorganisms play major roles as agents of soil fertility and in supporting domestic animal production. Many large-scale industrial processes, such as the production of antibiotics and human proteins, rely heavily on microorganisms. Thus both the detrimental and the beneficial aspects of microorganisms affect the everyday lives of humans.

The Importance of Microorganisms

In this book we will see that microorganisms play central roles in both human activities and the web of life on Earth. Although microorganisms are the smallest forms of life, collectively they constitute the largest mass of living material on Earth and carry out many chemical processes necessary for other organisms. In the absence of microorganisms, other life forms would never have arisen and could not now be sustained. Indeed, the very oxygen we breathe is the result of past microbial activity (Figure 1.1*b*). Moreover, we will see how humans, plants, and animals are intimately tied to microbial activities for the recycling of key nutrients and for degrading organic matter. No other life forms are as important as microorganisms for the support and maintenance of life on Earth.

Microorganisms existed on Earth for billions of years before plants and animals appeared (see Figure 1.6), and we will see in later chapters that the diversity of microbial life far exceeds that of the plants and animals. This huge diversity accounts for some of the spectacular properties of microorganisms. For example, we will see how microorganisms can live in places unsuitable for other organisms and how the diverse physiological capacities of microorganisms rank them as Earth's premier chemists. We will trace the evolutionary history of microorganisms and see that three huge groups of cells can be distinguished by their evolutionary relationships. And finally, we will see how microorganisms have established important relationships with other organisms, some beneficial and some harmful.

We begin our study of microbiology with a consideration of the cellular nature of microorganisms.

1.2 Microorganisms as Cells

The **cell** is the fundamental unit of life. A single cell is an entity, isolated from other cells by a membrane; many cells also contain a cell wall outside the membrane. A cell contains a variety of chemicals and subcellular structures (**Figure 1.2**). The membrane forms a compartment or "container" that is necessary to maintain the correct proportions of internal constituents in the cell and to protect it against outside forces. But the fact that a cell is a compartment does not mean that it is a *sealed* compartment. Instead, the membrane is semipermeable and thus the cell is an open, dynamic structure. Cells communicate and exchange materials with their environments, and they are constantly undergoing change. We study cell structure and function in detail in Chapters 2, 4, and 18.

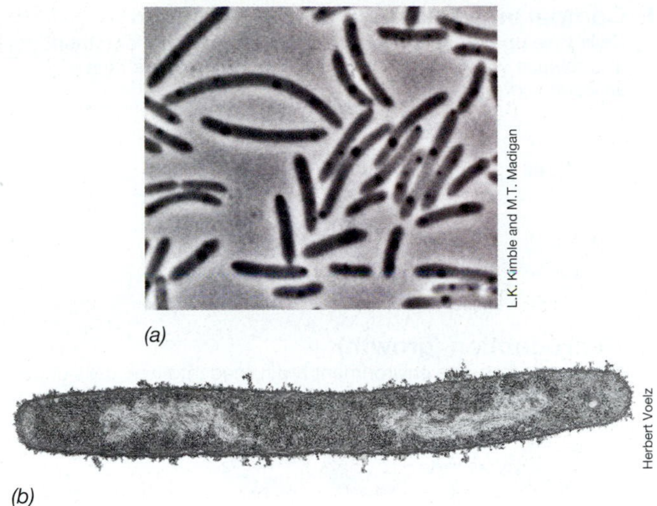

Figure 1.2 *(a)* Rod-shaped bacterial cells as seen in the light microscope; a single cell is about 1 μm in diameter. *(b)* Longitudinal section through a dividing bacterial cell as viewed with an electron microscope. The two lighter areas are nucleoids, regions in the cell containing aggregated DNA.

Cell Chemistry and Key Structures

Cells are highly organized structures that consist of an assortment of four chemical components: proteins, nucleic acids, lipids, and polysaccharides. These large molecules are called **macromolecules**, and, collectively, they make up greater than 95% of the dry weight of a cell (the remainder is a mixture of macromolecular precursors and inorganic ions). It is the precise chemistry and arrangement of macromolecules in different kinds of cells that make them distinct from one another. We can thus say that all cells have much in common but each different kind of cell is chemically unique.

There are a number of key structures in a cell. The **cytoplasmic membrane** (also called the cell membrane) is the barrier that separates the inside of the cell from the outside environment. Inside the cell membrane are various structures and chemicals suspended or dissolved in a fluid called the **cytoplasm**. The "machinery" for cell growth and function in the cytoplasm includes the nucleus or nucleoid, where the cells' **DNA** (the **genome**) is stored, and **ribosomes**, structures consisting of protein and **RNA** upon which new proteins are made in the cell. Most microbial cells also contain a cell wall. It is the cell wall rather than the cell membrane that confers structural strength on the cell and prevents it from osmotic bursting.

1. Compartmentalization and metabolism
Cells take up nutrients from the environment, transform them, and release wastes into the environment. The cell is thus an *open* system.

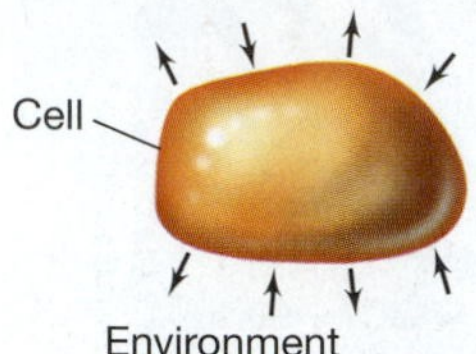

2. Reproduction (growth)
Chemicals from the environment are turned into new cells under the genetic direction of preexisting cells.

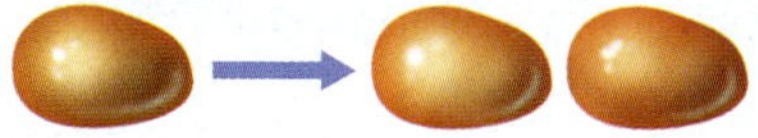

3. Differentiation
Some cells can form new cell structures such as a spore, usually as part of a cellular life cycle.

4. Communication
Cells *communicate* or *interact* by means of chemicals that are released or taken up.

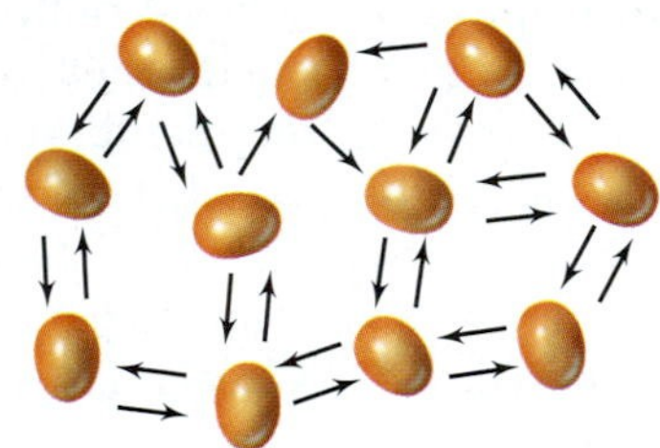

5. Movement
Some cells are capable of self-propulsion.

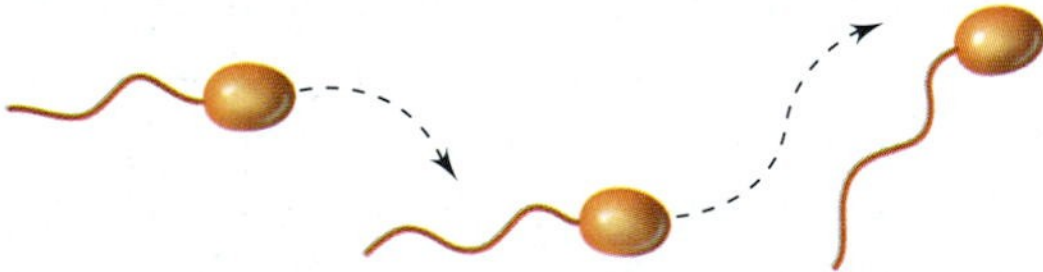

6. Evolution
Cells contain genes and *evolve* to display new biological properties. Phylogenetic trees show the evolutionary relationships between cells.

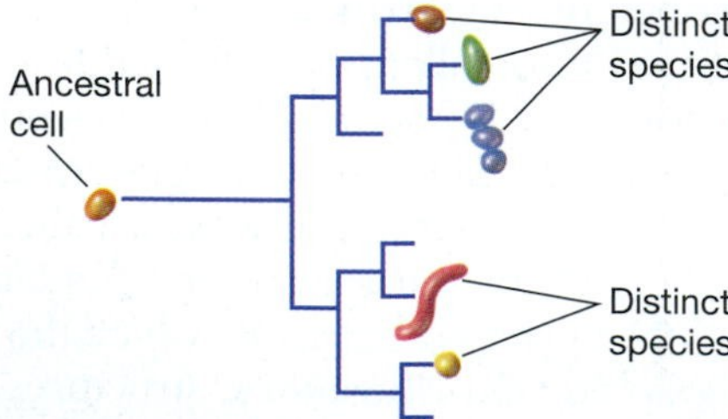

Figure 1.3 The characteristics of cellular life. Only certain cells undergo differentiation or are able to move.

Characteristics of Living Systems

What are the essential characteristics of life? What differentiates cells from inanimate objects? Our concept of a living organism is constrained by what we observe on Earth today and can deduce from the fossil record. But from our knowledge of biology thus far, we can identify several characteristics shared by most living systems. These characteristics of cellular organisms are summarized in **Figure 1.3**.

All cellular organisms show some form of **metabolism**. That is, within the framework of the physical container that defines a cell, nutrients are taken up from the environment and chemically transformed. During this process, energy is conserved and waste products eliminated. All cells show regeneration and *reproduction*. That is, a cell can repair and replace its components as needed and then accumulate multiple copies of each component before partitioning them to divide and form two cells. Some cells undergo *differentiation*, a process by which new substances or structures that modify the cell are synthesized. For example, cell differentiation is often part of a cellular life cycle in which cells form special structures, such as spores, involved in reproduction, dispersal, or survival.

Cells respond to chemical signals in their environment, including those produced by other cells. Cells "process" these signals in a variety of ways, which often trigger new activities. Cells thus exhibit *communication*. Many cells are capable of *movement* by self-propulsion; in the microbial world we will see several different mechanisms of motility. Finally, unlike nonliving structures, cells undergo *evolution;* over time, the characteristics of cells change and these changes are transmitted to their offspring.

Cells as Machines and as Coding Devices

The activities of cells can be viewed in two ways. On one hand, cells can be considered to be living machines that carry out chemical transformations. The catalysts of these chemical machines are **enzymes**, proteins that accelerate the rate of chemical reactions within the cell (**Figure 1.4**). On the other hand, cells can be considered to be coding devices, like computers. Just as computers store and process *digital* information, cells store and process *genetic* information (DNA) that is eventually passed on to offspring during reproduction (Figure 1.4). Replication and processing (transcription and translation) of the stored genetic information will be considered in detail in Chapter 7.

In reality, cells are both chemical machines and coding devices, and the link between these two cellular attributes is cell growth. Under proper conditions, a cell will grow larger and eventually divide to form two cells (Figure 1.4). In the events that lead up to cell division, all constituents in the cell double. This requires the chemical machinery of the cell to supply energy and precursors for biosynthesis of macromolecules. Also, when a cell divides, each of the two resulting cells must contain a copy of the genetic information. Thus, DNA must replicate during the growth process (Figure 1.4). The machine and coding functions of the cell must therefore be

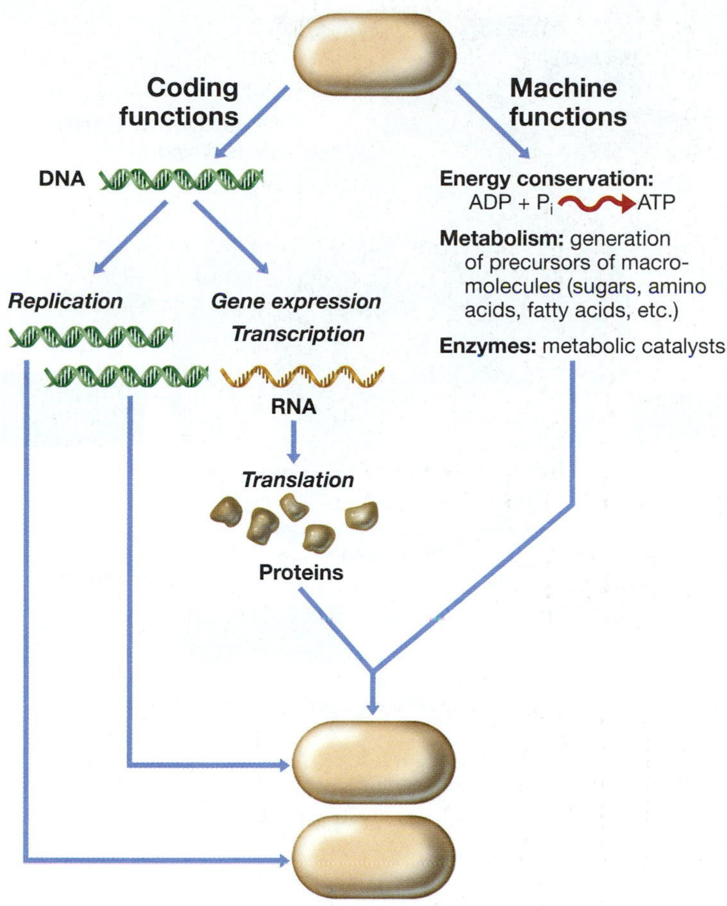

Figure 1.4 The machine and coding functions of the cell. For a cell to reproduce itself there must be energy and precursors for the synthesis of new macromolecules, the genetic instructions must be replicated such that upon division each cell receives a copy, and genes must be expressed to produce proteins and other macromolecules.

highly coordinated. Also, as we will see later, the machine and coding functions are subject to regulation that ensures substances are made in the proper order and concentrations to keep the cell optimally attuned to its environment.

1.1 and 1.2 MiniReview

Microorganisms include all microscopic organisms, including viruses. The cytoplasmic membrane is a barrier that partitions the cytoplasm away from the environment. Other major cell structures include the nucleus or nucleoid (genome), cytoplasm, ribosomes, and cell wall. Metabolism and reproduction are key features associated with the living state, and cells can be considered both chemical machines and coding devices.

- List the four classes of cellular macromolecules.
- List six characteristics of living organisms. Why might each characteristic be important to the survival of a cell?
- Compare the machine and coding functions of a microbial cell. Why is neither of value to a cell without the other?

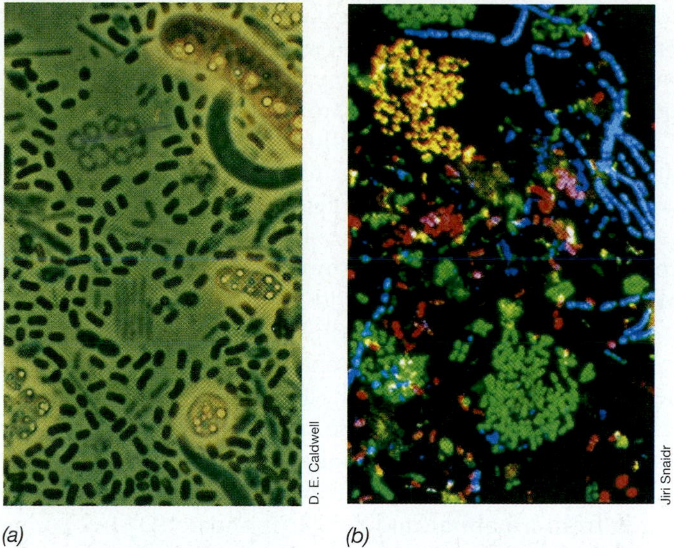

Figure 1.5 Microbial communities. *(a)* A bacterial community that developed in the depths of a small lake (Wintergreen Lake, Michigan), showing cells of various bacteria. *(b)* A bacterial community in a sewage sludge sample. The sample was stained with a series of dyes, each of which stained a specific bacterial group. From R. Amann, J. Snaidr, M. Wagner, W. Ludwig, and K.-H. Schleifer, 1996. *Journal of Bacteriology* 178: 3496–3500, Fig. 2b. © 1996 American Society for Microbiology.

1.3 Microorganisms and Their Natural Environments

In nature, microbial cells live in association with other cells in populations. Microbial populations are groups of cells derived from a single parent cell by successive cell divisions. The environment in which a microbial population lives is called its **habitat**. In microbial habitats, a population of cells rarely lives alone. Rather, cell populations live and interact with other populations in assemblages called *microbial communities* (**Figure 1.5**). The diversity and abundance of microorganisms in a microbial community is controlled by the resources (foods) and conditions (temperature, pH, oxygen content, and so on) that exist in the environment. The study of microorganisms in their natural environments is called **microbial ecology**, a major theme in this book.

Microbial Interactions

Microbial populations interact and cooperate in various ways, some beneficial and some harmful. For example, the waste products of the metabolic activities of some organisms can be nutrients for others. Microorganisms also interact with their physical and chemical environment. Habitats differ markedly in their characteristics, and a habitat that is favorable for the growth of one organism may actually be harmful for another. Collectively, we call all the living organisms, together with the physical and chemical constituents of their environment, an **ecosystem**. Major microbial ecosystems are found in aquatic environments (oceans, ponds, lakes, streams, ice, hot springs) and terrestrial environments (soil, deep subsurface environments), and in other organisms such as plants and animals.

An ecosystem is greatly influenced—even controlled—by microbial activities. Microorganisms carrying out metabolic processes remove nutrients from the ecosystem and use them to build new cells. At the same time, they excrete waste products back into the environment. Thus, over time, microbial ecosystems expand and contract, depending on the resources and conditions available; the metabolic activities of microorganisms gradually change these ecosystems, both chemically and physically. The habitat may thus change in significant ways. For example, molecular oxygen (O_2) is a vital nutrient for some microorganisms but a poison to others. If oxygen-consuming (aerobic) microorganisms remove oxygen from a habitat and render it anoxic (O_2-free), conditions may then favor the growth of anaerobic microorganisms that were present in the habitat but previously unable to grow. Thus, as resources and conditions in microbial habitats change, cell populations rise and fall, changing the habitat once again.

In later chapters, after we have learned about microbial structure and function, genetics, evolution, and diversity, we will return to a discussion of the ways in which microorganisms affect animals, plants, and the whole global ecosystem.

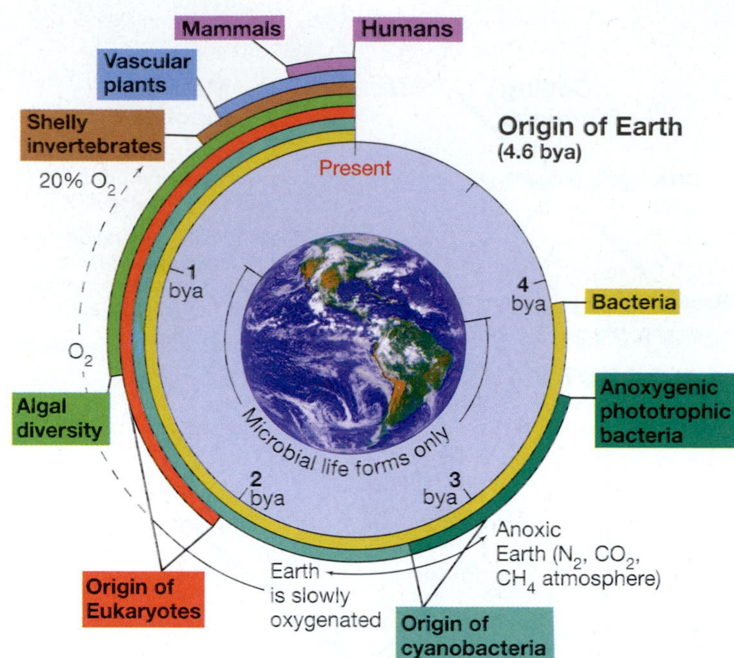

Figure 1.6 A summary of life on Earth through time. Cellular life was present on Earth about 3.8 billion years ago (bya). Cyanobacteria began the slow oxygenation of Earth about 3 bya, but current levels of oxygen in the atmosphere were not achieved until 500–800 million years ago. Eukaryotes are nucleated cells and include both microbial and multicellular organisms.

1.3 MiniReview

Microorganisms exist in nature in populations that interact with other populations to form microbial communities. The activities of microorganisms in microbial communities can greatly affect and rapidly change the chemical and physical properties of their habitats.

- What is a microbial habitat? How does a microbial community differ from a microbial population?
- How can microorganisms change the characteristics of their habitats?

1.4 The Antiquity and Extent of Microbial Life

Microorganisms were the first life forms on Earth to show the basic characteristics of living systems (Figure 1.3). We have already seen that cyanobacteria paved the way for the evolution of other life forms by producing the oxygen in Earth's atmosphere (Figure 1.1*b*). But long before cyanobacteria appeared on Earth, the planet was already teeming with life and diverse communities of microorganisms were widespread.

The First Cells

Where did the first cells come from? Were cells as we know them today the first self-replicating structures on Earth? Because all cells are constructed in similar ways, it is likely that all cells have descended from a common ancestral cell, the *universal ancestor* of all life. However, as we will see later in this book, the first self-replicating entities may not have been cells but instead small molecules of RNA. Eventually, though, evolution selected the cell as the best structural solution for supporting the fundamental characteristics of life. Once the first cells arose from nonliving materials, a process that occurred over hundreds of millions of years, their subsequent growth and division formed cell populations, and these began to interact as microbial communities. Evolution could then select for improvements and diversification of these early cells to yield the highly complex and diverse cells we see today. We will get a taste of this complexity and diversity in Chapter 2 and then consider the topic in detail in Chapters 4 and 15–18. We consider the topic of the origin of life in Chapter 14.

Life on Earth through the Ages

Earth is 4.6 billion years old. Scientists have evidence that cells first appeared on Earth between 3.8 and 3.9 billion years ago; these organisms were exclusively microbial. In fact, microorganisms were the only life on Earth for most of its history (**Figure 1.6**). Gradually, and over enormous periods of time, higher organisms appeared (throughout the book, for convenience, we will refer to life forms that evolved after microorganisms as "higher" organisms). What were some of the highlights along the way?

During the first 2 billion years or so of Earth's existence, the atmosphere was anoxic; oxygen was absent, and nitrogen (N_2), carbon dioxide (CO_2), and a few other gases were present. Only microorganisms capable of anaerobic metabolisms could survive under these conditions, but these included many different types of cells, including those that produce methane, called *methanogens*. The evolution of *phototrophic* microorganisms—organisms that harvest energy from sunlight—occurred within a billion years of the formation of Earth. The first phototrophs were relatively simple ones, such as the purple bacteria and their relatives, still widespread in anoxic habitats today (Figure 1.1*a*, *c*). Cyanobacteria evolved from these early phototrophs nearly a billion years later and began the long slow process of oxygenating the atmosphere (Figure 1.6). Triggered by increases in oxygen in the atmosphere, multicellular life forms eventually evolved culminating in the plants and animals we know today (Figure 1.6).

How do we know that events in the evolution of life occurred as depicted in Figure 1.6? The answer is that we will probably never know for sure. However, microbiologists have identified key chemical components in present-day organisms that are unique biomarkers for particular groups. Traces of many of these biomarkers can still be found in ancient rocks. The time line shown in Figure 1.6 pieces together what we know of molecular fossils from rocks of specific ages. Over time, microorganisms diversified and came to colonize every habitat on Earth that would support life, including many habitats unsuitable for other life forms. This brings us to the current distribution of microbial life on Earth. What is this picture like?

The Extent of Microbial Life

Microbial life is all around us. Examination of natural materials such as soil or water invariably reveals microbial cells. But unusual habitats such as boiling hot springs and glacial ice are also teeming with microorganisms. Although widespread on Earth, such tiny cells may seem inconsequential. But if we could count them all, what number would we reach?

Estimates of total microbial cell numbers on Earth are on the order of 5×10^{30} cells. The total amount of carbon present in this very large number of very small cells equals that of all plants on Earth (and plant carbon far exceeds animal carbon). But in addition, the collective contents of nitrogen and phosphorous in microbial cells is more than 10 times that in all plant biomass.

Thus, microbial cells, small as they are, constitute the major portion of biomass on Earth and are key reservoirs of essential nutrients for life. An equally startling revelation is that most microbial cells do not reside on Earth's surface but instead lie underground in the oceanic and terrestrial subsurfaces. Depths up to about 10 km under Earth's surface appear to be hospitable for microbial life. We will see later that these buried microbial habitats support diverse populations of microbial cells that make their livings in unusual ways and grow extremely slowly. However, because Earth's subsurface is a relatively unexplored frontier, there is much left for microbiologists to discover and understand about the life forms that dominate Earth's biology.

1.4 MiniReview

Diverse microbial populations were widespread on Earth for billions of years before higher organisms appeared, and today the cumulative microbial biomass on Earth exceeds that of higher organisms. Cyanobacteria in particular were important because they oxygenated the atmosphere.

- Were the earliest life forms cellular? Why is it that cellular life is the only form of life we see on Earth today?
- How old is Earth, and when did cellular life forms first appear? How can we use science to reconstruct the sequence of organisms that appeared on Earth?
- Where are most microbial cells located on Earth?

1.5 The Impact of Microorganisms on Humans

Microbiologists have been highly successful in discovering how microorganisms work, increasing their beneficial effects, and curtailing their harmful effects. Microbiology has thus greatly advanced human health and welfare. An overview of the impact of microorganisms on human affairs is shown in **Figure 1.7**.

Microorganisms as Disease Agents

The statistics summarized in **Figure 1.8** show the microbiologist's success in controlling microorganisms. These data compare the present causes of death in the United States with those of 100 years ago. At the beginning of the twentieth century, the major causes of death were infectious diseases, which are caused by microorganisms called **pathogens**. Children and the aged in particular succumbed in large numbers to microbial diseases. Today, however, infectious diseases are much less lethal, at least in developed countries. Control of infectious disease has come from increased understanding of disease processes, improved sanitary and public health practices, and the use of antimicrobial agents. As we will see later in this chapter, the science of microbiology had its roots in the study of infectious disease.

Although many infectious diseases can now be controlled, microorganisms can still be a major threat to survival, even in developed countries. Consider, for example, the individual dying slowly of a microbial infection as a consequence of acquired immunodeficiency syndrome (AIDS) or the individual infected with a multiple drug-resistant pathogen. In many developing countries, microbial diseases are still the major causes of death. Although the worldwide eradication of smallpox was a stunning triumph for medical science, millions still die yearly from other microbial diseases such as malaria, tuberculosis, cholera, African sleeping sickness, measles, pneumonia and other respiratory diseases, and diarrheal syndromes. In addition to these, humans worldwide are under threat from diseases that could emerge suddenly, such as

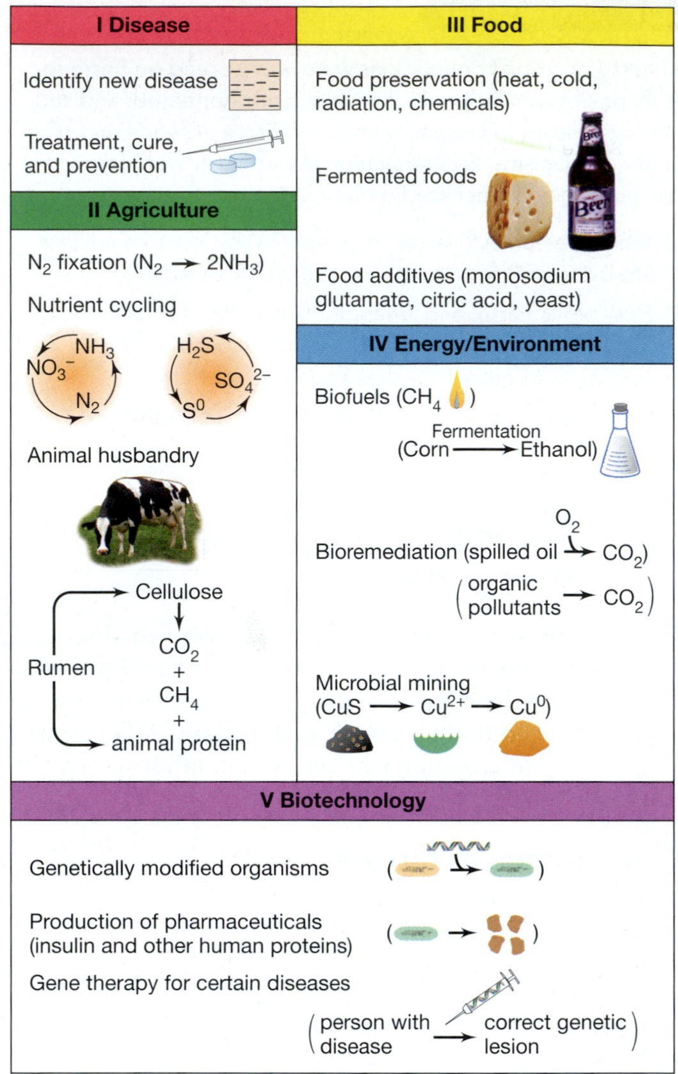

Figure 1.7 The impact of microorganisms on humans. Although many people think of microorganisms in the context of infectious diseases, few microorganisms actually cause disease. Microorganisms do, however, affect many other aspects of our lives as shown here.

bird flu, a viral disease of birds that has the potential to infect alternate hosts and spread through a population quickly. Exotic and rare diseases such as ebola hemorrhagic fever could also spread easily to developed countries because today global travel is so common. And if this weren't enough, consider the threat to humans from those who would deploy microbial bioterrorism agents. Clearly, microorganisms are still serious health threats to humans.

Although we should appreciate the powerful threat posed by microorganisms, in reality, most microorganisms are not harmful to humans. In fact, the vast majority of microorganisms cause no harm whatsoever to higher organisms but instead are beneficial—and in many cases even essential—to human welfare and the functioning of the planet. We consider these aspects of microorganisms now.

Microorganisms and Agriculture

Our whole system of intensive agriculture depends in many important ways on microbial activities (Figure 1.7). For example, a number of major crops are plants called *legumes*. Legumes live in close association with bacteria that form structures called *nodules* on their roots. In the root nodules, these bacteria convert atmospheric nitrogen (N_2) into fixed nitrogen (NH_3) that the plants use for growth. Thanks to the activities of these nitrogen-fixing bacteria, the legumes have no need for costly and polluting nitrogen fertilizers.

Also of major agricultural importance are the microorganisms that are essential for the digestive process in ruminant animals such as cattle and sheep. These important farm animals have a special digestive vessel called the *rumen* in which dense populations of microorganisms carry out the digestion of cellulose, the major component of plant cell walls. Without these microorganisms, cattle and sheep could not thrive on cellulose-rich but otherwise nutrient-poor substances such as grass and hay.

Microorganisms also play key roles in the cycling of important nutrients in plant nutrition, particularly those of carbon, nitrogen, and sulfur. Microbial activities in soil and water convert these elements to forms that are readily assimilated by plants. However, in addition to benefiting agriculture, microorganisms can also have negative effects; microbial diseases of plants and animals cause major economic losses in the agricultural industry. For example, diseases such as mad cow disease can have dramatic effects on the marketability of beef, and microbial diseases of crop plants can greatly reduce the yield of grain or other valuable plant products.

Microorganisms and Food

Once plants and animals for consumption are produced, they must be delivered in wholesome form to consumers. Microorganisms play important roles in the food industry (Figure 1.7). Food spoilage alone results in huge economic losses each year. Indeed, the canning, frozen-food, and dried-food industries emerged to preserve foods that would otherwise undergo microbial spoilage. Foodborne disease is also a consideration. Because food fit for human consumption can support the growth of microorganisms including pathogens, foods must be properly prepared and monitored to avoid transmission of disease-causing microorganisms.

However, not all microorganisms in foods have harmful effects on food products or those who eat them. For example, many dairy products depend on microbial transformations, including the fermentations that yield cheeses, yogurt, and buttermilk. Sauerkraut, pickles, and some sausages also owe their existence to microbial fermentations. Moreover, baked goods and alcoholic beverages rely on the fermentative activities of yeast, generating carbon dioxide (CO_2) to raise the dough and alcohol as a key ingredient, respectively. Many of these fermentations are discussed in Chapters 21, 25, and 37.

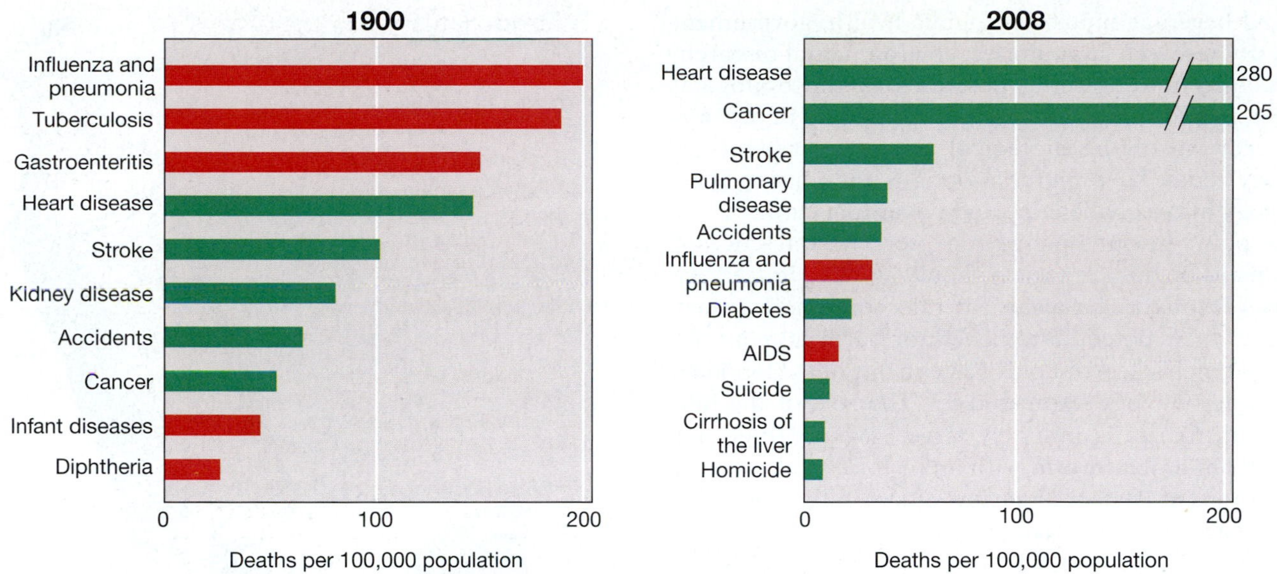

Figure 1.8 Death rates for the leading causes of death in the United States: 1900 and 2008 Infectious diseases were the leading causes of death in 1900, whereas today they are much less significant. Microbial diseases are shown in red, nonmicrobial causes of death in green. Data from the United States National Center for Health Statistics.

Microorganisms, Energy, and the Environment

Microorganisms play major roles in energy production (Figure 1.7). Natural gas (methane) is a product of microbial activity, arising from the metabolism of methanogenic microorganisms. Phototrophic microorganisms can harvest light energy for the production of biomass, energy stored in living organisms. Microbial biomass and waste materials, such as domestic refuse, surplus grain, and animal wastes, can be converted to biofuels, such as methane and ethanol, by the activities of microorganisms. Ethanol produced by the microbial fermentation of glucose from sugarcane or cornstarch is the major motor fuel in some countries, such as Brazil, and is becoming a more important component of fuels in the United States.

Microorganisms can also be used to help clean up pollution created by human activities, a process called *microbial bioremediation* (Figure 1.7). Microorganisms can be employed to consume spilled oil, solvents, pesticides, and other environmentally toxic pollutants. With time, if the input of pollutants ceases, polluted areas may be cleaned without human intervention through the activities of naturally occurring microorganisms. Bioremediation simply accelerates the natural cleanup process by introducing pollutant-consuming microorganisms or specific nutrients that help microorganisms degrade the pollutants. Within the huge diversity of microorganisms on Earth, vast genetic resources exist. Researchers are now tapping into these genes to develop new solutions to ever more challenging pollution problems.

Microorganisms and Their Genetic Resources

Besides cleaning up the environment, the genetic riches of the microbial world can be put to work to make products of commercial value. Microorganisms have been used for hundreds of years to produce fermented milk products, alcoholic beverages, and the like. More recently, microorganisms have been grown on massive scales for the production of antibiotics, specific enzymes, and various chemicals. But today we live in the world of "biotech." Biotechnology employs genetically modified microorganisms to synthesize products of high commercial value (Chapter 26). Biotechnology uses the tools of genetic engineering—the artificial manipulation of genes and gene products (Figure 1.7). Genes from any source can be manipulated and modified using microorganisms and their enzymes as molecular tools. For instance, human insulin, a hormone found in abnormally low amounts in people with the disease diabetes, is produced today by genetically engineered bacteria into which human insulin genes have been inserted. Using **genomics**, the sequencing and analysis of the genomes of organisms (Chapter 13), one can search the genome of any organism for genes encoding proteins of commercial interest. It is now a routine operation to clone a gene of interest into a suitable host and produce the protein it encodes on a scale to meet commercial demand.

Microbiology as a Career

The field of microbiology is ripe with opportunities for those who seek fulfilling and rewarding careers in science. For example, microbiologists are on the front lines of clinical medicine in developing and deploying methods for the diagnosis and treatment of infectious diseases. Research and development in pharmaceutical, chemical and biochemical, and biotechnology companies is often the foundation of these achievements and is itself fueled by the scientific activities of microbiologists. In addition, microbiologists play key roles in the food and beverage industries, public health, government and university research laboratories, environmental research, and in teaching and training positions in the biological sciences. So, in addition to grasping the principles of perhaps the most exciting of the biological sciences, students of microbiology today have multiple options for turning their love of biology into a challenging and rewarding career.

At this point in our opening chapter, the influence of microorganisms on human society should be clear. Indeed, we have many reasons to be aware of microorganisms, their activities, and their potential to benefit or harm humans. As the eminent French scientist Louis Pasteur, one of the founders of modern microbiology, expressed it: "The role of the infinitely small in nature is infinitely large." We continue our journey to the microbial world with an historical overview of the contributions of Pasteur and a few others to the science of microbiology as we know it today.

1.5 MiniReview

Microorganisms can be both beneficial and harmful to humans. Although we tend to emphasize harmful microorganisms (infectious disease agents), many more microorganisms in nature are beneficial than harmful.

- In what ways are microorganisms important in the food and agricultural industries?
- List two fuels that are made by microorganisms.
- What is biotechnology and how might it improve the lives of humans?

II PATHWAYS OF DISCOVERY IN MICROBIOLOGY

Like any science, microbiology owes much to its past. Although able to claim early roots, the science of microbiology didn't really develop until the nineteenth century. Since that time, the field has exploded and spawned several new but related fields. We retrace these pathways of discovery now.

1.6 The Historical Roots of Microbiology: Hooke, van Leeuwenhoek, and Cohn

Although the existence of creatures too small to be seen with the naked eye had long been suspected, their discovery was linked to the invention of the microscope. Robert Hooke (1635–1703), an English mathematician and natural historian, was also an excellent microscopist. In his famous book *Micrographia* (1665), the first book devoted to microscopic observations, Hooke illustrated, among many other things, the fruiting structures of molds (**Figure 1.9**). This was the first known description of microorganisms. The first person to see bacteria was the Dutch draper and amateur microscope

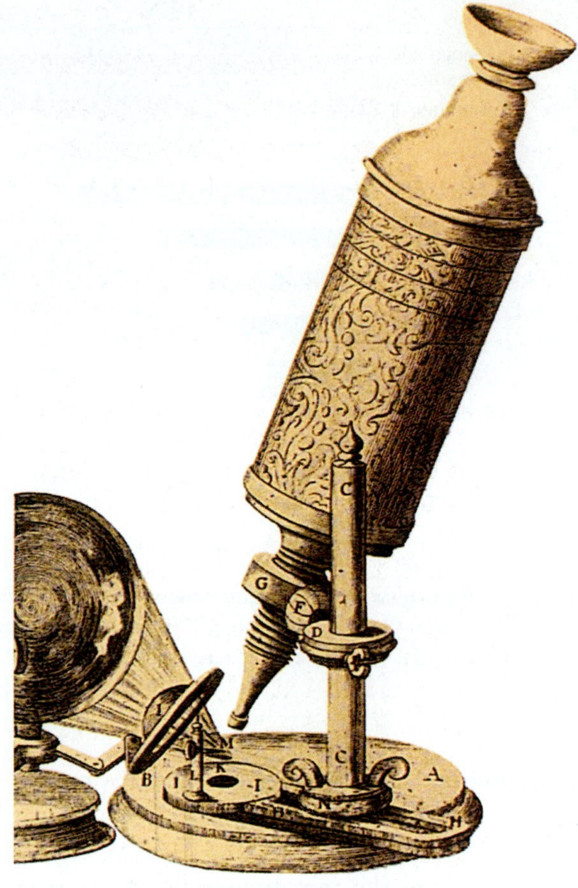

(a)

(b)

Figure 1.9 Robert Hooke and early microscopy. *(a)* A drawing of the microscope used by Robert Hooke in 1664. The objective lens was fitted at the end of an adjustable bellows (G), with illumination focused on the specimen by a single lens (1). *(b)* A drawing by Robert Hooke. This drawing, published in *Micrographia* in 1655, is the first description of a microorganism. The organism is a bluish-colored mold growing on the surface of leather. The round structures (sporangia) contain spores of the mold.

builder Antoni van Leeuwenhoek (1632–1723). In 1684, van Leeuwenhoek, who was aware of the work of Hooke, used extremely simple microscopes of his own construction (**Figure 1.10**) to examine the microbial content of a variety of natural substances.

Van Leeuwenhoek's microscopes were crude by today's standards, but by careful manipulation and focusing he was able to see bacteria, microorganisms considerably smaller than molds. He discovered bacteria in 1676 while studying pepper–water infusions. He reported his observations in a series of letters to the prestigious Royal Society of London, which published them in 1684 in English translation. Drawings of some of van Leeuwenhoek's "wee animalcules," as he referred to them, are shown in Figure 1.10*b*.

As years went by, van Leeuwenhoek's observations were confirmed by others, but progress in understanding the nature and importance of these tiny organisms remained slow for nearly the next 150 years. Only in the nineteenth century did improved microscopes become widely distributed, and about this time the extent and nature of microbial life forms became more apparent.

In the mid- to late nineteenth century major advances were made in the new science of microbiology, primarily because of the attention that was given to two major questions that pervaded biology and medicine at the time: (1) does spontaneous generation occur and (2) what is the nature of infectious disease. Answers to these penetrating questions emerged from the work of two giants in the fledgling field of microbiology: the French chemist Louis Pasteur and the German physician Robert Koch. But before we explore their work, let us briefly consider the groundbreaking work of a German botanist, Ferdinand Cohn, a contemporary of Pasteur and Koch and the founder of the field we now call *bacteriology*.

Ferdinand Cohn and the Science of Bacteriology

Ferdinand Cohn (1828–1898) was born in Breslau (now in Poland). He was trained as a botanist and became an excellent microscopist. His interests in microscopy naturally led him to the study of unicellular plants—the algae—and later to photosynthetic bacteria. Cohn believed that all bacteria, even those lacking photosynthetic pigments, were members of the plant kingdom, and his microscopic studies gradually drifted away from plants and algae to bacteria, including the large sulfur bacterium *Beggiatoa* (**Figure 1.11**).

Cohn was particularly interested in heat resistance in bacteria, which led him to discover the important group of bacteria that form endospores. We now know that bacterial endospores are extremely heat resistant. Cohn described the life cycle of the endospore-forming bacterium *Bacillus* (vegetative cell → endospore → vegetative cell) and discovered that vegetative cells of *Bacillus* but not their endospores were killed by boiling. Indeed, Cohn's discovery of endospores helped explain why his contemporaries, such as the Irish scientist John Tyndall, had found boiling to be an unreliable means of preventing fluid infusions from supporting microbial growth.

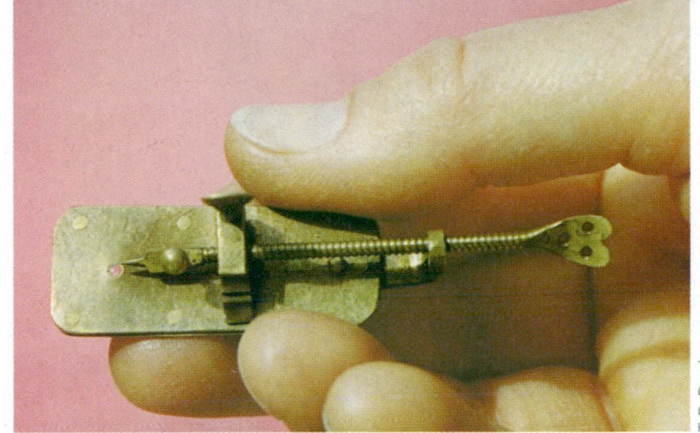

(a)

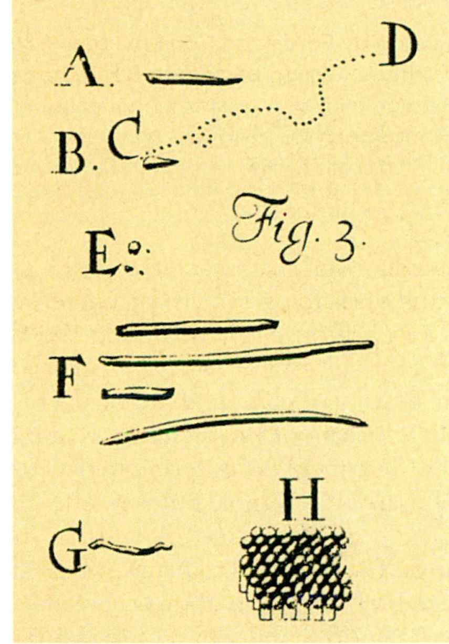

(b)

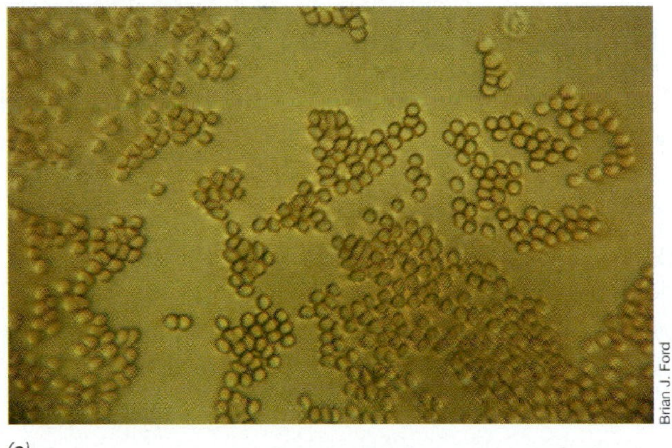

(c)

Figure 1.10 The van Leeuwenhoek microscope. *(a)* A replica of van Leeuwenhoek's microscope. The lens is mounted in the brass plate adjacent to the tip of the adjustable focusing screw. *(b)* Antoni van Leeuwenhoek's drawings of bacteria, published in 1684. Even from these relatively crude drawings we can recognize several shapes of common bacteria: A, C, F, and G, rod shaped; E, spherical or coccus-shaped; H, cocci packets. *(c)* Photomicrograph of a human blood smear taken through a van Leeuwenhoek microscope. Red blood cells are clearly apparent. A single red blood cell is about 6 μm in diameter.

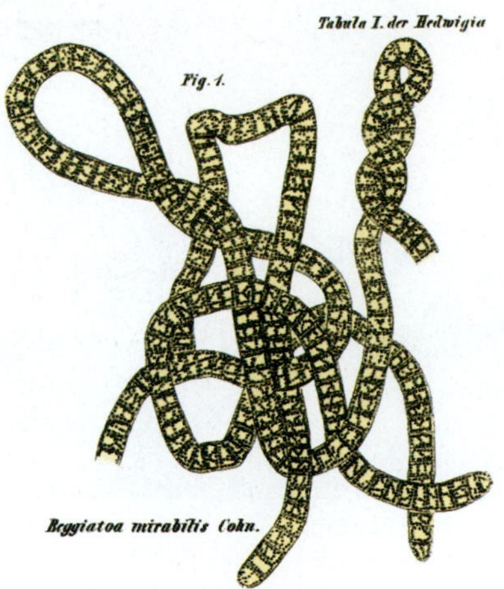

Figure 1.11 Drawing by Ferdinand Cohn made in 1866 of the large filamentous sulfur-oxidizing bacterium *Beggiatoa mirabilis*. The small granules inside the cell consist of elemental sulfur, produced from the oxidation of hydrogen sulfide (H_2S). Cohn was the first to identify the granules as sulfur. A cell of *B. mirabilis* is about 15 μm in diameter.

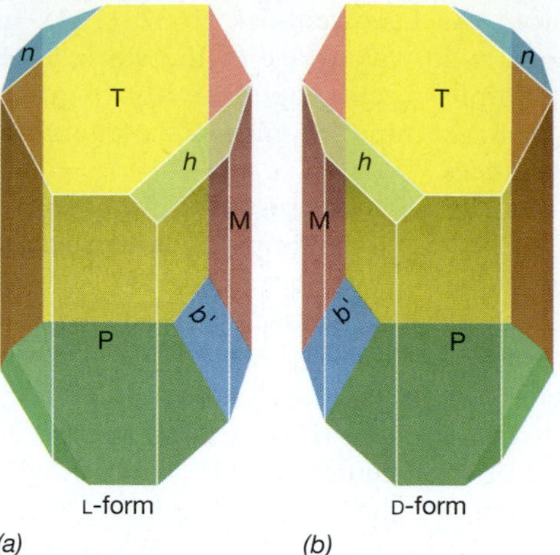

Figure 1.12 Louis Pasteur's drawings of tartaric acid ($C_4H_6O_6$) crystals that illustrated his famous paper on optical activity. *(a)* Left-handed crystal (L form). *(b)* Right-handed crystal (D form). Note that the two crystals are mirror images (Section 3.6). The letters on the faces of the crystals were Pasteur's way of labeling the mirror image faces of the two crystals. Color has been added to show the mirror image faces more clearly.

Cohn continued to work with bacteria until his retirement. He laid the groundwork for a system of bacterial classification, including an early attempt to define the nature of a bacterial species, an issue still unresolved today, and founded a major scientific journal of plant and microbial biology. Cohn was also a strong advocate of the techniques and research of Robert Koch, the first medical microbiologist. Cohn also is credited with helping devise simple but very effective methods for preventing the contamination of culture media, such as the use of cotton for closing flasks and tubes. These methods were later used by Koch and allowed him to make rapid progress in the isolation and characterization of several disease-causing bacteria (Section 1.8).

1.6 MiniReview

Robert Hooke was the first to describe microorganisms, and Antoni van Leeuwenhoek was the first to describe bacteria. Ferdinand Cohn founded the field of bacteriology and discovered bacterial endospores.

- What prevented the science of microbiology from developing before the era of van Leeuwenhoek?
- What major discovery emerged from Cohn's study of heat resistance in microorganisms?

1.7 Pasteur and the Defeat of Spontaneous Generation

The mid- to late nineteenth century saw the science of microbiology blossom. The concept of spontaneous generation was crushed and the science of pure culture microbiology emerged. Several scientific giants emerged in this era, and the first was the Frenchman Louis Pasteur (1822–1895), a contemporary of Cohn.

Optical Isomers and Fermentations

Pasteur was trained as a chemist and was one of the first scientists to recognize the significance of optical isomers (Section 3.6). A molecule is optically active if a pure solution or crystal of the molecule diffracts light in only one direction. Pasteur studied crystals of tartaric acid that he separated by hand into those that bent a beam of polarized light to the left and those that bent the beam to the right (**Figure 1.12**). Tartrate was chosen because it was an undesirable side product of the wine-making process, and so was readily available in France, and also because solutions of tartaric acid crystallize particularly well. Pasteur found that the mold *Aspergillis* metabolizes only D-tartrate and not its optical isomer L-tartrate. For Pasteur, the fact that a living organism discriminated between optical isomers was of profound significance. He began to see living processes as inherently asymmetric and nonliving chemical processes as not asymmetric. To Pasteur, only living things could selectively produce or consume optical isomers.

Pasteur's thoughts on the asymmetry of life carried over into his work on fermentations and, eventually, spontaneous generation. At the invitation of a local industrialist who was having problems making alcohol by the fermentation of beets, Pasteur began a detailed study of the mechanism of the alcoholic fermentation, at that time thought to be a strictly chemical process. The yeast cells in the fermenting broth were thought to be a complex chemical substance and a result, rather than a catalyst, of the fermentation. One of the side products of the beet fermentation is amyl alcohol, and Pasteur tested the fermenting juice and found the amyl alcohol to be optically active. Microscopic observations and other simple but rigorous experiments convinced Pasteur that the alcoholic fermentation was catalyzed by living yeast cells. Indeed, in Pasteur's own words: ". . . fermentation is associated with the life and structural integrity of the cells and not with their death and decay." From this foundation, Pasteur began a series of classic experiments on spontaneous generation, experiments that are forever linked to his name and to the science of microbiology.

Spontaneous Generation

The concept of **spontaneous generation** had existed since biblical times. The basic idea of spontaneous generation can easily be understood. For example, if food is allowed to stand for some time, it putrefies. When the putrefied material is examined microscopically, it is found to be teeming with bacteria and perhaps even higher organisms such as maggots and worms. Where do these organisms that are not apparent in the fresh food come from? Some people said they developed from seeds or germs that entered the food from air. Others said they arose spontaneously from nonliving materials, that is, spontaneous generation. Who was right? Keen insight was necessary to solve this controversy, and this was exactly the kind of problem that appealed to Louis Pasteur.

Pasteur was a powerful opponent of spontaneous generation. Following his discoveries about fermentation, Pasteur showed that microorganisms closely resembling those observed in putrefying materials could be found in air. Pasteur concluded that the organisms found in putrefying materials originated from microorganisms present in the air and on the surfaces of the containers that held the materials. He postulated that cells are constantly being deposited on all objects and that they grow when conditions are favorable. Furthermore, Pasteur reasoned that if food were treated in such a way as to destroy all living organisms contaminating it, that is, if it were rendered **sterile** and then protected from further contamination, it should not putrefy.

Pasteur used heat to eliminate contaminants. Other workers had shown that when a nutrient solution was sealed in a glass flask and heated to boiling for several minutes, it did not support microbial growth (of course, only if endospores were not present; see discussion of Cohn). Killing all the bacteria or other microorganisms in or on objects is a process we now call *sterilization*. Proponents of spontaneous generation criticized such experiments by declaring that "fresh air" was necessary for the phenomenon to occur. Boiling, so they claimed, in some way affected the air in the sealed flask so that it could no longer support spontaneous generation. In 1864 Pasteur countered this objection simply and brilliantly by constructing a swan-necked flask, now called a *Pasteur flask* (**Figure 1.13**). In such a flask nutrient solutions could be heated to boiling and sterilized. However, after the flask was cooled, air was allowed to reenter, but bends in the neck (the "swan neck" design) prevented particulate matter (containing microorganisms) from entering the main body of the flask and causing putrefaction.

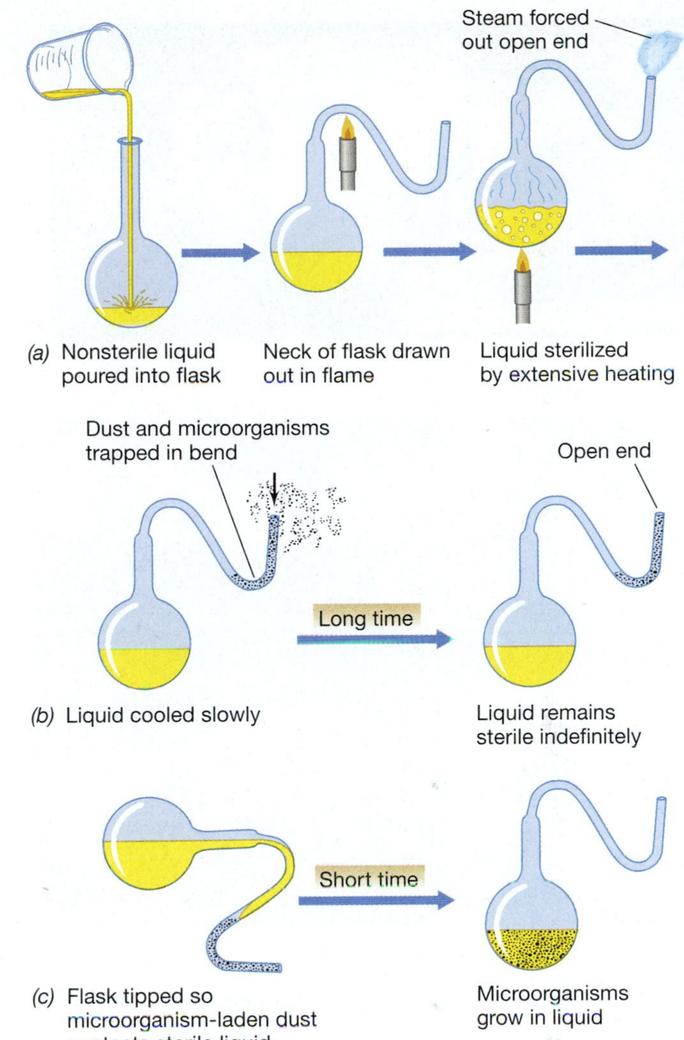

Figure 1.13 The defeat of spontaneous generation: Pasteur's experiment with the swan-necked flask. *(a)* Sterilizing the contents of the flask. *(b)* If the flask remained upright, no microbial growth occurred. *(c)* If microorganisms trapped in the neck reached the sterile liquid, microbial growth ensued.

Broth sterilized in a Pasteur flask did not putrefy, and microorganisms never appeared in the flask as long as the neck did not contact the sterile liquid. If, however, the flask was tipped to allow the sterile liquid to contact the contaminated neck of the flask (Figure 1.13*c*), putrefaction occurred and the liquid soon teemed with microorganisms. This simple experiment effectively settled the controversy surrounding spontaneous generation, and the science of microbiology was able to move ahead on firm footing. Incidentally, Pasteur's work also led to the development of effective sterilization procedures that were eventually refined and carried over into both basic and applied microbiological research. Food science also owes a debt to Pasteur, as his principles are applied today in the canning and preservation of milk and other foods (pasteurization). **www.microbiologyplace.com Online Tutorial 1.1: Pasteur's Experiment**

(a)

(b)

Figure 1.14 Louis Pasteur and symbols of his contributions to microbiology. *(a)* A French 5-franc note. The note contains a painting of Pasteur and symbols of his many scientific accomplishments. The shepherd boy Jean Baptiste Jupille is shown dispatching a rabid dog who had attacked a group of children. Pasteur's rabies vaccine saved Jupille's life. In France, the franc preceded the euro as a currency. *(b)* The Pasteur Institute, Paris France. Photo of the original structure built for Pasteur by the French government and opened in 1888. The Pasteur Institute today is a campus of several buildings. The Pasteur crypt and museum displaying Pasteur's original swan-neck flasks and other scientific equipment is located in the original building.

Other Accomplishments of Louis Pasteur

Pasteur went on to many other triumphs in microbiology and medicine beyond his seminal work on spontaneous generation. Some highlights include his development of vaccines for the diseases anthrax, fowl cholera, and rabies during a very scientifically productive period in his life from 1880 to 1890. Pasteur's work on rabies was his most famous success, culminating in July of 1885 with the first administration of a rabies vaccine to a human, a young French boy named Joseph Meister who had been bitten by a rabid dog. In those days, a bite from a rabid animal was akin to a death sentence. News of the success of Meister's vaccination, and that of a young shepherd boy, Jean Baptiste Jupille (**Figure 1.14a**), administered shortly thereafter, spread quickly, and within a year nearly 2500 people had come to Paris to be treated with Pasteur's rabies vaccine.

Pasteur's fame from his rabies research was legendary and led the French government to build the Pasteur Institute in Paris in 1888. Originally established as a clinical center for treatment of rabies and other contagious diseases, the Pasteur Institute is today a major biomedical research center focused on antiserum and vaccine production (Figure 1.14*b*). The medical and veterinary breakthroughs of Pasteur were not only highly significant in their own right but helped solidify the concept of the germ theory of disease, whose principles were being developed at about this same time by a second giant of the era, Robert Koch.

1.7 MiniReview

Louis Pasteur is best remembered for his ingenious experiments showing that living organisms were not spontaneously generated from nonliving matter. Pasteur's work in this area led to many of the basic techniques central to the science of microbiology, including the concept and practice of sterilization.

- Define the term sterile.
- How did Pasteur's swan-neck flask experiment show that the concept of spontaneous generation was invalid?

1.8 Koch, Infectious Disease, and the Rise of Pure Culture Microbiology

Proof that microorganisms could cause disease provided perhaps the greatest impetus for the development of the science of microbiology. Even in the sixteenth century it was thought that something that induced a disease could be transmitted from a diseased person to a healthy person. After the discovery of microorganisms, it was widely believed that they were responsible, but definitive proof was lacking. Improvements in sanitation by Ignaz Semmelweis and Joseph Lister provided indirect evidence for the importance of microorganisms in causing human diseases, but it was not until the work of a German physician, Robert Koch (1843–1910), that the concept of infectious disease was given experimental support.

The Germ Theory of Disease and Koch's Postulates

In his early work Koch studied anthrax, a disease of cattle and occasionally of humans. Anthrax is caused by an endospore-forming bacterium called *Bacillus anthracis*. By careful microscopy and by using special stains, Koch established that the bacteria were always present in the blood of an animal that was succumbing to the disease. However, Koch reasoned that mere association of the bacterium with the disease was not proof that it actually caused the disease. Instead, the bacterium might be a result of the disease. How could cause and effect be linked? With anthrax Koch sensed an opportunity to study cause and effect experimentally, and his results formed the standard by which infectious diseases have been studied ever since.

Koch used mice as experimental animals. Using all of the proper controls, Koch demonstrated that when a small

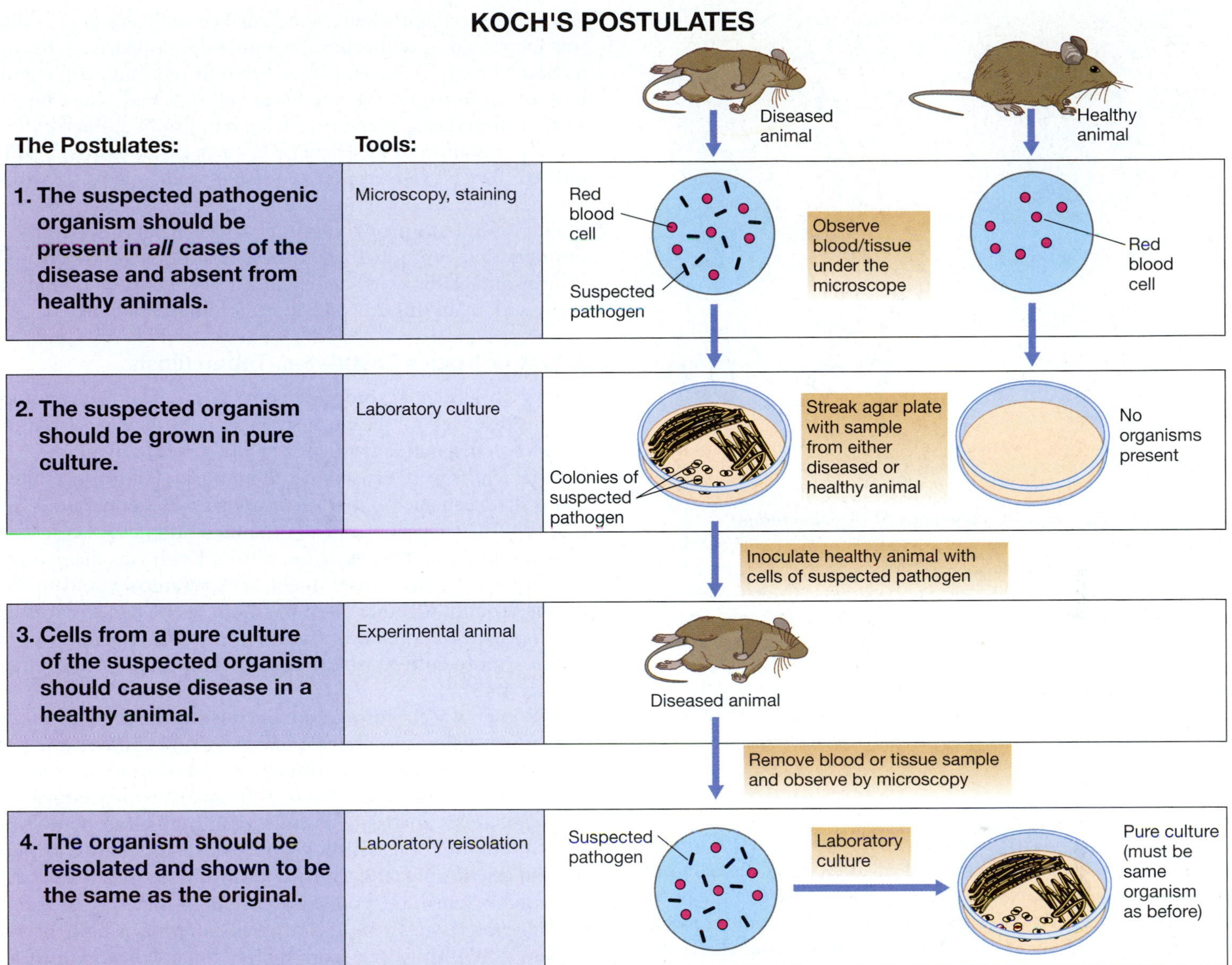
KOCH'S POSTULATES
Diseased animal
Healthy animal
The Postulates:
Tools:
1. The suspected pathogenic organism should be present in *all* cases of the disease and absent from healthy animals.
Microscopy, staining
Red blood cell
Suspected pathogen
Observe blood/tissue under the microscope
Red blood cell
2. The suspected organism should be grown in pure culture.
Laboratory culture
Colonies of suspected pathogen
Streak agar plate with sample from either diseased or healthy animal
No organisms present
Inoculate healthy animal with cells of suspected pathogen
3. Cells from a pure culture of the suspected organism should cause disease in a healthy animal.
Experimental animal
Diseased animal
Remove blood or tissue sample and observe by microscopy
4. The organism should be reisolated and shown to be the same as the original.
Laboratory reisolation
Suspected pathogen
Laboratory culture
Pure culture (must be same organism as before)

Figure 1.15 **Koch's postulates for proving that a specific microorganism causes a specific disease.** Note that following isolation of a pure culture of the suspected pathogen, a laboratory culture of the organism should both initiate the disease and be recovered from the diseased animal. Establishing the correct conditions for growing the pathogen is essential, otherwise it will be missed.

amount of blood from a diseased mouse was injected into a healthy mouse, the latter quickly developed anthrax. He took blood from this second animal, injected it into another, and again obtained the characteristic disease symptoms. However, Koch carried this experiment a critically important step further. He discovered that the anthrax bacteria could be grown in nutrient fluids outside the animal body and that even after many transfers in laboratory culture, the bacteria still caused the disease when inoculated into a healthy animal.

On the basis of these and related experiments carried out in his seminal work on the causative agent of tuberculosis, Koch formulated a set of rigorous criteria, now known as **Koch's postulates**, for definitively linking a specific microorganism to a specific disease:

1. The disease-causing organism must always be present in animals suffering from the disease and should not be present in healthy animals.
2. The organism must be cultivated in a pure culture away from the animal body.
3. The isolated organism must cause the disease when inoculated into a healthy susceptible animal.
4. The organism must be reisolated from these experimental animals and cultured again in the laboratory, after which it should still be the same as the original organism.

Koch's postulates are summarized in **Figure 1.15**. Koch's postulates were a monumental step forward in the study of

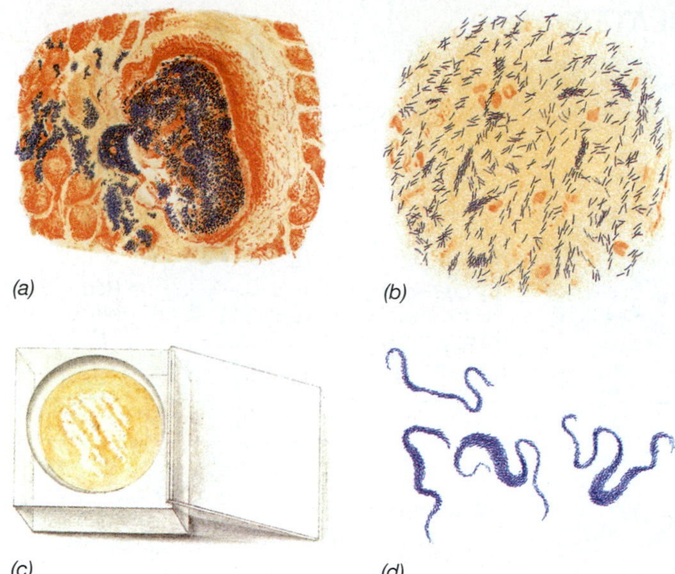

Figure 1.16 Robert Koch's drawings of *Mycobacterium tuberculosis.* Robert Koch was the first to isolate *M. tuberculosis* and to show it causes tuberculosis. *(a)* Section through a tubercle from lung tissue. Cells of *M. tuberculosis* stain blue, whereas the lung tissue stains brown. *(b)* Cells of *M. tuberculosis* in a sputum sample of a tuberculous patient. *(c)* Growth of *M. tuberculosis* on a glass plate of coagulated blood serum inside a glass box (lid open). *(d)* A colony of *M. tuberculosis* cells taken from the plate in *(c)* and observed microscopically at 700×; cells appear as long cordlike forms. Original drawings from Koch, R. 1884. "Die Aetiologie der Tuberkulose." *Mittheilungen aus dem Kaiserlichen Gesundheitsamte* 2:1–88.

infectious diseases. The postulates not only offered a means for linking the cause and effect of an infectious disease, but also stressed the importance of laboratory culture of the putative infectious agent. With these postulates as a guide, Koch, his students, and those that followed them discovered the causative agents of most of the important infectious diseases of humans and other animals. These discoveries led to the development of successful treatments for the prevention and cure of many of these diseases, thereby greatly improving the scientific basis of clinical medicine and human health and welfare (Figure 1.8).

Koch and Pure Cultures

To link a specific microorganism to a specific disease, the microorganism must first be isolated away from other microorganisms in laboratory culture; in microbiology we say that such a culture is *pure*. This concept was not lost on Robert Koch in formulating his famous postulates (Figure 1.15), and to accomplish this, he developed several simple but ingenious methods of obtaining bacteria in pure culture (see the Microbial Sidebar, "Solid Media, the Petri Plate, and Pure Cultures").

Koch started in a crude way by using solid nutrients such as a potato slice to culture bacteria. But he quickly developed more reliable methods, many of which are still in use today. Koch observed that when a solid surface such as a potato slice was incubated in air, bacterial colonies developed, each having a characteristic shape and color. He inferred that each colony had arisen from a single bacterial cell that had fallen on the surface, found suitable nutrients, and multiplied. Each colony was a population of identical cells, or in other words, a **pure culture**. Koch realized that the use of solid media provided a simple way of obtaining pure cultures. However, because not all organisms grow on potato slices, Koch devised more uniform and reproducible nutrient solutions solidified with gelatin and, later, with agar, laboratory techniques that remain with us to this day (see the Microbial Sidebar).

A Test of Koch's Postulates: Tuberculosis

Koch's crowning accomplishment in medical bacteriology was his discovery of the causative agent of tuberculosis. At the time Koch began this work (1881), one-seventh of all reported human deaths were caused by tuberculosis (Figure 1.8). There was a strong suspicion that tuberculosis was a contagious disease, but the suspected causal organism had never been seen, either in diseased tissues or in culture. Koch was determined to demonstrate the causal agent of tuberculosis, and to this end he brought together all of the methods he had so carefully developed in his previous studies with anthrax: microscopy, staining, pure culture isolation, and an animal model system (Figure 1.15).

As is now well known, the bacterium that causes tuberculosis, *Mycobacterium tuberculosis,* is very difficult to stain because of the large amounts of a waxy lipid present in its cell wall. But Koch devised a staining procedure for *M. tuberculosis* in tissue samples using alkaline methylene blue in conjunction with a second stain (Bismarck brown) that stained only the tissue. Using this method, Koch observed bright blue, rod-shaped cells of *M. tuberculosis* in tuberculous tissues, the tissue itself staining a light brown (**Figure 1.16**). However, from his previous work on anthrax, Koch fully realized that simply *identifying* an organism associated with tuberculosis was not enough. He knew he must *culture* the organism in order to prove that it was the specific cause of tuberculosis.

Obtaining cultures of *M. tuberculosis* was not easy, but eventually Koch was successful in growing colonies of this organism on a medium containing coagulated blood serum. Later he used agar, which had just been introduced as a solidifying agent (see the Microbial Sidebar). Under the best of conditions, *M. tuberculosis* grows slowly in culture, but Koch's persistence and patience eventually led to pure cultures of this organism from a variety of human and animal sources.

From here it was relatively easy for Koch to use his postulates (Figure 1.15) to obtain definitive proof that the organism he had isolated was the cause of the disease tuberculosis. Guinea pigs can be readily infected with *M. tuberculosis* and eventually succumb to systemic tuberculosis. Koch showed that diseased guinea pigs contained masses of *M. tuberculosis* cells in their tissues and that pure cultures obtained from

Microbial Sidebar

Solid Media, the Petri Plate, and Pure Cultures

Robert Koch was the first to grow bacteria on solid culture media. Koch's early use of potato slices as solid media was fraught with problems. Besides being rather selective in terms of which bacteria would grow on the slices, the slices were frequently overgrown with molds. Koch thus needed a more reliable and reproducible means of growing bacteria on solid media, and he found the answer in agar.

Koch initially employed gelatin as a solidifying agent for the various nutrient fluids he used to culture bacteria and developed a method for preparing horizontal slabs of solid media that were kept free of contamination by covering them with a bell jar or glass box (see Figure 1.16*c*). Nutrient gelatin was a good culture medium for the isolation and study of various bacteria, but it had several drawbacks, the most important being that it did not remain solid at 37°C, the optimum temperature for growth of most human pathogens. Thus, a different solidifying agent was needed.

Agar is a polysaccharide derived from red algae. It was used widely in the nineteenth century as a gelling agent. Walter Hesse, an associate of Koch, first used agar as a solidifying agent for bacteriological culture media (**Figure 1**). The actual suggestion that agar be used instead of gelatin was made by Hesse's wife, Fannie. She had used agar to solidify fruit jellies. When it was tried as a solidifying agent in microbial nutrient media, its superior gelling qualities were immediately evident. Hesse wrote to Koch about this discovery, and Koch quickly adapted agar to his own studies, including his classic studies on the isolation of the bacterium *Mycobacterium tuberculosis*, the cause of the disease tuberculosis (see text and Figure 1.16).

Agar has many other properties that make it desirable as a gelling agent for microbial culture media. In particular, agar remains solid at 37°C (human body temperature) and, after melting during the sterilization process, remains liquid to about 45°C, at which time it can be poured into sterile vessels. In addition, unlike gelatin, which many bacteria can degrade, causing the medium to liquify, agar is not degraded by most bacteria. Agar also renders most solid culture media transparent, making it easier to differentiate bacterial colonies from inanimate particulate matter suspended in the medium. Hence, agar found its place early in the annals of microbiology and is still used today for obtaining and maintaining pure cultures of bacteria.

In 1887 Richard Petri, a German bacteriologist, published a brief paper describing a modification of Koch's flat plate technique (Figure 1.16*c*). Petri's enhancement, which turned out to be amazingly useful, was the development of the transparent double-sided dishes that bear his name (**Figure 2**). The advantages of Petri dishes were immediately apparent. They could easily be stacked and sterilized separately from the medium, and, following the addition of molten culture medium to the smaller of the two dishes, the larger dish could be used as a cover to prevent contamination. Colonies that formed on the surface of the agar in the Petri dish remained fully exposed to air and could easily be manipulated for further study. The original idea of Petri has not been improved on to this day, and the Petri dish, made either of reusable glass and sterilized by heat or of disposable plastic and sterilized by ethylene oxide (a gaseous sterilant), is a mainstay of the microbiology laboratory.

Koch was keenly aware of the implications his pure culture methods had for the study of microbial systematics. He observed that different colonies (differing in color, morphology, size, and the like, see Figure 2) developed on solid media exposed to a contaminated object and that they bred true and could be distinguished from one another by their colony characteristics. Cells from different colonies also differed microscopically and often in their temperature or nutrient requirements as well. Koch realized that these differences among microorganisms met all the requirements that taxonomists had established for the classification of larger organisms, such as plant and animal species. In Koch's own words (translated

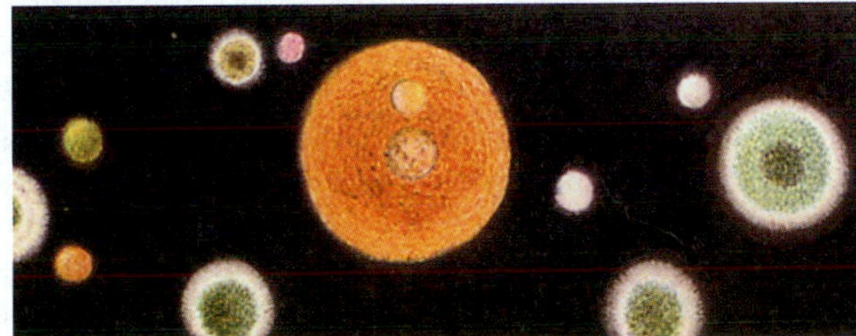

Figure 1 *A hand-colored photograph taken by Walter Hesse of colonies formed on agar. The colonies include those of fungi (molds) and bacteria and were obtained during studies Hesse initiated on the microbiological content of air in Berlin, Germany, in 1882.*

From Hesse, W. 1884. "Ueber quantitative Bestimmung der in der Luft enthaltenen Mikroorganismen," in Struck, H. (ed.), *Mittheilungen aus dem Kaiserlichen Gesundheitsamte*. August Hirschwald.

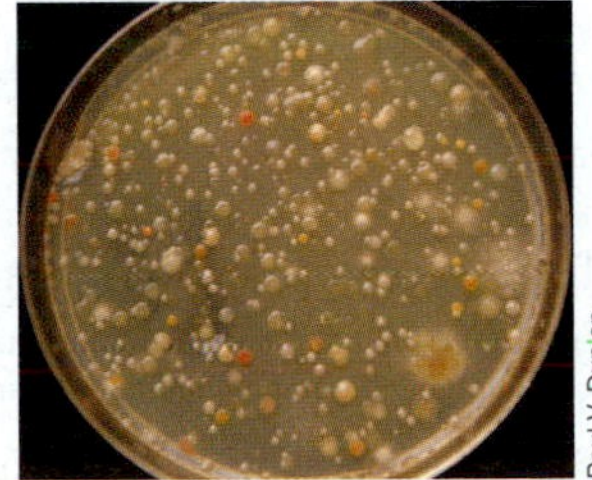

Paul V. Dunlap

Figure 2 *Photo of a Petri dish containing colonies of marine bacteria. Each colony contains billions of bacterial cells descended from a single cell.*

Microbial Sidebar

Solid Media, the Petri Plate, and Pure Cultures *(continued)*

from the German): "All bacteria which maintain the characteristics which differentiate one from another when they are cultured on the same medium and under the same conditions, should be designated as species, varieties, forms, or other suitable designation." Koch also realized from the study of pure cultures that one could show that specific organisms have specific effects, not only in causing disease, but in other capacities as well. Such insightful thinking was significant in the relatively rapid acceptance of microbiology as an independent biological science in the early twentieth century.

Koch's discovery of solid culture media and his emphasis on pure culture microbiology reached far beyond the realm of medical bacteriology. His discoveries supplied critically needed tools for development of the fields of bacterial taxonomy, genetics, and several other subdisciplines. Indeed, the entire field of microbiology owes much to Robert Koch and his associates for the intuition they displayed in grasping the significance of pure cultures and developing some of the most basic methods in microbiology.

such animals transmitted the disease to uninfected animals. Thus, Koch successfully satisfied all four of his postulates (Figure 1.15), and the cause of tuberculosis was understood. Koch announced his discovery of the cause of tuberculosis in 1882 and published a very thorough paper on the subject in 1884. It is in the latter that his postulates are most clearly stated. For his contributions on tuberculosis, Robert Koch was awarded the 1905 Nobel Prize for Physiology or Medicine.

Koch's Postulates Today

For diseases in which an animal model is available it is relatively easy to prove Koch's postulates. In modern clinical medicine, however, this is not always so easy. Even Koch had a hard time satisfying his postulates in some cases. Take cholera, for example. Today there is a suitable animal assay system for cholera, but in Koch's day this was not the case. And since only a fraction of human volunteers fed cells of *Vibrio cholerae,* the bacterium that causes cholera, contract cholera, obtaining definitive proof was difficult.

Even today it is sometimes impossible to satisfy Koch's postulates. For instance, the causative agents of several diseases of humans will not cause disease in any known experimental animals. These include many of the diseases associated with obligately intracellular bacteria, such as the rickettsias and chlamydias, and diseases caused by some viruses and protozoan parasites. Since for most of these diseases it would be unethical to use human volunteers to satisfy Koch's postulates, it is likely that cause and effect will never be unequivocally proven for them. However, for many of these diseases the clinical and epidemiological (disease tracking) evidence provides all but certain proof of the specific cause of the disease. Thus, although Koch's postulates remain the "gold standard" in medical microbiology, it has so far proven impossible to satisfy all of his postulates for every infectious disease.

1.8 MiniReview

Robert Koch developed criteria for the study of infectious microorganisms and developed the first methods for growth of pure cultures of microorganisms.

- How do Koch's postulates prove cause and effect in a disease?
- What advantages do solid media offer for the culture of microorganisms?
- What is a pure culture?

1.9 Microbial Diversity and the Rise of General Microbiology

As microbiology advanced out of the nineteenth and into the twentieth century, the initial focus on medical aspects of microbiology broadened to include studies of the microbial diversity of soil and water and the metabolic processes that organisms in these habitats carried out. This was the beginnings of *general microbiology,* a term that refers primarily to the nonmedical aspects of microbiology. Two giants of this era included the Dutchman Martinus Beijerinck and the Russian Sergei Winogradsky.

Martinus Beijerinck and the Enrichment Culture Technique

Martinus Beijerinck (1851–1931), a professor at the Delft Polytechnic School in Holland, was originally trained in botany; he began his career in microbiology studying plants. Beijerinck's greatest contribution to the field of microbiology was his clear formulation of the **enrichment culture technique**. In enrichment cultures microorganisms are isolated from natural samples in a highly selective fashion by manipulating nutrient and incubation conditions. Beijerinck's skill was aptly demonstrated when, following Winogradsky's discovery of the process of nitrogen fixation (see Figure 1.19), he enriched the aerobic nitrogen-fixing bacterium *Azotobacter* (**Figure 1.17**; Figure 22.1).

Using the enrichment culture technique, Beijerinck isolated the first pure cultures of many soil and aquatic microorganisms, including sulfate-reducing and sulfur-oxidizing bacteria, nitrogen-fixing root nodule bacteria, *Lactobacillus* species, green algae, various anaerobic bacteria, and many others. In his studies of tobacco mosaic disease, Beijerinck used selective filtering techniques to show that the infectious agent (a virus) was smaller than a bacterium and that it somehow became incorporated into cells of the living host plant. In this insightful work, Beijerinck not only described the first virus, but also the basic principles of virology, which we present later in Chapter 10.

Sergei Winogradsky and the Concept of Chemolithotrophy

Sergei Winogradsky (1856–1953) had scientific interests similar to Beijerinck's and was also successful in isolating or at least enriching several key bacteria from natural samples. Winogradsky was particularly interested in bacteria that cycle nitrogen and sulfur compounds, such as the nitrifying bacteria and purple sulfur bacteria (**Figure 1.18**). He showed in this work that specific bacteria are linked to specific biogeochemical transformations. For example, bacteria that cycle *nitrogen* compounds do not cycle *sulfur* compounds and vice versa. Moreover, Winogradsky's keen insight into the biology of these organisms revealed the metabolic significance of their biogeochemical transformations. From his studies of sulfur-oxidizing bacteria, for instance, Winogradsky proposed the concept of **chemolithotrophy**, the oxidation of *inorganic* compounds linked to energy conservation (**Figure 1.19*a***). And from his studies of the chemolithotrophic process of nitrification (the oxidation of ammonia to nitrate), Winogradsky showed that the organisms responsible—the nitrifying bacteria—obtained their carbon from CO_2. Winogradsky thus showed that, like phototrophic organisms, the nitrifying bacteria were autotrophs (Figure 1.19*a*).

Using an enrichment method, Winogradsky performed the first isolation of a nitrogen-fixing bacterium, the anaerobe *Clostridium pasteurianum,* thus formulating the concept of nitrogen fixation (Figure 1.19*b*). Beijerinck used this discovery to guide his isolation of aerobic nitrogen-fixing bacteria years later (Figure 1.17). Winogradsky lived to be almost 100, publishing many scientific papers and a major monograph, *Microbiologie du Sol (Soil Microbiology)*. This work, a milestone in microbiology, contains drawings of many of the organisms Winogradsky studied during his lengthy career (Figure 1.18).

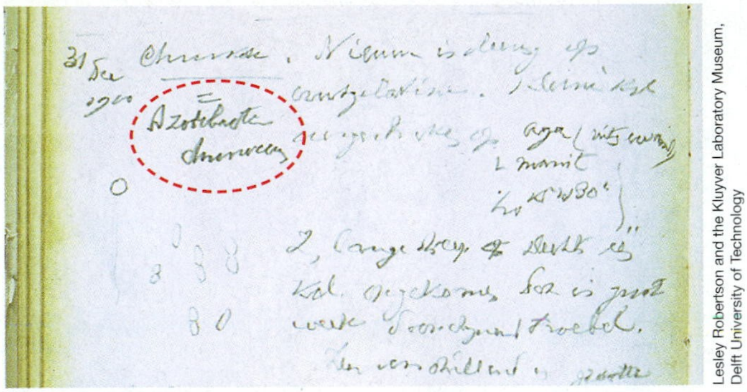

(a)

Lesley Robertson and the Kluyver Laboratory Museum, Delft University of Technology

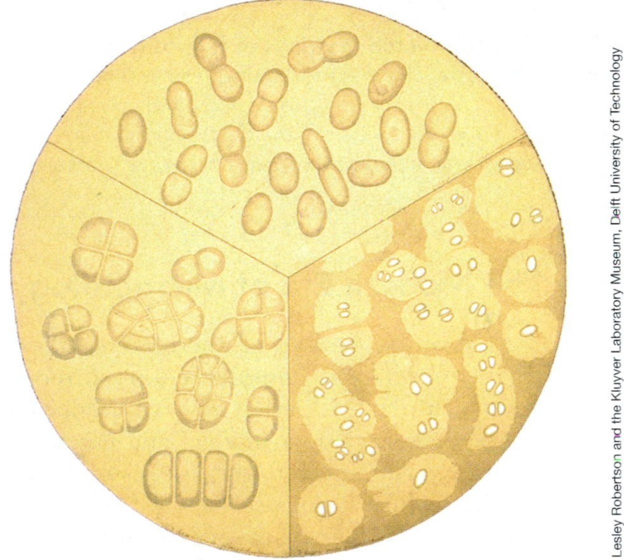

(b)

Lesley Robertson and the Kluyver Laboratory Museum, Delft University of Technology

Figure 1.17 Martinus Beijerinck and *Azotobacter*. *(a)* Portion of a page from the laboratory notebook of M. Beijerinck dated December 31, 1900, describing his observations on the aerobic nitrogen-fixing bacterium *Azotobacter chroococcum* (name circled in red). It is on this page that Beijerinck uses this name, which is still recognized today, for the first time. Compare Beijerinck's drawings of pairs of *A. chroococcum* cells with a photomicrograph of cells of *Azotobacter* shown in Figure 15.18*a*. *(b)* A painting by M. Beijerinck's sister, Henrëtte Beijerinck, showing cells of *Azotobacter chroococcum*. Beijerinck used such paintings to illustrate his lectures.

1.9 MiniReview

Beijerinck and Winogradsky studied bacteria inhabiting soil and water and developed the enrichment culture technique for the isolation of various microorganisms. Major new concepts in general microbiology emerged during this period, including enrichment cultures, chemolithotrophy, chemoautotrophy, and nitrogen fixation.

- What is the enrichment culture technique?
- What information in Figure 1.19 tells you that sulfur oxidation and nitrification are chemolithotrophic (energy-yielding) processes and that nitrogen fixation is not?

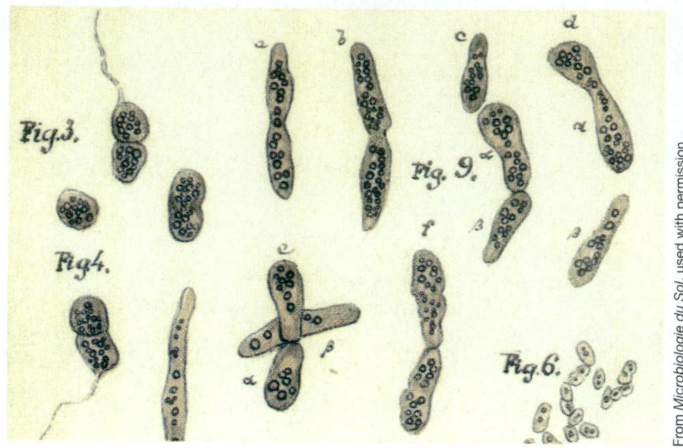

Figure 1.18 Hand-colored drawings of cells of purple sulfur phototrophic bacteria. The original drawings were made by Sergei Winogradsky about 1887 and then copied and hand-colored by his wife Hèléne. Figures 3 and 4 show cells of the genus *Chromatium*, such as *C. okenii*. Compare with a photomicrograph of cells of *C. okenii* in Figure 15.4*a*.

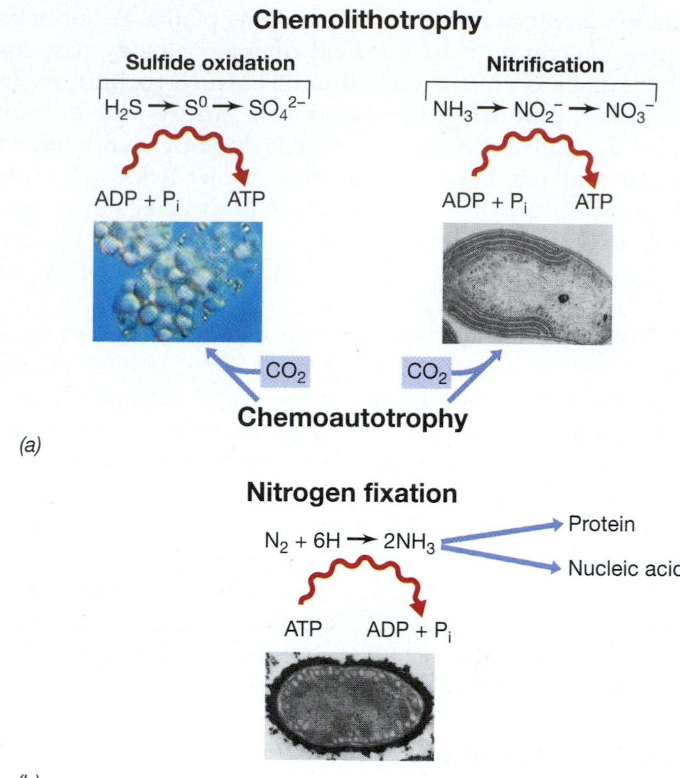

Figure 1.19 Major concepts conceived by Sergei Winogradsky. *(a)* Chemolithotrophy and chemoautotrophy. Oxidation of the sulfur or nitrogen compounds yields energy (ATP), and the cell obtains carbon from CO_2. Photos, left, the sulfur bacterium *Achromatium;* right, *Nitrobacter,* a bacterium that carries out the first step ($NH_3 \rightarrow NO_2^-$) in nitrification. *(b)* Nitrogen fixation. This process consumes ATP but allows the cell to use nitrogen gas (N_2) for all of its nitrogen needs. Photo, *Azotobacter,* an aerobic nitrogen-fixing bacterium (see also Figure 1.17).

1.10 The Modern Era of Microbiology

In the twentieth century, the field of microbiology developed rapidly in two different yet complementary directions—*applied* and *basic*. During this period new laboratory tools to study microorganisms became available, and the science of microbiology began to mature and spawn new subdisciplines. Few of these subdisciplines were purely applied or purely basic. Instead, most had both their discovery (basic) aspects and their problem-solving (applied) aspects. **Table 1.1** summarizes some of the key accomplishments during the first 300 years of microbiology; **Figure 1.20** shows a timeline of major accomplishments from 1975 to the present.

Origin of the Major Subdisciplines of Applied Microbiology

The advances of Robert Koch in understanding the nature of infectious diseases and culturing pathogens in the laboratory were catalysts for development of the fields of *medical microbiology* and *immunology,* both key subdisciplines of microbiology today. Work in these areas resulted in the discovery of many new bacterial pathogens of humans and other animals and elucidation of the mechanisms by which these pathogens infect the body or are resisted by the body's defenses. Other practical advances, bolstered by the discoveries of Beijerinck and Winogradsky, were in the field of *agricultural microbiology,* which began our understanding of microbial processes in the soil, such as nitrogen fixation (Figure 1.19*b*), that benefit plant growth. Later in the twentieth century, studies of soil microorganisms led to the discovery of antibiotics and other important chemicals. This spawned the field of *industrial microbiology,* the large-scale growth of microorganisms for the production of commercial products.

Advances in soil microbiology also provided the foundation for studies of microbial processes in lakes, rivers, and the oceans—*aquatic microbiology* and *marine microbiology.* One branch of aquatic microbiology deals with treating sewage and other wastewaters to render them harmless to humans and the environment and provide safe drinking water. Marine microbiology is presently enjoying great popularity because of the host of new tools available for studying the diversity and activities of marine microorganisms and the recognition that marine microorganisms likely control many important global parameters, including climate and atmospheric chemistry. As interest in the biodiversity and activities of microorganisms in their natural environments grew, the field of *microbial ecology* emerged in the 1960s and 1970s. Today microbial ecology is enjoying a "golden era" catalyzed by an influx of new molecular tools, in particular those of genomics. These powerful tools allow microbial ecologists to assess the microbial biodiversity and activities of even very complex microbial communities.

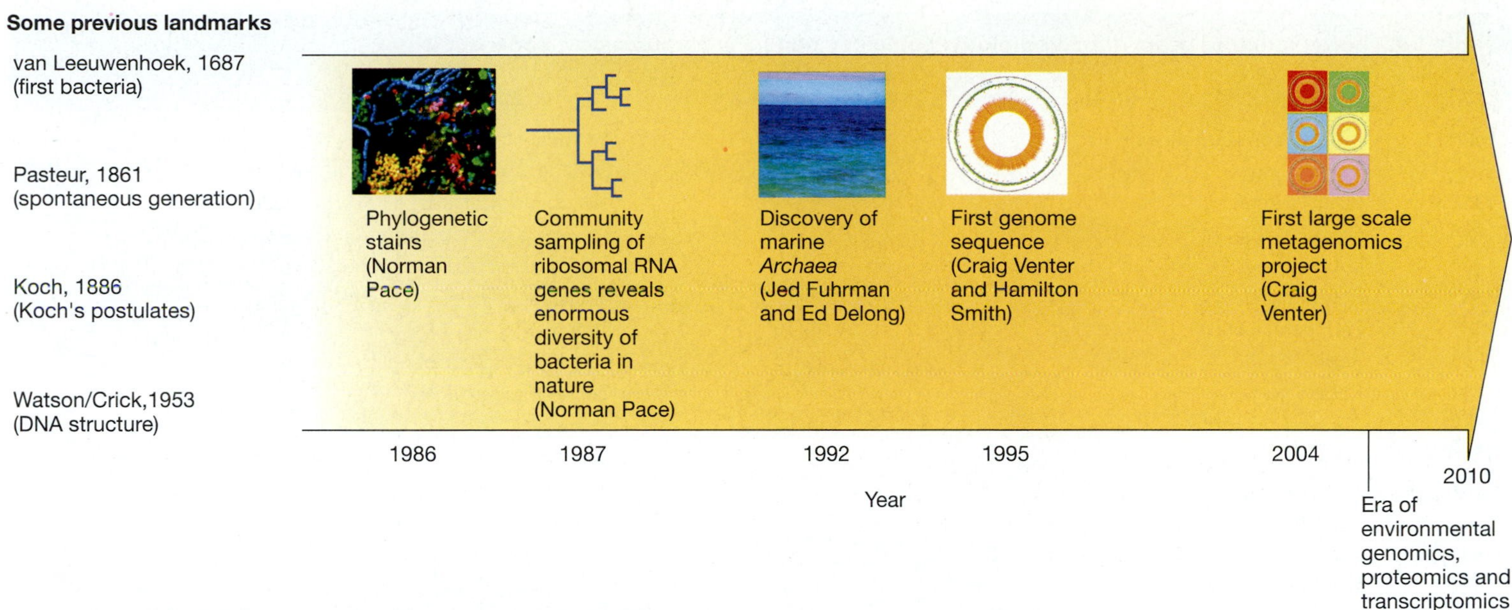

Figure 1.20 Some landmarks in molecular microbiology since 1985. The icons are representative of the discoveries. Not all contributors to a specific discovery could be listed.

Basic Science Subdisciplines in Microbiology

The twentieth century saw many new basic science subdisciplines develop in microbiology. Since the middle of the twentieth century, many new microorganisms have been cultured, resulting in considerable refinement of *microbial systematics,* the science of grouping and classifying microorganisms. This has culminated in the construction of a phylogenetic tree of life, as we discuss in the next chapter. Microbial ecology has fueled these advances by revealing, in most cases without actually culturing any organisms, that a vast world of unexplored microbial diversity awaits us in virtually any habitat one can examine. *Microbial physiology* studies the nutrients that microorganisms require for metabolism and growth and the products that they make from these nutrients. Enhanced understanding of the structure of microorganisms (*cytology*) and the discovery of microbial enzymes and the chemical reactions they carry out (*microbial biochemistry*) have also greatly influenced how microbiology is practiced today.

A key area of basic research that moved forward rapidly in the mid-twentieth century was the study of heredity and variation in bacteria, the subdiscipline of *bacterial genetics*. Although some aspects of bacterial genetics were known early in the twentieth century, it was not until the discovery of genetic exchange in bacteria around 1950 that bacterial genetics became a major field of study. Bacterial genetics, biochemistry, and physiology shared common roots during the 1950s and even today focus on many of the same scientific questions. By the early 1960s, these fields had provided an advanced understanding of DNA, RNA, and protein synthesis. *Molecular biology* arose to a great extent from studies of bacterial genetics (Figure 1.20).

Virology, the study of viruses, also blossomed in the twentieth century. Although Beijerinck discovered the first virus more than 100 years ago, it was not until the middle of the twentieth century that viruses were really understood. Much of the relevant work involved viruses that infect bacteria, called *bacteriophages*. Scientists realized that virus infection was a type of genetic transfer, and the relationship between viruses and cells was worked out primarily from research on bacteriophages. Today, animal and plant viruses occupy center stage in virology because of the ever-growing list of pathogenic viruses and the recognition that genetic diversity among these viruses is enormous.

The Era of Molecular Microbiology

By the 1970s, our knowledge of bacterial physiology, biochemistry, and genetics had advanced to the point that cellular genomes could be manipulated. DNA from one organism could be "transplanted" into a bacterium and the proteins encoded by the DNA harvested. This led to development of the field of *biotechnology*. At about this same time, nucleic acid sequencing techniques were developed, and the ramifications of this new technology were felt in all areas of biology. In microbiology, DNA sequencing revealed the phylogenetic (evolutionary) relationships among bacteria, which led to revolutionary new concepts in microbial systematics. DNA sequencing also gave birth in the mid-1990s to the field of *genomics*. The huge amounts of genomic information available today have fueled major advances in medicine, agriculture, biotechnology, and many other areas. And the fast-moving field of genomics has itself spawned new subdisciplines, such as *proteomics* and *metabolomics,* the patterns of protein and metabolic expression in cells, respectively.

Table 1.1 Three hundred years of microbiology: Some key papers in microbiology, 1684–2000[a]

Year	*Investigator(s)*	*Discovery*
1684	Antoni van Leeuwenhoek	Bacteria
1798	Edward Jenner	Smallpox vaccination
1857	Louis Pasteur	Microbiology of lactic acid fermentation
1860	Louis Pasteur	Role of yeast in alcoholic fermentation
1864	Louis Pasteur	Fallacy of spontaneous generation
1867	Robert Lister	Antiseptic principles in surgery
1876	Ferdinand Cohn	Endospores
1881	Robert Koch	Methods for study of bacteria in pure culture
1882	Robert Koch*	Cause of tuberculosis
1882	Élie Metchnikoff*	Phagocytosis
1884	Robert Koch	Cause of cholera; also first formulation of Koch's postulates
1884	Christian Gram	Gram-staining method
1885	Louis Pasteur	Rabies vaccine
1889	Sergei Winogradsky	Chemolithotrophy
1889	Martinus Beijerinck	Concept of a virus
1890	Emil von Behring* and Shibasaburo Kitasato	Diphtheria antitoxin
1890	Sergei Winogradsky	Autotrophy in chemolithotrophs
1901	Martinus Beijerinck	Enrichment culture method
1901	Karl Landsteiner*	Human blood groups
1908	Paul Ehrlich*	Chemotherapeutic agents
1911	Francis Rous*	First cancer virus
1915/1917	Frederick Twort and Felix d'Hérelle	Bacterial viruses (bacteriophage)
1928	Frederick Griffith	Pneumococcus transformation
1929	Alexander Fleming*	Penicillin
1931	Cornelius van Niel	H_2S (sulfide) as a photosynthetic electron donor
1935	Gerhard Domagk*	Sulfa drugs
1935	Wendall Stanley	Crystallization of tobacco mosaic virus
1941	George Beadle* and Edward Tatum*	One gene–one enzyme hypothesis
1943	Max Delbruck* and Salvador Luria*	Inheritance of genetic characteristics in bacteria
1944	Oswald Avery, Colin Macleod, Maclyn McCarty	DNA is genetic material

The concepts of genomics, proteomics, and metabolomics are all developed in more detail in Chapter 13.

New Frontiers

Not only has the 325 years of microbiology since the days of van Leeuwenhoek brought us startling insight into the biology of microorganisms, it has brought new challenges, both good and bad. On the one hand, new emerging diseases, such as SARS (severe acute respiratory syndrome) and bird flu, seem to appear without warning and challenge even our most sophisticated understanding of microbial diseases. On the other hand, new thrusts in microbiology, such as genomics, have given us an unprecedented understanding of how a cell works at the most fundamental level. Microbial research today is close to defining the minimalist genome—the minimum complement of genes necessary for a living cell. When such a genetic blueprint is available, scientists should be able to define precisely, at least in biochemical terms, all of the prerequisites for life. When this day arrives, can the laboratory creation of a living cell be that far off?

As the evolutionary biologist Stephen Jay Gould put it, we are living in the "age of bacteria." What an exciting time to be learning the science of microbiology! Stay tuned. Much more is in store!

1.10 MiniReview

In the middle to latter part of the twentieth century, basic and applied microbiology worked hand in hand to usher in the current era of molecular microbiology.

- List the subdisciplines of microbiology whose focus is the following: metabolism, enzymology, nucleic acid and protein synthesis, microorganisms and their natural environments, microbial classification, and microbial cell structure.

Table 1.1 (continued)

Year	Investigator(s)	Discovery
1944	Selman Waksman* and Albert Schatz	Streptomycin
1946	Edward Tatum and Joshua Lederberg*	Bacterial conjugation
1951	Barbara McClintock*	Transposable elements
1952	Joshua Lederberg and Norton Zinder	Bacterial transduction
1953	James Watson,* Francis Crick,* Rosalind Franklin, Maurice Wilkins*	Structure of DNA
1959	Arthur Pardee, François Jacob,* Jacques Monod,* Andre Lwoff*	Gene regulation by repressor proteins
1959	Rodney Porter*	Immunoglobulin structure
1959	F. Macfarlane Burnet	Clonal selection theory
1960	François Jacob, David Perrin, Carmon Sanchez, Jacques Monod	Concept of an operon
1960	Rosalyn Yalow* and Solomon Bernson	Radioimmunoassay (RIA)
1961	Sydney Brenner,* François Jacob,* Matthew Meselson	Messenger RNA and ribosomes as the site of protein synthesis
1966	Marshall Nirenberg* and H. Gobind Khorana*	Genetic code
1967	Thomas Brock	Bacteria inhabit boiling hot springs
1969	Howard Temin,* David Baltimore,* Renato Dulbecco*	Retroviruses/reverse transcriptase
1969	Thomas Brock and Hudson Freeze	*Thermus aquaticus*, source of *Taq* DNA polymerase
1970	Hamilton Smith* and David Nathans*	Restriction enzymes
1973	Stanley Cohen, Annie Chang, Robert Helling, Herbert Boyer, and Paul Berg*	Recombinant DNA technology
1975	Georges Kohler,* Cesar Milstein*	Monoclonal antibodies
1976	Susumu Tonegawa*	Rearrangement of immunoglobulin genes
1977	Carl Woese** and George Fox	*Archaea*
1977	Fred Sanger,* Steven Niklen, Alan Coulson	Methods for sequencing DNA
1981	Stanley Prusiner*	Prions
1982	Karl Stetter	First cultures of hyperthermophiles
1982	Barry Marshall* and Robin Warren*	Cause of peptic ulcers: *Helicobacter pylori*
1983	Luc Montagnier	Human immunodeficiency virus
1985	Kary Mullis*	Polymerase chain reaction (PCR)

[a]Major reference sources here include Brock, T. D. (1961), *Milestones in Microbiology*, Prentice Hall, Englewood Cliffs, NJ; Brock, T. D. (1990), *The Emergence of Bacterial Genetics*, Cold Spring Harbor Press, Cold Spring Harbor, NY. *Year* refers to the year in which the discovery was published.

*Nobel Laureates. The first Nobel Prizes were awarded in 1901 and Robert Koch received the Nobel Prize for Physiology or Medicine in 1905.

**Recipient of Crafoord Prize in Biosciences in 2003.

Review of Key Terms

Cell the fundamental unit of living matter

Chemolithotrophy a form of metabolism in which inorganic compounds are used to generate energy

Cytoplasm the fluid portion of a cell, bounded by the cell membrane

Cytoplasmic membrane a semipermeable barrier that separates the cell interior (cytoplasm) from environment

DNA deoxyribonucleic acid, the genetic material of cells and some viruses

Ecosystem organisms plus their nonliving environment

Enrichment culture technique a method for isolating specific microorganisms from nature using specific culture media and incubation conditions

Enzyme a protein (or in some cases an RNA) catalyst that functions to speed up chemical reactions

Genomics the identification and analysis of genomes

Genome an organism's full complement of genes

Habitat the environment in which a microbial population resides

Koch's postulates a set of criteria for proving that a given microorganism causes a given disease

Macromolecules the proteins, nucleic acids, lipids, and polysaccharides in a cell

Metabolism all biochemical reactions in a cell

Microbial ecology the study of microorganisms in their natural environments

Microorganism a microscopic organism consisting of a single cell or cell cluster, including the viruses

Pathogen a disease-causing microorganism

Pure culture a culture containing a single kind of microorganism

RNA ribonucleic acid, functions in protein synthesis as messenger RNA, transfer RNA, and ribosomal RNA

Ribosome structures composed of RNAs and proteins upon which new proteins are made

Spontaneous generation the hypothesis that living organisms can originate from nonliving matter

Sterile free of all living organisms and viruses

Review Questions

1. List six key properties associated with the living state. Which of these are characteristics of all cells? Which are characteristics of only some types of cells (Sections 1.1 and 1.2)?
2. Cells can be thought of as both machines and coding devices. Explain how these two attributes of a cell differ (Section 1.2).
3. What is needed for translation to occur in a cell? What is the product of the translational process (Section 1.2)?
4. What is an ecosystem? Do microorganisms live in pure cultures in an ecosystem? What effects can microorganisms have on their ecosystems (Section 1.3)?
5. Why did the evolution of cyanobacteria change Earth forever (Section 1.4)?
6. How would you convince a friend that microorganisms are much more than just agents of disease (Section 1.5)?
7. For what contributions are Hooke and van Leeuwenhoek remembered in microbiology? How did Ferdinand Cohn contribute to bacteriology (Section 1.6)?
8. Explain the principle behind the use of the Pasteur flask in studies on spontaneous generation (Section 1.7).
9. What is a pure culture and how can one be obtained? Why was knowledge of how to obtain a pure culture important for development of the science of microbiology (Section 1.8)?
10. What are Koch's postulates and how did they influence the development of microbiology? Why are they still relevant today (Section 1.8)?
11. Describe a major contribution to microbiology of the early microbiologist Martinus Beijerinck (Section 1.9).
12. What major concepts in microbiology do we owe to Sergei Winogradsky (Section 1.9)?
13. What major advances in microbiology have occurred in the past 60 years (Section 1.10)?

Application Questions

1. Pasteur's experiments on spontaneous generation were of enormous importance for the advance of microbiology, contributing to the methodology of microbiology, ideas on the origin of life, and techniques for the preservation of food, to name just a few. Explain briefly how Pasteur's experiments affected each of these topics.
2. Describe the lines of proof Robert Koch used to definitively associate the bacterium *Mycobacterium tuberculosis* with the disease tuberculosis. How would his proof have been flawed if any of the tools he developed for studying bacterial diseases had not been available for his study of tuberculosis?
3. Imagine that if by some action all microorganisms suddenly disappeared from Earth. From what you have learned in this chapter, why do you think that animals would eventually disappear from Earth? Why would plants disappear? If by contrast, all higher organisms suddenly disappeared, what aspect of Figure 1.6 tells you that a similar fate would not befall microorganisms?

2

A Brief Journey to the Microbial World

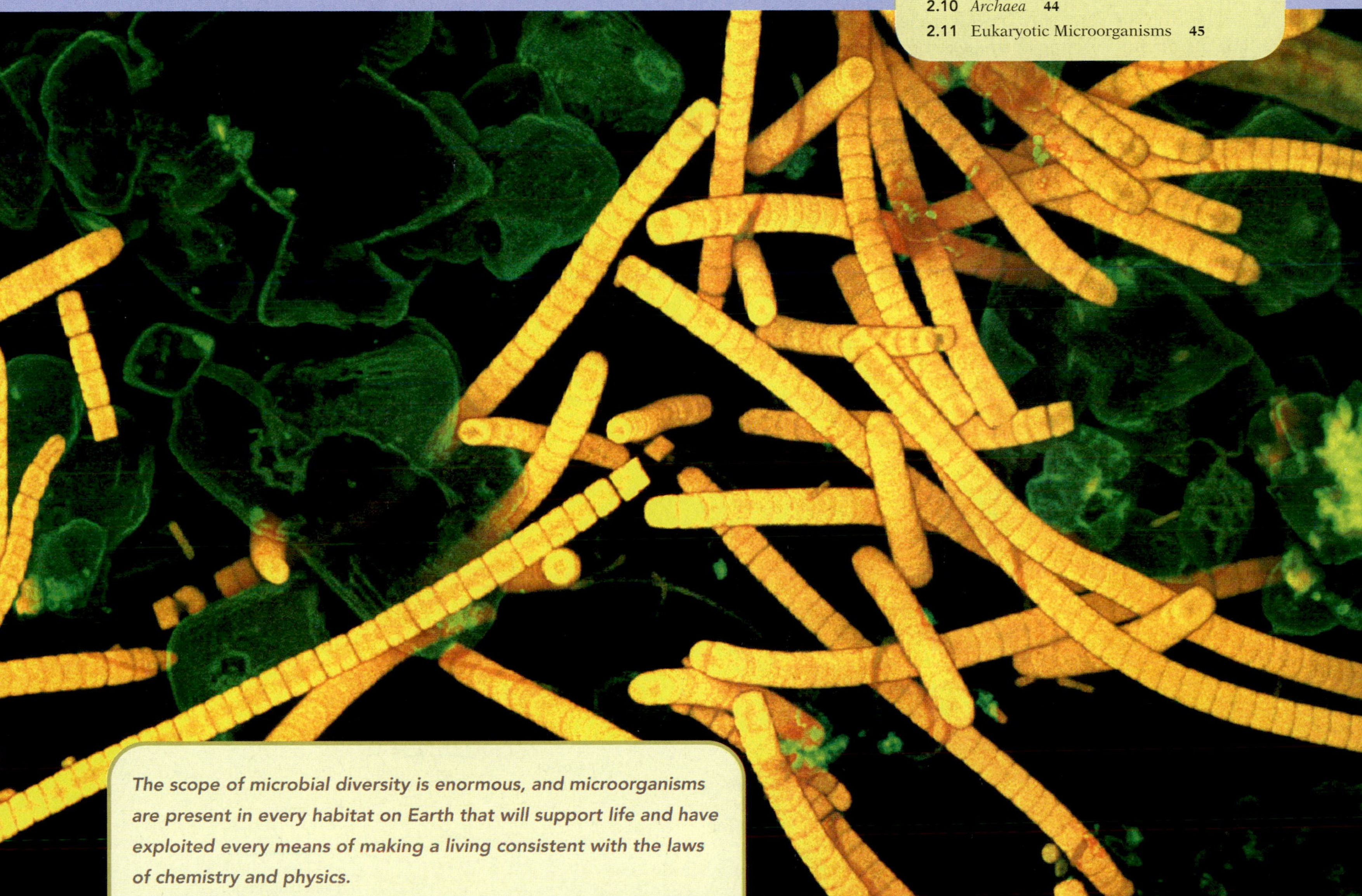

The scope of microbial diversity is enormous, and microorganisms are present in every habitat on Earth that will support life and have exploited every means of making a living consistent with the laws of chemistry and physics.

I SEEING THE VERY SMALL

Historically, the science of microbiology blossomed as the ability to see microorganisms improved. In other words, *microbiology* and *microscopy* advanced hand-in-hand. This is especially true for the main subject of this chapter—microbial diversity. The microscope is the microbiologist's most important tool, and the student of microbiology needs to have some background on how microscopes work and how microscopy is done. We begin our brief journey to the microbial world by considering different types of microscopes and the applications of microscopy to imaging microorganisms.

2.1 Some Principles of Light Microscopy

Visualization of microorganisms requires a microscope, either a *light* microscope or an *electron* microscope. In general, light microscopes are used to look at intact cells at relatively low magnifications, and electron microscopes are used to look at internal cell structure and the details of cell surfaces at very high magnification.

All microscopes employ lenses that magnify the original image. Equal in importance to magnification, however, is **resolution**, the ability to distinguish two adjacent objects as distinct and separate. Although magnification can be increased virtually without limit, resolution cannot because it is determined by the physical properties of light. It is thus resolution and not magnification that ultimately dictates what we can see with a microscope.

We begin with the light microscope, for which the limits of resolution are about 0.2 μm (micrometer, 10^{-6} m). We then proceed to the electron microscope, for which resolution is improved over that of the light microscope by about 1,000-fold.

The Compound Light Microscope

The light microscope uses visible light to illuminate cell structures. Several types of light microscopes are commonly used in microbiology: *bright-field, phase-contrast, dark-field,* and *fluorescence*.

With the bright-field microscope specimens are visualized because of the slight differences in contrast (density) that exist between them and their surrounding medium. Contrast differences arise because cells absorb or scatter light to varying degrees. The bright-field microscope is commonly used in laboratory courses in biology and microbiology and consists of two series of lenses (objective lens and ocular lens) that function in unison to form the image. The light source is focused on the specimen by a third lens, the condenser (**Figure 2.1**). Bacterial cells are typically difficult to see well with the bright-field microscope because they lack contrast with the surrounding medium. Pigmented microorganisms are an exception

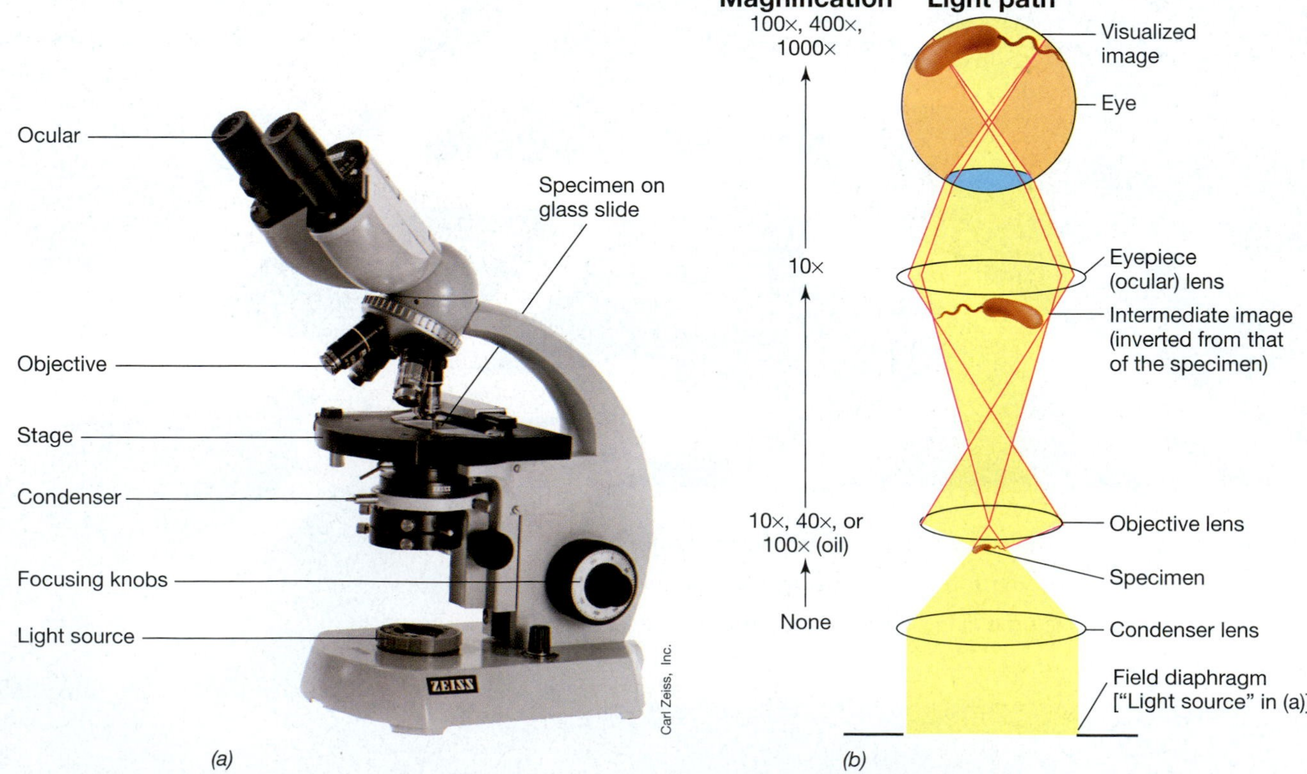

Figure 2.1 Microscopy. *(a)* A compound light microscope. *(b)* Path of light through a compound light microscope. Besides 10×, eyepieces (oculars) are available in 15–30×.

because the color of the organism itself adds contrast, thus improving visualization (**Figure 2.2**). For cells lacking pigments there are ways to boost contrast, and we consider these methods in the next section.

Magnification and Resolution

The total magnification of a compound microscope is the product of the magnification of its objective and ocular lenses (Figure 2.1*b*). Magnifications of about 1500× are the upper limit for a compound light microscope. Above this limit, resolution does not improve. Resolution is a function of the wavelength of light used and a characteristic of the objective lens known as its *numerical aperture* (a measure of light-gathering ability). There is a correlation between the magnification of a lens and its numerical aperture: Lenses with higher magnification typically have higher numerical apertures (the numerical aperture of a lens is stamped on the lens alongside the magnification). The diameter of the smallest object resolvable by any lens is equal to 0.5λ/numerical aperture, where λ is the wavelength of light used. Based on this formula, resolution is greatest when blue light is used to illuminate a specimen (blue light has a shorter wavelength than white or red light) and the objective used has a very high numerical aperture. Many light microscopes come fitted with a blue filter over the condenser lens to improve resolution.

As mentioned, the highest resolution possible in a compound light microscope is about 0.2 μm. What this means is that two objects that are closer together than 0.2 μm cannot be resolved as distinct and separate. Most microscopes used in microbiology have oculars that magnify 10–15× and objectives of 10–100× (Figure 2.1*b*). At 1000×, objects 0.2 μm in diameter can just be resolved. With a total of the 1000× objective, and with certain other objectives of very high numerical aperture, a special optical oil is placed between the specimen and the objective. Lenses on which oil is used are called *oil-immersion* lenses. Immersion oil increases the light-gathering ability of a lens by allowing rays emerging from the specimen at angles (that would otherwise be lost to the objective lens) to be collected and viewed.

2.1 MiniReview

Microscopes are essential for studying microorganisms. A bright-field microscope uses a series of lens to magnify and resolve the image.

- Define the term resolution.
- What is the upper limit of magnification for a bright-field microscope? Why is this so?

2.2 Improving and Adjusting Contrast in Light Microscopy

In microscopy, improving contrast improves the final image observed. Staining is an easy way to improve contrast but there are other ways to do so as well.

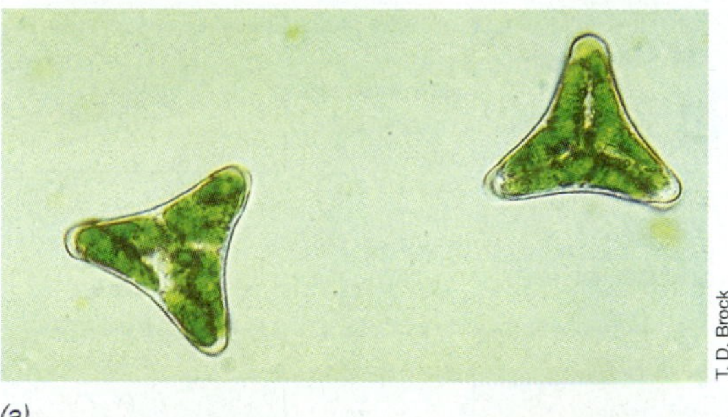

(a)

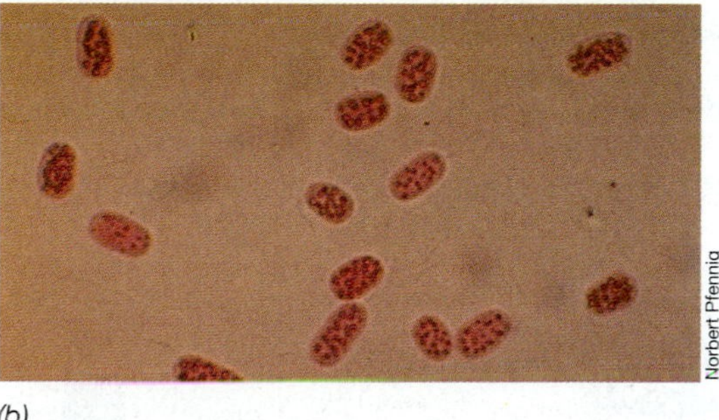

(b)

Figure 2.2 Bright-field photomicrographs of pigmented microorganisms. *(a)* A green alga (eukaryote). The green structures are chloroplasts. *(b)* Purple phototrophic bacteria (prokaryote). The alga cell is about 15 μm wide, and the bacterial cells are about 5 μm wide.

Staining: Increasing Contrast for Bright-Field Microscopy

Dyes can be used to stain cells and increase their contrast so that they can be more easily seen in the bright-field microscope. Dyes are organic compounds, and each class of dye has an affinity for specific cellular materials. Many dyes used in microbiology are positively charged dyes, called *basic dyes*, and bind strongly to negatively charged cellular constituents such as nucleic acids and acidic polysaccharides. Examples of basic dyes include methylene blue, crystal violet, and safranin. Because cell surfaces also tend to be negatively charged, these dyes combine with high affinity to structures on the surfaces of cells and hence are excellent general purpose stains.

To perform a simple stain one begins with dried preparations of cells (**Figure 2.3**). A glass slide containing a dried suspension of heat-fixed cells is flooded for a minute or two with a dilute solution of a dye, rinsed several times in water, and blotted dry. Because the cells are so small, it is common to observe dried, stained preparations of bacteria with a high-power (oil-immersion) lens (Figure 2.3).

Differential Stains: The Gram Stain

Stains that render different kinds of cells different colors are called *differential* stains. An important differential-staining procedure widely used in microbiology is the **Gram stain** (**Figure 2.4a**). On the basis of their reaction to the Gram stain, bacteria can be divided into two major groups: *gram positive* and *gram negative*. After Gram staining, **gram-positive bacteria**

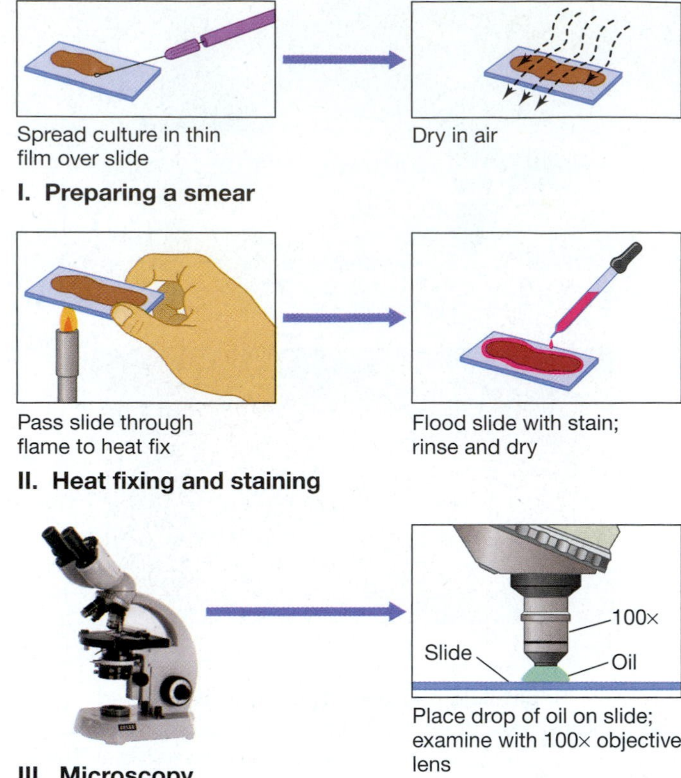
Spread culture in thin film over slide
Dry in air
I. Preparing a smear
Pass slide through flame to heat fix
Flood slide with stain; rinse and dry
II. Heat fixing and staining
100×
Slide
Oil
Place drop of oil on slide; examine with 100× objective lens
III. Microscopy

Figure 2.3 Staining cells for microscopic observation. Stains improve the contrast between cells and their background.

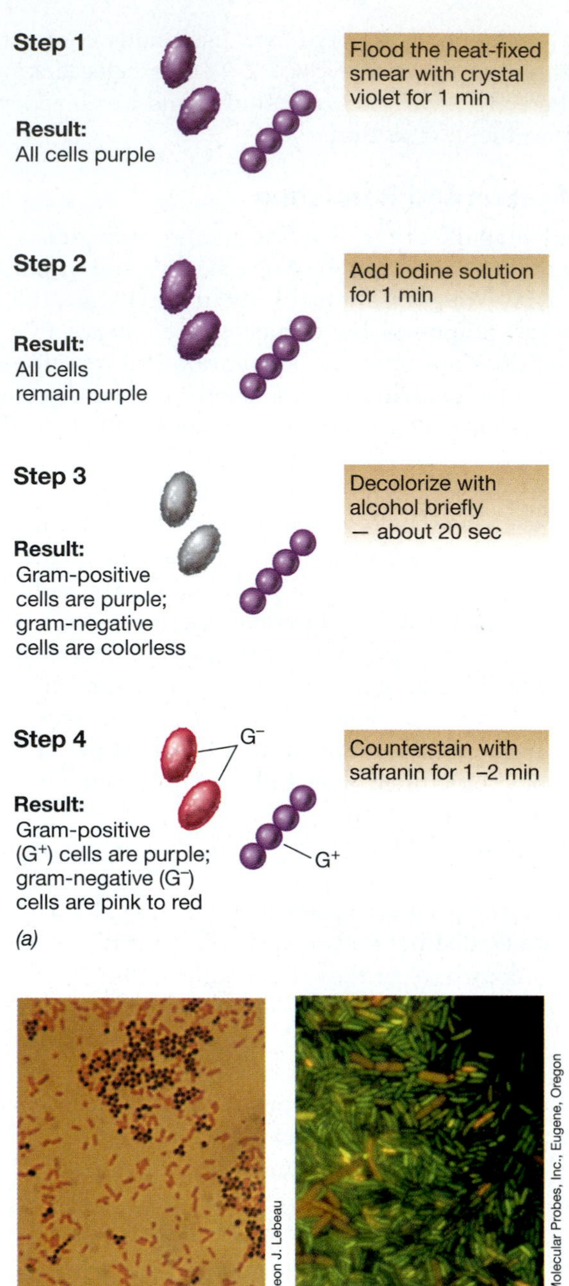
Step 1
Flood the heat-fixed smear with crystal violet for 1 min
Result:
All cells purple
Step 2
Add iodine solution for 1 min
Result:
All cells remain purple
Step 3
Decolorize with alcohol briefly — about 20 sec
Result:
Gram-positive cells are purple; gram-negative cells are colorless
Step 4
Counterstain with safranin for 1–2 min
G⁻
Result:
Gram-positive (G⁺) cells are purple; gram-negative (G⁻) cells are pink to red
G⁺
(a)
Leon J. Lebeau
Molecular Probes, Inc., Eugene, Oregon
(b)
(c)

Figure 2.4 The Gram stain. *(a)* Steps in the Gram-stain procedure. *(b)* Gram-stained *Bacteria* that are gram-positive (purple) and gram-negative (pink). The species are *Staphylococcus aureus* and *Escherichia coli*, respectively. *(c)* Cells of *Pseudomonas aeruginosa* (gram-negative, green) and *Bacillus cereus* (gram-positive, orange) stained with a one-step fluorescent staining method. This method allows for differentiating gram-positive from gram-negative cells in a single staining step.

appear purple and **gram-negative bacteria** appear pink (Figure 2.4*b*). This difference in reaction to the Gram stain arises because of differences in the cell wall structure of gram-positive and gram-negative cells (∞ Sections 4.6 and 4.7). After staining with a basic dye, typically crystal violet, treatment with ethanol decolorizes gram-negative but not gram-positive cells. Following counterstaining with a different-colored stain, the two cell types can be distinguished microscopically (Figure 2.4).

The Gram stain is one of the most useful staining procedures in microbiology. Typically, one begins the characterization of a new bacterium by determining whether it is gram positive or gram negative. If a fluorescent microscope, discussed below, is available, the Gram stain can be reduced to a one-step procedure in which gram-positive and gram-negative cells fluoresce different colors (Figure 2.4*c*).

Phase-Contrast and Dark-Field Microscopy

Staining, although a widely used procedure in light microscopy, kills cells and can distort their features. Two forms of light microscopy improve contrast without the use of stain. These are phase-contrast microscopy and dark-field microscopy (**Figure 2.5**). The phase-contrast microscope is widely used in research because it allows the observation of wet-mount (living) preparations.

Phase-contrast microscopy was invented in 1936 by Frits Zernike, a Dutch mathematical physicist. It is based on the principle that cells differ in refractive index (a factor by which light is slowed as it passes through a material) from their surroundings. Light passing through a cell thus differs in phase from light passing through its surroundings. This subtle difference is amplified by a device in the objective lens of the phase-contrast microscope called the *phase ring*, resulting in a dark image on a light background (Figure 2.5*b*). The ring

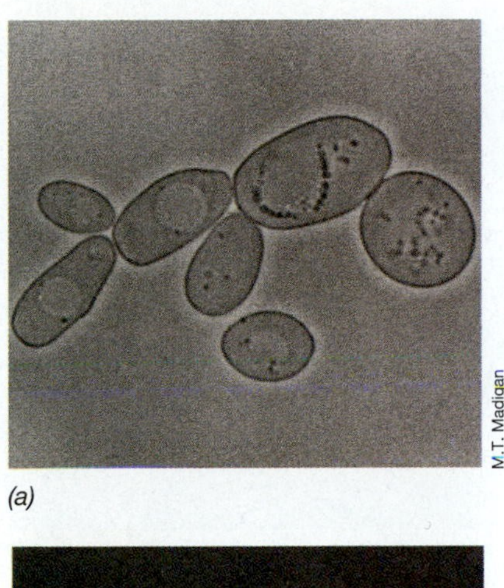

(a)

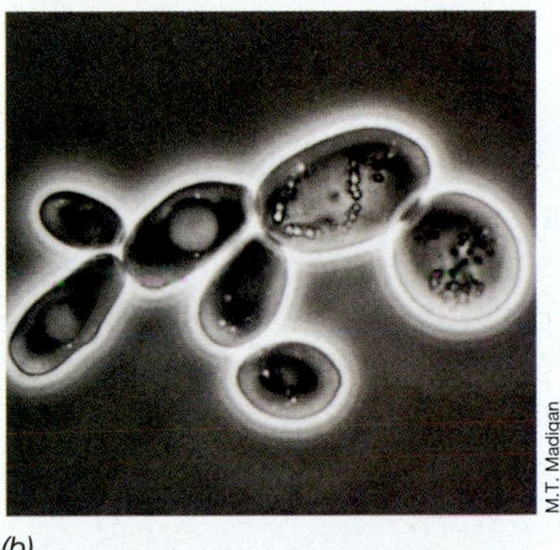

(b)

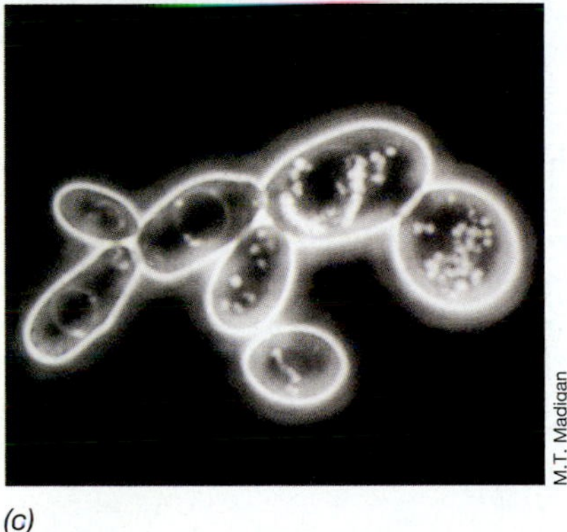

(c)

Figure 2.5 Cells of the baker's yeast *Saccharomyces cerevisiae* visualized by different types of light microscopy. *(a)* Bright-field microscopy. *(b)* Phase-contrast microscopy. *(c)* Dark-field microscopy. Cells average 8–10 μm wide.

consists of a phase plate—the key discovery of Zernike—that amplifies the minute variation in phase. Zernike's discovery of differences in contrast between cells and their background stimulated other innovations in microscopy, such as fluorescence and confocal microscopy (discussed below). For his invention of phase-contrast microscopy, Zernike was awarded the 1953 Nobel Prize in Physics.

The dark-field microscope is a light microscope in which the light reaches the specimen from the sides only. The only light reaching the lens is scattered by the specimen, and thus the specimen appears light on a dark background (Figure 2.5*c*). Resolution by dark-field microscopy is somewhat better than by light microscopy, and thus objects can often be resolved by dark-field that cannot be resolved by bright-field or even phase-contrast microscopes. Dark-field microscopy is also an excellent way to observe the motility of microorganisms, as bundles of flagella are often resolvable with this technique (∞ Figure 4.46*a*).

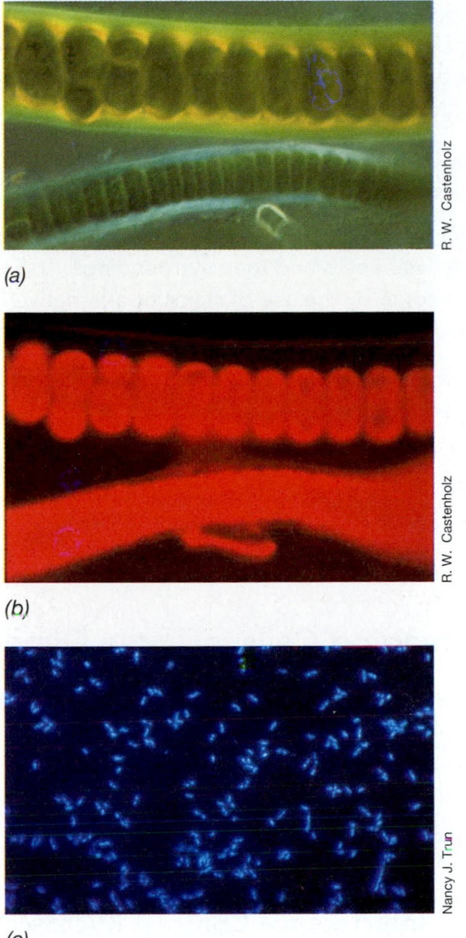

Figure 2.6 Fluorescence microscopy. *(a, b)* Cyanobacteria. *(a)* Cells observed by bright-field microscopy. *(b)* The same cells observed by fluorescence microscopy (cells exposed to light of 546 nm). The cells fluoresce red because they contain chlorophyll *a* and other pigments. *(c)* Fluorescence photomicrograph of cells of *Escherichia coli* made fluorescent by staining with the fluorescent dye, DAPI.

Fluorescence Microscopy

The fluorescence microscope is used to visualize specimens that fluoresce, that is, emit light of one color when light of another color shines upon them (**Figure 2.6**). Cells fluoresce either because they contain naturally fluorescent substances such as chlorophyll or other fluorescing components (autofluorescence) (see Figure 2.6*a, b*) or because the cells have been stained with a fluorescent dye (Figures 2.6*c*). DAPI (diamidino-2-phenylindole) is a widely used fluorescent dye, staining cells bright blue (Figure 2.6*c*). DAPI can be used to identify cells in a complex milieu, such as soil, water, food, or a clinical specimen. Fluorescence microscopy is widely used in clinical diagnostic microbiology and also in microbial ecology for enumerating bacteria in a natural environment or cell suspension (Figure 2.6*c*).

2.2 MiniReview

An inherent limitation of bright-field microscopy is its lack of contrast between cells and their surroundings. This problem can be overcome by the use of stains or alternative forms of light microscopy, such as phase contrast or dark field.

- What color will a gram-negative bacterium be after Gram staining by the conventional method?
- What major advantage does phase-contrast microscopy have over staining?
- How can cells be made fluorescent?

2.3 Imaging Cells in Three Dimensions

Up to now we have discussed forms of microscopy in which the images obtained are two-dimensional. How can this limitation be overcome? We will see in the next section that the scanning electron microscope offers one solution to this problem, but so can certain forms of light microscopy that we consider now.

Differential Interference Contrast Microscopy

Differential interference contrast (DIC) microscopy is light microscopy that employs a polarizer to produce polarized light. The polarized light then passes through a prism that generates two distinct beams. These beams traverse the specimen and enter the objective lens where they are recombined into one. Because the two beams pass through different substances with slightly different refractive indices, the combined beams are not totally in phase but instead create an interference effect. This effect intensifies subtle differences in cell structure. Thus, by DIC microscopy, structures such as the nucleus of eukaryotic cells (**Figure 2.7*a***), and endospores, vacuoles, and granules of prokaryotic cells, take on a three-dimensional appearance. DIC microscopy is particularly useful for observing unstained cells because it can reveal internal cell structures that are less apparent (or even invisible) by bright-field techniques (compare Figure 2.5*a* with Figure 2.7*a*).

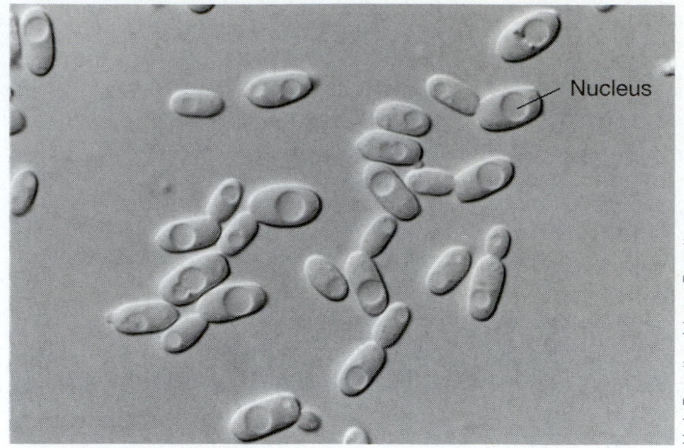

(a)

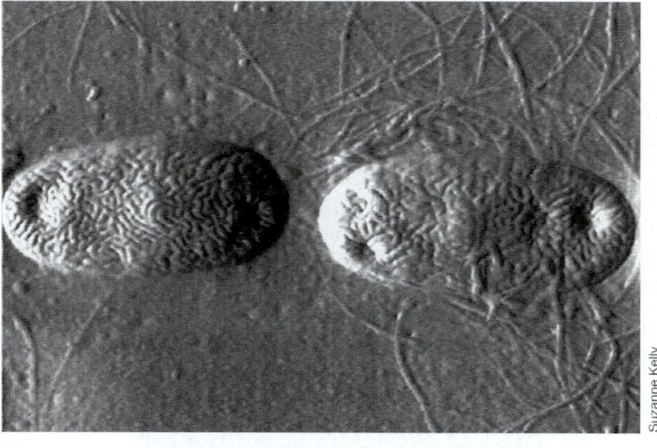

(b)

Figure 2.7 Three-dimensional imaging of cells. *(a)* Interference contrast microscopy and *(b)* atomic force microscopy. The yeast cells in *(a)* are about 8 μm wide. Note the clearly visible nucleus and compare Figure 2.5a. The bacterial cells in *(b)* are about 2.2 μm long and are from a natural biofilm that developed on the surface of a glass slide immersed for 24 h in a dog's water bowl. The slide was air dried before viewing with an atomic force microscope.

Atomic Force Microscopy

Another type of microscope useful for three-dimensional imaging of biological structures is the atomic force microscope (AFM). In atomic force microscopy, a tiny stylus is positioned extremely close to the specimen such that weak repulsive forces are established between the probe and atoms in the specimen. During scanning, the stylus rides up and down the "hills and valleys" of the specimen, continually recording its deviations from a flat surface. The pattern that is generated is processed by a series of detectors that feed the digital information into a computer, which outputs an image (Figure 2.7*b*).

Although the images obtained from an AFM appear similar to those from the scanning electron microscope (compare Figure 2.7*b* with Figure 2.10*b*), the AFM has the advantage that the specimen needn't be treated with fixatives or coatings. The AFM thus allows living specimens to be viewed, something that is generally not possible with electron microscopes.

Confocal Scanning Laser Microscopy

A confocal scanning laser microscope (CSLM) is a computerized microscope that couples a laser source to a light microscope. This generates a three-dimensional digital image of microorganisms and other biological specimens in the sample (**Figure 2.8**). The laser beam is precisely adjusted such that only a particular layer within a specimen is visible. By precisely illuminating only a single plane of focus, stray light from other focal planes is eliminated. Thus, when observing a

(a)

(b)

Figure 2.8 Confocal scanning laser microscopy. *(a)* Confocal image of a mixed microbial biofilm community cultivated in the laboratory. The green, rod-shaped cells are *Pseudomonas aeruginosa* experimentally introduced into the biofilm. Other cells of different colors are present at different depths in the biofilm. *(b)* Confocal image of a filamentous cyanobacterium growing in a soda lake. Cells are about 5 μm wide.

relatively thick specimen such as a microbial biofilm (Figure 2.8*a*), not only are cells on the surface of the biofilm apparent, as would be the case with conventional light microscopy, but cells in the various layers can also be observed by adjusting the laser beam. By illuminating a specimen with a laser whose intensity varies as a sine wave, it has been possible to improve on the 0.2-μm resolution of the compound light microscope to a limit of about 0.1 μm.

Cells in CSLM preparations are frequently stained with fluorescent dyes to make them more distinct (Figure 2.8). Alternatively, false-color images of unstained preparations can be generated such that different layers in the specimen take on different colors. The CLSM comes equipped with computer software that assembles digital images for subsequent image processing. Thus, images obtained from different layers can be stored and then digitally overlaid to reconstruct a three-dimensional image of the entire specimen (Figure 2.8*a*).

CSLM has found widespread use in microbial ecology, especially for identifying phylogenetically distinct populations of cells present in a microbial habitat (∞ Figure 1.5b) or for resolving the different components of a structured microbial habitat, such as a biofilm (Figure 2.8*a*). CSLM is particularly useful anywhere thick specimens need to be examined for their microbial content throughout their depth.

2.3 MiniReview

Differential interference contrast (DIC) and confocal scanning (CSLM) are forms of light microscopy that allow for greater three-dimensional imaging than other forms of light microscopy, and CSLM allows imaging through thick specimens. The atomic force microscope (AFM) can yield a detailed three-dimensional image of live preparations.

- What structure in eukaryotic cells is more easily seen in DIC than in bright-field microscopy? (*Hint:* Compare Figures 2.5*a* and 2.7*a*).
- How is CSLM able to view different layers in a thick preparation?

2.4 Electron Microscopy

Electron microscopes use electrons instead of photons to image cells or cell structures. In the *transmission electron microscope* (TEM), electromagnets function as lenses, and the whole system operates in a vacuum (**Figure 2.9**). Electron microscopes are fitted with cameras to allow a photograph, called an *electron micrograph*, to be taken.

The TEM is typically used to examine cell structure at very high magnification and resolution. The resolving power of the electron microscope is much greater than that of the

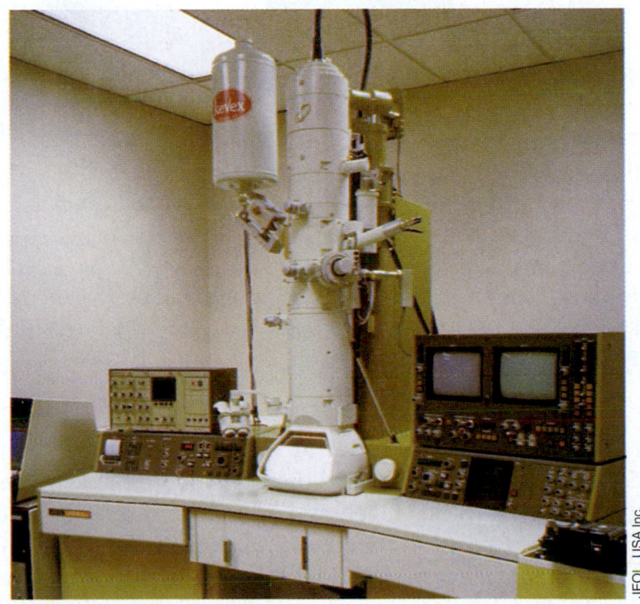

Figure 2.9 The electron microscope. This instrument encompasses both transmission and scanning electron microscope functions.

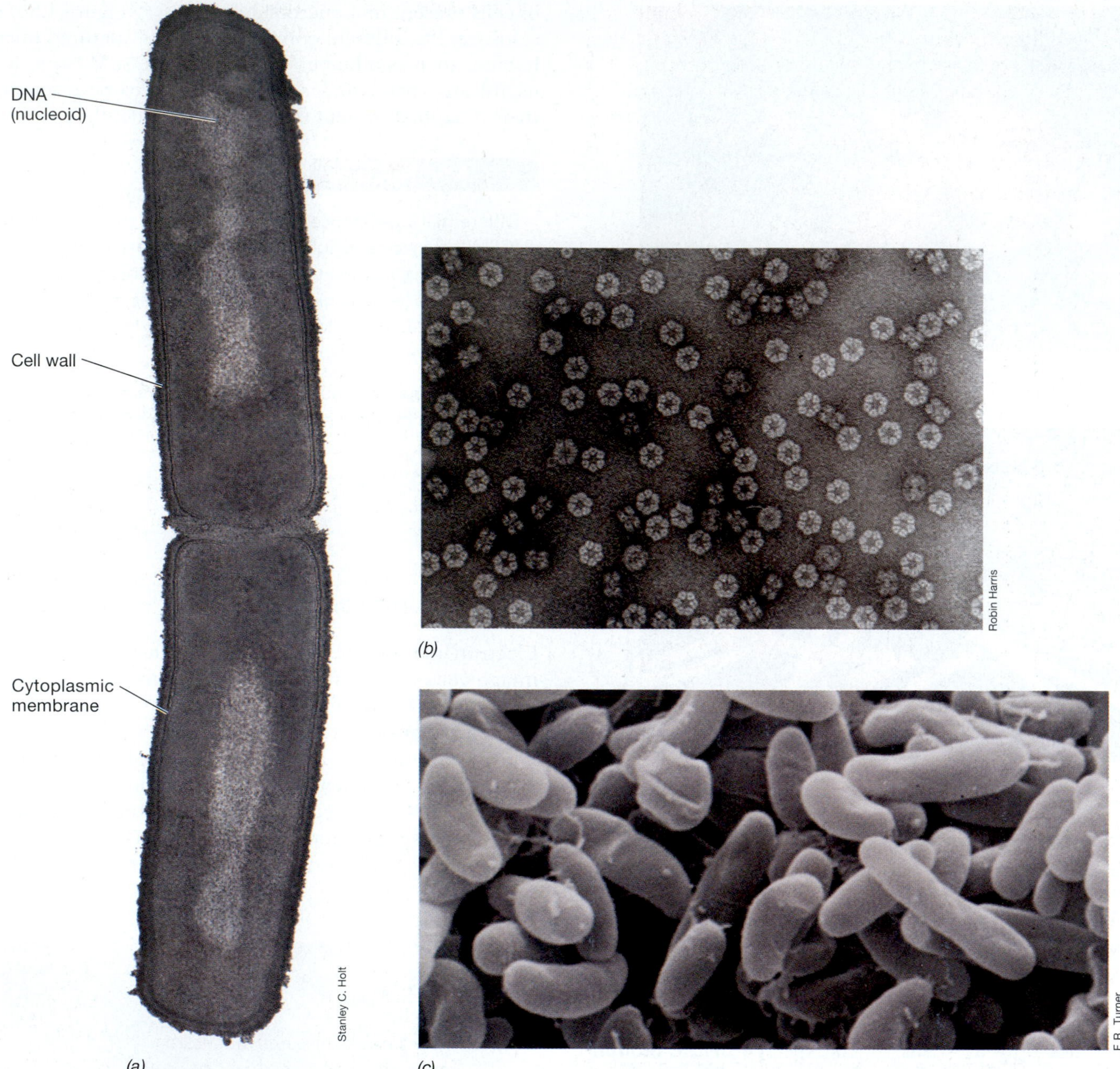

Figure 2.10 Electron micrographs. *(a)* Micrograph of a thin section of a dividing cell of the gram-positive bacterium, *Bacillus subtilis*, taken by transmission electron microscopy (TEM). Note the DNA forming the nucleoid. The cell is about 0.8 μm wide. *(b)* TEM of negatively stained molecules of hemoglobin from the marine worm *Nereis virens*. Each hexagonal-shaped molecule is about 25 nanometers (nm) in diameter and consists of two donut-shaped rings, a total of 15 nm wide. *(c)* Cells of the phototrophic bacterium *Rhodovibrio sodomensis*. A single cell is about 0.75 μm wide.

light microscope, enabling one to view structures at the molecular level. This is because the wavelength of electrons is much shorter than the wavelength of visible light, and wavelength affects resolution (Section 2.2). For example, whereas the resolving power of a high-quality light microscope is about 0.2 *micrometers*, the resolving power of a high quality TEM is about 0.2 *nanometers* (nm, 10^{-9}). Thus, even individual molecules, such as proteins and nucleic acids, can be visualized in the transmission electron microscope (**Figure 2.10*b***, and see Figure 2.14*b*).

Unlike visible light, however, electron beams do not penetrate very well; even a single cell is too thick to reveal its internal contents directly by TEM. Consequently, special techniques of thin sectioning are needed to prepare specimens before observing them. A single bacterial cell, for instance, is cut into many, very thin (20–60 nm) slices, which are then examined individually by TEM (Figure 2.10*a*). To obtain sufficient contrast, the preparations are treated with stains such as osmic acid, or permanganate, uranium, lanthanum, or lead salts. Because these substances are composed of atoms of high atomic weight, they scatter electrons well and thus improve contrast (Figure 2.10*b*).

Scanning Electron Microscopy

If only the external features of an organism need to be observed, thin sections are unnecessary. Intact cells or cell components can be observed directly by TEM with a technique called *negative staining* (Figure 2.10*b*). Alternatively, one can use the *scanning electron microscope* (SEM) (Figure 2.9).

In scanning electron microscopy, the specimen is coated with a thin film of a heavy metal such as gold. An electron beam from the SEM then scans back and forth across the specimen. Electrons scattered from the metal are collected, and they activate a viewing screen to produce an image (Figure 2.10*b*). In the SEM, even fairly large specimens can be observed, and the depth of field is extremely good. A wide range of magnifications can be obtained with the SEM, from as low as 15× up to about 100,000×, but only the surface of an object can be visualized.

Electron micrographs taken by either TEM or SEM are black and white images. Often false color is added to these images by manipulating the micrographs with a computer. But false color does not improve resolution of the micrograph; resolution is set by the magnification used to take the original micrograph.

2.4 MiniReview

Electron microscopes have far greater resolving power than do light microscopes, the limits of resolution being about 0.2 nm. Two major types of electron microscopy are performed: transmission electron microscopy, for observing internal cell structure down to the molecular level, and scanning electron microscopy, useful for three-dimensional imaging and for examining surfaces.

- What is an electron micrograph? How does an electron micrograph differ from a photomicrograph in terms of how the image is obtained? Why do electron micrographs have so much greater resolution than light micrographs?
- What type of electron microscope would be used to view a cluster of cells? What type would be used to observe the bacterial nucleoid?

II CELL STRUCTURE AND EVOLUTIONARY HISTORY

The next part of this chapter introduces concepts of microbial cell structure and diversity that underlie topics throughout the book. We first compare the internal architecture of microbial cells and differentiate cells from viruses. We then explore the evolutionary tree of life to set the stage for the introduction of the major groups of microorganisms that affect our lives and our planet.

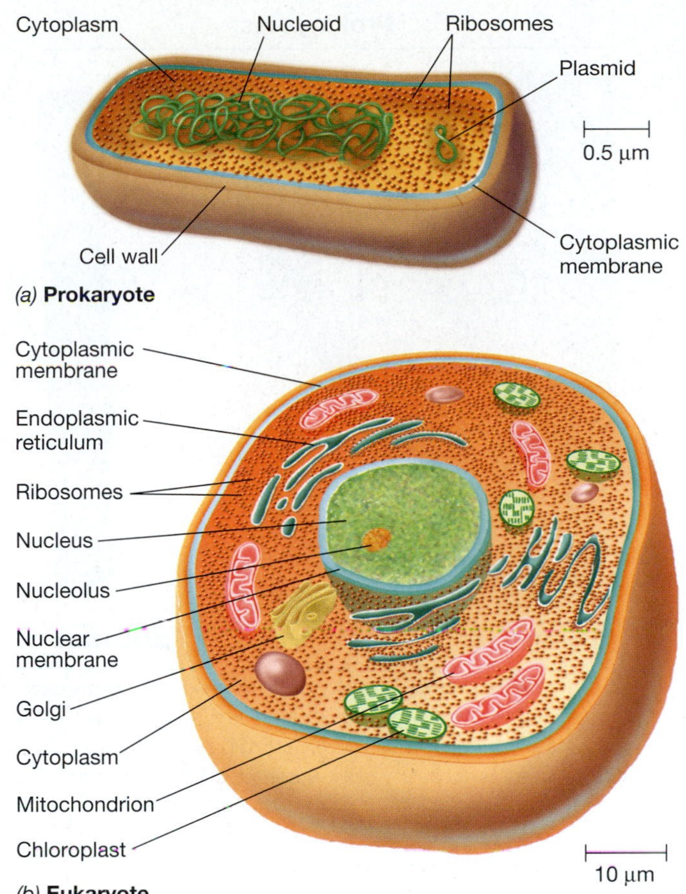

Figure 2.11 Internal structure of microbial cells. Note differences in scale and internal structure between the prokaryotic and eukaryotic cells.

2.5 Elements of Cell and Viral Structure

All cells have much in common and contain many of the same components. As we learned in Chapter 1, all cells have a permeability barrier called the **cytoplasmic membrane** that separates the inside of the cell, the **cytoplasm**, from the outside (**Figure 2.11**). Major components dissolved in the cytoplasm include macromolecules, small organic molecules (mainly precursors of macromolecules), various inorganic ions, and **ribosomes**—the cell's protein-synthesizing structures. Ribosomes interact with cytoplasmic proteins and messenger and transfer RNAs in the key process of protein synthesis (translation) (Figure 1.4).

The **cell wall** gives structural strength to a cell. The cell wall is relatively permeable and located outside the membrane (Figure 2.11*a*); it is a much stronger layer than the membrane itself. Plant cells and most microorganisms have cell walls, whereas animal cells, with rare exceptions, do not. In place of a cell wall, animal cells are reinforced by molecular scaffolding within the cytoplasm, called the *cytoskeleton*.

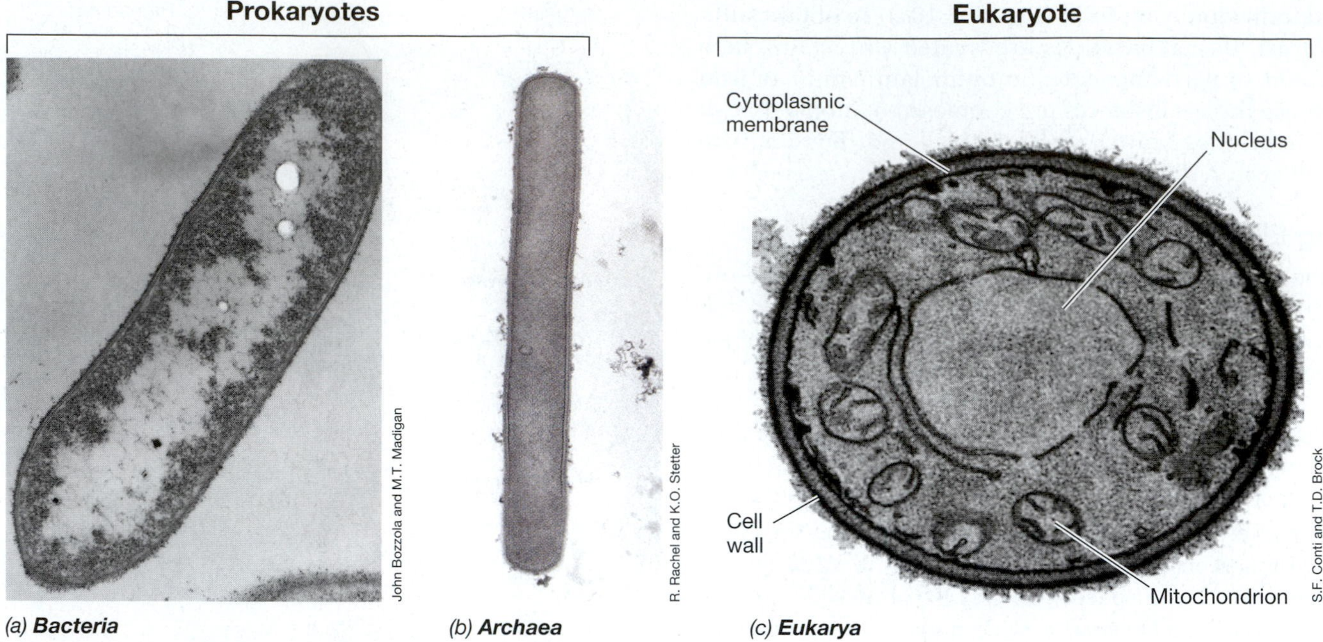

Figure 2.12 Electron micrographs of sectioned cells from each of the domains of living organisms. *(a) Heliobacterium modesticaldum*; the cell measures 1 × 3 μm. *(b) Methanopyrus kandleri*; the cell measures 0.5 × 4 μm. Reinhard Rachel and Karl O. Stetter, 1981. *Archives of Microbiology* 128:288–293. © Springer-Verlag GmbH & Co. KG. *(c) Saccharomyces cerevisiae*; the cell measures 8 μm in diameter.

Prokaryotic and Eukaryotic Cells

Examination of the internal structure and other features of cells reveal two patterns: **prokaryote** and **eukaryote** (Figure 2.11 and **Figure 2.12**). Eukaryotes have their DNA in a membrane-enclosed **nucleus** and are typically larger and structurally more complex than prokaryotic cells. In eukaryotic cells the key processes of transcription and translation (Figure 1.4) are partitioned; transcription occurs in the nucleus and translation in the cytoplasm. Eukaryotic microorganisms include algae, fungi, and protozoa. All multicellular plants and animals are also constructed of eukaryotic cells. We consider eukaryotic cells in detail in Chapter 18.

A major feature of eukaryotic cells is the presence of membrane-enclosed structures called **organelles**. These include, first and foremost, the nucleus, but also mitochondria and chloroplasts (the latter in photosynthetic cells only) (Figures 2.2*a* and 2.12*c*). As mentioned, the nucleus houses the genome and is also the site of transcription in eukaryotic cells. Mitochondria and chloroplasts contain their own very small genomes and play specific roles in energy generation by carrying out respiration and photosynthesis, respectively.

In contrast to eukaryotic cells, prokaryotic cells have a simpler internal structure that lacks membrane-enclosed organelles (Figures 2.11*a* and 2.12*a, b*). Prokaryotes differ from eukaryotes in many other ways as well. For example, prokaryotes couple transcription to translation directly within the cytoplasm because their DNA is not enclosed within a nucleus as it is in eukaryotes (Section 2.6). Moreover, in contrast to eukaryotes, most prokaryotes use their cytoplasmic membrane to drive energy-conserving reactions and have small, compact genomes consisting of circular DNA (Section 2.6).

Despite many clearcut differences between prokaryotes and eukaryotes, it is important that we not equate cellular organization with evolutionary relatedness. We will see in Section 2.7 that the prokaryotic world consists of two large and evolutionarily distinct groups. In Chapters 7 and 8 we compare and contrast the molecular biology of prokaryotes, highlighting their similarities and differences and relating them to molecular processes in eukaryotes.

Cell Size

In general, microbial cells are very small, particularly prokaryotes. For example, a typical rod-shaped prokaryote is 1–5 μm long and about 1 μm wide and thus is invisible to the naked eye. To conceive of how small a prokaryote is, consider that 500 prokaryotic cells each 1 μm long could be placed end-to-end across the period at the end of this sentence. Eukaryotic cells are typically much larger than prokaryotic cells, but the range of sizes in eukaryotic cells is quite large. Eukaryotic cells are known to have diameters as small as 0.8 μm or as large as several hundred micrometers. We revisit the subject of cell size in more detail later (Section 4.2).

Viruses

Viruses are a major class of microorganisms, but they are not cells (**Figure 2.13**). Viruses lack many of the attributes of cells (Figure 1.3), the most important of which is that they are not dynamic open systems. Instead, a virus particle is static, quite stable, and unable to change or replace its parts. Only when it infects a cell does a virus acquire the key attribute of a living system—replication. Unlike cells, viruses have no metabolic abilities of their own. Although they contain their

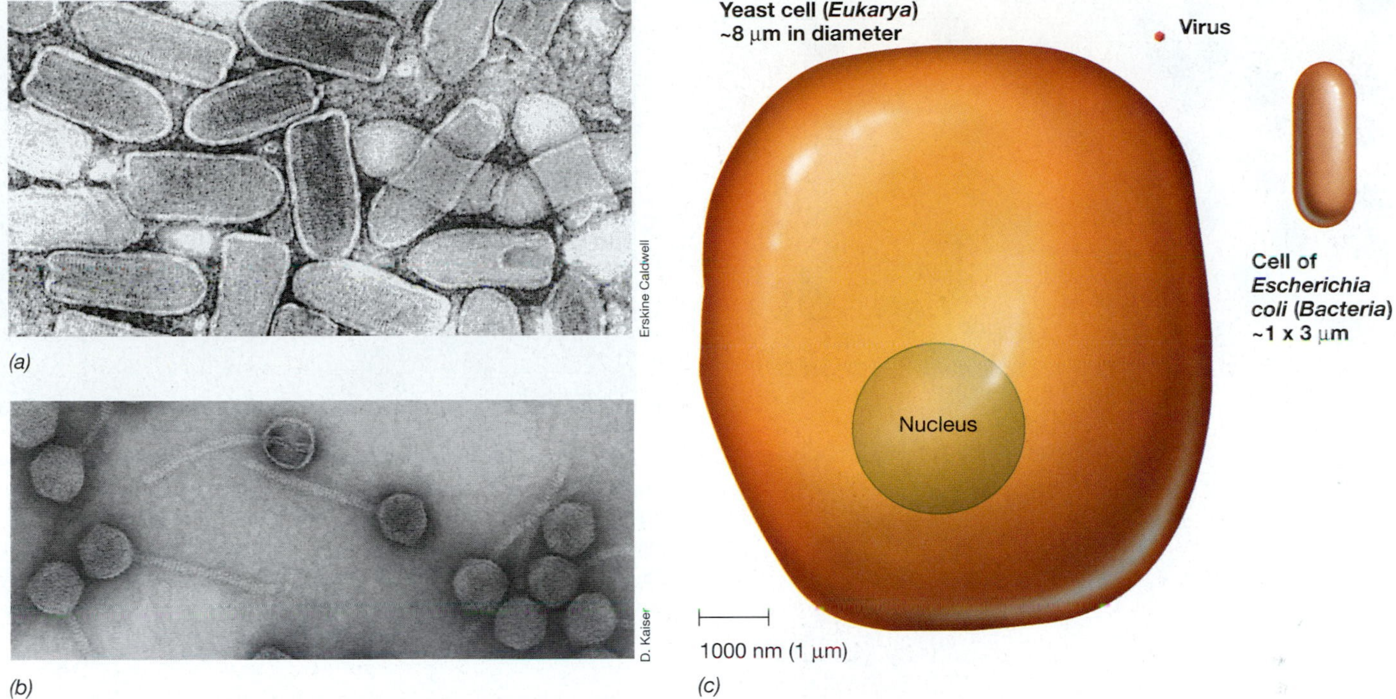

Figure 2.13 Virus structure and size comparisons of viruses and cells. *(a)* Particles of rhabdovirus (a virus that infects plants and animals). A single virus particle is about 65 nm (0.065 µm) wide. *(b)* Bacterial virus (bacteriophage) lambda. The head of each particle is about 65 nm wide. *(c)* The size of the viruses shown in *(a)* and *(b)* in comparison to a bacterial and eukaryotic cell.

own genomes, viruses lack ribosomes; to synthesize proteins they depend totally on the biosynthetic machinery of the cells they have infected. Moreover, unlike cells, viruses contain only a single form of nucleic acid, either DNA or RNA; thus, some viruses have RNA genomes. We discuss viruses in detail in Chapters 10 and 19.

Viruses are known to infect all types of cells, including microbial cells. Many viruses cause disease in the organisms they infect. However, viral infection can have many profound effects on cells other than disease, including genetic alterations that can actually improve the capabilities of the cell. Viruses are also much smaller than cells, even much smaller than prokaryotic cells (Figure 2.13). Viruses vary in size, with the smallest known viruses being only about 10 nm in diameter.

2.5 MiniReview

All microbial cells share certain basic structures such as the cytoplasmic membrane and ribosomes; most have a cell wall. Two cell structural patterns are recognized: the prokaryote and the eukaryote. Viruses are not cells but depend on cells for their replication.

- By looking inside a cell how could you tell if it was a prokaryote or a eukaryote?
- What important function do ribosomes play in cells?
- Why are viruses not cells?

2.6 Arrangement of DNA in Microbial Cells

The life processes of all cells are governed by their complement of genes, their **genome**. A gene can be defined as a segment of DNA that encodes a protein or an RNA molecule. In Chapter 13 we consider the rapid advances that have been made in sequencing and analyzing the genomes of organisms, from viruses through bacteria to humans. These advances have yielded detailed genetic blueprints of hundreds of different organisms and have allowed for extensive and rather revealing comparisons to be made. Here we consider only how genomes are organized in prokaryotic and eukaryotic cells and consider the number of genes and proteins present in a typical bacterial cell.

Nucleus versus Nucleoid

The genomes of prokaryotic and eukaryotic cells are organized differently. In prokaryotic cells, DNA is present in a large double-stranded molecule called the *chromosome*. The chromosome aggregates within the cell to form a mass visible in the electron microscope, called the **nucleoid** (**Figure 2.14**). We will see in Chapter 7 that the nucleoid is covalently closed and circular in most prokaryotes.

Most prokaryotes have only a single chromosome. Because of this, they typically contain only a single copy of each gene and are therefore genetically *haploid*. Many prokaryotes

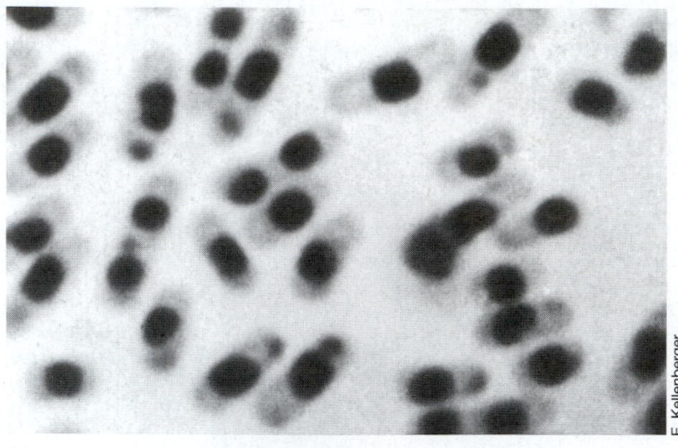

(a)

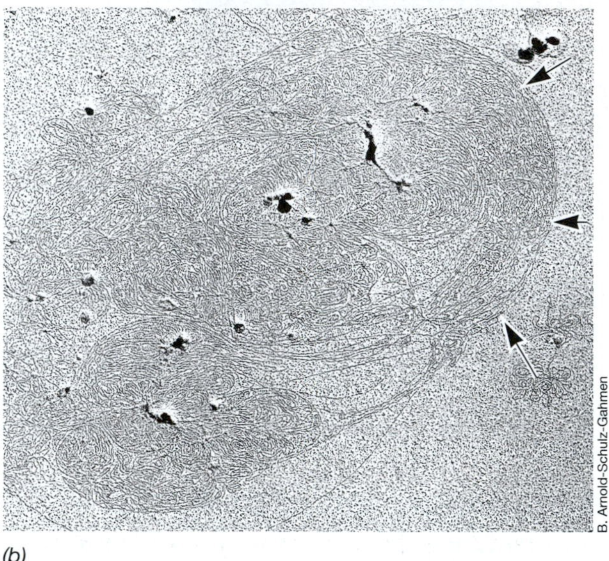

(b)

Figure 2.14 The nucleoid. *(a)* Photomicrograph of cells of *Escherichia coli* treated in such a way as to make the nucleoid visible. A single cell is about 3 μm in length. *(b)* Transmission electron micrograph of an isolated nucleoid released from a cell of *E. coli*. The cell was gently lysed to allow the highly compacted nucleoid to emerge intact. Arrows point to the edge of DNA strands.

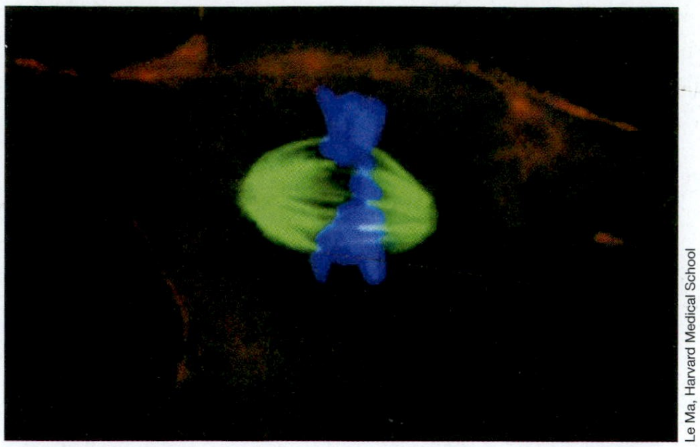

Figure 2.15 Mitosis in stained kangaroo rat cells. The cell was photographed while in the metaphase stage of mitotic division; only eukaryotic cells undergo mitosis. The green color stains a protein called tubulin, important in pulling chromosomes apart (∞ Section 18.1). The blue color is from a DNA-binding dye and shows the chromosomes.

also contain small amounts of circular extrachromosomal DNA called **plasmids**. Plasmids typically contain genes that confer special properties (such as unique metabolisms) on a cell. This is in contrast to essential ("housekeeping") genes, which are needed for basic survival and are located on the chromosome.

In eukaryotes, DNA is present in linear molecules within the membrane-enclosed nucleus; the DNA molecules are packaged with proteins and organized to form **chromosomes**. Chromosome number varies considerably. For example, the baker's yeast *Saccharomyces cerevisiae* contains 16 chromosomes arranged in 8 pairs, and human cells contain 46 chromosomes (23 pairs). Chromosomes in eukaryotes contain proteins that assist in folding and packing the DNA and other proteins that are required for gene expression (∞ Section 8.5). A key genetic difference between prokaryotes and eukaryotes is that eukaryotes typically contain two copies of each gene and are thus genetically *diploid*. During cell division in eukaryotic cells the nucleus divides (following a doubling of chromosome number) in the process called *mitosis* (**Figure 2.15**). Two identical daughter cells result, and each daughter cell receives a full complement of genes.

The diploid genome of eukaryotic cells is halved in the process of *meiosis* to form haploid gametes for sexual reproduction. Fusion of two gametes during zygote formation restores the cell to the diploid state. We discuss these processes in more detail in Chapter 8.

Genes, Genomes, and Proteins

How many genes and proteins does a cell have? The genome of *Escherichia coli*, a typical prokaryote, is a single circular chromosome of 4.68 million base pairs of DNA. Because the *E. coli* genome has been completely sequenced, we also know that it contains about 4,300 genes. The genomes of some prokaryotes have nearly three times this many genes, and the genomes of others contain fewer than one-eighth as many (∞ Table 13.1). Eukaryotic cells typically have much larger genomes than prokaryotes. A human cell, for example, contains over 1,000 times as much DNA as a cell of *E. coli* and about seven times as many genes.

A single cell of *E. coli* contains about 1,900 different kinds of proteins and a total of about 2.4 million protein molecules (∞ Table 3.2). However, some proteins in *E. coli* are very abundant, others are only moderately abundant, and some are present in only one or a very few copies per cell. Thus, *E. coli* has mechanisms for regulating its genes so that not all genes are *expressed* (transcribed and translated, ∞ Figure 1.4) at the same time or to the same extent. Gene regulation is an important mechanism in all cells, and we focus on the major mechanisms of gene regulation in Chapter 9.

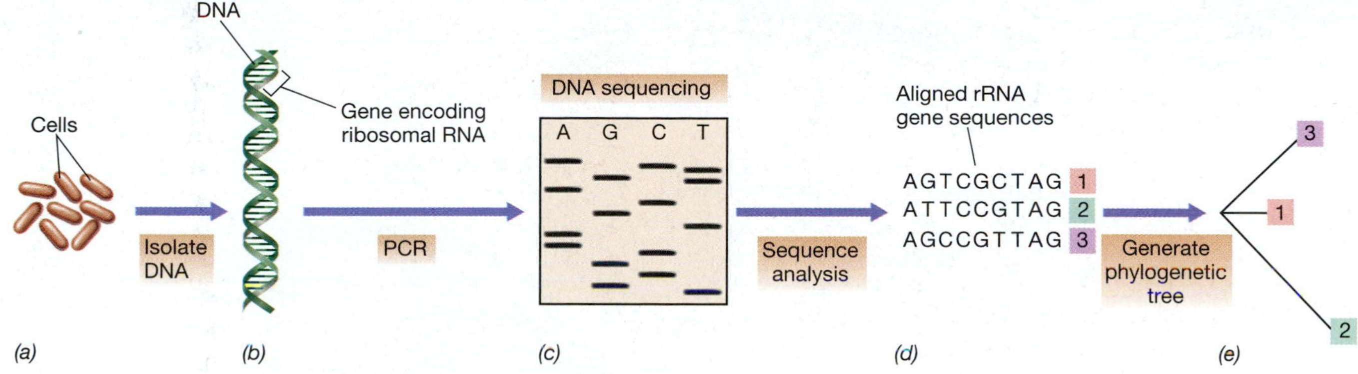

Figure 2.16 Ribosomal RNA (rRNA) gene sequencing and phylogeny. *(a)* Cells are broken open. *(b)* The gene-encoding rRNA is isolated, and many identical copies are made by the technique called the polymerase chain reaction (∞ Section 12.8). *(c, d)* The gene is sequenced (∞ Section 12.5), and the sequence obtained is aligned with other rRNA sequences. A computer algorithm makes pairwise comparisons and generates a phylogenetic tree *(e)* that depicts the differences in rRNA sequence between the organisms analyzed. In the example shown, the sequence differences are as follows: organism 1 versus organism 2, three differences; 1 versus 3, two differences; 2 versus 3, four differences. Thus organisms 1 and 3 are closer relatives than are 2 and 3 or 1 and 2.

2.6 MiniReview

Genes govern the properties of cells, and a cell's complement of genes is called its genome. DNA is arranged in cells to form chromosomes. Most prokaryotic species have a single circular chromosome; eukaryotic species have multiple linear chromosomes.

- Differentiate between the nucleus and the nucleoid.
- How do plasmids differ from chromosomes?
- Why does it make sense that a human cell would have more genes than a bacterial cell?

2.7 The Evolutionary Tree of Life

Evolution is the process of change in a line of descent over time that results in the appearance of new varieties and species of organisms. Evolution occurs in any self-replicating system in which variation occurs as the result of mutation and selection and differential fitness is a potential result. Thus, over time, all cells and viruses evolve.

Determining Evolutionary Relationships

The evolutionary relationships between organisms are the subject of **phylogeny**. Phylogenetic relationships between cells can be deduced by comparing the genetic information (nucleotide or amino acid sequences) that exists in their nucleic acids or proteins (∞ Chapter 14). Macromolecules that form the ribosome, in particular *ribosomal RNAs* (*rRNA*), turn out to be excellent tools for determining evolutionary relationships. Because all cells contain ribosomes (and thus rRNA), this molecule can and has been used to construct a phylogenetic tree of all cells, including microorganisms (see Figure 2.17). Viral phylogenies have also been determined, but because these microorganisms lack ribosomes, other molecules have been used as evolutionary barometers. Carl Woese, an American microbiologist, pioneered the use of rRNA as a barometer of microbial phylogeny and, in so doing, revolutionized our understanding of cellular evolution.

The steps in generating an RNA-based phylogenetic tree are outlined in **Figure 2.16.** In brief, genes encoding rRNA from two or more organisms are sequenced (that is, the precise order of nucleotides in the molecule are determined, ∞ Section 12.5), and the sequences are aligned and inspected, base-by-base, using a computer. The greater the rRNA gene sequence variation between any two organisms, the greater their evolutionary divergence. This divergence can then be depicted in a phylogenetic tree (Figure 2.16).

The Three Domains of Life

From comparative rRNA sequencing, three phylogenetically distinct lineages of cells have been identified. The lineages, called **domains**, are the ***Bacteria*** and the ***Archaea*** (both consisting of prokaryotes) and the ***Eukarya*** (eukaryotes) (**Figure 2.17**). The domains are thought to have diverged from a common ancestral organism or community of organisms early in the history of life on Earth.

The phylogenetic tree of life reveals two very important evolutionary facts: (1) as previously stated, all prokaryotes are *not* phylogenetically closely related and (2) *Archaea* are more closely related to *Eukarya* than to *Bacteria* (Figure 2.17). Thus, from the last universal common ancestor of all life, evolutionary diversification initially went in two directions: *Bacteria* and a second main lineage. This second lineage eventually diverged to yield the *Archaea,* which retained a prokaryotic cell structure, and the *Eukarya,* which did not. In this book when the word "bacteria" (lower case b and no italics) is used, the reference should be understood to be to some species of the domain *Bacteria*.

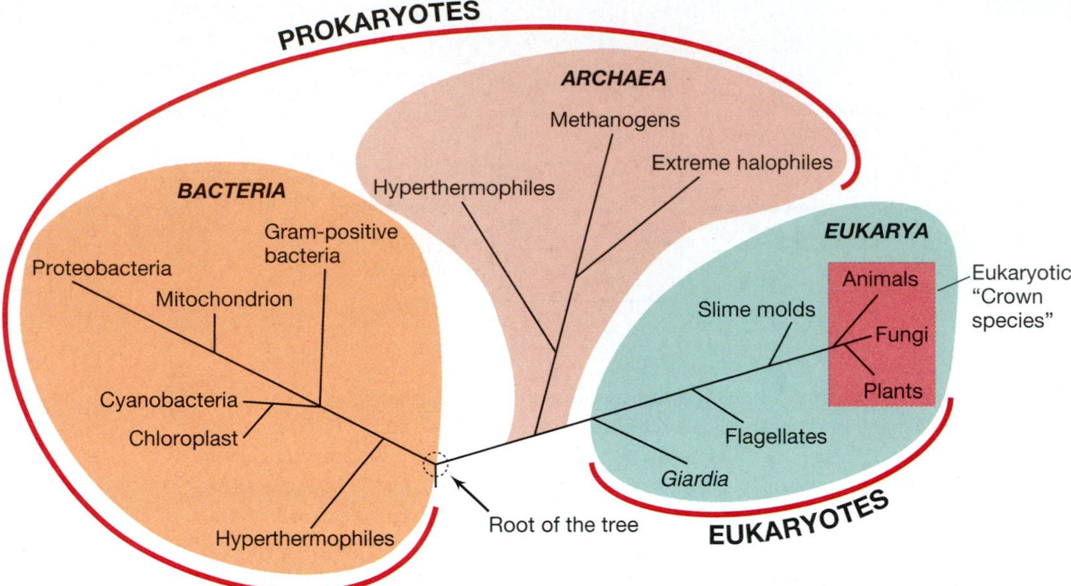

Figure 2.17 The phylogenetic tree of life as defined by comparative rRNA gene sequencing. The tree consists of three domains of organisms: the *Bacteria* and the *Archaea*, cells of which are prokaryotic, and the *Eukarya* (eukaryotes). Only a few of the groups of organisms within each domain are shown. Hyperthermophiles are prokaryotes that grow best at temperatures of 80°C or higher. The groups shaded in red are macroorganisms. All other organisms on the tree of life are microorganisms. Single domain phylogenetic trees can be found in Figures 2.19, 2.28, and 2.32.

Eukarya

Because the cells of animals and plants are all eukaryotic, it follows that eukaryotic microorganisms were the ancestors of multicellular organisms. The tree of life clearly bears this out. As expected, microbial eukaryotes branch off early on the eukaryotic lineage, and plants and animals branch near the tree's crown (Figure 2.17). However, molecular sequencing and other evidence have shown that eukaryotic cells contain genes from cells of two domains. In addition to the genome in the chromosomes of the nucleus, mitochondria and chloroplasts of eukaryotes contain their own genomes (DNA arranged in circular fashion, as in prokaryotes) and ribosomes. Using rRNA sequencing technology (Figure 2.16), these organelles have been shown to be highly derived ancestors of specific lineages of *Bacteria* (Figure 2.17 and Section 2.9). Mitochondria and chloroplasts were thus once free-living bacterial cells that took up an intracellular existence in cells of *Eukarya* eons ago. The process by which this stable arrangement developed is known as **endosymbiosis** and is discussed in later chapters (∞ Sections 14.4 and 18.4).

Contributions of Molecular Sequencing to Microbiology

Molecular phylogenies have confirmed the evolutionary connections among all cells. Application of molecular sequencing has also created an evolutionary framework for the prokaryotes, something that the science of microbiology had been without since its founding. In addition, RNA-based phylogenies have spawned new tools that have affected many subdisciplines of microbiology. These include, in particular, microbial classification, microbial ecology, and clinical diagnostics. In these areas molecular phylogeny has helped shape our concept of a bacterial species and has given microbial ecologists and clinical microbiologists the tools necessary to identify organisms without actually culturing them. This has greatly improved our picture of microbial diversity and has led to the staggering conclusion that most of the microbial diversity that exists on Earth has yet to be brought into laboratory culture.

2.7 MiniReivew

Comparative rRNA sequencing has defined the three domains of life: *Bacteria*, *Archaea*, and *Eukarya*. Molecular sequencing has also shown that the organelles of *Eukarya* have evolutionary roots in the *Bacteria* and has yielded new tools for microbial ecology and clinical microbiology.

- How can species of *Bacteria* and *Archaea* be distinguished using molecular biology?
- How did the process of endosymbiosis benefit eukaryotic cells?

II MICROBIAL DIVERSITY

Evolution has molded all life on Earth. The diversity we see in microbial cells today is the result of nearly 4 billion years of evolutionary change (∞ Figure 1.6). Microbial diversity can be seen in many forms, including cell size and cell morphology (shape), physiology, motility, mechanisms of cell division, pathogenicity, developmental biology, adaptation to environmental extremes, phylogeny, and so on. In the following sections we paint a picture of microbial diversity with a broad brush. We return to the theme of microbial diversity in more detail in Chapters 14–19.

We preface our discussion of *microbial* diversity by first considering *metabolic* diversity. The two are closely linked. Microorganisms have exploited every conceivable means of "making a living" consistent with the laws of chemistry and physics. This enormous versatility has allowed microorganisms to inhabit every conceivable habitat on and in planet Earth. Metabolic diversity will be covered in detail in Chapters 5, 6, 20, and 21.

2.8 Physiological Diversity of Microorganisms

All cells require energy and a means to conserve it for other uses. Energy can be obtained from three sources in nature: organic chemicals, inorganic chemicals, and light (**Figure 2.18**).

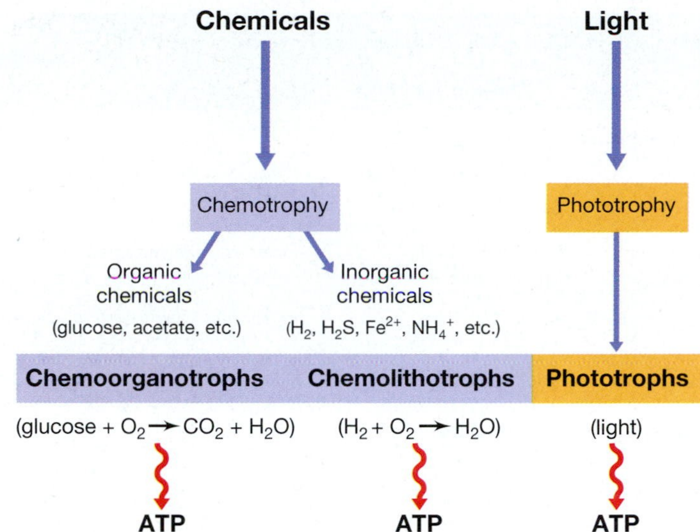

Figure 2.18 Metabolic options for conserving energy. The organic and inorganic chemicals listed here are just a few of the many different chemicals used by various chemotrophic organisms. Chemotrophic organisms oxidize organic or inorganic chemicals, which yields ATP. Phototrophic organisms convert solar energy to chemical energy in the form of ATP.

Chemoorganotrophs

Organisms that obtain energy from chemicals are called *chemotrophs*, and those that use organic chemicals are called **chemoorganotrophs** (Figure 2.18). Thousands of different organic chemicals can be used by one or another microorganism. Indeed, all natural and even most synthetic organic compounds can be metabolized. Energy is conserved from the oxidation of the compound and is stored in the cell as the energy-rich compound adenosine triphosphate (ATP).

Some microorganisms can extract energy from an organic compound only in the presence of oxygen; these organisms are called *aerobes*. Others can extract energy only in the absence of oxygen (*anaerobes*). Still others can break down organic compounds in either the presence or absence of oxygen. Most microorganisms that have been brought into laboratory culture are chemoorganotrophs.

Chemolithotrophs

Many prokaryotes can tap the energy available in inorganic compounds. This is a form of metabolism called *chemolithotrophy* (discovered by Winogradsky, ∞ Section 1.9) and is carried out by organisms called **chemolithotrophs** (Figure 2.18). Chemolithotrophy is a process found only in prokaryotes and is widely distributed among species of *Bacteria* and *Archaea*. The spectrum of different inorganic compounds used is quite broad, but typically, a particular group of prokaryotes specializes in the use of a related group of inorganic compounds.

It should be obvious why the capacity to conserve energy from the oxidation of inorganic chemicals is a good metabolic strategy—competition from chemoorganotrophs is not an issue. But in addition to this, many of the inorganic compounds oxidized by chemolithotrophs, for example H_2 and H_2S, are actually the waste products of chemoorganotrophs. Thus, chemolithotrophs have evolved strategies for exploiting resources that chemoorganotrophs are unable to use.

Phototrophs

Phototrophic microorganisms contain pigments that allow them to use light as an energy source, and thus their cells are colored (Figure 2.2). Unlike chemotrophic organisms, **phototrophs** do not require chemicals as a source of energy; they synthesize ATP from the energy of sunlight. This is a significant metabolic advantage because competition for energy sources with chemotrophic organisms is not an issue and light is available in a wide variety of microbial habitats.

Two major forms of phototrophy are known in prokaryotes. In one form, called *oxygenic* photosynthesis, oxygen (O_2) is produced. Among microorganisms, oxygenic photosynthesis is characteristic of cyanobacteria, algae, and their phylogenetic relatives. The other form, *anoxygenic* photosynthesis, occurs in the purple and green bacteria and does not result in O_2 production. Both groups of phototrophs use light to make ATP, however, and we will see later the great similarities in their mechanisms of ATP synthesis. We consider photosynthesis in more detail in Chapter 20.

Heterotrophs and Autotrophs

All cells require carbon as a major nutrient. Microbial cells are either **heterotrophs**, which require one or more organic compounds as their carbon source, or **autotrophs**, which use carbon dioxide (CO_2) as their carbon source. Chemoorganotrophs are by definition heterotrophs. By contrast, most

Table 2.1 Classes and examples of extremophiles[a]

Extreme	Descriptive term	Genus/species	Domain	Habitat	Minimum	Optimum	Maximum
Temperature							
High	Hyperthermophile	*Pyrolobus fumarii*	*Archaea*	Hot, undersea hydrothermal vents	90°C	**106°C**	113°C[b]
Low	Psychrophile	*Polaromonas vacuolata*	*Bacteria*	Sea ice	0°C	**48°C**	128°C
pH							
Low	Acidophile	*Picrophilus oshimae*	*Archaea*	Acidic hot springs	−0.06	**0.7**[c]	4
High	Alkaliphile	*Natronobacterium gregoryi*	*Archaea*	Soda lakes	8.5	**10**[d]	12
Pressure	Barophile	*Moritella yayanosii*[e]	*Bacteria*	Deep ocean sediments	500 atm	**700 atm**	>1000 atm
Salt (NaCl)	Halophile	*Halobacterium salinarum*	*Archaea*	Salterns	15%	**25%**	32% (saturation)

[a]The organisms listed are the current "record holders" for growth at a particular extreme condition.

[b]*Geogemma barossii*, a new species of hyperthermophilic *Archaea*, has been reported to grow at 121°C. However, *Pyrolobus* remains the best-characterized prokaryote growing above 110°C.

[c]*P. oshimae* is also a thermophile, growing optimally at 60°C.

[d]*N. gregoryi* is also an extreme halophile, growing optimally at 20% NaCl.

[e]*Moritella yayanosii* is also a psychrophile, growing optimally at about 4°C.

chemolithotrophs and virtually all phototrophs are autotrophs. Autotrophs are sometimes called *primary producers* because they synthesize organic matter from CO_2 for both their own benefit and that of chemoorganotrophs. The latter either feed directly on the primary producers or live off products they excrete. In one way or another, all organic matter on Earth has been synthesized by primary producers; in particular, by phototrophs.

Habitats and Extreme Environments

Microorganisms are present everywhere on Earth that will support life. These include habitats we are all familiar with—soil, water, animals, and plants—as well as virtually any structures made by humans. Indeed, sterility (the absence of life forms) in a natural sample of any sort is very rare.

Some microbial environments are those that we humans would find too extreme for life. Although these environments can pose challenges to microorganisms as well, extreme environments are often teeming with microbial life. Organisms inhabiting extreme environments are called **extremophiles**, a remarkable group of microorganisms that collectively define the physiochemical limits of life.

Extremophiles abound in such harsh environments as boiling hot springs; on or in the ice covering lakes, glaciers, or the polar seas; in extremely salty bodies of water; and in soils and waters having a pH as low as 0 or as high as 12. These prokaryotes do not just *tolerate* these extremes, but actually *require* the extreme condition in order to grow. That is why they are called extremophiles (the suffix *phile* means "loving"). **Table 2.1** summarizes the current "record holders" among extremophiles and lists the types of habitats in which they reside. We revisit many of these organisms in later chapters (Chapters 6 and 15–17).

2.8 MiniReview

Carbon and energy sources are needed by all cells. The terms chemoorganotroph, chemolithotroph, and phototroph refer to organisms that use organic chemicals, inorganic chemicals, or light, respectively, as their source of energy. Autotrophic organisms use CO_2 as their carbon source, and heterotrophs use organic carbon. Extremophiles thrive under environmental conditions that other organisms cannot tolerate.

- How might you distinguish a phototrophic microorganism from a chemotrophic one by simply looking at it under a microscope?
- What are extremophiles?

2.9 Bacteria

As we have seen, prokaryotes cluster into two phylogenetically distinct domains, the *Archaea* and the *Bacteria* (Figure 2.17). We begin with the *Bacteria,* because most of the known prokaryotes reside in this domain.

Proteobacteria, Gram-Positive Bacteria, and Cyanobacteria

The domain *Bacteria* contains an enormous variety of prokaryotes. All known disease-causing (pathogenic) prokaryotes are *Bacteria,* as are thousands of nonpathogenic species. There is a large variety of morphologies and physiologies in this domain. The ***Proteobacteria*** make up the largest division of *Bacteria* (**Figure 2.19**). Many chemoorganotrophic bacteria

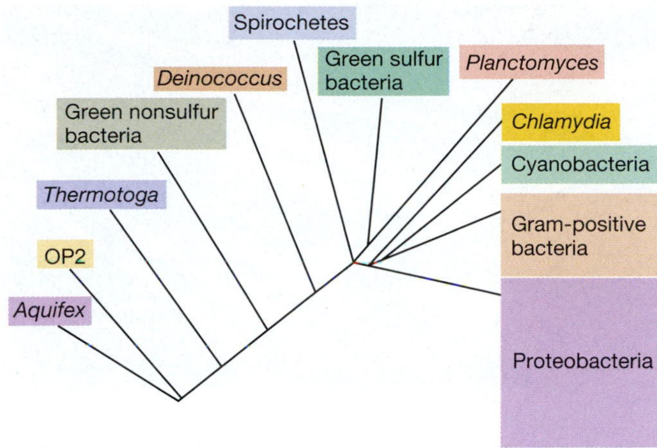

Figure 2.19 Phylogenetic tree of *Bacteria*. The relative sizes of the colored boxes reflect the number of known genera and species in each of the groups. The *Proteobacteria* are the largest group of *Bacteria* known. The lineage on the tree labeled OP2 does not represent a cultured organism but instead is a sequence of an rRNA gene isolated from an organism in a natural sample. In this example, the closest known relative of OP2 would be *Aquifex*. Although not shown on this tree, many thousands of other environmental sequences are known, and they branch all over the tree. Not all known groups of *Bacteria* are depicted on this tree.

are *Proteobacteria,* including *Escherichia coli*, the model organism of microbial physiology, biochemistry, and molecular biology. Several phototrophic and chemolithotrophic species are also Proteobacteria (**Figure 2.20**). Many of these use hydrogen sulfide (H_2S, the smell from rotten eggs) in their metabolism, producing elemental sulfur that is stored within or outside the cell (Figure 2.20). The sulfur is an oxidation product of H_2S and is further oxidized to sulfate (SO_4^{2-}). The sulfide and sulfur are oxidized to fuel important metabolic functions such as CO_2 fixation (autotrophy) or energy generation (Figure 2.18).

Several other common prokaryotes of soil and water, and species that live in or on plants and animals in both harmless and disease-causing ways, are members of the *Proteobacteria.* These include species of *Pseudomonas,* many of which can degrade complex and otherwise toxic natural and synthetic organic compounds, and *Azotobacter,* a nitrogen-fixing bacterium. A number of key pathogens are *Proteobacteria,* including *Salmonella, Rickettsia, Neisseria,* and many others. And finally, it is from within the *Proteobacteria* that the mitochondrion arose by endosymbiosis (Section 2.7), probably several times during the course of evolution (Figure 2.17).

As we learned in Section 2.2, bacteria can be distinguished by the Gram-staining procedure, a technique that stains cells either gram-positive or gram-negative. The gram-positive phylum of *Bacteria* (Figure 2.19) contains many organisms that are united by their common phylogeny and cell wall structure. Here we find the endospore-forming *Bacillus* (discovered by Ferdinand Cohn, ∞ Section 1.6) (**Figure 2.21*a***) and *Clostridium* and related spore-forming bacteria such as the antibiotic-producing *Streptomyces*. Also included

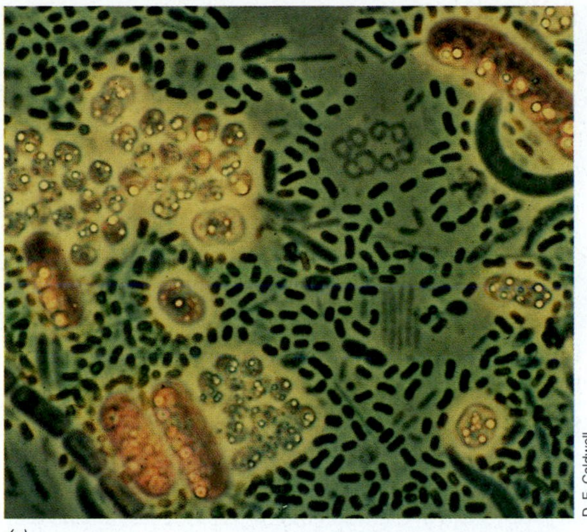

(a)

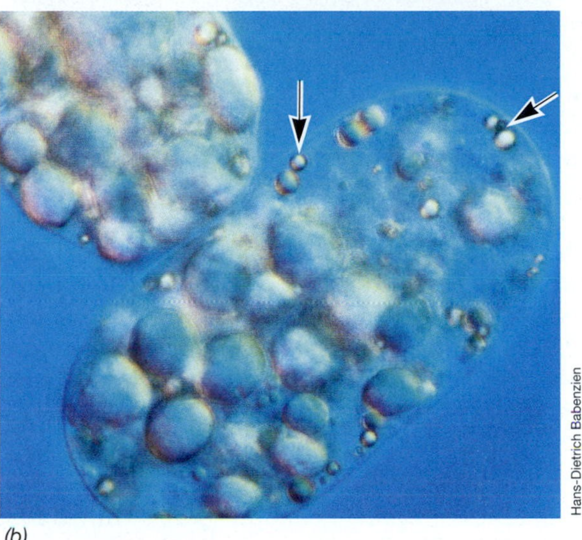

(b)

Figure 2.20 Phototrophic and chemolithotrophic *Proteobacteria*. *(a)* The phototrophic purple sulfur bacterium, *Chromatium* (the large, red-orange, rod-shaped cells in this photomicrograph of a natural microbial community). A cell is about 10 μm wide. (b) The large chemolithotrophic sulfur-oxidizing bacterium, *Achromatium*. A cell is about 20 μm wide. Globules of elemental sulfur can be seen in the cells (arrows). Both of these organisms oxidize hydrogen sulfide (H_2S).

here are the lactic acid bacteria, common inhabitants of decaying plant material and dairy products that include organisms such as *Streptococcus* (Figure 2.11*b*) and *Lactobacillus*. Other interesting bacteria that fall among the phylogenetic division of gram-positive bacteria are the mycoplasmas. These bacteria lack a cell wall and have very small genomes; many of them are pathogenic. *Mycoplasma* is a major genus of organisms in this medically important group (∞ Section 16.3).

The **cyanobacteria** are phylogenetic relatives of gram-positive bacteria (Figure 2.19) and are oxygenic phototrophs. The photosynthetic organelle of eukaryotic phototrophs, the chloroplast (Figure 2.2*a*), is related to the cyanobacteria (Figure 2.17). Cyanobacteria were critical in the evolution of life,

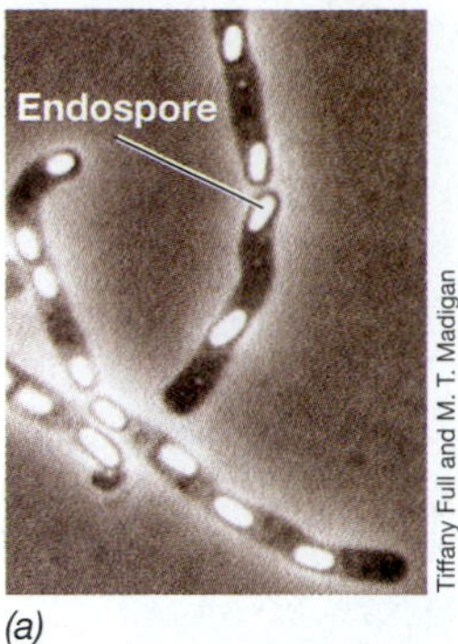

(a)

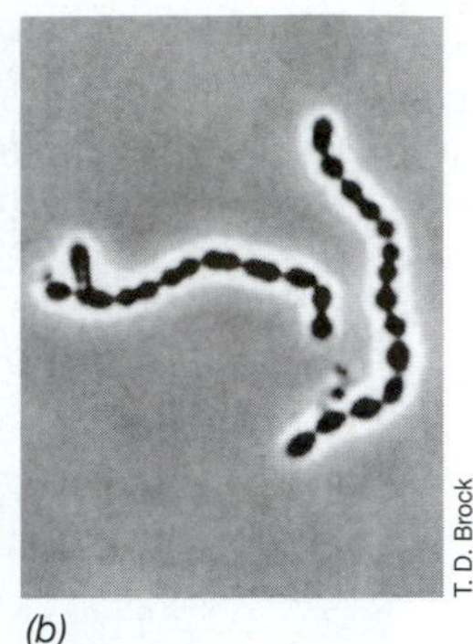

(b)

Figure 2.21 Gram-positive bacteria. *(a)* The rod-shaped endospore-forming bacterium *Bacillus,* here shown as cells in a chain. Note the presence of endospores (bright refractile structures) inside the cells. Endospores are extremely resistant to heat, chemicals, and radiation. Cells are about 16 μm in diameter. *(b) Streptococcus,* a spherical cell that exists in chains. Streptococci are widespread in dairy products, and some are potent pathogens. Cells are about 0.8 μm in diameter.

as they were the first oxygenic phototrophs to evolve on Earth. The production of O_2 on an originally anoxic Earth paved the way for the evolution of prokaryotes that could respire using oxygen. The development of higher organisms, such as the plants and animals, followed billions of years later when Earth had a more oxygen-rich environment (∞ Figure 1.6). Cells of some cyanobacteria join to form filaments (**Figure 2.22**). Many other morphological forms of cyanobacteria are known, including unicellular, colonial, and heterocystous cyanobacteria, which contain special structures called *heterocysts* that carry out nitrogen fixation (∞ Section 16.7).

Other Major Phyla of Bacteria

Several lineages of *Bacteria* contain species with unique morphologies. These include the aquatic Planctomyces group, characterized by cells with a distinct stalk that allows the organisms to attach to a solid substratum (**Figure 2.23**), and the helically shaped spirochetes (**Figure 2.24**). Several diseases, most notably syphilis and Lyme disease (∞ Sections 34.13 and 35.4, respectively), are caused by spirochetes.

Two other major lineages of *Bacteria* are phototrophic: the green sulfur bacteria and the green nonsulfur bacteria (*Chloroflexus* group) (**Figure 2.25**). Species in both of these lineages contain similar photosynthetic pigments and are also autotrophs. *Chloroflexus* is a filamentous prokaryote that inhabits hot springs and shallow marine bays and is often the dominant organism in stratified microbial mats, laminated microbial communities (∞ Section 22.7). *Chloroflexus* is also noteworthy because it is believed to be an important link in the evolution of photosynthesis (∞ Sections 16.18 and 20.7).

Other major lineages of *Bacteria* include the Chlamydia and Deinococcus groups (Figure 2.19). The genus *Chlamydia* harbors respiratory and sexually transmitted pathogens of humans (∞ Sections 16.9 and 34.14). Chlamydia are obligate intracellular parasites. By this it is meant that they live *inside* the cells of higher organisms, in this case, human cells.

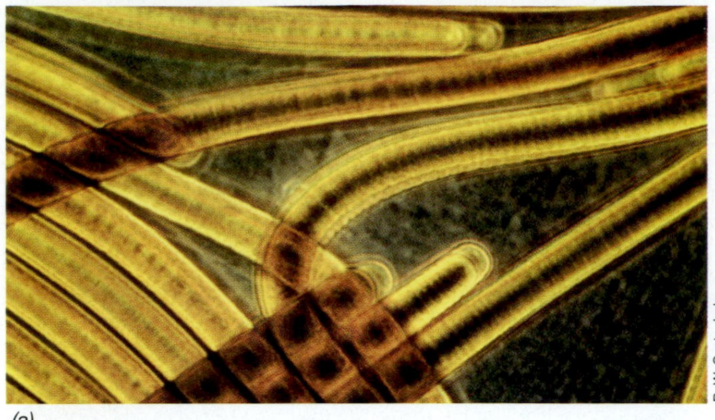

(a)

(b)

Figure 2.22 Filamentous cyanobacteria. *(a) Oscillatoria, (b) Spirulina.* Cells of both organisms are about 10 μm wide.

Several other pathogenic prokaryotes (for example, species of *Rickettsia,* a genus of *Proteobacteria* that causes diseases such as typhus and Rocky Mountain spotted fever, and *Mycobacterium tuberculosis,* a gram-positive bacterium that causes tuberculosis) are also intracellular pathogens. By living inside their host's cells, these pathogens avoid destruction by the host's immune response.

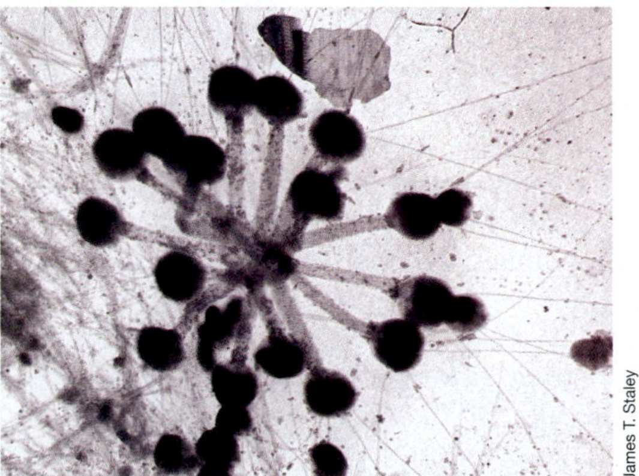

Figure 2.23 The morphologically unusual stalked bacterium *Planctomyces.* Shown are several cells attached by their stalks to form a rosette. Cells are about 1.4 μm wide.

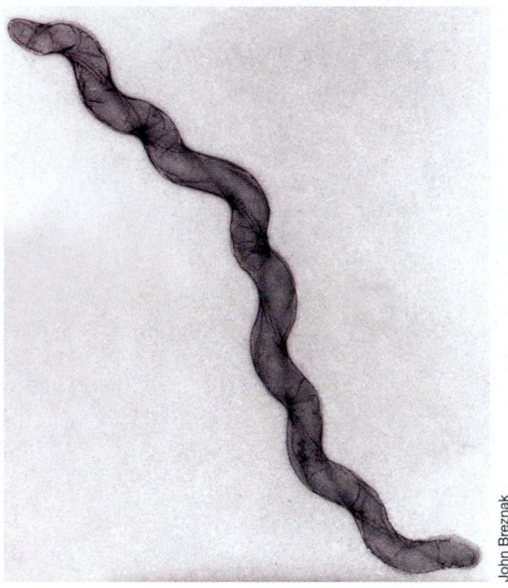

Figure 2.24 Spirochetes. Scanning electron micrograph of a cell of *Spirochaeta zuelzerae*. The cell is about 0.3 μm wide and tightly coiled.

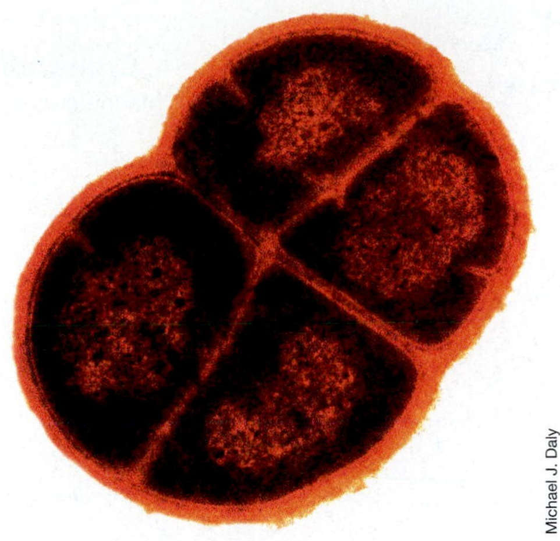

Figure 2.26 The highly radiation-resistant bacterium *Deinococcus radiodurans*. A single cell is about 2.5 μm wide.

The Deinococcus phylum contains species with unusual cell walls and an innate resistance to high levels of radiation; *Deinococcus radiodurans* (**Figure 2.26**) is a major species in this group. This organism can survive doses of radiation many-fold greater than that sufficient to kill animals and can even reassemble its chromosome after it has been shattered by radiation. We learn more about this amazing organism in Section 16.17.

Finally, several phyla of *Bacteria* branch off very early on the phylogenetic tree, very near the root (Figure 2.19). Although phylogenetically distinct from one another, these groups are unified by the common property of growth at very high temperature (*hyperthermophily*, Table 2.1). Organisms such as *Aquifex* (**Figure 2.27**) and *Thermotoga* grow in hot springs that are near the boiling point. The early phylogenetic branching of these phyla (Figure 2.19) is consistent with the hypothesis that the early Earth was much hotter than it is today. Assuming that early life forms were hyperthermophilic, it is not surprising that their closest living relatives would be hyperthermophiles themselves. Interestingly, the phylogenetic trees of both *Bacteria* and *Archaea* support this (Figure 2.19 and see Figure 2.28); organisms such as *Aquifex, Methanopyrus,* and *Pyrolobus* are thus modern descendants of very ancient cell lineages.

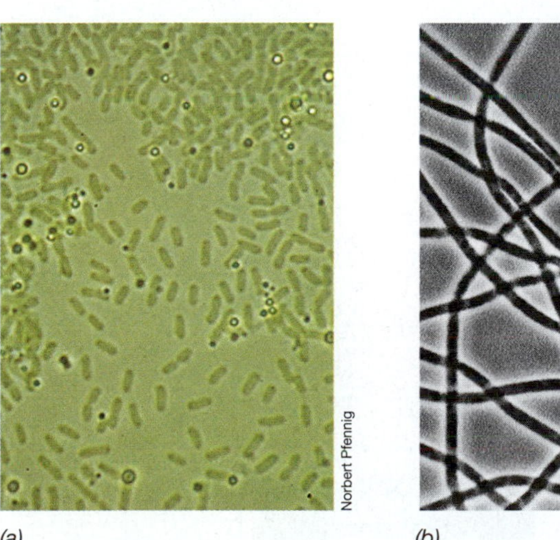

Figure 2.25 Phototrophic green bacteria. *(a) Chlorobium* (green sulfur bacteria). A single cell is about 0.8 μm wide. *(b) Chloroflexus* (green nonsulfur bacteria). A filament is about 1.3 μm wide. Despite sharing many features such as pigments and photosynthetic membrane structures, these two genera are phylogenetically distinct (Figure 2.19).

2.9 MiniReview

Several phyla of *Bacteria* are known, and an enormous diversity of cell morphologies and physiologies are represented.

- What is the largest phylum of *Bacteria*?
- Why can it be said that the cyanobacteria prepared Earth for the evolution of higher life forms?
- What is physiologically unique about *Deinococcus*?

Figure 2.27 The hyperthemophile *Aquifex*. Transmission electron micrograph using a technique called freeze-etching, where a frozen replica of the cell is made and then visualized. The cell is about 0.5 μm wide.

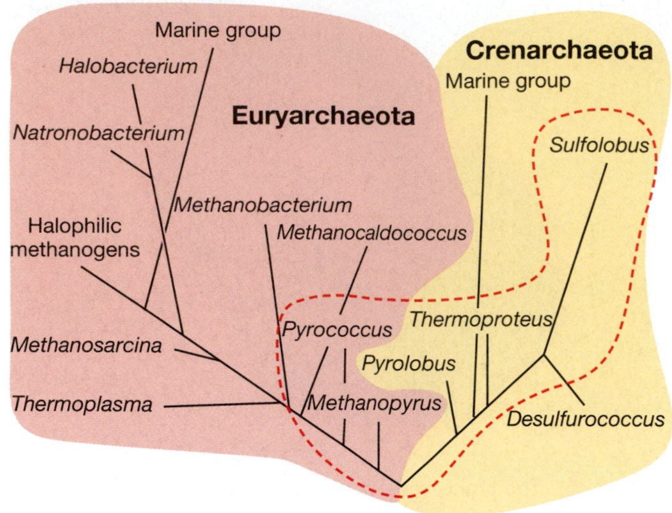

Figure 2.28 Phylogenetic tree of *Archaea*. The organisms circled in red are hyperthermophiles, growing at very high temperatures (*Crenarchaeota*). In pink are shown methanogens and the extreme halophiles and acidophiles (*Euryarchaeota*). The "marine group" sequences are environmental rRNA sequences from marine *Archaea* that are not yet cultured. Not all known groups of *Archaea* are depicted on this tree.

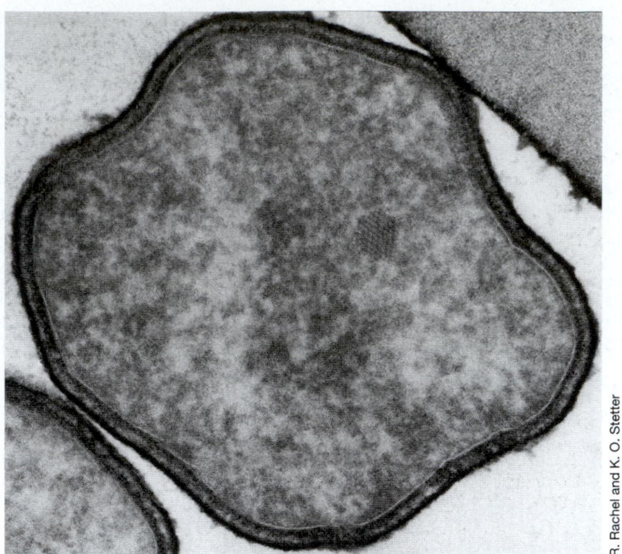

Figure 2.29 *Pyrolobus*. This hyperthermophile grows optimally above the boiling point of water. The cell is 1.4 μm wide.

2.10 *Archaea*

Two phyla exist in the domain *Archaea,* the *Euryarchaeota* and the *Crenarchaeota* (**Figure 2.28**). Each of these forms a major branch on the archaeal tree. Most cultured *Archaea* are extremophiles, with species capable of growth at the highest temperatures, salinities, and extremes of pH of all known microorganisms. The organism *Pyrolobus* (**Figure 2.29**), for example, is one of the most thermophilic of all known prokaryotes (Table 2.1).

Although all *Archaea* are chemotrophic, *Halobacterium* (to be discussed shortly) can use light to make ATP but in a way quite distinct from that of phototrophic organisms. Some *Archaea* use organic compounds in their energy metabolism. Many others are chemolithotrophs, with hydrogen gas (H_2) being a widely used energy source. Chemolithotrophy is particularly widespread among hyperthermophilic *Archaea*.

Euryarchaeota

The *Euryarchaeota* branch on the tree of *Archaea* (Figure 2.28) contains three groups of organisms that have dramatically different physiologies, the methanogens, the extreme halophiles, and the thermoacidophiles. Some of these require O_2 whereas others are killed by it, and some grow at the upper or lower extremes of pH (Table 2.1). Methanogens such as *Methanobacterium* are strict anaerobes. Their metabolism is unique in that energy is conserved during the production of methane (natural gas). Methanogens are important organisms in the anaerobic degradation of organic matter in nature, and most of the natural gas found on Earth is a result of their metabolism.

The extreme halophiles are relatives of the methanogens (Figure 2.28), but are physiologically distinct from them. Unlike methanogens, which are killed by oxygen, most extreme halophiles require oxygen, and all are unified by their requirement for very large amounts of salt (NaCl) for metabolism and reproduction. It is for this reason that these organisms are called *halophiles* (salt lovers). In fact, organisms like *Halobacterium* are so salt loving that they can actually grow on and within salt crystals (**Figure 2.30**).

As we have seen, many prokaryotes are phototrophic and can generate adenosine triphosphate (ATP) from light (Section 2.8). Although *Halobacterium* species do not produce chlorophyll as phototrophs do, they nevertheless contain

Figure 2.30 Extremely halophilic *Archaea*. A vial of brine with precipitated salt crystals containing cells of the extreme halophile, *Halobacterium*. The organism contains pigments that absorb light and lead to ATP production. Cells of *Halobacterium* can also live within salt crystals themselves (Microbial Sidebar, Chapter 4, How Long Can an Endospore Survive?).

light-sensitive pigments that can absorb light and trigger ATP synthesis (∞ Section 17.3). Extremely halophilic *Archaea* inhabit salt lakes, salterns, and other very salty environments. Some extreme halophiles, such as *Natronobacterium,* inhabit soda lakes, environments characterized by high levels of salt and high pH. Such organisms are *alkaliphilic* and grow at the highest pH of all known organisms (Table 2.1).

The third group of *Euryarchaeota* are the thermoacidophiles. These include *Thermoplasma* (**Figure 2.31**), an organism that, like *Mycoplasma* (Section 2.9), contains a cytoplasmic membrane but no cell wall. *Thermoplasma* grows best at moderately high temperatures and low pH. The thermoacidophiles also includes *Picrophilus,* the most acidophilic (acid-loving) of all known prokaryotes (Table 2.1).

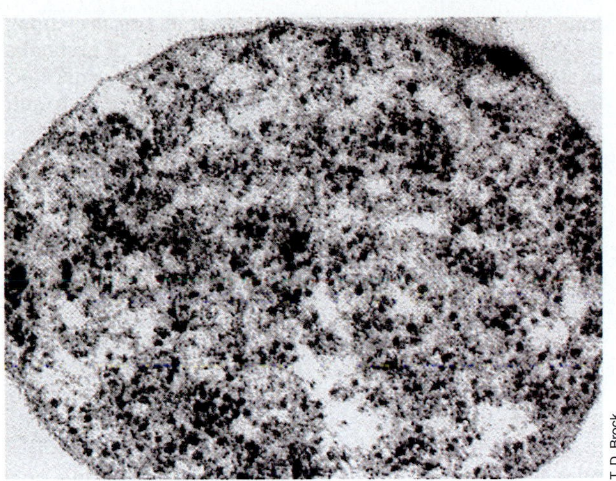

Figure 2.31 Extremely acidophilic *Archaea*. The organism *Thermoplasma* lacks a cell wall. The cell measures 1 μm wide.

Crenarchaeota

The vast majority of cultured *Crenarchaeota* are hyperthermophiles (Figure 2.29). These organisms are either chemolithotrophs or chemoorganotrophs and grow in such high-temperature environments as hot springs and hydrothermal vents (deep-sea hot springs). For the most part these organisms are anaerobes (because of the high temperature, their habitats are typically anoxic), and many of them use hydrogen gas (H_2) present in their geothermal habitats as an energy source.

Some *Crenarchaeota* inhabit environments that contrast dramatically with high-temperature environments. For example, many of the prokaryotes drifting in the open oceans are *Crenarchaeota:* Their environment is fully oxic and cold (~3°C). Some marine *Crenarchaeota* are chemolithotrophs that use ammonia (NH_3) as their energy source, but we know little about the metabolic activities of most marine *Archaea*. *Crenarchaeota* have also been detected in soil and freshwaters and so appear to be widely distributed in nature.

Phylogenetic Analyses of Natural Microbial Communities

Although microbiologists believe that thus far we have cultured only a small fraction of the *Archaea* and *Bacteria* that exist in nature, we still know a lot about their diversity. This is because it is possible to do phylogenetic analyses on rRNA genes present in a natural sample without first having to culture the organisms that contain them. If a sample of soil or water contains rRNA, it is because organisms that made that rRNA are present in the sample. Thus, if we isolate all of the different rRNA genes from a soil or water sample, we can use the techniques described in Figure 2.16 to order them on a phylogenetic tree. Conceptually, this is the same as first isolating pure cultures of all of the organisms in the sample and then extracting their rRNA genes. But these powerful techniques skip the culturing step—often the bottleneck in microbial diversity studies—and instead go right for the rRNA genes themselves.

From studies done using these methods of molecular microbial ecology, initially devised by the American microbiologist Norman Pace, it is clear that the extent of microbial diversity is far greater than what laboratory culturing has been able to reveal. A sampling of virtually any habitat typically shows that the vast majority of microorganisms in that habitat have never been in laboratory culture. The phylogeny of these uncultured organisms, known only from environmental rRNA sequences (Figures 2.19 and 2.28), is depicted in phylogenetic trees as branches labeled with letters or numbers instead of actual microbial names to identify lineages.

The results of molecular microbial ecology have given new impetus to innovative culturing techniques to grow the great "uncultured majority" of prokaryotes known to exist today. Genomic analyses of uncultured *Archaea* and *Bacteria* (using the techniques of environmental genomics, ∞ Sections 13.13 and 22.6) have helped in this regard. This is because knowledge of the full complement of genes in uncultured organisms often reveals secrets about their metabolic capacities and suggests ways to bring them into laboratory culture.

2.10 MiniReview

Two major phyla of *Archaea* are known, *Euryarchaeota* and *Crenarchaeota*. Retrieval and analysis of rRNA genes from cells in natural samples have shown that many phylogenetically distinct species of *Archaea* and *Bacteria* exist in nature but have not yet been cultured.

- What is unusual about the genus *Halobacterium*? What group of *Archaea* is responsible for producing the natural gas we use as a fuel? Chemically, what is natural gas?
- How can we know the microbial diversity of a natural habitat without first isolating and growing the organisms it contains?

2.11 Eukaryotic Microorganisms

Eukaryotic microorganisms are related by their distinct cell structure (Figure 2.11) and phylogenetic history (Figure 2.17). Inspection of the domain *Eukarya* (**Figure 2.32**) shows plants

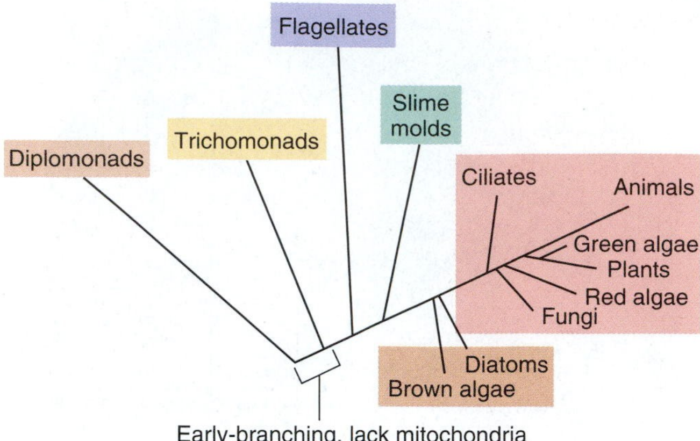

Figure 2.32 Phylogenetic tree of *Eukarya*. Some early-branching species of *Eukarya* lack organelles other than the nucleus. Note that plants and animals branch near the apex of the tree. Not all known lineages of *Eukarya* are depicted.

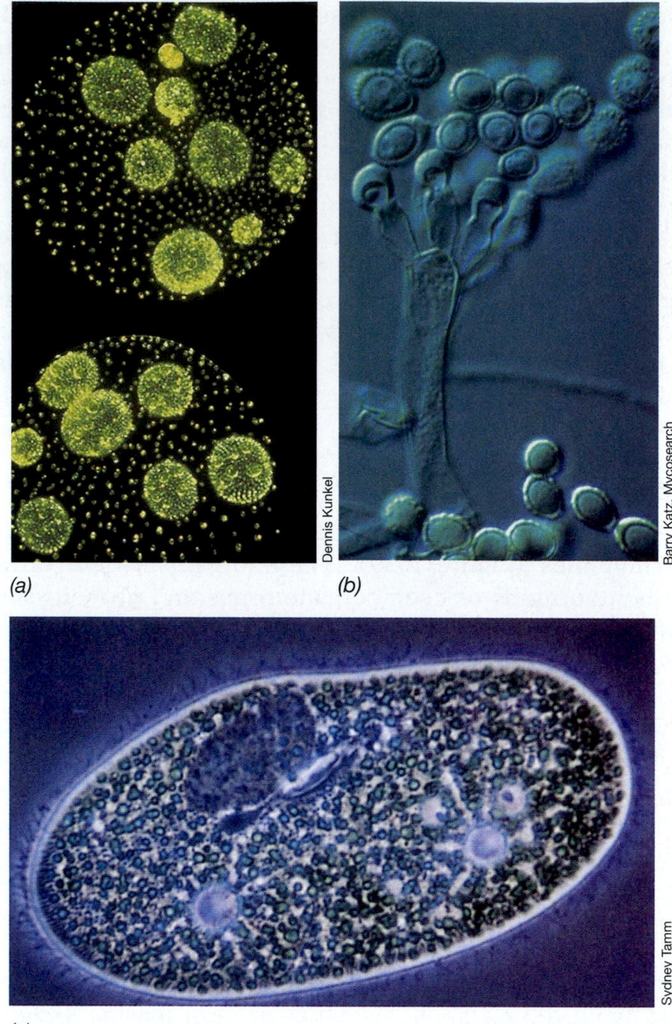

(a) (b) (c)

Figure 2.33 Microbial *Eukarya*. *(a)* Algae; the colonial green alga, *Volvox*. Each spherical cell contains several chloroplasts, the photosynthetic organelle of phototrophic eukaryotes. *(b)* Fungi; the spore-bearing structures of a typical mold. Each spore can give rise to a new filamentous fungus. *(c)* Protozoa; the ciliated protozoan *Paramecium*. Cilia function like oars in a boat, conferring motility on the cell. Microbial eukaryotes are covered in Chapter 18.

and animals to be farthest out on the branches of the tree; such late-branching groups are said to be the "most derived." Some of the early-branching *Eukarya* are structurally simple eukaryotes lacking mitochondria and some other organelles. These cells, such as the diplomonad *Giardia,* may be modern descendents of primitive eukaryotic cells that did not engage in endosymbiosis. Alternatively, such cells contained endosymbionts but for reasons that are not yet clear discarded them as they evolved to colonize the habitats we find them in today. Most of these early eukaryotes are parasites of humans and other animals, unable to live a free and independent existence.

Eukaryotic Microbial Diversity

A diverse array of eukaryotic microorganisms is known. Collectively, microbial eukaryotes are called *protists,* and major groups are algae, fungi, protozoa, and slime molds. Some protists, such as the algae (**Figure 2.33*a***), are phototrophic. Algae contain chloroplasts and can live in environments containing only a few minerals (for example, K, P, Mg, N, S), water, CO_2, and light. Algae inhabit both soil and aquatic habitats and are major primary producers in nature. Fungi (Figure 2.33*b*) lack photosynthetic pigments and are either unicellular (yeasts) or filamentous (molds). Fungi are major agents of biodegradation in nature and recycle much of the organic matter produced in soils and other ecosystems.

Cells of algae and fungi have cell walls, whereas the protozoa (Figure 2.33*c*) do not. Protozoans are typically motile, and different species are widespread in nature in aquatic habitats or as pathogens of humans and other animals. Protozoa are spread about the phylogenetic tree of *Eukarya*. Some, like the flagellates, are fairly early-branching species, whereas others, like the ciliates such as *Paramecium* (Figure 2.33*c*), appear later on the phylogenetic tree (Figure 2.32).

The slime molds resemble protozoa in that they are motile and lack cell walls. However, slime molds differ from protozoa in both their phylogeny and by the fact that their cells undergo a life cycle. During the slime mold life cycle, motile cells aggregate to form a multicellular structure called a *fruiting body* from which spores are produced that yield new motile cells. Slime molds are the earliest known protists to show the cellular cooperation needed to form multicellular structures.

Lichens are leaflike structures often found growing on rocks, trees, and other surfaces (**Figure 2.34**). Lichens are an example of microbial mutualism, a situation in which two organisms live together for mutual benefit. Lichens consist of a fungus and a phototrophic partner organism, either an alga (a eukaryote) or a cyanobacterium (a prokaryote). The phototrophic component is the primary producer and the fungus provides an anchor and protection from the elements. Lichens have thus evolved a successful strategy of mutualistic interaction between two quite different microorganisms.

(a)

(b)

Figure 2.34 Lichens. *(a)* An orange-pigmented lichen growing on a rock, and *(b)* a yellow-pigmented lichen growing on a dead tree stump, Yellowstone National Park, USA. The color of the lichen comes from the pigmented (algal) component of the lichen structure. Besides chlorophylls, the algal components of lichens contain carotenoid pigments, which can be yellow, orange, brown, red, green, or purple.

Final Remarks

Our tour of microbial diversity here is only an overview of the subject. The story is expanded in Chapters 14–18. The viruses were intentionally left out of this overview of diversity. This omission is because viruses are not cells (although they require cells for their replication as mentioned in Section 2.5). Cells in all domains of life have viral parasites, and we explore viral diversity in Chapters 10 and 19.

However, before we can proceed to a more detailed consideration of microbial diversity, we must become familiar with the molecular features of cells. We begin this in the next chapter where we examine the chemistry of cellular components. Our focus in Chapter 3 will be on the chemistry of macromolecules and how different kinds of cells are distinct because of the structure and function of their macromolecules. We proceed from there to study in successive chapters the structure, metabolism, growth, and genetics of microorganisms. By then we will be better prepared to revisit microbial diversity in a major way.

2.11 MiniReview

Microbial eukaryotes are a diverse group that includes algae, protozoa, fungi, and slime molds. Some algae and fungi have developed mutualistic associations called lichens.

- List at least two ways algae differ from cyanobacteria.
- List at least two ways algae differ from protozoa.
- How do each of the components of a lichen benefit each other?

Review of Key Terms

Archaea one of two known domains of prokaryotes; compare with *Bacteria*

Autotroph an organism able to grow on carbon dioxide (CO_2) as sole carbon source

Bacteria one of two known domains of prokaryotes; compare with *Archaea*

Cell wall a rigid layer present outside the cytoplasmic membrane that confers structural strength on the cell and protection from osmotic lysis

Chemolithotroph an organism that obtains its energy from the oxidation of inorganic compounds

Chemoorganotroph an organism that obtains its energy from the oxidation of organic compounds

Chromosome a genetic element containing genes essential to cell function

Cyanobacteria prokaryotic oxygenic phototrophs

Cytoplasm the fluid portion of a cell, bounded by the cytoplasmic membrane

Cytoplasmic membrane the cell's permeability barrier; encloses the cytoplasm

Domain the highest level of biological classification

Endosymbiosis the process by which mitochondria and chloroplasts originated from descendants of *Bacteria*

Eukarya the domain of life that includes all eukaryotic cells

Eukaryote a cell having a membrane-bound nucleus and usually other membrane-bound organelles

Evolution change in a line of descent over time leading to new species or varieties within a species

Extremophile an organism that grows optimally under one or more environmental extremes

Genome the complement of genes in an organism

Gram stain a differential staining technique in which cells stain either pink (gram-negative) or purple (gram-positive) depending upon their structural and phylogenetic makeup

Gram-positive bacteria a major phylogenetic lineage of prokaryotic cells that contain mainly peptidoglycan in their cell walls; stain purple in the Gram stain

Gram-negative bacteria prokaryotic cells that contain an outer membrane and peptidoglycan in their cell walls; stain pink in the Gram stain

Heterotroph an organism that requires organic carbon as its carbon source

Nucleoid the aggregated mass of DNA that constitutes the chromosome of cells of *Bacteria* and *Archaea*

Nucleus a membrane-enclosed structure that contains the chromosomes in eukaryotic cells

Organelle a unit membrane-enclosed structure such as a mitochondrion or chloroplast present in the cytoplasm of eukaryotic cells

Phototroph an organism that obtains its energy from light

Phylogeny the evolutionary relationships between organisms

Plasmid an extrachromosomal genetic element nonessential for growth

Prokaryote a cell that lacks a membrane-enclosed nucleus and other organelles

Proteobacteria a large phylum of *Bacteria* that includes many of the common gram-negative bacteria, such as *Escherichia coli*

Resolution in microbiology, the ability to distinguish two objects as distinct and separate under the microscope

Ribosome a cytoplasmic particle that functions in protein synthesis

Virus a genetic element that contains either DNA or RNA and replicates in cells; has an extracellular form

Review Questions

1. What is the function of staining in light microscopy? why are cationic dyes used for general staining purposes (Sections 2.1 and 2.2)?
2. What is the advantage of a differential interference contrast microscope over a bright-field microscope? A phase-contrast microscope over a bright-field microscope (Section 2.3)?
3. What is the major advantage of electron microscopes over light microscopes? What type of electron microscope would be used to view the three-dimensional features of a cell (Section 2.4)?
4. Why does a cell need a cytoplasmic membrane (Section 2.5)?
5. Which domains of life have a prokaryotic cell structure? Is prokaryotic cell structure a predictor of phylogenetic status (Section 2.5)?
6. How long is a cell of the bacterium *Escherichia coli*? How much larger are you than this single cell (Section 2.5)?
7. How do viruses resemble cells? How do they differ from cells (Section 2.5)?
8. What is meant by the word genome? How does the chromosome of prokaryotes differ from that of eukaryotes (Section 2.6)?
9. How many genes does an organism such as *Escherichia coli* have? How does this compare with the number of genes in one of your cells (Section 2.6)?
10. What is meant by the word endosymbiosis (Section 2.7)?
11. Molecular studies have shown that many macromolecules in species of *Archaea* resemble their counterparts in various eukaryotes more closely than those in species of *Bacteria*. Explain (Section 2.7).
12. From the standpoint of energy metabolism, how do chemoorganotrophs differ from chemolithotrophs? What carbon sources do members of each group use? Are they therefore heterotrophs or autotrophs (Section 2.8)?
13. What domain contains the phylum *Proteobacteria*? What is notable about the *Proteobacteria* (Section 2.9)?
14. What is unusual about the organism *Pyrolobus* (Sections 2.8 and 2.10)?
15. What similarities and differences exist between the following three organisms: *Pyrolobus, Halobacterium*, and *Thermoplasma* (Section 2.10)?
16. Examine Figure 2.28. What does the lineage "marine group" mean (Section 2.10)?
17. How does *Giardia* differ from a human cell, both structurally and phylogenetically (Section 2.11)?

Application Questions

1. Calculate the size of the smallest resolvable object if 600-nm light is used to observe a specimen with a 100× oil-immersion lens having a numerical aperture of 1.32. How could resolution be improved using this same lens?
2. Prokaryotic cells containing plasmids can often be "cured" of their plasmids (that is, the plasmids can be permanently removed) with no ill effects, whereas removal of the chromosome would be lethal. Explain.
3. It has been said that knowledge of the evolution of macroorganisms greatly preceded that of microorganisms. Why do you think that reconstruction of the evolutionary lineage of horses, for example, might have been an easier task than doing the same for any group of prokaryotes?
4. Examine the phylogenetic tree shown in Figure 2.16. Using the sequence data shown, describe why the tree would be incorrect if its branches remained the same but the positions of organisms 2 and 3 on the tree were switched?
5. Microbiologists have cultured a great diversity of microorganisms but know that an even greater diversity exists, despite the fact that they have never seen these organisms or grown them in the laboratory. Explain.
6. What data from this chapter could you use to convince your friend that extremophiles are not just organisms that were "hanging on" in their respective habitats?
7. Defend this statement: If cyanobacteria had never evolved, life on Earth would have remained strictly microbial.

3

Chemistry of Cellular Components

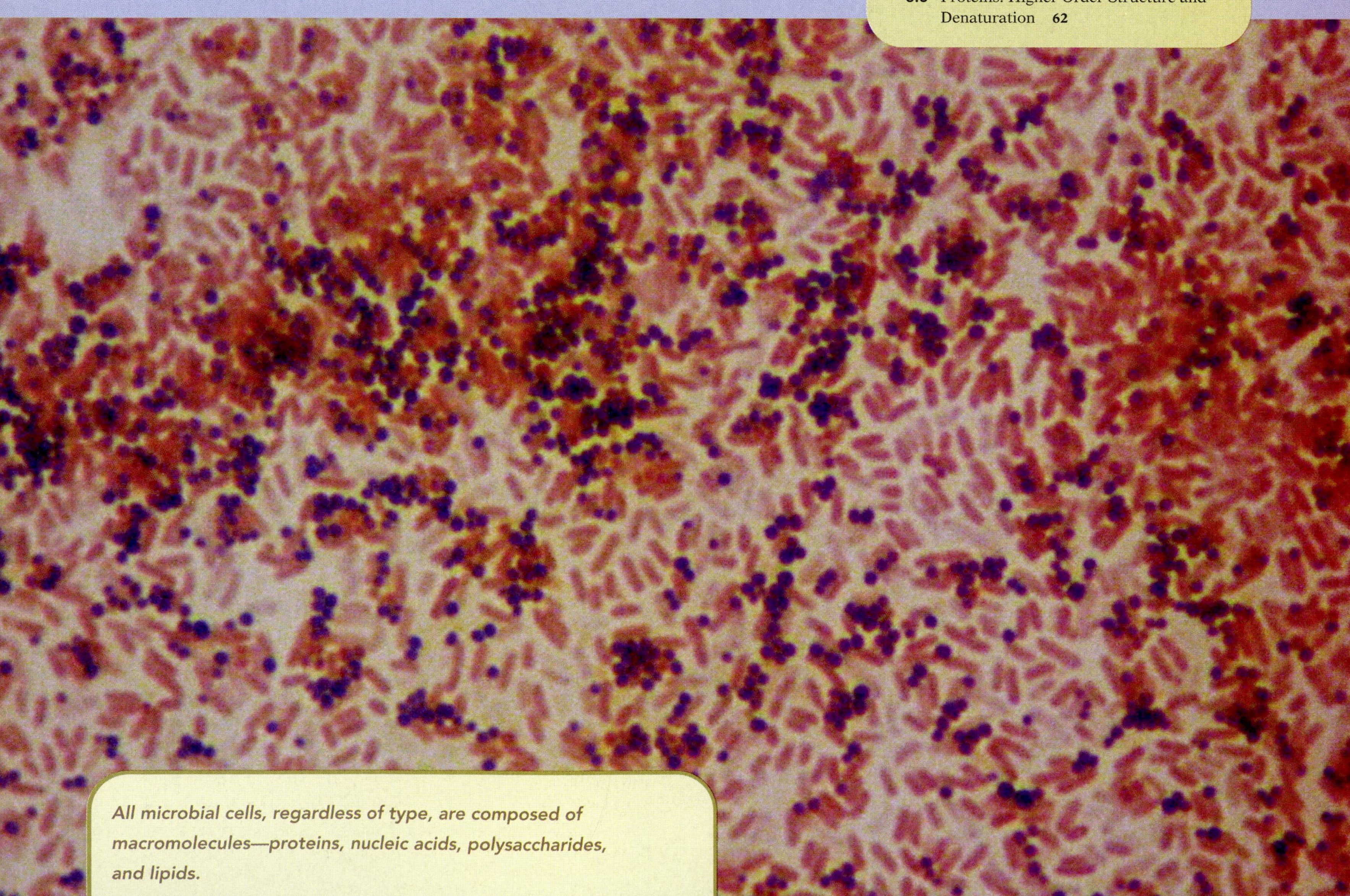

All microbial cells, regardless of type, are composed of macromolecules—proteins, nucleic acids, polysaccharides, and lipids.

I CHEMICAL BONDING, MACROMOLECULES, AND WATER

As we learned in the first two chapters, all cells have much in common. The heart of microbial diversity lies in the variations that cells display in the chemistry and arrangement of their cellular components. These variations confer special properties on each type of cell and allow the cell to carry out specific functions.

To really understand how a cell works, it is necessary to know the molecules that are present and the chemical processes that take place. Molecules, especially macromolecules, are the "guts" of the cell and are the subject of this chapter. It is assumed that the reader has some background in elementary chemistry, especially regarding the nature of atoms and atomic bonding. Here we will expand on this background with a primer on relevant biochemical bonds, followed by a discussion of the structure and function of the four classes of macromolecules.

3.1 Strong and Weak Chemical Bonds

The major chemical elements in living things include hydrogen, oxygen, carbon, nitrogen, phosphorus, and sulfur (Section 5.1). These elements bond in various ways to form the molecules of life. A **molecule** consists of two or more atoms chemically bonded to one another. Thus, two oxygen (O) atoms can combine to form a molecule of oxygen (O_2). Likewise, carbon (C), hydrogen (H), and O atoms can combine to form glucose, $C_6H_{12}O_6$, a hexose sugar (see Figure 3.4).

Covalent Bonds

In living things, chemical elements typically form strong bonds in which electrons are shared more or less equally between atoms. These are called **covalent bonds**. To envision a covalent bond, consider the formation of a molecule of water from the elements O and H:

$$\ddot{\text{O}} + 2\text{H}\cdot \longrightarrow \text{H}{:}\ddot{\text{O}}{:}\text{H}$$

Oxygen has six electrons in its outermost shell, and hydrogen contains only a single electron. When O and 2H combine to form H_2O, covalent bonds maintain the three atoms in tight association as a water molecule. In some compounds, double and even triple covalent bonds can form (**Figure 3.1**). The strength of these bonds increases dramatically with their number. In cells single and double covalent bonds are most common; triply bonded substances are rare.

Chemical elements bond in different combinations to form **monomers**, small molecules that in turn bond with each other to form larger molecules called **polymers**. Covalently bonded polymers in living things are called **macromolecules**. Thousands of different monomers are known but only a relatively small number play important roles in the four classes of macromolecules. To a large extent it is the chemical properties of monomers that give macromolecules their distinctive structure and function.

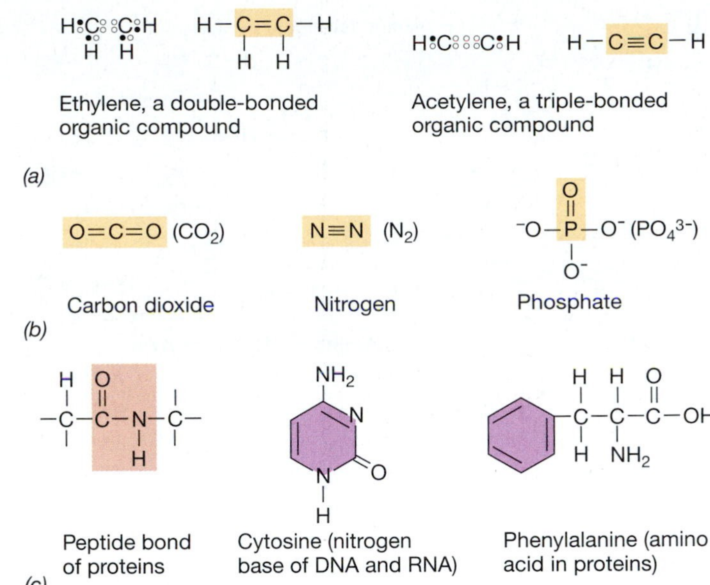

Figure 3.1 Covalent bonding of some molecules containing double or triple bonds. *(a)* For acetylene and ethylene, both the electronic configuration of the molecules and the conventional shorthand for bonds is shown. *(b)* Some inorganic compounds with double bonds. *(c)* Some organic compounds with double bonds.

Hydrogen Bonding and Polarity

In addition to covalent bonds, several weaker chemical bonds also play an important role in biological molecules. Foremost among these are hydrogen bonds. **Hydrogen bonds** form as the result of weak electrostatic interactions between hydrogen atoms and more electronegative (electron attracting) atoms, such as oxygen or nitrogen (**Figure 3.2**). For example, because an oxygen atom is electronegative but a hydrogen atom is not, in the covalent bond between oxygen and hydrogen the shared electrons orbit slightly nearer the oxygen nucleus than the hydrogen nucleus. Because electrons carry a negative charge, this creates a slight charge separation, oxygen slightly negative and hydrogen slightly positive; this bridge is the hydrogen bond. An individual hydrogen bond by itself is very weak. However, when many hydrogen bonds form within and between molecules, overall stability of the molecules can increase dramatically.

Water is a **polar** substance. Because of this, water molecules tend to associate with one another and remain apart from **nonpolar** (hydrophobic) molecules. Water is extensively hydrogen bonded. As water molecules orient themselves in solution, the slight positive charge on a hydrogen atom can bridge the negative charges on oxygen atoms (Figure 3.2*a*). Hydrogen bonds also form between atoms in macromolecules (Figure 3.2*b,c*). As these weak forces accumulate in a large molecule such as a protein, they increase the stability of the molecule and can also affect its overall structure. We will see in Sections 3.5, 3.7, and 3.8 that hydrogen bonds play major roles in the biological properties of proteins (Figure 3.2*b*) and nucleic acids (Figure 3.2*c*).

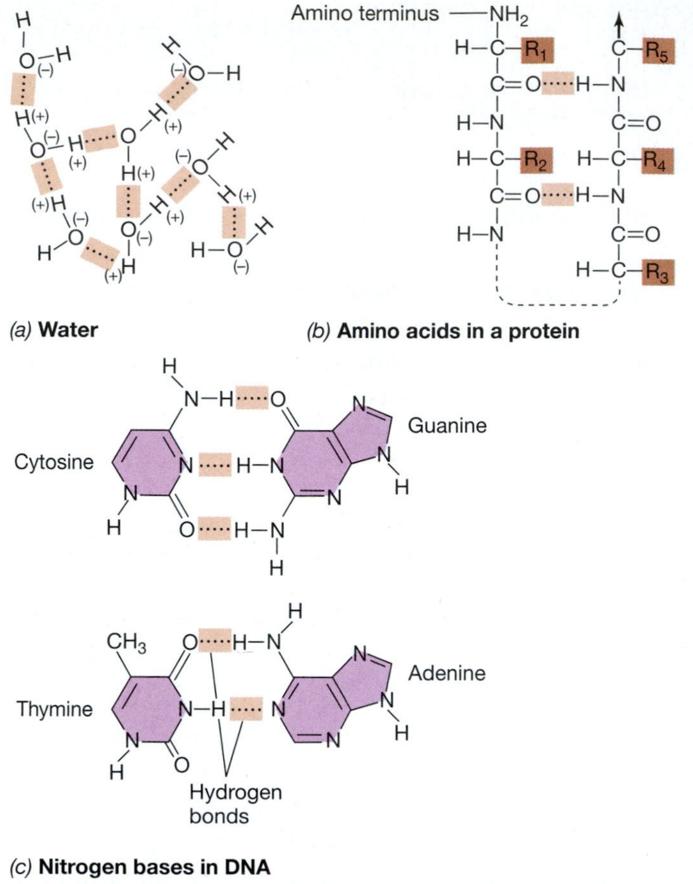

(a) **Water**

(b) **Amino acids in a protein**

(c) **Nitrogen bases in DNA**

Figure 3.2 Hydrogen bonding in water and organic compounds. Hydrogen bonds are shown as highlighted dotted lines. *(b)* Proteins. R represents the side chain of the amino acid (Figure 3.12). *(c)* Hydrogen bonds formed during complementary base pairing in DNA.

Other Weak Bonds

Weak interactions other than hydrogen bonds are also important in cells. For instance, *van der Waals forces* are weak attractive forces that occur between atoms when they become closer than about 3–4 angstroms (Å); van der Waals forces can play significant roles in the binding of substrates to enzymes (Section 5.5) and in protein–nucleic acid interactions.

Ionic bonds, such as that between Na^+ and Cl^- in NaCl, are weak electrostatic interactions that support ionization in aqueous solution. Many important biomolecules, such as carboxylic acids and phosphates (**Table 3.1**), are ionized at cytoplasmic pH (typically pH 6–8) and thus can be dissolved to high levels in the cytoplasm.

Hydrophobic interactions are also considered weak bonds. Hydrophobic interactions occur when nonpolar molecules or nonpolar regions of molecules associate tightly in a polar environment. Hydrophobic interactions can play major roles in controlling the folding of proteins (Sections 3.7 and 3.8). Like van der Waals forces, hydrophobic interactions help bind substrates to enzymes (Section 5.5). In addition, hydrophobic interactions often control how different subunits in a multisubunit protein associate with one another (quaternary structure, Section 3.8) to form the biologically active molecule, and they also help stabilize RNA and cytoplasmic membranes.

Bonding Patterns in Biological Molecules

The element carbon is a major component of all macromolecules. Carbon can bond not only with itself, but with many other elements as well, to yield large structures of considerable diversity and complexity. Different organic (carbon-containing) compounds have different bonding patterns. Each of these patterns, called *functional groups*, has unique chemical properties that are important in determining their biological role within the cell. An awareness of key functional groups will make our later discussion of macromolecular structure, cell physiology, and biosynthesis easier to follow. Table 3.1 lists several functional groups of biochemical importance and examples of molecules or macromolecules that contain them.

3.1 MiniReview

Covalent bonds are strong bonds that bind elements in macromolecules. Weak bonds, such as hydrogen bonds, van der Waals forces, and hydrophobic interactions, also affect macromolecular structure, but they do so through more subtle atomic interactions. Various functional groups are common in biomolecules.

- Why are covalent bonds stronger than hydrogen bonds?
- How can a hydrogen bond play a role in macromolecular structure?

3.2 An Overview of Macromolecules and Water as the Solvent of Life

If you were to chemically analyze a cell of the common intestinal bacterium *Escherichia coli*, what would you find? You would find water as the major constituent, but after removing the water, you would find large amounts of macromolecules, much smaller amounts of monomers, and a variety of inorganic ions (**Table 3.2**). About 95% of the dry weight of a cell consists of macromolecules, and of these, proteins are by far the most abundant class (Table 3.2).

Proteins are polymers of monomers called **amino acids**. Proteins are found throughout the cell, playing both structural and enzymatic roles (**Figure 3.3*a***). An average cell will have over a thousand different types of proteins and multiple copies of each (Table 3.2).

Nucleic acids are polymers of **nucleotides** and are found in the cell in two forms, RNA and DNA. After proteins, ribonucleic acids (RNAs) are the next most abundant macromolecule in an actively growing cell (Table 3.2 and Figure 3.3*b*). This is because there are thousands of ribosomes (the "machines" that make new proteins) in each cell, and

Table 3.1 Some functional groups of biochemical importance

Functional group	*Structure*[a]	*Biological relevance*	*Example*
Carboxylic acid	$-\mathrm{C(=O)-OH}$	Organic, amino, and fatty acids; lipids; proteins	Acetate[b]
Aldehyde	$-\mathrm{C(=O)-H}$	Functional group of reducing sugars such as glucose; aldehydes	Formaldehyde
Alcohol	$-\mathrm{C(H)(H)-OH}$	Lipids; carbohydrates	Glucose
Keto	$-\mathrm{C(=O)}-$	Citric acid cycle intermediates	α-ketoglutarate
Ester	$-\mathrm{C(=O)-O-C(H)(H)}-$	Triglycerides	Lipids of *Bacteria* and *Eukarya*
Phosphate ester	$^{-}\mathrm{O-P(O^{-})(=O)-O-C}-$	Nucleic acids	DNA, RNA
Thioester	$-\mathrm{C(=O)}\sim\mathrm{S}-$	Energy metabolism; biosynthesis of fatty acids	Acetyl-CoA
Ether	$-\mathrm{C(H)(H)-O-C(H)(H)}-$	Certain types of lipids	Lipids of *Archaea*
Acid anhydride	$-\mathrm{C(=O)}\sim\mathrm{O-P(O^{-})(O^{-})=O}$	Energy metabolism	Acetyl phosphate
Phosphoanhydride	$^{-}\mathrm{O-P(O^{-})(=O)}\sim\mathrm{O-P(O^{-})(=O)-O^{-}}$	Energy metabolism	Adenosine triphosphate (ATP)
Peptide	$\mathrm{R-C-C(=O)-N-C-R}$	Proteins	Cellular proteins

[a]A squiggle-type bond depiction (~) indicates an "energy-rich" bond (∞ Section 5.8).
[b]Acetate (H_3CCOO^-) is the ionized form of acetic acid (H_3CCOOH).

ribosomes are composed of a mixture of RNAs and protein. In addition, smaller amounts of RNA are present in the form of messenger and transfer RNAs, other key players in protein synthesis. In contrast to RNA, DNA makes up a small fraction of the bacterial cell (Table 3.2).

Lipids have both hydrophobic and hydrophilic properties and play crucial roles in the cell as the backbone of membranes and as storage depots for excess carbon (Figure 3.3*d*). **Polysaccharides** are polymers of sugars and are present in the cell, primarily in the cell wall. Like lipids, however, polysaccharides such as glycogen (discussed in the next section) can be major forms of carbon and energy storage in the cell (Figure 3.3*c*).

Water as a Biological Solvent

Macromolecules and all other molecules in cells are bathed in water. Water has several important features that make it an ideal biological solvent. Two key features are its polarity and cohesiveness.

The polar properties of water are important because many biologically important molecules (Table 3.2) are themselves polar and thus readily dissolve in water. As we will see in Chapter 4, dissolved substances are continually passing into and out of the cell through transport activities of the cytoplasmic membrane (∞ Sections 4.3 and 4.4). These substances include nutrients needed to build new cell material and waste products of metabolic processes.

Table 3.2 Chemical composition of a prokaryotic cell[a]

Molecule	Percent of dry weight[b]	Molecules per cell (different kinds)
Total macromolecules	96	24,610,000 (~2,500)
Protein	55	2,350,000 (~1,850)
Polysaccharide	5	4,300 (2)[c]
Lipid	9.1	22,000,000 (4)[d]
Lipopolysaccharide	3.4	1,430,000 (1)
DNA	3.1	2.1 (1)
RNA	20.5	255,500 (~660)
Total monomers	3.0	—[e](~350)
Amino acids and precursors	0.5	—(~100)
Sugars and precursors	2	—(~50)
Nucleotides and precursors	0.5	—(~200)
Inorganic ions	1	—(18)
Total	100%	—

[a]Data from Neidhardt, F.C., et al. (eds.), 1996. *Escherichia coli* and *Salmonella typhimurium—Cellular and Molecular Biology*, 2nd edition. American Society for Microbiology, Washington, DC.

[b]Dry weight of an actively growing cell of *E. coli* = 2.8×10^{-13}g; total weight (70% water) = 9.5×10^{-13} g.

[c]Assuming peptidoglycan and glycogen to be the major polysaccharides present.

[d]There are several classes of phospholipids, each of which exists in many kinds because of variability in fatty acid composition between species and because of different growth conditions.

[e]Reliable estimates of monomer and inorganic ion composition are lacking.

The polar properties of water also promote the stability of large molecules because of the increased opportunities for hydrogen bonding. Water forms three-dimensional networks, both with itself (Figure 3.2*a*) and within macromolecules. By so doing, water molecules help to position atoms within biomolecules for potential interaction. The high polarity of water is also beneficial to the cell because it forces nonpolar substances to aggregate and remain together. Membranes, for example, contain large amounts of lipids, which have major nonpolar (hydrophobic) components, and these aggregate in such a way as to prevent the unrestricted flow of polar molecules into and out of the cell.

The polar nature of water makes it highly cohesive. This means that water molecules have a high affinity for one another and form ordered arrangements in which hydrogen bonds (Figure 3.2*a*) are constantly forming, breaking, and re-forming. The cohesiveness of water is responsible for some of its biologically important properties, such as high surface tension and high specific heat (heat required to raise the temperature 1°C). Also, the fact that water expands on freezing to yield a less dense solid form (ice) has a profound effect on life in temperate and polar aquatic environments. In a lake, for example, ice on the surface insulates the water beneath the ice and prevents it from freezing, thus allowing aquatic organisms to survive under the overlying ice.

Life originated in water about 3.9 billion years ago (Figure 1.6), and virtually anywhere on Earth where liquid water exists, microorganisms are likely to be found. With

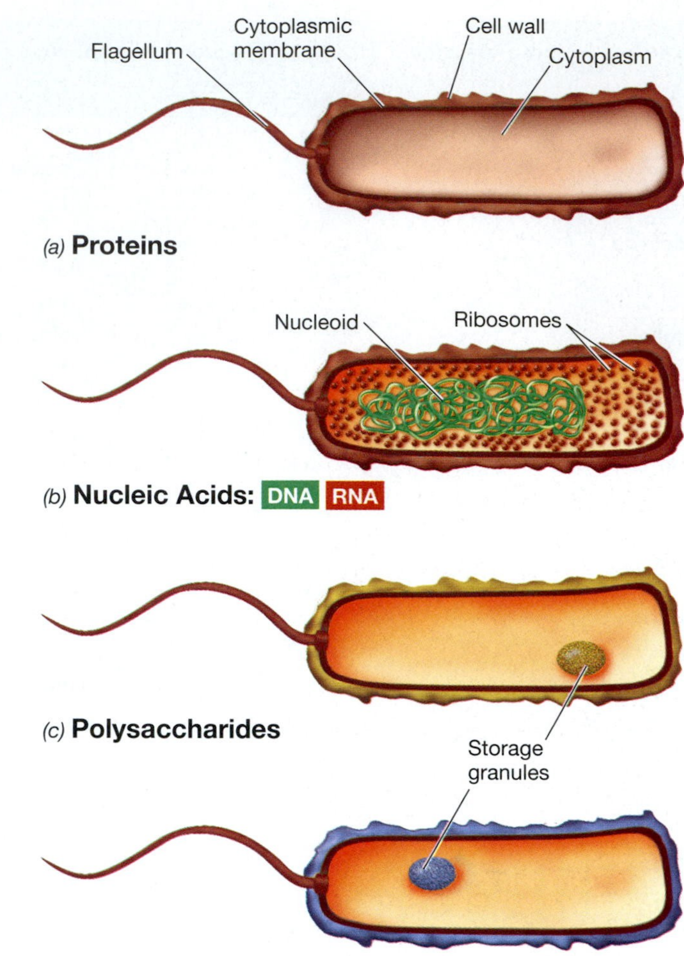

Figure 3.3 Macromolecules in the cell. *(a)* Proteins (brown) are found throughout the cell both as parts of cell structures and as enzymes. The flagellum is a structure involved in swimming motility. *(b)* Nucleic acids. DNA (green) is found in the nucleoid of prokaryotic cells and in the nucleus of eukaryotic cells. RNA (orange) is found in the cytoplasm (mRNA, tRNA) and in ribosomes (rRNA). *(c)* Polysaccharides (yellow) are located in the cell wall and occasionally in internal storage granules. *(d)* Lipids (blue) are found in the cytoplasmic membrane, the cell wall, and in storage granules.

these important properties of water in mind, we now consider the structure of the major macromolecules of life (Table 3.2 and Figure 3.3) in more detail.

3.2 MiniReview

Proteins are the most abundant class of macromolecule in the cell. Other macromolecules include the nucleic acids, lipids, and polysaccharides. Water is an excellent solvent for life because of its polarity and cohesiveness.

- Why do protein and RNA make up such a large proportion of an actively growing cell?
- Why does the high polarity of water make it useful as a biological solvent?

II NONINFORMATIONAL MACROMOLECULES

In this unit we examine the structure and function of noninformational macromolecules—polysaccharides and lipids. The sequence of monomers in these macromolecules does not carry genetic information, but the macromolecules themselves play important roles in the cell, primarily as structural or reserve materials.

3.3 Polysaccharides

Carbohydrates (sugars) are organic compounds that contain carbon, hydrogen, and oxygen in a ratio of 1:2:1. The structural formula for glucose, the most abundant sugar on Earth is $C_6H_{12}O_6$ (**Figure 3.4**). The most biologically relevant carbohydrates are those containing four, five, six, and seven carbon atoms (designated as C_4, C_5, C_6, and C_7). C_5 sugars (pentoses) are of special significance because of their role as structural backbones of nucleic acids. Likewise, C_6 sugars (hexoses) are the monomeric constituents of cell wall polymers and energy reserves. Figure 3.4 shows the structure of a few common sugars.

Derivatives of simple carbohydrates are common in cells. When other chemical species replace one or more of the hydroxyl groups on the sugar, derivatives are formed. For example, the important bacterial cell wall polymer *peptidoglycan* (∞ Section 4.6) contains the glucose derivative *N*-acetylglucosamine (**Figure 3.5**). Besides sugar derivatives, sugars having the same *structural* formula can differ in their *stereoisomeric* properties. For example, a polysaccharide composed of D-glucose differs from one containing L-glucose (Section 3.6). Hence, a large number of different sugars are available to the cell for the construction of polysaccharides.

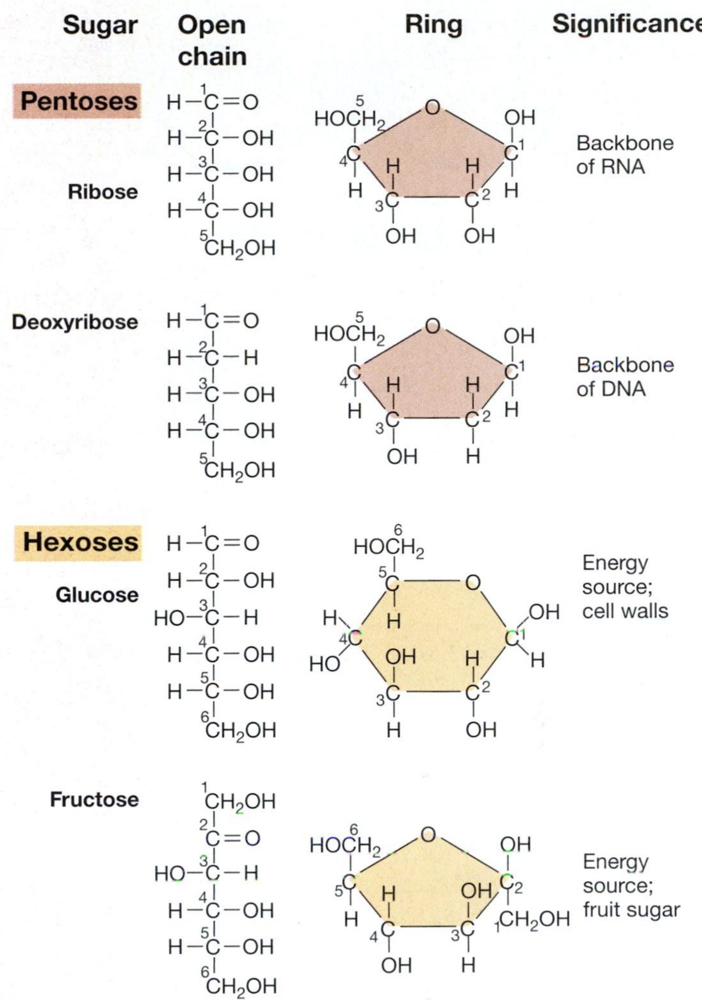

Figure 3.4 Structural formulas of a few common sugars. The formulas can be depicted in two alternate ways, open chain and ring. The open chain is easier to visualize, but the ring form is the commonly used structure. Note the numbering system on the ring. Glucose and fructose are isomers of one another; they have the same molecular composition but have different structures (Section 3.6).

The Glycosidic Bond

Polysaccharides are carbohydrates containing many (sometimes hundreds or even thousands) of monomeric units called *monosaccharides*. The latter are connected by covalent bonds called **glycosidic bonds** (**Figure 3.6**). If two monosaccharides are bonded by a glycosidic linkage, the resulting molecule is a disaccharide. The addition of one more monosaccharide yields a trisaccharide, and several more an oligosaccharide. An extremely long chain is then a *polysaccharide*.

Glycosidic bonds can form in two different geometric orientations, alpha (α) and beta (β) (Figure 3.6*a*). Polysaccharides with a repeating structure composed of glucose units bonded between carbons 1 and 4 in the alpha orientation (for example, glycogen and starch, Figure 3.6*b*) function as important carbon and energy reserves in bacteria, plants, and animals. Glucose units joined by β-1,4 linkages are present in cellulose (Figure 3.6*b*), a stiff plant and algal cell wall component. Thus, even though both starch and cellulose contain only D-glucose, their functional properties differ because of the different configurations, α or β, of their glycosidic bonds.

Complex Polysaccharides

Polysaccharides can also combine with other classes of macromolecules, such as proteins and lipids, to form complex polysaccharides—*glycoproteins* and *glycolipids*. These

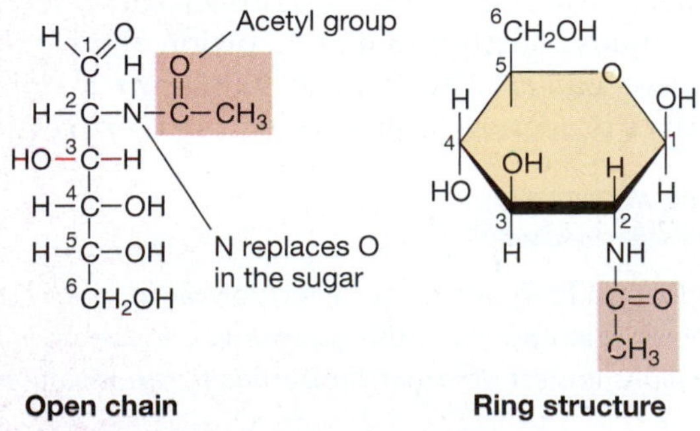

Acetyl group
N replaces O in the sugar
Open chain
Ring structure

Figure 3.5 ***N*-acetylglucosamine, a derivative of glucose.** The O on C-2 is replaced with an *N*-acetyl group.

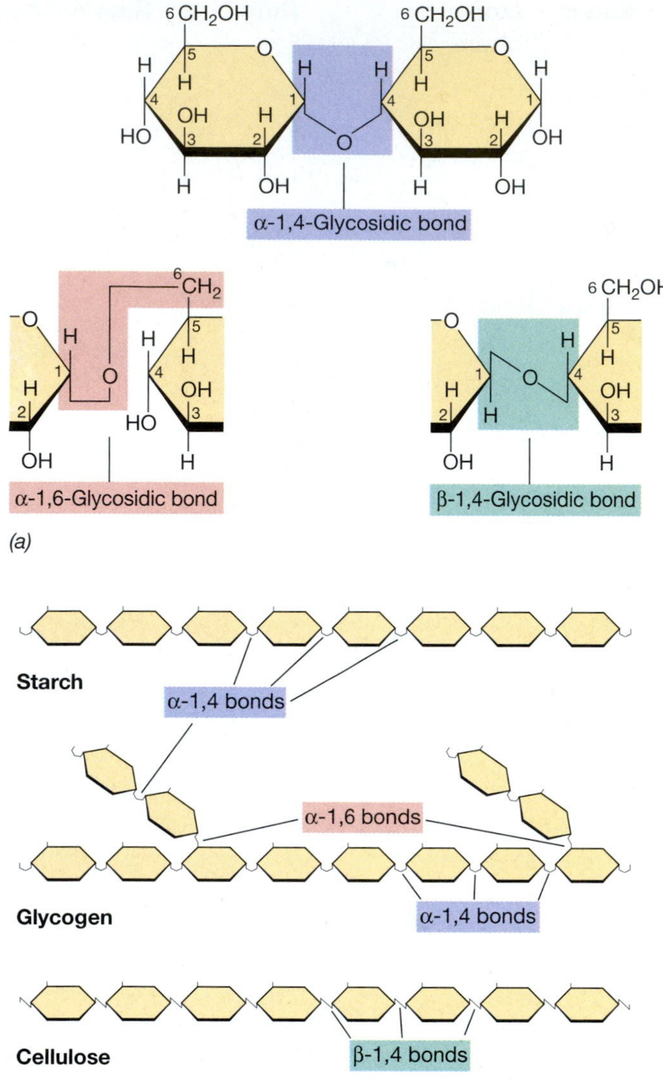

Figure 3.6 The glycosidic bond and polysaccharides. *(a)* Structure of different glycosidic bonds. Note that both the linkage (position on the ring of the carbon atoms bonded) and the geometry (α or β) of the linkage can vary about the glycosidic bond. *(b)* Structures of some common polysaccharides. Compare color coding to (a).

compounds play important roles in cells, in particular as cell-surface receptor molecules in cytoplasmic membranes. The compounds typically reside on the external surfaces of the membrane where they are in contact with the environment. Glycolipids constitute a major portion of the cell wall of gram-negative bacteria and, as such, impart a number of unique surface properties to these organisms (Section 4.9).

3.3 MiniReview

Sugars can form long polymers called polysaccharides. The two different orientations of the glycosidic bonds that link sugar residues impart different properties to the resultant molecules. Polysaccharides can also contain other molecules such as protein or lipid, forming complex polysaccharides.

- How can glycogen and cellulose differ so much in their physical properties when they both consist of 100% D-glucose?

3.4 Lipids

Lipids are essential components of cells and are *amphipathic* macromolecules, meaning that they show both hydrophilic and hydrophobic character. Lipid structure varies between the domains of life, and even within a domain many different lipids are known. **Fatty acids** are major constituents of *Bacteria* and *Eukarya* lipids. By contrast, lipids of *Archaea* contain a hydrocarbon (phytanyl) side chain not composed of fatty acids (Section 4.3).

Fatty acids contain both hydrophobic and hydrophilic components. Palmitate (the ionized form of palmitic acid), is a common fatty acid in membrane lipids. Palmitate is a 16-carbon fatty acid composed of a chain of 15 saturated (fully hydrogenated and thus highly hydrophobic) carbon atoms and a single carboxylic acid group (the hydrophilic portion) (**Figure 3.7**). Other common fatty acids in the lipids of *Bacteria* include saturated or monounsaturated forms, from C_{12} to C_{20} (Figure 3.7).

Triglycerides and Complex Lipids

Simple lipids (fats) consist of fatty acids (or phytanyl units in *Archaea*) bonded to the C_3 alcohol glycerol (Figure 3.7*a, b*). Simple lipids are also called *triglycerides* because three fatty acids are linked to the glycerol molecule. We will see when we consider membrane structure (Section 4.3) that the bond between glycerol and the hydrophobic side chain is an *ester* bond (Table 3.1) in cells of *Bacteria* and *Eukarya* but an *ether* bond (Table 3.1) in *Archaea*.

Complex lipids are simple lipids that contain additional elements such as phosphorus, nitrogen, or sulfur, or small hydrophilic organic compounds such as sugars (Figure 3.7*d*), ethanolamine (Figure 3.7*c*), serine, or choline. Lipids containing a phosphate group, called *phospholipids*, are an important class of complex lipids because they play a major structural role in the cytoplasmic membrane (Section 4.3).

The amphipathic property of lipids makes them ideal structural components of membranes. Lipids aggregate to form membranes; the hydrophilic (glycerol) portion is in contact with the cytoplasm and the external environment whereas the hydrophobic portion remains buried away inside the membrane (Section 4.3 and Figures 4.4 and 4.5). Because of this property, membranes are ideal permeability barriers. The inability of polar substances to flow through the hydrophobic region of the lipids renders the membrane impermeable and prevents leakage of cytoplasmic constituents. However, this also means that polar substances necessary for cell function do not leak *in*, either, but we reserve this story for the next chapter (transport, Section 4.5)

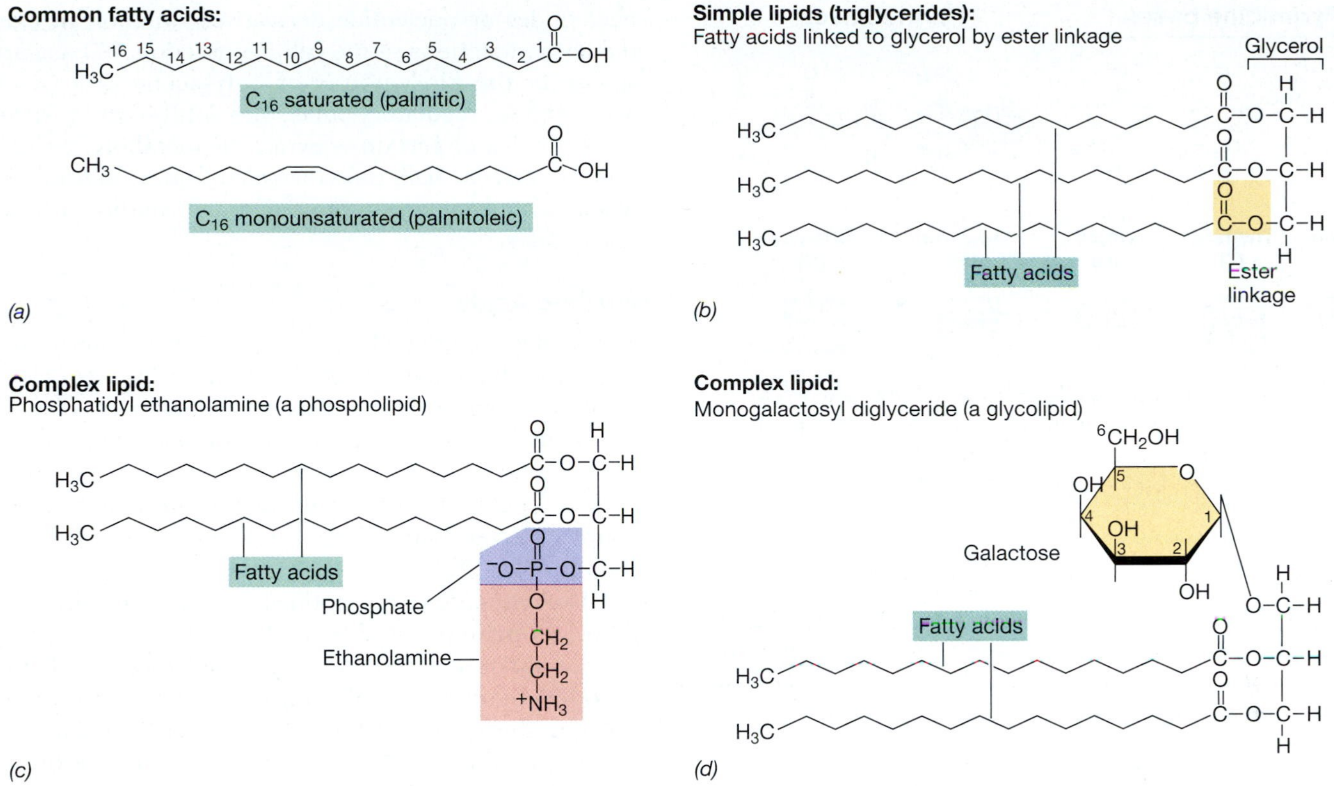

Common fatty acids:

C_{16} saturated (palmitic)

C_{16} monounsaturated (palmitoleic)

(a)

Simple lipids (triglycerides):
Fatty acids linked to glycerol by ester linkage

Glycerol

Fatty acids

Ester linkage

(b)

Complex lipid:
Phosphatidyl ethanolamine (a phospholipid)

Fatty acids

Phosphate

Ethanolamine

(c)

Complex lipid:
Monogalactosyl diglyceride (a glycolipid)

Galactose

Fatty acids

(d)

Figure 3.7 Lipids. *(a)* Fatty acids differ in length, in position, and in number of double bonds. *(b)* Simple lipids are formed by a dehydration reaction between fatty acids and glycerol to yield an ester linkage. The fatty acid composition of a cell varies with growth temperature. *(c, d)* Complex lipids are simple lipids containing other molecules.

3.4 MiniReview

Lipids contain both hydrophobic and hydrophilic components; their chemical properties make them ideal structural components for cytoplasmic membranes.

- What part of a fatty acid molecule is hydrophobic? Hydrophilic?
- How does a phospholipid differ from a triglyceride?
- Draw the chemical structure of butyrate, a C4 fully saturated fatty acid.

III INFORMATIONAL MACROMOLECULES

The sequence of monomers in nucleic acids carries genetic information, and the sequence of monomers in proteins carries structural and functional information. In contrast to polysaccharides and lipids, nucleic acids and proteins are thus *informational* macromolecules.

3.5 Nucleic Acids

The **nucleic acids** deoxyribonucleic acid, **DNA**, and ribonucleic acid, **RNA**, are macromolecules composed of monomers called *nucleotides*. Therefore, DNA and RNA are **polynucleotides**. As we already know, DNA carries the genetic blueprint for the cell and RNA is the intermediary molecule that converts the blueprint into defined amino acid sequences in proteins (∞ Figure 1.4).

A nucleotide is composed of three components: a pentose sugar, either ribose (in RNA) or deoxyribose (in DNA), a nitrogen base, and a molecule of phosphate, PO_4^{3-}. The general structure of nucleotides of both DNA and RNA is very similar (**Figure 3.8**).

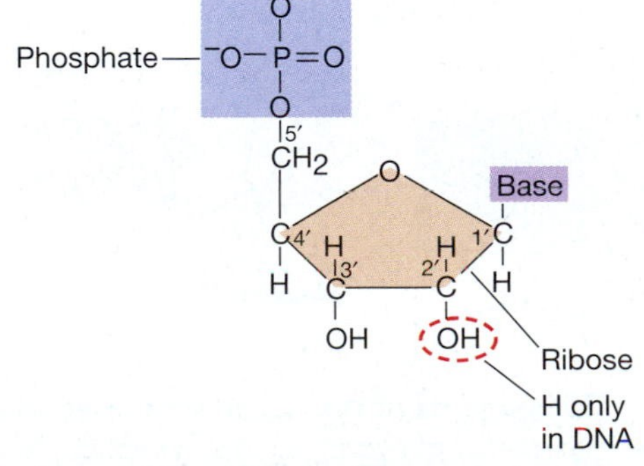

Figure 3.8 Nucleotides. The numbers on the sugar contain a prime (′) after them because the ring structure in the nitrogen base is also numbered (Figure 3.9).

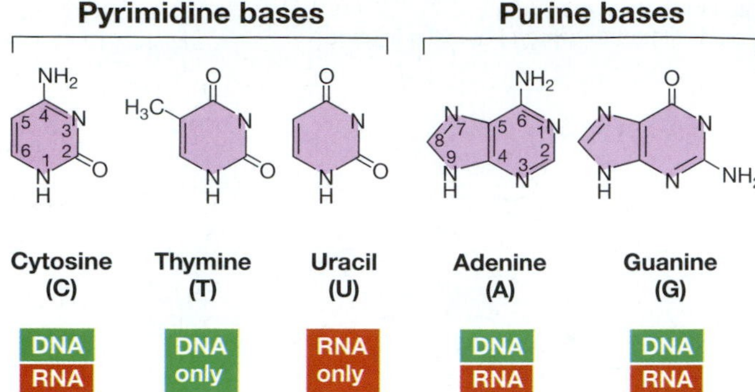

Figure 3.9 Structure of the nitrogen bases of DNA and RNA. Note the numbering system of the rings. In attaching itself to the 1′ carbon of the sugar phosphate shown in Figure 3.8, a pyrimidine base bonds through N-1 and a purine base bonds at N-9.

Nucleotides

The nitrogen bases of nucleic acids belong to one of two chemical classes. **Purine** bases—*adenine* and *guanine*—contain two fused heterocyclic rings (a heterocyclic ring contains more than one kind of atom). **Pyrimidine** bases—*thymine*, *cytosine*, and *uracil*—contain a single six-membered heterocyclic ring (**Figure 3.9**). Guanine, adenine, and cytosine are present in both DNA and RNA. Thymine is present (with minor exceptions) only in DNA, and uracil is present only in RNA.

Nucleotides consist of a nitrogen base attached to a pentose sugar by a glycosidic linkage between carbon atom 1 of the sugar and a nitrogen atom of the base, either the nitrogen atom labeled 1 (in a pyrimidine base) or 9 (in a purine base). Without the phosphate, a nitrogen base bonded to its sugar is called a *nucleoside*. Nucleotides are thus nucleosides containing one or more phosphates (**Figure 3.10**).

Nucleotides play other roles in the cell besides their major role as components of nucleic acids. Nucleotides, especially adenosine triphosphate (ATP) (Figure 3.10), are key forms of chemical energy within the cell, releasing sufficient energy during the hydrolysis of a phosphate bond to drive energy-requiring reactions in the cell (Section 5.8). Other

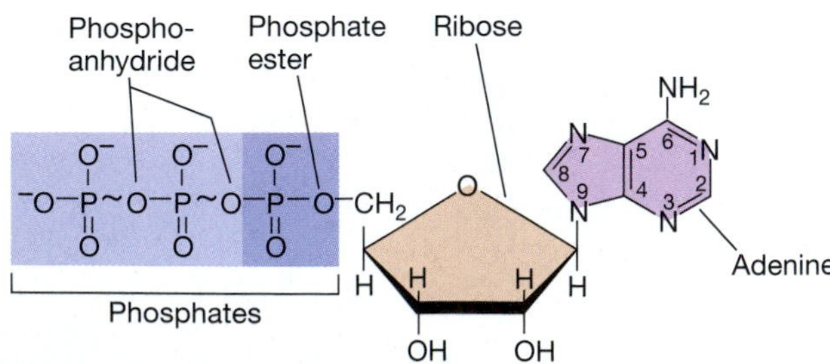

Figure 3.10 Components of the important nucleotide, adenosine triphosphate. The energy of hydrolysis of a phosphoanhydride bond (shown as squiggles) is greater than that of a phosphate ester bond, which is significant in bioenergetics (Section 5.8). With the phosphate group removed, the molecule would be the nucleoside adenosine.

nucleotides or nucleotide derivatives function in oxidation–reduction reactions in the cell (Section 5.7) as carriers of sugars in the biosynthesis of polysaccharides (Section 5.15) and as regulatory molecules inhibiting or stimulating the activities of certain enzymes or metabolic events. However, we discuss here only the role of nucleotides as building blocks of nucleic acids, the major informational function of nucleotides.

Nucleic Acids

The nucleic acid backbone is a polymer of alternating sugar and phosphate molecules. Polynucleotides consist of nucleotides covalently bonded via phosphate from carbon 3—called the 3′ (3 prime) carbon—of one sugar to the 5′ carbon of the adjacent sugar (**Figure 3.11*a***). The phosphate linkage is called a **phosphodiester bond** because a phosphate molecule connects two sugar molecules by ester linkage (Figure 3.11*a*; Table 3.1).

The sequence of nucleotides in a DNA or RNA molecule is called its **primary structure**. As we have discussed, the sequence of bases in a DNA or RNA molecule is informational, encoding the sequence of amino acids in proteins or encoding specific ribosomal or transfer RNAs. The replication of DNA and the synthesis of RNA are key events in the life of a cell (Section 1.2 and Figure 1.4). We will see later that a virtually error-free mechanism is employed to ensure the faithful transfer of genetic traits from one generation to another (Chapter 7).

DNA

In the genome of cells, DNA is *double-stranded*. Each chromosome consists of two strands of DNA, with each strand containing hundreds of thousands to several million nucleotides linked by phosphodiester bonds. The strands associate with one another by hydrogen bonds that form between the nitrogen bases in nucleotides of one strand and the nitrogen bases in nucleotides of the other strand. When positioned adjacent to one another, purine and pyrimidine bases can undergo hydrogen bonding (see Figure 3.2*c*).

Hydrogen bonding is most stable when guanine (G) bonds with cytosine (C) and adenine (A) bonds with thymine (T) (see Figure 3.2*c*). Specific base pairing, A with T and G with C, thus ensures that the two strands of DNA are *complementary* in base sequence; that is, wherever a G is found in one strand, a C is found in the other, and wherever a T is present in one strand, its complementary strand has an A (Figure 3.11*b*).

RNA

With a few exceptions, all RNAs are *single-stranded* molecules. However, RNAs typically fold back upon themselves in regions where complementary base pairing is possible to form folded structures. This pattern of folding in RNA is called its **secondary structure** (Figure 3.11*c*). In certain very large RNA molecules, such as ribosomal RNA (Sections 7.15 and 14.9), some parts of the molecule contain only primary

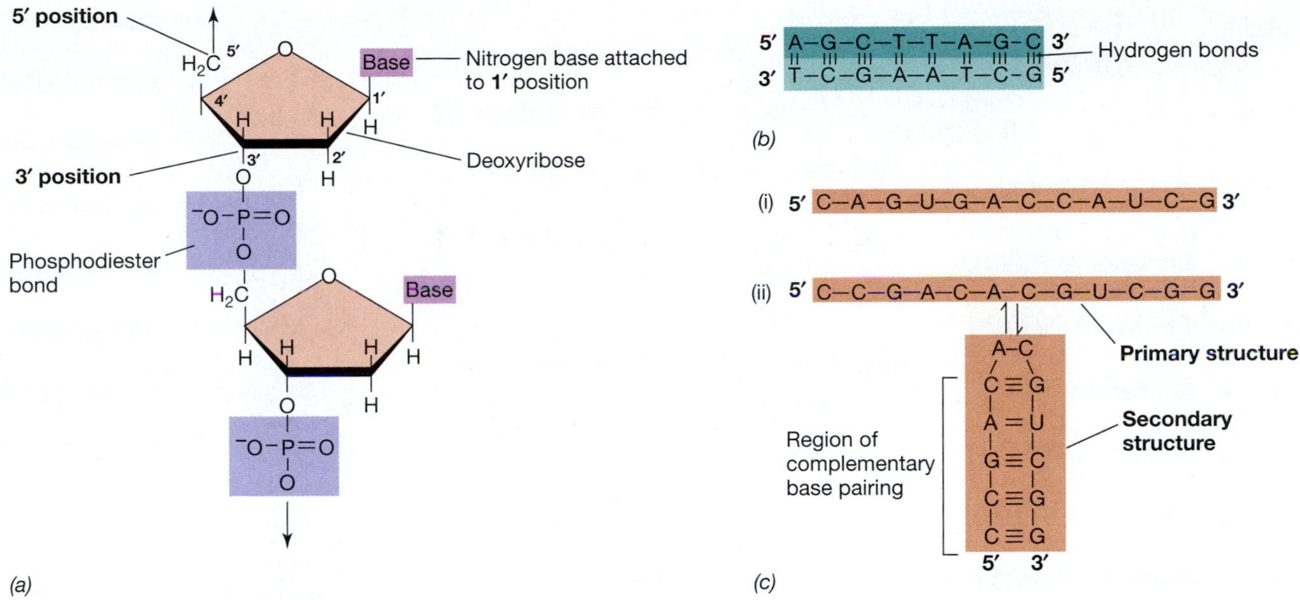

Figure 3.11 Nucleic acids: DNA and RNA. *(a)* Structure of part of a DNA chain. The nitrogen bases can be adenine, guanine, cytosine, or thymine. In RNA, an OH group is present on the 2′ carbon of the pentose sugar (see Figure 3.8) and uracil replaces thymine. *(b)* Simplified structure of DNA in which only the nitrogen bases are shown. The two strands are complementary in base sequence, with A joined to T by two hydrogen bonds and G joined to C by three hydrogen bonds (note that the hydrogen bonds are indicated by two and three lines rather than by dots as in Figure 3.2). *(c)* RNA: (i) a sequence showing only primary structure; (ii) a sequence that allows for secondary structure. In RNA, secondary structures form when opportunities for *intra*strand base pairing arise, as shown here.

structure but others contain both primary and secondary structure. This leads to highly folded and twisted molecules whose biological function is critically dependent on their final three-dimensional shape.

At least four classes of RNA exist in cells. *Messenger RNA* (mRNA) carries the genetic information of DNA in a single-stranded molecule complementary in base sequence to that of DNA. *Transfer RNAs* (tRNAs) convert the genetic information present in mRNA into the language of amino acids, the building blocks of proteins. *Ribosomal RNAs* (rRNAs), of which there are several types, are important structural and catalytic components of the ribosome, the protein-synthesizing system of the cell. In addition to these, a variety of *small RNAs* exist in cells. These RNAs function to regulate the production or activity of other RNAs. These various types of RNA are discussed in detail in Chapters 7, 9, and 11.

3.5 MiniReview

The informational content of a nucleic acid is determined by the sequence of nitrogen bases along the polynucleotide chain. Both RNA and DNA are informational macromolecules. RNA can fold into various configurations to obtain secondary structure.

- What components are found in a nucleotide?
- How does a nucleo*side* differ from a nucleo*tide*?
- Distinguish between the primary and secondary structure of RNA.

3.6 Amino Acids and the Peptide Bond

Amino acids are the monomers of **proteins**. Most amino acids consist of carbon, hydrogen, oxygen, and nitrogen only, but 2 of the 22 genetically encoded amino acids also contain sulfur, and 1 contains selenium. All amino acids contain two important functional groups, a carboxylic acid group (—COOH) and an amino group (—NH_2) (Table 3.1 and **Figure 3.12*a***). These groups are key to the structure of proteins because covalent bonds can form between the carboxyl carbon of one amino acid and the amino nitrogen of a second amino acid (with elimination of a molecule of water) to form the **peptide bond** (**Figure 3.13**).

Structure of Amino Acids

All amino acids have the general structure shown in Figure 3.12*a*. But each type of amino acid is unique because of its unique side group (abbreviated R in Figure 3.12*a*) attached to the α-carbon. The α-carbon is the carbon atom adjacent to the carboxylic acid group. The side chains vary considerably in structure, from as simple as a hydrogen atom in the amino acid glycine to aromatic rings in phenylalanine, tyrosine, and tryptophan (Figure 3.12*b*).

The chemical properties of an amino acid are governed by its side chain. Amino acids that have similar chemical properties are grouped into related amino acid "families" as shown in Figure 3.12*b*. For example, the side chain may contain a carboxylic acid group, such as in aspartic acid or glutamic

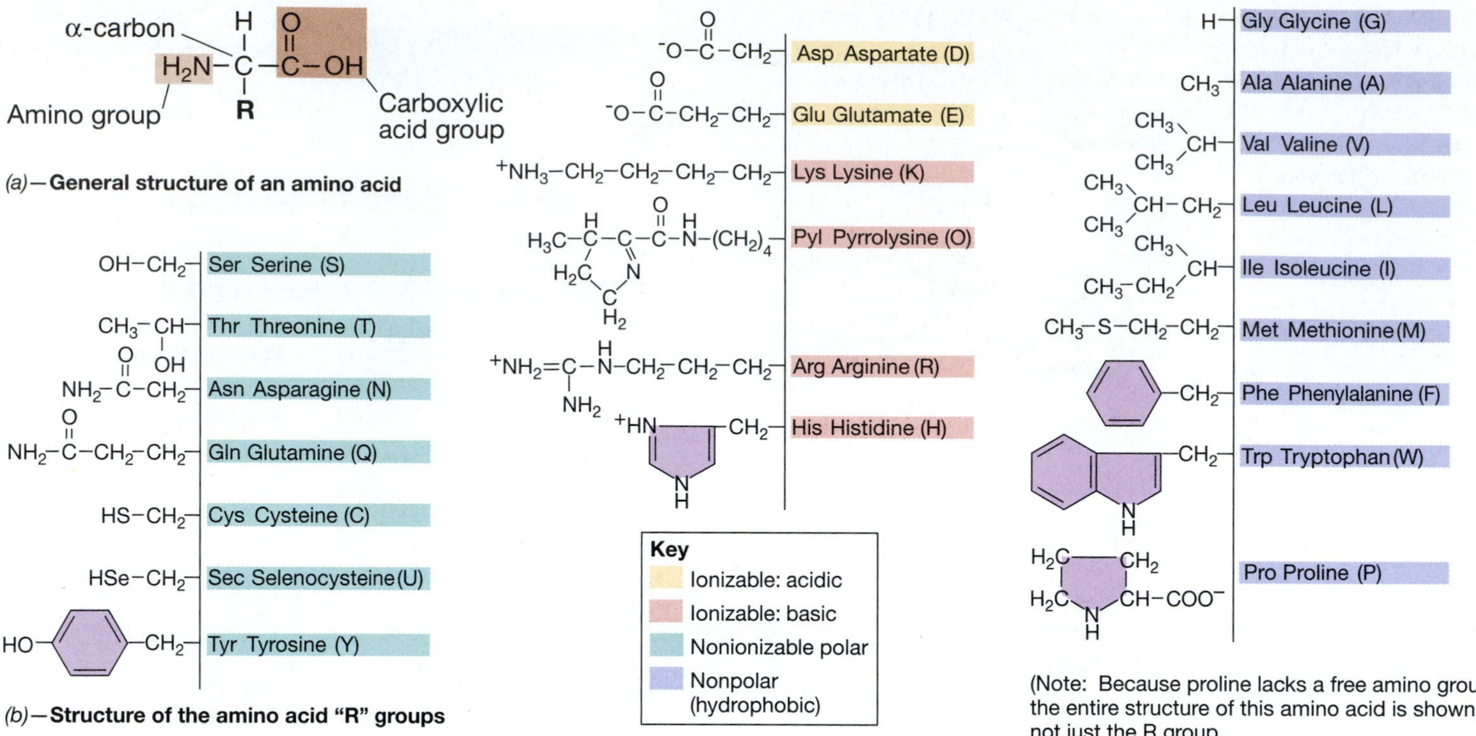

Figure 3.12 Structure of the 22 genetically encoded amino acids. *(a)* General structure. *(b)* R group structure. The three-letter codes for the amino acids are to the left of the names, and the one-letter codes are in parentheses to the right of the names. Pyrrolysine has thus far been found only in certain methanogenic *Archaea* (Sections 2.10 and 17.4).

acid, rendering the amino acid acidic. Others contain additional amino groups, rendering them basic. Alternatively, several amino acids contain hydrophobic side chains and are grouped together as nonpolar amino acids. The amino acid cysteine contains a sulfhydryl group (—SH). Sulfhydryl groups can connect one chain of amino acids to another by disulfide linkage (R—S—S—R) through two cysteine molecules, one from each chain.

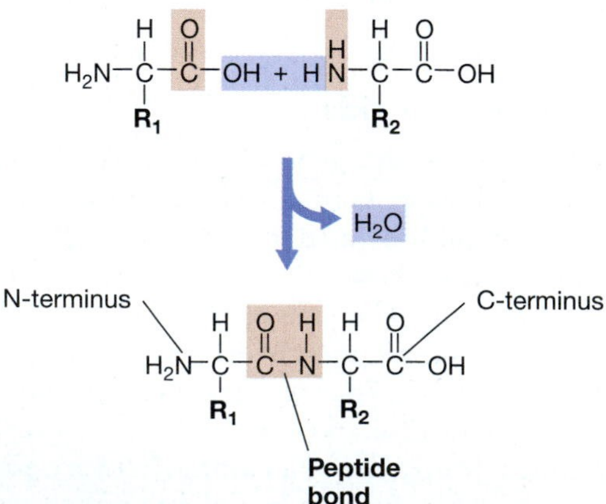

Figure 3.13 Peptide bond formation. R_1 and R_2 refer to the variable portions (side chains) of the amino acids (Figure 3.12). Note how, following peptide bond formation, a free OH group is present at the C-terminus for formation of the next peptide bond.

The diversity of chemically distinct amino acids makes possible an enormous number of unique proteins with widely different biochemical properties. For example, if one assumes that an average polypeptide contains 100 amino acids, there are 22^{100} different polypeptide sequences that are theoretically possible. No cell has anywhere near this many different proteins. However, a cell of *Escherichia coli* contains almost 2,000 different kinds of proteins (Table 3.2). These include different soluble and membrane-integrated enzymes, structural proteins, transport proteins, sensory proteins, and many others.

Isomers

Two molecules may have the same molecular formula but exist in different structural forms. These related but nonidentical molecules are called **isomers**. For example, the hexose sugars glucose and fructose (Figure 3.4) are isomers. Louis Pasteur, the famous early microbiologist who quashed the theory of spontaneous generation (Section 1.7), began his scientific career as a chemist studying a class of isomers called *optical* isomers. Optical isomers that have the same molecular and structural formulas, except that one is a "mirror image" of the other (just as the left hand is a mirror image of the right), are called **enantiomers**. The enantiomers of a given compound can never be superimposed one over the other and are designated as either D or L (**Figure 3.14**), depending on whether a pure solution rotates light to the right or left, respectively. Sugars of the D enantiomer predominate in biological systems.

Amino acids also have D or L enantiomers. However, in proteins cells employ the L-amino acid rather than the D form (Figure 3.14*c*). Nevertheless, D-amino acids are occasionally found in cells, most notably in the cell wall polymer peptidoglycan (∞ Section 4.6) and in certain peptide antibiotics (∞ Section 27.9). Cells can interconvert certain enantiomers by the activity of enzymes called *racemases*. For instance, some prokaryotes can grow on L-sugars or D-amino acids because they have racemases that can convert these forms into the opposite enantiomer before metabolizing them.

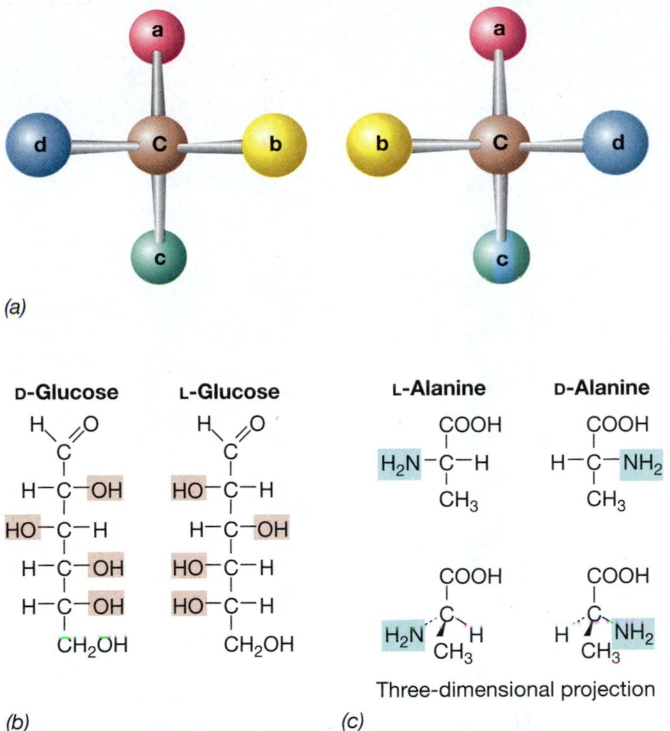

Figure 3.14 Isomers. *(a)* Ball-and-stick model showing mirror images. *(b)* Enantiomers of glucose. *(c)* Enantiomers of the amino acid alanine. In the three-dimensional projection the arrow should be understood as coming *toward* the viewer and the dashed line indicates a plane *away* from the viewer. Note that no matter how the three-dimensional views are rotated, the L and D forms can never be superimposed. This is a characteristic of enantiomers.

3.6 MiniReview

Twenty-two different amino acids are found in cells and can bond to each other via the peptide bond. Mirror image (enantiomeric) forms of sugars and amino acids exist, but only one optical isomer of each is found in most cell polysaccharides and proteins.

- Why can it be said that all amino acids are structurally similar yet different simultaneously?
- Draw the complete structure of a dipeptide containing the amino acids alanine and tyrosine. Outline the peptide bond.
- Which enantiomeric forms of sugars and amino acids are commonly found in living organisms? Why doesn't the amino acid glycine have different enantiomers? (*Hint:* Look carefully at Figure 3.14*c* and replace the alanine shown with glycine.)

3.7 Proteins: Primary and Secondary Structure

Proteins play several key roles in cell function. In essence, a cell is what it is and does what it does because of the kinds and amounts of proteins it contains; that is, every different type of cell has a different complement of proteins. An understanding of protein structure is therefore essential for understanding how cells work.

Two major classes of proteins are *catalytic* proteins (enzymes) and *structural* proteins. **Enzymes** are the catalysts for chemical reactions that occur in cells (∞ Chapters 5, 20, and 21). By contrast, structural proteins are integral parts of the major structures of the cell: membranes, walls, cytoplasmic components, and so on. However, all proteins show certain basic features in common, and we discuss these now.

Primary Structure

As we have said, proteins are polymers of amino acids covalently bonded by *peptide bonds* (Figure 3.13). Two amino acids bonded by peptide linkage constitute a dipeptide, three amino acids, a tripeptide, and so on. When many amino acids are covalently linked via peptide bonds, they form a **polypeptide**.

A protein consists of one or more polypeptides. The number of amino acids differs greatly from one protein to another; proteins containing as few as 15 or as many as 10,000 amino acids are known. Because proteins differ in their composition, sequence, and number of amino acids, it is obvious that enormous variation in protein structure (and thus function) is possible.

The linear array of amino acids in a polypeptide is called its **primary structure**. The primary structure of a polypeptide is critical to its final function because it is consistent with only certain types of folding patterns. And it is only the final, folded polypeptide that assumes biological activity. The two ends of a polypeptide are so designated by whether a free carboxylic acid group or a free amino group exists; the terms "C-terminus" and "N-terminus" are used to describe these two ends, respectively (Figure 3.2*b*).

Secondary Structure

Once formed, a polypeptide does not remain a linear structure. Instead it *folds* to form a more stable structure. Interactions of the R groups on the amino acids in a polypeptide force the molecule to twist and fold in a specific way. This forms the **secondary structure**. Hydrogen bonds, the weak noncovalent linkages discussed earlier (Section 3.1),

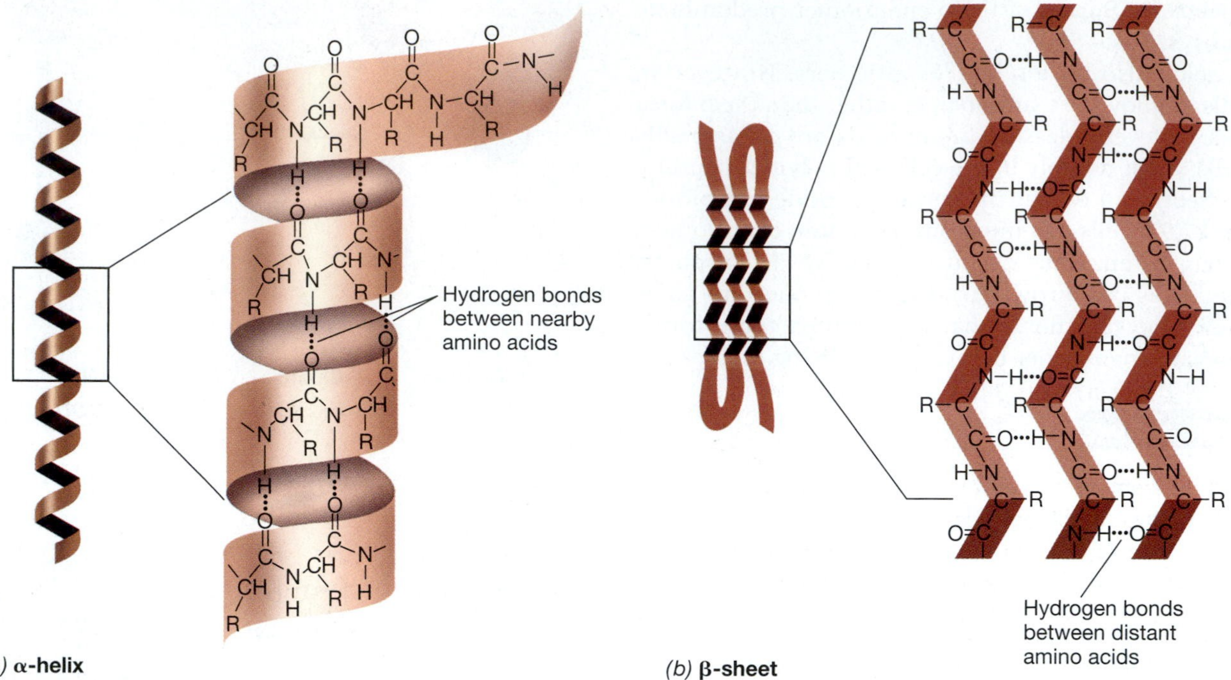

Figure 3.15 Secondary structure of polypeptides. *(a)* α-helix secondary structure. *(b)* β-sheet secondary structure. Note that the hydrogen bonding is between atoms in the peptide bonds and does not involve the R groups.

play important roles in polypeptide secondary structure. One common type of secondary structure is the *α-helix*. To envision an α-helix, imagine a linear polypeptide wound around a cylinder (**Figure 3.15a**). In this twisted structure, oxygen and nitrogen atoms from different amino acids become positioned close enough to allow hydrogen bonding. These hydrogen bonds give the α-helix its inherent stability (Figure 3.15*a*).

The primary structure of some polypeptides induces a different type of secondary structure, called a *β-sheet*. In the β-sheet, the chain of amino acids in the polypeptide folds back and forth upon itself instead of forming a helix. However, as in the α-helix, the folding in a β-sheet exposes hydrogen atoms that can undergo hydrogen bonding (Figure 3.15*b*). Typically, a β-sheet secondary structure yields a polypeptide that is rather rigid and α-helical secondary structures are more flexible. Thus, an enzyme, for example, whose activity may depend on its being rather flexible, may contain a high degree of α-helix secondary structure. By contrast, a structural protein that functions in cellular scaffolding may contain large regions of β-sheet secondary structure.

Many polypeptides contain regions of both α-helix *and* β-sheet secondary structure, the type of folding and its location in the molecule being determined by the primary structure and the available opportunities for hydrogen bonding and hydrophobic interactions (see Figure 3.16). A typical protein is thus made up of many **domains**, as they are called, regions of the protein that have a specific structure and function in the final, biologically active, molecule.

3.8 Proteins: Higher Order Structure and Denaturation

Once a polypeptide has achieved secondary structure it continues to fold to form an even more stable molecule. This folding results in a unique three-dimensional shape called the **tertiary structure** of the protein.

Like secondary structure, tertiary structure is ultimately determined by primary structure. However, tertiary structure is also governed to some extent by the secondary structure of the molecule because the side chain of each amino acid in the polypeptide is positioned in a specific way (Figure 3.15). If additional hydrogen bonds, covalent bonds, hydrophobic interactions, or other atomic interactions are able to form, the polypeptide will fold to accommodate them (**Figure 3.16**). The tertiary folds of the polypeptide ultimately form exposed regions or grooves in the molecule (Figure 3.16 and see Figure 3.17) that are important for binding other molecules (for example, in the binding of a substrate to an enzyme or the binding of DNA to a specific regulatory protein) (∞ Sections 5.5 and 9.2).

Frequently a polypeptide folds in such a way that adjacent sulfhydryl groups of cysteine residues are exposed. These free —SH groups can form a disulfide bond between the two amino acids. If the two cysteine residues are located in different polypeptides in a protein, the disulfide bond covalently links the two molecules (Figure 3.16*a*). In addition, a single polypeptide chain can fold and bond to itself if a disulfide bond can form within the molecule.

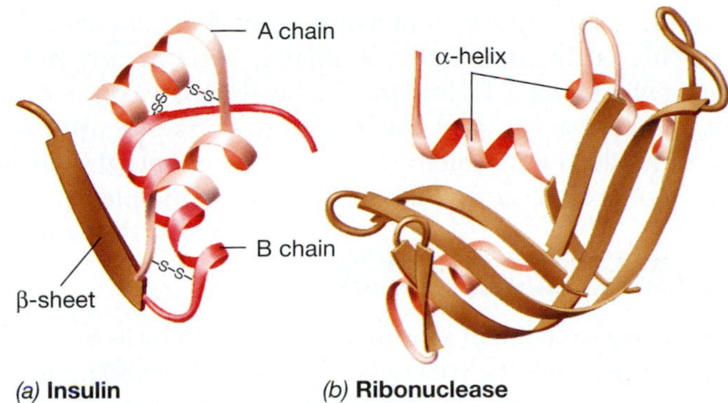

Figure 3.16 Tertiary structure of polypeptides. *(a)* Insulin, a protein containing two polypeptide chains; note how the B chain contains both α-helix and β-sheet secondary structure and how disulfide linkages (S–S) help in dictating folding patterns (tertiary structure). *(b)* Ribonuclease, a large protein with several regions of α-helix and β-sheet secondary structure.

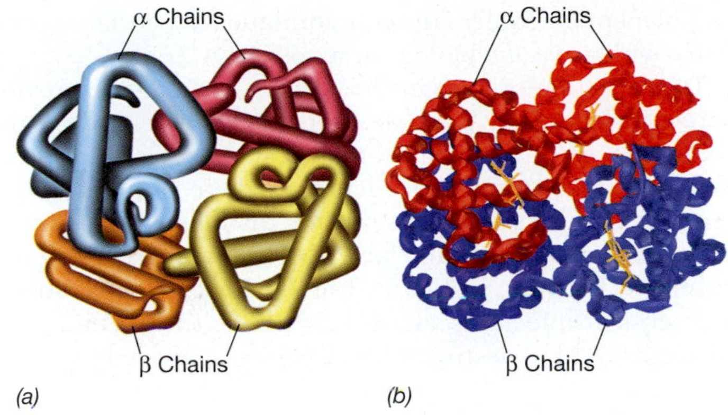

Figure 3.17 Quaternary structure of human hemoglobin. *(a)* There are two kinds of polypeptide in human hemoglobin, α chains (shown in blue and red) and β chains (shown in orange and yellow), but a total of four polypeptides in the final protein molecule (two α chains and two β chains). Separate colors are used to distinguish each chain. *(b)* Molecular structure of human hemoglobin as determined by X-ray crystallography. In this view each α chain is red and each β chain is blue.

Quaternary Structure

If a protein consists of two or more polypeptides, and many proteins do, the number and type of polypeptides that form the final protein molecule are referred to as its **quaternary structure** (**Figure 3.17**). In proteins showing quaternary structure, each polypeptide, called a *subunit*, contains primary, secondary, and tertiary structure. Some proteins contain multiple copies of a single subunit. A protein containing two identical subunits, for example, would be called a *homodimer*. Other proteins may contain nonidentical subunits, each present in one or more copies (a heterodimer, for example, contains one copy each of two different polypeptides). The subunits in multisubunit proteins are held together by noncovalent interactions (hydrogen bonding, van der Waals forces, and hydrophobic interactions) or by covalent linkages, typically disulfide bonds.

Denaturation

When proteins are exposed to extremes of heat or pH or to certain chemicals or metals that affect their folding, they may undergo **denaturation** (**Figure 3.18**). Denaturation causes the polypeptide chain to unfold, destroying the higher order (secondary, tertiary, and quaternary, if relevant) structure of the molecule. Depending on the severity of the denaturant or denaturing conditions, the polypeptide may refold after the denaturant is removed (Figure 3.18). Typically, however, denatured proteins unfold such that their hydrophobic regions become exposed and stick together to form protein aggregates that lack biological activity.

The biological properties of a protein are usually lost when it is denatured. Peptide bonds (Figure 3.13) are unaffected, however, and so a denatured molecule retains its primary structure. This shows that biological activity is not inherent in the primary structure of a protein but instead is a function of the uniquely folded form of the molecule as ultimately directed by primary structure. In other words, folding

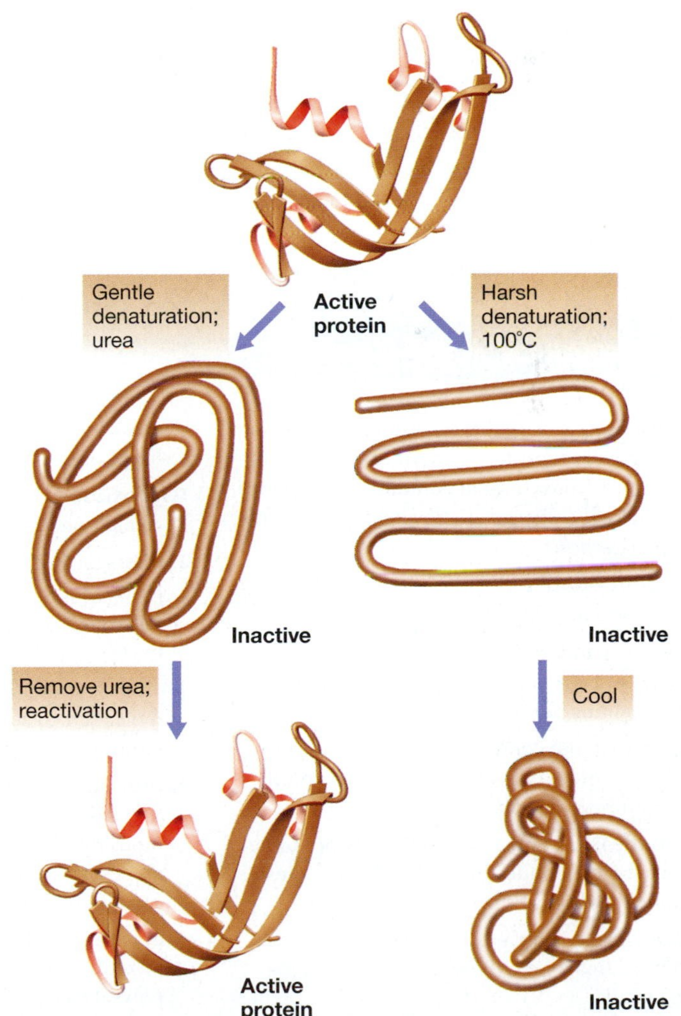

Figure 3.18 Denaturation of the protein ribonuclease. Ribonuclease structure was shown in Figure 3.16*b*. Note how harsh denaturation permanently destroys a molecule (from the standpoint of biological function) because of improper folding but primary structure is retained.

of a polypeptide confers upon it a unique shape that is compatible with a *specific* biological function.

Denaturation of proteins is a major means of destroying microorganisms. For example, alcohols such as phenol and ethanol are effective disinfectants because they readily penetrate cells and irreversibly denature their proteins. Such chemical agents are thus useful for disinfecting inanimate objects such as surfaces and have enormous practical value in household, hospital, and industrial disinfectant applications. We discuss disinfectants, along with other chemical and physical agents used to destroy microorganisms, in Chapter 27.

Moving On

Now that we have reviewed the chemistry of cellular components, we are in a better position to understand the structural details of cells. In the next chapter we will see how macromolecules come together to form major structures of the cell, such as the cytoplasmic membrane, the cell wall, and the flagellum. From there we will consider the basic metabolic properties of cells in Chapter 5. Metabolism, the *machine* function of a cell, drives the biosynthesis and assembly of new copies of macromolecules; these processes result in cell growth (Chapter 6). The metabolic events themselves are directed by the *coding* functions of the cell, the essential genetic events carried out by all cells. We discuss molecular biology in Chapters 7–9.

As the contemporary microbiologist Norman Pace has put it, "life is fundamentally *chemistry*." And as any microbiologist will attest, a feeling for the biochemistry of proteins, lipids, nucleic acids, and polysaccharides is essential to a grasp of modern microbiology and will accelerate the understanding of both basic and more advanced principles.

3.7 and 3.8 MiniReview

The primary structure of a protein is determined by its amino acid sequence, but the folding (higher order structure) of the polypeptide determines how the protein functions in the cell.

- Define the terms primary, secondary, and tertiary with respect to protein structure.
- How does a polypeptide differ from a protein?
- What secondary structural features tend to make β-sheet proteins more rigid than α-helices?
- Describe the number and kinds of polypeptides present in a homotetrameric protein.
- Describe the structural and biological effects of the denaturation of a protein. Of what practical value is knowledge of protein denaturation?

Review of Key Terms

Amino acid one of the 22 different monomers that make up proteins; chemically, a two-carbon carboxylic acid containing an amino group and a characteristic substituent on the alpha carbon

Covalent bond a chemical bond in which electrons are shared between two atoms

Denaturation destruction of the folding properties of a protein leading (usually) to protein aggregation and loss of biological activity

Domain in the context of proteins, a portion of the protein typically possessing a specific structure or function

DNA (deoxyribonucleic acid) a polymer of deoxyribonucleotides linked by phosphodiester bonds that carries genetic information

Enantiomer a form of a molecule that is the mirror image of another form of the same molecule

Enzyme a protein or an RNA that catalyzes a specific chemical reaction in a cell

Fatty acid an organic acid containing a carboxylic acid group and a hydrocarbon chain of various lengths; major components of lipids of *Bacteria* and *Eukarya*

Glycosidic bond a covalent bond linking sugars together in a polysaccharide

Hydrogen bond a weak chemical interaction between a hydrogen atom and a second, more electronegative element, usually an oxygen or nitrogen atom

Isomers two molecules with the same molecular formula but a difference in structure

Lipid a polar compound such as glycerol bonded to fatty acids or other hydrophobic molecules by ester or ether linkage, often also containing other groups, such as phosphate or sugars

Macromolecule a polymer of covalently linked monomeric units, such as DNA, RNA, polysaccharides, and lipids

Molecule two or more atoms chemically bonded to one another

Monomer a small molecule that is a building block for larger molecules

Nonpolar possessing hydrophobic (water-repelling) characteristics and not easily dissolved in water

Nucleic acid DNA or RNA

Nucleotide a monomer of a nucleic acid containing a nitrogen base (adenine, guanine, cytosine, thymine, or uracil), one or more molecules of phosphate, and a sugar, either ribose (in RNA) or deoxyribose (in DNA)

Peptide bond a type of covalent bond linking amino acids in a polypeptide

Phosphodiester bond a type of covalent bond linking nucleotides together in a polynucleotide

Polar possessing hydrophilic (water-loving) characteristics and generally water soluble

Polymer a large molecule made up of monomers

Polynucleotide a polymer of nucleotides bonded to one another by covalent bonds called phosphodiester bonds

Polypeptide a polymer of amino acids bonded to one another by peptide bonds

Polysaccharide a polymer of sugar units bonded to one another by glycosidic bonds

Primary structure in an informational macromolecule such as a polypeptide or a nucleic acid, the precise sequence of monomeric units

Protein a polypeptide or group of polypeptides forming a molecule of specific biological function

Purine one of the nitrogen bases of nucleic acids that contain two fused rings; adenine and guanine

Pyrimidine one of the nitrogen bases of nucleic acids that contain a single ring; cytosine, thymine, and uracil

Quaternary structure in proteins, the number and types of individual polypeptides in the final protein molecule

RNA (ribonucleic acid) a polymer of ribonucleotides linked by phosphodiester bonds that plays many roles in cells, in particular, during protein synthesis

Secondary structure the initial pattern of folding of a polypeptide or a polynucleotide, usually dictated by opportunities for hydrogen bonding

Tertiary structure the final folded structure of a polypeptide that has previously attained secondary structure

Review Questions

1. Which are the major elements found in living organisms? Why are oxygen and hydrogen particularly abundant in living organisms (Section 3.1)?
2. Define the word "molecule." How many atoms are in a molecule of hydrogen gas? How many atoms are in a molecule of glucose (Sections 3.1 and 3.3)?
3. Refer to the structure of the nitrogen base cytosine shown in Figure 3.1. Draw this structure and then label the positions of all single bonds and double bonds in the cytosine molecule (Section 3.1).
4. Compare and contrast the words "monomer" and "polymer." Give three examples of biologically important polymers and list the monomers of which they are composed. Which classes of macromolecules are most abundant (by weight) in a cell (Sections 3.1 and 3.2)?
5. List the components that would make up a simple lipid. How does a triglyceride differ from a complex lipid (Section 3.4)?
6. Examine the structures of the triglyceride and of phosphatidyl ethanolamine shown in Figure 3.7. How might the substitution of phosphate and ethanolamine for a fatty acid alter the chemical properties of the lipid (Section 3.4)?
7. RNA and DNA are similar types of macromolecules but show distinct differences as well. List three ways in which RNA differs chemically or physically from DNA. What is the cellular function of DNA and RNA (Section 3.5)?
8. Why are amino acids so named? Write a general structure for an amino acid. What is the importance of the R group to final protein structure? Why does the amino acid cysteine have special significance for protein structure (Section 3.6)?
9. What type of reaction between two amino acids leads to formation of the peptide bond (Section 3.6)?
10. Define the types of protein structure: primary, secondary, tertiary, and quaternary. Which of these structures are altered by denaturation (Sections 3.7 and 3.8)?
11. Fill in the blanks. A glycosidic bond is to a __________ as a __________ bond is to a polypeptide and a __________ is to a nucleic acid. All of these bonds are examples of __________ bonds, which are chemically much stronger than weak bonds, such as __________, __________, and __________

Application Questions

1. Observe the following nucleotide sequences of RNA: (*a*) GUCAAAGAC, (*b*) ACGAUAACC. Can either of these RNA molecules have secondary structure? If so, draw the potential secondary structure(s).
2. A few soluble (cytoplasmic) proteins contain a high content of hydrophobic amino acids. How would you predict these proteins would fold into their tertiary structure and why?
3. Cells of the genus *Halobacterium*, an organism that lives in very salty environments, contain over 5 molar (M) potassium (K^+). Because of this high K^+ content, many cytoplasmic proteins of *Halobacterium* cells are enriched in two specific amino acids that are present in much higher proportions in *Halobacterium* proteins than in functionally similar proteins from *Escherichia coli* (which has only very low levels of K^+ in its cytoplasm). Which amino acids are enriched in *Halobacterium* proteins and why? (*Hint:* Which amino acids could best neutralize the positive charges due to K^+?)
4. When a culture of the bacterium *Escherichia coli*, an inhabitant of the human gut, is placed in a beaker of boiling water, significant changes in the cells occur almost immediately. However, when a culture of *Pyrodictium*, a hypothermophile that grows optimally in boiling hot springs is put in the same beaker, similar changes do not occur. Explain.
5. Review Figure 3.6*b* and then describe the differences that make each of these polymers unique. If all of the glycosidic bonds in these polymers were hydrolyzed, what single molecule would remain?
6. Review Figure 3.12*b*. Of all the amino acids shaded in blue, what is it about their chemistry that unites them as a "family"?

4

Cell Structure and Function in *Bacteria* and *Archaea*

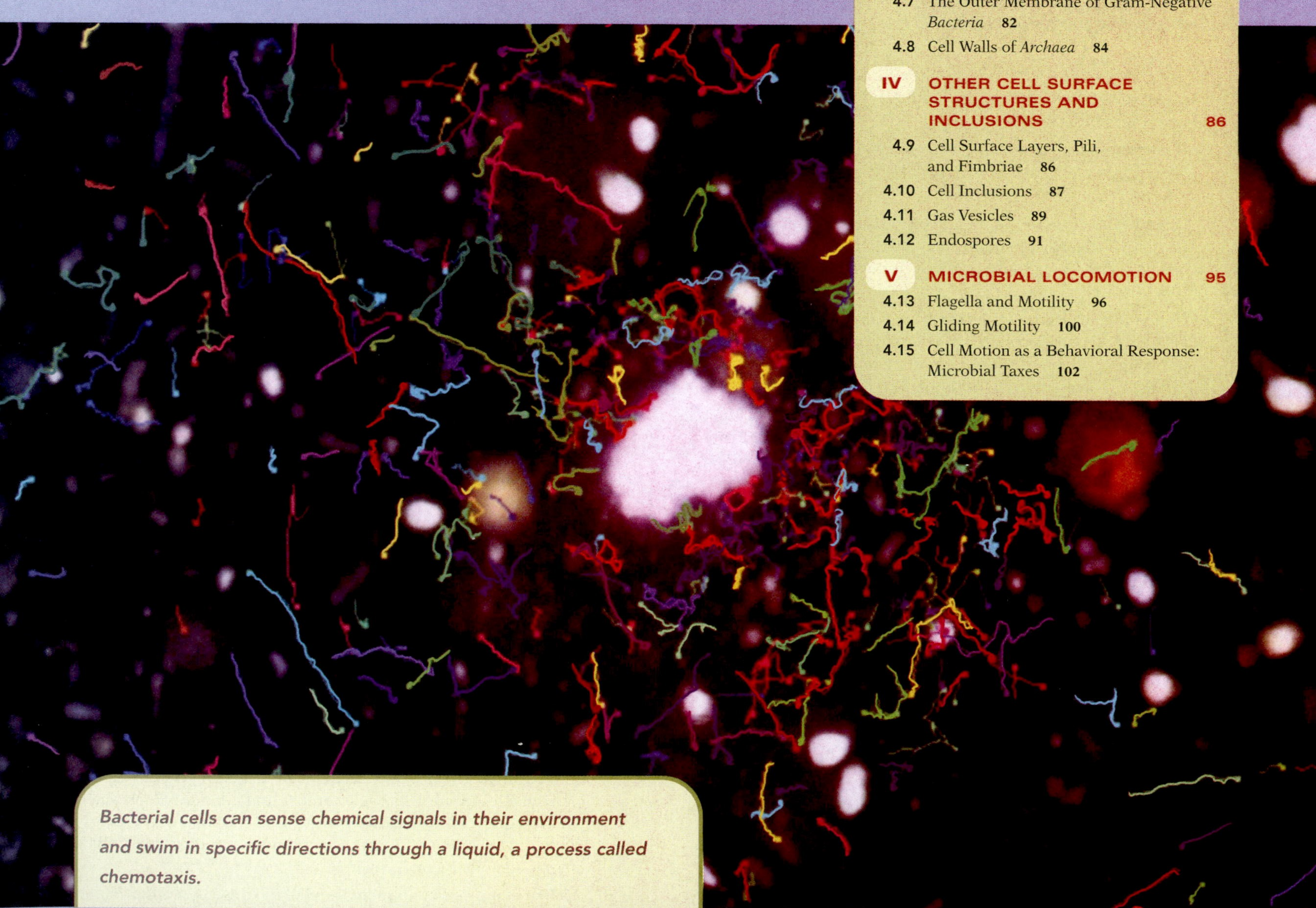

Bacterial cells can sense chemical signals in their environment and swim in specific directions through a liquid, a process called chemotaxis.

I CELL SHAPE AND SIZE

In this chapter we move on from considering the chemical structure of macromolecules to examining how macromolecules combine to form the key components of the cell: the cytoplasmic membrane, the cell wall, cell inclusions, the flagellum, and so on. Our theme in this chapter will be structure and function. There is an old saying in biology that "form follows function." That is, the structure (form) of a biological component evolved as it did because it was better than the alternatives at carrying out a specific function, either within the cell or within the whole organism. Microorganisms are no exception in this regard as the key structures of microbial cells evolved because they carry out specific functions effectively.

We begin this chapter by discussing two key features of prokaryotic cells—their shape and small size. Prokaryotes typically have defined shapes and are extremely small cells. Shape is useful for differentiating prokaryotic cells and size has profound effects on their biology.

4.1 Cell Morphology

In microbiology, the term **morphology** means cell shape. Several morphologies are known among prokaryotes, and the most common ones are described by terms that are part of the essential lexicon of the microbiologist.

Major Cell Morphologies

Examples of bacterial morphologies are shown in **Figure 4.1**. A bacterium that is spherical or ovoid in morphology is called a *coccus* (plural, *cocci*). A bacterium with a cylindrical shape is called a *rod*. Some rods twist into spiral shapes and are called *spirilla*. The cells of many prokaryotic species remain together in groups or clusters after cell division, and the arrangements are often characteristic of certain genera. For instance, some cocci form long chains (for example, the bacterium *Streptococcus*), others occur in three-dimensional cubes (*Sarcina*), and still others in grapelike clusters (*Staphylococcus*).

Several groups of bacteria are immediately recognizable by the unusual shapes of their individual cells. Examples include spirochetes, which are tightly coiled bacteria, appendaged bacteria, which possess extensions of their cells as long tubes or stalks, and filamentous bacteria, which form long, thin cells or chains of cells (Figure 4.1).

The cell shapes in Figure 4.1 should be examined with the understanding that they are *representative* morphologies for prokaryotic cells. Many variations of these basic morphological types are known. Thus there are fat rods, thin rods, short rods, and long rods, a rod simply being a cell that is longer in one dimension than in the other. As we will see, there are even square bacteria and star-shaped bacteria! Cell morphologies thus form a continuum, with some shapes very common and others more unusual.

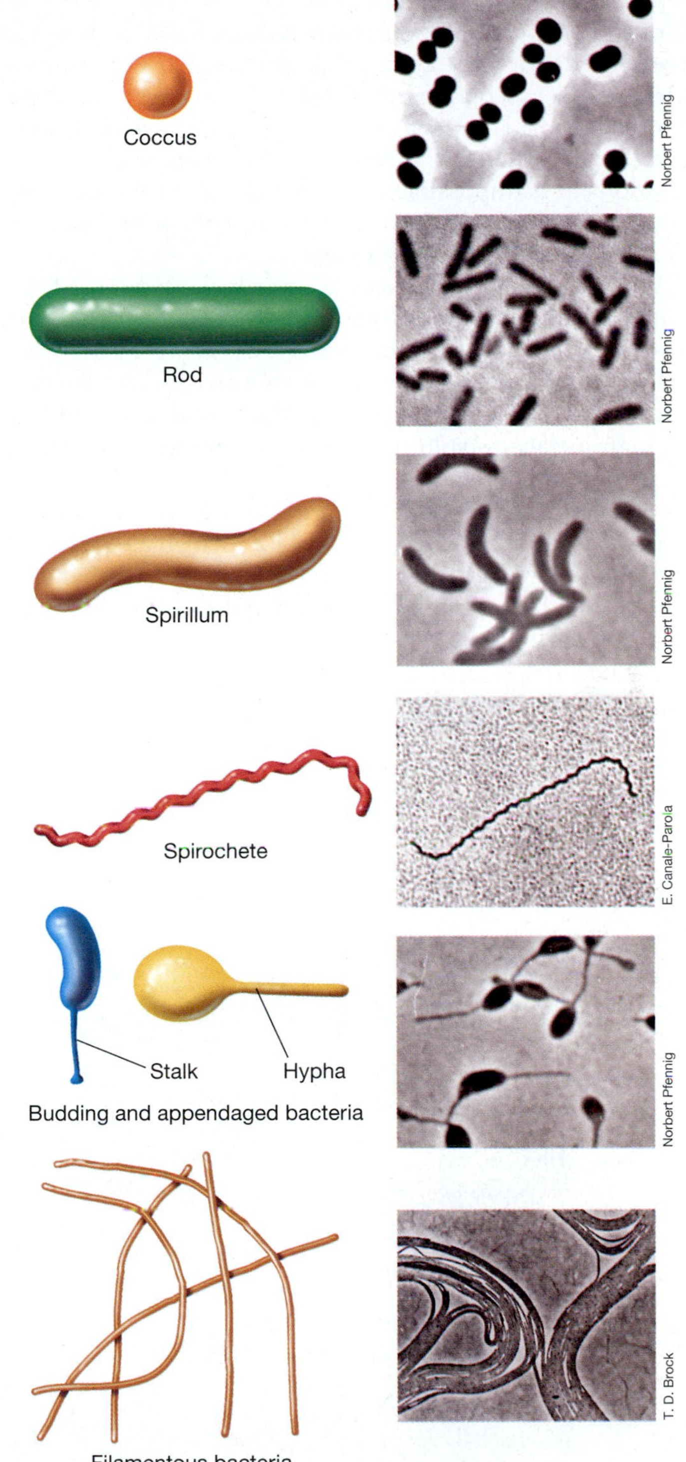

Figure 4.1 Representative cell morphologies of prokaryotes. Next to each drawing is a phase-contrast photomicrograph showing an example of that morphology. Organisms are coccus, *Thiocapsa roseopersicina* (diameter of a single cell = 1.5 μm); rod, *Desulfuromonas acetoxidans* (diameter = 1 μm); spirillum, *Rhodospirillum rubrum* (diameter = 1 μm); spirochete, *Spirochaeta stenostrepta* (diameter = 0.25 μm); budding and appendaged, *Rhodomicrobium vannielii* (diameter = 1.2 μm); filamentous, *Chloroflexus aurantiacus* (diameter = 0.8 μm).

Morphology and Biology

Although cell morphology is easily recognized, it is in general a poor predictor of a cell's other properties. For example, under the microscope many rod-shaped *Archaea* look identical to rod-shaped *Bacteria,* yet we know they are of different phylogenetic domains (Section 2.7). Thus, with very rare exceptions, it is impossible to predict the physiology, ecology, phylogeny, or virtually any other property of a prokaryotic cell by simply knowing its morphology.

What sets the morphology of a particular species? This is an interesting question to which we don't have firm answers. However, it is suspected that several selective forces are in play in setting the morphology of a given species. These include optimization for nutrient uptake (small cells and those with high surface-to-volume ratios), swimming motility in viscous environments or near surfaces (helical or spiral-shaped cells), gliding motility (filamentous bacteria), and so on. Thus morphology is not a trivial feature of a microbial cell. A cell's morphology is a genetically directed characteristic and has been selected by evolution to maximize fitness for the species in a particular habitat.

4.1 MiniReview

There are prokaryotic cells of many different shapes. Rods, cocci, and spirilla are common cell morphologies.

- Write a single sentence that describes how cocci and rods differ in morphology.
- Is cell morphology a good predictor of other properties of the cell?

4.2 Cell Size and the Significance of Smallness

Prokaryotes vary in size from cells as small as about 0.2 μm in diameter to those more than 700 μm in diameter (**Table 4.1**). The vast majority of rod-shaped prokaryotes that have been cultured in the laboratory are between 0.5 and 4 μm wide and less than 15 μm long, but a few very large prokaryotes, such as *Epulopiscium fishelsoni*, are huge, with cells being longer than 600 μm (0.6 millimeter). This bacterium, phylogenetically related to the endospore-forming bacterium *Clostridium* and found in the gut of the surgeonfish, is interesting not only because it is so large, but also because it has an unusual form of cell division and contains multiple copies of its genome. Multiple offspring are formed and are then released from the *Epulopiscium* "mother cell" (**Figure 4.2a** shows a mother cell). A mother cell of *Epulopiscium* contains several thousand genome copies, each of which is about the same size as the genome of *Escherichia coli* (4.6 megabase pairs, or Mbp). The many

Table 4.1 Cell size and volume of prokaryotic cells, from the largest to the smallest

Organism	*Characteristics*	*Morphology*	*Size*[a] *(μm)*	*Cell volume* (μm^3)	*E. coli volumes*
Thiomargarita namibiensis	Sulfur chemolithotroph	Cocci in chains	750	200,000,000	10^8
Epulopiscium fishelsoni	Chemoorganotroph	Rods with tapered ends	80 × 600	3,000,000	1.5×10^6
Beggiatoa sp.	Sulfur chemolithotroph	Filaments	50 × 160	1,000,000	5×10^5
Achromatium oxaliferum	Sulfur chemolithotroph	Cocci	35 × 95	80,000	4×10^4
Lyngbya majuscula	Cyanobacterium	Filaments	8 × 80	40,000	2×10^4
Prochloron sp.	Prochlorophyte	Cocci	30	14,000	7×10^3
Thiovulum majus	Sulfur chemolithotroph	Cocci	18	3,000	1.5×10^3
Staphylothermus marinus	Hyperthermophile	Cocci in irregular clusters	15	1,800	9×10^2
Titanospirillum velox	Sulfur chemolithotroph	Curved rods	5 × 30	600	3×10^2
Magnetobacterium bavaricum	Magnetotactic bacterium	Rods	2 × 10	30	15
Escherichia coli	Chemoorganotroph	Rods	1 × 2	2	1
Pelagibacter ubique	Marine chemoorganotroph	Rods	0.2 × 0.5	0.014	1.4×10^{-2}
Mycoplasma pneumoniae	Pathogenic bacterium	Pleomorphic[b]	0.2	0.005	2.5×10^{-3}

[a]Where only one number is given, this is the diameter of spherical cells. The values given are for the largest cell size observed in each species. For example, for *T. namibiensis*, an average cell is only about 200 μm in diameter. But on occasion, giant cells of 750 μm are observed. Likewise, an average cell of *S. marinus* is about 1 μm in diameter.

[b]*Mycoplasma* is a cell wall-less bacterium and can take on many shapes (*pleomorphic* means "many shapes").

Source: Data obtained from Schulz, H.N., and B.B. Jørgensen. 2001. *Ann. Rev. Microbiol. 55:* 105–137.

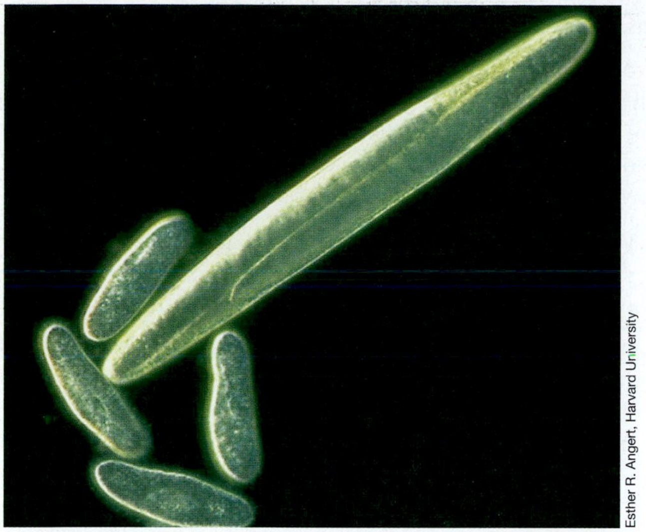

(a)

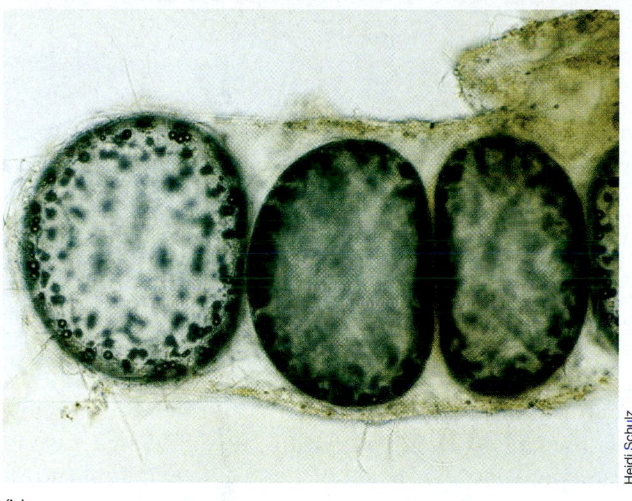

(b)

Figure 4.2 Some very large prokaryotes. *(a)* Dark-field photomicrograph of a giant prokaryote, the surgeonfish symbiont *Epulopiscium fishelsoni.* The rod-shaped *E. fishelsoni* cell in this field is about 600 μm (0.6 mm) long and 75 μm wide and is shown with three cells of the protist (eukaryote) *Paramecium,* each of which is about about 150 μm long. *E. fishelsoni* is a member of the *Bacteria* and phylogenetically related to *Clostridium* species. *(b) Thiomargarita namibiensis,* a large sulfur chemolithotroph (phylum *Proteobacteria* of the *Bacteria*) and currently the largest known prokaryote. Each ovoid-shaped cell is about 400 μm wide.

copies are apparently necessary because the cell volume of *Epulopiscium* is so large that a single copy of its genome wouldn't be sufficient to support the transcriptional and translational needs of the cell.

Cells of the largest known prokaryote, the sulfur chemolithotroph *Thiomargarita* (Figure 4.2*b*), can be 750 μm in diameter, nearly visible to the naked eye. Most very large prokaryotes are either sulfur chemolithotrophs or cyanobacteria (Table 4.1). Why these cells are so large is not well understood, although for sulfur bacteria large cell size may be a mechanism for storing sulfur (an energy source). It is thought that limitations in nutrient uptake ultimately dictate upper limits for the size of prokaryotic cells. The metabolic rate of a cell varies inversely with the square of its size. Thus for very large cells uptake processes eventually limit metabolism to the point that the cell is no longer competitive with smaller cells.

Very large cells are not the norm in the prokaryotic world. By contrast to *Thiomargarita* or *Epulopiscium* (Figure 4.2), the dimensions of an average rod-shaped prokaryote, the bacterium *E. coli,* for example, are about 1 × 2 μm; these dimensions are typical of prokaryotes. For comparison, average eukaryotic cells can be 10 to more than 200 μm in diameter. In general, then, prokaryotes are very small cells compared with eukaryotes.

Surface-to-Volume Ratios, Cell Growth Rates, and Evolution

There are significant advantages to being a small cell. Small cells contain more surface area relative to cell volume than do large cells; that is, they have a higher surface-to-volume ratio. Consider a spherical coccus. The volume of such a cell is a function of the cube of its radius ($V = 4/3\pi r^3$), and its surface area is a function of the square of the radius ($S = 4\pi r^2$). Therefore, the (S/V) ratio of a spherical coccus is $3/r$ (**Figure 4.3**). As a cell increases in size, its S/V ratio decreases. To illustrate this, consider the S/V ratio for some of the cells of different sizes listed in Table 4.1: *Pelagibacter ubique*, 22; *E. coli*, 4.5; and *E. fishelsoni*, 0.05.

The S/V ratio of a cell affects several aspects of its biology, including evolution. For instance, because a cell's growth rate depends, among other things, on the rate of nutrient exchange, the higher S/V ratio of smaller cells supports greater nutrient exchange per unit of cell volume

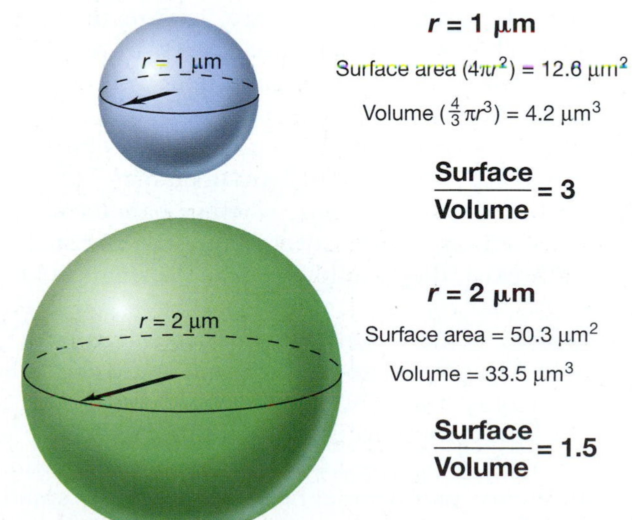

Figure 4.3 Surface area and volume relationships in cells. As a cell increases in size, its S/V ratio decreases.

in comparison with larger cells. Because of this, smaller cells, in general, grow faster than larger cells. In addition, a given amount of resources (the nutrients available to support growth) will support a larger population of small cells than of large cells. Why is this important? Each time a cell divides, its chromosome replicates as well. As DNA is replicated, occasional errors, called *mutations,* occur. Because mutation rates appear to be roughly the same in all cells, large or small, the more chromosome replications that occur, the greater the *total number* of mutations in the population. Mutations are the "raw materials" of evolution; the larger the pool of mutations, the greater the evolutionary possibilities. Thus, because prokaryotic cells are quite small and are also genetically haploid (allowing mutations to be expressed immediately), they have, in general, the capacity for more rapid growth and evolution than larger, genetically diploid cells.

In the larger diploid cells, not only is the S/V ratio smaller, but the effects of a mutation in one gene can be masked by a second unmutated gene copy. These fundamental differences between prokaryotic and eukaryotic cells support the observations that prokaryotes are often seen to adapt quite rapidly to changing environmental conditions and can more easily exploit new habitats than can eukaryotic cells. We will see this concept in action in later chapters when we consider, for example, the enormous metabolic diversity of prokaryotes or the spread of antibiotic resistance.

Lower Limits of Cell Size

From the discussion above, it may seem that smaller and smaller bacteria will have greater and greater selective advantages in nature. Obviously, there must be lower limits to cell size, but what are they? Some microbiologists have proposed from observations of soil and other habitats that very small bacteria exist in nature, cells called *nanobacteria* (the word "nano" comes from Greek, meaning "dwarf"). The sizes reported for nanobacteria are 0.1 μm in diameter and smaller. This is extremely small, even by prokaryotic standards (Table 4.1). If one considers the volume needed to house the essential biomolecules of a free-living cell (Chapter 3), a structure of 0.1 μm or less is insufficient to do the job, and structures 0.15 μm in diameter are on the borderline in this regard. Thus, whether nanobacteria are really living cells is an unanswered question. If such tiny cells actually exist, they would be by far the smallest known cells.

Regardless of the status of nanobacteria, many very small prokaryotic cells are known and many have been grown in the laboratory. The open oceans, for example, contain 10^4–10^5 prokaryotic cells per milliliter, and these tend to be very small cells, 0.2–0.4 μm in diameter. We will see later that many pathogenic bacteria are very small as well. Thus, very small cells are not uncommon, but cellular organisms smaller than 0.15 μm in diameter are unlikely to exist.

4.2 MiniReview

Prokaryotes are typically smaller in size than eukaryotes, although some very large and some very small prokaryotes are known. The typical small size of prokaryotic cells affects their physiology, growth rate, ecology, and evolution. The lower limit of cell diameter is probably about 0.15 μm.

- What physical property of cells increases as cells become smaller?
- Using the formula for volume, calculate the volume of a nanobacterium 0.1 μm in diameter (the cell radius is half of its diameter). How many nanobacteria this size could fit into a cell of *Mycoplasma pneumoniae,* the smallest known prokaryote (see Table 4.1)?

II THE CYTOPLASMIC MEMBRANE AND TRANSPORT

We now consider an extremely important cell structure, the cytoplasmic membrane, and review the major functions that the membrane has, in particular, in the transport of substances into and out of the cell.

4.3 The Cytoplasmic Membrane in *Bacteria* and *Archaea*

The **cytoplasmic membrane** is a thin structure that surrounds the cell. Although very thin, this vital structure is the barrier separating the inside of the cell (the cytoplasm) from its environment. If the membrane is broken, the integrity of the cell is destroyed, the cytoplasm leaks into the environment, and the cell dies. The cytoplasmic membrane is also a highly selective permeability barrier, enabling a cell to concentrate specific metabolites and excrete waste materials.

Composition of Membranes

The general structure of biological membranes is a phospholipid bilayer (**Figure 4.4**). As previously discussed (Section 3.4), phospholipids contain both hydrophobic (fatty acid) and hydrophilic (glycerol–phosphate) components and can exist in many different chemical forms as a result of variation in the groups attached to the glycerol backbone. As phospholipids aggregate in an aqueous solution, they naturally form bilayer structures. In a phospholipid membrane, the fatty acids point inward toward each other to form a hydrophobic environment, and the hydrophilic portions remain exposed to the external environment or the cytoplasm (Figure 4.4*a*).

The cytoplasmic membrane, which is 6–8 nanometers wide, can be seen with the electron microscope, where it appears as two light-colored lines separated by a darker area

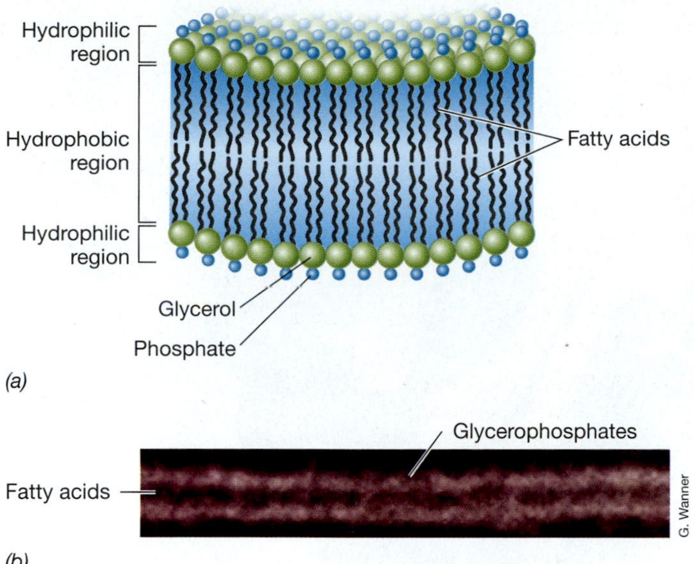

Figure 4.4 Structure of a phospholipid bilayer. *(a)* The general chemical structure of a phospholipid is shown in Figure 3.7*c*. *(b)* Transmission electron micrograph of a membrane from the bacterium *Halorhodospira halochloris*. The dark inner area is the hydrophobic region of the model membrane shown in *(a)*.

(Figure 4.4*b*). This *unit membrane,* as it is called (because each phospholipid leaf forms half of the "unit"), consists of a phospholipid bilayer with proteins embedded in it (**Figure 4.5**). The overall structure of the cytoplasmic membrane is stabilized by hydrogen bonds and hydrophobic interactions (Section 3.1). In addition, Mg^{2+} and Ca^{2+} help stabilize the membrane by forming ionic bonds with negative charges on the phospholipids.

Although in a diagram the cytoplasmic membrane may appear rather rigid (Figure 4.5), in reality it is somewhat fluid, having a viscosity approximating that of a light-grade oil. Membrane biologists used to think that membranes were highly fluid, with proteins free to float around within a "sea" of lipid. We now know that this model is incorrect; some movement in the membrane is likely, although how extensive this is and how important it is to membrane function is unknown.

Membrane Proteins

The major proteins of the cytoplasmic membrane have hydrophobic surfaces in their regions that span the membrane and hydrophilic surfaces in their regions that contact the environment and the cytoplasm (Figure 4.5). The outer surface of the cytoplasmic membrane faces the environment and in certain bacteria interacts with a variety of proteins that bind substrates or process large molecules for transport into the cell (periplasmic proteins, discussed in Section 4.7). The inner side of the cytoplasmic membrane faces the cytoplasm and interacts with proteins involved in energy-yielding reactions and other important cellular functions.

Many membrane proteins are firmly embedded in the membrane and are called *integral* membrane proteins. Other proteins have one portion anchored in the membrane and extramembrane regions that point into or out of the cell (Figure 4.5). Still other proteins, called *peripheral* membrane proteins, are not embedded in the membrane at all but are nevertheless firmly associated with membrane surfaces. Some of these peripheral membrane proteins are lipoproteins, proteins that contain a lipid tail that anchors the protein into the membrane. These proteins typically interact with integral membrane proteins in important cellular processes such as energy metabolism and transport.

Proteins in the cytoplasmic membrane are arranged in patches (Figure 4.5); instead of being distributed evenly, proteins are clustered, a strategy that allows the grouping of proteins that interact or that have similar function. The overall protein content of the membrane is also quite high membrane proteins are indeed rather crowded—and it is thought that the lipid bilayer varies in thickness from 6 to 8 nm to accommodate thicker and thinner patches of proteins.

Membrane-Strengthening Agents: Sterols and Hopanoids

One major eukaryotic–prokaryotic difference in cytoplasmic membrane composition is that eukaryotes have **sterols** in their membranes (**Figure 4.6*a***). Sterols are absent from the membranes of almost all prokaryotes (methanotrophic bacteria and mycoplasmas are exceptions, Sections 15.6 and 16.3, respectively). Depending on the cell type, sterols can make up as little as 5% or as much as 25% of the total lipids of eukaryotic membranes.

Sterols are rigid, planar molecules, whereas fatty acids are flexible. The presence of sterols in a membrane thus strengthens and stabilizes it and makes it less flexible. Molecules similar to sterols, called *hopanoids,* are present in the membranes of many *Bacteria* and likely play a role there similar to that of sterols in eukaryotic cells. One widely distributed hopanoid is the C_{30} hopanoid diploptene (Figure 4.6*b*). As far as is known, hopanoids are not present in *Archaea*.

Archaeal Membranes

The membrane lipids of *Archaea* differ from those of *Bacteria* and *Eukarya*. In contrast to the lipids of *Bacteria* and *Eukarya* in which *ester* linkages bond the fatty acids to glycerol (**Figure 4.7*a***; Section 3.4), the lipids of *Archaea* contain *ether* bonds between glycerol and their hydrophobic side chains (Figure 4.7*b*). In addition, archaeal lipids lack fatty acids. Instead, the side chains are composed of repeating units of the five-carbon hydrocarbon isoprene (Figure 4.7*c*). Despite these chemical differences, the fundamental construction of the cytoplasmic membrane of *Archaea*—inner and outer hydrophilic surfaces and a hydrophobic interior—is the same as that of membranes in *Bacteria* and *Eukarya*.

The major lipids of *Archaea* are glycerol diethers, which have 20-carbon side chains (the 20-C unit is called a *phytanyl*

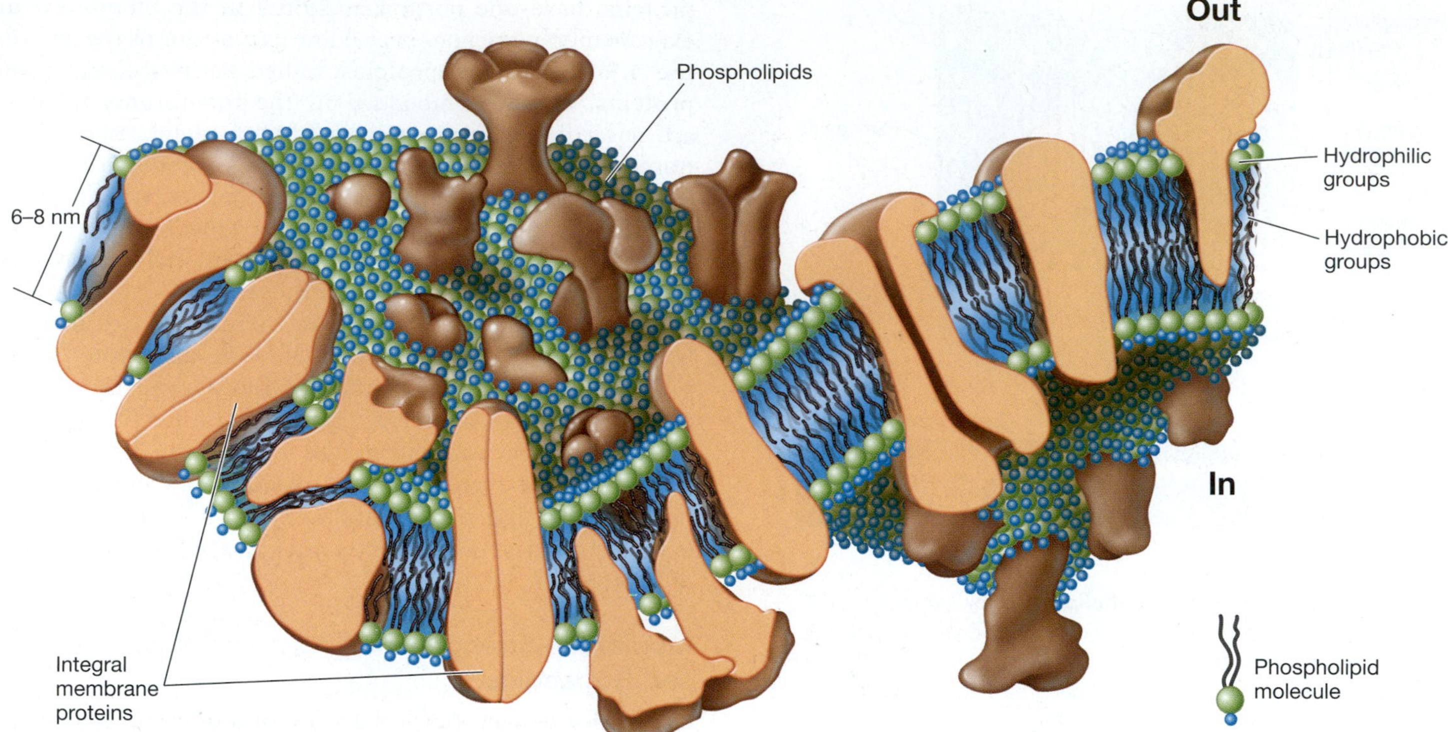

Figure 4.5 Structure of the cytoplasmic membrane. The inner surface (**In**) faces the cytoplasm and the outer surface (**Out**) faces the environment. Phospholipids compose the matrix of the cytoplasmic membrane, with the hydrophobic groups directed inward and the hydrophilic groups toward the outside, where they associate with water. Embedded in the matrix are proteins that are hydrophobic in the region that traverses the fatty acid bilayer. Hydrophilic proteins and other charged substances, such as metal ions, may attach to the hydrophilic surfaces. Although there are some chemical differences, the overall structure of the cytoplasmic membrane shown is similar in both prokaryotes and eukaryotes (but an exception to the bilayer design is shown in Figure 4.8*d*).

group), and diglycerol tetraethers, which have 40-carbon side chains (**Figure 4.8*a,b***). In the tetraether lipid, phytanyl side chains from each glycerol molecule are covalently linked (Figure 4.8*b*). Within a membrane this structure yields a lipid *monolayer* instead of a lipid *bilayer* membrane (Figure 4.8*c, d*). Unlike lipid bilayers, lipid monolayers are quite resistant to peeling apart. Not surprisingly, then, monolayer membranes are widespread among hyperthermophilic *Archaea*, prokaryotes that grow at temperatures so high (Sections 6.14 and 17.9–17.11) that the heat could peel apart lipid bilayers, causing cell lysis.

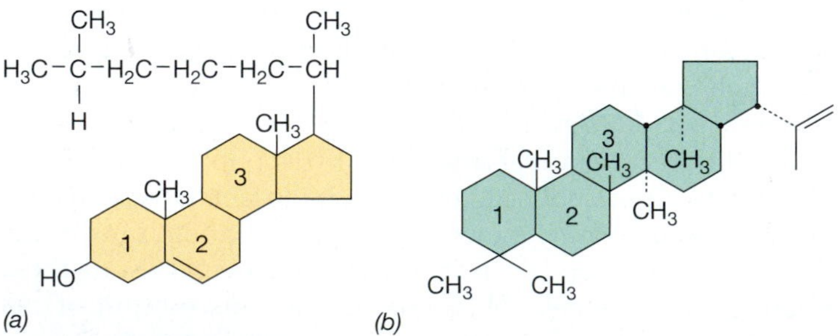

Figure 4.6 Sterols and hopanoids. *(a)* The structure of cholesterol, a typical sterol. *(b)* The structure of the hopanoid diploptene. Sterols are found in the membranes of eukaryotes and hopanoids in the membranes of some prokaryotes. The intraring labels 1, 2, and 3 highlight similarities in the parent structure of sterols and hopanoids.

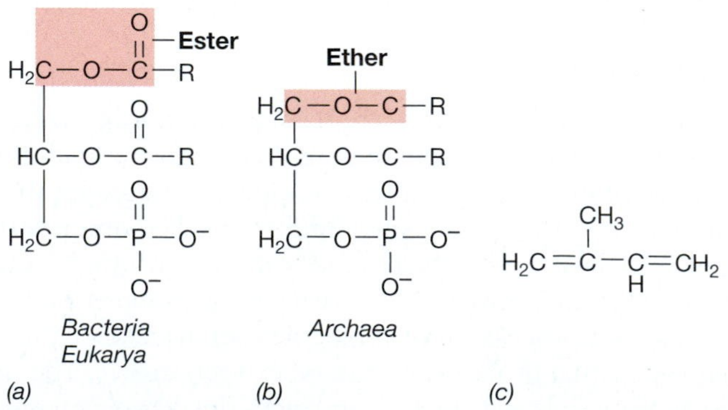

Ester
$H_2C{-}O{-}C(=O){-}R$
$HC{-}O{-}C(=O){-}R$
$H_2C{-}O{-}P(=O)(O^-){-}O^-$
Bacteria
Eukarya
(a)

Ether
$H_2C{-}O{-}C{-}R$
$HC{-}O{-}C{-}R$
$H_2C{-}O{-}P(=O)(O^-){-}O^-$
Archaea
(b)

$H_2C{=}C(CH_3){-}CH{=}CH_2$
(c)

Figure 4.7 General structure of lipids. *(a)* The ester linkage. *(b)* The ether linkage. *(c)* Isoprene, the parent structure of the hydrophobic side chains of archaeal lipids. By contrast, in lipids of *Bacteria* and *Eukarya*, the side chains are composed of fatty acids.

4.3 MiniReview

The cytoplasmic membrane is a highly selective permeability barrier constructed of lipids and proteins that form a bilayer with hydrophilic exteriors and a hydrophobic interior. Other molecules, such as sterols and hopanoids, may strengthen the membrane. Unlike *Bacteria* and *Eukarya*, *Archaea* contain ether-linked lipids, and some species have membranes of monolayer instead of bilayer construction.

- Draw the basic structure of a lipid bilayer.
- Why should compounds like sterols and hopanoids be good at stabilizing the cytoplasmic membrane?
- Contrast the linkage between glycerol and the hydrophobic portion of lipids in *Bacteria* and *Archaea*.

4.4 The Functions of Cytoplasmic Membranes

The cytoplasmic membrane is more than just a barrier separating the inside from the outside of the cell. The membrane plays critical roles in cell function. First and foremost, the membrane functions as a permeability barrier, preventing the passive leakage of substances into or out of the cell (**Figure 4.9**). Secondly, the membrane is an anchor for many proteins. Some of these are enzymes that catalyze bioenergetic reactions and others transport substances into and out of the cell. We will learn in Chapter 5 that the cytoplasmic membrane is also a major site of energy conservation in the cell. The membrane has an energetically charged form in which protons (H^+) are separated from hydroxyl ions (OH^-) across its surface (Figure 4.9). This charge separation is a form of energy, analogous to the potential energy present in a charged battery. This energy source, called the *proton motive force*, is responsible for driving many energy-requiring functions in the cell, including some forms of transport, motility, and biosynthesis of the cell's energy currency, ATP.

The Cytoplasmic Membrane as a Permeability Barrier

The interior of the cell (the cytoplasm) consists of an aqueous solution of salts, sugars, amino acids, nucleotides, vitamins, coenzymes, and other soluble materials. The hydrophobic internal portion of the cytoplasmic membrane (Figure 4.5) is a tight barrier to diffusion. Although some small hydrophobic molecules can pass through the membrane by diffusion, polar

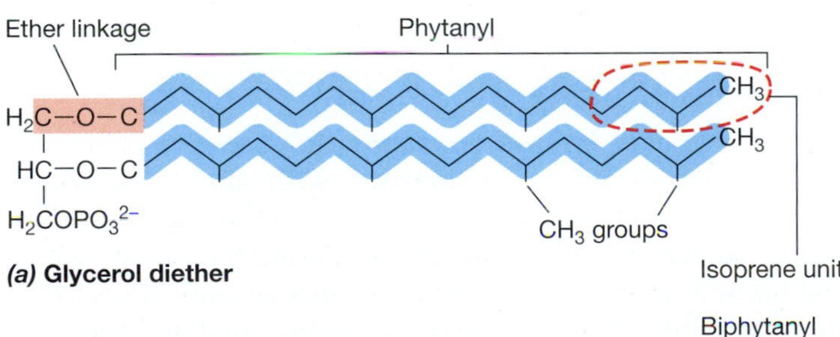

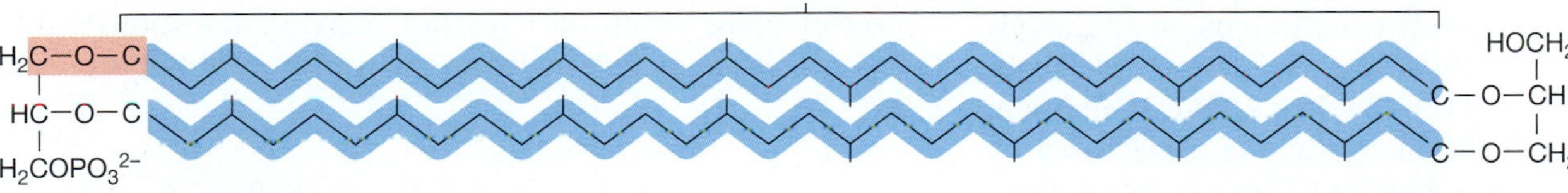

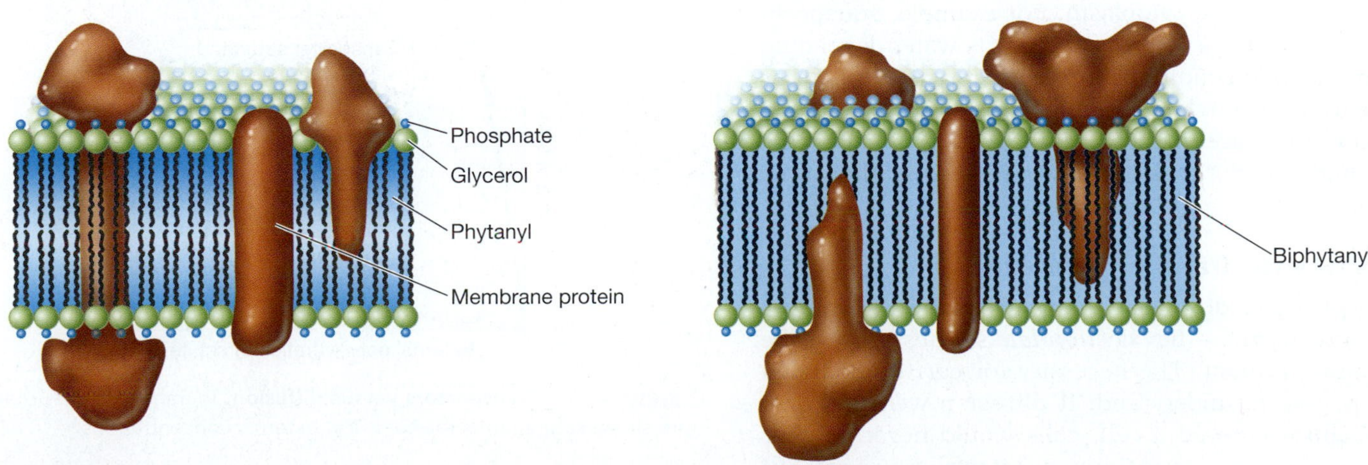

Figure 4.8 Major lipids of *Archaea* and the structure of archaeal cytoplasmic membranes. Note that in both *(a)* and *(b)*, the hydrocarbon of the lipid is attached to the glycerol by an ether linkage. The hydrocarbon in *(a)* is phytanyl (C_{20}) and in *(b)* biphytanyl (C_{40}). *(c, d)* Membrane structure in *Archaea*. The monolayer structure is the result of the tetraether composition of the membrane.

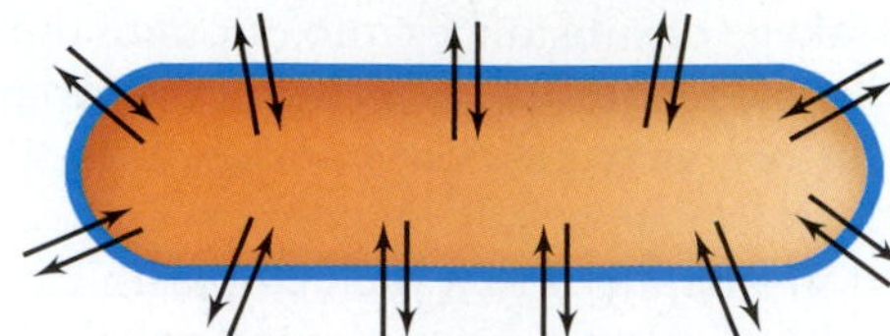

1. **Permeability Barrier** – Prevents leakage and functions as a gateway for transport of nutrients into and out of the cell

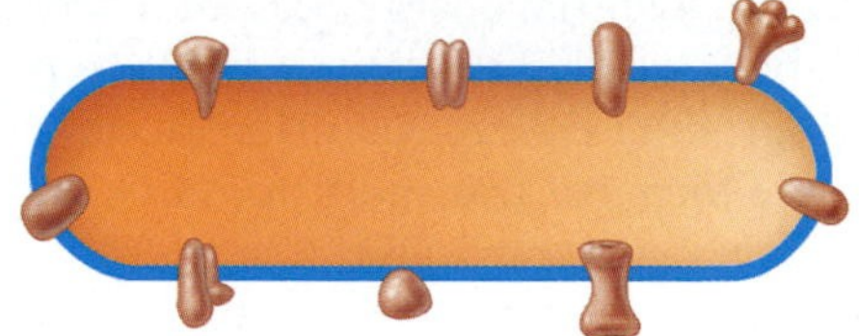

2. **Protein Anchor** – Site of many proteins involved in transport, bioenergetics, and chemotaxis

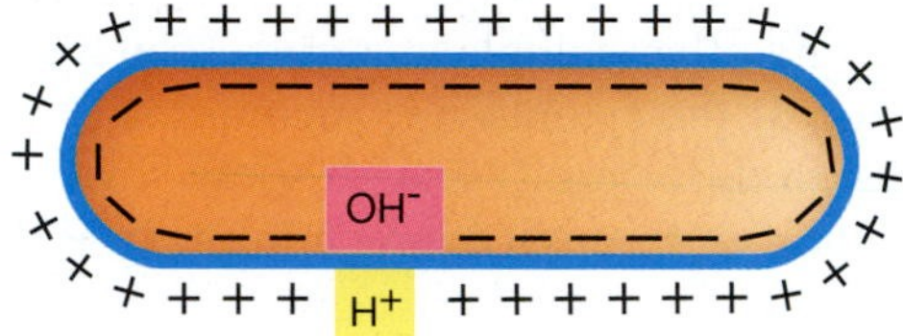

3. **Energy Conservation** – Site of generation and use of the proton motive force

Figure 4.9 The major functions of the cytoplasmic membrane. Although structurally weak, the cytoplasmic membrane has many important cellular functions.

and charged molecules do not pass through but instead must be specifically transported. Because the cytoplasmic membrane is charged, even a substance as small as a proton cannot diffuse across the cytoplasmic membrane.

One molecule that does freely penetrate the membrane is water, which is sufficiently small to pass between phospholipid molecules in the lipid bilayer (**Table 4.2**). But in addition, water movement through the membrane is accelerated by transport proteins called *aquaporins*. These proteins form membrane-spanning channels that specifically transport water into or out of the cytoplasm. For example, aquaporin AqpZ of *Escherichia coli* imports or exports water depending on whether osmotic conditions in the cytoplasm are high or low, respectively. The relative permeability of a few biologically relevant substances is shown in Table 4.2. As can be seen, most substances do not passively enter the cell and thus must be transported.

The Necessity for Transport Proteins

Transport proteins do more than just ferry substances across the membrane—they *accumulate* solutes against the concentration gradient. The necessity for carrier-mediated transport is easy to understand. If diffusion were the only way that solutes entered a cell, cells would never achieve the intracellular concentrations necessary to carry out biochemical reactions. This is for two reasons. First, as we have seen, few things diffuse across the cytoplasmic membrane. But secondly, even if solutes could diffuse across the membrane, their rate of uptake and their intracellular concentration would only be proportional to their external concentration (**Figure 4.10**), which in nature is often quite low. Hence, cells must have mechanisms for accumulating solutes, most of which are vital nutrients, to levels higher than those in their habitats, and this is the role of transport systems.

Table 4.2 Comparative permeability of membranes to various molecules

Substance	*Rate of permeability*[a]	*Potential for diffusion into a cell*
Water	100	Excellent
Glycerol	0.1	Good
Tryptophan	0.001	Fair/Poor
Glucose	0.001	Fair/Poor
Chloride ion (Cl^-)	0.000001	Very poor
Potassium ion (K^+)	0.0000001	Extremely poor
Sodium ion (Na^+)	0.00000001	Extremely poor

[a]Relative scale—permeability with respect to permeability of water given as 100. Permeability of the membrane to water may be affected by aquaporins (see text)

Properties of Transport Proteins

Carrier-mediated transport systems show several characteristic properties. First, in contrast with simple diffusion, transport systems show a *saturation effect*. If the concentration of substrate is high enough to saturate the carrier, which can occur at even the very low substrate concentrations found in nature, the rate of uptake becomes maximal and the addition of more substrate does not increase the rate (Figure 4.10). This characteristic feature of transport proteins greatly

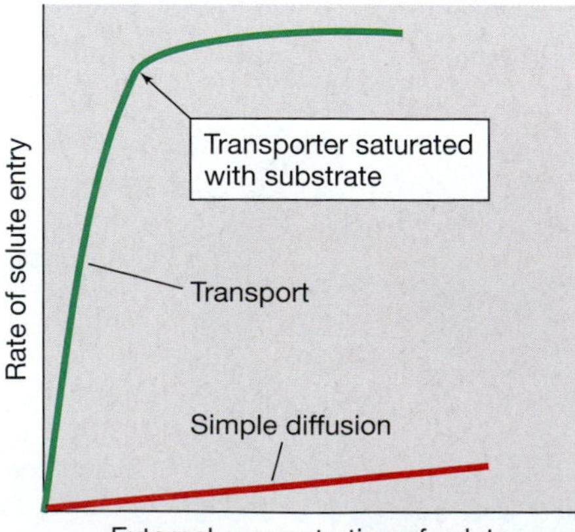

Figure 4.10 Transporters versus diffusion. In transport, the uptake rate shows saturation at relatively low external concentrations.

assists cells in concentrating nutrients from an often very dilute environment.

A second characteristic of carrier-mediated transport is the *high specificity* of the transport event. Many carrier proteins react only with a single molecule, and others show affinities for a closely related class of molecules such as sugars or amino acids. This economy in uptake reduces the need for separate transport proteins for each different amino acid or sugar.

Another major characteristic of transport systems is that biosynthesis of the transport proteins is typically *regulated* by the cell. That is, the specific complement of transporters present in the membrane at any given time is a function of both the nutrients present in the environment and their concentrations. Biosynthetic control is important because a particular nutrient may need to be transported by one transporter when the nutrient is at high concentration and by a different, higher-affinity transporter, when at low concentration.

4.4 MiniReview

The major functions of the cytoplasmic membrane are permeability, transport, and energy conservation. To accumulate nutrients against the concentration gradient, specific transport mechanisms are employed.

- List two reasons why a cell cannot depend on diffusion as a means of acquiring nutrients.
- Why is physical damage to the cytoplasmic membrane a more critical problem for the cell than damage to some other cell component?

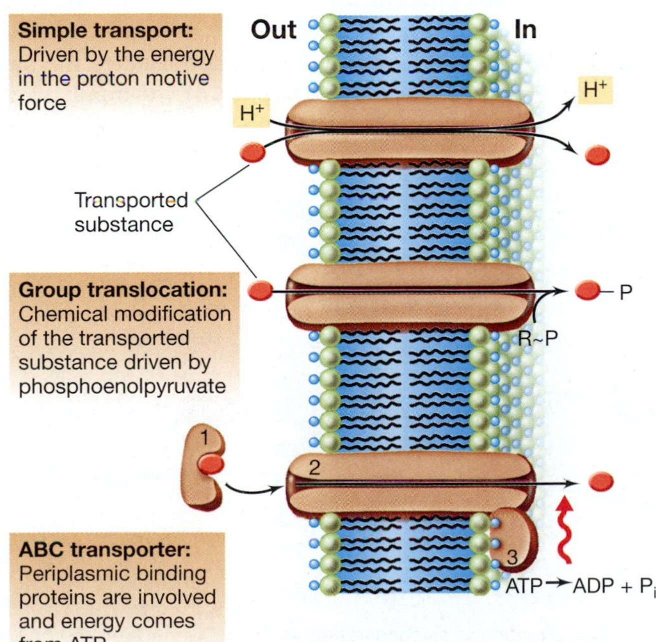

Figure 4.11 The three classes of membrane transport systems. Note how simple transporters and the ABC system transport substances without chemically modifying them, whereas group translocation results in the chemical modification (phosphorylation) of the transported substance. The three proteins of the ABC system are labeled 1, 2, and 3.

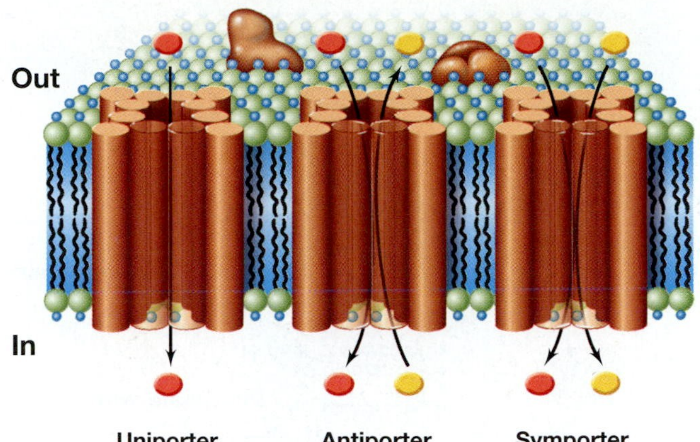

Figure 4.12 Structure of membrane-spanning transporters and types of transport events. In prokaryotes, membrane-spanning transporters typically contain 12 α-helices (each shown here as a cylinder) that aggregate to form a channel through the membrane. Shown here are three transporters with differing types of transport events. For antiporters and symporters, the cotransported substance is shown in yellow.

4.5 Transport and Transport Systems

Nutrient transport is a vital cellular event. Different mechanisms for transport exist in prokaryotes, each with its own unique features. We explore this subject here.

Structure and Function of Membrane Transport Proteins

At least three transport systems exist in prokaryotes: **simple transport**, **group translocation**, and the **ABC system**. Simple transport consists only of a membrane-spanning transport protein, group translocation involves a series of proteins in the transport event, and the ABC system consists of three components: a substrate-binding protein, a membrane-integrated transporter, and an ATP-hydrolyzing protein (**Figure 4.11**). All transport systems require energy in some form, either from the proton motive force or ATP (or some other energy-rich organic compound).

Figure 4.11 contrasts transport systems of prokaryotes. Regardless of the system, the membrane-spanning components typically show significant similarities in amino acid sequence, an indication of the common evolutionary roots of all transport systems. Membrane transporters contain 12 α-helix domains (∞ Section 3.7) that weave back and forth through the membrane to form a channel. It is through this channel that the solute is actually carried into the cell (**Figure 4.12**). The transport event involves a conformational change in the transport protein after it binds its solute. Like a gate swinging open, the conformational change then brings the solute into the cell.

Three transport events are possible: uniport, symport, and antiport (Figure 4.12). *Uniporters* are proteins that transport a molecule unidirectionally across the membrane. *Symporters* are proteins that function as cotransporters; they transport

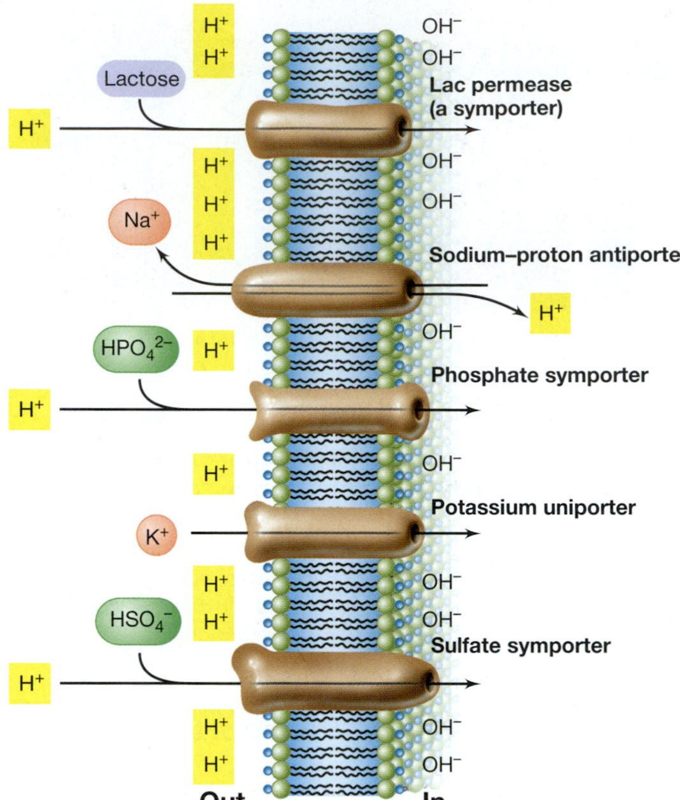

Figure 4.13 Function of the Lac permease symporter of *Escherichia coli* and several other well-characterized simple transporters. Although for simplicity the membrane-spanning proteins are drawn here in globular form, note that their structure is actually as depicted in Figure 4.12.

one molecule along with another substance, typically a proton. *Antiporters* are proteins that transport a molecule across the membrane while simultaneously transporting a second molecule in the opposite direction (Figure 4.12).

Simple Transport: Lac Permease of *Escherichia coli*

The bacterium *Escherichia coli* metabolizes the disaccharide sugar lactose. Lactose is transported into cells of *E. coli* by the activity of a simple transporter, *lac permease*, a symporter. This is shown in **Figure 4.13**, where the activity of lac permease is compared with that of some other simple transporters, including uniporters and antiporters. We will see later that lac permease is one of three proteins required to metabolize lactose in *E. coli* and that the synthesis of these proteins is highly regulated by the cell (∞ Section 9.9).

Activity of lac permease is energy driven. As each lactose molecule is transported into the cell, the energy in the proton motive force (Figure 4.9) is diminished by the cotransport of protons into the cytoplasm. The strength of the proton motive force is reestablished through energy-yielding reactions that we will describe in later chapters (∞ Chapters 5 and 21). The net result of the activity of lac permease is the accumulation of lactose within the cell coupled to the consumption of energy.

Group Translocation: The Phosphotransferase System

Group translocation is a form of transport in which the substance transported is chemically modified during its uptake across the membrane. The best-studied group translocation system transports the sugars glucose, mannose, and fructose in *E. coli*. These compounds are modified by phosphorylation during transport by the *phosphotransferase system*.

The phosphotransferase system consists of a family of proteins, five of which are necessary to transport a given sugar. Before the sugar is transported, the proteins in the phosphotransferase system are themselves alternately phosphorylated and dephosphorylated in cascading fashion until the actual transporter, Enzyme II_c, phosphorylates the sugar during the transport event (**Figure 4.14**). A small protein

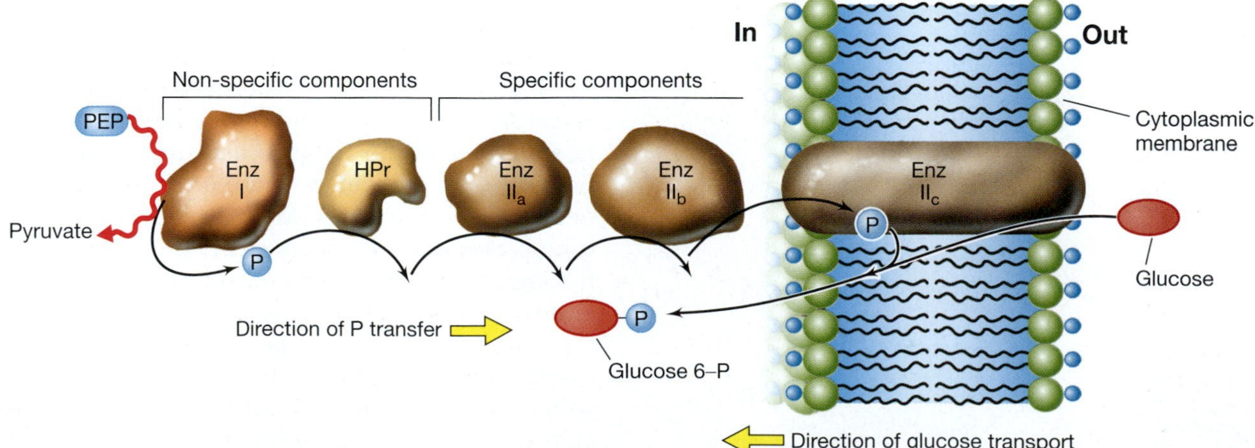

Figure 4.14 Mechanism of the phosphotransferase system of *Escherichia coli*. For glucose uptake, the system consists of five proteins: Enzyme (Enz) I, Enzymes II_a, II_b, and II_c, and HPr. Sequential phosphate transfer occurs from phosphoenolpyruvate (PEP) through the proteins shown to Enzyme II_c, which actually transports and phosphorylates the sugar. Proteins HPr and Enz I are nonspecific and will transport any sugar. The Enz II components are specific for each particular sugar.

called *HPr,* the enzyme that phosphorylates it (Enzyme I), and Enzyme II_a are all cytoplasmic proteins. By contrast, Enzyme II_b lies on the inner surface of the membrane and Enzyme II_c is an integral membrane protein (Figure 4.14). HPr and Enzyme I are nonspecific components of the phosphotransferase system and participate in the uptake of various sugars. A specific Enzyme II exists for each different sugar transported (Figure 4.14).

Energy for the phosphotransferase system comes from the energy-rich compound phosphoenolpyruvate (Section 5.8). Although energy in the form of one energy-rich phosphate bond is consumed in the process of transporting the glucose molecule (Figure 4.14), the phosphorylation of glucose to glucose-6-P is the first step in its intracellular metabolism anyway (glycolysis, Section 5.10). Thus, the phosphotransferase system prepares glucose for immediate entry into this major metabolic pathway.

Periplasmic-Binding Proteins and the ABC System

We will learn a bit later in this chapter that gram-negative bacteria contain a region called the *periplasm* that lies between the cytoplasmic membrane and a second membrane layer called the *outer membrane* (Section 4.7). The periplasm contains many different proteins, several of which function in transport and are called *periplasmic-binding proteins*. Transport systems that contain periplasmic-binding proteins along with a membrane transporter and ATP-hydrolyzing proteins are called *ABC transport systems,* the "ABC" being an acronym for ATP-*b*inding *c*assette, a structural feature of proteins that bind ATP (**Figure 4.15**). More than 200 different ABC transport systems have been identified in prokaryotes. ABC transporters exist for the uptake of organic compounds such as sugars and amino acids, inorganic nutrients such as sulfate and phosphate, and trace metals.

One of the characteristic properties of ABC transporters is the typically high substrate affinity of the periplasmic-binding proteins. These proteins can bind their substrate even when the substrate is present at extremely low concentration. For example, substrates present at concentrations as low as 1 micromolar (10^{-6} M) can easily be trapped by periplasmic-binding proteins. Once its substrate is bound, the periplasmic-binding protein interacts with its respective membrane-spanning transporter to transport the substrate driven by the energy of ATP hydrolysis (Figure 4.15).

Interestingly, even though gram-positive bacteria lack a periplasm, they have ABC systems. In gram-positive bacteria, however, specific substrate-binding proteins are anchored to the external surface of the cytoplasmic membrane. Nevertheless, as in gram-negative bacteria, once these proteins bind substrate, they interact with a membrane transporter to catalyze uptake of the substrate at the expense of ATP hydrolysis, just as they do in gram-negative bacteria (Figure 4.15).

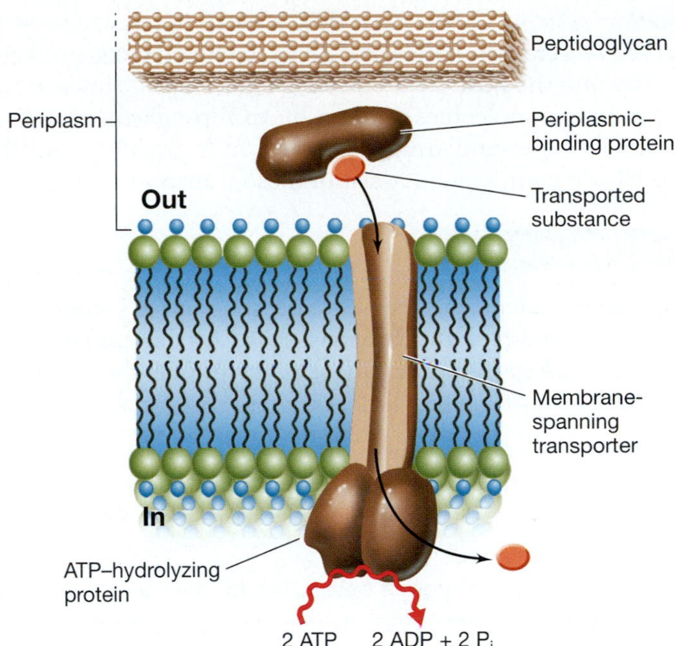

Figure 4.15 Mechanism of an ABC transporter. The periplasmic-binding protein has high affinity for substrate, the membrane-spanning proteins form the transport channel, and the cytoplasmic ATP-hydrolyzing proteins supply the energy for the transport event.

Protein Export

Thus far our discussion of transport has focused on small molecules. What about the transport of large molecules, such as proteins? To function properly, many proteins need to be either transported outside the cytoplasmic membrane or inserted into the membrane in a specific way. Proteins are exported through and inserted into prokaryotic membranes by the activities of proteins called *translocases,* a key one being the Sec (for secretory) system. The Sec system both exports proteins and inserts integral membrane proteins into the membrane. Proteins destined for transport are recognized by the Sec system because they are tagged in a specific way. We discuss this process later (Section 7.17).

The Sec translocase system consists of seven proteins, with SecYEG constituting the actual transmembrane transporter. Other components of the Sec system include SecA, an ATP-hydrolyzing enzyme that provides energy for the process; SecB, a protein that prevents folding of proteins in the cytoplasm; and Sec D and F, proteins that assist in the translocation process while consuming energy from the proton motive force.

Protein export is important to bacteria because many bacterial enzymes function outside the cell (exoenzymes). For example, hydrolytic exoenzymes such as amylase or cellulase are excreted directly into the environment where they cleave starch or cellulose (Figure 3.6*b*), respectively, into glucose, which is then used by the cell as a carbon and energy source. In gram-negative bacteria, many enzymes are localized to the periplasm, and these must get past the cytoplasmic membrane and into the periplasm in order to function. Moreover, many pathogenic bacteria excrete protein toxins or other harmful proteins into the host during infection. Many toxins are excreted by a second translocase system called the *type III* *secretion system*. This system differs from the Sec system in that the secreted protein is translocated from the bacterial cell directly into the host, for example, a human cell. However, all of these large molecules need to move through the cytoplasmic membrane, and translocases such as SecYEG and the type III secretion system assist in these transport events.

4.5 MiniReview

At least three functional classes of transporters are known: simple transporters, phosphotransferase-type transporters, and ABC systems. Transport requires energy from the proton motive force, ATP, or some other energy-rich substance.

- Contrast simple transporters, the phosphotransferase system, and ABC transporters in terms of (1) energy source, (2) chemical alterations of the substrate transported, and (3) number of proteins involved.
- Which transport system is best suited for the transport of nutrients present in the environment at extremely low levels, and why?
- How are proteins exported from the cell?

III CELL WALLS OF PROKARYOTES

4.6 The Cell Wall of *Bacteria*: Peptidoglycan

Because of the activities of transport systems, the cytoplasm of bacterial cells maintains a high concentration of dissolved solutes. This causes significant osmotic pressure to develop—about 2 atmospheres in a bacterium such as *Escherichia coli*. This is roughly the same as the pressure in an automobile tire. To withstand these pressures and prevent bursting—a process called *lysis*—bacteria have cell walls. Besides preventing osmotic lysis, cell walls also give shape and rigidity to the cell.

Species of bacteria can be divided into two major groups, called **gram-positive** and **gram-negative**. The distinction between gram-positive and gram-negative bacteria is based on the **Gram stain** reaction (∞ Section 2.2). But differences in cell wall structure are at the heart of the Gram-staining reaction. The appearance of the cell walls of gram-positive and gram-negative cells in the electron microscope differs markedly, as is shown in **Figure 4.16**. The gram-negative cell wall is a multilayered structure and quite complex, whereas the gram-positive cell wall is typically much thicker and consists almost entirely of a single type of molecule.

The focus of this section is on the polysaccharide component of the cell walls of *Bacteria,* both gram positive and gram negative. In Section 4.7 we describe the special wall components found in gram-negative *Bacteria*. And finally, in Section 4.8 we describe the various cell walls of *Archaea*.

Peptidoglycan

The cell walls of *Bacteria* have a rigid layer that is primarily responsible for the strength of the wall. In gram-negative bacteria, additional layers are present outside this rigid layer. The rigid layer, called **peptidoglycan**, is a polysaccharide composed of two sugar derivatives—*N-acetylglucosamine* and *N-acetylmuramic acid*—and a few amino acids, including L-alanine, D-alanine, D-glutamic acid, and either lysine or diaminopimelic acid (DAP) (**Figure 4.17**). These constituents are connected to form a repeating structure, the glycan tetrapeptide (**Figure 4.18**).

Long chains of peptidoglycan are biosynthesized adjacent to one another to form a sheet surrounding the cell (∞ Section 6.4). The chains are connected through cross-links of amino acids. The glycosidic bonds connecting the sugars in the glycan strands are covalent bonds, but these can provide rigidity to the structure in only one direction. Only after cross-linking is peptidoglycan strong in both the X and Y directions (**Figure 4.19**). Cross-linking occurs to different extents in different species of *Bacteria*, with greater rigidity the result of more extensive cross-linking.

In gram-negative bacteria, peptidoglycan cross-linkage occurs by peptide bond formation from the amino group of DAP of one glycan chain to the carboxyl group of the terminal D-alanine on the adjacent glycan chain (Figure 4.19). In gram-positive bacteria, cross-linkage occurs by way of a peptide interbridge, the kinds and numbers of amino acids in the interbridge varying from organism to organism. For example, in *Staphylococcus aureus,* a well-studied gram-positive bacterium, the interbridge peptide consists of five glycine residues (Figure 4.19*b*). The overall structure of a peptidoglycan molecule is shown in Figure 4.19*c*.

Diversity of Peptidoglycan

Peptidoglycan is present only in species of *Bacteria*—the sugar *N*-acetylmuramic acid and the amino acid DAP have never been found in the cell walls of *Archaea* or *Eukarya*. However, not all *Bacteria* examined have DAP in their peptidoglycan. This amino acid analog is present in peptidoglycan from all gram-negative bacteria and some gram-positive species; however, most gram-positive cocci contain lysine instead of DAP (Figure 4.19*b*), and a few other gram-positive bacteria have other amino acids. Another unusual feature of peptidoglycan is the presence of two amino acids that have the D configuration, D-alanine and D-glutamic acid. As we saw in Chapter 3, in cellular proteins amino acids are always of the L stereoisomer (∞ Section 3.6).

More than 100 different peptidoglycans are known, with the diversity focused on the chemistry of the peptide cross-links and interbridge. In each different peptidoglycan the glycan portion is constant; only the sugars *N*-acetylglucosamine and *N*-acetylmuramic are present. Moreover, these sugars are always connected in β-1,4 linkage (Figure 4.18). The tetrapeptide of the repeating unit shows major variation in only one amino acid, the lysine–DAP alternation. However, the D-glutamic acid at position 2 is hydroxylated in the

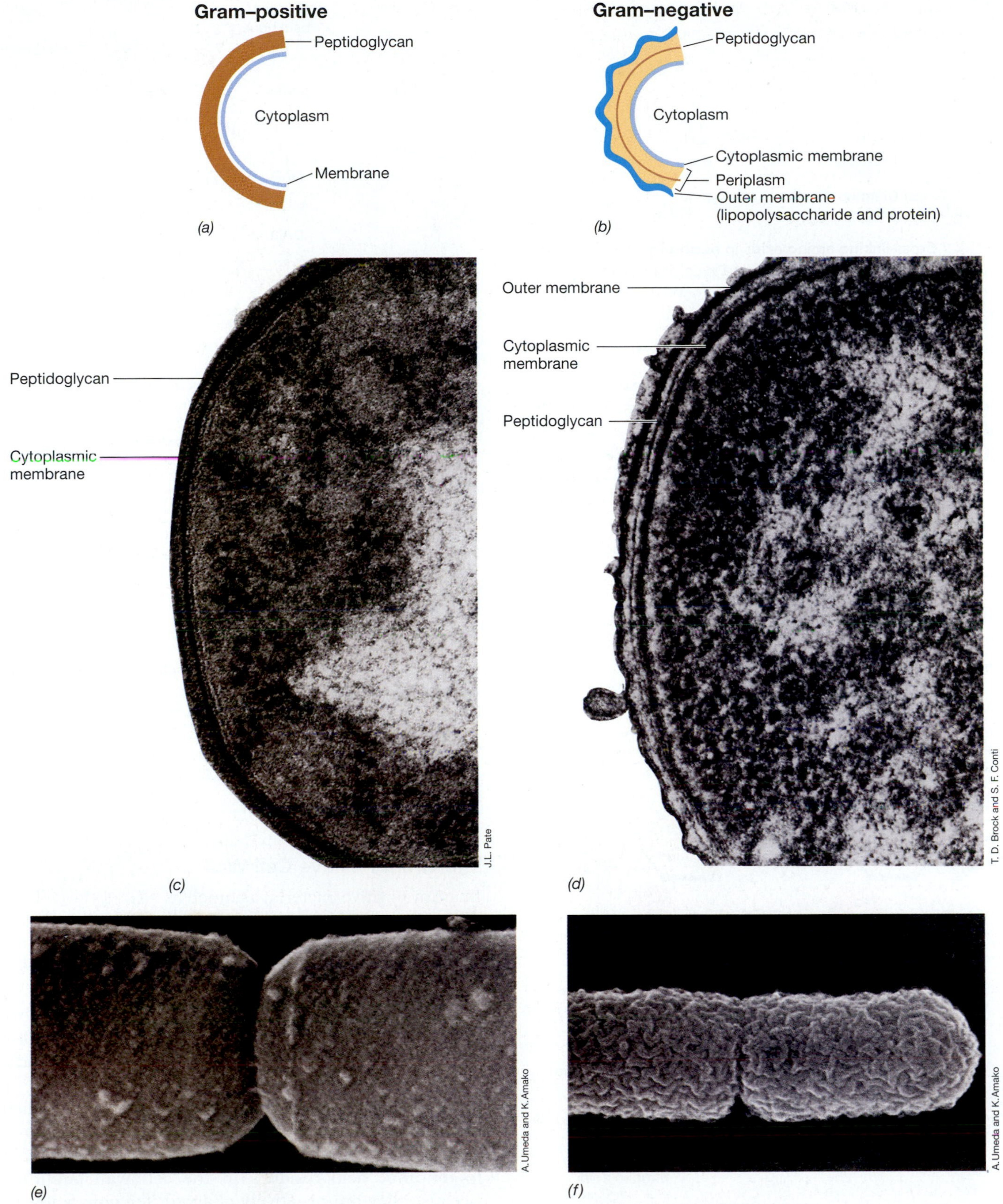

Figure 4.16 **Cell walls of *Bacteria*.** *(a, b)* Schematic diagrams of gram-positive and gram-negative cell walls. Transmission electron micrographs showing the cell wall of *(c)* a gram-positive bacterium, *Arthrobacter crystallopoietes*, and *(d)* a gram-negative bacterium, *Leucothrix mucor*. *(e, f)* Scanning electron micrographs of gram-positive (*Bacillus subtilis*) and gram-negative (*Escherichia coli*) bacteria. Note differences in the surface texture in the cells shown in *(e)* and *(f)*. A single cell of *B. subtilis* or *E. coli* is about 1 μm wide.

(a) Diaminopimelic acid (b) Lysine

Figure 4.17 Cross-linking amino acids in peptidoglycan. The only difference in the two molecules is highlighted in color. Besides these two amino acids, several other amino acids are found in peptidoglycan cross-links (Figure 4.18).

peptidoglycan of some organisms, and there are substitutions in amino acids at positions 1 and 3 in some others. Any of the amino acids present in the tetrapeptide can also occur in the interbridge. In addition, several other amino acids such as glycine, threonine, serine, and aspartic acid can be in the interbridge. However, branched-chain amino acids, aromatic amino acids, sulfur-containing amino acids, and histidine, arginine, and proline (Figure 3.12) have never been found in the interbridge.

Thus, although the peptide chemistry of peptidoglycan varies, the backbone of peptidoglycan—alternating repeats of N-acetylglucosamine and N-acetylmuramic acid—is the same in all species of *Bacteria*.

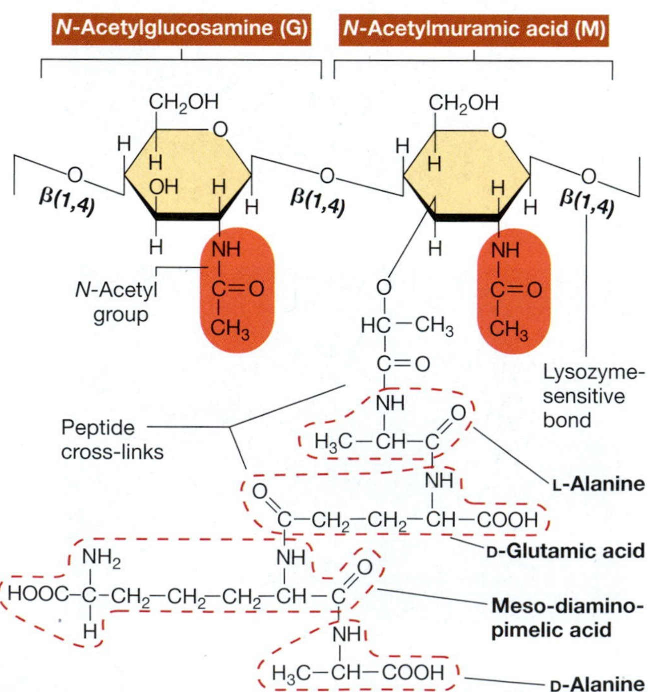

Figure 4.18 Structure of the repeating unit in peptidoglycan, the glycan tetrapeptide. The structure given is that found in *Escherichia coli* and most other gram-negative *Bacteria*. In some *Bacteria*, other amino acids are found.

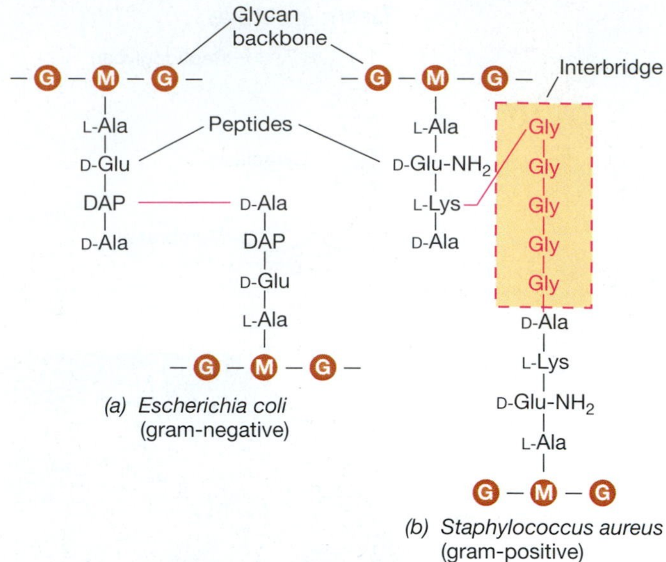

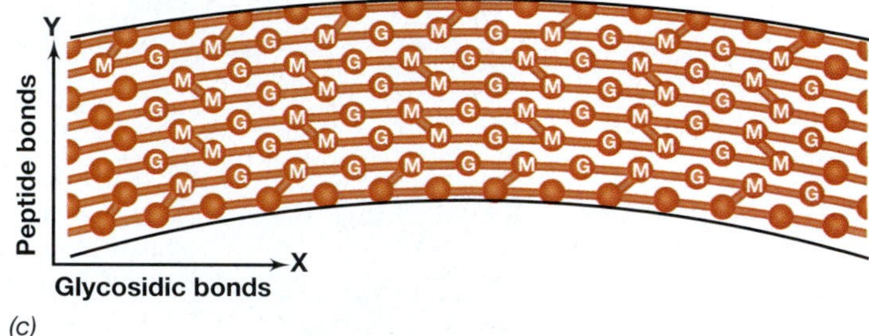

Figure 4.19 Peptidoglycan in *Escherichia coli* and *Staphylococcus aureus*. *(a)* No interbridge is present in *E. coli* and other gram-negative *Bacteria*. *(b)* The glycine interbridge in *S. aureus* (gram-positive). *(c)* Overall structure of peptidoglycan. G, *N*-acetylglucosamine; M, *N*-acetylmuramic acid. Note how glycosidic bonds confer strength to peptidoglycan in the X direction whereas peptide bonds confer strength in the Y direction.

The Gram-Positive Cell Wall

In gram-positive bacteria, as much as 90% of the cell wall consists of peptidoglycan. And, although some bacteria have only a single layer of peptidoglycan surrounding the cell, many bacteria, especially gram-positive bacteria, have several (up to about 25) sheets of peptidoglycan stacked one upon another.

Many gram-positive bacteria have acidic substances called **teichoic acids** embedded in their cell wall. Teichoic acids include all cell wall, cytoplasmic membrane, and capsular polymers containing glycerophosphate or ribitol phosphate residues. These polyalcohols are connected by phosphate esters and usually have other sugars and D-alanine attached (**Figure 4.20*a***). Teichoic acids are covalently bonded to muramic acid residues in the cell wall peptidoglycan. Because they are negatively charged, teichoic acids are partially responsible for the negative charge of the cell surface. Teichoic acids also function to bind Ca^{2+} and Mg^{2+} for eventual transport into the cell. Certain teichoic acids are covalently bound to membrane lipids; thus they have been called *lipoteichoic* acids.

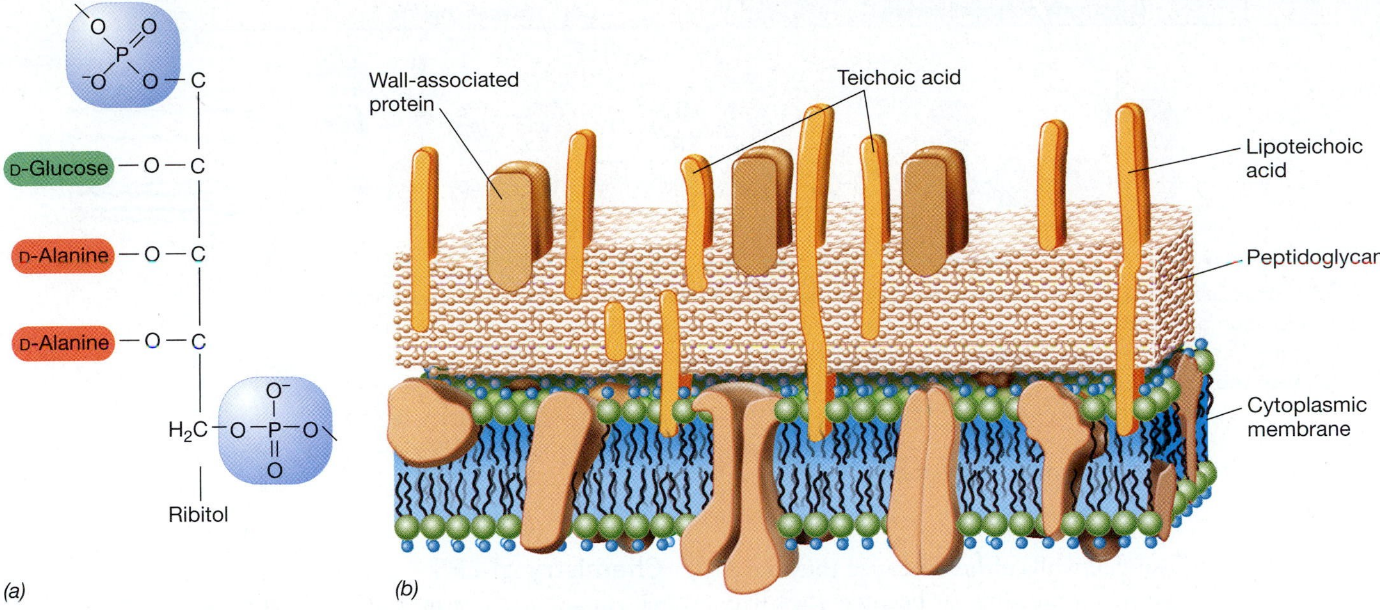

Figure 4.20 Teichoic acids and the overall structure of the gram-positive bacterial cell wall. *(a)* Structure of the ribitol teichoic acid of *Bacillus subtilis*. The teichoic acid is a polymer of the repeating ribitol units shown here. *(b)* Summary diagram of the gram-positive bacterial cell wall.

Figure 4.20*b* summarizes the structure of the cell wall of gram-positive *Bacteria* and shows how teichoic acids and lipoteichoic acids are arranged in the overall wall structure.

Lysozyme and Protoplasts

Peptidoglycan can be destroyed by certain agents. One such agent is the enzyme *lysozyme*, a protein that breaks the β-1,4-glycosidic bonds between *N*-acetylglucosamine and *N*-acetylmuramic acid in peptidoglycan (Figure 4.18), thereby weakening the wall. Water then enters the cell and the cell swells and eventually bursts (cell lysis) (**Figure 4.21*a***). Lysozyme is found in animal secretions including tears, saliva, and other body fluids, and functions as a major line of defense against bacterial infection (∞ Section 28.3).

If a solute that does not penetrate the cell, such as sucrose, is added to a cell suspension containing lysozyme, the solute concentration outside the cell balances the concentration inside (these conditions are called *isotonic*). Under isotonic conditions, if lysozyme is used to digest peptidoglycan, water does not enter the cell and lysis does not occur. Instead, a **protoplast** (a bacterium that has lost its cell wall) is formed (Figure 4.21*b*). If such sucrose-stabilized protoplasts are placed in water, they immediately lyse. The word *spheroplast* is often used as a synonym for protoplast, although the two words have slightly different meanings. Protoplasts are cells that are free of residual cell wall material, whereas spheroplasts contain pieces of wall material attached to the otherwise membrane-enclosed structure.

Cells That Lack Cell Walls

Although most prokaryotes cannot survive in nature without their cell walls, some are able to do so. These include the mycoplasmas, a group of pathogenic bacteria that causes a variety of infectious diseases in humans and other animals (∞ Section 16.3), and the *Thermoplasma* group, species of *Archaea* that naturally lack cell walls (∞ Section 17.5). These prokaryotes are essentially free-living protoplasts, and they

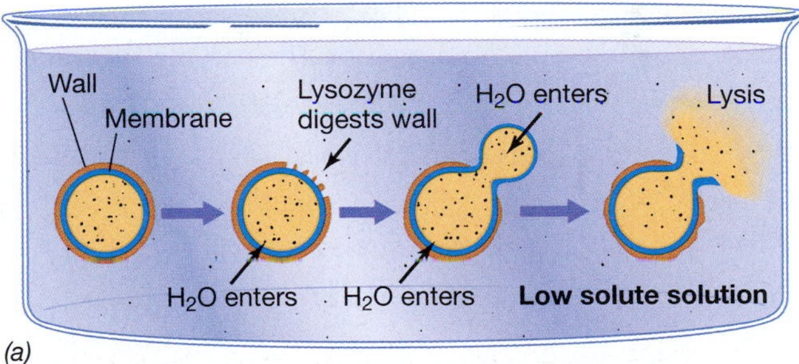

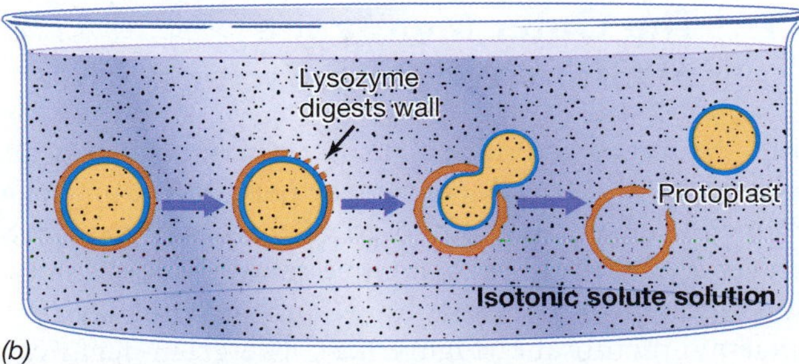

Figure 4.21 Protoplasts and their formation. Lysozyme breaks the β-1,4 glycosidic bonds in peptidoglycan (Figure 4.18). *(a)* In dilute solutions, breakdown of the cell wall is immediately followed by cell lysis because the cytoplasmic membrane is structurally very weak. *(b)* In a solution containing an isotonic concentration of a solute such as sucrose, water does not enter the protoplast and it remains stable.

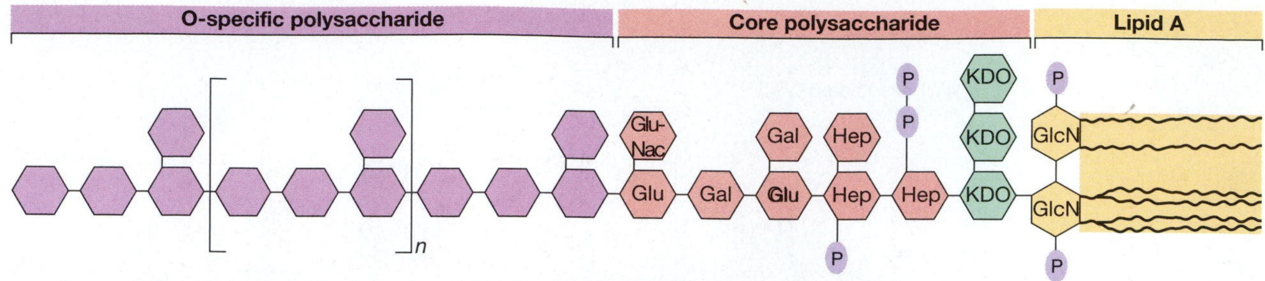

Figure 4.22 Structure of the lipopolysaccharide of gram-negative *Bacteria*. The chemistry of lipid A and the polysaccharide components varies among species of gram-negative *Bacteria*, but the major components (lipid A–KDO–core–O-specific) are typically the same. The O-specific polysaccharide varies greatly among species. KDO, ketodeoxyoctonate; Hep, heptose; Glu, glucose; Gal, galactose; GluNac, *N*-acetylglucosamine; GlcN, glucosamine; P, phosphate. Glucosamine and the lipid A fatty acids are linked through the amine groups. The lipid A portion of LPS can be toxic to animals and comprises the endotoxin complex. Compare this figure with Figure 4.23 and follow the LPS components by the color-coding.

are able to survive without cell walls either because they have unusually tough cytoplasmic membranes or because they live in osmotically protected habitats such as the animal body. Most mycoplasmas have sterols in their cytoplasmic membranes, and these probably function to add strength and rigidity to the membrane as they do in the cytoplasmic membranes of eukaryotic cells (Section 4.3).

4.6 MiniReview

The cell walls of *Bacteria* contain a polysaccharide called peptidoglycan. Peptidoglycan consists of alternating repeats of *N*-acetylglucosamine and *N*-acetylmuramic acid, which forms the glycan tetrapeptide with strands cross-linked by short peptides. One to several sheets of peptidoglycan can be present, depending on the organism. The enzyme lysozyme destroys peptidoglycan, leading to cell lysis.

- Why do bacterial cells need cell walls? Do all bacteria have cell walls?
- Why is peptidoglycan such a strong molecule?
- What does the enzyme lysozyme do?

4.7 The Outer Membrane of Gram-Negative *Bacteria*

In gram-negative bacteria such as *Escherichia coli* only about 10% of the total cell wall consists of peptidoglycan. Instead, most of the cell wall is composed of the **outer membrane**. This layer is effectively a second lipid bilayer, but it is not constructed solely of phospholipid and protein as is the cytoplasmic membrane (Figure 4.5). The gram-negative cell outer membrane also contains polysaccharide. The lipid and polysaccharide are linked in the outer membrane to form a complex. Because of this, the outer membrane is called the **lipopolysaccharide** layer, or simply **LPS**.

Chemistry of LPS

The chemistry of LPS from several bacteria is known. As seen in **Figure 4.22**, the polysaccharide portion of LPS consists of two components, the *core polysaccharide* and the *O-polysaccharide*. In *Salmonella* species, where LPS has been best studied, the core polysaccharide consists of ketodeoxyoctonate (KDO), seven-carbon sugars (heptoses), glucose, galactose, and *N*-acetylglucosamine. Connected to the core is the O-polysaccharide, which typically contains galactose, glucose, rhamnose, and mannose (all hexoses), as well as one or more unusual dideoxy sugars such as abequose, colitose, paratose, or tyvelose. These sugars are connected in four- or five-membered sequences, which often are branched. When the sequences repeat, the long O-polysaccharide is formed.

The relationship of the O-polysaccharide to the rest of the LPS is shown in **Figure 4.23**. The lipid portion of the LPS, called *lipid A*, is not a typical glycerol lipid (see Figure 4.7*a*), but instead the fatty acids are connected through the amine groups from a disaccharide composed of glucosamine phosphate (Figure 4.22). The disaccharide is attached to the core polysaccharide through KDO (Figure 4.22). Fatty acids commonly found in lipid A include caproic (C_6), lauric (C_{12}), myristic (C_{14}), palmitic (C_{16}), and stearic (C_{18}) acids.

LPS replaces most of the phospholipids in the outer half of the outer membrane; the structure of the inner half more closely resembles that of the cytoplasmic membrane. However, a lipoprotein complex is also present on the inner half of the outer membrane (Figure 4.23*a*). Lipoprotein functions as an anchor between the outer membrane and peptidoglycan. Thus, although the outer membrane is considered a lipid bilayer, its structure is distinct from that of the cytoplasmic membrane, especially in the outer half in contact with the environment (compare Figures 4.5 and 4.23*a*).

Endotoxin

Although the major function of the outer membrane is undoubtedly structural, one of its important biological properties is its toxicity to animals. Gram-negative bacteria that

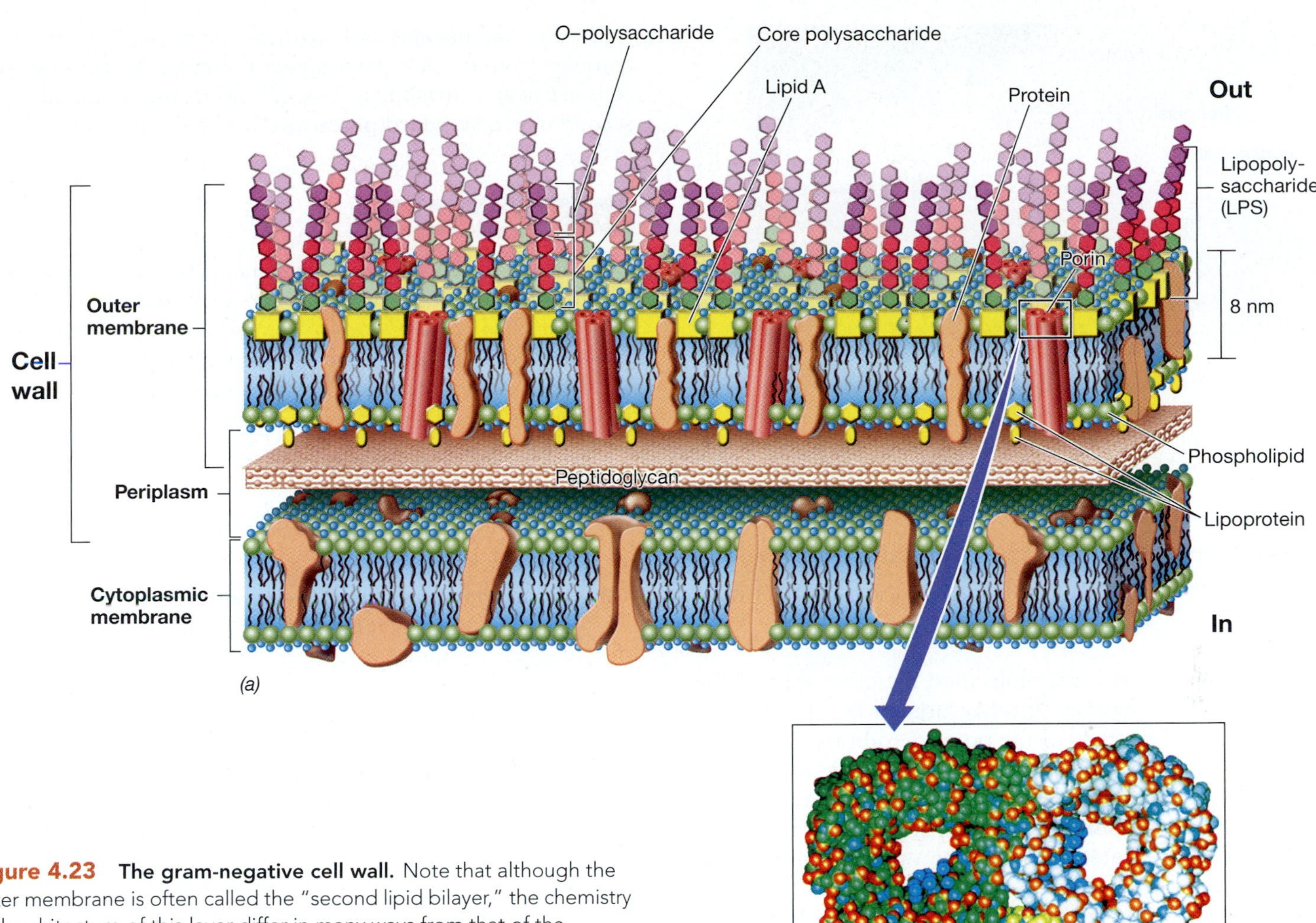

Figure 4.23 The gram-negative cell wall. Note that although the outer membrane is often called the "second lipid bilayer," the chemistry and architecture of this layer differ in many ways from that of the cytoplasmic membrane. *(a)* Arrangement of lipopolysaccharide, lipid A, phospholipid, porins, and lipoprotein in the outer membrane. See Figure 4.22 for details of the structure of LPS. *(b)* Molecular model of porin proteins. Note the four pores present, one within each of the proteins forming a porin molecule and a smaller central pore between the porin proteins. The view is perpendicular to the plane of the membrane. Model based on X-ray diffraction studies of *Rhodobacter blasticus* porin.

are pathogenic for humans and other mammals include species of *Salmonella, Shigella,* and *Escherichia,* among others, and some of the intestinal symptoms these pathogens typically elicit in their hosts are due to their toxic outer membrane.

The toxic properties are associated with the LPS layer, in particular, lipid A. The term *endotoxin* refers to this toxic component of LPS, as we discuss in Section 28.12. Some endotoxins cause violent symptoms in humans, including severe gastrointestinal distress (gas, diarrhea, vomiting). Endotoxins are responsible for a number of bacterial illnesses, including *Salmonella* food infection (∞ Section 37.7). Interestingly, LPS from several nonpathogenic bacteria have also been shown to have endotoxin activity. Thus, the organism itself need not be pathogenic to contain toxic outer membrane components.

Porins

Unlike the cytoplasmic membrane, the outer membrane of gram-negative bacteria is relatively permeable to small molecules even though it is basically a lipid bilayer. This is because proteins called *porins* are present in the outer membrane that function as channels for the entrance and exit of hydrophilic low-molecular-weight substances (Figure 4.23). Several different porins exist, including both specific and nonspecific classes.

Nonspecific porins form water-filled channels through which any small substance can pass. By contrast, specific porins contain a binding site for only one or a small group of structurally related substances. Porins are transmembrane proteins that contain three identical subunits (Figure 4.23*a*). Besides the channel present in each barrel of the porin, the barrels of the porin proteins associate in such a way that a small hole about 1 nm (10^{-9} m) in diameter is formed in the

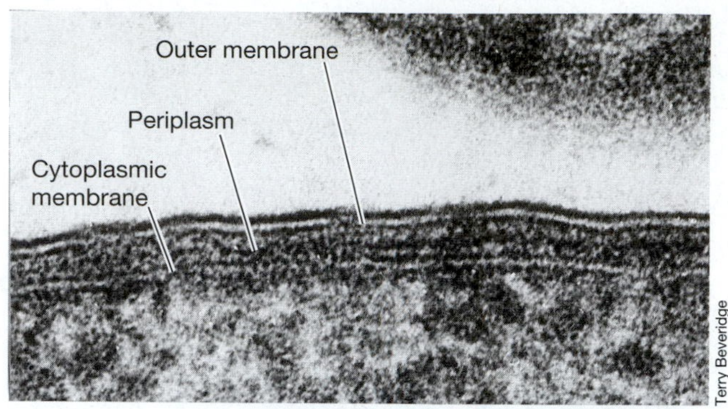

Figure 4.24 The cell wall of *Escherichia coli* as seen by the electron microscope. High-magnification thin section transmission electron micrograph of the cell envelope of *E. coli* showing the periplasm bounded by the outer and cytoplasmic membranes.

outer membrane through which very small substances can travel (Figure 4.23*b*).

The Periplasm

Although permeable to small molecules, the outer membrane is not permeable to enzymes or other large molecules. In fact, one of the major functions of the outer membrane is to keep proteins that are present outside the cytoplasmic membrane from diffusing away from the cell. These proteins are present in a region called the **periplasm** (see Figure 4.23 and Figure 4.24). This space, located between the outer surface of the cytoplasmic membrane and the inner surface of the outer membrane, is about 15 nm wide. The periplasm contents are gel-like in consistency because of the high concentration of proteins present there.

The periplasm can contain several different classes of proteins. These include hydrolytic enzymes, which function in the initial degradation of food molecules; binding proteins, which begin the process of transporting substrates (Section 4.5); and chemoreceptors, which are proteins involved in the chemotaxis response (Section 4.15 and ∞ Section 9.7). Most of these proteins reach the periplasm by way of the Sec protein exporting transport system in the cytoplasmic membrane (Section 4.5).

Relationship of Cell Wall Structure to the Gram Stain

The structural differences between the cell walls of gram-positive and gram-negative *Bacteria* are thought to be responsible for differences in the Gram stain reaction. In the Gram stain, an insoluble crystal violet–iodine complex forms inside the cell. This complex is extracted by alcohol from gram-negative but not from gram-positive bacteria (∞ Section 2.2). As we have seen, gram-positive bacteria have very thick cell walls consisting of several layers of peptidoglycan (Figure 4.20); these become dehydrated by the alcohol, causing the pores in the walls to close and preventing the insoluble crystal violet–iodine complex from escaping. By contrast, in gram-negative bacteria, alcohol readily penetrates the lipid-rich outer membrane and extracts the crystal violet–iodine complex from the cell. After alcohol treatment, gram-negative cells are nearly invisible unless they are counterstained with a second dye, a standard procedure in the Gram stain (∞ Figure 2.4).

4.7 MiniReview

In addition to peptidoglycan, gram-negative bacteria have an outer membrane consisting of LPS, protein, and lipoprotein. Proteins called porins allow for permeability across the outer membrane. The space between the outer and cytoplasmic membranes is the periplasm, which contains various proteins involved in important cellular functions.

- What components constitute the LPS layer of gram-negative bacteria?
- What is the function of porins and where are they located in a gram-negative cell wall?
- What component of the cell has endotoxin properties?
- Why does alcohol readily decolorize gram-negative but not gram-positive bacteria?

4.8 Cell Walls of *Archaea*

Peptidoglycan, the "signature" molecule in the cell walls of *Bacteria,* is absent from the cell walls of *Archaea*. An outer membrane is typically lacking in *Archaea* as well. Instead, a variety of chemistries are found in the cell walls of *Archaea,* including polysaccharides, proteins, and glycoproteins.

Pseudomurein

The cell walls of certain methanogenic *Archaea* (species that produce methane, natural gas), contain a polysaccharide that is very similar to peptidoglycan, a polysaccharide called *pseudomurein* (the term "murein" has Latin roots meaning "wall" and was an old term for peptidoglycan; Figure 4.25). The backbone of pseudomurein is composed of alternating repeats of *N*-acetylglucosamine (also found in peptidoglycan) and *N*-acetyltalosaminuronic acid; the latter replaces the *N*-acetylmuramic acid of peptidoglycan (compare Figures 4.18 and 4.25). Pseudomurein also differs from peptidoglycan in that the glycosidic bonds between the sugar derivatives are β-1,3 instead of β-1,4, and the amino acids are all of the L stereoisomer.

Peptidoglycan and pseudomurein are remarkably similar molecules, yet they are chemically distinct. It is likely that they arose either by convergent evolution after the two prokaryotic domains of life had separated or by divergence from a common polysaccharide present in the cell walls of prokaryotic cells before the divergence of the domain *Bacteria* from the domain *Archaea*.

Other Polysaccharide Cell Walls

Cell walls of some other *Archaea* lack pseudomurein and contain other polysaccharides. For example, *Methanosarcina*

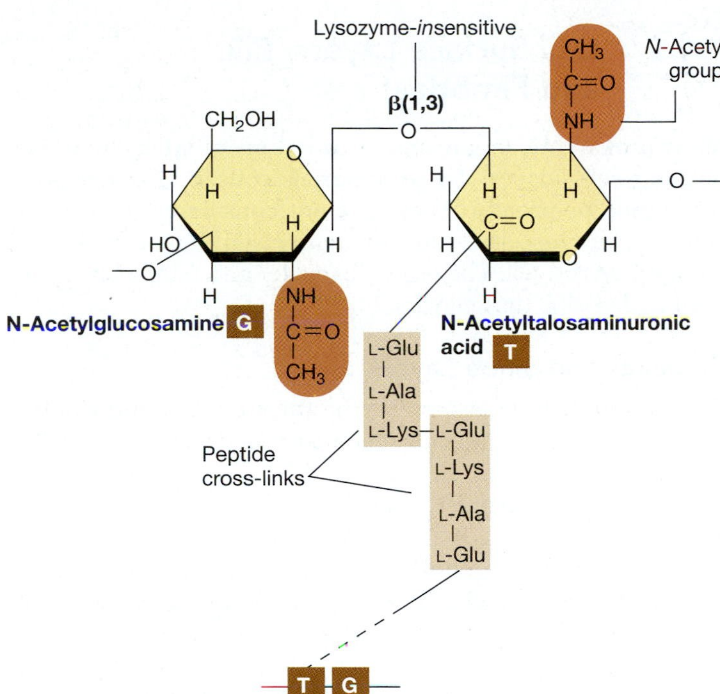

Figure 4.25 Pseudomurein. Structure of pseudomurein, the cell wall polymer of *Methanobacterium* species. Note the similarities and differences between pseudomurein and peptidoglycan (Figure 4.18).

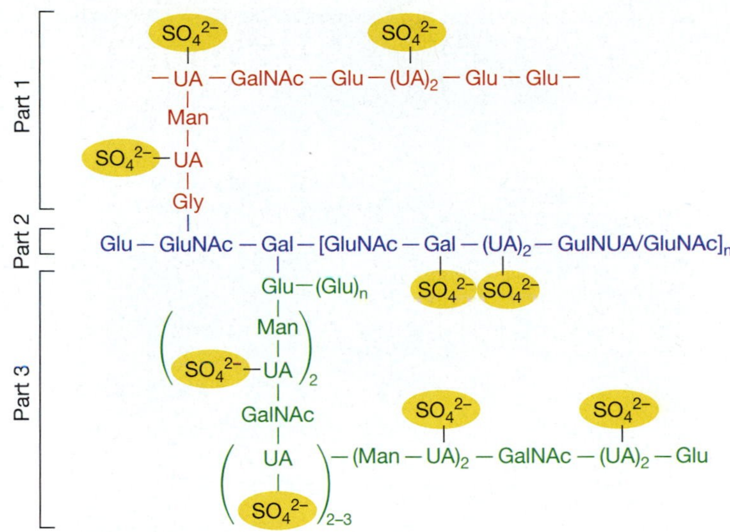

Figure 4.26 Polysaccharide cell walls of *Archaea*. Shown is the cell wall structure of *Halococcus*, an extreme halophile. The wall consists of a repeating three-part structure. UA, uronic acid; Glu, glucose; Gal, galactose; GluNAc, *N*-acetylglucosamine, GalNAc, *N*-acetylgalactosamine; Gly, glycine; GulNUA, *N*-acetylgulosaminuronic acid; Man, mannose.

species have thick polysaccharide walls composed of glucose, glucuronic acid, uronic acid galactosamine, and acetate. Extremely halophilic (salt-loving) *Archaea* such as *Halococcus* have cell walls similar to that of *Methanosarcina* but which contain, in addition, sulfate (SO_4^{2-}) (**Figure 4.26**). The negatively charged sulfate groups bind the high concentration of Na^+ present in the habitats of *Halococcus,* salt evaporation ponds and saline seas and lakes (⚭ Section 17.3), helping to stabilize the cell wall in such polar environments. We will see later that *Methanosarcina* and the extreme halophiles are phylogenetically closely related, so their similarities in cell wall structure are not surprising.

S-Layers

The most common cell wall type among the *Archaea* is the paracrystalline surface layer, or **S-layer**. S-layers consist of protein or glycoprotein and show an ordered appearance when viewed with the electron microscope (**Figure 4.27**). The paracrystalline structure of S-layers becomes arranged into various symmetries, such as hexagonal, tetragonal, or trimeric, depending upon the number and structure of the protein or glycoprotein subunits of which they are composed. S-layers have been found in representatives of all major groups of *Archaea,* the extreme halophiles, the methanogens, and the hyperthermophiles. Several species of *Bacteria* also have S-layers on their outer surfaces (Figure 4.27).

The cell walls of some *Archaea,* for example that of the methanogen *Methanococcus jannaschii,* consist of only a single S-layer. Thus, S-layers are themselves sufficiently strong to withstand osmotic bursting. However, in many organisms S-layers are present in addition to other cell wall components, usually polysaccharides. For example, in *Bacillus brevis,* a species of *Bacteria*, an S-layer is present along with peptidoglycan. When an S-layer is present along with other wall components, the S-layer is always the outermost wall layer. Besides serving as structural reinforcement, S-layers may have other functions. For example, as the interface between the cell and its environment, it is likely that the S-layer functions as a selective sieve, allowing the passage of low-molecular-weight substances and excluding large molecules and structures (such as viruses). The S-layer may also function to retain proteins near the cell surface, much as the outer membrane (Section 4.7) does in gram-negative bacteria.

Other Archaeal Cell Walls

Natronococcus is a haloalkaliphilic species of *Archaea* that thrives in highly alkaline and saline habitats, such as soda lakes. *Natronococcus* contains a glycoprotein cell wall that is not a typical S-layer. This is because the glycoprotein contains only a single type of amino acid, L-glutamate, as a backbone from which glucose and glucose derivatives are linked. Interestingly, polyglutamate protein surface layers are known from some *Bacteria* as well, most notably, that of *Bacillus anthracis,* the causative agent of anthrax. But a glycoprotein containing glutamic acid as the sole amino acid has been found only in *Natronococcus*.

In species of *Archaea* we thus see several cell wall chemistries, varying from molecules that closely resemble peptidoglycan to those lacking a polysaccharide component. But with rare exception, all *Archaea* contain a cell wall of some

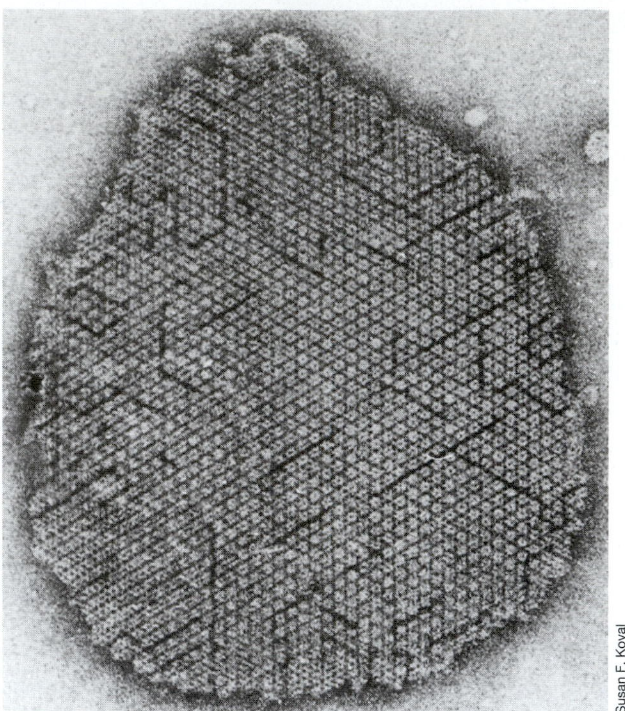

Figure 4.27 The S-Layer. Transmission electron micrograph of an S-layer showing the paracrystalline structure. Shown is the S-layer from *Aquaspirillum serpens* (a species of *Bacteria*); this S-layer shows hexagonal symmetry as is common in S-layers of *Archaea* as well.

sort, and as in *Bacteria*, the archaeal cell wall functions to prevent osmotic lysis and gives the cell its shape. In addition, because they lack peptidoglycan in their cell walls, *Archaea* are naturally resistant to the activity of lysozyme (Section 4.6) and the antibiotic penicillin, agents that either destroy peptidoglycan or prevent its proper synthesis (Section 6.2).

4.8 MiniReview

The cell walls of *Archaea* are of several types, including pseudomurein, various types of polysaccharides, S-layers, and other protein or glycoprotein cell walls.

- How does pseudomurein resemble peptidoglycan? How do the two molecules differ?
- What is the composition of an S-layer?
- Why are *Archaea* insensitive to penicillin?

IV OTHER CELL SURFACE STRUCTURES AND INCLUSIONS

In addition to cell walls, prokaryotic cells can have other outer layers or structures in contact with the environment. Moreover, cells often contain one or more types of cellular inclusions. We examine some of these structures here.

4.9 Cell Surface Layers, Pili, and Fimbriae

Many prokaryotic organisms secrete slimy or sticky materials on their cell surface. These materials consist of either polysaccharide or protein. These are not considered part of the cell wall because they do not confer significant structural strength on the cell. The terms "capsule" and "slime layer" are used to describe these layers.

Capsules and Slime Layers

Capsules and slime layers may be thick or thin and rigid or flexible, depending on their chemistry and degree of hydration. If the material is organized in a tight matrix that excludes small particles, such as India ink, it is called a **capsule** (**Figure 4.28**). If the material is more easily deformed, it will not exclude particles and is more difficult to see; this form is called a *slime layer*. In addition, capsules are

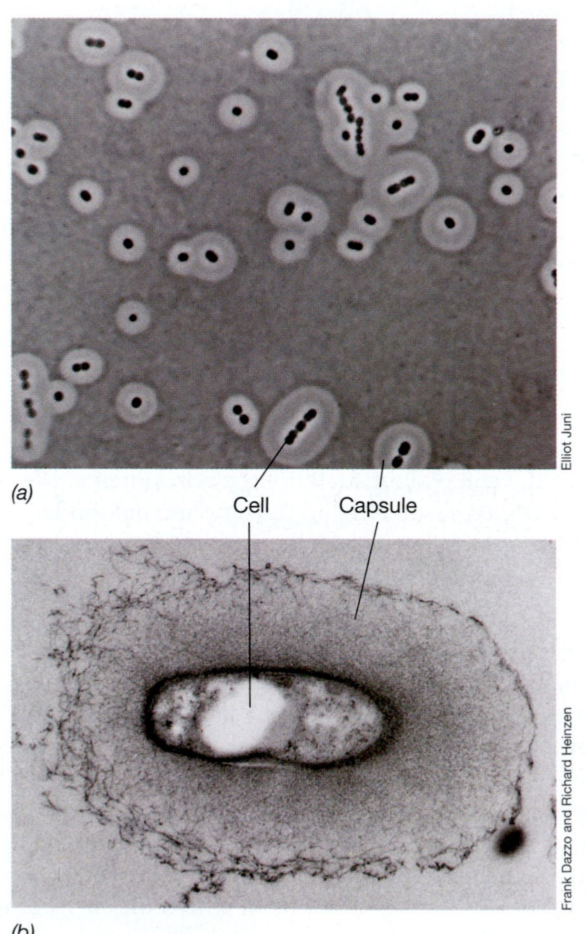

Figure 4.28 Bacterial capsules. *(a)* Capsules of *Acinetobacter* species observed by negative staining cells with India ink and phase-contrast microscopy. India ink does not penetrate the capsule and so the capsule appears as a light area surrounding the cell, which appears black. *(b)* Electron micrograph of a thin section of a cell of *Rhizobium trifolii* stained with ruthenium red to reveal the capsule. The diameter of the cell proper (not including the capsule) is about 0.7 μm.

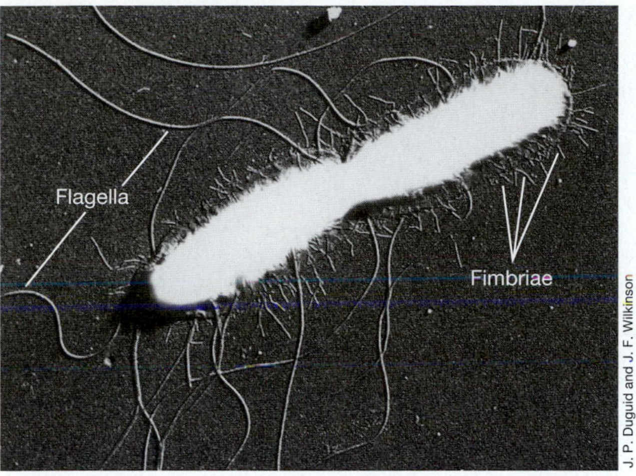

Figure 4.29 Fimbriae. Electron micrograph of a dividing cell of *Salmonella typhi*, showing flagella and fimbriae. A single cell is about 0.9 μm wide.

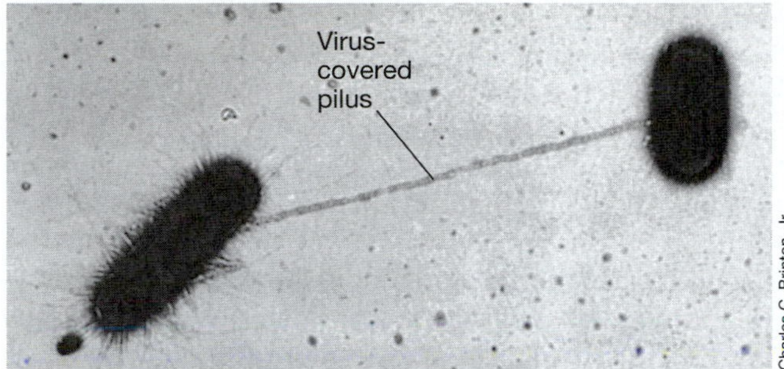

Figure 4.30 Pili. The pilus on an *Escherichia coli* cell that is undergoing genetic transfer with a second cell is revealed by the viruses that have adhered to it. The cells are about 0.8 μm wide.

typically firmly attached to the cell wall, whereas slime layers are loosely attached and can be lost from the cell surface.

Polysaccharide layers have several functions in bacteria. Surface polysaccharides assist in the attachment of microorganisms to solid surfaces. As we will see later, pathogenic microorganisms that enter the animal body by specific routes usually do so by first binding specifically to surface components of host tissues (⚭ Section 28.6). This binding is often mediated by surface polysaccharides on the bacterial cell. Many nonpathogenic bacteria also bind to solid surfaces in nature, sometimes forming a thick layer of cells called a *biofilm*. Extracellular polysaccharides play a key role in the development of biofilms (⚭ Microbial Sidebar in Chapter 6, "Microbial Growth in the Real World: Biofilms"; Sections 23.4 and 23.5).

Polysaccharide outer layers play other roles as well. For example, encapsulated pathogenic bacteria are typically more difficult for phagocytic cells of the immune system (⚭ Section 29.2) to recognize and subsequently destroy. In addition, because outer polysaccharide layers bind a significant amount of water, it is likely that these layers play some role in resistance of the cell to desiccation.

Fimbriae and Pili

Fimbriae and pili are filamentous structures composed of protein that extend from the surface of a cell and can have many functions. *Fimbriae* (**Figure 4.29**) enable organisms to stick to surfaces, including animal tissues in the case of some pathogenic bacteria, or to form pellicles (thin sheets of cells on a liquid surface) or biofilms on surfaces. Notorious among human pathogens in which these structures assist in the disease process include *Salmonella* species (salmonellosis), *Neisseria gonorrhoeae* (gonorrhea), and *Bordetella pertussis* (whooping cough).

Pili are similar to fimbriae but are typically longer structures, and only one or a few pili are present on the surface of a cell. Because pili can be receptors for certain types of viruses, they can best be seen under the electron microscope when they become coated with virus particles (**Figure 4.30**). Although they may attach to surfaces as do fimbriae, pili also have other functions. A very important function is facilitating genetic exchange between prokaryotic cells in the process of conjugation, as will be discussed in Section 11.12.

Many classes of pili are known, distinguished by their structure and function. One class, called *type IV pili*, performs an unusual form of cell motility called *twitching motility*. Type IV pili are 6 nm in diameter and can extend for several micrometers away from the cell surface. Twitching motility is a type of gliding motility, movement along a solid surface (Section 4.14). In twitching motility, extension of the pili followed by their retraction drag the cell along a solid surface. Energy for twitching motility is supplied by ATP hydrolysis. Certain species of *Pseudomonas* and *Moraxella* are well known for their twitching motility.

Unlike other pili, type IV pili are present only at the poles of rod-shaped cells. Type IV pili have been implicated as key colonization factors for certain human pathogens, including *Vibrio cholerae* (cholera) and *Neisseria gonorrhoeae*. The twitching motility of these pathogens presumably assists the organisms in their movement across host tissues. Type IV pili are also thought to mediate genetic transfer by the process of transformation (⚭ Section 11.10).

4.9 MiniReview

Many prokaryotic cells contain capsules, slime layers, pili, or fimbriae. These structures have several functions, including attachment, genetic exchange, and twitching motility.

- Could a bacterial cell dispense with a cell wall if it had a capsule? Why or why not?
- How do fimbriae differ from pili, both structurally and functionally?

4.10 Cell Inclusions

Granules or other inclusions are often present in prokaryotic cells. Inclusions function as energy reserves or as a reservoir of structural building blocks such as carbon. Inclusions can

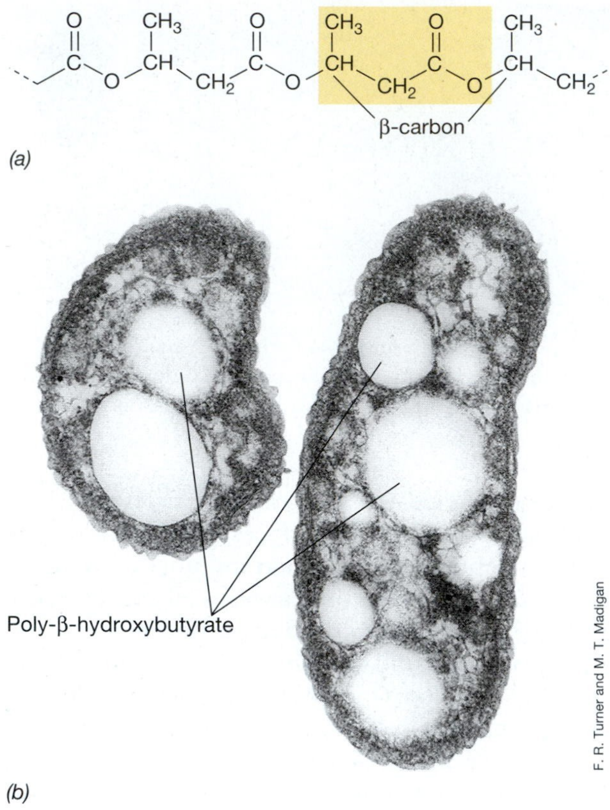

Figure 4.31 Poly-β-hydroxybutyrate (PHB). *(a)* Chemical structure of PHB, a common poly-β-hydroxyalkanoate. A monomeric unit is shown in color. Other alkanoate polymers are made by substituting longer-chain hydrocarbons for the $-CH_3$ group on the β carbon. *(b)* Electron micrograph of a thin section of cells of the phototrophic bacterium *Rhodovibrio sodomensis* containing granules of PHB.

often be seen directly with the light microscope and are typically enclosed by structurally atypical membranes that partition them off in the cell. Storing carbon or other substances in an insoluble form is advantageous to the cell because it reduces the osmotic stress that would be encountered if the same amount of the substance was stored in soluble form within the cytoplasm.

Carbon Storage Polymers

One of the most common inclusion bodies in prokaryotic organisms consists of **poly-β-hydroxybutyric acid (PHB)**, a lipid that is formed from β-hydroxbutyric acid units (**Figure 4.31a**). The monomers of PHB bond by ester linkage to form the long PHB polymer, and then the polymer aggregates into granules (Figure 4.31*b*). The length of the monomer in the polymer can vary considerably, from as short as C_4 to as long as C_{18}. Thus, the more generic term poly-β-hydroxyalkanoate (PHA) is used to describe this whole class of carbon- and energy-storage polymers. PHAs are synthesized when there is an excess of carbon and are broken down for use as carbon skeletons for biosynthesis or to make ATP when conditions warrant. Many prokaryotes, including species of both *Bacteria* and *Archaea,* produce PHAs.

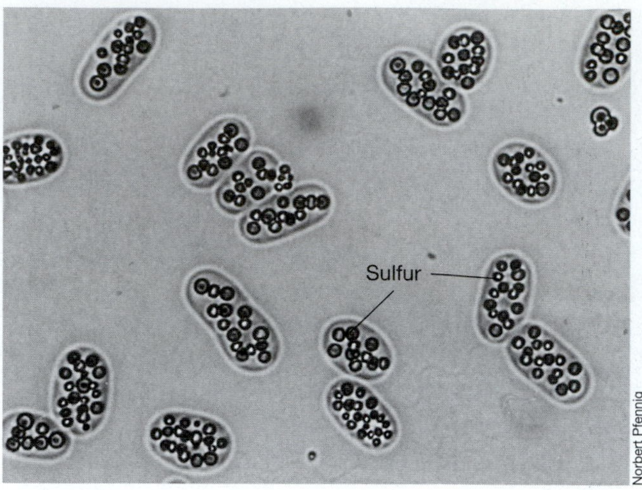

Figure 4.32 Sulfur globules. Bright-field photomicrograph of cells of the purple sulfur bacterium *Isochromatium buderi*. The intracellular inclusions are sulfur globules formed from the oxidation of hydrogen sulfide (H_2S). A single cell measures 4 × 7 μm.

Another storage product is *glycogen,* which is a polymer of glucose (Section 3.3). Like PHAs, glycogen is a storehouse of both carbon and energy. Glycogen is produced when carbon is in excess in the environment and is consumed when carbon is limited. Glycogen resembles starch, the major storage reserve of plants, but differs from starch in the manner in which the glucose units are linked together (Figure 3.6*b*).

Polyphosphate and Sulfur

Many microorganisms accumulate inorganic phosphate (PO_4^{2-}) in the form of granules of *polyphosphate*. These granules can be degraded and used as sources of phosphate for nucleic acid and phospholipid biosyntheses and in some organisms can be used to make the energy-rich compound ATP. Phosphate is often a limiting nutrient in natural environments. Thus if a cell happens upon an excess of phosphate, it is advantageous to store it as polyphosphate for future use.

Many gram-negative prokaryotes can oxidize reduced sulfur compounds, such as hydrogen sulfide (H_2S). The oxidation of sulfide is linked to either reactions of energy metabolism (chemolithotrophy, Sections 20.8 and 20.10) or phototrophic CO_2 fixation (autotrophy, Section 20.6). In either case, *elemental sulfur* may accumulate inside the cell in readily visible globules (**Figure 4.32**). These globules of elemental sulfur remain as long as the source of reduced sulfur is still present. However, as the reduced sulfur source becomes limiting, the sulfur in the granules is oxidized to sulfate (SO_4^{2-}), and the granules slowly disappear as this reaction proceeds.

It has been shown that the sulfur globules of sulfur bacteria actually form in the periplasm of the cell rather than in the cytoplasm. The periplasm expands outward to accommodate the globules as H_2S is oxidized to S^0. The periplasm then contracts inward as S^0 is oxidized to SO_4^{2-}. It is likely that some of the other "cytoplasmic" inclusions that form in gram-negative bacteria are actually periplasmic as well. For example, granules of PHA in gram-negative cells (Figure 4.31) are almost certainly located in the periplasm.

Magnetic Storage Inclusions: Magnetosomes

Some bacteria can orient themselves specifically within a magnetic field because they contain **magnetosomes**. These structures are intracellular particles of the iron mineral magnetite—Fe_3O_4 (**Figure 4.33**). Magnetosomes impart a magnetic dipole on a cell, allowing it to respond to a magnetic field. Bacteria that produce magnetosomes exhibit *magnetotaxis,* the process of orienting and migrating along Earth's magnetic field lines. Although the suffix "*-taxis*" is used in the word magnetotaxis, there is no evidence that magnetotactic bacteria employ the sensory systems of chemotactic or phototactic bacteria (Section 4.15). Instead, the alignment of magnetosomes in the cell simply imparts a magnetic moment upon it, which then orients the cell in a particular direction in its environment.

The major function of magnetosomes is unknown. However, magnetosomes have been found in several aquatic organisms that grow best in laboratory cultures at low O_2 concentrations. It has thus been hypothesized that one function of magnetosomes may be to guide these aquatic cells downward (the direction of Earth's magnetic field) toward the sediments where O_2 levels are lower.

Magnetosomes are surrounded by a membrane containing phospholipids, proteins, and glycoproteins (Figure 4.33*b*, *c*). This membrane is not a unit membrane, as is the cytoplasmic membrane (Figure 4.5), but instead is a nonunit membrane, like that surrounding granules of PHB (Figure 4.31). Magnetosome membrane proteins probably play a role in precipitating Fe^{3+} (brought into the cell in soluble form by chelating agents) as Fe_3O_4 in the developing magnetosome. The morphology of magnetosomes appears to be species specific, varying in shape from square to rectangular to spike-shaped in different species, forming into chains inside the cell (Figure 4.33).

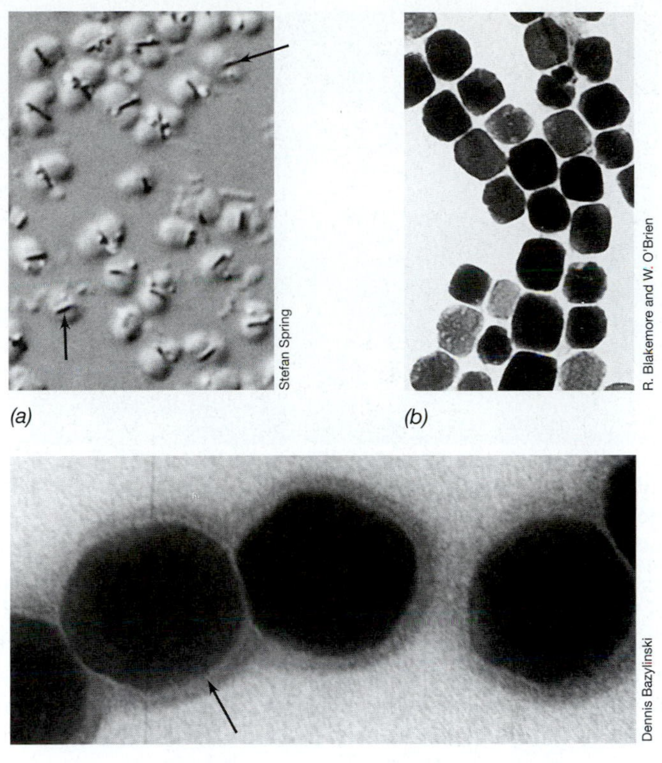

Figure 4.33 Magnetotactic bacteria and magnetosomes. *(a)* Differential interference contrast micrograph of coccoid magnetotactic bacteria. Note magnetosomes (arrows). A single cell is 2.2 μm wide. *(b)* Magnetosomes isolated from the magnetotactic bacterium *Magnetospirillum magnetotacticum*. Each particle is about 50 nm long. *(c)* Transmission electron micrograph of magnetosomes from a magnetic coccus. The arrow points to the membrane surrounding each magnetosome. A single magnetosome is about 90 nm wide.

4.10 MiniReview

Prokaryotic cells often contain inclusions of sulfur, polyphosphate, carbon polymers, or magnetosomes. These substances function as storage materials or in magnetotaxis.

- Under what growth conditions would you expect PHAs or glycogen to be produced?
- Why would it be impossible for gram-positive bacteria to store sulfur as sulfur-oxidizing chemolithotrophs can?
- What form of iron is present in magnetosomes?

4.11 Gas Vesicles

Some prokaryotes are *planktonic,* meaning that they live a floating existence within the water column of lakes and the oceans. These organisms can float because they contain **gas vesicles**. These structures confer buoyancy on cells, allowing them to position themselves in a water column in response to environmental factors.

The most dramatic examples of gas-vesiculate bacteria are cyanobacteria that form massive accumulations called *blooms* in lakes or other bodies of water (**Figure 4.34**). Gas-vesiculate cells rise to the surface of the lake and are blown by winds into dense masses. Certain purple and green phototrophic bacteria (∞ Sections 15.2 and 16.15) have gas vesicles, as do some nonphototrophic bacteria that live in lakes and ponds. Some *Archaea* also contain gas vesicles. Gas vesicles are absent from eukaryotes.

General Structure of Gas Vesicles

Gas vesicles are spindle-shaped gas-filled structures made of protein; they are hollow yet rigid and of variable length and diameter. Gas vesicles in different organisms vary in length from about 300 to more than 1000 nm and in width from 45 to 120 nm, but the vesicles of a given organism are more or less of constant size (**Figure 4.35**).

Figure 4.34 Buoyant cyanobacteria. Flotation of gas-vesiculate cyanobacteria in a bloom that formed in a nutrient-rich lake, Lake Mendota, Madison, Wisconsin, USA.

Gas vesicles may number from a few to hundreds per cell. The gas-vesicle membrane is composed of protein, is about 2 nm thick, and is impermeable to water and solutes but permeable to gases. The presence of gas vesicles in cells can be determined either by light microscopy, where clusters of vesicles, called *gas vacuoles,* appear as irregular bright inclusions (Figure 4.34), or by electron microscopy (**Figure 4.36**).

Molecular Structure of Gas Vesicles

A gas vesicle is composed of two different proteins (**Figure 4.37**). The major gas vesicle protein, called *GvpA*, is a small, highly hydrophobic and very rigid protein. The rigidity is essential for the structure to resist the pressures exerted on it from outside. GvpA makes up the gas vesicle shell and composes 97% of total gas vesicle protein. The minor protein, called *GvpC,* functions to strengthen the shell of the gas vesicle.

Gas vesicles are biosynthesized when copies of GvpA align to form parallel "ribs" yielding a watertight shell. The GvpA ribs are strengthened by GvpC protein. GvpC cross-links GvpA at an angle, binding several GvpA molecules together like a clamp (Figure 4.37). The final shape of the gas vesicle can vary in different organisms from long and thin to short and fat (compare Figures 4.35 and 4.36*b*) and is a function of how the GvpA and GvpC proteins are arranged to form the intact vesicle.

How do gas vesicles confer buoyancy and what ecological benefit does buoyancy confer? The composition and

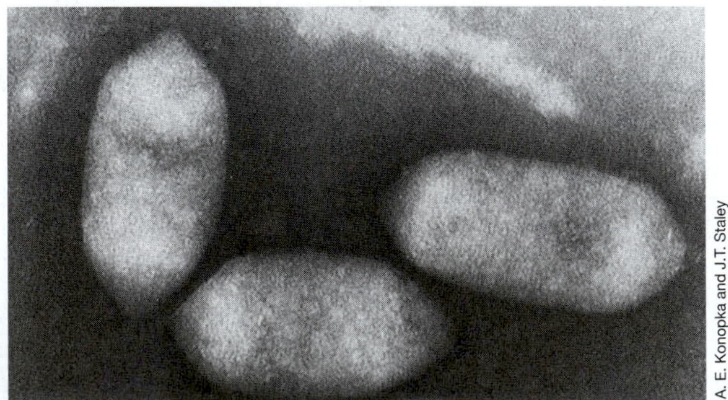

Figure 4.35 Gas vesicles. Transmission electron micrographs of gas vesicles purified from the bacterium *Ancyclobacter aquaticus* and examined in negatively stained preparations. A single gas vesicle is about 100 nm in diameter. [Reproduced with permission from *Archives of Microbiology* 112: 133–140 (1977).]

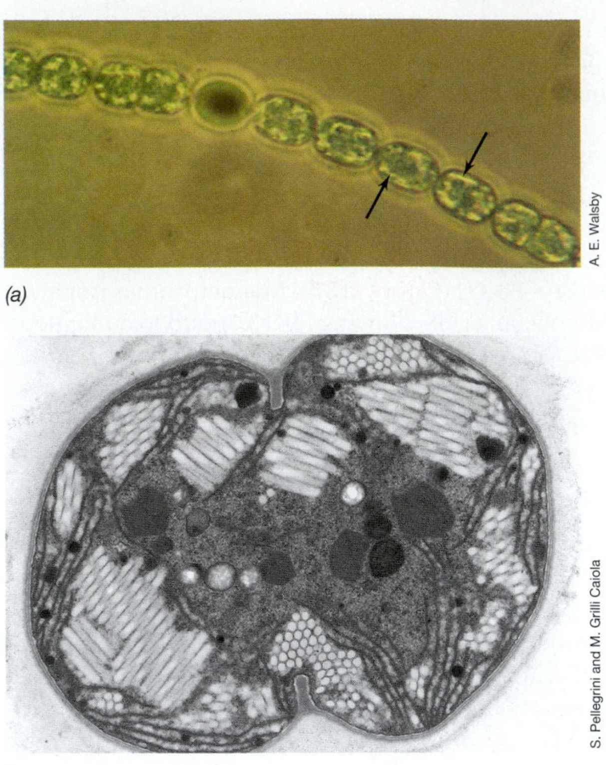

Figure 4.36 Gas vesicles of the cyanobacteria *Anabaena* and *Microcystis*. *(a) Anabaena flosaquae.* The dark cell in the center (a heterocyst) lacks gas vesicles. In the other cells, the vesicles group together as phase-bright clusters called gas vacuoles (arrows). *(b)* Transmission electron micrograph of the cyanobacterium *Microcystis.* Gas vesicles are arranged in bundles, here observable in both longitudinal and cross section.

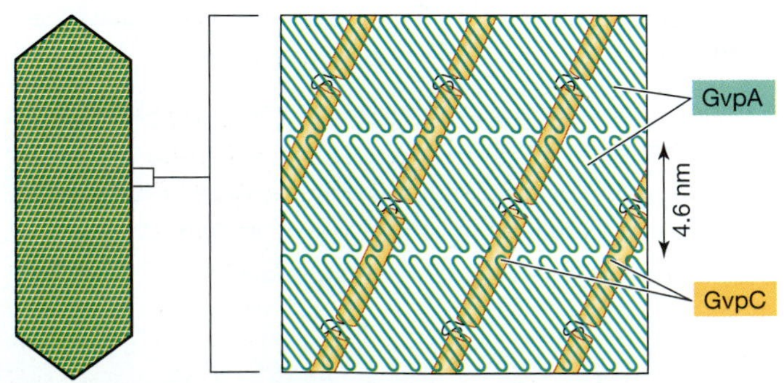

Figure 4.37 Gas vesicle proteins. Model of how the two proteins making up the gas vesicle, GvpA and GvpC, interact to form a watertight but gas-permeable structure. GvpA, a rigid β-sheet, makes up the rib, and GvpC, an α-helix structure, is the cross-linker.

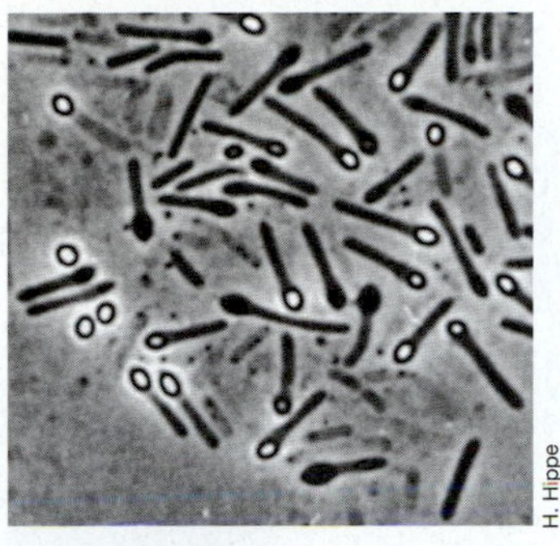

(*a*) **Terminal spores**

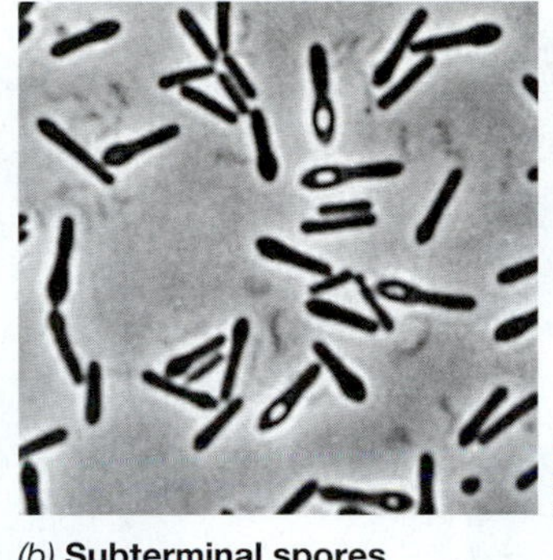

(*b*) **Subterminal spores**

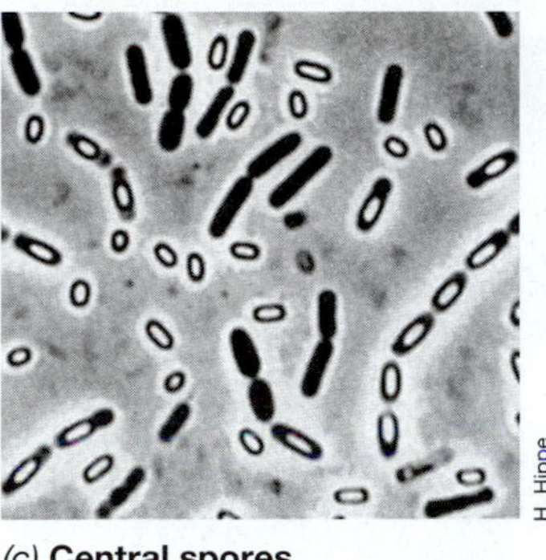

(*c*) **Central spores**

Figure 4.38 The bacterial endospore. Phase-contrast photomicrographs illustrating endospore morphologies and intracellular locations in different species of endospore-forming bacteria.

pressure of the gas inside a gas vesicle is the same as that of the gas in which the organism is suspended. However, because the inflated gas vesicle has a density of only some 5–20% of that of the cell proper, gas vesicles decrease the density of the cell, thereby increasing its buoyancy. Phototrophic organisms in particular can benefit from gas vesicles because they allow cells to adjust their position vertically in a water column to regions where the light intensity for photosynthesis is optimal.

4.11 MiniReview

Gas vesicles are intracellular gas-filled structures composed of proteins that function to confer buoyancy on cells. Gas vesicles contain two different proteins arranged to form a gas-permeable but watertight structure.

- What might be the benefit of gas vesicles to phototrophic cells? (*Hint*: What do phototrophs need to grow?).
- How are the two proteins that make up the gas vesicle, GvpA and GvpC, arranged to form such a water-impermeable structure?

4.12 Endospores

Certain species of *Bacteria* produce structures called **endospores** (**Figure 4.38**) during a process called *sporulation*. Endospores (the prefix "endo" means "within") are highly differentiated cells that are extremely resistant to heat, harsh chemicals, and radiation. Endospores function as survival structures and enable the organism to endure difficult times, including but not limited to extremes of temperature, drying, or nutrient depletion. Endospores can thus be thought of as the dormant stage of a bacterial life cycle: vegetative cell → endospore → vegetative cell. Endospores are also ideal structures for dispersal of an organism by wind, water, or through the animal gut. Endospore-forming bacteria are found most commonly in the soil, and the genera *Bacillus* and *Clostridium* are the best studied of endospore-forming bacteria.

Endospore Formation and Germination

During endospore formation, a vegetative cell is converted into a nongrowing, heat-resistant structure (**Figure 4.39**). Cells do not sporulate when they are actively growing but only when growth ceases owing to the exhaustion of an essential nutrient. Thus, cells of *Bacillus,* a typical endospore-forming bacterium, cease vegetative growth and begin sporulation when, for example, a key nutrient such as carbon or nitrogen becomes limiting.

An endospore can remain dormant for years (see the Microbial Sidebar, "How Long Can an Endospore Survive?" on page 94), but it can convert back to a vegetative cell relatively rapidly. This process involves three steps: *activation, germination*, and *outgrowth* (**Figure 4.40**). Activation is most easily accomplished by heating freshly formed endospores for several minutes at an elevated but sublethal temperature. Activated endospores are then conditioned to germinate when placed in the presence of specific nutrients, such as certain amino acids (alanine is a particularly good trigger of endospore

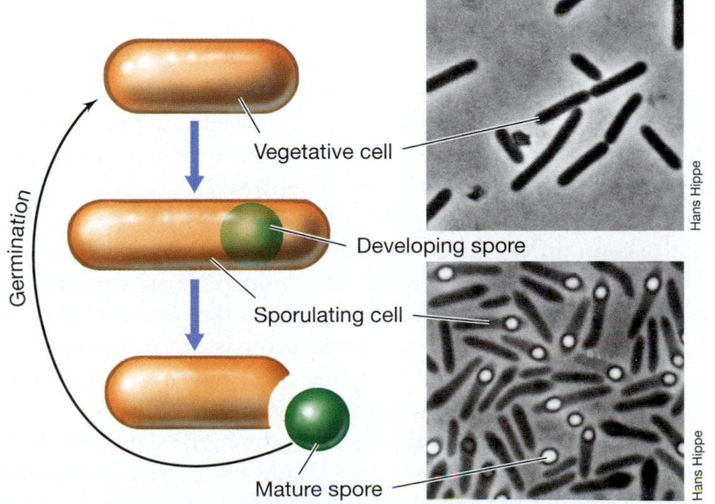

Figure 4.39 The life cycle of an endospore-forming bacterium. The phase-contrast photomicrographs are of cells of *Clostridium pascui*. A cell is about 0.8 μm wide.

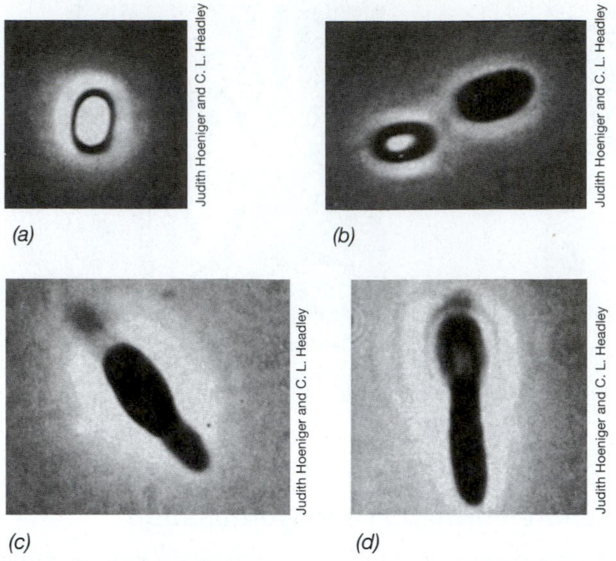

Figure 4.40 Endospore germination in *Bacillus*. Conversion of an endospore into a vegetative cell. The series of photomicrographs shows the sequence of events starting from a highly refractile mature endospore *(a)*. *(b)* (Activation) refractility is being lost. *(c)*, *(d)* The new vegetative cell is emerging (outgrowth).

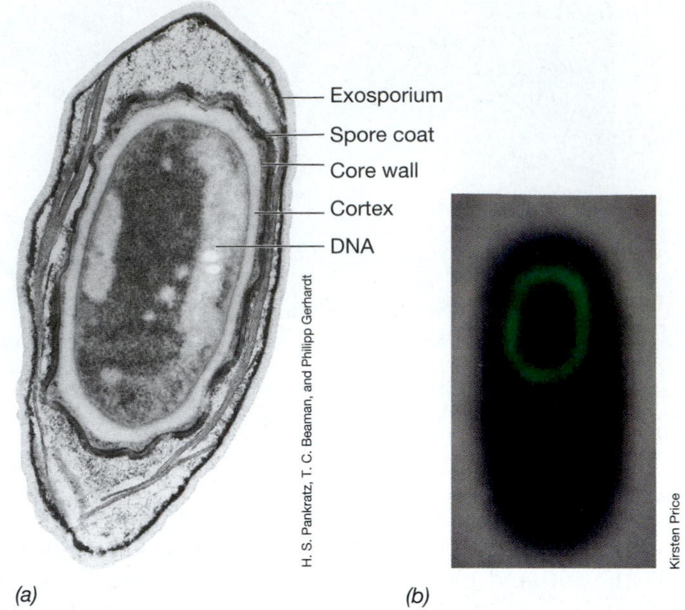

Figure 4.41 Structure of the bacterial endospore. *(a)* Transmission electron micrograph of a thin section through an endospore of *Bacillus megaterium*. *(b)* Fluorescent photomicrograph of a cell of *Bacillus subtilis* undergoing sporulation. The green area is due to a dye that specifically stains a sporulation protein in the spore coat.

germination). Germination, usually a rapid process (on the order of several minutes), involves loss of microscopic refractility of the endospore, increased ability to be stained by dyes, and loss of resistance to heat and chemicals. The final stage, outgrowth, involves visible swelling due to water uptake and synthesis of new RNA, proteins, and DNA. The cell emerges from the broken endospore and begins to grow (Figure 4.40). The cell then remains in vegetative growth until environmental signals once again trigger sporulation.

Endospore Structure

Endospores stand out under the light microscope as strongly refractile structures (see Figures 4.38–4.40). Endospores are impermeable to most dyes, so occasionally they are seen as unstained regions within cells that have been stained with basic dyes such as methylene blue. To stain endospores, special stains and procedures must be used. In the classical endospore-staining protocol, malachite green is used as a stain and is infused into the spore with steam.

The structure of the endospore as seen with the electron microscope differs distinctly from that of the vegetative cell (**Figure 4.41**). In particular, the endospore is structurally more complex in that it has many layers that are absent from the vegetative cell. The outermost layer is the *exosporium*, a thin protein covering. Within this are the *spore coats*, composed of layers of spore-specific proteins (Figure 4.41*b*). Below the spore coat is the *cortex*, which consists of loosely cross-linked peptidoglycan, and inside the cortex is the *core*, which contains the core wall, cytoplasmic membrane, cytoplasm, nucleoid, ribosomes, and other cellular essentials. Thus, the endospore differs structurally from the vegetative cell primarily in the kinds of structures found outside the core wall.

One substance that is characteristic of endospores but absent from vegetative cells is **dipicolinic acid** (**Figure 4.42*a***). This substance has been found in the endospores of all endospore-forming bacteria examined and is located in the core. Endospores are also enriched in calcium (Ca^{2+}), most of which is complexed with dipicolinic acid (Figure 4.42*b*). The calcium–dipicolinic acid complex of the core represents about 10% of the dry weight of the endospore. The complex functions to reduce water availability within the endospore, thus helping to dehydrate it. In addition, the complex intercalates (inserts between bases) in DNA, and in so doing stabilizes DNA to heat denaturation.

The Endospore Core and SASPs

Although both contain a copy of the chromosome and other essential cellular components, the core of a mature endospore differs greatly from the vegetative cell from which it was formed. Besides the high levels of calcium dipicolinate (Figure 4.42), which help reduce the water content of the core, the core becomes greatly dehydrated during the sporulation process. The core of a mature endospore contains only 10–25% of the water content of the vegetative cell, and thus the consistency of the core cytoplasm is that of a gel. Dehydration of the core greatly increases the heat resistance of macromolecules within the spore. Some bacterial endospores survive heating to temperatures as high as 150°C, although 121°C, the standard for microbiological sterilization (121°C is autoclave temperature, Section 27.1) kills the endospores

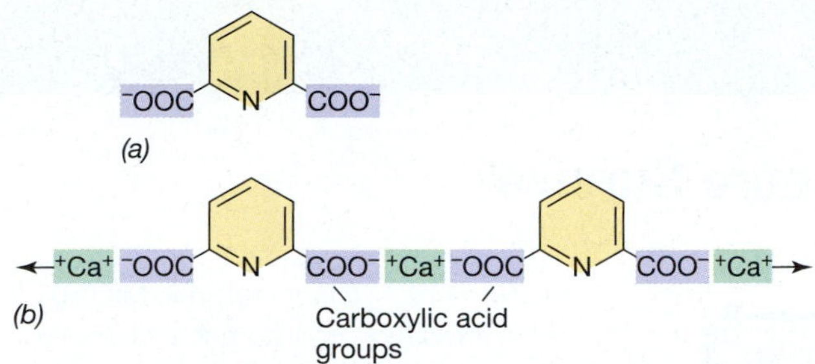

Figure 4.42 Dipicolinic acid (DPA). *(a)* Structure of DPA. *(b)* How Ca^{2+} cross-links DPA molecules to form a complex.

of most species. Boiling has essentially no negative effect on endospores. Dehydration has also been shown to confer resistance in the endospore to chemicals, such as hydrogen peroxide (H_2O_2) and causes enzymes remaining in the core to become inactive. In addition to the low water content of the endospore, the pH of the core is about one unit lower than that of the vegetative cell cytoplasm.

The endospore core contains high levels of *small acid-soluble proteins* (SASPs). These proteins are made during the sporulation process and have at least two functions. SASPs bind tightly to DNA in the core and protect it from potential damage from ultraviolet radiation, desiccation, and dry heat. Ultraviolet resistance is conferred when SASPs change the molecular structure of DNA from the normal "B" form to the more compact "A" form. A-form DNA is more resistant to the formation of pyrimidine dimers by UV radiation, a means of mutation (∞ Section 11.7), and to the denaturing effects of dry heat. In addition, SASPs function as a carbon and energy source for the outgrowth of a new vegetative cell from the endospore during germination.

The Sporulation Process

Sporulation is a complex series of events in cellular differentiation; many genetically directed changes in the cell underlie the conversion from vegetative growth to sporulation. The structural changes occurring in sporulating cells of *Bacillus* are shown in **Figure 4.43**. Sporulation can be divided into several stages. In *Bacillus subtilis*, where detailed studies have been done, the entire sporulation process takes about 8 hours and begins with asymmetric cell division (Figure 4.43). Genetic studies of mutants of *Bacillus*, each blocked at one of the stages of sporulation shown in Figure 4.43, indicate that more than 200 genes are specific to the process. Sporulation requires that the synthesis of many proteins needed for vegetative cell functions cease and that specific endospore proteins be made. This is accomplished by the activation of several families of endospore-specific genes in response to an environmental trigger to sporulate. The proteins encoded by these genes catalyze the series of events leading from a moist,

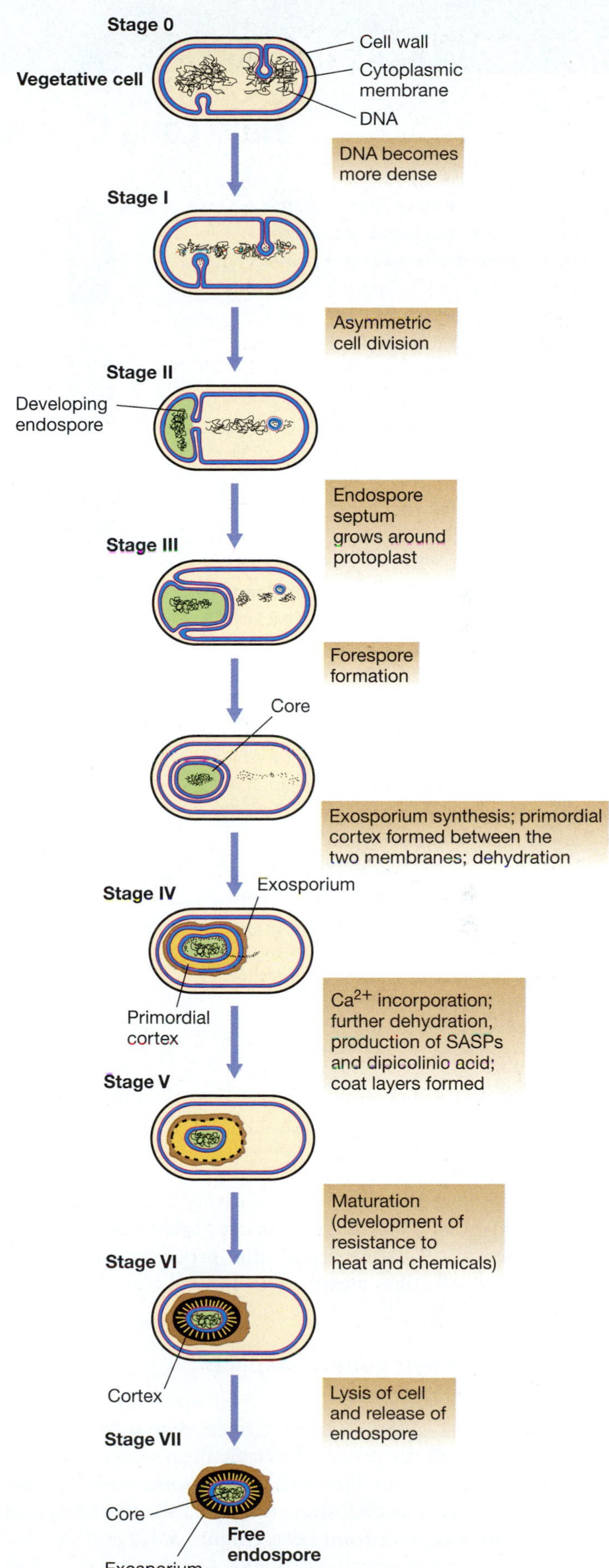

Figure 4.43 Stages in endospore formation. Stages 0 through VII are defined from genetic studies and microscopic analyses of sporulation in *Bacillus subtilis*.

Microbial Sidebar

How Long Can an Endospore Survive?

In this chapter we have discussed the dormancy and resistance properties of bacterial endospores and have pointed out that endospores can survive for long periods in a dormant state. But how long is *long*?

Evidence for endospore longevity suggests that these structures can remain viable (that is, capable of germination into vegetative cells) for at least several decades and probably for much longer than that. For example, a suspension of endospores of the bacterium *Clostridium aceticum* (**Figure 1a**) prepared in 1947 was placed in sterile growth medium in 1981, 34 years later, and in less than 12 h growth commenced, leading to a robust culture. *C. aceticum* was originally isolated by the Dutch scientist K.T. Wieringa in 1940 but was thought to have been lost until this vial of *C. aceticum* endospores was found in a storage room at the University of California at Berkeley and revived.[1] *C. aceticum* is an acetogen, a bacterium that makes acetate from $CO_2 + H_2$ or from glucose (Section 21.9).

Other more dramatic examples of endospore longevity have been documented. Bacteria of the genus *Thermoactinomyces* are thermophilic endospore formers that are widespread in nature in soil, plant litter, and fermenting plant material. Microbiological examination of a Roman archaeological site in the United Kingdom that was dated to more than 2,000 years ago yielded significant numbers of viable *Thermoactinomyces* endospores in various pieces of debris. Additionally, *Thermoactinomyces* endospores were recovered in fractions of sediment cores from a Minnesota lake known to be over 7,000 years old. Although contamination is always a possibility in such studies, samples in both of these cases were processed in such a way as to virtually rule out contamination with "recent" endospores.[2] Thus, endospores can probably last for several thousands of years, but is this the limit?

Figure 1 Longevity of endospores. (a) *A tube containing endospores from the bacterium* Clostridium aceticum *prepared on May 7, 1947. After remaining dormant for over 30 years, the endospores were suspended in a culture medium after which growth occurred within 12 h.* (b) *Halophilic bacteria trapped within salt crystals. These crystals (about 1 cm in diameter) were grown in the laboratory in the presence of* Halobacterium *cells (orange) that remain viable in the crystals. Crystals similar to these but of Permian age (≈250 million years old) were reported to contain viable halophilic endosporulating bacteria.*

What factors could limit the age of an endospore? Cosmic radiation has been considered a major factor because it can introduce mutations in DNA.[2] It has been hypothesized that over periods of thousands of years, the cumulative effects of cosmic radiation could introduce so many mutations into the genome of an organism that even highly radiation-resistant structures such as endospores would succumb to the genetic damage. However, extrapolations from experimental assessments of the effect of

metabolizing, vegetative cell to a relatively dry, metabolically inert, but extremely resistant endospore (**Table 4.3** and Figure 4.43). In Section 9.12 we revisit this process and examine some of the molecular events that control the sporulation process.

Diversity and Phylogenetic Aspects of Endospore Formation

Nearly 20 genera of *Bacteria* have been shown to form endospores, although the process has only been studied in detail in a few species of *Bacillus* and *Clostridium*. Nevertheless, many of the secrets to endospore survival, like the formation of calcium–dipicolinate complexes (Figure 4.42) and the production of endospore-specific proteins, seem universal. Thus, although some of the details of sporulation may vary from one organism to the next, the general principles are likely the same in all endospore-forming bacteria.

From a phylogenetic perspective, the capacity to produce endospores is found only in a particular sublineage of the gram-positive bacteria (Sections 2.9 and 16.2). Despite this, the physiologies of endospore-forming bacteria are highly diverse and include anaerobes, aerobes, phototrophs, and chemolithotrophs. Considering this physiological diversity, the actual triggers for endospore formation may vary with different species and could include signals other than simple nutrient starvation, the major trigger for endospore formation in *Bacillus* and *Clostridium* species. Interestingly, no species of *Archaea* have been shown to form endospores, suggesting

(continued)

natural radiation on endospores suggest that if suspensions of endospores were partially shielded from cosmic radiation, for example, by being embedded in layers of organic matter, such as in the Roman archaeological dig or the lake sediments described above, they could retain viability over periods as great as several hundred thousand years. Amazing, but is *this* the upper limit?

In 1995 a group of scientists reported the revival of bacterial endospores they claimed were 25–40 million years old.[3] The endospores were allegedly preserved in the gut of an extinct bee trapped in amber of known geological age. The presence of endospore-forming bacteria in these bees was previously suspected because electron microscopic studies of the insect gut showed endospore-like structures (Figure 4.41*a*) and because *Bacillus*-like DNA was recovered from the insect. Incredibly, samples of bee tissue incubated in a sterile culture medium quickly yielded endospore-forming bacteria. Rigorous precautions were taken to demonstrate that the endospore-forming bacterium revived from the amber-encased bee was not a modern-day contaminant. Subsequently, an even more spectacular claim was made that halophilic (salt-loving) endospore-forming bacteria had been isolated from fluid inclusions in salt crystals of Permian age, over 250 million years old.[4] Presumably, these cells were trapped in brines within the crystal (Figure 1*b*) as it formed eons ago and remained viable for more than a quarter billion years! Molecular experiments on even older material, 425 million-year-old halite, showed evidence for prokaryotic cells as well.[5]

If these astonishing claims are supported by repetition of the results in independent laboratories (and such confirmation is crucial in science for verifying controversial findings), then it appears that endospores stored under the proper conditions can remain viable indefinitely. This is remarkable testimony to a structure that probably evolved as a means of surviving relatively brief dormant periods or as a mechanism to withstand drying but that turned out to be so well designed that survival for millions of years may be possible.

[1] Braun, M., F. Mayer, and G. Gottschalk. 1981. *Clostridium aceticum* (Wieringa), a microorganism producing acetic acid from molecular hydrogen and carbon dioxide. *Arch. Microbiol. 128:* 288–293.

[2] Gest, H., and J. Mandelstam. 1987. Longevity of microorganisms in natural environments. *Microbiol. Sci. 4:* 69–71.

[3] Cano, R. J., and M. K. Borucki. 1995. Revival and identification of bacterial spores in 25- to 40-million-year-old Dominican amber. *Science 268:* 1060–1064.

[4] Vreeland, R.H., W.D. Rosenzweig, and D.W. Powers. 2000. Isolation of a 250 million-year-old halotolerant bacterium from a primary salt crystal. *Nature 407:* 897–900.

[5] Fish, S.A., T.J. Shepherd, T.J. McGenity, and W.D. Grant. 2002. Recovery of 16S ribosomal RNA gene fragments from ancient halite. *Nature 417:* 432–436.

that the capacity to produce endospores evolved sometime after the major prokaryotic lineages diverged billions of years ago (Section 2.3 and Figure 2.7).

4.12 MiniReview

The endospore is a highly resistant and differentiated bacterial cell produced by certain gram-positive *Bacteria.* Endospores are highly dehydrated and contain essential macromolecules and various protective substances, including calcium dipicolinate and small acid-soluble proteins, absent from vegetative cells. Endospores can remain dormant indefinitely but can germinate quickly when conditions warrant.

- What is dipicolinic acid and where is it found?
- What are SASPs and what is their function?
- What happens when an endospore germinates?

V MICROBIAL LOCOMOTION

We finish our survey of microbial structure and function by examining cell locomotion. Many cells can move under their own power. Motility allows cells to reach different parts of their environment. In the struggle for survival, movement to a new location may offer a cell new resources and opportunities and spell the difference between life and death.

Table 4.3 Differences between endospores and vegetative cells

Characteristic	*Vegetative cell*	*Endospore*
Structure	Typical gram-positive cell; a few gram-negative cells	Thick spore cortex; Spore coat; exosporium
Microscopic appearance	Nonrefractile	Refractile
Calcium content	Low	High
Dipicolinic acid	Absent	Present
Enzymatic activity	High	Low
Metabolism (O_2 uptake)	High	Low or absent
Macromolecular synthesis	Present	Absent
mRNA	Present	Low or absent
DNA and ribosomes	Present	Present
Heat resistance	Low	High
Radiations resistance	Low	High
Resistance to chemicals (for example, H_2O_2) and acids	Low	High
Stainability by dyes	Stainable	Stainable only with special methods
Action of lysozyme	Sensitive	Resistant
Water content	High, 80–90%	Low, 10–25% in core
Small acid-soluble proteins (product of *ssp* genes)	Absent	Present
Cytoplasmic pH	About pH 7	About pH 5.5–6.0 (in core)

We examine here the two major types of cell movement, *swimming* and *gliding*. We then consider how motile cells are able to move in a directed fashion toward or away from particular stimuli (phenomena called *taxes*) and present examples of these simple behavioral responses.

4.13 Flagella and Motility

Many prokaryotes are motile by swimming, and this function is typically due to a structure called the **flagellum** (plural, flagella) (**Figure 4.44**). We see here that the flagellum functions by rotation to push or pull the cell through a liquid medium.

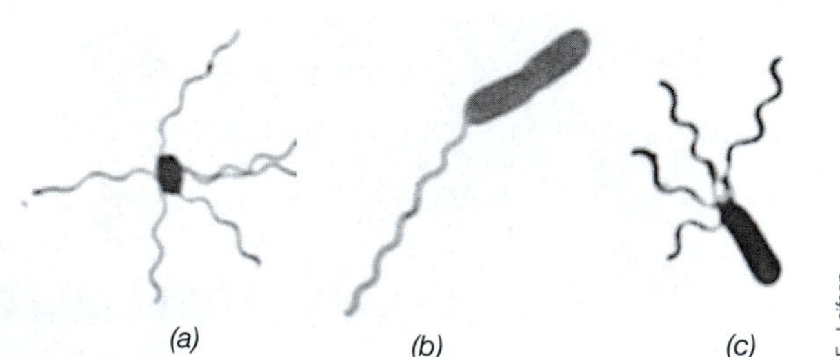

Figure 4.44 Bacterial flagella. Light photomicrographs of prokaryotes containing different arrangements of flagella. Cells are stained with Leifson flagella stain. *(a)* Peritrichous. *(b)* Polar. *(c)* Lophotrichous.

Flagella of *Bacteria*

Bacterial flagella are long, thin appendages free at one end and attached to the cell at the other end. Bacterial flagella are so thin (15–20 nm, depending on the species) that a single flagellum can be seen with the light microscope only after being stained with special stains that increase their diameter (Figure 4.44). However, flagella are easily seen with the electron microscope (**Figure 4.45**).

Flagella can be attached to cells in different patterns. In **polar flagellation**, the flagella are attached at one or both ends of the cell (Figures 4.44*b* and 4.45*a*). Occasionally a tuft (group) of flagella may arise at one end of the cell, a type of polar flagellation called *lophotrichous* ("lopho" means "tuft"; "trichous" means "hair") (Figure 4.44*c*). Tufts of flagella of this type can be seen in living cells by dark-field microscopy (**Figure 4.46a**), where the flagella appear light and are attached to light-colored cells against a dark background. In relatively large prokaryotes, tufts of flagella can also be observed by phase-contrast microscopy (Figure 4.46*b*). When a tuft of flagella emerges from both poles, flagellation is called *amphitrichous*. In **peritrichous flagellation** (Figures 4.44*a* and 4.45*b*), flagella are inserted at many locations around the cell surface (*"peri"* means "around"). The type of flagellation, polar or peritrichous, is a characteristic used in the classification of bacteria.

Flagellar Structure

Flagella are not straight but are helical. When flattened, flagella show a constant distance between adjacent curves, called

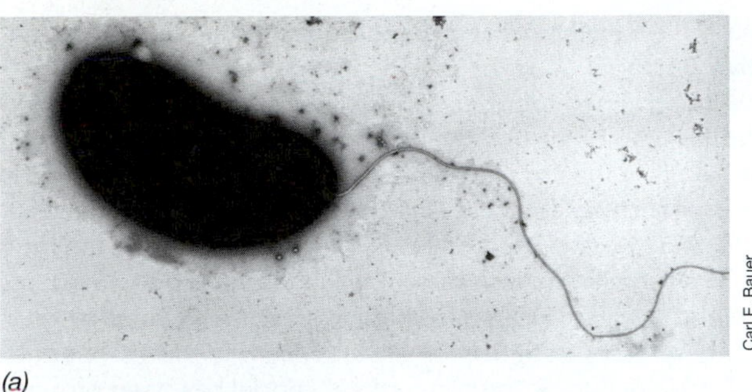

(a)

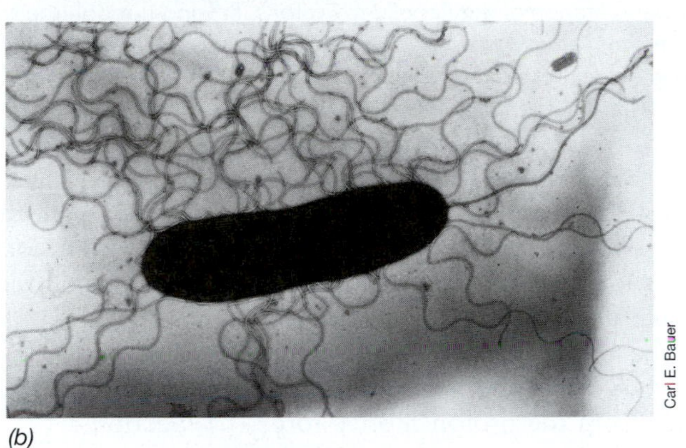

(b)

Figure 4.45 Bacterial flagella as observed by negative staining in the transmission electron microscope. *(a)* A single polar flagellum. *(b)* Peritrichous flagella. Both micrographs are of cells of the phototrophic bacterium *Rhodospirillum centenum*, which are about 1.5 μm wide. Cells of *R. centenum* are normally polarly flagellated but under certain growth conditions form peritrichously flagellated "swarmer" cells. See also Figure 4.55*b*.

the *wavelength*, and this wavelength is characteristic for the flagella of any given species (Figures 4.44–4.46). The filament of bacterial flagella is composed of many copies of a protein called *flagellin*. The shape and wavelength of the flagellum are in part determined by the structure of the flagellin protein and also to some extent by the direction of rotation of the filament. Flagellin is highly conserved in amino acid sequences in species of *Bacteria*, suggesting that flagellar motility has deep roots within this evolutionary domain.

A flagellum consists of several components and moves by rotation, much like a propeller of a boat motor. The base of the flagellum is structurally different from the filament (**Figure 4.47a**). There is a wider region at the base of the filament called the *hook*. The hook consists of a single type of protein and connects the filament to the motor portion in the base.

The motor is anchored in the cytoplasmic membrane and cell wall. The motor consists of a central rod that passes through a series of rings. In gram-negative bacteria, an outer ring, called the *L ring*, is anchored in the lipopolysaccharide layer. A second ring, called the *P ring*, is anchored in the peptidoglycan layer of the cell wall. A third set of rings, called the *MS* and *C rings*, are located within the cytoplasmic membrane and the cytoplasm, respectively (Figure 4.47*a*). In gram-positive bacteria, which lack an outer membrane, only the inner pair of rings is present. Surrounding the inner ring and anchored in the cytoplasmic membrane are a series of proteins called *Mot proteins*. A final set of proteins, called the *Fli proteins* (Figure 4.47*a*), function as the motor switch, reversing the direction of rotation of the flagella in response to intracellular signals.

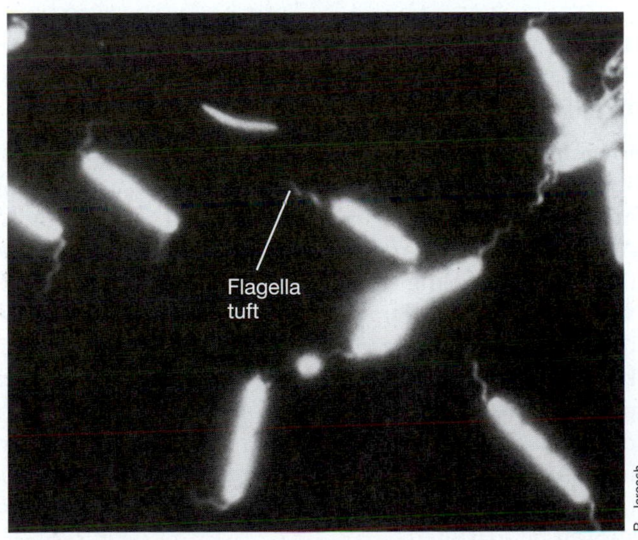

(a)

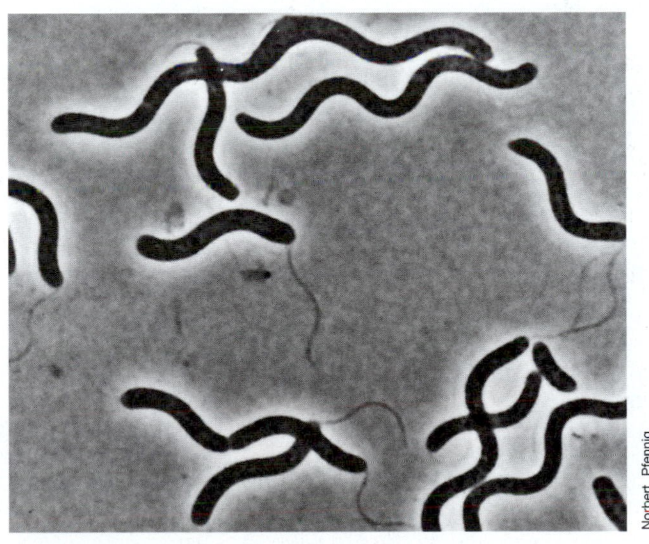

(b)

Figure 4.46 Bacterial flagella observed in living cells. *(a)* Dark-field photomicrograph of a group of large rod-shaped bacteria with flagellar tufts at each pole. A single cell is about 2 μm wide. *(b)* Phase-contrast photomicrograph of cells of the large phototrophic purple bacterium *Rhodospirillum photometricum* with lophotrichous flagella that emanate from one of the poles. A single cell measures about 3 × 30 μm.

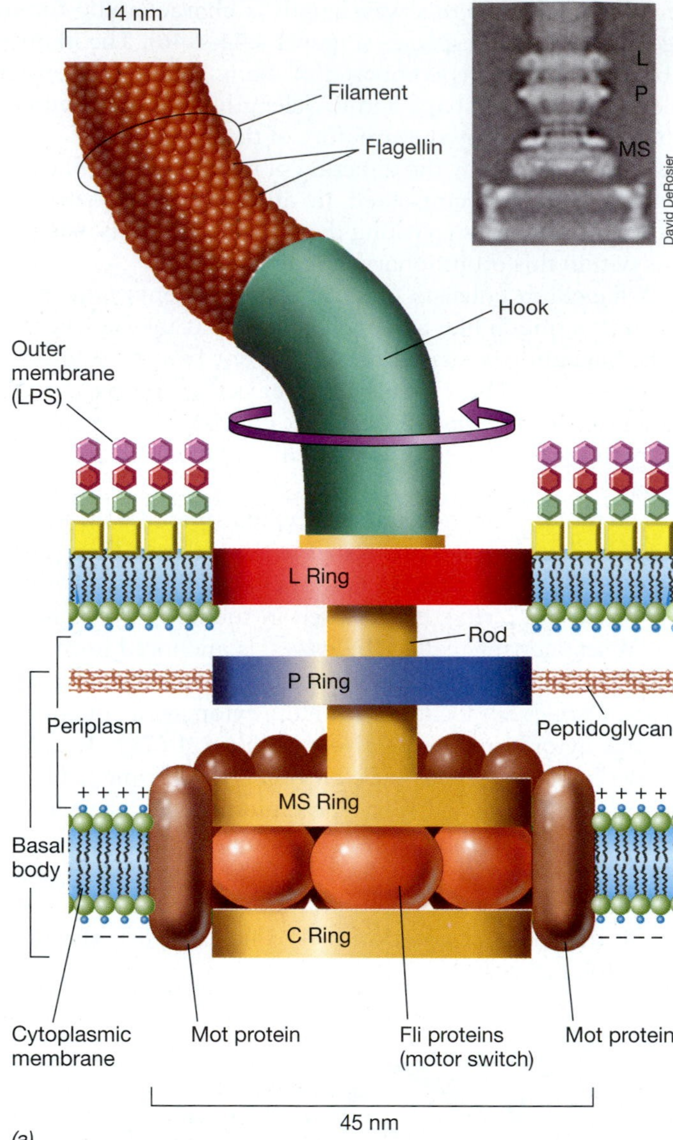

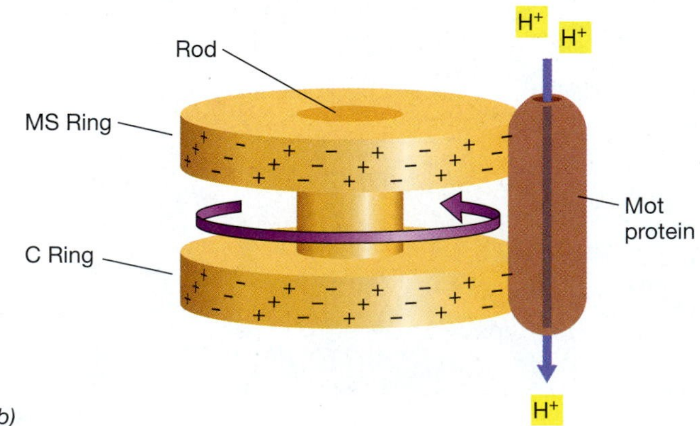

Figure 4.47 Structure and function of the flagellum in gram-negative *Bacteria*. *(a)* Structure. The L ring is embedded in the LPS and the P ring in peptidoglycan. The MS ring is embedded in the cytoplasmic membrane and the C ring in the cytoplasm. A narrow channel exists in the rod and filament through which flagellin molecules diffuse to reach the site of flagellar synthesis. The Mot proteins function as the flagellar motor, whereas the Fli proteins function as the motor switch. The flagellar motor rotates the filament to propel the cell through the medium. Inset: transmission electron micrograph of a flagellar basal body from *Salmonella enterica* with the various rings labeled. *(b)* Function. A "proton turbine" model has been proposed to explain rotation of the flagellum. Protons, flowing through the Mot proteins, may exert forces on charges present on the C and MS rings, thereby spinning the rotor. Electron micrograph from *J. Bacteriol. 183*: 6404–6412 (2001).

Flagellar Movement

The flagellum is a tiny rotary motor. How does this motor work? Rotary motors typically contain two main components: the *rotor* and the *stator*. In the flagellar motor, the rotor consists of the central rod and the L, P, C, and MS rings. Collectively, these structures make up the **basal body**. The stator consists of the Mot proteins that surround the basal body and function to generate torque.

The rotary motion of the flagellum is imparted by the basal body. The energy required for rotation of the flagellum comes from the proton motive force (∞ Section 5.12). Proton movement across the cytoplasmic membrane through the Mot complex (Figure 4.47) drives rotation of the flagellum; about 1000 protons are translocated per rotation of the flagellum. How this actually occurs is not yet known. However, a "proton turbine" model has been proposed to explain the results of experiments on flagellar function. In this model, protons flowing through channels in the stator exert electrostatic forces on helically arranged charges on the rotor proteins. Attractions between positive and negative charges cause the basal body to rotate as protons flow though the stator (Figure 4.47*b*). **www.microbiologyplace.com Online Tutorial 4.1: The Prokaryotic Flagellum**

Archaeal Flagella

Flagellar motility is widespread among *Archaea*, as representatives of all of the major physiological and phylogenetic groups (∞ Chapter 17) are capable of swimming motility. Archaeal flagella are significantly thinner than bacterial flagella, measuring only 10–14 nm in width (**Figure 4.48**). Archaeal flagella clearly rotate as they do in *Bacteria*, but the structure of the archaeal flagellar motor is not yet known. Recall that the components of the cell wall that play an important role in anchoring the flagellum in gram-negative *Bacteria*—peptidoglycan and the outer membrane in particular (Figure 4.47*a*)—are absent from *Archaea* (Section 4.8), and thus differences in the archaeal structure should be expected.

Unlike the situation in most *Bacteria* in which a single flagellin protein is present in the flagellar filament, several different flagellin proteins exist in *Archaea*. In addition, unlike those of most *Bacteria*, some archaeal flagellins are glycoproteins. The amino acid sequences of archaeal flagellins bear no relationship to that of bacterial flagellin. However, archaeal

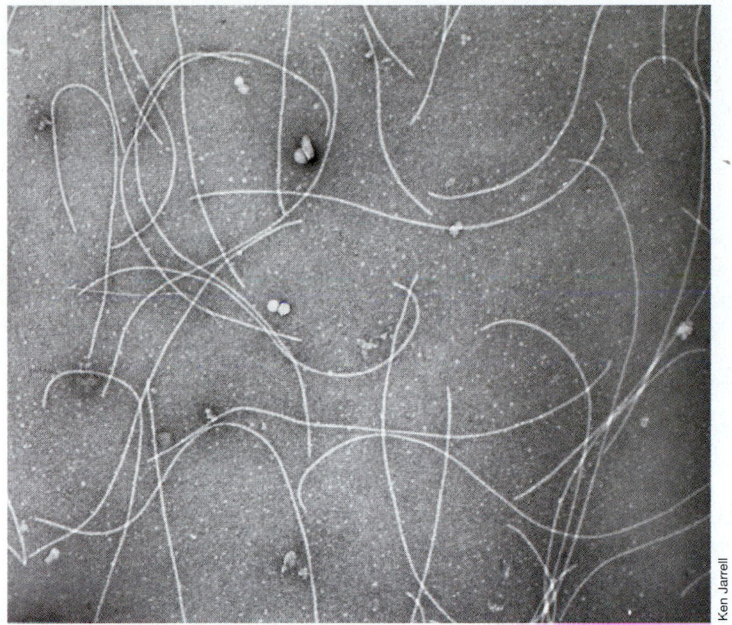

Figure 4.48 Archaeal flagella. Transmission electron micrograph of flagella isolated from cells of the methanogen *Methanococcus maripaludis.* A single flagellum is about 12 nm wide.

flagellins share certain molecular features with bacterial type IV pili, which function in other forms of motility (Sections 4.9 and 4.14). Studies of swimming cells of the extreme halophile *Halobacterium* show that they swim at speeds only about one-tenth that of cells of *Escherichia coli*. Whether this is typical for *Archaea* is unknown, but the significantly smaller diameter of the archaeal flagellum compared to the bacterial flagellum likely reduces the torque and, ultimately, the power of the flagellar motor such that slower swimming speeds would not be surprising.

Flagellar Synthesis

Several genes are required for flagellar synthesis and subsequent motility in *Bacteria*. In *E. coli* and *Salmonella typhimurium*, where studies have been most extensive, over 50 genes are involved in motility. These genes have several functions, including encoding structural proteins of the flagellar apparatus, export of flagellar components through the cytoplasmic membrane to the outside of the cell, and regulation of the many biochemical events surrounding the synthesis of new flagella.

A flagellar filament grows not from its base, as does an animal hair, but from its tip. The MS ring is synthesized first and inserted into the cytoplasmic membrane. Then other anchoring proteins are synthesized along with the hook before the filament forms (**Figure 4.49**). Flagellin molecules synthesized in the cytoplasm pass up through a 3-nm channel inside the filament and add on at the terminus to form the mature flagellum. At the end of the growing flagellum a protein "cap" exists. Cap proteins assist flagellin molecules that have diffused through the channel to organize at the flagellum termini to form new filament (Figure 4.49). Approximately 20,000 flagellin molecules are needed to make one filament. The flagellum grows more-or-less continuously until it reaches its final length. Broken flagella still rotate and can be repaired with new flagellin units passed through the filament channel to replace the lost ones.

Cell Speed and Motion

Flagella do not rotate at a constant speed but instead can increase or decrease their speed in relation to the strength of the proton motive force. They can rotate at up to 300 revolutions per second and can move cells through a liquid at speeds of up to 60 cell lengths/second (sec). Although this is only about 0.00017 kilometer/hour (km/h), when comparing this speed with that of higher organisms in terms of the number of lengths moved per second, it is extremely fast. The fastest animal, the cheetah, moves at a maximum rate of about 110 km/h, but this represents

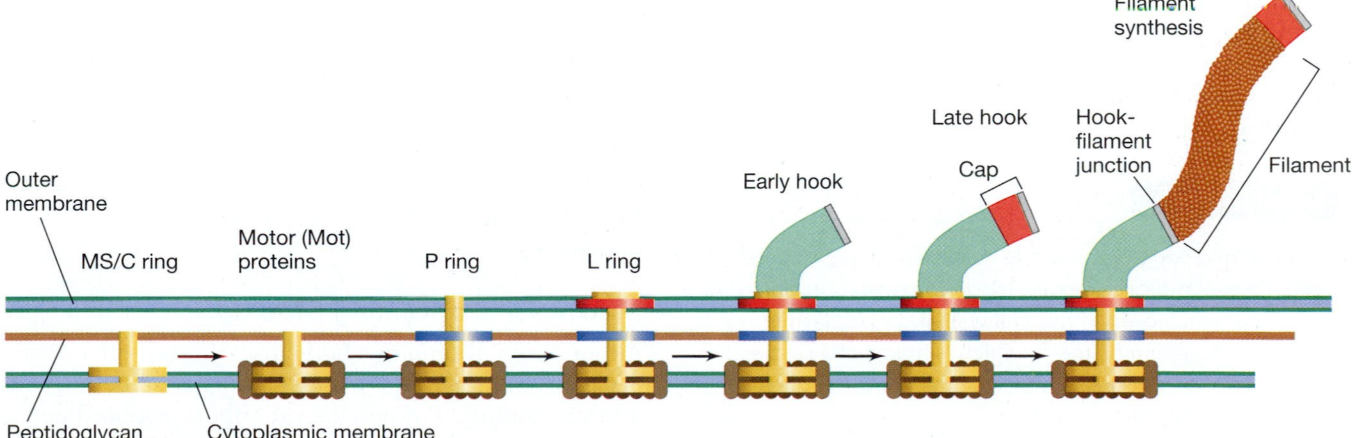

Figure 4.49 Flagella biosynthesis. Synthesis begins with assembly of MS and C rings in the cytoplasmic membrane. Then the other rings, the hook and the cap, form. At this point, flagellin protein flows through the hook to form the filament. Flagellin molecules are guided into position by cap proteins to ensure that the growing filament develops evenly.

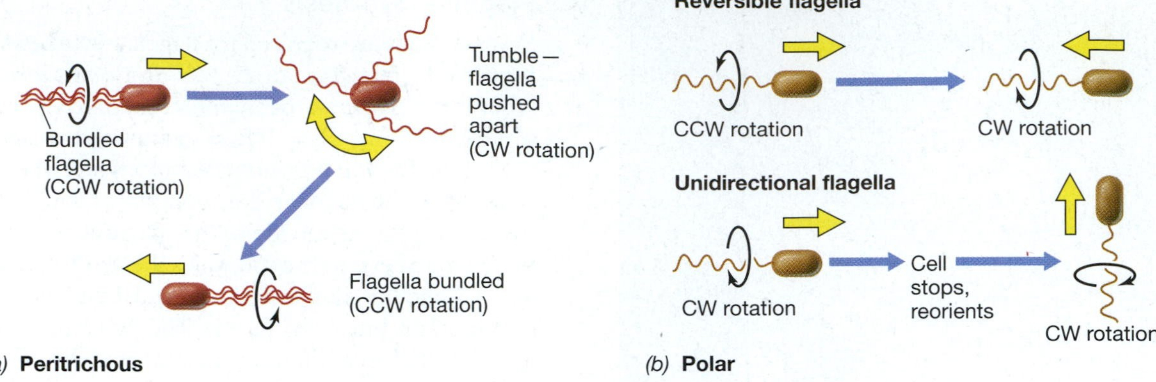

Figure 4.50 Movement in peritrichously and polarly flagellated prokaryotes. *(a)* Peritrichous: Forward motion is imparted by all flagella rotating counterclockwise (CCW) in a bundle. Clockwise (CW) rotation causes the cell to tumble, and then a return to counterclockwise rotation leads the cell off in a new direction. *(b)* Polar: Cells change direction by reversing flagellar rotation (thus pulling instead of pushing the cell) or, with unidirectional flagella, by stopping periodically to reorient, and then moving forward by clockwise rotation of its flagella. The yellow arrows show the direction the cell is traveling.

only about 25 body lengths/sec. Thus, when size is accounted for, prokaryotic cells swimming at 50–60 lengths/sec are actually moving faster than larger organisms!

The swimming motions of polarly and lophotrichously flagellated organisms differ from those of peritrichously flagellated organisms, and these can be distinguished by observing swimming cells under the microscope (**Figure 4.50**). Peritrichously flagellated organisms typically move in a straight line in a slow, deliberate fashion. Polarly flagellated organisms, on the other hand, move more rapidly, spinning around and seemingly dashing from place to place. The different behavior of flagella on polar and peritrichous organisms, including differences in reversibility of the flagellum, is illustrated in Figure 4.50.

Swimming speed is a genetically governed property because different motile species, even different species that have the same cell size, can swim at different maximum speeds. Moreover, when assessing the capacity of an organism for swimming motility and swimming speed, observations should only be made on young cultures. In old cultures, otherwise motile cells often stop swimming and the culture may appear to be nonmotile.

4.13 MiniReview

Motility in most microorganisms is due to flagella. In prokaryotes the flagellum is a complex structure made of several proteins, most of which are anchored in the cell wall and cytoplasmic membrane. The flagellum filament, which is made of a single kind of protein, rotates at the expense of the proton motive force, which drives the flagellar motor.

- How do flagella of *Bacteria* and *Archaea* compare?
- How does a bacterial flagellum move a cell forward? What is flagellin and where is it found?
- How does polar flagellation differ from peritrichous flagellation?

4.14 Gliding Motility

Some prokaryotes are motile but lack flagella. These nonswimming yet motile bacteria move across solid surfaces in a process called *gliding*. Unlike flagellar motility, in which cells stop and then start off in a different direction, gliding motility is a slower and smoother form of movement and typically occurs along the long axis of the cell. Gliding motility is widely distributed among *Bacteria* but has been well studied in only a few groups. The gliding movement itself—up to 10 μm/sec in some gliding bacteria—is considerably slower than propulsion by flagella but still offers the cell a means of moving about its habitat.

Gliding prokaryotes are filamentous or rod-shaped cells (**Figure 4.51**), and the gliding process requires that the cells be in contact with a solid surface. The morphology of colonies of a typical gliding bacterium (colonies are masses of bacterial cells that form from successive cell divisions of a single cell, Section 5.3) are distinctive, because cells glide out and move away from the center of the colony (Figure 4.51*c*). Perhaps the most well-known gliding bacteria are the filamentous cyanobacteria (Figure 4.51*a, b*; Section 16.7), certain gram-negative *Bacteria*, such as *Myxococcus* and other myxobacteria (Section 15.17), and species of *Cytophaga* and *Flavobacterium* (Figure 4.51*c, d*; Sections 16.12 and 16.14).

Mechanisms of Gliding Motility

Although no gliding mechanism is thoroughly understood, it is clear that more than one mechanism can be responsible for gliding motility. Cyanobacteria (phototrophic bacteria, Figure 4.51*a, b*) glide by secreting a polysaccharide slime on the outer surface of the cell. The slime contacts both the cell surface and the solid surface against which the cell moves. As the excreted slime adheres to the surface, the cell is pulled along. This mechanism is supported by the identification of

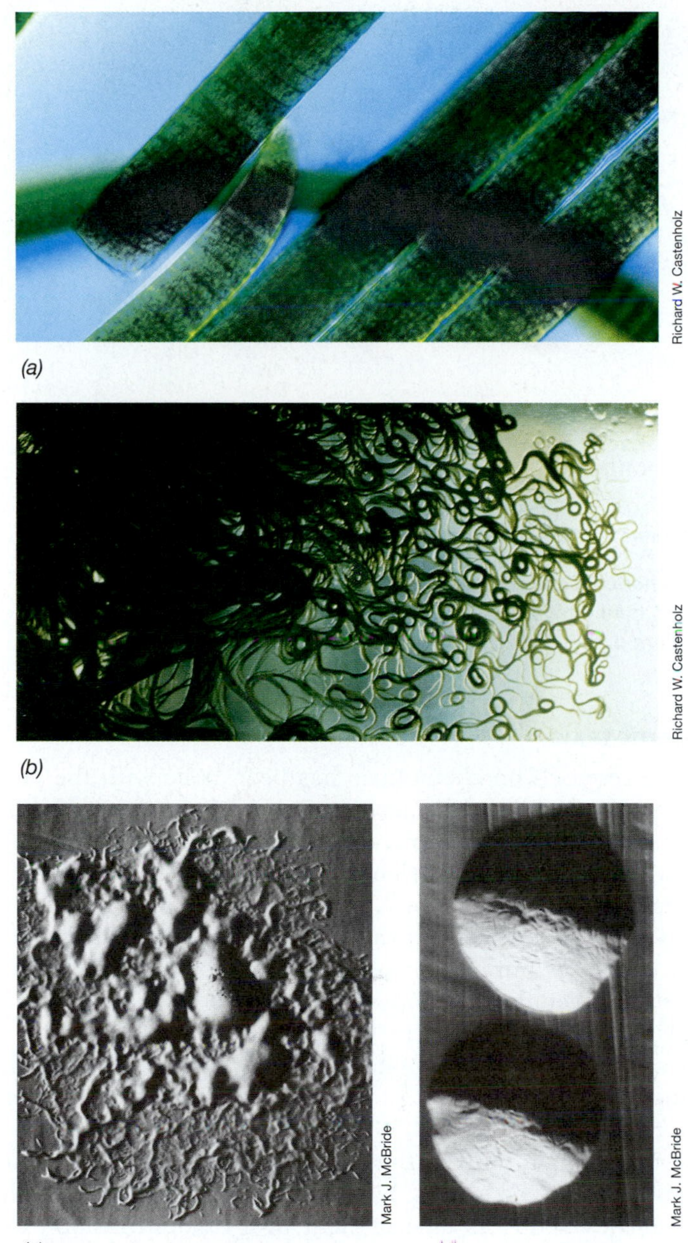

Figure 4.51 **Gliding bacteria.** *(a, b)* The large filamentous cyanobacterium *Oscillatoria princeps. (a)* A cell is about 35 μm wide. *(b)* Filaments of *O. princeps* gliding on an agar surface. *(c, d)* The gliding bacterium *Flavobacterium johnsoniae. (c)* Masses of cells gliding away from the center of the colony (the colony is about 2.7 mm wide). *(d)* Nongliding mutant strain showing typical colony morphology of nongliding bacteria (the colonies are 0.7–1 mm in diameter). See also Figure 4.52.

slime-excreting pores on the cell surface of gliding filamentous cyanobacteria. The nonphototrophic gliding bacterium *Cytophaga* also moves at the expense of slime excretion, rotating along its long axis as it does.

As previously mentioned, cells capable of "twitching motility," also display a form of gliding motility using a mechanism by which repeated extension and retraction of type IV pili propel the cell along a surface (Section 4.9). The gliding

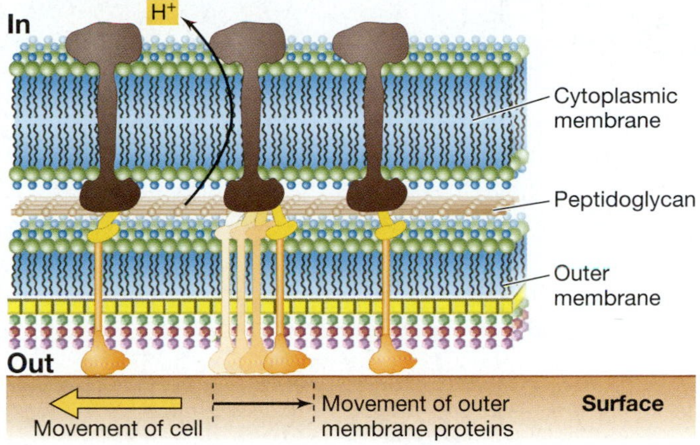

Figure 4.52 **Gliding motility in *Flavobacterium johnsoniae*.** Tracks (yellow) are thought to exist in the peptidoglycan that connect cytoplasmic proteins (brown) to outer membrane proteins (orange) and propel the outer membrane proteins along the solid surface. Note that the outer membrane proteins and the cell proper move in opposite directions.

myxobacterium *Myxococcus xanthus* has two forms of gliding motility. One form is driven by type IV pili whereas the other is distinct from either the type IV pili or the slime extrusion methods. In this form of *M. xanthus* motility a protein adhesion complex is formed at one pole of the rod-shaped cell and remains at a fixed position on the surface as the cell glides forward. This means that the adhesion complex moves in the direction opposite that of the cell, presumably fueled by some sort of cytoplasmic motility engine perhaps linked to the cell cytoskeleton (Section 6.3). These different forms of motility can be expressed at the same time and are somehow coordinated by the cell, presumably in response to various signals from the environment (Section 4.15).

Neither slime extrusion nor twitching is the mechanism of gliding in other gliding bacteria. In *Flavobacterium johnsoniae* (Figure 4.51*c*), for example, no slime is excreted and the cells lack type IV pili. Instead, the movement of proteins on the cell surface may be the mechanism of gliding in this organism. Specific motility proteins anchored in the cytoplasmic and outer membranes are thought to propel cells of *F. johnsoniae* forward by a ratcheting mechanism (**Figure 4.52**). Movement of gliding-specific proteins in the cytoplasmic membrane is driven by energy from the proton motive force (Section 5.12) that is somehow transmitted to gliding-specific proteins in the outer membrane. It is thought that movement of these proteins against the solid surface literally pulls the cell forward (Figure 4.52).

Like other forms of motility, gliding motility has significant ecological relevance. Gliding allows a cell to exploit new resources and to interact with other cells. In the latter regard, it is of interest that myxobacteria, such as *Myxococcus xanthus*, have a very social and cooperative lifestyle. In these bacteria gliding motility may play an important role in cell-to-cell interactions that are necessary to complete a very elaborate life cycle (Section 15.17).

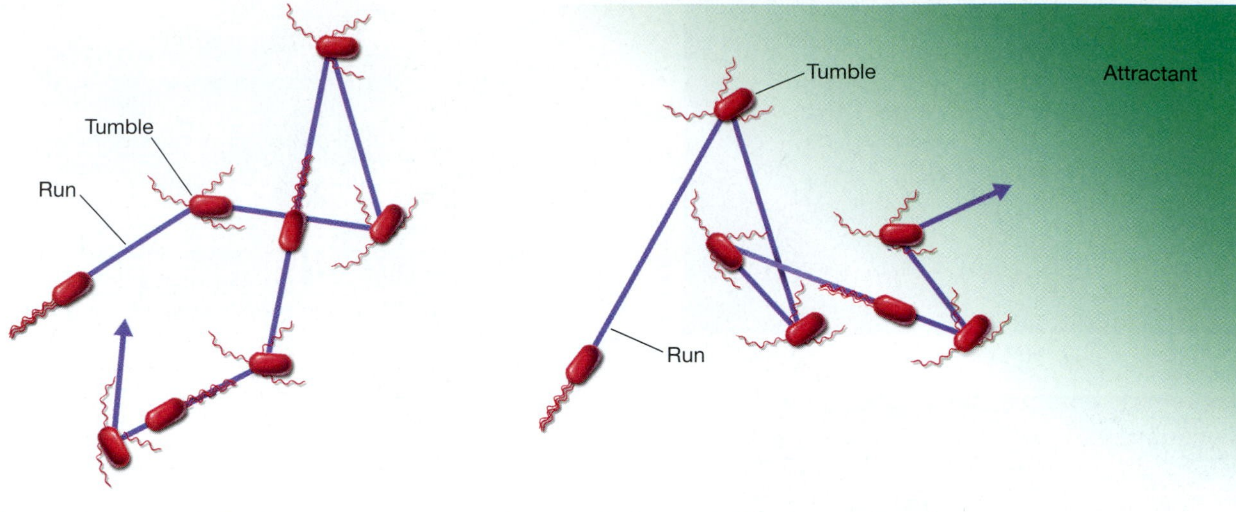

Figure 4.53 Chemotaxis in a peritrichously flagellated bacterium such as *Escherichia coli*. *(a)* In the absence of a chemical attractant the cell swims randomly in runs, changing direction during tumbles. *(b)* In the presence of an attractant runs become biased, and the cell moves up the gradient of the attractant.

4.14 MiniReview

Bacteria that move by gliding motility do not employ rotating flagella but instead creep along a solid surface by any of several possible mechanisms.

- How does gliding motility differ from swimming motility in both mechanism and requirements?
- Contrast the mechanism of gliding motility in a filamentous cyanobacterium and in *Flavobacterium*.

4.15 Cell Motion as a Behavioral Response: Microbial Taxes

Prokaryotes often encounter gradients of physical or chemical agents in nature and have evolved means to respond to these gradients by moving either toward or away from the agent. Such a directed movement is called a *taxis* (plural, taxes). **Chemotaxis**, a response to chemicals, and **phototaxis**, a response to light, are two well-studied taxes. Here we discuss these taxes in a general way. In Section 9.7 we examine the mechanism of chemotaxis and its regulation in *Escherichia coli* as a model for all prokaryotic taxes.

Chemotaxis has been well studied in swimming bacteria, and much is known at the genetic level concerning how the chemical state of the environment is communicated to the flagellar assembly. Our discussion here will thus deal solely with swimming bacteria. However, some gliding bacteria (Section 4.14) are also chemotactic, and there are phototactic movements in filamentous cyanobacteria (Figure 4.51*b*). In addition, although they reside in a different evolutionary domain, many species of *Archaea* are also chemotactic, and many of the proteins that control chemotaxis in *Bacteria* are present in these *Archaea* as well.

Chemotaxis

Much research on chemotaxis has been done with the peritrichously flagellated bacterium *E. coli*. To understand how chemotaxis affects the behavior of *E. coli,* consider the situation in which a cell experiences a gradient of some chemical in its environment (**Figure 4.53**). In the absence of a gradient, cells move in a random fashion that includes *runs,* in which the cell is swimming forward in a smooth fashion, and *tumbles,* when the cell stops and jiggles about (Figure 4.53*a*). During forward movement in a run, the flagellar motor rotates counterclockwise. When flagella rotate clockwise, the bundle of flagella pushes apart, forward motion ceases, and the cells tumble (Figure 4.50).

Following a tumble, the direction of the next run is random (Figure 4.53*a*). Thus, by means of runs and tumbles, the cell moves about its environment in a random fashion but does not really go anywhere. However, if a gradient of a chemical attractant is present, these random movements become biased. As the organism senses that it is moving toward higher concentrations of the attractant, runs become longer and tumbles are less frequent. The result of this behavioral response is that the organism moves up the concentration gradient of the attractant (Figure 4.53*b*). If the organism senses a repellent, the same general mechanism applies, although in this case it is the decrease in concentration of the repellent (rather than the increase in concentration of an attractant) that promotes runs.

Prokaryotic cells are too small to sense a gradient of a chemical along the length of a single cell. Instead, while moving, the cell monitors its environment, comparing its chemical or physical state with that sensed a few moments before. Bacterial cells are thus responding to *temporal* rather than *spatial* differences in the concentration of a chemical as they swim. This information is fed through an elaborate cascade of proteins that eventually affect the direction of rotation of the flagellar motor. The attractants and repellants are sensed by a series of membrane proteins called *chemoreceptors*. These proteins bind the chemicals and begin the process of sensory transduction to the flagellum (∞ Section 9.7). In a way, chemotaxis can thus be considered a sensory response system, analogous to sensory responses in the nervous system of animals.

Chemotaxis in Polarly Flagellated Bacteria

Chemotaxis in polarly flagellated cells shows similarities to and differences from that in peritrichously flagellated cells such as *E. coli*. Many polarly flagellated bacteria, such as *Pseudomonas* species, can reverse the direction of rotation of their flagella and in so doing reverse their direction of movement (Figure 4.50*b*). However, some polarly flagellated bacteria, such as the phototrophic bacterium *Rhodobacter sphaeroides,* have flagella that rotate only in a clockwise direction. How do such cells change direction, and are they chemotactic?

In cells of *R. sphaeroides,* which have only a single flagellum inserted subpolarly, rotation of the flagellum stops periodically. When it stops, the cell becomes reoriented in a random way by Brownian motion. As the flagellum begins to rotate again, the cell moves in a new direction. Nevertheless, cells of *R. sphaeroides* are strongly chemotactic to certain organic compounds and also show tactic responses to oxygen and light. *R. sphaeroides* cannot reverse its flagellar motor and tumble as *E. coli* can, but there is a similarity in that the cells maintain runs as long as they sense an increasing concentration of attractant; movement ceases if the cells sense a decreasing concentration of attractant. By random reorientation, a cell eventually finds a path of increasing attractant and maintains a run until either its chemoreceptors are saturated or it begins to sense a decrease in the level of attractant.

Measuring Chemotaxis

Bacterial chemotaxis can be demonstrated by immersing a small glass capillary tube containing an attractant in a suspension of motile bacteria that does not contain the attractant. From the tip of the capillary, a gradient forms into the surrounding medium, with the concentration of chemical gradually decreasing with distance from the tip (**Figure 4.54*a***). When an attractant is present, the bacteria will move toward it, forming a swarm around the open tip (Figure 4.54*c*) with many of the bacteria swimming into the capillary. Of course, because of random movements some bacteria will move into the capillary even if it contains a solution of the same composition as the medium (Figure 4.54*b*). However, when an attractant is present, movement becomes biased. The concentration of bacteria within the capillary can be many times higher than the external concentration. If the capillary is removed after a time period and the cells are counted and compared with a control, attractants and repellants can easily be identified (Figure 4.54*e*).

If the inserted capillary contains a repellent, the concentration of bacteria within the capillary will be considerably

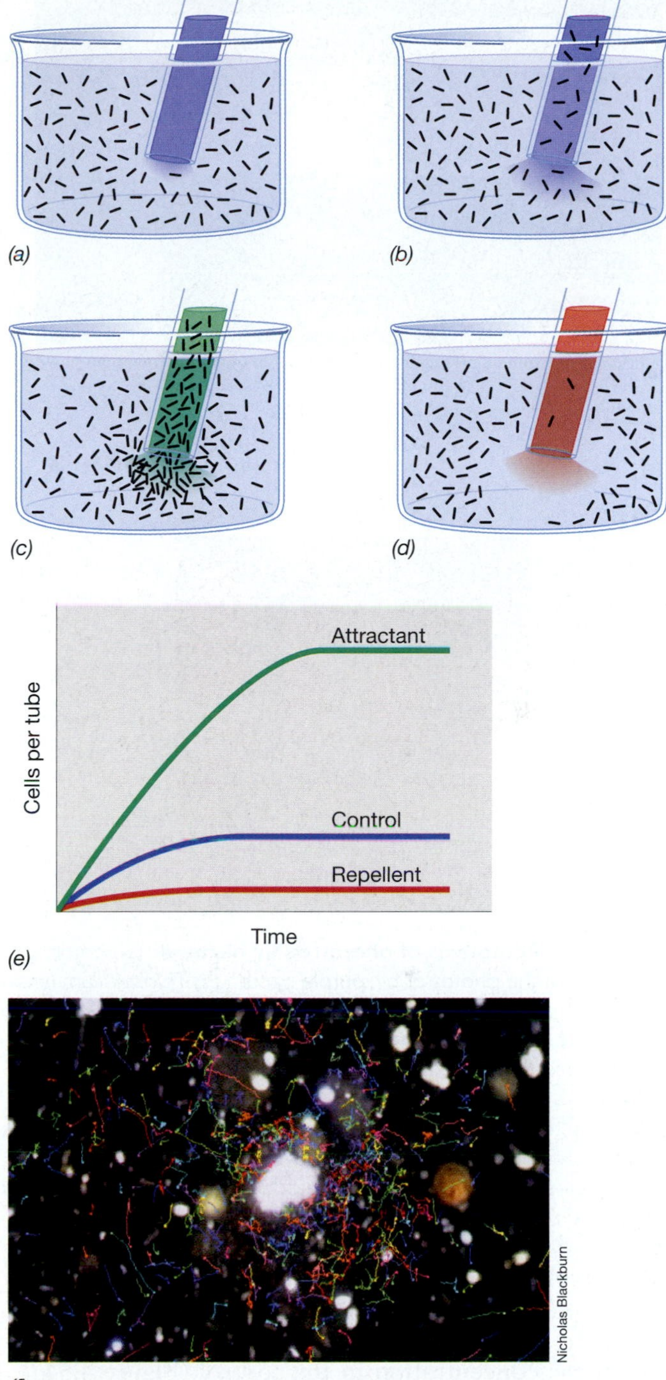

Figure 4.54 Measuring chemotaxis using a capillary tube assay. *(a)* Insertion of the capillary into a bacterial suspension. As the capillary is inserted, a gradient of the chemical begins to form. *(b)* Control capillary contains a salt solution that is neither an attractant nor a repellent. Cell concentration inside the capillary becomes the same as that outside. *(c)* Accumulation of bacteria in a capillary containing an attractant. *(d)* Repulsion of bacteria by a repellent. *(e)* Time course showing cell numbers in capillaries containing various chemicals. *(f)* Tracks of motile bacteria in seawater swarming around an algal cell (large white spot, center) photographed with a tracking video camera system attached to a microscope. Note how the bacterial cells are showing positive aerotaxis by moving toward the oxygen-producing algal cell. The average velocity of the cells was about 25 μm/sec. The alga is about 60 μm in diameter.

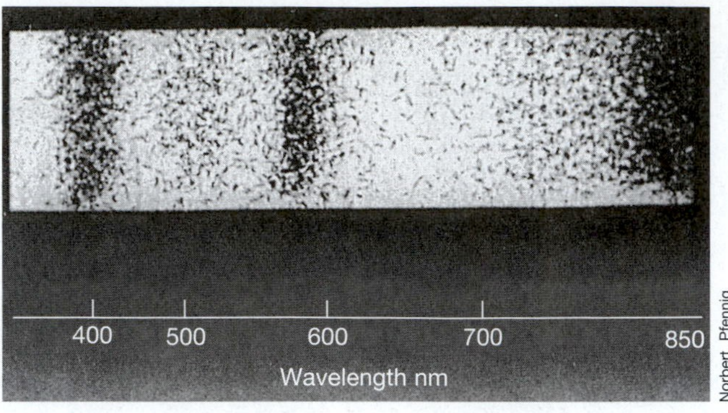

(a)

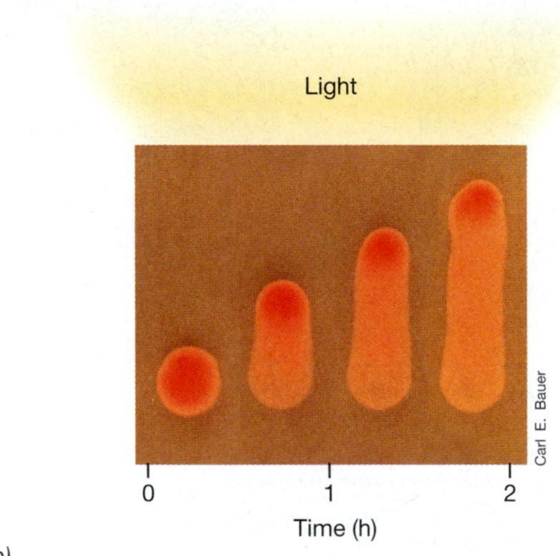

(b)

Figure 4.55 Phototaxis of phototrophic bacteria. *(a)* Scotophobic accumulation of the phototrophic purple bacterium *Thiospirillum jenense* at wavelengths of light at which its pigments absorb. A light spectrum was displayed on a microscope slide containing a dense suspension of the bacteria; after a period of time, the bacteria had accumulated selectively and the photomicrograph was taken. The wavelengths at which the bacteria accumulated are those at which the photosynthetic pigment bacteriochlorophyll *a* absorbs (compare with Figure 20.3b). *(b)* Phototaxis of an entire colony of the purple phototrophic bacterium *Rhodospirillum centenum*. These strongly phototactic cells move in unison toward the light source at the top. See Figure 4.45 for electron micrographs of *R. centenum* cells.

less than the concentration in the control (Figure 4.54*d*). In this case, the cell senses an increasing gradient of repellent, and the appropriate chemoreceptors affect flagellar rotation to move the cell away from the repellent (Figure 4.54*d*). Using the capillary method, it is possible to screen chemicals to see if they are attractants or repellents for a given bacterium.

Chemotaxis can also be observed visually under a microscope. Using a video camera that captures the position of bacterial cells with time and shows the motility tracks of each cell, it is possible to see the chemotactic movements of cells (Figure 4.54*f*). This method has been adapted to studies of chemotaxis of bacteria in natural environments. In nature it is thought that the major chemotactic agents for bacteria are nutrients excreted from larger microbial cells or from live or dead macroorganisms. Algae, for example, produce both organic compounds and oxygen (O_2, from photosynthesis) that can trigger chemotactic movements of bacteria toward an algal cell (Figure 4.54*f*).

Phototaxis

Many phototrophic microorganisms can move toward light, a process called *phototaxis*. The advantage of phototaxis for a phototrophic organism is that it allows the organism to orient itself most efficiently to receive light for photosynthesis. This can be seen if a light spectrum is spread across a microscope slide on which there are motile phototrophic bacteria. On such a slide the bacteria accumulate at wavelengths at which their photosynthetic pigments absorb (**Figure 4.55*a***; Sections 20.1–20.5 discuss photosynthetic pigments).

Two different light-mediated taxes are observed in phototrophic bacteria. One, called *scotophobotaxis*, can be observed only microscopically and occurs when a phototrophic bacterium happens to swim outside the illuminated field of view of the microscope into darkness. Entering darkness negatively affects the energy state of the cell and signals it to tumble, reverse direction, and once again swim in a run, thus reentering the light.

True phototaxis differs from scotophobotaxis. Phototaxis is analogous to chemotaxis except the attractant in this case is light instead of a chemical. In some species, such as the highly motile phototrophic organism *Rhodospirillum centenum* (Figure 4.45), entire colonies of cells show phototaxis and move in unison toward the light (Figure 4.55*b*).

Several components of the regulatory system that govern chemotaxis also control phototaxis. This conclusion has emerged from the study of mutants of phototrophic bacteria defective in phototaxis; such mutants show defective chemotaxis systems as well. A *photoreceptor*, a protein that functions similar to a chemoreceptor but senses a gradient of light instead of chemicals, is the initial sensor in the phototaxis response. The photoreceptor then interacts with the same cytoplasmic proteins that control flagellar rotation in chemotaxis, maintaining the cell in a run if it is swimming toward an increasing intensity of light. Thus, although the stimulus in chemotaxis and phototaxis differs—chemicals versus light—the same molecular machinery is employed in the cytoplasm to process both signals. We discuss this cytoplasmic machinery in detail in Section 9.7.

Other Taxes

Other bacterial taxes, such as movement toward or away from oxygen (*aerotaxis*, see Figure 4.54*f*) or toward or away from conditions of high ionic strength (*osmotaxis*), are known among various swimming prokaryotes. Like chemotaxis and phototaxis, these are simple forms of behavior, and it appears that a common molecular mechanism controls each of these behavioral responses. Cells periodically sample their environment for chemicals, light, oxygen, salt, or other substances, and then process the results through a network of proteins that ultimately control the direction of flagellar rotation. In some gliding cyanobacteria (Figure 4.51*a,b*), an unusual taxis, *hydrotaxis,* also occurs. This allows gliding cyanobacteria inhabiting dry environments, such as soils, to glide up a gradient of increasing hydration.

From our coverage of microbial taxes it should be clear that motile prokaryotes do not just swim around at random, but instead are keenly attuned to the chemical and physical state of their habitat. Indeed, regular and periodic "assays" of their environment are probably a key to ecological success for prokaryotes. By being able to move toward or away from various stimuli, prokaryotic cells have a better chance of competing successfully for resources and avoiding the harmful effects of substances that could damage or kill them.

4.15 MiniReview

Motile bacteria can respond to chemical and physical gradients in their environment. In swimming prokaryotes, movement of the cell can be biased either toward or away from a stimulus by control of runs and tumbles. Tumbles are controlled by the direction of rotation of the flagellum, which in turn is controlled by a network of sensory and response proteins.

- Define the word chemotaxis. How does chemotaxis differ from aerotaxis?
- What causes a run versus a tumble?
- How does scotophobotaxis differ from phototaxis?

Review of Key Terms

ABC (ATP-binding cassette) transport system a membrane transport system consisting of three proteins, one of which hydrolyzes ATP, which transports specific nutrients into the cell

Basal body the "motor" portion of the bacterial flagellum, embedded in the cytoplasmic membrane and wall

Capsule a polysaccharide or protein outermost layer, usually rather slimy, present on some bacteria

Chemotaxis directed movement of an organism toward (positive chemotaxis) or away from (negative chemotaxis) a chemical gradient

Cytoplasmic membrane the permeability barrier of the cell, separating the cytoplasm from the environment

Dipicolinic acid a substance unique to endospores that confers heat resistance on these structures

Endospore a highly heat-resistant, thick-walled, differentiated structure produced by certain gram-positive *Bacteria*

Flagellum a long, thin cellular appendage capable of rotation and responsible for swimming motility in prokaryotic cells

Gas vesicles gas-filled cytoplasmic structures bounded by protein and conferring buoyancy on cells

Gram-negative a prokaryotic cell whose cell wall contains small amounts of peptidoglycan, and an outer membrane, containing lipopolysaccharide, lipoprotein, and other complex macromolecules

Gram-positive a prokaryotic cell whose cell wall consists chiefly of peptidoglycan and lacks the outer membrane of gram-negative cells

Gram stain A differential staining procedure that stains cells either purple (gram-positive cells) or pink (gram-negative cells)

Group translocation an energy-dependent transport system in which the substance transported is chemically modified during the transport process

Lipopolysaccharide (LPS) a combination of lipid with polysaccharide and protein that forms the major portion of the outer membrane in gram-negative *Bacteria*

Magnetosome a particle of magnetite (Fe_3O_4) organized into a nonunit membrane-enclosed structure in the cytoplasm of magnetotactic *Bacteria*

Morphology the *shape* of a cell—rod, coccus, spirillum, and so on

Outer membrane a phospholipid- and polysaccharide-containing unit membrane that lies external to the peptidoglycan layer in cells of gram-negative *Bacteria*

Peptidoglycan a polysaccharide composed of alternating repeats of *N*-acetylglucosamine and *N*-acetylmuramic acid arranged in adjacent layers and cross-linked by short peptides

Periplasm a gel-like region between the outer surface of the cytoplasmic membrane and the inner surface of the lipopolysaccharide layer of gram-negative *Bacteria*

Peritrichous flagellation flagella located in many places around the surface of the cell

Phototaxis movement of an organism toward light

Pili thin, filamentous structures that extend from the surface of a cell and, depending on type, facilitate cell attachment, genetic exchange, or twitching motility

Polar flagellation having flagella emanating from one or both poles of the cell

Poly-β-hydroxybutyrate (PHB) a common storage material of prokaryotic cells consisting of a polymer of β-hydroxybutyrate or another β-alkanoic acid or mixtures of β-alkanoic acids

Protoplast an osmotically protected cell whose cell wall has been removed

S-layer an outermost cell surface layer composed of protein or glycoprotein present on some *Bacteria* and *Archaea*

Simple transport system a transporter that consists of only a membrane-spanning protein and typically driven by energy from the proton motive force

Sterol a hydrophobic heterocyclic ringed molecule that strengthens the cytoplasmic membrane of eukaryotic cells and a few prokaryotes

Teichoic acid a phosphorylated polyalcohol found in the cell wall of some gram-positive *Bacteria*

Review Questions

1. What are the major morphologies of prokaryotes? Draw cells for each morphology you list (Section 4.1).
2. How large can a prokaryote be? How small? Why is it that we likely know the lower limit more accurately than the upper limit? What are the dimensions of the rod-shaped bacterium *Escherichia coli* (Section 4.2)?
3. Describe in a single sentence the structure of a unit membrane (Section 4.3).
4. Describe a major chemical difference between membranes of *Bacteria* and *Archaea* (Section 4.3).
5. Explain in a single sentence why ionized molecules do not readily pass through the cytoplasmic membrane of a cell. How do such molecules get through the cytoplasmic membrane (Sections 4.4 and 4.5)?
6. Cells of *Escherichia coli* take up lactose via lac permease, glucose via the phosphotransferase system, and maltose via an ABC-type transporter. For each of these sugars describe: (1) the components of their transport system and (2) the source of energy that drives the transport event (Section 4.5).
7. Why is the bacterial cell wall rigid layer called peptidoglycan? What are the chemical reasons for the rigidity that is conferred on the cell wall by the peptidoglycan structure (Section 4.6)?
8. Why is sucrose able to stabilize bacterial cells from lysis by lysozyme (Section 4.6)?
9. List several functions of the outer membrane in gram-negative *Bacteria*. What is the chemical composition of the outer membrane (Section 4.7)?
10. What cell wall polysaccharide common in *Bacteria* is absent from *Archaea*? What is unusual about S-layers compared to other cell walls of prokaryotes ? What types of cell walls are found in *Archaea* (Section 4.8)?
11. What function(s) do non-cell wall polysaccharide layers have in prokaryotes (Section 4.9)?
12. What types of cytoplasmic inclusions are formed by prokaryotes? How does an inclusion of poly-β-hydroxybutyric acid differ from a magnetosome in composition and metabolic role (Section 4.10)?
13. What is the function of gas vesicles? How are these structures made such that they can remain gas tight (Section 4.11)?
14. In a few sentences, indicate how the bacterial endospore differs from the vegetative cell in structure, chemical composition, and ability to resist extreme environmental conditions (Section 4.12).
15. Define the following terms: mature endospore, vegetative cell, and germination (Section 4.12).
16. Describe the structure and function of a bacterial flagellum. What is the energy source for the flagellum (Section 4.13)? How do the flagella of *Bacteria* differ from those of *Archaea* in both size and composition?
17. How does the mechanism and energy requirements for motility in *Flavobacterium* differ from that in *Escherichia coli* (Sections 4.13 and 4.14)?
18. In a few sentences, write an explanation describing how a motile bacterium is able to sense the direction of an attractant and move toward it (Section 4.15).

Application Questions

1. Calculate the surface-to-volume ratio of a spherical cell 15 μm in diameter and a cell 2 μm in diameter. What are the consequences of these differences in surface-to-volume ratio for cell function?
2. Assume you are given two cultures, one of a species of gram-negative *Bacteria* and one of a species of *Archaea*. Other than by phylogenetic analyses, discuss at least four different ways you could tell which culture was which.
3. Calculate the amount of time it would take a cell of *Escherichia coli* (1 × 3 μm) swimming at maximum speed (60 cell lengths per second) to travel all the way up a 3-cm-long capillary tube containing a chemical attractant.
4. Assume you are given two cultures of rod-shaped bacteria, one gram positive and the other gram negative. How could you differentiate them using (a) light microscopy; (b) electron microscopy; (c) chemical analyses of cell walls; and (d) phylogenetic analyses?

5

Nutrition, Culture, and Metabolism of Microorganisms

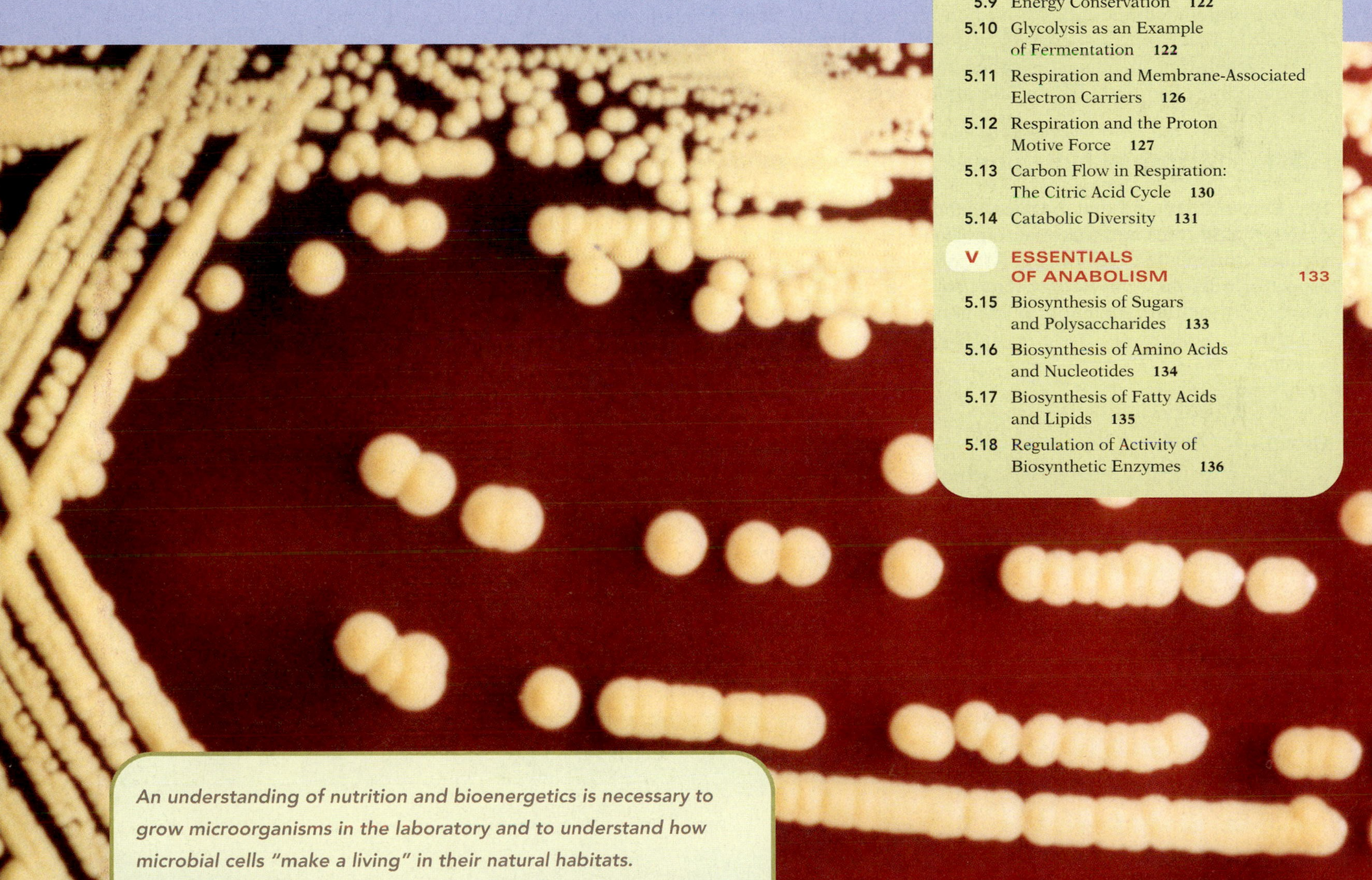

An understanding of nutrition and bioenergetics is necessary to grow microorganisms in the laboratory and to understand how microbial cells "make a living" in their natural habitats.

I NUTRITION AND CULTURE OF MICROORGANISMS

Before a cell can replicate, it must coordinate many different chemical reactions and organize molecules into specific structures. Collectively, these reactions are called **metabolism.** Metabolic reactions are either energy *releasing,* called **catabolic reactions (catabolism),** or energy *requiring,* called **anabolic reactions (anabolism).** Several catabolic and anabolic reactions occur in cells, and we will examine some of the key ones in this and future chapters. However, before we do, we consider how microorganisms are grown in the laboratory and the nutrients they need for growth. Indeed, most of what we know about the metabolism of microorganisms has emerged from the study of laboratory cultures. Our focus here will be on chemoorganotrophic microorganisms, also called *heterotrophs;* these are organisms that use organic compounds for carbon and energy (∞ Section 2.8). Later in this chapter we will briefly consider energy-generating mechanisms other than chemoorganotrophy, including chemolithotrophy and phototrophy.

5.1 Microbial Nutrition

Recall from Chapter 3 that, besides water, cells consist mainly of macromolecules and that macromolecules are polymers of smaller units called monomers. Microbial nutrition is that aspect of microbial physiology that deals with the supply of monomers (or the precursors of monomers) that cells need for growth. Collectively, these required substances are called *nutrients*.

Different organisms need different complements of nutrients, and not all nutrients are required in the same amounts. Some nutrients, called *macronutrients,* are required in large amounts, while others, called *micronutrients,* are required in trace amounts. All microbial nutrients originate from the chemical elements; however, just a handful of elements dominate biology (**Figure 5.1**). Although cells consist mainly of H, O, C, N, P, and S, at least 50 of the chemical elements are metabolized in some way by microorganisms (Figure 5.1). We begin our study of microbial nutrition with the key macronutrient elements, *carbon* and *nitrogen* (Figure 5.1 and **Table 5.1**).

Carbon and Nitrogen

All cells require carbon, and most prokaryotes require organic compounds as their source of carbon. On a dry weight basis, a typical cell is about 50% carbon, and carbon is the major element in all classes of macromolecules. Bacteria can assimilate organic compounds and use them to make new cell material. Amino acids, fatty acids, organic acids, sugars, nitrogen bases, aromatic compounds, and countless other organic compounds can be used by one or another bacterium. By contrast, some microorganisms are *autotrophs* and able to build all of their cellular structures from carbon dioxide (CO_2). The energy needed to support autotrophy is obtained from either light or inorganic chemicals.

Following carbon, the next most abundant element in the cell is nitrogen. A bacterial cell is about 12% nitrogen (by dry weight), and nitrogen is a key element in proteins, nucleic acids, and several other cell constituents. In nature, nitrogen is available in both organic and inorganic forms (Table 5.1).

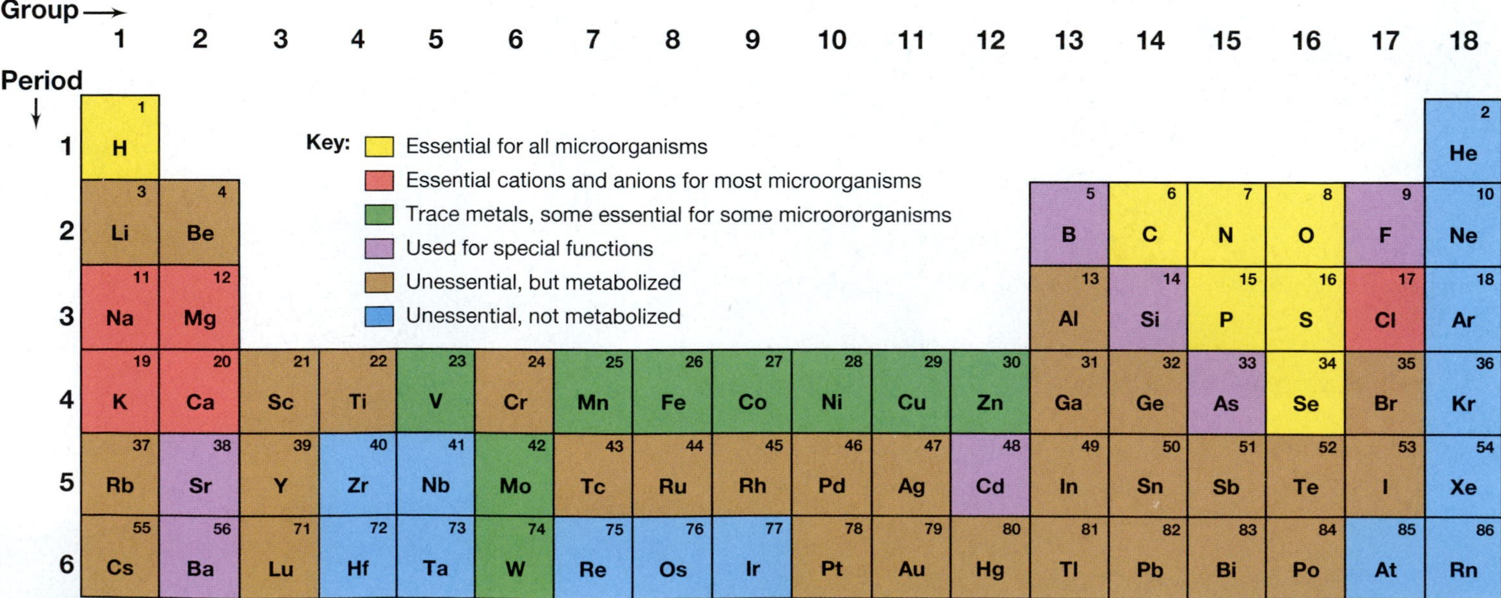

Figure 5.1 A microbial periodic table of the elements. With the exception of uranium, which can be metabolized by some prokaryotes, elements in period 7 or beyond in the complete periodic table of the elements are not metabolized. The atomic number of each element is shown in the upper right corner of each box.

Table 5.1 Macronutrients

Element	*Usual form found in the environment*
Carbon (C)	CO_2, organic compounds
Hydrogen (H)	H_2O, organic compounds
Oxygen (O)	H_2O, O_2, organic compounds
Nitrogen (N)	NH_3, NO_3^-, N_2, organic nitrogen compounds
Phosphorus (P)	PO_4^{3-}
Sulfur (S)	H_2S, SO_4^{2-}, organic S compounds, metal sulfides (FeS, CuS, ZnS, NiS, and so on)
Potassium (K)	K^+ in solution or as various K salts
Magnesium (Mg)	Mg^{2+} in solution or as various Mg salts
Sodium (Na)	Na^+ in solution or as NaCl or other Na salts
Calcium (Ca)	Ca^{2+} in solution or as $CaSO_4$ or other Ca salts
Iron (Fe)	Fe^{2+} or Fe^{3+} in solution or as FeS, $Fe(OH)_3$, or many other Fe salts

However, the bulk of available nitrogen is in inorganic form, either as ammonia (NH_3), nitrate (NO_3^-), or N_2. Virtually all bacteria can use ammonia as their nitrogen source, and many can also use nitrate. However, nitrogen gas (N_2) can satisfy the nitrogen needs of only a few bacteria, the nitrogen-fixing bacteria, discussed in later chapters.

Other Macronutrients: P, S, K, Mg, Ca, Na

After C and N, other essential elements are needed in smaller amounts. Phosphorus occurs in nature in the form of organic and inorganic phosphates and is required by the cell primarily for synthesis of nucleic acids and phospholipids. Sulfur is required because of its structural role in the amino acids cysteine and methionine and because it is present in several vitamins, including thiamine, biotin, and lipoic acid, as well as in coenzyme A. Most sulfur originates from inorganic sources in nature, either sulfate (SO_4^{2-}) or sulfide (HS^-) (Table 5.1).

Potassium is required by all organisms. Many enzymes, including some necessary for protein synthesis, require potassium for activity. Magnesium functions to stabilize ribosomes, membranes, and nucleic acids and is also required for the activity of many enzymes. Calcium helps stabilize cell walls in many microorganisms and plays a key role in the heat stability of endospores (∞ Section 4.12). Sodium is required by some but not all microorganisms and is typically a reflection of the habitat. For example, seawater contains relatively high levels of Na^+, and marine microorganisms usually require sodium for growth. By contrast, freshwater species are usually able to grow in the absence of sodium.

Iron and Trace Metals

Microorganisms require various metals for growth (Figure 5.1). Chief among these is iron, which plays a major role in cellular respiration. Iron is a key component of cytochromes and iron–sulfur proteins involved in electron transport reactions. Under anoxic conditions, iron is generally in the ferrous (Fe^{2+}) form and soluble. However, under oxic conditions, iron is typically in the ferric (Fe^{3+}) form in insoluble minerals. To obtain iron from such minerals, cells produce iron-binding agents called **siderophores** that bind iron and transport it into the cell. One major group of siderophores consists of derivatives of hydroxamic acid, which chelate ferric iron strongly (**Figure 5.2a**). Once the iron–hydroxamate complex has passed into the cell, the iron

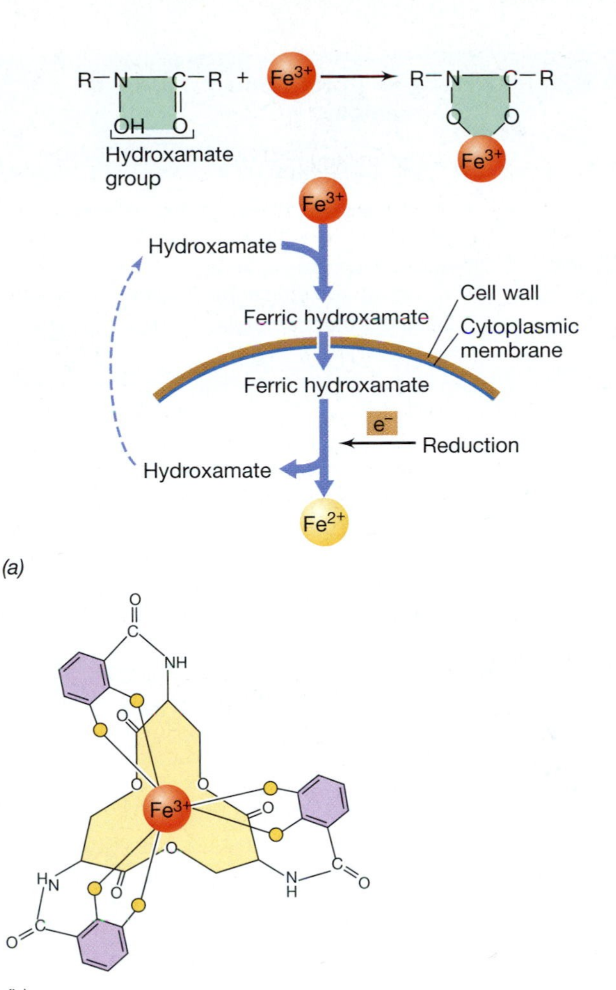

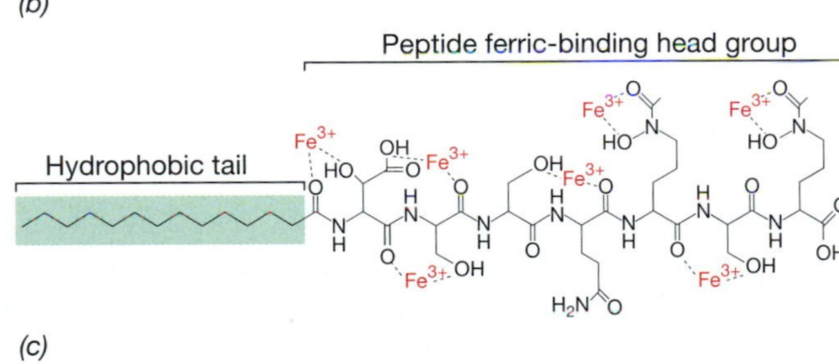

Figure 5.2 Iron-chelating agents produced by microorganisms. *(a)* Hydroxamate. Iron is bound as Fe^{3+} and released inside the cell as Fe^{2+}. The hydroxamate then exits the cell and repeats the cycle. *(b)* Ferric enterobactin of *Escherichia coli*. The oxygen atoms of each catechol molecule are shown in yellow. *(c)* The peptidic and tailed siderophore aquachelin, showing Fe^{3+} binding. The hydrophobic tail assists in threading aquachelin through the membrane into the cell.

Table 5.2 Micronutrients (trace elements) needed by microorganisms[a]

Element	*Cellular function*
Boron (B)	Present in an autoinducer for quorum sensing in bacteria; also found in some polyketide antibiotics
Chromium (Cr)	Required by mammals for glucose metabolism; microbial requirement possible but not proven
Cobalt (Co)	Vitamin B_{12}; transcarboxylase (propionic acid bacteria)
Copper (Cu)	Respiration, cytochrome *c* oxidase; photosynthesis, plastocyanin, some superoxide dismutases
Iron (Fe)[b]	Cytochromes; catalases; peroxidases; iron–sulfur proteins; oxygenases; all nitrogenases
Manganese (Mn)	Activator of many enzymes; present in certain superoxide dismutases and in the water-splitting enzyme in oxygenic phototrophs (photosystem II)
Molybdenum (Mo)	Certain flavin-containing enzymes; some nitrogenases, nitrate reductases, sulfite oxidases, DMSO-TMAO reductases; some formate dehydrogenases
Nickel (Ni)	Most hydrogenases; coenzyme F_{430} of methanogens; carbon monoxide dehydrogenase; urease
Selenium (Se)	Formate dehydrogenase; some hydrogenases; the amino acid selenocysteine
Tungsten (W)	Some formate dehydrogenases; oxotransferases of hyperthermophiles
Vanadium (V)	Vanadium nitrogenase; bromoperoxidase
Zinc (Zn)	Carbonic anhydrase; alcohol dehydrogenase; RNA and DNA polymerases; and many DNA-binding proteins

[a]Not every micronutrient listed is required by all cells; some metals listed are found in enzymes present in only specific microorganisms.
[b]Needed in greater amounts than other trace metals.

is released and the hydroxamate can be excreted and used again for iron transport.

Many other iron-binding agents are known. Some bacteria produce phenolic siderophores called *enterobactins* (Figure 5.2*b*). These siderophores are derivatives of the aromatic compound catechol and have an extremely high binding affinity for iron. Iron is present in the oceans at extremely low levels, often only a few picograms (1 picogram is 10^{-12} g) of iron per milliliter. To obtain this iron, many marine bacteria produce peptide siderophores such as aquachelin (Figure 5.2*c*) that can sequester iron present in these vanishingly small amounts. Peptide siderophores contain a peptide head group that complexes Fe^{3+} and a lipid tail that associates with the cytoplasmic membrane. Once aquachelin binds iron, the lipid tail helps transport it through the cytoplasmic membrane and into the cell.

As important as iron is for cells, some organisms can grow in the absence of iron. For example, the bacteria *Lactobacillus plantarum* and *Borrelia burgdorferii* (the latter is the causative agent of Lyme disease, ∞ Section 35.4) do not contain detectable iron. In these bacteria, Mn^{2+} substitutes for iron as the metal component of enzymes that normally contain Fe^{2+}.

Table 5.3 Growth factors: Vitamins and their functions

Vitamin	*Function*
p-Aminobenzoic acid	Precursor of folic acid
Folic acid	One-carbon metabolism; methyl group transfer
Biotin	Fatty acid biosynthesis; β-decarboxylations; some CO_2 fixation reactions
Cobalamin (B_{12})	Reduction of and transfer of single carbon fragments; synthesis of deoxyribose
Lipoic acid	Transfer of acyl groups in decarboxylation of pyruvate and α-ketoglutarate
Nicotinic acid (niacin)	Precursor of NAD^+ (see Figure 5.11); electron transfer in oxidation-reduction reactions
Pantothenic acid	Precursor of coenzyme A; activation of acetyl and other acyl derivatives
Riboflavin	Precursor of FMN (see Figure 5.16), FAD in flavoproteins involved in electron transport
Thiamine (B_1)	α-Decarboxylations; transketolase
Vitamins B_6 (pyridoxal-pyridoxamine group)	Amino acid and keto acid transformations
Vitamin K group; quinones	Electron transport; synthesis of sphingolipids
Hydroxamates	Iron-binding compounds; solubilization of iron and transport into cell

Many other metals are required or otherwise metabolized by microorganisms (Figure 5.1). Like iron, these micronutrients are called *trace elements*. Micronutrients typically play a role as components of enzymes, the cells' catalysts (Section 5.5). **Table 5.2** lists the major micronutrients and examples of enzymes in which each plays a role.

Growth Factors

Growth factors are organic compounds that share with trace metals the fact that they are required in only small amounts and then only by certain organisms. Growth factors include vitamins, amino acids, purines, and pyrimidines. Although most microorganisms are able to synthesize all of these compounds, some require one or more of them preformed from the environment and thus must be supplied with these compounds when cultured in the laboratory.

Vitamins are the most commonly required growth factors. Most vitamins function as coenzymes, and these are summarized in **Table 5.3**. Vitamin requirements vary among microorganisms, ranging from none to several. Lactic acid bacteria, which include the genera *Streptococcus*, *Lactobacillus*, and *Leuconostoc*, among others (∞ Section 16.1), are renowned for their many vitamin requirements, which are even more extensive than those of humans (see Table 5.4).

Table 5.4 Examples of culture media for microorganisms with simple and demanding nutritional requirements[a]

Defined culture medium for Escherichia coli	*Defined culture medium for Leuconostoc mesenteroides*	*Complex culture medium for either E. coli or L. mesenteroides*	*Defined culture medium for Thiobacillus thioparus*
K_2HPO_4 7 g KH_2PO_4 2 g $(NH_4)_2SO_4$ 1 g $MgSO_4$ 0.1 g $CaCl_2$ 0.02 g Glucose 4–10 g Trace elements (Fe, Co, Mn, Zn, Cu, Ni, Mo) 2–10 μg each Distilled water 1,000 ml pH 7	K_2HPO_4 0.6 g KH_2PO_4 0.6 g NH_4Cl 3 g $MgSO_4$ 0.1 g Glucose 25 g Sodium acetate 25 g Amino acids (alanine, arginine, asparagine, aspartate, cysteine, glutamate, glutamine, glycine, histidine, isoleucine, leucine, lysine, methionine, phenylalanine, proline, serine, threonine, tryptophan, tyrosine, valine) 100–200 μg of each Purines and pyrimidines (adenine, guanine, uracil, xanthine) 10 mg of each Vitamins (biotin, folate, nicotinic acid, pyridoxal, pyridoxamine, pyridoxine, riboflavin, thiamine, pantothenate, *p*-aminobenzoic acid) 0.01–1 mg of each Trace elements (see first column) 2–10 μg each Distilled water 1,000 ml pH 7	Glucose 15 g Yeast extract 5 g Peptone 5 g KH_2PO_4 2 g Distilled water 1,000 ml pH 7	KH_2PO_4 0.5 g NH_4Cl 0.5 g $MgSO_4$ 0.1 g $CaCl_2$ 0.05 g KCl 0.5 g $Na_2S_2O_3$ 2 g Trace elements (same as *E. coli* recipe) Distilled water 1,000 ml pH 7 Carbon source CO_2 from air
(a)		(b)	

[a]The photos are tubes of *(a)* the defined medium described, and *(b)* the complex medium described. Note how the complex medium is colored from the various organic extracts and digests that it contains. Photos courtesy of Cheryl L. Broadie and John Vercillo, Southern Illinois University at Carbondale.

5.1 MiniReview

Cells are made up primarily of H, O, C, N, P, and S; however, many other elements may be present as well. The hundreds of chemical compounds present inside a living cell are formed from available nutrients in the environment. Elements required in fairly large amounts are called macronutrients, whereas metals and organic compounds needed in very small amounts are called micronutrients (trace metals) and growth factors, respectively.

- What two classes of macromolecules contain the bulk of the nitrogen in a cell?
- Why is an element such as Co^{2+} considered a micronutrient whereas an element such as C is considered a macronutrient?
- What roles does iron play in cellular metabolism? How do cells sequester iron?

5.2 Culture Media

Culture media are the nutrient solutions used to grow microorganisms in the laboratory. Because laboratory culture is required for the detailed study of a microorganism, careful attention must be paid to both the selection and preparation of media for successful culture to take place.

Classes of Culture Media

Two broad classes of culture media are used in microbiology: **defined media** and **complex media** (Table 5.4). Defined media are prepared by adding precise amounts of highly purified inorganic or organic chemicals to distilled water. Therefore, the *exact* chemical composition of a defined medium is known. Of major importance in any culture medium is the carbon source, since all cells need large amounts of carbon to make new cell material. In a simple defined medium (Table 5.4), a single carbon source is usually present. The nature of the carbon source and its concentration depends on the organism to be cultured.

For growing many organisms, knowledge of the exact composition of a medium is not essential. In these instances complex media may suffice and may even be advantageous. Complex media employ digests of animal or plant products, such as casein (milk protein), beef (beef extract), soybeans (tryptic soy broth), yeast cells (yeast extract), or any of a number of other highly nutritious yet impure substances. These digests are commercially available in powdered form and can be quickly weighed and dissolved in distilled water to yield a medium. However, a major concession in using a complex medium is loss of control over its precise nutrient composition.

In particular situations, especially in clinical microbiology, culture media are often made to be selective or differential (or both). A *selective* medium contains compounds that selectively inhibit the growth of some microorganisms but not others. By contrast, a *differential* medium is one in which an indicator, typically a dye, is added that allows for the differentiation of particular chemical reactions that have occurred during growth. Differential media are quite useful for distinguishing between species of bacteria, some of which may carry out the particular reaction while others do not. Differential and selective media are further discussed in Chapter 32.

Nutritional Requirements and Biosynthetic Capacity

Table 5.4 lists four recipes for culture media, three defined and one complex. The complex medium is easiest to prepare and supports growth of both *Escherichia coli* and *Leuconostoc mesenteroides*. However, the simple defined medium shown in Table 5.4 supports growth of *E. coli* but not of *L. mesenteroides*. Growth of the latter organism, a nutritionally demanding bacterium, in defined medium requires the addition of several nutrients not needed by *E. coli* (Table 5.4). Which organism, *E. coli* or *L. mesenteroides*, has the greater biosynthetic capacity? Obviously, it is *E. coli*, because its ability to grow in a simple defined medium means that it can synthesize all of its organic constituents from a single carbon compound, in this case glucose (Table 5.4). By contrast, *L. mesenteroides* has multiple growth factor and other nutrient requirements, revealing its limited biosynthetic capacity. The nutritional needs of *L. mesenteroides* can be satisfied by preparing either a highly supplemented defined medium, a rather laborious undertaking because of all of the individual nutrients that need to be added (Table 5.4), or by preparing a complex medium, typically a much quicker operation.

Some pathogenic microorganisms have even more stringent nutritional requirements than those shown for *L. mesenteroides*. For culturing these organisms, an *enriched medium* may be needed. An enriched medium is a complex medium base to which additional nutrients, such as serum or whole blood, are added. These added nutrients better mimic conditions in the host and are required for the successful laboratory culture of some human pathogens such as *Streptococcus pyogenes* (strep throat) and *Neisseria gonorrhoeae* (gonorrhea).

The fourth medium shown in Table 5.4 is that for the sulfur chemolithotroph *Thiobacillus thioparus*. This organism is an autotroph and thus has no organic carbon requirements. *T. thioparus* derives all of its carbon from CO_2 and obtains its energy from the oxidation of the reduced sulfur compound thiosulfate ($Na_2S_2O_3$). Thus, *T. thioparus* has the greatest biosynthetic capacity of all of the organisms listed in Table 5.4, surpassing even *E. coli* in this regard.

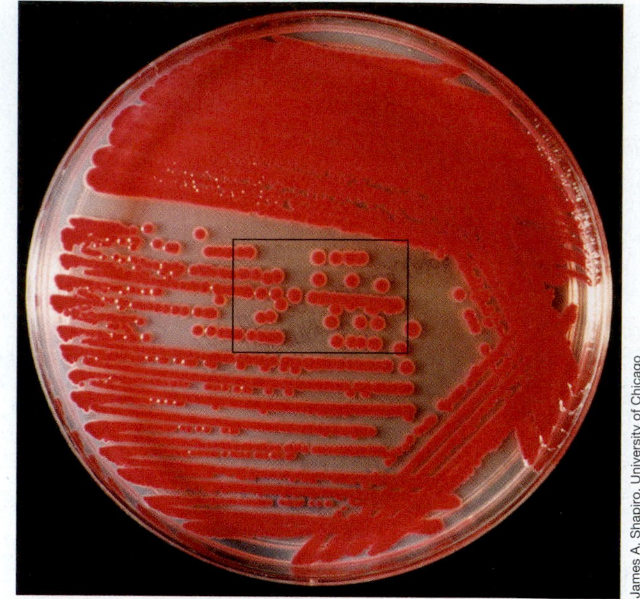

James A. Shapiro, University of Chicago

(a)

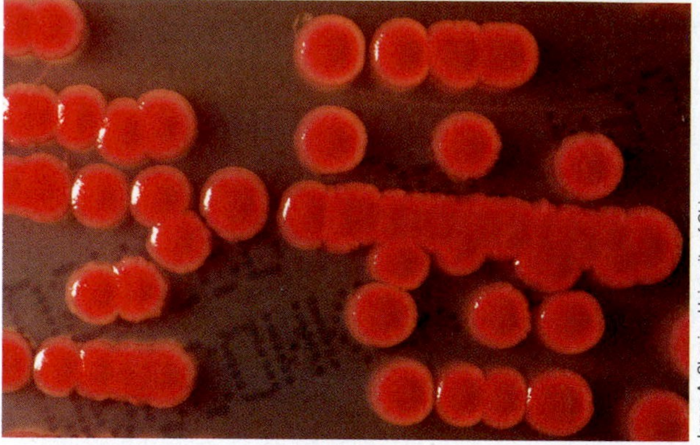

James A. Shapiro, University of Chicago

(b)

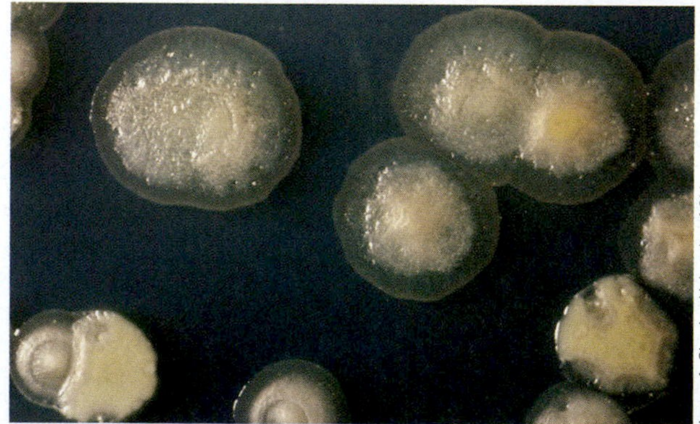

James A. Shapiro, University of Chicago

(c)

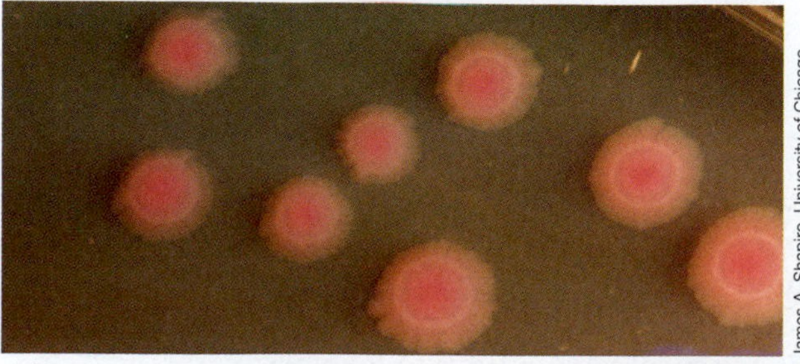

James A. Shapiro, University of Chicago

(d)

Figure 5.3 Bacterial colonies. Colonies are visible masses of cells formed from the subsequent division of one or a few cells and can contain over a billion (10^9) individual cells. *(a) Serratia marcescens*, grown on MacConkey agar. *(b)* Close-up of colonies outlined in *(a)*. *(c) Pseudomonas aeruginosa*, grown on Trypticase soy agar. *(d) Shigella flexneri*, grown on MacConkey agar.

What does Table 5.4 tell us? Simply put, it tells us that different microorganisms can have vastly different nutritional requirements. Thus, for successful culture of any microorganism it is necessary to understand its nutritional requirements and then supply the nutrients in the proper form and in the proper proportions in a culture medium. If care is taken in preparing culture media, it is fairly easy to culture many different types of microorganisms in the laboratory. We discuss the procedures for culturing microorganisms next.

5.2 MiniReview

Culture media supply the nutritional needs of microorganisms and are either chemically defined or undefined (complex). "Selective," "differential," and "enriched" are the terms that describe media used for the isolation of particular species or for comparative studies of microorganisms.

- Why would the routine culture of *Leuconostoc mesenteroides* be easier in a complex medium than in a chemically defined medium?
- In which medium shown in Table 5.4, defined or complex, do you think *Escherichia coli* would grow the fastest? Why? Although they are both simple defined media, why wouldn't *E. coli* grow in the medium described for *Thiobacillus thioparus*?

5.3 Laboratory Culture of Microorganisms

Once a culture medium has been prepared and made **sterile** to render it free of all microorganisms, organisms can be inoculated (added to it) and the culture incubated under conditions that will support growth. In a laboratory, inoculation will typically be of a **pure culture,** a culture containing only a single kind of microorganism.

It is essential to prevent other organisms from entering a pure culture. Such unwanted organisms, called *contaminants,* are ubiquitous (as Pasteur discovered over 125 years ago, Section 1.7), and microbiological techniques are designed to avoid contamination. A major method for obtaining pure cultures and for assessing the purity of a culture is the use of solid media, specifically, solid media prepared in the Petri plate, and we consider this now.

Solid versus Liquid Culture Media

Culture media are sometimes prepared in a semisolid form by the addition of a gelling agent to liquid media. Such solid culture media immobilize cells, allowing them to grow and form visible, isolated masses called *colonies* (**Figure 5.3**). Bacterial colonies can be of various shapes and sizes depending on the organism, the culture conditions, the nutrient supply, and several other physiological parameters. Some bacteria produce pigments that cause the colony to be colored (Figure 5.3). Colonies permit the microbiologist to visualize the purity of the culture. Plates that contain more than one colony type are indicative of a contaminated culture. The appearance and uniformity of colonies on a Petri plate has been used as one criterion of culture purity for over 100 years (Section 1.8).

Solid media are prepared the same way as for liquid media except that before sterilization, agar is added as a gelling agent, usually at a concentration of about 1.5%. The agar melts during the sterilization process, and the molten medium is then poured into sterile glass or plastic plates and allowed to solidify before use (Figure 5.3). **www.microbiologyplace.com Online Tutorial 5.1: Aseptic Transfer and the Streak Plate Method**

Aseptic Technique

Because microorganisms are everywhere, culture media must be sterilized before use. Sterilization is typically achieved with moist heat in a large pressure cooker called an *autoclave*. We discuss the operation and principles of the autoclave later, along with other methods of sterilization (Section 27.1).

Once a sterile culture medium has been prepared, it is ready to receive an inoculum from a previously grown pure culture to start the growth process once again. This manipulation requires the practice of **aseptic technique**, a series of steps to prevent contamination during manipulations of cultures and sterile culture media (**Figures 5.4** and **5.5**). A mastery of aseptic technique is required for success in the

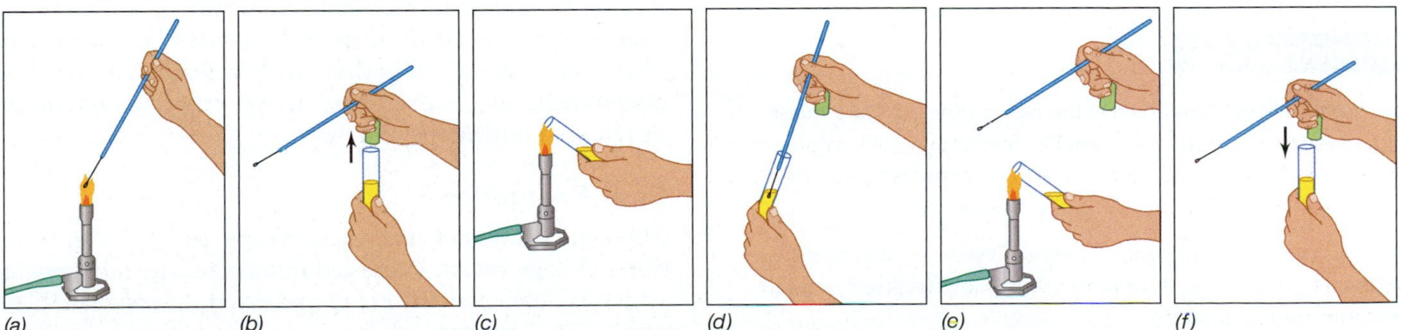

Figure 5.4 Aseptic transfer. *(a)* Loop is heated until red hot and cooled in air briefly. *(b)* Tube is uncapped. *(c)* Tip of tube is run through the flame. *(d)* Sample is removed on sterile loop and transferred to a sterile medium. *(e)* The tube is reflamed. *(f)* The tube is recapped. Loop is reheated before being taken out of service.

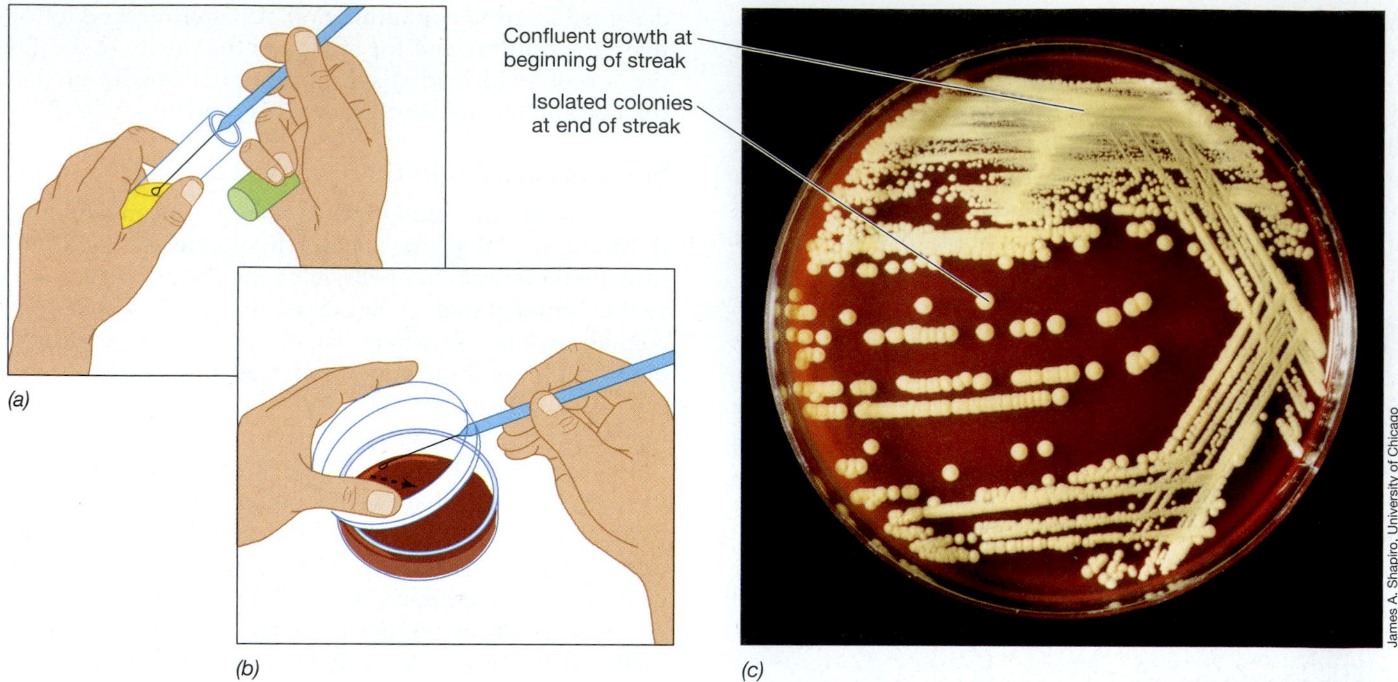

Figure 5.5 Method of making a streak plate to obtain pure cultures. *(a)* The loop is sterilized and a loopful of inoculum is removed from tube. *(b)* Streak is made and spread out on a sterile agar plate. Following the initial streak, subsequent streaks are made at angles to it, the loop being resterilized between streaks. *(c)* Appearance of a well-streaked plate after incubation. Colonies of the bacterium *Micrococcus luteus* on a blood agar plate. It is from such well-isolated colonies that pure cultures can usually be obtained.

microbiology laboratory, and it is one of the first methods learned by the novice microbiologist. Airborne contaminants are the most common problem because the dust in laboratory air contains microorganisms. When containers are opened, they must be handled in such a way that contaminant-laden air does not enter (Figures 5.4 and 5.5).

Aseptic transfer of a culture from one tube of medium to another is accomplished with an inoculating loop or needle that has previously been sterilized in a flame (Figure 5.4). Cells from liquid cultures can also be transferred to the surface of agar plates, where colonies develop from the growth and division of single cells (Figure 5.5). Picking an isolated colony and restreaking it is the main method for obtaining pure cultures from samples containing several different organisms.

5.3 MiniReview

Microorganisms can be grown in the laboratory in culture media containing the nutrients they require. Successful cultivation and maintenance of pure cultures of microorganisms can be done only if aseptic technique is practiced.

- What is meant by the word "sterile"? What would happen if freshly prepared culture media were not sterilized and left at room temperature?
- Why is aseptic technique necessary for successful cultivation of pure cultures in the laboratory?

II ENERGETICS AND ENZYMES

We learned in Chapter 2 of the different energy classes of microorganisms—*chemoorganotrophs, chemolithotrophs,* and *phototrophs*. However, regardless of how an organism makes a living, it must be able to conserve some of the energy released in its energy-yielding reactions. Here we discuss the principles of energy conservation, using some simple laws of chemistry and physics to guide our understanding. We then consider enzymes, the cell's biocatalysts.

5.4 Bioenergetics

Energy is defined as the ability to do work. In microbiology, energy is discussed in units of kilojoules (kJ), a measure of heat energy. We see now that all chemical reactions are associated with *changes* in energy, energy either being required for or released during the reaction.

Basic Energetics

Although in any chemical reaction some energy is lost as heat, in microbiology we are interested in **free energy** (abbreviated **G**), which is defined as the energy released that is available to do work. The change in free energy during a reaction is expressed as $\Delta G^{0\prime}$, where the symbol Δ should be read "change in." The "0" and "prime" superscripts mean that the free-energy value is for standard conditions: pH 7, 25°C,

1 atmosphere of pressure, and all reactants and products at 1 M concentration.

Consider the reaction:

$$A + B \rightarrow C + D$$

If $\Delta G^{0\prime}$ for this reaction is *negative* in arithmetic sign, then the reaction will proceed with the release of free energy, energy that the cell may conserve as ATP. Such energy-yielding reactions are called **exergonic**. However, if $\Delta G^{0\prime}$ is *positive,* the reaction requires energy in order to proceed. Such reactions are called **endergonic**. Thus, exergonic reactions *yield* energy while endergonic reactions *require* energy.

Free Energy of Formation and Calculating $\Delta G^{0\prime}$

To calculate the free energy yield of a reaction, one first needs to know the free energy of its reactants and products. This is the free energy of formation (G_f^0), the energy released or required during the formation of a given molecule from the elements. **Table 5.5** gives a few examples of G_f^0. By convention, the free energy of formation of the elements in their elemental and electrically neutral form (for instance, C, H_2, N_2) is zero. The free energies of formation of compounds, however, are not zero. If the formation of a compound from its elements proceeds exergonically, then the G_f^0 of the compound is negative (energy is released). If the reaction is endergonic, then the G_f^0 of the compound is positive (energy is required).

For most compounds G_f^0 is negative. This reflects the fact that compounds tend to form spontaneously (that is, with energy being released) from their elements. However, the positive G_f^0 for nitrous oxide (+104.2 kJ/mol, Table 5.5) indicates that this molecule will not form spontaneously. Instead, it will decompose spontaneously to nitrogen and oxygen. The free energies of formation of some compounds of microbiological interest are given in Appendix 1. A more complete list can be found in physical chemistry handbooks.

Using free energies of formation, it is possible to calculate a very useful value: the *change* in free energy ($\Delta G^{0\prime}$) that occurs in a given reaction. For the reaction A + B → C + D, $\Delta G^{0\prime}$ is calculated by subtracting the sum of the free energies of formation of the reactants (A + B) from that of the products (C + D). Thus:

$$\Delta G^{0\prime} = G_f^0[C + D] - G_f^0[A + B]$$

The $\Delta G^{0\prime}$ value obtained tells us whether the reaction is exergonic or endergonic. The phrase "products minus reactants" is a simple way to recall how to calculate changes in free energy during chemical reactions. However, it is first necessary to balance the reaction before free energy calculations can be made. Appendix 1 details the steps involved in balancing reactions and calculating free energies for any hypothetical reaction.

$\Delta G^{0\prime}$ versus ΔG

Although calculations of $\Delta G^{0\prime}$ are usually good estimates of actual free energy changes, under some circumstances they are not. We will see later in this book that the actual concentrations of products and reactants in nature, which are rarely at 1 M levels, can alter the bioenergetics of reactions, sometimes in significant ways. Thus, what may be most relevant to a bioenergetic calculation is not $\Delta G^{0\prime}$, but ΔG, the free energy change that occurs under the actual conditions in which the organism is growing. ΔG is a form of the free energy equation that takes into account the actual concentrations of reactants and products in the reaction, and it is calculated as:

$$\Delta G = \Delta G^{0\prime} + RT \ln K$$

where R and T are physical constants and K is the equilibrium constant for the reaction in question (Appendix 1). We distinguish between $\Delta G^{0\prime}$ and ΔG in Chapter 21, where we consider metabolic diversity in more detail, but for now, we only need to focus on the expression $\Delta G^{0\prime}$ and what it tells us about a chemical reaction catalyzed by a particular microorganism. Only reactions that are exergonic are capable of yielding energy and thus supporting growth.

Table 5.5 Free energy of formation for a few compounds of biological interest

Compound	Free energy of formation[a]
Water (H_2O)	−237.2
Carbon dioxide (CO_2)	−394.4
Hydrogen gas (H_2)	0
Oxygen gas (O_2)	0
Ammonium (NH_4^+)	−79.4
Nitrous oxide (N_2O)	+104.2
Acetate ($C_2H_3O_2^-$)	−369.4
Glucose ($C_6H_{12}O_6$)	−917.3
Methane (CH_4)	−50.8
Methanol (CH_3OH)	−175.4

[a]The free energy of formation values (G_f^0) are in kJ/mol. See Table A1.1 for a more complete list of free energies of formation.

5.4 MiniReview

The chemical reactions of the cell are accompanied by changes in energy, expressed in kilojoules. A chemical reaction can occur with the release of free energy (exergonic) or with the consumption of free energy (endergonic).

- What is free energy?
- In general, are catabolic reactions exergonic or endergonic?
- Using the data in Table 5.5, calculate $\Delta G^{0\prime}$ for the reaction $CH_4 + \frac{1}{2}O_2 \rightarrow CH_3OH$. How does $\Delta G^{0\prime}$ differ from ΔG?
- Does glucose formation from the elements release or require energy?

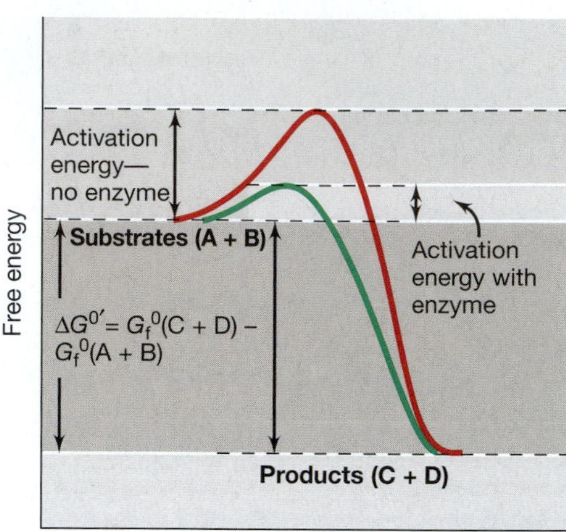

Figure 5.6 **Activation energy and catalysis.** Even chemical reactions that release energy may not proceed spontaneously because the reactants must first be activated. Once they are activated, the reaction proceeds spontaneously. Catalysts such as enzymes lower the required activation energy.

5.5 Catalysis and Enzymes

Free energy calculations only reveal whether energy is released or required in a given reaction. The value obtained says nothing about the *rate* of the reaction. Consider the formation of water from gaseous oxygen and hydrogen. The energetics of this reaction are quite favorable: $H_2 + \frac{1}{2}O_2 \rightarrow H_2O$ $\Delta G^{0\prime} = -237$ kJ. However, if we were to mix O_2 and H_2 together in a sealed bottle and leave it for years, no measurable formation of water would occur. This is because the bonding of oxygen and hydrogen atoms to form water requires that the chemical bonds of the reactants first be broken. The breaking of these bonds requires some energy, and this energy is called **activation energy**.

Activation energy is the energy required to bring all molecules in a chemical reaction into the reactive state. For a reaction that proceeds with a net release of free energy (that is, an exergonic reaction), the situation is as diagrammed in **Figure 5.6**. Although the activation energy barrier is virtually insurmountable in the absence of catalysis, in the presence of the proper catalyst it is much less formidable (Figure 5.6).

Enzymes

The concept of activation energy leads us to consider catalysis and enzymes. In biochemistry, a **catalyst** is a substance that lowers the activation energy of a reaction, thereby increasing the reaction rate. Catalysts facilitate reactions but are not consumed or transformed by them. Moreover, catalysts do not affect the energetics or the equilibrium of a reaction; catalysts affect only the rate at which reactions proceed.

Most cellular reactions would not proceed at appreciable rates without catalysis. Biological catalysts are called **enzymes**. Enzymes are proteins (or in a few cases, RNAs) that are highly specific in the reactions they catalyze. That is, each enzyme catalyzes only a single type of chemical reaction, or in the case of some enzymes, a single class of closely related reactions. This specificity is a function of the precise three-dimensional structure (as dictated by protein structure, ∞ Sections 3.7 and 3.8) of the enzyme molecule.

In an enzyme-catalyzed reaction, the enzyme combines with the reactant, called a *substrate* (S), forming an enzyme–substrate complex. Then, as the reaction proceeds, the *product* (P) is released and the enzyme (E) is returned to its original state:

$$E + S \leftrightarrow E{-}S \leftrightarrow E + P$$

The enzyme is generally much larger than the substrate(s), and binding of enzyme and substrate(s) typically relies on weak bonds, such as hydrogen bonds, van der Waals forces, and hydrophobic interactions (∞ Section 3.1). The portion of the enzyme to which substrate binds is called the *active site*, and the entire reaction from initial substrate binding to product release may take only a few milliseconds.

Enzyme Catalysis

The catalytic power of enzymes is remarkable. Enzymes increase the rate of chemical reactions anywhere from 10^8 to 10^{20} times the rate that would occur spontaneously. To catalyze a specific reaction, an enzyme must do two things: (1) bind its substrate and (2) position the substrate relative to the catalytically active amino acids in the enzyme's active site. The enzyme–substrate complex (**Figure 5.7**) aligns reactive groups and places strain on specific bonds in the substrate(s). The net result is a reduction in the activation energy required to make the reaction proceed from substrate(s) to product(s) (Figure 5.6). These steps are shown in Figure 5.7 for the glycolytic enzyme fructose bisphosphate aldolase (Section 5.10).

The particular reaction depicted in Figure 5.6 is exergonic because the free energy of formation of the substrates is greater than that of the products. However, enzymes can also catalyze reactions that require energy, converting energy-poor substrates into energy-rich products.

In these cases, however, not only must an activation energy barrier be overcome, but sufficient free energy must also be put into the reaction to raise the energy level of the substrates to that of the products. This is done by coupling the energy-*requiring* reaction to an energy-*yielding* one, such as the hydrolysis of ATP.

Theoretically, all enzymes are reversible in their activity. However, enzymes that catalyze highly exergonic or highly endergonic reactions are essentially unidirectional in their activity. If a particularly exergonic or endergonic reaction needs to be reversed, a different enzyme usually catalyzes the reverse reaction.

Structure and Nomenclature of Enzymes

Most enzymes are proteins, polymers of amino acids (∞ Sections 3.6–3.8), and every protein has a specific three-dimensional shape. A given protein thus assumes specific binding and physical properties. The structure of an enzyme may be

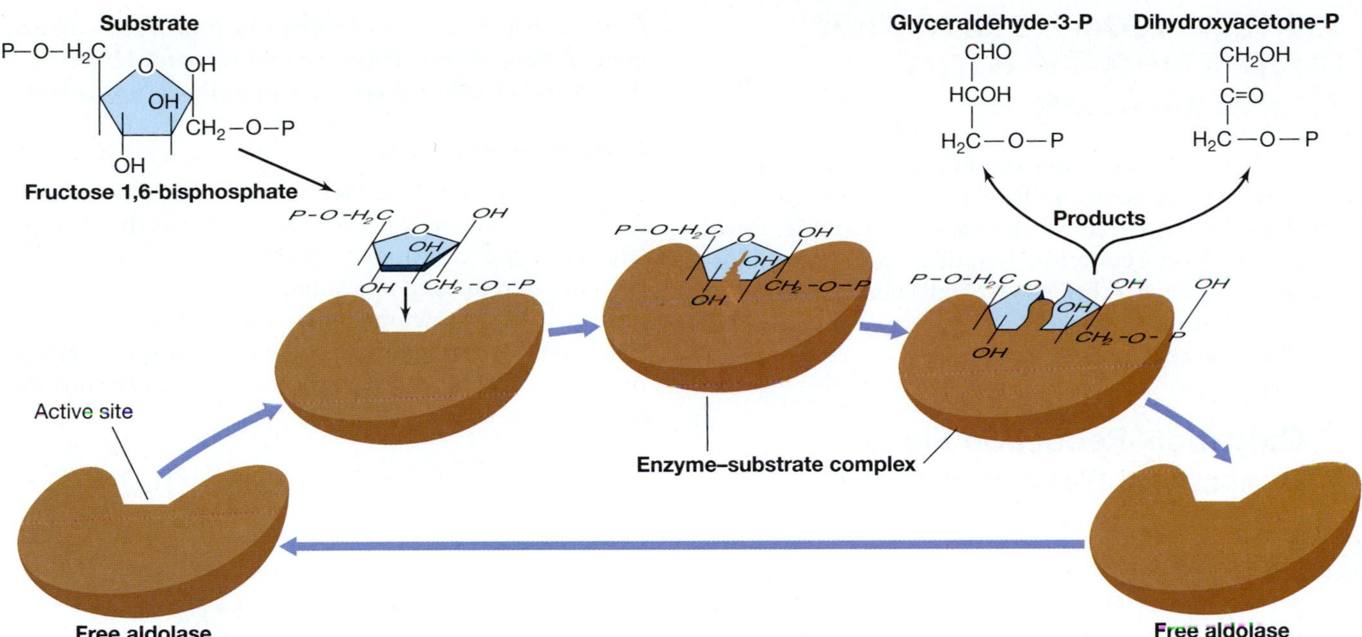

Figure 5.7 The catalytic cycle of an enzyme. The enzyme depicted here, fructose bisphosphate aldolase, catalyzes the reaction: fructose 1,6-bisphosphate → glyceraldehyde-3-phosphate + dihydroxyacetone phosphate in glycolysis. Following binding of fructose 1,6-bisphosphate in the formation of the enzyme–substrate complex, the conformation of the enzyme alters, placing strain on bonds of the substrate, which break and yield the two products.

visualized in a space-filling model (Figure 5.8). In this example of the peptidoglycan-cleaving enzyme lysozyme, the large cleft is the site where the substrate binds (the active site). The active site of lysozyme binds peptidoglycan but does not bind other polysaccharides, even those whose structure may closely resemble peptidoglycan.

Many enzymes contain small nonprotein molecules that participate in catalysis but are not themselves substrates. These small molecules can be divided into two classes based on the way they associate with the enzyme: *prosthetic groups* and *coenzymes*. Prosthetic groups are bound very tightly to their enzymes, usually covalently and permanently. The heme group present in cytochromes (Section 5.11) is an example of a prosthetic group. **Coenzymes**, by contrast, are loosely bound to enzymes, and a single coenzyme molecule may associate with a number of different enzymes. Most coenzymes are derivatives of vitamins, and NAD^+/NADH is a good example, being a derivative of the vitamin niacin (Table 5.3).

Enzymes are named for either the substrate they bind or for the chemical reaction they catalyze, by addition of the suffix "-ase." Thus, cellulase is an enzyme that attacks cellulose, glucose oxidase is an enzyme that catalyzes the oxidation of glucose, and ribonuclease is an enzyme that decomposes ribonucleic acid. A more formal enzyme nomenclature system exists, in which a numbering system is used to classify enzymes according to the type of reaction they catalyze. However, for our purposes, enzyme names ending in "ase" are sufficient and will be used exclusively in this book.

5.5 MiniReview

The reactants in a chemical reaction must first be activated before the reaction can take place, and this requires a catalyst. Enzymes are catalytic proteins that speed up the rate of biochemical reactions. Enzymes are highly specific in the reactions they catalyze, and this specificity resides in the three-dimensional structure of the polypeptide(s) in the protein.

- What is the function of a catalyst? What are enzymes made of?
- Where on an enzyme does the substrate bind?
- What is activation energy?

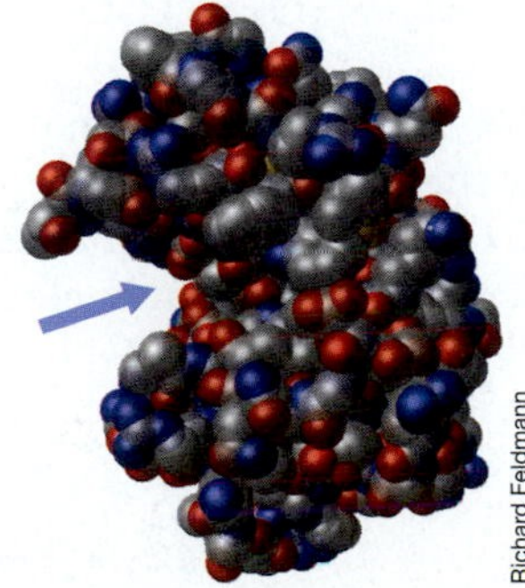

Figure 5.8 Space-filling model of the enzyme lysozyme. The substrate-binding site (the active site) is in the large cleft (arrow) on the left side of the model.

III OXIDATION–REDUCTION AND ENERGY-RICH COMPOUNDS

Energy is conserved in cells from oxidation–reduction (redox) reactions. The energy released in these reactions is conserved in the synthesis of energy-rich compounds, such as ATP. Here we consider oxidation–reduction reactions and the major electron carriers present in the cell. We will then examine the compounds that actually conserve the energy released in oxidation–reduction reactions.

5.6 Oxidation–Reduction Electron Donors and Electron Acceptors

An oxidation is the removal of an electron or electrons from a substance and a reduction is the addition of an electron or electrons to a substance. Oxidations and reductions can involve just electrons or an electron plus a proton (that is, a hydrogen atom).

Electron Donors and Electron Acceptors

Redox reactions occur in pairs. For example, hydrogen gas (H_2) can release electrons and protons and become oxidized:

$$H_2 \rightarrow 2\ e^- + 2\ H^+$$

However, electrons cannot exist alone in solution; they must be part of atoms or molecules. Thus, the equation as drawn does not itself represent an independent reaction. The reaction is only a *half reaction,* a term that implies the need for a second half reaction. This is because for any substance to be oxidized, another substance must be reduced. The oxidation of H_2 can be coupled to the reduction of many different substances, including O_2, in a second reaction:

$$\tfrac{1}{2}O_2 + 2\ e^- + 2\ H^+ \rightarrow H_2O$$

This half reaction, which is a reduction, when coupled to the oxidation of H_2, yields the following overall balanced reaction:

$$H_2 + \tfrac{1}{2}O_2 \rightarrow H_2O$$

In reactions of this type, we refer to the substance *oxidized,* in this case H_2, as the **electron donor**, and the substance *reduced,* in this case O_2, as the **electron acceptor** (**Figure 5.9**). The concept of electron donors and electron acceptors is very important in microbiology as it pervades all aspects of energy metabolism. Thus, we will refer to the terms electron donor and electron acceptor many times in this book.

Reduction Potentials

Substances vary in their tendency to become oxidized or reduced. This tendency is expressed as the **reduction potential** (E_0', standard conditions) of the half reaction, measured in volts (V) in reference to a standard substance, H_2. By convention, reduction potentials are expressed for half reactions as *reductions* with reactions at pH 7, because the cytoplasm of most cells is neutral, or nearly so. Using these conventions, at pH 7 the E_0' of

$$\tfrac{1}{2}O_2 + 2\ H^+ + 2\ e^- \rightarrow H_2O$$

is +0.816 volt (V), and the E_0' of

$$2\ H^+ + 2\ e^- \rightarrow H_2$$

is −0.421 V. We will see shortly that these E_0' values mean that O_2 is an excellent electron acceptor and that H_2 is an excellent electron donor.

Oxidation–Reduction Couples

Substances can be either electron donors or electron acceptors under different circumstances, depending on the substances with which they react. The chemical constituents on each side of the arrow in half reactions can be thought of as representing a *redox couple,* such as $2\ H^+/H_2$ or $\tfrac{1}{2}O_2/H_2O$. By convention, when writing a redox couple, the oxidized form of the couple is always placed on the left, before the slash mark.

In constructing redox reactions from their constituent half reactions, the reduced substance of a redox couple whose E_0' is more negative donates electrons to the oxidized substance of a redox couple whose E_0' is more positive. Thus, in the couple $2\ H^+/H_2$ (E_0' − 0.42 V), H_2 has a greater tendency to donate electrons than do protons to accept them. On the other hand, in the couple $\tfrac{1}{2}O_2/H_2O$ (E_0' + 0.82 V), H_2O has a very weak tendency to donate electrons, whereas O_2 has a great tendency to accept them. It follows then that in a reaction of H_2 and O_2, H_2 will be the electron donor and become oxidized, and O_2 will be the electron acceptor and become reduced (Figure 5.9).

As previously mentioned, by convention in electrochemistry all half reactions are written as *reductions*. However, in an actual redox reaction, the half reaction with the more negative E_0' proceeds as an oxidation and is therefore written in the

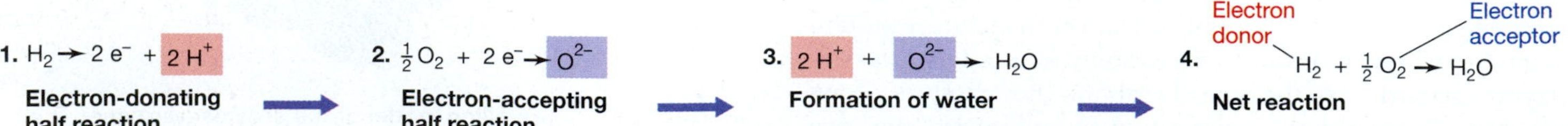

Figure 5.9 **Example of an oxidation–reduction reaction.** The formation of H_2O from the electron donor H_2 and the electron acceptor O_2.

opposite direction. Thus, in the reaction shown in Figure 5.9, the oxidation of H_2 to 2 H^+ + 2 e^- is reversed from its formal half reaction, written as a reduction.

The Redox Tower

A convenient way of viewing electron transfer reactions in biological systems and their importance to bioenergetics is to imagine a vertical tower (**Figure 5.10**). The tower represents the range of reduction potentials possible for redox couples in nature, from those with the most negative on the top to those with the most positive at the bottom; thus, we can call the tower a *redox tower*. The *reduced* substance in the redox couple at the top of the tower has the greatest tendency to donate electrons, whereas the *oxidized* substance in the redox couple at the bottom of the tower has the greatest tendency to accept electrons.

Using the tower analogy, imagine electrons from an electron donor near the top of the tower falling and being "caught" by electron acceptors at various levels. The difference in reduction potential between the donor and acceptor redox couples is expressed as $\Delta E_0'$. The farther the electrons drop from a donor before they are caught by an acceptor, the greater the amount of energy released. That is, $\Delta E_0'$ is proportional to $\Delta G^{0'}$ (Figure 5.10). Oxygen (O_2), at the bottom of the redox tower, is the strongest naturally occurring electron acceptor. In the middle of the redox tower, redox couples can be either electron donors or acceptors depending on which redox couples they react with. For instance, the 2 H^+/H_2 couple (−0.42 V) can react with the fumarate/succinate (+0.02 V), NO_3^-/NO_2^- (+0.42 V), or $\frac{1}{2}O_2/H_2O$ (+0.82 V) couples, with increasing amounts of energy being released, respectively (Figure 5.10).

Electron donors used in energy metabolism are also called *energy sources* because energy is released when they are oxidized (Figure 5.10). Many potential electron donors exist in nature, including a wide variety of organic and inorganic compounds. However, it should be understood that it is not the electron donor per se that contains energy but the chemical reaction in which the electron donor participates that actually releases energy. The presence of a suitable electron acceptor is just as important as the presence of a suitable electron donor. Lacking one or the other, the energy releasing reaction cannot proceed.

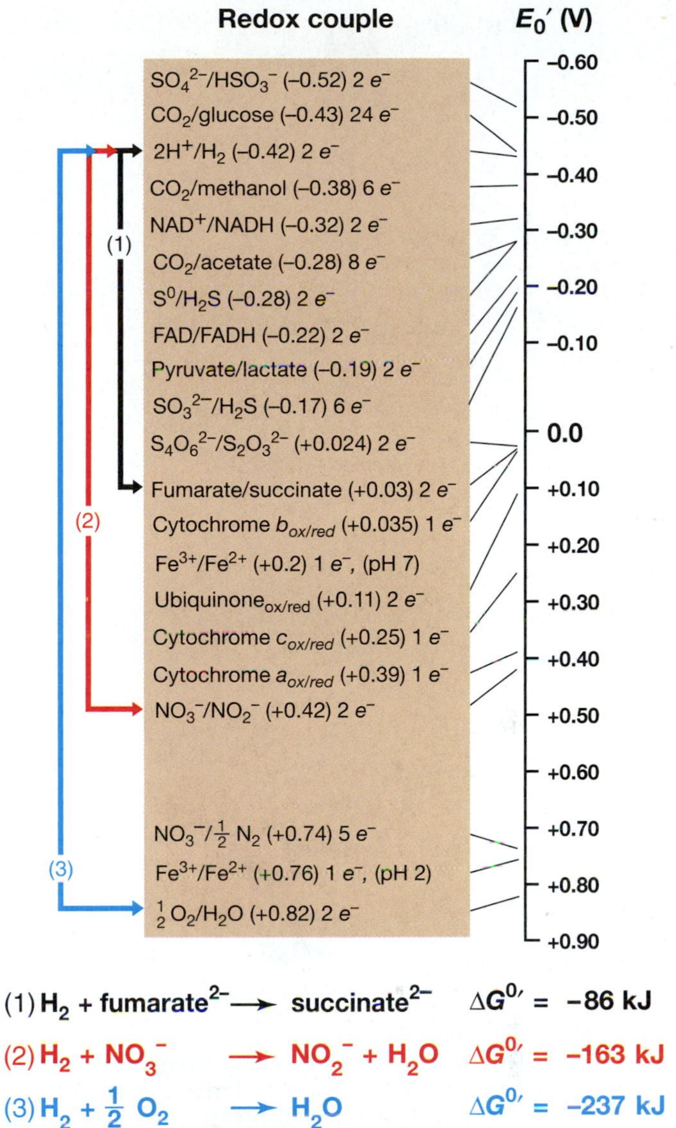

Figure 5.10 **The redox tower.** Redox couples are arranged from the strongest reductants (negative reduction potential) at the top to the strongest oxidants (positive reduction potentials) at the bottom. As electrons are donated from the top of the tower, they can be "caught" by acceptors at various levels. The farther the electrons fall before they are caught, the greater the difference in reduction potential between electron donor and electron acceptor and the more energy is released. As an example of this, the differences in energy released when a single electron donor, H_2, reacts with any of three different electron acceptors, fumarate, nitrate, and oxygen is shown.

5.6 MiniReview

Oxidation–reduction reactions require electron donors and electron acceptors. The tendency of a compound to accept or release electrons is expressed quantitatively by its reduction potential, E_0'.

- In the reaction $H_2 + \frac{1}{2}O_2 \rightarrow H_2O$, what is the electron donor and what is the electron acceptor?
- What is the E_0' of the 2 H^+/H_2 couple? Are protons good electron acceptors? Why or why not?
- Why is nitrate (NO_3^-) a better electron acceptor than fumarate?

5.7 NAD as a Redox Electron Carrier

Redox reactions in microbial cells typically involve reactions between one or more intermediates called *carriers*. Electron carriers can be divided into two classes: those that are freely diffusible (coenzymes) and those that are firmly attached to enzymes in the cytoplasmic membrane (prosthetic groups). The fixed carriers function in membrane-associated electron transport reactions and are discussed in Section 5.11. Common diffusible carriers include the coenzymes nicotinamide-adenine dinucleotide (NAD^+) and NAD-phosphate ($NADP^+$)

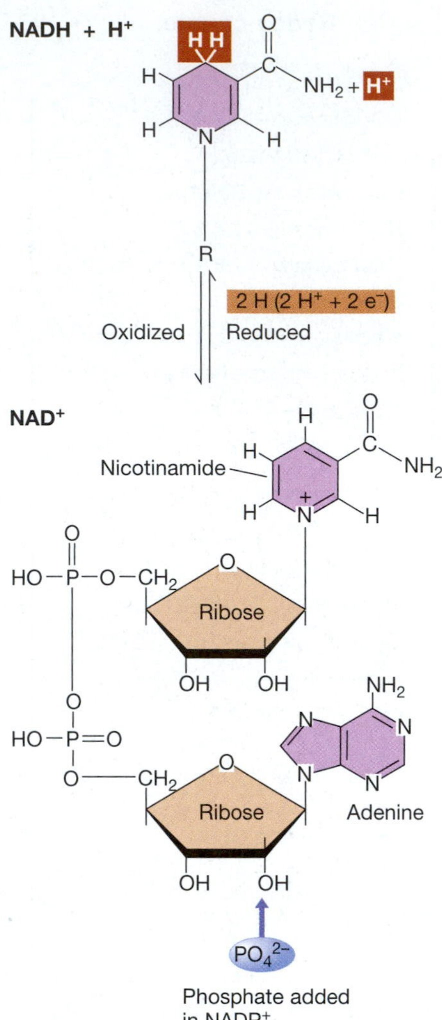

Figure 5.11 The oxidation–reduction coenzyme nicotinamide adenine dinucleotide (NAD^+). In $NADP^+$, a phosphate group is present, as indicated. Both NAD^+ and $NADP^+$ undergo oxidation–reduction as shown, are freely diffusible, and are $2e^- + 2H^+$ carriers. "R" in the top portion of the art is the adenine dinucleotide portion of NAD^+, shown in full in the bottom portion of the figure.

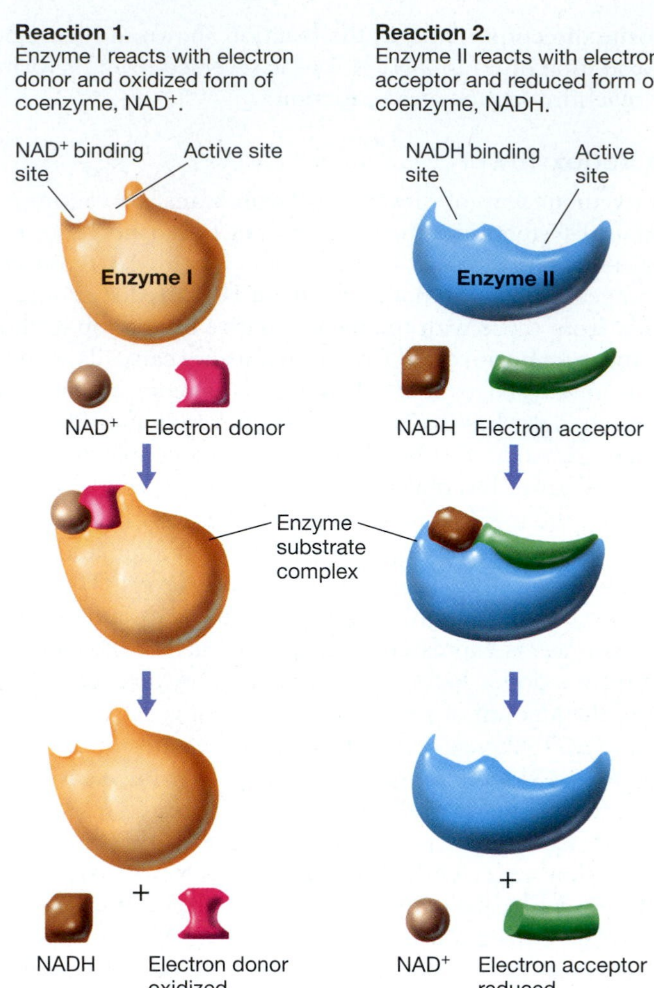

Figure 5.12 NAD^+/NADH cycling. A schematic example of a redox reaction requiring two different enzymes, linked by their use of a common coenzyme.

(**Figure 5.11**). NAD^+ and $NADP^+$ are electron plus proton carriers, transporting $2e^-$ and $2H^+$ at a time.

The reduction potential of the NAD^+/NADH (or $NADP^+$/NADPH) couple is −0.32 V, which places it fairly high on the electron tower; that is, NADH (or NADPH) is a good electron donor (Figure 5.10). However, although the NAD^+ and $NADP^+$ couples have the same reduction potentials, they typically function in different capacities in the cell. NAD^+/NADH play roles in energy-generating (catabolic) reactions, whereas $NADP^+$/NADPH play roles in biosynthetic (anabolic) reactions.

NAD/NADH Cycling

Coenzymes increase the diversity of redox reactions possible in a cell by allowing chemically dissimilar electron donors and acceptors to interact, the coenzyme acting as the intermediary. With NAD^+/NADH, for example, electrons removed from an electron donor can reduce NAD^+ to NADH, and the latter can be converted back to NAD^+ by donating electrons to the electron acceptor. **Figure 5.12** shows an example of electron shuttling by NAD^+/NADH. In the reaction, NAD^+ and NADH facilitate the redox reaction without being consumed in the process. Thus, unlike the primary electron donor (the substance that was oxidized to yield NADH) or acceptor (such as O_2), where relatively large amounts of each are required, the cell needs only a tiny amount of NAD^+ and NADH because they are constantly being recycled. All that is needed is an amount sufficient to service the redox enzymes in the cell that use these coenzymes in their reaction mechanisms (Figure 5.12).

5.7 MiniReview

The transfer of electrons from donor to acceptor in a cell typically involves electron carriers. Some electron carriers are membrane-bound, whereas others, such as NAD^+/NADH, are freely diffusible coenzymes.

- Review Figure 5.10. Is NADH a better electron donor than H_2? Is NAD^+ a better acceptor than H^+? How do you determine this?

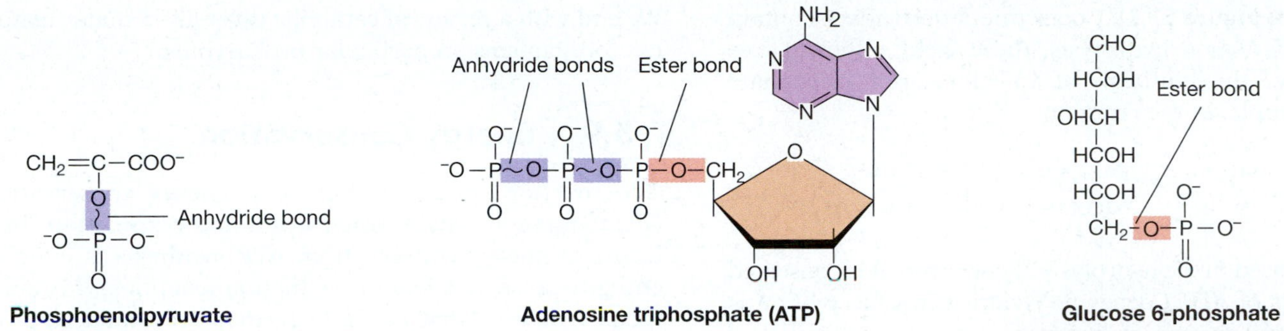

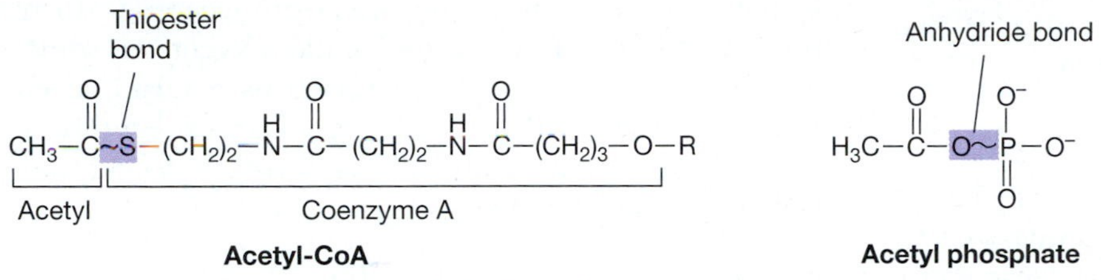

Compound	$G^{0'}$ kJ/mol
$\Delta G^{0'}$ > 30kJ	
Phosphoenolpyruvate	−51.6
1,3-Bisphosphoglycerate	−52.0
Acetyl phosphate	−44.8
ATP	−31.8
ADP	−31.8
Acetyl CoA	−31
$\Delta G^{0'}$ < 30kJ	
AMP	−14.2
Glucose 6-phosphate	−13.8

Figure 5.13 Some compounds important in energy transformations in cells. The table shows the free energy of hydrolysis of some of the key phosphate esters and anhydrides, indicating that some of the phosphate ester bonds are more energy-rich than others. Structures of four of the compounds show the position of ester and anhydride bonds. ATP contains three phosphates, but only two of them have free energies of hydrolysis >30 kJ. Also shown is the structure of the coenzyme acetyl-CoA. The thioester bond between C and S has a free energy of hydrolysis of >30 kJ. The "R" group of acetyl-CoA is a 3′ phospho ADP group.

5.8 Energy-Rich Compounds and Energy Storage

Energy released from redox reactions must be conserved by the cell if it is to be used to drive energy-requiring cell functions. In living organisms, chemical energy released in redox reactions is conserved primarily in the form of certain phosphorylated compounds that are energy rich. The free energy released upon hydrolysis of the phosphate in energy-rich compounds is significantly greater than that of the average covalent bond in the cell.

Phosphate can be bonded to organic compounds by either ester or anhydride bonds, as illustrated in **Figure 5.13**. However, not all phosphate bonds are energy-rich. As seen, the $\Delta G^{0'}$ of hydrolysis of the phosphate *ester* bond in glucose 6-phosphate is only −13.8 kJ/mol. By contrast, the $\Delta G^{0'}$ of hydrolysis of the phosphate *anhydride* bond in phosphoenolpyruvate is −51.6 kJ/mol, almost four times that of glucose 6-phosphate. Although either compound could be hydrolyzed to yield energy, cells typically use a small group of compounds whose $\Delta G^{0'}$ of hydrolysis is greater than −30 kJ/mol as energy "currencies" in the cell (Figure 5.13). Thus, phosphoenolpyruvate is energy-rich and glucose 6-phosphate is not (Figure 5.13).

Adenosine Triphosphate (ATP)

The most important energy-rich phosphate compound in cells is **adenosine triphosphate (ATP)**. ATP consists of the ribonucleoside adenosine to which three phosphate molecules are bonded in series. ATP is the prime energy currency in all cells, being generated during certain exergonic reactions and consumed in certain endergonic reactions. From the structure of ATP (Figure 5.13) it can be seen that two of the phosphate bonds are phosphoanhydrides and thus have free energies of hydrolysis greater than 30 kJ. Thus the reactions ATP → ADP + P_i and ADP → AMP + P_i each release roughly 32 kJ/mol of energy. By contrast, AMP is not energy-rich because its free energy of hydrolysis is only about half that of ADP or ATP (Figure 5.13).

Although the energy released in ATP hydrolysis is −32 kJ, a caveat must be introduced here to more precisely define the energy requirements for the synthesis of ATP. In an actively growing cell, the ratio of ATP to ADP is maintained near 1,000. This major deviation from equilibrium significantly affects the energy requirements for ATP synthesis. In such a cell, the actual energy expenditure (that is, the ΔG, Section 5.4) for the synthesis of one mole of ATP is on the order of −55 to −60 kJ. Nevertheless, for the purposes of learning and applying the basic principles of bioenergetics, we will consider reactions to conform to "standard conditions" ($\Delta G^{0'}$) and thus the energy required for synthesis or hydrolysis of ATP to be the same, 32 kJ/mol.

Coenzyme A

In addition to energy-rich phosphate compounds, cells can produce other energy-rich compounds. These include, in particular, derivatives of *coenzyme A* (for example, acetyl-CoA;

see structure in Figure 5.13). Coenzyme A derivatives contain thioester bonds. Upon hydrolysis, these yield sufficient free energy to drive the synthesis of an energy-rich phosphate bond. For example, in the reaction

$$\text{acetyl-S-CoA} + H_2O + \text{ADP} + P_i \rightarrow \text{acetate}^- + \text{HS-CoA} + \text{ATP} + H^+$$

the energy released in the hydrolysis of coenzyme A is conserved in the synthesis of ATP. Coenzyme A derivatives (acetyl-CoA is just one of many) are especially important to the energetics of anaerobic microorganisms, in particular those whose energy metabolism depends on fermentation. We will return to the importance of coenzyme A derivatives in Chapter 21.

Long-Term Energy Storage

ATP is a dynamic molecule in the cell; it is continuously being broken down to drive anabolic reactions and resynthesized at the expense of catabolic reactions. For longer-term energy storage, microorganisms typically produce insoluble polymers that can be oxidized later for the production of ATP.

Examples of energy storage polymers in prokaryotes include glycogen, poly-β-hydroxybutyrate and other polyhydroxyalkanoates, and elemental sulfur, stored from the oxidation of H_2S by sulfur chemolithotrophs. These polymers are deposited within the cell as large granules that can be seen with the light or electron microscope (∞ Section 4.10). In eukaryotic microorganisms, polyglucose in the form of starch (∞ Figure 3.6) or lipids in the form of simple fats (∞ Figure 3.7) are the major reserve materials. In the absence of an external energy source, cells can break down these polymers in order to make new cell material or to supply the very low amount of energy needed to maintain cell integrity, called *maintenance energy*, when in a nongrowing state.

5.8 MiniReview

The energy released in redox reactions is conserved in the formation of compounds that contain energy-rich phosphate or sulfur bonds. The most common of these compounds is ATP, the prime energy carrier in the cell. Longer-term storage of energy is linked to the formation of polymers, which can be consumed to yield ATP.

- How much energy is released per mole of ATP converted to ADP + P_i? Per mole of AMP converted to adenosine and P_i?
- During periods of nutrient abundance, how can cells prepare for periods of nutrient starvation?

IV ESSENTIALS OF CATABOLISM

We now consider some key catabolic reactions in the cell and how energy released in these reactions is conserved in microorganisms. We begin with fermentation and respiration and then examine how the proton motive force is generated. We end with a review of catabolic diversity, a major feature of microorganisms, in particular prokaryotes.

5.9 Energy Conservation

Two reaction series are linked to energy conservation in chemoorganotrophs: **fermentation** and **respiration**. In both forms of energy conservation, ATP synthesis is coupled to energy release in oxidation–reduction reactions. However, the redox reactions that occur in fermentation and respiration differ significantly. In fermentation the redox process occurs in the absence of exogenous electron acceptors. In respiration, on the other hand, molecular oxygen or some other exogenous electron acceptor serves as a terminal electron acceptor. As we will see, this lack of an exogenous electron acceptor in fermentation places severe constraints on the yield of energy obtained.

Substrate-Level Phosphorylation and Oxidative Phosphorylation

Fermentation and respiration differ in the mechanism by which ATP is synthesized. In fermentation, ATP is produced by **substrate-level phosphorylation**. In this process, ATP is synthesized directly from an energy-rich intermediate during steps in the catabolism of the fermentable compound (**Figure 5.14*a***). This is in contrast to **oxidative phosphorylation**, in which ATP is produced at the expense of the proton motive force (Figure 5.14*b*). A third form of ATP synthesis, **photophosphorylation**, occurs in phototrophic organisms. We will see later that both photophosphorylation and oxidative phosphorylation rely on the proton motive force to drive ATP synthesis. In fermentation, by contrast, the proton motive force is not involved.

5.9 MiniReview

Fermentation and respiration are the two means by which chemoorganotrophs can conserve energy from the oxidation of organic compounds. During these catabolic reactions, ATP is synthesized by either substrate-level phosphorylation (fermentation) or oxidative phosphorylation (respiration).

- Which form(s) of ATP synthesis requires cytoplasmic membrane participation? Why?
- How does substrate-level phosphorylation differ from oxidative phosphorylation?

5.10 Glycolysis as an Example of Fermentation

The substance fermented in a fermentation is both the electron donor and electron acceptor. Not all substances can be fermented; fatty acids, for example, are too reduced to be fermentable. However, many compounds can be fermented, and sugars, especially hexoses such as glucose, are excellent examples. A common pathway for the fermentation

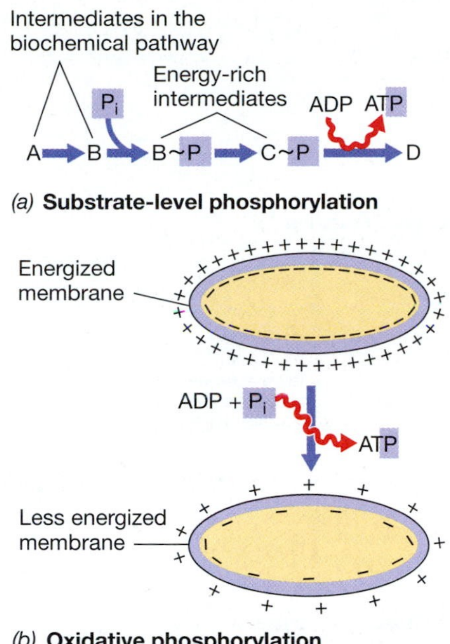

Figure 5.14 Energy conservation in fermentation and respiration. *(a)* In fermentation, substrate-level phosphorylation produces ATP. *(b)* In respiration, the cytoplasmic membrane, energized by the proton motive force, dissipates some of that energy in the formation of ATP from ADP plus inorganic phosphate (P_i) by oxidative phosphorylation. The proton motive force is coupled to ATP synthesis by way of ATP synthase (ATPase, see Figure 5.21).

of glucose is **glycolysis**, also called the *Embden–Meyerhof pathway* for its major discoverers. Several other microbial fermentations are known, and we discuss these in Chapter 21.

Glycolysis is an anaerobic process and can be divided into three stages, each involving a series of enzymatic reactions (**Figure 5.15**). Stage I comprises "preparatory" reactions; these are not redox reactions and do not release energy but lead to the production of two molecules of a key intermediate, glyceraldehyde 3-phosphate, from glucose. In Stage II, redox reactions occur, energy is conserved in the form of ATP, and two molecules of pyruvate are formed. In Stage III, redox reactions occur once again and *fermentation products* are formed (Figure 5.15).

Stage I: Preparatory Reactions

In Stage I, glucose is phosphorylated by ATP, yielding glucose 6-phosphate. The latter is then converted to fructose 6-phosphate. A second phosphorylation leads to the production of fructose 1,6-bisphosphate. The enzyme aldolase splits fructose 1,6-bisphosphate into two 3-carbon molecules, *glyceraldehyde 3-phosphate* and its isomer, *dihydroxyacetone phosphate,* which can be converted into glyceraldehyde 3-phosphate. Note that thus far, all of the reactions, including the consumption of ATP, have proceeded without redox reactions.

Stage II: The Production of NADH, ATP, and Pyruvate

The first redox reaction of glycolysis occurs in Stage II during the oxidation of glyceraldehyde 3-phosphate to 1,3-bisphosphoglyceric acid. In this reaction (which occurs twice, once for each of the two molecules of glyceraldehyde 3-phosphate produced from glucose), the enzyme glyceraldehyde-3-phosphate dehydrogenase reduces NAD^+ to NADH. Simultaneously, each glyceraldehyde-3-phosphate molecule is phosphorylated by the addition of a molecule of inorganic phosphate. This reaction, in which inorganic phosphate is converted to organic form, sets the stage for energy conservation. ATP formation is possible because 1,3-bisphosphoglyceric acid is an energy-rich compound (Figure 5.13). The synthesis of ATP occurs (1) when each molecule of 1,3-bisphosphoglyceric acid is converted to 3-phosphoglyceric acid, and (2) when each molecule of phosphoenolpyruvate is converted to pyruvate (Figure 5.15).

At this point in glycolysis, *two* ATP molecules have been consumed (in Stage I) and *four* ATP molecules have been synthesized in the reactions of Stage II (Figure 5.15). Thus, the net energy yield in glycolysis is *two* molecules of ATP per molecule of glucose fermented.

Stage III: Consumption of NADH and the Production of Fermentation Products

During the formation of two 1,3-bisphosphoglyceric acid, two NAD^+ are reduced to NADH (Figure 5.15). However, as previously discussed (Section 5.7 and Figure 5.12), NAD^+ is only an electron shuttle, not a terminal electron acceptor. Thus, the NADH produced from NAD^+ in glycolysis must be oxidized back to NAD^+ in order for glycolysis to continue, and this is accomplished by enzymes that reduce pyruvate to fermentation products (Figure 5.15). In the fermentation of yeast, pyruvate is reduced by NADH to ethanol with the subsequent production of CO_2. By contrast, in lactic acid bacteria pyruvate is reduced by NADH to lactate (Figure 5.15). Many other possibilities for pyruvate reduction in glycolysis are possible, but the net result is always the same: NADH is reoxidized to NAD^+ during the process.

In any energy-yielding process, oxidations must balance reductions, and there must be an electron acceptor for each electron removed. In this example, the reduction of NAD^+ at one enzymatic step in glycolysis is balanced by the oxidation of NADH at another. The final product(s) must also be in redox and atomic balance with the starting substrate, glucose. Hence, the fermentation products discussed here, ethanol plus CO_2, or lactate plus protons, are in both electrical and atomic balance with the starting substrate, glucose (Figure 5.15).

Glucose Fermentation: Net and Practical Results

During glycolysis glucose is consumed, two ATPs are made, and fermentation products are generated. For the organism the crucial product is ATP, which is used in a wide variety of energy-requiring reactions, and fermentation products are merely waste products. However, the latter substances are hardly considered waste products by the distiller, the brewer, the cheese maker, or the baker (see the Microbial Sidebar, "The Products of Yeast Fermentation and the Pasteur Effect"). Thus, fermentation is more than just an energy-yielding

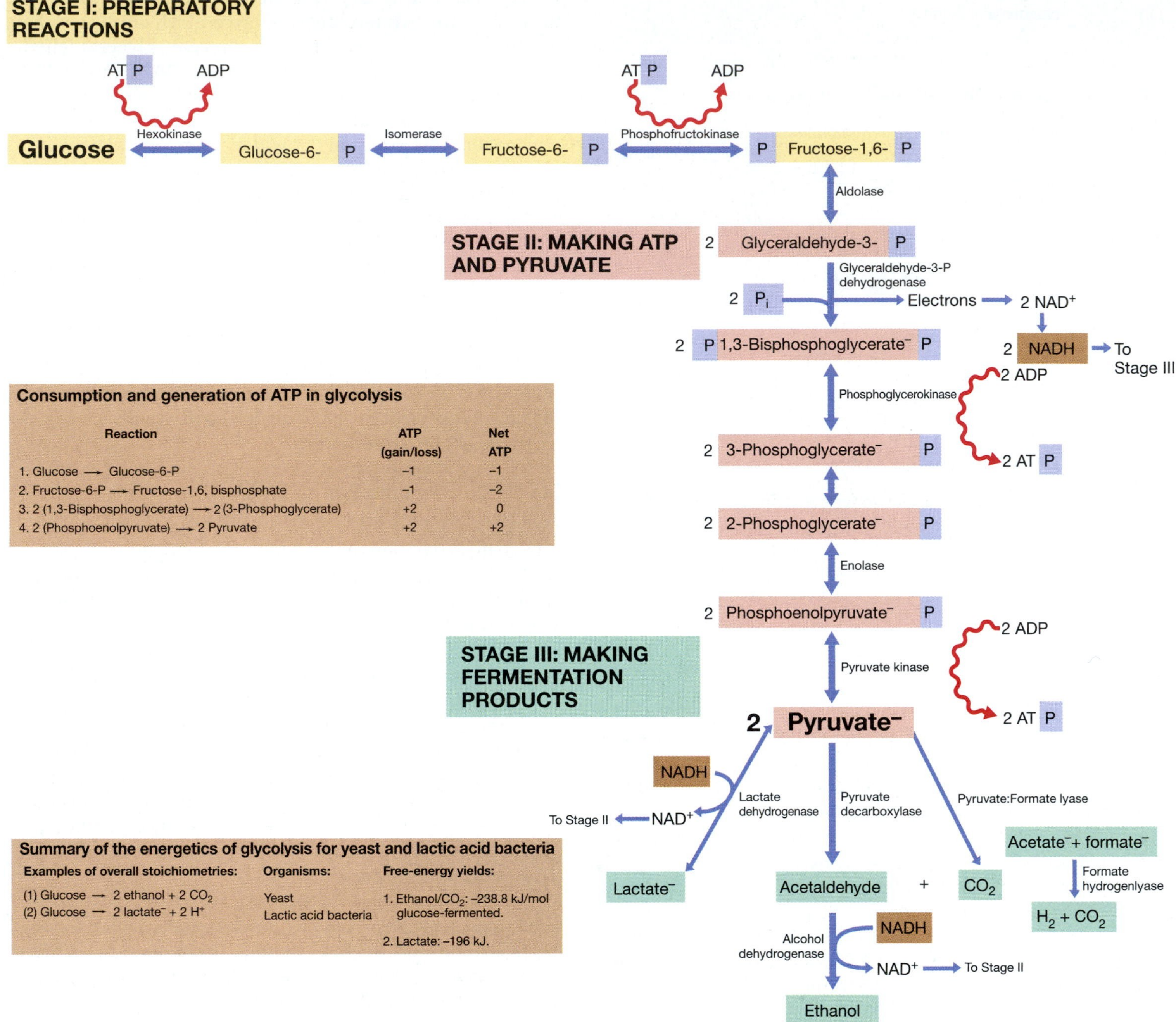

Consumption and generation of ATP in glycolysis

Reaction	ATP (gain/loss)	Net ATP
1. Glucose ⟶ Glucose-6-P	−1	−1
2. Fructose-6-P ⟶ Fructose-1,6, bisphosphate	−1	−2
3. 2 (1,3-Bisphosphoglycerate) ⟶ 2 (3-Phosphoglycerate)	+2	0
4. 2 (Phosphoenolpyruvate) ⟶ 2 Pyruvate	+2	+2

Summary of the energetics of glycolysis for yeast and lactic acid bacteria

Examples of overall stoichiometries:	Organisms:	Free-energy yields:
(1) Glucose ⟶ 2 ethanol + 2 CO_2	Yeast	1. Ethanol/CO_2: −238.8 kJ/mol glucose-fermented.
(2) Glucose ⟶ 2 lactate⁻ + 2 H^+	Lactic acid bacteria	2. Lactate: −196 kJ.

Figure 5.15 Embden–Meyerhof pathway (glycolysis). The sequence of reactions in the conversion of glucose to pyruvate and then to fermentation products (enzymes are shown in small type). The product of aldolase is actually glyceraldehyde 3-P and dihydroxyacetone P, but the latter is converted to glyceraldehyde 3-P. Note how pyruvate is the central "hub" of glycolysis—all fermentation products are made from pyruvate, and just a few common examples are given.

process. It is also a means of producing natural products useful to humans. We discuss industrial production of fermentation products in more detail in Chapter 25.

5.10 MiniReview

Glycolysis is a major pathway of fermentation and is a widespread means of anaerobic metabolism. The end result of glycolysis is the release of a small amount of energy that is conserved as ATP and the production of fermentation products. For each glucose consumed in glycolysis, two ATPs are produced.

- Which reaction(s) in glycolysis involve oxidations and reductions?
- What is the role of NAD^+/NADH in glycolysis?
- Why are fermentation products made during glycolysis?

Microbial Sidebar

The Products of Yeast Fermentation and the Pasteur Effect

Every home wine maker, brewer, and baker is an amateur microbiologist, perhaps without even realizing it. Indeed, the anaerobic processes of energy generation are at the heart of some of the most striking discoveries of the human race: fermented foods and beverages (**Figure 1**).

In the production of breads and most alcoholic beverages, the yeast *Saccharomyces cerevisiae* is exploited to produce ethanol plus CO_2. Found in various sugar-rich environments such as fruit juices and nectar, yeasts can carry out the two opposing modes of chemoorganotrophic metabolism discussed in this chapter, *fermentation* and *respiration*. When oxygen is present in high amounts, yeast grows efficiently on various sugars, making yeast cells and CO_2 (the latter from the citric acid cycle, Section 5.13) in the process. However, when conditions are anoxic, yeasts switch to fermentative metabolism. This results in a reduced cell yield but significant amounts of alcohol and CO_2.

Figure 1 *Major products in which fermentation by the yeast* Saccharomyces cerevisiae *is critical.*

The early microbiologist Louis Pasteur (Section 1.7) recognized this change in metabolic patterns of yeast during his studies on fermentation. Pasteur showed that the ratio of glucose consumed by a yeast suspension (which he called "the ferment") to the weight of cells produced varied with the concentration of oxygen supplied; the maximum ratio occured in the absence of O_2. In Pasteur's own words, "the ferment lost its fermentative abilities in proportion to the concentration of this gas."

Pasteur describes what has come to be known as the "Pasteur effect," a phenomenon that occurs in any organism (even humans) that can both ferment and respire glucose. The fermentation of glucose occurs maximally under anoxic conditions and is inhibited by O_2 because respiration yields much more energy per glucose than does fermentation.

The Pasteur effect occurs in the alcoholic beverage fermentation. When grapes are squeezed to make juice, called *must*, small numbers of yeast cells present on the grapes are transferred to the must. During the first several days of the wine-making process, yeast grow primarily by respiration and consume O_2, making the juice anoxic. The yeast respire the glucose in the juice rather than ferment it because more energy is available from the respiration of glucose than from its fermentation. However, as soon as the oxygen in the grape juice is depleted, fermentation begins along with alcohol formation. This switch from aerobic to anaerobic metabolism is crucial in wine making, and care must be taken to ensure that air is kept out of the fermentation vessel.

Wine is only one of many alcoholic products made with yeast. Others include beer (Microbial Sidebar, "Home Brew," Chapter 25) and distilled spirits such as brandy, whisky, vodka, and gin. In distilled spirits, the ethanol, produced in relatively low amounts (10–15% by volume) by the yeast, is concentrated by distilling to make a beverage containing 40–70% alcohol. Even alcohol for motor fuel is made with yeast in parts of the world where sugar is plentiful but petroleum is in short supply (such as Brazil). In the United States, ethyl alcohol for use as an industrial solvent and motor fuel is produced using corn starch as a source of glucose. Yeast also serves as the leavening agent in bread, although here it is not the alcohol that is important, but CO_2, the *other* product of the alcohol fermentation (Figure 5.15). We discuss yeast and yeast products in Chapter 25. The CO_2 raises the dough, and the alcohol produced along with it is volatilized during the baking process.

We should thus appreciate the impact that the yeast cell, forced to carry out a fermentative lifestyle because the oxygen it needs for respiration is absent, has had on the lives of humans. Substances that from a physiological standpoint are "waste products" of the glycolytic pathway—ethanol plus CO_2—are the basis of the alcoholic beverage and baking industries, respectively.

5.11 Respiration and Membrane-Associated Electron Carriers

Fermentation occurs anaerobically in the absence of usable external electron acceptors, and it releases only a relatively small amount of energy. Only a few ATP molecules are synthesized as a result. By contrast, if O_2 or some other usable terminal acceptor is present, the production of fermentation products is unnecessary and glucose can be oxidized completely to CO_2. When this occurs, a far higher yield of ATP is possible. Oxidation using O_2 as the terminal electron acceptor is called *aerobic respiration*.

Our discussion of aerobic respiration will deal with both the carbon transformations and the redox reactions, and will thus focus on two issues: (1) the way electrons are transferred from the organic compound to the terminal electron acceptor and (2) the pathway by which organic carbon is converted into CO_2. During the former, ATP synthesis occurs at the expense of the proton motive force (Figure 5.14*b*); thus we begin with a discussion of electron transport, the series of reactions that lead to the proton motive force.

Electron Transport Carriers

Electron transport systems are membrane associated. These systems have two basic functions. First, electron transport systems mediate the transfer of electrons from primary donor to terminal acceptor. And second, these systems conserve some of the energy released during this process and use it to synthesize ATP.

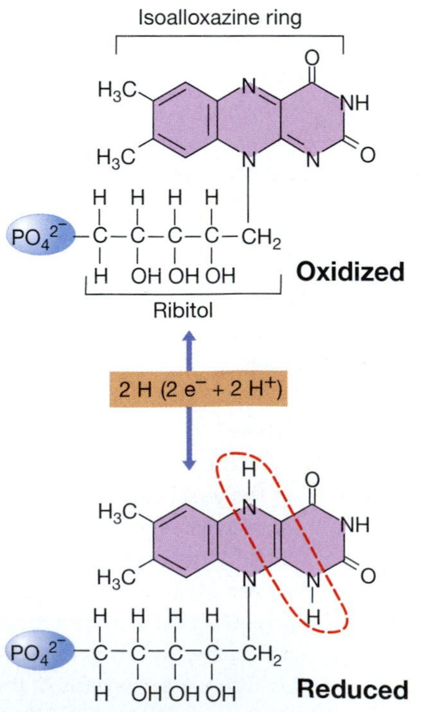

Figure 5.16 Flavin mononucleotide (FMN), a hydrogen atom carrier. The site of oxidation–reduction (dashed red circle) is the same in FMN and flavin-adenine dinucleotide (FAD).

Several types of oxidation–reduction enzymes participate in electron transport. These include *NADH dehydrogenases, flavoproteins, iron-sulfur proteins,* and *cytochromes*. In addition, a class of nonprotein electron carriers exists in the *quinones*. The carriers are arranged in the membrane in order of their increasingly positive reduction potentials, with NADH dehydrogenase first and the cytochromes last (see Figure 5.20).

NADH dehydrogenases are proteins bound to the inside surface of the cytoplasmic membrane. They have an active site that binds NADH and accepts $2\ e^- + 2\ H^+$ when NADH is converted to NAD^+ (Figures 5.11 and 5.12). The $2\ e^- + 2\ H^+$ are then passed on to flavoproteins, the next carrier in the chain.

Flavoproteins contain a derivative of the vitamin riboflavin (**Figure 5.16**). The flavin portion, which is bound to a protein, is a prosthetic group that is reduced as it accepts electrons plus protons and oxidized when electrons are passed on to the next carrier in the chain. Note that flavoproteins *accept* $2\ e^- + 2\ H^+$ but *donate* only electrons. We will consider what happens to the two protons later. Two flavins are common in cells, flavin mononucleotide (FMN) (Figure 5.16) and flavin-adenine dinucleotide (FAD). In the latter, FMN is bonded to ribose and adenine through a second phosphate. Riboflavin, also called vitamin B_2, is a source of the parent flavin molecule in flavoproteins and is a required growth factor for some organisms.

The cytochromes are proteins that contain heme prosthetic groups (**Figure 5.17**). Cytochromes undergo oxidation and reduction through loss or gain of a single electron by the iron atom in the heme of the cytochrome:

$$\text{Cytochrome} -Fe^{2+} \longleftrightarrow \text{Cytochrome} -Fe^{3+} + e^-$$

Several classes of cytochromes are known, differing widely in their reduction potentials (Figure 5.10). Different classes of cytochromes are designated by letters, such as cytochrome *a*, cytochrome *b*, cytochrome *c*, and so on, depending upon the type of heme they contain. The cytochromes of a given class in one organism may differ slightly from those of another, and so there are designations such as cytochromes a_1, a_2, a_3, and so on among cytochromes of the same class. Occasionally, cytochromes form complexes with other cytochromes or with iron-sulfur proteins. An example is the cytochrome bc_1 complex, which contains two different *b*-type cytochromes and one *c*-type cytochrome. The cytochrome bc_1 complex plays an important role in energy metabolism, as we will see later.

In addition to the cytochromes, in which iron is bound to heme, one or more proteins with iron not bound to heme are typically present in electron transport chains. These proteins contain clusters of iron and sulfur atoms, with Fe_2S_2 and Fe_4S_4 clusters being the most common. The iron atoms are bonded to free sulfur and to the protein via sulfur atoms from cysteine residues (**Figure 5.18**). Ferredoxin, a common iron-sulfur protein, has an Fe_2S_2 configuration.

The reduction potentials of iron-sulfur proteins vary over a wide range depending on the number of iron and sulfur atoms present and how the iron centers are embedded in the

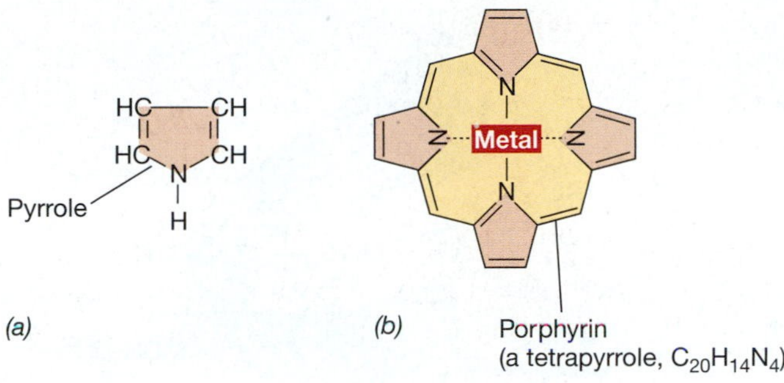

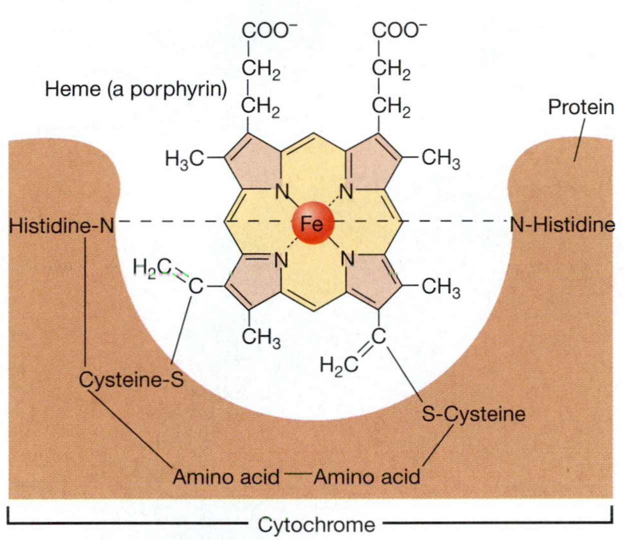

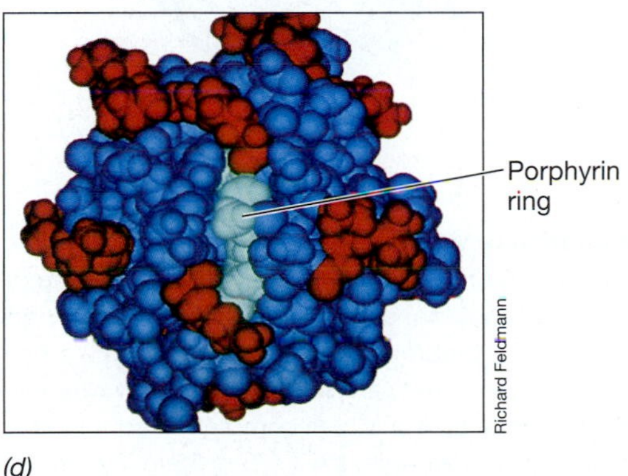

Figure 5.17 Cytochrome and its structure. *(a)* Structure of the pyrrole ring. *(b)* Four pyrrole rings are condensed to form the porphyrin ring. *(c)* In cytochrome *c* (shown), the porphyrin ring is covalently linked via disulfide bridges to cysteine molecules in the protein. *(d)* Computer-generated model of cytochrome *c*. The protein completely surrounds the porphyrin ring (light color) in the center. Cytochromes carry electrons only; the redox site is the iron atom, which can alternate between the Fe^{2+} and Fe^{3+} oxidation states.

protein. Thus, different iron-sulfur proteins can function at different locations in the electron transport process. Like cytochromes, iron-sulfur proteins carry electrons only.

Quinones (**Figure 5.19**) are hydrophobic nonprotein-containing molecules involved in electron transport. Some quinones found in bacteria are related to vitamin K, a growth

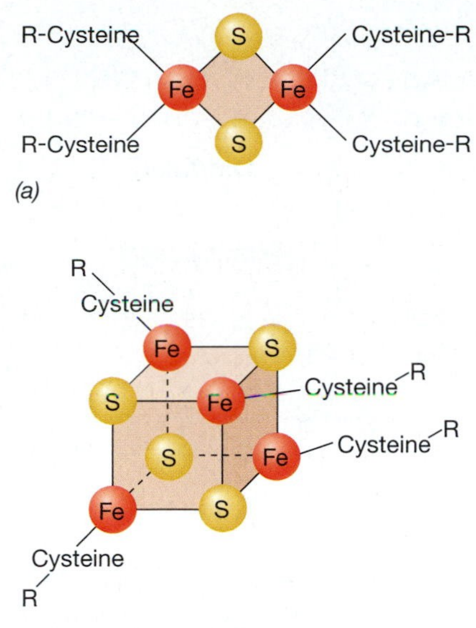

Figure 5.18 Arrangement of the iron-sulfur centers of nonheme iron-sulfur proteins. *(a)* Fe_2S_2 center. *(b)* Fe_4S_4 center. The cysteine linkages are from the protein portion of the molecule.

factor for higher animals. Like flavoproteins, quinones accept $2\ e^- + 2\ H^+$, but transfer only electrons to the next carrier in the chain.

5.11 MiniReview

Electron transport systems consist of a series of membrane-associated electron carriers that function in an integrated fashion to carry electrons from the primary electron donor to a terminal electron acceptor such as oxygen.

- In what major way do quinones differ from other electron carriers in the membrane?
- Which electron carriers described in this section accept $2\ e^- + 2\ H^+$? Which accept electrons only?

5.12 Respiration and the Proton Motive Force

The production of ATP by oxidative phosphorylation is linked to an energized state of the membrane. This energized state is established by electron transport reactions with the electron carriers we have just discussed; we see how this occurs now.

The Proton Motive Force

To understand how electron transport is linked to ATP synthesis, we must first understand how the electron transport system is oriented in the cytoplasmic membrane. Electron transport carriers are oriented in the membrane in such a way that, as electrons are transported, protons are separated from electrons. Two electrons plus two protons enter the electron

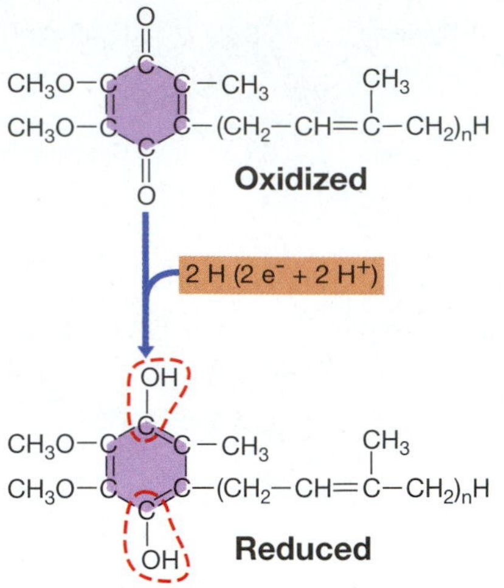

Figure 5.19 Structure of oxidized and reduced forms of coenzyme Q, a quinone. The five-carbon unit in the side chain (an isoprenoid) occurs in a number of multiples. In prokaryotes, the most common number is $n = 6$; in eukaryotes, $n = 10$. Note that oxidized quinone requires 2 e^- and 2 H^+ to become fully reduced (dashed red circles). An intermediate form, the semiquinone (one H more reduced than oxidized quinone), is formed during the reduction of a quinone.

transport chain from NADH to initiate the process. Carriers in the electron transport chain are arranged in the membrane in order of their increasingly positive reduction potential, with the final carrier in the chain donating the electrons plus protons to a terminal electron acceptor such as O_2 (**Figure 5.20**). During this process, several protons are released into the environment, causing a slight acidification of the external surface of the membrane. These protons originate from two sources: (1) NADH and (2) the dissociation of water into H^+ and OH^- in the cytoplasm. The extrusion of H^+ to the environment causes OH^- to accumulate on the inside of the membrane. Despite their small size, neither H^+ nor OH^- can diffuse through the membrane because they are charged (Section 4.3). Thus, equilibrium cannot be spontaneously restored.

The result of electron transport is the generation of a pH gradient and an electrochemical potential across the membrane (Figure 5.20). The *inside* surface of the membrane becomes electrically negative and alkaline and the *outside* surface of the membrane electrically positive and acidic. This pH gradient and electrochemical potential, collectively called the **proton motive force**, causes the membrane to be energized, much like a battery. Some of this energy is then conserved in the formation of ATP. Besides driving ATP synthesis, the energized state of the membrane can also be used to do work, such as ion transport, flagellar rotation, and many other energy-requiring reactions in the cell.

We now consider the individual electron transport reactions that lead to formation of the proton motive force.

Generation of the Proton Motive Force: Complexes I and II

The proton motive force develops from the activities of flavin enzymes, quinones, the cytochrome bc_1 complex, and the

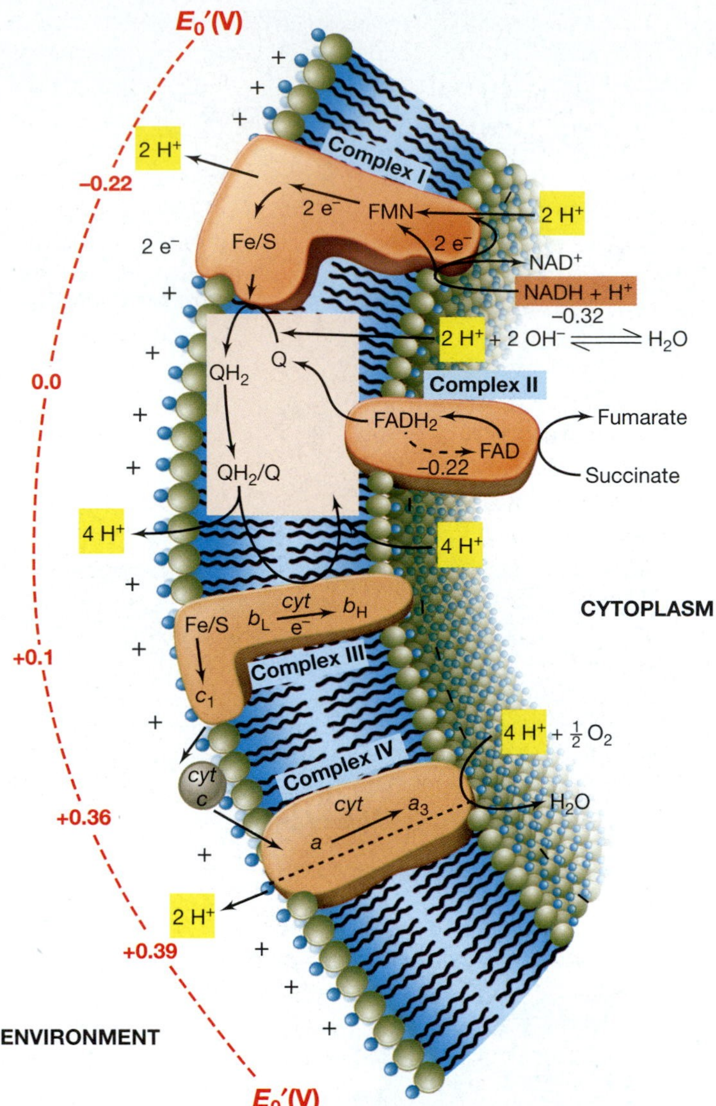

Figure 5.20 Generation of the proton motive force during aerobic respiration. The orientation of electron carriers in the membrane of *Paracoccus denitrificans*, a model for studies of respiration. The + and – charges at the edges of the membrane represent H^+ and OH^-, respectively. E_0' values for the major carriers are shown. Note how when a hydrogen atom carrier (for example, FMN in complex I) reduces an electron-accepting carrier (for example, the Fe/S protein in complex I) protons are extruded to the outer surface of the membrane. Abbreviations: FMN, flavoprotein; FAD, flavin adenine dinucleotide; Q, quinone; Fe/S, iron sulfur protein; cyt *a*, *b*, *c*, cytochromes (b_L and b_H, low and high potential *b*-type cytochromes, respectively). At the quinone site electrons are recycled during the "Q cycle." This is because electrons from QH_2 can be split in the bc_1 complex (Complex III) between the Fe/S protein and the *b*-type cytochromes. Electrons that travel through the cytochromes reduce Q (in two, one-electron steps) back to QH_2, thus increasing the number of protons pumped at the Q-bc_1 site. Electrons that travel to Fe/S proceed to reduce cytochrome c_1, then cytochrome *c*, and then *a*-type cytochromes in Complex IV, eventually reducing O_2 to H_2O (two electrons and two protons are required to reduce $\frac{1}{2}O_2$ to H_2O, and these come from electrons through cyt *c* and cytoplasmic protons, respectively). Complex II, the succinate dehydrogenase complex, bypasses Complex I and feeds electrons directly into the quinone pool.

terminal cytochrome oxidase. Following the donation of 2 e^- + 2 H^+ from NADH to FAD to form FADH, 2 H^+ are extruded to the outer surface of the membrane when FADH donates 2 e^- to a nonheme iron protein, part of membrane protein *Complex I,* shown in Figure 5.20. These electron carriers are referred to as complexes because each actually consists of several proteins. For example, Complex I in *Escherichia coli* contains 14 distinct proteins. Complex I is also called *NADH:quinone oxidoreductase,* because the reaction is one in which NADH is oxidized and quinone is reduced. Two protons are taken up from the dissociation of water in the cytoplasm when the iron protein of Complex I reduces coenzyme Q (Figure 5.20).

Complex II simply bypasses Complex I and feeds electrons and protons from FADH directly into the quinone pool. Complex II is also called the *succinate dehydrogenase complex* because of the specific substrate, succinate (a product of the citric acid cycle, Section 5.13), that it oxidizes. However, because Complex II bypasses Complex I, fewer protons are pumped per two electrons that enter the electron transport chain here than for two electrons that enter from NADH (Figure 5.20).

Complexes III and IV: *bc*$_1$ and *a*-Type Cytochromes

Reduced coenzyme Q passes electrons one at a time to the cytochrome bc_1 complex (*Complex III,* Figure 5.20. The cytochrome bc_1 complex is an association of several proteins containing hemes (Figure 5.17) or other metal centers. These include two *b* type hemes (b_L and b_H), one *c* type heme (c_1), and one iron-sulfur protein. The bc_1 complex is present in the electron transport chain of most organisms that can respire. It also plays a role in photosynthetic electron flow (Sections 20.4 and 20.5).

The major function of the cytochrome bc_1 complex is to transfer electrons from quinones to cytochrome *c* (Figure 5.20). Electrons travel from the bc_1 complex to a molecule of cytochrome *c*, located in the periplasm. Cytochrome *c* functions as a shuttle to transfer electrons to the high-potential cytochromes *a* and a_3 (*Complex IV,* Figure 5.20). Complex IV is the terminal oxidase and reduces O_2 to H_2O in the final step of the electron transport chain. Complex IV also pumps two H^+ to the outer surface of the membrane per each 2 e^- consumed in the reaction (Figure 5.20).

Besides transferring electrons to cytochrome *c*, the cytochrome bc_1 complex can also interact with the quinone pool in such a way that two additional protons are pumped at the Q-bc_1 site (Figure 5.20). This happens in a series of electron exchanges between cytochrome bc_1 and Q, called the *Q cycle.* Because quinone and bc_1 have roughly the same E_0' (near 0 V, Figure 5.20), different molecules of quinone can alternately oxidize and reduce one another using electrons fed back from the bc_1 complex. This mechanism allows on average a total of 4H^+ to be pumped to the outer surface of the membrane at the Q-bc_1 site for every 2 e^- that enter the chain in Complex I (Figure 5.20).

The electron transport chain shown in Figure 5.20 is one of many different sequences of electron carriers known from different organisms. However, three features are characteristic of all electron transport chains: (1) arrangement of membrane-associated electron carriers in order of increasingly more positive E_0', (2) alternation of electron-only and electron-plus-proton carriers in the chain, and (3) generation of a proton motive force (Figure 5.20).

As we will see now, it is this last characteristic, the proton motive force, that drives ATP synthesis.

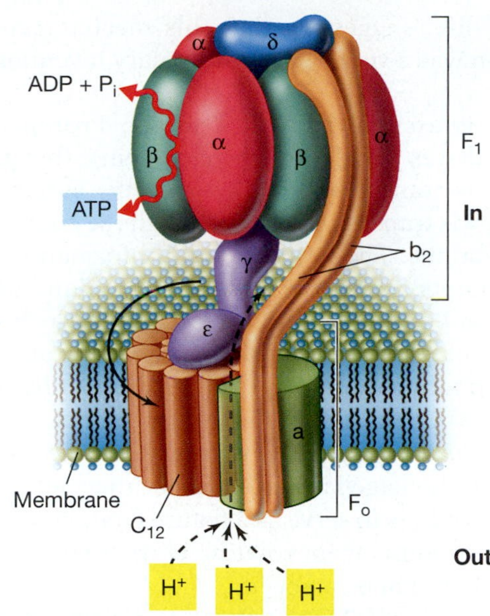

Figure 5.21 Structure and function of ATP synthase (ATPase). F_1 consists of five different polypeptides present as an $\alpha_3\beta_3\gamma\varepsilon\delta$ complex. F_1 is the catalytic complex responsible for the interconversion of ADP + P_i and ATP. F_o is integrated in the membrane and consists of three polypeptides in an ab_2c_{12} complex. Protons cross the membrane between the a and c_{12} subunits. Subunit b protrudes outside the membrane and forms, along with the b_2 and δ subunits, the stator. As protons enter, the dissipation of the proton motive force drives ATP synthesis. The ATPase is reversible in its action; that is, ATP hydrolysis can drive formation of a proton motive force.

ATPase

How does the proton motive force generated by electron transport (Figure 5.20) actually drive ATP synthesis? Interestingly, there is a strong parallel here between the mechanism of ATP synthesis and the mechanism of the motor that drives rotation of the bacterial flagellum (Section 4.13 and Figure 4.47). In a fashion similar to the way the proton motive force applies torque to the bacterial flagellum, it also applies force to a large protein complex that makes ATP. The complex that converts the proton motive force into ATP is called **ATP synthase,** or **ATPase** for short.

ATPases consist of two components, a multiprotein extramembrane complex called F_1 that faces into the cytoplasm and a proton-conducting intramembrane channel called F_o (**Figure 5.21**). ATPase catalyzes a reversible reaction between ATP and ADP + P_i as shown in Figure 5.21. The structure of ATPase proteins is highly conserved throughout all the domains of life, suggesting that this mechanism of energy conservation was a very early evolutionary invention (Section 14.2).

ATPase is a small biological motor. Proton movement through F_o drives rotation of the *c* proteins; this generates a torque that is transmitted to F_1 by the $\gamma\varepsilon$ subunits (Figure 5.21). Energy is transferred to F_1 through the coupled rotation of the $\gamma\varepsilon$ subunits. The latter causes conformational changes in the β subunits, and this is a form of potential energy that can be tapped to make ATP. This is possible because the conformational changes in the β subunits allow for the sequential binding of ADP + P_i to each subunit. ATP is synthesized when the β subunits return to their original conformation, releasing the energy needed to drive the process. In analogy to the flagellar motor, the primary function of the $b_2\delta$ subunits of F_1 is to serve as a fixture (stator). This prevents the α and β subunits from rotating with $\gamma\varepsilon$ such that conformational changes in β can occur.

ATPase-catalyzed ATP synthesis is called oxidative phosphorylation if the proton motive force originates from respiration reactions and photophosphorylation if it originates from photosynthetic reactions. Measurements of the stoichiometry between the number of protons consumed by ATPase per ATP produced yield a number of 3–4.

Reversibility of the ATPase

ATPase is reversible. The hydrolysis of ATP supplies torque for γ to rotate in the opposite direction from that in ATP synthesis, and this catalyzes the pumping of protons from the inside to the outside of the cell. This generates instead of dissipates the proton motive force. Reversibility of the ATPase explains why strictly fermentative organisms that lack electron transport chains and are unable to carry out oxidative phosphorylation still have ATPases. As we have said, many important reactions in the cell, such as motility and transport, require energy from the proton motive force rather than from ATP (Sections 4.5 and 4.13). Thus, ATPase in nonrespiratory organisms such as the strictly fermentative lactic acid bacteria, for example, functions unidirectionally to generate the proton motive force necessary to drive these cell functions.

Inhibitors and Uncouplers

Electron transport reactions have long been studied with chemicals that affect electron flow or the proton motive force. Two classes of such chemicals are known: *inhibitors* and *uncouplers*. Inhibitors block electron flow and, thus, establishment of the proton motive force. Examples include carbon monoxide (CO) and cyanide (CN^-), both of which bind tightly to *a*-type cytochromes (Figure 5.20) and prevent their functioning. By contrast, uncouplers prevent ATP synthesis without affecting electron transport. Uncouplers are lipid-soluble substances, such as dinitrophenol and dicumarol. These substances make membranes leaky, thereby destroying the proton motive force and its ability to drive ATP synthesis (Figure 5.21). Thus inhibitors and uncouplers both block ATP synthesis but for distinctly different reasons.

5.12 MiniReview

When electrons are transported through an electron transport chain, protons are extruded to the outside of the membrane forming the proton motive force. Key electron carriers include flavins, quinones, the cytochrome bc_1 complex, and other cytochromes, depending on the organism. The cell uses the proton motive force to make ATP through the activity of ATPase.

- How do electron transport reactions generate the proton motive force?
- What is the ratio of protons extruded per NADH oxidized through the electron transport chain of *Paracoccus* shown in Figure 5.20? At which sites in the chain is the proton motive force being established?
- What structure in the cell converts the proton motive force to ATP? How does it operate?

5.13 Carbon Flow in Respiration: The Citric Acid Cycle

Now that we have a grasp of how ATP is made in respiration, we need to consider the important reactions in carbon metabolism associated with formation of ATP. Our focus here is on the citric acid cycle, a key pathway found in virtually all cells.

The Respiration of Glucose

The early steps in the respiration of glucose are the same biochemical steps as those of glycolysis; all steps from glucose to pyruvate (Figure 5.15) are the same. However, whereas in fermentation pyruvate is reduced and converted into fermentation products that are subsequently excreted, in respiration pyruvate is oxidized to CO_2. The pathway by which pyruvate is completely oxidized to CO_2 is called the **citric acid cycle** (CAC), summarized in **Figure 5.22**.

Pyruvate is first decarboxylated, leading to the production of CO_2, NADH, and the energy-rich substance, *acetyl-CoA* (Figure 5.13). The acetyl group of acetyl-CoA then combines with the four-carbon compound oxalacetate, forming citric acid (Figure 5.22). A series of hydration, decarboxylation, and oxidation reactions follow, and two additional CO_2 molecules, three more NADH, and one FADH are formed. Ultimately, oxalacetate is regenerated and returns as an acetyl acceptor, thus completing the cycle (Figure 5.22).

CO_2 Release and Fuel for Electron Transport

For each pyruvate molecule oxidized through the citric acid cycle, three CO_2 molecules are released (Figure 5.22). Electrons released during the enzymatic oxidation of intermediates in the CAC are transferred to NAD^+ to form NADH or FAD to form FADH. Here's where respiration and fermentation differ in a major way. Instead of being used in the reduction of pyruvate as in fermentation (Figure 5.15), in respiration electrons from NADH and FADH are transferred to oxygen by way of

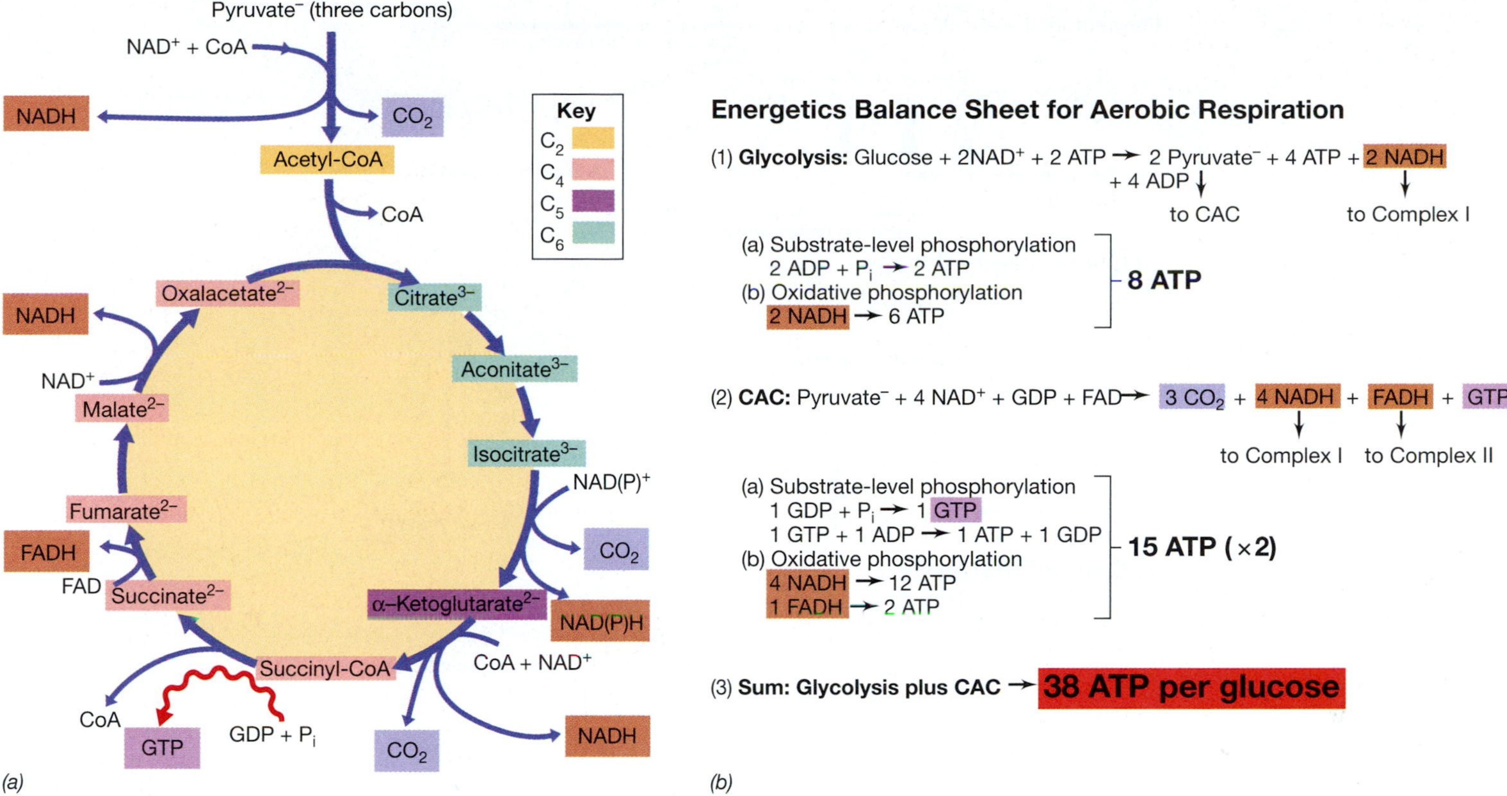

Figure 5.22 The citric acid cycle. *(a)* The citric acid cycle begins when the two-carbon compound acetyl-CoA condenses with the four-carbon compound oxalacetate to form the six-carbon compound citrate. Through a series of oxidations and transformations, this six-carbon compound is ultimately converted back to the four-carbon compound oxalacetate, which then begins another cycle with addition of the next molecule of acetyl-CoA. *(b)* The overall balance sheet of fuel (NADH/FADH) for the electron transport chain and CO_2 generated in the citric acid cycle. NADH and FADH feed into electron transport chain Complexes I and II, respectively (Figure 5.20).

the electron transport chain. Thus, unlike in fermentation, the presence of an electron acceptor (O_2) in respiration allows for the complete oxidation of glucose to CO_2 with a much greater yield of energy (Figure 5.22*b*). Thus while only *2 ATP* are produced per glucose fermented in the alcoholic or lactic acid fermentations (Figure 5.15), *38 ATP* can be made by respiring the same glucose molecule to CO_2 (Figure 5.22).

Biosynthesis and the Citric Acid Cycle

Besides playing a key role in catabolism, the citric acid cycle is vital to the cell for other reasons as well. This is because the cycle generates several key biochemical compounds that can be drawn off for biosynthetic purposes when needed. Particularly important in this regard are α-ketoglutarate and oxalacetate, which are precursors of a number of amino acids (Section 5.16), and succinyl-CoA, needed to form cytochromes, chlorophyll, and several other tetrapyrrole compounds (Figure 5.17). Oxalacetate is also important because it can be converted to phosphoenolpyruvate, a precursor of glucose. In addition, acetyl-CoA provides the starting material for fatty acid biosynthesis (Section 5.17). The citric acid cycle thus plays two major roles in the cell: *bioenergetic* and *biosynthetic*.

Much the same can be said about the glycolytic pathway, as certain intermediates from this pathway are drawn off for various biosynthetic needs as well (Section 5.16).

5.13 MiniReview

Respiration results in the complete oxidation of an organic compound with much greater energy release than occurs during fermentation. The citric acid cycle plays a major role in the respiration of organic compounds.

- How many molecules of CO_2 and pairs of electrons are released per pyruvate oxidized in the citric acid cycle?
- What two major roles do the citric acid cycle and glycolysis have in common?

5.14 Catabolic Diversity

Thus far in this chapter we have dealt only with the reactions of chemoorganotrophs. We now briefly consider catabolic diversity, some of the alternatives to the use of organic

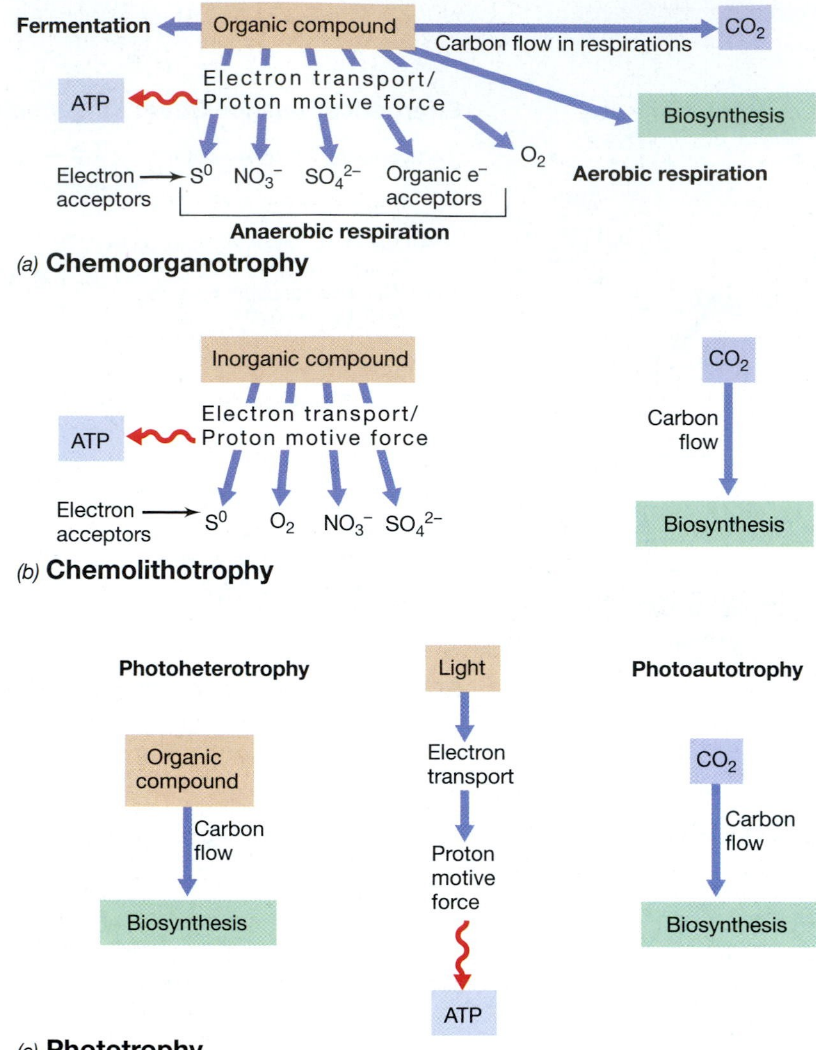

Figure 5.23 Catabolic diversity. *(a)* Chemoorganotrophic metabolism, *(b)* chemolithotrophic metabolism, and *(c)* phototrophic metabolism. Note how in phototrophic metabolism carbon for biosynthesis can come from CO_2 (photoautotrophy) or organic compounds (photoheterotrophy). Note also the importance of electron transport leading to proton motive force formation in each case.

compounds as energy sources, with an emphasis on electron and carbon flow. **Figure 5.23** summarizes the mechanisms by which cells generate energy other than by fermentation and aerobic respiration. These include anaerobic respiration, chemolithotrophy, and phototrophy.

Anaerobic Respiration

Under anoxic conditions, electron acceptors other than oxygen can be used to support respiration. These processes are called **anaerobic respiration**. Some of the electron acceptors used in anaerobic respiration include nitrate (NO_3^-), ferric iron (Fe^{3+}), sulfate (SO_4^{2-}), carbonate (CO_3^{2-}), and even certain organic compounds. Because of their positions on the redox tower (none of these acceptors has an E_0' as positive as the O_2/H_2O couple; Figure 5.10), less energy is released when these electron acceptors are used instead of oxygen (recall that $\Delta G^{0\prime}$ is proportional to $\Delta E_0'$; Section 5.6). Nevertheless, because oxygen is often limiting in anoxic environments, anaerobic respirations can be very important for supporting the growth of prokaryotes in such habitats. Like aerobic respiration, anaerobic respirations depend on electron transport, generation of a proton motive force, and the activity of ATPase.

Chemolithotrophy

Organisms able to use inorganic chemicals as electron donors are called **chemolithotrophs**. Examples of relevant inorganic electron donors include hydrogen sulfide (H_2S), hydrogen gas (H_2), ferrous iron (Fe^{2+}), and ammonia (NH_3).

Chemolithotrophic metabolism is typically aerobic, but begins with the oxidation of an inorganic rather than an organic electron donor (Figure 5.23). However, like chemoorganotrophs, chemolithotrophs have electron transport chains and form a proton motive force. However, one important distinction between chemolithotrophs and chemoorganotrophs, besides their electron donors, is their sources of carbon for biosynthesis. Chemoorganotrophs use organic compounds (glucose, acetate, and the like) as carbon sources. By contrast, chemolithotrophs use carbon dioxide (CO_2) as a carbon source and are therefore **autotrophs**. We consider many examples of chemolithotrophy in Chapter 20.

Phototrophy

Many microorganisms are **phototrophs**, using light as an energy source in the process of photosynthesis. The mechanisms by which light is used as an energy source are complex, but the end result is the same as in respiration: generation of a proton motive force that can be used in the synthesis of ATP. Light-mediated ATP synthesis is called **photophosphorylation**. Most phototrophs use energy conserved in ATP for the assimilation of carbon dioxide as the carbon source for biosynthesis; they are called *photoautotrophs*. However, some phototrophs use organic compounds as carbon sources with light as the energy source; these are the *photoheterotrophs* (Figure 5.23).

As we mentioned in Chapter 2, photosynthesis can be of two different types: oxygenic and anoxygenic. Oxygenic photosynthesis, carried out by cyanobacteria and their relatives, is similar to that of higher plants and results in O_2 evolution. Anoxygenic photosynthesis is a simpler process found in purple and green bacteria and in which O_2 evolution does not occur. The reactions leading to proton motive force formation in both forms of photosynthesis have strong parallels, as we see in Chapter 20.

The Proton Motive Force and Catabolic Diversity

Microorganisms show an amazing diversity of bioenergetic strategies. Thousands of organic compounds, many inorganic compounds, and light can be used by one or another microorganism as an energy source. However, with the exception of fermentations, where substrate-level phosphorylation occurs, ATP synthesis in respiration and photosynthesis requires generation of a proton motive force.

Whether electrons come from the oxidation of organic or inorganic chemicals or from phototrophic processes, in all forms of respiration and photosynthesis, electron transport occurs and energy conservation is linked to the proton motive force through ATPase (Figure 5.21). Considered in this way, respiration and anaerobic respiration are simply metabolic variations employing different electron acceptors. Likewise, chemoorganotrophy, chemolithotrophy, and photosynthesis are simply variations upon a theme of different electron donors. Electron transport and the proton motive force link all of these processes, bringing these seemingly quite different forms of metabolism into a common focus. We pick up on this theme in Chapters 20 and 21.

5.14 MiniReview

Electron acceptors other than O_2 can function as terminal electron acceptors for energy generation. Because O_2 is absent under these conditions, the process is called anaerobic respiration. Chemolithotrophs use inorganic compounds as electron donors, whereas phototrophs use light to form a proton motive force. The proton motive force operates in all forms of respiration and photosynthesis.

- In terms of their electron donor(s), how do chemoorganotrophs differ from chemolithotrophs?
- What is the carbon source for autotrophic organisms?
- How do photoautotrophs differ from photoheterotrophs?

V ESSENTIALS OF ANABOLISM

We close this chapter with a consideration of biosynthesis. Our focus here will be on biosynthesis of the individual units—sugars, amino acids, nucleotides, and fatty acids—that make up the four classes of macromolecules. Collectively, these biosynthetic processes are called **anabolism**. In Chapter 7 we consider synthesis of the macromolecules themselves, in particular, nucleic acids and proteins.

Many detailed biochemical pathways lie behind the metabolic patterns we present here, but we will keep our focus on the essential principles. We finish this unit with a brief look at how the enzyme activities in these biosynthetic processes are controlled by the cell. For a cell to be competitive, it must regulate its metabolism. This happens in several ways and at several levels, one of which, the control of enzyme activity, is relevant to our discussion here.

5.15 Biosynthesis of Sugars and Polysaccharides

Polysaccharides are key constituents of the cell walls of many organisms, and in *Bacteria*, the peptidoglycan cell wall (∞ Section 4.6) has a polysaccharide backbone. In addition, cells often store carbon and energy reserves in the form of the polysaccharides glycogen or starch (∞ Sections 3.3 and 4.10). The monomeric units of these polysaccharides are six-carbon sugars called *hexoses*, in particular, glucose or glucose derivatives. In addition to hexoses, five-carbon sugars called *pentoses* are common in the cell. Most notably, these include ribose and deoxyribose, present in the backbone of RNA and DNA, respectively.

In prokaryotes, polysaccharides are synthesized from either uridine diphosphoglucose (UDPG; **Figure 5.24a**) or adenosine-diphosphoglucose (ADPG), both of which are activated forms of glucose. ADPG is the precursor for the biosynthesis of glycogen. UDPG is the precursor of various glucose derivatives needed for the biosynthesis of other polysaccharides in the cell, such as *N*-acetylglucosamine and *N*-acetylmuramic acid in peptidoglycan or the lipopolysaccharide component of the gram-negative outer membrane (∞ Sections 4.6 and 4.7).

When a cell is growing on a hexose such as glucose, obtaining glucose for polysaccharide synthesis is obviously not a problem. But when the cell is growing on other carbon compounds, glucose must be synthesized. This process, called *gluconeogenesis*, uses phosphoenolpyruvate, one of the intermediates of glycolysis (Figure 5.15), as starting material.

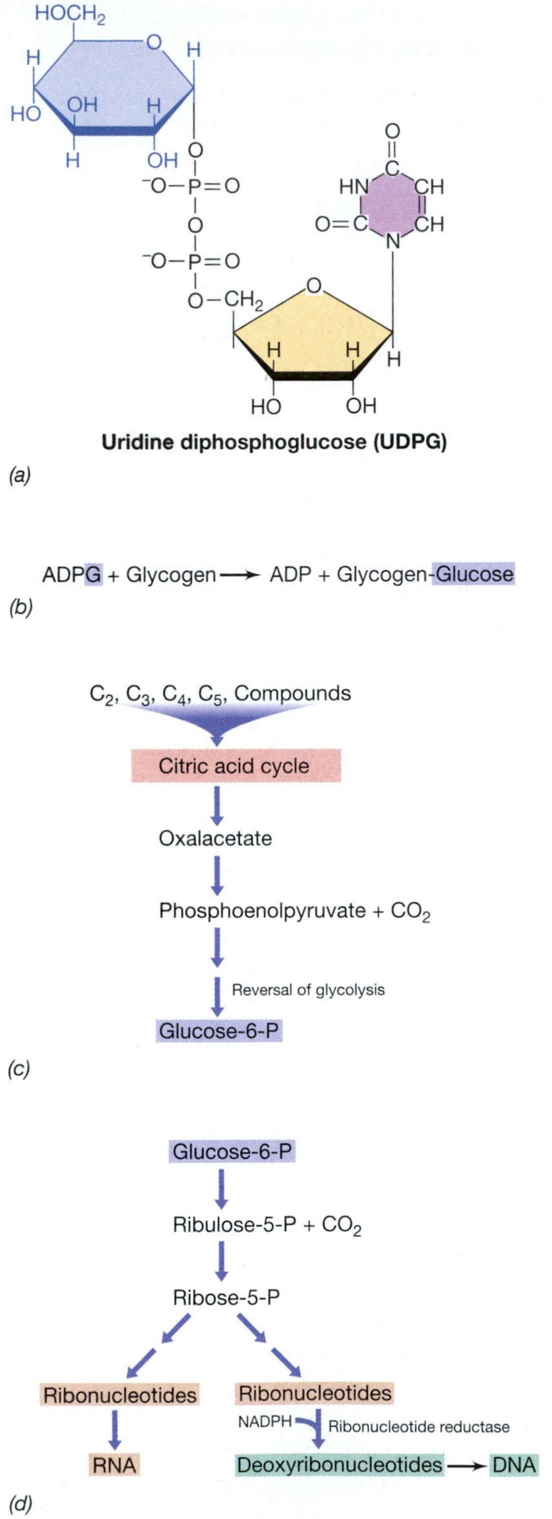

Uridine diphosphoglucose (UDPG)

(a)

ADPG + Glycogen ⟶ ADP + Glycogen-Glucose

(b)

C_2, C_3, C_4, C_5, Compounds → Citric acid cycle → Oxalacetate → Phosphoenolpyruvate + CO_2 → Reversal of glycolysis → Glucose-6-P

(c)

Glucose-6-P → Ribulose-5-P + CO_2 → Ribose-5-P → Ribonucleotides → RNA; Ribonucleotides → (NADPH, Ribonucleotide reductase) → Deoxyribonucleotides ⟶ DNA

(d)

Figure 5.24 Sugar metabolism. *(a)* Polysaccharides are synthesized from activated forms of hexoses such as UDPG. Glucose is shown here in blue. *(b)* Glycogen is biosynthesized from adenosine-diphosphoglucose (ADPG) by the sequential addition of glucose. *(c)* Gluconeogenesis. When glucose is needed, it can be biosynthesized from other carbon compounds, generally by the reversal of steps in glycolysis. *(d)* Pentoses for nucleic acid synthesis are formed by decarboxylation of hexoses like glucose-6-phosphate. Note how the precursors of DNA are produced from the precursors of RNA by the enzyme ribonucleotide reductase.

Phosphoenolpyruvate can be synthesized from oxalacetate, a citric acid cycle intermediate (Figure 5.22). An overview of gluconeogenesis is shown in Figure 5.24*c*.

Biosynthesis of Pentoses

Pentoses are formed by the removal of one carbon atom from a hexose, typically as CO_2. The pentoses needed for nucleic acid synthesis, ribose and deoxyribose, are formed as shown in Figure 5.24*d*. The enzyme ribonucleotide reductase converts ribose into deoxyribose by reduction of the 2′ carbon on the ring. Interestingly, this reaction occurs after, not before, synthesis of nucleotides. Thus, ribonucleotides are biosynthesized, and some of them are later reduced to *deoxy*ribonucleotides for use as precursors of DNA.

5.15 MiniReview

Polysaccharides are important structural components of cells and are biosynthesized from activated forms of their monomers. Gluconeogenesis is the production of glucose from nonsugar precursors.

- What form of glucose is used in the biosynthesis of glycogen?

5.16 Biosynthesis of Amino Acids and Nucleotides

The monomeric constituents of proteins and nucleic acids are amino acids and nucleotides, respectively. Their biosyntheses are often long, multistep pathways requiring many individual enzymes to complete. We approach their biosyntheses here by identifying the carbon skeletons needed to begin the biosynthetic pathways.

Monomers of Proteins: Amino Acids

Organisms that cannot obtain some or all of their amino acids preformed from the environment must synthesize them from other sources. Amino acids can be grouped into structurally related families that share biosynthetic steps. The carbon skeletons for amino acids come almost exclusively from intermediates of glycolysis or the citric acid cycle (**Figure 5.25**).

The amino group of amino acids is typically derived from some inorganic nitrogen source in the environment, such as ammonia (NH_3). Ammonia is most often incorporated through formation of the amino acids glutamate or glutamine by the enzymes *glutamate dehydrogenase* and *glutamine synthetase*, respectively (**Figure 5.26**). When ammonia is present at high levels, glutamate dehydrogenase or other amino acid dehydrogenases are used. However, when ammonia is present at low levels, glutamine synthetase, with its energy-consuming reaction mechanism (Figure 5.26*b*) and high affinity for substrate, is employed. We discuss control of the activity of glutamine synthetase in Section 5.18.

Once ammonia is incorporated, it can be used to form other nitrogenous compounds. For example, glutamate can

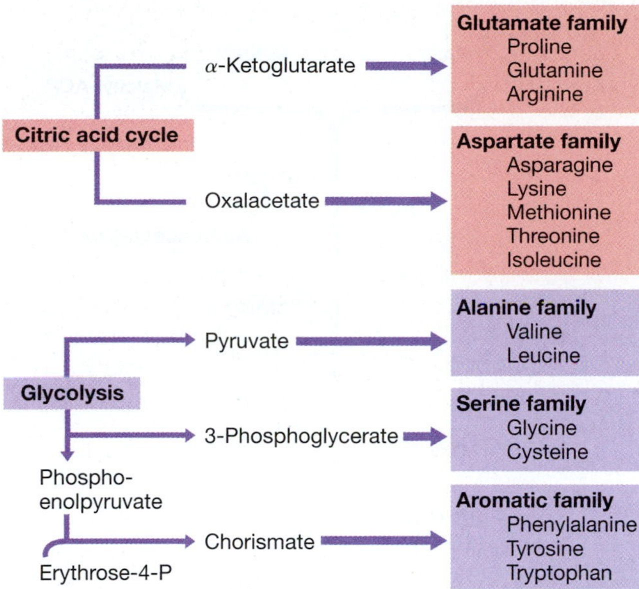

Figure 5.25 Amino acid families. Most amino acids are derived from either the citric acid cycle or from glycolysis. Synthesis of the various amino acids in a family frequently requires many separate enzymatically catalyzed steps starting with the parent amino acid (shown in bold).

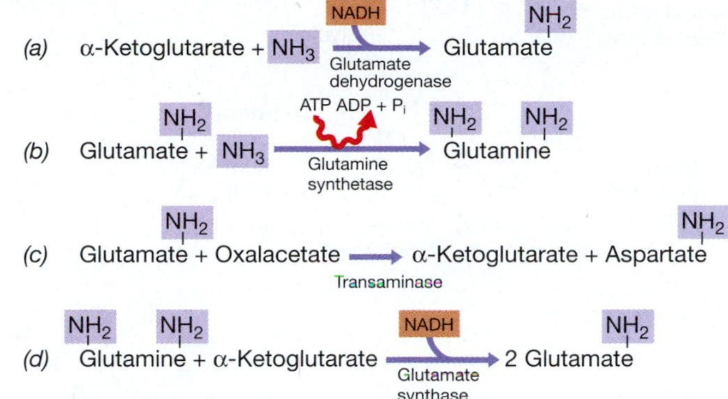

Figure 5.26 Ammonia incorporation in bacteria. To follow the flow of nitrogen, both free ammonia and the amino groups of all amino acids are shown in blue. Two major pathways for NH_3 assimilation in bacteria are those catalyzed by the enzymes *(a)* glutamate dehydrogenase and *(b)* glutamine synthetase. *(c)* Transaminase reactions transfer an amino group from an amino acid to an organic acid. *(d)* The enzyme glutamate synthase forms two glutamates from one glutamine and one α-ketoglutarate.

donate its amino group to oxalacetate in a transaminase reaction, producing α-ketoglutarate and aspartate (Figure 5.26*c*). Alternatively, glutamine can react with α-ketoglutarate to form two molecules of glutamate in an aminotransferase reaction (Figure 5.26*d*). The end result of these types of reactions is the shuttling of ammonia into various carbon skeletons from which further biosynthetic reactions can occur to form all the 22 amino acids (Figure 3.12) needed to make proteins.

Monomers of Nucleic Acids: Nucleotides

The biochemistry behind purine and pyrimidine biosynthesis is quite complex. Purines are constructed literally atom by atom from several distinct carbon and nitrogen sources, including CO_2 (**Figure 5.27*a***). The first key purine, inosinic acid (Figure 5.27*b*), is the precursor of the purine nucleotides adenine and guanine (Figure 3.9). Once these are synthesized (in their triphosphate forms) and have been attached to their correct pentose sugar, they are ready to be incorporated into DNA or RNA.

Like the purine ring, the pyrimidine ring is also constructed from several sources (Figure 5.27*c*). The first key pyrimidine is the compound uridylate, and from this the pyrimidines thymine, cytosine, and uracil (Figure 3.9) are derived (Figure 5.27*d*).

5.16 MiniReview

Amino acids are formed from carbon skeletons generated during catabolism whereas nucleotides are biosynthesized using carbon from several sources.

- What form of nitrogen is commonly used to form the amino group of amino acids?
- Which nitrogen bases are purines and which are pyrimidines?

5.17 Biosynthesis of Fatty Acids and Lipids

Lipids are important constituents of cells, as they are major structural components of membranes. Lipids can also be carbon and energy reserves. Other lipids function in and around the cell surface, including, in particular, the lipopolysaccharide layer (outer membrane) of gram-negative bacteria (Section 4.7). A cell produces many different types of lipids, some of which are produced only under certain conditions or have special functions in the cells. The biosynthesis of fatty acids is thus a major series of reactions in most cells.

Fatty Acid Biosynthesis

Fatty acids are biosynthesized two carbon atoms at a time with the help of a protein called *acyl carrier protein* (ACP). ACP holds the growing fatty acid as it is being synthesized and releases it once it has reached its final length (**Figure 5.28**; Figure 3.7). Interestingly, although fatty acids are constructed *two* carbons at a time, each two-carbon unit is donated from the *three*-carbon compound malonate, which is attached to the ACP to form malonyl-ACP. As each malonyl residue is donated, one molecule of CO_2 is released (Figure 5.28).

The fatty acid composition of cells varies from species to species and can also vary within a species due to differences in temperature (growth at low temperatures promotes the biosynthesis of shorter-chain fatty acids and growth at higher temperatures promotes longer-chain fatty acids). The most common fatty acids in lipids of *Bacteria* contain C_{12}–C_{20} fatty acids.

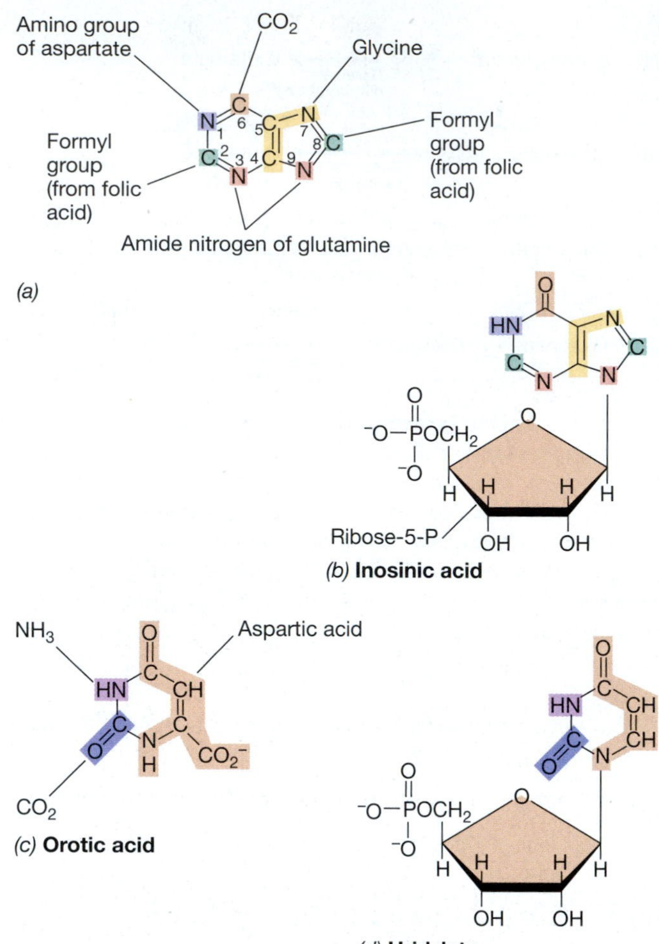

Figure 5.27 Biosynthesis of purines and pyrimidines. *(a)* The precursors of the purine skeleton. *(b)* Inosinic acid, the precursor of all purine nucleotides. *(c)* The precursor of the pyrimidine skeleton, orotic acid. *(d)* Uridylate, the precursor of all pyrimidine nucleotides. Uridylate is formed from orotate following a decarboxylation and the addition of ribose-5-phosphate.

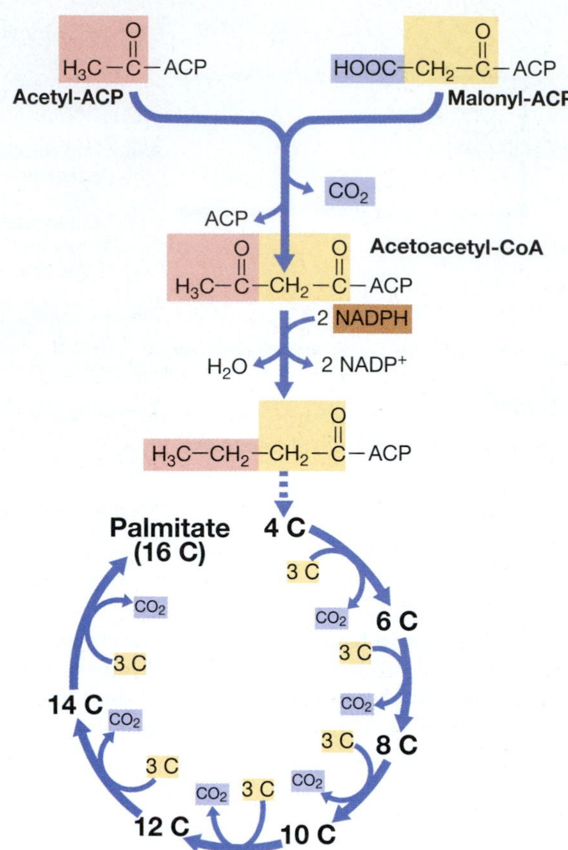

Figure 5.28 The biosynthesis of fatty acids. Shown is the biosynthesis of the C_{16} fatty acid palmitate (Figure 3.7). ACP, acyl carrier protein. The condensation of acetyl-ACP and malonyl-ACP forms acetoacetyl-CoA. Each successive addition of an acetyl unit comes from malonyl-ACP.

In addition to saturated, even-carbon-number fatty acids, fatty acids can also be unsaturated, branched, or have an odd number of carbon atoms. Unsaturated fatty acids contain one or more double bonds in the long hydrophobic portion of the molecule. The number and position of these double bonds is often species- or group-specific, and the double bonds are typically added by desaturating a saturated fatty acid. Branched chain and odd-carbon-number fatty acids are biosythesized using an initiating molecule that contains a branched chain fatty acid or a propionyl (C_3) group, respectively.

Lipids

In the final assembly of lipids in *Bacteria* and *Eukarya*, fatty acids are added to glycerol. For simple triglycerides (fats), all three glycerol carbons are esterified with fatty acids. In complex lipids, one of the glycerol carbon atoms contains a molecule of phosphate, ethanolamine, a sugar, or some other polar substance (Figure 3.7). In *Archaea*, lipids contain phytanyl side chains instead of fatty acid side chains (Section 4.3), and the biosynthesis of phytanyl is distinct from that described here for fatty acids. However, as for the lipids of *Bacteria* or *Eukarya*, the third carbon of the glycerol backbone in archaeal membrane lipids contains a polar group of some sort to allow the typical membrane structure to form: a hydrophobic interior with hydrophilic surfaces (Section 4.3).

5.17 MiniReview

Fatty acids are synthesized two carbons at a time and then attached to glycerol to form lipids. Only the lipids of *Bacteria* and *Eukarya* contain fatty acids.

- Explain why in fatty acid synthesis fatty acids are constructed two carbon atoms at a time even though the immediate donor for these carbons contains three carbon atoms.

5.18 Regulation of Activity of Biosynthetic Enzymes

We have just considered some of the key biosyntheses that occur in cells. There are hundreds of different enzymatic reactions that occur in these anabolic processes, and many of the en-

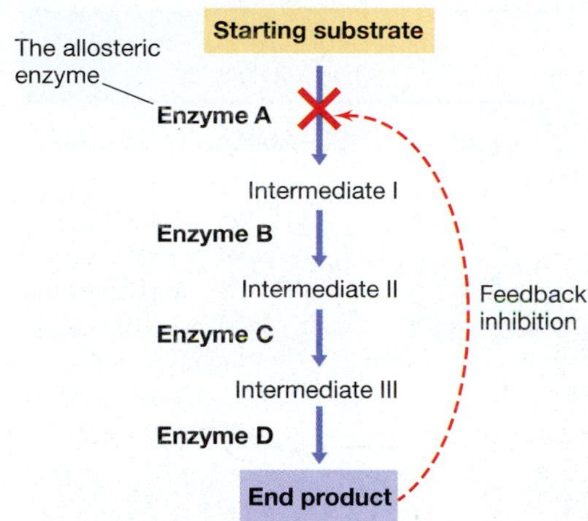

Figure 5.29 Feedback inhibition of enzyme activity. The activity of the first enzyme of the pathway is inhibited by the end product, thus shutting off the production of the end product.

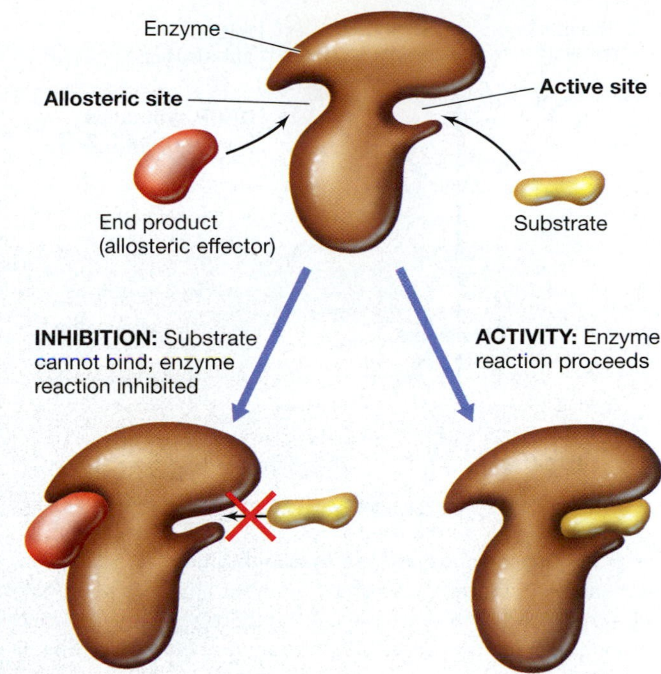

Figure 5.30 Allostery, the mechanism of enzyme inhibition by the end product of a pathway. When the end product combines with the allosteric site, the conformation of the enzyme is altered such that the substrate can no longer bind to the active site.

zymes that carry these out are highly regulated. The advantage to the cell of regulating enzymes of a biosynthetic pathway should be clear: Neither carbon nor energy is wasted as they would be if a metabolite already present in the environment in sufficient amount were instead biosynthesized by the cell.

There are two major modes of enzyme regulation, one that controls the *amount* (or even the complete presence or absence) of an enzyme in the cell and another that controls the *activity* of a preexisting enzyme by temporarily inactivating the protein. In prokaryotic cells, regulation of the number of copies (amount) of a given enzyme occurs at the gene level, and we reserve discussion of this until after we have considered the principles of molecular biology. Here we focus on what the cell can do to control the activity of enzymes that already exist in the cell.

Inhibition of an enzyme's activity is the result of either covalent or noncovalent changes in the enzyme structure. We begin with the simple case of feedback inhibition and isoenzymes, both examples of noncovalent interactions, and end with the classical case of covalent modification of the enzyme glutamine synthetase.

Feedback Inhibition

A major mechanism for the control of enzymatic activity is **feedback inhibition**. Feedback inhibition is a mechanism for turning off the reactions in an entire biosynthetic pathway, such as a biosynthetic pathway of an amino acid or nucleotide. The regulation in feedback inhibition occurs because an excess of the end product of the pathway inhibits activity of the *first* enzyme of the pathway. Inhibiting the first step effectively shuts down the entire pathway because no intermediates are generated for other enzymes further down the pathway (**Figure 5.29**). Feedback inhibition is reversible, however, because once levels of the end product become limiting, its synthesis resumes.

How can the end product of a pathway inhibit the activity of an enzyme whose substrate is quite unrelated to it? This is possible because the inhibited enzyme is an **allosteric enzyme**. Such enzymes have two binding sites, the *active site*, where the substrate binds (Section 5.5), and the *allosteric site*, where the end product of the pathway binds reversibly. When the end product binds at the allosteric site, the conformation of the enzyme changes such that the substrate can no longer bind at the active site (**Figure 5.30**). When the concentration of the end product begins to fall in the cytoplasm, the enzyme returns to its catalytic form and is once again active.

Isoenzymes

Some biosynthetic pathways controlled by feedback inhibition employ *isoenzymes* ("iso" means "same"). Isoenzymes are different enzymes that catalyze the same reaction but are subject to different regulatory controls. An example is the synthesis of the aromatic amino acids tyrosine, tryptophan, and phenylalanine in *Escherichia coli*.

The enzyme 3-deoxy-D-arabino-heptulosonate 7-phosphate (DAHP) synthase plays a central role in aromatic amino acid biosynthesis (**Figure 5.31**). In *E. coli*, three DAHP synthase isoenzymes catalyze the first reaction in this pathway, each regulated independently by only one of the end-product amino acids. However, unlike the example of feedback inhibition where an end product completely inhibited an enzyme activity, in the case of DAHP synthase enzyme activity is diminished in a stepwise fashion; enzyme activity falls to zero only when *all three* products are present in excess. Some organisms use only a single enzyme to accomplish the same thing. In these organisms *concerted* feedback inhibition occurs. Each end product feedback inhibits the enzyme's

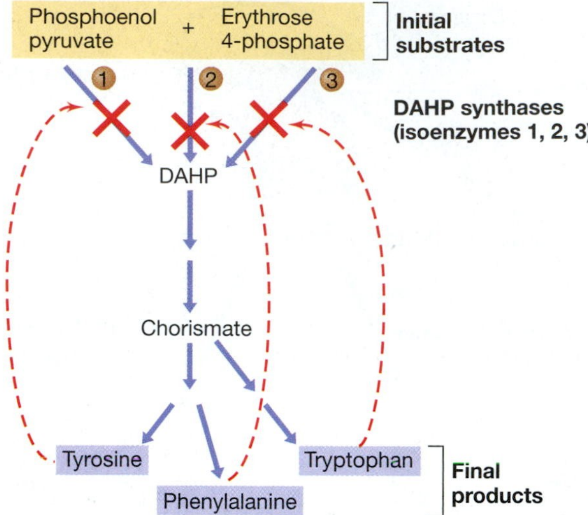

Figure 5.31 Isoenzymes and feedback inhibition. The common pathway leading to the synthesis of the aromatic amino acids contains three isoenzymes of DAHP synthase. Each of these enzymes is specifically feedback inhibited by one of the aromatic amino acids. Note how an excess of all three amino acids is required to completely shut off the synthesis of DAHP.

activity only partially; complete inhibition occurs only when all three end products are present in excess.

Enzyme Regulation by Covalent Modification

Some biosynthetic enzymes are regulated by covalent modification, typically by attaching or removing some small molecule to the protein. As for allosteric proteins, binding of the small molecule changes the conformation of the protein, inhibiting its catalytic activity. Removal of the molecule then returns the enzyme to an active state. Common modifiers include the nucleotides adenosine monophosphate (AMP) and adenosine diphosphate (ADP), inorganic phosphate (PO_4^{2-}), and methyl (CH_3) groups. We consider here the well-studied case of glutamine synthetase (GS), a key enzyme in ammonia assimilation (**Figure 5.32** and Figure 5.26*b*), whose activity is modulated by the addition of AMP, a process called *adenylylation*.

Each molecule of GS is composed of 12 identical subunits, and each subunit can be adenylylated. When the enzyme is fully adenylylated (that is, each molecule of GS contains 12 AMP groups), it is catalytically inactive. When it is partially adenylylated, it is partially active. As the glutamine pool in the cell increases, GS becomes more highly adenylylated and its activity diminishes. As glutamine levels diminish, GS becomes deadenylylated and its activity increases (Figure 5.32).

Other enzymes in the cell add and remove the AMP groups from GS, and these enzymes are themselves controlled, ultimately by levels of ammonia in the cell. Why should there be all of this elaborate regulation surrounding the enzyme GS? The activity of GS requires ATP (Figure 5.26*b*), and nitrogen assimilation is a major biosynthetic process in the cell. However, when ammonia is present at high levels in the cell, it can be assimilated into amino acids by enzymes that do not consume ATP (Figure 5.26*a*); under these conditions, GS remains inactive. When ammonia levels are very low, however, GS is needed, and it then becomes catalytically active. By using GS to assimilate ammonia present only at low levels, the cell conserves ATP that would be used unnecessarily if GS were active when ammonia was present in high amounts.

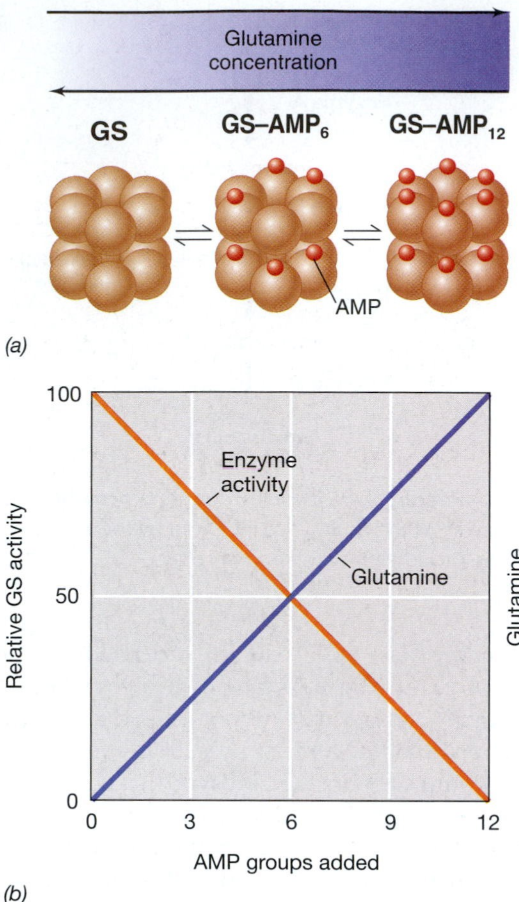

Figure 5.32 Regulation of glutamine synthetase by covalent modification. *(a)* When cells are grown in a medium rich in fixed nitrogen, glutamine synthetase (GS) is covalently modified by becoming progressively adenylylated; as many as 12 adenyl (AMP) groups can be added. When the medium becomes nitrogen poor, the groups are removed, forming ADP. *(b)* Adenylylated GS subunits are catalytically inactive, so the overall GS activity decreases as more subunits are adenylylated. See Figure 5.26*b* for the reaction carried out by glutamine synthetase.

5.18 MiniReview

One mechanism for modulating an enzyme's activity is feedback inhibition, in which the final product of a biosynthetic pathway inhibits the first enzyme unique to that pathway. Employing isoenzymes is a second way of doing this, and reversible covalent modification is yet another way of doing this.

- What is feedback inhibition?
- What is an allosteric enzyme?
- In glutamine synthetase, what does adenylylation do to enzyme activity?

Review of Key Terms

Activation energy the energy required to bring the substrate of an enzyme to the reactive state

Adenosine triphosphate (ATP) a nucleotide that is the primary form in which chemical energy is conserved and utilized in cells

Allosteric enzyme an enzyme containing an active site plus an allosteric site for binding an effector molecule

Anabolic reactions (Anabolism) the sum total of all biosynthetic reactions in the cell

Anaerobic respiration a form of respiration in which oxygen is absent and alternative electron acceptors are reduced

Aseptic technique manipulations to prevent contamination of sterile objects or microbial cultures during handling

ATPase (ATP synthase) a multiprotein enzyme complex embedded in the cytoplasmic membrane that catalyzes the synthesis of ATP coupled to dissipation of the proton motive force

Autotroph an organism capable of biosynthesizing all cell material from CO_2 as the sole carbon source

Catabolic reactions (Catabolism) biochemical reactions leading to energy conservation (usually as ATP) by the cell

Catalyst a substance that accelerates a chemical reaction but is not consumed in the reaction

Chemolithotrophs organisms that can grow with inorganic compounds as electron donors in energy metabolism

Citric acid cycle a cyclical series of reactions resulting in the conversion of acetate to two CO_2

Coenzyme a small and loosely bound nonprotein molecule that participates in a reaction as part of an enzyme

Complex medium a culture medium composed of digests of chemically undefined substances such as yeast and meat extracts

Culture medium an aqueous solution of various nutrients suitable for the growth of microorganisms

Defined medium a culture medium whose precise chemical composition is known

Electron acceptor a substance that can accept electrons from an electron donor, becoming reduced in the process

Electron donor a substance that can donate electrons to an electron acceptor, becoming oxidized in the process

Endergonic energy requiring

Enzyme a protein that can speed up (catalyze) a specific chemical reaction

Exergonic energy releasing

Feedback inhibition a process in which an excess of the end product of a multistep pathway inhibits activity of the first enzyme in the pathway

Fermentation anaerobic catabolism in which an organic compound is both an electron donor and an electron acceptor, and ATP is produced by substrate-level phosphorylation

Free energy (*G*) energy available to do work; $G^{0\prime}$ is free energy under standard conditions

Glycolysis a biochemical pathway in which glucose is fermented yielding ATP and various fermentation products; also called the Embden–Meyerhof pathway

Metabolism the sum total of all the chemical reactions that occur in a cell

Oxidative phosphorylation the production of ATP from a proton motive force formed by electron transport of electrons from organic or inorganic electron donors

Photophosphorylation the production of ATP from a proton motive force formed from light-driven electron transport

Phototrophs organisms that use light as their source of energy

Proton motive force a source of potential energy resulting from the separation of charge and protons from hydroxyl ions across the cytoplasmic membrane

Pure culture a culture that contains a single kind of microorganism

Reduction potential ($E_0{}'$) the inherent tendency, measured in volts under standard conditions, of a compound to donate electrons

Respiration the process in which a compound is oxidized with O_2 (or an O_2 substitute) as the terminal electron acceptor, usually accompanied by ATP production by oxidative phosphorylation

Siderophore an iron chelator that can bind iron present at very low concentrations

Substrate-level phosphorylation production of ATP by the direct transfer of an energy-rich phosphate molecule from a phosphorylated organic compound to ADP

Sterile absence of all microorganisms, including viruses

Review Questions

1. Why are carbon and nitrogen macronutrients while cobalt is a micronutrient (Section 5.1)?
2. What are siderophores and why are they necessary (Section 5.1)?
3. Why would the following medium not be considered a chemically defined medium: glucose, 5 grams (g); NH_4Cl, 1 g; KH_2PO_4, 1 g; $MgSO_4$, 0.3 g; yeast extract, 5 g; distilled water, 1 liter (Section 5.2)?
4. What is aseptic technique and why is it necessary (Section 5.3)?
5. Describe how you would calculate $\Delta G^{0\prime}$ for the reaction: glucose + 6 $O_2 \rightarrow$ 6 CO_2 + 6 H_2O. If you were told that this reaction is highly *exergonic,* what would be the arithmetic sign (negative or positive) of the $\Delta G^{0\prime}$ you would expect for this reaction (Section 5.4)?
6. Distinguish between $\Delta G^{0\prime}$, ΔG, and $G_f{}^0$ (Section 5.4).
7. Why are enzymes needed by the cell (Section 5.5)?
8. Describe the difference between a coenzyme and a prosthetic group (Section 5.5).
9. The following is a series of coupled electron donors and electron acceptors (written as donor/acceptor). Using just the data given in Figure 5.10, order this series from most energy-yielding to least energy-yielding. H_2/Fe^{3+}, H_2S/O_2, methanol/NO_3^- (producing NO_2^-), H_2/O_2, Fe^{2+}/O_2, NO_2^-/Fe^{3+}, H_2S/NO_3^- (Section 5.6).

10. What is the reduction potential of the NAD^+/NADH couple (Section 5.7)?
11. Why is acetyl phosphate considered an energy-rich compound while glucose 6-phosphate is not (Section 5.8)?
12. How is ATP made in fermentation and respiration (Section 5.9)?
13. Where in glycolysis is NADH produced? Where is NADH consumed (Section 5.10)?
14. What is needed to reduce NAD^+ to NADH? Cytochrome $bc_{1\text{-oxidized}}$ to $bc_{1\text{-reduced}}$ (Section 5.11)?
15. What is meant by the term proton motive force and why is this concept so important in biology (Section 5.12)?
16. How is rotational energy in the ATPase used to produce ATP (Section 5.12)?
17. The chemicals dinitrophenol and cyanide are both cellular poisons but act in quite different ways. Compare and contrast the modes of action of these two chemicals (Section 5.12).
18. Work through the energy balance sheets for fermentation and respiration, and account for all sites of ATP synthesis. Organisms can obtain nearly 20 times more ATP when growing aerobically on glucose than by fermenting it. Write one sentence that accounts for this difference (Section 5.13).
19. Why can it be said that the citric acid cycle plays two major roles in the cell (Section 5.13)?
20. What are the differences in electron donor and carbon source used by *Escherichia coli* and *Thiobacillus thioparus* (a sulfur chemolithotroph) (Section 5.14 and see Table 5.4)?
21. What two catabolic pathways supply carbon skeletons for sugar and amino acid biosyntheses (Sections 5.15 and 5.16)?
22. Describe the process by which a fatty acid such as palmitate (a C_{16} straight-chain saturated fatty acid) is synthesized in a cell (Section 5.17).
23. Contrast regulation of DAHP synthase and glutamine synthetase (Section 5.18).

Application Questions

1. Design a defined culture medium for an organism that can grow aerobically on acetate as a carbon and energy source. Make sure all the nutrient needs of the organism are accounted for and in the correct relative proportions.
2. *Desulfovibrio* can grow anaerobically with H_2 as electron donor and SO_4^{2-} as electron acceptor (which is reduced to H_2S). Based on this information and the data in Table A1.2 (Appendix 1), indicate which of the following components could not exist in the electron transport chain of this organism and why: cytochrome *c*, ubiquinone, cytochrome c_3, cytochrome aa_3, ferredoxin.
3. Again, using the data in Table A1.2, predict the sequence of electron carriers in the membrane of an organism growing aerobically and producing the following electron carriers: ubiquinone, cytochrome aa_3, cytochrome *b*, NADH, cytochrome *c*, FAD.
4. Explain the following observation in light of the redox tower: cells of *Escherichia coli* fermenting glucose grow faster when NO_3^- is supplied to the culture (NO_2^- is produced) and then grow even faster (and stop producing NO_2^-) when the culture is highly aerated.

6

Microbial Growth

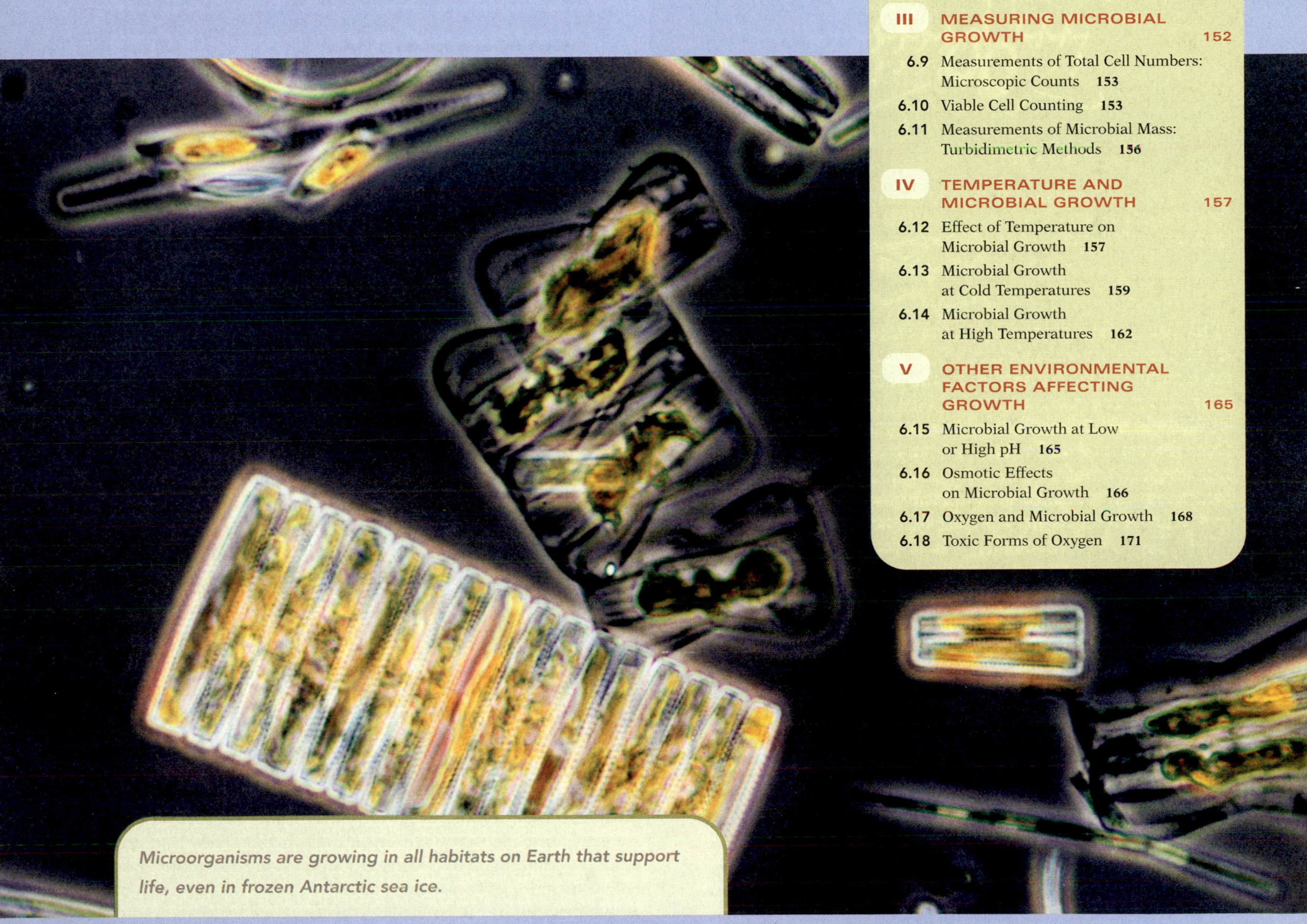

Microorganisms are growing in all habitats on Earth that support life, even in frozen Antarctic sea ice.

I BACTERIAL CELL DIVISION

In the past three chapters we have discussed the structure of cellular macromolecules (Chapter 3), cell structure and function (Chapter 4), and the principles of microbial nutrition and metabolism (Chapter 5). Before we begin our study of the biosynthesis of macromolecules in microorganisms (Chapters 7 and 8), we consider here microbial growth. Growth is the ultimate process in the life of a cell—one cell becoming two.

6.1 Cell Growth and Binary Fission

In microbiology, **growth** is defined as *an increase in the number of cells*. Microbial cells have a finite life span, and a species is maintained only as a result of continued growth of its population. Many practical situations call for the control of microbial growth. Knowledge of how microbial populations can rapidly expand is useful for designing methods to control microbial growth. We will study these control methods in Chapter 27.

Bacterial cell growth depends upon a large number of chemical reactions of a wide variety of types. Some of these reactions transform energy. Others synthesize small molecules—the building blocks of macromolecules. Still others provide the various cofactors and coenzymes needed for enzymatic reactions. However, the main reactions of cell synthesis are the polymerization reactions that make macromolecules from monomers. As macromolecules accumulate in the cytoplasm of a cell, they are assembled into new structures, such as the cell wall, cytoplasmic membrane, flagella, ribosomes, inclusion bodies, enzyme complexes, and so on, eventually leading to cell division.

Binary Fission

In a growing rod-shaped cell, elongation continues until the cell divides into two new cells. This process is called **binary fission** ("binary" to express the fact that two cells have arisen from one). In a growing culture of a rod-shaped bacterium such as *Escherichia coli,* cells elongate to approximately twice their original length and then form a partition that separates the cell into two daughter cells (**Figure 6.1**). This partition is called a *septum* and results from the inward growth of the cytoplasmic membrane and cell wall from opposing directions, which continues until the two daughter cells are pinched off. By definition, when one cell divides to form two, one *generation* has occurred, and the time required for this process is called the **generation time** (Figure 6.1).

In the period of one generation, all cellular constituents increase proportionally; cells are thus said to be in *balanced growth*. Each daughter cell receives a chromosome and sufficient copies of ribosomes and all other macromolecular complexes, monomers, and inorganic ions to exist as an independent cell. Partitioning of the replicated DNA molecule between the two daughter cells depends on the DNA remaining attached to the cytoplasmic membrane during division, with septum formation leading to separation of the chromosomes, one to each daughter cell (see Figure 6.3).

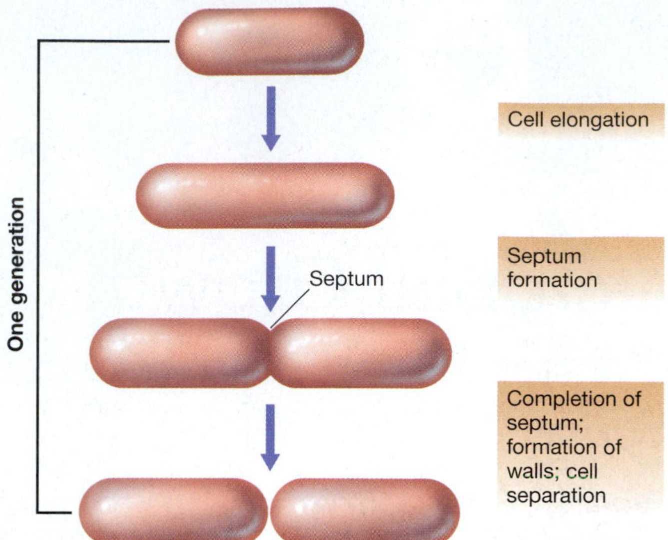

Figure 6.1 Binary fission in a rod-shaped prokaryote. Cell numbers double every generation.

The time required for a generation in a given bacterial species is highly variable and is dependent on both nutritional and genetic factors. Under the best nutritional conditions the generation time of a laboratory culture of *E. coli* is about 20 min. A few bacteria can grow even faster than this, but many grow much slower. In nature it is likely that microbial cells grow much slower than their maximum rate, because rarely are all conditions and resources necessary for optimal growth present at the same time.

6.1 MiniReview

Microbial growth involves all of the reactions necessary for doubling the amount of all cellular components, followed by cell division to form two daughter cells. Most microorganisms grow by binary fission.

- Define the term generation. What is meant by the term generation time?

6.2 Fts Proteins and Cell Division

A series of proteins present in all prokaryotes, called *Fts proteins,* are essential for cell division. The acronym *Fts* stands for *f*ilamentous *t*emperature *s*ensitive, which describes the properties of cells that have mutations in the genes that encode Fts proteins. Such cells do not divide normally but instead form long filamentous cells that fail to divide. **FtsZ**, a key Fts protein, has been well studied in *Escherichia coli* and several other bacteria, and much is known concerning its important role in cell division.

Fts proteins are universally distributed among prokaryotes, including the *Archaea,* and FtsZ-type proteins have even been found in mitochondria and chloroplasts, further emphasizing the evolutionary ties that these organelles have to the

Bacteria. Interestingly, the protein FtsZ is related to tubulin, the important cell division protein in eukaryotes (Section 18.5).

Fts Proteins and Cell Division

Fts proteins interact to form a cell division apparatus called the **divisome**. In rod-shaped cells, formation of the divisome begins with the attachment of molecules of FtsZ in a ring precisely around the center of the cell. This ring defines what will eventually become the cell-division plane. In a cell of *E. coli* about 10,000 FtsZ molecules polymerize to form the ring, and then the ring attracts other divisome proteins, including *FtsA* and *ZipA* (**Figure 6.2**). ZipA is an anchor that connects the FtsZ ring to the cytoplasmic membrane and stabilizes it (Figure 6.2). FtsA, a protein related to actin, also helps to connect the FtsZ ring to the cytoplasmic membrane and has an additional role in recruiting other divisome proteins. The divisome forms well after elongation of a newborn cell has already begun. For example, in cells of *E. coli* the divisome forms about three-quarters of the way into cell division. However, before the divisome forms, the cell is already elongating and DNA is replicating (see Figure 6.3).

The divisome also contains Fts proteins needed for peptidoglycan synthesis, such as FtsI (Figure 6.2). FtsI is one of several *penicillin-binding proteins* present in the cell. Penicillin-binding proteins are so named because their activity is blocked by the antibiotic penicillin (Section 6.4). The divisome orchestrates synthesis of new cytoplasmic membrane and cell wall material, called the *division septum*, at the center of a rod-shaped cell until it reaches twice its original length. Following this, the elongated cell divides, yielding two daughter cells (Figure 6.1).

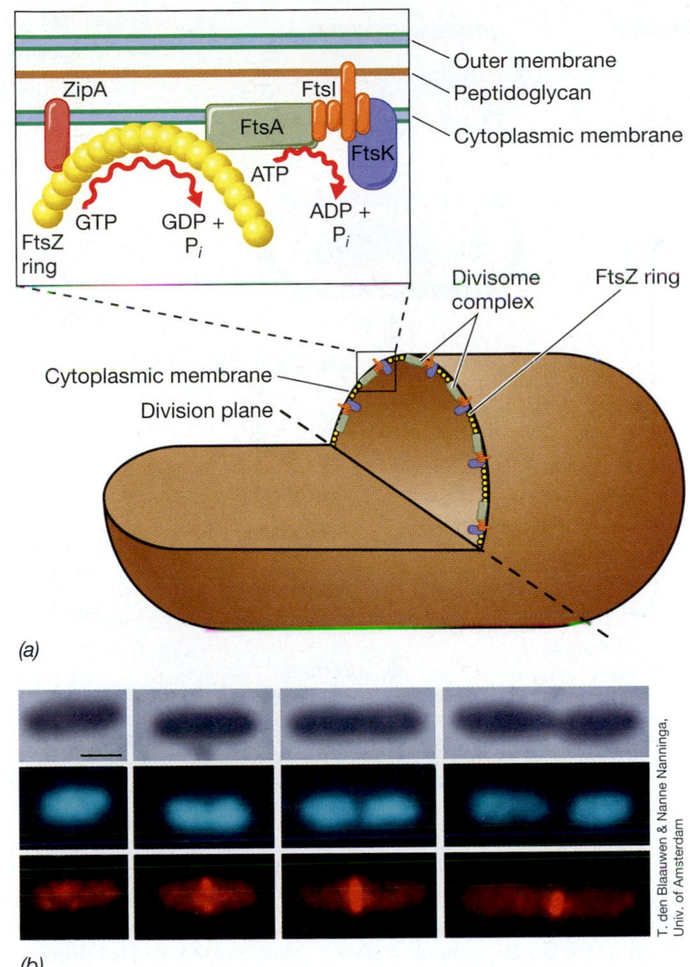

Figure 6.2 The FtsZ ring and cell division. *(a)* Cut-away view of a rod-shaped cell showing the ring of FtsZ molecules around the division plane. Blowup shows the arrangement of individual divisome proteins. ZipA is an FtsZ anchor, FtsI is a peptidoglycan biosynthesis protein, FtsK assists in chromosome separation, and FtsA is an ATPase. *(b)* Appearance and breakdown of the FtsZ ring during the cell cycle of *Escherichia coli*. Microscopy: upper row, phase-contrast; second row, nucleoid stain; third row, cells stained with a specific reagent against FtsZ. Cell division events: first column, FtsZ ring not yet formed; second column, FtsZ ring appears as nucleoids start to segregate; third column, full FtsZ ring forms as cell elongates; fourth column, breakdown of the FtsZ ring and cell division. Marker bar in upper left photo, 1 μm.

DNA Replication, Min Proteins, and Cell Division

As we noted, DNA replicates before the FtsZ ring forms (**Figure 6.3**). The ring forms in the space between the duplicated nucleoids, because before the nucleoids segregate, they effectively block formation of the FtsZ ring. Location of the actual cell midpoint by FtsZ is facilitated by a series of proteins called *Min* proteins, especially MinC, MinD, and MinE. MinD is present as a spiral structure on the surface of the cytoplasmic membrane and oscillates back and forth from pole to pole; MinD is also required to localize MinC to the cytoplasmic membrane. Together, MinC and D inhibit cell division by preventing the FtsZ ring from forming. MinE also oscillates from pole to pole, sweeping MinC and D aside as it moves along. Because MinC and MinD dwell longer at the poles than elsewhere in the cell during their oscillation, on average the center of the cell has the lowest concentration of these proteins. As a result, the cell center is the most permissive site for FtsZ ring assembly, and the FtsZ ring thus defines the division plane. In this way the Min proteins ensure that the divisome forms only at the cell center and not at the cell poles (Figure 6.3).

As cell elongation continues and septum formation begins, the two copies of the chromosome are pulled apart, each to its own daughter cell (Figure 6.3). The Fts protein *FtsK* and several other proteins assist in this process. As the cell constricts, the FtsZ ring begins to depolymerize, triggering the inward growth of wall materials to form the septum and seal off one daughter cell from the other. The enzymatic activity of FtsZ hydrolyzes guanosine triphosphate (GTP) to yield the energy necessary to fuel the polymerization and depolymerization of the FtsZ ring (Figures 6.2 and 6.3).

Properly functioning Fts proteins are essential for cell division. Much new information on cell division in *Bacteria* and *Archaea* has emerged in recent years, and genomic studies have confirmed that Fts proteins are highly conserved across broad phylogenetic lines. There is great interest in understanding bacterial cell division at the molecular level, not only for basic reasons, but also because such understanding could lead to the

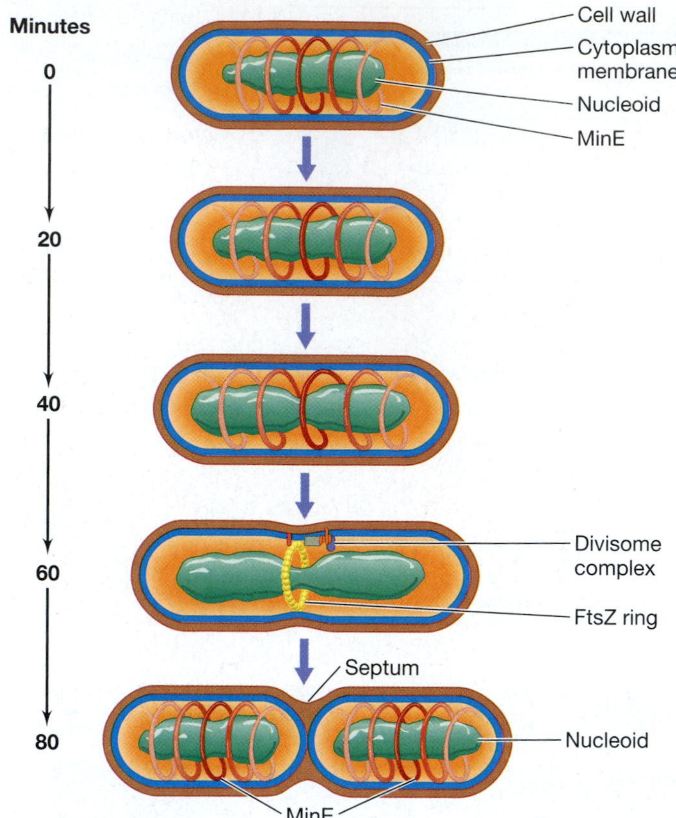

Figure 6.3 DNA replication and cell-division events. The protein MinE directs formation of the FtsZ ring and divisome complex at the cell-division plane. Shown is a schematic for cells of *Escherichia coli* growing with a doubling time of 80 min. MinC and MinD (not shown) are most abundant at the cell poles.

development of new drugs that target specific steps in the growth of pathogenic bacteria. Like penicillin (a drug that targets bacterial cell wall synthesis, Section 6.4), drugs that interfere with the function of specific Fts or other bacterial cell-division proteins could have applications in clinical medicine.

6.2 MiniReview

Cell division and chromosome replication are coordinately regulated, and the Fts proteins are keys to these processes. With the help of MinE, FtsZ defines the division plane in prokaryotes and orchestrates assembly of the divisome.

- When does the bacterial chromosome replicate in the binary fission process?
- How does FtsZ find the cell midpoint?

6.3 MreB and Determinants of Cell Morphology

Just as specific proteins exist in prokaryotes that direct cell division, specific proteins are present that define cell shape. Interestingly, these shape-determining proteins show significant homology to key cytoskeletal proteins in eukaryotic cells. As these proteins have become better understood, it has become clear that, like eukaryotes, prokaryotes also contain a cell cytoskeleton, one that is both dynamic and multifaceted.

Cell Shape and Actin-Like Proteins in Prokaryotes

The major shape-determining factor in prokaryotes is a protein called *MreB*. MreB forms a simple cytoskeleton in cells of *Bacteria* and probably in *Archaea* as well. MreB forms spiral-shaped bands around the inside of the cell, just underneath the cytoplasmic membrane (**Figure 6.4**). Presumably, the MreB cytoskeleton defines cell shape by recruiting other proteins that orchestrate cell wall growth in a specific pattern. Inactivation of the gene encoding MreB in rod-shaped bacteria causes the cells to become coccus-shaped. Interestingly, coccus-shaped bacteria lack the gene that encodes MreB and thus lack MreB. This indicates that the "default" shape for a bacterium is a sphere (coccus). Variations in the arrangement of MreB filaments in cells of nonspherical bacteria are probably responsible for the common morphologies of prokaryotic cells (∞ Figure 4.1).

Mechanism of MreB

How does MreB define a cell's shape? The answer is not entirely clear, but experiments on cell division and its link to cell wall synthesis have yielded two important clues. First, the helical structures formed by MreB (Figure 6.4) are not static, but instead can rotate within the cytoplasm of a growing cell. Second, newly synthesized peptidoglycan (Section 6.4) is associated with the MreB helices at points where the helices contact the cytoplasmic membrane (Figure 6.4). It thus appears that MreB functions to localize synthesis of new peptidoglycan and other cell wall components to specific locations along the cylinder of a rod-shaped cell during growth. This would explain the fact that new cell wall material in an elongated rod-shaped cell forms at several points along its long axis rather than from a single location at the FtsZ site outward, as in spherical bacteria (see Figure 6.5). By rotating within the cell cylinder and initiating cell wall synthesis where it contacts the cytoplasmic membrane, MreB would direct new wall synthesis in such a way that a rod-shaped cell would elongate only along its long axis.

Crescentin

Caulobacter crescentus, a vibrio-shaped species of *Bacteria* (∞ Section 15.16), produces a shape-determining protein called *crescentin* in addition to MreB. Copies of crescentin protein organize into filaments about 10 nm wide that localize onto the concave face of the curved cell. The arrangement and localization of crescentin filaments are thought to somehow impart the characteristic curved morphology on the *Caulobacter* cell (Figure 6.4*c*). *Caulobacter* is an aquatic bacterium that undergoes a life cycle in which swimming cells, called *swarmers,* eventually form a stalk and attach to surfaces. Attached cells then undergo cell division to form new swarmer cells that are released to colonize new habitats. The

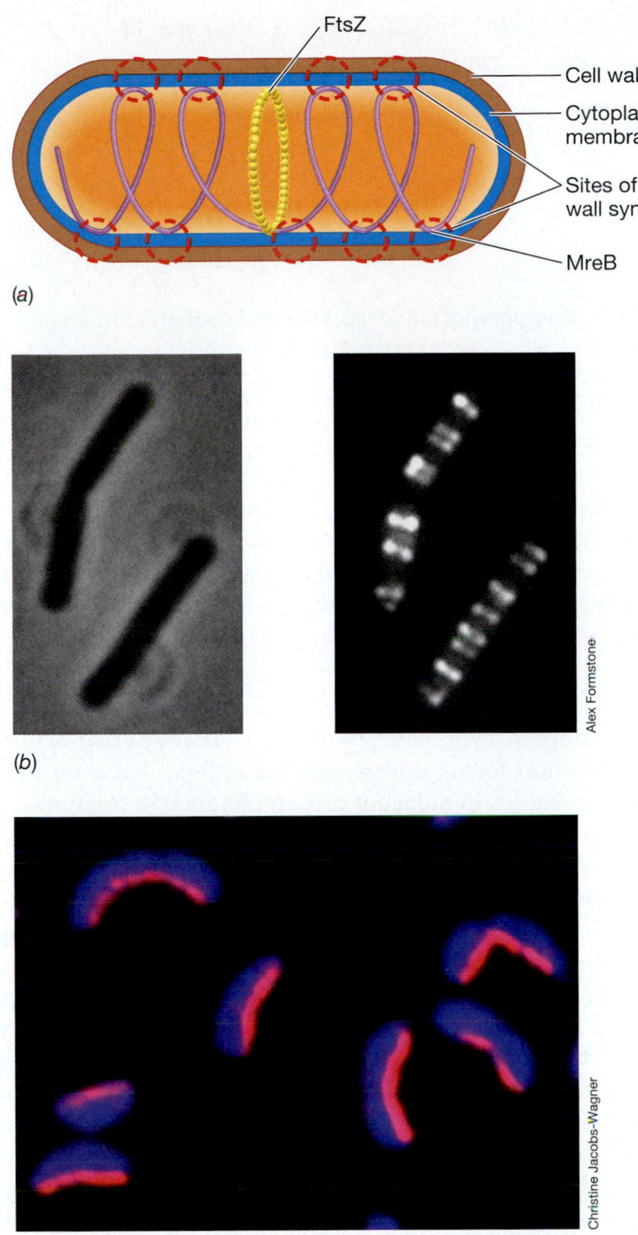

Figure 6.4 MreB and crescentin as determinants of cell morphology. *(a)* The cytoskeletal protein MreB is an actin analog that winds as a coil through the long axis of a rod-shaped cell, making contact with the cytoplasmic membrane in several locations (red dashed circles). These are sites of new cell wall synthesis. *(b)* Photomicrograph of the same cells of *Bacillus subtilis*. Left, phase-contrast; right, fluorescence. The cells contained a substance that makes the MreB protein fluoresce, shown here as bright white. *(c)* Cells of *Caulobacter crescentus*, a naturally curved (vibrio-shaped) cell. Cells are stained to show the shape-determining protein crescentin (red), which lies along the concave surface of the cell, and with DAPI, which stains DNA and thus the entire cell (blue).

steps in this life cycle are highly orchestrated at the genetic level, and *Caulobacter* has been used as a model system for the study of gene expression in cellular differentiation (⚭ Section 9.13). Although thus far crescentin has been found only in *Caulobacter*, proteins similar to crescentin have been found in other helically shaped cells, such as *Helicobacter*. This suggests that these proteins may be necessary for the formation of curved cells.

Archaeal Cell Morphology and the Evolution of Cell Division and Cell Shape

Although less is known about how cell morphology is controlled in *Archaea* than in *Bacteria*, the genomes of most *Archaea* contain genes that encode MreB-like proteins. Thus it is likely that these function in *Archaea* as they do in *Bacteria*. Along with the finding that FtsZ also exists in *Archaea*, it appears that there are strong parallels in cell-division processes and morphological determinants in all prokaryotes.

How do the determinants of cell shape and cell division in prokaryotes compare with those in eukaryotes? Interestingly, the protein MreB is structurally related to the eukaryotic protein actin, and FtsZ is related to the eukaryotic protein tubulin. In eukaryotic cells actin assembles into structures called *microfilaments* that function as scaffolding in the cell cytoskeleton and in cytokinesis, whereas tubulin forms *microtubules* that are important in eukaryotic mitosis and other processes (⚭ Sections 8.6, 18.1, and 18.5). In addition, the shape-determining protein crescentin in *Caulobacter* is related to the keratin proteins that make up *intermediate filaments* in eukaryotic cells. Intermediate filaments are also part of the eukaryotic cytoskeleton. Thus it appears that the proteins that control cell division and cell shape in eukaryotic cells have their evolutionary roots in prokaryotic cells, cells that preceded them on Earth by billions of years (⚭ Figure 1.6).

6.3 MiniReview

MreB protein helps define cell shape. In rod-shaped cells, MreB forms a coil that directs cell wall synthesis along the long axis of the cell. The protein crescentin probably plays an analogous role in *Caulobacter,* leading to formation of a curved cell. Shape and cell division proteins in eukaryotes have prokaryotic counterparts.

- What eukaryotic protein is related to MreB? What does this protein do in eukaryotic cells?
- What is crescentin and what does it do?

6.4 Peptidoglycan Synthesis and Cell Division

In the previous section we summarized the major events in binary fission and learned that a major feature of the cell division process is the production of new cell wall material. In cocci, cell walls grow in opposite directions outward from the FtsZ ring (**Figure 6.5**), whereas the walls of rod-shaped cells grow at several locations along the length of the cell (Figure 6.4). However, in both cases preexisting peptidoglycan has to be severed to allow newly synthesized peptidoglycan to be inserted. How does this occur?

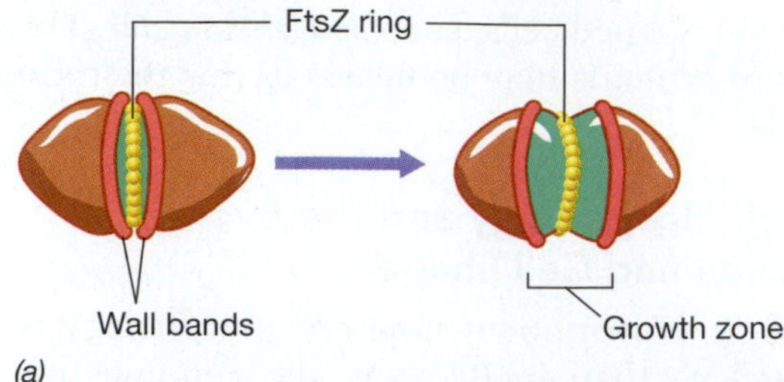

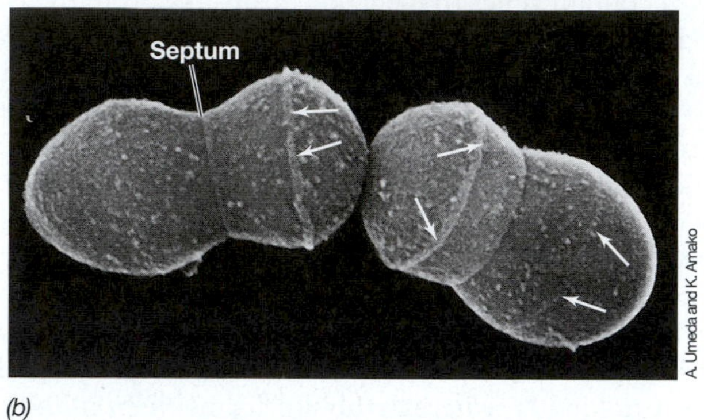

Figure 6.5 Cell-wall synthesis in gram-positive *Bacteria*. *(a)* Localization of cell wall synthesis during cell division. In cocci, wall synthesis (shown in green) is localized at only one point (compare with Figure 6.4). *(b)* Scanning electron micrograph of cells of *Streptococcus hemolyticus* showing wall bands (arrows). A single cell is about 1 μm in diameter.

$$H_3C-\overset{CH_3}{C}=CHCH_2(CH_2\overset{CH_3}{C}-CHCH_2)_9CH_2\overset{CH_3}{C}=CHCH_2-O-\underset{O^-}{\overset{O}{P}}-O-\underset{O^-}{\overset{O}{P}}-O-M-G$$

Figure 6.6 Bactoprenol (undecaprenol diphosphate). This highly hydrophobic molecule carries cell wall peptidoglycan precursors through the cytoplasmic membrane.

Beginning at the FtsZ ring (Figures 6.2 and 6.3), small openings in the wall are created by enzymes called *autolysins,* enzymes that function like lysozyme (∞ Section 4.6) to hydrolyze the β-1,4 glycosidic bonds that connect *N*-acetylglucosamine and *N*-acetylmuramic acid in the peptidoglycan backbone. New cell wall material is then added across the openings (Figure 6.5*a*). The junction between new and old peptidoglycan forms a ridge on the cell surface of gram-positive bacteria called a *wall band* (Figure 6.5*b*), analogous to a scar. It is essential in peptidoglycan synthesis that new cell-wall precursors (muramic acid/glucosamine/tetrapeptide units, see Figure 6.7) be spliced into existing peptidoglycan in a coordinated manner to prevent a breach in peptidoglycan integrity at the splice point. A breach could cause **autolysis** (spontaneous cell lysis).

Biosynthesis of Peptidoglycan

We discussed the general structure of peptidoglycan in Chapter 4 (∞ Section 4.6). The peptidoglycan layer can be thought of as a stress-bearing fabric, much like a thin sheet of rubber. Synthesis of new peptidoglycan during growth requires the controlled cutting of preexisting peptidoglycan by autolysins along with the simultaneous insertion of peptidoglycan precursors. A lipid carrier molecule called *bactoprenol* (**Figure 6.6**) plays a major role in this process.

Bactoprenol is a hydrophobic C_{55} alcohol that bonds to the *N*-acetyl glucosamine/*N*-acetylmuramic acid/pentapeptide peptidoglycan precursor (**Figure 6.7*a***). Bactoprenol transports peptidoglycan precursors across the cytoplasmic membrane by rendering them sufficiently hydrophobic to pass through the interior of the cytoplasmic membrane. Once in the periplasm, bactoprenol interacts with enzymes called *glycolases* that insert cell wall precursors into the growing point of the cell wall and catalyze glycosidic bond formation (Figure 6.7*b*).

Transpeptidation

The final step in cell wall synthesis is **transpeptidation**. Transpeptidation forms the peptide cross-links between muramic acid residues in adjacent glycan chains (∞ Section 4.6 and Figures 4.18 and 4.19). In gram-negative bacteria such as *Escherichia coli,* cross-links form between diaminopimelic acid on one peptide and D-alanine on the adjacent peptide (Figure 6.7*b*). Initially, there are two D-alanine residues at the end of the peptidoglycan precursor, but one D-alanine molecule is removed during the transpeptidation reaction (Figure 6.7*b*). This reaction, which is exergonic, supplies the energy necessary to drive the reaction forward (transpeptidation occurs outside the cytoplasmic membrane where ATP is unavailable). In *E. coli,* the protein FtsI (Figure 6.2*a*) is thought to be the key protein in transpeptidation. In gram-positive bacteria, where a glycine interbridge is common, cross-links occur across the interbridge, typically from an L-lysine of one peptide to a D-alanine on the other (∞ Section 4.6 and Figures 4.19 and 4.20).

Transpeptidation and Penicillin

Transpeptidation is medically noteworthy because it is the reaction inhibited by the antibiotic penicillin. Several penicillin-binding proteins have been identified in bacteria, including the previously mentioned FtsI (Figure 6.2*a*). When penicillin binds to penicillin-binding proteins, they lose their catalytic activity. In the absence of new wall synthesis, the continued activity of autolysins (Figure 6.7) weakens the cell wall and leads to osmotic bursting.

Penicillin has been such a successful drug in clinical medicine for at least two reasons. First, humans are *Eukarya* and therefore lack peptidoglycan; the drug can thus be administered in high doses and is typically nontoxic. And second, virtually all pathogenic bacteria contain peptidoglycan and are thus potential targets of the drug. However, in this regard, the continual and probably excessive use of penicillin since

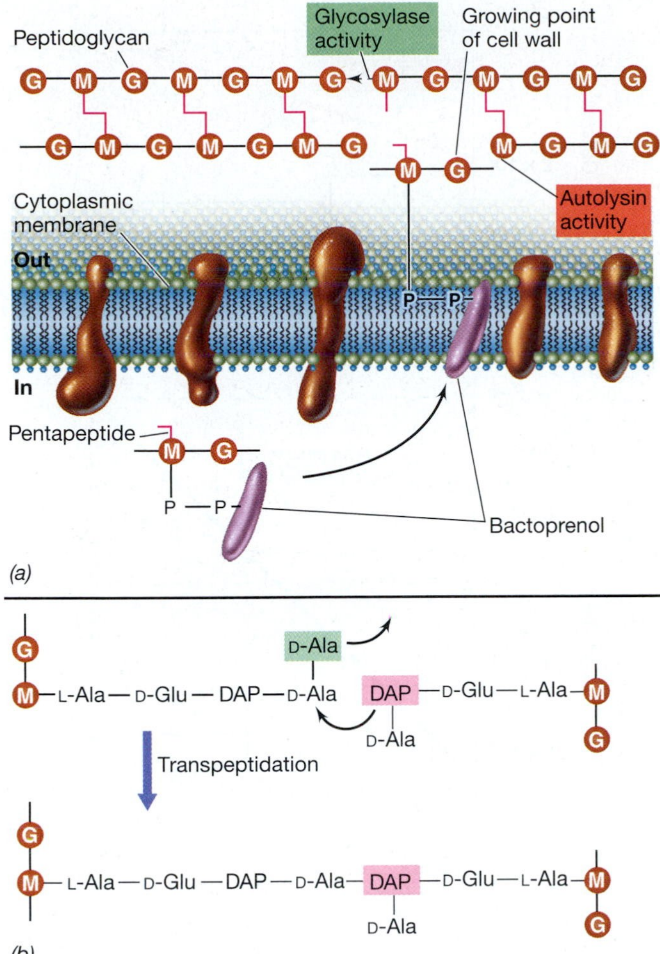

Figure 6.7 Peptidoglycan synthesis. *(a)* Transport of peptidoglycan precursors across the cytoplasmic membrane to the growing point of the cell wall. Autolysin breaks glycolytic bonds in preexisting peptidoglycan, while glycolase synthesizes them, linking old peptidoglycan with new. *(b)* The transpeptidation reaction that leads to the final cross-linking of two peptidoglycan chains. Penicillin inhibits this reaction.

the early 1940s has selected for resistant mutants of many common pathogens that have now become widespread in human and other animal populations. We engage this serious problem of antibiotic and other drug resistance in Chapter 27.

6.4 MiniReview

During bacterial growth new cell wall is synthesized by the insertion of new glycan tetrapeptide units into preexisting wall material. The hydrophobic alcohol bactoprenol facilitates transport of new glycan units through the cytoplasmic membrane to become part of the growing cell wall. Transpeptidation reactions complete the cross-linking of peptidoglycan chains after incorporation of new precursors.

- What are autolysins and why are they necessary?
- What is the function of bactoprenol?
- What is transpeptidation and why is it important?

II GROWTH OF BACTERIAL POPULATIONS

As we have mentioned, microbial growth is defined as an increase in the *number* of cells in a population. So we now move on from considering the growth and division of an individual cell to consider the dynamics of growth in bacterial populations.

6.5 Growth Terminology and the Concept of Exponential Growth

During cell division one cell becomes two. During the time that it takes for this to occur (the generation time), both total cell *number* and *mass* double. Generation times vary widely among microorganisms. In general, most bacteria have shorter generation times than do most microbial eukaryotes. The generation time of a given organism in culture is dependent on the growth medium and the incubation conditions used. Many bacteria have minimum generation times of 0.5–6 h under the best of growth conditions, but a few very rapidly growing organisms are known whose doubling times are less than 20 min and a few slow-growing organisms whose doubling times are as long as several days or even weeks.

Exponential Growth

A growth experiment beginning with a single cell having a doubling time of 30 min is presented in **Figure 6.8**. This pattern of population increase, where the number of cells doubles during a constant time interval, is called **exponential growth**. When the cell number from such an experiment is graphed on arithmetic (linear) coordinates as a function of time, one obtains a curve with a continuously increasing slope (Figure 6.8*b*).

By contrast, when the number of cells is plotted on a logarithmic ($\log_{10}$) scale and time is plotted arithmetically (a *semilogarithmic* graph), as shown in Figure 6.8*b*, the points fall on a straight line. This straight-line function reflects the fact that the cells are growing exponentially—the cell population is doubling in a constant time interval. Semilogarithmic graphs are also convenient and simple to use to estimate generation times from a set of growth data. The generation time may be read directly from the graph as shown in **Figure 6.9**.

The Consequences of Exponential Growth

During exponential growth, the increase in cell number is initially rather slow but increases at an ever faster rate. In the later stages of growth, this results in an explosive increase in cell numbers. For example, in the experiment in Figure 6.8, the rate of cell production in the first 30 min of growth is 1 cell per 30 min. However, between 4 and 4.5 h of growth, the rate of cell production is considerably higher, 256 cells per 30 min, and between 5.5 and 6 h of growth it is 2,048 cells per 30 min (Figure 6.8). Thus in an actively growing bacterial culture, cell numbers can get very large very quickly.

Time (h)	Total number of cells	Time (h)	Total number of cells
0	1	4	256 (2^8)
0.5	2	4.5	512 (2^9)
1	4	5	1,024 (2^{10})
1.5	8	5.5	2,048 (2^{11})
2	16	6	4,096 (2^{12})
2.5	32	.	.
3	64	.	.
3.5	128	10	1,048,576 (2^{19})

(a)

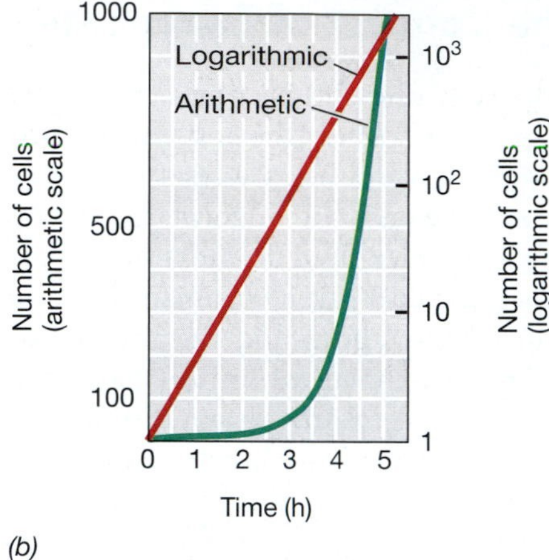

(b)

Figure 6.8 The rate of growth of a microbial culture. *(a)* Data for a population that doubles every 30 min. *(b)* Data plotted on arithmetic (left ordinate) and logarithmic (right ordinate) scales.

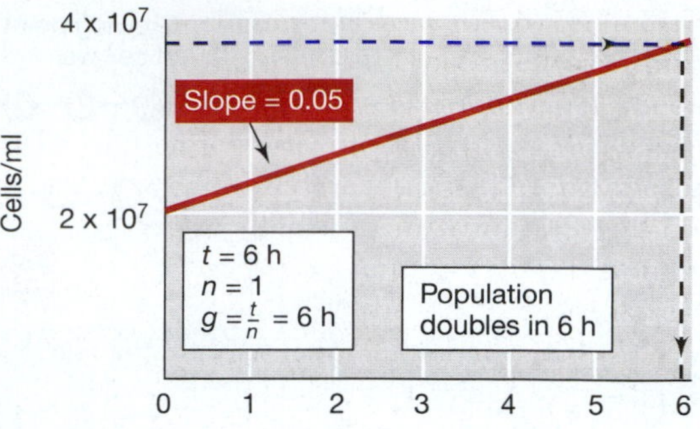

(a)

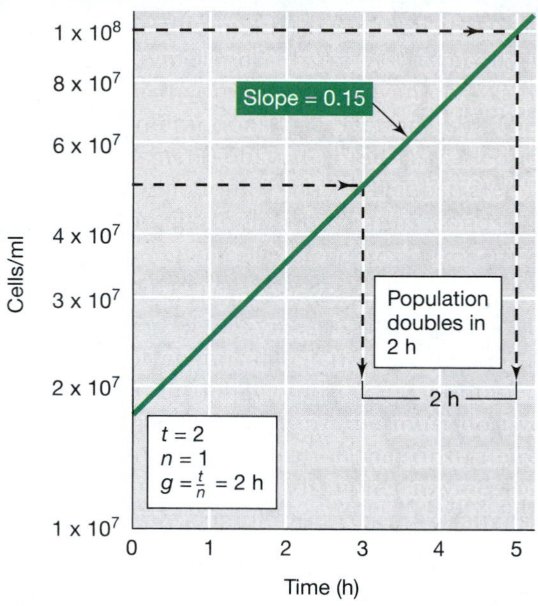

(b)

Figure 6.9 Calculating microbial growth parameters. Method of estimating the generation times (g) of exponentially growing populations with generation times of *(a)* 6 h and *(b)* 2 h from data plotted on semilogarithmic graphs. The slope of each line is equal to $0.301/g$, and n is the number of generations in the time t. All numbers are expressed in scientific notation; that is, 10,000,000 is 1×10^7, 60,000,000 is 6×10^7, and so on.

Consider the following practical implication of exponential growth. For a nonsterile and nutrient-rich food product such as milk to stand under ideal bacterial growth conditions for a few hours during the early stages of exponential growth, when total cell numbers are relatively low, is not detrimental. However, standing for the same length of time during the later stages of growth, when cell numbers are initially much higher, is disastrous. The lactic acid bacteria responsible for milk spoilage contaminate milk during its collection. These harmless organisms are held in check by refrigeration temperatures (~4°C), and only after several days of slow growth and lactic acid production are the effects of spoilage (rancid milk) noticeable. But at room temperature or above, rapid growth of the same organisms accelerates the spoilage process and overnight, the milk can be spoiled.

6.5 MiniReview

Microbial populations show a characteristic type of growth pattern called exponential growth, which is best seen by plotting the number of cells over time on a semilogarithmic graph.

- Why does exponential growth lead to large cell populations in so short a period of time?
- What is a *semilogarithmic* plot?

6.6 The Mathematics of Exponential Growth

The increase in cell number in an exponentially growing bacterial culture is a geometric progression of the number 2 (Figure 6.8*a*). As one cell divides to become two cells, we express this as $2^0 \rightarrow 2^1$. As two cells become four, we express this as $2^1 \rightarrow 2^2$, and so on (Figure 6.8*a*). A relationship exists between the number of cells present in a culture initially and the number present after a period of exponential growth that can be expressed mathematically:

$$N = N_0 2^n,$$

where N is the final cell number, N_0 is the initial cell number, and n is the number of generations during the period of exponential growth. The generation time (g) of the exponentially growing population is t/n, where t is the duration of exponential growth expressed in days, hours, or minutes. From a knowledge of the initial and final cell numbers in an exponentially growing cell population, it is possible to calculate n, and from n and knowledge of t, the generation time, g.

The Relationship of N and N_0 to n

The equation $N = N_0 2^n$ can be expressed in terms of n as follows:

$$N = N_0 2^n$$

$$\log N = \log N_0 + n \log 2$$

$$\log N - \log N_0 = n \log 2$$

$$n = \frac{\log N - \log N_0}{\log 2} = \frac{\log N - \log N_0}{0.301}$$

$$= 3.3\,(\log N - \log N_0)$$

With this formula, we can calculate generation times in terms of measurable quantities, N and N_0. As an example, consider the actual data from the graph in Figure 6.9*b*, in which $N = 10^8$, $N_0 = 5 \times 10^7$, and $t = 2$:

$$n = 3.3\,[\log(10^8) - \log(5 \times 10^7)] = 3.3(8 - 7.69)$$
$$= 3.3(0.301) = 1$$

Thus, in this example, $g = t/n = 2/1 = 2$ h. If exponential growth continued for another 2 h, the cell number would be 2×10^8. Two hours later the cell number would be 4×10^8, and so on.

Other Growth Expressions

The generation time of an exponentially growing culture can also be determined from the slope of the straight-line function obtained in a semilogarithmic plot of exponential growth. The slope is calculated as 0.301 n/t (0.301/g). In the above example, the slope would thus be 0.301/2, or 0.15. Since g is equal to 0.301/slope, we arrive at the same value of 2 for g. The term 0.301/g is called the *specific growth rate*, abbreviated k.

Another useful growth expression is the reciprocal of the generation time, called the *division rate*, abbreviated v. The division rate is equal to 1/g and has units of reciprocal hours (h^{-1}). Whereas the term g is a measure of the *time* it takes for a population to double in cell number, v is a measure of the *number of generations* per unit of time in an exponentially growing culture. The slope of the line relating log cell number to time (Figure 6.9) is equal to v/3.3.

Armed with knowledge of n and t, one can calculate g, k, and v for different microorganisms growing under different culture conditions. This is often useful for optimizing culture conditions for a particular organism and also for testing the positive or negative effect of some treatment on the bacterial culture. For example, compared with an unamended control, factors that stimulate or inhibit growth can be identified by measuring their effect on the various growth parameters discussed here.

6.6 MiniReview

From knowledge of the initial and final cell numbers and the time of exponential growth, the generation time and growth rate constant of a cell population can be calculated directly. Key parameters here are *n, g, v, k,* and *t*.

- Distinguish between the terms specific growth rate and generation time.
- If in 8 h an exponentially growing cell population increases from 5×10^6 cells/ml to 5×10^8 cells/ml, calculate *g, n, v,* and *k*.

6.7 The Microbial Growth Cycle

The data presented in Figures 6.8 and 6.9 reflect only part of the growth cycle of a microbial population, the part called *exponential growth.* For a culture growing in an enclosed vessel, such as a tube or a flask, a condition called a **batch culture**, exponential growth cannot continue indefinitely. Instead, a typical growth curve for a population of cells is obtained, as illustrated in **Figure 6.10**. The growth curve describes an entire growth cycle, including the lag phase, exponential phase, stationary phase, and death phase.

Lag Phase

When a microbial population is inoculated into a fresh medium, growth usually begins only after a period of time called the *lag phase.* This interval may be brief or extended, depending on the history of the inoculum and the growth conditions. If an exponentially growing culture is transferred into the same medium under the same conditions of growth, there is no lag and exponential growth begins immediately. However, if the inoculum is taken from an old (stationary phase) culture and transferred into the same medium, there is usually a lag even if all the cells in the inoculum are alive. This is because the cells are depleted of various essential constituents and time is required for their biosynthesis. A lag also ensues when the inoculum consists of cells that have been damaged (but not killed) by treatment with heat, radiation, or toxic chemicals because of the time required for the cells to repair the damage.

A lag is also observed when a microbial population is transferred from a rich culture medium to a poorer one; for example, from a complex medium to a defined medium (Section 5.2). To grow in any culture medium the cells must have a complete complement of enzymes for synthesis of the essential metabolites not present in that medium. Hence, upon transfer to a medium where essential metabolites

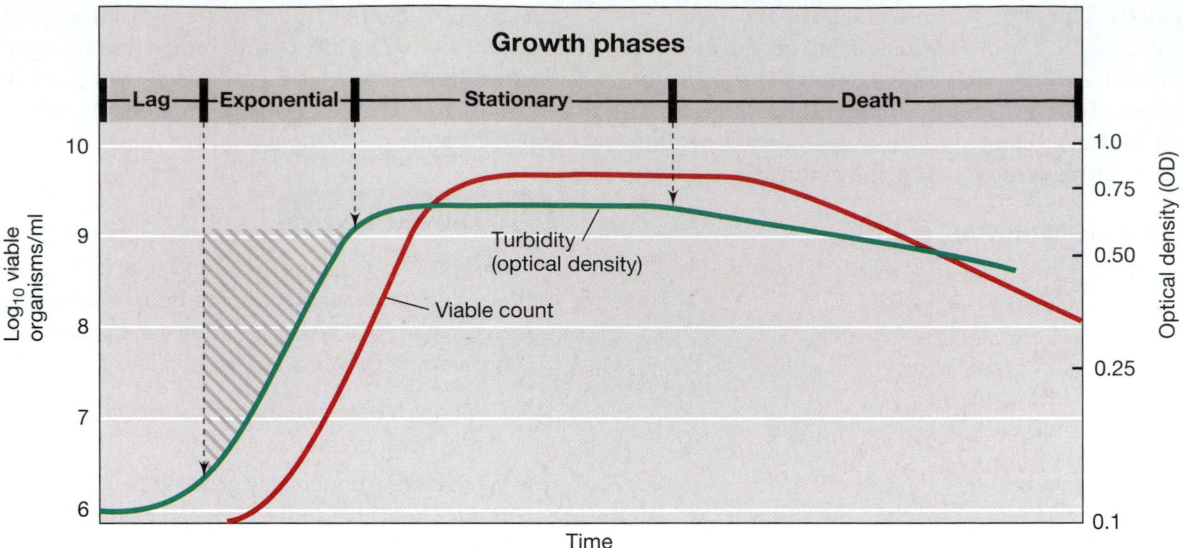

Figure 6.10 Typical growth curve for a bacterial population. A viable count measures the cells in the culture that are capable of reproducing. Optical density (turbidity), a quantitative measure of light scattering by a liquid culture, increases with the increase in cell number.

must be biosynthesized, time is needed for production of the new enzymes that will carry out these reactions.

Exponential Phase

As we saw in the previous section, during the *exponential phase* of growth each cell divides to form two cells, each of which also divides to form two more cells, and so on, for a brief or extended period, depending on the available resources and other factors. Cells in exponential growth are typically in their healthiest state and hence are most desirable for studies of their enzymes or other cell components.

Rates of exponential growth vary greatly. The rate of exponential growth is influenced by environmental conditions (temperature, composition of the culture medium), as well as by genetic characteristics of the organism itself. In general, prokaryotes grow faster than eukaryotic microorganisms, and small eukaryotes grow faster than large ones. This should remind us of the previously discussed concept of surface-to-volume ratio. Recall how small cells have an increased capacity for nutrient and waste exchange compared with larger cells, and this metabolic advantage can greatly affect their growth and other properties (Section 4.2).

Stationary Phase

In a batch culture, such as in a tube or a flask, exponential growth is limited. Consider the fact that a single bacterium with a 20-min generation time would produce, if allowed to grow exponentially in a batch culture for 48 h, a population of cells that weighed 4000 times the weight of Earth! This is particularly impressive when it is considered that a single bacterial cell weighs only about one-trillionth (10^{-12}) of a gram.

Obviously, this scenario is impossible. Something must happen to limit the growth of the population. Typically, either or both of two things limit growth: (1) an essential nutrient of the culture medium is used up, or (2) a waste product of the organism accumulates in the medium and inhibits growth. Either way, exponential growth ceases and the population reaches the *stationary phase*.

In the stationary phase, there is no net increase or decrease in cell number and thus the growth rate of the population is zero. Although the population may not grow during the stationary phase, many cell functions can continue, including energy metabolism and biosynthetic processes. In some cases there may be some cell division during the stationary phase but no net increase in cell number occurs. This is because some cells in the population grow, whereas others die, the two processes balancing each other out. This is a phenomenon called *cryptic growth*.

Death Phase

If incubation continues after a population reaches the stationary phase, the cells may remain alive and continue to metabolize, but they will eventually die. When this occurs, the population enters the *death phase* of the growth cycle. In some cases death is accompanied by actual cell lysis. Figure 6.10 indicates that the death phase of the growth cycle is also an exponential function. Typically, however, the rate of cell death is much slower than the rate of exponential growth.

The phases of bacterial growth shown in Figure 6.10 are reflections of the events in a *population* of cells, not in individual cells. Thus the terms lag phase, exponential phase, stationary phase, and death phase have no meaning with respect to individual cells but only to cell populations. Growth of an individual cell is a necessary prerequisite for population growth. But it is population growth that is most relevant to the ecology of microorganisms, because measurable microbial activities require microbial populations, not just an individual microbial cell.

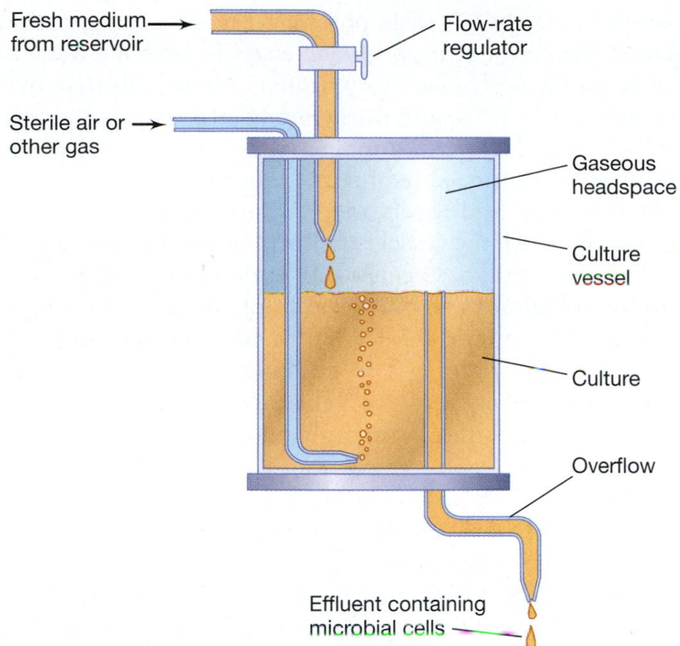

Figure 6.11 Schematic for a continuous culture device (chemostat). The population density is controlled by the concentration of limiting nutrient in the reservoir, and the growth rate is controlled by the flow rate. Both parameters can be set by the experimenter.

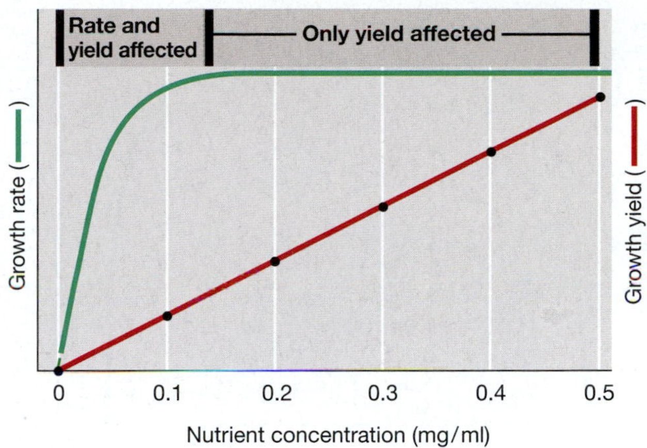

Figure 6.12 The effect of nutrients on growth. Relationship between nutrient concentration, growth rate (green curve), and growth yield (red curve) in a batch culture (closed system). Only at low nutrient concentrations are both growth rate and growth yield affected.

6.7 MiniReview

Microorganisms show a characteristic growth pattern when inoculated into a fresh culture medium. There is usually a lag phase, and then growth commences in an exponential fashion. As essential nutrients are depleted or toxic products build up, growth ceases and the population enters the stationary phase. If incubation continues, cells may begin to die.

- In what phase of the growth curve in Figure 6.10 are cells dividing in a regular and orderly process?
- Under what conditions does a lag phase not occur?
- Why do cells enter stationary phase?

6.8 Continuous Culture: The Chemostat

Our discussion of population growth thus far has been confined to batch cultures. A batch culture is continually being altered by the metabolic activities of growing organisms and is therefore a *closed* system. In the early stages of exponential growth in batch cultures, conditions may remain relatively constant, but in later stages when cell numbers become quite large, the chemical and physical composition of the culture medium changes drastically.

For many studies it is desirable to keep cultures in constant environments for long periods. For example, if one is studying a physiological process such as the synthesis of a particular enzyme, the ready availability of exponentially growing cells may be very useful. Such is only possible with a continuous culture device. Unlike a batch culture, a continuous culture is an *open* system. The continuous culture vessel maintains a constant volume to which fresh medium is added at a constant rate, and an equal volume of spent culture medium is removed at the same rate. Once such a system is in equilibrium, the chemostat volume, cell number, and nutrient status remain constant, and the system is said to be in *steady state*.

The Chemostat

The most common type of continuous culture device is the **chemostat** (**Figure 6.11**). In the chemostat both the growth rate and the population density of the culture can be controlled independently and simultaneously. Two factors are important in such control: (1) the dilution rate, the rate at which fresh medium is pumped in and spent medium is removed; and (2) the concentration of a limiting nutrient, such as a carbon or nitrogen source, present in the sterile medium entering the chemostat vessel.

In a batch culture, the nutrient concentration can affect both growth rate and growth yield (**Figure 6.12**). At very low concentrations of a given nutrient, the growth rate is submaximal because the nutrient cannot be transported into the cell fast enough to satisfy metabolic demand. At moderate or higher nutrient levels, however, the growth rate plateaus, but the final cell yield may continue to increase in proportion to the concentration of nutrients in the medium up to some fixed limit (Figure 6.12). In a chemostat, by contrast, growth rate and growth yield are controlled independently: The growth rate by the dilution rate and the growth yield (cell number/milliliter) by the limiting nutrient. Independent control of these two growth parameters is impossible in a batch culture because it is a closed system where growth conditions are constantly changing with time.

Varying Chemostat Parameters

The effects of varying dilution rate and concentration of the growth-limiting nutrient in a chemostat are shown

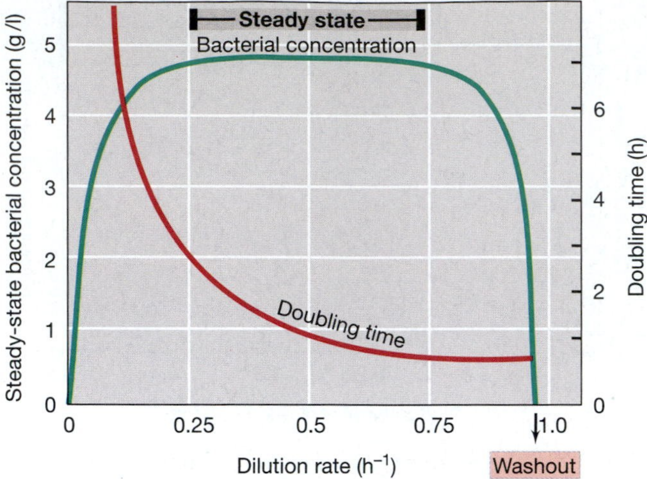

Figure 6.13 Steady-state relationships in the chemostat. The dilution rate is determined from the flow rate and the volume of the culture vessel. Thus, with a vessel of 1,000 ml and a flow rate through the vessel of 500 ml/h, the dilution rate would be 0.5 h^{-1}. Note that at high dilution rates, growth cannot balance dilution, and the population washes out. Note also that although the population density remains constant during steady state, the growth rate (doubling time) can vary over a wide range.

in **Figure 6.13**. As seen, there are rather wide limits over which the dilution rate controls growth rate, although at both very low and very high dilution rates the steady state breaks down. At too high a dilution rate, the organism cannot grow fast enough to keep up with its dilution and is washed out of the chemostat. By contrast, at too low a dilution rate, cells may die from starvation because the limiting nutrient is not being added fast enough to permit maintenance of cell metabolism. However, between these limits, different growth rates can be achieved by simply varying the dilution rate.

Cell density in a chemostat is controlled by a limiting nutrient, just as it is in a batch culture (Figure 6.12). If the concentration of this nutrient in the incoming medium is raised, with the dilution rate remaining constant, the cell density will increase while the growth rate remains the same. Thus, by adjusting the dilution rate and nutrient level accordingly, the experimenter can obtain dilute (for example, 10^5 cells/ml), moderate (for example, 10^7 cells/ml), or dense (for example, 10^9 cells/ml) cell populations growing at low, moderate, or high rates.

Experimental Uses of the Chemostat

A practical advantage to the chemostat is that a cell population may be maintained in the exponential growth phase for long periods, days and even weeks. Because exponential phase cells are usually most desirable for physiological experiments, with a chemostat such cells can be available at any time. Moreover, repetition of experiments can be done with the knowledge that each time the cell population will be as close to being the same as possible. For some applications, such as the study of a particular enzyme, enzyme activities may be significantly lower in stationary phase cells than in exponential phase cells, and thus chemostat-grown cultures are ideal. In practice, after a sample is removed from the chemostat, a period of time is required for the vessel to return to its original volume and for steady state to be reachieved. Once this has occurred, the vessel is ready to be sampled once again.

The chemostat has been used in microbial ecology as well as in microbial physiology. For example, because the chemostat can easily mimic the low substrate concentrations that often prevail in nature, it is possible to prepare mixed or pure bacterial populations in a chemostat and study the competitiveness of different organisms at particular nutrient concentrations. Using these methods together with the powerful tools of phylogenetic stains and gene tracking (∞ Chapters 14 and 22), experimenters can monitor changes in the microbial community in the chemostat as a function of different growth conditions. Such experiments often reveal interactions within the population that are not obvious from growth studies in batch culture.

Chemostats have also been used for enrichment and isolation of bacteria. From a natural sample, one can select a stable population under the nutrient and dilution rate conditions chosen and then slowly increase the dilution rate until a single organism remains. In this way, microbiologists studying the growth rates of various soil bacteria isolated an organism with a 6-minute doubling time—the fastest growing bacterium known!

6.8 MiniReview

Continuous culture devices (chemostats) are a means of maintaining cell populations in exponential growth for long periods. In a chemostat, the rate at which the culture is diluted governs the growth rate, and the cell density is governed by the concentration of the growth-limiting nutrient entering the vessel.

- How do microorganisms in a chemostat differ from microorganisms in a batch culture?
- What happens in a chemostat if the dilution rate exceeds the maximal growth rate of the organism?
- Do pure cultures have to be used in a chemostat?

III MEASURING MICROBIAL GROWTH

Population growth is measured by following changes in the number of cells or changes in the level of some cellular component. The latter could be protein, nucleic acids, or the dry weight of the cells themselves. We consider here two common measures of cell growth: cell counts and turbidity, the latter a measure of cell mass.

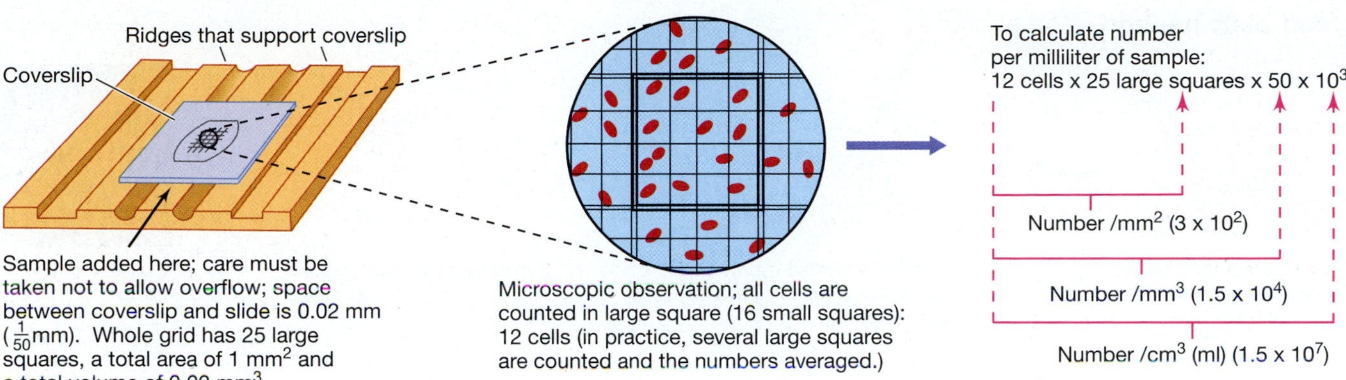

Figure 6.14 Direct microscopic counting procedure using the Petroff–Hausser counting chamber. A phase-contrast microscope is typically used to count the cells to avoid the necessity for staining.

6.9 Measurements of Total Cell Numbers: Microscopic Counts

A total count of microbial numbers can be achieved using a microscope to observe and enumerate the cells present in a culture or natural sample. The method is simple, but the results can be unreliable.

Total Cell Count

The most common total count method is the microscopic cell count. Microscopic counts can be done on either samples dried on slides or on samples in liquid. Dried samples can be stained to increase contrast between cells and their background (∞ Section 2.2). With liquid samples, specially designed counting chambers are used. In such a counting chamber, a grid with squares of known area is marked on the surface of a glass slide (**Figure 6.14**). Over each square on the grid is a volume of known amount, very small but precisely measured. The number of cells per unit area of grid can be counted under the microscope, giving a measure of the number of cells per small chamber volume. The number of cells per milliliter of suspension is calculated by employing a conversion factor based on the volume of the chamber sample (Figure 6.14).

A second method of enumerating cells in liquid samples is with a flow cytometer. This is a machine that employs a laser beam and complex electronics to count individual cells. Flow cytometry is rarely used for the routine counting of microbial cells but has applications in the medical field for counting and differentiating blood cells and other cell types from clinical samples (∞ Figure 32.21). It has also been used in microbial ecology to separate different types of cells for isolation purposes.

Microscopic counting is a quick and easy way of estimating microbial cell number. However, it has several limitations: (1) without special staining techniques (∞ Sections 2.2 and 22.3), dead cells cannot be distinguished from live cells; (2) small cells are difficult to see under the microscope, and some cells are inevitably missed; (3) precision is difficult to achieve; (4) a phase-contrast microscope is required if the sample is not stained; (5) cell suspensions of low density (less than about 10^6 cells/milliliter) have few if any bacteria in the microscope field unless a sample is first concentrated and resuspended in a small volume; (6) motile cells must be immobilized before counting; (7) debris in the sample may be mistaken for microbial cells.

In microbial ecology, total cell counts are often performed on natural samples using stains to visualize the cells. The stain DAPI stains all cells in a sample because it reacts with DNA. By contrast, fluorescent stains can be prepared that are highly specific for certain organisms or groups of organisms by attaching them to specific nucleic acid probes (∞ Section 22.4). If cells are present at low densities, for example in a sample of open ocean water, this problem can be overcome by first concentrating cells on a filter and then counting them after staining. Because they are easy to do and can yield useful information, total cell counts of one sort or another are common in microbial studies of natural environments.

www.microbiologyplace.com Online Tutorial 6.1: Direct Microscopic Counting Procedure

6.9 MiniReview

Cell counts done microscopically measure the total number of cells in either a laboratory culture or a natural sample.

- What are some of the problems that can arise when unstained preparations are used to make total cell counts of samples from natural environments?

6.10 Viable Cell Counting

A **viable** cell is one that is able to divide and form offspring, and in most cell-counting situations, these are the cells we are most interested in. For these purposes, we can use a viable counting method. Usually, we determine the number of cells in the sample capable of forming colonies on a suitable agar medium. For this reason, the viable count is also called a **plate count**. The assumption made in the viable counting procedure is that each viable cell can grow and divide to yield one colony. Thus, colony numbers and cell numbers are proportional.

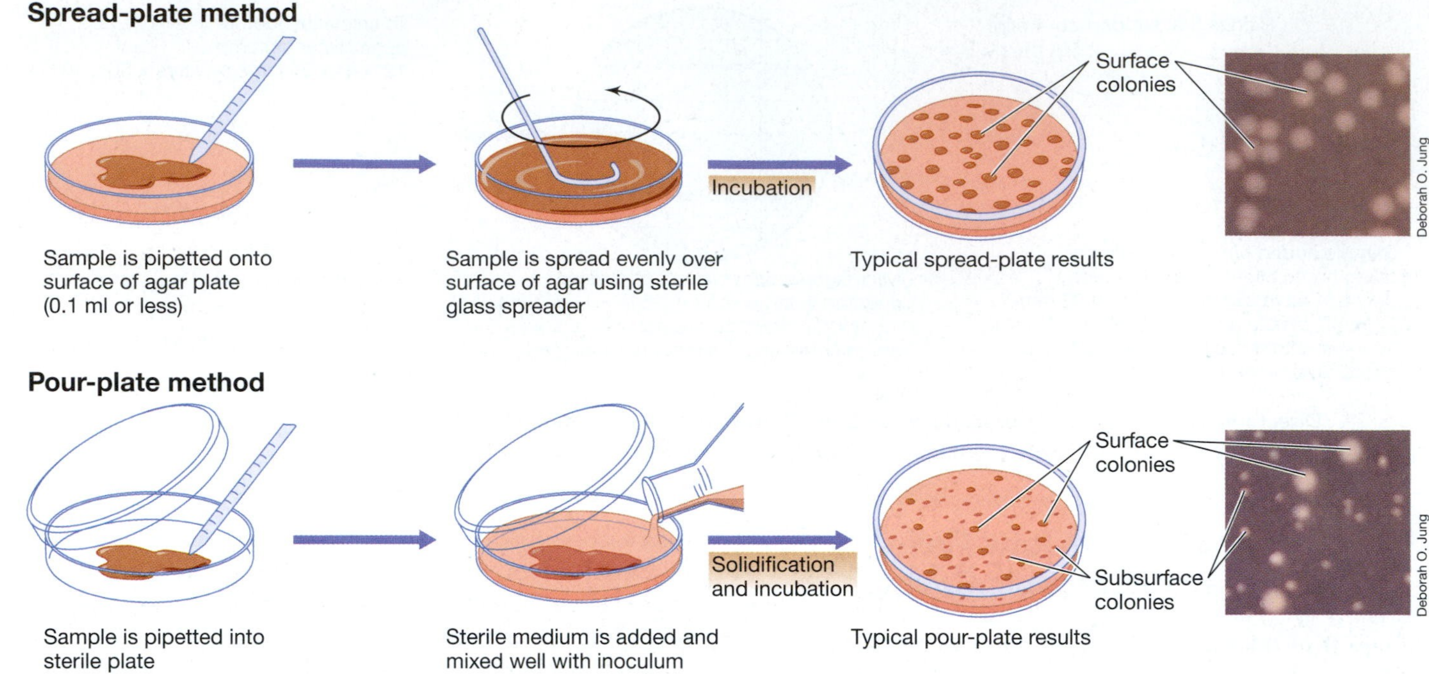
Spread-plate method
Incubation
Surface colonies
Deborah O. Jung
Sample is pipetted onto surface of agar plate (0.1 ml or less)
Sample is spread evenly over surface of agar using sterile glass spreader
Typical spread-plate results
Pour-plate method
Surface colonies
Solidification and incubation
Subsurface colonies
Deborah O. Jung
Sample is pipetted into sterile plate
Sterile medium is added and mixed well with inoculum
Typical pour-plate results

Figure 6.15 Two methods for the viable count. In the pour-plate method, colonies form within the agar as well as on the agar surface. In the fourth column is shown a photo of colonies of *Escherichia coli* formed from cells plated by the spread-plate method (top) or the pour-plate method (bottom).

There are at least two ways of performing a plate count: the *spread-plate method* and the *pour-plate method* (**Figure 6.15**). In the spread-plate method, a volume (usually 0.1 ml or less) of an appropriately diluted culture is spread over the surface of an agar plate using a sterile glass spreader. The plate is then incubated until colonies appear, and the number of colonies is counted. The surface of the plate must not be too moist so that the liquid soaks in and the cells remain stationary. Volumes greater than about 0.1 ml are avoided in this method because the excess liquid does not soak in and may cause the colonies to coalesce as they form, making them difficult to count.

In the pour-plate method (Figure 6.15), a known volume (usually 0.1–1.0 ml) of culture is pipetted into a sterile Petri plate. Melted agar medium is then added and mixed well by gently swirling the plate on the benchtop. Because the sample is mixed with the molten agar medium, a larger volume can be used than with the spread plate. However, with this method the organism to be counted must be able to withstand brief exposure to the temperature of molten agar (~45–50°C). Here, colonies form throughout the plate and not just on the agar surface as in the spread-plate method. The plate must therefore be examined closely to make sure all colonies are counted. If the pour-plate method is used to enumerate cells from a natural sample, another problem may arise; any debris in the sample must be distinguishable from actual bacterial colonies or the count will be erroneous.

Diluting Cell Suspensions before Plating

With both the spread-plate and pour-plate methods, it is important that the number of colonies developing on the plate not be too many or too few. On crowded plates some cells may not form colonies, and some colonies may fuse, leading to erroneous measurements. If the number of colonies is too small, the statistical significance of the calculated count will be low. The usual practice, which is statistically most valid, is to count colonies only on plates that have between 30 and 300 colonies.

To obtain the appropriate colony number, the sample to be counted must almost always be diluted. Because one may not know the approximate viable count ahead of time, it is usually necessary to make more than one dilution. Several 10-fold dilutions of the sample are commonly used (**Figure 6.16**). To make a 10-fold (10^{-1}) dilution, one can mix 0.5 ml of sample with 4.5 ml of diluent, or 1.0 ml sample with 9.0 ml diluent. If a 100-fold (10^{-2}) dilution is needed, 0.05 ml can be mixed with 4.95 ml diluent, or 0.1 ml with 9.9 ml diluent. Alternatively, a 10^{-2} dilution can be made by making two successive 10-fold dilutions. With dense cultures, such *serial* dilutions are needed to reach a suitable dilution. Thus, if a $10^{-6}(1/10^6)$ dilution is needed, it can be achieved by making three successive $10^{-2}(1/10^2)$ dilutions or six successive 10^{-1} dilutions (Figure 6.16).

Sources of Error in Plate Counting

The number of colonies obtained in a viable count experiment depends not only on the inoculum size and the viability of the culture, but also on the suitability of the culture medium and the incubation conditions. The colony number can also change with the length of incubation. For example, if a mixed culture is used, the cells deposited on the plate will not all form colonies at the same rate; if a short incubation time is used, fewer than the maximum number of colonies will be obtained. Furthermore, the size of colonies may vary. If some tiny colonies develop, they may be missed during the counting. With pure cultures, colony development is a more synchronous process and uniform colony morphology is the norm.

Viable counts can be subject to rather large errors for several reasons. These include pipetting inconsistencies, inhomogeneity of the sample, insufficient mixing, and other factors. Hence, if accurate counts are to be obtained, great care and consistency must be taken in sample preparation and pipetting, and replicate plates of key dilutions must be prepared. Note also that if two or more cells are in a clump, they will grow to form only a single colony. So if a sample contains many cell clumps, a viable count of that sample may be erroneously low. Data are often expressed as the number of *colony-forming units* obtained rather than the actual number of viable cells, because a colony-forming unit may contain one or more cells.

Despite the difficulties associated with viable counting, the procedure gives the best estimate of the number of viable cells in a sample and so is widely used in many areas of microbiology. For example, in food, dairy, medical, and aquatic microbiology, viable counts are employed routinely. The method has the virtue of high sensitivity, because as few as one viable cell per sample plated can be detected. This feature allows for the sensitive detection of microbial contamination of foods or other materials.

The use of highly selective culture media and growth conditions in viable counting procedures allows one to target only particular species, or in some cases even a single species, in a mixed population of microorganisms present in the sample. For example, in practical applications such as in the food industry, viable counting on both complex and selective media allows for both quantitative and qualitative assessments of the microorganisms present in a food product. That is, with a single sample one medium may be employed for a total count and a second medium used to target a particular organism, such as a specific pathogen. Targeted counting is common in wastewater and other water analyses (∞ Figures 36.1 and 36.2). For example, because enteric bacteria originate from fecal matter and are easy to target using selective media, if enteric bacteria are detected in a water sample from a swimming site, for example, their presence is a signal that the water is unsafe for swimming.

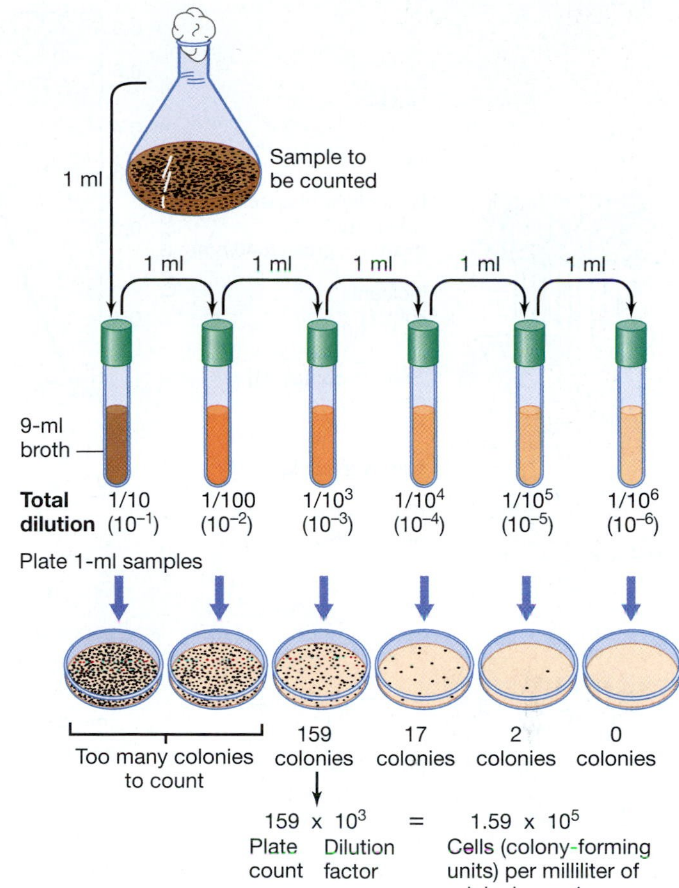

Figure 6.16 Procedure for viable counting using serial dilutions of the sample and the pour-plate method. The sterile liquid used for making dilutions can simply be water, but a balanced salt solution or growth medium may yield a higher recovery. The dilution factor is the reciprocal of the dilution.

The Great Plate Count Anomaly

Direct microscopic counts of natural samples typically reveal far more organisms than are recoverable on plates of any single culture medium. Thus, although a very sensitive technique, plate counts can be highly unreliable when used to assess total cell numbers of natural samples, such as soil and water. Some microbiologists have referred to this as "the great plate count anomaly."

Why do plate counts show lower numbers of cells than direct microscopic counts? One obvious factor is that microscopic methods count dead cells, whereas viable methods do not. More importantly, however, is the fact that different organisms, even those present in a very small natural sample, may have vastly different requirements for nutrients and growth conditions in laboratory culture. Thus, one medium and set of growth conditions can at best be expected to support growth of only a subset of the total microbial community. If this subset makes up, for example, 10^6 cells/g in a total viable community of 10^9 cells/g, the plate count will reveal only 0.1% of the total microbial population, a vast underestimation of the actual number of viable cells.

Plate count results can hence carry a large caveat. Targeted plate counts using highly selective media, as in, for example, the microbial analysis of sewage or food, can yield reliable data. By contrast, "total cell counts" of the same samples using a single medium and set of growth conditions may be, and usually are, underestimates of actual cell numbers by one to several orders of magnitude.

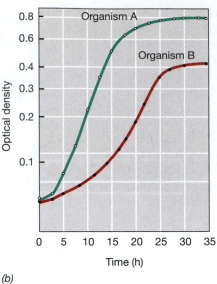

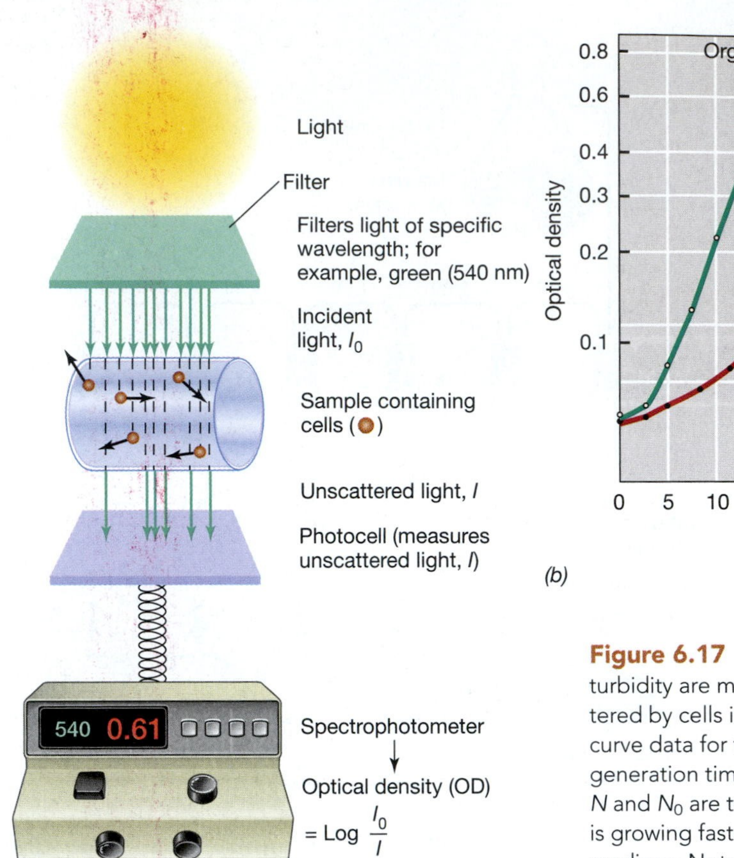

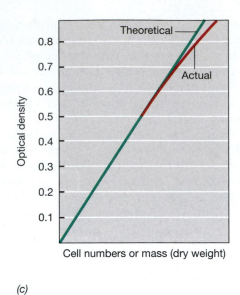

Figure 6.17 Turbidity measurements of microbial growth. *(a)* Measurements of turbidity are made in a spectrophotometer. The photocell measures incident light unscattered by cells in suspension and gives readings in optical density units. *(b)* Typical growth curve data for two organisms growing at different growth rates. For practice, calculate the generation time (g) of the two cultures using the formula $n = 3.3 (\log N - \log N_0)$ where N and N_0 are two different OD readings taken between a time interval t. Which organism is growing faster, A or B? *(c)* Relationship between cell number or dry weight and turbidity readings. Note that the one-to-one correspondence between these relationships breaks down at high turbidities. See Figure 1 of the Microbial Sidebar, page 158.

6.10 MiniReview

Viable cell counts (plate counts) measure only the living population present in the sample.

- Why is a viable count more sensitive than a microscopic count?
- What is the major assumption made in relating plate count results to cell number?
- Describe how you would dilute a bacterial culture by 10^{-7}.
- What is the "great plate count anomaly"?

6.11 Measurements of Microbial Mass: Turbidimetric Methods

During exponential growth, all cellular components increase in proportion to the increase in cell numbers. Thus, instead of measuring changes in cell number over time, we could instead measure the increase in protein, DNA, or dry weight of a culture as a barometer of growth. A rapid and quite useful method of estimating cell growth in this way is by *turbidity measurements*.

A suspension of cells looks cloudy (turbid) to the eye because cells scatter light passing through the suspension. The more cells that are present, the more light is scattered, and hence the more turbid the suspension. What is actually assessed in a turbidimetric measurement is total cell mass. However, because cell mass is proportional to cell number, we can use turbidity as a measure of cell numbers in a growing culture.

Optical Density

Turbidity is measured with a spectrophotometer, an instrument that passes light through a cell suspension and detects the unscattered light that emerges (**Figure 6.17**). A spectrophotometer employs a prism or diffraction grating to generate incident light of a specific wavelength (Figure 6.17*a*). Commonly used wavelengths for bacterial turbidity measurements include 480 nm (blue), 540 nm (green), 600 nm (orange), and 660 nm (red). Sensitivity is best at shorter wavelengths, but measurements of dense cell suspensions are more accurate at longer wavelengths. However, at any wavelength it is the decrease in unscattered light caused by turbidity that is the measured quantity. In a spectrophotometer the unit of turbidity measurement is *optical density (OD)* at the wavelength specified, for example, OD_{540} for optical density measurements at 540 nm (Figure 6.17).

Relating OD to Cell Numbers

For unicellular organisms, OD is proportional, within certain limits, to cell number. Turbidity readings can therefore be used as a substitute for total or viable counting methods. However, before this is done, a standard curve must be prepared that relates a cell number (microscopic or viable count), dry weight, or protein content, to turbidity. As can be seen in such a plot, proportionality only holds within limits (Figure 6.17*c*). Thus, at high cell concentrations, light scattered away from the spectrophotometer's photocell by one cell can be scattered back by another. To the photocell this makes it appear as if light had never been scattered in the first place. At such high cell densities, the correspondence between cell number and turbidity deviates from linearity (Figure 6.17*c*) and the OD measurements made are therefore less accurate. However, up to this limit, which varies for every organism, turbidity measurements can be accurate measures of cell number or dry weight.

Turbidity measurements have the virtue of being quick and easy to perform. Turbidity measurements can typically be made without destroying or significantly disturbing the sample. For these reasons, turbidity measurements are widely employed to monitor growth of microbial cultures. The same sample can be checked repeatedly and the measurements plotted on a semilogarithmic plot versus time. From these, it is easy to calculate the generation time and other parameters of the growing culture (Figure 6.17*b*).

Turbidity measurements are sometimes problematic. Although many microorganisms grow in even suspensions in liquid medium, many others do not. Some bacteria form small to large clumps, and in such instances, OD measurements may be quite inaccurate as a measure of total microbial mass. In addition, many bacteria grow in films on the sides of tubes or other growth vessels, mimicking in laboratory culture how they actually grow in nature (see the Microbial Sidebar "Microbial Growth in the Real World: Biofilms"). Thus for ODs to accurately reflect cell mass (and thus cell numbers) in a liquid culture, clumping and biofilms have to be minimized. This can often be accomplished by stirring, shaking, or in some way keeping the cells well mixed during the growth process to prevent the formation of cell aggregates and biofilms.

6.11 MiniReview

Turbidity measurements are an indirect but very rapid and useful method of measuring microbial growth. However, in order to relate a direct cell count to a turbidity value, a standard curve must first be established.

- List two advantages of using turbidity as a measure of cell growth.
- Describe how you could use a turbidity measurement to tell how many colonies you would expect from plating a culture of a given OD.

IV TEMPERATURE AND MICROBIAL GROWTH

The activities of microorganisms are greatly affected by the chemical and physical state of their environment. Many environmental factors can be considered. However, four key factors control the growth of all microorganisms: temperature, pH, water availability, and oxygen, and we consider these here. Some other factors can potentially affect the growth of microorganisms, such as pressure and radiation. These more specialized environmental factors will be considered later in this book when we encounter microbial habitats in which they play major roles.

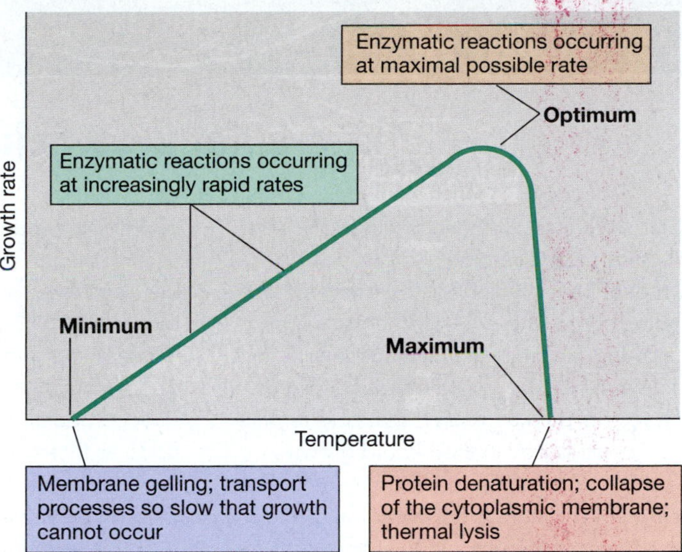

Figure 6.18 The cardinal temperatures: Minimum, optimum, and maximum. The actual values may vary greatly for different organisms (see Figure 6.19).

6.12 Effect of Temperature on Microbial Growth

Temperature is probably *the* most important environmental factor affecting the growth and survival of microorganisms. At either too cold or too hot a temperature, microorganisms will not be able to grow and may even die. The minimum and maximum temperatures for growth vary greatly among different microorganisms and usually reflect the temperature range and average temperature of their habitats.

Cardinal Temperatures

Temperature affects microorganisms in two opposing ways. As temperatures rise, chemical and enzymatic reactions in the cell proceed at more rapid rates and growth becomes faster; however, above a certain temperature, cell components may be irreversibly damaged. Thus, as the temperature is increased within a given range, growth and metabolic function increase up to a point where denaturation reactions set in. Above this point, cell functions fall to zero.

For every microorganism there is thus a minimum temperature below which growth is not possible, an optimum temperature at which growth is most rapid, and a maximum temperature above which growth is not possible (**Figure 6.18**). The optimum temperature is always nearer the maximum than the minimum. These three temperatures, called the **cardinal temperatures**, are characteristic for any given microorganism.

Microbial Sidebar

Microbial Growth in the Real World: Biofilms

In this chapter we have discussed several ways in which microbial growth can be measured, including microscopic methods, viable counts, and measurements of light scattering (turbidity) by cells suspended in a liquid culture. The turbidimetric measures of bacterial growth assume that cells remain evenly distributed in their liquid growth medium. Under these conditions the optical density of a culture is proportional to the log of the number of cells in suspension (**Figure 1**). This floating lifestyle, called *planktonic,* is the way some bacteria actually live in nature, for example, cells that inhabit the water column of a lake. However, many other microorganisms are *sessile,* meaning that they grow attached to a surface. These attached cells can then develop into **biofilms**.

Humans encounter bacterial biofilms on a daily basis, for example, when cleaning out a pet's water bowl that has been sitting unattended for a few days or when you sense with your tongue the "film" that develops on your unbrushed teeth.

A biofilm is an attached polysaccharide matrix containing bacterial cells. Biofilms form in stages: (1) reversible attachment of planktonic cells, (2) irreversible attachment of the same cells, (3) cell growth and production of polysaccharide, (4) further development to form the tenacious and nearly impenetrable mature biofilm. In the early stages of biofilm formation, the attachment of bacterial cells to a surface triggers biofilm-specific gene expression. Genes are turned on that direct production of cell surface polysaccharides; as these are produced, they facilitate the attachment of more cells. Eventually, through growth and recruitment, entire microbial communities develop within the slimy polysaccharide matrix.

Bacterial biofilms can dramatically affect humans. For example, bacterial infections are often linked to pathogens that develop in biofilms during the disease process. The disease cystic fibrosis (CF) is characterized by development of a biofilm containing *Pseudomonas aeruginosa* and other bacteria in the lungs of CF patients (**Figure 2**). The biofilm matrix, which contains alginate and other polysaccharides as well as bacterial DNA, greatly reduces the ability of antimicrobial agents, such as antibiotics, to penetrate, and thus bacteria within the biofilm may be unaffected by the drugs. Bacterial biofilms have also been implicated in difficult-to-treat infections of implanted medical devices, such as replacement heart valves and artificial joints.

Biofilms are also a major problem in industry. Microbial biofilms can cause fouling of equipment and the contamination of products, especially if the flowing liquid is a good microbial substrate, such as milk. Biofilms can also do long-term damage to water distribution facilities and other public utilities. Biofilms that develop in bulk storage containers, such as fuel storage tanks, can contaminate the fuel and cause souring from chemicals, such as hydrogen sulfide (H_2S), excreted by the biofilm bacteria.

Biofilms are a common form of bacterial growth in nature. Not only does the biofilm offer protection from harmful chemicals, the thick matrix of the biofilm provides a barrier to grazing by other microorganisms and prevents bacterial cells from being washed away into a less-favorable habitat. So, while optical densities give us a laboratory picture of the perfectly suspended bacterial culture, in the "real" world bacterial growth in the biofilm state may be the norm.

We examine biofilms in more detail in our focus on surfaces as microbial habitats in Chapter 23 (Sections 23.3–23.5).

Figure 1 *Liquid cultures of* Escherichia coli. *In these cultures cells are in a planktonic state and are evenly suspended in the medium. The increasing (left to right) optical density (OD_{540}) of each culture is shown below the tube.*

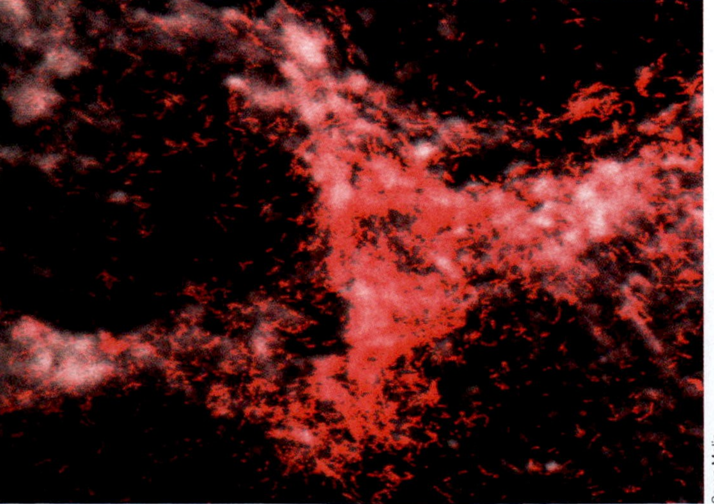

Figure 2 *Fluorescently stained cells of* Pseudomonas aeruginosa *from a sputum sample of a cystic fibrosis patient. The red cells are of* P. aeruginosa *and the white material is alginate, a polysaccharide-like material that is produced by* P. aeruginosa.

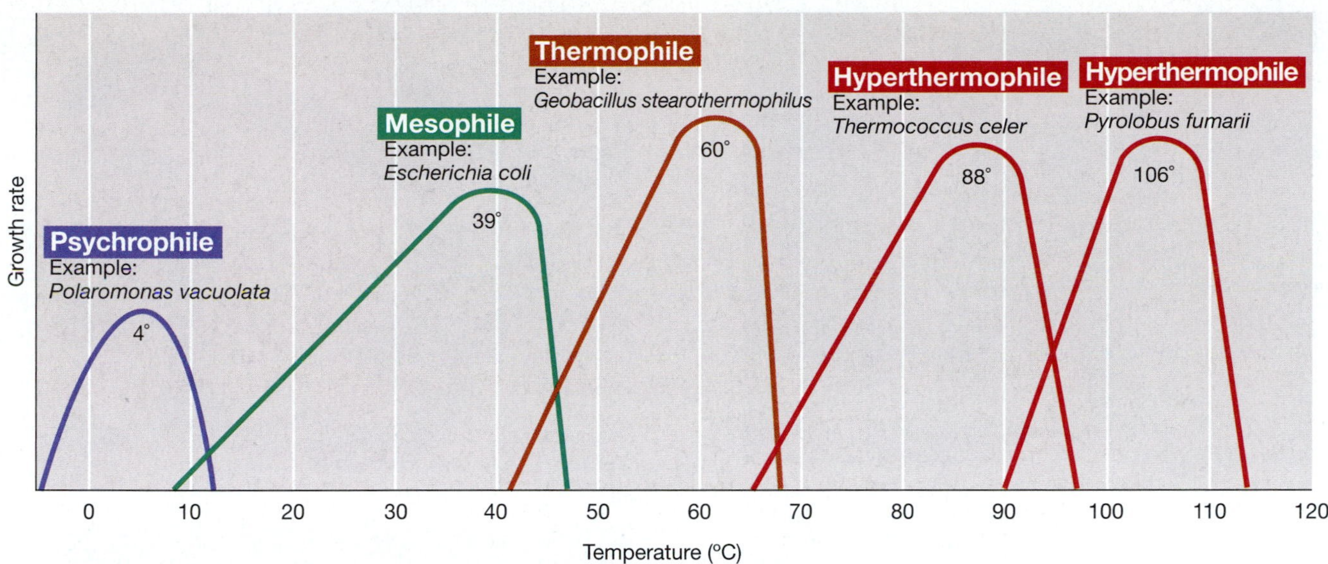

Figure 6.19 Temperature and growth relations in different temperature classes of microorganisms. The temperature optimum of each example organism is shown on the graph.

The cardinal temperatures of different microorganisms differ widely; some organisms have temperature optima as low as 4°C and some higher than 100°C. The temperature range throughout which microorganisms grow is even wider than this, from below freezing to well above the boiling point of water. However, no single organism can grow over this whole temperature range, as the range for any given organism is typically 25–40 degrees.

The maximum growth temperature of an organism reflects the temperature above which denaturation of one or more essential cell components, such as a key enzyme, occurs. The factors controlling an organism's minimum growth temperature are not as clear. However, as previously discussed (Section 4.5), the cytoplasmic membrane must be in a fluid state for transport and other important functions to take place. An organism's minimum temperature may well be governed by membrane functioning: If an organism's cytoplasmic membrane stiffens to the point that it no longer functions properly in nutrient transport or can no longer develop a proton motive force, the organism cannot grow. The growth temperature *optimum* reflects a state in which all or most cellular components are functioning at their maximum rate.

Temperature Classes of Organisms

Although there is a continuum of organisms, from those with very low temperature optima to those with high temperature optima, it is possible to distinguish at least four groups of microorganisms in relation to their growth temperature optima: **psychrophiles**, with low temperature optima; **mesophiles**, with midrange temperature optima; **thermophiles**, with high temperature optima; and **hyperthermophiles**, with very high temperature optima (**Figure 6.19**).

Mesophiles are widespread in nature. They are found in warm-blooded animals and in terrestrial and aquatic environments in temperate and tropical latitudes. Psychrophiles and thermophiles are found in unusually cold and unusually hot environments, respectively. Hyperthermophiles are found in extremely hot habitats such as hot springs, geysers, and deep-sea hydrothermal vents.

Escherichia coli is a typical mesophile, and its cardinal temperatures have been precisely defined. The optimum temperature for most strains of *E. coli* is 39°C, the maximum is 48°C, and the minimum is 8°C. Thus, the temperature *range* for *E. coli* is 40 degrees, near the high end for prokaryotes (Figure 6.19).

We now turn to the interesting cases of microorganisms growing at very low or very high temperatures—the extremophiles.

6.12 MiniReview

Temperature is a major environmental factor controlling microbial growth. The cardinal temperatures describe the minimum, optimum, and maximum temperatures at which each organism grows. Microorganisms can be grouped by the temperature ranges they require.

- What are the cardinal temperatures for *Escherichia coli?* To what temperature class does it belong?
- How does a hyperthermophile differ from a psychrophile?
- *Escherichia coli* can grow at a higher temperature in a complex medium than in a defined medium. Why?

6.13 Microbial Growth at Cold Temperatures

Because humans live and work on the surface of Earth where temperatures are generally moderate, it is natural to consider very hot and very cold environments as "extreme." However,

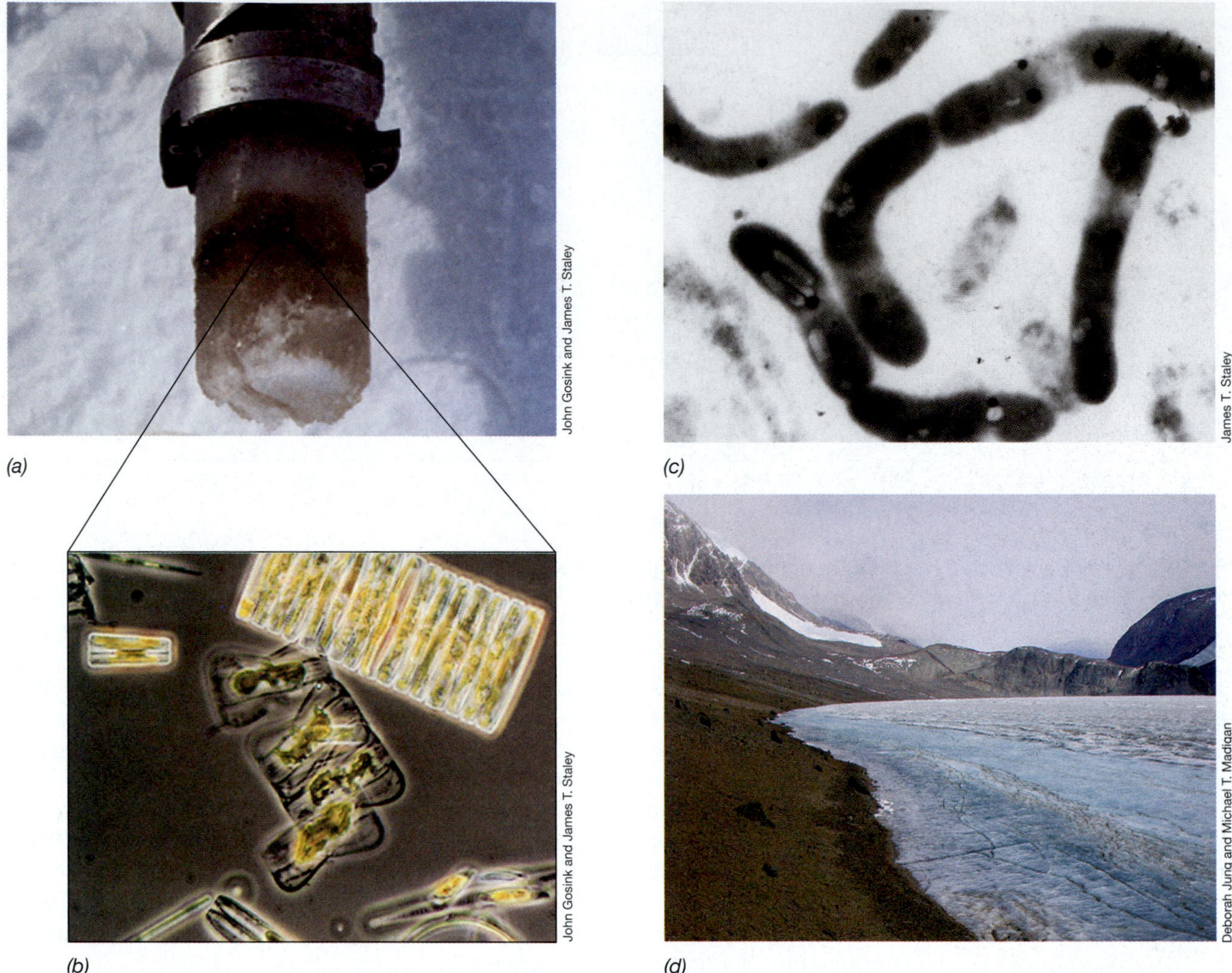

Figure 6.20 Antarctic microbial habitats and microorganisms. *(a)* A core of frozen seawater from McMurdo Sound, Antarctica. The core is about 8 cm wide. Note the dense coloration due to pigmented microorganisms. *(b)* Phase-contrast micrograph of phototrophic microorganisms from the core shown in *(a)*. Most organisms are either diatoms or green algae (both eukaryotic phototrophs). *(c)* Transmission electron micrograph of *Polaromonas*, a gas vesiculate bacterium that lives in sea ice and grows optimally at 4°C. *(d)* Photo of the surface of Lake Bonney, McMurdo Dry Valleys, Antarctica. Like many other Antarctic lakes, Lake Bonney, which is about 40 meters deep, remains permanently frozen and has an ice cover of about 5 meters. The water column of Lake Bonney remains near 0°C and contains both oxic and anoxic zones; thus both aerobic and anaerobic microorganisms inhabit the lake. However, no higher eukaryotic organisms inhabit Dry Valley lakes, making them uniquely microbial ecosystems.

many microbial habitats are very hot or very cold. The organisms that live in these environments are therefore called **extremophiles** (Section 2.4 and Table 2.1) and, interestingly, in most cases have evolved to grow optimally under their extreme condition. We consider the biology of these fascinating organisms here and in the next section.

Cold Environments

Much of Earth's surface is cold. The oceans, which make up over half of Earth's surface, have an average temperature of 5°C, and the depths of the open oceans have constant temperatures of 1–3°C. Vast land areas of the Arctic and Antarctic are permanently frozen or are unfrozen for only a few weeks in summer (**Figure 6.20**). These cold environments are far from sterile, as viable microorganisms will inhabit any low-temperature environment in which some liquid water remains. Salts and other solutes, for example, depress the freezing point of water and allow microbial growth to occur below the freezing point of pure water, 0°C. But even in frozen materials there are often small pockets of liquid water where solutes have concentrated and microorganisms can metabolize and grow. Within glaciers, for example, there exists a network of liquid water channels in which prokaryotes thrive and reproduce.

In considering cold environments, it is important to distinguish between environments that are constantly cold and those that are cold only in winter. The latter, characteristic of temperate climates, may have summer temperatures as high as 40°C. A temperate lake, for example, may have a period of ice cover in the winter, but the time that the water is at 0°C is relatively brief. Such highly variable environments are less favorable for cold-adapted organisms than the constantly cold environments characteristic of polar regions, high altitudes, and the depths of the oceans. For example, lakes in the Antarctic McMurdo Dry Valleys contain a permanent ice cover several meters thick (Figure 6.20*d*). The water column below the ice in these lakes remains at 0°C or colder year round and is thus an ideal habitat for cold-active microorganisms.

Psychrophilic Microorganisms

As noted earlier, organisms with low temperature optima are called *psychrophiles*. A psychrophile is defined as an organism with an optimal growth temperature of 15°C or lower, a maximum growth temperature below 20°C, and a minimal growth temperature at 0°C or lower. Organisms that grow at 0°C but have optima of 20–40°C are called **psychrotolerant**.

Psychrophiles are found in environments that are constantly cold and may be rapidly killed by warming, even to as little as 20°C. For this reason, their laboratory study requires that great care be taken to ensure that they never warm up during sampling, transport to the laboratory, isolation, or other manipulations. In open ocean waters, where temperatures remain constant at about 3°C, various cold-active *Bacteria* and *Archaea* are present, although only a relatively few have been isolated in laboratory culture. Temperate environments, which warm up in summer, cannot support the heat-sensitive psychrophiles because they cannot survive the warming.

Psychrophilic algae grow in dense masses within and under sea ice (frozen seawater that forms seasonally) in polar regions (Figures 6.20*a, b*) and are also often present on the surfaces of snowfields and glaciers at such densities that they impart a distinctive red or green coloration to the surface (**Figure 6.21*a***). The common snow alga *Chlamydomonas nivalis* is an example of this, its spores responsible for the brilliant red color of the snow surface (Figure 6.21*b*). This green alga grows within the snow as a green-pigmented vegetative cell and then sporulates. As the snow dissipates by melting, erosion, and ablation, the spores become concentrated on the surface. Related species of snow algae contain different carotenoid pigments, and thus, fields of snow algae can also be green, orange, brown, or purple.

In addition to snow algae, several psychrophilic bacteria have been isolated, most from either marine sediments or from Antarctica. Some of these, particularly isolates from sea ice such as *Polaromonas* (Figure 6.20*c*), have very low growth temperature optima and maxima (4°C and 12°C, respectively).

Psychrotolerant Microorganisms

Psychrotolerant microorganisms are more widely distributed in nature than are psychrophiles and can be isolated from soils and water in temperate climates as well as from meat, milk and other dairy products, cider, vegetables, and fruit

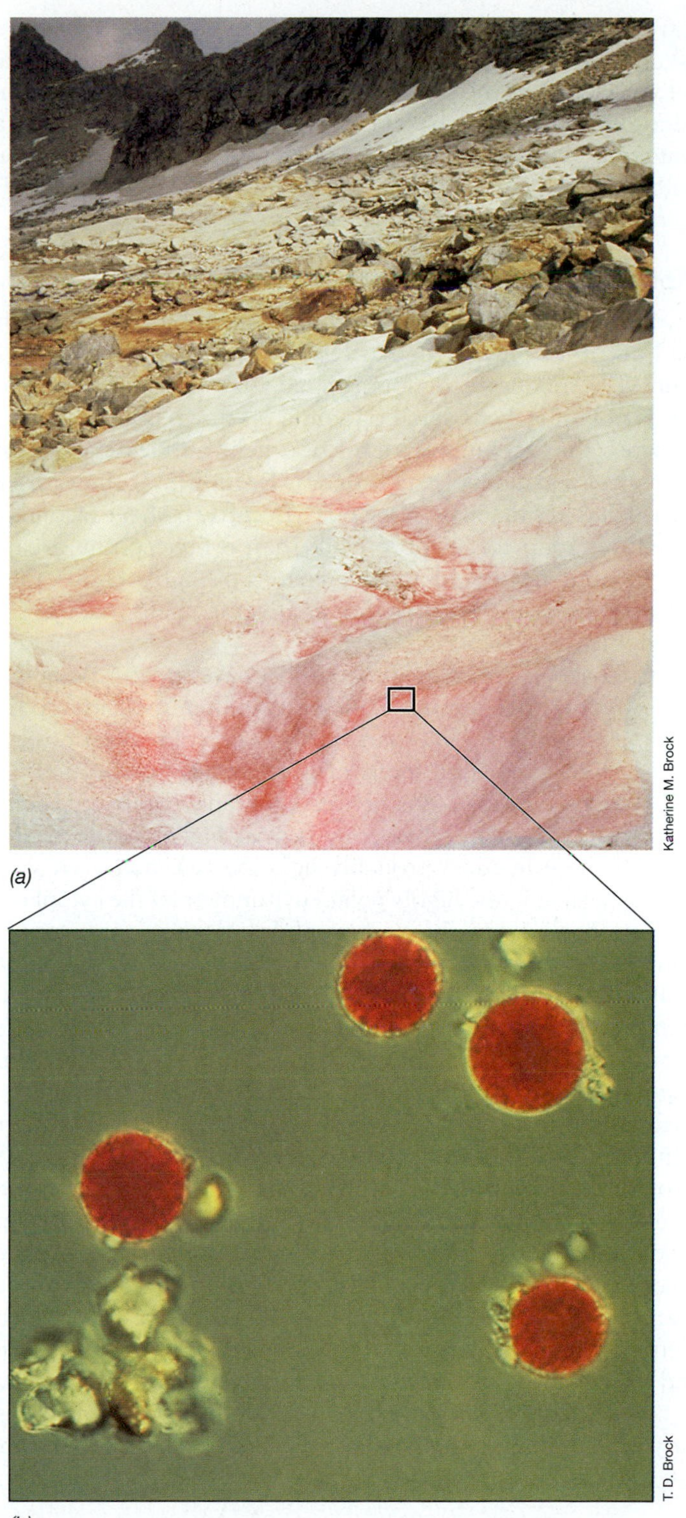

Figure 6.21 Snow algae. *(a)* Snow bank in the Sierra Nevada, California, with red coloration caused by the presence of snow algae. Pink snow such as this is common on summer snow banks at high altitudes throughout the world. *(b)* Photomicrograph of red-pigmented spores of the snow alga *Chlamydomonas nivalis*. The spores germinate to yield motile green algal cells. Some strains of snow algae are true psychrophiles but many are psychrotolerant, growing best at temperatures above 20°C.

stored at refrigeration temperatures (~4°C). As noted, psychrotolerant microorganisms grow best at a temperature between 20° and 40°C. Moreover, although psychrotolerant microorganisms do grow at 0°C, most do not grow very well at that temperature, and one must often wait several weeks before visible growth is seen in laboratory cultures. Various *Bacteria*, *Archaea*, and microbial eukaryotes are psychrotolerant.

Molecular Adaptations to Psychrophily

Psychrophiles produce enzymes that function optimally in the cold and that may be denatured or otherwise inactivated at even very moderate temperatures. The molecular basis for this is not entirely understood, but it is known that cold-active enzymes show greater amounts of α-helix and lesser amounts of β-sheet secondary structure (Section 3.7) than do enzymes that are inactive in the cold. Because β-sheet secondary structures tend to be more rigid than α-helices, the greater α-helix content of cold-active enzymes probably allows these proteins greater flexibility for catalyzing their reactions at cold temperatures.

Cold-active enzymes also tend to have greater polar and lesser hydrophobic amino acid contents than their mesophilic and thermophilic counterparts. Moreover, cold-active proteins tend to have decreased levels of weak bonds (Section 3.1) and decreased interactions between domains compared to proteins from organisms that grow best at higher temperatures. Both of these features probably help the cold-active enzymes remain flexible in the highly polar environment of the cytoplasm.

Another feature of psychrophiles is that compared to mesophiles, transport processes (Section 4.5) function optimally at low temperature, an indication that the cytoplasmic membranes of psychrophiles are modified such that low temperatures do not inhibit membrane functions. Cytoplasmic membranes from psychrophiles tend to have a higher content of unsaturated fatty acids (Section 5.17). This helps to maintain a semifluid state of the membrane at low temperatures (membranes composed of predominantly saturated fatty acids would become waxy at low temperatures). The lipids of some psychrophilic bacteria contain polyunsaturated fatty acids or other long-chain hydrocarbons with multiple double bonds. For example, a hydrocarbon with nine double bonds ($C_{31:9}$) has been identified from the membrane lipids of some Antarctic bacteria, and the psychrophilic bacterium *Psychroflexus* has been shown to contain fatty acids with four and five double bonds. These fatty acids remain more flexible at low temperatures than saturated or monounsaturated fatty acids.

Freezing

Despite the ability of some organisms to grow at low temperatures, there is a lower limit below which reproduction is impossible. Pure water freezes at 0°C and seawater at −2.5°C, but freezing is not a continuous process. Microscopic pockets of water continue to exist at these and even much lower temperatures. As long as liquid water is available, microbial growth is possible below the freezing point. Convincing evidence for microbial growth at −12°C has been demonstrated but not at temperatures below this.

Although freezing prevents microbial growth, it does not necessarily cause microbial death. Microbial cells can continue to metabolize at temperatures far beneath that which will support growth. For example, microbial respiration leading to CO_2 production has been shown in tundra soils at temperatures as low as −39°C. Thus, enzymes continue to function at temperatures far below those that allow for cell growth.

The medium in which cells are suspended also affects their sensitivity to freezing. If solutes such as glycerol or dimethylsulfoxide (DMSO) are present in the medium, this depresses the freezing point. The solutes penetrate the cells and protect them by reducing the severity of dehydration and preventing ice crystal formation. In fact, the addition of such *cryoprotectants* is a common way of preserving microbial cultures at very low temperatures. To freeze cells for long-term preservation, cells are typically suspended in growth medium containing ~10% DMSO or glycerol and quickly frozen at −70 to −196°C. Properly prepared frozen cells that are not allowed to thaw and refreeze can remain viable for decades or even longer.

We now travel to the other end of the thermometer and look at microorganisms growing at high temperatures.

6.13 MiniReview

Organisms with cold temperature optima are called psychrophiles, and the most extreme representatives inhabit permanently cold environments. Psychrophiles have evolved biomolecules that function best at cold temperatures but that can be unusually sensitive to warm temperatures.

- How do psychrotolerant organisms differ from psychrophilic organisms?
- What molecular adaptations to the cytoplasmic membrane are seen in psychrophiles, and why are they necessary?

6.14 Microbial Growth at High Temperatures

Microbial life flourishes in high-temperature environments, up to and including boiling water. Above about 65°C, only prokaryotic life forms are found. Even here, a huge diversity of both *Bacteria* and *Archaea* exist.

Thermal Environments

Organisms whose growth temperature optimum exceeds 45°C are called *thermophiles* and those whose optimum exceeds 80°C are called *hyperthermophiles* (Figure 6.19). Temperatures as high as these are found only in certain areas. For example, the surface of soils subject to full sunlight can be heated to above 50°C at midday, and some surface soils may become warmed to even 70°C. Fermenting materials such as compost piles and silage can also reach temperatures of 70°C. However, the most extensive and extreme high-temperature environments in nature are associated with volcanic phenomena. These include, in particular, hot springs.

Table 6.1 Presently known upper temperature limits for growth of living organisms

Group	Upper temperature limits (°C)
Macroorganisms	
Animals	
Fish and other aquatic vertebrates	38
Insects	45–50
Ostracods (crustaceans)	49–50
Plants	
Vascular plants	45
Mosses	50
Microorganisms	
Eukaryotic microorganisms	
Protozoa	56
Algae	55–60
Fungi	60–62
Prokaryotes	
Bacteria	
Cyanobacteria	73
Anoxygenic phototrophs	70–73
Chemoorganotrophic/chemolithotrophic *Bacteria*	95
Archaea	
Chemoorganotrophic/chemolithotrophic *Archaea*	121[a]

[a]The upper temperature limit for growth of the organism *Pyrolobus fumarii* is 113°C. A species of *Pyrodictium* can grow up to 121°C.

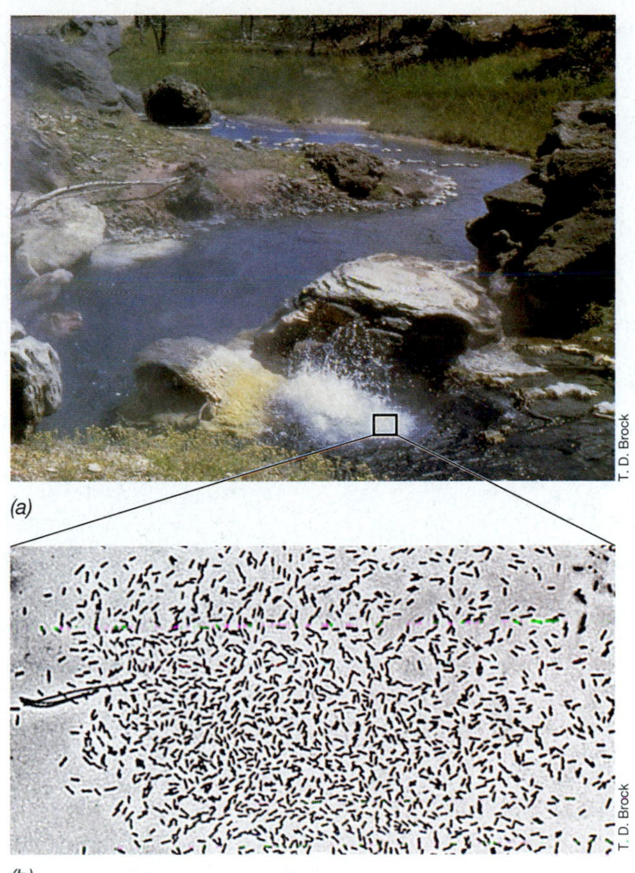

Figure 6.22 Growth of hyperthermophiles in boiling water. *(a)* Boulder Spring, a small boiling spring in Yellowstone National Park. This spring is superheated, having a temperature 1–2°C above the boiling point. The mineral deposits around the spring consist mainly of silica and sulfur. *(b)* Photomicrograph of a microcolony of prokaryotes that developed on a microscope slide immersed in a boiling spring such as that shown in *(a)*.

Many hot springs have temperatures at or near boiling, and steam vents (fumaroles) may reach 150–500°C. Hydrothermal vents in the bottom of the ocean have temperatures of 350°C or greater (Section 24.11). Hot springs exist throughout the world, but they are especially abundant in the western United States, New Zealand, Iceland, Japan, Italy, Indonesia, Central America, and central Africa. The world's largest single concentration of hot springs is in Yellowstone National Park, Wyoming (USA).

Although some hot springs vary widely in temperature, many are nearly constant, varying less than 1–2°C over many years. In addition, different springs have different chemical compositions and pH values. Above 65°C, only prokaryotes are present (**Table 6.1**), but the diversity of *Bacteria* and *Archaea* may be extensive.

Hyperthermophiles in Hot Springs

In boiling hot springs (**Figure 6.22**), a variety of hyperthermophiles are typically present, including both chemoorganotrophic and chemolithotrophic species. Growth of natural populations of hyperthermophiles can be studied by immersing a microscope slide into a spring and then retrieving it after a few days; microscopic examination reveals colonies of prokaryotes that have developed from single cells that attached to and grew on the glass surface (Figure 6.22*b*). Such ecological studies of organisms living in boiling springs have shown that growth rates are often rapid; doubling times as short as 1 h have been recorded.

Cultures of many hyperthermophiles have been obtained, and a variety of morphological and physiological types of both *Bacteria* and *Archaea* are known. Phylogenetic studies using ribosomal RNA sequencing have shown great evolutionary diversity among these hyperthermophiles as well. Some hyperthermophilic *Archaea* show growth-temperature optima greater than 100°C. Laboratory cultures of such organisms are grown in pressurized vessels that permit temperatures in the growth medium to rise above the boiling point. No species of *Bacteria* are known capable of growth above 95°C.

Thermophiles

Many thermophiles (optima 45–80°C) are also present in hot springs and other thermal environments. In hot springs, as

Figure 6.23 Growth of thermophilic cyanobacteria in a hot spring in Yellowstone National Park. Characteristic V-shaped pattern (shown by the dashed red lines) formed by cyanobacteria at the upper temperature for phototrophic life, 70–74°C, in the thermal gradient formed from a boiling hot spring. The pattern develops because the water cools more rapidly at the edges than in the center of the channel. The spring flows from the back of the picture toward the foreground. The light-green color is from a high-temperature strain of the cyanobacterium *Synechococcus*. As water flows down the gradient, the density of cells increases, less thermophilic strains enter, and the color becomes more intensely green.

boiling water overflows the edges of the spring and flows away from the source, it gradually cools, setting up a thermal gradient. Along this gradient, various microorganisms grow, with different species growing in the different temperature ranges (**Figure 6.23**). By studying the species distribution along such thermal gradients and by examining hot springs and other thermal habitats at different temperatures around the world, it has been possible to determine the upper temperature limits for each type of organism (Table 6.1). From this information we can conclude that (1) prokaryotic organisms are able to grow at far higher temperatures than are eukaryotes, (2) the most thermophilic of all prokaryotes are certain species of *Archaea,* and (3) nonphototrophic organisms can grow at higher temperatures than can phototrophic organisms.

Thermophilic prokaryotes have also been found in artificial thermal environments, such as hot water heaters. The domestic or industrial hot water heater has a temperature of 60–80°C and is therefore a favorable habitat for the growth of thermophilic prokaryotes. Organisms resembling *Thermus aquaticus,* a common hot spring thermophile (∞ Section 16.17), have been isolated from hot water heaters. Electric power plants, industrial hot water discharges, and other artificial thermal sources also provide sites where thermophiles can grow. Many of these organisms can be readily isolated using complex media incubated at the temperature of the habitat from which the sample originated.

Protein Stability at High Temperatures

How can thermophiles and hyperthermophiles survive at high temperature? First, their enzymes and other proteins are much more heat-stable than are those of mesophiles and actually function *optimally* at high temperatures. How is heat stability achieved? Amazingly, studies of several heat-stable enzymes have shown that they often differ very little in amino acid sequence from heat-sensitive forms of the enzymes that catalyze the same reaction in mesophiles. It appears that critical amino acid substitutions at only a few locations in the enzyme allow it to fold in a way that is consistent with heat stability.

Heat stability of proteins in hyperthermophiles is also bolstered by an increased number of ionic bonds between basic and acidic amino acids and their often highly hydrophobic interiors; the latter naturally resist unfolding in the aqueous cytoplasm. Finally, solutes such as di-inositol phosphate, diglycerol phosphate, and mannosylglycerate are produced at high levels in certain hyperthermophiles, and these may also help stabilize their proteins against thermal degradation.

Membrane Stability at High Temperatures

In addition to enzymes and other macromolecules in the cell, the cytoplasmic membranes of thermophiles and hyperthermophiles must be heat-stable. We mentioned earlier that psychrophiles have membrane lipids rich in unsaturated fatty acids, thus making the membranes semifluid and functional at low temperatures. Conversely, thermophiles typically have lipids rich in saturated fatty acids. This feature allows the membranes to remain stable and functional at high temperatures. Saturated fatty acids form a stronger hydrophobic environment than do unsaturated fatty acids, which helps account for membrane stability.

Hyperthermophiles, most of which are *Archaea*, do not contain fatty acids in their membranes but instead have C_{40} hydrocarbons composed of repeating units of isoprene (∞ Figures 4.7*c* and 4.8*b*) bonded by ether linkage to glycerol phosphate. In addition, the overall structure of the cytoplasmic membranes of hyperthermophiles forms a lipid *monolayer* rather than a lipid *bilayer* (Figure 4.8*d*). This structure improves the ability of the membrane to resist melting at the high growth temperatures of hyperthermophiles. We consider other aspects of heat stability in hyperthermophiles, including that of DNA, in Sections 17.13–17.15.

Thermophily and Biotechnology

Thermophiles and hyperthermophiles are interesting for more than just basic biological reasons. These organisms offer some major advantages for industrial and biotechnological processes, many of which can be run more rapidly and efficiently at high temperatures. For example, enzymes from thermophiles and hyperthermophiles are widely used in industrial microbiology. Such enzymes can catalyze biochemical reactions at high temperatures and are in general more stable than enzymes from mesophiles, thus prolonging the shelf life of purified enzyme preparations.

A classic example of a heat-stable enzyme of great importance to biology is the DNA polymerase isolated from *T. aquaticus*. *Taq polymerase*, as this enzyme is known, has been used to automate the repetitive steps in the polymerase chain reaction (PCR) technique (Section 12.8), an extremely important tool for biology. Several other uses of heat-stable enzymes and other heat-stable cell products are also known or are being developed for industrial applications.

6.14 MiniReview

Organisms with growth temperature optima between 45 and 80°C are called *thermophiles*, and those with optima greater than 80°C are called hyperthermophiles. These organisms inhabit hot environments that can have temperatures in excess of 100°C. Thermophiles and hyperthermophiles produce heat-stable macromolecules.

- Which domain of organisms includes species that can grow above 100°C?
- What is the structure of membranes of hyperthermophilic *Archaea*, and why might this structure be useful for growth at high temperature?
- What is Taq polymerase and why is it important?

V OTHER ENVIRONMENTAL FACTORS AFFECTING GROWTH

Temperature has a major effect on the growth of microorganisms. But many other factors do as well, chief among them pH, osmolarity, and oxygen.

6.15 Microbial Growth at Low or High pH

Acidity or alkalinity of a solution is expressed by its **pH** on a scale on which neutrality is pH 7 (**Figure 6.24**). pH values less than 7 are *acidic* and those greater than 7 are *alkaline*. It is important to remember that pH is a logarithmic function—a change of 1 pH unit corresponds to a tenfold change in hydrogen ion concentration. Thus, vinegar (pH near 2) and household ammonia (pH near 11) differ in hydrogen ion concentration by a billionfold.

Every microorganism has a pH range within which growth is possible and typically shows a well-defined growth pH optimum. Most organisms show a growth pH range of 2–3 units. Most natural environments have pH values between 4 and 9, and organisms with optima in this range are most commonly encountered. Only a few species can grow at pH values of lower than 3 or greater than 9.

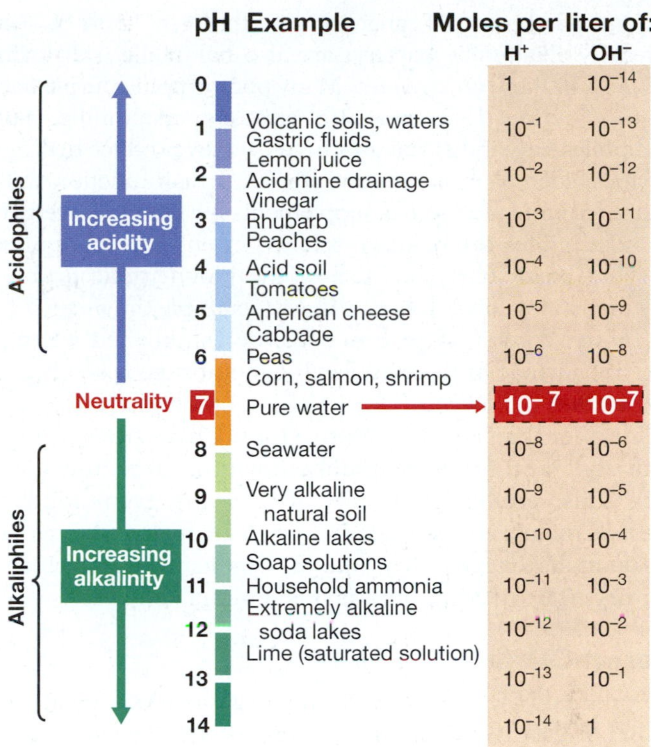

Figure 6.24 The pH scale. Although some microorganisms can live at very low or very high pH, the cell's internal pH remains near neutrality.

Acidophiles

Organisms that grow optimally at low pH, typically below pH 6, are called **acidophiles**. Fungi as a group tend to be more acid tolerant than bacteria. Many fungi grow best at pH 5 or below, and a few grow well at pH values as low as 2. Many prokaryotes are also acidophilic. In fact, some of these are *obligate* acidophiles, unable to grow at neutral pH. Obligately acidophilic bacteria include species of *Acidithiobacillus* and several genera of *Archaea*, including *Sulfolobus*, *Thermoplasma*, and *Ferroplasma*.

A critical factor governing acidophily is the stability of the cytoplasmic membrane. When the pH is raised to neutrality, the cytoplasmic membranes of strongly acidophilic bacteria are destroyed and the cells lyse. This indicates that these organisms are not just acid *tolerant* but that high concentrations of hydrogen ions are actually *required* for membrane stability. For example, the most acidophilic prokaryote known, *Picrophilus oshimae (Archaea)*, grows optimally at pH 0.7. Above pH 4, cells of *P. oshimae* spontaneously lyse. *P. oshimae* inhabits extremely acidic thermal soils associated with volcanic activity.

Alkaliphiles

A few extremophiles have very high pH optima for growth, sometimes as high as pH 11. Microorganisms showing growth pH optima of 9 or higher are called **alkaliphiles**. Alkaliphilic microorganisms are typically found in highly alkaline habitats, such as soda lakes and high-carbonate soils. The most well-studied alkaliphilic prokaryotes have been *Bacillus* species, such as *Bacillus firmus*. This organism is alkaliphilic but has an unusually broad pH range for growth, from 7.5 to 11. Some extremely alkaliphilic bacteria are also halophilic (salt loving), and most of these are *Archaea*. Many phototrophic purple bacteria (Section 15.2) are also strongly alkaliphilic. Some alkaliphiles have industrial uses because they produce hydrolytic enzymes, such as proteases and lipases, which function well at alkaline pH and are used as supplements for laundry detergents.

Alkaliphiles are of basic science interest for several reasons but particularly because of the bioenergetic problems they face living at such high pH. For example, how can a proton motive force (Section 5.12) be established when the external surface of the cytoplasmic membrane is so alkaline? In *B. firmus* it has been shown that a sodium (Na^+) motive force rather than a proton motive force drives transport reactions and motility. Remarkably, however, a proton motive force is also established in cells of *B. firmus*, even though the external membrane surface is awash in hydroxyl ions. The proton motive force in *B. firmus* is responsible for driving respiratory ATP synthesis, just as it is in nonalkaliphiles.

Internal Cell pH

The optimal pH for growth of any organism is a measure of the pH of the *extracellular* environment only. The *intracellular* pH must remain relatively close to neutrality to prevent destruction of macromolecules in the cell. For the majority of microorganisms whose pH optimum for growth is between pH 6 and 8, organisms called *neutrophiles*, the cytoplasm remains neutral or very nearly so (Figure 6.24). However, in acidophiles and alkaliphiles the internal pH can vary from neutrality. For example, in the previously mentioned acidophile *P. oshimae*, the internal pH has been measured at pH 4.6, and in extreme alkaliphiles an intracellular pH of as high as 9.5 has been measured. If these are not the lower and upper limits of cytoplasmic pH, respectively, they are extremely close to the limits. This is because DNA is acid-labile and RNA is alkaline-labile; if a cell cannot maintain these key macromolecules in a stable state, it obviously cannot survive.

Buffers

In a batch culture, the pH can change during growth as the result of metabolic reactions of microorganisms that consume or produce acidic or basic substances. Thus, buffers are frequently added to microbial culture media to keep the pH relatively constant. However, a given buffer works over only a narrow pH range. Hence, different buffers must be used at different pH values.

For near-neutral pH ranges, potassium phosphate (KH_2PO_4) and calcium carbonate ($CaCO_3$) are good buffers. Many other buffers for use in microbial growth media or for the assay of enzymes extracted from microbial cells are available, and the best buffering system for one organism or enzyme may be considerably different from that for another. Thus, the optimal buffer for use in a particular situation must usually be determined empirically. For assaying enzymes *in vitro*, though, a buffer that works well in an assay of the enzyme from one organism will usually work well for assaying the same enzyme from other organisms.

6.15 MiniReview

The acidity or alkalinity of an environment can greatly affect microbial growth. Some organisms have evolved to grow best at low or high pH, but most organisms grow best between pH 6 and 8. The internal pH of a cell must stay relatively close to neutral.

- What is the increase in concentration of protons when going from pH 7 to pH 3?
- What are buffers and why are they needed?

6.16 Osmotic Effects on Microbial Growth

Water is the solvent of life, and water availability is an important factor affecting the growth of microorganisms. Water availability not only depends on the absolute water content of an environment, that is, how moist or dry it is, but it is also a function of the concentration of solutes such as salts, sugars, or other substances that are dissolved in the water. Dissolved substances have an affinity for water, which makes the water associated with solutes less available to organisms.

Water Activity and Osmosis

Water availability is expressed in physical terms as **water activity**. Water activity, abbreviated a_w, is defined as the ratio of the vapor pressure of the air in equilibrium with a substance or solution to the vapor pressure of pure water. Thus, values of a_w vary between 0 and 1; some representative values are given in **Table 6.2**. Water activities in agricultural soils generally range between 0.90 and 1.

Table 6.2 Water activity of several substances

Water activity (a_w)	*Material*	*Example organisms*[a]
1.000	Pure water	*Caulobacter, Spirillum*
0.995	Human blood	*Streptococcus, Escherichia*
0.980	Seawater	*Pseudomonas, Vibrio*
0.950	Bread	Most gram-positive rods
0.900	Maple syrup, ham	Gram positive cocci such as *Staphylococcus*
0.850	Salami	*Saccharomyces rouxii* (yeast)
0.800	Fruit cake, jams	*Saccharomyces bailii, Penicillium* (fungus)
0.750	Salt lakes, salted fish	*Halobacterium, Halococcus*
0.700	Cereals, candy, dried fruit	*Xeromyces bisporus* and other xerophilic fungi

[a]Selected examples of prokaryotes or fungi capable of growth in culture media adjusted to the stated water activity.

Water diffuses from regions of high water concentration (low solute concentration) to regions of lower water concentration (higher solute concentration) in the process of *osmosis* (∞ Section 4.6 and Figure 4.21). The cytoplasm of a cell has a higher solute concentration than the environment, so water tends to diffuse into the cell. Under such conditions, the cell is said to be in *positive water balance*. However, when a cell finds itself in an environment where the solute concentration exceeds that of the cytoplasm, water will flow out of the cell. This can cause serious problems if a cell has no way to counteract it because a dehydrated cell cannot grow.

Halophiles and Related Organisms

In nature, osmotic effects are of interest mainly in habitats with high concentrations of salts. Seawater contains about 3% NaCl plus small amounts of many other minerals and elements. Marine microorganisms usually have a specific requirement for NaCl in addition to growing optimally at the water activity of seawater (**Figure 6.25**). Such organisms are called **halophiles**. The growth of halophiles requires at least some NaCl, but the optimum varies with the organism. For example, the terms *mild halophile* and *moderate halophile* are used to describe halophiles with low (1–6%) and moderate (7–15%) NaCl requirements, respectively (Figure 6.25).

Most microorganisms are unable to cope with environments of very low water activity and either die or become dehydrated and dormant under such conditions. **Halotolerant** organisms can tolerate some reduction in the a_w of their environment but grow best in the absence of the added solute (Figure 6.25). By contrast, some organisms thrive and indeed require low water activity for growth. These organisms are of interest not only from the standpoint of their adaptation to life under these conditions, but also from an applied standpoint, for example, in the food industry, where solutes such as salt and sucrose are commonly used as preservatives to inhibit microbial growth.

Organisms capable of growth in very salty environments are called **extreme halophiles** (Figure 6.25). These organisms require 15–30% NaCl, depending on the species, for optimum growth. Organisms able to live in environments high in sugar as a solute are called **osmophiles**, and those able to grow in very dry environments (made dry by lack of water rather than from dissolved solutes) are called **xerophiles**. Examples of these various classes of organisms are given in Table 6.2.

Compatible Solutes

When an organism grows in a medium with a low water activity, it can obtain water from its environment only by increasing its internal solute concentration. The internal solute concentration can be raised by either pumping inorganic ions into the cell from the environment or by synthesizing or concentrating an organic solute. Many organisms are known that employ one or the other of these strategies, and several examples are given in **Table 6.3**.

The solute used inside the cell for adjustment of cytoplasmic water activity must be noninhibitory to biochemical

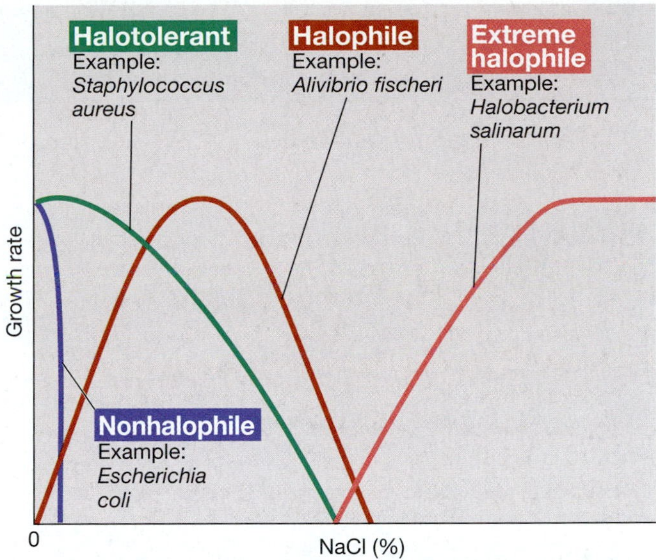

Figure 6.25 Effect of sodium chloride concentration on growth of microorganisms of different salt tolerances or requirements. The optimum NaCl concentration for marine microorganisms such as *V. fischeri* is about 3%; for extreme halophiles, it is between 15 and 30%, depending on the organism.

processes within the cell. Such compounds are called **compatible solutes**. Several different compatible solutes are known in microorganisms. These substances are all highly water-soluble sugars or sugar alcohols, other alcohols, or amino acids or their derivatives (Table 6.3 and **Figure 6.26**). The compatible solute of extremely halophilic *Archaea*, such as *Halobacterium*, and a very few extremely halophilic *Bacteria* is K^+ (∞ Section 17.3).

The concentration of compatible solute in a cell is a function of the level of solutes present in its environment; however, in any given organism the maximal amount of compatible solute is a genetically directed characteristic. As a result, different organisms can tolerate different ranges of water potential (Tables 6.2 and 6.3). Nonhalotolerant, halotolerant, halophilic, and extremely halophilic microorganisms (Figure 6.25) are to a major extent defined by their genetic capacity to produce or accumulate compatible solutes.

Gram-positive cocci of the genus *Staphylococcus* are notoriously halotolerant (in fact, a common isolation procedure for them is to use media containing 7.5–10% NaCl), and these organisms use the amino acid proline as a compatible solute. Glycine betaine is a derivative of the amino acid glycine in which hydrogen atoms on the amino group are replaced by methyl groups. This places a positive charge on the N atom (Figure 6.26) and greatly increases solubility. Glycine betaine is widely distributed as a compatible solute, especially among halophilic *Bacteria* and cyanobacteria (Table 6.3).

Some extremely halophilic bacteria produce the compatible solute ectoine (Figure 6.26), which is a cyclic derivative of the amino acid aspartate. Various glycosides and dimethylsulfoniopropionate (Figure 6.24) are produced

Table 6.3 Compatible solutes of microorganisms

Organism	Major solute(s) accumulated	Minimum a_w for growth
Bacteria, nonphototrophic	Glycine betaine, proline (mainly gram-positive), glutamate (mainly gram-negative)	0.97–0.90
Freshwater cyanobacteria	Sucrose, trehalose	0.98
Marine cyanobacteria	α-Glucosylglycerol	0.92
Marine algae	Mannitol, various glycosides, proline, dimethylsulfoniopropionate	0.92
Salt lake cyanobacteria	Glycine betaine	0.90–0.75
Halophilic anoxygenic phototrophic *Bacteria* (*Ectothiorhodospira/Halorhodospira* and *Rhodovibrio* species)	Glycine betaine, ectoine, trehalose	0.90–0.75
Extremely halophilic *Archaea* (for example, *Halobacterium*) and some *Bacteria* (for example, *Haloanaerobium*)	KCl	0.75
Dunaliella (halophilic green alga)	Glycerol	0.75
Xerophilic yeasts	Glycerol	0.83–0.62
Xerophilic filamentous fungi	Glycerol	0.72–0.61

by marine algae as compatible solutes, but with rare exception they accumulate in only low amounts because these organisms are rarely very halophilic. Xerophilic yeasts and halophilic green algae mainly produce glycerol as a compatible solute.

6.16 MiniReview

Water activity of a microbial habitat is controlled by the dissolved solute concentration. To survive in high-solute environments, organisms produce or accumulate compatible solutes to maintain the cell in positive water balance. Some microorganisms have evolved to grow best at reduced water potential, and some even require high levels of salts for growth.

- What is the a_w of pure water?
- What is a compatible solute and why is it needed?
- What is the compatible solute for *Halobacterium*?

6.17 Oxygen and Microbial Growth

Because animals require molecular oxygen (O_2), it is easy to assume that all organisms require oxygen. However, this is not true; many microorganisms can, and some must live in the total absence of oxygen.

Oxygen is poorly soluble in water, and because of the constant respiratory activities of microorganisms in aquatic or other moist habitats, O_2 can quickly become exhausted. Thus, anoxic microbial habitats are common in nature and include muds and other sediments, bogs and marshes, water-logged soils, intestinal tracts of animals, sewage sludge, the deep subsurface of Earth, and many other environments. In these anoxic habitats, microorganisms, particularly prokaryotes, thrive.

Oxygen Classes of Microorganisms

Microorganisms vary in their need for, or tolerance of, oxygen. In fact, microorganisms can be grouped according to how oxygen affects them, as outlined in **Table 6.4** on page 170. **Aerobes** can grow at full oxygen tensions (air is 21% O_2) and respire oxygen in their metabolism. Many aerobes can even tolerate elevated concentrations of oxygen (hyperbaric oxygen). **Microaerophiles**, by contrast, are aerobes that can use oxygen only when it is present at levels reduced from that in air (microoxic conditions). This is because of their limited capacity to respire or because they contain some oxygen-sensitive molecule such as an oxygen-labile enzyme. Many aerobes are **facultative**, meaning that under the appropriate nutrient and culture conditions they can grow under either oxic or anoxic conditions.

Some organisms cannot respire oxygen; such organisms are called **anaerobes**. There are two kinds of anaerobes: **aerotolerant anaerobes**, which can tolerate oxygen and grow in its presence even though they cannot use it, and **obligate anaerobes**, which are inhibited or even killed by O_2 (Table 6.4). The reason obligate anaerobes are killed by oxygen is unknown, but it is likely because they are unable to detoxify some of the products of oxygen metabolism (Section 6.18).

So far as is known, obligate anaerobiosis is found in only three groups of microorganisms: a wide variety of prokaryotes, a few fungi, and a few protozoa. The best-known group of obligately anaerobic *Bacteria* belongs to the genus *Clostridium*, a group of gram-positive endospore-forming rods. Clostridia are widespread in soil, lake sediments, and intestinal tracts and are often responsible for spoilage of

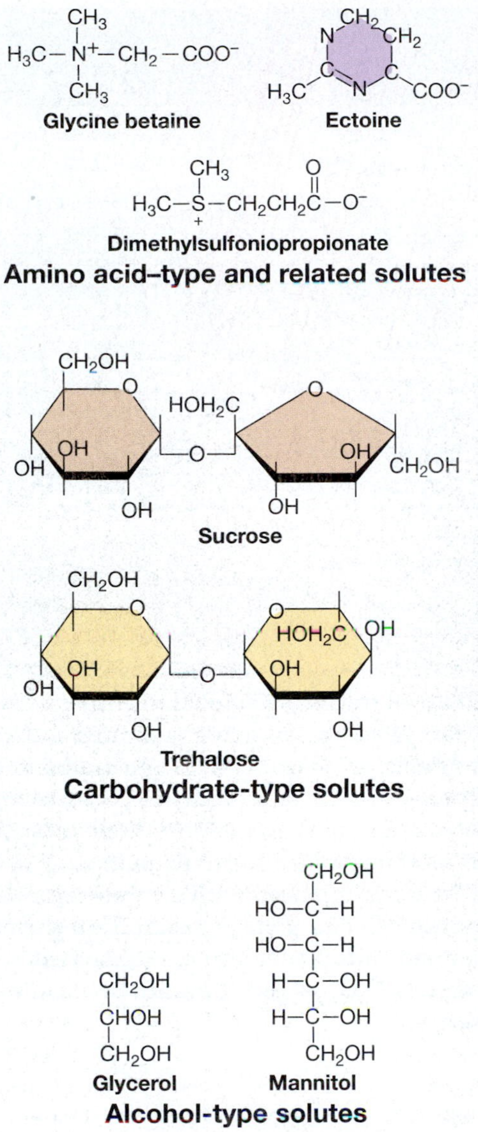

Glycine betaine

Ectoine

Dimethylsulfoniopropionate

Amino acid–type and related solutes

Sucrose

Trehalose

Carbohydrate-type solutes

Glycerol

Mannitol

Alcohol-type solutes

Figure 6.26 Structures of some common and highly soluble compatible solutes in microorganisms. The structures of glutamate and proline, other common solutes, were shown in Figure 3.12. The formal name of ectoine is 1,4,5,6-tetrahydro-2-methyl-4-pyrimidine carboxylate.

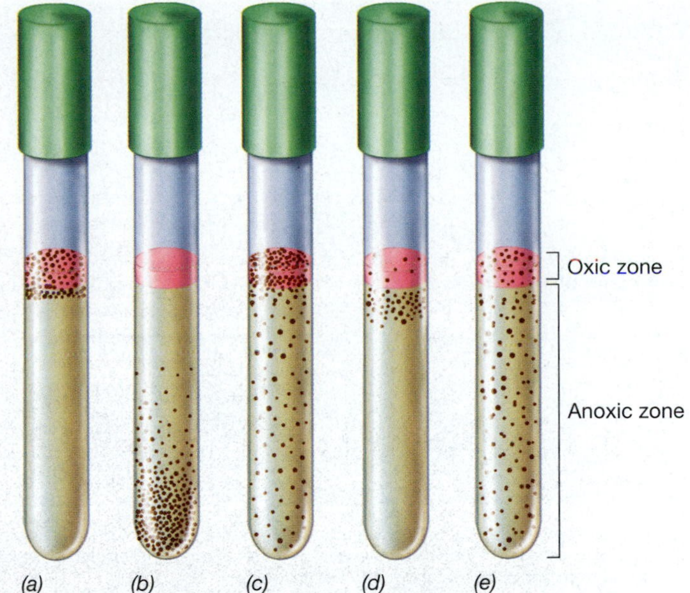

Figure 6.27 Growth versus oxygen concentration. *(a–e)* Aerobic, anaerobic, facultative, microaerophilic, and aerotolerant anaerobe growth, as revealed by the position of microbial colonies (depicted here as black dots) within tubes of thioglycolate broth culture medium. A small amount of agar has been added to keep the liquid from becoming disturbed, and the redox dye, resazurin, which is pink when oxidized and colorless when reduced, is added as a redox indicator. *(a)* Oxygen penetrates only a short distance into the tube, so obligate aerobes only grow close to the surface. *(b)* Anaerobes, being sensitive to oxygen, grow only away from the surface. *(c)* Facultative aerobes are able to grow in either the presence or the absence of oxygen and thus grow throughout the tube. However, growth is better near the surface because these organisms can respire. *(d)* Microaerophiles grow away from the most oxic zone. *(e)* Aerotolerant anaerobes grow throughout the tube. Growth is not better near the surface because these organisms can only ferment.

canned foods. Other obligately anaerobic organisms are the methanogens and many other *Archaea,* the sulfate-reducing and homoacetogenic bacteria, and many of the bacteria that inhabit the animal gut and oral cavity. Among obligate anaerobes, however, the sensitivity to oxygen varies greatly. Some species can tolerate traces of oxygen or even full exposure to oxygen, whereas others cannot.

Culture Techniques for Aerobes and Anaerobes

For the growth of many aerobes, it is necessary to provide extensive aeration. This is because the oxygen that is consumed by the organisms during growth is not replaced fast enough by simple diffusion from the air. Therefore, forced aeration of cultures is needed and can be achieved by either vigorously shaking the flask or tube on a shaker or by bubbling sterilized air into the medium through a fine glass tube or porous glass disc. Aerobes typically grow better with forced aeration than with oxygen supplied from simple diffusion.

For the culture of anaerobes, the problem is not to provide air, but to exclude it. Obligate anaerobes vary in their sensitivity to oxygen, and procedures are available for reducing the oxygen content of cultures. Some of these techniques are simple and suitable mainly for less oxygen-sensitive organisms; others are more complex, but necessary for growth of obligate anaerobes. Bottles or tubes filled completely to the top with culture medium and provided with tightly fitting stoppers provide suitably anoxic conditions for organisms that are not overly sensitive to small amounts of oxygen. A chemical called a *reducing agent* may be added to culture media; the reducing agent reacts with oxygen and reduces it to H_2O. An example is thioglycolate, which is added to thioglycolate broth, a medium commonly used to test an organism's requirements for oxygen (**Figure 6.27**).

Thioglycolate broth is a complex medium containing a small amount of agar, making the medium viscous but still fluid.

Table 6.4 Oxygen relationships of microorganisms

Group	*Relationship to O_2*	*Type of metabolism*	*Example*[a]	*Habitat*[b]
Aerobes				
Obligate	Required	Aerobic respiration	*Micrococcus luteus* (B)	Skin, dust
Facultative	Not required, but growth better with O_2	Aerobic respiration, anaerobic respiration, fermentation	*Escherichia coli* (B)	Mammalian large intestine
Microaerophilic	Required but at levels lower than atmospheric	Aerobic respiration	*Spirillum volutans* (B)	Lake water
Anaerobes				
Aerotolerant	Not required, and growth no better when O_2 present	Fermentation	*Streptococcus pyogenes* (B)	Upper respiratory tract
Obligate	Harmful or lethal	Fermentation or anaerobic respiration	*Methanobacterium* (A) *formicicum*	Sewage sludge digestors, anoxic lake sediments

[a]Letters in parentheses indicate phylogenetic status (B, *Bacteria;* A, *Archaea*). Representative of either domain of prokaryotes are known in each category. Most eukaryotes are obligate aerobes, but facultative aerobes (for example, yeast) and obligate anaerobes (for example, certain protozoa and fungi) are known.

[b]Listed are typical habitats of the example organism.

After thioglycolate reacts with oxygen throughout the tube, oxygen can penetrate only near the top of the tube where the medium contacts air. Obligate aerobes grow only at the top of such tubes. Facultative organisms grow throughout the tube but grow best near the top. Microaerophiles grow near the top but not right at the top. Anaerobes grow only near the bottom of the tube, where oxygen cannot penetrate. The redox indicator dye *resazurin* is added to the medium to differentiate oxic from anoxic regions; the dye is pink when oxidized and colorless when reduced and so gives a visual assessment of the degree of penetration of oxygen into the medium (Figure 6.27).

To remove all traces of O_2 for the culture of strict anaerobes, one can place an oxygen-consuming system in a jar holding the tubes or plates. One of the simplest devices for this is an anoxic jar, a glass or gas-impermeable plastic jar fitted with a gastight seal within which tubes, plates, or other containers are placed for incubation (**Figure 6.28a**). The air in the jar is replaced with a mixture of H_2 and CO_2, and in the presence of a palladium catalyst, the traces of O_2 left in the jar and culture medium are consumed in the formation of water ($H_2 + O_2 \rightarrow H_2O$), eventually leading to anoxic conditions.

(a)

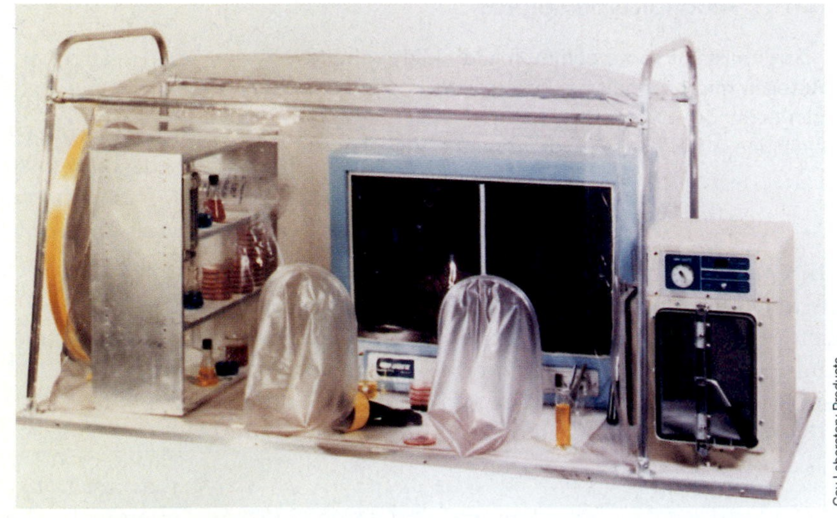

(b)

Figure 6.28 Incubation under anoxic conditions. *(a)* Anoxic jar. A chemical reaction in the envelope in the jar generates $H_2 + CO_2$. The H_2 reacts with O_2 in the jar on the surface of a palladium catalyst to yield H_2O; the final atmosphere contains N_2, H_2, and CO_2. *(b)* Anoxic glove bag for manipulating and incubating cultures under anoxic conditions. The airlock on the right, which can be evacuated and filled with O_2-free gas, serves as a port for adding and removing materials to and from the glove bag.

For obligate anaerobes it is usually necessary to not only remove all traces of O_2 but also to carry out all manipulations of cultures in a completely anoxic atmosphere. Strict anaerobes can be killed by even a brief exposure to oxygen. In these cases, a culture medium is first boiled to render it oxygen free, then a reducing agent such as H_2S is added, and the mixture is sealed under an oxygen-free gas. All manipulations are carried out under a jet of sterile oxygen-free hydrogen or nitrogen gas that is directed into the culture vessel when it is open, thus driving out any oxygen that might enter. For extensive research on anaerobes, special devices called *anoxic glove bags* permit work with open cultures in completely anoxic atmospheres (Figure 6.28*b*).

6.17 MiniReview

Aerobes require oxygen to live, whereas anaerobes do not and may even be killed by oxygen. Facultative organisms can live with or without oxygen. Special techniques are needed to grow aerobic and anaerobic microorganisms.

- What is a facultative aerobe?
- How does a reducing agent work? Give an example of a reducing agent.

6.18 Toxic Forms of Oxygen

Oxygen is a powerful oxidant and the best electron acceptor for respiration. But oxygen can also be a poison to obligate anaerobes. Why? It turns out that oxygen itself is not the poison, but instead it is the toxic derivatives of oxygen that can damage cells. We consider this topic here.

Oxygen Chemistry

Oxygen in its ground state is called *triplet* oxygen ($^3O_2$). However, other electronic configurations of oxygen are possible, and most are toxic to cells. One major form of toxic oxygen is *singlet* oxygen ($^1O_2$), a higher energy form of oxygen in which outer shell electrons surrounding the nucleus become highly reactive and can carry out spontaneous and undesirable oxidations within the cell. Singlet oxygen is produced both photochemically and biochemically, the latter through the

Reactants	Products	
$O_2 + e^- \rightarrow$	O_2^-	Superoxide
$O_2^- + e^- + 2\,H^+ \rightarrow$	H_2O_2	Hydrogen peroxide
$H_2O_2 + e^- + H^+ \rightarrow$	H_2O + $OH\bullet$	Hydroxyl
$OH\bullet + e^- + H^+ \rightarrow$	H_2O	Water

Outcome:
$O_2 + 4\,e^- + 4\,H^+ \rightarrow 2\,H_2O$

Figure 6.29 Four-electron reduction of O_2 to H_2O by stepwise addition of electrons. All the intermediates formed are reactive and toxic to cells, except for water, of course.

$H_2O_2 + H_2O_2 \rightarrow 2\,H_2O + O_2$

(*a*) **Catalase**

$H_2O_2 + NADH + H^+ \rightarrow 2\,H_2O + NAD^+$

(*b*) **Peroxidase**

$O_2^- + O_2^- + 2\,H^+ \rightarrow H_2O_2 + O_2$

(*c*) **Superoxide dismutase**

$4\,O_2^- + 4\,H^+ \rightarrow 2\,H_2O + 3\,O_2$

(*d*) **Superoxide dismutase/catalase in combination**

$O_2^- + 2\,H^+ + \text{rubredoxin}_{reduced} \rightarrow H_2O_2 + \text{rubredoxin}_{oxidized}$

(*e*) **Superoxide reductase**

Figure 6.30 Enzymes that destroy toxic oxygen species. (*a*) Catalases and (*b*) peroxidases are porphyrin-containing proteins, although some flavoproteins may consume toxic oxygen species as well. (*c*) Superoxide dismutases are metal-containing proteins, the metals being copper and zinc, manganese, or iron. (*d*) Combined reaction of superoxide dismutase and catalase. (*e*) Superoxide reductase catalyzes the one electron reduction of O_2^- to H_2O_2.

activity of various peroxidase enzymes. Organisms that frequently encounter singlet oxygen, such as airborne bacteria and phototrophic microorganisms, often contain pigments called *carotenoids,* which function to convert singlet oxygen to nontoxic forms.

Superoxide and Other Toxic Oxygen Species

Besides singlet oxygen, many other toxic forms of oxygen exist, including *superoxide anion* (O_2^-), *hydrogen peroxide* (H_2O_2), and *hydroxyl radical* ($OH\bullet$). All of these are produced as by-products of the reduction of O_2 to H_2O in respiration (**Figure 6.29**). Flavoproteins, quinones, and iron-sulfur proteins (Section 5.11), found in virtually all cells, can also catalyze the reduction of O_2 to O_2^-. Thus, whether or not it can respire oxygen (Table 6.3), a cell can be exposed to toxic oxygen species from time to time.

Superoxide and hydroxyl radical are strong oxidizing agents and can oxidize virtually any organic compound in the cell, including macromolecules. Peroxides such as H_2O_2 can also damage cell components but are not as toxic as superoxide or hydroxyl radical. The latter is the most reactive of all toxic oxygen species but is transient and quickly removed in other reactions. In later chapters we will see that toxic oxygen species are made by cells of the immune system specifically for the purpose of killing microbial invaders (Section 29.7).

Superoxide Dismutase and Other Enzymes That Destroy Toxic Oxygen

With so many toxic oxygen derivatives to deal with, it is not surprising that organisms have evolved enzymes that destroy these compounds (**Figure 6.30**). Superoxide and hydrogen peroxide are the most common toxic oxygen species so enzymes that destroy these compounds are widely distributed. The enzyme *catalase* attacks hydrogen peroxide forming

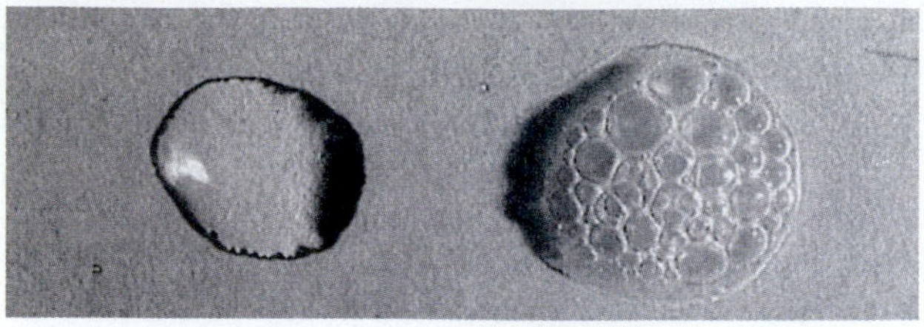

T. D. Brock

Figure 6.31 Method for testing a microbial culture for the presence of catalase. A heavy loopful of cells from an agar culture was mixed on a slide (right) with a drop of 30% hydrogen peroxide. The immediate appearance of bubbles is indicative of the presence of catalase. The bubbles are O_2 produced by the reaction $H_2O_2 + H_2O_2 \rightarrow 2\,H_2O + O_2$.

oxygen and water; its activity is illustrated in Figure 6.30*a* and **Figure 6.31**. Another enzyme that destroys hydrogen peroxide is *peroxidase* (Figure 6.30*b*), which differs from catalase in requiring a reductant, usually NADH, for activity and producing only H_2O as a product. Superoxide is destroyed by the enzyme *superoxide dismutase,* an enzyme that generates hydrogen peroxide and oxygen from two molecules of superoxide (Figure 6.30*c*). Superoxide dismutase and catalase thus work together to bring about the conversion of superoxide to oxygen and water (Figure 6.30*d*).

Aerobes and facultative aerobes typically contain both superoxide dismutase and catalase. Superoxide dismutase is an essential enzyme for aerobes, and the absence of this enzyme in obligate anaerobes was originally thought to explain why oxygen is toxic to them (but see next paragraph). Some aerotolerant anaerobes, such as the lactic acid bacteria, also lack superoxide dismutase, but they use protein-free manganese (Mn^{2+}) complexes to carry out the dismutation of O_2^- to H_2O_2 and O_2. Such a system is not as efficient as superoxide dismutase, but may have functioned as a primitive form of this enzyme in ancient anaerobic organisms faced with oxygen for the first time when cyanobacteria first appeared on Earth.

Superoxide Reductase

Another means of superoxide disposal is present in certain obligately anaerobic *Archaea*. In the hyperthermophile *Pyrococcus furiosus,* for example, superoxide dismutase is absent, but a unique enzyme, *superoxide reductase*, is present and functions to remove superoxide. However, unlike superoxide dismutase, superoxide reductase reduces superoxide to H_2O_2 without the production of O_2 (Figure 6.30*e*), thus avoiding exposure of the organism to O_2. The electron donor for superoxide reductase activity is rubredoxin, a low-potential iron–sulfur protein (Figure 6.30*e*). *P. furiosus* also lacks catalase, an enzyme that, like superoxide dismutase, also generates O_2 (Figure 6.30*a*). Instead, the H_2O_2 produced by superoxide reductase is removed by the activity of peroxidase-like enzymes that yield H_2O as a final product (Figure 6.30*b*).

Superoxide reductases are present in many other obligate anaerobes as well, such as sulfate-reducing bacteria (*Bacteria*) and methanogens (*Archaea*), as well as in certain microaerophilic species of *Bacteria*, such as *Treponema*. Thus these organisms, previously thought to be oxygen sensitive because they lacked superoxide dismutase, can indeed consume superoxide. The sensitivity of these organisms to oxygen may therefore be for entirely other and as yet unknown reasons.

Many obligately anaerobic hyperthermophiles such as *Pyrococcus* inhabit deep-sea hydrothermal vents (∞ Section 24.11) but are quite tolerant of cold, oxic conditions. Although they do not grow under these conditions, superoxide reductase presumably prevents their killing when they are exposed to O_2. It is thought that oxygen tolerance may be an important factor enabling transport of these organisms in fully oxic ocean water from one deep sea hydrothermal system to another.

6.18 MiniReview

Several toxic forms of oxygen can form in the cell, but enzymes are present that can neutralize most of them. Superoxide in particular seems to be a common toxic oxygen species.

- How does superoxide dismutase protect a cell from toxic oxygen?
- How does the activity of superoxide dismutase differ from that of superoxide reductase?

Review of Key Terms

Acidophile an organism that grows best at low pH; typically below pH 6

Aerobe an organism that can use oxygen (O_2) in respiration; some require oxygen

Aerotolerant anaerobe a microorganism unable to respire oxygen (O_2) but whose growth is unaffected by oxygen

Alkaliphile an organism that has a growth pH optimum of 9 or higher

Anaerobe an organism that cannot use O_2 in respiration and whose growth is typically inhibited by O_2

Autolysis spontaneous cell lysis, usually due to the activity of lytic proteins called autolysins

Batch culture a closed-system microbial culture of fixed volume

Binary fission cell division following enlargement of a cell to twice its minimum size

Biofilm an attached polysaccharide matrix containing bacterial cells

Cardinal temperatures the minimum, maximum, and optimum growth temperatures for a given organism

Chemostat a device that allows for the continuous culture of microorganisms with independent control of both growth rate and cell number

Compatible solute a molecule that is accumulated in the cytoplasm of a cell for adjustment of water activity but that does not inhibit biochemical processes

Divisome a complex of proteins that directs cell division processes in prokaryotes

Exponential growth growth of a microbial population in which cell numbers double within a specific time interval

Extreme halophile a microorganism that requires very large amounts of salt (NaCl), usually greater than 10% and in some cases near to saturation, for growth

Extremophile an organism that grows optimally under one or more chemical or physical extremes, such as high or low temperature or pH

Facultative with respect to oxygen, an organism that can grow in either its presence or absence

FtsZ a protein that forms a ring along the mid-cell-division plane to initiate cell division

Generation time the time required for a population of microbial cells to double

Growth an increase in cell number

Halophile a microorganism that requires NaCl for growth

Halotolerant not requiring NaCl for growth but able to grow in the presence of salt, in some cases, substantial levels of salt

Hyperthermophile a prokaryote that has a growth temperature optimum of 80°C or greater

Mesophile an organism that grows best at temperatures between 20 and 45°C

Microaerophile an aerobic organism that can grow only when oxygen tensions are reduced from that present in air

Obligate anaerobe an organism that cannot grow in the presence of O_2

Osmophile an organism that grows best in the presence of high levels of solute, typically a sugar

pH the negative logarithm of the hydrogen ion (H^+) concentration of a solution

Psychrophile an organism with a growth temperature optimum of 15°C or lower and a maximum growth temperature below 20°C

Plate count a viable counting method where the number of colonies on a plate is used as a measure of all numbers

Psychrotolerant capable of growing at low temperatures but having an optimum above 20°C

Thermophile an organism whose growth temperature optimum lies between 45 and 80°C

Transpeptidation formation of peptide cross-links between muramic acid residues in peptidoglycan synthesis

Viable capable of reproducing

Water activity the ratio of the vapor pressure of air in equilibrium with a solution to that of pure water

Xerophile an organism that is able to live, or that lives best, in very dry environments

Review Questions

1. Describe the key molecular processes that occur when a cell grows and divides. What proteins assist in this cell division process (Section 6.1)?
2. Describe the role that Fts proteins play in the cell division process (Section 6.2).
3. In what way do derivatives of the rod-shaped bacterium *Escherichia coli* that carry mutations that inactivate the protein MreB look different microscopically from wild-type (unmutated) cells? What is the reason for this (Section 6.3)?
4. How does the antibiotic penicillin kill bacterial cells, and why does it kill only growing cells (Section 6.4)?
5. What is the difference between the specific growth rate (k) of an organism and its generation time (g) (Sections 6.5 and 6.6)?
6. Describe the growth cycle of a population of bacterial cells from the time this population is first inoculated into fresh medium (Section 6.7).
7. How can a chemostat regulate growth rate and cell numbers independently (Section 6.8)?
8. What is the difference between a total cell count and a viable cell count (Sections 6.9 and 6.10)?
9. How can turbidity be used as a measure of cell numbers (Section 6.11)?
10. Examine the graph describing the relationship between growth rate and temperature (Figure 6.18). Give an explanation, in biochemical terms, of why the optimum temperature for an organism is usually closer to its maximum than its minimum (Section 6.12).
11. Describe a habitat where you would find a psychrophile; a hyperthermophile (Sections 6.13 and 6.14).
12. Concerning the pH of the environment and of the cell, in what ways are acidophiles and alkaliphiles different? In what ways are they similar (Section 6.15)?
13. Write an explanation in molecular terms for how a cell of a halophile is able to make water molecules flow into itself (Section 6.16).
14. Contrast an aerotolerant and an obligate anaerobe in terms of sensitivity to oxygen and ability to grow in the presence of oxygen (O_2). How does an aerotolerant anaerobe differ from a microaerophile (Section 6.17)?
15. Compare and contrast the enzymes catalase, superoxide dismutase, and superoxide reductase from the following points of view: substrates, oxygen products, organisms containing them, and role in oxygen tolerance of the cell (Section 6.18).

Application Questions

1. Starting with four bacterial cells per milliliter in a rich nutrient medium, with a 1-h lag phase and a 20-min generation time, how many cells will there be in 1 liter of this culture after 1 h? After 2 h? After 2 h if one of the initial four cells was dead?

2. Calculate g and k in a growth experiment in which a medium was inoculated with 5×10^6 cells/ml of *Escherichia coli* cells and, following a 1-h lag, grew exponentially for 5 h, after which the population was 5.4×10^9 cells/ml.

3. Return to Chapter 3 and locate a figure that best describes what happens to enzymes from a cell of a mesophile like *Escherichia coli* when placed in a culture medium at 80°C. Contrast this with a figure from Chapter 6 that best describes what would happen if cells of *Pyrolobus fumarii* were placed under the same conditions. Describe why neither organism would grow.

4. Would you expect to find a psychrophilic microorganism alive in a hot spring? Why? It is frequently possible to isolate hyperthermophilic microorganisms from cold-water environments. Supply an explanation for this.

5. In which direction (into or out of the cell) will water flow in cells of *Escherichia coli* (an organism found in your large intestine) suddenly suspended into a solution of 20% NaCl? What if the cells were suspended into distilled water? If growth nutrients were added to each cell suspension, which, if either, would be able to grow and why?

7

Essentials of Molecular Biology

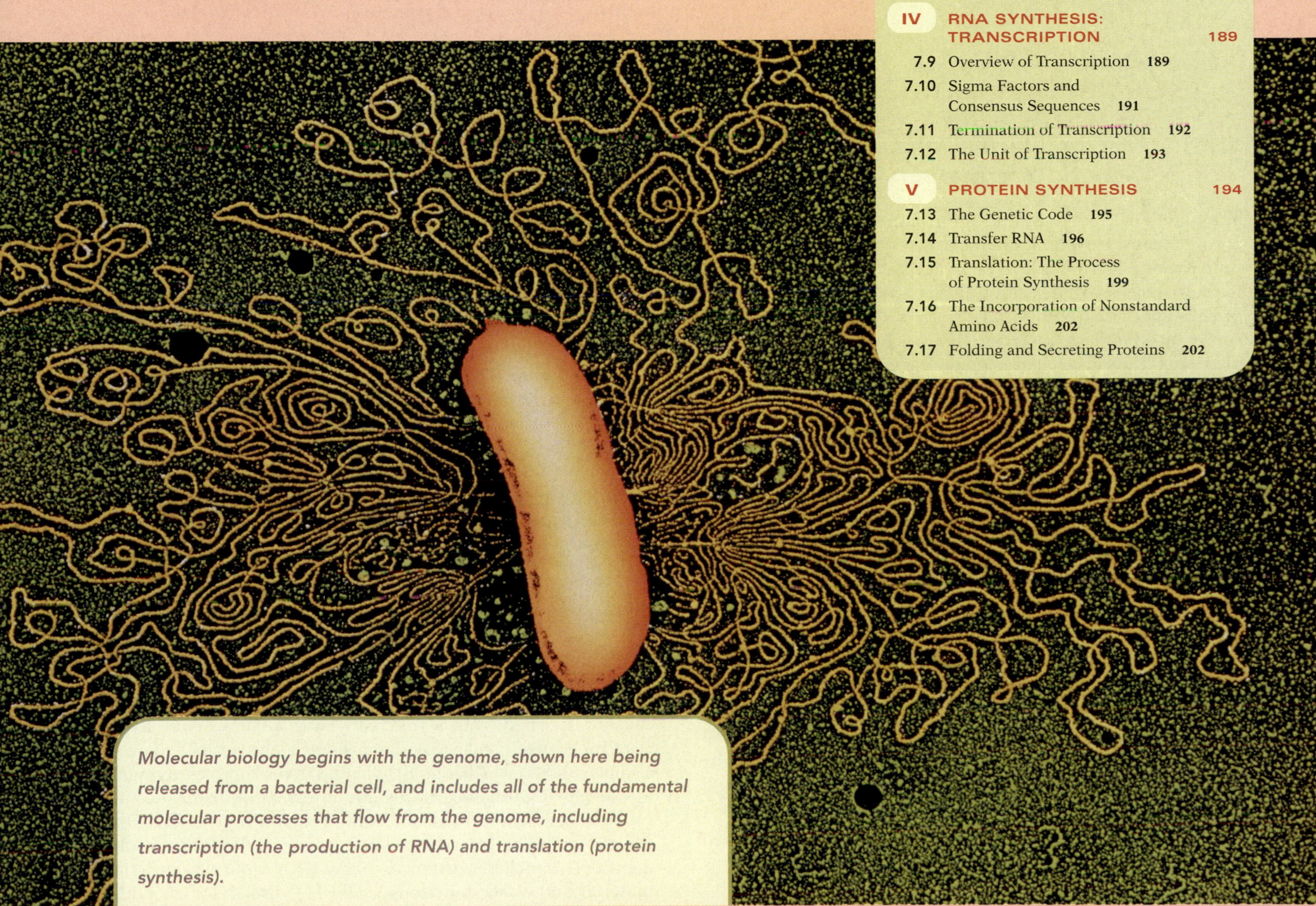

Molecular biology begins with the genome, shown here being released from a bacterial cell, and includes all of the fundamental molecular processes that flow from the genome, including transcription (the production of RNA) and translation (protein synthesis).

In Chapter 1 we characterized cells as chemical machines and coding devices. As chemical machines, cells accumulate and transform their vast array of macromolecules into new cells. As coding devices, they store, process, and use genes, the genetic information of the cell. Genes and gene expression are the subject of molecular biology. In particular, the review of molecular biology in this chapter includes the chemical nature of genes, the structure and function of DNA and RNA, and the replication of DNA. We then consider the biosynthesis of proteins, macromolecules that play important roles in both the structure and the functioning of the cell. Our focus in this chapter is on these processes as they occur in *Bacteria*. In particular, *Escherichia coli*, a member of the *Bacteria*, is the model organism for molecular biology and is the main example used.

I GENES AND GENE EXPRESSION

7.1 Macromolecules and Genetic Information

The functional unit of genetic information is the **gene**. All microorganisms, indeed all life forms, contain genes, and thus a fundamental understanding of what a gene is and what it does is important for understanding the structure and behavior of cells and viruses. Physically, genes are located on chromosomes or other large molecules known collectively as **genetic elements**. Nowadays, in the "genomics era," biology is tending to define cells in terms of their complement of genes. Thus, we must understand the process of biological information flow if we are to understand how microorganisms function.

Genes and the Steps in Information Flow

In all cells genes are composed of deoxyribonucleic acid (DNA). The information in the gene is present as the sequence of bases—purine (adenine and guanine) and pyrimidines (thymine and cytosine)—in the DNA. The information stored in DNA is transferred to ribonucleic acid (RNA). RNA may act as an informational intermediate (a messenger), as a structural component, or even as an enzyme. Finally, information carried by RNA is used to manufacture proteins. Because all three of these molecules, DNA, RNA, and protein, contain genetic information in their sequences, they are called **informational macromolecules** (∞ Section 3.2).

The molecular processes underlying genetic information flow can be divided into three stages (**Figure 7.1**).

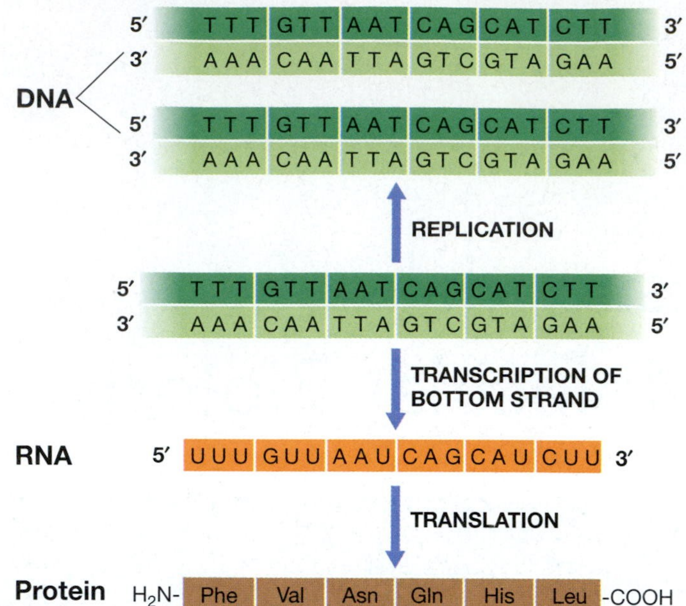

Figure 7.1 Synthesis of the three types of informational macromolecules. Note that for any particular gene only one of the two strands of the DNA double helix is transcribed.

1. **Replication**. The DNA molecule is a double helix of two long chains (∞ Section 3.5). During replication, DNA is duplicated, producing two double helices (Figure 7.1).
2. **Transcription**. DNA participates in protein synthesis through an RNA intermediate. Transfer of the information to RNA is called transcription, and the RNA molecule that encodes one or more polypeptides is called **messenger RNA** (mRNA). Some genes contain information for other types of RNA, in particular **transfer RNA** (tRNA) and **ribosomal RNA** (rRNA). These play roles in protein synthesis but do not themselves encode the genetic information for making proteins.
3. **Translation**. The sequence of amino acids in a polypeptide is determined by the specific sequence of bases in the mRNA. There is a linear correspondence between the base sequence of a gene and the amino acid sequence of a polypeptide (Figure 7.1). Each group of three bases on an mRNA molecule encodes a single amino acid, and each such triplet of bases is called a **codon**. This genetic code is translated into protein by means of the protein-synthesizing system. This system consists of ribosomes (which are themselves made up of proteins and rRNA), tRNA, and a number of proteins known as translation factors.

The three steps shown in Figure 7.1 are used in all cells and constitute the central dogma of molecular biology (DNA → RNA → protein). Note that many different RNA molecules are each transcribed from a relatively short region of the long DNA molecule. In eukaryotes, each gene is transcribed to give a single mRNA (∞ Chapter 8), whereas in prokaryotes, a single messenger RNA may carry genetic information from several genes, that is, more than one coding region (**Figure 7.2**). In Chapter 10 we will learn that some viruses violate the central dogma. This includes instances where RNA is the viral genetic material and either functions as mRNA directly or encodes mRNA. This also occurs in retroviruses such as HIV—the causative agent of AIDS—where an RNA genome encodes the production of DNA, a process called reverse transcription. Note that in both of these cases information transfer remains from nucleic acid to nucleic acid but not in the way originally stated by the central dogma.

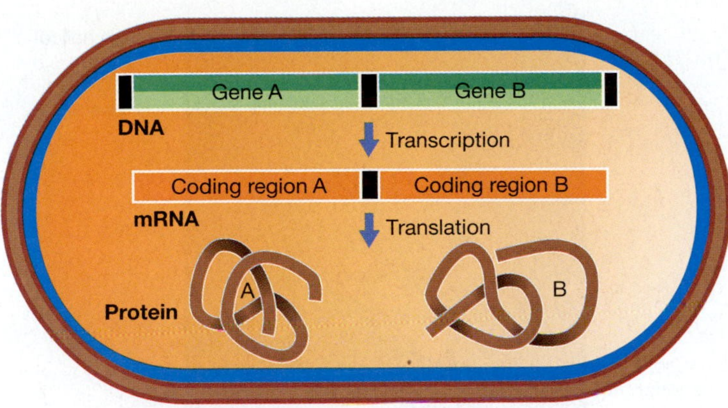

Figure 7.2 Transcription in prokaryotes. A single mRNA often contains more than one coding region.

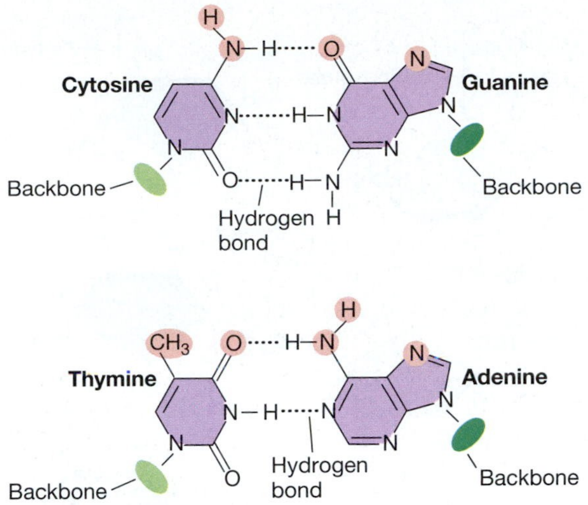

7.1 MiniReview

The three key processes of macromolecular synthesis are: (1) DNA replication; (2) transcription (the synthesis of RNA from a DNA template); and (3) translation (the synthesis of proteins using messenger RNA as template).

- What three informational macromolecules are involved in genetic information flow?
- In all cells there are three processes involved in genetic information flow. What are they?

II DNA STRUCTURE

We dealt with the structure of nucleic acids in a general way in Chapter 3. In the next few sections of this chapter we discuss the details of DNA structure, including the types of genetic elements containing DNA that are found in cells. With this information as a foundation, we can then discuss how DNA is replicated, transcribed into RNA, and translated into protein.

7.2 The Double Helix

Four nucleic acid bases are found in DNA: adenine (A), guanine (G), cytosine (C), and thymine (T). As already shown in Figure 3.11, the backbone of the DNA chain consists of alternating repeats of phosphate and the pentose sugar deoxyribose; connected to each sugar is one of the nucleic acid bases. Recall especially the numbering system for the positions of sugar and base. The phosphate connecting two sugars spans from the 3′-carbon of one sugar to the 5′-carbon of the adjacent sugar (see Figure 7.4). At one end of each DNA strand the sugar has a phosphate on the 5′-hydroxyl whereas at the other end the sugar has a free hydroxyl at the 3′-position.

DNA as a Double Helix

In all cells and many viruses, DNA exists as a double-stranded molecule with two polynucleotide strands whose base sequences are **complementary**. (As discussed in Chapter 10, the genomes of some viruses are single-stranded.) The complementarity of DNA arises because of the specific pairing of the purine and pyrimidine bases: Adenine always pairs with thymine, and guanine always pairs with cytosine (**Figure 7.3**). The two strands in the resulting double-stranded molecule are arranged in an **antiparallel** fashion (Figure 7.3 and **Figure 7.4**, distinguished as two shades of green). For example, in Figure 7.4 the strand on the left is arranged 5′ to 3′ (top to bottom), whereas the other strand is 5′ to 3′ (bottom to top).

Figure 7.3 Specific pairing between adenine (A) and thymine (T) and between guanine (G) and cytosine (C) via hydrogen bonds. These are the typical base pairs found in double-stranded DNA. Atoms that are found in the major groove of the double helix and that interact with proteins are highlighted in pink. The deoxyribose phosphate backbones of the two strands of DNA are also indicated. Note the different shades of green for the two strands of DNA, a convention used throughout this book.

The two strands of DNA are wrapped around each other to form a double helix (**Figure 7.5**). In the double helix, DNA forms two distinct grooves, the *major groove* and the *minor groove*. Of the many proteins that interact specifically with DNA (∞ Chapter 9), most bind in the major groove, where there is a considerable amount of space. Because of the regularity of the double helix, some atoms of each base are always exposed in the major groove (and some in the minor groove). Key regions of nucleotides that are important in interactions with proteins are shown in Figure 7.3.

Several double-helical structures are possible for DNA. The Watson and Crick double helix is known as the B-form or *B-DNA* to distinguish it from the A- and Z-forms. The A-form is shorter and fatter than the B-form of the double helix; has 11 base pairs per turn, and the major groove is narrower and deeper. Double-stranded RNA or hybrids of one RNA plus one DNA strand often form the A-helix. The Z-DNA double helix has 12 base pairs per turn and is left-handed. Its sugar phosphate backbone is a zigzag line rather than a smooth curve. Z-DNA is found in GC- or GT-rich regions, especially when negatively supercoiled. Occasional enzymes and regulatory proteins bind preferentially to Z-DNA.

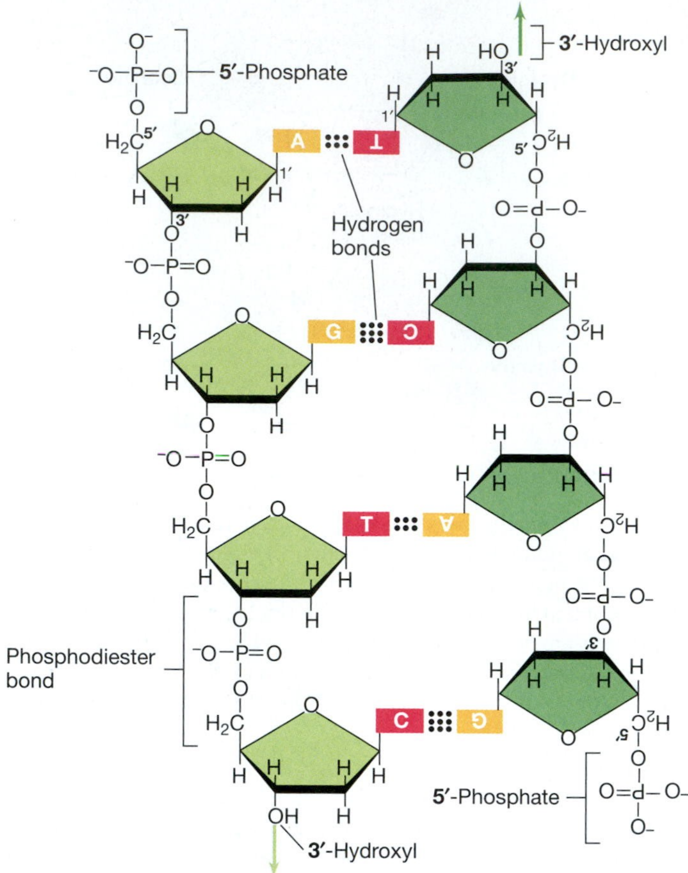

Figure 7.4 DNA structure. Complementary and antiparallel nature of DNA. Note that one chain ends in a 5′-phosphate group, whereas the other ends in a 3′-hydroxyl. The red bases represent the pyrimidines cytosine (C) and thymine (T), and the yellow bases represent the purines adenine (A) and guanine (G).

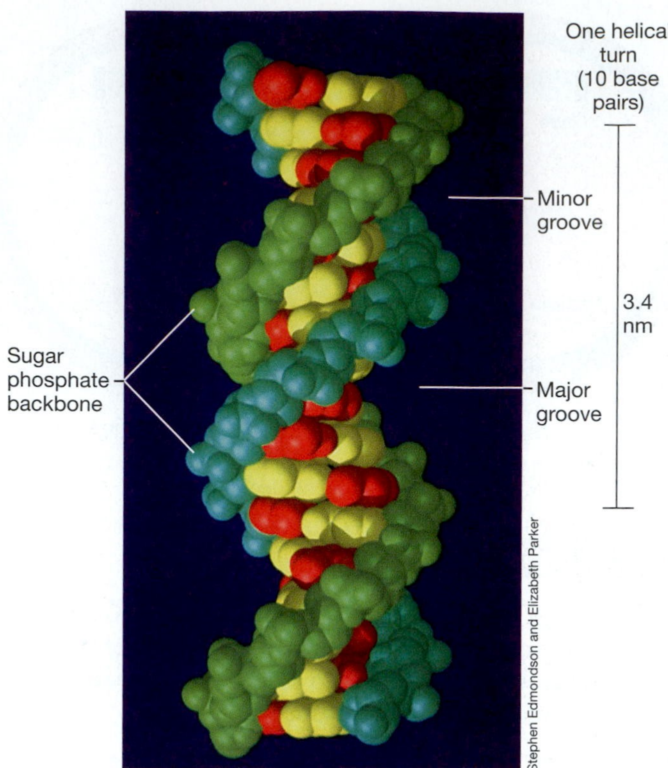

Figure 7.5 A computer model of a short segment of DNA showing the overall arrangement of the double helix. One of the sugar–phosphate backbones is shown in blue and the other in green. The pyrimidine bases are shown in red and the purines in yellow. Note the locations of the major and minor grooves (compare with Figure 7.3). One helical turn contains 10 base pairs.

Size of a DNA Molecule

The size of a DNA molecule can be expressed as the number of thousands of nucleotide bases or base pairs per molecule. Thus, a DNA molecule with 1000 bases is 1 kilobase (kb) of DNA. If the DNA is a double helix, then one speaks of *kilobase pairs* (kbp). Thus, a double helix 5000 base pairs in size would be 5 kbp. The bacterium *Escherichia coli* has about 4640 kbp of DNA in its chromosome. Since many genomes are quite large, it is easiest to speak of millions of base pairs, or *megabase pairs* (Mbp). The genome of *E. coli* is thus 4.64 Mbp.

In terms of actual length, each base pair takes up 0.34 nanometer (nm) in length along the helix, and each turn of the helix contains approximately 10 base pairs. Therefore, 1 kb of DNA has 100 turns of the helix and is 0.34 μm long. Calculations such as this can be very interesting, as we shall soon see.

Secondary Structure, Inverted Repeats, and Stem-Loops

Long DNA molecules are quite flexible, but stretches of DNA less than 100 base pairs are more rigid. Some short segments of DNA can be bent by proteins that interact with them. However, certain base sequences themselves cause DNA to bend. The sequences of this bent DNA often have several runs of five or six adenines (in the same strand), each separated by four or five other bases. In addition, some sequences contribute to the ability of DNA to bend when certain proteins interact with the DNA. DNA bending is one mechanism for the regulation of gene expression (the turning on and off of genes), as we shall discuss in Chapter 9.

Short, repeated sequences are often found in DNA molecules. Many proteins interact with regions of DNA containing repeated sequences that are arranged in inverse orientation (∞ Chapter 9). This type of repeat is called an inverted repeat and gives the DNA sequence a twofold symmetry. As shown in **Figure 7.6**, nearby inverted repeats can lead to the formation of stem-loop structures. The stems are short double-helical regions with normal base pairing and antiparallel strands. The loop consists of the unpaired bases between the two repeat units.

The production of stem-loop structures in DNA itself is relatively rare in cells. However, the production of stem-loop structures in the RNA produced from DNA following transcription is common. Such secondary structure (∞ Section 3.5) formed by base pairing within a single strand of nucleic acid is critical to the functioning of transfer RNA (Section 7.14)

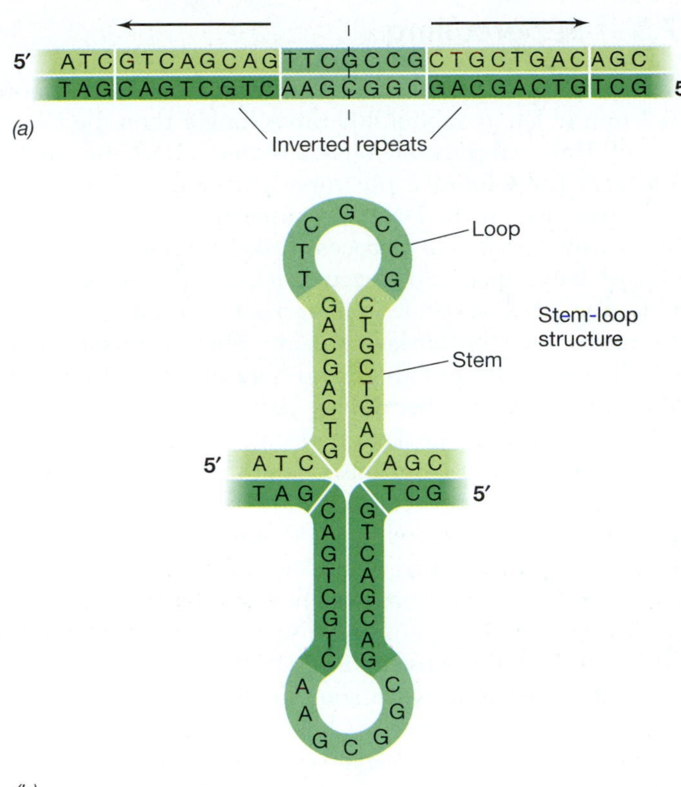

Figure 7.6 Inverted repeats and the formation of a stem-loop structure. *(a)* Nearby inverted repeats in DNA. The arrows indicate the symmetry around the imaginary axis (dashed line). *(b)* Formation of stem-loop structures by pairing of complementary bases on the same strand. Note how if RNA was formed from either strand of DNA, the RNA could form a single stem-loop (compare with Figure 7.23).

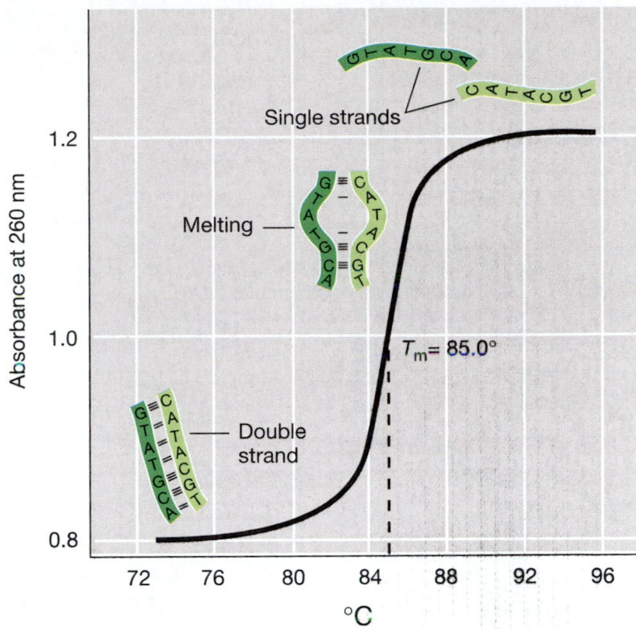

Figure 7.7 Thermal denaturation of DNA. DNA absorbs more ultraviolet radiation at 260 nm as the double helix is denatured. The transition is quite abrupt, and the temperature of the midpoint, T_m, is proportional to the GC content of the DNA. Although the denatured DNA can be renatured by slow cooling, the process does not follow a similar curve. Renaturation becomes progressively more complete at temperatures well below the T_m and then only after a considerable incubation time.

and ribosomal RNA (Section 7.15). Even if a stem-loop does not form, inverted repeats in DNA are often binding sites for specific DNA-binding proteins that regulate transcription (∞ Chapter 9).

The Effect of Temperature on DNA Structure

Although the hydrogen bonds between the base pairs are individually very weak (∞ Section 3.1), many such bonds in a long DNA molecule effectively hold the two strands together. There may be millions or even hundreds of millions of hydrogen bonds in a DNA molecule, depending on the number of base pairs present. Recall from Figure 7.3 that each adenine-thymine base pair has *two* hydrogen bonds, while each guanine-cytosine base pair has *three*. This makes GC pairs stronger than AT pairs.

When isolated from cells and kept near room temperature and at physiological salt concentrations, DNA remains in a double-stranded form. However, if the temperature is raised, the hydrogen bonds will break but the covalent bonds holding a chain together will not, and so the two DNA strands will separate. This process is called denaturation (melting) and can be measured experimentally because single-stranded and double-stranded nucleic acids differ in their ability to absorb ultraviolet radiation at 260 nm (**Figure 7.7**).

DNA with a high percentage of GC pairs melts at a higher temperature than a similar-sized molecule with more AT pairs. If the heated DNA is allowed to cool slowly, the double-stranded DNA can reform, a process called annealing. This can be used not only to reform native DNA but also to form hybrid molecules whose two strands come from different sources. Hybridization, the artificial construction of a double-stranded nucleic acid by complementary base pairing of two single strands, is a powerful technique in molecular biology (∞ Section 12.2).

7.2 MiniReview

DNA is a double-stranded molecule that forms a helix. Its length is measured in terms of numbers of base pairs. The two strands in the double helix are antiparallel, but inverted repeats allow for the formation of secondary structure. The strands of a double-helical DNA molecule can be denatured by heat and allowed to reassociate following cooling.

- Explain what the word antiparallel means in terms of the structure of double-stranded DNA.
- Define the term complementary when used to refer to two strands of DNA.
- Define the terms denaturation, reannealing, and hybridization as they apply to nucleic acids.

(a) Relaxed, covalently closed circular DNA

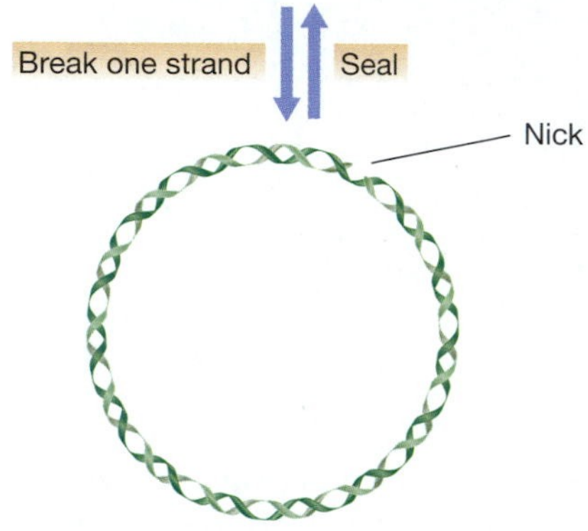

(b) Relaxed, nicked circular DNA

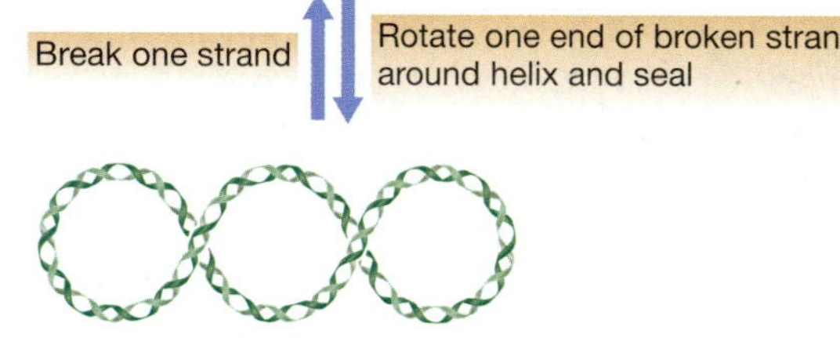

(c) Supercoiled circular DNA

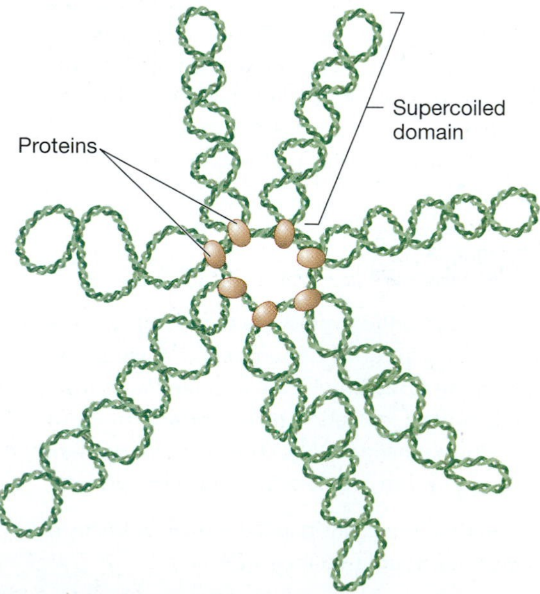

(d) Chromosomal DNA with supercoiled domains

7.3 Supercoiling

If linearized, the *Escherichia coli* chromosome would be more than 1 mm in length, about 400 times longer than the *E. coli* cell itself! How is it possible to pack so much DNA into such a little space? The solution is the imposition of a kind of "higher order" structure on the DNA, in which the double-stranded DNA is further twisted in a process called *supercoiling*. **Figure 7.8** shows how supercoiling occurs in a circular DNA duplex. If a circular DNA molecule is linearized, any supercoiling is lost and the DNA becomes "relaxed." Thus, a relaxed DNA molecule is one with exactly the number of turns of the helix predicted from the number of base pairs.

Supercoiling puts the DNA molecule under torsion, much like the added tension to a rubber band that occurs when it is twisted. DNA can be supercoiled in either a positive or negative manner. In positive supercoiling the double helix is overwound, whereas in negative supercoiling the double helix is underwound. Negative supercoiling results when the DNA is twisted about its axis in the opposite sense from the right-handed double helix. Negatively supercoiled DNA is the form predominantly found in nature, with some interesting exceptions in certain species of *Archaea* (∞ Chapter 8). In *Escherichia coli* more than 100 supercoiled domains are thought to exist, each of which is stabilized by binding to specific proteins.

Topoisomerases: DNA Gyrase

Supercoils may be introduced or removed by enzymes known as topoisomerases. Two major classes of topoisomerase exist and each functions by a different mechanism. Class I topoisomerases make a single-stranded break in the DNA that allows the rotation of one strand of the double helix around the other. Each such rotation adds or removes a single supercoil. After this, the nick is resealed. Class II topoisomerases make double-stranded breaks, pass the intact double helix through the break, and finally reseal the break. Each such operation adds or removes two supercoils. Insertion of supercoils into DNA requires ATP to provide energy, whereas releasing supercoils does not.

In *Bacteria* and most *Archaea,* there is an enzyme called **DNA gyrase** that introduces negative supercoils into DNA. DNA gyrase belongs to the group of type II topoisomerases. Its operation occurs in several stages. First, the circular DNA molecule is twisted; then, where the two chains touch, a double-stranded break is made in one chain. Next, the intact chain is passed through the break, and then the broken double helix is resealed (**Figure 7.9**). It is worth noting that some of the antibiotics that affect *Bacteria,* such as the quinolone nalidixic acid, the fluoroquinolone ciprofloxacin, and novobiocin, inhibit the activity of DNA gyrase. Novobiocin is also effective against several species of *Archaea,* where it also seems to inhibit DNA gyrase.

Figure 7.8 Supercoiled DNA. Parts *(a), (b),* and *(c)* show supercoiled circular DNA and relaxed, nicked circular DNA. A nick is a break in a phosphodiester bond of one strand. *(d)* In fact, the double-stranded DNA in the bacterial chromosome is arranged not in one supercoil but in several supercoiled domains, as shown here.

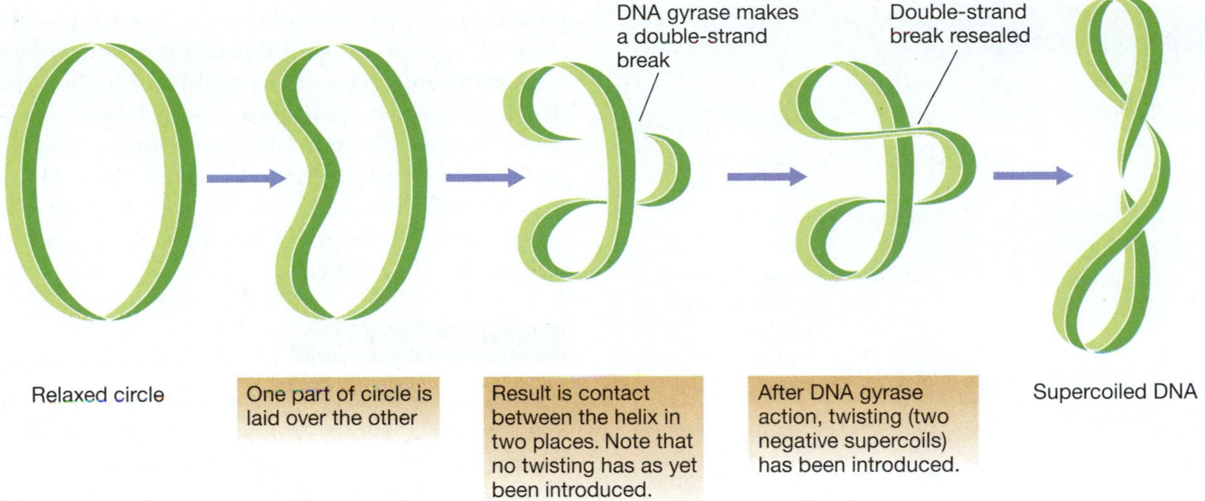
DNA gyrase makes a double-strand break
Double-strand break resealed
Relaxed circle
One part of circle is laid over the other
Result is contact between the helix in two places. Note that no twisting has as yet been introduced.
After DNA gyrase action, twisting (two negative supercoils) has been introduced.
Supercoiled DNA

Figure 7.9 DNA gyrase. Introduction of negative supercoiling into circular DNA by the activity of DNA gyrase (topoisomerase II), which makes double-strand breaks.

The removal of surplus supercoiling in DNA is generally done by topoisomerase I. This enzyme introduces a single-stranded break in the DNA and causes the rotation of one strand of the double helix around the other. As was shown in Figure 7.8, a break in the backbone (a nick) of either strand allows the DNA to return to the relaxed state. However, to prevent the entire bacterial chromosome from becoming relaxed every time a nick is made, the chromosome contains supercoiled domains as shown in Figure 7.8*d*. A nick in the DNA in one of these domains does not relax DNA in the others. It is unclear precisely how these domains are formed, although specific DNA-binding proteins are involved.

Through the activity of topoisomerases, the DNA molecule can be alternately supercoiled and relaxed. Because supercoiling is necessary for packing the DNA into the cell and relaxation is necessary for DNA replication and transcription, the complementary processes of supercoiling and relaxation play an important role in the biology of the cell. In most prokaryotes, the extent of negative supercoiling is the result of a balance between the activity of DNA gyrase and topoisomerase I. In addition to replication, however, supercoiling also affects gene expression. Certain genes are more actively transcribed when DNA is supercoiled, whereas transcription of other genes is inhibited by excessive supercoiling.

7.3 MiniReview

Very long DNA molecules can be packaged into cells because they are supercoiled. In prokaryotes this supercoiling is brought about by enzymes called topoisomerases. DNA gyrase is a key enzyme in prokaryotes and introduces negative supercoils into the DNA.

- Why is supercoiling important?
- What function do topoisomerases serve inside cells?

7.4 Chromosomes and Other Genetic Elements

Structures containing genetic material (DNA in most organisms, but RNA in some viruses) are called *genetic elements.* The **genome** is the total complement of genes in a cell or virus. Although the main genetic element in prokaryotes is the **chromosome**, other genetic elements are known and these play important roles in gene function in both prokaryotes and eukaryotes (**Table 7.1**). In Section 2.6 we noted that a typical prokaryote has a single circular chromosome containing all (or most) of the genes found inside the cell. Although a single chromosome is the rule among prokaryotes, there are exceptions. A few prokaryotes contain two chromosomes. Eukaryotes have multiple chromosomes making up their genome (∞ Section 8.5). Also, the DNA in all known eukaryotic chromosomes is linear in contrast to most prokaryotic chromosomes, which are circular DNA molecules.

Viruses and Plasmids

Besides the chromosome, a number of other genetic elements are known, including virus genomes, plasmids, organellar genomes, and transposable elements.

Viruses contain genomes, *either* DNA or RNA, that control their own replication and their transfer from cell to cell. Both linear and circular viral genomes are known. In addition, the nucleic acid in viral genomes may be single-stranded or double-stranded. Viruses are of special interest because they often cause disease. We discuss viruses in Chapters 10 and 19 and a variety of viral diseases in later chapters.

Plasmids are genetic elements that replicate separately from the chromosome. The great majority of plasmids are double-stranded DNA, and although most plasmids are circular, some are linear. Plasmids differ from viruses in two ways: (1) they do not cause cellular damage (generally they are

Table 7.1 Kinds of genetic elements

Organism	Element	Type of nucleic acid	Description
Prokaryote	Chromosome	Double-stranded DNA	Extremely long, usually circular
Eukaryote	Chromosome	Double-stranded DNA	Extremely long, linear
All organisms	Plasmid[a]	Double-stranded DNA	Relatively short circular or linear molecule, extrachromosomal
All organisms	Transposable element	Double-stranded DNA	Always found inserted into another DNA molecule
Mitochondrion or chloroplast	Genome	Double-stranded DNA	Medium length, usually circular
Virus	Genome	Single- or double-stranded DNA or RNA	Relatively short, circular or linear

[a]Plasmids are uncommon in eukaryotes.

beneficial), and (2) they do not have extracellular forms, whereas viruses do. Although only a few eukaryotes contain plasmids, one or more plasmids have been found in most prokaryotic species and can be of profound importance. We discuss plasmids in Chapter 11.

Some plasmids contain genes whose protein products confer important properties on the host cell, such as resistance to antibiotics. What is the difference, then, between a large plasmid and a chromosome? A chromosome is a genetic element that contains genes whose products are necessary for essential cellular functions. Such essential genes are called *housekeeping genes*. Some of these encode essential proteins, such as DNA and RNA polymerases, and others encode essential RNAs, such as ribosomal and transfer RNA. By contrast to the chromosome, plasmids are often expendable and rarely contain genes required for growth under all conditions. There are many genes on a chromosome that are unessential as well, but the presence of *essential* genes is necessary for a genetic element to be called a chromosome.

Transposable Elements

Transposable elements are segments of DNA that can move from one site on a DNA molecule to another site on the same molecule or on a different DNA molecule. Chromosomes, plasmids, virus genomes, and any other type of DNA molecule may act as host molecules for transposable elements. Transposable elements are not found as separate molecules of DNA but are inserted into other DNA molecules including chromosomes, plasmids, or virus genomes. Transposable elements are found in both prokaryotes and eukaryotes and play important roles in genetic variation. In prokaryotes there are three main types of transposable elements: insertion sequences, transposons, and some special viruses. Insertion sequences are the simplest type of transposable element and carry no genetic information other than that required for them to move about the chromosome. Transposons are larger and contain other genes. We discuss both of these in more detail in Chapter 11. In Chapter 19 we discuss a bacterial virus, Mu, that is itself a transposable element. The unique feature common to all transposable elements is that they replicate as part of some other molecule of DNA.

7.4 MiniReview

In addition to the chromosome, a number of other genetic elements exist in cells. Plasmids are DNA molecules that exist separately from the chromosome of the cell. Viruses contain a genome, either DNA or RNA, that controls their own replication. Transposable elements exist as a part of other genetic elements.

- What is a genome?
- What genetic material is found in all cellular chromosomes?
- What defines a chromosome in prokaryotes?

III DNA REPLICATION

The flow of biological information begins with DNA replication. DNA replication is necessary for cells to divide, whether to reproduce new organisms, as in unicellular microorganisms, or to produce new cells as part of a multicellular organism. The process of DNA replication must occur at sufficiently high accuracy so that the daughter cells are genetically identical to the mother cell (or nearly so). This involves a host of special enzymes and processes.

7.5 Templates and Enzymes

DNA exists in cells as a double helix with complementary base pairing (Figures 7.3 and 7.4). If the DNA double helix is opened up, a new strand can be synthesized as the complement of each of the parental strands. As shown in **Figure 7.10**, replication is **semiconservative**, meaning that the two resulting double helices consist of one newly made strand and one parental strand. The DNA strand that is copied is called the template, and in DNA replication each parental strand is a template for one newly synthesized (and complementary) daughter strand (Figure 7.10).

The chemistry of DNA, the nature of its precursors, and the activities of the enzymes involved in replication place some important restrictions on the manner in which new strands are synthesized. The precursor of each new nucleotide in the chain is a deoxynucleoside 5′-triphosphate, from which the two terminal phosphates are removed and the internal phosphate is attached covalently to deoxyribose of the growing chain (**Figure 7.11**). The addition of the nucleotide to the

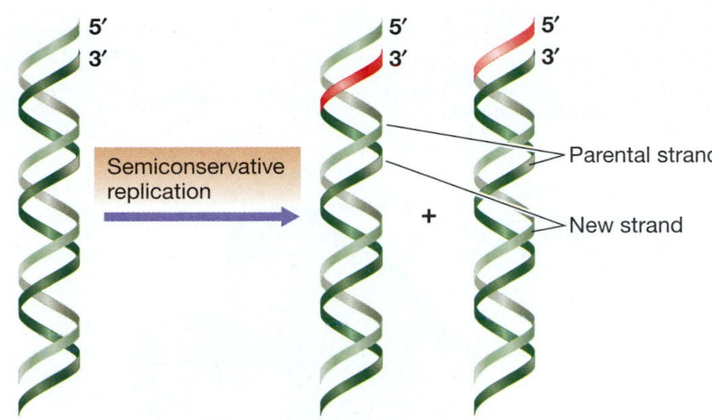

Figure 7.10 Overview of DNA replication. DNA replication is a semiconservative process in both prokaryotes and eukaryotes. Note that the new double helices each contain one new (shown topped in red) and one parental strand.

growing chain requires the presence of a free hydroxyl group, which is available only at the 3′ end of the molecule. This restriction leads to an important principle that is at the basis of DNA replication: DNA replication always proceeds from the 5′ end to the 3′ end, the 5′ -phosphate of the incoming nucleotide being attached to the 3′-hydroxyl of the previously added nucleotide.

DNA Polymerase and Primase

Enzymes that catalyze the addition of deoxynucleotides are called **DNA polymerases**. Several such enzymes exist, and each has a specific function. There are five different DNA polymerases in *Escherichia coli,* called DNA polymerases I, II, III, IV, and V. DNA polymerase III (Pol III) is the primary enzyme for replicating chromosomal DNA. DNA polymerase I (Pol I) is also involved in chromosomal replication, though to a lesser extent (see below). The other DNA polymerases help repair damaged DNA (∞ Section 11.7).

All known DNA polymerases synthesize DNA in the 5′ → 3′ direction. However, no known DNA polymerase can initiate a new chain; all of these enzymes can only add a nucleotide onto a preexisting 3′-OH group. Therefore, to start a new chain a **primer**, a nucleic acid molecule to which DNA polymerase can attach the first nucleotide, is required. In most cases this primer is a short stretch of RNA.

When the double helix is opened at the beginning of replication, an RNA-polymerizing enzyme makes the RNA primer. This enzyme, called *primase,* synthesizes a short stretch of RNA of around 11–12 nucleotides that is complementary in base pairing to the template DNA. At the growing end of this RNA primer is a 3′-OH group to which DNA polymerase can add the first deoxyribonucleotide. Continued extension of the molecule thus occurs as DNA rather than RNA. The newly synthesized molecule has a structure like that shown in Figure 7.12. The primer will eventually be removed and replaced with DNA, as described later. We will examine the details of DNA replication next.

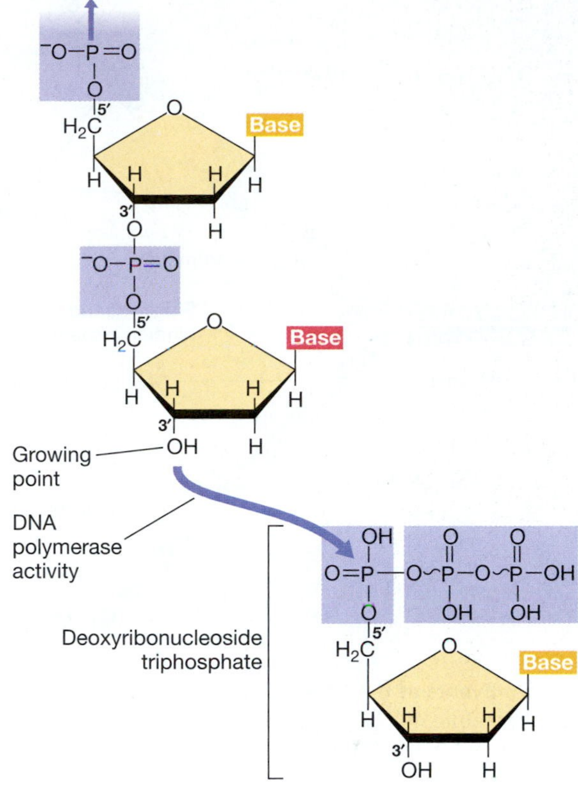

Figure 7.11 Structure of the DNA chain and mechanism of growth by addition of a deoxyribonucleoside triphosphate at the 3′-end of the chain. Growth proceeds from the 5′-phosphate to the 3′-hydroxyl end. The enzyme DNA polymerase catalyzes the addition reaction. The four precursors are deoxythymidine triphosphate (dTTP), deoxyadenosine triphosphate (dATP), deoxyguanosine triphosphate (dGTP), and deoxycytidine triphosphate (dCTP). Upon nucleotide insertion, the two terminal phosphates of the triphosphate are split off as pyrophosphate (PP_i). Thus, two energy-rich phosphate bonds are consumed in the addition of a single nucleotide.

7.5 MiniReview

Both strands of the DNA helix serve as templates for the synthesis of two new strands (semiconservative replication). The two progeny double helices each contain one parental strand and one new strand. The new strands are elongated by addition of deoxyribonucleotides to the 3′ end. DNA polymerases require a primer made of RNA.

- To which end (5′ end or 3′ end) of a newly synthesized strand of DNA does polymerase add a base?
- Why is a primer required for DNA replication?

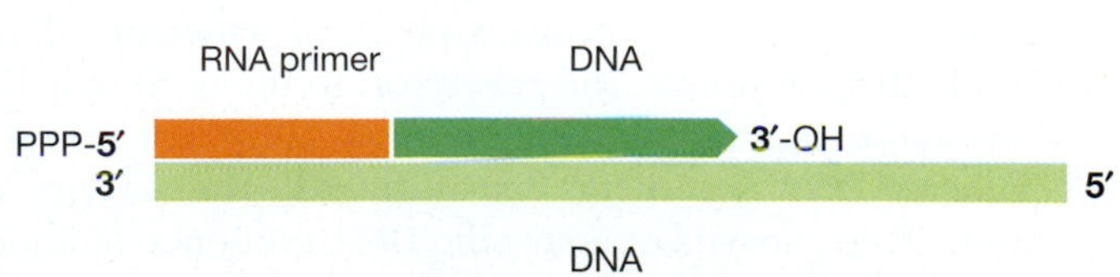

Figure 7.12 The RNA primer. Structure of the RNA–DNA combination formed at the initiation of DNA synthesis.

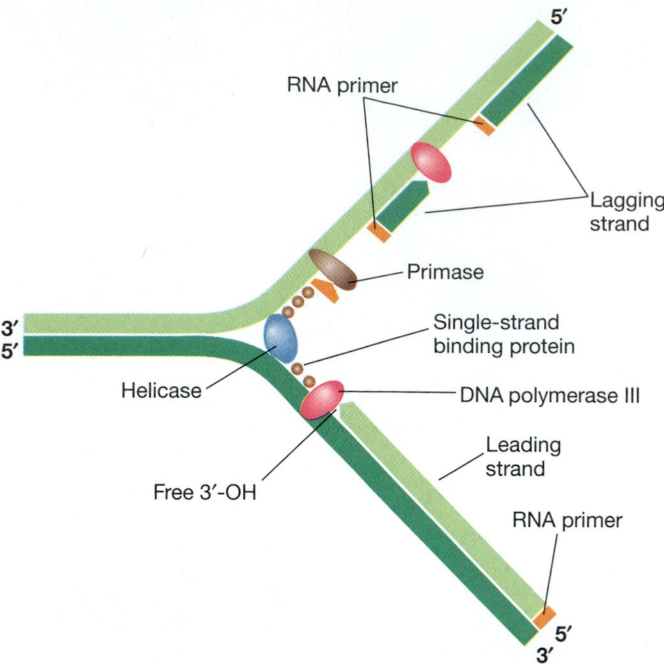

Figure 7.13 Events at the DNA replication fork. Note the polarity and antiparallel nature of the DNA strands.

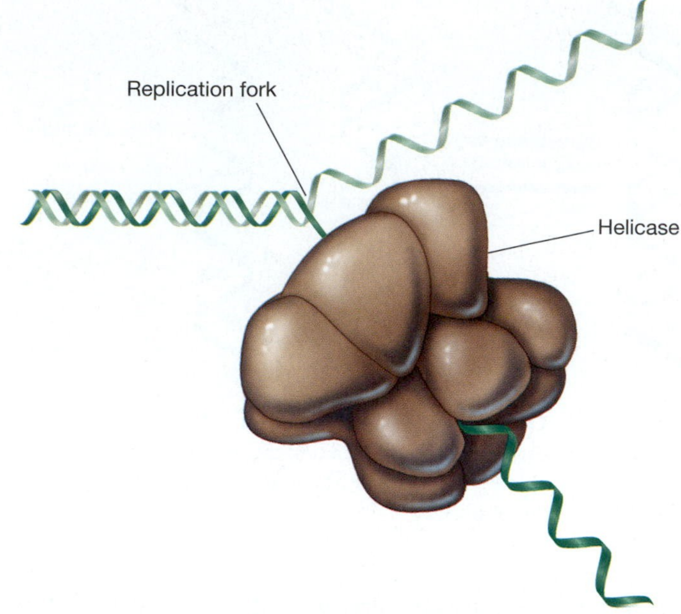

Figure 7.14 DNA helicase unwinding a double helix. In this figure, the protein and DNA molecules are drawn to scale. Simple diagrams are often used to give an idea of molecular events in the cell, but they may give the incorrect impression that most proteins are relatively small compared to DNA. Although DNA molecules are generally extremely long, they are relatively thin compared to many proteins.

7.6 The Replication Fork

Much of our understanding of the mechanism of DNA replication has been obtained from studying the bacterium *Escherichia coli,* and the following discussion deals primarily with this organism. However, the process of DNA replication is probably quite similar in all *Bacteria*. By contrast, although most species of *Archaea* have circular chromosomes, many events in DNA replication resemble those in eukaryotic cells more than those in *Bacteria* (Chapter 8). This again reflects the phylogenetic affiliation between *Archaea* and *Eukarya* (Section 2.7).

Initiation of DNA Synthesis

Before DNA polymerase can synthesize new DNA, the preexisting double helix must be unwound to expose the template strands. The zone of unwound DNA where replication occurs is known as the **replication fork**. An enzyme known as DNA helicase unwinds the DNA double helix and exposes a short single-stranded region (**Figures 7.13** and **7.14**). The energy required by helicases for unwinding DNA comes from the hydrolysis of ATP. Helicases move along the helix and separate the strands just in advance of the replication fork. The single-stranded region is complexed with a special protein, the single-strand binding protein, which stabilizes the single-stranded DNA, preventing the formation of intrastrand hydrogen bonds and reversion to the double helix.

In prokaryotes, there is a single location on the chromosome where DNA synthesis is initiated, the origin of replication. This consists of a specific DNA sequence of about 250 bases that is recognized by specific initiation proteins, in particular a protein called DnaA (**Table 7.2**), which binds to this region and opens up the double helix to expose the individual strands to the replication machinery. Next to assemble is the helicase (also known as DnaB), which is helped onto the DNA by the helicase loader protein (or DnaC). Two helicases are loaded, one onto each strand, facing in opposite directions. Next, two primase and then two DNA polymerase enzymes are loaded onto the DNA behind the helicases. Initiation of DNA replication then begins on the two single strands. As replication proceeds, the replication fork appears to move along the DNA (Figure 7.13). www.microbiologyplace.com **Online Tutorial 7.1: DNA Replication**

Leading and Lagging Strands

Figure 7.13 shows an important distinction in replication of the two DNA strands due to the fact that replication always proceeds from 5′ → 3′ (always adding a new nucleotide to the 3′-OH of the growing chain). On the strand growing from the 5′-PO_4^{2-} to the 3′-OH, called the **leading strand**, DNA synthesis occurs continuously because there is always a free 3′-OH at the replication fork to which a new nucleotide can be added. But on the opposite strand, called the **lagging strand**, DNA synthesis occurs discontinuously because there is no 3′-OH at the replication fork to which a new nucleotide can attach. Where is the 3′-OH on this strand? It is located at the opposite end, away from the replication fork. Therefore, on

Table 7.2 Major enzymes involved in DNA replication in *Bacteria*

Enzyme	*Encoding genes*	*Function*
DNA gyrase	*gyrAB*	Unwinds supercoils ahead of replisome
Origin-binding protein	*dnaA*	Binds origin of replication; aids DNA melting to open double helix
Helicase loader	*dnaC*	Loads helicase at origin
Helicase	*dnaB*	Unwinds double helix at replication fork
Single-strand binding protein	*ssb*	Prevents single strands from annealing
Primase	*dnaG*	Primes new strands of DNA
DNA polymerase III		Main polymerizing enzyme
Sliding clamp	*dnaN*	Holds Pol III on DNA
Clamp loader	*holA–E*	Loads Pol III onto sliding clamp (Figure 7.19)
Dimerization subunit	*dnaX*	Holds together the two core enzymes for the leading and lagging strands
Polymerase subunit	*dnaE*	Strand elongation
Proofreading subunit	*dnaQ*	Proofreading
DNA polymerase I	*polA*	Excises RNA primer and fills in gaps
DNA ligase	*ligA, ligB*	Seals nicks in DNA
Tus protein	*tus*	Binds terminus and blocks progress of the replication fork
Topoisomerase IV	*parCE*	Unlinking of interlocked circles

the lagging strand, RNA primers must be synthesized by primase multiple times to provide free 3′-OH groups. By contrast, the leading strand is primed only once, at the origin. As a result, the lagging strand is made in short segments, called *Okazaki fragments,* after their discoverer, Reiji Okazaki. These fragments are joined together later to give a continuous strand of DNA.

DNA Replication: Synthesis of the New DNA Strands

After synthesizing the RNA primer, primase is replaced by Pol III. This enzyme is a complex of several proteins (Table 7.2), including the polymerase core enzyme itself. Each polymerase is held on the DNA by a sliding clamp, which encircles and slides along the single template strands of DNA. Consequently, the replication fork contains two polymerase core enzymes and two sliding clamps, one set for each strand. However, there is only a single clamp-loader complex. This is needed to assemble the sliding clamps onto the DNA. After assembly on the lagging strand, the elongation component of

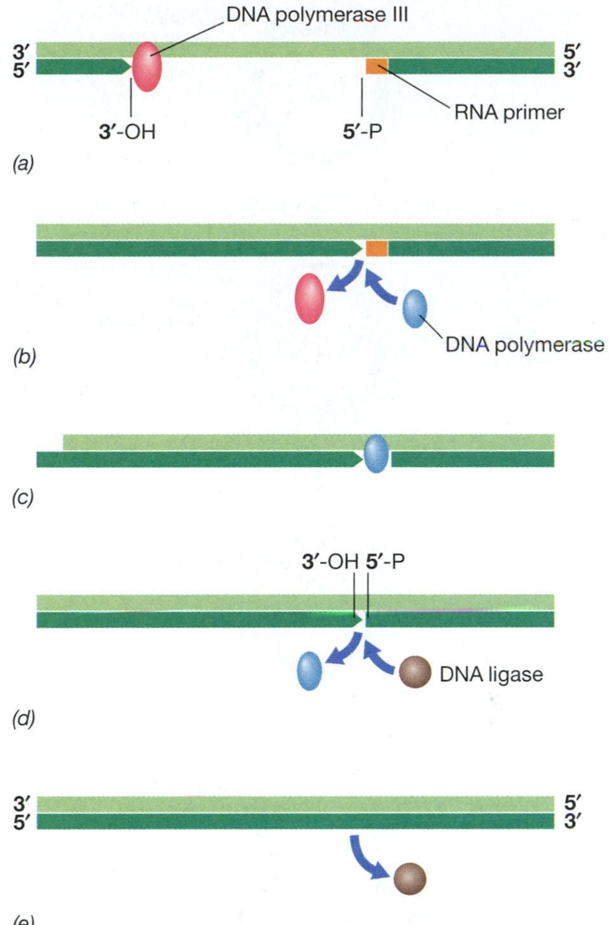

Figure 7.15 Sealing two fragments on the lagging strand. *(a)* DNA polymerase III is synthesizing DNA in the 5′ → 3′ direction toward the RNA primer of a previously synthesized fragment on the lagging strand. *(b)* On reaching the fragment, DNA polymerase III leaves and is replaced by DNA polymerase I. *(c)* DNA polymerase I continues synthesizing DNA while removing the RNA primer from the previous fragment. *(d)* DNA ligase replaces DNA polymerase I after the primer has been removed. *(e)* DNA ligase seals the two fragments together.

Pol III, DnaE, then adds deoxyribonucleotides until it reaches previously synthesized DNA (**Figure 7.15**). At this point, the activity of Pol III stops.

The next enzyme to take part, Pol I, has more than one enzymatic activity. Besides synthesizing DNA, Pol I has a 5′ → 3′ exonuclease activity that removes the RNA primer preceding it (Figure 7.15). When the primer has been removed and replaced with DNA, Pol I is released. The last phosphodiester bond is made by an enzyme called *DNA ligase.* This enzyme seals nicks in DNAs that have an adjacent 5′-PO_4^{2-} and 3′-OH (something that Pol III is unable to do), and along with Pol I, it also participates in DNA repair. DNA ligase also plays an important role in sealing genetically manipulated DNA in the process of molecular cloning (∞ Section 12.3).

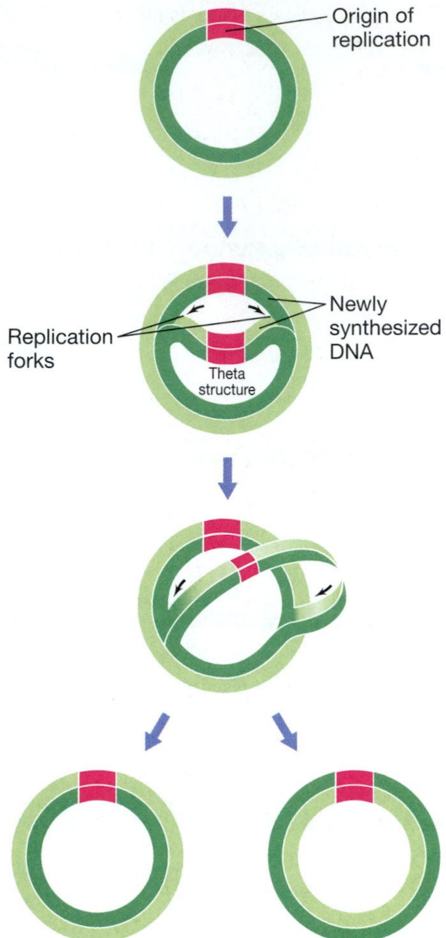

Figure 7.16 Replication of circular DNA: the theta structure. In circular DNA, bidirectional replication from an origin forms an intermediate structure resembling the Greek letter theta (θ).

7.6 MiniReview

DNA synthesis begins at a unique location called the origin of replication. The double helix is unwound by helicase and is stabilized by single-strand binding protein. Extension of the DNA occurs continuously on the leading strand but discontinuously on the lagging strand. The fragments of the lagging strand are joined together later.

- Why are there leading and lagging strands?
- What recognizes the origin of replication?
- What enzymes take part in joining the fragments of the lagging strand?

7.7 Bidirectional Replication and the Replisome

The circular nature of the chromosome of *Escherichia coli* and most other prokaryotes creates an opportunity for speeding up replication. Here we consider how this fits into the overall scheme of replication and cell division.

Bidirectional Chromosome Replication

In *E. coli,* and probably in all prokaryotes with circular chromosomes, replication is *bidirectional* from the origin of replication, as shown in **Figures 7.16** and **7.17**. There are thus two replication forks on each chromosome moving in opposite directions. In circular DNA, bidirectional replication leads to the formation of characteristic structures called theta structures (Figure 7.16). Most large DNA molecules, whether from prokaryotes or eukaryotes, have bidirectional replication from fixed origins. In fact, a single eukaryotic chromosome has many origins. Close inspection of the replicating DNA shows that during bidirectional replication, synthesis is occurring in both a leading and lagging fashion on each template strand (Figure 7.17).

Bidirectional DNA synthesis along a circular chromosome allows DNA to replicate as rapidly as possible. Even taking this into account and considering that Pol III can add nucleotides to a growing DNA strand at the rate of about 1,000 per second, chromosome replication in *E. coli* still takes about 40 min. Interestingly, under the best growth conditions, *E. coli* can grow with a doubling time of about 20 min. However, under these conditions, chromosome replication still takes 40 min. The solution to this conundrum is that cells of *E. coli* growing at doubling times shorter than 40 min contain multiple DNA replication forks. That is, a new round of DNA replication begins before the last round has been completed (**Figure 7.18**). Only in this way can a generation time shorter than the chromosome replication time be maintained.

The Replisome

While DNA is being synthesized at the replication fork, the coiling of the DNA is changing due to the activities of topoisomerases (Section 7.3). Unwinding is an essential feature of DNA replication, and because supercoiled DNA is under tension, it unwinds more easily than DNA that is not supercoiled. Thus, by regulating the degree of supercoiling, topoisomerases may regulate the process of replication (and also transcription, as discussed later).

Figure 7.13 shows the differences in replication of the leading and the lagging strands and the enzymes involved. From such a simplified drawing it would appear that each replication fork contains a host of different proteins all working independently. Actually, this is not the case. The highly dynamic proteins aggregate to form a large replication complex called the *replisome* (**Figure 7.19**). The lagging strand of DNA loops out to allow the replisome to move smoothly along both strands, and the replisome literally pulls the DNA template through it as replication occurs (Figure 7.18). Therefore, it is the DNA, and not DNA polymerase, that moves during replication. Note also in the replisome how helicase and primase form a subcomplex, called the *primosome,* which aids

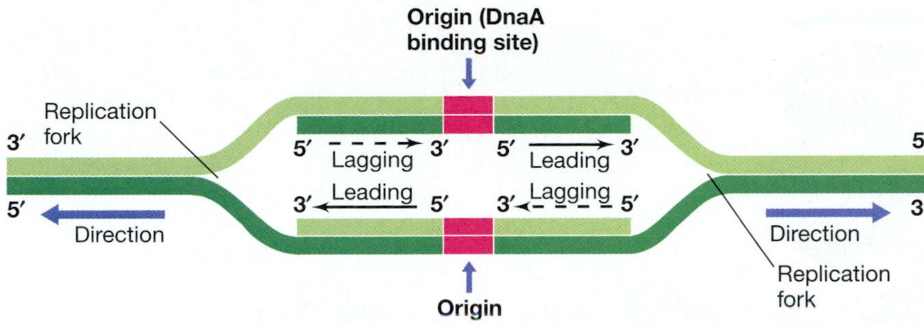

Figure 7.17 Dual replication forks in the circular chromosome. At an origin of replication that directs bidirectional replication, two replication forks must start. Therefore, two leading strands must be primed, one in each direction. In *Escherichia coli*, the origin of replication is identified by the binding of a specific protein, DnaA.

their working in close association during the replication process (Figure 7.19).

In summary, in addition to Pol III, the replisome contains several key replication proteins: (1) DNA gyrase, which removes supercoils; (2) DNA helicase and primase (the primosome), which unwind and prime the DNA; and (3) single-strand binding protein, which prevents the separated template strands from re-forming a double helix (Figure 7.19). Table 7.2 summarizes the properties of proteins essential for DNA replication.

7.7 MiniReview

Starting from a single origin, DNA synthesis proceeds in both directions around circular chromosomes. Therefore, there are two replication forks in operation simultaneously. The proteins at the replication fork form a large complex known as the replisome.

- What is the replisome and what are its components?
- How does *Escherichia coli* manage to carry out cell division in less time than it takes to duplicate its chromosome?

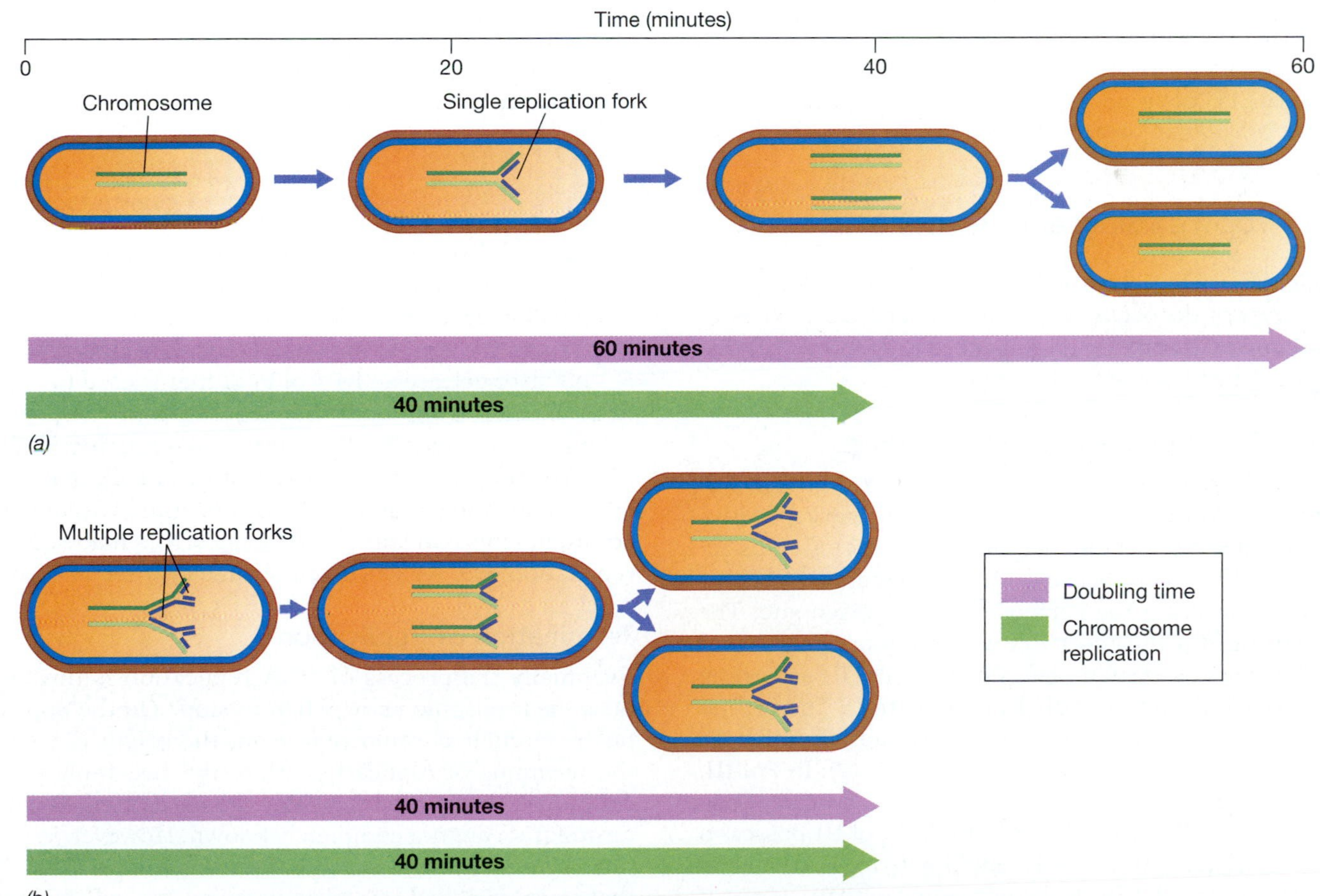

Figure 7.18 Cell division versus chromosome duplication. *(a)* Cells of *Escherichia coli* take approximately 40 min to replicate the chromosome and an additional 20 min for cell division. *(b)* When cells double in less than 60 min, a new round of chromosome replication must be initiated before the previous round is finished.

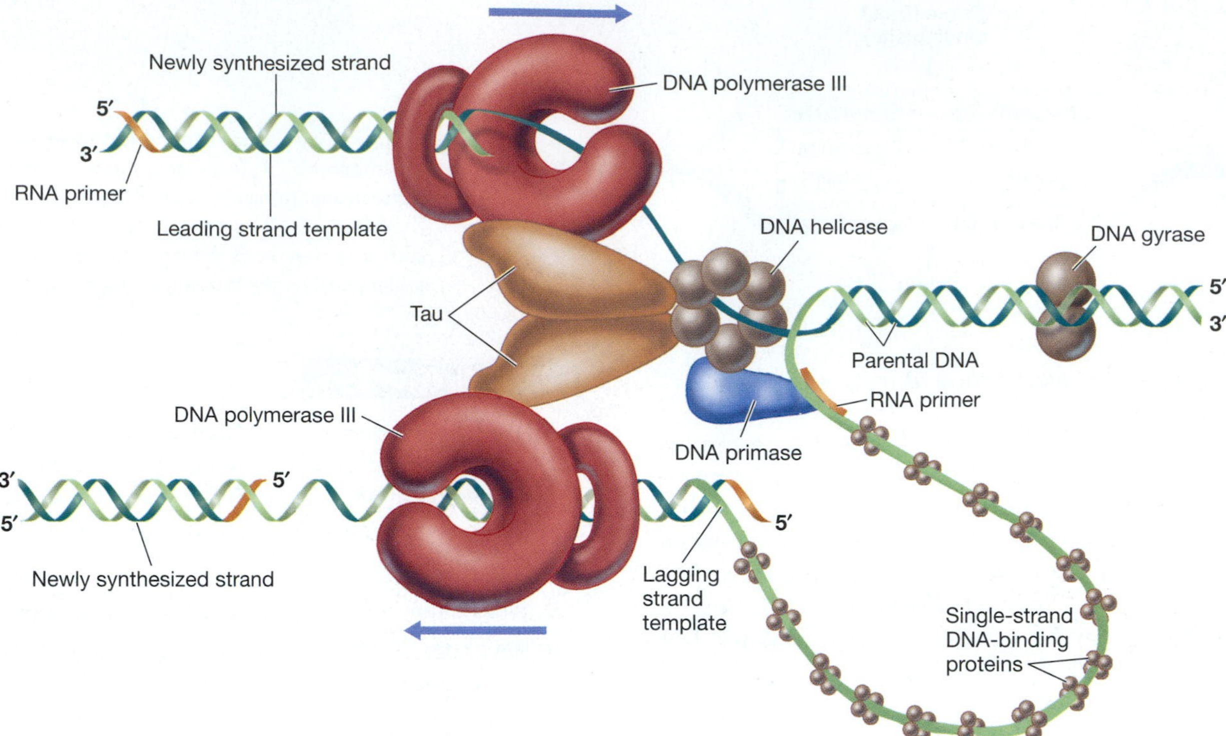

Figure 7.19 The replisome. The replisome consists of two copies of DNA polymerase III, plus helicase and primase (together forming the primosome), and many copies of single-strand DNA-binding protein. The tau subunits hold the two DNA polymerase assemblies and helicase together. Just upstream of the replisome, DNA gyrase removes supercoils in the DNA to be replicated. Note that the two polymerases are replicating the two individual strands of DNA in opposite directions. Consequently, the lagging-strand template loops around so that the whole replisome moves in the same direction along the chromosome.

7.8 Proofreading and Termination

DNA replicates with a remarkably low error rate. Nevertheless, when errors do occur, a back-up mechanism exists to detect and correct them. After considering this, we will discuss how DNA synthesis is terminated.

Fidelity of DNA Replication: Proofreading

Errors in DNA replication introduce mutations, changes in DNA sequence. Mutation rates in cells are remarkably low, between 10^{-8} and 10^{-11} errors per base pair inserted. This accuracy is possible partly because DNA polymerases get two chances to incorporate the correct base at a given site. The first chance follows the insertion of complementary bases opposite the bases on the template strand by Pol III according to the base-pairing rules, A with T and G with C. The second chance depends upon a second enzymatic activity of both Pol I and Pol III, called *proofreading* (**Figure 7.20**). In Pol III, a separate protein subunit, DnaQ, performs the proofreading.

How does proofreading work? Pol I and Pol III possess a $3' \rightarrow 5'$ exonuclease (exo means "end") activity that can remove a misinserted nucleotide. Proofreading activity occurs if an incorrect base has been inserted because this creates a mismatch in base pairing. The polymerase senses this because a misinserted nucleotide does not form the correct hydrogen-bonding pattern with its partner base in the complementary strand, and this causes a slight distortion in the double helix. After the removal of a mismatched nucleotide, Pol III gets a second chance to insert the correct nucleotide (Figure 7.20).

Proofreading exonuclease activity is distinct from the $5' \rightarrow 3'$ exonuclease activity of Pol I that is used to remove the RNA primer from both the leading and lagging strands (Figure 7.15). Only Pol I has this latter activity. Exonuclease proofreading occurs in prokaryotes, eukaryotes, and viral DNA replication systems. However, many organisms have additional mechanisms for reducing errors made during DNA replication, and we will discuss some of these in Chapter 11.

Termination of Replication

Eventually the process of DNA replication is finished. How does the replisome know when to stop? On the opposite side of the circular chromosome from the origin is a site called the terminus of replication. Here the two replication forks collide as the new circles of DNA are completed. The details of termination are not completely known. However, in the terminus region there are several DNA sequences called *Ter* sites that are recognized by a protein called Tus, whose function is to block progress of the replication forks. When replication of the circular chromosome is complete (Figure 7.16), the two circular molecules are linked together, much like the links of a

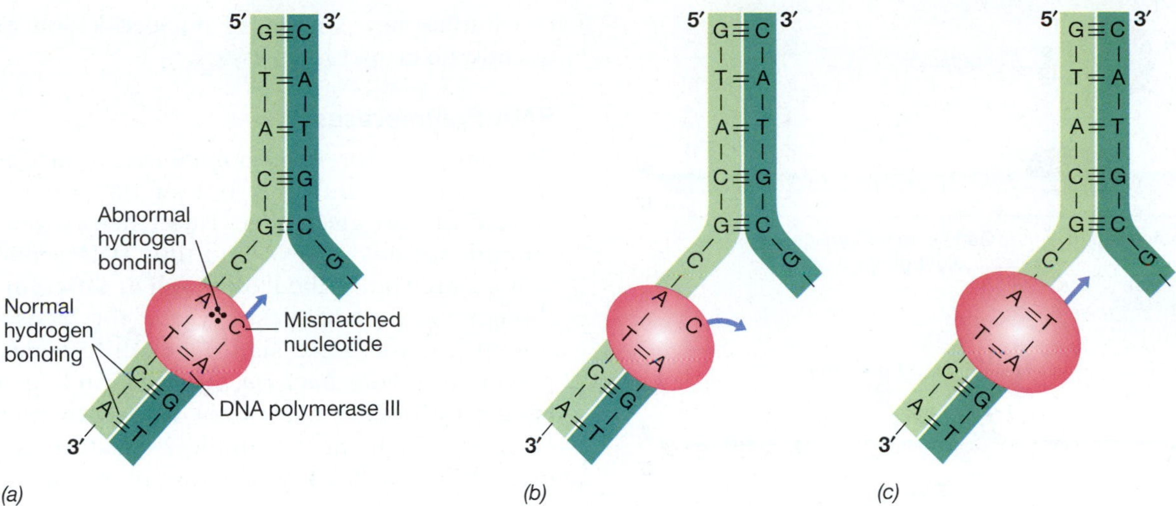

Figure 7.20 Proofreading by the 3′ → 5′ exonuclease activity of DNA polymerase III. A mismatch in base pairing at the terminal base pair *(a)* causes the polymerase to pause briefly. This is a signal for the proofreading activity to excise the mismatched nucleotide *(b)*, after which the correct base is incorporated by the polymerase activity *(c)*.

chain. They are unlinked by another enzyme, topoisomerase IV. In Chapter 6 we saw how cell wall synthesis and DNA replication are coupled with cell division (Section 6.2). Obviously, it is critical that, after DNA replication, the DNA is partitioned so that each daughter cell receives a copy of the chromosome. This process may be assisted by the important cell division protein FtsZ, which helps orchestrate several key events of cell division (Section 6.2).

7.8 MiniReview

Most errors in base pairing that occur during replication are corrected by the proofreading functions of DNA polymerases. Incorrect nucleotides are removed and replaced. Finally, DNA replication terminates when the replication forks meet at a special terminus region on the chromosome.

- How is proofreading carried out during DNA replication?
- What brings the replication forks to a halt in the terminus region of the chromosome?

IV RNA SYNTHESIS: TRANSCRIPTION

Ribonucleic acid (RNA) plays a number of important roles in the cell. There are three key differences in the chemistry of RNA and DNA: (1) RNA contains the sugar ribose instead of deoxyribose; (2) RNA contains the base uracil instead of thymine; and (3) except in certain viruses, RNA is not double-stranded. A change from deoxyribose to ribose affects the chemistry of a nucleic acid; enzymes that act on DNA usually have no effect on RNA, and vice versa. However, the change from thymine to uracil does not affect base pairing, as these two bases pair with adenine equally well.

Three major types of RNA are involved in protein synthesis: *messenger RNA (mRNA)*, *transfer RNA (tRNA)*, and *ribosomal RNA (rRNA)*. Several other types of RNA are also found that are mostly involved in regulation (Chapter 9). These RNA molecules are all products of the transcription of DNA. It should be emphasized that RNA plays a role at two levels, genetic and functional. At the genetic level, mRNA carries the genetic information from the genome to the ribosome. In contrast, rRNA has both a functional and a structural role in ribosomes and tRNA has an active role in carrying amino acids for protein synthesis. Some RNA even has catalytic (enzymatic) activity (ribozymes, Section 8.8). In this section we focus on how RNA is synthesized in species of *Bacteria*, using *Escherichia coli* as our model organism.

7.9 Overview of Transcription

Transcription of genetic information from DNA to RNA is carried out by the enzyme **RNA polymerase**. Like DNA polymerase, RNA polymerase catalyzes the formation of phosphodiester bonds but in this case between ribonucleotides rather than deoxyribonucleotides. RNA polymerase requires DNA as a template. The precursors of RNA are the ribonucleoside triphosphates ATP, GTP, UTP, and CTP.

The mechanism of RNA synthesis is much like that of DNA synthesis (Figure 7.11). That is, during elongation of an RNA chain, ribonucleoside triphosphates are added to the 3′-OH of the ribose of the preceding nucleotide. Polymerization is driven by the release of energy from the two energy-rich phosphate bonds of the incoming ribonucleoside triphosphates. In both DNA replication and RNA transcription the overall direction of chain growth is from the 5′ end to the 3′ end, thus the newly synthesized strand is antiparallel to the template strand. Unlike DNA polymerase, however, RNA polymerase

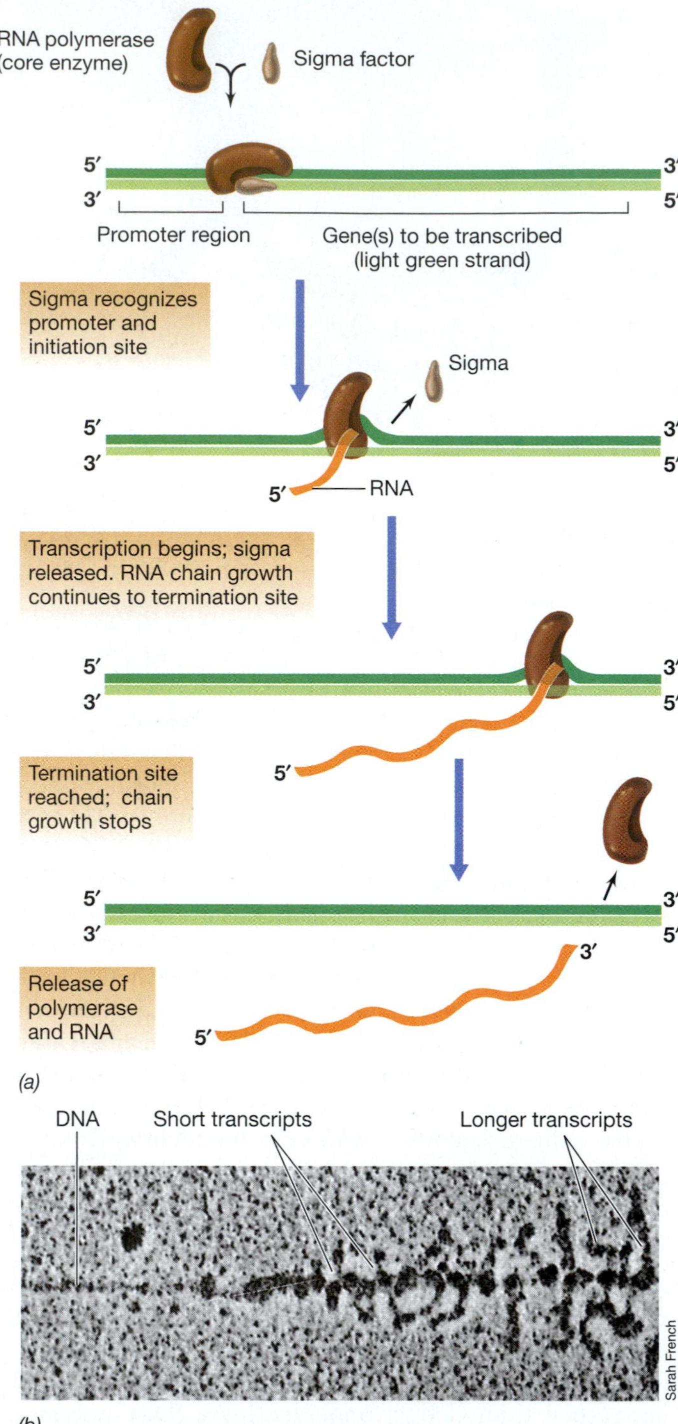

Figure 7.21 Transcription. *(a)* Steps in RNA synthesis. The initiation site (promoter) and termination site are specific nucleotide sequences on the DNA. RNA polymerase moves down the DNA chain, temporarily opening the double helix and transcribing one of the DNA strands. *(b)* Electron micrograph of transcription along a gene on the *Escherichia coli* chromosome. The region of active transcription is about 2 kb pairs of DNA. Transcription is proceeding from left to right, with the shorter transcripts on the left becoming longer as transcription proceeds.

can initiate new strands of nucleotides on its own; consequently, no primer is necessary.

RNA Polymerases

The template for RNA polymerase is a double-stranded DNA molecule, but only one of the two strands is transcribed for any given gene. Nevertheless, genes are present on both strands of DNA and thus DNA sequences on both strands are transcribed, although at different locations. Although these principles are true for transcription in all organisms, there are significant differences among RNA polymerase from *Bacteria, Archaea,* and *Eukarya.* The following discussion deals only with RNA polymerase from *Bacteria,* which has the simplest structure and about which most is known (RNA polymerase in *Archaea* and *Eukarya* is discussed in Chapter 8).

RNA polymerase from *Bacteria* has five different subunits, designated β, β', α, ω (omega) and σ (sigma), with α present in two copies. The β and β' (beta prime) subunits are similar but not identical. The subunits interact to form the active enzyme, called the RNA polymerase holoenzyme, but the sigma factor is not as tightly bound as the others and easily dissociates, leading to the formation of the RNA polymerase core enzyme, $\alpha_2\beta\beta'\omega$. The core enzyme alone synthesizes RNA, whereas the sigma factor recognizes the appropriate site on the DNA for RNA synthesis to begin. The omega subunit is needed for assembly of the core enzyme but is not required for RNA synthesis. RNA synthesis carried out by RNA polymerase and sigma is illustrated in **Figure 7.21**. www.microbiologyplace.com **Online Tutorial 7.2: Transcription**

Promoters

RNA polymerase is a large protein and makes contact with many bases of the DNA simultaneously. Proteins such as RNA polymerase can interact specifically with DNA because portions of the bases are exposed in the major groove (Figure 7.5). However, in order to initiate RNA synthesis correctly, RNA polymerase must first recognize the initiation sites on the DNA. These important sites, called **promoters,** are recognized by the sigma factor (**Figure 7.22**).

Once the RNA polymerase has bound to the promoter, transcription can proceed. In this process, the DNA double helix at the promoter is opened up by the RNA polymerase to form a transcription bubble. As the polymerase moves, it unwinds the DNA in short segments. This transient unwinding exposes the template strand and allows it to be copied into the RNA complement. Thus, promoters can be thought of as pointing RNA polymerase in one direction or the other along the DNA. If a region of DNA has two nearby promoters pointing in opposite directions, then transcription from one of the promoters will proceed in one direction (on one of the strands) while transcription from the other promoter will proceed in the opposite direction (on the other strand).

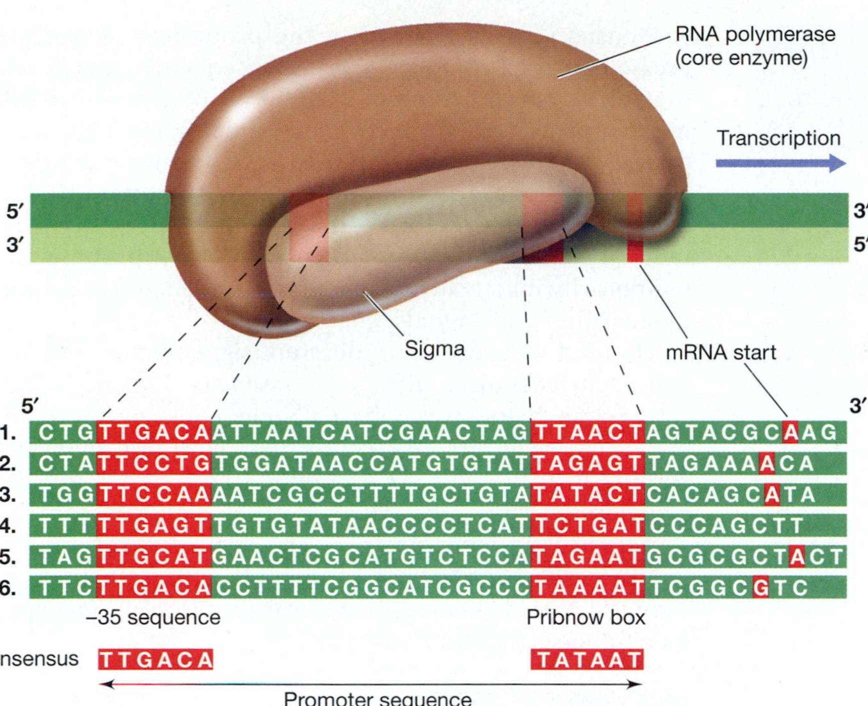

Figure 7.22 The interaction of RNA polymerase with the promoter. Shown below the RNA polymerase and DNA are six different promoter sequences identified in *Escherichia coli*, a species of *Bacteria*. The contacts of the RNA polymerase with the −35 sequence and the Pribnow box (−10 sequence) are shown. Transcription begins at a unique base just downstream from the Pribnow box. Below the actual sequences at the −35 and Pribnow box regions are consensus sequences derived from comparing many promoters. Note that although sigma recognizes the promoter sequences on the 5′ → 3′ (dark green) strand of DNA, the RNA polymerase core enzyme will actually transcribe the light green strand running 3′ → 5′ because core enzyme works only in a 5′ → 3′ direction.

Once a short stretch of RNA has been formed, the sigma factor dissociates. Elongation of the RNA molecule is then carried out by the core enzyme alone (Figure 7.21). Sigma is only involved in forming the initial RNA polymerase–DNA complex at the promoter. As the newly made RNA dissociates from the DNA, the opened DNA closes back into the original double helix. Transcription stops at specific sites called transcription terminators (Section 7.11).

Unlike DNA replication, which copies entire genomes, transcription involves much smaller units of DNA, often as little as a single gene. This system allows the cell to transcribe different genes at different frequencies, depending on the needs of the cell for different proteins. In other words, gene expression is regulated. As we shall see in Chapter 9, regulation of transcription is an important and elaborate process that uses many different mechanisms and is very efficient at controlling gene expression and conserving cell resources.

7.9 MiniReview

The three major types of RNA are messenger RNA (mRNA), transfer RNA (tRNA), and ribosomal RNA (rRNA). Transcription of RNA from DNA is due to the enzyme RNA polymerase, which adds nucleotides onto 3′ ends of growing chains. Unlike DNA polymerase, RNA polymerase needs no primer and recognizes a specific start site on the DNA called the promoter.

- In which direction (5′ → 3′ or 3′ → 5′) along the template strand does transcription occur?
- What is a promoter? What protein recognizes the promoters in *Escherichia coli*?

7.10 Sigma Factors and Consensus Sequences

As we have noted, the promoter plays a central role in transcription. Promoters are specific DNA sequences to which RNA polymerase binds. The sequences of a large number of promoters from many organisms have been determined, and Figure 7.22 shows the sequence of a few promoters from *Escherichia coli*.

All the DNA sequences in Figure 7.22 are recognized by the same sigma factor, the major sigma factor in *E. coli*, called σ^{70} (the superscript 70 indicates the size of this protein, 70 kilodaltons). If you compare the promoter DNA sequences, you will see that they are not identical. However, two shorter sequences within the promoter region are highly conserved between promoters, and it is these that sigma recognizes. Both sequences precede (are upstream of) the site where transcription starts. One is a region 10 bases before the start of transcription, the −10 region, called the *Pribnow box*. Notice that although each promoter is slightly different, most bases are the same within the −10 region. When comparing the −10 regions of all the promoters recognized by this sigma factor to determine which base occurs most often at each position, one arrives at the consensus sequence: TATAAT. In our example, each promoter has from three to five matches for these bases. The second region of conserved sequence in the promoter is about 35 bases from the start of transcription. The consensus sequence in the −35 region is TTGACA (Figure 7.22). Once again, most of the sequences are not exactly the same as the consensus sequence, but are very close.

Table 7.3 Sigma factors in *Escherichia coli*

Name[a]	Upstream consensus recognition sequence[b]	Function
σ^{70} RpoD	TTGACA	For most genes, major sigma factor during normal growth
σ^{54} RpoN	TTGGCACA	Nitrogen assimilation
σ^{38} RpoS	CCGGCG	Major sigma factor during stationary phase, also for genes involved in oxidative and osmotic responses
σ^{32} RpoH	TNTCNCCTTGAA[c]	Heat shock response
σ^{28} FliA	TAAA	For genes involved in flagella synthesis
σ^{24} RpoE	GAACTT	Response to misfolded proteins in periplasm
σ^{19} FecI	AAGGAAAAT	For certain genes in iron transport

[a]Superscript number in name indicates size of protein in kilodaltons. Many factors also have other names, for example, σ^{70} is also called σ^{D}.

[b]For a discussion of consensus sequences, see Sections 7.10 and 7.11 and Figure 7.30.

[c]N = any nucleotide.

Note that in Figure 7.22 the promoter sequence is shown for only one strand of the DNA. By convention, the strand shown is the one oriented with its 5′ end upstream (therefore, it is not the strand used as the template by RNA polymerase). Showing the sequence of only one strand is simply "shorthand" for writing DNA sequences. In reality, DNA is double-stranded, and RNA polymerase binds to double-stranded DNA and then unwinds it. A single strand of the unwound DNA, the transcribed strand is then used as template by the core RNA polymerase. Although it binds to both DNA strands, sigma recognizes the specific sequences in the −10 and −35 regions on the nontranscribed strand with which it makes most of its contacts.

Some sigma factors in other bacteria are much more specific as regards binding sequences than the example shown in Figure 7.22. In such cases, very little leeway is allowed in the critical bases that are recognized. In *E. coli*, promoters that are most like the consensus sequence are usually more effective in binding RNA polymerase. These more effective promoters are called strong promoters and are very useful in genetic engineering, as is discussed in Chapter 12.

Alternative Sigma Factors in *Escherichia coli*

Most genes in *E. coli* require the standard sigma factor, known as σ^{70}, for transcription and have promoters like those shown in Figure 7.22. However, several alternative sigma factors are known that recognize different consensus sequences (**Table 7.3**). Each of these alternative sigma factors is specific for a group of genes required under special circumstances. Thus σ^{38}, also known as RpoS, recognizes a consensus sequence found in the promoters of genes expressed during stationary phase. Consequently, it is possible to control the expression of each family of genes by regulating the level of the corresponding sigma factor. This may be done by changing either the rate of synthesis or the rate of degradation of the sigma factor. In addition, the activity of alternative sigma factors can be modulated by other proteins called *anti-sigma factors*. These proteins can temporarily inactivate a particular sigma factor in response to environmental signals.

In total there are seven different sigma factors in *E. coli*, and each recognizes different consensus sequences (Table 7.3). Sigma factors were originally named according to their molecular weight. More recently, they have been named according to their roles, for example, RpoN stands for "*R*NA *po*lymerase—*N*itrogen." Most of these sigma factors have counterparts in other *Bacteria*. The endospore-forming bacterium *Bacillus subtilis* has 14 sigma factors, with 4 different sigma factors dedicated to the transcription of endospore-specific genes (Section 4.12).

7.10 MiniReview

In *Bacteria*, promoters are recognized by the sigma subunit of RNA polymerase. Regions of DNA recognized by a particular DNA-binding protein have very similar sequences. Alternative sigma factors allow joint regulation of large families of genes in response to growth conditions.

- What is a consensus sequence?
- To what parts of the promoter region does sigma bind?
- How may families of genes required during specialized conditions be controlled as a group using sigma factors?

7.11 Termination of Transcription

Only those genes that need to be expressed should be transcribed. Therefore it is important to terminate transcription at the correct position. **Termination** of RNA synthesis is governed by specific base sequences on the DNA. In *Bacteria* a common termination signal on the DNA is a GC-rich sequence containing an inverted repeat with a central nonrepeating segment (see Section 7.2 and Figure 7.6 for an explanation of inverted repeats). When such a DNA sequence is transcribed, the RNA forms a stem-loop structure by intrastrand base pairing (**Figure 7.23**). Such stem-loop structures, followed by a run of adenosines in the DNA template and therefore a run of uridines in the mRNA, are effective transcription terminators. This is due to the formation of a stretch of U:A base pairs that holds the RNA and DNA template together. This structure is very weak as U:A base pairs have only two hydrogen bonds each. The RNA polymerase pauses at the stem-loop, and the

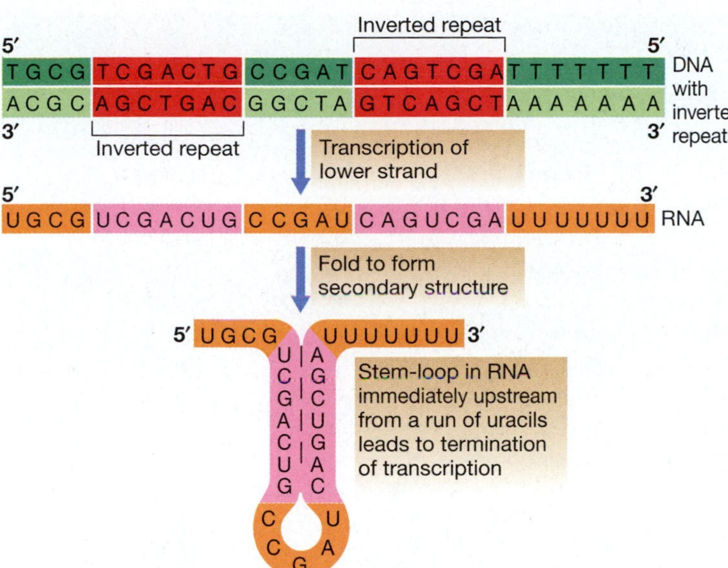

Figure 7.23 Inverted repeats and transcription termination. Inverted repeats in transcribed DNA form a stem-loop structure in the RNA that terminates transcription when followed by a run of uracils.

DNA and RNA come apart at the run of uridines. This terminates transcription. Sequence patterns that terminate transcription without addition of any extra factors are referred to as intrinsic terminators.

Rho-Dependent Termination

The other mechanism for transcription termination uses a specific protein factor, known in *Escherichia coli* as Rho. Rho does not bind to RNA polymerase or to the DNA, but binds tightly to RNA and moves down the chain toward the RNA polymerase–DNA complex. Once RNA polymerase has paused at a Rho-dependent termination site (a specific sequence on the DNA template), Rho causes both the RNA and RNA polymerase to be released from the DNA, thus terminating transcription. Although the termination sequences function at the level of RNA, remember that RNA is transcribed from DNA. Consequently, transcription termination is ultimately determined by specific nucleotide sequences on the DNA.

7.11 MiniReview

RNA polymerase stops transcription at specific sites called transcription terminators. Although encoded by DNA, these terminators function at the level of RNA. Some are intrinsic terminators and require no accessory proteins beyond RNA polymerase itself. In *Bacteria* these sequences are usually stem-loops followed by a run of uridines. Other terminators require proteins such as Rho.

- What is an intrinsic terminator?
- What is a stem-loop structure?

7.12 The Unit of Transcription

Chromosomes are organized into units that are bounded by sites where transcription of DNA into RNA is initiated and terminated: units of transcription. One might assume that each unit of transcription includes only a single gene. Although this is true in some instances, it is not always the case. Some units of transcription contain two or more genes. These genes are then said to be cotranscribed, yielding a single RNA molecule.

Ribosomal and Transfer RNAs and RNA Longevity

As we learned in Section 7.1, most genes encode proteins, but others encode RNAs that are not translated, such as ribosomal RNA (rRNA) and transfer RNA (tRNA). There are several different types of rRNA in an organism (with a ribosome having one copy of each type; Section 7.13). Prokaryotes have three types: 16S rRNA, 23S rRNA, and 5S rRNA (we discussed the importance of 16S rRNA in evolutionary studies of prokaryotes in Chapter 2). As shown in **Figure 7.24**, there are clusters containing one gene for each of these rRNAs, and the genes in such a cluster are cotranscribed. The situation is similar in eukaryotes. Therefore, in all organisms the unit of transcription for most rRNA is longer than a single gene. In prokaryotes tRNA genes are often cotranscribed with each other or even, as shown in Figure 7.24, with genes for rRNA. However, these cotranscribed transcripts must be processed (cleaved into individual units) to yield mature (functional) rRNAs or tRNAs (Figure 7.24). RNA processing is rare in

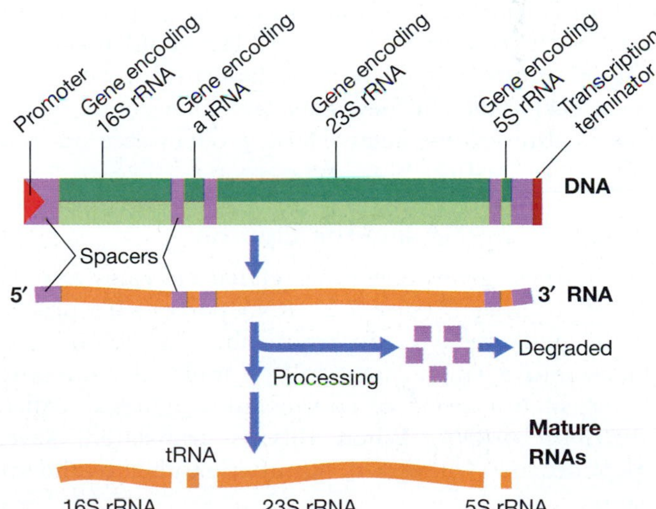

Figure 7.24 A ribosomal rRNA transcription unit from *Bacteria* and its subsequent processing. In *Bacteria* all rRNA transcription units have the genes in the order 16S rRNA, 23S rRNA, and 5S rRNA (shown approximately to scale). Note that in this particular transcription unit the "spacer" between the 16S and 23S rRNA genes contains a tRNA gene. In other transcription units this region may contain more than one tRNA gene. Often one or more tRNA genes also follow the 5S rRNA gene and are cotranscribed. *Escherichia coli* contains seven rRNA transcription units.

Table 7.4 The genetic code as expressed by triplet base sequences of mRNA

Codon	Amino acid	Codon	Amino acid	Codon	Amino acid	Codon	Amino acid
UUU	Phenylalanine	UCU	Serine	UAU	Tyrosine	UGU	Cysteine
UUC	Phenylalanine	UCC	Serine	UAC	Tyrosine	UGC	Cysteine
UUA	Leucine	UCA	Serine	UAA	None (stop signal)	UGA	None (stop signal)
UUG	Leucine	UCG	Serine	UAG	None (stop signal)	UGG	Tryptophan
CUU	Leucine	CCU	Proline	CAU	Histidine	CGU	Arginine
CUC	Leucine	CCC	Proline	CAC	Histidine	CGC	Arginine
CUA	Leucine	CCA	Proline	CAA	Glutamine	CGA	Arginine
CUG	Leucine	CCG	Proline	CAG	Glutamine	CGG	Arginine
AUU	Isoleucine	ACU	Threonine	AAU	Asparagine	AGU	Serine
AUC	Isoleucine	ACC	Threonine	AAC	Asparagine	AGC	Serine
AUA	Isoleucine	ACA	Threonine	AAA	Lysine	AGA	Arginine
AUG (start)[a]	Methionine	ACG	Threonine	AAG	Lysine	AGG	Arginine
GUU	Valine	GCU	Alanine	GAU	Aspartic acid	GGU	Glycine
GUC	Valine	GCC	Alanine	GAC	Aspartic acid	GGC	Glycine
GUA	Valine	GCA	Alanine	GAA	Glutamic acid	GGA	Glycine
GUG	Valine	GCG	Alanine	GAG	Glutamic acid	GGG	Glycine

[a]AUG encodes *N*-formylmethionine at the beginning of polypeptide chains of *Bacteria*.

prokaryotes but common in eukaryotes, as we will see later (Chapter 8).

In prokaryotes, most messenger RNAs have a short half-life (on the order of a few minutes), after which they are degraded by cellular ribonucleases. This is in contrast to rRNA and tRNA, which are stable RNAs. The stability occurs because tRNAs and rRNAs form highly folded structures that prevent them from being degraded by ribonucleases. By contrast, normal mRNA does not form such structures and is susceptible to ribonuclease attack. The rapid turnover of prokaryotic mRNAs is likely a mechanism that permits the cell to quickly adapt to new environmental conditions and halt translation of messages whose products are no longer needed.

Polycistronic mRNA and the Operon

In prokaryotes, genes encoding related enzymes are often clustered together (Figure 7.2). RNA polymerase proceeds through such clusters and transcribes the whole group of genes into a single, long mRNA molecule. An mRNA encoding such a group of cotranscribed genes is called a *polycistronic mRNA*. When this is translated, several polypeptides are synthesized, one after another, by the same ribosome.

A group of related genes that are transcribed together to give a single polycistronic mRNA is known as an **operon**. Assembling genes for the same biochemical pathway or genes needed under the same conditions into an operon allows their expression to be regulated in a coordinated manner. Despite this, eukaryotes do not have operons and polycistronic mRNA (Chapter 8). Often, transcription of an operon is controlled by a specific region of the DNA found just upstream of the protein-coding region of the operon. This is considered in more detail in Chapter 9.

7.12 MiniReview

The unit of transcription in prokaryotes often contains more than a single gene. Several genes are then transcribed into a single mRNA molecule that contains information for more than one polypeptide. A cluster of genes that are transcribed together from a single promoter constitute an operon. In all organisms, genes encoding rRNA are cotranscribed but then processed to form the final rRNA species.

- What is messenger RNA (mRNA)?
- What is a polycistronic mRNA?
- What are operons and why are they useful to prokaryotes?

V PROTEIN SYNTHESIS

In the first two steps in biological information transfer, replication and transcription, nucleic acids are synthesized on nucleic acid templates. In the last step, **translation**, there is synthesis on a nucleic acid template, but in this case the final product is a *protein* rather than a *nucleic acid.*

7.13 The Genetic Code

Before we describe the mechanism of translation, we need to discuss the heart of biological information transfer: the

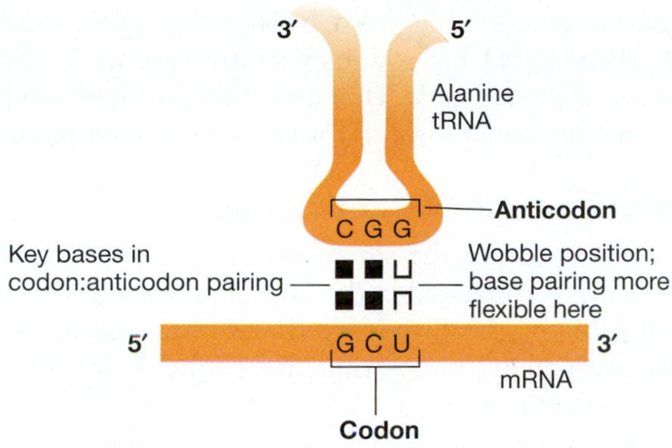

Figure 7.25 The wobble concept. Base pairing is more flexible for the third base of the codon than for the first two. Only a portion of the tRNA is shown (the structure of the whole tRNA is shown in Figure 7.27).

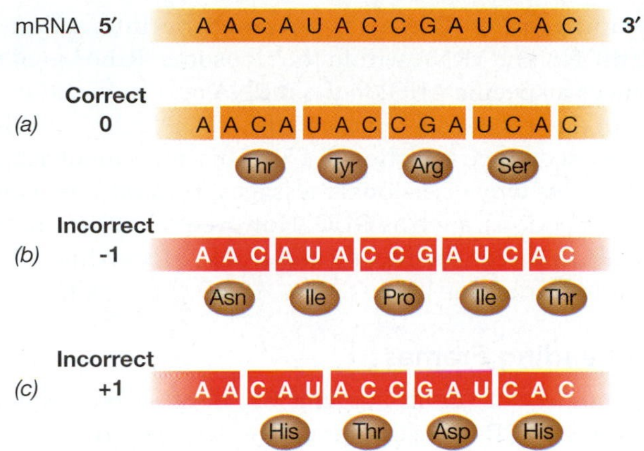

Figure 7.26 Possible reading frames in an mRNA. An interior sequence of an mRNA is shown. *(a)* The amino acids that would be encoded if the ribosome is in the correct reading frame (designated the "0" frame). *(b)* The amino acids that would be encoded by this region of the mRNA if the ribosome were in the −1 reading frame. *(c)* The amino acids that would be encoded if the ribosome were in the +1 reading frame.

correspondence between the nucleic acid template and the amino acid sequence of the polypeptide product. This is known as the **genetic code**. As we mentioned in Section 7.1, a triplet of three bases called a *codon* encodes a specific amino acid. The genetic code is written as mRNA rather than as DNA because it is mRNA that is translated. The 64 possible codons (four bases taken three at a time, 4^3) of mRNA are shown in **Table 7.4**. Note that in addition to the codons specifying the various amino acids, there are also specific codons for starting and stopping translation.

Properties of the Genetic Code

One of the most interesting features of the genetic code is that many amino acids are encoded by more than one codon. In these cases there is no one-to-one correspondence between the amino acid and the codon. Consequently, knowing the amino acid at a given location does not mean that the codon at that location is automatically known. A code such as this in which there is no one-to-one correspondence between word and code is called a degenerate code. The converse is true, however. Knowing the DNA sequence and the correct reading frame, one can specify the amino acid in the protein. This permits the determination of amino acid sequences from DNA base sequences and is at the heart of genomics (Chapter 13). Interestingly, in most cases where multiple codons encode the same amino acid, the multiple codons are closely related in RNA sequence (Table 7.4).

A codon is recognized by specific base-pairing with a complementary sequence of three bases called the **anticodon**, a sequence found on tRNAs. If this base-pairing were always the standard pairing of A with U and G with C, then at least one specific tRNA would exist for each codon. In some cases, this is true. For instance, there are six different tRNAs in *Escherichia coli* that carry the amino acid leucine, one for each codon (Table 7.4). By contrast, some tRNAs can recognize more than one codon. For instance, although there are two lysine codons in *E. coli*, there is only one lysyl tRNA whose anticodon can base-pair with either AAA or AAG (see Table 7.4). In these special cases, tRNA molecules form standard base pairs at only the first two positions of the codon while tolerating irregular base pairing at the third position. This phenomenon is called **wobble** and is illustrated in **Figure 7.25** where it can be seen that pairing between G and U (rather than G with C) is allowed at the wobble position.

Stop and Start Codons

A few codons do not encode an amino acid (Table 7.4). These codons (UAA, UAG, and UGA) are the **stop codons**, and they signal the termination of translation of a protein-coding sequence on the mRNA (Section 7.15). Stop codons are also called **nonsense codons**, because they interrupt the "sense" of the growing polypeptide when they terminate translation.

Messenger RNA is translated beginning with the **start codon (AUG)**, which encodes a chemically modified methionine, *N*-formylmethionine. Although AUG at the beginning of a coding region encodes *N*-formylmethionine, AUG within the coding region encodes methionine. Two different tRNAs are involved in this process (see below). With a triplet code it is critical for translation to begin at the correct nucleotide. If it does not, the whole reading frame of the mRNA will be shifted and thus an entirely different protein will be made. If the shift introduces a stop codon into the reading frame, the protein will terminate prematurely. By convention the reading frame that is translated to give the protein encoded by the gene is called the 0 frame. As can be seen in **Figure 7.26**, the other two possible reading frames (−1 and +1) do not encode the same amino acid sequence. Therefore it is essential that the ribosome finds the correct start codon to begin translation and, once it has, that it moves down the mRNA exactly three bases at a time. How is the correct reading frame ensured?

Reading frame fidelity is governed by interactions between mRNA and rRNA within the ribosome. Ribosomal RNA recognizes a specific AUG on the mRNA as a start codon with the aid of an upstream sequence in the mRNA called the Shine–Dalgarno sequence (Section 7.15). This alignment requirement explains why occasional messages from *Bacteria* can use other start codons, such as GUG. However, even these unusual start codons direct the incorporation of *N*-formylmethionine as the initiator amino acid.

Open Reading Frames

The genomes of many organisms have been sequenced (Chapter 13). But these mountains of sequence data would be useless unless scientists can determine the location of protein-encoding genes. One common method of identifying protein encoding genes is to examine each strand of the DNA sequence for **open reading frames (ORF)**. Remember that mRNA is transcribed from DNA, so that if one knows the sequence of DNA, one also knows the sequence of RNA that is transcribed from it. If an RNA can be translated, it contains an open reading frame: a start codon (typically AUG) followed by a number of codons and then a stop codon in the same reading frame as the start codon. In practice, only ORFs long enough to encode a protein of realistic length are accepted as true coding sequences. Although most functional proteins are at least 100 amino acids in length, a few protein hormones and regulatory peptides are much shorter. Consequently, it is not always possible to tell from sequence data alone whether a relatively short ORF is merely due to chance or encodes a genuine, albeit short, protein.

A computer can be programmed using the above guidelines to scan long DNA base sequences to look for open reading frames. In addition to looking for start and stop codons, the search may include promoters and Shine–Dalgarno ribosome-binding sequences as well. The search for ORFs is very important in genomics (Chapter 13). If an unknown piece of DNA has been isolated and sequenced, the presence of an open reading frame indicates that it can encode protein.

Codon Bias

Several amino acids are encoded by multiple codons. One might assume that such multiple codons would be used at equal frequencies. However, this is not the case, and sequence data show major **codon bias**. In other words, some codons are greatly preferred over others even though they encode the same amino acid. Moreover, this bias is organism specific. In *E. coli*, for instance, only about 1 out of 20 isoleucine residues in proteins is encoded by the isoleucine codon AUA, the other 19 being encoded by the other isoleucine codons, AUU and AUC (Table 7.4). Codon bias is correlated with a corresponding bias in the concentration of different tRNA molecules. Thus a tRNA corresponding to a rarely used codon will be in relatively short supply.

The origin of codon bias is unclear, but it is easily recognized and may be taken into account in practical uses of gene sequence information. For example, a gene from one organism whose codon usage differs dramatically from that of another may not be translated well if the gene is cloned into the latter using genetic engineering (Chapter 12). This is due to a shortage of the tRNA for codons that are rare in the host but frequent in the cloned gene. However, this problem can be corrected or at least compensated for by genetic manipulation.

Modifications to the Genetic Code

All cells appear to use the same genetic code. Therefore, the genetic code is a universal code. However, this view has been tempered a bit by the discovery that some organelles and a few cells use genetic codes that are slight variations of the "universal" genetic code.

Alternative genetic codes were first discovered in the genomes of animal mitochondria. These modified codes typically use nonsense codons as sense codons. For example, animal (but not plant) mitochondria use the codon UGA to encode tryptophan instead of using it as a stop codon (Table 7.4). Several organisms are known that also use slightly different genetic codes. For example, in the genus *Mycoplasma (Bacteria)* and the genus *Paramecium (Eukarya)*, certain nonsense codons encode amino acids. These organisms simply have fewer nonsense codons because one or two of them are used as sense codons. In a few rare cases, nonsense codons encode unusual amino acids rather than one of the 20 common amino acids (see below).

7.13 MiniReview

The genetic code is expressed as RNA, and a single amino acid may be encoded by several different but related codons. In addition to the nonsense codons, there is also a specific start codon that signals where the translation process should begin.

- Why is it important for the ribosome to read "in frame"?
- Describe an open reading frame. If you were given a nucleotide sequence, how would you find ORFs?

7.14 Transfer RNA

Recall from Section 7.13 that it is the anticodon portion of the tRNA that base-pairs with the codon. However, a tRNA is much more than simply an anticodon (**Figure 7.27**). A tRNA is specific for both a codon and its cognate amino acid (that is, the amino acid corresponding to the anticodon of the tRNA). The tRNA and its specific amino acid are brought together by specific enzymes that ensure that a particular tRNA receives its correct amino acid. These enzymes, called **aminoacyl-tRNA synthetases**, have the important function of recognizing both the cognate amino acid and the specific tRNA for that amino acid.

General Structure of tRNA

There are about 60 different tRNAs in bacterial cells and 100–110 in mammalian cells. Transfer RNA molecules are short, single-stranded molecules that contain extensive secondary structure and have lengths of 73–93 nucleotides.

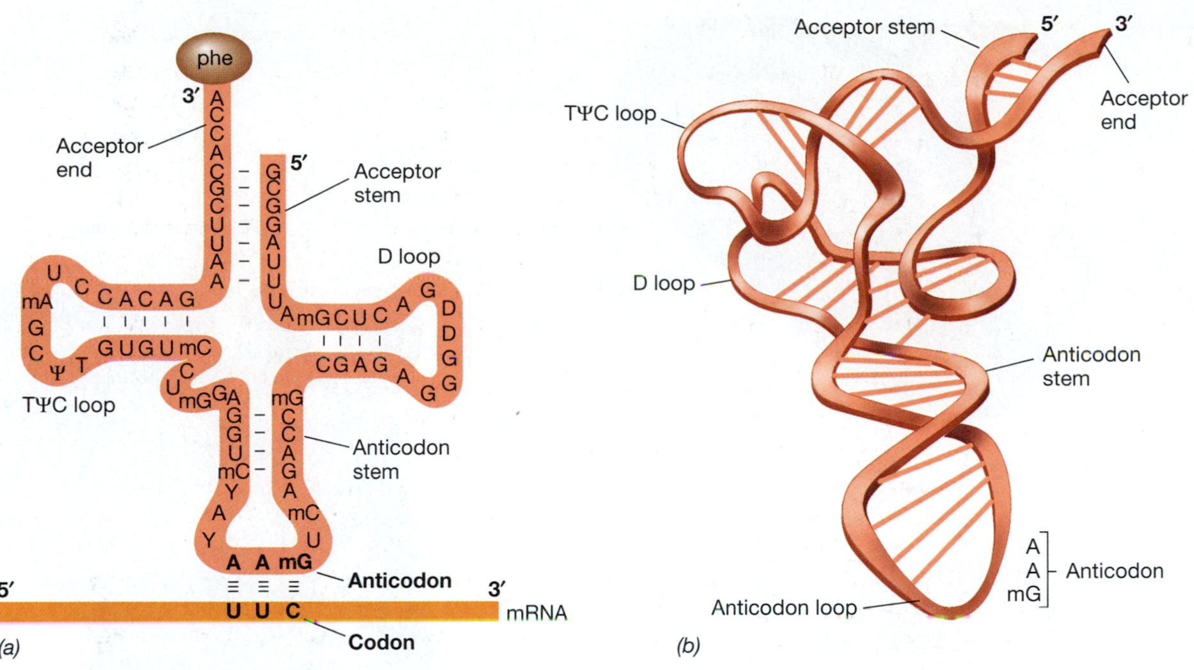

Figure 7.27 Structure of a transfer RNA. *(a)* The conventional cloverleaf structure of yeast phenylalanine tRNA. The amino acid is attached to the ribose of the terminal A at the acceptor end. A, adenine; C, cytosine; U, uracil; G, guanine; T, thymine; ψ, pseudouracil; D, dihydrouracil; m, methyl; Y, a modified purine. *(b)* In fact, the tRNA molecule folds so that the D loop and TψC loops are close together and associate by hydrophobic interactions.

Certain bases and secondary structures are constant for all tRNAs, whereas other parts are variable. Transfer RNA molecules also contain some purine and pyrimidine bases that differ slightly from the normal bases found in RNA because they are chemically modified. These modifications are made to the bases after transcription. Some of these unusual bases are pseudouridine, inosine, dihydrouridine, ribothymidine, methyl guanosine, dimethyl guanosine, and methyl inosine. The final mature and active tRNA not only contains unusual bases but also extensive double-stranded regions within the molecule. This secondary structure forms by internal base-pairing when the single-stranded molecule folds back on itself (Figure 7.27).

The structure of a tRNA can be drawn in a cloverleaf fashion, as in Figure 7.27*a*. Some regions of tRNA secondary structure are named after the bases most often found there (the TψC and D loops) or after their specific functions (anticodon loop and acceptor stem). The three-dimensional structure of a tRNA is shown in Figure 7.27*b*. Note that bases that appear widely separated in the cloverleaf model are actually closer together when viewed in three dimensions. This allows some of the bases in the loops to pair with bases in other loops.

The Anticodon and Amino Acid-Binding Site

One of the key variable parts of the tRNA molecule is the anticodon, the group of three bases that recognizes the codon on the mRNA. The anticodon is found in the anticodon loop (Figure 7.27). The three nucleotides of the anticodon recognize the codon by specifically pairing with its three bases (Section 7.13 and Figure 7.25). By contrast, other portions of the tRNA interact with both the rRNA and protein components of the ribosome, nonribosomal translation proteins, and the aminoacyl synthetase enzyme.

At the 3′-end, or acceptor stem, of all tRNAs, are three unpaired nucleotides. The sequence of these three nucleotides is always cytosine-cytosine-adenine (CCA), and they are absolutely essential for function. Curiously, in most organisms these three nucleotides are not encoded by the tRNA genes on the chromosome. Instead they are added, one after another, by an enzyme called the CCA-adding enzyme, using CTP and ATP as substrates. The cognate amino acid is covalently attached to the terminal adenosine of the CCA end by an ester linkage to the ribose sugar. As we shall see, from this location on the tRNA, the amino acid is incorporated into the growing polypeptide chain on the ribosome by a mechanism described in the next section.

Recognition, Activation, and Charging of tRNAs

Recognition of the correct tRNA by an aminoacyl-tRNA synthetase involves specific contacts between key regions of the tRNA and the synthetase (**Figure 7.28**). As might be expected because of its unique sequence, the anticodon of the tRNA is important in recognition by the synthetase. However, other contact sites between the tRNA and the synthetase are also important. Studies of tRNA binding to aminoacyl-tRNA synthetases, in which specific bases in the tRNA have been changed by mutation, have shown that only a small number of key nucleotides in a tRNA in addition to the anticodon region, are involved in recognition. These other key recognition

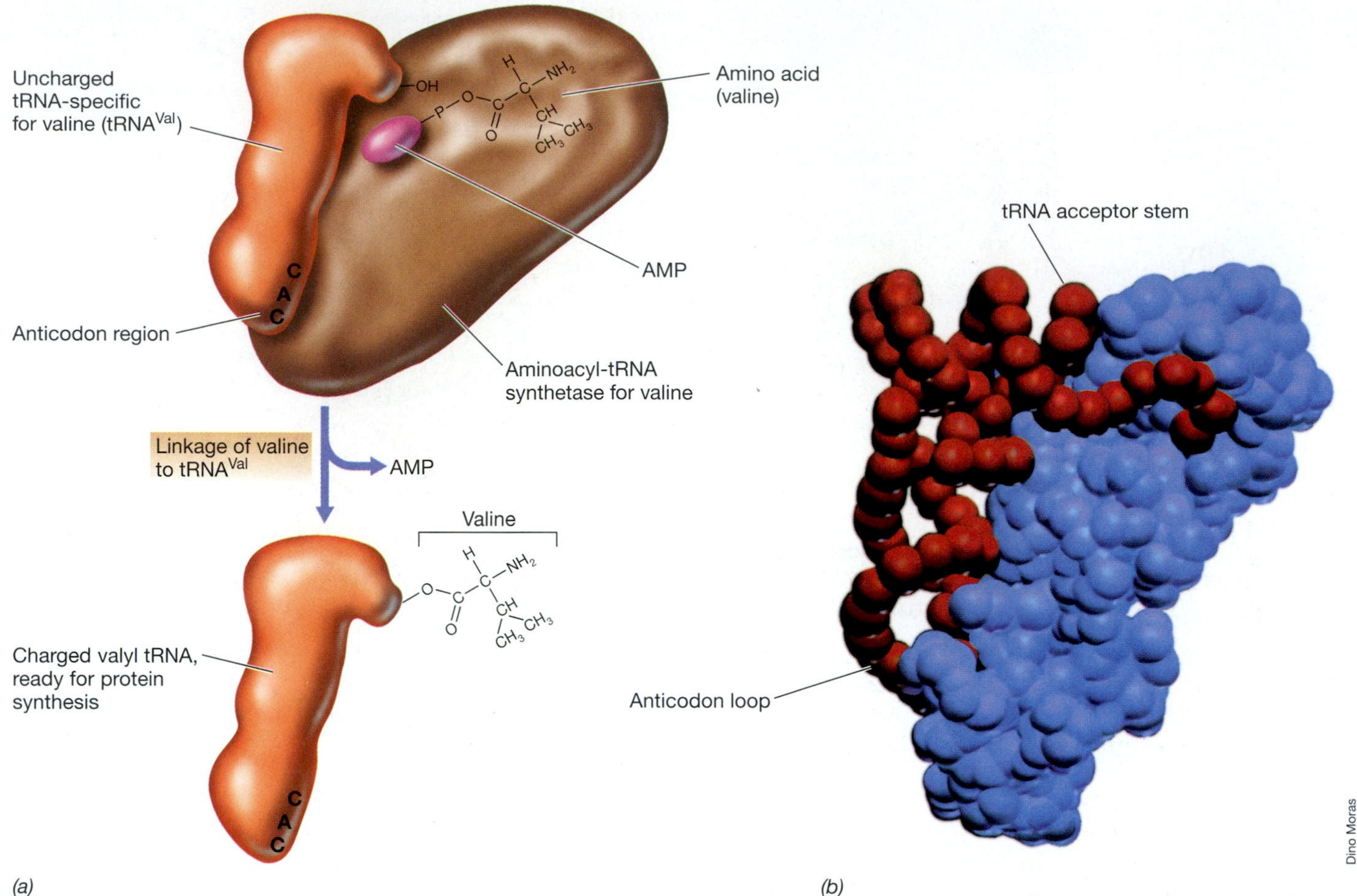

Figure 7.28 Aminoacyl-tRNA synthetase. *(a)* Mode of activity of an aminoacyl-tRNA synthetase. Recognition of the correct tRNA by a particular synthetase involves contacts between specific nucleic acid sequences in the D-loop and acceptor stem of the tRNA and specific amino acids of the synthetase. In this diagram, valyl-tRNA synthetase is shown catalyzing the final step of the reaction, where the valine in valyl-AMP is transferred to tRNA. *(b)* A computer model showing the interaction of glutaminyl-tRNA synthetase (blue) with its tRNA (red). Reprinted with permission from M. Ruff et al. 1991. *Science* 252: 1682–1689. © 1991, AAAS.

nucleotides are often part of the acceptor stem or D-loop of the tRNA molecule (Figure 7.27). It should be emphasized that the fidelity of this recognition process is crucial, for if the wrong amino acid is attached to the tRNA, it will be inserted into the growing polypeptide, likely leading to the synthesis of a faulty protein.

The specific reaction between amino acid and tRNA catalyzed by the aminoacyl-tRNA synthetase begins with activation of the amino acid by reaction with ATP:

$$\text{Amino acid} + \text{ATP} \longleftrightarrow \text{aminoacyl—AMP} + \text{P—P}$$

The aminoacyl-AMP intermediate formed normally remains bound to the enzyme until collision with the appropriate tRNA molecule. Then, as shown in Figure 7.28*a*, the activated amino acid is attached to the tRNA to form a charged tRNA:

$$\text{Aminoacyl—AMP} + \text{tRNA} \longleftrightarrow \text{aminoacyl—tRNA} + \text{AMP}$$

The pyrophosphate (PP_i) formed in the first reaction is split by a pyrophosphatase, giving two molecules of inorganic phosphate. Because ATP is used and AMP is formed in these reactions, a total of two energy-rich phosphate bonds are needed to charge a tRNA with its cognate amino acid. After activation and charging, the aminoacyl-tRNA leaves the synthetase and travels to the ribosome where the polypeptide is synthesized.

7.14 MiniReview

One or more tRNAs exist for each amino acid incorporated into proteins by the ribosome. Enzymes called aminoacyl-tRNA synthetases attach amino acids to their cognate tRNAs. Once the correct amino acid is attached to its tRNA, further specificity resides primarily in the codon–anticodon interaction.

- What is the function of the anticodon of a tRNA?
- What is the function of the acceptor stem of a tRNA?

7.15 Translation: The Process of Protein Synthesis

We learned in Chapter 3 that it is the amino acid sequence that determines the structure, and hence the function, of a protein. Thus, it is vital for proper functioning of proteins that the correct amino acids are inserted at the proper locations in the polypeptide chain. This is the task of the protein-synthesizing machinery, the ribosome.

Ribosomes

Ribosomes are the sites of protein synthesis. A cell may have many thousand ribosomes, the number being positively correlated with growth rate. Each ribosome is constructed of two subunits (**Figure 7.29*a***). Prokaryotes possess 30S and 50S ribosome subunits, yielding intact 70S ribosomes. The S-values are Svedberg units, which refer to the sedimentation coefficients of ribosome subunits (30S and 50S) or intact ribosomes (70S) when subjected to centrifugal force in an ultracentrifuge. (Although larger particles have larger S-values, the relationship is not linear and S-values cannot be added together.)

Each ribosomal subunit is a ribonucleoprotein complex made up of specific ribosomal RNAs and ribosomal proteins. The 30S subunit contains 16S rRNA and 21 proteins, and the 50S subunit contains 5S and 23S rRNA and 31 proteins. Thus, in *Escherichia coli,* there are 52 distinct ribosomal proteins, most present at one copy per ribosome. The ribosome is a dynamic structure whose subunits alternately associate and dissociate and also interact with many other proteins. There are several proteins that are essential for ribosome function and interact with the ribosome at various stages of translation (discussed below), but that are not considered "ribosomal proteins," per se.

Steps in Protein Synthesis

Protein is synthesized in a cycle in which the various ribosomal components play specific roles. Although protein synthesis is a continuous process, it can be broken down into a number of steps: initiation, elongation, and termination. In addition to mRNA, tRNA, and ribosomes, the process requires a number of proteins designated initiation, elongation, and termination factors. The energy-rich compound guanosine triphosphate (GTP) provides the necessary energy for the process. The key steps in protein synthesis are shown in Figure 7.29*b*.

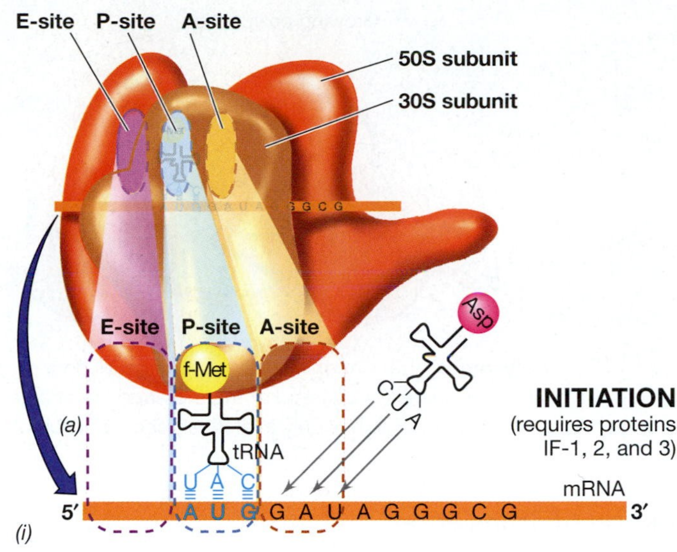

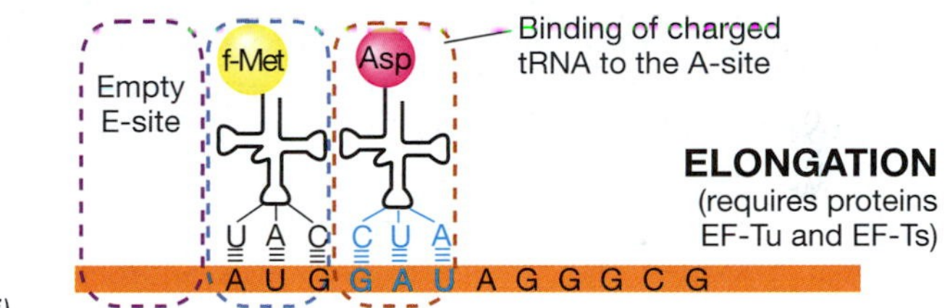

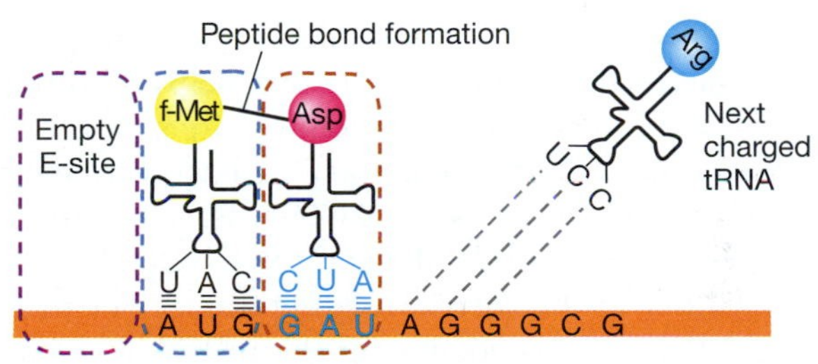

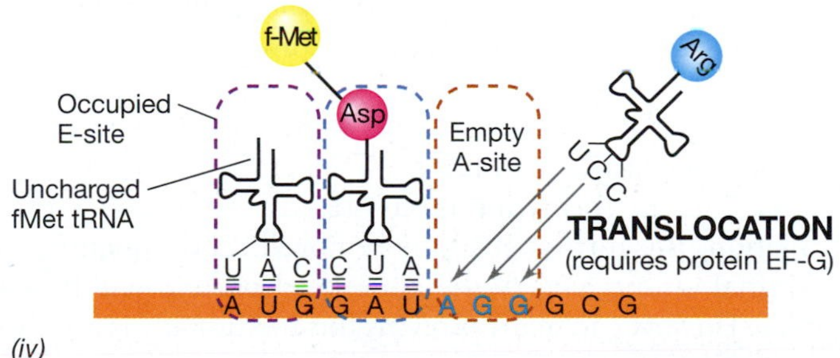

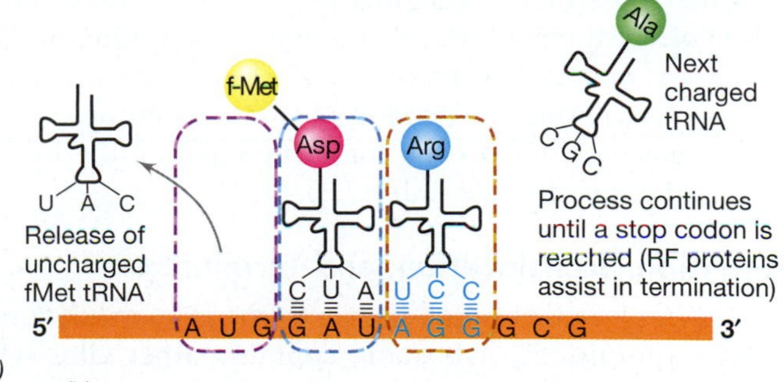

Figure 7.29 The ribosome and protein synthesis. *(a)* Structure of the ribosome, showing the position of the A (acceptor) site, the P (peptide) site, and the E (exit) site. *(b)* Translation. Initiation and elongation. (i, ii) Interaction between the codon and anticodon brings the correct charged tRNAs into position—in this case the initiator tRNA and the second charged tRNA. (iii) The formation of a peptide bond between amino acids on adjacent tRNA molecules completes elongation. (iv) Translocation of the ribosome from one codon to the next occurs with release of the tRNA from the E site. (v) The next charged tRNA binds to the A site.

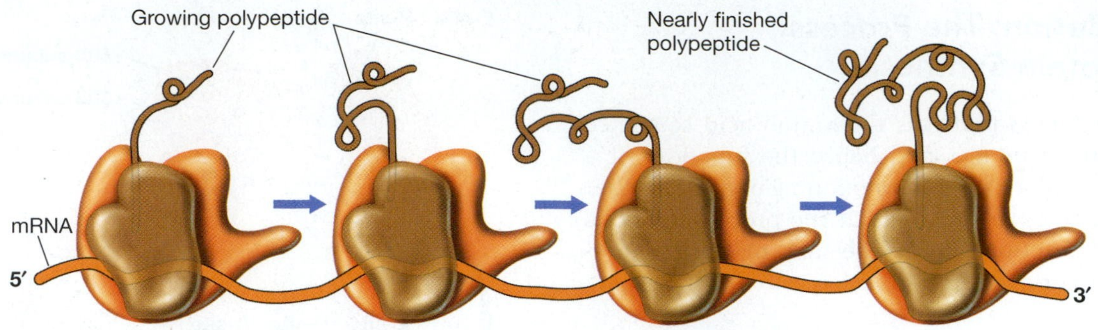

Figure 7.30 Polysomes. Translation by several ribosomes on a single messenger RNA forms the polysome. Note how the ribosomes nearest the 5′-end of the message are at an earlier stage in the translation process than ribosomes nearer the 3′-end, and thus only a relatively short portion of the final polypeptide has been made.

Initiation of Translation

In *Bacteria,* such as *E. coli,* initiation of protein synthesis always begins with a free 30S ribosomal subunit. From this, an initiation complex forms consisting of the 30S subunit, plus mRNA, formylmethionine tRNA, and several initiation proteins called IF1, IF2, and IF3. GTP is also required for this step. Next, a 50S ribosomal subunit is added to the initiation complex to form the active 70S ribosome. At the end of the translation process, the ribosome separates again into 30S and 50S subunits.

Just preceding the start codon on the mRNA is a sequence of three to nine nucleotides called the Shine–Dalgarno sequence or ribosome-binding site that helps bind the mRNA to the ribosome. The ribosome binding site is at the 5′-end of the mRNA and is complementary in base sequence to sequences in the 3′-end of the 16S rRNA. Base-pairing between these two molecules holds the ribosome–mRNA complex securely together in the correct reading frame. The presence of the Shine–Dalgarno sequence on the mRNA and its specific interaction with 16S rRNA allow bacterial ribosomes to translate polycistronic mRNA because the ribosome can find each initiation site within a message by binding to its Shine–Dalgarno site (Section 7.13).

Translational initiation always begins with a special initiator aminoacyl-tRNA binding to the start codon, AUG. In *Bacteria* this is formylmethionyl-tRNA. After polypeptide completion, the formyl group is removed. Consequently, the N-terminal amino acid of the completed protein will be methionine. However, in many proteins this methionine is removed by a specific protease. Because the Shine–Dalgarno sequences (and other possible interactions between the rRNA and the mRNA) direct the ribosome to the proper start site, prokaryotic mRNAs can use a start codon other than AUG. The most common alternative start codon is GUG. When used in this context, however, GUG calls for formylmethionine initiator tRNA (and not valine, see Table 7.3).

Elongation, Translocation, and Termination

The mRNA threads through the ribosome primarily bound to the 30S subunit. The ribosome contains other sites where the tRNAs interact. Two of these sites are located primarily on the 50S subunit, and they are termed the A site and the P site (Figure 7.29*b*). The A site, the acceptor site, is the site on the ribosome where the new charged tRNA first attaches. Travel and attachment of a tRNA to the A site is assisted by one of a series of elongation factor (EF) proteins called EF-Tu.

The P site, the peptide site, is the site where the growing polypeptide is held by a tRNA. During peptide bond formation, the growing polypeptide chain moves to the tRNA at the A site as a new peptide bond is formed. Several nonribosomal proteins are required for elongation, especially the elongation factors, EF-Tu and EF-Ts, as well as more GTP (to simplify Figure 7.29*b*, the elongation factors are omitted and only a portion of the ribosome is shown). Following elongation, the tRNA holding the polypeptide is translocated (moved) from the A site to the P site, thus opening up the A site for another charged tRNA (Figure 7.29*b*).

Translocation requires a specific EF protein called EF-G and one molecule of GTP for each translocation event. At each translocation step the ribosome advances three nucleotides, exposing a new codon at the A site. Translocation pushes the now empty tRNA to a third site, called the E site. It is from this exit site that the tRNA is actually released from the ribosome (Figure 7.29*b*). The precision of the translocation step is critical to the accuracy of protein synthesis. The ribosome must move exactly one codon at each step. Although during this process mRNA appears to be moving through the ribosome complex, in reality, the ribosome is moving along the mRNA. Thus, the three sites on the ribosome that we have identified in Figure 7.29 are not static locations but instead are moving parts of a complex biomolecular machine.

Several ribosomes can simultaneously translate a single mRNA molecule, forming a complex called a polysome (**Figure 7.30**). Polysomes increase the speed and efficiency of translation, and because the activity of each ribosome is independent of that of its neighbors, each ribosome in a polysome complex makes a complete polypeptide. Note in Figure 7.30 how ribosomes closest to the 5′-end (the beginning) of the mRNA molecule have short polypeptides attached to them because only a few codons have been read, while ribosomes closest to the 3′-end of the mRNA have nearly finished polypeptides.

Protein synthesis terminates when the ribosome reaches a nonsense codon (stop codon). No tRNA binds to a nonsense codon. Instead, specific proteins called *release factors* (RFs) recognize the stop codon and cleave the attached polypeptide from the final tRNA, releasing the finished product. Following this, the ribosome subunits dissociate, and the 30S and 50S subunits are then free to form new initiation complexes and repeat the process.

Role of Ribosomal RNA in Protein Synthesis

Ribosomal RNA plays a vital role in all stages of protein synthesis, from initiation to termination. The role of the many proteins present in the ribosome, although less clear, may be to act as a scaffold to position key sequences in the ribosomal RNAs.

As already discussed, in prokaryotes it is clear that 16S rRNA is involved in initiation through base-pairing with the Shine–Dalgarno sequence on the mRNA. There are also other mRNA–rRNA interactions during elongation. On either side of the codons in the A and P sites, the mRNA is held in position by binding to 16S rRNA and ribosomal proteins. Ribosomal RNA also plays a role in ribosome subunit association, as well as in positioning tRNA in the A and P sites on the ribosome (Figure 7.29*b*). Although charged tRNAs that enter the ribosome recognize the correct codon by codon–anticodon base pairing, they are also bound to the ribosome by interactions of the anticodon stem-loop of the tRNA with specific sequences within 16S rRNA. Moreover, the acceptor end of the tRNA (Figure 7.27) base-pairs with sequences in the 23S rRNA.

In addition to all of this, the actual formation of peptide bonds is catalyzed by rRNA. This happens on the 50S subunit of the ribosome during the peptidyl transferase reaction. This reaction is catalyzed by the 23S rRNA itself, rather than by any of the ribosomal proteins. The 23S rRNA also plays a role in translocation, and in this regard, the EF proteins are known to interact specifically with 23S rRNA. Thus, besides its role as the structural backbone of the ribosome, ribosomal RNA plays a major catalytic role in the translation process as well. **www.microbiologyplace.com Online Tutorial 7.3: Translation**

Freeing Trapped Ribosomes

A defective mRNA that lacks a stop codon causes a problem in translation. Such a defect may arise, for example, from a mutation that removed the stop codon, defective synthesis of the mRNA, or partial degradation of the mRNA. If a ribosome reaches the end of an mRNA molecule and there is no stop codon, release factor cannot bind and the ribosome cannot be released from the mRNA. The ribosome is trapped.

Bacterial cells contain a small RNA molecule, called *tmRNA*, that frees stalled ribosomes (**Figure 7.31**). The "tm" in its name refers to the fact that tmRNA mimics both tRNA, in that it carries the amino acid alanine, and mRNA, in that it contains a short stretch of RNA that can be translated. When tmRNA collides with a stalled ribosome, it binds alongside the defective mRNA. Protein synthesis can then proceed, first by adding the alanine on the tmRNA and then by translating the short tmRNA message. Finally, tmRNA contains a stop codon that allows release factor to bind and disassemble the ribosome. The protein made as a result of this rescue operation is defective and is subsequently degraded. The short sequence of amino acids, encoded by tmRNA and added to the end of the defective protein, is a signal for a specific protease to degrade the protein. Thus, through the activity of tmRNA, stalled ribosomes are freed up to participate in protein synthesis once again.

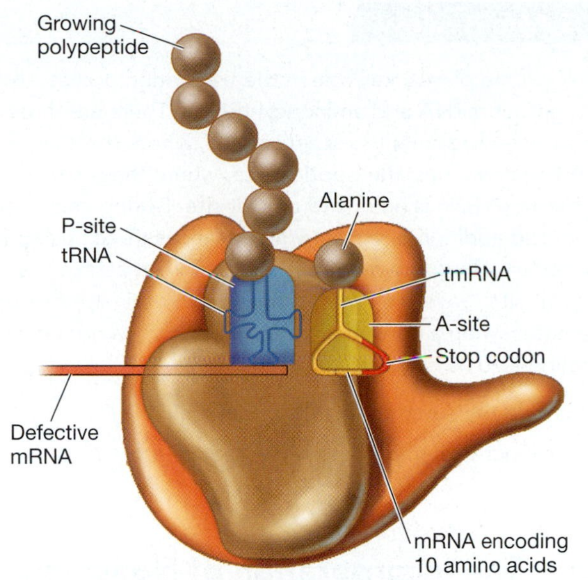

Figure 7.31 Freeing of a stalled ribosome by tmRNA. A defective mRNA lacking a stop codon stalls a ribosome that has a partly synthesized polypeptide attached to a tRNA (blue) in the P site. Binding of tmRNA (yellow) in the A site releases the polypeptide. Translation then continues up to the stop codon provided by the tmRNA.

Effect of Antibiotics on Protein Synthesis

A large number of antibiotics inhibit translation by interacting with the ribosome. These interactions are quite specific, and many have been shown to involve rRNA. Several of these antibiotics are clinically useful, and several are also effective research tools because they are specific for different steps in protein synthesis. For instance, streptomycin inhibits initiation. By contrast, puromycin, chloramphenicol, cycloheximide, and tetracycline inhibit elongation. Adding to their clinical usefulness is the fact that many antibiotics specifically inhibit ribosomes of organisms from only one or two of the phylogenetic domains. Of the antibiotics just listed, for example, chloramphenicol and streptomycin are specific for the ribosomes of *Bacteria* and cycloheximide for ribosomes of *Eukarya*. The mode of action of these and other antibiotics will be discussed in Chapter 27.

7.15 MiniReview

The ribosome plays a key role in the translation process, bringing together mRNA and aminoacyl tRNAs. There are three sites on the ribosome: the acceptor site, where the charged tRNA first combines; the peptide site, where the growing polypeptide chain is held; and an exit site. During each step of amino acid addition, the ribosome advances three nucleotides (one codon) along the mRNA and the tRNA that is in the acceptor site moves to the peptide site. Protein synthesis terminates when a nonsense codon, which does not encode an amino acid, is reached.

- What are the components of a ribosome?
- What functional roles does rRNA play in protein synthesis?

7.16 The Incorporation of Nonstandard Amino Acids

The Variety of Amino Acids

The universal genetic code has codons for 20 amino acids (Table 7.4). However, many proteins contain other amino acids. In fact, more than 100 different amino acids have been found in various proteins. Most of these are made by modifying one of the standard amino acids after it is incorporated into a protein molecule, a process called *posttranslational modification*. However, at least two nonstandard amino acids are inserted into proteins during protein synthesis itself. These exceptions are selenocysteine and pyrrolysine, the 21st and 22nd genetically encoded amino acids (Figure 3.12).

Selenocysteine and Pyrrolysine

Selenocysteine has the same structure as cysteine except it contains a selenium rather than a sulfur atom. It is formed by modifying a serine after it has been attached to selenocysteine tRNA by the aminoacyl-tRNA synthetase. Pyrrolysine is a lysine derivative with an extra aromatic ring. In this case, pyrrolysine is fully synthesized and only then joined to pyrrolysyl tRNA by the corresponding aminoacyl-tRNA synthetase.

Both selenocysteine and pyrrolysine are encoded by stop codons (UGA and UAG, respectively). Both have their own tRNAs that contain anticodons that read these stop codons. Both selenocysteine and pyrrolysine also have specific aminoacyl-tRNA synthetases to charge the tRNA with the amino acids. Note that these amino acids are used by organisms that use the universal genetic code; in other words, most stop codons in these organisms do indeed indicate stop. However, occasional stop codons are recognized as encoding selenocysteine or pyrrolysine. How can a codon sometimes be a nonsense codon and sometimes a sense codon? The answer in the case of selenocysteine is well understood and depends on a recognition sequence just downstream of the selenocysteine-encoding UGA codon. This forms a stem-loop structure that binds a special protein factor, the SelB protein. The SelB protein also binds charged selenocysteine tRNA and brings it to the ribosome when needed during translation. Similarly, pyrrolysine incorporation relies on a recognition sequence just downstream of the pyrrolysine-encoding UAG codon.

Selenocysteine and pyrrolysine are both relatively rare. *Escherichia coli* makes only a handful of proteins with selenocysteine, including two different formate dehydrogenase enzymes. It was sequencing the genes for these enzymes that led to the discovery of selenocysteine and its incorporation mechanism. Most organisms, including plants and animals, have a small number of proteins that contain selenocysteine. Pyrrolysine is rarer still. It has been found in certain *Archaea* and *Bacteria* but was first discovered in species of methanogenic *Archaea*, organisms whose metabolism generates natural gas (methane) (Section 21.10). In certain methanogens the enzyme methylamine methyltransferase contains a pyrrolysine residue. Whether there are still other genetically encoded amino acids that use reassigned nonsense codons, or perhaps even reassigned sense codons, remains a possibility.

7.16 MiniReview

Many nonstandard amino acids are found in proteins as a result of posttranslational modification. In contrast, the two rare amino acids selenocysteine and pyrrolysine are inserted into growing polypeptide chains during protein synthesis. They are both encoded by special stop codons that have a nearby recognition sequence that is specific for insertion of selenocysteine or pyrrolysine.

- Explain the term posttranslational modification.
- What specific components (apart from a ribosome and a stop codon) are needed for the insertion of selenocysteine into a growing polypeptide chain?

7.17 Folding and Secreting Proteins

Thus far we have described how genetic information present in the sequence of bases in DNA is replicated into an identical copy, transcribed into a sequence of bases in RNA, and translated into a sequence of amino acids in a polypeptide chain. For a protein to function, however, it must be folded correctly (Sections 3.6–3.8), and it must also end up in the correct location in the cell. Here we briefly discuss these two processes.

Protein Folding

Most polypeptides fold spontaneously into their active form while they are being synthesized. However, some do not and require assistance from other proteins called **molecular chaperones** or **chaperonins** for proper folding or for assembly into larger complexes. The chaperonins themselves do not become part of the assembled proteins but only assist in folding. Indeed, one important function of chaperones is to prevent improper aggregation of proteins.

There are several different kinds of molecular chaperones. Some help newly synthesized proteins fold correctly. Other chaperonins are very abundant in the cell, especially under growth conditions that put protein stability at risk (for example, high temperatures). Chaperonins are widespread in all domains of life, and their sequences are highly conserved among all organisms.

Four key chaperones in *Escherichia coli* are the proteins DnaK, DnaJ, GroEL, and GroES. DnaK and DnaJ are ATP-dependent enzymes that bind to newly formed polypeptides and keep them from folding too abruptly, a process that increases the risk of improper folding (**Figure 7.32**). Slower folding thus improves the chances of correct folding. If the DnaKJ complex is unable to fold the protein properly, it may transfer the partially folded protein to the two multi-subunit proteins, GroEL and GroES. The protein first enters GroEL, a large barrel-shaped protein that, using the energy of ATP hydrolysis, properly folds the protein. GroES assists in this (Figure 7.32). It is estimated that only about 100 of the several thousand proteins of *E. coli* need help in folding from the GroEL/GroES complex and of these approximately a dozen are essential for survival of the bacteria.

In addition to folding newly synthesized proteins, chaperones can also refold proteins that have partially denatured in the cell. A protein may denature for many reasons, but often it is because the organism has temporarily experienced high temperatures. Chaperones are thus one type of *heat shock protein,* and their synthesis is greatly accelerated when a cell is stressed by excessive heat (∞ Section 9.11). The heat shock response is an attempt by the cell to refold its partially denatured proteins for re-use before proteases recognize them as improperly folded and destroy them. Refolding is not always successful, and cells contain proteases whose function is to specifically target and destroy misfolded proteins, which then frees their amino acids to make new proteins.

Protein Secretion and the Signal Recognition Particle

Many proteins carry out their function in the cytoplasmic membrane, in the periplasm of gram-negative cells (∞ Section 4.9), or even outside the cell proper. Such proteins must get from their site of synthesis on ribosomes into or through the cytoplasmic membrane. How is it possible for a cell to selectively transfer some proteins across a membrane while leaving most proteins in the cytoplasm?

Most proteins that must be transported into or through membranes are synthesized with an amino acid sequence of about 15–20 residues, called the **signal sequence**, at the beginning of the protein molecule. Signal sequences are quite variable, but typically have a few positively charged residues at the beginning, a central region of hydrophobic residues, and then a more polar region. The signal sequence "signals" the cell's secretory system that this particular protein is to be exported and also helps prevent the protein from completely folding, a process that could interfere with its secretion.

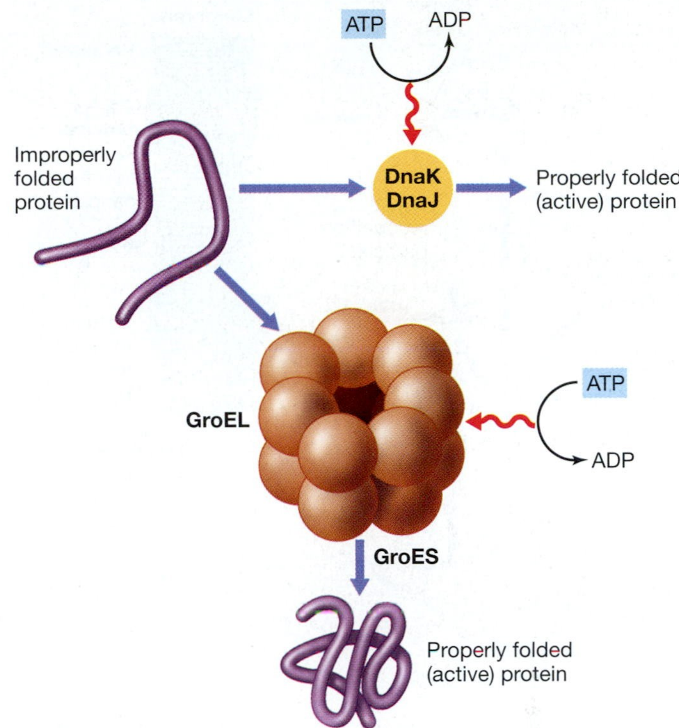

Figure 7.32 The activity of molecular chaperones. An improperly folded protein can be refolded by either the DnaKJ complex or by the GroEL/ES complex. In both cases, energy for refolding comes from ATP.

Because the signal sequence is the first part of the protein to be synthesized, the early steps in export may actually begin before the protein is completely synthesized (**Figure 7.33**).

Proteins to be exported are identified by their signal sequences either by the *SecA protein* or the *signal recognition particle (SRP)* (Figure 7.33). Generally, SecA binds proteins that are fully exported across the membrane into the periplasm whereas the SRP binds proteins that are inserted into the membrane but are not released on the other side. SRPs are found in all cells. In *Bacteria,* they contain a single protein and a small noncoding RNA molecule (4.5S RNA). Both SecA and the SRP deliver proteins to be secreted to the membrane secretion complex. In bacteria this is normally the Sec system, whose channel consists of the three proteins SecYEG (∞ Section 4.5). The protein is exported across the cytoplasmic membrane through this channel. It may then either remain in the membrane or be released into the periplasm or the environment (Figure 7.33). After crossing the membrane, the signal sequence is removed by a protease.

Secretion of Folded Proteins: The Tat System

In the Sec system of protein transport, the transported proteins are threaded through the cytoplasmic membrane in an unfolded state and only fold afterwards (Figure 7.33). However, there are a small number of proteins that must be transported outside the cell after they have already achieved their final folded structure. Usually this is because they contain small cofactors that must be inserted into the protein

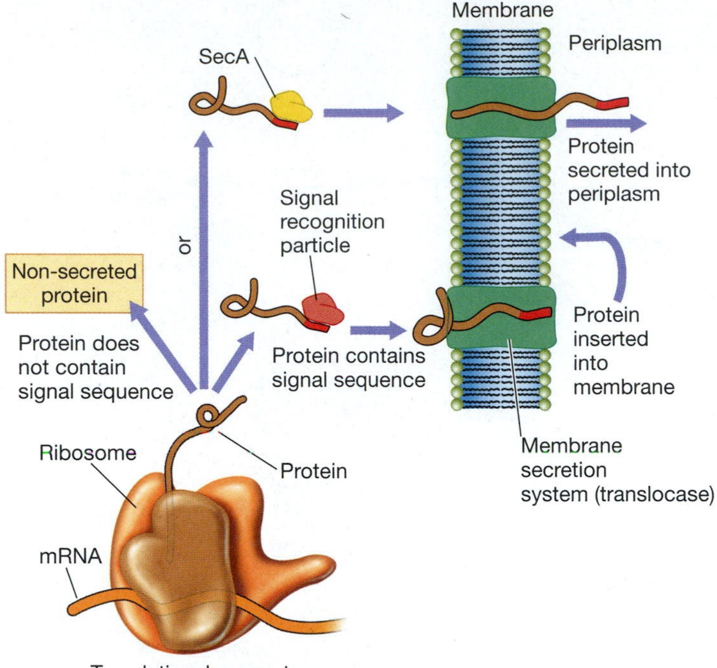

Figure 7.33 Export of proteins via the major secretory system. The signal sequence is recognized either by SecA or by the signal recognition particle, which carries the protein to the membrane secretion system. The signal recognition particle binds proteins that are inserted into the membrane whereas SecA binds proteins that are secreted across the membrane. In gram-negative bacteria, the latter proteins will be secreted into the periplasmic space (∞ Section 4.9).

as it folds into its final form. Such proteins fold in the cytoplasm and then are exported by a transport system distinct from Sec, called the *Tat protein export system.*

The acronym Tat stands for "twin arginine translocase" because the transported proteins contain a short signal sequence containing a pair of arginine residues. This signal sequence on a folded protein is recognized by the TatBC proteins, which carries the protein to TatA, the membrane transporter. Energy is required for the actual transport event, and this is supplied by the proton motive force (∞ Sections 4.4 and 5.12). A wide variety of proteins are transported by the Tat system, especially proteins required for energy metabolism that function in the periplasm. This includes iron-sulfur proteins and several other "redox" type proteins. In addition, the Tat pathway transports proteins needed for outer membrane (∞ Section 4.9) biosynthesis and a few proteins that do not contain cofactors but only fold properly within the cytoplasm.

7.17 MiniReview

Proteins must be properly folded in order to function correctly. Folding may occur spontaneously but may also involve other proteins called molecular chaperones. Many proteins also need to be transported into or through membranes. Such proteins are synthesized with a signal sequence that is recognized by the cellular export apparatus and is removed either during or after export.

- What is a molecular chaperone?
- Why do some proteins have a signal sequence?
- What is a signal recognition particle?

In this chapter we have covered the essentials of the key molecular processes that occur in bacteria. We next consider how archaeal and eukaryotic cells carry out the same processes. There are many similarities but also some major differences, in replication, transcription, and translation, among organisms in the three domains of life.

Review of Key Terms

Aminoacyl-tRNA synthetase an enzyme that catalyzes attachment of an amino acid to its cognate tRNA

Anticodon a sequence of three bases in a tRNA molecule that base-pairs with a codon during protein synthesis

Antiparallel in reference to double-stranded DNA, the two strands run in opposite directions (one runs 5′ → 3′ and the complementary strand 3′ → 5′)

Chaperonin or molecular chaperone a protein that helps other proteins fold or refold from a partly denatured state

Chromosome a genetic element, usually circular in prokaryotes, carrying genes essential to cellular function

Codon a sequence of three bases in mRNA that encodes an amino acid

Codon bias nonrandom usage of multiple codons encoding the same amino acid

Complementary nucleic acid sequences that can base-pair with each other

DNA gyrase an enzyme found in most prokaryotes that introduces negative supercoils in DNA

DNA polymerase an enzyme that synthesizes a new strand of DNA in the 5′ → 3′ direction using an antiparallel DNA strand as a template

Gene a segment of DNA specifying a protein (via mRNA), a tRNA, an rRNA, or any other non-coding RNA

Genetic code correspondence between nucleic acid sequence and amino acid sequence of proteins

Genetic element a structure that carries genetic information, such as a chromosome, a plasmid, or a virus genome

Genome the total complement of genes contained in a cell or virus

Informational macromolecule any large polymeric molecule that carries genetic information, including DNA, RNA, and protein

Lagging strand the new strand of DNA that is synthesized in short pieces during DNA replication and then joined together later

Leading strand the new strand of DNA that is synthesized continuously during DNA replication

Messenger RNA (mRNA) an RNA molecule that contains the genetic information to encode one or more polypeptides

Nonsense codon another name for a stop codon

Open reading frame (ORF) a sequence of DNA or RNA that could be translated to give a polypeptide

Operon a cluster of genes that are cotranscibed to give a single messenger RNA

Primer an oligonucleotide to which DNA polymerase can attach the first deoxyribonucleotide during DNA replication

Promoter a site on DNA to which RNA polymerase binds to commence transcription

Replication synthesis of DNA using DNA as a template

Replication fork the site on the chromosome where DNA replication occurs and where the enzymes replicating the DNA are bound to untwisted, single-stranded DNA

Ribosomal RNA (rRNA) types of RNA found in the ribosome; some participate actively in the process of protein synthesis

Ribosome a cytoplasmic particle composed of ribosomal RNA and protein and whose function is to synthesize proteins

RNA polymerase an enzyme that synthesizes RNA in the 5′ → 3′ direction using a complementary and antiparallel DNA strand as a template

Semiconservative replication DNA synthesis yielding new double helices, each consisting of one parental and one progeny strand

Signal sequence a special N-terminal sequence of approximately 20 amino acids that signals that a protein should be exported across the cytoplasmic membrane

Start codon a special codon, usually AUG, that signals the start of a protein

Stop codon a codon that signals the end of a protein

Termination stopping the elongation of an RNA molecule at a specific site

Transcription the synthesis of RNA using a DNA template

Transfer RNA (tRNA) a small RNA molecule used in translation that possesses an anticodon at one end and has the corresponding amino acid attached to the other end

Translation the synthesis of protein using the genetic information in RNA as a template

Wobble less rigid form of base pairing allowed only in codon–anticodon pairing

UNIT 2

Review Questions

1. Describe the central dogma of molecular biology (Section 7.1).
2. Genes were discovered before their chemical nature was known. Define a gene without mentioning its chemical nature. Of what is a gene composed (Section 7.1)?
3. Inverted repeats can give rise to stem-loops. Show this by giving the sequence of a double-stranded DNA containing an inverted repeat and show how the transcript from this region can form a stem-loop (Section 7.2).
4. Is the sequence 5′-GCACGGCACG-3′ an inverted repeat? Explain your answer (Section 7.2).
5. DNA molecules that are AT-rich separate into two strands more easily when the temperature is raised than do DNA molecules that are GC-rich. Explain this observation based on the properties of AT and GC base pairing (Section 7.2).
6. Describe how DNA, which when linearized, is many times the length of a cell, fits into the cell (Section 7.3).
7. List the major genetic elements known in microorganisms (Section 7.4).
8. A structure commonly seen in circular DNA during replication is the theta structure. Draw a diagram of the replication process and show how a theta structure could arise (Sections 7.5–7.7).
9. Why are errors in DNA replication so rare? What enzymatic activity, in addition to polymerization, is associated with DNA polymerase III and how does it reduce errors (Section 7.8)?
10. Do genes for tRNAs have promoters? Do they have start codons? Explain (Sections 7.12 and 7.13).
11. The start and stop sites for mRNA synthesis (on the DNA) are different from the start and stop sites for protein synthesis (on the mRNA). Explain (Sections 7.10 and 7.13).
12. What is "wobble" and what makes it necessary in protein synthesis (Section 7.13 and 7.14)?
13. What are aminoacyl-tRNA synthetases and what types of reactions do they carry out? Approximately how many different types of these enzymes are present in the cell? How does a synthetase recognize its correct substrates (Section 7.14)?
14. The activity that forms peptide bonds on the ribosome is called peptidyl transferase. What catalyzes this reaction (Section 7.15)?
15. Sometimes misfolded proteins can be correctly refolded, but sometimes they cannot and are destroyed. What kinds of proteins are involved in refolding misfolded proteins? What kinds of enzymes are involved in destroying misfolded proteins (Section 7.17)?
16. How does a cell know which of its proteins are designed to function outside of the cell (Section 7.17)?

Application Questions

1. The genome of the bacterium *Neisseria gonorrhoeae* consists of one double-stranded DNA molecule that contains 2220 kilobase pairs. Calculate the length of this DNA molecule in centimeters. If 85% of this DNA molecule is made up of the open reading frames of genes encoding proteins, and the average protein is 300 amino acids long, how many protein-encoding genes does *Neisseria* have? What kind of information do you think might be present in the other 15% of the DNA?
2. Compare and contrast the activity of DNA and RNA polymerases. What is the function of each? What are the substrates of each? What is the main difference in the behavior of the two polymerases?
3. What would be the result (in terms of protein synthesis) if RNA polymerase initiated transcription one base upstream of its normal starting point? Why? What would be the result (in terms of protein synthesis) if translation began one base downstream of its normal starting point? Why?
4. In Chapter 10 we will learn about mutations, inheritable changes in the sequence of nucleotides in the genome. By inspecting Table 7.4, discuss how the genetic code has evolved to help minimize the impact of mutations.

8

Archaeal and Eukaryotic Molecular Biology

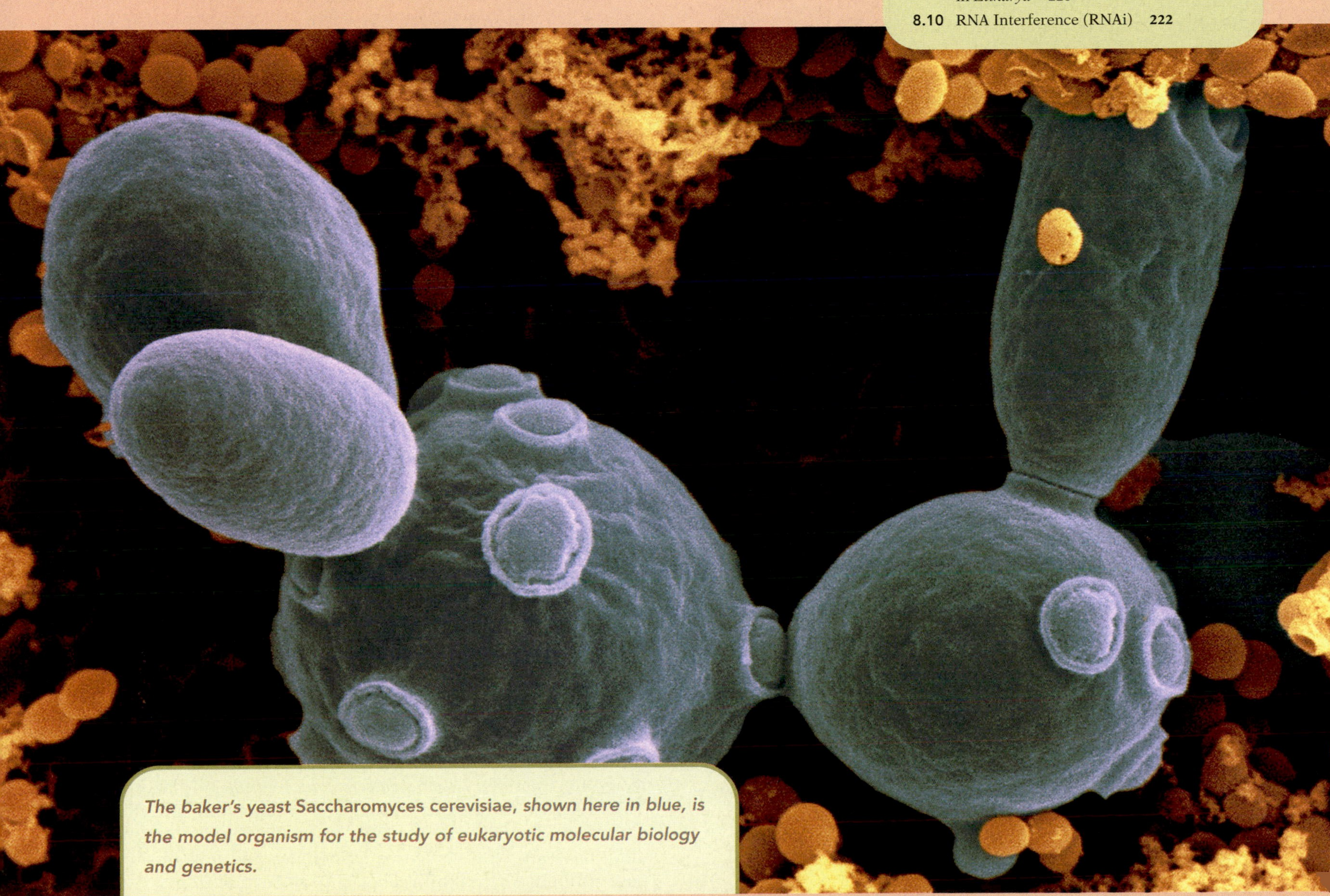

The baker's yeast Saccharomyces cerevisiae*, shown here in blue, is the model organism for the study of eukaryotic molecular biology and genetics.*

We emphasized in Chapter 2 how the two domains of prokaryotes—*Bacteria* and *Archaea*—are phylogenetically distinct, despite sharing a number of properties. This distinction is also obvious in many aspects of the molecular biology of these organisms. In Chapter 7 we covered the essentials of molecular biology as illustrated by the model prokaryotic organism, *Escherichia coli*, a species of *Bacteria*. In both *Archaea* and higher organisms the overall flow of information, from DNA to RNA to protein, as outlined in the central dogma of molecular biology, is the same as in the *Bacteria*. However, some of the details are different, and in eukaryotic cells there are complications due to the existence of the nucleus as a separate compartment.

Recent investigations have shown that, despite lacking a nucleus, *Archaea* are more similar in many of their properties to eukaryotes than to *Bacteria*. Indeed, the *Archaea* are being used as model organisms to investigate mechanisms they share with the eukaryotes that are more difficult to study in complex eukaryotic cells. Here we present an overview of key elements of the molecular biology of the *Archaea*, in particular the elements that show similarities to those of the eukaryotes.

I MOLECULAR BIOLOGY OF *ARCHAEA*

The *Archaea* were originally regarded as aberrant members of the *Bacteria* because both groups share the same overall prokaryotic cell design in which there are no membrane-bound compartments and, in particular, no nucleus. Moreover, both groups of prokaryotes typically contain a single circular chromosome and often transcribe several genes onto the same polycistronic mRNA. However, comparative rRNA sequencing has revealed a closer genetic relationship between *Archaea* and *Eukarya* than between *Bacteria* and *Archaea* (Section 2.7). This is reflected in such areas as the use of histones for DNA packaging and the detailed mechanism of translation. We start by comparing the genes and chromosomes of the *Archaea* with those of the other two domains.

8.1 Chromosomes and DNA Replication in *Archaea*

The chromosomes of *Archaea* resemble those of *Bacteria* in that they are circular and carry from around 500 to a few thousand genes. However, DNA packaging and chromosome replication reveal greater similarities with the eukaryotes.

DNA Packaging in *Archaea*

In all organisms, long DNA molecules are packaged by supercoiling, although the mechanisms vary among the three domains. The *Bacteria* use DNA gyrase (Section 7.3) whereas the eukaryotes wind their DNA around proteins known as histones (Section 8.5). The *Archaea* possess both DNA gyrase and histones. Thus the DNA of *Archaea* is either condensed by supercoiling mediated by DNA gyrase as in *Bacteria* or by binding to histones as in eukaryotes.

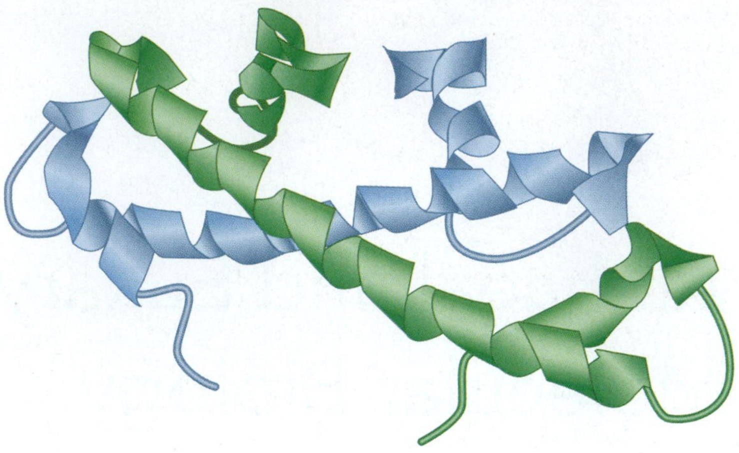

Figure 8.1 Histones in the *Archaea*. Although archaeal histones are shorter than eukaryotic histones, they contain the same histone fold structure. Shown here is a dimer of the histone HPhA from the hyperthermophilic archaeon *Pyrococcus horikoshii*. Adapted from structure 1KU5 of the Protein Data Bank.

Unlike *Bacteria*, both *Archaea* and *Eukarya* possess **histones** (Figure 8.1). Histones are present in all eukaryotes and in at least some members of all archaeal phyla, although not in every archaeal species. Histones are positively charged proteins that neutralize the negative charge of DNA (resulting from the phosphate groups). The DNA is wound around clusters of histones, forming structures known as **nucleosomes**, which are spaced along the DNA double helix at regular intervals. Archaeal histones form clusters of four (sometimes referred to as tetrasomes) rather than the eight found in eukaryotic histones, and accommodate approximately 80 base pairs (bp) of DNA. Archaeal histones are shorter than eukaryotic histones but are homologous in amino acid sequence and similar in their three-dimensional structure. Archaeal and eukaryotic histones share the so-called histone fold, the central region of the histone protein that is necessary for forming nucleosomes. Eukaryotic histones have extra N- and C-terminal domains that are not necessary for nucleosome assembly. Although *Bacteria* do not possess genuine histones, they do contain histone-like proteins that bind to DNA. These histone-like proteins are not homologous in sequence to true histones nor do they form nucleosomes.

Despite the presence of true histones in *Archaea*, some species, for example, *Thermoplasma acidophilum*, rely largely on DNA gyrase to package their DNA by supercoiling. Such *Archaea* are consequently sensitive to antibiotics that act by inhibiting DNA gyrase, such as quinolones and novobiocin. These antibiotics also inhibit *Bacteria* (Section 27.6).

A few *Archaea* that grow at extremely high temperatures (hyperthermophiles; Chapter 17), contain an enzyme called **reverse gyrase**. This topoisomerase introduces positive supercoils into DNA. Positive supercoiling probably protects DNA against denaturation by heat. Supporting this hypothesis is the fact that with rare exceptions reverse gyrase is present only in hyperthermophiles. However, as in all cells, to be of use, the genetic information in DNA must be accessible to the cell's replication and transcriptional machinery. Therefore, whether supercoiling is controlled by the activity of DNA gyrase or reverse gyrase, the structure of DNA within the cell likely remains quite dynamic.

Replication of Chromosomes in *Archaea*

The circular archaeal chromosome is structurally similar to and replicates by bidirectional synthesis in a mode similar to that of *Bacteria* (∞ Section 7.7). Nonetheless, the machinery of chromosomal replication in *Archaea* shows greater similarity to that of eukaryotes.

It is generally thought that the multiple origins of replication seen in eukaryotic chromosomes are necessary to replicate the very long chromosomes within a reasonable time period. One curious observation is that several *Archaea* are known whose single circular chromosome, despite being relatively short compared with those of eukaryotes, also has multiple origins of replication (Table 8.1). For example, *Halobacterium* appears to have two origins of replication on its chromosome and *Sulfolobus* has three.

The proteins of *Archaea* and *Eukarya* that recognize the origin of replication and help synthesize DNA show much greater similarity to each other than to functionally equivalent proteins of *Bacteria* (**Table 8.1**). In some cases, such as with DNA helicase or the origin recognition complex (ORC), the *Eukarya* have enzyme complexes consisting of multiple different (but related) protein subunits. The *Archaea* have only a single protein that forms equivalent complexes. For example, the eukaryotic helicase protein forms rings consisting of six different protein subunits, whereas the archaeal helicase forms rings of six identical subunits. Overall, then, the *Archaea* seem to have a simplified version of the eukaryotic replication apparatus.

All organisms possess multiple DNA polymerases specialized for different roles such as replication and DNA repair. There are three structural families of DNA polymerase, referred to as A, B, and C. Organisms in different domains use members of these three families for different roles. For example, *Bacteria* use DNA polymerases of family C (for example, Pol III of *Escherichia coli*) as their main replicative enzymes, whereas DNA polymerases of family A and B are used mostly in DNA repair. In contrast, the *Archaea* and *Eukarya* use DNA polymerases of family B as their main replicative enzymes and use DNA polymerases of family A and C for repair (Table 8.1).

Table 8.1 Three-domain comparison of chromosomes and their replication[a]

	Bacteria	*Archaea*	*Eukarya*
Chromosome topology	Circular	Circular	Linear
Chromosome number	One	One	Multiple
Origins per chromosome	One	One, two, or three	Many
Origin recognition	DnaA protein	Cdc6/Orc[b] protein	ORC[b]
Helicase	DnaB protein	MCM protein	MCM protein
Helicase loader	DnaC protein	Cdc6/Orc[b] protein	Cdc6 protein[b]
Sliding clamp	DnaN protein	PCNA protein	PCNA protein
Primase	DnaG protein	PriS and PriL	Pri1 and Pri2
Main DNA polymerase	PolC family	PolB family	PolB family

[a]Relatively uncommon exceptions, such as bacteria with linear chromosomes are omitted from this table.

[b]The Cdc6/Orc protein of *Archaea* is similar to both the Cdc6 and some ORC (origin recognition complex) subunits of eukaryotes and may be bifunctional. Some *Archaea* have more than one Cdc6-type protein.

8.1 MiniReview

The packaging of very long DNA molecules relies on some form of supercoiling. In *Bacteria* this is due to DNA gyrase, whereas in eukaryotes DNA is wound around nucleosomes that consist of proteins called histones. *Archaea* possess both DNA gyrase and histones. Some *Archaea* possess reverse gyrase, which introduces positive supercoiling. Despite resembling *Bacteria* in having single circular chromosomes, the DNA replication apparatus of *Archaea* is much more similar to that of eukaryotes.

- What similarities are there between the histones of *Archaea* and *Eukarya*?
- How do the activities of DNA gyrase and reverse gyrase differ?
- How many origins of replication are found on chromosomes from *Bacteria*, *Archaea* and *Eukarya*?

8.2 Transcription and RNA Processing in *Archaea*

The fundamental genetic relationship of the *Archaea* and the *Eukarya* was originally revealed by sequence comparisons of ribosomal RNA (∞ Section 2.7). Further comparisons of the molecular features of *Archaea* and *Eukarya* have shown that transcription and translation in the two domains share many structural and mechanistic features, indicating that these two domains are more closely related to each other than either is to *Bacteria*. In this section we present an overview of transcription in the *Archaea*.

Transcription in *Archaea*

Similar to chromosome replication, *Archaea* seem to have a simplified version of the eukaryotic transcription apparatus. Both the sequences of archaeal promoters and the structure and activity of RNA polymerase resemble those of eukaryotes.

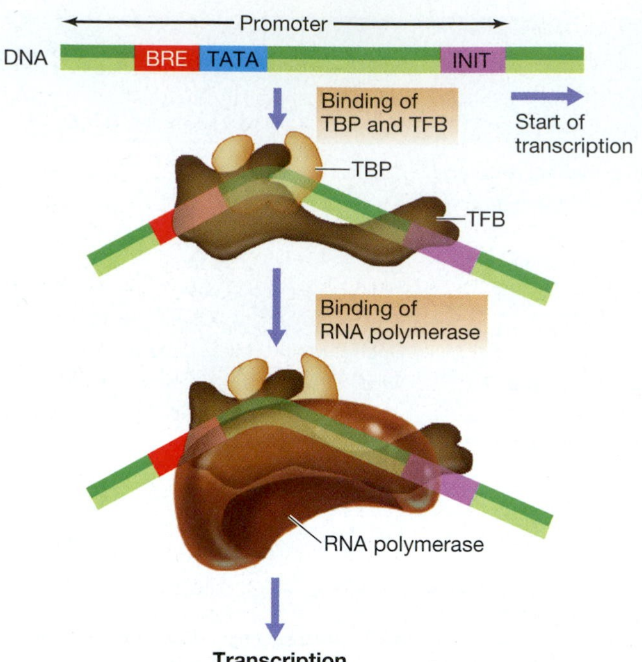

Figure 8.2 Promoter architecture and transcription in the *Archaea*. Three promoter elements are critical for promoter recognition in the *Archaea*: the initiator element (INIT), the TATA box, and the B recognition element (BRE). The TATA-binding protein (TBP) binds the TATA box; transcription factor B (TFB) binds to both BRE and INIT. Once both TBP and TFB are in place RNA, polymerase binds.

Conversely, the regulation of transcription in *Archaea* shares major similarities with *Bacteria*, perhaps because of the overall small genome size and lack of a nucleus. Regulation of gene expression of both *Bacteria* and *Archaea* is covered in Chapter 9 (∞ Section 9.8 for *Archaea*).

The *Archaea* contain only a single RNA polymerase, which most closely resembles eukaryotic RNA polymerase II (Section 8.9). The archaeal RNA polymerase typically has eight subunits, and eukaryotic RNA polymerase II has a dozen or more. RNA polymerase from *Bacteria* has only four subunits (excluding the sigma recognition subunit) (∞ Section 7.9). The antibiotic rifampicin inhibits bacterial RNA polymerase, but does not inhibit either the eukaryotic or archaeal enzymes. Furthermore, the structure of archaeal promoters resembles that of eukaryotic promoters recognized by eukaryotic RNA polymerase II. Three main recognition sequences are part of the promoters in both domains, and these sequences are recognized by a series of proteins called *transcription factors* that are similar in *Eukarya* and *Archaea*.

The most important recognition sequence in archaeal and eukaryotic promoters is a 6- to 8-base pair TATA box, located 18–27 nucleotides upstream of the transcriptional start site (**Figure 8.2**). This is recognized by TBP (*TATA-binding* protein). Upstream of this is the BRE (*B* recognition *element*) sequence that is recognized by TFB (*transcription factor B*). In addition, the initiator element, located at the start of transcription, is also important. Once TBP has bound to the TATA box and TFB has bound to the BRE, then archaeal RNA polymerase can bind and initiate transcription. This process is similar in eukaryotes except that more transcription factors are required (Section 8.9).

Less is known about the transcription termination signals in the *Archaea*. Some archaeal genes have inverted repeats followed by an AT-rich sequence, sequences very similar to those found in many bacterial transcription terminators (∞ Section 7.11). However, such termination sequences are not found in other archaeal genes. One other type of suspected transcription terminator lacks inverted repeats but contains repeated runs of Ts. In some way, this signals the archaeal termination machinery to terminate the transcription process. No Rho-like proteins have been found in *Archaea*. (Rho is required for termination of some genes in *Bacteria*, ∞ Section 7.11.)

Intervening Sequences in *Archaea*

As in *Bacteria*, intervening sequences in genes that encode proteins are extremely rare in *Archaea*. This is in contrast to *Eukarya* where many such genes are split into two or more coding regions separated by noncoding regions (Section 8.8). The segments of coding sequence are called **exons**, and the intervening noncoding regions, **introns**. The term **primary transcript** refers to the RNA molecule that is originally transcribed before the introns are removed to generate the final mRNA, consisting solely of the exons.

However, several tRNA- and rRNA-encoding genes of *Archaea* do possess introns that must be removed after transcription to generate the mature tRNA or rRNA. These introns were named archaeal introns because they are not processed by the spliceosome that processes typical eukaryotic introns (Section 8.8). Instead, archaeal introns are excised by a specific endoribonuclease that recognizes exon–intron junctions (**Figure 8.3**).

Archaeal introns are also found in the nuclear tRNA genes of eukaryotes. Furthermore, the archaeal endoribonuclease that splices introns is homologous to two of the subunits of the enzyme complex that removes introns from eukaryotic nuclear tRNA. The overall frequency of intervening gene sequences is similar in *Bacteria* and *Archaea*, possibly because they both contain small compact genomes. On the other hand, the splicing mechanism of archaeal introns is shared with a small subset of introns in the nucleus of eukaryotes.

8.2 MiniReview

The transcription apparatus and the promoter architecture of *Archaea* and *Eukarya* have many features in common, although the components are usually relatively simpler in the *Archaea*. A few unusual intervening sequences are found in *Archaea* that are similar to those found in eukaryotic nuclear tRNA.

- What three major components make up an archaeal promoter?
- What are archaeal introns?

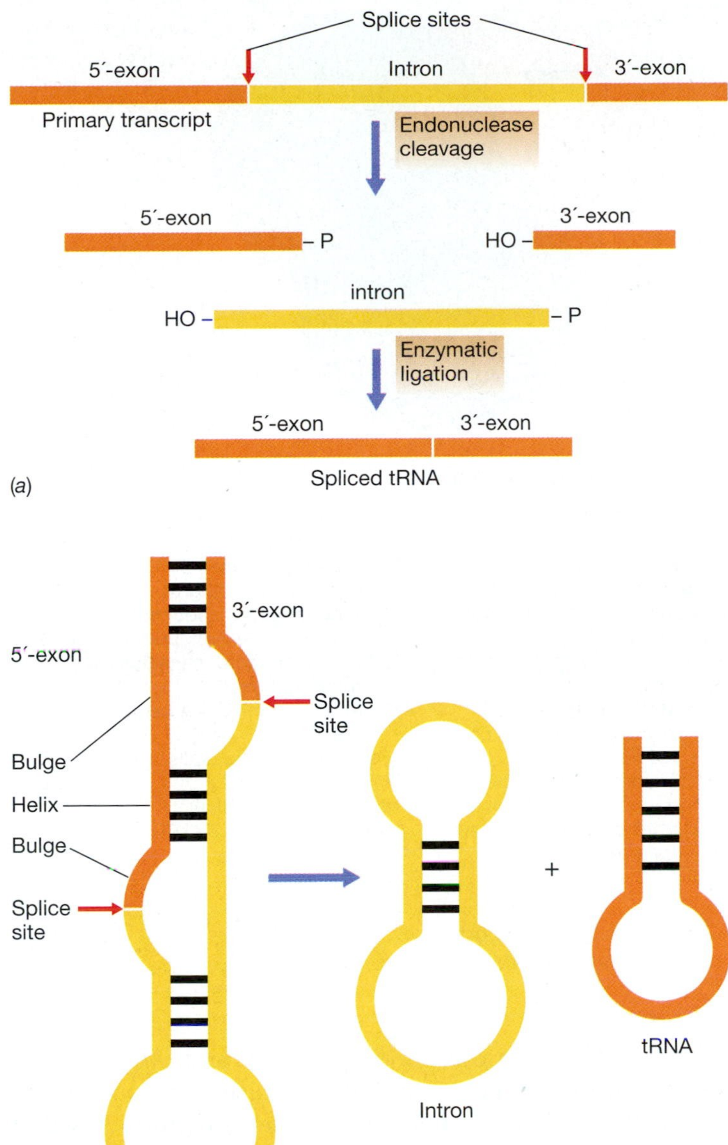

Figure 8.3 Splicing of archaeal introns. *(a)* Reaction scheme. Removal of archaeal introns is a two-step reaction. In the first step a specific endonuclease excises the intron. In the second step a ligase joins the 5′ exon to the 3′ exon, generating the mature, spliced tRNA. *(b)* Folding of the tRNA precursor. The two splice sites (red arrows) are recognized by their characteristic "bulge-helix-bulge" motifs. The products of the reaction are the tRNA and a circular intron.

8.3 Protein Synthesis in *Archaea*

As previously stated, the genetic relationship of *Archaea* to the eukaryotic nuclear genome was originally discovered by comparative sequencing of rRNA, which is a key component of the translation machinery (Section 7.15). Analyses of most other components of the translation machinery support the close relationship of the *Archaea* to the eukaryotes, and we consider these here.

Translation in *Archaea*

The ribosome of eukaryotes contains several rRNA molecules plus 78 proteins. Of these proteins, 34 are common to all three domains of life, another 33 are shared by the *Archaea* and *Eukarya*, and another 11 are unique to eukaryotes—none are common to only *Bacteria* and *Eukarya*. A similar situation is seen with the various translation factors. Some are common to all three domains, and others are shared by the *Archaea* and *Eukarya*. In addition to the 11 uniquely eukaryotic ribosomal proteins, eukaryotes have one extra rRNA, the 5.8S rRNA, that is not found in either *Bacteria* or *Archaea*.

Translation needs not only a functional ribosome, but also several translation factors (initiation, elongation, and release factors; Section 7.15). Eukaryotes and *Archaea* have approximately twice as many translation factors as *Bacteria*. Those of eukaryotes tend to have more subunits than those of *Archaea*, as with many proteins involved in replication (Section 8.1). However, the translation factors of *Archaea* and *Eukarya* show a high degree of homology.

Overall, the components of the translation machinery are more closely related in *Archaea* and *Eukarya* than they are to those of *Bacteria*. Despite this, the mechanism of protein synthesis in *Archaea* shares some features with both *Bacteria* and *Eukarya*, and as a result, the situation is rather complicated. For example, the ribosomes of both the *Bacteria* and *Archaea* recognize mRNA by means of Shine–Dalgarno sequences (Section 7.15), which are absent from the *Eukarya*. Conversely, both the *Archaea* and *Eukarya* insert methionine as the first amino acid, in contrast to *Bacteria*, which use *N*-formylmethionine (Section 7.13). At present little experimental data are available on the detailed mechanism of translation or the order in which the various components assemble for the *Archaea*. Overall, sequence comparisons underscore the similarity between the *Eukarya* and *Archaea* of the RNA and proteins that make up the translation machinery.

8.3 MiniReview

The translation apparatus of *Archaea* and *Eukarya* share many features, although the number of ribosomal components and translation factors is fewer in the *Archaea*. However, the *Archaea* show some similarities with *Bacteria*, such as the use of Shine–Dalgarno sequences.

- How does translation in *Archaea* resemble that in *Bacteria*?
- How does translation in *Archaea* resemble that in *Eukarya*?

8.4 Shared Features of *Bacteria* and *Archaea*

Despite the closer genetic relationship of the *Archaea* to the eukaryotes, organisms in the two prokaryotic domains, the *Bacteria* and the *Archaea*, nonetheless share several fundamental properties that are absent from eukaryotes. **Figure 8.4** summarizes how the genetic features of the three domains are

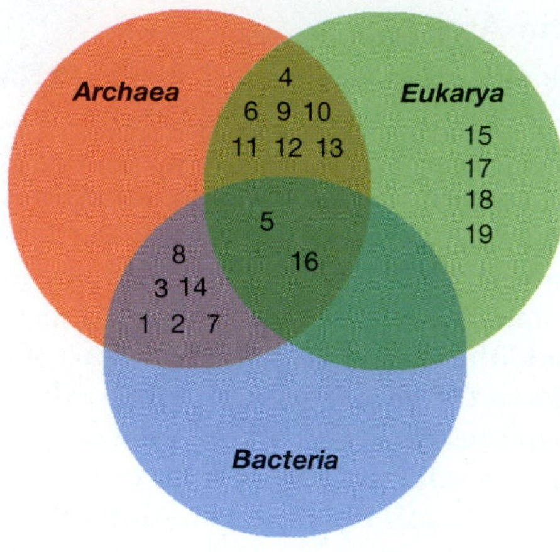

Genome

1 Chromosome circular versus linear
2 Single chromosome versus multiple chromosomes
3 Introns rare
4 Archaeal type introns
5 Inteins
6 Histones
7 DNA gyrase
8 Reverse gyrase
9 Multiple chromosomal origins
10 Eukaryotic origin recognition complex
11 Eukaryotic type helicase
12 B family DNA polymerase is major replicative enzyme
13 Eukaryotic type sliding clamp
14 Restriction enzymes
15 RNAi
16 Genome of double stranded DNA
17 Multiple retroelements in genome
18 Centromeres
19 Telomeres and telomerase

(a)

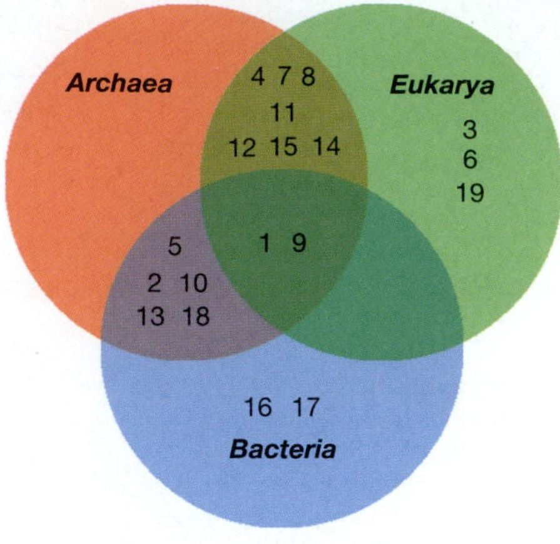

Transcription and Translation

1 RNA used as a genetic messenger
2 Polycistronic mRNA
3 Cap and tail on mRNA
4 TATA box and BRE sequence in promoter
5 Repressors binding directly to DNA in promoter
6 Multiple RNA polymerases
7 RNA polymerase II with 8 or more subunits
8 Multiple transcription factors needed
9 Ribosomes synthesize proteins
10 70S versus 80S ribosomes
11 Ribosomal RNA sequence homologies
12 Ribosomal protein sequence homologies
13 Shine Dalgarno sequences
14 Multiple translation factors
15 Elongation factor sensitive to diphtheria toxin
16 N-formyl methionine versus methionine
17 tmRNA rescues stalled ribosomes
18 16S and 23S rRNA
19 18S, 28S and 5.8S rRNA

(b)

Figure 8.4 Molecular features of the three domains. Venn diagrams show which features are shared by the domains and which are unique. *(a)* Genomic features. *(b)* Features of transcription and translation.

distributed. Both groups of prokaryotes are typically single-celled microorganisms, most of which divide by binary fission, although some divide by budding. Neither *Bacteria* nor *Archaea* possess a nucleus or membrane-bound organelles. Consequently, *Bacteria* and *Archaea* share several features such as coupled transcription and translation (⚭ Section 7.15) that are impossible in eukaryotes due to the fact that transcription occurs inside the nucleus. Other shared features, such as the near absence of intervening sequences in both *Bacteria* and *Archaea,* may be a result of a "prokaryotic lifestyle." In other words, relatively rapid growth as single cells creates pressure for a small, compact genome with little surplus noncoding DNA.

Neither *Bacteria* nor *Archaea* possess the large eukaryotic type of flagellum that uses ATP as an energy source. Instead, both *Bacteria* and *Archaea* use flagella with rotary bases that are energized by the proton motive force and whose shaft is a single helix of protein subunits (⚭ Section 4.13). However, the protein that makes up the flagellar shaft in *Archaea* is related to the pilus protein of *Bacteria* (⚭ Section 4.9), not to their flagellin.

Bacteria and *Archaea* share many other properties. Both possess a single circular chromosome and transcribe clusters of genes, controlled as a unit (an operon), into single polycistronic mRNA molecules that are translated to give several proteins. Furthermore, both *Bacteria* and *Archaea* use Shine–Dalgarno sequences to indicate to the ribosome where to start translation, and both lack the cap structure typical of eukaryotic mRNA (Section 8.8). In addition, regulation of transcription in *Archaea* is largely bacterial in nature, relying on DNA-binding proteins, both repressors and activators, which recognize sites within the promoter region (⚭ Section 9.8).

In eukaryotes, reproduction and genetic exchange are coupled during sexual reproduction, in which haploid gametes from two parents fuse to create a new diploid individual. Thus, eukaryotes alternate between haploid and diploid states. In *Bacteria* and *Archaea,* reproduction and gene exchange are two distinct processes. There is no diploid phase, and reproduction is equivalent to cell division. In both prokaryotic domains, several different mechanisms are responsible for genetic transfer but all are unidirectional, and DNA is transferred from a donor organism to a recipient in processes that do not involve cell division (⚭ Chapter 11).

RNA interference (RNAi), a major mechanism for protection against infection by RNA viruses, is found only in eukaryotes (Section 8.10). Conversely, both *Bacteria* and *Archaea* produce restriction endonucleases (⚭ Section 12.1) that function to destroy incoming foreign DNA, whereas eukaryotes lack restriction enzyme systems.

8.4 MiniReview

The *Bacteria* and *Archaea* share many properties related to their simple cell structure and relatively small genome size. In addition, they share several fundamental genetic features, including circular chromosomes, polycistronic mRNA, Shine–Dalgarno sequences, bacterial style gene regulation, restriction enzymes, and flagella with rotary bases driven by the proton motive force.

- What shared features of *Bacteria* and *Archaea* are likely due to the lack of a nucleus and membrane-bound organelles?
- What shared features of *Bacteria* and *Archaea* affect transcription and translation?

II EUKARYOTIC MOLECULAR BIOLOGY

The fact that eukaryotes have a nucleus has several profound consequences. These include the necessary disassembling of the nucleus during cell division as seen in mitosis and meiosis and the physical separation of transcription (nuclear) and translation (cytoplasmic). Eukaryotes are also largely unique in the replication of linear (as opposed to circular) chromosomes and in the processing of mRNA—both of which occur inside the nucleus. We briefly explore each of these topics here. A summary of molecular events in eukaryotes is given in **Figure 8.5**.

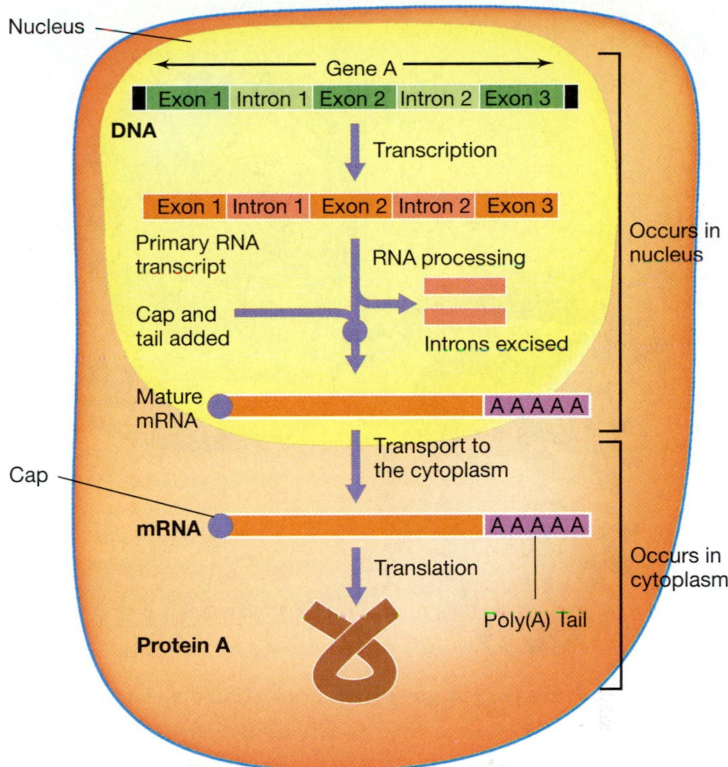

Figure 8.5 Information transfer in eukaryotes. Prior to translation, noncoding regions (introns) are removed from the primary RNA transcript, leaving exons to be joined to form the mature mRNA. In addition, a cap and a poly(A) tail are added before the mRNA is allowed to exit the nucleus.

8.5 Genes and Chromosomes in the *Eukarya*

In eukaryotes, protein-encoding genes are often split into two or more coding regions, known as exons, with noncoding regions, known as introns, separating them. Both introns and exons are transcribed into the primary transcript. From this, the mature (functional) mRNA is formed by excision of the introns and splicing together the exons (Section 8.8 and Figure 8.5). Only after this occurs does the mRNA exit the nucleus to be translated by the ribosomes in the cytoplasm. As we have seen (Section 8.2), a few genes in prokaryotes contain introns, but the vast majority do not.

DNA Content

Eukaryotes typically contain much more DNA than is needed to encode all the proteins required for cell function. For instance, in the human genome only about 3% of the total DNA encodes protein. In contrast, in prokaryotes this fraction often exceeds 90%. The "extra" DNA in eukaryotes is present as introns or repetitive sequences, some repeated hundreds or thousands of times. Eukaryotic microorganisms have fewer introns than do higher eukaryotes. For example, about 70% of DNA in the yeast *Saccharomyces cerevisiae* encodes protein. Additionally, eukaryotes often have multiple copies of certain genes, such as those that encode tRNAs and rRNAs. The genes for rRNA are found repeated in some prokaryotes, but usually only a few copies are present.

In contrast to the single circular prokaryotic chromosome in the cytoplasm, a eukaryote has multiple linear chromosomes in the nucleus. Because these chromosomes are in the nucleus and the ribosomes are in the cytoplasm, transcription and translation are spatially separated processes. Signals for gene expression that originate outside the cell or within the cytoplasm must be transmitted into the nucleus in order to take effect.

In eukaryotes, large amounts of protein are bound to the DNA in a very regular fashion. The linear DNA molecules that make up eukaryotic chromosomes are wound around proteins called histones to form structures called *nucleosomes* (**Figure 8.6**). In eukaryotic chromosomes, the formation of the nucleosome introduces negative supercoils into the DNA. The nucleosomes are spaced along the double helix at very regular intervals, and each nucleosome contains approximately 200 bp of DNA plus nine histone proteins, two each of the four core histones (H2A, H2B, H3, and H4) and one linker histone, H1 (Figure 8.6). As noted above (Section 8.1), histones are also present in *Archaea*. However, archaeal histones are shorter than eukaryotic histones and assemble into tetramers.

Histones are positively charged proteins that neutralize the negative charge of DNA caused by the presence of phosphate groups. Of all known proteins, eukaryotic core histones

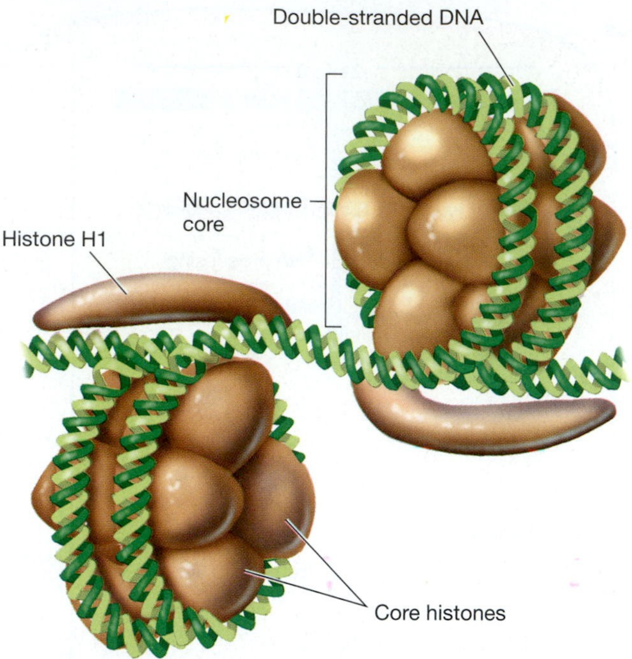

Figure 8.6 Packaging of eukaryotic DNA around a histone core to form a nucleosome. Nucleosomes are arranged along the DNA strand somewhat like beads on a string. In eukaryotes, nucleosomes consist of a core with eight proteins, two copies each of histones H2A, H2B, H3, and H4, plus a linker with one copy of histone H1.

have been the most highly conserved during evolution. For example, only 2 amino acids out of 102 are different between histone H4 of cows and peas.

This complex of DNA plus histones is called chromatin. This can be further compacted by folding and looping to eventually form very dense structures. Highly condensed chromatin is known as heterochromatin and cannot be transcribed because it cannot be accessed by RNA polymerase. During eukaryotic cell division the DNA is condensed into heterochromatin and the chromosomes become much more compact. It is these compacted chromosomes that are visualized in eukaryotic cells during cell division.

8.5 MiniReview

Due to the presence of a nucleus, the organization of genetic information is more complex in eukaryotes. Most eukaryotic genes have both coding regions (exons) and noncoding regions (introns).

- What effect does the presence of a nucleus have on genetic information flow in eukaryotes?

8.6 Overview of Eukaryotic Cell Division

Eukaryotic cells divide by a process in which the chromosomes are replicated, the nucleus is disassembled, the chromosomes are segregated into two sets, and a nucleus is

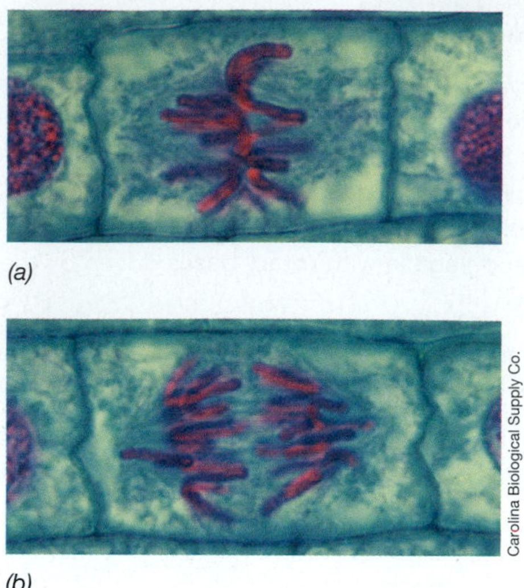

Figure 8.7 Mitosis, as seen in the light microscope. These are onion root tip cells that have been stained to reveal nucleic acid and chromosomes. *(a)* Metaphase. Chromosomes are paired in the center of the cell. *(b)* Anaphase. Chromosomes are separating.

finally reassembled in each daughter cell. Many eukaryotes also undergo sexual reproduction, during which male and female gametes are formed and then fuse to form the zygote. This is fundamentally different from the unidirectional mating processes of prokaryotes (Chapter 11).

Mitosis and Meiosis

From a genetic standpoint, eukaryotic cells can exist in two forms: *haploid* or *diploid*. Diploid cells have two copies of each chromosome whereas haploid cells have only one. Cells of many single-celled eukaryotes, such as the brewer's yeast, *Saccharomyces cerevisiae*, can exist indefinitely in the haploid stage (containing 16 chromosomes) as well as in the diploid stage (32 chromosomes). Occasionally, two haploid yeast cells will fuse (mate) to yield a diploid cell (Section 18.18). This process should not be confused with the situation in cells of multicellular plants and animals. In these organisms the diploid phase is present in the organism proper, and the haploid phase occurs only transiently, in the gametes.

Mitosis is the process following DNA replication in a eukaryotic cell. During mitosis, the chromosomes condense, divide, and are separated into two sets, one for each daughter cell (**Figure 8.7**). Diploid cells of both yeasts and higher eukaryotes undergo mitosis before each cell division to maintain the diploid number of chromosomes per cell. Similarly, haploid *S. cerevisiae* cells undergo mitosis before each cell division to maintain the haploid number of 16 chromosomes per cell (Section 18.18).

Meiosis is the process of conversion from the diploid to the haploid stage. Meiosis consists of two cell divisions. In the first meiotic division, homologous chromosomes segregate into separate cells, changing the genetic state from diploid to haploid. The second meiotic division is essentially the same as mitosis, as the two haploid cells divide to form a total of four haploid gametes. In higher organisms these are the eggs and sperm; in eukaryotic microorganisms, they are spores or related structures.

8.6 MiniReview

Eukaryotic microorganisms can mate and exchange DNA during sexual reproduction. Mitosis ensures appropriate segregation of the chromosomes during asexual cell division. Meiosis generates haploid cells known as gametes that fuse to form a diploid zygote.

- If the diploid number of human chromosomes is 46, how many chromosomes are there in a human sperm cell?
- What are the important differences between mitosis and meiosis?

8.7 Replication of Linear DNA

Most prokaryotic chromosomes are circular, as are most plasmids, some virus genomes, and the genomes of most organelles (∞ Section 13.4). In contrast to prokaryotes, the nucleus of eukaryotic cells contains linear DNA. Almost all the steps in DNA replication are the same whether the chromosome is linear or circular. However, there is a key problem with the replication of linear genetic elements that is not an issue with circular ones: replication of DNA at the extreme 5′ end of each strand.

To understand the nature of the problem, first review Figure 7.12. Imagine that the left end of the DNA in this diagram is actually one end of a linear chromosome. Even if the RNA primer is very short and there is a special enzyme to remove it, no DNA polymerase can replace it with DNA because all known DNA polymerases require a primer. Therefore, if nothing were done about this problem, the DNA molecule would become shorter each time it was replicated. The replication of linear DNA thus requires special attention, and there are at least two solutions to this problem.

Replication of Linear DNA Using a Protein Primer

Viruses that contain linear DNA genomes (this includes most viruses that infect eukaryotes) and many linear plasmids solve the problem of replicating linear DNA by using a protein primer instead of an RNA primer. Although all DNA polymerases must add each nucleotide to a free hydroxyl (OH) group, some DNA polymerases can add the first base onto a hydroxyl group present on specific proteins that bind to the ends of linear chromosomes (**Figure 8.8**). These proteins are encoded by the plasmid or virus, and they recognize and bind the ends of the chromosomes. These protein primers are not removed, so these plasmids and viruses have proteins permanently attached to the 5′ ends of their DNA. Protein primers are also the means by which some linear chromosomes of *Bacteria* are replicated.

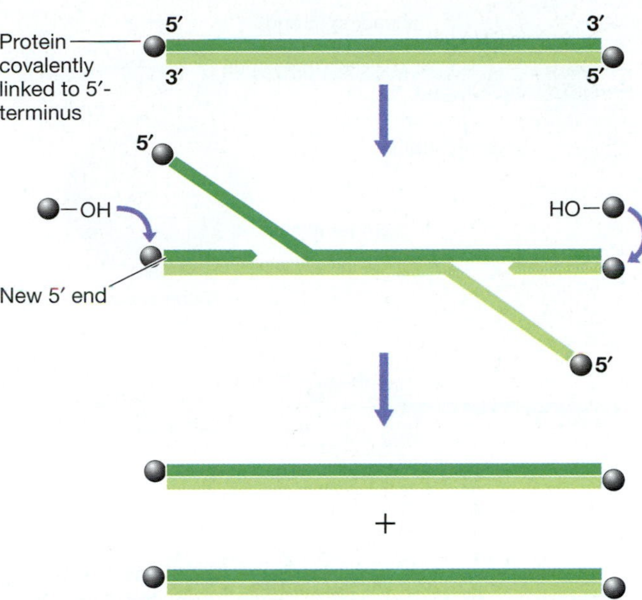

Figure 8.8 Replication of linear DNA using protein primers. New strands of DNA are primed by proteins covalently attached to their 5′ ends. Note the free OH group on the protein. DNA polymerase III can add a nucleotide to this OH group.

Telomeres and Telomerase

A special method is used to replicate the ends of eukaryotic chromosomes, which are called *telomeres*. Telomeres contain repetitive DNA—a short sequence (often 6 bp) tandemly repeated from 20 to several hundred times. The sequences from the telomeres of different eukaryotes are closely related, and one strand always has several guanines. During replication, this guanine-rich sequence is present on the leading strand of the DNA duplex and can base-pair with the 3′ end of a complementary RNA present in an enzyme called **telomerase** (**Figure 8.9**).

Telomerase becomes active once bound to the 3′ end of the G-rich overhang of telomeric DNA. Telomerase does not need a template to begin DNA synthesis because the enzyme itself contains a small RNA template as a cofactor. Technically, telomerase is an unusual type of reverse transcriptase (∞ Section 10.12), making DNA using an RNA template. Telomerase works repetitively to make a long extension of the leading strand. Once this extension is made, the other strand (the lagging strand) can be primed with an RNA primer in the normal fashion and the DNA replicated (Figure 8.9*b*). Telomeres do not need to be a precise number of repeats long, but just long enough to ensure that no portion of a gene becomes lost during DNA replication.

Centromeres and Kinetochores

In addition to telomeres, eukaryotic chromosomes contain *centromeres*. These are regions that provide an attachment site for the spindle fibers that pull the pairs of homologous chromosomes apart during mitosis (**Figure 8.10**). Centromeres cannot be situated at the very ends of eukaryotic

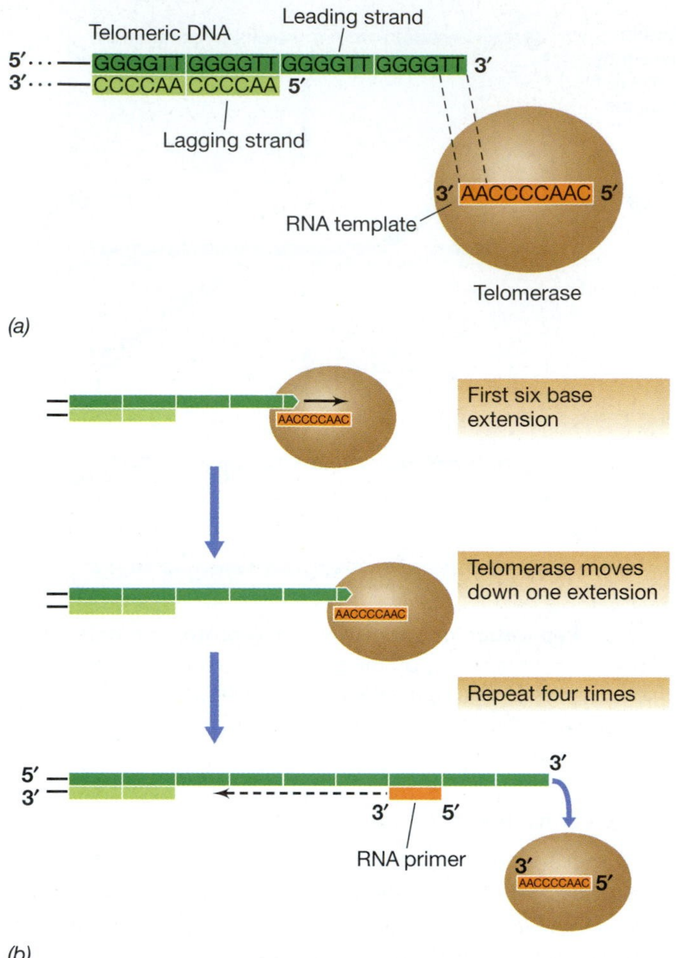

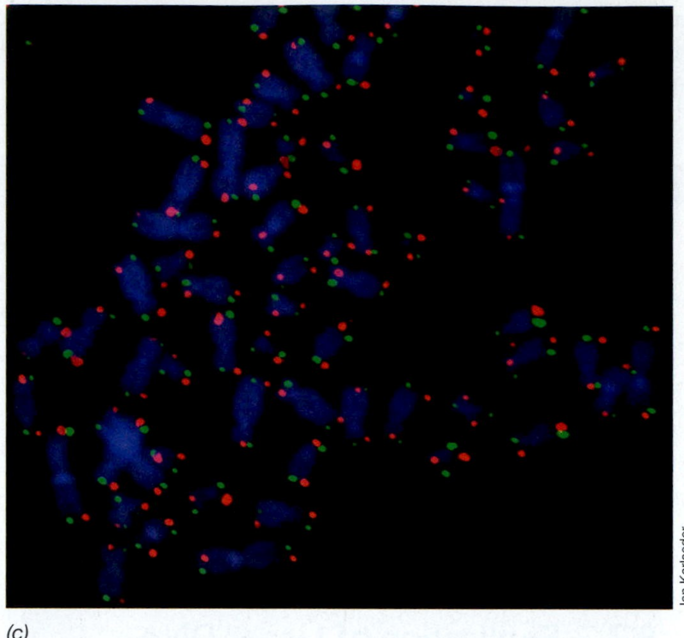

Figure 8.9 Model for the activity of telomerase at one end of a eukaryotic chromosome. *(a)* A diagram of the sequence of the end of the DNA in a telomere, with four of the guanine-rich repeats and the enzyme telomerase, which contains a short RNA template. *(b)* Steps in elongation of the guanine-rich strand catalyzed by telomerase. After telomerase finishes, the lagging strand can be primed with an RNA primer by primase followed by completion of the lagging strand by DNA polymerase and ligase. *(c)* A preparation of HeLa cell chromosomes stained with fluorescent dyes. The red dots are leading strand telomeres and the green dots are lagging strand telomeres.

chromosomes and are often more or less centrally located. The term kinetochore refers to the complex assemblage of proteins that links the DNA of the centromere region to the spindle fibers.

Unlike telomeres, centromeres vary greatly in sequence and length from one eukaryotic organism to another. In higher animals such as humans, the centromere consists of multiple repeat sequences spread over about 1 Mbp, whereas in *Saccharomyces cerevisiae*, the total centromere sequence is 125 bp long. The short centromere sequence of yeast is fortunate from the viewpoint of genetic engineering, because stable maintenance and replication of plasmids or other cloning vectors in eukaryotes requires the presence of a centromere sequence (Section 12.13). A short sequence reduces the space needed on the vector devoted to the centromere.

8.7 MiniReview

The ends of linear DNA present a problem to the replication machinery that circular DNA molecules do not. Some linear DNA molecules of prokaryotes and viruses solve this problem using a protein primer. Eukaryotes solve the problem using a special enzyme called telomerase to extend one strand of the DNA.

- How can a protein prime DNA for replication?
- What is telomerase and what is its function?
- Why are centromeres important?

8.8 RNA Processing

In Chapter 7 we learned that in the process of transcription RNA is formed from a DNA template (Section 7.9). The RNA product of transcription is known as the primary transcript. However, many RNA molecules need alterations—known as **RNA processing**—before they are mature, that is, ready to carry out their role in the cell.

As we have seen, many genes in eukaryotes contain intervening sequences, the introns, between the protein-coding regions, the exons. These intervening sequences are removed from the primary transcript. The RNA is cleaved to remove the introns, and the exons are joined to form a contiguous protein-coding sequence in the mature mRNA (Figure 8.5). The process by which introns are removed and exons are joined is called *splicing*. Occasional introns are found in prokaryotes, but the mechanism of removal is different. (See also the Microbial Sidebar, "Inteins and Protein Splicing.")

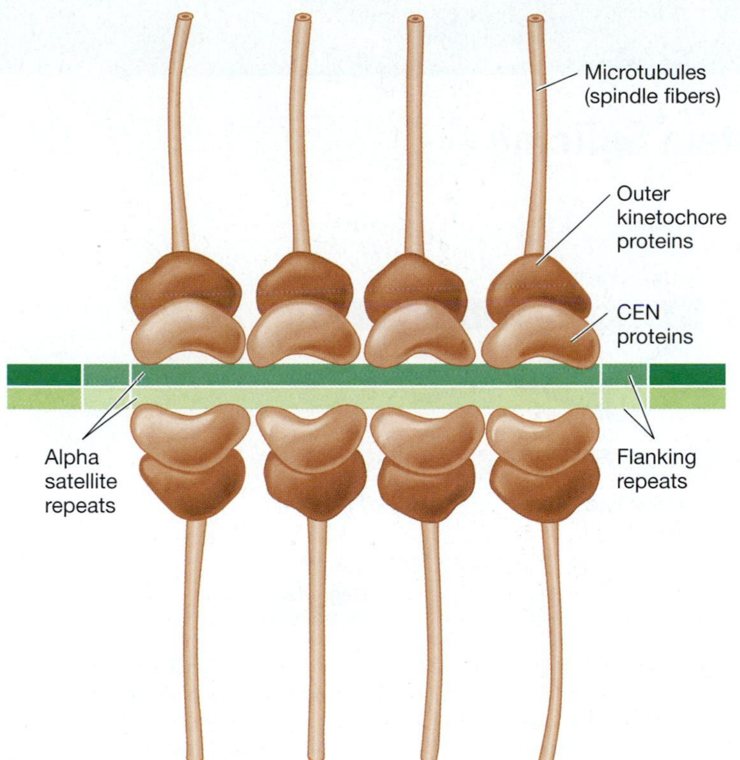

Figure 8.10 The eukaryotic centromere. DNA at the centromere of human chromosomes consists of multiple repeats of 171 bp flanked by other repeats, both of which are highly condensed into heterochromatin. The CEN proteins bind directly to the centromere DNA, and other proteins forming the kinetochore complex assemble onto the CEN proteins. The microtubules that make up the spindle attach to the kinetochore.

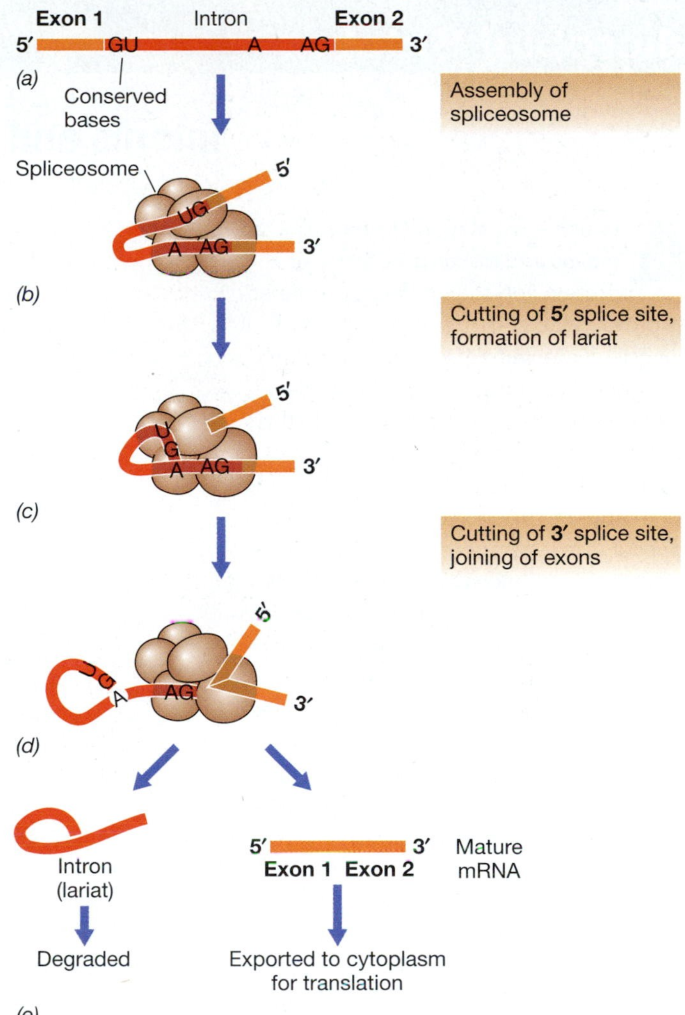

Figure 8.11 Activity of the spliceosome. Removal of an intron from the transcript of a protein-coding gene in a eukaryote. *(a)* A primary transcript containing a single intron. The sequence GU is conserved at the 5′ splice site and AG at the 3′ splice site. There is also an interior A that serves as a branch point. *(b)* Several small ribonucleoprotein particles (shown in brown) assemble on the RNA to form a spliceosome. Each of these particles contains distinct small RNA molecules that act in the splicing mechanism. *(c)* The 5′ splice site has been cut with the simultaneous formation of a branch point. *(d)* The 3′ splice site has been cut and the two exons have been joined. Note that overall, two phosphodiester bonds were broken, but two others were formed. *(e)* The final products are the joined exons (the mRNA) and the released intron.

The Spliceosome

RNA splicing takes place in the nucleus. Splicing is done by a large macromolecular complex about the size of a ribosome, called the **spliceosome**. The spliceosome removes introns and joins adjacent exons to form mature mRNA (**Figure 8.11**). The spliceosome contains four large RNA–protein complexes, called small nuclear ribonucleoproteins (snRNPs), together with many protein factors; indeed, over 100 proteins participate in its activity. The snRNPs each contain noncoding RNA molecules known as small nuclear RNA (snRNA) that recognize sequences on the primary transcript by base pairing. In particular, some of the snRNA molecules recognize conserved sequences at the splice junctions.

Once the splice junctions at both ends of an intron have been located by the snRNA, the proteins of the spliceosome cut out the intron and join the flanking exons together. The intron is removed as a lariat structure (Figure 8.11) that is then degraded, releasing free nucleotides. Many genes, especially in higher animals and plants, have multiple introns, so it is clearly important that they should all be recognized and removed by the spliceosome to generate the final mature mRNA.

RNA Capping and the Poly(A) Tail

There are two other unique steps in the processing of eukaryotic mRNA. Both steps take place in the nucleus prior to splicing. The first, called *capping,* occurs before transcription is complete. Capping is the addition of a methylated guanine nucleotide at the 5′-phosphate end of the mRNA. The cap nucleotide is added in reverse orientation relative to the rest of the mRNA molecule. Occasionally, other nucleotides next to the 5′ end of the eukaryotic mRNA are modified by methylation during capping. The guanosine cap is needed for translation and promotes the formation of the initiation complex between the mRNA and the ribosome through specific cap-binding proteins (Section 8.9).

Microbial Sidebar

Inteins and Protein Splicing

A rather unusual type of processing removes and discards portions of a protein and then reconnects the active protein domains. The splicing out of noncoding intervening sequences that interrupt genes is normally done at the level of RNA, as discussed in Section 8.8. In this case, the introns are removed during processing of the primary transcript to yield the final mature mRNA, consisting of the exons. However, a few instances are known where intervening sequences are removed at the level of the protein instead of the RNA. This process is called **protein splicing**, the peptide removed is called an **intein**, and the final mature protein consists of the **exteins**.

Protein splicing is rare overall, but is found in proteins from *Archaea*, *Bacteria*, chloroplasts, and lower *Eukarya*. Relative to the number of described organisms, inteins are actually most frequent in the *Archaea*. Usually there is just a single intein per protein, but occasional examples are known in which two inteins are inserted into the same host protein.

Figure 1 shows how protein splicing works in the production of DNA gyrase in the bacterium *Mycobacterium leprae,* the causative agent of leprosy (Section 34.5). The A subunit of the *M. leprae* DNA gyrase is the protein GyrA, which is encoded by the *gyrA* gene (Figure 1). Note that the flanking sequences in the GyrA polypeptide, the exteins, are ligated together to form the final active gyrase protein, while the inteins in the precursor GyrA are discarded (Figure 1). Interestingly, inteins are *self-splicing* entities and thus have the enzymatic activity of a highly specific protease.

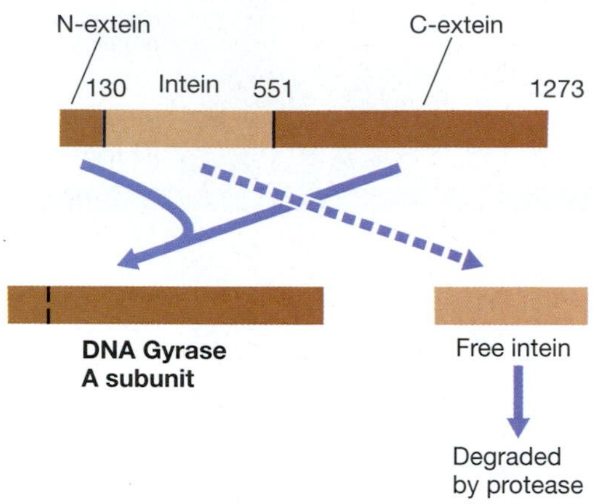

Figure 1 Protein splicing. *The protein synthesized from the gyrA mRNA in* Mycobacterium leprae *is 1,273 amino acid residues in length. The N-extein is the amino terminal extein and the C-extein is the carboxyl terminal extein. Residues 131 to 550 make up an intein, which removes itself in a self-splicing reaction that generates the free intein and the DNA gyrase A subunit.*

In addition to joining two exteins, the spliced-out intein polypeptide has a second enzyme activity; it acts as a site-specific deoxyribonuclease (DNase). Its function is to protect the existence of the intein. If a mutation occurred that deleted the intein DNA sequence from the middle of the host gene, the host cell would not be harmed, but the intein would be lost. However, the intein DNase will cut the DNA of such a host cell in the middle of the host gene. This is likely to kill any cell that deletes the useless intein. Only cells that keep the intein survive. This cutting occurs at the site where the intein is normally inserted; thus host genes with an intein are not cut because the nuclease recognition site is disrupted by the presence of the intein DNA sequence. Inteins are unnecessary for cell survival, and intein-encoding DNA may therefore be regarded as a form of selfish DNA (that is, DNA that promotes its own survival but confers no benefit on the host cell). The origin of inteins is obscure.

Inteins have found uses in biotechnology. Carrier proteins fused to a target protein are often used to facilitate protein purification (Section 26.2). However, after purification, the carrier must be removed. To accomplish this, inteins have been engineered as carrier proteins. Once the target protein has been purified, the intein is triggered to cut itself out. This releases the purified target protein without the need for additional and potentially complicated steps.

The second processing step consists of trimming the 3′ end of the precursor mRNA and adding 100–200 adenylate residues as the *poly(A) tail*. Precursor mRNA molecules have a tail recognition sequence, AAUAAA, close to the 3′ end and beyond the stop codon of the protein encoded by the mRNA. [Thus the poly(A) tail is not translated.] The poly(A) tail stabilizes mRNA and must be removed before the mRNA can be degraded. The poly(A) tail is also required for translation; it indicates to the translation machinery that the RNA is mRNA rather than some other form of RNA and that it is ready for translation. The three steps leading to the formation of mature eukaryotic mRNA are summarized in **Figure 8.12**.

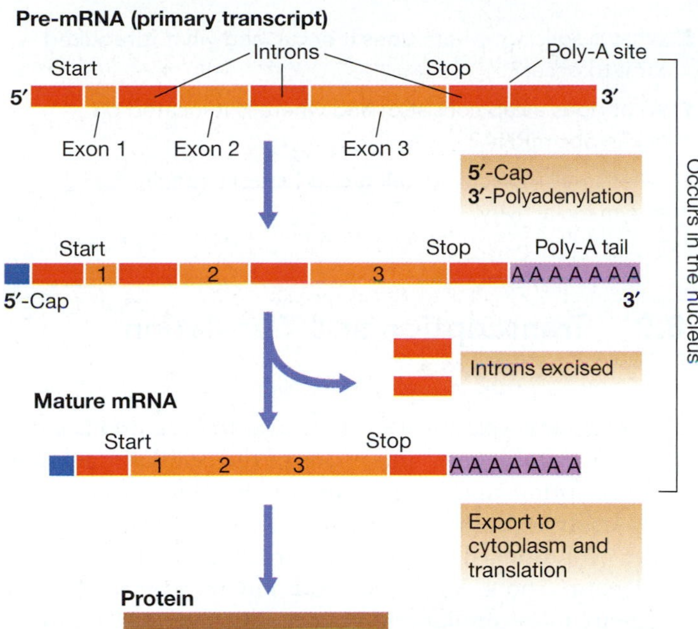

Figure 8.12 An overview of the processing of pre-mRNA into mature mRNA in eukaryotes. The processing steps include adding a cap at the 5′ end, removing the introns, and clipping the 3′ end of the transcript while adding a poly (A) tail. All these steps are carried out in the nucleus. The location of the start and stop codons to be used during translation are indicated.

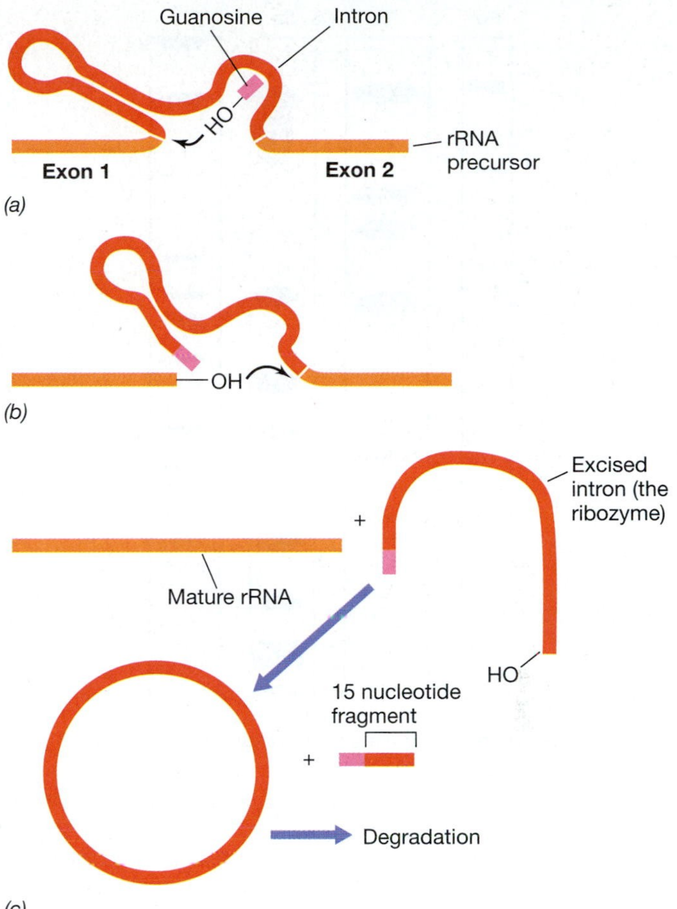

Figure 8.13 Self-splicing ribozymal intron of the protozoan *Tetrahymena*. There is considerable secondary structure in such molecules, which is critical for the splicing reaction. *(a)* An rRNA precursor contains a 413-nucleotide intron. *(b)* Following the addition of the nucleoside guanosine, the intron splices itself out and joins the two exons. *(c)* The intron is spliced out and then circularizes with the loss of a 15-nucleotide fragment.

Only when all three are complete is the mature mRNA transported into the cytoplasm for translation.

Some bacterial mRNAs are also polyadenylated, although the tails are relatively short (10–40 bases) and the enzyme that adds the poly(A) tail is associated with the ribosomes. In addition, the role of the poly(A) tail on bacterial mRNA is quite different from that on eukaryotic mRNA, as the poly(A) tail triggers self-degradation. Similarly, adding a poly(A) tail to chloroplast mRNA promotes its degradation, and this correlates with the prokaryotic ancestry of these organelles (∞ Section 18.4).

Self-Splicing Introns

Certain RNAs have enzymatic activity. These catalytic RNAs, called **ribozymes**, participate in several cellular reactions, the most important being polypeptide synthesis (∞ Section 7.15). Ribozymes work like protein enzymes in that they have an "active site" that binds the substrate and catalyzes formation of a product (∞ Section 5.5).

Self-splicing introns are introns that fold up to generate three-dimensional structures with ribozyme activity. This enzymatic activity allows them to excise themselves from an RNA molecule while joining adjacent exons together. Most self-splicing introns are found in the genes of mitochondria and chloroplasts. Others are present in the rRNA (but not mRNA) of lower eukaryotes, such as the single-celled ciliate *Tetrahymena*. In this case, a 413-nucleotide intron splices itself out of a longer precursor rRNA and joins two adjacent exons to form the mature rRNA (**Figure 8.13**). This *Tetrahymena* ribozyme is thus a sequence-specific endoribonuclease that carries out a reaction analogous to that of the spliceosome (Figure 8.11). Once removed from the precursor RNA, the intron circularizes following the removal of a short oligonucleotide fragment from itself (Figure 8.13). The circular fragment of the *Tetrahymena* intron is eventually degraded by the cell.

Two major classes of self-splicing introns are known. Those that require a free guanosine in the splicing reaction, such as the *Tetrahymena* intron in Figure 8.13, are group I introns. In contrast, group II introns use an internal adenosine residue to initiate splicing and generate a lariat product similar to that produced by spliceosomes (Figure 8.11).

Although catalyzing a specific reaction just as protein enzymes do, the ribozymes of self-splicing introns differ from protein enzymes in a key way. Unlike protein enzymes, a self-splicing ribozyme catalyzes its reaction *only once*. (Note that this is not true of ribozymes as a whole. Some ribozymes do indeed catalyze multiple reactions, just as protein enzymes do.)

UNIT 2

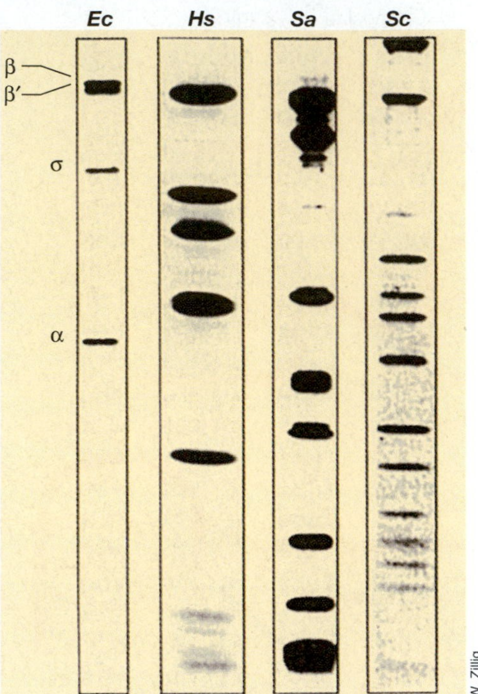

Figure 8.14 RNA polymerase and phylogeny. RNA polymerases from representatives of the three domains: Ec, *Escherichia coli* (*Bacteria*), Hs, *Halobacterium salinarum* (*Archaea*), Sa, *Sulfolobus acidocaldarius* (*Archaea*), and Sc, *Saccharomyces cerevisiae* (*Eukarya*). The purified components of the RNA polymerases have been separated by electrophoresis on a polyacrylamide gel. The largest polypeptide subunits are on the top, and the smallest subunits are on the bottom. Only members of the *Bacteria* contain the simple (four-polypeptide) RNA polymerase.

Why do ribozymes exist in a world in which protein enzymes dominate? It has been proposed that ribozymes are the vestigial remains of a simpler form of life, "RNA life," which may have predated the era in which proteins are the cell's major catalysts. Indeed, it is possible that an RNA world existed long before cellular structures evolved. We discuss this concept further in Chapter 14. Although proteins have replaced RNA enzymes in most areas of biocatalysis, a few reactions catalyzed by RNA remain. Either these are the last "holdouts" from the RNA world, or alternatively, they may be reactions that protein enzymes catalyze only very poorly and thus RNA catalysis is required to get the job done.

8.8 MiniReview

The processing of eukaryotic precursor mRNAs is unique and has three distinct steps: splicing, capping, and tailing. Only after processing can eukaryotic mRNA exit the nucleus to be translated. Introns in some transcripts are self-splicing, and the RNA itself catalyzes the reaction. RNA molecules with catalytic activity are called ribozymes and play an important role in the cell.

- What is splicing, where does it occur, and what is required for it to occur?
- What does a cap consist of and where is it located on eukaryotic mRNA?
- What small molecule do all group I introns require for enzymatic activity?

8.9 Transcription and Translation in *Eukarya*

The close genetic relationships between *Archaea* and *Eukarya* are most noticeable in comparisons of the cellular machinery for transcription and translation in the three domains of life. RNA polymerases from all sources contain some subunits that are evolutionarily conserved. Nonetheless, true to their phylogenetic roots, archaeal and eukaryotic RNA polymerases are more similar and structurally more complex than those of *Bacteria*. Similarly, protein synthesis in *Archaea* and *Eukarya* shares several features that contrast with *Bacteria* (Section 8.3).

Transcription in Eukaryotes

Eukaryotes have multiple RNA polymerases, unlike species of *Bacteria* and *Archaea*, which have just one. The eukaryotic nucleus contains three RNA polymerases, RNA polymerases I, II, and III, that transcribe different categories of nuclear genes. The single RNA polymerase of *Archaea* most closely resembles eukaryotic RNA polymerase II. Whereas the *Bacteria* contain a relatively simple (four-polypeptide) RNA polymerase, the RNA polymerase of *Archaea* and *Eukarya* has eight or more subunits (**Figure 8.14**). The membrane-bound organelles, mitochondria and chloroplasts, also possess RNA polymerase, but this resembles the bacterial enzyme.

Eukaryotic RNA polymerase I transcribes the genes for the two large rRNA molecules, 18S and 28S. RNA polymerase III transcribes genes for tRNA, 5S rRNA, and other small RNA molecules. Protein-encoding genes are transcribed by RNA polymerase II, which is therefore responsible for making all of the cell's mRNA. Each RNA polymerase recognizes a distinct class of promoters. In contrast, in prokaryotes the promoter for a gene encoding a protein could well be identical to a promoter for a gene encoding a tRNA.

A depiction of RNA polymerase II binding to a promoter on the DNA is shown in **Figure 8.15**. This promoter contains a TATA box, a conserved sequence resembling, in some respects, the Pribnow box in the promoters of *Bacteria* (Figure 7.22). The promoter also contains an initiator element very near the transcription start site. These two regions are key elements of eukaryotic promoters, although there may also be other sequences in any given promoter.

As for *Archaea*, eukaryotic RNA polymerases require transcription factors to recognize specific promoters. However, these proteins (and those of the *Archaea*) recognize the

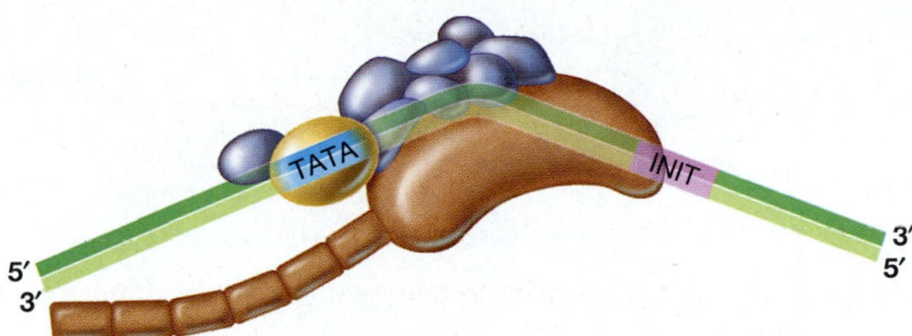

Figure 8.15 The interaction of eukaryotic RNA polymerase II with a promoter. The polymerase itself (brown) is positioned at the initiator element (INIT) of the promoter. A TATA box-binding protein (yellow) is shown bound at the TATA box. The polymerase has a repetitive amino acid sequence at one end (shown as a tail-like structure) that can be phosphorylated, affecting activity of the polymerase. The other proteins shown in blue are a few of the large number of accessory factors required for initiation of transcription in eukaryotes.

promoter independently, not as part of a polymerase holoenzyme as in *Bacteria* (∞ Figure 7.22). General transcription factors are needed for the functioning of all promoters recognized by a particular RNA polymerase. Specific transcription factors are needed for expression of certain genes under specific circumstances.

Translation in Eukaryotes

Protein synthesis by eukaryotic ribosomes is generally more complex than in *Bacteria*. The cytoplasmic ribosomes of eukaryotic cells (80S ribosomes) are larger than bacterial ribosomes (70S ribosomes) and contain more rRNA and protein molecules. In particular, the large ribosomal subunit contains three rRNA molecules, 5S, 5.8S, and 28S. The 5.8S rRNA is homologous to the 5′ end of bacterial 23S rRNA, so the eukaryotic 5.8S and 28S rRNAs taken together may be regarded as equivalent to the 23S rRNA of bacteria. Eukaryotes also have more initiation factors and a more complex initiation procedure. (Note that eukaryotic cells also contain bacterial-type ribosomes in their mitochondria and chloroplasts. The term "eukaryotic ribosomes" refers to those found in the eukaryotic *cytoplasm* and whose components are encoded by nuclear genes.) These differences are summarized in **Table 8.2**.

In prokaryotes, both *Bacteria* and *Archaea,* mRNA is polycistronic and may be translated to give several proteins. In eukaryotes, mRNA carries only a single gene that is translated into a single protein. That is, eukaryotic mRNA is monocistronic (Figure 8.5).

Initiation of protein synthesis differs significantly among the three domains. All organisms have a special initiator tRNA that recognizes the start codon and inserts the first amino acid. *Bacteria* start proteins with *N*-formylmethionine whereas eukaryotes and *Archaea* use methionine. However, unlike mRNA in both *Bacteria* and *Archaea,* eukaryotic mRNA has no ribosome-binding site (Shine–Dalgarno sequence). Instead, the mRNA is recognized by its cap (Section 8.8). A specific protein, cap-binding protein, binds both the mRNA cap and the ribosome. The first AUG to be found on the mRNA is normally used as the start codon in eukaryotes. In addition, the order of assembly of the ribosomal initiation complex is different: In eukaryotes the initiator Met-tRNA

Table 8.2 Comparison of translation

Bacteria	*Archaea*	*Eukarya (cytoplasm)*
Coupled transcription and translation	Coupled transcription and translation	No coupled transcription and translation for nuclear genes
Polycistronic mRNA	Polycistronic mRNA	Monocistronic mRNA
No cap on mRNA	No cap on mRNA	5′ end of mRNA recognized by cap
Start codon is next AUG after Shine–Dalgarno sequence	Start codon is often next AUG after Shine–Dalgarno sequence	First AUG in mRNA is used No ribosome-binding site
First amino acid is formylmethionine	First methionine is unmodified	First methionine is unmodified
70S ribosomes with 30S and 50S subunits	70S ribosomes with 30S and 50S subunits	80S ribosomes with 40S and 60S subunits
Small subunit rRNA: 16S (1,500 nucleotides)	Small subunit rRNA: 16S (1,500 nucleotides)	Small subunit rRNA: 18S (2,300 nucleotides)
Large subunit rRNA: 23S (2,900 nucleotides) and 5S (120 nucleotides)	Large subunit rRNA: 23S (2,900 nucleotides) and 5S (120 nucleotides)	Large subunit rRNA: 28S (4,200 nucleotides), 5.8S (160 nucleotides), and 5S (120 nucleotides)
Protein homologies: bacterial	Protein homologies: eukaryotic	Protein homologies: eukaryotic
rRNA homologies: bacterial	rRNA homologies: eukaryotic	rRNA homologies: eukaryotic
Not inhibited by diphtheria toxin	Inhibited by diphtheria toxin	Inhibited by diphtheria toxin
Two elongation factors	Multiple elongation factors	Multiple elongation factors
Three initiation factors	Multiple initiation factors	Multiple initiation factors

UNIT 2

binds to the small ribosomal subunit before the mRNA (the opposite is true in *Bacteria*).

Finally, translation is inhibited in eukaryotic cells by a protein toxin made by the bacterium *Corynebacterium diphtheriae*, the causative agent of the disease diphtheria. This toxin can actually kill cells and is a major factor in the pathology of the disease (Section 34.3). In line with the many similarities in the translational mechanisms of *Archaea* and eukaryotes, diphtheria toxin also inhibits translation in cells of *Archaea* but has no effect on the process in *Bacteria* (Table 8.2).

8.9 MiniReview

In the *Eukarya*, the major classes of RNA are transcribed by different RNA polymerases, with RNA polymerase II responsible for transcribing mRNA. Translation in the *Eukarya* involves larger ribosomes and more initiation factors. Eukaryotes begin each polypeptide chain with an unmodified methionine.

- Which type of eukaryotic RNA polymerase resembles most closely the RNA polymerase of *Archaea*?
- In eukaryotes, which type of genes are transcribed by RNA polymerases I and III?
- How is mRNA recognized by the ribosome in eukaryotes?

8.10 RNA Interference (RNAi)

Healthy cells contain double-stranded DNA and single-stranded RNA but not double-stranded RNA (dsRNA). The presence of dsRNA usually indicates the presence of an RNA virus genome within the cell. Note that even if the virion of an RNA virus contains single-stranded RNA, the virus genome must go through a double-stranded form (the replicative form) during replication. **RNA interference** (RNAi), a mechanism for defending against RNA viruses, both cleaves the double-stranded RNA and destroys any ssRNA (usually mRNA) that corresponds to the targeted dsRNA sequence. RNA interference is found only in eukaryotes, including protozoa, animals, and plants. It is not found in *Bacteria* or *Archaea*. (In *Bacteria*, ribonuclease III rapidly degrades dsRNA molecules, even those as short as 12 bp.)

RNAi is triggered by dsRNA of greater than 20 bp in length. Longer molecules of dsRNA are cleaved into fragments of 21–23 bp by a dsRNA-specific nuclease known as Dicer. These short dsRNA fragments are known as **short interfering RNA** (siRNA) and are bound by the RNA-induced silencing complex (RISC). The RISC complex recognizes and destroys single-stranded RNA that corresponds in sequence to siRNA (**Figure 8.16**). In practice, such single-stranded RNA will usually be mRNA made by the infecting RNA virus. To achieve this, the RISC complex separates the siRNA into its two strands. Any longer single-stranded RNA in the cell that base-pairs with these fragments is then cleaved by the nuclease Slicer, which is part of RISC.

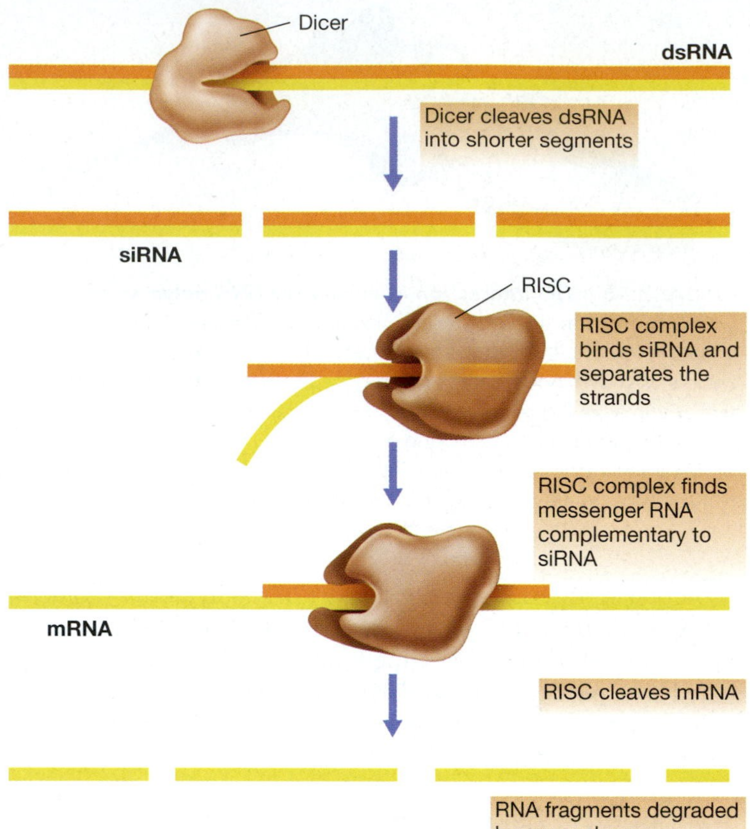

Figure 8.16 RNA interference. The nuclease Dicer cleaves double-stranded RNA into segments of 21–23 base pairs known as short interfering RNA (siRNA). This is recognized by RISC (RNA-induced silencing complex), which separates the strands of the siRNA. Finally, RISC cleaves target RNA that hybridizes to the siRNA.

The RNAi effect can spread from cell to cell and may travel quite a long way through an organism. This is especially noticeable in plants. The spread is due to copying of siRNA by an enzyme known as RNA-dependent RNA polymerase (RdRP). The siRNA then travels from cell to cell via intercellular connections called plasmodesmata. In lower animals, such as *Caenorhabditis elegans*, the RNAi effect can be "inherited" for several generations. However, mammals do not possess the RdRP needed to amplify RNAi, so the effect remains localized and cannot be transmitted.

8.10 MiniReview

In the *Eukarya* the mechanism of RNA interference (RNAi) destroys alien mRNAs. RNAi is triggered by the presence of double-stranded RNA and protects against infection by RNA viruses.

- What triggers RNA interference and what is the result?
- How does the RNAi effect spread throughout a plant or lower animal?

Review of Key Terms

Exon the coding DNA sequences in a split gene (contrast with intron)

Extein the portion of a protein that remains and has biological activity after the splicing out of any inteins

Histones basic proteins that protect and compact the DNA in eukaryotes and *Archaea*

Intein an intervening sequence in a protein; a segment of a protein that can splice itself out

Intron the intervening noncoding DNA sequences in a split gene (contrast with *exon*)

Meiosis specialized form of nuclear division that halves the diploid number of chromosomes to the haploid number for gametes of eukaryotic cells

Mitosis normal form of nuclear division in eukaryotic cells in which chromosomes are replicated and partitioned into two daughter nuclei

Nucleosome spherical complex of eukaryotic DNA plus histones

Primary transcript an unprocessed RNA molecule that is the direct product of transcription

Protein splicing removal of intervening sequences from a protein

Reverse gyrase an enzyme that introduces positive supercoils into DNA

Ribozyme catalytic RNA

RNA interference (RNAi) a response that is triggered by the presence of double-stranded RNA and results in the degradation of ssRNA homologous to the inducing dsRNA

RNA processing the conversion of a precursor RNA to its mature form

Self-splicing intron an intron that possesses ribozyme activity and splices itself out

Short interfering RNA (siRNA) short double-stranded RNA molecules that trigger RNA interference

Spliceosome a complex of ribonucleoproteins that catalyze the removal of introns from RNA primary transcripts

Telomerase an enzyme complex that replicates DNA at the end of eukaryotic chromosomes

Review Questions

1. List at least five features of *Archaea* that clearly differentiate the members of this domain from *Bacteria* (Sections 8.1–8.3).
2. Describe the common features of promoter sequences in *Archaea* and *Eukarya* (Section 8.2).
3. Compare translation in the *Archaea* with translation in the other two domains (Section 8.3 and 8.9).
4. List at least five features of eukaryotic cells that clearly differentiate them from both types of prokaryotic cells (Section 8.4).
5. Why are most eukaryotic genes much longer than corresponding genes in *Bacteria* (Section 8.5)?
6. Compare and contrast the processes of mitosis and meiosis. Which process is absolutely necessary for growth of *Saccharomyces cerevisiae* and why (Section 8.6)?
7. Why would genes eventually be lost from eukaryotic DNA in the absence of telomerase activity (Section 8.7)?
8. Why do eukaryotic mRNAs have to be "processed," whereas most prokaryotic RNAs do not (Section 8.8)?
9. How many RNA polymerases does a eukaryotic cell have (Section 8.9)? Which eukaryotic RNA polymerase is similar to the RNA polymerase of *Archaea*?
10. What is the purpose of RNA interference?

Application Questions

1. *Archaea* and *Bacteria* both have single circular chromosomes. Suggest possible reasons for this similarity.
2. Eukaryotic cells usually have much larger genomes than prokaryotic cells. List three reasons why this is not surprising.
3. Eukaryotic cells often have interrupted genes that contain introns whereas prokaryotes do not. Suggest possible reasons for this difference.

Regulation of Gene Expression

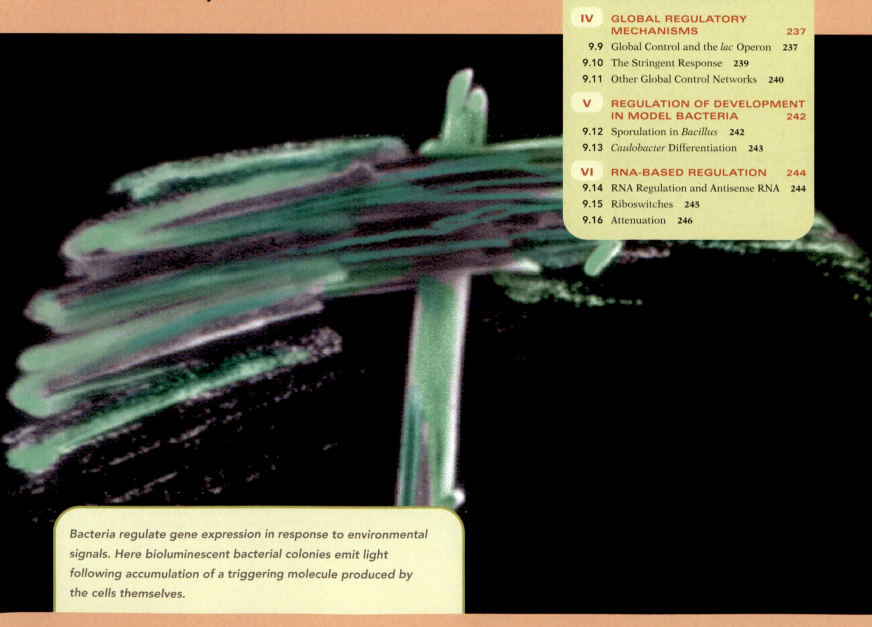

Bacteria regulate gene expression in response to environmental signals. Here bioluminescent bacterial colonies emit light following accumulation of a triggering molecule produced by the cells themselves.

In Chapters 7 and 8 we saw how the genetic information stored as a sequence of nucleotides in a gene could be transcribed into RNA in cells of the three domains of life. This information is then translated to yield a specific polypeptide. Collectively, these processes are called **gene expression**. Most proteins are enzymes (Section 5.5) that carry out the hundreds of different biochemical reactions needed for cell growth. However, to succeed in nature, microorganisms must respond rapidly to changes in their environment. To efficiently orchestrate the numerous reactions that occur in a cell and make maximal use of available resources, cells need to *regulate* the kinds and amounts of macromolecules they make. Such metabolic regulation is the focus of this chapter.

I OVERVIEW OF REGULATION

Some proteins and RNA molecules are needed in the cell at about the same level under all growth conditions. The expression of these molecules is said to be *constitutive*. However, it is more common for a particular macromolecule to be needed under some conditions but not others. For instance, enzymes required for the catabolism of the sugar lactose are useful to the cell only if lactose is present in its environment. Microbial genomes encode many more proteins than are actually present in the cell under any particular condition. Thus, regulation is a major process in all cells and a mechanism for conserving energy and resources.

9.1 Major Modes of Regulation

There are two major levels of regulation in the cell. One controls the *activity* of preexisting enzymes, and one controls the *amount* of an enzyme (**Figure 9.1**). The activity of a protein can be regulated only after it has been synthesized (that is, posttranslationally). By contrast, the amount of protein synthesized can be regulated at either the level of transcription, by varying the amount of messenger RNA (mRNA) made, or at the level of translation, by translating or not translating the mRNA.

Regulation of enzyme *activity* in prokaryotes is typically a very rapid process (occurring in seconds or less), whereas the regulation of enzyme *synthesis* is a relatively slow process (taking a few minutes). If a new enzyme needs to be synthesized, it will take some time before it is present in the cell in amounts sufficient to affect metabolism. Alternatively, if synthesis of an enzyme is stopped, a considerable amount of time may elapse before the existing enzyme is diluted sufficiently to no longer affect metabolism. However, working together, regulating enzyme activity and enzyme synthesis efficiently control cell metabolism.

Systems that control the level of expression of particular genes are varied, and genes can be regulated by more than one system. The processes that regulate the activity of preformed enzymes have already been discussed (Section 5.18). Here we consider how the synthesis of RNA and proteins is controlled.

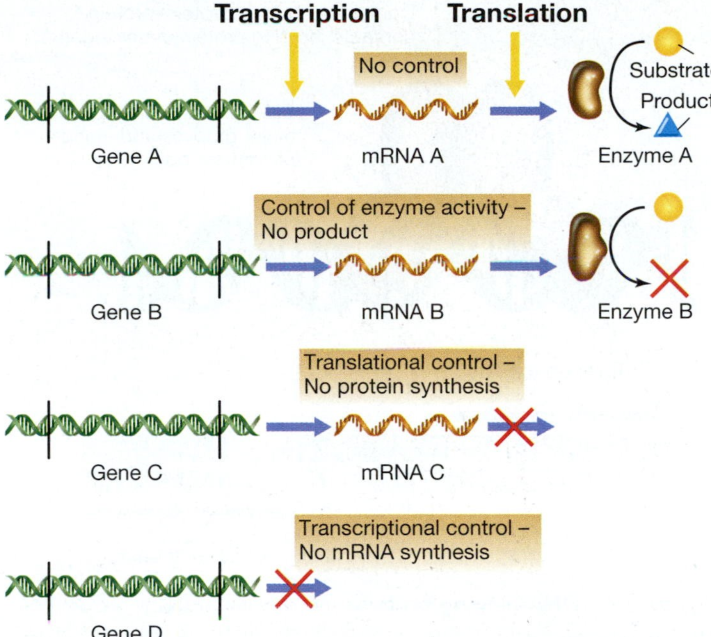

Figure 9.1 An overview of mechanisms of regulation. The product of gene A is enzyme A, which in this example is synthesized constitutively and carries out its reaction. Enzyme B is also synthesized constitutively, but its activity can be inhibited. The synthesis of the gene C product can be prevented by control at the level of translation. The synthesis of the gene D product can be prevented by control at the level of transcription.

9.1 MiniReview

Most genes encode proteins, and most proteins are enzymes. Expression of an enzyme-encoding gene is regulated by controlling the activity of the enzyme or controlling the amount of enzyme produced.

- What steps in the synthesis of protein might be subject to regulation?
- Which is likely to be more rapid, the regulation of activity or the regulation of synthesis? Why?

II DNA-BINDING PROTEINS AND REGULATION OF TRANSCRIPTION

The amount of a protein present in a cell may be controlled at the level of transcription, at the level of translation, or, occasionally, by protein degradation. Our discussion focuses on control at the level of transcription because this is the major means of regulation in prokaryotes.

Recall that the half-life of a typical mRNA in prokaryotes is short, only a few minutes at best. Because the half-life of a mRNA is so short, prokaryotes can exploit this to respond quickly to changing environmental parameters. Although there are energy costs in resynthesizing mRNAs that have been translated only a few times before being degraded, there

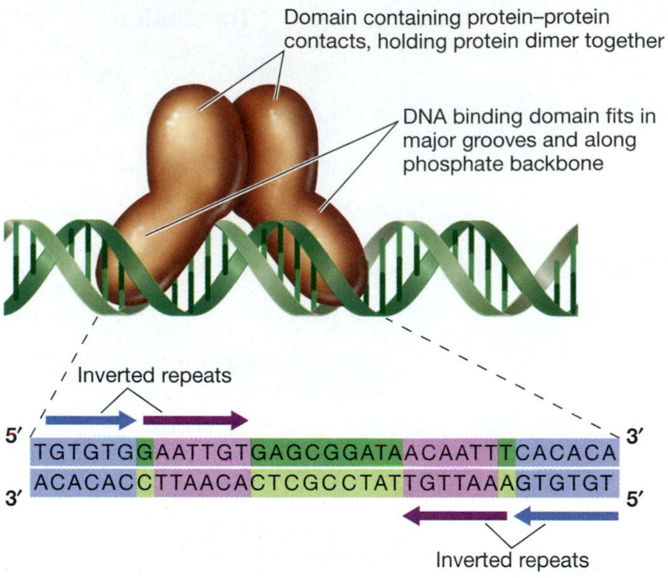

Figure 9.2 DNA-binding proteins. Many such proteins are dimers that combine specifically with two sites on the DNA. The specific DNA sequences that interact with the protein are inverted repeats. The nucleotide sequence of the operator gene of the lactose operon is shown, and the inverted repeats, which are sites at which the *lac* repressor makes contact with the DNA, are shown in shaded boxes.

are benefits from removing mRNAs rapidly when they are no longer needed, as this prevents the production of unneeded proteins. In the growing cell, transcription and mRNA degradation are thus coexisting processes.

For a gene to be transcribed, RNA polymerase must recognize a specific promoter on the DNA and begin functioning (Section 7.10). Regulation of transcription, both the turning on and turning off, typically requires proteins that can bind to DNA. Thus, before discussing specific regulatory mechanisms, we must consider DNA-binding proteins.

9.2 DNA-Binding Proteins

Small molecules are often involved in the regulation of transcription. However, they rarely do so directly. Instead, they typically influence the binding of certain proteins, called *regulatory proteins,* to specific sites on the DNA. It is these proteins that actually regulate transcription.

Interaction of Proteins with Nucleic Acids

Interactions between proteins and nucleic acids are central to replication, transcription, and translation, as well as to the regulation of these processes. Protein–nucleic acid interactions may be nonspecific or specific, depending on whether the protein attaches anywhere along the nucleic acid or whether the interaction is sequence specific. Histones (Section 8.5) are good examples of nonspecific binding proteins. Histones are universally present in *Eukarya* and are also present in many *Archaea.* Because they are positively charged, histones combine strongly and relatively nonspecifically with negatively charged DNA. If the DNA is covered with histones, RNA polymerase is unable to bind and transcription cannot occur. However, removal of histones does not automatically lead to transcription, but simply leaves genes in a position to be activated by other factors. For example, even after binding and beginning transcription, eukaryotic RNA polymerases need several protein factors to help them elongate through the histone-coated DNA.

Most DNA-binding proteins interact with DNA in a sequence-specific manner. Specificity is provided by interactions between specific amino acid side chains of the proteins and specific chemical groups on the bases and the sugar–phosphate backbone of the DNA. Because of its size, the *major groove* of DNA is the main site of protein binding. Figure 7.3 identified atoms of the bases in the major groove that are known to interact with proteins. To achieve high specificity, the binding protein must interact simultaneously with more than one nucleotide, frequently several. In practice, this means that a specific binding protein binds only to DNA containing a specific base sequence.

We have already described a structure in DNA called an *inverted repeat* (Figure 7.6). Such inverted repeats are frequently the locations at which regulatory proteins bind specifically to DNA (**Figure 9.2**). Note that this interaction does not involve the formation of stem-loop structures in the DNA, as was shown in Figure 7.6. DNA-binding proteins are typically homodimeric, meaning they are composed of two identical polypeptides. On each polypeptide chain is a domain that interacts specifically with a region of DNA in the major groove. When protein dimers interact with inverted repeats on DNA, each of the polypeptides binds to one of the inverted repeat regions and, in doing so, combines with each of the DNA strands (Figure 9.2). The DNA-binding protein does not recognize base sequences, per se, but instead recognizes molecular contacts associated with specific base sequences.

Structure of DNA-Binding Proteins

Studies of the structure of several DNA-binding proteins from both prokaryotes and eukaryotes have revealed several classes of protein domains that are critical for proper binding of these proteins to DNA. One of these is called the *helix-turn-helix motif* (**Figure 9.3**). The helix-turn-helix-motif consists of two segments of polypeptide chain with an α-helix secondary structure connected by a short sequence forming the "turn." The first helix is the *recognition helix* that interacts specifically with DNA. The second helix, the *stabilizing helix,* stabilizes the first helix by interacting hydrophobically with it. The turn linking the two helices consists of three amino acid residues, the first of which is typically a glycine (Figure 9.3*a*). Sequences are recognized by noncovalent interactions, including hydrogen bonds and van der Waals contacts (Section 3.1), between the recognition helix of the protein and specific chemical groups in the sequence of base pairs on the DNA.

Many different DNA-binding proteins from *Bacteria* contain the helix-turn-helix structure. These include many repressor proteins, such as the *lac* and *trp* repressors of *Escherichia coli* (Section 9.3), and some bacteriophage proteins, such as the phage lambda repressor (Figure 9.3*b*).

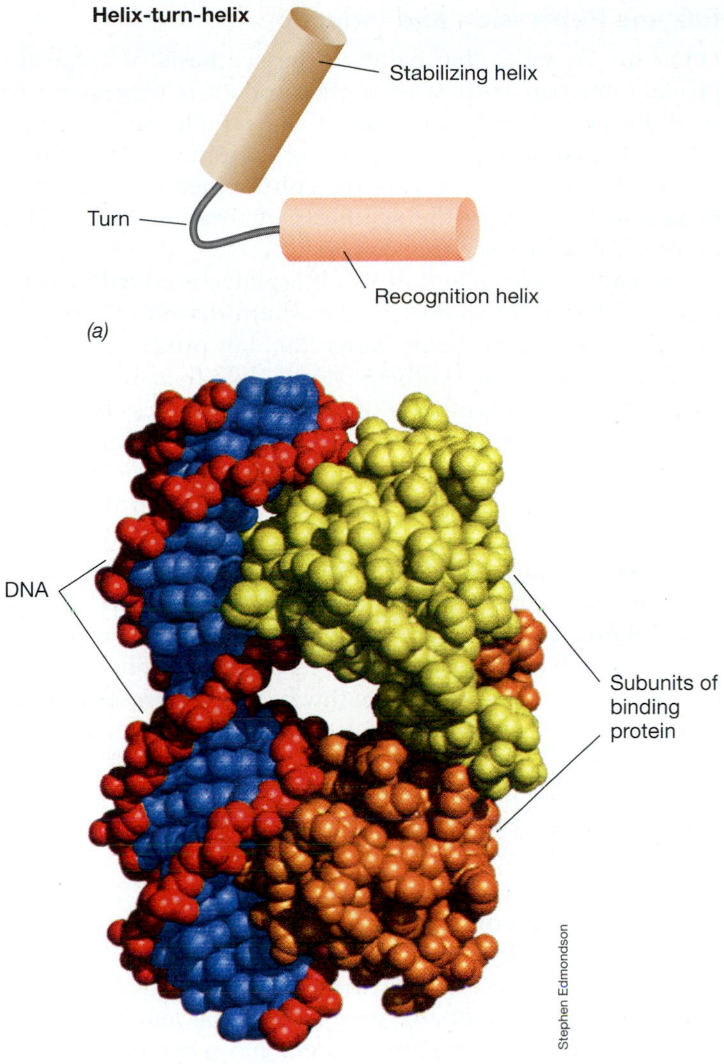

Figure 9.3 The helix-turn-helix structure of some DNA-binding proteins. *(a)* A simple model of the helix-turn-helix components. *(b)* A computer model of the bacteriophage lambda repressor, a typical helix-turn-helix protein, bound to its operator. DNA is red and blue, and one subunit of the dimeric repressor is brown and the other yellow. One subunit of the dimeric repressor is shown in dark brown and the other in dark yellow. Each subunit contains a helix-turn-helix structure. The coordinates used to generate this image were downloaded from the Protein Data Base, Brookhaven, NY (**www.pdb.org/pdb/home/**).

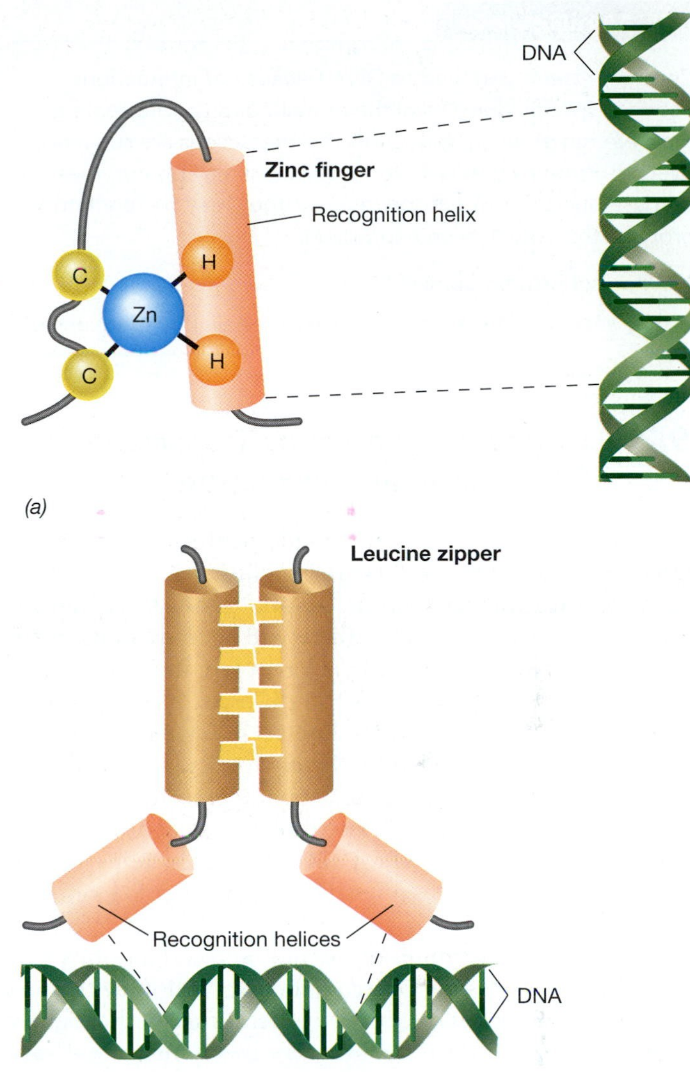

Figure 9.4 Simple models of protein substructures found in eukaryotic DNA-binding proteins. α-Helices are represented by cylinders. Recognition helices are the domains that bind DNA. *(a)* The zinc finger structure. The amino acids holding the Zn^{2+} ion always include at least two cysteine residues (C), with the other residues being histidine (H). *(b)* The leucine zipper structure. The leucine residues (yellow) are spaced exactly every seven amino acids. The interaction of the leucine side chains helps hold the two helices together.

Indeed, there are over 250 different proteins known with this motif that bind to DNA to regulate transcription in *E. coli.*

Two other types of protein domains are commonly found in DNA-binding proteins. One of these, the *zinc finger,* is frequently found in regulatory proteins in eukaryotes. The zinc finger is a protein structure that, as its name implies, binds a zinc ion (**Figure 9.4a**). Part of the "finger" of amino acids that is created forms an α-helix, and this recognition helix interacts with DNA in the major groove. There are typically at least two or three zinc fingers on proteins that use them for DNA binding.

The other protein domain commonly found in DNA-binding proteins is the *leucine zipper*. These proteins contain regions in which leucine residues are spaced every seven amino acids, somewhat resembling a zipper. Unlike the helix-turn-helix structure and the zinc finger, the leucine zipper does not interact with DNA itself but functions to hold two recognition helices in the correct orientation to bind DNA (Figure 9.4*b*).

Once a protein binds at a specific site on the DNA, a number of outcomes are possible. In some cases, the DNA-binding protein is an enzyme that catalyzes a specific reaction on the DNA, such as transcription by RNA polymerase. In other cases, however, the binding event can either block transcription (negative regulation, Section 9.3) or activate it (positive regulation, Section 9.4).

9.2 MiniReview

Certain proteins can bind to DNA because of interactions between specific domains of the proteins and specific regions of the DNA molecule. In most cases the interactions are sequence-specific. Proteins that bind to nucleic acid may be enzymes that use nucleic acid as substrates, or they may be regulatory proteins that affect gene expression.

- What is a protein domain?
- Why are some interactions specific to certain DNA sequences?

9.3 Negative Control of Transcription: Repression and Induction

Transcription is the first step in biological information flow (Figure 9.1); because of this, it is relatively easy to affect gene expression at this point. If one gene is transcribed more frequently than another, there will be more of its mRNA available for translation and therefore a greater amount of its protein product in the cell. Several different mechanisms for controlling gene expression are known in bacteria, and all of them are greatly influenced by the environment in which the organism is growing, in particular by the presence or absence of specific small molecules. These molecules can interact with specific proteins such as the DNA-binding proteins just described. The result is the control of transcription or, more rarely, translation.

We begin by describing repression and induction, simple forms of regulation that govern gene expression at the level of transcription. In this section we deal only with **negative control** of transcription, a regulatory mechanism that stops transcription (hence, the term *negative*).

Enzyme Repression and Induction

Often the enzymes that catalyze the synthesis of a specific product are not synthesized if the product is present in the medium in sufficient amounts. For example, the enzymes needed to synthesize the amino acid arginine are made only when arginine is absent from the culture medium; an excess of arginine represses the synthesis of these enzymes. This event is called enzyme **repression**.

As can be seen in Figure 9.5, if arginine is added to a culture growing exponentially in a medium devoid of arginine, growth continues at the previous rate, but production of the enzymes for arginine synthesis stops. Note that this is a specific effect, as the synthesis of all other enzymes in the cell continues at the previous rate. This is because the enzymes affected by a particular repression event make up only a tiny fraction of the entire complement of proteins in the cell at that time.

Enzyme repression is widespread in bacteria as a means of controlling the synthesis of enzymes required for the biosynthesis of amino acids and the nucleotide precursors purines and pyrimidines. In most cases, it is the final product of a particular biosynthetic pathway that represses the enzymes of the pathway. In these cases repression is quite specific and usually has no effect on the synthesis of enzymes other than those in the specific biosynthetic pathway (Figure 9.5). The benefit of enzyme repression should be obvious. It ensures that the organism does not waste energy and nutrients synthesizing unneeded enzymes.

Enzyme **induction** is conceptually the opposite of enzyme repression. In enzyme induction an enzyme is made only when its substrate is present. Enzyme repression typically affects biosynthetic (anabolic) enzymes. In contrast, enzyme induction usually affects degradative (catabolic) enzymes.

Consider, for example, the utilization of the sugar lactose as a carbon and energy source by *Escherichia coli*. Figure 9.6

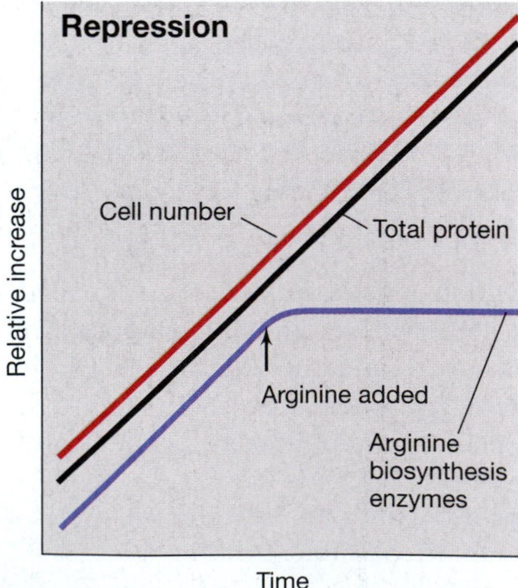

Figure 9.5 Enzyme repression. The addition of arginine to the medium specifically represses production of enzymes needed to make arginine. Net protein synthesis is unaffected.

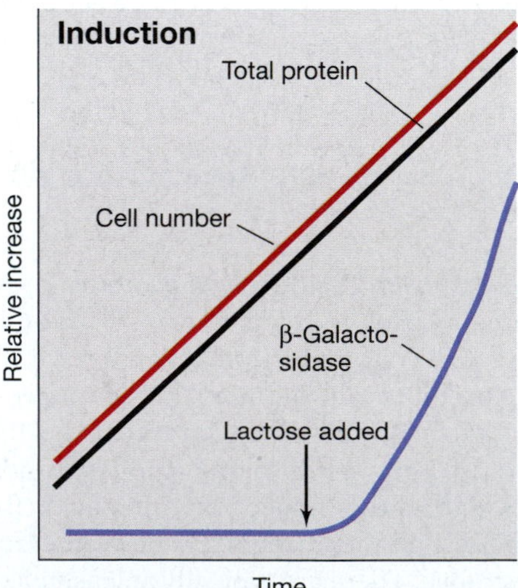

Figure 9.6 Enzyme induction. The addition of lactose to the medium specifically induces synthesis of the enzyme β-galactosidase. Net protein synthesis is unaffected.

shows the induction of β-galactosidase, the enzyme that cleaves lactose into glucose and galactose. This enzyme is required for *E. coli* to grow on lactose. If lactose is absent from the medium, the enzyme is not synthesized, but synthesis begins almost immediately after lactose is added. The three genes in the *lac* operon encode three proteins, including β-galactosidase, that are induced simultaneously upon adding lactose. One can immediately see the value to the organism of such a control mechanism; it provides a means to synthesize specific enzymes only when they are needed.

Inducers and Corepressors

The substance that induces enzyme synthesis is called an *inducer*, and a substance that represses enzyme synthesis is called a *corepressor*. These substances, which are normally small molecules, are collectively called *effectors*. Interestingly, not all inducers and corepressors are actual substrates or end products of the enzymes involved. For example, structural analogs may induce or repress even though they are not substrates of the enzyme. Isopropyl-thiogalactoside (IPTG), for instance, is an inducer of β-galactosidase even though IPTG cannot be hydrolyzed by this enzyme. In nature, however, inducers and corepressors are probably normal cell metabolites. However, detailed studies of lactose utilization in *E. coli* have shown that the actual inducer of β-galactosidase, is not lactose, but instead the structurally similar compound *allolactose*, a derivative made by the cell from lactose. www.microbiologyplace.com **Online Tutorial 9.1: Negative Control of Transcription and the *lac* Operon**

Mechanism of Repression and Induction

How can inducers and corepressors affect transcription in such a specific manner? They do this indirectly by binding to specific DNA-binding proteins, which, in turn, affect transcription. In the case of a repressible enzyme (Figure 9.5), the corepressor (in this case, arginine) binds to a specific **repressor protein**, the arginine repressor, which is present in the cell (**Figure 9.7**). The repressor protein is itself an allosteric protein (∞ Section 5.18); that is, its conformation is altered when the corepressor binds to it.

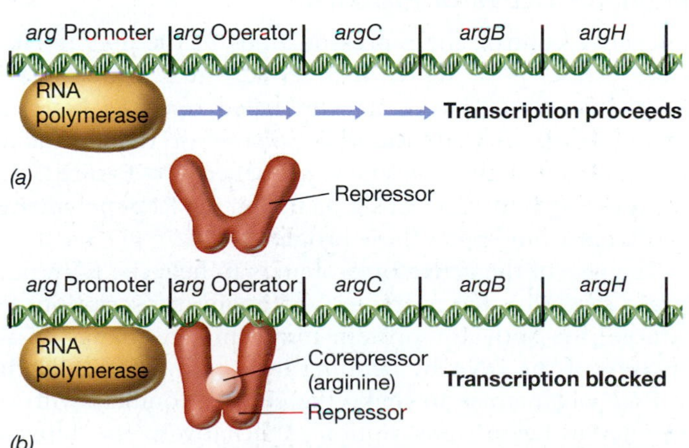

Figure 9.7 The process of enzyme repression using the arginine operon as an example. *(a)* The operon is transcribed because the repressor is unable to bind to the operator. *(b)* After a corepressor (small molecule) binds to the repressor, the repressor binds to the operator and blocks transcription; mRNA and the proteins it encodes are not made. For the *argCBH* operon, the amino acid arginine is the corepressor that binds to the arginine repressor.

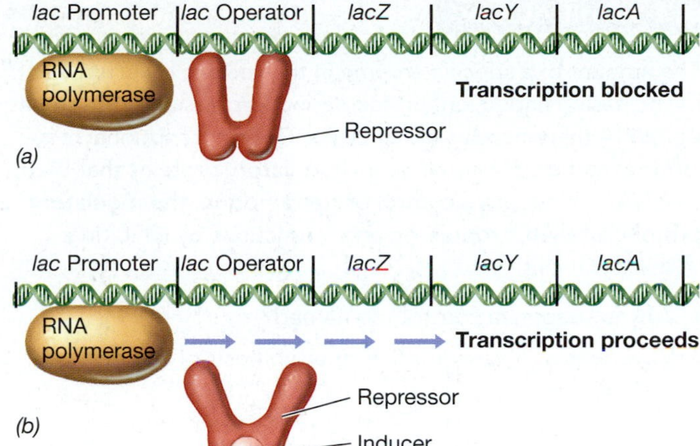

Figure 9.8 The process of enzyme induction using the lactose operon as an example. *(a)* A repressor protein binds to the operator region and blocks the binding of RNA polymerase. *(b)* An inducer molecule binds to the repressor and inactivates it so that it no longer can bind to the operator. RNA polymerase then transcribes the DNA and makes an mRNA for that operon. For the *lac* operon, the sugar allolactose is the inducer that binds to the lactose repressor.

By binding its effector, the repressor protein becomes active and can then bind to a specific region of the DNA near the promoter of the gene, the *operator*. This region gave its name to the **operon**, a cluster of genes arranged in a linear and consecutive fashion whose expression is under the control of a single operator (∞ Section 7.12). All of the genes in an operon are transcribed as a single unit yielding a single mRNA. The operator is located downstream of the promoter where synthesis of mRNA is initiated (Figure 9.7). If the repressor binds to the operator, transcription is physically blocked because RNA polymerase can neither bind nor proceed. Hence, the polypeptides encoded by the genes in the operon cannot be synthesized. If the mRNA is polycistronic (∞ Section 7.12), all the polypeptides encoded by this mRNA will be repressed.

Enzyme induction may also be controlled by a repressor. In this case, the repressor protein is active in the absence of the inducer, completely blocking transcription. When the inducer is added, it combines with the repressor protein and inactivates it; inhibition is overcome and transcription can proceed (**Figure 9.8**).

All regulatory systems employing repressors have the same underlying mechanism: inhibition of mRNA synthesis by the activity of specific repressor proteins that are themselves under the control of specific small effector molecules. And, as previously noted, because the repressor's role is inhibitory, regulation by repressors is called *negative control*.

UNIT 2

9.3 MiniReview

The amount of a specific enzyme in the cell can be controlled by increasing (induction) or decreasing (repression) the amount of mRNA that encodes the enzyme. This transcriptional regulation is carried out by allosteric regulatory proteins that bind to DNA. For negative control of transcription, the regulatory protein is called a repressor, and it functions by inhibiting mRNA synthesis.

- Why is "negative control" so named?
- How does a repressor inhibit the synthesis of a specific mRNA?

9.4 Positive Control of Transcription

In negative control, the controlling element—the repressor protein—brings about repression of mRNA synthesis. By contrast, in **positive control** of transcription, a regulator protein activates the binding of RNA polymerase to DNA. An excellent example of positive regulation is the catabolism of the disaccharide sugar maltose in *Escherichia coli.*

Maltose Catabolism in *Escherichia coli*

The enzymes for maltose catabolism in *E. coli* are synthesized only after the addition of maltose to the medium. The expression of these enzymes thus follows the pattern shown for β-galactosidase in Figure 9.6 except that maltose rather than lactose is required to affect gene expression. However, control of the synthesis of maltose-degrading enzymes is not under negative control as in the *lac* operon, but under positive control; transcription requires the binding of an **activator protein** to the DNA.

The maltose activator protein cannot bind to the DNA unless this protein first binds maltose, the inducer (also called an effector). When the maltose activator protein binds to DNA, it allows RNA polymerase to begin transcription (**Figure 9.9**). Like repressor proteins, activator proteins bind specifically only to certain sequences on the DNA. However, the region on the DNA that is the site of the activator is not called an operator (Figures 9.7 and 9.8), but instead an *activator-binding site* (Figure 9.9). Nevertheless, the genes controlled by this activator-binding site are still called an operon.

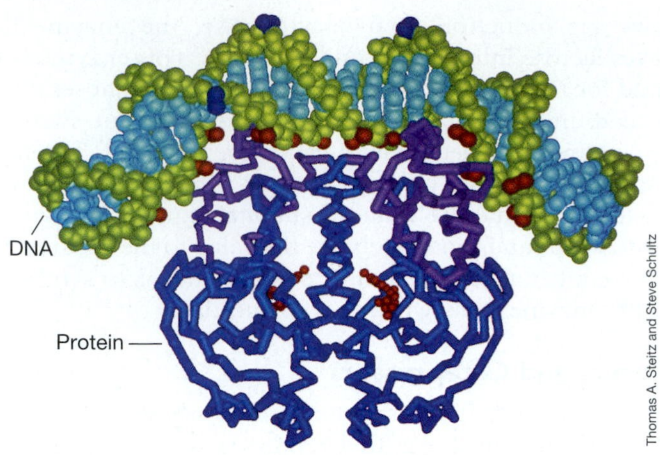

Figure 9.10 Computer model of the interaction of a positive regulatory protein with DNA. This model shows the cyclic AMP-binding protein (CRP protein; Section 9.9), a regulatory protein that helps control several operons. The α-carbon backbone of this protein is shown in blue and purple. The protein is binding to a DNA double helix (green and light blue). Note that binding of the CRP protein to DNA has bent the DNA.

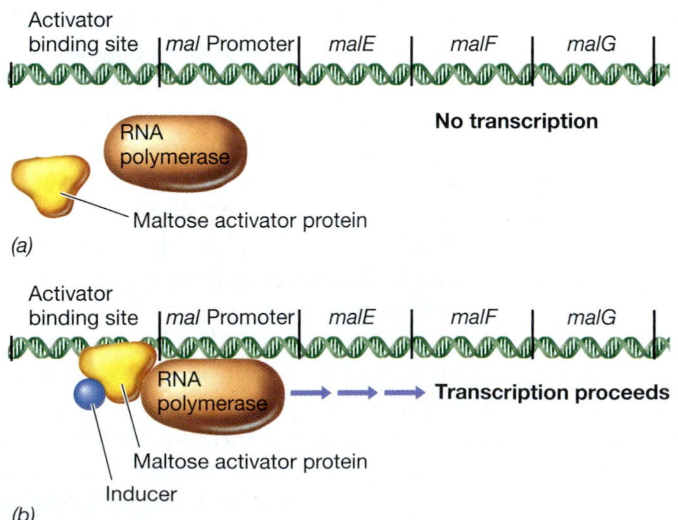

Figure 9.9 Positive control of enzyme induction using the maltose operon as an example. *(a)* In the absence of an inducer, neither the activator protein nor the RNA polymerase can bind to the DNA. *(b)* An inducer molecule (for the *malEFG* operon it is the sugar maltose) binds to the activator protein (MalT), which in turn binds to the activator-binding site. This allows RNA polymerase to bind to the promoter and begin transcription.

Binding of Activator Proteins

In negative control, the repressor binds to the operator and blocks transcription. How does an activator protein work? The promoters of positively controlled operons have nucleotide sequences that bind RNA polymerase weakly and are poor matches to the consensus sequence (Figure 7.22). Thus, even with the correct sigma factor, the RNA polymerase has difficulty binding to these promoters.

The role of the activator protein is to help the RNA polymerase recognize the promoter and begin transcription. For example, the activator protein may cause a change in the structure of the DNA by bending it (**Figure 9.10**), allowing the RNA polymerase to make the correct contacts with the promoter to begin transcription. Alternatively, the activator protein may interact directly with the RNA polymerase. This can happen either when the activator-binding site is close to the promoter (**Figure 9.11*a***) or when it is several hundred base pairs away from the promoter, a situation in which DNA looping is required to make the necessary contacts (Figure 9.11*b*).

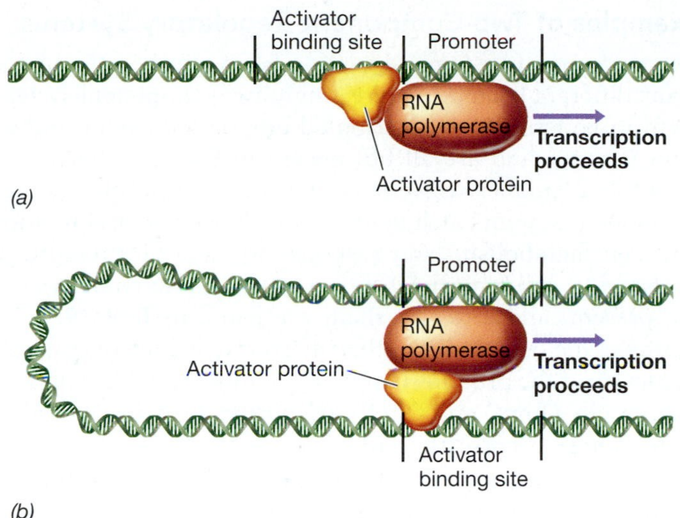

Figure 9.11 Activator protein interactions with RNA polymerase. *(a)* The activator-binding site is near the promoter. *(b)* The activator-binding site is several hundred base pairs from the promoter. In this case, the DNA must be looped to allow the activator and the RNA polymerase to contact.

Many genes in *E. coli* have promoters under positive control, and many have promoters under negative control. In addition, many operons have a promoter with multiple types of control and some have more than one promoter, each with its own control system! Thus, the rather simple picture outlined above is not typical of all operons. Multiple control features are common in the operons of virtually all prokaryotes, and thus their overall regulation can be very complex.

Operons versus Regulons

In *E. coli*, the genes required for maltose utilization are spread out over the chromosome in several operons, each of which has an activator-binding site to which a copy of the maltose activator protein can bind. Therefore, the maltose activator protein actually controls the transcription of more than one operon. When more than one operon is under the control of a single regulatory protein, these operons are collectively called a **regulon**. Therefore, the enzymes for maltose utilization are encoded by the maltose regulon.

Regulons are also known for operons under negative control as well. For example, the arginine biosynthetic enzymes (Section 9.3) are encoded by the arginine regulon, whose operons are all under the control of the arginine repressor protein (only one of the arginine operons was shown in Figure 9.7). However, in regulon control a specific DNA-binding protein binds only at those operons it controls regardless of whether it is functioning as an activator or repressor; other operons are not affected.

9.4 MiniReview

Positive regulators of transcription are called activator proteins. They bind to activator-binding sites on the DNA and stimulate transcription. As for repressors, the activity of activating proteins is modified by inducers. For positive control of enzyme induction, the inducer promotes the binding of the activator protein and thus stimulates transcription.

- Compare and contrast the activities of an activator protein and a repressor protein.
- Distinguish between an operon and a regulon.

UNIT 2

III SENSING AND SIGNAL TRANSDUCTION

9.5 Two-Component Regulatory Systems

Prokaryotes regulate cell metabolism in response to environmental fluctuations, including temperature changes, changes in pH and oxygen availability, changes in the availability of nutrients, and even changes in the number of cells present. Therefore, there must be mechanisms by which cells receive signals from the environment and transmit them to the specific target to be regulated.

We have seen in preceding sections that some signals can be small molecules that enter the cell and function as effectors. However, in many cases the external signal is not transmitted directly to the regulatory protein but instead is detected by a sensor that transmits it to the rest of the regulatory machinery, a process called **signal transduction**.

Sensor Kinases and Response Regulators

Most signal transduction systems contain two parts and are thus called **two-component regulatory systems**. Characteristically, such systems include two different proteins: a specific **sensor kinase protein** located in the cytoplasmic membrane and a partner **response regulator protein** present in the cytoplasm.

A kinase is an enzyme that phosphorylates compounds, typically using phosphate from ATP. Sensor kinases detect a signal from the environment and phosphorylate themselves (a process called autophosphorylation) at a specific histidine residue on their cytoplasmic surface (**Figure 9.12**). Sensor kinases are thus also called *histidine kinases*. The phosphoryl group is then transferred to another protein inside the cell, the response regulator. The response regulator is typically a DNA-binding protein that regulates transcription, either in a positive or negative fashion, depending on the system (Figure 9.12).

A balanced regulatory system must have a feedback loop, that is, a way to complete the regulatory circuit and terminate the initial signal. This resets the system for another cycle. This feedback loop involves a phosphatase, an enzyme activity that removes the phosphoryl group from the response regulator protein at a constant rate. In many systems this reaction is carried out by the response regulator

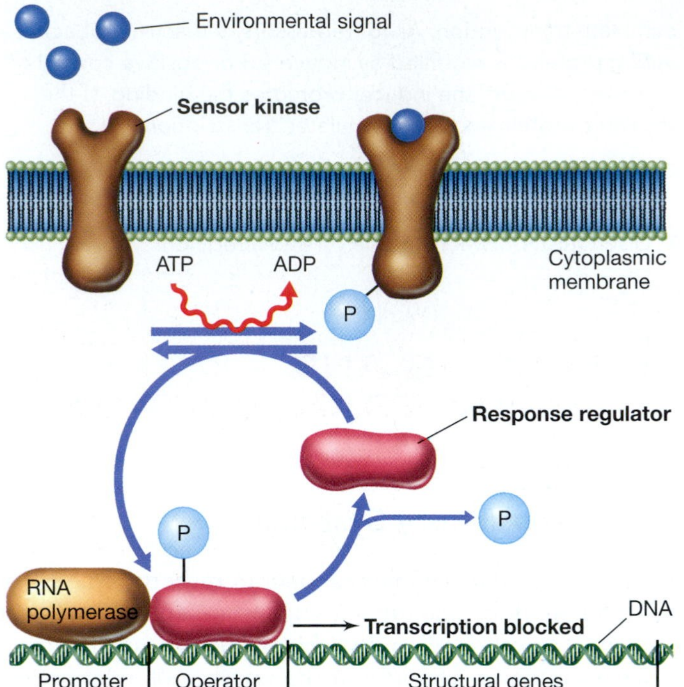

Figure 9.12 The control of gene expression by a two-component regulatory system. The main components of the system include a sensor kinase in the cytoplasmic membrane that phosphorylates itself in response to an environmental signal. The phosphoryl group is then transferred to the other main component, a response regulator. In the system shown here the phosphorylated response regulator is a repressor protein. The phosphatase activity of the response regulator slowly releases the phosphate from the response regulator and closes the circuit.

itself, although in many other cases separate proteins are present that stimulate this reaction (Figure 9.12). Phosphatase activity is typically a slower process than response regulator phosphorylation. However, if phosphorylation ceases due to reduced sensor kinase activity, phosphatase activity eventually returns the response regulator to the fully unphosphorylated state.

Examples of Two-Component Regulatory Systems

Two-component systems regulate a large number of genes in many different bacteria. Interestingly, two-component systems seem to be less widely distributed among species of *Archaea* than *Bacteria* and are all but absent in bacteria that live as parasites of higher organisms. A few key examples of two-component systems include those for phosphate assimilation, nitrogen metabolism, and response to osmotic pressure. In *Escherichia coli* almost 50 different two-component systems are present, and a few of them are listed in **Table 9.1**. For example, the osmolarity of the environment controls the relative levels of the proteins OmpC and OmpF in the *E. coli* outer membrane. OmpC and OmpF are porins, proteins that allow metabolites to cross the outer membrane of gram-negative bacteria (Section 4.7). If osmotic pressure is low, the synthesis of OmpF, a porin with a larger pore, increases; if osmotic pressure is higher, OmpC, a porin with a smaller pore, is made in larger amounts. The response regulator of this system is OmpR. When OmpR is phosphorylated, it activates transcription of the *ompC* gene and represses transcription of the *ompF* gene. The *ompF* gene in *E. coli* is also controlled by antisense RNA (Section 9.14).

Some signal transduction systems have multiple regulatory elements. For instance, in the Ntr regulatory system, which regulates nitrogen assimilation in many *Bacteria*, including *E. coli*, the response regulator is the activator protein NRI (nitrogen regulator I). NRI activates transcription from promoters recognized by RNA polymerase using σ^{54} (RpoN), an alternative sigma factor (Section 7.10 and Table 7.3). The sensor kinase in the Ntr system, NRII (nitrogen regulator II), fills a dual role as both kinase and phosphatase. The activity of NRII is in turn regulated by the addition or removal of uridine monophosphate groups from another protein, PII.

The Nar regulatory system (Table 9.1) controls a set of genes that allow the use of nitrate and/or nitrite as alternative electron acceptors during anaerobic respiration (Sections 21.6 and 21.7). The Nar system contains two different sensor kinases and two different response regulators. In addition, all

Table 9.1 Examples of two-component regulatory systems that regulate transcription in *Escherichia coli*

System	*Environmental signal*	*Sensor kinase*	*Response regulator*	*Activity of response regulator*[a]
Arc system	Oxygen	ArcB	ArcA	Repressor/activator
Nitrate and nitrite respiration (Nar)	Nitrate and nitrite	NarX	NarL	Activator/repressor
		NarQ	NarP	Activator/repressor
Nitrogen utilization (Ntr)	Shortage of organic nitrogen	NRII (=GlnL)	NRI (=GlnG)	Activator of promoters requiring RpoN/σ^{54}
Pho regulon	Inorganic phosphate	PhoR	PhoB	Activator
Porin regulation	Osmotic pressure	EnvZ	OmpR	Activator/repressor

[a]Note that many response regulator proteins act as both activators and repressors depending on the genes being regulated. Although ArcA can function as either an activator or a repressor, it functions as a repressor on most operons that it regulates.

of the genes regulated by this system are also controlled by the FNR protein, a global regulator for genes involved in anaerobic respiration (see Table 9.3). This type of multiple regulation is not uncommon in systems of central importance to the metabolism of a cell.

Genomic analyses allow for the ready detection of genes encoding two-component regulatory systems because the histidine kinases show significant amino acid sequence conservation. Some species of *Archaea,* in particular the methanogens (∞ Section 17.4), also possess two-component regulatory systems genes. Two-component systems closely related to those in *Bacteria* are also present in microbial eukaryotes, such as the yeast *Saccharomyces cerevisiae*, and even in plants. However, most eukaryotic signal transduction pathways rely on phosphorylation of serine, threonine and tyrosine residues of proteins that are unrelated to those of bacterial two-component systems.

9.5 MiniReview

Signal transduction systems transmit environmental signals to the cell. In prokaryotes signal transduction is typically carried out by a two-component regulatory system that includes a membrane-integrated sensor kinase and a cytoplasmic response regulator. The activity of the response regulator depends on its state of phosphorylation.

- What are kinases and what is their role in two-component regulatory systems?
- After depletion of an environmental signal, how are the two components reset for a second stimulus response?

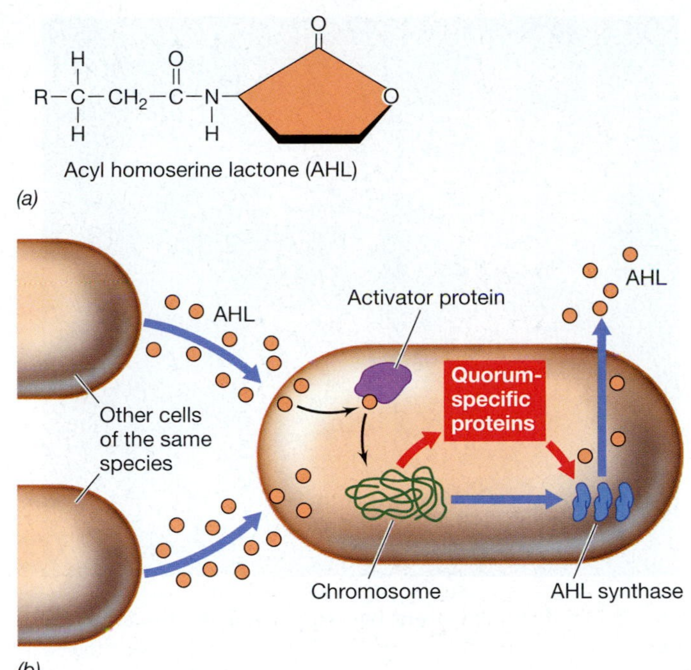

Figure 9.13 Quorum sensing. *(a)* General structure of an acyl homoserine lactone (AHL). Different AHLs are variants of this parent structure. R = alkyl group (C_1–C_{17}); the carbon next to the R group is often modified to a keto group (C=O). *(b)* A cell capable of quorum sensing expresses acyl homoserine lactone synthase at basal levels. This enzyme makes the cell's specific AHL. When cells of the same species reach a certain density, the concentration of AHL rises sufficiently to bind to the activator protein, which activates transcription of quorum-specific genes.

9.6 Quorum Sensing

As previously mentioned, organisms regulate genes in response to signals in the environment. One potential signal that prokaryotes can respond to is the presence of other cells of the same species in their surroundings. Some prokaryotes have regulatory pathways that are controlled by the density of cells of their own kind. This type of control is called **quorum sensing** (the word "quorum" in this sense means "sufficient numbers").

Mechanism of Quorum Sensing

Quorum sensing is a mechanism by which bacteria assess their population density. This is of practical use to ensure that sufficient cell numbers of a given species are present before initiating a response that requires a certain cell density to have an effect. For example, a pathogenic (disease-causing) bacterium that secretes a toxin can have no effect as a single cell; production of the toxin by that cell alone would be a waste of resources. However, if there is a sufficiently high population of cells present, the coordinated expression of the toxin may successfully initiate disease.

Quorum sensing is widespread among gram-negative bacteria but is also found in gram-positive bacteria. Each species that employs quorum sensing synthesizes a specific signal molecule called an **autoinducer**. This molecule diffuses freely across the cell envelope in either direction. Because of this, the autoinducer reaches high concentrations inside the cell only if there are many cells nearby, each making the same autoinducer. Inside the cell, the autoinducer binds to a specific activator protein and in so doing triggers transcription of specific genes (**Figure 9.13*b***).

There are several different classes of autoinducers (**Table 9.2**). The first to be identified were the *acyl homoserine lactones* (AHLs) (Figure 9.13*a*). Several different AHLs, with acyl groups of different lengths, are found in different species of gram-negative bacteria. In addition, many gram-negative bacteria make AI-2 (*auto*i*nducer 2*; a cyclic furan derivative). This is apparently used as a common autoinducer between many species of bacteria. Among gram-positive bacteria, certain short peptides function as autoinducers.

Quorum sensing was first discovered as the mechanism of regulating light emission in bioluminescent bacteria. Several bacterial species can emit light, including the marine bacterium *Aliivibro fischeri* (∞ Section 15.12). **Figure 9.14** shows bioluminescent colonies of *A. fischeri.* The light is due to the activity of an enzyme called luciferase. The *lux* operons, which encode the proteins needed for bioluminescence, are under control of the activator protein LuxR and are induced when the concentration of the specific *A. fischeri* AHL,

Figure 9.14 Bioluminescent bacteria producing the enzyme luciferase. Cells of the bacterium *Aliivibrio fischeri* (Section 15.12) were streaked on nutrient agar in a Petri dish and allowed to grow overnight. The photograph was taken in a darkened room using only the light generated by the bacteria.

N-3-oxohexanoyl homoserine lactone, becomes high enough. This AHL is synthesized by the enzyme encoded by the *luxI* gene.

Examples of Quorum Sensing

A variety of genes are controlled by a quorum-sensing system, including some in pathogenic bacteria. For example, pseudomonads use 4-hydroxy-alkyl quinolines as autoinducers to induce genes involved in virulence and related activities. In *Pseudomonas aeruginosa,* for instance, quorum sensing triggers the expression of a large number of unrelated genes when the population density becomes sufficiently high. These genes assist cells of *P. aeruginosa* in the transition from growing freely suspended in liquid to growing in a semisolid matrix called a *biofilm*. The biofilm, formed from specific polysaccharides produced by *P. aeruginosa,* increases the pathogenicity of this organism and prevents the penetration of antibiotics (Sections 23.4 and 23.5).

The pathogenesis of *Staphylococcus aureus* (Sections 34.10 and 37.5) involves, among many other things, the production and secretion of small cell-surface and extracellular peptides that damage host cells or that interfere with the immune system. The genes encoding these virulence factors are under the control of a quorum-sensing system that responds to a peptide produced by the organism. The regulation of these genes is quite complex and requires a regulatory RNA molecule discussed below. The *S. aureus* quorum-sensing system also uses regulatory proteins that are part of a signal-transduction system (Section 9.5).

Quorum sensing probably exists in *Archaea,* in particular haloalkaliphilic *Archaea,* organisms that inhabit alkaline saline habitats (Section 17.3), although the reports are preliminary and the details are unclear. However, quorum sensing also occurs in microbial eukaryotes. For example, in the eukaryotic budding yeast, *Saccharomyces cerevisiae,* specific aromatic alcohols are produced as autoinducers and control the transition between growth of *S. cerevisiae* as single cells and as elongated filaments. Similar transitions are seen in other fungi, some of which, such as *Candida,* whose quorum sensing is mediated by the long chain alcohol farnesol, cause disease in humans.

9.6 MiniReview

Quorum sensing allows cells to monitor their environment for cells of their own kind. Quorum sensing depends on the sharing of specific small molecules known as autoinducers. Once a sufficient concentration of the autoinducer is present, specific gene expression is triggered.

- What are the properties required for a molecule to function as an autoinducer?
- In terms of autoinducers, how does quorum sensing differ between gram-negative and gram-positive bacteria?

Table 9.2 Examples of quorum sensing and autoinducers

Organism	*Autoinducer*	*Receptor*[a]	*Process regulated*
Proteobacteria	Acyl homoserine lactones	LuxR protein	Diverse processes
Many diverse bacteria	AI-2 (furanone ± borate)[b]	LuxQ protein	Diverse processes
Pseudomonads	4-Hydroxy alkyl quinolines	?	Virulence; biofilms
Streptomyces	Gamma-butyrolactones	ArpA repressor	Antibiotic synthesis; sporulation
Gram-positive bacteria	Oligopeptides (linear or cyclic)	Two-component systems	Diverse processes
Yeast	Aromatic alcohols	?	Filamentation

[a]The details of how many of the quorum-sensing systems operate is complex. Some involve two-component regulatory systems (Section 9.5).

[b]The AI-2 autoinducer exists in several slightly different structures, some of which have an attached borate group.

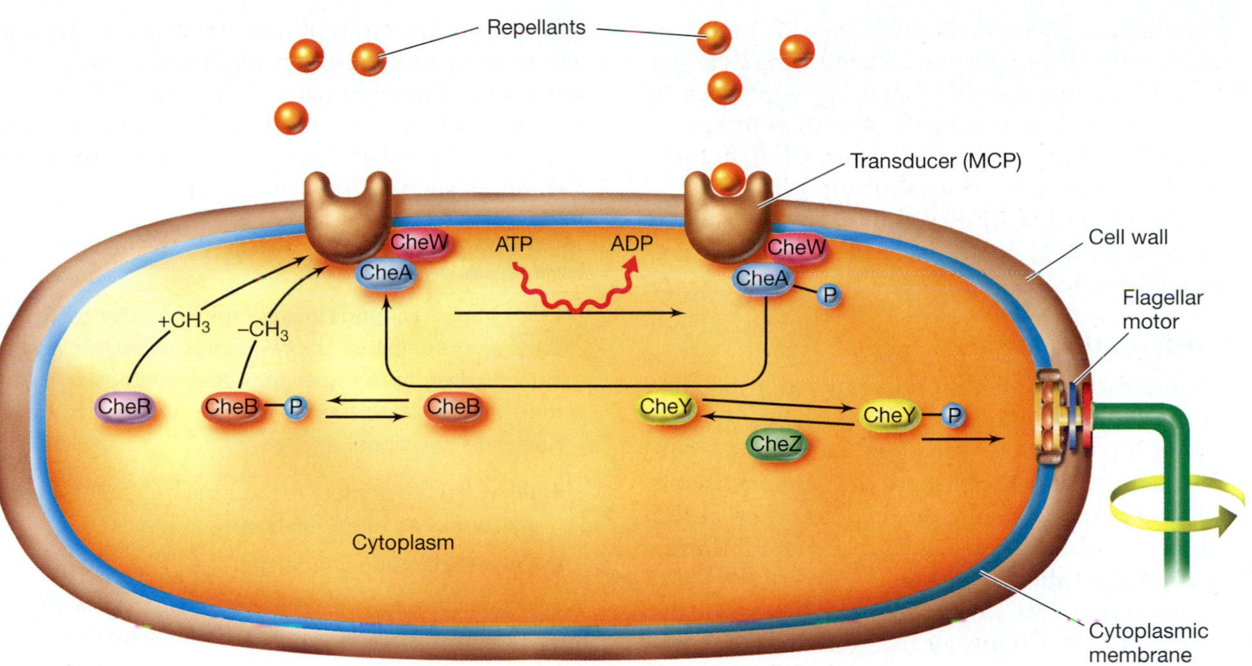

Figure 9.15 Interactions of MCPs, Che proteins, and the flagellar motor in bacterial chemotaxis. The MCP forms a complex with the sensor kinase CheA and the coupling protein CheW. This combination results in a signal-regulated autophosphorylation of CheA to CheA-P. CheA-P can then phosphorylate the response regulators CheB and CheY. Phosphorylated CheY (CheY-P) interacts directly with the flagellar motor switch. CheZ dephosphorylates CheY-P. CheR continually adds methyl groups to the transducer. CheB-P (but not CheB) removes them. The degree of methylation of the MCPs controls their ability to respond to attractants and repellents and leads to adaptation.

9.7 Regulation of Chemotaxis

We have previously seen how prokaryotes can move toward attractants or away from repellents, a process called *chemotaxis* (∞ Section 4.15). We noted that prokaryotes are too small to actually sense spatial gradients of a chemical, but they can respond to temporal gradients. That is, they can sense the *change* in concentration of a chemical over time rather than the absolute concentration of the chemical stimulus. Prokaryotes use a modified two-component system to sense temporal changes in attractants or repellents and process this information to regulate flagellar rotation. Note that in the case of chemotaxis we have a two-component system that directly regulates the activity of preexisting flagella not the transcription of the genes encoding the flagella.

Step One: Response to Signal

The mechanism of chemotaxis is complex and depends upon a variety of different proteins. Several sensory proteins reside in the cytoplasmic membrane and sense the presence of attractants and repellents. These sensor proteins are not themselves sensor kinases but interact with cytoplasmic sensor kinases. These sensory proteins allow the cell to monitor the concentration of various substances over time.

The sensory proteins are called *methyl-accepting chemotaxis proteins* (*MCPs*). In *Escherichia coli,* five different MCPs have been identified, and each is a transmembrane protein (**Figure 9.15**). Each MCP can sense a variety of compounds. For example, the Tar MCP of *E. coli* can sense the attractants aspartate and maltose as well as repellents such as the heavy metals cobalt and nickel.

MCPs bind attractants or repellents directly or in some cases indirectly through interactions with periplasmic binding proteins. Binding of an attractant or repellent initiates a series of interactions with cytoplasmic proteins that eventually affects flagellar rotation. Recall that if rotation of the flagellum is counterclockwise, the cell will continue to move in a run, whereas if the flagellum rotates clockwise, the cell will tumble (∞ Section 4.13).

MCPs are in contact with the cytoplasmic proteins CheW and CheA (Figure 9.15). CheA is the sensor kinase in chemotaxis. When an MCP binds a chemical, it changes conformation and (with help from CheW) affects the autophosphorylation of CheA to form CheA-P. Attractants *decrease* the rate of autophosphorylation, whereas repellents *increase* this rate. CheA-P then passes the phosphate to CheY (forming CheY-P); this is the response regulator. CheA-P can also pass the phosphate to CheB, another response regulator, but this is a much slower reaction than the phosphorylation of CheY. We will discuss the activity of CheB-P later.

Step Two: Controlling Flagellar Rotation

CheY is a key protein in the system because it governs the direction of rotation of the flagellum. CheY-P interacts with the flagellar motor to induce clockwise flagellar rotation and tumbling (∞ Section 4.13). If nonphosphorylated, CheY cannot bind to the flagellar motor and the flagellum continues to rotate counterclockwise; this causes the cell to continue to run. Another protein, CheZ, dephosphorylates CheY, returning it to a form that allows runs instead of tumbles. Because repellents increase the level of CheY-P, they lead to tumbling, whereas attractants lead to a lower level of CheY-P and smooth swimming (runs).

Step Three: Adaptation

In addition to processing a signal and regulating flagellar rotation, the system must be able to detect a change in concentration of the attractant or repellent with time. This problem is solved by yet another aspect of chemotaxis, the process of *adaptation*. Adaptation is the feedback loop necessary to reset the system. This involves covalent modification of the MCPs by CheB, mentioned earlier.

As their name implies, MCPs can be methylated. The cytoplasmic protein, CheR (Figure 9.15), continually adds methyl groups to the MCPs at a slow rate using *S*-adenosylmethionine as a methyl donor. The response regulator CheB is a demethylase that removes methyl groups from the MCPs. Phosphorylation of CheB greatly increases its rate of activity. The changes in level of methylation of the MCPs results in conformational changes similar to those caused by binding of attractant or repellant. Methylation thus allows adaptation to sensory signals as outlined below.

If the level of an attractant remains *high*, the level of phosphorylation of CheA (and, therefore, of CheY and CheB) remains *low*, and the cell swims smoothly. The level of methylation of the MCPs increases during this period because CheB-P is not present to demethylate them rapidly. However, MCPs no longer respond to the attractant when they are fully methylated. Therefore, even if the level of attractant remains high, the level of CheA-P and CheB-P increases and the cell begins to tumble. However, now the MCPs can be demethylated by CheB-P, and when this happens, the receptors are "reset" and can once again respond to further increase or decrease in level of attractants.

The course of events is just the opposite for repellents. Fully methylated MCPs respond best to an increasing gradient of repellents and send a signal for cell tumbling to begin. The cell then moves off in a random direction while MCPs are slowly demethylated. With this mechanism for adaptation, chemotaxis successfully achieves the ability to monitor small changes in the concentrations of both attractants and repellents over time.

Other Taxes

In addition to chemotaxis, several other microbial taxes are known, for example, *phototaxis* (movement toward light), *aerotaxis* (movement toward oxygen), and other types of directed movements (∞ Section 4.15). Interestingly, many of the cytoplasmic Che proteins that function in chemotaxis also play a role in these other taxes. In phototaxis, for example, a light sensor protein replaces the MCPs of chemotaxis, and in aerotaxis, a redox protein monitors levels of oxygen. These sensors then interact with cytoplasmic Che proteins to direct runs or tumbles in response to these other signals. It thus appears that several different prokaryotic taxes can be driven by a common set of cytoplasmic proteins.

9.7 MiniReview

Chemotaxis responds in a complex manner to both attractants and repellents. The regulation of swimming affects the activity of proteins rather than their synthesis. Adaptation by methylation allows the system to reset itself to the continued presence of a signal.

- What are the primary response regulator and the primary sensor kinase for regulating chemotaxis?
- Why is adaptation important?
- What is the major difference in response by the chemotaxis system to an attractant versus a repellent?

9.8 Control of Transcription in *Archaea*

All organisms regulate expression of their genes. There are two alternative approaches to regulating the activity of RNA polymerase. One strategy, predominately used by *Bacteria* and *Archaea,* is to use DNA-binding proteins that either prevent RNA polymerase activity (repressor proteins) or promote RNA polymerase activity (activator proteins) (Section 9.3). The alternative, used most often by eukaryotes, is to transmit signals to the protein subunits of the RNA polymerase itself. It is perhaps surprising, then, that despite the greater overall similarity between the *mechanisms* of replication and transcription in *Archaea* and *Eukarya* (∞ Chapter 8), the *regulation* of transcription in *Archaea* more closely resembles that of *Bacteria*.

Few repressor or activator proteins from *Archaea* have yet been characterized in detail, but it is clear that both types of regulatory proteins exist. Archaeal repressor proteins function either by blocking the binding of RNA polymerase itself or by blocking binding of the TBP and TFB proteins required for promoter recognition and RNA polymerase binding in *Archaea* (∞ Section 8.2). At least some archaeal activator proteins function in just the opposite way, by recruiting TBP to the promotor, thereby facilitating transcription.

Repressor Proteins in *Archaea*

A good example of an archaeal repressor is the NrpR protein from the methanogen *Methanococcus maripaludis;* this protein represses genes involved in nitrogen metabolism (**Figure 9.16**) such as those involved in nitrogen fixation (∞ Section 20.14) or glutamine synthesis (∞ Section 5.16). When levels of organic nitrogen are plentiful in the cell, NrpR represses genes involved in nitrogen fixation and glutamine biosynthesis. However, if the level of nitrogen compounds

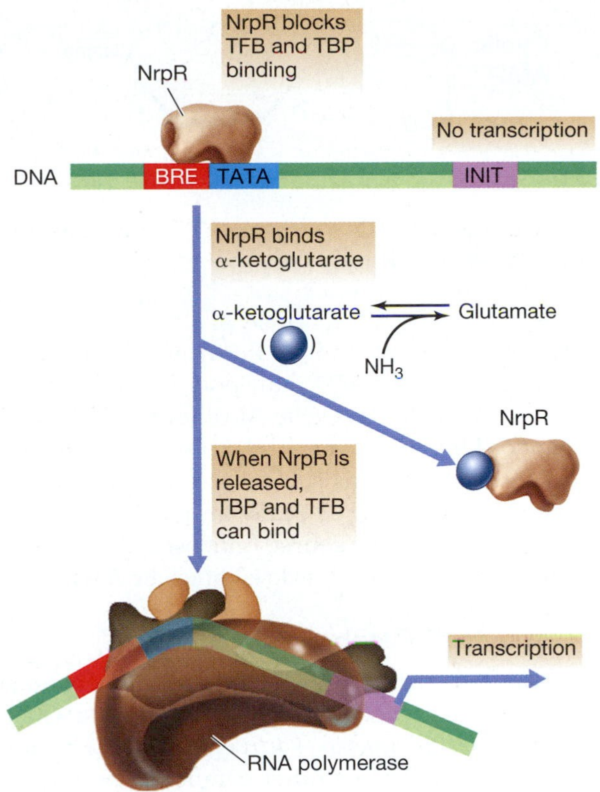

Figure 9.16 Repression of genes for nitrogen metabolism in *Archaea*. The NrpR protein of *Methanococcus maripaludis* acts as a repressor. It blocks the binding of the TFB and TBP proteins, which are required for promoter recognition, to the BRE site and TATA box, respectively. If there is a shortage of ammonia, α-ketoglutarate is not converted to glutamate. The α-ketoglutarate accumulates and binds to NrpR, releasing it from the DNA. Now TBP and TFB can bind. This in turn allows RNA polymerase to bind and transcribe the operon, starting from a point within the initiator box.

becomes limiting in the cell, this is a signal to the cell that more nitrogen is needed and that the repressor must be inactivated.

Interestingly, the signal that the level of nitrogen is limiting comes from elevated levels of α-ketoglutarate, a citric acid cycle intermediate, in the cell. This compound is converted to the amino acid glutamate by the addition of ammonia. When levels of α-ketoglutarate rise, this is a signal that ammonia is limiting and that additional means for obtaining ammonia, such as nitrogen fixation or the high-affinity nitrogen assimilation enzyme glutamine synthetase, need to be synthesized. Elevated levels of α-ketoglutarate function as an inducer by binding to the NrpR protein, which is then released from the DNA. This allows RNA polymerase to bind and transcribe the genes. Here the mechanism of inactivation of a repressor protein by a small molecule is analogous to the mechanism of induction considered earlier in *Bacteria*, where the presence of an inducer, such as lactose in the *lac* operon of *Escherichia coli*, triggered the transcription of genes involved in lactose catabolism (Section 9.3).

9.8 MiniReview

Archaea resemble *Bacteria* in using DNA-binding activator and repressor proteins to regulate gene expression at the level of transcription.

- What is the major difference between transcriptional regulation in *Archaea* versus eukaryotes?

UNIT 2

IV GLOBAL REGULATORY MECHANISMS

An organism often needs to regulate many unrelated genes simultaneously in response to a change in its environment. Regulatory mechanisms that respond to environmental signals by regulating the expression of many different genes are called *global control systems*. Both the lactose operon and the maltose regulon respond to global controls in addition to their own specific regulation discussed in Sections 9.3 and 9.4. We begin our consideration of global regulation by revisiting the *lac* operon.

9.9 Global Control and the *lac* Operon

Our discussions of the lactose and maltose regulatory systems did not consider the possibility that the cell's environment might contain several different carbon sources that the bacteria could use. For example, *Escherichia coli* can use many different sugars. When faced with several sugars, including glucose, do cells of *E. coli* use them simultaneously or one at a time? The answer is that glucose is always used first. Indeed, it would be wasteful to induce enzymes for the metabolism of other sugars if glucose were available, because *E. coli* grows faster on glucose than on any other carbon source. Thus, one mechanism of global control, catabolite repression, ensures that glucose is used first if available.

Catabolite Repression

In **catabolite repression** the syntheses of unrelated, primarily catabolic enzymes are repressed when cells are grown in a medium that contains glucose. Catabolite repression is also called the "glucose effect" because glucose was the first substance shown to initiate this response. In some organisms, however, carbon sources other than glucose cause catabolite repression. The key point is that the substrate that represses the use of other substrates is a better carbon and energy source. In this way, catabolite repression ensures that the organism uses the *best* available carbon and energy source first.

One consequence of catabolite repression is that it leads to two exponential growth phases, a situation called *diauxic growth*. If two usable energy sources are available, the cells grow first on the better energy source. Then, following a lag period, growth resumes on the other energy source. Diauxic

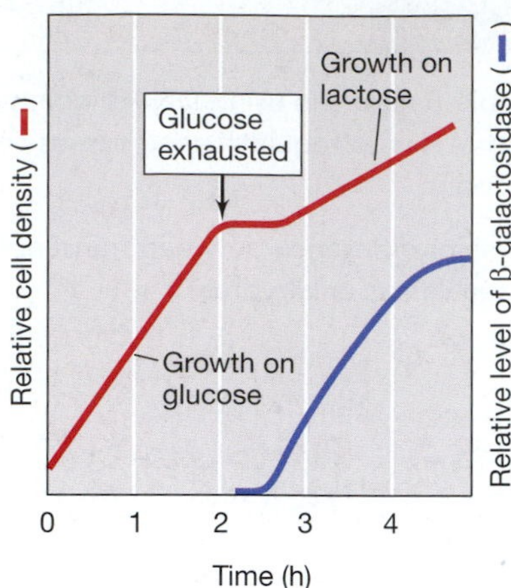

Figure 9.17 Diauxic growth of *Escherichia coli* on a mixture of glucose and lactose. Glucose represses the synthesis of β-galactosidase. After glucose is exhausted, a lag occurs until β-galactosidase is synthesized, and then growth resumes on lactose but at a slower rate.

growth is illustrated in **Figure 9.17** for the growth of *E. coli* on a mixture of glucose and lactose.

As we have seen, the proteins of the *lac* operon, including the enzyme β-galactosidase, are required for using lactose and are induced in its presence (Figures 9.6 and 9.8). But, in addition, their synthesis is subject to catabolite repression. As long as glucose is present, the *lac* operon is not expressed and lactose is not used. However, when glucose is exhausted, catabolite repression is abolished, and after a brief lag period, the *lac* operon is expressed and the cells grow on lactose. As Figure 9.17 shows, the cells grow more rapidly on glucose than on lactose. Even though glucose and lactose are both excellent energy sources for *E. coli*, glucose is a better carbon source for this organism, and growth is faster on this substrate.

Cyclic AMP and CRP

Despite its name, in catabolite repression transcription is controlled by an activator protein and is actually a form of positive control (Section 9.4). The activator protein is called the *cyclic AMP receptor protein* (*CRP*) (also called the catabolite activator protein, or CAP). A gene that encodes a catabolite-repressible enzyme is expressed only if CRP protein binds to DNA in the promoter region. This then allows RNA polymerase to bind to the promoter. CRP is an allosteric protein and binds to DNA only if it has first bound a small molecule called *cyclic adenosine monophosphate* (*cyclic AMP* or *cAMP*) (**Figure 9.18**). As for most DNA-binding proteins (Section 9.2), CRP binds to DNA as a dimer.

Cyclic AMP is a key molecule in many metabolic control systems, both in prokaryotes and eukaryotes. Because it is derived from a nucleic acid precursor, it is a **regulatory nucleotide**. Other regulatory nucleotides include cyclic GMP (cyclic guanosine monophosphate; important mostly in eukaryotes), cyclic di-GMP, and ppGpp (Section 9.10). Cyclic AMP is synthesized from ATP by an enzyme called *adenylate*

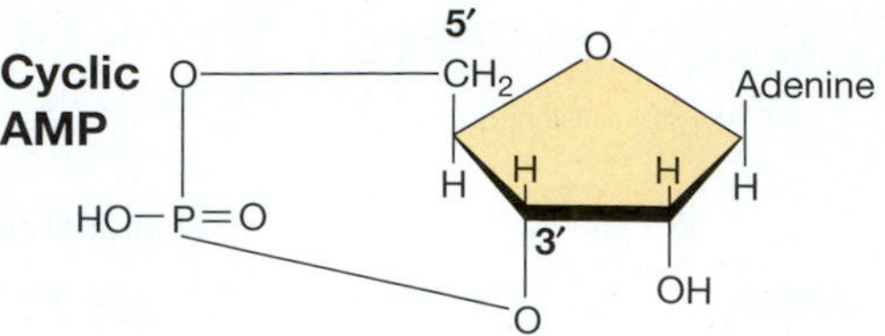

Figure 9.18 Cyclic AMP. Cyclic adenosine monophosphate is made from ATP by the enzyme adenylate cyclase.

cyclase. However, glucose inhibits the synthesis of cyclic AMP and also stimulates cyclic AMP transport out of the cell. When glucose enters the cell, the cyclic AMP level is lowered, CRP protein cannot bind DNA, and RNA polymerase fails to bind to the promoters of operons subject to catabolite repression. Thus, catabolite repression is really an indirect result of the presence of a better energy source (glucose). The direct cause of catabolite repression is a low level of cyclic AMP.

Global Aspects of Catabolite Repression

Why is catabolite repression considered a mechanism of *global* control? In *E. coli* and other organisms for which glucose is the preferred energy source, catabolite repression prevents expression of most other catabolic operons as long as glucose is present. Dozens of catabolic operons are affected, including those for lactose, maltose, a host of other sugars, and most of the other commonly used carbon and energy sources of *E. coli*. In addition, genes for the synthesis of flagella are controlled by catabolite repression—if bacteria have a good carbon source available, there is no need to swim around in search of nutrients.

Let us return to the *lac* operon to put catabolite repression into the context of the entire regulatory picture. To show this, the entire regulatory region of the *lac* operon is diagrammed in **Figure 9.19**. For *lac* genes to be transcribed, two requirements must be met: (1) the level of cyclic AMP must be high enough that the CRP protein binds to the CRP-binding site (positive control), and (2) lactose must be present so that the lactose repressor does not block transcription by binding to the operator (negative control). If these two conditions are met, the cell is signaled that glucose is absent and lactose is present; then and only then does transcription of the *lac* operon begin.

9.9 MiniReview

Global control systems regulate the expression of many genes simultaneously. Catabolite repression is a global control system that helps cells make the most efficient use of available carbon sources. The *lac* operon is under the control of catabolite repression as well as its own specific negative regulatory system.

- Explain how catabolite repression depends on an activator protein.
- Explain how the *lac* operon is both positively and negatively controlled.

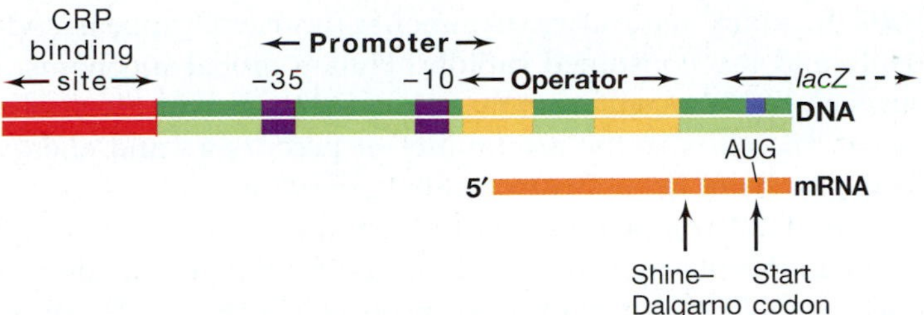

Figure 9.19 Overall regulation of the lactose operon. The first structural gene in this operon, *lacZ,* encodes the enzyme β-galactosidase, which breaks down lactose. The operon contains two other genes that also take part in lactose metabolism. The two halves of the operator (where the repressor binds) and the CRP-binding site are almost perfect inverted repeats. The transcriptional start site is located on the DNA exactly at the 5′-end of the mRNA. The location of the −35 sequence and the −10 sequence, which are part of the promoter (Figure 7.22), are also shown. In addition, the location of the Shine–Dalgarno sequence and the start codon are also given. These two sequences are critical for translation of the mRNA (Section 7.15).

9.10 The Stringent Response

Nutrient levels in the natural environments of bacterial cells often change significantly, even if only briefly. Such conditions can easily be simulated in the laboratory, and much work has been done with *Escherichia coli* and other prokaryotes on the regulation of gene expression following a "shift down" or "shift up" in nutrient status. These include, in particular, the regulatory events triggered by starvation for amino acids or energy.

As a result of a shift down from amino acid excess to limitation—as would occur when a culture is transferred from a rich complex medium to a defined medium with a single carbon source—the synthesis of rRNA and tRNA ceases almost immediately. No new ribosomes are produced. Protein and DNA synthesis is curtailed, but the biosynthesis of new amino acids is activated (**Figure 9.20*a***). Following such a shift, new proteins must be made to synthesize the amino acids no longer available in the environment; these are made by existing ribosomes. After a while, rRNA synthesis (and hence, the production of new ribosomes) begins again but at a new rate commensurate with the cell's reduced growth rate (Figure 9.20*a*). This course of events is called the **stringent response** (or stringent control) and is another example of global control.

ppGpp and the Mechanism of the Stringent Response

The stringent response is triggered by a mixture of two regulatory nucleotides, *guanosine tetraphosphate* (ppGpp) and *guanosine pentaphosphate* (pppGpp); this mixture is often written as (p)ppGpp (Figure 9.20*b*). In *E. coli,* these nucleotides, which are also called *alarmones,* rapidly accumulate during a shift down from amino acid excess to amino acid starvation. Alarmones are synthesized by a specific protein, called RelA, using ATP as a phosphate donor (Figure 9.20*b,c*). RelA adds two phosphate groups from ATP to GTP or GDP,

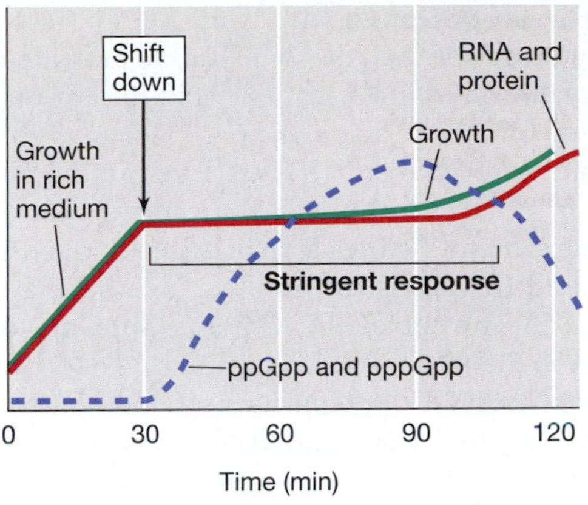

(a)

(b)

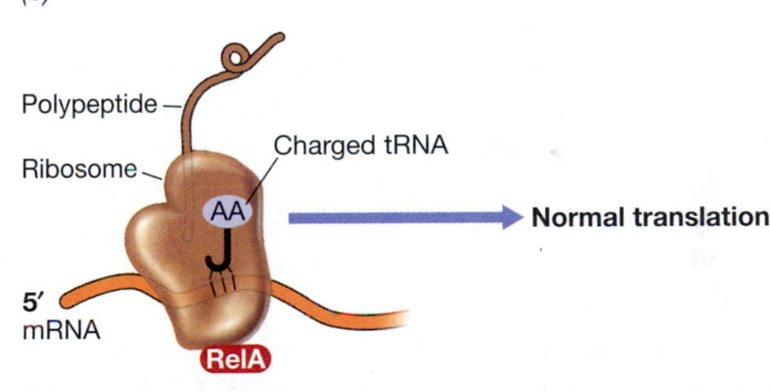

(c)

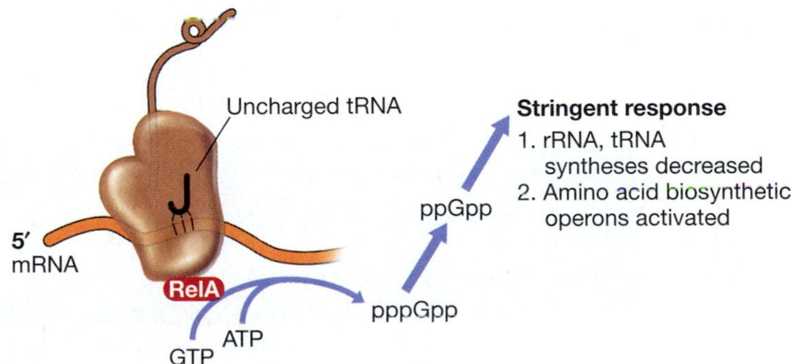

(d)

Figure 9.20 The stringent response. *(a)* Upon nutrient downshift, rRNA, tRNA, and protein syntheses temporarily cease. Some time later growth resumes at a new (decreased) rate. *(b)* Structure of guanosine tetraphosphate (ppGpp), a trigger of the stringent response. *(c)* Normal translation, which requires charged tRNAs. *(d)* Synthesis of ppGpp. When cells are starved for amino acids, an uncharged tRNA can bind to the ribosome, which stops ribosome activity. This event triggers the RelA protein to synthesize a mixture of pppGpp and ppGpp.

UNIT 2

thus producing pppGpp or ppGpp, respectively. RelA is associated with the 50S subunit of the ribosome and is activated by a signal from the ribosome during amino acid limitation. When the cell is limited for amino acids, the pool of *uncharged* tRNAs increase relative to *charged* tRNAs. Eventually, an uncharged tRNA is inserted into the ribosome instead of a charged tRNA during protein synthesis. When this happens, the ribosome stalls, and this leads to (p)ppGpp synthesis by RelA (Figure 9.20*c*). The protein Gpp converts pppGpp to ppGpp so that ppGpp is the major overall product.

The alarmones ppGpp and pppGpp have global control effects. They strongly inhibit rRNA and tRNA synthesis by binding to RNA polymerase and preventing initiation of transcription of genes for these RNAs. On the other hand, alarmones activate the biosynthetic operons for certain amino acids as well as catabolic operons that yield precursors for amino acid synthesis. By contrast, operons that encode biosynthetic proteins whose amino acid products are present in sufficient amounts remain shut down. The stringent response also inhibits the initiation of new rounds of DNA synthesis and cell division and slows down the synthesis of cell envelope components, such as membrane lipids. Efficient binding of (p)ppGpp to RNA polymerase requires the protein DksA, which is needed to position the (p)ppGpp correctly in the channel that normally allows substrates (that is nucleoside triphosphates) into the RNA polymerase active site.

In addition to RelA, another protein, SpoT, helps trigger the stringent response. The SpoT protein can either make (p)ppGpp or degrade it. Under most conditions, SpoT is responsible for degrading (p)ppGpp; however, SpoT synthesizes (p)ppGpp in response to certain stresses or when there is a shortage of energy. Thus the stringent response is the result of not only the absence of precursors for protein synthesis but also the lack of energy for biosynthesis.

The stringent response can be thought of as a mechanism for adjusting the cell's biosynthetic machinery to the availability of the required precursors and energy. By so doing, the cell achieves a new balance between anabolism and catabolism. In many natural environments, nutrients appear suddenly and are consumed rapidly. Thus a global mechanism such as the stringent response that balances the metabolic state of a cell with the availability of precursors and energy likely improves its ability to compete in nature.

The RelA/(p)ppGpp system is found only in *Bacteria* and in the chloroplasts of plants. *Archaea* and eukaryotes do not make (p)ppGpp in response to resource shortages. Although *Archaea* display an overall response similar to the stringent response of *Bacteria* when faced with carbon and energy shortages, they use regulatory mechanisms other than those described here to deal with these nutritional situations.

9.10 MiniReview

The stringent response is a global control mechanism triggered by amino acid starvation. The alarmones ppGpp and pppGpp are produced by RelA, a protein that monitors ribosome activity. Within the cell the stringent response achieves balance between protein production and amino acid requirements.

- Which genes are activated during the stringent response, and why?
- Which genes are repressed during the stringent response, and why?
- How are the alarmones ppGpp and pppGpp synthesized?

9.11 Other Global Control Networks

In Section 9.9 we discussed catabolite repression in *Escherichia coli*, and in Section 9.10 we discussed the stringent response. Both are examples of global control. There are several other global control systems in *E. coli* (and probably in all prokaryotes), and a few of these are shown in **Table 9.3**.

Global control systems regulate more than one regulon (Section 9.4). Global control networks may include activators, repressors, signal molecules, two-component regulatory

Table 9.3 Examples of global control systems known in *Escherichia coli*[a]

System	*Signal*	*Primary activity of regulatory protein*	*Number of genes regulated*
Aerobic respiration	Presence of O_2	Repressor (ArcA)	>50
Anaerobic respiration	Lack of O_2	Activator (FNR)	>70
Catabolite repression	Cyclic AMP level	Activator (CRP)	>300
Heat shock	Temperature	Alternative sigmas (RpoH and RpoE)	36
Nitrogen utilization	NH_3 limitation	Activator (NR_I)/alternative sigma RpoN	>12
Oxidative stress	Oxidizing agents	Activator (OxyR)	>30
SOS response	Damaged DNA	Repressor (LexA)	>20

[a]For many of the global control systems, regulation is complex. A single regulatory protein can play more than one role. For instance, the regulatory protein for aerobic respiration is a repressor for many promoters but an activator for others, whereas the regulatory protein for anaerobic respiration is an activator protein for many promoters but a repressor for others. Regulation can also be indirect or require more than one regulatory protein. Some of the regulatory proteins involved are members of two-component systems (Section 9.5). Many genes are regulated by more than one global system.

systems (Section 9.5), regulatory RNA (Sections 9.13 and 9.14), and alternative sigma factors (Section 7.10). An example of a global response that is widespread in *Bacteria* is the response to high temperature. In many bacteria, this **heat shock response** is largely controlled by alternative sigma factors.

Heat Shock Proteins

Most proteins are relatively stable. Once made, they continue to perform their functions and are passed along at cell division. However, some proteins are less stable at elevated temperatures and tend to unfold. Such improperly folded proteins are recognized by protease enzymes in the cell and are degraded. Consequently, cells that are heat stressed induce the synthesis of a set of proteins, the **heat shock proteins**, which help counteract the damage. Heat shock proteins assist the cell in recovering from stress. They are not only induced by heat, but by several other stress factors that the cell can encounter. These include exposure to high levels of certain chemicals—such as ethanol—and exposure to high doses of ultraviolet radiation.

In *E. coli* and in most prokaryotes examined, there are three major classes of heat shock protein, Hsp70, Hsp60, and Hsp10. We have encountered these proteins before, although not by these names. The Hsp70 protein of *E. coli* is DnaK, which prevents aggregation of newly synthesized proteins and stabilizes unfolded proteins (Section 7.17). Major representatives of the Hsp60 and Hsp10 families in *E. coli* are the proteins GroEL and GroES, respectively. These are molecular chaperones that catalyze the correct refolding of misfolded proteins (Section 7.17 and Figure 7.40). Another class of heat shock proteins includes various proteases that function in the cell to remove denatured or irreversibly aggregated proteins.

The heat shock proteins are apparently very ancient and are highly conserved. Molecular sequencing of heat shock proteins, especially Hsp70, has been used to help unravel the phylogeny of eukaryotes. Heat shock proteins are present in all cells although the regulatory system that controls their expression varies greatly in different groups of organisms.

Heat Shock Response

In many bacteria, such as *E. coli*, the heat shock response is controlled by the alternative sigma factors RpoH (σ^{32}) and RpoE (**Figure 9.21**). The RpoH sigma factor controls expression of heat shock proteins in the cytoplasm, and RpoE (about which less is known) regulates the expression of a different set of heat shock proteins in the periplasm and cell envelope. RpoH is normally degraded within a minute or two of its synthesis. However, when cells suffer a heat shock, degradation of RpoH is inhibited and its level therefore increases. Consequently, transcription of those operons whose promoters are recognized by RpoH increases too. The rate of degradation of RpoH depends on the level of free DnaK protein, which inactivates RpoH. In unstressed cells the level of free DnaK is relatively high and the level of intact RpoH is correspondingly low. However, if heat stress unfolds proteins, DnaK binds preferentially to the unfolded proteins and so is no longer free to promote degradation of RpoH. Thus the more denatured proteins there are, the lower the level of free DnaK and the higher the level of RpoH; the result is heat shock gene expression.

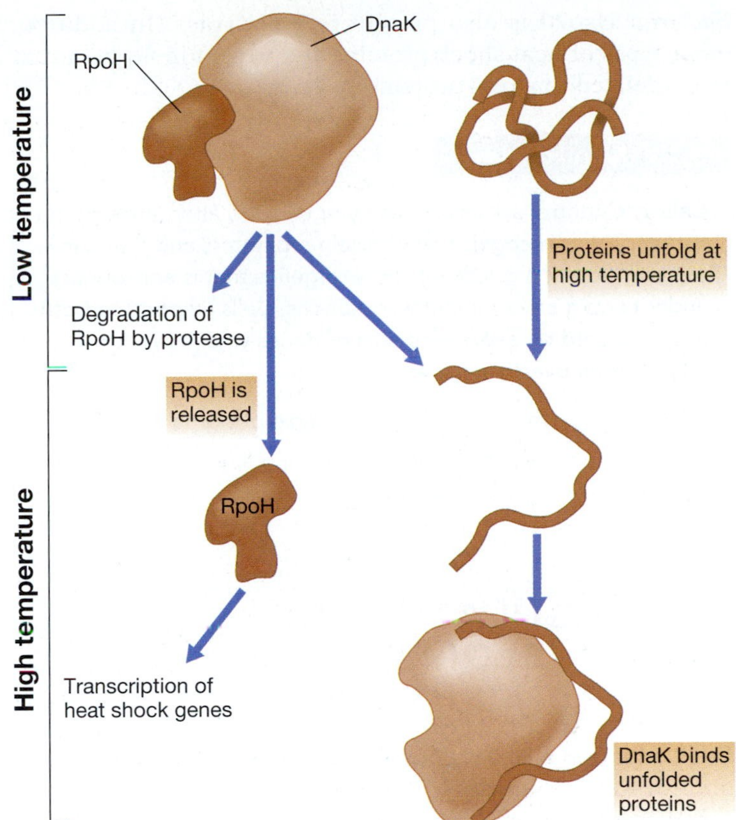

Figure 9.21 Control of heat shock in *Escherichia coli*. The RpoH alternative sigma factor is broken down rapidly by proteases at normal temperatures. This is stimulated by binding of the DnaK chaperonin to RpoH. At high temperatures, some proteins are denatured, and DnaK recognizes and binds to the unfolded polypeptide chains. This removes DnaK from RpoH, which slows the degradation rate. The level of RpoH rises, and the heat shock genes are transcribed.

When the stress situation has passed, for example, upon a temperature downshift, RpoH is rapidly inactivated by DnaK, and the synthesis of heat shock proteins is greatly reduced. Because heat shock proteins perform vital functions in the cell, there is always a low level of these proteins present, even when cells are growing under optimal conditions. However, the rapid synthesis of heat shock proteins in stressed cells emphasizes how important they are in surviving excessive heat, chemicals, or physical agents. Heat or other stresses can generate large amounts of inactive proteins that need to be refolded (and in the process, reactivated) or degraded to release free amino acids for the biosynthesis of new proteins.

A heat-shock response also occurs in *Archaea*, even in species that grow best at very high temperatures. An analog of the bacterial Hsp70 is found in many *Archaea* and is structurally quite similar to those found in gram-positive species of

Bacteria. Hsp70 is also present in eukaryotes. In addition, other types of heat shock proteins are present in *Archaea* that are unrelated to stress proteins of *Bacteria*.

9.11 MiniReview

Cells can control sets of genes by employing alternative sigma factors. These recognize only certain promoters and thus allow transcription of a select category of genes that is appropriate under certain environmental conditions. Cells respond to both heat and cold by expressing sets of genes whose products help the cell overcome stress.

- Why do cells have more than one type of sigma factor?
- Why might the proteins induced during a heat shock not be needed during a cold shock?

V REGULATION OF DEVELOPMENT IN MODEL BACTERIA

Differentiation and development are largely characteristics of multicellular organisms. Because most prokaryotic microorganisms grow as single cells, few show differentiation. Nonetheless, there are occasional examples among single-celled prokaryotes that illustrate the basic principle of differentiation, namely that one cell gives rise to two genetically identical descendents that perform different roles and must therefore express different sets of genes. Here we discuss two well-studied examples, the formation of endospores in the gram-positive bacterium *Bacillus* and the formation of two cell types, mobile and stationary, in the gram-negative bacterium *Caulobacter*.

Although forming just two different cell types may seem superficially simple, the regulatory systems that control these processes are highly complex. There are three major phases for the regulation of differentiation: (1) triggering the response, (2) asymmetric development of two sister cells and (3) reciprocal communication between the two differentiating cells.

9.12 Sporulation in *Bacillus*

Many microorganisms, both prokaryotic and eukaryotic, respond to adverse conditions by forming spores. Once good conditions return, the spore germinates and the microorganism returns to its normal lifestyle (Section 4.12). Among the *Bacteria*, the genus *Bacillus* is well known for the formation of endospores, that is, spores formed inside a mother cell. Prior to endospore formation, the cell divides asymmetrically. The smaller cell develops into the endospore, which is surrounded by the larger mother cell (Section 4.12). Once development is complete, the mother cell bursts, releasing the endospore.

Endospore formation in *Bacillus subtilis* is triggered by adverse external conditions, such as starvation, desiccation, or growth-inhibitory temperatures. In fact, multiple aspects of the environment are monitored by a group of five sensor kinases. These function via a phosphotransfer relay system whose mechanism resembles that of a two-component regulatory system (Section 9.5) but is considerably more complex (**Figure 9.22**). The net result of multiple adverse conditions is the successive phosphorylation of several proteins called *sporulation factors,* culminating with the sporulation factor Spo0A (Figure 9.22a). When Spo0A is highly phosphorylated, sporulation events proceed.

Development of the Endospore

Once triggered, endospore development is controlled by four different sigma factors, two of which (σF and σG) activate genes needed inside the developing endospore itself and two (σE and σK) of which activate genes needed in the mother cell that surrounds the endospore (Figure 9.22b). The sporulation signal, transmitted via Spo0A, activates σF in the smaller cell that is destined to become the endospore. σF is already present but is inactive, as it is bound by an anti-sigma factor. The signal from Spo0A activates a protein that binds to the anti-sigma factor. This inactivates the anti-sigma factor and liberates σF. Once free, σF binds to RNA polymerase and allows transcription (inside the spore) of genes whose products are needed for the next stage of sporulation. These include the gene for the sigma factor σG and the genes for proteins that cross into the mother cell and activate σE. Active σE is required for transcription inside the mother cell of yet more genes, including the gene for the sigma factor σK. The sigma factors σG (in the endospore) and σK (in the mother cell) are required for transcription of genes needed even later in the sporulation process.

One fascinating aspect of endospore formation is that it is preceded by what is in effect cellular cannibalism. Those cells in which Spo0A has already become activated secrete a protein that lyses nearby cells of the same species whose Spo0A protein has not yet become activated. This toxic protein is accompanied by a second protein that delays sporulation of neighboring cells. Cells committed to sporulation also make an antitoxin protein to protect themselves against the effects of their own toxin. Their sacrificed sister cells are used as a source of nutrients for developing endospores. Shortages of certain nutrients, such as phosphate, increase the expression level of the toxin-encoding gene.

9.12 MiniReview

Sporulation during adverse conditions is triggered via a complex phosphorelay system that monitors multiple aspects of the environment. The sporulation factor Spo0A then sets in motion a cascade of regulatory responses under the control of several alternative sigma factors.

- How are different sets of genes expressed in the developing endospore and the mother cell?
- What is an anti-sigma factor and how can its effect be overcome?

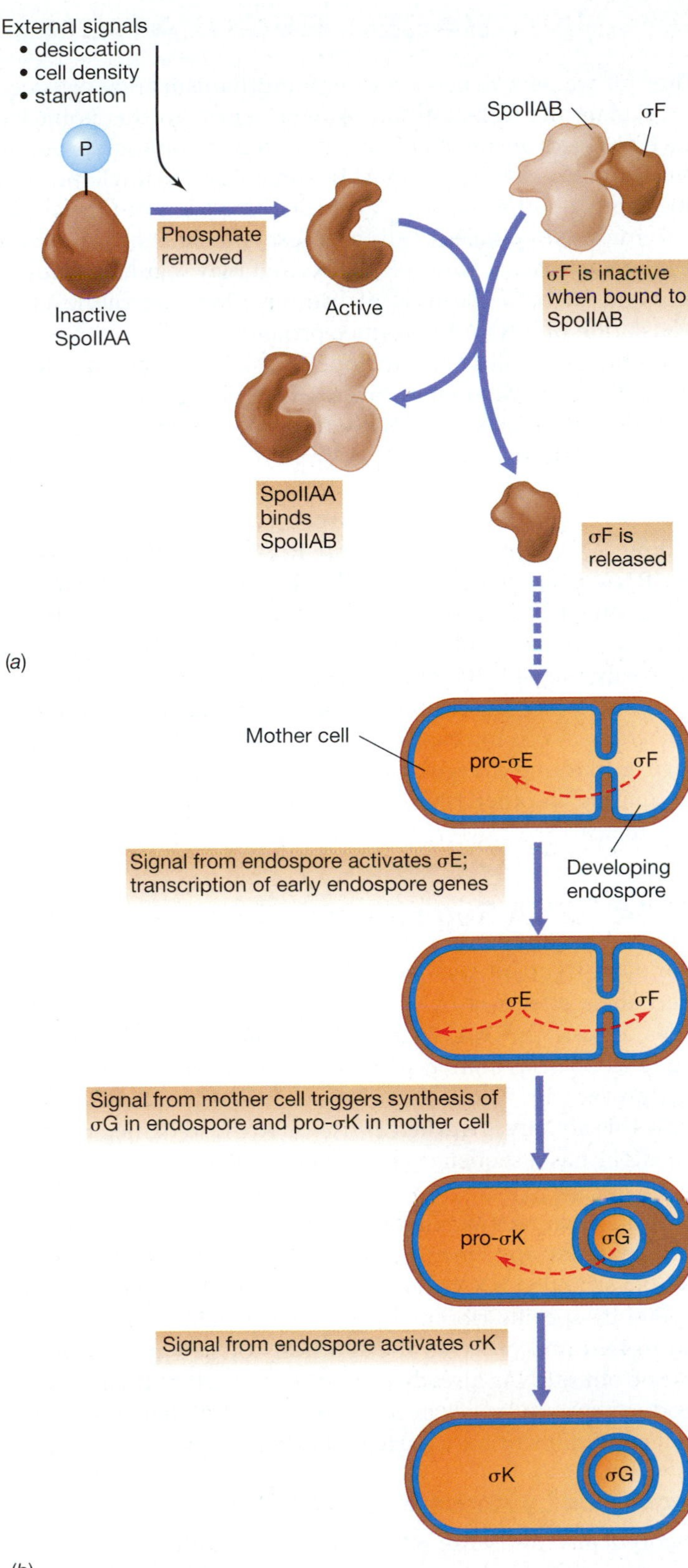

Figure 9.22 Control of endospore formation in *Bacillus*. After an external signal is received, a cascade of sigma factors controls differentiation. *(a)* Active SpoIIAA binds the anti-sigma factor SpoIIAB, thus liberating the first sigma factor, σF. *(b)* Sigma factor σF initiates a cascade of sigma factors, some of which already exist and need to be activated, others of which are not yet present and whose genes need to be expressed. These sigma factors then promote transcription of genes needed for endospore development.

9.13 *Caulobacter* Differentiation

Caulobacter provides another example in which a cell divides into two genetically identical daughter cells that perform different roles and express different sets of genes. *Caulobacter* is a species of *Proteobacteria* that is common in aquatic environments, typically in waters that are very nutrient poor (oligotrophic) (Section 15.16). In the *Caulobacter* life cycle, free-swimming (swarmer) cells alternate with cells that lack flagella and are attached to surfaces by a stalk with a holdfast at its end. The role of the swarmer cells is dispersal, as swarmers cannot divide or replicate their DNA. Conversely, the role of the stalked cell is reproduction.

The *Caulobacter* Life Cycle

The *Caulobacter* cell cycle is controlled by three major regulatory proteins whose concentrations oscillate in succession. Two of these are the transcriptional regulators, GcrA and CtrA. The third is DnaA, a protein that functions both in its normal role in DNA replication and also as a transcriptional regulator. Each of these regulators is active at a specific stage in the cell cycle, and each controls many other genes that are needed at that particular stage in the cycle (**Figure 9.23**).

CtrA is activated by phosphorylation in response to external signals. Once phosphorylated, CtrA-P activates genes needed for the synthesis of the flagella and other functions in swarmer cells. Conversely, CtrA-P represses the synthesis of GcrA and also inhibits the initiation of DNA replication by binding to and blocking the origin of replication (Figure 9.23). As the cell cycle proceeds, CtrA is degraded by a specific protease; as a consequence, levels of DnaA rise. The absence of CtrA-P allows access to the chromosomal origin of replication, and, as in all *Bacteria*, DnaA binds to the origin and triggers the initiation of DNA replication (Section 7.6). In addition, in *Caulobacter*, DnaA activates several other genes needed for chromosomal replication. The level of DnaA then falls due to protease degradation, and the level of GcrA rises. The GcrA regulator promotes the elongation phase of chromosome replication, cell division, and the growth of the stalk on the immobile daughter cell. Eventually, GcrA levels fall and high levels of CtrA reappear (in the daughter cell destined to swim away) (Figure 9.23).

Many of the details of the regulation of the *Caulobacter* cell cycle are still uncertain. Both external stimuli and internal factors such as nutrient and metabolite pools are known to affect the cycle but how this information is integrated into the overall control system is only partly understood.

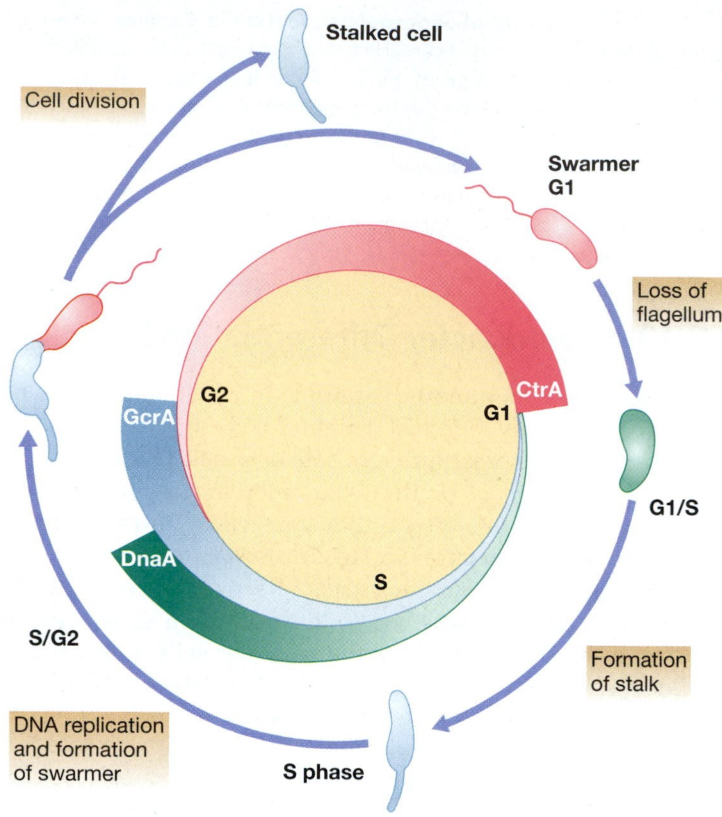

Figure 9.23 Cell cycle regulation in *Caulobacter*. Three global regulators CtrA, DnaA, and GcrA, oscillate in levels through the cycle as shown. In G1 swarmer cells, CtrA represses initiation of DNA replication and expression of GcrA. At the G1/S transition, CtrA is degraded and DnaA levels rise. DnaA binds to the origin of replication and initiates replication. GcrA also rises and activates genes for cell division and DNA synthesis. At the S/G2 transition, CtrA levels begin to rise again and shut down GcrA expression. GcrA levels slowly decline in the stalked cell but are rapidly degraded in the swarmer. CtrA is degraded in the stalked cell.

However, since its genome has been sequenced and good genetic systems are available, differentiation in *Caulobacter* has been used as a model system for studying cell developmental processes.

9.13 MiniReview

Differentiation in *Caulobacter* consists of the alternation between motile cells and those that are attached to surfaces. Three major regulatory proteins CtrA, GcrA, and DnaA, act in succession to control the three phases of the cell cycle. Each in turn controls many other genes needed at specific times in the cell cycle.

- Why are the levels of DnaA protein controlled during the *Caulobacter* cell cycle?
- When do the regulators CtrA and GcrA carry out their main roles during the *Caulobacter* life cycle?

VI RNA-BASED REGULATION

Thus far we have discussed several mechanisms that cells use to regulate the transcription of their genes. To this point we have focused on mechanisms in which regulatory proteins sense signals or bind to DNA. In some cases a single protein does both; in other cases separate proteins carry out these two activities. Nonetheless, all of these mechanisms rely on *regulatory proteins*. However, RNA itself may regulate gene expression, at both the level of transcription and the level of translation of mRNA to produce proteins.

RNA molecules that are not translated to give proteins are collectively known as **noncoding RNA**. This category includes the rRNA and tRNA molecules that take part in protein synthesis and the RNA present in the signal recognition particle, a structure that identifies newly synthesized proteins for secretion from the cell (Section 7.17). Noncoding RNA also includes small RNA molecules necessary for RNA processing, especially the splicing of mRNA in eukaryotes (Section 8.8). In addition, a variety of small RNA (sRNA) molecules are known. Small RNAs range from approximately 40–400 nucleotides long and function to regulate gene expression in both prokaryotes and eukaryotes. In *Escherichia coli,* for example, a number of sRNA molecules have been found to regulate various aspects of cell physiology by binding to other RNAs or even to other small molecules in some cases.

9.14 RNA Regulation and Antisense RNA

The most frequent way in which regulatory RNA molecules exert their effects is by base-pairing with other RNA molecules, usually mRNA, that have regions of complementary sequence. These double-stranded regions tie up the mRNA and prevent its translation (**Figure 9.24**). Small RNAs that show this activity are called *antisense RNA*, so named because the sRNA has a sequence complementary to the coding sense of the mRNA.

Antisense sRNAs survey the cytoplasm for their mRNA complements and form double-stranded RNAs with them. Double-stranded RNAs cannot be translated and are soon degraded by specific ribonucleases (Figure 9.24). This removes the mRNA and prevents the synthesis of new protein molecules from mRNAs already present in the cell but whose gene products are no longer needed because of a change in nutrional conditions or for a variety of other reasons.

Regulation by Antisense RNA

Theoretically, antisense RNA could be made by transcribing the non-template strand of the same gene that yielded a specific mRNA. Instead, a distinct "anti-gene" is used to form the antisense RNA. Only a relatively short piece of antisense RNA is needed to block transcription of a mRNA, and therefore the "anti-gene" that encodes the antisense RNA is much shorter

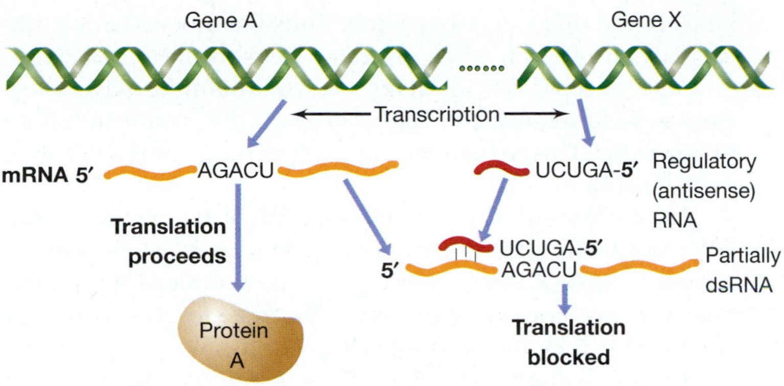

Figure 9.24 Regulation by Antisense RNA. Gene A is transcribed from its promoter to yield an mRNA that can be translated to form protein A. Gene X is a small gene with a sequence identical to that of part of Gene A but with its promoter at the opposite end. Therefore, if it is transcribed, the resulting RNA will be complementary to the mRNA of gene A. If these two RNAs base pair, translation will be blocked.

than the gene that encodes the original message. Typically, antisense RNAs are around 100 nucleotides long and bind to a target region approximately 30 nucleotides long. In addition, however, each antisense RNA can regulate several different mRNAs, all of which share the same target sequence for antisense RNA binding.

Transcription of antisense RNA is enhanced under conditions in which its target gene(s) needs to be turned off. For example, the RyhB antisense RNA of *E. coli* is transcribed when iron is limiting for growth. RyhB RNA binds to a dozen or more target mRNAs that encode proteins needed for iron metabolism or that use iron as cofactors. The base-paired RyhB/mRNA molecules are then degraded by ribonucleases, in particular, ribonuclease E. This forms part of the mechanism by which *E. coli* and related bacteria respond to a shortage of iron. Other responses to iron limitation in *E. coil* include transcriptional controls involving repressor and activator proteins (Sections 9.3 and 9.4) that function to increase the capacity of cells to take up iron and to tap into intracellular iron stores.

Although antisense RNA usually blocks translation of mRNA, occasional examples are known in which antisense RNA does just the opposite and actually enhances the translation of its target mRNA. It is hypothesized that in these cases the native mRNA attains a secondary structure that prevents translation. The antisense RNA is then thought to bind to a short region of the mRNA and unfold it, thereby allowing access to the ribosome.

9.14 MiniReview

Cells can control genes in several ways by employing regulatory RNA molecules. One way is to take advantage of base pairing and use antisense RNA to form a double-stranded RNA that cannot be translated.

- Why are antisense RNAs much shorter than the mRNA molecules to which they bind?
- How do cells synthesize antisense RNA molecules?
- What happens to mRNA molecules following binding of their antisense RNAs?

9.15 Riboswitches

One of the most interesting findings in molecular biology has been the discovery that RNA can carry out many roles once thought to be the purview of proteins only. In particular, RNA can specifically recognize and bind other molecules, including low-molecular-weight metabolites. It is important to emphasize that such binding does not involve complementary base pairing (as does antisense RNA described in the previous section) but occurs as a result of a folding of the RNA into a specific three-dimensional structure that recognizes the target molecule, much as a protein enzyme recognizes its substrate. Some of these RNA molecules are called *ribozymes* because they are catalytically active like enzymes. We discussed ribozymes in Section 8.8. Other RNA molecules resemble repressors and activators in binding metabolites such as amino acids or vitamins and regulating gene expression; these are the **riboswitches**.

Certain mRNAs contain regions upstream of the coding sequences that can fold into specific three-dimensional structures that bind small molecules (**Figure 9.25**). These recognition domains are riboswitches and exist as two alternative structures, one with the small molecule bound and the other without. Alternation between the two forms of the riboswitch thus depends on the presence or absence of the small molecule and in turn controls expression of the mRNA. Riboswitches have been found that control the synthesis of enzymes in biosynthetic pathways for various enzymatic cofactors, such as the vitamins thiamine, riboflavin, and cobalamin (B_{12}), for a few amino acids, for the purine bases adenine and guanine, and for glucosamine 6-phosphate, a precursor in peptidoglycan synthesis.

Mechanism of Riboswitches

The riboswitch control mechanism is analogous to one we have seen before. Early in this chapter we discussed the regulation of enzyme synthesis by negative control of transcription (Section 9.3).

In this case, the presence of a specific metabolite shuts down the transcription of genes encoding enzymes for the corresponding biosynthetic pathway. This is achieved by use of a protein repressor, such as the arginine repressor in

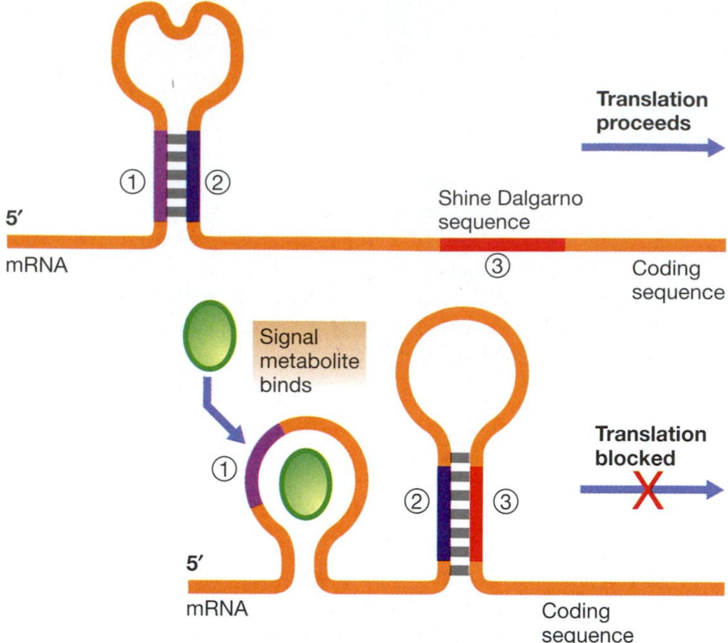

Figure 9.25 Regulation by a Riboswitch. Binding of a specific metabolite alters the secondary structure of the riboswitch domain, which is located in the 5′-untranslated region of the mRNA, preventing translation. The Shine–Dalgarno site is where the ribosome binds the RNA.

the case of the arginine biosynthetic operon (Figure 9.5). In the case of a riboswitch, there is no regulatory protein. Instead, the metabolite binds directly to the riboswitch domain at the 5′ end of the mRNA. Riboswitches usually exert their control after the mRNA has already been synthesized. Therefore, most riboswitches function to control *translation* of the mRNA, rather than its *transcription.*

The metabolite that is bound by the riboswitch is typically the product of a biosynthetic pathway whose constituent enzymes are encoded by the mRNAs that carry the corresponding riboswitches. For example, the thiamine riboswitch that binds thiamine pyrophosphate is upstream of the coding sequences for enzymes that participate in the thiamine biosynthetic pathway. When the pool of thiamine pyrophosphate is sufficient in the cell, this metabolite binds to its specific riboswitch mRNA. The new secondary structure of the riboswitch blocks the Shine–Dalgarno ribosome-binding sequence on the mRNA (Section 7.15) and prevents the message from binding to the ribosome; this prevents translation (Figure 9.25). If the concentration of thiamine pyrophosphate drops sufficiently low, this molecule can dissociate from its riboswitch mRNA. This unfolds the message and exposes the Shine–Dalgarno site, allowing the mRNA to bind to the ribosome and become translated.

The thiamine analog pyrithiamine blocks the synthesis of thiamine and hence, inhibits bacterial growth. Until the discovery of riboswitches, the site of action of pyrithiamine remained mysterious. It now appears that pyrithiamine is converted by cells to pyrithiamine pyrophosphate, which then binds to the thiamine riboswitch. Thus the biosynthetic pathway is shut off even when no thiamine is available. Bacterial mutants selected for resistance to pyrithiamine have alterations in the sequence of the riboswitch that result in failure to bind both pyrithiamine pyrophosphate and thiamine pyrophosphate.

In *Bacillus subtilis*, where about 2% of the genes are under riboswitch control, the same riboswitch is present on several mRNAs that together encode the proteins for a particular pathway. For example, over a dozen genes in six operons are controlled by the thiamine riboswitch.

Despite being part of the mRNA, some riboswitches nevertheless do control transcription. The mechanism is similar to that seen in attenuation—a conformational change in the riboswitch causes premature termination of the synthesis of the mRNA that carries it (Section 9.16).

Riboswitches and Evolution

How widespread are riboswitches and how did they evolve? Thus far riboswitches have been found only in some bacteria and a few plants and fungi. Some scientists believe that riboswitches are remnants of the RNA world, a period eons ago before cells, DNA, and protein, when it is hypothesized that catalytic RNAs were the only self-replicating life forms (Section 14.2). In such an environment, riboswitches may have been a primitive mechanism of metabolic control—a simple means by which RNA life forms could have controlled the synthesis of other RNAs. As proteins evolved, riboswitches might have been the first control mechanisms for their synthesis, as well. If true, the riboswitches that remain today may be the last vestiges of this simple form of control, because, as we have seen in this chapter, metabolic regulation is almost exclusively carried out by way of regulatory proteins.

9.15 MiniReview

Riboswitches are RNA domains at the 5′-ends of mRNA that recognize small molecules and respond by changing their three-dimensional structure. This in turn affects the translation of the mRNA or, sometimes, its premature termination. Riboswitches are mostly used to control biosynthetic pathways for amino acids, purines, and a few other metabolites.

- What happens when a riboswitch binds the small molecule it recognizes?
- What are the major differences between using a repressor protein versus a riboswitch to control gene expression?

9.16 Attenuation

Attenuation is a form of transcriptional control that functions by premature termination of mRNA synthesis. That is, in attenuation, control is exerted *after* the initiation of transcription but *before* its completion. Consequently, the number of completed transcripts from an operon is reduced, even though the number of initiated transcripts is not.

The basic principle of attenuation is that the first part of the mRNA to be made, called the *leader region,* can fold up into two alternative secondary structures. In this respect, the mechanism of attenuation resembles that of riboswitches. In attenuation, one mRNA secondary structure allows continued synthesis of the mRNA, whereas the other secondary structure causes premature termination. Folding of the mRNA depends on either events at the ribosome or the activity of regulatory proteins, depending on the organism. The best examples of attenuation involve regulation of genes controlling the biosynthesis of certain amino acids in gram-negative *Bacteria.* The first such system to be described was in the tryptophan operon in *Escherichia coli,* and we focus on it here. Attenuation control has been documented in several other species of *Bacteria,* and genomic analyses of *Archaea* suggest that the mechanism is present in this domain as well. However, because the processes of transcription and translation are spatially separated in eukaryotes, attenuation control is absent from *Eukarya.*

Attenuation and the Tryptophan Operon

The tryptophan operon contains structural genes for five proteins of the tryptophan biosynthetic pathway plus the usual promoter and regulatory sequences at the beginning of the operon (**Figure 9.26**). Like many operons, the tryptophan operon has more than one type of regulation. The first enzyme in the pathway, anthranilate synthase (a multisubunit enzyme encoded by *trpD* and *trpE*) is subject to feedback inhibition by tryptophan (Section 5.18). Transcription of the entire tryptophan operon is also under negative control (Section 9.3). However, in addition to the promoter (P) and operator (O) regions needed for negative control, there is a sequence in the operon called the *leader sequence*. The leader encodes a short polypeptide, the *leader peptide,* that contains tandem tryptophan codons near its terminus and functions as an attenuator (Figure 9.26).

The basis of control of the tryptophan attenuator is as follows. If tryptophan is plentiful in the cell, there will be a sufficient pool of charged tryptophan tRNAs and the leader peptide will be synthesized. On the other hand, if tryptophan is in short supply, the tryptophan-rich leader peptide will not be synthesized. Synthesis of the leader peptide results in termination of transcription of the remainder of the *trp* operon, which includes the structural genes for the biosynthetic enzymes. By contrast, if synthesis of the leader peptide is blocked by tryptophan deficiency, the rest of the operon is transcribed. **www.microbiologyplace.com Online Tutorial 9.2: Attenuation and the Tryptophan Operon**

Mechanism of Attenuation

How does translation of the leader peptide regulate transcription of the tryptophan genes downstream? Consider that in prokaryotic cells transcription and translation are simultaneous processes; as mRNA is released from the DNA, the ribosome binds to it and translation begins (Figure 7.29). That is, while transcription of downstream DNA sequences is still proceeding, translation of sequences transcribed has already begun (**Figure 9.27**).

Attenuation occurs (transcription stops) because a portion of the newly formed mRNA folds into a unique stem-loop that causes cessation of RNA polymerase activity. The stem-loop structure forms in the mRNA because two stretches of nucleotides near each other are complementary and can thus base-pair. If tryptophan is plentiful, the ribosome will translate the leader sequence until it comes to the leader stop codon. The remainder of the leader RNA then assumes the stem-loop, a transcription pause site, which is followed by a uracil-rich sequence that actually causes termination (Figure 9.27).

If tryptophan is in short supply, transcription of genes encoding tryptophan biosynthetic enzymes is obviously desirable. During transcription of the leader, the ribosome pauses at a tryptophan codon because of a shortage of charged tryptophan tRNAs. The presence of the stalled ribosome at this position allows a stem-loop to form that differs from the terminator stem-loop (sites 2 and 3 in Figure 9.27). This alternative stem-loop is not a transcription termination signal. Instead, it prevents the terminator stem-loop (sites 3 and 4 in Figure 9.27) from forming. This allows RNA polymerase to move past the termination site and begin

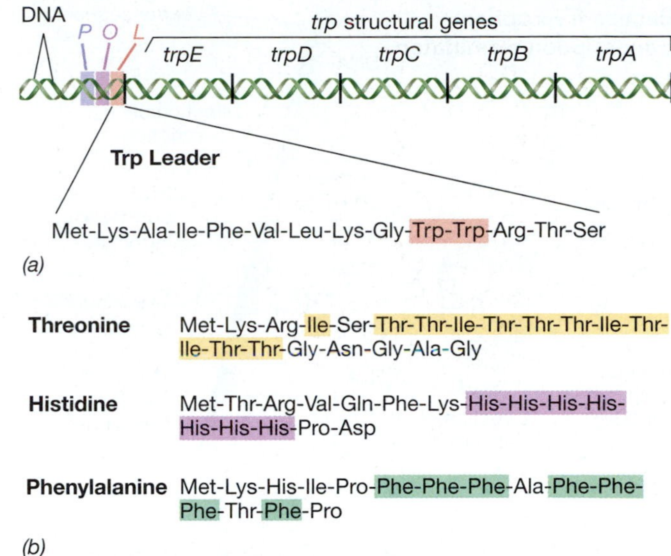

Figure 9.26 Attenuation and the leader peptide. Structure of the tryptophan operon and of tryptophan and other leader peptides in *Escherichia coli. (a)* Arrangement of the tryptophan operon. Note that the leader (*L*) encodes a short peptide containing two tryptophan residues near its terminus (there is a stop codon following the Ser codon). The promoter is labeled *P,* and the operator is labeled *O.* The genes labeled *trpE* through *trpA* encode the enzymes needed for tryptophan synthesis. *(b)* Amino acid sequences of leader peptides of some other amino acid synthetic operons. Because isoleucine is made from threonine, it is an important constituent of the threonine leader peptide.

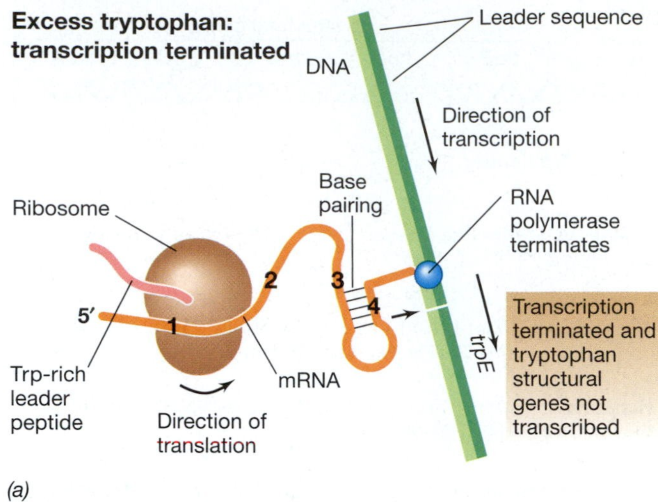

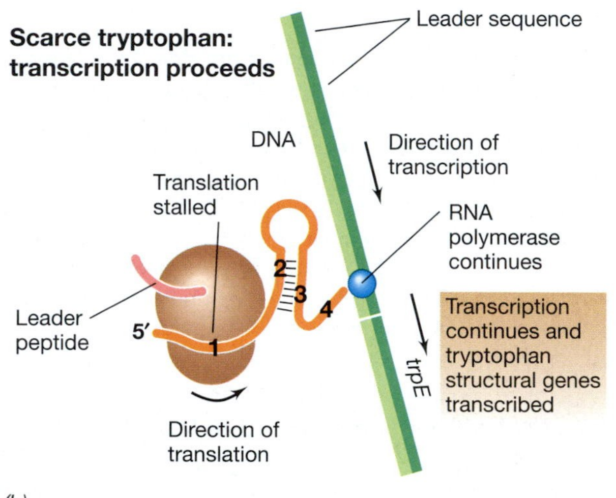

Figure 9.27 Mechanism of attenuation. Control of transcription of tryptophan operon structural genes by attenuation in *Escherichia coli*. The leader peptide is encoded by regions 1 and 2 of the mRNA. Two regions of the growing mRNA chain are able to form double-stranded loops, shown as 3:4 and 2:3. *(a)* When there is excess tryptophan, the ribosome translates the complete leader peptide, and so region 2 cannot pair with region 3. Regions 3 and 4 then pair to form a loop that terminates RNA polymerase. *(b)* If translation is stalled because of tryptophan starvation, a loop forms by pairing of region 2 with region 3, loop 3:4 does not form, and transcription proceeds past the leader sequence.

transcription of tryptophan structural genes. Thus, in attenuation control, the rate of transcription is influenced by the rate of translation.

Attenuation also occurs in *Escherichia coli* in the biosynthetic pathways for histidine, threonine-isoleucine, phenylalanine, and several other amino acids and essential metabolites. As shown in Figure 9.26*b*, the leader peptide for each of these amino acid biosynthetic operons is rich in that particular amino acid. The *his* operon is dramatic in this regard because its leader contains seven histidines in a row near the end of the peptide (Figure 9.26*b*). This longer stretch of regulatory codons gives attenuation a major effect in regulation, which may compensate for the fact that unlike the *trp* operon, the *his* operon in *E. coli* is not under negative control.

Translation-Independent Attenuation Mechanisms

Gram-positive *Bacteria,* such as *Bacillus,* also use attenuation of transcription to regulate certain amino acid biosynthetic operons. And, as in gram-negative *Bacteria,* the mechanism relies on alternative mRNA secondary structures, which in one configuration lead to termination. However, the mechanism is independent of translation and requires an RNA-binding protein.

In the *Bacillus subtilis* tryptophan operon, the binding protein is called the *trp* attenuation protein. In the presence of sufficient amounts of the amino acid tryptophan, this regulatory protein binds to the leader mRNA and causes transcription termination. By contrast, if tryptophan is limiting, the protein does not bind to the mRNA. This allows the favorable secondary structure to form and transcription proceeds.

Attenuation also occurs with genes unrelated to amino acid biosynthesis. These mechanisms obviously do not rely on amino acid levels. Some of the operons for pyrimidine biosynthesis (the *pyr* operons) in *E. coli* are regulated by attenuation, and the same is true for *Bacillus.* The mechanisms in the two organisms are, however, quite different, although each employs a system to assess the level of pyrimidines in the cell. In *E. coli* the mechanism monitors the rate of transcription, not translation. If pyrimidines are plentiful, RNA polymerase moves along and transcribes the leader DNA at a normal rate; this allows a stem-loop that signals termination to form in the mRNA. By contrast, if pyrimidines are scarce, the polymerase pauses at pyrimidine-rich sequences, which leads to formation of a nonterminator stem-loop that allows further transcription.

In *Bacillus,* a different mechanism is employed. For *pyr* attenuation, an RNA-binding protein controls the alternative stem-loop structures of the *pyr* mRNA, terminating transcription when pyrimidines are in excess. In this way the cell can maintain levels of pyrimidines—compounds that require significant cell resources to biosynthesize (∞ Section 5.16)—at levels needed to balance biosynthetic needs.

9.16 MiniReview

Attenuation is a mechanism whereby gene expression is controlled after initiation of mRNA synthesis. Attenuation mechanisms depend upon alternative stem-loop structures in the mRNA.

- Explain how the formation of one stem-loop in the RNA can block the formation of another.
- How does attenuation of the tryptophan operon differ between *Escherichia coli* and *Bacillus subtilis*?

Review of Key Terms

Activator protein a regulatory protein that binds to specific sites on DNA and stimulates transcription; involved in positive control

Attenuation a mechanism for controlling gene expression; typically transcription is terminated after initiation but before a full-length mRNA is produced

Autoinducer small signal molecule that takes part in quorum sensing

Catabolite repression the suppression of alternative catabolic pathways by a preferred source of carbon and energy

Cyclic AMP a regulatory nucleotide that participates in catabolite repression

Gene expression transcription of a gene followed by translation of the resulting mRNA into protein(s)

Heat shock proteins proteins induced by high temperature (or certain other stresses) that protect against high temperature, especially by refolding partially denatured proteins or by degrading them

Heat shock response response to high temperature that includes the synthesis of heat shock proteins together with other changes in gene expression

Induction production of an enzyme in response to a signal (often the presence of the substrate for the enzyme)

Negative control a mechanism for regulating gene expression in which a repressor protein prevents transcription of genes

Noncoding RNA an RNA molecule that is not translated into protein

Operon one or more genes transcribed into a single RNA and under the control of a single regulatory site

Positive control a mechanism for regulating gene expression in which an activator protein functions to promote transcription of genes

Quorum sensing a regulatory system that monitors the population level and controls gene expression based on cell density

Regulatory nucleotide a nucleotide that functions as a signal rather than being incorporated into RNA or DNA

Regulon a series of operons controlled as a unit

Repression preventing the synthesis of an enzyme in response to a signal

Repressor protein a regulatory protein that binds to specific sites on DNA and blocks transcription; involved in negative control

Response regulator protein one of the members of a two-component system; a protein that is phosphorylated by a sensor kinase and then acts as a regulator, often by binding to DNA

Riboswitch an RNA domain, usually in an mRNA molecule, that can bind a specific small molecule and alter its secondary structure; this in turn controls translation of the mRNA

Sensor kinase protein one of the members of a two-component system; a protein that phosphorylates itself in response to an external signal and then transfers the phosphoryl group to a response regulator protein

Signal transduction *see* two-component regulatory system.

Stringent response a global regulatory control that is activated by amino acid starvation or energy deficiency

Two-component regulatory system a regulatory system consisting of two proteins: a sensor kinase and a response regulator

Review Questions

1. Describe why a protein that binds to a specific sequence of double-stranded DNA is unlikely to bind to the same sequence if the DNA is single-stranded (Section 9.2).
2. Most biosynthetic operons need only be under negative control for effective regulation, whereas most catabolic operons need to be under both negative and positive control. Why (Sections 9.4 and 9.5)?
3. What are the two components that give their name to a signal transduction system in prokaryotes? What is the function of each of the components (Section 9.5)?
4. How can quorum sensing be considered a regulatory mechanism for conserving cell resources (Section 9.6)?
5. Adaptation allows the mechanism controlling flagellar rotation to be reset. How is this achieved (Section 9.7)?
6. What is the difference between an operon and a regulon (Section 9.9)?
7. Describe the mechanism by which cAMP receptor protein (CRP), the regulatory protein for catabolite repression, functions using the lactose operon as an example (Section 9.9).
8. What events trigger the stringent response? Why are the events that occur in the stringent response a logical consequence of the trigger of the response (Section 9.10)?
9. Describe the proteins produced when cells of *Escherichia coli* experience a heat shock. Of what value are they to the cell (Section 9.11)?
10. How does regulation by antisense RNA differ from that of riboswitches (Section 9.14 and 9.15)?
11. Describe how transcriptional attenuation works. What is actually being “attenuated” (Section 9.16)?
12. Why hasn't the type of attenuation that controls several different amino acid biosynthetic pathways in *Escherichia coli* also been found in eukaryotes (Section 9.16)?

Application Questions

1. What would happen to regulation from a promoter under negative control if the region where the regulatory protein binds were deleted? What if the promoter were under positive control?
2. Promoters from *Escherichia coli* under positive control are not close matches to the DNA consensus sequence for *E. coli* (Section 7.10). Why?
3. The attenuation control of some of the pyrimidine biosynthetic pathway genes in *Escherichia coli* actually involves coupled transcription and translation. Can you describe a mechanism whereby the cell could somehow make use of translation to help it measure the level of pyrimidine nucleotides?
4. Most of the regulatory systems described in this chapter involve regulatory proteins. However, regulatory RNA is also important. Describe how one could achieve negative control of the *lac* operon using either of two different types of regulatory RNA.
5. Many amino acid biosynthetic operons under attenuation control are also under negative control. Considering that the environment of a bacterium can be highly dynamic, what advantage could be conferred by having attenuation as a second layer of control?

10

Overview of Viruses and Virology

Viruses that infect bacterial cells are widespread in nature. Infection of a lawn of bacteria growing on agar results in cleared zones called plaques where the cells have been lysed.

Viruses are genetic elements that cannot replicate independently of a living cell, called the **host cell**. Viruses are therefore obligate intracellular parasites that rely on entering a suitable living cell to carry out their replication cycle. However, viruses do possess their own genetic information and are thus independent of the host cell's genome (Section 7.4). However, unlike genetic elements such as plasmids (Section 11.2), viruses have an extracellular form, the virus particle, that enables them to exist outside the host for long periods and that facilitates transmission from one host cell to another. To multiply, viruses must enter a cell in which they can replicate, a process called *infection*.

Viruses, like plasmids and some other genetic elements, exploit the metabolic machinery of the cell. Like these other genetic elements, viruses can confer important new properties on their host cell. These properties will be inherited when the host cell divides if each new cell also inherits the viral genome. These changes are not always harmful and may even be beneficial. Viruses can replicate in a way that is destructive to the host cell, and this accounts for the fact that some viruses are agents of disease. We cover a number of human diseases caused by viruses in Chapters 34 and 35.

The study of viruses is called *virology*, and we introduce the essentials of the field in this chapter, which is divided into four parts. The first part introduces basic concepts of virus structure, infection of the host cell, and how viruses can be detected and quantified. The second part deals with the basic molecular biology of virus replication. The third part provides an overview of some key viruses that infect bacteria and animals; further coverage of viral diversity can be found in Chapter 19. The fourth part deals with subviral entities.

Viruses are the most numerous microorganisms on our planet and infect all types of cellular organisms. Therefore, they are interesting in their own right. However, scientists also study viruses for what they can tell us about the genetics and biochemistry of cellular processes and, for many viruses, the development of disease. Furthermore, as we shall see in Chapters 11 and 12, viruses are also important tools in microbial genetics and genetic engineering.

I VIRUS STRUCTURE AND GROWTH

10.1 General Properties of Viruses

Although viruses are not cells and thus are nonliving, they nonetheless possess a genome encoding the information they need in order to replicate. However, viruses rely on host cells to provide the energy and materials needed for replicating their genomes and synthesizing their proteins. Consequently, they cannot replicate unless they have entered a suitable host cell.

Viruses can exist in either extracellular or intracellular forms. In its extracellular form, a virus is a microscopic particle containing nucleic acid surrounded by protein and sometimes, depending on the specific virus, other macromolecules. In this form, the virus particle, called the **virion**, is metabolically inert and does not carry out respiration or biosynthesis. The virion is the structure by which the virus genome moves from the cell in which it was produced to another cell. Once in the new cell, the intracellular state begins and the virus replicates: New copies of the virus genome are produced, and the components that make up the virus coat are synthesized. Certain animal viruses (such as polio and respiratory syncytial virus) may skip the extracellular stage when moving from cell to cell within the same organism. Instead, they mediate the fusion of infected cells with uninfected cells and transfer the virus in this way. However, when moving from one organism to another they are truly extracellular.

Viral genomes are usually very small, and they encode primarily those functions that viruses cannot adapt from their hosts. Therefore, during replication inside a cell, viruses depend heavily on host cell structural and metabolic components. The virus redirects host metabolic functions to support virus replication and the assembly of new virions. Eventually, new viral particles are released, and the process can repeat itself.

Viral Genomes

All cells contain double-stranded DNA genomes. By contrast, viruses have either DNA or RNA genomes. (One group of viruses does use both DNA and RNA as their genetic material but at different stages of their replication cycle.) Virus genomes can be classified according to whether the nucleic acid in the virion is DNA or RNA and further subdivided according to whether the nucleic acid is single- or double-stranded, linear, or circular (**Figure 10.1**). Some viral genomes are circular, but most are linear.

Although those viruses whose genome consists of DNA follow the central dogma of molecular biology (DNA → RNA → protein, Section 7.1), RNA viruses are exceptions to this rule. Nonetheless, genetic information still flows from nucleic acid to protein. Moreover, all viruses use the cell's translational machinery, and so regardless of the genome structure of the virus, messenger RNA (mRNA) must be generated that can be translated on the host cell ribosomes.

Viral Hosts and Taxonomy

Viruses can be classified on the basis of the hosts they infect as well as by their genomes. Thus, we have bacterial viruses, animal viruses, plant viruses, and viruses that infect other kinds of eukaryotic cells. Bacterial viruses, sometimes called **bacteriophages** (or phage for short; from the Greek *phagein*, meaning "to eat"), have been intensively studied as model systems for the molecular biology and genetics of virus replication. Species of both *Bacteria* and *Archaea* are infected by specific bacteriophages. Indeed, many of the basic concepts of virology were first worked out with bacterial viruses and subsequently applied to viruses of higher organisms. Because of their frequent medical importance, animal viruses also have been extensively studied, whereas plant viruses,

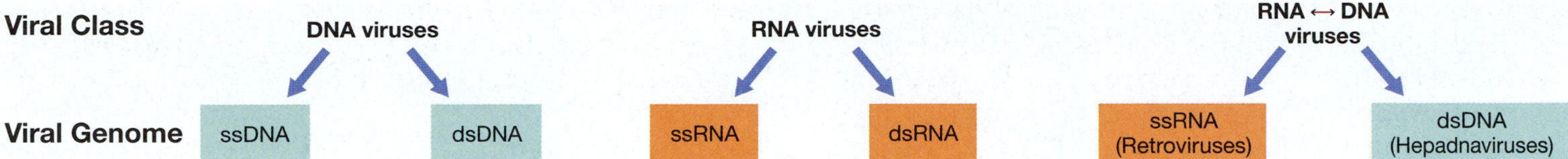

Figure 10.1 Viral genomes. The genomes of viruses can be either DNA or RNA, and some use both as their genomic material at different stages in their replication cycle. However, only one type of nucleic acid is found in the virion of any particular type of virus. This can be single-stranded (ss), double-stranded (ds), or in the hepadnaviruses, partially double-stranded. Some viral genomes are circular, but most are linear.

although of enormous importance to modern agriculture, have been the least studied of all viruses.

A formal system of viral classification exists that groups viruses into various taxa, such as orders, families, and even genus and species. The family taxon seems particularly useful. Members of a family of viruses all have a similar virion morphology, genome structure, and strategy of replication. Virus families have names that include the suffix *-viridae* (as in *Poxviridae*). We discuss a few of these in Chapter 19.

10.1 MiniReview

A virus is an obligate intracellular parasite that cannot replicate without a suitable host cell. A virion is the extracellular form of a virus and contains either an RNA or a DNA genome inside a protein shell. The virus genome may be introduced into a new host cell by infection. The virus redirects the host metabolism to support virus replication. Viruses are classified by their nucleic acids and type of host.

- How does a virus differ from a plasmid?
- How does a virion differ from a cell?
- What is a bacteriophage?

10.2 Nature of the Virion

Virions come in many sizes and shapes. Most viruses are smaller than prokaryotic cells, ranging in size from 0.02 to 0.3 μm (20–300 nm). A common unit of measure for viruses is the nanometer, which is one-thousandth of a micrometer. Smallpox virus, one of the largest viruses, is about 200 nm in diameter (about the size of the smallest cells of *Bacteria*). Poliovirus, one of the smallest viruses, is only 28 nm in diameter (about the size of a ribosome). Consequently, viruses could not be properly characterized until the invention of the electron microscope, part way through the 20th century.

Viral genomes are smaller than those of most cells. Most bacterial genomes are between 1,000 and 5,000 kilobase pairs (kbp) of DNA, with the smallest known being about 500 kbp. (Interestingly, *Bacteria* with the smallest genomes are, like viruses, parasites that replicate in other cells; Sections 13.2, 15.13, and 16.3). The largest known viral genome, that of *Mimivirus*, consists of 1.18 Mbp of dsDNA (double-stranded DNA). This virus, which infects protists such as *Amoeba*, is one of only a few viruses currently known whose genome is larger than some cellular genomes. More typical virus genome sizes are listed in **Table 10.1**. Some viruses have genomes so small they contain fewer than five genes. Also, as can be seen in the table, the genome of some viruses, such as reovirus or influenza virus, is segmented into more than one molecule of nucleic acid.

Viral Structure

The structures of virions are quite diverse, varying widely in size, shape, and chemical composition. The nucleic acid of the virion is always located within the particle, surrounded by a protein shell called the **capsid**. This protein coat is composed of a number of individual protein molecules, which are arranged in a precise and highly repetitive pattern around the nucleic acid (**Figure 10.2**).

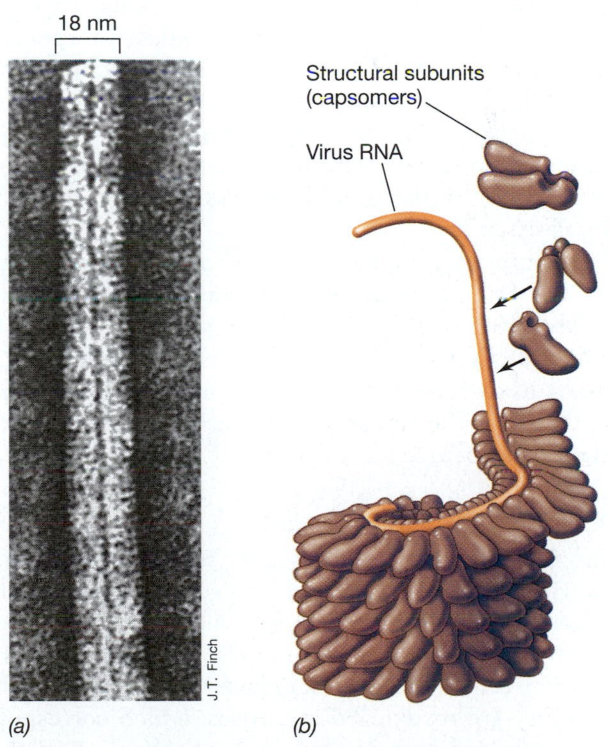

Figure 10.2 The arrangement of nucleic acid and protein coat in a simple virus, tobacco mosaic virus. *(a)* A high-resolution electron micrograph of a portion of the virus particle. *(b)* Assembly of the tobacco mosaic virus virion. The RNA assumes a helical configuration surrounded by the protein capsid. The center of the particle is hollow.

Table 10.1 Some types of viral genomes[a]

Virus	Host	DNA or RNA	Viral genome: Single- or double-stranded	Structure	Number of molecules	Size[a]
H-1 parvovirus	Animals	DNA	Single-stranded	Linear	1	5,176
ϕX174	*Bacteria*	DNA	Single-stranded	Circular	1	5,386
Simian virus 40 (SV40)	Animals	DNA	Double-stranded	Circular	1	5,243
Poliovirus	Animals	RNA	Single-stranded	Linear	1	7,433
Cauliflower mosaic virus	Plants	DNA	Double-stranded	Circular	1	8,025
Cowpea mosaic virus	Plants	RNA	Single-stranded	Linear	2 different	9,370 (total)
Reovirus type 3	Animals	RNA	Double-stranded	Linear	10 different	23,549 (total)
Bacteriophage lambda	*Bacteria*	DNA	Double-stranded	Linear	1	48,514[b]
Herpes simplex virus type I	Animals	DNA	Double-stranded	Linear	1	152,260
Bacteriophage T4	*Bacteria*	DNA	Double-stranded	Linear	1	168,903
Human cytomegalovirus	Animals	DNA	Double-stranded	Linear	1	229,351

[a]The sizes of the viral genomes chosen for this table are known accurately because they have been sequenced. However, this accuracy can be misleading because only a particular strain or isolate of a virus was sequenced. Therefore, the sequence and exact number of bases for other isolates may be slightly different. No attempt has been made to choose the largest and smallest viruses known, but rather to give a fairly representative sampling of the sizes and structures of the genomes of viruses containing both single- and double-stranded RNA and DNA. The size is in bases or base pairs depending on whether the virus is single or double stranded.
[b]This includes single-stranded extensions of 12 nucleotides at either end of the linear form of the DNA (see Section 10.10).

The small genome size of most viruses restricts the number of different viral proteins. A few viruses have only a single kind of protein in their capsid, but most viruses have several chemically distinct proteins that are themselves associated in specific ways to form larger assemblies called **capsomers** (Figure 10.2). The capsomer is the smallest morphological unit that can be seen with the electron microscope. A single virion can have a large number of capsomers. The information for proper folding and aggregation of the proteins into capsomers is contained within the structure of the proteins themselves; hence, the overall process of virion assembly is called *self-assembly*.

The complete complex of nucleic acid and protein packaged in the virion is called the virus **nucleocapsid**. Inside the virion are often one or more virus-specific enzymes. Such enzymes play a role during the infection and replication processes, as discussed later in this chapter. Some viruses are *naked*, whereas others possess layers around the nucleocapsid called an *envelope* (**Figure 10.3**).

Virus Symmetry

The nucleocapsids of viruses are constructed in highly symmetric ways. Symmetry refers to the way in which the capsomers are arranged in the virus capsid. When a symmetric structure is rotated around an axis, the same form is seen again after a certain number of degrees of rotation. Two kinds of symmetry are recognized in viruses, which correspond to the two primary shapes, rod and spherical. Rod-shaped viruses have *helical* symmetry, and spherical viruses have *icosahedral* symmetry. In all cases, the characteristic structure of the virus is determined by the structure of the protein subunits of which it is constructed.

A typical virus with helical symmetry is the tobacco mosaic virus (TMV) illustrated in Figure 10.2. It is an RNA virus in which the 2130 identical capsomers are arranged in a helix. The overall dimensions of the TMV virion are 18 × 300 nm. The lengths of helical viruses are determined by the length of the nucleic acid, but the width of the helical virion is determined by the size and packaging of the protein subunits.

An **icosahedron** is a symmetric structure containing 20 faces and 12 vertices and is roughly spherical in shape (**Figure 10.4**). Icosahedral symmetry is the most efficient arrangement of subunits in a closed shell because it uses the smallest number of units to build the shell. The simplest arrangement of capsomers is three per face, for a total of 60 units per virion. Most viruses have more nucleic acid than can be packed into a shell made of just 60 morphological units. The next possible structure that permits close packing

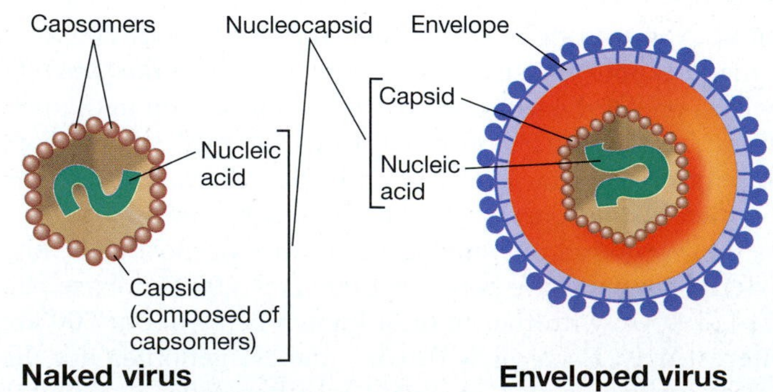

Figure 10.3 Comparison of naked and enveloped virus particles.

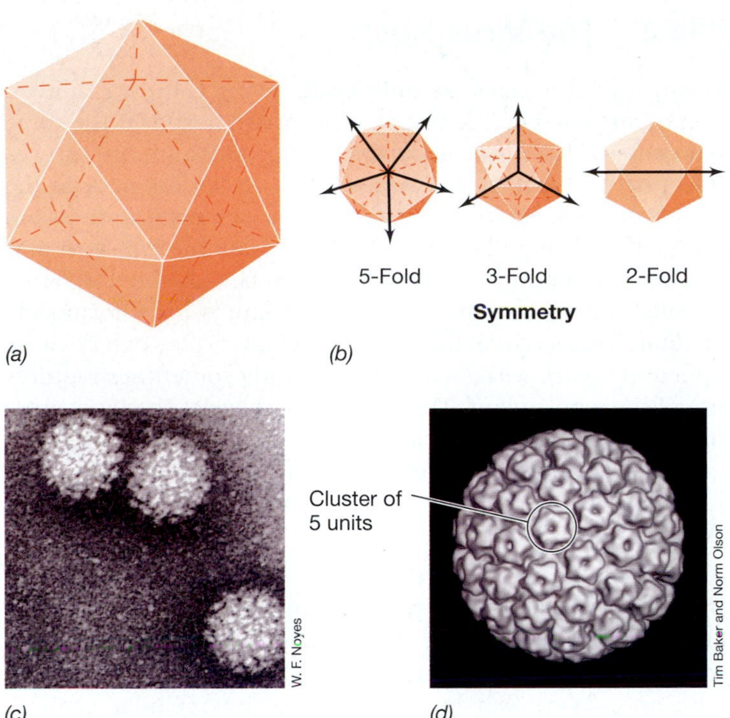

Figure 10.4 Icosahedral symmetry. *(a)* A model of an icosahedron. *(b)* Three views of an icosahedron showing the 5-3-2 symmetry. *(c)* Electron micrograph of human papillomavirus, a virus with icosahedral symmetry. A virion is about 55 nm in diameter. *(d)* Three-dimensional reconstruction of human papillomavirus calculated from images of frozen hydrated virions. The virus contains 360 units arranged in 72 clusters of 5 each.

contains 180 units, and many viruses have shells with this configuration. Other common configurations contain 240 units and 420 units.

Figure 10.4*a* shows a model of an icosahedron. Figure 10.4*b* shows the same icosahedron viewed from three different angles to illustrate its complex 5-3-2 symmetry. The axes of symmetry divide the icosahedron into segments (5, 3, or 2) of identical size and shape. Figure 10.4*c* shows an electron micrograph of a typical icosahedral virus, human papillomavirus; this virus contains 360 morphological units clustered into groups of five. Figure 10.4*d* shows a computer model of the same virus, where the five-unit clusters are more easily seen.

Enveloped Viruses

Enveloped viruses contain a membrane surrounding the nucleocapsid (**Figure 10.5*a***). Many viruses are enveloped, and most infect animal cells (for example, influenza virus), although a few enveloped bacterial and plant viruses are also known. The viral envelope consists of a lipid bilayer with proteins, usually glycoproteins, embedded in it. The lipids of the viral membrane are derived from the membranes of the host cell, but viral membrane proteins are encoded by viral genes. These virus-specific proteins are critical for attachment of the virion to the host cell during infection or for release of the virion from the host cell after replication. The symmetry of enveloped viruses is not expressed by the virion as a whole but by the nucleocapsid present inside the virus envelope.

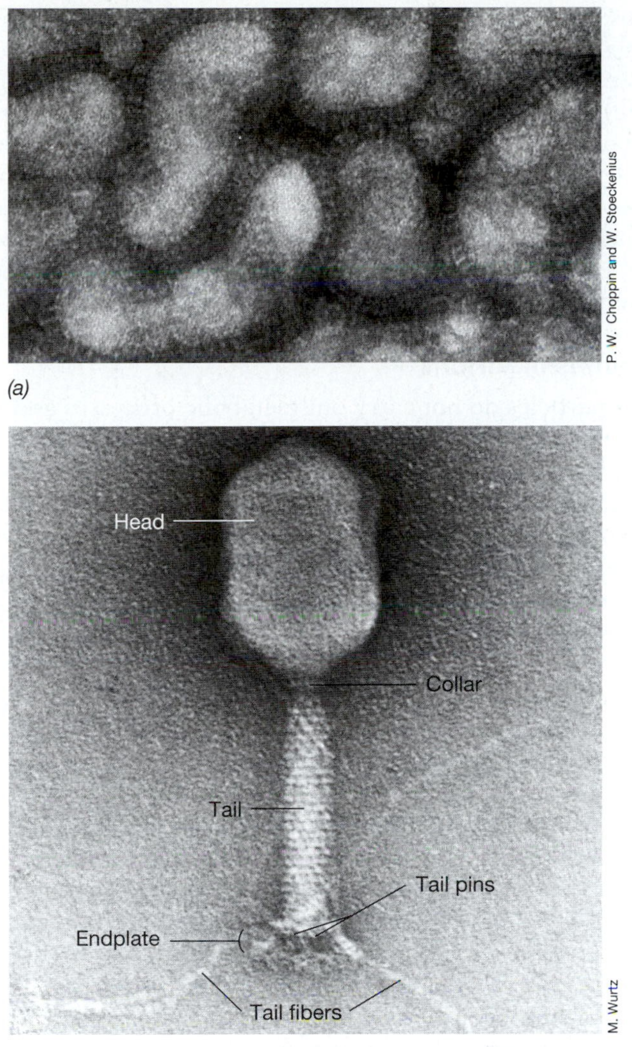

Figure 10.5 Electron micrographs of animal and bacterial viruses. *(a)* Influenza virus, an enveloped virus. The virions are about 80 nm in diameter, but have no defined shape (Section 19.8 and Figure 19.16). *(b)* Bacteriophage T4 of *Escherichia coli*. The tail components function in attachment of the virion to the host and injection of the nucleic acid (Figure 10.10). The head is about 85 nm in diameter.

What is the function of the envelope in a virus particle? We discuss this in detail later, but note that the envelope is the structural component of the virus particle that makes initial contact with the host cell. The specificity of virus infection and some aspects of virus penetration are thus controlled in part by characteristics of virus envelopes.

Complex Viruses

Some virions are even more complex than anything discussed so far, being composed of several parts, each with separate shapes and symmetries. The most complicated viruses in terms of structure are some of the bacterial viruses, which possess icosahedral heads plus helical tails. In some bacterial viruses, such as bacteriophage T4 of *Escherichia coli* (Figure 10.5*b*), the tail itself has a complex structure. The complete T4 tail has almost 20 different proteins, and the T4 head has several more proteins. In such complex viruses, assembly is also quite involved. For instance, in T4 the complete tail is formed as a subassembly, and then the tail is added to the DNA-containing head. Finally, tail fibers formed from another protein are added to make the mature, infectious virion.

Enzymes in Virions

Virus particles do not carry out metabolic processes and thus a virus is metabolically inert outside a host cell. However, some virions do contain enzymes that play important roles in infection. Some of these enzymes are required for very early events in the infection process. For example, some bacteriophages contain the enzyme lysozyme (Section 4.6), which they use to make a small hole in the bacterial cell wall. This allows the viral nucleic acid to enter. Lysozyme is again produced in large amounts in the later stages of infection, causing lysis of the host cell and release of the virions.

Many viruses contain their own nucleic acid polymerases for replication of the viral genome and for transcription of virus-specific RNA. For example, retroviruses are RNA viruses that replicate via DNA intermediates. These viruses possess an RNA-dependent DNA polymerase called *reverse transcriptase* that transcribes the viral RNA to form a DNA intermediate. Other viruses contain RNA genomes and require their own RNA polymerase. These virion enzymes are necessary because cells cannot make DNA or RNA from an RNA template (Sections 7.5 and 7.9).

Some viruses contain enzymes that aid in their release from the host. For example, certain animal viruses contain surface proteins called *neuraminadases,* enzymes that cleave glycosidic bonds in glycoproteins and glycolipids of animal cell connective tissue, liberating the virions. Thus, although most virus particles lack their own enzymes, those that contain them do so for good reason: The host cell would not be able to produce virions in the absence of these extra enzymes.

10.2 MiniReview

In the virion of the naked virus, only nucleic acid and protein are present, with the nucleic acid on the inside; the whole unit is called the nucleocapsid. Enveloped viruses have one or more lipoprotein layers surrounding the nucleocapsid. The nucleocapsid is arranged in a symmetric fashion, with a precise number and arrangement of structural subunits surrounding the virus nucleic acid. Although viruses are metabolically inert, in some viruses one or more key enzymes are present within the virion.

- What is the difference between a naked virus and an enveloped virus?
- What kinds of enzymes can be found within the virions of specific viruses?

10.3 The Virus Host

Because viruses replicate only inside living cells, the cultivation of viruses requires use of appropriate hosts. Of the three types of viral hosts—prokaryotes, plants, and animals—viruses infecting prokaryotes are typically the easiest to grow in the laboratory. For the study of bacterial viruses, pure cultures are used either in liquid or on semisolid (agar) media. Most animal viruses and many plant viruses can be cultivated in tissue or cell cultures, and the use of such cultures has enormously facilitated research on these viruses. Plant viruses can be more difficult to work with, because their study sometimes requires use of the whole plant. This is a problem because plants grow much slower than bacteria, and plant viruses also often require a break in the thick plant cell wall in order to infect.

Animal Cell Cultures

Animal cell cultures are derived from cells originally taken from an organ of an experimental animal. Unless blood cells are used, cell cultures are usually obtained by aseptically removing pieces of tissue, dissociating the cells by treatment with an enzyme that breaks apart the intercellular cement, and spreading the resulting suspension over a flat surface, such as the bottom of a culture flask or a Petri dish. The thin layer of cells adhering to the glass or plastic dish, called a *monolayer,* is overlaid with a suitable culture medium and incubated at a suitable temperature. The culture media used for cell cultures are typically quite complex, containing a number of amino acids and vitamins, salts, glucose, and a bicarbonate buffer system. To obtain the best growth, addition of a small amount of blood serum is usually necessary, and several antibiotics are added to prevent bacterial contamination.

Some cell cultures prepared in this way can be subcultured and grown indefinitely as *permanent cell lines.* Cell lines are convenient for virus research because cell material is continuously available. In many cases, a culture will not grow indefinitely but may remain alive for a number of days. Such cultures, called *primary cell cultures,* may still be useful for growing a virus, although new cultures need to be prepared from fresh sources from time to time, an expensive and time-consuming process. In some cases, primary or permanent cell lines cannot be obtained, but whole organs or pieces of organs can successfully replicate the virus. Such organ cultures may still be useful in virus research because they permit growth of viruses under more- or less-controlled laboratory conditions.

10.3 MiniReview

Viruses can replicate only in certain types of cells or in whole organisms. Bacterial viruses have proved useful as model systems because the host cells are easy to grow and manipulate in culture. Many animal and plant viruses can be grown in cultured cells.

- In virology, what is a host?
- Why is it helpful to use cell culture for viral research?

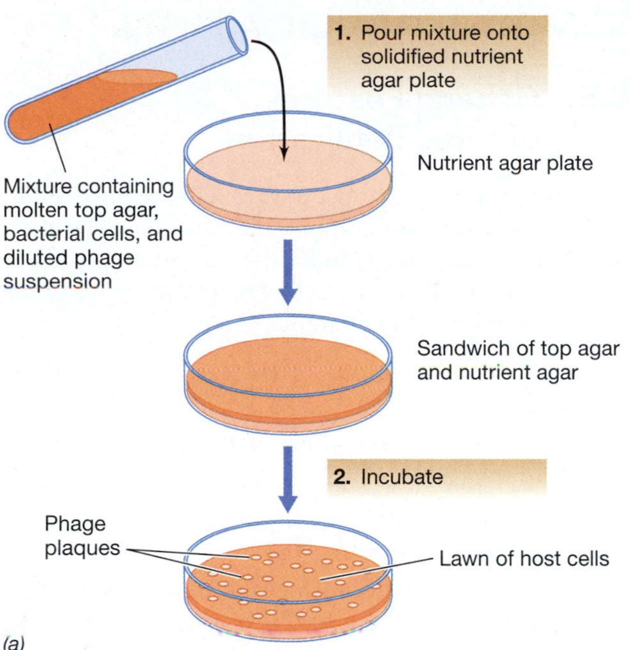

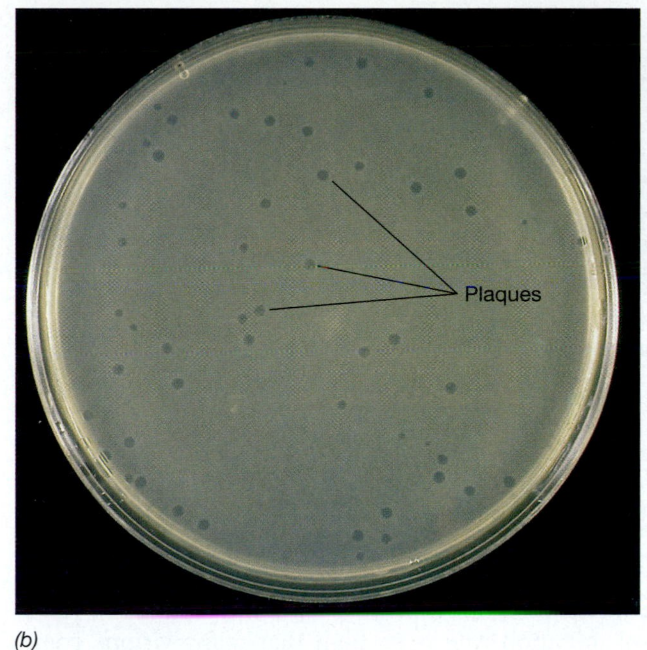

Figure 10.6 Quantification of bacterial virus by plaque assay using the agar overlay technique. *(a)* A dilution of a suspension containing the virus is mixed in a small amount of melted agar with the sensitive host bacteria. The mixture is poured on the surface of an agar plate of the appropriate medium. The host bacteria, which have been spread uniformly throughout the top agar layer, begin to grow, and after overnight incubation form a lawn of confluent growth. Virion-infected cells are lysed, forming plaques in the lawn. The size of the plaque depends on the virus, the host, and conditions of culture. *(b)* Photograph of a plate showing plaques formed by a bacteriophage on a lawn of sensitive bacteria. The plaques shown are about 1–2 mm in diameter.

10.4 Quantification of Viruses

For many purposes, in virology it is necessary to quantify the number of virus particles in a suspension. Although one can enumerate virions using an electron microscope (Figure 10.4), the number of virions in a suspension can be more easily quantified by measuring effects on the host. Using such a method, we see that a virus infectious unit is the smallest unit that causes a detectable effect when added to a susceptible host. This can be as few as one virion, although more often a larger inoculum is required. By determining the number of infectious units per volume of fluid, a measure of virus quantity, called a *titer,* can be obtained.

Plaque Assay

When a virion initiates an infection on a layer of host cells growing on a flat surface, a zone of lysis may be seen as a clear area in the lawn of growing host cells. This clearing is called a **plaque**, and it is assumed that each plaque originated from the replication of a single virion (**Figure 10.6**).

Plaques are essentially "windows" in the lawn of confluent cell growth. With bacteriophages, plaques may be obtained when virus particles are mixed into a thin layer of host bacteria that is spread out in an agar overlay on the surface of an agar medium (Figure 10.6*a*). During incubation of the culture, the bacteria grow and form a turbid layer that is visible to the naked eye. However, wherever a successful viral infection has been initiated, cells are lysed forming a plaque (Figure 10.6*b*). By counting the number of *plaque-forming units,* one can calculate the number of virus infectious units present in the original sample.

The plaque procedure also permits the isolation of pure virus strains. This is because if a plaque has arisen from a single virion, all the viruses in this plaque should be genetically identical. Some of the virions from this plaque can be picked and inoculated into a fresh bacterial culture to establish a pure virus line. The development of the plaque assay technique was as important for the advancement of virology as Koch's development of solid media (∞ Section 1.8) was for pure culture microbiology.

Plaques may be obtained for animal viruses by using animal cell culture systems as hosts. A monolayer of cultured animal cells is prepared on a plate or flat bottle and the virus suspension overlaid. Plaques are revealed by zones of destruction of the animal cells, and from the number of plaques produced, an estimation of the virus titer can be made (**Figure 10.7**).

Efficiency of Plating

The concept of *efficiency of plating* is important in quantitative virology. In any given viral system, the number of plaque-forming units is always lower than counts of the viral suspension made with an electron microscope. The efficiency with which virions infect host cells is thus rarely 100% and may often be considerably less. Virions that fail to cause infection are often inactive, although this is not always the case. Some viruses produce many incomplete virus particles

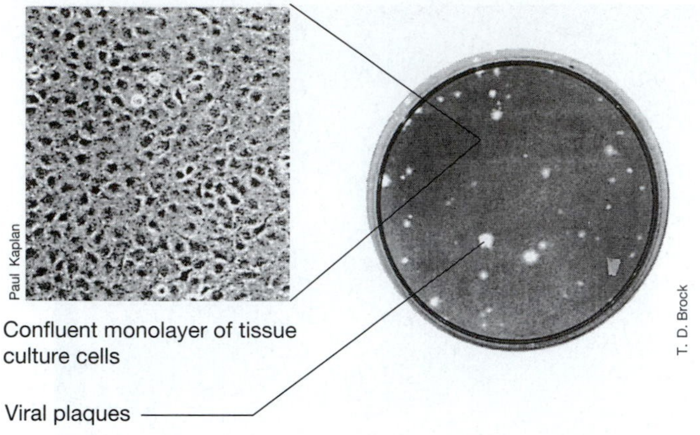

Figure 10.7 Cell cultures in monolayers grown on a Petri plate. Note the presence of plaques. Also shown is a photomicrograph of a cell culture.

during infection. In other cases, especially with RNA viruses, the viral mutation rate is so high that many virions contain defective genomes. However, sometimes a low efficiency of plating merely means that under the conditions used, some virions did not successfully infect cells. Although with bacterial viruses efficiency of plating is often higher than 50%, with many animal viruses it may be much lower, 0.1 or 1%. Knowledge of plating efficiency is useful in cultivating viruses because it allows one to estimate how concentrated a viral suspension needs to be (that is, its titer) to yield a certain number of plaques.

Intact Animal Methods

Some viruses do not cause recognizable effects in cell cultures yet cause death in whole animals. In such cases, quantification can be done only by titration in infected animals. The general procedure is to carry out a serial dilution of the virus sample (Section 6.10), generally at tenfold dilutions, and to inject samples of each dilution into several sensitive animals. After a suitable incubation period, the fraction of dead and live animals at each dilution is tabulated and an end point dilution is calculated. This is the dilution at which, for example, half of the injected animals die (the LD_{50}, Section 28.8). Although using whole animals is much more cumbersome and much less accurate than cell culture methods, it may be essential for the study of certain types of viruses.

10.4 MiniReview

Although it requires only a single virion to initiate an infectious cycle, not all virions are equally infectious. One of the most accurate ways of measuring virus infectivity is by the plaque assay. Plaques are clear zones that develop on lawns of host cells. Theoretically, each plaque is due to infection by a single virus particle. The virus plaque is analogous to the bacterial colony.

- Give a definition of efficiency of plating.
- What is a plaque-forming unit?

II VIRAL REPLICATION

10.5 General Features of Virus Replication

For a virus to replicate it must induce a living host cell to synthesize all the essential components needed to make more virions. These components must then be assembled into new virions that escape from the cell. The phases of viral replication can be divided into five steps (**Figure 10.8**).

1. *Attachment* (adsorption) of the virion to a susceptible host cell.
2. *Penetration* (entry, injection) of the virion or its nucleic acid into the cell.
3. *Synthesis* of virus nucleic acid and protein by cell metabolism as redirected by the virus.
4. *Assembly* of capsids (and membrane components in enveloped viruses) and *packaging* of viral genomes into new virions. This whole process is called *maturation*.
5. *Release* of mature virions from the cell.

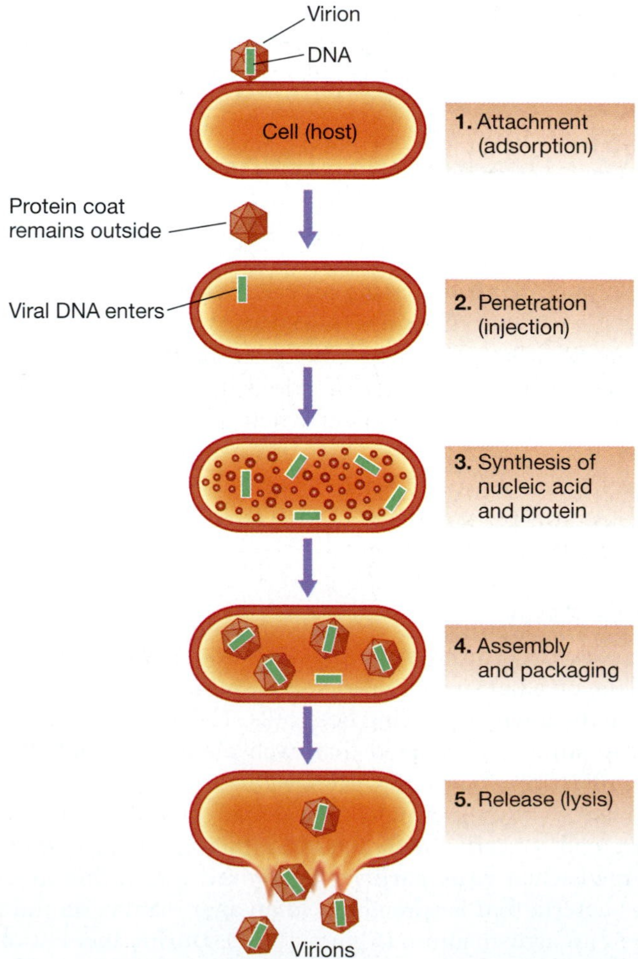

Figure 10.8 The replication cycle of a bacterial virus. Note that the viruses and cell are not drawn to scale.

These stages of virus replication are illustrated in **Figure 10.9**. In the first few minutes after infection the virus is said to undergo an *eclipse*. During this period infectious particles cannot be detected in the culture medium. The eclipse begins as soon as infectious particles are removed from the environment by adsorbing to host cells. Once attached, the virions are no longer available to infect other cells. This is followed by the entry of viral nucleic acid (or intact virion) into the host cell. If the infected cell breaks open at this point, the virion no longer exists as an infectious entity since the viral genome is no longer inside its capsid. *Maturation* begins as the newly synthesized nucleic acid molecules become packaged inside protein coats. During the maturation phase, the titer of active virions inside the cell rises dramatically. However, the new virus particles still cannot be detected unless the cells are artificially lysed to release them. Because newly synthesized virions have not yet appeared outside the cell, the eclipse and maturation periods together are called the latent period.

At the end of maturation, mature virions are released, either as a result of cell lysis or by budding or excretion, depending on the virus. The number of virions released, called the burst size, varies with the particular virus and the particular host cell and can range from a few to a few thousand. The timing of this overall virus replication cycle varies from 20–60 min (in many bacterial viruses) to 8–40 h (in most animal viruses). Because the release of virions is more or less simultaneous, virus replication is typically characterized by a *one-step growth curve* (Figure 10.9). In the next two sections we consider a few key steps of the virus replication cycle in more detail.

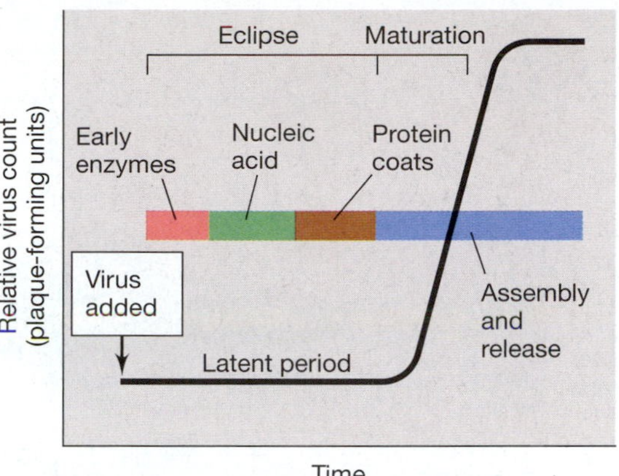

Figure 10.9 The one-step growth curve of virus replication. This graph displays the results of a single round of viral replication in a population of cells. Following adsorption, the infectivity of the virus particles disappears, a phenomenon called eclipse. This is due to the uncoating of the virus particles. During the latent period, viral nucleic acid replicates and protein synthesis occurs. The maturation period follows, when virus nucleic acid and protein are assembled into mature virus particles. Finally, the virions are released, either with or without cell lysis. This general picture is amplified for bacteriophage T4 in Figure 10.15.

10.5 MiniReview

The virus replication cycle can be divided into five stages: attachment (adsorption), penetration (injection), protein and nucleic acid synthesis, assembly and packaging, and virion release.

- What is packaged into the virions?
- Explain the term maturation.
- What events happen during the latent period of viral replication?

10.6 Viral Attachment and Penetration

In this section we focus on virus attachment and penetration, the first steps in replication. In addition, we consider the mechanism by which some bacteria react to penetration by bacteriophage DNA.

Attachment

The most common basis for the host specificity of a virus depends upon attachment. The virion itself (whether it is naked or enveloped) has one or more proteins on the outside that interact with specific cell surface components called receptors. These receptors are normal surface components of the host, such as proteins, carbohydrates, glycoproteins, lipids, lipoproteins, or complexes of these, to which the virion attaches. The receptors carry out normal functions for the cell. For example, the receptor for bacteriophage T1 is an iron-uptake protein and that for bacteriophage lambda is normally involved in maltose uptake. Animal virus receptors may include macromolecules needed for cell–cell contact or by the immune system. For example, the receptors for poliovirus and for HIV are normally used in interactions between human cells.

In the absence of its specific receptor, the virus cannot adsorb and hence cannot infect. Moreover, if the receptor site is altered, for example, by mutation, the host may become resistant to virus infection. However, mutants of the virus can also arise that gain the ability to adsorb to previously resistant hosts. In addition, some animal viruses may be able to use more than one receptor, so the loss of one may not necessarily prevent attachment. Thus, the host range of a particular virus is, to some extent, determined by the availability of a suitable receptor that the virus can recognize. In multicellular organisms, cells in different tissues or organs often express different proteins on their cell surfaces. Consequently, viruses that infect animals often infect only cells of certain tissues. For example, many viruses that cause coughs and colds infect only cells of the upper respiratory tract.

Penetration

The attachment of a virus to its host cell results in changes to both the virus and the cell surface that result in penetration. Viruses must replicate within cells. Therefore, at a minimum, the viral genome must enter the cell (Figure 10.8). However,

UNIT 2

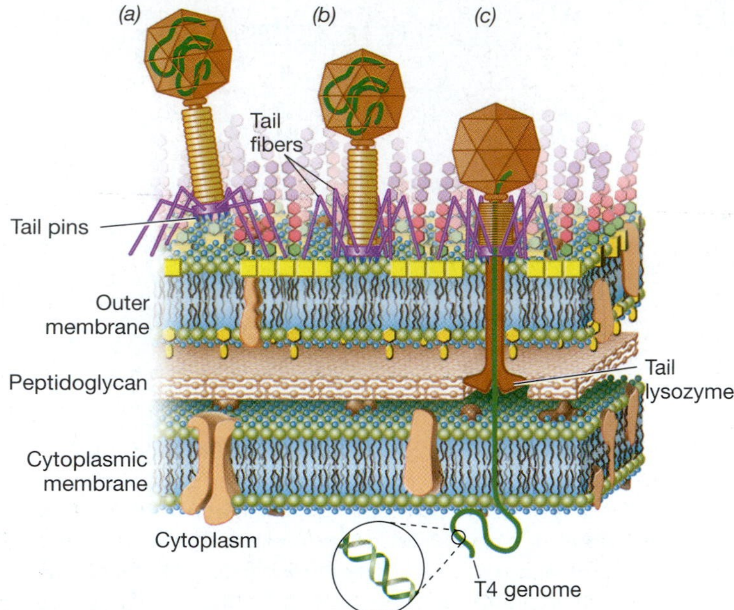

Figure 10.10 Attachment of bacteriophage T4 to the cell wall of *Escherichia coli* and injection of DNA. *(a)* Attachment of a T4 virion to the cell wall by the long tail fibers interacting with core lipopolysaccharide. *(b)* Contact of cell wall by the tail pins. *(c)* Contraction of the tail sheath and injection of the T4 genome. The tail tube penetrates the outer membrane, and the tail lysozyme digests a small opening through the peptidoglycan.

we mentioned (Section 10.2) that for some viruses to replicate, certain proteins must also enter the host cell. It is important to note that attachment to and even penetration of a susceptible cell will not lead to virus replication if the information in the viral genome cannot be read. A cell that allows the complete replication cycle of a virus to take place is said to be *permissive* for that virus.

Different viruses have different strategies for penetration. Some enveloped animal viruses are uncoated at the cytoplasmic membrane, releasing the virion contents into the cytoplasm. However, the entire virion of naked animal viruses and many enveloped animal viruses enter the cell via endocytosis. In such cases the virus must be uncoated inside the host cell so that the genome is exposed and replication can proceed. Some enveloped viruses are uncoated in the cytoplasm and others (such as influenza) are uncoated at the nuclear membrane.

Tailed Bacteriophage Attachment and Penetration

Cells that have cell walls, such as most bacteria, are infected in a manner different from animal cells, which lack a cell wall. The most complex penetration mechanisms have been found in viruses that infect bacteria. The bacteriophage T4, which infects *Escherichia coli,* is a good example.

The structure of bacteriophage T4 was shown in Figure 10.5*b*. The virion has a head, within which the viral linear double-stranded DNA is folded, and a long, fairly complex tail, at the end of which is a series of tail fibers and tail pins. The virions first attach to cells by means of the tail fibers (**Figure 10.10**). The ends of the fibers interact specifically with polysaccharides that are part of the outer layer of the gram-negative cell envelope (∞ Section 4.7). These tail fibers then retract, and the core of the tail makes contact with the cell wall of the bacterium through a series of fine tail pins at the end of the tail. The activity of a lysozyme-like enzyme forms a small pore in the peptidoglycan. The tail sheath then contracts, and the viral DNA passes into the cytoplasm through a hole in the tip of the phage tail, with the majority of the coat protein remaining outside (Figure 10.10).

Virus Restriction and Modification by the Host

Animals can often eliminate invading viruses by immune defense mechanisms before the viral infection becomes widespread or sometimes even before the virus has penetrated target cells. In addition, eukaryotes, including animals and plants, possess an antiviral mechanism known as RNA interference (∞ Section 8.10). Prokaryotes lack both these defenses. However, prokaryotes do have specific DNA destruction systems that can destroy double-stranded viral DNA after it has been injected. This destruction is brought about by host restriction endonucleases (∞ Section 12.1), enzymes that cleave the viral DNA at one or several places, thus preventing its replication. This phenomenon is called *restriction* and is part of a general host mechanism to prevent the invasion of foreign nucleic acid. For such a system to be effective the host must have a mechanism for protecting its own DNA. This is accomplished by specific modification of its DNA at the sites where the restriction enzymes cut (∞ Section 12.1).

Restriction enzymes are specific for double-stranded DNA, and thus single-stranded DNA viruses and all RNA viruses are unaffected by restriction systems. Although host restriction systems confer significant protection, some DNA viruses have overcome their host's restriction mechanisms by modifying their own nucleic acids so that they are no longer subject to restriction enzyme attack. Two patterns of chemical modification of viral DNA are known: glucosylation and methylation. For instance, the T-even bacteriophages (T2, T4, and T6) have their DNA glucosylated to varying degrees, which prevents endonuclease attack. Many other viral nucleic acids can be modified by methylation. However, whether glucosylated or methylated, viral nucleic acids are modified after genomic replication has occurred by modification proteins encoded by the virus.

Other viruses, such as the bacteriophages T3 and T7, avoid destruction by host restriction enzymes by encoding proteins that inhibit the host restriction systems. To counter this, some hosts have multiple restriction and methylation systems that help prevent infection by viruses that can circumvent only one of them. Hosts also contain other DNA methylases in addition to those that protect them from their own restriction enzymes. Some of these methylases take part in DNA repair or in gene regulation, but others protect the host DNA from foreign endonucleases. This is necessary because some viruses encode restriction systems themselves, systems that are designed to destroy host DNA! It is thus clear that viruses and hosts have responded to each other's defense mechanisms by continuing to evolve their own mechanisms to better their chances of infection or survival, respectively.

10.6 MiniReview

The attachment of a virion to a host cell is a highly specific process requiring complementary receptors on the surface of a susceptible host cell and its infecting virus. Resistance of the host to infection by the virus can involve restriction–modification systems that recognize and destroy foreign double-stranded DNA.

- How does the attachment process contribute to virus–host specificity?
- Why do some viruses need to be uncoated after penetration and others do not?

10.7 Production of Viral Nucleic Acid and Protein

Once a host has been infected, new copies of the viral genome must be made and virus-specific proteins must be synthesized in order for the virus to replicate. Typically, the production of at least some viral proteins begins very early after the viral genome has entered the cell. The synthesis of these proteins requires viral mRNA. For certain types of RNA viruses, the genome itself is the mRNA. For most viruses, however, the mRNA must first be transcribed from the DNA or RNA genome and then the genome must be replicated. We consider these important events here.

The Baltimore Classification Scheme and DNA Viruses

The virologist David Baltimore, who along with Howard Temin and Renato Dulbecco shared the Nobel Prize for Physiology or Medicine in 1975 for the discovery of retroviruses and reverse transcriptase, developed a classification scheme for viruses based on the relationship of the viral genome to its mRNA. In the Baltimore Classification scheme (**Table 10.2**), seven classes of viruses are recognized. Double-stranded (ds) DNA viruses are in Class I. The mechanism of mRNA production and genome replication of Class I viruses is the same as that used by the host cell genome, although different viruses use different strategies to ensure that viral mRNA is expressed in preference to host mRNA.

Class II viruses are single-stranded (ss) DNA viruses. Before mRNA can be produced from such viruses, a complementary DNA strand must be synthesized because RNA polymerase uses double-stranded DNA as a template (Section 7.9). These viruses form a double-stranded DNA intermediate during replication, and it is this intermediate that is used for transcription (**Figure 10.11**). The synthesis of the dsDNA intermediate and its subsequent transcription can be carried out by cellular enzymes (although viral proteins may also be required). The double-stranded intermediate is also used to generate the viral genome; one strand becomes the genome while the other is discarded (Figure 10.11). Until recently, all known ssDNA viruses contained DNA with the same sequence as their mRNA (positive-strand viruses, see later). However, in 1997 a novel virus with circular ssDNA of negative polarity was discovered. Torque Teno virus (TTV) is widespread in humans and other animals but causes no obvious disease symptoms. The mode of replication of TTV has not yet been fully investigated.

Positive- and Negative-Strand RNA Viruses

The production of mRNA and genome replication is different for RNA viruses (Classes III–VI). However, before discussing these viruses, some nomenclature related to nucleic acid strand orientation is needed. Recall that mRNA is complementary in base sequence to the template strand of DNA. By convention in virology, mRNA is said to be of the plus (+) configuration. Its complement is thus of the minus (−) configuration. This convention is used to describe the configuration of the genome of a single-stranded virus, whether its genome contains RNA or DNA (Figure 10.11). For example, a virus that has a single-stranded RNA genome with the same orientation as its mRNA is a **positive-strand RNA virus**. A virus whose single-stranded RNA genome is complementary to its mRNA is a **negative-strand RNA virus**. How do these viruses replicate?

Cellular RNA polymerases do not normally catalyze the formation of RNA from an RNA template, but instead require a DNA template. Therefore, RNA viruses, whether positive, negative, or double-stranded, require a specific RNA-dependent

Table 10.2 The Baltimore classification system of viruses

		Examples	
Class	*Description of genome and replication strategy*	*Bacterial viruses*	*Animal viruses*
I	Double-stranded DNA genome	Lambda, T4	Herpesvirus, pox virus
II	Single-stranded DNA genome	ϕX174	Chicken anemia virus
III	Double-stranded RNA genome	ϕ6	Reoviruses (Section 19.10)
IV	Single-stranded RNA genome of plus sense	MS2	Poliovirus
V	Single-stranded RNA genome of minus sense		Influenza virus, rabies virus
VI	Single-stranded RNA genome that replicates with DNA intermediate		Retroviruses
VII	Double-stranded DNA genome that replicates with RNA intermediate		Hepatitis B virus

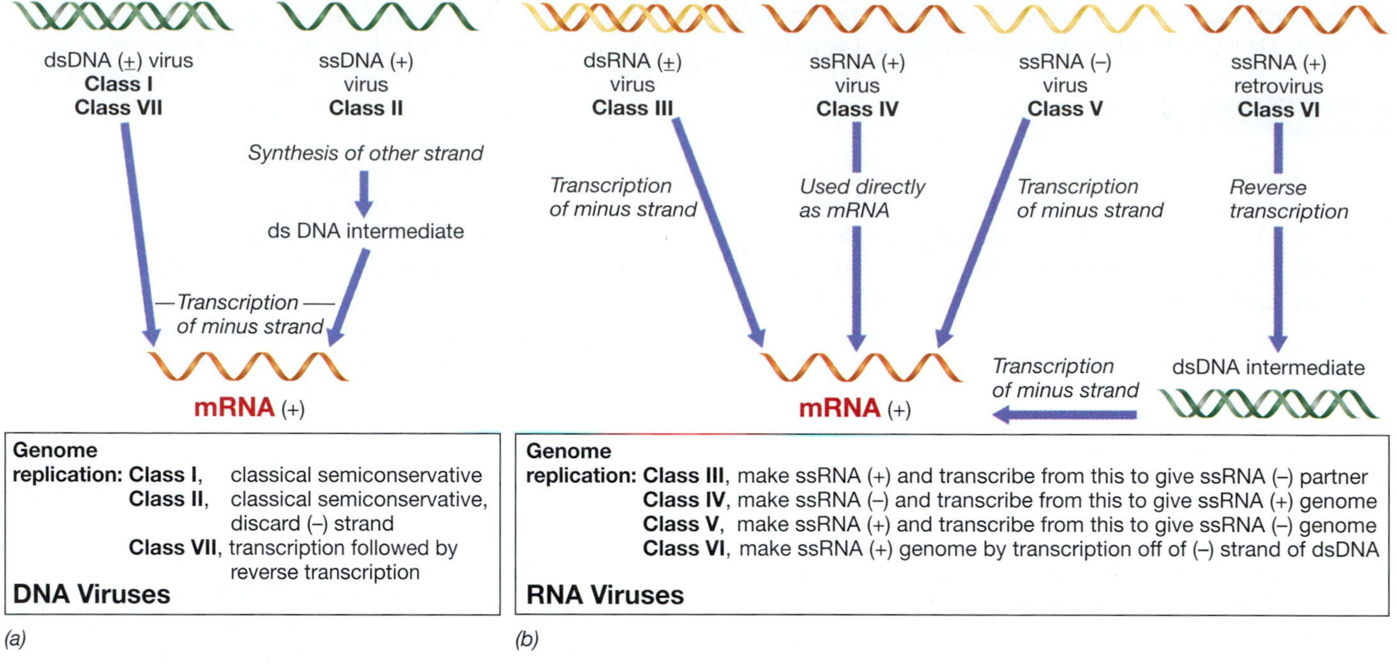

Figure 10.11 Formation of mRNA and new genomes in *(a)* DNA viruses, and *(b)* RNA viruses. By convention, the mRNA is always considered to be of the plus (+) orientation. Examples of each class of virus are given in Table 10.2.

RNA polymerase. The simplest case is the positive-strand RNA viruses (Class IV) in which the viral genome is of the plus configuration and hence can function directly as mRNA (Figure 10.11). In addition to other required proteins, this mRNA encodes a virus-specific RNA-dependent RNA polymerase (also called RNA replicase). Once synthesized, this polymerase first makes complementary minus strands of RNA and then uses them as templates to make more plus strands. The plus strands produced can either be translated as mRNA or packaged as the genome in newly synthesized virions (Figure 10.11).

For negative-strand RNA viruses (Class V) the situation is more complex. The incoming RNA is the wrong polarity to serve as mRNA, and therefore mRNA must be synthesized first. Because cells do not have an RNA polymerase capable of this, these viruses carry some of this enzyme in their virions, and the enzyme enters the cell along with the genomic RNA. The complementary plus strand of RNA is synthesized by this RNA-dependent RNA polymerase and is then used as mRNA. This plus-strand mRNA is also used as a template to make more negative-strand genomes (Figure 10.11). The double-stranded RNA viruses (Class III) face a similar problem. Although the virion does contain plus-strand RNA, this is part of the double-stranded RNA genome and cannot be released to act as mRNA. Consequently, the virions of dsRNA viruses also contain RNA polymerases that transcribe the dsRNA genome to produce plus-strand mRNA upon entry into the host cell.

Retroviruses

The **retroviruses** are animal viruses that are responsible for causing certain kinds of cancers and acquired immunodeficiency syndrome, AIDS. Retroviruses have single-stranded RNA in their virions but replicate through a dsDNA intermediate (Class VI). The process of copying the information found in RNA into DNA is called **reverse transcription**, and thus these viruses require an enzyme called *reverse transcriptase.* Although the incoming RNA of retroviruses is the plus strand, it is not used as message, and therefore these viruses must carry reverse transcriptase in their virions. After infection, the virion ssRNA is converted to dsDNA via a hybrid RNA–DNA intermediate. The dsDNA is then the template for mRNA synthesis by normal cellular enzymes.

Finally, Class VII viruses are those that have double-stranded DNA in their virions but replicate through an RNA intermediate. These unusual viruses also use reverse transcriptase (Section 19.15). The strategy these viruses use to produce mRNA is the same as that of Class I viruses (Figure 10.11), although their DNA replication is very unusual because, as we will see later, the genome is only partially double-stranded (Section 19.15).

While the Baltimore scheme might seem to cover all possibilities, there are exceptions. For example, ambiviruses contain a single-stranded RNA genome, half of which is in the plus orientation (and can thus be used as mRNA) and half in the minus configuration (which cannot). A complementary strand must be synthesized from the latter half before the genes there can be translated. Evolution has clearly pushed viral genome diversity to the limits!

Viral Proteins

Once viral mRNA is made (Figure 10.11), viral proteins can be synthesized. These proteins can be grouped into two broad categories:

1. Proteins synthesized soon after infection, called **early proteins**, which are necessary for the replication of virus nucleic acid, and
2. Proteins synthesized later, called **late proteins**, which include the proteins of the virus coat.

Generally, both the timing and amount of virus proteins are highly regulated. Early proteins are typically enzymes that act catalytically and are therefore synthesized in smaller amounts. By contrast, late proteins are typically structural components of the virion and are made in much larger amounts.

Virus infection upsets the regulatory mechanisms of the host because there is a marked overproduction of viral nucleic acid and protein in the infected cell. In some cases, virus infection causes a complete shutdown of host macromolecular synthesis, whereas in other cases, host synthesis proceeds concurrently with virus synthesis. In either case, regulation of virus synthesis is under the control of the virus rather than the host. Several elements of this control resemble the regulatory mechanisms discussed in Chapter 9, but there are also some uniquely viral regulatory mechanisms. We discuss these regulatory mechanisms next when we consider some well-studied viruses.

10.7 MiniReview

Before viral nucleic acid can replicate, new virus proteins are needed, and these are encoded by mRNA transcribed from the virus genome. In some RNA viruses, the viral genomic RNA is also the mRNA. In others, the virus genome is a template for the formation of viral mRNA, and in certain cases, essential transcriptional enzymes are contained in the virion.

- Why must some types of virus contain enzymes in the virion in order for mRNA to be produced?
- Distinguish between a positive-strand RNA virus and a negative-strand RNA virus.
- Both positive-strand RNA viruses and retroviruses contain plus configuration RNA genomes. Contrast mRNA production in these two classes of viruses.

III VIRAL DIVERSITY

10.8 Overview of Bacterial Viruses

Bacteriophages are quite diverse, and examples of the various classes are illustrated in **Figure 10.12**. Most of the bacterial viruses that have been investigated in detail infect well-studied bacteria, such as *Escherichia coli* and *Salmonella enterica*. However, viruses are known that infect various prokaryotes, both *Bacteria* and *Archaea*.

Most known bacteriophages contain double-stranded DNA genomes, and this type of bacteriophage is thought to be the most common type in nature. However, there are many other kinds known, including those with single-stranded RNA genomes, double-stranded RNA genomes, and single-stranded DNA genomes (Figure 10.11).

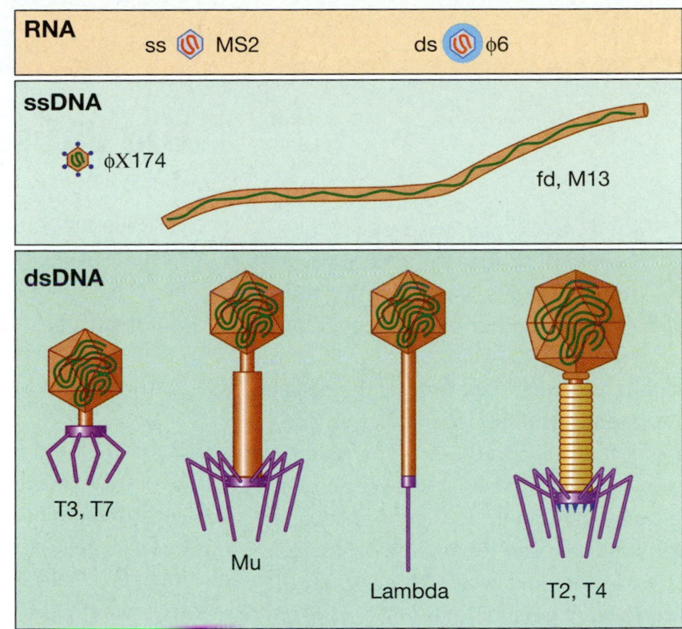

Figure 10.12 Schematic representations of the main types of bacterial viruses. Sizes are to approximate scale. The nucleocapsid of ϕ6 is surrounded by a membrane.

A few bacterial viruses have lipid envelopes, but most are naked. However, many bacterial viruses are structurally complex. All examples of bacteriophages with double-stranded DNA genomes shown in Figure 10.12 have heads and tails. The tails of bacteriophages T2, T4, and Mu are contractile and function in nucleic acid penetration of the host (Figure 10.10). By contrast, the tail of phage lambda is flexible.

Although tailed bacterial viruses were first studied as model systems for understanding general features of virus replication, some of them are now used as convenient tools for genetic engineering. Understanding bacterial viruses is not only valuable as background for the discussion of animal viruses but also is essential for the material presented in the chapters on microbial genetics (Chapter 11) and genetic engineering (Chapter 12).

In the next two sections we examine two contrasting viral life cycles: *virulent* and *temperate*. In the virulent mode viruses lyse or kill their hosts after infection, whereas in the temperate mode viruses replicate their genomes in tandem with the host genome and without killing their hosts.

10.8 MiniReview

Bacterial viruses, or bacteriophages, are very diverse. The best-studied bacteriophages infect bacteria such as *Escherichia coli* and are structurally quite complex, containing heads, tails, and other components.

- What type of nucleic acid is thought to be most common in bacteriophage genomes?

UNIT 2

Microbial Sidebar

Did Viruses Invent DNA?

The three-domain theory of cellular evolution divides living cells into three lineages, *Bacteria*, *Archaea*, and *Eukarya*, based on the sequence of their ribosomal RNA (Section 14.8). In addition, molecular analyses of the cellular components required for translation and transcription support this scheme rather well. However, when molecular analyses of the components required for DNA replication, recombination, and repair are considered, the three-domain scheme does not hold up so well. For example, type II topoisomerases of the *Archaea* are more closely related to those of the *Bacteria* than to those of the *Eukarya*. In addition, viral DNA-processing enzymes show erratic relationships to those of cellular organisms. For example, the DNA polymerase of bacteriophage T4 is more closely related to the DNA polymerases of eukaryotes than to those of its bacterial hosts.

Another major problem in microbial evolution is where viruses fit into the universal tree of life. Did they emerge relatively late as rogue genetic elements escaping from cellular genomes, or were they around at the same time as the very earliest cells? A related issue is when DNA entered the evolutionary scene and took over from RNA as the genetic material. The scenario of the RNA world (Section 14.2) proposes that RNA was the original genetic material of cells and that DNA took over relatively early because it was a more stable molecule than RNA. Hence, in this scheme, the last universal common ancestor (LUCA) to the three domains of life was a DNA-containing cell. But how did LUCA obtain its DNA?

Recently, Patrick Forterre of the Institut Pasteur (Figure 1.14) has suggested a novel evolutionary scenario for how cells obtained DNA that also explains how the cellular machinery that deals with DNA originated in cells in the first place. Forterre argues that minor improvements in genetic stability would not have been sufficiently beneficial to select for the upheaval of converting an entire cellular genome from RNA to DNA. Instead, he suggests that viruses invented DNA as a modification mechanism to protect their genomes from host cell enzymes designed to destroy them (**Figure 1**). Viruses are known today that contain genomes of RNA, DNA, DNA containing uracil instead of thymine, and DNA containing hydroxymethylcytosine in place of cytosine (Figure 1*a*). Moreover, modern cells of all three domains contain systems designed to destroy incoming foreign DNA or RNA.

Forterre's hypothesis starts with an RNA world consisting of cells with RNA genomes plus viruses with RNA genomes. Viruses with DNA genomes were then selected because this protected them from degradation by cellular nucleases. This would have occurred before the LUCA (also containing an RNA genome) split into the three domains (Figure 1*b*). Then three nonvirulent DNA viruses ("founder viruses") infected the ancestors of the three domains. The three DNA viruses replicated inside their host cells as DNA plasmids, much as a P1 prophage replicates inside *Escherichia coli* today. Furthermore, two of the founder viruses were more closely related to each other (and these infected the ancestors of today's *Archaea* and *Eukarya*) than to the third founder virus (which infected the ancestor of *Bacteria*). Gradually, cells converted their genes from RNA into DNA due to its greater stability. Reverse transcriptase is believed to be an enzyme of very ancient origin, and it is conceivable that it was involved in the conversion of RNA genes to DNA, as occurs in retroviruses today.

To recap the hypothesis, the LUCA diverged into the three cellular ancestors to the three domains of life, and this laid the groundwork for the transcription and translation machinery in cells—that is, those functions that involve RNA (but not DNA). However, the use of DNA as a storage system for genetic information—now a universal property of cells—was provided by a family of DNA viruses that infected cells eons ago. Because DNA is a more stable molecule than RNA, cells with RNA genomes that were not infected by DNA viruses never became DNA-based cells and eventually became extinct (Figure 1*b*).

The Forterre model explains the origin of DNA in cells and provides a mechanism for the gradual replacement of RNA genomes with DNA. And, importantly, it also explains the noncongruence of the DNA replication, recombination, and repair machinery of cells of the different domains as compared with the transcription and translation machinery. Although this hypothesis does not wholly explain the origin of viruses, it does explain their diversity of replication systems and the very ancient structural similarities between certain families of DNA and RNA viruses.

(continued)

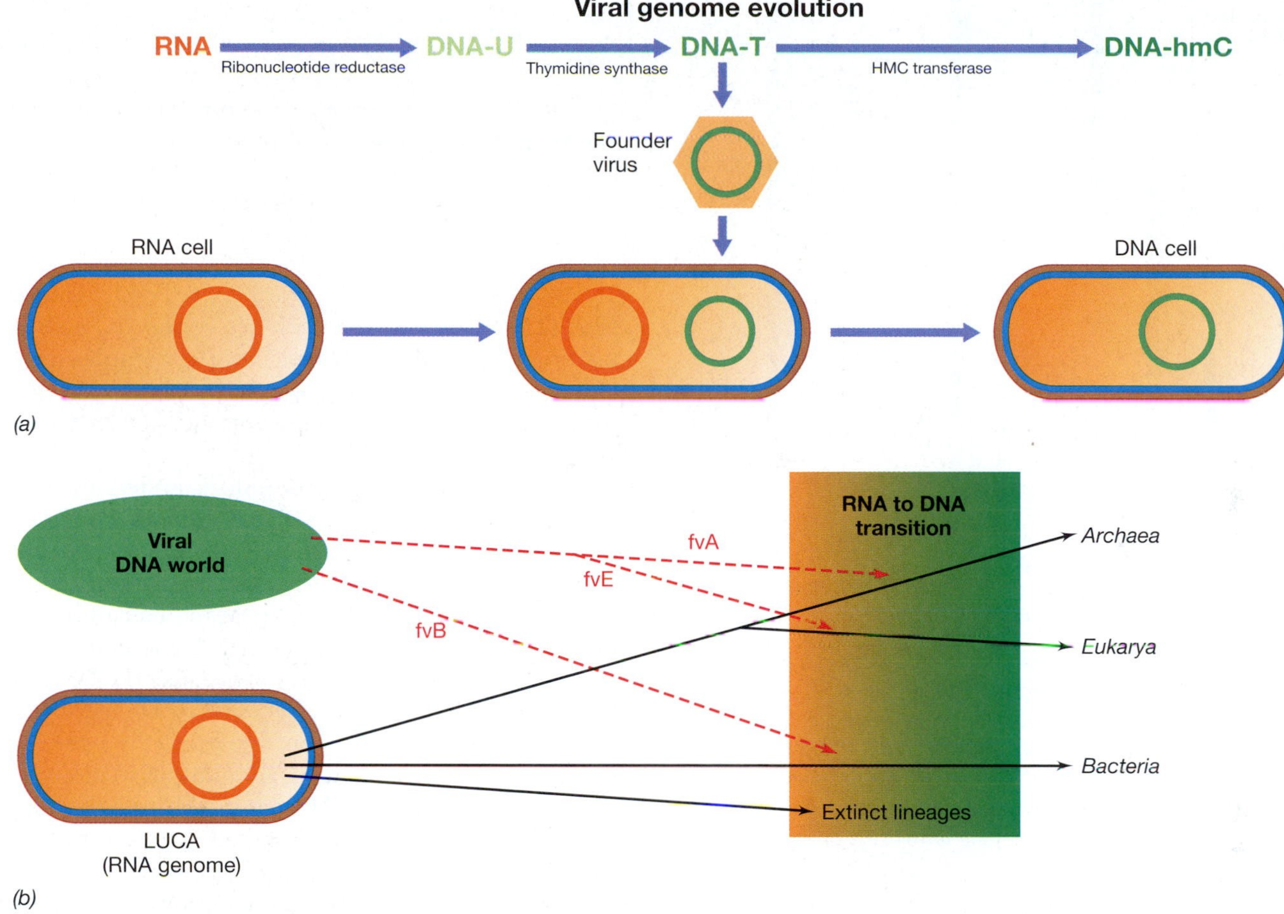

Figure 1 Viral origin of DNA. (a) *Several successive cycles of mutation and selection resulted in the appearance of viral nucleic acids more resistant to degradation by the host cell: DNA-U, DNA with uracil; DNA-T, DNA with thymine (i.e., normal DNA); DNA-hmC, DNA with 5-hydroxymethyl cytosine. All four types of nucleic acid are found in present-day viruses, although DNA-U and DNA-hmC are rare. Conversion of RNA cellular genomes to DNA postulates lysogeny by a DNA "founder virus" followed by movement of host genes onto the DNA genome.* (b) *Three founder viruses, fvB, fvA, fvE, are hypothesized to have infected the ancestors of the Bacteria, Archaea, and Eukarya, respectively. Note that viruses fvA and fvE are more closely related to each other than to fvB. As a result of viral infection the genomes of these three ancestral lines were eventually converted from RNA to DNA. Other cellular lineages derived from the last universal common ancestor (LUCA) that retained RNA genomes are presumably extinct.*

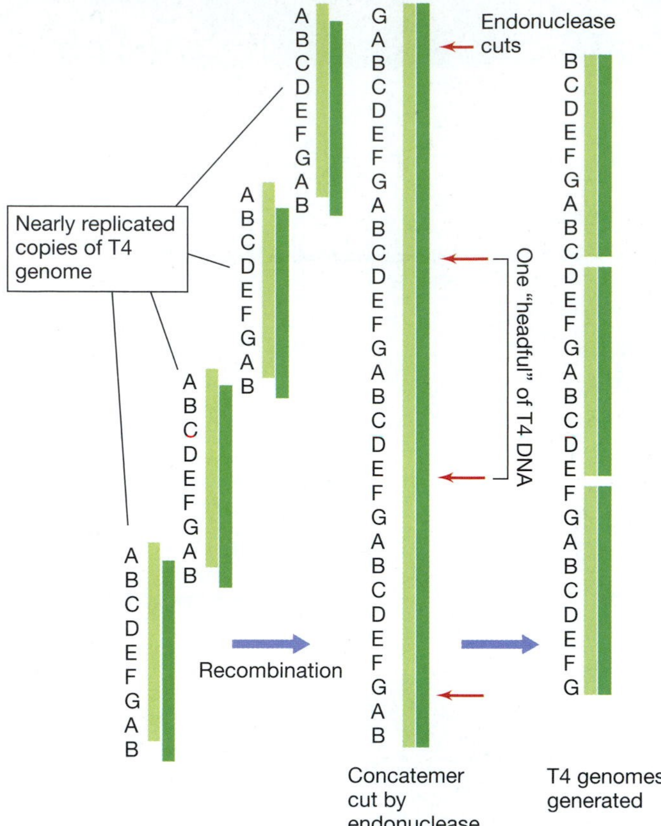

Figure 10.13 Circular permutation. Generation of virus-length T4 DNA molecules with permuted sequences by an endonuclease that cuts off constant lengths of DNA irrespective of the sequence. Left: nearly replicated copies of infecting T4 genome are recombined to form a concatamer. Middle: red arrows, sites of endonuclease cuts. Right: genome molecules generated. Note how each of the T4 genomes formed on the right contains genes A–G, but that the termini are unique in each molecule.

10.9 Virulent Bacteriophages and T4

Virulent viruses kill their hosts after infection. The first such viruses to be studied in detail were bacteriophages with linear, double-stranded DNA genomes that infect *Escherichia coli* and a number of related *Bacteria*. Virologists began isolating and studying these viruses as model systems for virus replication and used them to establish many of the fundamental principles of molecular biology and genetics. These phages were designated T1, T2, and so on, up to T7, with the "T" referring to the tail these phages contain. Already in this chapter we have briefly mentioned how one of these viruses, T4, attaches to its host and how its DNA penetrates the host (Section 10.6 and Figure 10.10). Here we consider this virus in more detail to illustrate the replication cycle of virulent viruses.

The Genome of T-Even Bacteriophages

Bacteriophages T2, T4, and T6 are closely related viruses, but T4 is the most extensively studied. The virion of phage T4 is structurally complex (Figure 10.5*b*). It consists of an elongated icosahedral head whose overall dimensions are 85 × 110 nm.

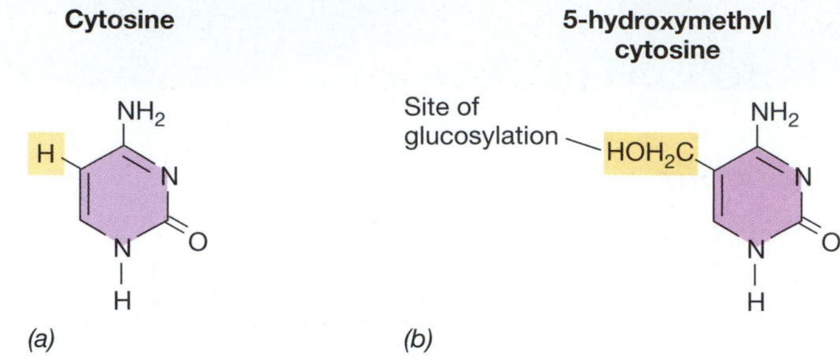

Figure 10.14 The unique base in the DNA of the T-even bacteriophages, 5-hydroxymethylcytosine. *(a)* Cytosine. *(b)* 5-Hydroxymethylcytosine. DNA containing glucosylated 5-hydroxymethylcytosine is resistant to cutting by restriction enzymes.

To this head is attached a complex tail consisting of a helical tube (25 × 110 nm) to which are connected a sheath, a connecting "neck" with "collar," and a complex end plate, to which are attached long, jointed tail fibers (Figure 10.5*b*). Altogether, the virus particle contains over 25 distinct types of structural proteins.

The genome of T4 is a double-stranded linear DNA molecule of 168,903 base pairs. The T4 genome encodes over 250 different proteins, and although no known virus encodes its own translational apparatus, T4 does encode several of its own tRNAs. Although the T4 genome has a unique linear sequence, the actual genomic DNA molecules in different virions are not identical. This is because the DNA of phage T4 is circularly permuted. Molecules that are circularly permuted appear to have been linearized by opening identical circles at different locations. In addition to circular permutation, the DNA in each T4 virion has repeated sequences of about 3–6 kbp at each end, called terminal repeats. Both of these factors affect genome packaging.

The terminally redundant T4 DNA infecting a single host cell is first replicated as a unit, and then several genomic units are recombined end-to-end to form a long DNA molecule called a *concatemer* (**Figure 10.13**). During the packaging of T4 DNA, the DNA is not cut at a specific sequence. Instead, a segment of DNA long enough to fill a phage head is cut from the concatemer. Because the T4 head holds slightly more than a genome length, this "headful mechanism" leads to circular permutation and terminal redundancy. T4 DNA contains the modified base *5-hydroxymethylcytosine* in place of cytosine (**Figure 10.14**). It is these residues that are glucosylated (Section 10.6), and DNA with this modification is resistant to virtually all known restriction enzymes. Consequently, the incoming T4 DNA is protected from host defenses.

Events During T4 Infection

Things happen rapidly in a T4 infection. Early in infection T4 directs the synthesis of its own RNA and also begins to replicate its unique DNA. About 1 min after attachment and penetration of the host by T4 DNA, the synthesis of host DNA and RNA ceases and transcription of specific phage genes begins. Translation of viral mRNA begins soon after, and within 4 min of infection, phage DNA replication has begun.

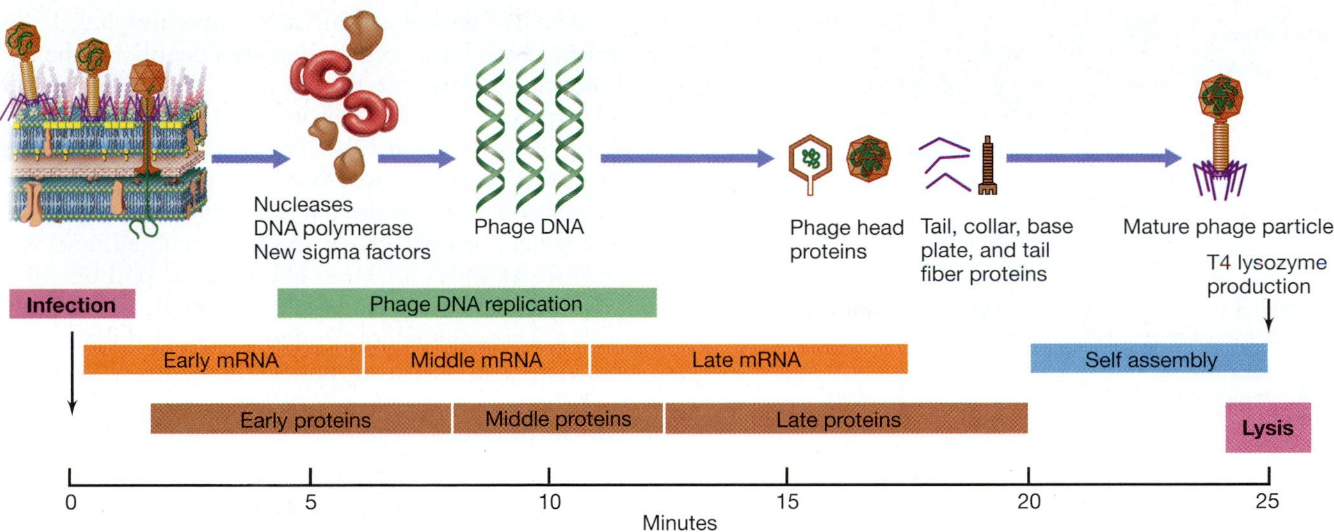

Figure 10.15 Time course of events in phage T4 infection. Following injection of DNA, early and middle mRNA is produced that codes for nucleases, DNA polymerase, new phage-specific sigma factors, and other proteins needed for DNA replication. Late mRNA codes for structural proteins of the phage virion and for T4 lysozyme, which is needed to lyse the cell and release new phage particles.

The T4 genome can be divided into three parts, encoding *early proteins, middle proteins,* and *late proteins,* respectively (**Figure 10.15**). The early and middle proteins are primarily enzymes needed for DNA replication and transcription, whereas the late proteins are the head and tail proteins and the enzymes required to liberate the mature phage particles from the cell. The time course of events during T4 infection is shown in Figure 10.15.

Although T4 has a very large genome for a virus, it does not encode its own RNA polymerase. The control of T4 mRNA synthesis requires the production of proteins that sequentially modify the specificity of the host RNA polymerase so that it recognizes phage promoters. The early promoters are read directly by the host RNA polymerase and require the host sigma factor. Host transcription is shut down shortly thereafter by a phage-encoded anti-sigma factor that binds to host σ^{70} (Section 7.10) and interferes with its recognition of host promoters.

Phage-specific proteins encoded by the early genes also covalently modify the host RNA polymerase α-subunits (Section 7.9), and a few phage-encoded proteins also bind to the RNA polymerase. These modifications change the specificity of the polymerase so that it now recognizes T4 middle promoters. One of the T4 early proteins, MotA, recognizes a particular DNA sequence in middle promoters and guides RNA polymerase to these sites. Transcription from the late promoters requires a new T4-encoded sigma factor. Sequential modification of host cell RNA polymerase as described here for phage T4 is used to regulate gene expression by many other bacteriophages as well.

T4 encodes over 20 new proteins that are synthesized early after infection. These include enzymes for the synthesis of the unusual base 5-hydroxymethylcytosine (Figure 10.14) and for its glucosylation, as well as an enzyme that degrades the normal DNA precursor deoxycytidine triphosphate. In addition, T4 encodes a number of enzymes that have functions similar to those of host enzymes in DNA replication but that are formed in larger amounts, thus permitting faster synthesis of T4-specific DNA. Additional early proteins include those involved in the processing of newly replicated phage DNA (Figure 10.13).

Most of the late genes encode structural proteins for the virion, including those for the head and tail. The assembly of heads and tails is independent. The DNA is actively pumped into the head until the internal pressure reaches the required level—which is over ten times that of bottled champagne! The tail and tail fibers are added after the head has been filled (Figure 10.15). The virus exits when the cell is lysed. The phage encodes an enzyme, *T4 lysozyme,* which attacks the peptidoglycan of the host cell. After each replication cycle, which takes only about 25 min (Figure 10.15), over 100 new virions are released from each host cell, which itself has now been almost completely destroyed.

10.9 MiniReview

After a virion of T4 attaches to a host cell and the DNA penetrates into the cytoplasm, the expression of viral genes is regulated so as to redirect the host synthetic machinery to the production of viral nucleic acid and protein. New virions are then assembled and are released by lysis of the cell. T4 has a double-stranded DNA genome that is circularly permuted and terminally redundant.

- What does it mean that the bacteriophage T4 genome is both circularly permuted and terminally redundant?
- Give an example of a mechanism used by T4 to ensure that its genes, rather than those of the host, are transcribed.

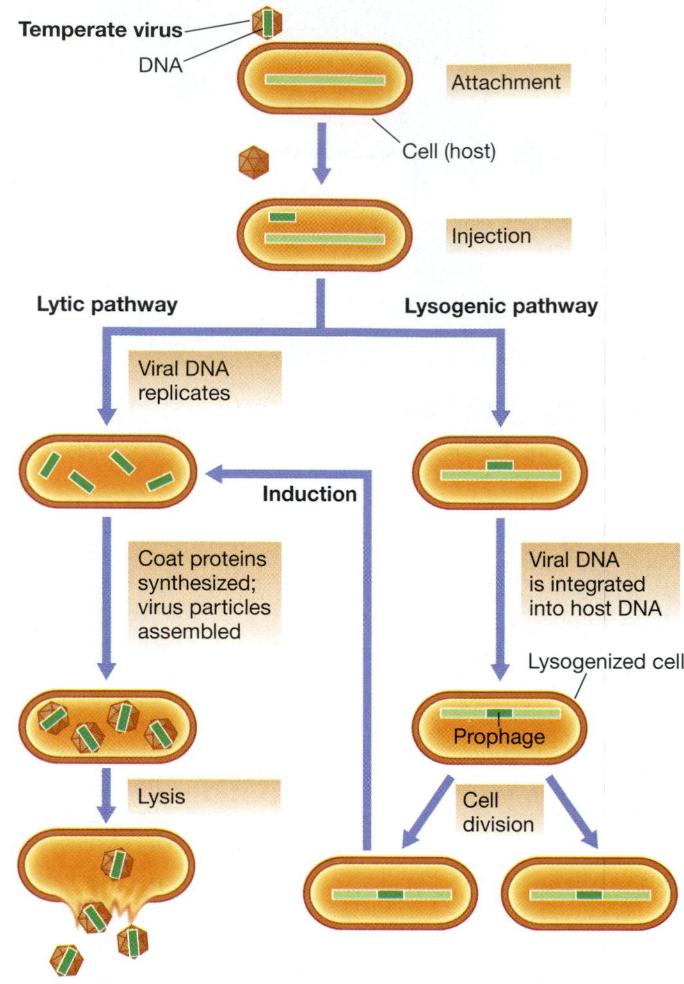

Figure 10.16 The consequences of infection by a temperate bacteriophage. The alternatives upon infection are replication and release of mature virus (lysis) or lysogenization, often, as shown here, by integration of the virus DNA into the host DNA. The lysogenic cell can be induced to produce mature virus and lyse.

10.10 Temperate Bacteriophages, Lambda and P1

Bacteriophage T4 is virulent. However, some other viruses, although able to kill cells through a lytic cycle, can also undergo a different life cycle resulting in a stable genetic relationship with the host. Such viruses are called **temperate viruses**. Such viruses can enter into a state called **lysogeny**, where most virus genes are not expressed and the virus genome, called a **prophage**, is replicated in synchrony with the host chromosome. It is expression of the viral genome that harms the host cell, not the mere presence of viral DNA. Consequently, host cells can harbor viral genomes without harm provided that the expression of genes controlling lytic functions of the virus is kept in check. In cells that harbor a temperate virus, called **lysogens**, the phage genome is replicated along with the host genome and, during cell division, is passed from one generation to the next. Under certain conditions lysogenic viruses may revert to the **lytic pathway** and begin to produce virions.

The two best-characterized temperate phages are *lambda* and *P1*. Both have contributed significantly to the advance of molecular genetics and are important tools in bacterial genetics (phage P1, Section 11.11) and molecular cloning (lambda, Section 12.14). Lysogeny is also of ecological importance, because most bacteria isolated from nature are lysogens for one or more bacteriophages. Lysogeny can confer new genetic properties on the bacterial cell, and we will see several examples in later chapters of pathogenic bacteria whose virulence depends on the lysogenic bacteriophage they harbor. Lysogeny is not limited to bacteriophages. Many animal viruses establish similar relationships with their hosts.

The Replication Cycle of a Temperate Phage

Temperate phages may enter the lytic mode after infection or they may establish lysogeny. An overall view of the life cycle of a temperate bacteriophage is shown in **Figure 10.16**. During lysogeny, the temperate virus does not exist as a virus particle inside the cell. Instead, the virus genome is either integrated into the bacterial chromosome (e.g., bacteriophage lambda) or exists in the cytoplasm in plasmid form (e.g., bacteriophage P1). In either case, it replicates in step with the host cell as long as the genes activating its lytic pathway are not expressed. These nonlytic forms are known as prophages. Typically, this control is due to a phage-encoded repressor protein (clearly, the gene encoding the repressor protein must be expressed). The virus repressor protein not only controls the lytic genes on the prophage, but also prevents the expression of any incoming genes of the same virus. This results in the lysogens having immunity to infection by the same type of virus.

If the phage repressor is inactivated or if its synthesis is prevented, the prophage is induced (Figure 10.16). New virions are produced, and the host cell is lysed. In some cases (e.g., bacteriophage lambda) altered conditions, especially damage to the host cell DNA, induce the lytic pathway. If the virus loses the ability to leave the host genome (because of mutation), it becomes a cryptic virus. Genomic studies have shown that many bacterial chromosomes contain DNA sequences that were clearly once part of a viral genome. Thus the establishment and breakdown of the lysogenic state is likely a dynamic process in prokaryotes. www.microbiologyplace.com **Online Tutorial 10.1: A Temperate Bacteriophage**

Bacteriophage Lambda

Bacteriophage lambda, which infects *Escherichia coli*, has been studied in great detail. As with other temperate viruses, both the lytic and the lysogenic pathways are possible (Figure 10.16). Lambda virions resemble those of other tailed bacteriophages, although no tail fibers are present (**Figure 10.17** and Figure 10.12). The lambda genome consists of linear dsDNA. However, at the 5′-terminus of each of the strands is a single-stranded region 12 nucleotides long. These single-stranded cohesive ends are complementary, and when lambda DNA enters the host cell, they base-pair, forming the *cos* site (**Figure 10.18**). The DNA is then ligated, forming a double-stranded circle.

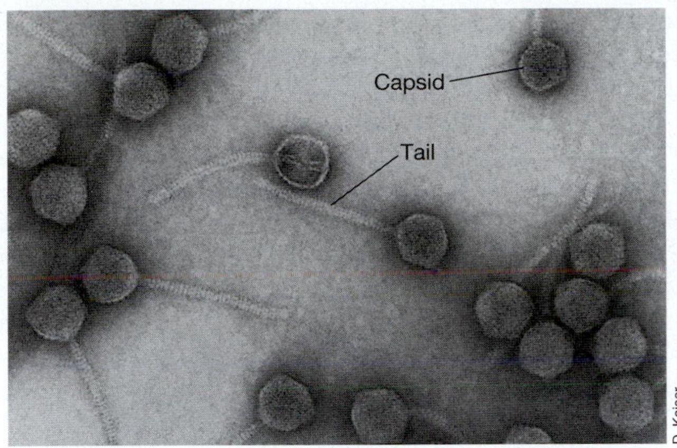

Figure 10.17 Bacteriophage lambda. Electron micrograph by negative staining of phage lambda virions. The head of each virion is about 65 nm in diameter and contains linear dsDNA.

When lambda is lysogenic, it integrates into the *E. coli* chromosome at a unique site known as the lambda attachment *site, attλ*. Integration requires the enzyme lambda integrase, which recognizes the phage and bacterial attachment sites (labeled *att* in Figure 10.18) and facilitates their union. The integrated lambda DNA is then replicated along with the rest of the host genome and transmitted to progeny cells.

When lambda enters the lytic pathway, it synthesizes long, linear concatamers of DNA by rolling circle replication (**Figure 10.19**). In contrast to semiconservative replication, this mechanism is asymmetrical and occurs in two stages. In the first stage, one strand of the circular lambda genome is nicked. Then a long single-stranded concatamer is made, using the unbroken strand as a template. In the second stage, a second strand is made, using the single-stranded concatemer as a template. Finally, the double-stranded concatemer is cut into genome-sized lengths at the *cos* sites, giving cohesive ends. The linear genomes are then packaged into phage heads, and the tails are added; the cell is then lysed by phage-encoded enzymes. Many DNA and RNA viruses and some plasmids (∞ Section 11.2) use variants of rolling circle replication. In some cases, single-stranded concatamers are cut and packaged; in other cases, the complementary strand is made before packaging, as in lambda.

Lambda: Lysis or Lysogeny?

Whether lysis or lysogeny occurs during lambda infection depends on an exceedingly complex genetic switch. The key elements are two repressor proteins, the lambda repressor, or *cI protein* (∞ Figure 9.3*b*), and the *Cro repressor*. To establish lysogeny, two events must happen: (1) The production of late proteins must be prevented; and (2) a copy of the lambda genome must be integrated into the host chromosome. If cI is made, it represses the synthesis of all other lambda-encoded proteins and lysogeny is established. Conversely, Cro represses the expression of the cII and cIII proteins, which are needed to maintain lysogeny by inducing synthesis of the lambda repressor (cI). Thus, when Cro is made in high amounts, lambda is

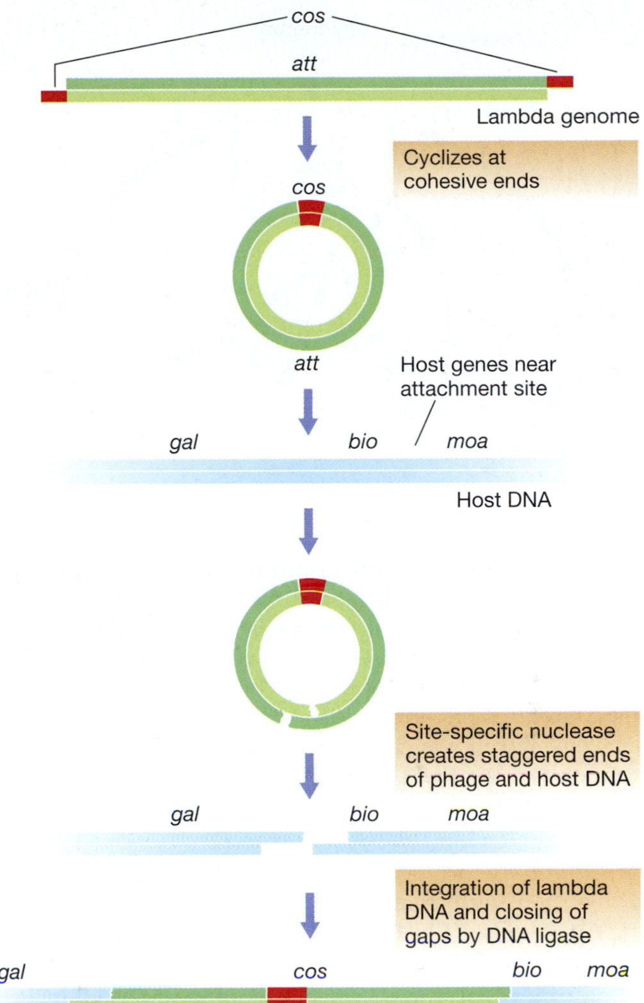

Figure 10.18 Integration of lambda DNA into the host. Integration always occurs at specific attachment sites (*att* sites) on both the host DNA and the phage. Some host genes near the attachment site are given: *gal* operon, galactose utilization; *bio* operon, biotin synthesis; *moa* operon, molybdenum cofactor synthesis. A site-specific enzyme (integrase) is required, and specific pairing of the complementary ends results in integration of phage DNA.

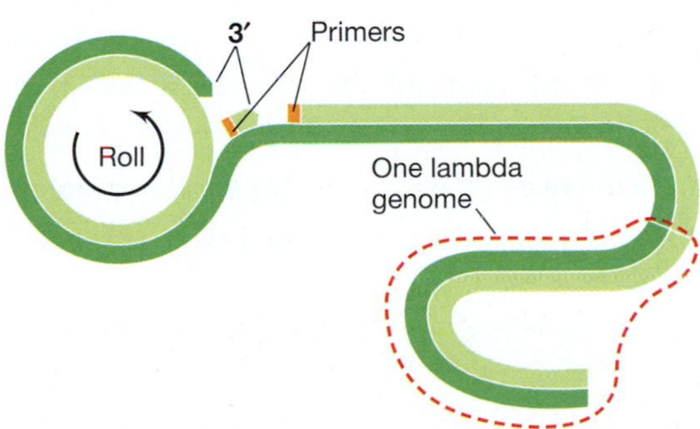

Figure 10.19 Rolling circle replication of the lambda genome. As the dark green strand rolls out, it is being replicated at its opposite end. Note that this synthesis is *asymmetric* because one of the parental strands continues to serve as a template and the other is used only once.

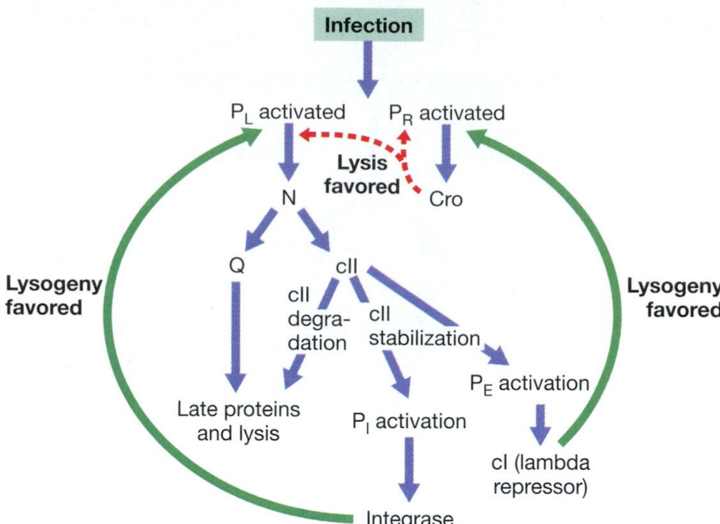

Figure 10.20 Summary of the steps in lambda infection. Lysis versus lysogeny is governed by whether or not the lambda repressor (cI) is made. High Cro activity prevents sufficient cII protein from being made, which in turn prevents synthesis of cI protein, and the result is lysis. In addition, the level of cII depends on its degradation by host proteases versus its protection by lambda cIII protein. If sufficient cII is present, the promoters for cI and for integrase are activated and both cI and integrase are made. This results in integration and lysogeny. P_L is the lambda leftward promoter and P_R is the lambda rightward promoter.

committed to the lytic pathway. The degradation of cII by a host cell protease (FtsH protein) is also critical. The lambda cIII protein protects cII against protease attack and stabilizes it. A summary of the steps controlling lambda lysis and lysogeny is shown in **Figure 10.20**. The final outcome is determined by whether Cro protein or cI dominates in a given infection. If Cro dominates regulatory events, the outcome is lysis, whereas if cI dominates, lysogeny will occur.

10.10 MiniReview

Lysogeny is a state in which lytic events are repressed. Viruses capable of entering the lysogenic state are called temperate viruses. In lysogeny, the virus genome becomes a prophage, either by integration into the host chromosome or by replicating like a plasmid in step with the host cell. However, lytic events can be induced by certain environmental stimuli.

- What are the two pathways available to a temperate virus?
- What events need to happen for lambda to become a prophage?
- What might be the advantage to hosts of temperate viruses?

10.11 Overview of Animal Viruses

The first few sections of this chapter were devoted to general properties of viruses, and little was said about animal viruses. Here we consider some of the attributes of animal viruses, setting the stage for a discussion of retroviruses, one of which causes acquired immunodeficiency syndrome (AIDS).

In our discussions of viral reproduction in Sections 10.5–10.7, we dealt with the virus "host" in a general way. However, it is important to remember that the bacteriophage host is a bacterial cell, whereas the host of an animal virus is a eukaryotic cell. We will enlarge on these important differences in Chapter 19, where we discuss several types of animal virus in more detail. However, the key points are that (1) unlike in prokaryotes, the entire virion enters the animal cell, and (2) eukaryotic cells contain a nucleus, where many animal viruses replicate. In addition, some details of eukaryotic RNA processing affect the replication of animal viruses, but these need not concern us here.

Classification of Animal Viruses

Various types of animal viruses are illustrated in **Figure 10.21**. We discussed the principles of virus classification in Section 10.7. Animal viruses are classified according to the same schemes as bacterial viruses, including the Baltimore Classification System (Table 10.2), which classifies viruses on the basis of genome type and reproductive strategy. Animal viruses are known in all replication categories, and an example of each is discussed in Chapter 19. Most of the animal viruses that have been studied in detail are those that can replicate in cell cultures (Section 10.3).

Note that there are many more kinds of enveloped animal viruses than enveloped bacterial viruses (Section 10.7). This likely relates to the differences in host cell exteriors. Unlike prokaryotic cells, animal cells lack a cell wall, and thus viruses are more easily secreted from the cell. As they exit, they remove part of the animal cell's lipid bilayer as they pass through the membrane.

Consequences of Virus Infection in Animal Cells

Viruses can have several different effects on animal cells. Lytic infection results in the destruction of the host cell (**Figure 10.22**). With enveloped viruses, however, release of virions, which occurs by a kind of budding process, may be slow, and the host cell may not be lysed. The infected cell may therefore remain alive and continue to produce virus indefinitely. Such infections are called persistent infections (Figure 10.22).

Viruses may also cause latent infection of a host. In a latent infection, there is a delay between infection by the virus and lytic events. Fever blisters (cold sores), caused by the herpes simplex virus (Section 19.12), are a typical example of a latent viral infection; the symptoms (the result of lysed cells) reappear sporadically as the virus emerges from latency. The latent stage in viral infection of an animal cell is not usually due to integration of the viral genome into the host genome, as often happens with latent (lysogenic) infections by temperate bacteriophages. Instead, herpesviruses exist in a relatively inactive state within nerve cells. A low level of transcription continues, but the viral DNA does not replicate.

Some enveloped viruses promote fusion between multiple animal cells, creating giant cells with several nuclei (Figure 10.22). Not surprisingly, such fused cells fail to develop cor-

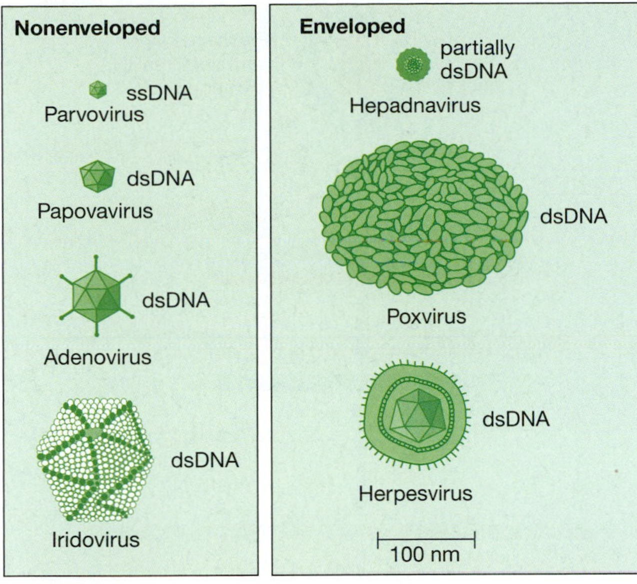

(a) **DNA viruses**

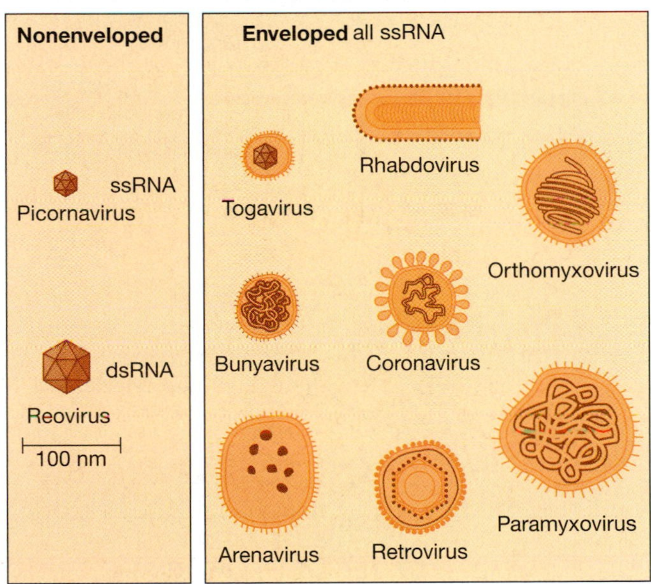

(b) **RNA viruses**

Figure 10.21 Diversity of animal viruses. The shapes and relative sizes of the major groups of vertebrate viruses. The hepadnavirus genome has one complete DNA strand and part of the complementary strand (Section 19.15).

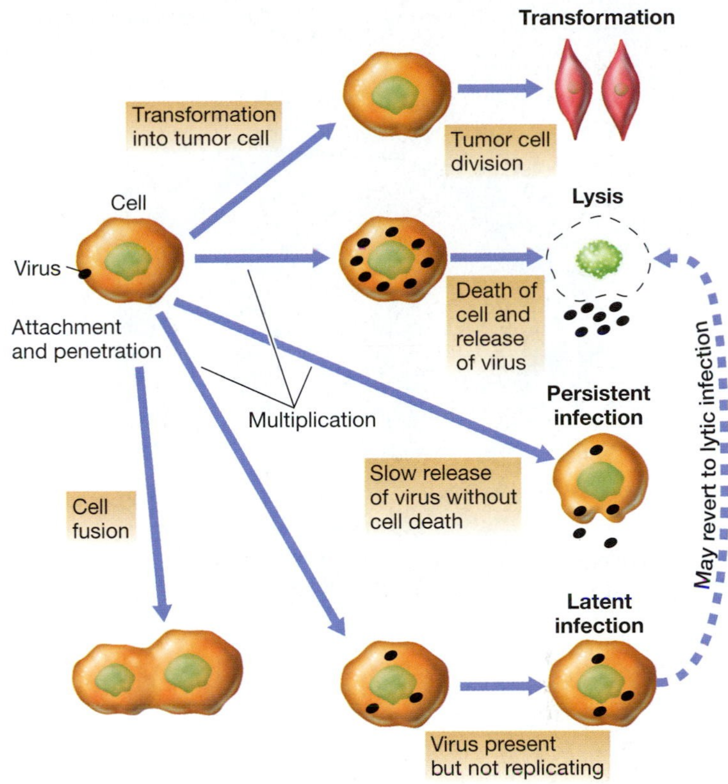

Figure 10.22 Possible effects that animal viruses may have on cells they infect. Most viruses are lytic, and only very few are known to cause cancer.

rectly and are short-lived. In addition, cell fusion allows viruses to avoid exposure to the immune system by moving between host cell nuclei without emerging from the host cells. Finally, certain animal viruses can convert a normal cell into a tumor cell, a process called **transformation**. We discuss cancer-causing viruses in Sections 19.11 and 19.12.

Many different viruses are known. But of all the viruses listed in Figure 10.21, one group stands out as having an absolutely unique mode of replication. These are the retroviruses. We explore them next as an example of a complex and highly unusual animal virus with significant medical implications.

10.11 MiniReview

There are animal viruses with all known modes of viral genome replication. Many animal viruses are enveloped, picking up portions of host membrane as they leave the cell. Not all infections of animal host cells result in cell lysis or death; latent or persistent infections are common, and a few animal viruses can cause cancer.

- Differentiate between a persistent and a latent viral infection.
- Contrast the mechanisms by which animal viruses enter cells with those used by bacterial viruses.

10.12 Retroviruses

Retroviruses contain an RNA genome that is replicated through a DNA intermediate (Section 10.1 and Figure 10.11). The term *retro* means "backward," and the name retrovirus is derived from the fact that these viruses transfer information backward from RNA to DNA. Retroviruses employ the enzyme *reverse transcriptase* to carry out this interesting process. The use of reverse transcriptase is not restricted to the retroviruses. Hepatitis B virus (a human virus) and cauliflower mosaic virus (a plant virus) also use reverse transcription in their replication processes (Section 19.15). However,

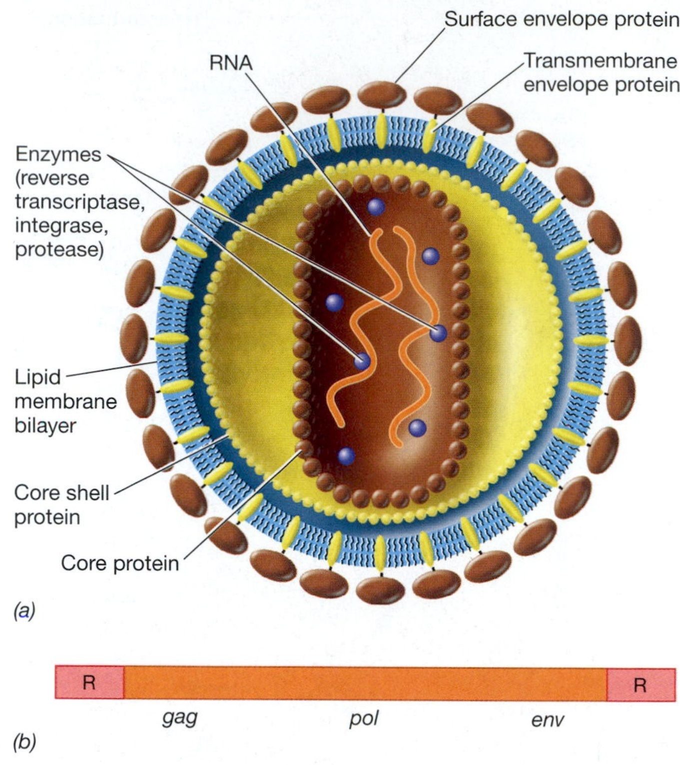

Figure 10.23 Retrovirus structure and function. *(a)* Structure of a retrovirus. *(b)* Genetic map of a typical retrovirus genome. Each end of the genomic RNA contains direct repeats (R).

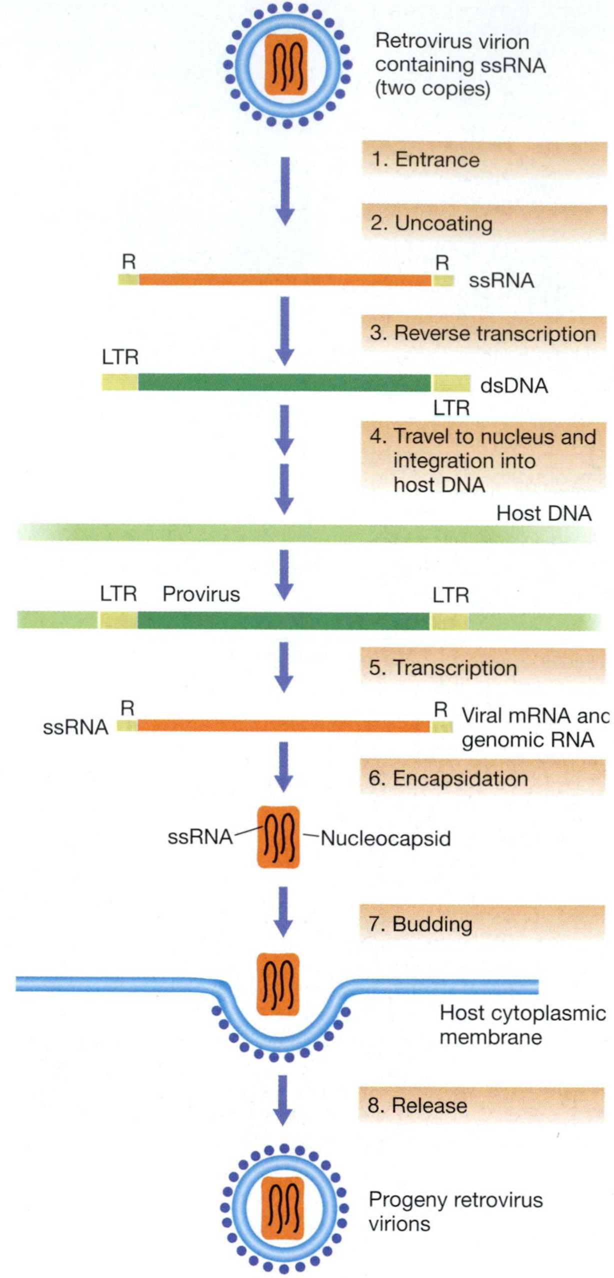

Figure 10.24 Replication process of a retrovirus. R, direct repeats; LTR, long terminal repeats. For more detail on conversion of RNA to DNA (step 3) refer to Figure 19.25.

unlike the retroviruses, these other viruses carry the DNA version of their genome in the virion rather than RNA.

Retroviruses are interesting for several other reasons. For example, they were the first viruses shown to cause cancer and have been studied most extensively for their carcinogenic characteristics. Also, one retrovirus, human immunodeficiency virus (HIV), causes acquired immunodeficiency syndrome (AIDS). This virus infects a specific kind of T lymphocyte (T-helper cell) in humans that is vital for proper functioning of the immune system. In later chapters we discuss the medical and immunological aspects of AIDS (∞ Section 34.15).

Retroviruses are enveloped viruses (**Figure 10.23*a***). There are several proteins in the virus coat and typically seven internal proteins, four of which are structural and three of which are enzymatic. The enzymes found in the virion are reverse transcriptase, integrase, and a protease. The virion also contains specific cellular tRNA molecules used in replication (discussed later in this section).

Features of Retroviral Genomes and Replication

The genome of the retrovirus is unique. It consists of two identical single-stranded RNA molecules of the plus (+) orientation. A genetic map of a typical retrovirus genome is shown in Figure 10.23*b*. Although there are differences between the genetic maps of different retroviruses, all contain the following genes and are arranged in the same order: *gag*, encoding structural proteins; *pol*, encoding reverse transcriptase and integrase; and *env*, encoding envelope proteins. Some retroviruses, such as Rous sarcoma virus, carry a fourth gene downstream from *env* that is active in cellular transformation and cancer. The terminal repeats shown on the map are essential for the viral replication process.

The overall process of replication of a retrovirus can be summarized in the following steps (**Figure 10.24**):

1. Entrance into the cell by fusion with the cytoplasmic membrane at sites of specific receptors.
2. Removal of the virion envelope at the membrane, but leaving the genome and enzymes in the virus core.

3. Reverse transcription of one of the two RNA genomes into a single-stranded DNA that is subsequently converted by reverse transcriptase to a linear double-stranded DNA, which then enters the nucleus.
4. Integration of retroviral DNA into the host genome.
5. Transcription of retroviral DNA, leading to the formation of viral mRNAs and viral genomic RNA.
6. Assembly and packaging of the genomic RNA into nucleocapsids in the cytoplasm.
7. Budding of enveloped virions at the cytoplasmic membrane and release from the cell.

Activity of Reverse Transcriptase

A very early step after the entry of the RNA genome into the cell is **reverse transcription**: conversion of RNA into a DNA copy using the enzyme reverse transcriptase present in the virion. The DNA formed is a linear double-stranded molecule and is synthesized in the cytoplasm within an uncoated viral core particle. Details of this process can be found in Section 19.15. Here we note only that reverse transcriptase is a type of DNA polymerase and, like all DNA polymerases, must have a primer (∞ Section 7.5). The primer for retrovirus reverse transcription is a specific transfer RNA (tRNA) encoded by the host cell. The type of tRNA used as primer depends on the virus and is packaged into the virion from the previous host cell.

The overall process of reverse transcription results in a product that has long terminal repeats (LTRs, Figure 10.24) that are longer than the terminal repeats on the RNA genome itself (Figure 10.23). This entire dsDNA molecule enters the nucleus along with the integrase protein; here the viral DNA is integrated into the host DNA. The LTRs contain strong promoters of transcription and participate in the integration process. The integration of the viral DNA into the host genome is analogous to the integration of phage DNA into a bacterial genome to form the lysogenic state (Section 10.10). Viral DNA can be integrated anywhere in the cellular DNA, and once integrated, the element, now called a **provirus**, is a stable genetic element and can remain so indefinitely.

If the promoters in the right LTR are activated, the integrated proviral DNA is transcribed by a host cell RNA polymerase into transcripts. These RNA transcripts may either be packaged into virus particles (as the genome) or may be translated into virus proteins. Some virus proteins are made initially as a large primary *gag* protein, which is split by proteolysis into the capsid proteins. Occasionally, read-through past the *gag* region occurs (due either to inserting an amino acid at a stop codon or a shift in reading frame by the ribosome). This leads to the translation of *pol,* the reverse transcriptase gene, which yields reverse transcriptase for insertion into virions.

When virus structural proteins have accumulated in sufficient amounts, nucleocapsids are assembled. Encapsidation of the RNA genome leads to the formation of mature nucleocapsids, which move to the cytoplasmic membrane for final assembly into the enveloped virus particles. As nucleocapsids bud through the membrane they are sealed and then released and may infect neighboring cells (Figure 10.24).

10.12 MiniReview

Retroviruses are RNA viruses that replicate through a DNA intermediate. The retrovirus human immunodeficiency virus (HIV) causes AIDS. The retrovirus virion contains an enzyme, reverse transcriptase, that copies the information from its RNA genome into DNA. The DNA becomes integrated into the host chromosome in the manner of a temperate virus. The retrovirus DNA can be transcribed to yield mRNA (and new genomic RNA) or may remain in a latent state.

- Why are these viruses called retroviruses?
- How does the replication cycle of a temperate bacteriophage differ from that of a retrovirus?

IV SUBVIRAL ENTITIES

We have defined a virus as a genetic element that subverts normal cellular processes for its own replication and that has an infectious extracellular form. There are a few infectious agents that resemble viruses but whose properties are at odds with this definition, thus are not considered viruses. Defective viruses are clearly derived from viruses but have become dependent on other, complete, viruses to supply certain gene products. In contrast, two of the most important subviral entities, viroids and prions, are not merely incomplete viruses, but differ in fundamental ways from viruses. They both illustrate the unusual ways that genetic elements can replicate and the unexpected ways they can subvert the host cells. However, prions stand out among all the genetic elements we have considered in this chapter because the infectious transmissible agent lacks nucleic acid.

10.13 Defective Viruses

Some viruses are parasitic on other viruses, known as **helper viruses**. Some of these so-called **defective viruses** merely rely on intact viruses of the same type to provide necessary functions. Far more interesting are those defective viruses—referred to as *satellite viruses*—for which no intact version exists; these defective viruses rely on unrelated viruses as helpers.

Examples of Defective and Satellite Viruses

Many defective viruses are known. For example, bacteriophage P4 of *Escherichia coli* can replicate, but its genome does not encode any capsid proteins. Instead, it relies on the related phage P2 as a helper to provide capsid proteins for assembly of the phage particle.

Satellite viruses are found in both animals and plants. For example, adeno-associated virus (AAV) is a satellite virus of humans that depends on the adenovirus as helper. The adeno-associated virus and adenovirus belong to two quite different virus families. Thus, AAV is not just a defective mutant of adenovirus, but an unrelated virus that inhabits the same host cells. However, because it causes little or no damage to the host, AAV is now being used as a eukaryotic cloning vector in

Figure 10.25 Viroids and plant diseases. Photograph of healthy (left) and potato spindle tuber viroid (PSTV)-infected (right) tomato plants. The host range of most viroids is quite restricted. However, PSTV infects tomatoes as well as potatoes, causing growth stunting, a flat top, and premature plant death (right).

gene therapy (Section 26.9). In this system, AAV can be used to carry replacement genes to specific host tissues without causing disease itself.

10.13 MiniReview

Defective viruses are parasites of complete viruses. The helper viruses supply proteins that the defective virus no longer encodes. Some defective viruses rely on closely related but intact helper viruses. However, satellite viruses rely on unrelated intact viruses that infect the same host cells to complete replication events.

- What is a helper virus?
- What is a satellite virus?

10.14 Viroids

Viroids are infectious RNA molecules that differ from viruses in that they lack a protein coat. Despite this they do have a reasonably stable extracellular form that travels from one host cell to another. Viroids are small, circular, single-stranded RNA molecules that are the smallest known pathogens. They range in size from 246 to 399 nucleotides and show a considerable degree of sequence homology, suggesting that they have common evolutionary roots. Viroids cause a number of important plant diseases and can have a severe agricultural impact (**Figure 10.25**). A few well-studied viroids include coconut cadang-cadang viroid (246 nucleotides), citrus exocortis viroid (375 nucleotides), and potato spindle tuber viroid (359 nucleotides). No viroids are known that infect animals or prokaryotes.

Figure 10.26 Viroid structure. Viroids consist of single-stranded circular RNA that forms a seemingly double-stranded structure by intra-strand base pairing.

Viroid Structure and Function

The extracellular form of the viroid is naked RNA—there is no protein capsid of any kind. Although the viroid RNA is a single-stranded covalently closed circle, there is so much secondary structure that it resembles a short double-stranded molecule with closed ends (**Figure 10.26**). This apparently makes the viroid sufficiently stable to exist outside the host cell. Because it lacks a protein coat, the viroid does not use a receptor to enter the host cell. Instead, the viroid enters the plant through a wound, as from insect or other mechanical damage. Once inside, viroids move from cell to cell via the plasmodesmata—thin strands of cytoplasm that link plant cells (**Figure 10.27**).

Even more curious, the RNA molecule contains no protein-encoding genes, and therefore the viroid is almost totally dependent on host function for its replication. The viroid is replicated in the host cell nucleus or chloroplast by one of the plant RNA polymerases. The result is a multimeric RNA consisting of many viroid units joined end to end. The viroid does contribute one function to its own replication—part of the viroid itself has ribozyme activity (Section 8.8). This ribozyme activity is used for self-cleavage of the multimeric RNA, which releases individual viroids.

Viroid Disease

Viroid-infected plants can be symptomless or develop symptoms that range from mild to lethal, depending on the viroid (Figure 10.25). Most severe symptoms are growth related, suggesting that viroids mimic or interfere in some way with small regulatory RNAs (Section 9.14), examples of which are widely known in plants. Thus, viroids could themselves be regulatory RNAs that have evolved away from carrying out beneficial roles in the cell to inducing destructive events. No viroid diseases of animals are known, and the mechanisms by which viroids cause plant diseases remain unclear.

10.14 MiniReview

Viroids are circular single-stranded RNA molecules that do not encode proteins and are dependent on host-encoded enzymes, except for the ribozyme activity of the viroid molecule itself. Viroids are the smallest known pathogens that contain nucleic acids.

- If viroids are circular molecules, why are they usually drawn as compact rods?
- In what part of the host cell are viroids replicated?

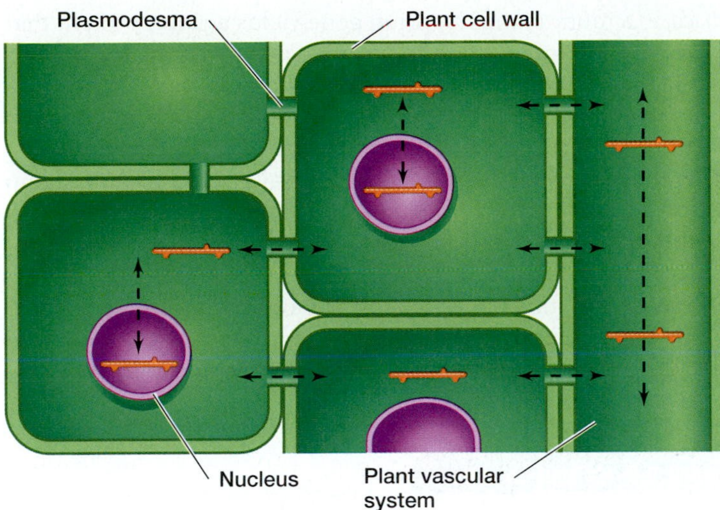

Figure 10.27 Viroid movement inside plants. After entry into a plant cell, viroids (orange) replicate either in the nucleus (shown here in purple) or in the chloroplast (not shown). Viroids can move between plant cells via the plasmodesmata (thin threads of cytoplasm that penetrate the walls and connect plant cells). In addition, on a larger scale, viroids can move around the plant via the plant vascular system.

10.15 Prions

Prions represent the other extreme from viroids. They have a distinct extracellular form, which consists entirely of protein. The prion particle contains neither DNA nor RNA. Nonetheless, it is infectious, and prions are known to cause diseases in animals, such as scrapie in sheep, bovine spongiform encephalopathy (BSE or "mad cow disease") in cattle, chronic wasting disease in deer and elk, and kuru and Creutzfeldt-Jakob disease (CJD) in humans. No prion diseases of plants are known, although prions have been found in yeast. Collectively, animal prion diseases are known as **transmissible spongiform encephalopathies (TSEs)**. In 1997 the American scientist Stanley B. Prusiner won the Nobel Prize for Physiology or Medicine for his pioneering work with these diseases and with the prion proteins.

In 1996 it became clear from disease tracking in England that the prion that causes BSE in cattle can also infect humans, resulting in a novel type of CJD, called variant CJD (vCJD). Because transmission was from contaminated beef products, vCJD quickly became a worldwide health concern, with a major impact on the animal husbandry industry (∞ Section 37.11). Most such instances of BSE occurred in the United Kingdom or other European Union (EU) countries and were linked to improper feeding practices in which protein supplements containing rendered cattle and sheep (including nervous tissues) were used to feed uninfected animals. Since 1994 this practice has been banned in all EU countries, and cases of BSE have dropped dramatically. Thus far, TSE transmission via other domesticated animals, such as swine, chicken, or fish, has not been found.

Forms of the Prion Protein

As prions lack nucleic acid, how is the protein they consist of encoded? The host cell contains a gene, *Prnp*, which encodes the native form of the prion protein, known as PrP^{C} (prion protein cellular), which is found in healthy animals, mostly in neurons, especially in the brain. The pathogenic form of the prion protein is designated PrP^{Sc} (prion protein scrapie), because scrapie in sheep was the first prion disease to be discovered. The pathogenic form of the prion protein is identical in amino acid sequence to PrP^{C} from the same species but has a different conformation. Prion proteins from different species of mammals are very similar but not identical in sequence. Susceptibility to infection depends on the protein sequence in a manner not fully understood. For example, pathological cow prions can infect humans, although at a very low frequency. However, pathological sheep prions have never been observed to infect people. Healthy prions consist largely of α-helical segments, whereas rogue prions have less α-helix and more β-sheet regions instead. This causes the prion protein to lose its normal function, to become partially resistant to proteases, and to become insoluble, leading to aggregation within the neural cell (**Figure 10.28**). In this state, prion protein accumulates and neurological symptoms commence.

Prion Diseases and the Prion Infectious Cycle

When a pathogenic prion enters a host cell that is expressing normal prion protein, it promotes the conversion of the PrP^{C}

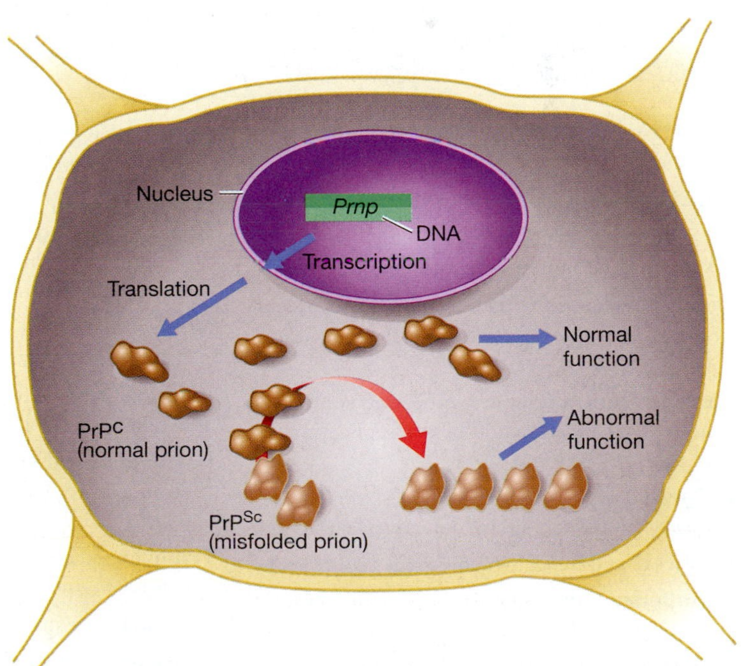

Figure 10.28 Mechanism of prion misfolding. Neuronal cells produce the normal form of the prion protein. Abnormally folded prion protein catalyzes the refolding of normal prions into the abnormal form. The abnormal form is protease resistant, insoluble, and forms aggregates in neural cells. This eventually leads to destruction of neural tissues and neurological symptoms.

protein into PrP^{Sc}. Thus the rogue prion does not subvert host enzymes or genes as a virus does; rather, it "replicates" by converting normal prion proteins that already exist in the host cell into the pathogenic form. As the pathogenic prions accumulate, they form insoluble aggregates in the brain cells (Figure 10.28). This leads to disease symptoms that are invariably neurological and in most cases are due to destruction of brain or related nervous tissue (Section 37.11). Whether the destruction of brain tissue is directly due to the accumulation of aggregated prion protein is uncertain. Normal prion protein functions in the cell as a cytoplasmic membrane glycoprotein, and it has been shown that membrane attachment of pathogenic prions is necessary for disease symptoms to commence. Mutant versions of the pathogenic prion protein that can no longer attach to nerve cell cytoplasmic membranes may still aggregate, but no longer cause disease.

Prion disease occurs by three distinct mechanisms, although all lead to the same result. In *infectious prion disease,* as described above, the pathogenic form of the prion protein is transmitted between animals or humans. In *sporadic prion disease,* random misfolding of a PrP^{c} molecule occurs in a normal, uninfected individual. This change is propagated as for infectious prion disease, and eventually the pathogenic form of the prion protein accumulates until symptoms appear. In humans this occurs in about one person in a million. In *inherited prion disease,* a mutation in the prion gene yields a prion protein that changes more often into the disease-causing form. Several different mutations are known whose symptoms vary slightly.

What happens if an incoming PrP^{sc} protein finds no healthy prions to alter? The answer is that no disease results. This may seem surprising, but is logical given the mechanism of prion replication. Mice that have been engineered with both copies of the *Prnp* gene disrupted and thus do not produce PrP^{c} are resistant to infection with pathogenic prions. Interestingly, such mice also live for a normal time and do not show any obvious behavioral abnormalities. This leaves wide open the puzzling question of what role the normal prion protein plays in brain cells.

10.15 MiniReview

Prions consist of protein but have no nucleic acid. Prions exist in two conformations, the original cellular form and the pathogenic form. The pathogenic form "replicates" itself by converting healthy prion proteins into the pathogenic conformation.

- What is the difference between the healthy and pathogenic forms of the prion?
- How does sporadic prion disease differ from the transmitted form?

Review of Key Terms

Bacteriophage a virus that infects prokaryotic cells

Capsid the protein shell that surrounds the genome of a virus particle

Capsomer subunit of a capsid

Defective virus a virus that relies on another virus, the helper virus, to provide some of its components

Early protein a protein synthesized soon after virus infection and before replication of the virus genome

Helper virus a virus that provides some necessary components for a defective virus

Host cell cell inside which a virus replicates

Icosahedron three-dimensional figure with 20 triangular faces

Late protein a protein synthesized later in virus infection, after replication of the virus genome

Lysogen a bacterium containing a provirus

Lysogeny a state following virus infection in which the viral genome is replicated as a provirus along with the genome of the host

Lytic pathway a series of steps after virus infection that leads to virus replication and the destruction (lysis) of the host cell

Negative-strand virus a virus with a single-stranded genome that has the opposite sense to (is complementary to) the viral mRNA

Nucleocapsid the complex of nucleic acid and proteins of a virus

Plaque a zone of lysis or cell inhibition caused by virus infection of a lawn of sensitive cells

Positive-strand virus a virus with a single-stranded genome that has the same complementarity as the viral mRNA

Prion an infectious protein whose extracellular form contains no nucleic acid

Prophage *see* provirus

Provirus (prophage) the genome of a temperate virus when it is replicating in step with, and often integrated into, the host chromosome

Retrovirus a virus whose RNA genome has a DNA intermediate as part of its replication cycle

Reverse transcription the process of copying information found in RNA into DNA by the enzyme reverse transcriptase

Temperate virus a virus whose genome is able to replicate along with that of its host without causing cell death in a state called lysogeny

Transformation in eukaryotes, a process by which a normal cell becomes a cancer cell

Transmissible spongiform encephalopathy (TSE) degenerative disease of the brain caused by prion infection

Virion the infectious virus particle; the nucleic acid genome surrounded by a protein coat and in some cases other layers of material

Virulent virus a virus that lyses or kills the host cell after infection; a nontemperate virus

Virus a genetic element containing either RNA or DNA surrounded by a protein capsid and that replicates only inside host cells

Viroid small, circular, single-stranded RNA that causes certain plant diseases

Review Questions

1. In what ways do viral genomes differ from those of cells (Section 10.1)?
2. Define virus. What are the minimal features needed to fit your definition (Section 10.2)?
3. Define the term "host" as it relates to viruses (Section 10.3).
4. Describe the events that occur on an agar plate containing a bacterial lawn when a single bacteriophage particle causes the formation of a bacteriophage plaque (Section 10.4).
5. Under some conditions, it is possible to obtain nucleic acid-free protein coats (capsids) of certain viruses. Under the electron microscope, these capsids look very similar to complete virions. What does this tell you about the role of the virus nucleic acid in the virus assembly process? Would you expect such virions to be infectious? Why (Section 10.5)?
6. Describe how a restriction endonuclease might play a role in resistance to bacteriophage infection. Why could a restriction endonuclease play such a role whereas a generalized DNase could not (Section 10.5)?
7. One can divide the replication process of a virus into five steps. Describe the events associated with each of these steps (Sections 10.6 and 10.7).
8. Specifically, why are both the life cycle and the virion of a positive-strand RNA virus likely to be simpler than those of a negative-strand RNA virus (Section 10.8)?
9. In terms of structure, how does the genome of bacteriophage T4 resemble and differ from that of *Escherichia coli* (Section 10.9)?
10. Many of the viruses we have considered have early genes and late genes. What is meant by these two classifications? What types of proteins tend to be encoded by early genes? What type of proteins by late genes? For bacteriophage T4 describe how expression of the late genes is controlled (Section 10.9).
11. Define the following: virulent, lysogeny, prophage (Section 10.10).
12. A strain of *Escherichia coli* that is missing the outer membrane protein responsible for maltose uptake is resistant to lambda infection. A lambda lysogen is immune to lambda infection. Describe the difference between resistance and immunity (Section 10.10).
13. Describe and differentiate the effects animal virus infection can have on an animal (Section 10.11).
14. Typically, tRNA is used in translation. However, it also plays a role in the replication of retroviral nucleic acid. Explain this role (Section 10.12).
15. What does a helper virus provide a satellite virus that allows it to replicate (Section 10.13)?
16. What are the similarities and differences between viruses and viroids (Section 10.14)?
17. What are the similarities and differences between prions and viruses (Section 10.15)?

Application Questions

1. What causes the viral plaques that appear on a bacterial lawn to stop growing larger?
2. The promoters for mRNA encoding early proteins in viruses like T4 have a different sequence than the promoters for mRNA encoding late proteins in the same virus. Explain why this might be true.
3. One characteristic of temperate bacteriophages is that they cause turbid rather than clear plaques on bacterial lawns. Can you think why this might be? (Remember the process by which a bacterial plaque develops.)
4. Suggest possible reasons why viroids infect only plants and not animals or bacteria.
5. Contrast the enzyme(s) present in the virions of a retrovirus and a positive-strand RNA bacteriophage. Why do they differ if each has plus complementarity single-stranded RNA as their genome?
6. Since viral infection leads to more viral particles being formed, explain why the "growth curve" for viruses is stepped rather than a smooth curve (as seen with bacterial multiplication).

11

Principles of Bacterial Genetics

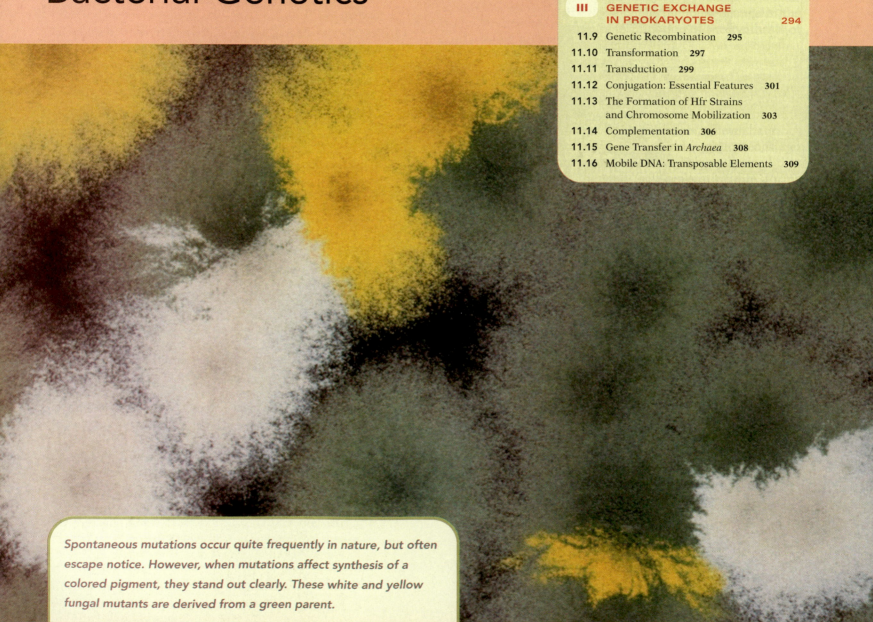

Spontaneous mutations occur quite frequently in nature, but often escape notice. However, when mutations affect synthesis of a colored pigment, they stand out clearly. These white and yellow fungal mutants are derived from a green parent.

In this chapter we present the basic principles and techniques of bacterial genetics. First we discuss the physical structures on which the genes of bacteria are located, chromosomes and plasmids. Next we consider how alterations in the genetic material arise. Finally, we consider how genes can be transferred from one organism to another.

I BACTERIAL CHROMOSOMES AND PLASMIDS

Today, many bacterial genomes have been completely sequenced, thus revealing the number and location of the genes they possess. Cloning and sequencing has revolutionized the study of genome structure and function in all organisms, and we consider the field of genomics in Chapter 13. However, in the early days of molecular biology, other genetic analyses were necessary. In particular, the three mechanisms of genetic exchange described later in this chapter, transformation, transduction, and conjugation, were used to map the locations of genes on a bacterial chromosome.

11.1 Genetic Map of the *Escherichia coli* Chromosome

The genes of *Escherichia coli* were initially mapped using conjugation (Section 11.12). However, conjugation does not permit ordering genes that are closely linked. Therefore, generalized transduction by bacteriophage P1 (Section 11.11) was used for detailed mapping of genes on the *E. coli* chromosome. The genetic techniques used to map the chromosomes of other prokaryotes were dictated by the efficiency with which genetic transfer occurs in the particular organism. For example, transformation is very inefficient in *E. coli* but is very efficient in gram-positive bacteria (Section 11.10). Thus, transformation proved an effective tool for mapping genes in *Bacillus*, a model gram-positive bacterium.

The genetic map of *E. coli* strain K-12 is shown in **Figure 11.1**. The map distances are given in "minutes" of transfer (sometimes called centisomes), with the entire chromosome containing 100 minutes. Zero is arbitrarily set at *thrABC* (the threonine operon), because the *thrABC* genes were the first shown to be transferred by conjugation in *E. coli*. The genetic map in Figure 11.1 shows only a few of the several thousand genes in the *E. coli* chromosome and also shows the location of a few restriction enzyme recognition sites. The size of the chromosome is given in both minutes (the original units determined by conjugation studies) and in kilobase pairs.

The *E. coli* chromosome, like that of many other prokaryotes, has been completely sequenced. Sequencing of relatively small bacterial chromosomes of 3 or 4 megabases is now done routinely by automated sequencing of random, or shotgun, clones (∞ Section 12.6). Because of the enormous amount of data, the information from genomic analyses can only be managed through computer databases. This situation has spawned a whole new field of molecular biology called *bioinformatics* (∞ Chapter 13). Database analyses using powerful gene search software has placed the field of bioinformatics in a central position in genomic studies.

Escherichia coli as a Model Prokaryote

Many factors have favored the use of *E. coli* as the workhorse for studies of biochemistry, genetics, and bacterial physiology. As emphasized in Chapters 10 and 19, even *E. coli* viruses have served as model systems of study. Although the chromosome of *E. coli* was not the first prokaryotic chromosome to be sequenced, this organism remains the best characterized of any organism, prokaryote or eukaryote. *E. coli* continues to be the organism of choice for both research and genetic engineering (∞ Chapter 26).

The strain of *E. coli* whose chromosome was originally sequenced, strain MG1655, is a derivative of *E. coli* K-12, the traditional strain used for genetics. Wild-type *E. coli* K-12 is lysogenic for bacteriophage lambda (∞ Section 10.10) and also contains the F plasmid (Section 11.3). However, strain MG1655 had been "cured" of lambda (by radiation) and of the F plasmid (by acridine treatment, Section 11.3) before sequencing began. The chromosome of strain MG1655 contains 4,639,221 bp. Analysis of the genomic sequence showed 4,288 possibly functional open reading frames (ORF, ∞ Section 7.13), which account for about 88% of the genome. Approximately 1% of the genome is composed of genes encoding tRNAs and rRNAs, and about 0.5% is noncoding, repetitive sequences. The remaining 10% of the genome contains all of the regulatory sequences: promoters, operators, origin and terminus of DNA replication, and so on.

Arrangement and Expression of Genes on the *Escherichia coli* Chromosome

Early mapping experiments and studies on the regulation of the genes that control the enzymes of a single biochemical pathway in *E. coli* had shown that these genes were often clustered. On the genetic map in Figure 11.1, a few such clusters are shown. Notice, for instance, the *gal* gene cluster at 18 min, the *trp* gene cluster at about 28 min, and the *his* cluster at 44 min. Each of these gene clusters constitutes an *operon* that is transcribed as a single polycistronic mRNA (∞ Sections 7.12 and 9.3).

Genes for some biochemical pathways in *E. coli* are not clustered. For example, genes involved in arginine biosynthesis (*arg* genes) are scattered throughout the chromosome as part of the *arg* regulon, a series of operons controlled as a unit (∞ Section 9.3). Because of the early discovery of multigene operons and their use in studying gene regulation (for example, the *lac* operon; ∞ Sections 9.3 and 9.9), the impression may be given that such operons are the rule in prokaryotes. However, sequence analysis of the *E. coli* chromosome has shown that over 70% of the 2,584 predicted or known transcriptional units contain only a single gene. Only about 6% of the operons have four or more genes.

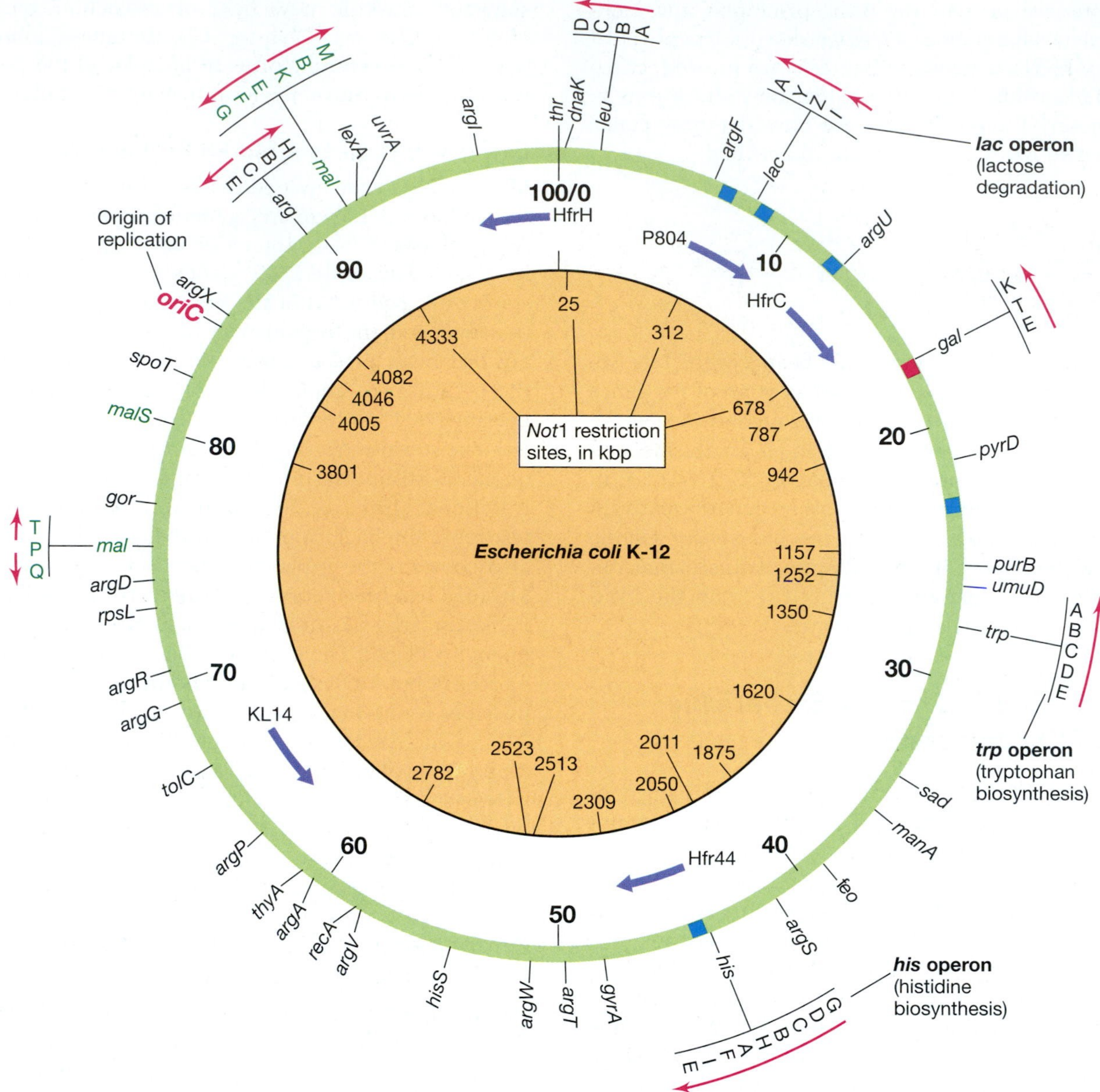

Figure 11.1 Circular linkage map of the chromosome of *Escherichia coli* strain K-12. The *E. coli* chromosome contains 4,639,221 base pairs and 4,288 ORFs. On the outer edge of the map, the locations of a few genes are indicated. A few operons are also shown, with their directions of transcription. Along the inner edge of the map, the numbers from 0 to 100 refer to map position in minutes. Note that 0 (in both minutes and kilobase pairs [kbp]) is located by convention at the *thr* locus. Replication proceeds bidirectionally from the origin of DNA replication, *oriC*, at 84.3 min (∞ Section 7.7 and Figure 7.16). The inner circle shows the locations, in kilobase pairs, of the sites where the restriction enzyme *Not*I cuts. The origins and directions of transfer of a few Hfr strains are also shown (arrows). The locations of five copies of the transposable element IS*3* found in a particular strain are shown in blue. The site where bacteriophage lambda integrates is shown in red. If the lambda prophage were present, it would add an extra 48.5 kbp (slightly over 1 min) to the map. The genes of the maltose regulon, which includes several operons, are shown in green. Although most maltose genes are abbreviated *mal*, note that the one exception is *lamB*. This gene encodes an outer membrane protein involved in maltose uptake, but that is also the receptor for bacteriophage lambda. The gene *rpsL* (73 min) encodes a ribosomal protein. This gene was once called *str* because mutations in this gene lead to streptomycin resistance.

In *E. coli* the transcription of some operons proceeds clockwise around the chromosome, whereas transcription of others proceeds counterclockwise. This means that some coding sequences are on one strand of the chromosome and others are on the opposite strand. Genomic analysis has shown that there are about equal numbers of operons on both strands. The direction of transcription of a few multigene operons is shown by the arrows in Figure 11.1. Many genes that are highly expressed in *E. coli* are oriented on the chromosome so that they are transcribed in the same direction that the DNA replication fork moves through them. The two replication forks start at the origin, *oriC* located at about 84 min on the map shown in Figure 11.1, and move bidirectionally toward the terminus, which is located at approximately 34 min on the map. Genomic sequencing has confirmed that all seven of the rRNA operons of *E. coli* and 53 of its 86 tRNA genes are transcribed in the same direction as replication. Presumably, this arrangement for the transcription of highly expressed genes allows RNA polymerase to avoid collision with the replication fork, because this would be moving in the same direction as the RNA polymerase.

Almost 2,000 *E. coli* proteins, or genes encoding proteins, had been identified by classical genetic analyses before the chromosome was completely sequenced. From sequence analyses we know that as many as 4,288 different proteins can be encoded by the *E. coli* chromosome. Approximately 38% of these proteins are of unknown function or are simply hypothetical ORFs. The "average" *E. coli* protein contains slightly more than 300 amino acid residues, but many proteins are smaller and many are much larger. The largest gene in *E. coli* encodes a protein of 2,383 amino acids, but this protein has not yet been characterized. However, from comparative genomic analyses, this giant protein shows similarities to proteins found in pathogenic enteric bacteria closely related to *E. coli*. This large protein may thus play some role in infection.

Although *E. coli* has very few duplicated genes, computer analyses of its genome have shown that many of the genes that encode proteins have arisen by gene duplication during evolutionary history (Section 11.5). There are also some large gene families in the *E. coli* genome, groups of genes with related sequences encoding products that have related functions (∞ Section 13.9). For example, there is a family of 70 genes whose products are all membrane transport proteins. Gene families are common, not only within a species, but across broad taxonomic lines. Thus gene duplication is a powerful force that affects the genomic composition of prokaryotes.

Insertions within the *Escherichia coli* Chromosome and Lateral Gene Transfer

The *E. coli* genome contains many other genetic elements that replicate as part of it. For example, there are multiple copies of several different insertion sequences (IS elements), including seven copies of IS*2* and five copies of IS*3* (Section 11.16). Both of these IS elements are also found on the F plasmid, and both take part in the formation of Hfr strains (Section 11.13 and Figure 11.22). There are also several defective prophages (three of which are related to lambda) in the *E. coli* genome and partial genomes from several more. It is thus likely that lysogenic bacteriophages have been a part of the biology of *E. coli* for a long time.

Genomic analyses have shown that *E. coli* obtained at least a portion of its genome by *lateral* (*horizontal*) gene transfer from other organisms. Lateral transfer contrasts with *vertical* gene transfer in which genes move from mother cell to daughter cell. In fact, it has been estimated that nearly 20% of the *E. coli* genome originated from lateral transfers. Laterally transferred genes can often be detected by their having significantly different GC ratios or a significantly different distribution of codons (codon bias, ∞ Section 7.13) from those of the host organism.

Lateral gene transfer may cause large-scale changes in an organism's genome. For example, strains of *E. coli* are known that contain virulence genes located on large, unstable regions of the chromosome called *pathogenicity islands* that can be acquired by lateral transfer (∞ Section 13.12 and Figure 13.13). Lateral gene transfer does not necessarily result in an ever-larger genome size. Many of the genes acquired in this way provide no selective advantage and so are lost by deletion. This keeps the chromosome of a given species at roughly the same size over time. For example, comparisons of genome sizes of several strains of *E. coli* have shown them all to be about 4.5–5.5 Mbp, despite the fact that prokaryotic genomes can vary from under 0.5 to over 10 Mbp. Genome size is therefore a species-specific trait.

Deeper Analysis of Prokaryotic Genomes

The molecular genetics of *E. coli* have been intensively explored for several decades, and the genome has now been sequenced. Does this mean that future studies will be exclusively bioinformatic and that traditional genetic and biochemical studies of *E. coli* are over? Absolutely not! Although computer analysis of a sequence can yield much information, to understand *gene function*, particularly of regulatory sequences, it is still necessary to isolate mutants, map their mutations, and use biochemical and physiological analyses to determine their effects on the organism. This is especially true of the 20–40% (depending on the species) of genes that show up in all genomic analyses (∞ Section 13.3) as encoding proteins of unknown function.

This huge repository of hypothetical proteins almost certainly holds new biochemical secrets that, collectively, will expand the extensive metabolic capabilities of prokaryotes already known. But in addition, because many prokaryotic genes have homologues in eukaryotic organisms including humans, understanding gene function in prokaryotes has practical consequences in terms of better understanding human genetics. Many human diseases have a genetic link. Because of this, the more we understand what proteins do in bacteria, the more we will understand their function in humans. If history is any indication, this will lead to, among other things, the development of new drugs and other treatments to control diseases.

11.1 MiniReview

The *Escherichia coli* chromosome has been mapped using conjugation, transduction, molecular cloning, and sequencing. *E. coli* has been a useful model organism, and a considerable amount of information has been obtained from it, not only about gene structure but also gene function and regulation.

- Genetic maps of prokaryotic chromosomes are now typically made using only molecular cloning and DNA sequencing. Why were other methods also used for *E. coli*?
- How large is an average bacterial protein?
- Approximately how large is the *E. coli* genome in base pairs? How many genes does it contain?

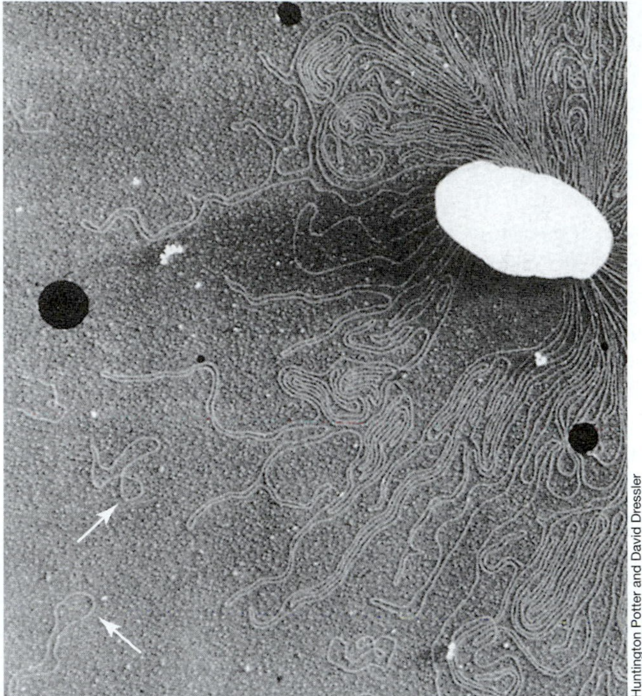

Figure 11.2 The bacterial chromosome and bacterial plasmids, as seen in the electron microscope. The plasmids (arrows) are the circular structures and are much smaller than the main chromosomal DNA. The cell (large, white structure) was broken gently so the DNA would remain intact.

11.2 Plasmids: General Principles

In addition to the chromosome, many prokaryotic cells also contain other genetic elements, in particular, **plasmids**. Plasmids are genetic elements that replicate independently of the host chromosome. Unlike viruses, plasmids do not have an extracellular form and exist inside cells as free and typically circular DNA. Plasmids and chromosomes can be differentiated in that plasmids carry only unessential (but often very helpful) genes. Essential genes reside on chromosomes. Literally thousands of different plasmids are known. Indeed, over 300 different naturally occurring plasmids have been isolated from strains of *Escherichia coli* alone. In this section we discuss the properties of a few of them.

Physical Nature and Replication of Plasmids

Almost all known plasmids consist of double-stranded DNA. Most plasmids are circular, but many linear plasmids are also known. Naturally occurring plasmids vary in size from approximately 1 kbp to more than 1 Mbp. The typical plasmid is a circular double-stranded DNA molecule less than 5% the size of the chromosome (**Figure 11.2**). Most of the plasmid DNA isolated from cells is in the supercoiled configuration, which is the most compact form for DNA to exist within the cell (∞ Figure 7.8).

The enzymes needed to replicate plasmids are normal cell enzymes. Therefore, the genes carried by the plasmid itself are concerned primarily with control of the initiation of replication and with partition of the replicated plasmids between daughter cells. Also, different plasmids are present in cells in different numbers; this is called the *copy number*. Some plasmids are present in the cell in only 1–3 copies, whereas others may be present in over 100 copies. Copy number is controlled by genes on the plasmid and by interactions between the host and the plasmid.

Most plasmids in gram-negative *Bacteria* replicate in a manner similar to that already described for the chromosome (∞ Section 7.7). This involves initiation at an origin of replication and bidirectional replication around the circle, giving a theta intermediate (∞ Figure 7.16). However, some plasmids have unidirectional replication, with just a single replication fork. Because of the small size of plasmid DNA relative to the chromosome, the plasmid replicates very quickly, perhaps in a tenth or less of the total time of the cell division cycle.

Most plasmids of gram-positive *Bacteria*, plus a few from gram-negative *Bacteria* and *Archaea*, replicate by a rolling circle mechanism similar to that used by the phage ϕX174 (∞ Section 19.2 and Figure 19.4). This mechanism proceeds via a single-stranded intermediate. Most linear plasmids replicate by using a protein bound to the 5′ end of each strand for priming DNA synthesis (∞ Section 8.7).

Plasmid Incompatibility and Plasmid Curing

Some bacteria may contain several different types of plasmids. For example, *Borrelia burgdorferi* (the Lyme disease pathogen, ∞ Section 35.4) contains 17 different circular and linear plasmids! The ability of two different plasmids to each replicate in the same cell is controlled by plasmid genes that regulate DNA replication. When a plasmid is transferred into a cell that already carries another plasmid, a common observation is that the second plasmid may be lost during subsequent cell replication. If this occurs, the two plasmids are said to be *incompatible*. A number of incompatibility (Inc) groups have been recognized. Plasmids belonging to the same Inc group exclude each other from replicating in the same cell but can coexist with plasmids from other groups. Plasmids of each Inc group share a common mechanism of regulating their replication and are thus related to one another. Therefore, although a bacterial cell may contain different kinds of plasmids, each is genetically distinct.

Some plasmids, called *episomes*, can integrate into the chromosome, and under such conditions their replication comes under control of the chromosome. This situation is remarkably similar to that of several viruses whose genomes can become incorporated into the host genome (prophages, ∞ Section 10.10).

Plasmids can sometimes be eliminated from host cells by various treatments. This removal, called *curing*, results from inhibition of plasmid replication without parallel inhibition of chromosome replication. As a result of cell division, the plasmid is diluted out. Curing may occur spontaneously, but it is greatly increased by treatments with certain chemicals such as acridine dyes, which become inserted into DNA, or other treatments that seem to interfere more with plasmid replication than with chromosome replication.

Cell-to-Cell Transfer of Plasmids

How do plasmids manage to infect new host cells? Some prokaryotic cells can take up free DNA from the environment (Section 11.10). Consequently, plasmids released by the death and lysis of their previous host cell (however this may have occurred) may be taken up by a new host. However, few bacterial species have this ability, and it is unlikely to account for much cell-to-cell plasmid transfer. The main mechanism of cell-to-cell plasmid transfer is *conjugation,* a function encoded by some plasmids themselves. Conjugation involves cell-to-cell contact and is discussed in Section 11.12.

Plasmids that govern their own transfer by cell-to-cell contact are called *conjugative*. Not all plasmids are conjugative. Transfer by conjugation is controlled by a set of genes on the plasmid called the *tra* (for transfer) region. The *tra* region contains genes encoding proteins that function in DNA transfer and replication and others that function in mating pair formation. The presence of a *tra* region in a plasmid can have another important consequence if the plasmid becomes integrated into the chromosome. In that case, the plasmid can mobilize the chromosomal DNA, which may then be transferred from one cell to another. The use of conjugation to transfer host genes is discussed further in Section 11.13.

Some conjugative plasmids from *Pseudomonas* have a broad host range. This means that they are transferable to a wide variety of other gram-negative *Bacteria*. Such plasmids can transfer genes between distantly related organisms. Conjugative plasmids have been shown to transfer between gram-negative and gram-positive *Bacteria,* between *Bacteria* and plant cells, and between *Bacteria* and fungi. Even if the plasmid cannot replicate independently in the new host, transfer of the plasmid itself could have important evolutionary consequences if genes from the plasmid recombine with the genome of the new host.

11.2 MiniReview

Plasmids are small circular or linear DNA molecules that carry nonessential genes. Although a cell can contain more than one plasmid, these cannot be closely related genetically. Although they have no extracellular form, plasmids can be transferred by the process of conjugation.

- How can a large plasmid be differentiated from a small chromosome?
- What function do the *tra* genes of the F plasmid carry out?

11.3 Types of Plasmids and Their Biological Significance

Clearly, all plasmids must carry genes that ensure their own replication. As we have seen, some plasmids also carry genes necessary for conjugation. Although plasmids do not carry genes that are essential to the host, plasmids may carry genes that have a profound influence on host cell phenotype.

Table 11.1 Examples of phenotypes conferred by plasmids in prokaryotes

Phenotype class	*Organisms*[a]
Antibiotic production	*Streptomyces*
Conjugation	Wide range of bacteria
Metabolic functions	
Degradation of octane, camphor, naphthalene	*Pseudomonas*
Degradation of herbicides	*Alcaligenes*
Formation of acetone and butanol	*Clostridium*
Lactose, sucrose, citrate, or urea utilization	Enteric bacteria
Pigment production	*Erwinia, Staphylococcus*
Gas vesicle production	*Halobacterium*
Resistance	
Antibiotic resistance	Wide range of bacteria
Resistance to toxic metals	Wide range of bacteria
Virulence	
Tumor production in plants	*Agrobacterium*
Nodulation and symbiotic nitrogen fixation	*Rhizobium*
Bacteriocin production and resistance	Wide range of bacteria
Animal cell invasion	*Salmonella, Shigella, Yersinia*
Coagulase, hemolysin, enterotoxin	*Staphylococcus*
Toxins and capsule	*Bacillus anthracis*
Enterotoxin, K antigen	*Escherichia*

[a]All of the organisms in the list are *Bacteria* except for *Halobacterium*, which is a member of the *Archaea*.

In some cases plasmids encode properties fundamental to the ecology of the bacterium. For example, the ability of *Rhizobium* to interact with plants and form nitrogen-fixing root nodules requires certain plasmid functions (∞ Section 24.15). Other plasmids confer special metabolic properties on bacterial cells, such as the ability to degrade toxic pollutants. Indeed, plasmids seem to be a major mechanism for conferring special properties on bacteria and for mobilizing these properties by lateral gene flow. A few special properties conferred on prokaryotes by plasmids are summarized in **Table 11.1**.

Resistance Plasmids

Among the most widespread and well-studied groups of plasmids are the resistance plasmids, usually just called *R plasmids,* which confer resistance to antibiotics and various other growth inhibitors. Several antibiotic resistance genes can be carried by a single R plasmid, or, alternatively, a cell may contain several R plasmids. In either case, the result is multiple resistance. R plasmids were first discovered in Japan in the 1950s in strains of enteric bacteria that had acquired resistance to sulfonamide antibiotics. Since then they have been found throughout the world. The emergence of bacteria resistant to antibiotics is of considerable medical significance

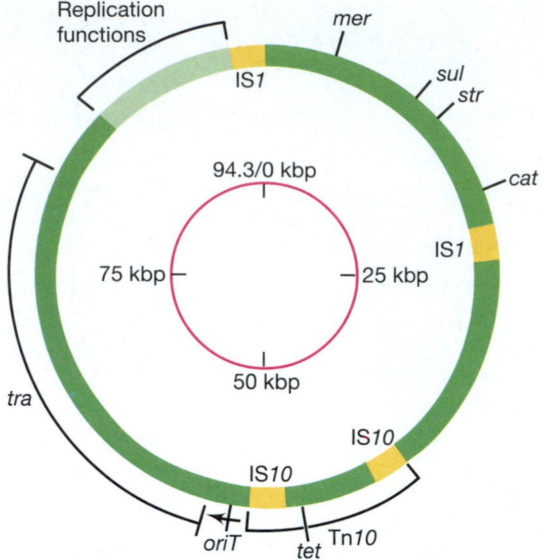

Figure 11.3 Genetic map of the resistance plasmid R100. The inner circle shows the size of the plasmid in kilobase pairs. The outer circle shows the location of major antibiotic resistance genes and other key functions: *cat*, chloramphenicol resistance; *oriT*, origin of conjugative transfer; *mer*, mercuric ion resistance; *sul*, sulfonamide resistance; *str*, streptomycin resistance; *tet*, tetracycline resistance; *tra*, transfer functions. The locations of insertion sequences (IS) and the transposon Tn*10* are also shown. Several genes related to plasmid replication are found in the region from 88 to 92 kbp.

and is correlated with the increasing use of antibiotics for the treatment of infectious diseases (⚭ Section 27.12). Soon after these resistant strains were isolated, it was shown that they could transfer resistance to sensitive strains via cell-to-cell contact. The infectious nature of the conjugative R plasmids permitted their rapid spread through cell populations. Resistance plasmids are now a major problem in clinical medicine.

In general, resistance genes encode proteins that either inactivate the antibiotic or protect the cell by some other mechanism. Plasmid R100, for example, is a 94.3-kbp plasmid (**Figure 11.3**) that carries genes encoding resistance to sulfonamides, streptomycin, spectinomycin, fusidic acid, chloramphenicol, and tetracycline. Plasmid R100 also carries several genes conferring resistance to mercury. Plasmid R100 can be transferred between enteric bacteria of the genera *Escherichia, Klebsiella, Proteus, Salmonella,* and *Shigella,* but does not transfer to gram-negative bacteria outside the enteric group. Different R plasmids with genes for resistance to most antibiotics are known. Many drug-resistant modules on R plasmids, such as those on R100, are also transposable elements (⚭ Section 11.16), and this, combined with the fact that many of these plasmids are conjugative, have made them a serious threat to traditional antibiotic therapy.

Plasmids Encoding Toxins and Other Virulence Characteristics

We will discuss in Chapter 28 the characteristics of pathogenic microorganisms that enable them to colonize hosts and establish infections. Here we merely note the two major characteristics of the virulence (disease-causing ability) of pathogens: (1) the ability of the pathogen to attach to and colonize specific host tissue and (2) the production of substances (toxins, enzymes, and other molecules) that cause damage to the host.

In several pathogenic bacteria these virulence characteristics are encoded by plasmid genes. For example, enteropathogenic strains of *Escherichia coli* are characterized by an ability to colonize the small intestine and to produce a toxin that causes diarrhea. Colonization requires the presence of a cell-surface protein called *colonization factor antigen,* encoded by a plasmid. This protein confers on bacterial cells the ability to attach to epithelial cells of the intestine. At least two toxins in enteropathogenic *E. coli* are known to be encoded by plasmids: the hemolysin, which lyses red blood cells, and the enterotoxin, which induces extensive secretion of water and salts into the bowel. It is the enterotoxin that is responsible for diarrhea (⚭ Section 28.11).

Some virulence factors are encoded on plasmids. Other virulence factors are encoded by other mobile genetic elements, such as transposons and bacteriophages. Some virulence factors are chromosomal. Several examples are known in which multiple virulence genes are present on different genetic elements within the same cell. For instance, the genes encoding the virulence determinants of shigatoxin-producing strains of *E. coli* (⚭ Section 37.8) are distributed among the chromosome, a bacteriophage, and a plasmid.

Bacteriocins

Many bacteria produce proteins that inhibit or kill closely related species or even different strains of the same species. These agents, called *bacteriocins* to distinguish them from antibiotics, have a more narrow spectrum of activity than antibiotics. The genes encoding bacteriocins and the proteins needed for processing and transporting them (and for conferring immunity on the producing organism) are often carried on a plasmid or a transposon. Bacteriocins are often named after the species of organism that produces them. Thus, *E. coli* produces *colicins*; *Yersinia pestis* produces *pesticins,* and so on.

The Col plasmids of *E. coli* encode various colicins. Col plasmids can be either conjugative or nonconjugative. Colicins released from the producer cell bind to specific receptors on the surface of susceptible cells. The receptors for colicins are proteins whose normal function is to transport some substance, frequently a growth factor or micronutrient, through the outer membrane of the cell. Colicins kill cells by disrupting some critical cell function. For example, many colicins form channels in the cell membrane that allow potassium ions and protons to leak out, leading to a loss of the cell's ability to generate energy. Another major group of colicins are nucleases and degrade DNA or RNA. For example, colicin E2 is a DNA endonuclease that can cleave DNA, and colicin E3 is a ribonuclease that cuts at a specific site in 16S rRNA and therefore inactivates ribosomes.

The bacteriocins or bacteriocin-like agents of gram-positive bacteria are quite different from the colicins but are also often encoded by plasmids; some even have commercial value. For instance, lactic acid bacteria produce the bacteriocin Nisin A, which strongly inhibits the growth of a wide range of gram-positive bacteria and is used as a preservative in the food industry.

Engineered Plasmids

Plasmids have been widely exploited in genetic engineering. Indeed, countless new, artificial plasmids have been constructed in the laboratory. Genes from a wide variety of sources have been incorporated into such plasmids, thus allowing their transfer across any species barrier. The only requirements for artificial plasmids are that they contain genes controlling their own replication and are stably maintained in the host of choice. This topic is discussed further in Chapter 12.

11.3 MiniReview

The genetic information that plasmids carry is not essential for cell function under all conditions but may confer a selective growth advantage under certain conditions. Examples include antibiotic resistance, enzymes for degradation of unusual organic compounds, and special metabolic pathways. Virulence factors of many pathogenic bacteria are plasmid encoded.

- How does an R plasmid differ from the F plasmid? How are they similar?
- How do bacteriocins differ from antibiotics?

II MUTATION

All organisms contain a specific sequence of nucleotide bases in their genome, their genetic blueprint. A **mutation** is a heritable change in the base sequence of that genome. Mutations can lead to changes—some good, some bad—in an organism. Genetic alterations can also be brought about by recombination (Section 11.9), the physical exchange of DNA between genetic elements. Recombination creates new combinations of genes even in the absence of mutation. Whereas mutation usually brings about only a very small amount of genetic change in a cell, genetic recombination typically generates much larger changes. Entire genes, sets of genes, or even larger segments of DNA can be transferred between chromosomes or other genetic elements. Taken together, mutation and recombination fuel the evolutionary process.

Unlike most eukaryotes, prokaryotes do not reproduce sexually. However, prokaryotes possess mechanisms of lateral genetic exchange that allow for both gene transfer and recombination. To detect genetic exchange between two prokaryotes, it is therefore necessary to employ genetic markers whose transfer can be detected. The term "marker" refers to any gene whose presence is monitored during a genetics experiment. If possible, markers are chosen that are relatively easy to detect. Genetically altered strains are used in gene transfer experiments, the alteration(s) being due to one or more mutations in their DNA. These mutations may involve changes in only one or a few base pairs or even the insertion or deletion of entire genes. Before discussing genetic exchange, we will therefore consider the molecular mechanism of mutation and the properties of mutant microorganisms.

11.4 Mutations and Mutants

As previously mentioned, a mutation is a heritable change in the base sequence of the nucleic acid in the genome of an organism (or a virus or other genetic element). In all cells the genome consists of double-stranded DNA (Section 7.1). In viruses, by contrast, the genome may consist of single- or double-stranded DNA or RNA. A strain of any cell or virus carrying a change in nucleotide sequence is called a **mutant**. A mutant by definition differs from its parental strain in **genotype**, the nucleotide sequence of the genome. But in addition, the observable properties of the mutant—its **phenotype**—may also be altered relative to the parental strain. This altered phenotype is called a *mutant phenotype*. It is common to refer to a strain isolated from nature as a **wild-type strain**. Mutant derivatives can be obtained either directly from wild-type strains or from other strains previously derived from the wild type, for example, another mutant.

Genotype Versus Phenotype

Depending on the mutation, a mutant strain may or may not differ in phenotype from its parent. By convention in bacterial genetics, the genotype of an organism is designated by three lowercase letters followed by a capital letter (all in italics) indicating a particular gene. For example, the *hisC* gene of *Escherichia coli* encodes a protein called HisC that functions in biosynthesis of the amino acid histidine. Mutations in the *hisC* gene would be designated as *hisC1*, *hisC2*, and so on, the numbers referring to the order of isolation of the mutant strains. Each *hisC* mutation would be different, and each *hisC* mutation might affect the HisC protein in different ways.

The phenotype of an organism is designated by a capital letter followed by two lowercase letters, with either a plus or minus superscript to indicate the presence or absence of that property. For example, a His^+ strain of *E. coli* is capable of making its own histidine, whereas a His^- strain is not. The His^- strain would require a histidine supplement for growth. Returning to our previous example, any mutation in the *hisC* gene may lead to a His^- phenotype if it eliminates the function of the HisC protein.

Isolation of Mutants: Screening Versus Selection

Virtually any characteristic of an organism can be changed by mutation. However, some mutations are selectable, conferring some type of advantage on organisms possessing them, whereas others are nonselectable, even though they may lead to a very clear change in the phenotype of an organism.

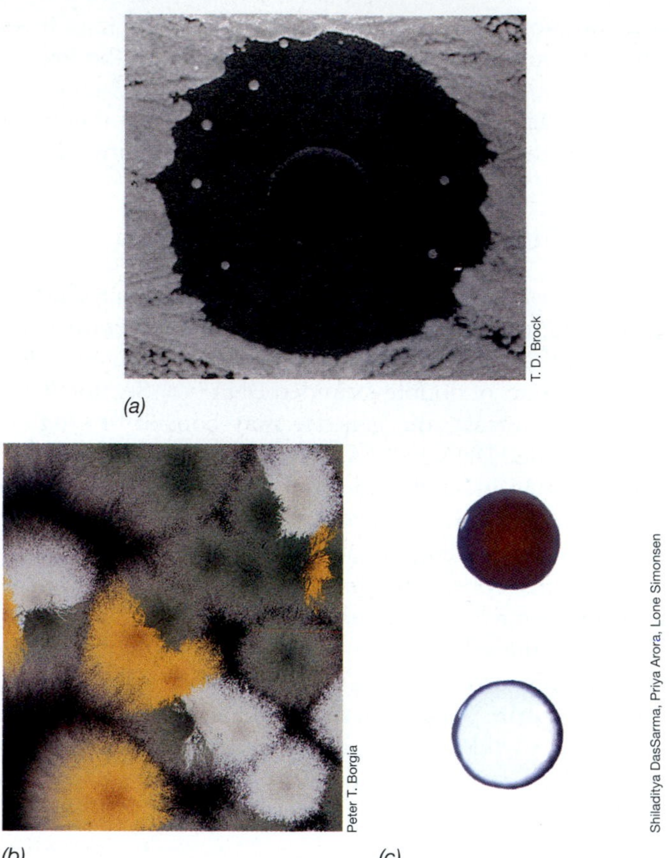

Figure 11.4 Selectable and nonselectable mutations. *(a)* Development of antibiotic-resistant mutants, a type of easily selectable mutation, within the inhibition zone of an antibiotic assay disc. *(b)* Nonselectable mutations. Spontaneous pigmented and nonpigmented mutants of the fungus *Aspergillus nidulans*. The wild type has a green pigment. The white or colorless mutants make no pigment, whereas the yellow mutants cannot convert the yellow pigment precursor to the normal (green) color. *(c)* Colonies of mutants of a species of *Halobacterium*, a member of the *Archaea*. The wild-type colonies are white. The orangish brown colonies are mutants that lack gas vesicles (Section 4.11). The gas vesicles scatter light and mask the color of the colony.

A selectable mutation confers a clear advantage on the mutant strain under certain environmental conditions, so the progeny of the mutant cell are able to outgrow and replace the parent. A good example of a selectable mutation is drug resistance: An antibiotic-resistant mutant can grow in the presence of antibiotic concentrations that inhibit or kill the parent (**Figure 11.4a**) and is thus selected for under these conditions. It is relatively easy to detect and isolate selectable mutants by choosing the appropriate environmental conditions. **Selection** is therefore an extremely powerful genetic tool, allowing the isolation of a single mutant from a population containing millions or even billions of parental organisms.

An example of a nonselectable mutation is color loss in a pigmented organism (Figure 11.4*b*). Nonpigmented cells usually have neither an advantage nor a disadvantage over the pigmented parent cells when grown on agar plates, although pigmented organisms may have a selective advantage in nature. We can detect such mutations only by examining large numbers of colonies and looking for the "different" ones, a process called **screening**.

Isolation of Nutritional Auxotrophs and Penicillin Selection

Although screening is always more tedious than selection, methods are available for screening large numbers of colonies for certain types of mutations. For instance, nutritionally defective mutants can be detected by the technique of *replica plating* (**Figure 11.5**). With the use of sterile velveteen cloth or filter paper, an imprint of colonies from a master plate is made onto an agar plate lacking the nutrient. Parental colonies will grow normally, whereas those of the mutant will not. Thus, the inability of a colony to grow on the replica plate containing minimal medium signals that it is a mutant (Figure 11.5). The colony on the master plate corresponding to the vacant spot on the replica plate can then be picked, purified, and characterized. A mutant with a nutritional requirement for growth is called an **auxotroph**, and the parent from which it was derived is called a *prototroph*. (A prototroph may or may not be the wild type. An auxotroph may be derived from the wild type or from a mutant derivative of the wild type.) For instance, mutants of *Escherichia coli* with a His$^-$ phenotype are histidine auxotrophs. Although of great utility, replica plating is nevertheless a screening process, and it can be laborious to isolate mutants by screening.

An ingenious method widely used to isolate auxotrophs is the penicillin-selection method. Ordinarily, mutants that require specific nutrients are at a disadvantage in competition with the parent cells and so there is no direct way of isolating them. For instance, histidine mutants are rare in a mutagenized culture, and it could take a great deal of time to obtain them by replica plating alone. However, penicillin selection can be used to enrich a population of mutagenized cells in His$^-$ mutants, after which replica plating can be much more effective. How does penicillin selection work?

Penicillin is an antibiotic that kills only growing cells. If penicillin is added to a population of cells growing in a medium lacking the nutrient required by the desired mutant, the parent cells will be killed, whereas any nongrowing mutant cells will be unaffected. Thus, after preliminary incubation in the absence of the nutrient in a penicillin-containing medium, the population is washed free of the penicillin and transferred to plates containing the nutrient. The colonies that appear include some wild-type cells that escaped penicillin killing, but also include some of the desired mutants. Penicillin selection is thus a kind of negative selection; the selection is not *for* the mutant but instead *against* the parental type. Penicillin selection is often used as a prelude to replica plating to increase the chances of obtaining auxotrophic mutants.

Examples of common classes of mutants and the means by which they are detected are listed in **Table 11.2**. **www.microbiologyplace.com Online Tutorial 11.1: Replica Plating**

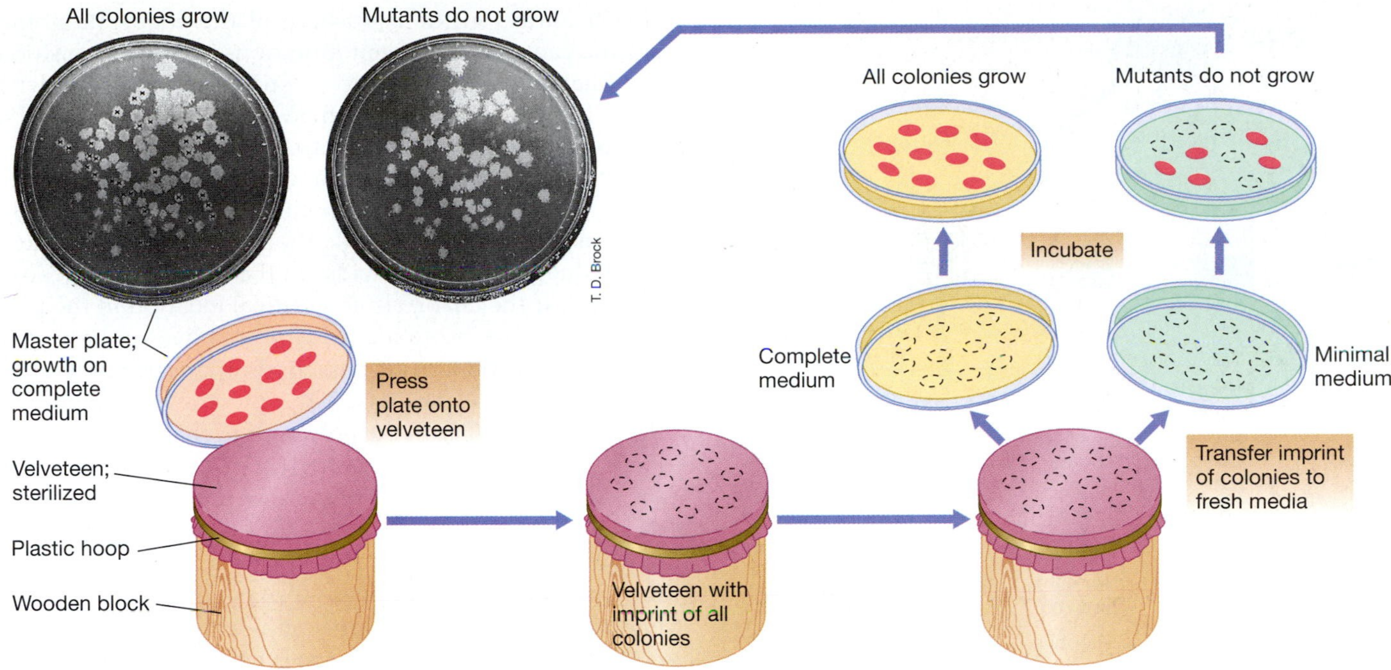

Figure 11.5 Screening for nutritional auxotrophs. The replica plating method can be used for the detection of nutritional mutants. Photos: The photograph on the left shows the master plate. The colonies not appearing on the replica plate are marked with an X. The replica plate (right) lacked one nutrient (leucine) present in the master plate. Therefore, the colonies marked with an X on the master plate are leucine auxotrophs.

11.4 MiniReview

Mutation is a heritable change in DNA sequence that can lead to a change in phenotype. Selectable mutations are those that give the mutant a growth advantage under certain environmental conditions and are especially useful in genetic research. If selection is not possible, mutants must be identified by screening.

- Distinguish between the words "mutation" and "mutant."
- Distinguish between the words "screening" and "selection."

11.5 Molecular Basis of Mutation

Mutations can be either spontaneous or induced. **Induced mutations** are those that are made deliberately. **Spontaneous mutations** are those that occur without human intervention. They can result from exposure to natural radiation (cosmic rays, and so on) that alters the structure of bases in the DNA. Also, oxygen radicals (∞ Section 6.18) can affect DNA structure by chemically modifying DNA. For example, oxygen radicals can convert guanine into 8-hydroxyguanine, and this causes spontaneous mutations. However, the bulk of

Table 11.2 Kinds of mutants

Phenotype	*Nature of change*	*Detection of mutant*
Auxotroph	Loss of enzyme in biosynthetic pathway	Inability to grow on medium lacking the nutrient
Temperature-sensitive	Alteration of an essential protein so it is more heat-sensitive	Inability to grow at a high temperature (for example, 40°C) that normally supports growth
Cold-sensitive	Alteration of an essential protein so it is inactivated at low temperature	Inability to grow at a low temperature (for example, 20°C) that normally supports growth
Drug-resistant	Detoxification of drug or alteration of drug target or permeability to drug	Growth on medium containing a normally inhibitory concentration of the drug
Rough colony	Loss or change in lipopolysaccharide layer	Granular, irregular colonies instead of smooth, glistening colonies
Nonencapsulated	Loss or modification of surface capsule	Small, rough colonies instead of larger, smooth colonies
Nonmotile	Loss of flagella or nonfunctional flagella	Compact instead of flat, spreading colonies
Pigmentless	Loss of enzyme in biosynthetic pathway leading to loss of one or more pigments	Presence of different color or lack of color
Sugar fermentation	Loss of enzyme in degradative pathway	Lack of color change on agar containing sugar and a pH indicator
Virus-resistant	Loss of virus receptor	Growth in presence of large amounts of virus

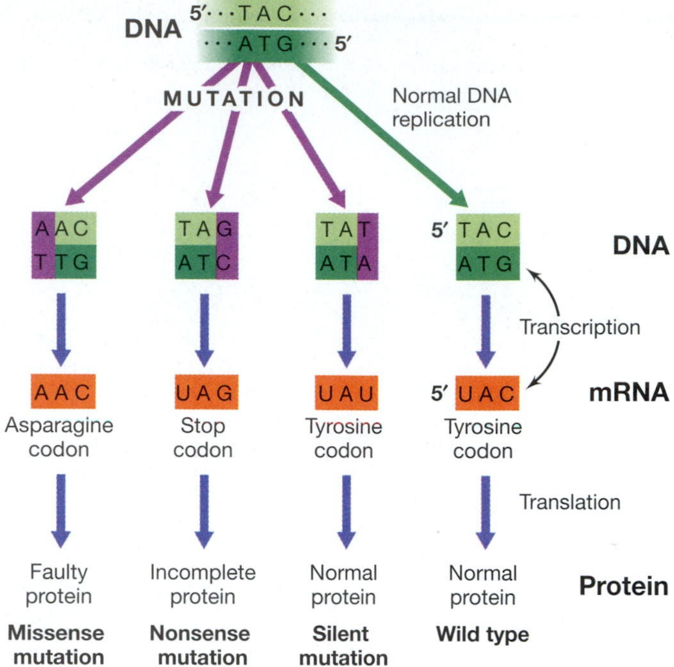

Figure 11.6 Possible effects of base-pair substitution in a gene encoding a protein. Three different protein products are possible from changes in the DNA for a single codon.

spontaneous mutations result from errors in the pairing of bases during DNA replication.

Mutations that change only one base pair are called **point mutations**. Point mutations are caused by base-pair substitutions in the DNA or by the loss or gain of a single base pair. As is the case with all mutations, the phenotypic change that results from a point mutation depends on exactly where the mutation occurs in the gene, what the nucleotide change is, and what product the gene encodes.

Base-Pair Substitutions

If a point mutation is within the coding region of a gene that encodes a polypeptide, any change in the phenotype of the cell is almost certainly the result of a change in the amino acid sequence of the polypeptide. The error in the DNA is transcribed into mRNA, and the erroneous mRNA in turn is translated to yield a polypeptide. **Figure 11.6** shows the consequences of various base-pair substitutions.

In interpreting the results of a mutation, we must first recall that the genetic code is degenerate (∞ Section 7.13 and Table 7.4). Because of this degeneracy, not all mutations in the base sequence encoding a polypeptide will change the polypeptide. This is illustrated in Figure 11.6, which shows several possible results when the DNA that encodes a single tyrosine codon in a polypeptide is mutated. First, a change in the RNA from UAC to UA*U* would have no apparent effect because UAU is also a tyrosine codon. Although they do not affect the sequence of the encoded polypeptide, such changes in the DNA are indeed still mutations. They are one type of **silent mutation**, that is, a mutation that does not affect the phenotype of the cell. Note that silent mutations in coding regions are almost always in the third base of the codon (arginine and leucine can also have silent mutations in the first position).

Changes in the first or second base of the triplet more often lead to significant changes in the polypeptide. For instance, a single-base change from *U*AC to *A*AC (Figure 11.6) results in an amino acid change within the polypeptide from tyrosine to asparagine at a specific site. This is referred to as a **missense mutation** because the informational "sense" (precise sequence of amino acids) in the ensuing polypeptide has changed. If the change is at a critical location in the polypeptide chain, the protein could be inactive or have reduced activity. However, not all missense mutations necessarily lead to nonfunctional proteins. The outcome depends on where the substitution lies in the polypeptide chain and on how it affects protein folding and activity. For example, mutations in the active site of an enzyme are more likely to destroy activity than mutations in other regions of the protein.

Another possible outcome of a base-pair substitution is the formation of a nonsense (stop) codon. This results in premature termination of translation, leading to an incomplete polypeptide that would almost certainly not be functional (Figure 11.6). Mutations of this type are called **nonsense mutations** because the change is from a codon for an amino acid (sense codon) to a nonsense codon (∞ Table 7.4). Unless the nonsense mutation is very near the end of the gene, the incomplete product will be completely inactive.

The terms "transition" and "transversion" are used to describe the type of base substitution in a point mutation. **Transitions** are mutations in which one purine base (A or G) is substituted for another purine, or one pyrimidine base (C or T) is substituted for another pyrimidine. **Transversions** are point mutations in which a purine base is substituted for a pyrimidine base or vice versa.

Frameshifts and Other Insertions or Deletions

Because the genetic code is read from one end of the nucleic acid in consecutive blocks of three bases (that is, as codons), any deletion or insertion of a single base pair results in a shift in the reading frame. These frameshift mutations can have serious consequences. Single-base insertions or deletions change the primary sequence of the encoded polypeptide, typically in a major way (**Figure 11.7**). Such microinsertions or microdeletions can result from replication errors. Insertion or deletion of two base pairs also causes a frameshift; however, insertion or deletion of three base pairs adds or removes a whole codon. This results in addition or deletion of a single amino acid in the polypeptide sequence. Although this may well be deleterious to protein function, it is usually not as bad as a frameshift, which scrambles the entire polypeptide sequence after the mutation point.

Insertions or deletions can also result in the gain or loss of hundreds or even thousands of base pairs. Such changes inevitably result in complete loss of gene function. Some deletions are so large that they may include several genes. If any of the deleted genes are essential, the mutation will be lethal. Such deletions cannot be restored through further mutations, but only through genetic recombination. Indeed, one way in which large deletions are distinguished from point mutations is that the latter are reversible through further mutations, whereas the former are not. Larger insertions and deletions may arise as a result of errors during genetic recombination. In addition, many insertion mutations are due to the insertion of specific identifiable DNA sequences 700–1400 bp in length called *insertion sequences*, a type of transposable element (Section 11.16). The effect of transposable elements on the evolution of bacterial genomes is discussed further in Section 13.10.

Other types of large-scale mutations are rearrangements brought about by errors in recombination. These include translocations, in which a large section of chromosomal DNA is moved to a new location (and in eukaryotes often to a different chromosome), and inversions, in which the orientation of a particular segment of DNA is reversed relative to the surrounding DNA. **www.microbiologyplace.com Online Tutorial 11.2: The Molecular Basis for Mutations**

Site-Directed Mutagenesis and Transposons

The mutations that we have considered thus far have been random, that is, not directed at any particular gene. However, recombinant DNA technology and the use of synthetic DNA make it possible to induce specific mutations in specific genes. The procedures for carrying out mutagenesis of specific sites in the genome are called *site-directed mutagenesis* and are discussed in Chapter 12. Mutations can also be deliberately introduced by transposon mutagenesis (Section 11.16). If a transposable element inserts within a gene, loss of gene function generally results. Because transposable elements can insert into the chromosome at various locations, transposons are widely used to generate mutations.

Back Mutations or Reversions

Point mutations are typically reversible, a process known as **reversion**. A revertant is a strain in which the original phenotype that was changed in the mutant is restored. Revertants can be of two types. In same-site revertants, the mutation that restores activity is at the same site as the original mutation. If the back mutation is not only at the same site but also restores the original sequence, it is called a *true revertant*.

In second-site revertants, the mutation is at a different site in the DNA. Second-site mutations can cause restoration of a wild-type phenotype if they function as suppressor mutations—mutations that compensate for the effect of the original mutation and restore the original phenotype. Several classes of suppressor mutations are known. These include (1) a mutation somewhere else in the same gene that restores enzyme function, such as a second frameshift mutation near the first that restores the original reading frame; (2) a mutation in another gene that restores the function of the original mutated gene; and (3) a mutation in another gene that results in the production of an enzyme that can replace the mutated one.

An interesting subclass of suppressor mutations are those due to alterations in tRNA. Nonsense mutations can be suppressed by changing the anticodon sequence of a tRNA molecule so that it now recognizes a stop codon. Such an altered tRNA is known as a suppressor tRNA and will insert its cognate amino acid at the stop codon that it now reads. Suppressor tRNA mutations would be lethal unless a cell has more than one tRNA for a particular codon. One tRNA may then be mutated into a suppressor, and the other performs the original function. Most cells have multiple tRNAs and so suppressor mutations are reasonably common, at least in microorganisms. Sometimes the amino acid inserted by the suppressor tRNA is identical to the original amino acid and the protein is fully restored. In other cases, a different amino acid is inserted and a partially active protein may be produced.

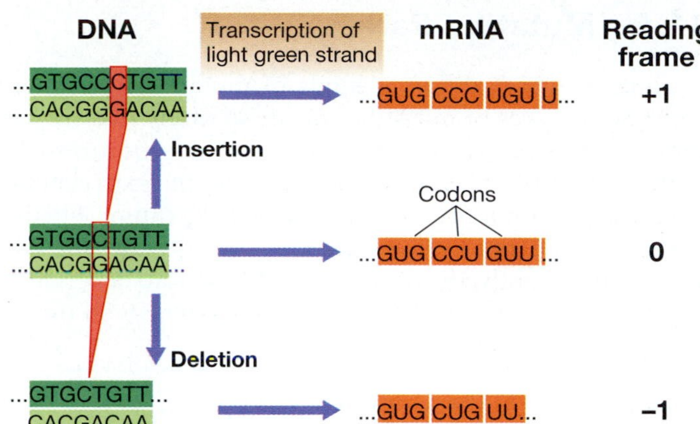

Figure 11.7 Shifts in the reading frame of mRNA caused by insertions or deletions. The reading frame in mRNA is established by the ribosome that begins at the 5′ end (toward the left in the figure) and proceeds by units of three bases (codons, Section 7.13). The normal reading frame is referred to as the 0 frame, that missing a base the −1 frame, and that with an extra base the +1 frame. To determine the effects of a frameshift, translate the codons using Table 7.4.

Unlike point mutations, large-scale deletions do not revert. By contrast, large-scale insertions can revert as the result of a subsequent deletion that removes the insertion. Typically, frameshift mutations of any magnitude are difficult to restore to the wild type, and mutants carrying frameshift mutations are therefore genetically quite stable. For this reason, geneticists often use them in genetic crosses to avoid accidental reversion of mutant strains during the course of a genetic study.

11.5 MiniReview

Mutations, which can be either spontaneous or induced, arise because of changes in the base sequence of the nucleic acid of an organism's genome. A point mutation, which is due to a change in a single base pair, can lead to a single amino acid change in a polypeptide or to no change at all, depending on the particular codon. In a nonsense mutation, the codon becomes a stop codon and an incomplete polypeptide is made. Deletions and insertions cause more dramatic changes in the DNA, including frameshift mutations that often result in complete loss of gene function.

- What does it mean to say that point mutations can spontaneously revert?
- Do missense mutations occur in genes encoding tRNA? Why or why not?

11.6 Mutation Rates

The rates at which different kinds of mutations occur vary widely. Some types of mutations occur so rarely that they are almost impossible to detect, whereas others occur so frequently that they present difficulties for an experimenter trying to maintain a genetically stable stock culture. Furthermore, all organisms possess a variety of systems for DNA repair. Consequently, the observed mutation rate depends not only on the frequency of DNA alterations but also on the efficiency of DNA repair.

Spontaneous Mutation Frequencies

For most microorganisms, errors in DNA replication occur at a frequency of 10^{-6} to 10^{-7} per kilobase pair during a single round of replication. A typical gene has about 1,000 base pairs. Therefore, the frequency of a mutation *in a given gene* is also in the range of 10^{-6} to 10^{-7} per generation. For instance, in a bacterial culture having 10^8 cells/ml, there are likely to be a number of different mutants for any given gene in each milliliter of culture. Higher organisms with very large genomes tend to have replication error rates about tenfold lower than typical bacteria, whereas DNA viruses, especially those with very small genomes, may have error rates 100-fold to 1,000-fold higher than those of cellular organisms. RNA viruses have even higher error rates.

Single-base errors during DNA replication are more likely to lead to missense mutations than to nonsense mutations because most single-base substitutions yield codons that encode other amino acids (∞ Table 7.4). The next most frequent type of codon change caused by a single-base change leads to a silent mutation. This is because for the most part alternate codons for a given amino acid differ from each other by a single-base change in the "silent" third position (∞ Table 7.4). A given codon can be changed to any of 27 other codons by a single-base substitution, and on average, about two of these will be silent mutations, about one a nonsense mutation, and the rest will be missense mutations. There are also some DNA sequences, typically areas containing short repeats, that are hot spots for mutations because the error frequency of DNA polymerase is relatively high there. The error rate at hot spots is affected by the base sequence in the vicinity.

Unless a mutation can be selected for, its experimental detection is difficult, and much of the skill of the microbial geneticist involves increasing the efficiency of mutation detection. As we see in the next section, it is possible to significantly increase the mutation rate by the use of mutagenic treatments. In addition, the mutation rate may change in certain situations, such as under high-stress conditions.

Mutations in RNA Genomes

Whereas all cells have DNA as their genetic material, some viruses have RNA genomes (∞ Section 10.1). These genomes can also undergo mutation. Interestingly, the mutation rate in RNA genomes is about 1,000-fold higher than in DNA genomes. Why should this be so?

Some RNA polymerases have proofreading activities like those of DNA polymerases (∞ Section 7.8), thus limiting the total number of polymerase errors. However, although there are several repair systems for DNA that can correct changes before they become fixed in the genome as mutations (Section 11.7), comparable RNA repair mechanisms do not exist. This leads to heightened mutation rates for RNA genomes. This high mutation rate in RNA viruses has dramatic consequences. For example, the RNA genomes of viruses that cause disease can mutate very rapidly, presenting a constantly changing and evolving population of viruses. Such changes are one of many challenges to human medicine posed by the AIDS virus, HIV, an RNA virus with a notorious ability to undergo genetic changes that affect its virulence (∞ Section 19.15).

11.6 MiniReview

Different types of mutations occur at different frequencies. For a typical bacterium, mutation rates of 10^{-6} to 10^{-9} per kilobase pair are generally seen. Although RNA and DNA polymerases make errors at about the same rate, RNA genomes typically accumulate mutations at much higher frequencies than DNA genomes.

- Which class of mutation, missense or nonsense, is more common, and why?
- Why are RNA viruses genetically highly variable?

11.7 Mutagenesis

The spontaneous rate of mutation is very low, but there are a variety of chemical, physical, and biological agents that can increase the mutation rate and are therefore said to induce mutations. These agents are called **mutagens**. We discuss some of the major categories of mutagens and their activities here.

Chemical Mutagens

An overview of some of the major chemical mutagens and their modes of action is given in **Table 11.3**. Several classes of chemical mutagens exist. One class is the *nucleotide base analogs,* molecules that resemble the purine and pyrimidine bases of DNA in structure yet display faulty pairing properties (**Figure 11.8**). If one of these base analogs is incorporated into DNA in place of the natural base, the DNA may replicate normally most of the time. However, DNA replication errors occur at higher frequencies at these sites, due to incorrect base pairing. The result is the incorporation of a wrong base into the new strand of DNA and thus introduction of a mutation. During subsequent segregation of this strand in cell division, the mutation is revealed.

Other chemical mutagens induce *chemical modifications* in one base or another, resulting in faulty base pairing or related changes (Table 11.3). For example, alkylating agents (chemicals that react with amino, carboxyl, and hydroxyl groups in proteins and nucleic acids, substituting them with

Table 11.3 Chemical and physical mutagens and their modes of action

Agent	*Action*	*Result*
Base analogs		
5-Bromouracil	Incorporated like T; occasional faulty pairing with G	AT → GC and occasionally GC → AT
2-Aminopurine	Incorporated like A; faulty pairing with C	AT → GC and occasionally GC → AT
Chemicals reacting with DNA		
Nitrous acid (HNO_2)	Deaminates A and C	AT → GC and GC → AT
Hydroxylamine (NH_2OH)	Reacts with C	GC → AT
Alkylating agents		
Monofunctional (for example, ethyl methane sulfonate)	Puts methyl on G; faulty pairing with T	GC → AT
Bifunctional (for example, mitomycin, nitrogen mustards, nitrosoguanidine)	Cross-links DNA strands; faulty region excised by DNase	Both point mutations and deletions
Intercalating dyes		
Acridines, ethidium bromide	Inserts between two base pairs	Microinsertions and microdeletions
Radiation		
Ultraviolet	Pyrimidine dimer formation	Repair may lead to error or deletion
Ionizing radiation (for example, X-rays)	Free-radical attack on DNA, breaking chain	Repair may lead to error or deletion

alkyl groups) such as nitrosoguanidine, are powerful mutagens and generally induce mutations at higher frequency than base analogs. Unlike base analogs, which have an effect only when incorporated during DNA replication, alkylating agents are able to introduce changes even in nonreplicating DNA. Both base analogs and alkylating agents tend to induce base-pair substitutions (Section 11.5).

Another group of chemical mutagens, the acridines, are planar molecules that function as *intercalating agents*. These mutagens become inserted between two DNA base pairs and in the process push them apart. During replication, this abnormal conformation can lead to insertions or deletions in acridine-containing DNA. Thus, acridines typically induce frameshift mutations (Section 11.5). Ethidium bromide, which is often used to detect DNA in electrophoresis, is also an intercalating agent and therefore a mutagen.

Radiation

Several forms of radiation are highly mutagenic. We can divide mutagenic electromagnetic radiation into two main categories, nonionizing and ionizing (**Figure 11.9**). Although both kinds of radiation are used in microbial genetics to generate mutations, nonionizing radiation such as ultraviolet (UV) radiation has the widest use.

The purine and pyrimidine bases of nucleic acids absorb UV radiation strongly, and the absorption maximum for DNA and RNA is at 260 nm (Figure 7.7). Killing of cells by UV radiation is due primarily to its effect on DNA. Although several effects are known, one well-established effect is the production of pyrimidine dimers, in which two adjacent pyrimidine bases (cytosine or thymine) on the same strand of DNA become covalently bonded to one another. This results either in impeding DNA polymerase or in a greatly increased probability of DNA polymerase misreading the sequence at this point.

The UV radiation source most commonly used for mutagenesis is the germicidal lamp, which emits UV radiation in the 260-nm region. A dose of UV radiation is used that kills about 50–90% of the cell population, and mutants are then selected or screened for among the survivors. If much higher doses of radiation are used, the number of viable cells is too low. If lower

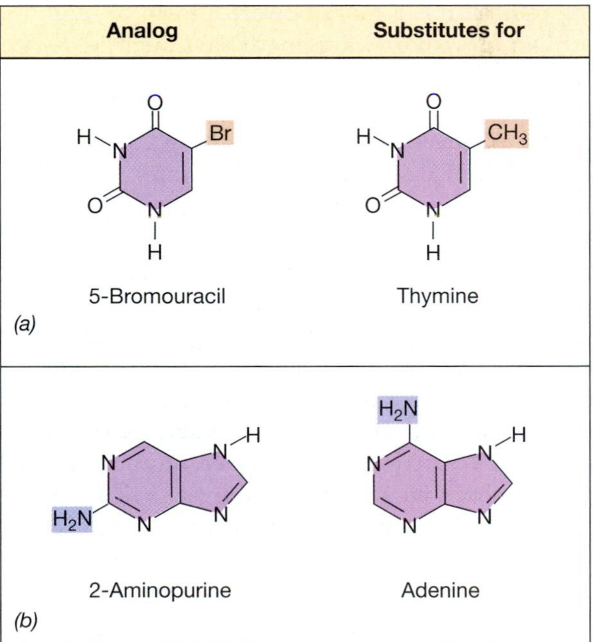

Figure 11.8 Nucleotide base analogs. Structure of two common nucleotide base analogs used to induce mutations and the normal nucleic acid bases for which they substitute. *(a)* 5-Bromouracil can base-pair with guanine, causing AT to GC substitutions. *(b)* 2-Aminopurine can base-pair with cytosine, causing AT to GC substitutions.

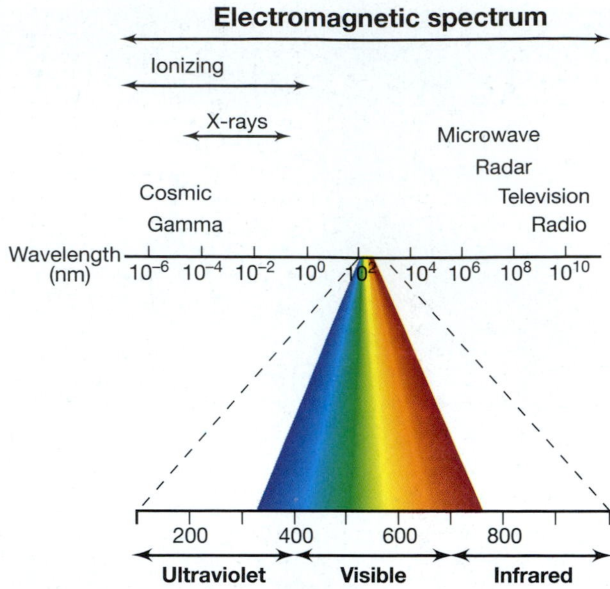

Figure 11.9 Wavelengths of radiation. Note that ultraviolet radiation consists of wavelengths just shorter than visible light. For any electromagnetic radiation, the shorter the wavelength, the higher the energy. DNA absorbs strongly at 260 nm.

doses are used, damage to DNA is insufficient to make recovery of the desired mutants possible. When used at the correct dose, UV radiation is a very convenient tool for isolating mutants and avoids the necessity of handling toxic chemicals.

Ionizing Radiation

Ionizing radiation is a more powerful form of radiation than UV and includes short-wavelength rays such as X-rays, cosmic rays, and gamma rays (Figure 11.9). These rays cause water and other substances to ionize, and mutagenic effects are brought about indirectly through this ionization. Among the potent chemical species formed by ionizing radiation are chemical free radicals, the most important being the hydroxyl radical, OH• (∞ Section 6.18).

Free radicals react with and damage macromolecules in the cell, of which the most important is DNA. At low doses of ionizing radiation only a few "hits" on DNA occur, but at higher doses, multiple hits occur, leading to the death of the cell. In contrast to UV radiation, ionizing radiation penetrates readily through glass and other materials. Therefore, ionizing radiation is used frequently to induce mutations in animals and plants (where its penetrating power makes it possible to reach the gamete-producing cells of these organisms). However, because ionizing radiation is more dangerous to use and is less readily available than UV, it finds less use in microbial genetics.

DNA Repair Systems

Recall that a mutation is a heritable change in the genetic material. Therefore, if damaged DNA can be corrected before the cell divides, no mutation will occur. Most cells have a variety of different DNA repair processes to correct mistakes or repair damage. Most of these DNA repair systems are virtually error-free. However, some are error-prone and the repair process itself introduces the mutation. DNA repair processes may be grouped into three categories: direct reversal, repair of single-strand damage, and repair of double-strand damage.

Direct reversal applies to bases that have been chemically altered but whose identity is still recognizable. No base pairing (that is, no template strand) is needed. For example, some alkylated bases are repaired by direct chemical removal of the alkyl group. Another direct repair system is photoreactivation, which cleaves pyrimidine dimers generated by UV radiation. The enzyme photolyase absorbs blue light and uses the energy to drive the cleavage reaction.

Several systems exist that repair single-strand DNA damage. In these cases, the damaged DNA is removed from only one strand. Then the opposite (undamaged) strand is used as a template for replacing the missing nucleotides. In base excision repair, a single damaged base is removed and replaced. In nucleotide excision repair and mismatch repair, a short stretch of single-stranded DNA containing the damage is removed and replaced. Double-strand damage, including both cross-strand links and double-stranded breaks, is especially dangerous. These lesions are repaired by recombinational mechanisms and may require error-prone repair.

Mutations That Arise from DNA Repair: The SOS System

Some types of DNA damage, especially large-scale damage from highly mutagenic chemicals or large doses of radiation, may interfere with replication. If this cannot be successfully repaired by the error-free repair systems, then the cell must use a second type of repair system, but one more error-prone. This allows replication to proceed and the cell to divide but does so at the cost of introducing mutations itself. This mechanism, called the *SOS regulatory system* is activated by some types of DNA damage and initiates a number of DNA repair processes. However, in the SOS system, DNA repair can occur without a template, that is, without base pairing; expectedly, this results in many errors and hence many mutations.

The SOS system is a **regulon**, that is, a set of related genes that are coordinately regulated although they are transcribed separately. The SOS system is regulated by two proteins, LexA and RecA. LexA is a repressor protein that normally prevents expression of the SOS regulon. The RecA protein, which normally functions in genetic recombination (Section 11.9), is activated by the presence of DNA damage (Figure 11.10). The activated form of RecA stimulates LexA to inactivate itself by self-cleavage. This leads to derepression of the SOS system and results in the coordinate expression of a number of proteins that take part in DNA repair. Because some of the DNA repair mechanisms of the SOS system are inherently error-prone, many mutations arise. Once the DNA damage has been repaired, the SOS regulon is repressed and further mutagenesis ceases.

When DNA replication stalls because of serious DNA damage, the cell replaces the normal DNA polymerase with special repair polymerases that can move past DNA damage—a process known as translesion synthesis. Even if no template is available to allow insertion of the correct bases, it is less dangerous to fill the gap than leave broken or damaged DNA in place. Consequently, translesion synthesis generates many errors. In *E. coli,* in which the process of mutagenesis has been studied in great detail, the two error-prone repair polymerases are DNA polymerase V, an enzyme encoded by the *umuCD* genes (Figure 11.10), and DNA polymerase IV, encoded by *dinB*. Both are induced as part of the SOS response.

Changes in Mutation Rate

High fidelity (low error frequency) in DNA replication is essential if organisms are to remain genetically stable. On the other hand, perfect fidelity is counterproductive because it would prevent evolution. Therefore, a mutation rate has evolved in cells that is very low yet detectable. This allows organisms to balance the need for genetic stability with that for evolutionary improvement.

The fact that organisms as phylogenetically disparate as hyperthermophilic *Archaea* and *E. coli* have about the same mutation rate might make one believe that evolutionary pressure has selected organisms with the lowest possible mutation rates. However, this is not so. The mutation rate in an organism is subject to change. For example, mutants of some organisms that are hyperaccurate in DNA replication and repair have been selected in the laboratory. However, in these strains, the improved proofreading and repair mechanisms have a significant metabolic cost; thus, hyperaccurate mutants might well be at a disadvantage in the natural environment. On the other hand, some organisms seem to benefit from enhanced DNA repair systems that enable them to occupy particular niches in nature. A good example is the bacterium *Deinococcus radiodurans* (∞ Section 16.17). This organism is 20 times more resistant to UV radiation and 200 times more resistant to ionizing radiation than is *E. coli*. This resistance, dependent in part upon redundant DNA repair systems and on a mechanism for exporting damaged nucleotides, allows the organism to survive in environments in which other organisms cannot, such as near concentrated sources of radiation or on the surfaces of dust particles exposed to intense sunlight.

In contrast to hyperaccuracy, some organisms actually benefit from an increased mutation rate. DNA repair systems are themselves genetically encoded and thus subject to mutation. For example, the protein subunit of DNA polymerase III involved in proofreading (∞ Section 7.8) is encoded by the gene *dnaQ*. Certain mutations in *dnaQ* lead to mutants that are still viable but have an increased rate of mutation. These are known as hypermutable or **mutator strains**. Mutations leading to a mutator phenotype are known in several other DNA repair systems as well. The mutator phenotype is apparently selected for in complex and changing environments, because strains of bacteria with mutator phenotypes appear to be more abundant under these conditions. Presumably, whatever disadvantage an increased mutation rate may have in such environments is offset by the ability to generate greater numbers of useful mutations. These mutations ultimately increase evolutionary fitness of the population and make the organism more successful in its ecological niche.

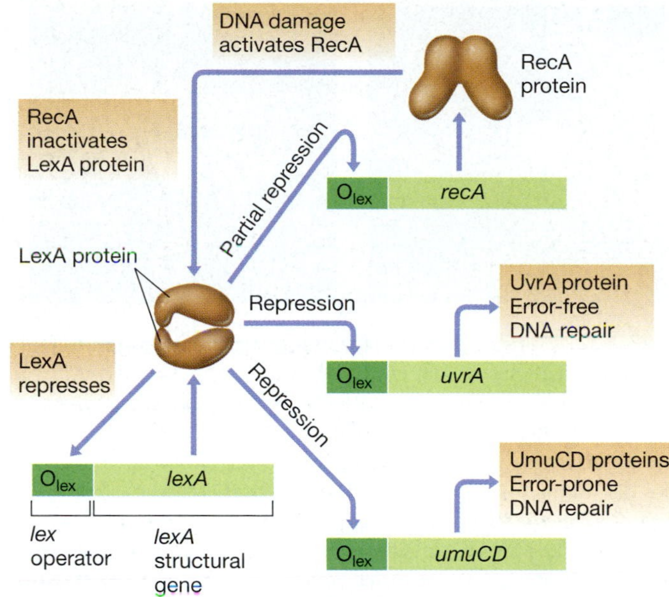

Figure 11.10 Mechanism of the SOS response. DNA damage activates RecA protein which in turn activates the protease activity of LexA. The LexA protein then cleaves itself. LexA protein normally represses the activities of the *recA* gene and the DNA repair genes *uvrA* and *umuCD* (the UmuCD proteins are part of DNA polymerase V). However, repression is not complete. Some RecA protein is produced even in the presence of LexA protein. With LexA inactivated, these genes become highly active.

As indicated earlier, a mutator phenotype may be induced in wild-type strains by stress situations. For instance, the SOS response induces error-prone repair. Therefore, when the SOS response is activated, the mutation rate increases. In some cases this is merely an inevitable by-product of DNA repair, but in other cases, the increased mutation rate may itself be of selective value to the organism for survival purposes.

11.7 MiniReview

Mutagens are chemical, physical, or biological agents that increase the mutation rate. Mutagens can alter DNA in many different ways. However, alterations in DNA are not mutations unless they are inherited. Some DNA damage can lead to cell death if not repaired, and both error-prone and high-fidelity DNA repair systems exist.

- How do mutagens work?
- Why might a mutator phenotype be successful in an environment experiencing rapid changes?

11.8 Mutagenesis and Carcinogenesis: The Ames Test

The Ames test makes practical use of bacterial mutations for detecting potentially hazardous chemicals in the environment.

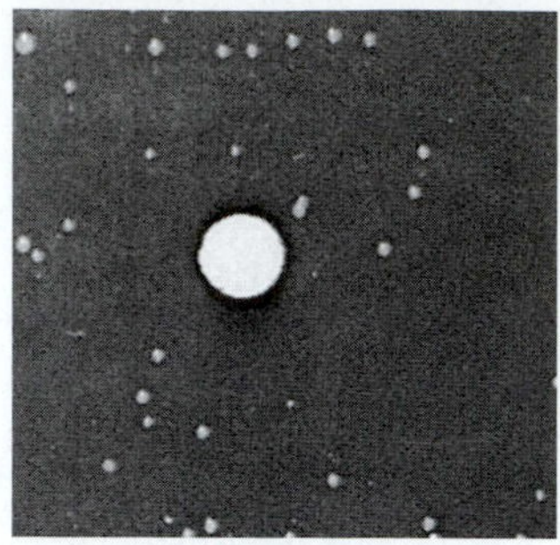

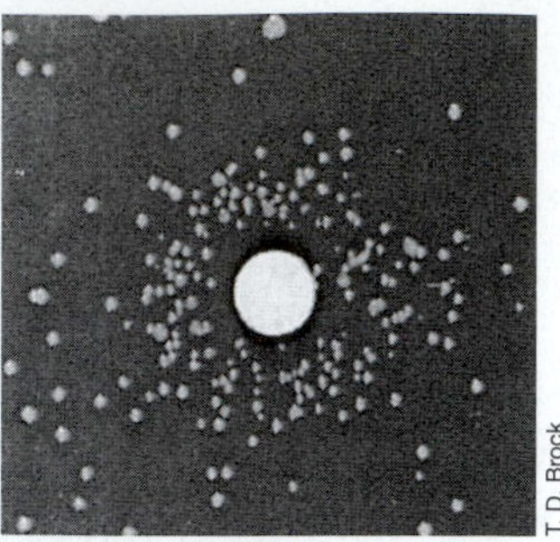

T. D. Brock

Figure 11.11 The Ames test for assessing the mutagenicity of a chemical. Two plates were inoculated with a culture of a histidine-requiring mutant of *Salmonella enterica*. The medium does not contain histidine, so only cells that revert back to wild type can grow. Spontaneous revertants appear on both plates, but the chemical on the filter paper disc in the test plate (right) has caused an increase in the mutation rate, as shown by the large number of colonies surrounding the disc. Revertants are not seen very close to the test disc because the concentration of mutagen is lethally high there. The plate on the left was the negative control; its filter paper disc had only water added.

Because selectable mutants can be detected in large populations of bacteria with very high sensitivity, bacteria can be used to screen chemicals for potential mutagenicity. This is relevant because many mutagenic chemicals are also carcinogenic, capable of causing cancer in humans or other animals.

The variety of chemicals, both natural and artificial, that humans encounter through agricultural and industrial exposure is enormous. There is good evidence that some human cancers have environmental causes, most likely from various chemicals, making the detection of chemical carcinogens an important matter. It does not necessarily follow that because a compound is mutagenic it is also carcinogenic. The correlation, however, is high, and knowing that a compound is mutagenic to bacteria is a warning of possible danger. The development of bacterial tests for carcinogenic screening was carried out primarily by Bruce Ames and colleagues at the University of California in Berkeley and consequently, the mutagenicity test for carcinogens is known as the *Ames test* (**Figure 11.11**).

Protocol for an Ames Test

The standard way to test chemicals for mutagenesis is to look for an increase in the rate of back mutation (reversion) in auxotrophic strains of bacteria in the presence of the suspected mutagen. It is important that the auxotrophic strain carry a point mutation because the reversion rate in such a strain is measurable. Cells of such an auxotroph do not grow on a medium lacking the required nutrient (for example, an amino acid), and even very large populations of cells can be spread on the plate without formation of visible colonies. However, if back mutants (revertants) are present, those cells form colonies. Thus, if 10^8 cells are spread on the surface of a single plate, even as few as 10–20 revertants can be detected by the 10–20 colonies they form (Figure 11.11, left photo). However, if the reversion rate has been increased by the presence of a chemical mutagen, the number of revertant colonies is even greater. Histidine auxotrophs of *Salmonella enterica* (Figure 11.11) and tryptophan auxotrophs of *E. coli* have been the major tools of the Ames test.

Two additional elements have been introduced in the Ames test to make it much more powerful. The first of these is to use test strains that almost exclusively use error-prone pathways to repair DNA damage; normal repair mechanisms are thus thwarted (Section 11.7). The second important element in the Ames test is the addition of liver enzyme preparations to convert the chemicals to be tested into their active mutagenic (and potentially carcinogenic) forms. It has been well established that many carcinogens are not directly carcinogenic or mutagenic themselves but undergo modifications in the human body that convert them into active substances. These changes take place primarily in the liver, where enzymes called mixed-function oxygenases, whose normal function is detoxification, generate activated forms of the compounds that are highly reactive (and thus mutagenic) toward DNA.

In the Ames test, a preparation of enzymes from rat liver is first used to activate the test compound. The activated complex is then soaked into a filter-paper disk, which is placed in the center of a plate on which the proper bacterial strain has been overlaid. After overnight incubation, the mutagenicity of the compound can be detected by looking for a halo of back mutations in the area around the paper disk (Figure 11.11). It is always necessary to carry out this test with several different concentrations of the compound and with appropriate positive (known mutagens) and negative (no mutagen) controls, because compounds vary in their mutagenic activity and may be lethal at higher levels. A wide variety of chemicals have been subjected to the Ames test, and it has become one of the most useful screens for determining the potential carcinogenicity of a compound.

11.8 MiniReview

The Ames test employs a sensitive bacterial assay system for detecting chemical mutagens in the environment.

- Why does the Ames test measure the rate of back mutation rather than the rate of forward mutation?
- Of what significance is the detection of mutagens to the prevention of cancer?

III GENETIC EXCHANGE IN PROKARYOTES

For genetic analyses, the microbial geneticist must cross strains of an organism that have different genotypes (and phenotypes) and look for recombinants. Three mechanisms of genetic exchange are known in prokaryotes: (1) transformation, in which free DNA released from one cell is taken up by another (Section 11.10); (2) transduction, in which DNA transfer is mediated by a virus (Section 11.11); and (3) conjugation, in which DNA transfer involves cell-to-cell contact and a conjugative plasmid in the donor cell (Sections 11.12 and 11.13). These processes are contrasted in **Figure 11.12**.

Before discussing the mechanisms of transfer further, we must consider the fate of the transferred DNA. Whether it was transferred by transformation, transduction, or conjugation, the incoming DNA faces three possible fates: It may be degraded by restriction enzymes; it may replicate by itself (but only if it possesses its own origin of replication such as a plasmid or phage genome); or it may recombine with the host chromosome.

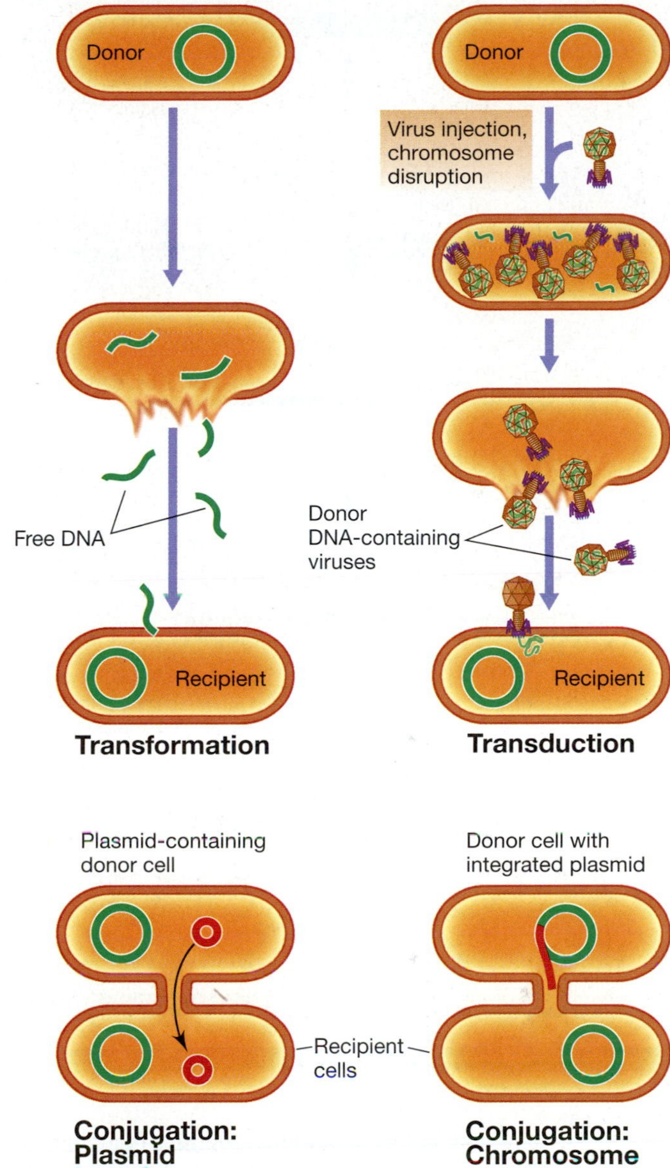

Figure 11.12 Processes by which DNA is transferred from donor to recipient bacterial cell. Just the initial steps in transfer are shown. For details of how the DNA is integrated into the recipient, see Figures 11.13, 11.16, 11.17, and 11.21.

11.9 Genetic Recombination

Recombination is the physical exchange of DNA between genetic elements. In this section we focus on *homologous* recombination, a process that results in genetic exchange between homologous DNA sequences from two different sources. Homologous DNA sequences are those that have nearly the same sequence; therefore, bases can pair over an extended length of the two DNA molecules. This type of recombination is involved in the process referred to as "crossing over" in classical genetics.

Molecular Events in Homologous Recombination

The RecA protein, previously mentioned in regard to the error-prone SOS repair system (Section 11.7), is the key to homologous recombination. RecA is essential in nearly every homologous recombination pathway. RecA-like proteins have been identified in all prokaryotes examined, including the *Archaea,* as well as in yeast and in higher *Eukarya*.

A molecular mechanism for homologous recombination is shown in **Figure 11.13**. An endonuclease begins the process by nicking one strand of one of the DNA molecules. This nicked strand must be displaced from the other strand by proteins having helicase activity (∞ Section 7.6). In some pathways specialized enzymes, such as the RecBCD enzyme of *E. coli,* have both endonuclease and helicase activities. Single-stranded binding protein (∞ Section 7.6) then binds to the resulting single-stranded segment. Next, the RecA protein binds to the single-stranded region, forming a complex that facilitates annealing with the complementary sequence in the second DNA duplex, simultaneously displacing the resident strand (Figure 11.13). This process is called *strand invasion* and leads to the pairing of DNA molecules over long stretches. The exchange of strands leads to the formation of recombination intermediates containing extensive **heteroduplex** regions, where each strand has originated from a different chromosome. These structures are called Holliday junctions (after Robin Holliday who proposed this model in 1964) and can migrate along the DNA; this migration is energized by a complex of several other proteins. Finally, the linked molecules are separated or "resolved" by resolvases that cut and rejoin the second (previously unbroken) strands. In *E. coli,* the RecG and RuvC proteins both function as resolvases, and their activity generates two recombined DNA molecules. Depending on the orientation of the Holliday junction during resolution, two types of products, patches or splices, are formed that differ in the conformation of the heteroduplex regions remaining after resolution (Figure 11.13).

Effect of Homologous Recombination on Genotype

For homologous recombination to generate new genotypes, it is essential that the two homologous sequences be related but genetically distinct. This is obviously the case in a diploid eukaryotic cell, which has two sets of chromosomes, one from each parent. In prokaryotes, genetically distinct but

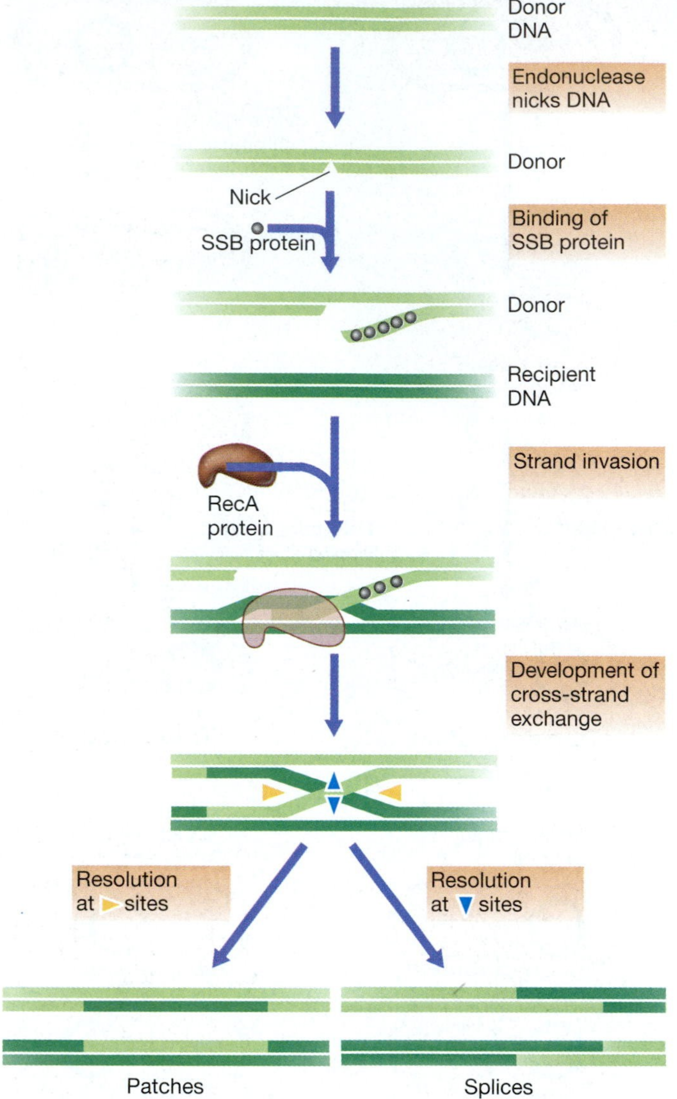

Figure 11.13 A simplified version of homologous recombination. Homologous DNA molecules pair and exchange DNA segments. The mechanism involves breakage and reunion of paired segments. Two of the proteins involved, single-stranded binding (SSB) protein and the RecA protein, are shown. The other proteins involved are not shown. The diagram is not to scale: Pairing may occur over hundreds or thousands of bases. Resolution occurs by cutting and rejoining the cross-linked DNA molecules. Note that there are two possible outcomes, patches or splices, depending on where strands are cut during the resolution process.

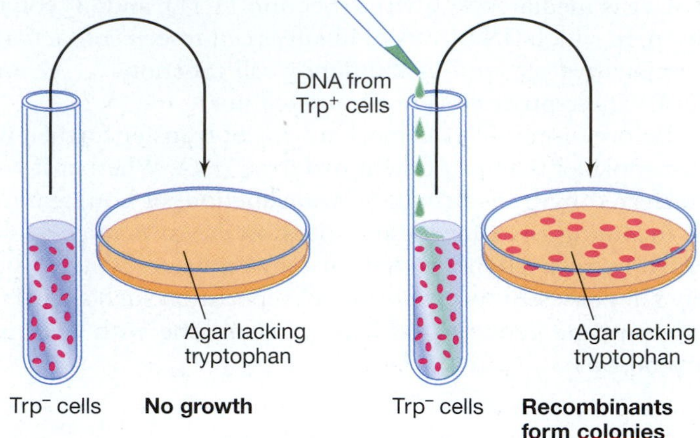

Figure 11.14 Using a selective medium to detect rare genetic recombinants. On the selective medium only the rare recombinants form colonies even though a very large population of bacteria was plated. Procedures such as this, which offer high resolution for genetic analyses, can ordinarily be used only with microorganisms. The type of genetic exchange being illustrated is transformation.

homologous DNA molecules are brought together in different ways, but the process of genetic recombination is no less important. Genetic recombination in prokaryotes occurs after fragments of homologous DNA from a donor chromosome are transferred to a recipient cell by transformation, transduction, or conjugation. It is only after the transfer event, when the DNA fragment from the host is in the recipient cell, that homologous recombination occurs. In prokaryotes only a chromosomal fragment is transferred; therefore, if recombination does not occur, the DNA fragment will be lost because it cannot replicate independently. Thus, in prokaryotes, transfer is just the first step in generating recombinant organisms.

Detection of Recombination

To detect physical exchange of DNA segments, the cells resulting from recombination must be phenotypically different from both parents. Genetic crosses usually depend on using recipient strains that lack some selectable character that the recombinants will gain. For instance, the recipient may be unable to grow on a particular medium, and genetic recombinants are selected that can. Various kinds of selectable markers, such as drug resistance and nutritional requirements, were discussed in Section 11.4.

The exceedingly great sensitivity of the selection process allows even a few recombinant cells to be detected in a large population of nonrecombinant cells (**Figure 11.14**). The only requirement for effective detection of recombination is that the reverse mutation rate for the selected characteristic should be low, because revertants will also form colonies. This problem can often be overcome by using double mutants—strains that carry two different mutations—in genetic crosses because it is very unlikely that two back mutations will occur in the same cell. Alternatively, frameshift mutants can be used, because their reversion rates are typically extremely low.

Much of the skill of the bacterial geneticist lies in the choice of proper mutants and selective media for efficient detection of genetic recombination. Because selection is so powerful and because crosses can be made using billions of individual cells, recombinational analysis following gene transfer is an important tool for the microbial geneticist (Sections 11.10–11.12).

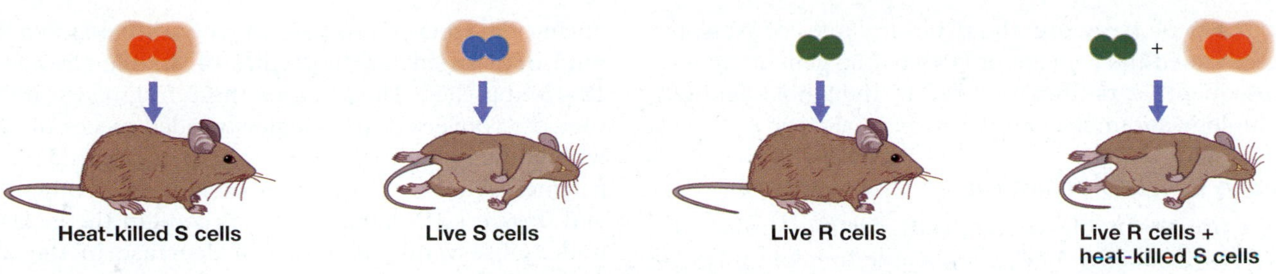

Figure 11.15 Griffith's experiments with pneumococcus. Live smooth (S) cells contain a capsule and kill mice because immune cells cannot kill the encapsulated bacteria; the cells proliferate in the lung and cause a fatal pneumonia. Rough (R) cells have no capsule and are not pathogenic. But a combination of live R and dead S cells kill mice, and live S cells can be isolated from the animals. DNA carrying genes for capsule production is released from dead S cells and taken up by R cells, thus transforming them into S cells.

UNIT 2

11.9 MiniReview

Homologous recombination arises when closely related DNA sequences from two distinct genetic elements are combined together in a single element. Recombination is an important evolutionary process, and cells have specific mechanisms for ensuring that recombination takes place.

- Which protein, found in all prokaryotes, facilitates the pairing required for homologous recombination?
- In eukaryotes, recombination involves entire chromosomes, but this is not true in prokaryotes. Explain.

11.10 Transformation

Transformation is a genetic transfer process by which free DNA is incorporated into a recipient cell and brings about genetic change. Several prokaryotes are naturally transformable, including certain species of both gram-negative and gram-positive *Bacteria* and also some species of *Archaea* (Section 11.15). Because the DNA of prokaryotes is present in the cell as a large single molecule, when the cell is gently lysed, the DNA pours out. Because of their extreme length (1700 μm in *Bacillus subtilis,* for example), bacterial chromosomes break easily. Even after gentle extraction, the *B. subtilis* chromosome of 4.2 Mbp is converted to fragments about 10 kbp each. Because the DNA that corresponds to an average gene is about 1,000 nucleotides, each of the fragments of *B. subtilis* DNA therefore contains about ten genes. This is a typical transformable size. A single cell usually incorporates only one or a few DNA fragments, so only a small proportion of the genes of one cell can be transferred to another by a single transformation event.

Transformation in the History of Molecular Biology

Although genetic recombination in eukaryotes had been known for a long time, the discovery of genetic recombination in bacteria is more recent. However, the discovery of transformation was one of the key events in biology, as it led to experiments demonstrating that DNA was the genetic material (**Figure 11.15**). This discovery became a cornerstone of molecular biology and modern genetics, and so we review it here.

The British scientist Frederick Griffith obtained the first evidence of bacterial transformation in the late 1920s. Griffith was working with *Streptococcus pneumoniae* (pneumococcus), a bacterium that owes its ability to invade the body in part to the presence of a polysaccharide capsule (Section 4.9). Mutants can be isolated that lack this capsule and thus cannot cause infection. Such mutants are called *R strains* because their colonies appear rough on agar, in contrast to the smooth appearance of encapsulated strains, called *S strains*. A mouse infected with only a few cells of an S strain succumbs in a day or two to a massive pneumococcus infection. By contrast, even large numbers of R cells do not cause death when injected. Griffith showed that if heat-killed S cells were injected along with living R cells, the mouse developed a fatal infection and the bacteria isolated from the dead mouse were of the S type (Figure 11.15). Because the S cells isolated in such an experiment always had the capsule type of the heat-killed S cells, Griffith concluded that the R cells had been transformed into a new type. This process set the stage for the discovery of DNA.

Oswald T. Avery and his associates at the Rockefeller Institute in New York provided the molecular explanation for the transformation of pneumococcus in a series of studies carried out during the 1930s and 1940s. Avery and his coworkers showed that transformation could be carried out in the test tube instead of the mouse and that a cell-free extract of heat-killed cells could induce transformation. In a series of painstaking biochemical experiments, the active fraction was purified from cell-free extracts and was shown to be DNA. The transforming activity of purified DNA preparations was very high, and only a very small amount of material was necessary. Subsequently, others showed that transformation in pneumococcus not only applied to capsular characteristics but for other genetic characteristics as well, such as antibiotic resistance and sugar fermentation.

In 1953, James Watson and Francis Crick published their model of the structure of DNA, providing a theoretical framework for how DNA could serve as genetic material. Thus, three types of studies, the bacteriological ones of Griffith, the biochemical ones of Avery, and the structural ones of Watson and Crick, solidified the concept of DNA as the genetic material. In subsequent years, this work led to the whole field of molecular biology and molecular genetics.

Competence in Transformation

Even within transformable genera, only certain strains or species are transformable. A cell that is able to take up DNA and be transformed is said to be *competent,* and this capacity is genetically determined.

Competence in most naturally transformable bacteria is regulated, and special proteins play a role in the uptake and processing of DNA. These competence-specific proteins include a membrane-associated DNA-binding protein, a cell wall autolysin, and various nucleases. One pathway of natural competence in *B. subtilis*—an easily transformed species—is regulated by a quorum-sensing system (a regulatory system that responds to cell density; Section 9.6). Cells produce and excrete a small peptide during growth, and the accumulation of this peptide to high concentrations induces the cells to become competent. In *Bacillus,* roughly 20% of the cells in a culture become competent and stay that way for several hours. However, in *Streptococcus,* 100% of the cells can become competent, but only for a brief period during the growth cycle.

High-efficiency natural transformation is known in only a few *Bacteria*. For example, *Acinetobacter, Bacillus, Streptococcus, Haemophilus, Neisseria,* and *Thermus* are naturally competent and easily transformable. By contrast, many prokaryotes are poorly transformed, if at all, under natural conditions. *E. coli* and many other gram-negative bacteria fall into this category. However, if cells of *E. coli* are treated with high concentrations of calcium ions and then chilled for several minutes, they become adequately competent. Cells of *E. coli* treated in this manner take up double-stranded DNA, and therefore transformation of this organism by plasmid DNA is relatively efficient. This is important because getting DNA into *E. coli*—the workhorse of genetic engineering—is critical for biotechnology, as we will see in Chapter 26.

Electroporation is a physical technique that is used to get DNA into organisms that are difficult to transform, especially those with thick cell walls. In electroporation, cells are mixed with DNA and then exposed to brief high-voltage electrical pulses. This makes the cell envelope permeable and allows entry of the DNA. Electroporation is a quick process and works for most types of cells, including *E. coli,* most other *Bacteria,* some members of the *Archaea,* and even yeast and certain plant cells.

Uptake of DNA in Transformation

During natural transformation, competent bacteria reversibly bind DNA. Soon, however, the binding becomes irreversible. Competent cells bind much more DNA than do noncompetent cells—as much as 1,000 times more. As noted earlier, the sizes of the transforming fragments are much smaller than that of the whole genome, and the fragments are further degraded during the uptake process. In *S. pneumoniae* each cell can bind only about ten molecules of double-stranded DNA of 10–15 kbp each. However, as these fragments are taken up, they are converted into single-stranded pieces of about 8 kb, with the complementary strand being degraded. The DNA fragments in the mixture compete with each other for uptake, and if excess DNA that does not contain the genetic marker under observation is added, a decrease in the number of transformants occurs.

In preparations of transforming DNA, typically only about 1 out of 100–300 DNA fragments contains the genetic marker being studied. Thus, at high concentrations of DNA, the competition between DNA molecules results in saturation of the system, so even under the best of conditions it is impossible to transform all the cells in a population for a given marker. The maximum frequency of transformation that has so far been obtained is about 20% of the population; the usual values obtained being between 0.1% and 1.0%. But when recipient population sizes are very high, even this low frequency is easy to detect. The minimum concentration of DNA yielding detectable transformants is about 0.01 ng/ml, which is so low that it is chemically undetectable.

Interestingly, transformation in *Haemophilus influenzae* requires the DNA fragment to have a particular 11-bp sequence for irreversible binding and uptake to occur. This sequence is found at an unexpectedly high frequency in the *Haemophilus* genome. Evidence such as this, and the fact that certain bacteria have been shown to become competent in their natural environment, suggests that transformation is not a laboratory artifact but plays an important role in lateral gene transfer in nature. By promoting new combinations of genes, naturally transformable bacteria increase the diversity and fitness of the microbial community as a whole.

Integration of Transforming DNA

Transforming DNA is bound at the cell surface by a DNA-binding protein. Following this, either the entire double-stranded fragment is taken up, or a nuclease degrades one strand and the remaining strand is taken up, depending on the organism (**Figure 11.16**). After uptake, the DNA becomes attached to a competence-specific protein. This protects the DNA from nuclease attack until it reaches the chromosome, where the RecA protein takes over. The DNA is integrated into the genome of the recipient by recombination (Figures 11.16 and 11.13). If single-stranded DNA is integrated, a heteroduplex DNA is formed. During the next round of chromosomal replication, one parental and one recombinant DNA molecule are generated. On segregation at cell division, the recombinant molecule is present in the transformed cell, which is now genetically altered compared to its parent. The preceding only pertains to small pieces of linear DNA. Many naturally transformable *Bacteria* are transformed only poorly by plasmid DNA because the plasmid must remain double-stranded and circular in order to replicate.

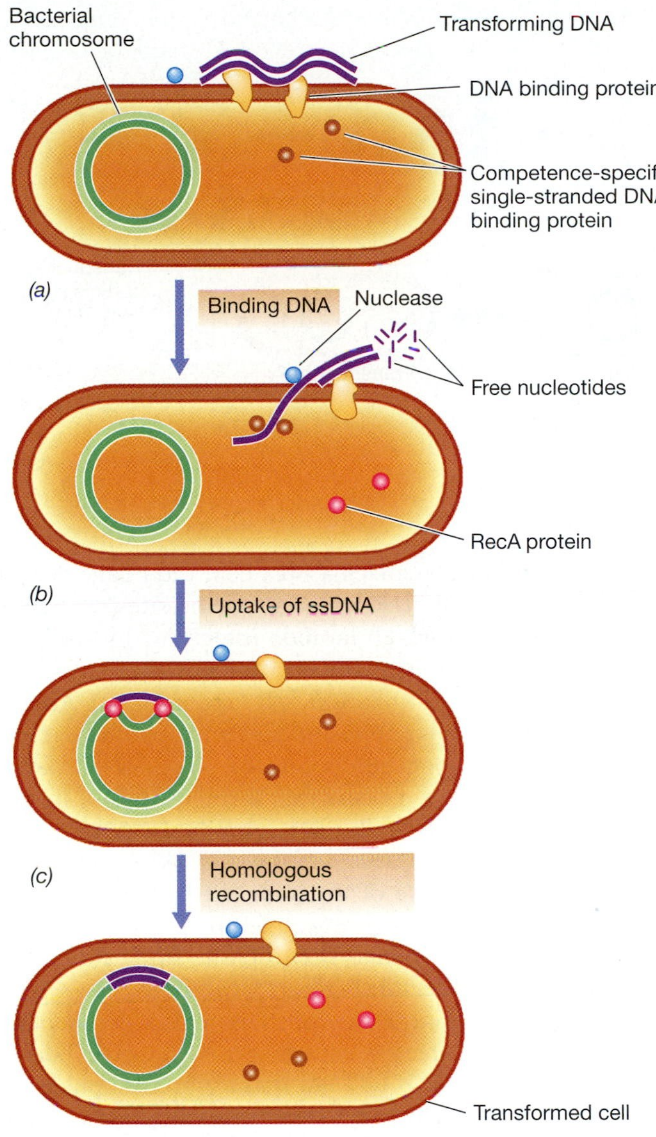

Figure 11.16 Mechanism of transformation in a gram-positive bacterium. *(a)* Binding of double-stranded DNA by a membrane-bound DNA-binding protein. *(b)* Passage of one of the two strands into the cell while nuclease activity degrades the other strand. *(c)* The single strand in the cell is bound by specific proteins, and recombination with homologous regions of the bacterial chromosome is mediated by RecA protein. *(d)* Transformed cell.

Transfection

Bacteria can be transformed with DNA extracted from a bacterial virus rather than from another bacterium. This process is called *transfection*. If the DNA is from a lytic bacteriophage, transfection leads to virus production and can be measured by the standard phage plaque assay (∞ Section 10.4). Transfection is useful for studying the mechanisms of transformation and recombination because the small size of phage genomes allows the isolation of a nearly homogeneous population of DNA molecules. By contrast, in conventional transformation the transforming DNA is typically a random assortment of chromosomal DNA fragments of various lengths, and this tends to complicate experiments designed to study the mechanism of transformation.

UNIT 2

11.10 MiniReview

Certain prokaryotes exhibit competence, a state in which cells are able to take up free DNA released by other bacteria. Incorporation of donor DNA into a recipient cell requires the activity of single-stranded binding protein, RecA protein, and several other enzymes. Only competent cells are transformable.

- The donor bacterial cell in a transformation is probably dead. Explain.
- Even in naturally transformable cells, competency is usually inducible. What does this mean?

11.11 Transduction

In **transduction**, a bacterial virus (bacteriophage) transfers DNA from one cell to another. Viruses can transfer host genes in two ways. In the first, called *generalized transduction,* DNA derived from virtually any portion of the host genome is packaged inside the mature virion in place of the virus genome. In the second, called *specialized transduction,* DNA from a specific region of the host chromosome is integrated directly into the virus genome—usually replacing some of the virus genes. This only occurs with certain temperate viruses (∞ Section 10.10). The transducing bacteriophage in both generalized and specialized transduction is usually noninfectious because bacterial genes have replaced all or some necessary viral genes.

In generalized transduction, the donor genes cannot replicate independently and are not part of a viral genome. Unless the donor genes recombine with the recipient bacterial chromosome, they will be lost. In specialized transduction, homologous recombination may also occur. However, since the donor bacterial DNA is actually a part of a temperate phage genome, it may be integrated into the host chromosome during lysogeny (∞ Section 10.10).

Transduction occurs in a variety of *Bacteria*, including the genera *Desulfovibrio, Escherichia, Pseudomonas, Rhodococcus, Rhodobacter, Salmonella, Staphylococcus,* and *Xanthobacter,* as well as *Methanothermobacter thermoautotrophicus*, a species of *Archaea*. Not all phages can transduce, and not all bacteria are transducible, but the phenomenon is sufficiently widespread that it likely plays an important role in gene transfer in nature.

Generalized Transduction

In generalized transduction, virtually any gene on the donor chromosome can be transferred to the recipient. Generalized transduction was first discovered and extensively studied in the bacterium *Salmonella enterica* with phage P22 and has also been studied with phage P1 in *E. coli*. An example of how

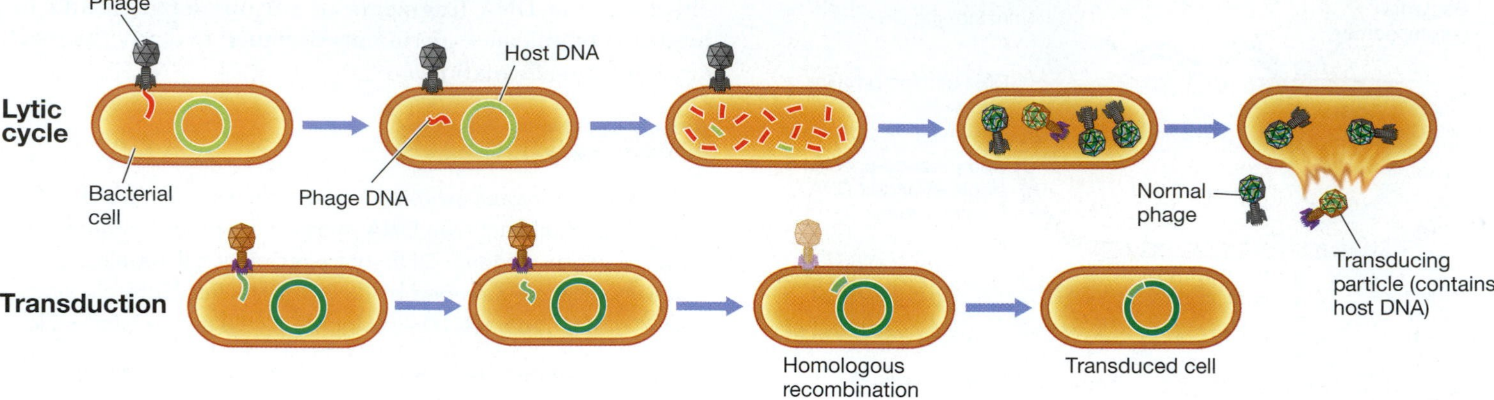

Figure 11.17 Generalized transduction. Note that "normal" virions contain phage genes, whereas a transducing particle contains host genes.

transducing particles are formed is given in **Figure 11.17**. When a bacterial cell is infected with a phage, the lytic cycle may be initiated. However, during the lytic infection, the enzymes responsible for packaging viral DNA into the bacteriophage sometimes package host DNA accidentally. The resulting virion is called a *transducing particle*. Because transducing particles cannot lead to a viral infection (they contain no viral DNA), they are said to be *defective*. On lysis of the cell, these particles are released along with normal virions (that is, those containing the virus genome). Consequently, the lysate contains a mixture of normal virions and transducing particles.

When this lysate is used to infect a population of recipient cells, most of the cells become infected with normal virus. However, a small proportion of the population receives transducing particles that inject the DNA they packaged from the previous host bacterium. Although this DNA cannot replicate, it can undergo genetic recombination with the DNA of the new host. Because only a small proportion of the particles in the lysate are defective, and each of these contains only a small fragment of donor DNA, the probability of a given transducing particle containing a particular gene is quite low. Typically, only about 1 cell in 10^6 to 10^8 is transduced for a given marker.

Phages that form transducing particles can be either temperate or virulent, the main requirements being that they have a DNA-packaging mechanism that accepts host DNA and that DNA packaging occurs before the host genome is completely degraded. Transduction is most likely when the ratio of input phage to recipient bacteria is low so that cells are infected with only a single phage particle; with multiple infection, the cell is likely to be killed by the normal virions in the lysate.

Phage Lambda and Specialized Transduction

Generalized transduction allows the transfer of any gene from one bacterium to another, but at a low frequency. In contrast, specialized transduction allows extremely efficient transfer but is selective and transfers only a small region of the bacterial chromosome. In the first case of specialized transduction to be discovered, the galactose genes were transduced by the temperate phage lambda of *E. coli*.

When lambda lysogenizes a host cell, the phage genome is integrated into the host DNA at a specific site (∞ Section 10.10). The region in which lambda integrates in the *E. coli* chromosome is next to the cluster of genes that encode the enzymes for galactose utilization (∞ Figure 10.18). After insertion, viral DNA replication is under control of the bacterial host chromosome. Upon induction, the viral DNA separates from the host DNA by a process that is the reverse of integration (**Figure 11.18**). Usually, the lambda DNA is excised precisely as a unit, but occasionally, the phage genome is excised incorrectly. Some of the adjacent bacterial genes to one side of the prophage (for example, the galactose operon) are excised along with phage DNA. At the same time, some phage genes are left behind (Figure 11.18).

One type of altered phage particle, called *lambda dgal* (λ*dgal; dgal* means "defective galactose"), is defective because of the phage genes lost. It will not make a mature phage in a subsequent infection. However, a viable lambda virion known as a helper phage can provide those functions missing in the defective particle. When cells are coinfected with λ*dgal* and the helper phage, the culture lysate contains a few λ*dgal* particles mixed in with a large number of normal lambda virions. When a galactose-negative (Gal^-) bacterial culture is infected with such a lysate and Gal^+ transductants selected, many are double lysogens carrying both lambda and λ*dgal*. When such a double lysogen is induced, the lysate contains large numbers of λ*dgal* virions and can transduce at high efficiency, although only for the restricted group of *gal* genes.

For a lambda virion to be viable, there is a limit to the amount of phage DNA that can be replaced with host DNA. Sufficient phage DNA must be retained to encode the phage protein coat and other phage proteins needed for lysis and lysogeny. However, if a helper phage is used together with a defective phage in a mixed infection, then far fewer phage-specific genes are needed in the defective phage. Only the *att* (attachment) region, the *cos* site (cohesive ends, for packaging), and the replication origin of the lambda genome are absolutely needed for production of a transducing particle when a helper phage is used. By deleting the normal chromosomal *att* site and forcing lambda to integrate at other locations, specialized transducing phages covering many specific regions of the *E. coli* genome have been isolated. In addition, lambda-transducing phages can be constructed by the techniques of genetic engineering to contain genes from any organism (∞ Section 12.14).

Phage Conversion

Alteration of the phenotype of a host cell by lysogenization is called *phage conversion*. When a normal (that is, nondefective) temperate phage lysogenizes a cell and becomes a prophage, the cell becomes immune to further infection by the same type of phage. Such immunity may itself be regarded as a change in phenotype. However, other phenotypic alterations unrelated to phage immunity are often observed in lysogenized cells.

Two cases of phage conversion have been especially well studied. One involves a change in structure of a polysaccharide on the cell surface of *Salmonella anatum* on lysogenization with bacteriophage ε^{15}. The second involves the conversion of nontoxin-producing strains of *Corynebacterium diphtheriae* (the bacterium that causes the disease diphtheria) to toxin-producing (pathogenic) strains following lysogeny with phage β (∞ Section 34.3). In both cases, the genes responsible for the changes are an integral part of the phage genome and hence are automatically transferred upon infection by the phage and lysogenization.

Lysogeny probably carries a strong selective value for the host cell because it confers resistance to infection by viruses of the same type. Phage conversion may also be of considerable evolutionary significance because it results in efficient genetic alteration of host cells. Many bacteria isolated from nature are natural lysogens. It seems reasonable to conclude that lysogeny is common and may often be essential for survival of the host cells in nature.

11.11 MiniReview

Transduction is the transfer of host genes from one bacterium to another by a bacterial virus. In generalized transduction, defective virus particles randomly incorporate fragments of the cell's chromosomal DNA, but the transducing efficiency is low. In specialized transduction, the DNA of a temperate virus excises incorrectly and takes adjacent host genes along with it; the transducing efficiency here may be very high.

- What is the major difference between generalized transduction and transformation?
- In specialized transduction, the donor DNA can replicate inside the recipient cell without homologous recombination taking place, but this is not true in generalized transduction. Explain.

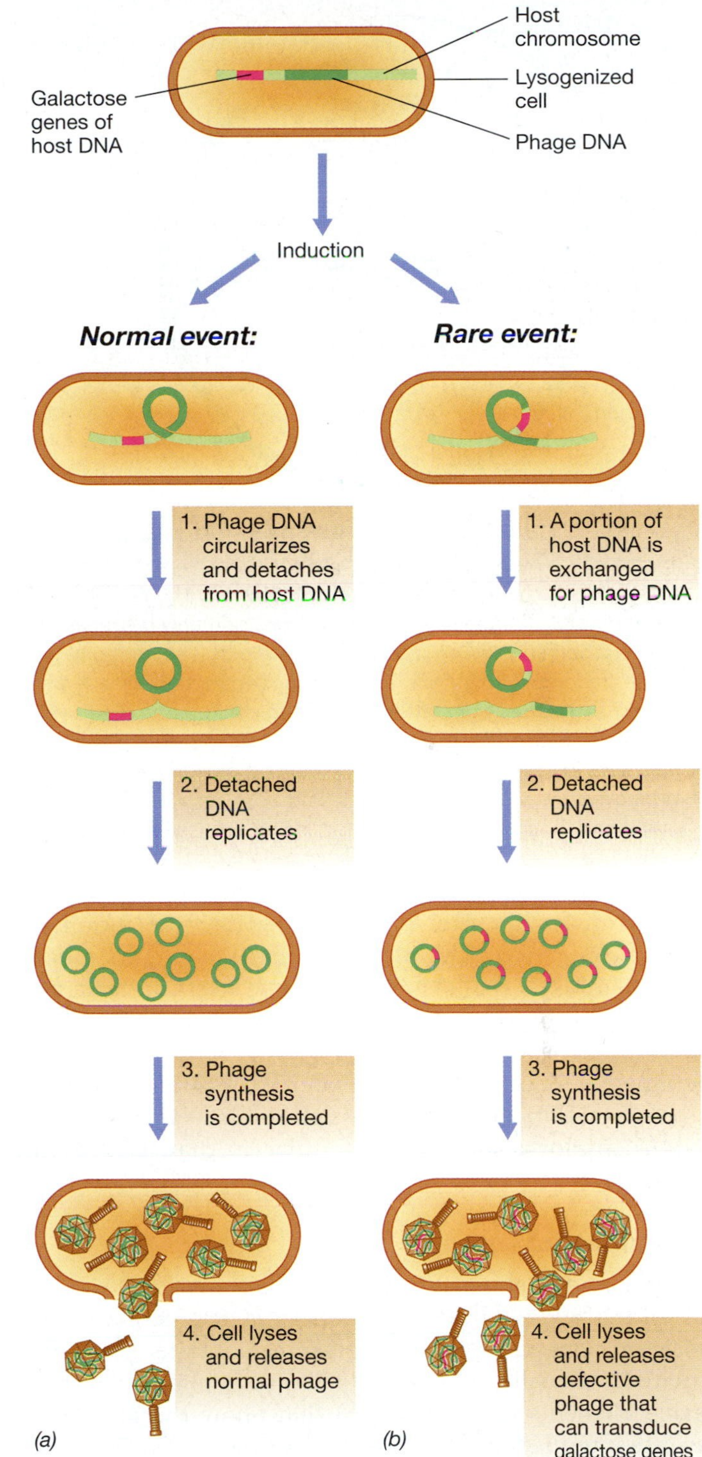

Figure 11.18 Specialized transduction. *(a)* Normal lytic events, and *(b)* the production of particles transducing the galactose genes in an *Escherichia coli* cell containing a lambda prophage.

11.12 Conjugation: Essential Features

Bacterial **conjugation** (mating) is a mechanism of genetic transfer that involves cell-to-cell contact. Conjugation is a plasmid-encoded mechanism. Conjugative plasmids use this mechanism to transfer copies of themselves to new host cells. Thus the process of conjugation involves a *donor* cell, which contains the conjugative plasmid, and a *recipient* cell, which does not. In addition, some genetic elements that

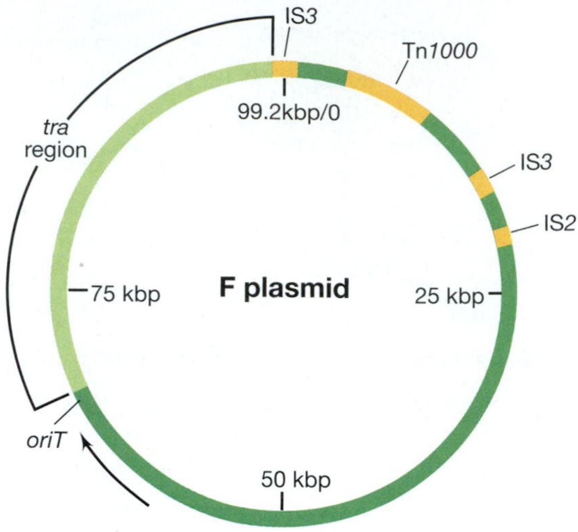

Figure 11.19 Genetic map of the F (fertility) plasmid of *Escherichia coli*. The numbers on the interior show the size of the plasmid in kilobase pairs (the exact size is 99,159 bp). The region in dark green at the bottom of the map contains genes primarily responsible for the replication and segregation of the F plasmid. The light green region, the *tra* region, contains the genes needed for conjugative transfer. The *oriT* site is the origin of transfer during conjugation. The arrow indicates the direction of transfer (the *tra* region is transferred last). The regions shown in yellow are insertion sequences. These may recombine with identical elements on the bacterial chromosome, which leads to integration and the formation of different Hfr strains (see Figure 11.25).

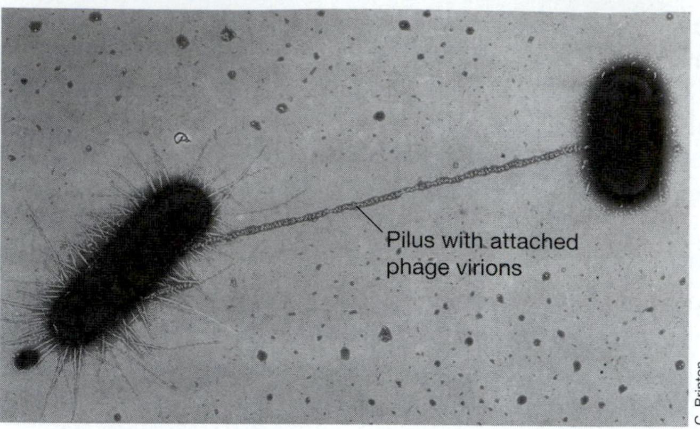

Figure 11.20 Formation of a mating pair. Direct contact between two conjugating bacteria is first made via a pilus. The cells are then drawn together to form a mating pair by retraction of the pilus, which is achieved by depolymerization. Certain small phages (F-specific bacteriophages; Section 19.1) use the sex pilus as receptor and can be seen here attached to the pilus.

cannot transfer themselves can sometimes be mobilized during conjugation. These other genetic elements can be other plasmids or the host chromosome itself. Indeed, conjugation was discovered because the F plasmid of *E. coli* can mobilize the host chromosome (see Figure 11.24). Mechanisms of conjugative transfer may differ depending on the plasmid involved, but most plasmids in gram-negative *Bacteria* employ a mechanism similar to that used by the F plasmid.

F Plasmid

The F plasmid (F stands for "fertility") is a circular DNA molecule of 99,159 bp. **Figure 11.19** shows a genetic map of the F plasmid. One region of the plasmid contains genes that regulate DNA replication. It also contains a number of transposable elements (Section 11.16) that allow the plasmid to integrate into the host chromosome. Moreover, the F plasmid has a large region of DNA, the *tra* region, containing genes that encode transfer functions. Many genes in the *tra* region are involved in mating pair formation, and most of these have to do with the synthesis of a surface structure, the sex pilus (Section 4.9). Only donor cells produce these pili. Different conjugative plasmids may have slightly different *tra* regions, and the pili may vary somewhat in structure. The F plasmid and its relatives encode F pili.

Pili allow specific pairing to take place between the donor and recipient cells. All conjugation in gram-negative *Bacteria* is thought to depend on cell pairing brought about by pili. The pilus makes specific contact with a receptor on the recipient cell and then is retracted by disassembling its subunits. This pulls the two cells together (**Figure 11.20**). Following this process, donor and recipient cells remain in contact by binding proteins located in the outer membrane of each cell. DNA is then transferred from donor to recipient cell through this conjugation junction.

Mechanism of DNA Transfer During Conjugation

DNA synthesis is necessary for DNA transfer by conjugation. This DNA is synthesized not by normal semiconservative replication (Section 7.6) but by **rolling circle replication**, a mechanism also used by some viruses (Section 10.10) and shown in **Figure 11.21**. DNA transfer is triggered by cell-to-cell contact, at which time one strand of the circular plasmid DNA is nicked and is transferred to the recipient. The nicking enzyme required to initiate the process, TraI, is encoded by the *tra* operon of the F plasmid. This protein also has helicase activity and thus also unwinds the strand to be transferred. As this transfer occurs, DNA synthesis by the rolling circle mechanism replaces the transferred strand in the donor, while a complementary DNA strand is being made in the recipient. Therefore, at the end of the process, both donor and recipient possess complete plasmids. For transfer of the F plasmid, if an F-containing donor cell, which is designated F^+, mates with a recipient cell lacking the plasmid, designated F^-, the result is two F^+ cells (Figure 11.21).

Transfer of plasmid DNA is efficient and rapid; under favorable conditions virtually every recipient cell that pairs with a donor acquires a plasmid. Transfer of the F plasmid, comprising approximately 100 kbp of DNA, takes about 5 minutes. If the plasmid genes can be expressed in the recipient, the recipient itself becomes a donor and can transfer the plasmid to other recipients. In this fashion, conjugative plasmids can spread rapidly among bacterial populations, behaving much

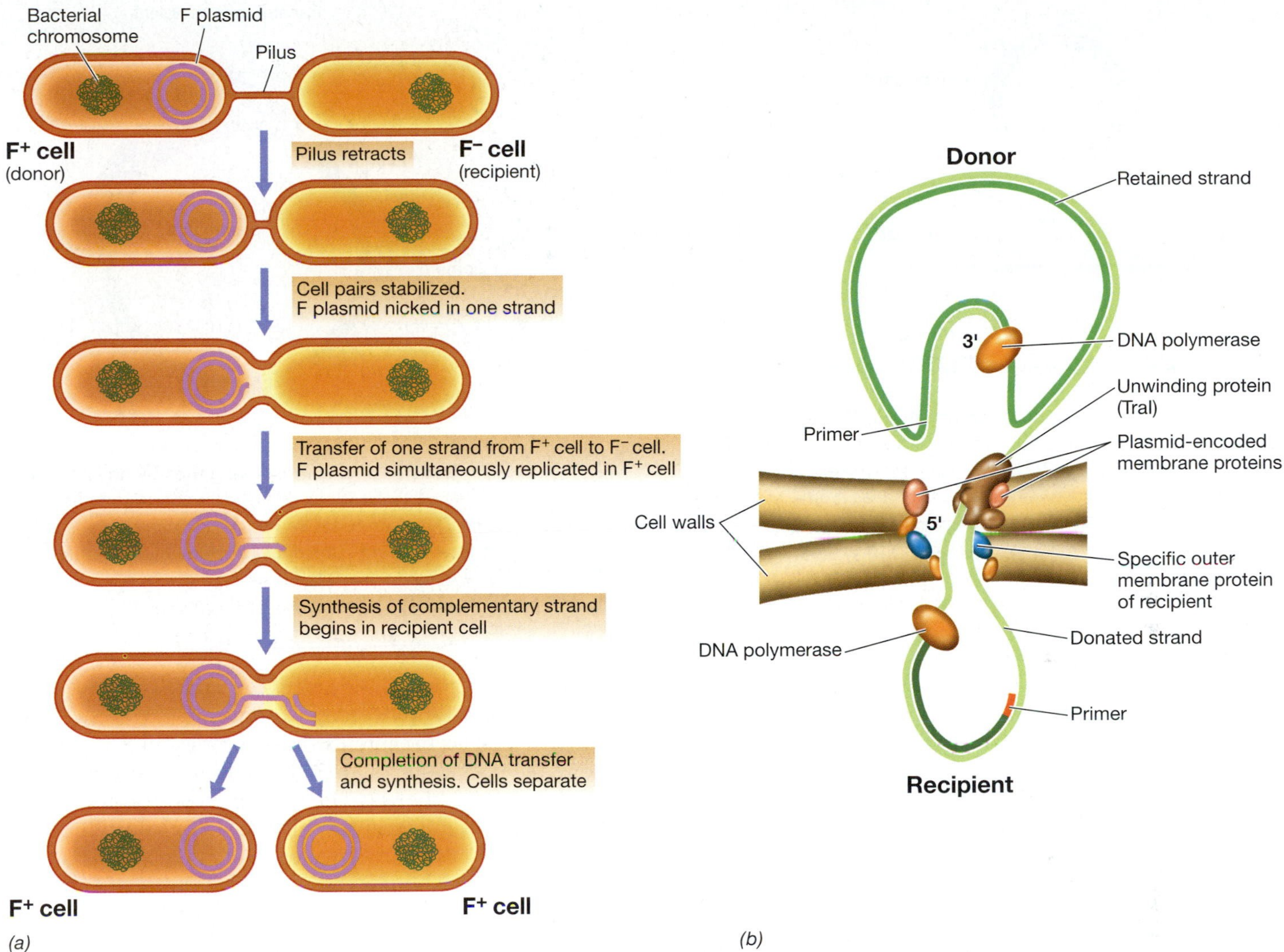

Figure 11.21 Transfer of plasmid DNA by conjugation. *(a)* The transfer of the F plasmid converts an F^- recipient cell into an F^+ cell. Note the mechanism of rolling circle replication (Figure 10.19). *(b)* Details of the replication and transfer process.

like infectious agents. This is of major ecological significance because a few plasmid-containing cells introduced into a population of recipients can convert the entire population into plasmid-bearing (and thus donating) cells in a short time.

Plasmids can be lost from a cell by curing. This may happen spontaneously in natural populations when there is no selection pressure to maintain the plasmid. For example, plasmids conferring antibiotic resistance can be lost without affecting cell viability if there are no antibiotics in the cells' environment.

11.12 MiniReview

Conjugation is a mechanism of DNA transfer in prokaryotes that requires cell-to-cell contact. Conjugation is controlled by genes carried by certain plasmids (such as the F plasmid) and involves transfer of the plasmid from a donor cell to a recipient cell. Plasmid DNA transfer involves replication via the rolling circle mechanism.

- In conjugation, how are donor and recipient cells brought into contact with each other?
- Explain how rolling circle DNA replication allows both donor and recipient to end up with a complete copy of plasmids transferred by conjugation.

11.13 The Formation of Hfr Strains and Chromosome Mobilization

Chromosomal genes can be transferred by conjugation. As mentioned above, the F plasmid of *Escherichia coli* can, under certain circumstances, mobilize the chromosome for transfer during cell-to-cell contact. The F plasmid is actually an episome,

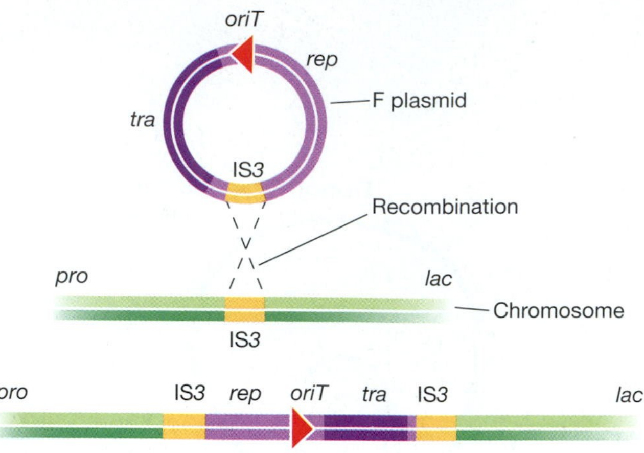

Figure 11.22 The formation of an Hfr strain. Integration of the F plasmid into the chromosome may occur at a variety of specific sites where IS elements are located. The example shown here is an *IS3* located between the chromosomal genes *pro* and *lac*. Some of the genes on the F plasmid are shown. The arrow indicates the origin of transfer, *oriT*, with the arrow as the leading end. Thus, in this Hfr *pro* would be the first chromosomal gene to be transferred and *lac* would be among the last.

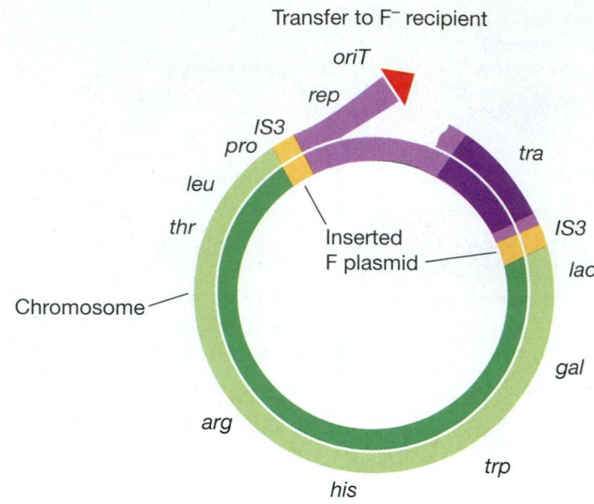

Figure 11.23 Transfer of chromosomal genes by an Hfr strain. The Hfr chromosome breaks at the origin of transfer within the integrated F plasmid. The transfer of DNA to the recipient begins at this point. DNA replicates during transfer as for a free F plasmid (Figure 11.21). This figure is not to scale; the inserted F plasmid is actually less than 3% of the size of the *Escherichia coli* chromosome.

a plasmid that can integrate into the host chromosome, and when the F plasmid is integrated, chromosomal genes can be transferred. Following genetic recombination between donor and recipient, lateral gene transfer by this mechanism can be very extensive in terms of the number and different kinds of chromosomal genes that can be transferred.

Cells possessing a nonintegrated F plasmid are called F^+. Those with an F plasmid integrated into the chromosome are called **Hfr** (for *h*igh *f*requency of *r*ecombination) **cells**. This term refers to the high rates of genetic recombination between genes on the donor chromosome and that of the recipient. Both F^+ and Hfr cells are donors, but unlike conjugation between an F^+ and an F^-, conjugation between an Hfr donor and an F^- leads to transfer of genes from the host chromosome. This is because the chromosome and plasmid now form a single molecule of DNA. Consequently, when rolling circle replication is initiated by the F plasmid, replication continues on into the chromosome. Thus, the chromosome is also replicated and transferred. Hence, integration of a conjugative plasmid provides a mechanism for mobilizing a cell's genetic resources.

Overall, the presence of the F plasmid results in three distinct alterations in the properties of a cell: (1) the ability to synthesize the F pilus, (2) the mobilization of DNA for transfer to another cell, and (3) the alteration of surface receptors so the cell can no longer act as a recipient in conjugation and is unable to take up a second copy of the F plasmid or genetically related plasmids.

Integration of F and Chromosome Mobilization

The F plasmid and the chromosome of *E. coli* both carry several copies of mobile elements called *insertion sequences* (IS; Section 11.16). IS provide regions of sequence homology between chromosomal and F plasmid DNA. Consequently, homologous recombination between an IS on the F plasmid and a corresponding IS on the chromosome results in integration of the F plasmid into the host chromosome. **Figure 11.22** shows the integration of an F plasmid into a chromosomal IS. Once integrated, the plasmid no longer controls its own replication, but the *tra* operon still functions normally and the strain synthesizes pili. When a recipient is encountered, conjugation is triggered just as in an F^+ cell, and DNA transfer is initiated at the *oriT* (origin of transfer) site. However, because the plasmid is now part of the chromosome, after part of the plasmid DNA is transferred, chromosomal genes begin to be transferred (**Figure 11.23**). As is the case of conjugation with just the F plasmid itself (Figure 11.21), chromosomal DNA transfer also involves replication. After transfer, the Hfr strain remains Hfr because it retains a copy of the integrated F plasmid. However, the recipient does not become Hfr because only part of the integrated F plasmid is transferred (**Figure 11.24**).

Because a number of distinct insertion sequences are present on the chromosome, a number of distinct Hfr strains are possible. A given Hfr strain always donates genes in the same order, beginning with the same position. However, Hfr strains that differ in the integration position of the F plasmid in the chromosome transfer genes in different orders (**Figure 11.25**). At some insertion sites, the F plasmid is integrated with its origin pointing in one direction, whereas at other sites the origin points in the opposite direction. The orientation of the F plasmid determines which chromosomal genes enter the recipient first (Figure 11.25). By using various Hfr strains in mating experiments, it was possible to determine the arrangement and orientation of virtually all of the genes in the *E. coli* chromosome long before it was sequenced.

Because the DNA strand typically breaks during transfer, only part of the donor chromosome is transferred. Because a

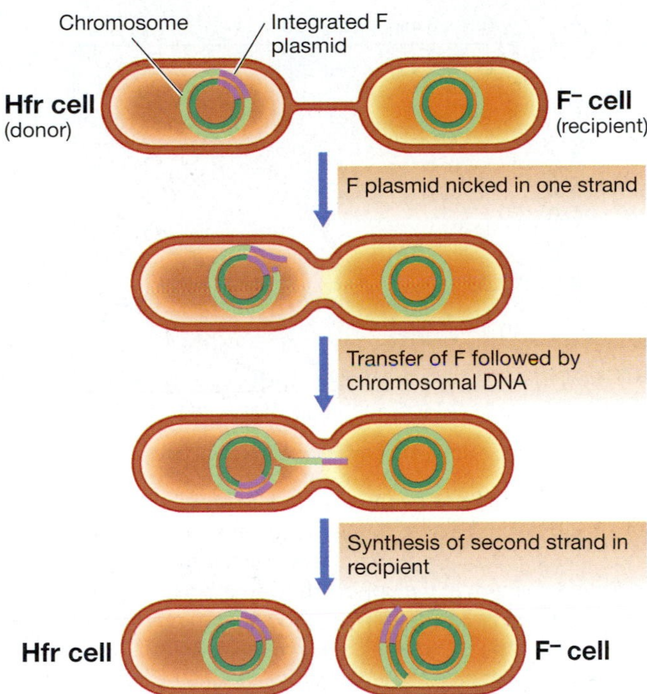

Figure 11.24 Transfer of chromosomal DNA by conjugation. Transfer of the integrated F plasmid from an Hfr strain results in the co-transfer of chromosomal DNA because this is linked to the plasmid. The steps in transfer are similar to those in Figure 11.21 *(a)*. However, the recipient remains F⁻ and receives a linear fragment of donor chromosome attached to part of the F plasmid. For donor DNA to survive, it must be recombined into the recipient chromosome after transfer (not shown).

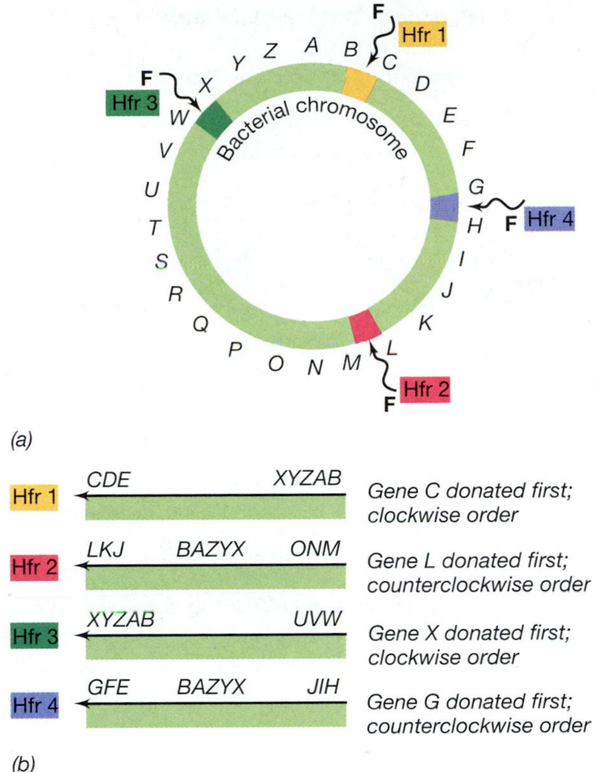

Figure 11.25 Formation of different Hfr strains. Different Hfr strains donate genes in different orders and from different origins. *(a)* F plasmids can be inserted into various insertion sequences on the bacterial chromosome, forming different Hfr strains. *(b)* Order of gene transfer for different Hfr strains.

partial chromosome cannot replicate, for incoming donor DNA to survive, it must recombine with the recipient chromosome. Following recombination, the recipient cell may express a new phenotype due to incorporation of donor genes. Although Hfr strains transmit chromosomal genes at high frequency, they generally do not convert F⁻ cells to F⁺ or Hfr because the entire F plasmid is rarely transferred. Instead, an Hfr × F⁻ cross yields the original Hfr and an F⁻ cell that now has a new genotype. As in transformation and transduction, genetic recombination between Hfr genes and F⁻ genes involves homologous recombination in the recipient cell.

Use of Hfr Strains in Genetic Crosses

As for any system of bacterial gene transfer, the experimenter selects recombinants from conjugation. However, in contrast to transformation and transduction, both the donor and recipient cells are viable during conjugation. It is thus necessary to choose selection conditions in which the desired recombinants can grow but where neither of the parental strains can grow. Typically, a recipient is used that is resistant to an antibiotic but is auxotrophic for some nutrient, and a donor is used that is sensitive to the antibiotic but is prototrophic for the same nutrient. Thus, on minimal medium containing the antibiotic, only recombinant cells will grow following the mating event.

For instance, in the experiment shown in **Figure 11.26**, an Hfr donor that is sensitive to streptomycin (Str^s) and is wild type for synthesis of the amino acids threonine and leucine (Thr^+ and Leu^+) and for utilization of lactose (Lac^+), is mated with a recipient cell that cannot make these amino acids or use lactose, but that is resistant to streptomycin (Str^r). The selective minimal medium contains streptomycin so that only recombinant cells can grow. The composition of each selective medium is varied depending on which genotypic characteristics are desired in the recombinant, as shown in Figure 11.26. The frequency of gene transfer is measured by counting the colonies grown on the selective medium.

The order in which genes are present on the donor chromosome can also be determined by following the kinetics of transfer of individual markers. For example, in the process called *interrupted mating*, conjugating cells are separated by agitation in a mixer or blender. If mixtures of Hfr and F⁻ cells are agitated at various times after mixing and the genetic recombinants scored, it is found that the longer the time between pairing and agitation, the greater the number of genes from the Hfr that are found in the recombinant. As shown in **Figure 11.27**, genes present closer to the origin of transfer enter the recipient cell first and are present in a higher percentage of the recombinants than genes that are transferred later. In addition to showing that gene transfer from donor to recipient occurs sequentially, experiments of this kind provide

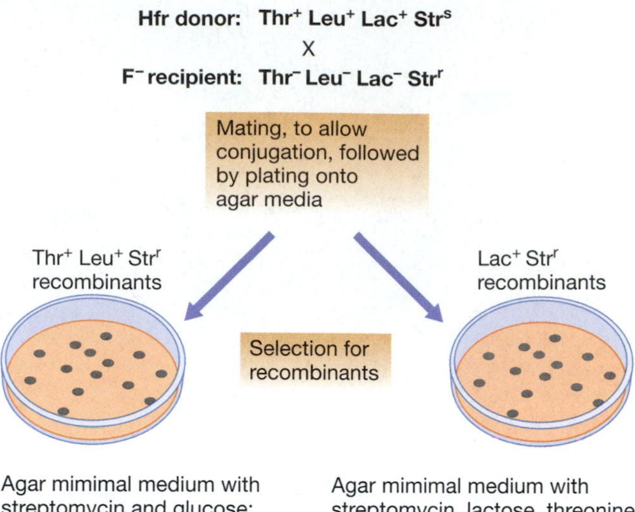

Figure 11.26 Example experiment for the detection of conjugation. Thr, threonine; Leu, leucine; Lac, lactose; Str, streptomycin. Note that each medium selects for specific classes of recombinants. The controls for the experiment are made by plating samples of the donor and the recipient before they are mixed. Neither should be able to grow on the selective media used.

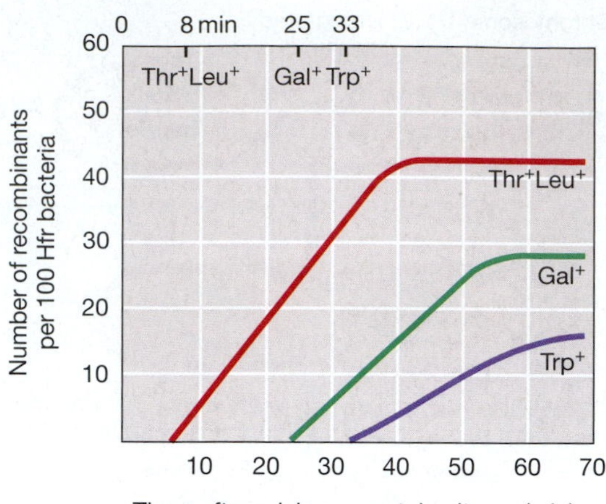

Figure 11.27 Time of gene entry in a mating culture. The rate of appearance of recombinants containing different genes after mating Hfr and F⁻ bacteria is shown. The location of the genes along the Hfr chromosome is shown at the upper left. Genes closest to the origin (0 min) are the first to be transferred. The experiment is done by mixing Hfr and F⁻ cells under conditions in which most Hfr cells find recipients. At various times, samples of the mixture are shaken violently to separate mating pairs and plated onto selective medium on which only recombinants can form colonies.

a method to determine the order of the genes on the bacterial chromosome.

Transfer of Chromosomal Genes to the F Plasmid

Occasionally integrated F plasmids may be excised from the chromosome. During excision, chromosomal genes may sometimes be incorporated into the liberated F plasmid. This can happen because both the F plasmid and the chromosome contain multiple identical insertion sequences where recombination can occur (Figure 11.23). F plasmids containing chromosomal genes are called *F′ plasmids*. These differ from normal F plasmids in containing identifiable chromosomal genes. When F′ plasmids promote conjugation, they transfer these chromosomal genes at high frequency to the recipients. F′-mediated transfer resembles specialized transduction (Section 11.11) in that only a restricted group of chromosomal genes is transferred by any given F′ plasmid. Transferring a known F′ into a recipient allows one to establish diploids (two copies of each gene) for a limited region of the chromosome. Such partial diploids are important for doing complementation tests, as we will see in the next section.

Other Conjugation Systems

Although we have discussed conjugation almost exclusively as it occurs in *E. coli*, conjugative plasmids have been found in many other gram-negative *Bacteria*. Conjugative plasmids of the incompatibility group IncP can be maintained in virtually all gram-negative *Bacteria* and even transferred between different genera. Conjugative plasmids are also known in gram-positive *Bacteria* (for example, in *Streptococcus* and *Staphylococcus*). A process of genetic transfer similar to conjugation in *Bacteria* also occurs in some *Archaea* (Section 11.15).

11.13 MiniReview

The donor cell chromosome can be mobilized for transfer to a recipient cell. This requires an F plasmid to integrate into the chromosome to form the Hfr phenotype. Transfer of the host chromosome is rarely complete but can be used to map the order of the genes on the chromosome. F′ plasmids are previously integrated F plasmids that have excised and captured some chromosomal genes.

- In conjugation involving the F plasmid of *Escherichia coli*, how is the host chromosome mobilized?
- Why does an Hfr × F⁻ mating not yield two Hfr cells?
- At which sites in the chromosome can the F plasmid integrate?

11.14 Complementation

In all three methods of bacterial gene transfer, only a portion of the donor chromosome enters the recipient. Therefore, unless recombination takes place with the recipient chromosome, the donor DNA will be lost because it cannot replicate independently in the recipient. Nonetheless, it is possible to stably maintain a state of partial diploidy for use in bacterial genetic analysis, and we consider this now.

Merodiploids and Complementation

A bacterial strain that carries two copies of any particular chromosomal segment is known as a partial diploid or *merodiploid*. In general, one copy is present on the chromosome itself and the second copy on another genetic element, such as a plasmid or a bacteriophage. Because it is possible to create a specialized transducing phage or specific plasmids using recombinant DNA techniques (∞ Chapter 12), it is possible to put any portion of the bacterial chromosome onto a phage or plasmid.

Consequently, if the chromosomal copy of a gene is defective due to a mutation, it is possible to supply a functional copy of the gene on a plasmid or phage. For example, if one of the genes for tryptophan biosynthesis has been inactivated, this will give a Trp$^-$ phenotype. That is, the mutant strain will be a tryptophan auxotroph and will require the amino acid tryptophan for growth. However, if a copy of the wild-type gene is inserted into the same cell on a plasmid or viral genome, this gene will encode the necessary protein and restore the wild-type phenotype. This process is called *complementation* because the wild-type gene is said to complement the mutation, in this case converting the Trp$^-$ cell into Trp$^+$.

Complementation Tests and the Cistron

When two mutant strains are genetically crossed (whether by conjugation, transduction, or transformation), homologous recombination can yield wild-type recombinants unless both mutations affect exactly the same base pairs. For example, if two different Trp$^-$ *Escherichia coli* mutants are crossed and Trp$^+$ recombinants are obtained, it is obvious that the mutations in the two strains were not in the same base pairs. However, this kind of experiment cannot determine whether two mutations are in two different genes that both affect tryptophan synthesis or in different regions of the same gene. This can be determined by a complementation test.

To perform a complementation test, two copies of the region of DNA under investigation must be present, carried on two different molecules of DNA. One copy is normally present on the chromosome, and the other is carried on a second DNA molecule, typically a plasmid. For example, if we are analyzing mutants in tryptophan biosynthesis, then two copies of the whole tryptophan operon must be present.

Suppose that we wish to know if two Trp$^-$ strains have a mutation in the same gene. To do this we must arrange for one mutation to be present on the chromosome and the other on a plasmid. The mutations are then referred to as being in *trans* with respect to one another. If the two mutations are in the *same* gene, the recombinant cell will have two defective copies of the same gene and will display the negative phenotype. Conversely, if the two mutations are in *different* genes, the recipient cell will have one unmutated copy of each gene (one on the chromosome and the other on the plasmid) and be able to synthesize tryptophan. The possible combinations are shown diagrammatically in **Figure 11.28**. If one DNA molecule carries both mutations (that is, the mutations are in *cis*) a second DNA molecule can complement if it is wild type for the given gene. Having the mutations in *cis* serves as a positive control in a complementation experiment. This type of complementation test is therefore called a *cis-trans* test.

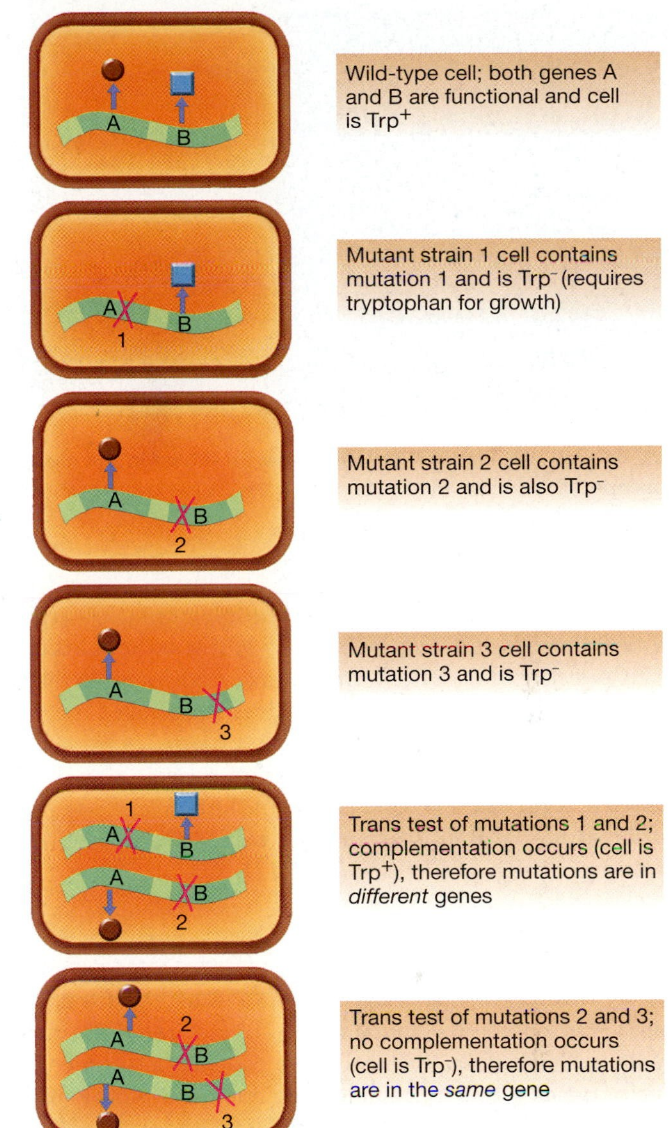

Figure 11.28 Complementation analysis. In this example, the protein products of both genes A and B are required to synthesize tryptophan. Mutations 1, 2, and 3 each lead to the same phenotype, a requirement for tryptophan (Trp$^-$). Complementation analysis indicates that mutations 2 and 3 are in the same gene and mutation 1 is in a separate gene.

A gene as defined by the *cis-trans* test is called a **cistron** and is equivalent to defining a structural gene as a segment of DNA that encodes a single polypeptide chain. If two mutations occur in genes encoding different enzymes, or even different protein subunits of the same enzyme, complementation of the two mutations is possible, and the mutations are therefore not in the same cistron (Figure 11.28). It is important to note that complementation does not rely on recombination; the two genes in question remain on separate genetic elements.

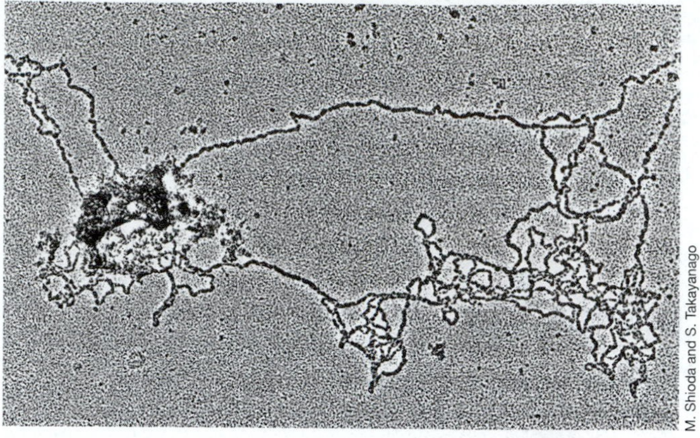

Figure 11.29 An archaeal chromosome, as shown in the electron microscope. The circular chromosome is from the hyperthermophile *Sulfolobus*, a member of the *Archaea*.

Although genetic crosses to test complementation are still done in bacterial genetics, it is often easier to sequence the gene in question to identify the nature and location of any mutations. This is especially true if the sequence of the wild-type gene is already known. The word "cistron" is now rarely used in microbial genetics except when describing whether an mRNA has the genetic information from one gene (*monocistronic* mRNA) or from more than one gene (*polycistronic* mRNA) (∞ Section 7.12).

11.14 MiniReview

A defective copy of a gene may be complemented by the presence of a second, unmutated copy of that gene. The construction of merodiploids carrying two copies of a specific gene or genes allows for complementation tests to determine if two mutations are in the same or different genes. This is necessary when mutations in different genes in the same pathway yield the same phenotype. Recombination does not occur in complementation tests.

- What is a merodiploid?
- Complementation tests have been referred to as *cis-trans* tests. Explain.

11.15 Gene Transfer in *Archaea*

Although *Archaea* contain a single circular chromosome like most *Bacteria* (**Figure 11.29**) and the genomes of several members of the *Archaea* have been entirely sequenced, the development of gene transfer systems lags far behind that for *Bacteria*. Practical problems include the need to grow many *Archaea* under extreme conditions. Thus, the temperatures necessary to culture some hyperthermophiles will melt agar, and alternative materials are required to form solid media and obtain colonies.

Another problem is that most antibiotics do not affect *Archaea*. For example, penicillins do not affect *Archaea* because their cell walls lack peptidoglycan. The choice of selectable markers for genetic crosses is therefore often limited. However, novobiocin (a DNA gyrase inhibitor) and mevinolin (an inhibitor of isoprenoid biosynthesis) are used to inhibit halophiles, and puromycin and neomycin (both protein synthesis inhibitors) inhibit methanogens.

Archaeal Genetic Systems

No single species of *Archaea* has become a model organism for archaeal genetics, although more genetic work has probably been done on select species of extreme halophiles (*Halobacterium, Haloferax,* ∞ Section 17.3) than on any other *Archaea*. Instead, individual mechanisms for gene transfer have been found scattered among a range of *Archaea*. Examples of transformation, viral transduction, and conjugation are known. In addition, several plasmids have been isolated from *Archaea* and some have been used to construct cloning vectors, allowing genetic analysis through cloning and sequencing rather than traditional genetic crosses. Transposon mutagenesis has been well developed in certain methanogen species including *Methanococcus* and *Methanosarcina*, and other tools such as shuttle vectors and other in vitro methods of genetic analysis have been developed for study of the highly unusual biochemistry of the methanogens (∞ Section 17.4).

Transformation works reasonably well in several *Archaea*. Transformation procedures vary in detail from organism to organism. One approach involves removal of divalent metal ions, which in turn results in the disassembly of the glycoprotein cell wall layer surrounding many archaeal cells and hence allows access by DNA. However, *Archaea* with rigid cell walls have proven difficult to transform, although electroporation sometimes works. One exception is in *Methanosarcina* species, organisms with a thick cell wall, for which high-efficiency transformation systems have been developed that employ DNA-loaded lipid preparations (liposomes) to deliver DNA into the cell.

Although viruses that infect *Archaea* are plentiful, viral transduction is extremely rare. Only one archaeal virus, which infects the thermophilic methanogen *Methanobacterium thermoautotrophicum*, has been shown to transduce the genes of its host. Unfortunately the low burst size (about six phages liberated per cell) makes using this system for gene transfer impractical.

Two types of conjugation have been detected in *Archaea*. Some strains of *Sulfolobus solfataricus* (∞ Section 17.10) contain plasmids that promote conjugation between two cells in a manner similar to that seen in *Bacteria*. In this process, cell pairing occurs before plasmid transfer, and DNA transfer is unidirectional. However, most of the genes encoding these functions seem to have little similarity to those in gram-negative *Bacteria*. The exception is a gene similar to *traG* from the F plasmid, whose protein product is involved in stabilizing mating pairs. It thus seems likely that the actual mechanism of conjugation in *Archaea* is quite different from that in *Bacteria*.

Some halobacteria, in contrast, perform a novel form of conjugation. No fertility plasmids are involved, and DNA transfer is bidirectional. Cytoplasmic bridges form between the mating cells and appear to be used for DNA transfer. Neither type of conjugation has been developed to the point of being used for routine gene transfer or genetic analysis. However, these genetic resources will likely be useful for developing facile genetic systems in these organisms.

11.15 MiniReview

Archaea lag behind *Bacteria* in the development of systems for gene transfer. Many antibiotics are ineffective against *Archaea*, making it difficult to select recombinants effectively. Moreover the unusual growth conditions needed by many *Archaea* make genetic experimentation difficult. Nevertheless, the genetic transfer systems of *Bacteria*—transformation, transduction, and conjugation—are all known in *Archaea*.

- Why is it usually more difficult to select recombinants with *Archaea* than with *Bacteria*?
- Why do penicillins not kill members of the *Archaea*?

11.16 Mobile DNA: Transposable Elements

As we have seen, molecules of DNA may move from one cell to another, but to a geneticist, "mobile DNA" has a specialized meaning. Mobile DNA refers to discrete segments of DNA that move as units from one location to another *within* other DNA molecules. Although the DNA of certain viruses can be inserted into and excised from the genome of the host cell, most mobile DNA consists of **transposable elements**. These are stretches of DNA that can move from one site to another. However, transposable elements are always found inserted into another DNA molecule such as a plasmid, a chromosome, or a viral genome. Transposable elements do not possess their own origin of replication. Instead, they are replicated when the host DNA molecule into which they are inserted is replicated.

Transposable elements move by a process called *transposition* that is important both in evolution and in genetic analysis. The frequency of transposition is extremely variable, and ranges from 1 in 1,000 to 1 in 10,000,000 per transposable element per cell generation, depending on both the transposable element and the organism. Transposition was originally observed in corn (maize) in the 1940s by Barbara McClintock before the DNA double helix was even discovered! McClintock later received the Nobel Prize for this discovery. The molecular details of transposition were revealed using *Bacteria* due to the powerful genetic analyses possible in these organisms. Transposable elements are widespread in nature and can be found in the genomes of all three domains of life as well as in many viruses and plasmids.

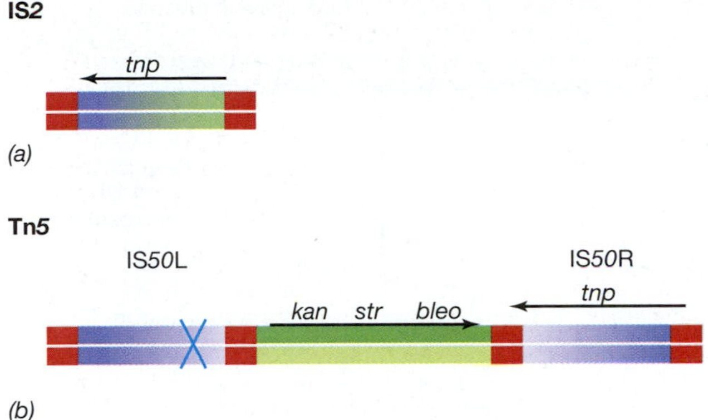

Figure 11.30 Maps of the transposable elements IS*2* and Tn*5*. Inverted repeats are shown in red. The arrows above the maps show the direction of transcription of any genes on the elements. The gene encoding the transposase is *tnp*. (*a*) IS*2* is an insertion sequence of 1327 bp with inverted repeats of 41 bp at its ends. (*b*) Tn*5* is a composite transposon of 5.7 kbp containing the insertion sequences IS*50*L and IS*50*R at its left and right ends, respectively. IS*50*L is not capable of independent transposition because there is a nonsense mutation, marked by a blue cross, in its transposase gene. Otherwise, the two IS*50* elements are almost identical. The genes *kan*, *str*, and *bleo* confer resistance to the antibiotics kanamycin (and neomycin), streptomycin, and bleomycin. Tn*5* is commonly used to generate mutants in *Escherichia coli* and other gram-negative bacteria.

Transposons and Insertion Sequences

The two major types of transposable elements in *Bacteria* are *insertion sequences* and *transposons*. Both elements have two important features in common: They carry genes encoding transposase, the enzyme necessary for transposition, and they have short inverted terminal repeats at their ends that are also needed for transposition. Note that the ends of transposable elements are not free but are continuous with the host DNA molecule into which the transposable element has inserted. Figure 11.30 shows genetic maps of the insertion element IS*2* and of the transposon Tn*5*.

Insertion sequences (IS) are the simplest type of transposable element. They are short DNA segments, about 1,000 nucleotides long and typically contain inverted repeats of 10–50bp. Each different IS has a specific number of base pairs in its terminal repeats. The only gene they possess is for the transposase. Several hundred distinct IS elements have been characterized. Insertion sequences are found in the chromosomes and plasmids of both *Bacteria* and *Archaea,* as well as in certain bacteriophages. Individual strains of the same bacterial species vary in the number and location of the IS elements they harbor. For instance, one strain of *Escherichia coli* has five copies of IS*2* and five copies of IS*3*. Many plasmids, such as the F plasmid, also carry insertion sequences. Indeed, integration of the F plasmid into the *E. coli* chromosome is due to homologous recombination between identical insertion sequences on the F plasmid and the chromosome (Section 11.13).

Transposons are larger than insertion sequences but have the same two essential components: inverted repeats at both ends and a gene that encodes transposase. The transposase

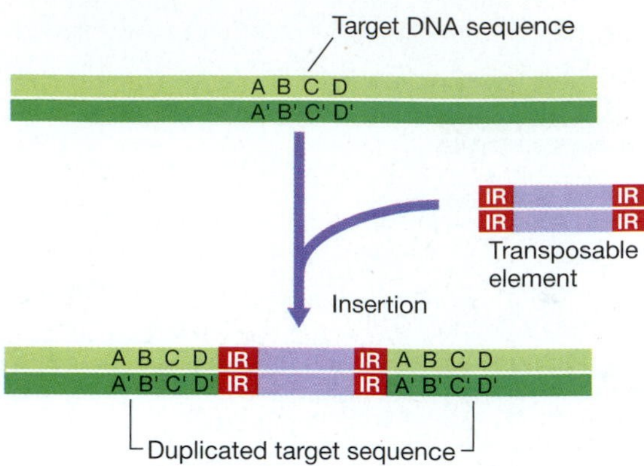

Figure 11.31 Transposition. Insertion of a transposable element generates a duplication of the target sequence. Note the presence of inverted repeats (IR) at the ends of the transposable element.

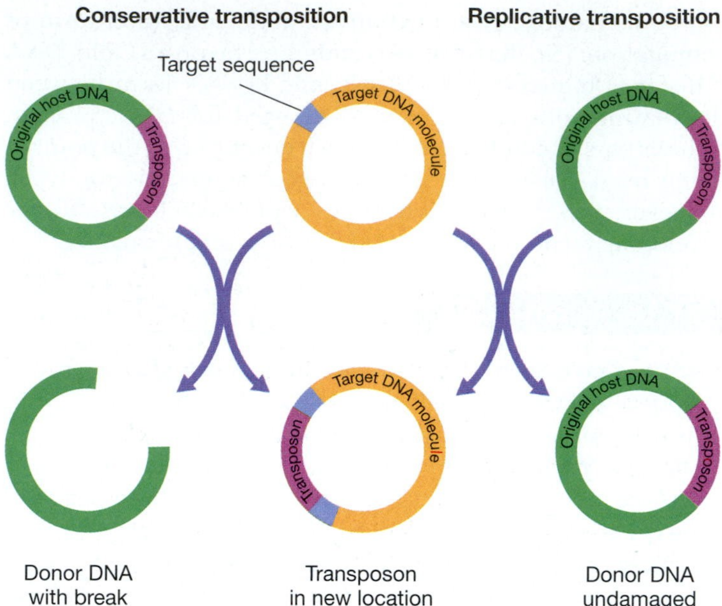

Figure 11.32 Two mechanisms of transposition. Donor DNA (carrying the transposon) is shown in light green, and recipient DNA carrying the target sequence is shown in orange. In both conservative and replicative transposition the transposase inserts the transposon (purple) into the target site (blue) on the recipient DNA. During this process, the target site is duplicated. In conservative transposition, the donor DNA is left with a double-stranded break at the previous location of the transposon. In contrast, after replicative transposition, both donor and recipient DNA possess a copy of the transposon.

recognizes the inverted repeats and moves the segment of DNA flanked by them from one site to another. Consequently, any DNA that lies between the two inverted repeats is moved and is, in effect, part of the transposon. Genes included inside transposons vary widely. Some of these genes, such as antibiotic resistance genes, confer important new properties on the organism harboring the transposon. Because antibiotic resistance is both important and easy to detect, most highly investigated transposons have antibiotic resistance genes as selectable markers. Examples include transposon Tn*5*, which carries kanamycin resistance (Figure 11.30) and Tn*10*, with tetracycline resistance.

Because any genes lying between the inverted repeats become part of a transposon, it is possible to get hybrid transposons that display complex behavior. For example, conjugative transposons contain *tra* genes and can move between bacterial species by conjugation as well as transpose from place to place within a single bacterial genome. Even more complex is bacteriophage Mu, which is both a virus and a transposon (∞ Section 19.4). In this case a complete virus genome is contained within a transposon. Other composite genetic elements consist of a segment of DNA lying between two identical insertion sequences. This whole structure can move as a unit and is called a *composite transposon*. The behavior of composite transposons indicates that novel transposons likely arise periodically in cells that contain insertion sequences located close to one another.

Mechanisms of Transposition

Both the inverted repeats found at the ends of transposable elements and transposase are essential for transposition. The transposase recognizes, cuts, and ligates the DNA during transposition. When a transposable element is inserted into target DNA, a short sequence in the target DNA at the site of integration is duplicated during the insertion process (**Figure 11.31**). The duplication arises because single-stranded DNA breaks are made by the transposase. The transposable element is then attached to the single-stranded ends that have been generated. Finally, enzymes of the host cell repair the single-strand portions, which results in the duplication.

Two mechanisms of transposition are known: *conservative* and *replicative* (**Figure 11.32**). In conservative transposition, as occurs with the transposon Tn*5*, the transposon is excised from one location and is reinserted at a second location. The copy number of a conservative transposon therefore remains at one. By contrast, during replicative transposition, a new copy is produced and is inserted at the second location. Thus, after a replicative transposition event, one copy of the transposon remains at the original site, and there is a second copy at the new site.

Transposition is a type of recombination called *site-specific* recombination, because specific DNA sequences (the inverted repeats and the target sequence) are recognized by a protein (the transposase). This contrasts with *homologous* recombination (Section 11.9) in which homologous DNA sequences recognize each other by base pairing.

Mutagenesis with Transposons

When a transposon inserts itself within a gene, a mutation occurs in that particular gene (**Figure 11.33**). Mutations due to transposon insertion do occur naturally. However, use of transposons to generate mutations is a convenient way to create bacterial mutants in the laboratory. Typically, transposons carrying antibiotic resistance genes are used. The transposon is introduced into the target cell on a phage or plasmid that cannot replicate in that particular host. Consequently,

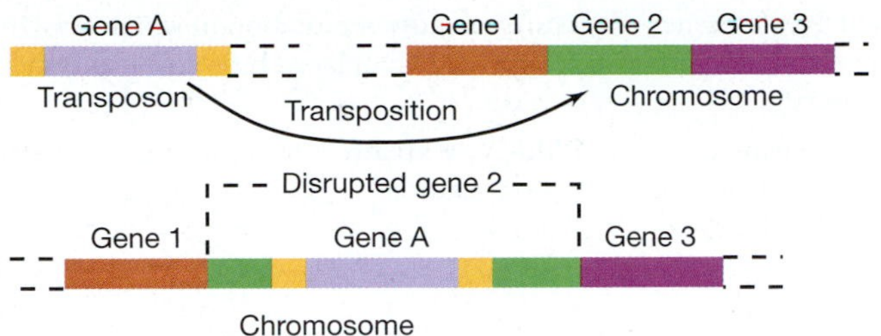

Figure 11.33 Transposon mutagenesis. The transposon moves into the middle of gene 2. Gene 2 is now disrupted by the transposon and is inactivated. Gene A in the transposon is expressed in both locations.

antibiotic-resistant colonies will mostly be due to insertion of the transposon into the bacterial genome.

Because bacterial genomes contain relatively little non-coding DNA, most transposon insertions will occur in genes that encode proteins. If inserted into a gene encoding an essential protein, the mutation may be lethal under certain growth conditions and become suitable for genetic selection. For example, if transposon insertions are selected on rich medium on which all auxotrophs can grow, they can subsequently be screened on minimal medium supplemented with various nutrients to determine if a nutrient is required. Further analyses can then be performed to reveal which gene the transposon has disrupted. Auxotrophic mutations due to transposon insertions are very useful in bacterial genetics. Normally, auxotrophic recombinants cannot be isolated by positive selection. However, the presence of a transposon with an antibiotic resistance marker allows for positive selection.

Two transposons widely used for mutagenesis of *E. coli* and related bacteria are Tn*5* (Figure 11.30), which confers neomycin and kanamycin resistance, and Tn*10*, which confers tetracycline resistance. Many *Bacteria*, a few *Archaea*, and the yeast *Saccharomyces cerevisiae*, have all been mutagenized using transposon mutagenesis. More recently, transposons have even been used to isolate mutations in animals, including the mouse.

UNIT 2

11.16 MiniReview

Transposons and insertion sequences are genetic elements that can move from one location on a host DNA molecule to another by transposition, a type of site-specific recombination. Transposition can be either replicative or conservative. Transposons often carry genes encoding antibiotic resistance and can be used as biological mutagens.

- Which features do insertion sequences and transposons have in common?
- What is the significance of the terminal inverted repeats of transposons?

Review of Key Terms

Auxotroph an organism that has developed a nutritional requirement, often as a result of mutation

Cistron a gene as defined by the *cis-trans* test; a segment of DNA (or RNA) that encodes a single polypeptide chain

Conjugation transfer of genes from one prokaryotic cell to another by a mechanism involving cell-to-cell contact

Genotype the complete genetic makeup of an organism; the complete description of a cell's genetic information

Heteroduplex a DNA double helix composed of single strands from two different DNA molecules

Hfr cell a cell with the F plasmid integrated into the chromosome

Induced mutation a mutation caused by external agents such as mutagenic chemicals or radiation

Insertion sequence (IS) the simplest type of transposable element, which carries only genes involved in transposition

Missense mutation a mutation in which a single codon is altered so that one amino acid in a protein is replaced with a different amino acid

Mutagen an agent that causes mutation

Mutant an organism whose genome carries a mutation

Mutation a heritable change in the base sequence of the genome of an organism

Mutator strain a mutant strain in which the rate of mutation is increased

Nonsense mutation a mutation in which the codon for an amino acid is changed to a stop codon

Phenotype the observable characteristics of an organism

Plasmid an extrachromosomal genetic element that has no extracellular form

Point mutation a mutation that involves a single base pair

Recombination the process by which DNA molecules from two separate sources exchange sections or are brought together into a single DNA molecule

Regulon a set of genes or operons that are transcribed separately but are coordinately controlled by the same regulatory protein

Reversion alteration in DNA that reverses the effects of a prior mutation

Rolling circle replication mechanism of replicating double-stranded circular DNA that starts by nicking and unrolling one strand and using the other (still circular) strand as a template for DNA synthesis

Screening a procedure that permits the identification of organisms by phenotype or genotype, but does not inhibit or enhance the growth of particular phenotypes or genotypes

Selection placing organisms under conditions where the growth of those with a particular phenotype or genotype will be favored or inhibited

Silent mutation a change in DNA sequence that has no effect on the phenotype

Spontaneous mutation a mutation that occurs "naturally" without the help of mutagenic chemicals or radiation

Transduction transfer of host cell genes from one cell to another by a virus

Transformation transfer of bacterial genes involving free DNA (but see alternative usage in Chapter 10)

Transition a mutation in which a pyrimidine base is replaced by another pyrimidine or a purine is replaced by another purine

Transposable element a genetic element with the ability to move (transpose) from one site to another on host DNA molecules

Transposon a type of transposable element that carries genes in addition to those involved in transposition

Transversion a mutation in which a pyrimidine base is replaced by a purine or vice versa

Wild-type strain a bacterial strain isolated from nature

Review Questions

1. What is the size of the *Escherichia coli* chromosome? About how many proteins can it encode? How much noncoding DNA is present in the *E. coli* chromosome (Section 11.1)?
2. How do plasmids replicate and how does this differ from chromosomal replication (Section 11.2)?
3. What are R factors and why are they of medical concern (Section 11.3)?
4. Write a one-sentence definition of the term "genotype." Do the same for the term "phenotype." Does the phenotype of an organism automatically change when a change in genotype occurs? Why or why not? Can phenotype change without a change in genotype? In both cases, give examples to support your answer (Section 11.4).
5. Explain why an *E. coli* strain that is His^- is an auxotroph and one that is Lac^- is not. (*Hint:* Think about what *E. coli* does with histidine and lactose.) (Section 11.4).
6. What are silent mutations? From your knowledge of the genetic code, why do you think most silent mutations affect the third position in a codon (Section 11.5)?
7. Microinsertions that occur in promoters are not frameshift mutations. Define the terms microinsertion, frameshift, mutation, and promoter (∞ Section 7.10). Explain how the statement can be true (Section 11.5).
8. Explain how it is possible for a frameshift mutation early in a gene to be corrected by another frameshift mutation farther along the gene (Section 11.5).
9. What is the average rate of mutation in a cell? Can this rate change (Section 11.6)?
10. Give an example of one biological, one chemical, and one physical mutagen and describe the mechanism by which each causes a mutation (Section 11.7).
11. How can the Ames test, an assay using bacteria, have any relevance to human cancer (Section 11.8)?
12. How does homologous recombination differ from site-specific recombination (Section 11.9)?
13. Why is it difficult in a single experiment using transformation or transduction to transfer a large number of genes to a cell (Section 11.10 and 11.11)?
14. Explain why in generalized transduction one always refers to a transducing particle but in specialized transduction one refers to a transducing virus (or transducing phage) (Section 11.11).
15. What is a sex pilus and which cell type, F^- or F^+, would produce this structure (Section 11.12)?
16. What does an F^+ cell need to do before it can transfer chromosomal genes (Section 11.13)?
17. What does it mean to complement a mutation "in *trans*" (Section 11.14)?
18. Explain why performing genetic selections is difficult when studying *Archaea*. Give examples of some selective agents that work well with *Archaea* (Section 11.15).
19. What are the major differences between insertion sequences and transposons (Section 11.16)?
20. The most useful transposons for isolating a variety of bacterial mutants are transposons containing antibiotic resistance genes. Why are such transposons so useful for this purpose (Section 11.16)?

Application Questions

1. A constitutive mutant is a strain that continuously makes a protein that is inducible in the wild type. Describe two ways in which a change in a DNA molecule could lead to the emergence of a constitutive mutant. How could these two types of constitutive mutants be distinguished genetically?
2. In Chapter 7, we saw that it was critical for the ribosome to translocate with great accuracy in order to maintain the proper reading frame. However, sometimes ribosomes make frameshift errors. Compare the impact on the cell of a ribosome periodically making a frameshift error in the mRNA from a particular gene with the impact of a frameshift mutation in the same gene.
3. Although a large number of mutagenic chemicals are known, none is known that induces mutations in only a single gene (gene-specific mutagenesis). From what you know about mutagens, explain why it is unlikely that a gene-specific chemical mutagen will be found. How then is site-specific mutagenesis accomplished?
4. Transposable elements cause mutations when inserted within a gene. These elements disrupt the continuity of a gene. Introns also disrupt the continuity of a gene, yet the gene is still functional. Explain why the presence of an intron in a gene does not inactivate that gene but insertion of a transposable element does.

12

Genetic Engineering

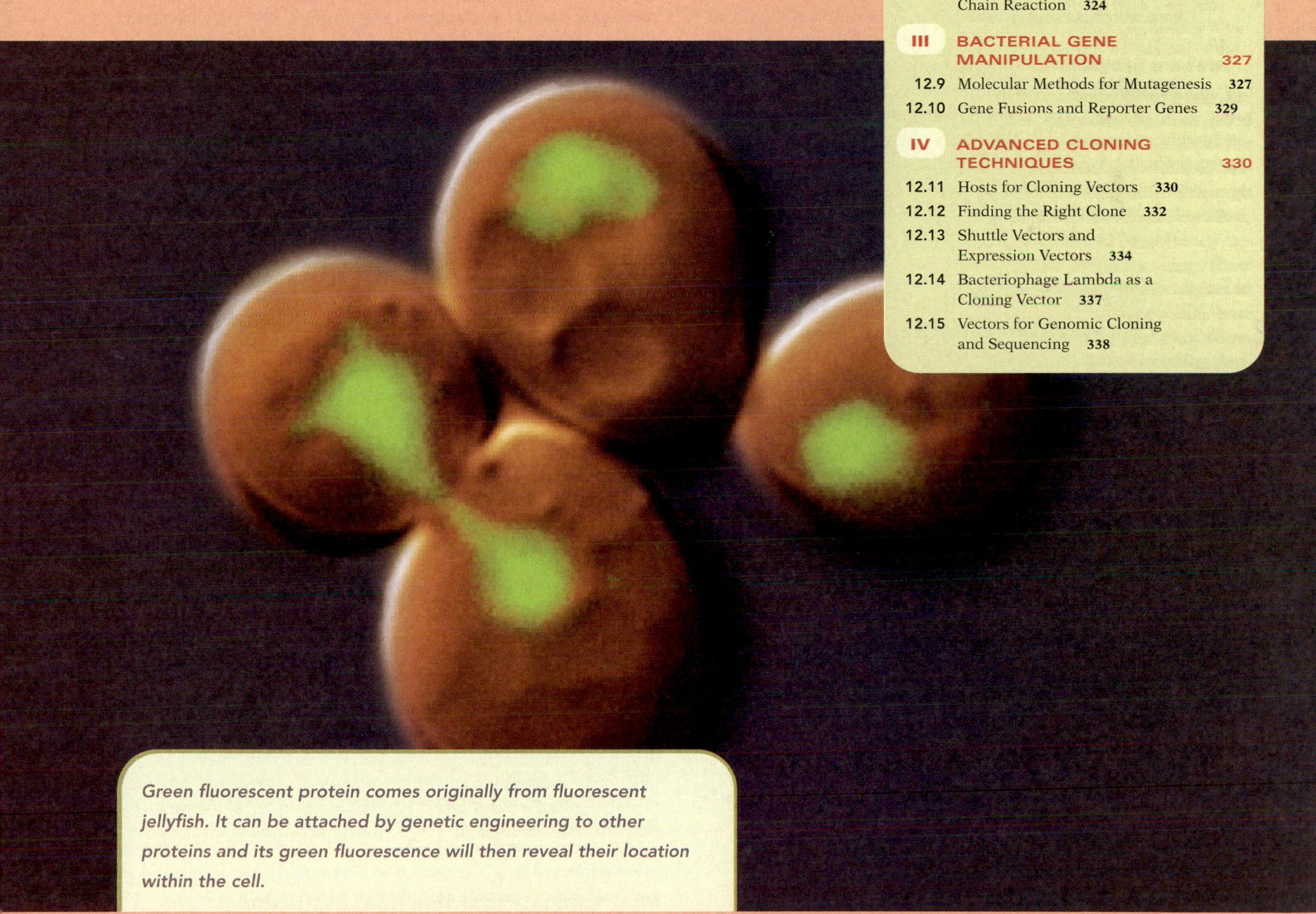

Green fluorescent protein comes originally from fluorescent jellyfish. It can be attached by genetic engineering to other proteins and its green fluorescence will then reveal their location within the cell.

In this chapter we discuss techniques used in genetic engineering, in particular those used to clone genes, alter genes, and to express them efficiently in host organisms. The mechanisms of genetic exchange discussed in Chapter 11 (transformation, transduction, and conjugation) allow us to move genes among bacteria. However, performing genetics only *in vivo* (in living bacteria) has many limitations that can be overcome by manipulating DNA *in vitro* (in a test tube).

I TOOLS AND TECHNIQUES OF GENETIC ENGINEERING

Genetic engineering refers to the use of *in vitro* techniques to alter genetic material in the laboratory. Such altered genetic material may be reinserted into the original source organism or into some other organism. The basic techniques of genetic engineering include the ability to cut the DNA of interest into specific fragments and to purify these fragments for further manipulation. We begin this chapter by considering some of the basic tools of genetic engineering, including restriction enzymes, the separation of nucleic acids by electrophoresis, nucleic acid hybridization, nucleic acid probes, molecular cloning, and cloning vectors.

12.1 Restriction and Modification Enzymes

All cells contain enzymes that can chemically modify DNA in one way or another. One major class of such enzymes is the restriction endonucleases, or **restriction enzymes** for short. Restriction enzymes recognize specific base sequences within DNA and cut the DNA. Although they are widespread among prokaryotes (both *Bacteria* and *Archaea*), they are very rare in eukaryotes. *In vivo,* restriction enzymes protect prokaryotes from hostile foreign DNA such as virus genomes. However, restriction enzymes are also essential for *in vitro* DNA manipulation, and their discovery gave birth to the field of genetic engineering.

Mechanism of Restriction Enzymes

Restriction endonucleases are divided into three major classes. The type I and III restriction enzymes bind to the DNA at their recognition sequences but cut the DNA a considerable distance away. In contrast, the type II restriction enzymes cleave the DNA within their recognition sequences, and thus this group of enzymes is much more useful for the specific manipulation of DNA. **Figure 12.1** shows the 6-bp sequence that is recognized and cleaved by the restriction enzyme from *Escherichia coli* called *Eco*RI (this acronym stands for *Escherichia coli,* strain RY13, restriction enzyme I). The cleavage sites are indicated by arrows and the axis of symmetry by a dashed line. Note that the two strands of the recognition sequence have the same sequence if one is read from the left and the other from the right (or if both are read 5′ → 3′). Such inverted repeat sequences are called *palindromes* (the term palindrome is derived from the Greek, meaning "to run back again").

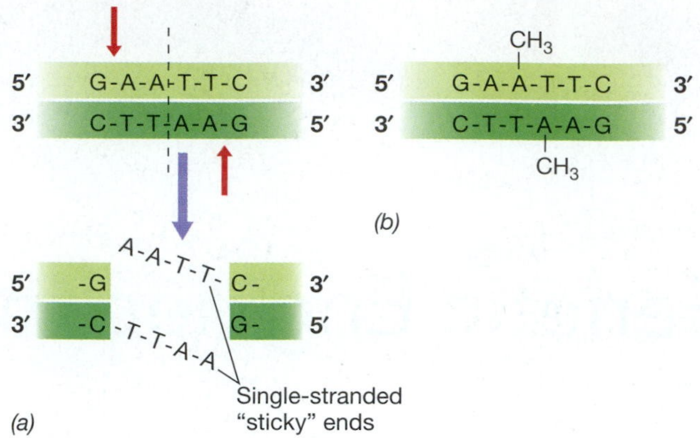

Figure 12.1 Restriction and modification of DNA. *(a)* (Top panel) The sequence of DNA recognized by the restriction endonuclease *Eco*RI. The red arrows indicate the bonds cleaved by the enzyme. The dashed line indicates the axis of symmetry of the sequence. (Bottom panel) Appearance of DNA after cutting with restriction enzyme *Eco*R1. Note the single-stranded "sticky ends." *(b)* The same sequence after modification by the *Eco*RI methylase. The methyl groups added by this enzyme are shown, and these protect the restriction site from cutting by *Eco*R1.

Most restriction enzymes are homodimeric proteins; that is, they are composed of two identical polypeptide subunits, and each subunit recognizes and cuts the DNA on one of its two strands. This therefore results in a double-stranded break. Many type II restriction enzymes make staggered cuts, leaving short, single-stranded overhangs known as "sticky ends" at the ends of the two fragments (Figure 12.1). As explained below, such fragments with defined termini have many uses, especially in molecular cloning. Other restriction enzymes cut both strands of the DNA directly opposite each other, yielding blunt ends.

Most of the DNA sequences recognized by type II restriction enzymes are short inverted repeats of from 4 to 8 bp. Consider the enzyme *Eco*RI, which recognizes a specific 6-bp sequence (Figure 12.1). Any specific 6-base sequence should appear in a strand of DNA about once every 4,096 nucleotides (4 bases taken 6 at a time, 4^6), assuming that the DNA has a "random" sequence. Thus, several *Eco*RI cut sites should be present in a lengthy DNA molecule, including the chromosome of *E. coli* (see later). The recognition sequences and cut sites for several restriction enzymes are given in **Table 12.1**. Several thousand restriction enzymes with several hundred different specificities are known; some leave "sticky ends" with a 5′ overhang and others with a 3′ overhang, whereas others generate blunt ends. Restriction enzymes are such important tools in modern molecular genetic research that they have become widely available commercially.

Modification: Protection from Restriction

The major function of restriction enzymes in prokaryotes is probably to protect the cell from invasion by foreign DNA, for example, viral DNA. If foreign DNA enters the cell, the cell's restriction enzymes will destroy it. However, a cell must protect its own DNA from inadvertent destruction by its

Table 12.1 Recognition sequences of a few restriction endonucleases

Organism	Enzyme designation[a]	Recognition sequence[b]
Bacillus globigii	*Bgl*II	A↓GATCT
Bacillus subtilis	*Bsu*RI	GG↓CC
Brevibacterium albidum	*Bal*I	TGG↓CCA
Escherichia coli	*Eco*RI	G↓AATTC[c]
Haemophilus haemolyticus	*Hha*I	GCG↓C
Haemophilus influenzae	*Hind*II	GTPy↓PuAC
Haemophilus influenzae	*Hind*III	A↓AGCTT
Klebsiella pneumoniae	*Kpn*I	GGTAC↓C
Nocardia otitidiscaviarum	*Not*I	GC↓GGCCGC
Proteus vulgaris	*Pvu*I	CGAT↓CG
Serratia marcescens	*Sma*I	CCC↓GGG
Thermus aquaticus	*Taq*I	T↓CGA

[a]Nomenclature: The first letter of the three-letter abbreviation of a restriction endonuclease designates the genus from which the enzyme originates, the second two letters, the species. The roman numeral designates the order of discovery of enzymes in that particular organism, and any additional letters are strain designations.

[b]Arrows indicate the sites of enzymatic attack. Asterisks indicate the site of methylation (modification). G, guanine; C, cytosine; A, adenine; T, thymine; Pu, any purine; Py, any pyrimidine. Only the 5′ → 3′ sequence is shown.

[c]See Figure 12.1*a*

own restriction enzymes. Such protection is conferred by **modification enzymes**, which chemically modify specific nucleotides in the restriction recognition sequences in the cell's own DNA. The modified sequences can no longer be cut by the cell's own restriction enzymes. Each restriction enzyme is partnered with a corresponding modification enzyme that shares the same recognition sequence. Typically, modification consists of methylating specific bases within the recognition sequence so that the restriction endonuclease can no longer bind. For example, the sequence recognized by the *Eco*RI restriction enzyme (Figure 12.1*a*) can be modified by methylation of the two most interior adenines (Figure 12.1*b*). The enzyme that performs this modification is *Eco*RI methylase. If even a single strand is modified, the sequence is no longer a substrate for the restriction enzyme *Eco*RI.

Gel Electrophoresis: Separation of DNA Molecules

Because the base sequences recognized by many restriction enzymes are four to six nucleotides long (Table 12.1), they cut DNA molecules into segments that range in length from a few hundred to a few thousand base pairs. After cleaving the DNA, the fragments generated can be separated from each other by **gel electrophoresis** and analyzed.

Electrophoresis is a procedure that separates charged molecules migrating in an electrical field. The rate of migration is determined by the charge on the molecule and by its size and shape. In gel electrophoresis (**Figure 12.2a**) the molecules are separated in a porous gel, rather than in free solution. Gels made of *agarose*, a polysaccharide, are used for separating DNA fragments. When an electrical current is applied, nucleic acids move through the gel toward the positive electrode due to their negatively charged phosphate groups. The presence of the gel meshwork hinders the progress of the DNA, and small or compact molecules migrate more rapidly than large molecules. After a certain period of time, the gel can be stained with a compound that binds to DNA, such as *ethidium bromide* (Section 11.7), and the DNA will then fluoresce orange under ultraviolet light (Figure 12.2*b*). DNA fragments can be purified from gels and used for a variety of purposes.

Elizabeth Parker

(a)

–
A B C D
Size in base pairs
5000 –
4000 –
3000 –
2000 –
1800 –
1000 –
500 –
+
Jack Parker

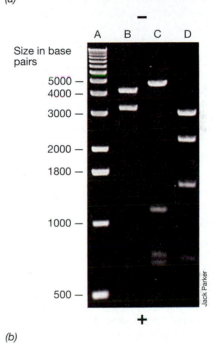

(b)

Figure 12.2 Agarose gel electrophoresis of DNA. *(a)* DNA samples are loaded into wells in a submerged agarose gel. *(b)* A photograph of a stained agarose gel. The DNA has been loaded into wells toward the top of the gel (negative pole) as shown, and the positive pole of the electrical field is at the bottom. The standard sample in lane A has fragments of known size. Using these standards, one can determine the sizes of the fragments in the other lanes. Bands stain less intensely at the bottom of the gel because the fragments are smaller, and thus there is less DNA to stain.

UNIT 2

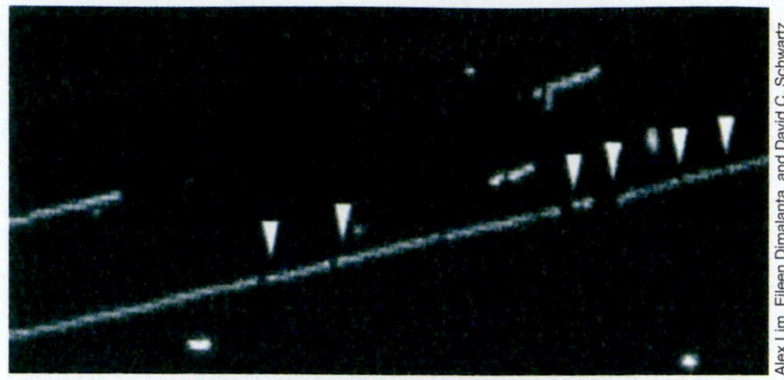

Figure 12.3 Optical mapping of restriction fragments. A digital fluorescence micrograph of a portion of the *Escherichia coli* chromosome digested with the restriction enzyme *XhoI* (isolated from *Xanthomonas holica*). Arrows indicate the sites of cutting by the restriction enzyme. Although the DNA strand itself is too small to be seen by light microscopy, fluorescence of the DNA is visible. The length of the DNA shown here is about 260 kbp.

Restriction Analyses of DNA

A typical agarose gel is shown in Figure 12.2*b*. Lane A contains a standard DNA sample consisting of DNA fragments of known sizes. The other lanes contain purified DNA (in this case a plasmid) cut by one or more restriction enzymes. Because a given restriction enzyme always cuts at the same site, the banding pattern of DNA molecules with the same sequence will always be the same. The size of the fragments can therefore be determined by comparison with the standard. This approach can be used to generate a **restriction map** of the DNA. Combining restriction enzyme digestion of a single DNA molecule with fluorescent microscopy allows a restriction map to be made directly from the pattern of cuts along the DNA (**Figure 12.3**). This is called *optical mapping* and has been used to create restriction maps of the chromosomes of *E. coli* and other organisms such as *Deinococcus radiodurans*.

Restriction analyses have many uses in molecular biology. They are useful in molecular cloning where they provide a guide as to which restriction enzyme and cut sites to use. Restriction analyses are also used to compare different but related DNA sequences for studies on classification of microorganisms. For example, the banding patterns generated from restriction analyses of either whole chromosomes or specific genes from a series of organisms can yield an indication of their genetic relationships. Multiple restriction digests, using different enzymes either one at a time or in combination, can generate patterns that allow the relative relationships between different DNA molecules to be assessed.

12.1 MiniReview

Restriction enzymes recognize specific short sequences in DNA and make cuts in the DNA. The products of restriction enzyme digestion can be separated using gel electrophoresis.

- Why are restriction enzymes useful to the molecular biologist?
- What is the basis for separating molecules by electrophoresis?

12.2 Nucleic Acid Hybridization and the Southern Blot

When DNA is denatured (that is, the two strands are separated), the single strands can form double-stranded molecules with other single-stranded DNA (or RNA) molecules to form hybrid molecules having complementary (or almost complementary) base sequences (∞ Section 7.2). This is called *nucleic acid hybridization*, or **hybridization** for short, and is widely used in detecting, characterizing, and identifying segments of DNA. Segments of single-stranded nucleic acids whose identity is already known and that are used in hybridization are called **nucleic acid probes** or, simply, probes. To allow detection, probes can be made radioactive or labeled with chemicals that are colored or yield fluorescent products.

Hybridization can be very useful for finding related sequences in different chromosomes or other genetic elements or to find the location of a specific gene. In Southern blotting, named for its inventor, E.M. Southern, probes of known sequence are hybridized to DNA fragments that have been separated by gel electrophoresis. The hybridization procedure when DNA is in the gel and RNA or DNA is the probe is called a **Southern blot**. By contrast, when RNA is in the gel and DNA or RNA is the probe, the procedure is called a **Northern blot**.

In a Southern blot the DNA fragments in the gel are denatured to yield single strands and transferred to a synthetic membrane. The membrane is then exposed to a labeled probe. If the probe is complementary to any of the fragments, hybrids form, and the probe attaches to the membrane at the locations of the complementary fragments. Hybridization can be detected by monitoring the label that has bound to the membrane. **Figure 12.4** shows how a Southern blot can be used to identify fragments of DNA containing sequences that hybridize to the probe.

12.2 MiniReview

Complementary nucleic acid sequences may be detected by hybridization. Probes composed of single-stranded DNA or RNA and labeled with radioactivity or a fluorescent dye are hybridized to target DNA or RNA sequences.

- What is a Southern blot and what does it tell you?
- What is the difference between a Northern blot and a Southern blot?

12.3 Essentials of Molecular Cloning

In **molecular cloning** a fragment of DNA is isolated and replicated. The basic strategy of molecular cloning is to isolate the desired gene (or other segment of DNA) from its original location and move it to a small, simple genetic element, such as a plasmid or virus that is called a **vector** (**Figure 12.5**). When the vector replicates, the cloned DNA

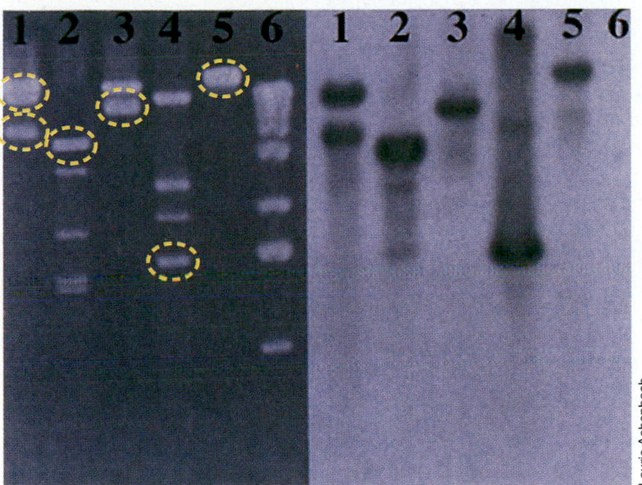

Figure 12.4 Southern blotting. (Left panel) Purified molecules of DNA from several different plasmids were treated with restriction enzymes and then subjected to agarose gel electrophoresis. (Right panel) Southern blot of the DNA gel shown to the left. After blotting, DNA in the gel was hybridized to a radioactive probe. The positions of the bands were visualized by X-ray autoradiography. Note that only some of the DNA fragments (circled in yellow) have sequences complementary to the labeled probe. Lane 6 contained DNA used as a size marker, and none of the bands hybridized to the probe.

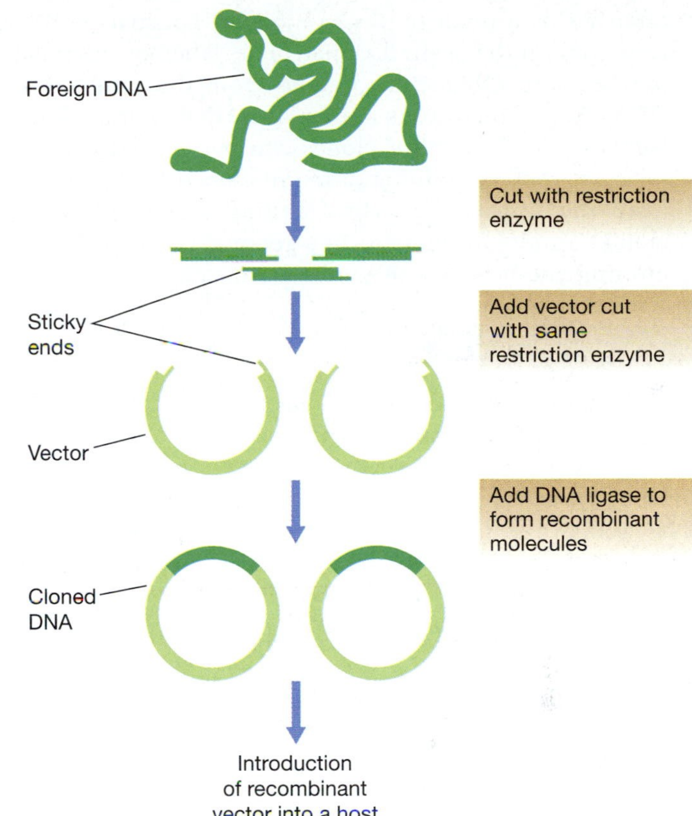

Figure 12.5 Major steps in gene cloning. The vector can be a plasmid or a viral genome. By cutting the foreign DNA and the vector DNA with the same restriction enzyme, complementary sticky ends are generated that allow foreign DNA to be inserted into the vector.

that it carries is also replicated. Molecular cloning thus includes locating the gene of interest, obtaining and purifying a copy of the gene, and placing it into a convenient vector. Once successfully cloned, the gene of interest can be manipulated in various ways and may eventually be put back into a living cell. This approach provides the foundation for much of genetic engineering and has greatly facilitated the detailed analysis of genomes.

The objective of gene cloning is to isolate copies of specific genes in pure form. Consider the nature of the problem. For a genetically "simple" organism such as *Escherichia coli,* an average gene is encoded by 1–2 kbp of DNA out of a genome of over 4,600 kbp. An average *E. coli* gene is thus less than 0.05% of the total DNA in the cell. In human DNA the problem is even more complicated because the coding regions of average genes are not much larger than in *E. coli,* genes are typically split into pieces, and the genome is almost 1,000 times larger! How then can a specific gene be obtained? Fortunately, our knowledge of DNA chemistry and enzymology allows us to break and join and replicate DNA molecules *in vitro*. Consequently, restriction enzymes, DNA ligase, the polymerase chain reaction, and synthetic DNA are important tools used in molecular cloning.

Steps in Gene Cloning: A Summary

1. **Isolation and fragmentation of the source DNA.** The source DNA can be total genomic DNA from an organism of interest, DNA synthesized from an RNA template by reverse transcriptase (⚭ Section 19.15), a gene or genes amplified by the polymerase chain reaction (Section 12.8), or even wholly synthetic DNA made *in vitro* (Section 12.7). If genomic DNA is the source, it is first cut with restriction enzymes (Section 12.1) to give a mixture of fragments of manageable size (Figure 12.5).
2. **Inserting the DNA fragment into a cloning vector.** Cloning vectors are small, independently replicating genetic elements used to replicate genes. Most vectors are derived from plasmids or viruses. Cloning vectors are typically designed to allow *in vitro* insertion of foreign DNA at a restriction site that cuts the vector without affecting its replication (Figure 12.5). If the source DNA and the vector are cut with a restriction enzyme that yields sticky ends, joining of the two molecules is greatly assisted by annealing of the sticky ends (Figure 12.5). Blunt ends generated by some restriction enzymes can be joined by direct ligation or by using synthetic DNA linkers or adapters. In either case, the strands are joined by DNA ligase, an enzyme that covalently links both strands of the vector and the inserted DNA.
3. **Introduction of the cloned DNA in a host organism.** Recombinant DNA molecules made in the test tube are introduced into suitable host organisms where they can replicate. Transformation (⚭ Section 11.10) is often used to get recombinant DNA into cells. In practice this

often yields a mixture of recombinant constructs. Some cells contain the desired cloned gene, whereas other cells may contain other cloned genes from the same source DNA. Such a mixture is known as a **DNA (gene) library** because many different clones can be purified from the mixture, each containing different cloned DNA segments from the source organism. Making a gene library by cloning random fragments of a genome is called **shotgun cloning** and is widely used in genomic analyses.

12.3 MiniReview

The isolation of a specific gene or region of a chromosome by molecular cloning is done using a plasmid or virus as the cloning vector. Restriction enzymes and DNA ligase are used *in vitro* to produce a chimeric DNA molecule composed of DNA from two or more sources. Once introduced into a suitable host, the cloned DNA can be produced in large amounts under the control of the cloning vector.

- What is the purpose of molecular cloning?
- What are the roles of a cloning vector, restriction enzymes, and DNA ligase in molecular cloning?

12.4 Plasmids as Cloning Vectors

Plasmids replicate independently of the host chromosome. In addition to carrying genes required for their own replication, most plasmids are natural vectors because they often carry other genes that confer important properties on their hosts. Plasmids have very useful properties as cloning vectors. These include (1) small size, which makes the DNA easy to isolate and manipulate; (2) independent origin of replication, so plasmid replication in the cell proceeds independently of direct chromosomal control; (3) multiple copy number, so they are present in the cell in several copies, thus giving both high yields of DNA and high-level expression of cloned genes; and (4) presence of selectable markers such as antibiotic resistance genes, which makes detection and selection of plasmid-borne clones easier.

Although conjugative plasmids are transferred by cell-to-cell contact in nature, most plasmid cloning vectors have been genetically modified to abolish conjugative transfer. This prevents unwanted movement of the vector into other organisms. However, vector transfer in the laboratory can be facilitated by chemically mediated transformation or electroporation (Section 11.10). Depending on the host–plasmid system, replication of the plasmid may be under tight cellular control, in which case only a few copies are made, or under relaxed cellular control, in which case a large number of copies is made. Obtaining a high copy number is often important in gene cloning, and by proper selection of the host–plasmid system and manipulation of cellular macromolecule synthesis, plasmid copy numbers of several thousand per cell can be obtained.

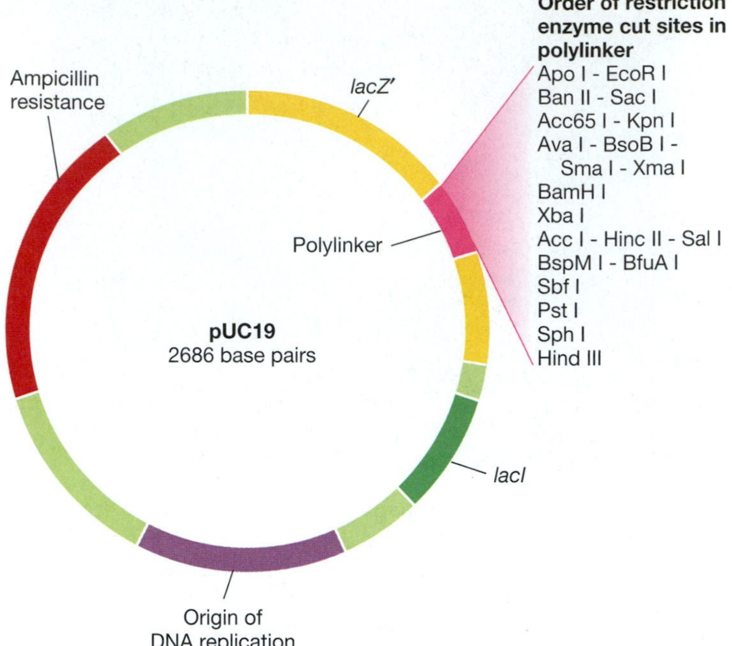

Figure 12.6 Cloning vector plasmid pUC19. Essential features include an ampicillin resistance marker and the polylinker with multiple restriction enzyme cut sites. Insertion of cloned DNA within the polylinker inactivates the *lacZ'* gene that encodes β-galactosidase and allows for easy identification of transformants by blue–white screening (see Figure 12.7).

An Example of a Cloning Vector: The Plasmid pUC19

The first plasmid cloning vectors used were natural isolates. In particular, the ColE plasmids of *Escherichia coli* that encode colicin E were used because they are relatively small and are naturally present in multiple copies, making DNA isolation easier. However, these were soon replaced by plasmids that were themselves the result of *in vitro* manipulations. A widely used plasmid cloning vector is pUC19 (**Figure 12.6**). This was derived in several steps from the ColE1 plasmid by removing the colicin genes and inserting genes for ampicillin resistance and for a blue–white color-screening system (see below). A segment of artificial DNA with cut sites for many restriction enzymes, called a *polylinker* or *multiple cloning site*, is inserted into the *lacZ* gene, the gene that encodes the lactose-degrading enzyme, β-galactosidase (Section 9.3). The presence of this short polylinker does not inactivate *lacZ*. Cut sites for restriction enzymes present in the polylinker are absent from the rest of the vector. Consequently, treatment with each of these enzymes opens the vector at a unique location but does not cut the vector into multiple pieces. Plasmid pUC19 has a number of characteristics that make it suitable as a cloning vehicle:

1. It is relatively small, only 2,686 bp.
2. It is stably maintained in its host (*E. coli*) in relatively high copy number, about 50 copies per cell.

3. It can be amplified to a very high number (1,000–3,000 copies per cell, about 40% of the cellular DNA) by inhibiting protein synthesis with the antibiotic chloramphenicol.
4. It is easy to isolate in the supercoiled form using routine techniques.
5. Moderate amounts of foreign DNA can be inserted, although inserts of more than 10 kbp lead to plasmid instability.
6. The complete base sequence of the plasmid is known, allowing identification of all restriction enzyme cut sites.
7. The polylinker contains single cut sites for a dozen restriction enzymes, such as *Eco*RI, *Sal*I, *Bam*HI, *Pst*I, and *Hin*dIII. It is crucial that only a single cut site for any given restriction enzyme should exist on a cloning vector so that treatment with that enzyme linearizes the vector but does not cut it into pieces. Single cut sites for multiple restriction enzymes increases the versatility of the vector.
8. It has a gene conferring ampicillin resistance on its host. This permits ready selection of hosts containing the plasmid because such hosts gain resistance to the antibiotic.
9. It can be inserted into cells easily by transformation.
10. Insertion of foreign DNA into the polylinker can be detected by blue–white screening because of *lacZ*.

Cloning Genes into Plasmid Vectors

The use of plasmid vectors in gene cloning is shown in **Figure 12.7**. A suitable restriction enzyme with a cut site within the polylinker is chosen. Both the vector and the foreign DNA to be cloned are cut with this enzyme. The vector is linearized. Segments of the foreign DNA are inserted into the open cut site and ligated into position with DNA ligase. This disrupts the *lacZ* gene, a phenomenon called *insertional inactivation*. This may be used to detect the presence of foreign DNA within the vector. When the colorless reagent Xgal is added to the medium, β-galactosidase cleaves it, generating a blue product. Thus, cells containing the vector *without* cloned DNA form blue colonies, whereas cells containing the vector *with* an insert of cloned DNA do not form β-galactosidase and are therefore white. Thus, after DNA ligation, the resulting plasmids are transformed into cells of *E. coli*. The colonies are selected on media containing both ampicillin, to select for the presence of the plasmid, plus Xgal, to test for β-galactosidase activity. Those colonies that are *blue* contain the plasmid without any inserted foreign DNA (i.e., the plasmid merely recyclized without picking up foreign DNA), whereas those colonies that are *white* contain plasmid with inserted foreign DNA and are picked for further analyses (see Figure 12.24 for a related example of the blue–white selection system).

Many subsequent vectors include features similar to those of pUC19 listed above. Sometimes insertional inactivation can be detected by selection, rather than by screening. For example, in some vectors, the gene carrying the polylinker normally produces a protein that is lethal to the host cell.

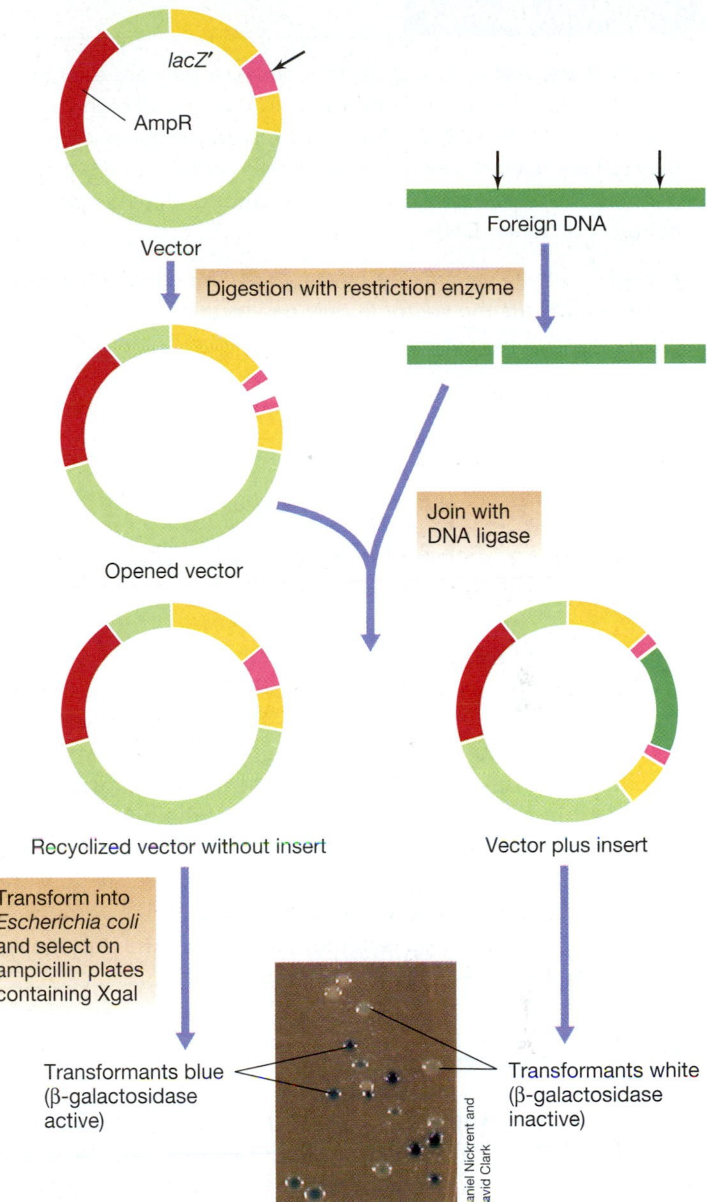

Figure 12.7 Cloning into the plasmid vector pUC19. The cloning vector is opened by cutting with a suitable restriction enzyme in the polylinker. Insertion of DNA inactivates β-galactosidase allowing blue–white screening for the presence of the insert. The photo on the bottom shows colonies of *Escherichia coli* on an Xgal plate. The enzyme β-galactosidase can cleave the normally colorless Xgal to form a blue product.

Therefore, only cells containing a plasmid in which this gene has been inactivated can grow.

Cloning using plasmid vectors is versatile and widely used in genetic engineering, particularly when the fragment to be cloned is fairly small. Also, plasmids are often used as cloning vectors if expression of the cloned gene is desired, since regulatory genes can be engineered into the plasmid to obtain expression of the cloned genes under specific conditions (Section 12.13).

12.4 MiniReview

Plasmids are useful cloning vectors because they are easy to isolate and purify and are often able to multiply to high copy numbers in bacterial cells. Antibiotic resistance genes on the plasmid are used to select bacterial cells containing the plasmid, and color-screening systems are used to identify colonies containing cloned DNA.

- Explain why in cloning it is necessary to use a restriction enzyme that cuts the vector in only one location.
- What is insertional inactivation?
- What is a polylinker?

II SEQUENCING, SYNTHESIS, AND AMPLIFICATION OF DNA

Restriction enzyme and hybridization analyses are useful, but full characterization of a segment of DNA requires that the complete nucleotide sequence be known. Once some sequence information is available, segments of DNA may be artificially synthesized. These may be used as primers or probes in further analyses or to provide altered versions of parts of genes or regulatory regions. When handling DNA segments, both for sequencing or construction, it is often necessary to replicate the DNA to obtain sufficient quantities for experimentation. This is achieved by using the polymerase chain reaction.

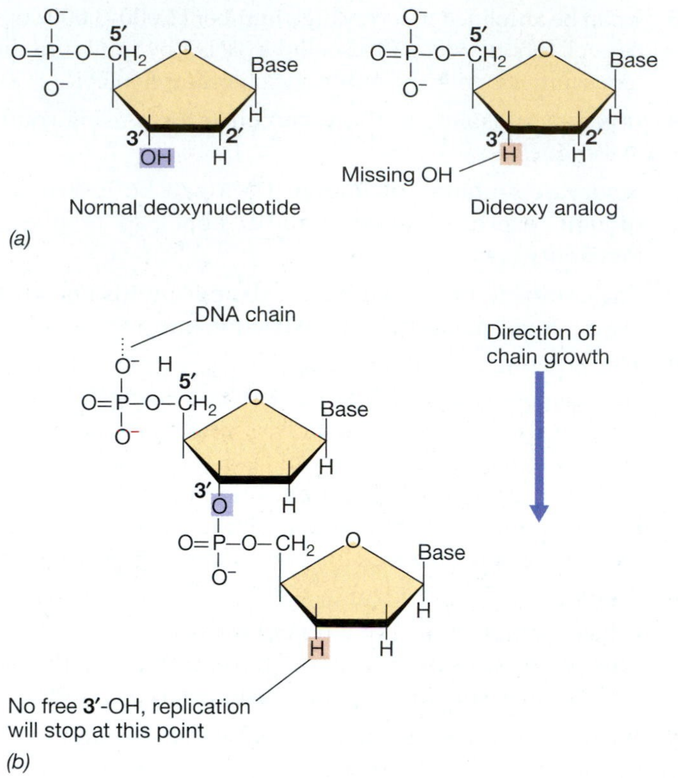

Figure 12.8 Dideoxynucleotides and Sanger sequencing. *(a)* A normal deoxynucleotide has a hydroxyl group on the 3′-carbon and a dideoxynucleotide does not. *(b)* Elongation of the chain terminates where a dideoxynucleotide is incorporated. Compare this figure with Figure 7.11.

12.5 Sequencing DNA

The term **sequencing** refers to determining the precise order of nucleotides in a DNA (or RNA) molecule. For use in sequencing, short fragments of DNA with defined sequences, called *oligonucleotides* (typically 10–20 nucleotides), are artificially synthesized. These DNA molecules are used as **primers**. Primers are short lengths of DNA or RNA used to initiate the synthesis of new strands of nucleic acid. During DNA replication *in vivo*, RNA primers are used (∞ Section 7.5), but in biotechnology, both in DNA synthesis and in the polymerase chain reaction (Section 12.8), primers made of DNA are used because they are more stable than RNA primers. Nucleic acid sequencing is a key tool for molecular biology, and we describe the essentials of the sequencing process here.

DNA Sequencing: The Sanger Dideoxy Method

Most DNA sequencing today is done by the dideoxy method invented by the British scientist Fred Sanger, who won a Nobel Prize for this important technique. This method generates DNA fragments of different lengths that terminate at each of the four bases and that are labeled with either radioactivity or a fluorescent dye. These fragments are then separated by gel electrophoresis such that molecules that differ by only one nucleotide in length are clearly resolved. The procedure requires four separate reactions (and four separate gel lanes) for each sequence determination, one for fragments ending at each of the four bases in DNA: adenine, guanine, cytosine, and thymine. The positions of the fragments are located by autoradiography (the darkening of photographic film by exposure to radiation) or fluorescence, and the sequences can then be read off the gel.

In the Sanger procedure the sequence is actually determined by making a *copy* of the single-stranded DNA using the enzyme DNA polymerase. As we have previously seen, this enzyme uses deoxyribonucleoside triphosphates as substrates and adds them to a primer. In each of the four incubation mixtures are small amounts of a different *dideoxy analog* of the deoxyribonucleoside triphosphates (**Figure 12.8**). Because the dideoxy sugar lacks the 3′-hydroxyl, when it is inserted, elongation of the chain cannot continue. The dideoxy analog thus functions as a specific chain-termination reagent. Oligonucleotide fragments of variable length are obtained, depending on the incubation conditions. Electrophoresis of these fragments is then carried out, and the positions of the bands are determined by exposing X-ray film or by fluorescence. By aligning the four dideoxynucleotide lanes and noting the vertical position of each fragment relative to its neighbor, the sequence of the DNA copy can be read directly from the gel (**Figure 12.9**).

The Sanger method can be used to sequence RNA as well as DNA. To sequence RNA, a single-stranded DNA copy is first made (using the RNA as the template) by the enzyme reverse transcriptase (∞ Section 10.12). The single-stranded DNA is then sequenced by the Sanger dideoxy method as described above, and the RNA sequence is deduced by base-pairing rules from the DNA sequence.

Automated Sequencing

The demands of large-scale sequencing projects have led to the development of automated DNA-sequencing systems. With such systems, the sequencing reactions are still based on the dideoxy method, but fluorescent dye-labeled primers (or nucleotides) are used instead of radioactivity. The products are separated by automated electrophoresis and the bands detected by fluorescence spectroscopy. Each of the four different reactions uses a fluorescent label of a different color, and the lanes are scanned by a fluorescence-detecting laser. In this way, all four reactions can be run on a single lane. The results are analyzed by computer and a sequence generated, with each of the four bases being distinguished by a separate color (Figure 12.9*c*).

454 Pyrosequencing

Recent technical advances have revolutionized DNA sequencing. In particular, the system developed by the 454 Life Sciences corporation has the potential to generate sequence data 100 times faster than previous methods. The 454 system relies on two major advances, *massively parallel liquid handling* and *pyrosequencing* instead of dideoxy sequencing.

In the 454 system, the DNA sample is broken into single-stranded segments of about 100 bases each, and each fragment is immobilized to a microscopic bead. The DNA is amplified by the polymerase chain reaction (Section 12.8), resulting in each bead carrying several identical copies of the DNA strand. The beads are then put into a fiber-optic plate with over a million wells, each of which holds just one bead, and the four nucleotides are flowed sequentially over the plate in a fixed order; sequencing reactions then occur simultaneously across the entire plate.

As for Sanger sequencing, the principle of 454 sequencing involves synthesis of a complementary strand by DNA polymerase. However, instead of chain termination that drives the dideoxy method (Figure 12.9), in the 454 method each time a nucleotide is incorporated in the complementary strand, a molecule of pyrophosphate is released, which provides the energy needed for the release of light by the enzyme luciferase (∞ Section 9.6), also incorporated into the system. Thus, each light pulse signals that the sequence being analyzed incorporated a base complementary to the one present at the site of insertion. By base-pairing rules, the identity of the nucleotide in the sample DNA at this site is also revealed.

Although 454 sequencing is remarkably fast because the sample DNA in each well is being analyzed at the same time

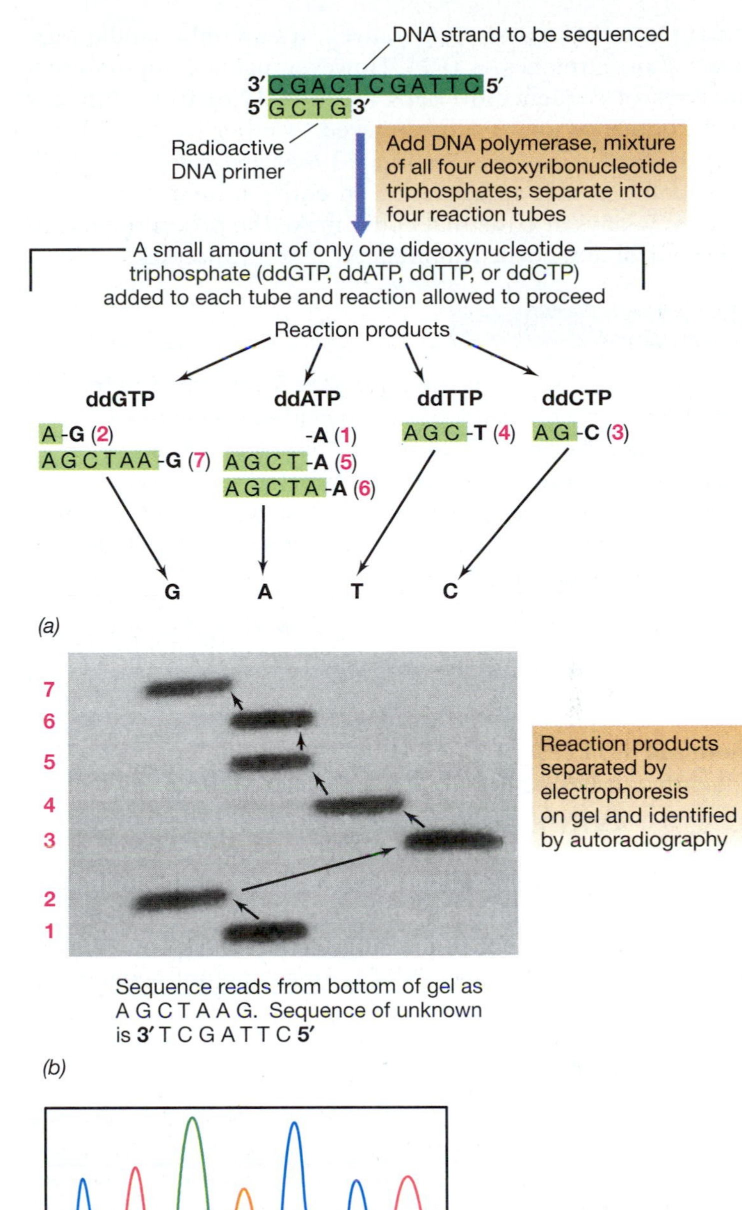

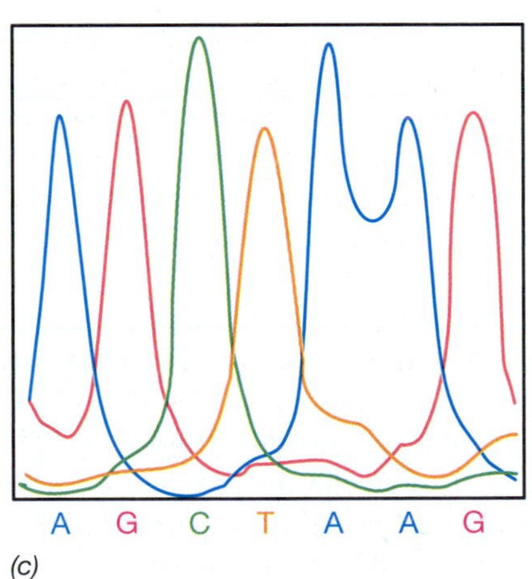

Figure 12.9 DNA sequencing using the Sanger method. *(a)* Note that four different reactions must be run, one with each dideoxynucleotide. Because these reactions are run *in vitro*, the primer for DNA synthesis need not be RNA, and for convenience is DNA. *(b)* A portion of a gel containing the reaction products from *(a)*. *(c)* Results of sequencing the same DNA as shown in *(a)* and *(b)*, but using an automated sequencer and fluorescent labels.

UNIT 2

(thus the term "massively parallel"), it can only handle relatively short stretches of DNA. However, using computational analyses of sequence overlaps (Section 12.6) from different wells, entire genomes can be pieced together by the 454 system. In fact, it has been estimated that 454 sequencing will eventually be able to sequence an entire human genome in just a few days at costs that could make the procedure useful for medical diagnostics and related genetic analyses.

12.5 MiniReview

DNA sequencing can be done using the Sanger chain termination method that employs dideoxynucleotides to block elongation of growing DNA chains. The use of radioactive labeling has largely given way to automated sequencing that relies on fluorescent labeling. Advances in technology, such as the 454 system, have greatly increased the speed by which DNA can be sequenced.

- How do dideoxynucleotides block chain elongation?
- Why are DNA primers necessary for DNA sequencing?

12.6 Sequencing and Annotating Entire Genomes

The analysis of a genome begins with the formation of a genomic library—the molecular cloning of all of the DNA fragments of the genome generated from restriction enzyme cuts (Section 12.3). Once this is done, actual sequencing begins.

Shotgun Sequencing

Virtually all genomic sequencing projects today employ shotgun sequencing. This technique is made possible by high-throughput sequencing capacity, robotics, and powerful computational capacities. Shotgun sequencing has been used on chromosomes from prokaryotes as well as eukaryotes, including the privately funded version of the human genome project.

In the whole-genome shotgun approach the entire genome is cloned in a random fashion, and the resultant clones are then sequenced. That is, the clones are sequenced without knowing the order or orientation of the DNA fragments. The sequences are then analyzed by a computer that searches for overlapping sequences. The overlaps allow the computer to assemble the sequenced fragments in the correct order.

Much of the sequencing in the shotgun method is redundant. To ensure that a complete genome sequence is obtained it is necessary to sequence a very large number of clones, many of which will be identical or nearly identical. Typically, there will be 7–10 replicate sequences for any given part of the genome. This seven- to tenfold coverage greatly reduces errors in the sequence because the redundancy in sequencing allows for a consensus nucleotide to be selected at any point in the sequence where ambiguity may exist.

For shotgun sequencing to work effectively, the cloning itself must be efficient (one needs a large number of clones) and, in so far as possible, the DNA segments cloned should be randomly generated. Restriction sites are not random, but by cutting the genomic DNA with an enzyme that recognizes a short common sequence, one can approach random digestion. To obtain more truly random fragments, DNA can be mechanically sheared by forcing it through a nebulizer. This is a device with a small opening similar to a nozzle that reduces the DNA solution to a spray; in the process, the DNA is sheared. The DNA fragments can be purified by size using gel electrophoresis (Section 12.1) and then cloned into a vector and transformed into a host.

Genome Assembly

Once shotgun sequencing is completed, all of the fragments must be ordered, a process called *assembly*. This puts the fragments in the correct order, eliminating overlaps, and generates a genome suitable for annotation, the process of identifying genes and other functional regions.

Sometimes shotgun sequencing and assembly does not yield a complete genome sequence and there are gaps left in the sequence. In such situations, individual clones can be sought to cover the gap. One method of doing this is to perform polymerase chain reactions (Section 12.8) using specific primers complementary to the known sequences that flank a given gap. Note that these additional clones are targeted, not random, as in the shotgun method. However, the sequences in these clones are likely to contain overlapping sequences sufficient to close the gap.

Some genome projects have the goal of obtaining a *closed genome,* meaning that the entire genome sequence is determined. Other projects stop at the *draft stage,* dispensing with sequencing of the small gaps. Since shotgun sequencing and assembly are heavily automated procedures, but gap closure is not, a closed genome is much more expensive and time consuming to generate than a draft genome sequence.

Annotating the Genome

Once sequencing and assembly is completed, the next step in genomic analysis is *annotation,* the conversion of raw sequence data into a list of the genes present in the genome. The great majority of genes of any organism encode proteins, and in most microbial genomes, the great majority of the genome consists of coding sequences. Because the genomes of microbial eukaryotes typically have fewer introns (∞ Section 8.5) than plant and animal genomes, and prokaryotes have almost none, microbial genomes essentially consist of hundreds to thousands of open reading frames (ORFs) (∞ Section 7.13) separated by short regulatory regions and transcriptional terminators.

How Does the Computer Find an ORF?

A *functional ORF* is one that actually encodes a protein in the cell. Thus, the simplest way to locate protein-encoding genes, or potential protein-encoding genes, is to have a computer search the sequence of the genome for ORFs. In the cell, ribosomes establish a reading frame by initiating translation at a start codon, usually an AUG. The ribosome then proceeds until it reaches an in-frame stop codon (Section 7.13). Therefore, the first step in finding an ORF is to look for these signals in the sequence.

However, the identification of possible functional ORFs is typically more complex than just searching for in-frame start and stop codons, because these will appear randomly with reasonable frequency. Thus, other clues are sought. One hint that an ORF is functional is its size. The majority of cellular proteins contain 100 or more amino acids, so most functional ORFs are longer than 100 codons (300 nucleotides). However, simply programming the computer to ignore ORFs shorter than 100 codons will miss some genuine but short genes. So other factors such as **codon bias** (**codon usage**) must be considered. More than one codon exists for most of the 20 amino acids (Table 7.4), and some codons are used more frequently than others. Codon bias differs between organisms. If the codon usage in a given ORF is considerably different from the consensus codon usage, the ORF may not be functional or may be functional but obtained by lateral gene transfer (Section 13.11). Furthermore, prokaryotic ribosomes start translation, not at the first possible start codon, but at one immediately downstream of a Shine–Dalgarno (ribosome-binding) sequence on the mRNA (Section 7.15). Therefore, searching the DNA sequence of a prokaryotic genome for potential Shine–Dalgarno sequences can help establish both whether an ORF is functional and which start codon is actually used.

Although any given gene is always transcribed from a single strand, in all but the smallest plasmid or viral genomes, both strands are transcribed in some part of the genome. Thus, computer inspection of both strands of DNA is required. In addition, some genes encode tRNAs and rRNAs that are not recognized by programs that search only for ORFs. However, genes for tRNAs and rRNAs can usually be located easily because the sequences of these RNAs are highly conserved (Section 14.9).

Locating Putative Genes: Genomic Comparisons

An ORF is also likely to be functional if its sequence is similar to the sequences of ORFs in the genomes of other organisms (regardless of whether they encode known proteins) or if some part of the ORF has a sequence known to encode a protein functional domain. This is because proteins with similar functions in different cells tend to be *homologous*; that is, they are related in an evolutionary sense and typically share sequence and structural features. Computers are used to search for such sequence similarities in databases such as GenBank. This database contains over 100 billion base pairs of information and can be searched online at http://www.ncbi.nlm.nih.gov/Genbank/index.html. The most widely used database search tool is BLAST (Basic Local Alignment Search Tool), which has several variants depending on whether nucleic acid or protein sequences are used for searching. For example, the tool BLASTn searches nucleic acid databases using a nucleotide query, whereas BLASTp searches protein databases using a protein query.

It would be almost impossible to assemble even a very small genome and to locate genes and other important functional DNA sequences without the availability of sophisticated computational tools to handle large databases. Using computers to do analyses of this type is called working *in silico,* a term analogous to the terms *in vivo* and *in vitro* (*in silico* refers to the silicon processor chips present in the computer).

UNIT 2

12.6 MiniReview

Shotgun techniques employ random cloning and sequencing of relatively small genome fragments followed by computer-generated assembly of the genome using overlaps as a guide to the final sequence. After major sequencing is finished, computers search for ORFs and genes encoding protein homologues as part of the annotation process.

- Why is shotgun sequencing considered to give a very accurate genome sequence?
- What is done during genome assembly?
- What is an open reading frame (ORF)?
- How can protein homology assist in the annotation process?

12.7 Synthesizing DNA

DNA of defined sequence is necessary for many molecular procedures. Systems for synthesizing DNA are thus available and are completely automated such that an oligonucleotide of 30–35 bases can be made routinely and oligonucleotides of over 100 bases in length can be made if necessary. For the synthesis of longer polynucleotides, individual oligonucleotide fragments can be joined enzymatically using DNA ligase (Section 12.1).

Oligonucleotide Synthesis

DNA is synthesized *in vitro* in a solid-phase procedure in which the first nucleotide in the chain is fastened to an insoluble support (such as porous glass beads about 50 μm in size). The overall procedure is shown in **Figure 12.10**. Several steps are needed for the addition of each nucleotide, and the chemistry is intricate. After each step is completed, the reaction mixture is flushed out of the solid support and the series of reactions repeated for the addition of the next nucleotide. Once the oligonucleotide is the desired length, it is cleaved from the solid-phase support by a specific reagent and purified to eliminate by-products and contaminants.

Synthetic DNA molecules are widely used for various purposes. In addition to their use as primers for DNA sequencing, they are used as primers for the polymerase chain reaction (Section 12.8). They are also used as probes to detect, through nucleic acid hybridization, specific DNA or RNA sequences in

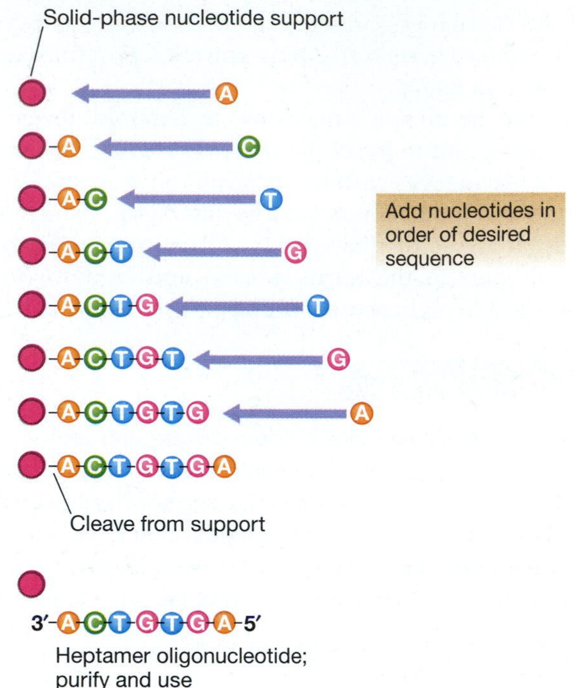

Figure 12.10 Solid-phase procedure to synthesize DNA. Chemical synthesis proceeds by adding one nucleotide at a time to the growing chain. Note that artificial synthesis proceeds from 3′ to 5′ (the opposite of natural DNA replication).

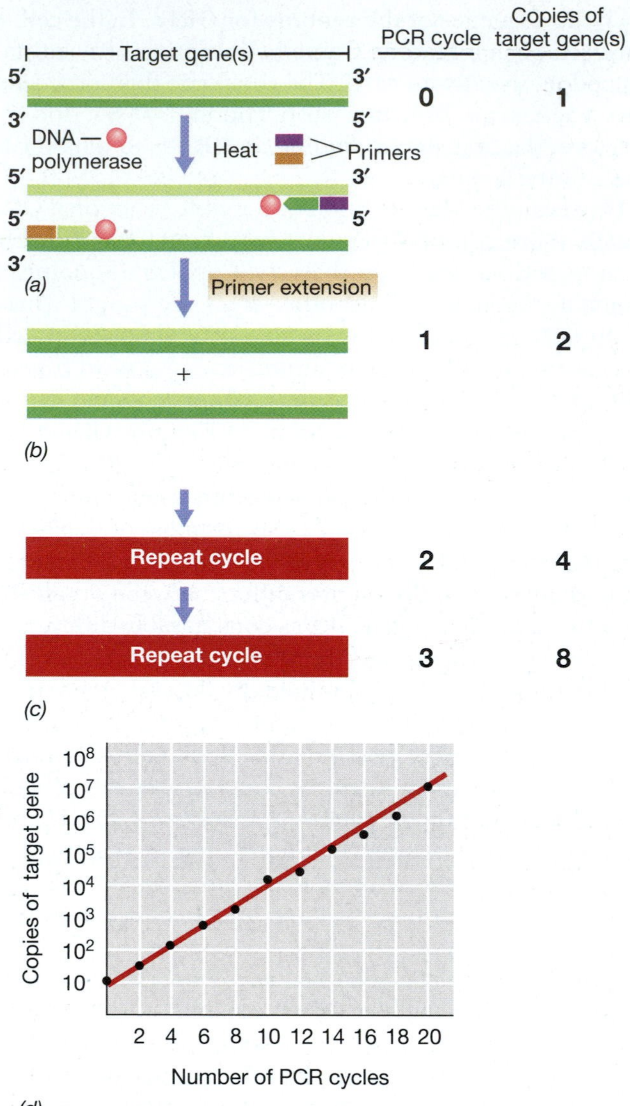

Figure 12.11 The polymerase chain reaction (PCR). The PCR amplifies specific DNA sequences. *(a)* Target DNA is heated to separate the strands, and a large excess of two oligonucleotide primers, one complementary to the target strand and one to the complementary strand, is added along with DNA polymerase. *(b)* Following primer annealing, primer extension yields a copy of the original double-stranded DNA. *(c)* Two additional PCR cycles yield four and eight copies, respectively, of the original DNA sequence. *(d)* Effect of running 20 PCR cycles on a DNA preparation originally containing ten copies of a target gene. Note that the plot is semilogarithmic.

Southern or Northern blots, respectively (Section 12.2). Synthetic DNA is also used in site-directed mutagenesis to create mutations of interest at specific locations within a gene (Section 12.9).

12.7 MiniReview

Moderate lengths of DNA can be synthesized chemically by an automated procedure. Synthetic DNA is used as primers and probes for several techniques, as well as in the artificial creation of gene segments.

- What techniques use short DNA primers and probes?
- Why is a solid support used during chemical synthesis of DNA?

12.8 Amplifying DNA: The Polymerase Chain Reaction

Copies of a specific DNA sequence are necessary for many molecular procedures, and the **polymerase chain reaction (PCR)** is the method by which copies are made *in vitro*. The polymerase chain reaction can copy segments of DNA by up to a billionfold in the test tube, a process called *amplification*, yielding large amounts of specific genes or other DNA segments for a host of applications in molecular biology. PCR makes use of the enzyme DNA polymerase, which naturally copies DNA molecules (Section 7.6). PCR does not actually copy whole DNA molecules but amplifies stretches of up to a few thousand base pairs (the *target*) from within a larger DNA molecule (the *template*). PCR was conceived by Kary Mullis, who received a Nobel Prize for this achievement.

The steps in PCR amplification of DNA can be summarized as follows:

1. Two DNA oligonucleotide primers flanking the target DNA (**Figure 12.11a**) are made on an oligonucleotide synthesizer (Section 12.7) and added in excess to heat-denatured template DNA.

2. As the mixture cools, the excess primers relative to the template DNA ensures that most template strands anneal to a primer and not to each other (Figure 12.11*a*).
3. DNA polymerase then extends the primers using the original DNA as the template (Figure 12.11*b*).
4. After an appropriate incubation period, the mixture is heated again to separate the strands. The mixture is then cooled to allow the primers to hybridize with complementary regions of newly synthesized DNA, and the whole process is repeated (Figure 12.11*c*).

The power of PCR is that the products of one primer extension are templates for the next cycle such that each cycle doubles the amount of the original target DNA. In practice, 20–30 cycles are usually run, yielding a 10^6-fold to 10^9-fold increase in the target sequence (Figure 12.11*d*). Because the technique consists of several highly repetitive steps, PCR machines, called *thermocyclers*, are available that run through the heating and cooling cycles automatically. Because each cycle requires only about 5 minutes, the automated procedure allows for large amplifications in only a few hours.

PCR at High Temperature

The original PCR technique employed the DNA polymerase *Escherichia coli* Pol III, but because of the high temperatures needed to denature the double-stranded copies of DNA being made, the enzyme was also denatured and had to be replenished every cycle. This problem was solved by employing a thermostable DNA polymerase isolated from the thermophilic hot spring bacterium *Thermus aquaticus*, an organism isolated by Thomas Brock, an American microbiologist. DNA polymerase from *T. aquaticus*, called *Taq polymerase*, is stable to 95°C and thus is unaffected by the denaturation step employed in the PCR reaction. The use of *Taq* DNA polymerase also increased the specificity of the PCR reaction because the DNA is copied at 72°C rather than 37°C. At such high temperatures, nonspecific hybridization of primers to nontarget DNA is rare, thus making the product of *Taq* PCR more homogeneous than that obtained using the *E. coli* enzyme. On the other hand, the primer hybridization step is often carried out at lower temperatures, which may allow some nonspecific binding.

DNA polymerase from *Pyrococcus furiosus*, a hyperthermophile with a growth temperature optimum of 100°C (Section 17.11) is called *Pfu polymerase* and is even more thermostable than *Taq* polymerase. Moreover, unlike *Taq* polymerase, *Pfu* polymerase has proofreading activity (Section 7.8), making it a particularly good enzyme when high accuracy is crucial. Thus, the error rate for *Taq* polymerase under standard conditions is 8.0×10^{-6} (per base duplicated), whereas for *Pfu* polymerase it is only 1.3×10^{-6}. To supply the demand for thermostable DNA polymerases in the PCR and DNA sequencing markets, the genes for these enzymes have been cloned into *E. coli* and produced commercially in large quantities. **www.microbiologyplace.com Online Tutorial 12.1: Polymerase Chain Reaction (PCR)**

Applications and Sensitivity of PCR

PCR is a powerful tool. It is easy to perform, extremely sensitive and specific, and highly efficient. During each round of amplification the amount of product doubles, leading to an exponential increase in the DNA (Figure 12.11) This means not only that a large amount of amplified DNA can be produced in just a few hours, but that only a few molecules of target DNA need be present in the sample to start the reaction. The reaction is so specific that, with primers of 15 or so nucleotides and high annealing temperatures, there is almost no "false priming," and therefore the PCR product is virtually homogeneous.

PCR is extremely valuable for obtaining DNA for cloning genes or for sequencing purposes because the gene or genes of interest can easily be amplified if flanking sequences are known. PCR is also used routinely in comparative or phylogenetic studies to amplify genes from various sources. In these cases the primers are made for regions of the gene that are conserved in sequence across a wide variety of organisms. For example, because 16S rRNA, a molecule used for phylogenetic analyses (Section 2.7), has both highly conserved and highly variable regions, primers specific for the 16S rRNA gene from *Bacteria* or *Archaea* can be synthesized and used to survey these organisms in a specific habitat. Moreover, if primers with greater specificity are used, only certain subgroups within each domain are targeted. This technique is in widespread use in microbial ecology and has revealed the enormous diversity of the microbial world, much of it not yet cultured (Section 22.5).

Because it is such a sensitive technique, PCR can be used to amplify very small quantities of DNA. For example, PCR has been used to amplify and clone DNA from sources as varied as mummified human remains and fossilized plants and animals. The ability of PCR to amplify and analyze DNA from cell mixtures has also made it a common tool of diagnostic microbiology. For example, if a clinical sample shows evidence of a gene specific to a particular pathogen, then it can be assumed that the pathogen was present in the sample. Treatment of the patient can then begin without the need to culture the organism, a time-consuming and often fruitless process. PCR has also been used in *DNA fingerprinting*, a powerful forensic tool that permits identification of human individuals, or relationships between individuals, from very small samples of their DNA (see the Microbial Sidebar, "DNA Fingerprinting").

Reverse Transcriptase and Real-Time PCR

The power of PCR has been extended by the development of related techniques such as *reverse transcriptase PCR*, *real-time PCR*, and *quantitative PCR*. These techniques are routinely applied to the analysis of environmental samples (Section 22.5) or to the diagnosis of pathogens in clinical samples (Section 32.12).

The first step in reverse transcriptase PCR (RT-PCR) is to use the enzyme reverse transcriptase (Section 19.15) to make a complementary DNA (cDNA) copy of an RNA sample.

UNIT 2

Microbial Sidebar

DNA Fingerprinting

Because every individual (except for identical twins) has a unique genomic DNA sequence, DNA samples can be used for identification. The use of DNA technology to differentiate between very closely related organisms, including the identification of individual humans, is called **DNA fingerprinting**.

DNA fingerprinting is greatly helped by the fact that higher organisms contain many repetitive DNA sequences, some of which vary in number and arrangement in different individuals. Some repeats exist as families of related sequences scattered around the genome. One class of very useful repetitive sequences contains those with identical sequences but varying numbers of repeats at a single site on a particular chromosome. These sequences are called *variable number tandem repeats* (*VNTR*), and several have been identified.

The use of VNTRs in DNA fingerprinting is illustrated in **Figure 1a**. This shows two different alleles (alternative forms of the same gene) of a eukaryotic chromosome that differ

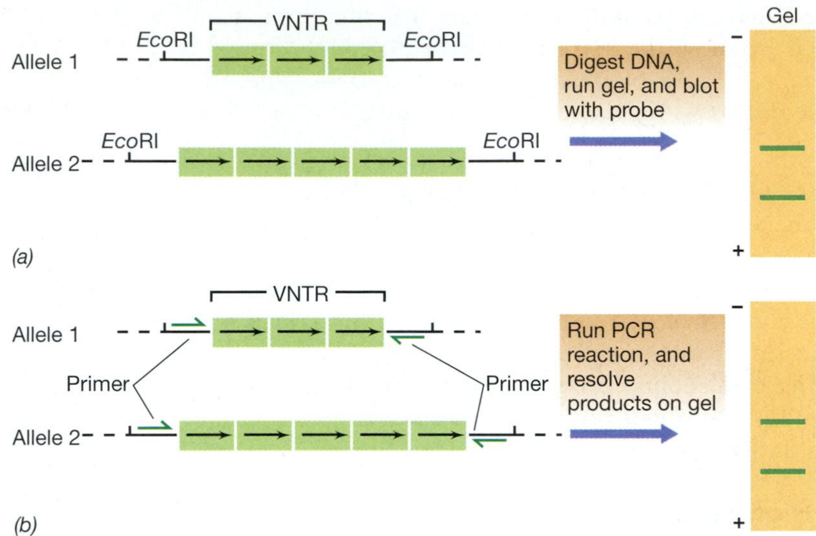

Figure 1 DNA fingerprinting. (a) *Two different alleles of a region of a single chromosome. The alleles differ only in the number of repeats in the VNTR. DNA from cells containing these chromosomes can be cut with the restriction enzyme EcoRI (which does not cut within the VNTR) and the fragments separated on an agarose gel (smaller fragments migrate farther in a gel than do larger fragments). The fragments containing the VNTR are then identified after Southern blotting by hybridization with a probe specific to the VNTR. The figure shows only the result from individuals whose two chromosomes have different alleles at this site. If an individual had the same allele on both chromosomes, only a single band would be observed.* (b) *The same alleles, but using specific primers to amplify the VNTR segments by PCR. The products of the PCR reaction can be loaded directly onto the gel without restriction digestion.*

Next, PCR is used to amplify the cDNA, which can have several important applications. For example, one could monitor microbial gene expression in an environmental sample by isolating RNA and employing RT-PCR to make DNA copies of the corresponding gene(s). The DNA can then be sequenced or probed to identify the genes. Another common application is for cloning eukaryotic genes; the mRNAs are amplified using RT-PCR. This avoids the need to remove the introns which must be done if the genes are cloned directly from chromosomal DNA (Section 26.2).

In real-time PCR the accumulation of target DNA is monitored during the PCR process. This is achieved by adding to the PCR reaction mixture fluorescent probes whose fluorescence increases upon binding to DNA. As the target DNA is amplified, the level of fluorescence increases proportionally. The fluorescent probes may be specific for the target DNA sequence or nonspecific. For example, the dye SYBR Green binds nonspecifically, although only to double-stranded DNA, and becomes fluorescent only when bound. Specific probes are made by attaching a fluorescent dye to a short DNA probe that matches the target sequence being amplified. These fluoresce only when DNA of the correct sequence accumulates.

The advantage of real-time PCR is that amplification can be monitored continuously, thus running a gel afterwards to confirm amplification isn't necessary. For example, if you wanted to know whether a gene diagnostic of a particular pathogen were present in a DNA preparation obtained from a clinical sample, you would know this in an hour or two instead of the usual overnight processing required by standard PCR. Moreover, by monitoring the *rate* of fluorescence increase in the PCR reaction, it is possible to accurately determine the *amount* of target DNA present in the original sample. This is known as quantitative PCR, or quantitative real-time PCR (qrt-PCR). This form of PCR also has many uses, in particular, environmental DNA analyses. For instance, qrt-PCR can be used to assess the abundance of an

(continued)

only in how many copies of the repeated sequence are present. Because the sequence of the VNTR is known, restriction enzymes whose sites are not present in the VNTR are chosen to digest the DNA. Digestion thus releases the intact VNTR. When DNA from two chromosomes with a different number of repeats at a particular locus is digested, the fragments of DNA differ in size. These are separated by gel electrophoresis and the VNTR-containing fragments are detected by Southern blotting (Section 12.2) using a probe made from the cloned VNTR sequence. Even if multiple alleles are present in a population (rather than just two as in the example), differences at a single site are insufficient to unequivocally identify an individual. Many individuals in a population would be expected to have the same pattern at any given site. However, it is possible to probe the digested DNA simultaneously for several different VNTR markers. With such methods, the probability of identifying a particular individual by comparing two different DNA samples is extremely high.

The polymerase chain reaction (PCR) can simplify the identification process. PCR can be used to amplify segments of DNA carrying VNTRs (Figure 1*b*). Although PCR is not needed for DNA fingerprinting, *per se*, it is essential when the amount of DNA in the sample is very small—such as that found in the cells of a single hair. To use PCR, it is necessary to know the sequences flanking the VNTR so that specific primers can be synthesized. With enough cycles of amplification, it is possible to detect the PCR-generated bands by simply staining the gels rather than using a hybridization probe; the bands run to different regions of the gel based on their size (number of repeats). Because PCR amplifies only the DNA between the primers, the DNA does not need to be cut with restriction enzymes before being run on a gel.

Some VNTRs have 100–200 different variants, making them very useful for analysis. By testing enough different VNTRs, a unique pattern can be found for any individual. The STR (short tandem repeat) is a subcategory of VNTR in which the repeat is 2–6 nucleotides long. Many STRs have 10–20 variants. Currently, individual identification is normally done by using PCR and a set of about a dozen STRs. One of the most fascinating applications of STR technology was the identification of the partial remains of the last Russian Czar, Nicholas II, and his family, who were executed by the Bolsheviks in 1918. Their remains were reburied in 1998.

VNTRs are not restricted to higher organisms. The genome of *Bacillus anthracis*, the organism that causes the disease anthrax, contains several different VNTRs. This allowed strains of *B. anthracis* used as biological weapons to be identified and tracked. This helps to identify the source of *B. anthracis* strains or can be used as evidence that a particular strain was used in a bioterrorism incident (Sections 33.11 and 33.12).

organism in a sample by quantifying a gene characteristic of that particular organism.

12.8 MiniReview

The polymerase chain reaction is a procedure for amplifying DNA *in vitro* and employs heat-stable DNA polymerases. Heat is used to denature the DNA into two single-stranded molecules, each of which is copied by the polymerase. After each cycle, the newly formed DNA is denatured and a new round of copying proceeds. After each cycle, the amount of target DNA doubles.

- Why is a primer needed at each end of the DNA segment being amplified?
- From which organisms are thermostable DNA polymerases obtained?
- How does real-time PCR differ from standard PCR?

III BACTERIAL GENE MANIPULATION

When methods for *in vitro* DNA manipulation were developed, they provided new approaches in many areas, including both the generation of mutants and the analysis of gene expression. Both of these fields have been revolutionized by the ability to change one or a few bases at will or to link distinct segments of DNA in a predetermined order.

12.9 Molecular Methods for Mutagenesis

As we have seen, conventional mutagens introduce mutations *at random* in the intact organism (Section 11.7). In contrast, *in vitro* mutagenesis, better known as **site-directed mutagenesis**, uses synthetic DNA plus DNA-cloning techniques to introduce mutations into genes *at precisely determined sites*. In addition to changing just a few bases,

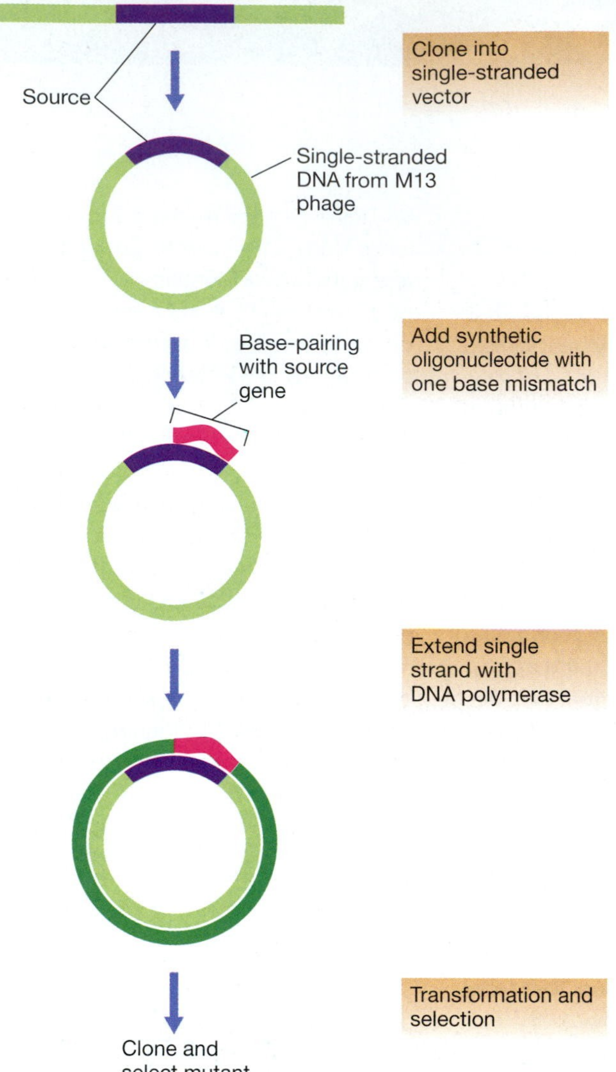

Figure 12.12 Site-directed mutagenesis using synthetic DNA. Short synthetic oligodeoxyribonucleotides may be used to generate mutations. Cloning into the genome of bacteriophage M13 yields the single-stranded DNA needed for site-directed mutagenesis to work.

mutations may also be engineered by inserting large segments of DNA at precisely determined locations.

Site-Directed Mutagenesis

Site-directed mutagenesis is a powerful tool, as it allows for a change to any base pair in a specific gene and thus has many uses in genetics. The basic procedure is to synthesize a short DNA oligonucleotide primer containing the desired base change (mutation) and to allow this to base-pair with single-stranded DNA containing the target gene. Pairing will be complete except for the short region of mismatch. Then the synthetic oligonucleotide is extended using DNA polymerase, thus copying the rest of the gene. The double-stranded molecule obtained is inserted into a host cell by transformation. Mutants are often selected by some sort of positive selection, such as antibiotic resistance; in this case, the modified DNA would also carry a nearby antibiotic resistance marker.

One procedure for site-directed mutagenesis is illustrated in **Figure 12.12**. The process starts with cloning the target gene into a single-stranded DNA vector. A widely used vector for this purpose is bacteriophage M13, a single-stranded DNA phage whose genome is easy to manipulate and that replicates in *Escherichia coli* (∞ Section 19.2). Several cloning methods are available based on M13 (Section 12.15). The mutagenized DNA can then bind by complementary base pairing with the target gene (Figure 12.12). This gives a DNA molecule with a mismatch. After cell division and vector DNA replication, both daughter molecules will be fully base-paired, but one daughter molecule will carry the mutation and the other will be wild type. Progeny bacteria are then screened for those carrying the mutation.

The basic scheme of site-directed mutagenesis may also be carried out using PCR. In this case, the short DNA oligonucleotide with the required mutation is used as a PCR primer. The mutagenic primer is designed to anneal to the target with the mismatch in the middle and must have enough matching nucleotides on both sides for binding to be stable during the PCR reaction. The mutagenic primer is paired with a normal primer. When the PCR reaction amplifies the target DNA, it incorporates the mutations in the mutagenic primer.

Applications of Site-Directed Mutagenesis

Site-directed mutagenesis can be used to assess the activity of proteins containing known amino acid substitutions. Suppose one was studying the importance of particular amino acids in the active site of an enzyme. Site-directed mutagenesis could be used to change a specific amino acid in the enzyme, and the modified enzyme could then be assayed and compared to the wild-type enzyme. In such an experiment, the vector encoding the mutant enzyme would be inserted into a mutant host strain unable to make the original enzyme. Consequently, the activity measured would be due to the mutant version of the enzyme alone.

Using site-directed mutagenesis, enzymologists can link virtually any aspect of an enzyme's activity, such as catalysis, resistance or susceptibility to chemical or physical agents, or interactions with other proteins, to specific amino acids in the protein. In particular, site-directed mutagenesis has given detailed information to structural biologists interested in which amino acids are critical for an enzyme's function and significant insight into the mechanisms behind enzyme-catalyzed reactions.

Cassette Mutagenesis and Gene Disruption

Because of the large number of restriction enzymes commercially available and therefore the large number of different DNA sequences that can be cut, it is usually possible to find several different restriction sites within any target gene. However, if sites for the appropriate restriction enzyme are not present at the required location, they can be inserted by site-directed mutagenesis as shown in Figure 12.12. If restriction sites are reasonably close together, the intervening DNA fragment can be excised and replaced by a synthetic DNA fragment in which one or more of the nucleotides have been changed. These synthetic fragments are called **DNA cassettes** (or cartridges), and the process is known as **cassette mutagenesis**. When using cassettes to replace sections of genes, the cassettes are typically the same size as the wild-type DNA fragments they replace.

Another type of cassette mutagenesis is called **gene disruption**. In this technique, cassettes are inserted into the middle of a gene, thus disrupting the coding sequence. Cassettes used for making insertion mutations can be almost any size and can even carry an entire gene. To facilitate the selection process, cassettes that encode antibiotic resistance are commonly used. The process of gene disruption is illustrated in **Figure 12.13**. In this case, a DNA cassette carrying a gene conferring kanamycin resistance, the Kan cassette, is inserted at a restriction site in a cloned gene. The vector carrying this mutant gene is then linearized by cutting with a different restriction enzyme. Finally, the linear DNA is transformed into the host, and kanamycin resistance is selected. The linearized plasmid cannot replicate, and so resistant cells arise mostly by homologous recombination (∞ Section 11.9) between the mutated gene on the plasmid and the wild-type gene on the chromosome.

Note that when a cassette is inserted, the cells not only gain antibiotic resistance but also lose the function of the gene into which the cassette is inserted. Such mutations are called *knockout mutations*. These are similar to insertion mutations made by transposons (∞ Section 13.10), but here the experimenter chooses which gene will be mutated. Knockout mutations in haploid organisms (such as prokaryotes), yield viable cells only if the disrupted gene is unessential. Indeed, gene knockouts are a convenient means of determining whether a given gene is an essential one.

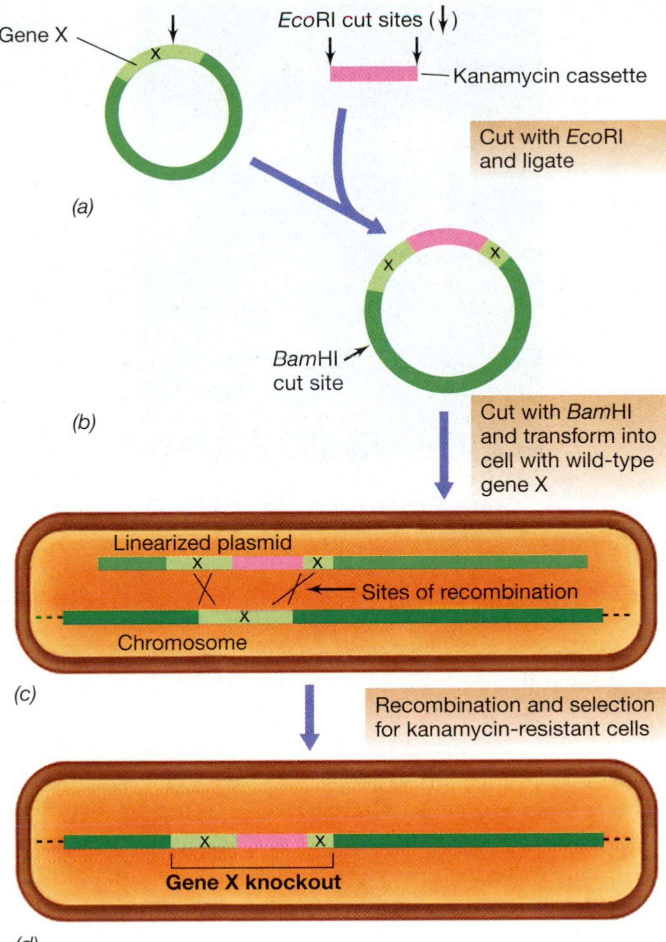

Figure 12.13 Gene disruption by cassette mutagenesis. *(a)* A cloned wild-type copy of gene X, carried on a plasmid is cut with *Eco*RI and mixed with the kanamycin cassette. *(b)* The cut plasmid and the cassette are ligated, creating a plasmid with the kanamycin cassette as an insertion mutation within gene X. This new plasmid is cut with *Bam*H1 and transformed into a cell. *(c)* The transformed cell contains the linearized plasmid with a disrupted gene X and its own chromosome with a wild-type copy of the gene. *(d)* In some cells, homologous recombination occurs between the wild-type and mutant forms of gene X. Cells that can grow in the presence of kanamycin have only a single disrupted copy of gene X.

12.9 MiniReview

Synthetic DNA molecules of desired sequence can be made *in vitro* and used to construct a mutated gene directly or to change specific base pairs within a gene by site-directed mutagenesis. Also, genes can be disrupted by inserting DNA fragments, called cassettes, into them, generating knockout mutants.

- How can site-directed mutagenesis be useful to enzymologists?
- What are knockout mutations?

12.10 Gene Fusions and Reporter Genes

Not only did *in vitro* DNA manipulation allow new approaches to generating mutations, it also revolutionized the study of gene regulation. Constructs may be engineered in which a coding sequence from one source is placed next to a regulatory region from another source. This may be used to increase expression of a desired gene product as discussed below (Section 12.13) or to analyze the response of a regulatory region to diverse conditions.

Reporter Genes

We discussed above (Section 12.4) the benefit of locating a cloning site within a gene whose activity is easy to assay. That system employed *lacZ*, the gene encoding β-galactosidase, an enzyme whose activity was easily detected on indicator plates (Figure 12.7 and see Figure 12.24) or by simple liquid assays. In this application, the *lacZ* gene was being used as a **reporter gene**.

The key property of a reporter gene is that it encodes a protein that is easy to detect and assay. Reporter genes are used for a variety of purposes. They may be used to signal the presence or absence of a particular genetic element (such as a plasmid) or DNA inserted within a vector. They can also be fused to other genes or to the promoter of other genes so that expression can be studied.

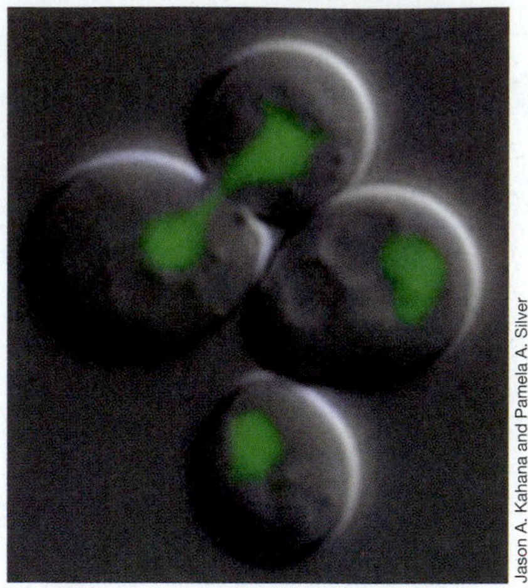

Figure 12.14 Green fluorescent protein (GFP). GFP can be used as a tag for protein localization *in vivo*. In this example, the gene encoding Pho2, a DNA-binding protein from the yeast *Saccharomyces cerevisiae*, was fused to the gene encoding GFP and photographed by fluorescence microscopy. The recombinant gene was transformed into budding yeast cells. These expressed the fluorescent fusion protein localized in the nucleus.

There are many choices for reporter genes besides *lacZ*. The luciferase enzyme, which we discussed earlier, makes cells expressing it luminescent (Section 9.6). Colonies containing this reporter system can be detected on agar plates by their luminescence against a large background of other colonies. However, the expression of luciferase depends on more than one gene because several accessory factors are needed. By contrast, the **green fluorescent protein** (GFP) needs no accessory factors and is widely used as a reporter (**Figure 12.14**). Although the gene for GFP was originally cloned from the jellyfish *Aequorea victoria*, the GFP protein may be expressed in most cells. It is stable and causes little or no disruption of host cell metabolism. If expression of a cloned gene is linked to that of GFP, expression of GFP signals that the cloned gene has also been expressed (Figure 12.14).

Gene Fusions

It is possible to engineer constructs that consist of segments from two different genes. Such constructs are known as **gene fusions**. If the promoter that controls a coding sequence is removed, the coding sequence can be fused to a different regulatory region to place the gene under the control of a different promoter. Alternatively, the promoter can be fused to a gene whose product is easy to assay. This approach is often used in studying gene regulation, especially where measuring the levels of the natural gene product is difficult or time consuming. The regulatory region of the gene of interest is fused to the coding sequence for a reporter gene, such as that for β-galactosidase or GFP. The reporter is then made under the conditions that would trigger expression of the target gene (**Figure 12.15**). The expression of the reporter can then be assayed under a variety of conditions to determine how the gene of interest is regulated.

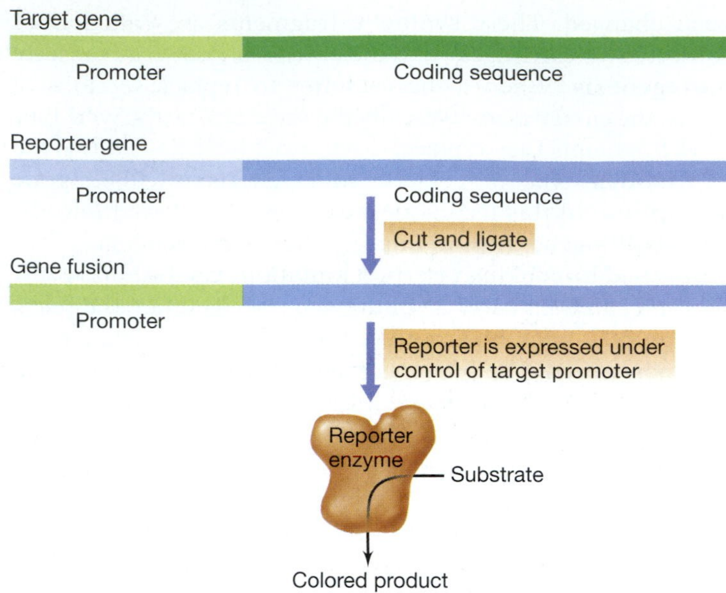

Figure 12.15 Construction and use of gene fusions. The promoter of the target gene is fused to the reporter coding sequence, and the reporter gene is expressed under those conditions where the target gene would normally be expressed. The reporter shown here is an enzyme (such as β-galactosidase) that converts a substrate to a colored product that is easy to detect. This approach greatly facilitates the investigation of regulatory mechanisms.

Gene fusions may also be used to test for the effects of regulatory genes. Mutations that affect regulatory genes are introduced into cells carrying gene fusions, and expression is measured and compared to cells lacking the regulatory mutations. This allows the rapid screening of multiple regulatory genes that are suspected of controlling the target gene.

12.10 MiniReview

Reporter genes are genes whose products, such as β-galactosidase or GFP, are easy to assay or detect. They are used to simplify and increase the speed of genetic analysis. In gene fusions, segments from two different genes, one of which is usually a reporter gene, are spliced together.

- Why are gene fusions useful in studying gene regulation?
- What is a reporter gene?

IV ADVANCED CLONING TECHNIQUES

12.11 Hosts for Cloning Vectors

To obtain large amounts of cloned DNA, an ideal host should grow rapidly in an inexpensive culture medium. The host should also be nonpathogenic, be capable of incorporating DNA, be genetically stable in culture, and have the appropriate enzymes to allow replication of the vector. It is also helpful if considerable background information on the host and a wealth of tools for its genetic manipulation exists.

The most useful hosts for cloning are microorganisms that are easily grown and for which we have much information. These include the bacteria *Escherichia coli* and *Bacillus subtilis* and the yeast *Saccharomyces cerevisiae* (**Figure 12.16**). Complete genome sequences are available for all of these organisms, and they are widely used as cloning hosts. However, in some cases other hosts and specialized vectors may be necessary to get the DNA properly cloned and expressed.

Prokaryotic Hosts

Although most molecular cloning has been done in *E. coli* (Figure 12.16), this host has a few disadvantages. *E. coli* is an excellent choice for initial cloning work but is problematic as an expression host because it is found in the human intestine and some wild-type strains are potentially pathogenic. However, several modified *E. coli* strains have been developed for cloning purposes, and thus *E. coli* remains the organism of choice for most molecular cloning.

The gram-positive organism *B. subtilis* is also used as a cloning host (Figure 12.16). Although the technology for cloning in *B. subtilis* is less advanced than for *E. coli,* several plasmids and phages suitable for cloning have been developed, and transformation is a well-developed procedure in *B. subtilis*. The main disadvantage of using *B. subtilis* as a cloning host is plasmid instability. It is often difficult to maintain plasmid replication over many subcultures of the organism. Also, foreign DNA is not as well maintained in *B. subtilis* as in *E. coli,* thus the cloned DNA is often unexpectedly lost.

Often host organisms for cloning must have specific genotypes to be effective. For instance, if the cloning vector uses the *lacZ* gene for screening (Section 12.10), then the host must carry a mutation disabling this gene. Because the bacteriophage M13 infects only bacteria with F pili (Section 12.15), hosts used with M13-derived vectors must contain and express genes on the F plasmid. These types of considerations and others, such as the ease of selection of transformants, must be taken into account when choosing a cloning host.

Eukaryotic Hosts

Cloning in eukaryotic microorganisms has focused on the yeast *S. cerevisiae* (Figure 12.16). Plasmid vectors as well as artificial chromosomes (Section 12.15) have been developed for yeast. One important advantage of eukaryotic cells as hosts for cloning vectors is that they already possess the complex RNA and posttranslational processing systems required for the production of eukaryotic proteins. Thus these systems do not have to be engineered into the vector or host cells as would be required if eukaryotic DNA was cloned and to be expressed in a prokaryotic cloning host.

For many applications, gene cloning in mammalian cells has been done. Mammalian cell culture systems can be handled in some ways like microbial cultures, and they find wide use in research on human genetics, cancer, infectious disease, and physiology. A disadvantage of mammalian cells as hosts is that they are expensive and difficult to produce under large-scale conditions. Insect cell lines are simpler to grow, and vectors have been developed from an insect DNA virus, the baculovirus (Section 12.15). For some applications, in particular for plant agriculture, the cloning host can be a plant cell tissue culture line or even an entire plant. Indeed, genetic engineering has many applications in plant agriculture (Section 26.10). However, regardless of eukaryotic host type, it is necessary to get the vector DNA into the host cells, and we consider this now.

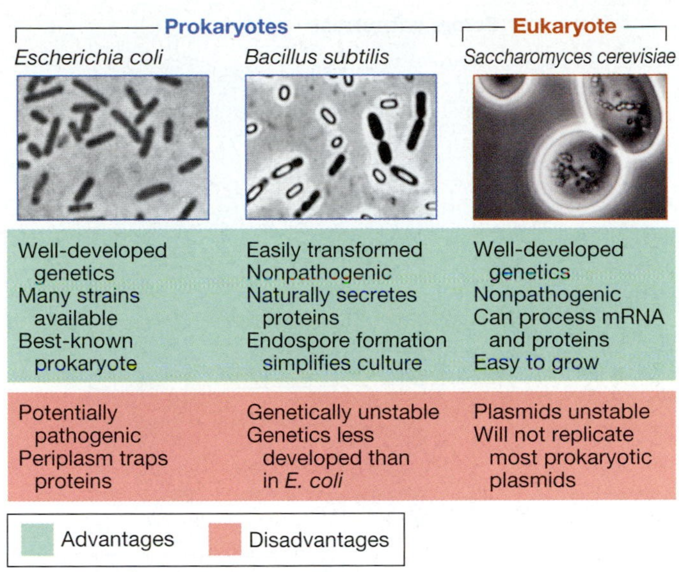

Prokaryotes
Eukaryote
Escherichia coli
Bacillus subtilis
Saccharomyces cerevisiae
Well-developed genetics
Many strains available
Best-known prokaryote
Easily transformed
Nonpathogenic
Naturally secretes proteins
Endospore formation simplifies culture
Well-developed genetics
Nonpathogenic
Can process mRNA and proteins
Easy to grow
Potentially pathogenic
Periplasm traps proteins
Genetically unstable
Genetics less developed than in E. coli
Plasmids unstable
Will not replicate most prokaryotic plasmids
Advantages
Disadvantages

Figure 12.16 Hosts for molecular cloning. A summary of the advantages and disadvantages of some common cloning hosts.

Transfection of Eukaryotic Cells

Eukaryotic microorganisms and animal and plant cells can take up DNA in a process that resembles bacterial transformation (Section 11.10). Because the word "transformation" is used to describe the conversion of mammalian cells to a cancerous state, the introduction of DNA into mammalian cells is usually called *transfection*.

Transfection of cultured animal cells was originally accomplished by precipitating DNA in such a way that the cells would take it up by phagocytosis, which is possible because animal cells do not have cell walls. In animal cells, DNA can also be injected directly into the nucleus using micropipets, a technique called *microinjection*.

In yeast, which is an important organism for genetic engineering, transfection at low efficiencies can be mediated by various methods. However, a technique called *electroporation* is widely applicable for eukaryotic cells and can be used whether or not the cell wall of the organism is removed. Electroporation exposes host cells to pulsed electrical fields in the presence of cloned DNA. The electrical treatment opens small pores in the cytoplasmic membrane through which the DNA can enter. The pores generated are insufficient to cause major cell damage or lysis, but sufficient to allow cloned DNA to enter.

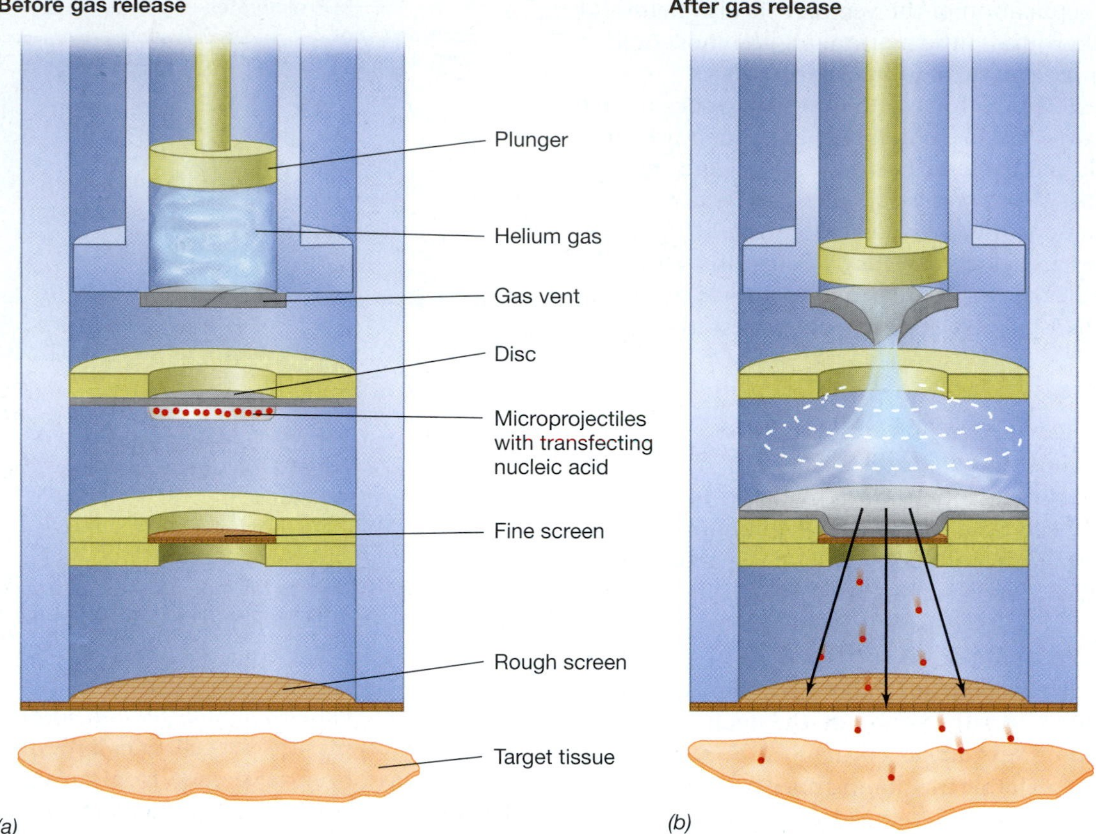

Figure 12.17 DNA gun for transfection of eukaryotic cells. The inner workings of the gun show how metal pellets coated with nucleic acids (microprojectiles) are projected at target cells. *(a)* Before firing and *(b)* after firing. A shock wave due to gas release throws the disc carrying the microprojectiles against the fine screen. The microprojectiles continue on into the target tissue.

In addition to electroporation, a high-velocity microprojectile "gun" can be used to get DNA into cells. The particle gun or gene gun operates by using a propellant such as pressurized helium to fire DNA-coated particles through a small steel cylinder at target cells (**Figure 12.17**). The particles bombard the cell, piercing the cell wall and cytoplasmic membrane without actually killing the cells. The nucleic acid entering the cells can then recombine with host DNA. The particle gun has been used successfully to transfect yeast, algae, and higher plant cells. Moreover, unlike electroporation, the particle gun can be used to get DNA into intact tissues, such as plant seeds, and into mitochondria and chloroplasts.

12.11 MiniReview

The choice of a cloning host depends on the final application. In many cases the host can be a prokaryote, but in others, it is essential that the host be a eukaryote. Any host must be able to take up DNA, and there are a variety of techniques by which this can be accomplished, both natural and artificial.

- Why does molecular cloning require a host?
- Describe three mechanisms by which cells can take up DNA.

12.12 Finding the Right Clone

Genetic engineering begins with the isolation of a clone containing a gene of interest. Various manipulations can then be performed on this gene, depending on the application. But first it is necessary to identify the host containing the correct clone. Methods to clone DNA include making gene libraries from total genomic DNA (Section 12.3) or cloning a DNA fragment made by PCR (Section 12.8). PCR products are usually used when the gene of interest has already been identified and the goal is to obtain a single clone. However, in a gene library there may be thousands or tens of thousands of clones, and often only one or a few may contain the genes of interest. How is the desired clone identified?

The Initial Screen

One can isolate host cells containing a plasmid vector by selecting for a vector marker, such as antibiotic resistance, so that only these cells form colonies (Section 12.4). For host cells containing a viral vector, one simply looks for plaques (Section 12.14). These colonies or plaques can also be screened for vectors that contain foreign DNA inserts by looking for the inactivation of a vector gene (Figure 12.7).

When cloning a single DNA fragment generated by PCR or purified by some other means, such simple selections or screenings are usually sufficient. However, with a heterogeneous collection of DNA fragments, as in a gene library, identifying cells carrying cloned DNA is only the first step. The biggest challenge remains finding the clone that has the gene of interest. One must examine colonies of bacteria or plaques from viral-infected cells growing on agar plates and detect those few that contain the gene of interest. Sometimes the cloned gene is expressed and the protein is synthesized in the cloning host and may be detected. Often, however, the cloned gene is not expressed. Then the only option is to look for the DNA itself.

Foreign Genes Expressed in the Cloning Host

If the foreign gene is expressed in the cloning host, the encoded protein can be searched for in colonies of the host cell. Obviously, for this to work the host itself must not produce the protein being studied. Then when the foreign gene is incorporated, its expression can be detected. This makes selection of cells containing cloned genes relatively simple, especially if expression of the cloned genes is easy to observe. A striking example is the cloning of luciferase genes (∞ Figure 9.14) into other organisms, which allows the host organism to glow in the dark.

Using Antibodies to Detect Proteins

Antibodies can be used as detection reagents for a protein of interest. Produced by an animal, antibodies are serum proteins that bind in a highly specific way to a target molecule, the antigen (∞ Section 29.5). In the application of antibodies for detection, the protein encoded by the cloned gene is the antigen and is used to produce an antibody in an experimental animal. Because the antibody combines specifically with the antigen, when the antigen is present in one or more colonies on the plate, the locations of these colonies can be determined by observing the binding of the antibody.

Because only a small amount of the protein (antigen) is present in each colony, only a small amount of antibody is bound, and so a highly sensitive procedure for detecting bound antibody must be available. In practice, detection relies on radioisotopes, chemiluminescent chemicals, or enzymes. These and other extremely sensitive techniques for detecting antigens are discussed in Chapter 32.

This method of detection depends on screening, not selection, and so thousands of clones may need to be examined. These can be colonies containing plasmids or plaques containing viruses that produce the cloned product. The procedure using radioactive detection is outlined in **Figure 12.18*a***. As shown, the replica-plating procedure (∞ Figure 11.5) is used to make a duplicate of the master plate. However, the duplication is done onto a synthetic membrane filter, and all further manipulations are done with this filter. The duplicate colonies are lysed to release the protein (antigen) of interest. (Screening using virus vectors eliminates this step because the bacteria are already lysed.) The antibody is then added, and reacts with the antigen. Unbound antibody is washed off, and a radioactive

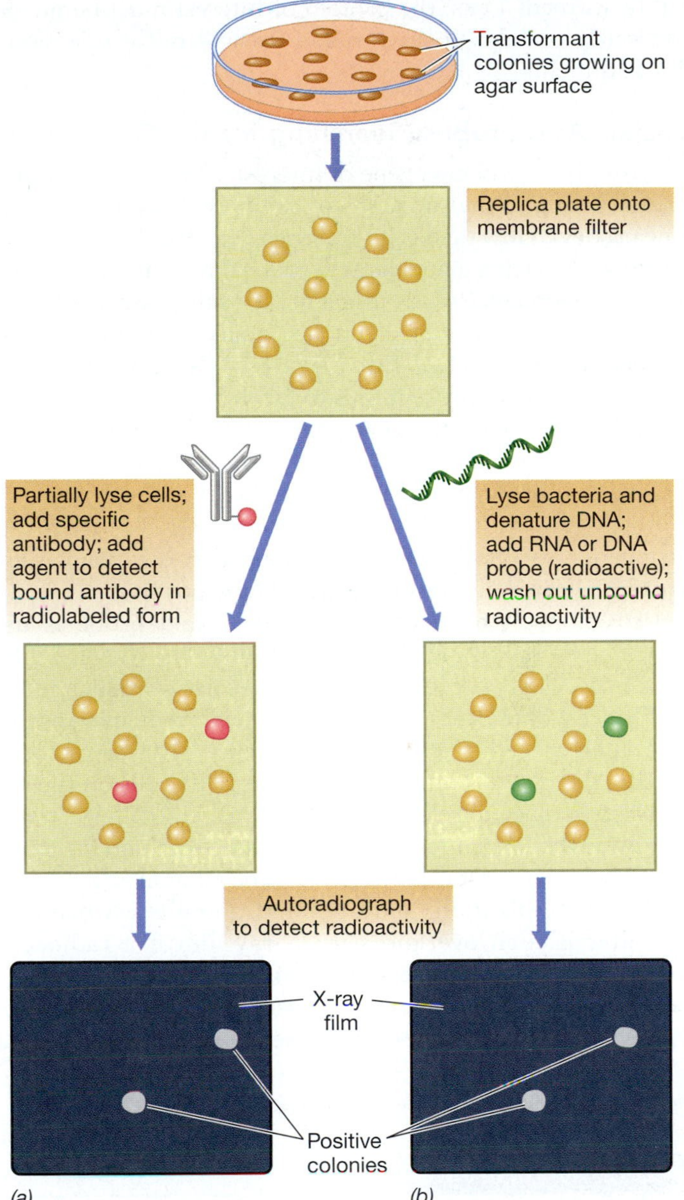

Figure 12.18 Finding the right clone. *(a)* Method for detecting production of protein by using a specific antibody. *(b)* Method for detecting recombinant clones by colony hybridization with a radioactive nucleic acid probe. Formation of a DNA duplex binds the DNA probe to a particular spot on the membrane.

reagent is added that is specific for the antibody. A sheet of X-ray film is placed over the filter and exposed. If a radioactive colony is present, a spot is observed on the X-ray film after it is developed (Figure 12.18*a*). The location of this spot on the film corresponds to the location of a colony on the master plate that produces the protein. This colony can then be picked from the master plate and subcultured.

One limitation of this procedure is that an antibody must be available that is specific for the protein in question. Antibodies are produced by injecting the specific protein antigen into an animal. But to be successful, the protein injected must be pure; otherwise, antibodies of several specificities will be formed. Thus, the protein of interest must be purified previously or false-positive reactions will make selection of clones difficult.

Nucleic Acid Probes: Searching for the Gene Itself

Suppose that the cloned gene of interest is not expressed in the cloning host and that no assay or antibody is available for the gene product. How then is its presence detected? This is done by hybridization using a nucleic acid probe that contains a key part of the base sequence of the target gene. As discussed, hybridization can be used to detect nucleic acids containing specific sequences (Section 12.2). Either DNA or RNA can be used as a probe. The general procedure is to label the nucleic acid probe, typically with radioactive phosphate ($^{32}PO_4^{\ 2-}$) or a fluorescent chemical group. Following this, the single-stranded probe is hybridized to single-stranded DNA from the clone. By using appropriate hybridization conditions, it is possible to adjust the "stringency" of the hybridization such that complementarity must be nearly exact; this helps avoid false positives.

In colony hybridization, a nucleic acid probe is used to detect recombinant DNA in colonies, as shown in Figure 12.18*b*. This procedure again makes use of replica plating to produce a duplicate of the master plate on a membrane filter. (The same procedure can be carried out with virus vectors by blotting the plaques onto a membrane.) The cells on the filter are lysed in place to release their DNA, and the filter is treated to convert the DNA into a single-stranded form and fix it to the filter. This filter is then exposed to a labeled nucleic acid probe to allow hybridization, and unbound probe is washed away. The filter is then overlaid with X-ray film if a radioactive probe was used. After development, the X-ray film is examined for spots (Figure 12.18*b*). These correspond to locations on the membrane where the radioactive probe hybridized the DNA from a particular colony. Colonies corresponding to these spots are then chosen and studied further.

12.12 MiniReview

Special procedures are needed to detect a foreign gene in the cloning host. If the gene is expressed, the presence of the encoded protein, as detected either by its activity or by reaction with specific antibodies, is evidence that the gene is present. However, if the gene is not expressed, it may be detected by hybridization to a nucleic acid probe.

- Does use of nucleic acid probes depend on gene expression? Explain.
- Why is it necessary to lyse cells containing plasmids in order to detect the product of the cloned gene?

12.13 Shuttle Vectors and Expression Vectors

Once a gene has been cloned, it may be used for a variety of purposes. Specialized vectors have been engineered for use in different situations. Here we consider how to move a cloned gene between organisms of different species and how to optimize expression of the cloned gene. Two classes of vectors are appropriate, *shuttle* vectors and *expression* vectors.

Shuttle Vectors

Vectors that can replicate and are stably maintained in two (or more) unrelated host organisms are called **shuttle vectors**. Genes carried by a shuttle vector can thus be moved between unrelated organisms. Shuttle vectors have been developed that replicate in both *Escherichia coli* and *Bacillus subtilis*, *E. coli* and yeast, and *E. coli* and mammalian cells, as well as in many other pairs of organisms. The importance of a shuttle vector is that DNA cloned in one organism can be replicated in a second host without modifying the vector in any way to do so.

Many shuttle vectors have been designed to move genes between *E. coli* and yeast. Bacterial plasmid vectors were the starting point and were modified to function in yeast as well. Because bacterial origins of replication do not function in eukaryotes, it is necessary to provide a yeast replication origin. One bonus is that DNA sequences of replication origins are similar in different eukaryotes, so the yeast origin functions in other higher organisms. When eukaryotic cells divide, the duplicated chromosomes are pulled apart by microtubules ("spindle fibers") attached to their centromeres (∞ Section 8.6). Consequently, shuttle vectors for eukaryotes must contain a segment of DNA from the centromere in order to be properly distributed at cell division (**Figure 12.19**). Luckily, the yeast centromere recognition sequence, the CEN sequence, is relatively short and easy to insert into shuttle vectors.

Another requirement is a convenient marker to select for the plasmid in yeast. Unfortunately, yeast is not susceptible to most antibiotics that are effective against bacteria. In practice, yeast host strains that are defective in making a particular amino acid or purine or pyrimidine base are used. A functional copy of the biosynthetic gene that is defective in the host is inserted into the shuttle vector. For example, if the *URA3* gene, needed for synthesis of uracil, is used, the yeast will not grow in the absence of uracil unless it gains a copy of the shuttle vector.

Expression Vectors

Organisms have complex regulatory systems, and cloned genes are often expressed poorly or not at all in a foreign host cell. This problem is tackled by using **expression vectors**, vectors that are designed to allow the experimenter to control the expression of cloned genes. Generally, the objective is to obtain high levels of expression, especially in biotechnological applications. However, when dealing with potentially toxic gene products, a low but strictly controlled level may be appropriate.

Control over expression is achieved because expression vectors contain regulatory sequences that allow manipulation of gene expression. Usually the control is transcriptional

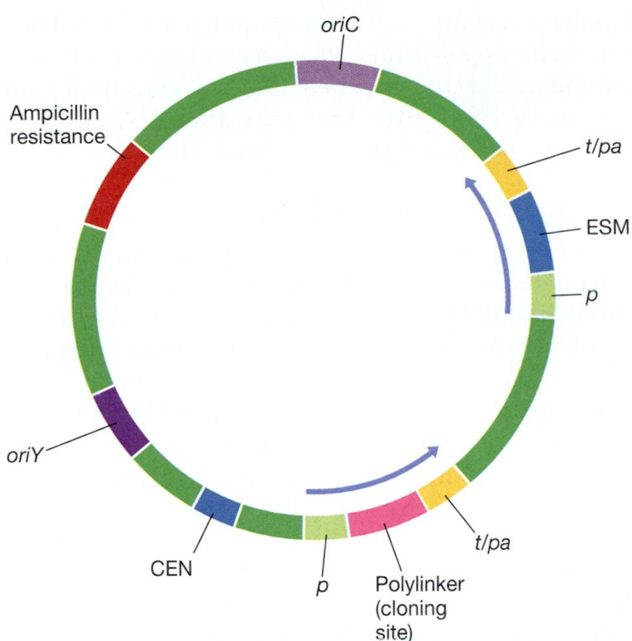

Figure 12.19 Genetic map of a shuttle vector used in yeast. The vector contains components that allow it to shuttle between *Escherichia coli* and yeast and be selected in each organism: *oriC*, origin of replication in *E. coli*; *oriY*, origin of replication in yeast; ESM, eukaryotic selectable marker; CEN, yeast centromere sequence; *p*, promoter; *t/pa*, transcription termination/polyadenylation signals. Arrows indicate the direction of transcription.

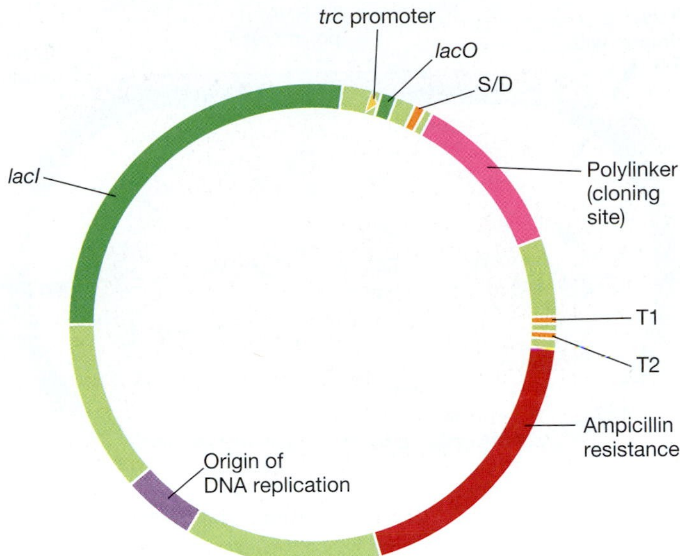

Figure 12.20 Genetic map of the expression vector pSE420. This vector was developed by Invitrogen Corp., a biotechnology company. The polylinker contains many different restriction enzyme recognition sequences to facilitate cloning. This region, plus the inserted cloned gene, are transcribed by the *trc* promoter, which is immediately upstream of the *lac* operator (*lacO*). Immediately upstream of the polylinker is a sequence that encodes a Shine–Dalgarno (S/D) ribosome-binding site on the resulting mRNA. Downstream of the polylinker are two transcription terminators (T1 and T2). The plasmid also contains the *lacI* gene, which encodes the *lac* repressor, and a gene conferring resistance to the antibiotic ampicillin. These two genes are under the control of their own promoters, which are not shown.

because for high levels of expression it is essential to produce high levels of mRNA. In practice, high levels of transcription require strong promoters that bind RNA polymerase efficiently (∞ Section 7.10). However, the native promoter of a cloned gene may work poorly in the new host. For example, promoters from eukaryotes or even from other bacteria function poorly or not at all in *E. coli*. Indeed, even some *E. coli* promoters function at low levels in *E. coli* because their sequences match the promoter consensus poorly and bind RNA polymerase inefficiently (∞ Section 7.10).

For this reason, expression vectors must contain a promoter that functions efficiently in the host and one that is correctly positioned to drive transcription of the cloned gene. Promoters from *E. coli* that are used in expression vectors include *lac* (the *lac* operon promoter), *trp* (the *trp* operon promoter), *tac* and *trc* (synthetic hybrids of the *trp* and *lac* promoters), and lambda P_L (the leftward lambda promoter; ∞ Section 10.10). These are all "strong" promoters in *E. coli* and ones that can be specifically regulated.

Regulation of Transcription from Expression Vectors

It is important to regulate the expression of cloned genes. Although one typically wants to produce very high levels of mRNA (and have it translated), it is rarely desirable to transcribe cloned genes at high levels at all times. Ideally, the culture containing the expression vector should grow until a large population of cells is obtained, each containing many copies of the vector. Expression of the desired gene(s) can then be triggered by a genetic switch.

The regulation of transcription by a repressor protein (∞ Section 9.3) is a useful way to control a cloned gene. A strong repressor can completely block the synthesis of the proteins under its control by binding to the operator region. Furthermore, when gene expression is required, the repressor can be removed by adding the inducer, thus allowing transcription of the regulated genes. For a repressor–operator system to control the production of a foreign protein, the expression vector is designed such that the cloned gene is inserted just downstream from the chosen promoter and operator region. A strong ribosome-binding site is often included between the promoter and cloned gene. This results in control of the cloned gene by the chosen promoter together with efficient transcription and translation. The operator and promoter usually correspond to each other (for instance, the *lac* operator is used with the *lac* promoter), but this need not be so. For example, hybrid regulatory regions are sometimes used (for example, fusing the *trp* promoter to the *lac* operator to form the *trc* regulatory element).

Figure 12.20 shows an expression vector controlled by *trc*. This plasmid also contains a copy of the *lacI* gene that encodes the *lac* repressor. The level of repressor in a cell containing this plasmid is sufficient to prevent transcription

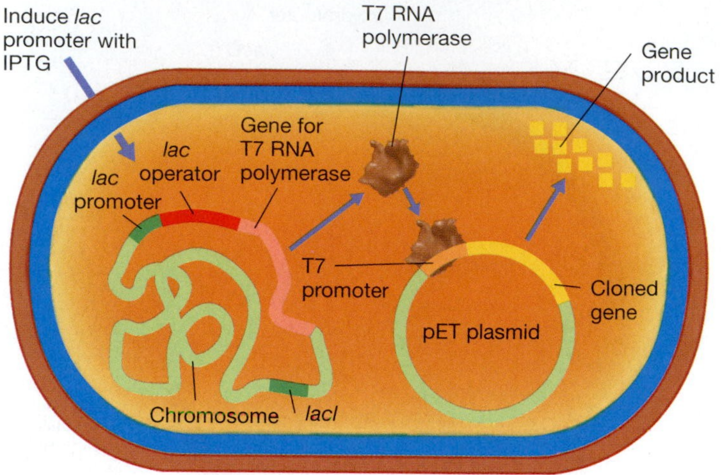

Figure 12.21 The T7 expression system. The gene for T7 RNA polymerase is in a gene fusion under control of the *lac* promoter and is inserted into the chromosome of a special host strain of *Escherichia coli*. Addition of IPTG induces the *lac* promoter, causing expression of T7 RNA polymerase. This transcribes the cloned gene, which is under control of the T7 promoter and is carried by the pET plasmid.

from the *trc* promoter until inducer is added. Addition of lactose or related *lac* inducers triggers transcription of cloned DNA (Figure 12.20). In addition to a strong and easily regulated promoter, most expression vectors contain an effective transcription terminator (∞ Section 7.11). This prevents transcription from the strong cloning promoter continuing on into other genes on the vector, which would interfere with vector stability. The expression vector shown in Figure 12.20 has strong transcription terminators to halt transcription immediately downstream from the cloned gene.

Regulating Expression with Bacteriophage T7 Control Elements

In some cases the transcriptional control system may not be a normal part of the host at all. An example of this is the use of the bacteriophage T7 promoter and T7 RNA polymerase to regulate expression. When T7 infects *E. coli*, it encodes its own RNA polymerase that recognizes only T7 promoters, thus effectively shutting down host transcription (∞ Section 19.3). In T7 expression vectors, cloned genes are placed under control of the T7 promoter, thereby limiting transcription to the cloned genes. To achieve this, the gene for T7 RNA polymerase must also be present in the cell under the control of an easily regulated system, such as *lac* (**Figure 12.21**). This is usually done by integrating the gene for T7 RNA polymerase with a *lac* promoter into the chromosome of a specialized host strain.

The BL21 series of host strains are especially designed to work with the pET series of T7 expression vectors. The cloned genes are expressed shortly after T7 RNA polymerase transcription has been switched on by a *lac* inducer, such as IPTG. Because it recognizes only T7 promoters, the T7 RNA polymerase transcribes only the cloned genes. The T7 RNA polymerase is so highly active that it uses most of the RNA precursors; consequently, host genes that require host RNA polymerase are for the most part not transcribed, and thus the cells stop growing. Protein synthesis in such cells is then dominated by the protein of interest. Thus the T7 control system is very effective for generating extremely large amounts of a particular protein of interest.

Translation of the Cloned Gene

Expression vectors must also be designed to ensure that the mRNA produced is efficiently translated. To synthesize protein from an mRNA, it is essential that the ribosomes bind at the correct site and begin reading in the correct frame. In bacteria this is accomplished by having a ribosome-binding site (Shine–Dalgarno sequence, ∞ Section 7.15) and a nearby start codon on the mRNA. Bacterial ribosome-binding sites are not found in eukaryotic genes and must be engineered into the vector if high levels of eukaryotic gene expression are to be obtained. Once again, the vector shown in Figure 12.20 has such a site.

Often, adjustments have to be made to ensure high efficiency translation after the gene has been cloned. For example, codon usage can be an obstacle. Codon usage is related to the concentration of the appropriate tRNA in the cell. Therefore, if a cloned gene has a very different codon usage from its expression host, it will probably be translated inefficiently in that host. Site-directed mutagenesis (Sections 12.7 and 12.9) can be used to change select codons in the gene, making it more amenable to the codon usage pattern of the host.

Finally, if the cloned gene contains introns, as eukaryotic genes typically do (∞ Section 8.8), the correct protein product will not be made if the host is a prokaryote. This problem can also be corrected by using synthetic DNA. However, the usual method to create an intron-free gene is to obtain the mRNA (from which the introns have already been removed) and use reverse transcriptase (∞ Section 26.2) to generate a cDNA copy of this.

Eukaryotic Vectors

It is often desirable to clone and express genes directly in eukaryotes, and vectors are available for cloning into yeast. Yeast is one of the few eukaryotes that contain a plasmid, called the two-micron circle, and most yeast vectors are based on this. The use of yeast artificial chromosomes for cloning very large fragments of DNA is discussed below (Section 12.15).

Virus vectors are commonly used in multicellular eukaryotes. For example, the DNA virus SV40 (∞ Section 19.11), which causes tumors in primates, has been engineered as a cloning vector for cultured human cells. Derivatives of SV40 that do not induce tumors have been developed for cloning and expressing mammalian genes. Other mammalian vectors use adenovirus (∞ Section 19.14) or vaccinia virus (∞ Section 19.13). Vaccinia virus vectors in particular have been used to develop recombinant vaccines (∞ Section 26.5). Vectors derived from baculovirus, a DNA virus that replicates in insect cells, can be used to make large quantities of the products of cloned genes.

Other expression vectors have been developed specifically to stably maintain and express cloned genes in an organism or tissue. These **integrating vectors** are maintained at low copy number (typically one copy per genome) by integrating into the host chromosome. They have been developed in eukaryotes ranging from yeast to mammals, as well as in certain bacteria. Integrating vectors have uses in both basic science and in applications such as gene therapy (Section 26.9). In particular, retroviruses may be used to introduce genes into mammalian cells because these viruses replicate via a DNA form that is integrated into the host chromosome (Section 19.15).

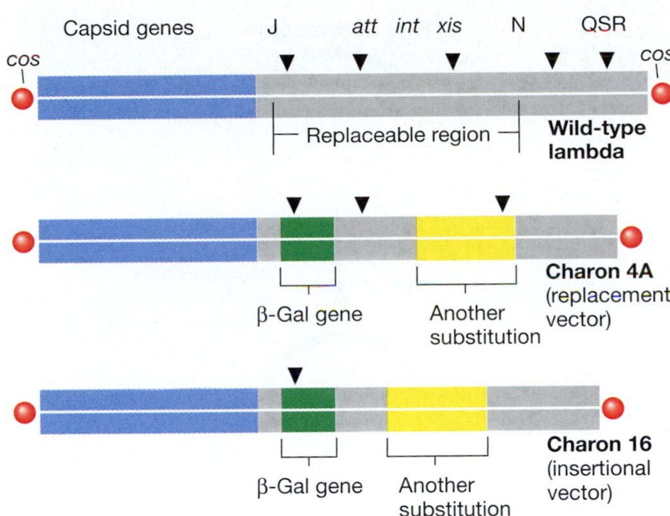

Figure 12.22 Lambda cloning vectors. Abbreviated genetic map of bacteriophage lambda showing the cohesive ends as spheres. Charon 4A and 16 are both derivatives of lambda with various substitutions and deletions in the nonessential region. Each has the *lacZ* gene, encoding the enzyme β-galactosidase, which permits detection of phage containing inserted clones. Whereas the wild-type lambda genome is 48.5 kbp, that for Charon 4A is 45.4 and for Charon 16 is 41.7 kbp. The arrowheads above the maps of each phage indicate the sites recognized by the restriction enzyme *Eco*RI.

12.13 MiniReview

Many cloned genes are not expressed efficiently in a foreign host. Expression vectors have been developed for both prokaryotic and eukaryotic hosts that contain genes or regulatory sequences that both increase transcription of the cloned gene and control the level of transcription. Signals to improve the efficiency of translation may also be present in the expression vector.

- Describe some of the components of an expression vector that improve expression of the cloned gene.
- Describe the components needed for an efficient shuttle vector.

12.14 Bacteriophage Lambda as a Cloning Vector

Recall that during the process of specialized transduction some host genes become incorporated into a bacteriophage genome and that bacteriophage lambda is the most studied specialized transducing phage (Section 11.11). During specialized transduction, lambda acts as a vector, but, recombination occurs in the cell, not in a test tube. Lambda can also be used as a cloning vector for *in vitro* recombination.

Lambda is a particularly useful cloning vector because its biology is well understood, it can hold larger amounts of DNA than most plasmids, and DNA can be efficiently packaged into phage particles *in vitro*. These can be used to infect suitable host cells, and infection is much more efficient than transformation (transfection). Phage lambda has a large number of genes; however the central third of the lambda genome, the region between genes J and N (**Figure 12.22**), is unessential for infectivity and can be replaced with foreign DNA. This allows relatively large DNA fragments, up to about 20 kbp, to be cloned into lambda. This is twice the cloning capacity of typical small plasmid vectors, such as pUC19 (Section 12.4).

Modified Lambda Phages

Wild-type lambda is not suitable as a cloning vector because its genome has too many restriction enzyme sites. To avoid this difficulty, modified lambda phages have been constructed especially for cloning. In one set of modified lambda phages, called *Charon phages*, unwanted restriction enzyme sites have been removed by mutation. Foreign DNA can be inserted into variants that have only a single restriction site, such as Charon 16. By contrast, in variants with two sites, such as Charon 4A, foreign DNA can replace a specific segment of the lambda DNA (Figure 12.22). Such *replacement vectors* are especially useful in cloning large DNA fragments. When Charon 4A is used as a replacement vector, the two small interior fragments are cut out and discarded.

Steps in Cloning with Lambda

Cloning with lambda replacement vectors involves the following steps (**Figure 12.23**):

1. Isolating vector DNA from phage particles and cutting it into two fragments with the appropriate restriction enzyme.
2. Connecting the two lambda fragments to fragments of foreign DNA using DNA ligase. Conditions are chosen so molecules are formed of a length suitable for packaging into phage particles.
3. Packaging of the DNA by adding cell extracts containing the head and tail proteins and allowing the formation of viable phage particles to occur spontaneously.
4. Infecting *E. coli* cells and isolating phage clones by picking plaques on a host strain.
5. Checking recombinant phage for the presence of the desired foreign DNA sequence using nucleic acid hybridization procedures, DNA sequencing, or observation of genetic properties.

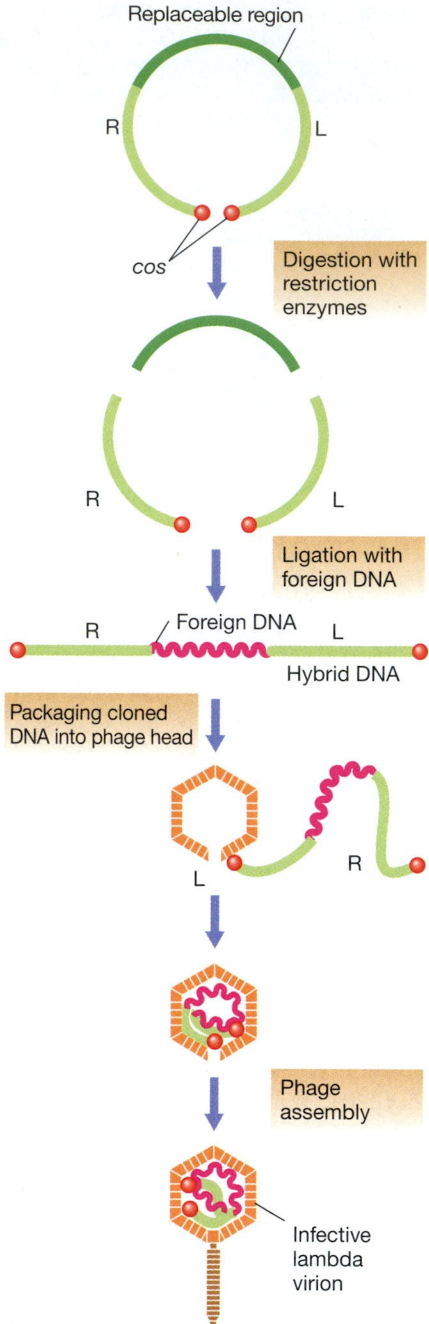

Figure 12.23 Bacteriophage lambda as a cloning vector. The maximum size of inserted DNA is about 20 kbp.

Selection of recombinants is less of a problem with lambda replacement vectors (such as Charon 4A) than with plasmids because (1) the efficiency of transfer of recombinant DNA into the cell by lambda is very high and (2) lambda fragments that have not received new DNA are too small to be incorporated into phage particles. Thus, every viable phage virion should contain cloned DNA.

Both Charon vectors are also engineered to contain reporter genes, such as the gene for β-galactosidase previously discussed. When the vectors replicate in a lactose-negative (Lac^-) strain of *E. coli*, β-galactosidase is synthesized from the phage gene, and the presence of lactose-positive (Lac^+) plaques can be detected by using a color indicator agar (Section 12.4). However, if a foreign gene is inserted into the β-galactosidase gene, the Lac^+ character is lost. Such Lac^- plaques can be readily detected as colorless plaques among a background of colored plaques (see Figure 12.24b).

Cosmid Vectors

Like replacement vectors, cosmid vectors employ specific lambda genes and are packaged into lambda virions. *Cosmids* are plasmid vectors containing the *cos* site from the lambda genome, which yields cohesive ends when cut (∞ Section 10.10). The *cos* site is required for packaging DNA into lambda virions. Cosmids are constructed from plasmids by ligating the lambda *cos* region to the plasmid DNA. Foreign DNA is then ligated into the vector. The modified plasmid, plus cloned DNA, can then be packaged into lambda virions *in vitro* as described previously and the phage particles used to transduce *E. coli*.

One major advantage of cosmids is that they can be used to clone large fragments of DNA, with inserts as large as 50 kbp accepted by the system. Therefore, with bigger inserts, fewer clones are needed to cover a whole genome. Using cosmids avoids the necessity of having to transform *E. coli*, which is especially inefficient with larger plasmids. Cosmids also permit storage of the DNA in phage particles instead of as plasmids. Phage particles are more stable than plasmids, so the recombinant DNA can be kept for long periods of time.

12.14 MiniReview

Bacteriophages such as lambda have been modified to make useful cloning vectors. Larger amounts of foreign DNA can be cloned with lambda than with many plasmids. In addition, the recombinant DNA can be packaged *in vitro* for efficient transfer to a host cell. Plasmid vectors containing the lambda *cos* sites are called cosmids, and they can carry large fragments of foreign DNA.

- Why is the ability to package recombinant DNA in a test tube useful?
- What is a replacement vector?

12.15 Vectors for Genomic Cloning and Sequencing

Most of the techniques discussed in this section are variants of the *in vitro* techniques discussed above. The principles remain the same, but now the emphasis is on the entire genome of an organism, rather than on individual genes. Plasmid vectors and the Charon derivatives of bacteriophage lambda, for example, have been extensively used for cloning and sequencing, including for genomic analysis. Other, more specialized vectors for genome analysis include bacteriophage M13 and bacterial and yeast artificial chromosomes, and we focus on these here.

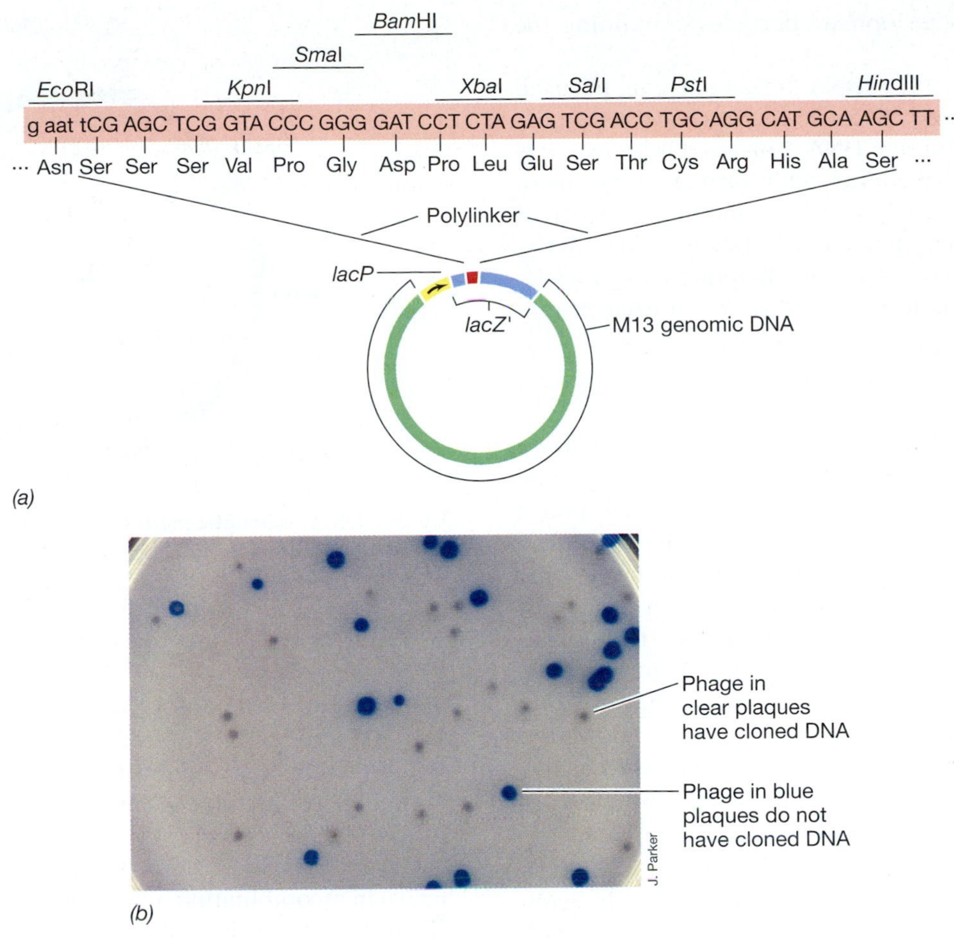

Figure 12.24 Cloning using bacteriophage M13. *(a)* A partial map of M13mp18, a derivative of M13 constructed for use as a cloning vector. The vector contains the *lac* promoter (*lacP*) and truncated gene, lacZ′, which encodes a functional part of β-galactosidase. At the beginning of this gene is a polylinker that contains several restriction sites but maintains the proper reading frame. The amino acids encoded by the polylinker are shown. Most DNA fragments cloned into the polylinker disrupt the lacZ′ gene and abolish β-galactosidase activity. *(b)* Portion of an Xgal-containing Petri plate showing white plaques formed by M13mp18 containing cloned DNA and blue plaques formed from phage lacking cloned DNA (refer to Figure 12.7 for Xgal information). Unless a PCR product is cloned, one must still screen the white plaques to find the ones containing the cloned gene of interest.

Vectors Derived from Bacteriophage M13

M13 is a filamentous bacteriophage with single-stranded DNA that replicates without killing its host (Section 19.2). Mature particles of M13 are released from host cells without lysing by a budding process, and infected cultures can provide continuous sources of phage DNA. Most of the genome of phage M13 contains genes essential for virus replication. However, a small region called the *intergenic sequence* can be used as a cloning site. Variable lengths of foreign DNA, up to about 5 kbp, can be cloned without affecting phage viability. As the genome gets larger, the virion simply grows longer.

Phage M13mp18 is a derivative of M13 in which the intergenic region has been modified to facilitate cloning (**Figure 12.24a**). One useful modification is the insertion of a functional fragment of *lacZ*, the *E. coli* gene that encodes the enzyme β-galactosidase. Cells infected with M13mp18 can be detected easily by their color on indicator plates. This is achieved using the artificial substrate, Xgal, which is cleaved by β-galactosidase to yield a blue color (Figure 12.24*b*). The *lacZ* gene has itself been modified to contain a 54-bp polylinker, which contains several restriction enzyme cut sites absent from the original M13 genome. The polylinker is inserted into the beginning of the coding portion of the *lacZ* gene but does not affect enzyme activity. However, inserting additional cloned DNA into the polylinker inactivates the gene. Phages with such DNA inserts yield colorless plaques (no β-galactosidase activity), and can therefore be easily identified (Figure 12.24*b, c*).

Use of M13 in Molecular Cloning

To clone DNA in M13 vectors, double-stranded replicative form DNA (Section 19.2) is isolated from the infected host and treated with a restriction enzyme. The foreign DNA is then treated with the same restriction enzyme. On ligation, double-stranded M13 molecules are obtained that contain the foreign DNA. When these molecules are introduced into the cell by transformation, they are replicated and produce single-stranded DNA bacteriophage particles containing the cloned DNA.

The single-stranded M13 DNA produced can be used directly in DNA sequencing. Because the base sequence next to the cut site where the foreign DNA is inserted is known, it is possible to construct an oligonucleotide primer complementary to this region and use this to determine the sequence of the DNA downstream from this point. In this way, M13 derivatives have proven extremely useful in sequencing foreign DNA, even rather long molecules, and have featured prominently in the sequencing of several genomes.

Artificial Chromosomes

Vectors such as M13, or plasmid vectors that hold about 2–10 kbp of cloned DNA, are adequate for making gene libraries for sequencing prokaryotic genomes. Bacteriophage lambda vectors, which hold 20 kbp or more, are also widely used in genomics projects. However, as the size of the genome to be sequenced increases, so does the number of clones needed to obtain a complete sequence. Therefore, for making libraries of DNA from eukaryotic microorganisms or from higher eukaryotes such as humans, it is useful to have vectors that can carry very large segments of DNA. This allows the size of the initial library to be manageable. Such vectors have been developed and are called **artificial chromosomes**.

Bacterial Artificial Chromosomes: BACs

Many bacteria contain large plasmids that are stably replicated within the cell, for example, the F plasmid of *E. coli* (∞ Section 11.12). Naturally occurring derivatives of the F plasmid, called F′ plasmids, are known that may carry large amounts of chromosomal DNA (∞ Section 11.13). Because of these desirable properties, the F plasmid has been used to construct cloning vectors called **bacterial artificial chromosomes** (BACs).

Figure 12.25 shows the structure of a BAC based on the F plasmid. The vector is only 6.7 kb compared to the 99.2 kb of F itself. The BAC contains only a few genes from F, including *oriS* and *repE*, which are necessary for replication, and *sopA* and *sopB*, which keep the copy number very low. The plasmid also contains the *cat* gene, which confers chloramphenicol resistance on the host, and a polylinker that includes several restriction sites for cloning DNA. Foreign DNA of more than 300 kbp can be inserted and stably maintained in a BAC vector such as this. The host for a BAC is typically a mutant strain of *E. coli* that lacks the normal restriction and modification systems of the wild type (Section 12.1). This prevents the BAC from being destroyed. The host strain is usually also defective in recombination, which prevents recombination of the cloned DNA from the BAC into the host cell chromosome.

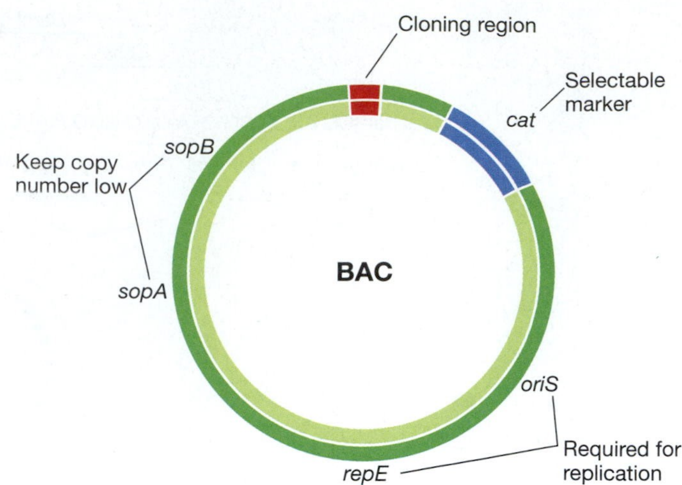

Figure 12.25 Genetic map of a bacterial artificial chromosome. The BAC diagrammed is 6.7 kbp. The cloning region has several unique restriction enzyme sites. This BAC contains only a small fraction of the 99.2-kbp F plasmid.

Eukaryotic Artificial Chromosomes

Historically, the first artificial chromosomes were **yeast artificial chromosomes** (YACs). These vectors replicate in yeast like normal chromosomes, but they have sites where very large fragments of DNA can be inserted. To function like normal eukaryotic chromosomes, YACs must have (1) an origin of DNA replication, (2) telomeres for replicating DNA at the ends of the chromosome (∞ Section 8.7), and (3) a centromere for segregation during mitosis (∞ Section 8.6). They must also contain a cloning site and a gene for selection following transformation into the host, typically the yeast *Saccharomyces cerevisiae*. **Figure 12.26** shows a diagram of a

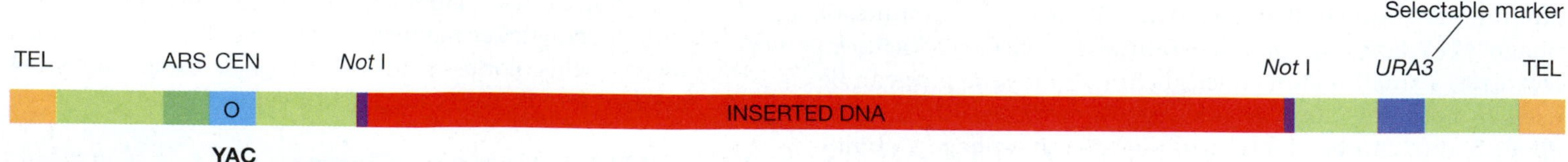

Figure 12.26 A yeast artificial chromosome containing foreign DNA. The foreign DNA was cloned into the vector at a *Not*I restriction site. The telomeres are labeled TEL and the centromere CEN. The origin of replication is labeled ARS (for autonomous replication sequence). The gene used for selection is *URA3.* The host into which the clone is transformed has a mutation in *URA3* and requires uracil for growth (Ura$^-$). Host cells containing this YAC become Ura$^+$. The diagram is not drawn to scale; vector DNA is only 10 kbp whereas cloned DNA can be up to 800 kbp.

YAC vector into which foreign DNA has been cloned. YAC vectors are themselves only about 10 kbp, but they can have 200–800 kbp of cloned DNA inserted.

After confirming that a particular fragment of DNA has been cloned on a BAC or a YAC, this segment can be subcloned into a plasmid or bacteriophage vector for more detailed analysis or sequencing. Although YACs can hold more DNA than BACs, there is a greater problem with recombination and rearrangement of the cloned DNA within yeast than within *E. coli*. For this reason BACs are now more widely used in genomic cloning than are YACs.

Human artificial chromosomes (HACs) have also been developed and are similar to YACs in overall structure. Oddly enough, either linear DNA molecules with telomeres or circular DNA will both replicate in cultured human cells, so most HACs are actually circular. They are constructed by adding segments of human DNA to BACs. The critical issue is to provide an efficient centromere. In mammals the centromere consists of long stretches of repeated sequences (Section 8.6), and HACs must have long arrays of these repeats to be inherited stably. It is hoped that HACs will find future use in gene therapy to carry large human genes with their natural regulatory sequences.

12.15 MiniReview

Specialized cloning vectors have been constructed for the sequencing and assembly of genomes. Some, like the M13 derivatives, are useful for both cloning and for direct DNA sequencing. Others, like artificial chromosomes, are useful for cloning very large fragments of DNA, fragments approaching a megabase in size.

- Compare the capacity for cloning foreign DNA in M13, lambda, BACs, and YACs.
- The yeast artificial chromosome behaves like a chromosome in a yeast cell. What makes this possible?

Review of Key Terms

Artificial chromosome a single copy vector that can carry extremely long inserts of DNA and is widely used for cloning segments of large genomes

Bacterial artificial chromosome (BAC) circular artificial chromosome with bacterial origin of replication

Cassette mutagenesis creating mutations by the insertion of a DNA cassette

Codon bias the relative proportions of different codons encoding the same amino acid; it varies in different organisms. Same as codon usage

DNA cassette an artificially designed segment of DNA that usually carries a gene for resistance to an antibiotic or some other convenient marker and is flanked by convenient restriction sites

DNA fingerprinting use of DNA technology to determine the origin of DNA in a sample of tissue

DNA library (also called a gene library) a collection of cloned DNA segments that is big enough to contain at least one copy of every gene from a particular organism; same as gene library

Expression vector a cloning vector that contains the necessary regulatory sequences to allow transcription and translation of cloned genes

Gel electrophoresis technique for separation of nucleic acid molecules by passing an electric current through a gel made of agarose or polyacrylamide

Gene disruption (also called cassette mutagenesis or gene knockout) inactivation of a gene by inserting into it a DNA fragment containing a selectable marker. The inserted fragment is called a DNA cassette

Gene fusion a structure created by joining together segments of two separate genes, in particular when the regulatory region of one gene is joined to the coding region of a reporter gene

Green fluorescent protein (GFP) a protein that fluoresces green and is widely used in genetic analysis

Genetic engineering the use of *in vitro* techniques in the isolation, alteration, and expression of DNA and in the development of genetically modified organisms

Human artificial chromosome (HAC) artificial chromosome with human centromere sequence array

Hybridization base pairing of single strands of DNA or RNA from two different (but related) sources to give a hybrid double helix

Integrating vector a cloning vector that becomes integrated into a host chromosome

Modification enzyme an enzyme that chemically modifies bases within a restriction enzyme recognition site and thus prevents the site from being cut

Molecular cloning isolation and incorporation of a fragment of DNA into a vector where it can be replicated

Northern blot a hybridization procedure where RNA is in the gel and DNA or RNA is the probe

Nucleic acid probe a strand of nucleic acid that can be labeled and used to hybridize to a complementary molecule from a mixture of other nucleic acids

Polymerase chain reaction (PCR) artificial amplification of a DNA sequence by repeated cycles of strand separation and replication

Primer a short length of DNA or RNA used to initiate synthesis of a new DNA strand

Reporter gene a gene used in genetic analysis because the product it encodes is easy to detect

Restriction enzyme an enzyme that recognizes a specific DNA sequence and then cuts the DNA; also known as a restriction endonuclease

Restriction map a map showing the location of restriction enzyme cut sites on a segment of DNA

Sequencing deducing the sequence of a DNA or RNA molecule by a series of chemical reactions

Shotgun cloning making a gene library by random cloning of DNA fragments

Shuttle vector a cloning vector that can replicate in two or more dissimilar hosts

Site-directed mutagenesis a technique whereby a gene with a specific mutation can be constructed *in vitro*

Southern blot a hybridization procedure where DNA is in the gel and RNA or DNA is the probe

Vector (as in cloning vector) a self-replicating DNA molecule that is used to carry cloned genes or other DNA segments for genetic engineering

Yeast artificial chromosome (YAC) artificial chromosome with yeast origin of replication and CEN sequence

Review Questions

1. What are restriction enzymes? What is the probable function of a restriction enzyme in the cell that produces it? Why does the presence of a restriction enzyme in a cell not cause the degradation of that cell's DNA (Section 12.1)?
2. How could you detect a colony containing a cloned gene if you already knew the sequence of the gene (Section 12.2)?
3. Genetic engineering depends on vectors. Describe the properties needed in a well-designed plasmid cloning vector (Section 12.4).
4. How does the insertional inactivation of β-galactosidase allow the presence of foreign DNA in a plasmid vector such as pUC19 to be detected (Section 12.4)?
5. Why do dideoxynucleotides function as chain terminators (Section 12.5)?
6. What characteristics are used to identify open reading frames using sequence data (Section 12.6)?
7. What are the major uses for artificially synthesized DNA (Section 12.7)?
8. Describe the basic principles of gene amplification using the polymerase chain reaction (PCR). How have thermophilic and hyperthermophilic prokaryotes simplified the use of PCR (Section 12.8)?
9. How has DNA technology allowed identification of humans from forensic samples (Section 12.8, sidebar)?
10. What is a VNTR (Section 12.8, sidebar)?
11. What does site-directed mutagenesis allow you to do that normal mutagenesis does not (Section 12.9)?
12. What is a reporter gene? Describe two widely used reporter genes (Section 12.10).
13. How are gene fusions used to investigate gene regulation (Section 12.10)?
14. Describe two prokaryotic cloning hosts and the beneficial and detrimental features of each (Section 12.11).
15. How could you detect a colony containing a cloned gene if you didn't know the gene sequence but had available purified protein encoded by the gene (Section 12.12)?
16. Describe the similarities and differences between expression vectors and shuttle vectors (Section 12.13).
17. How has bacteriophage T7 been used in expressing foreign genes in *Escherichia coli* and what desirable features does this regulatory system possess (Section 12.13)?
18. What advantages are there to using a lambda-based cloning vector rather than a plasmid vector (Section 12.14)?
19. What are the essential characteristics of an artificial chromosome? What is the difference between a BAC and a YAC? What characteristics of the F plasmid make it less useful for use *in vitro* (Section 12.15)?

Application Questions

1. Suppose you are given the task of constructing a plasmid expression vector suitable for molecular cloning in an organism of industrial interest. List the characteristics such a plasmid should have. List the steps you would use to create such a plasmid.
2. Suppose you have just determined the DNA base sequence for an especially strong promoter in *Escherichia coli* and you are interested in incorporating this sequence into an expression vector. Describe the steps you would use. What precautions are necessary to be sure that this promoter actually works as expected in its new location?
3. Many genetic systems use the *lacZ* gene encoding β-galactosidase as a reporter. What advantages or problems would there be if (a) luciferase or (b) green fluorescent protein were used as reporters instead of β-galactosidase?

13

Microbial Genomics

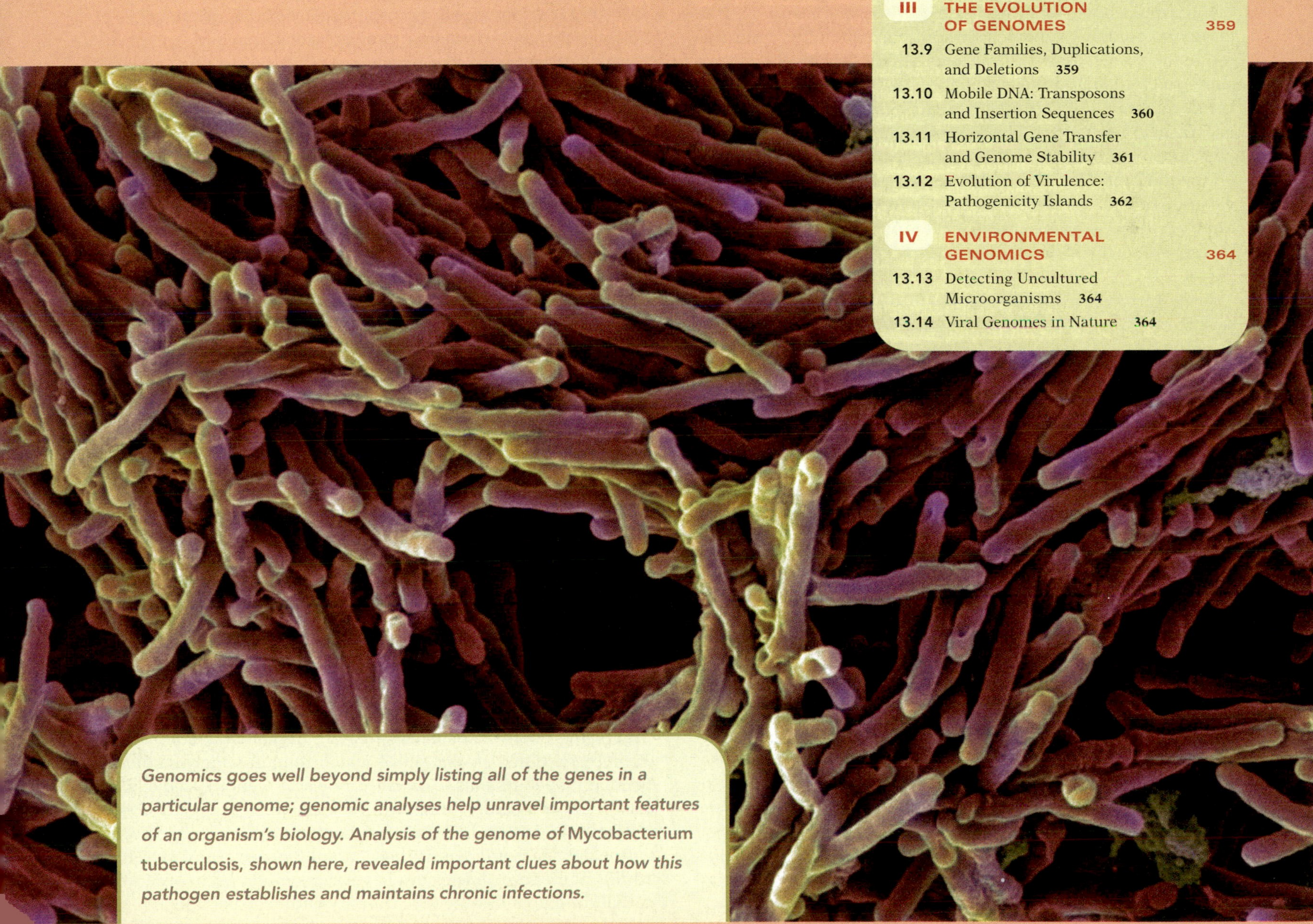

Genomics goes well beyond simply listing all of the genes in a particular genome; genomic analyses help unravel important features of an organism's biology. Analysis of the genome of Mycobacterium tuberculosis*, shown here, revealed important clues about how this pathogen establishes and maintains chronic infections.*

An organism's **genome** is its entire complement of genetic information, including the genes themselves together with regulatory sequences and noncoding DNA. Genome analyses make up the field of genomics, the subject of this chapter. Knowledge of the genome sequence of an organism reveals not only its genes, but also yields important clues to how the organism functions and its evolutionary history. Genomic sequences also aid the study of gene expression, the transcription and translation of the genetic information in the genome. The traditional approach to studying gene expression was to focus on a single gene or group of related genes. Today, expression of the entire complement of an organism's genes can be examined in a single experiment.

It would have been impossible to sequence and analyze large and complex genomes without corresponding improvements in molecular technology. Major advances include shotgun cloning and the automation of DNA sequencing, and the development of powerful computational methods for DNA and protein sequences. The technology behind DNA analyses continues to improve as new advances drive down the cost and increase the speed at which genomes are sequenced. Here we focus on microbial genomes, some techniques used to analyze these genomes, and what microbial genomics has revealed thus far.

I MICROBIAL GENOMES

The word **genomics** refers to the discipline of mapping, sequencing, analyzing, and comparing genomes. Several hundred microbial genomes, mostly prokaryotic, have now been sequenced, and hundreds more projects are ongoing. Here we discuss only a few of them and explore what analysis of these genomes tells us.

13.1 A Short History of Genomics

The first genome to be sequenced was the 3569-nucleotide RNA genome of the *virus* MS2 (Section 19.1) in 1976. The first DNA genome to be sequenced was the 5386-nucleotide sequence of the small, single-stranded DNA virus, ϕX174 (Section 19.2) in 1977. This feat was accomplished by a group led by the British scientist Fred Sanger and debuted the dideoxynucleotide technique for DNA sequencing (Section 12.5).

The First Cellular Genome

The first cellular genome to be sequenced was the 1,830,137-bp chromosome of *Haemophilus influenzae* described in 1995 by Hamilton O. Smith, J. Craig Venter, and their colleagues at The Institute for Genomic Research in Maryland. At present well over 2,000 genomes of prokaryotes have either been sequenced or are in progress. Because new advances in DNA sequencing are introduced quite frequently, it is likely that the number of sequenced prokaryotic genomes will grow rapidly.

In addition, we now have genome sequences of several higher animals, including the haploid human genome, which contains about 3 billion bp yet only around 25,000 genes. The largest genomes so far sequenced, in terms of the number of genes, are those of the higher plants, rice and black cottonwood (a species of poplar), and the protozoans, *Paramecium* and *Trichomonas,* all of which have significantly more genes than humans. Information from genome sequences has provided new insight on topics as diverse as clinical medicine and microbial evolution.

13.1 MiniReview

Small viruses were the first organisms subjected to genomic sequencing, but now many prokaryotic and eukaryotic genomes have been sequenced.

- Do humans have the largest genome?

13.2 Prokaryotic Genomes: Sizes and ORF Contents

The DNA sequences of several hundred prokaryotic genomes are now available in public databases (for an up-to-date list of genome sequencing projects search http://www.genomesonline.org/), and several thousand prokaryotic genome sequences have either been sequenced or are in progress. **Table 13.1** lists some representative examples. These include many species of *Bacteria* and *Archaea* and representatives containing circular as well as linear genomes. Although rare, linear chromosomes are present in several *Bacteria,* including *Borrelia burgdorferi,* the causative agent of Lyme disease, and the important antibiotic-producing genus *Streptomyces*.

Note that the list of prokaryotic genomes in Table 13.1 contains several pathogens. Naturally, such organisms have high priority for sequencing. The hyperthermophiles on the list may have important uses in biotechnology because the enzymes in these organisms are heat-stable. Indeed, the needs of the biomedical and biotechnology industries have been important drivers for selecting organisms for genome sequencing. However, the list in Table 13.1 also includes organisms such as *Bacillus subtilis, Escherichia coli,* and *Pseudomonas aeruginosa,* all of which remain widely studied genetic model systems. Currently, however, genome sequencing has become so routine and inexpensive that sequencing projects are no longer driven by medical or biotechnological rationales. In some cases the genomes of several different *strains* of the same bacterium have been sequenced in order to reveal the extent of genetic variability within a species.

Size Range of Prokaryotic Genomes

Genomes of species in both prokaryotic domains, *Bacteria* and *Archaea,* show a strong correlation between genome size

Table 13.1 Select prokaryotic genomes[a]

Organism	*Cell type*[b]	*Size (base pairs)*	*ORFs*[c]	*Comments*
Bacteria				
Carsonella ruddii	E	159,662	182	Degenerate aphid endosymbiont
Buchnera aphidicola BCc	E	422,434	362	Primary aphid endosymbiont
Mycoplasma genitalium	P	580,070	470	Smallest nonsymbiotic bacterial genome
Borrelia burgdorferi	P	910,725	853	Spirochete, linear chromosome, causes Lyme disease
Chlamydia trachomatis	P	1,042,519	894	Obligate intracellular parasite, common human pathogen
Rickettsia prowazekii	P	1,111,523	834	Obligate intracellular parasite, causes epidemic typhus
Treponema pallidum	P	1,138,006	1041	Spirochete, causes syphilis
Pelagibacter ubique	FL	1,308,759	1354	Marine heterotroph
Aquifex aeolicus	FL	1,551,335	1544	Hyperthermophile, autotroph
Prochlorococcus marinus	FL	1,657,990	1716	Most abundant marine oxygenic phototroph
Streptococcus pyogenes	FL	1,852,442	1752	Causes strep throat and scarlet fever
Thermotoga maritima	FL	1,860,725	1877	Hyperthermophile
Chlorobaculum tepidum	FL	2,154,946	2288	Model green phototrophic bacterium
Deinococcus radiodurans	FL	3,284,156	2185	Radiation resistant, multiple chromosomes
Synechocystis sp.	FL	3,573,470	3168	Model cyanobacterium
Bdellovibrio bacteriovorus	FL	3,782,950	3584	Predator of other prokaryotes
Caulobacter crescentus	FL	4,016,942	3767	Complex life cycle
Bacillus subtilis	FL	4,214,810	4100	Gram-positive genetic model
Mycobacterium tuberculosis	P	4,411,529	3924	Causes tuberculosis
Escherichia coli K12	FL	4,639,221	4288	Gram-negative genetic model
Escherichia coli O157:H7	FL	5,594,477	5361	Enteropathogenic strain of *E. coli*
Bacillus anthracis	FL	5,227,293	5738	Pathogen, biowarfare agent
Rhodopseudomonas palustris	FL	5,459,213	4836	Metabolically versatile anoxygenic phototroph
Pseudomonas aeruginosa	FL	6,264,403	5570	Metabolically versatile opportunistic pathogen
Streptomyces coelicolor	FL	8,667,507	7825	Linear chromosome, produces antibiotics
Bradyrhizobium japonicum	FL	9,105,828	8317	Nitrogen fixation, nodulates soybeans
Archaea				
Nanoarchaeum equitans	P	490,885	552	Smallest nonsymbiotic cellular genome
Thermoplasma acidophilum	FL	1,564,905	1509	Thermophile, acidophile
Methanocaldococcus jannaschii	FL	1,664,976	1738	Methanogen, hyperthermophile
Aeropyrum pernix	FL	1,669,695	1841	Hyperthermophile
Pyrococcus horikoshii	FL	1,738,505	2061	Hyperthermophile
Methanothermobacter thermautotrophicus	FL	1,751,377	1855	Methanogen
Archaeoglobus fulgidus	FL	2,178,400	2436	Hyperthermophile
Halobacterium salinarum	FL	2,571,010	2630	Extreme halophile, bacteriorhodopsin
Sulfolobus solfataricus	FL	2,992,245	2977	Hyperthermophile, sulfur chemolithotroph
Haloarcula marismortui	FL	4,274,642	4242	Extreme halophile, bacteriorhodopsin
Methanosarcina acetivorans	FL	5,751,000	4252	Acetotrophic methanogen

[a]Information on these and hundreds of other prokaryotic genomes can be found in the TIGR Database (**www.tigr.org/tdb**), a web site maintained by The Institute for Genomic Research (TIGR), Rockville, MD, a not-for-profit research institute, and at **http://www.genomesonline.org**. Links are listed there to other relevant web sites.

[b]E, endosymbiont; P, parasite; FL, free-living.

[c]Open reading frames. The purpose of reporting ORFs is to predict the total number of proteins that an organism might encode. Of course, genes encoding known proteins are included, as are all ORFs that could encode a protein greater than 100 amino acid residues. Smaller ORFs are typically not included unless they show similarity to a gene from another organism or unless the codon usage is typical of the organism being studied.

and ORF content (**Figure 13.1**). Regardless of the organism, each megabase of prokaryotic DNA encodes about 1,000 open-reading frames (ORFs, ∞ Section 7.13). As prokaryotic genomes increase in size, they also increase proportionally in gene number. This contrasts with eukaryotes, in which non-coding DNA can be a large fraction of the genome, especially in organisms with large genomes (see Table 13.4).

Analyzing genomic sequences can yield answers to fundamental biological questions. For example, how many genes are necessary for life to exist? For free-living cells, the smallest genomes found so far encode approximately 1,400 genes.

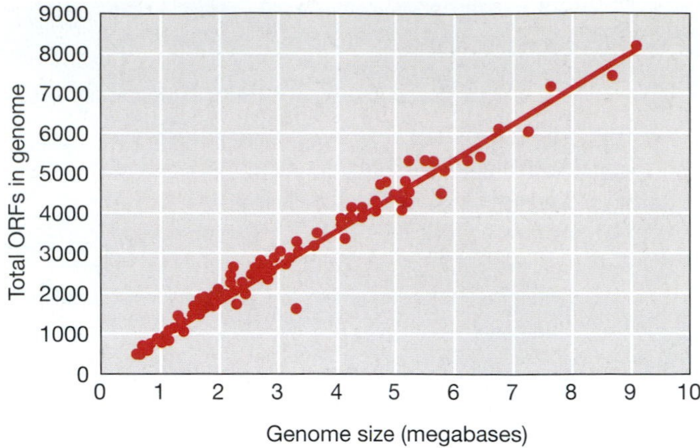

Figure 13.1 Correlation between genome size and ORF content in prokaryotes. These data are from analyses of 115 completed prokaryotic genomes and include species of both *Bacteria* and *Archaea*. Data from *Proc. Natl. Acad. Sci. (USA)* 101: 3160–3165 (2004).

Both free-living *Bacteria* and *Archaea* are known that have genomes in this range. The record at present is held by *Pelagibacter ubique,* a marine heterotroph that has approximately 1,350 genes (Section 23.9). This organism is extremely efficient in its use of DNA; it has no introns, inteins, or transposons and has the shortest intergenic spaces yet recorded. The largest genomes of prokaryotes contain over 10,000 genes and are primarily from soil organisms, such as the myxobacteria, prokaryotes that undergo a complex life cycle (Section 15.17).

Perhaps surprisingly, genomic analyses have shown that autotrophic organisms need only a few more genes than heterotrophs. For example, *Methanocaldococcus jannaschii* (*Archaea*) is an autotroph and contains only 1,738 ORFs. This enables it to be not only free living, but also to rely on carbon dioxide as its sole carbon source. *Aquifex aeolicus* (*Bacteria*) is also an autotroph and contains the smallest known genome of any autotroph at just 1.5 **Megabase pair (Mbp**, one million base pairs) (Table 13.1). Both *Methanocaldococcus* and *Aquifex* are also hyperthermophiles, growing optimally at temperatures above 80°C. Thus, a large genome is not necessary to support an extreme lifestyle, including that of autotrophs living in near-boiling water.

Small Genomes

The smallest cellular genomes belong to prokaryotes that are parasitic or endosymbiotic. A disproportionate number of such small genomes have been sequenced, partly because many are of medical importance and partly because smaller genomes are easier to sequence. Genome sizes for prokaryotes that are obligate parasites range from 490 kbp, for *Nanoarchaeum equitans* (*Archaea*), to 4,400 kbp, for *Mycobacterium tuberculosis* (*Bacteria*). The genomes of *N. equitans* and several other bacteria, including *Mycoplasma, Chlamydia,* and *Rickettsia,* are smaller than the largest known *viral* genome, that of Mimivirus with 1.2 Mbp (Section 19.10, Microbial Sidebar).

Excluding endosymbionts, the smallest prokaryotic genome is that of the archaeon *N. equitans,* which is some 90 kbp smaller that that of *Mycoplasma genitalium* (Table 13.1). However, the genome of *N. equitans* actually contains more ORFs than the larger genome of *M. genitalium.* This is because the *N. equitans* genome is extremely compact and contains virtually no noncoding DNA. *N. equitans* is a hyperthermophile and a parasite of a second hyperthermophile, the archaeon *Ignicoccus* (Section 17.8). Analyses of the gene content of *N. equitans* show it to be free of virtually all genes that encode proteins for anabolism or catabolism.

Using *Mycoplasma,* which has around 500 genes, as a starting point, several investigators have estimated that around 250–300 genes are the minimum for a viable cell. These estimates rely partly on comparisons with other small genomes. In addition, systematic mutagenesis has been performed to identify those *Mycoplasma* genes that are essential. However, these estimates are compromised by the fact that *Mycoplasma* is not a free-living organism and cannot make many vital components, such as purines and pyrimidines.

Smaller still than the genomes of prokaryotic parasites are those of some endosymbionts. Endosymbiotic bacteria live inside the cells of their hosts and are relatively common among insects. *Buchnera aphidicola,* for example, is an endosymbiont of aphids that supplies its host with essential amino acids in exchange for a protected habitat. Different strains of *B. aphidicola* that live in different species of aphids have genomes that vary considerably in size, from around 600 kbp down to 420 kbp. Even smaller is the remarkable genome of the insect endosymbiont *Carsonella ruddii.* This genome has a mere 160 kbp making it by far the smallest bacterial genome yet characterized. This genome has many genes of reduced length and many overlapping genes. Although *Carsonella* retains many genes for amino acid biosynthesis, it lacks many other genes previously regarded as essential for life. It has been suggested that *Carsonella* may be degenerating into an organelle-like status.

Large Genomes

Some prokaryotes have very large genomes that are comparable in size to those of eukaryotic microorganisms. Because eukaryotes tend to have significant amounts of noncoding DNA and prokaryotes do not, some prokaryotic genomes actually have more genes than microbial eukaryotes, despite having less DNA. For example, *Bradyrhizobium japonicum,* which forms nitrogen-fixing root nodules on soybeans, has 9.1 Mbp of DNA and 8,300 ORFs, whereas the yeast *Saccharomyces cerevisiae,* a eukaryote, has 12.1 Mbp of DNA but only 5,800 ORFs (see Table 13.4). *Myxococcus xanthus* also has 9.1 Mbp of DNA, whereas its relatives in the Delta proteobacteria have genomes approximately half this size (Section 15.17). It has been suggested that there were multiple duplication events of substantial segments of DNA in the evolutionary history of *M. xanthus.*

The largest prokaryotic genome known at present is the 12.3-Mbp chromosome of *Sorangium cellulosum,* a gliding bacterium, which has not yet been completely sequenced and annotated. In contrast to *Bacteria,* the largest genomes found in *Archaea* thus far are around 5 Mbp (Table 13.1). Overall, prokaryotic genomes thus range in size from those of large viruses to those of eukaryotic microorganisms.

13.2 MiniReview

Sequenced prokaryotic genomes range in size from 0.16 Mbp to 9.1 Mbp, and even larger genomes are known. The smallest prokaryotic genomes are smaller than those of the largest viruses. The largest prokaryotic genomes have more genes than some eukaryotes. In prokaryotes, gene content is typically proportional to genome size.

- What is the lifestyle of prokaryotic organisms that have genomes smaller than that of certain viruses?
- Approximately how many protein-encoding genes will a bacterial genome of 4 Mbp contain?
- Which organism is likely to have more genes, a prokaryote with 8 Mbp of DNA or a eukaryote with 10 Mbp of DNA? Explain.

13.3 Prokaryotic Genomes: Bioinformatic Analyses and Gene Distributions

Following assembly and annotation of a genome, a typical genomic analysis proceeds to the gene comparison stage. The basic question here is how does the complement of genes in this organism compare with that of other organisms? Such analyses put the just-sequenced genome in perspective when considering the genomes of similar as well as more distantly related organisms. These activities are a major part of the field of **bioinformatics**, the science that applies powerful computational tools to DNA and protein sequences for the purpose of analyzing, storing, and accessing the sequences for comparative purposes.

Gene Content of Prokaryotic Genomes

The complement of genes in a particular organism defines its biology. Conversely, genomes are molded by an organism's lifestyle. One might imagine, for instance, that obligate parasites such as *Treponema pallidum* (Section 16.16) would require relatively few genes for amino acid biosynthesis because the amino acids they need can be supplied by their hosts. This is indeed the case, as the *T. pallidum* genome lacks recognizable genes for amino acid biosyntheses, although genes are found encoding several proteases, enzymes that can convert peptides taken up from the host into free amino acids. In contrast, the free-living bacterium *Escherichia coli* has 131 genes for amino acid biosynthesis and metabolism, and the soil bacterium *Bacillus subtilis* has over 200.

Comparative analyses are useful in the search for genes that encode enzymes that almost certainly must exist because of the known properties of an organism. *Thermotoga maritima* (*Bacteria*), for example, is a hyperthermophile found in hot marine sediments, and laboratory studies have shown it to be able to catabolize a large number of sugars. **Figure 13.2** summarizes some of the metabolic pathways and transport systems of *T. maritima* that have been deduced from analysis of its genome. As expected, its genome is rich in genes for transport, particularly for carbohydrates and amino acids. In fact, a large portion of the *T. maritima* genome (7%) consists of genes that encode proteins for the metabolism of simple and complex sugars. All this suggests that *T. maritima* exists in an environment rich in organic material.

Table 13.2 Gene function in bacterial genomes

Functional categories	Percentage of genes on chromosome in that category: *Escherichia coli* (4.64 Mbp)[a]	*Haemophilus influenzae* (1.83 Mbp)[a]	*Mycoplasma genitalium* (0.58 Mbp)[a]
Metabolism	21.0	19.0	14.6
Structural	5.5	4.7	3.6
Transport	10.0	7.0	7.3
Regulation	8.5	6.6	6.0
Translation	4.5	8.0	21.6
Transcription	1.3	1.5	2.6
Replication	2.7	4.9	6.8
Other, known	8.5	5.2	5.8
Unknown	38.1	43.0	32.0

[a]Chromosome size, in megabase pairs. Each organism listed contains only a single circular chromosome.

A functional analysis of genes and their activities in several prokaryotes is given in **Table 13.2**. Similar data are assembled when each new genome is published. Thus far a distinct pattern has emerged of gene distribution in prokaryotes. Metabolic genes are typically the most abundant class of genes in prokaryotic genomes, although genes for protein synthesis overtake metabolic genes on a percentage basis as genome size decreases (Table 13.2, and see Figure 13.3). Interestingly, as vital as they are, genes for DNA replication and transcription make up only a minor fraction of the typical prokaryotic genome.

In addition to protein-encoding genes, most organisms have a substantial number of genes that encode nontranslated RNA. These include genes for ribosomal RNA, which are often present in multiple copies, tRNAs, and a variety of small regulatory RNAs (Section 9.14).

Uncharacterized ORFs

Although there are differences from organism to organism, in most cases the number of genes whose role can be clearly identified in a given genome is 70% or less of the total number of ORFs detected. Uncharacterized ORFs are said to encode *hypothetical proteins,* proteins that likely exist although their function is unknown. An uncharacterized ORF shares with a known protein-encoding gene an uninterrupted reading frame of reasonable length and start and stop codons, but it encodes a protein lacking sufficient amino acid sequence homology with any known protein and thus cannot be readily identified as such.

Uncharacterized ORFs reveal that there is still much we don't know about the function of prokaryotic genes. On the

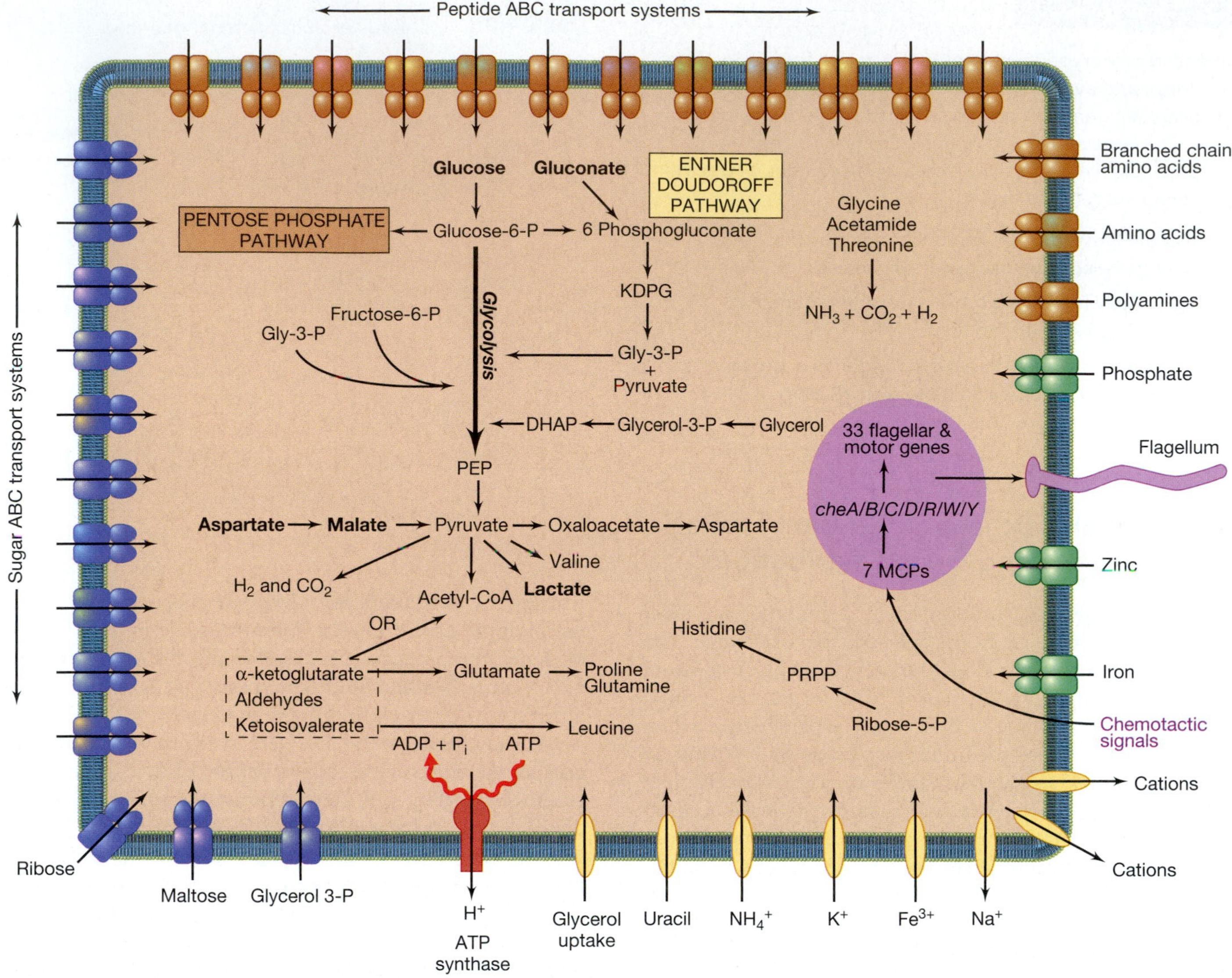

Figure 13.2 Overview of metabolism and transport in *Thermotoga maritima*. The figure shows a blueprint for the metabolic capabilities of this organism. These include some of the pathways for energy production and the metabolism of organic compounds, including transport proteins that were identified from analysis of the genomic sequence. Gene names are not shown. The genome contains several ABC-type transport systems, 12 for carbohydrates, 14 for peptides and amino acids, and still others for ions. These are shown as multisubunit structures in the figure. Other types of transport proteins have also been identified and are shown as simple ovals. The flagellum is shown, and this organism has seven transducers (MCPs) and several chemotaxis (*che*) genes and genes required for flagellar assembly. A few aspects of sugar metabolism are also shown. This figure is adapted from one published by The Institute for Genomic Research.

other hand, as gene functions are identified in one organism, homologous ORFs in other organisms can be assigned functions. Even in the world's best-understood organism, *E. coli*, functions still have not been assigned to over 1,000 of its almost 4,300 genes. However, most of the genes for macromolecular syntheses and central metabolism essential for growth of *E. coli* have been identified. Therefore, as the functions of the remaining ORFs are identified, it is likely that most of them will encode nonessential proteins.

As progress is made in identifying hypothetical proteins, the relative percentage of *E. coli* genes involved in macromolecular syntheses or central metabolism will probably decrease. Many of the unidentified genes in *E. coli* are predicted to encode regulatory or redundant proteins; these might include proteins needed only under special conditions or as "backup" systems for key enzymes. However, it must be remembered that the precise function of many genes, even in well-studied organisms such as *E. coli*, are often unpredictable. Some gene identifications merely assign a given gene to a family or to a general function (such as "transporter"). By contrast, other genes are completely unknown and have only been predicted using bioinformatics, whereas some are actually incorrect; it has been estimated that as many as 10% of genes in databases are incorrectly annotated.

Gene Categories as a Function of Genome Size

As the data of Table 13.2 show, the percentage of an organism's genes devoted to one or another cell function is to some degree a function of genome size. This is summarized for a large number of bacterial genomes in **Figure 13.3**. Core cellular processes, such as protein synthesis, DNA replication and energy production, show only minor variations in gene number with genome size (Figure 13.3). Consequently, the relative percentage of genes devoted to protein synthesis, for example, rises dramatically in small genome organisms. By contrast, genes devoted to regulation of transcription increase significantly in large genome organisms.

The data summarized in Figure 13.3 suggest that although many genes can be dispensed with, genes that encode the protein-synthesizing apparatus cannot. Thus, the smaller the genome, the greater the percentage of its genes encode translational processes (Table 13.2). Large genomes, on the other hand, contain more genes for regulation than small genomes. These additional regulatory systems allow the cell to be more flexible in diverse environmental situations by controlling gene expression accordingly. In small genome prokaryotes these regulatory processes are dispensable because such organisms are often parasitic and obtain much of what they need from their hosts.

Large genome organisms can afford to encode many regulatory and specialized metabolic genes. This likely makes these organisms more competitive in their habitats, which, for prokaryotes with very large genomes, is often soil. Soil is a habitat in which carbon and energy sources are often scarce or available only intermittently, and vary greatly (Section 23.7). A large genome encoding multiple metabolic options would thus be strongly selected for in such a habitat. Interestingly, all of the prokaryotes listed in Table 13.1 whose genomes are in excess of 6 Mbp inhabit soil.

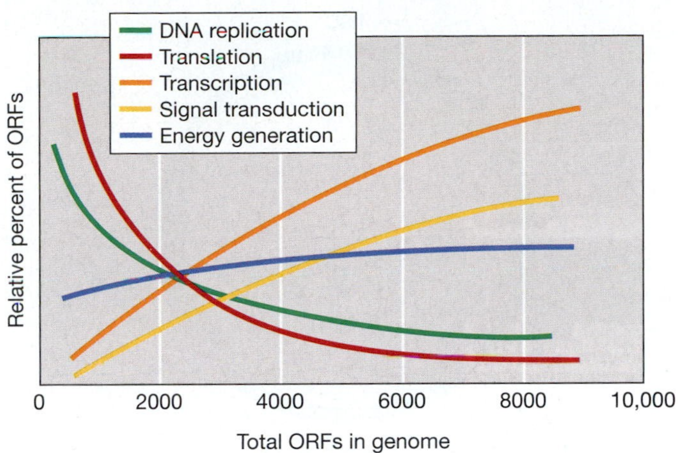

Figure 13.3 Functional category of genes as a percentage of the genome. Note how genes encoding products for translation or DNA replication increase in percentage in small genome organisms, whereas transcriptional regulatory genes increase in percentage in large genome organisms. Data from *Proc. Natl. Acad. Sci. (USA)* 101: 3160–3165 (2004).

Gene Distribution in *Bacteria* and *Archaea*

Analyses of gene categories have been done on several *Bacteria* and *Archaea* and the results are compared in **Figure 13.4**. Note that these data reflect the average gene content for several separate genomes. On average, species of *Archaea* devote a higher percentage of their genomes to energy and coenzyme production than do *Bacteria* (these results are undoubtedly skewed a bit due to the large number of novel coenzymes produced by methanogenic *Archaea*, Section 21.10). On the other hand, *Archaea* appear to contain fewer genes for carbohydrate metabolism or cytoplasmic membrane functions, such as transport and membrane biosynthesis, than do *Bacteria*. However, this finding may be skewed because the corresponding pathways are less well studied in

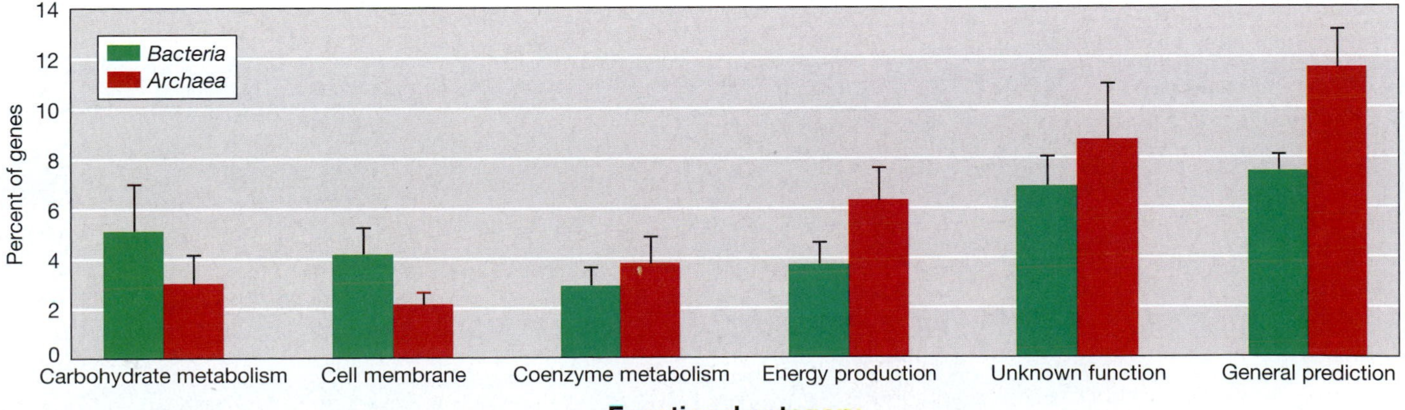

Figure 13.4 Variations in gene category in *Bacteria* and *Archaea*. Data are averages from 34 species of *Bacteria* and 12 species of *Archaea*. "Unknown function" represents genes known to encode proteins but whose functions are unknown. Genes labeled as "general prediction" encode hypothetical proteins that may or may not exist. Data from *Proc. Natl. Acad. Sci. (USA)* 101: 3160–3165 (2004).

UNIT 2

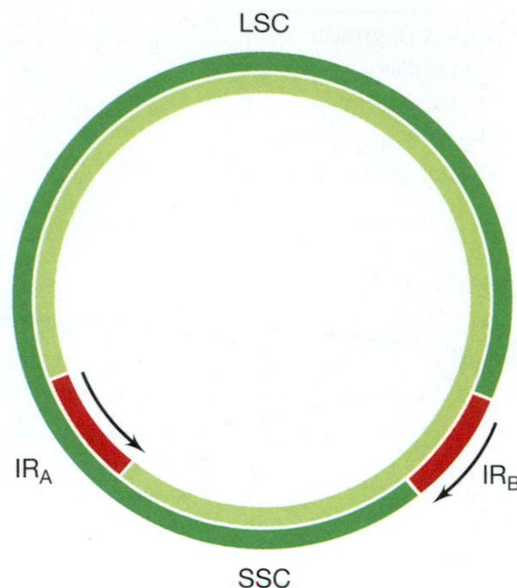

Figure 13.5 Map of a typical chloroplast genome. The genomes of chloroplasts are circular double-stranded DNA molecules. Most contain two inverted repeat regions (IR_A and IR_B), which form the borders of a small single copy region (SSC) and a large single copy region (LSC).

Archaea than in *Bacteria,* and many of the corresponding archaeal genes are probably still unknown.

Both domains of prokaryotes have relatively large numbers of genes whose functions are either unknown or that encode only hypothetical proteins, although in both categories more uncertainty exists among the *Archaea* than the *Bacteria* (Figure 13.4). However, this may be an artifact due to the availability of fewer genome sequences from the *Archaea* than from the *Bacteria*.

13.3 MiniReview

Many genes can be identified by their sequence similarity to genes found in other organisms. However, a significant percentage of sequenced genes are of unknown function. On average, the gene complements of *Bacteria* and *Archaea* are related but distinct. Bioinformatics plays an important role in genomic analyses.

- Does every functional ORF encode a protein?
- What is a hypothetical protein?
- What category of genes do prokaryotes contain the most of on a percentage basis?

13.4 The Genomes of Eukaryotic Organelles

Eukaryotic cells contain a variety of membrane-bound organelles in addition to the nucleus (Sections 18.1–18.5). The mitochondrion and the chloroplast each contain a small genome. In addition, both organelles contain the machinery necessary for protein synthesis, including ribosomes, transfer RNAs, and all the other components necessary for translation and formation of functional proteins. Indeed, these organelles share many traits in common with prokaryotic cells, to which they are phylogenetically related (Sections 14.4 and 18.4).

The Chloroplast Genome

Green plant cells contain chloroplasts, the organelles responsible for photosynthesis. Known chloroplast genomes are all circular DNA molecules. Although there are several copies of the genome in each chloroplast, they are identical. The typical chloroplast genome is about 120–160 kbp and contains two inverted repeats of 6–76 kbp (**Figure 13.5**). Several chloroplast genomes have been completely sequenced, and a few of these are summarized in **Table 13.3**. The flagellated protozoan *Mesostigma viride* belongs to the earliest diverging green plant lineage. Its chloroplast contains more protein-encoding genes and tRNA genes than any other so far known and has the typical genome structure illustrated in Figure 13.5.

Many of the chloroplast genes encode proteins for photosynthesis and autotrophy. However, the chloroplast genome also encodes rRNA used in chloroplast ribosomes, tRNA used in translation, several proteins used in transcription and translation, as well as some other proteins. Some proteins that

Table 13.3 Some chloroplast genomes[a]

Organism		Size (bp)	Genes encoding: Proteins[b]	tRNA	rRNA[c]	Inverted repeats[d]
Chlorella vulgaris	Green alga	150,613	77	31	1	Absent
Euglena gracilis	Protozoan	143,170	67	27	3	Absent
Mesostigma viride	Protozoan	118,360	92	37	2	Present
Pinus thunbergii	Black pine	119,707	72	32	1	Present[e]
Oryza sativa	Rice	134,525	70	30	2	Present
Zea mays	Corn	140,387	70	30	2	Present

[a]All chloroplast genomes are circular, double-stranded DNA (Sections 14.4 and 18.4).
[b]These include genes encoding proteins of known function and ORFs that might be functional.
[c]Each unit is an rRNA operon, containing genes for each of the rRNAs (Section 7.12).
[d]See Figure 13.5.
[e]Although the inverted repeats are present, they are greatly truncated.

function in the chloroplast are encoded by nuclear genes. These are thought to be genes that migrated to the nucleus as the chloroplast evolved from an endosymbiont into a photosynthetic organelle. Unlike free-living prokaryotes, introns are common in chloroplast genes, and they are primarily of the self-splicing type (Section 8.8).

Analyses of chloroplast genomes firmly support the endosymbiotic hypothesis (Sections 14.4 and 18.4). For example, chloroplast genomes contain genes that are homologs of those in *Escherichia coli*, cyanobacteria, and other *Bacteria*. These include genes encoding proteins for cell division, suggesting that the mechanism of chloroplast division resembles that of bacterial cells. In addition, chloroplast genes for protein transport through membranes are highly related to those of *Bacteria*.

Mitochondrial Genomes

Mitochondria are responsible for energy production by respiration and are found in most eukaryotic organisms. Mitochondrial genomes primarily encode proteins for oxidative phosphorylation and, as do chloroplast genomes, also encode rRNAs, tRNAs, and proteins involved in protein synthesis. However, most mitochondrial genomes encode many fewer proteins than do those of chloroplasts.

Several hundred mitochondrial genomes have been sequenced. The largest mitochondrial genome has 62 protein-encoding genes, but others encode as few as 3 proteins. The mitochondria of almost all mammals, including humans, encode only 13 proteins plus 22 tRNAs and 2 rRNAs. **Figure 13.6** shows a map of the 16,569-bp human mitochondrial genome. The mitochondrial genome of the yeast *Saccharomyces cerevisiae* is larger (85,779 bp), but has only 8 protein-encoding genes. Besides the genes encoding the RNA and proteins, the genome of yeast mitochondria contains large stretches of extremely AT-rich DNA that has no apparent function.

Whereas chloroplasts use the "universal" genetic code, mitochondria use slightly different, simplified genetic codes (Section 7.13). These seem to have arisen from selection pressure for smaller genomes. For example, the 22 tRNAs produced in mitochondria are insufficient to read the entire genetic code, even with the "standard" wobble pairing taken into consideration. Therefore, base pairing between the anticodon and the codon (Figure 7.25) is even more flexible in mitochondria than it is in cells.

Unlike chloroplast genomes, which are all single, circular DNA molecules, the genomes of mitochondria, are quite diverse. For example, some mitochondrial genomes are linear, including those of some species of algae, protozoans, and fungi. In other cases, such as in the bakers and brewer's yeast *S. cerevisiae,* although genetic analyses indicate that the mitochondrial genome is circular, it seems that the major *in vivo* form is linear. (Recall that bacteriophage T4 has a genetically circular genome but is physically linear, Section 10.9.)

Finally, it should be noted that small plasmids exist in the mitochondria of several organisms, complicating mitochondrial genome analysis. An additional problem in analyzing some

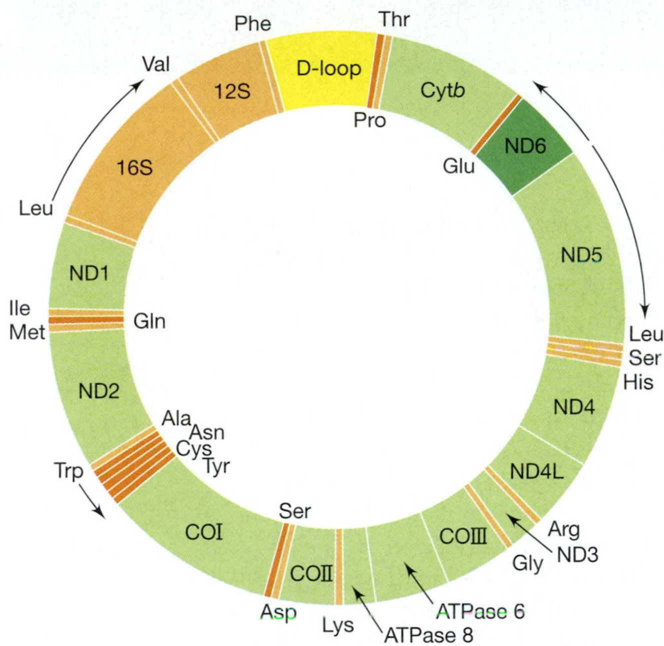

Figure 13.6 Map of the human mitochondrial genome. The circular genome of the human mitochondrion contains 16,569 bp. The genome encodes the 16S and 12S rRNA (corresponding to the prokaryotic 23S and 16S rRNAs) and 22 tRNAs. Genes that are transcribed counterclockwise (CCW) are in dark orange, and those transcribed clockwise (CW) in light orange. The amino acid designations for tRNA genes are on the outside for CCW-transcribed genes and on the inside of the map for CW genes. The 13 protein-encoding genes are shown in green (dark green, transcribed CCW; light green, transcribed CW). Cyt*b*, cytochrome *b*; ND1-6, components of the NADH dehydrogenase complex; COI-III, subunits of the cytochrome oxidase complex; ATPase 6 and 8, polypeptides of the mitochondrial ATPase complex. The two promoters are in the region called the D-loop, a region also involved in DNA replication.

organelle genomes is that it is sometimes difficult to find the gene for a particular protein even when the sequence of the protein and the organelle DNA are both known. This is because of **RNA editing** (see the Microbial Sidebar, "RNA Editing").

Organelles and the Nuclear Genome

Chloroplasts and mitochondria require many more proteins than they encode. For example, far more proteins are needed for translation in organelles than are encoded by the organelle genome. Thus, many organelle functions are encoded by nuclear genes.

It is estimated that the yeast mitochondrion contains over 400 different proteins; however, only 8 of them are encoded by the yeast mitochondrial genome, the remaining proteins being encoded by nuclear genes. Although one might predict that proteins that function in specific processes in the eukaryotic nucleus and cytoplasm could be put to the same use in organelles, this is not the case. Although the genes for many organelle proteins are present in the nucleus, transcribed there, and translated on the 80S ribosomes in the eukaryotic cytoplasm, the proteins are used specifically by the organelles and must be transported into them.

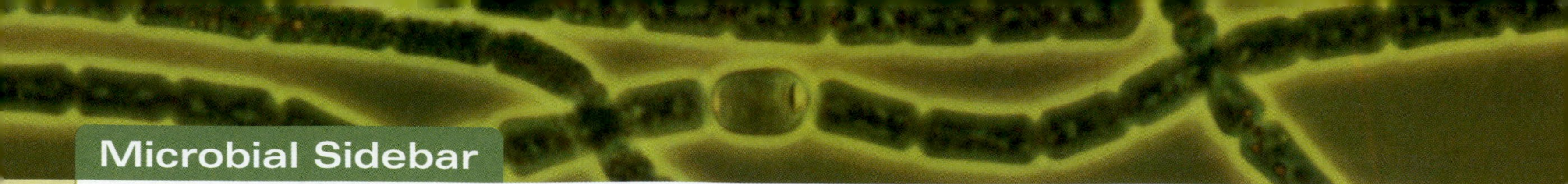

Microbial Sidebar

RNA Editing

RNA editing is the process of altering the base sequence of a messenger RNA *after* transcription. Consequently, the nucleotide sequence of the final mRNA does not correspond exactly to the sequence of the DNA from which it was transcribed. There are two forms of RNA editing. In one, nucleotides are either inserted or deleted in the mRNA. In the other, a base in the mRNA is chemically modified in a way that changes its identity. In either case, RNA editing can alter the coding sequence of a mRNA with the consequence that the amino acid sequence of the resulting polypeptide differs from that predicted from its gene sequence.

RNA editing is very rare in most organisms, especially higher animals. Mammals and plants only employ editing by chemical modification. One important human example is the adenosine to inosine editing of mRNA for some neurotransmitter receptor proteins in the nervous system. In this case, adenosine is converted into inosine by deamination and the inosine generated functions as a guanosine during translation. Defects in this editing can cause severe neurological symptoms.

RNA editing is more common in the mitochondria and chloroplasts of higher plants. At specific sites in some mRNAs, a C will be converted to a U by oxidative deamination (the opposite modification is more rare). There are at least 25 sites of C to U conversion in the maize chloroplast. Depending on the location of the editing, a new codon may be formed, leading to an altered protein sequence.

Protein ...Leu Cys Phe Trp Phe Arg Phe Phe Cys...

mRNA ...uuG uGu UUU UGG uuu AGG uuu uuu uGu...

DNA ... G G TTT TCC AGG G ...

... C C AAA AGG TCC C ...

Figure 1 RNA editing. *The first row shows a portion of the amino acid sequence of subunit III of the enzyme cytochrome oxidase from the protozoan* Trypanosoma brucei. *This protein is encoded by the mitochondrial genome. The sequence of the mRNA for this region is shown beneath the amino acid sequence. The bases in uppercase letters are those transcribed from the gene, which is shown below in green. The bases in the mRNA (orange) in lowercase are inserted into the transcript by RNA editing. The spaces shown between the DNA base pairs are simply to aid in visualization; there are no actual gaps in the DNA molecule itself.*

Editing of mRNA by the insertion or deletion of nucleotides occurs in certain protozoa, notably the trypanosomes and their relatives, and more often in mitochondrial genes than in nuclear genes. Some mitochondrial transcripts are edited such that large numbers (hundreds in some cases) of uridines are added or, more rarely, deleted. An example of this type of RNA editing is shown in **Figure 1**. In such cases, the coding sequences in the DNA usually have incorrect reading frames. If the mRNA were not edited, the resulting proteins would be frame-shifted and defective.

RNA editing is precisely controlled by short sequences in the mRNA that "guide" editing enzymes. In the case of insertional editing, the sequences in the mRNA are recognized by short guide RNA molecules that are complementary to the mRNA except that they have an extra A. The U residues are inserted into the mRNA opposite the extra A on the guide RNA. Obviously, this process must be very precisely controlled. Inserting too many or too few bases would generate frameshift errors that would yield dysfunctional proteins.

RNA editing, although a curious phenomenon, was not a significant obstacle in analyzing organellar genomes. This is because the number of proteins they encode is small and the proteins highly conserved. By contrast, if RNA editing had been widespread in nuclear genomes or in the genomes of prokaryotic cells, comparative genomics would be an even more formidable task.

The function and origin of RNA editing is unknown. But some scientists have suggested that this process may be yet another remnant, along with ribozymes, of an era on Earth before the evolution of cells or DNA, the RNA world (Sections 8.8 and 14.2).

Those nuclear-encoded mitochondrial proteins involved in translation and energy generation are closely related to counterparts in the *Bacteria,* not to those that function in the eukaryotic cytoplasm. Thus, it initially appeared that most genes encoding mitochondrial proteins had been transferred from the mitochondrion to the nucleus during maturation of the endosymbiotic process. However, what was needed to confirm such a hypothesis was both the nuclear and mitochondrial genome sequences of a eukaryote, the genome sequence of a species of *Bacteria* phylogenetically closely related to the mitochondrial genome, and the genomes of other *Bacteria* for comparative purposes. All these requirements were met in the case of the yeast *S. cerevisiae* and the necessary *Bacteria,* and the analysis has been revealing.

Surprisingly, of the 400 nuclear genes encoding mitochondrial proteins, only about 50 were closely related to the phylogenetic lineage in *Bacteria* that led to mitochondria (the *Alpha proteobacteria*, Section 15.1). Another 150 were clearly related to proteins of *Bacteria*, but not necessarily *Alpha proteobacteria*. These *Bacteria*-like proteins were mostly needed for energy conversions, translation, and biosynthesis. However, the remaining 200 or so mitochondrial proteins were encoded by genes that have no identifiable homologues among known genes of *Bacteria*. These proteins were mostly required for membranes, regulation, and transport. Thus, although the mitochondrion shows many of signs of having originated from endosymbiotic events (Sections 14.4 and 18.4), genomic analyses have shown that its genetic history is more complicated than previously thought.

13.4 MiniReview

All eukaryotic cells (except for a few parasites) contain mitochondria. In addition, plant cells contain chloroplasts. Both organelles contain circular DNA genomes that encode rRNAs, tRNAs, and a few proteins involved in energy metabolism. Although the genomes of the organelles are independent of the nuclear genome, the organelles themselves are not. Many genes in the nucleus encode proteins required for organelle function.

- What is unusual about the genes that encode mitochondrial functions in yeast?
- How are genome size and gene content correlated in yeast and human mitochondria?
- What is RNA editing? How does it differ from RNA processing?

13.5 Eukaryotic Microbial Genomes

A large number of eukaryotes are known, and the genomes of several microbial and higher eukaryotes have now been sequenced (**Table 13.4**). The genomes of mammals, including human, mouse, and rat, have around 25,000 genes—about

Table 13.4 Some eukaryotic nuclear genomes[a]

Organism	Comments	Organism/ Cell type[b]	Genome size (Mbp)	Haploid chromosome number	Protein-encoding genes[c]
Nucleomorph of *Bigelowiella natans*	Degenerate endosymbiotic nucleus	E	0.37	3	331
Encephalitozoon cuniculi	Smallest known eukaryotic genome; human pathogen	P	2.9	11	2,000
Cryptosporidium parvum	Parasitic protozoan	P	9.1	8	3,800
Plasmodium falciparum	Malignant malaria	P	23	14	5,300
Saccharomyces cerevisiae	Yeast, a model eukaryote	FL	13	16	5,800
Ostreococcus tauri	Marine green alga; smallest free-living eukaryote	FL	12.6	20	8,200
Aspergillus nidulans	Filamentous fungus	FL	30	8	9,500
Giardia lamblia	Flagellated protozoan; causes acute gastroenteritis	P	12	5	9,700
Dictyostelium discoideum	Social amoeba	FL	34	6	12,500
Drosophila melanogaster	Fruit fly; model organism for genetic studies	FL	180	4	13,600
Caenorhabditis elegans	Roundworm; model organism for animal development	FL	97	6	19,100
Arabidopsis thaliana	Model plant for genetic studies	FL	125	5	26,000
Mus musculus	Mouse, a model mammal	FL	2,500	23	25,000
Homo sapiens	Human	FL	2,850	23	25,000
Oryza sativa	Rice; the world's most important crop plant	FL	390	12	38,000
Paramecium tetraaurelia	Ciliate protozoan	FL	72	>50	40,000
Populus trichocarpa	Black poplar, a tree	FL	500	19	45,000
Trichomonas vaginalis	Flagellated protozoan; human pathogen	P	160	6	60,000

[a]All data are for the haploid nuclear genomes of these organisms.

[b]E, endosymbiont; P, parasite; FL, free-living.

[c]The number of protein-encoding genes is in all cases an estimate based on the number of known genes and sequences that seem likely to encode functional proteins.

UNIT 2

twice the number found in insects and four times that of yeast. However, the genomes of higher plants, such as rice and black poplar, contain even more genes, approaching twice that of humans. It is thought that sequencing of corn (maize) and other large plant genomes presently in progress will reveal even higher gene numbers.

Interestingly, both single-celled protozoans *Paramecium* (40,000 genes) and *Trichomonas,* (60,000 genes) have significantly more genes than humans do (Table 13.4). Indeed, *Trichomonas* presently holds the record for gene number of any organism. This is puzzling because *Trichomonas* is a human parasite, and as we have seen, such organisms typically have small genomes relative to comparable free-living organisms.

Of single-celled eukaryotes, the yeast *Saccharomyces cerevisiae* is most widely used as a model organism as well as being extensively used in industry, and so we focus on it here.

The Yeast Genome

The haploid yeast genome contains 16 chromosomes ranging in size from 220 kbp to about 2,352 kbp. The total yeast nuclear genome (excluding the mitochondria and some plasmid and viruslike genetic elements) is approximately 13,392 kbp. Why are the words "about" and "approximately" used to describe this genome when it has been completely sequenced? Yeast, like many other eukaryotes, has a large amount of repetitive DNA (Section 8.5). When the yeast genome was published in 1997, not all of the "identical" repeats had been sequenced. It is difficult to sequence a very long run of identical or nearly identical sequences and then assemble the data into a coherent framework. For example, yeast chromosome XII contains a stretch of approximately 1,260 kbp containing 100–200 repeats of yeast rRNA genes. Another repeated sequence follows this long series of rRNA gene repeats. Because of such identical repeats, the sizes of eukaryotic genomes are inevitably only close approximations.

In addition to having multiple copies of the rRNA genes, the yeast nuclear genome has around 300 genes for tRNAs (only a few are identical) and nearly 100 genes for other types of noncoding RNA. As with other eukaryotic genomes, the number of predicted ORFs in yeast has changed somewhat as sequence analysis is refined. As of 2006, approximately 5,800 ORFs plus another 800 possible ORFs have been identified in the yeast genome; this is fewer ORFs than in some prokaryotic genomes (Tables 13.1 and 13.4). Of the yeast ORFs, about 3,500 encode proteins whose functions are known. The wide variety of genetic and biochemical techniques available for studying this organism have resulted in significant advances in the understanding of the function of the remaining proteins as well (Section 13.7).

Minimal Gene Complement of Yeast

How many of the known yeast genes are actually essential? This question can be approached by systematically inactivating each gene in turn with knockout mutations (mutations that render a gene nonfunctional, Section 12.9). Knockout mutations cannot normally be obtained in genes essential for cell viability in a haploid organism. However, yeast can be grown in both diploid and haploid states (Section 18.18). By generating knockout mutations in diploid cells and then investigating whether they can also exist in haploid cells, it is possible to determine whether a particular gene is essential for cell viability.

Using knockout mutations, it has been shown that at least 877 yeast ORFs are essential, whereas 3,121 clearly are not. Note that this number of essential genes is much greater than the approximately 300 genes (or possibly fewer than 200 genes, Section 13.3) predicted to be the minimal number required in prokaryotes. However, because eukaryotes are more complex than prokaryotes, a larger minimal gene complement would be expected.

Yeast Introns

Yeast is a eukaryote and contains introns (Section 8.8). However, the total number of introns in the protein-encoding genes of yeast is a mere 225. Most yeast genes with introns have a single small intron near the 5′ end of the gene. This situation differs greatly from that seen in higher eukaryotes (Table 13.4). In the worm *Caenorhabditis elegans,* for example, the average gene has 5 introns, and in the fruit fly *Drosophila,* the average gene has 4 introns. Introns are also very common in the genes of higher plants. Thus the plant *Arabidopsis* averages 5 introns per gene, and over 75% of *Arabidopsis* genes have introns. In humans almost all protein-encoding genes have introns, and it is not uncommon for a single gene to have 10 or more. Moreover, human introns are typically much larger than human exons. Indeed, exons make up only about 1% of the human genome, whereas introns account for 24%.

Other Eukaryotic Microorganisms

The genomes of several other eukaryotic microorganisms, mostly those of medical importance, have been sequenced. The smallest eukaryotic cellular genome known belongs to *Encephalitozoon cuniculi,* an intracellular pathogen of humans and other animals that causes lung infections. *E. cuniculi* lacks mitochondria, and although its haploid genome contains 11 chromosomes, the genome size is only 2.9 Mbp with approximately 2,000 genes (Table 13.4); this is smaller than many prokaryotic genomes (Table 13.1). As is also true in the prokaryotes, the smallest eukaryotic genome belongs to an endosymbiont (Table 13.4). Known as a *nucleomorph,* this is the degenerate remains of a eukaryotic endosymbiont found in certain green algae that have acquired photosynthesis by secondary endosymbioses (Section 18.21). Nucleomorph genomes range from about 0.45 to 0.85 Mbp.

As previously mentioned, the largest eukaryotic genome belongs to *Trichomonas,* which has about 60,000 genes despite its parasitic existence (Table 13.4). The free-living ciliate *Paramecium* has about 40,000 genes, and the free-living social amoeba, *Dictyostelium,* has about 12,500 genes (but note that *Dictyostelium* has both single-celled and multicellular phases in its life cycle, Section 18.12). For comparison, the pathogenic amoeba *Entamoeba histolytica,* the causative agent of amebic dysentery, has approximately 10,000 genes.

Apart from the strange case of *Trichomonas,* parasitic eukaryotic microorganisms have genomes containing 10–30 Mbp of DNA and between 4,000 and 11,000 genes. For example, the genome of the trypanosome *Trypanosoma brucei*, the agent of African sleeping sickness, has 11 chromosomes, 35 Mbp of DNA, and almost 11,000 genes. The most important eukaryotic parasite is *Plasmodium*, which causes malaria (Section 35.5). The 25-Mbp genome of *Plasmodium falciparum* consists of 14 chromosomes ranging in size from 0.7 to 3.4 Mbp. The estimated number of genes for *P. falciparum*, which infects humans, is 5,300, and for the related species *Plasmodium yoelli*, which infects rodents, is 5,900.

13.5 MiniReview

The complete genomic sequence of the yeast *Saccharomyces cerevisiae* and that of many other microbial eukaryotes has been determined. Yeast may encode up to 5,800 proteins, of which only about 900 appear essential for viability. Relatively few of the protein-encoding genes of yeast contain introns. The number of genes in single-celled eukaryotes ranges from 2,000 (less than many bacteria) to 60,000 (more than twice as many as humans).

- How can you show whether a gene is essential?
- What is unusual about the genome of the eukaryote *Encephalitozoon*?

II GENOME FUNCTION AND REGULATION

Despite the major effort required to generate an annotated genome sequence, in some ways the net result is simply a "list of parts." To understand how a cell functions, we need to know more than which genes are present. It is also necessary to investigate both gene expression (transcription) and the function of the final gene product.

We focus here on gene expression. In analogy to the term "genome," the entire complement of RNA produced under a given set of conditions is known as the **transcriptome**.

13.6 Microarrays and the Transcriptome

Knowing the conditions under which a gene is transcribed may reveal a gene's function. We have already discussed how nucleic acid hybridization reveals the location of genes on specific fragments of DNA (Section 12.2). Hybridization techniques can also be used in conjunction with genomic sequence data to measure gene expression by hybridizing mRNA to specific DNA fragments. This technique has been radically enhanced with the development of microarrays, or *gene chips* as they are also called.

Microarrays and the DNA Silica Chip

Microarrays are small solid-state supports to which genes or more often, portions of genes, are fixed and arrayed spatially in a known pattern. The gene segments are synthesized by the polymerase chain reaction (PCR), or, alternatively, oligonucleotides are designed for each gene based on the genomic sequence. Once attached to the solid support, these DNA segments can be hybridized with mRNA from cells grown under specific conditions and scanned and analyzed by computer. Hybridization between a specific mRNA and the corresponding DNA on the chip indicates that the gene has been transcribed.

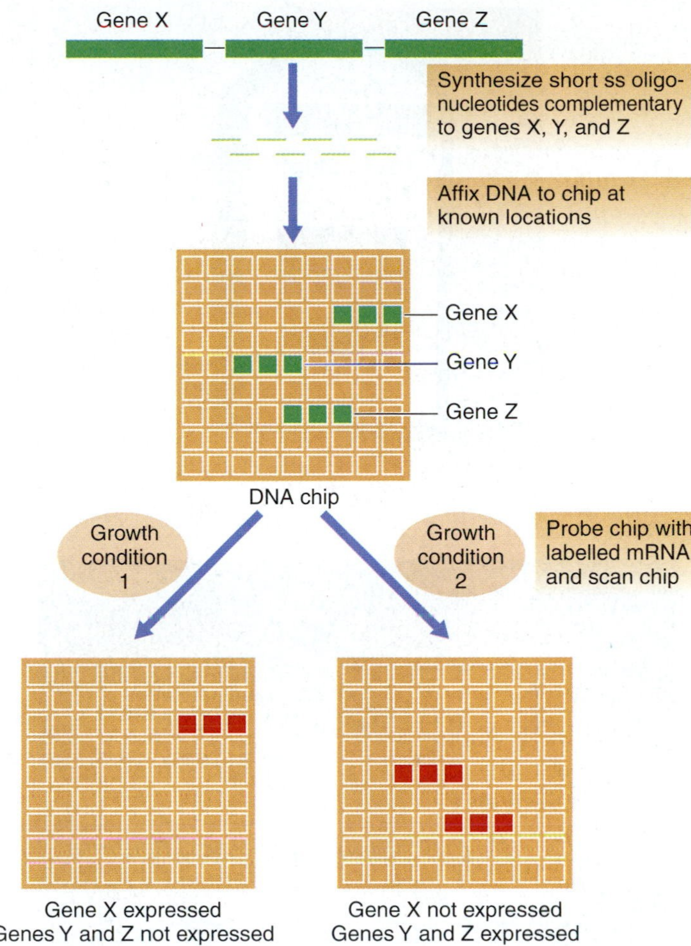

Figure 13.7 Making and using DNA chips. Short single-stranded oligonucleotides corresponding to all the genes of an organism are synthesized individually and affixed at known locations to make a DNA chip (microarray). The DNA chip is assayed by hybridizing fluorescently labeled mRNA obtained from cells grown under a specific condition to the DNA probes on the chip and then scanning the chip with a laser.

In practice, mRNA is present in amounts too low for direct use. Consequently, the mRNA sequences must first be amplified. Reverse transcriptase (RT) is used to generate complementary DNA (cDNA) from the mRNA. The cDNA may then be amplified by PCR (these two steps constitute the RT-PCR procedure; Section 12.8). Alternatively, the cDNA may be used as a template by T7 RNA polymerase, which generates multiple RNA copies called *cRNA*. The cDNA or cRNA is then applied to the array.

A method for making and using microarrays is shown in Figure 13.7. Photolithography, a process used to produce computer chips, has been adapted to produce silica microarray

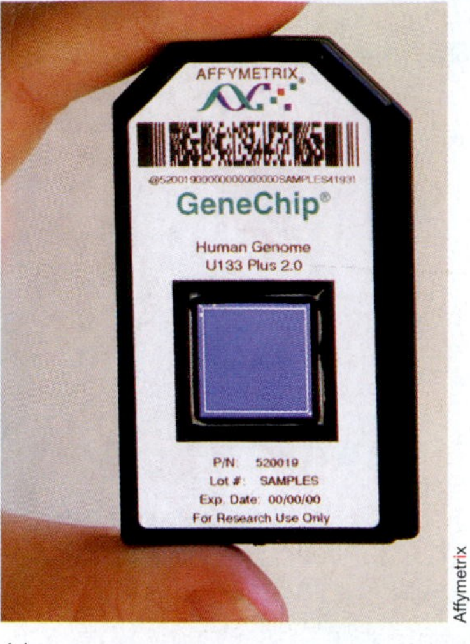

(a)

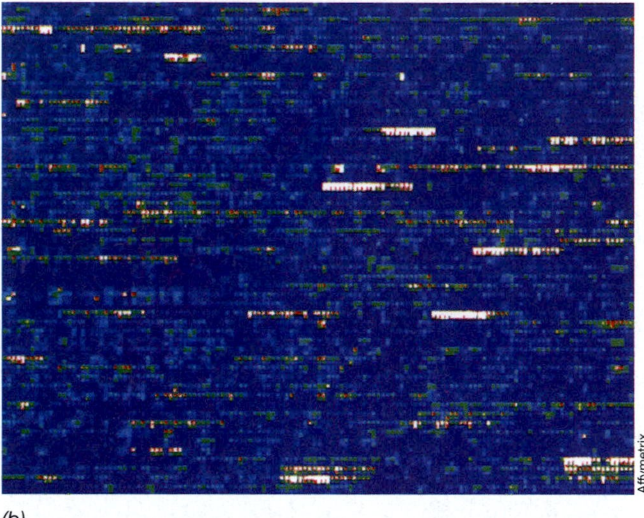

(b)

Figure 13.8 Using DNA chips to assay gene expression. (*a*) The human genome chip containing over 40,000 gene fragments. (*b*) A hybridized yeast chip. The photo shows fragments from one-fourth of the entire genome of baker's yeast, *Saccharomyces cerevisiae*, affixed to a gene chip. Each gene is present in several copies and has been probed with fluorescently labeled cDNA derived from the mRNA extracted from yeast cells grown under a specific condition. The background of the chip is blue. Locations where the cDNA has hybridized are indicated by a gradation of colors up to maximum hybridization, which shows as white. Because the location of each gene on the chip is known, when the chip is scanned, it reveals which genes were expressed.

chips of 1 to 2 cm in size, each of which can hold thousands of different DNA fragments. Whole genome arrays contain DNA segments representing the entire genome of an organism. In practice, each gene is often represented more than once in the array to provide increased reliability. For example, one company markets a human genome chip that contains the entire human genome (**Figure 13.8a**). This single chip can analyze over 47,000 human transcripts and has room for 6,500 additional oligonucleotides for use in clinical diagnostics.

Figure 13.8*b* shows a part of a chip used to assay expression of the *Saccharomyces cerevisiae* genome. This chip easily holds the 5,600 protein-encoding genes of *S. cerevisiae* (Section 13.5) so that global gene expression in this organism can be measured in a single experiment. To do this, the chip is hybridized with cRNA or cDNA derived from mRNA obtained from yeast cells grown under specific conditions. Any particular cRNA/cDNA binds only to the DNA on the chip that is complementary in sequence. To visualize binding, the cRNA/cDNA is tagged with a fluorescent dye, and the chip is scanned with a laser fluorescence detector. The signals are then analyzed by computer. A distinct pattern of hybridization is observed, depending upon which DNA sequences correspond to which mRNAs (Figures 13.7 and 13.8*b*). The intensity of the fluorescence gives a quantitative measure of gene expression (Figure 13.8*b*). This allows the computer to make a list of which genes were expressed and to what extent. Thus, using gene chips, the transcriptome of the organism of interest grown under specified conditions is revealed from the pattern and intensity of the fluorescent spots generated.

Applications of DNA Chips: Gene Expression

Gene chips may be used in several ways, depending on the genes attached to the chip. Global gene expression is monitored by affixing to the array an oligonucleotide complementary to each gene in the genome and then using the entire population of mRNA as the test sample (Figure 13.8*b*). Alternatively, one can compare expression of specific groups of genes under different growth conditions. The ability to analyze the simultaneous expression of thousands of genes has tremendous potential for unraveling the complexities of metabolism and regulation. This is true both for organisms as "simple" as bacteria or as complex as higher eukaryotes.

The *S. cerevisiae* gene chip (Figure 13.8*b*) has been used to study metabolic control in this important industrial organism (Sections 25.11 and 25.14). Yeast can grow by fermentation and by respiration. Transcriptome analysis can reveal which genes are shut down and which are turned on when yeast cells are switched from fermentative (anaerobic) to respiratory metabolism or vice versa. Transcriptome analyses of such gene expression show that yeast undergoes a major metabolic "reprogramming" during the switch from anaerobic to aerobic growth. A number of genes that control production of ethanol (a key fermentation product) are strongly repressed, whereas citric acid cycle functions (needed for aerobic growth) are strongly activated by the switch. Overall, over 700 genes are turned on and over 1,000 turned off during this metabolic transition. Moreover, by using a microarray, the expression pattern of genes of unknown function are also monitored during the fermentative to respiratory switch, yielding clues to their possible role. At present, no other available method can give as much information about gene expression as microarrays.

Another important use of DNA microarrays is the comparison of genes in closely related organisms. For example, this approach has been used to follow the evolution of pathogenic bacteria from their harmless relatives. In human medicine microarrays are used to monitor gene loss or duplication in cancer cells. In environmental microbiology microarrays have been used to assess microbial diversity. Phylochips, as they are called, contain oligonucleotides complementary to the 16S rRNA sequences of different bacterial species. After extracting bulk DNA or RNA from an environment, the presence or absence of each species can be assessed by the presence or absence of hybridization on the chip (∞ Section 22.5 and Figure 22.15). **www.microbiologyplace.com Online Tutorial 13.1: DNA Chips**

Applications in Identification

Besides probing gene expression, microarrays can be used to specifically identify microorganisms. In this case the array contains a set of characteristic DNA sequences from each of a variety of organisms or viruses. Such an approach can be used to differentiate between closely related strains by differences in their hybridization patterns. This allows very rapid identification of pathogenic viruses or bacteria from clinical samples or detection of these organisms in various other substances, such as food. For example, identification (ID) chips have been used in the food industry to detect particular pathogens, for example *Escherichia coli* O157:H7.

DNA chips are also available to identify higher organisms as well. A commercially available chip called the FoodExpert-ID contains 88,000 gene fragments from vertebrate animals and is used in the food industry to ensure food purity. For example, the chip can confirm the presence of the meat listed on a food label and can also detect foreign animal meats that may have been added as supplements to or substitutes for the official ingredients. The eventual goal is to have each meat product receive an "identity card" listing all the animal species whose tissues were detected in it. This is intended to give consumers more confidence in the wholesomeness of their food products. The FoodExpert-ID can also be used to detect vertebrate by-products in animal feed, a growing concern with the advent of prion-mediated diseases such as mad cow disease (∞ Section 10.15).

13.6 MiniReview

Microarrays consist of genes or gene fragments attached to a solid support in a known pattern. These arrays are used to hybridize to mRNA and then analyzed to determine patterns of gene expression. The arrays are large enough and dense enough that the transcription pattern of an entire genome (the transcriptome) can be analyzed.

- What do microarrays tell you that studying gene expression by assaying a particular enzyme cannot?
- Why is it useful to know how gene expression of the entire genome responds to a particular condition?

13.7 Proteomics

A further aim of genomic studies is to determine which expressed genes actually yield protein products and then to determine the function of these proteins. The genomewide study of the structure, function, and regulation of an organism's proteins is called **proteomics**.

The number and types of proteins present in a cell are subject to change in response to an organism's environment or other factors, such as developmental cycles. As a result, the term **proteome** has unfortunately become ambiguous. In its wider sense, a proteome refers to *all* the proteins encoded by an organism's genome. In its narrower sense, however, it refers to those proteins present in a cell *at any given time*.

Methods in Proteomics

The first major approach to proteomics was developed decades ago with the advent of two-dimensional (2D) polyacrylamide gel electrophoresis. This technique can separate, identify, and measure all the proteins present in a cell sample. A 2D gel separation of proteins from *Escherichia coli* is shown in **Figure 13.9**. In the first dimension (the horizontal dimension in the figure), the proteins are separated by differences in their isoelectric points, the pH at which the net charge on each protein reaches zero. In the second dimension, the proteins are denatured in a way that gives each amino acid residue a fixed charge. The proteins are then separated by size (in much the same way as for DNA molecules; ∞ Section 12.1).

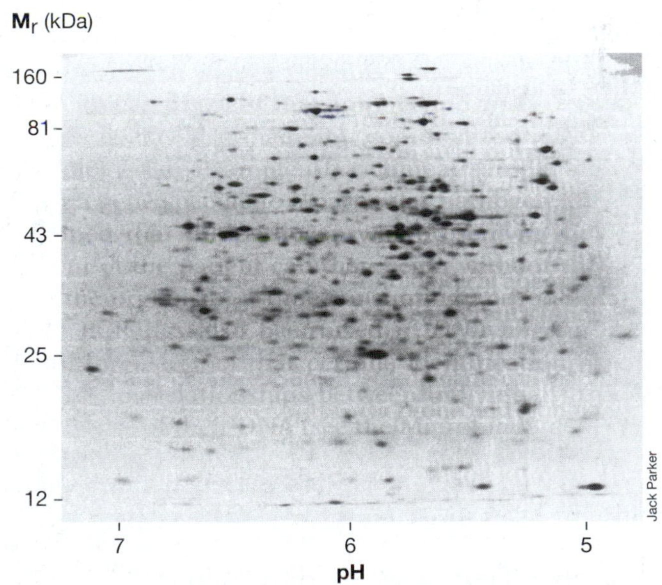

Figure 13.9 Two-dimensional polyacrylamide gel electrophoresis of proteins. Autoradiogram of the proteins of cells of *Escherichia coli*. Each spot on the gel is a different protein. The proteins are radioactively labeled to allow for visualization and quantification. The proteins were separated in the first dimension (X-direction) by isoelectric focusing under denaturing conditions. The second dimension (Y-direction) separates denatured proteins by their mass (M_r; in kilodaltons), with the largest proteins being toward the top of the gel.

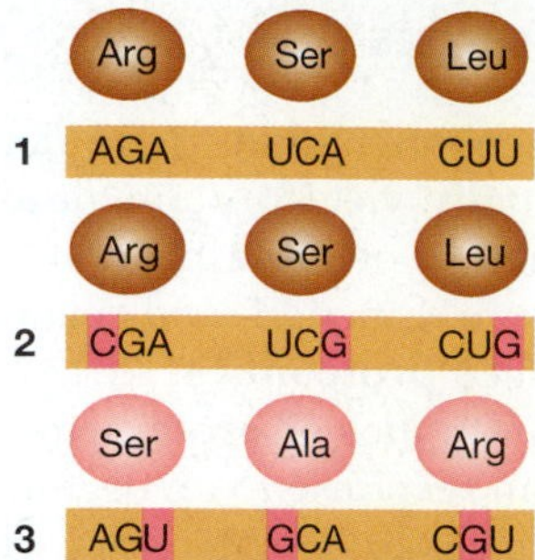

Figure 13.10 Comparison of nucleic acid and amino acid sequence similarities. Three different nucleotide sequences are shown (for convenience RNA is shown). Both sequence 2 and sequence 3 differ from sequence 1 in only three positions. However, the amino acid sequence encoded by 1 and 2 are identical, whereas that encoded by sequence 3 is unrelated to the other two.

In studies of *E. coli* and a few other organisms, hundreds of proteins separated in 2D gels have been identified by biochemical or genetic means, and their regulation has been studied under various conditions. Using 2D gels, the presence of a particular protein under different growth conditions can be measured and related to environmental signals. One method of connecting an unknown protein with a particular gene using the 2D gel system is to elute the protein from the gel and sequence a portion of it, usually from its N-terminal end. More recently, eluted proteins have been identified by a technique called *mass spectrometry* (Section 13.8), usually after preliminary digestion to give a characteristic set of peptides. This sequence information may be sufficient to completely identify the protein. Alternatively, partial sequence data may allow for the design of oligonucleotide probes or primers to locate the gene encoding the protein from genomic DNA by hybridization or polymerase chain reaction. Then, following sequencing of the DNA, the gene's identity may be determined.

Today, liquid chromatography is increasingly used to separate protein mixtures. In high-pressure liquid chromatography (HPLC), the sample is dissolved in a suitable liquid and forced under pressure through a column packed with a stationary phase material that separates proteins by variations in their chemical properties, such as size, ionic charge, or hydrophobicity. As the mixture travels through the column, it is separated by interaction of the proteins with the stationary phase. Fractions are collected at the column exit. The proteins in each fraction are digested by proteases and the peptides are identified by mass spectrometry.

Comparative Genomics and Proteomics

Although proteomics often requires intensive experimentation, *in silico* techniques can also be quite useful. Once the sequence of an organism's genome is obtained, it can be compared to that of other organisms to locate and identify genes that are similar to those already known. The sequence that is most important here is the amino acid sequence of the encoded proteins. Because the genetic code is degenerate (∞ Section 7.13), differences in DNA sequence may not necessarily lead to differences in the amino acid sequence (**Figure 13.10**).

Proteins with greater than 50% sequence identity frequently have similar functions. Proteins with identities above 70% are almost certain to have similar functions. Many proteins consist of distinct structural modules, called *protein domains,* each with characteristic functions. Such regions include metal-binding domains, nucleotide-binding domains, or domains for certain classes of enzyme activity, such as helicase or nuclease. Identification of domains of known function within a protein may reveal much about its role, even in the absence of complete sequence homology.

Structural proteomics refers to the proteome-wide determination of the three-dimensional (3D) structures of proteins. At present, it is not possible to directly predict the 3D structure of proteins from their amino acid sequences. However, structures of unknown proteins can often be modeled if the 3D structure is available for a protein with 30% or greater identity in amino acid sequence.

Coupling proteomics with genomics is yielding important clues to how gene expression in different organisms correlates with environmental stimuli. Not only does such information have important basic science benefits, but it also has potential applications. These include advances in medicine, the environment, and agriculture. In all of these areas, understanding the link between the genome and the proteome and how it is regulated could give humans unprecedented control in fighting disease and pollution, as well as unprecedented benefits for agricultural productivity.

13.7 MiniReview

Proteomics is the analysis of all the proteins present in an organism. The ultimate aim of proteomics is to understand the structure, function, and regulation of these proteins.

- Why is the term "proteome" ambiguous, whereas the term "genome" is not?
- What are the most common experimental methods used to survey the proteome?

13.8 Metabolomics

By analogy with the terms *genome* and *proteome,* the **metabolome** is the complete set of metabolic intermediates and other small molecules produced in an organism. Metabolomics has lagged behind other "omics" in large part due to the immense chemical diversity of small metabolites. This makes systematic screening technically challenging. Early attempts used nuclear magnetic resonance (NMR) analysis of extracts from cells labeled with ^{13}C-glucose. However, this method is limited in sensitivity, and the number of compounds that can be simultaneously identified in a mixture is too low for resolution of complete cell extracts.

Mass Spectrometric Analyses of the Metabolome

The most promising approach to metabolomics is the use of newly developed techniques of mass spectrometry. This approach is not limited to particular classes of molecules and can be extremely sensitive. The mass of carbon-12 is defined as exactly 12 molecular mass units (Daltons). However, the masses of other atoms, such as nitrogen-14 or oxygen-16 are not exact integers. Mass spectrometry using extremely high mass resolution, which is now possible in special instruments, allows the unambiguous determination of the molecular formula of any small molecule. Clearly, isomers will have the same molecular formula, but they may be distinguished by their different fragmentation patterns during mass spectrometry.

Metabolome analysis is especially useful in the study of plants, many of which produce several thousand different metabolites—more than most other types of organism. This is because plants make many *secondary metabolites,* such as scents, flavors, alkaloids, and pigments, many of which are commercially important. Metabolomic investigations have monitored the levels of several hundred metabolites in the model plant *Arabidopsis,* and significant changes were observed in the levels of many of these metabolites in response to changes in temperature. Future directions for metabolomics, presently under development, include assessing the effect of disease on the metabolome of various human organs and tissues. Such results should greatly improve our understanding of how the human body fights off infectious and noninfectious disease.

13.8 MiniReview

The metabolome is the complete set of metabolic intermediates produced by an organism.

- What techniques are used to monitor the metabolome?

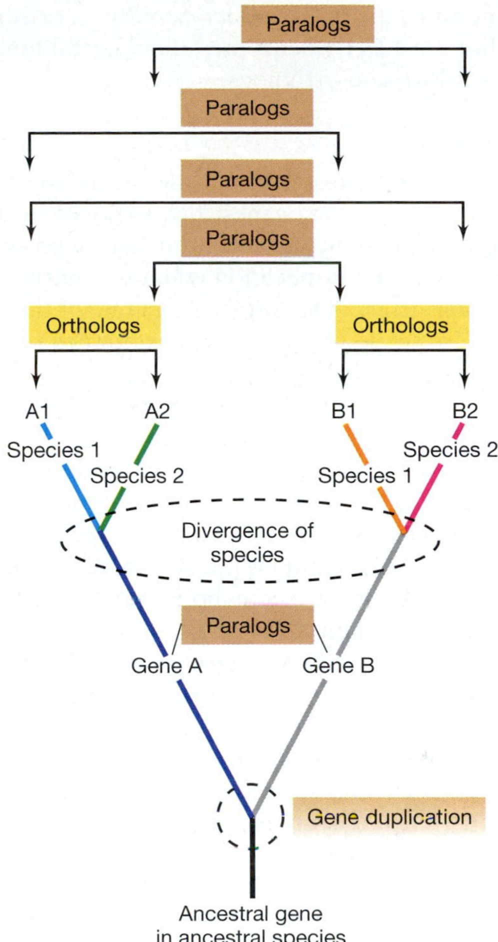

Figure 13.11 Orthologs and paralogs. This family tree depicts an ancestral gene that duplicated and diverged into two paralogous genes, A and B. Next, the ancestral species diverged into species 1 and species 2, both of which have genes for A and B (designated A1 and B1 and A2 and B2 respectively). Each such pair are paralogs. However, because species 1 and 2 are now separate species, A1 is an ortholog of A2 and B1 is an ortholog of B2.

III THE EVOLUTION OF GENOMES

In addition to understanding how genes function and organisms interact with the environment, comparative genomics can also reveal evolutionary relationships between organisms. Reconstructing evolutionary relationships from genome sequences helps to distinguish between primitive and derived characteristics and can resolve ambiguities in phylogenetic trees based on analyses of a single gene, such as small subunit rRNA (Sections 14.6–14.9). Genomics is also a link to understanding early life forms and, eventually, may answer the most fundamental of all questions in biology: How did life first arise?

13.9 Gene Families, Duplications, and Deletions

Genomes from both prokaryotic and eukaryotic sources often contain multiple copies of genes that are related in sequence due to shared evolutionary ancestry; such genes are called **homologous** genes. Groups of gene homologs are called **gene families**. Not surprisingly, larger genomes tend to contain more individual members from a particular gene family.

Paralogs and Orthologs

Comparative genomics has shown that many genes have arisen by duplication of other genes. Such homologs may be subdivided, depending on their origins. Genes whose similarity is the result of gene duplication at some time in the evolution of an organism are called **paralogs**. Genes found in one organism that are similar to genes in another organism, but that differ because of speciation or more distant evolutionary events, are called **orthologs** (**Figure 13.11**). An example of paralogous genes are those encoding several different lactate dehydrogenase (LDH) isoenzymes in humans. These enzymes are structurally distinct yet are all highly related and carry out the same enzymatic reaction. By contrast, the corresponding LDH from *Lactobacillus* is orthologous to all of the human LDH isoenzymes. Thus, gene families contain both paralogs and orthologs.

Gene Duplication

It is widely thought that gene duplication is the mechanism for the evolution of most new genes. If a segment of duplicated DNA is long enough to include an entire gene or group of genes, the organism with the duplication contains multiple copies of these particular genes. After duplication, one of the duplicates is free to evolve while the other copy continues to supply the cell with the original function. In this way, evolution can "experiment" with one copy of the gene. Such gene-duplication events, followed by diversification of one copy, are thought to be the major events that fuel microbial evolution. Genomic analyses have revealed numerous examples of protein-encoding genes that were clearly derived from gene duplication.

Duplications that occur in genetic material may include just a handful of bases or even whole genomes. For example, comparison of the genomes of the yeast *Saccharomyces cerevisiae* and other fungi suggests that the ancestor of *Saccharomyces* duplicated its entire genome. This was followed by extensive deletions that eliminated much of the duplicated genetic material. Analysis of the genome of the model plant *Arabidopsis* suggests that there was one or more whole genome duplications in the ancestor of the flowering plants.

Did bacterial genomes evolve by whole genome duplication? The distribution of duplicated genes and gene families in the genomes of bacteria suggests that many frequent but relatively small duplications have occurred. For example, among the *Deltaproteobacteria,* the soil bacterium *Myxococcus* has a genome of 9.1 Mbp. This is approximately twice that of the genomes of other typical *Deltaproteobacteria,* which range from 4 to 5 Mbp. Among the *Alphaproteobacteria,* genome sizes range from 1.1 to 1.5 Mbp for parasitic members to 4 Mbp for free-living *Caulobacter,* and up to 7–9 Mbp for plant-associated bacteria. However, in all of these cases gene distribution analysis points to frequent small-scale duplications rather than whole genome duplications. Conversely, in bacteria that are parasitic, frequent successive deletions have eliminated genes no longer needed for a parasitic lifestyle, leading to their unusually small genomes (Section 13.2).

Gene Analysis in Different Domains

The comparison of genes and gene families is a major task in comparative genomics. Because chromosomes from many different microorganisms have already been sequenced, such comparisons can be easily done, and the results are often surprising. For instance, genes in *Archaea* involved in DNA replication, transcription, and translation are more similar to those in *Eukarya* than to those in *Bacteria*. Unexpectedly, however, many other genes in *Archaea,* for example, those encoding metabolic functions other than information processing, are more similar to those in *Bacteria* than those in *Eukarya*. The powerful analytical tools of bioinformatics allow genetic relationships between any organisms to be deduced very quickly and at the single gene, gene group, or entire genome level. The results obtained thus far lend further support to the phylogenetic picture of life deduced originally by comparative ribosomal RNA sequence analysis (Sections 2.7 and 14.8) and suggest that many genes in all organisms have common evolutionary roots. However, such analyses have also revealed instances of horizontal gene flow, an important issue to which we now turn.

13.9 MiniReview

Genomics can be used to study the evolutionary history of an organism. Organisms contain gene families, genes with related sequences. If these arose because of gene duplication, the genes are said to be paralogs; if they arose by speciation, they are called orthologs.

- What is a gene family?
- Contrast gene paralogs with gene orthologs.

13.10 Mobile DNA: Transposons and Insertion Sequences

Evolution is based on the transfer of genetic traits from one generation to the next. However, in prokaryotes, **horizontal (lateral) gene transfer** also occurs, and it can complicate evolutionary studies, especially those of entire genomes. Horizontal gene transfer occurs whenever genes are transferred from one cell to another other than by the usual inheritance process, from mother cell to daughter cell. In prokaryotes, at least three mechanisms for horizontal gene transfer are known: *transformation, transduction,* and *conjugation* (Chapter 11).

Horizontal gene flow may be extensive in nature and may sometimes cross even phylogenetic domain boundaries. However, to be detectable by comparative genomics, the difference between the organisms must be rather large. For example, several genes with eukaryotic origins have been found in *Chlamydia* and *Rickettsia,* both human pathogens. In particular, two genes encoding histone H1-like proteins have been found in the *Chlamydia trachomatis* genome, suggesting horizontal transfer from a eukaryotic source, possibly even its human host. Note that this is opposite the situation in which genes from the ancestor of the mitochondrion were transferred to the eukaryotic nucleus (Section 13.4).

Detecting Horizontal Gene Flow

Horizontal gene transfers can be spotted in genomes once the genes have been annotated. The presence of genes that encode proteins typically found only in distantly related species is one signal that the genes originated from horizontal transfer. However, another clue to horizontally transferred genes is the presence of a stretch of DNA whose GC content or codon bias differs significantly from the rest of the genome. Using these clues, many likely examples of horizontal transfer have been documented in the genomes of various prokaryotes. A classic example exists with the organism *Thermotoga maritime*, a species of *Bacteria*, which was shown to contain over 400 genes (greater than 20% of its genome) of archaeal origin. Of these genes, 81 were found in discrete clusters. This strongly suggests that they were obtained by horizontal gene transfer, presumably from thermophilic *Archaea* that share the hot environments inhabited by *Thermotoga*.

Horizontally transferred genes typically encode metabolic functions other than the core molecular processes of DNA replication, transcription, and translation, and may account for the previously mentioned similarities of metabolic genes in *Archaea* and *Bacteria* (Section 13.9). In addition, there are several examples of virulence genes of pathogens having been transferred by horizontal means. It is obvious that prokaryotes are exchanging genes in nature, and the process likely functions to "fine-tune" an organism's genome to a particular situation or habitat.

Caveat to Horizontal Flow

It is necessary to be cautious when invoking horizontal gene transfer to explain the distribution of genes. When the human genome was first sequenced, a couple hundred genes were identified as being horizontal transfers from prokaryotes. However, when more eukaryotic genomes became available for examination, homologs were found for most of these genes in many eukaryotic lineages. Consequently, it now seems that most of these genes are in fact of eukaryotic origin. Only about a dozen human genes are now accepted as strong candidates for having relatively recent prokaryotic origins. The phrase "relatively recent" here refers to genes transferred from prokaryotes after separation of the major eukaryotic lineages, not to genes of possible ancient prokaryotic origin that are shared by eukaryotes as a whole.

13.10 MiniReview

Organisms may acquire genes from other organisms in their environment by a process called horizontal gene transfer. Such transfer may cross the domain boundaries between *Bacteria*, *Archaea*, and *Eukarya*.

- Which class of genes is rarely transferred horizontally? Why?
- List the major mechanisms by which horizontal gene transfer occurs in prokaryotes.

13.11 Horizontal Gene Transfer and Genome Stability

As described in Section 11.16, mobile DNA refers to segments of DNA that move from one location to another within host DNA molecules. Most mobile DNA consists of transposable elements, but integrated virus genomes and integrons are also found. All of these mobile elements play important roles in genome evolution.

Genome Evolution and Transposons

Transposons may move between different host DNA molecules, including chromosomes, plasmids, and viruses. In doing so they may pick up and horizontally transfer genes for a variety of characteristics, including resistance to antibiotics or production of toxins. However, transposons may also mediate a variety of large-scale chromosomal changes. Bacteria that are undergoing rapid evolutionary change often contain relatively large numbers of mobile elements, especially insertion sequences. Recombination among identical elements generates chromosomal rearrangements such as deletions, inversions, or translocations. This is thought to provide a source of genome diversity upon which selection can act. Thus, chromosomal rearrangements that accumulate in bacteria during stressful growth conditions are often flanked by repeats or insertion sequences.

Conversely, once a species settles into a stable evolutionary niche, most mobile elements are apparently lost. For example, genomes of species of *Sulfolobus* (*Archaea*) have unusually high numbers of insertion sequences and show a high frequency of gene translocations. By contrast, *Pyrococcus* (*Archaea*) shows an almost complete lack of insertion sequences and a correspondingly low number of gene translocations. This suggests that for whatever reason, perhaps because of fluctuations in conditions in their habitats, the genome of *Sulfolobus* is more dynamic than the more stable genome of *Pyrococcus*.

Insertion Sequences

Chromosomal rearrangements due to insertion sequences have apparently contributed to the evolution of several bacterial pathogens. In *Bordetella*, *Yersinia*, and *Shigella*, the more highly pathogenic species show a much greater frequency of insertion sequences. For example, *Bordetella bronchiseptica* has a genome of 5.34 Mbp but carries no known insertion sequences. Its more pathogenic relative, *Bordetella pertussis*, has a smaller genome (4.1 Mb) but has more than 260 insertion sequences. Comparison of these genomes suggests that the insertion sequences are responsible for substantial genome rearrangement, including deletions responsible for the reduction of genome size in *B. pertussis*.

Insertion sequences also play a role in assembling genetic modules to generate novel plasmids. Thus 46% of the 220-kbp virulence megaplasmid of *Shigella flexneri* consists of insertion sequence DNA! In addition to full-length insertion sequences, there are many fragments in this plasmid that imply multiple ancestral rearrangements.

Integrons and Super-Integrons

Integrons are genetic elements that collect and express genes carried on mobile segments of DNA, called *cassettes*. Gene cassettes suitable for integration consist of a coding sequence lacking a promoter, but containing an integrase recognition

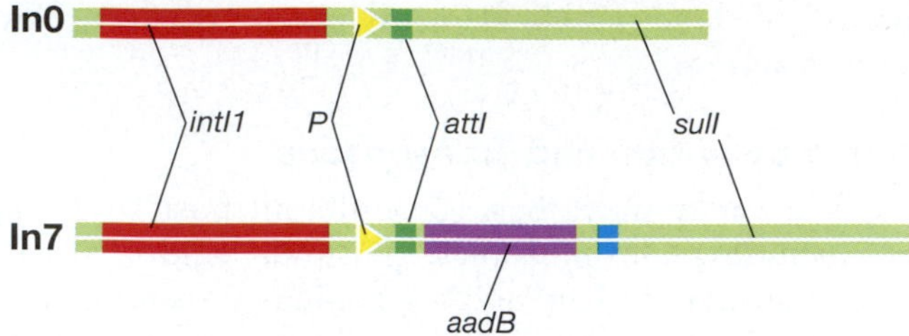

Figure 13.12 Structure of two naturally occurring integrons from *Pseudomonas*. Integron In0 has the basic set of genes: *intI1*, integrase; *attI*, the integration site; P, promoter; and *sulI*, a gene conferring sulfonamide resistance. Integron In7 contains all of these plus an integrated gene cassette. All cassettes contain a site (blue) for site-specific recombination. This cassette contains *aadB*, which confers resistance to certain aminoglycoside antibiotics.

site, the *attC* site. The integrons themselves contain a corresponding integration site, the *attI* site, into which gene cassettes may be integrated. The integron also possesses a gene encoding the **integrase**, the enzyme responsible for inserting cassettes. Integration occurs by recombination between the *attI* site and the *attC* site. Once a gene cassette has been inserted into an integron, the gene it carries may be expressed from a promoter that is provided by the integron.

Neither integrons nor gene cassettes are transposable elements (they do not have terminal inverted repeats, nor do they transpose). However, gene cassettes may exist transiently as free nonreplicating circular DNA incapable of gene expression, or they may be found integrated into the *attI* site of an integron. Thus the gene cassettes are a form of mobile DNA that may move from one integron to another. Most integrons are found on plasmids or in transposons and may collect multiple gene cassettes. A few integrons are found on bacterial chromosomes and may collect hundreds of gene cassettes, whereupon they are called *superintegrons*. For example, the second chromosome of *Vibrio cholerae* (causative agent of cholera) has a super-integron with approximately 200 genes, mostly of unknown function.

Most known integrons carry genes for antibiotic resistance. However, this is probably due to a bias in observation, since antibiotic resistance is of clinical importance. Over 40 different antibiotic resistance genes have been identified on integron gene cassettes, as have some genes associated with virulence in certain pathogenic bacteria. **Figure 13.12** shows the structure of two integrons from *Pseudomonas aeruginosa,* a potentially serious pathogen. Integrons have been found in various species of *Bacteria,* often in clinical isolates, and their selection by horizontal gene transfer in such antibiotic-rich environments as hospitals and clinics is obvious. What is less obvious is the origin of the gene cassettes themselves. These are not simply random genes, as they must possess specific DNA sequences that are recognized by the integrase; they are incapable of expression until they become part of an integron and can be transcribed from the integron's promoter.

13.11 MiniReview

Mobile DNA elements, including transposons, integrons, and viruses are important in genome evolution. Mobile DNA often carries genes encoding antibiotic resistance or virulence factors. Integrons are genetic structures that collect and express promoterless genes carried on gene cassettes.

- Why are transposons especially important in the evolution of pathogenic bacteria?
- What are integrons and how do they differ from transposons?

13.12 Evolution of Virulence: Pathogenicity Islands

Comparison of the genomes of pathogenic bacteria with those of their harmless close relatives often reveals extra blocks of genetic material that contain genes that encode *virulence factors,* special proteins or other molecules or structures necessary to cause disease (Section 28.9).

Some virulence genes are carried on plasmids or lysogenic bacteriophages (Section 10.10). However, many others are clustered in chromosomal regions called **pathogenicity islands**. For example, the identity and chromosomal location of most genes of pathogenic strains of *Escherichia coli* correspond to those of the harmless laboratory strain *E. coli* K-12, as would be expected. However, most pathogenic strains contain pathogenicity islands of considerable size that are absent from *E. coli* K-12 (**Figure 13.13**). Consequently, two strains of the same bacterial species may show significant differences in genome size. For example, as shown in Table 13.1, the enterohemorrhagic strain *E. coli* O157:H7 contains 20% more DNA and genes than the *E. coli* strain K12.

Chromosomal Islands

Pathogenicity islands are merely the best-known case of **chromosomal islands**. Such islands are presumed to have a "foreign" origin based on several observations. First, these extra regions are often flanked by inverted repeats, implying that the whole region was inserted into the chromosome by transposition at some period in the recent evolutionary past. Second, the base composition and codon usage in chromosomal islands often differ significantly from that of the rest of the genome. Third, chromosomal islands are often found in some strains of a particular species but not in others.

Some chromosomal islands carry a gene for an integrase and are thought to move in a manner analogous to conjugative transposons (Section 11.16). Chromosomal islands are typically inserted into a gene for a tRNA; however, because the target site is duplicated upon insertion, an intact tRNA gene is regenerated during the insertion process. In a few cases, transfer of a whole chromosomal island between related bacteria has been demonstrated in the laboratory;

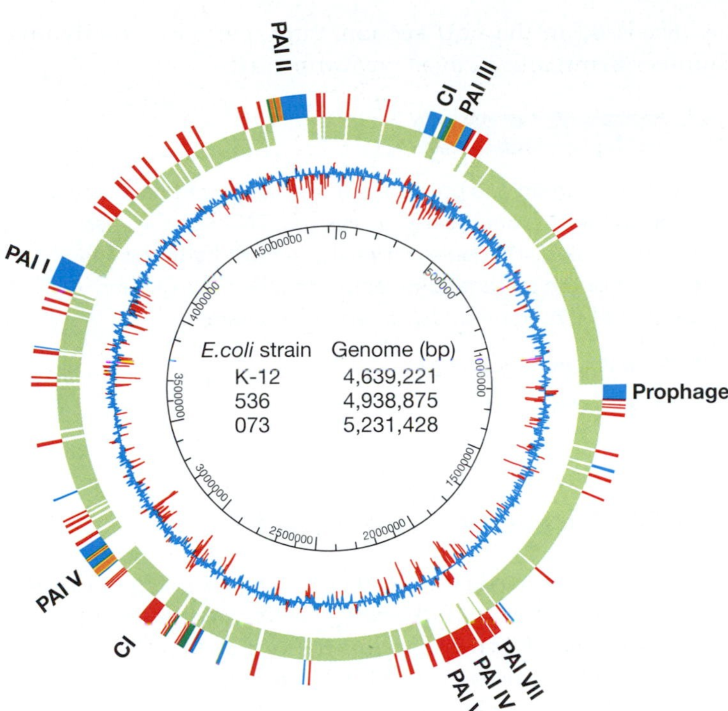

Figure 13.13 Pathogenicity islands in *Escherichia coli*. Genetic map of *E. coli* strain 536, a urinary tract pathogen, compared with a second pathogenic strain (073) and the wild-type strain K-12. The pathogenic strains contain pathogenicity islands, and thus their chromosomes are larger than that of K-12. Inner circle, nucleotide base pairs. Jagged circle, DNA GC distribution; regions where GC content varies dramatically from the genome average are in red. Outermost circle, three-way genomic comparison: green, genes common to all strains; red, genes present in the pathogenic strains only; blue, genes found only in strain 536; orange, genes of strain 536 present in a different location in strain 073. Some very small inserts deleted for clarity. PAI, pathogenicity islands; CI, chromosomal island. Prophage, DNA from a temperate bacteriophage. Note the correlation between genomic islands and skewed GC content. Data adapted from *Proc. Natl. Acad. Sci. (USA)* 103: 12879–12884 (2006).

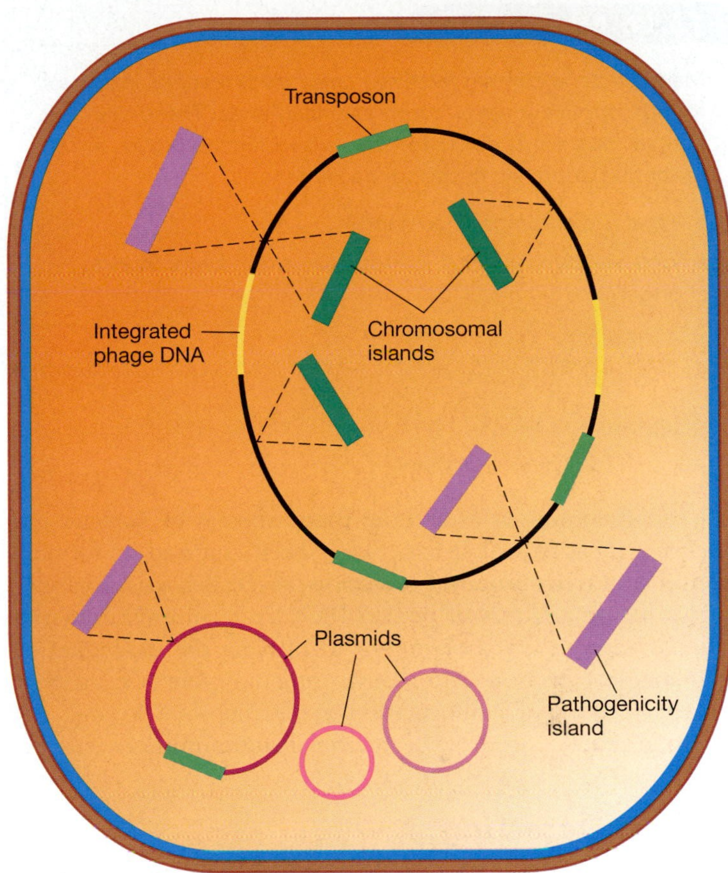

Figure 13.14 Pan genome versus core genome. The core genome is represented by the black regions of the chromosome and is present in all strains of a species. The pan genome includes elements that are present in one or more strains but not in all strains. Each colored wedge indicates a single insertion. Where two wedges emerge from the same location, they represent alternative islands that can insert at that site. However, only one insertion can be present at a given location.

transfer can presumably occur by any of the mechanisms of lateral transfer previously discussed: transformation, transduction, and conjugation. It is thought that after insertion into the genome of a new host cell, chromosomal islands gradually accumulate mutations—both point mutations and small deletions. Thus, over many generations, chromosomal islands tend to lose their ability to move.

Chromosomal islands contribute specialized functions that are not needed for simple survival. Not surprisingly, pathogenicity islands with their clinical relevance have drawn the most interest. However, chromosomal islands are also known that carry genes for the biodegradation of various substrates derived from human activity, such as aromatic hydrocarbons and herbicides. In addition, many of the genes essential for the symbiotic relationship of rhizobia with plants in the root nodule symbiosis (⚭ Section 24.15) are carried in symbiosis islands inserted into the genome of these bacteria. Perhaps the most unique chromosomal island is the magnetosome island of the bacterium *Magnetospirillum*; this DNA fragment carries the genes needed for the formation of magnetosomes, intracellular magnetic particles used to orient the organism in a magnetic field and influence the direction of its motility; (⚭ Sections 4.10 and 15.14).

The presence or absence of chromosomal islands, transposable elements, integrated virus genomes, and plasmids means that there may be major differences in the total amount of DNA and the suite of accessory capabilities (virulence, symbiosis, biodegradation) between strains of a single bacterial species. This has led to the concept that the genome of a bacterial species consists of two components, the *core* genome and the *pan* genome. The core genome is shared by all strains of the species, whereas the pan-genome includes all of the optional extras present in some but not all strains of the species (**Figure 13.14**). In other words, one could say that the core genome is typical of the species as a whole, whereas the pan genome is unique to particular strains within a species.

13.12 MiniReview

Many bacteria contain relatively large chromosomal inserts of foreign origin known as chromosomal islands. These islands contain clusters of genes for specialized functions such as biodegradation, symbiosis, or pathogenesis.

- What is a chromosomal island?
- Why are chromosomal islands believed to be of foreign origin?

IV ENVIRONMENTAL GENOMICS

Microbial communities contain many species of *Bacteria* and *Archaea*, most of which have never been cultured or formally identified. **Environmental genomics**, also called *metagenomics*, refers to the analysis of pooled DNA from an environmental sample without first isolating or identifying the individual organisms (Section 22.6). Environmental genomic analyses have yielded valuable information on the microbial diversity and the total genetic resources of microbial communities.

13.13 Detecting Uncultured Microorganisms

Most microorganisms present in nature have never been brought into laboratory culture (Sections 22.5 and 22.6). Nevertheless, it is possible to obtain information on the organisms and their activities by carrying out analyses of DNA or RNA directly extracted from the environment. Just as the total gene content of an organism is its genome, so the total gene content of the organisms inhabiting an environment is known as its **metagenome**.

Several environments have been surveyed by large-scale metagenome sequencing projects. Extreme environments, such as acid mine run-off waters tend to have limited species diversity. In such environments it has been possible to isolate community DNA and assemble much of it into nearly complete genomes. Conversely, complex environments such as fertile soil yield too much sequence data to allow successful assembly at present.

In addition to metagenome analyses, microarrays (Section 13.6) have been used to explore the patterns of gene expression in natural microbial communities. These techniques have empowered microbiologists to ask new and very complex questions about the structure and function of microbial communities and to measure the contributions of specific members of a microbial community to overall ecosystem function.

One curious recent revelation is that most DNA in natural habitats does not belong to living cells. First, about 50–60% of the DNA in the oceans is extracellular DNA found in deep-sea sediments. This is deposited when dead organisms from the upper layers of the ocean sink to the bottom and disintegrate. Because nucleic acids are repositories of phosphate, this DNA is a major contributor to the global phosphorus cycle. Second, as discussed in the next section, virus particles greatly outnumber living cells in most environments.

13.13 MiniReview

Most microorganisms present in the environment have never been cultured. Nonetheless, analysis of DNA samples has revealed colossal sequence diversity in most habitats. The concept of the metagenome embraces the total genetic content of the organisms in a particular habitat.

- What is a metagenome?
- How is a metagenome analyzed?

13.14 Viral Genomes in Nature

The number of prokaryotic cells on Earth is far greater than the total number of eukaryotic cells; estimates of total prokaryotic cell numbers are on the order of 10^{30}. However, the number of viruses on Earth is even greater—an estimated 10^{31}. The best estimates of both cell and virus numbers in the environment come from the analysis of seawater, as this is much easier to analyze than soil or sediments.

Viral Metagenomics

There are millions of bacteria, and approximately ten times as many viruses, present in every milliliter of seawater (**Figure 13.15**). Not surprisingly, most of these viruses are bacteriophages, and these populations turn over rapidly. It has been

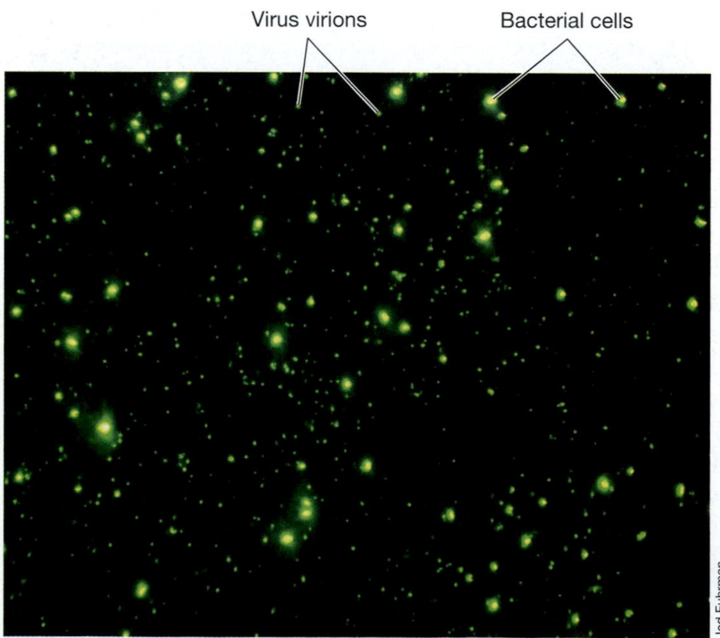

Figure 13.15 Viruses and bacteria in seawater. An epifluorescence photomicrograph of seawater stained with the dye SYBR Green to reveal prokaryotic cells and viruses. Although viruses are too small to be seen with the light microscope, fluorescence from a stained virus is visible.

estimated that every day from 5–50% of the bacteria in seawater are killed by bacteriophages, and most of the others are eaten by protozoa.

Overall, most of the genetic diversity on Earth is thought to reside in viruses, mostly bacteriophages. As discussed in Chapter 10, many bacteriophages can integrate into the genomes of their bacterial hosts and can transfer bacterial genes from one bacterium to another. Thus, viruses have a massive influence on bacterial evolution, partly by culling the large bacterial populations on a daily basis and partly by catalyzing horizontal gene transfer.

The *viral metagenome* is the sum total of all the virus genes in a particular environment. Several viral metagenomic studies have been undertaken, and they invariably show that immense viral diversity exists on Earth. For example, approximately 75% of the gene sequences found in viral metagenomic studies show no similarity to any genes presently in viral or cellular gene databases. By comparison, surveys of bacterial metagenomes typically reveal approximately 10% unknown genes. Thus, most viruses await discovery and most viral genes have unknown functions.

13.14 MiniReview

Viruses outnumber cells about tenfold in most habitats. Consequently, most global genetic information is carried by virus genomes. Most virus genes are presently uncharacterized and many are unlike those found in cells of any of the domains of life.

▮ How do viruses affect the evolution of bacterial genomes?

Review of Key Terms

Bioinformatics the use of computational tools to acquire, analyze, store, and access DNA and protein sequences

Chromosomal island region of bacterial chromosome of foreign origin that contains clustered genes for some extra property such as virulence or symbiosis

Environmental genomics genomic analysis of pooled DNA from an environmental sample without first isolating or identifying the individual organisms

Gene family genes that are related in sequence to each other as the result of a common evolutionary origin

Genome the total complement of genetic information of a cell or a virus

Genomics the discipline that maps, sequences, analyzes, and compares genomes

Homologous related in sequence to an extent that implies common genetic ancestry; includes both orthologs and paralogs

Horizontal gene transfer the transfer of genetic information between organisms as opposed to its vertical inheritance from parental organism(s)

Integrase the enzyme that inserts cassettes into an integron

Integron a genetic element that accumulates and expresses genes carried by cassettes

Lateral gene transfer the same as horizontal gene transfer

Megabase pair (Mbp) one million base pairs

Metabolome the total complement of small molecules and metabolic intermediates of a cell or organism

Metagenome the total genetic complement of all the cells present in a particular environment

Microarray small solid-state supports to which genes or portions of genes are affixed and arrayed spatially in a known pattern (also called *gene chips*)

Ortholog a gene found in one organism that is similar to that in another organism but differs because of speciation (see also *Paralog*)

Paralog a gene whose similarity to one or more other genes in the same organism is the result of gene duplication (see also *Ortholog*)

Pathogenicity island region of bacterial chromosome of foreign origin that contains clustered genes for virulence

Proteome the total set of proteins encoded by a genome or the total protein complement of an organism

Proteomics the genomewide study of the structure, function, and regulation of the proteins of an organism

RNA editing changing the coding sequence of an RNA molecule by altering, adding, or removing bases

Transcriptome the complement of mRNAs produced in an organism under a specific set of conditions

Review Questions

1. What is the relationship between genome size and ORF content of prokaryotic genomes (Section 13.2)?
2. As a proportion of the total genome, which class of genes predominates in small genome organisms? In large genome organisms (Section 13.3)?
3. Whose genomes are larger, those of chloroplasts or those of mitochondria? Describe one unusual feature about a chloroplast and a mitochondrial genome (Section 13.4).
4. How much larger is your genome than that of yeast? How many more genes do you have than yeast (Section 13.5)?

5. In *Bacteria* and *Archaea* the acronym ORF is almost a synonym for the word "gene." However, in eukaryotes this is not, strictly speaking, true. Explain (Section 13.5).
6. Distinguish between the terms *genome, proteome,* and *transcriptome* (Sections 13.6 and 13.7).
7. Why is the term *proteome* ambiguous (Section 13.7)?
8. What does a 2D protein gel show? How can the results of such a gel be tied to protein function (Section 13.7)?
9. Why is investigation of the metabolome lagging behind that of the proteome (Section 13.8)?
10. What is the major difference in how duplications have contributed to the evolution of prokaryote versus eukaryote genomes (Section 13.9)?
11. Explain how horizontally transferred genes can be detected in a genome (Section 13.11).
12. Explain how chromosomal islands might be expected to move between different bacterial hosts (Section 13.12).
13. How can gene expression be measured in uncultured bacteria (Section 13.13)?
14. Most of the genetic information on our planet does not belong to cellular organisms. Discuss (Section 13.14).

Application Questions

1. Although the sequence of the yeast nuclear genome was published, the entire sequence was never actually completely determined. Describe the practical difficulties that were encountered in the sequencing.
2. Describe how one might determine which proteins in *Escherichia coli* are repressed when a culture is shifted from a minimal medium (which contains only a single carbon source) to a rich medium, containing a large number of amino acids, bases, and vitamins? Describe how one might study which genes are expressed during each growth condition.
3. The gene encoding the beta subunit of RNA polymerase from *Escherichia coli* is said to be orthologous to the *rpoB* gene of *Bacillus subtilis*. What does that mean about the relationship between the two genes? What protein do you suppose the *rpoB* gene of *Bacillus subtilis* encodes? The genes for the different sigma factors of *Escherichia coli* are paralogous. What does that say about the relationship between these genes?

14

Microbial Evolution and Systematics

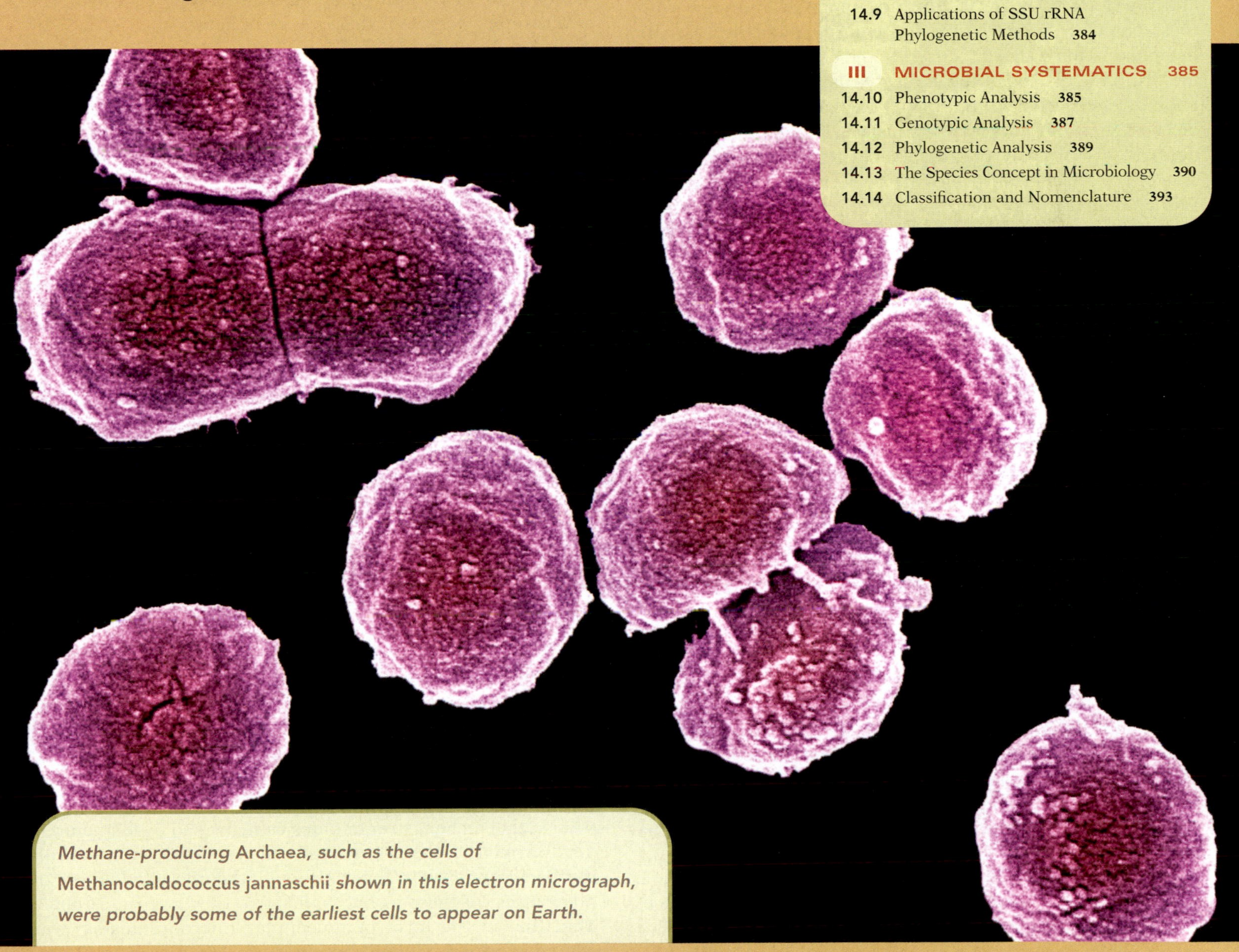

Methane-producing Archaea, *such as the cells of* Methanocaldococcus jannaschii *shown in this electron micrograph, were probably some of the earliest cells to appear on Earth.*

I EARLY EARTH AND THE ORIGIN AND DIVERSIFICATION OF LIFE

A theme that unifies all of biology is **evolution**, DNA sequence change and the inheritance of that change, often under the selective pressures of a changing environment. From the time of its origin about 4.5 billion years ago, Earth has undergone a continual process of physical and geological change. These changes created conditions leading to the origin of life about 4 billion years ago, and they have presented living organisms with new opportunities and challenges from that time to the present. In turn, as microbial metabolisms and physiologies arose to take advantage of and cope with these changes, microbial life has continually altered Earth's biosphere. This chapter focuses on the evolution of microbial life, from the origins of the earliest cells and microbial metabolic diversification to the origin of eukaryotes. Methods for discerning evolutionary relationships among modern-day descendents of early microbial lineages and the emerging concept of a bacterial species are also discussed. The goal of this chapter is to provide an evolutionary and systematic foundation for our examination of the diversity of microbial life to follow in the next four chapters.

14.1 Formation and Early History of Earth

In these first few sections, we consider the possible conditions under which life arose, the processes that might have given rise to the first cellular life, its divergence into two evolutionary lineages, ***Bacteria*** and ***Archaea***, and the later formation, through endosymbiosis, of a third lineage, the ***Eukarya***. Although much about these events and processes remains speculative, geological and molecular evidence is providing an increasingly clear view of how life might have arisen and diversified.

Origin of Earth

Earth is thought to have formed about 4.5 billion years ago, based on data from slowly decaying radioactive isotopes. Our planet and the other planets of our solar system arose from materials making up a disc-shaped nebular cloud of dust and gases released by the supernova of a massive old star. As a new star—our sun—formed within this cloud, it began to compact, undergo nuclear fusion, and release large amounts of heat and light. Materials left in the nebular cloud began to clump and fuse due to collisions and gravitational pull, forming tiny accretions that gradually grew larger to form clumps that eventually coalesced into planets. Energy released in this process heated the emerging Earth as it formed, as did energy released by radioactive decay within the condensing materials, transforming Earth into a planet of fiery hot magma. As Earth cooled over time, a metallic core, rocky mantle, and a thin lower-density surface crust formed.

The fiery, inhospitable conditions of early Earth, characterized by a molten surface under intense bombardment from space by masses of accreted materials, are thought to have persisted for over 500 million years. Water on Earth originated from innumerable collisions with icy comets and asteroids and from volcanic outgassing of the planet's interior. At this time, water would have been present as vapor due to the heat. No rocks dating to the origin of our planet have yet been discovered, presumably because they have undergone geological metamorphosis. Ancient sedimentary rocks, which form under liquid water, have been found in several locations on Earth. Some of the oldest sedimentary rocks discovered thus far are in the Itsaq Gneiss Complex, in southwestern Greenland, which date to about 3.86 billion years ago. The sedimentary nature of these rocks indicates that at least by that time Earth had cooled sufficiently for the water vapor to have condensed and formed the early oceans.

Even more ancient materials, crystals of the mineral zircon ($ZrSiO_4$), however, have been discovered, and these materials give us a glimpse of even earlier conditions on Earth. Impurities trapped in the crystals and the mineral's isotopic ratios of oxygen (Section 22.8) indicate that Earth cooled much earlier than previously believed, with solid crust forming and water condensing into oceans perhaps as early as 4.4 to 4.3 billion years ago. The presence of liquid water implies that conditions might have been compatible with life within a few hundred million years after our planet formed.

Evidence for Microbial Life on Early Earth

The fossilized remains of cells and the isotopically "light" carbon abundant in these rocks (we discuss the use of isotopic analyses of carbon and sulfur as an indication of living processes in Section 22.8) provide evidence for early microbial life. Some ancient rocks contain what appear to be bacteria-like microfossils, typically as simple rods or cocci (**Figure 14.1**).

In rocks of 3.5 billion years old or younger, microbial formations called **stromatolites** are abundant. Stromatolites are fossilized microbial mats consisting of layers of filamentous prokaryotes and trapped sediment (**Figure 14.2**); we discuss some characteristics of microbial mats in Section 22.7. What kind of organisms were these ancient stromatolitic bacteria? By comparing ancient stromatolites with modern stromatolites growing in shallow marine basins (Figure 14.2*c* and *e*) or in hot springs (Figure 14.2*d*; Figure 22.19*b*), it has been concluded that ancient stromatolites were formed by filamentous phototrophic bacteria, perhaps relatives of the green nonsulfur bacterium *Chloroflexus*. Although the microbial nature of the earliest of these fossils is debated, they give an estimate that life, in the form of unicellular microorganisms, was abundant by 3.5 billion years ago.

Figure 14.3 shows photomicrographs of thin sections of more recent rocks containing cell-like structures remarkably similar to modern filamentous bacteria and green algae (Section 18.21). In the oldest stromatolites these organisms

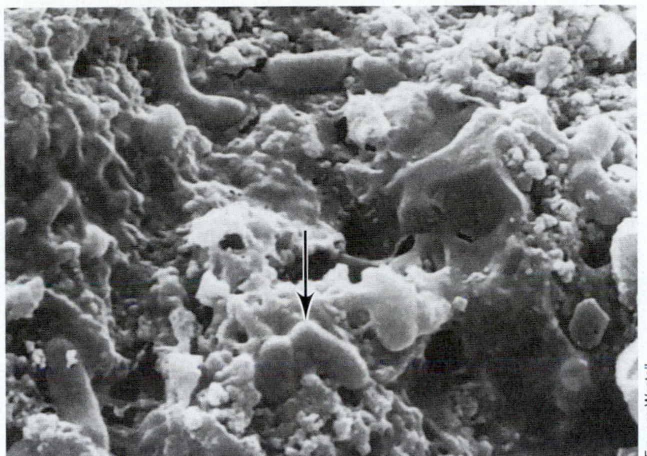

Figure 14.1 Ancient microbial life. Scanning electron micrograph of microfossil bacteria from 3.45 billion-year-old rocks of the Barberton Greenstone Belt, South Africa. Note the rod-shaped bacteria (arrow) attached to particles of mineral matter. The cells are about 0.7 μm in diameter.

Figure 14.2 Ancient and modern stromatolites. *(a)* The oldest known stromatolite, found in a rock about 3.5 billion years old, from the Warrawoona Group in Western Australia. Shown is a vertical section through the laminated structure preserved in the rock. Arrows point to the laminated layers. *(b)* Stromatolites of conical shape from 1.6 billion-year-old dolomite rock of the McArthur basin of the Northern Territory of Australia. *(c)* Modern stromatolites in a warm marine bay, Shark Bay, Western Australia. *(d)* Modern stromatolites composed of thermophilic cyanobacteria growing in a thermal pool in Yellowstone National Park. Each structure is about 2 cm high. *(e)* Another view of modern and very large stromatolites from Shark Bay. Individual structures are 0.5–1 m in diameter.

were likely anoxygenic (nonoxygen-evolving) phototrophic bacteria rather than the O_2-evolving cyanobacteria that dominate modern stromatolites (Figure 14.2*c, e*). In summary, it seems likely that *Bacteria* and *Archaea* evolved an impressive morphological and metabolic diversity very early in Earth's history.

14.1 MiniReview

Planet Earth is approximately 4.5 billion years old. The first evidence for microbial life can be found in rocks 3.86 billion years old. In rocks 3.5 billion years old or younger, microbial formations called stromatolites are abundant.

- Why have rocks dating to the origin of Earth not been found?
- What do crystals of the mineral zircon tell us about when the first oceans may have formed?
- What does the possible presence of liquid water on Earth 4.4 to 4.3 billion years ago suggest about the origin of life?

14.2 Origin of Cellular Life

Here, we examine some possible ways in which life might have originated, developing from abiotic materials into self-replicating cells. Because all life is cellular, from single-celled bacteria to multicellular animals and plants, our focus here is on the following questions: How might the first cells have arisen? and What might those early cells have been like?

Surface Origin Hypothesis

One hypothesis for the origin of life holds that the first membrane-enclosed, self-replicating cells arose out of a primordial soup rich in organic and inorganic compounds in a "warm little pond." Although there is experimental evidence that organic precursors to living cells can form spontaneously under certain conditions, surface conditions of early Earth are now thought to have been hostile to life and to its inorganic and organic precursors. The dramatic temperature fluctuations and mixing resulting from meteor impacts, dust clouds, and storms, together with the highly oxidizing atmosphere present at the time, make a surface origin for life unlikely.

Subsurface Origin Hypothesis

An alternative hypothesis is that life originated at hydrothermal springs on the ocean floor, well below Earth's surface, where conditions would have been much less hostile and more stable. A steady and abundant supply of energy in the form of reduced inorganic compounds, for example, H_2 and H_2S, may have been available at these spring sites (**Figure 14.4**).

When this very warm (90–100°C), alkaline, hydrothermal water flowed up through the crust and mixed with the cooler,

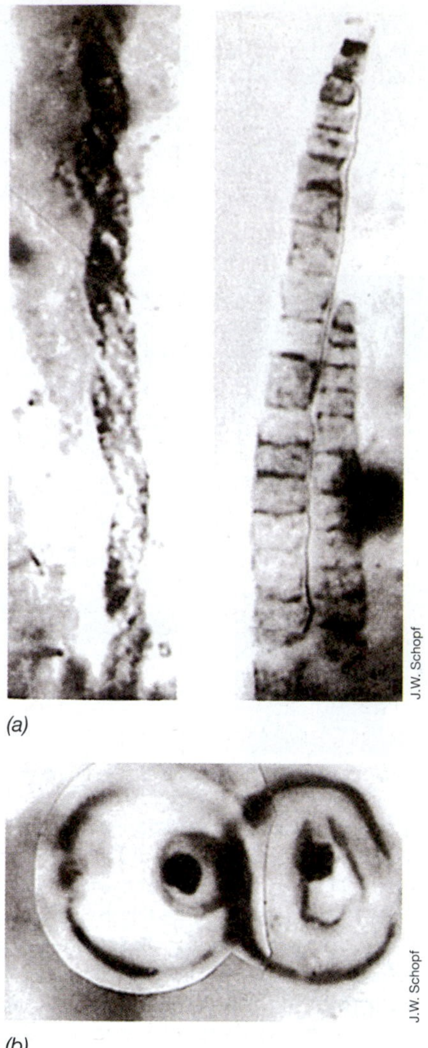

Figure 14.3 More recent fossil bacteria and eukaryotes. The two photographs in *(a)* show fossil bacteria from the Bitter Springs Formation, a rock formation in central Australia about 1 billion years old. These forms bear a striking resemblance to modern filamentous cyanobacteria, anoxygenic phototrophs, or filamentous sulfur chemolithotrophs (Chapters 15 and 16). Cell diameters, 5–7 μm. *(b)* Microfossils of eukaryotic cells from the same rock formation. The cellular structure is remarkably similar to that of certain modern green algae, such as *Chlorella* species. Cell diameter, about 15 μm.

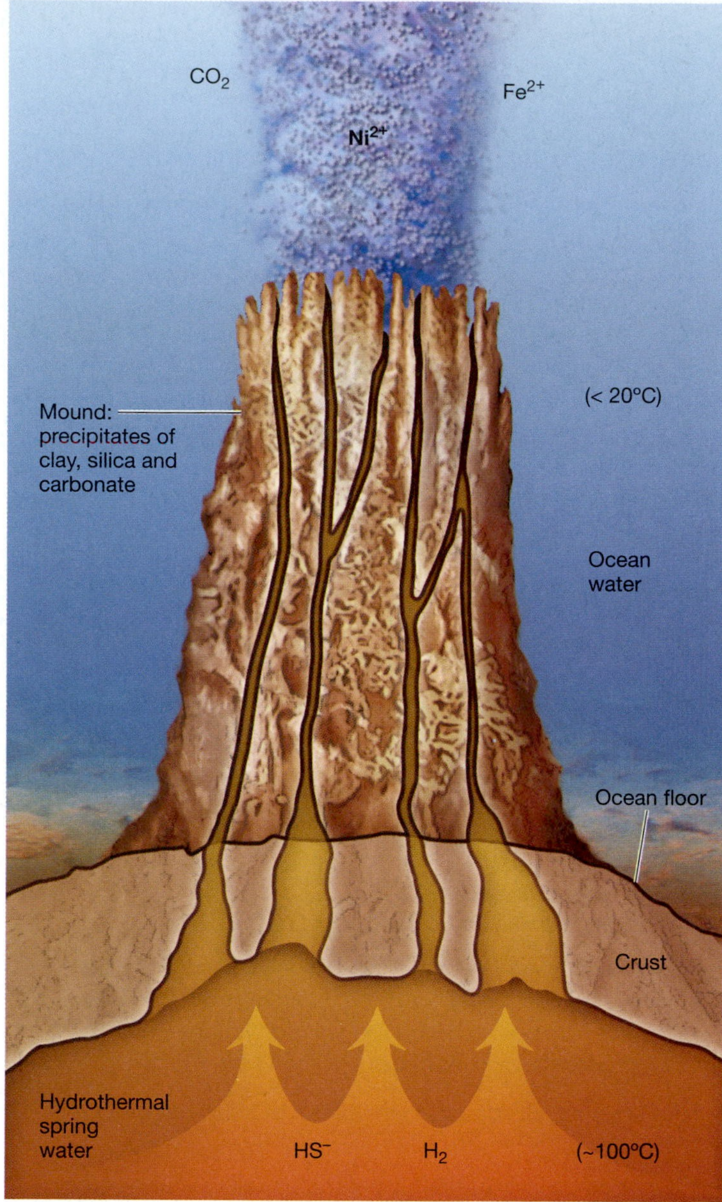

Figure 14.4 Submarine mound formed at a hydrothermal spring in Earth's early ocean. Hot reduced alkaline hydrothermal fluid mixes with cooler, more oxidized, more acidic ocean water, forming precipitates. The mound is made up of precipitates of Fe and S compounds, clays, silicates, and carbonates. Reaction of H_2 and CO_2 as an early prebiotic metabolism may have led to the generation of organic compounds.

slightly acidic, iron-containing, more oxidized oceanic waters, precipitates of colloidal pyrite (FeS), silicates, carbonates, and magnesium-containing montmorillonite clays formed. These precipitates built up into structured mounds of gel-like adsorptive surfaces containing pore-filled semipermeable enclosures (**Figure 14.5**). The surfaces and pores were rich in minerals such as Fe and Ni sulfides, which catalyzed formation of amino acids, simple peptides, sugars, and nitrogenous bases and trapped and concentrated these compounds. With phosphate from seawater, nucleotides such as AMP and ATP were formed, with their polymerization into RNA catalyzed by montmorillonite clay, which has been shown to catalyze various chemical reactions. The flow of reduced inorganic compounds from the crust provided steady sources of electrons for this prebiotic chemistry, which was fed from ocean water by carbon dioxide, phosphate, iron, and other minerals. It was powered by redox and pH gradients developed across the semipermeable FeS membrane-like surfaces, providing a prebiotic proton motive force (Section 5.12).

An RNA World and Protein Synthesis

The synthesis and concentration of organic compounds by this prebiotic chemistry set the stage for self-replicating systems, the precursors to cellular life. How might self-replicating

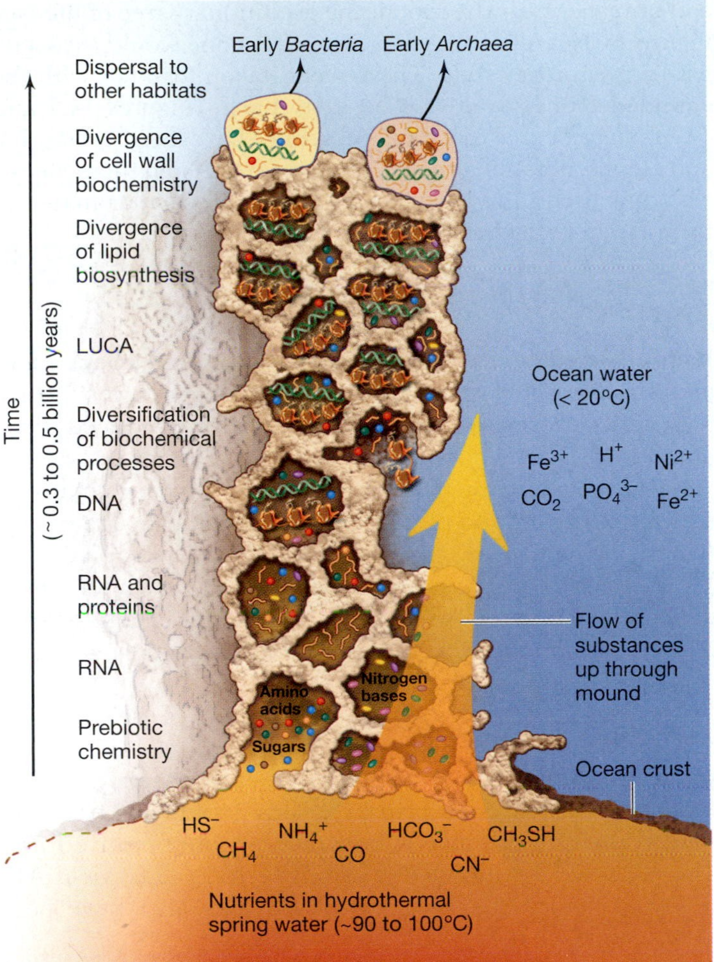

Figure 14.5 A model for the origin of cellular life and its divergence into early *Bacteria* and early *Archaea*. A portion of the submarine mound from Figure 14.4 is shown. Transitions from prebiotic chemistry to cellular life are depicted. Key features of the model are self-replicating RNA, enzymatic activity of proteins, and DNA taking on the genetic coding function, leading to early cellular life. This was followed by evolution of biochemical pathways and divergence in lipid biosynthesis and cell wall biochemistry, giving rise to early *Bacteria* and *Archaea*. LUCA, Last Universal Common Ancestor.

systems have arisen? One possibility is that there was an early *RNA world*, in which the first self-replicating systems were molecules of RNA (Figure 14.5). Although fragile, RNA could have survived in the cooler temperatures of the gel-like precipitates forming at ocean floor springs. RNA can bind small molecules, such as ATP and other nucleotides, and has catalytic activity (ribozymes, Section 8.8), so RNA might have catalyzed its own synthesis from the available sugars, bases and phosphate.

RNA also can bind other molecules, such as amino acids, catalyzing the synthesis of primitive proteins. As different proteins were made and accumulated on early Earth they may have coated the inner surfaces of the structured FeS mounds and taken over some of the function of a semipermeable membrane. Later, as different proteins emerged, they took over RNA's catalytic role (Figure 14.5). Eventually, DNA, more stable than

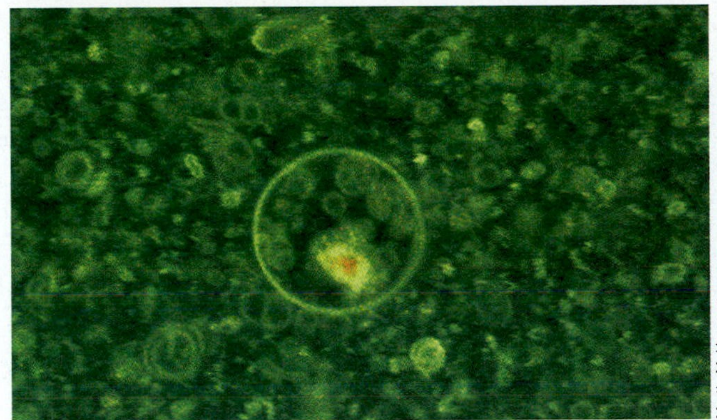

Figure 14.6 Lipid vesicles made in the laboratory from the fatty acid myristic acid and RNA. The vesicle itself stains green, and the RNA complexed inside the vesicle stains red. Vesicle synthesis is catalyzed by the surfaces of montmorillonite clay particles.

RNA and therefore a better repository of genetic (coding) information, arose and assumed the template role for RNA synthesis. This three-part system—DNA, RNA, and protein—became fixed early on in cellular evolution as the best solution to biological information processing (Figure 14.5). Following these steps, one can envision a time of extensive biochemical innovation and experimentation in which much of the structural and functional machinery of these earliest self-replicating systems was invented and refined under natural selection.

Lipid Membranes and Cellular Life

Other important steps in the emergence of cellular life were the buildup of lipids and the synthesis of phospholipid membrane vesicles that enclosed the cell's biochemical and replication machinery. Proteins embedded in the lipids would have maintained a semipermeable state for the vesicles, shuttling nutrients and wastes across the lipid membrane, setting the stage for evolution of energy-conserving processes and ATP synthesis. By entrapping RNA and DNA, these lipoprotein vesicles, which may have been similar to montmorillonite clay vesicles (**Figure 14.6**), may have formed the first self-replicating cells, thereby gaining a measure of independence from the structured FeS precipitate mounds that nurtured them.

From this population of early cells, considered to be the last universal common ancestor (LUCA), cellular life may then have evolved differences in lipid biosynthesis and cell wall biochemistry, diverging into the ancestors of modern day *Bacteria* and *Archaea* (Figure 14.5). An indication of that early divergence is retained in the stereochemistry of the glycerol-phosphate backbone of phospholipids. In *Bacteria*, phospholipids contain glycerol-3-phosphate, whereas in *Archaea* it is glycerol-1-phosphate, and the enzymes for the synthesis of these different glycerol-phosphates appear to have arisen independently in ancestral *Bacteria* and *Archaea*.

Early Metabolism

From the time of its formation, the early ocean and all of Earth was anoxic. Molecular oxygen did not appear in any

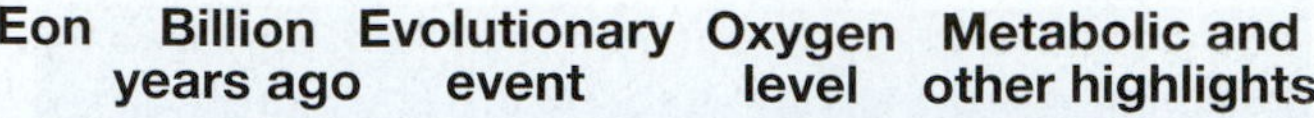

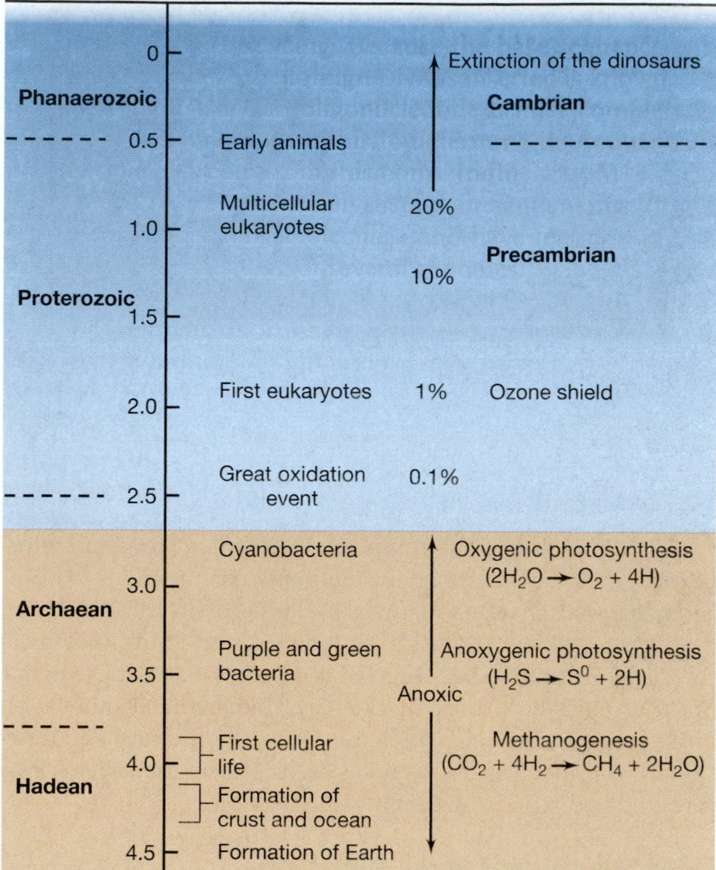

Figure 14.7 Major landmarks in biological evolution, Earth's changing geochemistry, and microbial metabolic diversification. The maximum time for the origin of life is fixed by the time of the origin of Earth, and the minimum time for the origin of oxygenic photosynthesis is fixed by the Great Oxidation Event, about 2.3 billion years ago. Note how the oxygenation of the atmosphere due to cyanobacterial metabolism was a gradual process, occurring over a period of about 2 billion years. Although full (20%) oxygen levels are required for animals and most other higher organisms, this is not true of *Bacteria*, as many are facultative aerobes or microaerophiles (Section 6.17 and Table 6.4). Thus, *Bacteria* respiring at reduced O_2 levels may have dominated Earth for a period of a billion years or so before Earth's atmosphere reached current levels of oxygen. Compare this figure wih the introduction to the antiquity of life on Earth, shown in Figure 1.6.

significant quantities until much later, following evolution of oxygenic photosynthesis by cyanobacteria (**Figure 14.7**). Thus, the energy-generating metabolism of primitive cells would have been exclusively anaerobic. It may also have been autotrophic, because any consumption of abiotically formed organic compounds as sources of carbon for cellular material probably would have exhausted these compounds relatively quickly. The possibility that autotrophy was an early physiological lifestyle is supported also by the discovery of autotrophic *Bacteria* such as *Aquifex*. This member of the *Bacteria* is a hyperthermophile, contains a very small genome, and branches near the root of the evolutionary tree of life (see Figure 14.16). All of these are properties one would associate with a "primitive" state. These ideas, taken together with the abundance of H_2 and CO_2 on early Earth (Figures 14.4 and 14.5), suggest that the first cells may have been anaerobic autotrophs, obtaining their carbon from CO_2 and their energy, electrons used to reduce carbon dioxide to cellular material, from H_2, in accord with the formula:

$$2\ H_2 + CO_2 \rightarrow H_2O + [CH_2O]$$

If life emerged in this way, it would have required an abundant and steady supply of H_2, as well as nitrogen, phosphorus, and various minerals, and it would have produced large amounts of waste, possibly in the form of acetate. A similar situation exists with hyperthermophilic *Archaea*, many of which oxidize H_2 and reduce S^0 and branch near the root of the phylogenetic tree.

Reactions with the substrate ferrous iron, which was abundant on early Earth, also have been proposed as energy-yielding reactions for primitive organisms. For example, the reaction

$$FeS + H_2S \rightarrow FeS_2 + H_2 \qquad \Delta G^{0'} = -42\text{kJ}$$

proceeds exergonically with the release of energy. This reaction yields H_2, and primitive cells could have used this powerful electron donor to generate a proton motive force. The H_2 could have fueled a primitive ATPase to yield ATP (**Figure 14.8**). Other sources of H_2 may also have been

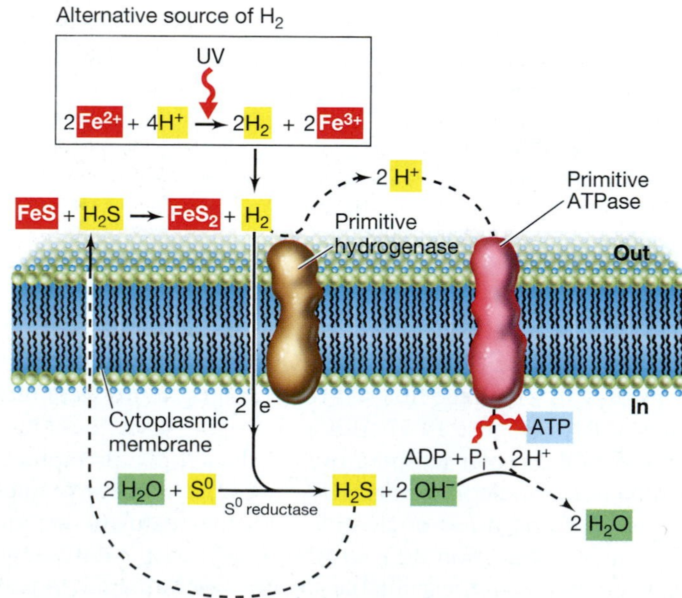

Figure 14.8 A possible energy-generating scheme for primitive cells. Formation of pyrite leads to H_2 production and S^0 reduction, which fuels a primitive ATPase. Note how H_2S plays only a catalytic role; the net substrates would be FeS and S^0. Also note how few different proteins would be required. The $\Delta G^{0'}$ of the reaction $FeS + H_2S \rightarrow FeS_2 + H_2 = -42$ kJ. An alternative source of H_2 could have been the UV-catalyzed reduction of H^+ by Fe^{2+} as shown.

available from ferrous iron on early Earth. For example, Fe^{2+} can reduce protons to hydrogen in the presence of ultraviolet radiation as an energy source (Figure 14.8). Early oceans were thought to be loaded with ferrous iron, and their surface waters were subject to intense UV radiation.

With H_2 as electron donor, an electron acceptor would also have been required to form a redox pair (Section 5.6); this could have been elemental sulfur (S^0). As shown in Figure 14.8, the oxidation of H_2 with the production of H_2S would have required few enzymes. Moreover, because of the abundance of H_2 and sulfur compounds on early Earth, cells would have had a nearly limitless supply of energy.

These early forms of chemolithotrophic metabolism would have supported the production of large amounts of organic compounds from autotrophic CO_2 fixation. Over time, these organic materials would have accumulated and provided abundant, diverse, and continually renewed sources of reduced organic carbon, triggering the evolution of various chemoorganotrophic bacteria whose metabolic strategies employ organic compounds as electron donors in energy metabolism.

14.2 MiniReview

Life may have arisen at hydrothermal springs on the early ocean floor. The first life forms may have been self-replicating RNAs that catalyzed synthesis of proteins. Eventually, DNA took over the template role for RNA synthesis, and the three-part system of DNA, RNA, and protein became universal in early cells. Early microbial metabolism was anaerobic and likely chemolithotrophic, exploiting the abundant sources of H_2, FeS, and H_2S present. Carbon metabolism initially may have been autotrophy, followed by heterotrophy as various organic compounds were formed and accumulated.

- What roles did the mounds of mineral-rich materials at warm hydrothermal springs play in the origin of life?
- What important cell structure was necessary for life to proceed from a possible RNA world to cellular life?
- How could cells have obtained energy from FeS + H_2S?

14.3 Microbial Diversification: Consequences for Earth's Biosphere

Following the origin of cells and the development of early forms of energy and carbon metabolism, microbial life underwent a long process of metabolic diversification, taking advantage of the various and abundant energy resources available on Earth. Undoubtedly, as individual kinds of resources were consumed and became limiting, competition and natural selection drove the evolution of more efficient and new metabolisms. Also, microbial life, through its metabolic activity, altered the chemical environment, depleting some resources and creating others through the production of waste products and cellular material. These changes and new resources in turn presented both challenges to survival and opportunities for utilization. In the following sections, we examine the overall scope of metabolic diversification and focus on one metabolic waste product, oxygen, which had a profound influence on the course of evolution.

Metabolic Diversification

Geological and molecular biological data allow us to look back in time to gain insight into microbial diversification. Molecular evidence, in contrast to geological materials, which can be examined directly, is indirect; phylogenies are based on comparison of DNA sequences that allow us to estimate when the ancestors of modern bacteria may have existed. In Section 14.5 we describe molecular clocks and DNA sequence-based analysis, but here we use information from DNA to estimate a timescale for when some of the major metabolic groups of bacteria appeared on Earth.

LUCA, the last universal common ancestor, may have existed as early as 4.25 billion years ago (Figure 14.7). Molecular evidence, however, suggests that ancestors of modern day *Bacteria* and *Archaea* had diverged by 4.1–3.9 billion years ago. As these two lineages diverged, they developed distinct metabolisms. Early *Bacteria* may have used H_2 and CO_2 to produce acetate, or ferrous iron compounds, for energy generation, as noted above. At the same time, early *Archaea* developed the ability to use H_2 and CO_2, or possibly acetate as it accumulated, as substrates for methanogenesis (the production of methane), according to the following formulas:

$$4\,H_2 + CO_2 \rightarrow CH_4 + 2\,H_2O$$

$$H_3C\text{—}COO^- + H_2O \rightarrow CH_4 + HCO_3^-$$

Phototrophy (Sections 20.1–20.5) arose somewhat later, about 3.2 billion years ago, and only in *Bacteria*. The ability to use solar radiation as an energy source allowed phototrophs to diversify extensively. With the exception of the early-branching hyperthermophilic *Bacteria*, *Aquifex*, and *Thermotoga*, the common ancestor of all other *Bacteria* appears to be an anaerobic phototroph, possibly similar to *Chloroflexus*. About 2.7 billion years ago, the cyanobacteria lineage developed a photosystem that could use H_2O in place of H_2S for photosynthetic reduction of CO_2, releasing O_2 instead of S^0 as a waste product (Section 20.5). The development of oxygenic photosynthesis dramatically changed the course of evolution.

The Rise of Oxygen: Banded Iron Formations

Molecular and chemical evidence indicates that oxygen-generating cyanobacteria first appeared on Earth about 2.7 billion years ago. Geological evidence, however, indicates that the accumulation of a significant level of oxygen in the atmosphere took another 300 million years. By 2.4 billion years ago, oxygen levels had risen to one part per million, a tiny amount, but enough to initiate what has come to be called the *Great Oxidation Event* (Figure 14.7). What delayed the buildup of oxygen for so long?

UNIT 3

Figure 14.9 Banded iron formations. An exposed cliff about 10 m in height containing layers of iron oxides interspersed with layers containing iron silicates and other silica materials. Brockman Iron Formation, Hammersley Basin, Western Australia. The iron oxides contain iron in the ferric (Fe^{3+}) form produced from ferrous iron (Fe^{2+}) primarily by the oxygen released by cyanobacterial photosynthesis.

The O_2 that cyanobacteria produced did not begin to accumulate in the atmosphere until it reacted with reduced materials, especially iron (such as FeS and FeS_2) in the oceans; these materials react spontaneously with O_2 to produce H_2O. The Fe^{3+} produced in pyrite (FeS_2) oxidation became a prominent marker in the geological record. Much of the iron in rocks of Precambrian origin (>0.5 billion years ago, see Figure 14.7) exists in banded iron formations (**Figure 14.9**), laminated sedimentary rocks formed in deep-water deposits of alternating layers of iron-rich minerals and iron-poor, silica-rich material. The metabolism of cyanobacteria yielded O_2 that oxidized Fe^{2+} to Fe^{3+}. The ferric iron formed various iron oxides that accumulated as banded iron formations (Figure 14.9). Once the abundant Fe^{2+} on Earth was consumed, the stage was set for O_2 to accumulate in the atmosphere, a major trigger of evolutionary events (Figure 14.7).

New Metabolisms and the Ozone Shield

The evolution of oxygenic photosynthesis had enormous consequences for Earth because, as O_2 accumulated, the atmosphere gradually changed from anoxic to oxic (Figure 14.7). *Bacteria* and *Archaea* unable to adapt to this change were increasingly restricted to anoxic habitats because of the toxicity of oxygen and because it oxidized the reduced substances upon which their metabolisms were dependent. The oxic atmosphere, however, also created conditions that led to the evolution of various new metabolic pathways, such as sulfate reduction, nitrification, and iron oxidation (Sections 21.8, 20.12, and 20.11, respectively). *Bacteria* that developed the ability to respire oxygen, either facultatively or exclusively, gained a tremendous energetic advantage and diversified rapidly thereafter, because with the oxidation of organic compounds they were able to obtain more energy than could anaerobes. With the development of larger cell populations, chances increased for the evolution of new types of organisms and metabolic schemes.

During this time, organelle-containing eukaryotic microorganisms evolved (Section 14.4), and the rise in oxygen also spurred their rapid evolution. The oldest known eukaryotic microfossil is about 2 billion years old. Multicellular and increasing complex microfossils of algae are evident from 1.9 to 1.4 billion years ago. By 0.6 billion years ago, oxygen was near present levels, and large multicellular organisms, the Ediacaran fauna, were present in the sea (Figure 14.7). In a relatively short time, multicellular eukaryotes diversified into the ancestors of modern-day algae, plants, fungi, and animals (Section 14.8).

An important consequence of O_2 for the evolution of life was the formation of ozone, a substance that provides a barrier preventing much of the intense ultraviolet radiation of the sun from reaching the Earth. When O_2 is subject to UV radiation, it is converted to O_3, which strongly absorbs wavelengths up to 300 nm. Until an ozone shield developed in Earth's upper atmosphere, evolution could have continued only beneath the ocean surface and in protected terrestrial environments where organisms were not exposed to the lethal DNA damage by intense UV radiation from the sun. However, after the photosynthetic production of O_2 and subsequent development of an ozone shield, organisms could then range over the surface of Earth, exploiting new habitats and evolving greater diversity. Figure 14.7 summarizes some major events in biological evolution and changes in the Earth's geochemistry from a highly reducing to a highly oxidizing planet.

14.3 MiniReview

Early *Bacteria* and *Archaea* may have diverged from a common ancestor 4.1–3.9 billion years ago. Microbial metabolism diversified on early Earth, with the invention of methanogenesis and anoxygenic photosynthesis. Oxygenic photosynthesis led to development of an oxic environment, banded iron formations, and great bursts of biological evolution.

- Why is the advent of cyanobacteria considered a critical step in evolution?
- In what form is iron present in banded iron formations?
- What role did ozone play in biological evolution, and how did cyanobacteria make the production of ozone possible?

14.4 Endosymbiotic Origin of Eukaryotes

The divergence of the lineages that gave rise to *Bacteria* and *Archaea* was followed by many millions of years of metabolic diversification as the members of these lineages evolved and diversified. Major changes in Earth's biosphere occurred as a consequence of this diversification, most notably the generation of an increasingly oxygenated atmosphere. Up to that point in the evolution of life, all cells apparently lacked a membrane-enclosed nucleus and organelles, as do modern-day *Bacteria* and *Archaea*. Eukaryotes (*Eukarya*), however, are characterized by the presence of a membrane-enclosed nucleus and organelles. Chapter 2 gave a brief tour of eukaryotic microbial diversity (∞ Section 2.11), and Chapter 18 will examine eukaryotic microorganisms in more detail. Here we probe the origin of *Eukarya,* asking when and how this third domain of life, bearing a membrane-enclosed nucleus and organelles, emerged.

Origin of Eukaryotes

Geological evidence from microfossils (Figure 14.3) indicates that unicellular eukaryotes arose on Earth about 2 billion years ago (Figure 14.7). Thus, the lineages that gave rise to modern-day *Bacteria* and *Archaea* had existed as the only life forms on our planet for about 2 billion years before eukaryotes evolved. This timing tells us that the origin of eukaryotes came after the rise in atmospheric oxygen, the invention of respiratory metabolism in *Bacteria,* and the development of enzymes such as superoxide dismutase (∞ Section 6.18), by which cells could detoxify oxygen radicals generated as a by-product of aerobic respiration. How might *Eukarya* have arisen and in what ways did the availability of oxygen influence evolution?

Endosymbiosis

A well-supported hypothesis for the origin of the eukaryotic cells is that of **endosymbiosis**, which states that the mitochondria and chloroplasts of modern-day eukaryotes arose from the stable incorporation into another type of cell of a chemoorganotrophic bacterium, which carried out facultatively aerobic metabolism, and a cyanobacterium, which carried out oxygenic photosynthesis. Oxygen was a factor in endosymbiosis through its consumption in energy metabolism by the ancestor of the mitochondrion and its production in photosynthesis by the ancestor of the chloroplast. The greater amounts of energy released by aerobic respiration undoubtedly contributed to rapid evolution of eukaryotes, as did the ability to exploit sunlight for energy.

The overall physiology and metabolism of mitochondria and chloroplasts and the sequence and structures of their genomes support the endosymbiosis hypothesis. For example, both mitochondria and chloroplasts contain ribosomes of the prokaryotic type (70S, ∞ Table 7.4) and show 16S ribosomal RNA gene sequences (Section 14.6) characteristic of certain *Bacteria*. Moreover, the same antibiotics that affect ribosome function in free-living *Bacteria* inhibit ribosome function in these organelles. Mitochondria and chloroplasts also contain small amounts of DNA arranged in a covalently closed, circular form, typical of *Bacteria* (∞ Section 2.6). Indeed, many telltale signs of *Bacteria* are present in organelles from modern eukaryotic cells (∞ Section 18.4). However, there is more to the question of how the eukaryotic cell arose, including the nature of the cell that acquired mitochondria and later chloroplasts and how the nuclear membrane formed.

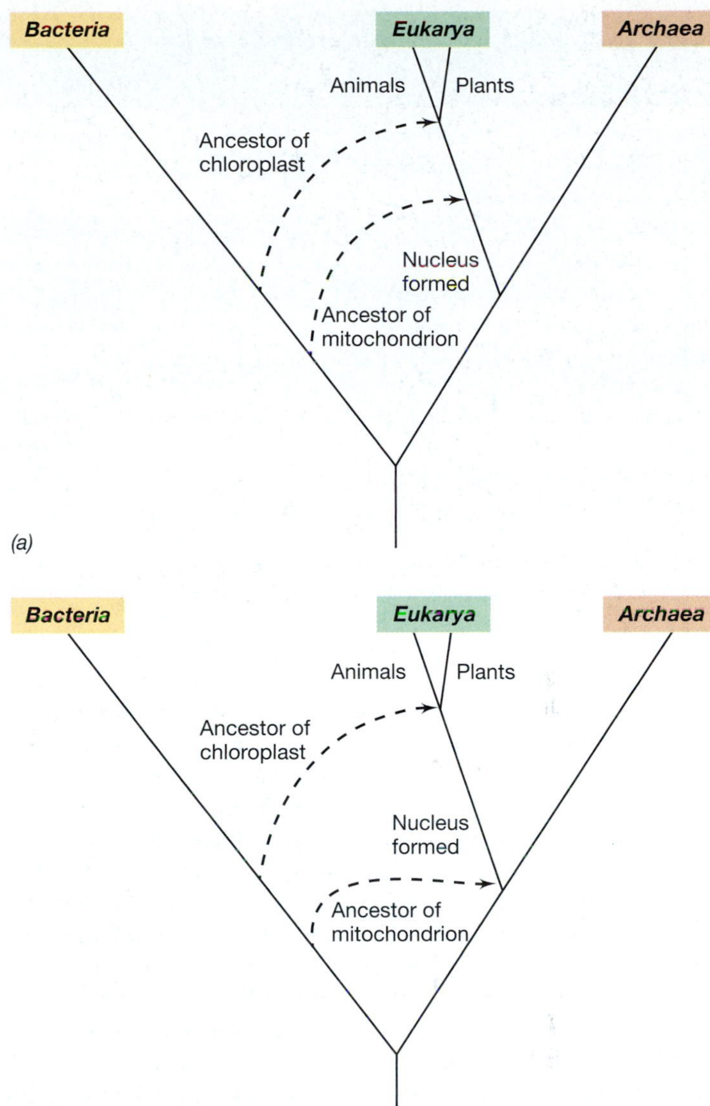

Figure 14.10 Models for the origin of the eukaryotic cell. *(a)* The nucleated line diverged from the archaeal line and later acquired by endosymbiosis the bacterial ancestor of the mitochondrion and then the cyanobacterial ancestor of the chloroplast, at which point the nucleated line diverged into the lineages giving rise to plants and animals. *(b)* In contrast, the bacterial ancestor of the mitochondrion was taken up endosymbiotically by a member of *Archaea*, and from this association, the nucleus later emerged, with a later endosymbiotic acquisition of the cyanobacterial ancestor of the chloroplast. Note the position of the mitochondrion and chloroplast on the universal phylogenetic tree shown in Figure 14.16.

Formation of the Eukaryotic Cell

Two hypotheses have been put forward to explain the formation of the eukaryotic cell (**Figure 14.10**). In one, eukaryotes began as a nucleus-bearing lineage that later acquired mitochondria and chloroplasts by endosymbiosis (Figure 14.10a). In this scenario, the nucleus-bearing cell line arose from an early kind of cell, and the nucleus is thought to have arisen spontaneously, probably in response to the increasing genome size of the early eukaryote.

UNIT 3

Table 14.1 Major features that group *Bacteria* or *Archaea* with *Eukarya*[a]

Characteristic	Bacteria	Archaea	Eukarya
Membrane lipids	Ester-linked	Ether-linked	Ester-linked
Chlorophyll-based photosynthesis	Yes	No	Yes (in chloroplasts)
Histone proteins present	No[b]	Yes	Yes
Cell wall	Muramic acid present	Muramic acid absent	Muramic acid absent
Initiator tRNA	Formylmethionine	Methionine	Methionine
Ribosome sensitivity to diphtheria toxin[c]	No	Yes	Yes
RNA polymerases (Figure 8.14)	One (4 subunits)	One (8–12 subunits each)	Three (12–14 subunits each)
Transcription factors required (Sections 7.10, 8.2, and 8.9)	No	Yes	Yes
Promoter structure (Sections 7.9, 7.10, 8.2, and 8.9)	−10 and −35 sequences (Pribnow box)	TATA box	TATA box
Sensitivity to chloramphenicol, streptomycin, and kanamycin	Yes	No	No

[a]See also Table 14.2 for additional features of the three domains.
[b]DNA-binding protein HU is present in *Bacteria* (Section 17.5).
[c]Sections 8.9 and 34.3.

The second hypothesis, called the *hydrogen hypothesis,* proposes that the eukaryotic cell arose from an intracellular association between an oxygen-consuming, hydrogen-producing member of *Bacteria,* the symbiont, which gave rise to the mitochondrion, and a member of *Archaea,* the host (Figure 14.10*b*). The host's dependence on H_2 as an energy source, produced as a waste product of the symbiont's metabolism is thought to have led to this association. In this scenario, the nucleus arose after these two kinds of cells had formed a stable association and genes for lipid synthesis were transferred from the symbiont to the host chromosome. The transfer of these genes to the host chromosome may then have led to synthesis of bacterial (symbiont) lipids by the host, eventually forming an internal membrane system, the endoplasmic reticulum (Section 18.5), and the beginnings of the eukaryotic nucleus. Increasing size of the host genome led to compartmentalizing and sequestering the genetic coding information within a membrane away from the cytoplasm to protect it and to permit more efficient replication and gene expression. Later, this mitochondrion-containing, nucleated cell line acquired chloroplasts by endosymbiosis (Figure 14.10*b*).

Chimeric Nature of the Eukaryotic Cell

Both hypotheses to explain the origin of eukaryotes suggest that the eukaryotic cell is a chimera, a cell made up of attributes of both *Bacteria* and *Archaea*. Indeed, this appears to be the case. Eukaryotes have the type of lipids found in *Bacteria,* but they have a transcription and translation apparatus more similar to that of *Archaea* than that of *Bacteria* (**Table 14.1**; Chapter 8). However, in energy metabolism—that is, ATP-producing pathways in mitochondria, hydrogenosomes (degenerate mitochondria, Section 18.2), and the cytoplasm, as well as glycolytic enzymes in the cytoplasm—they are more similar to *Bacteria* than to *Archaea*.

The development of the eukaryotic cell was a major step in evolution, creating complex, genetically chimeric cells with new capabilities, powered by oxygen-respiring mitochondria and phototrophic, oxygen-generating chloroplasts. Like the origin of the first cells from abiotic materials and the diversification of *Bacteria* and *Archaea,* the evolutionary process by which the eukaryotic cell and its individual components arose took long periods of time and was likely to have undergone many false starts and evolutionary dead ends. Each step along the way, however, created new opportunities for variation to arise and natural selection to work, discarding some functions and attributes while refining others, and inventing wholly new capabilities as a consequence of the progressive genetic integration between symbiont and host. The origin of the eukaryotic cell set the stage for an explosion in biological diversity that built on and followed the massive diversification of *Bacteria* and *Archaea*. The period from about 2 billion years ago to the present saw the rise and diversification of unicellular eukaryotic microorganisms, the origin of multicellularity, and the appearance of structurally complex plant, animal, and fungal life (Figures 14.7, 14.10, and see 14.16).

14.4 MiniReview

Unicellular eukaryotes arose about 2 billion years ago. The eukaryotic cell may have arisen from the endosymbiosis between a facultatively aerobic H_2-producing *Bacteria* and a H_2-consuming *Archaea.* The eukaryotic cell is believed to be a chimera made up of properties from both *Bacteria* and *Archaea*.

- What evidence supports the idea that the eukaryotic mitochondrion and chloroplast were once free-living members of the domain *Bacteria*?
- How might the eukaryotic nucleus have arisen?
- In what ways are modern eukaryotes a combination of attributes of *Bacteria* and *Archaea*?

II MICROBIAL EVOLUTION

14.5 The Evolutionary Process

Evolution is descent with modification, a change in the sequence of an organism's genomic DNA and the inheritance of that change by the next generation. In this explicitly Darwinian view of life, all organisms are related through descent from an ancestor that lived in the past. We have outlined an hypothesis for the origin of the most distant of those ancestors, LUCA (Section 14.2). Since that time, about 4 billion years ago, life has undergone an extensive process of change as new kinds of organisms arose from other kinds existing in the past. Evolution has also led to the loss of life forms, with unsuccessful organisms becoming extinct over time. Evolution accounts not only for the tremendous diversity we see today, but also for the high level of complexity in modern organisms. Indeed, no organism living today is primitive. All extant life forms are modern organisms, well adapted to and successful in their ecological niches, having arisen by the process of evolutionary change under the pressure of natural selection.

Mutations and Selection

DNA sequence variation can arise in a number of different ways. Most important for the way organisms evolve are mutations, changes in the nucleotide sequence of an organism's genome (∞ Sections 11.4–11.8). Mutations occur due to errors in the fidelity of replication, to UV radiation, and other factors, and they are essential in order for life to be able to change and adapt through natural selection. Adaptive mutations are those that improve the **fitness** of an organism, increasing its survival or reproduction compared to competing organisms. By contrast, harmful mutations lower an organism's fitness. Most mutations, however, are neutral, neither benefiting nor causing harm to the cell, and over time these mutations can accumulate in an organism's genome.

Other processes also bring about heritable changes in the sequence of an organism's genome. These include gene duplication, which can set the stage for the evolutionary development of new functions as the sequence of the duplicated gene changes over time; **horizontal gene transfer** (∞ Section 13.11), which can bring in genes from distantly related lineages; and gene loss, which can lead to the genome reduction often seen in obligate symbionts and parasites. Also, recombination of homologous DNA that has undergone divergence from closely related strains within a species can contribute to variation.

Regardless of whether a mutation or other genetic change is neutral, beneficial, or harmful, these changes provide the opportunity for selection of organisms whose genomes carry those changes. As environments change and as new habitats arise, cells are presented with new conditions under which they must either survive and successfully compete for nutrients or become extinct. The sequence variation present in a population provides the raw material by which the reproduction of those individuals bearing mutations beneficial under the new circumstances is favored. We will return to these ideas later, in a discussion of bacterial speciation, but here we examine how to track the evolutionary history of an organism.

14.5 MiniReview

Evolution is descent with modification. Natural selection works by favoring the survival and reproduction of organisms that by chance have mutations that improve fitness under new environmental conditions.

- According to the Darwinian view of life, all organisms are related by ________ .
- What class of mutation—beneficial, neutral, or harmful—is most common?
- How does the accumulation of mutations set the stage for selection?

UNIT 3

14.6 Evolutionary Analysis: Theoretical Aspects

The evolutionary history of a group of organisms is called its **phylogeny**, and a major goal of evolutionary analysis is to understand this history. Because we do not have direct knowledge of the path of evolution, except for certain experimental populations of bacteria in the laboratory, phylogeny is inferred indirectly from nucleotide sequence data. Our premises are that bacteria (and all organisms) are related by descent and that the sequence of a bacterium's genome is an explicit, though sometimes fuzzy, record of the bacterium's ancestry. Because evolution is a process of inherited nucleotide sequence change, analyzing DNA sequence differences among bacteria allows us to reconstruct their phylogenetic history. Here, we examine some of the ways in which this is carried out.

Genes Employed in Phylogenetic Analysis

Various genes are used in molecular phylogenetic studies of microorganisms. Most widely used and useful for defining relationships are the genes encoding 16S rRNA (**Figure 14.11**) and its counterpart in eukaryotes, 18S rRNA. These **small subunit ribosomal RNA (SSU rRNA)** genes have been used extensively for sequence-based evolutionary analysis because they are (1) universally distributed, (2) functionally constant, (3) sufficiently conserved (that is, slowly changing), and (4) of adequate length, such that they can provide a view of evolution encompassing all living organisms.

Carl Woese at the University of Illinois pioneered the use of SSU rRNA for phylogenetic studies in the early 1970s. His work established the presence of three domains of life, *Bacteria, Archaea,* and *Eukarya,* and provided for the first time a unified phylogenetic framework for bacteria. For his accomplishments, Woese received the 2003 Crafoord Prize, the highest recognition for scientific achievement in biology, from the Royal Swedish Academy of Sciences.

A large and growing database of rRNA gene sequences exists. For example, the **Ribosomal Database Project II**

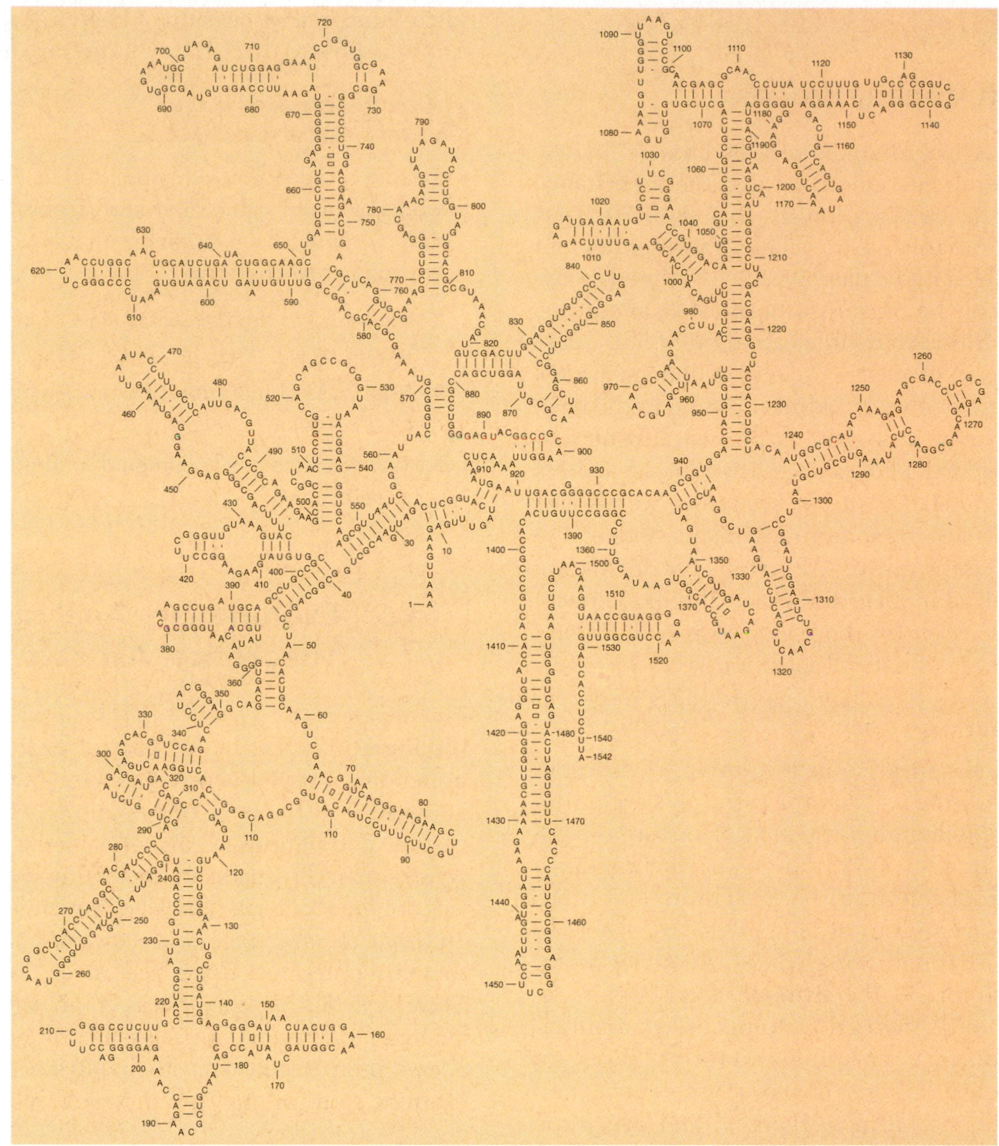

Figure 14.11 Ribosomal RNA. Primary and secondary structure of 16S rRNA from *Escherichia coli* (*Bacteria*). The16S rRNA from *Archaea* is similar in overall secondary structure (folding) but has numerous differences in primary structure (sequence). The counterpart in *Eukarya* to bacterial 16S rRNA is 18S rRNA, which is present in cytoplasmic ribosomes.

(RDP-II; http://rdp.cme.msu.edu) contains a collection of such sequences, now numbering over 440,000, and provides a variety of analytical programs. The 23S large-subunit rRNA (LSU-rRNA) gene is also phylogenetically highly informative, with its longer sequence providing additional information, though its length makes it more costly and time consuming to sequence. Along with these genes, those for several highly conserved proteins have been used effectively in phylogenetic analysis, including protein synthesis elongation factor Tu (∞ Section 7.15), heat-shock protein Hsp60 (∞ Section 9.11), and several tRNA synthetases (∞ Section 7.15).

The highly conserved SSU and LSU genes have changed slowly and provide a view of evolution that is deep enough to encompass all living organisms. This strength, however, is also a limitation, in that the essential functions of rRNAs apparently have limited the amount these genes can change over evolutionary time. Consequently, the amount of variation present in SSU rRNA gene sequences may not be sufficient to provide good discrimination between bacterial species. Another possible problem with using a single gene to study bacterial evolutionary relationships is that a given gene might not be present in all organisms. An example is *recA*, which encodes a recombinase enzyme that facilitates genetic recombination in *Bacteria* (∞ Section 11.9). Genes homologous to *recA*, that is, having a shared ancestry with *recA*, are not present in *Archaea* and *Eukarya*, so *recA* would not be suitable for evolutionary studies extending beyond *Bacteria*. We discuss later (Section 14.8) ways of bypassing these problems in phylogeny and taxonomy by using genes whose sequences have diverged more than the 16S rRNA gene, consequently revealing distinctions between closely related bacteria, and by using multiple genes for evolutionary analysis.

Molecular Clocks

An unresolved question in phylogenetics is whether DNA (and protein) sequences change at a constant rate. If they do, then the amount of change between two homologous sequences would serve as an approximate **molecular clock**, or chronometer, of evolution, allowing the time in the past when the two sequences diverged from a common ancestral sequence to be estimated. Major assumptions of the molecular clock approach are that nucleotide changes accumulate in a sequence in proportion to time, that such changes generally are neutral and do not interfere with gene function, and that they are random. The molecular clock approach has been used to estimate the time of divergence of *Escherichia coli* and *Salmonella typhimurium*, closely related members of the *Gammaproteobacteria* (∞ Section 15.11), from a common ancestor about 120–140 million years ago. These data have also been combined with evidence from the geological record on isotopes and specific biological markers to approximate when different metabolic patterns emerged in bacteria (Section 14.3).

The main problem with the molecular clock approach, however, is that DNA sequences do change at different rates, which means that direct and reliable correlations to a timescale will be difficult to make. However, much of phylogenetic analysis is concerned with relative relationships among organisms, shown by branching order on phylogenetic trees. These relationships are generally discernible regardless of whether different sequences change at similar rates, so the accuracy of the molecular clock approach is not a major concern.

14.6 MiniReview

The evolutionary history of life, also called phylogeny, can be reconstructed through analysis of DNA sequences. The work of Carl Woese established the presence of three domains of life. Two of the domains, *Bacteria* and *Archaea*, lack a nucleus, whereas the third domain, *Eukarya*, has a nucleus.

- Why are SSU rRNA genes suitable for phylogenetic analysis?
- What information does the RDP-II provide?
- What value do molecular clocks have in phylogenetic analysis?

14.7 Evolutionary Analysis: Analytical Methods

Modern phylogenetics is based on nucleotide sequence comparisons, for which specific methods have been developed. We consider these methods here.

Obtaining DNA Sequences

Phylogenetic analysis using DNA sequences relies heavily on the polymerase chain reaction (PCR) to obtain sufficient copies of a gene for efficient sequencing (∞ Section 12.8). Specific oligonucleotide primers are designed that bind to the ends of the gene of interest, or to DNA flanking the gene, allowing DNA polymerase to bind and copy the gene. The source of DNA bearing a gene of interest typically is genomic DNA purified from individual bacterial strains. The PCR product is then visualized by agarose gel electrophoresis (**Figure 14.12**), excised from the gel, extracted and purified from the agarose, and then sequenced, often using the same oligonucleotides as primers for the sequencing reactions.

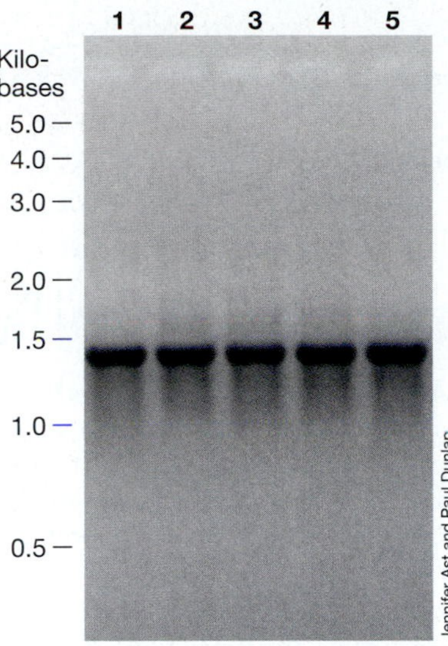

Figure 14.12 PCR-amplification of the 16S rRNA gene. Standard primers complementary to the ends of the 16S rRNA gene 27f (5′-AGAGTTTGATCCTGGCTCAG-3′) and 1492r (5′-TACGGYTACCTTGTTACGACTT-3′) were used to PCR-amplify the 16S rRNA gene from genomic DNA of five different unknown bacterial strains (lanes 1 through 5), and the products were separated by agarose gel electrophoresis (in the primer shown, Y can be either a C or a T). The bands of amplified DNA are approximately 1,465 nucleotides in length. Positions of DNA kilobase size markers are indicated at the left. Excision from the gel and purification of these PCR products is followed by sequencing (∞ Sections 12.5 and 22.6) and analysis to identify the bacteria.

More information on PCR, gel electrophoresis, and DNA sequencing is presented in Chapter 12. An important aspect of PCR amplification is primer design, deciding what sequence of primers to use to amplify a specific gene. Standard primers exist for many highly conserved genes, such as primers 27f and 1492r, among others, for the 16S rRNA gene of *Bacteria* (Figures 14.11 and 14.12). These standard primers allow the 16S rRNA genes of many different *Bacteria* to be amplified for sequencing. Amplification of newly characterized or more highly divergent genes, however, often requires using a computer program, along with some trial and error, to identify primers that will effectively and specifically amplify the gene of interest.

Sequence Alignment

Phylogenetic analysis is based on homology, that is, analysis of DNA sequences that are related by common ancestry. Homologous genes in different organisms may be either

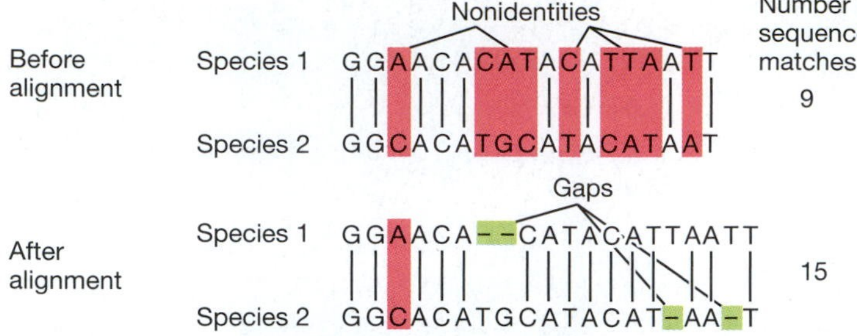

Figure 14.13 Alignment of DNA sequences. Sequences for a hypothetical region of a gene are shown for two organisms, before alignment and after the insertion of gaps to improve the matchup of nucleotides, indicated by the vertical lines showing identical nucleotides in the two sequences. The insertion of gaps in both sequences substantially improves the alignment.

orthologs, which differ because of sequence divergence as the organisms followed different evolutionary paths, or *paralogs,* which arise through gene duplication (Figure 13.11 and Section 13.9 for a further explanation of these terms). Once the DNA sequence of a gene is obtained, the next step in developing a phylogeny is to align that sequence with sequences of homologous (orthologous) genes from other strains or species. This way, differences between the sequences, that is, nucleotide mismatches and insertions and deletions (gaps), some of which can be phylogenetically informative, can be pinpointed.

Figure 14.13 gives an example of sequence alignment. The web-based BLAST (*B*asic *L*ocal *A*lignment *S*earch *T*ool) of the National Institutes of Health (http://www.ncbi.nlm.nih.gov/BLAST) does this automatically and can help identify genes homologous to a new sequence from among the many thousands of genes already sequenced. Homologous sequences are then downloaded from GenBank (http://www.ncbi.nih.gov/Genbank), an annotated collection of all publicly available DNA sequences, and aligned. The alignment is critical to phylogenetic analysis, because the assignment of mismatches and gaps is an explicit hypothesis of how the sequences have diverged from a common ancestral sequence. Protein-coding genes usually are aligned with the aid of their inferred amino acid sequences. Other genes, such as those encoding 16S rRNA, can be aligned by eye or through the use of computer programs designed to minimize the number of mismatches and gaps. Secondary structure, for example, the folding of the **16S ribosomal RNA** (Figure 14.11), is also helpful in making accurate gene alignments.

Phylogenetic Trees

Reconstructing evolutionary history from observed nucleotide sequence differences involves construction of a *phylogenetic tree,* which is a graphic illustration of the relationships among sequences of the organisms under study, much like a family tree. A phylogenetic tree is composed of nodes and branches (**Figure 14.14**). The internal nodes represent ancestors, and the tips of the branches, also called nodes, are individual strains of bacterial species that exist now and from which the sequence data were obtained. The internal nodes are points in evolution where an ancestor diverged into two new entities, each of which then began to accumulate

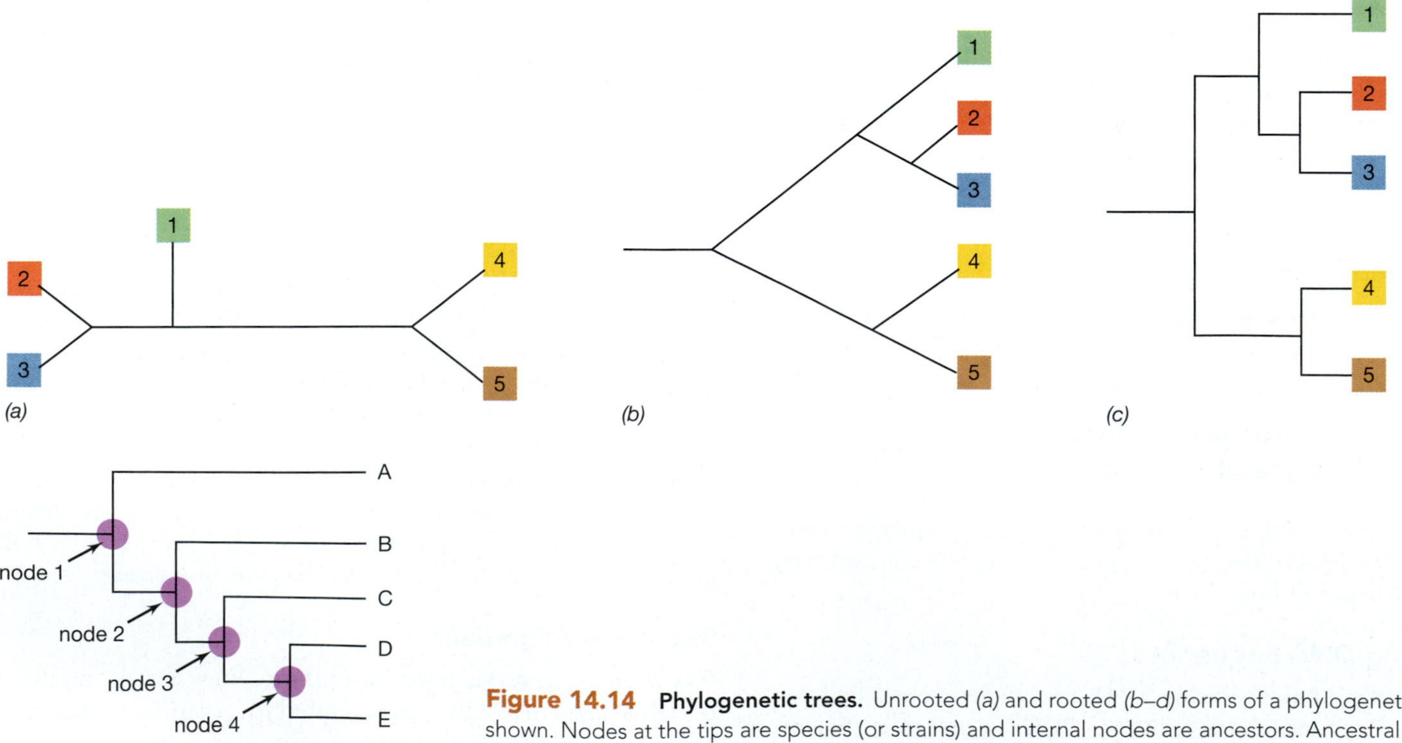

Figure 14.14 Phylogenetic trees. Unrooted *(a)* and rooted *(b–d)* forms of a phylogenetic tree are shown. Nodes at the tips are species (or strains) and internal nodes are ancestors. Ancestral relationships are revealed by the branching order in rooted trees *(d)*.

differences during its subsequent independent evolution. The branches define (1) the order of descent and (2) the ancestry of the nodes, and the branch length represents the number of changes that have occurred along that branch. Trees can be either *unrooted,* showing the relationships among the bacterial strains under study but not the evolutionary path leading from an ancestor to a strain (Figure 14.14*a*), or *rooted,* in which case the unique path from an ancestor (internal node) to each strain is defined (Figure 14.14*b–d*). Trees are rooted by the inclusion in the analysis of an *outgroup,* a bacterium that is less closely related to the organisms under study than the organisms are to each other, but that shares with them homologs of the gene or genes under study.

A phylogenetic tree is a depiction of lines of descent, and the relationship between two organisms therefore should be read in terms of common ancestry. That is, the more recently two species share a common ancestor, the more closely related they are. The tree in Figure 14.14*d* illustrates this point. Species B is more closely related to species C than it is to species A, because B and C share a more recent common ancestor (node 2) than do B and A (node 1).

Tree Reconstruction

Modern evolutionary analysis uses character-state methods, also called *cladistics,* for tree reconstruction. Character-state methods define phylogenetic relationships by examining changes in nucleotides at individual positions in the sequence, using those characters that are *phylogenetically informative*. These are characters that define a **monophyletic** group. **Figure 14.15** describes how phylogenetically informative characters are recognized. Computer-based analysis of these changes generates a phylogenetic tree, or *cladogram*.

A widely used cladistic method is *parsimony,* which is based on the assumption that evolution is most likely to have proceeded by the path requiring fewest changes. Computer algorithms based on parsimony provide a way of identifying the tree with the smallest number of character changes. Other cladistic methods, *maximum likelihood* and *Bayesian analysis,* proceed like parsimony, but they differ by assuming a model of evolution, for example, that certain kinds of nucleotide changes are more common than others (for example, an A changing into a C happens more often than a change from A to T or G). Inexpensive computer applications, such as PAUP* (*P*hylogenetic *A*nalysis *U*nder *P*arsimony, and Other Methods), guidebooks, and web-accessible tutorials are available for learning the basic procedures of cladistic analysis and tree reconstruction.

14.7 MiniReview

Analytical methods for evolutionary analysis include sequence alignment and reconstructing phylogenetic trees, which should be read in terms of common ancestry. Character-state methods such as parsimony are used for tree reconstruction.

- How are DNA sequences obtained for phylogenetic analysis?
- What does a phylogenetic tree depict?
- Why is sequence alignment critical to phylogenetic analysis?

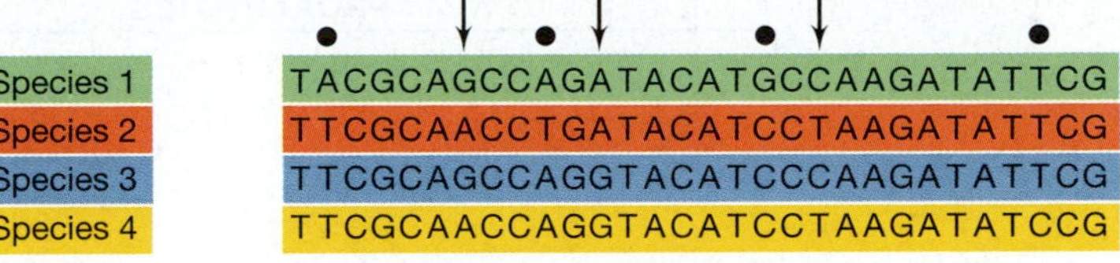

Figure 14.15 Identification of phylogenetically informative sites. Aligned sequences for four species are shown. Invariant sites are unmarked, and phylogenetically neutral sites are indicated by dots. Phylogenetically informative sites, varying in at least two of the sequences, are marked with an arrow.

14.8 Microbial Phylogeny

Biologists previously grouped living organisms into five kingdoms: plants, animals, fungi, protists, and bacteria. DNA sequence-based phylogenetic analysis, on the other hand, has revealed that the five kingdoms do not represent five primary evolutionary lines. Instead, as previously outlined in Chapter 2, cellular life on Earth has evolved along three primary lineages, called **domains**, the *Bacteria,* the *Archaea,* and the *Eukarya*. Two of these domains, the *Bacteria* and the *Archaea,* are exclusively microbial and are composed only of cells that lack a membrane-enclosed nucleus (that is, prokaryotic cells). The third lineage contains the eukaryotes (**Figure 14.16**) and is primarily microbial (that is, unicellular), and includes all of the original five kingdoms except for bacteria. The terms *Bacteria, Archaea,* and *Eukarya* designate the three domains of life, the domain being the highest biological taxon. Thus, plants, animals, fungi, and protists are all kingdoms within domain *Eukarya*.

A SSU rRNA Gene-Based Phylogeny of Life

The **universal phylogenetic tree** based on SSU rRNA genes (Figure 14.16) is a genealogy of all life on Earth. It depicts the evolutionary history of the cells of all organisms and clearly reveals the three domains. The root of the universal tree represents a point in evolutionary history when all extant life on Earth shared a common ancestor, LUCA, the *l*ast *u*niversal *c*ommon *a*ncestor.

Whole genome DNA sequences have confirmed the concept of the *Archaea,* species of which contain a large complement of genes that have no counterparts in *Bacteria* or *Eukarya*. The three-domain concept is also supported by analysis of specific genes shared among all organisms. Analysis of over 30 genes present in more than 190 species of *Bacteria, Archaea,* and *Eukarya* whose genomes have been completely sequenced confirms the distinct and unequivocal separation between these three lines of descent. Although branching orders and relationships among some lineages within *Bacteria, Archaea,* and *Eukarya* will be subject to revision as more is learned, analysis of multiple genes from genomic studies supports the basic structure of life proposed by Woese based on sequence analysis of SSU rRNA genes.

The presence of genes in common to *Bacteria, Archaea,* and *Eukarya* raises an interesting question. If these lineages diverged from each other so long ago from a common ancestor, how is it they share so many genes? One hypothesis is that early

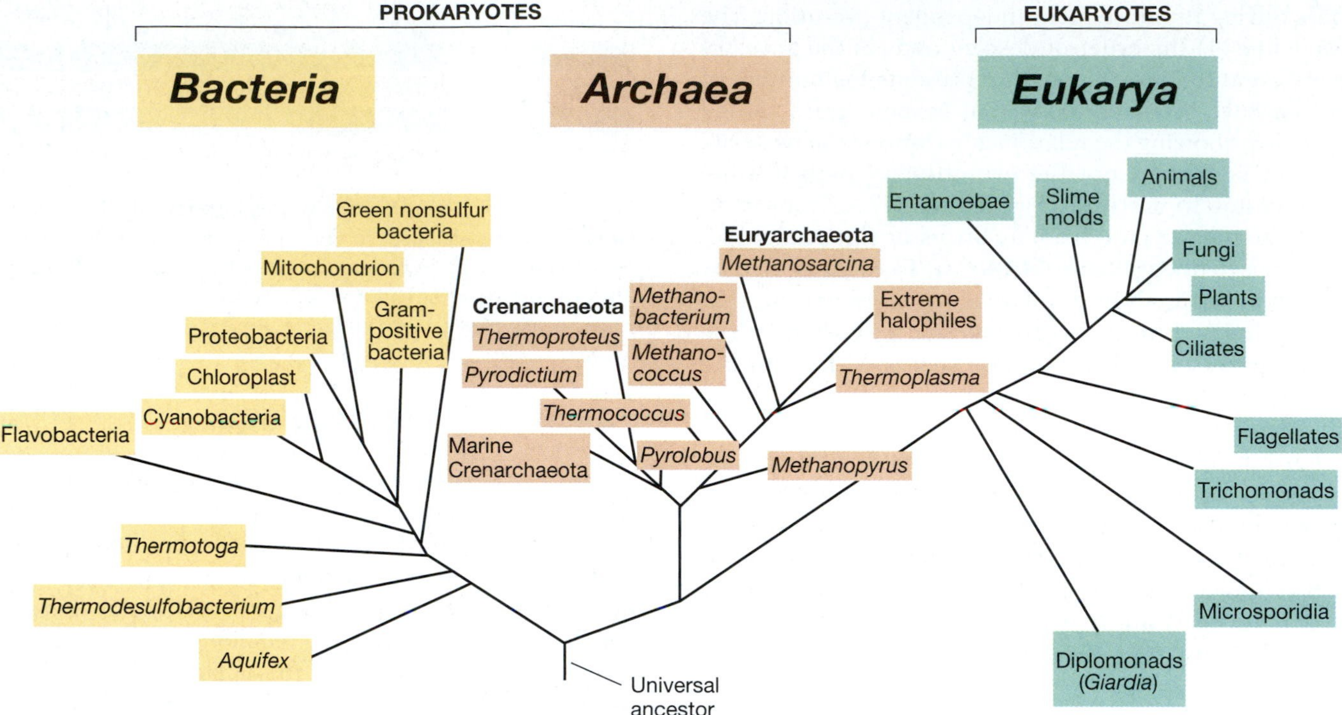

Figure 14.16 Universal phylogenetic tree as determined from comparative rRNA gene sequence analysis. Only a few key organisms or lineages are shown in each domain. Of the three domains, two (*Bacteria* and *Archaea*) contain only organisms lacking a membrane-enclosed nucleus (prokaryotic cells, ⚭ Section 2.5).

in the history of life, before the primary domains had diverged, horizontal gene transfer (⚭ Section 13.11) was extensive. During this time, genes encoding proteins that conferred exceptional fitness, for example, genes for the core cellular functions of transcription and translation, were promiscuously transferred among a population of primitive organisms derived from a common ancestral cell. If true, this would explain why, as genome analyses have shown, all cells regardless of domain have many core functional genes in common, as would be expected if all cells shared a common ancestor (Figure 14.16).

But what about the genetic differences observed from complete genome sequences? It is further hypothesized that over time barriers to unrestricted horizontal transfer evolved, perhaps from the selective colonization of habitats (thereby generating reproductive isolation) or as the result of structural or enzymatic (for example, restriction endonuclease) barriers that in some way prevented free genetic exchange. As a result, the previously genetically promiscuous population slowly began to sort out into the primary lines of evolutionary descent, the *Bacteria* and *Archaea* (Figure 14.16). As each lineage continued to evolve, certain unique biological traits became fixed within each group. Today, after some 4 billion years of microbial evolution (Figure 14.7), we see the grand result: three domains of cellular life that on the one hand share many common features but on the other hand display distinctive evolutionary histories of their own.

Bacteria

Among *Bacteria*, at least 80 major evolutionary groups (called phyla, singular **phylum**, or divisions) have been discovered thus far; several key ones are shown in the universal tree in Figure 14.16. Many groups have been defined from environmental sequences only (Section 14.9). Some of the lineages in the domain *Bacteria* are those previously distinguished by some phenotypic property, such as morphology or physiology; the spirochetes (⚭ Section 16.16) and the cyanobacteria (⚭ Section 16.7), respectively, are good examples of this. But most major groups of *Bacteria* consist of species that, although specifically related from a phylogenetic standpoint, lack strong phenotypic cohesiveness. The **Proteobacteria** are a good example here; the mixture of physiologies present within this group includes all known forms of microbial physiology. This clearly indicates that physiology and phylogeny are not necessarily linked.

Eukaryotic organelles clearly originated from within the domain *Bacteria*. As shown on the universal tree, mitochondria arose from the *Proteobacteria,* a major group of *Bacteria* (Figure 14.16), specifically from a relative of *Rhizobium* and the rickettsias. Interestingly, like mitochondria, these organisms live intracellularly, within the cells of either plants (⚭ Sections 24.14 and 24.15) or animals (⚭ Sections 15.13 and 35.3). The chloroplast arose from within the cyanobacterial phylum (Figures 14.10 and 14.16), as could be predicted because both cyanobacteria and the chloroplast carry out oxygenic photosynthesis (⚭ Section 20.5).

Archaea

From a phylogenetic perspective, the domain *Archaea* consists of two groups, the *Crenarchaeota* and *Euryarchaeota*

(Figure 14.16). Branching close to the root of the universal tree are hyperthermophilic *Crenarchaeota,* such as *Thermoproteus, Pyrolobus,* and *Pyrodictium* (Figure 14.16). They are followed by the *Euryarchaeota,* the methane-producing (methanogenic) *Archaea* and the extreme halophiles; *Thermoplasma,* an acidophilic, thermophilic member of *Archaea* lacking a cell wall is loosely related to this latter group (Figure 14.16).

There are some branches in the *Crenarchaeota* lineage (Figure 14.16, and Figure 17.1) that are known only from environmental sampling of ribosomal genes from the environment (Sections 14.9, 13.13, and 22.6). Interestingly, however, these sequences have originated from organisms that inhabit the open oceans, including Antarctic marine waters where temperatures are far colder than in the hot-spring or deep sea hydrothermal vent habitats of known *Crenarchaeota* (Section 17.9). Other *Crenarchaeota* sequences have emerged from environmental sampling of soil and lake water. We discuss these cold-adapted *Crenarchaeota* in more detail in Section 17.12.

Eukarya

Phylogenetic trees of species in the domain *Eukarya* are generated from comparative sequence analysis of the 18S rRNA gene, the functional equivalent of the 16S rRNA gene. *Eukarya* include a wide diversity of organisms. At one extreme are the single-celled microsporidia and diplomonads, obligate parasites that live in association with representatives of various groups of eukaryotes (for example, the pathogen *Giardia,* discussed in Chapter 18, is a member of the diplomonad group) and that bear degenerate mitochondria (hydrogenosomes and mitosomes, Section 18.4). At another extreme are the multicellular organisms, including the largest and most structurally complex of the *Eukarya,* the plants and animals.

When the fossil record is compared with the phylogenetic tree of *Eukarya,* the beginnings of a rapid evolutionary radiation can be traced to about 2 billion years ago. Geochemical evidence suggests that this is the period in Earth's history in which significant oxygen levels had accumulated in the atmosphere (Figure 14.7). It is thus likely that the onset of oxic conditions and subsequent development of an ozone shield (which would have greatly expanded the number of surface habitats available for colonization) was a major trigger for the rapid diversification of *Eukarya*. We consider the major groups of microbial *Eukarya* and their biology in Chapter 18.

UNIT 3

Distinguishing Characteristics of the Domains of Life

Although the primary domains (*Bacteria, Archaea,* and *Eukarya*) were founded on the basis of comparative ribosomal RNA gene sequencing—genetic criteria—each domain can also be characterized by various phenotypic properties. Table 14.1 listed traits linking *Eukarya* with either *Bacteria* or *Archaea*. Several other traits distinguish *Eukarya* from *Bacteria* and *Archaea,* and these are summarized in **Table 14.2**. Chapters 4, 8, and 18 provide additional information on these properties.

Table 14.2 Major features distinguishing *Bacteria* and *Archaea* from *Eukarya*[a]

Characteristic	Bacteria	Archaea	Eukarya
Morphological and Genetic			
Prokaryotic cell structure	Yes	Yes	No
DNA present in covalently closed and circular form	Yes	Yes	No
Membrane-enclosed nucleus	Absent	Absent	Present
Ribosomes (mass)	70S	70S	80S
Introns in most genes	No	No	Yes
Operons	Yes	Yes	No
Capping and poly(A) tailing of mRNA	No	No	Yes
Plasmids	Yes	Yes	Rare
Physiological/Special Structures			
Dissimilative reduction of S^0 or SO_4^{2-} to H_2S, or Fe^{3+} to Fe^{2+}	Yes	Yes	No
Nitrification	Yes	No[b]	No
Denitrification	Yes	Yes	No
Nitrogen fixation	Yes	Yes	No
Rhodopsin-based energy metabolism	Yes	Yes	No
Chemolithotrophy (Fe, S, H_2)	Yes	Yes	No
Gas vesicles	Yes	Yes	No
Synthesis of carbon storage granules composed of poly-β-hydroxyalkanoates	Yes	Yes	No
Growth above 70°C	Yes	Yes	No

[a]Note that for many features only particular representatives within a domain show the property.
[b]Environmental genomics studies of *Bacteria* and *Archaea* in marine waters strongly suggest that nitrifying *Archaea* exist (Section 22.6).

14.8 MiniReview

Life on Earth evolved along three major lines, domains *Bacteria*, *Archaea*, and *Eukarya*. Each domain contains several major evolutionary groups.

- How does the SSU rRNA tree of life differ from the grouping of life based on five kingdoms?
- What kinds of evidence support the three-domain concept of life?
- How does the universal tree support the hypothesis of endosymbiosis (Figure 14.10)?

14.9 Applications of SSU rRNA Phylogenetic Methods

SSU ribosomal RNA gene sequencing has spawned many research tools that employ this technology. These include signature sequences and rRNA probes for use in microbial ecology and diagnostic medicine, and methods for typing bacteria based on the SSU rRNA gene.

Signature Sequences

Computer analyses of ribosomal rRNA sequences have revealed **signature sequences**, short oligonucleotides unique to certain groups of organisms, for example, for each of the domains of cellular life. Moreover, signatures defining a specific group within a domain or, in some cases, a particular genus or even a single species, are also known or can be determined by computer inspection of aligned sequences. Because of their exclusivity, signature sequences are very useful. For example, signature sequences can help place newly isolated or previously misclassified organisms into their correct phylogenetic group. However, the most common use of signature sequences is for the design of specific nucleic acid probes.

Phylogenetic Probes and FISH

Recall that a probe is a strand of nucleic acid that can be labeled and used to hybridize to a complementary nucleic acid from a mixture (Section 12.2). Probes can be general or specific. For example, universal SSU rRNA probes are designed to bind to conserved sequences in the rRNA of all organisms, regardless of domain. By contrast, specific probes can be designed that will react only with species of the domain *Bacteria* because of unique signatures found in their RNA. Likewise, archaeal-specific and eukaryl-specific probes will react only with species of *Archaea* or *Eukarya*, respectively. **Phylogenetic probes** can be designed also to target groups within a domain, such as members of individual families, genera, or species.

The binding of probes to cellular ribosomes can be seen microscopically when a fluorescent dye is attached to the probe. By treating cells with the appropriate reagents, membranes become permeable and allow penetration of the probe–dye mixture. After hybridization of the probe directly to rRNA in ribosomes, the cells become uniformly fluorescent and can be observed under a fluorescent microscope (**Figure 14.17**). This technique has been nicknamed **FISH**, for *f*luorescent *i*n *s*itu *h*ybridization, and can be applied directly to cells in culture or in a natural environment (the term *in situ* means "in the environment"). In essence, FISH is a *phylogenetic stain*.

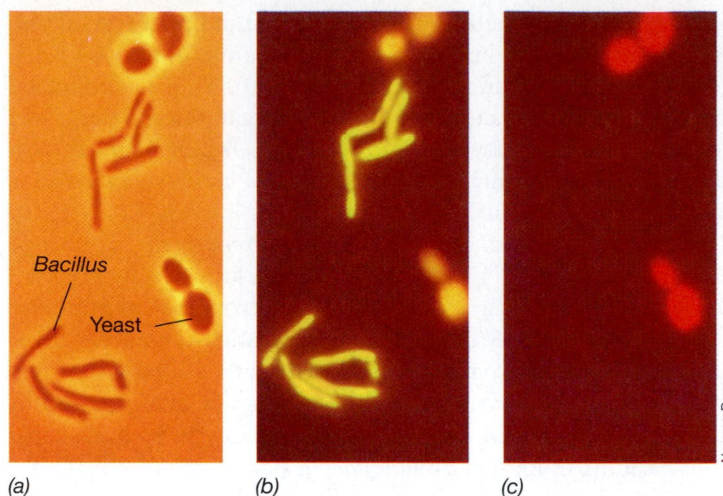

Figure 14.17 Fluorescently labeled rRNA probes: Phylogenetic stains. *(a)* Phase-contrast photomicrograph of cells of *Bacillus megaterium* (rod, member of the *Bacteria*) and the yeast *Saccharomyces cerevisiae* (oval cells, *Eukarya*). *(b)* Same field; cells stained with a yellow-green universal rRNA probe (this probe reacts with species from any domain). *(c)* Same field; cells stained with a eukaryal probe (only cells of *S. cerevisiae* react). Cells of *B. megaterium* are about 1.5 μm in diameter, and cells of *S. cerevisiae* are about 6 μm in diameter.

FISH technology is widely used in microbial ecology and clinical diagnostics. In ecology, FISH can be used for the microscopic identification and tracking of organisms directly in the environment. FISH also offers a method for assessing the composition of microbial communities directly by microscopy (Section 22.4). In clinical diagnostics, FISH has been used for the rapid identification of specific pathogens from patient specimens. The technique circumvents the need to grow an organism in culture. Instead, microscopic examination of a specimen can confirm the presence of a specific pathogen. In clinical diagnostics this facilitates the rapid diagnosis and treatment of a patient suffering from a microbial infection. By contrast, isolation and identification techniques typically take a minimum of 24 hours to complete and can take much longer.

Microbial Community Analysis

PCR-amplified rRNA genes do not need to originate from a pure culture grown in the laboratory. Using methods described in detail in Chapter 22, a phylogenetic snapshot of a natural microbial community can be taken using PCR to amplify the genes encoding SSU rRNA from members of that community. Such genes can easily be sorted out, sequenced, and aligned. From these data, a phylogenetic tree can be generated from these "environmental" sequences that shows the different rRNA genes present in the community. From this tree, specific organisms can be inferred even though none of them were actually cultivated or otherwise identified. Such microbial community analyses, as they have come to be called, are a major thrust of microbial ecology research today and have revealed many key features of microbial community structure and microbial interactions.

Ribotyping

Information from rRNA-based phylogenetic analyses also finds application in a technique for bacterial identification called **ribotyping**. Unlike comparative sequencing methods, however, ribotyping does not involve sequencing. Instead, it reports the specific pattern of bands, or DNA fingerprint, that is generated when DNA from an organism is digested by a restriction enzyme and the fragments are separated and probed with an rRNA gene probe (**Figure 14.18**). Differences between organisms in the sequence of their 16S rRNA genes translate into the presence or absence of cut sites recognized by different restriction enzymes (∞ Section 12.1 for a discussion of restriction enzymes). The DNA banding pattern, or *ribotype*, of a particular bacterial species may therefore be unique and diagnostic, allowing discrimination between species (Figure 14.18) and between different strains of a species, if there are differences in their SSU rRNA gene sequences.

In practice, ribotyping involves digestion of the bacterium's DNA with one or more restriction enzymes and separation of the DNA fragments by gel electrophoresis. The DNA fragments are then transferred from the gel onto nylon membranes and are hybridized with a labeled rRNA gene probe (∞ Section 12.2). The pattern generated from the fragments of DNA on the gel is then digitized, and a computer used to make comparisons of this pattern with patterns from reference organisms available from a database (Figure 14.18). Ribotyping is both highly specific and rapid (because it bypasses the actual sequencing, sequence alignment, and requirements for phylogenetic analysis). For these reasons, ribotyping has found many applications for bacterial identification in clinical diagnostics and for the microbial analyses of food, water, and beverages.

14.9 MiniReview

Phylogenetic analyses of SSU rRNA genes have led to the development of research tools useful in ecology and medicine.

- What are signature sequences?
- How can oligonucleotide probes be made visible under the microscope? What is this technology called?
- What kinds of questions can be addressed using microbial community analysis?
- How is ribotyping able to distinguish between different bacteria?

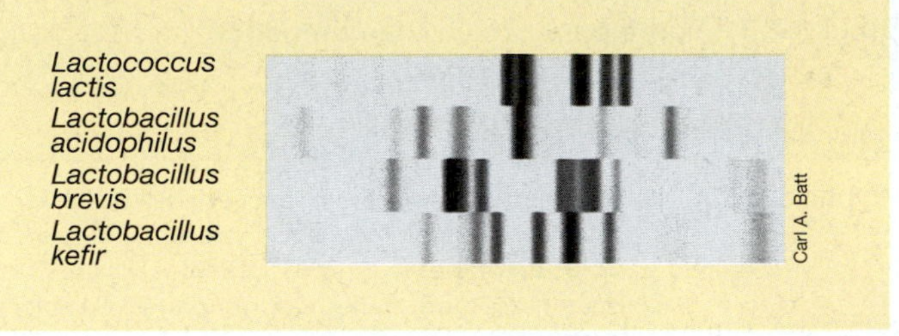

Figure 14.18 Ribotyping. Ribotype results from four different lactic acid bacteria. For each species there is a unique pattern of DNA fragments generated from restriction enzyme digestion of DNA taken from a colony of each bacterium and then probed with a 16S rRNA gene probe. Variations in both position and intensity of the bands are important in identification.

III MICROBIAL SYSTEMATICS

Systematics is the study of the diversity of organisms and their relationships. It links together phylogeny, just discussed, with **taxonomy**, in which organisms are characterized, named, and placed into groups according to their natural relationships. Bacterial taxonomy traditionally has focused on practical aspects of identification and description, activities that have relied heavily on phenotypic comparisons. At present, the growing use of genetic information, especially DNA sequence data, is allowing taxonomy to increasingly reflect phylogenetic relationships. Ideally, as Charles Darwin stated in his famous monograph, *On the Origin of Species*, "Our classifications will come to be, as far as they can be so made, genealogies." In the sections that follow, we consider basic elements of bacterial taxonomy—the identification of bacteria, species concepts, and bacterial classification and nomenclature.

Bacterial taxonomy has changed substantially in the past few decades, incorporating new methods for the identification of bacteria and additional criteria for the description of new species. This *polyphasic* approach to taxonomy uses three kinds of methods, phenotypic, genotypic, and phylogenetic, for the identification and description of bacteria. Phenotypic analysis examines the morphological, metabolic, physiological, and chemical characteristics of the cell. Genotypic analysis considers comparative aspects of cells at the level of the genome. These two kinds of analysis group organisms on the basis of similarities. They are complemented by phylogenetic analysis, which seeks to place organisms in a framework of evolutionary relationships. There is also growing recognition of the importance of information on the bacterium's habitat and ecology in polyphasic taxonomy.

14.10 Phenotypic Analysis

The observable characteristics of a bacterium provide many traits that can be used to differentiate between species. Typically, in describing a new species, and also to identify a bacterium, several of these traits are determined for the strain or strains of interest. The results are then compared with known organisms as controls, either examined in parallel

Table 14.3 Some phenotypic characteristics of taxonomic value

Major category	*Components*
Morphology	Colony morphology; Gram reaction; cell size and shape; pattern of flagellation; presence of spores, inclusion bodies (e.g., PHB[a] granules, gas vesicles, magnetosomes); stalks or appendages; fruiting-body formation
Motility	Nonmotile; gliding motility; swimming (flagellar) motility; swarming; motile by gas vesicles
Metabolism	Mechanism of energy conservation (phototroph, chemoorganotroph, chemolithotroph); utilization of individual carbon, nitrogen, or sulfur compounds; fermentation of sugars; nitrogen fixation; growth factor requirements
Physiology	Temperature, pH, and salt ranges for growth; response to oxygen (aerobic, facultative, anaerobic); presence of catalase or oxidase; production of extracellular enzymes
Cell chemistry	Fatty acids; polar lipids; respiratory quinones
Other traits	Pigments; luminescence; antibiotic sensitivity; serotype

[a]PHB, poly-β-hydroxybutyric acid (Section 4.10).

with the unknowns or from published information. The specific traits used depend on the kind of organism, and which traits are chosen for testing may arise from substantial prior knowledge of the bacterial group to which the new organism likely belongs as well as on the investigator's purpose. For example, in applied situations, such as in clinical diagnostic microbiology, where identification may be an end in itself and speed may be critical, a well-defined subset of traits often is used that quickly discriminates among likely or possible identifications. **Table 14.3** lists general categories and examples of some phenotypic traits used in identifications and species descriptions, and we examine one of these examples here.

Fatty Acid Analyses: FAME

The types and proportions of fatty acids present in cytoplasmic membrane and outer membrane (gram-negative bacteria, Sections 4.3 and 4.7) lipids of cells are major phenotypic traits. The technique for determining these fatty acids has been nicknamed **FAME**, for *f*atty *a*cid *m*ethyl *e*ster and is in widespread use in clinical, public health, and food and water inspection laboratories where the identification of pathogens or other bacterial hazards needs to be done on a routine basis. It is widely used in the characterization of new species of bacteria.

The fatty acid composition of *Bacteria* can be highly variable, including differences in chain length, the presence or absence of double bonds, rings, branched chains, or hydroxy groups (**Figure 14.19a**). Hence, a fatty acid profile can often identify a particular bacterial species. For the analyses, fatty acids, extracted from cell hydrolysates of a culture grown under standardized conditions, are chemically derivatized to form their corresponding methyl esters. These now volatile derivatives are then identified by gas chromatography. A chromatogram showing the types and amounts of fatty acids from

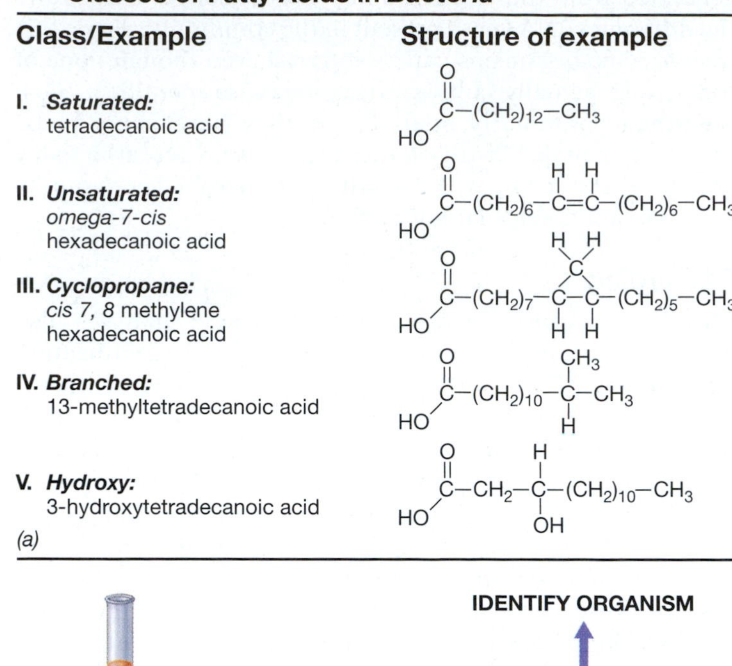

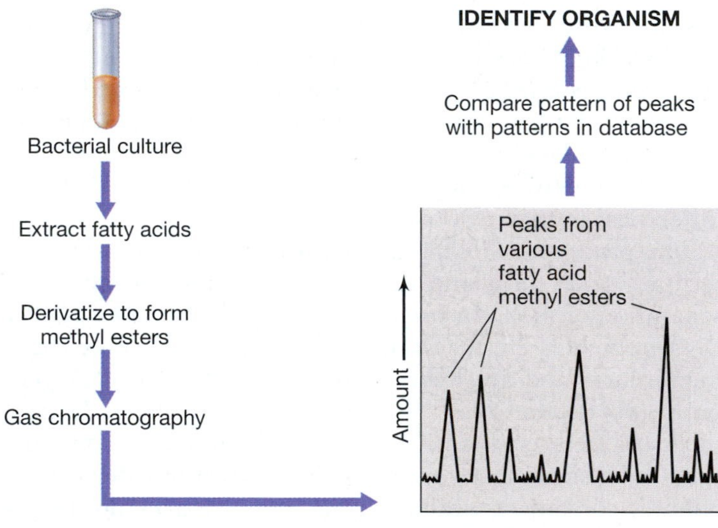

Figure 14.19 Fatty acid methyl ester (FAME) analysis in bacterial identification. *(a)* Classes of fatty acids in *Bacteria.* Only a single example is given of each class, but in fact, more than 200 different fatty acids have been discovered from bacterial sources. A methyl ester contains a methyl group (CH_3) in place of the proton on the carboxylic acid group (COOH) of the fatty acid. *(b)* Procedure. Each peak from the gas chromatograph is due to one particular fatty acid methyl ester, and the peak height is proportional to the amount.

the unknown bacterium is then compared with a database containing the fatty acid profiles of thousands of reference bacteria grown under the same conditions. The best matches to that of the unknown are then selected (Figure 14.19*b*).

As a phenotypic trait for species identification and description, FAME does have some drawbacks. In particular, FAME analyses require rigid standardization because fatty acid profiles of an organism, like many phenotypic traits, can vary as a function of temperature, growth phase (exponential versus stationary), and to a lesser extent, growth medium. Thus, for consistent results, it is necessary to grow the unknown organism

Table 14.4 Some genotypic methods used in bacterial taxonomy

Method	Description/application
DNA–DNA hybridization	A genomewide comparison of sequence similarity. Useful to distinguish species within a genus
DNA profiling	Ribotyping (Section 14.9), AFLP, rep-PCR. Rapid method to distinguish between species and strains within a species
MLST	Strain typing using DNA sequences of multiple genes. High resolution, useful for distinguishing even very closely related strains within a species
GC ratio	Percentage of guanine plus cytosine base pairs in the genome. Less commonly used in taxonomy because of poor resolution. If the GC ratio of two organisms differs by more than about 5%, they cannot be closely related, and organisms with similar or even identical GC ratios may be unrelated

on a specific medium and at a specific temperature in order to compare its fatty acid profile with those of organisms from the database that have been grown in the same way. For many organisms this is impossible, of course, and thus FAME analyses are limited to those organisms that can be grown under the specified conditions. In addition, the extent of variation in FAME profiles among strains of a species, a necessary consideration in studies to discriminate between species, is not yet well documented.

14.10 MiniReview

Systematics is the study of the diversity and relationships of living organisms. Polyphasic taxonomy is based on phenotypic, genotypic, and phylogenetic information. Phenotypic traits useful in taxonomy include morphology, motility, metabolism, and cell chemistry, especially lipid analyses.

- What is FAME analysis?
- What are some of the drawbacks of FAME analysis?
- Do these drawbacks apply to other kinds of phenotypic traits?

14.11 Genotypic Analysis

Comparative analysis of the genome also provides many traits for discriminating between species of bacteria. Genotypic analysis has particular appeal in microbial taxonomy because of the insights it provides at the DNA level. Several methods of genotypic analysis are used (as listed in **Table 14.4**) depending on the questions posed, with DNA–DNA hybridization and DNA profiling among the more commonly used in microbial taxonomy.

DNA–DNA Hybridization

When two organisms share many highly similar (or identical) genes, their DNAs would be expected to hybridize to one another in approximate proportion to the similarities in their gene sequences. In this way, measurement of **DNA–DNA hybridization** between the genomes of two organisms provides

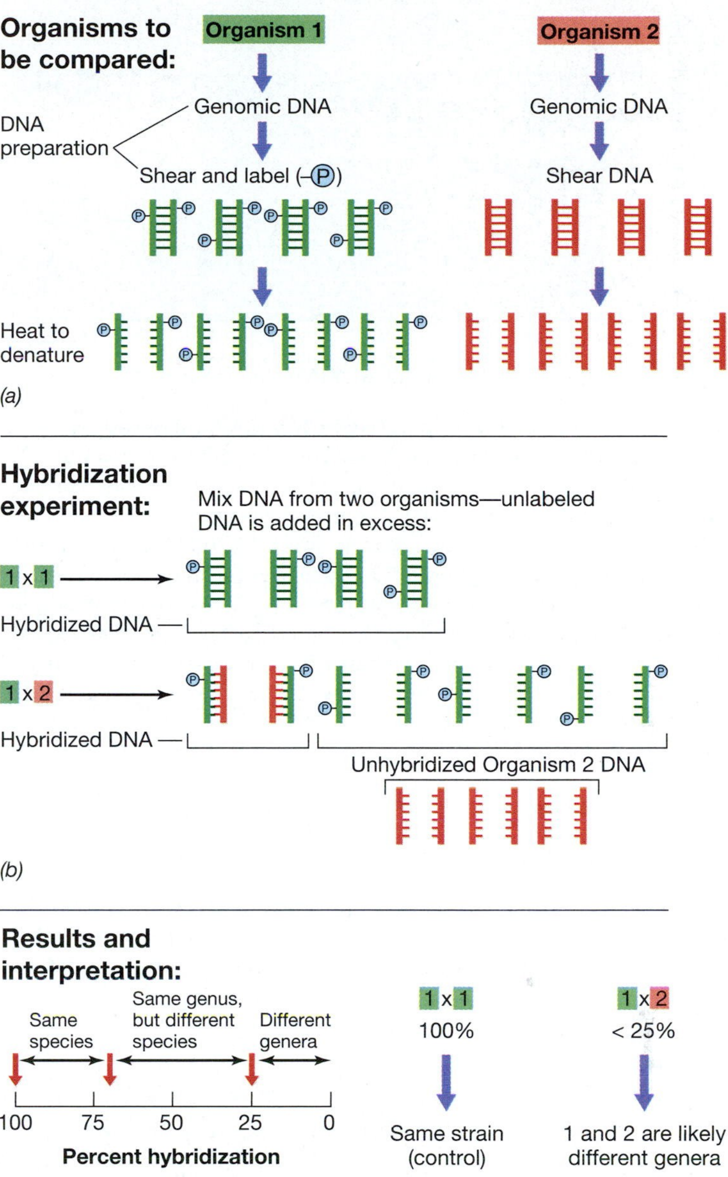

Figure 14.20 Genomic hybridization as a taxonomic tool. *(a)* Genomic DNA is isolated from test organisms. One of the DNAs is labeled (shown here as radioactive phosphate in the DNA of Organism 1). *(b)* Excess unlabeled DNA is added to prevent labeled DNA from reannealing with itself. Following hybridization, hybridized DNA is separated from unhybridized DNA before measuring radioactivity in the hybridized DNA only. *(c)* Radioactivity in the control (Organism 1 DNA hybridizing to itself) is taken as the 100% hybridization value.

a rough index of their similarity to each other. DNA–DNA hybridization therefore is useful for differentiating between organisms as a complement to SSU rRNA gene sequencing.

We discussed the theory and methodology of nucleic acid hybridization in Sections 7.2 and 12.2. In a hybridization experiment, genomic DNA isolated from one organism is made radioactive with ^{32}P or ^{3}H, sheared to a relatively small size, heated to separate the two strands, and mixed with an excess of unlabeled DNA prepared in the same way from a second organism (**Figure 14.20**). The DNA mixture is then cooled to

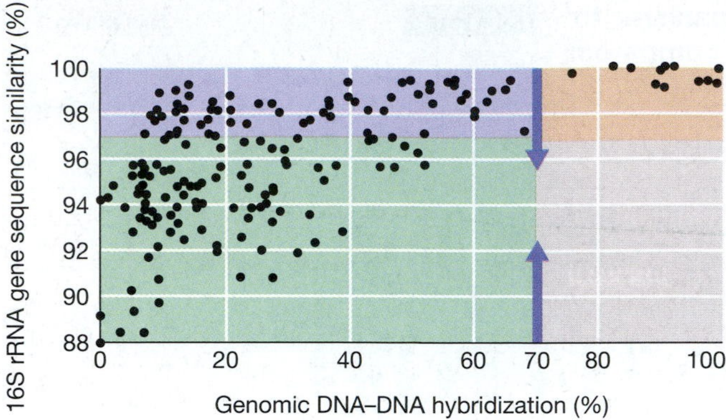

Figure 14.21 Relationship between 16S rRNA gene sequence similarity and genomic DNA-hybridization for different pairs of organisms. These data are the results from several independent experiments with various species of the domain *Bacteria*. Points in the orange box represent combinations for which 16S rRNA gene sequence similarity and genomic hybridization were both very high; thus, in each case, the two organisms tested were clearly the same species. Points in the green box represent combinations that detected different species, and both methods show this. The blue box shows examples of organisms that seemed to be different species as measured by genomic DNA–DNA hybridization but not by 16S rRNA gene sequence. Note that above 70% DNA–DNA hybridization, no 16S rRNA gene similarities were found that were less than 97%.

Data from Rosselló-Mora, R., and R. Amann. 2001. *FEMS Microbiol. Revs. 25*:39–67.

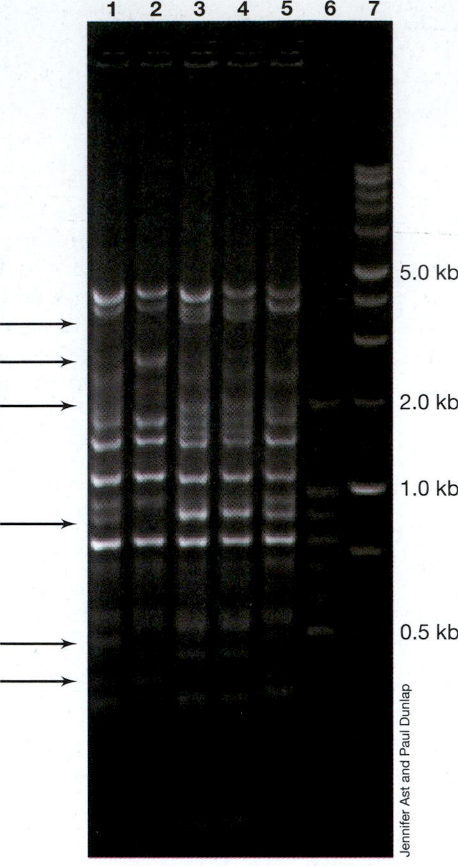

Figure 14.22 DNA fingerprinting with rep-PCR. Genomic DNAs from five strains (1–5) of a single species of bacteria were PCR-amplified using REP (repetitive extragenic palidromic) primers, REP1R-I (5′-IIIICG-ICGICATCIGGC-3′) and REP2-I (5′-ICGICTTATCIGGCCTAC-3′), and the PCR products were separated by gel electrophoresis on the basis of size to generate DNA fingerprints (in the primers shown, "I" can be any nucleotide). Arrows indicate some of the different bands in each strain. Strains 3 and 4 have very similar DNA profiles. Lanes 6 and 7 are 100-bp and 1-kbp DNA size ladders, respectively, used in estimating sizes of DNA fragments.

allow the single strands to reanneal. The double-stranded DNA is separated from any remaining unhybridized DNA. Following this, the amount of radioactivity in the hybridized DNA is determined and compared with the control, which is taken as 100% (Figure 14.20).

DNA–DNA hybridization is a sensitive method for revealing subtle differences in the genomes of two organisms and is often therefore useful for differentiating very similar organisms. Although there is no fixed convention as to how much hybridization between two DNAs is necessary to assign two organisms to the same taxonomic rank, hybridization values of 70% or greater are recommended for considering two isolates to be of the same species. By contrast, values of at least 25% are required to argue that two organisms should reside in the same genus. DNAs from more distantly related organisms, for example, *Clostridium* (gram-positive) and *Salmonella* (gram-negative), hybridize at only background levels, 10% or less (**Figure 14.21**).

DNA Profiling Methods

Several methods can be used to generate DNA fragment patterns for analysis of genotypic similarity among bacterial strains. One of these DNA-profiling methods, ribotyping, was described in Section 14.9 on applications of SSU sequence analysis. Other commonly used methods for rapid genotyping of bacteria include *repetitive extragenic palindromic PCR (rep-PCR)* and *amplified fragment length polymorphism (AFLP)*. In contrast to ribotyping, which focuses on a single gene, rep-PCR and AFLP assay for variations in DNA sequence throughout the genome.

The rep-PCR method is based on the presence of highly conserved repetitive DNA elements interspersed randomly around the bacterial chromosome. The number and positions of these elements differ between strains of a species that have diverged in genome sequence. Oligonucleotide primers, designed to be complementary to these elements, enable PCR amplification of different genomic fragments, generating a pattern of bands from genomic DNA. The pattern of bands differs for different strains, giving a strain-specific DNA "fingerprint" (**Figure 14.22**).

AFLP is based on the digestion of genomic DNA with one or two restriction enzymes and selective PCR-amplification of resulting fragments, which are separated by agarose gel electrophoresis. Strain-specific banding patterns similar to that of rep-PCR or other DNA fingerprinting methods are generated, with the large number of bands giving a high degree of discrimination between strains within a species. A similar technique to AFLP called T-RFLP is widely used in phylogenetic analyses of natural microbial communities (Section 22.5).

Multilocus Sequence Typing

One of the limitations of both ribosomal RNA gene sequence analysis and ribotyping (but not of strain typing with rep-PCR or AFLP) is that these analyses focus on only a single gene, which may not provide sufficient information for unequivocal discrimination between bacterial strains. **Multilocus sequence typing (MLST)** circumvents this problem and is a powerful technique for characterizing strains within a species.

MLST involves sequencing several different "housekeeping genes" from an organism and comparing their sequences with sequences of the same genes from different strains of the same organism. Recall that housekeeping genes encode essential functions in cells and are located on the chromosome rather than on a plasmid (∞ Section 7.4). For each gene, an approximately 450-bp sequence is amplified using PCR and is then sequenced. Each nucleotide along the sequence is compared and differences are noted. Each difference, or sequence variant, is considered an **allele** and is assigned a number. A given strain will then be assigned a series of numbers, its allelic profile, or multilocus sequence type. In MLST, strains with identical sequences for a given gene will have the same allele number for that gene, and two strains with identical sequences for all the genes would have the same allelic profile (and would be considered identical by this method). The relatedness between each allelic profile is expressed in a dendrogram of linkage distances that vary from 0 (strains are identical) to 1 (strains are only distantly related) (Figure 14.23).

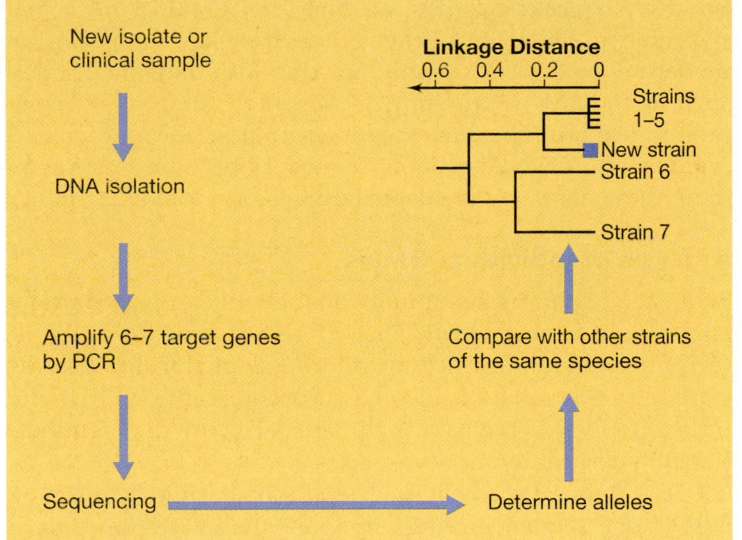

Figure 14.23 Multilocus sequence typing. Steps in MLST leading to a similarity phenogram are shown. Strains 1–5 are virtually identical, whereas strains 6 and 7 are both distinct from one another and from strains 1–5.

MLST has sufficient resolving power to distinguish even very closely related strains. In practice, strains can be discriminated on the basis of a single nucleotide change in just one of the several analyzed genes. MLST is not useful, however, for comparing organisms above the species level; its resolution is too sensitive to yield meaningful information for higher order groups, that is, genera or families.

MLST has made a variety of impacts in microbiology. In clinical microbiology, it has been effective for differentiating strains of a particular pathogen. This is important because some strains within a species—*Escherichia coli* K-12, for example—may be harmless, whereas others, such as strain O157:H7, can cause serious and even fatal infections (∞ Section 37.8). MLST is also quite useful for epidemiological studies, to track a virulent strain of a bacterial pathogen as it moves through a population, and for environmental studies, to define the geographic distributions of strains.

MLST analyses also have revealed some interesting genetic patterns in bacteria. For example, some *Bacteria,* such as *Staphylococcus aureus,* show very little MLST variation, whereas other organisms, such as *Neisseria meningitidis,* show great variability by MLST analyses. This difference implies that organisms such as *N. meningitidis* have undergone much greater horizontal gene flow (∞ Section 13.11) in the past than has *S. aureus,* or that their genes accumulate mutations at a faster rate.

GC Ratios

Another method used to compare and describe bacteria is their **GC ratios**. The GC ratio is the percentage of guanine plus cytosine in an organism's genomic DNA. GC ratios vary over a wide range, with values as low as 20% and as high as nearly 80% among *Bacteria* and *Archaea,* a range that is somewhat broader than for eukaryotes. It is generally considered that if two organisms' GC ratios differ by more than about 5%, they will share few DNA sequences in common and are therefore unlikely to be closely related. However, two organisms can have identical GC ratios and yet turn out to be quite unrelated, because very different nucleotide sequences are possible from DNA of the same overall base composition. In this case, the identical GC ratios are taxonomically misleading. Because gene sequence data are increasingly easy to obtain, GC ratios may find less application in bacterial taxonomy in the future.

14.11 MiniReview

Genotypic analysis examines traits of the genome. Bacterial species can be distinguished genotypically on the basis of DNA–DNA hybridization and DNA profiling.

- What is DNA fingerprinting and how is it useful for distinguishing bacteria?
- Hybridization of less than 25% between two organisms' DNA indicates that they are ______.
- How do AFLP and MLST differ from ribotyping?

14.12 Phylogenetic Analysis

Phylogenetic data are being used increasingly in microbial taxonomy to complement phenotypic and genotypic information. In addition, phylogenetic analysis permits organisms to be placed in a unified classification system structured on their evolutionary relationships, an important goal of microbial systematics. Thus, phylogenetic data are a key component of modern polyphasic taxonomy. Starting with individual genes, such as the SSU rRNA gene, phylogenetic analysis is beginning to use multiple gene sequences to describe and identify organisms. As whole-genome sequences become more common, these data, too, are contributing to microbial taxonomy.

Analysis of Individual Genes

Sequence comparisons of individual genes can provide valuable insight for taxonomy. The 16S rRNA gene, for example, the importance of which in microbial phylogeny was described in Section 14.6, also has proven exceptionally useful in taxonomy, serving as a "gold standard" for the identification and description of new species.

It is generally considered that a bacterium whose 16S rRNA gene sequence differs by more than 3% from that of other organisms should be considered a new species. This proposal is supported by the observation that genomic DNA from two microorganisms whose 16S rRNA gene sequences are less than 97% identical typically hybridize to less than 70%, a minimal value considered evidence for two organisms being of the same species (Figure 14.20). The relationship between these methods is depicted in Figure 14.21. The proposed cutoff value serves as a guide for discriminating between strains at the species level, and it provides widely used data for the practical description of a new species. What constitutes a genus, the next level in taxonomic hierarchy (Section 14.11), is more a matter of judgment than for species, but 16S rRNA gene sequence differences of more than 5% from other organisms have been taken as support for an organism being a new genus. Other highly conserved genes, such as *recA,* which encodes a recombinase protein, and *gyrB,* which encodes a DNA gyrase protein, also can be useful for distinguishing bacteria at the species level.

Reliance in taxonomy, as well as in phylogeny, on the sequence of a single gene, however, can have pitfalls. The lack of divergence of the 16S rRNA gene, for example, can limit its effectiveness in discriminating between bacteria at the species level. This problem can be overcome by analysis of genes that are more highly divergent but that still reliably reflect systematic relationships, when such genes can be identified for the bacterial group under study. Individual genes also may be subject to horizontal gene transfer, however, which would invalidate taxonomic conclusions based on their sequences. For these reasons, a *multi-gene* approach is being used more frequently now in bacterial taxonomy and phylogeny.

Multi-Gene Sequence Analysis

The use of multiple genes for the identification and description of bacteria can avoid problems associated with reliance on individual genes. Multi-gene sequence analysis is similar to MLST, except that complete gene sequences are obtained when possible and comparisons are made using cladistic methods. By sequencing a few to several functionally unrelated genes, one can obtain a sampling of the genome that is more representative than a single gene, and instances of horizontal gene transfer can be detected and those genes excluded from further consideration. Taxonomic descriptions based on multiple genes therefore can provide a more reliable measure of the bacterium's systematic relationships.

Whole-Genome Sequence-Based Analysis

Rapid progress in sequencing the whole genomes of *Bacteria, Archaea,* and *Eukarya* has the potential of providing new tools for bacterial taxonomy. At this time, more than 670 genomes have been completely sequenced, over 550 of which are *Bacteria* (∞ Sections 13.2 and 13.5). Approximately 2,300 additional genome sequencing projects are in progress. When the genome of a bacterium is sequenced, computational methods can be used to compare its genes with those of any and all other sequenced organisms.

Whole-genome sequences provide many traits for comparative genotypic analysis. For example, major physical differences between species in genome structure, including size and number of chromosomes, their GC content, and whether the chromosomes are linear or circular may have taxonomic significance. Comparative analyis of gene content (the presence or absence of genes) and the order of those genes in the genome also will undoubtedly provide insights with implications for taxonomy, as will MLST and phylogenetic analysis based on identification of genes found to be universally present in the organisms of interest. One can envision a time in the not too distant future when descriptions of new bacterial species will include, along with phenotypic, genotypic, phylogenetic, and ecological data, the sequence of the bacterium's genome.

14.12 MiniReview

Phylogenetic approaches to bacterial taxonomy can be based on sequence analysis of individual genes, multiple genes, or potentially on the whole genome. Phylogenetic analysis is an important aspect of modern polyphasic taxonomy.

- What limits the value of the 16S rRNA gene (or any gene) in taxonomy?
- Why is the *recA* gene not suitable for phylogenetic analysis in *Archaea*?
- What distinguishes multi-gene phylogenetic analysis from MLST?

14.13 The Species Concept in Microbiology

At present, there is no universally accepted concept of **species** for prokaryotes. We have seen that microbial taxonomy combines phenotypic, genotypic, and sequence-based phylogenetic data, and that, as will be discussed in the following section, certain standards and guidelines are in place for distinguishing bacteria as separate species. These standards and guidelines provide an effective framework for the practice of describing and identifying prokaryotes, but they leave unaddressed the question of what constitutes a prokaryotic

Table 14.5 Taxonomic hierarchy for the purple sulfur bacterium *Allochromatium warmingii*

Taxonomic division	Name	Properties	Confirmed by
Domain	*Bacteria*	Bacterial cells; ribosomal RNA gene sequences typical of *Bacteria*	Microscopy; 16S rRNA gene sequence analysis; presence of unique biomarkers, for example, peptidoglycan
Phylum	*Proteobacteria*	rRNA gene sequence typical of *Proteobacteria*	16S rRNA gene sequence analysis
Class	*Gammaproteobacteria*	Gram-negative bacteria; rRNA sequence typical of *Gammaproteobacteria*	Gram-staining, microscopy
Order	*Chromatiales*	Phototrophic purple bacteria	Characterizing pigments (Figure 20.3)
Family	*Chromatiaceae*	Purple sulfur bacteria	Ability to oxidize H_2S and store S^0 within cells; observe culture microscopically for presence of S^0 (see photo)
Genus	*Allochromatium*	Rod-shaped purple sulfur bacteria	Microscopy (see photo)
Species	*warmingii*	Cells 3.5–4.0 μm × 5–11 μm; store sulfur mainly in poles of cell (see photo)	Measure cells in microscope using a micrometer; look for position of S^0 globules in cells (see photo)

Sulfur (S^0) globules

Norbert Pfennig

Photomicrograph of cells of the purple sulfur bacterium Allochromatium warmingii.

species. Because species are the fundamental units of biological diversity, this question has importance for our perception of and interaction with the microbial world. How the concept of species is defined in microbiology determines how we distinguish and classify the units of diversity that make up the microbial world.

Current Definition of Prokaryotic Species

A prokaryotic species presently is defined operationally as a collection of strains sharing a high degree of similarity in several independent traits. In polyphasic taxonomy, phenotypic, genotypic, and phylogenetic data are used to describe a new species. Traits currently considered most important for grouping strains together as a species include 70% or greater DNA–DNA hybridization and 97% or greater 16S rRNA gene sequence identity (Sections 14.10 and 14.11). **Table 14.5** gives an example of species definition in practice, listing relevant traits for the hierarchical classification of the phototrophic bacterium *Allochromatium warmingii*. Based on these kinds of criteria, nearly 7,000 species of *Bacteria* and *Archaea* have been formally recognized.

There is substantial interest in microbiology, however, in moving beyond a reliance on practical methods for identification to develop a prokaryotic species concept that is grounded in fundamental principles of biology and evolution. The biological species concept, which states that a species is an interbreeding population of organisms that are reproductively isolated from other interbreeding populations, is widely accepted as effective for grouping many eukaryotic organisms in a biologically meaningful way. However, the biological species concept is not effective or meaningful for the vast majority of life's diversity—the *Bacteria* and *Archaea*, which are haploid and do not undergo sexual reproduction. Even if we factor in acquisition of homologous DNA by recombination and acquisition of nonhomologous DNA by horizontal gene transfer and illegitimate recombination, descent in prokaryotes is predominantly vertical, and there is no true sexual reproduction, the fusion of haploid gametes to form a diploid cell, in bacteria. For these organisms, one haploid cell reproduces to form two new haploid cells that are identical genetically or essentially identical except for mutations. Therefore, a species definition based on sexual reproduction and reproductive isolation is not suitable for *Bacteria* and *Archaea*.

An alternative to the biological species concept that seems suitable for haploid organisms is the genealogical species concept (also referred to as the phylogenetic species concept). According to the genealogical species concept, a prokaryotic species is a group of strains that—based on DNA sequences of multiple genes—cluster closely with each other phylogenetically and are distinct from other groups of strains. In other

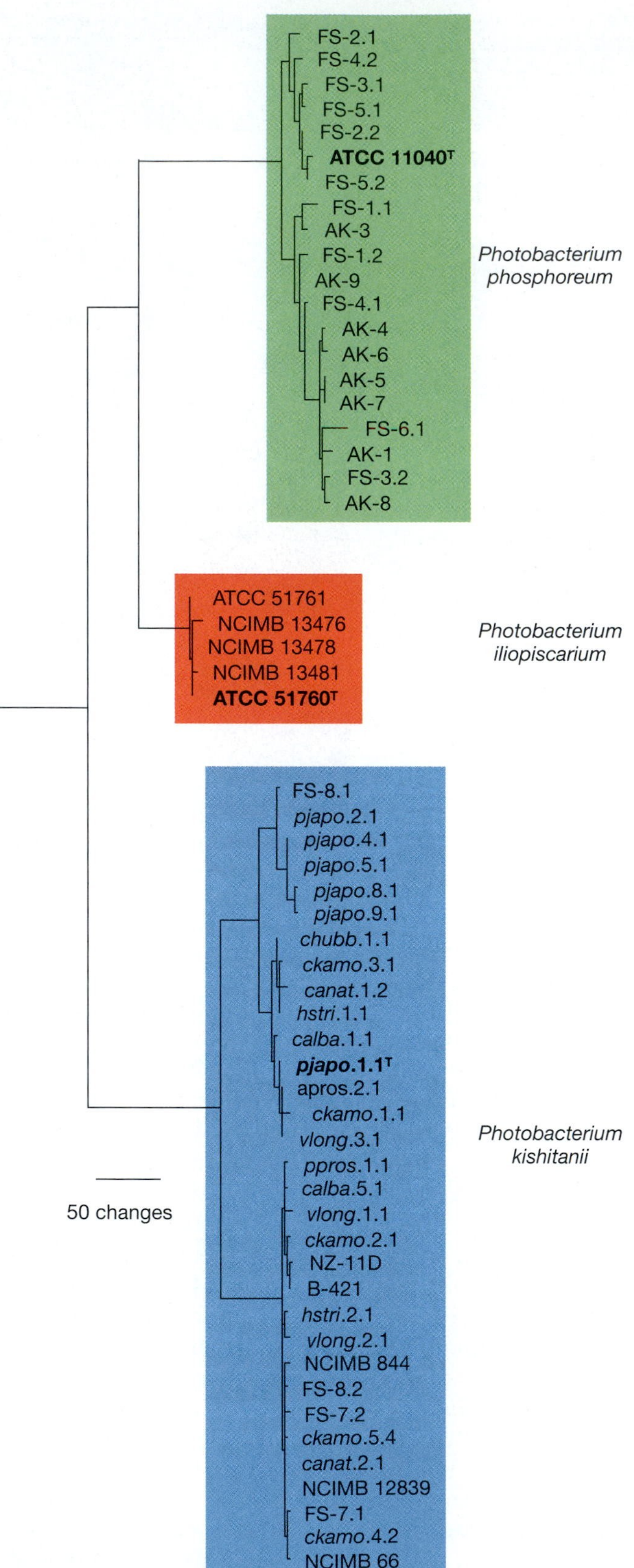

Figure 14.24 Multi-gene phylogenetic analysis. A phylogeny is shown for strains in the genus *Photobacterium* (*Gammaproteobacteria*) based on combined parsimony analysis of the 16S rRNA gene, *gyrB* (a housekeeping gene that encodes DNA gyrase subunit B), and the *luxABFE* genes (encoding the light-emitting enzyme luciferase and other luminescence enzymes). Although 16S rRNA gene sequences alone do not distinguish among these bacteria, due to the lack of divergence in the sequence of this gene, the combined analysis, due to the greater sequence divergences of the *gyrB* and *luxABFE* genes, resolves them as three evolutionarily distinct clades (that is, species), *P. phosphoreum*, *P. iliopiscarium*, and *P. kishitanii*. The scale bar indicates the branch length equal to 50 nucleotide changes. Type strains of each species, designated with a superscript "T" appended to their strain designations, are shown in bold.

words, each of these groups of strains is monophyletic, having descended from a different common ancestor. An example of the use multiple gene sequences to define separate species is shown in **Figure 14.24**. The analysis shown here uses the 16S rRNA gene, the sequence of which has diverged very little and which therefore provides good family and genus level resolution, but not good species level resolution. For good species level resolution, the *gyrB* gene, which encodes DNA gyrase subunit B, and the *luxABFE* genes, which encode luminescence enzymes, are included in the analysis. The *gyrB* and *luxABFE* genes are less constrained functionally; that is, their sequences can change more than that of the 16S rRNA gene without a loss of function of the proteins they encode. The combined analysis (Figure 14.24) separates closely related, phenotypically similar strains of *Photobacterium*, a genus of *Bacteria*, into three evolutionarily distinct clades (that is, *species*). In this way, through its reliance on multiple gene sequence analysis, the genealogical species concept has the potential to provide a strong evolutionary framework for species descriptions that are based on polyphasic taxonomy.

Speciation in Prokaryotes

How might new prokaryotic species arise? One possibility is by the process of periodic selection. Imagine a population of bacteria that originated from a single cell and that occupies a particular niche in a habitat. If cells in this population share a particular resource (for example, a key nutrient), this population of cells can be called an **ecotype**. Different ecotypes can coexist in a habitat, but each is only successful within its prime niche in the habitat.

Within each ecotype population, random mutations will occur over time in the genomes of the cells as they reproduce. Most of the mutations are neutral and have no effect. If there is a beneficial mutation, one that increases fitness, in a cell in one of the ecotypes, that cell may reproduce better than other kinds, so its progeny, which carry the mutation, will become dominant. In time, this may purge the population of the less well-adapted cells. Repeated rounds of mutation and selection in this ecotype may lead it to become more and more distinct genetically from the other ecotypes. Given enough time and sufficient changes, cells of this lineage would likely be identified by various traits as a new species (**Figure 14.25**).

Selection of strains bearing beneficial mutations can proceed gradually, or it can occur quite suddenly due to rapid environmental change. Note that this series of events within an ecotype has no effect on other ecotypes, because different ecotypes do not compete for the same resources (Figure 14.25). It is also possible that a new genetic capability in an ecotype may arise from genes obtained from cells of another ecotype by horizontal gene transfer, rather than from mutation and selection.

The extent of horizontal gene transfer among bacteria is variable. Genome sequence analyses have revealed examples in which horizontal tranfer of genes has apparently been frequent and others in which it has been rare (Section 13.11). In addition, MLST (Section 14.11) has revealed that genetic exchange within some species, called intraspecific recombination, is widespread, but virtually nonexistent within other species. Despite the impact of horizontal gene transfer, speciation in *Bacteria* and *Archaea* is thought to be driven primarily by mutation and periodic selection (Figure 14.25) rather than by horizontal gene transfer. This is because horizontal gene transfer occurs rarely in most bacteria and may confer only temporary benefits; the transferred genes can readily be lost if the selective pressure to retain them decreases.

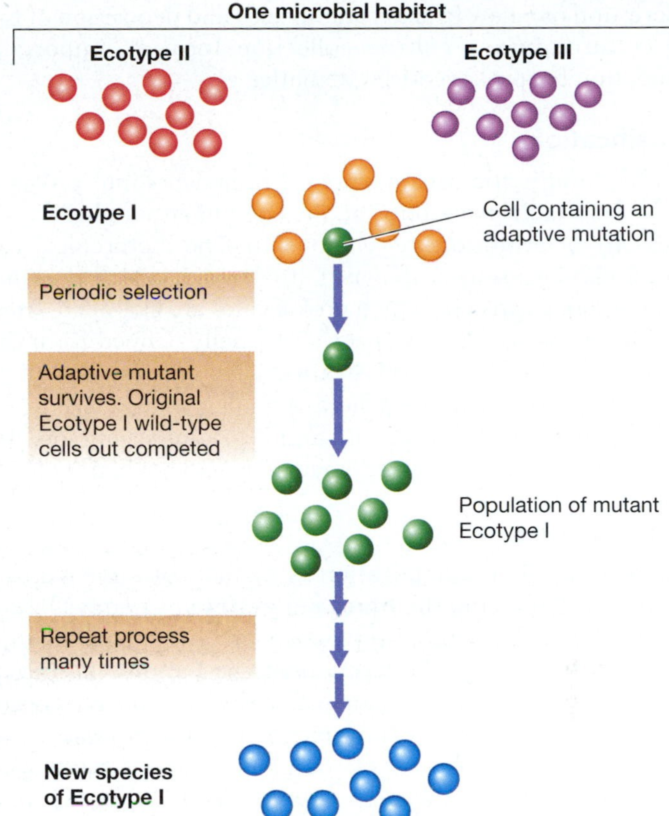

Figure 14.25 A model for bacterial speciation. Several ecotypes can coexist in a single microbial habitat, each occupying its own prime ecological niche. A cell within an ecotype that has a beneficial mutation may become a population that eventually replaces the original ecotype. As this occurs repeatedly within a given ecotype, a genetically distinct population of cells arises that represents a new species. Because other ecotypes do not compete for the same resources, they are unaffected by genetic and selection events outside of their prime niche.

How Many Prokaryotic Species Are There?

The result of over 4 billion years of evolution (Figure 14.7) is the prokaryotic world we see today. Microbial taxonomists agree that no firm estimate of the number of prokaryotic species can be given at present, in part because of uncertainty about what delimits a species. However, they also agree that in the final analysis, this number will be very large. Nearly 7,000 species of *Bacteria* and *Archaea* are already known, and thousands more, perhaps as many as 100,000–1,000,000 in total, are thought likely to exist. If we factor in application of the genealogical species concept, which provides better resolution of evolutionarily distinct monophyletic groups (species) (Figure 14.24), the estimate increases many fold.

It is important to remember that microbial community analyses (Section 14.9 and Sections 22.5 and 22.6) indicate that we have only scratched the surface in our ability to culture the diversity of *Bacteria* and *Archaea* in nature. With more exacting tools, both molecular and cultural, for revealing diversity, it is possible that the large number of species already predicted will be an underestimate. The reality today is that an accurate estimate of prokaryotic species is simply out of reach of current understanding and technology. But as with other things in microbiology, this will likely change. The area of microbial systematics is just beginning to open up, and many exciting discoveries will be made over the next several years.

14.13 MiniReview

A prokaryotic species presently is defined operationally. The biological species concept is not suitable for *Bacteria* and *Archaea*, because the members of these domains do not undergo true sexual reproduction. Prokaryotes may speciate as a result of repeated periodic selection for favorable traits within an ecotype.

- How does the genealogical species concept compare to an operational definition of a bacterial species?
- What is meant by the term monophyletic?
- What is an ecotype?
- How many species of *Bacteria* and *Archaea* are already known? How many likely exist?

14.14 Classification and Nomenclature

We conclude this chapter with a brief description of how *Bacteria* and *Archaea* are classified and named. Information is presented also on culture collections, which serve as repositories for scientific deposition of cultures, on some key taxonomic resources available for microbiology, and on some of the procedures for naming new species. The formal description of a new prokaryotic species and deposition of living cultures into a culture collection form an important foundation for prokaryotic systematics.

Classification

Classification is the organization of organisms into progressively more inclusive groups on the basis of either phenotypic similarity or evolutionary relationship. The hierarchical nature of classification is shown in Table 14.5, which presents the taxonomic ranks by which prokaryotes are classified, from domain to species. A new species is usually defined from the characterization of several strains, with groups of similar species then collected into genera (singular, genus). Groups of similar genera are collected into families, families into orders, orders into classes, up to the domain, the taxon of highest level.

Nomenclature

Nomenclature is the application of formal rules for naming organisms. Following the **binomial system** of nomenclature used throughout biology, prokaryotes are given genus names and species epithets. The terms used are Latin or Latinized Greek derivations, often of some descriptive property appropriate for the organism, and are set in print in *italics*. For example, over 100 species of the genus *Bacillus* have been described, including *Bacillus subtilis, Bacillus cereus,* and *Bacillus megaterium*. These species epithets mean "slender," "waxen," and "big beast," respectively, and refer to key morphological, physiological, or ecological traits characteristic of each organism. By classifying organisms into groups and naming them, we order the natural word and make it possible to communicate effectively about all aspects of individual organisms, including their behavior, ecology, physiology, pathogenesis, and evolutionary relationships.

In contrast to classification, nomenclature is subject to specific rules. The assignment of names for species and higher groups of *Bacteria* and *Archaea* is regulated by the Bacteriological Code—*The International Code of Nomenclature of Bacteria*. This source presents the formal framework by which *Bacteria* and *Archaea* are to be officially named and the procedures by which existing names can be changed, for example, when new data warrants taxonomic rearrangements. There are even rules for rejecting names if errors were made in the original naming process or a name has otherwise become invalid. The Bacteriological Code deals only with procedures for assigning names to organisms, not with issues of taxonomic methods or interpretation.

Bergey's Manual and *The Prokaryotes*

Because taxonomy is largely a matter of scientific judgment, there is no "official" classification of *Bacteria* and *Archaea*. Presently, the classification system most widely accepted by microbiologists is the "Taxonomic Outline of the Prokaryotes" arising out of the second edition of *Bergey's Manual of Systematic Bacteriology,* a major taxonomic treatment of *Bacteria* and *Archaea*. See Appendix 2 for a list of genera and higher order groups from *Bergey's Manual*. Widely used, *Bergey's Manual* has served the community of microbiologists since 1923 and is a compendium of information on all recognized species of bacteria. Each chapter, written by experts, contains tables, figures, and other systematic information useful for identification purposes. Volume I of the second edition of *Bergey's Manual* appeared in 2001, Volume 2 in 2005, and three additional volumes in 2007. The second edition of *Bergey's Manual* has incorporated many of the concepts that have emerged from SSU rRNA gene sequencing and genomic studies and blends this with a wealth of phenotypic information as well.

A second major reference in bacterial diversity is *The Prokaryotes,* which provides detailed information on the enrichment, isolation, and culture of the many groups of *Bacteria* and *Archaea* assembled by experts on each microbial group. This work, with more than 4,100 pages (four volumes) in its second edition (1992), is now available in a third edition (2006) of seven volumes. Collectively, *Bergey's Manual* and *The Prokaryotes* offer microbiologists the foundations as well as the details of the taxonomy and phylogeny of *Bacteria* and *Archaea* as we know it today. Typically they are the primary resources for microbiologists characterizing newly isolated organisms.

Culture Collections

National microbial culture collections are an important foundation of microbial systematics (as well as industrial microbiology, ∞ Section 25.1). These permanent collections catalog and store microorganisms and provide them upon request, usually for a fee, to researchers in academia, medicine, and industry. They play an important role as repositories for the natural diversity of bacteria, through the deposition of new kinds of microorganisms by the scientists who discover them. In this way, these collections serve to maintain and protect microbial biodiversity, much in the same way that museums maintain plant and animal specimens for future study. However, microbial culture collections store the deposited microorganisms, not as chemically preserved or dried, dead specimens, but as *viable* cultures, typically frozen at very low temperatures or as a freeze-dried culture. These storage methods maintain the bacteria essentially indefinitely as viable cultures in their original state.

A related and key role of culture collections is as repositories for *type strains*. When a new species of bacteria is described in a scientific journal, a strain is designated as the nomenclatural type of the taxon for future taxonomic comparison with other strains of that species. Deposition of this type strain in the national culture collections of at least two countries, thereby making the strain publicly available, is a prerequisite for validation of the new species name. Some of the large national culture collections are listed in **Table 14.6**. Their web sites contain searchable databases of strain holdings, together with information on the environmental sources of strains and publications on them.

Describing New Species

When a new prokaryote is isolated from nature and thought to be unique, a decision must be made as to whether it is

Table 14.6 Some national microbial culture collections

Collection	*Name*	*Location*	*Web address*
ATCC	American Type Culture Collection	Manassas, Virginia	http://www.atcc.org
BCCM/LMG	Belgium Coordinated Collection of Microorganisms	Ghent, Belgium	http://bccm.belspo.be
CIP	Collection de l'Institut Pasteur	Paris, France	http://www.pasteur.fr
DSMZ	Deutsche Sammlung von Mikroorganismen und Zellkulturen	Braunschweig, Germany	http://www.dsmz.de
JCM	Japan Collection of Microorganisms	Saitama, Japan	http://www.jcm.riken.go.jp
NCCB	Netherlands Culture Collection of Bacteria	Utrecht, The Netherlands	http://www.cbs.knaw.nl/nccb
NCIMB	National Collection of Industrial, Marine and Food Bacteria	Aberdeen, Scotland	http://www.ncimb.com

sufficiently different from other species to be described as novel, or perhaps even sufficiently different from all described genera to warrant description as a new genus (in which case a new species name is automatically created). To achieve formal validation of taxonomic standing as a new genus or species, a detailed description of the organism's characteristics and distinguishing traits, along with its proposed name, is published, and viable cultures of the organism are deposited in at least two international culture collections. The description and new name should be published in the *International Journal of Systematic and Evolutionary Microbiology (IJSEM)*, the official publication of record for the taxonomy and classification of *Bacteria* and *Archaea*. Alternatively, if the new organism is published in a different journal, a copy of the publication and additional supporting information can be submitted to *IJSEM* to request validation of the name and formal acceptance as a new taxon. In each issue, the *IJSEM* publishes an approved list of newly validated names. By providing validation of newly proposed names, publication in *IJSEM* paves the way for their inclusion in *Bergey's Manual of Systematic Bacteriology*. Two web sites provide listings of valid, approved bacterial names: List of Prokaryotic Names with Standing in Nomenclature (**http:// www.bacterio.clct.fr**), and Bacterial Nomenclature Up-to-Date (**http://www.dsmz.de/bactnom/bactname.htm**).

The International Committee on Systematics of Prokaryotes (ICSP) is responsible for overseeing nomenclature and taxonomy of *Bacteria* and *Archaea*. The ICSP oversees the publication of *IJSEM* and the Bacteriological Code, and it guides several subcommittees tasked with setting up minimal standards for the description of new species in the different groups.

14.14 MiniReview

Formal recognition of a new prokaryotic species requires depositing a sample of the organism in culture collections and official publication of the new species name and description. *Bergey's Manual of Systematic Bacteriology* and *The Prokaryotes* are major taxonomic compilations of *Bacteria* and *Archaea*.

- What roles do culture collections play in microbial systematics?
- What is the *IJSEM* and what taxonomic function does it fulfill?
- Why might viable cell cultures be of more use in microbial taxonomy than preserved specimens?

Review of Key Terms

Allele a sequence variant of a given gene

Archaea a group of phylogenetically related prokaryotes distinct from *Bacteria*

Bacteria a group of phylogenetically related prokaryotes distinct from *Archaea*

Binomial system the system devised by Linnaeus for naming living organisms by which an organism is given a genus name and a species epithet

Cladistics phylogenetic methods that group organisms by their evolutionary relationships, not by their phenotypic similarities

Domain in a taxonomic sense, the highest level of biological classification

DNA–DNA hybridization the experimental determination of genomic similarity by measuring the extent of hybridization of DNA from the genome of one organism with that of another

Ecotype a population of genetically identical cells sharing a particular resource within an ecological niche

Endosymbiosis the theory that a chemoorganotrophic bacterium and a cyanobacterium, respectively, were stably incorporated into another cell type to give rise to the mitochondria and chloroplasts of modern-day eukaryotes

Eukarya all eukaryotes: algae, protists, fungi, slime molds, plants, and animals

Evolution descent with modification; DNA sequence variation and the inheritance of that variation

FAME fatty acid methyl ester

FISH fluorescent *in-situ* hybridization

Fitness the capacity of an organism to survive and reproduce as compared to competing organisms

GC ratio in DNA from an organism, the percentage of the total nucleic acid that consists of guanine and cytosine bases

Horizontal gene transfer the transfer of DNA from one cell to another, often distantly related, cell

Molecular clock a gene, such as for ribosomal RNA, whose DNA sequence can be used as a comparative temporal measure of evolutionary divergence

Monophyletic in phylogeny, a group descended from one ancestor

Multilocus sequence typing (MLST) a taxonomic tool for classifying organisms on the basis of gene sequence variations in several housekeeping genes

Phylogenetic probe an oligonucleotide, sometimes made fluorescent by attachment of a dye, complementary in sequence to some ribosomal RNA signature sequence

Phylogeny the evolutionary history of an organism

Phylum a major lineage of cells in one of the three domains of life

Proteobacteria a large group of phylogenetically related gram-negative *Bacteria*

Ribosomal Database Project (RDP) a large database of small subunit ribosomal RNA sequences that can be retrieved electronically and used in comparative ribosomal RNA sequence studies

Ribotyping a means of identifying microorganisms from analysis of DNA fragments generated from restriction enzyme digestion of genes encoding their 16S rRNA

Signature sequence short oligonucleotides of defined sequence in SSU ribosomal RNA characteristic of specific organisms or a group of phylogenetically related organisms; useful for constructing probes

16S ribosomal RNA a large polynucleotide (~1,500 bases) that functions as part of the small subunit of the ribosome of *Bacteria* and *Archaea* and from whose gene sequence evolutionary information can be obtained; eukaryotic counterpart, 18S rRNA

Small subunit (SSU) RNA ribosomal RNA from the 30S ribosomal subunit of *Bacteria* and *Archaea* or the 40S ribosomal subunit of eukaryotes, that is 16S or 18S ribosomal RNA, respectively

Species defined in microbiology as a collection of strains that all share the same major properties and differ in one or more significant properties from other collections of strains; defined phylogenetically as monophyletic, exclusive groups based on DNA sequence

Stromatolite a laminated microbial mat, typically built from layers of filamentous *Bacteria* and other microorganisms, which can become fossilized

Systematics the study of the diversity organisms and their relationships; includes taxonomy and phylogeny

Taxonomy the science of identification, classification, and nomenclature

Universal phylogenetic tree a tree that shows the position of representatives of all domains of living organisms

Review Questions

1. What is the age of planet Earth? When did the oceans form, and what is the age of the earliest known microfossils (Section 14.1)?
2. Under what conditions did life likely originate? What were the steps leading from prebiotic chemistry to living cells? (Section 14.2)?
3. What kind of energy and carbon metabolisms likely characterized early cellular life (Section 14.2)?
4. Why was the evolution of cyanobacteria of such importance to the further evolution of life on Earth? What component of the geological record is used to date the evolution of cyanobacteria (Section 14.3)?
5. What is the hydrogen hypothesis and how does it relate to the endosymbiotic origin of the eukaryotic cell (Section 14.4)?
6. What does the phrase "descent with modification" imply about natural relationships among living organisms (Secton 14.5)?
7. Why are SSU rRNA genes good choices for phylogenetic studies, and what are their limitations (Section 14.6)?
8. What is the fundamental basis for character-state methods of phylogenetic analysis (Section 14.7)?
9. What major evolutionary finding has emerged from the study of rRNA sequences? How did this modify the classic view of evolution? How has this discovery supported previous beliefs on the origin of eukaryotic organisms (Section 14.8)?
10. What major physiological and biochemical properties do *Archaea* share with *Eukarya*? With *Bacteria* (Section 14.8)?
11. What are signature sequences and of what phylogenetic value are they? How are signature sequences discerned (Section 14.9)?
12. What is FISH technology? Give an example of how it would be used (Section 14.9).
13. What major phenotypic and genotypic properties are used to classify organisms in bacterial taxonomy (Sections 14.10 and 14.11)?
14. What is measured in FAME analyses (Section 14.10)?
15. How does 16S rRNA gene sequence analysis differ from multilocus sequence typing as an identification tool (Section 14.11)?
16. How is multi-gene phylogenetic analysis an improvement over analyses based on individual genes (Section 14.12)?
17. How is it thought that new bacterial species arise? How many bacterial species are there? Why don't we know this number more precisely (Section 14.13)?
18. What roles do microbial culture collections play in microbial systematics (Section 14.14)?

Application Questions

1. Compare and contrast the physical and chemical conditions on Earth at the time life first arose with conditions today. From a physiological standpoint, discuss at least two reasons why *animals* could not have existed on early Earth.
2. Why is it highly unlikely that life could originate today as it did billions of years ago?
3. Imagine that you are debating someone who is arguing against the theory of endosymbiosis. What evidence would you use to support the view that endosymbiosis did occur? (You may wish to review Section 14.4 before writing your answer.)
4. In what ways has microbial metabolism altered Earth's biosphere? How might life on Earth be different if oxygenic photosynthesis had not evolved?
5. For the following sequences, identify the phylogenetically informative sites. Identify also the phylogenetically neutral sites and those that are invariant.

   Taxon 1: TCCGTACGTTA
   Taxon 2: TCCCCACGGTT
   Taxon 3: TCGGTACCGTA
   Taxon 4: TCGGTACCGTA
6. Imagine that you are doing lipid analyses of two microorganisms along the lines shown in Figure 14.22. Your results on culture A show an abundance of short-chain unsaturated fatty acids. Analyses of cells in culture B show ether-linked phytanyl lipids to be present. Based on this information, to which phylogenetic domain do organisms A and B belong? Also, if you were told that these organisms were both extremophiles (Section 2.8) and that one originated from a boiling hot spring and one from polar sea ice, which would be which? Finally, based on your lipid analyses and knowledge of where most lipid is located in a cell (Section 3.4 and Figure 3.3), describe how the substances you detected might benefit each organism in thriving in its extreme environment. (You will likely need to review material in Sections 6.12–6.14 before answering.)
7. Imagine that you have been given several bacterial strains from various countries around the world and that all the strains are thought to cause the same gastrointestinal disease and to be genetically identical. Upon carrying out a DNA fingerprint analysis of the strains, you find that four different strain types are present. What methods could you use to test whether the different strains are actually members of the same species?
8. What reference resource(s) would you check for information on the taxonomy and phylogeny of *Bacteria* and *Archaea*? On enrichment, isolation, and culture? If your library has these sources, compare their tables of contents. Which has the greater emphasis on classification and nomenclature?
9. Imagine that you have discovered a new form of microbial life, one that appears to represent a fourth domain. How would you go about characterizing the new organism and determining if it actually is evolutionarily distinct from *Bacteria, Archaea,* and *Eukarya*?

15

Bacteria: The *Proteobacteria*

The large cells of this filamentous sulfur-oxidizing chemolithotroph, Thioploca*, occur in bundles surrounded by a common sheath. Thick mats of* Thioploca *occur on the ocean floor, where they contribute to the ecological cycling of sulfur and nitrogen compounds.*

I THE PHYLOGENY OF *BACTERIA*

In Chapter 14 we examined evolutionary relationships among microorganisms. In this and the next three chapters we expand on these concepts with a discussion of properties of members of the major microbial groups. We begin this tour of the diversity of microorganisms with a focus on *Proteobacteria,* a major group within the *Bacteria* that contains many of the commonly encountered bacteria. Then, in the three chapters that follow, we examine representatives of other major lineages of *Bacteria,* of *Archaea,* and of microbial *Eukarya,* respectively.

With several thousand species of bacteria known, obviously we cannot consider them all. Therefore, using a phylogenetic tree to focus our discussion, we will explore some of the best-known species, particularly ones for which much phenotypic information is available. For more detailed information on prokaryotic diversity the reader is encouraged to review the two major reference sources for these organisms: *Bergey's Manual of Systematic Bacteriology* and *The Prokaryotes* (∞ Section 14.14).

15.1 Phylogenetic Overview of Bacteria

Many major lineages of *Bacteria,* called *phyla,* are known from the study of laboratory cultures, and many others have been identified from retrieval and sequencing of ribosomal RNA (rRNA) genes from microbial communities in natural habitats. **Figure 15.1** gives an overview of the evolutionary relationships of major phyla of *Bacteria* for which laboratory cultures have been obtained. When one includes phyla of *Bacteria* known only from 16S rRNA sequences retrieved from the environment (∞ Section 22.5), well over 80 phyla can be distinguished. However, since little phenotypic information is available on species in which cultures have not yet been obtained, we focus here on phyla with cultured species.

As Figure 15.1 clearly shows, the most phylogenetically ancient (least derived) phylum contains the genus *Aquifex* and relatives, all of which are hyperthermophilic H_2-oxidizing chemolithotrophs. Other "early" phyla such as *Thermodesulfobacterium, Thermotoga,* and the green nonsulfur bacteria (the *Chloroflexus* group), also contain thermophilic species.

Continuing past the green nonsulfur bacteria, we see the deinococci and relatives, the morphologically unique spirochetes, the phototrophic green sulfur bacteria, the chemoorganotrophic *Flavobacterium* and *Cytophaga* groups, the budding *Planctomyces-Pirellula* and the *Verrucomicrobium*

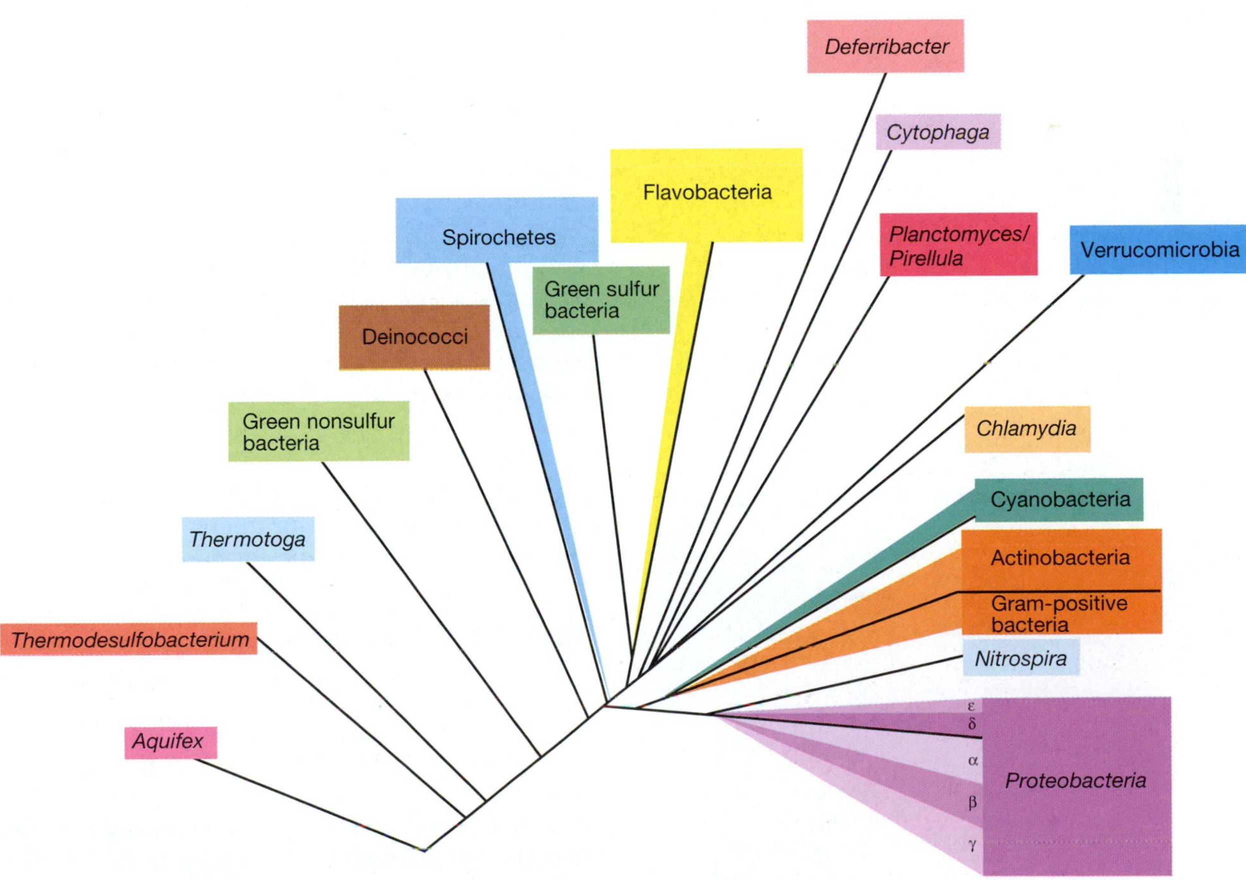

Figure 15.1 Major lineages (phyla) of *Bacteria* based on 16S ribosomal RNA gene sequence comparisons. Over 80 phyla of *Bacteria* are currently known, including many phyla known only from environmental sequences obtained in community sampling.

Table 15.1 Major genera of *Proteobacteria*[a]

Subdivision	Genera	
Alpha	*Acetobacter*	*Nitrobacter*
	Agrobacterium	*Paracoccus*
	Alcaligenes	*Rhodospirillum*
	Azospirillum	*Rhodopseudomonas*
	Beijerinckia	*Rhodobacter*
	Bradyrhizobium	*Rhodomicrobium*
	Brucella	*Rhodovulum*
	Caulobacter	*Rhodopila*
	Ehrlichia	*Rhizobium*
	Gluconobacter	*Rickettsia*
	Hyphomicrobium	*Sphingomonas*
	Methylocystis	*Zymomonas*
Beta	*Aquaspirillum*	*Oxalobacter*
	Bordetella	*Polaromonas*
	Burkholderia	*Ralstonia*
	Chromobacterium	*Rhodocyclus*
	Dechloromonas	*Rhodoferax*
	Gallionella	*Sphaerotilus*
	Leptothrix	*Spirillum*
	Methylophilus	*Thiobacillus*
	Neisseria	*Zoogloea*
	Nitrosomonas	
Gamma	*Acetobacter*	*Photobacterium*
	Acinetobacter	*Pseudomonas*
	Aliivibrio	*Methylococcus*
	Azotobacter	*Methylobacter*
	Chromatium	*Nitrosococcus*
	Escherichia	*Nitrococcus*
	Ectothiorhodospira	*Thermochromatium*
	Erwinia	*Thiomicrospira*
	Francisella	*Thiospirillum* and other purple sulfur bacteria
	Halomonas	*Salmonella* and other enteric bacteria
	Halorhodospira	*Vibrio*
	Halothiobacillus	*Xanthomonas*
	Legionella	
	Leucothrix	
	Methylomonas	
	Oceanospirillum	
Delta	*Acinetobacter*	*Geobacter*
	Aeromonas	*Halomonas*
	Bdellovibrio	*Moraxella*
	Desulfuromonas	*Myxococcus* and other myxobacteria
	Desulfovibrio and most other sulfate-reducing bacteria	*Pelobacter*
	Francisella	*Syntrophobacter*
Epsilon	*Campylobacter*	*Thiovulum*
	Helicobacter	*Wolinella*

[a]This table is not meant to be inclusive but only lists some well-described genera of *Proteobacteria*. For a complete list of genera of *Proteobacteria* and genera of other lineages of *Bacteria*, see Appendix 2.

groups, the *Chlamydia*, and the genera *Nitrospira* and *Deferribacter* (Figure 15.1). Other major groups include the gram-positive bacteria and the cyanobacteria. The gram-positive bacteria are a large group of primarily chemoorganotrophic *Bacteria* and are discussed in detail in Sections 16.1–16.6. They can be separated into two subgroups, called low GC and high GC (*Actinobacteria*), the terms referring to the fact that the species tend to have DNA GC base ratios (Section 14.11) either well below or well above 50%, respectively. The cyanobacteria are oxygenic phototrophic bacteria with evolutionary roots near those of the gram-positive *Bacteria*; these organisms are covered in Sections 16.7 and 16.8.

The *Proteobacteria*

The remaining phylum of cultured *Bacteria*, the ***Proteobacteria*** (Figure 15.1), is by far the largest and most metabolically diverse of all *Bacteria*. *Proteobacteria* constitute the majority of known bacteria of medical, industrial, and agricultural significance. **Table 15.1** lists some key genera. As a group, the *Proteobacteria* are all gram-negative. They show an exceptionally wide diversity of energy-generating mechanisms, with chemolithotrophic, chemoorganotrophic, and phototrophic species. Indeed, we will see in Chapters 20 and 21 the great diversity of energy-generating metabolisms used by representatives of this group. The *Proteobacteria* are equally diverse physiologically, with anaerobic, microaerophilic, and facultatively aerobic forms, for example. Morphologically, also, they exhibit a wide range of cell shapes, including straight and curved rods, cocci, spirilla, and budding and appendaged forms.

Phylogenetically, based on 16S rRNA gene sequences, the phylum *Proteobacteria* is divided into five classes, *Alphaproteobacteria, Betaproteobacteria, Gammaproteobacteria, Deltaproteobacteria,* and *Epsilonproteobacteria* (Figure 15.1 and Table 15.1), each containing many genera and species. Despite this phylogenetic division, members of the different classes of *Proteobacteria* often have physiological traits in common. For example, phototrophy occurs in representatives of three different classes of *Proteobacteria,* and the nitrifying bacteria span four different classes of *Proteobacteria* plus an additional genus that makes up a separate phylum of *Bacteria*! The overlap of these traits should thus serve to remind us that phenotype and phylogeny often give different perspectives on prokaryotes.

We now consider the major groups of *Proteobacteria,* grouping them along some common phenotypic themes.

II PHOTOTROPHIC, CHEMOLITHOTROPHIC, AND METHANOTROPHIC *PROTEOBACTERIA*

Our first group of *Proteobacteria* include prokaryotes able to carry out anoxygenic photosynthesis (Section 20.4), sulfur-, iron-, hydrogen-, or nitrogen-dependent chemolithotrophy (Section 20.4), or the oxidation of methane (CH_4) (Section 21.16). We begin with purple bacteria, classic examples of the phototrophic lifestyle.

Figure 15.2 Photograph of liquid cultures of phototrophic purple bacteria showing the color of species with various carotenoid pigments. The blue culture is a carotenoidless mutant derivative of *Rhodospirillum rubrum* showing that bacteriochlorophyll *a* is blue. The bottle on the far right (*Rhodobacter sphaeroides* strain G) lacks one of the carotenoids of the wild type and thus is greener.

15.2 Purple Phototrophic Bacteria

Key Genera: *Chromatium, Ectothiorhodospira, Rhodobacter, Rhodospirillum*

The purple phototrophic bacteria carry out anoxygenic photosynthesis. Thus, unlike the cyanobacteria (∞ Section 16.7), no O_2 is released. The purple bacteria are a morphologically diverse group, and the classification of these organisms has been established along phylogenetic, morphological, and physiological lines. Different genera fall within the *Alpha-*, *Beta-*, or *Gammaproteobacteria*.

Purple bacteria contain bacteriochlorophylls and carotenoid pigments. Together, these pigments give purple bacteria their spectacular colors, usually purple, red, or brown (**Figure 15.2**). We examine the structure of these pigments and learn how they function in light-mediated energy generation (photophosphorylation) in Section 20.4.

Purple bacteria produce intracytoplasmic photosynthetic membrane systems into which their pigments are inserted. These membranes can be of various morphologies (**Figure 15.3**) but in all cases originate from invaginations of the cytoplasmic membrane. These internal membranes allow purple bacteria to increase the amount of pigment they contain and to thus better utilize the available light. When cells are grown at high light intensities, internal membranes are few and pigment contents are low. By contrast, at low light intensities, the cells are packed with membranes and photopigments.

Purple Sulfur Bacteria

Purple bacteria that utilize hydrogen sulfide (H_2S) as an electron donor for CO_2 reduction in photosynthesis are called **purple sulfur bacteria** (**Table 15.2**). The sulfide is oxidized to elemental sulfur (S^0) that is stored in globules inside the cells (**Figure 15.4**); the sulfur later disappears as it is oxidized to sulfate (SO_4^{2-}). Many purple sulfur bacteria can also use other reduced sulfur compounds as photosynthetic electron

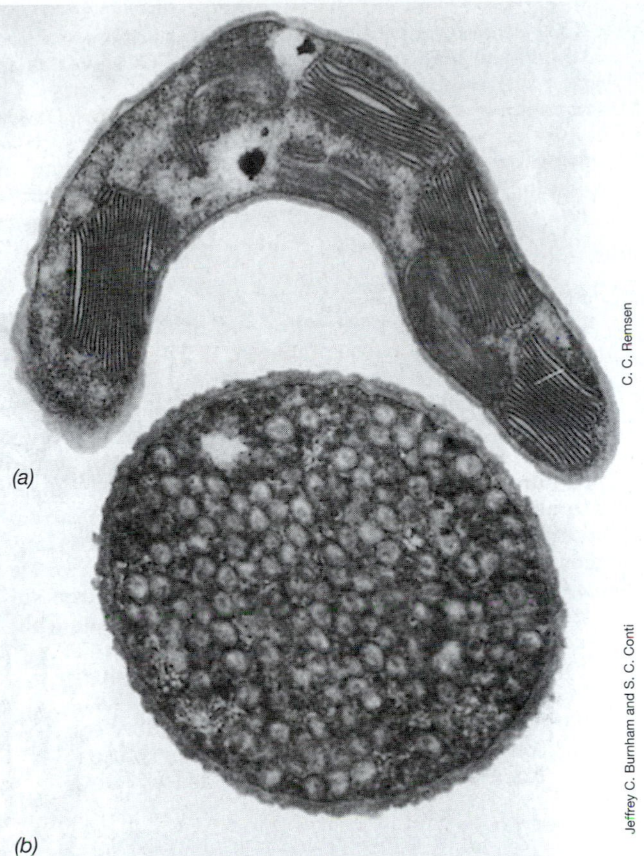

Figure 15.3 Membrane systems of phototrophic purple bacteria as revealed by the electron microscope. *(a) Ectothiorhodospira mobilis*, showing the photosynthetic membranes in flat sheets (lamellae). *(b) Allochromatium vinosum*, showing the membranes as individual, spherical-shaped vesicles.

donors, thiosulfate ($S_2O_3^{2-}$) being a key one commonly used to grow laboratory cultures. All purple sulfur bacteria discovered thus far are *Gammaproteobacteria*.

Purple sulfur bacteria are generally found in illuminated anoxic zones of lakes and other aquatic habitats where H_2S accumulates and also in "sulfur springs," where geochemically or biologically produced H_2S can trigger the formation of blooms of purple sulfur bacteria (**Figure 15.5**). The most favorable lakes for development of purple sulfur bacteria are meromictic (permanently stratified) lakes. Meromictic lakes stratify because they have denser (usually saline) water in the bottom and less dense (usually freshwater) nearer the surface. If sufficient sulfate is present to support sulfate reduction, the sulfide, produced in the sediments, diffuses upward into the anoxic bottom waters, and here purple sulfur bacteria can form dense cell masses, called *blooms,* usually in association with green phototrophic bacteria (Figure 15.5*c*).

The genera *Ectothiorhodospira* and *Halorhodospira* are of special interest. Unlike other purple sulfur bacteria, these organisms oxidize H_2S but produce S^0 outside the cell rather than inside the cell (Figure 15.4*d*). These genera are also interesting because many species are extremely halophilic (salt-loving) or alkaliphilic and are among the most extreme in

UNIT 3

Table 15.2 Genera and characteristics of purple sulfur bacteria[a]

Characteristics	*Genus*
Sulfur deposited externally:	
Spirilla, polar flagella	*Ectothiorhodospira*
Spirilla, extreme alkaliphiles	*Thiorhodospira*
Spirilla, extreme halophiles	*Halorhodospira*
Sulfur deposited internally:	
Do not contain gas vesicles	
Ovals or rods, polar flagella	*Chromatium* *Allochromatium; Halochromatium; Rhabdochromatium; Thermochromatium; Isochromatium; Marichromatium*
Spheres, alkaliphilic	*Thioalkalicoccus*
Spheres, contain bacteriochlorophyll *b*	*Thioflavicoccus*
Spheres	*Thiorhodococcus*
Spheres, diplococci, tetrads, nonmotile; cells 1.2–3 μm in diameter	*Thiocapsa*
Spheres or ovals, polar flagella; cells 2.5-3 μm in diameter	*Thiocystis*
Spheres, 1.5–2.5 μm in diameter	*Thiohalocapsa*
Spheres, 1–2 μm in diameter	*Thiorhodococcus*
Spheres, 1.2–1.5 μm in diameter	*Thiococcus*
Large spirilla, polar flagella	*Thiospirillum*
Small spirilla	*Thiorhodovibrio*
Contain gas vesicles	
Irregular spheres forming platelets of 4–16 cells	*Thiolamprovum*
Rods	*Lamprobacter*
Spheres, ovals, polar flagella	*Lamprocystis*
Rods, nonmotile; forming irregular network	*Thiodictyon*
Spheres, nonmotile; forming flat sheets of tetrads	*Thiopedia*

[a]From a phylogenetic standpoint, all are members of the *Gammaproteobacteria*.

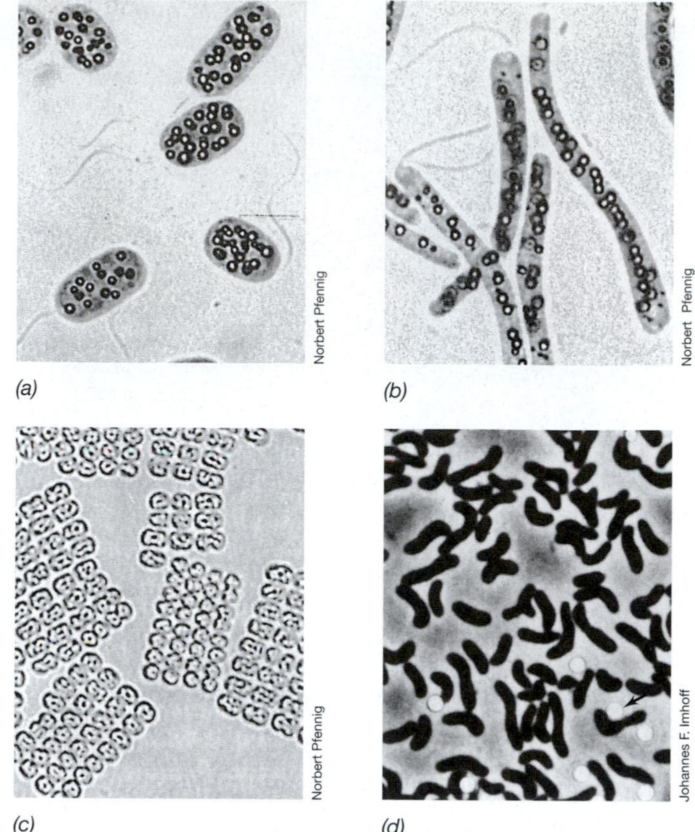

Figure 15.4 Bright-field and phase-contrast photomicrographs of purple sulfur bacteria. *(a) Chromatium okenii*; cells are about 5 μm wide. Note the globules of elemental sulfur inside the cells. *(b) Thiospirillum jenense*, a very large, polarly flagellated spiral; cells are about 30 μm long. Note the sulfur globules. *(c) Thiopedia rosea*; cells are about 1.5 μm wide. *(d)* Phase micrograph of cells of *Ectothiorhodospira mobilis*. Cells are about 0.8 μm wide. Note external sulfur globules (arrow). Compare the photo of *Chromatium okenii* with the drawings of purple sulfur bacteria made by the Russian microbiologist, Sergei Winogradsky, over 120 years ago, shown in Figure 1.18.

these characteristics of all known bacteria. These organisms are typically found in saline lakes, soda lakes, and salterns.

Purple Nonsulfur Bacteria

Some purple bacteria are called **purple nonsulfur bacteria** because it was originally thought that they were unable to use sulfide as an electron donor for the reduction of CO_2 to cell material. In fact, sulfide can be used by most species in this group, although the levels of sulfide ideal for purple sulfur bacteria (1–3 mM) are often toxic to most purple nonsulfur bacteria. Some purple nonsulfur bacteria can also grow anaerobically in the dark using fermentative or anaerobic respiratory metabolism, and most can grow aerobically in darkness by respiration. Under the latter conditions, synthesis of the photosynthetic machinery is repressed by oxygen, and the electron donor can be an organic compound or in some species even an inorganic compound, such as H_2.

However, it is the capacity of this group for *photoheterotrophy* (a condition where light is the energy source and an organic compound is the carbon source) that accounts for their competitive success in nature. Purple nonsulfur bacteria are typically nutritionally diverse, using fatty, organic, or amino acids; sugars; alcohols; or even aromatic compounds like benzoate or toluene as carbon sources. Most species can also grow photoautotrophically with $CO_2 + H_2$ or CO_2 + low levels of H_2S.

Enrichment and isolation of purple nonsulfur bacteria is easy using a mineral salts medium supplemented with an organic acid as carbon source. Such media, inoculated with a mud, lake water, or sewage sample and incubated anoxically in the light, invariably select for purple nonsulfur bacteria. Enrichment cultures can be made even more selective by omitting fixed nitrogen sources (for example, NH_4^+) or organic

(a)

(b)

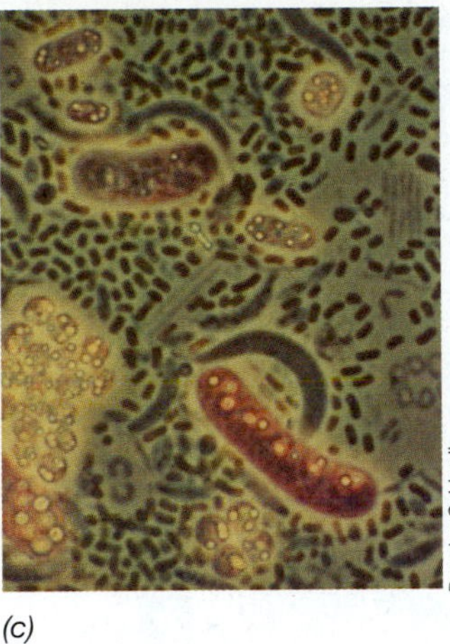

(c)

Figure 15.5 Blooms of purple sulfur bacteria. *(a) Lamprocystis roseopersicina*, in a sulfide spring in Madison, Wisconsin. The bacteria grow near the bottom of the spring pool and float to the top (by virtue of their gas vesicles) when disturbed. The green color is from cells of the eukaryotic alga *Spirogyra* (⚭ Figure 18.38*d*). *(b)* Sample of water from a depth of 7 m in Lake Mahoney, British Columbia. The major organism is *Amoebobacter purpureus*. *(c)* Phase-contrast photomicrograph of layers of purple sulfur bacteria from a small, stratified lake in Michigan. The purple sulfur bacteria include *Chromatium* species (large rods) and *Thiocystis* (small cocci).

nitrogen sources (for example, yeast extract or peptone) from the medium and supplying a gaseous headspace of N_2; virtually all purple nonsulfur bacteria can fix N_2 (⚭ Section 20.14) and will thrive under such conditions, typically outcompeting other bacteria.

The morphological diversity of purple nonsulfur bacteria is typical of that of purple bacteria (**Table 15.3** and **Figure 15.6**), and it is clearly a heterogeneous group in this regard. All purple nonsulfur bacteria isolated thus far are either *Alpha*- or *Betaproteobacteria* (Figure 15.1).

15.2 MiniReview

Purple bacteria are anoxygenic phototrophs that grow phototrophically, obtaining carbon from CO_2 + H_2S (purple sulfur bacteria) or organic compounds (purple nonsulfur bacteria). Purple nonsulfur bacteria are physiologically diverse, and most can grow as chemoorganotrophs in darkness. The purple bacteria reside in the *Alpha*, *Beta*, and *Gamma* classes of the *Proteobacteria*.

- What is meant by the term anoxygenic?
- Give a major reason why purple nonsulfur bacteria do not carry out photosynthesis under aerobic conditions.
- Can purple bacteria grow in the absence of light?

15.3 The Nitrifying Bacteria

Key Genera: *Nitrosomonas, Nitrobacter*

Many species of *Bacteria* are **chemolithotrophs**. Chemolithotrophic bacteria are physiologically united by their ability to utilize *inorganic* electron donors as energy sources (we discuss the conceptual basis of chemolithotrophy in Section 20.8). Most chemolithotrophs are also capable of autotrophic growth and in this way share a major physiological trait with phototrophic bacteria and cyanobacteria. We focus here on the best-studied chemolithotrophs: those capable of oxidizing reduced sulfur or nitrogen compounds or H_2.

Table 15.3 Genera and characteristics of purple nonsulfur bacteria[a]

Characteristics	*Genus*
Alphaproteobacteria	
Spirilla, polarly flagellated	*Rhodospirillum* *Phaeospirillum;* *Rhodovibrio;* *Rhodothalassium;* *Roseospira;* *Rhodospira;* *Roseospirillum*
Rods, polarly flagellated; divide by budding	*Rhodopseudomonas* *Rhodoplanes* *Rhodobium*
Rods; divide by binary fission	*Rhodobacter*
Ovoid to rod-shaped cells	*Rhodovulum*
Ovals, peritrichously flagellated; growth by budding and hypha formation	*Rhodomicrobium*
Large spheres, acidophilic (pH 5 optimum)	*Rhodopila*
Small spheres, alkaliphilic (pH 9 optimum)	*Rhodobaca*
Betaproteobacteria	
Ring-shaped or spirilla	*Rhodocyclus*
Curved rods	*Rubrivivax*
Curved rods	*Rhodoferax*

[a]All are members of the *Proteobacteria* (see Figure 15.1 and Table 15.1).

UNIT 3

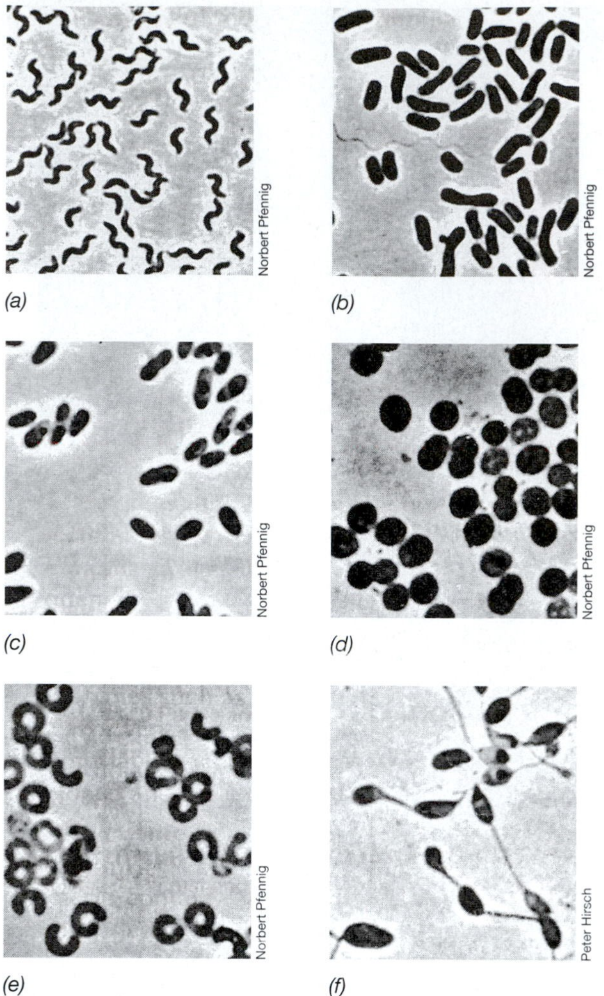

Figure 15.6 Representatives of several genera of purple nonsulfur bacteria. *(a) Phaeospirillum fulvum*; cells are about 3 μm long. *(b) Rhodoblastus acidophilus*; cells are about 4 μm long. *(c) Rhodobacter sphaeroides*; cells are about 1.5 μm wide. *(d) Rhodopila globiformis*; cells are about 1.6 μm wide. *(e) Rhodocyclus purpureus*; cells are about 0.7 μm in diameter. *(f) Rhodomicrobium vannielii*; cells are about 1.2 μm wide. See also Table 15.3.

Ammonia- and Nitrite-Oxidizers

Bacteria able to grow chemolithotrophically at the expense of reduced inorganic nitrogen compounds are called **nitrifying bacteria**. Several genera are recognized on the basis of morphology and phylogeny and the particular steps in the oxidation sequences that they carry out (**Table 15.4**). Phylogenetically, the nitrifying bacteria are scattered among four of the *Proteobacteria* classes: *Alpha*, *Beta*, *Gamma*, and *Delta*. The genus *Nitrospira* forms its own phylum of *Bacteria* (Figure 15.1 and see Section 15.38) and is related to other nitrifying bacteria in a metabolic sense only.

No chemolithotroph is known that carries out the complete oxidation of ammonia to nitrate. Thus, nitrification in nature results from the sequential action of two separate groups of organisms, the ammonia-oxidizing bacteria, sometimes called the *nitrosifyers* (**Figure 15.7**), and the nitrite-oxidizing bacteria, the actual nitrate-producing bacteria (**Figure 15.8**). Nitrosifying bacteria typically have genus names beginning in "Nitroso," while genus names of nitrate producers begin with "Nitro." For example, *Nitrosomonas* and *Nitrobacter* are major genera of the two groups of nitrifying bacteria, respectively (Table 15.4). Historically, the nitrifying bacteria were the first organisms to be shown to grow chemolithotrophically. Winogradsky showed that they were able to produce organic matter and cell mass when provided with CO_2 as sole carbon source (Microbial Sidebar, Chapter 20).

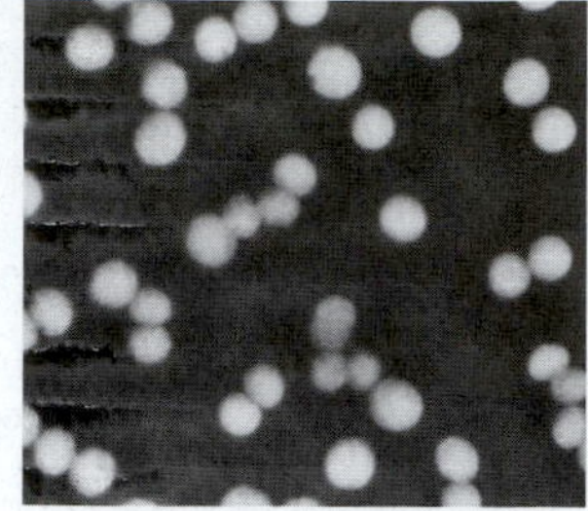

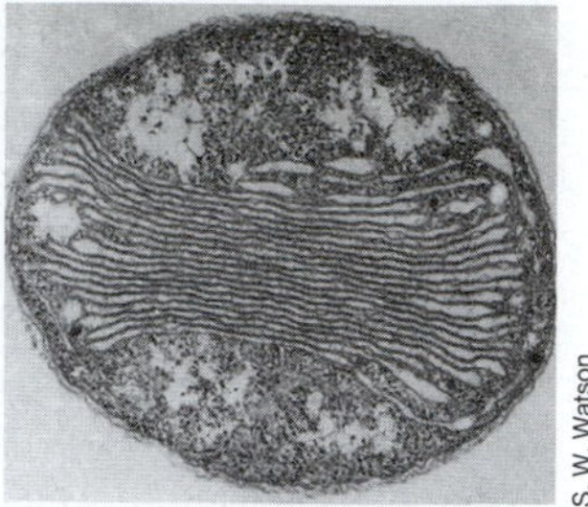

Figure 15.7 Phase-contrast photomicrograph (left) and electron micrograph (right) of the nitrosifying bacterium *Nitrosococcus oceani*. A single cell is about 2 μm in diameter.

Many species of nitrifying bacteria have internal membrane systems that closely resemble the photosynthetic membranes found in their close phylogenetic relatives, the purple phototrophs (Section 15.2) and the methane-oxidizing (methanotrophic) bacteria (Section 15.6). The membranes are the location of key enzymes in nitrification: ammonia monooxygenase, which oxidizes NH_3 (ammonia) to NH_2OH (hydroxylamine), and nitrite oxidase, which oxidizes NO_2^- (nitrite) to NO_3^-. Hydroxylamine is oxidized to NO_2^- by the nitrosifying bacteria, generating the substrate for the nitrite-oxidizing bacteria (Section 20.12).

Ecology, Isolation, and Culture

The nitrifying bacteria are widespread in soil and water. They are present in highest numbers in habitats where considerable amounts of ammonia are present, such as sites with extensive protein decomposition (ammonification), and also in sewage treatment facilities (Section 36.2). Nitrifying bacteria develop especially well in lakes and streams that

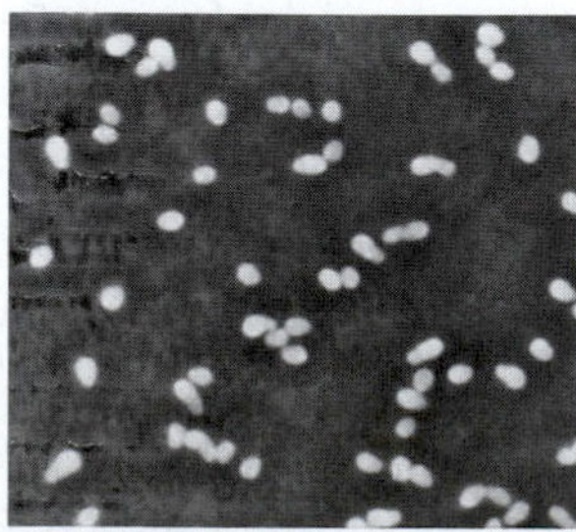

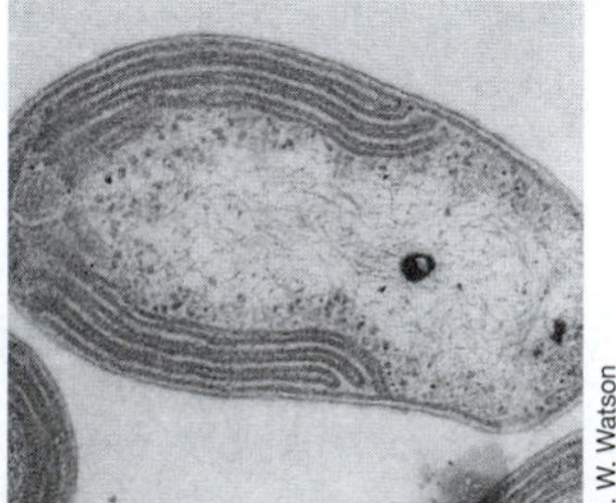

Figure 15.8 Phase-contrast photomicrograph (left) and electron micrograph (right) of the nitrifying bacterium *Nitrobacter winogradskyi*. A cell is about 0.7 μm in diameter.

Table 15.4 Characteristics of the nitrifying bacteria

Characteristics	*Genus*	*Phylogenetic group*[a]	*Habitats*
Oxidize ammonia:			
Gram-negative short to long rods, motile (polar flagella) or nonmotile; peripheral membrane systems	*Nitrosomonas*	*Beta*	Soil, sewage, freshwater, marine
Large cocci, motile; vesicular or peripheral membranes	*Nitrosococcus*	*Gamma*	Freshwater, marine
Spirals, motile (peritrichous flagella); no obvious membrane system	*Nitrosospira*	*Beta*	Soil, freshwater
Pleomorphic, lobular, compartmented cells; motile (peritrichous flagella)	*Nitrosolobus*	*Beta*	Soil
Oxidize nitrite:			
Short rods, reproduce by budding, occasionally motile (single subterminal flagellum); membrane system arranged as a polar cap	*Nitrobacter*	*Alpha*	Soil, freshwater, marine
Long, slender rods, nonmotile; no obvious membrane system	*Nitrospina*	*Delta*	Marine
Large cocci, motile (one or two subterminal flagella); membrane system randomly arranged in tubes	*Nitrococcus*	*Gamma*	Marine
Helical to vibrioid-shaped cells, nonmotile; no internal membranes	*Nitrospira*	*Nitrospira* group	Marine, soil

[a]Phylogenetically, all nitrifying bacteria thus far examined are *Proteobacteria*, except for *Nitrospira*, which constitutes its own phylogenetic lineage (Figure 15.1), or certain marine crenarchaeotes (⚭ Section 17.12).

receive inputs of sewage or other wastewaters because these are frequently high in ammonia (⚭ Figure 23.9).

Enrichment cultures of nitrifying bacteria can be achieved using mineral salts media containing ammonia or nitrite as the electron donor and bicarbonate (HCO_3^-) as the sole carbon source. Because of the inefficient growth of these organisms (⚭ Section 20.12), visible turbidity may not develop even after extensive nitrification has occurred. An easy means of monitoring growth is thus to assay for the production of nitrite (with ammonia as electron donor) or the disappearance of nitrite or the production of nitrate (with nitrite as electron donor). Most of the nitrifying bacteria are obligate chemolithotrophs and obligate aerobes. Species of *Nitrobacter* are an exception and are able to grow chemoorganotrophically on acetate or pyruvate as the sole carbon and energy source. One group, the *anammox* organisms (⚭ Section 20.13), are phylogenetically distinct and oxidize ammonia anaerobically.

15.4 Sulfur- and Iron-Oxidizing Bacteria

Key Genera: *Thiobacillus, Acidithiobacillus, Achromatium, Beggiatoa*

The ability to grow chemolithotrophically on reduced sulfur compounds is a property of a diverse group of *Proteobacteria* (**Table 15.5**). Two broad ecological classes of sulfur-oxidizing bacteria exist, those living at neutral pH and those living at acidic pH. Some of the acidophiles also have the ability to grow chemolithotrophically using ferrous iron (Fe^{2+}) as an electron donor. We discuss the biogeochemistry of acidophilic sulfur- and iron-oxidizing bacteria in Sections 24.4–24.6 and the biochemistry of these processes in Sections 20.10 and 20.11.

Thiobacillus and *Achromatium*

The genus *Thiobacillus* and close relatives contain several gram-negative, rod-shaped bacteria, indistinguishable morphologically from most other gram-negative rods (**Figure 15.9*a***); they are the best studied of the sulfur chemolithotrophs. Phylogenetically, species of thiobacilli are scattered among the *Proteobacteria*, with different species residing in the *Alpha*, *Beta*, and *Gamma* classes (Table 15.5). The sulfur compounds most commonly used as electron donors in chemolithotrophic metabolism of thiobacilli are H_2S, S^0, and $S_2O_3^{2-}$. Large amounts of energy are available from the reactions of these compounds, and we will see in Section 20.10 that some of this energy can be trapped as ATP from electron transport reactions that lead to a proton motive force. Moreover, the use of these substrates generates large amounts of sulfuric acid, and thus several thiobacilli are acidophilic. One acidophilic species, *Acidithiobacillus ferrooxidans*, can also grow chemolithotrophically by the oxidation of ferrous iron and is a major biological agent for the oxidation of this metal. Iron pyrite (FeS_2) is a major source of Fe^{2+} as well as sulfide. The oxidation of FeS_2, especially in mining operations, can be both beneficial (because leaching of the ore releases the iron from the sulfide mineral) and ecologically disastrous (the environment is acidified and contaminated with other heavy metals associated with the pyrite) (⚭ Sections 24.5 and 24.6).

Achromatium is a spherical sulfur-oxidizing chemolithotroph that is common in freshwater sediments containing sulfide. Cells of *Achromatium* are large cocci that can have diameters of 10–100 μm (Figure 15.9*b*). Phylogenetic analyses of natural populations of *Achromatium* have shown that several species likely exist (probably each of distinct size), although pure cultures of this organism have not yet been

UNIT 3

Table 15.5 Physiological characteristics of sulfur-oxidizing chemolithotrophic prokaryotes

Genus and species	*Inorganic electron donor*	*Range of pH for growth*	*Phylogenetic group*[a]
Species growing poorly if at all in organic media:			
Thiobacillus thioparus	H_2S, sulfides, S^0, $S_2O_3^{2-}$	6–8	*Beta*
Thiobacillus denitrificans[b]	H_2S, S^0, $S_2O_3^{2-}$	6–8	*Beta*
Halothiobacillus neapolitanus	S^0, $S_2O_3^{2-}$	6–8	*Gamma*
Acidithiobacillus thiooxidans	S^0	2–4	*Gamma*
Acidithiobacillus ferrooxidans	S^0, metal sulfides, Fe^{2+}	2–4	*Gamma*
Species growing well in organic media:			
Starkeya novella	$S_2O_3^{2-}$	6–8	*Alpha*
Thiomonas intermedia	$S_2O_3^{2-}$	3–7	*Beta*
Filamentous sulfur chemolithotrophs:			
Beggiatoa	H_2S, $S_2O_3^{2-}$	6–8	*Gamma*
Thiothrix	H_2S	6–8	*Gamma*
Thioploca[c]	H_2S, S^0	—	*Gamma*
Other genera:			
Achromatium	H_2S	—	*Gamma*
Thiomicrospira	$S_2O_3^{2-}$, H_2S	6–8	*Gamma*
Thiosphaera[d]	H_2S, $S_2O_3^{2-}$, H_2	6–8	*Alpha*
Thermothrix	H_2S, $S_2O_3^{2-}$, SO_3^{-}	6.5–7.5	*Beta*
Thiovulum	H_2S, S^0	6–8	*Epsilon*

[a]All are *Proteobacteria*.
[b]Facultative aerobes; use NO_3^- as electron acceptor anaerobically.
[c]Pure cultures not yet available.
[d]*Thiosphaera pantotropha* has the same 16S rRNA gene sequence as *Paracoccus denitrificans*.

achieved. Phylogenetically, *Achromatium* belongs to the *Gammaproteobacteria* and is specifically related to phototrophic purple bacteria, such as its phototrophic counterpart *Chromatium* (Section 15.2 and Figure 15.4*a*). Like *Chromatium*, cells of *Achromatium* store elemental sulfur internally (Figure 15.9*b*); the granules later disappear as sulfur is oxidized to sulfate. Cells of *Achromatium* also store large granules of calcite ($CaCO_3$) (Figure 15.9*b*), possibly as a carbon source for autotrophic growth.

Culture

Some sulfur chemolithotrophs are obligate chemolithotrophs, locked into a lifestyle of using inorganic instead of organic compounds as electron donors. When growing in this fashion, they are also autotrophs, converting CO_2 into cell material by reactions of the Calvin cycle (∞ Section 20.6). **Carboxysomes** are often present in cells of obligate chemolithotrophs (Figure 15.9*a*). These structures contain high levels of Calvin cycle enzymes and probably increase the rate at which these organisms can fix CO_2.

Other sulfur chemolithotrophs are facultative chemolithotrophs, facultative in the sense that they can grow either chemolithotrophically (and thus, also as autotrophs) or chemoorganotrophically (Table 15.5). Most species of *Beggiatoa*, however, can obtain energy from the oxidation of inorganic sulfur compounds but lack enzymes of the Calvin cycle. They thus require organic compounds as carbon sources. Such a nutritional lifestyle is called **mixotrophy**.

Beggiatoa

Organisms of this genus are filamentous, gliding, sulfur-oxidizing bacteria. Filaments of *Beggiatoa* are usually large in diameter and long, consisting of many short cells attached end to end (**Figure 15.10**). Filaments then flex and twist so that many filaments may become intertwined to form a complex tuft.

Beggiatoa is found in nature primarily in habitats rich in H_2S, such as sulfur springs (Figure 15.10*b*), decaying seaweed beds, mud layers of lakes, and waters polluted with sewage. In such environments, filaments of *Beggiatoa* are typically filled with sulfur granules (Figure 15.10*a*). *Beggiatoa* are also common inhabitants of hydrothermal vents (∞ Section 24.11). It was with *Beggiatoa* that Winogradsky first demonstrated that a living organism could oxidize H_2S to S^0 and then to SO_4^{2-}, leading him to formulate the concept of chemolithotrophy (∞ Microbial Sidebar, Chapter 20). Although a few strains of *Beggiatoa* are truly chemolithotrophic autotrophs, most grow best mixotrophically with reduced sulfur compounds as electron donors and organic compounds as carbon sources.

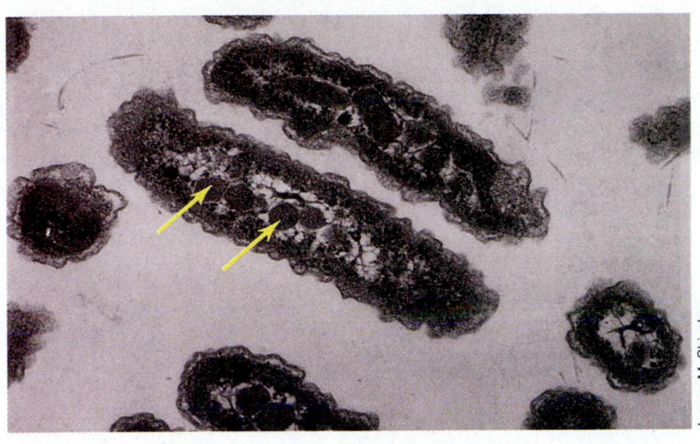

(a)

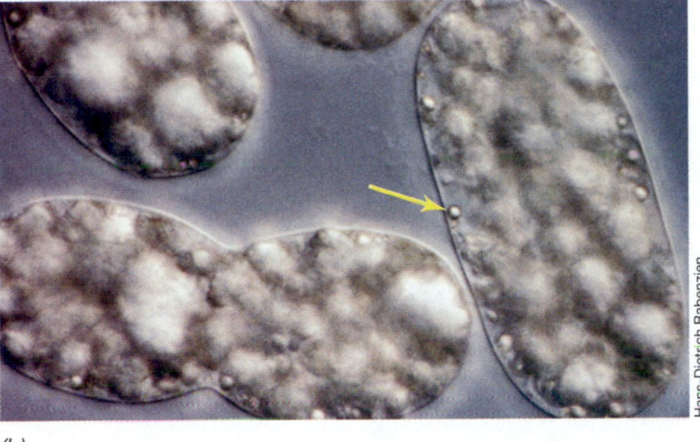

(b)

Figure 15.9 Nonfilamentous sulfur chemolithotrophs. *(a)* Transmission electron micrograph of cells of the chemolithotrophic sulfur oxidizer *Halothiobacillus neapolitanus*. A single cell is about 0.5 μm in diameter. Note the polyhedral bodies (carboxysomes) distributed throughout the cell (arrows). *(b) Achromatium.* Cells isolated from a small lake in Germany and photographed by Nomarski light microscopy. The small globular structures near the periphery of the cells (arrow) are elemental sulfur, and the large granules consist of calcium carbonate. A single *Achromatium* cell is about 25 μm in diameter.

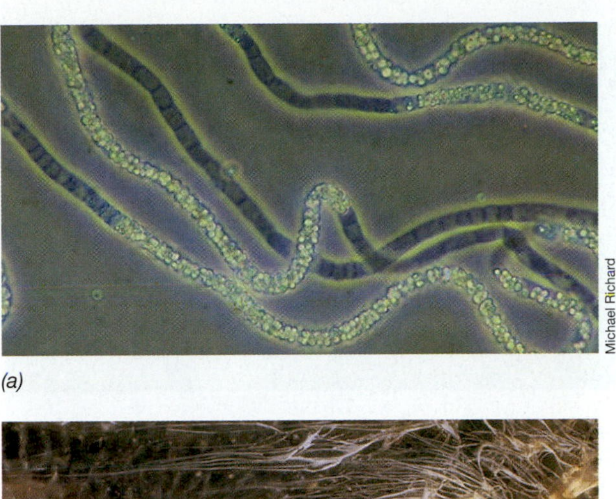

(a)

(b)

Figure 15.10 Filamentous sulfur-oxidizing bacteria. *(a)* Phase-contrast photomicrograph of a *Beggiatoa* species isolated from a sewage treatment plant. Note the abundant elemental sulfur granules in some of the cells. *(b)* Sulfur-oxidizing bacteria in the outflow of a small sulfide spring. The filamentous cells twist together to form thick streamers, and the white color is due to their abundant elemental sulfur content.

An interesting habitat of *Beggiatoa* is the rhizosphere of plants (rice, cattails, and other swamp plants) living in flooded, and hence anoxic, soils. Such plants pump oxygen down into their roots so a sharply defined oxic–anoxic boundary develops between the root and the soil. *Beggiatoa* (and probably other sulfur bacteria) develops at this boundary and plays a beneficial role for the plant by oxidizing (and thus detoxifying) the H_2S.

Beggiatoa and other filamentous bacteria such as *Sphaerotilus* (Section 15.14) can cause major settling problems in sewage treatment facilities and in industrial waste lagoons from canning, paper pulping, brewing, or milling operations. This condition is called *bulking* and occurs when filamentous bacteria such as *Beggiatoa* overgrow the normal flora of the waste system, producing a loose detrital floc instead of the normal and more easily settling tight floc containing organisms such as *Zoogloea* and related sewage bacteria (∞ Section 36.2). If bulking occurs, the wastewater remains improperly treated because the effluent discharged remains high in organic matter and inorganic nutrients (∞ Section 36.2).

Thioploca and *Thiothrix*

Other filamentous sulfur-oxidizing bacteria include *Thioploca* and *Thiothrix*. *Thioploca* is a large, filamentous sulfur-oxidizing chemolithotroph that forms cell bundles surrounded by a common sheath (**Figure 15.11**). Thick mats of a marine *Thioploca* species have been found on the ocean floor off the coast of Chile and Peru. Studies on the ecology of these organisms have shown that they carry out the anoxic oxidation of H_2S coupled with the reduction of nitrate (NO_3^-) to ammonium (NH_4^+) (∞ Section 21.7). Cells of *Thioploca* can accumulate huge amounts of nitrate intracellularly, and this nitrate can then support extended periods of anaerobic respiration with H_2S as electron donor. It is thought that these *Thioploca* mats fix substantial amounts of CO_2 and also play a major role in sulfur and nitrogen cycling in the marine environment. Microbial mats consisting primarily of *Beggiatoa* are also found near hydrothermal vents (∞ Section 24.11); however, the connection with nitrate respiration here is not as well established.

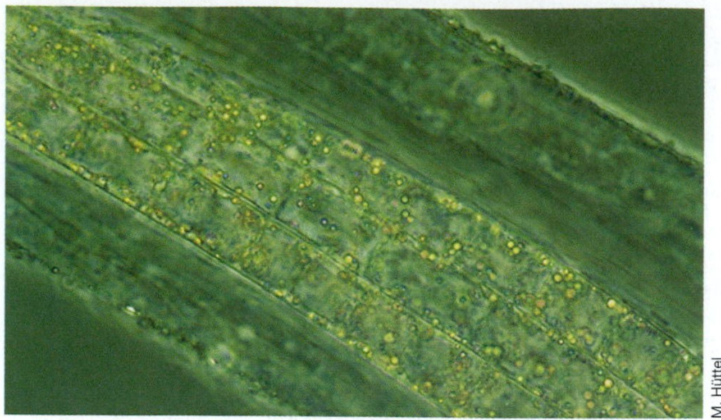

Figure 15.11 Cells of a large marine *Thioploca* species. Cells contain sulfur granules (yellow) and are about 40–50 μm wide.

Thiothrix is a filamentous sulfur-oxidizing organism in which the filaments group together at their ends by way of a holdfast to form cell arrangements called rosettes (**Figure 15.12**). Physiologically, *Thiothrix* is an obligately aerobic mixotroph, and in this and most other respects it resembles *Beggiatoa*.

15.5 Hydrogen-Oxidizing Bacteria

Key Genera: *Ralstonia, Paracoccus*

Many bacteria can grow with H_2 as sole electron donor and O_2 as electron acceptor using the "knallgas" reaction, the reduction of O_2 with H_2, as their energy metabolism:

$$H_2 + \tfrac{1}{2}O_2 \rightarrow H_2O \qquad \Delta G^{0\prime} = -237\,\text{kJ}$$

Most of these organisms can also grow autotrophically (using reactions of the Calvin cycle to incorporate CO_2) and are grouped together here as the chemolithotrophic hydrogen-oxidizing bacteria. Both gram-positive and gram-negative hydrogen bacteria are known, with the best-studied representatives classified in the genera *Ralstonia* (**Figure 15.13**), *Pseudomonas,* and *Paracoccus* (**Table 15.6**). Different hydrogen bacteria are scattered among the *Alpha, Beta,* and *Gamma* classes of *Proteobacteria*. All hydrogen-oxidizing bacteria contain one or more hydrogenase enzymes that function to bind H_2 and use it to either produce ATP (Section 20.9) or for reducing power for autotrophic growth (Table 15.6).

Almost all hydrogen bacteria are facultative chemolithotrophs, meaning that they can also grow chemoorganotrophically with organic compounds as energy sources. This is a major distinction between hydrogen chemolithotrophs and many sulfur chemolithotrophs or nitrifying bacteria. Most representatives of these groups are *obligate* chemolithotrophs—they cannot grow in the absence of the inorganic energy source. By contrast, hydrogen chemolithotrophs can switch between chemolithotrophic and chemoorganotrophic modes of metabolism as nutritional conditions in their habitats warrant.

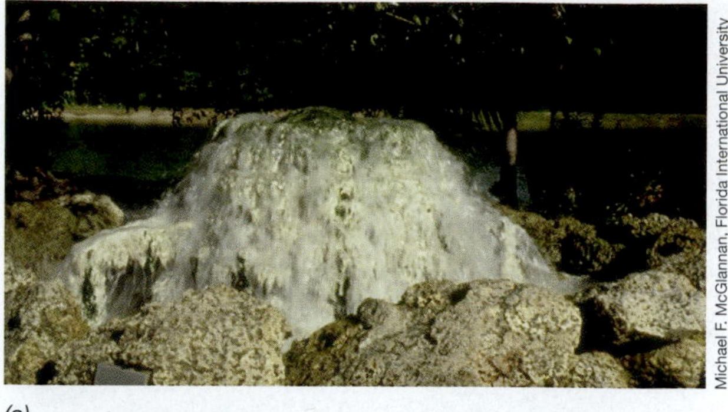

(a)

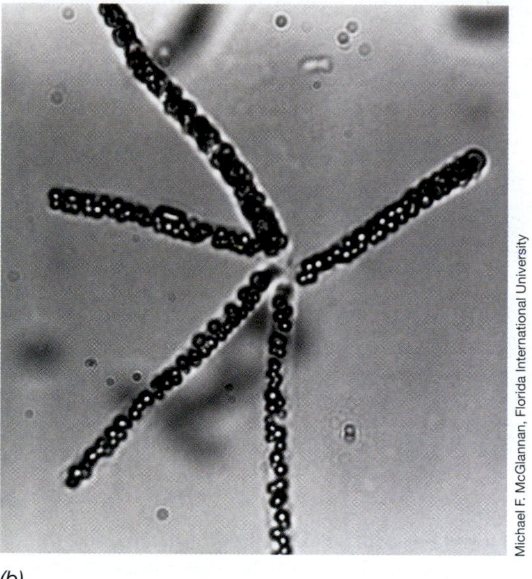

(b)

Figure 15.12 ***Thiothrix.*** *(a)* A sulfide-containing artesian spring in Florida (USA). The outside of the spring is coated with a mat of *Thiothrix*. The mat is about 1.5 m in diameter. *(b)* Phase-contrast photomicrograph of a rosette of cells of *Thiothrix* isolated from the spring and grown in pure culture. Note the internal sulfur globules produced from the oxidation of sulfide. Each filament is about 4 μm in diameter.

Physiology and Ecology of Hydrogen Bacteria

When growing chemolithotrophically on H_2, most hydrogen bacteria grow best under microoxic (5–10% O_2) conditions because hydrogenases are typically oxygen-sensitive. The element nickel must be present in the medium for chemolithotrophic growth of hydrogen bacteria because virtually all hydrogenases contain Ni^{2+} as a key metal cofactor. A few hydrogen bacteria also fix molecular nitrogen, making possible their culture in a mineral salts medium supplied with only H_2, O_2, CO_2, and N_2 as carbon, energy, and nitrogen sources!

Hydrogen-oxidizing bacteria can be enriched if a small amount of mineral salts medium containing trace metals (especially Ni^{2+} and Fe^{2+}) is inoculated with soil or water and incubated in a large, sealed flask containing a head space of 5% O_2, 10% CO_2, and 85% H_2. When the liquid becomes turbid, plates of the same medium are streaked and incubated in a glass jar containing the same gas mixture (one must exercise care here, as mixtures of O_2 and H_2 are potentially explosive).

CO Oxidation

Some hydrogen bacteria can grow on carbon monoxide, CO, as electron donor. Electrons from the oxidation of CO to CO_2 enter the electron transport chain and form a proton motive force. CO-oxidizing bacteria, called *carboxydotrophic* bacteria, grow autotrophically using the Calvin cycle (Section 20.6) to fix CO_2 generated from the oxidation of CO. Carbon monoxide is oxidized to CO_2 by the enzyme carbon monoxide dehydrogenase, which is a molybdenum-containing enzyme. The molybdenum in CO dehydrogenase is bound to a small cofactor consisting of a multiringed structure called a *pterin*, a situation similar to that of the enzyme nitrate reductase (Section 21.7).

CO consumption by carboxydotrophic bacteria in nature is a significant ecological process. Although much CO is generated from human and other sources, CO levels in air have not risen significantly over many years. Because the most significant releases of CO (primarily from automobile exhaust, incomplete combustion of fossil fuels, and the catabolism of lignin) are in oxic environments, carboxydotrophic bacteria in the upper layers of soil probably represent the most significant sink for CO in nature.

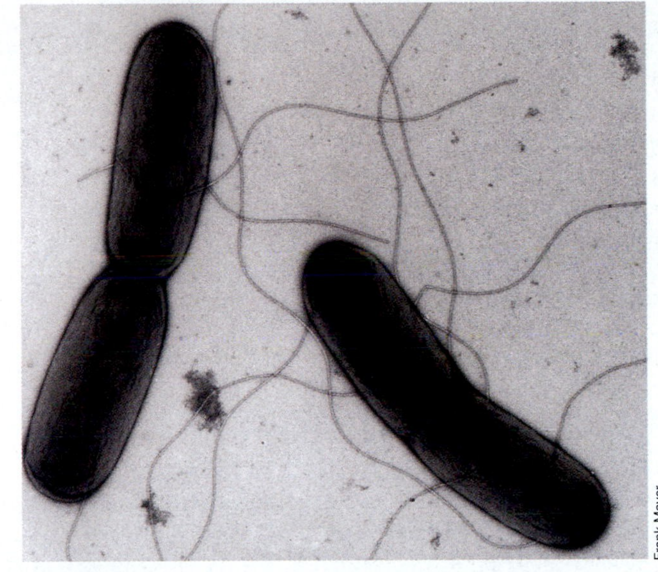

Figure 15.13 Hydrogen bacteria. Transmission electron micrograph of negatively stained cells of the hydrogen-oxidizing chemolithotroph *Ralstonia eutropha*. A cell is about 0.6 μm in diameter and contains several flagella.

Table 15.6 Differential characteristics of a few common species of hydrogen-oxidizing bacteria

Genus and species	*Denitrification*	*Growth on fructose*	*Motility*	*Phylogenetic group*[a]	*Other characteristics*
Gram-negative					
Acidovorax facilis	−	+	+	Beta	Membrane-bound hydrogenase
Ralstonia eutropha	+	+	+	Beta	Membrane-bound and cytoplasmic hydrogenases
Achromobacter xylosoxidans	−	+	+	Beta	Membrane-bound and cytoplasmic hydrogenases
Aquaspirillum autotrophicum	−	−	+	Beta	Membrane-bound hydrogenase
Pseudomonas carboxydovorans	−	−	+	Gamma	Membrane-bound hydrogenase; also oxidizes CO
Hydrogenophaga flava	−	+	+	Beta	Colonies are bright yellow
Paracoccus denitrificans	+	+	−	Alpha	Membrane-bound hydrogenase; strong denitrifier
Aquifex pyrophilus	+	−	+	*Aquifex* group[b]	Hyperthermophile, grows microaerophilically or anaerobically (with NO_3^-), obligate chemolithotroph; also uses S^0 or $S_2O_3^{2-}$ as electron donor
Hydrogenobacter thermophilus	−	−	−	*Aquifex* group[b]	As for *Aquifex*, but obligate aerobe (microaerophile)
Gram-positive					
Bacillus schlegelii	−	−	+	Low GC gram-positive[c]	Produces endospores; thermophile; also uses CO or $S_2O_3^{2-}$ as electron donor
Arthrobacter sp.	−	+	−	Actinobacteria[d]	Membrane-bound hydrogenase
Mycobacterium gordonae	−	?	−	Actinobacteria[d,e]	Acid-fast; colonies yellow to orange

[a]Aerobic hydrogen bacteria are *Proteobacteria* except as indicated.
[b]See Section 16.20.
[c]See Section 16.2.
[d]See Section 16.4.
[e]See Section 16.5.

Table 15.7 Substrates used by methylotrophic bacteria[a]

I. Substrates used for growth	
Methane, CH_4[b]	Formate, $HCOO^-$
Methanol, CH_3OH	Formamide, $HCONH_2$
Methylamine, CH_3NH_2	Carbon monoxide, CO
Dimethylamine, $(CH_3)_2NH$	Dimethyl ether, $(CH_3)_2O$
Trimethylamine, $(CH_3)_3N$	Dimethyl carbonate, $CH_3OCOOCH_3$
Tetramethylammonium, $(CH_3)_4N^+$	Dimethyl sulfoxide, $(CH_3)_2SO$
Trimethylamine N-oxide, $(CH_3)_3NO$	Dimethylsulfide, $(CH_3)_2S$
Trimethylsulfonium, $(CH_3)_3S^+$	
II. Substrates oxidized but not used for growth	
Ammonium, NH_4^+	Bromomethane, CH_3Br
Ethylene, $H_2C{=}CH_2$	Higher hydrocarbons (ethane, propane)
Chloromethane, CH_3Cl	

[a]A single isolate does not use all of the above, but at least one methylotrophic bacterium has been reported to oxidize each of the listed compounds.
[b]Methylotrophs able to oxidize methane are called *methanotrophs*.

At least one carboxydobacterium can grow on CO anaerobically with nitrate as electron acceptor, but this does not seem to be a widespread property of the group. Like the hydrogen bacteria, virtually all isolates of carboxydotrophic bacteria also grow chemoorganotrophically on organic compounds as well as on CO.

15.3–15.5 MiniReview

Chemolithotrophs are bacteria that can oxidize inorganic electron donors and in many cases use CO_2 as sole carbon source.

- Compare and contrast the nitrifying bacteria with the sulfur, iron, and hydrogen bacteria in terms of inorganic electron donors and carbon source used, as well as their major habitats.
- What common pathway is present for assimilation of CO_2 in many chemolithotrophs?

15.6 Methanotrophs and Methylotrophs

Key Genera: *Methylomonas, Methylobacter*

Methane (CH_4) is found extensively in nature. It is produced in anoxic environments by methanogenic *Archaea* (Sections 17.4 and 21.10) and is a major gas of anoxic muds, marshes (Figure 17.5), anoxic zones of lakes, the rumen, and the mammalian intestinal tract. Methane is the major constituent of natural gas and is also present in many coal formations.

Methane is a chemically stable molecule, but some bacteria, the **methanotrophs**, oxidize it readily. These bacteria use methane and a few other one-carbon compounds as electron donors for energy generation and as sole sources of carbon. These bacteria are all aerobes and are widespread in nature in soil and water. They exhibit diverse morphologies and are related both phylogenetically and in their ability to oxidize methane. Methane is scarce in marine sediments, where the major form of respiration is anaerobic sulfate reduction, not methanogenesis (Section 24.2).

C_1 Metabolism

A list of compounds catabolized by methanotrophs is given in **Table 15.7**. From a biochemical viewpoint, these compounds share a key characteristic: they lack carbon–carbon bonds. Thus, all organic compounds in the cell must be synthesized from C_1 precursors. Organisms that can grow using carbon compounds that lack C—C bonds are called **methylotrophs**. Many, but not all methylotrophs are also methanotrophs. However, methanotrophs are unique in that they can grow not only on some of the more oxidized one-carbon compounds (Table 15.7), but also on methane.

Methanotrophs possess a key enzyme, methane monooxygenase, that incorporates an atom of oxygen from O_2 into methane, forming methanol (Section 21.16). The requirement for O_2 as a reactant in the initial oxygenation of methane thus explains why these methanotrophs are obligate aerobes (we will see in Section 21.13 that methane can also be oxidized under anoxic conditions by a consortium of special organisms). All methanotrophs are obligate C_1 utilizers, unable to use compounds containing carbon–carbon bonds. By contrast, many nonmethanotrophic methylotrophs can use organic acids, ethanol, and sugars.

Methane-oxidizing bacteria are unique among bacteria in possessing relatively large amounts of sterols. As we noted in Section 4.3, sterols are found in eukaryotes as a functional part of the membrane system but are absent from most bacteria. We will see that in methanotrophs, sterols may be an essential part of the complex internal membrane system for methane oxidation. The only other group of bacteria in which sterols are extensively distributed is in the cell wall-less mycoplasmas (Section 16.3).

Classification of Methanotrophs

Table 15.8 gives a taxonomic overview of the methanotrophs. These bacteria were initially distinguished on the basis of morphology and formation of resting stages. However, now they are classified into two major groups, based on their internal cell structure, phylogeny, and carbon assimilation pathway. *Type I methanotrophs* assimilate one-carbon compounds via the ribulose monophosphate cycle and are phylogenetically *Gammaproteobacteria*. By contrast, *type II methanotrophs* assimilate C_1 intermediates via the serine pathway and are phylogenetically *Alphaproteobacteria* (Table 15.8). We discuss the biochemical details of these pathways in Section 21.16.

Both groups of methanotrophs contain extensive internal membrane systems for methane oxidation. Membranes in type I methanotrophs are arranged as bundles of disc-shaped vesicles distributed throughout the cell (**Figure 15.14*b***).

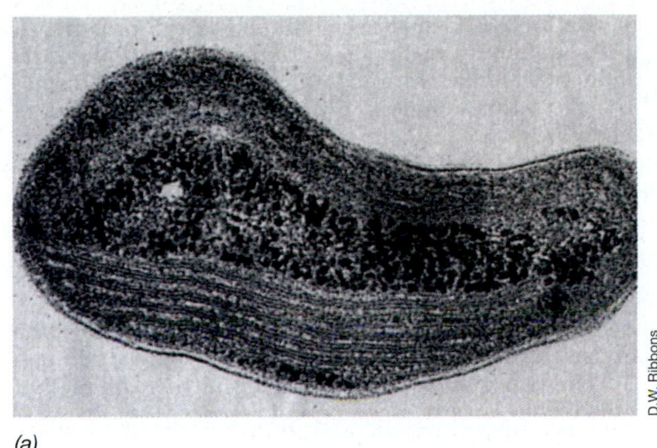

(a)

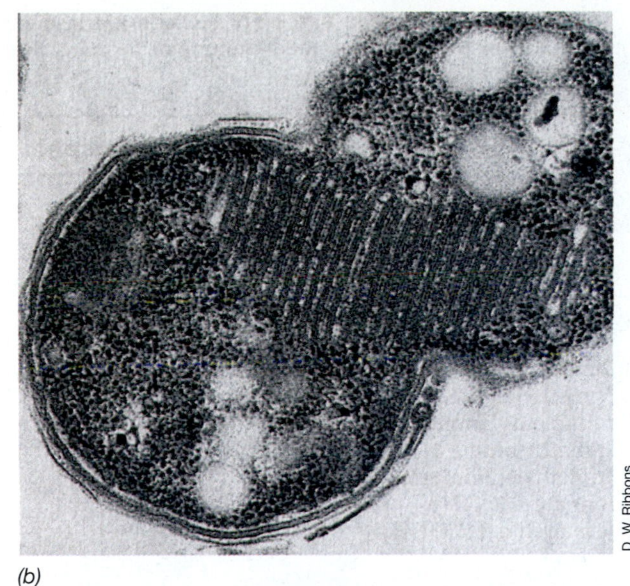

(b)

Figure 15.14 Electron micrographs of methanotrophs. (*a*) A *Methylosinus* species, illustrating a type II membrane system. Cells are about 0.6 μm in diameter. (*b*) *Methylococcus capsulatus*, illustrating a type I membrane system. Cells are about 1 μm in diameter.

Type II species possess paired membranes running along the periphery of the cell (Figure 15.14*a*). The key enzyme methane monooxygenase is located in these membranes.

Type I methanotrophs are also characterized by a lack of a complete citric acid cycle (the enzyme α-ketoglutarate dehydrogenase is absent), whereas type II organisms possess a complete cycle. Absence of a complete citric acid cycle (⚭ Figure 5.22) greatly diminishes the ability of an organism to grow chemoorganotrophically and prevents growth at the expense of many organic compounds, because reactions of the cycle are important for generating NADH. This, then, could be a major reason why type I methanotrophs are obligate methylotrophs.

Ecology and Isolation

Methanotrophs are widespread in aquatic and terrestrial environments, being found wherever stable sources of methane are present. Methane produced in the anoxic regions of lakes

Table 15.8 Some characteristics of methanotrophic bacteria

Organism	Morphology	Phylogenetic group[a]	Resting stage	Internal membranes[b]	Citric acid cycle[c]	Carbon assimilation pathway[d]	N_2 fixation
Methylomonas	Rod	*Gamma*	Cystlike body	I	Incomplete	Ribulose monophosphate	No
Methylomicrobium	Rod	*Gamma*	None	I	Incomplete	Ribulose monophosphate	No
Methylobacter	Coccus to ellipsoid	*Gamma*	Cystlike body	I	Incomplete	Ribulose monophosphate	No
Methylococcus	Coccus	*Gamma*	Cystlike body	I	Incomplete	Ribulose monophosphate	Yes
Methylosinus	Rod or vibrioid	*Alpha*	Exospore	II	Complete	Serine	Yes
Methylocystis	Rod	*Alpha*	Exospore	II	Complete	Serine	Yes
Methylocella[e]	Rod	*Alpha*	Exospore	II	Complete	Serine	Yes

[a]All are *Proteobacteria*.

[b]Internal membranes: Type I, bundles of disc-shaped vesicles distributed throughout the organism; type II, paired membranes running along the periphery of the cell. See Figure 15.15.

[c]Organisms with an incomplete citric acid cycle lack the enzyme α-ketoglutarate dehydrogenase and thus cannot oxidize acetate to CO_2.

[d]See Figures 21.33 and 21.34. Unlike other methylotrophs, *Methylococcus* species contain Calvin cycle enzymes.

[e]Acidophilic, growth optimal at pH 5.

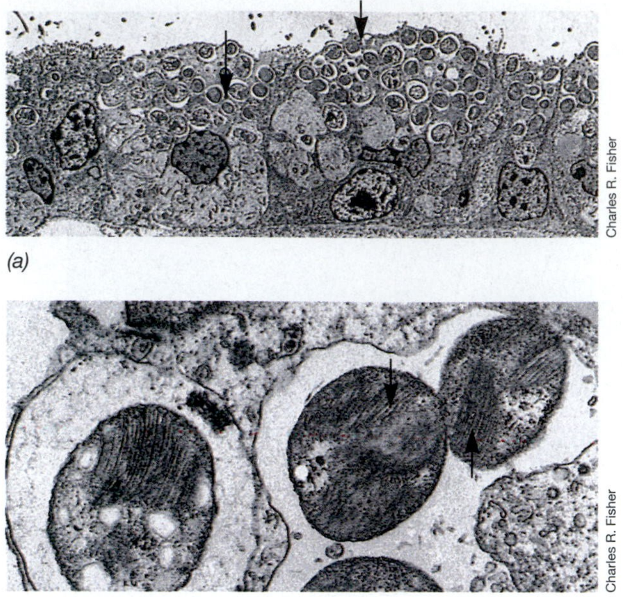

Figure 15.15 Methanotrophic symbionts of marine mussels. (*a*) Electron micrograph of a thin section at low magnification of gill tissue of a marine mussel living near hydrocarbon seeps in the Gulf of Mexico. Note the symbiotic methanotrophs (arrows) in the tissues. (*b*) High magnification view of gill tissue showing type I methanotrophs. Note membrane bundles (arrows). The methanotrophs are about 1 μm in diameter. Compare with Figure 15.14*b*.

rises through the water column, and methanotrophs are often concentrated in a narrow band at the zone where methane and oxygen meet. Methane-oxidizing bacteria therefore play an important role in the carbon cycle, converting methane derived from anoxic decomposition back into cell material and CO_2.

For the enrichment of methanotrophs, all that is needed is a mineral salts medium with an atmosphere of 80% methane and 20% air. Once good growth is obtained, purification can be achieved by repeated streaking on mineral salts agar plates incubated in a jar containing a methane–air mixture. Colonies appearing on the plates are typically of two types: nonmethanotrophic chemoorganotrophs growing on traces of organic matter in the medium, which appear in 1–2 days, and methanotrophs, which appear after about a week. The colonies of many methanotrophs are pink from the production of various carotenoid pigments, and this feature can help in their isolation.

Methanotrophs and Nitrosifying Bacteria

Methanotrophs are able to oxidize ammonia, although they cannot grow chemolithotrophically using ammonia as sole electron donor. In addition to methane oxidation, methane monooxygenase also functions to oxidize ammonia, and a competitive interaction between the two substrates exists. For this reason, ammonia is generally toxic to methanotrophs, and the preferred nitrogen source is nitrate.

It has been speculated that methanotrophic bacteria evolved from the nitrosifying bacteria via mutations that converted an ammonia monooxygenase into a methane monooxygenase. The fact that both groups of bacteria have elaborate internal membrane systems (Section 15.3) and are phylogenetically closely related supports this theory. In addition, however, methanotrophic bacteria contain some of the same genes and make some of the same proteins as methanogenic (methane-producing) bacteria, all of which are phylogenetically *Archaea*. Such results suggest that there has been horizontal (lateral) gene transfer between these groups as the biochemistry of methane metabolism evolved eons ago. We will see how the contrasting processes of methanogenesis and methanotrophy are related in this regard in Sections 21.10 and 21.16.

Methanotrophic Symbionts of Animals

Methanotrophic bacteria and certain marine mussels and sponges develop symbiotic relationships. Some marine mussels live in the vicinity of hydrocarbon seeps on the seafloor, places where methane is released in substantial amounts. Isolated mussel gill tissues consume methane at high rates in the presence of O_2. In these tissues, coccoid-shaped bacteria are present in high numbers (**Figure 15.15*a***). The bacterial symbionts contain intracytoplasmic membranes typical of type I methanotrophs (Figure 15.15*b*). The symbionts are found in vacuoles within animal cells near the gill surface, which probably ensures an effective gas exchange with seawater. That these symbionts are indeed methanotrophs has been demonstrated by phylogenetic analyses.

Methane assimilated by the methanotrophs in the mussels is distributed throughout the animal by the excretion of organic compounds by the methanotrophs. Methanotrophic symbioses are therefore conceptually similar to those that develop between sulfide-oxidizing chemolithotrophs and hydrothermal vent tube worms and giant clams (Section 24.11).

15.6 MiniReview

Methylotrophs are bacteria able to grow on carbon compounds that lack carbon–carbon bonds. Some methylotrophs are also methanotrophs, organisms able to catabolize methane. Two classes of methanotrophs are known, each having characteristic structural and biochemical properties. Methanotrophs reside in water and soil and can also exist as symbionts of marine animals.

- What is the difference between a methanotroph and a methylotroph?
- What features differentiate type I from type II methanotrophs?
- What types of animals harbor methanotrophic symbionts and where does the methane the symbionts need come from?

III AEROBIC AND FACULTATIVELY AEROBIC CHEMOORGANOTROPHIC *PROTEOBACTERIA*

The next few groups to be considered are the classic examples of chemoorganotrophic bacteria that carry out respiratory metabolisms. Here we meet the pseudomonads, the enteric bacteria, the aerobic nitrogen-fixing bacteria and many of their close relatives.

15.7 *Pseudomonas* and the Pseudomonads

Key Genera: *Pseudomonas, Burkholderia, Zymomonas, Xanthomonas*

All the genera in this group are straight or slightly curved chemoorganotrophic aerobic rods with *polar* flagella (**Figure 15.16*a*,*b***). The key genera are *Pseudomonas, Commamonas, Ralstonia,* and *Burkholderia,* discussed in some detail here. Other genera include *Xanthomonas, Zoogloea,* and *Gluconobacter*. *Xanthomonas* is primarily a plant pathogen that is responsible for a number of necrotic plant lesions and that is characterized by its yellow-colored pigments. *Zoogloea* is characterized by its formation of an extracellular fibrillar polymer that causes the cells to aggregate into distinctive flocs (this organism is a major component of activated sewage sludge) (Section 36.2). *Gluconobacter* is characterized by its incomplete oxidation of sugars or alcohols to acids, such as the oxidation of glucose to gluconic acid or ethanol to acetic acid (this organism is discussed briefly with the other acetic acid bacteria in Section 15.8). Phylogenetically, the pseudomonads scatter within the *Proteobacteria,* primarily within the *Alpha* and *Beta* classes (Table 15.1).

(a)

James Shapiro

(b)

Arthur Kelman

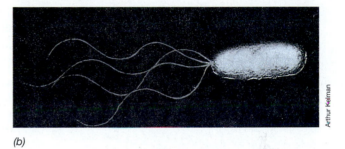

Figure 15.16 Typical pseudomonad colonies and cell morphology common in pseudomonads. *(a)* Photograph of colonies of *Burkholderia cepacia* on an agar plate. *(b)* Shadow-cast transmission electron micrograph of a *Pseudomonas* species. The cell measures about 1 μm in diameter.

Characteristics of Pseudomonads

The distinguishing characteristics of the pseudomonad group are given in **Table 15.9**. Also listed in this table are the minimal characteristics needed to identify an organism as a pseudomonad. Key identifying characteristics are the absence of gas formation from glucose and the positive oxidase test, both of which help to distinguish pseudomonads from enteric bacteria (Section 15.11).

Species of the genus *Pseudomonas* and related genera are defined on the basis of phylogeny and various physiological characteristics, as outlined in **Tables 15.10** and **15.11**. Pseudomonads have very simple nutritional requirements and grow chemoorganotrophically at neutral pH and at temperatures in the mesophilic range.

One of the striking properties of pseudomonads is their ability to use many different organic compounds as carbon and energy sources. Some species utilize over 100 different compounds, and only a few species utilize fewer than 20. As an example of this versatility, a single strain of *Burkholderia cepacia* can use many different sugars, fatty acids, dicarboxylic acids, tricarboxylic acids, alcohols, polyalcohols, glycols, aromatic compounds, amino acids and amines, plus miscellaneous organic compounds not fitting into any of these categories. On the other hand, pseudomonads generally lack the hydrolytic enzymes necessary to break down polymers into their component monomers. These nutritionally versatile pseudomonads

Table 15.9 Characteristics of pseudomonads

General characteristics:
Straight or curved rods but not vibrioid; size 0.5–1.0 μm by 1.5–4.0 μm; no spores; gram-negative; polar flagella: single or multiple; no sheaths, appendages, or buds; respiratory metabolism, never fermentative, although may produce small amounts of acid from glucose aerobically; use low-molecular-weight organic compounds, not polymers; some are chemolithotrophic, using H_2 or CO as sole electron donor; some can use nitrate as electron acceptor anaerobically; some can use arginine as energy source anaerobically
Minimal characteristics for identification:
Gram-negative, straight or slightly curved; no spores; motile (always); polar flagella (flagellar stain); oxidative-fermentative medium with glucose: tube open, acid produced; tube sealed, acid not produced; gas not produced from glucose (distinguishes them easily from enteric bacteria and *Aeromonas*); oxidase, almost always positive (enterics are oxidase-negative); catalase always positive; photosynthetic pigments absent (distinguishes them from purple nonsulfur bacteria); indole-negative; methyl red-negative; Voges-Proskauer-negative (for discussion of many of these biochemical tests, see Section 32.2)

UNIT 3

Table 15.10 Subgroups and characteristics of pseudomonads

Group	Phylogenetic group[a]	Characteristics
Fluorescent subgroup	*Gamma*	**Most produce water-soluble, yellow-green fluorescent pigments; do not form poly-β-hydroxybutyrate; single DNA homology group**
Pseudomonas aeruginosa		Pyocyanin production; growth at up to 43°C; single polar flagellum; capable of denitrification
Pseudomonas fluorescens		Does not produce pyocyanin or grow at 43°C; tuft of polar flagella
Pseudomonas putida		Similar to *P. fluorescens* but does not liquefy gelatin and does grow on benzylamine
Pseudomonas syringae		Lacks arginine dihydrolase; oxidase-negative; pathogenic to plants
Pseudomonas stutzeri		Soil saprophyte; strong denitrifyer and nonfluorescent
Acidovorans subgroup	*Beta*	**Nonpigmented; form poly-β-hydroxybutyrate; tuft of polar flagella; do not use carbohydrates; single DNA homology group**
Commamonas acidovorans		Uses muconic acid as sole carbon source and electron donor
Commamonas testosteroni		Uses testosterone as sole carbon source
Pseudomallei-cepacia subgroup	*Beta*	**No fluorescent pigments; tuft of polar flagella; forms poly-β-hydroxybutyrate; single DNA homology group**
Burkholderia cepacia		Extreme nutritional versatility; some strains pathogenic to plants; human pathogen
Burkholderia pseudomallei		Causes melioidosis in animals; nutritionally versatile
Burkholderia mallei		Causes glanders in animals; nonmotile; nutritionally restricted
Diminuta-vesicularis subgroup	*Alpha*	**Single flagellum of very short wavelength; require vitamins (pantothenate, biotin, B_{12})**
Brevundimonas diminuta		Nonpigmented; does not use sugars
Brevundimonas vesicularis		Carotenoid pigment; uses sugars
Ralstonia subgroup	*Beta*	
Ralstonia solanacearum		Plant pathogen
Ralstonia saccharophila		Grows chemolithotrophically with H_2; digests starch
Stenotrophomonas maltophilia	*Gamma*	Requires methionine; does not use NO_3^- as N source; oxidase-negative

[a]All pseudomonads are members of the *Proteobacteria* (see Table 15.1).

Table 15.11 Pathogenic pseudomonads

Species	Relationship to disease
Animal pathogens	
Pseudomonas aeruginosa	Opportunistic pathogen, especially in hospitals; in patients with metabolic, hematologic, and malignant diseases; hospital-acquired (nosocomial) infections from catheterizations, tracheostomies, lumbar punctures, and intravenous infusions; in patients given prolonged treatment with immunosuppressive agents, corticosteroids, antibiotics, and radiation; may contaminate surgical wounds, abscesses, burns, ear infections, lungs of patients treated with antibiotics; cystic fibrosis; primarily a soil organism
Pseudomonas fluorescens	Rarely pathogenic, as does not grow well at 37°C; may grow in and contaminate blood and blood products under refrigeration
Stenotrophomonas maltophilia	A ubiquitous, free-living organism that is a common nosocomial pathogen
Burkholderia cepacia	Causes onion bulb rot; has also been isolated from humans and from environmental sources of medical importance
Burkholderia pseudomallei	Causes melioidosis, a disease endemic in animals and humans in Southeast Asia
Burkholderia mallei	Causes glanders, a disease of horses that is occasionally transmitted to humans
Pseudomonas stutzeri	Often isolated from humans and environmental sources; may live saprophytically in the body
Plant pathogens	
Ralstonia solanacearum	Causes wilts of many cultivated plants (for example, potato, tomato, tobacco, peanut)
Pseudomonas syringae	Attacks foliage, causing chlorosis and necrotic lesions on leaves; rarely found free in soil
Pseudomonas marginalis	Causes soft rot of various plants; active pectinolytic species
Xanthomonas campestris	Causes necrotic lesions on foliage, stems, fruits; also causes wilts and tissue rots; rarely found free in soil

contain numerous inducible operons (Section 9.3) because the catabolism of unusual organic substrates often requires the activity of several different enzymes.

The pseudomonads are ecologically important organisms in soil and water and are probably responsible for the degradation of many soluble compounds derived from the breakdown of plant and animal materials in oxic habitats. They are also capable of breaking down many xenobiotic (not naturally occurring) compounds, such as pesticides and other toxic chemicals, and are thus important agents of bioremediation in the environment.

Most pseudomonads metabolize glucose via the Entner–Doudoroff pathway, a variation of the glycolytic pathway in which early steps in glucose catabolism differ from the glycolytic pattern (Section 21.2). A survey for the presence of the Entner–Doudoroff pathway has shown it to be absent from gram-positive *Bacteria* but present in bacteria of the genera *Pseudomonas, Rhizobium, Agrobacterium, Zymomonas,* and several other gram-negative *Bacteria*.

Pathogenic Pseudomonads

A number of pseudomonads are pathogenic (Table 15.11). Among the fluorescent pseudomonads, the species *Pseudomonas aeruginosa* is frequently associated with infections of the urinary and respiratory tracts in humans. *P. aeruginosa* infections are also common in patients receiving treatment for severe burns or other traumatic skin damage and in people suffering from cystic fibrosis. *P. aeruginosa* is not an obligate pathogen. Instead, the organism is an opportunist, initiating infections in individuals whose resistance is low. In addition to urinary tract infections, it can also cause systemic infections, usually in individuals who have experienced extensive skin damage. Other pseudomonads are also human pathogens, such as *Burkholderia cepacia,* which like *P. aeruginosa* can infect the lungs of patients with cystic fibrosis.

P. aeruginosa is naturally resistant to many of the widely used antibiotics, so chemotherapy is often difficult. Resistance is due to a resistance transfer plasmid (R plasmid) (Sections 11.2 and 27.12), which is a plasmid carrying genes encoding proteins that detoxify various antibiotics, and to the presence of multidrug efflux systems that pump antibiotics out of the cell (Section 27.12). *P. aeruginosa* is commonly found in the hospital environment and can easily infect patients receiving treatment for other illnesses (nosocomial infections, Section 33.7). Polymyxin, an antibiotic not ordinarily used in human therapy because of its toxicity, is effective against *P. aeruginosa* and is used in certain medical situations.

Certain species of *Pseudomonas, Ralstonia,* and *Burkholderia* and the genus *Xanthomonas* are well-known plant pathogens (phytopathogens) (Table 15.11). In many cases these organisms are so highly adapted to the plant environment that they are rarely isolated from other habitats, including soil. Phytopathogens frequently inhabit nonhost plants (in which disease symptoms are not apparent) and from them are transmitted to host plants and initiate infection. Disease symptoms vary considerably, depending on the particular phytopathogen and host plant. The pathogen releases plant toxins, lytic enzymes, plant growth factors, and other substances that destroy or distort plant tissue. In many cases the disease symptoms help identify the phytopathogen. Thus, *Pseudomonas syringae* is typically isolated from leaves showing chlorotic (yellowing) lesions, whereas *P. marginalis* is a typical "soft-rot" pathogen, infecting stems and shoots, but rarely leaves.

Zymomonas

The genus *Zymomonas* consists of large, gram-negative rods that carry out a vigorous fermentation of sugars to ethanol. Although strictly fermentative, *Zymomonas* shows phylogenetic affiliation with the pseudomonads and employs the Entner–Doudoroff pathway (Section 21.2).

Zymomonas is a common organism that carries out alcoholic fermentations of various plant saps, and in many tropical areas of South and Central America, Africa, and Asia it occupies a position in the fermented beverage industry similar to that of *Saccharomyces cerevisiae* (yeast) in North America and Europe. For example, *Zymomonas* participates in the alcoholic fermentation of agave in Mexico to form the drink *pulque* and the fermentation of palm sap in many other tropical areas. It also carries out an alcoholic fermentation of sugarcane juice and honey. Although *Zymomonas* is rarely the sole organism responsible for these alcoholic fermentations, it is often the dominant organism and is probably responsible for the production of most of the ethanol in these beverages. *Zymomonas* is also responsible for spoilage of fruit juices, such as apple and pear ciders, and is also a constituent of the bacterial flora of spoiled beer.

Zymomonas is distinguished from *Pseudomonas* by its fermentative metabolism, microaerophilic to anaerobic nature, oxidase negativity, and other molecular taxonomic characteristics. It also phenotypically resembles the acetic acid bacteria (Section 15.8) and is often found associated with these organisms in nature. This is of interest because, like yeast, *Zymomonas* ferments glucose to ethanol, whereas the acetic acid bacteria oxidize ethanol to acetic acid. Thus the acetic acid bacteria depend on the activity of yeast and *Zymomonas* for the production of their growth substrate, ethanol.

15.8 Acetic Acid Bacteria

Key Genera: *Acetobacter, Gluconobacter*

As originally defined, the acetic acid bacteria comprised gram-negative, aerobic, motile rods that carry out the incomplete oxidation of alcohols and sugars, leading to the accumulation of organic acids as end products. With ethanol as a substrate, acetic acid is produced, which gives acetic acid bacteria their name.

Acetic acid bacteria have a relatively high tolerance to acidic conditions; most strains can grow well at pH values lower than 5. This acid tolerance should, of course, be essential for an organism producing large amounts of acid. The acetic acid bacteria are a heterogeneous assemblage, comprising both peritrichously and polarly flagellated organisms. The polarly flagellated organisms are classified in the genus

UNIT 3

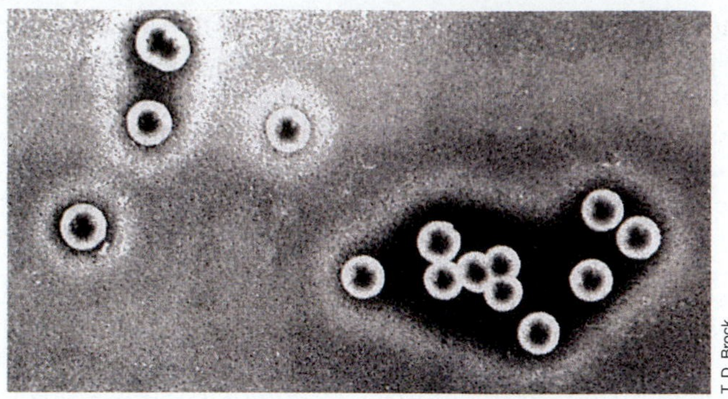

Figure 15.17 Colonies of *Acetobacter aceti* on calcium carbonate agar containing ethanol as energy source. Note the clearing around the colonies due to the dissolution of calcium carbonate ($CaCO_3$) by the acetic acid produced by the bacteria.

Gluconobacter, and the peritrichously flagellated species are grouped into the genus *Acetobacter*. All known acetic acid bacteria group phylogenetically with the *Alphaproteobacteria* (Table 15.1).

In addition to flagellation, *Acetobacter* differs from *Gluconobacter* in being able to further oxidize the acetic acid it forms to CO_2. This difference in ability to oxidize acetic acid is related to the presence of a complete citric acid cycle. *Gluconobacter,* which lacks a complete citric acid cycle, is unable to oxidize acetic acid, whereas *Acetobacter,* which has all enzymes of the cycle, can oxidize it.

Ecology and Industrial Uses

The acetic acid bacteria are commonly found in alcoholic juices. Acetic acid bacteria can often be isolated from an alcoholic fruit juice such as hard cider or wine, or from beer. Colonies of acetic acid bacteria can be recognized on $CaCO_3$ agar plates containing ethanol, because the acetic acid produced dissolves and causes a clearing of the otherwise insoluble $CaCO_3$ (**Figure 15.17**). Cultures of acetic acid bacteria are used in the commercial production of vinegar (∞ Section 25.12).

In addition to ethanol, the acetic acid bacteria carry out an incomplete oxidation of such organic compounds as higher alcohols and sugars. For instance, glucose is oxidized only to gluconic acid, galactose to galactonic acid, arabinose to arabonic acid, and so on. This property of "underoxidation" is exploited in the manufacture of ascorbic acid (vitamin C). Ascorbic acid can be formed from sorbose, but sorbose is difficult to synthesize chemically. It is, however, conveniently obtainable from acetic acid bacteria, which oxidize sorbitol (a readily available sugar alcohol) to sorbose, a process called *biotransformation* (∞ Section 25.8).

Another interesting property of some acetic acid bacteria is their ability to synthesize cellulose. The cellulose formed does not differ significantly from that of plant cellulose, with the exception that it is pure and not mixed in with other polymers like the hemicelluloses, pectin, or lignins of plants. Cellulose from acetic acid bacteria is formed as a matrix outside the cell wall and causes cells to become embedded in a tangled mass of cellulose microfibrils. When these species of acetic acid bacteria grow in an unshaken vessel, they form a surface pellicle of cellulose in which the bacteria develop. Because these bacteria are obligate aerobes, the ability to form such a pellicle may be a means by which the organisms remain at the surface of the liquid where oxygen is readily available.

Table 15.12 Genera of free-living aerobic nitrogen-fixing bacteria

Characteristics	*Genus*
Gammaproteobacteria	
Large rod; produces cysts; primarily found in neutral to alkaline soils	*Azotobacter*
Large rod; no cysts; primarily aquatic	*Azomonas*
Alphaproteobacteria	
Microaerophilic rod; associates with plants	*Azospirillum*
Pear-shaped rod with large lipid bodies at each end; produces extensive slime; inhabits acidic soils	*Beijerinckia*
Betaproteobacteria	
Small curved cells; no cysts	*Azoarcus*
Very thin curved cells; no cysts	*Azovibrio*
Very thin rods to vibrios; no cysts	*Azospira*
Cells form coils up to 50 μm long; no cysts	*Azonexus*
Rods; form coarse, wrinkled colonies	*Derxia*

15.9 Free-Living Aerobic Nitrogen-Fixing Bacteria

Key Genera: *Azotobacter, Azomonas*

A variety of organisms inhabit soil and are capable of fixing N_2 *aerobically* (**Table 15.12**). The genus *Azotobacter* comprises large, gram-negative, obligately aerobic rods capable of fixing N_2 nonsymbiotically (**Figure 15.18**). The first species of this genus was discovered by the Dutch microbiologist Martinus Beijerinck early in the twentieth century (∞ Section 1.9). Beijerinck used an aerobic enrichment culture technique with a medium containing N_2 (air) but devoid of a combined nitrogen source (∞ Section 22.1 and Figure 22.1). Phylogenetically, most free-living nitrogen-fixing bacteria are *Alpha-*, *Beta-*, or *Gammaproteobacteria* (Table 15.12).

Taxonomy

The major free-living nitrogen-fixing bacteria that have been studied include *Azotobacter, Azospirillum,* and *Beijerinckia*. *Azotobacter* cells are large, many isolates being almost the size of yeasts, with diameters of 2–4 μm or more. Pleomorphism is

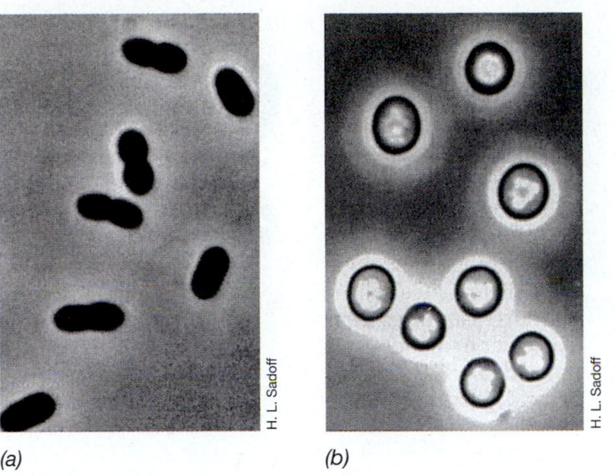

Figure 15.18 ***Azotobacter vinelandii.*** *(a)* Vegetative cells and *(b)* cysts visualized by phase-contrast microscopy. A cell measures about 2 μm in diameter and a cyst about 3 μm. Compare with Figure 1.17*b*.

common, and a variety of cell shapes and sizes have been described. Some strains are motile by peritrichous flagella. Extensive capsules or slime layers are produced by free-living N_2-fixing bacteria when they are fixing nitrogen on carbohydrate-containing media (**Figure 15.19**; Figure 20.37).

Despite the fact that *Azotobacter* is an obligate aerobe, its nitrogenase, the enzyme that catalyzes biological N_2 fixation (Section 20.14), is O_2-sensitive. It is thought that the high respiratory rate characteristic of *Azotobacter* and the abundant capsular slime help protect nitrogenase from O_2. *Azotobacter* is able to grow on many different carbohydrates, alcohols, and organic acids. The metabolism of carbon compounds is strictly oxidative, and acids or other fermentation products are rarely produced. All members fix nitrogen, but can also grow on simple forms of combined nitrogen: ammonia, urea, and nitrate.

Azotobacter can form resting structures called *cysts*. Like bacterial endospores, *Azotobacter* cysts (Figure 15.18*b*) show negligible endogenous respiration and are resistant to desiccation, mechanical disintegration, and ultraviolet and ionizing radiation. In contrast to endospores, however, cysts are not especially heat-resistant, and they are not completely dormant because they rapidly oxidize carbon sources if supplied.

The remaining key genera of free-living N_2 fixers include *Azomonas*, a genus of large coccus to rod-shaped bacteria that resemble *Azotobacter* except that they do not produce cysts and are primarily aquatic, and *Beijerinckia* (**Figures** 15.19*b* and **15.20a**) and *Derxia* (Figure 15.20*b*), two genera that grow well in acidic soils. *Azospirillum*, a rod- to spirillum-shaped nitrogen-fixing bacterium that forms nonspecific symbiotic associations with plants, in particular, corn (Table 15.12), rounds out this group.

Azotobacter and Alternative Nitrogenases

We study the important process of biological N_2 fixation in Section 20.14. There we will learn of the central importance of the metals molybdenum (Mo) and iron (Fe) to the enzyme nitrogenase. The species *Azotobacter chroococcum* was the first nitrogen-fixing bacterium shown capable of growth on N_2 in the absence of molybdenum. In *A. chroococcum* it was first shown that in place of the Mo nitrogenase, two "alternative nitrogenases" containing the metals vanadium (V) plus Fe or Fe can be formed only under certain conditions. Subsequent

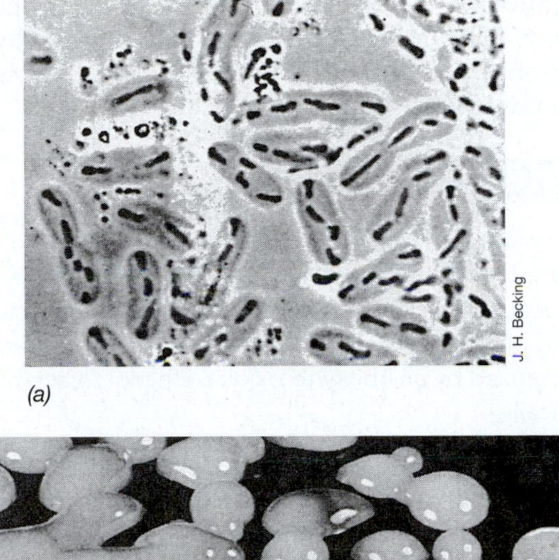

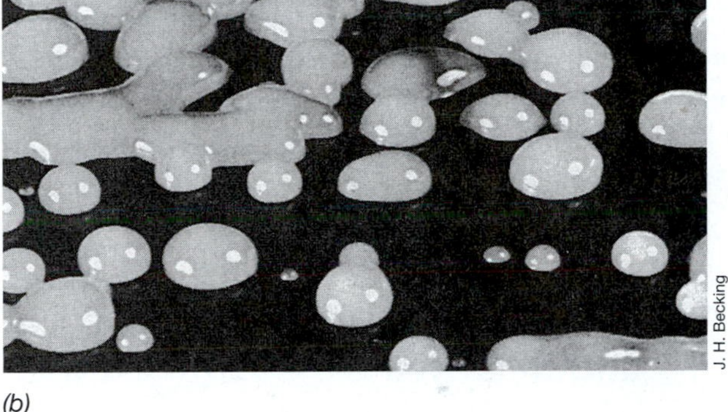

Figure 15.19 **Examples of slime production by free-living N_2-fixing bacteria.** *(a)* Cells of *Derxia gummosa* encased in slime. Cells are about 1–1.2 μm wide. *(b)* Colonies of *Beijerinckia* species growing on a carbohydrate-containing medium. Note the raised, glistening appearance of the colonies due to abundant capsular slime.

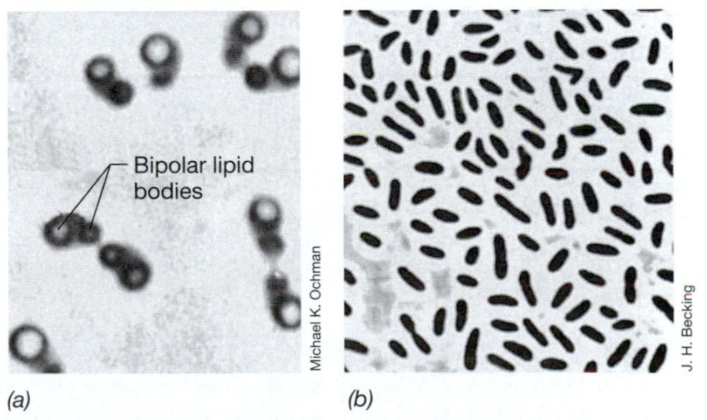

Figure 15.20 **Phase-contrast photomicrographs of two genera of acid-tolerant, free-living N_2-fixing bacteria.** *(a) Beijerinckia indica.* The cells are roughly pear-shaped, about 0.8 μm in diameter, and contain a large globule of poly-β-hydroxybutyrate at each end. *(b) Derxia gummosa*. Cells are about 1 μm in diameter.

UNIT 3

investigations of other nitrogen-fixing bacteria have shown that these "backup" nitrogenase systems are widely distributed among nitrogen-fixing bacteria, including *Archaea,* of which a few species fix N_2.

15.7–15.9 MiniReview

Pseudomonads include many gram-negative chemoorganotrophic aerobic rods; many N_2-fixing species are phylogenetically closely related. The acetic acid bacteria are also phylogenetically related to pseudomonads and are characterized by an ability to oxidize ethanol to acetate aerobically.

- Compare and contrast the pseudomonads, *Azotobacter,* and the acetic acid bacteria in terms of O_2 and nitrogen requirements, electron donors, pathogenicity, and habitats.
- Compare and contrast the organisms *Acetobacter* and *Gluconobacter.*

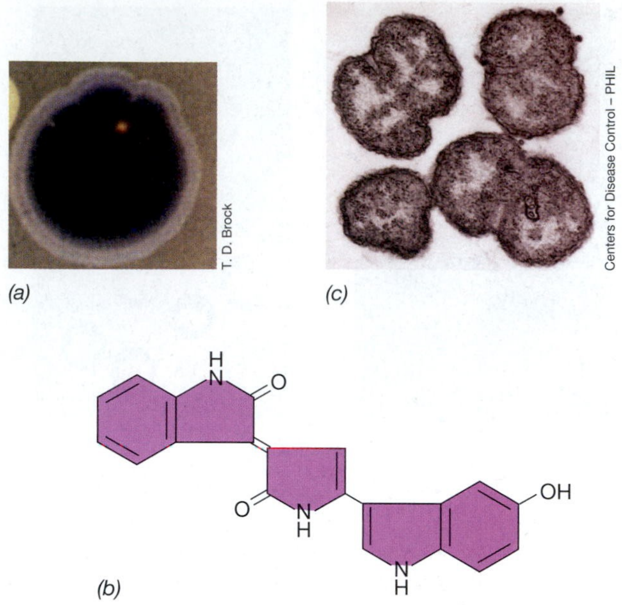

Figure 15.21 ***Chromobacterium* and *Neisseria*.** *(a)* A large colony of *Chromobacterium violaceum*. The purple pigment is an aromatic compound called violacein, the structural formula of which is shown in *(b)*. *(c)* Transmission electron micrograph of cells of *Neisseria gonorrhoeae* showing the typical diplococcus cell arrangements.

15.10 *Neisseria, Chromobacterium,* and Relatives

Key Genera: *Neisseria, Chromobacterium*

This group of *Beta-* and *Gammaproteobacteria* comprises a diverse collection of organisms, related phylogenetically as well as by Gram stain, morphology, lack of swimming motility, and aerobic metabolism. The genera *Neisseria, Moraxella, Branhamella, Kingella,* and *Acinetobacter* are distinguished as outlined in **Table 15.13**.

In the genus *Neisseria,* the cells are always cocci (**Figures 15.21c** and Figure 34.29), whereas cells of the other genera are rod-shaped, becoming coccoid only in the stationary phase of growth. This has led to designation of these organisms as coccobacilli. Organisms of the genera *Neisseria, Kingella,* and *Moraxella* are commonly isolated from animals, and some of them are pathogenic. We discuss the clinical microbiology of *Neisseria gonorrhoeae,* the causative agent of the disease gonorrhea, in Section 32.1 and the pathogenesis of gonorrhea itself in Section 34.13.

Table 15.13 Characteristics of the genera of gram-negative cocci[a]

Characteristics	*Genus*
I. Oxidase-positive, penicillin-sensitive:	
Cocci; complex nutrition, utilize carbohydrates, obligate aerobes; *Betaproteobacteria* or *Gammaproteobacteria*	*Neisseria* *Moraxella*
Rods or cocci; generally no growth factor requirements, generally do not utilize carbohydrates; do not contain flagella, but some species exhibit twitching motility; many are commensals or pathogens of animals; *Betaproteobacteria*	*Branhamella* *Kingella*
II. Oxidase-negative, penicillin-resistant:	
Some strains can utilize a restricted range of sugars, and some exhibit twitching motility; saprophytes in soil, water, and sewage; *Gammaproteobacteria*	*Acinetobacter*

[a]All are *Proteobacteria.*

Species of *Acinetobacter* are common soil and water organisms, although they are occasionally found as parasites of animals and have been implicated in some nosocomial (hospital-associated) infections. Some strains of *Moraxella* and *Acinetobacter* possess the interesting property of twitching motility, exhibited as brief translocative movements or "jumps" covering distances of about 1–5 μm. Twitching bacteria contain special force-generating pili (Section 4.9) that facilitate their movement.

Chromobacterium is a close phylogenetic relative of *Neisseria* but is rod-shaped in morphology, resembling the pseudomonads or enteric bacteria. The best-known *Chromobacterium* species is *C. violaceum,* a purple-pigmented organism (Figure 15.21*a*) found in soil and water and occasionally in pus-forming infections of humans and other animals. *C. violaceum* and a few other chromobacteria produce the purple pigment *violacein* (Figure 15.21*b*), a water-insoluble pigment that has antibiotic-like properties and is produced only in media containing the amino acid tryptophan. Like enteric bacteria, *Chromobacterium* is a facultative aerobe, growing fermentatively on sugars and aerobically on various carbon sources.

15.10 MiniReview

Neisseria, Chromobacterium, and their relatives can be isolated from animals, and some species of this group are pathogenic.

- What species causes the disease gonorrhea?
- What is twitching motility and what cellular structure is involved?
- What is the name of the water-soluble, purple pigment produced by *Chromobacterium violaceum*?

15.11 Enteric Bacteria

Key Genera: *Escherichia, Salmonella, Proteus, Enterobacter*

The **enteric bacteria** comprise a relatively homogeneous phylogenetic group within the *Gammaproteobacteria* and consist of facultatively aerobic, gram-negative, nonsporulating rods that are either nonmotile or motile by peritrichous flagella (**Figure 15.22**). Enteric bacteria are also oxidase-negative, have relatively simple nutritional requirements, and ferment sugars to a variety of end products. The phenotypic characteristics that separate the enteric bacteria from other bacteria of similar morphology and physiology are given in **Table 15.14**.

Among the enteric bacteria are many species pathogenic to humans, other animals, or plants, as well as other species of industrial importance. *Escherichia coli,* the best known of all organisms, is the classic example of an enteric bacterium. Because of the medical importance of the enteric bacteria, an extremely large number of isolates have been studied and characterized, and numerous genera have been defined. Despite the fact that there is marked genetic relatedness among many of the enteric bacteria, as shown by high levels of genomic DNA hybridization (Section 14.11) and widespread genetic recombination, separate genera are maintained, largely for practical reasons, such as identification in clinical microbiology. Because these organisms are frequently cultured from diseased hosts, some means of identifying them is necessary to facilitate rapid treatment, and thus phenotypic characteristics have traditionally been very important in distinguishing between genera of enteric bacteria.

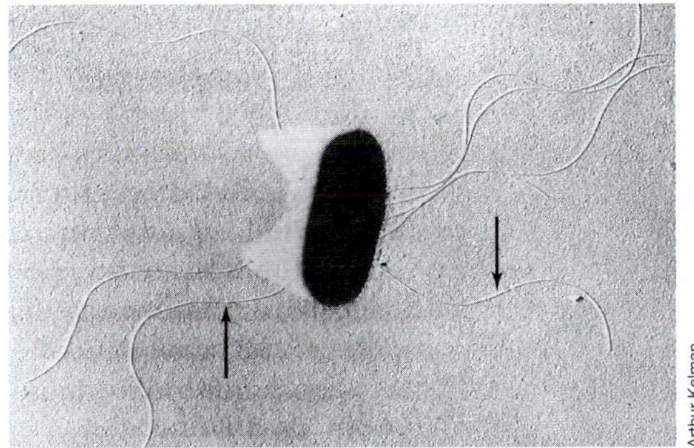

Figure 15.22 Butanediol-producing bacterium. Electron micrograph of a shadow-cast preparation of cells of the butanediol-producing enteric bacterium *Erwinia carotovora.* The cell is about 0.8 μm wide. Note the peritrichously arranged flagella (arrows).

Table 15.14 Defining characteristics of the enteric bacteria

General characteristics:
Gram-negative straight rods; motile by peritrichous flagella, or nonmotile; nonsporulating; facultative aerobes, producing acid from glucose; sodium neither required nor stimulatory; catalase-positive; oxidase-negative; usually reduce nitrate to nitrite (not to N_2); 16S rRNA gene of *Gammaproteobacteria* (see Table 15.1)
Key tests to distinguish enteric bacteria from other bacteria of similar morphology[a]:
Oxidase test, enterics always negative—separates enterics from oxidase-positive bacteria of genera *Pseudomonas, Aeromonas, Vibrio, Alcaligenes, Achromobacter, Flavobacterium, Cardiobacterium,* which may have similar morphology; nitrate reduced only to nitrite (assay for nitrite after growth)—distinguishes enteric bacteria from bacteria that reduce nitrate to N_2 (gas formation detected), such as *Pseudomonas* and many other oxidase-positive bacteria; ability to ferment glucose—distinguishes enterics from obligately aerobic bacteria

[a]See Section 32.2 and Figure 32.7.

Fermentation Patterns in Enteric Bacteria

One important taxonomic characteristic separating the various genera of enteric bacteria is the type and proportion of fermentation products produced by anaerobic fermentation of glucose. Two broad patterns are recognized, the *mixed-acid* fermentation and the *2,3-butanediol* fermentation (**Figure 15.23**).

In the mixed-acid fermentation, three acids are formed in significant amounts—acetic, lactic, and succinic; ethanol, CO_2, and H_2 are also formed, but not butanediol. In the butanediol fermentation, smaller amounts of acids are formed, and butanediol, ethanol, CO_2, and H_2 are the main products. As a result of a mixed-acid fermentation, equal amounts of CO_2 and H_2 are produced, whereas in a butanediol fermentation, considerably more CO_2 than H_2 is produced. This is because mixed-acid fermenters produce CO_2 only from formic acid by means of the enzyme system formate hydrogen lyase

$$HCOOH \rightarrow H_2 + CO_2,$$

and this reaction results in equal amounts of CO_2 and H_2. The butanediol fermenters also produce CO_2 and H_2 from formic acid, but they produce two additional molecules of CO_2 during the formation of each molecule of butanediol (Figure 15.23*b*).

Identification of Enteric Bacteria

Diagnostic tests and differential media are used to separate the various genera in the two broad groups of enteric bacteria.

UNIT 3

Glucose —Glycolysis→ Pyruvate → Lactate
CO_2
Succinate
Acetyl~CoA → Ethanol, Acetate
+
Formate → CO_2, H_2

Typical products **(molar amounts)**

Acidic : neutral
4 : 1
CO_2 : H_2
1 : 1

(*a*) **Mixed acid fermentation** (for example, *Escherichia coli*)

Typical products **(molar amounts)**

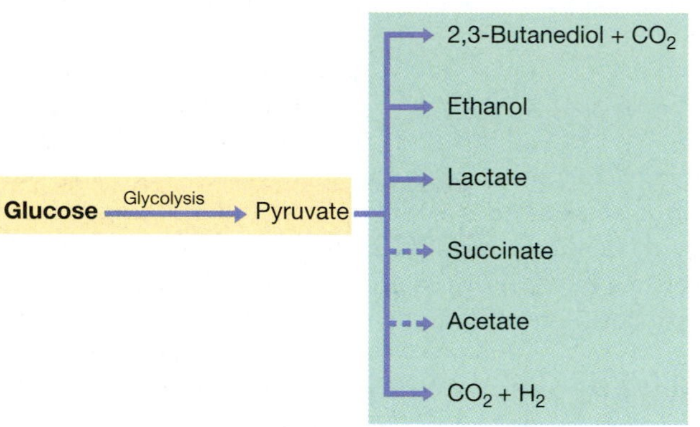

Acidic : neutral
1 : 6
CO_2 : H_2
5 : 1

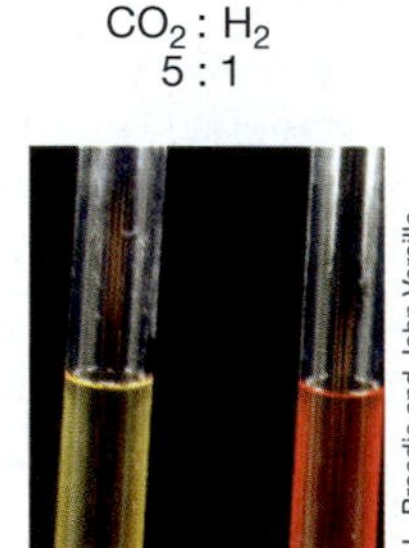

(*b*) **Butanediol fermentation** (for example, *Enterobacter*)

Figure 15.23 Enteric fermentations. Distinction between (*a*) mixed acid and (*b*) butanediol fermentation in enteric bacteria. The bold arrows indicate reactions leading to major products. Dashed arrows indicate minor products. The upper photo shows the production of acid (yellow) and gas (in the inverted Durham tube) in a culture of *Escherichia coli* (purple tube was uninoculated). The bottom photo shows the pink-red color in the Voges–Proskauer (VP) test, which indicates butanediol production, following growth of *Enterobacter aerogenes*. The left (yellow) tube was not inoculated. Note the major difference in CO_2 production in the two pathways, with butanediol production leading to substantially greater CO_2 yields. Because the production of one molecule of butanediol from two pyruvates consumes only one NADH, 0.5 molecules of ethanol must be made for each butanediol produced to consume the second NADH generated in glycolysis.

On the basis of these tests, genera can be defined as outlined in **Tables 15.15** and 15.16. Because enteric bacteria are genetically very closely related, their positive identification often presents considerable difficulty. In clinical laboratories, identification is often based on computer analysis of a large number of diagnostic tests carried out using miniaturized rapid diagnostic media kits and immunological and nucleic acid probes (∞ Figure 32.1), with consideration being given for variable reactions of exceptional strains. The separation of genera given in Tables 15.15 and 15.16 must only be considered as approximate, for it is always possible to isolate a strain that does not possess one or another characteristic normally considered positive for the genus as a whole. With these limitations in mind, an even more simplified separation of the key genera is found in **Figure 15.24**. This key permits a quick decision on the likely genus in which to place a new enteric isolate. We now consider some of the properties of key genera.

Molecular methods for the identification of enteric bacteria (∞ Sections 32.5–32.12) are in use and under development. Examples include genus- and species-specific probes for individual genes using PCR (∞ Section 32.12), tests for specific genes involved in pathogenicity, and the use of partial or complete 16S rRNA gene sequencing and 16S rRNA gene ribotyping (∞ Section 14.9; Figure 14.18).

The Genus *Escherichia*

Members of the genus *Escherichia* are almost universal inhabitants of the intestinal tract of humans and other warm-blooded animals, although they are by no means the dominant organisms in this habitat. *Escherichia* may play a nutritional role in the intestinal tract by synthesizing vitamins, particularly vitamin K. As a facultative aerobe, this organism probably also helps consume oxygen, thus rendering the large intestine anoxic. Wild-type *Escherichia* strains rarely show any growth-factor requirements and are able to grow on a wide variety of carbon and energy sources such as sugars, amino acids, organic acids, and so on.

Some strains of *Escherichia* are pathogenic. These have been implicated in diarrhea in infants, a major public health problem in developing countries. *Escherichia* is a major cause of urinary tract infections in women. Enteropathogenic *E. coli* (abbreviated as EPEC) are becoming more frequently implicated in gastrointestinal infections and generalized fevers. As noted in Sections 28.6, 28.11, and 37.8, certain of these strains form a structure called *K antigen*, permitting attachment and colonization of the small intestine, and enterotoxin, responsible for the symptoms of diarrhea. Some strains, such as enterohaemorrhagic *E. coli* (abbreviated as EHEC), an important representative of which is strain O157:H7, can cause sporadic outbreaks of severe foodborne disease. Infection

Table 15.15 Key diagnostic reactions used to separate some key genera of enteric bacteria[a]

Genus	H_2S (TSI)	Urease	VP[b]	Indole	Motility	Gas from glucose[b]	β-Galactosidase
Escherichia	−	−	−	+	+ or −	+	+
Enterobacter	−	−	+	−	+	+	+
Shigella	−	−	−	+ or −	−	−	+ or −
Edwardsiella	+	−	−	+	+	+	−
Salmonella	+	−	−	−	+	+	+ or −
Klebsiella	−	+	+ or −	−	−	+	+
Citrobacter	+ or −	−	−	−	+	+	+
Proteus	+ or −	+	−	+ or −	+	+ or −	−
Providencia	−	−	−	+	+	−	−
Yersinia	−	+	−	−	+[c]	−	+
Hafnia	−	−	+	−	+	+	+ or −

Genus	KCN	Citrate	Mucate utilization	Phenyl-methyl red	Tartrate utilization	Alanine deaminase
Escherichia	−	−	+	+	+	−
Enterobacter	+	+	+	−	−	−
Shigella	−	−	−	+	−	−
Edwardsiella	−	−	−	+ or −	−	−
Salmonella	−	+ or −	+ or −	+	+ or −	−
Klebsiella	+	+	+	−	+ or −	−
Citrobacter	+ or −	+	+	+	+	−
Proteus	+	+ or −	−	+	+	+
Providencia	+	+	−	+	+	+
Yersinia	−	−	−	+	−	−
Hafnia	+	+	−	+	−	−

[a]See Table 32.3 for the procedures for these diagnostic reactions.
[b]See Figure 15.23 for a photo of this reaction.
[c]Motile when grown at room temperature; nonmotile at 37°C.

occurs primarily through consumption of contaminated foods, such as raw or undercooked ground beef, unpasteurized milk, or contaminated water. In a small percentage of cases, *E. coli* O157:H7 causes a life-threatening complication related to its production of enterotoxin.

Salmonella and *Shigella*

Salmonella and *Escherichia* are quite closely related, the two genera showing about 50% genomic hybridization (∞ Section 14.11). However, in contrast to most *Escherichia,* members of the genus *Salmonella* are usually pathogenic, either to humans or to other warm-blooded animals; *Salmonella* also is found in intestines of poikilothermic (cold-blooded) animals, such as turtles and lizards. In humans the most common diseases caused by salmonellas are typhoid fever and gastroenteritis (∞ Sections 36.8 and 37.7). The salmonellas are characterized immunologically on the basis of three cell surface antigens, the O, or cell wall (somatic) antigen; the H, or flagellar, antigen; and the Vi (outer polysaccharide layer) antigen, found primarily in strains of *Salmonella* causing typhoid fever.

The O antigens are part of the lipopolysaccharides that constitute the outermost layer of the outer membrane of these organisms (∞ Sections 4.7 and 28.12). We discussed the chemical structure of lipopolysaccharides in Section 4.7 (∞ Figures 4.22 and 4.23). The genus *Salmonella* contains over 1,000 distinct serotypes having different antigenic specificities in their O antigens. Additional antigenic subdivisions are based on the antigenic specificities of the flagellar (H) antigens. There is little or no correlation between the antigenic type of *Salmonella* and the disease symptoms elicited, but immunological typing permits tracking a single strain type in an epidemic.

The shigellas are also genetically very closely related to *Escherichia*. Tests for DNA hybridization show that strains of *Shigella* have 70% or even higher genomic hybridization with *E. coli* and therefore probably form a single species (∞ Section 14.11). In contrast to *Escherichia,* however, *Shigella* is commonly pathogenic to humans, causing a rather severe gastroenteritis called bacillary dysentery, although some strains of *E. coli* can also cause a dysentery-like diarrhea. *Shigella dysenteriae* is transmitted by food- and waterborne routes and is capable of invading intestinal epithelial cells (∞ Section 37.11). This organism produces both an endotoxin and a neurotoxin that exhibits enterotoxic (gastrointestinal) effects.

UNIT 3

	Diagnostic test	Go to number
1	MR+; VP – (mixed-acid fermenters)	2
	MR –; VP + (butanediol producers)	7
2	Urease +	*Proteus*
	Urease –	3
3	H_2S (TSI) +	4
	H_2S (TSI) –	6
4	KCN +	*Citrobacter*
	KCN –	5
5	Indole +; citrate –	*Edwardsiella*
	Indole –; citrate +	*Salmonella*
6	Gas from glucose	*Escherichia*
	No gas from glucose	*Shigella*
7	Nonmotile; ornithine –	*Klebsiella*
	Motile; ornithine +	8
8	Gelatin+; DNAse +	*Serratia* (red pigment)
	Gelatin slow; DNAse –	*Enterobacter*

Key
- Mixed-acid fermenters
- Butanediol producers

Figure 15.24 A simple key to the main genera of enteric bacteria. Only the most common genera are given. See text for precautions in the use of this key. Diagnostic tests for use with this figure are given in Table 32.3. Other characteristics of the genera are given in Tables 15.14–15.16.

Commensal forms of *E. coli* (and *Shigella*) have become pathogenic, as in EPEC strains, through acquisition of virulence factors from other bacterial species. Indeed, whole genome sequence analysis of *E. coli* suggests that approximately 17% of this bacterium's genes were acquired by horizontal gene transfer (Chapter 13).

Proteus

The genus *Proteus* is characterized by rapid motility (**Figure 15.25**) and by production of the enzyme *urease*. By the extent of genomic DNA hybridization, it shows only a distant relationship to *E. coli*. *Proteus* is a frequent cause of urinary tract infections in humans and probably benefits in this regard from its ready ability to degrade urea. Because of the rapid motility of *Proteus* cells, colonies growing on agar plates often exhibit a characteristic swarming phenotype (Figure 15.25*b*). Cells at the edge of the growing colony are more rapidly motile than those in the center of the colony. The former move a short distance away from the colony in a mass and then undergo a reduction in motility, settle down, and divide, forming a new population of motile cells that again swarm. As a result, the mature colony appears as a series of concentric rings, with higher concentrations of cells alternating with lower concentrations (Figure 15.25*b*).

Butanediol Fermenters: *Enterobacter*, *Klebsiella*, and *Serratia*

The butanediol fermenters are genetically more closely related to each other than to the mixed-acid fermenters, a finding that

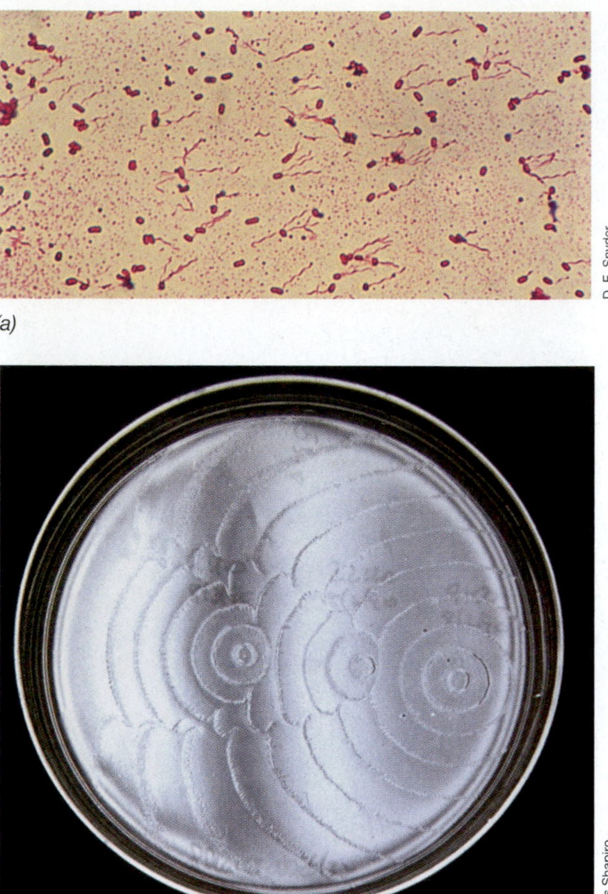
D. E. Snyder

(a)

James Shapiro

(b)

Figure 15.25 Swarming in *Proteus*. *(a)* Cells of *Proteus mirabilis* stained with a flagella stain; the peritrichous flagella of each cell group into a bundle. *(b)* Photo of a swarming colony of *Proteus vulgaris*. Note the concentric rings.

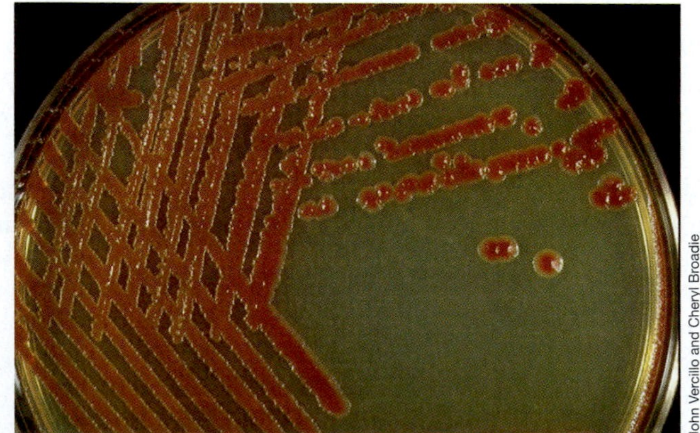
John Vercillo and Cheryl Broadie

Figure 15.26 Colonies of *Serratia marcescens*. The orange-red pigmentation is due to the pyrrole-containing pigment prodigiosin.

Table 15.16 Key diagnostic reactions used to separate the various genera of 2,3-butanediol producers[a]

Genus	Ornithine decarboxylase	Gelatin hydrolysis	Temperature optimum (°C)	Pigmentation	Motility	Lactose	DNase	Sorbitol
Klebsiella	–	–	37–40	None	–	+	–	+
Enterobacter	+	Slow	37–40	Yellow (or none)	+	+	–	+
Serratia	+	+	37–40	Red (or none)	+	–	+	–
Erwinia[b]	–	+ or –	27–30	Yellow (or none)	+	+ or –	–	+
Hafnia	+	–	35	None	+	–	–	–

[a]See Table 32.3 for a description of these diagnostic tests.
[b]See Figure 15.22.

is in agreement with the observed physiological differences. A classification of this group is outlined in **Table 15.16**.

Enterobacter aerogenes is a common species in water and sewage as well as the intestinal tract of warm-blooded animals and is an occasional pathogen in urinary tract infections. One species of *Klebsiella, K. pneumoniae,* occasionally causes pneumonia in humans, but klebsiellas are most commonly found in soil and water. Most *Klebsiella* strains also fix N_2 (⚭ Section 20.14), a property not found among other enteric bacteria.

The genus *Serratia* forms a series of red pyrrole-containing pigments called *prodigiosins* (**Figure 15.26**). Prodigiosin is produced in stationary phase as a secondary metabolite (⚭ Section 25.2) and is of interest because it contains the pyrrole ring also found in the pigments for energy transfer: porphyrins, chlorophylls, and phycobilins (⚭ Sections 20.2 and 20.3). However, there is no evidence that prodigiosin plays any role in energy transfer, and its exact function is unknown. Species of *Serratia* can be isolated from water and soil as well as from the gut of various insects and vertebrates and occasionally from the intestines of humans.

15.12 *Vibrio, Aliivibrio,* and *Photobacterium*

Key Genera: *Vibrio, Aliivibrio, Photobacterium*

The *Vibrio* group, family *Vibrionaceae,* contains gram-negative, facultatively aerobic rods and curved rods that possess a fermentative metabolism. Most species of *Vibrio* are polarly flagellated, although some are peritrichously flagellated. One key difference between the *Vibrio* group and enteric bacteria is that members of the former are oxidase-*positive* (⚭ Table 32.3), whereas members of the latter are oxidase-*negative*. Although *Pseudomonas* species are also polarly flagellated and oxidase-positive, they are not fermentative and so are clearly distinct from *Vibrio* species. The best known genera of the *Vibrio* group are *Vibrio, Aliivibrio,* and *Photobacterium*.

Most vibrios and related bacteria are aquatic, found in marine, brackish, or freshwater habitats. *Vibrio cholerae* is the specific cause of the disease cholera in humans (⚭ Sections 28.11 and 36.5); the organism does not normally cause disease in other hosts. Cholera is one of the most common infectious human diseases in underdeveloped countries and one that has had a long history (the famous early medical microbiologist Robert Koch first isolated *V. cholerae* in 1884). The organism is transmitted almost exclusively via water, and studies on its distribution in the nineteenth century played a major role in demonstrating the importance of water purification in urban areas. We discuss the pathogenesis of *V. cholerae* in Section 28.11.

Vibrio parahaemolyticus is a marine organism. It is a major cause of gastroenteritis in Japan (where raw fish is widely consumed) and has also been implicated in outbreaks of gastroenteritis in other parts of the world, including the United States. The organism can be frequently isolated from seawater or from shellfish and crustaceans, and its primary habitat is probably marine animals, with human infection being a secondary development (⚭ Section 36.1).

Bacterial Bioluminescence

Several species of bacteria can emit light, a process called **bioluminescence** (**Figure 15.27**). Most of these bacteria have been classified in the genera *Photobacterium, Aliivibrio,* and *Vibrio,* but a few luminous species are found also in *Shewanella,* a genus of primarily marine bacteria, and in *Photorhabdus,* a genus of terrestrial bacteria. Most bioluminescent bacteria inhabit the marine environment and can be isolated from seawater, marine sediments, and the external surfaces and gut tracts of marine animals. Some species also colonize specialized organs of certain marine fishes and squids (⚭ Section 24.12), called *light organs,* producing light that the animal uses for signaling, avoiding predators, and attracting prey (Figure 15.27*c–f*; ⚭ Section 24.12). When living symbiotically in light organs of fish and squids, or saprophytically, for example, on the skin of a dead fish or parasitically in the body of a crustacean, luminous bacteria can be recognized by the light they produce. (To see bioluminescence readily, one should observe the material in a completely dark room after the eyes have become adapted to the dark, Figure 15.27*a,b*). Because some pathogenic strains of *Vibrio* are also luminous, such as certain strains of *V. cholerae* and *V. vulnificus,* care should always be taken when handling luminous bacteria.

Although *Photobacterium, Aliivibrio,* and *Vibrio* isolates are facultative aerobes, they are bioluminescent only when O_2 is present. Luminescence in bacteria is catalyzed by bacterial luciferase, which uses O_2, a long-chain aliphatic aldehyde (for

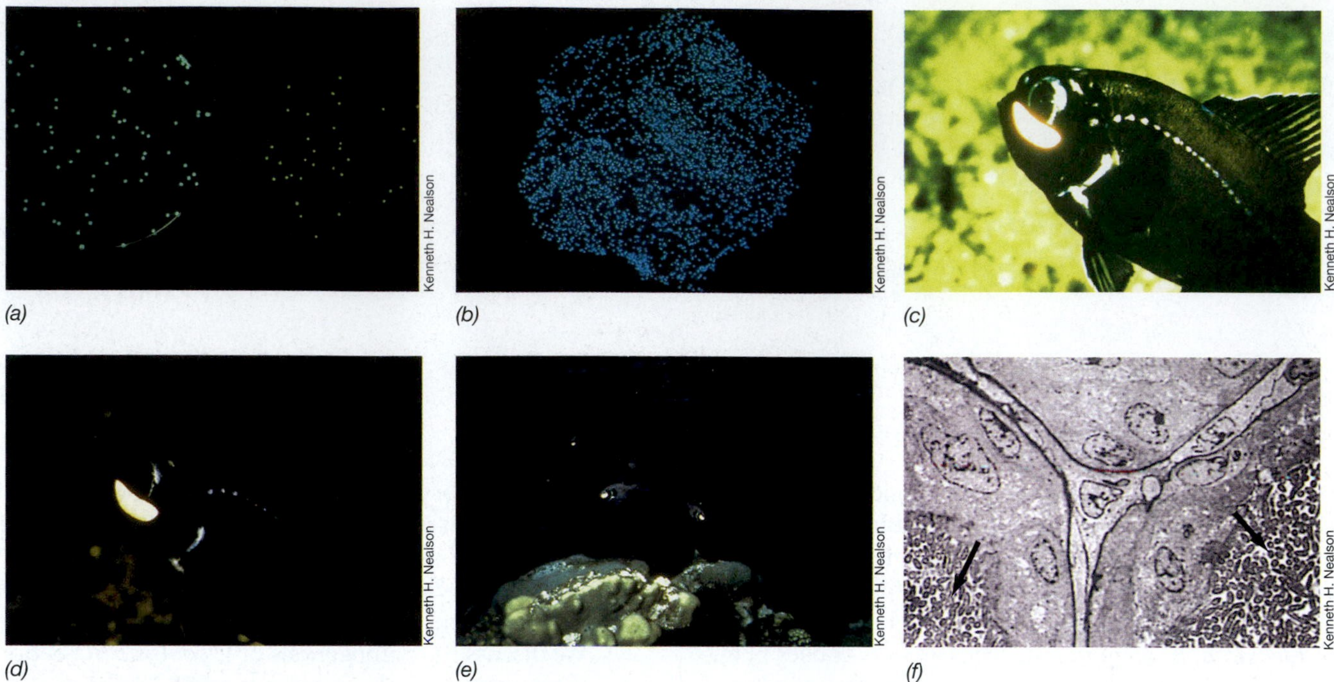

Figure 15.27 Bioluminescent bacteria and their role as light organ symbionts in the flashlight fish. (*a*) Two Petri plates of luminous bacteria photographed by their own light. Note the different colors. Left, *Aliivibrio fischeri* strain MJ-1, blue light, and right, strain Y-1, green light. (*b*) Colonies of *Photobacterium phosphoreum* photographed by their own light. (*c*) The flashlight fish *Photoblepharon palpebratus*; the bright area is the light organ containing bioluminescent bacteria. (*d*) Same fish photographed by its own light. (*e*) Underwater photograph taken at night of *P. palpebratus* in coral reefs in the Gulf of Eilat. (*f*) Electron micrograph of a thin section through the light-emitting organ of *P. palpebratus*, showing the dense array of bioluminescent bacteria (arrows).

example, tetradecanal), and reduced flavin mononucleotide ($FMNH_2$) as substrates. The primary electron donor is NADH, and the bacterial luciferase reaction is as follows:

$$FMNH_2 + O_2 + RCHO \xrightarrow{\text{Luciferase}} FMN + RCOOH + H_2O + \text{Light}$$

The light-generating system constitutes a bypass route for shunting electrons from $FMNH_2$ to O_2, without involving other electron carriers such as quinones and cytochromes. The evolutionary origins of bacterial bioluminescence are obscure, but one possibility is that a primitive luciferase arose in a lineage of bacteria that gave rise to modern day species of *Vibrionaceae* as a means of detoxifying oxygen as Earth's atmosphere became progressively more oxic (Section 14.3). However, many species and strains of *Vibrio*, *Aliivibrio*, and *Photobacterium* lack *luxCDABE*, the genes necessary for producing luciferase and the enzymes for synthesis of the long-chain aldehyde. Apparently, these genes have been lost from many *Vibrio*, *Aliivibrio*, and *Photobacterium* lineages over evolutionary time.

Regulation of Bioluminescence

The expression of luminescence in many luminous bacteria is induced at high population density. The enzyme luciferase and other proteins of the bacterial luminescence system exhibit a population density-responsive induction, called **autoinduction**, in which transcription of the *luxCDABE* genes is controlled in *Aliivibrio fischeri* and related species by a regulatory protein, LuxR, and an inducer molecule, acyl-homoserine lactone (Section 9.6).

During growth, cells produce autoinducer, which can rapidly cross the cytoplasmic membrane in either direction, diffusing in and out of cells. Under conditions in which a high local population density of cells is attained, as in a test tube, colony on a plate, or in the light organ of a fish or squid (Section 24.12), autoinducer can accumulate. When it reaches a threshold concentration in the cell (equal to the concentration in the cell's local environment), the autoinducer interacts with LuxR, forming a complex that activates transcription of *luxCDABE*, and cells become strongly luminous (Figure 15.27; Figure 1.1*c*). In contrast, when cells of luminous bacteria are at low population densities, such as in seawater where the autoinducer can diffuse away, luminescence is not induced. The primary autoinducer in *Aliivibrio fischeri* has been identified as *N*-3-oxo-hexanoylhomoserine lactone; other luminous species produce other kinds of homoserine lactones and even different classes of compounds, resulting in highly specific autoinduction mechanisms. This gene regulatory mechanism is also called *quorum sensing* because of the population density-dependent nature of the phenomenon (Section 9.6).

In saprophytic, parasitic, and symbiotic habitats (Figure 15.27*c–f*), a rationale for population density-responsive induction of luminescence may be that luminescence develops only when sufficiently high population densities are reached to allow the light produced to be visible to animals. The bacterial light can then attract animals to feed on the luminous material, thereby bringing the bacteria into the animal's nutrient-rich gut tract for further growth, or serve as the source of light in symbiotic, light organ associations. The genetics of quorum sensing using bioluminescence as a model experimental system has been actively explored; this form of regulation has also been found to be a general feature of many different nonluminous bacteria, including several animal and plant pathogens. Quorum sensing in these bacteria controls activities such as the production of extracellular enzymes and expression of virulence factors for which a high population density is thought to be beneficial for the bacteria (Section 9.6).

15.11 and 15.12 MiniReview

The enteric bacteria are a large group of facultative aerobic rods of medical and molecular biological significance. *Vibrio*, *Aliivibrio*, and *Photobacterium* species are marine organisms; some species are pathogenic and some are bioluminescent.

- How is *Escherichia coli* distinguished from *Enterobacter aerogenes* on physiological grounds?
- Describe two major properties of *Proteus* species that distinguish them from other enteric bacteria.
- What is required for an organism such as *Aliivibrio* to give off visible light?

15.13 Rickettsias

Key Genera: *Rickettsia*, *Wolbachia*

The rickettsias are small, gram-negative, coccoid or rod-shaped *Proteobacteria* in the size range of 0.3–0.7 μm wide and 1–2 μm long. They are, with one exception, obligate intracellular parasites and have not yet been cultivated in the absence of host cells (**Figure 15.28*a***). Rickettsias are the causative agents of several human diseases, including typhus (any one of several similar diseases caused by louseborne bacteria), Rocky Mountain spotted fever, and Q fever (Section 35.3).

Electron micrographs of thin sections of rickettsial cells show a typical prokaryotic morphology (Figure 15.28*b*); both cell wall and cytoplasmic membrane are clearly present. The rickettsial cell wall contains muramic acid and diaminopimelic acid, and the cells divide by binary fission with doubling times of about 8 h. The penetration of a host cell by a rickettsial cell is an active process, requiring both host and parasite to be alive and metabolically active. Once inside the host phagocytic cell, the bacteria multiply primarily in the cytoplasm and continue replicating until the host cell is loaded

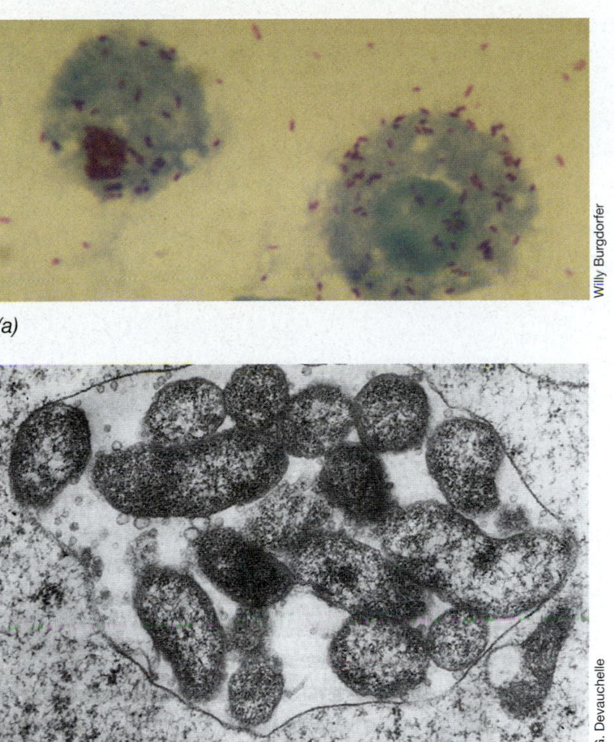

Figure 15.28 Rickettsias growing within host cells. *(a) Rickettsia rickettsii* in tunica vaginalis cells of the vole, *Microtus pennsylvanicus*. Cells are about 0.3 μm in diameter. *(b)* Electron micrograph of cells of *Rickettsiella popilliae* within a blood cell of its host, the beetle *Melolontha melolontha*. Notice that the bacteria are growing in a vacuole within the host cell.

with parasites (Figures 15.28; Figure 35.6). The host cell then bursts and liberates the bacterial cells into the surrounding fluid. Several genera of rickettsias are known, and the properties of four key genera are shown in **Table 15.17**.

Metabolism and Pathogenesis

Much research has been done on the metabolic activities and biochemical pathways of rickettsias in an attempt to explain why they are obligate intracellular parasites. Many rickettsias possess a highly specific energy metabolism: They can oxidize only glutamate or glutamine and cannot oxidize glucose or organic acids. However, *Coxiella burnetii*, the causative agent of the disease Q fever, is able to utilize both glucose and pyruvate as electron donors. Rickettsias possess a respiratory chain complete with cytochromes and are able to carry out electron transport phosphorylation, using NADH as electron donor. They are also able to synthesize at least some of the small molecules needed for macromolecular synthesis and growth, and they obtain the rest of their nutrients from the host cell. Thus, although parasites, rickettsias maintain a number of independent metabolic functions.

Rickettsias do not survive long outside their hosts, and this may explain why they must be transmitted from animal to animal by arthropod vectors. When the arthropod obtains a

Table 15.17 Characteristics of rickettsias

Genus and Species	Rickettsial group	Alternate host	Cellular location	Phylogenetic group[a]	DNA hybridization to R. rickettsii DNA (%)[b]
Rickettsia					
R. rickettsii	Spotted fever	Tick	Cytoplasm and nucleus	*Alpha*	100
R. prowazekii[c]	Typhus	Louse	Cytoplasm		53
R. typhi	Typhus	Flea	Cytoplasm		36
Rochalimaea					
R. quintana	Trench fever	Louse	Epicellular	*Alpha*	30
R. vinsonii	—	Vole	Epicellular		30
Coxiella					
C. burnetii	Q fever	Tick	Vacuoles	*Gamma*	—
Ehrlichia					
E. chaffeensis	Ehrlichiosis (humans)	Tick or domestic animals	Mononuclear leukocytes	*Alpha*	—
E. equi	Potomac fever (horses)	Tick	Granulocyte	*Alpha*	—
Wolbachia[d]					
W. pipientis	—	Arthropods	Cytoplasm	*Alpha*	—

[a]All are *Proteobacteria*.
[b]For discussion of DNA:DNA hybridization, see Section 14.11.
[c]The genome of this organism has been sequenced and shows several similarities to the mitochondrial genome.
[d]Not a pathogen of humans or other animals.

blood meal from an infected animal, rickettsias present in the blood are ingested by the arthropod, where they penetrate to the epithelial cells of the gastrointestinal tract, multiply, and appear later in the feces. When the arthropod feeds on an uninfected individual, it then transmits the rickettsias either directly with its mouthparts or by contaminating the bite with its feces. However, *C. burnetii* (Section 35.3) can also be transmitted to the respiratory system by aerosols. *C. burnetii* is the most resistant of the rickettsias to physical damage, probably because it produces a resistant, sporelike form. This probably explains the ability of *C. burnetii* to survive in air.

Rochalimaea is an atypical rickettsia because it can be grown in culture and is thus not an obligate intracellular parasite. In addition, when growing in tissue culture, cells of *Rochalimaea* grow on the outside surface of the eukaryotic host cells rather than within the cytoplasm or the nucleus. *Rochalimaea quintana* is the causative agent of trench fever, a disease that decimated troops in World War I. Species of the genus *Ehrlichia* cause disease in humans and other animals, two of which, ehrlichiosis in humans and Potomac fever in horses, can be quite debilitating (Section 35.3).

Wolbachia

The genus *Wolbachia* contains species of rod-shaped *Alphaproteobacteria* that are intracellular parasites of arthropod insects (**Figure 15.29**). *Wolbachia* are phylogenetically related to the rickettsias and can have any of several effects on their insect hosts. These include inducing parthenogenesis (development of unfertilized eggs), the killing of males, and feminization (the conversion of male insects into females).

Wolbachia pipientis is the best-studied species in the genus. Cells colonize the insect egg (Figure 15.29), where they multiply in vacuoles of host cells surrounded by a membrane of host origin. Cells of *W. pipientis* are passed from an infected female to her offspring through this egg infection. *Wolbachia*-induced parthenogenesis occurs in a number of species of wasps. In these insects, males normally arise from unfertilized eggs (which contain only one set of chromosomes), while females arise from fertilized eggs (which contain two sets of chromosomes). However, in unfertilized eggs infected with *Wolbachia*, the organism somehow triggers a doubling of the chromosome number, thus yielding only females. Predictably, if female insects are fed antibiotics that kill *Wolbachia*, parthenogenesis ceases.

Like parasitic bacteria in general, *W. pipientis* has a small genome (about 1.5 Mbp), which has been sequenced. Although this parasite does not cause disease in either vertebrates or its invertebrate hosts, its various reproductive effects raise the interesting question of the extent it has affected evolution and speciation in a major class of insects (arthropods make up 70% of all known insects). In some insects *Wolbachia* infection may actually be essential for survival. For example, in the nematode worms that cause the diseases elephantiasis and river blindness, antibiotic treatment kills the worms, apparently by killing their *Wolbachia* symbionts.

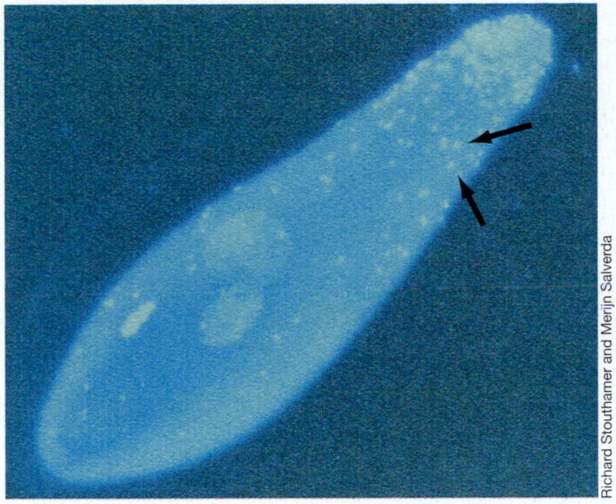

Richard Stouthamer and Merijn Salverda

Figure 15.29 ***Wolbachia.*** Photomicrograph of a 4′,6-diamidine-2′ phenylindole dihydrochloride-stained (DAPI) egg of the parasitoid wasp, *Trichogramma kaykai* infected with *Wolbachia pipientis,* which induces parthenogenesis. The *W. pipientis* cells are primarily located in the narrow end of the egg (arrows).

15.13 MiniReview

The rickettsias are obligate intracellular parasites, many of which cause disease. Rickettsias are deficient in many metabolic functions and obtain key metabolites from their hosts.

- Name a disease caused by a *Rickettsia* species.
- What is meant by the term "obligate intracellular parasite"?
- What effects can *Wolbachia* have on its insect hosts?

IV MORPHOLOGICALLY UNUSUAL *PROTEOBACTERIA*

Some *Proteobacteria* have unusual morphologies or undergo fascinating life cycles. We consider some of these here, focusing on common curved and spiral-shaped bacteria, a few filamentous bacteria that encase their cells in a sheath, and the morphologically unusual prosthecate and stalked bacteria. In the latter group, we will focus on the important model bacterium for the study of cell differentiation, *Caulobacter*. Physiologically, all of these organisms are chemoorganotrophs, consuming organic matter in a wide variety of primarily aquatic habitats.

15.14 Spirilla

Key Genera: *Spirillum, Bdellovibrio, Campylobacter*

The **spirilla** are gram-negative, motile, spiral-shaped *Proteobacteria* that show a wide variety of physiological attributes. Some of the key taxonomic criteria used are cell shape, size, number of polar flagellation (single or multiple), relation to oxygen (obligately aerobic, microaerophilic, facultative), relationship to plants (as symbionts or plant pathogens) or animals (as pathogens), fermentative ability, and certain other physiological characteristics (for example, nitrogen-fixing ability, halophilic nature, thermophilic nature). The genera to be covered here are given in **Table 15.18**, where it can be seen that spirilla are found in each of the five classes of *Proteobacteria*.

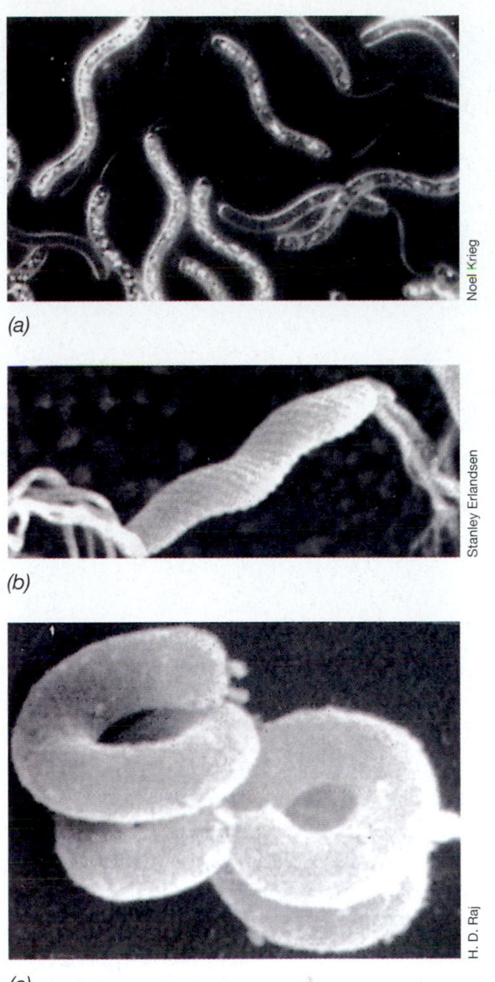

(a) Noel Krieg; (b) Stanley Erlandsen; (c) H. D. Raj

Figure 15.30 **Spirilla.** *(a) Spirillum volutans,* visualized by dark-field microscopy, showing flagellar bundles and volutin (polyphosphate) granules. Cells are about 1.5 × 25 μm. *(b)* Scanning electron micrograph of an intestinal spirillum. Note the polar flagellar tufts and the spiral structure of the cell surface. *(c)* Scanning electron micrograph of cells of *Ancyclobacter aquaticus.* Cells are about 0.5 μm in diameter.

Spirillum, Aquaspirillum, Oceanospirillum, and *Azospirillum*

The spirilla, which are helically curved rods, are motile by means of polar flagella, usually tufts at both poles (**Figure 15.30*b***). The number of turns in the helix may vary from less than one complete turn (in which case the organism looks like a vibrio) to many turns. Spirilla with many turns can superficially resemble spirochetes (∞ Section 16.16) but differ

Table 15.18 Characteristics of the genera of spiral-shaped bacteria[a]

Genus	Phylogenetic group[b]	Characteristics
Spirillum	Beta	Cell diameter 1.7 μm; microaerophilic; freshwater
Aquaspirillum	Beta	Cell diameter 0.2–1.5 μm; aerobic; freshwater
Magnetospirillum	Alpha	Vibrio to spirillum-shaped; cell diameter about 0.3 μm; contains magnetosomes; microaerophilic
Oceanospirillum	Gamma	Cell diameter 0.3–1.2 μm; aerobic; marine (require 3% NaCl)
Azospirillum	Alpha	Cell diameter 1 μm; microaerophilic; soil and rhizosphere; fixes N_2
Herbaspirillum	Beta	Cell diameter 0.6–0.7 μm; microaerophilic; soil and rhizosphere; fixes N_2
Campylobacter	Epsilon	Cell diameter 0.2–0.8 μm; microaerophilic to anaerobic; pathogenic or commensal in humans and other animals; single polar flagellum
Helicobacter	Epsilon	Cell diameter 0.5–1 μm; tuft of polar flagella; associated with pyloric ulcers in humans
Bdellovibrio	Delta	Cell diameter 0.25–0.4 μm; aerobic; predatory on other bacteria; single polar sheathed flagellum
Ancyclobacter	Alpha	Cell diameter 0.5 μm; curved rods forming rings; nonmotile, aerobic; sometimes gas-vesiculate

[a]All are gram-negative and respiratory but never fermentative.
[b]All are *Proteobacteria*.

distinctly from the latter phylogenetically. In addition, spirilla do not have the outer sheath and endoflagella of spirochetes, but instead contain typical bacterial flagella (Section 4.13).

The cells of some spirilla are very large and were seen by early microscopists. For example, it is likely that Antoni van Leeuwenhoek first observed *Spirillum* species in the 1670s (Section 1.6). The organism seen by van Leeuwenhoek was probably *Spirillum volutans,* a rather large prokaryote (Figure 15.30*a*). A phototrophic organism resembling *S. volutans* is the purple bacterium *Thiospirillum,* considered earlier (Figure 15.4*b*). *S. volutans* is microaerophilic, requiring O_2, but is inhibited by O_2 at normal levels. Another prominent characteristic of cells of *S. volutans* is the formation of granules (volutin granules) consisting of polyphosphate (Figure 15.30*a*; Section 4.10).

Azospirillum lipoferum is a nitrogen-fixing organism, which was originally described and named *Spirillum lipoferum* by Beijerinck in 1925. *A. lipoferum* is of considerable interest because it enters into a symbiotic relationship with tropical grasses and grain crops. Although not as intimate an association as that between root nodule bacteria of the genus *Rhizobium* and leguminous plants (Section 24.15), the *A. lipoferum*–corn association clearly benefits the corn plant by nitrogen fixation.

The small-diameter spirilla (which are not microaerophilic) have been separated into two genera, *Aquaspirillum* and *Oceanospirillum*. The former includes freshwater species, and the latter includes species that inhabit seawater and require NaCl for growth (Table 15.18). Numerous species of *Aquaspirillum* and *Oceanospirillum* have been described, the various species being separated on physiological and phylogenetic grounds. These organisms undoubtedly play an important role in the recycling of organic matter in aquatic environments.

Magnetotactic Spirilla

Highly motile microaerophilic magnetic spirilla have been isolated from freshwater habitats. These organisms demonstrate a dramatic directed movement in a magnetic field called *magnetotaxis*. In an artificial magnetic field, magnetic spirilla quickly orient their long axis along the north–south magnetic moment of the field. Within the cells are chains of 5–40 magnetic particles called *magnetosomes,* consisting of magnetite (Fe_3O_4) and greigite (Fe_3S_4) (Section 4.10 and Figure 4.33). Magnetosomes function as internal magnets that orient the cells along a specific magnetic field.

Magnetic bacteria display one of two magnetic polarities depending on the orientation of magnetosomes within the cell. Cells in the Northern Hemisphere have the north-seeking pole of their magnetosomes forward with respect to their flagella and thus move in a northward direction. Cells in the Southern Hemisphere have the opposite polarity and move southward. Although the ecological role of bacterial magnets is unclear, the ability to orient in a magnetic field may be of selective advantage in maintaining these microaerophilic organisms in zones of low O_2 concentration near the oxic–anoxic interface. The spirillum *Magnetospirillum magnetotacticum* (**Figure 15.31** and Table 15.18) is a major organism in this group.

Bdellovibrio

These small curved organisms have the unusual property of preying on other bacteria, using as nutrients the cytoplasmic constituents of their hosts. These bacterial predators are small, highly motile cells that stick to the surfaces of their prey cells. Because of this, they have been given the name *Bdellovibrio* (*bdello* is a prefix meaning "leech"). Other predatory bacteria have been isolated and given such suggestive genus names as *Vampirovibrio*. However, *Bdellovibrio* has a unique mode of attack and develops intraperiplasmically.

After attachment of a *Bdellovibrio* cell to its prey, the predator penetrates through the prey wall and replicates in the periplasmic space, eventually forming a spherical structure called a *bdelloplast*. The stages of attachment and penetration are shown in electron micrographs in **Figure 15.32** and

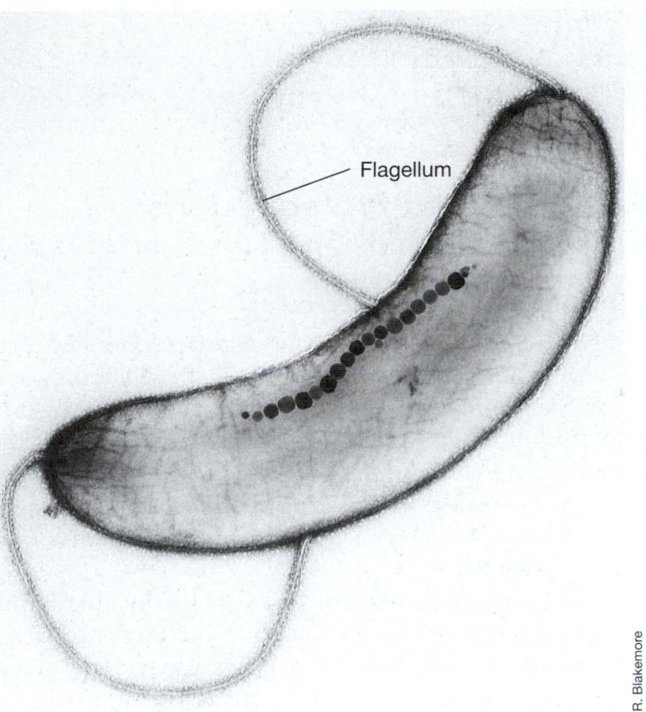

Figure 15.31 Negatively stained electron micrograph of a magnetotactic spirillum, *Magnetospirillum magnetotacticum*. A cell measures 0.3 × 2 μm. This bacterium contains particles of Fe_3O_4 (magnetite), called magnetosomes (∞ Figure 4.33), arranged in a chain; the particles align the cell along geomagnetic lines. The organism was isolated from a water treatment plant in Durham, New Hampshire.

diagrammatically in **Figure 15.33**. A wide variety of gram-negative bacteria can be attacked by a single *Bdellovibrio* species; gram-positive cells are not attacked.

Bdellovibrio is an obligate aerobe, obtaining its energy from the oxidation of amino acids and acetate. In addition, *Bdellovibrio* assimilates nucleotides, fatty acids, peptides, and even some intact proteins, directly from its host without first breaking them down. Thus it is clear that interesting and unusual biochemical processes have developed as part of the predatory lifestyle of *Bdellovibrio*. Prey-independent derivatives of predatory strains of *Bdellovibrio* can be isolated, however, showing that predation is not obligatory. Prey-independent strains can be grown on complex media containing yeast extract and peptone.

Phylogenetically, bdellovibrios are members of the *Deltaproteobacteria* (Figure 15.1) and contain a genome about half the size of that of *Escherichia coli*. Taxonomically, two species of *Bdellovibrio* and one species of the genus *Bacteriovorax* are recognized and supported by the results of genomic DNA–DNA hybridization (∞ Section 14.11).

Bdellovibrio is widespread in soil and water, including the marine environment. Their detection and isolation require methods reminiscent of those used in the study of bacterial viruses (∞ Section 10.4). Prey bacteria are spread on the surface of an agar plate to form a lawn, and the surface is inoculated with a small amount of soil suspension that has been filtered through a membrane filter; the latter retains most bacteria but allows the small *Bdellovibrio* cells to pass. On incubation of the agar plate, plaques analogous to bacteriophage plaques (∞ Figure 10.6*b*) are formed at locations where *Bdellovibrio* cells are growing. However, unlike phage plaques, which continue to enlarge only while the bacterial host is still growing, *Bdellovibrio* plaques continue to enlarge even after their prey cells have stopped growing. This results in ever enlarging plaques on the agar surface. Pure cultures of *Bdellovibrio* can then be isolated from these plaques. *Bdellovibrio* cultures have been obtained from many soils and are thus widespread in distribution.

Ancyclobacter

Members of the genus *Ancyclobacter* are ring-shaped, nonmotile, extremely nutritionally diverse chemoorganotrophic bacteria (Figure 15.30*c*). They resemble very tightly coiled vibrios and are widely distributed in aquatic environments. A phototrophic counterpart to *Ancyclobacter* is the purple

UNIT 3

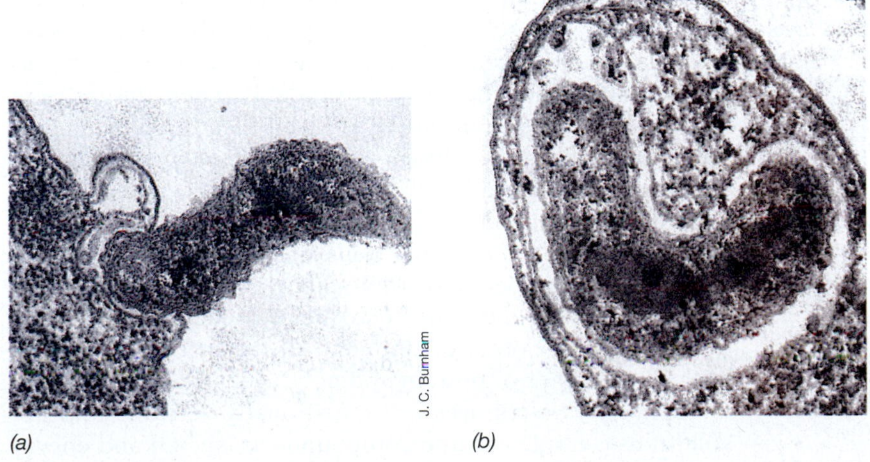

Figure 15.32 Stages of attachment and penetration of a prey cell by *Bdellovibrio*. A *Bdellovibrio* cell measures about 0.3 μm in diameter. *(a, b)* Electron micrographs of thin sections of *Bdellovibrio* attacking *Escherichia coli*: *(a)* early penetration; *(b)* complete penetration. The *Bdellovibrio* cell is enclosed in a membranous infolding of the prey cell (the bdelloplast) and replicates in the periplasmic space between the wall and the cytoplasmic membrane.

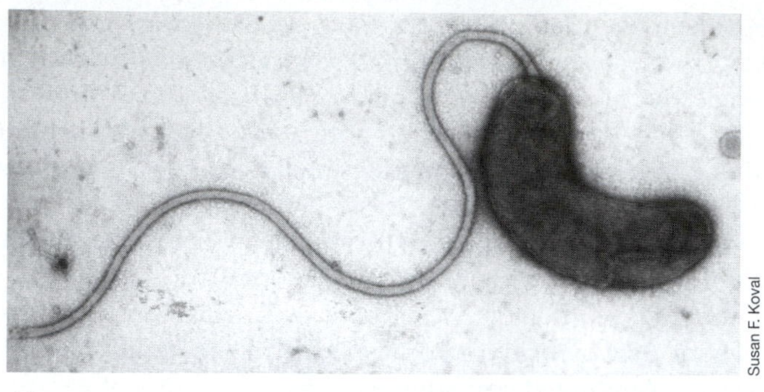

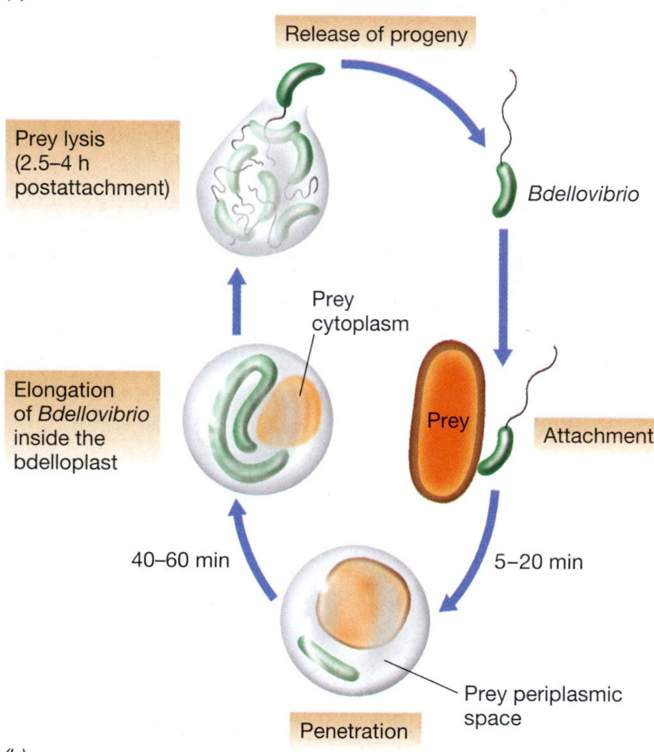

Figure 15.33 Developmental cycle of the bacterial predator *Bdellovibrio bacteriovorus*. *(a)* Electron micrograph of a cell of *Bdellovibrio bacteriovorus*. Note the very thick flagellum present on this organism (compare with Figure 4.45). *(b)* Events in predation. Following primary contact with a gram-negative bacterium, a highly motile *Bdellovibrio* cell attaches to and penetrates into the prey periplasmic space. Once inside, *Bdellovibrio* elongates and within 4 h progeny cells are released. The number of progeny cells released varies with the size of the prey bacterium. For example, 5–6 bdellovibrios are released from each infected *Escherichia coli* cell and 20–30 for a larger cell, such as a species of *Aquaspirillum*.

nonsulfur bacterium *Rhodocyclus purpureus*, considered earlier (Section 15.2 and Figure 15.6*e*).

15.14 MiniReview

Spirilla are spiral-shaped, chemoorganotrophic bacteria widespread in the aquatic environment. The genera *Helicobacter* and *Campylobacter* are pathogenic spirilla. Spirilla are distributed among all five classes of the *Proteobacteria*.

- What is a volutin granule?
- What is unique about the spirilla *Bdellovibrio* and *Magnetospirillum*?

15.15 Sheathed *Proteobacteria*: *Sphaerotilus* and *Leptothrix*

Key Genera: *Sphaerotilus*, *Leptothrix*

Sheathed bacteria are filamentous *Betaproteobacteria* (Table 15.1) with a unique life cycle in which flagellated swarmer cells form within a long tube or sheath. Under unfavorable growth conditions, the swarmer cells move out and become dispersed to new environments, leaving behind the empty sheath. Under favorable conditions, the cells grow vegetatively within the filament, leading to the formation of long, cell-packed sheaths.

Sheathed bacteria are common in freshwater habitats that are rich in organic matter, such as polluted streams. Because they are typically found in flowing waters, they are also abundant in trickling filters and activated sludge digestors in sewage treatment plants (Section 36.3). In habitats in which reduced iron (Fe^{2+}) or manganese (Mn^{2+}) is present, the sheaths may become coated with ferric hydroxide or manganese oxide (Figure 20.29). Iron precipitation is probably due to chemical reactions, but some sheathed bacteria have the biochemical ability to oxidize manganous ions to manganese oxide. Two genera are the major organisms here: *Sphaerotilus*, whose members do not oxidize manganese, and *Leptothrix*, whose members do.

Sphaerotilus

The *Sphaerotilus* filament is composed of a chain of rod-shaped cells enclosed in a closely fitting sheath. This thin, transparent sheath is difficult to see when it is filled with cells, but when it is partially empty, the sheath can easily be seen by phase-contrast microscopy (**Figure 15.34*a***) or by staining. Individual cells are 1–2 μm wide and 3–8 μm long and stain gram negatively.

The cells within the sheath divide by binary fission (Figure 15.34*b*), and the new cells pushed out at the end synthesize new sheath material. Thus, the sheath is always formed at the tips of the filaments. Eventually swarmer cells are liberated from the sheaths, probably when the nutrient supply is low. These cells are actively motile, the flagella being arranged lophotrichously (in a bundle at one pole) (Figure 15.34*c*). Probably the flagella are synthesized before the cells leave the sheath and, if so, may even aid in their liberation. It is thought that the swarmer cells then migrate, attach to a solid surface, and begin to grow, each swarmer being the forerunner of a new filament. The sheath, which is devoid of muramic acid or other components of the peptidoglycan cell wall, is a complex of protein and polysaccharide.

Sphaerotilus cultures are nutritionally versatile and are able to use simple organic compounds as carbon and energy

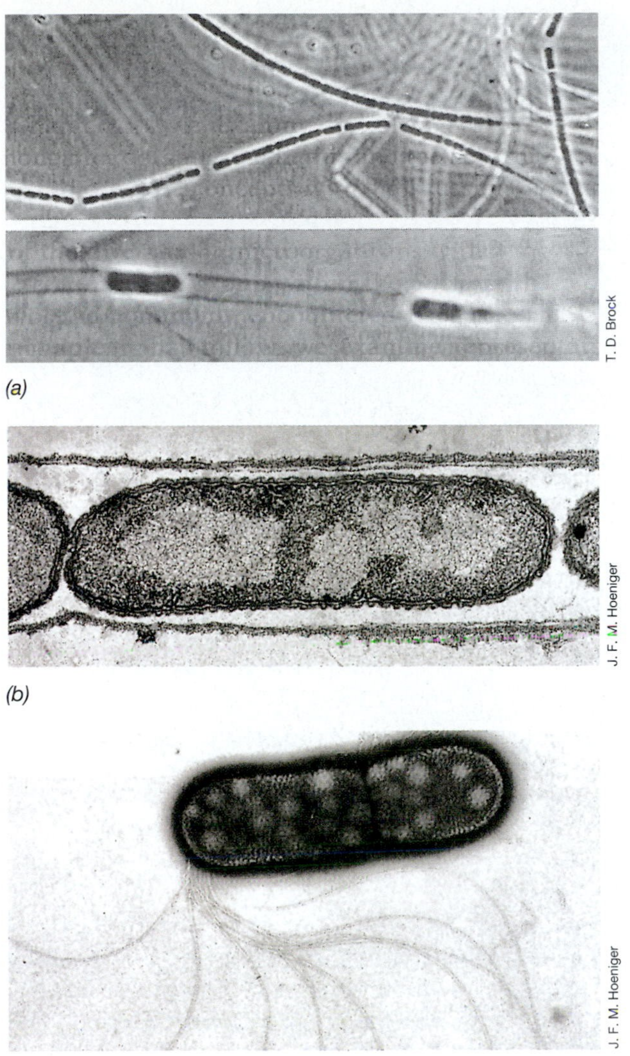

Figure 15.34 ***Sphaerotilus natans.*** A single cell is about 2 μm wide. *(a)* Phase-contrast photomicrographs of material collected from a polluted stream. Active growth stage (above) and swarmer cells leaving the sheath. *(b)* Electron micrograph of a thin section through a filament. *(c)* Electron micrograph of a negatively stained swarmer cell. Notice the polar flagellar tuft.

sources, along with inorganic nitrogen sources. Befitting its habitat in flowing waters, *Sphaerotilus* is an obligate aerobe. *Sphaerotilus* blooms often occur in the fall of the year in streams and brooks when leaf litter causes a temporary increase in the organic content of the water. Its filaments are the main component of a microbial complex that wastewater engineers call "sewage fungus," a filamentous slime found on the rocks in streams receiving sewage pollution. In activated sludge of sewage treatment plants (Section 36.2), *Sphaerotilus* is often responsible for bulking, the detrimental condition mentioned earlier in connection with *Beggiatoa* (Section 15.4). The tangled masses of *Sphaerotilus* filaments so increase the bulk of the sludge that it does not settle properly. This has a negative effect on the oxidation of organic matter and the recycling of inorganic nutrients.

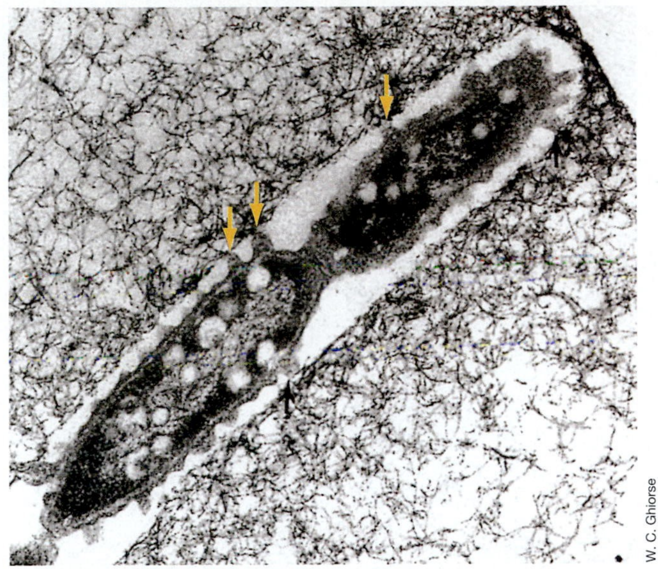

Figure 15.35 ***Leptothrix* and iron precipitation.** Transmission electron micrograph of a thin section of *Leptothrix* in a sample from a ferromanganese film in a swamp in Ithaca, New York. A single cell measures about 0.9 μm in diameter. Note the protuberances of the cell envelope that contact the sheath (arrows).

Leptothrix

The ability of *Sphaerotilus* and *Leptothrix* to precipitate iron oxides on their sheaths is well established, and when sheaths become iron-encrusted, as occurs in iron-rich waters, they can frequently be seen microscopically (**Figure 15.35**). Iron precipitates when iron, chelated to organic materials such as humic or tannic acids, is metabolized to iron oxides. The iron binds to the sheath and the organic constituents are taken up and used as a carbon or energy source.

Besides Fe^{2+} oxidation, *Leptothrix* can oxidize Mn^{2+} to Mn^{4+}. The gene encoding the manganese-oxidizing protein of *Leptothrix* has been isolated, and biochemical studies have shown that the protein resides in the sheath, not inside the cells. Thus the sheath of *Leptothrix* is critical for energy metabolism in this bacterium.

15.16 Budding and Prosthecate/Stalked Bacteria

Key Genera: *Hyphomicrobium, Caulobacter*

This large and heterogeneous group of primarily *Alphaproteobacteria* contains organisms that form various kinds of cytoplasmic extrusions: *stalks, hyphae,* or *appendages* (**Table 15.19**). Extrusions of these kinds, which are smaller in diameter than the mature cell, contain cytoplasm, and are bounded by the cell wall, are collectively called **prosthecae** (**Figure 15.36**).

Budding Division

Budding bacteria divide as a result of unequal cell growth. In contrast to binary fission that forms two equivalent cells

Table 15.19 Characteristics of stalked, appendaged (prosthecate), and budding bacteria

Characteristics	*Genus*	*Phylogenetic group*[a]
Stalked bacteria:		
Stalk an extension of the cytoplasm and involved in cell division	*Caulobacter*	*Alpha*
Stalked, fusiform-shaped cells	*Prosthecobacter*	*Verrucomicrobiaceae*[b]
Stalked, but stalk an excretory product not containing cytoplasm:		
Stalk depositing iron, cell vibrioid	*Gallionella*	*Beta*
Laterally excreted gelatinous stalk not depositing iron	*Nevskia*	*Gamma*
Appendaged (prosthecate) bacteria:		
Single or double prosthecae	*Asticcacaulis*	*Alpha*
Multiple prosthecae		
Short prosthecae, multiply by fission, some with gas vesicles	*Prosthecomicrobium*	*Alpha*
Flat, star-shaped cells, some with gas vesicles	*Stella*	*Alpha*
Long prosthecae, multiply by budding, some with gas vesicles	*Ancalomicrobium*	*Alpha*
Budding bacteria:		
Phototrophic, produce hyphae	*Rhodomicrobium*	*Alpha*
Phototrophic, budding without hyphae	*Rhodopseudomonas*	*Alpha*
Chemoorganotrophic, rod-shaped cells	*Blastobacter*	*Alpha*
Chemoorganotrophic, buds on tips of slender hyphae		
Single hyphae from parent cell	*Hyphomicrobium*	*Alpha*
Multiple hyphae from parent cell	*Pedomicrobium*	*Alpha*

[a]All but *Prosthecobacter* are *Proteobacteria*.
[b]See Section 16.11.

(Figure 6.1), cell division in stalked and budding bacteria forms a totally new daughter cell, with the mother cell retaining its original identity (**Figure 15.37**). The major genera of budding bacteria and a summary of their cell division processes are shown in Figure 15.37.

A fundamental difference between these bacteria and conventional bacteria is not the formation of buds or stalks but the formation of new cell wall material from a single point (polar growth) rather than throughout the whole cell (intercalary growth) as in binary fission (Sections 6.1–6.4). Several genera not normally considered to be budding bacteria show polar growth without differentiation of cell size (Figure 15.37). An important consequence of polar growth is that internal structures, such as membrane complexes, are not involved in the cell division process and must be formed *de novo*. However, this has an advantage in that more complex internal structures can be formed in budding cells than in cells that divide by binary fission, since the latter cells would have to partition these structures between the two newly forming daughter cells. Many budding bacteria, particularly phototrophic and chemolithotrophic species, contain extensive internal membrane systems. **www.microbiologyplace.com Online Tutorial 15.1: Cell Division in Conventional, Budding, and Stalked *Bacteria***

Budding Bacteria: *Hyphomicrobium*

Two well-studied budding bacteria are closely related phylogenetically: *Hyphomicrobium,* which is chemoorganotrophic, and *Rhodomicrobium,* which is phototrophic. These organisms release buds from the ends of long, thin hyphae. The hypha is a direct cellular extension of the mother cell and contains cell wall, cytoplasmic membrane, ribosomes, and occasionally DNA.

Figure 15.38 shows the process of reproduction in *Hyphomicrobium*. The mother cell, which is often attached by its base to a solid substrate, forms a thin outgrowth that lengthens to become a hypha. At the end of the hypha, a bud forms. This bud enlarges, forms a flagellum, breaks loose from the mother cell, and swims away. Later, the daughter cell loses its flagellum and after a period of maturation forms a hypha and buds. More buds can also form at the hyphal tip of the mother cell leading to arrays of cells connected by hyphae. In some cases, a bud begins to form directly from the mother cell without the intervening formation of a hypha, whereas in other cases a single cell forms hyphae from each end (**Figure 15.39a**).

Nucleoid replication events during the budding cycle are of interest (Figure 15.38). First, the DNA located in the mother cell replicates. Then, once the bud has formed, a copy of the circular chromosome moves down the length of the hypha and into the bud. A cross-septum then forms, separating the still-developing bud from the hypha and mother cell.

Hyphomicrobium is a methylotrophic bacterium (Section 15.6) and is widespread in freshwater, marine, and terrestrial habitats. Preferred carbon sources are one-carbon compounds such as methanol, methylamine, formaldehyde, and formate. Enrichment cultures of *Hyphomicrobium* can be prepared, taking advantage of the ability of this organism to grow in very dilute conditions. All that is needed is a mineral salts

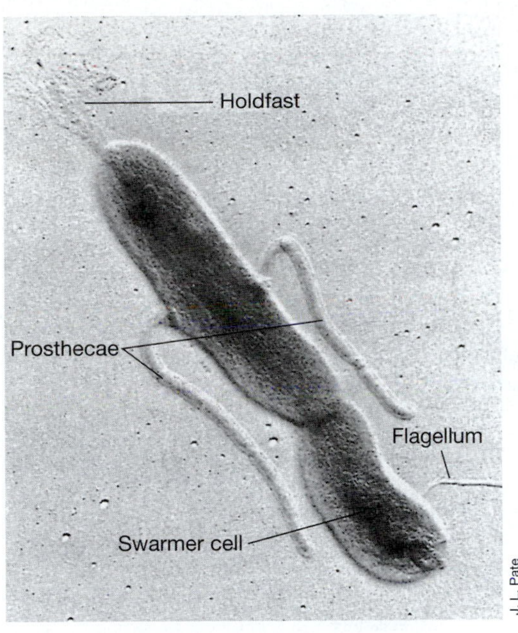

(a)

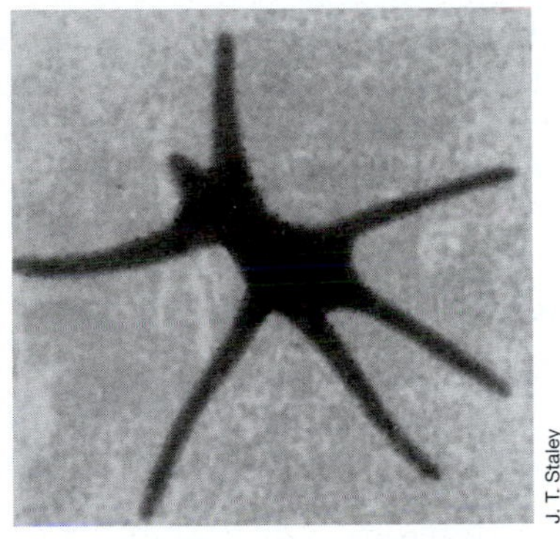

(b)

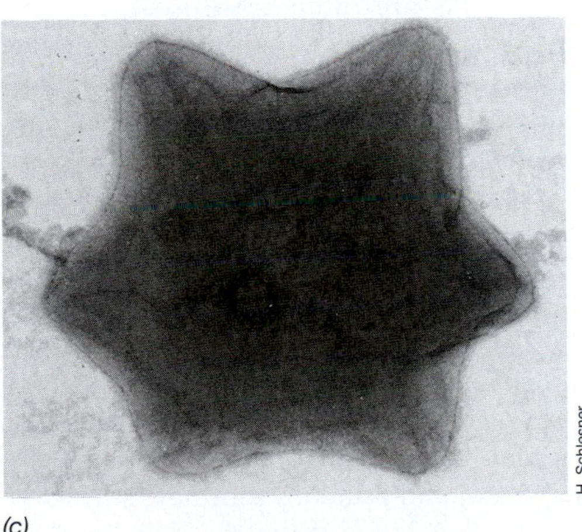

(c)

Figure 15.36 Prosthecate bacteria. *(a)* Electron micrograph of a shadow-cast preparation of *Asticcacaulis biprosthecum*, illustrating the location and arrangement of the prosthecae. Cells are about 0.6 μm wide. Note also the holdfast material and the swarmer cell in the process of differentiation. *(b)* Electron micrograph of a negatively stained preparation of a cell of the prosthecate bacterium *Ancalomicrobium adetum*. The appendages are cellular (prosthecae) because they are bounded by the cell wall, contain cytoplasm, and are about 0.2 μm in diameter. *(c)* Electron micrograph of the star-shaped bacterium *Stella*. Cells are about 0.8 μm in diameter.

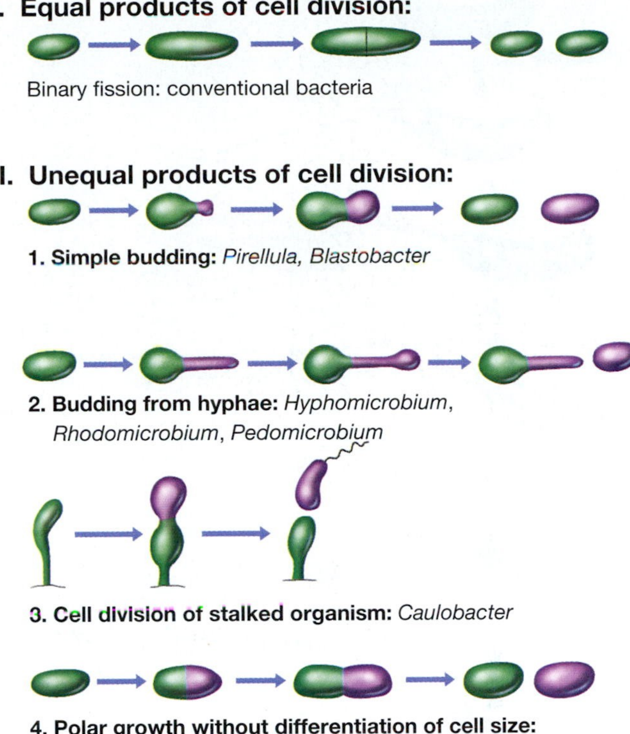

Figure 15.37 Cell division. Contrast between cell division in conventional bacteria and in budding and stalked bacteria.

medium lacking organic carbon and nitrogen to which an inoculum of soil, mud, or water is added. After several weeks of incubation, a surface film develops and is streaked out on agar medium containing methylamine or methanol as a sole carbon source. Colonies are then checked microscopically for the characteristic *Hyphomicrobium* cellular morphology (Figure 15.39). A fairly specific enrichment procedure for *Hyphomicrobium* uses methanol as electron donor with nitrate as electron acceptor under anoxic conditions. Virtually the only denitrifying (∞ Section 21.7) bacterium that uses methanol is *Hyphomicrobium*, and so this procedure can select this organism out of a wide variety of environments.

Prosthecate and Stalked Bacteria

Prosthecate and stalked bacteria (**Figures** 15.36 and **15.40**) are appendaged bacteria that attach to particulate matter, plant material, or other microorganisms in aquatic habitats. Although a major function of these appendages is undoubtedly attachment, they also significantly increase the surface-to-volume ratio of the cells (prosthecae have large surface areas but almost no volume). Recall from our earlier discussions that the high surface-to-volume ratio of prokaryotic cells confers an increased ability to take up nutrients and expel wastes (∞ Section 4.2). The unusual morphology of appendaged bacteria (Figure 15.36) carries this theme to an extreme and may be an evolutionary adaptation to life in the oligotrophic (nutrient-poor) waters where these organisms are most commonly found.

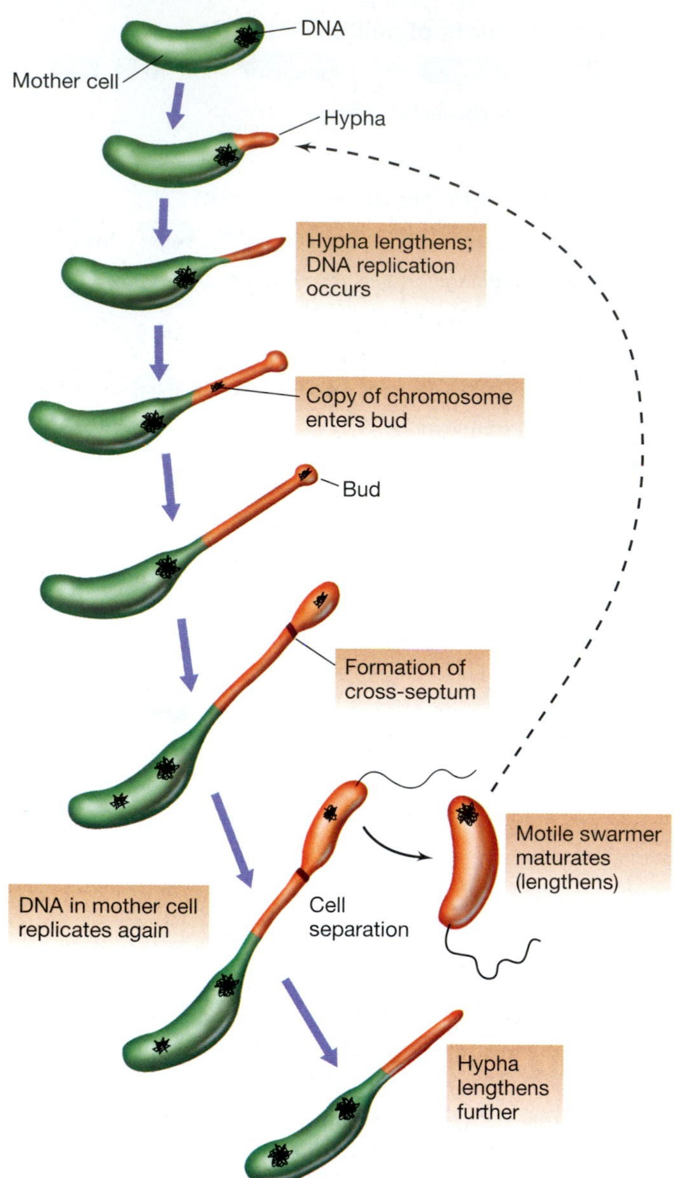

Figure 15.38 Stages in the *Hyphomicrobium* cell cycle. The single chromosome of *Hyphomicrobium* is circular.

Another function of prosthecae may be to decrease a cell's sedimentation rate in aquatic environments. It has been shown that centrifugation of prosthecate cells in the laboratory requires greater centrifugal force than for nonprosthecate cells. This inherent tendency to resist sinking is likely due to the prosthecae and might be an advantage for prosthecate cells in their natural habitats. Because these organisms are in general strict aerobes, prosthecae may keep cells from sinking into sediments or other anoxic zones where they would be unable to respire.

Selective isolation of prosthecate/stalked bacteria can be achieved by mixing an inoculum with a very dilute nutrient solution, such as 0.01% peptone, and leaving it to sit undisturbed. Within a few days, a surface film often develops containing prosthecae and stalked bacteria. Isolation can then be achieved by streaking on agar plates of the same, very dilute, media.

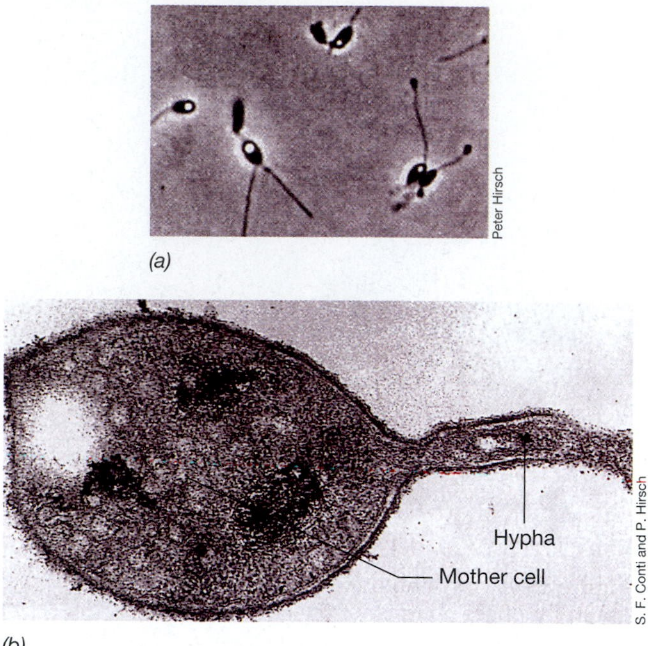

Figure 15.39 Morphology of *Hyphomicrobium*. *(a)* Phase micrograph of cells of *Hyphomicrobium*. Cells are about 0.7 μm wide. *(b)* Electron micrograph of a thin section of a single *Hyphomicrobium* cell. The hypha is about 0.2 μm wide.

Caulobacter and *Gallionella*

Two common stalked bacteria are *Caulobacter* (Figure 15.40) and *Gallionella* (see Figure 15.42). The former is a chemoorganotroph that produces a cytoplasm-filled stalk, that is, a prostheca, while the latter is a chemolithotrophic iron-oxidizing bacterium whose stalk is composed of ferric hydroxide.

Caulobacter cells are often seen on surfaces in aquatic environments with the stalks of several cells attached to form *rosettes* (Figure 15.40*a*). At the end of the stalk is a structure called a *holdfast* by which the stalk attaches the cell to a surface. The *Caulobacter* cell division cycle (**Figure 15.41**; ∞ Figure 9.23) is of special interest because it involves unequal binary fission.

Much molecular research has been done on *Caulobacter* as a model system for cell division and developmental events (∞ Section 9.13). A stalked cell of *Caulobacter* divides by elongation of the cell followed by fission, a single flagellum forming at the pole opposite the stalk. The flagellated cell so formed, called a *swarmer,* separates from the nonflagellated mother cell, swims around, and attaches to a new surface, forming a new stalk at the flagellated pole; the flagellum is then lost (Figure 15.41). Stalk formation is a necessary precursor of cell division and is coordinated with DNA synthesis (Figure 15.41; ∞ Figure 9.23). The cell division cycle in *Caulobacter* is thus more complex than simple binary fission or budding division because the stalked and swarmer cells are structurally different and the growth cycle must include both forms.

Gallionella forms a twisted stalklike structure containing ferric hydroxide [$Fe(OH)_3$] (**Figure 15.42**). However, the stalk

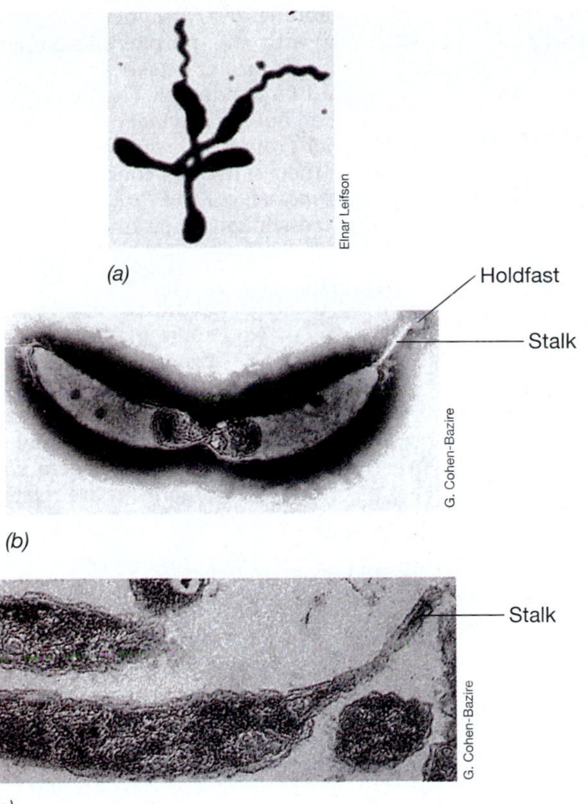

Figure 15.40 Stalked bacteria. *(a)* A *Caulobacter* rosette. A single cell is about 0.5 μm wide. The five cells are attached by their stalks (prosthecae). Two of the cells have divided, and the daughter cells have formed flagella. *(b, c)* Electron micrographs of *Caulobacter* cells. *(b)* Negatively stained preparation of a cell in division. *(c)* A thin section. Notice that cytoplasmic constituents are present in the stalk region.

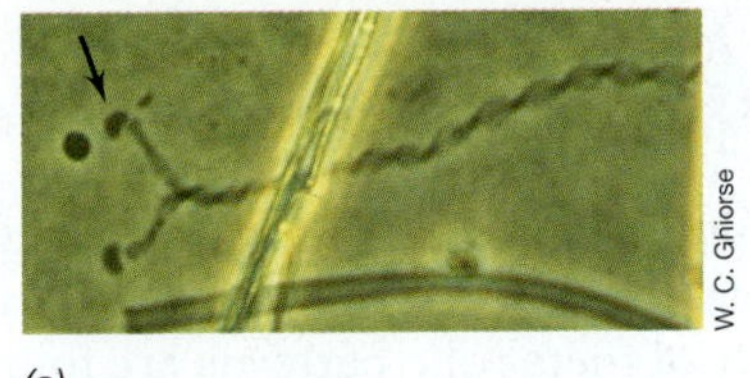

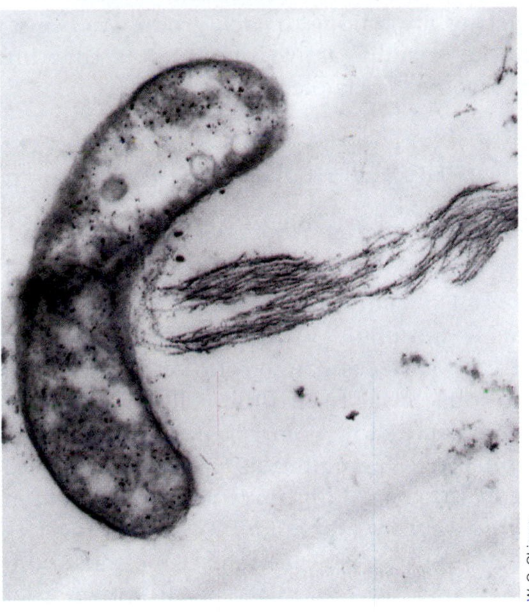

Figure 15.42 The neutrophilic ferrous iron oxidizer, *Gallionella ferruginea*. *(a)* Photomicrograph of cells from an iron seep near Ithaca, New York. Notice the twisted stalk leading to the two bean-shaped cells (arrow). *(b)* Transmission electron micrograph of a thin section of a cell. Cells are about 0.6 μm wide. Note the twisted stalk of ferric hydroxide emanating from the center of the cell in both photos.

of *Gallionella* is not an integral part of the cell but is simply excreted from the cell surface. It contains an organic matrix on which the ferric hydroxide accumulates. *Gallionella* is common in the waters draining bogs, iron springs, and other habitats where ferrous iron (Fe^{2+}) is present, usually in association with sheathed bacteria such as *Sphaerotilus*. *Gallionella* is an autotrophic chemolithotroph containing enzymes of the Calvin cycle (∞ Section 20.6) by which CO_2 is incorporated into cell material with Fe^{2+} as electron donor.

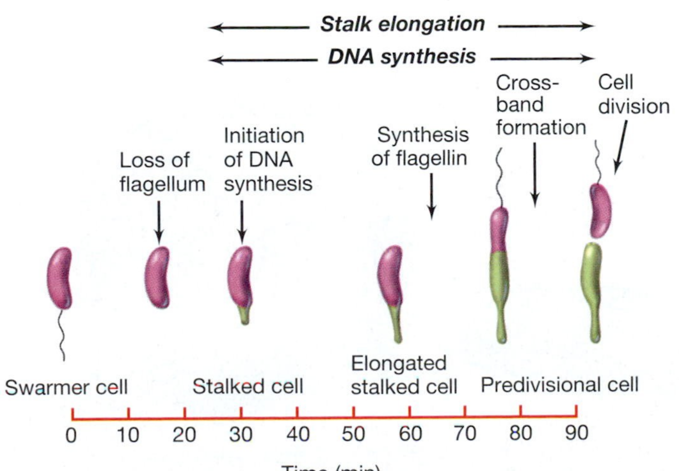

Figure 15.41 Growth of *Caulobacter*. Stages in the *Caulobacter* cell cycle, beginning with a swarmer cell. Compare with Figure 9.23.

15.15–15.16 MiniReview

Sheathed bacteria are filamentous *Proteobacteria* in which individual cells form chains within an outer layer called the sheath. Budding and prosthecate bacteria are appendaged cells that form stalks or prosthecae used for attachment or nutrient absorption and are primarily aquatic.

- Physiologically, what is unique about the sheathed bacterium *Leptothrix*?
- How does budding division differ from binary fission? How does binary fission differ from the division process in *Caulobacter*?
- What advantage might a prosthecate organism have in a very nutrient-poor environment?

UNIT 3

V DELTA- AND EPSILONPROTEOBACTERIA

To round out the *Proteobacteria* we now consider some *Delta-* and *Epsilonproteobacteria*. As is typical of *Proteobacteria* in general, a variety of metabolic patterns are found in species of these two proteobacterial classes, and several other phenotypic properties are unique to each group. We begin with the myxobacteria, socially active prokaryotes that interact to form macroscopically visible masses of cells.

15.17 Gliding Myxobacteria

Key Genera: *Myxococcus, Stigmatella*

Some bacteria exhibit a form of motility called *gliding* (Section 4.14). Gliding bacteria, typically either long rods or filaments in morphology, lack flagella but can move when in contact with surfaces. One group of gliding bacteria, the fruiting myxobacteria, form multicellular structures called *fruiting bodies* and show complex developmental life cycles involving intercellular communication (see Figure 15.46). Phylogenetically, the gliding myxobacteria are members of the *Deltaproteobacteria* (**Table 15.20**).

The fruiting myxobacteria exhibit the most complex behavioral patterns and life cycles of all known bacteria. To encode this complexity, the chromosome of some myxobacteria is very large. *Myxococcus xanthus*, for example, has a single circular chromosome of 9.2 Mbp, twice that of *Escherichia coli* (Section 13.2). Indeed, this is two-thirds the size of the entire genome of yeast, a eukaryote, which is distributed across 16 chromosomes (Section 13.5). The vegetative cells of the fruiting myxobacteria are simple, nonflagellated, gram-negative rods (**Figure 15.43*a***) that glide across surfaces and obtain their nutrients primarily by lysing other bacteria and utilizing the released nutrients. Under appropriate conditions, a swarm of vegetative cells aggregate and construct fruiting bodies, within which some of the cells become converted to resting structures called *myxospores* (Figure 15.43*b*).

Table 15.20 Classification of the fruiting myxobacteria[a]

Characteristics	*Genus*
Vegetative cells tapered:	
Spherical or oval myxospores, fruiting bodies usually soft and slimy without well-defined sporangia or stalks	*Myxococcus*
Rod-shaped myxospores: Myxospores not contained in sporangia, fruiting bodies without stalks	*Archangium*
Myxospores embedded in slime envelope:	
Fruiting bodies without stalks	*Cystobacter*
Stalked fruiting bodies, single sporangia	*Melittangium*
Stalked fruiting bodies, multiple sporangia	*Stigmatella*
Fruiting bodies are dark-brown clusters consisting of tiny spherical or disclike sporangia with an outer wall	*Angiococcus*
Vegetative cells not tapered (blunt, rounded ends); myxospores resemble vegetative cells; sporangia always produced:	
Fruiting bodies without stalks; myxospores rod-shaped	*Polyangium*
Fruiting bodies without stalks; myxospores oval; highly cellulolytic	*Sorangium*
Fruiting bodies without stalks; myxospores coccoid	*Nannocystis*
Stalked fruiting bodies	*Chondromyces*

[a]Phylogenetically, those species examined are members of the *Deltaproteobacteria* (see Table 15.1).

Fruiting Bodies

The fruiting bodies of the myxobacteria vary from simple globular masses of myxospores in loose slime to complex forms with a fruiting body wall and a stalk (**Figure 15.44**). The fruiting bodies are often strikingly colored and morphologically elaborate (**Figures** 15.44 and **15.45**). They can usually be seen with a hand lens or dissecting microscope on moist pieces of decaying wood or plant material. Fruiting

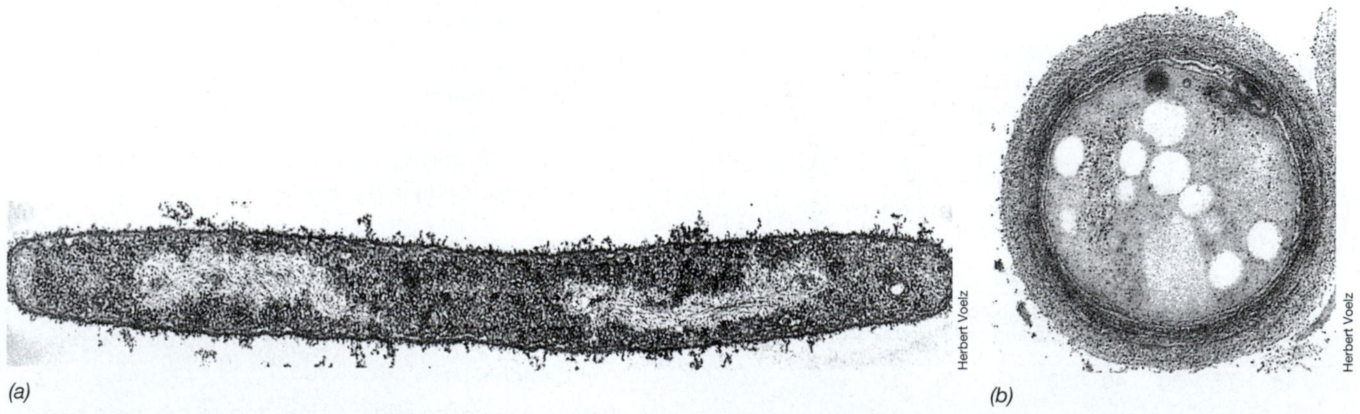

Figure 15.43 Myxococcus. *(a)* Electron micrograph of a thin section of a vegetative cell of *Myxococcus xanthus*. A cell measures about 0.75 μm wide. *(b)* Myxospore of *M. xanthus*, showing the multilayered outer wall. Myxospores measure about 2 μm in diameter.

(a)

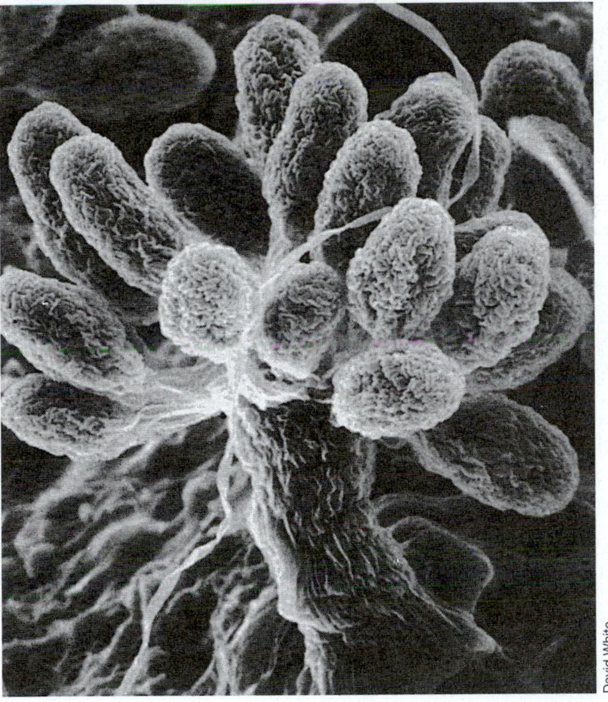

(b)

Figure 15.44 ***Stigmatella aurantiaca.*** *(a)* Color photo of a single fruiting body. The structure is about 150 μm high. *(b)* Scanning electron micrograph of a fruiting body growing on a piece of wood. Note the individual cells visible in each fruiting structure. The color of the fruiting body shown in *(a)* is due to the production of carotenoid pigments.

bodies of myxobacteria often develop on dung pellets (for example, rabbit pellets) after they have been incubated for a few days in a moist chamber.

Another method for isolating fruiting myxobacteria is to prepare Petri plates of water agar (1.5% agar in distilled water with no added nutrients) onto which is spread a heavy suspension of virtually any bacterium. In the center of the plate a small amount of soil, decaying bark, or other natural material is placed. Myxobacteria in the inoculum lyse the bacterial cells and use their liberated products as nutrients; as they grow, they swarm out across the plate from the inoculum site. After several days to a week, the plates are examined under a dissecting microscope for myxobacterial swarms or fruiting bodies, and pure cultures are obtained by transfer of cells from the fruiting bodies or from the edge of the swarm to organic media. Many myxobacteria can be grown in the laboratory on complex media containing peptone or casein hydrolysate, which provides amino acids or small peptides as nutrients, and most are obligate aerobes. A notable exception is the facultative organism *Anaeromyxobacter,* cells of which can grow by anaerobic respiration coupled to the reduction of nitrate, fumarate, or several chlorinated compounds.

Life Cycle of a Fruiting Myxobacterium

The life cycle of a typical fruiting myxobacterium is shown in **Figure 15.46**. A vegetative cell excretes slime, and as it moves across a solid surface it leaves a slime trail behind (**Figure 15.47**). This trail is then used by other cells in the swarm such that a characteristic radiating pattern is soon created with cells migrating along established slime trails (Figure 15.47). The fruiting body ultimately formed (Figures 15.44 and 15.45) is a complex structure produced by the differentiation of cells in the stalk region and in the myxospore-bearing head.

Fruiting bodies do not form if adequate nutrients for vegetative growth are present, but upon nutrient exhaustion, the vegetative swarms begin to fruit. Cells aggregate, likely through a chemotactic response (∞ Sections 4.14 and 4.15),

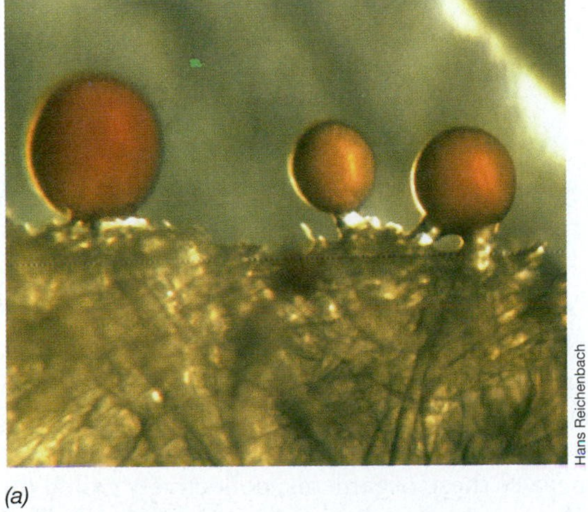

(a)

(b)

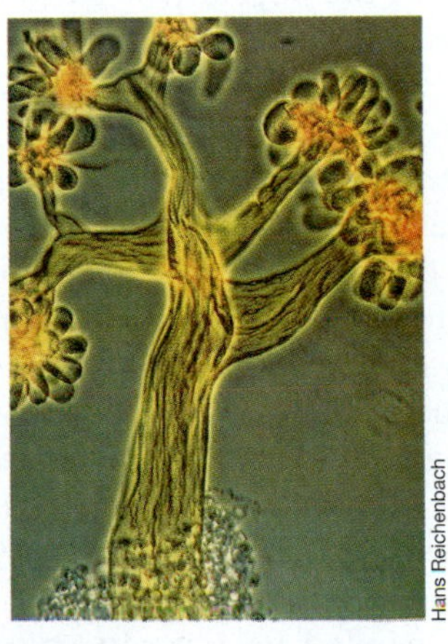

(c)

Figure 15.45 **Fruiting bodies of three species of fruiting myxobacteria.** *(a) Myxococcus fulvus* (125 μm high). *(b) Myxococcus stipitatus* (170 μm high). *(c) Chondromyces crocatus* (560 μm high).

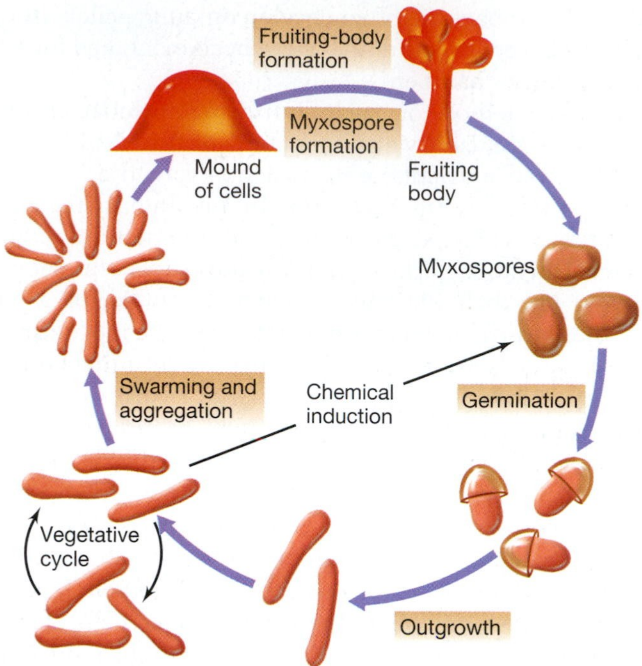

Figure 15.46 Life cycle of *Myxococcus xanthus*. Aggregation assembles vegetative cells that then undergo fruiting body formation, within which some vegetative cells undergo morphogenesis to resting cells called myxospores. The myxospores germinate under favorable nutritional and physical conditions to yield vegetative cells.

with the cells migrating toward each other and forming mounds or heaps (**Figure 15.48**); a single fruiting body may have 10^9 or more cells. As the cell masses become higher, the differentiation of the fruiting body into stalk and head begins (Figure 15.48*b* and *c*). Figure 15.48*d* clearly illustrates the differentiation of the fruiting body into stalk and head. The stalk is composed of slime, within which a few cells may be trapped. The majority of the cells accumulate in the fruiting body head, where they undergo differentiation into myxospores (Figures 15.43–15.46). In some genera, the myxospores are enclosed in large walled structures called *cysts*.

Compared to the vegetative cell, the myxospore is more resistant to drying, sonic vibration, UV radiation, and heat, but the degree of heat resistance is much less than that of the bacterial endospore (Section 4.12). The main function of encysted myxospores is therefore to enable the organism to survive desiccation during dispersal or during drying of the habitat. Upon dissemination to a suitable habitat or restoration of adequate growth conditions, the myxospore eventually germinates by a localized rupture of the capsule, with the growth and emergence of a new vegetative cell.

Myxobacteria are usually pigmented by carotenoid pigments (Figures 15.44*a* and 15.45), and the main pigments are carotenoid glycosides. Pigment formation is promoted by light, and at least one function of the pigment is photoprotection, which is needed because in nature the myxobacteria usually form fruiting bodies in the light. In the genus *Stigmatella* (Figure 15.44), light greatly stimulates fruiting

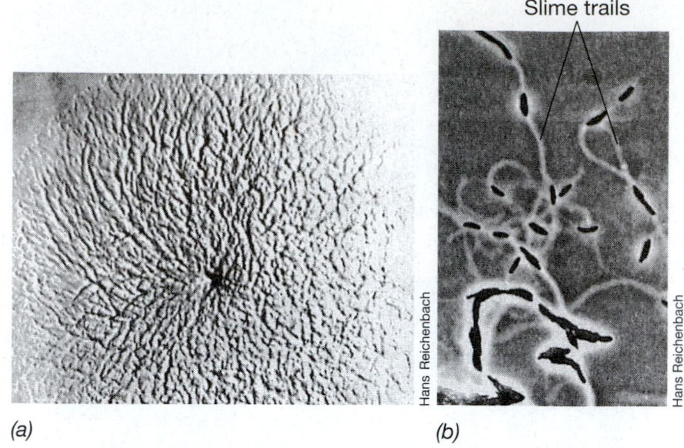

Figure 15.47 Swarming in *Myxococcus*. (*a*) Photomicrograph of a swarming colony (5-mm radius) of *Myxococcus xanthus* on agar. (*b*) Single cells of *Myxococcus fulvus* from an actively gliding culture, showing the characteristic slime trails on the agar. A cell of *M. fulvus* is about 0.8 μm in diameter.

body formation and catalyzes production of the lipid pheromone 2,5,8-trimethyl-8-hydroxy-nonane-4-one. This substance initiates the aggregation step. The fruiting myxobacteria are classified on morphological grounds using characteristics of the vegetative cells, the myxospores, and fruiting body structure (Table 15.20), and on phylogenetic grounds, using 16S rRNA gene sequence analyses.

15.17 MiniReview

The fruiting myxobacteria are rod-shaped, gliding bacteria that aggregate to form complex masses of cells called fruiting bodies. Myxobacteria are chemoorganotrophic soil bacteria that live by consuming dead organic matter or other bacterial cells.

- What environmental conditions trigger fruiting body formation in myxobacteria?
- What is a myxospore and how does it compare with an endospore?
- To what specific phylogenetic group do the myxobacteria belong?

15.18 Sulfate- and Sulfur-Reducing *Proteobacteria*

Key Genera: *Desulfovibrio, Desulfobacter, Desulfuromonas*

Sulfate (SO_4^{2-}) and sulfur (S^0) are electron acceptors for a large group of anaerobic *Deltaproteobacteria* that utilize organic compounds or H_2 as electron donors. Hydrogen sulfide (H_2S) is the product of both SO_4^{2-} and S^0 reduction. Over 40 genera of these organisms, collectively called the dissimilative **sulfate-reducing bacteria** and **sulfur-reducing bacteria**,

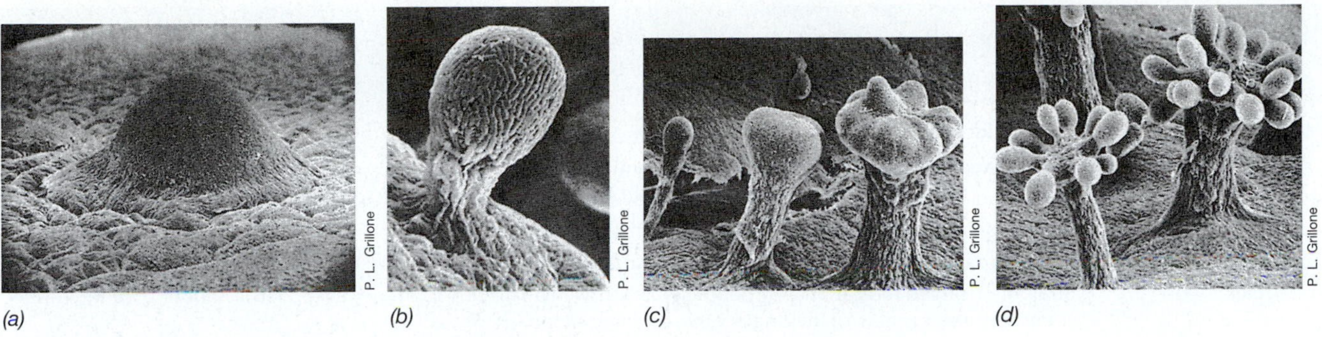

Figure 15.48 Scanning electron micrographs of fruiting body formation in *Chondromyces crocatus*. *(a)* Early stage, showing aggregation and mound formation. *(b)* Initial stage of stalk formation. Slime formation in the head has not yet begun and so the cells that compose the head are still visible. *(c)* Three stages in head formation. Note that the diameter of the stalk also increases. *(d)* Mature fruiting bodies. The entire fruiting structure is about 600 μm in height (compare with Figure 15.45*c*).

UNIT 3

are known, and some of the key ones are shown in **Table 15.21**. The word *dissimilative* refers to the use of sulfate or sulfur as electron acceptors in energy generation instead of their assimilation as biosynthetic sources of sulfur (∞ Section 21.8).

General Properties

The genera of sulfate-reducing bacteria that form group I, including *Desulfovibrio* (**Figure 15.49a**), *Desulfomonas, Desulfotomaculum,* and *Desulfobulbus* (Figure 15.49*c*), utilize lactate, pyruvate, ethanol, or certain fatty acids as electron donors, reducing sulfate to hydrogen sulfide; they are unable to catabolize acetate. The genera that form group II, such as *Desulfobacter* (Figure 15.49*d*), *Desulfococcus, Desulfosarcina* (Figure 15.49*e*), and *Desulfonema* (Figure 15.49*b*), specialize in the oxidation of fatty acids, particularly acetate, reducing sulfate to sulfide. The sulfate-reducing bacteria are, for the most part, obligate anaerobes, and strict anoxic techniques must be used in their cultivation (Figure 15.49*g*).

Sulfate-reducing bacteria are widespread in aquatic and terrestrial environments that become anoxic as a result of microbial decomposition processes. The best-studied genus is *Desulfovibrio* (Figure 15.49*a*), which is common in aquatic habitats or waterlogged soils containing abundant organic material and sufficient levels of sulfate. *Desulfotomaculum,* phylogenetically a member of the gram-positive *Bacteria,* consists of endospore-forming rods found primarily in soil. Growth and reduction of sulfate by *Desulfotomaculum* in certain canned foods leads to a type of spoilage called *sulfide stinker*. The remaining genera of sulfate reducers are indigenous to anoxic freshwater (group I) or marine environments (groups I and II) and can occasionally be isolated from the mammalian intestine.

Sulfur Reduction

The dissimilative sulfur-reducing bacteria can reduce elemental sulfur to sulfide but are unable to reduce sulfate to sulfide. Species of *Desulfuromonas* (Figure 15.49*f*) grow anaerobically by coupling the oxidation of acetate or ethanol to the reduction of sulfur to sulfide. However, the ability to reduce elemental sulfur, as well as other sulfur compounds such as thiosulfate, sulfite, or dimethyl sulfoxide (DMSO), is widespread in a number of chemoorganotrophic, generally facultatively aerobic, bacteria (for example, *Proteus, Campylobacter, Pseudomonas,* and *Salmonella*). *Desulfuromonas* differs from these aerobic bacteria in that it is an obligate anaerobe and utilizes only sulfur as an electron acceptor (Table 15.21).

Dissimilative sulfur-reducing bacteria reside in many of the same habitats as sulfate-reducing bacteria and often form associations with bacteria that oxidize H_2S to S^0, such as green sulfur bacteria (∞ Section 16.15). The sulfur produced from sulfide oxidation is then reduced back to H_2S during metabolism of the sulfur reducer, completing an anoxic sulfur cycle (∞ Section 24.4).

Physiology of Sulfate-Reducing Bacteria

The biochemistry of sulfate reduction is discussed in Section 21.8. Here we simply touch on some of the general physiological properties of this group. The range of electron donors used by sulfate-reducing bacteria is fairly broad. Hydrogen, lactate, and pyruvate are almost universally used, and many group I species also utilize malate, sulfonates, and certain primary alcohols (for example, ethanol, propanol, and butanol) as electron donors. Some strains of *Desulfotomaculum* utilize glucose, but this is rare among sulfate reducers. Group I sulfate reducers oxidize lactate, pyruvate, or ethanol to acetate and then excrete this fatty acid as an end product.

Group II sulfate reducers oxidize fatty acids (including acetate), lactate, succinate, and even benzoate in some species, completely to CO_2. *Desulfosarcina, Desulfonema, Desulfococcus, Desulfobacterium, Desulfotomaculum,* and certain species of *Desulfovibrio,* are unique among sulfate reducers in their ability to grow chemolithotrophically and autotrophically with H_2 as electron donor, SO_4^{2-} as electron acceptor, and CO_2 as sole carbon source. A few sulfate reducers can use hydrocarbons, even crude oil itself, as electron donors. This process is noteworthy because until such

Table 15.21 Characteristics of some key genera of sulfate- and sulfur-reducing bacteria[a]

Genus	*Characteristics*
Group I sulfate reducers: Nonacetate oxidizers	
Desulfovibrio	Polarly flagellated, curved rods, no spores; gram-negative; contain desulfoviridin; one thermophilic
Desulfomicrobium	Motile rods, no spores; gram-negative; desulfoviridin absent
Desulfobotulus	Vibrios; gram-negative; motile; desulfoviridin absent
Desulfofustis	Motile rods, specializes in the degradation of glycolate and glyoxalate
Desulfotomaculum	Straight or curved rods; motile by peritrichous or polar flagellation; gram-negative; desulfoviridin absent; produce endospores; capable of utilizing acetate as energy source
Desulfomonile	Rod; capable of reductive dechlorination of 3-chlorobenzoate to benzoate (Section 21.12)
Desulfobacula	Oval to coccoid cells, marine; can oxidize various aromatic compounds including the aromatic hydrocarbon toluene, to CO_2
Archaeoglobus	Archaeon; hyperthermophile, temperature optimum, 83°C; contains some unique coenzymes of methanogenic bacteria, makes small amount of methane during growth; H_2, formate, glucose, lactate, and pyruvate are electron donors, SO_4^{2-}, $S_2O_3^{2-}$, or SO_3^{2-}, electron acceptors (Section 17.7)
Desulfobulbus	Ovoid or lemon-shaped cells; no spores; gram-negative; desulfoviridin absent; if motile, by single polar flagellum; utilizes propionate as electron donor with acetate + CO_2 as products
Desulforhopalus	Curved rods, gas vacuolate, psychrophile; uses propionate, lactate, or alcohols as electron donor
Thermodesulfobacterium	Small, gram-negative rods; desulfoviridin present; thermophilic, optimum growth at 70°C; a member of the *Bacteria* but contains ether-linked lipids (Section 16.19)
Group II sulfate reducers: Acetate oxidizers	
Desulfobacter	Rods; no spores, gram-negative; desulfoviridin absent; if motile, by single polar flagellum; utilizes only acetate as electron donor and oxidizes it to CO_2 via the citric acid cycle
Desulfobacterium	Rods, some with gas vesicles, marine; capable of autotrophic growth via the acetyl-CoA pathway
Desulfococcus	Spherical cells; nonmotile; gram-negative; desulfoviridin present, no spores; utilizes C_1 to C_{14} fatty acids as electron donor with complete oxidation to CO_2; capable of autotrophic growth via the acetyl-CoA pathway
Desulfonema	Large, filamentous gliding bacteria; gram-positive, no spores; desulfoviridin present or absent; utilizes C_2 to C_{12} fatty acids as electron donor with complete oxidation to CO_2; capable of autotrophic growth via the acetyl-CoA pathway (H_2 as electron donor)
Desulfosarcina	Cells in packets (sarcina arrangement); gram-negative; no spores; desulfoviridin absent; utilizes C_2 to C_{14} fatty acids as electron donor with complete oxidation to CO_2; capable of autotrophic growth via the acetyl-CoA pathway (H_2 as electron donor)
Desulfarculus	Vibrios; gram-negative; motile; desulfoviridin absent; utilizes only C_1 to C_{18} fatty acids as electron donor
Desulfacinum	Cocci to oval-shaped cells; gram-negative; utilizes C_1 to C_{18} fatty acids, very nutritionally diverse, capable of autotrophic growth; thermophilic
Desulforhabdus	Rods; no spores; gram-negative; nonmotile; utilizes fatty acids with complete oxidation to CO_2
Thermodesulforhabdus	Gram-negative motile rods; thermophilic; uses fatty acids up to C_{18}
Dissimilative sulfur reducers	
Desulfuromonas	Straight rods, single lateral flagellum; no spores; gram-negative; does not reduce sulfate; acetate, succinate, ethanol, or propanol used as electron donor; obligate anaerobe; one species is capable of the reductive dechlorination of trichloroethylene (Section 21.12)
Desulfurella	Motile short rods; gram-negative; requires acetate; thermophilic
Sulfurospirillum	Small vibrios, reduces S^0 with H_2 or formate as electron donors
Campylobacter	Curved, vibrio-shaped rods; polar flagella; gram-negative; no spores; unable to reduce sulfate but can reduce sulfur, sulfite, thiosulfate, nitrate, or fumarate anaerobically with acetate or a variety of other carbon or electron donor sources; facultative aerobe; microaerophilic

[a]Phylogenetically, most sulfate- and sulfur-reducing bacteria are members of the *Deltaproteobacteria*. *Archaeoglobus* is a member of domain *Archaea* (Chapter 17); *Thermodesulfobacterium* is a deeply-branching hyperthermophilic bacterium (Section 16.19); and *Sulfurospirillum* and *Campylobacter* are members of *Epsilonproteobacteria* (Section 15.19 and see Table 15.22).

organisms were recognized, it was thought that hydrocarbons could only be oxidized under oxic conditions (Sections 21.13 and 21.15).

In addition to using sulfate as an electron acceptor, many sulfate-reducing bacteria can reduce nitrate to NH_3 or reduce sulfonates, such as isethionate ($HO—CH_2—CH_2—SO_3^-$), to sulfide. Elemental sulfur can also be reduced to sulfide by most sulfate-reducing bacteria.

Certain organic compounds can also be fermented by sulfate-reducing bacteria. The most common fermentable compound is pyruvate, which is converted via the phosphoroclastic reaction to acetate, CO_2, and H_2 (Figure 21.3).

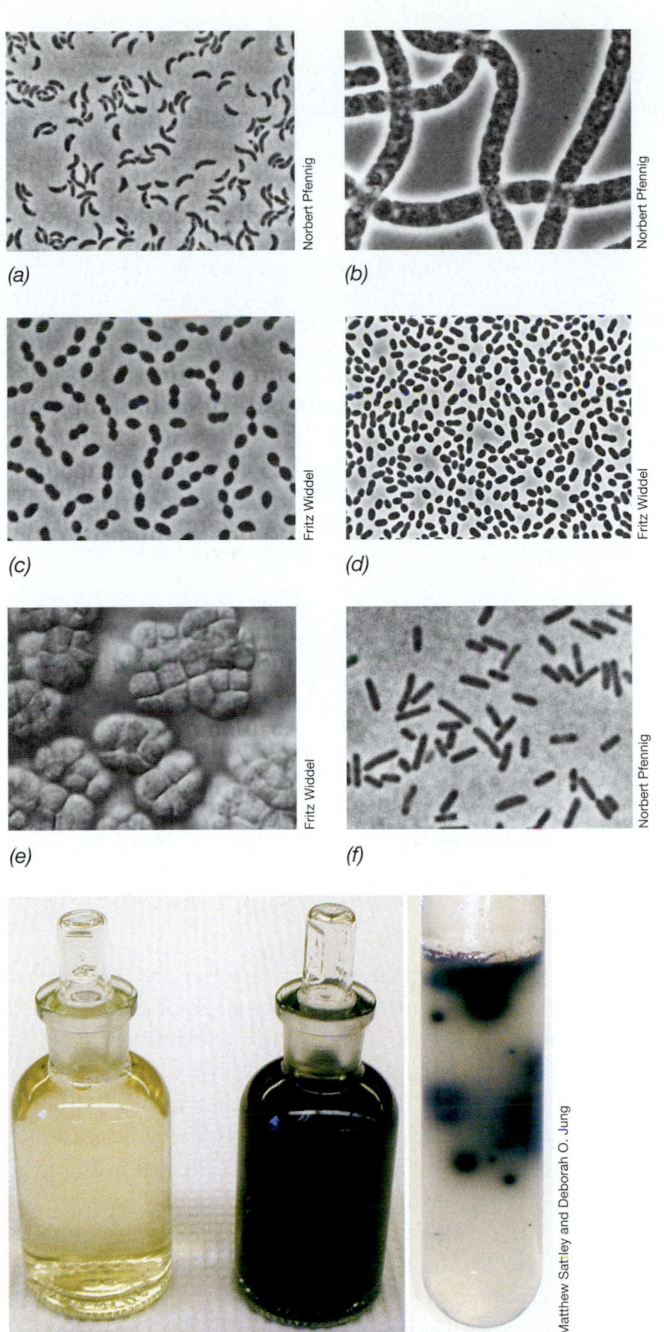

Figure 15.49 Representative sulfate-reducing and sulfur-reducing bacteria. *(a) Desulfovibrio desulfuricans*; cell diameter about 0.7 μm. *(b) Desulfonema limicola*; cell diameter 3 μm. *(c) Desulfobulbus propionicus*; cell diameter about 1.2 μm. *(d) Desulfobacter postgatei*; cell diameter about 1.5 μm. *(e) Desulfosarcina variabilis*; cell diameter about 1.25 μm. *(f) Desulfuromonas acetoxidans*; cell diameter about 0.6 μm. *(g)* Enrichment culture of sulfate-reducing bacteria. Left, sterile medium; center, a positive enrichment showing black ferrous sulfide (FeS); right, colonies of sulfate-reducing bacteria in a dilution tube. *(a–d, f)* Phase-contrast photomicrographs; *(e)* interference contrast micrograph.

Moreover, although thought to be obligate anaerobes, certain sulfate-reducing bacteria, primarily strains isolated from microbial mats where they coexist with O_2-producing cyanobacteria, are quite oxygen-tolerant and can respire with O_2 as electron acceptor. At least one species, *Desulfovibrio oxyclinae,* can actually grow with O_2 as electron acceptor under microoxic conditions.

Isolation

The enrichment of *Desulfovibrio* species is easy on an anoxic lactate-sulfate medium containing ferrous iron. A reducing agent, such as thioglycolate or ascorbate, is required to achieve a low E_0'. When sulfate-reducing bacteria grow, the sulfide formed from sulfate reduction combines with the ferrous iron to form black, insoluble ferrous sulfide (Figure 15.49*g*). This blackening not only indicates sulfate reduction, but the iron also binds and detoxifies the sulfide, making possible growth to higher cell yields.

After some growth has occurred in an enrichment as evidenced by blackening of the medium, purification can be accomplished by streaking onto a tube coated on the inside surface with a thin layer of agar, called *roll tubes,* or on Petri plates in an anoxic glove box (∞ Figure 6.28). Alternatively, agar dilution tubes can be used for purification purposes. In the shake tube method a small amount of liquid from the original enrichment is added to a tube of molten agar growth medium, mixed thoroughly, and sequentially diluted through a series of molten agar tubes (∞ Section 22.2 and Figure 22.3). On solidification, individual cells of sulfate-reducing bacteria become distributed throughout the agar and grow to form black colonies (Figure 15.49*g*) that can be removed aseptically to yield pure cultures.

UNIT 3

15.18 MiniReview

Sulfate- and sulfur-reducing bacteria are a large group of *Deltaproteobacteria* unified physiologically by their ability to reduce SO_4^{2-} or S^0 to H_2S under anoxic conditions. Two subgroups of sulfate-reducing bacteria are known: group I, which is incapable of oxidizing acetate to CO_2, and group II, which can.

- What organic substrate would you use to enrich and isolate a group II sulfate reducer from nature?
- For sulfate-reducing bacteria capable of chemolithotrophic and autotrophic growth: (1) What is the electron donor? (2) What is the electron acceptor? (3) What is the source of cell carbon?
- Physiologically, how does *Desulfuromonas* differ from *Desulfovibrio*?

15.19 The *Epsilonproteobacteria*

The *Epsilonproteobacteria,* the fifth class of *Proteobacteria,* was initially defined by certain pathogenic bacteria, in particular *Campylobacter* and *Helicobacter*. However, environmental studies of marine and terrestrial habitats have shown that a diversity of *Epsilonproteobacteria* can be found in a variety of habitats in nature, where their numbers and metabolic capabilities suggest they play important ecological roles.

Table 15.22 Characteristics of key genera of *Epsilonproteobacteria*

Genus	*Habitat*	*Descriptive characters*	*Physiology and metabolism*
Campylobacter	Reproductive organs, oral cavity, and intestinal tract of humans and other animals; pathogenic	Slender, spirally curved rods; corkscrew-like motility by single polar flagellum	Microaerophilic; chemoorganotrophic
Arcobacter	Diverse habitats (freshwater, sewage, saline environments, animal reproductive tract, plants); some species pathogenic for humans and other animals	Slender, curved rods; motile by single polar flagellum	Microaerophilic; aerotolerant or aerobic; chemoorganotrophic; oxidation of sulfide to elemental sulfur (S^0) by some species; nitrogen fixation in one species
Helicobacter	Intestinal tract and oral cavity of humans and other animals; pathogenic	Rods to tightly spiral; some species with tightly coiled periplasmic fibers	Microaerophilic, chemoorganotrophic; produce high levels of urease (nitrogen assimilation)
Sulfurospirillum	Freshwater and marine habitats containing sulfur	Vibrioid to spiral-shaped cells; motile by polar flagella	Microaerophilic; reduces elemental sulfur (S^0)
Thiovulvum	Freshwater and marine habitats containing sulfur; not yet in pure culture	Cells contain orthorhombic sulfur granules; rapid motility by peritrichous flagella	Microaerophilic; chemolithotrophic oxidizing reduced sulfur compounds (H_2S)
Wolinella	Bovine rumen	Rapidly motile by polar flagellum; single species, *W. succinogenes*, known	Anaerobe; anaerobic respiration using fumarate, nitrate, or other compounds as terminal electron acceptor, and with H_2 or formate as electron donor

Epsilonproteobacteria are especially abundant at oxic–anoxic interfaces in sulfur-rich environments, such as around hydrothermal vents (⚭ Section 24.11), where they catalyze metabolic transformations of sulfur and live in association with animals that live in these vent areas. Many of these bacteria are autotrophs and use H_2, formate, sulfide, or thiosulphate as electron donor, with nitrite, oxygen, or elemental sulfur as electron acceptor, depending on the species. In this section, we describe major groups of cultured *Epsilonproteobacteria* and comment on the environmental diversity of uncultured members of this class.

Campylobacter and *Helicobacter*

These two genera are key representatives of the *Epsilonproteobacteria*. Although they represent separate families in this class, *Campylobacteraceae* and *Helicobacteraceae,* they share a number of characteristics. They are all gram-negative, motile spirilla, and most species are pathogenic to humans or other animals (**Table 15.22**). *Campylobacter* and *Helicobacter* species are also microaerophilic (⚭ Section 6.17) and are therefore cultured from clinical specimens in media incubated at low (3–15%) O_2 and high (3–10%) CO_2.

Campylobacter species, over a dozen of which have been described, cause acute enteritis leading to (usually) bloody diarrhea, and pathogenesis is due to several factors, including an enterotoxin that is related to cholera toxin (⚭ Section 28.11). *Helicobacter pylori,* also a pathogen, causes both chronic and acute gastritis, leading to the formation of peptic ulcers (⚭ Figure 34.25). The discovery of *H. pylori* and its role in gastritis and peptic ulcer disease earned Barry Marshall and Robin Warren the 2005 Nobel Prize in Physiology or Medicine. We discuss the diseases caused by *Campylobacter* and *Helicobacter,* including modes of transmission of the organisms and clinical symptoms, in more detail in Sections 34.11 and 37.9.

Arcobacter

The genus *Arcobacter,* another member of *Campylobacteraceae,* is unusual among *Epsilonproteobacteria* in that its various species show an unusually wide diversity of habitats. Some species are pathogenic, infecting the reproductive and intestinal tracts of humans and other animals and causing reproductive failures in animals, diarrhea-like diseases in a wide range of animals, and gastroenteritis and appendicitis in humans. *Arcobacter* also have been detected in sewage and water reservoirs (Table 15.22), so a fecal-to-water-to-oral route may be typical in gastrointestinal infections of *Arcobacter*. One species, *Arcobacter nitrofigilis,* is associated with sediment and the roots of the salt marsh plant *Spartina* and can fix nitrogen.

Sulfurospirillum and *Thiovulvum*

Sulfurospirillum, family *Campylobacteraceae,* are nonpathogenic, free-living microaerophiles found in freshwater and marine habitats. They reduce elemental sulfur (S^0), and some species can also use selenate or arsenate as electron acceptors (⚭ Section 21.12). *Thiovulvum,* family *Helicobacteraceae,* also is microaerophilic and is found in freshwater and marine habitats in which sulfide-rich muds interface with oxgyen-containing waters, where it oxidizes sulfide, forming large internal sulfur granules that displace much of the cytoplasmic contents of the cell. When motile, *Thiovulvum* cells, which are peritrichously flagellated, swim at exceptionally high speed. Members of this genus, which have not yet been brought into pure culture, also secrete a slime stalk that attaches cells to solid surfaces.

Wolinella

Only one species, *Wolinella succinogenes,* has been described for the genus *Wolinella*, an anaerobic bacterium isolated from the bovine rumen (Table 15.22). Unlike other *Epsilonproteobacteria, W. succinogenes* grows best as an anaerobe and only poorly at very low ($< 2\%$) oxygen tensions. However, this bacterium can carry out anaerobic respiration using fumarate or nitrate as electron acceptors with H_2 or formate as electron donors.

Although thus far found only in the rumen, the genome of *W. succinogenes,* which for the most part shows strong homologies to that of *Campylobacter* and *Helicobacter,* contains additional genes that encode nitrogen fixation, extensive cell signaling mechanisms, and virtually complete metabolic pathways, absent from the genomes of its close relatives. This suggests that *Wolinella* inhabits other diverse environments outside of the rumen.

Environmental *Epsilonproteobacteria*

In addition to cultured representatives of the genera mentioned above, and many additional species and genera not mentioned here, there are large groups within this class that are known only from 16S rRNA gene sequences obtained from the environment. Through environmental sequencing studies and ongoing cultivation efforts, members of *Epsilonproteobacteria* are now becoming recognized as ubiquitous in marine and terrestrial environments where sulfur is present, particularly deep-sea hydrothermal vent habitats where sulfide-rich and oxygenated waters mix. Also, as epibionts of animals living in proximity to hydrothermal vents, such as the tube worm *Alvinella* and the shrimp *Rimicaris,* filamentous and as-yet uncultured *Epsilonproteobacteria* may, through their sulfur metabolism, detoxify sulfide that would be deleterious to their animal hosts, allowing the animals to thrive in an otherwise hostile habitat (∞ Section 24.11). Studies of the phylogenetic placement, metabolic activities, and ecological roles of the bacteria forming this fifth class of *Proteobacteria* provide an exciting new area of prokaryotic diversity.

UNIT 3

15.19 MiniReview

Epsilonproteobacteria contain both pathogens and nonpathogens that are abundant in certain environments. Many of the nonpathogenic forms are found at the oxic–anoxic interface of sulfide-rich environments.

- What discovery led to the 2005 Nobel Prize in Physiology or Medicine?
- From what habitat would one likely obtain additional representatives of *Wolinella*?
- From what kinds of habitats would one likely find undescribed species of *Epsilonproteobacteria*?

Review of Key Terms

Autoinduction a gene regulatory mechanism involving small, diffusible signal molecules

Bioluminescence the enzymatic production of visible light by living organisms

Carboxysome a polyhedral cellular inclusion of crystalline ribulose bisphosphate carboxylase (RubisCO), the key enzyme of the Calvin cycle

Chemolithotroph an organism able to oxidize inorganic compounds (such as H_2, Fe^{2+}, S^0, or NH_4^+) as energy sources (electron donors)

Enteric bacteria a large group of gram-negative rod-shaped *Bacteria* characterized by a facultatively aerobic metabolism and commonly found in the intestines of animals

Methanotroph an organism capable of oxidizing methane (CH_4) as an electron donor in energy metabolism

Methylotroph an organism capable of oxidizing organic compounds that do not contain carbon–carbon bonds; if able to oxidize CH_4, also a methanotroph

Mixotroph an organism that can conserve energy from the oxidation of inorganic compounds but requires organic compounds as a carbon source

Nitrifying bacteria chemolithotrophs capable of carrying out the transformation $NH_3 \rightarrow NO_2^-$, or $NO_2^- \rightarrow NO_3^-$

Prosthecae extrusions of cytoplasm, often forming distinct appendages, bounded by the cell wall

Proteobacteria a major lineage of *Bacteria* that contains a large number of gram-negative rods and cocci

Purple nonsulfur bacteria a group of phototrophic bacteria containing bacteriochlorophyll *a* or *b* that grows best as photoheterotrophs and has a relatively low tolerance for H_2S

Purple sulfur bacteria a group of phototrophic bacteria containing bacteriochlorophylls *a* or *b* and characterized by the ability to oxidize H_2S and store elemental sulfur inside the cells (or in the genera *Ectothiorhodospira* and *Halorhodospira,* outside the cell)

Spirilla spiral-shaped cells (singular, spirillum)

Sulfate-reducing and **sulfur-reducing bacteria** two groups of anaerobic *Bacteria* that respire anaerobically with SO_4^{2-} and S^0, respectively, as electron acceptors, producing H_2S

Review Questions

1. Compare and contrast the metabolism, morphology, and phylogeny of purple sulfur and purple nonsulfur bacteria.
2. In what classes of *Proteobacteria* are nitrogen-fixing species found?
3. Compare and contrast the nitrogen metabolism of nitrosifying bacteria with that of nitrifying bacteria.
4. Compare and contrast the phylogeny, ecology, and sulfur metabolism of *Thiobacillus* and *Desulfovibrio*.
5. Give examples from this chapter of mixotrophic bacteria. What other traits do these bacteria share?
6. *Pseudomonas* and *Vibrio* are polarly flagellated and oxidase-positive bacteria typically found in aquatic environments. What physiological trait distinguishes these bacteria and how is that trait tested?
7. Of the five classes of *Proteobacteria,* which one has the most members associated with animals as commensals, pathogens, or symbionts?
8. List an electron donor for energy metabolism for each of the following *Proteobacteria* and state whether the organism is an aerobe or an anaerobe: *Thiobacillus, Nitrosomonas, Ralstonia eutropha, Methylomonas, Acetobacter,* and *Gallionella*.
9. What morphologically similar attribute unites *Caulobacter* with *Hyphomicrobium*? How would you distinguish between these organisms morphologically?
10. Compare and contrast the life cycle of *Myxococcus* with that of *Bdellovibrio*.
11. Describe a key physiological feature of the following *Proteobacteria* that would differentiate each from the others: *Acetobacter, Methylococcus, Azotobacter, Photobacterium, Desulfovibrio*.
12. What physiological trait unites the majority of cultured *Epsilonproteobacteria*?

Application Questions

1. Defend the following statement using phylogenetic, ecological, and physiological arguments: *Wolbachia pipientis* is a more highly evolved bacterium than *Escherichia coli*.
2. Defend or refute the following statements, using examples from this chapter: (a) Cell morphology has phylogenetic predictive value. (b) Major physiological differences between *Proteobacteria* correlate with the different classes in this phylum.

16

Bacteria: Gram-Positive and Other Bacteria

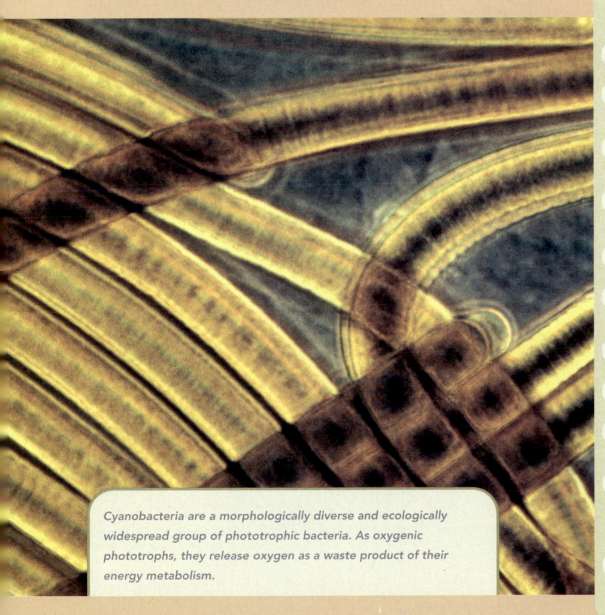

Cyanobacteria are a morphologically diverse and ecologically widespread group of phototrophic bacteria. As oxygenic phototrophs, they release oxygen as a waste product of their energy metabolism.

I OVERVIEW OF GRAM-POSITIVE AND OTHER BACTERIA

In Chapter 15, we examined the *Proteobacteria*, a large and diverse phylum of *Bacteria*. In this chapter, we focus on several of the many other phyla of *Bacteria* (Figure 15.1), exploring well-known cultured species with emphasis on those for which substantial information is available. These phyla include the gram-positive bacteria, a large group of mostly chemoorganotrophic bacteria; the cyanobacteria, a morphologically and ecologically diverse group of oxygenic phototrophs that have evolutionary roots near those of the gram-positive bacteria; phylogenetically early-branching phyla such as *Aquifex*; and other morphologically, evolutionarily distinct groups. Again we mention the resources to be found in *Bergey's Manual of Systematic Bacteriology* and *The Prokaryotes* (Section 14.14) for more detailed information on these bacterial groups and species. We start with an examination of gram-positive bacteria.

II GRAM-POSITIVE BACTERIA AND ACTINOBACTERIA

The gram-positive bacteria are a large and diverse group. Previously, the gram-positive bacteria were divided into two groups, called low GC and high GC (Actinobacteria), based on whether their DNA GC base ratios (Section 14.11) fell either below 50% (low GC) or above it (high GC). Although this grouping is finding less use today with the advent of DNA sequence analysis, and is not emphasized here, the major differences in GC ratios between gram-positive bacteria and actinobacteria serve as a reminder of their major phylogenetic differences.

16.1 Nonsporulating Gram-Positive Bacteria

Key Genera: *Staphylococcus, Micrococcus, Streptococcus, Lactobacillus, Sarcina*

In this section we consider genera of nonsporulating gram-positive bacteria, formerly grouped with endospore-forming gram-positive bacteria (Section 16.2) and mycoplasmas (Section 16.3) as *low-GC gram-positive* bacteria. This group is also called the *Firmicutes* and is composed of the lactic acid bacteria, which are classical nonsporulating gram-positive rods and cocci, and their various relatives. We also consider here *Micrococcus*, because it is morphologically quite similar to *Staphylococcus*, although it is actually a member of the other major subdivision, the Actinobacteria (**Table 16.1**).

Staphylococcus and *Micrococcus*

Staphylococcus (**Figure 16.1**) and *Micrococcus* are both aerobic organisms with a typical respiratory metabolism. They are catalase-positive, and this permits their distinction from *Streptococcus* and some other genera of gram-positive cocci. Gram-positive cocci are relatively resistant to reduced water potential and tolerate drying and high salt fairly well. Their ability to grow in media with high salt provides a selective means for isolation. For example, if an appropriate inoculum is spread on a rich medium agar plate containing 7.5% NaCl and the plate is incubated aerobically, gram-positive cocci

Table 16.1 Distinguishing features of major gram-positive cocci

Genus	*Motility*	*Arrangement of cells*	*Growth by fermentation*	*Phylogenetic group*[a]	*Other characteristics*
Micrococcus	−	Clusters, tetrads	−	*Actinobacteria*	Strict aerobe
Staphylococcus	−	Clusters, pairs	+	Low GC	Only genus of this group to contain teichoic acid in cell wall
Stomatococcus	−	Clusters, pairs	+	*Actinobacteria*	Only genus of this group containing a capsule
Planococcus	+	Pairs, tetrads	−	Low GC	Primarily marine
Sarcina	−	Cuboidal packets of eight or more cells	+	Low GC	Extremely acid-tolerant; cellulose in cell wall
Ruminococcus	+	Pairs, chains	+	Low GC	Obligate anaerobe; inhabits rumen, cecum, and large intestine of many animals
Peptococcus	−	Clusters, pairs	+	Low GC	Obligate anaerobe; ferments peptone but not sugars
Peptostreptococcus	−	Clumps, short chains	+	Low GC	Obligate anaerobe; ferments peptone; common member of human normal flora, skin, intestine, vagina; also isolated from vaginal and purulent discharges

[a]All are members of the gram-positive *Bacteria* (Figure 15.1).

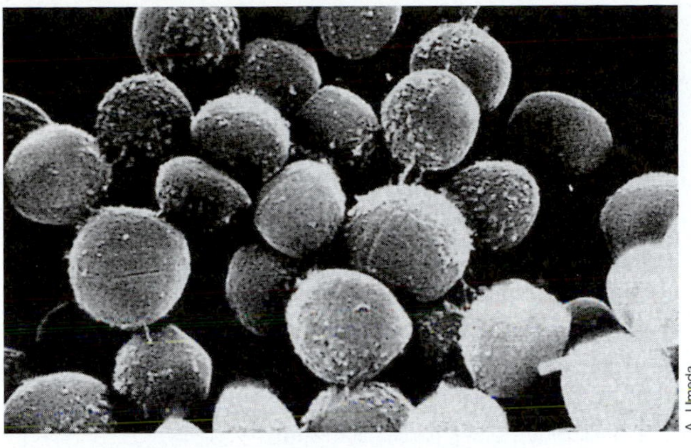

(a)

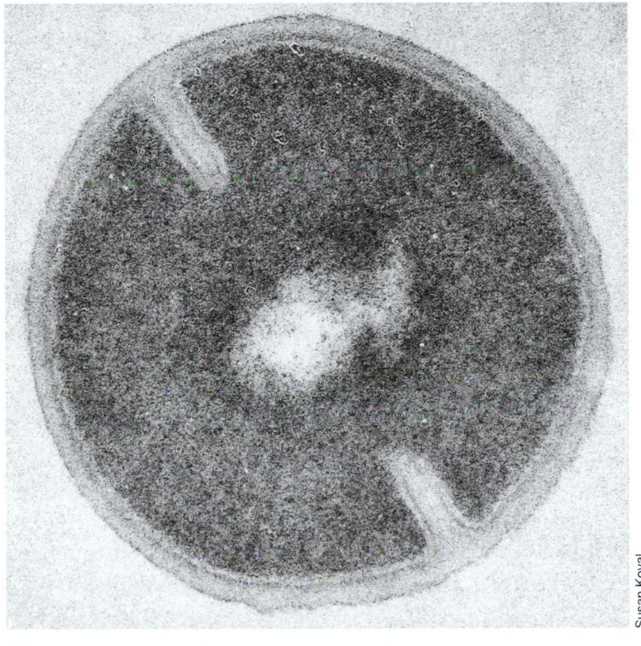

(b)

Figure 16.1 ***Staphylococcus.*** *(a)* Scanning electron micrograph of typical *Staphylococcus aureus* cells, showing the irregular arrangement of the cell clusters. Individual cells are about 0.8 μm in diameter. *(b)* Transmission electron micrograph of a dividing cell of *S. aureus.* Note the thick gram-positive cell wall.

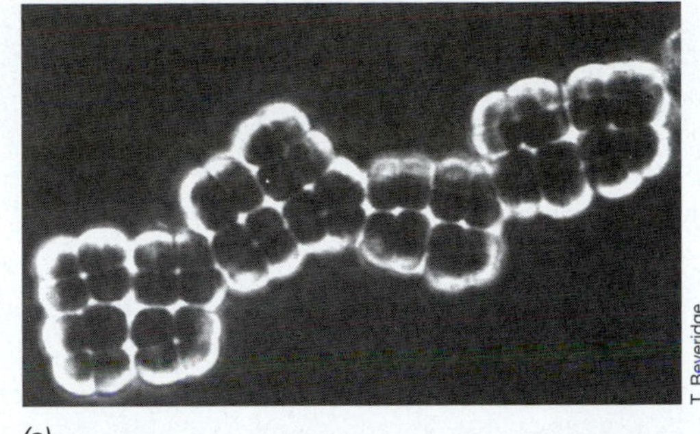

(a)

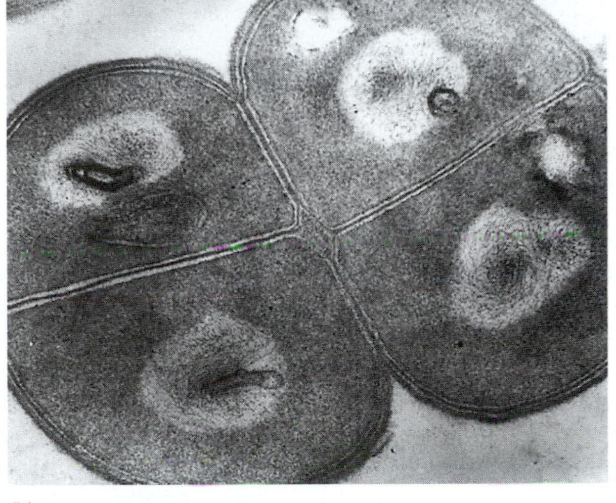

(b)

Figure 16.2 ***Sarcina.*** *(a)* Phase-contrast photomicrograph of cells of a typical gram-positive coccus *Sarcina.* A single cell is about 2 μm in diameter. *(b)* Electron micrograph of a thin section. The outermost layer of the cell consists of cellulose.

often form the predominant colonies. Many species are pigmented, and this provides an additional aid in selecting gram-positive cocci.

The genera *Micrococcus* and *Staphylococcus* can easily be separated based on the oxidation–fermentation (O/F) (Table 32.3) test. *Micrococcus* is an obligate aerobe and produces acid from glucose only under aerobic conditions, whereas *Staphylococcus* is a facultative aerobe and produces acid from glucose both aerobically and anaerobically. *Staphylococcus* also typically forms cell clusters (Figure 16.1*a*), whereas *Micrococcus* does not.

Staphylococci are common commensals and parasites of humans and animals, and they occasionally cause serious infections. In humans, there are two major species, *Staphylococcus epidermidis*, a nonpigmented, nonpathogenic organism usually found on the skin or mucous membranes, and *Staphylococcus aureus* (Figure 16.1), a yellow-pigmented species that is most commonly associated with pathological conditions, including boils, pimples, pneumonia, osteomyelitis, meningitis, and arthritis. We discuss the pathogenesis of *S. aureus* in Section 28.2 and staphylococcal diseases in Sections 34.10 and 37.5. *Micrococcus* species can also be isolated from skin but are much more common on surfaces of inanimate objects, on dust particles, and in soil.

Sarcina

The genus *Sarcina* contains species of bacteria that divide in three perpendicular planes to yield packets of eight or more cells (**Figure 16.2*a***). *Sarcina* are obligate anaerobes and are extremely acid-tolerant, being able to ferment sugars and grow in environments at a pH as low as 2. Cells of one species, *Sarcina ventriculi*, contain a thick fibrous layer of cellulose surrounding the cell wall (Figure 16.2*b*). The cellulose layers of adjacent cells become attached, and this functions as a cementing material to hold together packets of *S. ventriculi* cells.

Sarcina species can be isolated from soil, mud, feces, and stomach contents. Because of its extreme acid tolerance, *S. ventriculi* is one of only a few bacteria that can inhabit and

Table 16.2 Differentiation of the principal genera of lactic acid bacteria[a]

Cell form and arrangement	*Genus*
Cocci in chains or tetrads	
Homofermentative	*Streptococcus*
	Enterococcus
	Lactococcus
	Pediococcus
Heterofermentative	*Leuconostoc*
Rods, typically in chains	
Homofermentative	*Lactobacillus*
Heterofermentative	*Lactobacillus*

[a]For information on homofermentative and heterofermentative lactic acid bacteria, refer to Section 21.2 and Figure 21.4.

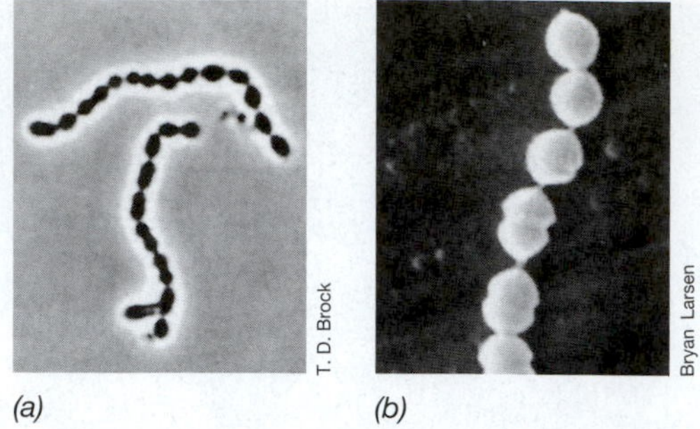

Figure 16.3 Gram-positive cocci. *(a) Lactococcus lactis*, phase-contrast. (*b*) *Streptococcus* sp., scanning electron micrograph. Cells in both (*a*) and (*b*) are 0.5–1 μm in diameter.

grow in the stomach of humans and other monogastric animals. Rapid growth of *S. ventriculi* is observed in the stomach of humans suffering from certain gastrointestinal disorders, such as pyloric ulcerations. These pathological conditions retard the flow of food to the intestine and often require surgery to correct.

Lactic Acid Bacteria and Lactic Acid Fermentations

The lactic acid bacteria are gram-positive rods and cocci that produce lactic acid as a major or sole fermentation product. Members of this group lack porphyrins and cytochromes, do not carry out oxidative phosphorylation, and hence obtain energy only by substrate-level phosphorylation. All lactic acid bacteria grow anaerobically. Unlike many anaerobes, however, most lactic acid bacteria are not sensitive to O_2 and can grow in its presence; thus they are called *aerotolerant anaerobes*. Most lactic acid bacteria obtain energy only from the metabolism of sugars and therefore are usually restricted to habitats in which sugars are present. They typically have limited biosynthetic abilities, and their complex nutritional requirements include needs for amino acids, vitamins, purines, and pyrimidines (Table 5.4).

One important difference between subgroups of the lactic acid bacteria lies in the pattern of products formed from the fermentation of sugars. One group, called **homofermentative**, produces a single fermentation product, *lactic acid*. The other group, called **heterofermentative**, produces other products, mainly ethanol and CO_2, as well as lactate (**Table 16.2**). Section 21.2 provides additional information on homofermentative and heterofermentative pathways in these bacteria. The various genera of lactic acid bacteria have been defined on the basis of cell morphology, phylogeny, and type of fermentative metabolism, as shown in Table 16.2.

Streptococcus and Other Cocci

The genus *Streptococcus* (**Figure 16.3*b***) contains homofermentative species with quite distinct habitats and activities that are of considerable practical importance to humans. Some species are pathogenic to humans and animals (Section 34.2). As producers of lactic acid, other streptococci play important roles in the production of buttermilk, silage, and other fermented products (Section 37.2), and certain species play a major role in the formation of dental caries (Section 28.3). To distinguish nonpathogenic from human pathogenic streptococci, two genera are recognized. The genus *Lactococcus* contains those streptococci of dairy significance, whereas the genus *Enterococcus* includes streptococci that are primarily of fecal origin.

Organisms in the genus *Streptococcus* have been divided into two groups of related species on the basis of characteristics enumerated in **Table 16.3**. Hemolysis on blood agar is of considerable importance in the subdivision of the genus. Colonies of those strains producing streptolysin O or S are surrounded by a large zone of complete red blood cell hemolysis when plated on blood agar, a condition called *β-hemolysis* (Figure 28.18*a*). On the other hand, many streptococci, lactococci, and enterococci do not produce hemolysins, but instead cause the formation of a greenish or brownish zone around colonies on blood agar. This is not due to true hemolysis, but to discoloration and loss of potassium from the red cells. This type of reaction is called *α-hemolysis*.

Streptococci and related cocci are also divided into immunological groups based on the presence of specific carbohydrate antigens (antigens are substances that elicit an immune response). These antigenic groups (or *Lancefield groups* as they are commonly known, named for Rebecca Lancefield, a pioneer in *Streptococcus* taxonomy) are designated by letters, A through O. Those β-hemolytic streptococci found in humans usually contain the group A antigen, whereas enterococci contain the group D antigen. Streptococci with group B antigen, usually found in association with animals, are a cause of mastitis (inflammation of the udder) in cows and have also been implicated in certain human infections. Lactococci are of antigen group N and are not pathogenic.

Heterofermentative cocci are placed in the genus *Leuconostoc*. Strains of *Leuconostoc* also produce the flavoring ingredients diacetyl and acetoin from the catabolism of citrate; they have been used as starter cultures in dairy fermentations. Some strains of *Leuconostoc* produce large

Table 16.3 Differential characteristics of streptococci, lactococci, and enterococci

Group	Antigenic (Lancefield) groups	Representative species	Type of hemolysis on blood agar	Good growth at 10°C	Good growth at 45°C	Survive 60°C for 30 min	Growth in Milk with 0.1% methylene blue	Growth in Broth with 40% bile	Habitat
Streptococci									
Pyogenes subgroup	A,B,C,F,G	*Streptococcus pyogenes*	Lysis (β)	–	–	–	–	–	Respiratory tract, systemic
Viridans subgroup	–	*Streptococcus mutans*	Greening (α)	–	+	–	–	–	Mouth, intestine
Enterococci	D	*Enterococcus faecalis*	Lysis (β), greening (α), or none	+	+	+	+	+	Intestine, vagina, plants
Lactococci	N	*Lactococcus lactis* (Figure 16.3a)	None	+	–	+	+	+	Plants, dairy products

amounts of dextran polysaccharides (α-1,6-glucan) when cultured on sucrose (Figure 21.38), and some of these have found medical use as plasma extenders in blood transfusions. Other strains of *Leuconostoc* produce other polysaccharide polymers such as fructose polymers (levans).

Lactobacillus

Lactobacilli are typically rod-shaped, varying from long and slender to short, bent rods (**Figure 16.4**). Most species are homofermentative, but some are heterofermentative (Table 16.2). Lactobacilli are common in dairy products, and some strains are used in the preparation of fermented milk products. For instance, *Lactobacillus acidophilus* (Figure 16.4*a*) is used in the production of acidophilus milk; *L. delbrueckii* (Figure 16.4*c*) is used in the preparation of yogurt; and other species are used in the production of sauerkraut, silage, and pickles (Section 37.2). The lactobacilli are usually more resistant to acidic conditions than are the other lactic acid bacteria and are able to grow well at pH values as low as 4. Because of this, they can be selectively isolated from natural materials by use of an acidic rich carbohydrate-containing medium such as tomato juice–peptone agar.

The acid resistance of the lactobacilli enables them to continue growing during natural lactic fermentations, even when the pH value has dropped too low for other lactic acid bacteria to grow. The lactobacilli are therefore responsible for the final stages of most lactic acid fermentations. They are rarely, if ever, pathogenic.

Listeria

Listeria are gram-positive coccobacilli that tend to form chains of three to five cells (Figure 37.9). *Listeria* is phylogenetically related to *Lactobacillus* species, and like homofermentative lactic acid bacteria, they produce acid but not gas from glucose. True lactic acid bacteria, however, are

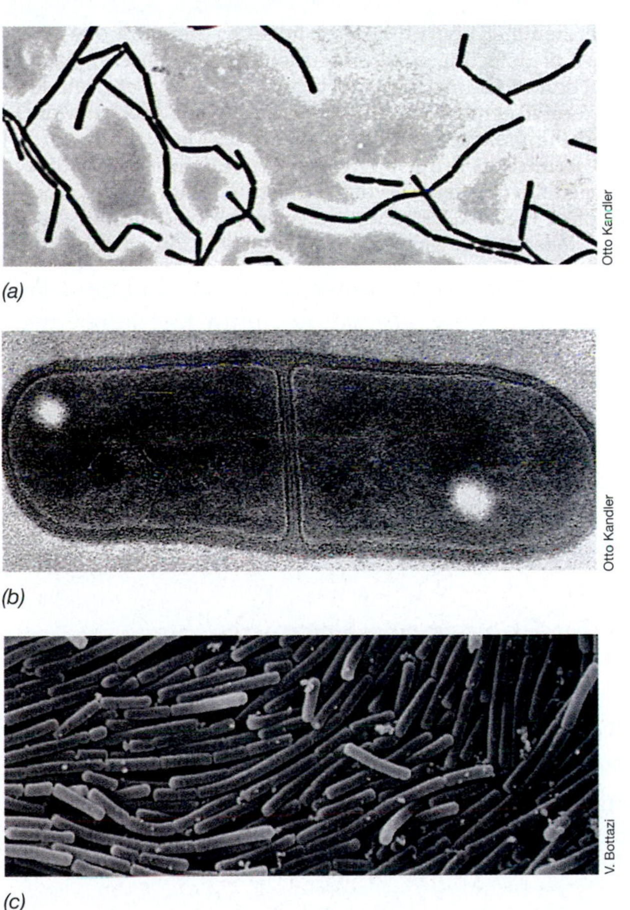

Figure 16.4 ***Lactobacillus* species.** *(a) Lactobacillus acidophilus*, phase-contrast. Cells are about 0.75 μm wide. *(b) Lactobacillus brevis*, transmission electron micrograph. Cells measure about 0.8 × 2 μm. *(c) Lactobacillus delbrueckii*, scanning electron micrograph. Cells are about 0.7 μm in diameter.

UNIT 3

Table 16.4 Major genera of endospore-forming bacteria[a]

Characteristics	*Genus*
Rods	
Aerobic or facultative, catalase produced	*Bacillus*
	Paenibacillus
Microaerophilic, no catalase; homofermentative lactic acid producer	*Sporolactobacillus*
Anaerobic:	
Sulfate-reducing	*Desulfotomaculum*
Does not reduce sulfate, fermentative	*Clostridium* (see Figure 16.5)
Thermophilic, temperature optimum 65–70°C, fermentative	*Thermoanaerobacter*
Gram-negative; can grow as homoacetogen on $H_2 + CO_2$	*Sporomusa*
Halophile, isolated from the Dead Sea	*Sporohalobacter*
Produces up to five spores per cell; fixes N_2	*Anaerobacter*
Other rod-shaped endospore formers	
Acidophile, pH optimum 3	*Alicyclobacillus*
Alkaliphile, pH optimum 9	*Amphibacillus*
Phototrophic	*Heliobacterium, Heliophilum, Heliorestis*
Syntrophic, degrades fatty acids but only in coculture with a H_2-utilizing bacterium (Section 21.5)	*Syntrophospora*
Reductively dechlorinates chlorophenols (Section 21.12)	*Desulfitobacterium*
Cocci (usually arranged in tetrads or packets), aerobic	
	Sporosarcina (see Figure 16.7)

[a]Phylogentiically, all organism are members of the low-GC subdivision of the gram-positive *Bacteria*.

capable of growth under strictly anoxic conditions and lack the enzyme catalase. *Listeria*, in contrast, requires microoxic or fully oxic conditions for growth and produces catalase.

Although several species of *Listeria* are known, the species *L. monocytogenes* is most noteworthy because it causes a major foodborne illness, *listeriosis* (Section 37.10). The organism is transmitted in contaminated, usually ready-to-eat, foods (commonly, cheese) and can cause anything from a mild illness to a fatal form of meningitis.

16.2 Endospore-Forming Gram-Positive Bacteria

Key Genera: *Bacillus, Clostridium, Sporosarcina, Heliobacterium*

Several genera of endospore-forming bacteria have been recognized (**Table 16.4**) and are distinguished on the basis of cell morphology, shape, and cellular position of the endospore (**Figure 16.5**), relationship to O_2, and energy metabolism. The structure and heat resistance of the bacterial endospore along with the process of endospore formation itself was discussed in Section 4.12. The two genera about which most is known are *Bacillus*, species of which are aerobic or facultatively aerobic, and *Clostridium*, which contains fermentative species. One group of endospore-formers, the heliobacteria, are phototrophic (the word *helio* comes from the Greek word for sun).

Considerable genetic heterogeneity exists among endospore formers, yet all endospore-forming bacteria are ecologically related because they are found in nature primarily in soil. Even those species that are pathogenic to humans or other animals are primarily saprophytic soil organisms and infect animals only incidentally. Indeed, the ability to produce endospores should be advantageous for a soil microorganism because soil is a highly variable environment in terms of nutrient levels, temperature, and water activity. Thus, a heat- and

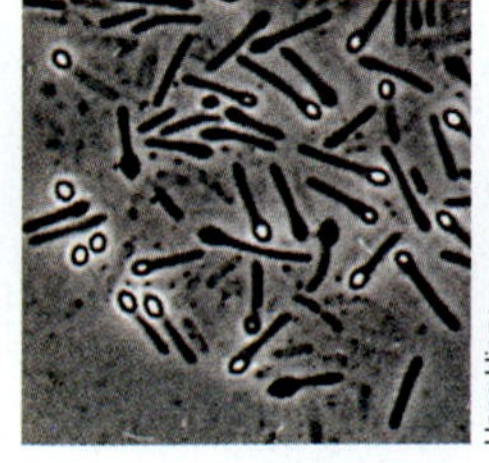

(a)

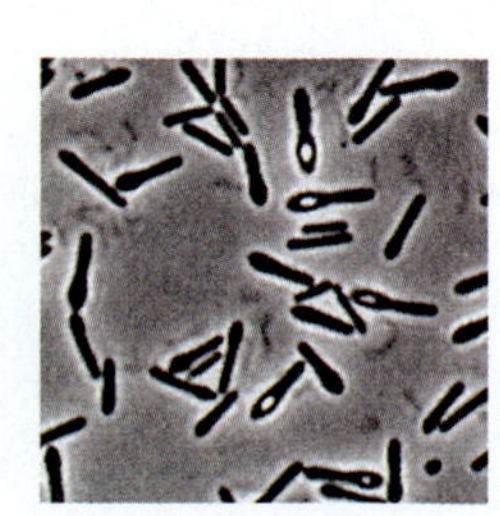

(b)

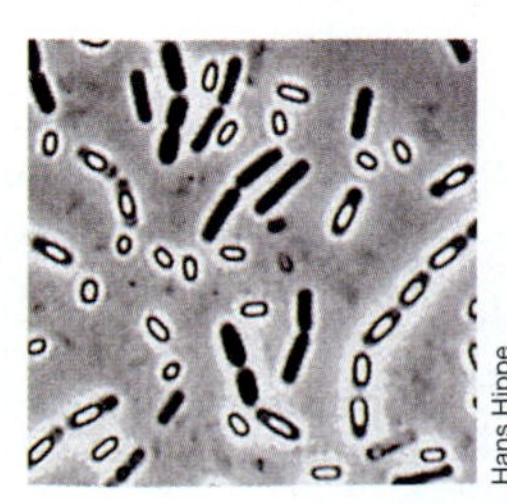

(c)

Figure 16.5 ***Clostridium*** **species and endospore location.** *(a) Clostridium cadaveris*, terminal endospores. Cells are about 0.9 μm wide. *(b) Clostridium sporogenes*, subterminal endospores. Cells are about 1 μm wide. *(c) Clostridium bifermentans*, central endospores. All phase-contrast micrographs. Cells are about 1.2 μm wide.

Table 16.5 Characteristics of representative species of bacilli

Characteristics	*Genus/Species*	*Endospore position*
I. Endospores oval or cylindrical, facultative aerobes, casein and starch hydrolyzed; sporangia not swollen, endospore wall thin		
Thermophiles and acidophiles	*Bacillus coagulans*	Central or terminal
	Alicyclobacillus acidocaldarius	Terminal
Mesophiles	*Bacillus licheniformis*	Central
	Bacillus cereus	Central
	Bacillus anthracis	Central
	Bacillus megaterium	Central
	Bacillus subtilis	Central
Insect pathogen	*Bacillus thuringiensis*	Central
Sporangia distinctly swollen, spore wall thick		
Thermophile	*Geobacillus stearothermophilus*	Terminal
Mesophiles	*Paenibacillus polymyxa*	Terminal
	Bacillus macerans	Terminal
	Bacillus circulans	Central or terminal
Insect pathogens	*Paenibacillus larvae*	Central or terminal
	Paenibacillus popilliae	Central
II. Endospores spherical, obligate aerobes, casein and starch not hydrolyzed		
Sporangia swollen	*Bacillus sphaericus*	Terminal
Sporangia not swollen	*Sporosarcina pasteurii*	Terminal

UNIT 3

desiccation-resistant structure capable of remaining dormant for long periods (perhaps even millions of years, ∞ Section 4.12) should offer considerable survival value in nature.

Endospore formers can be selectively isolated from soil, food, dust, and other materials by heating the sample to 80°C for 10 min, a treatment that effectively kills vegetative cells while the endospores present remain viable. Streaking heat-treated samples on plates of the appropriate medium and incubating either aerobically or anaerobically selectively yield species of *Bacillus* or *Clostridium*, respectively.

Bacillus and *Paenibacillus*

A list of representatives in the *Bacillus* group is shown in **Table 16.5**. Species of *Bacillus* and *Paenibacillus* grow well on defined media containing any of a number of carbon sources. Many bacilli produce extracellular hydrolytic enzymes that break down complex polymers such as polysaccharides (∞ Figure 21.37), nucleic acids, and lipids, permitting the organisms to use these products as carbon sources and electron donors. Many bacilli produce antibiotics, including bacitracin, polymyxin, tyrocidin, gramicidin, and circulin. In most cases the antibiotics are released during sporulation, when the culture enters the stationary phase of growth and after it is committed to sporulation.

Several bacilli, most notably *Paenibacillus popilliae* and *B. thuringiensis*, produce insect larvicides. *P. popilliae* causes a fatal condition called milky disease in Japanese beetle larvae and larvae of closely related beetles of the family *Scarabaeidae*. *Bacillus thuringiensis* causes a fatal disease of larvae of many different groups of insects, although individual strains are specific as to the host affected. Some strains are specific for lepidopterans, such as the silkworm, the cabbageworm, the tent caterpillar, and the gypsy moth. Other strains kill dipterans such as mosquitoes and black flies. Still others kill coleopterans such as Colorado potato beetles. Strains of *B. thuringiensis* have also been discovered that are toxic to Japanese beetles. Endospore preparations derived from *B. thuringiensis* and *P. popilliae* are commercially available as biological insecticides.

The disease caused by *P. popilliae* is a septicemia, whereas the disease caused by *B. thuringiensis* is essentially an intoxication. Both of these insect pathogens form a crystalline protein during sporulation called the *parasporal body*, which is deposited within the sporangium but outside the endospore proper (**Figure 16.6**). In *B. thuringiensis*, the parasporal body is a protoxin that is converted to a toxin by proteolytic cleavage in the larval gut. The toxin binds to intestinal epithelial cells and induces pore formation that causes leakage of the host cell cytoplasm followed by lysis.

Genes encoding crystal proteins from several *B. thuringiensis* strains have been isolated. The genes for the *B. thuringiensis* crystal protein (known commercially as "Bt-toxin") have been introduced into plants to render the plants "naturally" resistant to insects. Genetically altered Bt-toxins have also been developed by genetic engineering to help increase toxicity and reduce resistance (∞ Section 26.10).

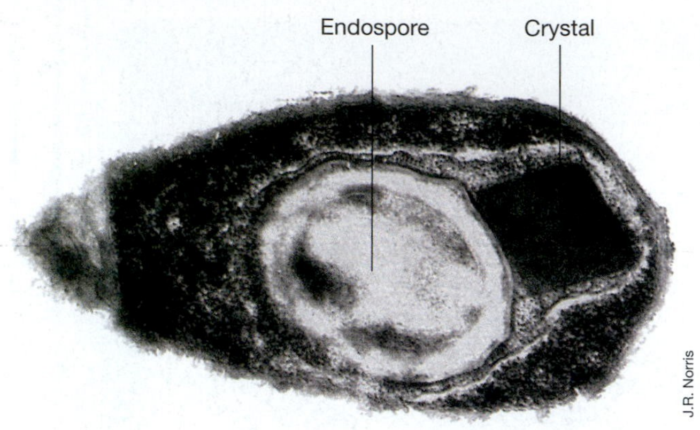

Figure 16.6 The toxic parasporal crystal in the insect pathogen *Bacillus thuringiensis*. Electron micrograph of a thin section of a sporulating cell. The crystalline protein (Bt-toxin) is toxic to certain insects by causing lysis of intestinal cells.

Clostridium

The clostridia lack a respiratory chain; unlike *Bacillus* species, they obtain ATP only by substrate-level phosphorylation. Many anaerobic energy-yielding mechanisms are known in the clostridia (clostridial fermentations are discussed in Section 21.3). Indeed, the separation of the genus *Clostridium* into subgroups is based primarily on these properties and on the fermentable substrate used (**Table 16.6**).

A number of clostridia ferment sugars, producing butyric acid as a major end product. Some of these also produce acetone and butanol, and at one time acetone–butanol fermentation by clostridia was industrially important as the main commercial source of these products. Some clostridia of the acetone–butanol type fix nitrogen. The most vigorous N_2 fixer is *Clostridium pasteurianum*, which probably is responsible for most anaerobic nitrogen fixation in the soil. One group of clostridia ferments cellulose with the formation of acids and alcohols, and these are likely the major organisms decomposing cellulose anaerobically in soil.

Another group of clostridia obtains its energy by fermenting amino acids. Some species ferment individual amino acids, but others ferment only amino acid pairs. The amino acids that can be fermented singly are alanine, cysteine, glutamate, glycine, histidine, serine, and threonine. The products are generally acetate, butyrate, CO_2, and H_2. Coupled catabolism of a mixture of amino acids is known as the **Stickland reaction**; for example, *Clostridium sporogenes* couples the use

Table 16.6 Characteristics of some groups of clostridia

Key characteristics	*Other characteristics*	*Species*
I. Ferment carbohydrates		
Ferment cellulose	Fermentation products: acetate, lactate, succinate, ethanol, CO_2, H_2	*C. cellobioparum*[a] *C. thermocellum*
Ferment sugars, starch, and pectin; some ferment cellulose	Fermentation products: acetone, butanol, ethanol, isopropanol, butyrate, acetate, propionate, succinate, CO_2, H_2; some fix N_2	*C. butyricum* *C. cellobioparum* *C. acetobutylicum* *C. pasteurianum* *C. perfringens*
Ferment sugars primarily to acetic acid	Total synthesis of acetate from CO_2; cytochromes present in some species	*C. aceticum* *Moorella thermoacetica* *C. formicaceticum*
Ferments only pentoses or methylpentoses	Ring-shaped cells form left-handed, helical chains; fermentation products: acetate, propionate, *n*-propanol, CO_2, H_2	*C. methylpentosum*
II. Ferment amino acids	Fermentation products: acetate, other fatty acids, NH_3, CO_2, sometimes H_2; some also ferment sugars to butyrate and acetate; may produce exotoxins (Sections 28.8, 28.10, and 37.5)	*C. sporogenes* *C. histolyticum* *C. putrefaciens* *C. tetani* *C. botulinum* *C. tetanomorphum*
	Ferments three-carbon amino acids (for example, alanine) to propionate, acetate, and CO_2	*C. propionicum*
III. Ferments carbohydrates or amino acids	Fermentation products from glucose: acetate, formate, small amounts of isobutyrate and isovalerate	*C. bifermentans*
IV. Purine fermenters	Ferments uric acid and other purines, forming acetate, CO_2, NH_3	*C. acidurici*
V. Ethanol fermentation to fatty acids	Produces butyrate, caproate, and H_2; requires acetate as electron acceptor; does not use sugars, amino acids, or purines	*C. kluyveri*

[a]All listings beginning with a "C." are species of the genus *Clostridium*.

of glycine and alanine. In the Stickland reaction, one amino acid functions as the electron donor and is oxidized, whereas the other is the electron acceptor and is reduced (Figure 21.7). Many of the products of amino acid fermentation by clostridia are foul-smelling substances, and the odor that results from putrefaction is a result mainly of clostridial action. In addition to butyric acid, other odoriferous compounds produced are isobutyric acid, isovaleric acid, caproic acid, hydrogen sulfide, methylmercaptan (from sulfur amino acids), cadaverine (from lysine), putrescine (from ornithine), and ammonia.

The main habitat of clostridia is the soil, where they live primarily in anoxic "pockets," made anoxic by facultative organisms metabolizing organic compounds. In addition, a number of clostridia inhabit the anoxic environment of the mammalian intestinal tract. Several clostridia are capable of causing severe disease in humans under specialized conditions, as is discussed in Section 28.10. Botulism is caused by *Clostridium botulinum*, tetanus by *Clostridium tetani*, and gas gangrene by *Clostridium perfringens* and a number of other clostridia, both sugar and amino acid fermenters. These pathogenic clostridia seem in no way unusual metabolically but are distinct in that they produce specific toxins or, in those causing gas gangrene, a group of toxins. *C. perfringens* and related species can also cause gastroenteritis in humans and domestic animals (Section 37.6), and botulism occurs in sheep and ducks and a variety of other animals. A major unsolved ecological problem is what role these extremely powerful toxins play in soil, the natural habitat of the organism.

Sporosarcina

The genus *Sporosarcina* is unique among endospore formers because cells are cocci instead of rods. *Sporosarcina* consists of strictly aerobic spherical to oval cells that divide in two or three perpendicular planes to form tetrads or packets of eight or more cells (**Figure 16.7**). The major species is *Sporosarcina ureae*. This organism (Figure 16.7) can be enriched from soil by plating dilutions of a pasteurized soil sample on alkaline nutrient agar supplemented with 8% urea and incubating in air. Most soil bacteria are strongly inhibited by as little as 2% urea. However, *S. ureae* actively decomposes even high levels of urea to CO_2 and NH_3, dramatically raising the pH.

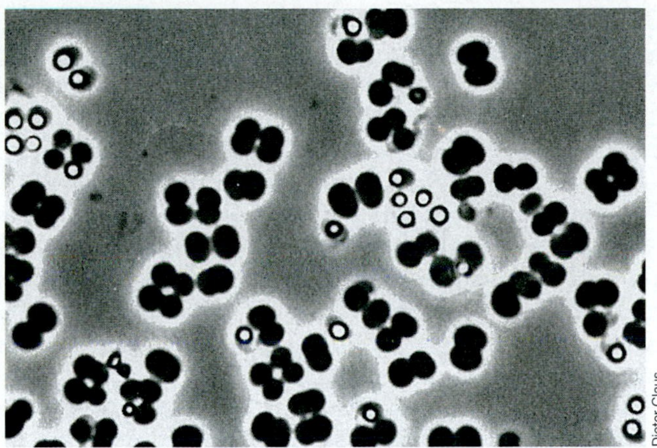

Figure 16.7 ***Sporosarcina ureae.*** Phase micrograph. A single cell is about 2 μm wide. Note bright refractile endospores. Most cell packets contain eight cells.

Sporosarcina ureae is remarkably alkaline-tolerant and can be grown in media up to pH 10. This feature can be used to advantage in its enrichment from soil.

Sporosarcina ureae is common in soils, and studies on its distribution suggest that numbers of *S. ureae* are greatest in soils that receive inputs of urine (a source of urea), such as soils in which animals periodically urinate. Because many soil organisms are quite sensitive to urea, these results suggest that *S. ureae* is ecologically important as a major urea degrader in nature.

Heliobacteria

Heliobacteria are phototrophic gram-positive *Bacteria*. **Heliobacteria** are anoxygenic phototrophs and produce a unique form of bacteriochlorophyll, bacteriochlorophyll *g* (Section 20.2). The group contains four genera: *Heliobacterium, Heliophilum, Heliorestis*, and *Heliobacillus*. All known heliobacteria are rod-shaped, either short or long rods, frequently with pointed ends (**Figure 16.8**). *Heliophilum* is morphologically interesting because its rod-shaped cells form bundles (Figure 16.8*b*) that are motile as a unit.

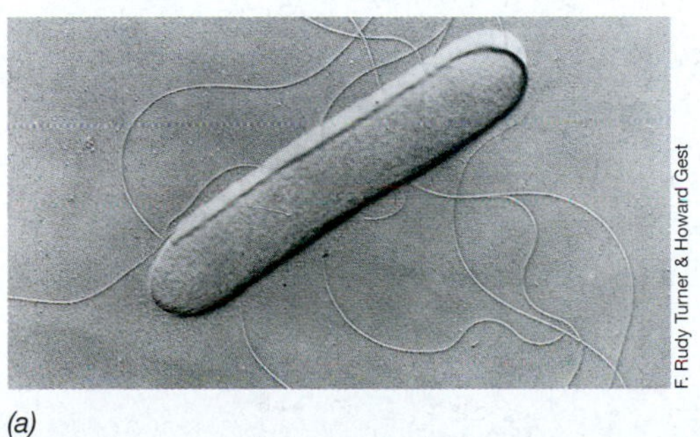

(a)

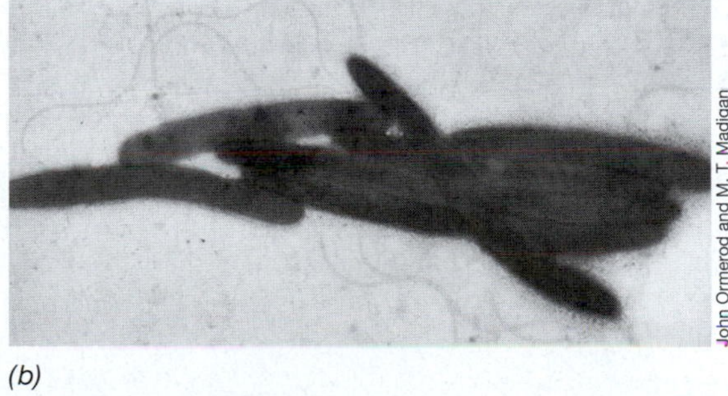

(b)

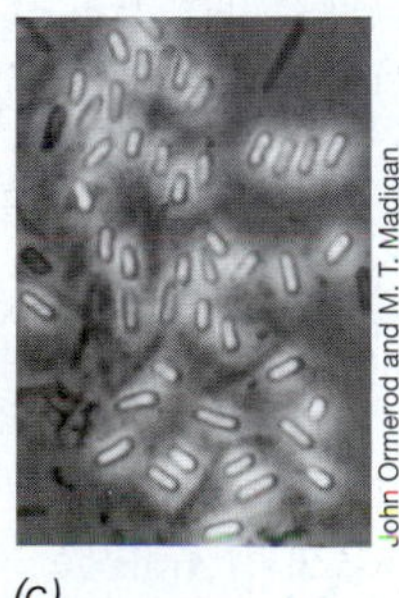

(c)

Figure 16.8 **Cells and endospores of heliobacteria.** *(a)* Electron micrograph of *Heliobacillus mobilis*, a peritrichously flagellated species. *(b) Heliophilum fasciatum* bundles as observed by electron microscopy. *(c)* Phase micrograph of endospores from *Heliobacterium gestii*.

UNIT 3

Heliobacteria are strict anaerobes, but in addition to anaerobic phototrophic growth they can grow chemotrophically in darkness by pyruvate fermentation (as can many clostridia, close relatives of the heliobacteria). Like the endospores of *Bacillus* or *Clostridium* species, the endospores of heliobacteria (Figure 16.8*c*) contain elevated Ca^{2+} levels and the signature molecule of the endospore, dipicolinic acid (∞ Section 4.12). Heliobacteria reside in soil, especially paddy field soils, where their N_2-fixation activities may benefit rice productivity. A large diversity of heliobacteria have also been found in highly alkaline environments, such as soda lakes and their surrounding alkaline soils.

16.1–16.2 MiniReview

Gram-positive *Bacteria (Firmicutes)* previously grouped in the low-GC subdivision are a large, phylogenetically related group that contains rods and cocci, sporulating and nonsporulating species. Production of endospores is a hallmark of the key genera *Bacillus* and *Clostridium*. Gram-positive bacteria are major agents for the degradation of organic matter in soil, and a few species are pathogenic.

- What are the major features that differentiate *Staphylococcus* from *Bacillus*?
- How could you distinguish between a heterofermentative and a homofermentative lactic acid bacterium?
- What is the major physiological distinction between *Bacillus* and *Clostridium* species?
- Among endospore-producing genera, what is unique about the heliobacteria?

16.3 Cell Wall-Less Gram-Positive Bacteria: The Mycoplasmas

Key Genera: *Mycoplasma, Spiroplasma*

The mycoplasmas are unusual prokaryotes because they lack cell walls and are one of the smallest organisms capable of autonomous growth. They are also of special evolutionary interest because of their simple cell structure and small genomes. Although they do not stain as gram-positive (because they lack cell walls), the mycoplasmas are phylogenetically related to nonsporulating and endospore-forming gram-positive bacteria. They are organisms that likely once had cell walls but lost the need for them because of their special habitats. Mycoplasmas are parasitic, inhabiting animal and plant hosts.

Properties of Mycoplasmas

The absence of cell walls in the mycoplasmas is readily observed by electron microscopy. The absence of cell walls has been confirmed by chemical analysis as well, which shows that the key components of peptidoglycan, muramic acid, and diaminopimelic acid, are missing. In Section 4.6 we discussed cell walls and how protoplasts can form when cells are treated in an osmotically protected medium. When the osmotic stabilizer is removed, protoplasts take up water, swell, and burst (∞ Figure 4.21). Mycoplasmas resemble protoplasts, but they are more resistant to osmotic lysis and are able to survive conditions under which protoplasts lyse. This ability to resist osmotic lysis is at least partially determined by the presence of sterols (∞ Section 4.3), which make the cytoplasmic membranes of mycoplasmas more stable than that of other bacteria. Indeed, some mycoplasmas require sterols in their growth media, and this sterol requirement is a basis for separating the mycoplasmas into two groups (**Table 16.7**).

Table 16.7 Major characteristics of mycoplasmas

Genus	Properties	Genome size (kilobase pairs)	Presence of lipoglycans
Require sterols			
Mycoplasma	Many pathogenic; facultative aerobes (see Figure 16.9)	600–1350	+
Anaeroplasma	May or may not require sterols; obligate anaerobes; degrade starch, producing acetic, lactic, and formic acids plus ethanol and CO_2; inhibited by thallium acetate; found in the bovine and ovine rumen	1500–1600	+
Spiroplasma	Spiral to corkscrew-shaped cells; associated with various phytopathogenic (plant disease) conditions (see Figure 16.11)	940–2200	–
Ureaplasma	Coccoid cells; occasional clusters and short chains; growth optimal at pH 6; strong urease reaction; associated with certain urinary tract infections in humans; inhibited by thallium acetate	750	–
Entomoplasma	Facultative aerobe; associated with insects and plants	790–1140	?
Do not require sterols			
Acholeplasma	Facultative aerobes	1500	+
Asteroleplasma	Obligate anaerobe; isolated from the bovine or ovine rumen	1500	+
Mesoplasma	Phylogenetically and ecologically related to *Entomoplasma*	870–1100	?

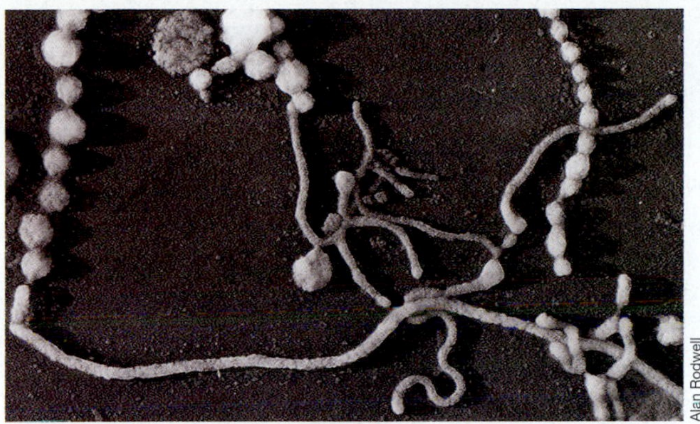

Figure 16.9 ***Mycoplasma mycoides.*** Metal-shadowed transmission electron micrograph. Note the coccoid and hyphalike elements. The average diameter of cells in chains is about 0.5 μm.

In addition to sterols, certain mycoplasmas contain compounds called *lipoglycans* (Table 16.7). Lipoglycans are long-chain heteropolysaccharides covalently linked to membrane lipids and embedded in the cytoplasmic membrane of many mycoplasmas. Lipoglycans resemble the lipopolysaccharides of gram-negative bacteria (Section 4.7) except they lack the lipid A backbone and the phosphate typical of bacterial lipopolysaccharides. Lipoglycans also function to help stabilize the cytoplasmic membrane and have also been identified as facilitating attachment of mycoplasmas to cell surface receptors of animal cells. Like lipopolysaccharides, lipoglycans stimulate antibody production when injected into experimental animals.

Growth of Mycoplasmas

Mycoplasma cells are small and pleomorphic. A single culture may exhibit small coccoid elements, larger, swollen forms, and filamentous forms of variable lengths, often highly branched (**Figure 16.9**). The small coccoid elements (0.2–0.3 μm in size) are the smallest mycoplasma units capable of independent growth. Derivatives of these cells with diameters close to 0.1 μm are occasionally seen in mycoplasma cultures, but these are not viable. Even so, the minimum reproductive unit of 0.2–0.3 μm probably represents the smallest free-living cell (Section 4.2).

The genomes of mycoplasmas are also smaller than those of most bacteria, between 500 and 1,100 kbp of DNA in most cases (Table 16.7). This is comparable to the genome size of the obligately parasitic chlamydia and rickettsia and about one-fifth to one-fourth that of *Escherichia coli*. For example, the genome of *Mycoplasma genitalium* contains only 580 kbp (Section 13.2). When the *M. genitalium* genome was sequenced in 1995 its very small genome was thought to be near the lower limit for a cell to carry out life processes. However, since then, several prokaryotic genomes have been found to be smaller, some radically so (Section 13.2).

The mode of growth of mycoplasmas differs in liquid and agar cultures. On agar, there is a tendency for the organisms to grow so that they become embedded in the medium. These colonies show a characteristic "fried-egg" appearance: a dense central core, which penetrates downward into the agar, surrounded by a circular spreading area that is lighter in color

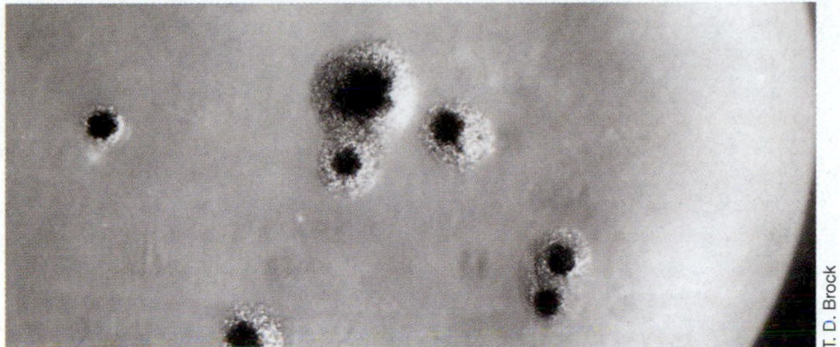

Figure 16.10 **Typical "fried egg" appearance of mycoplasma colonies on agar.** The colonies are about 0.5 mm in diameter.

(**Figure 16.10**). As would be expected of cells lacking cell walls, growth of mycoplasmas is not inhibited by penicillin, vancomycin, or other antibiotics that inhibit cell wall synthesis. However, mycoplasmas are as sensitive as other *Bacteria* to antibiotics whose targets are other than the cell wall.

Media for the culture of mycoplasmas are typically quite complex. For many species, growth is poor or absent even in complex yeast extract–peptone–beef heart infusion media. Fresh serum or ascitic fluid (peritoneal fluids) is needed as well to provide unsaturated fatty acids and sterols. Some mycoplasmas can be cultivated on relatively simple culture media, however, and defined media have been developed for some species. Most mycoplasmas use carbohydrates as carbon and energy sources and require vitamins, amino acids, purines, and pyrimidines as growth factors. The energy metabolism of mycoplasmas varies, with some species being strictly respiratory whereas others are facultative or even obligate anaerobes (Table 16.7).

Spiroplasma

The genus *Spiroplasma* consists of helical or spiral-shaped wall-less cells. Although they lack a cell wall and flagella, they are motile by means of a rotary (screw) motion or a slow undulation. Intracellular fibrils that are thought to play a role in motility have been demonstrated. The organism has been isolated from ticks, the hemolymph (**Figure 16.11**) and gut of

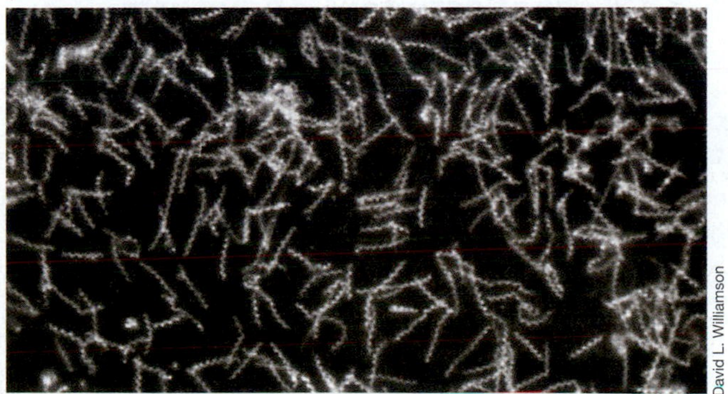

Figure 16.11 **"Sex ratio" spiroplasma from the hemolymph of the fly *Drosophila pseudoobscura*.** Dark-field micrograph. Female flies infected with the sex ratio spiroplasma bear only female progeny. Individual spiroplasma cells are about 0.15 μm in diameter. Dark-field microscopy makes these extremely thin cells visible.

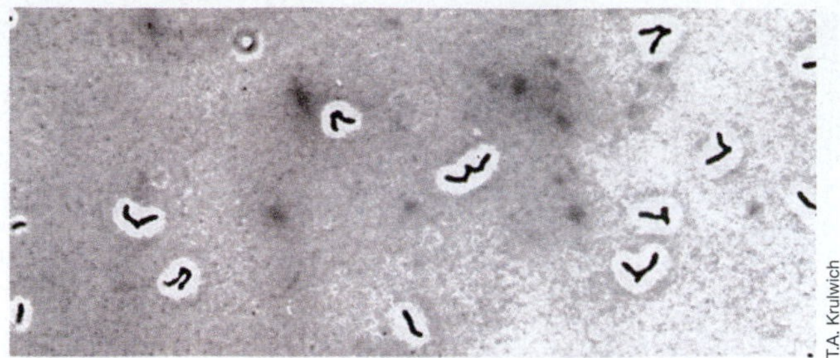

Figure 16.12 Snapping division in *Arthrobacter*. Phase micrograph of characteristic V-shaped cell groups in *Arthrobacter crystallopoietes* resulting from snapping division. Cells are about 0.9 μm in diameter.

insects, vascular plant fluids and insects that feed on fluids, and the surfaces of flowers and other plant parts.

Spiroplasma citri has been isolated from the leaves of citrus plants, where it causes a disease called citrus stubborn disease, and from corn plants suffering from corn stunt disease. A number of other mycoplasma-like bodies have been detected in diseased plants by electron microscopy, which indicates that a large group of plant-associated mycoplasmas may exist. Some species of *Spiroplasma* are known that cause insect diseases, such as honeybee spiroplasmosis and lethargy disease of the beetle *Melolontha*.

16.3 MiniReview

The mycoplasma group contains organisms that lack cell walls and have a very small genome. Many species require sterols to strengthen their cytoplasmic membranes, and several are pathogenic for humans, other animals, and plants.

- Why do mycoplasmas need to have stronger cytoplasmic membranes than other bacteria?
- Where do the mycoplasmas group phylogenetically?
- Why couldn't motile spiroplasmas contain a normal bacterial flagellum?

16.4 Actinobacteria: Coryneform and Propionic Acid Bacteria

Key Genera: *Corynebacterium, Arthrobacter, Propionibacterium*

A second major group of gram-positive bacteria, previously called the high-GC gram-positive bacteria, is the actinobacteria, which form their own phylum. This group is very large, consisting of over 30 taxonomic families of organisms (Section 14.14). Species of actinobacteria are typically rod-shaped to filamentous, primarily aerobic bacteria that are common inhabitants of the soil and of plant materials. For the most part they are harmless commensals, species of *Mycobacterium* (for example, *Mycobacterium tuberculosis*) being notable exceptions. Some are of great economic value in either the production of antibiotics or certain fermented dairy products. We begin with the rod-shaped representatives.

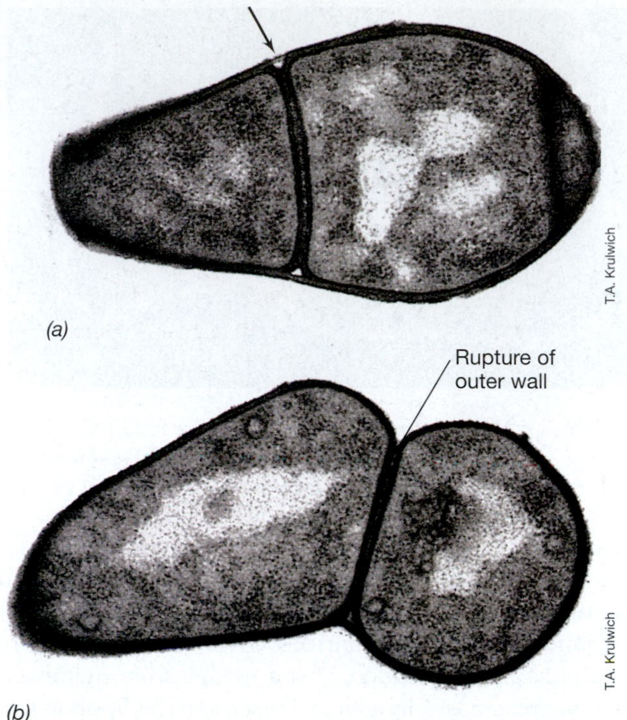

Figure 16.13 Cell division in *Arthrobacter*. Transmission electron micrograph of cell division in *Arthrobacter crystallopoietes*, illustrating how snapping division and V-shaped cell groups arise. *(a)* Before rupture of the outer cell wall layer (arrow). *(b)* After rupture of the outer layer on one side. Cells are 0.9–1 μm in diameter.

Corynebacteria

The coryneform bacteria are gram-positive, aerobic, nonmotile, rod-shaped organisms with the characteristic of forming irregular-shaped, club-shaped, or V-shaped cell arrangements during normal growth. V-shaped cell groups arise as a result of a snapping movement that occurs just after cell division (called postfission snapping movement or, simply, snapping division) (**Figure 16.12**). Snapping division occurs because the cell wall consists of two layers; only the inner layer participates in cross-wall formation, and so after the cross-wall is formed, the two daughter cells remain attached by the outer layer of the cell wall. Localized rupture of this outer layer on one side results in a bending of the two cells away from the ruptured side (**Figure 16.13**) and thus development of V-shaped forms.

The main genera of coryneform bacteria are *Corynebacterium* and *Arthrobacter*. The genus *Corynebacterium* consists of an extremely diverse group of bacteria, including animal and plant pathogens and saprophytes. Some species, such as *Corynebacterium diphtheriae*, are pathogenic (diphtheria, Section 34.3). The genus *Arthrobacter*, consisting primarily of soil organisms, is distinguished from *Corynebacterium* on the basis of a developmental cycle involving

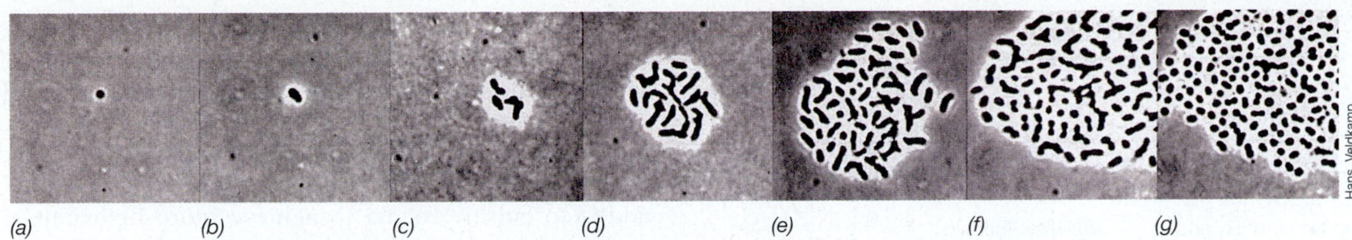

Figure 16.14 Stages in the life cycle of *Arthrobacter globiformis* as observed in slide culture. *(a)* Single coccoid element; *(b–e)* conversion to rod and growth of a microcolony consisting predominantly of rods; *(f–g)* conversion of rods to coccoid forms. Cells are about 0.9 μm in diameter.

conversion from rod to coccus and back to rod again (**Figure 16.14**). However, some corynebacteria are pleomorphic and form coccoid cells during growth, and so the distinction between the two genera on the basis of life cycle is not absolute. The *Corynebacterium* cell frequently has a swollen end, so it has a club-shaped appearance (hence the name of the genus: *koryne* is the Greek word for "club"), whereas *Arthrobacter* species are less commonly club shaped.

Species of *Arthrobacter* are among the most common of all soil bacteria. They are remarkably resistant to desiccation and starvation, despite the fact that they do not form spores or other resting cells. Arthrobacters are a heterogeneous group that have considerable nutritional versatility, and strains have been isolated that decompose herbicides, caffeine, nicotine, phenols, and other unusual organic compounds.

Propionic Acid Bacteria

The propionic acid bacteria (genus *Propionibacterium*) were first discovered in Swiss (Emmentaler) cheese, where their fermentative production of CO_2 produces the characteristic holes. Moreover, the propionic acid they produce is at least partly responsible for the unique flavor of the cheese. Although some other bacteria produce this acid, its production in large amounts by the propionic acid bacteria is a distinguishing characteristic of the genus. The bacteria in this group are gram-positive anaerobes that ferment lactic acid, carbohydrates, and polyhydroxy alcohols, producing primarily propionic acid, acetic acid, and CO_2 (∞ Section 21.3 and Figure 21.8). Their nutritional requirements are complex, and they usually grow fairly slowly.

The anaerobic fermentation of lactic acid to propionate is of interest because lactic acid itself is an end product of fermentation for many bacteria (Section 16.1). The starter culture in Swiss cheese manufacture consists of a mixture of homofermentative streptococci and lactobacilli, plus propionic acid bacteria. The homofermentative organisms carry out the initial fermentation of lactose to lactic acid during formation of the curd (protein and fat). After the curd has been drained, the propionic acid bacteria develop rapidly. The eyes (or holes) characteristic of Swiss cheese are formed by the accumulation of CO_2, the gas diffusing through the curd and gathering at weak points. In the fermentation, lactate is oxidized to pyruvate, from which it is converted to propionate (∞ Figure 21.8). The propionic acid bacteria are thus able to obtain energy anaerobically from a fermentation product that other bacteria have produced. This metabolic strategy is called a *secondary fermentation*.

Propionate is also formed in the fermentation of succinate by the bacterium *Propionigenium*. This organism is phylogenetically and ecologically unrelated to *Propionibacterium*, but energetic aspects of its fermentation are of considerable interest. We discuss the mechanism of the *Propionigenium* fermentation in Section 21.4.

16.5 Actinobacteria: *Mycobacterium*

Key Genus: *Mycobacterium*

The genus *Mycobacterium* consists of rod-shaped organisms that at some stage of their growth cycle possess the distinctive staining property called **acid-fastness**. This property is due to the presence on the surface of the mycobacterial cell of unique lipids called *mycolic acids*, found only in the genus *Mycobacterium*. First discovered by Robert Koch during his pioneering investigations on tuberculosis (∞ Section 1.8), acid-fast staining permitted the identification of the organism in tuberculous lesions. It has subsequently proved to be of great taxonomic use in defining the genus *Mycobacterium*.

Acid-Fast Ziehl–Neelsen Stain

A mixture of the dye basic fuchsin and phenol is used in acid-fast Ziehl–Neelsen stain. The stain is driven into the cells by slow heating of the smear on the microscope slide to the steaming point for 2–3 min. The role of the phenol is to enhance penetration of the fuchsin into the lipids. After washing in distilled water, the preparation is decolorized with acid–alcohol. After another wash, a final counterstain of methylene blue is used. Acid-fast organisms in the final preparation appear red, whereas the background and non-acid-fast organisms appear blue (∞ Figure 34.9).

As noted, the key component necessary for acid-fastness is mycolic acid. This substance is actually a group of complex branched-chain hydroxy lipids (the general structure is shown in **Figure 16.15*a***). In the acid-fast stain, the carboxylic acid group of the mycolic acid reacts with the fuchsin dye (Figure 16.15*b*). The mycolic acid is covalently bound to peptidoglycan in the mycobacterial wall, and this complex leaves the cell surface with a waxy, hydrophobic consistency. Mycobacteria

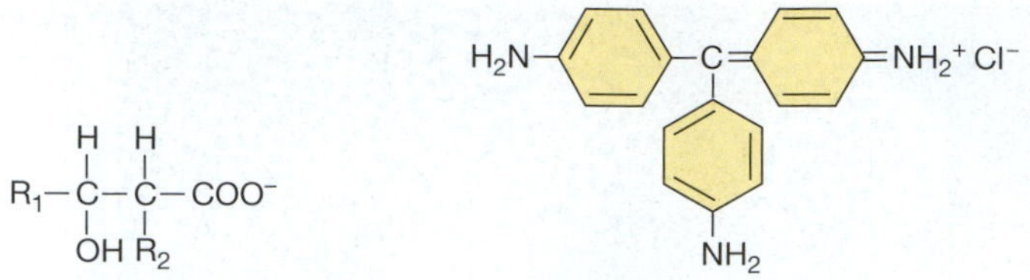

(a) Mycolic acid; R_1 and R_2 are long-chain aliphatic hydrocarbons

(b) Basic fuchsin

Figure 16.15 Acid-fast staining. Structure of *(a)* mycolic acid and *(b)* basic fuchsin, the dye used in the acid-fast stain. The fuchsin dye combines with the mycolic acid via ionic bonds between COO^- and NH_2^+.

are not readily stained by the Gram stain method because of the high surface-lipid content. If the lipoidal portion of the cell is removed with alkaline ethanol, however, the intact cell remaining is non-acid-fast but instead is gram-positive.

Characteristics of Mycobacteria

Mycobacteria are somewhat pleomorphic and may undergo branching or filamentous growth. However, in contrast to the filaments of the actinomycetes (Section 16.6), the filaments of the mycobacteria become fragmented into rods or coccoid elements upon slight disturbance; a true mycelium is not formed. In general, mycobacteria can be separated into two major groups, slow growers and fast growers (**Table 16.8**). *Mycobacterium tuberculosis* is a typical slow grower, and visible colonies are produced from dilute inoculum only after days to weeks of incubation. (Koch was successful in first isolating *M. tuberculosis* because he waited long enough after inoculating media; Section 1.8.) When growing on solid media, mycobacteria form tight, compact, often wrinkled colonies (**Figure 16.16*a***; Figure 1.16). This colony morphology is probably due to the high lipid content and hydrophobic nature of the cell surface that make cells stick together.

For the most part, mycobacteria have relatively simple nutritional requirements. They can grow in a simple mineral salts medium with ammonium as nitrogen source and glycerol or acetate as sole carbon source and electron donor incubated in air. Growth of *M. tuberculosis* is stimulated by lipids and fatty acids, and egg yolk (a good source of lipids) is often added to culture media to achieve more luxuriant growth. A glycerol-whole egg medium (Lowenstein–Jensen medium) is often used in primary isolation of *M. tuberculosis* from pathological materials.

Perhaps because of the high lipid content of its cell walls, *M. tuberculosis* is able to resist such chemical agents as alkali and phenol for considerable periods of time, and this property is used in the selective isolation of the organism from patient sputum and other materials that are grossly contaminated. The sputum is first treated with 1 M NaOH for 30 min and then neutralized and streaked onto an isolation medium.

A characteristic of many mycobacteria is their ability to form yellow carotenoid pigments (Figure 16.16*c*). Based on pigmentation, the mycobacteria can be classified into three groups: nonpigmented (including *Mycobacterium tuberculosis, Myobacterium bovis*); forming pigment only when cultured in the light, a property called photochromogenesis (including *Mycobacterium kansasii, Mycobacterium marinum*); and forming pigment even when cultured in the dark, a property called scotochromogenesis (for example, *Mycobacterium gordonae;* Table 16.8). Photoinduction of carotenoid formation requires short-wavelength (blue) light and occurs only in the presence of O_2. The evidence indicates that the critical event in photoinduction is a light-catalyzed oxidation event, and it appears that one of the early enzymes in carotenoid biosynthesis is photoinduced. As with other carotenoid-containing bacteria, it has been suggested that carotenoids protect mycobacteria against oxidative damage involving singlet oxygen (Section 6.18).

The virulence of *M. tuberculosis* cultures has been correlated with the formation of long, cordlike structures (Figure 16.16*b*) on agar or in liquid medium, due to side-to-side

Table 16.8 Some characteristics of representative mycobacteria

Species	Growth in 5% NaCl	Nitrate reduction	Growth at 45°C	Human pathogen	Pigmentation
Slow-growing species					
Mycobacterium tuberculosis	−	+	−	+	None
Mycobacterium avium	−	−	−	+	Old colonies pigmented (see Figure 16.16c)
Mycobacterium bovis	−	−	+	+	None
Mycobacterium kansasii	−	+	−	+	Photochromogenic
Fast-growing species					
Mycobacterium smegmatis	+	+	+	−	None
Mycobacterium phlei	+	+	+	−	Pigmented
Mycobacterium chelonae	+	−	−	+	None
Mycobacterium parafortuitum	+	+	−	−	Photochromogenic

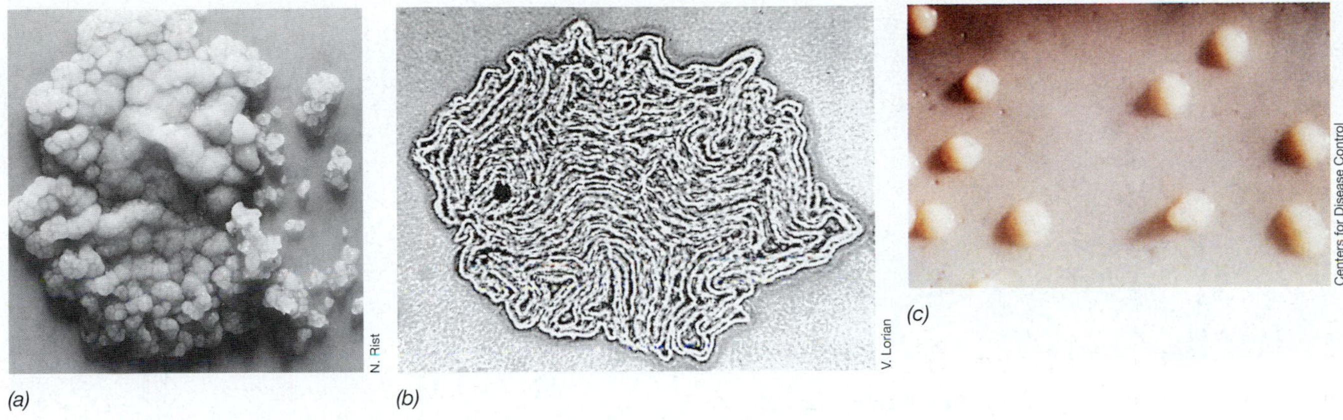

Figure 16.16 Characteristic colony morphology of mycobacteria. *(a) Mycobacterium tuberculosis*, showing the compact, wrinkled appearance of the colony. The colony is about 7 mm in diameter. *(b)* A colony of virulent *M. tuberculosis* at an early stage, showing the characteristic cordlike growth. Individual cells are about 0.5 μm in diameter. (See also the historic drawings of *M. tuberculosis* cells made by Robert Koch, Figure 1.16). *(c)* Colonies of *Mycobacterium avium* from a strain of this organism isolated as an opportunistic pathogen from an AIDS patient.

aggregation and intertwining of long chains of bacteria. Growth in cords reflects the presence of a characteristic glycolipid, the cord factor, on the cell surface (**Figure 16.17**). The pathogenesis of tuberculosis, along with the related disease leprosy, is discussed in Section 34.5.

16.4–16.5 MiniReview

Members of the *Actinobacteria* include such organisms as *Corynebacterium, Arthrobacter, Propionibacterium*, and *Mycobacterium*. They are mainly harmless soil saprophytes but *M. tuberculosis* is the causative agent of the disease tuberculosis. *Mycobacterium tuberculosis* cells have a lipid-rich, waxy outer surface layer that requires special staining procedures (the acid-fast stain) in order to observe the cells microscopically.

- What is snapping division and what organism exhibits it?
- What organism is involved in the production of Swiss cheese and what products does it make that help to flavor the cheese and make the holes?
- What is mycolic acid, what organism produces it, and what properties does this substance confer on cells that make it?

16.6 Filamentous Actinobacteria: *Streptomyces* and Relatives

Key Genera: *Streptomyces, Actinomyces*

The actinomycetes are a large group of filamentous, gram-positive *Bacteria* that form branching filaments. As a result of successful growth and branching, a ramifying network of filaments is formed, called a *mycelium* (**Figure 16.18**). Although it is of bacterial dimensions, the mycelium is analogous to the mycelium formed by the filamentous fungi (Figure 18.26).

Most actinomycetes form spores; the manner of spore formation varies and is used in separating subgroups, as outlined in **Table 16.9**. Phylogenetically, the filamentous actinomycetes form a coherent group. Thus, the mycelial spore-forming habit is of both phylogenetic and taxonomic importance. We focus here on the genus *Streptomyces*, the most important genus in this group.

Streptomyces

Over 500 species of *Streptomyces* are recognized. *Streptomyces* filaments are typically 0.5–1.0 μm in diameter, are of indefinite length, and often lack cross-walls in the vegetative phase.

Streptomyces grow at the tips of the filaments, often with branching. Thus, the vegetative phase consists of a complex,

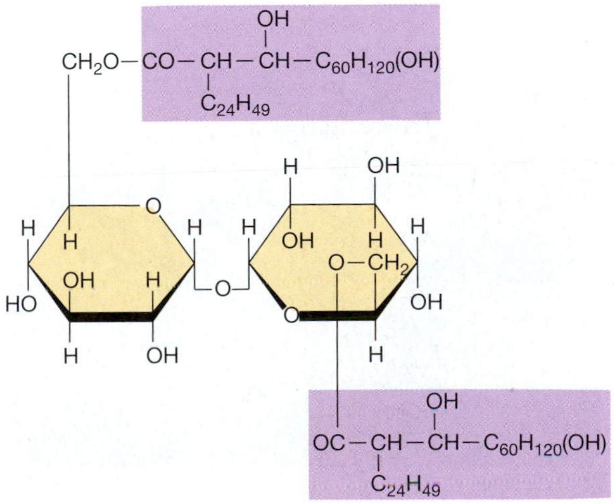

Figure 16.17 Structure of cord factor, a mycobacterial glycolipid: 6,6′-dimycolyltrehalose. The two identical long-chain dialcohol groups are shown in purple.

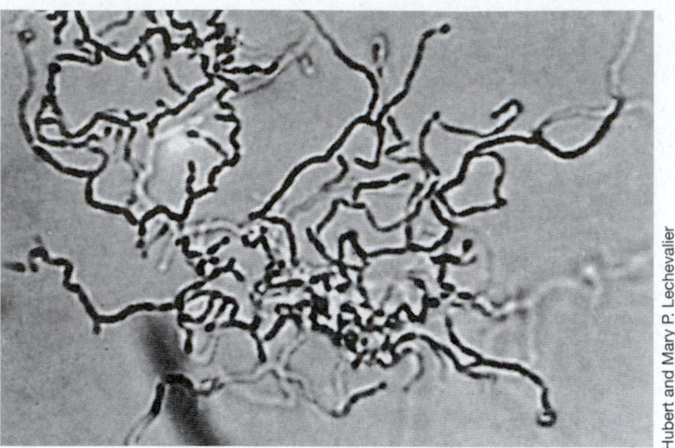

Figure 16.18 ***Nocardia.*** A young colony of an actinomycete of the genus *Nocardia*, showing typical filamentous cellular structure (mycelium). Each filament is about 0.8–1 μm in diameter.

tightly woven matrix, resulting in a compact, convoluted mycelium and subsequent colony. As the colony ages, characteristic aerial filaments called *sporophores* are formed, which project above the surface of the colony and give rise to spores (**Figure 16.19**). *Streptomyces* spores, called *conidia*, are quite distinct from the endospores of *Bacillus* and *Clostridium*. Unlike the elaborate cellular differentiation that leads to the formation of an endospore, streptomycete spores are produced by the formation of cross-walls in the multinucleate sporophores followed by separation of the individual cells directly into spores (**Figure 16.20**).

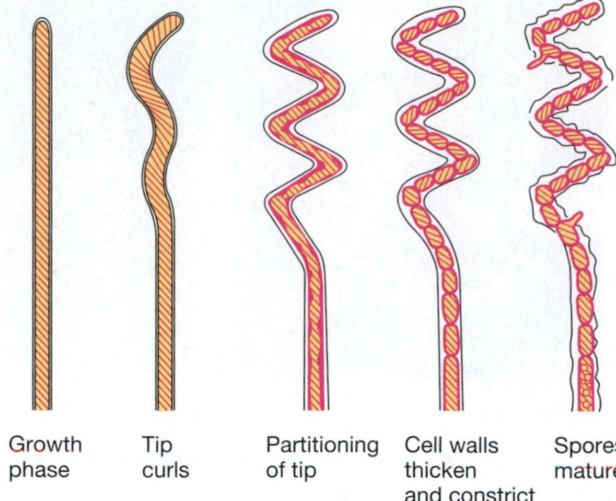

Figure 16.20 **Spore formation in *Streptomyces.*** Diagram of stages in the conversion of a streptomycete's aerial hypha (sporophores) into spores (conidia).

Differences in the shape and arrangement of aerial filaments and spore-bearing structures of various species are among the fundamental features used in classifying the *Streptomyces* species (**Figure 16.21**). The conidia and sporophores are often pigmented and contribute a characteristic

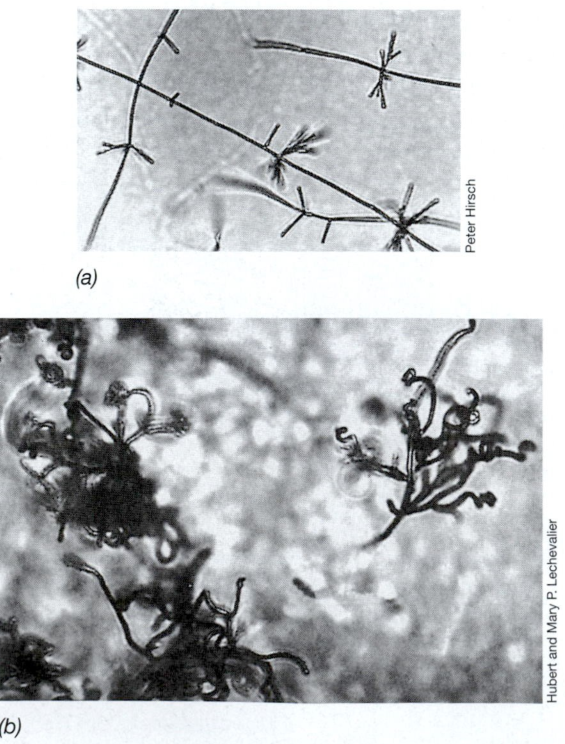

Figure 16.19 **Spore-bearing structures of actinomycetes.** Phase micrographs. (*a*) *Streptomyces*, a monoverticillate type. (*b*) *Streptomyces*, a closed spiral type. Filaments are about 0.8 μm wide in both types. Compare these photos with the art in Figure 16.21.

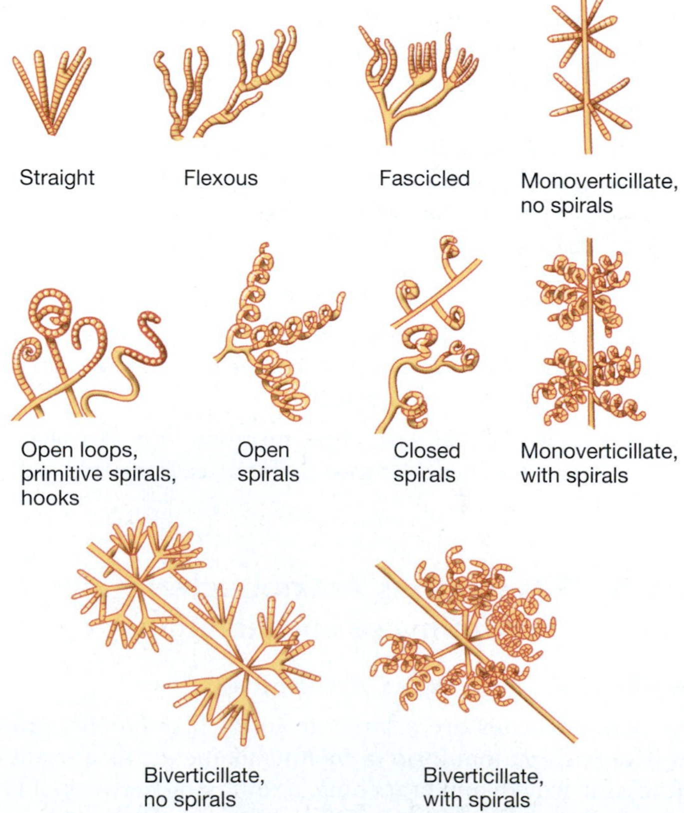

Figure 16.21 **Morphologies of spore-bearing structures in the streptomycetes.** A given species of *Streptomyces* produces only one morphological type of spore-bearing structure.

Table 16.9 **Representative *actinomycetes* and related genera of *actinobacteria* (all gram-positive)[a]**

Major groups
Coryneform group of bacteria: rods, often club-shaped, morphologically variable; not acid-fast or filamentous; snapping cell division
Corynebacterium: irregularly staining segments, sometimes granules; club-shaped swelling frequent; animal and plant pathogens, also soil saprophytes
Arthrobacter: coccus-rod morphogenesis; soil organisms
Cellulomonas: coryneform morphology; cellulose digested; facultative aerobe
Kurthia: rods with rounded ends occurring in chains; coccoid later
Brevibacterium: coccus-rod morphogenesis; cheese, skin
Propionic acid bacteria: anaerobic to aerotolerant; rods or filaments, branching
Propionibacterium: nonmotile; anaerobic to aerotolerant; produce propionic acid and acetic acid; dairy products (Swiss cheese); skin, may be pathogenic
Eubacterium: obligate anaerobes; produce mixture of organic acids, including butyric, acetic, formic, and lactic; intestine, infections of soft tissue, soil; may be pathogenic; probably the predominant member of the intestinal flora
Obligate anaerobes
Bifidobacterium: smooth microcolony, no filaments; coryneform cells common; found in intestinal tract of breast-fed infants
Acetobacterium: homoacetogen; sediments and sewage
Butyrivibrio: curved rods; rumen
Thermoanaerobacter: rods, thermophilic, found in hot springs
Actinomycetes: filamentous, often branching; highly diverse
Group I. Actinomycetes: not acid-fast; facultatively aerobic; mycelium not formed; branching filaments may be produced; rod, coccoid, or coryneform cells
Actinomyces: anaerobic to facultatively aerobic; filamentous microcolony, but filaments transitory and fragment into coryneform cells; may be pathogenic for humans or other animals; found in oral cavity
Other genera: *Arachnia, Bacterionema, Rothia, Agromyces*
Group II. Mycobacteria: acid-fast, filaments transitory
Mycobacterium: pathogens, saprophytes; obligate aerobes; lipid content of cells and cell walls high; waxes, mycolic acids; simple nutrition; growth slow; tuberculosis, leprosy, granulomas, avian tuberculosis; also soil organisms; hydrocarbon oxidizers
Group III. Nitrogen-fixing actinomycetes: nitrogen-fixing symbionts of plants; true mycelium produced
Frankia: forms nodules of two types on various plant roots; probably microaerophilic; grows slowly; fixes N_2
Group IV. Actinoplanes: true mycelium produced; spores formed, borne inside sporangia
Actinoplanes, Streptosporangium
Group V. Dermatophilus group: mycelial filaments divide transversely, and in at least two longitudinal planes, to form masses of motile, coccoid elements; aerial mycelium absent; occasionally responsible for epidermal infections
Dermatophilus, Geodermatophilus
Group VI. Nocardias: mycelial filaments commonly fragment to form coccoid or elongate elements; aerial spores occasionally produced; sometimes acid-fast; lipid content of cells and cell wall very high
Nocardia: common soil organisms; obligate aerobes; many hydrocarbon utilizers
Rhodococcus: soil saprophytes, also common in gut of various insects; utilize hydrocarbons
Group VII. Streptomycetes: mycelium remains intact, abundant aerial mycelium and long spore chains
Streptomyces: Nearly 500 recognized species, many produce antibiotics
Other genera (differentiated morphologically): *Streptoverticillium, Sporichthya, Kitasatoa*
Group VIII. Micromonosporas group: mycelium remains intact; spores formed singly, in pairs, or short chains; several thermophilic; saprophytes found in soil, rotting plant debris; one species produces endospores
Micromonospora, Microbispora, Thermobispora, Thermoactinomyces, Thermomonospora

[a]Phylogenetically, all species (except for *Acetobacterium, Butyrivibrio,* and *Thermoanaerobacter*) are members of the *Actinobacteria*.

color to the mature colony (**Figure 16.22a**). The dusty appearance of the mature colony, its compact nature, and its color make detection of *Streptomyces* colonies on agar plates relatively easy (Figure 16.22*b*).

Ecology and Isolation of *Streptomyces*

Although a few streptomycetes can be found in aquatic habitats, they are primarily soil organisms. In fact, the characteristic earthy odor of soil is caused by the production of a series of streptomycete metabolites called *geosmins*. These substances are sesquiterpenoid compounds—unsaturated ring compounds of carbon, oxygen, and hydrogen. A common geosmin is trans-1,10-dimethyl-trans-9-decalol. Geosmins are also produced by some cyanobacteria (Section 16.7).

Alkaline to neutral soils are more favorable for the development of *Streptomyces* than are acid soils. Moreover, higher numbers of *Streptomyces* are found in well-drained soils (such

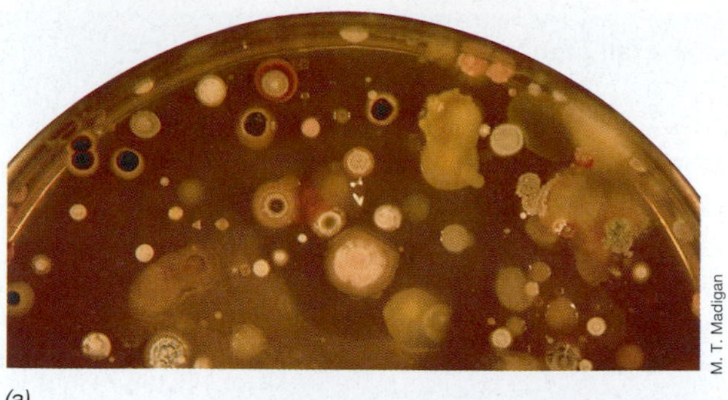

(a)

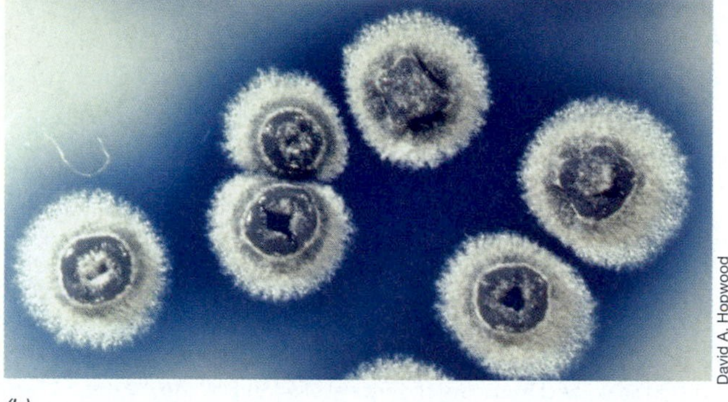

(b)

Figure 16.22 Streptomyces. *(a)* Colonies of *Streptomyces* and other soil bacteria derived from spreading a soil dilution on a casein–starch agar plate. The *Streptomyces* colonies are of various colors (several black *Streptomyces* colonies are in the foreground) but can easily be identified by their opaque, rough, nonspreading morphology. *(b)* Close-up photo of colonies of *Streptomyces coelicolor*.

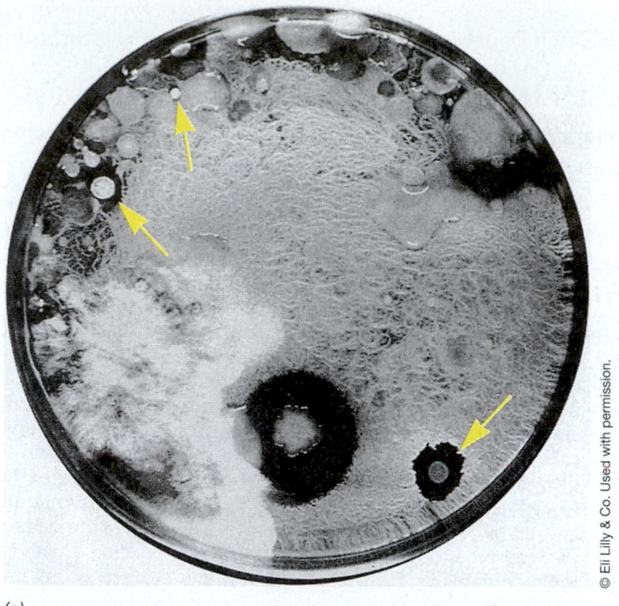

(a)

(b)

Figure 16.23 Antibiotics from *Streptomyces*. *(a)* Antibiotic action of soil microorganisms on a crowded plate. The smaller colonies surrounded by inhibition zones (arrows) are streptomycetes; the larger, spreading colonies are *Bacillus* species, some of which are also producing antibiotics. *(b)* The red-colored antibiotic undecylprodigiosin is being excreted by colonies of *Streptomyces coelicolor*.

as sandy loams or soils covering limestone), where conditions are more likely to be aerobic than in waterlogged soils. Isolation of *Streptomyces* from soil is relatively easy: A suspension of soil in sterile water is diluted and spread on selective agar medium, and the plates are incubated at 25°C (Figure 25.7). Media selective for *Streptomyces* contain mineral salts plus polymeric substances such as starch or casein as organic nutrients. After incubation for 5–7 days in air, the plates are examined for the presence of the characteristic *Streptomyces* colonies (**Figures** 16.22 and **16.23**), and spores from colonies can be restreaked to isolate pure cultures.

Nutritionally, the streptomycetes are quite versatile. Growth-factor requirements are rare, and a wide variety of carbon sources, such as sugars, alcohols, organic acids, amino acids, and some aromatic compounds, can be utilized. Most streptomycetes produce extracellular hydrolytic enzymes that permit utilization of polysaccharides (starch, cellulose, and hemicellulose), proteins, and fats, and some strains can use hydrocarbons, lignin, tannin, or even rubber. A single isolate may be able to break down over 50 distinct carbon sources.

Streptomycetes are strict aerobes whose growth in liquid culture is usually markedly stimulated by forced aeration. Sporulation usually does not take place in liquid culture, but only when the organism is growing on the surface of agar or another solid substrate; it can occur, however, when organisms form a pellicle on the surface of an unshaken liquid culture.

Antibiotics of *Streptomyces*

Perhaps the most striking property of the streptomycetes is the extent to which they produce *antibiotics* (**Table 16.10**). Evidence for antibiotic production is often seen on the agar plates used in the initial isolation of *Streptomyces*: Adjacent colonies of other bacteria show zones of inhibition (Figures 16.22*a* and 16.23*a*; Figure 25.7). About 50% of all *Streptomyces* isolated have been found to be antibiotic producers.

Because of the great economic and medical importance of many streptomycete antibiotics, an enormous amount of

Table 16.10 Some common antibiotics synthesized by species of *Streptomyces*

Chemical class	Common name	Produced by	Active against[a]
Aminoglycosides	Streptomycin	*S. griseus*[b]	Most gram-negative *Bacteria*
	Spectinomycin	*Streptomyces* spp.	*M. tuberculosis*, penicillinase-producing *N. gonorrhoeae*
	Neomycin	*S. fradiae*	Broad spectrum, usually used in topical applications because of toxicity
Tetracyclines	Tetracycline	*S. aureofaciens*	Broad spectrum, gram-positive and gram-negative *Bacteria*, rickettsias and chlamydias, *Mycoplasma*
	Chlortetracycline	*S. aureofaciens*	As for tetracycline
Macrolides	Erythromycin	*Saccharopolyspora erythraea*	Most gram-positive *Bacteria*, frequently used in place of penicillin, *Legionella*
	Clindamycin	*S. lincolnensis*	Effective against obligate anaerobes, especially *Bacteroides fragilis*
Polyenes	Nystatin	*S. noursei*	Fungi, especially *Candida* infections
	Amphoteracin B	*S. nodosus*	Fungi
None	Chloramphenicol	*S. venezuelae*	Broad spectrum; drug of choice for typhoid fever

[a]Most antibiotics are effective against several different *Bacteria*. The entries in this column refer to the common clinical application of a given antibiotic. The structures and mode of action of many of these antibiotics are discussed in Sections 27.6–27.9.

[b]All listings beginning with an "*S.*" are species of the genus *Streptomyces*.

research has been done on these producers. Over 500 distinct antibiotics have been shown to be produced by streptomycetes and many more are suspected (Sections 25.5 and 25.6); most of these have been identified chemically (Figure 16.23*b*). Some organisms produce more than one antibiotic, and often the several kinds produced by one organism are not chemically related molecules. Also, the same antibiotic may be formed by different species found in widely scattered parts of the world. And, although an antibiotic-producing organism is resistant to its own antibiotics, it usually remains sensitive to antibiotics produced by other streptomycetes. Many genes are required to encode the enzymes for antibiotic synthesis, and because of this, the genomes of *Streptomyces* species are typically quite large (8 Mbp and larger; Table 13.1)

More than 60 streptomycete antibiotics have found practical application in human and veterinary medicine, agriculture, and industry. Some of the more common antibiotics of *Streptomyces* origin are listed in Table 16.10. They are grouped into classes based on the chemical structure of the parent molecule. The search for new streptomycete antibiotics continues because existing antibiotics still do not adequately control many infectious diseases. Also, the development of antibiotic-resistant pathogens requires the continual discovery of new agents.

Ironically, despite the extensive practical studies on antibiotic-producing streptomycetes by the antibiotic industry and the fact that antibiotics from *Streptomyces* species are a multi-billion-dollar a year industry, the ecology of *Streptomyces* remains poorly understood. The interactions of these organisms with other bacteria and the ecological rationale for why antibiotics are produced in the first place is still an area about which we know very little. One hypothesis for why *Streptomyces* species produce antibiotics is that antibiotic production, which is linked to sporulation (a process itself triggered by nutrient depletion), might be a mechanism to inhibit the growth of other organisms competing with *Streptomyces* cells for limiting nutrients. This would allow the *Streptomyces* to complete the sporulation process and form a dormant structure that would have increased chances of survival.

16.6 MiniReview

The streptomycetes are a large group of filamentous, gram-positive bacteria that form spores at the end of aerial filaments. Many clinically useful antibiotics such as tetracycline and neomycin have come from *Streptomyces* species.

- How do spores and the process of sporulation in a *Streptomyces* species differ from that in a *Bacillus* species?
- What energy class of organism is a *Streptomyces* and from what types of compounds do these organisms obtain their energy?
- Why might antibiotic production be of advantage to streptomycetes?

III CYANOBACTERIA AND PROCHLOROPHYTES

16.7 Cyanobacteria

Key Genera: *Synechococcus, Oscillatoria, Nostoc*

Cyanobacteria comprise a large, morphologically, and ecologically heterogeneous group of phototrophic *Bacteria*. Cyanobacteria differ in fundamental ways from purple and green bacteria (Sections 15.2 and 16.15), most notably in that they are oxygenic phototrophs. Cyanobacteria represent one of the major phyla of *Bacteria* and show a distant relationship

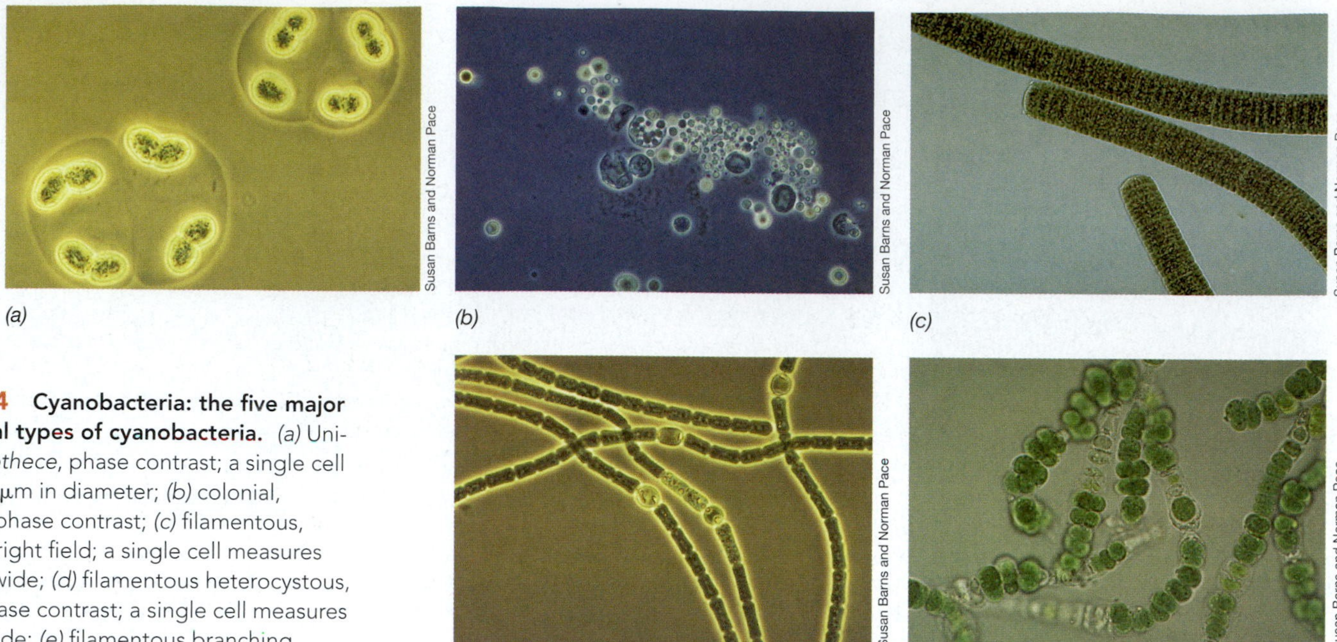

Figure 16.24 Cyanobacteria: the five major morphological types of cyanobacteria. *(a)* Unicellular, *Gloeothece*, phase contrast; a single cell measures 5–6 μm in diameter; *(b)* colonial, *Dermocarpa*, phase contrast; *(c)* filamentous, *Oscillatoria*, bright field; a single cell measures about 15 μm wide; *(d)* filamentous heterocystous, *Anabaena*, phase contrast; a single cell measures about 5 μm wide; *(e)* filamentous branching, *Fischerella*, bright field.

to gram-positive bacteria (Figure 15.1). As we saw in Section 14.3, these organisms were the first oxygen-evolving phototrophic organisms on Earth and were responsible for the conversion of the atmosphere of the Earth from anoxic to oxic.

Structure and Classification of Cyanobacteria

The morphological diversity of the cyanobacteria is impressive (**Figure 16.24**). Both unicellular and filamentous forms are known, and there is considerable variation within these morphological types. Still, cyanobacteria can be divided into just five morphological groups: (1) unicellular, dividing by binary fission (Figure 16.24*a*); (2) unicellular, dividing by multiple fission (colonial) (Figure 16.24*b*); (3) filamentous, containing differentiated cells called heterocysts that function in nitrogen fixation (Figure 16.24*d*, and see Figure 16.26); (4) filamentous nonheterocystous forms (Figure 16.24*c*); and (5) branching filamentous species (Figure 16.24*e*). **Table 16.11** lists the major genera currently recognized in each group. Cyanobacterial cells range in size from those of typical prokaryotes (0.5–1 μm in diameter) to cells as large as 40 μm in diameter (in the species *Oscillatoria princeps*, ∞ Figure 4.51*a*). Classification attempts based on 16S rRNA gene sequences and various characteristics of morphology, physiology, and ecology reveal the cyanobacteria to be a remarkably diverse group.

The structure of the cell wall of cyanobacteria is similar to that of gram-negative bacteria, and peptidoglycan is present in the walls. Many cyanobacteria produce extensive mucilaginous envelopes, or sheaths, that bind groups of cells or filaments together (Figure 16.24*a*). The photosynthetic membrane system is often complex and multilayered (∞ Figure 20.10*b*), although in some of the simpler cyanobacteria the thylakoid membranes are regularly arranged in concentric circles around the periphery of the cytoplasm (**Figure 16.25**).

Cyanobacteria have only one form of chlorophyll, chlorophyll *a*, and all of them also have characteristic biliprotein pigments, **phycobilins** (∞ Figure 20.10*a*), which function as accessory pigments in photosynthesis. One class of phycobilins, phycocyanins, are blue and, together with the green chlorophyll *a*, are responsible for the blue-green color of the bacteria. However, some cyanobacteria produce phycoerythrin, a red phycobilin, and species possessing phycoerythrin are red or brown.

Table 16.11 Genera and grouping of cyanobacteria

Group	*Genera*
Group I. Unicellular: single cells or cell aggregates	*Gloeothece* (Figure 16.24a), *Gloeobacter, Synechococcus, Cyanothece, Gloeocapsa, Synechocystis, Chamaesiphon, Merismopedia*
Group II. Pleurocapsalean: reproduce by formation of small spherical cells called baeocytes produced through multiple fission	*Dermocarpa* (Figure 16.24b), *Xenococcus, Dermocarpella, Pleurocapsa, Myxosarcina, Chroococcidiopsis*
Group III. Oscillatorian: filamentous cells that divide by binary fission in a single plane	*Oscillatoria* (Figure 16.24c), *Spirulina, Arthrospira, Lyngbya, Microcoleus, Pseudanabaena*
Group IV. Nostocalean: filamentous cells that produce heterocysts	*Anabaena* (Figure 16.24d), *Nostoc, Calothrix, Nodularia, Cylindrospermum, Scytonema*
Group V. Branching: cells divide to form branches	*Fischerella* (Figure 16.24e), *Stigonema, Chlorogloeopsis, Hapalosiphon*

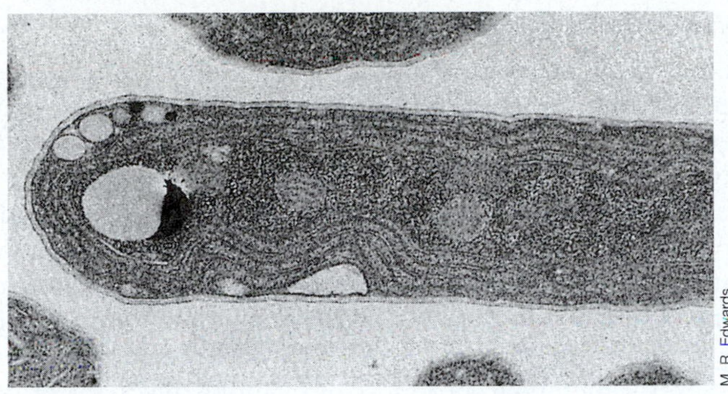

Figure 16.25 Thylakoids in cyanobacteria. Electron micrograph of a thin section of the cyanobacterium *Synechococcus lividus.* A cell is about 5 μm in diameter. Note thylakoid membranes running parallel to the cell wall.

Structural Variations: Gas Vesicles and Heterocysts

Among the cytoplasmic structures seen in many cyanobacteria are gas vesicles (Section 4.11), which are common in species that live in open waters (planktonic species). The function of gas vesicles is to regulate cell buoyancy such that cells can remain in a position in the water column where light intensity is optimal for photosynthesis. Some filamentous cyanobacteria form specialized cells called *heterocysts,* which are rounded and usually enlarged cells distributed regularly along a filament or located at one end of a filament (**Figure 16.26*a***). Heterocysts arise from differentiation of vegetative cells and are the sole sites of nitrogen fixation in heterocystous cyanobacteria. In *Anabaena,* a well-studied heterocystous cyanobacterium, complex gene rearrangements within the heterocyst yield a contiguous cluster of *nif* genes that can be expressed as a unit (*nif* genes encode nitrogenase, Section 20.15).

Heterocysts have intercellular connections with adjacent vegetative cells, and there is mutual exchange of materials between these cells. The products of photosynthesis move from vegetative cells to heterocysts, and products of nitrogen fixation move from heterocysts to vegetative cells (Figure 16.26*b*). Heterocysts are low in phycobilin pigments and lack photosystem II, the oxygen-evolving photosystem that generates reducing power from H_2O (Section 20.5). Without photosystem II, heterocysts are unable to fix CO_2 and thus lack the necessary electron donor to reduce N_2 to NH_3; fixed carbon imported to the heterocyst from an adjacent vegetative cell solves this problem (Figure 16.26*b*).

Heterocysts are surrounded by a thickened cell wall containing large amounts of glycolipid, which serves to slow the diffusion of O_2 into the cell. Because of the oxygen lability of the enzyme nitrogenase, the heterocyst maintains an anoxic environment, and by doing so stabilizes the nitrogen-fixing system in organisms that are not only aerobic but also oxygen producing. Indeed, some nonheterocystous filamentous cyanobacteria produce nitrogenase and fix nitrogen in normal vegetative cells if they are grown anaerobically by vigorous bubbling with N_2 to remove O_2.

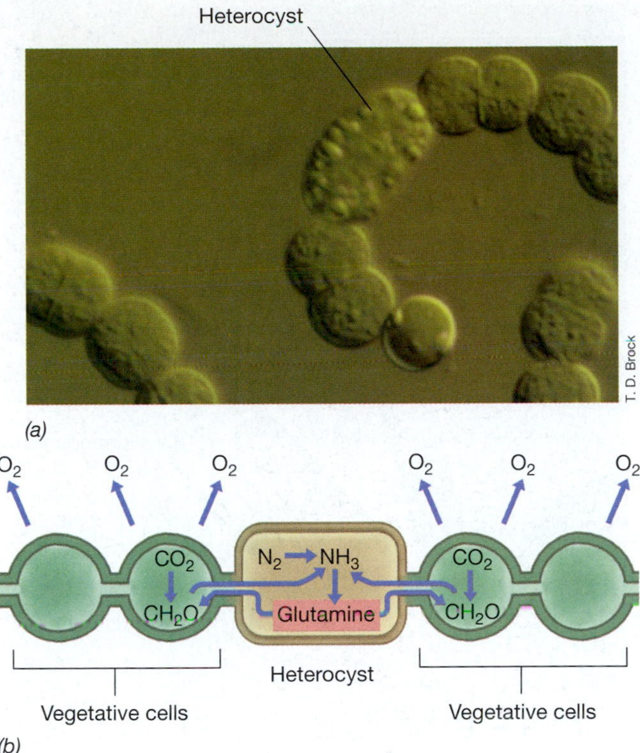

Figure 16.26 Heterocysts. *(a)* Heterocysts in the cyanobacterium *Anabaena.* Heterocysts are the sole site of nitrogen fixation in heterocystous cyanobacteria. *(b)* Model for the operation of a heterocyst. The heterocyst lacks oxygen-producing ability and obtains the needed reductant for nitrogen fixation from organic matter produced by adjacent vegetative cells. Glutamine is the form of fixed nitrogen transported from heterocysts to vegetative cells.

Cyanophycin and Other Structures

A structure called *cyanophycin* can be seen in electron micrographs of many cyanobacteria. This structure is a copolymer of aspartic acid and arginine and can constitute up to 10% of the cell mass. Cyanophycin is a nitrogen storage product, and when nitrogen in the environment becomes deficient, this polymer is broken down and used as a cellular nitrogen source. Cyanophycin is also an energy reserve in cyanobacteria. Arginine, derived from cyanophycin, can be hydrolyzed to yield ornithine, with the production of ATP through the activity of the enzyme *arginine dihydrolase,* with carbamyl phosphate (Section 21.1) occurring as an intermediate:

$$\text{Arginine} + \text{ADP} + \text{P}_i + \text{H}_2\text{O} \rightarrow \text{Ornithine} + 2\ \text{NH}_3 + \text{CO}_2 + \text{ATP}$$

Arginine dihydrolase is present in many cyanobacteria and may function as a source of ATP for maintenance purposes during dark periods.

Many cyanobacteria exhibit gliding motility. We discussed gliding motility in Section 4.14. Gliding occurs only when the cell or filament is in contact with a solid surface or

UNIT 3

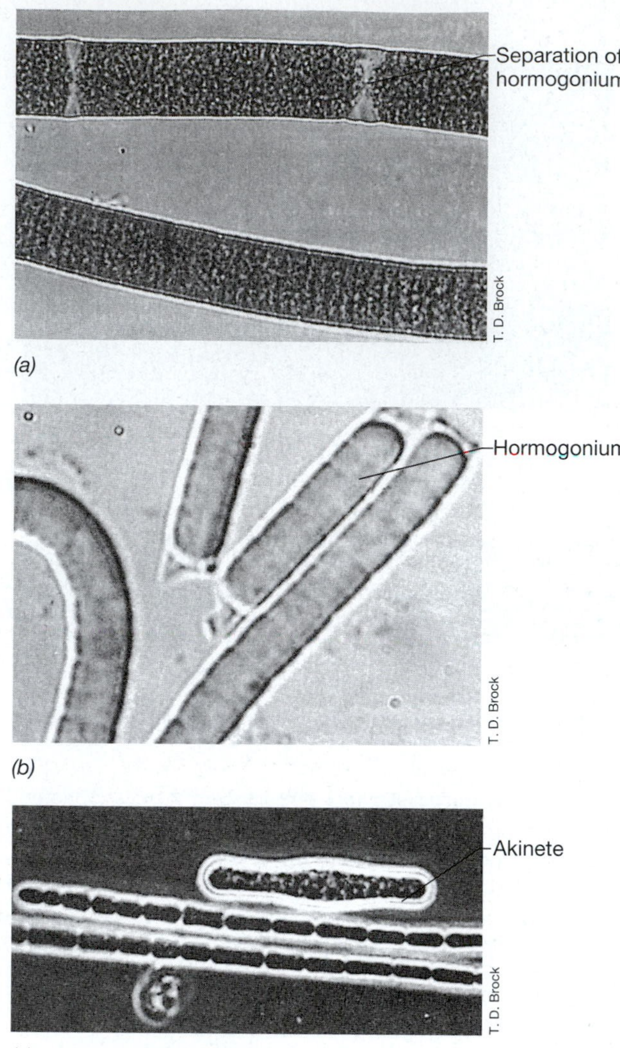

Figure 16.27 Structural differentiation in filamentous cyanobacteria. *(a)* Initial stage of hormogonium formation in *Oscillatoria*. Notice the empty spaces where the hormogonium is separating from the filament. *(b)* Hormogonium of a smaller *Oscillatoria* species. Notice that the cells at both ends are rounded. Nomarski interference-contrast microscopy. *(c)* Akinete (resting spore) of *Anabaena* by phase contrast.

with another cell or filament. In some cyanobacteria gliding is not a simple translational movement but is accompanied by rotations, reversals, and flexings of filaments. Most gliding species exhibit directional movement toward light (phototaxis), and chemotaxis (⚭ Section 4.15) may occur as well.

Among the filamentous cyanobacteria, the filaments can fragment to form small pieces called *hormogonia* (**Figure 16.27a,b**); these break away from the filaments and glide off. In some species, resting spores or akinetes (Figure 16.27*c*) form, which protect the organism during periods of darkness, drying, or freezing. These are cells with thickened outer walls; they germinate through the breakdown of the outer wall and outgrowth of a new vegetative filament. However, even the vegetative cells of many cyanobacteria are relatively resistant to drying or low temperatures.

Physiology of Cyanobacteria

The nutrition of cyanobacteria is simple. Vitamins are not required, and nitrate or ammonia is used as nitrogen source. Nitrogen-fixing species are common. Most species tested are obligate phototrophs, being unable to grow in the dark on organic compounds. However, some cyanobacteria can assimilate simple organic compounds such as glucose and acetate if light is present (photoheterotrophy). A few cyanobacteria, mainly filamentous species, can grow in the dark on glucose or sucrose, using the sugar as both carbon and energy source.

Several metabolic products of cyanobacteria are of considerable practical importance. Many cyanobacteria produce potent neurotoxins, and during water blooms when massive accumulations of cyanobacteria develop, animals ingesting such water may be killed. Many cyanobacteria are also responsible for the production of earthy odors and flavors in fresh waters, and if such waters are used as drinking water sources, aesthetic problems may arise. The major compound produced is geosmin, a substance also produced by many actinomycetes (Section 16.6).

Ecology and Phylogeny of Cyanobacteria

Cyanobacteria are widely distributed in nature in terrestrial, freshwater, and marine habitats. In general, they are more tolerant of environmental extremes than are algae and are often the dominant or sole oxygenic phototrophic organisms in hot springs (⚭ Table 6.1), saline lakes, and other extreme environments. Many species are found on the surfaces of rocks or soil and occasionally even within rocks themselves (⚭ Figure 18.37). In desert soils subject to intense sunlight, cyanobacteria often form extensive crusts over the surface, remaining dormant during most of the year and growing during the brief winter and spring rains. In shallow marine bays, where relatively warm seawater temperatures exist, cyanobacterial mats of considerable thickness may form. Freshwater lakes, especially those that are rich in nutrients, may develop blooms of cyanobacteria (⚭ Figure 23.9*b*). A few cyanobacteria are symbionts of liverworts, ferns, and cycads; a number are found as the phototrophic component of lichens. It has been shown that the cyanobacterial endophyte (a species of *Anabaena*) of the water fern *Azolla* (⚭ Sections 20.14 and 24.15) fixes nitrogen that becomes available to the plant. *Synechococcus* and the related *Prochlorococcus* (Section 16.8), small unicellular, generally nonmotile cyanobacteria, are found in abundance in the photic zones of the world's oceans, where they carry out photosynthesis that is responsible for a significant percentage of the CO_2 fixed globally.

Phylogenetically, cyanobacteria group along morphological lines in most cases. Filamentous heterocystous and nonheterocystous species form distinct groups, as do the branching forms. However, unicellular cyanobacteria are phylogenetically highly diverse, with different representatives showing phylogenetic relationships to different morphological groups.

16.8 Prochlorophytes

Key Genera: *Prochloron, Prochlorothrix, and Prochlorococcus*

Prochlorophytes are oxygenic phototrophs that contain chlorophyll *a* and *b* but do not contain phycobilins. Prochlorophytes therefore resemble both cyanobacteria (because they are prokaryotes and have chlorophyll *a*) and the green plant/green alga chloroplast (because they contain chlorophyll *b* instead of phycobilins). Phylogenetically, prochlorophytes form several separate groups in cyanobacteria.

Prochloron

Prochloron was the first prochlorophyte discovered. It is found in nature as a symbiont of marine invertebrates (didemnid ascidians), but it has not been cultured in the laboratory. Cells of *Prochloron* expressed from the cavities of didemnid tissue are roughly spherical (**Figure 16.28**) and 8–10 μm in diameter.

Electron micrographs of thin sections of *Prochloron* (Figure 16.28) show an extensive thylakoid membrane system similar to that of the chloroplast (Figure 18.5). Further evidence that *Prochloron* is phylogenetically a member of the *Bacteria* is the presence of muramic acid in the cell walls, indicating that peptidoglycan is present (Section 4.6). The carotenoids of *Prochloron* are similar to those of cyanobacteria, predominantly β-carotene and zeaxanthin. Substantial genetic heterogeneity exists between *Prochloron* obtained from different ascidians. Thus, different species of *Prochloron* probably exist, but confirmation must await laboratory culture and study of pure cultures.

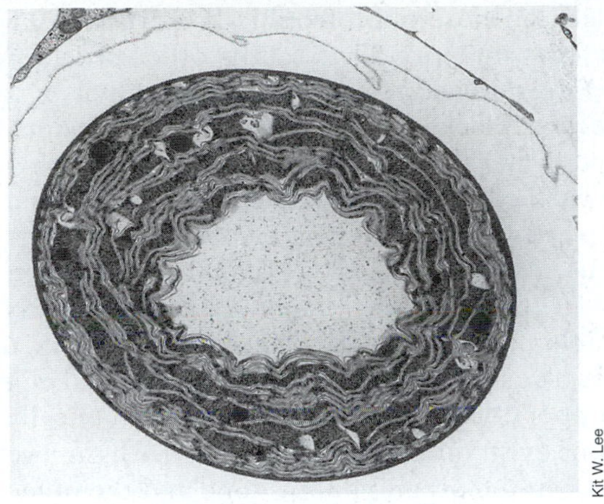

Figure 16.28 Electron micrograph of the prochlorophyte *Prochloron*. Note the extensive thylakoid membranes. Cells are about 10 μm in diameter.

Prochlorothrix and *Prochlorococcus*

Prochlorothrix is a filamentous prochlorophyte (**Figure 16.29**) that can be grown in pure culture. Like *Prochloron*, *Prochlorothrix* contains chlorophylls *a* and *b* and lacks phycobilins, although the thylakoid membranes are less well developed than in *Prochloron* (compare Figures 16.28 and 16.29*b*).

Prochlorococcus, a unicellular phototroph, inhabits the euphotic zone of the open oceans and may be the most abundant and perhaps the smallest photosynthetic organism on

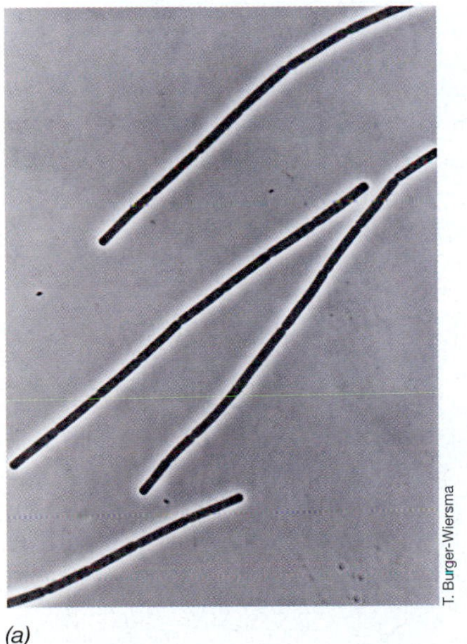

(a)

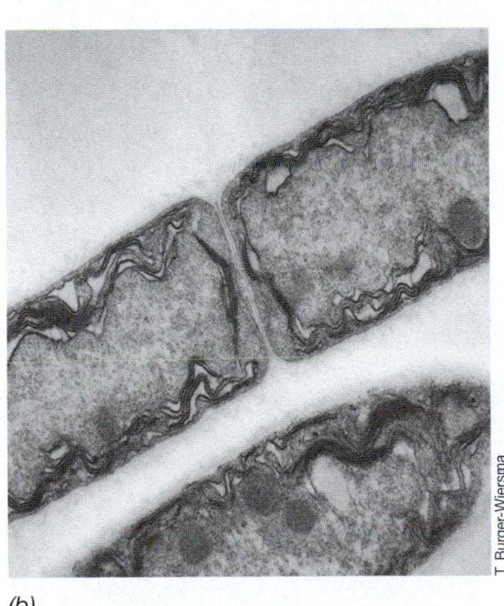

(b)

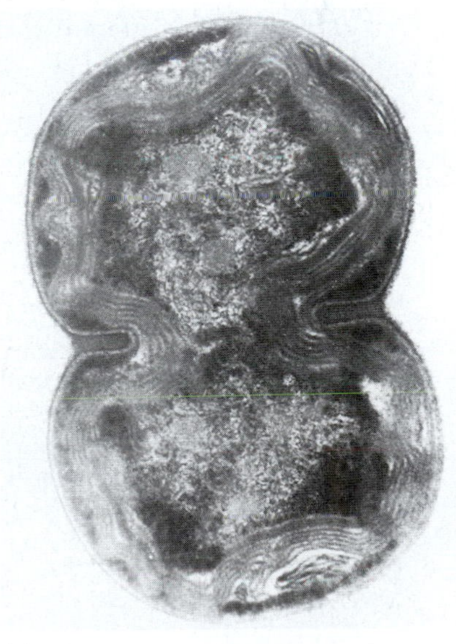

(c)

Figure 16.29 The prochlorophytes *Prochlorothrix* and *Acaryochloris*. *(a)* Phase contrast. *(b)* Electron micrograph of thin section showing arrangement of membranes in the filamentous cells. The diameter of cells is about 2 μm. *(c) Acaryochloris*, thin section. This prochlorophyte contains chlorophyll *d* as its main chlorophyll pigment. A cell is about 1.5 μm in diameter.

Earth. Cells of these phototrophic bacteria are 0.5–0.8 μm in diameter (∞ Figure 23.15). Like other prochlorophytes, cells of *Prochlorococcus* contain chlorophyll *b*. However, *Prochlorococcus* lacks true chlorophyll *a* and produces instead a modified form of chlorophyll *a*, divinyl chlorophyll *a*. Cells of *Prochlorococcus* also contain α- instead of β-carotene, a pigment previously unknown in bacteria.

Because its numbers in the oceans are very large, up to as many as 10^5 per milliliter of seawater, *Prochlorococcus* has considerable ecological significance as a primary producer. *Prochlorococcus* is found in temperate to tropical oligotrophic (nutrient-poor) open ocean waters. Two genetically distinct populations exist, one adapted to the lower light levels occurring deeper in the ocean's photic zone and the other adapted to higher light levels nearer the surface. Similarly, two ecotypes of the related unicellular phototroph *Synechococcus* exist, one better adapted to open ocean waters and the other better adapted to coastal and continental shelf waters. The adaptations relate to the content and ratio of chlorophylls in *Prochloron* and to the predominant chromophore, phycourobilin or phycoerythrobilin, associated with the photosynthetic accessory pigment phycoerythrin in *Synechococcus*. Other prochlorophytes have been isolated, including *Acaryochloris* (Figure 16.29*c*), which contains chlorophyll *d* as its major pigment.

16.7–16.8 MiniReview

Cyanobacteria and prochlorophytes are oxygenic phototrophic bacteria. Prochlorophytes differ most clearly from cyanobacteria in that prochlorophytes contain chlorophyll *b* or *d* and lack phycobilins. Oxygen in Earth's atmosphere is thought to have originated from cyanobacterial photosynthesis.

- Describe at least three ways in which cyanobacteria differ from purple bacteria.
- What is a heterocyst and what is its function?
- How are cyanobacteria, prochlorophytes, and the chloroplasts of corn plants similar and how do they differ?
- Of what ecological significance is *Prochlorococcus*?

IV CHLAMYDIA

16.9 The Chlamydia

Key Genera: *Chlamydia, Chlamydophila*

Organisms of the genera *Chlamydia* and *Chlamydophila* are obligately parasitic bacteria with poor metabolic capacities; they constitute the phylum *Chlamydia* (∞ Figure 15.1). Several key species are recognized (**Table 16.12**): *Chlamydophila psittaci*, the causative agent of the disease psittacosis; *Chlamydia trachomatis*, the causative agent of trachoma and a variety of other human diseases; and *Chlamydophila pneumoniae*, the cause of respiratory syndromes (Table 16.12).

Psittacosis is an epidemic disease of birds that is occasionally transmitted to humans and causes pneumonia-like symptoms. Trachoma is a debilitating disease of the eye

Table 16.12 Differential characteristics of species of the genera *Chlamydia* and *Chlamydophila*

Characteristic	Chlamydia trachomatis	Chlamydophila psittaci	Chlamydophila pneumoniae
Hosts	Humans	Birds, mammals, occasionally humans	Humans
Usual site of infection	Mucous membrane	Multiple sites	Respiratory mucosa
Human-to-human transmission	Common	Rare	Probable
Mol % GC of DNA	42–45	39–43	40
Percent homology to *C. trachomatis* DNA by DNA–DNA hybridization[a]	100	10	10
DNA, kilobase pairs/genome (*Escherichia coli* = 4639)	1000	550	~1000
Human diseases	Trachoma, otitis media, nongonococcal urethritis (males), urethral inflammation (females), lymphogranuloma venereum, cervicitis	Psittacosis	Respiratory syndromes
Domesticated animal diseases	—	Avian chlamydiosis (parrots, parakeets), pneumonia, synovial tissue arthritis, or conjunctivitis (kittens, lambs, calves, piglets, foals)	—

[a]For discussion of DNA:DNA hybridization, see Section 14.11.

characterized by vascularization and scarring of the cornea. Trachoma is the leading cause of blindness in humans. Other strains of *C. trachomatis* infect the genitourinary tract, and chlamydial infections are currently one of the leading sexually transmitted diseases (Section 34.14). A comparison of the properties of *C. psittaci, C. trachomatis*, and *C. pneumoniae* and the diseases they cause is shown in Table 16.12.

Molecular and Metabolic Properties

Besides being disease entities, the chlamydias are intriguing because of the biological, evolutionary, and metabolic problems they pose. Biochemical studies show that the chlamydias have gram-negative-type cell walls, and they have both DNA and RNA; that is, they are clearly cellular. Electron microscopy of thin sections of infected cells shows cells dividing by binary fission (**Figure 16.30**).

The biosynthetic capacities of the chlamydias are much more limited than even the rickettsias, the other group of obligate intracellular parasites known among the *Bacteria* (Section 15.13). Indeed, for some time it was thought that chlamydias were "energy parasites," obtaining not only biosynthetic intermediates from their hosts, as do the rickettsias, but also ATP. However, this hypothesis has been questioned following the sequencing of the genome of *C. trachomatis*. The approximately 1-Mbp chromosome of *C. trachomatis* contains genes encoding proteins for ATP synthesis and even contains a complement of genes encoding peptidoglycan biosynthetic functions. This suggests that this organism may well contain peptidoglycan even though chemical analyses of chlamydial cells have been negative.

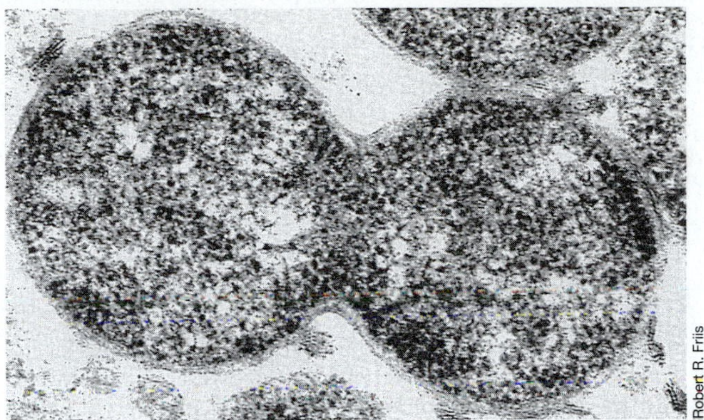

Figure 16.30 ***Chlamydia.*** Electron micrograph of a thin section of a dividing reticulate body of *Chlamydia psittaci*, a member of the psittacosis group, within a mouse tissue culture cell. A single chlamydial cell is about 1 μm in diameter.

Nevertheless, the chlamydias still probably have one of the simplest biochemical capacities of all known *Bacteria*, and this is summarized in **Table 16.13**.

Interestingly, the *C. trachomatis* genome lacks a gene encoding the protein FtsZ, a key protein in septum formation during cell division (Section 6.2). This protein was previously thought to be indispensable for growth of all prokaryotes. Moreover, genes are present in *C. trachomatis* that have a distinct "eukaryotic look" to them, suggesting that *C. trachomatis* has picked up some host genes that may encode functions that assist it in its pathogenic lifestyle (Table 16.12; Sections 28.7 and 34.14).

Table 16.13 Comparison of obligate intracellular parasites: rickettsias, chlamydias, and viruses

Property	*Rickettsias*	*Chlamydias*	*Viruses*
Structural			
Nucleic acid	RNA and DNA	RNA and DNA	Either RNA or DNA (single- or double-stranded), never both
Ribosomes	Present	Present	Absent
Cell wall	Peptidoglycan present	Peptidoglycan present[a]	No wall
Structural integrity during multiplication	Maintained	Maintained	Lost
Metabolic capacities			
Macromolecular synthesis	Carried out	Carried out	Only with use of host machinery
ATP-generating system	Present	Present[a]	Absent
Capable of oxidizing glutamate	Yes	No	No
Sensitivity to antibacterial antibiotics	Sensitive	Sensitive (except for penicillin)	Resistant
Phylogeny	*Alphaproteobacteria*	Chlamydial phylum	Not cells

[a]The genome of *Chlamydia trachomatis* (and several other chlamydial species) has been sequenced (Section 13.2), and genes for peptidoglycan synthesis and ATP synthesis are present. However, the lack of penicillin sensitivity of the chlamydia raises doubt whether peptidoglycan is synthesized.

UNIT 3

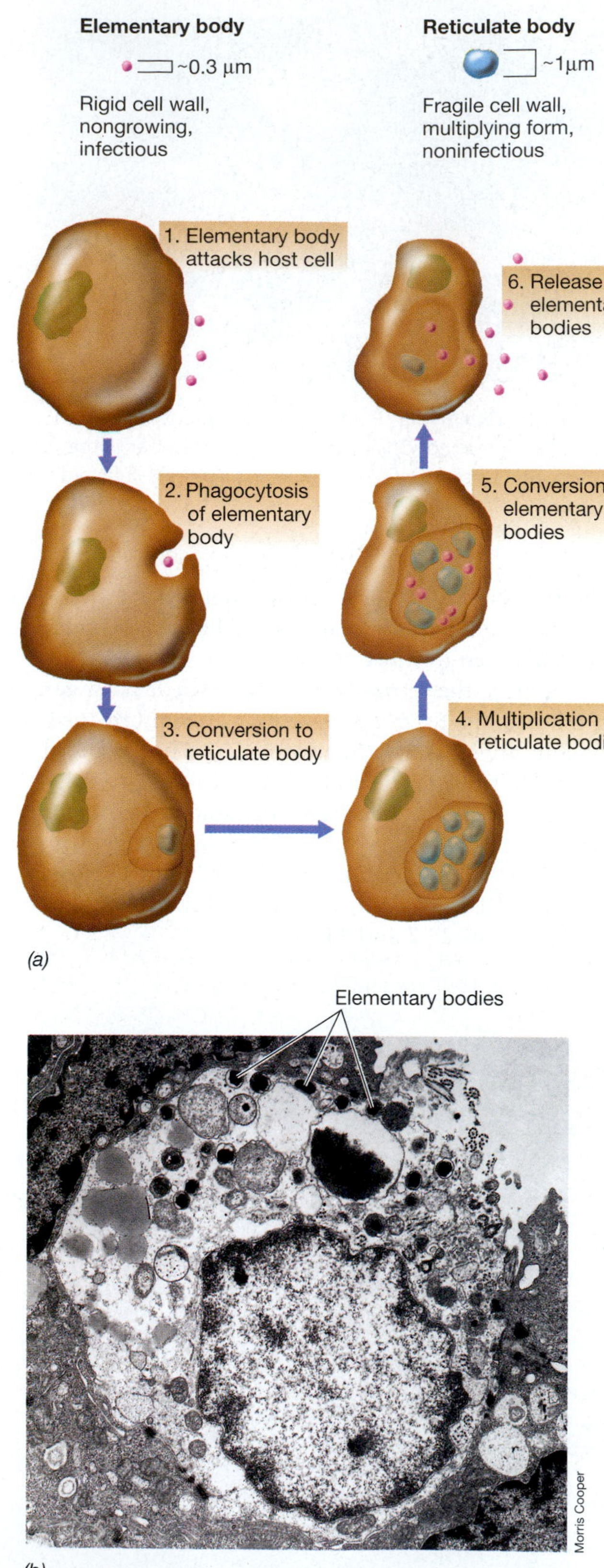

Figure 16.31 The infection cycle of chlamydia. *(a)* Schematic diagram of the cycle: the entire cycle takes about 48 h. *(b)* Human chlamydial infection. An infected fallopian tube cell is bursting, releasing mature elementary bodies.

Life Cycle of *Chlamydia*

The life cycle of a typical chlamydial species is shown in **Figure 16.31**. Two cellular types are seen in the life cycle: (1) a small, dense cell, called an *elementary body*, which is relatively resistant to drying and is the means of dispersal, and (2) a larger, less dense cell, called a *reticulate body*, which divides by binary fission and is the vegetative form.

Elementary bodies are nonmultiplying cells specialized for infectious transmission. By contrast, reticulate bodies are noninfectious forms that function only to multiply inside host cells to form a large inoculum for transmission. Unlike the rickettsias (∞ Section 15.13), the chlamydias are not transmitted by arthropods but are primarily airborne invaders of the respiratory system—hence the significance of resistance to drying of the elementary bodies. A dividing reticulate body can be seen in Figure 16.30. After a number of cell divisions, these vegetative cells are converted into elementary bodies that are released when the host cell disintegrates and can then infect other cells. Generation times of 2–3 h have been measured for reticulate bodies, which are considerably faster than those found for the rickettsias.

In sum, the chlamydias appear to have evolved an efficient and effective survival strategy including parasitizing the resources of the host (Table 16.13) and the production of resistant cell forms for transmission. It is thus not surprising that chlamydias have been associated with so many different disease syndromes (Table 16.12; ∞ Section 34.14).

16.9 MiniReview

Chlamydia are extremely small parasitic bacteria that cause human diseases. Chlamydia have a very small genome and are apparently deficient in many metabolic functions.

- Using the data of Table 16.13 as a guide, how can chlamydias be differentiated from rickettsias? From viruses?
- What is the difference between an elementary body and a reticulate body?
- What surprises have emerged from sequencing of the chlamydial genome?

V *PLANCTOMYCES/PIRELLULA*

16.10 *Planctomyces*: A Phylogenetically Unique Stalked Bacterium

Key Genera: *Planctomyces, Pirellula, Gemmata*

The *Planctomyces/Pirellula* group contains a number of morphologically unique bacteria including the genera *Planctomyces, Pirellula, Gemmata*, and *Isosphaera*. The best studied of these is *Planctomyces* (**Figure 16.32**). In Section 15.16 we considered stalked bacteria such as *Caulobacter*. *Planctomyces* is also a stalked bacterium. However, unlike

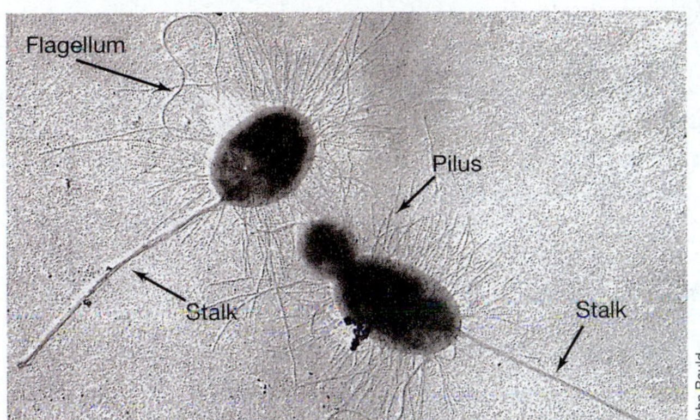

Figure 16.32 ***Planctomyces maris.*** Metal-shadowed transmission electron micrograph. A single cell is about 1–1.5 μm long. Note the fibrillar nature of the stalk. Pili are also abundant. Note also the flagella (curly appendages) on each cell and the bud that is developing from the nonstalked pole of one cell.

Caulobacter, the stalk of *Planctomyces* consists of protein and does not contain a cell wall or cytoplasm (compare Figure 16.32 with Figure 15.40). The *Planctomyces* stalk presumably functions in attachment, but it is a much narrower and finer structure than the prosthecal stalk of *Caulobacter*.

Other Features of the *Planctomyces* Group

Planctomyces and relatives are also of interest because they lack peptidoglycan and their cell walls are of an S-layer type (∞ Sections 4.6 and 4.9), consisting of protein containing large amounts of cysteine (as cystine) and proline. As would be expected of organisms lacking peptidoglycan, these organisms are resistant to antibiotics such as penicillin and cephalosporin that disrupt peptidoglycan synthesis.

Like *Caulobacter* (∞ Figures 9.23 and 15.41), *Planctomyces* is a budding bacterium with a life cycle wherein motile swarmer cells attach to a surface, grow a stalk from the attachment point, and generate a new cell from the opposite pole by budding. This daughter cell grows a flagellum, breaks away from the attached mother cell, and begins the cycle anew. Physiologically, *Planctomyces* species are facultatively aerobic chemoorganotrophs, growing either by fermentation or respiration of sugars.

The habitat of *Planctomyces* is primarily aquatic, both freshwater and marine, and the genus *Isosphaera* is a filamentous, gliding hot spring bacterium. The isolation of *Planctomyces* and relatives, like that of *Caulobacter*, requires dilute media, and because all known members of this group lack peptidoglycan, enrichments can be made even more selective by the addition of penicillin.

We learned in Section 2.5 of the major structural differences between prokaryotic and eukaryotic cells. In particular, eukaryotes have a membrane-enclosed nucleus. However, members of the phylum *Planctomycetes* are unique among all known prokaryotes in that they show extensive cell compartmentalization, including a membrane-enclosed nuclear structure. For example, in the bacterium *Gemmata* (**Figure 16.33**), the nucleoid is surrounded by a nuclear envelope. DNA in

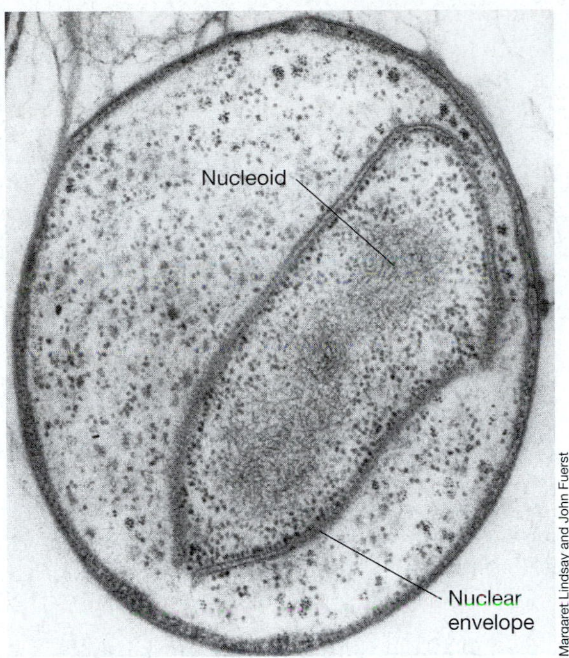

Figure 16.33 ***Gemmata:* a nucleated bacterium.** Thin-section transmission electron micrograph of a cell of *Gemmata obscuriglobus* showing the nucleoid surrounded by a nuclear envelope. The cell is about 1.5 μm in diameter.

Gemmata remains in a covalently closed, circular, and supercoiled form, typical of prokaryotes (∞ Section 7.3), but it is highly condensed and remains partitioned from the remaining cytoplasm by a true unit membrane (Figure 16.33). Another interesting compartment is the anammoxosome of *Brocadia anammoxidans*, a relative of *Planctomyces*. This bacterium carries out the anaerobic oxidation of ammonia within the enclosed anammoxosome structure (∞ Section 20.13).

Thus far, all species of *Planctomycetes* have been found to contain internal cell compartments. Some compartments lack DNA and thus have other functions (for example, metabolic functions, ∞ Section 20.13). In no other known bacteria do internal compartments so closely resemble those of the eukaryotic cell. Indeed, the unique compartmentalization of cells of planctomycetes tends to blur the structural distinction between prokaryotes and eukaryotes. Nevertheless, planctomycetes are distinct in many ways other than cell structure, including, in particular, the specific phylogenetic position they occupy within the heart of the domain *Bacteria* (∞ Figure 15.1).

VI THE VERRUCOMICROBIA

16.11 *Verrucomicrobium* and *Prosthecobacter*

Key Genera: *Verrucomicrobium*

This group of bacteria shares with prosthecate *Proteobacteria* (∞ Section 15.16) the formation of cytoplasmic appendages called *prosthecae*. The genera *Verrucomicrobium* and

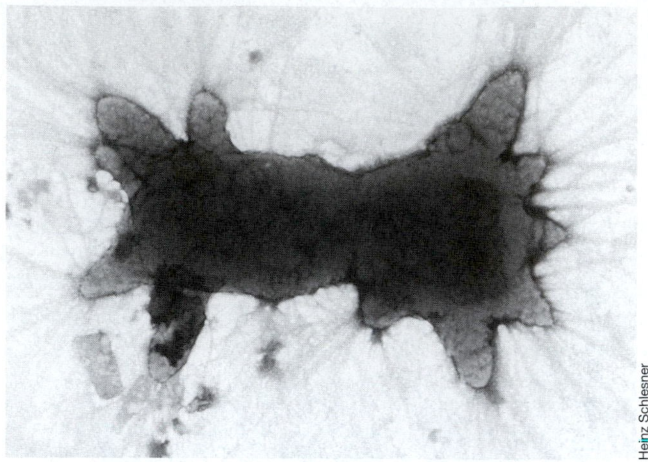

Figure 16.34 ***Verrucomicrobium spinosum.*** Negatively stained transmission electron micrograph. Note the wartlike prosthecae. A single cell is about 1 μm in diameter.

Prosthecobacter produce two to several prosthecae per cell (**Figure 16.34**). Also, unlike cells of *Caulobacter* (Figures 9.23 and 15.41), which contain a single prostheca and produce flagellated and nonprosthecate swarmer cells, *Verrucomicrobium* and *Prosthecobacter* divide symmetrically, and both mother and daughter cells contain prosthecae at the time of cell division. The genus name *Verrucomicrobium* derives from Greek roots meaning "warty," which is an appropriate description of cells of *Verrucomicrobium spinosum* with their multiple projecting prosthecae (Figure 16.34).

Species of verrucomicrobia share with other prosthecate bacteria the presence of peptidoglycan in their cell walls and are aerobic to facultatively aerobic bacteria, capable of fermenting various sugars. Verrucomicrobia are widespread in nature, inhabiting freshwater and marine environments as well as forest and agricultural soils. From a phylogenetic standpoint, verrucomicrobia are distinct from all known *Bacteria* (Figure 15.1). The group shows some phylogenetic affiliation with the *Planctomyces* and *Chlamydia* phyla (Figure 15.1), but is clearly sufficiently distinct from both of these groups to form its own separate lineage.

Prosthecobacter

Species of the genus *Prosthecobacter* contain two genes that show significant homology to the genes that encode tubulin in eukaryotic cells. Tubulin is the key protein that makes up the cytoskeleton of eukaryotic cells (Section 18.5). Although the important cell division protein FtsZ (Section 6.2) is also a tubulin homologue, the *Prosthecobacter* proteins are structurally more similar to eukaryotic tubulin than is FtsZ. The role of the tubulin proteins in *Prosthecobacter* is unknown; a eukaryotic-like cytoskeleton has not been observed in these organisms. However, these genes likely signal that a specific relationship exists between verrucomicrobia and eukaryotic cells, either in terms of shared ancestry or from horizontal gene transfer between cells of the two domains.

VII THE FLAVOBACTERIA

Flavobacteria range from obligate aerobes to obligate anaerobes, unified by a common phylogenetic thread. The organisms inhabit many different types of environments, and we focus next on two main genera in the group.

16.12 *Bacteroides* and *Flavobacterium*

Key Genera: *Bacteroides, Flavobacterium*

The genus *Bacteroides* contains obligately anaerobic, nonsporulating species that are saccharolytic, fermenting sugars or proteins (depending on the species) to primarily acetate and succinate as fermentation products. *Bacteroides* are normally commensals, found in the intestinal tract of humans and other animals (Section 28.4). In fact, *Bacteroides* species are the numerically dominant bacteria in the human large intestine, where measurements have shown that over 10^{10} prokaryotic cells are present per gram of human feces. However, species of *Bacteroides* can also be pathogens and are the most important anaerobic bacteria associated with human infections.

Species of *Bacteroides* are unusual among *Bacteria* in that they are one of the few groups of organisms to synthesize a special type of lipid called *sphingolipid,* a heterogeneous collection of lipid characterized by the long-chain amino alcohol sphingosine in place of glycerol (**Figure 16.35**). Sphingolipids such as sphingomyelin, cerebrosides, and gangliosides are common in mammalian tissues, especially in the brain and other nervous tissues.

In contrast to *Bacteroides, Flavobacterium* species are primarily found in aquatic habitats, both freshwater and marine, as well as in foods and food-processing plants. Colonies of *Flavobacterium* are frequently yellow-pigmented. Physiologically these organisms are aerobes and rather nutritionally restricted, using glucose as carbon and energy source but very few other carbon compounds. Flavobacteria are rarely pathogenic; however, one species, *Flavobacterium meningosepticum*, has been associated with cases of infant meningitis.

Other important genera in this group are psychrophilic or psychrotolerant (Section 6.13). These include, in particular, the genera *Polaribacter* and *Psychroflexus*. Many other genera in the group are also capable of good growth below 20°C.

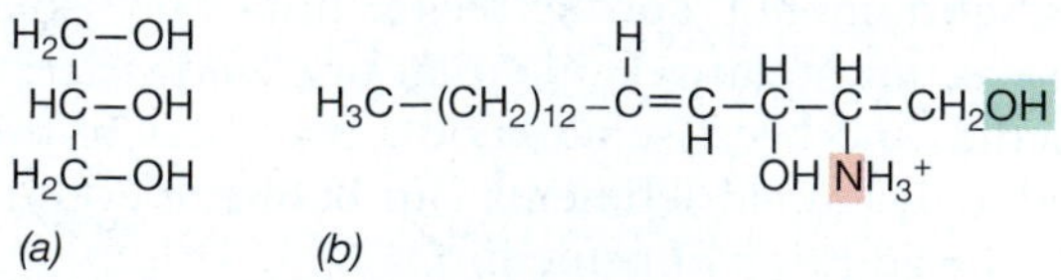

(a) $H_2C{-}OH$ / $HC{-}OH$ / $H_2C{-}OH$

(b) $H_3C{-}(CH_2)_{12}{-}CH{=}CH{-}CH(OH){-}CH(NH_3^+){-}CH_2OH$

Figure 16.35 Sphingolipids. Comparison of *(a)* glycerol with *(b)* sphingosine. In sphingolipids, characteristic of *Bacteroides* species, sphingosine is the esterifying alcohol; a fatty acid is bonded by peptide linkage through the N atom (shown in red), and the terminal —OH group (shown in green) can be any of a number of compounds including phosphatidyl choline (sphingomyelin) or various sugars (cerebrosides and gangliosides).

16.13 Acidobacteria

Key Genera: *Acidobacterium, Geothrix, Holophaga*

The acidobacteria form the heart of a phylum of *Bacteria* common in soils and other habitats. Three species have been characterized in detail, *Acidobacterium capsulatum*, *Geothrix fermentans*, and *Holophaga foetida*; all are gram-negative chemoorganotrophs. *A. capsulatum* is an acidophilic, encapsulated, aerobic bacterium isolated from acid mine drainage; it utilizes various sugars and organic acids. *G. fermentans*, a strict anaerobe, oxidizes simple organic acids (acetate, propionate, lactate, fumarate) to CO_2, with Fe^{3+} as the electron acceptor, and can also ferment citrate to acetate plus succinate. *H. foetida*, also a strictly anaerobic bacterium, can degrade methylated aromatic compounds to acetate.

Other acidobacteria are known, but most are not yet cultured and are known only from environmental 16S rRNA gene sequences. Evidence for as many as six major groups has been obtained based on variation in 16S rRNA gene sequences, indicating substantial phylogenetic and metabolic diversity of the members making up this phylum. Based on environmental 16S rRNA gene sequences, acidobacteria are abundant in soils, freshwater habitats, hot spring microbial mats, wastewater treatment reactors, and sewage sludge. Their abundance, widespread distribution, and likely metabolic diversity indicate they play important ecological roles comparable to that of other major bacterial groups.

VIII THE CYTOPHAGA GROUP

Key Genera: *Cytophaga, Flexibacter, Rhodothermus, Salinibacter*

16.14 *Cytophaga* and Relatives

Organisms of the *Cytophaga* group are long, slender, gram-negative rods, often containing pointed ends, that move by gliding (**Figure 16.36**). The related genus, *Sporocytophaga*, is similar to *Cytophaga* in morphology and physiology, but the cells form resting spherical structures called *microcysts* (Figure 16.36*d*), similar to those produced by some fruiting myxobacteria (∞ Section 15.17). They are widespread in soil and water, often in great abundance.

Many cytophagas digest polysaccharides such as cellulose (Figure 16.36*c*), agar (Figure 16.36*a*), or chitin. The cellulose decomposers can be easily isolated by placing small crumbs of soil on pieces of cellulose filter paper laid on the surface of mineral salts agar. The bacteria attach to and digest the cellulose fibers, forming spreading colonies (Figure 16.36*c*). The cytophagas do not produce soluble, extracellular, cellulose-digesting enzymes (cellulases). Instead, their cellulases remain attached to the cell envelope, which probably explains why the cells must adhere to cellulose fibrils in order to digest them.

(a)

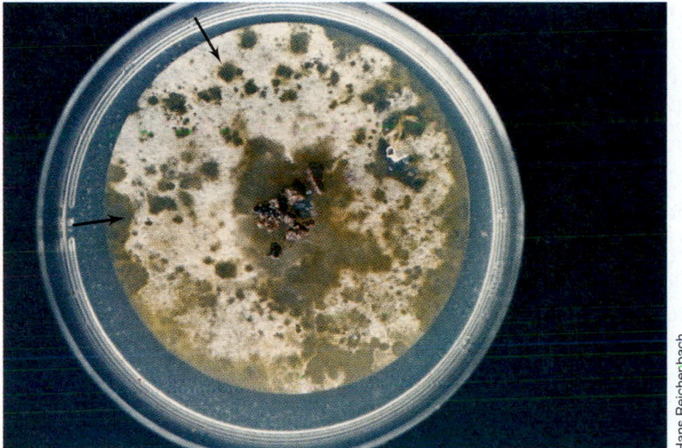

(b)

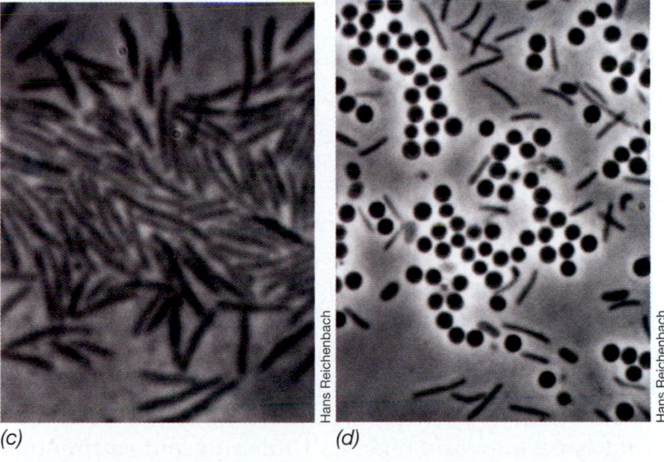

(c) (d)

Figure 16.36 ***Cytophaga* and *Sporocytophaga*.** *(a)* Streak of an agarolytic marine *Cytophaga* species hydrolyzing agar in the Petri dish. *(b)* Colonies of *Sporocytophaga* growing on cellulose. Note the clearing zones (arrows) where the cellulose has been degraded. *(c)* Phase-contrast photomicrograph of cells of *Cytophaga hutchinsonii* grown on cellulose filter paper (cells are about 1.5 μm in diameter). *(d)* Phase-contrast photomicrograph of the rod-shaped cells and spherical microcysts of *Sporocytophaga myxococcoides* (cells are about 0.5 μm and microcysts about 1.5 μm in diameter).

UNIT 3

In pure culture, *Cytophaga* can be cultured on agar containing embedded cellulose fibers, the presence of the organism being indicated by the clearing that occurs as the cellulose is digested (Figure 16.36*c* and Figure 21.36).

Species of *Cytophaga* and *Sporocytophaga* are obligately aerobic and probably account for much of the cellulose digestion by bacteria in oxic environments in nature. A number of *Cytophaga* species are fish pathogens and can cause serious problems in the cultivated fish industry. Two of the most important diseases are columnaris disease, caused by *Cytophaga columnaris*, and cold-water disease, caused by *Cytophaga psychrophila*. Both diseases preferentially affect stressed fish, such as those living in waters receiving pollutant discharges or living in high-density confinement situations such as fish hatcheries and aquaculture facilities. Infected fish show tissue destruction, frequently around the gills. This may stem from the fact that *Cytophaga* species isolated from infected fish are typically strongly proteolytic.

The genus *Flexibacter* differs from the cytophagas in that species of this genus usually require complex media for good growth and are not cellulolytic. Cells of some *Flexibacter* species also undergo changes in cell morphology from long, gliding, threadlike filaments lacking cross-walls to short, nonmotile rods. Many species are pigmented due to carotenoids located in the cytoplasmic membrane, or related pigments called flexirubins, located in the gram-negative outer membrane. *Flexibacter* species are common soil and freshwater saprophytes, and none have been identified as pathogens.

Rhodothermus and *Salinibacter*

The genera *Rhodothermus* and *Salinibacter* are in the *Cytophaga* group, but are only distant relatives. *Rhodothermus* and *Salinibacter* are gram-negative, red- or yellow-pigmented, obligatory aerobic chemoorganotrophic bacteria. *Rhodothermus* is thermophilic, with a temperature optimum near 60°C. *Rhodothermus* grows best on sugars or on simple and complex polysaccharides. The organism inhabits shallow-water submarine hot springs and has also been detected in terrestrial hot springs. *Rhodothermus* produces thermal stable hydrolytic enzymes of biotechnological interest, including an amylase (that degrades starch), a cellulase (that degrades cellulose), and a xylanase (that degrades hemicelluloses, abundant in plant cell walls), among many others.

Salinibacter is a genus of extremely halophilic red bacteria that is perhaps the most salt-tolerant and salt-requiring of all *Bacteria*. In fact, its salt requirements rival those of extremely halophilic *Archaea*, such as *Halobacterium* (Section 17.3). *Salinibacter ruber*, the only known species, lives in saltern crystallization ponds and related highly saline environments. *Salinibacter* shares with *Halobacterium* the use of K^+ as a compatible solute, a property that is rarely found among halophiles of the domain *Bacteria* (most halophilic *Bacteria* either synthesize or accumulate organic solutes to maintain water balance in their salty environments, Section 6.16). In contrast to the highly hydrolytic *Rhodothermus*, *Salinibacter* grows best with amino acids as electron donors, similar to *Halobacterium*. Thus, although phylogenetically unique, we see in *Halobacterium* and *Salinibacter* a likely case of convergent evolution, where physiological strategies have converged along the same pattern, probably because of the unique demands of the extreme environment that these organisms share.

16.10–16.14 MiniReview

The *Planctomyces* group contains stalked, budding bacteria, and the flavobacteria are gram-negative bacteria associated with animals, food, and soil that are motile by either flagella or by gliding. Species of verrucomicrobia are distinguished by their multiple prosthecate cells. The *Cytophaga* group includes obligately aerobic chemoorganotrophic bacteria that live in soil, water, hot springs, and thermal environments.

- What is unique about the cell wall and arrangement of DNA in *Planctomyces*?
- How does the stalk of *Planctomyces* differ from the stalk of *Caulobacter*?
- Where might you find large numbers of *Bacteroides* cells in nature?
- In what kinds of habitats would you expect to find *Acidobacterium*?
- Describe a method for isolating *Cytophaga* species from nature.
- Contrast the habitat and physiology of *Rhodothermus* and *Salinibacter*.

IX GREEN SULFUR BACTERIA

16.15 *Chlorobium* and Other Green Sulfur Bacteria

Key Genera: *Chlorobium, Chlorobaculum, Prosthecochloris, "Chlorochromatium"*

Green sulfur bacteria are a phylogenetically distinct group of nonmotile anoxygenic phototrophic bacteria that contain only obligately anaerobic and phototrophic species among cultured isolates. The group is morphologically restricted and includes short to long rods (**Table 16.14** and **Figure 16.37**).

Like purple sulfur bacteria, green sulfur bacteria utilize H_2S as an electron donor, oxidizing it first to S^0 and then to SO_4^{2-}. But unlike purple sulfur bacteria, the sulfur produced by green sulfur bacteria is deposited outside the cell (Figures 16.37*a* and Figure 20.17*b*). Most species can also assimilate a few organic compounds in the light (that is, photoheterotrophy, Section 20.4). Autotrophy, however, is supported not by the reactions of the Calvin cycle as in purple bacteria, but instead by a reversal of steps in the citric acid cycle (Section 20.7), a unique means of autotrophy.

Table 16.14 Genera and characteristics of phototrophic greeen sulfur bacteria

Characteristics	*Genus*
Straight or curved rods, some branching; nonmotile; color green or brown; some contain gas vesicles	*Chlorobium* *Chlorobaculum*
Spheres and ovals, nonmotile; forming prosthecae; green or brown	*Prosthecochloris*
Rods, motile by gliding; green	*Chloroherpeton*

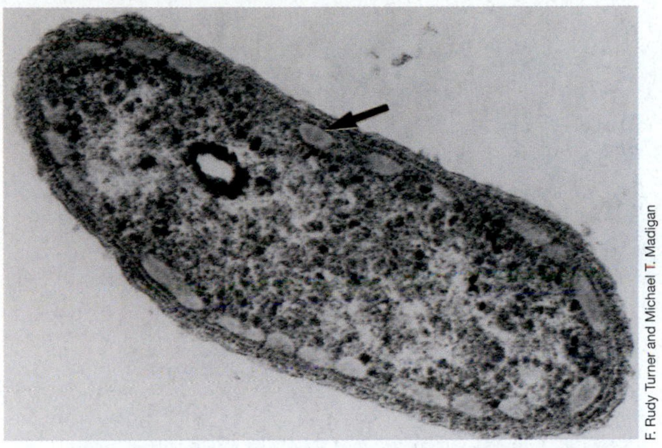

Figure 16.38 The thermophilic green sulfur bacterium *Chlorobaculum tepidum.* Thin section transmission electron micrograph. Note chlorosomes (arrow) in the cell periphery. A cell is about 0.7 μm wide.

Pigments and Ecology

The bacteriochlorophylls found in green sulfur bacteria include bacteriochlorophyll *a*, and either bacteriochlorophylls *c*, *d*, or *e*. The latter pigments function only in light-harvesting reactions (∞ Section 20.2) and are located in unique structures called **chlorosomes** (**Figure 16.38**). Chlorosomes are oblong bacteriochlorophyll-rich bodies bounded by a thin, nonunit membrane and attached to the cytoplasmic membrane in the periphery of the cell (Figure 16.38 and ∞ Figure 20.7).

Studies of energy transfer in green sulfur bacteria have shown that light energy absorbed by bacteriochlorophylls *c*, *d*, or *e* in the chlorosome is funneled to bacteriochlorophyll *a*, which resides in the cytoplasmic membrane. It is here where photosynthetic energy is converted and ATP is synthesized. Both green- and brown-colored species of green sulfur bacteria are known, the brown-colored species containing bacteriochlorophyll *e* and carotenoids that make cell suspensions brown (**Figure 16.39**; ∞ Figure 20.9).

Like purple sulfur bacteria (∞ Section 15.2), green sulfur bacteria live in anoxic aquatic environments, especially ones in which H_2S is abundant (in general, green sulfur bacteria are more tolerant of sulfide than are purple bacteria). Because the chlorosome is such an efficient light-harvesting structure, little light is required to support the photosynthetic activities of green sulfur bacteria, and they are thus typically found in lakes at the greatest depths of any phototrophic organism.

One species, *Chlorobaculum tepidum* (Figure 16.38), is thermophilic and forms dense microbial mats in high-sulfide hot springs. *C. tepidum* is also notable because its genome (2.1 Mbp) has been completely sequenced—the first genome

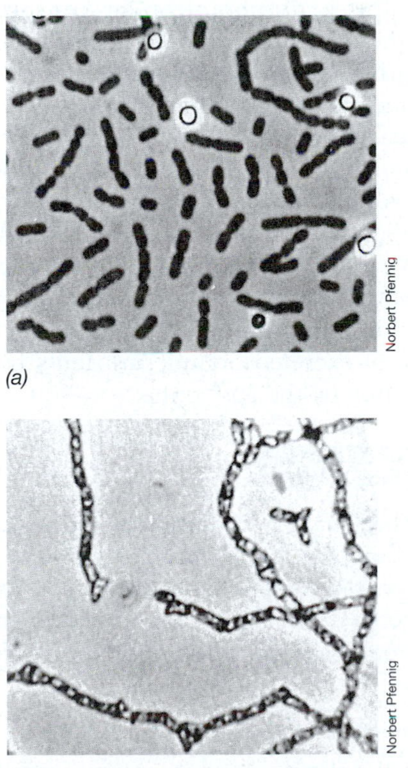

Figure 16.37 Phototrophic green sulfur bacteria. *(a) Chlorobium limicola*; cells are about 0.8 μm wide. Note the sulfur granules deposited extracellularly. *(b) Chlorobium clathratiforme*, a bacterium forming a three-dimensional network; cells are about 0.8 μm wide.

Figure 16.39 Green and brown chlorobia. Tube cultures of *(a) Chlorobaculum tepidum* and *(b) Chlorobaculum phaeobacteroides*. Cells of *C. tepidum* contain bacteriochlorophyll *c* and green carotenoids, and cells of *C. phaeobacteroides* contain bacteriochlorophyll *e* and isorenieratene, a brown carotenoid.

UNIT 3

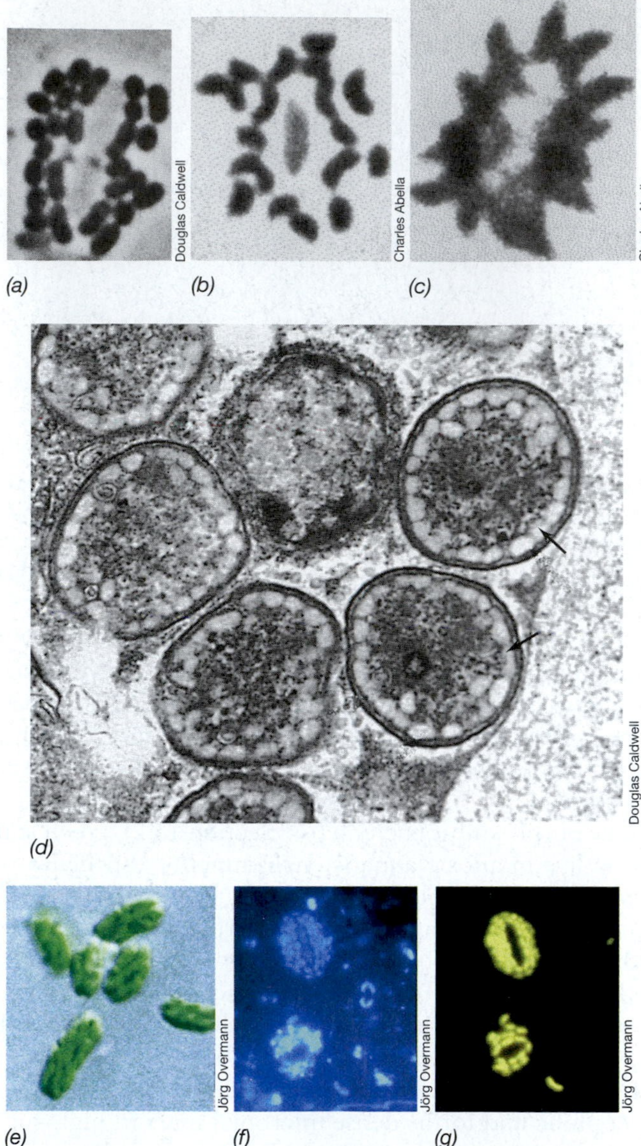

Figure 16.40 Green sulfur bacteria consortia. *(a–c)* Phase-contrast micrographs and *(d)* transmission electron micrograph. *(a–c)* The nonphototrophic central organism is much lighter in color than the pigmented phototrophic bacteria. *(d)* Note the chlorosomes (arrows). The entire consortium is about 3 μm in diameter. (e) Differential contrast micrograph of cells of "*Chlorochromatium aggregatum.*" *(f, g)* Fluorescence micrographs. *(f)* DAPI-stained cells (stains DNA). *(g)* Same preparation as in *(f)* stained with a phylogenetic probe for green sulfur bacteria (yellow).

of any anoxygenic phototroph. *C. tepidum* also grows rapidly and is amenable to genetic manipulation by both conjugation and transformation. It has become the model organism for studying the molecular biology of green sulfur bacteria.

Green Sulfur Bacteria Consortia

Certain green sulfur bacteria can form an intimate two-membered association with a chemoorganotrophic bacterium called a **consortium**. In the consortium, each organism benefits, and thus a variety of such consortia containing different phototrophic and chemotrophic components probably exist in nature. The phototrophic component, called the *epibiont*, is physically attached to the nonphototrophic cell (**Figure 16.40**), although the mechanism of attachment is not clear.

The name "*Chlorochromatium aggregatum*" has been used to describe a commonly observed consortium, although the name has no basis in formal taxonomy because it refers to two separate organisms rather than one. The "*C. aggregatum*" consortium is green because the epibionts, which surround a central nonphototrophic cell, are green sulfur bacteria that contain bacteriochlorophyll *c* or *d* and green-colored carotenoids (Figure 16.40*e*). A structurally similar organism called "*Pelochromatium roseum*" is brown. In other consortia, the morphology of the epibiont cells is vibroid (Figure 16.40*b,c*).

Some green sulfur bacterial consortia have been grown in laboratory culture. On average, the "*C. aggregatum*" consortium (Figure 16.40) contains about 12 epibionts per central cell, and the "*P. roseum*" consortium contains about 20. Evidence that the epibionts are indeed green sulfur bacteria comes not only from pigment analyses, but also from the fact that chlorosomes are visible in thin sections of the epibionts (Figure 16.40*d*) and also from molecular evidence. Treatment of the consortium with a fluorescent oligonucleotide probe specific for green sulfur bacterial 16S rRNA (FISH; ∞ Sections 14.9 and 22.4) causes the epibionts, but not the central cell, to fluoresce (Figure 16.40*f,g*). Studies of laboratory cultures of phototrophic consortia have also shown that the central cell and the epibiont divide in synchrony, indicating that the two components of the consortium must have some means of communication.

It is unclear why and how phototrophic consortia form. However, laboratory and field studies suggest that the epibionts are adapted to a very narrow regimen of light intensities and sulfide concentrations. The central cell may thus allow the otherwise nonmotile phototrophs to position themselves in a water column for optimal photosynthesis. The nonphototrophic central cell in the consortium likely benefits by the uptake of excreted organic nutrients from photosynthetic CO_2 fixation by the epibionts.

16.15 MiniReview

Green sulfur bacteria are obligately anaerobic anoxygenic phototrophs that produce chlorosomes. These organisms can grow at very low light intensities and oxidize H_2S to S^0 and SO_4^{2-}. Consortia containing phototrophic green bacteria and a nonphototrophic central cell are common in sulfidic aquatic environments.

- What pigments are found in the chlorosome?
- What is unique about autotrophy in green sulfur bacteria (∞ Section 20.7)?
- What evidence supports the idea that the epibionts of green bacterial consortia are truly green sulfur bacteria?

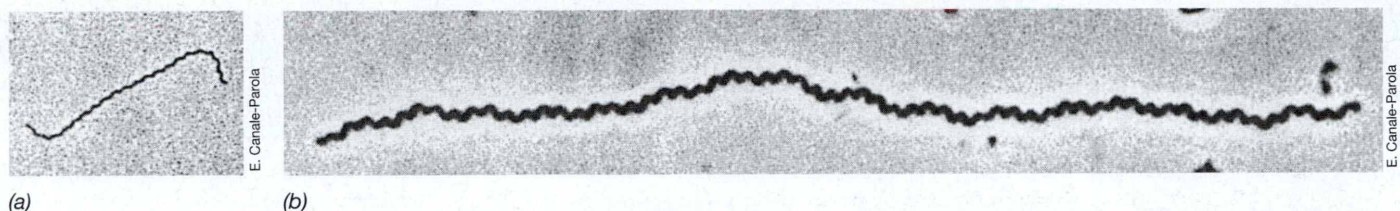

Figure 16.41 Morphology of spirochetes. Two spirochetes at the same magnification, showing the wide size range in the group. *(a) Spirochaeta stenostrepta*, by phase-contrast microscopy. A single cell is 0.25 μm in diameter. *(b) Spirochaeta plicatilis*. A single cell is 0.75 μm in diameter and can be up to 250 μm (0.25 mm) in length.

X THE SPIROCHETES

16.16 Spirochetes

Key Genera: *Spirochaeta, Treponema, Cristispira, Leptospira, Borrelia*

Spirochetes are gram-negative, motile, tightly coiled *Bacteria*, typically slender and flexuous in shape (**Figure 16.41**). These morphologically unique bacteria form a major phylogenetic lineage of *Bacteria* (∞ Figure 15.1). Spirochetes are widespread in aquatic environments and in animals. Some cause diseases, including syphilis, an important human sexually transmitted disease (∞ Section 34.13).

The spirochete cell is made up of a protoplasmic cylinder, consisting of the regions enclosed by the cell wall and cytoplasmic membrane. Motility is conferred by single to many flagella that emerge from each pole. However, unlike typical bacteria flagella (∞ Section 4.13), spirochete flagella fold back from each pole on to the protoplasmic cylinder and remain located in the periplasm of the cell; thus, they have also been called *endoflagella* (**Figure 16.42**). Both the endoflagella and the protoplasmic cylinder are surrounded by a multilayered but flexible membrane called the *outer sheath* (Figure 16.42*b*).

Motility of Spirochetes

Spirochetes have an unusual mode of motility. Each endoflagellum is anchored at one end and extends about two-thirds of the length of the cell. Endoflagella rotate, as do typical bacterial flagella (∞ Section 4.13). However, because the protoplasmic cylinder is rigid whereas the outer sheath is flexible, when both endoflagella rotate in the same direction, the protoplasmic cylinder rotates in the opposite direction, placing torsion on the cell as illustrated in Figure 16.42*b*. This causes

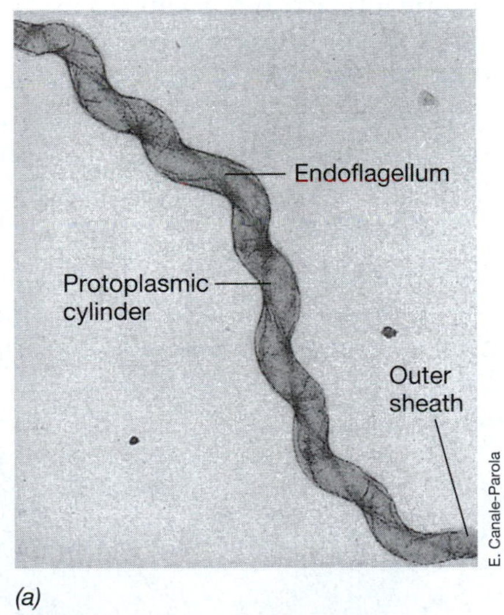

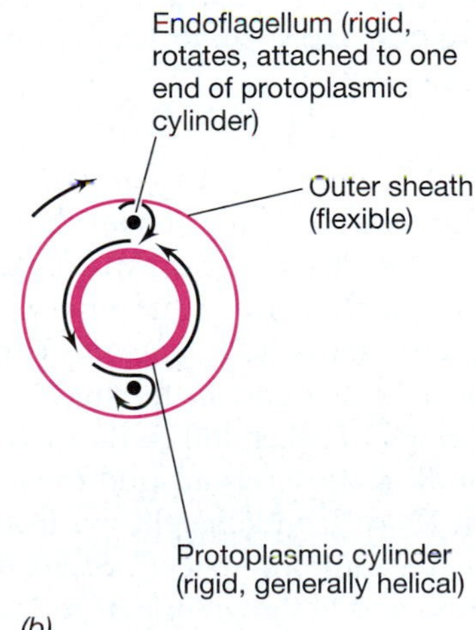

Figure 16.42 Motility in spirochetes. *(a)* Electron micrograph of a negatively stained preparation of *Spirochaeta zuelzerae*, showing the position of the endoflagellum. A single cell is about 0.3 μm in diameter. *(b)* Cross section of a spirochete cell, showing the arrangement of the protoplasmic cylinder, endoflagella, and external sheath. It also shows the manner in which the rotation of the rigid endoflagellum can generate rotation of the protoplasmic cylinder and (in the opposite direction) rotation of the external sheath. If the sheath is free, the cell will rotate about its longitudinal axis and move along it. If the sheath is in contact with a solid surface, the cell will creep forward.

Table 16.15 Genera of spirochetes and their characteristics

Genus	Dimensions (μm)	General characteristics	Number of endoflagella	Habitat	Diseases
Cristispira	30–150 × 0.5–3.0	3–10 complete coils; bundle of endoflagella visible by phase contrast microscopy	>100	Digestive tract of molluscs; has not been cultured	None known
Spirochaeta	5–250 × 0.2–0.75	Anaerobic or facultatively aerobic; tightly or loosely coiled	2–40	Aquatic, free-living, freshwater and marine	None known
Treponema	5–15 × 0.1–0.4	Microaerophilic or anaerobic; helical or flattened coil amplitude up to 0.5 μm	2–32	Commensal or parasitic in humans, other animals	Syphilis, yaws, swine dysentery, pinta
Borrelia	8–30 × 0.2–0.5	Microaerophilic; 5–7 coils of approx. 1 μm amplitude	7–20	Humans and other mammals, arthropods	Relapsing fever, Lyme disease, ovine and bovine borreliosis
Leptospira	6–20 × 0.1	Aerobic, tightly coiled, with bent or hooked ends; requires long-chain fatty acids	2	Free-living or parasitic in humans, other mammals	Leptospirosis
Leptonema	6–20 × 0.1	Aerobic; does not require long-chain fatty acids	2	Free-living	None known
Brachyspira	7–10 × 0.35–0.45	Anaerobe	8–28	Intestine of warm-blooded animals	Causes diarrhea in chickens and swine
Brevinema	4–5 × 0.2–0.3	Microaerophile, by 16S rRNA sequence analysis, forms deep branch in spirochete lineage (Figure 15.1)	2	Blood and tissue of mice and shrews	Infectious for laboratory mice

the spirochete cell to move by flexing or lashing motions due to torque exerted at the ends of the protoplasmic cylinder by the rotating endoflagella (Figure 16.42*b*). Thus, despite the fact that endoflagella do not extend away from the cell, their rotation provides motility, albeit of a more irregular and jerky form than motility provided by the flagella of other bacteria.

Classification

Spirochetes are classified into eight genera primarily on the basis of habitat, pathogenicity, phylogeny, and morphological and physiological characteristics. **Table 16.15** lists the major genera and their characteristics.

Spirochaeta and *Cristispira*

The genus *Spirochaeta* includes free-living, anaerobic, and facultatively aerobic spirochetes. These organisms are common in aquatic environments, such as the water and mud of rivers, ponds, lakes, and oceans. *Spirochaeta plicatilis* (Figure 16.41*b*) is a fairly large spirochete found in freshwater and marine H_2S-containing habitats. The endoflagella of *S. plicatilis* are arranged in a bundle that winds around the coiled protoplasmic cylinder. From 18 to 20 endoflagella are inserted at each pole of this spirochete. Another species, *Spirochaeta stenostrepta*, has been cultured and is shown in Figure 16.41*a*. It is an obligate anaerobe commonly found in H_2S-rich black muds. It ferments sugars via the glycolytic pathway to ethanol, acetate, lactate, CO_2, and H_2. The species *Spirochaeta aurantia* is an orange-pigmented facultative aerobe, fermenting sugars via the glycolytic pathway under anaerobic conditions and oxidizing sugars aerobically mainly to CO_2 and acetate.

The genus *Cristispira* (**Figure 16.43**) contains organisms with a unique distribution, being found in nature primarily in the crystalline style of certain molluscs, such as clams and oysters. The crystalline style is a flexible, semisolid rod seated in a sac and rotated against a hard surface of the digestive tract, thereby mixing with and grinding the small particles of food. Being large spirochetes, the cristispiras can easily be seen microscopically within the mollusc style as they rapidly rotate forward and backward in corkscrew fashion. *Cristispira* lives in both freshwater and marine molluscs, but not all species of molluscs possess them. Unfortunately, *Cristispira* has not been cultured, and so the physiological rationale for its restriction to this unique habitat is unknown.

Treponema

Anaerobic host-associated spirochetes that are commensals or parasites of humans and animals reside in the genus *Treponema*. *Treponema pallidum*, the causal agent of syphilis

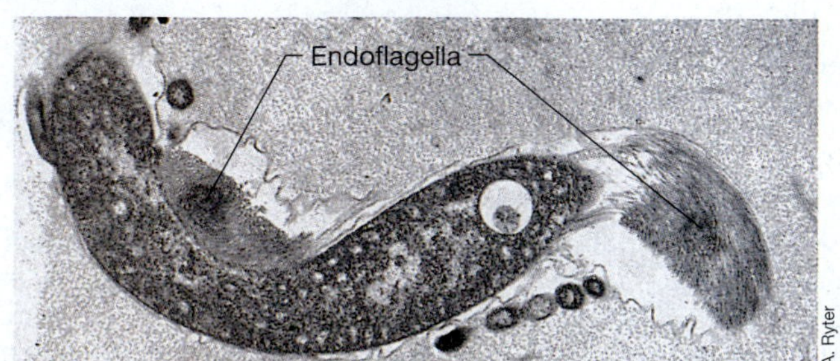

Figure 16.43 ***Cristispira.*** Electron micrograph of a thin section of a cell of *Cristispira*, a large spirochete. The cell measures about 2 μm in diameter. Notice the numerous endoflagella.

(∞ Section 34.13), is the best-known species of *Treponema*. It differs in morphology from other spirochetes; the cell is not helical but is flat and wavy. The *T. pallidum* cell is remarkably thin, measuring only 0.2 μm in diameter. Because of this, dark-field microscopy has long been used to examine exudates from suspected syphilitic lesions (∞ Figure 34.30).

In nature, *T. pallidum* appears restricted to humans, although artificial infections have been established in rabbits and monkeys. Although never grown in laboratory culture, it has been established from animal studies that virulent *T. pallidum* cells (purified from infected rabbits) contain a cytochrome system and are in fact microaerophiles. Such cells have also yielded sufficient DNA for the genome of *T. pallidum* (1.14 Mbp) to be sequenced.

Other species of the genus *Treponema* are commensal organisms in the oral cavity of humans and can typically be seen in material scraped from between the teeth and from the narrow space between the gums and the teeth. *Treponema denticola* is a major oral treponeme and ferments amino acids such as cysteine and serine, forming acetate as the major fermentation acid, as well as CO_2, NH_3, and H_2S.

Metabolically unusual species of the genus *Treponema* have been isolated from the hindgut of the termite. This environment is highly cellulolytic, as termites live mostly on wood and wood products. H_2 and CO_2 are produced from the fermentation of glucose released from the cellulose. *Treponema primitia* converts this H_2 plus CO_2 to acetate; that is, it is an acetogen (acetogenesis is discussed in Section 21.9). This is the first instance of this form of energy metabolism in a phylogenetic group of *Bacteria* outside of the clostridia and their relatives. The hindgut spirochete *Treponema azotonutricium* fixes molecular nitrogen (N_2 fixation; ∞ Section 20.14), a property also not previously found in spirochetes.

Spirochetes are also found in the rumen, the digestive organ of ruminant animals (∞ Section 24.10). *Treponema saccharophilum* (**Figure 16.44**) is a large, pectinolytic spirochete found in the bovine rumen. *T. saccharophilum* is an obligate anaerobe that ferments pectin, starch, inulin, and other plant polysaccharides. This and other spirochetes may play an important role in the conversion of plant polysaccharides to volatile fatty acids usable as energy sources by the ruminant (∞ Section 24.10). Although the genus *Treponema* is phylogenetically a unit, differences in GC base ratios between *T. pallidum* and other *Treponema* species indicate substantial genetic heterogeneity among these organisms.

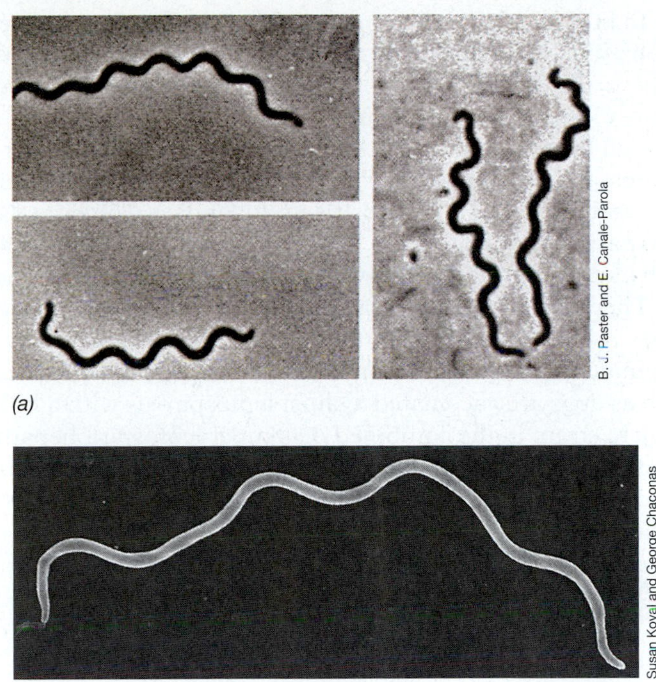

Figure 16.44 ***Treponema* and *Borrelia*.** *(a)* Phase-contrast photomicrographs of *Treponema saccharophilum*, a large pectinolytic spirochete from the bovine rumen. A cell measures about 0.4 μm in diameter. Left, regularly coiled cells; right, irregularly coiled cells. *(b)* A scanning electron micrograph of *Borrelia burgdorferii*, the causative agent of Lyme disease.

Borrelia

The majority of species in the genus *Borrelia* are animal or human pathogens. *Borrelia recurrentis* is the causative agent of relapsing fever in humans and is transmitted via an insect vector, usually by the human body louse. Relapsing fever is characterized by a high fever and generalized muscular pain that lasts for 3–7 days, followed by a recovery period of 7–9 days. Left untreated, the fever returns in two to three more cycles (hence the name, relapsing fever) and causes death from hemorrhaging and organ failure in up to 40% of those infected. Fortunately, the organism is quite sensitive to tetracycline, and if the disease is correctly diagnosed, treatment is straightforward. Other species of *Borrelia* are of veterinary importance, causing diseases in cattle, sheep, horses, and birds. In most of these diseases the bacterium is transmitted by ticks.

Borrelia burgdorferi (Figure 16.44*b*) is the causative agent of the tickborne disease called *Lyme disease*, which infects humans and animals. Lyme disease is discussed in Section 35.4. *B. burgdorferi* is also of interest because it is as yet one of the few known bacteria that has a linear (as opposed to a circular) chromosome. The rather small genome of *B. burgdorferi* (1.44 Mbp) has been completely sequenced.

Leptospira and *Leptonema*

The genera *Leptospira* and *Leptonema* contain strictly aerobic spirochetes that use long-chain fatty acids (for example, oleic acid) as electron donor and carbon sources. With few exceptions, these are the only substrates utilized by leptospiras for growth. The leptospira cell is thin, finely coiled, and usually bent at each end into a semicircular hook. At present, several species are recognized in this group, some free-living and many parasitic. Two major species are *Leptospira interrogans* (parasitic) and *Leptospira biflexa* (free-living). Strains of *L. interrogans* are parasitic for humans and animals. Rodents are the natural hosts of most leptospiras, although dogs and pigs are also important carriers of certain strains.

In humans the most common leptospiral syndrome is leptospirosis, a disorder in which the organism localizes in the kidneys and can cause renal failure and death. Leptospiras ordinarily enter the body through the mucous membranes or through breaks in the skin. After a transient multiplication in various parts of the body, the organism localizes in the kidneys and liver, causing nephritis and jaundice. The organism then passes out of the body in the urine; infection of another individual is most commonly by contact with infected urine.

Therapy with penicillin, streptomycin, or the tetracyclines is possible, but it may take extended courses to eliminate the organism from the kidney. Domestic animals such as dogs are vaccinated against leptospirosis with a killed virulent strain in the combined distemper-leptospira-hepatitis vaccine. In humans, prevention is effected primarily by elimination of the disease from animals, thereby eliminating the disease reservoir.

16.16 MiniReview

Spirochetes are tightly coiled, motile, helical bacteria that include both free-living and pathogenic species.

- How do the endoflagella of spirochetes compare with the flagella of *Escherichia coli* in structure and function?
- Name two diseases of humans caused by spirochetes.
- What is the habitat of the spirochete *Cristispira*?

XI DEINOCOCCI

16.17 *Deinococcus* and *Thermus*

Key Genera: *Deinococcus, Thermus*

The deinococci group contains only a few genera, the best studied being *Deinococcus* and *Thermus*. The latter contains thermophilic chemoorganotrophic bacteria including *Thermus aquaticus*, the organism from which *Taq* DNA polymerase is obtained. Because it is so heat-stable, this enzyme is the major one used in the polymerase chain reaction (PCR) technique for amplifying DNA, as was described in Section 12.8.

Deinococci stain gram-positively and contain an unusual form of peptidoglycan in which ornithine is present in place of diaminopimelic acid in the muramic acid cross-bridges (Section 4.6). A number of species of *Thermus* and *Deinococcus* have been described, and all grow aerobically by catabolism of sugars, amino and organic acids, or various complex mixtures. We focus the rest of the discussion here on *Deinococcus*.

The genus *Deinococcus* contains four species of gram-positive cocci; *Deinococcus radiodurans* is the best-studied species. The *D. radiodurans* cell wall is structurally complex and consists of several layers, including an outer membrane (**Figure 16.45**), normally present only in gram-negative bacteria (Section 4.7). However, unlike the outer membrane of

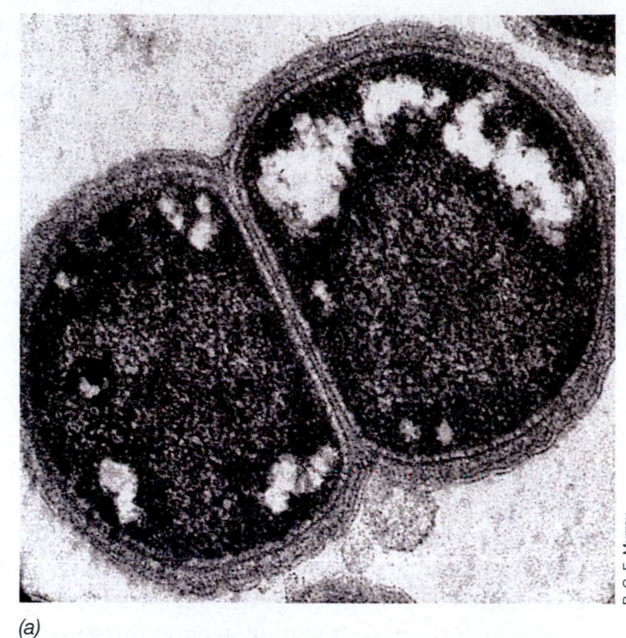

(a)

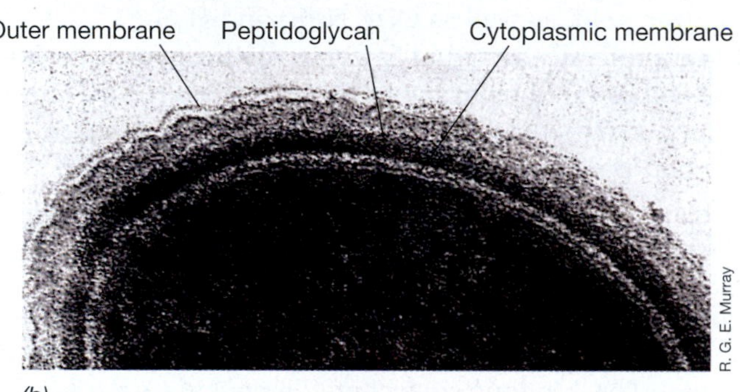

(b)

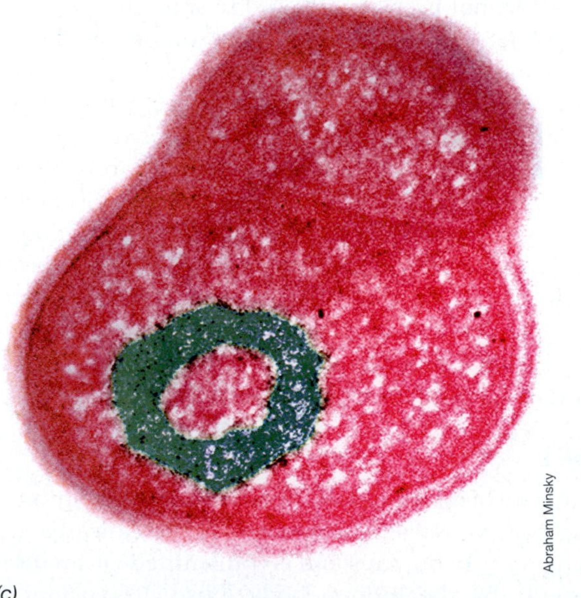

(c)

Figure 16.45 The radiation-resistant coccus *Deinococcus radiodurans*. An individual cell is about 2.5 μm in diameter. *(a)* Transmission electron micrograph of *D. radiodurans*. Note the outer membrane layer. *(b)* High-magnification micrograph of wall layer. *(c)* Transmission electron micrograph of cells of *D. radiodurans* colored to show the toroidal morphology of the nucleoid (green).

gram-negative bacteria such as *Escherichia coli*, the outer membrane of *D. radiodurans* lacks lipid A. The organism primarily inhabits soil and can also be isolated from dust particles.

Radiation Resistance of *Deinococcus radiodurans*

Most deinococci are red or pink due to carotenoids, and many strains are highly resistant to UV radiation and to desiccation. Resistance to radiation can be used to advantage in isolating deinococci. These remarkable organisms can be selectively isolated from soil, ground meat, dust, and filtered air following exposure of the sample to intense UV (or even gamma) radiation and plating on a rich medium containing tryptone and yeast extract. Because many strains of *Deinococcus radiodurans* are even more resistant to radiation than are bacterial endospores, treatment of a sample with strong doses of radiation effectively sterilizes the sample except for cells of *D. radiodurans*, making isolation of deinococci relatively straightforward. For example, *D. radiodurans* cells can survive exposure to up to 15,000 grays (Gy) of ionizing radiation (1 Gy = 100 rad). This is sufficient to shatter the organism's chromosome into hundreds of fragments (by contrast, a human can be killed by exposure to less than 10 Gy).

In addition to impressive radiation resistance, *D. radiodurans* is resistant to the mutagenic effects of many other mutagenic agents. The only chemical mutagens that seem to work on *D. radiodurans* are agents such as nitrosoguanidine, which can induce deletions in DNA. Deletions are apparently not repaired as efficiently as point mutations in this organism, and mutants of *D. radiodurans* can be isolated in this way.

DNA Repair in *Deinococcus radiodurans*

Studies of *D. radiodurans* have shown that it is highly efficient in repairing damaged DNA. Several different DNA repair enzymes exist in *D. radiodurans*. In addition to the DNA repair enzyme RecA (∞ Section 11.9), several RecA-independent DNA systems exist in *D. radiodurans* that can repair breaks in single- or double-stranded DNA and excise and repair misincorporated bases. In fact, repair processes are so effective, that the chromosome can even be reassembled from a fragmented state.

It is also thought that the unique arrangement of DNA in *D. radiodurans* cells plays a role in radiation resistance. Cells of *D. radiodurans* always exist as pairs or tetrads (Figure 16.45*a*). Instead of scattering DNA within the cell as in a typical nucleoid, DNA in *D. radiodurans* is ordered into a toroidal ringlike structure (Figure 16.45*c*). Repair is then facilitated by the fusion of nucleoids from adjacent compartments, because this provides a platform for homologous recombination. From this extensive recombination, a single repaired chromosome emerges, and the cell containing this chromosome can then grow and divide.

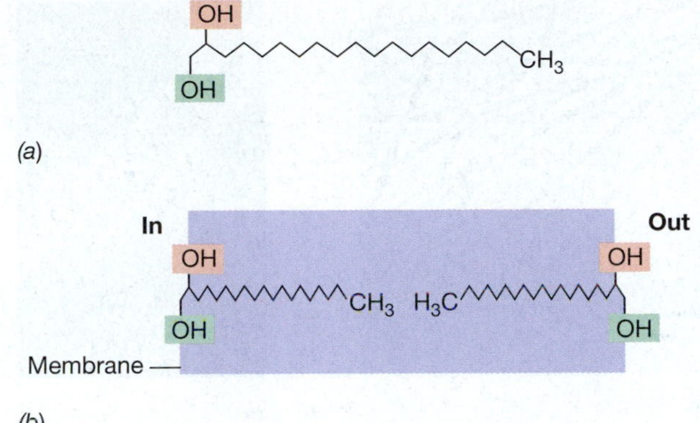

Figure 16.46 The unusual lipids of *Thermomicrobium*. *(a)* Membrane lipids from *Thermomicrobium roseum* contain long-chain diols like the one shown here (13-methyl-1,2 nonadecanediol). Note that unlike the lipids of other *Bacteria* or of *Archaea*, neither ester- nor ether-linked side chains are present. *(b)* To form a bilayer membrane, dialcohol molecules presumably oppose each other at the methyl groups, with the OH groups being the inner and outer hydrophilic surfaces. Small amounts of the diols have fatty acids esterified to the secondary —OH group (shown in red), whereas the primary —OH group (shown in green) can bond a hydrophilic molecule like phosphate.

XII THE GREEN NONSULFUR BACTERIA

The green nonsulfur bacteria are phylogenetically distinct from other *Bacteria* and contain just a few genera, the best known being the anoxygenic phototroph *Chloroflexus*; all cultured representatives are thermophilic. *Thermomicrobium* is a chemotrophic member of this group and is a strictly aerobic, gram-negative rod, growing optimally in complex media at 75°C. Besides its phylogenetic novelty, *Thermomicrobium* is also of interest because of its membrane lipids. Recall that the lipids of *Bacteria* and *Eukarya* contain fatty acids esterified to glycerol (∞ Section 3.4 and 4.3). By contrast, the lipids of *Thermomicrobium* are formed on 1,2-dialcohols instead of glycerol, and have neither ester *nor* ether linkages (**Figure 16.46**). In addition, cells of *Thermomicrobium* are unusual for species of *Bacteria* in that they lack peptidoglycan, the signature cell wall polymer of this domain (∞ Sections 4.6 and 14.8).

16.18 *Chloroflexus* and Relatives

Key Genera: *Chloroflexus, Heliothrix, Roseiflexus*

Chloroflexus and most other green nonsulfur bacteria (so named because of the physiology of *Chloroflexus*, as discussed later) are thermophilic filamentous bacteria that, along with cyanobacteria, form thick microbial mats in neutral to alkaline hot springs (**Figure 16.47**; ∞ Figure 22.18). Nonthermophilic *Chloroflexus*-like organisms have also been found in marine microbial mats.

Although an anoxygenic phototroph, *Chloroflexus* is a "hybrid" phototroph in the sense that its photosynthetic

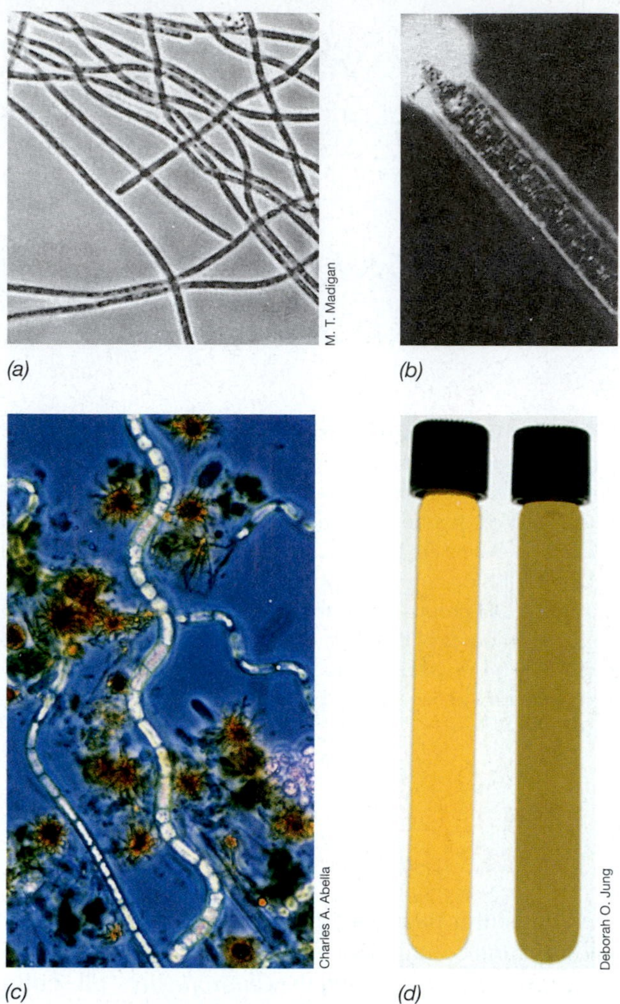

Figure 16.47 Green nonsulfur bacteria. *(a)* Phase photomicrograph of the thermophilic phototroph, *Chloroflexus aurantiacus.* Cells are about 1 μm in diameter. *(b)* Phase-contrast micrograph of the large phototroph *Oscillochloris.* Cells are about 5 μm wide. The brightly contrasting material is a holdfast, used for attachment. *(c)* Color photomicrograph of filaments of a *Chloronema* species growing in a stratified Michigan lake. These cells of *Chloronema* are wavy filaments and about 2.5 μm in diameter. *(d)* Tube cultures of *C. aurantiacus* (right) and *Roseiflexus* (left). *Roseiflexus* is yellow because it lacks bacteriochlorophyll *c* and chlorosomes.

mechanism shows features of both green sulfur bacteria (Section 16.15) and phototrophic purple bacteria. Like green sulfur bacteria, *Chloroflexus* contains bacteriochlorophyll *c* and chlorosomes (Figure 16.38). However, the bacteriochlorophyll *a* located in the cytoplasmic membrane of cells of *Chloroflexus* is arranged to form a photosynthetic reaction center structurally similar to those of purple bacteria (by contrast, the reaction center of green sulfur bacteria is structurally quite different; ∞ Figures 20.7 and 20.18).

Because of its phylogeny and unique mechanism of autotrophy (discussed below), it is possible that *Chloroflexus* may be a modern relative of early phototrophs that perhaps first evolved a photosynthetic reaction center and then received chlorosome-specific genes later by horizontal transfer. From a phylogenetic standpoint, *Chloroflexus* is clearly the earliest known phototrophic bacterium.

Physiology of *Chloroflexus*

Physiologically, *Chloroflexus*, like purple nonsulfur bacteria, can grow by photoautotrophy; however, they grow best phototrophically when organic compounds are available as carbon sources (photoheterotrophy). *Chloroflexus* also grows well in the dark as a chemoorganotroph by aerobic respiration. Interestingly, and this should be considered in light of the evolutionary position of *Chloroflexus* as the most ancient of anoxygenic phototrophs, autotrophy in *Chloroflexus* is based on the hydroxypropionate pathway of CO_2 incorporation, which is unique to this and a few other phylogenetically ancient organisms. We consider the biochemistry of this novel autotrophic pathway in Section 20.7.

Other Green Nonsulfur Bacteria

In addition to *Chloroflexus*, other phototrophic green nonsulfur bacteria include the thermophile *Heliothrix* and the large-celled mesophiles *Oscillochloris* (Figure 16.47*b*) and *Chloronema* (Figure 16.47*c*). *Oscillochloris* and *Chloronema* are unusual because they are rather large cells, 2–5 μm wide and up to several hundred micrometers long (Figure 16.47*c*). Members of both genera inhabit freshwater lakes containing low levels of sulfide, together with other species of green sulfur and purple sulfur bacteria.

Two genera in the green nonsulfur phylum, *Roseiflexus* and *Heliothrix*, differ quite dramatically from *Chloroflexus*. These phototrophs lack bacteriochlorophyll *c* and chlorosomes and thus more closely resemble phototrophic purple bacteria (∞ Section 15.2) than *Chloroflexus*. Despite this clear difference in pigment complement, *Chloroflexus, Roseiflexus*, and *Heliothrix* have many other traits in common. These include a filamentous morphology and thermophilic lifestyle. And if *Chloroflexus* obtained chlorosomes and bacteriochlorophyll *c* biosynthesis genes by horizontal transfer from green sulfur bacteria, as is suspected, it is possible that *Chloroflexus* and *Roseiflexus* were at one point phenotypically indistinguishable. *Heliothrix* is not available in pure culture, but *Roseiflexus* is. Cells of *Roseiflexus* are filamentous, and cultures are yellow-orange from their extensive carotenoid pigments and lack of bacteriochlorophyll *c* (Figure 16.47*d*).

16.17–16.18 MiniReview

Deinococcus and *Chloroflexus* are each key genera in separate major lineages of *Bacteria. Deinococcus radiodurans* is the most radiation resistant of all known organisms, and *Chloroflexus* is an anoxygenic phototroph that shows photosynthetic properties characteristic of both purple bacteria and green sulfur bacteria.

- How does *Deinococcus radiodurans* avoid being killed by high levels of radiation?
- In what ways do *Chloroflexus* and *Roseiflexus* resemble *Chlorobium*? *Rhodobacter*?
- What is unique about *Thermomicrobium*?

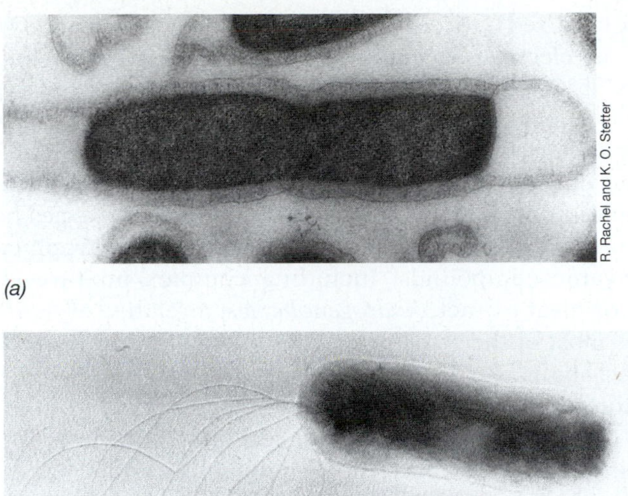

Figure 16.48 **Hyperthermophilic *Bacteria.*** *(a) Thermotoga maritima*—temperature optimum, 80°C. Note the outer covering, the toga. *(b) Aquifex pyrophilus*—temperature optimum, 85°C. Cells of *Thermotoga* (thin section) measure 0.6 × 3.5 μm; cells of *Aquifex* (freeze-fracture micrograph) measure 0.5 × 2.5 μm.

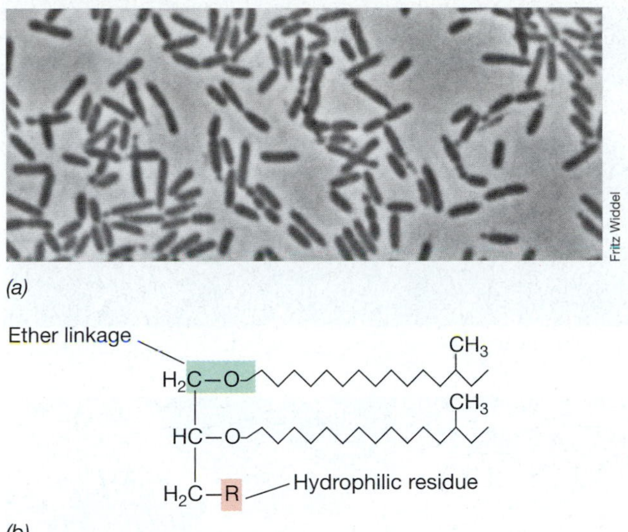

Figure 16.49 ***Thermodesulfobacterium.*** *(a)* Phase photomicrograph of cells of *Thermodesulfobacterium mobile.* *(b)* Structure of one of the lipids of *T. mobile.* Note that although the two hydrophobic side chains are ether-linked, they are not phytanyl units as in *Archaea.* The designation "R" is for a hydrophilic residue, such as a phosphate group.

XIII HYPERTHERMOPHILIC BACTERIA

Three groups of hyperthermophilic bacteria cluster deep in the phylogenetic tree of *Bacteria*, near the hypothetical root (Figures 14.16 and 15.1). Each group consists of one or two major genera, and a key physiological feature of most of the members of these genera is hyperthermophily—optimal growth at temperatures above 80°C (Sections 6.12 and 6.14).

16.19 *Thermotoga* and *Thermodesulfobacterium*

Key Genera: *Thermotoga, Thermodesulfobacterium*

Thermotoga is a rod-shaped hyperthermophile capable of growth at temperatures as high as 90°C (optimum, 80°C). Cells of *Thermotoga* contain a sheathlike envelope called a *toga* (**Figure 16.48a**), stain gram-negatively, and are nonsporulating. *Thermotoga* is an anaerobic, fermentative chemoorganotroph, catabolizing sugars and polymers such as starch, and producing lactate, acetate, CO_2, and H_2 as fermentation products. The organism can also grow by anaerobic respiration using H_2 as electron donor and ferric iron (Fe^{3+}) as electron acceptor. Species of *Thermotoga* have been isolated from terrestrial hot springs as well as marine hydrothermal vents.

The genome of *Thermotoga* has been completely sequenced, and interestingly, this organism contains many genes that show strong homology to genes from hyperthermophilic *Archaea*. In fact, over 20% of the genes of *Thermotoga* probably originated from hyperthermophilic species of *Archaea* by horizontal transfers (Section 13.11). It is hypothesized that in very hot habitats where the two coexisted, *Thermotoga* underwent extensive horizontal gene exchange with species of *Archaea*, perhaps picking up several genes useful for survival under such extreme conditions. Although a few archaeal-like genes have been identified in the genomes of other *Bacteria* and vice versa, thus far, only in *Thermotoga* has such large-scale horizontal transfer of genes between domains been detected.

Thermodesulfobacterium

Thermodesulfobacterium (**Figure 16.49a**) is a thermophilic, sulfate-reducing bacterium, positioned on the phylogenetic tree as a separate phylum between *Thermotoga* and *Aquifex* (Figure 15.1). Although not a true hyperthermophile because its growth temperature optimum is only 70°C, *Thermodesulfobacterium* is the most thermophilic of all known sulfate-reducing bacteria (the archaeal sulfate reducer *Archaeoglobus* is a true hyperthermophile, Section 17.7). Like the mesophilic sulfate reducer *Desulfovibrio* (Section 15.18), *Thermodesulfobacterium* is a strict anaerobe and cannot utilize acetate as an electron donor in its energy metabolism. Instead, *Thermodesulfobacterium* uses compounds such as lactate, pyruvate, and ethanol as electron donors, reducing SO_4^{2-} to H_2S.

An unusual biochemical feature of species of *Thermodesulfobacterium* is the presence of ether-linked lipids. Recall that such lipids are a hallmark of *Archaea* and that a polyisoprenoid C_{20} hydrocarbon (phytanyl) replaces fatty acid as the

UNIT 3

(a)

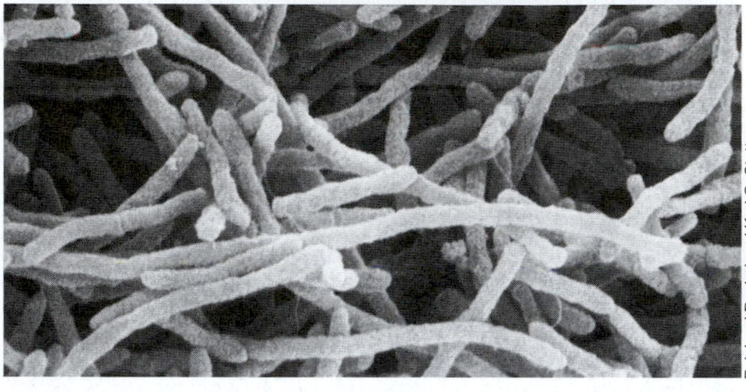

(b)

Figure 16.50 ***Thermocrinis.*** *(a)* Cells of *Thermocrinis ruber* growing as filamentous streamers (arrow) attached to siliceous sinter in the outflow (85°C) of Octopus Spring, Lower Geyser Basin, Yellowstone National Park. The pink color is due to a carotenoid pigment. *(b)* Scanning electron micrograph of rod-shaped cells of *T. ruber* grown on a silicon-coated cover glass. The hairlike structures are silicon. A single cell of *T. ruber* is about 0.4 μm in diameter and from 1 to 3 μm long.

side chain in archaeal lipids (Sections 4.3 and 14.8). The ether-linked lipids in *Thermodesulfobacterium* are unusual because the glycerol side chains are not phytanyl groups, as they are in *Archaea*, but instead are composed of a unique C_{17} hydrocarbon along with some fatty acids (Figure 16.49*b*).

Thus we see in *Thermodesulfobacterium* both a deep phylogenetic lineage (Figure 15.1) and a lipid profile that combines features of both the *Archaea* and the *Bacteria* (Figure 16.49*b*). Perhaps in *Thermotoga*, this is a case of lateral gene flow from *Archaea* to *Bacteria*. However, the bacterium *Ammonifex*, a gram-positive thermophilic member of *Bacteria* that grows anaerobically by H_2 oxidation coupled to the reduction of NO_3^- to NH_3, also contains lipids like those of *Thermodesulfobacterium*. Thus, ether-linked lipids may be more common among *Bacteria* than previously thought.

16.20 *Aquifex, Thermocrinis,* and Relatives

Key Genera: *Aquifex, Thermocrinis*

The genus *Aquifex* (Figure 16.48b) is an obligately chemolithotrophic and autotrophic hyperthermophile and is the most thermophilic of all known *Bacteria*. Various *Aquifex* species utilize H_2, S^0, or $S_2O_3^{2-}$ as electron donors and O_2 or NO_3^- as electron acceptors; cells inhabit environments as hot as 95°C (optimum, 85°C). *Aquifex* can tolerate only very low O_2 concentrations, but it remains (along with a few species of *Archaea*, Sections 17.5 and 17.9) one of the few aerobic (or more exactly, microaerophilic) hyperthermophiles known. Nutritional studies of *Aquifex* species have shown them totally unable to grow chemoorganotrophically on organic compounds, including complex mixtures like yeast or meat extract. *Hydrogenobacter*, a relative of *Aquifex*, shows most of the same properties as *Aquifex* but is an obligate aerobe.

Autotrophy in *Aquifex* is supported by enzymes of the reverse citric acid cycle, a series of reactions previously found only in green sulfur bacteria (Section 16.15 and Section 20.7) within the domain *Bacteria*. The complete genome sequence of *Aquifex aeolicus* has been determined, and its entirely chemolithotrophic and autotrophic lifestyle is supported by a very small genome of only 1.55 Mbp (one-third the size of the *Escherichia coli* genome).

The discovery that so many hyperthermophilic species of *Archaea* and *Bacteria*, like *Aquifex*, are H_2 chemolithotrophs, coupled with the finding that they branch as very early lineages on their respective phylogenetic trees (Figures 14.16, 15.1, and 17.1), suggests that H_2 was a key electron donor for primitive organisms under early Earth conditions (Sections 14.23 and 17.15). A simple model for the utilization of H_2 to generate a proton motive force and hence, usable energy as ATP, was shown in Figure 14.8.

Thermocrinis

Thermocrinis (**Figure 16.50**) is an interesting relative of *Aquifex* and *Hydrogenobacter*. This organism is a hyperthermophilic (temperature optimum, 80°C) chemolithotroph capable of oxidizing H_2, thiosulfate ($S_2O_3^{2-}$), or sulfur (S^0) as electron donors, with O_2 as electron acceptor.

Thermocrinis ruber, the only known species, grows in the outflow of certain hot springs in Yellowstone National Park, where it forms pink "streamers" consisting of a filamentous form of the cells attached to siliceous sinter (Figure 16.50*a*). In static culture, cells of *T. ruber* grow as individual rod-shaped cells (Figure 16.50*b*). However, when cultured in a flowing system in which growth medium is trickled over a solid glass surface to which cells can attach, *Thermocrinis* assumes the streamer morphology it forms in its constantly flowing habitat in nature (Figure 16.50*a*).

T. ruber is of historical significance in microbiology because it was one of the organisms studied in the 1960s by Thomas Brock, a pioneer in the field of thermal microbiology. The discovery by Brock that the pink streamers (Figure 16.50*a*) contained protein and nucleic acids clearly indicated that they were living organisms and not just mineral debris. Moreover, the presence of streamers in the outflow of hot springs at 80–90°C but not at lower temperatures supported Brock's hypothesis that hot spring microorganisms actually required high temperatures for growth and were likely to be present in even boiling water. Both of these conclusions were subsequently supported by the discovery by Brock and others of literally dozens of genera of hyperthermophilic bacteria inhabiting hot springs, hydrothermal vents, and other thermal environments. There is more coverage of hyperthermophiles in Sections 6.12, 6.14, 17.5–17.15, and 24.11.

XIV NITROSPIRA AND DEFERRIBACTER

16.21 *Nitrospira, Deferribacter,* and Relatives

Key Genera: *Nitrospira, Deferribacter*

Several other phyla of *Bacteria* have been identified by 16S rRNA gene sequence analysis of cultures, although relatively little is known about them. Two such phyla contain the organisms *Nitrospira* and *Deferribacter* (Figure 15.1). Physiologically, these organisms are either chemolithotrophs or chemoorganotrophs and are mesophiles to thermophiles.

Like nitrite-oxidizing *Proteobacteria* (Sections 15.3 and 20.12), *Nitrospira* oxidizes NO_2^- to NO_3^- and grows autotrophically. Despite this close physiological relationship to the classical nitrifying bacteria, *Nitrospira* is phylogenetically quite distinct from them. Also, *Nitrospira* lacks the extensive internal membranes found in species of nitrifying *Proteobacteria* (Figures 15.7 and 15.8). Nevertheless, *Nitrospira* inhabits many of the same environments as nitrite-oxidizing *Proteobacteria* such as *Nitrobacter*, so it has been suggested that its physiological capacities may have been transferred to it by horizontal gene flow from nitrifying *Proteobacteria* (or vice versa). As we know, this mechanism for acquiring physiological traits has been widely exploited in the bacterial world (Section 13.11). However, environmental surveys for the presence of nitrifying bacteria in nature have shown *Nitrospira* to be much more common than *Nitrobacter*; thus most of the NO_2^- oxidized in natural environments where nitrification is a significant process, such as in wastewater treatment plants, is probably due to the activities of *Nitrospira*.

Other genera in the *Nitrospira* group include *Leptospirillum*, an iron-oxidizing chemolithotroph responsible for much of the acid mine drainage associated with the mining of coal and iron (Section 24.6), and *Thermodesulfovibrio*, a thermophilic sulfate-reducing bacterium that inhabits hot spring microbial mats. Note that although similar in name and physiology, *Thermodesulfovibrio* and *Thermodesulfobacterium* (Section 16.19) are phylogenetically distinct organisms.

Deferribacter

The genus *Deferribacter* forms its own distinct lineage (Figure 15.1) and is composed of species that specialize in anaerobic energy metabolism. Other genera in this group include *Geovibrio* and *Flexistipes*; the latter genus is an obligately anaerobic and fermentative bacterium.

Deferribacter and *Geovibrio* are extremely versatile at anaerobic respiration using a variety of electron acceptors, including the metals Fe^{3+} and Mn^{4+}. We discuss anaerobic respiration in Sections 21.6–21.12 and show that the process can be linked to many different terminal electron acceptors. Members of the *Deferribacter* group appear to be unusual in the vast number of alternative electron acceptors they can use and in the fact that they are obligate anaerobes. By contrast, most organisms capable of growing by anaerobic respiration with nitrate or metals as electron acceptors are facultative aerobes, which are able to grow fully aerobically as well as by anaerobic respiration (Section 21.6).

UNIT 3

16.19–16.21 MiniReview

Thermotoga, Thermodesulfobacterium, and *Aquifex* grow at high temperature, and each represents a major lineage of *Bacteria. Aquifex* and *Thermocrinis* are H_2-oxidizing chemolithotrophs, and *Thermotoga* and *Thermodesulfobacterium* are both anaerobic chemoorganotrophs. *Nitrospira* and *Deferribacter* each form their own phylum of *Bacteria.*

- Compare the catabolic metabolism of *Thermotoga* and *Thermodesulfobacterium.*
- What is unusual about the lipids of *Thermodesulfobacterium*?
- From a genomic perspective, why does the fact that *Aquifex* is able to grow on the gases $H_2 + CO_2 + O_2$ seem surprising?
- Contrast the metabolic features of *Nitrospira* and *Deferribacter.*

Review of Key Terms

Acid-fastness a property of *Mycobacterium* species in which cells stained with the dye basic fuchsin resist decolorization with acidic alcohol

Chlorosome a cigar-shaped structure bounded by a nonunit membrane and containing the light-harvesting bacteriochlorophyll (*c, d,* or *e*) in green sulfur bacteria and *Chloroflexus*

Consortium two- or more-membered association of bacteria, usually living in an intimate symbiotic fashion

Cyanobacteria bacterial oxygenic phototrophs that contain chlorophyll *a* and phycobilins but not chlorophyll *b*

Green sulfur bacteria anoxygenic phototrophs containing chlorosomes and bacteriochlorophyll *c*, *d*, or *e* as light-harvesting chlorophyll

Heliobacteria anoxygenic phototrophs containing bacteriochlorophyll *g*

Heterofermentative in reference to lactic acid bacteria, capable of making more than one fermentation product

Homofermentative in reference to lactic acid bacteria, producing only lactic acid as a fermentation product

Phycobilin the light-capturing open chain tetrapyrole component of phycobiliproteins

Prochlorophyte a bacterial oxygenic phototroph that contains chlorophylls *a* and *b* but lacks phycobilins

Spirochete a slender, tightly coiled gram-negative bacterium characterized by possession of endoflagella used for motility

Stickland reaction fermentation of an amino acid pair in which one amino acid serves as an electron donor and a second serves as an electron acceptor

Review Questions

1. Describe key physiological and morphological features that would differentiate *Staphylococcus* from *Micrococcus*.
2. What morphological and physiological features distinguish *Sporosarcina* from *Clostridium*?
3. What key features could be used to differentiate the following genera of gram-positive bacteria: *Bacillus, Mycoplasma*, and *Mycobacterium*?
4. Compare and contrast the morphology of *Nocardia* with that of *Anabaena*.
5. Describe the kinds of oxygenic phototrophic bacteria that occur in the open ocean. Are their numbers low or high, and how significant is their contribution to the primary productivity (i.e., fixation of carbon dioxide) of the ocean?
6. Describe a key morphological feature that would differentiate each of the following *Bacteria: Streptococcus, Streptomyces, Verrucomicrobium, Gemmata*, and *Spirochaeta*.
7. Describe a key physiological feature of the following *Bacteria* that would differentiate each from the others: *Lactobacillus, Nitrobacter*, and *Oscillatoria*.
8. What do species in the *Planctomyces* phylum have in common with *Archaea*? With Eukarya?
9. Compare and contrast photosynthesis in cyanobacteria and green sulfur bacteria.
10. What traits do the chlamydia and the rickettsias (*Alphaproteobacteria*) have in common? In what ways do they differ? What is the function of each of the two types of cell formed by *Chlamydia*?
11. Describe the possible benefits to each member of a consortium composed of a green sulfur bacterium and a chemoorganotroph.
12. What major physiological property unites species of *Thermotoga, Aquifex*, and *Thermocrinis*?

Application Questions

1. Defend or refute the claim that convergent evolution accounts for the presence of phototrophic representatives in several different phyla of *Bacteria*.
2. Nitrogen fixation is found in representatives of various phyla of *Bacteria*. Discuss the conditions that would select for organisms with this capability and the reasons that might explain why nitrogen fixation is a strictly bacterial activity.

17

Archaea

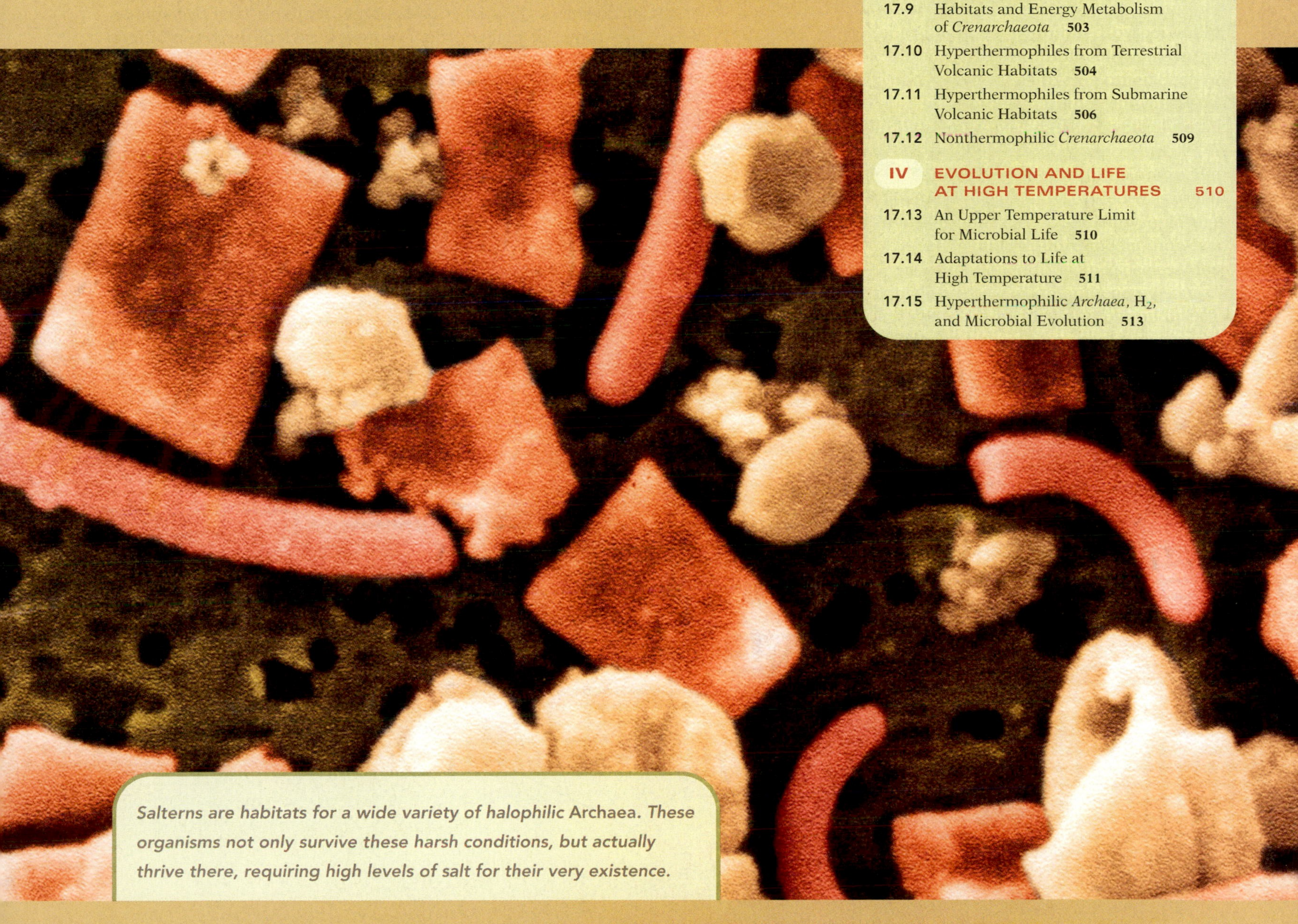

Salterns are habitats for a wide variety of halophilic Archaea. These organisms not only survive these harsh conditions, but actually thrive there, requiring high levels of salt for their very existence.

We now consider the domain *Archaea*. In Chapter 8 we discussed the major phenotypic similarities and differences between *Archaea* and *Bacteria*—the two domains of prokaryotic cells. In Chapter 14 we emphasized the profound phylogenetic differences between *Bacteria* and *Archaea*. Here we examine the organisms themselves. Some major characteristics of the *Archaea* include the absence of peptidoglycan in cell walls and the presence of ether-linked lipids and complex RNA polymerases. But beyond this, *Archaea* show enormous phenotypic diversity, and we consider this diversity here. As in Chapter 15 on *Bacteria*, we begin this chapter on *Archaea* with a phylogenetic overview to show the evolutionary relationships within the domain.

I PHYLOGENY AND GENERAL METABOLISM

The *Archaea* are one of the three domains of life. They share many characteristics with the other two domains, *Bacteria* and *Eukarya*, but are evolutionarily distinct from them. For example, like the *Bacteria*, *Archaea* have a prokaryotic cellular structure; however, their evolutionary history is separate from that of *Bacteria*. We begin this chapter with an overview of the phylogeny and metabolism of *Archaea* and then proceed to a discussion of the *Euryarchaeota* and *Crenarchaeota*, the two phyla of *Archaea*.

17.1 Phylogenetic Overview of *Archaea*

A phylogenetic tree of *Archaea* is shown in **Figure 17.1**. The tree, based on sequence of 16S rRNA genes, reveals a major evolutionary split of *Archaea* into two groups, the *Crenarchaeota* and the *Euryarchaeota*. The separation between these groups is supported by analysis of completely sequenced genomes. These groups may be phyla, the grouping used in this book, similar in phylogenetic divergence and complexity to phyla of *Bacteria*, or they might be higher order groupings, each of them containing multiple phyla. Either way, environmental sampling and culture-based studies are rapidly advancing our understanding of the diversity and phylogeny of *Archaea*.

Crenarchaeota

Among *Archaea* in laboratory culture, the ***Crenarchaeota*** contain mostly **hyperthermophiles**—organisms whose growth

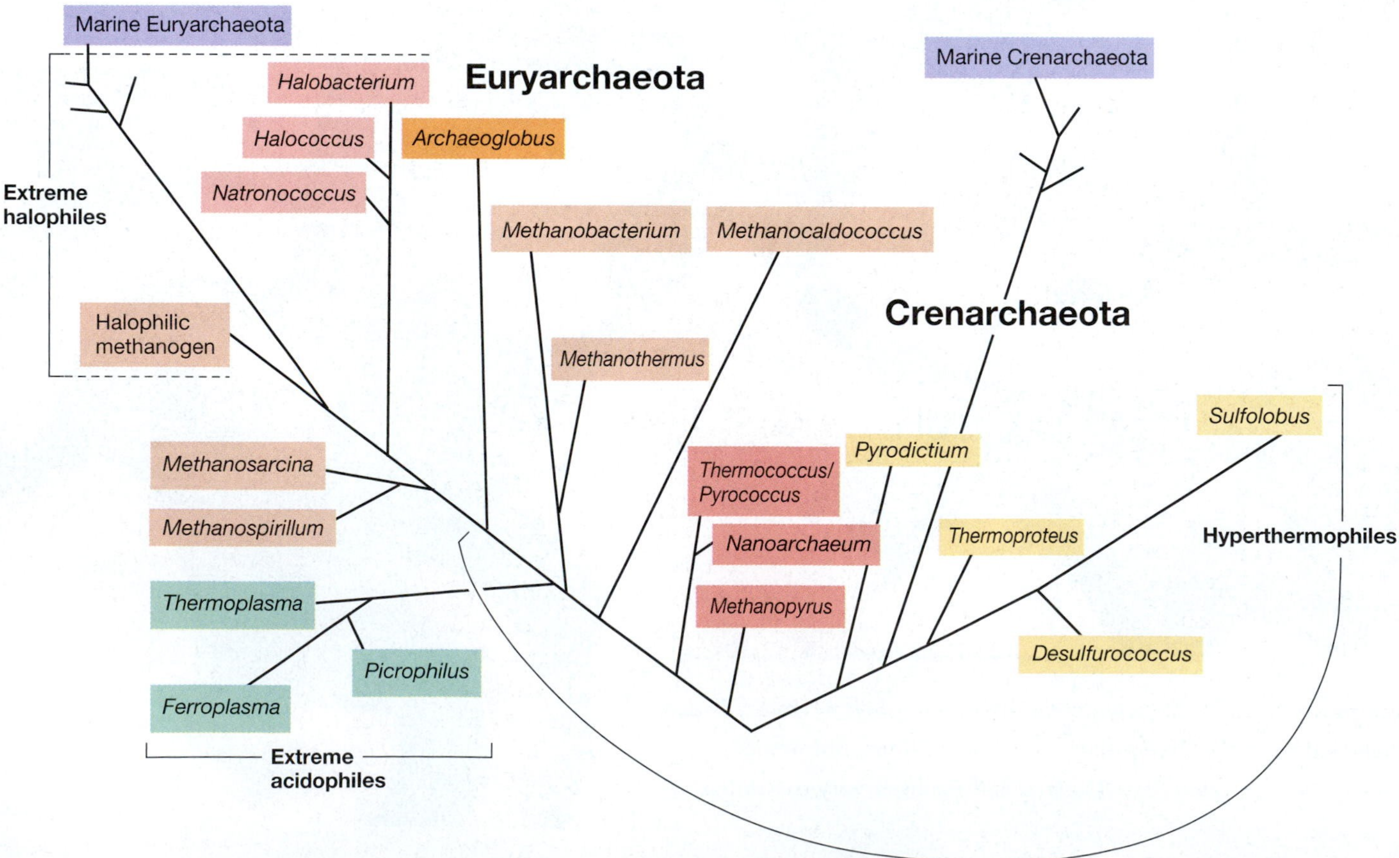

Figure 17.1 Detailed phylogenetic tree of the *Archaea* based on 16S rRNA gene sequence comparisons. Several members of the marine *Euryarchaeota* and marine *Crenarchaeota*, previously known only from community sampling, have been isolated and characterized recently. Not all *Methanobacterium* species are thermophiles.

temperature optimum is greater than 80°C—including those able to grow at the highest temperatures of all known organisms. Interestingly, however, several nonthermophilic *Crenarchaeota* related to hyperthermophilic species inhabit aquatic and terrestrial environments. Many hyperthermophiles are chemolithotrophic autotrophs, and because no phototrophs can survive such temperatures, these organisms are the sole primary producers in these habitats.

Hyperthermophilic species of *Crenarchaeota* tend to cluster closely together and occupy short branches on the 16S rRNA gene-based tree of life (∞ Figures 14.16 and 17.1). These organisms are therefore thought to be more slowly evolving than other lineages in the domain. If true, hyperthermophilic *Crenarchaeota* should be good models for "early" *Archaea* and perhaps early life forms in general; we return to this theme at the end of this chapter (Section 17.15). By contrast, cold-dwelling relatives of hyperthermophilic crenarchaeotes have been identified by community analysis of rRNA genes (∞ Sections 22.5 and 23.9) from *Archaea* in the oceans and a variety of other temperate and polar environments, and some of these organisms have been cultured. From a phylogenetic perspective, these species have longer branches on the tree and therefore might be more rapidly evolving. We consider the *Crenarchaeota* in more detail in Sections 17.9–17.12.

Euryarchaeota

Euryarchaeota comprise a physiologically diverse group of *Archaea*. Like crenarchaeotes, many inhabit extreme environments of one kind or the other. This phylum includes methanogens—organisms whose metabolism produces methane (CH_4)—and several genera of extremely halophilic *Archaea*, the "halobacteria" (Figure 17.1). As a study in physiological contrasts, these two groups are remarkable. Methanogens are the strictest of anaerobes, while extreme halophiles are for the most part obligate aerobes.

Other groups of euryarchaeotes include the hyperthermophiles *Thermococcus* and *Pyrococcus* and the methanogen *Methanopyrus*, all of which branch near the root (Figure 17.1), and the cell wall-less *Thermoplasma*, an organism phenotypically similar to the mycoplasmas (∞ Section 16.3). Finally, in parallel to the *Crenarchaeota*, a large group of as yet uncultured euryarchaeotes exists in the marine environment and forms long branches near the top of the archaeal tree (Figure 17.1). Many euryarchaeotes also inhabit freshwater and terrestrial habitats. We consider *Euryarchaeota* in more detail in Sections 17.3–17.8.

With this overview of archaeal phylogeny, we proceed to a brief description of the metabolic features of *Archaea*. From there we will describe major cultured representatives.

17.2 Energy Conservation and Autotrophy in *Archaea*

Energy metabolism in methanogens, a major group of *Archaea*, is unlike that of any other microbial group, *Bacteria* or *Archaea*. We reserve for later our discussion of the many novel features of methanogenesis (∞ Section 21.10) and focus here on the metabolism of nonmethanogenic *Archaea*, which more closely resembles the metabolism of *Bacteria*.

Chemoorganotrophy and Chemolithotrophy in *Archaea*

Several *Archaea* are chemoorganotrophic and thus use organic compounds as electron donors for energy metabolism. Catabolism of glucose in *Archaea* proceeds via slight modifications of the Entner–Doudoroff (∞ Section 15.7) or glycolytic (∞ Section 5.10) pathways. Oxidation of acetate to CO_2 in *Archaea* proceeds through the citric acid cycle (∞ Section 5.13) or some slight variation of this reaction series, or by the **acetyl-CoA pathway** (∞ Section 21.9). Little is known concerning biosynthesis of amino acids and other macromolecular precursors in *Archaea*; however, key monomers are likely produced from the central biosynthetic intermediates discussed previously for *Bacteria* (∞ Sections 5.15–5.17), and many of the same key enzymes in amino acid metabolism have counterparts in *Archaea*.

There are electron transport chains including cytochromes of the *a*, *b*, and *c* types in some *Archaea*. Employing these and other electron carriers, chemoorganotrophic metabolism in most *Archaea* proceeds by introduction of electrons from organic electron donors into an electron transport chain, leading to the reduction of O_2, S^0, or some other electron acceptor. Electron transport reactions drive the synthesis of a proton motive force that couples to ATP synthesis through membrane-bound ATPases (ATPase function is described in Section 5.12). Chemolithotrophy is also well established in the *Archaea*, with H_2 being a common electron donor (Section 17.15). We examine chemolithotrophic metabolism of hyperthermophilic *Archaea* in Section 17.9.

Autotrophy in *Archaea*

Autotrophy is widespread in the *Archaea* and proceeds by several different pathways. In hyperthermophilic methanogens and certain members of the *Archaeoglobales*, CO_2 is fixed in the acetyl-CoA pathway, or some modification of it. Most chemolithotrophic hyperthermophiles fix CO_2 via the reverse (reductive) citric acid cycle, a reaction series that also functions as the autotrophic pathway in the green sulfur bacteria, or by the 3-hydroxypropionate cycle (∞ Section 20.7), or possibly by the Calvin cycle, the most widespread autotrophic pathway in *Bacteria* and eukaryotes (∞ Section 20.6). Genes encoding functional and very thermostable RubisCO enzymes (RubisCO catalyzes the first step in the Calvin cycle) have been characterized from the methanogen *Methanocaldococcus jannaschii* and from a *Pyrococcus* species, both hyperthermophiles.

We can thus conclude that many of the catabolic and anabolic pathways in the *Archaea* are similar to those in *Bacteria*, a good reminder that metabolism has a long evolutionary history. With this as background and the phylogeny of *Archaea* (Figure 17.1) firmly in mind, we now consider the organismal diversity of this fascinating domain of life.

17.1–17.2 MiniReview

Archaea form two major groups, the *Euryarchaeota* and the *Crenarchaeota*. With the exception of methanogenesis, bioenergetics and intermediary metabolism in species of *Archaea* exhibit many similarities to these processes in various species of *Bacteria*.

- In which group of *Archaea* do the majority of cultured hyperthermophiles reside?
- What autotrophic pathways are found in *Archaea*?

II EURYARCHAEOTA

Euryarchaeotes comprise a physiologically diverse group of *Archaea*, and many species inhabit extreme environments. Included in *Euryarchaeota* are the extreme halophiles, the methanogens, and several thermophilic and extremely acidophilic species. Although some euryarchaeotes share traits such as hyperthermophily with crenarchaeotes, these two phyla are phylogenetically distinct (Figure 17.1). We start with a description of the extreme halophiles and then proceed to methanogens and other groups.

17.3 Extremely Halophilic *Archaea*

Key Genera: *Halobacterium, Haloferax, Natronobacterium*

Extremely halophilic *Archaea* (Figure 17.1), sometimes called *haloarchaea*, are a diverse group of prokaryotes that inhabits natural environments high in salt, such as solar salt evaporation ponds and salt lakes, or artificial saline habitats such as the surfaces of heavily salted foods, like certain fish and meats. Such salt habitats are called *hypersaline*. The term **extreme halophile** is used to indicate that these organisms are not only halophilic, but that their requirement for salt is very high, in some cases at levels near saturation.

An organism is considered an extreme halophile if it requires at least 1.5 M (about 9%) NaCl for growth. Most species of extreme halophiles require 2–4 M NaCl (12–23%) for optimal growth. Virtually all extreme halophiles can grow at 5.5 M NaCl (32%, the limit of saturation for NaCl), although some species grow very slowly at this salinity. Some phylogenetic relatives of extremely halophilic *Archaea*, for example, species of *Haloferax* and *Natronobacterium*, are able to grow at much lower salinities, such as at or near that of seawater (about 2.5% NaCl); nevertheless, these organisms are grouped with other extreme halophiles. The isolation of these "low-salinity" haloarchaea highlights a theme emerging from environmental sampling and culture-based studies of *Archaea*, which is the presence of substantial previously unknown ecological, metabolic, and physiological diversity (Section 17.12).

Hypersaline Environments: Chemistry and Productivity

Hypersaline habitats are common throughout the world, but extremely hypersaline habitats are rare. Most such environments are in hot, dry areas of the world. Salt lakes can vary considerably in ionic composition. The predominant ions in a hypersaline lake depend on the surrounding topography, geology, and general climatic conditions.

Great Salt Lake in Utah (USA) (**Figure 17.2a**), for example, is essentially concentrated seawater; the relative proportions of the various ions are those of seawater, although the overall concentration of ions is much higher. Sodium is the predominant cation in Great Salt Lake, whereas chloride is the predominant anion; significant levels of sulfate are also present at a slightly alkaline pH (**Table 17.1**). By contrast, another hypersaline basin, the Dead Sea, is relatively low in sodium but contains high levels of magnesium (Table 17.1).

Soda lakes are highly alkaline hypersaline environments. The water chemistry of soda lakes resembles that of hypersaline lakes such as Great Salt Lake, but because high levels of carbonate minerals are also present in the surrounding strata, the pH of soda lakes is quite high. Waters of pH 10–12 are not uncommon in these environments (Table 17.1 and Figure 17.2*c*). In addition, Ca^{2+} and Mg^{2+} are virtually absent from soda lakes because they precipitate out at high pH and carbonate concentrations (Table 17.1).

The diverse chemistries of hypersaline habitats has selected for a large diversity of halophilic microorganisms. Some organisms are only known from one environment and others are widespread in several habitats. Moreover, despite what may seem like rather harsh conditions, salt lakes can be highly productive ecosystems (the word *productive* here means high levels of CO_2 fixation). *Archaea* are not the only microorganisms present. The eukaryotic alga *Dunaliella* is the major, if not sole, oxygenic phototroph in most salt lakes. In highly alkaline soda lakes where *Dunaliella* is absent, anoxygenic phototrophic purple bacteria of the genera *Ectothiorhodospira* and *Halorhodospira* (∞ Section 15.2) predominate. Organic matter originating from primary production by oxygenic or anoxygenic phototrophs sets the stage for growth of haloarchaea, all of which are chemoorganotrophic. In addition, a few extremely halophilic anaerobic chemoorganotrophic *Bacteria*, such as *Halanaerobium*, *Halobacteroides*, and *Salinibacter*, thrive in such environments.

Marine salterns are also habitats for extreme halophiles. Marine salterns are small, enclosed basins filled with seawater that are left to evaporate, yielding solar sea salt (Figure 17.2*b, d*). As salterns approach the minimum salinity limits for extreme halophiles, the waters turn a reddish purple color due to the massive growth—called a *bloom*—of halophilic *Archaea* (the red coloration apparent in Figures 17.2*b* and *c* is due to carotenoids and other pigments discussed later). Morphologically unusual *Archaea* are often present in salterns, even including species with a square morphology (Figure 17.2*d*). Extreme halophiles are also present in highly salted

Figure 17.2 Hypersaline habitats for halophilic *Archaea*. (*a*) Great Salt Lake, Utah, a hypersaline lake in which the ratio of ions is similar to that in seawater but in which absolute concentrations of ions are about 10 times that of seawater. The green color is primarily from cells of cyanobacteria and green algae. (*b*) Aerial view near San Francisco Bay, California, of a series of seawater evaporating ponds where solar salt is prepared. The red-purple color is predominantly due to bacterioruberins and bacteriorhodopsin in cells of *Halobacterium*. (*c*) Lake Hamara, Wadi El Natroun, Egypt. A bloom of pigmented haloalkaliphiles is growing in this pH 10 soda lake. Note the deposits of trona ($NaHCO_3 \cdot Na_2CO_3 \cdot 2\ H_2O$) around the edge of the lake. (*d*) Scanning electron micrograph of halophilic bacteria including square bacteria present in a saltern in Spain.

foods, such as certain types of sausages, marine fish, and salted pork.

Taxonomy and Physiology of Extremely Halophilic *Archaea*

Table 17.2 lists many of the 27 currently recognized genera of extremely halophilic *Archaea*. Besides the term haloarchaea, these *Archaea* are commonly called "halobacteria," because the genus *Halobacterium* (**Figure 17.3**) was the first in this group to be described and is still the best-studied representative of the group. *Natronobacterium*, *Natronomonas*, and their relatives differ from other extreme halophiles in being extremely alkaliphilic as well as halophilic. As befits their soda lake habitat (Table 17.1 and Figure 17.2*c*), growth of natronobacteria is optimal at very low Mg^{2+} concentrations and high pH (9–11).

Extremely halophilic *Archaea* stain gram negatively, reproduce by binary fission, and do not form resting stages or spores. Most halobacteria are nonmotile, but a few strains are weakly motile by flagella or are able to float by gas vesicles (Section 4.11). The genomes of *Halobacterium* and *Halococcus* are unusual in that large plasmids containing up to 25–30% of the total cellular DNA are often present and the GC base ratio of these plasmids (57–60% GC) differs significantly from that of chromosomal DNA (66–68% GC). Plasmids from extreme halophiles are among the largest naturally occurring plasmids known and might actually be small chromosomes (Section 11.2).

Table 17.1 Ionic composition of some highly saline environments[a]

Ion	Concentration (g/l) Great Salt Lake[b]	Dead Sea	Lake Zugm[c]
Na^+	105	40.1	142
K^+	6.7	7.7	2.3
Mg^{2+}	11	44	<0.1
Ca^{2+}	0.3	17.2	<0.1
Cl^-	181	225	155
Br^-	0.2	5.3	—
SO_4^{2-}	27	0.5	23
HCO_3^-	0.7	0.2	67
pH	7.7	6.1	11

[a]For comparison, seawater contains (grams per liter): Na^+, 10.6; K^+, 0.38; Mg^{2+}, 1.27; Ca^{2+}, 0.4; Cl^-, 19; Br^-, 0.065; SO_4^{2-}, 2.65; HCO_3^-, 0.14; pH 7.8.

[b]See Figure 17.2*a*.

[c]Wadi El Natroun, Egypt (see Figure 17.2*c*).

Most species of extremely halophilic *Archaea* are obligate aerobes. Most halobacteria use amino acids or organic acids as energy sources and require a number of growth factors (mainly vitamins, (Section 5.1) for optimal growth. A few *Halobacterium* species oxidize carbohydrates, but this ability is rare. Electron transport chains containing cytochromes of the *a*, *b*, and *c* types are present in *Halobacterium*, and energy is conserved during aerobic growth via a proton motive force arising from electron transport. Some halophilic *Archaea* have been shown to grow anaerobically. Growth under anoxic conditions at the expense of sugar fermentation and by anaerobic respiration (Section 21.6) linked to the reduction of nitrate or fumarate has been demonstrated in certain species.

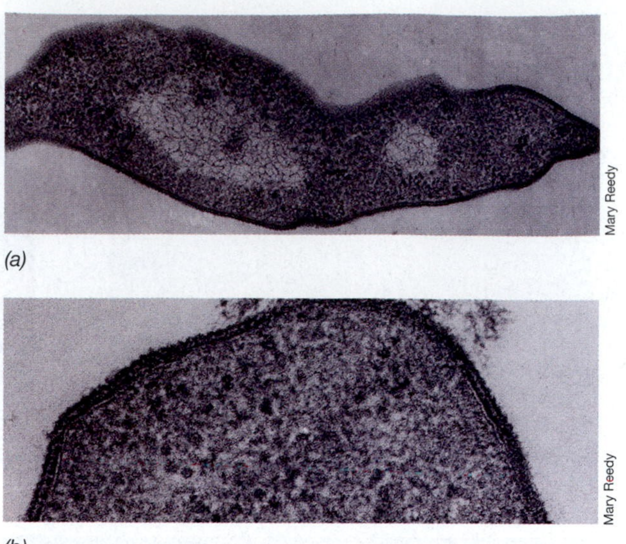

Figure 17.3 Electron micrographs of thin sections of the extreme halophile *Halobacterium salinarum*. A cell is about 0.8 μm in diameter. (*a*) Longitudinal section of a dividing cell showing the nucleoids. (*b*) High-magnification electron micrograph showing the glycoprotein subunit structure of the cell wall.

Table 17.2 Some genera of extremely halophilic *Archaea*

Genus	Morphology	Habitat
Extreme Halophiles		
Halobacterium	Rods	Salted fish; hides; hypersaline lakes; salterns
Halorubrum	Rods	Dead Sea; salterns
Halobaculum	Rods	Dead Sea
Haloferax	Flattened discs	Dead Sea; salterns
Haloarcula	Irregular discs	Salt pools, Death Valley, CA; marine salterns
Halococcus	Cocci	Salted fish; salterns
Halogeometricum	Rods	Solar salterns
Haloterrigena	Rods, ovals	Saline soil
Haloalkaliphiles		
Natronobacterium	Rods	Highly saline soda lakes
Natrinema	Rods	Salted fish; hides
Natrialba	Rods	Soda lakes; beach sand
Natronomonas	Rods	Soda lakes
Natronococcus	Cocci	Soda lakes
Natronorubrum	Flattened cells	Soda lakes

Water Balance in Extreme Halophiles

Extremely halophilic *Archaea* require large amounts of sodium for growth, typically supplied as NaCl. Detailed salinity studies of *Halobacterium* have shown that the requirement for Na^+ cannot be satisfied by any other ion, even the chemically related ion K^+. However, cells of *Halobacterium* need *both* Na^+ and K^+ for growth, because each plays an important role in maintaining osmotic balance.

As we learned in Section 6.16, microbial cells must withstand the osmotic forces that accompany life. To do so in a high-solute environment, such as the salt-rich habitats of *Halobacterium*, organisms must either accumulate or synthesize solutes intracellularly. These solutes are called **compatible solutes**. These compounds counteract the tendency of the cell to become dehydrated under conditions of high osmotic strength by placing the cell in positive water balance with its surroundings (Section 6.16). Cells of *Halobacterium*, however, do not synthesize or accumulate organic compounds but instead pump large amounts of K^+ from the environment into the cytoplasm. This ensures that the concentration of K^+ *inside* the cell is even greater than the concentration of Na^+ *outside* the cell (**Table 17.3**). This ionic condition maintains positive water balance.

The *Halobacterium* cell wall (Figure 17.3*b*) is composed of glycoprotein and stabilized by Na^+. Sodium ions bind to the outer surface of the *Halobacterium* wall and are absolutely essential for maintaining cellular integrity. When insufficient Na^+ is present, the cell wall breaks apart and the cell lyses. This is a consequence of the exceptionally high content of the

Table 17.3 Concentration of ions in cells of *Halobacterium salinarum*

Ion	Concentration in medium (M)	Concentration in cells (M)
Na^+	3.3	0.8
K^+	0.05	5.3
Mg^{2+}	0.13	0.12
Cl^-	3.3	3.3

acidic (negatively charged) amino acids aspartate and glutamate (∞ Figure 3.12) in the glycoprotein of the *Halobacterium* cell wall. The negative charges contributed by the carboxyl groups of these amino acids are shielded by Na^+. When Na^+ is diluted away, the negatively charged parts of the proteins actively repel each other, leading to cell lysis.

Halophilic Cytoplasmic Components

Like cell wall proteins, cytoplasmic proteins of *Halobacterium* are highly acidic, but it is K^+, not Na^+, that is required for activity. This, of course, makes sense because K^+ is the predominate cation in the cytoplasm of cells of *Halobacterium* (Table 17.3). Besides a high acidic amino acid composition, halobacterial cytoplasmic proteins typically contain lower levels of hydrophobic amino acids and lysine, a positively charged (basic) amino acid, than proteins of nonhalophiles. This is also to be expected because in a highly ionic cytoplasm, polar proteins would tend to remain in solution whereas nonpolar proteins would tend to cluster and perhaps lose activity. The ribosomes of *Halobacterium* also require high K^+ levels for stability, whereas ribosomes of nonhalophiles have no K^+ requirement.

Extremely halophilic *Archaea* are thus well adapted, both internally and externally, to life in a highly ionic environment. Cellular components exposed to the external environment require high Na^+ for stability, whereas internal components require high K^+. With the exception of a few extremely halophilic members of the *Bacteria* that also use K^+ as a compatible solute, in no other group of bacteria do we find this unique requirement for such high amounts of specific cations.

Bacteriorhodopsin and Light-Mediated ATP Synthesis in Halobacteria

Certain species of haloarchaea can carry out a light-driven synthesis of ATP. This occurs without chlorophyll pigments, so it is not photosynthesis. However, other light-sensitive pigments are present, including red and orange carotenoids—primarily C_{50} pigments called *bacterioruberins*—and inducible pigments involved in energy conservation as discussed now.

Under conditions of low aeration, *Halobacterium salinarum* and some other extreme halophiles synthesize and insert a protein called **bacteriorhodopsin** into their cytoplasmic membranes. Bacteriorhodopsin was so named because of its structural and functional similarity to rhodopsin, the visual pigment of the eye. Conjugated to bacteriorhodopsin is a

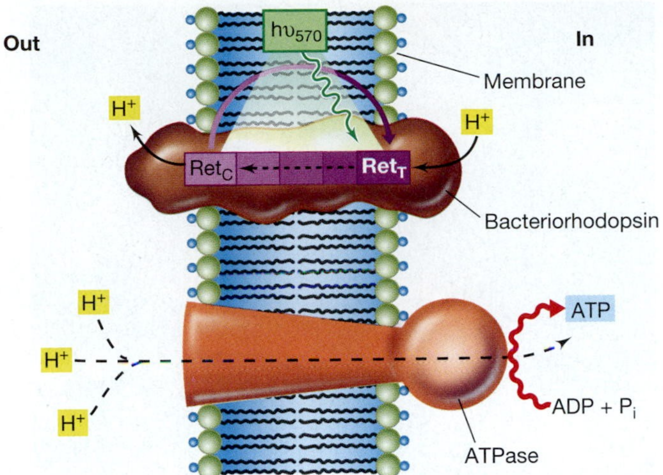

Figure 17.4 Model for the mechanism of bacteriorhodopsin. Light of 570 nm ($h\nu_{570}$) converts the protonated retinal of bacteriorhodopsin from the *trans* form (Ret_T) to the *cis* form (Ret_C), along with translocation of a proton to the outer surface of the cytoplasmic membrane, thus establishing a proton motive force. ATPase activity is driven by the proton motive force.

molecule of retinal, a carotenoid-like molecule that can absorb light energy and pump a proton across the cytoplasmic membrane. The retinal gives bacteriorhodopsin a purple hue. Thus cells of *Halobacterium* switched from growth under high aeration conditions to oxygen-limiting growth conditions (a trigger of bacteriorhodopsin synthesis) gradually change color from orange-red to purple-red as they synthesize bacteriorhodopsin and insert it into their cytoplasmic membranes.

Bacteriorhodopsin absorbs green light, with an absorption maximum at 570 nm. Following absorption, the retinal of bacteriorhodopsin, which normally exists in a *trans* configuration (Ret_T), becomes excited and converts to the *cis* (Ret_C) form (**Figure 17.4**). This transformation is coupled to the translocation of a proton across the cytoplasmic membrane. The retinal molecule then decays to the *trans* isomer along with the uptake of a proton from the cytoplasm, and this completes the cycle. The proton pump is then ready to repeat the cycle (Figure 17.4). As protons accumulate on the outer surface of the membrane, a proton motive force is generated that is coupled to ATP synthesis through the activity of a proton-translocating ATPase (∞ Section 5.12) (Figure 17.4).

Light-mediated ATP production in *H. salinarum* supports slow growth of this organism under anoxic conditions. The light-stimulated proton pump of *H. salinarum* also functions to pump Na^+ out of the cell by activity of a Na^+/H^+ antiport system (∞ Section 4.5) and also drives the uptake of nutrients, including the K^+ needed for osmotic balance. Amino acid uptake by *H. salinarum* is indirectly driven by light because amino acids are cotransported into the cell with Na^+ by an amino acid–Na^+ symporter (∞ Section 4.5); removal of Na^+ from the cell occurs by way of the light-driven Na^+/H^+ antiporter. **www.microbiologyplace.com Online Tutorial 17.1 Bacteriorhodopsin and Light-Mediated ATP Synthesis**

UNIT 3

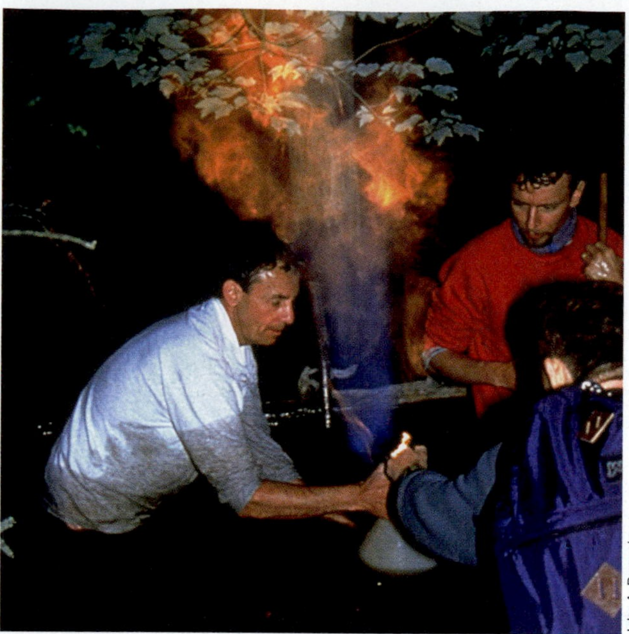

Figure 17.5 The Volta experiment as demonstrated in the microbial diversity summer course in Woods Hole, Massachusetts. A large inverted funnel was placed over freshwater sediments in Cedar Swamp, Woods Hole, where anaerobic decomposition was occurring. After water displaced the air in the funnel, the funnel was capped and the sediments were mixed with a stick, allowing trapped bubbles of methane to collect in the inverted funnel. Immediately after uncapping the funnel, a flame was held near the funnel port, thus igniting the methane. This experiment was performed over 200 years ago by the Italian physicist Alessandro Volta, which led him to describe methane as "combustible air."

John A. Breznak

Other Rhodopsins

Besides bacteriorhodopsin, at least three other rhodopsins are present in the cytoplasmic membrane of *H. salinarum*. **Halorhodopsin** is a light-driven pump that pumps chloride (Cl^-) into the cell as an anion for K^+. The retinal of halorhodopsin binds Cl^- and transports it into the cell. Two other light sensors, called *sensory rhodopsins*, are present in *H. salinarum*. These light sensors control phototaxis (movement toward light, ⚭ Section 4.15) by the organism. Through the interaction of a cascade of proteins similar to those in chemotaxis (⚭ Sections 4.15 and 9.7), sensory rhodopsins affect flagellar rotation, moving cells of *H. salinarum* toward light where bacteriorhodopsin can function to make ATP (Figure 17.4).

We will learn when we consider marine microbiology (⚭ Section 23.9) that several *Bacteria* that contain bacteriorhodopsin-like proteins called *proteorhodopsins* inhabit the upper layers of the ocean. As far as is known, proteorhodopsin functions like bacteriorhodopsin, except that several different spectral forms exist, each tuned to the absorption of different wavelengths of light. Proteorhodopsin as a mechanism for energy conservation in marine bacteria makes good ecological sense, because levels of dissolved organic matter in the open oceans are typically very low (⚭ Section 23.9), and thus a strictly chemoorganotrophic lifestyle would be difficult.

17.3 MiniReview

Extremely halophilic *Archaea* require large amounts of Na^+ for growth and accumulate large levels of K^+ in their cytoplasm as a compatible solute. These salts affect cell wall stability and enzyme activity. The light-mediated proton pump bacteriorhodopsin helps extreme halophiles make ATP.

- What is the major physiological difference between *Halobacterium* and *Natronobacterium*?
- If cells of *Halobacterium* require high levels of Na^+, why is this not true for cytoplasmic enzymes of *Halobacterium* (Table 17.3)?
- What benefit does bacteriorhodopsin confer on a cell of *Halobacterium salinarum*? Halorhodopsin?

17.4 Methane-Producing *Archaea*: Methanogens

Key Genera: *Methanobacterium, Methanocaldococcus, Methanosarcina*

Many *Euryarchaeota* are **methanogens**, microorganisms that produce methane (CH_4) as an integral part of their energy metabolism (methane production is called *methanogenesis*). Methane was described as "combustible air" by the Italian physicist Alessandro Volta; Volta collected gas from marsh sediments and showed that it was flammable. The Volta experiment, as it has come to be known, is easily reproduced if methane trapped in freshwater sediments is collected and carefully ignited (**Figure 17.5**).

In a later chapter we will study the biochemically unique and amazingly complex process of methane formation (⚭ Section 21.10). Moreover, we will learn how methanogenesis is the terminal step in the biodegradation of organic matter in many anoxic habitats in nature, such as the swamp shown in Figure 17.5. **Table 17.4** lists the habitats of

Table 17.4 Habitats of methanogens

I. Anoxic sediments: marsh, swamp (Figure 17.5), and lake sediments, paddy fields, moist landfills (⚭ Section 21.10)
II. Animal digestive tracts:
(a) Rumen of ruminant animals such as cattle, sheep, elk, deer, and camels (⚭ Section 24.10)
(b) Cecum of cecal animals such as horses and rabbits (⚭ Section 24.10)
(c) Large intestine of monogastric animals such as humans, swine, and dogs (⚭ Section 28.4)
(d) Hindgut of cellulolytic insects (for example, termites) (⚭ Section 24.2)
III. Geothermal sources of $H_2 + CO_2$: hydrothermal vents (⚭ Section 24.11)
IV. Artificial biodegradation facilities: sewage sludge digestors (⚭ Section 36.2)
V. Endosymbionts of various anaerobic protozoa (⚭ Figure 24.4)

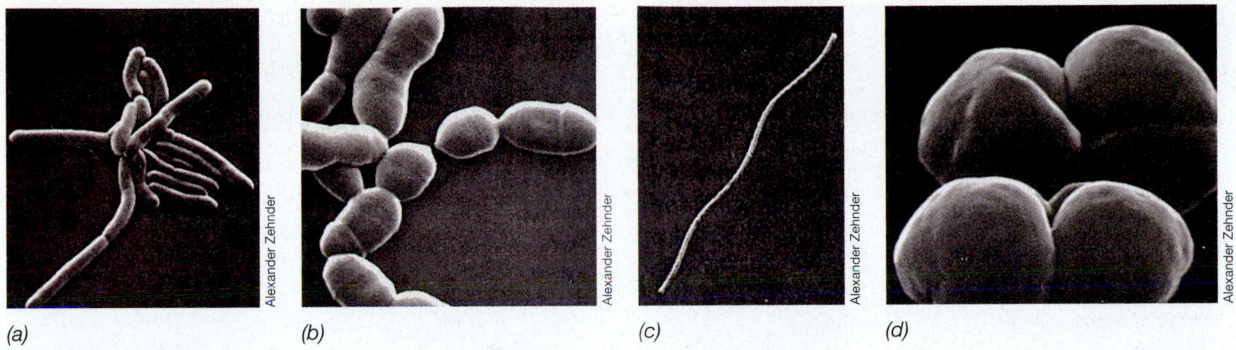

Figure 17.6 Scanning electron micrographs of cells of methanogenic *Archaea,* showing the considerable morphological diversity. (*a*) *Methanobrevibacter ruminantium.* A cell is about 0.7 μm in diameter. (*b*) *Methanobrevibacter arboriphilus.* A cell is about 1 μm in diameter. (*c*) *Methanospirillum hungatei.* A cell is about 0.4 μm in diameter. (*d*) *Methanosarcina barkeri.* A cell is about 1.7 μm wide.

methanogens; these habitats are major sources of biogenic methane in nature.

Diversity and Physiology of Methanogens

Methanogens show a variety of morphologies (**Figure 17.6** and **Table 17.5**). Their taxonomy is based on both phenotypic and phylogenetic analyses (Table 17.5), with several taxonomic orders being recognized (in taxonomy, an order contains groups of related families, each of which contains one or more genera ∞ Section 14.14).

Methanogens show a diversity of cell wall chemistries. These include the pseudomurein walls of *Methanobacterium* species and relatives (**Figure 17.7*a***); walls composed of methanochondroitin (so named because of its structural resemblance to chondroitin, the connective tissue polymer of vertebrate animals) in *Methanosarcina* and relatives (Figure 17.7*b*); the protein or glycoprotein walls of *Methanocaldococcus* (**Figure 17.8*a***) and *Methanoplanus* species, respectively; and the S-layer walls of *Methanospirillum* (Figure 17.6; ∞ Section 4.8).

Physiologically, methanogens are obligate anaerobes, and strict anoxic techniques are necessary to culture them. Depending on the species, cultures of methanogens can be established in a mineral salts medium under an atmosphere of H_2 plus CO_2 (in the ratio of 4:1, see the reaction shown later), or in complex media. Most known methanogens are mesophilic and nonhalophilic, although "extremophilic" species growing optimally at very high (Figure 17.8) or very low temperatures, or at very high salt concentrations, or at extremes of pH, have also been described.

Substrates for Methanogenesis

Eleven substrates can be converted to methane by pure cultures of methanogens (**Table 17.6**). Interestingly, these substrates do not include such common compounds as glucose and organic or fatty acids (other than acetate and pyruvate). Compounds such as glucose can be converted to methane, but only in reactions in which methanogens and other anaerobic bacteria cooperate. With the right mixture of organisms, virtually any

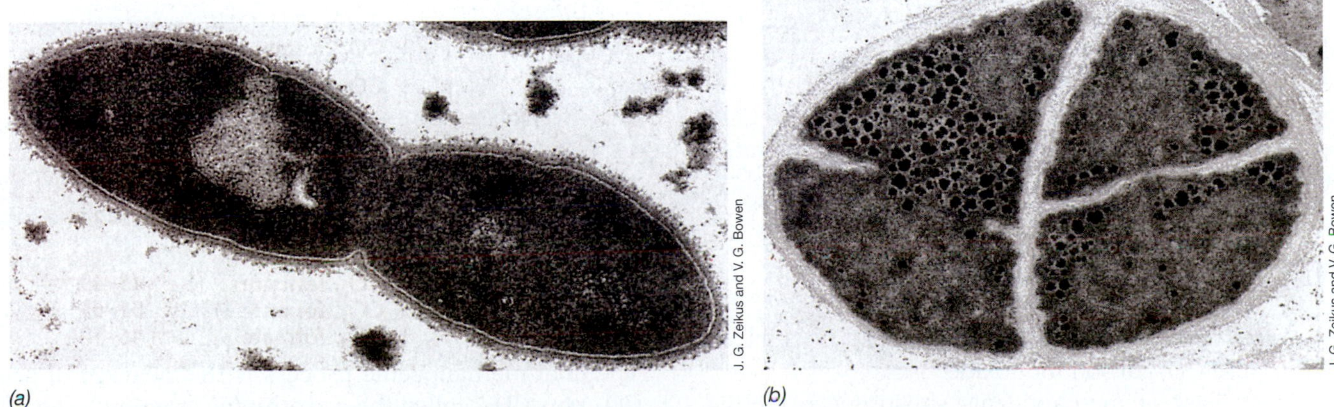

Figure 17.7 Transmission electron micrographs of thin sections of methanogenic *Archaea.* (*a*) *Methanobrevibacter ruminantium.* A cell is 0.7 μm in diameter. (*b*) *Methanosarcina barkeri,* showing the thick cell wall and the manner of cell segmentation and cross-wall formation. A cell is 1.7 μm in diameter.

Table 17.5 Characteristics of some methanogenic *Archaea*[a]

Order/Genus	*Morphology*	*Substrates for methanogenesis*
Methanobacteriales		
Methanobacterium	Long rods	$H_2 + CO_2$, formate
Methanobrevibacter	Short rods	$H_2 + CO_2$, formate
Methanosphaera	Cocci	Methanol + H_2 (both needed)
Methanothermus	Rods	$H_2 + CO_2$
Methanothermobacter	Rods	$H_2 + CO_2$, formate
Methanococcales		
Methanococcus	Irregular cocci	$H_2 + CO_2$, pyruvate + CO_2, formate
Methanothermococcus	Cocci	$H_2 + CO_2$, formate
Methanocaldococcus	Cocci	$H_2 + CO_2$
Methanotorris	Cocci	$H_2 + CO_2$
Methanomicrobiales		
Methanomicrobium	Short rods	$H_2 + CO_2$, formate
Methanogenium	Irregular cocci	$H_2 + CO_2$, formate
Methanospirillum	Spirilla	$H_2 + CO_2$, formate
Methanoplanus	Plate-shaped cells—occurring as thin plates with sharp edges	$H_2 + CO_2$, formate
Methanocorpusculum	Irregular cocci	$H_2 + CO_2$, formate, alcohols
Methanoculleus	Irregular cocci	$H_2 + CO_2$, alcohols, formate
Methanofollis	Irregular cocci	$H_2 + CO_2$, formate
Methanolacinia	Irregular rods	$H_2 + CO_2$, alcohols
Methanosarcinales		
Methanosarcina	Large irregular cocci in packets	$H_2 + CO_2$, methanol, methylamines, acetate
Methanolobus	Irregular cocci in aggregates	Methanol, methylamines
Methanohalobium	Irregular cocci	Methanol, methylamines
Methanococcoides	Irregular cocci	Methanol, methylamines
Methanohalophilus	Irregular cocci	Methanol, methylamines, methyl sulfides
Methanosaeta	Long rods to filaments	Acetate
Methanosalsum	Irregular cocci	Methanol, methylamines, dimethylsulfide
Methanopyrales		
Methanopyrus	Rods in chains	$H_2 + CO_2$

[a]Taxonomic orders are listed in bold.

organic compound, even hydrocarbons, can be converted to methane plus CO_2 (Section 21.10).

Three classes of compounds make up the list of methanogenic substrates shown in Table 17.6. These are CO_2-type substrates, methyl substrates, and acetotrophic substrates. CO_2-type substrates naturally include CO_2 itself, which is reduced to methane using H_2 as electron donor:

$$CO_2 + 4\,H_2 \rightarrow CH_4 + 2\,H_2O \quad \Delta G^{0\prime} = -131\text{ kJ}$$

Other substrates here include formate (which is $CO_2 + H_2$ in organic form) and CO, carbon monoxide.

The second class of methanogenic substrates is methylated substrates (Table 17.6). With methanol (CH_3OH) as a model methyl substrate, the formation of CH_4 can be formed in either of two ways. First, CH_3OH can be reduced using an external electron donor such as H_2:

$$CH_3OH + H_2 \rightarrow CH_4 + H_2O \quad \Delta G^{0\prime} = -113\text{ kJ}$$

Alternatively, in the absence of H_2, some CH_3OH can be oxidized to CO_2 to generate the electrons needed to reduce other molecules of CH_3OH to CH_4:

$$4\,CH_3OH \rightarrow 3\,CH_4 + CO_2 + 2\,H_2O \quad \Delta G^{0\prime} = -319\text{ kJ}$$

The final methanogenic process is the cleavage of acetate to CO_2 plus CH_4, called the *acetotrophic* reaction:

$$CH_3COO^- + H_2O \rightarrow CH_4 + HCO_3^- \quad \Delta G^{0\prime} = -31\text{ kJ}$$

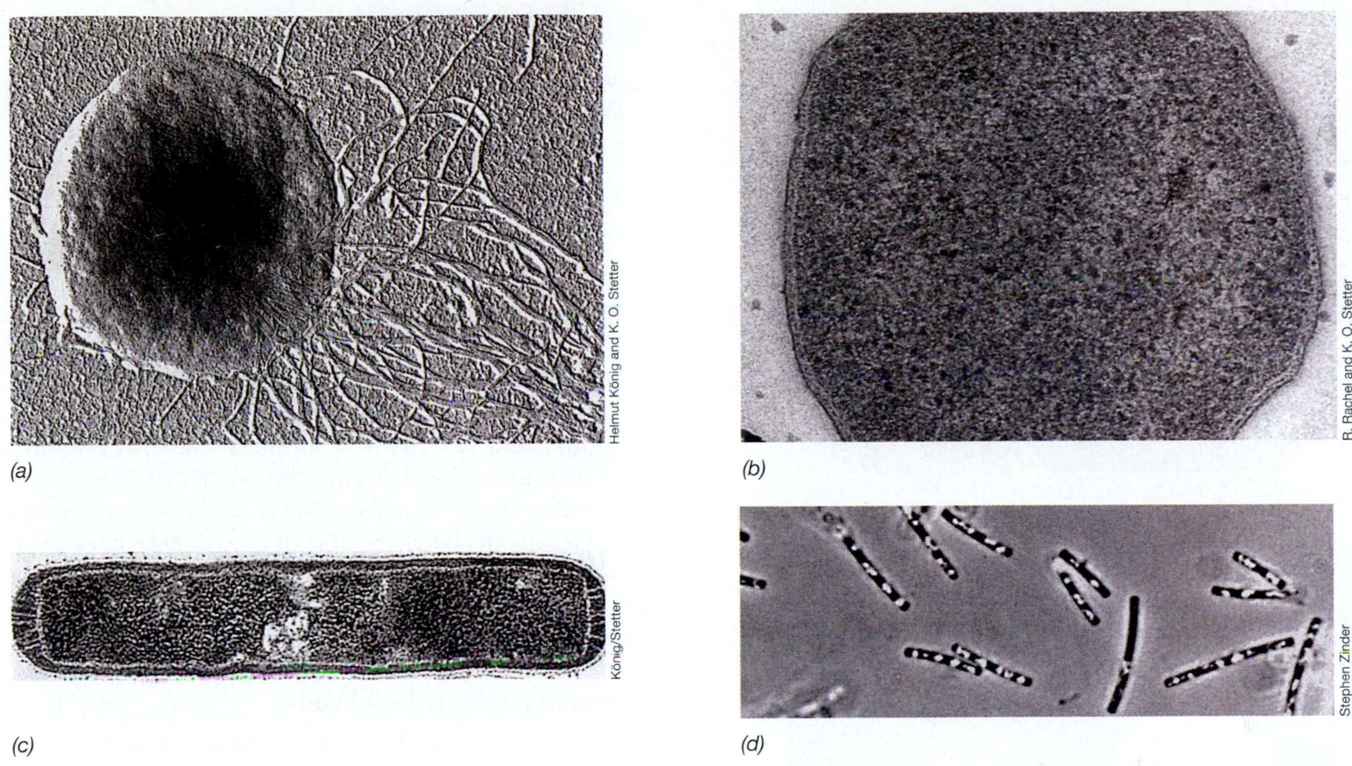

Figure 17.8 Hyperthermophilic and thermophilic methanogens. *(a) Methanocaldococcus jannaschii* (temperature optimum, 85°C), shadowed preparation electron micrograph. A cell is about 1 μm in diameter. *(b) Methanotorris igneus* (temperature optimum, 88°C), thin section. A cell is about 1 μm in diameter. *(c) Methanothermus fervidus* (temperature optimum, 88°C), thin-sectioned electron micrograph. A cell is about 0.4 μm in diameter. *(d) Methanosaeta thermophila* (temperature optimum, 60°C), phase contrast micrograph. A cell is about 1 μm in diameter. The refractile bodies inside the cells are gas vesicles.

Only a very few methanogens are acetotrophic (Table 17.5). However, measurements of methanogenesis in some methanogenic habitats, such as sewage sludge (Section 36.2), have shown that about two-thirds of the methane formed originates from acetate and one-third from $H_2 + CO_2$. Thus, although apparently lacking in terms of known diversity, acetotrophic methanogens are very important ecologically.

As can be seen by inspection of each of the above reactions, all are energy-releasing (exergonic) and can thus be used to synthesize ATP. The biochemical details of methanogenesis will be developed in Section 21.10. We only note here that the process is coupled to formation of a proton motive force, and ATP synthesis occurs through the activity of ATPases (Section 5.12). Substrate-level phosphorylation, typical of fermentative bacteria, apparently does not occur in methanogens.

Methanocaldococcus jannaschii as a Model Methanogen

As described in Section 13.2, the genome of the hyperthermophilic methanogen *Methanocaldococcus jannaschii* (Figure 17.8*a*) and that of several other methanogens have been sequenced. The 1.66-Mbp circular genome of *M. jannaschii* contains about 1700 genes, and genes encoding enzymes of methanogenesis and several other key cell functions have been identified. Interestingly, the majority of *M. jannaschii* genes encoding functions such as central metabolic pathways and cell division are similar to those in *Bacteria*. By contrast, most of the *M. jannaschii* genes encoding core molecular

Table 17.6 Substrates converted to methane by various methanogenic *Archaea*

Substrates
I. CO_2-type substrates
Carbon dioxide, CO_2 (with electrons derived from H_2, certain alcohols, or pyruvate)
Formate, $HCOO^-$
Carbon monoxide, CO
II. Methyl substrates
Methanol, CH_3OH
Methylamine, $CH_3NH_3^+$
Dimethylamine, $(CH_3)_2NH_2^+$
Trimethylamine, $(CH_3)_3NH^+$
Methylmercaptan, CH_3SH
Dimethylsulfide, $(CH_3)_2S$
III. Acetotrophic substrates
Acetate, CH_3COO^-
Pyruvate, CH_3COCOO^-

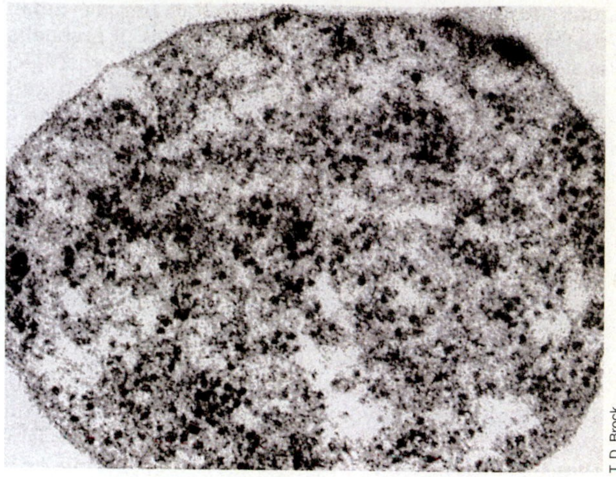

(a)

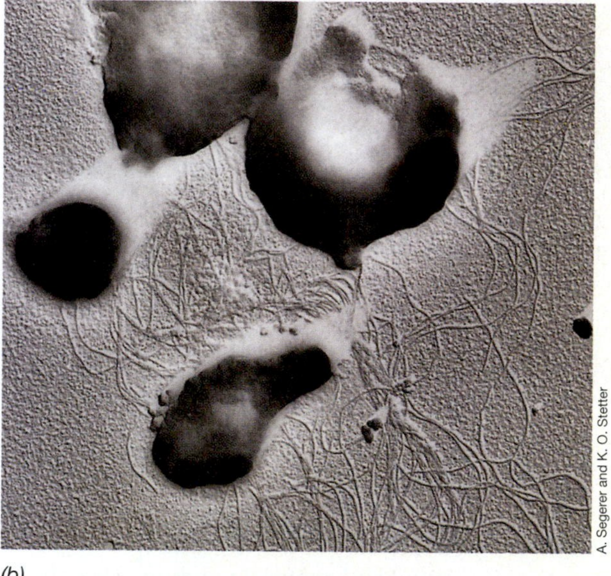

(b)

Figure 17.9 ***Thermoplasma* species.** *(a) Thermoplasma acidophilum*, an acidophilic, thermophilic mycoplasma-like archaeon. Electron micrograph of a thin section. The diameter of cells is highly variable from 0.2 to 5 μm. The cell shown is about 1 μm in diameter. *(b)* Shadowed preparation of cells of *Thermoplasma volcanium* isolated from hot springs. Cells are 1–2 μm in diameter. Notice the abundant flagella and irregular cell morphology.

processes such as transcription and translation more closely resemble those of eukaryotes. These findings reflect the various traits shared by *Archaea* and *Eukarya* and by *Bacteria* and *Eukarya* (∞ Tables 14.1 and 14.2), and they may reflect the hypothesized endosymbiotic origin of eukaryotic cells (∞ Section 14.4). However, analyses of the *M. jannaschii* genome also show that nearly 50% of its genes have no counterparts in known genes from any group of organisms. This suggests that there are either many new cellular functions encoded in archaeal DNA that have yet to be discovered or that many existing functions are encoded by redundant (and quite distinct) sets of genes.

17.4 MiniReview

Methanogenic *Archaea* are strict anaerobes whose metabolism is tied to the production of methane (CH_4).

- What does the Volta experiment demonstrate?
- What are the major substrates for methanogenesis?

17.5 *Thermoplasmatales*

Key Genera: *Thermoplasma, Picrophilus, Ferroplasma*

A phylogenetically distinct line of *Archaea* contains thermophilic and extremely acidophilic genera: *Thermoplasma, Ferroplasma*, and *Picrophilus* (Figure 17.1). These prokaryotes are among the most acidophilic of all known microorganisms, with *Picrophilus* being capable of growth even below pH 0. They also form their own taxonomic order within the *Euryarchaeota*, the *Thermoplasmatales*. We begin with a description of the mycoplasma-like organisms *Thermoplasma* and *Ferroplasma*.

Archaea Lacking Cell Walls

Thermoplasma and *Ferroplasma* lack cell walls, and in this respect they resemble the mycoplasmas (∞ Section 16.3). *Thermoplasma* (**Figure 17.9*a***) is a chemoorganotroph that grows optimally at 55°C and pH 2 in complex media. Two species of *Thermoplasma* have been described, *Thermoplasma acidophilum* and *Thermoplasma volcanium*. Species of *Thermoplasma* are facultative aerobes, growing either aerobically or anaerobically by sulfur respiration (∞ Section 21.8). Most strains of *T. acidophilum* have been obtained from self-heating coal refuse piles. Coal refuse contains coal fragments, pyrite (FeS_2), and other organic materials extracted from coal. When dumped into piles in surface-mining operations, coal refuse heats by spontaneous combustion (**Figure 17.10**). This sets the stage for growth of *Thermoplasma*, which apparently metabolizes organic compounds leached from the hot coal refuse. A second species, *Thermoplasma volcanium*, has

Figure 17.10 A typical self-heating coal refuse pile, habitat of *Thermoplasma*. The pile containing coal debris, pyrite, and other microbial substrates, self-heats from microbial metabolism.

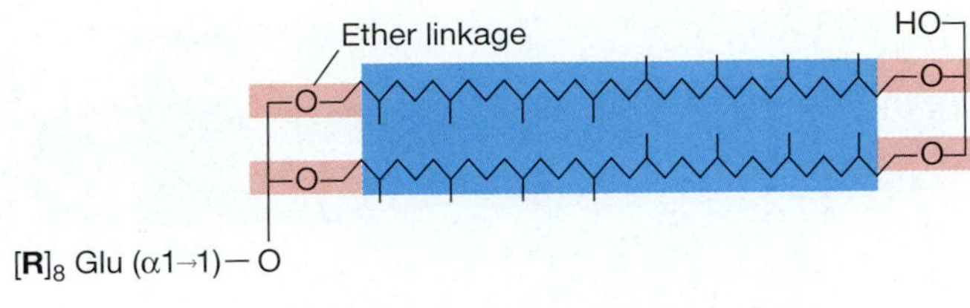

Figure 17.11 Structure of the tetraether lipoglycan of *Thermoplasma acidophilum*. Glu, Glucose; Man, mannose. Note the ether linkages (shown in pink) and the fact that this lipid would form a monolayer rather than a bilayer membrane (compare with Figure 4.8*b*).

been isolated in hot acidic soils throughout the world and is highly motile by multiple flagella (Figure 17.9*b*).

To survive the osmotic stresses of life without a cell wall and to withstand the dual environmental extremes of low pH and high temperature, *Thermoplasma* has evolved a unique cytoplasmic membrane structure. The membrane contains a lipopolysaccharide-like material called *lipoglycan* (Section 16.3). This substance consists of a tetraether lipid monolayer membrane with mannose and glucose (**Figure 17.11**). This molecule constitutes a major fraction of the total lipid composition of *Thermoplasma*. The membrane also contains glycoproteins but not sterols. These molecules render the *Thermoplasma* membrane stable to hot acidic conditions.

The genome of *Thermoplasma* is of interest. Like mycoplasmas (Sections 16.3 and 13.2), *Thermoplasma* contains a small genome (1.5 Mbp). In addition, *Thermoplasma* DNA is complexed with a highly basic DNA-binding protein that organizes the DNA into globular particles resembling the nucleosomes of eukaryotic cells (Section 8.5 discusses the arrangement of DNA in eukaryotes). This protein is homologous to the histone-like DNA-binding protein HU of *Bacteria*, which plays an important role in organization of the DNA in the cell. In contrast, several other *Euryarchaeota* contain basic proteins homologous to the DNA-binding histone proteins of eukaryotic cells (Section 17.14).

Ferroplasma

Ferroplasma is a chemolithotrophic relative of *Thermoplasma*. *Ferroplasma* is a strong acidophile; however, it is not a thermophile, as it grows optimally at 35°C. *Ferroplasma* oxidizes Fe^{2+} to Fe^{3+} to obtain energy (this reaction generates acid, see Figure 17.16*d*) and uses CO_2 as its carbon source (autotrophy).

Ferroplasma grows in mine tailings containing pyrite (FeS), which is its energy source. The extreme acidophily of *Ferroplasma* allows it to drive the pH of its habitat down to extremely acidic values. After moderate acidity is generated from Fe^{2+} oxidation by acidophilic organisms such as *Acidithiobacillus ferrooxidans* and *Leptospirillum ferrooxidans*, *Ferroplasma* becomes active and subsequently generates the very low pH values typical of acid mine drainage. Acidic waters at pH 0 can be generated by the activities of *Ferroplasma* (Figure 24.12 and Sections 24.5 and 24.6).

Picrophilus

A phylogenetic relative of *Thermoplasma* and *Ferroplasma* is *Picrophilus*. Although *Thermoplasma* and *Ferroplasma* are extreme acidophiles, *Picrophilus* is even more so, growing optimally at pH 0.7 and capable of growth at pH values lower than 0. *Picrophilus* also has a cell wall (an S-layer; Section 4.9) and a much lower DNA GC base ratio than does *Thermoplasma* or *Ferroplasma*. Although phylogenetically related, *Thermoplasma*, *Ferroplasma*, and *Picrophilus* have quite distinct genomes. Two species of *Picrophilus* have been isolated from acidic Japanese solfataras, and like *Thermoplasma*, both grow heterotrophically on complex media.

The physiology of *Picrophilus* is of interest as a model for extreme acid tolerance. Studies of its cytoplasmic membrane point to an unusual arrangement of lipids that forms a highly acid impermeable membrane at very low pH. By contrast, at moderate acidities such as pH 4, the membranes of cells of *Picrophilus* become leaky and disintegrate. Obviously, this organism has evolved to survive only in highly acidic habitats.

UNIT 3

17.5 MiniReview

Thermoplasma, *Ferroplasma*, and *Picrophilus* are extremely acidophilic thermophiles that form their own phylogenetic family of *Archaea* and inhabit coal refuse piles and highly acidic solfataras. Cells of *Thermoplasma* and *Ferroplasma* lack cell walls and thus resemble the mycoplasmas in this regard.

- In what ways are *Thermoplasma* and *Picrophilus* similar? In what ways do they differ?
- How does *Thermoplasma* strengthen its cytoplasmic membrane to survive life in the absence of a cell wall?
- How does *Ferroplasma* obtain energy for growth?

17.6 Thermococcales and Methanopyrus

Key Genera: *Thermococcus*, *Pyrococcus*, *Methanopyrus*

A few euryarchaeotes thrive in thermal environments, and some are hyperthermophiles. We consider here three hyperthermophilic euryarchaeotes that branch on the archaeal tree (Figure 17.1) very near the root. Two of these organisms, *Thermococcus* and *Pyrococcus*, show phenotypic properties very similar to those of hyperthermophilic crenarchaeotes (to be discussed in Sections 17.9–17.11). *Thermococcus* and *Pyrococcus* form a distinct order: the *Thermococcales* (Table 17.9). The third organism, *Methanopyrus*, is a methanogen that closely resembles other methanogens (Section 17.4 and Table 17.5) in its basic physiology but is unusual in its hyperthermophily and phylogenetic position (Figure 17.1).

Thermococcus and Pyrococcus

Thermococcus is a spherical hyperthermophilic euryarchaeote indigenous to anoxic thermal waters in various locations throughout the world. The spherical cells contain a tuft of

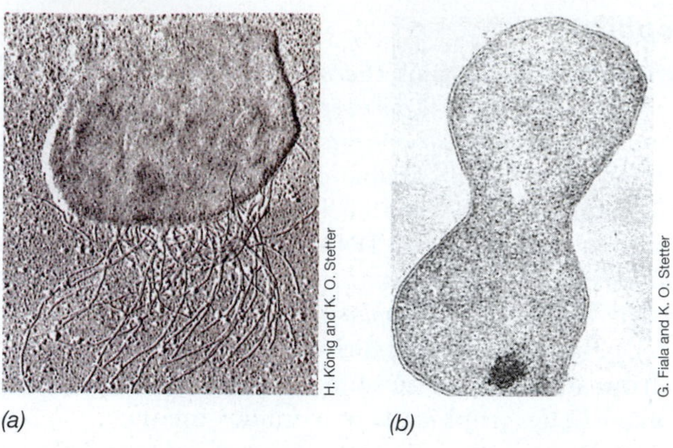

Figure 17.12 Spherical hyperthermophilic *Archaea* from submarine volcanic areas. *(a) Thermococcus celer.* Electron micrograph of shadowed cells (note tuft of flagella). *(b)* Dividing cell of *Pyrococcus furiosus*. Electron micrograph of thin section. Cells of both organisms are about 0.8 μm in diameter.

polar flagella and are thus highly motile (**Figure 17.12*a***). *Thermococcus* is an obligately anaerobic chemoorganotroph that metabolizes proteins and other complex organic mixtures (including some sugars) with S^0 as electron acceptor at temperatures from 55 to 95°C.

Pyrococcus is morphologically similar to *Thermococcus* (Figure 17.12*b*). *Pyrococcus* (the Latin "pyro" literally means "fire") differs from *Thermococcus* primarily by its higher temperature requirements; *Pyrococcus* grows between 70 and 106°C with an optimum of 100°C. *Thermococcus* and *Pyrococcus* are also metabolically quite similar. Proteins, starch, or maltose are oxidized as electron donors, and S^0 is the terminal acceptor and is reduced to H_2S. Both *Thermococcus* and *Pyrococcus* form H_2S when S^0 is present, but form H_2 when S^0 is absent. More properties of these organisms are given later in Tables 17.8 and 17.9.

Methanopyrus

Methanopyrus is a rod-shaped hyperthermophilic methanogen (**Figure 17.13**). *Methanopyrus* was isolated from hot sediments near submarine hydrothermal vents and from the walls of "black smoker" hydrothermal vent chimneys (Section 17.11; ∞ Section 24.11). The phylogenetic position of *Methanopyrus* is not entirely clear at this time. It appears to lie near the base of the archaeal tree (Figure 17.1), and it shares phenotypic properties with both the hyperthermophiles (the growth temperature maximum of *Methanopyrus* is 110°C) and the methanogens.

Methanopyrus only produces methane from $H_2 + CO_2$ and grows rapidly for an autotrophic organism (generation time less than 1 h at its temperature optimum of 100°C). Unlike mesophilic methanogens, however, cells of *Methanopyrus* contain large levels of the glycolysis derivative cyclic 2,3-diphosphoglycerate dissolved in the cytoplasm. This compound, present at more than 1 *M* concentration in *Methanopyrus*, is thought to function as a thermostabilizing agent to prevent denaturation of enzymes and DNA inside the cell (Section 17.14).

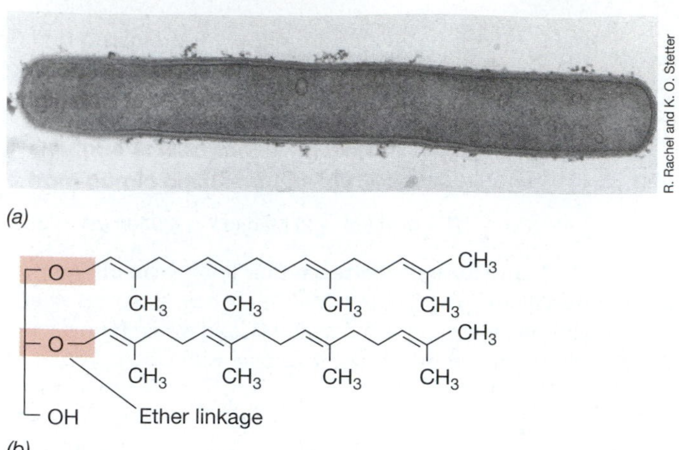

Figure 17.13 *Methanopyrus.* *Methanopyrus* grows optimally at 100°C and can make CH_4 only from $CO_2 + H_2$. *(a)* Electron micrograph of a cell of *Methanopyrus kandleri*, the most thermophilic of all known methanogens (upper temperature limit, 110°C). This cell measures 0.5 × 8 μm. *(b)* Structure of the novel lipid of *M. kandleri*. This is the normal ether-linked lipid of the *Archaea*, with the exception that the side chains are an unsaturated form of phytanyl called geranylgeraniol.

Methanopyrus is also unusual because it contains membrane lipids found in no other known organism. Recall that in the lipids of *Archaea*, the glycerol side chains contain **phytanyl** rather than fatty acids bonded in ether linkage to the glycerol (∞ Section 4.3). In *Methanopyrus*, this ether-linked lipid is an unsaturated form of the otherwise saturated dibiphytanyl tetraethers found in other hyperthermophilic *Archaea* (Figure 17.13*b*; ∞ Section 4.3 and Figure 4.8). From knowledge of the biochemistry of its biosynthetic pathway, this lipid is thought to be a primitive characteristic. The genome of *Methanopyrus kandleri* has been sequenced and found to consist of a single circular chromosome of approximately 1.7 Mbp that is similar in gene content and gene order to that of other methanogens. Along with its phylogenetic position, hyperthermophily, and anaerobic metabolism, the primitive lipids of *Methanopyrus* make this organism an intriguing model for one of Earth's earliest life forms (∞ Sections 14.2–14.4, and 17.15).

The discovery of *Methanopyrus* offers a possible explanation for the source of methane in hot oceanic sediments previously thought to be too hot to support biogenic methanogenesis. At the depth at which *Methanopyrus* was found—approximately 2000 m—water remains liquid at temperatures up to 350°C. This leaves open the possibility that other hyperthermophilic methanogens capable of growth at 110°C or at even higher temperatures also exist.

17.7 *Archaeoglobales*

Key Genera: *Archaeoglobus, Ferroglobus*

We will see later that a number of hyperthermophilic *Crenarchaeota* catalyze anaerobic respirations in which elemental sulfur (S^0) is used as an electron acceptor, being reduced to H_2S (see Table 17.8). Curiously, none of these crenarchaeotes are sulfate reducers, that is, capable of reducing SO_4^{2-} to H_2S during energy metabolism. However, one hyperthermophilic euryarchaeote, *Archaeoglobus*, can reduce SO_4^{2-} and forms a phylogenetically distinct lineage within the *Euryarchaeota* (Figure 17.1).

Archaeoglobus and Its Genome

Archaeoglobus was isolated from hot marine sediments near hydrothermal vents. In its metabolism, *Archaeoglobus* couples the oxidation of H_2, lactate, pyruvate, glucose, or complex organic compounds to the reduction of sulfate to sulfide. Cells of *Archaeoglobus* are irregular cocci (**Figure 17.14*a***), and cultures grow optimally at 83°C.

Archaeoglobus and methanogens share some characteristics. We will learn in Section 21.10 about the unique biochemistry of methanogenesis. Briefly, this process requires a series of novel coenzymes. With rare exceptions, these coenzymes have only been found in methanogens (Section 21.10). Surprisingly, however, *Archaeoglobus* also contains many of these coenzymes, and cultures of this organism can actually produce small amounts of methane. Thus, *Archaeoglobus*, which also shows a rather close phylogenetic relationship to methanogens (Figure 17.1), may be a metabolically intermediate type of organism, bridging the energy-conserving processes of methanogenesis and other forms of respiration.

The genome sequence of *Archaeoglobus* has revealed a conundrum concerning methanogenesis. Although *Archaeoglobus* can make methane, it lacks genes for a key enzyme of methanogenesis, methyl-CoM reductase (Section 21.10). Thus, the methane that is made by *Archaeoglobus* does not come from the activity of this key enzyme, and its source is unknown. The genome of *Archaeoglobus* contains about 2,400 genes and shares a number of genes with methanogens (Section 17.4). However, the *Archaeoglobus* genome also contains a number of genes unique to this organism, an example of the broad genetic diversity in *Archaea*. As in species of *Bacteria*, roughly one-third of the genes in every archaeal genome are unique, not present in any other organism (Section 13.2).

Ferroglobus

Ferroglobus (Figure 17.14*b*) is related to *Archaeoglobus* but is not a sulfate reducer. Instead, *Ferroglobus* is an iron-oxidizing chemolithotroph, conserving energy from the oxidation of Fe^{2+} to Fe^{3+} coupled to the reduction of NO_3^- to NO_2^- plus NO (see Table 17.8). *Ferroglobus* grows autotrophically and can also use H_2 or H_2S as electron donors in its energy metabolism. *Ferroglobus* was isolated from a shallow marine hydrothermal vent and grows optimally at 85°C.

Ferroglobus is interesting for several reasons, but especially for its ability to oxidize Fe^{2+} to Fe^{3+} under anoxic conditions. The process might help explain the abundance of Fe^{3+} in ancient rocks, such as the banded iron formations (Section 14.3) previously thought to have originated from the oxidation of Fe^{2+} by O_2 produced by cyanobacteria (oxygenic photosynthesis). The metabolism of *Ferroglobus* thus has implications for dating the origin of cyanobacteria and the subsequent oxygenation of Earth. Certain anoxygenic phototrophic bacteria can also oxidize Fe^{2+} under anoxic conditions (Section 20.11). So, several anaerobic routes to ancient Fe^{3+} obviously exist, making it a challenge to accurately estimate when cyanobacteria first evolved (Figure 14.7).

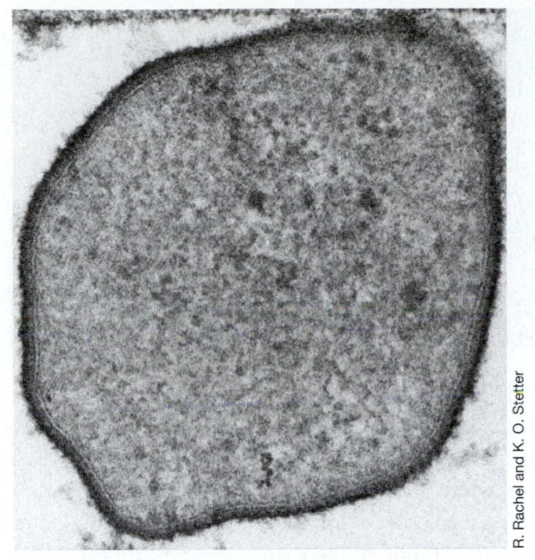

(a)

R. Rachel and K. O. Stetter

(b)

Figure 17.14 Archaeoglobales. *(a)* Transmission electron micrograph of the sulfate-reducing hyperthermophile *Archaeoglobus fulgidus*. The cell measures 0.7 μm in diameter. *(b)* Freeze-etched electron micrograph of *Ferroglobus placidus*, a ferrous iron-oxidizing, nitrate-reducing hyperthermophile. The cell measures about 0.8 μm in diameter.

17.8 *Nanoarchaeum* and *Aciduliprofundum*

Key Genera: *Nanoarchaeum and Aciduliprofundum*

Two other members of *Euryarchaeota* deserve mention here, *Nanoarchaeum* and *Aciduliprofundum*. *Nanoarchaeum equitans* is a very unusual prokaryote. As one of the smallest cellular organisms known and with one of the smallest genomes of any cell, it lives as an obligate symbiont of the crenarchaeote *Ignicoccus* (Section 17.11). The coccoid cells of *Nanoarchaeum* are very small, about 0.4 μm in diameter, having only about 1% of the volume of an *Escherichia coli* cell. They cannot grow in pure culture and replicate only when

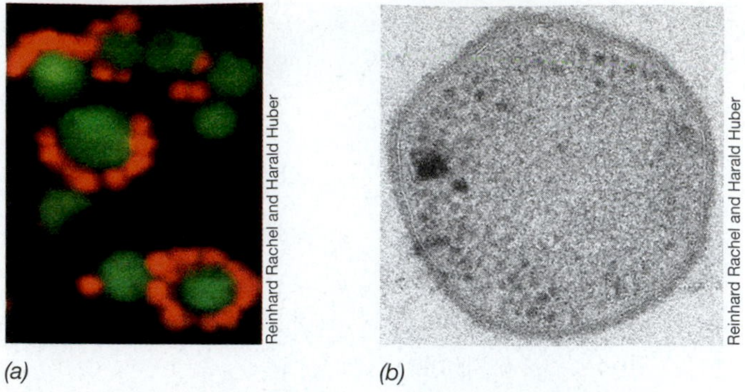

Figure 17.15 Nanoarchaeum. (a) Fluorescence micrograph of cells of *Nanoarchaeum* (red) attached to cells of *Ignicoccus* (green). Cells were stained by fluorescent *in-situ* hybridization (Section 14.9) using specific nucleic acid probes targeted to each organism. (b) Transmission electron micrograph of a thin section of a cell of *Nanoarchaeum*. Note the distinct cell wall. Cells of *Nanoarchaeum* are about 0.4 μm in diameter.

attached to the surface of *Ignicoccus* cells in densities up to 10 or more cells per *Ignicoccus* cell (**Figure 17.15a**). Whether large numbers of associated *Nanoarchaeum* cells might slow the growth of *Ignicoccus* and thereby act as a parasite of its host is not yet known. Thus, it remains to be seen if this is the first parasitic association identified in *Archaea*. Regardless, the appearance of *Nanoarchaeum* cells is typical of *Archaea*, with a cell wall consisting of an S-layer (Section 4.9) that overlays what appears to be a periplasmic space (Figure 17.15*b*).

N. equitans and its host *Ignicoccus* were first isolated from a submarine hydrothermal vent off the coast of Iceland. Environmental sampling of 16S rRNA genes (Section 22.5) indicates that there are organisms phylogenetically similar to *Nanoarchaeum* in other submarine hydrothermal vents and terrestrial hot springs; *Archaea* of this kind are probably distributed worldwide in suitable hot habitats. Like its host *Ignicoccus*, *Nanoarchaeum* grows at temperatures ranging from 70 to 98°C and optimally at about 90°C.

The metabolism of *Nanoarchaeum* is not yet well understood, but it appears to depend on its host for many of its metabolic functions; *Ignicoccus* is an autotroph that uses H_2 as electron donor and S^0 as electron acceptor. For example, *Nanoarchaeum* lacks the ability to metabolize hydrogen and sulfur for energy, and whether it generates ATP from compounds obtained from *Ignicoccus* or obtains its ATP directly from its host is not yet known.

Phylogeny and Genomics of *Nanoarchaeum*

Although the sequence of its 16S rRNA gene clearly places *Nanoarchaeum* in the domain *Archaea*, the sequence differs at many sites from 16S rRNA gene sequences of other *Archaea*, even in regions of the molecule that are highly conserved among *Archaea*. These differences led initially to placing *Nanoarchaeum* in a separate, new phylum. However, more detailed phylogenetic analyses, using genes for ribosomal proteins and other key cellular proteins, suggest that *Nanoarchaeum* is most likely a genus of *Euryarchaeota* and most closely related to members of the *Thermococcales* (Figure 17.1).

The sequence of the *N. equitans* genome provides insight into this organism's obligately symbiotic lifestyle. Its single, circular genome is only 490,885 nucleotides long, one of the smallest nonviral genomes yet sequenced. *Escherichia coli*, a member of the *Gammaproteobacteria*, for example, has a genome size more than nine times larger, at 4,639,000 nucleotides. Genes for several important metabolic functions appear to be missing in the *N. equitans* genome, including those for the biosynthesis of amino acids, nucleotides, coenzymes, and lipids. Also missing are genes encoding proteins for catabolic pathways, such as glycolysis. Presumably, all of these functions are carried out for *Nanoarchaeum* by its *Ignicoccus* host, with transfer of needed substances from *Ignicoccus* to the associated cells of *Nanoarchaeum*. *Nanoarchaeum* also contains some but not all of the genes necessary to encode ATPase, which suggests that it may not synthesize a functional ATPase. If no ATPase is present and substrate-level phosphorylation does not occur (due to the lack of glycolytic enzymes), then *Nanoarchaeum* may be dependent on its *Ignicoccus* host for both ATP and carbon. Like obligate intracellular parasitic or symbiotic *Bacteria*, such as *Rickettsia* and *Chlamydia*, which infect human hosts, *Nanoarchaeum* appears to be dependent on *Ignicoccus* for most of its metabolic needs.

With so many genes missing, which genes remain in the *Nanoarchaeum* genome? *Nanoarchaeum* contains recognizable genes for the key enzymes for DNA replication, transcription, and translation, as well as genes for DNA repair enzymes. In addition to its small size, the genome of *Nanoarchaeum* is among the most compact (gene dense) of any organism known; over 95% of the *Nanoarchaeum* chromosome encodes proteins, 552 proteins compared to 4,243 for *E. coli*. Clearly, *Nanoarchaeum* is living near limits of life in terms of cell volume and genetic capacity. The only known organism with a smaller genome and fewer genes than *Nanoarchaeum* is *Carsonella*, a bacterial endosymbiont of certain insects whose genome is only 159,000 nucleotides and encodes only 182 proteins.

Genome reduction, common in obligately parasitic or endosymbiotic organisms (Section 13.4), can arise by the elimination of unneeded genes during adaptation to and genetic integration with a host over evolutionary time. At present, the most likely scenario for the evolution of *Nanoarchaeum* is that it once was a free-living hyperthermophilic species of *Euryarchaeota* that entered a parasitic lifestyle and has since undergone an extensive genome reduction as it integrated genetically with its *Ignicoccus* host. An interesting and related question is whether free-living relatives of *Nanoarchaeum* exist. Thus far, none have been isolated. But since community sampling of SSU rRNA genes has shown that extensive uncultured archaeal diversity exists in high-temperature habitats, the eventual discovery of free-living *Nanoarchaeum* species would not be surprising.

Aciduliprofundum

Another remarkable *Euryarchaeote* is *Aciduliprofundum*. This thermophile has a growth temperature range of 55–75°C. It is particularly notable because it is an acidophile, with a pH range for growth of 3.3–5.8, living in sulfide deposits in hydrothermal vents. Before its discovery, only neutrophilic or acid-tolerant microorganisms had been isolated from these deposits, which had been predicted to provide a suitable habitat for acidophiles. *Aciduliprofundum*, a chemoorganotroph that uses elemental sulfur or ferric iron as electron acceptors, constitutes up to 15% of the *Archaea* present in these deposits, and it therefore may play a major role in the sulfur and iron cycles in these deep-sea hydrothermal vent sulfur deposits.

17.6–17.8 MiniReview

Hyperthermophilic *Euryarchaeota* include the *Thermococcales*, *Archaeoglobales*, *Methanopyrus*, and *Nanoarchaeum* and *Aciduliprofundum*. All organisms in this group have growth temperature optima above 80°C. *Nanoarchaeum* is a small, parasitic, member of the *Archaea*. Its genome is one of the smallest of all known organisms. *Nanoarchaeum* lacks genes for all but core molecular processes and thus depends on its host, *Ignicoccus*, for most of its cellular needs.

- How does the metabolism of *Thermococcus* differ from that of *Methanopyrus*?
- Which of the hyperthermophilic *Euryarchaeota* is a true sulfate reducer?
- How does the metabolism of *Ferroglobus* differ from that of *Archaeoglobus*?
- What aspects of *Nanoarchaeum* make it especially interesting from an evolutionary perspective?
- How do we know that *Nanoarchaeum* is a hyperthermophile?

III CRENARCHAEOTA

Crenarchaeotes are phylogenetically distinct from euryarchaeotes (Figure 17.1) and inhabit both ends of nature's temperature extremes: boiling water and freezing water. Most cultured crenarchaeotes are hyperthermophiles and some actually have optima above the boiling point of water. We start with an overview of the habitats and energy metabolism of crenarchaeotes and describe the properties of some key thermophilic genera. We then conclude this section with a discussion of *Crenarchaeota* from temperate and cold environments.

17.9 Habitats and Energy Metabolism of *Crenarchaeota*

A summary of the *Crenarchaeota* is shown in **Table 17.7.** Most hyperthermophilic *Archaea* have been isolated from geothermally heated soils or waters containing elemental sulfur and sulfides, and most species metabolize sulfur in one way or another. In terrestrial environments, sulfur-rich springs, boiling mud, and soils may have temperatures up to 100°C and are mildly to extremely acidic owing to the production of sulfuric acid (H_2SO_4) from the biological oxidation of H_2S and S^0 (∞ Sections 20.10 and 24.4). Such hot, sulfur-rich environments, called **solfataras**, are found throughout the world (**Figure 17.16**), including Italy, Iceland, New Zealand, and Yellowstone National Park in Wyoming (USA).

Depending on the surrounding geology, solfataras can be mildly acidic to slightly alkaline (pH 5–8) or extremely acidic, with pH values below 1. Hyperthermophilic crenarchaeotes have been obtained from both acidic and alkaline environments, but the majority of these organisms inhabit neutral or mildly acidic hot habitats. In addition to these natural habitats, hyperthermophilic *Archaea* also thrive within artificially heated environments, in particular the boiling outflows of geothermal power plants.

Hyperthermophilic *Crenarchaeota* also inhabit undersea hot springs called **hydrothermal vents**. We discuss the geology and microbiology of these habitats in Section 24.11. Here it is only necessary to note that submarine waters can be much hotter than surface waters because the water is under pressure. Indeed, all hyperthermophiles with growth temperature optima above 100°C have come from submarine sources. The latter include both shallow (2–10 m depth) vents such as those off the coast of Vulcano, Italy, to deep (2,000–4,000 m depth) vents near ocean-spreading centers (∞ Section 23.9). Deep hydrothermal vents are the hottest habitats so far known to yield prokaryotes.

Table 17.7 Habitats of *Crenarchaeota*

	Thermal area[a]		
Characteristic	***Terrestrial***	***Marine***	***Nonthermal area***[b]
Locations	Solfataras (hot springs, fumaroles, mudpots, steam-heated soils); geothermal power plants; deep in Earth's crust	Submarine hot springs, hot sediments and vents ("black smokers"); deep oil reservoirs	Planktonic in oceans worldwide; near-shore and deep Antarctic waters; sea ice; symbionts of marine sponges
Temperature	Surface to 100°C; subsurface, above 100°C	Up to 400°C (smokers)	−2 to +4°C
Salinity/pH	Usually less than 1% NaCl; pH 0.5–9	Moderate, about 3% NaCl; pH 5–9	3–8% NaCl; pH 7–9
Gases and other nutrients	CO_2, CO, CH_4, H_2, H_2S, S^0, $S_2O_3^{2-}$, SO_4^{2-}, NH_4^+, N_2	Same as for terrestrial	CO_2, N_2, O_2; for chemolithotrophs, inorganic substrates such as NH_4^+

[a]See Figures 17.10 and 17.16 (∞ Figures 24.32 and 24.33).
[b]See Figure 17.22. Some species of *Euryarchaeota* are also present in marine waters.

UNIT 3

(a)

(b)

(c)

(d)

Figure 17.16 Habitats of hyperthermophilic *Archaea*. *(a)* A typical solfatara in Yellowstone National Park. Steam rich in hydrogen sulfide rises to the surface of the earth. Because of the heat and acidity, only prokaryotes are present. *(b)* Sulfur-rich hot spring, a habitat containing dense populations of *Sulfolobus*. The acidity in solfataras and sulfur springs comes from the oxidation of H_2S and S^0 to H_2SO_4 (sulfuric acid) by *Sulfolobus* and related prokaryotes. *(c)* A typical boiling spring of neutral pH in Yellowstone Park; Imperial Geyser. Many different species of hyperthermophilic *Archaea* may reside in such a habitat. *(d)* An acidic iron-rich geothermal spring, another *Sulfolobus* habitat. Here the oxidation of Fe^{2+} to Fe^{3+} (rather than S^0 to SO_4^{2-}) causes acidic conditions ($Fe^{3+} + 3\,H_2O \rightarrow Fe(OH)_3 + 3\,H^+$).

Energy Metabolism

With a few exceptions, hyperthermophilic *Crenarchaeota* are obligate anaerobes. Their energy-yielding metabolism is either chemoorganotrophic or chemolithotrophic (or both, for example, in *Sulfolobus*) and is dependent on diverse electron donors and acceptors (**Table 17.8**). Fermentation is rare, and most bioenergetic strategies involve anaerobic respirations (Table 17.8). Energy is conserved during these respiratory processes by the same general mechanism widespread in *Bacteria*: electron transfer within the cytoplasmic membrane leading to the formation of a proton motive force from which ATP is made by way of proton-translocating ATPases (∞ Section 5.12).

Many hyperthermophilic crenarchaeotes can grow chemolithotrophically under anoxic conditions, with H_2 as electron donor and S^0 or NO_3^- as electron acceptor; a few can also oxidize H_2 aerobically (Table 17.8). H_2 respiration with ferric iron (Fe^{3+}) as electron acceptor is the process in several hyperthermophiles (Table 17.8). Other chemolithotrophic lifestyles include the oxidation of S^0 and Fe^{2+} aerobically or Fe^{2+} anaerobically with NO_3^- as acceptor (Table 17.8). Only one sulfate-reducing hyperthermophile is known (the euryarchaeote *Archaeoglobus*, Section 17.7). The only bioenergetic option not possible is photosynthesis (the most thermophilic phototrophic organism known can grow up to 73°C (∞ Table 6.1, and see Figure 17.26), far too cold for most hyperthermophiles (**Table 17.9**).

With this basic understanding of the habitats and energy metabolism of hyperthermophilic crenarchaeotes, let us now look at some representative organisms.

17.10 Hyperthermophiles from Terrestrial Volcanic Habitats

Key Genera: *Sulfolobus, Acidianus, Thermoproteus, Pyrobaculum*

Terrestrial volcanic habitats can have temperatures as high as 100°C and are thus suitable for hyperthermophilic *Archaea*. Two phylogenetically related organisms isolated from these environments are *Sulfolobus* and *Acidianus*. These genera form the heart of an order called the *Sulfolobales* (Table 17.9).

Table 17.8 Energy-yielding reactions of hyperthermophilic *Archaea*

Nutritional class	*Energy-yielding reaction*	*Metabolic[a] type*	*Example*
Chemoorganotrophic	Organic compound + $S^0 \rightarrow H_2S + CO_2$	AnR	*Thermoproteus, Thermococcus, Desulfurococcus, Thermofilum, Pyrococcus*
	Organic compound + $SO_4^{2-} \rightarrow H_2S + CO_2$	AnR	*Archaeoglobus*
	Organic compound + $O_2 \rightarrow H_2O + CO_2$	AeR	*Sulfolobus*
	Organic compound $\rightarrow CO_2 + H_2$ + fatty acids	AnR	*Staphylothermus, Pyrodictium*
	Organic compound + $Fe^{3+} \rightarrow CO_2 + Fe^{2+}$	AnR	*Pyrodictium*
	Pyruvate $\rightarrow CO_2 + H_2$ + acetate	AnR	*Pyrococcus*
	Peptides ($H_2 + S^0 \rightarrow H_2S$ stimulates growth)	F	*Hyperthermus*
Chemolithotrophic	$H_2 + S^0 \rightarrow H_2S$	AnR	*Acidianus, Pyrodictium, Thermoproteus, Stygiolobus, Ignicoccus*
	$H_2 + NO_3^- \rightarrow NO_2^- + H_2O$ (NO_2^- is reduced to N_2 by some species)	AnR	*Pyrobaculum*
	$4\,H_2 + NO_3^- + H^+ \rightarrow NH_4^+ + 2\,H_2O + OH^-$	AnR	*Pyrolobus*
	$H_2 + 2\,Fe^{3+} \rightarrow 2\,Fe^{2+} + 2\,H^+$	AnR	*Pyrobaculum, Pyrodictium, Archaeoglobus*
	$2\,H_2 + O_2 \rightarrow 2\,H_2O$	AeR	*Acidianus, Sulfolobus, Pyrobaculum*
	$2\,S^0 + 3\,O_2 + 2\,H_2O \rightarrow 2\,H_2SO_4$	AeR	*Sulfolobus, Acidianus*
	$2\,FeS_2 + 7\,O_2 + 2\,H_2O \rightarrow 2\,FeSO_4 + 2\,H_2SO_4$	AeR	*Sulfolobus, Acidianus, Metallosphaera*
	$2\,FeCO_3 + NO_3^- + 6\,H_2O \rightarrow 2\,Fe(OH)_3 + NO_2^- + 2\,HCO_3^- + 2\,H^+ + H_2O$	AnR	*Ferroglobus*
	$4\,H_2 + SO_4^{2-} + 2\,H^+ \rightarrow 4\,H_2O + H_2S$	AnR	*Archaeoglobus*
	$4\,H_2 + CO_2 \rightarrow CH_4 + 2\,H_2O$	AnR	*Methanopyrus, Methanocaldococcus, Methanothermus*

[a]AnR, anaerobic respiration; AeR, aerobic respiration; F, fermentation.

Sulfolobales

Sulfolobus grows in sulfur-rich acidic hot springs (Figure 17.16) at temperatures up to 90°C and at pH values of 1–5. *Sulfolobus* is an aerobic chemolithotroph that oxidizes H_2S or S^0 to H_2SO_4 and fixes CO_2 as carbon source. *Sulfolobus* can also grow chemoorganotrophically. Cells of *Sulfolobus* are more or less spherical but contain distinct lobes (**Figure 17.17a**). Cells adhere tightly to sulfur crystals where they can be seen with a microscope after preparation with fluorescent dyes (∞ Figure 20.26*b*). Besides the aerobic respiration of sulfur or organic compounds, *Sulfolobus* can also oxidize Fe^{2+} to Fe^{3+}, and this has been applied in the high-temperature leaching of iron and copper ores (∞ Section 24.6).

A facultative aerobe resembling *Sulfolobus* also lives in acidic solfataric springs. This organism, named *Acidianus* (Figure 17.17*b*), differs from *Sulfolobus* most clearly by its ability to grow using elemental sulfur both anaerobically as well as aerobically. Under aerobic conditions the organism uses S^0 as an electron *donor*, oxidizing S^0 to H_2SO_4, with O_2 as electron acceptor. Anaerobically, *Acidianus* uses S^0 as an electron *acceptor* with H_2 as electron donor, forming H_2S as the reduced product. Thus, the metabolic fate of S^0 in cultures of *Acidianus* depends on the presence or absence of O_2. Like *Sulfolobus, Acidianus* is roughly spherical in shape but is not as lobed (Figure 17.17*b*). It grows at temperatures from 65°C up to a maximum of 95°C, with an optimum of about 90°C.

Thermoproteales

Key genera within the Thermoproteales are *Thermoproteus*, *Thermofilum*, and *Pyrobaculum*. The genera *Thermoproteus* and *Thermofilum* consist of rod-shaped cells that inhabit neutral or slightly acidic hot springs. Cells of *Thermoproteus* are rigid rods about 0.5 μm in diameter and highly variable in length, ranging from short cells of 1–2 μm (**Figure 17.18a**) up to filaments 70–80 μm long. Filaments of *Thermofilum* are thinner, some 0.17–0.35 μm wide, with filament lengths ranging up to 100 μm (Figure 17.18*b*).

Both *Thermoproteus* and *Thermofilum* are strict anaerobes that carry out an S^0-based anaerobic respiration (Table 17.8). Unlike most hyperthermophiles, the oxygen sensitivity of *Thermoproteus* and *Thermofilum* is extreme, comparable to that of the methanogens (Section 17.4). Thus, strict precautions must be taken in their culture. Most *Thermoproteus* isolates can grow chemolithotrophically on H_2 or chemoorganotrophically on complex carbon substrates such as yeast extract, small peptides, starch, glucose, ethanol, malate, fumarate, or formate (Table 17.8). *Thermoproteus* and *Thermofilum* have similar GC base ratios (56–58% GC), but their genomes are distinct as determined by nucleic acid hybridization analyses (∞ Section 14.11).

Pyrobaculum (Figure 17.18*c*) is a rod-shaped hyperthermophile but is physiologically unique from other *Thermoproteales* in at least one sense: Some species of

Table 17.9 Properties of some hyperthermophilic *Crenarchaeota*

Order/Genus[a]	Morphology	Relationship to O_2[b]	Temperature Minimum	Temperature Optimum	Temperature Maximum	Optimum pH
Sulfolobales						
Sulfolobus	Lobed coccus	Ae	55	75	87	2–3
Acidianus	Coccus	Fac	60	88	95	2
Metallosphaera	Coccus	Ae	50	75	80	2
Stygiolobus	Lobed coccus	An	57	80	89	3
Sulfurisphaera	Coccus	Fac	63	84	92	2
Sulfurococcus	Coccus	Ae	40	75	85	2.5
Thermoproteales						
Thermoproteus	Rod	An	60	88	96	6
Thermofilum	Rod	An	70	88	95	5.5
Pyrobaculum	Rod	Fac	74	100	102	6
Caldivirga	Rod	An	60	85	92	4
Thermocladium	Rod	An	60	75	80	4.2
Desulfurococcales						
Desulfurococcus	Coccus	An	70	85	95	6
Aeropyrum	Coccus	Ae	70	95	100	7
Staphylothermus	Cocci in clusters	An	65	92	98	6–7
Pyrodictium	Disc-shaped with filaments	An	82	105	110[c]	6
Pyrolobus	Lobed coccus	Fac	90	106	113	5.5
Thermodiscus	Disc-shaped	An	75	90	98	5.5
Ignicoccus	Irregular coccus	An	65	90	103	5
Hyperthermus	Irregular coccus	An	75	102	108	7
Stetteria	Coccus	An	68	95	102	6
Sulfophobococcus	Disc-shaped	An	70	85	95	7.5
Thermosphaera	Coccus	An	67	85	90	7
Strain 121	Coccus	An	85	106	121	7
Archaeoglobales[c]						
Archaeoglobus	Coccus	An	64	83	95	7
Ferroglobus	Irregular coccus	An	65	85	95	7
Thermococcales[d]						
Thermococcus	Coccus	An	70	88	98	6–7
Pyrococcus	Coccus	An	70	100	106	6–8

[a]The group names ending in "ales" are order names (Section 14.14).
[b]Ae, aerobe; An, anaerobe; Fac, facultative.
[c]One species may grow up to 121°C.
[d]Phylogenetically, genera in this order of hyperthermophiles are members of the *Euryarchaeota* (see Sections 17.6 and 17.7).

Pyrobaculum are capable of aerobic respiration. In addition, cultures of *Pyrobaculum* can carry out anaerobic respirations with NO_3^-, Fe^{3+}, or S^0 as electron acceptors and H_2 as electron donor (that is, they can grow chemolithotrophically and autotrophically). Other species of *Pyrobaculum* can grow anaerobically on organic electron donors, reducing S^0 to H_2S. The growth temperature optimum of *Pyrobaculum* is 100°C, and species of this organism have been isolated from terrestrial hot springs and from hydrothermal vents.

17.11 Hyperthermophiles from Submarine Volcanic Habitats

Key Genera: *Pyrodictium, Pyrolobus, Ignicoccus, Staphylothermus*

We now turn to the microbiology of submarine volcanic habitats, homes to the most thermophilic of all known *Archaea*. These habitats include both shallow water thermal springs and deep-sea hydrothermal vents. We discuss the geology of these fascinating microbial habitats in Section 24.11. The

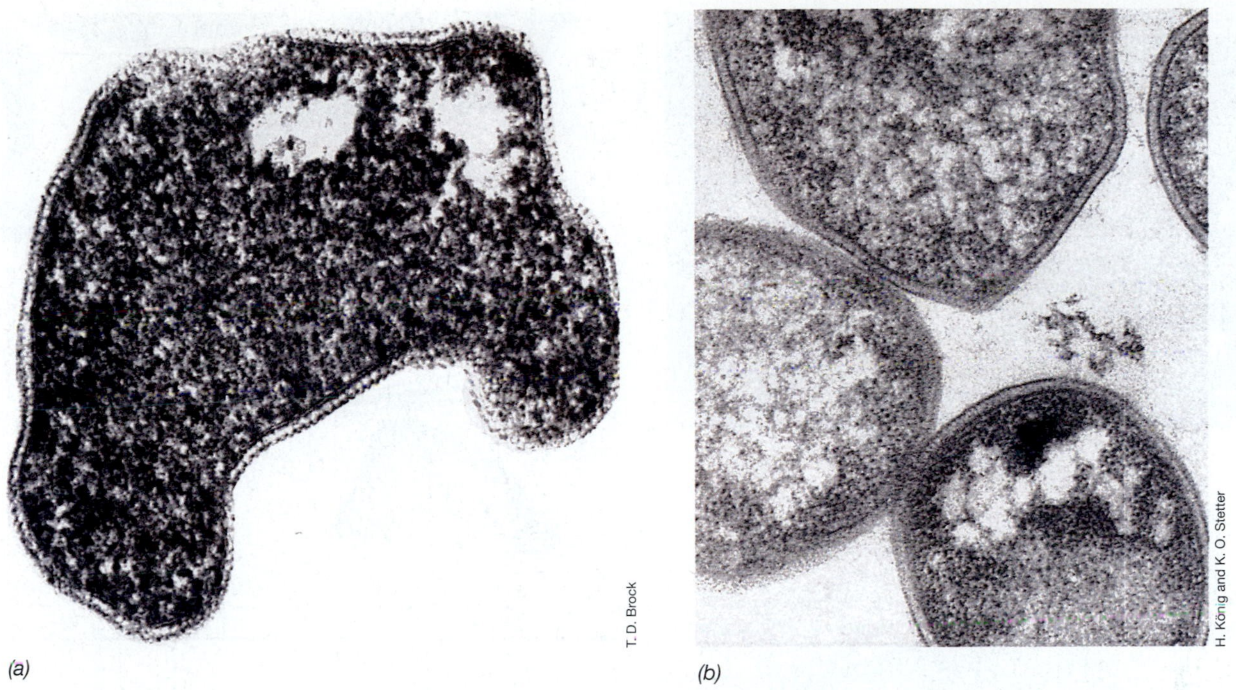

Figure 17.17 **Acidophilic hyperthermophilic *Archaea*, the *Sulfolobales*.** (*a*) *Sulfolobus acidocaldarius*. Electron micrograph of a thin section. (*b*) *Acidianus infernus*. Electron micrograph of a thin section. Cells of both organisms vary from 0.8 to 2 μm in diameter.

organisms to be described here constitute an order of *Archaea* called the *Desulfurococcales* (Table 17.9).

Pyrodictium and *Pyrolobus*

Pyrodictium and *Pyrolobus* are examples of microorganisms whose growth temperature optimum lies above 100°C; the optimum for *Pyrodictium* is 105°C and for *Pyrolobus* is 106°C. Cells of *Pyrodictium* are irregularly disc-shaped and grow in culture in a mycelium-like layer attached to crystals of elemental sulfur. The cell mass consists of a network of fibers to which individual cells are attached (**Figure 17.19*a,b***). The fibers are hollow and consist of protein arranged in a fashion similar to that of bacterial flagella (∞ Section 4.13). However, the filaments do not function in motility but instead as organs of attachment. The cell walls of *Pyrodictium* are composed of glycoprotein. Physiologically, *Pyrodictium* is a strict anaerobe that grows chemolithotrophically on H_2 with S^0 as electron acceptor or chemoorganotrophically on complex mixtures of organic compounds (Table 17.8).

Pyrolobus fumarii (Figure 17.19*c*) is one of the most thermophilic of the hyperthermophiles. Its growth temperature maximum is 113°C (Table 17.9). *Pyrolobus* lives in the walls of "black smoker" hydrothermal vent chimneys (∞ Section 24.11 and Figure 24.33), where its autotrophic abilities contribute organic carbon to this otherwise inorganic environment. *Pyrolobus* cells are coccoid-shaped (Figure 17.19*c*), and the cell wall is composed of protein. The organism is an obligate H_2 chemolithotroph, growing by the oxidation of H_2 coupled to the reduction of NO_3^- (to NH_4^+),

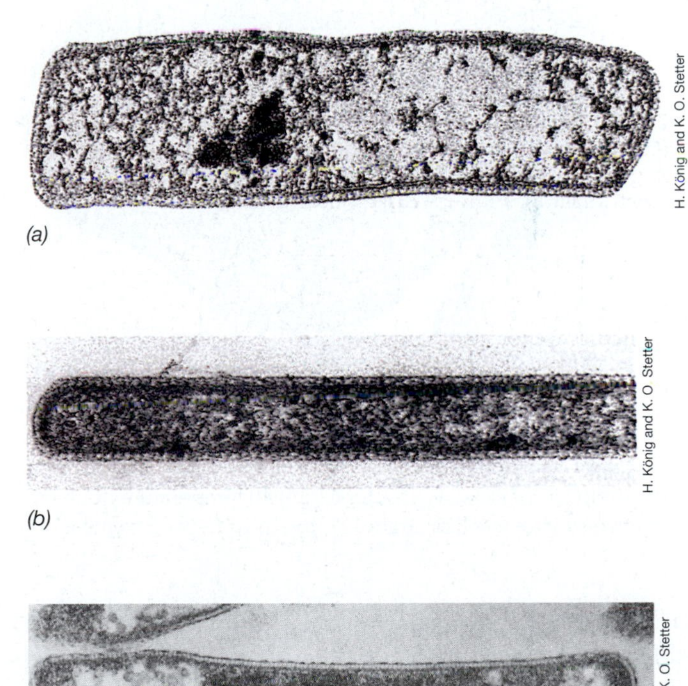

Figure 17.18 **Rod-shaped hyperthermophilic *Archaea*, the *Thermoproteales*.** (*a*) *Thermoproteus neutrophilus*. Electron micrograph of a thin section. A cell is about 0.5 μm in diameter. (*b*) *Thermofilum librum*. A cell is about 0.25 μm in diameter. (*c*) *Pyrobaculum aerophilum*. Transmission electron micrograph of a thin section; a cell measures 0.5 × 3.5 μm.

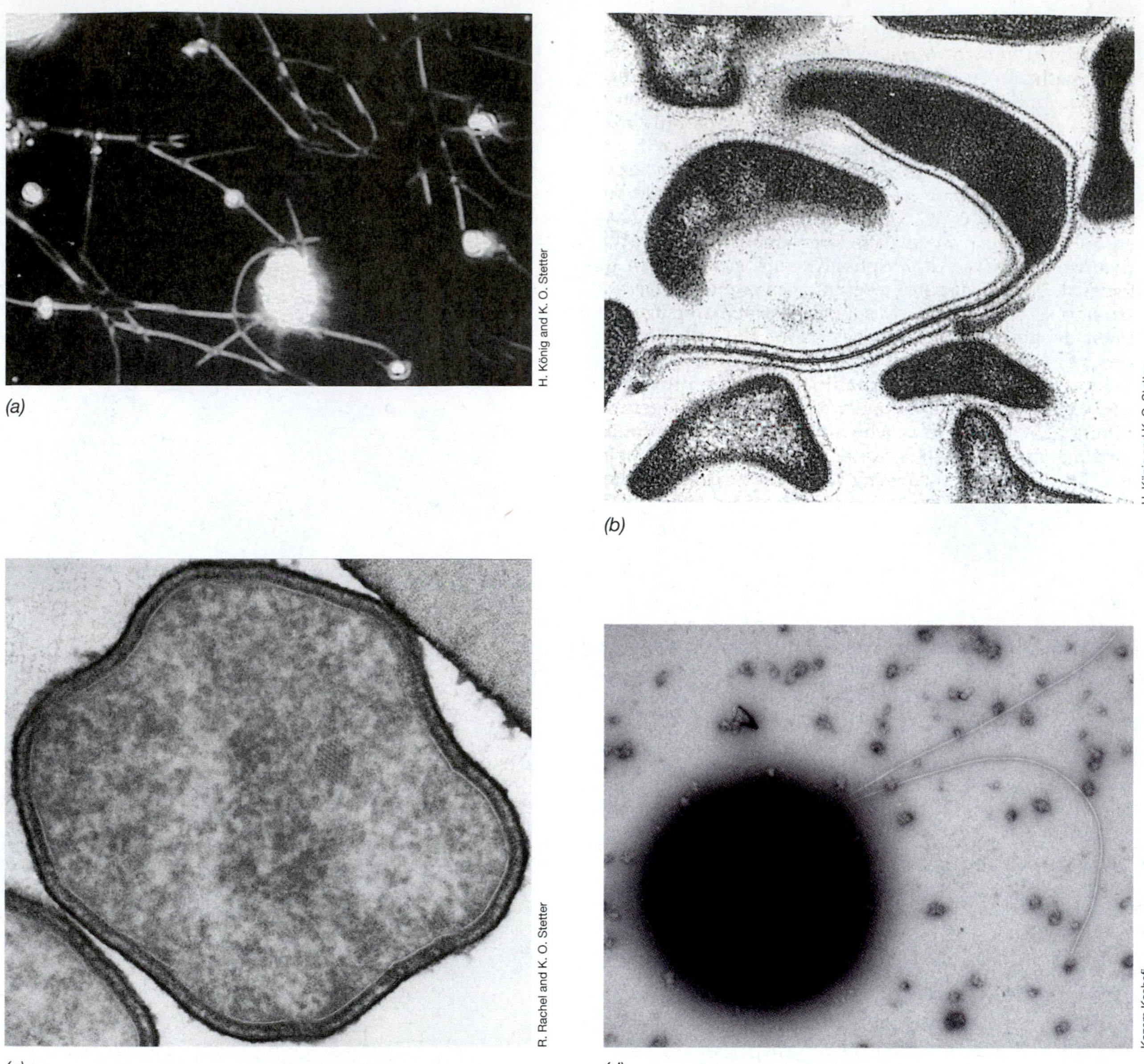

Figure 17.19 ***Desulfurococcales*** **with growth temperature optima >100°C.** *(a) Pyrodictium occultum* (growth temperature optimum, 105°C), dark-field micrograph. *(b)* Thin-section electron micrograph of *P. occultum.* Cells are highly variable in diameter from 0.3 to 2.5 μm. *(c)* Thin section of a cell of *Pyrolobus fumarii*, one of the most thermophilic of all known bacteria (growth temperature optimum, 106°C); a cell is about 1.4 μm in diameter. *(d)* Negative stain of a cell of strain 121, the most heat-loving of all known *Archaea*; a cell is about 1 μm wide.

$S_2O_3^{2-}$ (to H_2S), or very low concentrations of O_2 (to H_2O). Besides its extremely thermophilic nature, *Pyrolobus* is also resistant to temperatures substantially above its growth temperature maximum. For example, cultures of *P. fumarii* survive autoclaving (121°C) for 1 h, a condition that even bacterial endospores (Section 4.12) cannot withstand.

Strain 121

An organism related to *Pyrodictium* and given the provisional name strain 121 shares with *Pyrolobus* a growth temperature optimum of 106°C. However, strain 121 can grow up to 121°C, the temperature at which microbiological materials and media typically are sterilized by autoclaving, (Section 27.1). Strain 121 is thus the most thermophilic prokaryote in laboratory culture. Remarkably, cells of strain 121 survive even temperatures higher than 121°C, remaining viable after 2 h at 130°C.

Strain 121, which consists of coccoid, lophotrichously flagellated cells (Figure 17.19*d*), is a strict anaerobe and was isolated from a hydrothermal vent in the northeast Pacific Ocean using Fe^{3+} as the electron acceptor. Strain 121 grows chemolithotrophically and autotrophically, with Fe^{3+} oxide as electron acceptor and formate or H_2 as electron donor. Phylogenetic analysis based on the SSU rRNA gene sequencing places strain 121 in the family *Pyrodictiaceae*, along with *Pyrodictium* and *Pyrolobus*, probably as a new genus of this

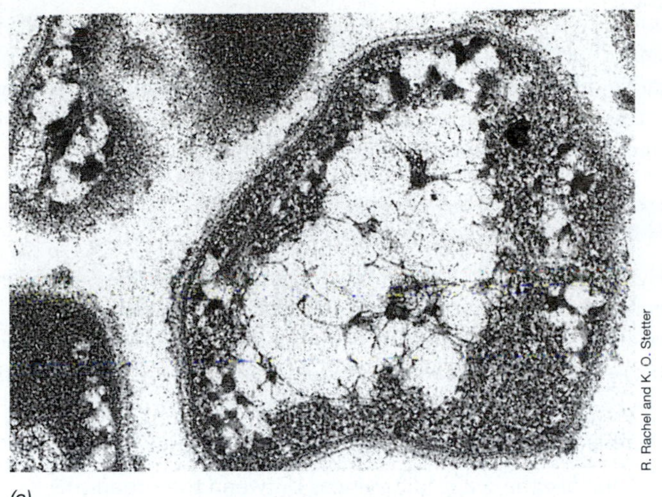

(a)

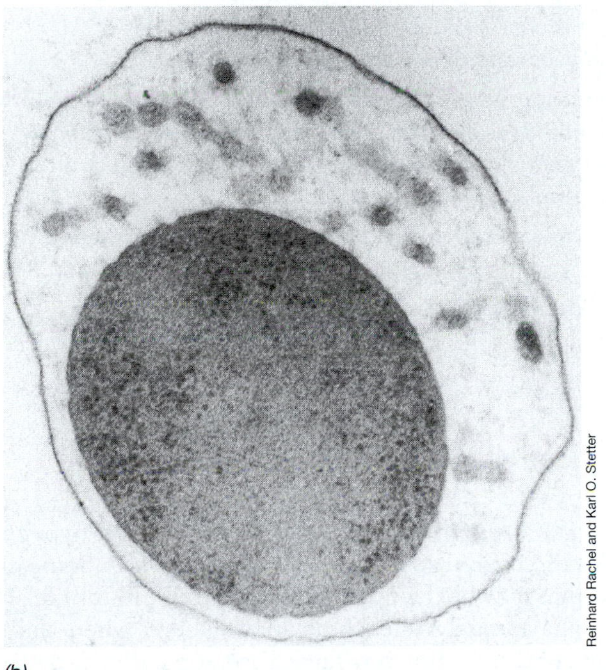

(b)

Figure 17.20 ***Desulfurococcales*** **with growth temperature optima below the boiling point.** *(a)* Thin section of a cell of *Desulfurococcus saccharovorans*; a cell is 0.7 μm in diameter. *(b)* Thin section of a cell of *Ignicoccus islandicus*. The cell proper is surrounded by an extremely large periplasm. The cell itself measures about 1 μm in diameter, and the cell plus periplasm measures 1.4 μm.

family. Thus this family of *Archaea* collectively contains the most hyperthermophilic of all known prokaryotes.

Desulfurococcus and *Ignicoccus*

Other notable members of the *Desulfurococcales* include *Desulfurococcus*, the genus for which the order is named (**Figure 17.20a**), and *Ignicoccus*. *Desulfurococcus* is a strictly anaerobic S^0-reducing organism like *Pyrodictium* but differs from this organism in its phylogeny and the fact that it is much less thermophilic, growing optimally at about 85°C.

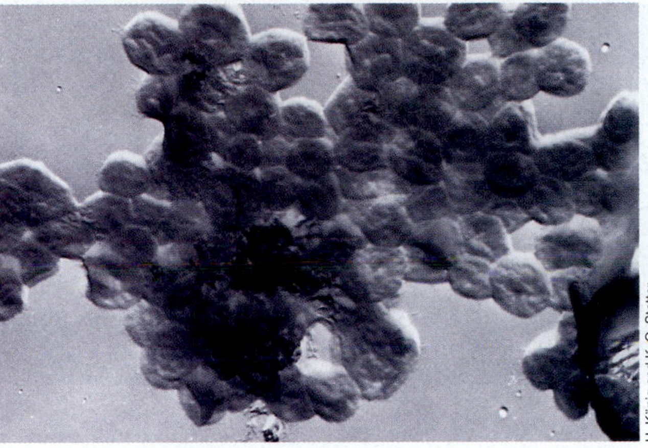

Figure 17.21 The hyperthermophile *Staphylothermus marinus* (growth temperature optimum, 92°C). Electron micrograph of shadowed cells. A single cell is about 1 μm in diameter.

Ignicoccus grows optimally at 90°C, and its metabolism is H_2/S^0 based, as is that of so many hyperthermophilic *Archaea* (Table 17.8). *Ignicoccus* (Figure 17.20*b*) is a novel hyperthermophile because it contains an *outer membrane* similar to that of gram-negative *Bacteria* (∞ Section 4.7). The outer membrane of *Ignicoccus* is unusual, however, in that it is present at some distance from the cytoplasm of the cell. This arrangement allows for an unusually large periplasm to form (Figure 17.20*b*). Indeed, the volume of the periplasm of *Ignicoccus* is some 2–3 times that of its cytoplasm, in contrast to that of gram-negative *Bacteria*, where periplasmic volume is about 25% that of the cytoplasm. The periplasm of *Ignicoccus* also contains membrane-bound vesicles (Figure 17.20*b*) that may function in exporting substances outside the cell. In addition, however, some *Ignicoccus* species are hosts to the small, parasitic bacterium, *Nanoarchaeum*, as discussed in Section 17.8.

Staphylothermus

A morphologically unusual member of the order *Desulfurococcales* is the genus *Staphylothermus* (**Figure 17.21**). Cells of *Staphylothermus* are spherical, about 1 μm in diameter, and form aggregates of up to 100 cells. *Staphylothermus* is not a chemolithotroph like so many of its hyperthermophilic relatives, but instead is a chemoorganotroph that grows optimally at 92°C. Energy is obtained from the fermentation of peptides, producing the fatty acids acetate and isovalerate as fermentation products (Table 17.8).

Isolates of *Staphylothermus* have been obtained from both shallow marine hydrothermal vents and very hot black smokers (∞ Section 24.11). This organism is thus widely distributed in submarine thermal areas where it is likely a significant consumer of proteins released from dead organisms.

17.12 Nonthermophilic *Crenarchaeota*

In contrast to hyperthermophiles, nonthermophilic crenarchaeotes (Table 17.7) have been identified from community sampling of SSU rRNA genes from many cool or cold marine

UNIT 3

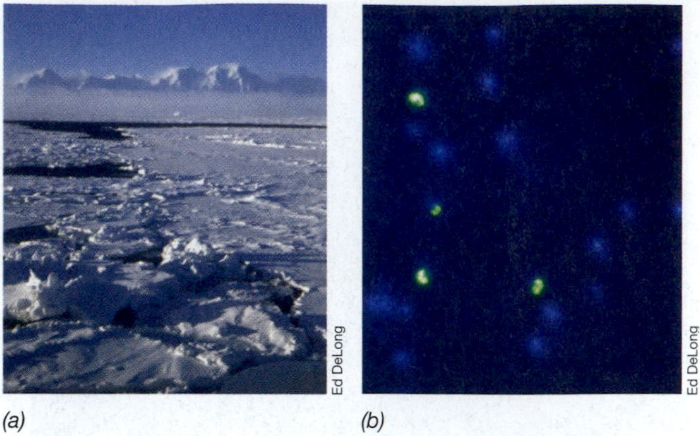

Figure 17.22 Cold-dwelling *Crenarchaeota*. *(a)* Photo of the Antarctic Peninsula taken from shipboard. The frigid waters that lie under the surface ice shown here are habitats for cold-dwelling crenarchaeotes. *(b)* Fluorescence photomicrograph of seawater treated with a FISH probe (Section 14.9) specific for species of *Crenarchaeota* (green cells). Blue cells are stained with DAPI, a stain that stains all cells.

and terrestrial environments. By using fluorescent phylogenetic probes (Sections 14.9 and 22.3), microbiologists have found crenarchaeotes in oxic marine waters worldwide. In stark contrast to the hyperthermophiles, marine crenarchaeotes thrive even in frigid waters and sea ice, such as those near Antarctica (**Figure 17.22**). These organisms are planktonic (suspended freely or attached to suspended particles in the water column, Figure 17.22*b*) and present in significant numbers (~10^4/ml) in waters that are both nutrient poor and very cold (2–4°C in seawater and less than 0°C in sea ice). Marine crenarchaeotes can account for up to 40% of the prokaryotes of deep ocean waters (Section 23.9 and Figure 23.19). Lipid analyses of marine crenarchaeotes filtered from seawater have shown that they contain ether-linked lipids, the hallmark of the *Archaea* (Section 4.3 and Table 14.1). Their high numbers, together with the ability to fix inorganic carbon, demonstrated through analysis of the carbon-stable isotope (Section 22.8) content of archaeal lipids, suggest that marine *Crenarchaeota* play a major role in the global carbon cycle.

Nonthermophilic species of *Euryarchaeota* are also present in marine environments. From studies thus far, it appears that these organisms are more abundant in temperate surface waters, whereas crenarchaeotes seem to be more abundant in cold (Figure 17.22*a*) waters and in waters of great depth.

Nitrification in *Archaea*

The physiology of marine crenarchaeotes has remained a mystery until the recent discovery of nitrification in some of these organisms. A crenarchaeote tentatively named *Nitrosopumilus maritimus* was isolated from a saltwater aquarium and shown to grow chemolithotrophically by aerobically oxidizing ammonia to nitrite, the first demonstration of nitrification in *Archaea* (Section 20.12). This organism also grows with CO_2 as the sole carbon source (autotrophy), as do nitrifying *Bacteria*. Extensive environmental sampling of oceanic waters for archaeal genes required for ammonia oxidation (for example, *amoA*), together with the isolation of other marine nitrifying crenarchaeotes, have revealed that marine *Crenarchaeota* play a more dominant role than do nitrifying *Bacteria* in this aspect of the ocean's nitrogen cycle. The ecology of marine *Crenarchaeota* is discussed in Section 23.9.

17.9–17.12 MiniReview

Hyperthermophilic *Crenarchaeota* inhabit the hottest habitats currently known to support life. Cold-dwelling phylogenetic relatives of these organisms are abundant in the oceans and in terrestrial habitats. Various morphological types of *Crenarchaeota* are known, as are several different metabolic strategies used to support growth.

- What are the major differences between the organisms *Sulfolobus* and *Pyrolobus*? *Staphylothermus* and *Ignicoccus*?
- What is unusual about the metabolic properties of *Acidianus* regarding elemental sulfur (S^0)?
- What energy class of organisms do we call those organisms that use H_2 as electron donor? List the genera of *Crenarchaeota* that can use H_2 and what they use as electron acceptors.
- In what way is strain 121 unique among prokaryotes?
- What are the major physiological capabilities of marine crenarchaeotes?

IV EVOLUTION AND LIFE AT HIGH TEMPERATURES

As we have seen in this chapter, most of the hyperthermophiles discovered so far are species of *Archaea*. Some grow near to what may be the upper temperature limit for life. They therefore have much to teach us about where, when, and under what conditions life may have arisen on a hot, early Earth (Section 14.2). Defining the upper temperature limit for life also may help identify the depths at which life might exist in Earth's subsurface and provide insight into the possibility that life has arisen on other planets that formed in ways similar to Earth. In this section, we consider the major factors that likely define the upper temperature limit for life and the biological adaptations of hyperthermophiles that permit them to exist at the exceptionally high temperatures of 100°C and higher. We end this section with a discussion of hyperthermophily and hydrogen metabolism.

17.13 An Upper Temperature Limit for Microbial Life

Habitats that contain liquid water, a prerequisite for cellular life, and that are at temperatures higher than 100°C are found at many places in the world's oceans where geothermally heated water flows out of vents or rifts in the ocean floor (Figures 24.30, 24.32, and 24.33). The hydrostatic pressure over these habitats keeps the water from boiling, allowing it to reach temperatures well above boiling, 350°C or even higher. In contrast, surface hot springs can boil, releasing heat, and therefore only attain temperatures near 100°C. We therefore would expect to find hyperthermophiles with the highest growth temperatures in hydrothermal vents on the sea floor at substantial depth, where habitats with superheated waters exist. Indeed, hydrothermal vent black smokers (∞ Figures 24.32 and 24.33) have been rich sources of exceptional hyperthermophilic *Archaea* (Table 17.9), including *Methanopyrus*, *Pyrodictium*, *Pyrolobus*, and strain 121 (Sections 17.6–17.11).

Black smokers emit hydrothermal vent fluid at 250–350°C or higher, forming upright metallic structures, called *chimneys*, that are formed by the metal sulfides precipitating out of the fluid as it mixes with the surrounding, much cooler seawater (approximately 2–4°C). The superheated vent water itself is sterile. However, several hyperthermophiles have been isolated from smoker chimney walls, where they presumably inhabit regions of the walls along the temperature gradient, from 250 to 350°C inside to 2 to 4°C outside, in zones where the temperature has decreased to levels that permit their survival and growth.

What's the Upper Limit?

How high a temperature can hyperthermophiles withstand? Over the past several decades, the known upper temperature limit of microbial existence has been pushed higher and higher with the isolation and characterization of new species of thermophiles and hyperthermophiles (**Figure 17.23**). Until recently, the record holder was *Pyrolobus fumarii*, with an upper temperature limit for growth of 113°C (Figure 17.19*c*). The current record holder, strain 121 (Section 17.11, Figure 17.19*d*), however, has pushed the limit somewhat higher, with the ability to grow at 121°C and to survive substantial periods at 130°C. Given the trend over the past several years (Figure 17.23), one can predict that *Archaea* even more hyperthermophilic than strain 121 may inhabit hydrothermal environments but have yet to be isolated. Thus, the upper temperature limit for prokaryotic growth might even exceed 130°C, with the maximum for survival some degrees higher.

Is there a temperature at which life as we know it is unlikely to exist? The answer to this is clearly yes, but we don't yet know what the upper limit is. However, laboratory experiments on the stability of biomolecules suggest that the upper limit lies somewhere between 140 and 150°C. Above 150°C, organisms would almost certainly be unable to overcome the heat lability of the essential molecules of life. For example, ATP is degraded instantly at 150°C and still very quickly at slightly cooler temperatures. Thus, the upper temperature limit for survival may well exceed 130°C, but it is almost certainly below 150°C.

17.14 Adaptations to Life at High Temperature

Because all cellular structures and activities are affected by heat, hyperthermophiles are likely to exhibit multiple adaptations to the exceptionally high temperatures of their habitats.

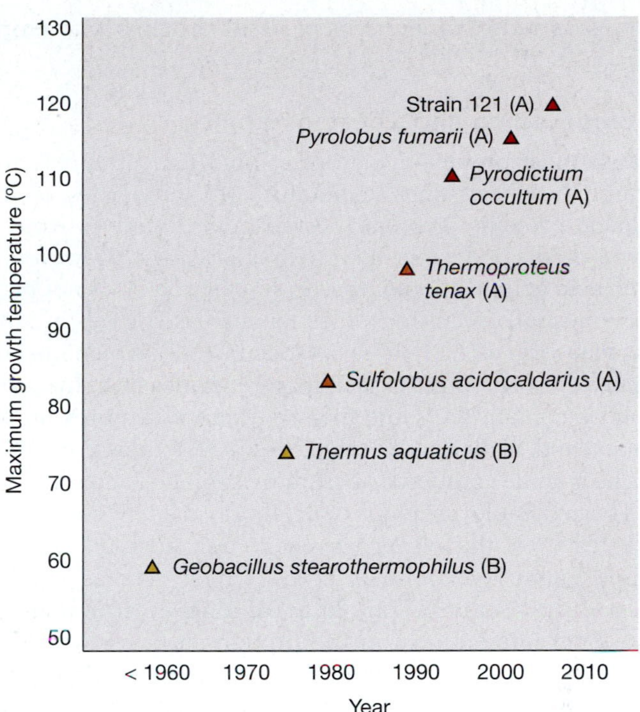

Figure 17.23 Thermophilic and hyperthermophilic prokaryotes. The graph gives the species that were, in turn, the record holders for growing at the highest temperature, from before 1960 to the present. The studies reporting these new species repeatedly extended the known upper temperature limit for life. B, *Bacteria*; A, *Archaea*. Modified from D.A. Cowan, *TRENDS in Microbiology* (2004) 12: 58–60.

In this section, we briefly examine the adaptations that may allow hyperthermophilic organisms to survive and grow at high temperatures. Knowledge of these adaptations may expand quickly in the future through studies of the cellular structures, metabolism, and the physiology of strain 121 and *Pyrolobus fumarii*, species that can grow in laboratory culture at the highest known temperatures.

Stability of Monomers

As mentioned above, the thermal lability of small molecules such as ATP may be critical in determining the upper temperature limits for growth and survival of hyperthermophiles. Even at temperatures as low as 120°C, some important small molecules are destroyed at significant rates. For example, two key molecules in energy metabolism, ATP and NAD^+ (∞ Sections 5.7 and 5.8), hydrolyze rapidly at these temperatures; the half-life of ATP or NAD^+ *in vitro* is less than 30 minutes at 120°C and shortens dramatically at temperatures above this, at least as measured in water. The growth of strain 121 at 121°C tells us, however, that the stability of molecules such as these must be significantly greater under the conditions of the cell's cytoplasm. The high concentrations of cytoplasmic solutes, such as salts, sugars, and other small molecules, likely have a protective effect on ATP and other key molecules. Alternatively, more heat-stable molecules might replace the function of certain of these compounds, for example, using nonheme iron proteins as redox proteins in place of proteins that employ NAD and NADH.

Protein Folding and Thermostability

Because most proteins denature at high temperatures, much research has been done to identify the properties of thermostable proteins. Protein thermostability derives from the folding of the molecule. Perhaps surprisingly, however, the amino acid composition of thermostable proteins is not particularly unusual, except for a bias for increased levels of amino acids that form alpha helices (∞ Section 3.7), which may help stabilize these proteins. In fact, many enzymes from hyperthermophiles contain the same major structural features, in both primary and higher-order structure (∞ Sections 3.6–3.8), as their heat labile counterparts from mesophilic bacteria.

Thermostable proteins typically do display some structural features that improve their thermostability. These include highly hydrophobic cores, which decrease the tendency of the protein to unfold in an ionic environment, and more ionic interactions on the protein surfaces, which also help hold the protein together and work against unfolding. Ultimately, it is the *folding* of the protein that most affects its heat stability, and noncovalent, ionic bonds called *salt bridges* on a protein's surface likely play a major role in maintaining the biologically active structure. But, as previously stated, many of these changes are possible with only minimal changes in primary structure when thermal stable and thermal labile forms of the same protein are compared.

Chaperonins: Assisting Proteins to Remain in Their Native State

We discussed earlier a class of proteins called *chaperonins* (heat-shock proteins; ∞ Sections 7.17 and 9.11) that function to refold partially denatured proteins. Hyperthermophilic *Archaea* produce special classes of chaperonins that function only at the highest growth temperatures. In cells of *Pyrodictium abyssi* (**Figure 17.24**), for example, a major chaperonin is the protein complex called the **thermosome**. This complex is thought to keep the cell's other proteins properly folded and functional at high temperature and help cells survive, even at temperatures above their maximal growth temperature. For example, cells of *P. abyssi* grown near its maximum temperature (110°C) contain large levels of the thermosome. Possibly because of this, the cells can remain viable following a heat shock, even a 1-h treatment in an autoclave (121°C). In cells experiencing such a treatment and then returned to the optimum temperature, the thermosome, which is itself quite heat resistant, is thought to refold sufficient copies of key denatured proteins that *P. abyssi* can once again begin to grow and divide. Thus, due to chaperonin activity, the upper temperature limit at which many hyperthermophiles can survive is higher than the upper temperature at which they can grow. The "safety net" of chaperonin activity probably ensures that cells that experience brief heat treatment above their growth temperature maximum are not killed by the exposure.

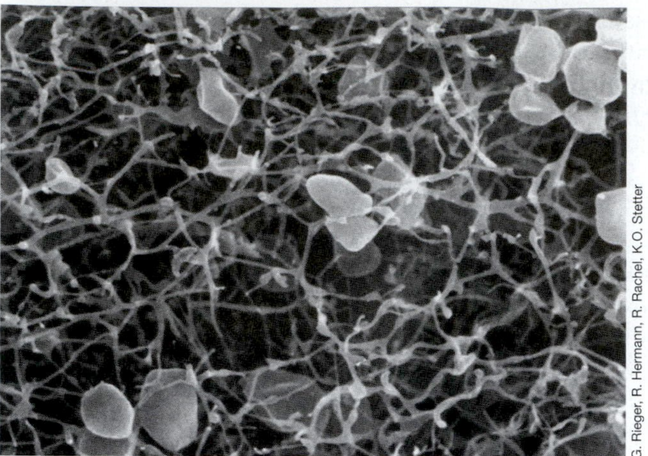
G. Rieger, R. Hermann, R. Rachel, K.O. Stetter

Figure 17.24 ***Pyrodictium abyssi*, scanning electron micrograph.** *Pyrodictium* has been studied as a model of macromolecular stability at high temperatures. Cells are enmeshed in a sticky glycoprotein matrix that binds them together.

DNA Stability at High Temperatures: Solutes and Reverse Gyrase

How does DNA stay intact and keep from melting at high temperatures? Various mechanisms may contribute. One such mechanism increases cellular solute levels, as ions, particularly potassium, and compatible organic solutes affect DNA stability. For example, as previously mentioned, the cytoplasm of the hyperthermophilic methanogen *Methanopyrus* (Section 17.6) contains molar levels of potassium cyclic 2,3-diphosphoglycerate. This solute prevents chemical damage to DNA, such as depurination or depyrimidization (loss of a nucleotide base through hydrolysis of the glycosidic bond) from high temperatures. These chemical changes can lead to mutation (∞ Section 11.4). This compound and other compatible solutes, such as potassium di-*myo*-inositol phosphate, that protect against osmotic stress are produced by thermophilic and hyperthermophilic prokaryotes in response to elevated growth temperatures. These compounds therefore are likely to play protective roles in the cell, maintaining both DNA and other macromolecules, such as proteins, in active forms.

A unique protein found *only* in hyperthermophiles is the likely major reason why DNA does not melt in these organisms. All hyperthermophiles produce a DNA topoisomerase called **reverse DNA gyrase**. Reverse gyrase introduces positive supercoils into DNA (in contrast to the negative supercoils introduced by DNA gyrase found in nonhyperthermophilic microorganisms; ∞ Section 7.3). Positive supercoiling stabilizes DNA to heat and thereby prevents the DNA helix from denaturing. Indeed, the absence of reverse gyrase in prokaryotes whose growth temperature optima are below about 80°C suggests a specific role for reverse gyrase in DNA stability at high temperatures.

Polyamines also play a role in DNA stability and in the stability of other macromolecules. These organic cations, examples of which are putrescine and spermidine, are present at high concentrations in most hyperthermophiles. Together with Mg^{2+}, polyamines function to stabilize RNA and DNA, and in thermophilic *Archaea* such as *Sulfolobus*, polyamines also help stabilize ribosomes, thereby facilitating protein synthesis at high temperatures.

DNA Stability: DNA-Binding Proteins

In addition to ions such as potassium, compatible solutes, and the enzyme reverse DNA gyrase, other proteins in hyperthermophiles may also function to maintain the integrity of duplex DNA. For example, a small heat-stable DNA-binding protein present in cells of *Sulfolobus* binds to the minor groove of DNA (Figure 7.5) in a nonspecific manner and increases its melting temperature by some 40°C. This protein, called Sac7d, sharply kinks the DNA and thus may also be involved in gene regulation (bent DNA; Sections 9.2 and 9.4).

Species of *Euryarchaeota* also contain DNA-binding proteins. But unlike the protein Sac7d, these are highly basic (positively charged) proteins that are remarkably similar in amino acid sequence and folding properties to the core histones of the *Eukarya* (Section 8.5). Archaeal histones from the hyperthermophilic methanogen *Methanothermus fervidus* (Figure 17.8*c*) have been particularly well studied. Histones from this organism wind and compact DNA into nucleosome-like structures (**Figure 17.25**) (Figure 8.6). These structures maintain the DNA in a double-stranded form at very high temperatures. Archaeal histones are found in most *Euryarchaeota*, including extremely halophilic *Archaea*, such as *Halobacterium*. However, because the extreme halophiles are not thermophiles, archaeal histones may have other functions besides assisting in DNA stability, such as in regulating gene expression.

Lipid Stability

What about cellular lipids? How do hyperthermophiles prevent their cytoplasmic membranes from separating at high temperatures? Virtually all hyperthermophilic *Archaea* synthesize lipids of the dibiphytanyl tetraether type (Section 4.3). Dibiphytanyl tetraether lipids are naturally heat resistant because the covalent bond between phytanyl units forms a *lipid monolayer* membrane structure instead of the normal lipid bilayer (Figure 4.8). This structure, supported by covalent bonds, resists the tendency of heat to pull apart a lipid bilayer constructed of fatty acids.

SSU rRNA Stability

A final point on adaptations to life at high temperatures is that of the base composition of 16S rRNAs. Hyperthermophiles, both *Bacteria* and *Archaea*, show as much as a 15% greater proportion of GC base pairs in their SSU rRNAs compared to other organisms. GC pairs, with three hydrogen bonds compared to the two of AU base pairs, confer greater thermal stability and therefore might be an adaptation to high temperatures. In contrast, the GC ratios (Section 14.11) of these organisms' genomes do not show an overall enrichment in the proportion of GC pairs, which suggests that the thermal stability of SSU rRNA specifically might be critical to life under hyperthermophilic conditions.

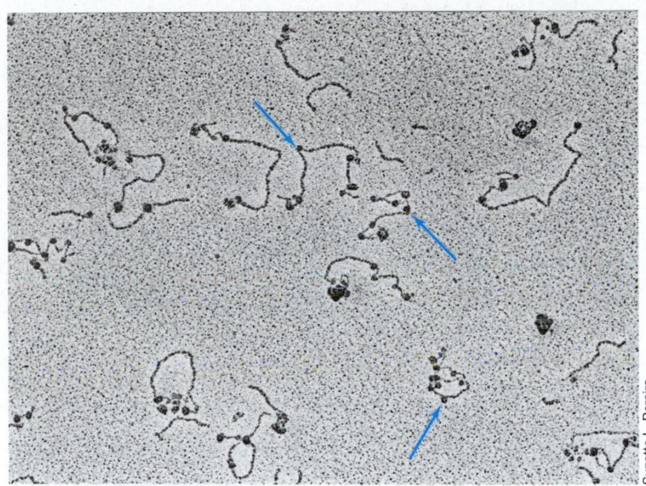

Figure 17.25 Archaeal histones and nucleosomes. Electron micrograph of linearized plasmid DNA wrapped around copies of archaeal histone Hmf (from the hyperthermophilic methanogen *Methanothermus fervidus*) to form the roughly spherical, darkly stained nucleosome structures (arrows). Compare this micrograph with an artist's depiction of the histones and nucleosomes of *Eukarya* shown in Figure 8.6.

UNIT 3

17.15 Hyperthermophilic *Archaea*, H_2, and Microbial Evolution

At the time that cellular life arose, approximately 4 billion years ago, it is almost certain that Earth was far hotter than it is today (Section 14.1), although the actual temperatures at which life arose remain controversial, even today. Prokaryotic life forms are thought to have arisen under conditions of very high temperatures, and for millions of years Earth may have been suitable only for hyperthermophiles. Given the discussion above on the temperature limits to life, it has been hypothesized that biological molecules, biochemical processes, and the first cells arose on Earth around hydrothermal springs and vents on the seafloor as they cooled to temperatures below 150°C. The phylogeny of modern hyperthermophiles, as well as the similarities in their habitats and metabolism to that of early Earth, suggests that these organisms, more so than any other prokaryotic life forms, may be the closest descendents of ancient hyperthermophilic cells. The detailed study of modern hyperthermophiles, especially of species with the highest growth temperatures, therefore might provide a window into the distant past of microbial life and evolution.

Hyperthermophiles and Evolutionary Constraints

Comparative analysis of sequences of the SSU rRNA gene place hyperthermophilic *Archaea* and hyperthermophilic *Bacteria* (such as *Thermotoga* and *Aquifex*) on the deepest, shortest branches of the phylogenetic tree (Figure 17.1; Figures 14.16, 15.1). This suggests that the 16S rRNA genes of hyperthermophiles are subject to unusually strong constraints, compared with those of nonhyperthermophilic organisms, in the amount of sequence change that can be

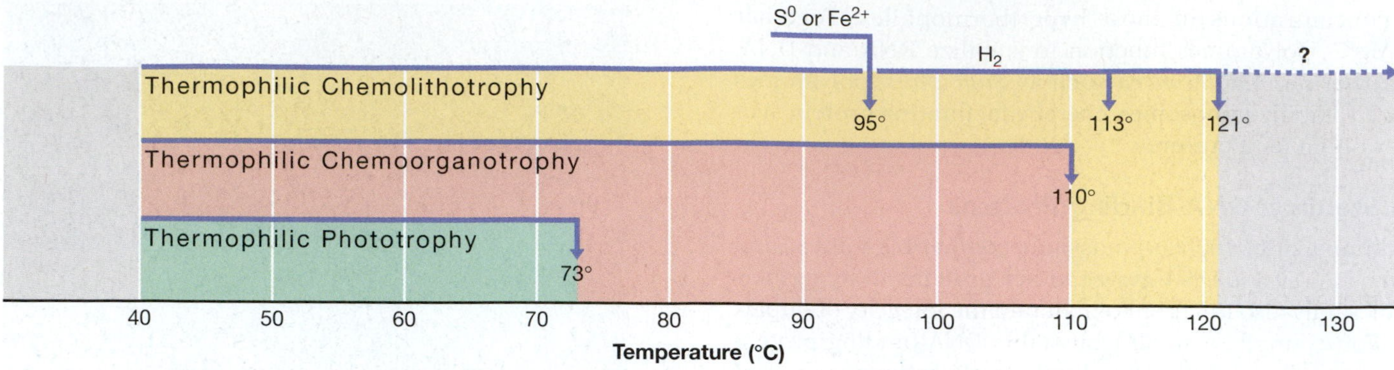

Figure 17.26 Upper temperature limits for energy metabolism. Phototrophy, *Synechococcus lividus* (*Bacteria*, cyanobacteria); chemoorganotrophy, *Pyrodictium occultum* (*Archaea*); chemolithotrophy, S^0 as electron donor, *Acidianus infernus* (*Archaea*); chemolithotrophy, Fe^{2+} as electron donor, *Ferroglobus placidus* (*Archaea*); chemolithotrophy, H_2 as electron donor, *Pyrolobus fumarii* (*Archaea*, 113°C); strain 121 (*Archaea*, 121°C). Strain 121 also grows on formate, but this compound may be metabolized through $H_2 + CO_2$.

tolerated in the SSU rRNA. These constraints are likely related to the exceptionally high temperatures at which these organisms live, with only a small amount of variation possible before heat stability and function become compromised. Indeed, the higher percentage of GC base pairs in the SSU rRNAs of hyperthermophiles may be a reflection of these constraints. Because this bias in GC base pairs is shared by both hyperthermophilic *Archaea* and hyperthermophilic *Bacteria*, it appears to be an example of evolutionary convergence rather than an indication of close relationship. Nonetheless, it is possible that if the allowable divergence in the sequence of the SSU rRNA of hyperthermophiles reached saturation early on in evolution, then the sequences of the 16S rRNA gene of modern hyperthermophiles may thereafter have diverged relatively little from those of early cells. Much the same might be said for other key macromolecules in the cell, such as proteins. Indeed it is possible that the biology of modern day hyperthermophiles is very similar to that of hyperthermophiles that lived billions of years ago.

Hyperthermophilic Habitats and Hydrogen (H_2) as an Energy Source

The oxidation of hydrogen with the reduction of Fe^{3+} is a feature common to many hyperthermophiles (**Figure 17.26**). This activity is consistent with the presence of H_2 and Fe^{3+} oxides in hydrothermal vents and other hot environments. These environments contain H_2 and an abundance of other inorganic electron donors and acceptors. Indeed, many hyperthermophiles grow anaerobically with H_2 as an electron donor and one or more of the electron acceptors S^0, NO_3^-, Fe^{3+}, or O_2 (Table 17.8). The diversity of H_2-oxidizing hyperthermophilic *Archaea* (and *Bacteria*; ∞ Sections 15.5 and 16.20) known today attests to the importance of H_2 oxidation as a key mechanism for energy conservation.

H_2 metabolism may have evolved in primitive organisms because of the ready availability of H_2 and suitable inorganic electron acceptors in their primordial environments, and also because H_2 catabolism requires few proteins (∞ Figure 14.8 and Section 21.11). The relationship between hyperthermophiles and H_2 metabolism is emphasized by comparison of the known upper temperature limits for growth based on the three energy-conserving processes of chemoorganotrophy, chemolithotrophy, and phototrophy.

Above 110°C, only chemolithotrophy may be possible, since only H_2-oxidizing *Archaea* are known, including *Pyrolobus* and strain 121 (Table 17.9 and Figure 17.26). Chemoorganotrophy occurs up to at least 110°C, as this is the upper temperature limit for growth of *Pyrodictium occultum*, an organism that can ferment certain organic compounds as well as grow chemolithotrophically on H_2 with S^0 as electron acceptor (Table 17.9). Photosynthesis is the least heat tolerant of these processes, with no hyperthermophilic representatives known.

Comparisons of this kind point to the H_2-oxidizing hyperthermophiles discussed in this chapter (and ∞ Section 16.20) as the best-known extant examples of Earth's earliest cellular life forms. More so than any other prokaryotes, these organisms retain the metabolic and physiological traits one would predict were necessary for existence in hyperthermophilic habitats on the early Earth.

17.13–17.15 MiniReview

Although hyperthermophiles live at very high temperatures, in some cases above the boiling point of water, there are temperature limits beyond which no living organism can survive. This limit is likely 140–150°C. Hydrogen (H_2) catabolism may have been the first energy-yielding metabolism of cells.

- List at least two reasons why an upper temperature limit to life undoubtedly exists.
- How do hyperthermophiles keep important macromolecules such as proteins and DNA from being destroyed by high heat?
- What phylogenetic and physiological evidence suggests that extant hyperthermophiles may exhibit similarities to early microorganisms?

Review of Key Terms

Acetyl-CoA pathway a pathway of autotrophic CO_2 fixation widespread in obligate anaerobes including methanogens, homoacetogens, and sulfate-reducing bacteria

Bacteriorhodopsin a membrane protein containing retinal produced by certain extreme halophiles and capable of light-mediated proton motive force formation

Compatible solute an organic or inorganic substance accumulated in the cytoplasm of a halophilic organism that maintains osmotic pressure

Crenarchaeota a phylum of *Archaea* that contains both hyperthermophilic and cold-dwelling organisms

Euryarchaeota a phylum of *Archaea* that contains primarily methanogens, the extreme halophiles, *Thermoplasma*, and some marine hyperthermophiles

Extreme halophile an organism whose growth is dependent on large concentrations (generally 10% or more) of NaCl

Halorhodopsin a light-driven chloride pump that accumulates Cl^- within the cytoplasm

Hydrothermal vent a deep-sea hot spring emitting warm (~20°C) to superheated (>300°C) water

Hyperthermophile an organism with a growth temperature optimum of 80°C or greater

Methanogen a methane-producing organism; CH_4 is produced by either reduction of CO_2 with H_2 or from certain organic compounds

Phytanyl a branched-chain hydrocarbon containing 20 carbon atoms and commonly found in the lipids of *Archaea*

Reverse DNA gyrase a protein universally present in hyperthermophiles that introduces positive supercoils into circular DNA

Solfatara a hot, sulfur-rich, generally acidic environment commonly inhabited by hyperthermophilic *Archaea*

Thermosome a heat-shock (chaperonin) protein complex that functions to refold partially heat-denatured proteins in hyperthermophiles

UNIT 3

Review Questions

1. What are some features that all *Archaea* have in common (Section 17.1)?
2. Which organism, *Pyrodictium*, *Thermoplasma*, or *Methanosarcina*, is the closest relative of the extreme halophile *Halobacterium* (*Hint*: The answer lies in Figure 17.1.)? What form of energy metabolism does each of these organisms have (Sections 17.1 and 17.2)?
3. How can organisms such as *Halobacterium* survive in a high-salt environment whereas an organism such as *Escherichia coli* cannot (Section 17.3)?
4. Contrast the roles of bacteriorhodopsin, halorhodopsin, and sensory rhodopsin in *Halobacterium salinarum* (Section 17.3).
5. What is the electron donor for methanogenesis when CO_2 is reduced to CH_4 (Section 17.4)?
6. What one major physiological feature unifies all members of the *Thermoplasmatales*? Why does this allow some of them to successfully colonize mine tailings (Section 17.5)?
7. What is physiologically unique about *Methanopyrus* compared with another methanogen such as *Methanobacterium* (Section 17.6)? What is physiologically unique about *Archaeoglobus* (Section 17.7)?
8. How is *Nanoarchaeum* similar to other *Archaea*? How does it differ (Section 17.8)?
9. What forms of energy metabolism are present in *Crenarchaeota*? What form is absent (Section 17.9)?
10. Why is the organism *Sulfolobus* of historical interest in the field of microbiology (Section 17.10)?
11. What is unusual about the organism *Pyrolobus fumarii* (Section 17.11)?
12. What is physiologically unusual about the marine crenarchaeote *Nitrosopumilus maritimus* (Section 17.12)?
13. What organism is the current record holder for the upper temperature limit for growth (Section 17.13)?
14. What is reverse gyrase and why is it important to hyperthermophiles (Section 17.14)?
15. Why might hydrogen metabolism have evolved in early organisms (Section 17.15)?

Application Questions

1. Using the information in Figure 17.1 as a guide, discuss why bacteriorhodopsin may have been a late evolutionary invention.
2. Defend or refute the following statement: The upper temperature limit to life is unrelated to the stability of proteins or nucleic acids.

18

Eukaryotic Cell Biology and Eukaryotic Microorganisms

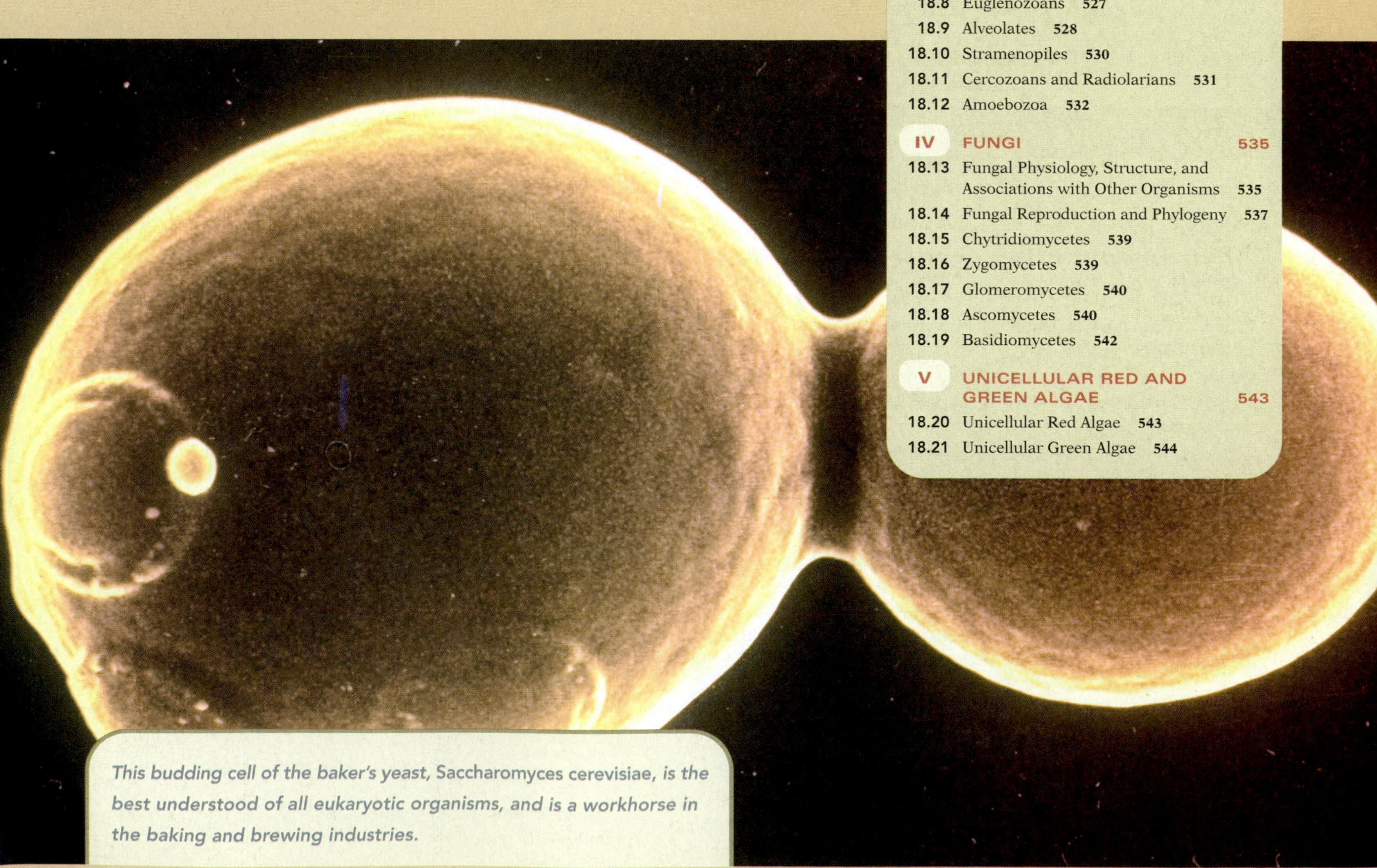

This budding cell of the baker's yeast, Saccharomyces cerevisiae, *is the best understood of all eukaryotic organisms, and is a workhorse in the baking and brewing industries.*

In this chapter we consider the cell structure, phylogeny, and diversity of eukaryotic microorganisms. The unicellular eukaryotes are diverse, have a long evolutionary history, and are of major ecological importance. Several of them also directly impact humans, either favorably through, for example, beneficial fermentations, or unfavorably, as human pathogens.

I EUKARYOTIC CELL STRUCTURE AND FUNCTION

The five sections that follow examine the structure of the eukaryotic cell and the ancestral endosymbiotic link between eukaryotic organelles and *Bacteria*.

18.1 Eukaryotic Cell Structure and the Nucleus

A typical eukaryotic cell is shown in **Figure 18.1**. In contrast to prokaryotes, **eukaryotes** contain a membrane-enclosed nucleus and, depending on the organism, several other organelles. For example, mitochondria are nearly universal among eukaryotic cells, but the pigmented chloroplasts are found only in phototrophic cells. Other internal structures typically include the Golgi complex, peroxisomes, lysosomes, endoplasmic reticula, and microtubules and microfilaments (Figure 18.1). These internal structures compartmentalize activities of the cell for efficient function. Some eukaryotic cells have flagella and cilia—organelles of motility—while others do not. Eukaryotic cells also may have extracellular components, such as a cell wall in fungi, algae, or plant cells (not found in animal cells or most protists), or an extracellular matrix in animal cells. Also, many eukaryotes, even many eukaryotic microorganisms, form multicellular structures.

Nucleus

The **nucleus** contains the chromosomes of the eukaryotic cell (**Figure 18.2**). In eukaryotes, DNA within the nucleus is wound around positively charged proteins called **histones**, which help tightly pack the negatively charged DNA to form nucleosomes and from them, chromosomes (∞ Section 8.6, Figure 8.7). In many eukaryotic cells the nucleus is many micrometers in diameter and is easily visible with the light microscope, even without staining (∞ Figure 2.5*a*). In

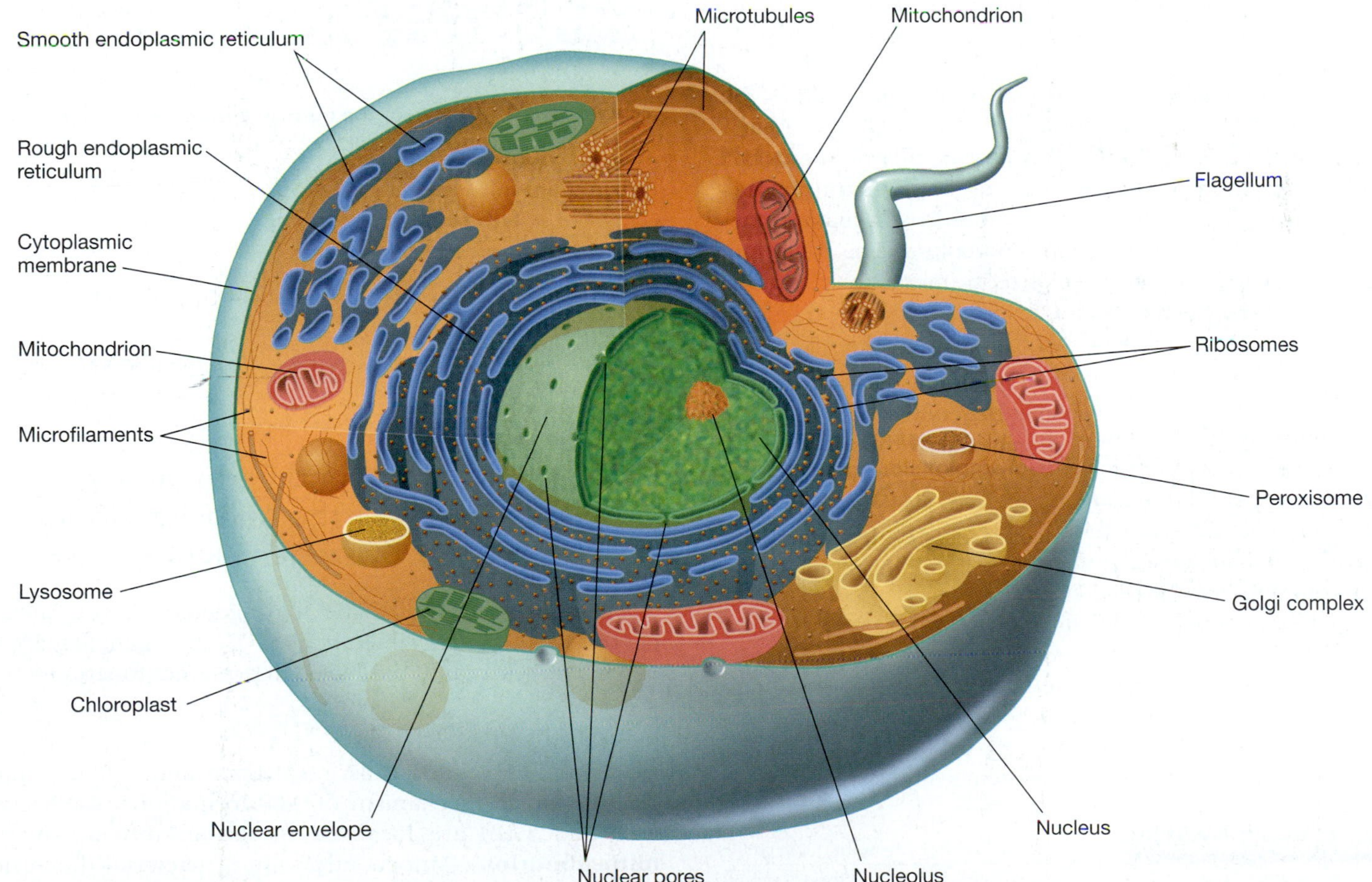

Figure 18.1 Schematic, cut-away view of a eukaryotic cell. Although all eukaryotic cells contain a nucleus, not all organelles and other structures shown are present in all eukaryotic cells. Not shown is the cell wall, found in fungi, algae, plants, and some protists.

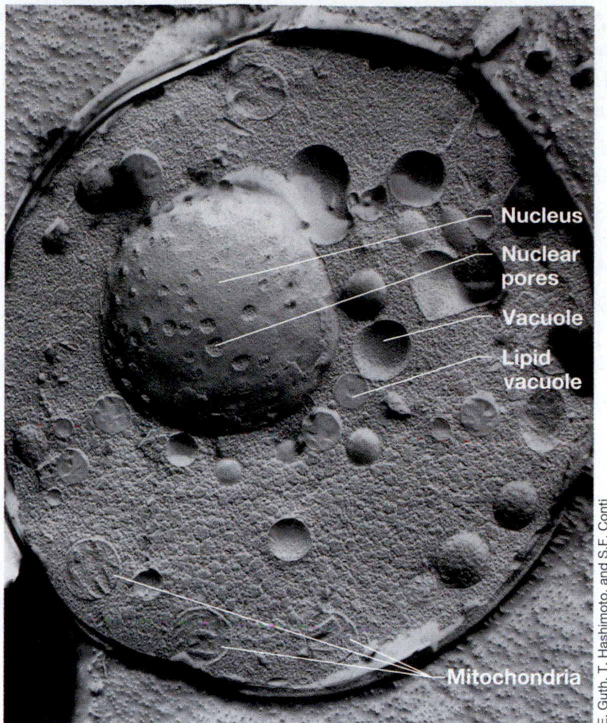

Figure 18.2 The nucleus. Electron micrograph of a yeast cell prepared by freeze-etching, showing a surface view of the nucleus. The cell is about 8 μm wide and the nucleus about 1.5 μm wide.

smaller eukaryotes, however, special staining procedures are often required to see the nucleus.

The nucleus is enclosed by a pair of membranes, each with its own function, separated by a space. The inner membrane is a simple sac; the outer membrane is in many places continuous with the endoplasmic reticulum. The inner and outer nuclear membranes specialize in interactions with the nucleoplasm and the cytoplasm, respectively.

The nuclear membrane contains pores (Figure 18.2), formed from holes where the inner and outer membranes are joined. The pores allow a complex of proteins to import and export other proteins and nucleic acids into and out of the nucleus, a process called nuclear transport. Nuclear transport requires energy, and this comes from hydrolysis of the energy-rich compound guanosine triphosphate (GTP).

Within the nucleus, the *nucleolus* (Figure 18.1) is the site of ribosomal RNA synthesis. The nucleolus is rich in RNA and often visible under the light microscope. Ribosomal proteins synthesized in the cytoplasm are transported into the nucleolus and combine with ribosomal RNA to form the small and large subunits of the eukaryotic ribosome. These are then exported to the cytoplasm, where they associate to form the intact ribosome and function in protein synthesis.

18.1 MiniReview

The nucleus contains the cell's chromosomes and is surrounded by two membranes. The nucleolus is the site of ribosomal RNA synthesis.

- What proteins is DNA wound around in eukaryotic chromosomes?
- What is the role of pores in the nuclear membrane?

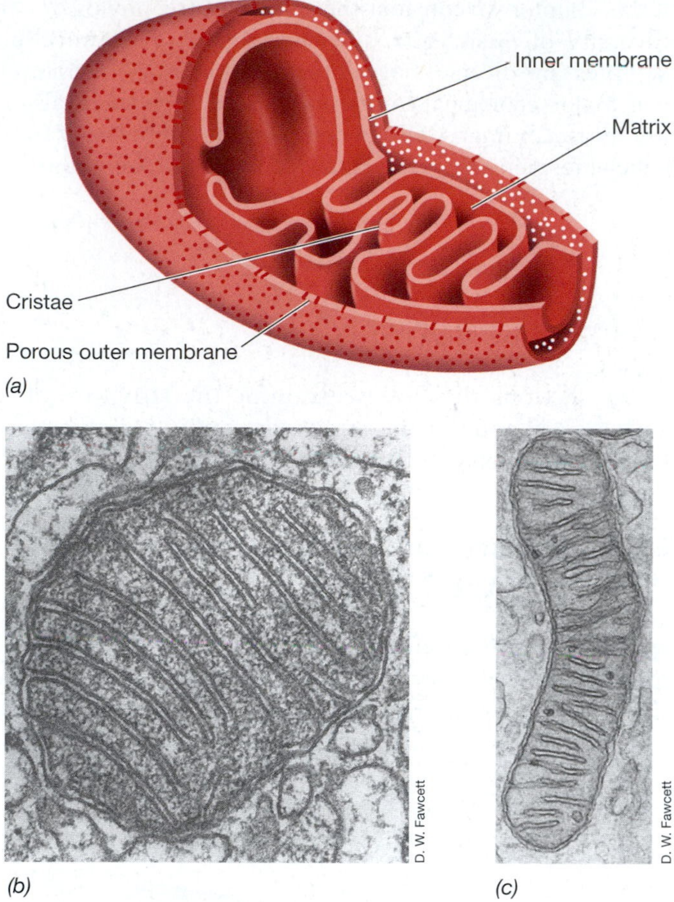

Figure 18.3 Structure of the mitochondrion. *(a)* Diagram showing the overall structure of the mitochondrion. Note inner and outer membranes. *(b, c)* Transmission electron micrographs of mitochondria from rat tissue, showing the variability in morphology. Note the cristae.

18.2 Respiratory and Fermentative Organelles: The Mitochondrion and the Hydrogenosome

The mitochondrion and the hydrogenosome specialize in chemotrophic energy metabolism. Both organelles are enclosed by membranes but have quite distinct functions.

Mitochondria

In aerobic eukaryotic cells, respiration and oxidative phosphorylation (a mechanism of ATP formation) (Sections 5.11 and 5.12) are localized in **mitochondria** (singular, **mitochondrion**). Mitochondria are of bacterial dimensions and can be rod-shaped or nearly spherical (**Figure 18.3**). A typical animal cell can contain over 1,000 mitochondria, but the number per cell depends somewhat on the cell type and size. A yeast cell may have many fewer mitochondria per cell (Figure 18.2). Mitochondria are surrounded by two membranes. The outer membrane, composed of an equal mixture of protein and lipid, is relatively permeable and contains numerous minute channels that allow passage of ions and small organic molecules. The inner membrane is more protein rich than the outer membrane and is also less permeable. Because mitochondrial membranes lack sterols (Section 4.3), they are much less rigid than the eukaryotic cytoplasmic membrane, which does contain sterols. Mitochondria can thus take on highly varied shapes (Figure 18.3*b,c*).

Mitochondria also possess a series of folded internal membranes called **cristae**. These membranes, formed by invagination of the inner membrane, are the sites of enzymes for respiration and ATP production. Cristae also contain specific transport proteins that regulate the passage of metabolites, in particular ATP, into and out of the *matrix* of the mitochondrion (Figure 18.3*a*). The matrix contains enzymes for the oxidation of organic compounds—in particular, enzymes of the citric acid cycle (Section 5.13).

The Hydrogenosome

Some anaerobic eukaryotic microorganisms lack mitochondria and instead contain **hydrogenosomes** (**Figure 18.4*a***). Although similar in size to a mitochondrion, the hydrogenosome lacks the citric acid cycle enzymes and usually also lacks cristae. Various microbial eukaryotes contain hydrogenosomes, and all are either obligate or aerotolerant anaerobes whose metabolism is strictly fermentative. Examples include human parasites such as the flagellate *Trichomonas* (Section 18.10) and various ciliated protists that inhabit the rumen of ruminant animals (Section 24.10) or anoxic muds and sediments.

The major biochemical reactions in the hydrogenosome are the oxidation of pyruvate to H_2, CO_2, and acetate (Figure 18.4*b*). Pyruvate is oxidized to the energy-rich compound acetyl-CoA (from which additional ATP is made during the formation of acetate, Section 21.9), along with H_2 plus CO_2 (Figure 18.4*b*). The key enzymes of the hydrogenosome are *pyruvate: ferredoxin reductase* and *hydrogenase*. Some anaerobic eukaryotes have H_2-consuming symbiotic bacteria such as methanogens residing in their cytoplasm (Figure 24.4*b,c*). The symbionts consume the H_2 and CO_2 produced by the hydrogenosome (Figure 18.4*b*), yielding methane. Because hydrogenosomes lack an electron transport chain and usually lack the citric acid cycle, they cannot oxidize the acetate produced from pyruvate catabolism as mitochondria do. Acetate is therefore excreted from the hydrogenosome into the cytoplasm of the host cell (Figure 18.4*b*).

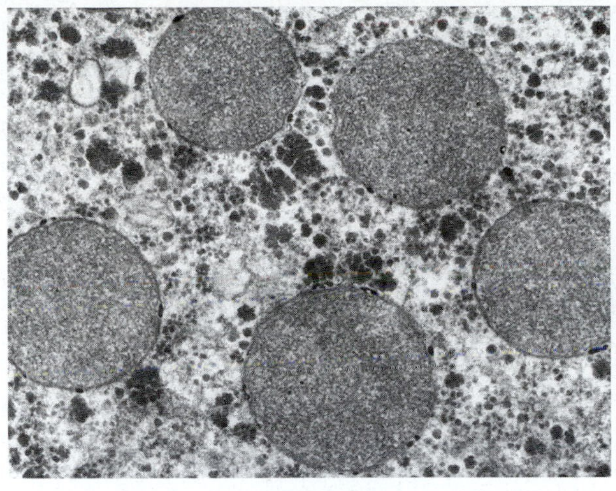

(a)

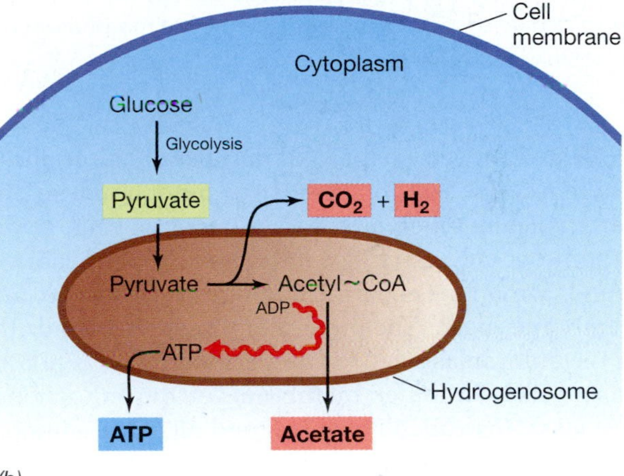

(b)

Figure 18.4 The hydrogenosome. *(a)* Electron micrograph of a thin section through a cell of the anaerobic parabasalid, *Trichomonas vaginalis*, showing five hydrogenosomes in cross-section. Compare their internal structure with that of mitochondria in Figure 18.3. *(b)* Biochemistry of the hydrogenosome. Pyruvate is taken up by the hydrogenosome, and H_2, CO_2, acetate, and ATP (all shown in red) are the products.

18.2 MiniReview

The mitochondrion and the hydrogenosome are energy-generating organelles of eukaryotic cells. Mitochondria carry out aerobic respiration, whereas hydrogenosomes, found only in certain anaerobic eukaryotes, ferment pyruvate to yield H_2 plus CO_2, acetate, and ATP.

- What key reactions occur in the mitochondrion? What key product does the mitochondrion make for the cell proper?
- Compare and contrast the metabolic fate of pyruvate in mitochondria and hydrogenosomes.

18.3 Photosynthetic Organelle: The Chloroplast

Chloroplasts are chlorophyll-containing organelles found in phototrophic eukaryotes—plants, unicellular and multicellular **algae**, and certain unicellular organisms grouped as protists. Chloroplasts of many protists and algae are relatively large and readily visible with the light microscope

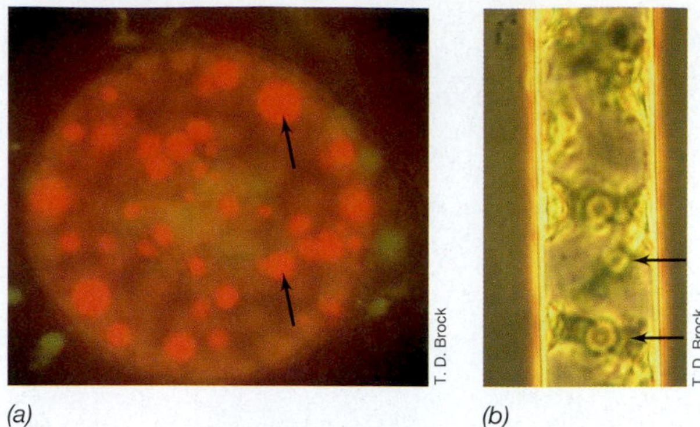

Figure 18.5 Photomicrographs of protist and green alga cells showing chloroplasts. *(a)* Fluorescence photomicrograph of the diatom *Stephanodiscus*. The chlorophyll in the chloroplasts (arrows) absorbs light and fluoresces red. The cell is about 40 μm wide. *(b)* Phase-contrast photomicrograph of the green alga *Spirogyra* showing the characteristic spiral-shaped chloroplasts (arrows) of this phototroph. A cell is about 20 μm wide.

(**Figure 18.5**). The size, shape, and number of chloroplasts per cell vary markedly, and in contrast to mitochondria, chloroplasts are typically much larger than bacterial cells.

Like mitochondria, chloroplasts have a permeable outer membrane, a much less-permeable inner membrane, and an intermembrane space. The inner membrane surrounds the lumen of the chloroplast, called the **stroma**, but it is not folded into cristae like the inner membrane of the mitochondrion (Figure 18.3*a*). Instead, chlorophyll and all other components needed for photosynthesis are located in a series of flattened membrane discs called **thylakoids** (**Figure 18.6**). The thylakoid membrane is highly impermeable to ions and other metabolites because its function is to establish the proton motive force necessary for ATP synthesis (∞ Section 5.12). In green algae and green plants, thylakoids are typically stacked into discrete structural units called *grana* (∞ Figure 20.5).

The chloroplast stroma contains large amounts of the enzyme ribulose bisphosphate carboxylase (RubisCO). RubisCO is a key catalyst of the **Calvin cycle**, the series of biosynthetic reactions by which most photosynthetic organisms convert CO_2 to organic compounds (∞ Section 20.6). RubisCO makes up over 50% of the total chloroplast protein and catalyzes the formation of phosphoglyceric acid, a key compound in the biosynthesis of glucose (∞ Sections 5.10 and 5.15). The permeability of the outer chloroplast membrane allows glucose and ATP produced during photosynthesis to diffuse into the cytoplasm where they can be used to build new cell material.

18.3 MiniReview

Chloroplasts are the site of photosynthetic energy production and CO_2 fixation in eukaryotic phototrophs.

- Differentiate the stroma from thylakoids.
- What is the function of RubisCO and where is it found?

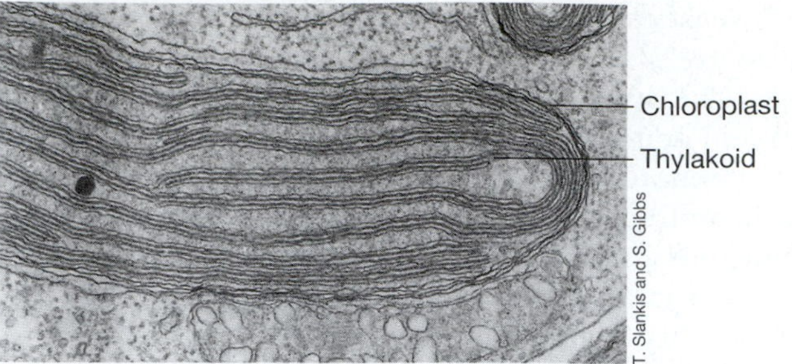

Figure 18.6 The chloroplast. Transmission electron micrograph showing a chloroplast of the stramenopile *Ochromonas danica*. Note the thylakoids.

18.4 Endosymbiosis: Relationships of Mitochondria and Chloroplasts to *Bacteria*

On the basis of their relative autonomy, size, and morphological resemblance to bacteria, it was suggested long ago that mitochondria and chloroplasts are descendants of ancient prokaryotic cells. More recent evidence indicates that the ancestor of mitochondria was a free-living, facultatively aerobic alphaproteobacterium, acquired by another cell and giving rise to the eukaryotic cell (∞ Section 14.4 and Figure 14.10). In a similar manner, the ancestor of the chloroplast was a cyanobacterium, acquired approximately 1.5 billion years ago by a heterotrophic eukaryote, sometime after eukaryotic cells had arisen. These two evolutionary events, engulfment by cells of the ancestors of the mitochondrion and the chloroplast, are called *primary* endosymbiosis to distinguish them from *secondary* endosymbiosis, which is discussed later. Several lines of molecular evidence support **endosymbiosis**—the engulfment of one cell type by another cell type and the subsequent stable association of the two cells—as the origin of eukaryotic cells. We consider here the molecular evidence for **primary endosymbiosis**.

1. **Mitochondria and chloroplasts contain DNA.** Although most of the functions of mitochondria and chloroplasts are encoded by nuclear DNA, a few of their components are encoded by a small genome within the organelle itself. These include ribosomal RNA, transfer RNAs, and certain proteins of the respiratory chain (mitochondria) and photosynthetic apparatus (chloroplast). Thus, nonphototrophic eukaryotic cells are genetic chimeras containing DNA from two different sources, the endosymbiont and the host cell nucleus. Phototrophic eukaryotes—algae and plants—contain DNA from three different sources, the mitochondrial and chloroplast endosymbionts, and the nucleus. Most mitochondrial DNA and all chloroplast DNA are of a covalently closed circular form like that of most *Bacteria* (∞ Sections 2.6, 7.2, and 7.3). Mitochondrial DNA can be seen in cells by using special staining methods (**Figure 18.7**). We discuss other features of organellar genomes in Section 13.4.

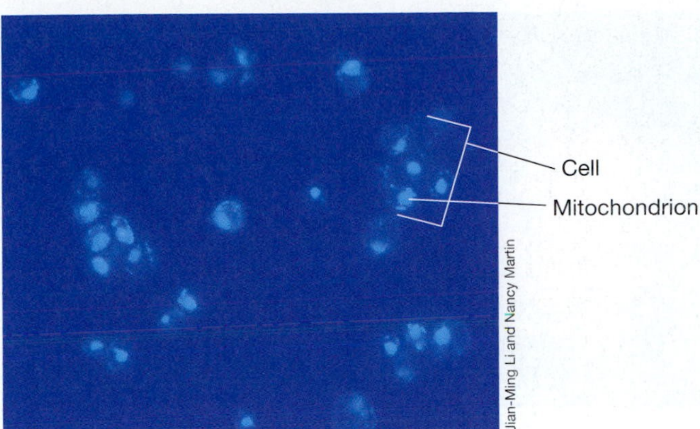

Figure 18.7 Cells of the ascomycete yeast *Saccharomyces cerevisiae*. The cells have been stained with 4′6-diamidine-2′-phenylindole dihydrochloride (DAPI) (∞ Section 22.3) to show mitochondrial DNA. Each mitochondrion has two to four circular chromosomes that stain blue with this fluorescent dye.

2. **The eukaryotic nucleus contains genes derived from bacteria.** Genomic sequencing and other genetic studies have clearly shown that several nuclear genes encode properties unique to mitochondria and chloroplasts. Because the sequences of these genes more closely resemble those of *Bacteria* than those of *Archaea* or *Eukarya,* it is concluded that these genes were transferred to the nucleus from endosymbionts that were oxygen-consuming *Bacteria* (∞ Section 14.4) during the evolutionary transition from engulfed cell to organelle.
3. **Mitochondria and chloroplasts contain their own ribosomes.** Ribosomes, cell structures that function in protein synthesis (∞ Section 7.15), exist in either a large form (80S), typical of the cytoplasm of eukaryotic cells, or in a smaller form (70S), present in *Bacteria* and *Archaea*. Mitochondria and chloroplasts also contain ribosomes, and they are 70S like those of prokaryotes.
4. **Antibiotic specificity.** Several antibiotics (streptomycin is one example) kill or inhibit *Bacteria* by specifically interfering with 70S-ribosome function. These same antibiotics also inhibit protein synthesis in mitochondria and chloroplasts.
5. **Molecular phylogeny.** Phylogenetic studies using comparative ribosomal RNA gene-sequencing methods (∞ Sections 14.5–14.9) and organellar genome studies (∞ Section 13.4) strongly support a model in which the chloroplast and mitochondrion originated from the domain *Bacteria*. According to this model, the modern eukaryotic cell thus arose from an association of at least two quite distinct organisms by the process of endosymbiosis (∞ Section 14.4 and Figure 14.10).

The evidence that hydrogenosomes are endosymbionts is also strong. For example, in the obligately anaerobic ciliated protist *Nyctotherus ovalis* that lives in the hindgut of termites (∞ Section 24.2), mitochondria are absent and hydrogenosomes that contain DNA and ribosomes have been identified. Moreover, the nucleus of hydrogenosome-containing eukaryotes contains genes encoding proteins that are close homologs of proteins from *Bacteria*. Thus it appears that the hydrogenosome originated by endosymbiosis, just as the mitochondrion did. In fact, hydrogenosomes are thought to be metabolically degenerate mitochondria that dispensed with respiration to exploit pyruvate fermentation as a means of energy conservation in a host cell with an anaerobic lifestyle. Although not equal to the energy available from respiration in the mitochondrion, acetate production in the hydrogenosome supplies cells with more energy than would be obtained if the fermentation products were lactate or ethanol (Figure 18.4; ∞ Section 5.10). The mitochondrion and the hydrogenosome can thus be viewed as functionally related organelles that specialize in different metabolic strategies for making ATP. Other structures called *mitosomes* are present in some eukaryotic cells and are probably even more degenerate mitochondria, having lost virtually all energy-related functions altogether. The evolutionary implications of the apparent absence of mitochondria and the presence of hydrogenosomes and mitosomes are discussed in Section 18.6.

These various forms of evidence suggest the following model (see Figure 18.12). Mitochondria and hydrogenosomes arose from the endosymbiosis of an alphaproteobacterium by another cell in forming the early eukaryotic cell. This was followed by the later endosymbiotic acquisition of a cyanobacterium as the ancestor of the chloroplast in eukaryotes that became photosynthetic. Through these associations, host cells obtained permanent partners specializing in energy generation, and the symbionts received a stable and supportive growth environment.

Secondary Endosymbiosis

Eukaryotes acquired the ability to carry out photosynthesis approximately 1.5 billion years ago through the capture by a heterotrophic eukaryote of a cyanobacterium, giving rise to the chloroplast. This instance of primary endosymbiosis just discussed gave rise to the chloroplast in the common ancestor of green algae, red algae, and plants (see Figure 18.12). However, following this primary endosymbiosis event, several groups of nonphototrophic eukaryotes acquired chloroplasts by **secondary endosymbiosis**, the process of engulfing a green algal cell or a red algal cell, retaining its chloroplast, and becoming phototrophic. Separate secondary endosymbioses with green algae account for chloroplasts in euglenids and chlorarachniophytes (Sections 18.8 and 18.11; see Figure 18.12). Alveolates (ciliates, apicomplexans, and dinoflagellates; Section 18.9) and Stramenopiles (Section 18.10) obtained their chloroplasts through secondary endosymbiosis with red algae. The ancestral red algal chloroplasts were lost from some lineages, such as the ciliates, or became reduced in others, such as the apicomplexans, or were replaced at various times, as in dinoflagellates, with a chloroplast from a different alga, including green algae.

In addition to secondary endosymbiotic events, molecular evidence indicates that several instances of *tertiary* endosymbiosis have occurred, in which ancestral chloroplasts in several organisms have been replaced with those from organisms that had acquired the chloroplast by secondary endosymbiosis. These many examples serve to highlight the importance of endosymbiosis in the evolution and diversification of eukaryotes.

18.4 MiniReview

Key metabolic organelles of eukaryotes are the chloroplast, which functions in photosynthesis, and the mitochondrion or hydrogenosome, which function in respiration or fermentation. These organelles were originally *Bacteria* that established permanent residence inside other cells (endosymbiosis).

- Summarize the molecular evidence that supports the relationship of organelles to *Bacteria*.
- Why might *Nyctotherus* be better off with hydrogenosomes than with mitochondria?
- Distinguish between primary and secondary endosymbiosis.

18.5 Other Organelles and Eukaryotic Cell Structures

Other cytoplasmic structures are typically present in eukaryotic cells. These include the endoplasmic reticulum, ribosomes, Golgi complex, lysosomes, peroxisomes, and the organelles of motility—flagella and cilia. An extracellular matrix also can be observed surrounding animal cells. In contrast to mitochondria and chloroplasts, however, these additional cytoplasmic structures lack DNA and are not of endosymbiotic origin.

Endoplasmic Reticulum, Ribosomes, and Golgi Complex

The endoplasmic reticulum (ER) is a network of membranes continuous with the nuclear membrane. Two types of endoplasmic reticulum are recognized: *rough,* which contains attached ribosomes, and *smooth,* which does not (Figure 18.1). Smooth ER participates in the synthesis of lipids and in some aspects of carbohydrate metabolism. Rough ER, through the activity of its ribosomes, is a major producer of glycoproteins and also produces new membrane material that is transported throughout the cell to enlarge the various membrane systems (Figure 18.1) before cell division.

The Golgi complex consists of a stack of membranes distinct from the ER (**Figures** 18.1 and **18.8**), but that functions in concert with the ER. In the Golgi complex, products of the ER are chemically modified and sorted into those destined for secretion—for example, hormones or digestive enzymes—and those that function in other membranous structures in the cell. Golgi arise from the division of preexisting Golgi and contain various enzymes that modify secretory and membrane proteins differently, depending on their final destination

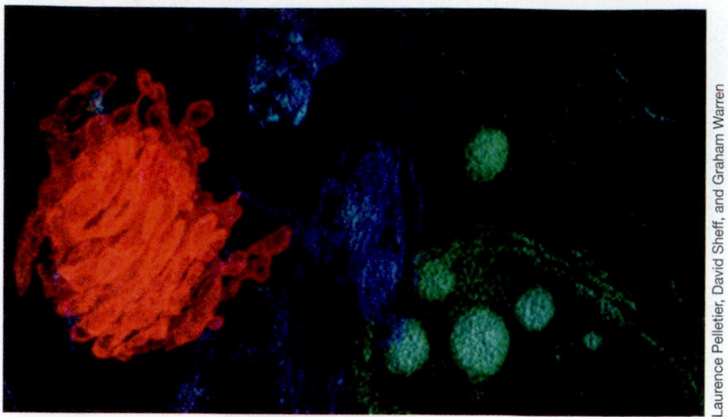

Figure 18.8 The Golgi complex. Transmission electron micrograph of a portion of a cell of the apicomplexan protist *Toxoplasma gondii.* The Golgi complex is colored in red. Note the multiple folded membranes of which it is composed (the membrane stacks are 0.5–1.0 μm in diameter). Other structures such as cytoplasmic granules are shown in other colors. *T. gondii* is a model system for the study of the Golgi complex because each cell contains only one such structure.

in the cell. Many of the modifications are glycosylations (addition of sugar residues) that convert the proteins into specific glycoproteins.

Lysosomes and Peroxisomes

Lysosomes (Figure 18.1) are membrane-enclosed compartments made from proteins and lipids transported from the Golgi complex; they also receive proteins and lipids from the cytoplasmic membrane during the process of endocytosis. Lysosomes contain various digestive enzymes that the cell uses to hydrolyze macromolecules, such as proteins, fats, and polysaccharides, to carry out intracellular digestion. The lysosome fuses with food vacuoles, releasing its digestive enzymes, which break down these macromolecules for use in cellular biosynthesis and energy generation. Lysosomes also function in hydrolyzing damaged cellular components and recycling these materials for new biosyntheses. The internal pH of the lysosome is about 5, two units lower than that of the cytoplasm; the hydrolytic enzymes within the lysosome function optimally at this pH. These hydrolytic enzymes are nonspecific in their activity and could potentially destroy key cellular macromolecules if not contained. Thus, the lysosome allows for lytic activities to be partitioned away from the cytoplasm proper. Following hydrolysis of macromolecules in the lysosome, the resulting monomers pass from the lysosome into the cytoplasm as nutrients for the cell.

The **peroxisome** is a specialized membrane-enclosed metabolic compartment (Figure 18.1). Peroxisomes originate in the cell by incorporating proteins and lipids from the cytoplasm, eventually becoming membrane-enclosed entities that can enlarge and divide in synchrony with the cell. The function of the peroxisome is to oxidize various compounds, such as alcohols and long-chain fatty acids, breaking them down into smaller molecules that are then used by the mitochondrion for energy generation. Peroxisomes also function to oxidize toxic compounds in the cell. The enzymes of the peroxisome

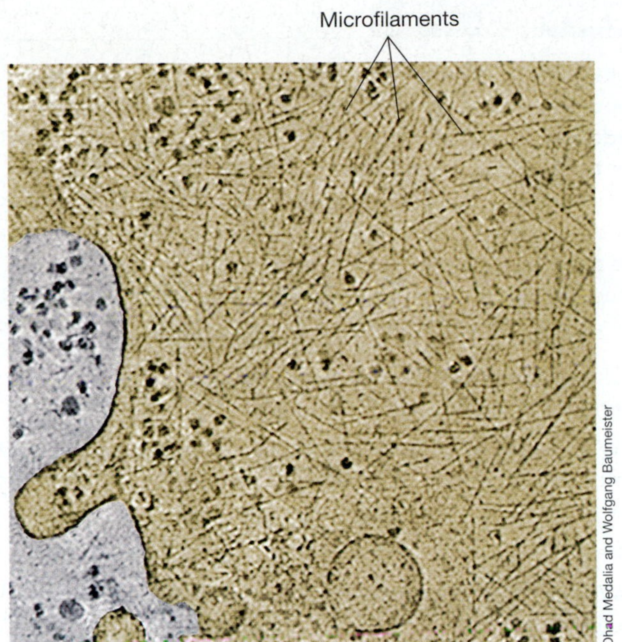

Figure 18.9 Microfilaments and eukaryotic cell architecture. An electron tomographic image of a cell of the cellular slime mold *Dictyostelium discoideum*, a protist, showing the network of actin microfilaments that along with microtubules functions as the cell cytoskeleton. Microfilaments are about 7 nm in diameter. Electron tomography is a method for three-dimensional reconstruction of cells from a series of images taken with a transmission electron microscope.

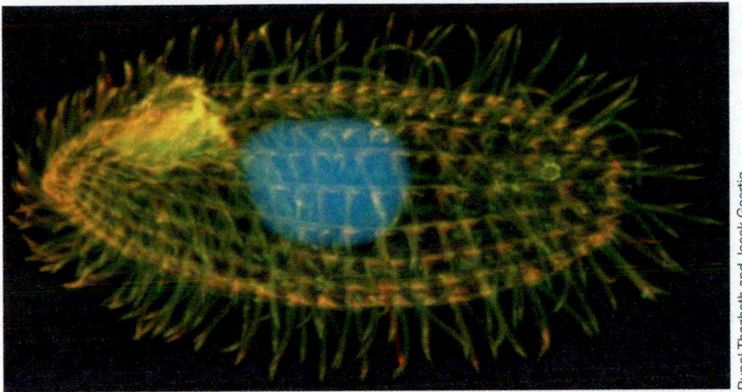

Figure 18.10 Tubulin of *Tetrahymena thermophila*. Fluorescence photomicrograph of a cell of this protist labeled with two types of anti-tubulin antibodies (red/green) and with DAPI, which stains DNA (blue, nucleus). A cell is about 10 μm wide.

transfer hydrogen from these compounds to O_2, producing hydrogen peroxide (H_2O_2) as a by-product. The H_2O_2 produced in the peroxisome, which is itself toxic, is degraded to H_2O and O_2 by the enzyme catalase (Section 6.18). Peroxisomes play other roles as well, such as synthesizing bile salts that aid in the absorption and digestion of fats.

Microtubules, Microfilaments, and Intermediate Filaments

Just as buildings are supported by structural reinforcement, the large size of eukaryotic cells and their ability to move requires structural reinforcement. This internal structural network comes from proteins that form filamentous structures called *microtubules, microfilaments,* and *intermediate filaments*. Together, these structures form the cell **cytoskeleton** (Figures 18.1 and 18.9).

Microtubules are tubes about 25 nm in diameter containing a hollow core about 15 nm wide and are composed of the proteins *α-tubulin* and *β-tubulin*. Microtubules function in maintaining cell shape, in cell motility by cilia and flagella (Figure 18.10), in chromosome movement during cell division, and in movement of organelles. **Microfilaments** are smaller filaments, about 7 nm in diameter and are polymers of two intertwined strands of the protein actin. Microfilaments function in maintaining and changing cell shape, in cell motility by pseudopodia, and during cell division. **Intermediate filaments** are fibrous keratin proteins supercoiled into thicker fibers 8–12 nm in diameter that function in maintaining cell shape and positioning organelles in the cell.

We saw in Chapter 6 that prokaryotic cells produce structural and functional homologs of actin and tubulin in the form of the proteins MreB and FtsZ, respectively (Sections 6.2 and 6.3). This indicates that although prokaryotes are typically much smaller than eukaryotes, they still require some minimal level of internal scaffolding.

Flagella and Cilia

Flagella and cilia are present on many eukaryotic microorganisms. Flagella and cilia are organelles of motility, allowing cells to move by swimming. Cilia are essentially short flagella that beat in synchrony to propel the cell—usually quite rapidly—through the medium. Flagella are long appendages present singly or in groups that propel the cell along—typically more slowly than by cilia—through a whiplike motion (Figure 18.11*a*). The flagella of eukaryotic cells are structurally

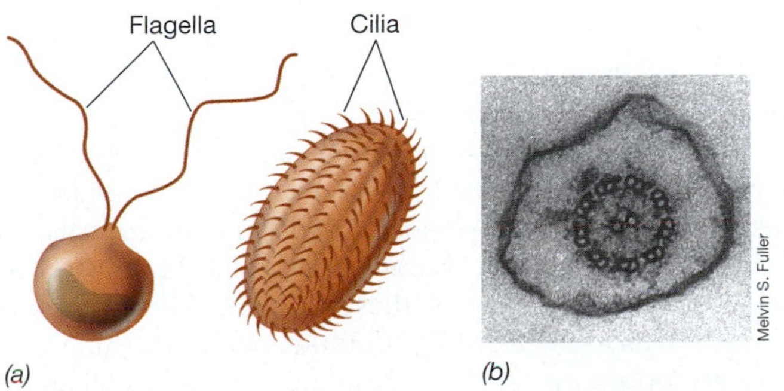

Figure 18.11 Flagella and cilia, motility organelles in eukaryotic cells. *(a)* Flagella can be present as single or multiple filaments. Cilia are structurally very similar to flagella but much shorter. Eukaryotic flagella move in a whiplike motion. *(b)* Cross-section through a flagellum of the chytrid fungus *Blastocladiella emersonii* showing the outer sheath, the outer nine pairs of microtubules, and the central pair of microtubules.

quite distinct from the flagella of prokaryotes (⚭ Section 4.13) and do not rotate.

In cross-section, cilia and flagella are very similar. Each contains a bundle of nine pairs of microtubules composed of tubulins, surrounding a central pair of microtubules called the *axoneme* (Figure 18.11*b*). A second protein, called *dynein,* is attached to the tubulin and functions as an ATPase, hydrolyzing ATP to yield the energy necessary to drive motility. Movement of flagella and cilia is similar. In both cases, movement involves the coordinated sliding of axonemal microtubules (Figure 18.11*b*) against one another in a direction toward or away from the base of the cell. This movement confers the whiplike action on the flagellum or cilium that results in cell propulsion.

A clear distinction can thus be made between the bacterial and eukaryotic flagellum. The filament of the bacterial flagellum is made from a helical array of a single protein, flagellin, and the structure itself is firmly anchored in a rotary motor complex embedded in the cell wall and cytoplasmic membrane (⚭ Figure 4.47). In addition, the bacterial flagellum functions as a propeller, its rotation being driven by the energy of the proton motive force (⚭ Sections 4.13 and 5.12). The eukaryotic flagellum, by contrast, propels the cell by the whip-like motion of sliding microtubules driven by the energy of ATP. Nevertheless, in all motile cells, motility likely has survival value; the ability to move allows motile organisms to move to and exploit new resources and move away from toxic substances or harmful conditions.

Extracellular Components: The Cell Wall and Extracellular Matrix

Cell walls are present in fungi, algae, and plant cells. Animal cells and most protists lack cell walls. The cell wall functions to provide shape to the cell, protect it from the environment, and limit the uptake of water. In multicellular organisms such as plants, the structural support provided by the cell wall helps the plant to withstand gravity. The composition of the plant wall varies with the organism, but the polysaccharide cellulose, together with other polysaccharides and proteins, is used, forming a strong matrix ranging from 0.1 to a few millimeters in thickness, much thicker than the cytoplasmic membrane that it surrounds.

In animal cells, some fungi, and some protists, an **extracellular matrix (ECM)** in which cells are embedded is present outside the cytoplasmic membrane. In animals, where the ECM has been best characterized, different glycoproteins make up the ECM. One of these proteins, called *fibronectin,* connects cells to the ECM by simultaneously attaching to other ECM glycoproteins, such as proteoglycans and collagen, and also by way of *integrins,* proteins embedded in the cytoplasmic membrane. The ECM, through its connection to integrin proteins, provides a means for cells to integrate changes occurring outside and inside the cell and to coordinate the activities of adjacent cells. An ECM is also found in the fruiting bodies formed by some fungi (mushrooms) and slime molds (*Dictyostelium discoideum*) and in biofilms (⚭ Sections 23.4 and 23.5) formed by some yeasts (*Candida albicans*).

18.5 MiniReview

Eukaryotes contain several different organelles and other important structures in the cytoplasm. These include the endoplasmic reticulum, the Golgi complex, lysosomes, and the peroxisome. In addition to all of this, structures called microtubules, microfilaments, and intermediate filaments are present, which form the cell's cytoskeleton. Flagella and cilia are organelles of motility and have extensive microtubular structure. The cell wall and the extracellular matrix connect the cells to the environment or to other cells.

- How does smooth endoplasmic reticulum differ from rough endoplasmic reticulum?
- Why are the activities in the lysosome best partitioned away from the cytoplasm proper?
- Besides scaffolding, what other functions do microtubules have?
- What functional roles does the extracellular matrix of animal cells play?

II EUKARYOTIC MICROBIAL DIVERSITY

We now examine the diversity of microbial ***Eukarya***. These include the *protists,* the *fungi,* and the *unicellular red* and *green algae*. We begin our tour with an overview of the phylogeny of these organisms and then proceed to consider individual groups. Here we learn that microbial eukaryotes have a complex and intriguing evolutionary history and that they are remarkably diverse, varying in many aspects of their biology.

18.6 Phylogeny of the *Eukarya*

From the universal phylogenetic tree of life (⚭ Figure 14.16) we learned that *Eukarya* form their own phylogenetic domain and that *Eukarya* is more closely related to *Archaea* than to *Bacteria*. *Eukarya*'s phylogeny was inferred from sequences of the 18S ribosomal RNA gene, which encodes the small subunit (SSU) RNA of cytoplasmic ribosomes (⚭ Section 14.6).

Problems with the SSU Tree of *Eukarya*

Phylogenetic trees based on comparative sequencing—particularly of SSU rRNA genes—are generally believed to be the best trees available and are thought to provide a realistic picture of microbial evolutionary relationships. However, agreement in this regard is much less strong for the phylogeny of the eukaryotes than for prokaryotes, the *Bacteria* and *Archaea*. Indeed, it now appears that a phylogeny of *Eukarya* based on SSU rRNA genes may not be an accurate picture.

The SSU rRNA view of eukaryotic phylogeny distinguishes certain organisms, such as the diplomonad *Giardia,* the microsporidian *Encephalitozoon,* and the parabasilid

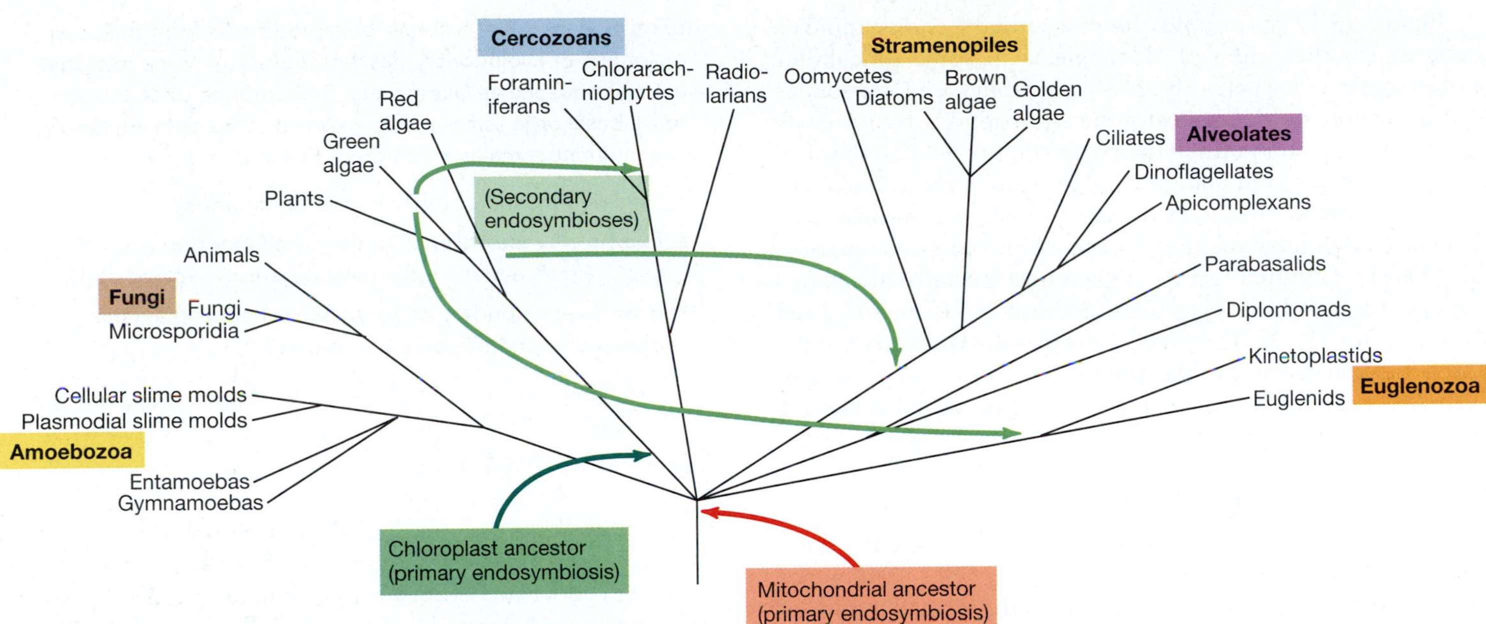

Figure 18.12 Phylogenetic tree of *Eukarya*. This composite generalized phylogenetic tree, based on sequences of various genes and proteins, depicts the relationships among the major groups of eukaryotic organisms. Thick arrows indicate primary endosymbiotic events for the acquisition of the mitochondrion (red) and the chloroplast (green). Thin arrows indicate secondary endosymbiotic acquisition of chloroplasts from red and green algae by various protists.

Trichomonas, as having diverged long ago, well before other eukaryotes, such as animals and plants (Figure 14.16). In support of this view, representatives of these "early branching" eukaryotic groups initially appeared to lack mitochondria and therefore to have arisen before the primary endosymbiotic event(s) that led to the eukaryotic cell containing mitochondria that we know today. However, "amitochondriate" eukaryotes such as *Trichomonas* (Section 18.7) have been found to contain hydrogenosomes, which are metabolically degenerate mitochondria (Section 18.2 and Figure 18.4). Furthermore, sensitive nucleic acid identification methods and genomic sequence data have shown that many amitochondriate eukaryotes contain nuclear genes that originated from *Bacteria*. These genes are analogous to mitochondrial genes that remain in the nucleus of aerobic eukaryotes today (Sections 18.4 and 13.4). In addition, some of these organisms, for example, *Giardia* (Section 18.7), make relic mitochondria—mitochondrial-like proteins that cluster together within the cell and are surrounded by tiny double membrane sacs. These relic mitochondria, called *mitosomes*, have also been found in several other eukaryotes, including the protist *Entamoeba* (Section 18.12), an amitochondriate, but phylogenetically more derived organism. The presence of hydrogenosomes, mitochondrial-like genes in the nucleus, and mitosomes suggests that none of the extant eukaryotic microorganisms that were thought to be amitochondriate were ever really so.

The New Picture of Eukaryotic Phylogeny

Since the development of the SSU phylogenetic tree of eukaryotes, molecular sequencing of several other eukaryotic genes, as well as several proteins, has been used to build alternative trees. These other phylogenetic markers include the amino acid sequences of tubulin proteins, RNA polymerase, and ATPase subunits, among many others. Some of the most revealing information has emerged from protein sequences of heat shock proteins and related chaperonins (Sections 7.17 and 9.11), proteins that are thought to be excellent phylogenetic markers.

Phylogenies based on these alternative markers reveal several differences from the SSU rRNA gene-based tree of *Eukarya*. First, it appears that a major phylogenetic radiation took place as an early event in eukaryote evolution. This radiation included evolution of the ancestors of all, or essentially all, modern-day eukaryotic organisms. A tree representing this view (**Figure 18.12**) shows the diplomonads and parabasilids, amitochondriate organisms once thought to be **basal** (early evolving), instead as **derived** organisms, arising later in eukaryote evolution. Second, animals and fungi appear to be closely related (Figure 18.12). Third, the microsporidia, which branched very early in a SSU rRNA gene tree (Figure 14.16), are revealed to be a highly derived group and close relatives of fungi, a highly derived group themselves (Figure 18.12). Interestingly, microsporidia are transmitted by small infectious sporelike structures similar in some respects to the spores of fungi. The study of heat shock genes has solidified the fungal–microsporidium connection, since bacterial heat shock genes (which are similar to those in the mitochondrion) have been detected in *Encephalitozoon*. This implies that *Encephalitozoon* once had mitochondria, but for some reason dispensed with them, and now contains only traces of these structures.

Figure 18.12 also shows how secondary endosymbiosis accounts for the origin of chloroplasts in some unicellular phototrophic eukaryotes. Following primary endosymbiosis of the cyanobacterial ancestor of chloroplasts by an early mitochondrion-containing eukaryote (Figure 18.12), that lineage diverged into red and green algae (and later into plants). Then, in separate secondary endosymbiosis events, ancestors of certain euglenozoans (Section 18.8) and cercozoans (Section 18.11) engulfed green algae, and certain alveolates (Section 18.9) and stramenopiles (Section 18.10) engulfed red algae (Figure 18.12). These secondary endosymbiosis events, which help account for the phylogenetic diversity of phototrophic eukaryotes, are likely to have occurred relatively recently in evolutionary time.

Eukaryotic Evolution: The Big Picture

What can we conclude from this emerging view of eukaryotic phylogeny? Although phylogenies based on SSU rRNA genes and on other genes and proteins confirm the three domains of life, *Bacteria, Archaea,* and *Eukarya,* our overall view of eukaryotic evolution has changed dramatically with the new information from the other genes and proteins. Also, new aspects of eukaryotic biology, including the finding of hydrogenosomes or gene and protein remnants of mitochondria in organisms previously thought to have never contained them, coupled with recent morphological and sequence-based evidence of secondary endosymbiosis, have shifted our thinking. Taken together, these data reveal that certain eukaryotic groups once thought to have arisen early in evolution are more likely to have arisen within a more recent, major evolutionary radiation. It follows that the origin of the mitochondrion may have predated this major radiation, as all extant *Eukarya* either bear this organelle (or hydrogenosomes) or contain some macromolecular traces of mitochondrial-like bodies.

An intriguing hypothesis then, is that endosymbiotic acquisition of the ancestor of the mitochondrion, which likely provided the early eukaryotic cell with dramatic new metabolic capabilities (Section 14.4), was the trigger for the radiation of *Eukarya*. Later in evolutionary time the ancestor of all phototrophic eukaryotes acquired the ancestor of the chloroplast in a primary endosymbiotic event. Eukaryotic phototrophic diversity arose later through secondary endosymbioses of chloroplast-containing red and green algae.

Despite our advances in understanding eukaryotic evolution, many aspects of the phylogeny of these organisms remain unknown. Thus, the tree shown in Figure 18.12 should not be considered the final word. As more comparative sequencing results come to hand and as new studies (such as the discovery of mitosomes) reveal previously unsuspected aspects of eukaryotic biology, the true phylogeny of the eukaryotes should emerge.

18.6 MiniReview

Based on phylogenetic trees from comparative SSU rRNA gene sequences and on trees from comparative sequencing of other genes and proteins, eukaryotic cells form their own major line of evolutionary descent (*Eukarya*). Some microbial eukaryotes, such as *Giardia* and *Trichomonas*, once thought to be basal organisms, are now known to be derived, having arisen during a major radiation of *Eukarya*.

- How do mitosomes differ from mitochondria?
- Explain why the absence of the mitochondrion could be false evidence that a eukaryotic organism evolved early.
- How does secondary endosymbiosis help explain the diversity of phototrophic eukaryotes?

III PROTISTS

Now that we have the overall phylogeny of *Eukarya* in mind, we proceed to examine the major groups of eukaryotic microorganisms. We begin with organisms traditionally grouped as protists, followed by the fungi and by the unicellular red and green algae.

Protists (**Figure 18.13**), also called protozoa, a word meaning "first" or "original" animals, were first seen over 325 years ago by van Leeuwenhoek (Section 1.6). Traditionally, **protists** have been grouped together based on general similarities. They can be described, for example, as unicellular eukaryotes that lack a cell wall and that are colorless and motile, although there are many exceptions to this description. Many are phototrophic and thus colored, for example, and some have cell walls. They exhibit a wide range of morphologies and inhabit many different kinds of habitats. Many of them also play major roles in human society and health. Recent studies of these organisms indicate they represent a tremendous phylogenetic diversity. Indeed, protists represent much of the diversity found in domain *Eukarya* (Figure 18.12). Our consideration of these organisms follows the major phylogenetic groupings.

18.7 Diplomonads and Parabasalids

Key Genera: *Giardia, Trichomonas*

Diplomonads and parabasalids are unicellular, flagellated protists that lack chloroplasts. They live in anoxic habitats, such as animal intestines, either symbiotically or as parasites, using fermentation for energy generation.

Diplomonads

Diplomonads, which have two nuclei of equal size, contain mitosomes, much reduced mitochondria lacking electron transport proteins and enzymes of the citric acid cycle (Section 18.6). The diplomonad *Giardia intestinalis,* also known as *Giardia lamblia,* causes giardiasis, one of the most common waterborne diarrheal diseases in the United States (Section 36.6 and Figure 36.13).

Parabasalids

Parabasalids contain a *parabasal body* that, among other functions, gives structural support to the Golgi complex. They lack

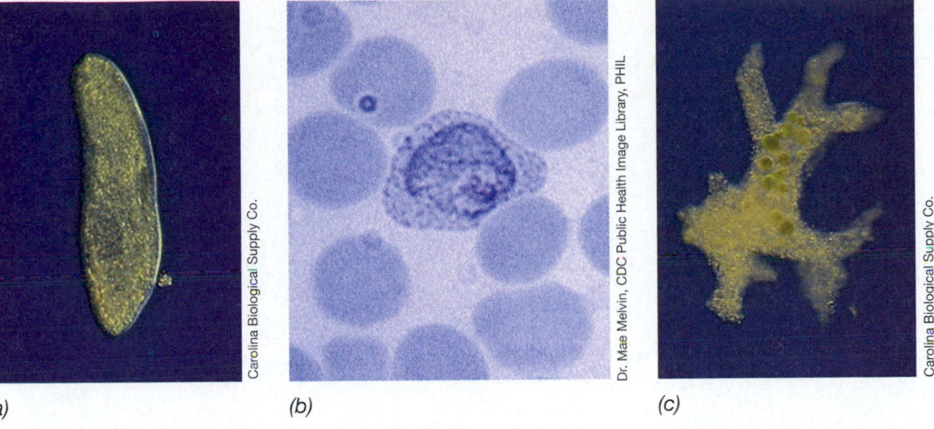

Figure 18.13 Typical protists. *(a)* A typical ciliate, *Paramecium*. A cell is about 60 μm wide. *(b) Plasmodium vivax*, an apicomplexan, growing in a human red blood cell. A cell is about 1.5 μm wide. *(c) Amoeba*, an amoebozoan. The cell is about 60 μm wide.

mitochondria but contain hydrogenosomes for anaerobic metabolism (Section 18.2). Parabasalids live in the intestinal and urogenital tract of vertebrates and invertebrates as parasites or as commensal symbionts (Section 23.1). The parabasalid *Trichomonas vaginalis* (Section 34.14) causes a sexually transmitted disease in humans.

The genomes of parabasalids are unique among eukaryotes in that most of them lack introns, the noncoding sequences characteristic of eukaryotic genes (Sections 8.5 and 13.5). In addition, the genome of *T. vaginalis* is huge for a parasitic organism, about 160 megabase pairs, and shows evidence of genes acquired from bacteria by lateral gene transfer. Much of the genome of *T. vaginalis* contains repetitive sequences, but the organism still contains nearly 60,000 genes, about twice that of the human genome, and near the upper limit observed thus far for eukaryotic genomes.

18.7 MiniReview

Diplomonads are unicellular, flagellated, nonphototrophic protists. Parabasalids such as *Trichomonas* contain huge genomes that lack introns.

- How do diplomonads obtain energy?
- In what kinds of environments do parabasalids inhabit?

18.8 Euglenozoans

Key Genera: *Trypanosoma, Euglena*

Euglenozoans are a diverse assemblage of unicellular, flagellated eukaryotes that are distinguished from other protists in part by the presence of a crystalline rod in their flagella. The function of this morphological feature is not known. Some euglenozoans are parasitic, whereas others are free-living phototrophs or chemoorganotrophs. The euglenozoan group includes the kinetoplastids and euglenids.

Kinetoplastids

Kinetoplastids are a well-studied group of Euglenozoans and are named for the presence of the *kinetoplast,* a mass of DNA present in their single, large mitochondrion. Kinetoplastids live primarily in aquatic habitats, where they feed on bacteria. Some species, however, are parasites of animals and cause a number of serious diseases in humans and vertebrate animals. In *Trypanosoma,* a genus infecting humans, the cells are small, about 20-μm-long, thin, and crescent-shaped. Trypanosomes have a single flagellum that originates in a basal body and folds back laterally across the cell where it is enclosed by a flap of cytoplasmic membrane (**Figure 18.14**). Both the flagellum and the membrane participate in propelling the organism, making effective movement possible even in viscous liquids, such as blood.

Trypanosoma brucei (Figure 18.14) is the species that causes *African sleeping sickness,* a chronic and usually fatal disease. In humans, this parasite lives and grows primarily in the bloodstream, but in the later stages of the disease it invades the central nervous system, causing an inflammation of the brain and spinal cord that is responsible for the characteristic neurological symptoms of the disease. The parasite is transmitted from host to host by the tsetse fly, *Glossina* sp., a bloodsucking fly found only in certain parts of Africa. After

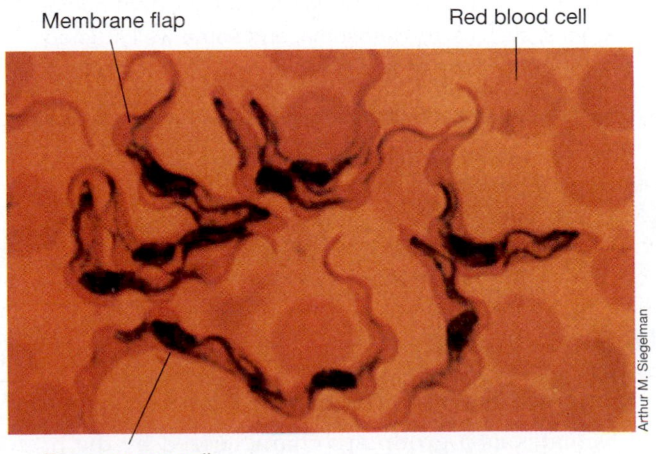

Figure 18.14 Trypanosomes. Photomicrograph of the flagellated euglenozoan, *Trypanosoma brucei*, the causative agent of African sleeping sickness, from a blood smear. A cell is about 3 μm wide.

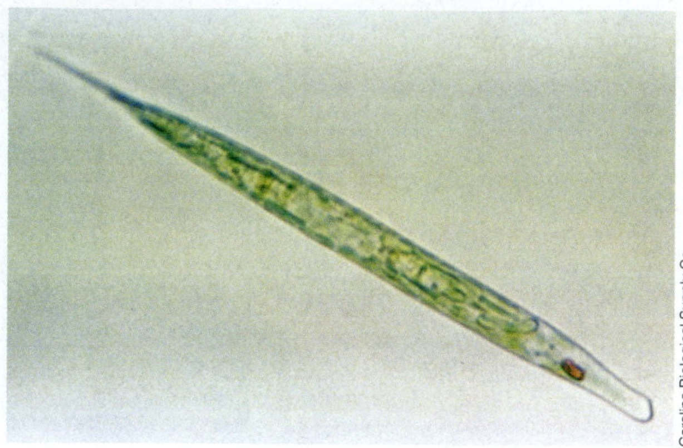

Figure 18.15 ***Euglena*, a euglenozoan.** This phototrophic protist, like other euglenozoans, is not pathogenic. A cell is about 15 μm wide.

moving from the human to the fly via blood, the parasite proliferates in the intestinal tract of the fly and invades the insect's salivary glands and mouthparts, from which it is transferred to a new human host by a fly bite.

Euglenids

Another well-studied group of euglenozoans are the euglenids. Unlike trypanosomes, these organisms are nonpathogenic and phototrophic. They live exclusively in aquatic habitats and contain chloroplasts, which allow for phototrophic growth (**Figure 18.15**). In darkness, however, cells of *Euglena,* a typical euglenid, can lose their chloroplasts and exist as completely heterotrophic organisms. Many euglenids can also feed on bacteria via **phagocytosis**, a process of surrounding a food particle with a portion of their flexible cytoplasmic membrane to engulf the particle and bring it into the cell where it is digested.

18.8 MiniReview

Euglenozoans are unicellular, flagellated protists. Some are phototrophic. This group includes some important human pathogens, such as *Trypanosoma*, and some well-studied nonpathogens, such as *Euglena*.

- How are euglenozoans distinguished from other protists?
- How do cells of *Trypanosoma brucei* get from one human host to another?
- What are the different ways that euglenids obtain nutrition?

18.9 Alveolates

Key Genera: *Gonyaulax, Plasmodium, Paramecium*

The alveolates as a group are characterized by the presence of *alveoli,* sacs present under the cytoplasmic membrane. Although the function of the alveoli is unknown, they may help the cell maintain osmotic balance. Three phylogenetically distinct, though somewhat closely related, kinds of organisms are placed in the alveolate group (Figure 18.12): the *ciliates,* which use cilia for motility; the *dinoflagellates,* which are motile by means of a flagellum; and the *apicomplexans,* which are animal parasites.

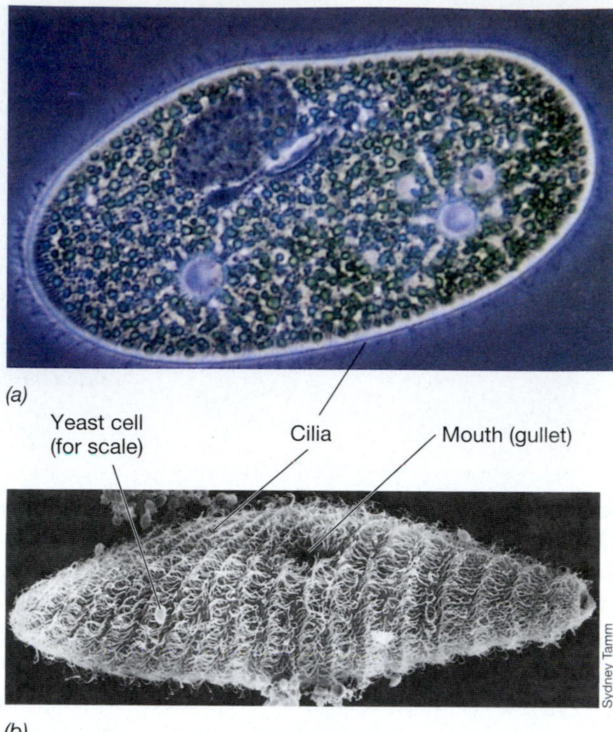

Figure 18.16 ***Paramecium*, a ciliate protist.** *(a)* Phase photomicrograph. *(b)* Scanning electron micrograph. Note the cilia in both micrographs. A single *Paramecium* cell is about 60 μm in diameter.

Ciliates

Ciliates at some stage of their life cycle possess *cilia* (**Figure 18.16**; see also Figure 18.13*a*), structures that function in motility (Section 18.5). Cilia may cover the cell or form tufts or rows, depending on the species. Probably the best-known and most widely distributed ciliates are those of the genus *Paramecium* (Figure 18.16). Like many other ciliates, *Paramecium* uses cilia not only for motility but also to obtain food, by ingesting particulate materials such as bacterial cells through a distinctive funnel-shaped oral groove. Cilia that line the oral groove move food down the groove to the cell mouth (Figure 18.16*b*). There, it is enclosed in a food vacuole by phagocytosis. Digestive enzymes secreted into the food vacuole then break down the food.

Ciliates are unique among protists in having two kinds of nuclei, *micronuclei* and *macronuclei*. Genes in the macronucleus regulate basic cellular functions, such as growth and feeding, whereas those of the micronucleus are involved in sexual reproduction, which occurs through conjugation, a partial fusion of two *Paramecium* cells and exchange of micronuclei.

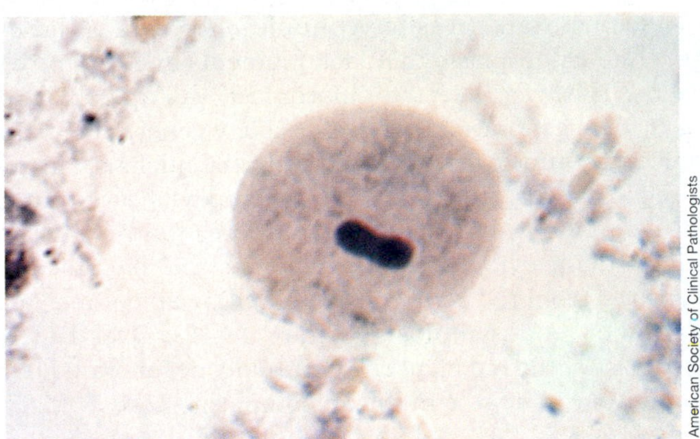

Figure 18.17 ***Balantidium coli*, a ciliate protist that causes a dysentery-like disease in humans.** The dark blue-stained structure is the macronucleus. The cell is about 55 μm wide.

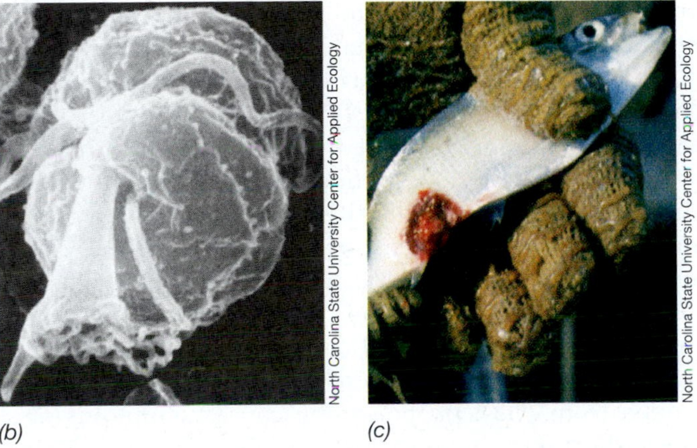

Figure 18.19 **Toxic dinoflagellates (members of the alveolates).** *(a)* Photograph of a "red tide" caused by massive growth of toxin-producing dinoflagellates, such as *Gonyaulax*. The toxin is excreted into the water and also accumulates in shellfish that feed on the dinoflagellates. The toxin is harmless to the shellfish but can poison fish and humans that eat infected shellfish. *(b–c)* Toxic *Pfiesteria*. *(b)* Scanning electron micrograph of a toxic spore of *P. piscicida*; the cell is about 12 μm wide. *(c)* Fish killed by *P. piscicida*. Note the lesions of decaying flesh.

Many *Paramecium* species (as well as many other protists) are hosts for endosymbiotic prokaryotes that reside in the cytoplasm or sometimes in the macronucleus. These organisms may play a nutritional role, synthesizing vitamins or other growth factors used by the host cell. Ciliated protists that are commensal in the termite hindgut carry endosymbiotic methanogens (*Archaea*). These organisms metabolize H_2 produced from pyruvate oxidation in the hydrogenosome (Figure 18.4) to yield methane, which is released to the atmosphere (∞ Section 24.2 and Figure 24.4). Moreover, ciliates themselves can be symbiotic, as obligately anaerobic ciliates are present in the rumen, the forestomach of ruminant animals. Rumen protists play a beneficial role in the digestive and fermentative processes of the animal (∞ Section 24.10).

In contrast to these examples of symbiosis, some ciliates are animal parasites, although this lifestyle is less common in ciliates than in some other groups of protists. The species *Balantidium coli* (**Figure 18.17**), for example, is primarily an intestinal parasite of domestic animals but occasionally infects the intestinal tract of humans, producing dysentery-like symptoms similar to those caused by *Entamoeba histolytica*.

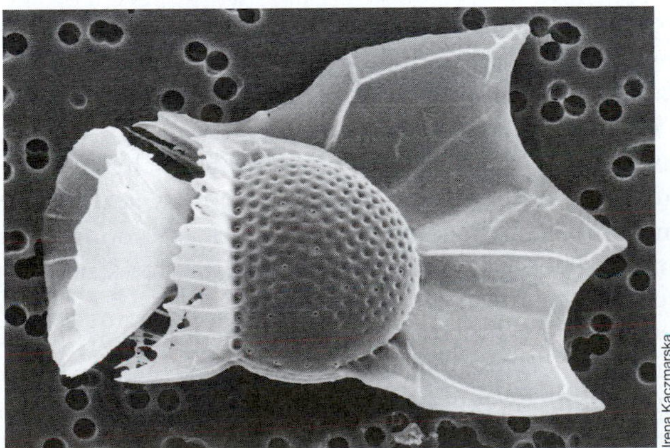

Figure 18.18 **The marine dinoflagellate, *Ornithocercus magnificus* (an alveolate).** The cell proper is the globular central structure; the attached ornate structures are called lists. A cell is about 30 μm wide.

Dinoflagellates

A second group of alveolates are the dinoflagellates, common and very diverse marine and freshwater phototrophic organisms (**Figures 18.18** and **18.19**). Flagella encircling the cell impart spinning movements that give dinoflagellates their name (*dinos* is Greek for "whirling"). Dinoflagellates have two flagella of different lengths and with different points of insertion into the cell, transverse and longitudinal. The transverse flagellum is attached laterally, whereas the longitudinal flagellum originates from the lateral groove of the cell and extends lengthwise (Figure 18.19*b*). Some dinoflagellates are free living (Figure 18.18), whereas others live a symbiotic existence with animals that make up coral reefs, obtaining a sheltered and protected habitat in exchange for supplying phototrophically fixed carbon as a food source for the reef.

In coastal marine waters rich in nutrients, dinoflagellates can reach high numbers. Several species of *Gonyaulax* are bioluminescent (producing visible light from biochemical reactions) and can create spectacular displays of luminescence when they are stimulated by shear forces, as by a wave crashing on shore or a boat or swimmer moving through the water. *Gonyaulax* also can be highly toxic. Dense suspensions of these cells, called "red tides" (Figure 18.19*a*) due to the bright red-colored xanthophyll pigments in this organism, can form in warm, usually polluted, coastal waters. Such blooms are often associated with fish kills and poisoning in humans following consumption of contaminated mussels that accumulate *Gonyaulax* through filter feeding. *Gonyaulax* toxicity results from its production of a substance called *saxitoxin,* a potent neurotoxin that can cause paralytic shellfish poisoning (PSP). Symptoms of PSP include numbness of the lips, dizziness, and difficulty breathing. Death can result from respiratory failure.

Pfiesteria is another genus of toxic dinoflagellates. Toxic spores of *Pfiesteria piscicida* (Figure 18.19*b*) infect fish and eventually kill them due to neurotoxins that affect movement and destroy skin. Lesions form on areas of the fish where the skin has been removed, and this allows the growth of opportunistic bacterial pathogens (Figure 18.19*c*). *P. piscicida* can cause massive fish kills. For example, over a billion fish were killed by a *Pfiesteria* outbreak in the Neuse Estuary, North Carolina (USA), in 1991, and other outbreaks along the eastern coast of the United States have been suspected. Human toxemia from *Pfiesteria* poisoning causes symptoms of skin rashes, short-term memory loss, and respiratory problems.

Apicomplexans

The third group of alveolates, the apicomplexans, are obligate parasites of animals (Figure 18.13*b*). Apicomplexans cause severe diseases such as malaria (*Plasmodium* species), toxoplasmosis (*Toxoplasma*), and coccidiosis (*Eimeria*). These organisms are characterized by nonmotile adult stages, and their food is taken up in soluble form across the cytoplasmic membrane, as in prokaryotes and in fungi. Despite the name *sporozoa,* by which these organisms also have been called, these organisms do not form true resting spores like those of bacteria, algae, and fungi, but instead produce analogous structures called *sporozoites,* which function in transmission of the parasite to a new host. The name apicomplexan derives from the presence at one apex of the sporozoite of a complex of organelles that penetrate host cells.

Numerous vertebrates and invertebrates are hosts for apicomplexans. In some cases, an alternation of hosts takes place, with some stages of the life cycle linked to one host and some to another. Important apicomplexans are the coccidia, typically bird parasites, and members of the genus *Plasmodium* (malaria parasites) (Figure 18.13*b*), which infect birds and mammals, including humans. Because malaria is a major disease of humans, especially in developing countries, we devote a considerable discussion to this disease and the properties of malarial parasites, including alternation of hosts in completing their life cycle, in Section 35.5.

Apicomplexan parasites contain *apicoplasts*. These are degenerate chloroplasts that lack pigments and phototrophic capacity. However, like the chloroplast, the apicoplast contains its own genome and expresses a few genes. The majority of apicoplast functions are encoded in the nucleus, similar in this regard to the mitochondrion. Apicoplasts carry out fatty acid, isoprenoid, and heme biosyntheses, and export their products to the cytoplasm. It is hypothesized that apicoplasts are derived from red algal cells engulfed by apicomplexans in a secondary endosymbiosis (Figure 18.12). Over time, the chloroplast of the red algal cell eventually degenerated to play a nonphototrophic role in the apicomplexan cell.

18.9 MiniReview

Three groups make up the alveolates: ciliates, dinoflagellates, and apicomplexans. Ciliates and dinoflagellates are free-living organisms, whereas apicomplexans are obligate parasites of animals.

- What is one possible function of alveoli?
- Describe how the ciliate *Paramecium* feeds.
- What is an apicoplast?

18.10 Stramenopiles

Key Genera: *Phytophthora, Nitzschia, Dinobryon*

The *stramenopiles* include both chemoorganotrophic and phototrophic organisms. Members of this group bear flagella with many short, hairlike extensions, and this morphological feature gives the group its name (from Latin *stramen* for "straw" and *pilos* for "hair"). The oomycetes, the diatoms, and the golden algae (Figure 18.12) make up unicellular and filamentous stramenopiles. A fourth member, the brown algae, or phaeophytes, which includes many seaweeds, all are multicellular organisms and are not discussed here.

Oomycetes

The oomycetes ("egg fungi"), also called *water molds,* were previously grouped with fungi based on their filamentous growth and the presence of **coenocytic** (that is, multinucleate) hyphae, morphological traits characteristic of fungi. Their life cycle, unlike that of many fungi, includes a diploid, asexually reproducing phase and a diploid, sexually reproducing phase. Phylogenetically, however, the oomycetes are distant from fungi and are closely related to other stramenopiles (Figure 18.12).

Oomycetes differ from fungi in other fundamental ways, as well. For example, the cell walls of oomycetes are typically made of cellulose, not chitin, as for fungi, and they have flagellated cells, which are lacking in all but a few fungi. Moreover, the diploid phase, which in most fungi is reduced, is dominant in oomycetes. Nonetheless, oomycetes are ecologically similar to fungi in that they grow as a mass of hyphae decomposing dead plant and animal material in aquatic habitats. Oomycetes have had a major impact on human society. The oomycete *Phytophthora infestans,* which causes late blight disease of potatoes, contributed to massive famines in Ireland in the seventeenth century. The famines led to the death of a million Irish and to the migration of at least a million more to the United States, Australia, Canada, and other countries.

Diatoms

Diatoms consist of over 100,000 species of unicellular marine and freshwater phototrophic organisms. They produce a specialized, extremely crush-resistant cell wall made of silica, to which protein and polysaccharide are added. The wall, which protects them against predation, exhibits widely different shapes in different species (**Figure 18.20**). The external structure formed by this wall, called a *frustule,* often remains after the cell dies and the organic materials have disappeared. Diatom frustules typically show morphological symmetry, including *pennate symmetry* (having similar parts arranged on opposite sides of an axis) (Figure 18.20*a*), as in the common diatom *Nitzschia,* and radial symmetry (Figure 18.20*b,c*). Because the diatom frustules are resistant to decay, they remain intact for long periods of time and constitute some of the best unicellular eukaryotic fossils known. From this excellent fossil record, it is known that diatoms first appeared on Earth about 200 million years ago.

Golden Algae

Golden algae, also called *chrysophytes,* are primarily unicellular marine and freshwater phototrophs. Some species are chemoorganotrophs and feed by either phagocytosis or by transporting soluble organic compounds across the cytoplasmic membrane. Some golden algae, such as *Dinobryon,* found in freshwater, are colonial. However, most golden algae are unicellular and motile by the activity of two flagella of unequal length.

Golden algae are so named because of their golden-brown color. This is due to chloroplast pigments dominated by the brown-colored carotenoid fucoxanthin. Also, the major chlorophyll pigment in golden algae is chlorophyll *c* rather than chlorophyll *a*, and they lack the phycobilins present in red algal chloroplasts (Section 18.20). Cells of the unicellular golden alga *Ochromonas,* the best-studied genus of this group, have only 1–2 chloroplasts.

18.10 MiniReview

Stramenopiles are protists that bear a flagellum with fine, hairlike extensions. They include oomycetes, diatoms, golden algae, and brown algae.

- In what ways do oomycetes resemble fungi?
- In what ways do oomycetes differ from fungi?
- What structure of diatoms accounts for their excellent fossil record?

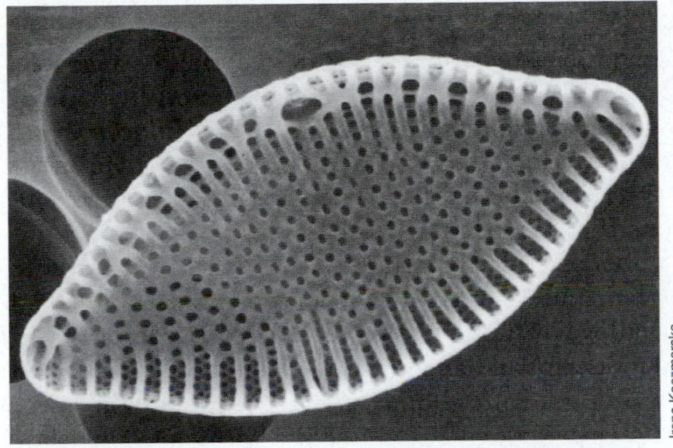

(a)

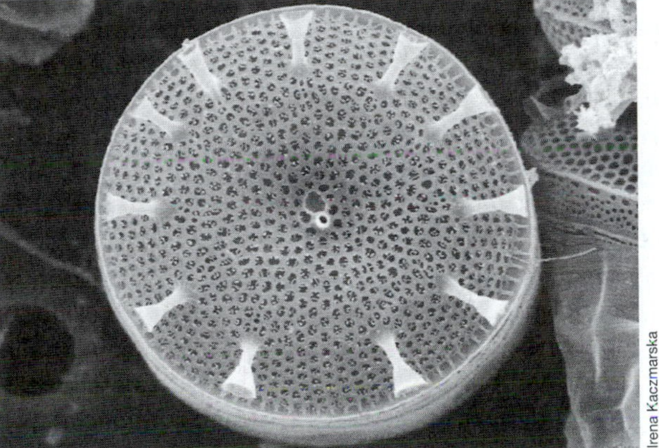

(b)

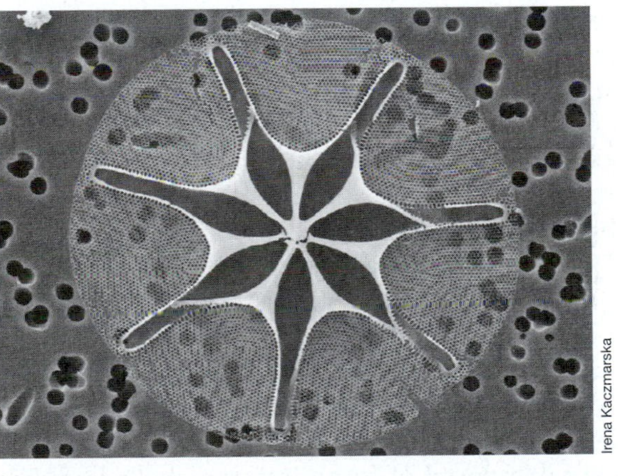

(c)

Figure 18.20 Diatom frustules. Scanning electron micrographs. A frustule is the silica enclosure or wall surrounding the diatom cell proper. *(a)* Frustule of the marine diatom *Nitzschia*, showing pennate symmetry. A cell is about 10 μm wide. *(b)* Frustule of the marine diatom *Thalassiosira*, showing radial symmetry. A cell is about 5 μm in diameter. *(c)* Frustule of the marine diatom *Asteriolampra*, showing radial symmetry. A cell is about 5 μm wide.

18.11 Cercozoans and Radiolarians

Cercozoans and radiolarians are distinguished from other protists by their threadlike pseudopodia, by which they move and feed. Cercozoans were previously called amoeba because

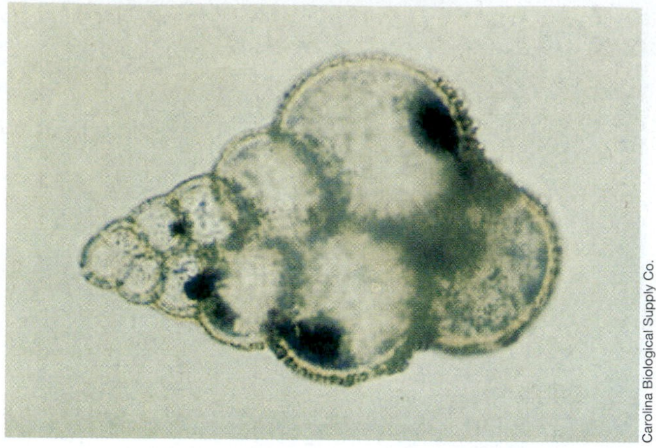

Figure 18.21 A foraminiferan (a cercozoan). Note the ornate and multilobed test. The test is about 1 mm wide.

of their pseudopodia, but it is now known that many phylogenetically diverse organisms use pseudopodia.

Cercozoans

Cercozoa include the chlorarachniophytes and foraminiferans, among other groups, and they are closely related to radiolarians, which also have threadlike pseudopodia. Chlorarachniophytes are phototrophic amoeba-like organisms that develop a flagellum for dispersal; their acquisition of chloroplasts is an example of a secondary symbiosis (Figure 18.12).

Foraminifera are exclusively marine organisms, living primarily in coastal waters. They form shell-like structures called *tests,* which have distinctive characteristics and are often quite ornate (**Figure 18.21**). Tests are typically made of organic materials reinforced with minerals such as calcium carbonate. The cell is not firmly attached to the test, and the amoeba-like cell may extend partway out of the shell during feeding. However, because of the weight of the test, the cell usually sinks to the bottom, and it is thought that the organisms feed on particulate deposits in the sediments, primarily bacteria and the remains of dead organisms. Foraminiferan tests are relatively resistant to decay and hence readily become fossilized (the White Cliffs of Dover, England, for example, are composed of foraminiferan shells laid down in an ancient sea). Because these organisms have left an excellent fossil record, we have a better idea of their distribution through geological time than for virtually any other protist.

Radiolarians

Radiolarians are mostly marine, heterotrophic organisms, and like cercozoans, they also have threadlike pseudopodia. The name radiolarian comes from the radial symmetry of their tests, which generally are made of silica and exist in one fused piece. The tests of marine radiolarians settle to the ocean floor when the cell dies and can build up there over time into thick layers of decaying cell material.

18.11 MiniReview

Cercozoans and radiolarians are two closely related protist groups. The cercozoans, which were previously grouped with amoebas, include chlorarachniophytes and foraminiferans. The tests of radiolarians exhibit radial symmetry.

- What structure distinguishes cercozoans and radiolarians from other protists?
- How are chlorarachniophytes thought to have acquired the ability to photosynthesize?
- Where do foraminiferans live?

18.12 Amoebozoa

Key Genera: *Amoeba, Entamoeba, Physarum, Dictyostelium*

The amoebozoa are a diverse group of terrestrial and aquatic protists that use lobe-shaped pseudopodia for movement and feeding, in contrast to the threadlike pseudopodia of cercozoans and radiolarians. The major groups of amoebozoa are the gymnamoebas, the entamoebas, and the plasmodial and cellular slime molds.

Gymnamoebas

The gymnamoebas are free-living protists that inhabit aquatic and soil environments. They use pseudopodia to move by amoeboid movement (**Figure 18.22**) and feed by phagocytosis on bacteria, other protists, and organic materials. **Amoeboid movement** results from streaming of the cytoplasm as it flows forward at the less contracted and viscous cell tip, taking the path of least resistance. Cytoplasmic streaming is facilitated by microfilaments, which exist in a thin layer just

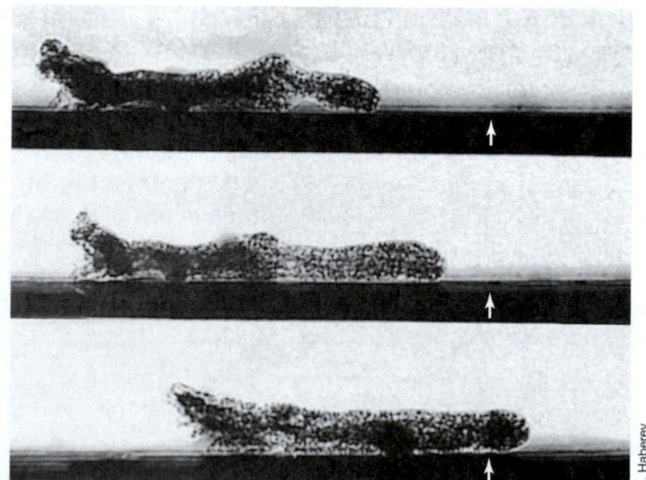

Figure 18.22 Side view of the amoebozoan, *Amoeba proteus*, taken from a film. The time interval from top to bottom is about 6 sec. The arrows point to a fixed spot on the surface. A single cell is about 80 μm wide.

beneath the cytoplasmic membrane in eukaryotic cells (Section 18.5 and Figures 18.1 and 18.9). *Amoeba* (Figure 18.13*c*) is a common freshwater genus, often present in pond water, with species varying in size from 15 μm in diameter—clearly microsocopic—to over 750 μm—almost visible with the naked eye.

Entamoebas

In contrast to gymnamoebas, the entamoebas are parasites of vertebrates and invertebrates. Their usual habitat is the oral cavity or intestinal tract of animals. For example, several species of *Entamoeba* infect humans, though without causing disease. One species, however, *Entamoeba histolytica* (Figure 36.17) is pathogenic in humans and can cause amoebic dysentery, an ulceration of the intestinal tract that results in a bloody diarrhea. This parasite is transmitted from person to person in the cyst form by fecal contamination of water, food, and eating utensils. In Section 36.8 we discuss the etiology and pathogenesis of amoebic dysentery, an important cause of death due to parasites in humans.

Slime Molds

The **slime molds** previously were grouped with fungi since they undergo a similar life cycle (Section 18.13) and produce fruiting bodies with spores for dispersal. As protists, however, slime molds are motile and can move across a solid surface fairly quickly (Figures 18.23–18.25). Phylogenetic analysis places the slime molds in the amoebozoa (Figure 18.12); they appear to be descended from the gymnamoebas.

The slime molds are divided into two groups, *plasmodial* slime molds, also called acellular slime molds, whose vegetative forms are masses of protoplasm of indefinite size and shape called plasmodia (**Figure 18.23**), and *cellular* slime molds, whose vegetative forms are single amoebae. Slime molds live primarily on decaying plant matter, such as leaf litter, logs, and soil. Their food consists mainly of other microorganisms, especially bacteria, which they ingest by phagocytosis. Slime molds can maintain themselves in a vegetative state for long periods but eventually form differentiated sporelike structures that can remain dormant and then germinate later to once again generate the active amoeboid state.

Plasmodial slime molds, such as *Physarum*, exist in the vegetative phase as an expanding single mass of protoplasm called the *plasmodium* that contains many diploid nuclei (Figure 18.23). The plasmodium is actively motile by amoeboid movement, the plasmodium flowing over the surface of the substratum, engulfing food particles as it moves. Cytoplasmic streaming within the plasmodial mass during amoeboid movement serves to distribute nutrients. The plasmodium is genetically diploid. From it, a sporangium and haploid spores can be produced. Under favorable conditions, the spores germinate to yield haploid flagellated swarm cells. The fusion of two swarm cells then regenerates a diploid plasmodium.

Figure 18.23 Slime mold. Plasmodia of the plasmodial slime mold *Physarum* growing on an agar surface. The slime is about 5 cm long and 3.5 cm wide.

UNIT 3

Cellular Slime Molds

Cellular slime molds differ in various ways from plasmodial slime molds. For example, cellular slime molds are haploid and form diploid macrocysts (described below) only under certain conditions. In addition, instead of the single plasmodial mass containing many nuclei seen in plasmodial slime molds, the moving and feeding stage of cellular slime molds are individual, independent amoeboid cells. When the available food is consumed, the cells then aggregate as a pseudoplasmodium (a slug), from which a fruiting body forms, although the individual cells remain separated by their cytoplasmic membranes.

The cellular slime mold *Dictyostelium discoideum* has been used for some time as a model for the study of intercellular communication in eukaryotic cell development. This communal organism undergoes a remarkable life cycle in which vegetative cells aggregate, migrate as a cell mass, and eventually produce fruiting bodies in which cells differentiate and form spores (**Figures 18.24** and **18.25**). When cells of *Dictyostelium* become starved, they aggregate and form a pseudoplasmodium; in this stage cells lose their individuality but do not fuse. Aggregation is triggered by the production of cyclic adenosine monophosphate (cAMP), which functions as a chemotactic agent (we discussed the involvement of cAMP in various regulatory systems in *Bacteria* in Section 9.9). The first cells of *Dictyostelium* that produce this compound become centers for the attraction of neighboring amoeboid cells and trigger their aggregation into slugs.

Fruiting body formation begins when the slug ceases to migrate and becomes vertically oriented (Figures 18.24 and 18.25). The fruiting body then becomes differentiated into a stalk and a head; cells in the anterior end of the slug become stalk cells and those in the posterior end become spores in the head. Cells that form stalk cells begin to secrete cellulose, which provides the rigidity of the stalk. Cells from the rear of the slug swarm up the stalk to the tip and form the head. Most of these posterior cells differentiate into spores. On maturation of the head, the spores are released and dispersed.

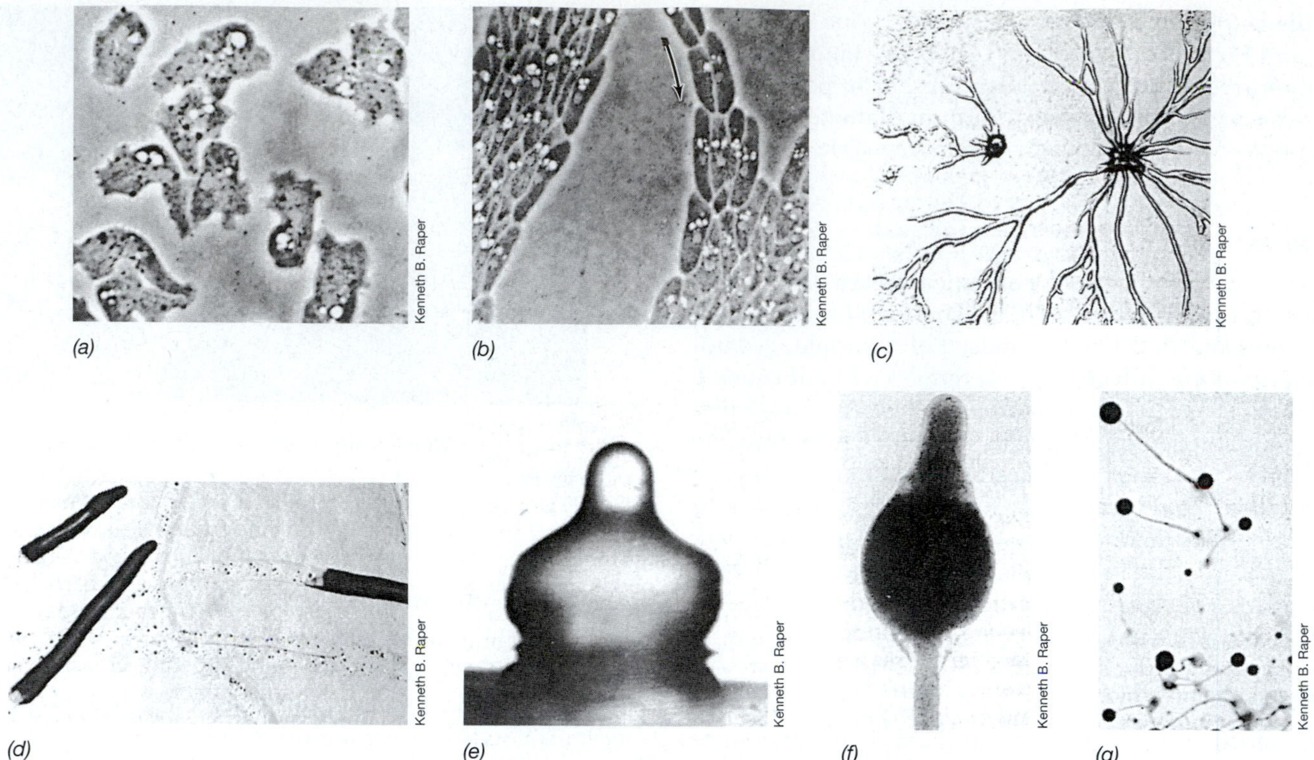

Figure 18.24 Photomicrographs of various stages in the life cycle of the cellular slime mold *Dictyostelium discoideum*. *(a)* Amoebae in preaggregation stage. Note irregular shape and lack of orientation. *(b)* Aggregating amoebae. Amoebae are about 300 μm in diameter. Notice the regular shape and orientation. The cells are moving in streams in one direction. *(c)* Low-power view of aggregating amoebae. *(d)* Migrating pseudoplasmodia (slugs) moving on an agar surface and leaving trails of slime in their wake. *(e, f)* Early stage of fruiting body. *(g)* Mature fruiting bodies. Figure 18.25 shows the sizes of these structures.

Each spore then germinates and releases an independent amoeba.

The cycle of fruiting body and spore formation in *Dictyostelium* is an asexual process. However, sexual spores called *macrocysts* can also be produced. Macrocysts are formed from aggregates of amoebae that become enclosed in a cellulose wall. Following the conjugation of two amoebae, a single large amoeba develops and proceeds to phagocytize the remaining amoebae. At this point a thick cellulose wall forms around this giant amoeba, forming the mature macrocyst; this can remain dormant for long periods. Eventually, the diploid nucleus undergoes meiosis to form haploid nuclei that become integrated into new amoebae that can initiate the asexual cycle (Figures 18.24 and 18.25).

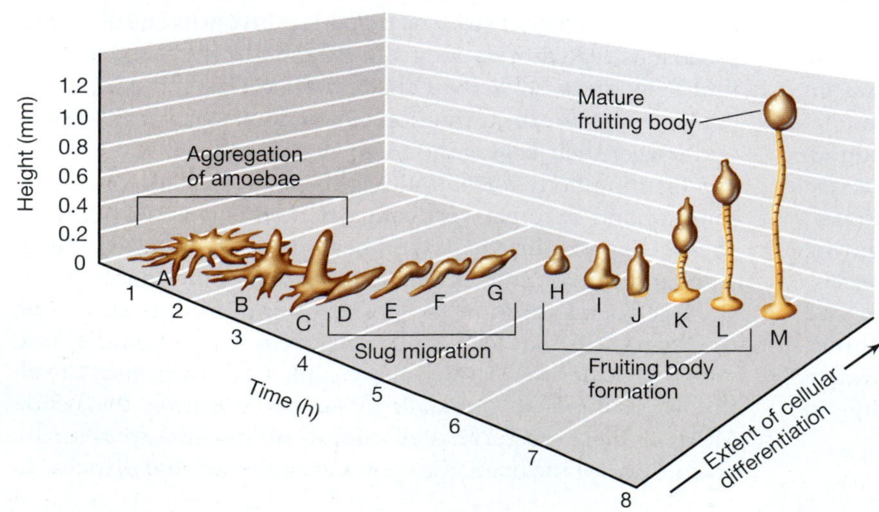

Figure 18.25 Stages in fruiting body formation in the cellular slime mold *Dictyostelium discoideum*. *(a–c)* Aggregation of amoebae. *(d–g)* Migration of the slug formed from aggregated amoebae. *(h–l)* Culmination of migration and formation of the fruiting body. *(m)* Mature fruiting body composed of stalk and head. Cells from the rear of the slug form the head and become spores. The fruiting body contains spores that can become amoebae and begin the life cycle anew.

18.12 MiniReview

Amoebozoa are protists that use pseudopodia for movement and feeding. Within amoebozoa are gymnamoebas, entamoebas, and slime molds. There are two groups of slime molds; the plasmodial slime molds form masses of motile protoplasm, whereas cellular slime molds are masses of individual cells that aggregate to form fruiting bodies from which spores are released.

- How are amoebozoans distinguished from cercozoans and radiolarians?
- Compare and contrast the lifestyles of gymnamoebas and entamoebas.
- Describe the major steps in the life cycle of *Dictyostelium discoideum*.

IV FUNGI

The **fungi** are a large, diverse, and widespread group of organisms, the *molds, mushrooms,* and *yeasts*. Approximately 100,000 species of fungi have been described, with estimates being that as many as 1.5 million species may exist. Fungi form a phylogenetic cluster distinct from other organisms and are most closely related to animals (Figure 18.12).

The habitats of fungi are quite diverse. Some fungi are aquatic, living primarily in freshwater, and a few marine fungi are also known. Most fungi, however, are terrestrial. They inhabit soil or dead plant matter and play crucial roles in the mineralization of organic carbon. A large number of fungi are plant parasites. Indeed, fungi cause many of the economically significant diseases of crop plants. A few fungi cause disease in animals, including humans, although in general fungi are less important as animal pathogens than are other microorganisms. Fungi also establish symbiotic associations with many plants, facilitating the plant's acquisition of minerals from soil (Section 24.13), and many fungi contribute beneficially to human life through fermentation and the synthesis of antibiotics. The following sections describe some of the general features of fungi as a group and then address the diversity of fungi from a phylogenetic perspective.

18.13 Fungal Physiology, Structure, and Associations with Other Organisms

In this section we describe some general features of fungi, including their physiology and ecology, cell structure, and the associations they develop with plants and animals. In the following section we examine fungal reproduction and phylogeny.

Nutrition and Physiology

Fungi are chemoorganotrophs, typically with simple nutritional requirements, and most are aerobes. They feed by secreting extracellular enzymes that digest complex organic materials, such as polysaccharides or proteins, into their monomeric constituents, sugars, peptides, amino acids, and so on. These are then taken up by the fungal cell as sources of energy, carbon, and other nutrients. As decomposers, fungi digest, and thereby recycle, nonliving organic material, such as leaf litter, fallen logs, and the bodies of dead plants and animals. As parasites of plants or animals, fungi use the same mode of nutrition but take up their nutrients from the living cells of the plants and animals they infect and invade.

A major ecological activity of fungi, especially basidiomycetes, is the decomposition of wood, paper, cloth, and other products derived from these natural sources. Fungi that attack these materials can use the cellulose or lignin present as carbon and energy sources. Lignin is a complex polymer in which the building blocks are phenolic compounds. Lignin is an important constituent of woody plants, and in association with cellulose, it confers rigidity on them. Lignin is decomposed in nature almost exclusively through the activities of certain basidiomycetes called *wood-rotting fungi*. Two types of wood rot are known: *brown rot,* in which the cellulose is attacked preferentially and the lignin left unmetabolized, and *white rot,* in which both cellulose and lignin are decomposed. The white rot fungi are of major ecological importance because they play such a key role in decomposing woody material in forests.

Many fungi can grow at environmental extremes of low pH or high temperature (up to 62°C; Table 6.1) and this, coupled with the ease of dispersal of fungal spores, makes these organisms common contaminants of food products, microbial culture media, and surfaces of all sorts.

Fungal Morphology and Spores

Most fungi are multicellular, forming a network of filaments called *hyphae* (singular, hypha). Hyphae are tubular cell walls that surround the cytoplasmic membrane. Fungal hyphae are often septate, with cross-walls dividing each hyphae into separate cells. In some cases, however, the vegetative cell of a fungal hypha contains more than one nucleus (coenocytic); often hundreds of nuclei are present due to repeated division without the formation of cross-walls. Each hyphal filament grows mainly at the tip, by extension of the terminal cell (**Figure 18.26**).

Hyphae typically grow together across a surface and form compact tufts, collectively called a *mycelium,* or mold, which can be seen easily without a microscope (**Figure 18.27**). The mycelium arises because the individual hyphae branch as they grow over and into the organic material on which the fungus is feeding, and these branches intertwine, forming a compact mat. From the mycelial mat, hyphal branches may reach up into the air above the surface, and on these aerial branches spores called **conidia** are formed (Figure 18.26). Conidia are asexual spores (their formation does not involve the fusion of gametes or meiosis), and they are often pigmented (Figure 18.27; Figure 11.4) and resistant to drying. Conidia function in the dispersal of the fungus to new habitats. When conidia form, the white color of the mycelium changes, taking on the color of the conidia, which may be black, blue green, red, yellow, or brown. The presence of these spores gives the mycelial mat a dusty appearance (Figure 18.27*a*). Some fungi

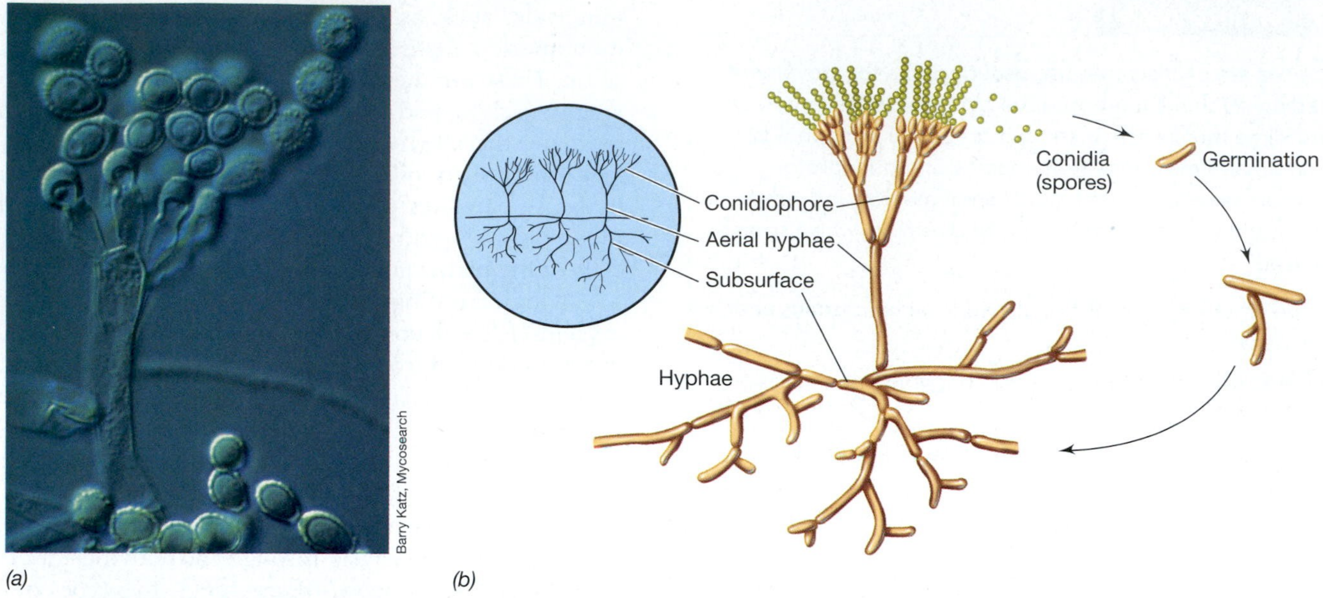

Figure 18.26 Fungal structure and growth. *(a)* Photomicrograph of a typical fungus. Spherical structures at the ends of aerial hyphae are spores. *(b)* Diagram of a mold life cycle. The conidia can be dispersed by either wind or animals and are about 2 μm wide.

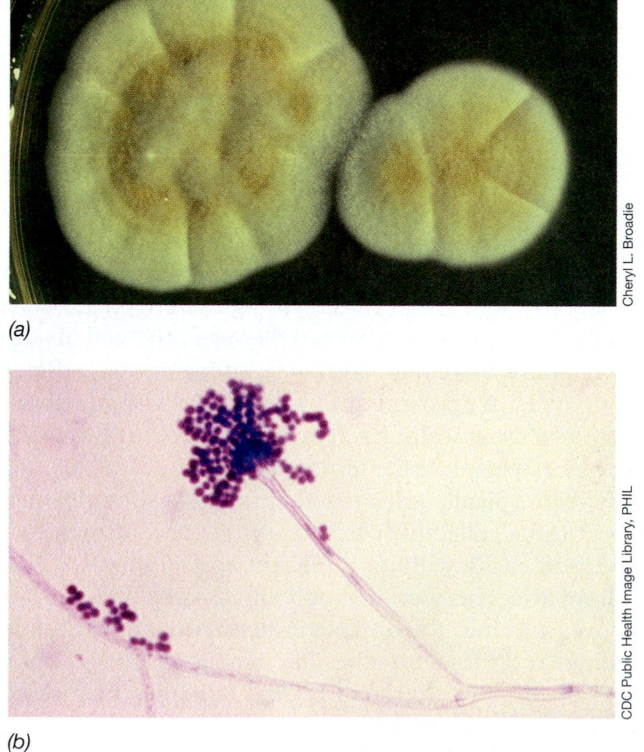

Figure 18.27 Fungi. *(a)* Colonies of an *Aspergillus* species, an ascomycete, growing on an agar plate. Note the appearance of the masses of filamentous cells (the mycelium) and asexual spores that give the colonies a dusty, matted appearance. *(b)* Conidiophore and conidia of *Aspergillus fumigatus*. The conidiophore is about 300 μm long and the condia about 3 μm wide.

form macroscopic reproductive structures called *fruiting bodies* (**mushrooms** or puff balls, for example), in which spores are produced and from which they can be dispersed (**Figure 18.28**). The fruiting bodies can release millions of spores that are spread by wind, water, or animals to new habitats where the spores can then germinate. Some fungi grow as single-celled forms; these are the **yeasts**.

Fungal Cell Walls

Fungal cell walls resemble plant cell walls in basic structure, but not in their chemistry. Although the plant cell wall polysaccharide cellulose is present in the walls of certain fungi, most fungi contain **chitin**, a polymer of the glucose derivative, *N*-acetylglucosamine (∞ Figure 3.3), in their cell walls. Chitin is arranged in the walls in microfibrillar bundles, as is cellulose in plant cell walls. Other polysaccharides such as mannans, galactosans, and chitosans replace chitin in some fungal cell walls. Fungal cell walls are typically 80–90% polysaccharide, with proteins, lipids, polyphosphates, and inorganic ions making up the wall-cementing matrix. An understanding of fungal cell wall chemistry is important because of the extensive biotechnological uses of fungi. In addition, the chemistry of the fungal cell wall has been used in classifying fungi for research and industrial purposes.

Fungal Symbioses

Most plants are dependent on fungi to facilitate their uptake of minerals from soil. The fungi form symbiotic associations with the plant roots called *mycorrhizae* (the word means, literally, "fungus roots"). Mycorrhizal fungi establish close physical contact with the roots and help the plant obtain phosphate and other minerals and also water from the soil. In return, the fungi obtain nutrients such as sugars from the plant root.

There are different kinds of mycorrhizal associations. One kind is *ectomycorrhizae,* typically formed between basidiomycete fungi and the roots of woody plants. In ectomycorrhizal associations the fungal hyphae form a sheath around the plant root but do not penetrate the root extensively. Another kind is *endomycorrhizae,* which form between glomeromycete fungi and many herbaceous (nonwoody) plants. In endomycorrhizae the fungal hyphae embed deeply in the plant root tissue, forming swollen vesicles or branching invaginations called arbuscules.

Fungi also form symbiotic associations with cyanobacteria or green algae. These are called *lichens* and appear as colorful, crusty growths on the surfaces of trees and rocks (Figure 24.35). Section 24.13 has additional discussion of the mycorrhizal and lichen symbioses.

Pathogenesis

Fungi can invade and cause disease in plants and animals. Plant-pathogenic fungi cause widespread crop and plant damage worldwide. For example, the ascomycete *Ophiostoma ulmi* devastated the Dutch elm tree in the United States. Fruit and grain crops suffer significant yearly losses due to fungal infection. Many plant-pathogenic fungi form specialized hyphae, called *haustoria,* that penetrate the plant cell wall and make the cell's cytoplasm available for fungal nutrition. In humans, fungal diseases, called *mycoses,* range from relatively minor and easily cured conditions, such as athlete's foot, to serious, life-threatening systemic mycoses, such as histoplasmosis. Section 35.8 describes some human diseases caused by fungi.

18.13 MiniReview

Fungi include the molds, mushrooms, and yeasts. Other than phylogeny, fungi primarily differ from protists by their rigid cell wall, production of spores, and lack of motility.

- What is chitin and what is its function in fungi?
- What are conidia? How does a conidium differ from a hypha? A mycelium?
- How do endomycorrhizal and ectomycorrhizal associations differ?

18.14 Fungal Reproduction and Phylogeny

Most fungi reproduce primarily by asexual means. Fungi reproduce asexually in three ways: by the growth and spread of hyphal filaments, by asexual production of spores (Figures 18.26 and 18.27), or by simple cell division as in budding yeasts (**Figure 18.29**). Fungi that do not have sexually produced spores were formerly grouped as deuteromycetes ("imperfect fungi"), whereas those using sexual reproduction were the "perfect fungi." With the advent of molecular systematics, this grouping is no longer widely used.

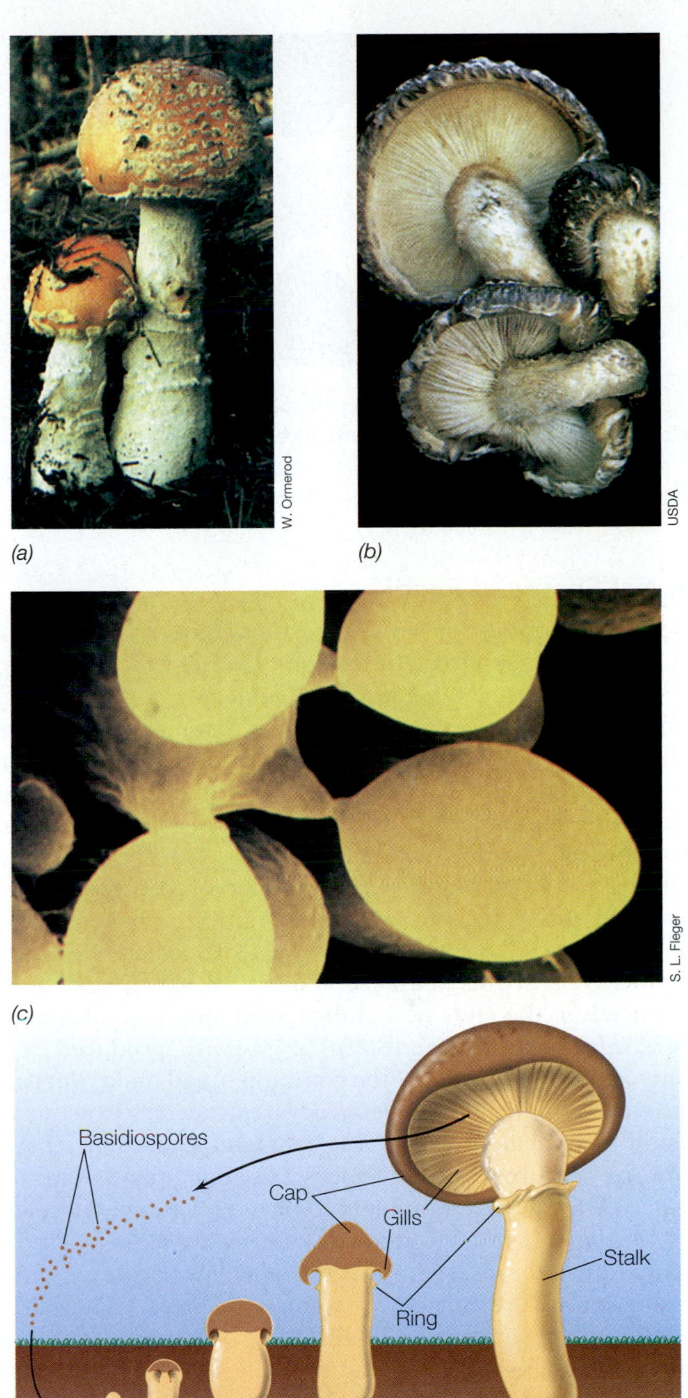

Figure 18.28 Mushrooms. *(a)* The basidiomycete *Amanita,* a highly poisonous mushroom. *(b)* Gills on the underside of the mushroom fruiting body contain the spore-bearing basidia. *(c)* Scanning electron micrograph of basidiospores released from mushroom basidia. *(d)* Life cycle of a typical mushroom. Mushrooms typically develop underground and then emerge on the surface rather suddenly (usually overnight), triggered by an influx of moisture.

UNIT 3

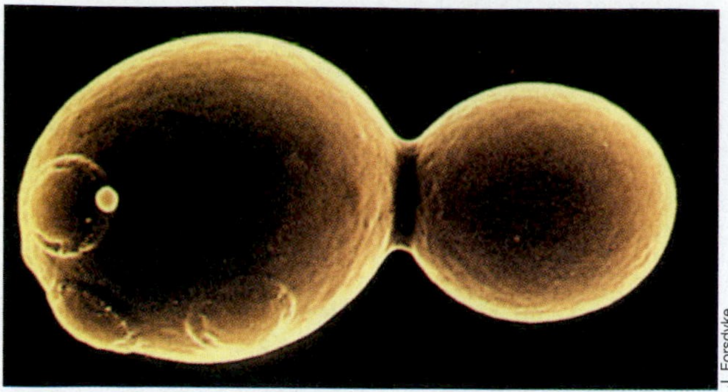

Figure 18.29 Scanning electron micrograph of the common baker's and brewer's yeast *Saccharomyces cerevisiae* (an ascomycete). Note the budding division and scars from previous budding. A single large cell is about 6 μm in diameter.

Sexual Spores of Fungi

Some fungi produce spores as a result of sexual reproduction. The spores develop from the fusion of either unicellular gametes or specialized hyphae called *gametangia*. Alternatively, sexual spores can originate from the fusion of two haploid cells to yield a diploid cell; this then undergoes meiosis and mitosis (∞ Section 8.6) to yield individual haploid spores. Depending on the group, different types of sexual spores are produced. Spores formed within an enclosed sac (ascus) are called *ascospores*. Many yeasts produce ascospores, and we consider this situation with the common baker's yeast, *Saccharomyces cerevisiae,* in Section 18.18, where we outline the life cycle of this yeast (see Figure 18.33). Sexual spores produced on the ends of a club-shaped structure (basidium) are *basidiospores* (Figure 18.28*c*). *Zygospores,* produced by zygomycetous fungi, such as the common bread mold *Rhizopus,* are macroscopically visible structures that result from the fusion of hyphae and genetic exchange. Eventually the zygospore matures and produces asexual spores that are dispersed by air and germinate to form new fungal mycelia.

Sexual spores of fungi are typically resistant to drying, heating, freezing, and some chemical agents. However, fungal sexual spores are not as resistant to heat as bacterial endospores (∞ Section 4.12). Either an asexual or a sexual spore of a fungus can germinate and develop into a new hypha and mycelium.

The Phylogeny of Fungi

Fungi share a more recent common ancestor with animals than with any other group of eukaryotic organisms (Figure 18.12); phylogenetically, fungi and animal cells are therefore *sister groups*. Fungi and animals are thought to have diverged approximately 1.5 billion years ago. The earliest fungal lineage is thought to be the chytridiomycetes, an unusual group in which cells produce flagellated spores (zoospores, Section 18.15). Thus the lack of flagella in most fungi indicates that motility is a characteristic that has been lost at various times in different fungal lineages.

A more detailed picture of fungal phylogeny than that shown in Figure 18.12 is depicted in the evolutionary tree in

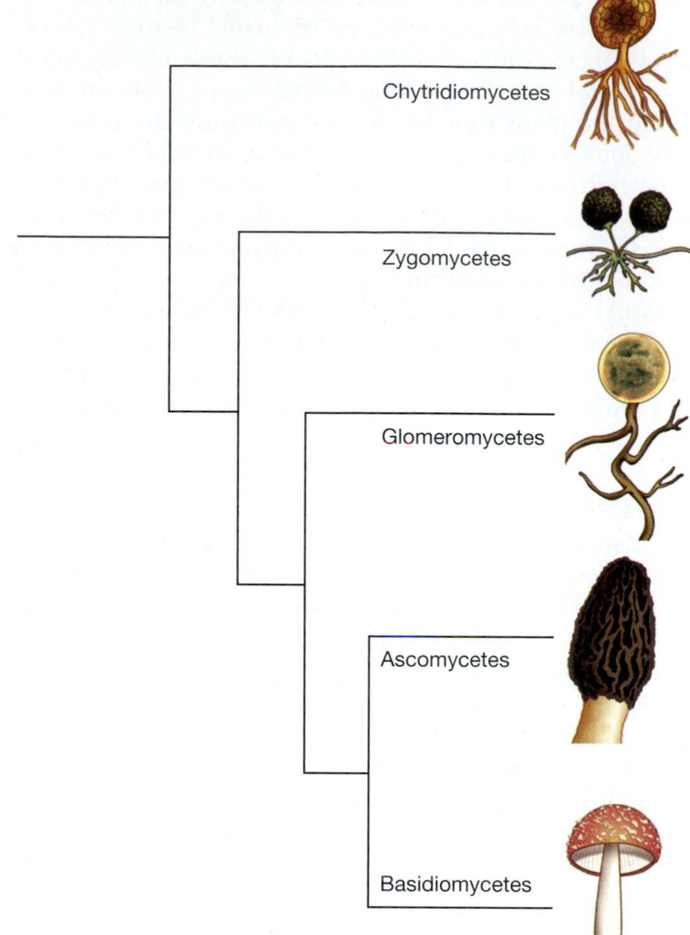

Figure 18.30 Phylogeny of fungi. This generalized phylogenetic tree based on 18S rRNA gene sequences depicts the relationships among the major groups (phyla) of fungi.

Figure 18.30. The phylogeny shown in this figure, based on comparative sequencing of SSU rRNA (these can be used to resolve fairly close, but not distant, relationships among eukaryotes, see Section 18.6), defines several distinct fungal groups: the chytridiomycetes, zygomycetes, glomeromycetes, ascomycetes, and basidiomycetes. Figure 18.30 also lends support to the idea that the chytridiomycetes are phylogenetically basal to all other fungal groups and that the most derived groups of fungi are the basidiomycetes and the ascomycetes.

18.14 MiniReview

A variety of sexual spores are produced by fungi, including ascospores, basidiospores, and zygospores. From a phylogenetic standpoint, fungi are most closely related to animal cells, and chytridiomycetes are the earliest lineage of fungi.

- Ascomycetes can produce both ascospores and conidia. Describe the basic differences between these two types of spores.
- How long ago are fungi and animals thought to have diverged?

18.15 Chytridiomycetes

Key Genera: *Allomyces, Batrachochytrium*

Chytridiomycetes, or *chytrids,* are the earliest diverging lineage of fungi (Figure 18.30). Their name refers to the structure of the fruiting body ("little pot"), which contains zoospores. The cell wall, as in other fungi, contains chitin. The chytrids differ from many other fungi, however, in producing motile, flagellated spores called *zoospores,* a possible remnant of their adaptation to aquatic environments, mostly freshwater and moist soils, where they are commonly found.

Many species of chytrids are known, and some exist as single cells, whereas others form colonies with hyphae. They include both free-living forms that degrade organic material, such as *Allomyces,* and parasites of animals, plants, and protists. The chytrid *Batrachochytrium dendrobatidis* causes chytridiomycosis of frogs (**Figure 18.31**), a condition in which the organism infects the frog's epidermis and interferes with the ability of the frog to respire across the skin. Chytrid fungi have been implicated in the massive die-off of amphibians worldwide, possibly in response to increases in environmental temperature associated with global warming. It is thought that warmer temperatures have improved growth conditions for chytrids and thus frog–chytrid encounters are more likely to occur. Other factors may also contribute to frog mortality, including increased susceptibility of the amphibians to chytrid infection due to habitat loss and aquatic pollution.

Unresolved aspects of the phylogeny of chytrids suggest that this group is not monophyletic. Some organisms currently classified as chytrids may actually be more closely related to species of other fungal groups, such as the zygomycetes. As is true for protists, much about the evolution of fungi remains to be learned.

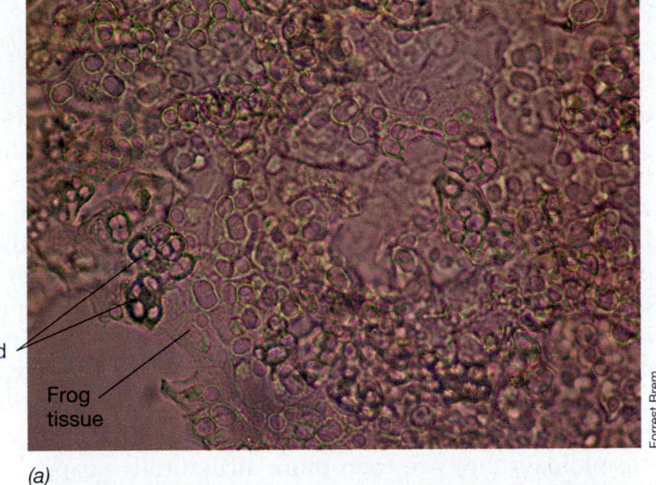

(a)

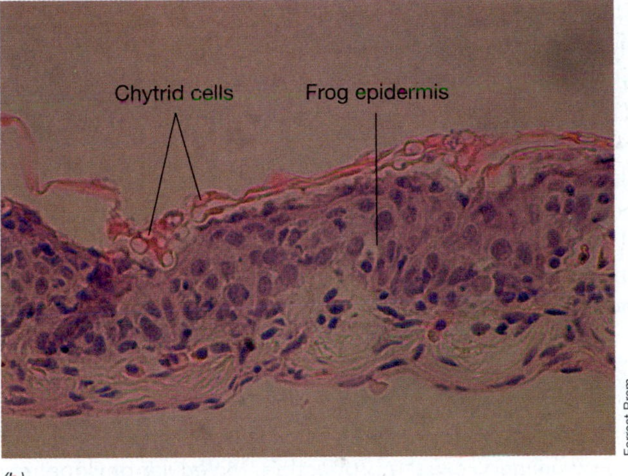

(b)

Figure 18.31 Light micrographs of a chytridiomycete. *(a)* Cells of the chytrid *Batrachochytrium dendrobatidis* growing on the surface of the skin of a frog and viewed from above. Chytrid cells range from 20–75 μm in diameter. *(b)* In this stained section viewed from the side, the chytrid cells can be seen covering the surface of the frog epidermis.

18.15 MiniReview

Chytrids are thought to be the earliest diverging lineage of fungi, and some species are amphibian pathogens.

- What is one feature of chytrids that links them with other fungi?
- What is one feature of chytrids that distinguishes them from other fungi?
- How do chytrids cause disease in frogs?

18.16 Zygomycetes

Key Genera: *Rhizopus, Encephalitozoon*

Zygomycetes, or zygote fungi, are known primarily for their role in food spoilage. They are commonly found in soil and on decaying plant material. All are coenocytic (multinucleate), and a unifying characteristic is the formation of zygospores (Section 18.14).

The black bread mold, *Rhizopus stolonifer,* is representative of the zygomycetes. This organism undergoes a complex life cycle that includes both asexual and sexual reproduction. In the asexual phase, the mycelia form sporangia, within which haploid spores are produced. When released, genetically identical spores can disperse and germinate, giving rise to vegetatively growing mycelia. In the sexual phase of reproduction there are mycelia of different mating compatibility types, with different chemical markers. When hyphae of different mating types come into close proximity, they form hyphal extensions (gametangia, Section 18.14) that contain several nuclei. The fusion of the two gametangia mixes the haploid nuclei together and initiates formation of a heterokaryotic (different nuclei) zygosporangium, which can remain dormant and resist dryness and other unfavorable conditions. When conditions are favorable, the different haploid nuclei fuse to become diploid, and meiosis then leads to production of haploid spores as the zoosporangium germinates into a sporangium. As in the asexual phase, the release of the spores, in this case genetically nonidentical, from the sporangium disperses the organism for vegetative, hyphal growth.

UNIT 3

Microsporidia

Closely related to or perhaps belonging within zygomycetes phylogenetically are the microsporidia, tiny (2–5 μm), unicellular, obligate parasites of animals and protists. For example, microsporidia are often one of the opportunistic pathogens that infect immune-compromised individuals, such as those who have AIDS. Based on SSU rRNA gene sequencing and their lack of mitochondria, microsporidia were once thought to form an early branching lineage of *Eukarya* (Section 18.6). Recent evidence, however, of the presence of genes typical of *Bacteria* and mitosome-like mitochondrial remnants in these organisms places them as close relatives of zygomycetes.

Microsporidia have adapted to a parasitic lifestyle through the elimination or loss of many key aspects of eukaryotic biology; they are even more structurally stripped down than other amitochondriate eukaryotes. The microsporidium *Encephalitozoon,* for example, lacks all organelles, including the Golgi complex and hydrogenosomes (whether mitosomes are present is unknown), and contain very small genomes. The genome of *Encephalitozoon* is only 2.9 Mbp and contains only about 2000 genes. This is 1.5 Mbp and 2600 genes smaller than that of the bacterium *Escherichia coli*! The *Encephalitozoon* genome lacks genes for seminal metabolic pathways, such as the citric acid cycle, meaning that this pathogen must be highly dependent on its host for even the most basic of metabolic processes.

18.16 MiniReview

Zygomycetes form coenocytic hyphae and undergo both asexual and sexual reproduction. Molecular phylogenetic analysis places microsporidia, once thought to be an early branching lineage of *Eukarya*, as closely related to or belonging within the zygomycetes.

- What feature of zygomycetes gives this group its name?
- In what ways have microsporidia adapted to a parasitic lifestyle?

18.17 Glomeromycetes

The glomeromycetes are a relatively small group of fungi that nonetheless have major ecological importance. In contrast to chytrids and zygomycetes, in which several hundred to a thousand species have been described, and to the ascomycetes and basidiomycetes, each with an excess of 30,000 species, only about 160 species of glomeromycetes are currently known. All known species of glomeromycetes form endomycorrhizae, also called arbuscular mycorrhizas (Sections 18.13 and 24.13), typically with the roots of herbaceous plants, but in some cases also with woody plants. As many as 70% or more of angiosperm plants form endomycorrhizal associations in which the fungal hyphae enter the plant cell walls and produce swollen vesicles or arbuscules. The increase in surface contact between the hyphae and the plant cell cytoplasm resulting from the arbuscule structure aids the plant's acquisition of minerals from the soil.

As plant symbionts, glomeromycetes are thought to have played an important role in the ability of early vascular plants to colonize land. So far, none of them have been grown independent of a plant. They reproduce only asexually and are mostly coenocytic in their morphology.

18.17 MiniReview

Glomeromycetes are fungi that form arbuscular mycorrhizal associations with plants.

- How do arbuscular mycorrhizae aid the acquisition of nutrients by plants?

18.18 Ascomycetes

Key Genera: *Saccharomyces, Candida, Neurospora*

The ascomycetes are a large and highly diverse group of fungi. Ascomycetes range from primarily single-celled species, such as the yeast *Saccharomyces* that is used in the baking and brewing industries, to species that grow as filaments, such as the bread mold *Neurospora crassa*.

The group ascomycetes, representatives of which are found in aquatic and terrestrial environments, takes its name from the production of asci (singular, ascus), cells in which two haploid nuclei from different mating types come together and fuse, forming a diploid nucleus that then undergoes meiosis to form haploid ascospores. In some ascomycetes, the asci are formed within a fruiting body, called an ascocarp. Ascomycetes reproduce asexually by the production of conidia that form by mitosis at the tips of specialized hyphae called *conidiophores*. The ecological role of ascomycetes is primarily that of decomposers of dead plant material, but many ascomycete species are ectomycorrhizal with forest trees, and a large number are partners in lichen symbioses together with cyanobacteria or green algae. We focus here on yeasts and especially *Saccharomyces* as a representative ascomycete.

Saccharomyces cerevisiae

The cells of *Saccharomyces* and other single-celled ascomycetes are typically spherical, oval, or cylindrical, and cell division typically takes place by budding. In the budding process, a new cell forms as a small outgrowth of the old cell; the bud gradually enlarges and then separates from the parent cell (**Figures** 18.29 and **18.32**). Although most ascomycete yeasts apparently reproduce only as single cells, some, including *Saccharomyces cerevisiae,* can form filaments *Saccharomyces* in response to certain environmental conditions. In these species, specific cellular characteristics may be expressed only in the filamentous form. For example, the filamentous phase is essential for pathogenicity in the normally unicellular *Candida albicans,* a yeast that can cause vaginal, oral, lung, or systemic infections in AIDS patients and other immunosuppressed patients (Section 34.15).

Yeast cells are typically much larger than bacterial cells and can be distinguished microscopically from prokaryotes by their larger size and by the obvious presence of internal cell structures, such as the nucleus or cytoplasmic vacuoles (Figure 18.32). Yeasts flourish in habitats where sugars are present, such as fruits, flowers, and the bark of trees. Most yeasts are facultative aerobes, capable of fully aerobic as well as fermentative metabolism. A number of yeast species live symbiotically with animals, especially insects, and a few species are pathogenic for animals and humans (Section 35.8). The most important commercial yeasts are the baker's and brewer's yeasts, which are members of the genus *Saccharomyces*. The original habitats of these yeasts were undoubtedly fruits and fruit juices. However, the commercial yeasts of today are probably quite different from wild strains because they have been greatly improved through the years by careful selection and genetic manipulation by industrial microbiologists. The yeast *S. cerevisiae* has been studied as a model eukaryote for many years (Section 18.1) and was the first eukaryote to have its genome completely sequenced (Section 13.5). **www.microbiologyplace.com Online Tutorial 18.1: Life Cycle and Mating Type in a Typical Yeast**

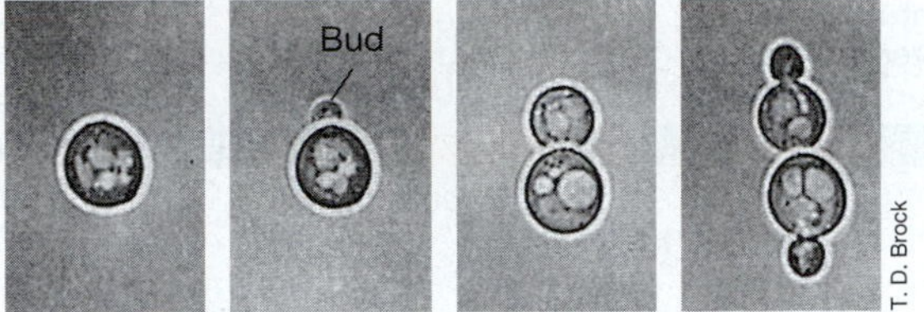

Figure 18.32 Growth by budding division in *Saccharomyces cerevisiae*. Shown is a time-lapse series of phase-contrast micrographs of the budding division starting from a single cell. Note the pronounced nucleus. A single cell of *S. cerevisiae* is about 5 μm in diameter.

Sexual Reproduction in Yeasts

Some yeasts exhibit sexual reproduction by a process called *mating*, in which two yeast cells fuse. Within the fused cell, called a zygote, ascospores are eventually formed. The sexual cycle of the model yeast *S. cerevisiae*, including the important property of *mating types*, is described here (**Figure 18.33**).

Cells of *S. cerevisiae* can exist in the haploid (vegetative) stage (containing 16 chromosomes), but in nature, two haploid yeast cells can fuse (mate) to yield a diploid cell (32 chromosomes). *Saccharomyces* grows vegetatively as a diploid. *S. cerevisiae* has two different forms of haploid cells called mating types; these can be considered analogous to male and female gametes. The mating of opposite mating types forms a diploid cell. From a single diploid cell growing under nutrient-poor conditions, an ascus forms that contains four ascospores two of each mating type (Figure 18.31).

The two mating types of *S. cerevisiae* are designated α and *a*. Cells of type *a* mate only with cells of type α, and whether a cell is mating type *a* or mating type α is genetically determined. Some haploid strains of *S. cerevisiae* remain *a* or α, but other strains are able to switch their mating type from one to the other. This switch in mating type occurs when the active mating gene is replaced with one of two otherwise silent genes (**Figure 18.34**).

There is a single location on one of the *S. cerevisiae* chromosomes called the MAT (for *mating type*) locus, at which either gene *a* or gene α can be inserted. At this locus, the MAT promoter controls transcription of whichever gene is present. If gene *a* is at that locus, then the cell is mating type *a*, whereas if gene α is at that locus, the cell is mating type α. Elsewhere in the yeast genome are copies of genes *a* and α that are not expressed. These silent copies are the source of the inserted gene. In the switch (Figure 18.34), the appropriate gene, *a* or α, is copied from its silent site and inserted into the MAT location, replacing the gene already present. The old mating-type gene is excised and discarded, and the new gene

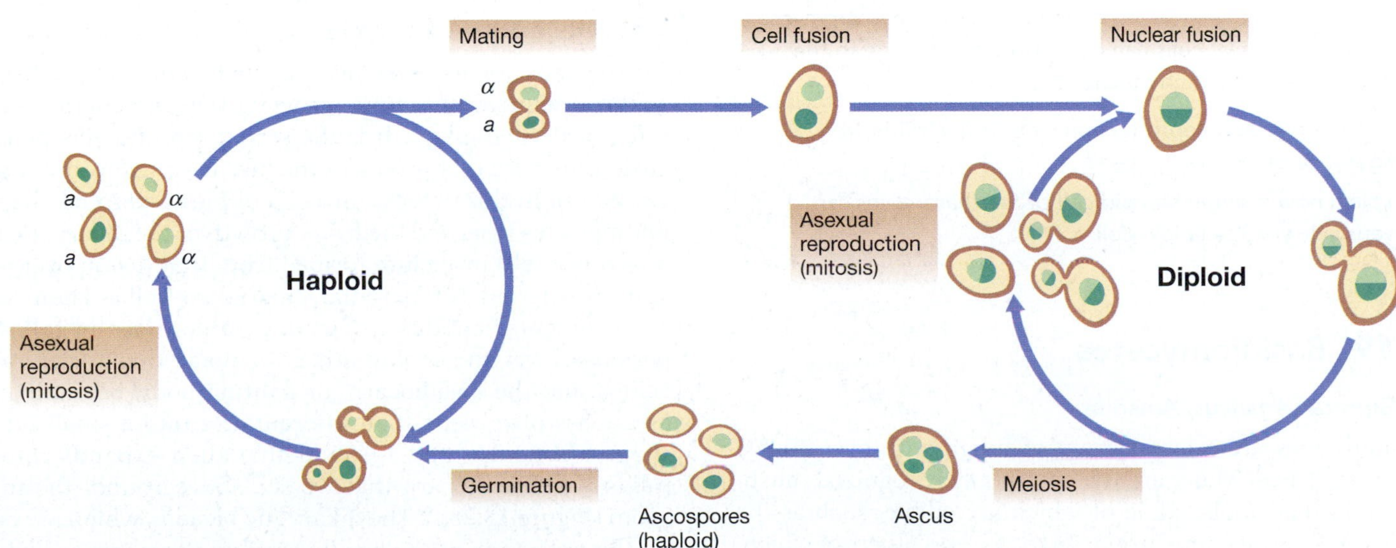

Figure 18.33 Life cycle of a typical ascomycete yeast, *Saccharomyces cerevisiae*.

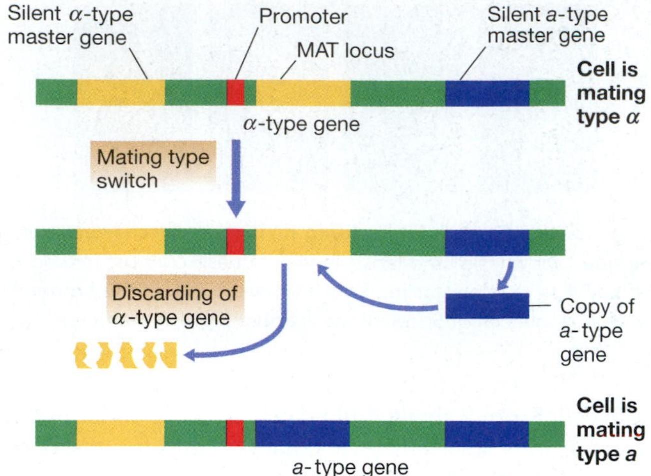

Figure 18.34 The cassette mechanism that switches an ascomycete yeast from mating type α to *a*. The cassette inserted at the active locus (reading head) determines the mating type. The process shown is reversible, so type *a* can also revert to type α.

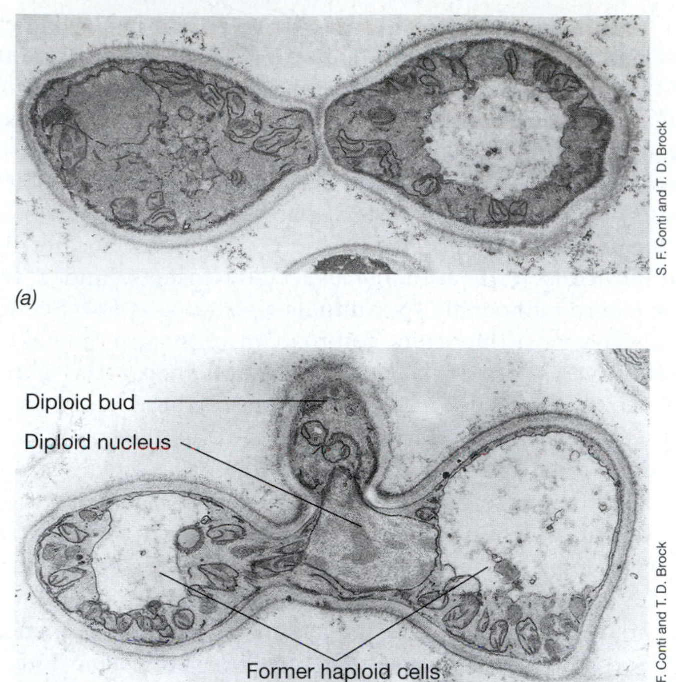

Figure 18.35 Electron micrographs of the mating process in an ascomycete yeast, *Hansenula wingei*. *(a)* Two cells have fused at the point of contact and have sent out protuberances toward each other. *(b)* Late stage of mating. The nuclei of the two cells have fused, and a diploid bud has formed at a right angle to the mating cells. This bud eventually separates and becomes the progenitor of a diploid cell line. A single cell of *Hansenula* is about 10 μm in diameter.

is inserted. Whichever gene is inserted in the MAT locus is the one that will be transcribed.

The α and *a* genes of *S. cerevisiae* are regulatory genes. Among other things, they regulate the production of the peptide hormones α factor or *a* factor, which are excreted by yeast cells undergoing mating. These hormones bind to cells of the opposite mating type and bring about changes in their cell surfaces that enable the cells to fuse (**Figure 18.35**). Once mating has occurred, the nuclei fuse, forming a diploid zygote (Figure 18.35). The zygote grows vegetatively by budding (Figure 18.35*b*). Under starvation conditions, *S. cerevisiae* will undergo meiosis and generate ascospores once again.

18.18 MiniReview

The ascomycetes are a large and diverse group of saprophytic fungi. Some, such as *Candida albicans*, can be pathogenic in humans. There are two mating types in the yeast *Saccharomyces cerevisiae*, and yeast cells can convert from one type to the other by a genetic switch mechanism.

- In what ways does reproduction by budding differ from vegetative growth of hyphae?
- Explain how a *single* haploid cell of *Saccharomyces* can eventually yield a diploid cell.

18.19 Basidiomycetes

Key Genera: *Agaricus, Amanita*

Basidiomycetes are a large group of fungi, with over 30,000 species described. Many are the commonly recognized mushrooms and toadstools, some of which are edible, such as the commercially grown mushroom *Agaricus*, and others of which, such as *Amanita* (Figure 18.28*a*) are highly poisonous. We will discuss in Sections 25.15 the commercial growth of these and other fungi as a food source. In addition to these familiar forms, there are yeasts and plant and human pathogens in this group. The unifying characteristic of the basidiomycetes is the basidium (plural, basidia), usually a single-celled structure, in which basidiospores are formed by meiosis. The basidium ("little pedestal") gives the group its name.

The Mushroom Life Cycle

During most of its existence, a mushroom fungus lives as a simple haploid mycelium, growing vegetatively in soil, leaf litter, or decaying logs. It is the sexual reproductive phase of basidiomycetes that produces the familiar mushroom (Figure 18.28*a, b*). In this process, mycelia of plus and minus haploid mating types fuse, and the faster growth of the dikaryotic (two nuclei per cell) mycelium formed from that fusion overgrows and crowds out the parental haploid mycelia. Then, when environmental conditions are favorable, usually following periods of wet and cool weather, the dikaryotic mycelium develops into the basidiocarp, or fruiting body, beginning first as a mass of mycelia that differentiates into a small button-shaped structure underground that then expands into the full-grown basidiocarp that we see aboveground, the mushroom (Figure 18.28*a*). The dikaryotic basidia, which are borne on the underside of the basidiocarp on flat plates called gills attached to the cap of the mushroom (Figure 18.28*b, d*), then undergo a fusion of the two nuclei, forming basidia with diploid nuclei. The two rounds of meiotic division generate four haploid nuclei in the basidia, and each of the nuclei becomes a basidiospore (Figure 18.28*c*). The genetically different basidiospores can then be dispersed by wind to new habitats to begin the cycle again, germinating under favorable conditions and growing as haploid mycelia (Figure 18.28*d*).

18.19 MiniReview

Basidiomycetes include the familiar mushrooms, some of which are highly poisonous and others of which are edible. Members of this fungal group undergo both vegetative reproduction as haploid mycelia and sexual reproduction via fusion of mating types and formation of haploid basidiospores.

- What is the technical term for the aboveground structure mushroom of Basidiomycetes?
- In what structure are basidiospores formed?

V UNICELLULAR RED AND GREEN ALGAE

We conclude our tour of eukaryotic microbial diversity with the **algae**. Phototrophic eukaryotes originated from a primary endosymbiosis, the engulfment and retention by a eukaryotic, heterotrophic cell of a cyanobacterium (Figure 18.12). Because cyanobacteria carry out oxygen-releasing (oxygenic) photosynthesis, phototrophic eukaryotes also carry out oxygenic photosynthesis, using water as a photosynthetic electron donor (Section 20.5). Later, this early phototrophic protist diverged into the ancestors of the red and the green algae, and the lineage leading to the green algae later gave rise to the ancestor of plants (Figure 18.12).

As discussed in Section 18.4, various secondary endosymbioses, in which cells of red and green algae were engulfed by other eukaryotic protists and their chloroplasts retained, led to a further diversification of phototrophic eukaryotes. Here we focus on the red and green algae, a large and diverse group of eukaryotic organisms that contain chlorophyll and carry out oxygenic photosynthesis.

18.20 Unicellular Red Algae

Key Genus: *Cyanidioschyzon*

The red algae, also called *rhodophytes*, are found mostly in marine habitats, but a few species inhabit freshwater and terrestrial habitats. They are phototrophic and contain chlorophyll *a*; they are noteworthy among the algae in that their chloroplasts lack chlorophyll *b* and contain phycobiliproteins, the major light-harvesting pigments of the cyanobacteria (Section 20.3). The reddish color (**Figure 18.36**) of many red algae results from phycoerythrin, an accessory pigment that masks the green color of chlorophyll. At greater depths in aquatic habitats, where less light penetrates, cells produce more phycoerythrin and are a darker red, whereas shallower-dwelling species often have less phycoerythrin and can be green in color.

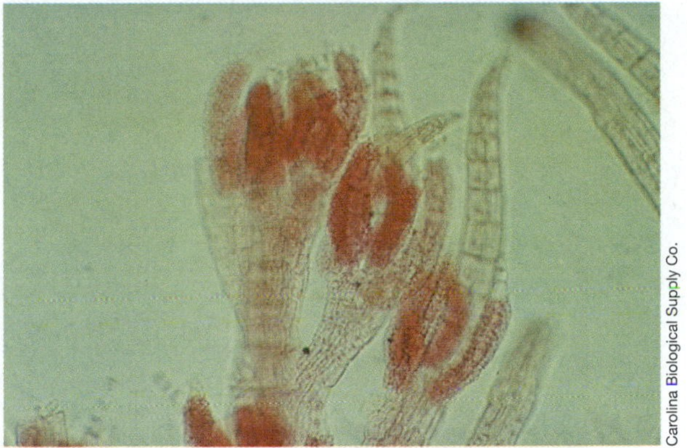

Figure 18.36 ***Polysiphonia*, a marine red alga.** Light micrograph. *Polysiphonia* grows attached to the surfaces of marine plants. Cells are about 150 μm wide.

Most species of red algae are multicellular and lack flagella. Some are included in the seaweeds and serve as the source of agar, the polysaccharide solidifying agent used in bacteriological media. Different species of red algae are filamentous, leafy, or, if they deposit calcium carbonate, *corraline* (coral like) in morphology, and often exhibit an alteration of multicellular diploid and haploid forms.

Cyanidium and Relatives

Unicellular species of red algae are also known. One such group, members of the *Cyanidiales* that includes the genera *Cyanidium, Cyanidioschyzon,* and *Galdieria* (**Figure 18.37**),

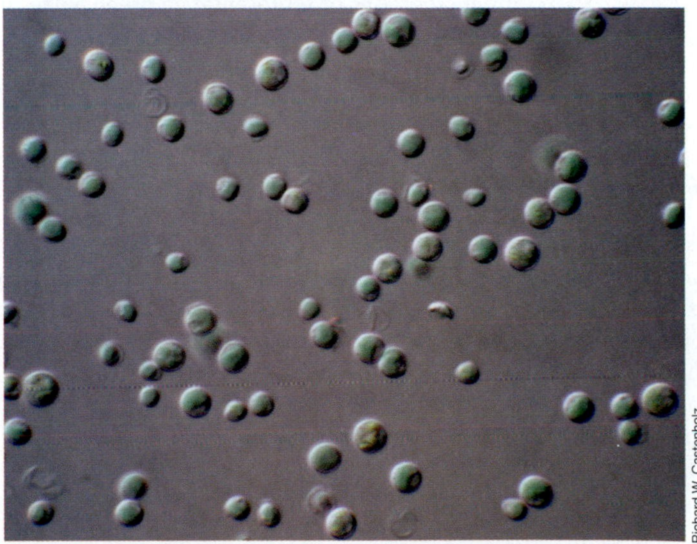

Figure 18.37 ***Galdieria*, a red alga.** These small unicellular phototrophic protists exist at low pH and high temperature in hot springs. The cells are about 25 μm in diameter and are green in color because, as a shallow-dwelling species, *Galdieria* has less phycoerythrin, a reddish accessory pigment that masks the green of chlorophyll in deeper-dwelling species.

UNIT 3

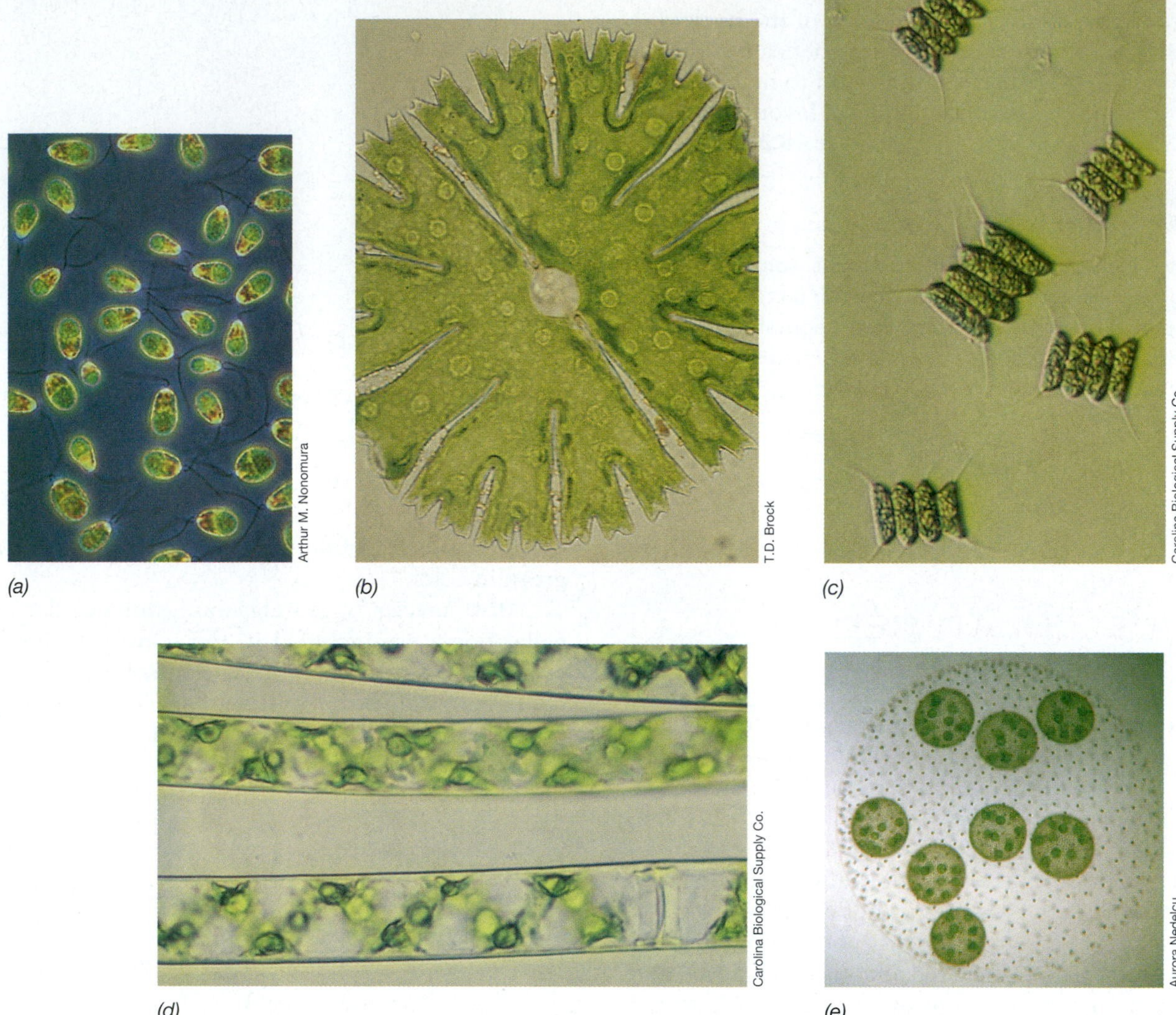

Figure 18.38 Light micrographs of representative green algae. *(a)* A single-celled, flagellated green alga, *Dunaliella*. A cell is about 5 μm wide. *(b) Micrasterias*. A single cell about 100 μm wide. *(c) Scenedesmus*, showing packets of four cells each. *(d) Spirogyra*, a filamentous alga with cells about 20 μm wide. Note the green spiral-shaped chloroplasts. *(e) Volvox carteri* colony with eight daughter colonies. Daughter colonies are about 50 μm wide. The colony is made up of several hundred flagellated somatic cells (the small dots on the surface of the colony and on the surfaces of the smaller green daughter colonies). The daughter colonies contain reproductive cells called gonidia (the dark green dots within the daughter colonies).

live in acidic hot springs at temperatures from about 35° to 56°C and at pH values from 0.5 to 4.0; under these extreme conditions, no other phototrophic microorganisms can exist. The unicellular red algae are unusual in other ways as well. For example, cells of *Cyanidioschyzon merolae* are unusually small (1–2 μm in diameter) for eukaryotes, and the genome of this species, approximately 16.5 Mbp, is one of the smallest genomes known for a phototrophic eukaryote.

18.20 MiniReview

Red algae are mostly marine and range from unicellular to multicellular. Their reddish color is due to the synthesis of the pigment phycoerythrin.

- What traits link cyanobacteria and red algae?
- How is the photosynthesis system of red algae different from that of plants?

18.21 Unicellular Green Algae

Key Genera: *Chlamydomonas, Volvox*

The green algae, also called *chlorophytes*, bear chloroplasts containing chlorophylls *a* and *b* giving them their characteristic green color, but lack phycobilins. In the composition of their photosynthetic pigments, they are similar to plants, to which they are closely related phylogenetically. There are two main groups of green algae, the chlorophytes, examples of which are the unicellular *Chlamydomonas* (∞ Figure 6.21) and *Dunaliella* (**Figure 18.38a**), and the charophyceans, the group that is actually most closely related to land plants.

Most green algae inhabit freshwater, although some are marine, and others are found in moist soil or growing in snow, to which they impart a pink color, or as symbionts in lichens. The morphology of chlorophytes ranges from unicellular (Figure 18.38*a, c*); to filamentous, with individual cells arranged end-to-end (Figure 18.36*d*); to colonial, as aggregates of cells (Figure 18.36*e*); and multicellular, as in the seaweed *Ulva*. Most have a complex life cycle, with both sexual and asexual reproductive stages.

One of the smallest eukaryotes known is the green alga *Ostreococcus tauri,* a common unicellular member of marine phytoplankton (∞ Section 23.9 and Figure 23.16*b*), with a cell diameter of approximately 2 μm. *O. tauri* also is notable in that it has the smallest genome of any known phototrophic eukaryote, approximately 12.6 Mbp. *Ostreococcus* thus provides a model organism for research into the evolution of genome reduction and specialization in eukaryotes. At the **colonial** level of organization in green algae is *Volvox* (Figure 18.38*d*). This alga forms colonies composed of several hundred flagellated cells involved in motility and photosynthesis and several other cells that specialize in reproduction. *Volvox* has been a model organism for research into the genetic mechanisms controlling multicellularity and the distribution of functions among cells in multicellular organisms.

Endolithic Phototrophs

Some green algae grow inside rocks. These *endolithic* (*endo* means "inside") phototrophs inhabit porous rocks, for example, those containing quartz, and are typically found in layers near the rock surface. Endolithic phototrophic communities are most common in dry environments such as deserts or cold dry environments such as the Antarctic. For example, in the Antarctic Dry Valleys, where temperatures and humidity are very low, life within a rock has its advantages. Rocks are heated by the sun and water from snow melt can be absorbed and retained for relatively long periods, supplying moisture needed for growth. Moreover, water absorbed by a porous rock makes the rock more transparent, thus funneling more light to the algal layers.

A wide variety of phototrophs can form endolithic communities, including cyanobacteria (**Figure 18.39**) and various green algae. In addition to being free-living phototrophs, green algae and cyanobacteria coexist with fungi in endolithic lichen communities (∞ Section 24.13 for discussion of the lichen symbiosis). Metabolism and growth of these internal rock communities slowly weathers the rock, allowing gaps to develop where water can enter, freeze and thaw, and eventually crack the rock, producing new habitats for microbial colonization.

(a)

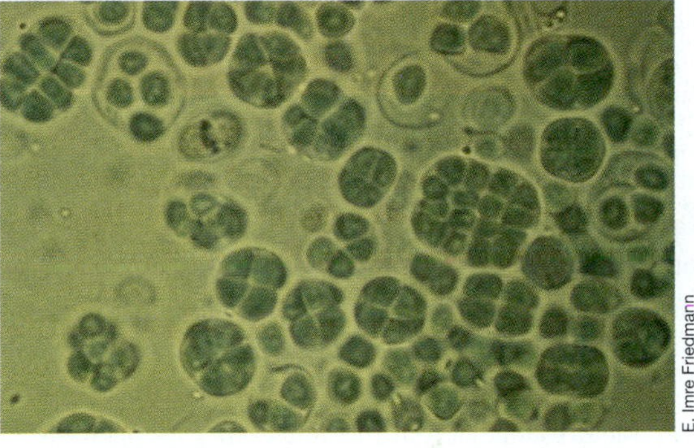

(b)

Figure 18.39 Endolithic phototrophs. *(a)* Photograph of a limestone rock from Makhtesh Gadol, Negev Desert, Israel, showing a layer of cells of the cyanobacterium *Chroococcidiopsis*. Cells are about 10 μm in width. *(b)* Photomicrograph of cells of *Chroococcidiopsis* isolated from a sandstone rock in the Negev Desert.

18.21 MiniReview

Green algae are common in aquatic environments and are also found growing in moist soils and even snow. They exist as unicellular, filamentous, colonial, and multicellular species. A unicellular green alga, *Ostreococcus*, has the smallest genome known for a phototrophic eukaryote. The green alga *Volvox* is a model multicellular phototroph.

- What phototrophic properties link green algae and plants?
- In what kinds of habitats would one most likely find endolithic green algae?

Review of Key Terms

Algae phototrophic eukaryotes, both microorganisms and macroorganisms

Amoeboid movement a type of motility in which cytoplasmic streaming via pseudopodia moves the organism forward

Basal an organism or group of organisms evolving early and giving rise to other groups

Calvin cycle the series of biosynthetic reactions by which most photosynthetic organisms convert CO_2 to organic compounds

Chitin a polymer of *N*-acetylglucosamine commonly found in the cell walls of fungi

Chloroplast the photosynthetic organelle of eukaryotic phototrophs

Ciliate protist characterized in part by rapid motility driven by numerous short appendages called cilia

Coenocytic the presence of multiple nuclei in fungal hyphae without septa

Colonial the growth form of certain protists and green algae in which several cells live together and cooperate for feeding, motility or reproduction; an early form of multicellularity

Conidia asexual spores of fungi

Cristae the internal membranes of a mitochondrion

Cytoskeleton cellular scaffolding typical of eukaryotic cells, in which microtubules, microfilaments, and intermediate filaments define the cell's shape

Derived an organism or group of organisms evolving later, from a more basal group

Endosymbiosis the engulfment of one cell type by another cell type and the subsequent stable association of the two cells

Eukarya eukaryotic organisms

Eukaryote a member of *Eukarya*

Extracellular matrix (ECM) proteins and polysaccharides that surround an animal cell and in which the cell is embedded

Fungi nonphototrophic eukaryotic microorganisms with rigid cell walls

Histones positively charged proteins that package eukaryotic DNA in nucleosomes

Hydrogenosome an organelle of endosymbiotic origin present in certain anaerobic eukaryotic microorganisms that functions to oxidize pyruvate to H_2, CO_2, and acetate along with the production of one molecule ATP

Intermediate filament a filamentous polymer of fibrous keratin proteins, supercoiled into thicker fibers, that functions in maintaining cell shape and the positioning of certain organelles in the eukaryotic cell

Lysosome an organelle containing digestive enzymes for hydrolysis of proteins, fats, and polysaccharides

Microfilament a filamentous polymer of the protein actin that helps maintain the shape of a eukaryotic cell

Microtubule a filamentous polymer of the proteins α-tubulin and β-tubulin that functions in eukaryotic cell shape and motility

Mitochondrion the respiratory organelle of eukaryotic organisms

Mushroom the aboveground fruiting body, or basidiocarp, of basidiomycete fungi

Nucleus the organelle that contains the eukaryotic cell's chromosomes

Peroxisome organelles that function to rid the cell of toxic substances such as peroxides, alcohols, and fatty acids

Phagocytosis a mechanism for ingesting particulate food in which a portion of the cytoplasmic membrane surrounds the particle and brings it into the cell

Primary endosymbiosis acquisition of the alphaproteobacterial ancestor of the mitochondrion or of the cyanobacterial ancestor of the chloroplast by another kind of cell

Protists unicellular eukaryotic microorganisms; may be flagellated or aflagellate, phototrophic or nonphototrophic, and most lack cell walls; also referred to as protozoa

Secondary endosymbiosis acquisition by a mitochondrion-containing eukaryotic cell of a red or green algal cell

Slime mold a nonphototrophic protist that lacks cell walls and that aggregates to form fruiting structures (cellular slime molds) or masses of protoplasm (acellular slime molds)

Stroma the inner membrane surrounding the lumen of the chloroplast

Thylakoid a membrane layer containing the photosynthetic pigments in chloroplasts

Yeast the single-celled growth form of various fungi

Review Questions

1. List at least three features of eukaryotic cells that clearly differentiate them from prokaryotic cells (Sections 18.1).
2. How are the mitochondrion and the hydrogenosome similar structurally? How do they differ? How do they differ metabolically (Section 18.2)?
3. What major physiological processes dealing with energy and carbon occur in the chloroplast (Section 18.3)?
4. What does knowing that an antibiotic such as streptomycin blocks protein synthesis in organelles tell us about their relationship to *Bacteria* (Section 18.4)?
5. What are the functions of the following eukaryotic cell structures: endoplasmic reticulum, Golgi complex, lysosome, and peroxisome (Section 18.5)?
6. Explain how a highly derived eukaryote might be mistakenly thought to be more basal evolutionarily (Section 18.6)?
7. In what ways do diplomonads and parabasalids differ from each other (Section 18.7)?
8. What morphological feature unites kinetoplastids and euglenids (Section 18.8)?

9. What organism causes "red tides" and why is this organism considered toxic (Section 18.9)?
10. In what ways are oomycetes similar and different from fungi (Section 18.10)?
11. What morphological trait links cercozoans and radiolarians and distinguishes them from other protists (Section 18.11)?
12. Why were slime molds once grouped with fungi (Section 18.12)?
13. What is the difference between fungal endomycorrhizal and ectomycorrhizal associations (Section 18.13)?
14. What was the common ancestor of fungi and animals likely to have been like (Section 18.14)?
15. In what way do chytrids differ from other fungi (Section 18.15)?
16. Is the formation of zygospores in zygomycetes a sexual or an asexual process (Section 18.16)?
17. What benefit do endomycorrhizal fungal relationships provide to the plants with which they associate (Section 18.17)?
18. How is the mating type of a yeast cell determined (Section 18.18)?
19. What morphological feature unites the basidiomycetes, and where is this feature found (Section 18.19)?
20. In what kinds of habitats would one likely find red algae (Section 18.20)?
21. What traits link green algae and plants (Section 18.21)?

Application Questions

1. If organelles such as the mitochondrion and chloroplast had originally been free-living eukaryotic cells, how would the molecular properties of organelles listed in Section 18.4 differ?
2. Explain why the process of endosymbiosis can be viewed as both an ancient event and a more recent event.
3. What does the presence of the apicoplast in apicomplexans reveal about their evolutionary history?

19

Viral Diversity

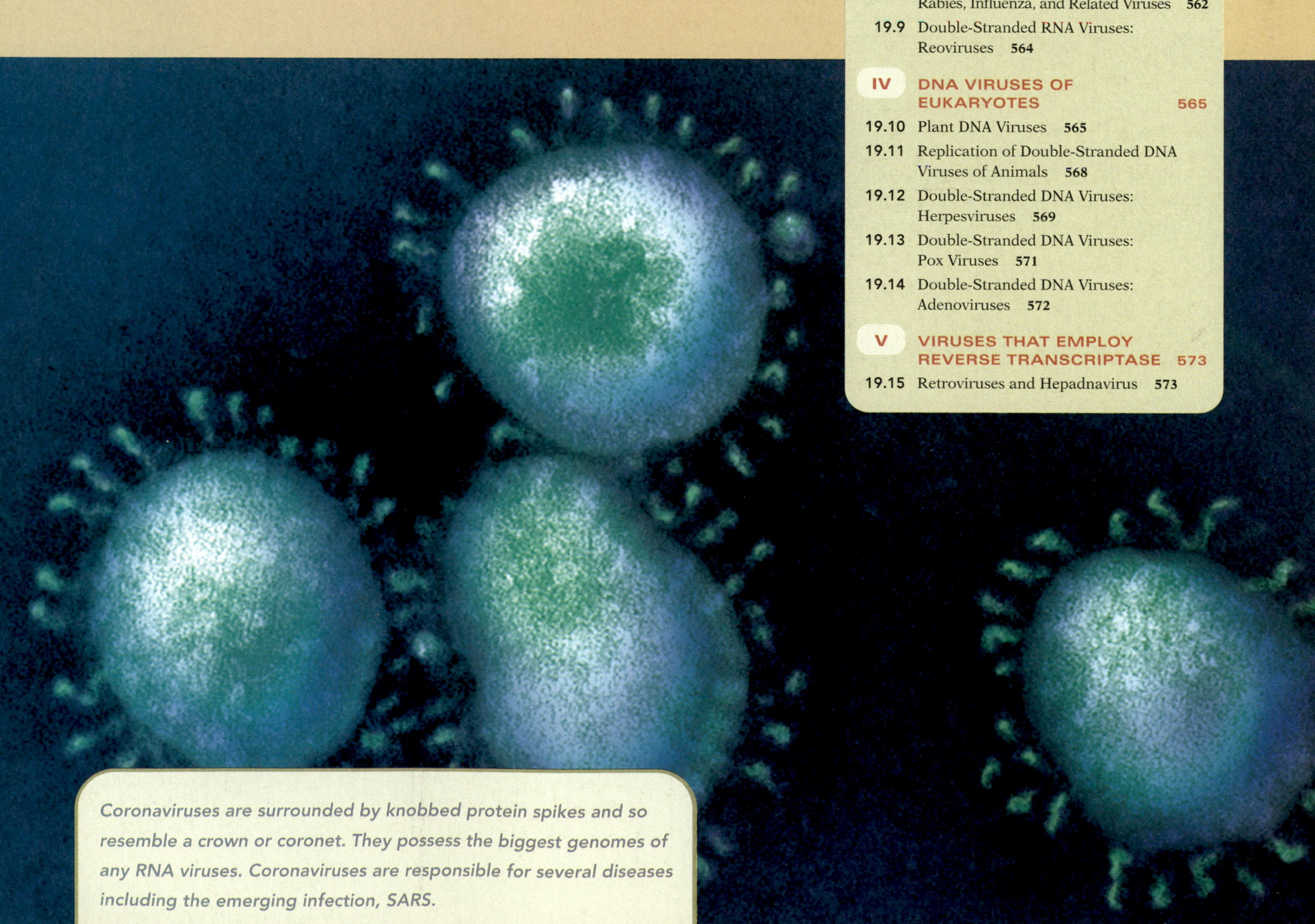

Coronaviruses are surrounded by knobbed protein spikes and so resemble a crown or coronet. They possess the biggest genomes of any RNA viruses. Coronaviruses are responsible for several diseases including the emerging infection, SARS.

The previous chapters in this unit have explored the enormous diversity of microbial cells: *Bacteria*, *Archaea*, and *Eukarya*. Here we focus on the diversity of viruses and their genomes. This chapter extends Chapter 10, which introduced the principles of virology.

All cells contain double-stranded DNA as their genetic material. By contrast, viruses are known that have single-stranded RNA, double-stranded RNA, single-stranded DNA, or double-stranded DNA as their genetic material (Section 10.1). This makes for interesting schemes of replication and gene expression. In this chapter we illustrate viral diversity by exploring the replication processes of viruses in each of these groups. We separate our discussion by the type of host cell infected, because the differences in cell structure between the three domains of life place certain constraints on the life cycles of the viruses that infect them.

I VIRUSES OF *BACTERIA*

Many viruses are known that infect *Bacteria*. As mentioned in Chapter 10 most known bacterial viruses, or **bacteriophages** as they are called, have double-stranded DNA genomes (Section 10.8). Even so, many bacteriophages have other types of genomes. **Table 19.1** lists several well-characterized bacteriophages that infect *Escherichia coli*. The simplest are those with RNA genomes, and we begin with these.

19.1 RNA Bacteriophages

Many bacteriophages contain RNA genomes of the plus (+) configuration. In such viruses the viral genome—**positive-strand RNA**—and the mRNA are of the same complementarity (Section 10.7). Interestingly, RNA viruses of the enteric bacteria infect only bacterial cells that contain a type of plasmid, called a conjugative plasmid, which allows the bacterial cell to function as a donor in the process of conjugation (Section 11.12). This restriction arises because these RNA viruses infect bacteria by first attaching to pili (**Figure 19.1**), which are encoded by the plasmid and present only on the donor cell.

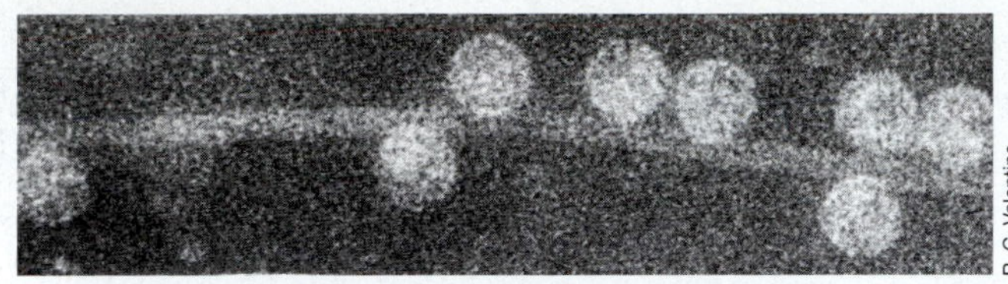

Figure 19.1 Small RNA Bacteriophages. Electron micrograph of the pilus of a donor bacterial cell of *Escherichia coli* showing virions of a small RNA phage attached to the pilus.

Phage MS2

The bacterial RNA viruses are all quite small, about 25 nm in diameter, and they are all icosahedral (Section 10.2) with 180 copies of coat protein per virion. The complete nucleotide sequences of several RNA phage genomes are known. For example, the genome of the RNA phage MS2, which infects *Escherichia coli*, is 3569 nucleotides long.

The genetic map of phage MS2 is shown in **Figure 19.2*a***, and the flow of events of MS2 replication is shown in Figure 19.2*b*. The small MS2 genome encodes only four proteins. These are the maturation protein (present in the mature virus particle as a single copy), coat protein, lysis protein (required for lysis of bacteria, which releases mature virus particles), and a subunit of **RNA replicase**, the enzyme that replicates the viral RNA. RNA replicase is a composite protein, composed partly of a virus-encoded polypeptide and partly of host polypeptides. The maturation protein acts as a protease to process some viral proteins necessary for producing infective virions.

Because the genome of phage MS2 is positive-strand RNA, it can be translated directly upon entry into the cell. After RNA replicase is synthesized, it can synthesize negative-strand RNA using the genomic RNA as a template (Figure 10.11). After negative-strand RNA has been synthesized, more positive-strand RNA is made using this negative-strand RNA as a template. The newly made positive strands are translated for continued synthesis of the viral proteins.

Phage MS2 RNA is folded into a complex form with extensive secondary structure. Of the four AUG translational start

Table 19.1 Some bacteriophages of *Escherichia coli*

Bacteriophage	*Virion structure*	*DNA or RNA*	*Single- or Double-Stranded*	*Structure of genome*	*Size of genome*[a]
MS2	Icosahedral	RNA	Single-stranded	Linear	3,569
ϕX174	Icosahedral	DNA	Single-stranded	Circular	5,386
M13, f1, and fd	Filamentous	DNA	Single-stranded	Circular	6,408
Lambda	Head & tail	DNA	Double-stranded	Linear	48,514[b]
T7 and T3	Head & tail	DNA	Double-stranded	Linear	39,936
T4	Head & tail	DNA	Double-stranded	Linear	168,903
Mu	Head & tail	DNA	Double-stranded	Linear	39,000[c]

[a]The sizes of some viral genomes are known accurately because they have been sequenced. However, the sequence and exact number of bases for related isolates may be slightly different. The size is in bases or base pairs depending on whether the virus is single- or double-stranded, respectively.

[b]This includes single-stranded extensions of 12 nucleotides at either end of the linear DNA.

[c]This includes 1,800 bp of host DNA that are attached to the ends of Mu DNA in the virion.

UNIT 3

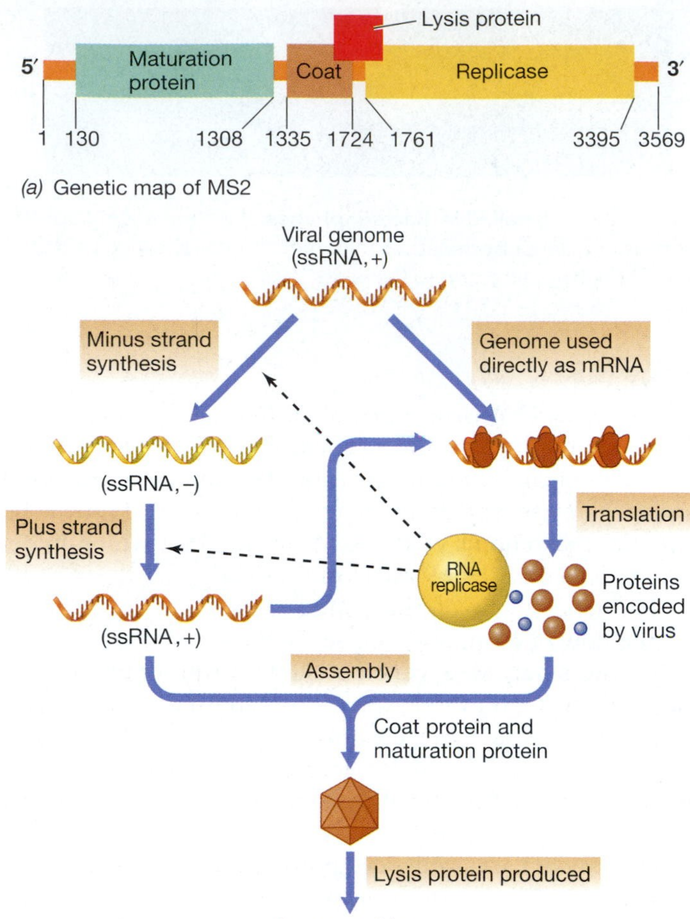

Figure 19.2 Bacteriophage MS2 *(a)* Note on the genetic map how the lysis protein gene overlaps with both the coat protein and replicase genes. The numbers in *(a)* refer to the nucleotide positions on the RNA. *(b)* Flow of events during multiplication.

sites on the mRNA, the most accessible to the translation machinery is that for the coat protein, and translation begins there very early in virus infection. The replicase mRNA is also translated early. As coat protein molecules increase in number in the cell, they bind to the RNA around the AUG start site for the replicase protein, effectively turning off synthesis of replicase. Although the gene for the maturation protein is at the 5′ end of the RNA, the extensive folding of the RNA limits access to the maturation protein start site. Consequently, only a few copies are synthesized. Thus, due to control of translation by RNA folding, the virus proteins are made in the relative amounts needed for virus assembly. In particular, the major virus protein synthesized is coat protein, which is needed in the largest amounts.

Overlapping Genes and Assembly of MS2

An interesting feature of bacteriophage MS2 is that the lysis protein is encoded by a gene that overlaps both the coat protein gene and the replicase gene (Figure 19.2*a*). This phenomenon of **overlapping genes** is quite common in very small viral genomes (Section 19.2) and makes small genomes more efficient. The start codon of the MS2 lysis gene is not easily accessible to ribosomes because of the secondary structure found in the RNA. However, when the ribosome terminates synthesis of the coat protein gene, the secondary structure in this region of the RNA is disrupted, and sometimes this disruption allows a ribosome to begin reading the lysis gene. This restriction of the efficiency of translation prevents premature lysis of the cell. Only after sufficient copies of coat protein are available for the assembly of mature virions does cell lysis commence.

Ultimately, self-assembly of MS2 virions takes place, and virions are released from the cell as a result of cell lysis. The features of replication of small RNA viruses such as MS2 are thus fairly simple. The viral RNA itself functions as an mRNA, and regulation is based on controlling access of ribosomes to the appropriate translation start sites on the viral RNA.

Although several positive-strand RNA bacteriophages are known that are similar in their biology to that of MS2, no bacteriophage has yet been discovered that contains negative-strand RNA, a type of genome that is fairly common among eukaryotic viruses (Section 19.8). However, a few bacteriophages are known that have segmented, double-stranded RNA as their genetic material. The best studied of these is ϕ6, whose host is *Pseudomonas syringae*. This virus, which is **enveloped** (enclosed by a lipid membrane; ∞ Figure 10.12), seems very closely related to the reoviruses (Section 19.9), which infect eukaryotes.

19.1 MiniReview

Many RNA viruses that infect bacteria are known. The small RNA genome of these bacterial viruses is translated directly and encodes only a few proteins.

- What is the difference between the genomes of positive-strand RNA viruses and negative-strand RNA viruses?
- Describe what is meant by the term overlapping genes.

19.2 Single-Stranded DNA Bacteriophages

Some bacteriophages contain a single-stranded DNA genome of the plus complementarity (there are no known negative-strand DNA bacteriophages). Before such a genome can be transcribed, a complementary strand of DNA must be synthesized, forming a double-stranded molecule, the **replicative form (RF)** (∞ Section 10.7 and Figure 10.11). These bacteriophages replicate this double-stranded DNA but then only package the positive strand of DNA into their virions. In this section we discuss two well-known single-stranded DNA phages that infect *Escherichia coli*, the icosahedral phage ϕX174 and the filamentous phage M13.

The Genome of Phage ϕX174

Bacteriophage ϕX174 contains a circular single-stranded DNA genome inside an icosahedral virion. This virus is very small—about 25 nm in diameter—and the principal building block of the coat is a single protein present in 60 copies, the minimum number of protein subunits possible in an icosahedral virus. Attached at the vertices of the icosahedron are several other proteins that make up spike-like structures (⚭ Figure 10.12). These small DNA viruses possess only a few genes, and the host cell DNA replication machinery is used exclusively in the replication of viral DNA.

The genome of ϕX174 consists of a circular single-stranded DNA molecule of 5,386 nucleotides. The DNA of ϕX174 was the first DNA molecule to be completely sequenced, a remarkable achievement when it was accomplished by Frederick Sanger and colleagues in 1977. Phage ϕX174 is also of special interest because it was the first genetic element shown to have *overlapping genes,* a condition already described for the RNA phage MS2 (Section 19.1). For example, in the ϕX174 genome gene B resides within gene A and gene K resides within both genes A and C (**Figure 19.3**). In very small viruses such as ϕX174, there is insufficient DNA to encode all viral-specific proteins unless parts of the genome are read more than once in different reading frames (Figure 19.3).

As seen in the genetic map of ϕX174, genes D and E also overlap, gene E being contained completely within gene D. Also, the termination codon of gene D overlaps the initiation codon of gene J (Figure 19.3*a*). In addition to overlapping genes, a small protein in ϕX174 called A* protein that functions in shutting down host DNA synthesis is synthesized by the reinitiation of *translation* (not transcription) within the mRNA for gene A. The A* protein is read and terminated from the same mRNA reading frame as A protein but has a different in-frame start codon and is thus a shorter protein.

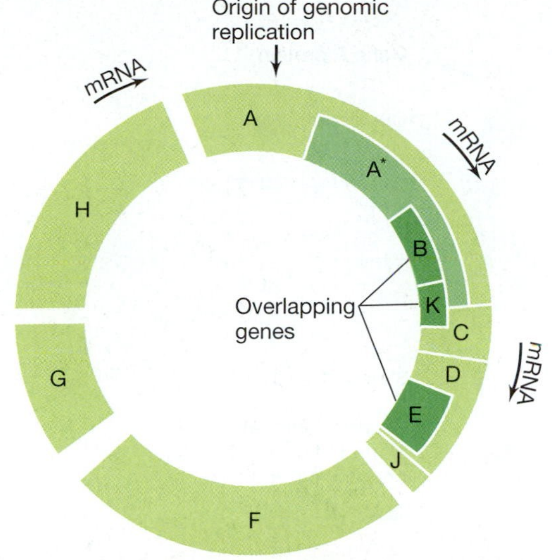

A	Replicative form DNA synthesis	E	Host cell lysis
A*	Shut off of host DNA synthesis	F	Major capsid protein
B	Formation of capsid precursors	G	Major spike protein
C	DNA maturation	H	Minor spike protein
D	Capsid assembly	J	DNA packaging protein
		K	Function unknown

(a) Genetic map of ϕX174

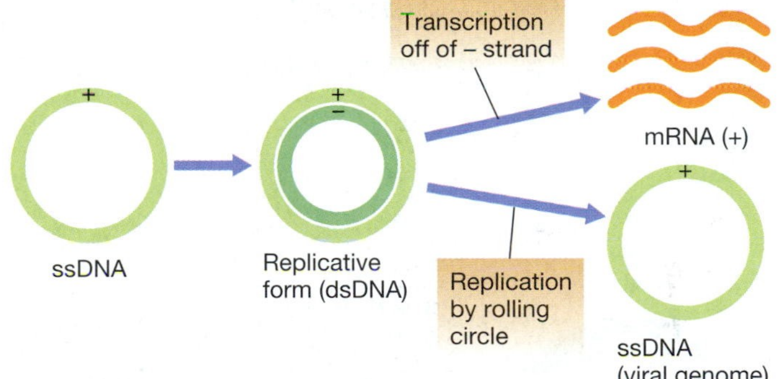

(b) Flow of events during ϕX174 replication

Figure 19.3 Bacteriophage ϕX174, a single-stranded DNA phage. *(a)* Note the regions of gene overlap in the genetic map. Intergenic regions are not colored. Protein A* is formed using only part of the coding sequence of gene A by reinitiation of translation. *(b)* Flow of events in ϕX174 replication. Progeny ssDNA is produced from replicative form dsDNA by rolling circle replication as shown in more detail in Figure 19.4.

DNA Replication by the Rolling Circle Mechanism

Several plasmids and viruses replicate by a mechanism different from the semiconservative replication of cellular DNA (⚭ Section 7.5). This alternative mechanism, **rolling circle replication,** is used by some viruses with a single-stranded DNA genome, including ϕX174. Upon infection, their positive-strand viral DNA becomes separated from the protein coat. Entry into the cell is accompanied by the conversion of this single-stranded DNA into the double-stranded replicative form (Figure 19.3*b*). Cell-encoded enzymes that help convert viral DNA to RF DNA include primase, DNA polymerase, ligase, and gyrase (⚭ Section 7.5).

In cells, replication of the lagging strand requires the synthesis of short RNA primers by primase (⚭ Section 7.5). These RNA primers are made at intervals along the lagging strand and are later removed and replaced with DNA by the enzyme DNA polymerase I (⚭ Figure 7.15). The situation with ϕX174 DNA is similar, except that the DNA in this case is a single-stranded closed circle. To begin replication of this DNA, primase synthesizes a short RNA primer at one or more specific initiation sites. DNA is then synthesized by DNA polymerase III, and the primer is removed and replaced by DNA using DNA polymerase I, exactly as in the case of lagging strand synthesis. This results in the formation of the complete, circular, double-stranded replicative form. Once the replicative form is completed, copies are made by conventional semiconservative replication with theta-form intermediates (⚭ Figure 7.16).

In the formation of single-stranded viral genomes, the rolling circle arises because the positive strand of the replicative form is cut and the 3′ end of the exposed DNA is used to prime synthesis of a new strand (**Figure 19.4**). Cutting

UNIT 3

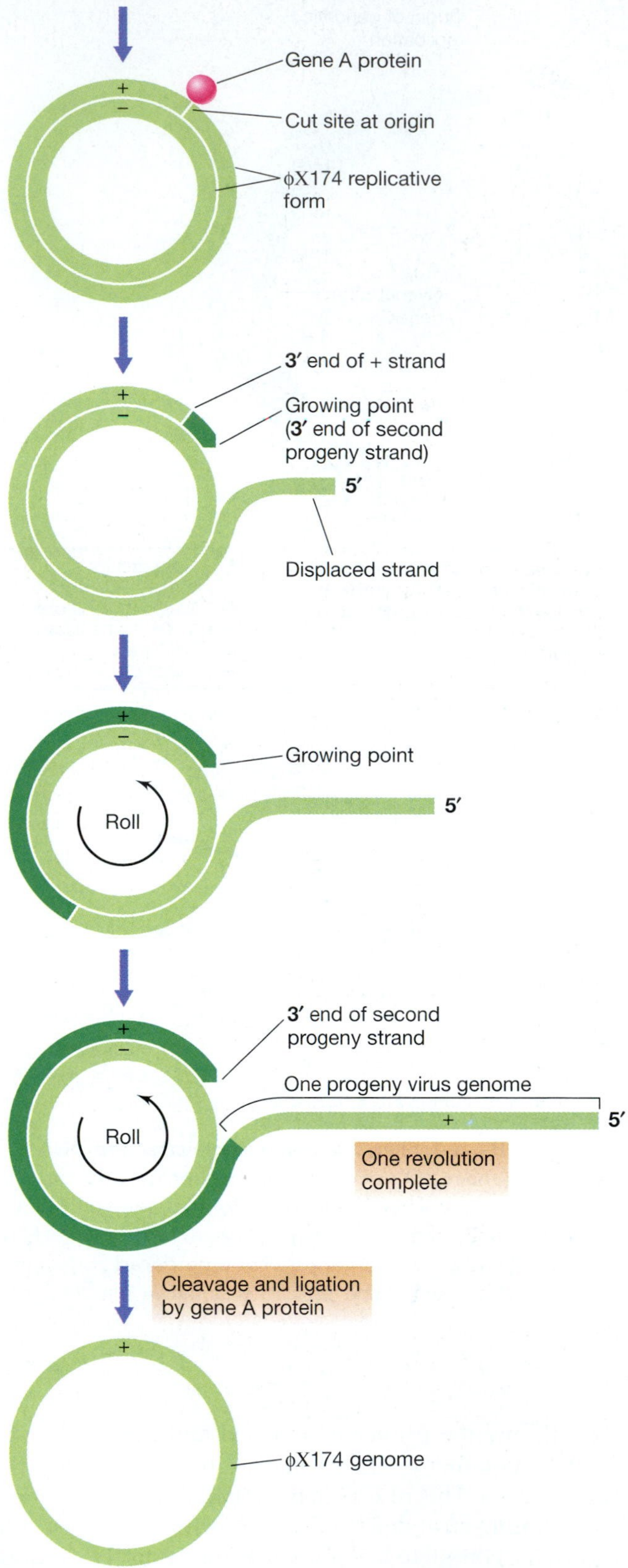

Figure 19.4 Rolling circle replication in phage ϕX174. Replication begins at the origin of the double-stranded replicative form with the cutting of the plus strand of DNA by gene A protein (both strands of DNA are shown in light green here to simplify the diagram.). After one new progeny strand has been synthesized (one revolution of the circle), the gene A protein cleaves the new strand and ligates its two ends.

of the plus strand is accomplished by the A protein. Continued rotation of the circle leads to the synthesis of a linear, single-stranded structure, the ϕX174 genome. Note that synthesis is asymmetric because only one of the strands (the negative strand) serves as a template. When the growing viral strand reaches unit length (5,386 residues for ϕX174), the A protein cleaves and then ligates the two ends of the newly synthesized single strand to give a single-stranded DNA circle.

Transcription and Translation in ϕX174

The replicative form is also used as a template for transcription of ϕX174 mRNA. Transcription of mRNA begins at several promoters and terminates at a number of sites. The polycistronic mRNA molecules are then translated into the various phage proteins. As already noted, several proteins are made from mRNA transcripts formed from different reading frames of the same DNA sequences (overlapping genes, Figure 19.3*a*). The efficiency with which such a small genome as that of ϕX174 can encode so many different proteins is impressive.

Ultimately, assembly of mature ϕX174 virions occurs. Release of virions from the cell takes place as a result of cell lysis, due to the E protein. Interestingly, protein E catalyzes lysis by inhibiting the activity of one of the enzymes involved in peptidoglycan synthesis in the cell wall (∞ Section 6.4). Because of the resulting weakness in newly synthesized cell wall material, the cell eventually lyses, releasing the viral particles.

Filamentous Single-Stranded DNA Bacteriophages

The filamentous DNA phages have helical rather than icosahedral symmetry. The most studied member of this group is phage M13, which infects *E. coli*, but related phages include f1 and fd. As with the small RNA bacteriophages, these filamentous DNA phages infect only conjugational donor cells, entering after attachment to the pilus (Figure 19.1). Even though these phages are linear (filamentous) in shape, they possess circular single-stranded DNA.

Bacteriophage M13

M13 is the model filamentous bacteriophage. It has found extensive use as a cloning and DNA-sequencing vector in genetic engineering (∞ Section 12.15). The virion of phage M13 is only 6 nm in diameter but is 860 nm long. These filamentous DNA phages have the additional interesting property of being released without lysing the host cell. Thus, a cell infected with phage M13 can continue to grow, all the while releasing M13 virions. Virus infection causes a slowing of cell growth, but otherwise a cell is able to coexist with its virus. Typical plaques are thus not observed; instead, only areas of reduced turbidity are seen within the bacterial lawn.

Many aspects of DNA replication in filamentous phages are similar to those of ϕX174 (Figure 19.4). The virions are released without killing the cell by a process called *budding*. In this mechanism, the end of the virion containing several copies of a protein known as the A protein is released first, with the remainder of the virion following (**Figure 19.5**). With phage M13 there is no accumulation of intracellular virions as with typical bacteriophages. Instead, mature M13 virions are assembled on the inner surface of the cytoplasmic membrane, and virus assembly is coupled with the budding process.

Using M13 in Genetic Engineering

Several features of phage M13 make it useful as a cloning and DNA sequencing vehicle. First, it has single-stranded DNA, which means that sequencing can easily be carried out by the Sanger dideoxynucleotide method (Section 12.5). Second, a double-stranded form of genomic DNA essential for cloning purposes is produced naturally when the phage produces its replicative form. Third, as long as infected cells are kept growing, they can be maintained indefinitely, so a continuous source of the cloned DNA is available. And finally, like phage lambda (Section 12.14), there is an intergenic space in the genome of phage M13 that does not encode proteins and can be replaced by variable amounts of foreign DNA (Section 12.15). For these and other reasons, phage M13 is an important part of the biotechnologist's toolbox.

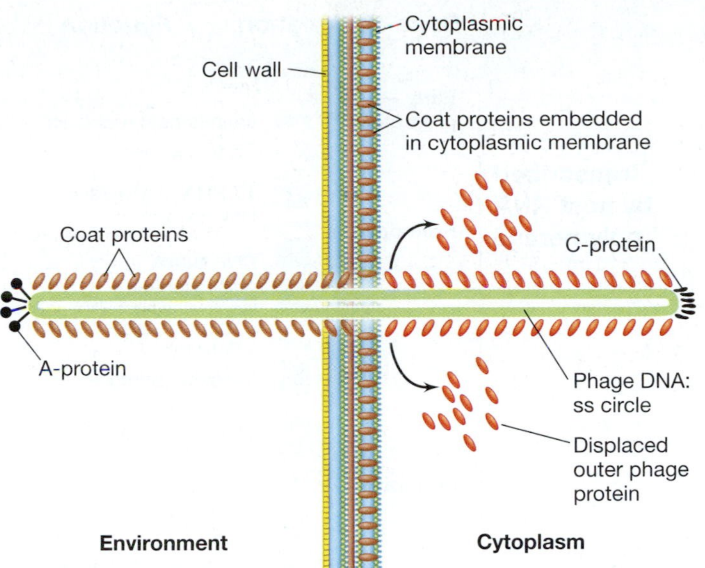

Figure 19.5 Release of filamentous phages. The virions of filamentous single-stranded phages (such as M13 or fd) exit from infected cells without lysis. The A protein passes through the cytoplasmic membrane first at a site where coat protein molecules have already become embedded into the cytoplasmic membrane. The intracellular circular DNA is coated with dimers of outer phage protein, which is displaced by coat protein as the DNA passes through the intact cytoplasmic membrane.

19.2 MiniReview

The single-stranded DNA genome of the virus ϕX174 is so small that genes overlap to encode all of its essential proteins. The viral DNA replicates by a rolling circle mechanism. Some single-stranded DNA viruses, such as M13, have filamentous virions that are released without cell lysis. These viruses are useful tools for DNA sequencing and genetic engineering.

- Why can't the genome of ϕX174, which is single-stranded and in the plus configuration, be used directly as mRNA as in phage MS2?
- How does the replicative form of ϕX174 nucleic acid differ from the form found in the virion?
- Describe the genome of phage M13. How does the genome relate to the mRNA produced by this phage?
- How can M13 virions be released without killing the infected host cell?

19.3 Double-Stranded DNA Bacteriophages

The double-stranded DNA bacteriophages are among the best studied of all viruses, and we have already discussed two important ones, T4 and lambda, in Chapter 10 (Sections 10.9 and 10.10). Because of their importance as models for understanding molecular biology and gene regulation, we consider two more such viruses, T7 in this section and Mu in the next section.

Replication of Bacteriophage T7: Early Events

Bacteriophage T7 and its close relative T3 are relatively small DNA viruses that infect *Escherichia coli* and a few related enteric bacteria, notably *Shigella*. The virion has an icosahedral head and a very short tail. The T7 genome is a linear double-stranded DNA molecule of 39,936 bp. The genetic map of T7 is shown in **Figure 19.6** and includes some overlapping genes (Section 19.2).

The order of the genes on the T7 chromosome influences the regulation of virus replication. When the virion attaches to the bacterial cell, the DNA is injected with the genes at the "left end" of the genetic map entering the cell first. Several genes at this end of the T7 genome are quickly transcribed by host RNA polymerase and then translated. One of these early proteins inhibits the host restriction system, a mechanism for protecting the cell from foreign DNA (Section 12.1). This occurs very rapidly, as the anti-restriction protein is synthesized before the entire T7 genome even enters the cell!

Another early protein is a viral RNA polymerase, called *T7 RNA polymerase*. Two other early mRNA molecules encode proteins that inhibit host RNA polymerase, thus turning off the transcription of the early genes as well as the transcription of host genes. Host RNA polymerase is thus used just to transcribe the first few genes. T7 RNA polymerase then takes over and carries out the transcription of most phage genes.

Phage T7 RNA polymerase recognizes only T7 promoters distributed along the T7 genome (Figure 19.6). These T7 promoters have a sequence unrelated to typical *E. coli* promoters. Because T7 RNA polymerase is extremely efficient, genetic engineers have used it to express cloned genes at very high levels (Section 12.13). Note that this strategy differs from that of

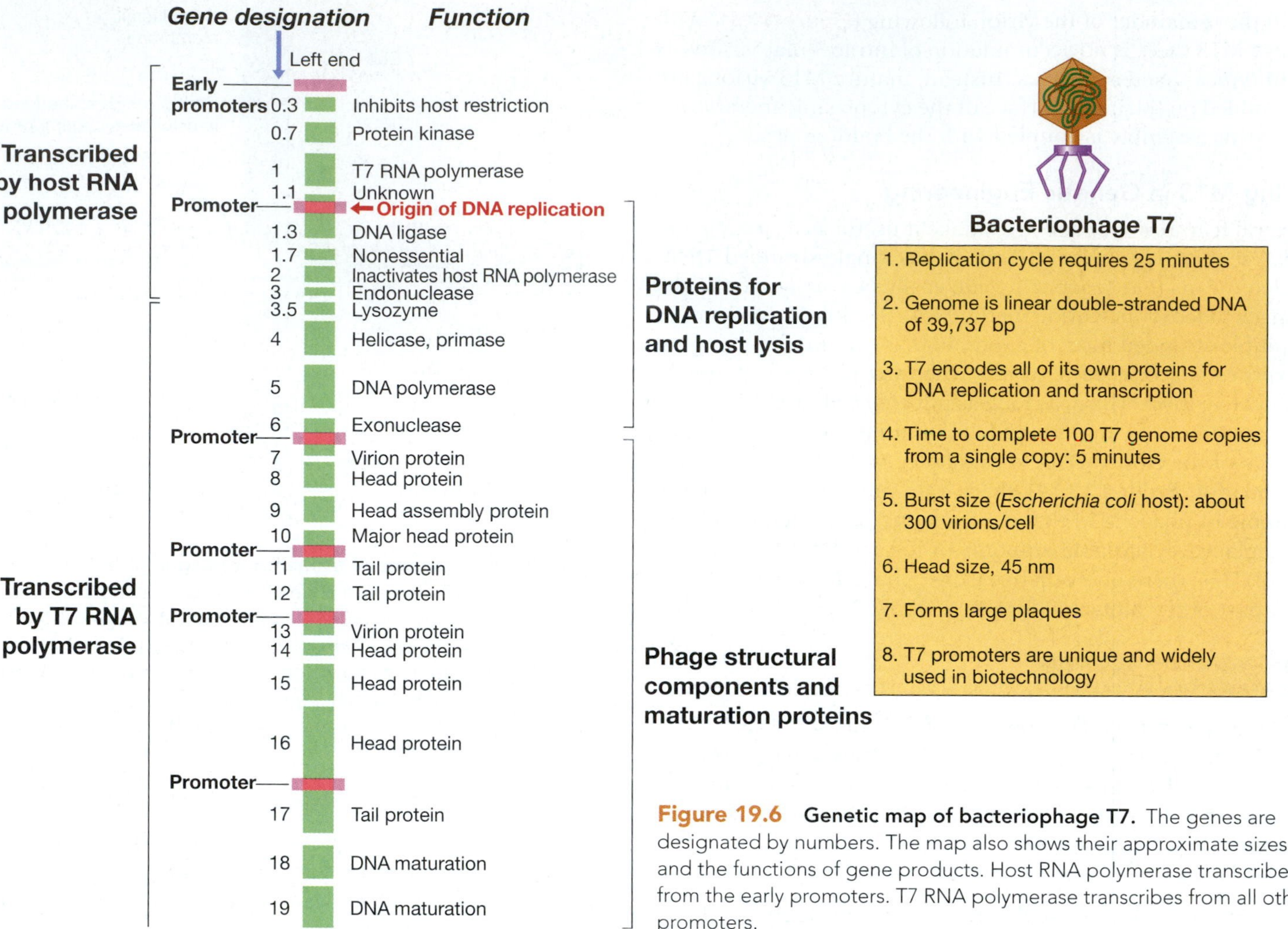

Figure 19.6 Genetic map of bacteriophage T7. The genes are designated by numbers. The map also shows their approximate sizes and the functions of gene products. Host RNA polymerase transcribes from the early promoters. T7 RNA polymerase transcribes from all other promoters.

phage T4. Phage T4 uses the host RNA polymerase but directs it only to phage genes using a T4-specific sigma factor (Section 10.9).

Genome Replication in T7

DNA replication in T7 begins at a single origin of replication (shown in Figure 19.6) and proceeds bidirectionally from this origin (**Figure 19.7**). Replicating molecules of T7 DNA can be recognized under the electron microscope by their characteristic structures. Because the origin of replication is near the left end, Y-shaped molecules are typically seen in transmission electron micrographs of replicating T7 DNA. Earlier in replication, bubble-shaped molecules can appear (Figure 19.7). Several virus-encoded proteins, including a T7-specific DNA polymerase, are required for T7 DNA replication. This differs from the use of host-encoded proteins by phage T4 (Section 10.9) and ϕX174 (Section 19.2).

Replication of linear DNA molecules must overcome the problem of the DNA shortening each round of replication due to removal of the RNA primer used to initiate DNA synthesis. Different solutions to this problem exist (Section 8.7). The solution employed by T7 resembles that used by T4 and relies on repeated sequences (Section 10.9). T7 DNA has a direct terminal repeat of 160 bp at both ends of the molecule. To replicate DNA near the 5′ terminus, RNA primer molecules must be removed before replication is complete. This leaves an unreplicated segment of the T7 DNA at the 5′ terminus of each strand (lower part of Figure 19.7*a*). The opposite single 3′ strands on two separate DNA molecules, being complementary, can pair with these 5′ strands, forming a DNA molecule twice as long as the original T7 DNA (Figure 19.7*b*). The unreplicated portions of this structure are then completed through the activity of T7 DNA polymerase and ligase, resulting in a linear double molecule called a **concatemer**.

Continued replication and recombination can lead to concatemers of considerable length, but ultimately a phage-encoded endonuclease cuts each concatemer at a specific site, resulting in the formation of virus-sized linear DNA molecules with terminal repeats (Figure 19.7*c*). Because T7 cuts the concatemer at specific sequences, the DNA sequence in each T7 virion is identical. Recall that this is not the case in phage T4, which processes DNA using a "headful mechanism." Consequently, T4 DNA is not only terminally redundant as is T7 DNA, but is also circularly permuted (Section 10.9).

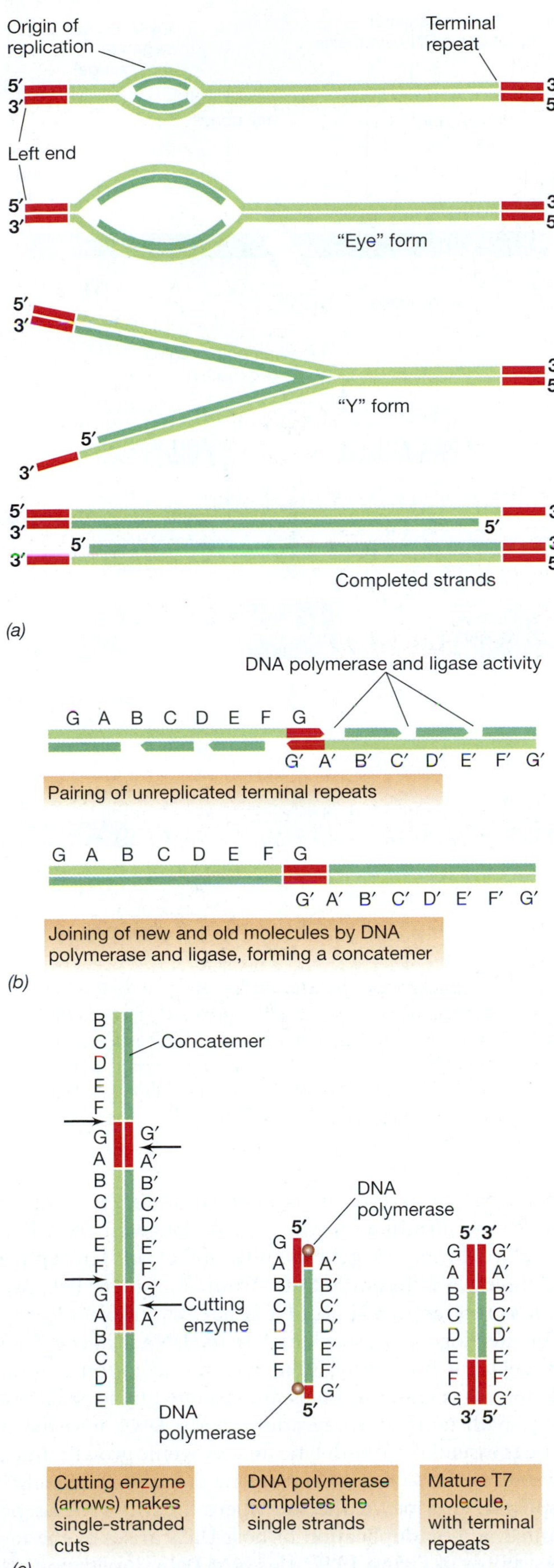

Figure 19.7 Replication of the bacteriophage T7 genome. *(a)* The linear, double-stranded DNA undergoes bidirectional replication giving rise to intermediate "eye" and "Y" forms (for simplicity, both template strands are shown in light green and both newly synthesized strands in dark green). *(b)* Formation of concatemers by joining DNA molecules at the unreplicated terminal ends. The designation of the genes is arbitrary. *(c)* Production of mature viral DNA molecules from T7 concatemers by activity of the cutting enzyme, an endonuclease.

19.3 MiniReview

The bacteriophage T7 double-stranded DNA genome always enters the host cell in the same orientation. The late genes in T7 are transcribed by a virus-encoded RNA polymerase. Replication of the T7 genome employs T7 DNA polymerase and involves terminal repeats and the formation of concatemers.

- Of what significance is it that the T7 genome enters the cell in only one orientation?
- What is meant by terminal repeats?
- In what ways is DNA replication similar and in what ways is it different in phages T4 and T7?

UNIT 3

19.4 Mu: A Double-Stranded Transposable DNA Bacteriophage

The bacteriophage Mu is temperate, like lambda (Section 10.10), but has the unusual property of replicating by *transposition* (Section 11.16). Transposable elements are sequences of DNA that can move from one location on their host genome to another as discrete genetic units. They are found in both prokaryotes and eukaryotes and play important roles in genetic variation. Although a bacteriophage, Mu is at the same time a very large transposable element that replicates its DNA by transposition.

This phage was named Mu because it is a *mutator* phage, inducing mutations in a host genome into which it becomes integrated. The mutagenic property of Mu arises because the genome of the virus can be inserted within host genes, thus interrupting coding sequences. Hence, a host cell that is infected with Mu will often gain a mutant phenotype. Mu is a useful phage in bacterial genetics because it can be used to easily generate bacterial mutants.

Basic Properties of Phage Mu

Bacteriophage Mu is a large virus with an icosahedral head, a helical tail, and six tail fibers (**Figure 19.8**). The genome of Mu consists of linear double-stranded DNA, and its genetic map is shown in **Figure 19.9*a***. Most Mu genes are involved in the synthesis of head and tail proteins, and important genes at each end of the genome are involved in replication and host range.

The Mu virion contains approximately 39 kbp of DNA, but only 37.2 kbp constitute the actual Mu genome. The additional length is host DNA attached to the ends of the Mu genome,

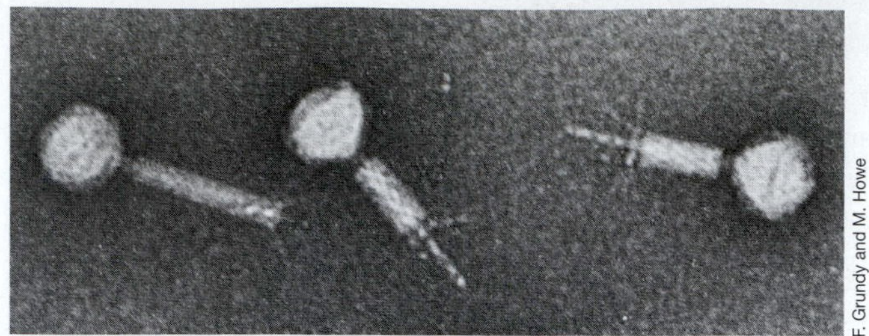

Figure 19.8 Bacteriophage Mu virions. Electron micrograph of virions of the double-stranded DNA phage Mu, the mutator phage.

50–150 bp at the left end and 1–2 kbp at the right end. These host DNA sequences are not unique but merely represent DNA adjacent to the location where Mu was inserted into the genome of its previous host. How does this happen?

When a Mu phage virion is formed, a length of DNA containing the Mu genome, just large enough to fill the phage head, is excised from the host. The DNA is packaged until the head is full, but the place at the right end where the DNA is cut varies from one virion to another. For that reason, as shown on the genetic map, there is a variable sequence of host DNA at the right-hand end of the phage (right of the *attR* site). Thus, each virion arising from a single infected cell is genetically unique in having different flanking host DNA.

Mu and the Invertible G Region

As shown on the genetic map (Figure 19.9*a*), a specific segment of the Mu genome called *G* is invertible, being present in the genome either in the orientation designated G^+ or in the inverted orientation G^-. The orientation of this segment determines the kind of tail fibers that are present on the phage. Because adsorption to the host cell is controlled by molecular interactions between the tail fibers and the cell surface, the host range of Mu is determined by the orientation of the G segment in the phage. For example, if the G segment is in the G^+ orientation then the phage will make tail fibers that allow it to infect *Escherichia coli* strain K12. By contrast, if the G segment is in the G^- orientation, then the phage will infect *E. coli* strain C or several other species of enteric bacteria. The two alternative tail fiber proteins are encoded on opposite strands within this small G segment.

Left of the G segment is a promoter that directs transcription into the G segment. In the orientation G^+, the promoter for genes S and U is active, whereas in the orientation G^-, a different promoter directs transcription of genes S′ and U′ on the opposite strand. Inversion of the G region is a rare event and is under the control of a gene adjacent to the G region. We thus see in this inversion phenomenon a simple mechanism the phage has evolved for attacking a variety of different host cells.

Replication of Mu

Note that the bacteriophages we have discussed that have double-stranded DNA genomes, T4 (Section 10.9), T7 (Section 19.3), lambda (Section 10.10), and Mu, all have *linear* genomes. Nonetheless, these viruses use three quite

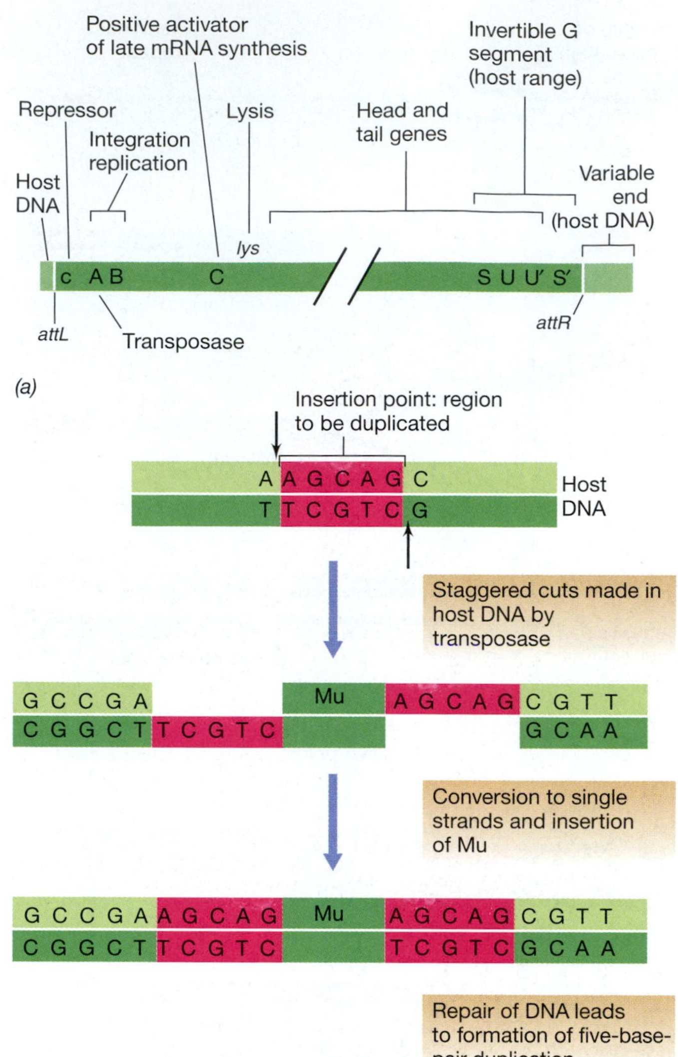

Figure 19.9 Bacteriophage Mu. *(a)* Genetic map of the Mu genome. Note that there is a lowercase *c* gene, which encodes a repressor protein, and an uppercase *C* gene, which encodes an activator protein. The region encoding the head and tail genes is not drawn to scale. *(b)* Integration of Mu into the host DNA, showing the generation of a 5-bp duplication of host DNA.

distinct schemes to replicate the ends of their genomes. Both T4 and T7 use terminal repeats to form concatemers, whereas lambda circularizes its genome after infection. Mu replicates in a completely different manner from all of these because its genome is replicated as part of a larger DNA molecule.

On infection of a host cell by Mu, the DNA is injected and is protected from host restriction by a modification system in which several adenine residues are modified by acetylation. In contrast with lambda, integration of Mu DNA into the host genome is essential for both lytic and lysogenic growth. Integration requires the activity of the gene A product, which is a **transposase** enzyme. At the site where the Mu DNA becomes integrated, a 5-bp duplication of host DNA arises at the target site. As shown in Figure 19.9*b*, this host DNA duplication arises because staggered cuts are made at the point in the host genome where Mu is inserted. The resulting single-stranded segments are converted to the double-stranded form as part of the Mu integration process. Duplication of short stretches of host DNA is typical of transposable element insertion (∞ Section 11.16).

Mu can enter the lytic pathway either upon initial infection if the Mu repressor (the product of the *c* gene) is not made or by induction of a lysogen. In either case, Mu DNA is replicated by repeated transposition of Mu to multiple sites on the host genome. Initially, only the early genes of Mu are transcribed, but after the C protein is expressed (C is a positive activator of late transcription), the Mu head and tail proteins are synthesized. Eventually, the cell is lysed and mature phage particles are released. The lysogenic state in Mu requires the sufficient accumulation of repressor protein to prevent transcription of integrated Mu DNA.

19.4 MiniReview

Bacteriophage Mu is a temperate virus that is also a transposable element. In either the lytic or lysogenic pathway, its genome is integrated into the host chromosome by the activity of a transposase. Even in the lytic pathway, its genome is replicated as part of a larger DNA molecule. The genome is packaged into the virion with short sequences of host DNA at either end.

- What mechanism is used to change the host range of Mu?
- What mechanism does Mu use to ensure that the ends of its linear genome are completely replicated?

II VIRUSES OF *ARCHAEA*

19.5 Viruses of *Archaea*

Several DNA viruses have been discovered whose hosts are species of *Archaea*, including representatives of both the *Euryarchaeota* and *Crenarchaeota* (∞ Chapter 17). Most viruses that infect species of *Euryarchaeota*, including both methanogenic and halophilic *Archaea*, are of the "head and tail" type, resembling phages that infect enteric bacteria, such as phage T4 (∞ Section 10.9). In fact, certain tailed viruses of halophilic and haloalkaliphilic *Archaea*, such as phage ΦH of *Halobacterium salinarum* and phage ΦCh1 of *Natrialba magadii*, have linear double-stranded DNA genomes that are circularly permuted and terminally redundant, as in phage T4. These viruses likely package DNA using the same "headful" mechanism of phage T4 (∞ Section 10.9). In contrast to this familiar group of viruses, many other DNA viruses that infect members of the *Crenarchaeota* are extremely unusual. Thus far, no RNA viruses infecting any member of the *Archaea* have been found.

Viruses of Hyperthermophiles

The most diverse and morphologically unusual archaeal viruses infect hyperthermophiles that are members of the *Crenarchaeota* (∞ Chapter 17), and so they are our focus here. For example, the sulfur chemolithotroph *Sulfolobus* is host to several structurally unusual viruses. Considering the habitat of *Sulfolobus*—hot, acidic soils and hot springs (∞ Section 17.9), these viruses must be remarkably resistant to heat and acid denaturation.

One of the viruses that infects *Sulfolobus* species, nicknamed SSV for *Sulfolobus* spindle-shaped virus, forms spindle-shaped virions that often cluster in rosettes (**Figure 19.10*a***).

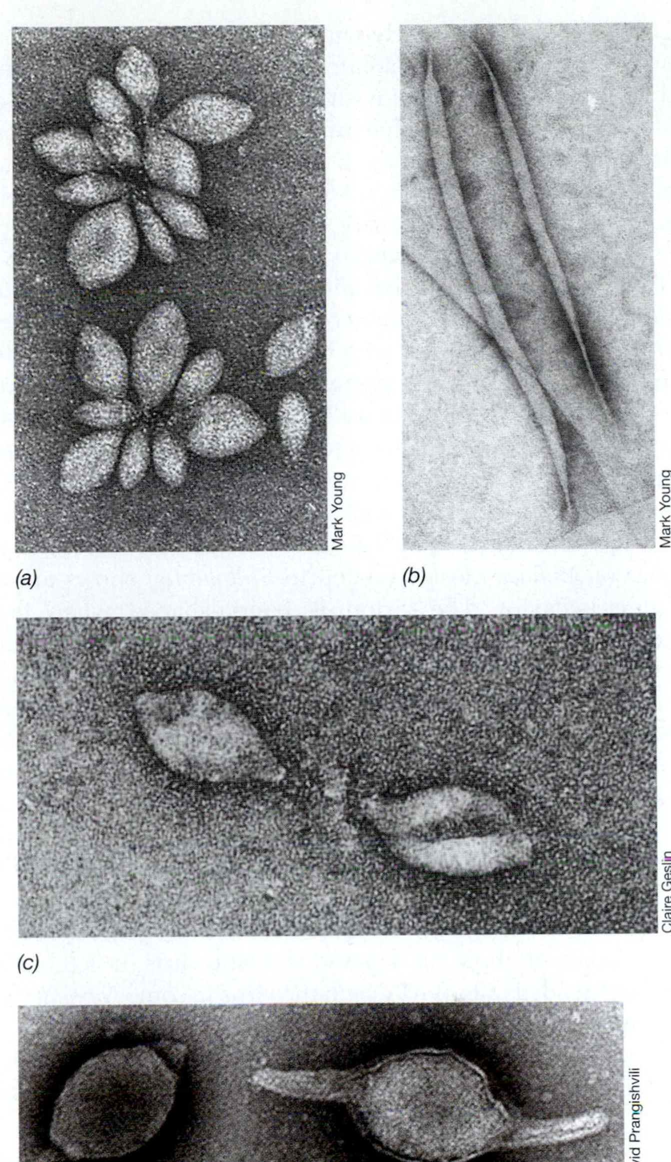

Figure 19.10 Archaeal viruses. Electron micrographs of *(a, b, d)* viruses of *Crenarchaeota*, and *(c)* a virus of a Euryarchaeote. *(a)* Spindle-shaped virus SSV1 that infects *Sulfolobus solfataricus*. The viral dimensions are 40 × 80 nm. *(b)* Filamentous virus SIFV that infects *S. solfataricus*. The viral dimensions are 50 × 900–1500 nm. *(c)* Spindle-shaped virus PAV1 that infects *Pyrococcus abyssi*. The viral dimensions are 80 × 120 nm. *(d)* ATV, the virus that infects the hyperthermophile *Acidianus convivator*. When released from the cell the virions are lemon-shaped (left virion) but proceed to grow appendages on both ends (right virion) that can grow to over three times that shown here. The virions are about 100 nm in diameter.

Such viruses are apparently widespread in very hot environments, as they have been isolated from hot springs in Iceland and Japan, as well as Yellowstone National Park. Virions of SSV contain circular double-stranded DNA about 15 kbp long, considerably smaller than the linear genome of a tailed bacteriophage like T4 (~168 kbp; no circular double-stranded DNA viruses of *Bacteria* are known). A second morphological type of *Sulfolobus* bacteriophage is a rigid, helical rod (Figure 19.10*b*). Viruses in this class, nicknamed SIFV for *Sulfolobuc islandicus f*ilamentous *v*irus, contain linear double-stranded DNA genomes of about 35 kbp. Many variations on the spindle- and rod-shaped patterns have been seen in viral isolation studies. These include very long rods similar in morphology to filamentous bacteriophages (Section 19.2) and spindle-shaped viruses that contain a large spindle-shaped center with appendages at each end (Figure 19.10*d*).

A spindle-shaped virus that infects *Acidianus convivator* (a species of *Archaea* closely related to *Sulfolobus*) shows an intriguing behavior. The virion is lemon-shaped when first released from the host cells. Following release from its host, the virion develops long thin tails, one at each end, and was thus named ATV for "*Acidianus* two-tailed virus" (Figure 19.10*d*). At 85–90°C, the optimum growth temperature for the host cell, the tails form in about an hour. When lemon-shaped ATV virions were stored at low temperature, tails did not develop. When such particles were returned to high temperatures, the tails formed. What is remarkable about the biology of ATV is that it is the first example of virus development in the complete absence of host cell contact. ATV contains a double-stranded circular DNA genome of about 68 kbp and is a lytic virus. It is thought that the extended tails of ATV help the virus in some way survive in the hot, acidic (pH 1.5) environment of *A. convivator*.

A spindle-shaped virus also infects *Pyrococcus*, a species that falls within the archaeal phylum *Euryarchaeota* (∞ Section 17.6). This virus, named PAV1 for *Pyrococcus abyssi virus 1*, resembles SSV but is larger and contains a very short tail (Figure 19.10*c*). Phage PAV1 has a circular double-stranded DNA genome of about 18 kbp and, interestingly, is released from host cells without cell lysis, probably by a budding mechanism similar to that of the filamentous *Escherichia coli* phage M13 (Section 19.2). *Pyrococcus* is a hyperthermophile with a growth temperature optimum of about 100°C, meaning that PAV1 virions must be extremely heat-stable. Despite their similar morphologies, genomic comparisons of PAV1 and SSV-type viruses show little sequence similarity, indicating that the two types of viruses do not have common evolutionary roots.

Replication and Evolution of Archaeal Viruses

Replication studies of archaeal viruses still need to be done to define the major events in genome replication and virion assembly. However, considering that the genomes of these viruses are all double-stranded DNA, it is unlikely that any major novel modes of replication will emerge from such studies. However, many molecular details, such as the extent to which viral rather than host polymerases and related enzymes are used in replication events, await further work on these relatively newly discovered viruses.

Spindle-shaped viruses are unknown among species of *Bacteria*. This, along with limited genomic similarity to viruses of *Bacteria*, suggests that these archaeal viruses do not share a common ancestor with known bacteriophages. By contrast, structural and genomic studies have shown that tailed viruses of *Archaea* may well have shared a common ancestry with tailed phages of *Bacteria*. This could be because these viruses share common evolutionary roots, or it could be the result of lateral gene transfer (∞ Chapter 13.11) of phage genes from *Bacteria* to *Archaea* (or vice versa). Both temperate viruses and transducing viruses (∞ Section 11.15) exist within the *Archaea*, although currently, examples are rare and have yet to be developed to the point of playing a major role in the laboratory genetics of these organisms.

19.5 MiniReview

Several viruses are known to infect *Archaea*. Many of these have double-stranded circular DNA genomes not known in bacteriophages of *Bacteria*. Although head- and tail-type viruses are known, most archaeal viruses have an unusual spindle-shaped morphology.

- In which groups of *Archaea* have head- and tail-type phages been discovered? Are they related to the tailed phages of *Bacteria*?
- Compare the shapes of viruses that infect methanogenic and halophilic *Archaea* with those that grow at high temperatures.

III RNA VIRUSES OF EUKARYOTES

The differences between prokaryotic cells and eukaryotic cells place some constraints on the mechanisms used by the viruses that infect them. For example, in prokaryotes transcription and translation can be coupled processes. In eukaryotes, by contrast, DNA is replicated and transcribed in the nucleus, whereas proteins are synthesized in the cytoplasm. In addition, mRNAs in eukaryotes are capped and have poly(A) tails (∞ Section 8.8). We will see that these details are relevant to viruses that replicate in eukaryotes.

We discussed some animal viruses in Chapter 10 (∞ Section 10.11). Our interest in animal viruses springs mostly from their role in human disease, and we mainly focus on animal viruses in the remainder of this chapter. However, plants also suffer from infectious diseases, some of which are due to viruses. Indeed, several virus diseases of crop plants are of major economic importance.

19.6 Plant RNA Viruses

Plant cell walls are extremely thick and strong. Yet there are a considerable number of viruses that infect plants and, in multicellular plants, spread from the infected cell to neighboring cells. The great majority of known plant viruses are positive-strand RNA viruses, perhaps because these small genomes can be transferred easily from cell to cell within the plant.

Tobacco Mosaic Virus: General Properties

In 1892, the Russian scientist Dmitri Ivanovsky showed that the causative agent of tobacco mosaic disease could pass through filters that retain bacteria. In 1898, the famous Dutch microbiologist Martinus Beijerinck (∞ Section 1.9) showed that this agent was not only filterable, but that it had many of the properties of a living organism. This agent, the first virus of any type to be recognized, was tobacco mosaic virus (TMV).

TMV has a rod-shaped virion with helical symmetry that contains 2,130 copies of a coat protein plus a single copy of the positive-strand RNA genome (∞ Section 10.2). It was TMV that was first used to show that RNA could be the genetic material of viruses, just as DNA is in other viruses and in cells. TMV remains a serious agricultural problem because it infects tomato plants as well as tobacco. TMV infection of a plant requires damage to plant cell walls through which the virion enters. Uncoating takes place in the cell.

Genome and Replication of TMV

The RNA genome of TMV contains 6,395 nucleotides, and a map of the genome is shown in **Figure 19.11**. Like the bacteriophage MS2 (Section 19.1), TMV encodes only four proteins. The genome has a 5′ cap (∞ Section 8.8), thus it can be used directly as an mRNA and translated in the plant cell. The 3′ end of the TMV genome folds into a transfer RNA-like structure (Figure 19.11). In bacteria, the genome of a positive-strand RNA virus may act as the viral mRNA (∞ Section 10.7). However, eukaryotes cannot translate polycistronic mRNA (∞ Section 8.9), so the expression of TMV genes differs from that of bacterial RNA viruses.

The first gene in TMV encodes an enzyme called MTH that has two enzymatic activities, a methyltransferase that caps RNA and an RNA helicase. The next gene encodes the RNA-dependent RNA polymerase (RNA replicase) that the virus must synthesize to replicate a negative-strand RNA copy from which it can then make more copies of the genomic RNA. This protein is synthesized as part of a *polyprotein* (Section 19.8) including the MTH domains and is only synthesized when a ribosome accidentally reads through the stop codon at the end of the MTH gene. Because this happens rather infrequently, this longer protein is made only at low levels.

The remaining two genes encode the movement protein and the coat protein. These two proteins are translated from small monocistronic mRNAs that are transcribed from the negative-strand RNA. Like coat proteins of other viruses, that of TMV is essential to the formation of the virion, which is essential for infecting new plants. However, it is the movement protein that enables TMV to infect neighboring cells in an already infected plant. Plant cells are interconnected by cytoplasmic strands called *plasmodesmata*, and these strands connect the cytoplasm of neighboring cells (∞ Section 10.14). Plasmodesmata have very narrow channels, so narrow

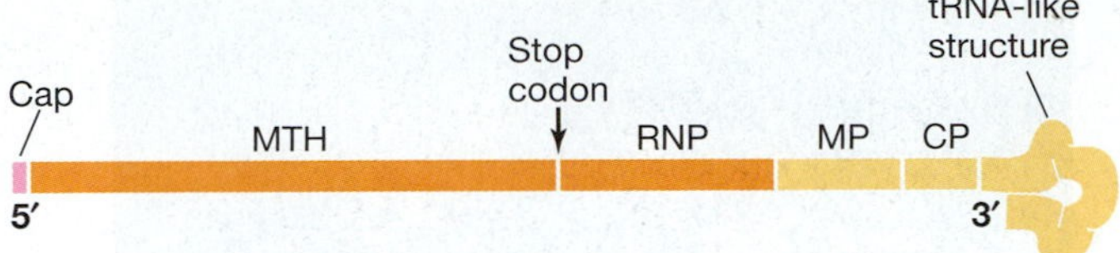

Figure 19.11 Genetic map of tobacco mosaic virus. The genome is a positive-strand RNA that is capped at its 5′ end and has a tRNA-like structure (drawn larger than scale) at its 3′ end.

in fact that neither the TMV virion nor free RNA can easily traverse these openings. The TMV movement protein binds to the new genomic positive-strand RNA and forms a complex that is extremely thin (about 2.5 nm) and that can move through the plasmodesmata and infect neighboring cells.

Many positive-strand RNA viruses are known that infect plants as well as those that infect *Bacteria* (Section 19.1) and animals (Section 19.7). Genomic analyses indicate that many of these viruses are closely related despite their quite different hosts and that all of their RNA replicases are related. Nonetheless, they have slightly different replication modes, usually related to differences in the host cells.

19.6 MiniReview

Most plant viruses have positive-strand RNA genomes, and one example is tobacco mosaic virus (TMV), the first virus discovered. The genomes of these viruses can move within the plant through intercellular connections that span the cell walls.

- Most plant RNA viruses encode a movement protein. What is its role in infection?
- Although the TMV genome is used as a messenger RNA, not all the proteins encoded by the virus can be translated from it. Explain.

19.7 Positive-Strand RNA Viruses of Animals: Poliovirus and Coronaviruses

Several positive-strand RNA animal viruses cause disease in humans and other animals, including the polioviruses, the rhinoviruses that cause the common cold, the coronaviruses that cause respiratory syndromes, including severe acute respiratory syndrome (SARS), and the hepatitis A virus. The first animal virus discovered, foot-and-mouth disease virus, the causative agent of a debilitating and eventually fatal disease of cloven-hoofed (ruminant) animals, is also in this group. With the exception of the coronaviruses, which are much larger, these viruses have been called picornaviruses because they are typically very small (about 30 nm in diameter; *pico* means "small") and contain single-stranded RNA. We focus in this section on poliovirus and the coronaviruses.

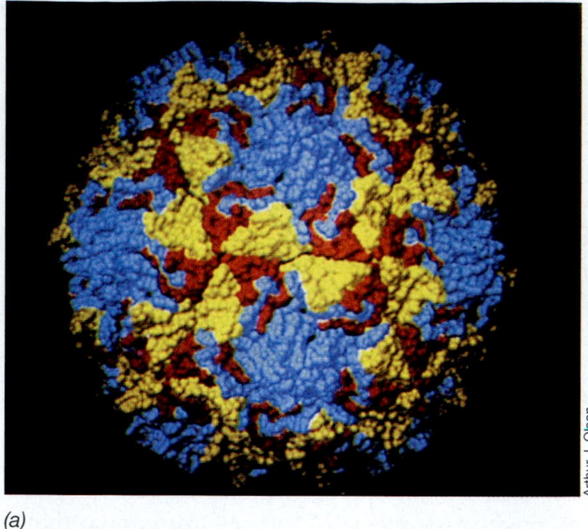

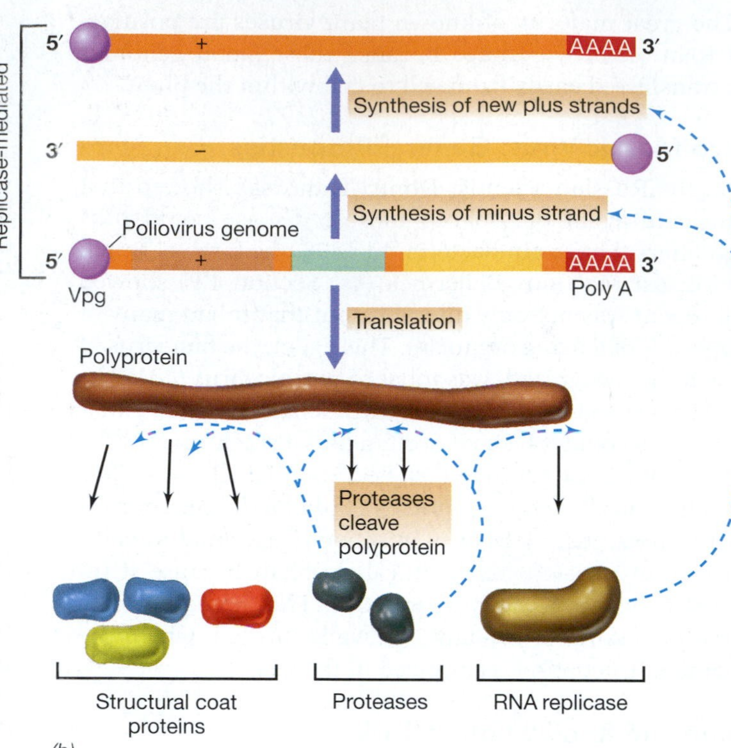

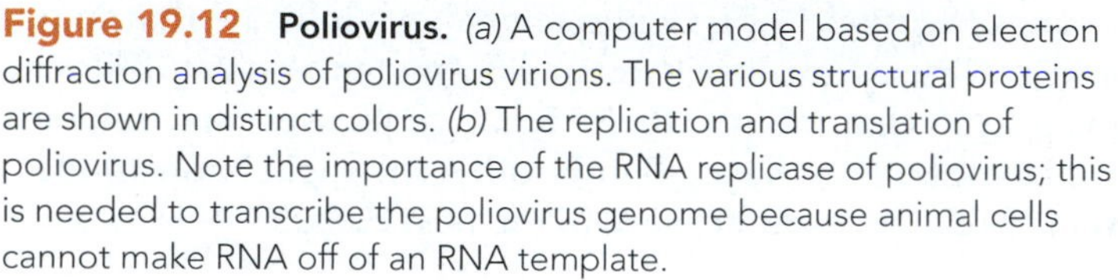
Figure 19.12 Poliovirus. *(a)* A computer model based on electron diffraction analysis of poliovirus virions. The various structural proteins are shown in distinct colors. *(b)* The replication and translation of poliovirus. Note the importance of the RNA replicase of poliovirus; this is needed to transcribe the poliovirus genome because animal cells cannot make RNA off of an RNA template.

An understanding of the biology of poliovirus is very important. At one time, polio was a major infectious disease of humans, but the development of an effective vaccine has brought the disease almost completely under control. The World Health Organization (WHO) has a vaccination program intended to eradicate the disease worldwide, and at present there are reports of the virus in only a few countries in Africa and Asia.

Poliovirus: General Features

The virion of poliovirus has an icosahedral structure with 60 morphological units per virion, each unit consisting of four distinct proteins. The genome of poliovirus is a linear single-stranded RNA molecule of 7,433 bases. At the 5′ terminus of the viral RNA is a protein, called the VPg protein, that is attached covalently to the RNA. At the 3′ terminus of the RNA is a poly(A) tail. The RNA genome of the virus is also the mRNA, even though the RNA is not capped, normally a prerequisite for translation in eukaryotes (∞ Section 8.8). Instead, the 5′ end of poliovirus RNA has a long sequence that can fold into several stem-loops. The VPg protein and the stem-loops mimic the cap-binding complex, and this permits binding of the poliovirus mRNA to the ribosome.

Although the viral RNA is monocistronic and has only a single start codon, it nonetheless encodes all the proteins of the virus. This trick is accomplished by making a single protein called a **polyprotein**. This giant protein (about 2,200 amino acid residues) then undergoes self-cleavage into about 20 smaller proteins (including cleavage intermediates), among which are the four different structural proteins of the virion. Other proteins include the RNA-linked VPg protein, an RNA replicase responsible for synthesis of both minus-strand and plus-strand RNA, and at least one virus-encoded protease, which carries out the polyprotein cleavage. This process, called posttranslational cleavage, occurs in many animal viruses as well as in normal cell metabolism in animal cells. Translation of a positive-strand mRNA to yield a single polyprotein that is subsequently cleaved is a scheme also used by some other viruses (Section 19.15).

Replication of Poliovirus RNA

An overview of poliovirus replication is illustrated in **Figure 19.12**. The whole poliovirus replication process occurs in the cell cytoplasm. To initiate infection, the poliovirus virion attaches to a specific receptor on the surface of a sensitive cell and enters the cell. Once inside the cell, the virus particle is uncoated, and the free RNA associates with ribosomes. The viral RNA is then translated to yield the large polyprotein as described above.

Replication of viral RNA begins within a short time after infection and is catalyzed by the RNA replicase released by cleavage of the polyprotein. This RNA replicase uses the positive-strand viral RNA as a template to synthesize a complementary negative-strand RNA. This negative strand is then the template for repeated synthesis of progeny positive strands, catalyzed by the same virus-encoded RNA replicase. Some of the progeny positive strands may again be used to make more negative strands, and as many as 1,000 negative strands may eventually be present in the cell. From these, as many as a million positive-strand copies may ultimately be formed. Both the positive and negative strands become covalently linked to the tiny VPg protein (only 22 amino acids long), which functions as a primer for RNA synthesis.

Once poliovirus replication begins, host RNA and protein syntheses are inhibited. Host protein synthesis is inhibited as a result of the destruction of an important host protein, the cap-binding protein, required for translation of capped mRNAs (Section 8.8). Poliovirus mRNA itself circumvents this limitation as previously mentioned through the extensive secondary structure of the 5′ end of its genome that allows binding to the ribosome. **www.microbiologyplace.com Online Tutorial 19.1: Replication of Poliovirus**

Coronaviruses and SARS

Coronaviruses are single-stranded plus-strand RNA viruses that, like poliovirus, replicate in the cytoplasm, but differ from poliovirus in their larger size and details of replication. Coronaviruses cause respiratory infections in humans and other animals, including about 15% of common colds (Section 34.8). In 2003, a novel coronavirus caused several outbreaks of SARS, an occasionally fatal pneumonia of the lower respiratory tract in humans.

Coronavirus virions are enveloped and more or less spherical, 60–220 nm in diameter, and contain club-shaped glycoprotein spikes on their surfaces (**Figure 19.13**). These give the virus the appearance of having a "crown" (the term *corona* is Latin for crown). Coronavirus genomes are noteworthy because they are the largest of any known RNA viruses (27–31 kb; strains of SARS virus average 29,700 nucleotides in length).

The coronavirus genome has a 5′ methylated cap and a 3′ poly(A) tail and so can function directly in the animal as mRNA. However, most viral proteins are not made from the translation of genomic RNA. Instead, upon infection, only a portion of the genome is translated, yielding a viral RNA replicase. The replicase uses the genomic RNA as a template to produce a full-length negative RNA strand from which several monocistronic mRNAs are transcribed. The latter are then translated to produce viral proteins. Progeny genomes are also made by transcription from the negative-strand RNA. The virions are assembled within the Golgi apparatus, a major secretory organelle in eukaryotic cells (Section 18.5), with fully assembled virions being released at the cell surface.

Coronavirus thus differs from poliovirus in terms of virion size, genome size and capping, lack of the VPg protein, and the absence of polyprotein formation and cleavage. Nevertheless, like poliovirus, coronaviruses can cause severe disease, including a fatal acute infection that is rarely seen in polio. Fatality rates from SARS range from 13% in those under age 60 to nearly 45% in those above 60 (Microbial Sidebar, "SARS as a Model of Epidemiological Success," Chapter 33).

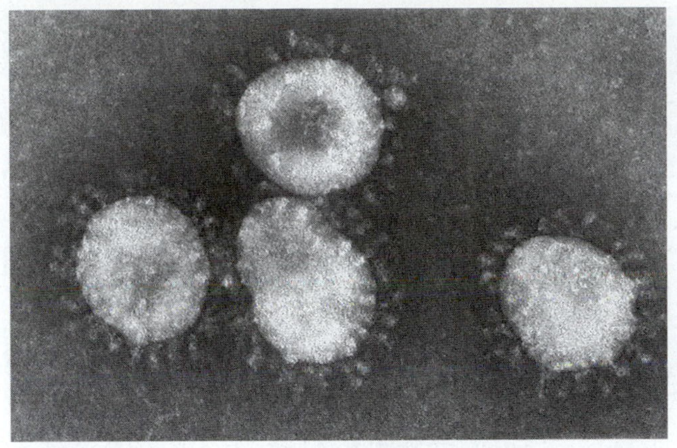

(a)

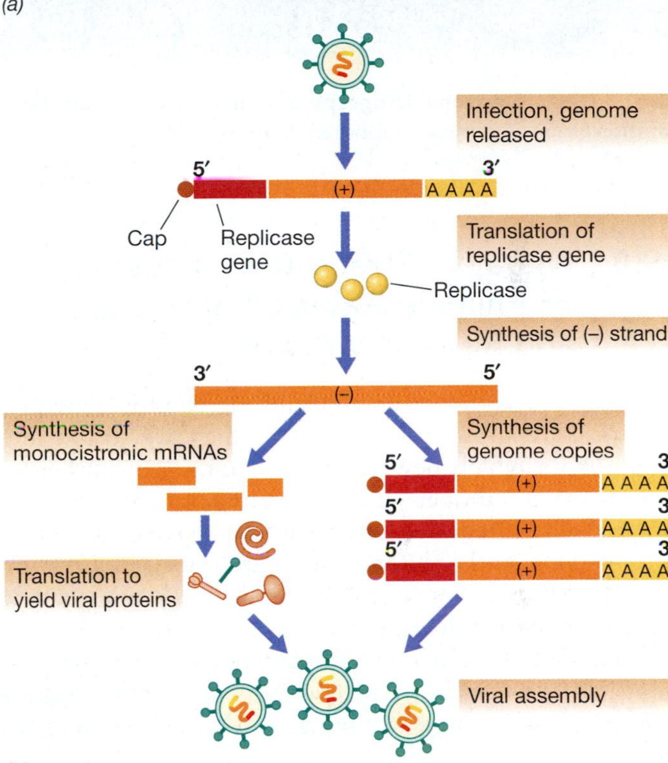

(b)

Figure 19.13 Coronaviruses. *(a)* Electron micrograph of a coronavirus. *(b)* Steps in coronavirus replication. The mRNA encoding viral proteins is transcribed from the negative strand made by the RNA replicase using the viral genome as template.

19.7 MiniReview

In small RNA viruses such as poliovirus, the viral RNA is translated directly, producing a long polyprotein that is broken down by enzymes into the numerous small proteins necessary for nucleic acid replication and virus assembly. Coronavirus is a large single-stranded RNA virus that resembles poliovirus in some but not all of its replication features.

- How can poliovirus RNA be synthesized in the cytoplasm whereas host RNA must be synthesized in the nucleus?
- In what ways are protein synthesis and genomic replication processes similar or different in poliovirus and the SARS virus?

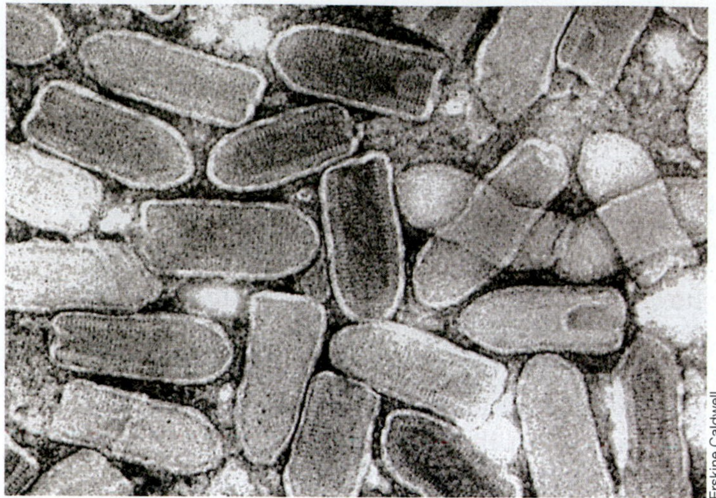

Figure 19.14 Electron micrograph of a rhabdovirus (vesicular stomatitis virus). A particle is about 65 nm in diameter.

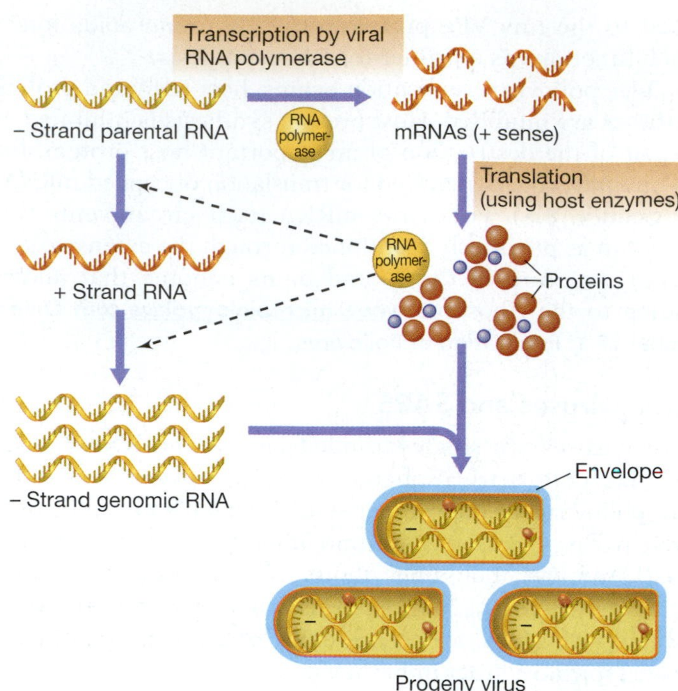

Figure 19.15 Flow of events during replication of a negative-strand RNA virus. Note the importance of the viral RNA polymerase, carried in the virion. This is critical because animal cells cannot make RNA using an RNA template.

19.8 Negative-Strand RNA Viruses of Animals: Rabies, Influenza, and Related Viruses

Poliovirus and coronavirus replication requires conversion of the positive-strand genome into a negative-strand intermediate from which new positive strands are synthesized. However, in a number of RNA animal viruses the RNA genome itself is **negative-strand RNA**—that is, RNA that is *complementary* to the mRNA. These are thus called *negative-strand RNA viruses*. We discuss here two important examples: rhabdoviruses, including rabies virus, and orthomyxoviruses, including influenza virus. The Ebola virus, a human pathogen responsible for an emerging infectious disease, is also a negative-strand RNA virus. There are no known negative-strand RNA bacteriophages or archaeal viruses.

Rhabdoviruses: General Features

One of the most important negative-strand RNA viral pathogens is the rabies virus, which causes the disease rabies in animals and humans. Worldwide, there are over 30,000 (mostly fatal) cases of rabies each year in humans and countless more in domesticated and wild animals (∞ Section 35.1). Rabies virus is called a rhabdovirus, from *rhabdo* meaning "rod," which refers to the shape of the virus particle. Another rhabdovirus that has been extensively studied is vesicular stomatitis virus (VSV) (**Figure 19.14**), a virus that causes the disease vesicular stomatitis in cattle, pigs, horses, and sometimes humans. Many rhabdoviruses, such as potato yellow dwarf virus, infect both insects and plants and can cause major losses of agricultural products.

The rhabdoviruses are enveloped viruses, with an extensive and rather complex lipid envelope surrounding the nucleocapsid. In rhabdoviruses the virion is bullet-shaped, about 70 nm in diameter and 175 nm long (Figure 19.14). The nucleocapsid is helically symmetric and makes up only a small part of the virus particle weight (about 2–3% of the virion is RNA).

Replication of Rhabdoviruses

The rhabdovirus virion contains several enzymes that are essential for the infection process. One of these is an RNA-dependent RNA polymerase (RNA replicase). As previously discussed (∞ Section 10.7), the presence of such an enzyme is essential because the genome of negative-strand viruses cannot be translated directly but must first be converted into the positive strand; host enzymes that transcribe RNA from an RNA template are not available.

The RNA of rhabdoviruses is transcribed in the cytoplasm into two distinct classes of RNAs (**Figure 19.15**). The first is a series of mRNAs encoding the structural genes of the virus (for example, VSV has five genes). Each mRNA is monocistronic, encoding a single protein. The second type of RNA is a positive-strand RNA that is a copy of the complete viral genome (the VSV genome is 11,162 nucleotides long). These full-length plus-strand RNAs are templates for the synthesis of negative-strand genomic RNA molecules for progeny virions. Once mRNA encoding the virus RNA polymerase is made in the primary transcription process, synthesis of more copies of the virus RNA polymerase occurs, leading to the formation of many plus-strand RNA molecules, both mRNAs and genomic RNA templates (Figure 19.15).

Assembly of Rhabdoviruses

Translation of viral mRNAs leads to the synthesis of viral coat proteins. Assembly of an enveloped virus is considerably more complex than assembly of a naked virion. Two kinds of coat proteins are formed, nucleocapsid proteins and envelope proteins. The **nucleocapsid** is formed first by association of the nucleocapsid protein molecules around the viral RNA.

The envelope proteins possess hydrophobic amino acid leader sequences at their amino-terminal ends (∞ Section 7.17). As these proteins are synthesized, sugar residues are added, leading to the formation of glycoproteins. Such viral glycoproteins, characteristic of membrane-associated proteins, migrate to the cytoplasmic membrane where the leader sequences are removed; here they replace host membrane proteins. Nucleocapsids then migrate to the areas on the cytoplasmic membrane where these virus-specific glycoproteins exist, recognizing the virus glycoproteins with great specificity. The nucleocapsids then become aligned with the glycoproteins and bud through them, becoming coated by the glycoproteins in the process.

The final result is an enveloped virion with a nucleocapsid center and a surrounding membrane whose lipid is derived from the host cytoplasmic membrane but whose membrane proteins are encoded by the virus. The budding process itself does not cause detectable damage to the cell, which may continue to release virions in this way for a considerable period of time. Host damage does eventually occur, of course, but is brought about by factors other than just the production of virus.

Influenza and Other Orthomyxoviruses

Another group of negative-strand viruses is the orthomyxoviruses, which contain the important human pathogen, *influenza virus*. The term "myxo" refers to the mucus or slime of cell surfaces with which these viruses interact. In the case of influenza virus, this mucus is on the mucous membrane of the respiratory tract; thus influenza virus is primarily transmitted by the respiratory route (∞ Section 34.9). The term "ortho" distinguishes this group from another group of negative-strand viruses, the paramyxovirus group. The paramyxoviruses, which include such important human pathogens as mumps and measles viruses, are quite similar to rhabdoviruses in their molecular biology.

The orthomyxoviruses have been extensively studied over many years, beginning with early work during the 1918 influenza pandemic that caused the deaths of millions of people worldwide. The orthomyxoviruses are enveloped viruses in which the viral RNA is present in the virion in a number of separate pieces. The genome of the orthomyxoviruses is thus segmented. In the case of influenza A virus, the genome is segmented into eight linear single-stranded molecules ranging in size from 890 to 2,341 nucleotides. The influenza virus nucleocapsid is of helical symmetry, about 6–9 nm in diameter and about 60 nm long. This nucleocapsid is embedded in an envelope that has a number of virus-specific proteins as well as lipid derived from the host (**Figure 19.16**).

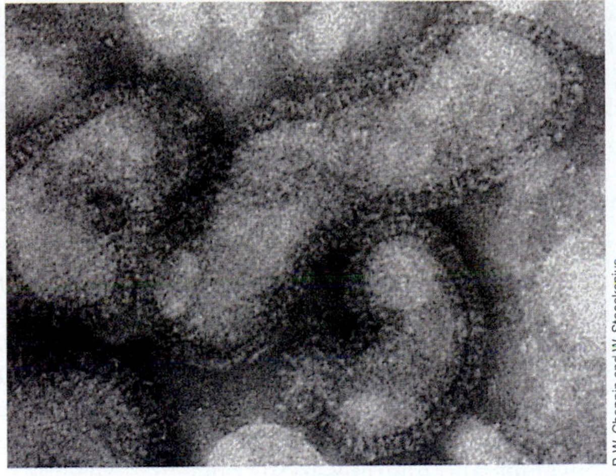

(a)

Neuraminidase
Hemagglutinin
Viral RNA polymerase
RNA endonuclease
Envelope
RNA genome (eight segments)

(b)

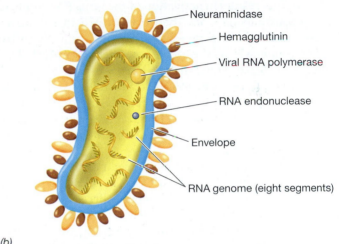

Figure 19.16 Influenza virus. *(a)* Electron micrograph of human influenza virions. *(b)* Diagram, showing some of the components including the segmented genome.

Because of the way influenza virus buds as it leaves the cell, the virus has no defined shape and is said to be polymorphic (Figure 19.16*a*). There are proteins on the outside of the envelope that interact with the host cell surface. One of these is called hemagglutinin, so named because it causes agglutination (clumping) of red blood cells. The red blood cell is not the type of host cell the virus normally infects but has on its surface the same type of membrane component, sialic acid, as the mucous membrane cells of the respiratory tract. Thus, the red blood cell is merely a convenient cell type for assaying agglutination activity. An important feature of the influenza virus hemagglutinin is that antibodies against this hemagglutinin prevent the virus from infecting a cell. Thus, antibody directed against the hemagglutinin neutralizes the virus, and this is the mechanism by which immunity to influenza is brought about during immunization (∞ Section 34.9).

A second type of protein on the influenza virus surface is an enzyme called neuraminidase (Figure 19.16*b*). Neuraminidase breaks down the sialic acid component of the host cytoplasmic membrane, which is a derivative of neuraminic acid. Neuraminidase appears to function primarily in the virus assembly process, destroying host membrane sialic acid that would otherwise block assembly or become incorporated into the mature virus particle.

Replication of Influenza Virus

In addition to neuraminidase, influenza virions possess two other key enzymes, an RNA-dependent RNA polymerase (RNA replicase), which converts the negative-strand genome into a positive strand (as already discussed for the rhabdoviruses), and an RNA endonuclease, which cuts a primer from the host's capped mRNA precursors. The virion enters the cell, after which the nucleocapsid becomes separated from the envelope and migrates to the nucleus. The viral nucleic acid then replicates in the nucleus.

Uncoating results in activation of the virus RNA replicase. The mRNA molecules are then transcribed in the nucleus from the virus RNA, using oligonucleotide primers cut from the 5′ ends of newly synthesized capped cellular mRNAs by the viral endonuclease. Thus, the viral mRNAs have 5′ caps. The poly(A) tails of the viral mRNAs are added, and the virus mRNA molecules move to the cytoplasm for translation. Thus, although influenza virus RNA replicates in the nucleus, influenza virus proteins, like all viral proteins, are synthesized in the cytoplasm.

Ten proteins are encoded by the eight segments of the influenza virus genome. The mRNAs transcribed from six segments each encode a single protein, whereas the other two segments encode two proteins each. The latter are not expressed by using true polycistronic mRNA as in prokaryotes because eukaryotic ribosomes recognize only the AUG closest to the 5′ end of the mRNA as a start codon (∞ Section 8.9). Therefore, they can make only one protein from a given RNA. The original full-length mRNAs transcribed from these two segments are each translated to give one protein. But in each case, an additional protein is translated from these messages following processing of the message by the host's RNA splicing machinery.

Some of the proteins made are needed for influenza virus RNA replication, whereas others are structural proteins of the virion. The overall pattern of genomic RNA synthesis resembles that of the rhabdoviruses, with primary RNA synthesis resulting in the formation of positive-strand RNA templates that are then used for the formation of progeny negative-strand RNA molecules. The complete enveloped virion forms by budding, as for the rhabdoviruses.

Antigenic Shift versus Antigenic Drift in Influenza

The segmented genome of the influenza virus has important practical consequences. Influenza virus and other viruses of this family exhibit a phenomenon called **antigenic shift** in which portions of the RNA genome from two genetically distinct strains of virus infecting the same cell are reassorted. This generates virions that express a set of surface proteins significantly different from that of both original viruses. Surface proteins are the major target of antibodies that form as a result of immunization by artificial means or through natural infection. However, after an antigenic shift, previous immunity to either original influenza virus is insufficient to ward off infection by the genetically "new" viruses. Antigenic shift is thought to bring about major pandemics and epidemics of influenza because immunity to the new forms of the virus is essentially absent from the population.

Antigenic shift can be contrasted with **antigenic drift**. In the latter, the structure of the neuraminidase and hemagglutinin proteins on the surface of the influenza virus virion is altered, usually in a subtle way, by mutation in the genes encoding them. The alterations change the surface properties of influenza virus sufficiently that antibodies that recognized the virus previously no longer do so, or do so less effectively. Thus, for effective immunity to be achieved, new antibodies must be produced. This is a major reason why influenza vaccines rarely confer protection for more than 1 year (∞ Section 34.9). Genetic drift during the previous year yields slightly modified forms of influenza virus for which new vaccines must be made.

19.8 MiniReview

In negative-strand viruses the virus RNA is not the mRNA, but it is copied into mRNA by an enzyme present in the virion. Important negative-strand viruses include rabies virus and influenza virus.

- Why is it essential that negative-strand viruses carry an enzyme in their virions?
- What is a segmented genome?
- What is the difference between antigenic shift and antigenic drift?

19.9 Double-Stranded RNA Viruses: Reoviruses

Reoviruses are an important family of animal viruses that have double-stranded RNA genomes. Rotavirus is a typical reovirus and is the most common cause of diarrhea in infants from 6 to 24 months of age. There are also reoviruses known that cause respiratory infections, and others that infect plants; we previously mentioned that the bacteriophage $\phi 6$ seems closely related to the reoviruses (Section 19.1). Reovirus virions consist of a nonenveloped nucleocapsid 60–80 nm in diameter, with a double shell of icosahedral symmetry (**Figure 19.17**). Predictably, the virions of these double-stranded RNA viruses contain virus-encoded enzymes necessary to synthesize mRNA and new RNA genomes.

Replication of Reoviruses

The genome of reoviruses is segmented into 10–12 molecules of *linear* double-stranded RNA. Replication occurs exclusively in the cytoplasm of the host. The double-stranded RNA is inactive as mRNA, and the first step in reovirus replication is transcription by a viral-encoded RNA-dependent RNA polymerase, using the minus strand as a template, to make plus-sense mRNA. The mRNA is then capped and methylated by viral enzymes and then translated.

Generally, each molecule of RNA in the genome encodes a single protein, although in a few cases the protein formed is cleaved to yield the final product. However, one of the reovirus mRNAs actually encodes two proteins, but the RNA does not have to be processed in order to translate both of these. Instead, a ribosome sometimes "misses" the start codon for the first gene in this message and travels on to the start codon of the second gene. Therefore, there are occasional exceptions to the generalization that eukaryotic ribosomes initiate at the first AUG codon in an mRNA.

In the initial infection process, the reovirus virion binds to a cellular receptor protein. The attached virus then enters the cell and is transported into lysosomes (Section 18.5), where normally it would be destroyed. However, within the lysosome the outer shell of the virus particle is modified by removal of two proteins and cleavage of another by lysosomal enzymes. The viral core, surrounded only by the inner protein shell, is then released into the cytoplasm of the host cell.

This uncoating process activates the viral RNA polymerase and hence initiates virus replication that occurs within the viral core, called the subviral particle, which remains intact in the cell. Each of the ten capped, single-stranded plus-strand RNAs serve as templates for the synthesis of complementary minus-strand RNA. This eventually yields progeny double-stranded viral genomic RNA that is encapsidated. When enough viral capsid proteins are present, mature virions are assembled and released by cell lysis.

19.9 MiniReview

Reoviruses contain segmented linear double-stranded RNA genomes. Like negative-strand RNA viruses, reoviruses contain an RNA-dependent RNA polymerase within the virion.

- How does reovirus genome replication resemble that of influenza virus, and how does it differ?

IV DNA VIRUSES OF EUKARYOTES

DNA viruses of eukaryotes are fewer in number than RNA viruses. Both single-strand and double-strand DNA viruses are known for both animals and plants. The dsDNA viruses include several important viruses that cause human diseases and some interesting large viruses of plants and protozoa.

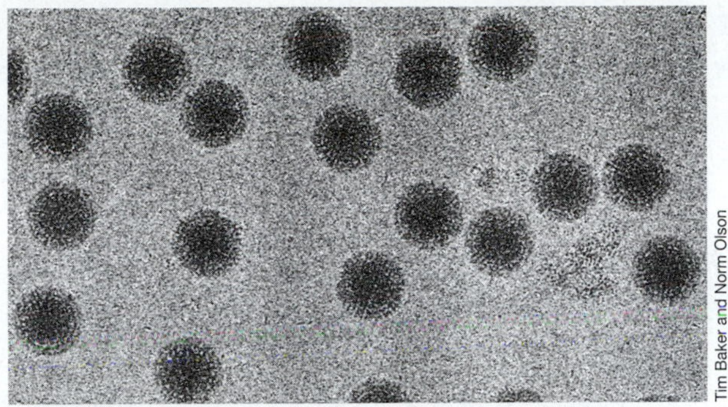

(a)

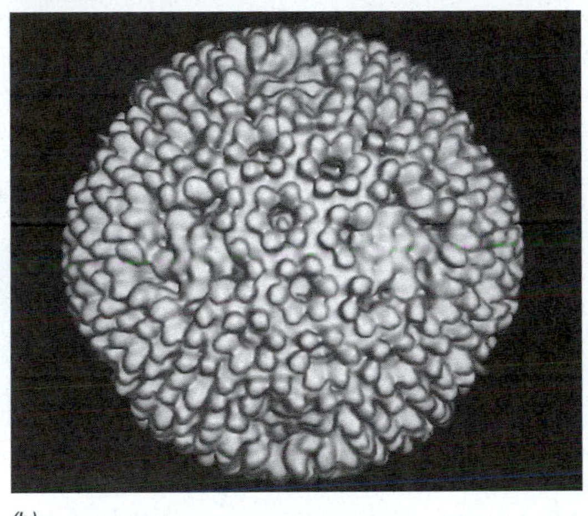

(b)

Figure 19.17 Double-stranded RNA viruses: The reoviruses. *(a)* An electron micrograph showing reovirus virions (each with a diameter of about 70 nm). *(b)* Three-dimensional reconstruction of a reovirus virion calculated from electron micrographs of frozen-hydrated virions.

19.10 Plant DNA Viruses

DNA viruses of plants are especially rare. However, some unusually large viruses are known that infect single-celled plants. Related viruses are known that infect certain protozoa (Microbial Sidebar, "Mimivirus and Viral Evolution").

DNA Plant Viruses: *Chlorella* Viruses

Green algae are green plants, and *Chlorella* is a widely distributed genus of microscopic, single-celled green alga. Most species of *Chlorella* are free living, but some *Chlorella*-like algae are endosymbionts of freshwater or marine animals, including protozoa such as *Paramecium* (Section 18.21). Many of these endosymbionts can be grown in the laboratory independently of the organism in which they reside, and some of these are hosts to viruses; the best studied of these is *Paramecium bursaria Chlorella* virus 1 (PBCV-1). PBCV-1 belongs to a large family of viruses known as phycodnaviruses, which are quite widespread in nature. Members of this family infect both *Chlorella* strains found as endosymbionts in other cells and also many free-living species of single-celled algae.

UNIT 3

Microbial Sidebar

Mimivirus and Viral Evolution

Mimivirus is the largest virus presently known, both in terms of its physical size and in genome size. When first discovered, it was misidentified as a gram-positive coccus because it stained with the Gram stain and is as large as some small bacteria. Mimivirus has an icosahedral capsid of about 0.5 μm diameter surrounded by filaments of 125 nm, giving a total diameter of about 0.75 μm (**Figure 1**). The capsid shows three layers of dense matter under the electron microscope, probably consisting of two lipid membranes inside a protein shell.

Mimivirus contains 1.2 Mbp of double-stranded DNA that encodes an estimated 911 proteins. Its genome is thus over twice as large as that of the next largest known virus, bacteriophage G of *Bacillus subtilis* (497,513 bp) and three times that of the *Paramecium bursaria Chlorella* virus 1 (330,743 bp) discussed in Section 19.10. In addition, the Mimivirus genome is larger than that of several cellular organisms, notably the mycoplasmas. Of the 911 Mimivirus proteins, only about 300 have functions predicted by homology. However, of the 600 "unknown/unidentified" proteins with no close relatives in any sequence database, nearly 50 are found in the virus particle. Thus, at least these 50 are genuine functional proteins, not merely "genetic junk" as is sometimes suggested for the many unidentified genes and proteins found in viruses.

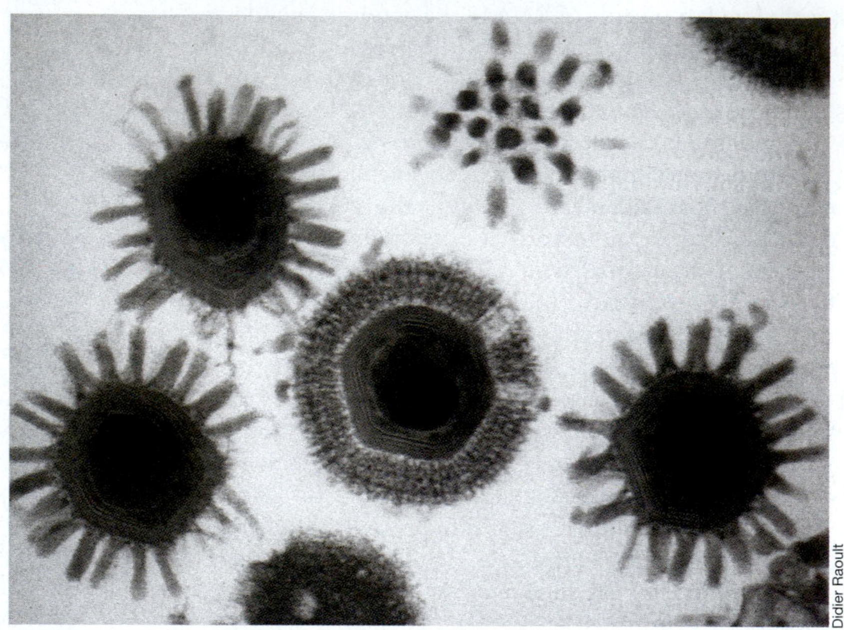

Figure 1 Mimivirus: The largest known virus. *Mimivirus virions are shown in this thin-section transmission electron micrograph. Each virion is approximately 0.75 μm in diameter, roughly three-quarters the diameter of an* Escherichia coli *cell.*

Mimivirus normally infects the amoeba *Acanthamoeba polyphaga*. It may also cause pneumonia in humans, although the evidence is largely indirect and consists of finding antibodies to Mimivirus in certain pneumonia patients. However, a laboratory technician working with Mimivirus developed pneumonia that was almost certainly due to infection by Mimivirus because virus particles were isolated from the technician's blood.

Mimivirus belongs to a group of large viruses that contain large genomes known as the *nucleocytoplasmic large DNA viruses* (NCLDV) (**Figure 2**). The NCLDV comprises several virus families, including the pox viruses (Section 19.13) and phycodnaviruses (Section 19.10). These viruses all share a set

PBCV-1 has large icosahedral virions (**Figure 19.18**) and a linear double-stranded DNA genome. The virions have a lipid component that is essential for infectivity but it is inside the capsid; therefore, the virion is not enveloped. The genomes of the *Chlorella* viruses are extremely large; all are over 300 kbp (for a comparison with other viruses see Table 10.1). In many cases the DNA is also extensively modified by methylation. The genome of PBCV-1 has been completely sequenced. This 330,742-bp genome encodes over 370 different proteins and 10 tRNAs. The ends of the genome are incompletely base-paired hairpin loops very similar to those found in poxviruses (Section 19.14).

Replication of *Chlorella* Viruses

PBCV-1 enters cells somewhat like bacteriophages do. The virion binds specifically to the cell wall of the host, and then

(continued)

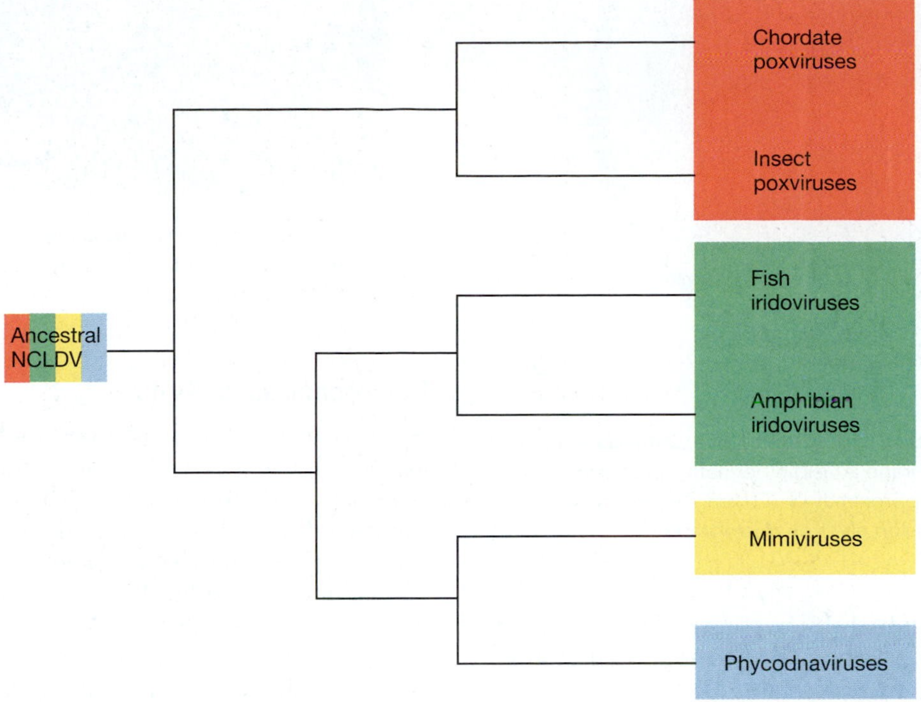

Figure 2 Nucleocytoplasmic large DNA viruses (NCLDV). *The phylogenetic tree shows the genomic relationships between the poxviruses, iridoviruses (large viruses of fish and amphibians), phycodnaviruses (large double-stranded DNA viruses of plants and some green algae), and Mimivirus.*

of highly homologous proteins, mostly involved in DNA replication. All NCLDV also share an internal lipid layer surrounding the central core with a protein shell outside the lipid. Thus, these viruses are not truly enveloped. All of these large viruses contain not only certain enzymes but also mRNA inside the virion. In particular, they contain mRNA encoding virus DNA polymerase.

Database searches of DNA sequences present in environmental samples implies that unisolated large viruses related to Mimivirus are reasonably frequent in nature. In particular, genes related to those of Mimivirus have been found among DNA sequences from marine environments.

at least five different enzymes carried by the virion digest away the cell wall at the point of contact. Viral DNA is then released into the cell leaving the empty virion behind. As is the case for most double-stranded DNA viruses of eukaryotes, the DNA of PBCV-1 is replicated in the nucleus, and RNA is also synthesized there. PBCV-1 encodes several enzymes for DNA replication, including a DNA polymerase. However, although PBCV-1 encodes some transcription factors, it does not encode its own RNA polymerase. A few of the genes of PBCV-1 contain introns that must be removed (Section 8.8). The virus mRNA is capped and some of the enzymes involved are virus-encoded. Early mRNA, but not late mRNA, has poly(A) tails. Like bacteriophage T4 and some other large DNA viruses, PBCV-1 encodes several of its own tRNAs.

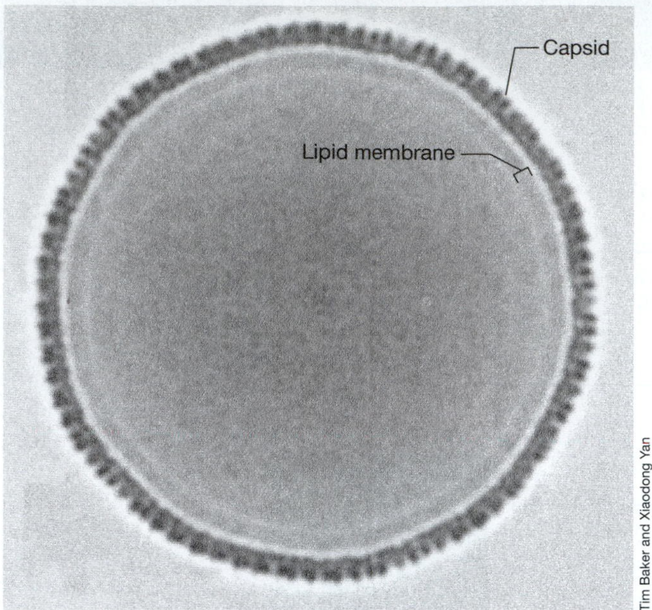

Figure 19.18 **A cross-section of a virion of the *Chlorella* virus PBCV-1.** A lipid bilayer membrane is visible beneath the capsid shell. The virion has a diameter of approximately 170 nm. The image is reconstructed from several transmission electron micrographs.

There is one other extremely interesting fact to note about the *Chlorella* viruses: Their genomes encode several restriction and modification enzyme systems (∞ Section 12.1). Indeed, they are the only source of restriction enzymes outside of the prokaryotes and a few bacteriophages. Thus we see features in *Chlorella* viruses typical of both prokaryotic and eukaryotic genetic elements.

19.10 MiniReview

Some of the largest known viruses infect single-celled algae. The *Chlorella* viruses have very large double-stranded DNA genomes that encode several hundred proteins.

- Which features of *Chlorella* viruses are prokaryote-like?
- Which features of *Chlorella* viruses are eukaryote-like?
- What is unusual about the Mimivirus genome?

19.11 Replication of Double-Stranded DNA Viruses of Animals

A large number of double-stranded DNA viruses of animals are known, among them several viruses that are human pathogens. These include the polyoma and papilloma viruses, the herpesviruses, the pox viruses, and the adenoviruses. The genomes of all of these replicate in the nucleus, except for the pox viruses, which replicate in the cytoplasm. In the following sections, we briefly discuss the replication of each of these groups of viruses.

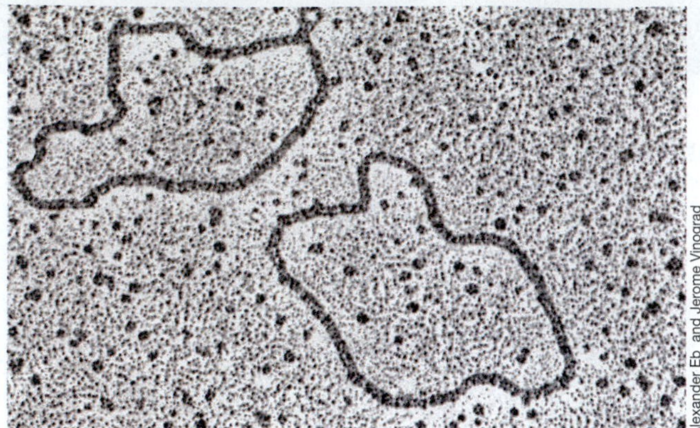

Figure 19.19 **Polyomaviruses.** Electron micrograph of relaxed (nonsupercoiled) circular DNA from a tumor virus. The contour length of each circle is about 1.5 μm.

Polyomaviruses: SV40

Some viruses of the polyomavirus family induce tumors in animals; indeed, the suffix "oma" means tumor. One of these DNA tumor viruses was first isolated from monkeys, and it was therefore called *s*imian *v*irus 40 or SV40. Virus SV40 was one of the first viruses to be studied by genetic engineering techniques and has been extensively used as a vector for moving genes into eukaryotic cells (∞ Section 12.13).

The SV40 virion is a nonenveloped particle 45 nm in diameter with an icosahedral head containing 72 protein subunits. Unlike RNA viruses, there are no enzymes in the virion. The genome of SV40 consists of one molecule of double-stranded DNA of 5,243 bp. The DNA is circular (**Figure 19.19**) and exists in a supercoiled configuration within the virion. The complete base sequence of SV40 is known, and a genetic map is shown in **Figure 19.20**.

SV40 nucleic acid is replicated in the nucleus, and the proteins are synthesized in the cytoplasm. The virion is assembled in the nucleus. The replication of these viruses can be divided into two distinct stages, early and late. During the early stage, the early region of the viral DNA is transcribed (Figure 19.20). A single RNA molecule—the primary transcript—is made by the cellular RNA polymerase, but it is then processed into two mRNAs, a large one and a small one, both of which are capped. Introns are present in the SV40 genome, so they are excised out of the primary RNA transcript. In the cytoplasm, the mRNAs are translated to yield two proteins. One of these proteins, the T antigen, binds to the site on the virus DNA that is the origin of replication; this initiates viral genome synthesis.

The genome of SV40 is too small to encode its own DNA polymerase, so host DNA polymerases are used. DNA is replicated in a bidirectional fashion (∞ Figure 7.16) from a single origin of replication. The process involves the same events that have already been described for host cell DNA replication (∞ Sections 7.5 and 7.6).

Late SV40 mRNA molecules are synthesized using the strand complementary to that used for early mRNA synthesis

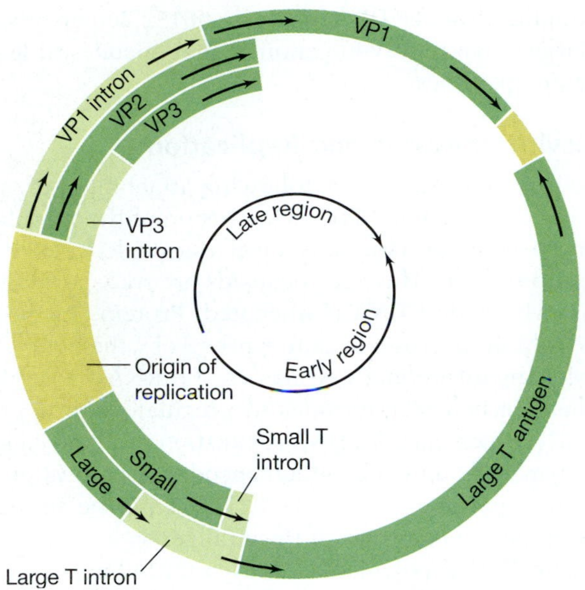

Figure 19.20 Genetic map of the polyomavirus SV40. VP1, VP2, and VP3 are the genes encoding the three proteins that make up the coat of SV40. The arrows show the direction of transcription. Note how genes encoding VP1, VP2, and VP3 overlap.

(Figure 19.20). Transcription begins at a promoter near the origin of replication. This late RNA is then processed by splicing, capping, and polyadenylation to yield mRNA corresponding to the three coat proteins: VP1, VP2, and VP3. The genes for these proteins overlap (Figure 19.20), a phenomenon seen in several other small viruses (Sections 19.1, 19.2, and 19.4). SV40 coat protein mRNAs are transported to the cytoplasm and translated into the viral coat proteins. The latter are then transported back into the nucleus where virion assembly takes place. New SV40 virions are released by cell lysis.

Some polyomaviruses cause cancer. When a virus of the polyomavirus group infects a host cell, one of two modes of replication can occur, depending on the type of host cell. In some types of host cells, known as permissive cells, virus infection results in the usual formation of new virions and the lysis of the host cell. In other types of host cells, known as nonpermissive, efficient replication does not occur. Instead, the virus DNA becomes integrated into host DNA, analogous to a prophage (∞ Section 10.10), genetically altering the cells in the process (**Figure 19.21**). Such cells can show loss of growth inhibition and become tumor cells, a process called transformation (∞ Figure 10.22). As in certain tumor-causing retroviruses (∞ Section 10.12), expression of specific polyomavirus genes converts cells to the transformed state (Figure 19.21).

19.11 MiniReview

Most double-stranded DNA animal viruses, such as SV40, replicate in the nucleus. SV40 has a tiny genome and employs the strategy of overlapping genes to boost its genetic-coding potential. Some of these viruses cause cancer.

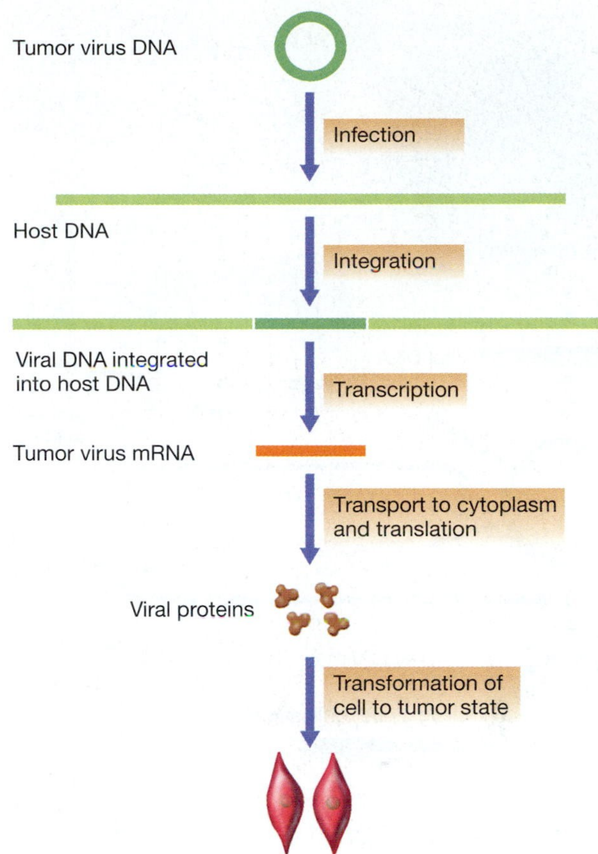

Figure 19.21 Events in cell transformation by a polyomavirus such as SV40. Either all or portions of the viral DNA are incorporated into host cell DNA. The viral genes responsible for cell transformation are transcribed and processed to viral mRNA molecules, which are transported to the cytoplasm. Here they are translated to form proteins that transform host cells into tumor cells (∞ Figure 10.22).

- Why doesn't a virus like SV40 need to carry enzymes in the virion as influenza virus does?
- How can one transcript yield more than one mRNA in SV40?

19.12 Double-Stranded DNA Viruses: Herpesviruses

The herpesviruses are a large group of double-stranded DNA viruses that cause diseases in humans and animals, including fever blisters (cold sores), venereal herpes, chickenpox, shingles, and infectious mononucleosis. Some of these diseases are discussed in Chapter 34. Herpesviruses are able to remain latent in the body for long periods of time, becoming active only under conditions of stress. Both **herpes simplex**, the virus that causes fever blisters and genital herpes, and varicella-zoster virus, the cause of chickenpox and shingles, are able to remain latent in the neurons of the sensory ganglia, from which they emerge periodically to cause infections of the skin.

An important group of herpesviruses are tumorigenic, causing clinical forms of cancer. For example, Epstein-Barr

UNIT 3

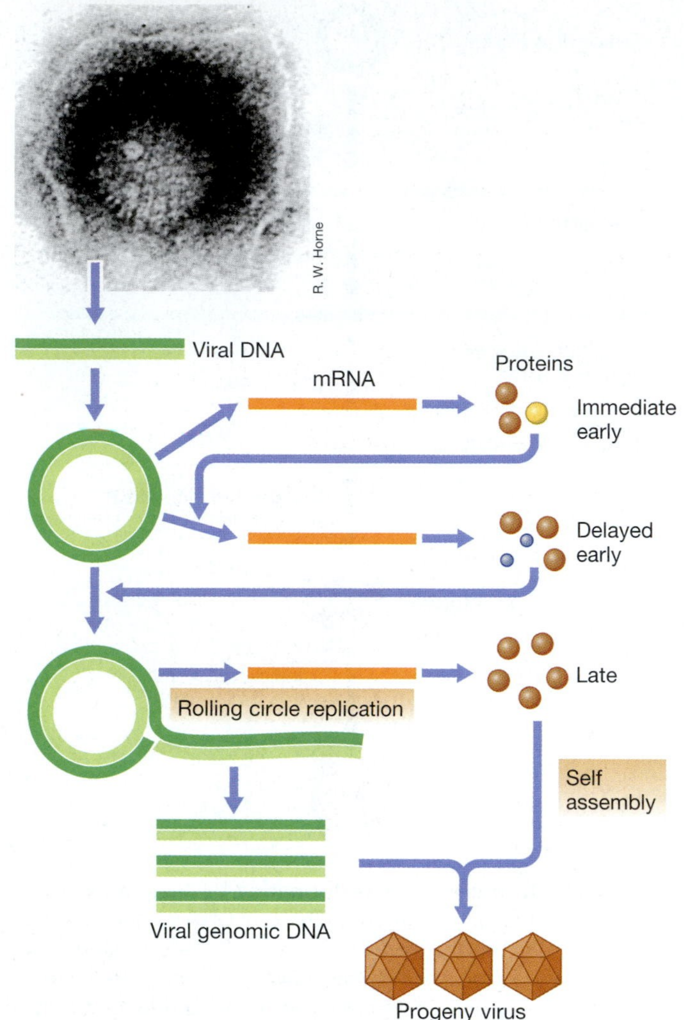

Figure 19.22 Herpesvirus. Flow of events in replication of herpes simplex virus starting from an electron micrograph of a herpes virion (diameter about 150 nm).

virus causes Burkitt's lymphoma, a tumor common among children in Central Africa and New Guinea. Burkitt's lymphoma was among the first human cancers to be linked to virus infection. Epstein-Barr virus can also cause a nonmalignant general malaise called infectious mononucleosis in humans.

Herpesviruses: General Features

The herpesvirus particle is structurally complex, consisting of four distinct morphologic units. Herpes simplex type I is an enveloped virus about 150 nm in diameter. The center of the virus, called the core, consists of linear double-stranded DNA. The herpesvirus nucleocapsid is of icosahedral symmetry and consists of 162 capsomers, each of which is composed of a number of distinct proteins. Outside the nucleocapsid is an amorphous layer called the *tegument*, a fibrous structure unique to the herpesviruses. Surrounding the tegument is an envelope whose outer surface contains many small spikes.

A large number of separate proteins are present within the virion, but not all of them have been characterized. The genome of herpes simplex type I virus consists of one large linear double-stranded DNA molecule of 152,260 bp (about 30 times larger than the SV40 genome) that encodes at least 84 distinct polypeptides.

Herpesvirus Infection and Replication

Herpesvirus infection occurs following attachment of virions to specific cell receptors. Following fusion of the cytoplasmic membrane with the virus envelope, the nucleocapsid is released into the cell. The nucleocapsids are transported to the nucleus, where viral DNA is uncoated. Proteins in the virus particle inhibit macromolecular synthesis by the host.

Following infection, three classes of mRNA are produced: immediate early, which encodes five regulatory proteins; delayed early, which encodes DNA replication proteins including DNA polymerase; and late, which encodes structural proteins of the virus particle (**Figure 19.22**). During the immediate early stage, about one-third of the viral genome is transcribed by a host cell RNA polymerase. Early mRNA encodes certain regulatory proteins that stimulate the synthesis of the delayed early proteins. The delayed early proteins appear only after the immediate early proteins have been made. During this stage, about 40% of the viral genome is transcribed. Among the many key proteins synthesized during the delayed early stage are a viral-specific DNA polymerase, enzymes for synthesis of deoxyribonucleotides, and a DNA-binding protein. These enzymes are all needed for viral DNA replication.

Herpesvirus DNA synthesis itself takes place in the nucleus. After infection, the herpesvirus genome circularizes (remarkably like bacteriophage lambda; Section 10.10) and replicates by a rolling circle mechanism (Figure 10.19). However, there seem to be three origins of replication. Long concatemers are formed that become processed into virus-length genomic DNA during the assembly process in a manner similar to that described for DNA bacteriophages (Sections 10.9 and 19.3). Viral nucleocapsids are assembled in the cell nucleus, and the viral envelope is added via a budding process through the inner membrane of the nucleus. Mature virions are subsequently released through the endoplasmic reticulum to the outside of the cell. Thus, the assembly of herpesvirus differs from that of the enveloped RNA viruses, which are assembled on the cytoplasmic membrane instead of the nuclear membrane.

Cytomegalovirus

A very widespread herpesvirus is cytomegalovirus (CMV). CMV replicates as a typical herpesvirus and is present in 50–85% of all adults in the United States by 40 years of age. For healthy individuals, infection with CMV comes with no apparent symptoms or long-term health consequences. However, CMV infection can cause serious disease in immune-compromised individuals, such as those receiving immunosuppressant drugs (for example, those who have received organ transplants and some cancer and dialysis patients) or those infected with the AIDS virus, HIV. In such individuals, CMV can cause pneumonia, retinitis (an eye condition), and gastrointestinal disease. These conditions can be serious and occasionally cause death.

In the healthy host the immune system keeps CMV in check. However, CMV remains dormant within cells of an infected individual and can become reactivated whenever the immune system is compromised, leading to symptoms whose severity depends on the degree to which the immune system is suppressed. Like the Epstein-Barr virus, CMV also causes some cases of infectious mononucleosis.

19.12 MiniReview

Herpesviruses are large double-stranded DNA viruses. The viral DNA circularizes and is replicated by a rolling circle mechanism. Herpesviruses cause a variety of disease syndromes and can maintain themselves in a latent state in the host indefinitely, initiating viral replication periodically.

- Of what use is it to the herpesviruses to circularize their DNA prior to replication?
- Where is the herpesvirus nucleocapsid assembled and how does this affect the protein content of its envelope?

19.13 Double-Stranded DNA Viruses: Pox Viruses

Pox viruses are among the most complex and largest animal viruses known (**Figure 19.23**). These viruses are also unique in being DNA viruses that replicate in the cytoplasm. Thus, a host cell infected with a pox virus exhibits DNA synthesis outside the nucleus, something that otherwise occurs in eukaryotic cells only in organelles.

General Properties of Pox Viruses

Pox viruses have been important medically as well as historically. Smallpox was the first virus to be studied in any detail and was the first virus for which a vaccine was developed (by Edward Jenner in 1798). By diligent application of this vaccine on a worldwide basis, the disease smallpox has been eradicated in the wild, the first infectious disease to be eliminated in this fashion. Other pox viruses of importance are cowpox and rabbit myxomatosis virus, an important infectious agent of rabbits that was intentionally introduced in Australia to control the Australian rabbit population (∞ Section 33.5). Some pox viruses also cause tumors.

The pox viruses are very large, so large in fact, that they can actually be seen under the light microscope. Most research on pox viruses has been done with vaccinia virus, a close relative of smallpox virus and the virus used as a smallpox vaccine. The vaccinia virion is a brick-shaped structure about 400 × 240 × 200 nm. The virion lacks an envelope but is covered on its outer surface with protein tubules arranged in a lattice-like pattern (Figure 19.23). Within the virion there are two lateral bodies of unknown composition and a nucleocapsid, which contains DNA bounded by a layer of protein subunits.

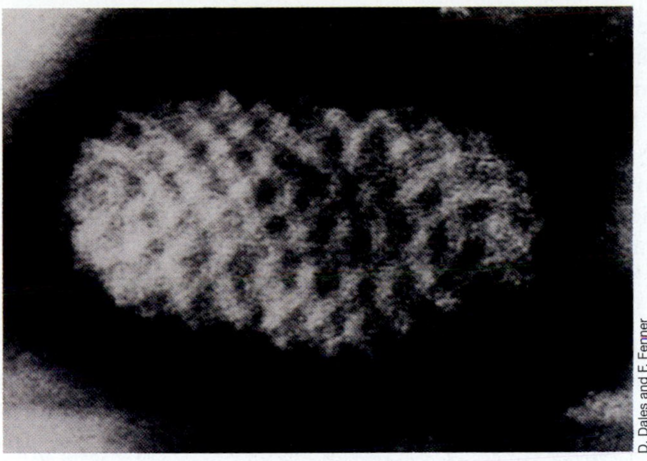

Figure 19.23 Pox viruses. Electron micrograph of a negatively stained vaccinia virus virion. The virion is approximately 400 nm (0.4 μm) long. Compare the size of pox virus with that of Mimivirus in Section 19.10.

The pox virus genome consists of linear double-stranded DNA. The vaccinia virus genome has about 185 kbp and about 180 genes. Pox virus DNA is unique because the two strands of the double helix are cross-linked at their termini as a result of phosphodiester bonds between adjacent strands. The ends of pox virus DNA are therefore very similar to those of the large *Chlorella* viruses discussed previously (Section 19.10).

Replication of Pox Viruses

Vaccinia virions are taken up into cells and the nucleocapsids liberated in the cytoplasm. The uncoating of the viral genome requires the activity of a viral protein that is synthesized after infection. The viral gene encoding this protein is transcribed by a viral-encoded RNA polymerase contained within the virion. In addition to this uncoating gene, a number of other viral genes are transcribed. The primary transcripts are turned into mRNAs by capping and polyadenylation while they are still inside the nucleocapsid. Copies of the genome are then made by a viral-encoded DNA polymerase.

Once the vaccinia DNA is fully uncoated, the formation of inclusion bodies within the cytoplasm begins. Within these inclusion bodies, the DNA is transcribed, replicated, and encapsidated into progeny virions. Each infecting virion initiates its own inclusion body, so the number of inclusions depends on the multiplicity of infection. Progeny DNA molecules form a pool from which individual molecules are incorporated into virions. Mature virions accumulate in the cytoplasm. There seems to be no specific release mechanism, and most virions are released only when the infected cell disintegrates.

Pox Viruses and Recombinant Vaccines

Vaccinia virus has been used as a host for genetically altered proteins of other viruses, permitting the construction of

UNIT 3

genetically engineered vaccines (∞ Section 26.5). As we will see in Section 30.5, a vaccine is a substance capable of eliciting an immune response in an animal and serves to protect the animal from future infection with the same agent. Vaccinia virus causes no serious health effects in humans but is highly immunogenic. Therefore, as a carrier of proteins from pathogenic viruses, vaccinia virus is a safe and effective tool for stimulating the immune response.

Molecular cloning methods (∞ Chapter 12) have been used to express key viral proteins of various viral pathogens (including influenza virus, rabies virus, herpes simplex type I virus, and hepatitis B virus) in vaccinia virions, which have then been used to develop a vaccine against that pathogen (∞ Section 26.5). A similar vaccine delivery system using adenovirus (discussed in the next section) as a vehicle has been developed, because, like vaccinia virus, adenoviruses are also of minor health consequence to humans.

19.13 MiniReview

The pox viruses, unlike the other DNA viruses we have discussed, are very large viruses that replicate entirely in the cytoplasm. These viruses also are responsible for several human diseases, but a vaccination campaign has eradicated the smallpox virus in the wild.

- Why is it notable that pox viruses replicate their DNA in the cytoplasm?
- How are poxviruses of use to genetic engineering?

19.14 Double-Stranded DNA Viruses: Adenoviruses

The adenoviruses are a major group of icosahedral linear double-stranded DNA viruses. The term *adeno* is derived from the Latin for "gland" and refers to the fact that these viruses were first isolated from the tonsils and adenoid glands of humans. Adenoviruses cause mild respiratory infections in humans, and a number of such viruses can be isolated, even from healthy individuals.

The genomes of the adenoviruses consist of linear double-stranded DNA of about 36 kbp. Attached in covalent linkage to the 5′ end of the DNA is a protein called the terminal protein, essential for replication of the DNA. The DNA also has inverted terminal repeats of 100–1,800 bp (the number varies with the virus strain) that are important in the replication process.

Replication of Adenoviruses: Early Events

Adenoviral DNA replicates in the nucleus. After the virus particle has been transported to the nucleus, the nucleocapsid is released and converted to a viral DNA-histone complex. Early transcription is carried out by a host RNA polymerase, and a number of primary transcripts are made. The transcripts contain introns and so must first be spliced

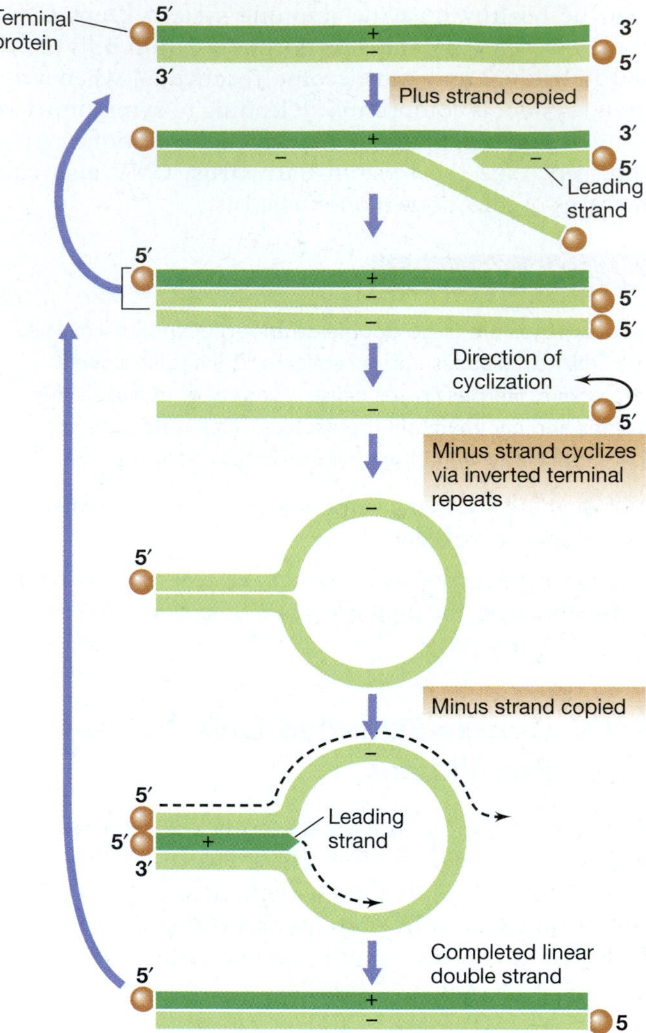

Figure 19.24 Replication of adenovirus DNA. Note how because of loop formation (cyclization), there is no lagging strand; DNA synthesis is leading on both strands.

to yield mature transcripts that are capped and polyadenylated before translation.

As is typical for many viruses, early adenoviral proteins regulate DNA replication and later proteins are structural. Viral DNA replication uses the terminal protein as a primer and another virus-encoded protein as DNA polymerase. The terminal protein contains a covalently bound cytosine residue from which DNA replication proceeds.

Replication of Adenoviruses: Genome Replication

Replication of the adenoviral genome begins at either end, the two strands being replicated asynchronously (**Figure 19.24**). The products of a round of replication are a double-stranded and a single-stranded molecule. However, then a unique replication mechanism proceeds. The single strand cyclizes by means of its inverted terminal repeats, and a new complementary strand is synthesized beginning from the 5′ end (Figure 19.24). This mechanism of replication is noteworthy because it does not require the formation of a lagging strand as conventional DNA replication does (∞ Section 7.6). Instead, DNA synthesis proceeds in leading fashion on both newly synthesized DNA strands (Figure 19.24).

19.14 MiniReview

Different double-stranded DNA animal viruses have different genome replication mechanisms. That of the adenoviruses requires protein primers and a mode of replication that avoids the synthesis of a lagging strand and that occurs within the nucleus.

- Describe how adenovirus replicates its double-stranded DNA genome without the synthesis of a lagging strand.

V VIRUSES THAT EMPLOY REVERSE TRANSCRIPTASE

19.15 Retroviruses and Hepadnaviruses

In Section 10.12 we discussed the **retroviruses**, a group of viruses that replicates through reverse transcription using the enzyme reverse transcriptase. There are two different types of viruses that use reverse transcriptase, and they differ in the type of nucleic acid in their genomes. The retroviruses have RNA genomes, whereas the hepadnaviruses have DNA genomes. In this section we deal with examples of both replication patterns.

Retroviruses: General Principles

Recall that retroviruses have enveloped virions that contain two copies of the RNA genome (∞ Figure 10.24). The virion also contains several enzymes, including reverse transcriptase, and also a specific tRNA. The reason that enzymes for retrovirus replication are carried in the virion is because although the retroviral genome is of the plus sense and is capped and tailed, it is not used directly as mRNA. Instead, one of the copies of the genome is converted to DNA by reverse transcriptase and is integrated into the host genome. The DNA that is eventually formed is a linear double-stranded molecule and is synthesized in the cytoplasm within an uncoated viral core particle. An outline of the steps in reverse transcription is given in **Figure 19.25**.

Activity of Reverse Transcriptase

Reverse transcriptase is essentially a DNA polymerase, but it actually possesses three enzymatic activities: (1) **reverse transcription** (the synthesis of DNA from an RNA template), (2) synthesis of DNA from a DNA template, and (3) ribonuclease H activity (an enzymatic activity that degrades the RNA strand of an RNA:DNA hybrid). Like all DNA polymerases, reverse transcriptase needs a primer for DNA synthesis. The primer for retrovirus reverse transcription is a specific cellular tRNA (Figure 19.25). This is packaged in the virion during its assembly in the previous host cell.

Using the tRNA primer, the 100 or so nucleotides at the 5′ terminus of the RNA are reverse-transcribed into DNA. Once reverse transcription reaches the 5′ end of the RNA, the process stops. To copy the remaining RNA, which is the bulk of the RNA of the virus, a different mechanism comes into play. First, terminally redundant RNA sequences at the 5′ end of the molecule are removed by the ribonuclease H activity of reverse transcriptase. This leads to the formation of a small, single-stranded DNA that is complementary to the RNA segment at the other end of the viral RNA. This short, single-stranded piece of DNA then hybridizes with the other end of the viral RNA molecule, where copying of the viral RNA sequences continues.

As summarized in Figure 19.25, continued reverse transcription and ribonuclease H activities lead to the formation of a double-stranded DNA molecule with long terminal repeats (LTRs) at each end. These LTRs contain strong transcriptional promoters and are involved in the integration process. The integration of the viral DNA into the host genome is analogous to the integration of virus Mu DNA (Section 19.4). Integration can occur anywhere in the DNA of the host cell chromosomes, and once integrated, the retroviral DNA, now called a provirus, is a stable genetic element (Figure 19.25).

Retroviral Gene Expression, Processing, and Virion Assembly

As a provirus, the retroviral genome may be expressed, or it may remain in a latent state and not be expressed. If the promoters in the right LTR are activated, the integrated proviral DNA is transcribed by a cellular RNA polymerase into transcripts that are capped and polyadenylated. These RNA transcripts may be either encapsidated into virions or processed and translated into virus proteins.

The translation and processing of such an mRNA from a retrovirus is shown in **Figure 19.26**. All retroviruses have the three genes *gag*, *pol*, and *env*, arranged in that order in the genome. The *gag* gene at the 5′ end of the mRNA actually encodes several small viral structural proteins. These are first synthesized as a polyprotein that is subsequently processed by a protease (which itself is a part of the polyprotein). The structural proteins make up the capsid, and the protease becomes packaged in the virion.

Next, the *pol* gene is translated into a large polyprotein that also contains the *gag* proteins (Figure 19.26). Compared to structural proteins, the *pol* products are required in only small amounts. This is achieved because their synthesis requires that the ribosome make an error. To produce *pol* gene products the ribosome must either read through a stop codon at the end of the *gag* gene or make a precise switch to a different reading frame in this region. These are rather rare events.

Once produced, the *pol* gene product is processed in two ways. First, it is removed from the *gag* proteins, and second, reverse transcriptase is cleaved from integrase, a protein required for DNA integration that is packaged in the virion along with reverse transcriptase (Figure 19.26). For the *env*

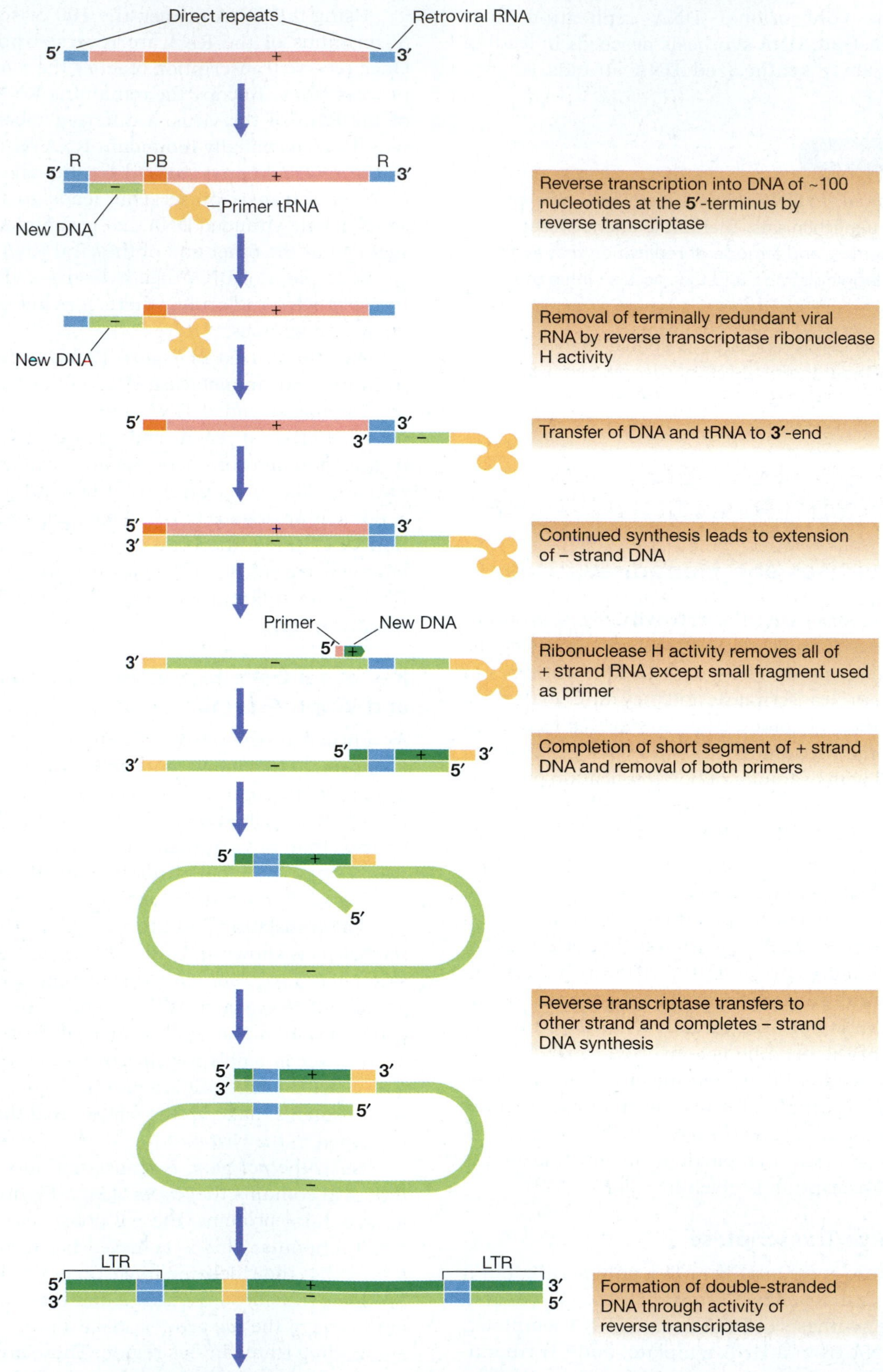

Figure 19.25 Formation of double-stranded DNA from retrovirus single-stranded RNA. The sequences labeled R on the RNA are direct repeats found at either end. The sequence labeled PB is where the primer (tRNA) binds. Note that DNA synthesis has yielded longer direct repeats on the DNA than were originally on the RNA. These are called long terminal repeats (LTRs).

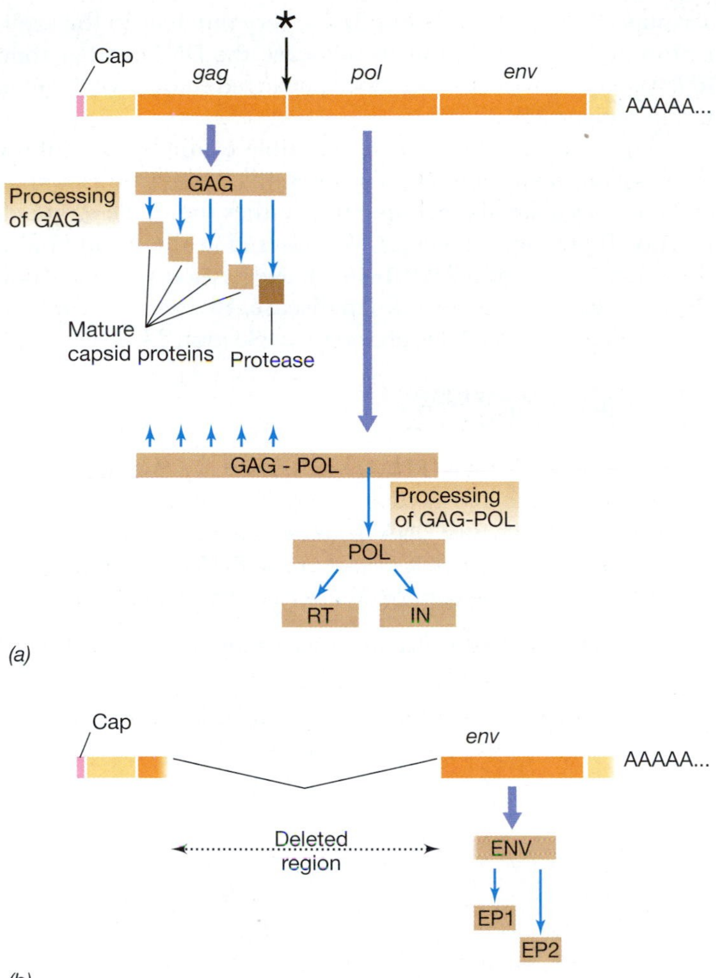

Figure 19.26 Translation of retrovirus mRNA and processing of the proteins. *(a)* The full-length mRNA with the three genes *gag*, *pol*, and *env*, is shown at the top. The asterisk shows the site where a ribosome must read through a stop codon or do a precise shift of reading frame to synthesize the GAG-POL polyprotein. The thick blue arrows indicate translation, and the thin blue arrows indicate protein-processing events. One of the *gag* gene products is a protease. The POL product is processed to give reverse transcriptase (RT) and integrase (IN). *(b)* The mRNA has been processed to remove most of the *gag-pol* region. This shortened message is translated to give the ENV polyprotein, which is cleaved into two envelope proteins (EP), EP1 and EP2.

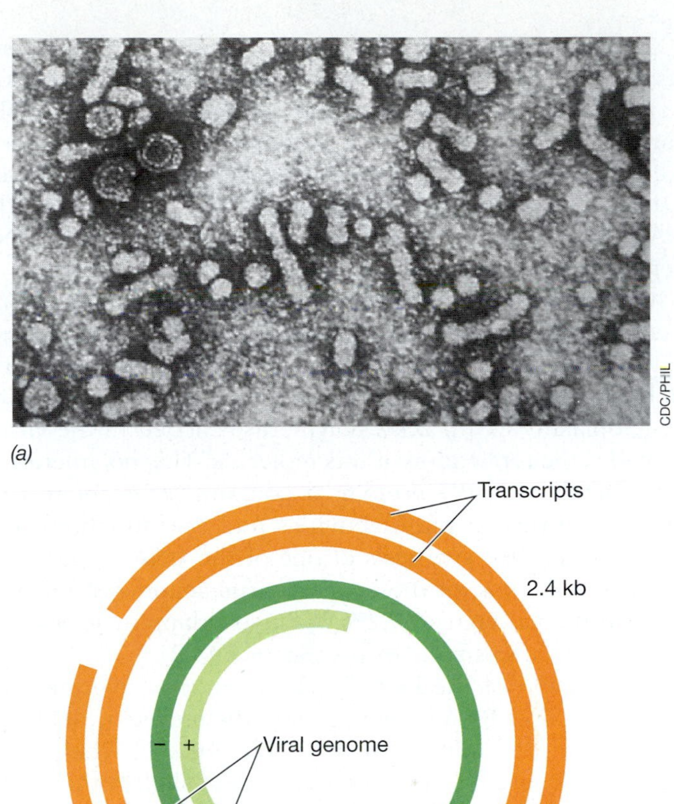

Figure 19.27 Hepadnaviruses. *(a)* Electron micrograph of hepatitis B virions. *(b)* Hepatitis B genome. The partially double-stranded genome is shown in green. Note that the positive strand is incomplete. The sizes of the transcripts are also shown. All of the genes in the hepatitis B virus overlap, and collectively, they cover every base in the genome. Reverse transcriptase produces the DNA genome from a single genome-length mRNA made by host RNA polymerase.

gene to be translated, the full-length mRNA is first processed to remove the *gag* and *pol* regions. The *env* product is made and then processed into two distinct envelope proteins (Figure 19.26).

HIV as a Retrovirus

Although the replication pattern of retroviruses described above is complex, this is actually the pattern typical of a rather "simple" retrovirus. By contrast, the genome of the retrovirus human immunodeficiency virus 1 (HIV-1), the causative agent of AIDS, is more complex than this and includes several additional small genes. Its expression not only requires extensive protein processing by proteases but also complex patterns of alternative splicing of introns. However, a hallmark of retroviruses is the many protease steps required to form mature proteins. In this regard, specific protease inhibitors have been developed for use in treating retroviral infections, including HIV-1 infections (∞ Section 34.15).

DNA Reverse-Transcribing Viruses: Hepadnaviruses

We have seen that the life cycles of viruses show a variety of unexpected genome structures and replication schemes. But none could be more unusual than the **hepadnaviruses**, such as human hepatitis B virus, a serious bloodborne pathogen (∞ Section 34.12). The term "hepadnavirus" comes from the fact that the virus infects the liver (thus "hepa") and the genome consists of DNA (thus "dna"). The virions of hepatitis B virus are small, irregular, rod-shaped particles (**Figure 19.27a**).

The genomes of hepadnaviruses are unusual for several reasons, including the fact they are among the smallest known of any viruses, some 3–4 kb, and are only *partially* double-stranded. The virus life cycle is also unique and very complex. Like the retroviruses, hepadnaviruses use reverse transcriptase in their replication cycle. However, unlike retroviruses, the DNA genome of hepadnaviruses is replicated through an RNA intermediate, the opposite of the pattern in retroviruses.

One strand of the genome of hepadnaviruses is incomplete, and both strands have gaps (Figure 19.27*b*). Nevertheless, the two strands are held together in a circular form by hydrogen bonding between complementary base pairs. On entering the cytoplasm, a viral DNA polymerase carried in the virion completes the replication of this molecule. This polymerase is an extremely versatile protein. It contains DNA polymerase plus reverse transcriptase activities and also functions as a protein primer for synthesis of one of the DNA strands! Despite the small size of the hepadnavirus genome, it encodes several proteins (Figure 19.27*b*) using overlapping genes as is frequently seen in small viruses (Section 19.2).

Replication of the hepadnavirus genome involves transcription by host RNA polymerase (in the nucleus), yielding a transcript with terminal repeats (Figure 19.27*b*). The repeats are made because the polymerase proceeds slightly more than once around the circular molecule. The viral reverse transcriptase then copies this into DNA, very much as in the replication of retroviruses, but in this case the DNA (rather than RNA as in retroviruses) becomes packaged into newly formed virions (Figure 19.27*b*).

Hepadnaviruses are thus incredible examples of making the most of a small genome. The hepatitis B virus, for example, at 3.4 kb has a smaller genome than either the single-stranded RNA or DNA bacteriophages we explored at the beginning of this chapter (Sections 19.1–19.3). Nevertheless, this small virus can cause a serious human disease. We examine the hepatitis syndromes in a later chapter (∞ Section 34.12).

19.15 MiniReview

The retroviruses contain RNA genomes and use reverse transcriptase to make a DNA copy during their life cycle. The hepadnaviruses contain DNA genomes and use reverse transcriptase to make genomic DNA from an RNA copy. Both types of viruses have complex patterns of gene expression.

- Why might protease inhibitors be an effective treatment for human AIDS?
- Describe the different role played by reverse transcriptase in the replication cycle of retroviruses and hepadnaviruses.

Review of Key Terms

Antigenic drift in influenza virus, minor changes in viral proteins (antigens) due to gene mutation

Antigenic shift in influenza virus, major changes in viral proteins (antigens) due to gene reassortment

Bacteriophage a virus that infects cells of *Bacteria*

Concatemer two or more identical linear nucleic acid molecules in tandem

Enveloped in reference to a virus, having a lipoprotein membrane surrounding the virion

Hepadnavirus a virus whose DNA genome replicates by way of an RNA intermediate

Herpes simplex the virus that causes both genital herpes and cold sores

Negative strand a nucleic acid strand that has the opposite sense to (is complementary to) the mRNA

Nucleocapsid the complete complex of nucleic acid and protein packaged in a virus particle

Overlapping genes two or more genes in which part or all of one gene is embedded in the other

Positive strand a nucleic acid strand that has the same sense as the mRNA

Polyprotein a large protein expressed from a single gene and subsequently cleaved to form several individual proteins

Replicative form a double-stranded DNA molecule that is an intermediate in the replication of single-stranded DNA viruses

Retrovirus a virus whose RNA genome has a DNA intermediate as part of its replication cycle

Reverse transcription the process of copying genetic information found in RNA into DNA

RNA replicase an enzyme that can produce RNA from an RNA template

Rolling circle replication a mechanism used by some plasmids and viruses of replicating circular DNA, which starts by nicking and unrolling one strand and using the other, still circular strand as a template for DNA synthesis

Transposase an enzyme that catalyzes the insertion of DNA segments into other DNA molecules

Review Questions

1. Describe the types of genomes found in viruses. Give an example of at least one virus (or group of viruses) with each type of genome (entire chapter).

2. What are overlapping genes? Give examples of viruses that have overlapping genes (Sections 19.1, 19.2, 19.3, 19.12, and 19.15).

3. Many bacteriophages have single-stranded DNA genomes. Describe how the genes carried by these viruses can be transcribed and translated (Sections 19.1 and 19.2).

4. Positive-strand RNA viruses are known for *Bacteria*, for plants, and for animals. What are the important distinctions between gene expression in these viruses, particularly between those that infect *Bacteria* and those that infect eukaryotes (Sections 19.1 and 19.7)?

5. One end of the T7 genome always enters the cell first during infection. This is necessary for the infection process to proceed. Explain (Section 19.3).

6. Why is bacteriophage Mu mutagenic? What features are necessary for Mu to insert into DNA (Section 19.4)?

7. What is unusual about the genome of archaeal viruses such as PAV-1 compared to the genomes of double-stranded DNA viruses of bacteriophages (Section 19.5)?

8. What type of genome does tobacco mosaic virus contain? How does it compare in terms of structure and size to that of phycodnaviruses that infect *Chlorella* (Sections 19.6 and 19.10)?

9. What is the function of the VPg protein of poliovirus, and how, since its genome is also plus-sense RNA, can coronaviruses replicate without a VPg protein (Section 19.7)?

10. The rhabdoviruses have an unusual genome structure. What is the nature of their genome and why *must* the virion contain an enzyme involved in its replication (Section 19.8)?

11. The reovirus genome is unique in all of biology. Explain (Section 19.9).

12. Several animal viruses can set up latent infections. Describe any two such types of animal viruses (Sections 19.12 and 19.15).

13. Although the genomes of herpesviruses and adenoviruses both contain double-stranded DNA, their replication features are quite distinct. Explain this (Sections 19.12 and 19.14).

14. Of all the double-stranded DNA animal viruses, pox viruses stand out concerning one unique aspect of their DNA replication process. What is this unique aspect and how can this be accomplished without special enzymes being packaged in the virion (Section 19.13)?

15. What is there about expression of retroviral genes that makes virus replication sensitive to protease inhibitors (Section 19.15)?

16. What reactions does the reverse transcriptase of retroviruses carry out? What molecule does it use as a primer (Section 19.15)?

Application Questions

1. Not all proteins are made from the RNA genome of bacteriophage MS2 in the same amounts. Can you explain why? One of the proteins functions very much like a repressor, but it functions at the translational level. Which protein is it and how does it function?

2. Give a mechanistic explanation of why the genomes of double-stranded RNA viruses might be segmented.

3. The mechanism of replication of both strands of DNA in some viruses, such as adenoviruses, is continuous (leading). Show how this can be without violating the "rule" learned in Chapter 7 that all DNA synthesis occurs in the overall direction of $5' \rightarrow 3'$.

4. Most genetic elements that express reverse transcriptase make it in small amounts and/or as part of a polyprotein. Can you think of any reason(s) why this might be so?

20

Metabolic Diversity: Phototrophy, Autotrophy, Chemolithotrophy, and Nitrogen Fixation

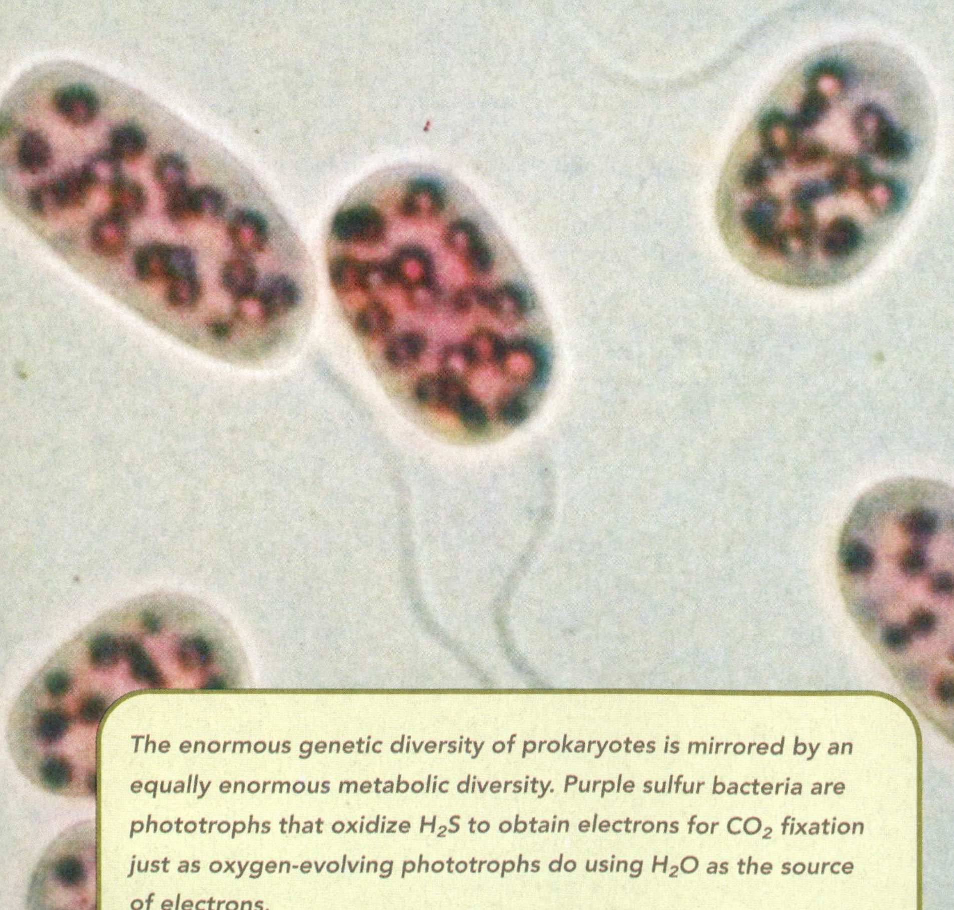

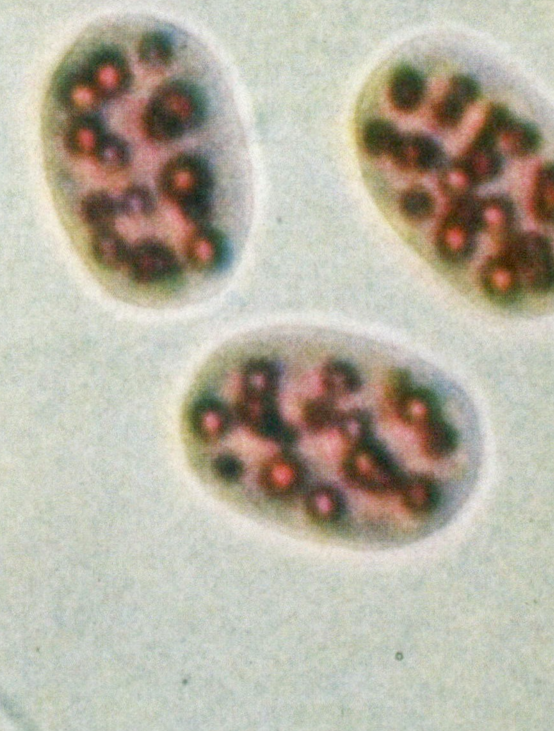

The enormous genetic diversity of prokaryotes is mirrored by an equally enormous metabolic diversity. Purple sulfur bacteria are phototrophs that oxidize H_2S to obtain electrons for CO_2 fixation just as oxygen-evolving phototrophs do using H_2O as the source of electrons.

The preceding five chapters considered the great *phylogenetic* diversity of microbial life on Earth. In this unit we focus on their *metabolic* diversity, with special emphasis on the biochemical processes behind this diversity. After this, we will consider microbial ecology—the interactions of microorganisms with each other and their environments.

Metabolic diversity and microbial ecology go hand in hand; metabolic diversity to be described in this chapter and the next fuels the nutrient cycles and other microbial activities that we will discuss in Chapters 23–24. Indeed, as we will see, microorganisms and their metabolic reactions play key roles in maintaining the web of life on Earth and are of crucial importance for supporting many different aspects of human affairs (Section 1.4).

In this chapter we focus on the metabolic diversity of phototrophs and chemolithotrophs, organisms that use light or inorganic compounds, respectively, as their sources of energy. In addition we discuss two important biosyntheses: *autotrophy*, the fixation of CO_2 into cell material, and *nitrogen fixation*, the reduction of atmospheric nitrogen (N_2) to ammonia to supply the cell's nitrogen requirements. In Chapter 21 we consider the metabolic diversity of chemoorganotrophs, especially the many forms of anaerobic metabolism in which prokaryotes excel.

I THE PHOTOTROPHIC WAY OF LIFE

Phototrophy, the use of *light* as an energy source, is widespread in the microbial world. In the first five sections of this chapter we consider the major forms of phototrophy, including that which forms the very oxygen we breathe, and see how light energy is converted into the energy of ATP.

20.1 Photosynthesis

The most important biological process on Earth is **photosynthesis**, the conversion of light energy to chemical energy. Organisms that carry out photosynthesis are called **phototrophs** (Figure 20.1). Most phototrophic organisms are also **autotrophs**, capable of growing with CO_2 as the sole carbon source. Energy from light is used in the reduction of CO_2 to organic compounds (*photoautotrophy*). However, some phototrophs use organic carbon as their carbon source; this lifestyle is called *photoheterotrophy* (Figure 20.1).

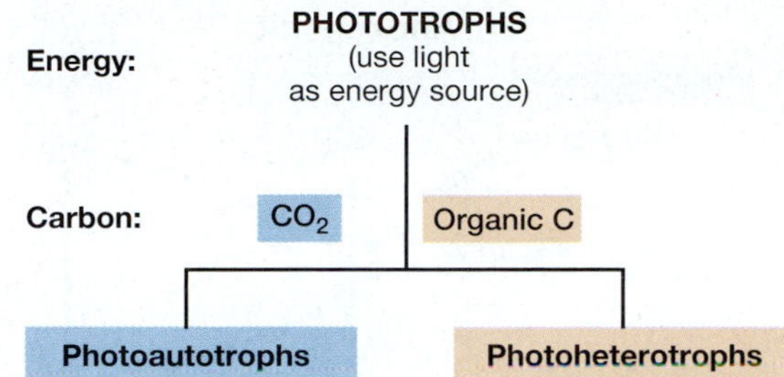

Figure 20.1 Classification of phototrophic organisms in terms of energy and carbon sources.

Photosynthesis requires light-sensitive pigments, the *chlorophylls*, found in plants, algae, and several groups of prokaryotes. Sunlight reaches phototrophic organisms in distinct units of energy called *quanta*. Absorption of light energy by chlorophylls begins the process of photosynthetic energy conversion, and the net result is chemical energy, ATP.

Energy Production and CO_2 Assimilation

Photoautotrophy requires two distinct sets of reactions: (1) ATP production and (2) CO_2 reduction to cell material. For autotrophic growth, energy is supplied from ATP, and electrons for the reduction of CO_2 come from NADH (or NADPH). These are produced by the reduction of NAD^+ (or $NADP^+$) by electrons originating from various electron donors. Some phototrophic bacteria obtain reducing power from electron donors in their environment, such as reduced sulfur sources H_2S, S^0, $S_2O_3^{2-}$ or from H_2. By contrast, green plants, algae, and cyanobacteria use H_2O, an inherently poor electron donor (Figure 5.10), as reducing power.

The oxidation of H_2O produces molecular oxygen O_2 as a by-product. Because oxygen is produced, photosynthesis in these organisms is called **oxygenic**. However, in many phototrophic bacteria water is not oxidized and oxygen is not produced, and thus the process is called **anoxygenic photosynthesis** (Figure 20.2). The production of NADH from substances such as H_2S by anoxygenic phototrophs may or may not be directly driven by light, depending on the organism. By contrast, the oxidation of H_2O to O_2 by oxygenic phototrophs is driven by light (Figure 20.2). Oxygenic phototrophs thus require light for both reducing power and energy conservation.

20.1 MiniReview

Phototrophs obtain their energy from light. In photosynthetic reactions ATP is generated from light and then is consumed in the reduction of CO_2 by NADH.

- How are photoautotrophs and photoheterotrophs similar and how do they differ?
- What is the fundamental difference between an oxygenic and an anoxygenic phototroph?

20.2 Chlorophylls and Bacteriochlorophylls

The only organisms that can perform photosynthesis are ones that produce some form of **chlorophyll** (oxygenic phototrophs) or **bacteriochlorophyll** (anoxygenic phototrophs). Chlorophyll is a porphyrin, as are the cytochromes (Section 5.11). But unlike the cytochromes, chlorophyll contains magnesium instead of iron at the center of the porphyrin ring. Chlorophyll also contains specific side chains on the porphyrin

UNIT 4

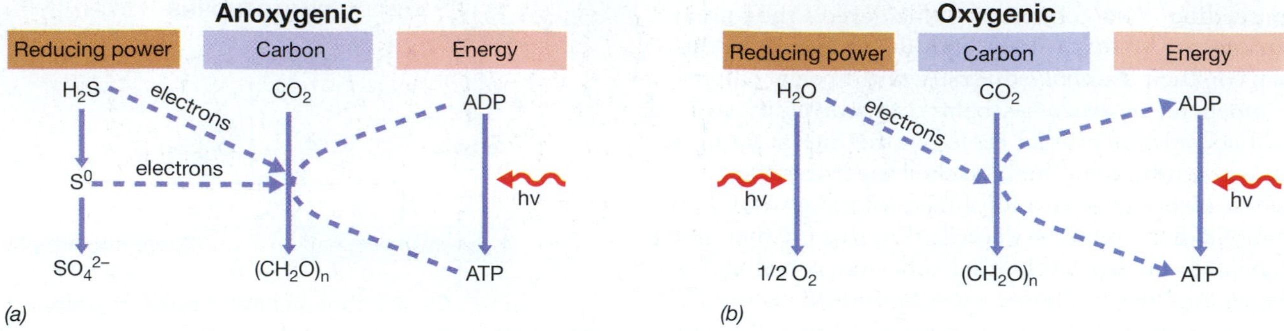

Figure 20.2 Patterns of photosynthesis. Energy and reducing power synthesis in *(a)* anoxygenic and *(b)* oxygenic phototrophs. Although both types of phototrophs obtain their energy from light (hv), in oxygenic phototrophs light also drives the oxidation of water to oxygen.

ring, as well as a hydrophobic alcohol side chain that helps to anchor the chlorophyll into the photosynthetic membranes.

The structure of chlorophyll *a*, the principal chlorophyll of higher plants, most algae, and the cyanobacteria, is shown in **Figure 20.3a**. Chlorophyll *a* is green because it *absorbs* red and blue light preferentially and *transmits* green light. The spectral properties of any pigment can best be expressed by its *absorption spectrum,* a measure of the absorbance of the pigment at different wavelengths. The absorption spectrum of cells containing chlorophyll *a* shows strong absorption of red light (maximum absorption at a wavelength of 680 nm) and blue light (maximum at 430 nm) (Figure 20.3*b*).

Chlorophyll Diversity

There are a number of different chlorophylls and bacteriochlorophylls, and each is distinguished by its unique absorption spectrum. Chlorophyll *b*, for instance, absorbs maximally at 660 nm rather than at 680 nm. Many plants have more than one chlorophyll, but the most common are chlorophylls *a* and *b*.

Among prokaryotes, cyanobacteria produce chlorophyll *a*, but anoxygenic phototrophs, such as the purple and green bacteria, can have any of a number of bacteriochlorophylls (**Figures** 20.3 and **20.4**). Bacteriochlorophyll *a* (Figure 20.3), present in most purple bacteria (Section 15.2), absorbs maximally between 800 and 925 nm, depending on the species of phototroph. Different species have different pigment-binding proteins, and the absorption maxima of bacteriochlorophyll *a* depends to some degree on the nature of these proteins and how they are arranged in photocomplexes. Other bacteriochlorophylls, whose distribution runs along phylogenetic lines, absorb in other regions of the visible and infrared spectrum (Figure 20.4).

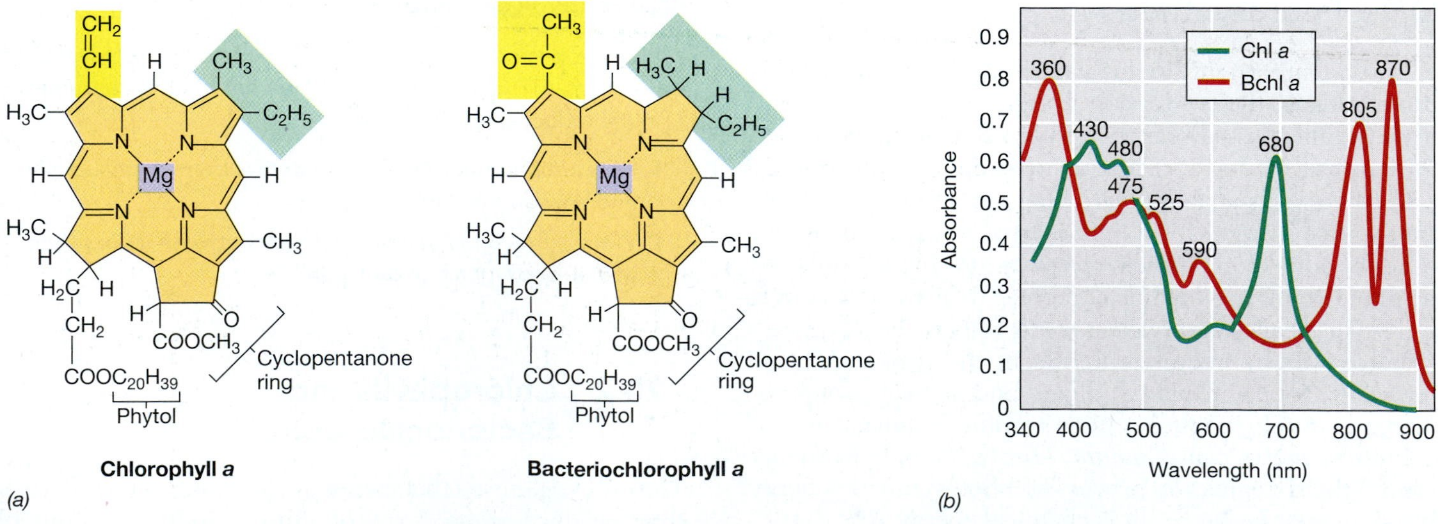

Figure 20.3 Structures and spectra of chlorophyll *a* and bacteriochlorophyll *a*. *(a)* The two molecules are identical except for those portions contrasted in yellow and green. *(b)* Absorption spectrum (green curve) of cells of the green alga *Chlamydomonas*. The peaks at 680 and 430 nm are due to chlorophyll *a*, and the peak at 480 nm is due to carotenoids. Absorption spectrum (red curve) of cells of the phototrophic purple bacterium *Rhodopseudomonas palustris*. Peaks at 870, 805, 590, and 360 nm are due to bacteriochlorophyll *a*, and peaks at 525 and 475 nm are due to carotenoids.

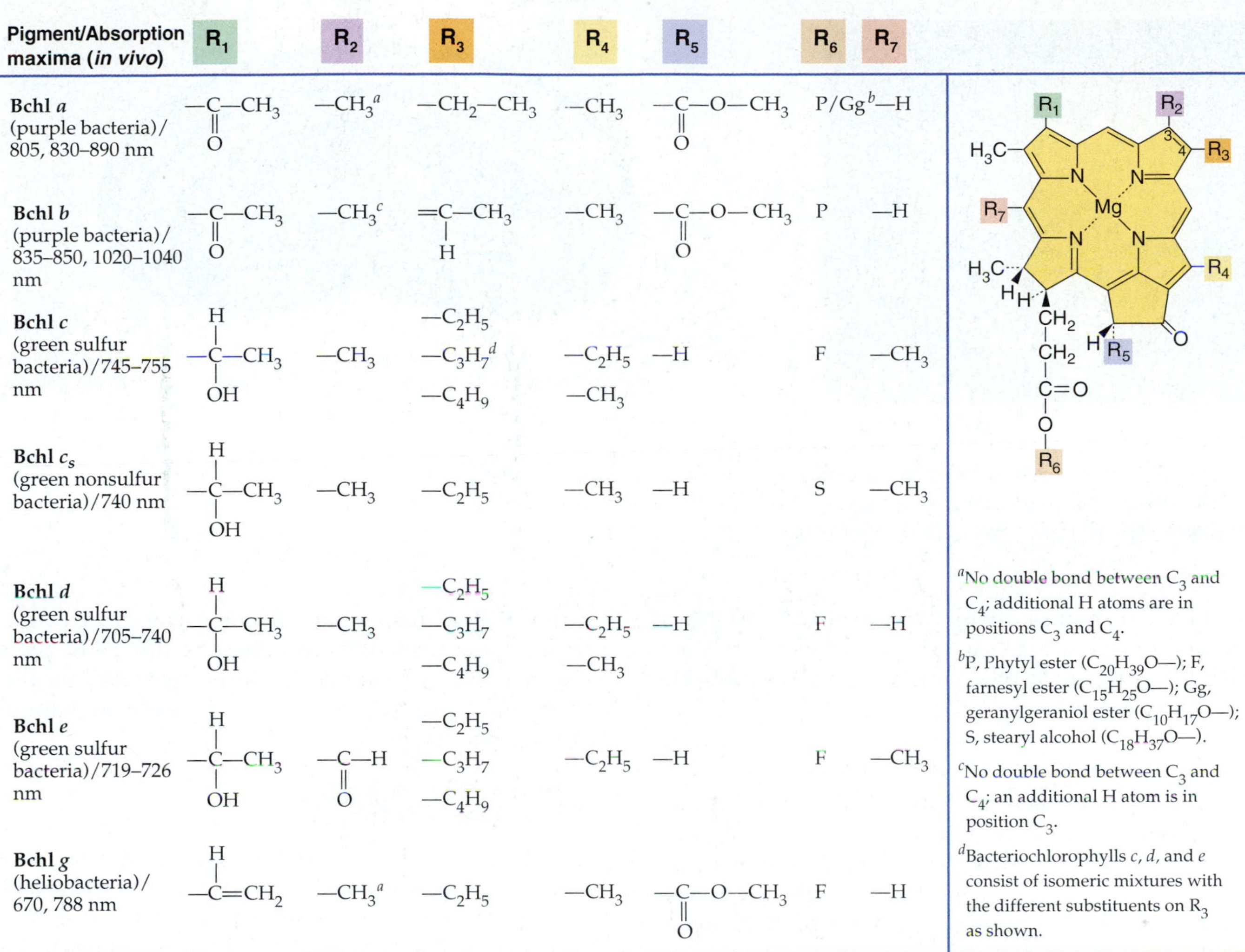

Pigment/Absorption maxima (*in vivo*)	R_1	R_2	R_3	R_4	R_5	R_6	R_7
Bchl *a* (purple bacteria)/ 805, 830–890 nm	$-C(=O)-CH_3$	$-CH_3$[a]	$-CH_2-CH_3$	$-CH_3$	$-C(=O)-O-CH_3$	P/Gg[b]	$-H$
Bchl *b* (purple bacteria)/ 835–850, 1020–1040 nm	$-C(=O)-CH_3$	$-CH_3$[c]	$=C(H)-CH_3$	$-CH_3$	$-C(=O)-O-CH_3$	P	$-H$
Bchl *c* (green sulfur bacteria)/745–755 nm	$-C(H)(OH)-CH_3$	$-CH_3$	$-C_2H_5$ $-C_3H_7$[d] $-C_4H_9$	$-C_2H_5$ $-CH_3$	$-H$	F	$-CH_3$
Bchl c_s (green nonsulfur bacteria)/740 nm	$-C(H)(OH)-CH_3$	$-CH_3$	$-C_2H_5$	$-CH_3$	$-H$	S	$-CH_3$
Bchl *d* (green sulfur bacteria)/705–740 nm	$-C(H)(OH)-CH_3$	$-CH_3$	$-C_2H_5$ $-C_3H_7$ $-C_4H_9$	$-C_2H_5$ $-CH_3$	$-H$	F	$-H$
Bchl *e* (green sulfur bacteria)/719–726 nm	$-C(H)(OH)-CH_3$	$-C(=O)-H$	$-C_2H_5$ $-C_3H_7$ $-C_4H_9$	$-C_2H_5$	$-H$	F	$-CH_3$
Bchl *g* (heliobacteria)/ 670, 788 nm	$-C(H)=CH_2$	$-CH_3$[a]	$-C_2H_5$	$-CH_3$	$-C(=O)-O-CH_3$	F	$-H$

[a]No double bond between C_3 and C_4; additional H atoms are in positions C_3 and C_4.

[b]P, Phytyl ester ($C_{20}H_{39}O-$); F, farnesyl ester ($C_{15}H_{25}O-$); Gg, geranylgeraniol ester ($C_{10}H_{17}O-$); S, stearyl alcohol ($C_{18}H_{37}O-$).

[c]No double bond between C_3 and C_4; an additional H atom is in position C_3.

[d]Bacteriochlorophylls *c*, *d*, and *e* consist of isomeric mixtures with the different substituents on R_3 as shown.

Figure 20.4 Structure of all known bacteriochlorophylls (Bchl). The different substituents present in the positions R_1 to R_7 in the accompanying structure are listed. Absorption properties can be determined by suspending intact cells in a viscous liquid such as 60% sucrose (this reduces light scattering and smoothes out spectra) and running absorption spectra as shown in Figure 20.3. *In vivo* absorption maxima are the physiologically relevant absorption peaks. The spectrum of bacteriochlorophylls dissolved in organic solvents is often quite different.

Why do various organisms have different forms of chlorophyll that absorb light of different wavelengths? This allows phototrophs to make better use of the available energy in the electromagnetic spectrum. Only light energy that is *absorbed* is useful for energy conservation. By having different pigments, different organisms can coexist in an illuminated habitat, each using wavelengths of light that others are not using. Thus, pigment diversity has ecological significance.

Photosynthetic Membranes and Chloroplasts

The chlorophyll pigments and all the other components of the light-gathering apparatus are present within special membrane systems, the photosynthetic membranes. The location of the photosynthetic membranes differs between prokaryotic and eukaryotic microorganisms. In eukaryotes, photosynthesis is associated with intracellular organelles, the chloroplasts (**Figure 20.5*a***; ∞ Section 18.3). As Figure 20.5*b* shows, the chlorophylls are attached to sheetlike membranes within the chloroplast. These photosynthetic membrane systems are called **thylakoids**; stacks of thylakoids form grana. The thylakoids are so arranged that the chloroplast is divided into two regions, the matrix space that surrounds the thylakoids and the inner space within the thylakoid array. This arrangement makes possible the development of a light-driven proton motive force that is used to synthesize ATP, as will be described in Section 20.5.

Prokaryotes do not have chloroplasts. Their photosynthetic pigments are integrated into internal membrane systems. These systems arise (1) from invagination of the cytoplasmic membrane (purple bacteria) (see Figure 20.12); (2) from the cytoplasmic membrane itself (heliobacteria; ∞ Section 16.2); (3) in both the cytoplasmic membrane

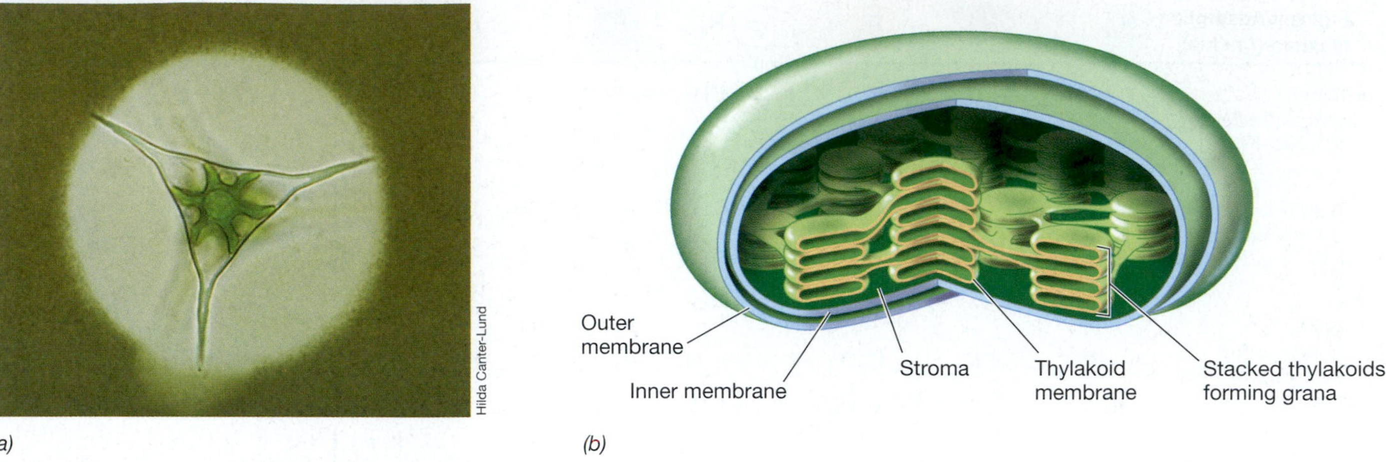

Figure 20.5 The chloroplast. *(a)* Photomicrograph of an algal cell showing chloroplasts. *(b)* Details of chloroplast structure, showing how the convolutions of the thylakoid membranes define an inner space called the stroma and form membrane stacks called grana.

and specialized non-unit membrane-enclosed structures called chlorosomes (green bacteria; see Figure 20.7 and Section 16.15); or (4) in thylakoid membranes (cyanobacteria, Section 16.7).

Reaction Centers and Antenna Pigments

Chlorophyll or bacteriochlorophyll molecules within a photosynthetic membrane are attached to proteins to form complexes consisting of anywhere from 50 to 300 molecules (**Figure 20.6**). Only a small number of these pigment molecules, called **reaction centers**, participate directly in the conversion of light energy to ATP (Figure 20.6). Reaction center chlorophylls and bacteriochlorophylls are surrounded by more numerous light-harvesting (antenna) chlorophylls/bacteriochlorophylls. The **antenna pigments** harvest light and funnel its energy to the reaction center (Figure 20.6). At the low light intensities that often prevail in nature, this arrangement of pigment molecules allows reaction centers to capture photons they would not otherwise be able to use.

Chlorosomes

The ultimate in capturing light at low intensity is the **chlorosome** (**Figure 20.7**). Chlorosomes are found in green sulfur bacteria, such as *Chlorobium* (Section 16.15) and *Chloroflexus* (Section 16.18). Chlorosomes are massive antenna systems. But unlike the antenna pigments of other phototrophs, bacteriochlorophyll molecules in the chlorosome are not attached to proteins. Chlorosomes contain bacteriochlorophyll *c, d,* or *e* (Figure 20.4) arranged in dense rod-like arrays running along the long axis of the structure. Light absorbed by these antenna bacteriochlorophylls transfers energy to bacteriochlorophyll *a* located in the reaction center in the cytoplasmic membrane (Figure 20.7). This arrangement of pigments is highly efficient for absorbing light at low intensities; light energy collected by antenna pigments is forwarded to the reaction center where ATP is made. Because of this, green sulfur bacteria can grow at the lowest light intensities of all known phototrophs. Thus, green bacteria are typically found in the deepest waters of lakes, inland seas, and other aquatic habitats where anoxygenic phototrophs are found.

20.2 MiniReview

Chlorophylls and bacteriochlorophylls are located in photosynthetic membranes where the light reactions of photosynthesis are carried out. Antenna chlorophyll molecules harvest light energy and transfer it on to reaction center chlorophylls. In green bacteria, chlorosomes function as massive antenna systems.

- Considering their function, why is it necessary for chlorophyll pigments to be located in membranes?
- What is the difference between the numbers of antenna and reaction center chlorophyll molecules in a photosynthetic complex, and why?
- What pigments are found within the chlorosome?

20.3 Carotenoids and Phycobilins

Although chlorophyll or bacteriochlorophyll is required for photosynthesis, phototrophic organisms contain an assortment of accessory pigments, as well. These include carotenoids and phycobilins. Carotenoids primarily play a photoprotective role in both anoxygenic and oxygenic phototrophs, while phycobilins function in light-harvesting in cyanobacteria.

Carotenoids

The most widespread accessory pigments are the **carotenoids**, which are always found in phototrophic organisms. Carotenoids

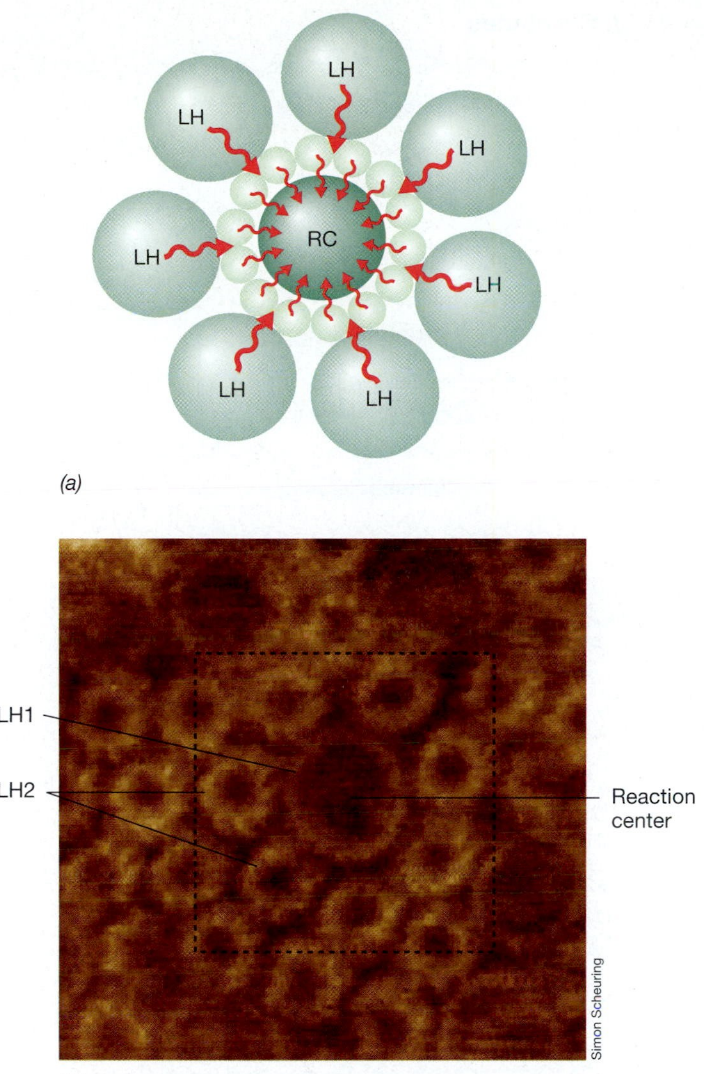

Figure 20.6 Arrangement of light-harvesting chlorophylls/bacteriochlorophylls and reaction centers within a photosynthetic membrane. *(a)* Light energy, absorbed by light-harvesting (LH) molecules (light green), is transferred to the reaction centers (dark green, RC) where photosynthetic electron transport reactions begin. Pigment molecules are secured within the membrane by specific pigment-binding proteins. Compare this figure to Figures 20.13 and 20.15. *(b)* Atomic force micrograph of photocomplexes of the purple bacterium, *Phaeospirillum molischianum*. This organism has two types of light-harvesting complexes, LH1 and LH2. LH2 complexes transfer energy to LH1 complexes, and these transfer energy to the reaction center.

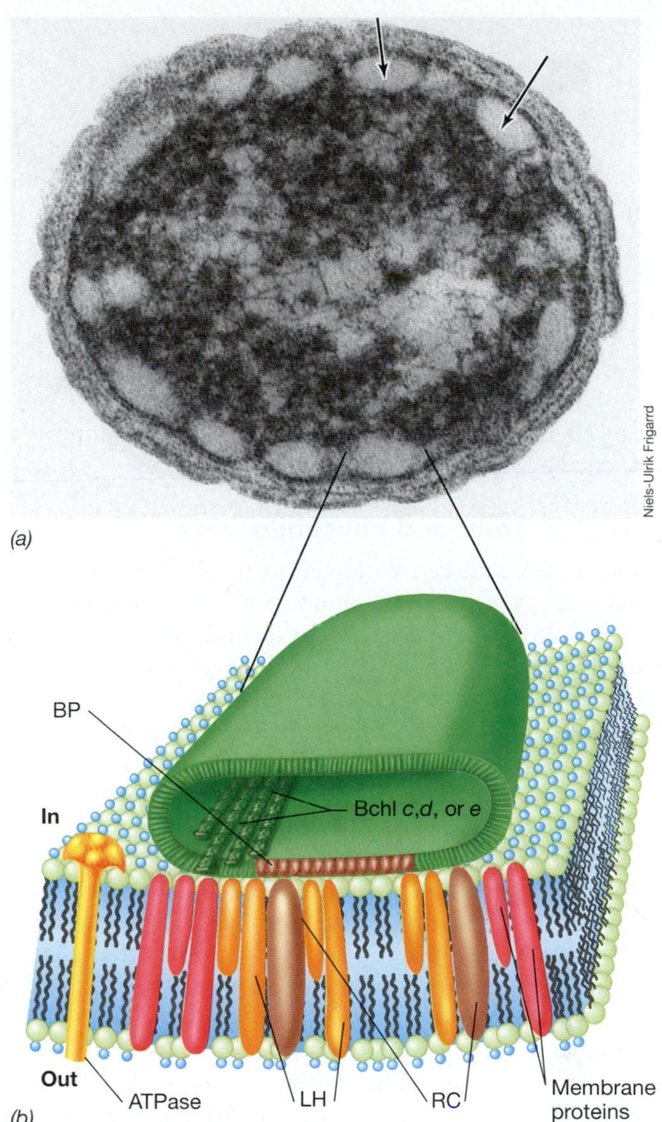

Figure 20.7 The chlorosome of green sulfur and green nonsulfur bacteria. *(a)* Electron micrograph of a cell of the green sulfur bacterium *Chlorobaculum tepidum*. Note the chlorosomes (arrows). *(b)* Model of chlorosome structure. The chlorosome (green) lies appressed to the inside surface of the cytoplasmic membrane. Antenna bacteriochlorophyll (Bchl) molecules are arranged in tubelike arrays inside the chlorosome, and energy is transferred from these bacteriochlorophylls through light-harvesting Bchl *a* molecules (LH) to reaction center (RC) Bchl *a* in the cytoplasmic membrane (blue). Base plate (BP) proteins function as connectors between the chlorosome and the cytoplasmic membrane.

are hydrophobic light-sensitive pigments that are firmly embedded in the photosynthetic membrane. **Figure 20.8** shows the structure of a typical carotenoid, β-carotene. Carotenoids are composed of branched isoprenoic units with conjugated double bonds (alternating C—C and C=C bonds).

Carotenoids are typically yellow, red, brown, or green (∞ Figure 15.2) and absorb light in the blue region of the spectrum (Figure 20.3). The major carotenoids of anoxygenic phototrophs are shown in **Figure 20.9**. These pigments are responsible for the brilliant colors of red, purple, pink, green, yellow, or brown that are observed in different species of anoxygenic phototrophs (see Figure 20.17).

Carotenoids are closely associated with chlorophyll or bacteriochlorophyll in photosynthetic complexes but do not function directly in ATP synthesis. However, energy absorbed by carotenoids can be transferred to the reaction center, and this energy can be used to make ATP. But carotenoids function primarily as photoprotective agents. Bright light can be harmful to cells in that it catalyzes photooxidation reactions that can lead to the production of toxic forms of oxygen, such as singlet

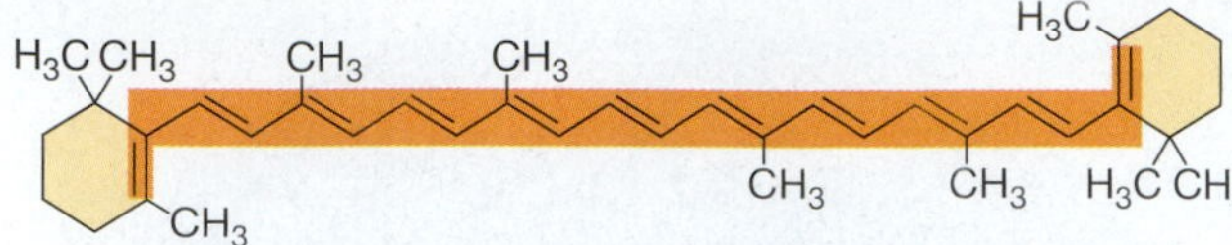

Figure 20.8 Structure of β-carotene, a typical carotenoid. The conjugated double-bond system is highlighted in orange.

oxygen ($^1O_2$) (∞ Section 6.18). Singlet oxygen can oxidize and thereby damage components of the photosynthetic apparatus itself. Carotenoids quench toxic oxygen species and absorb much of this harmful light. Because phototrophic organisms must by their very nature live in the light, the photoprotective role of carotenoids is thus an obvious advantage.

Phycobiliproteins and Phycobilisomes

Cyanobacteria and red algal chloroplasts, which are close phylogenetic relatives of cyanobacteria (∞ Section 18.20), contain **phycobiliproteins**. These are the main antenna pigments of these organisms. Phycobiliproteins are red or blue open-chain tetrapyrroles coupled to proteins (**Figure 20.10**). The red pigment, called *phycoerythrin*, absorbs light most strongly at wavelengths around 550 nm, whereas the blue pigment, *phycocyanin* (Figure 20.10*a*), absorbs most strongly at 620 nm (**Figure 20.11**). A third pigment, called *allophycocyanin*, absorbs at about 650 nm.

Phycobiliproteins form into aggregates called **phycobilisomes** that attach to the photosynthetic membranes (Figure 20.10*b,c*). Phycobilisomes are constructed such that the allophycocyanin molecules are in direct contact with the photosynthetic membrane. Allophycocyanin is surrounded by molecules of phycocyanin or phycoerythrin (or both, depending on the organism). The latter pigments absorb light of shorter wavelengths (higher energy) and transfer the energy to allophycocyanin, which is positioned closest to the reaction center chlorophyll and transfers energy to it (Figure 20.10*b*). Phycobilisomes thus accomplish efficient energy transfer that allows cyanobacteria to grow at fairly low light intensities. In this regard, the phycobilisome content of a cyanobacterium *increases* as light intensity *decreases*.

The light-gathering function of accessory pigments like carotenoids and phycobiliproteins is an obvious advantage for phototrophic organisms. Light from the sun is distributed over the whole visible range, and chlorophylls absorb in only part of this spectrum. Accessory pigments allow the organism to capture more of the available light, and this increases photosynthetic efficiency.

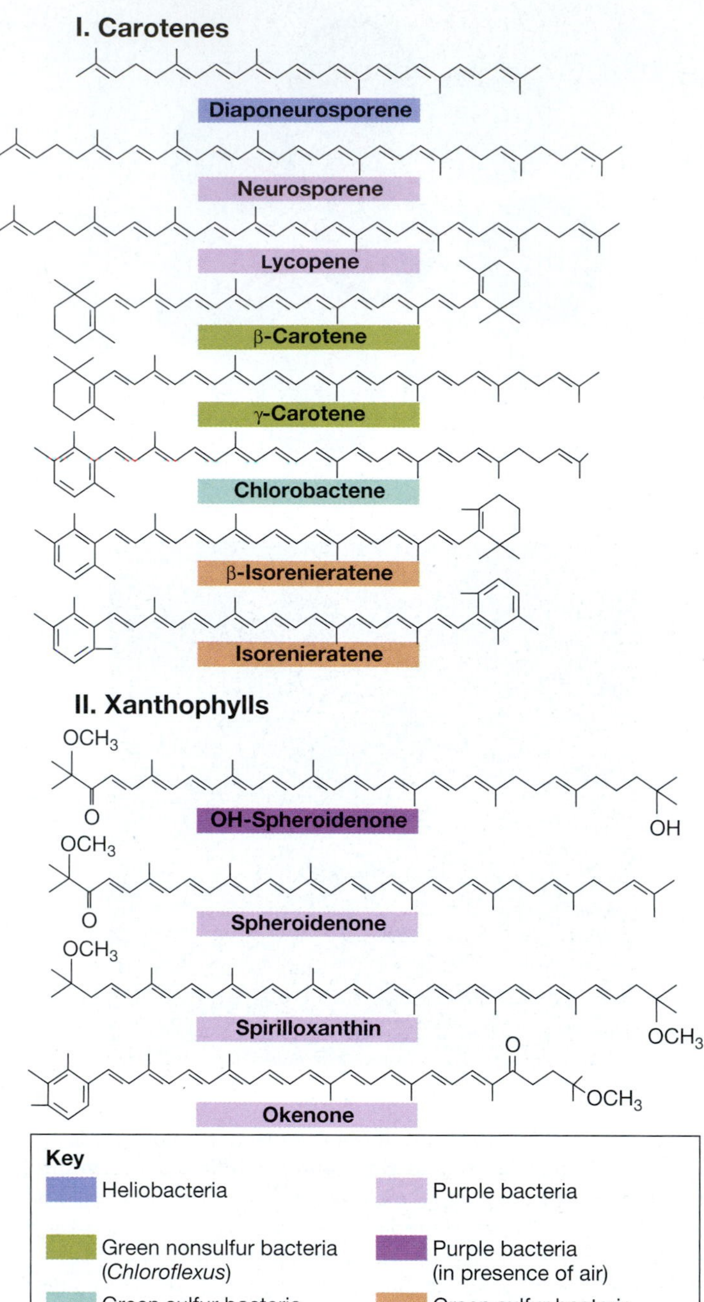

Figure 20.9 Structures of some common carotenoids found in anoxygenic phototrophs. Carotenes are hydrocarbon carotenoids and xanthophylls are oxygenated carotenoids. Compare the structure of β-carotene shown in Figure 20.8 with how it is drawn here. For simplicity in the structures shown here, methyl (CH_3) groups are designated by bond only.

20.3 MiniReview

Accessory pigments such as carotenoids and phycobilins absorb light and transfer the energy to reaction center chlorophyll, thus broadening the wavelengths of light usable in photosynthesis. Carotenoids also play an important photoprotective role in preventing photooxidative damage to cells.

- In what group of phototrophs are phycobiliproteins found?
- How does the structure of a phycobilin compare with that of a chlorophyll?
- Phycocyanin is blue-green. What are the wavelengths of light it is absorbing and why?

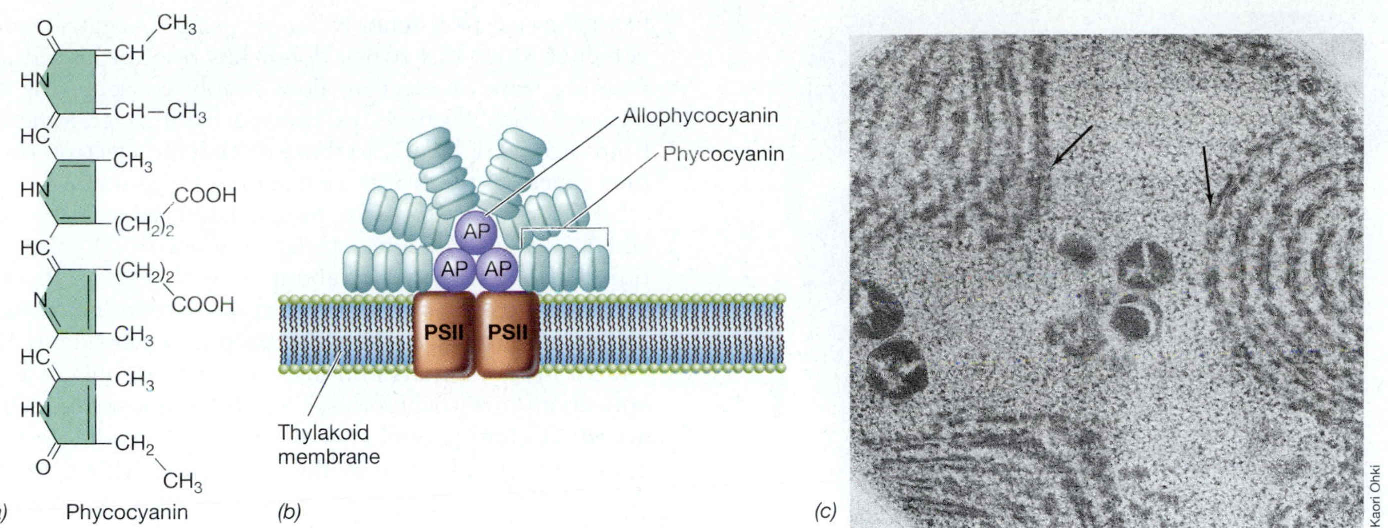

Figure 20.10 Phycobiliproteins and phycobilisomes. *(a)* A typical phycobilin. This compound is an open-chain tetrapyrrole derived biosynthetically from a closed porphyrin ring by loss of one carbon atom as carbon monoxide. The structure shown is the prosthetic group of phycocyanin, found in cyanobacteria (Section 16.7) and red algae (Section 18.20). *(b)* Structure of a phycobilisome. Phycocyanin absorbs at higher energies (shorter wavelengths) than allophycocyanin. Chlorophyll *a* absorbs at longer wavelengths (lower energies) than allophycocyanin. Energy flow is thus phycocyanin → allophycocyanin → chlorophyll *a* of photosystem II. *(c)* Electron micrograph of a thin section of the cyanobacterium *Synechocystis*. Note the darkly staining ball-like phycobilisomes (arrows) attached to the lamellar membranes.

20.4 Anoxygenic Photosynthesis

In light-mediated ATP synthesis in phototrophic organisms, electrons are transported through a series of electron carriers arranged in the photosynthetic complex in order of increasingly more electropositive reduction potentials (the symbol for reduction potential is E_0', Section 5.6). This generates a proton motive force and, subsequently, ATP. Anoxygenic photosynthesis is found in at least four phylogenetically defined phyla of *Bacteria:* the proteobacteria (purple bacteria); green sulfur bacteria; green nonsulfur bacteria; and the gram-positive bacteria (heliobacteria).

Photosynthetic Reaction Centers

The photosynthetic apparatus of purple phototrophic bacteria is contained in intracytoplasmic membrane systems of various morphologies. Membrane vesicles, sometimes called *chromatophores*, or membrane stacks called *lamellae* are common membrane morphologies (**Figure 20.12**). Reaction centers of purple bacteria consist of three polypeptides, designated L, M, and H. These proteins are firmly embedded in the photosynthetic membrane and traverse the membrane several times (**Figure 20.13*b***).

The L, M, and H polypeptides bind pigments in the reaction center photocomplex. This photocomplex consists of two molecules of bacteriochlorophyll *a*, called the *special pair*; two additional bacteriochlorophyll *a* molecules whose function is unknown; two molecules of bacteriopheophytin (bacteriochlorophyll *a* minus its magnesium atom); two molecules of quinone; and two molecules of a carotenoid pigment (Figure 20.13*a*). All components of the reaction center are integrated in such a way that they can interact in very fast electron transfer reactions that ultimately result in ATP production.

Photosynthetic Electron Flow in Purple Bacteria

Recall that photosynthetic reaction centers are surrounded by antenna pigments that function to funnel light energy to the reaction center (Figure 20.6). The energy in photons is transferred

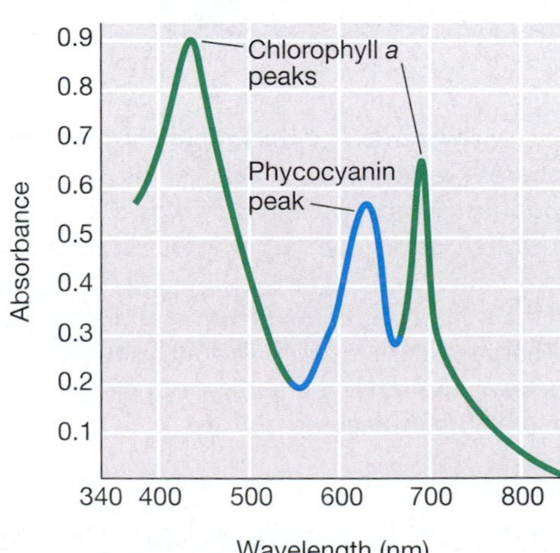

Figure 20.11 Absorption spectrum of a cyanobacterium that has a phycocyanin as an accessory pigment. Note how the presence of phycocyanin broadens the wavelengths of usable light energy (between 600 and 700 nm). Compare with Figure 20.3*b*.

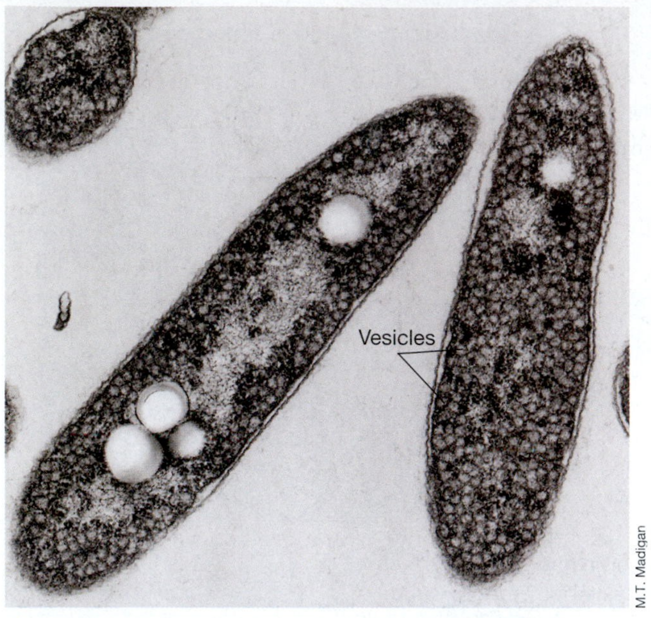

(a)

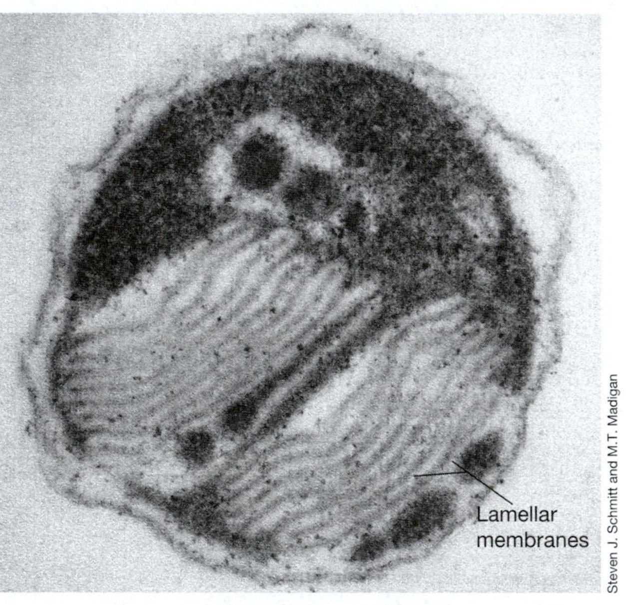

(b)

Figure 20.12 Membranes in anoxygenic phototrophs. *(a)* Chromatophores. Section through a cell of the phototrophic purple bacterium *Rhodobacter capsulatus* showing vesicular photosynthetic membranes. The vesicles are continuous with and arise by invagination of the cytoplasmic membrane. A cell is about 1 μm wide. *(b)* Lamellar membranes in the purple bacterium *Ectothiorhodospira.* A cell is about 1.5 μm wide. These membranes are also continuous with and arise from invagination of the cytoplasmic membrane, but instead of forming vesicles, they form membrane stacks.

from the antenna to the reaction center in packets called *excitons*, mobile forms of energy that migrate at high efficiency through the antenna pigments to the reaction center.

Photosynthesis begins when exciton energy strikes the special pair of bacteriochlorophyll *a* molecules (Figure 20.13*a*). The absorption of energy excites the special pair, converting it to a strong electron donor with a very electronegative reduction potential. Once this strong donor has been produced, the remaining steps in electron flow simply conserve the energy released when electrons are transported through a membrane from carriers of low E_0' to those of high E_0'; electron transport then generates the proton motive force (**Figure 20.14**).

Before excitation, the purple bacterial reaction center, which is called *P870*, has an E_0' of about +0.5V; after excitation it has a potential of about −1.0 V (Figure 20.14). The excited electron within P870 proceeds to reduce a molecule of bacteriopheophytin *a* within the reaction center (Figures 20.13*a* and 20.14). This transition takes place incredibly fast, taking only about three-trillionths (3×10^{-12}) of a second. Once reduced, bacteriopheophytin *a* reduces several intermediate quinone molecules within the membrane. This transition is also very fast, taking less than one-billionth of a second (**Figures** 20.14 and **20.15**). Relative to what happens in the reaction center, further electron transport reactions proceed rather slowly, on the order of microseconds to milliseconds.

From the quinone, electrons are transported in the membrane through a series of iron–sulfur proteins and cytochromes (Figures 20.14 and 20.15), eventually returning to the reaction center. Key electron transport proteins include cytochrome bc_1 and cytochrome c_2 (Figure 20.14). Cytochrome c_2 is a periplasmic cytochrome that is an electron shuttle between the membrane-bound bc_1 complex and the reaction center (Figures 20.14, and 20.15).

Photophosphorylation

ATP is synthesized during photosynthetic electron flow from the activity of ATPase that couples the proton motive force to ATP formation (∞ Section 5.12). Electron flow is completed when cytochrome c_2 donates an electron to the special pair (Figure 20.13); this returns these bacteriochlorophyll molecules to their original ground state potential (E_0' = +0.5 V). The reaction center is then capable of absorbing new energy and repeating the process.

This mechanism of ATP synthesis is called **photophosphorylation**, specifically *cyclic photophosphorylation*, because electrons move within a closed loop. Cyclic photophosphorylation resembles respiration in that electron flow through the membrane establishes a proton motive force. However, unlike in respiration, there is no net input or consumption of electrons in cyclic photophosphorylation; electrons simply travel a circuitous route.

The spatial relationship of the electron transport components in the purple bacterial photosynthetic membrane is illustrated in Figure 20.15. Note that as in respiratory electron flow (∞ Section 5.11 and Figure 5.20), the cytochrome bc_1 complex interacts with the quinone pool during photosynthetic electron flow as a major means of establishing the proton motive force used to drive ATP synthesis (Figure 20.15).

Genetics of Bacterial Photosynthesis

Purple phototrophic bacteria are gram-negative prokaryotes (∞ Section 15.2), and certain species are readily

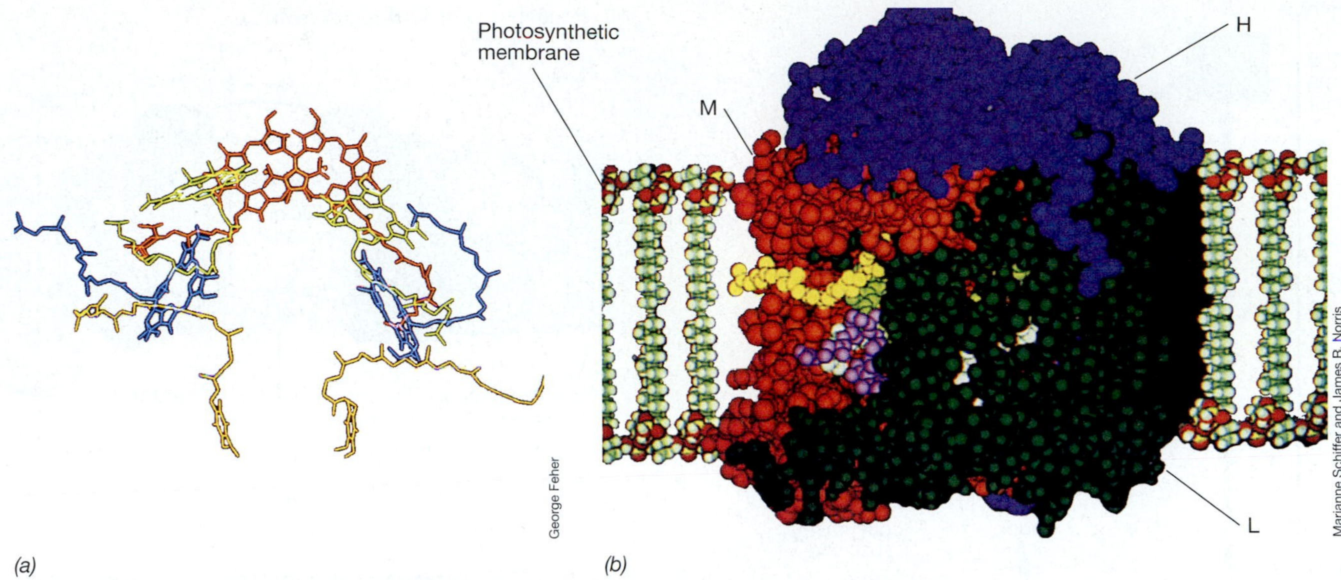

Figure 20.13 Structure of the reaction center of purple phototrophic bacteria. *(a)* Arrangement of pigment molecules in the reaction center. The "special pair" of bacteriochlorophyll molecules overlap and are shown in red at the top; quinones are in dark yellow and are at the bottom of the figure. The accessory bacteriochlorophylls are in lighter yellow near the special pair, and the bacteriopheophytin molecules are shown in blue. *(b)* Molecular model of the protein structure of the reaction center. The pigments discussed in *(a)* are bound to membranes by protein H (blue), protein M (red), and protein L (green). The reaction center pigment–protein complex is integrated into the lipid bilayer.

UNIT 4

amenable to genetic manipulation. Species of the genus *Rhodobacter*, especially *R. capsulatus* and *R. sphaeroides*, have been the main focus of genetic research in bacterial photosynthesis. The genomes of both of these purple bacteria have been sequenced. In *Rhodobacter*, most of the genes encoding proteins that participate in photosynthesis are clustered in several operons that span a 50-kbp region of the chromosome called the *photosynthetic gene cluster* (**Figure 20.16**). Genes in the photosynthetic gene cluster encode proteins involved in (1) bacteriochlorophyll biosynthesis (*bch* genes); (2) carotenoid biosynthesis (*crt* genes); and (3) reaction center and light-harvesting photocomplexes (*puf* and *puh* genes) (Figure 20.16).

The gene cluster allows for the coordinated synthesis of bacteriochlorophyll, carotenoids, and pigment-binding proteins in the correct proportions for final assembly into photocomplexes. The master regulatory signal governing transcription of the photosynthetic gene cluster in anoxygenic phototrophs is O_2. Molecular oxygen represses pigment synthesis in these organisms such that photosynthesis occurs only under anoxic conditions. In this connection, the protein CrtJ (encoded by *crtJ*) is a global regulatory protein (Sections 9.9 and 9.11), repressing the synthesis of both bacteriochlorophyll and carotenoids when O_2 levels exceed a certain threshold.

Genetic analysis of photosynthesis has been greatly assisted by the bioenergetic diversity of *Rhodobacter* species. In addition to their capacity for photosynthesis, these organisms can grow in darkness by respiration in the presence or absence of oxygen. Thus, mutants unable to photosynthesize but able to grow by respiration can be easily obtained and have been used in genetic crosses to characterize the number, organization, relatedness, and expression of photosynthesis genes (Figure 20.16).

Autotrophy in Purple Bacteria: Electron Donors and Reverse Electron Flow

For a purple bacterium to grow autotrophically, the formation of ATP is not enough. Reducing power (NADH) is also necessary so that CO_2 can be reduced to the level of cell material. As previously mentioned, for purple sulfur bacteria reducing power originates as H_2S, although S^0, $S_2O_3^{2-}$, and even Fe^{2+} or NO_2^- can be used by various species. When H_2S is the electron donor in purple sulfur bacteria, globules of S^0 are stored inside the cells (**Figure 20.17*a***).

Reduced substances such as H_2S are oxidized by cytochromes of the *c* type, and electrons from them eventually end up in the "quinone pool" of the photosynthetic membrane (Figure 20.14). However, the E_0' of quinone (about 0 volts) is insufficiently electronegative to reduce NAD^+ (− 0.32 V) directly. Instead, electrons from the quinone pool are forced to travel backward, against the thermodynamic gradient, to eventually reduce NAD^+ to NADH (Figure 20.14). This energy-requiring process, called **reverse electron transport**, is driven by the energy of the proton motive force. Reverse electron flow is also the mechanism by which chemolithotrophs reduce NAD^+ to NADH, in many cases from electron donors of extremely positive E_0' (Sections 20.8–20.12).

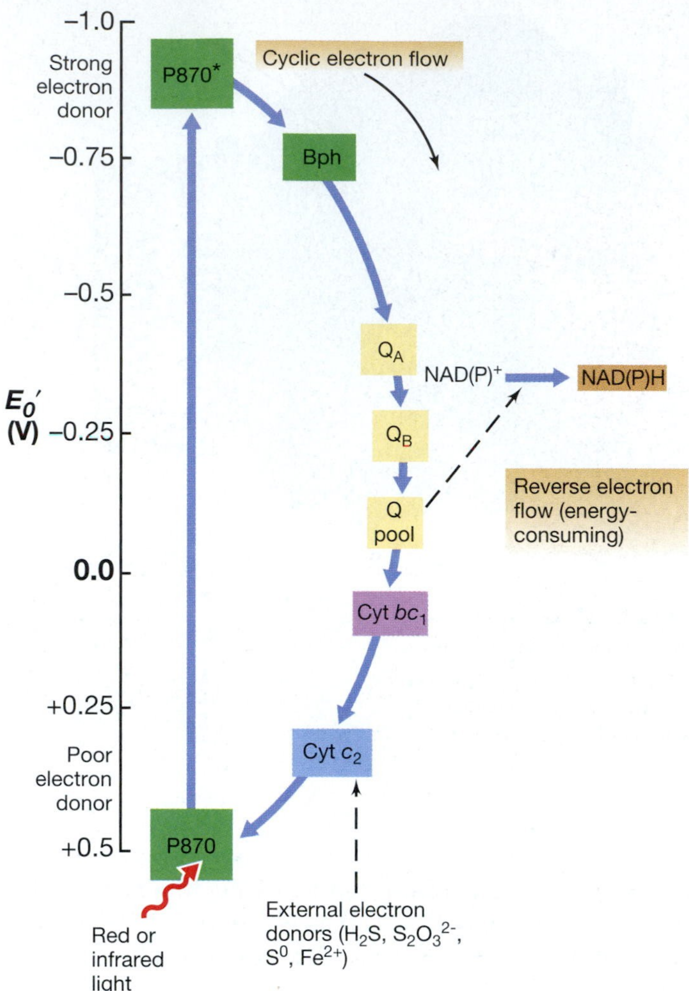

Figure 20.14 Electron flow in anoxygenic photosynthesis in a purple bacterium. Only a single light reaction occurs. Note how light energy converts a weak electron donor, P870, into a very strong electron donor, P870*, and that following this event, the remaining steps in photosynthetic electron flow are much the same as those of respiratory electron flow (Figure 5.20). Bph, bacteriopheophytin; Q_A, Q_B, intermediate quinones; Q pool, quinone pool in membrane; Cyt, cytochrome. Compare this figure with Figures 20.15 and 20.19.

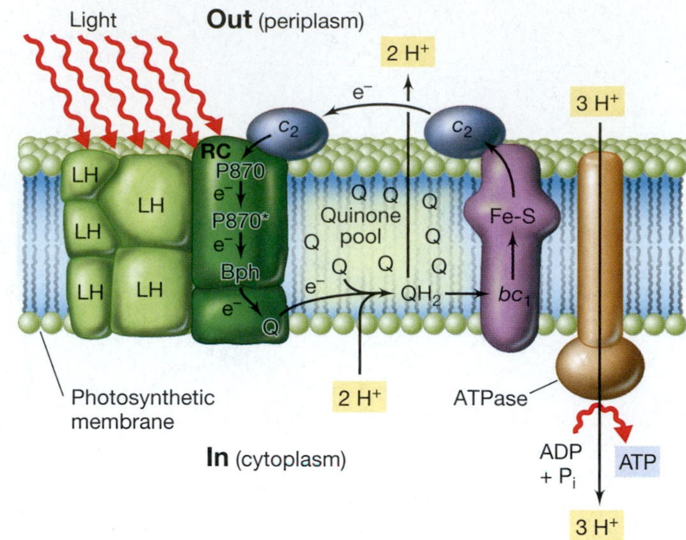

Figure 20.15 Arrangement of protein complexes in the purple bacterium reaction center. The light-generated proton gradient is used in the synthesis of ATP by the ATP synthase (ATPase). LH, Light-harvesting bacteriochlorophyll complexes; RC, reaction center; Bph, bacteriopheophytin; Q, quinone; FeS, iron-sulfur protein; bc_1, cytochrome bc_1 complex; c_2, cytochrome c_2. For a description of ATPase function, see Figure 5.21.

Photosynthesis in Other Anoxygenic Phototrophs

Our discussion of photosynthetic electron flow has thus far focused on purple bacteria. Although similar membrane components drive photophosphorylation in other anoxygenic phototrophs, there are differences in certain photochemical reactions that affect the biosynthesis of reducing power.

Figure 20.18 contrasts photosynthetic electron flow of purple and green bacteria and the heliobacteria. Note that in green bacteria and heliobacteria the excited state of the reaction center bacteriochlorophylls resides at a significantly more electronegative E_0' than in purple bacteria and that actual *chlorophyll a* (green bacteria) or a structurally modified form of chlorophyll *a*, called *hydroxychlorophyll a* (heliobacteria), is present in the reaction center. Thus, unlike in purple bacteria, where the first stable acceptor molecule (quinone) has an E_0' of about 0 V (Figure 20.14), the acceptors in green bacteria and heliobacteria (FeS proteins) have a much more electronegative E_0' than does NADH. This has a major effect on NADH synthesis in these organisms.

In green bacteria, a protein called *ferredoxin* (reduced by an FeS protein, Figure 20.18) is the direct electron donor for CO_2 fixation reactions in the reverse citric acid cycle (Figure 20.24*a*). Thus, as in oxygenic phototrophs (to be discussed next), in green bacteria both ATP and reducing power are direct products of the light reactions. A similar situation exists in the heliobacteria, but here the picture is complicated by the fact that autotrophic growth has so far not been demonstrated (heliobacteria thus appear to be obligate photoheterotrophs). When sulfide is the electron donor for the synthesis of reducing power in green bacteria, globules of S^0 are produced as in purple bacteria, but the globules are formed outside, rather than inside the cells (Figure 20.17*b*).

20.4 MiniReview

A series of electron transport reactions occur in the photosynthetic reaction center of anoxygenic phototrophs, resulting in the formation of a proton motive force and the synthesis of ATP. Reducing power for CO_2 fixation comes from reductants present in the environment and requires reverse electron transport for NADH production in purple phototrophs.

- How does photophosphorylation compare with electron transport phosphorylation in respiration?
- What is reverse electron flow and why is it necessary? Which phototrophs need to use reverse electron flow?

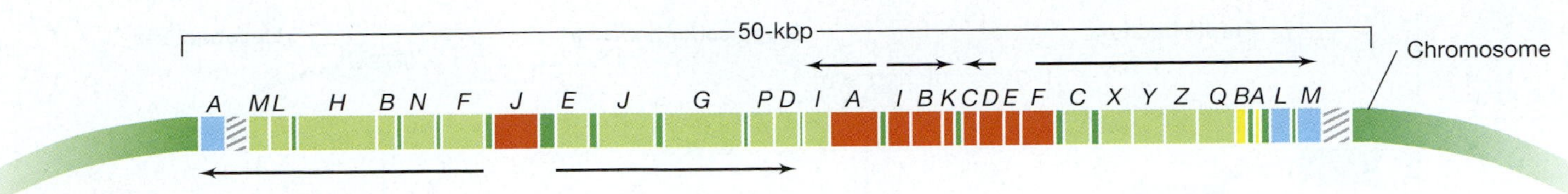

Figure 20.16 Map of the photosynthetic gene cluster of the purple phototrophic bacterium, *Rhodobacter capsulatus*. The *bch* genes, which encode bacteriochlorophyll synthesis proteins, are shown in green; *crt* genes, which encode proteins that synthesize carotenoids, are shown in red (*crtJ* is an exception in that it is a regulatory protein). Genes encoding reaction center polypeptides (*puh* and *puf* genes) are shown in blue, and genes encoding light-harvesting I polypeptides (B870 complex; *puf* genes) in yellow. Genes shown by diagonal lines are of unknown function. Arrows indicate direction of transcription.

20.5 Oxygenic Photosynthesis

In contrast to electron flow in *anoxygenic* phototrophs, electron flow in *oxygenic* phototrophs proceeds through two distinct, but interconnected, photochemical reactions. Oxygenic phototrophs use light to generate both ATP and NADPH, the electrons for the latter arising from the splitting of water into oxygen and electrons (Figure 20.2). The two light reactions are called *photosystem I* and *photosystem II*, each photosystem having a spectrally distinct form of reaction center chlorophyll *a*. Photosystem (PS) I chlorophyll, called P700, absorbs light at long wavelengths (far red light), whereas PS II chlorophyll, called P680, absorbs light at shorter wavelengths (near red light).

As in anoxygenic photosynthesis, oxygenic photochemical reactions occur in membranes. In eukaryotic cells, these membranes are found in the chloroplast (Figure 20.5), whereas in cyanobacteria, photosynthetic membranes are arranged in stacks within the cytoplasm (Figure 20.6). In both groups of phototrophs the membranes are arranged in a similar way, and the two forms of chlorophyll *a* are attached to specific proteins in the membrane and interact as shown in **Figure 20.19**.

Electron Flow in Oxygenic Photosynthesis

The path of electron flow in oxygenic phototrophs resembles the letter Z turned on its side, and Figure 20.19 outlines this so-called "Z scheme" of photosynthesis. The reduction potential of the P680 chlorophyll *a* molecule in PS II is very electropositive, even more positive than that of the O_2/H_2O couple. This facilitates the first step in oxygenic electron flow, the splitting of water into oxygen and hydrogen atoms (Figure 20.19), a thermodynamically very unfavorable reaction. An electron from water is donated to the oxidized P680 molecule following the absorption of a quantum of light near 680 nm. Light energy converts P680 into a moderately strong reductant, capable of reducing a molecule with an E_0' of about −0.5 V. The nature of this molecule is uncertain, but it is probably pheophytin *a* (chlorophyll *a* minus its magnesium atom).

From here the excited electron travels through several membrane carriers, including quinones, cytochromes, and a copper-containing protein called *plastocyanin* that donates the electron to PS I. The electron is accepted by the reaction center chlorophyll of PS I, P700, which has previously absorbed light quanta and begun the steps that lead to the reduction of $NADP^+$. These transfer electrons through several carriers of increasingly electropositive E_0', terminating with the reduction of $NADP^+$ to NADPH (Figure 20.19).

ATP Synthesis in Oxygenic Photosynthesis

Besides the net synthesis of reducing power (that is, NADPH), other important events take place while electrons flow in the photosynthetic membrane from one photosystem to the other. During the transfer of an electron from the acceptor in PS II to the reaction center chlorophyll molecule in PS I, the electron moves in a thermodynamically favorable (negative-to-positive) direction. This generates a proton motive force from which ATP can be produced. This mechanism for ATP synthesis is called *noncyclic photophosphorylation*. This is because electrons here do not cycle back to reduce the oxidized P680, but instead are used in the reduction of $NADP^+$.

When sufficient reducing power is present, ATP can also be produced in oxygenic phototrophs by *cyclic* photophosphorylation. This occurs when electrons travel from ferredoxin to the cytochrome *bf* complex (the *bf* complex is analogous to the bc_1 complex of anoxygenic phototrophs, Figure 20.14) from

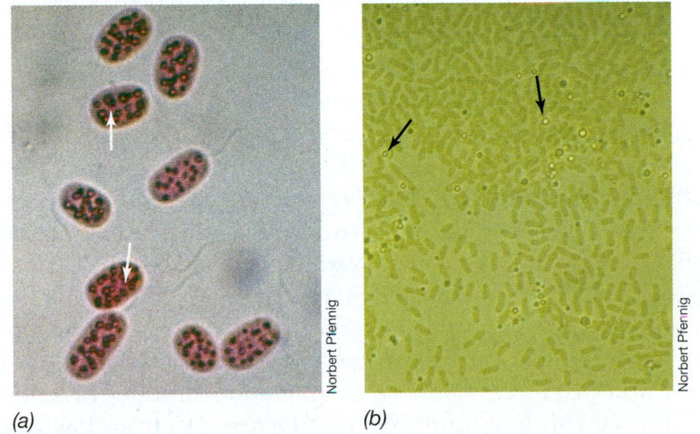

Figure 20.17 Phototrophic purple and green sulfur bacteria. *(a)* Purple bacterium, *Chromatium okenii*. Notice the sulfur granules deposited inside the cell (arrows). *(b)* Green bacterium, *Chlorobium limicola*. The refractile bodies are sulfur granules deposited outside the cell (arrows). In both cases the sulfur granules arise from the oxidation of H_2S to obtain reducing power. Cells of *C. okenii* are about 5 μm in diameter, and cells of *C. limicola* are about 0.9 μm in diameter. Both micrographs are bright-field images.

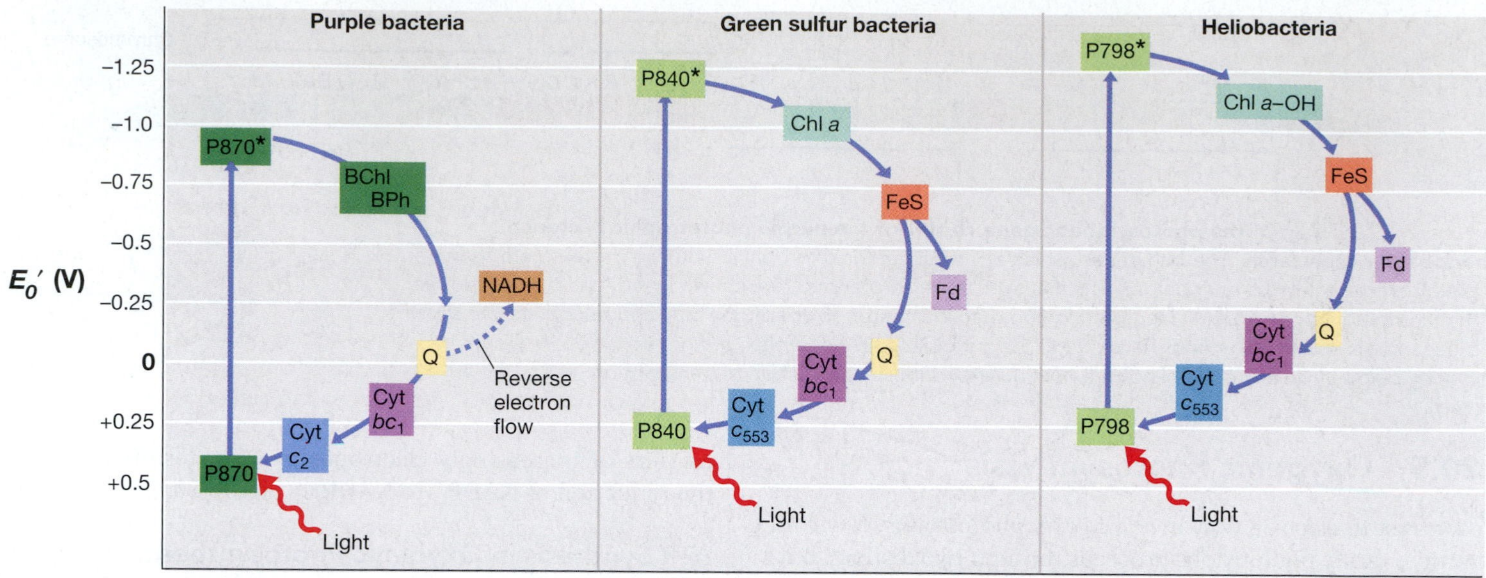

Figure 20.18 Electron flow in purple bacteria, green sulfur bacteria, and heliobacteria. Note that reverse electron flow in purple bacteria is necessary to produce NADH because the primary acceptor (quinone, Q) is more positive in potential than the $NAD^+/NADH$ couple. In green and heliobacteria ferredoxin (Fd), whose E_0' is more negative than that of NADH, is produced by light-driven reactions. In green bacteria, ferredoxin instead of NADH is the reducing power for autotrophy (Figure 20.24*a*). Bchl, Bacteriochlorophyll; BPh, bacteriopheophytin. P870 and P840 are reaction centers of purple and green bacteria, respectively, and consist of Bchl *a*. The reaction center of heliobacteria (P798) contains Bchl *g*, and the reaction center of *Chloroflexus* is of the purple bacterial type. Note chlorophyll *a* in the reaction centers of green bacteria and heliobacteria.

which electron transport returns the electron to P700. This flow creates a proton motive force and synthesis of additional ATP (dashed line in Figure 20.19).

Anoxygenic Photosynthesis in Oxygenic Phototrophs and the Evolution of Photosynthesis

Photosystems I and II normally function in tandem in oxygenic photosynthesis. However, under certain conditions, for example if PSII activity were blocked, some algae and cyanobacteria can carry out cyclic photophosphorylation using only PS I, obtaining reducing power for CO_2 reduction from sources other than water. This is, in effect, anoxygenic photosynthesis.

Many cyanobacteria can use H_2S as an electron donor for anoxygenic photosynthesis, whereas many green algae can use H_2. When H_2S is used, it is oxidized to elemental sulfur (S^0), and sulfur granules similar to those produced by green sulfur bacteria (Figure 20.17*b*) are deposited outside the cyanobacterial cells. **Figure 20.20** shows an example of this with the filamentous cyanobacterium *Oscillatoria limnetica*. This organism lives in sulfide-rich saline ponds where it carries out anoxygenic photosynthesis along with photosynthetic green and purple bacteria and produces sulfur as an oxidation product of sulfide (Figure 20.20). In cultures of *O. limnetica*, electron flow from PS II is strongly inhibited by H_2S, thus necessitating anoxygenic photosynthesis if the organism is to survive.

From an evolutionary point of view, the existence of cyclic photophosphorylation in both oxygenic and anoxygenic phototrophs is one of many indications of their close relationship. Oxygenic phototrophs such as *O. limnetica* have acquired PS II and hence the ability to split water. However, *O. limnetica* still retains the ability under certain conditions to use photosystem I alone, just as anoxygenic phototrophs do during phototrophic growth.

Further evidence of evolutionary relationships among phototrophs has been the discovery that the structure of the purple bacterial photosynthetic reaction center resembles that of PS II, whereas the structure of reaction center of green sulfur bacteria and heliobacteria resembles that of PS I. Because various lines of evidence show that purple and green bacteria preceded cyanobacteria on Earth by perhaps as many as 0.5 billion year (Section 14.3), it is evident that anoxygenic photosynthesis was the first form of photosynthesis on Earth. Cyanobacteria appeared later by borrowing key features from their anoxygenic predecessors and evolving the important new process of using water as an electron donor.

20.5 MiniReview

In oxygenic photosynthesis, water donates electrons to drive autotrophy, and oxygen is produced as a by-product. There are two separate photocomplexes, PS I and PS II. PS I resembles the system in anoxygenic photosynthesis. PS II splits H_2O to yield O_2 and electrons.

- Why is the term noncyclic electron flow used in reference to oxygenic photosynthesis?
- What are the major differences between the two reaction center chlorophyll molecules in photosystems I and II regarding their absorption properties and ground and excited state E_0'?

The Z Scheme:
PSII PSI

E_0' (V)
−1.25
−1.0
−0.75
−0.5
−0.25
0.0
+0.25
+0.5
+0.75
+1.0

P680*
Ph
PQ_A
PQ_B
PQ pool
Cyt *bf*
PC
P700
P700*
Chl a_0
Q_A
FeS
Fd
Fp
$NAD(P)^+$
NAD(P)H
Light
Photosystem I
Cyclic electron flow (generates proton motive force)
Noncyclic electron flow (generates proton motive force)
e^-
H_2O
$\frac{1}{2}O_2 + 2H$
P680
Light
Photosystem II

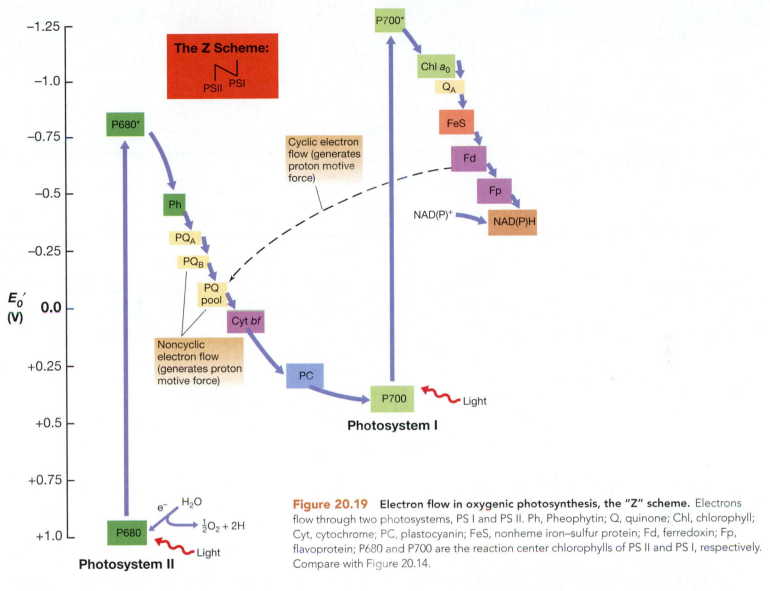

Figure 20.19 Electron flow in oxygenic photosynthesis, the "Z" scheme. Electrons flow through two photosystems, PS I and PS II. Ph, Pheophytin; Q, quinone; Chl, chlorophyll; Cyt, cytochrome; PC, plastocyanin; FeS, nonheme iron–sulfur protein; Fd, ferredoxin; Fp, flavoprotein; P680 and P700 are the reaction center chlorophylls of PS II and PS I, respectively. Compare with Figure 20.14.

UNIT 4

II AUTOTROPHY

Autotrophy is the process by which CO_2 is assimilated as carbon source. Many microorganisms are autotrophic, as are phototrophic macroorganisms—green plants. And, when we consider anaerobic metabolism in the next chapter we will encounter autotrophy again, but proceeding by a distinctly different mechanism from those described here. We focus here on autotrophy in phototrophic microorganisms, of which at least three pathways are known. One pathway, the Calvin cycle, is widely distributed, so we begin here.

20.6 The Calvin Cycle

Several biochemical mechanisms are known for the fixation of CO_2 into cell material for autotrophic growth. The **Calvin cycle**, named for its discoverer, Melvin Calvin, requires NAD(P)H, ATP, and two key enzymes, *ribulose bisphosphate*

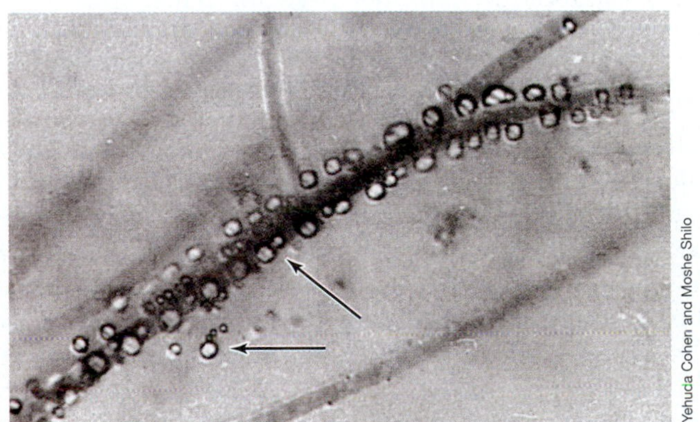

Figure 20.20 Oxidation of H_2S by *Oscillatoria limnetica*. Note the globules of elemental sulfur (arrows), the oxidation product of sulfide, formed outside the cells. *O. limnetica* can carry out oxygenic photosynthesis but sulfide inhibits this process and so cells convert to anoxygenic photosynthesis with sulfide as electron donor for CO_2 fixation.

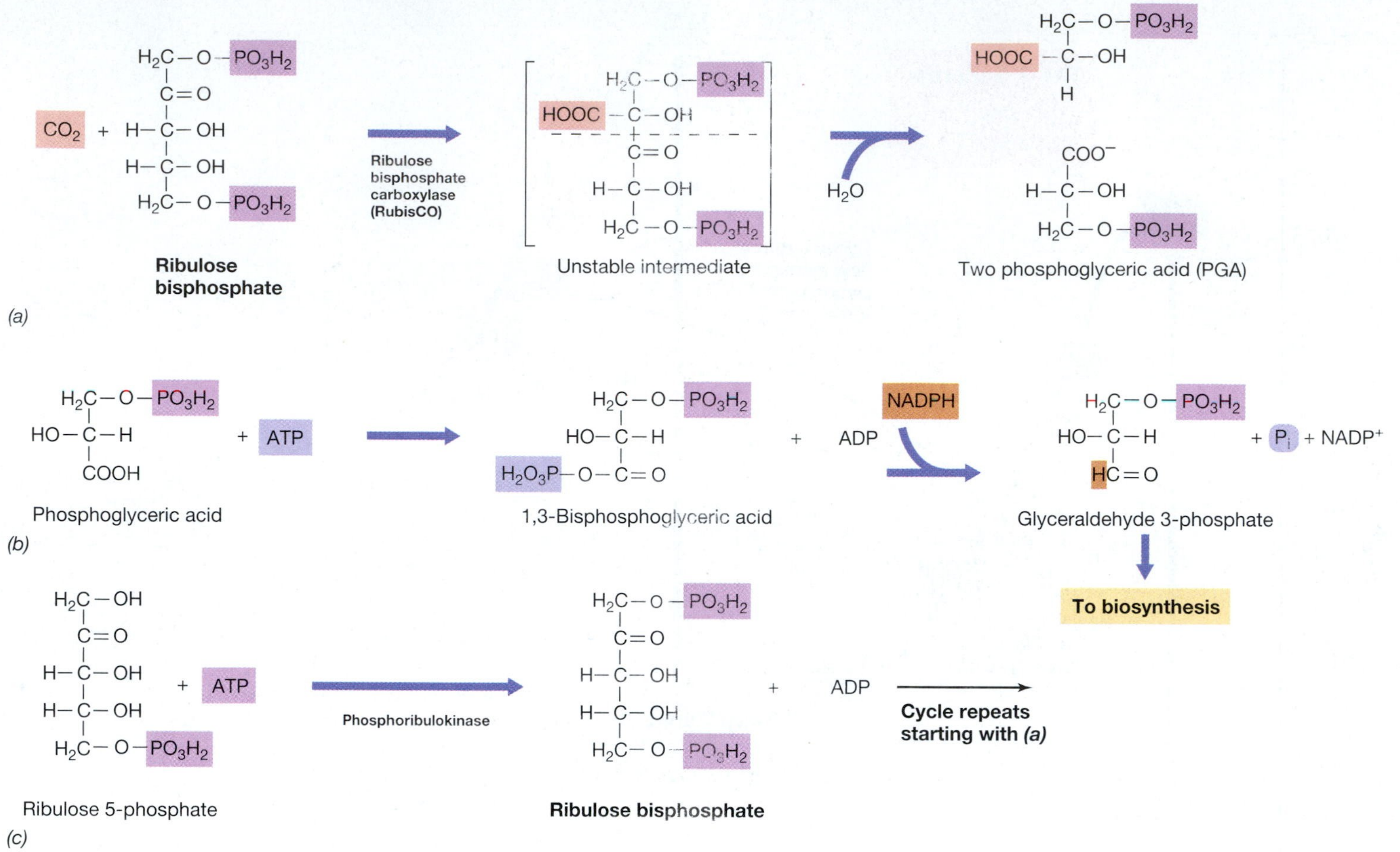

(a)
CO_2 + Ribulose bisphosphate → Ribulose bisphosphate carboxylase (RubisCO) → Unstable intermediate → H_2O → Two phosphoglyceric acid (PGA)

(b)
Phosphoglyceric acid + ATP → 1,3-Bisphosphoglyceric acid + ADP; NADPH → Glyceraldehyde 3-phosphate + P_i + $NADP^+$
To biosynthesis

(c)
Ribulose 5-phosphate + ATP → Phosphoribulokinase → Ribulose bisphosphate + ADP
Cycle repeats starting with *(a)*

Figure 20.21 Key reactions of the Calvin cycle. *(a)* Reaction of the enzyme ribulose bisphosphate carboxylase. *(b)* Steps in the conversion of 3-phosphoglyceric acid (PGA) to glyceraldehyde 3-phosphate. Note that both ATP and NADPH are required. *(c)* Conversion of ribulose 5-phosphate to the CO_2 acceptor molecule ribulose bisphosphate by the enzyme phosphoribulokinase.

carboxylase and *phosphoribulokinase*. The remainder of the cycle is catalyzed by enzymes used for other purposes in a wide variety of organisms.

RubisCO and the Formation of PGA

The first step in the Calvin cycle is catalyzed by the enzyme ribulose bisphosphate carboxylase, **RubisCO** for short. The Calvin cycle is present in purple bacteria, cyanobacteria, algae, green plants, most chemolithotrophic *Bacteria*, and even in some halophilic and hyperthermophilic *Archaea*. RubisCO catalyzes the formation of two molecules of 3-phosphoglyceric acid (PGA) from ribulose bisphosphate and CO_2 as shown in **Figure 20.21**. The PGA is then phosphorylated and reduced to a key intermediate of glycolysis, glyceraldehyde 3-phosphate. From here, glucose can be formed by reversal of the early steps in glycolysis (∞ Section 5.10 and Figure 5.15).

Stoichiometry of the Calvin Cycle

It is convenient to consider reactions of the Calvin cycle based on the incorporation of 6 molecules of CO_2, as this is what is required to make 1 molecule of glucose ($C_6H_{12}O_6$). For RubisCO to incorporate 6 molecules of CO_2, 6 molecules of ribulose bisphosphate (total, 30 carbons) are required; carboxylation of each yields 12 molecules of PGA (total, 36 carbon atoms) (**Figure 20.22**). These then form the carbon skeletons for 6 molecules of ribulose bisphosphate (total, 30 carbons) and one hexose molecule (6 carbons), the latter to be used for cell biosynthesis (Figure 20.22).

A series of rearrangements involving C_3, C_4, C_5, C_6, and C_7 sugars follow, yielding 6 molecules of ribulose 5-phosphate (total, 30 carbons). The final step is the phosphorylation of ribulose 5-phosphate with ATP by the enzyme phosphoribulokinase (∞ Figure 17.21*c*) to regenerate the acceptor molecule, ribulose bisphosphate. Like RubisCO, phosphoribulokinase is unique to the Calvin cycle.

In the overall stoichiometry of the Calvin cycle (Figure 20.22), 12 molecules each of ATP and NADPH are required for the reduction of 12 molecules of PGA to 12 molecules of glyceraldehyde 3-phosphate. Six more ATP molecules are required for the conversion of 6 ribulose phosphates to 6 ribulose

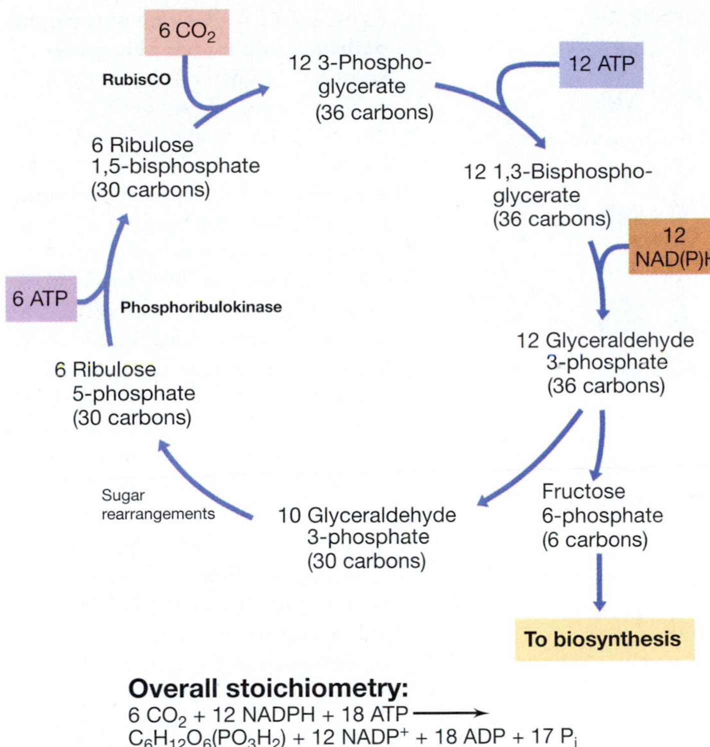

Figure 20.22 **The Calvin cycle.** Shown is the production of one hexose molecule from CO_2. For each six molecules of CO_2 incorporated, one fructose 6-phosphate is produced. In phototrophs, ATP comes from photophosphorylation and NAD(P)H from light or reverse electron flow. Use the color-coding here to follow the biochemical reactions in Figure 20.21.

bisphosphates. Thus, 12 NADPH and 18 ATP are required to synthesize 1 hexose from 6 CO_2 by the Calvin cycle. The hexose molecules can then be converted to storage polymers such as glycogen, starch, or poly-β-hydroxyalkanoates (∞ Section 4.10) and used later to build new cell material.

Carboxysomes

Several autotrophic prokaryotes that use the Calvin cycle for CO_2 fixation produce polyhedral cell inclusions called **carboxysomes**. The inclusions, about 100 nm in diameter, are surrounded by a thin, nonunit membrane and consist of a crystalline array of RubisCO molecules (**Figure 20.23**; ∞ Figure 15.9*a*). Carboxysomes are a mechanism for increasing the total amount of RubisCO in the cell to allow for rapid CO_2 fixation without the increase in osmolarity of the cytoplasm that would occur if the RubisCO molecules in the carboxysome were dissolved in the cytoplasm.

Carboxysomes have been detected in obligately chemolithotrophic sulfur-oxidizing bacteria, the nitrifying bacteria, and cyanobacteria and prochlorophytes. They are not present in facultative autotrophs (organisms that can grow either as autotrophs or as heterotrophs), such as purple anoxygenic phototrophs. Thus, the carboxysome appears to be an evolutionary adaptation to life under strictly autotrophic conditions.

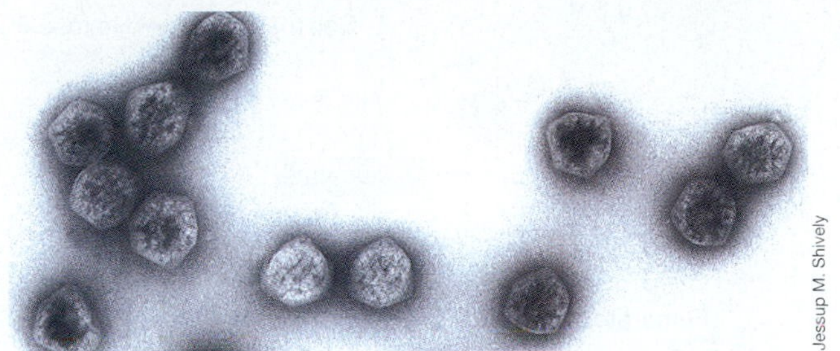

Figure 20.23 **Crystalline Calvin cycle enzymes: Carboxysomes.** Electron micrograph of carboxysomes purified from the chemolithotrophic sulfur oxidizer *Halothiobacillus neapolitanus*. The structures are about 100 nm in diameter. Carboxysomes are present in a wide variety of obligately autotrophic prokaryotes.

20.6 MiniReview

CO_2 is fixed by most phototrophic and other autotrophic organisms by the Calvin cycle, in which the enzyme RubisCO plays a key role. The Calvin cycle is an energy-demanding process in which CO_2 is converted into sugar.

- What reaction does the enzyme RubisCO carry out?
- Why is reducing power needed for autotrophic growth?
- What is a carboxysome?

20.7 Other Autotrophic Pathways in Phototrophs

Routes for autotrophic CO_2 fixation differ between green sulfur bacteria and green nonsulfur bacteria. The green sulfur bacteria *Chlorobaculum* and *Chlorobium* (Figures 20.7*a* and 20.17*b*) fix CO_2 by a reversal of steps in the citric acid cycle (∞ Figure 5.21), a pathway called the **reverse citric acid cycle** (**Figure 20.24a**). Green sulfur bacteria contain two ferredoxin-linked enzymes that catalyze the reductive fixation of CO_2 into intermediates of the citric acid cycle. Ferredoxin is an electron donor with a very electronegative E_0', about −0.4 V. The two ferredoxin-linked reactions catalyze the carboxylation of succinyl-CoA to α-ketoglutarate and the carboxylation of acetyl-CoA to pyruvate (Figure 20.24*a*). Most of the other reactions of the reverse citric acid cycle are catalyzed by enzymes working in reverse of the normal oxidative direction of the cycle. One exception is *citrate lyase*, an ATP-dependent enzyme that cleaves citrate into acetyl-CoA and oxaloacetate in green sulfur bacteria (Figure 20.24*a*). In the oxidative direction of the cycle, citrate is produced from these same precursors by the enzyme *citrate synthase*.

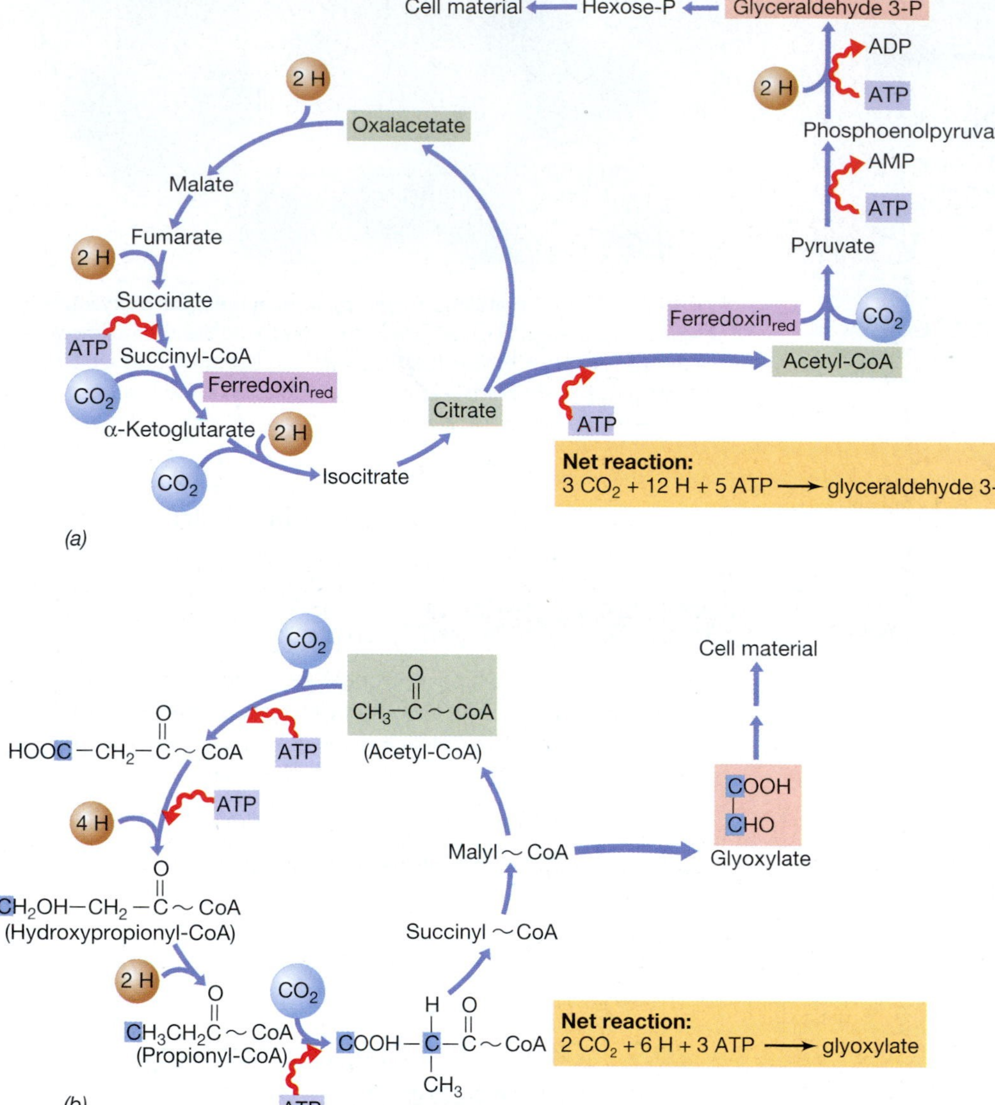

Figure 20.24 Unique autotrophic pathways in phototrophic green bacteria. *(a)* The reverse citric acid cycle is the mechanism of CO_2 fixation in green sulfur bacteria. Ferredoxin$_{red}$ indicates carboxylation reactions requiring reduced ferredoxin (2 H each). Reduced ferredoxin is generated in *Chlorobium* by light-driven reactions (Figure 20.18). Starting from oxalacetate, each turn of the cycle results in three molecules of CO_2 being incorporated and pyruvate as the product. The cleavage of citrate by the ATP-dependent enzyme citrate lyase regenerates the C_4 acceptor oxalacetate and produces acetyl-CoA for biosynthesis. The conversion of pyruvate to phosphoenolpyruvate consumes two high-energy phosphate bond equivalents. NADH or FADH supply the other reducing power needs of the cycle *(b)* The hydroxypropionate pathway is the autotrophic pathway in the green nonsulfur bacterium *Chloroflexus* (∞ Section 16.18). Acetyl-CoA is carboxylated twice to yield methylmalonyl-CoA. This intermediate is rearranged to yield a new acetyl-CoA acceptor molecule and a molecule of glyoxylate, which is converted to cell material through an amino acid intermediate. The source of reducing power in the hydroxypropionate pathway is NADPH.

The reverse citric acid cycle as a mechanism of autotrophy has also been found in some nonphototrophic prokaryotes as well. For example, the hyperthermophiles *Thermoproteus* and *Sulfolobus*, both *Archaea* (∞ Section 17.10), and *Aquifex* (*Bacteria*; ∞ Section 16.20), employ the reverse citric acid cycle for autotrophic growth, as do certain sulfur chemolithotrophic bacteria such as *Thiomicrospira*. Thus, this pathway, originally discovered in green sulfur bacteria and thought to be present only in these phototrophs, is probably distributed among several groups of autotrophic prokaryotes.

Autotrophy in *Chloroflexus*

The green nonsulfur phototroph *Chloroflexus* (∞ Section 16.18) grows autotrophically with either H_2 or H_2S as electron donor. However, neither the Calvin cycle nor the reverse citric acid cycle operates in this organism. Instead, two molecules of CO_2 are reduced to glyoxylate by a unique cyclic pathway, the **hydroxypropionate pathway**. This pathway is so named because hydroxypropionate, a three-carbon compound, is a key intermediate (Figure 20.24*b*).

In phototrophic bacteria, the hydroxypropionate pathway has been confirmed only in *Chloroflexus*, the earliest branching anoxygenic phototroph on the tree of *Bacteria* (∞ Figure 15.1). This suggests that the hydroxypropionate pathway was the first attempt at autotrophy in anoxygenic phototrophs. But in addition to *Chloroflexus*, the hydroxypropionate pathway functions in several hyperthermophilic *Archaea*, including *Metallosphaera*, *Acidianus*, and *Sulfolobus* (∞ Section 17.10). These are all nonphototrophic prokaryotes and lie near the base of the archaeal domain. The roots of the hydroxypropionate pathway may thus be very deep; it was perhaps nature's first attempt at autotrophy in any prokaryotic organism.

20.7 MiniReview

The reverse citric acid cycle and the hydroxypropionate cycle are pathways of CO_2 fixation in green sulfur and green nonsulfur bacteria, respectively, as well as among other prokaryotes.

- Including the route of CO_2 fixation, discuss at least three ways you could distinguish a purple sulfur bacterium from a green sulfur bacterium.
- Including the route of CO_2 fixation, what similarities and differences exist between green sulfur and green nonsulfur bacteria?

III CHEMOLITHOTROPHY

The next four sections focus on the chemolithotrophs, highlighting the strategies, problems, and advantages of a lifestyle geared toward the use of inorganic chemicals as energy sources.

20.8 The Energetics of Chemolithotrophy

Organisms that obtain energy from the oxidation of *inorganic* compounds are called **chemolithotrophs**. Most chemolithotrophic bacteria also obtain their carbon from CO_2, so they are also autotrophs. As we have noted, for growth on CO_2 as sole carbon source an organism needs (1) energy (ATP) and (2) reducing power (NADH or reduced ferredoxin). Some chemolithotrophs are **mixotrophs**, meaning that although they are able to obtain energy from the oxidation of an inorganic compound, they require an organic compound as carbon source.

ATP generation in chemolithotrophs is similar to that in chemoorganotrophs, except that the electron donor is inorganic rather than organic. Electron transport chains are present and ATP synthesis occurs by way of ATPases. Reducing power in chemolithotrophs is obtained in either of two ways: directly from the inorganic compound, if it has a sufficiently low reduction potential, such as H_2, or by reverse electron transport reactions (as discussed in Section 20.4 for phototrophic purple bacteria), if the electron donor is more electropositive than NADH. As we will see, the latter is typically the case.

Sources of Inorganic Electron Donors

Chemolithotrophs have many sources of inorganic electron donors. They may be geological, biological, or anthropogenic (the result of human activities). Volcanic activity is a major source of reduced sulfur compounds, primarily H_2S. Agricultural and mining operations add inorganic electron donors to the environment, especially reduced nitrogen and iron compounds, as does the burning of fossil fuels and the input of industrial wastes. Biological sources are also quite extensive, especially the production of H_2S, H_2, and NH_3. The ecological success and metabolic diversity of chemolithotrophs underscores the diversity of sources and abundance of inorganic electron donors available in nature.

Energetics of Chemolithotrophy

A review of reduction potentials listed in Table A1.2 reveals that a number of inorganic compounds can provide sufficient energy for ATP synthesis when oxidized with O_2 as electron acceptor. Recall from Chapter 5 that the further apart two half reactions are in terms of E_0', the greater the amount of energy released. For instance, the difference in reduction potential between the $2H^+/H_2$ couple and the $\frac{1}{2}O_2/H_2O$ couple is −1.23 V, which is equivalent to a free-energy yield of −237 kJ/mol (Appendix 1 gives the calculations). On the other hand, the potential difference between the $2H^+/H_2$ couple and the NO_3^-/NO_2^- couple is less, −0.84 V, equivalent to a free-energy yield of −163 kJ/mol. This is still quite sufficient for the production of ATP (the energy-rich phosphate bond of ATP has a free energy of −31.8 kJ/mol). However, a similar calculation shows that there is insufficient energy available from, for example, the oxidation of H_2S to S^0 using CO_2 as electron acceptor and CH_4 as the product (Appendix 1).

Energy calculations make it possible to predict the kinds of chemolithotrophs that should exist in nature. Because organisms must obey the laws of thermodynamics, only reactions that are thermodynamically favorable are potential energy-yielding reactions. **Table 20.1** summarizes energy yields for some chemolithotrophic reactions. The organisms themselves were considered in Chapters 15–17. We examine ecological aspects of chemolithotrophy in Chapter 23, where we will see that chemolithotrophic reactions form the heart of most nutrient cycles.

20.8 MiniReview

Chemolithotrophs oxidize inorganic chemicals as their sources of energy and reducing power. Most chemolithotrophs can also grow autotrophically.

- For what two purposes are inorganic compounds used by chemolithotrophs?
- Why does the oxidation of H_2 yield more energy with O_2 as electron acceptor than with SO_4^{2-} as electron acceptor?

20.9 Hydrogen Oxidation

Hydrogen (H_2) is a common product of microbial metabolism, and a number of chemolithotrophs are able to use it as an electron donor in energy metabolism. Anaerobic H_2-oxidizing *Bacteria* and *Archaea* are known, which differ in the electron acceptor they use (for example, nitrate, sulfate, ferric iron, and others), and these organisms are discussed in the next chapter. Here we consider only the *aerobic* H_2-oxidizing bacteria.

Energetics of H_2 Oxidation

Generation of ATP during H_2 oxidation by O_2 results from the formation of a proton motive force. The overall reaction

$$H_2 + \tfrac{1}{2}O_2 \longrightarrow H_2O \quad \Delta G^{0\prime} = -237 \text{ kJ}$$

UNIT 4

Table 20.1 Energy yields from the oxidation of various inorganic electron donors[a]

Electron donor	Chemolithotrophic reaction	Group of chemolithotrophs	E_0' of couple (V)	$\Delta G^{0\prime}$ (kJ/reaction)	Number of electrons/reaction	$\Delta G^{0\prime}$ (kJ/2e$^-$)
Phosphite[b]	$4\ HPO_3^{2-} + SO_4^{2-} + H^+ \rightarrow 4\ HPO_4^{2-} + HS^-$	Phosphite bacteria	−0.69	−91	2	−91
Hydrogen	$H_2 + \frac{1}{2}O_2 \rightarrow H_2O$	Hydrogen bacteria	−0.42	−237.2	2	−237.2
Sulfide	$HS^- + H^+ + \frac{1}{2}O_2 \rightarrow S^0 + H_2O$	Sulfur bacteria	−0.27	−209.4	2	−209.4
Sulfur	$S^0 + 1\frac{1}{2}O_2 + H_2O \rightarrow SO_4^{2-} + 2\ H^+$	Sulfur bacteria	−0.20	−587.1	6	−195.7
Ammonium[c]	$NH_4^+ + 1\frac{1}{2}O_2 \rightarrow NO_2^- + 2\ H^+ + H_2O$	Nitrifying bacteria	+0.34	−274.7	6	−91.6
Nitrite	$NO_2^- + \frac{1}{2}O_2 \rightarrow NO_3^-$	Nitrifying bacteria	+0.43	−74.1	2	−74.1
Ferrous iron	$Fe^{2+} + H^+ + \frac{1}{4}O_2 \rightarrow Fe^{3+} + \frac{1}{2}H_2O$	Iron bacteria	+0.77	−32.9	1	−65.8

[a]Data calculated from E_0' values in Appendix 1; values for Fe^{2+} are for pH 2, and others are for pH 7. At pH 7 the value for the Fe^{3+}/Fe^{2+} couple is about +0.2 V.

[b]Except for phosphite, all reactions are shown coupled to O_2 as electron acceptor. The only known phosphite oxidizer couples to SO_4^{2-} as electron acceptor.

[c]Ammonium can also be oxidized with NO_2^- as electron acceptor by anammox organisms (Section 20.13).

is highly exergonic and can thus support the synthesis of ATP. In this reaction, which is catalyzed by the enzyme **hydrogenase**, the electrons from H_2 are initially transferred to a quinone acceptor. From there electrons pass through a series of cytochromes to eventually reduce O_2 to water (**Figure 20.25**).

Some hydrogen bacteria have two distinct hydrogenases, one cytoplasmic and one membrane-integrated. The membrane-bound enzyme participates in energetics, whereas the soluble hydrogenase assimilates H_2 and reduces NAD^+ to NADH directly (the reduction potential of H_2, −0.42 V, is so low that reverse electron flow as a means of making reducing power is unnecessary; Figure 20.25). The organism *Ralstonia eutropha* has been a model for studying aerobic H_2 oxidation, and we discussed some of the properties of this organism in Section 15.5.

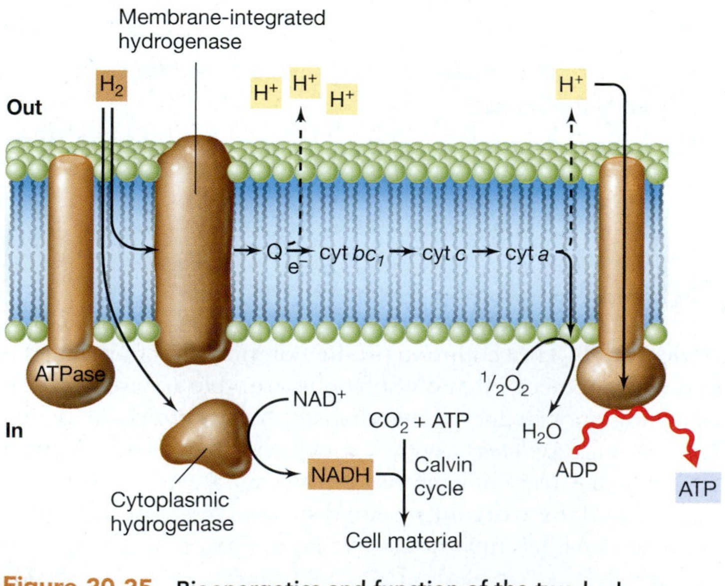

Figure 20.25 Bioenergetics and function of the two hydrogenases of aerobic H_2 bacteria. In *Ralstonia eutropha* two hydrogenases are present; the membrane-bound hydrogenase participates in energetics, whereas the cytoplasmic hydrogenase makes NADH for the Calvin cycle. Some H_2 bacteria have only the membrane-bound hydrogenase, and in these organisms reducing power is synthesized by reverse electron flow from Q back to NAD^+ to form NADH. Cyt, cytochrome; Q, quinone.

Autotrophy in H_2 Bacteria

Although most hydrogen bacteria can also grow as chemoorganotrophs, when growing chemolithotrophically, they fix CO_2 by the Calvin cycle (Section 20.6). The stoichiometry observed is

$$6\ H_2 + 2\ O_2 + CO_2 \longrightarrow (CH_2O) + 5\ H_2O$$

where (CH_2O) represents cell material. However, when readily useable organic compounds such as glucose are present, synthesis of Calvin cycle and hydrogenase enzymes by H_2 bacteria is repressed. In nature, H_2 levels in oxic environments are transient and low at best; most biological H_2 production is the result of fermentations, which are anoxic processes. Aerobic hydrogen bacteria must thus closely regulate their catabolic enzymes. These organisms likely shift their metabolism between chemoorganotrophy and chemolithotrophy, depending on levels of useable organic compounds and H_2 in their habitats. Moreover, many aerobic H_2 bacteria grow best microaerobically. It is therefore likely that these organisms are most successful as H_2 chemolithotrophs in oxic–anoxic interfaces where H_2 from fermentative metabolism is in greater and more continuous supply than in highly oxic habitats.

20.10 Oxidation of Reduced Sulfur Compounds

Many reduced sulfur compounds are used as electron donors by the colorless sulfur bacteria, called colorless to distinguish them from the bacteriochlorophyll-containing (pigmented) green and purple sulfur bacteria discussed earlier in this chapter (Figure 20.17). Historically, the concept of chemolithotrophy emerged from studies of the sulfur bacteria by the great Russian microbiologist Sergei Winogradsky (see the Microbial Sidebar, "Winogradsky and Chemolithoautotrophy").

Microbial Sidebar

Winogradsky and Chemolithoautotrophy

The concept of chemolithotrophic autotrophy was first conceived by the Russian microbiologist Sergei Winogradsky (Section 1.9). Winogradsky studied sulfur bacteria because certain colorless sulfur bacteria (*Beggiatoa, Thiothrix*) are very large (**Figure 1**) and easy to investigate, even in the absence of pure cultures. Springs with waters rich in H_2S are fairly common around the world. In the Bernese Oberland district of Switzerland, vast populations of *Beggiatoa* and *Thiothrix* develop in the outflow channels of sulfur springs. It was here that Winogradsky found suitable material for microscopic and physiological studies by merely lifting up the white filamentous masses of cells (Section 15.4 and Figures 15.10 and 15.12) and performing experiments directly in the field.

Winogradsky first showed that colorless sulfur bacteria were only present in water containing H_2S. As the water flowed away from the source, the H_2S gradually dissipated, and the sulfur bacteria disappeared. This suggested to him that their development depended on the presence of H_2S. Winogradsky then showed that when *Beggiatoa* filaments were starved for sulfide, they lost their sulfur granules. He found, however, that the granules were rapidly restored if a small amount of H_2S was added (Figure 1). He thus concluded that H_2S was being oxidized to elemental sulfur.

But what happened to the sulfur granules when filaments were starved of H_2S? Winogradsky showed by some clever microchemical tests that when the sulfur granules disappeared, sulfate appeared in the medium. He concluded that *Beggiatoa* (and by inference other colorless sulfur bacteria) oxidized H_2S to elemental sulfur and subsequently to sulfate ($H_2S \rightarrow S^0 \rightarrow SO_4^{2-}$). Because this organism seemed to require H_2S for development in the springs, he further postulated that this oxidation was the principal source of *energy* for these organisms.

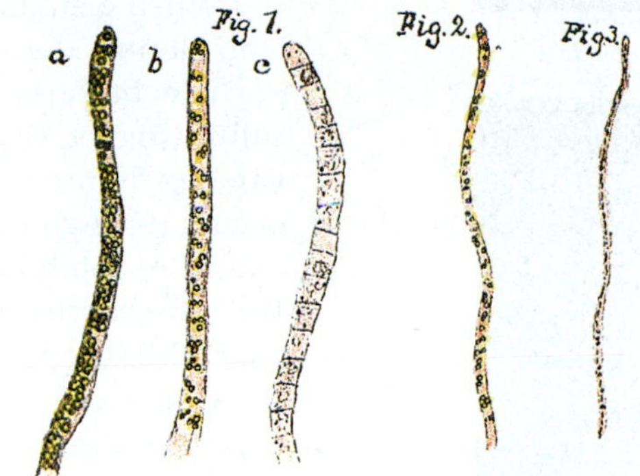

Figure 1 ***Beggiatoa* and chemolithotrophy.** *Drawings made by Winogradsky of Beggiatoa and translation (from the French) of the legend accompanying these figures. "Fig. 1. The tip of a filament of* Beggiatoa alba: *(a) in sulfurous [sulfide-containing] water, (b) after 24 h in water nearly depleted in H_2S, (c) after 48 h in water without H_2S [note depletion of sulfur globules with time]. Fig. 2. The tip of a filament of* Beggiatoa media. *Fig. 3. The tip of a filament of* Beggiatoa minima.*" From Winogradsky, S. 1949.* Microbiologie du Sol. *Masson, Paris.*

Winogradsky's studies on *Beggiatoa* thus provided the first evidence that an organism could oxidize an *inorganic* substance as an energy source. This was the origin of the concept of chemolithotrophy. From these beginnings, Winogradsky turned to a study of the nitrifying bacteria, and it was with this group that he first showed that autotrophic fixation of CO_2 was coupled to the oxidation of an inorganic compound. The process of nitrification had been known before Winogradsky's work from studies on the fate of sewage added to soil. For example, ammonia-rich sewage that passed through a soil column was converted to nitrate. Winogradsky proceeded to isolate nitrifying bacteria using completely mineral media in which CO_2 was the sole carbon source and ammonia was the sole electron donor. Because ammonia is chemically stable, it was easy to show that the oxidation of ammonia to nitrite, and subsequently to nitrate, was a strictly bacterial process.

Winogradsky further showed that nitrification was a *two-step process*, with one group of organisms converting NH_4^+ to NO_2^- and a second converting NO_2^- to NO_3^- (Section 20.12). Because no organic materials were present in the medium, it was also possible to show that organic matter (the bacterial cell material) was formed from CO_2: When the ammonia or nitrite was left out of the medium, the cells did not grow. Careful chemical analyses showed that the amount of organic matter formed by the bacteria was proportional to the amount of ammonia or nitrite they oxidized. Winogradsky concluded, "This [process] is contradictory to that fundamental doctrine of physiology which states that a complete synthesis of organic matter cannot take place in nature except through chlorophyll-containing plants by the action of light." The concept of chemolithoautotrophy was thus born.

We now know that at least in one way autotrophy in most chemolithotrophs and phototrophs is similar. In both groups, the pathway of CO_2 fixation follows the same biochemical steps (the Calvin cycle), which require the enzyme ribulose bisphosphate carboxylase (Section 20.6). Some other chemolithotrophs use alternative autotrophic pathways, such as the reverse citric acid cycle or the hydroxypropionate cycle, mechanisms first discovered in anoxygenic phototrophs.

As we will see in this chapter, many *Bacteria* and *Archaea* are chemolithotrophs, but all employ the same basic metabolic strategy: Oxidize the inorganic compound by an electron transport chain that generates a proton motive force. The various forms of chemolithotrophy are simply variations on a common metabolic theme first discovered by Winogradsky with the sulfur bacteria.

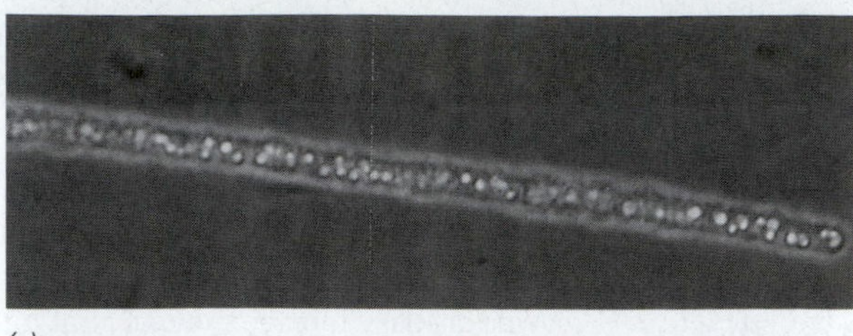

(a)

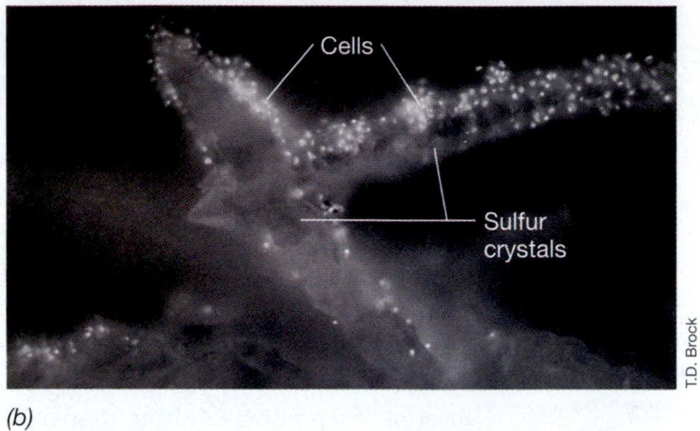

(b)

Figure 20.26 Sulfur bacteria. *(a)* Internal sulfur granules in *Beggiatoa.* *(b)* Attachment of cells of the sulfur-oxidizing archaeon *Sulfolobus acidocaldarius* to a crystal of elemental sulfur. Cells are visualized by fluorescence microscopy after being stained with the dye acridine orange. The sulfur crystal does not fluoresce.

Energetics of Sulfur Oxidation

The most common sulfur compounds used as electron donors are hydrogen sulfide (H_2S), elemental sulfur (S^0), and thiosulfate ($S_2O_3^{2-}$). In most cases the final product of sulfur oxidation is sulfate (SO_4^{2-}), and the total number of electrons generated between H_2S (oxidation state, −2) and sulfate (oxidation state, +6) is eight. Less energy is available when one of the intermediate sulfur oxidation states is used:

$$H_2S + 2\,O_2 \longrightarrow SO_4^{2-} + 2\,H^+$$

(final product, sulfate)

$$\Delta G^{0\prime} = -798.2 \text{ kJ/reaction } (-99.75 \text{ kJ/e}^-)$$

$$HS^- + \tfrac{1}{2}\,O_2 + H^+ \longrightarrow S^0 + H_2O$$

(final product, sulfur)

$$\Delta G^{0\prime} = -209.4 \text{ kJ/reaction } (-104.7 \text{ kJ/e}^-)$$

$$S^0 + H_2O + 1\tfrac{1}{2}\,O_2 \longrightarrow SO_4^{2-} + 2\,H^+$$

(final product, sulfate)

$$\Delta G^{0\prime} = -587.1 \text{ kJ/reaction } (-97.85 \text{ kJ/e}^-)$$

$$S_2O_3^{2-} + H_2O + 2\,O_2 \longrightarrow 2\,SO_4^{2-} + 2\,H^+$$

(final product, sulfate)

$$\Delta G^{0\prime} = -818.3 \text{ kJ/reaction } (-102 \text{ kJ/e}^-)$$

H_2S oxidation occurs in stages, with the first oxidation step yielding elemental sulfur, S^0. Some H_2S-oxidizing bacteria deposit this elemental sulfur inside the cell (**Figure 20.26*a***), where the sulfur forms an energy reserve. When the supply of H_2S has been depleted, additional energy can be obtained from the oxidation of sulfur to sulfate.

When elemental sulfur is provided externally as an electron donor, the organism must attach itself to the sulfur particle because of the extreme insolubility of elemental sulfur (Figure 20.26*b*). By adhering to the particle, the organism can remove sulfur atoms for oxidation to sulfate. This occurs through the activity of membrane or periplasmic proteins that solubilize the sulfur, probably by reduction of S^0 to HS^-, from which it is transported into the cell and enters chemolithotrophic metabolism (**Figure 20.27**).

One of the products of reduced sulfur oxidation reactions is H^+, and the production of protons lowers the pH. Consequently, one result of the oxidation of reduced sulfur compounds is the acidification of the medium. The acid formed by the sulfur bacteria is a strong acid, sulfuric acid (H_2SO_4), and because of this, sulfur bacteria are often able to bring about a marked reduction in the pH of the medium. Some sulfur bacteria are quite acid-tolerant or even acidophilic. *Acidithiobacillus thiooxidans,* for example, grows best at a pH below 3.

Biochemistry of Sulfur Oxidation

The biochemical steps in the oxidation of various sulfur compounds are summarized in Figure 20.27. Several oxidation systems are known in sulfur chemolithotrophs. In two of the systems, the starting substrate, H_2S, $S_2O_3^{2-}$, or S^0, is first oxidized to sulfite (SO_3^{2-}); starting with sulfide, six electrons are released. Then the sulfite is oxidized to sulfate. This can occur in either of two ways. The most widespread system is that employing the enzyme sulfite oxidase. Sulfite oxidase transfers electrons from SO_3^{2-} directly to cytochrome *c*, and ATP is made from this during electron transport and proton motive force formation (Figure 20.27*b*). By contrast, some sulfur chemolithotrophs oxidize SO_3^{2-} to SO_4^{2-} via a reversal of the activity of adenosine phosphosulfate (APS) reductase, an enzyme essential for the metabolism of sulfate-reducing bacteria (∞ Section 21.8). This reaction, run in the direction of SO_4^{2-} production by sulfur chemolithotrophs, yields one high-energy phosphate bond when AMP is converted to ADP (Figure 20.27*a*). When thiosulfate is the electron donor for sulfur chemolithotrophs, it is first split into S^0 and SO_3^{2-}, both of which are eventually oxidized to SO_4^{2-}.

Sox

A functionally distinct sulfide and thiosulfate oxidation system is present in *Paracoccus pantotrophus* and many other sulfur bacteria. This system, called the *Sox* (for *s*ulfur *ox*idation) *system*, oxidizes reduced sulfur compounds directly to sulfate without the intermediate formation of sulfite (Figure 20.27*a*). The Sox system comprises 15 genes that encode various cytochromes and other proteins necessary for

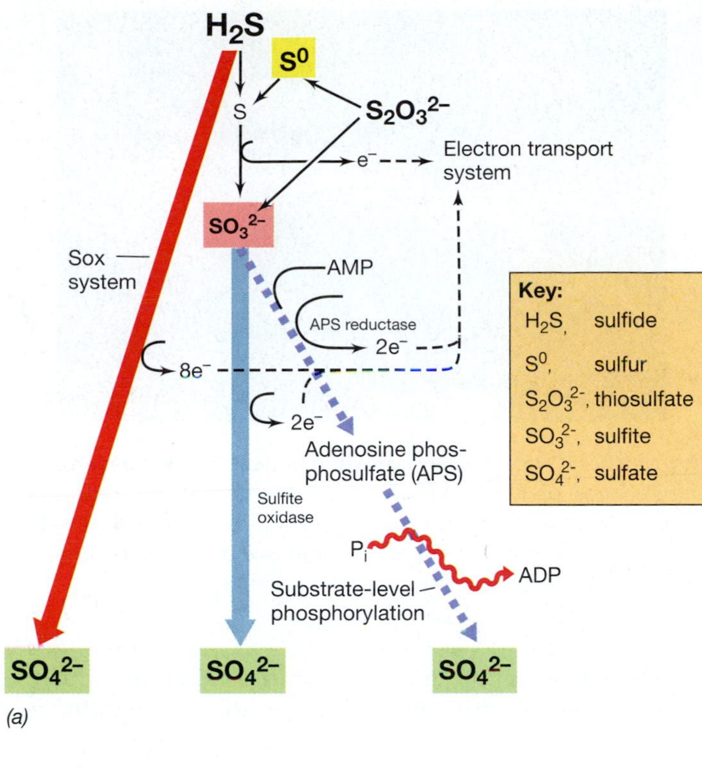

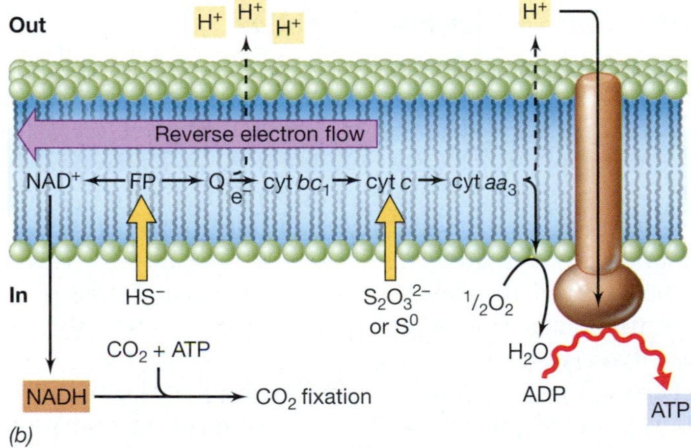

Figure 20.27 Oxidation of reduced sulfur compounds by sulfur chemolithotrophs. *(a)* Steps in the oxidation of different compounds. Three different pathways are known. *(b)* Electrons from sulfur compounds feed into the electron transport chain to drive a proton motive force; electrons from thiosulfate and elemental sulfur enter at the level of cytochrome *c*. NADH must be made by energy-consuming reactions of reverse electron flow because the electron donors have a more electropositive E_0' than does $NAD^+/NADH$. Cyt, cytochrome; FP, flavoprotein; Q, quinone. For the structure of APS, see Figure 21.15.

the oxidation of reduced sulfur compounds directly to sulfate. The Sox system is present in several sulfur chemolithotrophs and is also present in some phototrophic sulfur bacteria that oxidize H_2S to obtain reducing power for CO_2 fixation (Figure 20.17). Thus this same biochemical system for sulfide oxidation is distributed among prokaryotes that oxidize sulfide for inherently different reasons, an indication that the genes that encode it may have been transferred between species by horizontal gene flow.

Energy from Sulfur Oxidation

All the electrons from reduced sulfur compounds eventually reach the electron transport system as shown in Figure 20.27*b*. Depending on the E_0' of the couple, electrons enter at either the flavoprotein ($E_0' = \sim -0.2$) or cytochrome *c* ($E_0' = +0.3$) level and are transported through the chain to O_2, generating a proton motive force that leads to ATP synthesis by ATPase. Electrons for autotrophic CO_2 fixation come from reverse electron flow (Section 20.4), eventually yielding NADH. Autotrophy is supported by reactions of the Calvin cycle (Figure 20.27*b*) or some other autotrophic pathway. Although the sulfur chemolithotrophs are primarily an aerobic group (∞ Section 15.4), some species can grow anaerobically using nitrate as an electron acceptor; *Thiobacillus denitrificans* is a classic example of this lifestyle.

UNIT 4

20.9–20.10 MiniReview

Hydrogen (H_2) and reduced sulfur compounds such as H_2S, $S_2O_3^{2-}$, and S^0 are excellent electron donors for energy metabolism in chemolithotrophs. Electrons from these substances enter electron transport chains, yielding a proton motive force. Sulfur and hydrogen chemolithotrophs are also autotrophs and fix CO_2 by the Calvin cycle.

- What special enzyme is needed for growth on H_2?
- How many electrons are available from the oxidation of H_2S if S^0 is the final product? If SO_4^{2-} is the final product?
- How does the Sox system for oxidizing H_2S differ from other systems for oxidizing H_2S?

20.11 Iron Oxidation

The aerobic oxidation of iron from the ferrous (Fe^{2+}) to the ferric (Fe^{3+}) state is a chemolithotrophic reaction for some prokaryotes. At acidic pH, only a small amount of energy is available from this oxidation (Table 20.1), and for this reason the iron bacteria must oxidize large amounts of iron in order to grow. The ferric iron produced forms insoluble ferric hydroxide [$Fe(OH)_3$] precipitates in water (**Figure 20.28**). This is in part because at neutral pH ferrous iron rapidly oxidizes spontaneously to the ferric state. Fe^{2+} is thus stable for long periods only under anoxic conditions. At acidic pH, however, ferrous iron is stable under oxic conditions. This explains why most iron-oxidizing bacteria are obligately acidophilic.

Iron-Oxidizing Bacteria

The best-known iron-oxidizing bacteria, *Acidithiobacillus ferrooxidans* and *Leptospirillum ferrooxidans*, can both grow autotrophically using ferrous iron (Figure 20.28*b*) as electron donor. Both organisms can grow at pH values below pH 1, with growth optimal at pH 2–3. These organisms are common

(a)

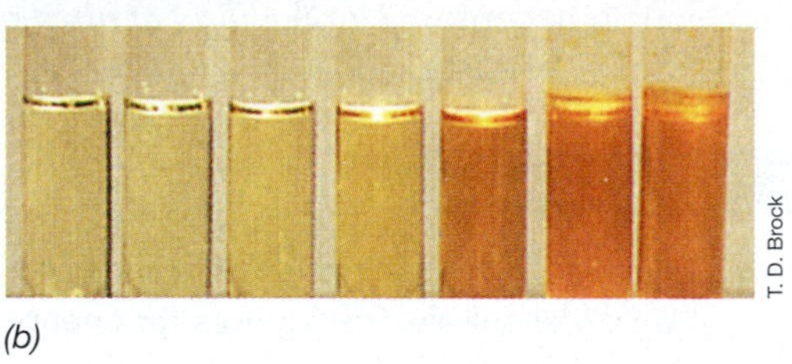

(b)

Figure 20.28 Iron-oxidizing bacteria. *(a)* Acid mine drainage, showing the confluence of a normal river and a creek draining a coal-mining area. The acidic creek is very high in ferrous iron (Fe^{2+}). At low pH values, ferrous iron does not oxidize spontaneously in air, but *Acidithiobacillus ferrooxidans* carries out the oxidation. Insoluble ferric hydroxide and complex ferric salts precipitate, forming the precipitate called "yellow boy" by coal miners. *(b)* Cultures of *A. ferrooxidans*. Shown is a dilution series, with no growth in the tube on the left and increasing amounts of growth from left to right. Growth is evident from the production of Fe^{3+} from Fe^{2+}, which readily forms $Fe(OH)_3$ and acidity, leading to the yellow-orange color ($Fe^{3+} + 3H_2O \rightarrow Fe(OH)_3 + 3H^+$).

Bill Strode

T. D. Brock

in acid-polluted environments such as coal-mining dumps (Figure 20.28*a*). *Ferroplasma*, a member of the *Archaea*, is an extremely acidophilic iron oxidizer and can even grow at pH values below 0 (∞ Section 17.5). We will discuss the role of all of these organisms in acid-mine pollution and mineral oxidation in Section 24.6.

Despite the instability of Fe^{2+} at neutral pH, there are some iron-oxidizing bacteria that thrive in such environments, but only in situations where ferrous iron is transitioning from anoxic to oxic conditions. For example, anoxic groundwater can contain Fe^{2+}, and when it is released, as in an iron spring, it becomes exposed to O_2. At such interfaces, iron bacteria oxidize Fe^{2+} as it emerges from the spring and before it oxidizes spontaneously. *Gallionella ferruginea* and *Sphaerotilus natans* are examples of organisms that live at these interfaces. They are typically seen mixed in with the characteristic ferric iron deposits they form (**Figure 20.29**; ∞ Figure 15.42). Iron bacteria that develop at neutral pH have an energetic advantage over their acidophilic counterparts. At neutral pH, the E_0' of the Fe^{3+}/Fe^{2+} couple is +0.2 V. This means that Fe^{2+} is a stronger electron donor at neutral pH than at acidic

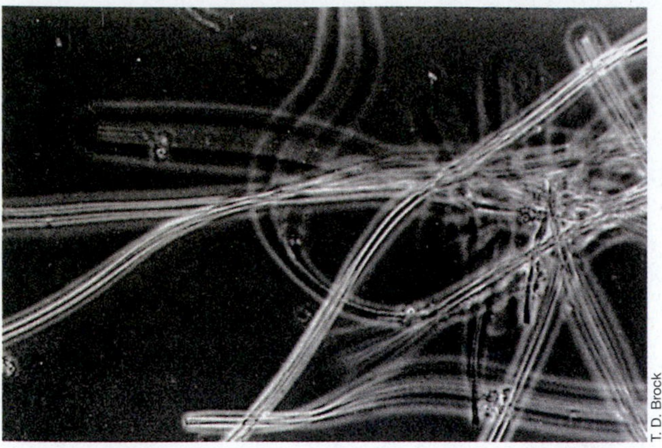

T. D. Brock

Figure 20.29 Iron bacteria growing at neutral pH: *Sphaerotilus*. Phase-contrast photomicrograph of empty iron-encrusted sheaths of *Sphaerotilus* collected from iron seepage at the edge of a small swamp. *Sphaerotilus* is a typical interface organism, living in habitats where anoxic ferrous-containing water meets oxic zones.

pH. This can affect the energetics of iron oxidation, which, as we see now, is quite marginal for the acidophilic iron oxidizers.

Energy from Ferrous Iron Oxidation

The bioenergetics of iron oxidation by *Acidithiobacillus ferrooxidans* and other acidophilic iron-oxidizers are of interest because of the very electropositive reduction potential of the Fe^{3+}/Fe^{2+} couple at acidic pH (+0.77 V at pH 2). The respiratory chain of *A. ferrooxidans* contains cytochromes of the *c* and *a* types (∞ Section 5.11) and a periplasmic copper-containing protein called *rusticyanin* (**Figure 20.30**). Because the reduction potential of the Fe^{3+}/Fe^{2+} couple is so high, the route of electron transport to oxygen ($\frac{1}{2}$ O_2/H_2O, $E_0' = +0.82$ V) is by necessity very short.

Ferrous iron oxidation begins in the periplasm, where rusticyanin oxidizes Fe^{2+} to Fe^{3+}, a one-electron transition. Rusticyanin then reduces cytochrome *c*, and this subsequently reduces cytochrome *a*. The latter interacts directly with O_2 to form H_2O (Figure 20.30). ATP is then synthesized from proton-translocating ATPases in the cytoplasmic membrane. However, here is where the energetics of acidophilic iron oxidizers takes a unique twist. In an environment that is pH 2, a large gradient of protons naturally exists across the *A. ferrooxidans* membrane (the periplasm is pH 1–2, whereas the cytoplasm is pH 5.5–6). But, the organisms cannot make ATP from this "natural proton motive force" for free. Protons entering the cytoplasm via the ATPase must be consumed in order to maintain the internal pH within acceptable limits. The protons are consumed during the production of H_2O in the electron transport chain, and this reaction requires electrons; these come from the oxidation of Fe^{2+} to Fe^{3+} (Figure 20.30). Thus, as long as *A. ferrooxidans* has Fe^{2+} available as an electron donor, ATP can be synthesized at the expense of the proton motive force that preexists across the cytoplasmic membrane (Figure 20.30).

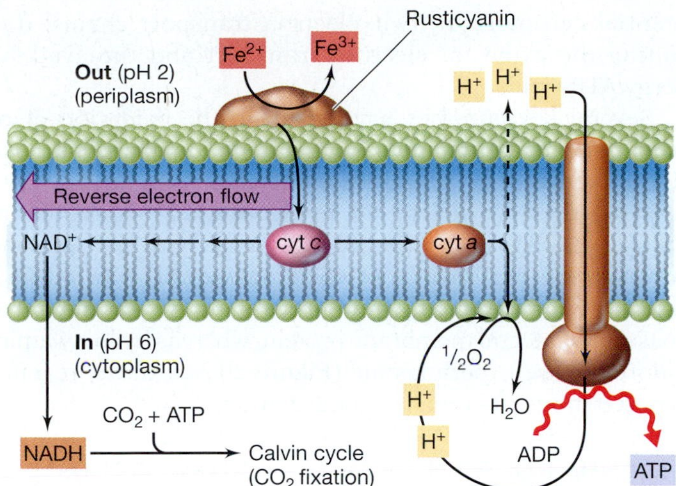

Figure 20.30 Electron flow during Fe^{2+} oxidation by the acidophile *Acidithiobacillus ferrooxidans*. The periplasmic copper-containing protein rusticyanin is the immediate acceptor of electrons from Fe^{2+}. From here, electrons travel a short electron transport chain resulting in the reduction of O_2 to H_2O. Reducing power to drive the Calvin cycle comes from reverse electron flow. Note the steep pH gradient (~4 units) across the membrane.

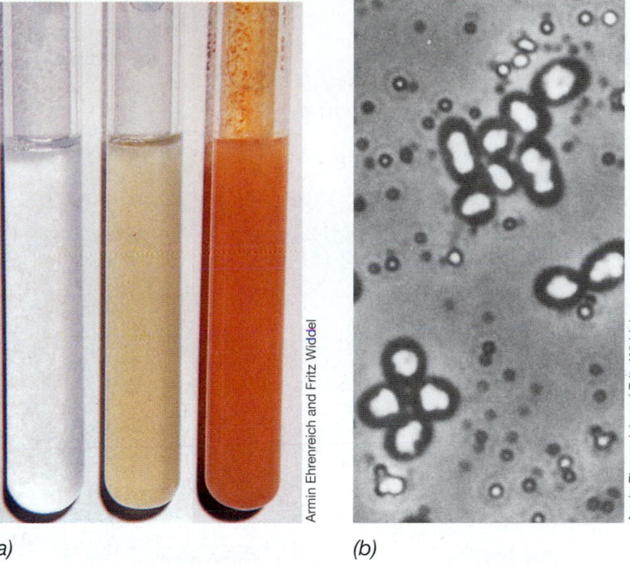

Figure 20.31 Ferrous iron oxidation by anoxygenic phototrophic bacteria. *(a)* Fe^{2+} oxidation in anoxic tube cultures. Left to right: Sterile medium, inoculated medium, a growing culture. The brown-red color is mainly due to $Fe(OH)_3$ precipitate. *(b)* Phase-contrast photomicrograph of an iron-oxidizing purple bacterium. The bright refractile areas within cells are gas vesicles (Section 4.11). The granules outside the cells are iron precipitates. This organism is phylogenetically related to the purple sulfur bacterium *Chromatium* (Section 15.2).

Autotrophy in *A. ferrooxidans* is driven by the Calvin cycle, and because of the high potential of the electron donor, Fe^{2+}, much energy is consumed in reverse electron flow reactions to obtain the reducing power (NADH) necessary to drive CO_2 fixation. Thus, a relatively poor energetic yield coupled with large energetic demands for biosynthesis means that *A. ferrooxidans* must oxidize large amounts of Fe^{2+} to produce even a very small amount of cell material. Because of this, in environments where acidophilic Fe^{2+}-oxidizing bacteria thrive, their presence is signaled not by the formation of much cell material but by the presence of large amounts of ferric iron precipitates (Figure 20.28; Figure 24.11). We consider the important ecological processes connected with the iron-oxidizing bacteria in Sections 24.5 and 24.6.

Ferrous Iron Oxidation by Anoxygenic Phototrophs

Ferrous iron can be oxidized under anoxic conditions by certain anoxygenic phototrophic bacteria (**Figure 20.31**). The ferrous iron is used in this case not as an electron donor in energy metabolism, but as an electron donor for CO_2 reduction (autotrophy). At neutral pH where these organisms thrive, the Fe^{3+}/Fe^{2+} couple is 0.2 V, and thus, electrons from Fe^{2+} can reduce cytochrome *c* in the photosystem of purple bacteria (anoxygenic photosynthesis, Section 20.4). The ferrous iron-oxidizing organisms, which are species of phototrophic purple bacteria (Figure 20.31*b*), can also use FeS as electron donor; under these conditions both Fe^{2+} and S^{2-} are oxidized as electron donors, Fe^{2+} to Fe^{3+}, and S^{2-} to SO_4^{2-} (Figure 20.27).

Certain phototrophic green sulfur bacteria (genus *Chlorobium*; Section 16.15) can also use Fe^{2+} as an electron donor for autotrophic growth. Moreover, various chemotrophic denitrifying bacteria have been isolated that can couple the oxidation of Fe^{2+} to the reduction of NO_3^- to N_2 and grow under anoxic conditions. However, like the aerobic iron bacteria, Fe^{2+} is an electron donor for both energy and reducing power needs in these organisms.

The discovery of Fe^{2+}-oxidizing phototrophs has important implications for both understanding the evolution of photosynthesis and explaining the large deposits of ferric iron found in ancient sediments on Earth. Ancient ferric iron was assumed to have formed from the oxidation of Fe^{2+} by O_2 produced by oxygenic phototrophs. However, because of the age of these sediments, which in many cases date to a time before the appearance of cyanobacteria on Earth (Section 14.2), it is more likely that the ferric iron was formed by anoxygenic phototrophs oxidizing Fe^{2+} in iron-rich anoxic environments.

20.11 MiniReview

The iron bacteria are chemolithotrophs that oxidize ferrous iron (Fe^{2+}) as electron donor. Most iron bacteria grow only at acidic pH and are often associated with acidic pollution from mineral and coal mining. Some phototrophic purple bacteria can oxidize Fe^{2+} to Fe^{3+} anaerobically.

- Why is only a very small amount of energy available from the oxidation of Fe^{2+} to Fe^{3+} at acidic pH?
- What is the function of rusticyanin and where is it found in the cell?
- How can Fe^{2+} be oxidized under anoxic conditions?

UNIT 4

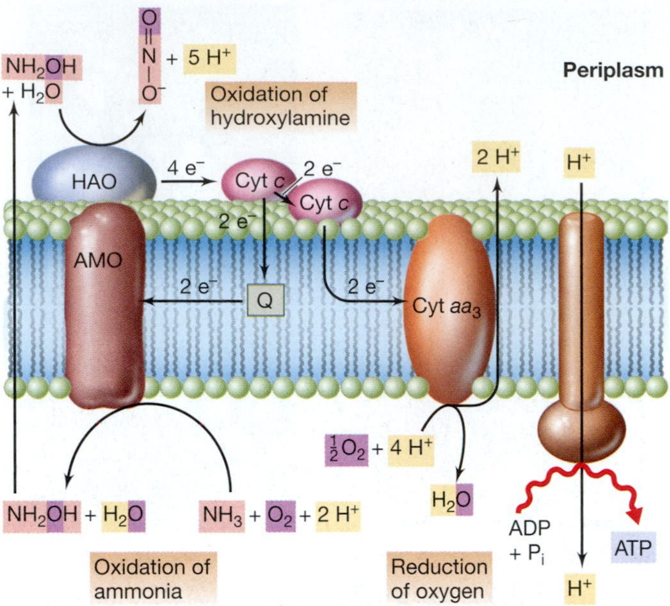

Figure 20.32 Oxidation of ammonia and electron flow in ammonia-oxidizing bacteria. The reactants and the products of this reaction series are highlighted. The cytochrome c (Cyt c) in the periplasm is a different form of Cyt c than that in the membrane. AMO, ammonia monooxygenase; HAO, hydroxylamine oxidoreductase; Q, ubiquinone.

20.12 Nitrification

The inorganic nitrogen compounds ammonia (NH_3) and nitrite (NO_2^-) are chemolithotrophic substrates. These compounds are oxidized aerobically by *nitrifying bacteria* (Section 15.3) during the process of **nitrification**. The nitrifying bacteria are widely distributed in soil and water. One group, the *nitrosifyers* (*Nitrosomonas* is an important genus, Section 15.3), oxidizes ammonia to nitrite, and another group (*Nitrospira* is an important genus, Section 16.21) oxidizes nitrite to nitrate. The complete oxidation of ammonia to nitrate, an eight-electron transfer, is thus carried out by two groups of organisms acting in concert, a process discovered by the Russian microbiologist, Winogradsky (as seen in the Microbial Sidebar).

Bioenergetics and Enzymology of Nitrification

The bioenergetics of nitrification is based on the same principles that govern other chemolithotrophic reactions: electrons from reduced inorganic substrates (in this case, reduced nitrogen compounds) enter an electron transport chain, and electron flow establishes a proton motive force that drives ATP synthesis.

Nitrifying bacteria are faced with bioenergetic problems similar to those of most other chemolithotrophs. The E_0' of the NO_2^-/NH_3 couple (the first step in the oxidation of NH_3) is high, +0.34 V. The E_0' of the NO_3^-/NO_2^- couple is even higher, about +0.43 V. These relatively high reduction potentials force nitrifying bacteria to donate electrons to rather high potential acceptors in their electron transport chains, thus limiting the extent of electron transport and the available energy/ATP yield.

Several key enzymes participate in the oxidation of reduced nitrogen compounds. In ammonia-oxidizing bacteria, NH_3 is oxidized by *ammonia monooxygenase* (monooxygenase enzymes are discussed in Section 21.14) producing NH_2OH and H_2O (**Figure 20.32**). A second key enzyme, *hydroxylamine oxidoreductase*, then oxidizes NH_2OH to NO_2^-, removing four electrons in the process. Ammonia monooxygenase is an integral membrane protein, whereas hydroxylamine oxidoreductase is periplasmic (Figure 20.32). In the reaction carried out by ammonia monooxygenase,

$$NH_3 + O_2 + 2\,H^+ + 2\,e^- \longrightarrow NH_2OH + H_2O,$$

there is a need for two exogenously supplied electrons plus two protons to reduce one atom of oxygen to water. These electrons originate from the oxidation of hydroxylamine and are supplied to ammonia monooxygenase from hydroxylamine oxidoreductase via cytochrome *c* and ubiquinone (Figure 20.32). Thus, for every *four* electrons generated from the oxidation of NH_3 to NO_2^-, only *two* actually reach the terminal oxidase (cytochrome aa_3, Figure 20.32) and can yield energy.

Nitrite-oxidizing bacteria employ the enzyme *nitrite oxidoreductase* to oxidize nitrite to nitrate, with electrons traveling a very short electron transport chain (because of the high potential of the NO_3^-/NO_2^- couple) to the terminal oxidase (**Figure 20.33**). Cytochromes of the *a* and *c* types are present in the electron transport chain of nitrite oxidizers, and the activity of cytochromes aa_3 generates a proton motive force (Figure 20.33). As in the iron oxidation reaction (Section 20.11), only small amounts of energy are available in this reaction. Thus, growth yields of nitrifying bacteria (grams of cells produced per mole of substrate oxidized) are low.

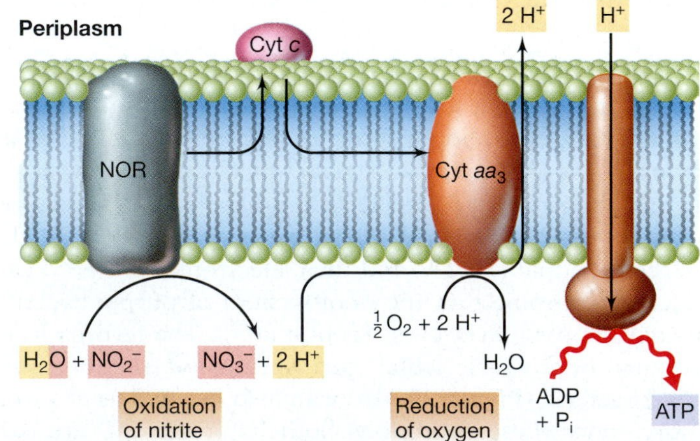

Figure 20.33 Oxidation of nitrite to nitrate by nitrifying bacteria. The reactants and products of this reaction series are highlighted to follow the reaction. NOR, nitrite oxidoreductase.

Other Nitrifying Prokaryotes

From a phylogenetic standpoint, all nitrifying bacteria discussed thus far have been *Bacteria*. However, we now know that at least one species of *Archaea* is a nitrifyer. This organism, *Nitrosopumilus*, is an autotrophic ammonia-oxidizing chemolithotroph and member of the archaeal phylum *Crenarchaeota* (∞ Section 17.12). *Nitrosopumilus* also contains genes related to those that encode ammonia monooxygenase in bacterial nitrifyers such as *Nitrosomonas*, and thus it is likely that the physiology of ammonia oxidation in *Bacteria* and *Archaea* is similar.

Thus far, nitrite-oxidizing *Archaea* are unknown. However, nitrite is an electron donor for certain purple anoxygenic phototrophic bacteria. In this case, however, nitrite is oxidized to nitrate under *anoxic* conditions because anoxygenic photosynthesis is an anoxic process (Section 20.4). Moreover, the electrons derived from nitrite oxidation by these purple bacteria are not used to obtain energy as they are in nitrifying bacteria, but instead are used as a source of reducing power for autotrophic CO_2 fixation (Calvin cycle, Section 20.6).

Carbon Metabolism and Ecology of Nitrifying Bacteria

Like sulfur- and iron-oxidizing chemolithotrophs, aerobic nitrifying bacteria employ the Calvin cycle for CO_2 fixation. The ATP and reducing power requirements of the Calvin cycle place additional burdens on an already relatively low-yielding energy-generating system (NADH to drive the Calvin cycle in nitrifyers is formed by reverse electron flow). The energetic constraints are particularly severe for nitrite oxidizers, and it is perhaps for this reason that most nitrite oxidizers have alternative energy-conserving mechanisms, growing chemoorganotrophically on glucose and certain other organic substrates (∞ Section 15.3).

Nitrifying bacteria play key ecological roles in the nitrogen cycle, converting ammonia into nitrate, a key plant nutrient. Nitrifying bacteria are also important in sewage and wastewater treatment, removing toxic amines and ammonia and releasing less toxic nitrogen compounds. Nitrifying bacteria play a similar role in the water column of lakes, where ammonia produced in the sediments from the decomposition of organic nitrogenous compounds is converted into nitrate for use by algae and cyanobacteria.

20.12 MiniReview

Ammonia and nitrite can be used as electron donors by the nitrifying bacteria. The ammonia-oxidizing bacteria produce nitrite, which is then oxidized by the nitrite-oxidizing bacteria to nitrate.

- What is the inorganic electron donor for *Nitrosomonas*? For *Nitrospira*?
- What are the substrates for the enzyme ammonia monooxygenase?
- What do nitrifying bacteria use as a carbon source?

20.13 Anammox

Although the classical nitrifying bacteria just discussed are *strict aerobes*, ammonia can also be oxidized under *anoxic* conditions. This process, known as **anammox** (for *an*oxic *amm*onia *ox*idation), is exergonic and is catalyzed by a unusual group of obligately anaerobic *Bacteria*.

In anammox ammonia is oxidized with nitrite as the electron acceptor to yield nitrogen gas:

$$NH_4^+ + NO_2^- \longrightarrow N_2 + 2\,H_2O \qquad \Delta G^{0\prime} = -357 \text{ kJ}$$

The first anammox organism discovered, *Brocadia anammoxidans*, is a member of the *Planctomycetes* phylum of *Bacteria* (∞ Section 16.10) (**Figure 20.34**). *Planctomycetes* are unusual *Bacteria*, lacking peptidoglycan and containing membrane-enclosed compartments inside the cell. In cells of *B. anammoxidans* one such compartment is the *anammoxosome*, and it is within this structure that the anammox reaction occurs (Figure 20.34*c*). In addition to *Brocadia*, several other genera of anammox bacteria are known, including *Kuenenia* and *Scalindua*, both of which also contain an anammoxosome.

The Anammoxosome

The anammoxosome is a unit membrane-enclosed structure (Figure 20.34*b*) and in this respect can be considered an organelle in the eukaryotic sense of the term (∞ Sections 18.2 and 18.3). However, lipids that make up the anammoxosome membrane are not the typical lipids of *Bacteria*. Anammoxosome lipids consist of fatty acids containing multiple four-membered (cyclobutane) rings that are connected to glycerol by both ester and ether bonds. The lipids aggregate to form an unusually dense membrane structure that is highly resistant to diffusion. The strong anammoxosome membrane is likely required to protect the cell from the toxic intermediates produced during the anammox reactions. These include, in particular, the compound *hydrazine*, a very strong reductant. In the anammox reaction hydrazine as well as hydroxylamine are produced in a cyclical series of reactions (Figure 20.34*c*). Of these, hydrazine is the most toxic, and so it is trapped in the anammoxosome until it is degraded to harmless N_2; this prevents it from reducing macromolecules indiscriminantly in the cytoplasm.

Autotrophy in Anammox Bacteria

Like classical nitrifying bacteria, anammox bacteria are also autotrophs. Anammox organisms grow with CO_2 as their sole carbon source and use nitrite as electron donor to produce cell material:

$$CO_2 + 2\,NO_2^- + H_2O \longrightarrow CH_2O + 2\,NO_3^-$$

Although they are autotrophs, anammox bacteria lack Calvin cycle enzymes, and the mechanism of CO_2 fixation is instead the acetyl-CoA pathway, an autotrophic pathway widespread

UNIT 4

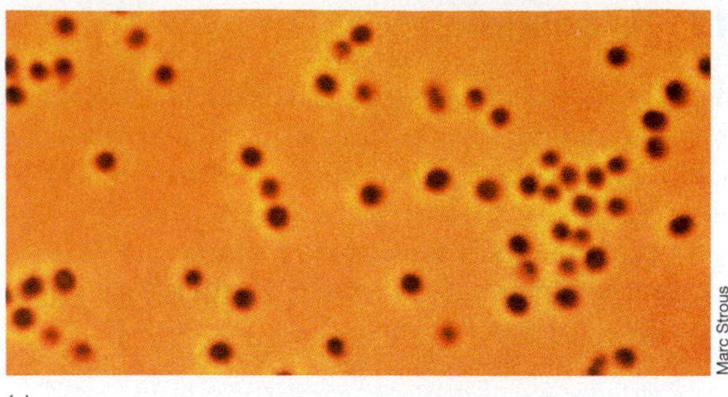

(a)

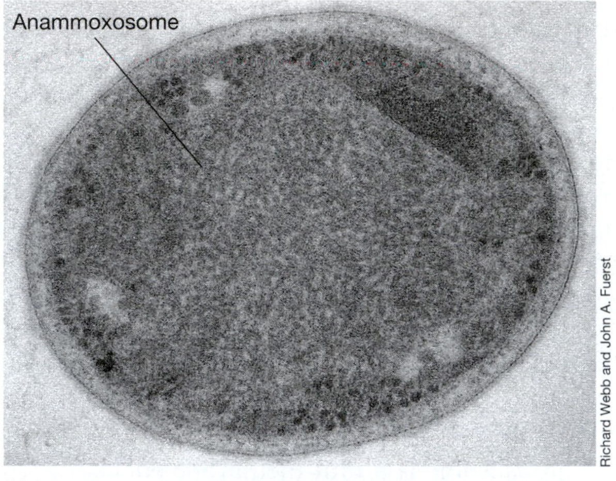

(b)

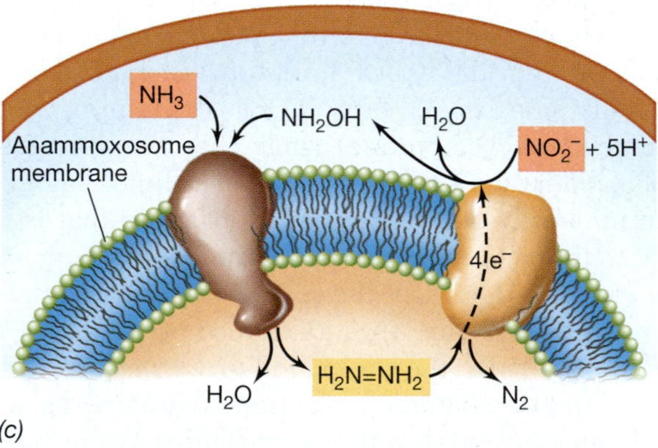

(c)

Figure 20.34 Anammox. *(a)* Phase-contrast photomicrograph of cells of *Brocadia anammoxidans*. A single cell is about 1 μm in diameter. *(b)* Transmission electron micrograph of a cell; note the membrane-enclosed compartments including the large fibrillar anammoxosome. *(c)* Reactions in the anammoxosome. Anammox substrates are shown in red; hydrazine in yellow.

among obligately anaerobic bacteria (Section 21.9). Interestingly, however, the high potential of the NO_2^-/NO_3^- couple precludes the use of NO_2^- directly as electron donor for the acetyl-CoA pathway, because the latter requires an electron donor of E_0' near −0.4 V. It is thought that anammox organisms overcome this problem by using hydrazine (N_2H_4) as an electron donor for the reductant needed for autotrophic growth. The oxidation of hydrazine to N_2 in the anammoxosome (Figure 20.34*c*) generates electrons at an extremely negative potential, about −0.75 V. These electrons can reduce ferredoxin (Section 20.7), which has an E_0' of −0.4 V, sufficient for the acetyl-CoA pathway.

Ecology of Anammox

The source of NO_2^- in the anammox reaction is the product of ammonia oxidation by aerobic nitrifying bacteria (Figure 20.32). The two groups of ammonia oxidizers, aerobic (for example, *Nitrosomonas*) and anaerobic (*Brocadia*), live together in ammonia-rich habitats such as sewage and other wastewaters. In these environments suspended particles containing both oxic and anoxic zones form where the two groups of ammonia oxidizers coexist. In mixed laboratory cultures, high levels of oxygen inhibit anammox and favor classic nitrification, and thus it is likely that the extent of anammox-mediated ammonia oxidation in nature is governed by the concentration of O_2 in the system.

Before the anammox process was discovered, it was thought that ammonia was stable under anoxic conditions, but now we know otherwise. From an environmental standpoint, anammox is a very beneficial process in the treatment of wastewaters. The anoxic removal of ammonia/amines along with the production of gaseous nitrogen helps reduce the fixed nitrogen pollution of rivers and streams from treated wastewater disposal, thereby maintaining high water quality. Ecological studies have shown that organisms similar to *Brocadia* carry out anammox in marine sediments. This discovery has helped account for a significant fraction of ammonia that was known to be lost from marine environments but previously unaccounted for. Indeed, anammox is likely to occur in any anoxic environment in which ammonia and nitrite coexist.

20.13 MiniReview

Anoxic ammonia oxidation is called anammox and consumes both ammonia and nitrite, forming N_2. The anammox reaction occurs within a membrane-enclosed compartment within the cell, called the anammoxosome.

- In what fundamental ways does anammox differ from nitrification?
- Why are anammox reactions carried out in a special structure within the cell?
- What is the carbon source for anammox organisms?

IV NITROGEN FIXATION

We have just covered chemolithotrophic transformations of nitrogen. We conclude this chapter with a different consideration of nitrogen, not as an electron donor in energy metabolism but as a source of nitrogen for biosynthetic needs.

Our focus is on nitrogen gas, N_2. The utilization of N_2 as a source of cell nitrogen is called **nitrogen fixation**. The ability to fix N_2 frees an organism from dependence on fixed forms of nitrogen, such as ammonia or nitrate. Because fixed nitrogen is in high demand in microbial ecosystems, the capacity to fix N_2 confers a significant ecological advantage on those cells capable of the process. Moreover, certain forms of nitrogen fixation are of enormous agricultural importance, supporting the nitrogen needs of key crops, such as soybeans.

20.14 Nitrogenase and Nitrogen Fixation

Only certain prokaryotes can fix N_2, and an abbreviated list of nitrogen-fixing organisms is given in **Table 20.2**. Some nitrogen-fixing bacteria are *free-living* and require no host in order to carry out the process. By contrast, others are *symbiotic* and fix nitrogen only in association with certain plants (∞ Section 24.15). But it is the bacterium, not the plant, that fixes the N_2; no eukaryotic organisms are known that fix nitrogen. Many different types of prokaryotes can fix nitrogen, including several extremophilic species. For example, nitrogen fixation has been recorded at temperatures as low as 0°C and as high as 92°C, suggesting that few nitrogen-limited environments on Earth would be off limits to nitrogen fixing bacteria.

Table 20.2 Some nitrogen-fixing organisms[a]

Free-living aerobes/Facultative aerobes		
Chemoorganotrophs	**Phototrophs**	**Chemolithotrophs**
Azotobacter	Cyanobacteria	*Alcaligenes*
Azomonas		*Thiobacillus*
Agrobacterium		*Acidithiobacillus*
Klebsiella[b]		*Streptomyces thermoautotrophicus*
Beijerinckia		
Bacillus polymyxa		
Mycobacterium flavum		
Azospirillum lipoferum		
Citrobacter freundii		
Acetobactor diazotrophicus		
Methylomonas		
Methylococcus		
Methylosinus		
Pseudomonas		
Free-living anaerobes		
Chemoorganotrophs	**Phototrophs**	**Chemolithotrophs**[c]
Clostridium	*Chromatium*	*Methanosarcina*
Desulfovibrio	*Ectothiorhodospira*	*Methanococcus*
Desulfobacter	*Thiocapsa*	*Methanobacterium*
Desulfotomaculum	*Chlorobium*	*Methanospirillum*
	Chlorobaculum	*Mathanolobus*
	Rhodospirillum	*Methanocaldococcus*
	Rhodopseudomonas	
	Rhodomicrobium	
	Rhodopila	
	Rhodobacter	
	Heliobacterium	
	Heliobacillus	
	Heliophilum	
	Heliorestis	

Symbiotic	
Leguminous plants	**Nonleguminous plants**
Soybeans, peas, clover, locust, and so on, in association with a bacterium of the genus *Rhizobium, Bradyrhizobium, Sinorhizobium*, or *Azorhizobium*	*Alnus, Myrica, Ceanothus, Comptonia, Casuarina*; in association with actinomycetes of the genus *Frankia*; *Anabaena* (a cyanobacterium) with the water fern *Azolla*.

[a]For some genera listed, N_2 fixation occurs in only one or a few species.
[b]N_2 fixation occurs only under anoxic conditions.
[c]All are *Archaea*.

Nitrogenase

In nitrogen fixation N_2 is reduced to ammonia and the ammonia converted to organic form. The reduction is catalyzed by a large enzyme complex called **nitrogenase**. Nitrogenase consists of two distinct proteins, *dinitrogenase* and *dinitrogenase reductase*. Both proteins contain iron, and dinitrogenase contains molybdenum as well. The iron and molybdenum in dinitrogenase are contained within a cofactor called *FeMo-co* (**Figure 20.35**), and the actual reduction of N_2 occurs on this iron–molybdenum center. The composition of FeMo-co is $MoFe_7S_8 \cdot$ homocitrate (Figure 20.35).

Owing to the stability of the dinitrogen triple bond, N_2 is extremely inert; its activation is therefore a very energy-demanding process. Six electrons must be transferred to reduce N_2 to NH_3; the three successive reduction steps occur directly on nitrogenase with no free intermediates accumulating (**Figure 20.36**).

Nitrogen fixation is inhibited by oxygen because dinitrogenase reductase is rapidly and irreversibly inactivated by O_2; this is true even if this enzyme is isolated from aerobic nitrogen fixers. In aerobic nitrogen-fixing bacteria, N_2 is fixed in the presence of O_2 in cells but not in purified enzyme preparations. Nitrogenase in such organisms is protected from O_2 inactivation by one of several different mechanisms, including the rapid removal of O_2 by respiration; the production of O_2-retarding slime layers (**Figure 20.37**), or, in certain cyanobacteria, by compartmentalization of nitrogenase in a special type of cell, the heterocyst (∞ Section 16.7). In addition, although N_2 is not fixed in cell extracts exposed to oxygen, in aerobic nitrogen fixers such as *Azotobacter*, nitrogenase is protected from oxygen inactivation by complexing with a specific protein; this process is called conformational protection and is reversible. When oxygen is no longer at inhibitory levels, the conformationally protected nitrogenase can once again become active.

Electron Flow in Nitrogen Fixation

The sequence of electron transfer in nitrogenase is as follows: electron donor → dinitrogenase reductase → dinitrogenase → N_2, with NH_3 being the final product (Figure 20.36). The electrons for nitrogen reduction are transferred to dinitrogenase reductase from ferredoxin or flavodoxin, both of which are low-potential iron–sulfur proteins (∞ Section 5.11). In

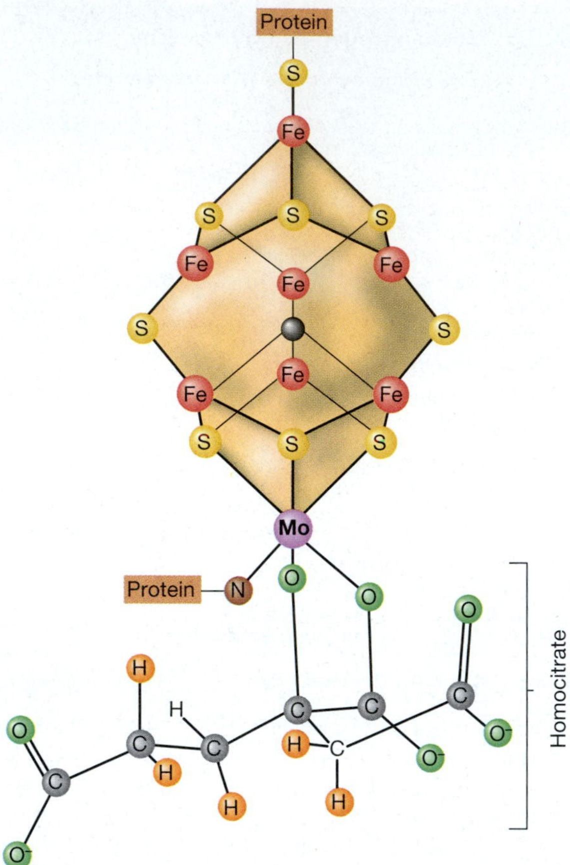

Figure 20.35 FeMo-co, the iron–molybdenum cofactor from nitrogenase. On the top is the Fe_7S_8 cube that binds to Mo along with oxygen atoms from homocitrate (bottom, all oxygen atoms shown in green) and N and S atoms from dinitrogenase. The central atom shown in black is unknown, but could be a carbon, oxygen, or nitrogen atom.

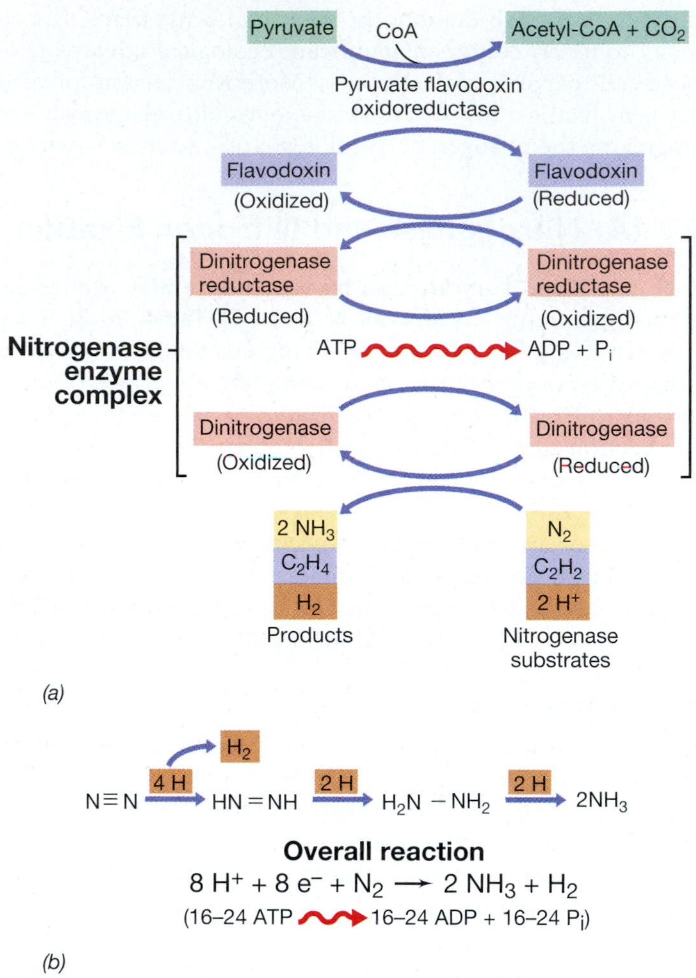

Figure 20.36 Nitrogenase function. *(a)* Shown are the steps in N_2 fixation starting from pyruvate. Electrons are supplied from dinitrogenase reductase to dinitrogenase one at a time, and each electron supplied is associated with the hydrolysis of 2–3 ATPs. *(b)* Hypothetical steps in N_2 reduction showing the H_2 evolution step and a summary of nitrogenase activity.

Clostridium pasteurianum ferredoxin is the electron donor and is reduced by the oxidation of pyruvate to acetyl-CoA + CO_2. In addition to reduced ferredoxin or flavodoxin, ATP is required for N_2 fixation. In each cycle of electron transfer, dinitrogenase reductase is reduced by ferredoxin or flavodoxin and binds two molecules of ATP. ATP binding alters the conformation of dinitrogenase reductase and lowers its reduction potential, allowing it to interact with dinitrogenase. Upon electron transfer to dinitrogenase, the ATP is hydrolyzed and dinitrogenase reductase dissociates from dinitrogenase and begins another cycle of reduction and ATP binding (Figure 20.36). When fully reduced, dinitrogenase reduces N_2 to NH_3, with the actual reduction occurring at the FeMo-co center (Figure 20.36).

Although only *six* electrons are necessary to reduce N_2 to two NH_3, *eight* electrons are actually consumed in the process, two electrons being lost as H_2 for each mole of N_2 reduced (Figure 20.36). The reason for this reducing power wastage is unknown, but it is clear that H_2 evolution is an intimate part of the reaction mechanism of nitrogenase and cannot be eliminated without inactivating nitrogenase.

Alternative Nitrogenases

Some nitrogen-fixing *Bacteria* and *Archaea* produce nonmolybdenum nitrogenases under conditions in which Mo is absent or limiting. These *alternative nitrogenases,* as they are called, contain either vanadium (V) or iron in place of Mo. Cofactors similar to FeMo-co are present in these alternative nitrogenases—FeVa-co in the vanadium nitrogenase and an iron–sulfur cluster resembling FeMo-co and FeVa-co but lacking both Mo and V in the iron nitrogenase. Alternative nitrogenases are not synthesized when sufficient molybdenum is present, as the molybdenum nitrogenase is the main nitrogenase in the cell. Instead, alternative nitrogenases seem to function as a "backup system" to support N_2 fixation when molybdenum is limiting in the habitat.

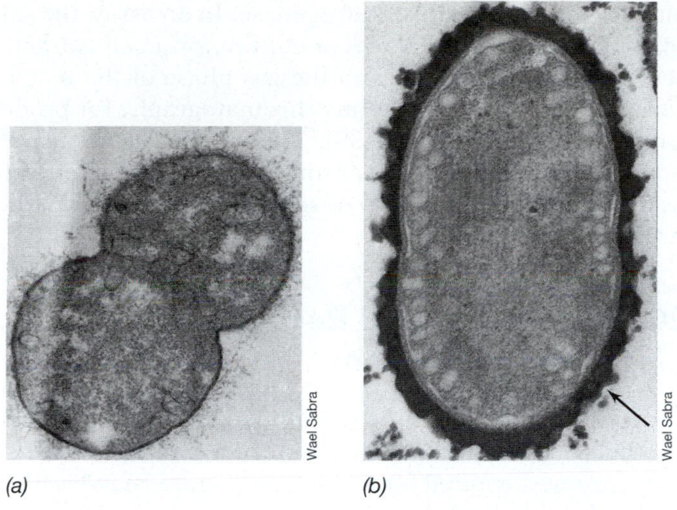

Figure 20.37 Induction of slime formation by O_2 in nitrogen-fixing cells of *Azotobacter vinelandii*. *(a)* Transmission electron micrograph of cells grown under micro-oxic conditions on 2.5% O_2; very little slime is evident. *(b)* Cells grown in air (21% O_2). Note the extensive darkly staining slime layer (arrow). The slime retards diffusion of O_2 into the cell, thus preventing nitrogenase inactivation by oxygen. A single cell of *A. vinelandii* is about 2 μm in diameter.

The Unique Nitrogenase of *Streptomyces thermoautotrophicus*

A structurally and functionally novel molybdenum nitrogenase is present in the streptomycete, *Streptomyces thermoautotrophicus*. This organism is a thermophilic (optimum temperature 65°C), filamentous member of the *Actinobacteria* (∞ Section 16.6). *S. thermoautotrophicus* is a hydrogen bacterium that uses carbon monoxide (CO) as an electron donor in energy metabolism and is also an autotroph.

S. thermoautotrophicus fixes nitrogen, and its nitrogenase contains Mo. But unlike classic Mo nitrogenase, the *S. thermoautotrophicus* nitrogenase is completely *insensitive* to O_2. The dinitrogenase component of the *S. thermoautotrophicus* nitrogenase, called *Str1*, contains three different polypeptides that show some structural similarity to dinitrogenase polypeptides from other nitrogen fixers. In contrast, the dinitrogenase reductase component, called *Str2*, shows no similarity to other dinitrogenase reductases. Str2 is, however, highly similar to manganese-containing enzymes called superoxide dismutases (∞ Section 6.18), and this is its role in this unusual nitrogenase (**Figure 20.38**).

There is a pattern to electron flow in the *S. thermoautotrophicus* nitrogenase system that mimics that of classical nitrogenase. Str2 supplies electrons to Str1 in the *S. thermoautotrophicus* nitrogenase. The source of the electrons is superoxide (O_2^-), and the O_2^- is formed from the reduction of O_2 by a CO dehydrogenase (Figure 20.38). Thus, in analogy to the pyruvate → flavodoxin → dinitrogenase reductase → dinitrogenase reaction sequence in classical nitrogen fixation (Figure 20.36), the *S. thermoautotrophicus* sequence is CO →

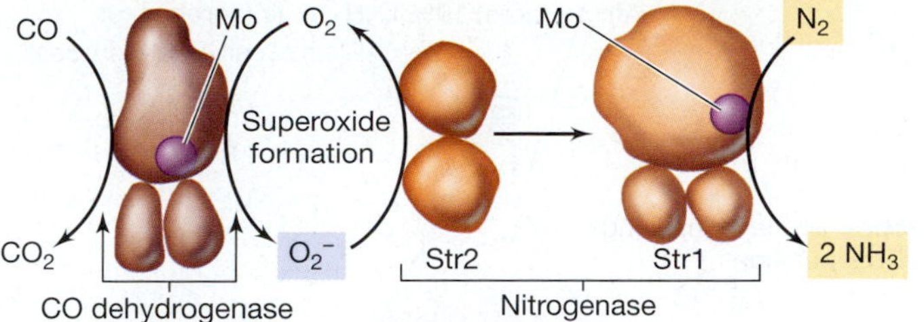

Figure 20.38 Reactions of nitrogen fixation in *Streptomyces thermoautotrophicus*. Although Str2 and Str1 are quite different from dinitrogenase reductase and dinitrogenase, they are functionally equivalent, respectively, to these proteins.

O_2^- → Str2 → Str1. And remarkably, instead of oxygen *inhibiting* nitrogenase (as it does in every classical nitrogenase that has ever been examined), oxygen is actually *required* in the reaction mechanism of the *S. thermoautotrophicus* nitrogenase.

Other unique properties of the *S. thermoautotrophicus* nitrogenase include the fact that the enzyme consumes less than half of the ATP of classical nitrogenases and does not reduce other triply bonded compounds, such as acetylene (see Figure 20.39). Collectively, these properties, especially that of oxygen insensitivity, lend hope to plant biotechnologists trying to engineer nitrogen fixation into crop plants such as corn (maize) that do not harbor known nitrogen-fixing symbionts. In plants (oxygenic phototrophs), of course, only an oxygen-insensitive nitrogenase could fix N_2.

Assaying Nitrogenase: Acetylene Reduction

Classic nitrogenases are not entirely specific for N_2 because they also reduce other triply bonded compounds, such as cyanide (CN^-) and acetylene ($HC{\equiv}CH$). The reduction of acetylene by nitrogenase is only a two-electron process, and *ethylene* ($H_2C{=}CH_2$) is produced. The reduction of acetylene to ethylene provides a simple and rapid method for measuring the activity of nitrogen-fixing systems by gas chromatography (**Figure 20.39**). This technique, known as the *acetylene reduction assay,* is widely used to detect and quantify nitrogen fixation.

Definitive proof for N_2 fixation is obtained using an isotope of nitrogen, ^{15}N, as a tracer. (^{15}N is not a radioisotope but a stable isotope. It is detected with a mass spectrometer.) The gas phase of a culture is enriched with $^{15}N_2$, and after an incubation period, the cells and medium are digested to release ammonia from all cellular nitrogenous compounds, and the ammonia produced is assayed for its ^{15}N content. If there has been a significant production of ^{15}N-labeled NH_3, it is proof of nitrogen fixation.

Although $^{15}N_2$ assimilation is occasionally used to demonstrate nitrogen fixation, the acetylene reduction method is a more rapid and sensitive method for measuring this process. Therefore, biological acetylene reduction is most frequently used for studying nitrogen fixation, and is taken as strong

UNIT 4

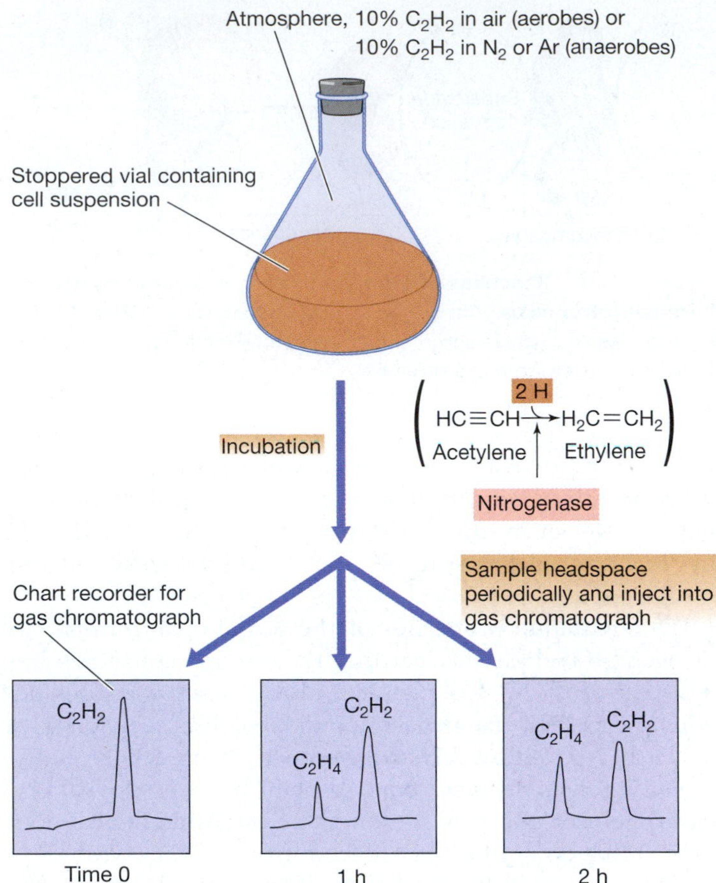

Figure 20.39 The acetylene reduction assay for nitrogenase activity. The results show no ethylene (C_2H_4) when the experiment begins (time 0), but increasing production of C_2H_4 as the assay proceeds. Note how as C_2H_4 is produced, acetylene (C_2H_2) is consumed. If the vial contained an enzyme extract, conditions would be anoxic, even if nitrogenase was from an aerobic bacterium.

evidence for the activity of nitrogenase. In an assay, the sample, which may be soil, water, a culture, or a cell extract, is incubated with acetylene, and the gas phase of the reaction mixture is later analyzed by gas chromatography for production of ethylene (Figure 20.39). This method is far simpler and faster than other methods and can easily be adapted for use in ecological studies of N_2-fixing bacteria directly in their habitats.

20.15 Genetics and Regulation of N_2 Fixation

Because the process of N_2 fixation is highly energy demanding, the synthesis and activity of nitrogenase and the many other enzymes required for N_2 fixation (Figure 20.36) is highly regulated.

Genetics of Nitrogen Fixation

The genes for dinitrogenase and dinitrogenase reductase in *Klebsiella pneumoniae*, a well-studied nitrogen fixer, are part of a complex regulon (a large network of operons, Section 9.9) called the *nif regulon*. The *K. pneumoniae nif* regulon spans 24 kbp of DNA and contains 20 genes arranged in several transcriptional units (**Figure 20.40**). In addition to nitrogenase structural genes, the genes for FeMo-co synthesis, genes controlling the electron transport proteins, and a number of regulatory genes are also present in the *nif* regulon.

Dinitrogenase is composed of two subunits, α (the product of *nifD*) and β (the product of *nifK*), each of which is present in two copies in the nitrogenase enzyme complex.

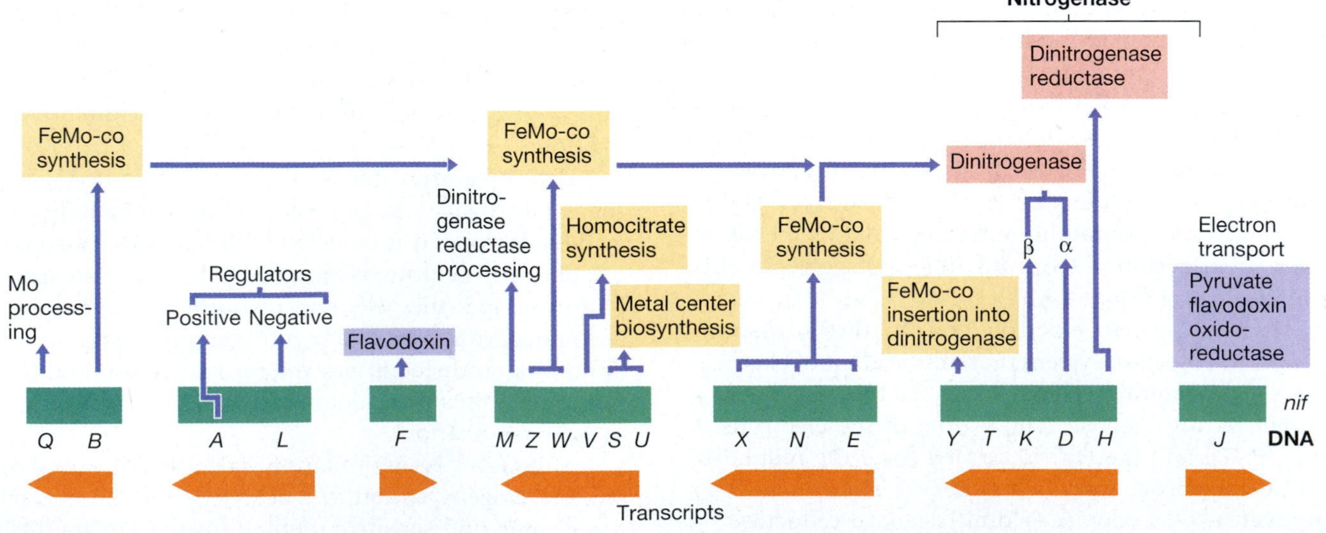

Figure 20.40 The *nif* regulon in *Klebsiella pneumoniae*, the best-studied nitrogen-fixing bacterium. The function of the *nifT* gene product is unknown. The mRNA transcripts are shown below the genes; arrows indicate the direction of transcription. Proteins that catalyze FeMo-co synthesis are shown in yellow. Other colors match those of Figure 20.36.

Dinitrogenase reductase is a protein dimer consisting of two identical subunits, the product of *nifH*. FeMo-co is synthesized by enzymes encoded by several genes, including *nifN, V, Z, W, E*, and *B*, as well as *nif Q*, which encodes a molybdenum processing enzyme. The *nifA* gene encodes a positive regulatory protein that activates transcription of other *nif* genes (Figure 20.40).

Nitrogenase is a highly conserved protein, and the *nifHDK* genes that encode it have been used as molecular probes to screen DNA from various prokaryotes for homologous genes, signaling their ability to fix nitrogen. In all nitrogen fixers examined (except for the highly unusual *Streptomyces thermoautotrophicus*, Section 20.14), *nifHDK*-like genes are present, suggesting that the genetic requirements for nitrogenase are rather specific. Alternative nitrogenases (see Section 20.14) are encoded by their own sets of genes, *vnfHDK* for the vanadium system and *anfHDK* for the iron-only system, but these genes show significant sequence similarity to *nifHDK*.

Regulation of Nitrogenase Synthesis

Nitrogenase is subject to strict regulatory controls. Nitrogen fixation is repressed by O_2 and by fixed forms of nitrogen, including NH_3, NO_3^-, and certain amino acids. A major part of this regulation is at the level of transcription, as shown in Figure 20.40. Transcription of *nif* structural genes is activated by the NifA protein (positive regulation, ∞ Section 9.4). By contrast, NifL is a negative regulator of *nif* gene expression and contains a molecule of FAD (recall that FAD is a redox coenzyme for flavoproteins, ∞ Section 5.11) that is involved in O_2-sensing. In the presence of sufficient O_2, NifL represses synthesis of other *nif* genes, which blocks synthesis of the oxygen-labile nitrogenase.

Ammonia represses N_2 fixation through a second protein, called *NtrC*, whose activity is regulated by the nitrogen status of the cell. When ammonia is limiting, NtrC is active and promotes transcription of *nifA*. This produces NifA, the nitrogen fixation activator protein, and *nif* transcription begins.

The ammonia produced by nitrogenase does not block enzyme synthesis because it is incorporated into amino acids (∞ Section 5.16) and used in biosynthesis as soon as it is made. But when ammonia is in excess (as in natural environments or culture media high in ammonia), nitrogenase synthesis is repressed. In this way, ATP is not wasted in making ammonia when it is already available in ample amounts.

Regulation of Nitrogenase Activity

Besides repressing the *synthesis* of nitrogenase, nitrogenase *activity* is also regulated by ammonia in many nitrogen-fixing bacteria; that is, nitrogenase proteins that already exist in the cell can be turned on or off. Although there are likely different mechanisms for shutting off the activity of nitrogenase, they all seem to function by shutting down electron flow to dinitrogenase in response to excess ammonia, thus making nitrogenase inactive.

The mechanism of nitrogenase shut down by ammonia is known in many phototrophic and some chemotrophic nitrogen-fixing *Proteobacteria* and is called the *ammonia switch-off effect*. In this mechanism, excess ammonia causes a covalent modification of dinitrogenase reductase, which results in a loss of enzyme activity. When ammonia becomes limiting, the modified dinitrogenase reductase is converted to its active form and N_2 fixation can resume. Ammonia switch-off is thus a rapid and reversible method of controlling ATP consumption by nitrogenase.

Ammonia switch-off has also been observed in nitrogen-fixing *Archaea*. In *Methanococcus* species, for example, ammonia quickly inhibits the activity of nitrogenase. Here, however, covalent modification of dinitrogenase reductase is not involved. Instead, it appears that an ammonia-sensing protein exists in the cell that can bind to nitrogenase or in some other way inactivate it when ammonia is in excess in the cell. Interestingly, the ammonia-sensing system appears to function not by detecting ammonia directly, but by detecting a shortage in the carbon compound α-ketoglutarate, a direct precursor of the amino acid glutamate (∞ Section 5.16). In this model for nitrogenase regulation, a shortage of α-ketoglutarate is sensed as an excess of glutamate, and thus nitrogenase activity would be unnecessary because it would produce more ammonia and lead to the synthesis of more glutamate (∞ Section 9.8 and Figure 9.16). This interesting regulatory system also controls nitrogenase activity in some species of *Bacteria*, in particular nitrogen-fixing species of the fermentative anaerobe *Clostridium* (∞ Section 16.2).

UNIT 4

20.14–20.15 MiniReview

Nitrogen fixation is the reduction of N_2 to NH_3 and requires the enzyme complex nitrogenase. Most nitrogenases contain molybdenum or vanadium plus iron as metal cofactors, and the process of nitrogen fixation is highly energy demanding. Nitrogenase and most associated regulatory proteins are encoded by the *nif* regulon. Certain substances that are structurally similar to N_2, such as acetylene or cyanide, are also reduced by nitrogenase. Nitrogenase and nitrogen fixation are both highly regulated, with O_2 and NH_3 being the two main regulatory effectors.

- Write a balanced equation for the reaction carried out by the enzyme nitrogenase.
- What is FeMo-co and what metals does it contain?
- How is C_2H_2 useful for studies of nitrogen fixation?
- What chemical and physical factors affect the synthesis or activity of nitrogenase? How does the *Streptomyces thermoautotrophicus* nitrogenase system differ in this regard?
- How do alternative nitrogenases differ from classical nitrogenase?

Review of Key Terms

Anammox anoxic ammonia oxidation

Anoxygenic photosynthesis photosynthesis in which O_2 is not produced

Antenna pigments light-harvesting chlorophylls or bacteriochlorophylls in photocomplexes that funnel energy to the reaction center

Autotroph an organism that uses CO_2 as sole carbon source

Bacteriochlorophyll the chlorophyll pigment of anoxygenic phototrophs

Calvin cycle the biochemical pathway for CO_2 fixation in many autotrophic organisms

Carboxysomes crystalline inclusions of RubisCO

Carotenoid a hydrophobic accessory pigment present along with chlorophyll in photosynthetic membranes

Chemolithotroph a microorganism that oxidizes inorganic compounds as electron donors in energy metabolism

Chlorophyll a light-sensitive, Mg-containing porphyrin of phototrophic organisms that initiates the process of photophosphorylation

Chlorosome cigar-shaped structures present in the periphery of cells of green sulfur and green nonsulfur bacteria and that contain the antenna bacteriochlorophylls (*c*, *d*, or *e*)

Hydrogenase an enzyme, widely distributed in anaerobic microorganisms, capable of taking up or evolving H_2

Hydroxypropionate pathway an autotrophic pathway found in *Chloroflexus* and a few *Archaea*

Mixotroph an organism in which an inorganic compound serves as electron donor in energy metabolism and organic compounds serve as carbon source

Nitrification the microbial conversion of NH_3 to NO_3^-

Nitrogenase an enzyme capable of reducing N_2 to NH_3 in the process of nitrogen fixation

Nitrogen fixation the biological reduction of N_2 to NH_3 by nitrogenase

Oxygenic photosynthesis photosynthesis carried out by cyanobacteria and green plants in which O_2 is evolved

Photophosphorylation the production of ATP in photosynthesis

Photosynthesis the series of reactions in which ATP is synthesized by light-driven reactions and CO_2 is fixed into cell material

Phototroph an organism that uses light as an energy source

Phycobiliprotein the accessory pigment complex in cyanobacteria that contains phycocyanin and allophycocyanin or phycoerythrin coupled to proteins

Phycobilisome aggregates of phycobiliproteins

Reaction center a photosynthetic complex containing chlorophyll or bacteriochlorophyll and several other components within which occurs the initial electron transfer reactions of photosynthetic electron flow

Reverse citric acid cycle a mechanism for autotrophy in green sulfur bacteria and a few other phototrophs

Reverse electron transport the energy-dependent movement of electrons against the thermodynamic gradient to form a strong reductant from a weaker electron donor

RubisCO the acronym for ribulose bisphosphate carboxylase, a key enzyme of the Calvin cycle

Thylakoids membrane stacks in cyanobacteria or in the chloroplast of eukaryotic phototrophs

Review Questions

1. What are the major differences between oxygenic and anoxygenic phototrophs (Sections 20.1–20.5)?
2. What are the functions of light-harvesting and reaction center chlorophylls? Why would a mutant incapable of making light-harvesting chlorophylls (such mutants can be readily isolated in the laboratory) probably not be a successful competitor in nature (Section 20.2)?
3. Where are the photosynthetic pigments located in a purple bacterium? A cyanobacterium? A green alga? Considering the function of chlorophyll pigments, why can't they be located elsewhere in the cell, for example, in the cytoplasm or in the cell wall (Section 20.3)?
4. What accessory pigments are present in phototrophs and what are their functions (Section 20.3)?
5. How does light result in ATP production in an anoxygenic phototroph? In what ways are photosynthetic and respiratory electron flow similar? In what ways do they differ (Section 20.4)?
6. How is reducing power for autotrophic growth in a purple bacterium made? In a cyanobacterium (Section 20.4)?
7. How does the reduction potential of chlorophyll *a* in PS I and PS II differ? Why must the reduction potential of photosystem II chlorophyll be so highly electropositive (Section 20.5)?
8. What two enzymes are unique to organisms that carry out the Calvin cycle? What reactions do these enzymes carry out? What would be the consequences if a mutant arose that lacked either of these enzymes (Section 20.6)?
9. What organisms employ the hydroxypropionate or reverse citric acid cycles as autotrophic pathways (Section 20.7)?
10. Compare and contrast the utilization of H_2S by a purple phototrophic bacterium and by a colorless sulfur bacterium such as *Beggiatoa*. What role does H_2S play in the metabolism of each organism (Section 20.8)?
11. Which inorganic electron donors are used by the organisms *Ralstonia* and *Thiobacillus* (Sections 20.9 and 20.10)?
12. What is unusual about the environment of *Acidithiobacillus ferrooxidans* that affects the energetics of its metabolism (Section 20.11)?

13. Contrast classical nitrification with anammox in terms of oxygen requirements, organisms involved, and the need for monooxygenases (Sections 20.12 and 20.13).

14. Write out the reaction catalyzed by the enzyme *nitrogenase*. How many electrons are required in this reaction? How many are actually used? Explain (Section 20.14).

15. How does the *Streptomyces thermoautotrophicus* nitrogenase differ from that of *Azotobacter* (Section 20.14)?

16. Distinguish between ammonia repression of nitrogenase and the ammonia "switch-off" effect. What is happening under each condition (Section 20.15)?

Application Questions

1. Compare and contrast the absorption spectrum of chlorophyll *a* and bacteriochlorophyll *a*. What wavelengths are preferentially absorbed by each pigment and how do the absorption properties of these molecules compare with the regions of the spectrum visible to our eye? Why are most plants green?

2. The growth rate of the phototrophic purple bacterium *Rhodobacter* is about twice as fast when the organism is grown *phototrophically* in a medium containing malate as carbon source as when it is grown with CO_2 as carbon source (with H_2 as electron donor). Discuss the reasons why this is true and list the nutritional class in which we would place *Rhodobacter* when growing under each of the two different conditions.

3. Although physiologically distinct, chemolithotrophs and chemoorganotrophs share a number of features with respect to the production of ATP. Discuss these common features along with reasons why the growth yield (grams of cells per mole of substrate) of a chemoorganotroph respiring glucose is so much higher than for a chemolithotroph respiring sulfur.

4. Employing biotechnology, you would like to genetically engineer corn (maize) to fix nitrogen. Discuss what type of nitrogenase you would try to engineer into the corn plant and why this would be the most suitable enzyme for the purpose.

21

Metabolic Diversity: Catabolism of Organic Compounds

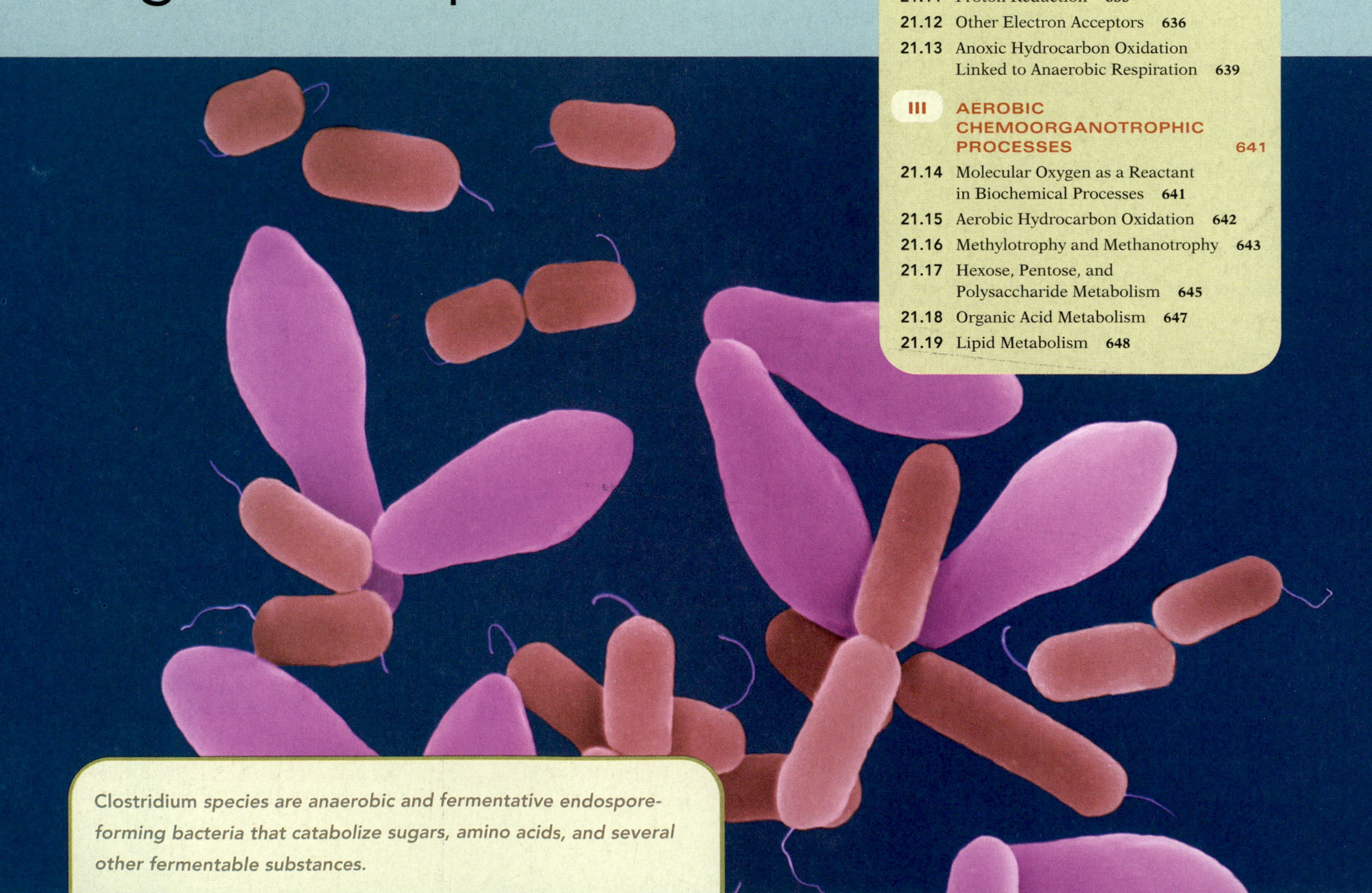

Clostridium species are anaerobic and fermentative endospore-forming bacteria that catabolize sugars, amino acids, and several other fermentable substances.

In Chapter 20 we considered bioenergetic processes other than those having organic compounds as reactants. In this chapter we focus on *organic compounds* as electron donors and the many ways in which chemoorganotrophic microorganisms conserve energy. A major focus is on anaerobic forms of metabolism, as strategies for growth under anoxic conditions are a hallmark of prokaryotic diversity. We end the chapter with a consideration of the aerobic catabolism of key organic compounds, primarily monomers released from the degradation of macromolecules.

I FERMENTATIONS

Two mechanisms for the catabolism of organic compounds are *fermentation* and *respiration*. These processes differ fundamentally in terms of redox considerations and mechanism of ATP synthesis. In respiration, whether aerobic or anaerobic, exogenous electron acceptors are present to accept electrons generated from the oxidation of electron donors. In fermentation, this is not the case. Thus in respiration but not fermentation we will see a common theme of electron transport and generation of a proton motive force. In addition, compared with respirations, fermentations are typically energetically marginal. However, we will see that a little free energy can go a long way and that fermentative diversity in the prokaryotic world is staggering.

21.1 Fermentations: Energetic and Redox Considerations

Many microbial habitats are **anoxic** (oxygen-free). In such environments, decomposition of organic material occurs anaerobically. If adequate supplies of electron acceptors such as SO_4^{2-}, NO_3^-, Fe^{3+}, and others to be considered later are not available in anoxic microbial habitats, organic compounds are catabolized by **fermentation** (**Figure 21.1**). We discussed the overall process of fermentation in Sections 5.9 and 5.10. There we saw how fermentation is an internally balanced redox process in which the fermentable substrate becomes both oxidized and reduced.

An organism faces two major problems if it is to catabolize organic compounds in energy-yielding metabolism: (1) energy conservation and (2) redox balance. In fermentation, ATP is typically synthesized by *substrate-level phosphorylation*. This is the mechanism in which energy-rich phosphate bonds from phosphorylated organic intermediates are transferred directly to ADP to form ATP (∞ Section 5.9). The second problem, redox balance, is solved by the production and subsequent excretion of fermentation products generated from the original substrate (**Figures** 21.1 and **21.2**).

Energy-Rich Compounds and Substrate-Level Phosphorylation

Energy can be conserved by substrate-level phosphorylation in many different but highly related ways. However, central to this mechanism of ATP synthesis is the production of *energy-rich compounds*. These are organic compounds that contain an energy-rich phosphate bond or a molecule of coenzyme-A; the hydrolysis of either of these is highly exergonic. **Table 21.1** lists some energy-rich intermediates formed during biochemical processes. The hydrolysis of most of the compounds listed can be coupled to ATP synthesis (−31.8 kJ/mol). Thus, if an organism can form one of these compounds during fermentative metabolism, it can make ATP by substrate-level phosphorylation. Several pathways for the anaerobic breakdown of fermentable substances to energy-rich intermediates are summarized in Figure 21.2.

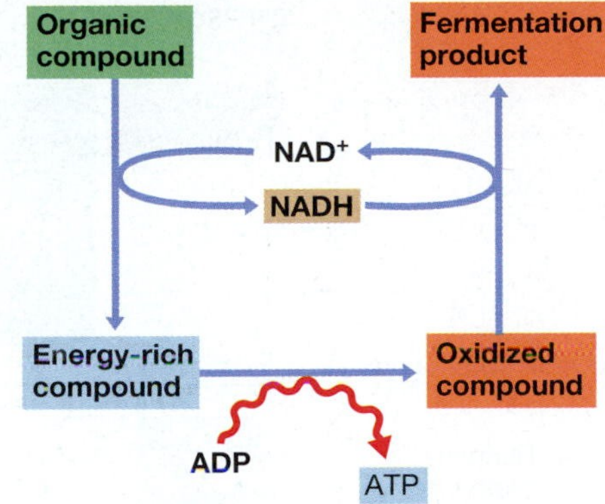

Figure 21.1 The essentials of fermentation. The fermentation product is excreted from the cell, and only a relatively small amount of the original organic compound is used for biosynthesis.

Oxidation–Reduction Balance, H_2, and Acetate Production

In any fermentation there must be oxidation–reduction (redox) balance; the total number of electrons in the products on the right side of the equation must balance the number in the substrates on the left side of the equation. When fermentations are studied experimentally in the laboratory, it is possible to calculate a fermentation balance to ensure that no products have been missed. The fermentation balance can also be calculated theoretically from the oxidation states of the substrates and products (Appendix 1 explains the procedure for calculating oxidation states).

In a number of fermentations, electron balance is maintained by the production of molecular hydrogen, H_2. The production of H_2 is associated with the activity of an iron-sulfur protein called *ferredoxin*, a very low potential electron carrier. The transfer of electrons from ferredoxin to H^+ is catalyzed by the enzyme **hydrogenase**, as illustrated in **Figure 21.3**. No energy is conserved from this reaction, and thus H_2 production functions as a "safety valve" to maintain redox balance.

Many anaerobic bacteria produce acetate as a major or minor fermentation product. The production of acetate and certain other fatty acids (Table 21.1) is energy conserving because it allows the organism to make ATP by substrate-level phosphorylation. The key intermediate generated in acetate production is acetyl-CoA (Table 21.1 and Figure 21.2), an energy-rich compound. Acetyl-CoA can be converted to acetylphosphate (Table 21.1) and the phosphate group of

UNIT 4

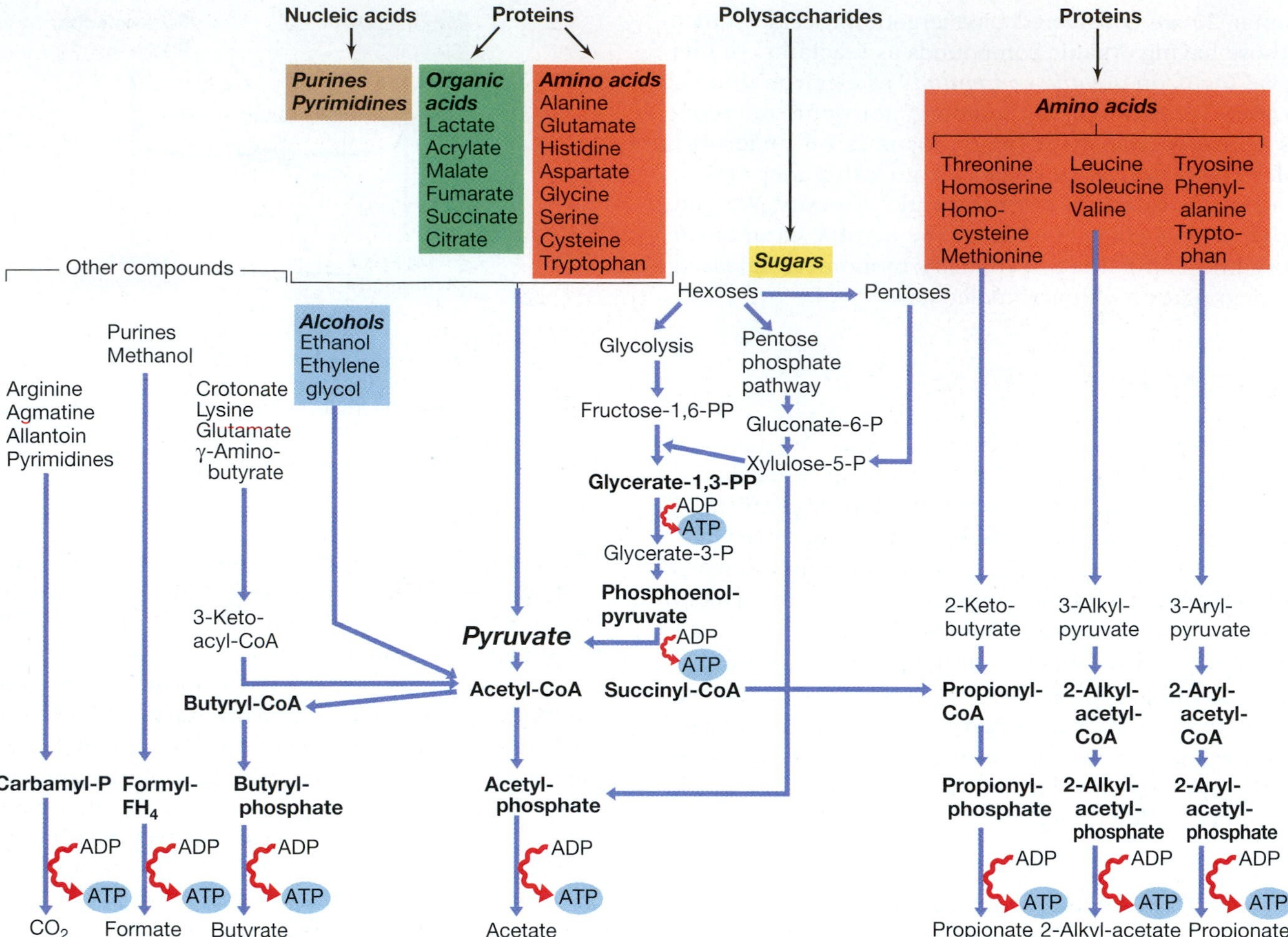

Figure 21.2 Routes of the anaerobic breakdown of major fermentable substances. The sites of substrate-level phosphorylation are shown. Energy-rich CoA derivatives and other energy-rich compounds are highlighted in bold. Italicized bold terms are group names. Refer to Table 21.1 if needed.

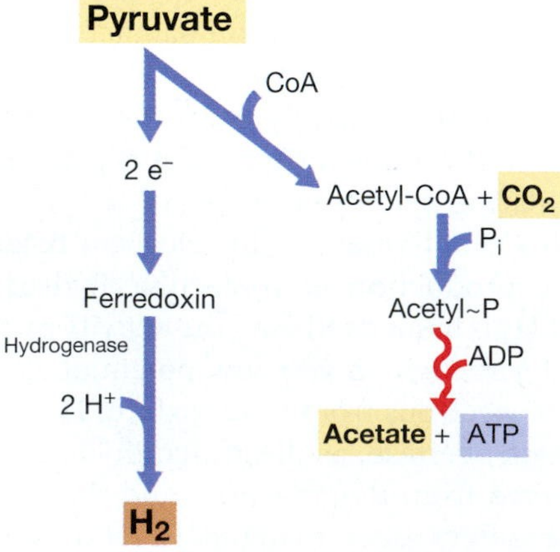

Figure 21.3 Production of molecular hydrogen (H_2) and acetate from pyruvate. Note how in the production of acetate, ATP is synthesized from the hydrolysis of the energy-rich intermediate, acetylphosphate (Table 21.1).

acetylphosphate subsequently transferred to ADP by the enzyme acetate kinase, yielding ATP. One of the main precursors of acetyl-CoA is pyruvate, a major product of glycolysis. The conversion of pyruvate to acetyl-CoA is a key oxidation reaction, and the electrons generated must either be used to form fermentation products or released as H_2 (Figure 21.3).

21.1 MiniReview

In the absence of external electron acceptors, organic compounds can be catabolized anaerobically only by fermentation. Only certain compounds are inherently fermentable, and a requirement for most fermentations is that an energy-rich organic intermediate be generated that can yield ATP by substrate-level phosphorylation. Redox balance must also be achieved in fermentations, and H_2 production is one means of disposing of excess electrons.

- What is substrate-level phosphorylation?
- Why is acetate formation in fermentation energetically beneficial?

Table 21.1 Energy-rich compounds involved in substrate-level phosphorylation[a]

Compound	*Free energy of hydrolysis, $\Delta G^{0\prime}$ (kJ/mol)*[b]
Acetyl-CoA	−35.7
Propionyl-CoA	−35.6
Butyryl-CoA	−35.6
Caproyl-CoA	−35.6
Succinyl-CoA	−35.1
Acetylphosphate	−44.8
Butyrylphosphate	−44.8
1,3-Bisphosphoglycerate	−51.9
Carbamyl phosphate	−39.3
Phosphoenolpyruvate	−51.6
Adenosine-phosphosulfate (APS)	−88
N^{10}-formyltetrahydrofolate	−23.4
Energy of hydrolysis of ATP (ATP → ADP + P_i)	−31.8

[a]Data from Thauer, R. K., K. Jungermann, and K. Decker, 1977. Energy conservation in chemotrophic anaerobic bacteria. *Bacteriol. Rev.* 41: 100–180.

[b]The $\Delta G^{0\prime}$ values shown here are for "standard conditions," which are not necessarily those of cells. Including heat loss, the energy costs of making an ATP are more like 60 kJ than 32 kJ, and the energy of hydrolysis of the energy-rich compounds shown here is thus likely higher. But for simplicity and comparative purposes, the values in this table will be taken as the actual energy released per reaction.

21.2 Fermentative Diversity: Lactic and Mixed-Acid Fermentations

Many different fermentations are known and are classified by either the substrate fermented or the fermentation products formed. **Table 21.2** summarizes some of the main types of fermentations classified on the basis of products formed. Note some of the broad categories, such as alcohol, lactic acid, propionic acid, mixed acid, butyric acid, and acetogenic acid. By contrast, a number of fermentations are classified on the basis of the substrate fermented rather than the fermentation product. For instance, some of the endospore-forming anaerobic bacteria (genus *Clostridium*) ferment amino acids, whereas others ferment purines and pyrimidines. Other anaerobes ferment aromatic compounds (**Table 21.3**). Clearly, a wide variety of organic compounds can be fermented.

Certain fermentations are carried out by only a very restricted group of anaerobes; in some cases this may be only a single known bacterium. A few examples are listed in Table 21.3. Many of these bacteria can be considered metabolic specialists, having evolved biochemical capabilities to catabolize a substrate or substrates not catabolized by other bacteria. However, for all fermentations, catabolism of the substrates requires that the organism be able to synthesize an energy-rich intermediate of the type listed in Table 21.1, which conserves some of the energy released as ATP.

We now consider two very common fermentations of sugars in which lactic acid is a major product.

Lactic Acid Fermentation

The lactic acid bacteria are gram-positive organisms that produce lactic acid as a major or sole fermentation product (∞ Section 16.1). Two fermentative patterns are observed. One, called **homofermentative**, yields a single fermentation product, lactic acid. The other, called **heterofermentative**, yields products in addition to lactate, mainly ethanol plus CO_2.

Figure 21.4 summarizes pathways for the fermentation of glucose by homofermentative and heterofermentative lactic acid bacteria. The differences observed in the fermentation patterns can be traced to the presence or absence of the enzyme *aldolase*, a key enzyme of glycolysis (∞ Figure 5.15). Homofermentative lactic acid bacteria contain aldolase and produce two lactates from glucose by the glycolytic pathway

Table 21.2 Common bacterial fermentations and some of the organisms carrying them out

Type	*Reaction*	*Organisms*
Alcoholic	Hexose → 2 Ethanol + 2 CO_2	Yeast, *Zymomonas*
Homolactic	Hexose → 2 $Lactate^-$ + 2 H^+	*Streptococcus*, some *Lactobacillus*
Heterolactic	Hexose → $Lactate^-$ + Ethanol + CO_2 + H^+	*Leuconostoc*, some *Lactobacillus*
Propionic acid	3 $Lactate^-$ → 2 $Propionate^-$ + $Acetate^-$ + CO_2 + H_2O	*Propionibacterium, Clostridium propionicum*
Mixed acid[a]	Hexose → Ethanol + 2,3-Butanediol + $Succinate^{2-}$ + $Lactate^-$ + $Acetate^-$ + $Formate^-$ + H_2 + CO_2	Enteric bacteria[b] *Escherichia, Salmonella, Shigella, Klebsiella, Enterobacter*
Butyric acid[b]	Hexose → $Butyrate^-$ + 2 H_2 + 2 CO_2 + H^+	*Clostridium butyricum*
Butanol[b]	2 Hexose → Butanol + Acetone + 5 CO_2 + 4 H_2	*Clostridium acetobutylicum*
Caproate/Butyrate	6 Ethanol + 3 $Acetate^-$ → 3 $Butyrate^-$ + $Caproate^-$ + 2 H_2 + 4 H_2O + H^+	*Clostridium kluyveri*
Acetogenic	Fructose → 3 $Acetate^-$ + 3 H^+	*Clostridium aceticum*

[a]Not all organisms produce all products. In particular, butanediol production is limited to only certain enteric bacteria. Reaction not balanced.

[b]Stoichiometry shows major products. Other products include some acetate and a small amount of ethanol (butanol fermention only).

UNIT 4

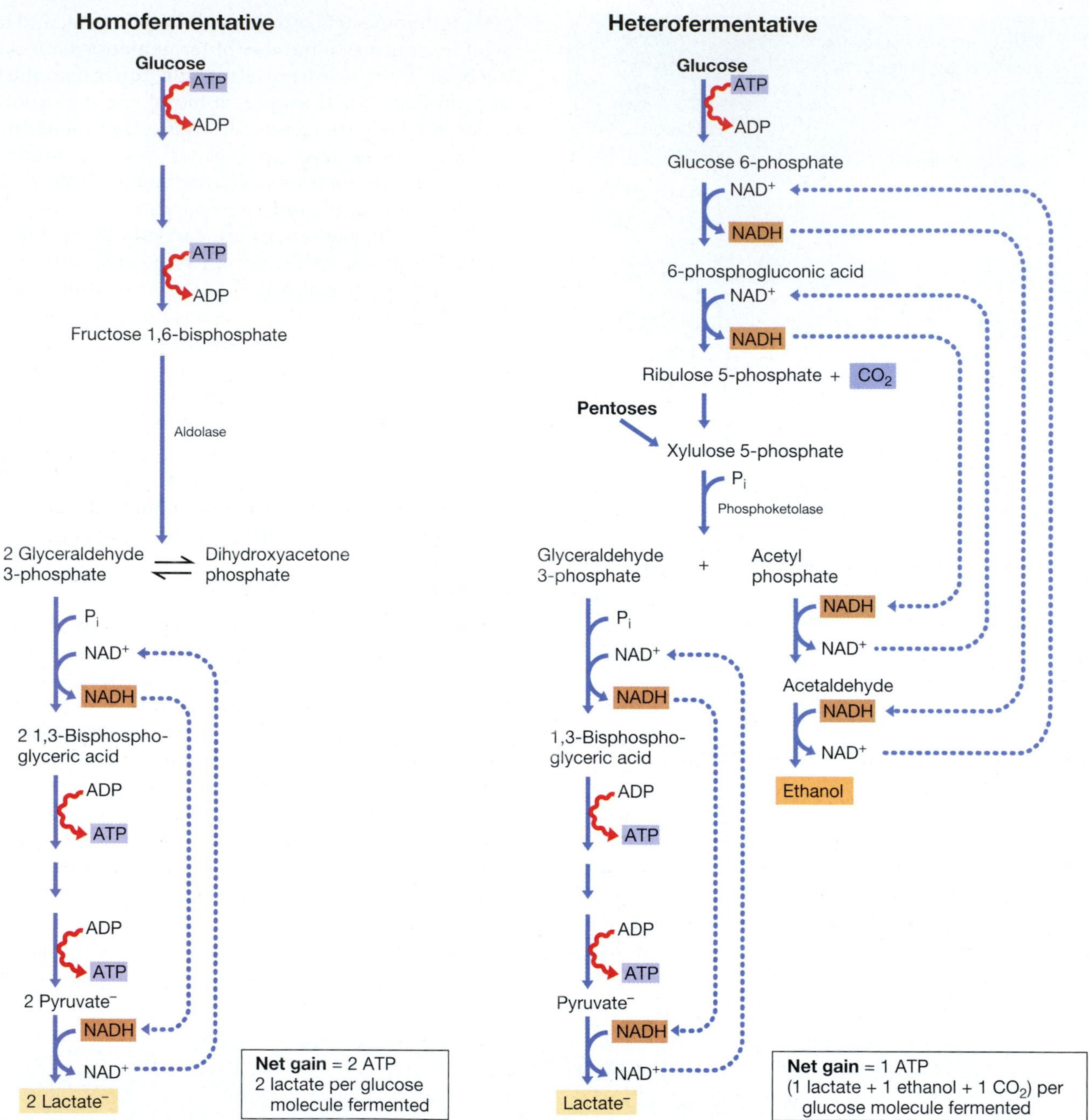

Figure 21.4 The fermentation of glucose in homofermentative and heterofermentative lactic acid bacteria. Note that no ATP is made in reactions leading to ethanol formation in heterofermentative organisms.

(Figure 21.4*a*). The heterofermenters lack aldolase and thus cannot break down fructose bisphosphate to triose phosphate. Instead, they oxidize glucose 6-phosphate to 6-phosphogluconate and then decarboxylate this to pentose phosphate. The pentose phosphate is converted to triose phosphate and acetylphosphate by the key enzyme *phosphoketolase* (Figure 21.4*b*). The early steps in catabolism by heterofermentative lactic acid bacteria are those of the pentose phosphate pathway (see Figure 21.39).

In heterofermenters, triose phosphate is converted ultimately to lactic acid with the production of ATP (Figure 21.4). However, to achieve redox balance, the acetylphosphate produced is reduced by NADH (generated during the production of pentose phosphate) and converted to ethanol. This occurs without ATP synthesis because the energy-rich CoA bond is lost during this reduction. Because of this, heterofermenters produce only *one* ATP/glucose instead of the *two* ATP/glucose produced by homofermenters. In addition, because heterofermenters decarboxylate 6-phosphogluconate, they produce CO_2 as a fermentation product; homofermenters do not produce CO_2. Thus a simple way of detecting a heterofermenter is to observe for the production of CO_2 in laboratory cultures.

Table 21.3 Some unusual bacterial fermentations

Type	*Reaction*	*Organisms*
Acetylene	$2\ C_2H_2 + 3\ H_2O \rightarrow$ Ethanol + Acetate$^-$ + H^+	*Pelobacter acetylenicus*
Glycerol	4 Glycerol + 2 $HCO_3^- \rightarrow$ 7 Acetate$^-$ + 5 H^+ + 4 H_2O	*Acetobacterium* spp.
Resorcinol (aromatic)	$2\ C_6H_4(OH)_2 + 6\ H_2O \rightarrow$ 4 Acetate$^-$ + Butyrate$^-$ + 5 H^+	*Clostridium* spp.
Phloroglucinol (aromatic)	$C_6H_6O_3 + 3\ H_2O \rightarrow$ 3 Acetate$^-$ + 3 H^+	*Pelobacter massiliensis* *Pelobacter acidigallici*
Putrescine	$10\ C_4H_{12}N_2 + 26\ H_2O \rightarrow$ 6 Acetate$^-$ + 7 Butyrate$^-$ + 20 NH_4^+ + 16 H_2 + 13 H^+	Unclassified gram-positive nonsporing anaerobes
Citrate	Citrate^{3-} + 2 $H_2O \rightarrow$ Formate$^-$ + 2 Acetate$^-$ + HCO_3^- + H^+	*Bacteroides* spp.
Aconitate	Aconitate^{3-} + H^+ + 2 $H_2O \rightarrow$ 2 CO_2 + 2 Acetate$^-$ + H_2	*Acidaminococcus fermentans*
Glyoxylate	4 Glyoxylate$^-$ + 3 H^+ + 3 $H_2O \rightarrow$ 6 CO_2 + 5 H_2 + Glycolate$^-$	Unclassified gram-negative bacterium
Benzoate	2 Benzoate$^- \rightarrow$ Cyclohexane carboxylate$^-$ + 3 Acetate$^-$ + HCO_3^- + 3 H^+	*Syntrophus aciditrophicus*

Entner–Doudoroff Pathway

A variant of the glycolytic pathway, called the *Entner–Doudoroff pathway,* is a widely distributed pathway for sugar catabolism in bacteria, especially among species of the pseudomonad group. In this pathway, glucose-6-phosphate is oxidized to 6-phosphogluconic acid and NADPH; the 6-phosphogluconic acid is dehydrated and split into pyruvate and glyceraldehyde 3-phosphate (G-3-P), a key intermediate of the glycolytic pathway. G-3-P is then catabolized as in glycolysis, generating NADH and 2 ATP, and used as an electron acceptor to balance redox reactions (Figure 21.4*a*).

Interestingly, because pyruvate is formed directly in the Entner–Doudoroff pathway and cannot yield ATP as can G-3-P (Figure 21.4), the Entner–Doudoroff pathway yields only half the ATP of the glycolytic pathway. Organisms using the Entner–Doudoroff pathway therefore share this physiological characteristic with heterofermentative lactic acid bacteria that also use a variant of the glycolytic pathway (Figure 21.4*b*). *Zymomonas,* an obligately fermentative pseudomonad, and *Pseudomonas,* a nonfermentative bacterium, are major genera that employ the Entner–Doudoroff pathway (∞ Section 15.7).

Mixed-Acid Fermentations

In the mixed-acid fermentation, characteristic of enteric bacteria (∞ Section 15.11), three diffferent acids are formed in significant amounts from the fermentation of glucose or other sugars—*acetic, lactic,* and *succinic*. Ethanol, CO_2, and H_2 are also formed in various amounts. Glycolysis is the pathway used by mixed-acid fermenters, such as *Escherichia coli,* and we outlined the steps in that pathway in Figure 5.15.

Some enteric bacteria produce acidic products in lower amounts than *E. coli* and balance their fermentations by producing larger amounts of neutral products. One key neutral product is the four-carbon alcohol *butanediol*. In this variation of the mixed-acid fermentation, butanediol, ethanol, CO_2, and H_2 are the main products observed (**Figure 21.5**). In the mixed-acid fermentation of *E. coli,* equal amounts of CO_2 and H_2 are produced, whereas in a butanediol fermentation,

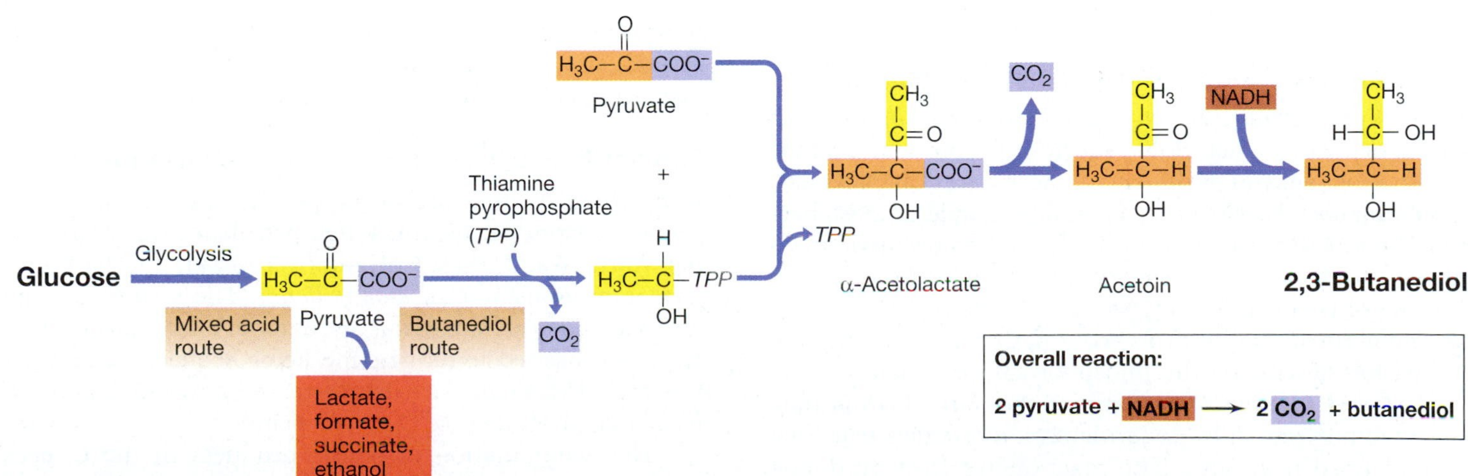

Figure 21.5 **Butanediol production in mixed-acid fermentations.** Pathway for the formation of butanediol from two molecules of pyruvate. Note how only one NADH but two pyruvates are required to make one butanediol.

considerably more CO_2 than H_2 is produced. This is because mixed-acid fermenters produce CO_2 only from formic acid by means of the enzyme formate hydrogen lyase:

$$HCOOH \longrightarrow H_2 + CO_2$$

By contrast, butanediol producers, such as *Enterobacter aerogenes,* produce CO_2 and H_2 from formic acid but also produce two additional molecules of CO_2 during the formation of each molecule of butanediol (Figure 21.5). Because they produce fewer acidic products, butanediol fermenters do not acidify their environment as extensively as do mixed-acid fermenters, and this is presumably a reflection of differences in acid tolerance in the two groups of organisms that have significance for their competitive success in nature.

21.2 MiniReview

The lactic acid fermentation is carried out by homofermentative and heterofermentative species. The mixed-acid fermentation results in either acids or acids plus neutral products, depending on the organism.

- How can CO_2 production differentiate homo- and heterofermentative lactic acid bacteria?
- Butanediol production leads to greater ethanol production than in a mixed-acid fermentation by *Escherichia coli.* Why?

21.3 Fermentative Diversity: Clostridial and Propionic Acid Fermentations

Species of the genus *Clostridium* are classical fermentative anaerobes (Section 16.2). Different clostridia ferment sugars, amino acids, purines and pyrimidines, and a few other compounds. In all cases, ATP synthesis is linked to substrate-level phosphorylations either in the glycolytic pathway or from the hydrolysis of a CoA intermediate (Table 21.1 and Figure 21.2). We begin with sugar-fermenting (saccharolytic) clostridia.

Sugar Fermentation by *Clostridium* Species

A number of clostridia ferment sugars, producing *butyric acid* as a major end product. Some species also produce acetone and butanol, neutral products, as fermentation products. *Clostridium acetobutylicum* is a classic example of these bacteria. The biochemical steps in the formation of butyric acid and neutral products from sugars are shown in **Figure 21.6**.

Glucose is converted to pyruvate via the glycolytic pathway, and pyruvate is split to acetyl-CoA, CO_2, and H_2 (through reduced ferredoxin) by the phosphoroclastic reaction (Section 21.1 and Figure 21.2). Some of the acetyl-CoA is then reduced to butyrate or other fermentation products using the NADH derived from glycolytic reactions as electron donor. The fermentation products observed are influenced by the duration and the conditions of the fermentation. During the

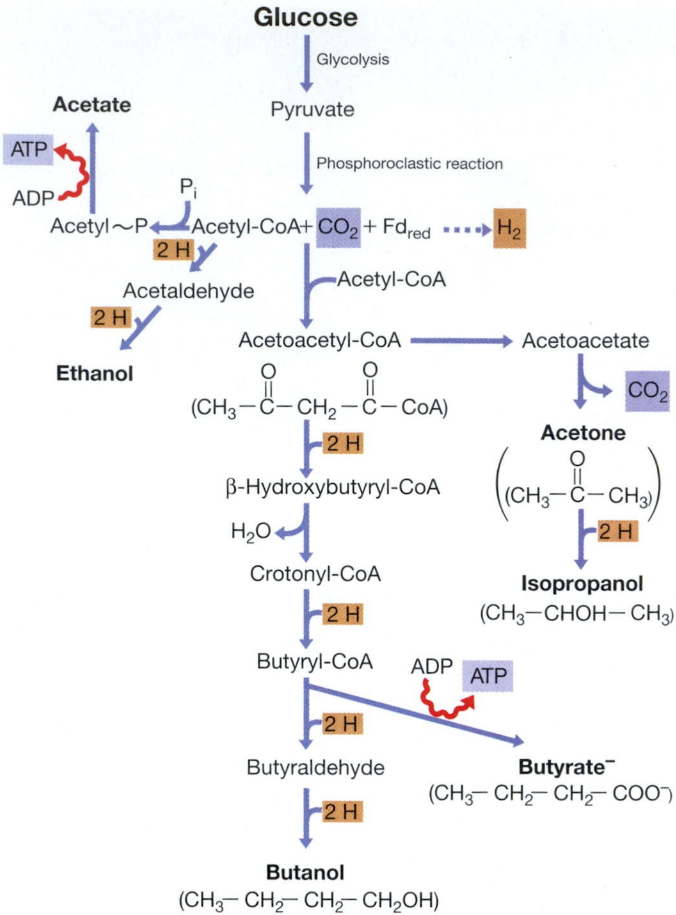

Figure 21.6 The butyric acid and butanol/acetone fermentation. All fermentation products from glucose are shown in bold. Note how the production of acetate and butyrate lead to additional ATP by substrate-level phosphorylation. By contrast, formation of butanol and acetone reduces the ATP yield because the butyryl-CoA step is bypassed. 2H, NADH; Fd, ferredoxin.

early stages of the butyric fermentation, butyrate and a small amount of acetate are produced. But as the pH of the medium drops, synthesis of acids ceases and the neutral products acetone and butanol begin to accumulate. However, if the pH of the medium is kept neutral with buffers, there is very little formation of neutral products and the fermentation proceeds primarily with the production of butyric acid.

Solvent Production and Energetic Consequences

Butanol and acetone are major industrial solvents, currently produced from petroleum. When petroleum was scarce during World War I, these solvents were made by the butanol fermentation using *C. acetobutylicum*. The historical importance of *C. acetobutylicum* as a model solvent-producing organism has led to its genome being sequenced, and from this and physiological experiments we know much about the fermentative strategy of this organism.

The accumulation of acidic products in the *C. acetobutylicum* fermentation lowers the pH, and this triggers derepression of genes responsible for solvent production. The

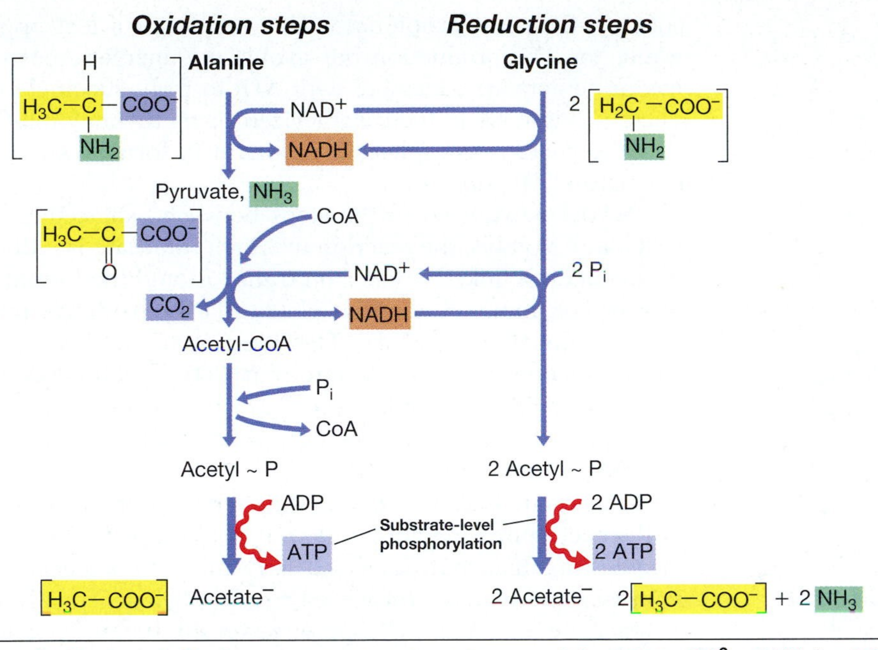

Amino acids participating in coupled fermentations (Stickland reaction)	
Amino acids oxidized:	**Amino acids reduced:**
Alanine	Glycine
Leucine	Proline
Isoleucine	Hydroxyproline
Valine	Tryptophan
Histidine	Arginine

Figure 21.7 The Stickland reaction. This example shows the cocatabolism of the amino acids alanine and glycine. The structures of key substrates, intermediates, and products are shown in brackets to allow the chemistry of the reaction to be followed. Note how in the reaction shown, alanine is the electron donor and glycine is the electron acceptor.

UNIT 4

production of butanol is actually a consequence of the production of acetone. For each acetone that is made, two NADH produced during glycolysis fail to be reoxidized, as they would be if butyrate was produced (Figure 21.6). Because redox balance is necessary for any fermentation to proceed, the cell uses butyrate as an electron acceptor. Butanol and acetone are therefore produced in equal amounts. Although neutral product formation helps the organism keep its environment from becoming too acidic, there is an energetic price to pay for this. In producing butanol, the cell loses the opportunity to convert butyryl-CoA to butyrate and ATP (Figure 21.6).

Amino Acid Fermentation by *Clostridium* Species and the Stickland Reaction

Some *Clostridium* species obtain their energy by fermenting amino acids. These are the "proteolytic" clostridia that degrade proteins, releasing and then catabolizing their amino acids. Some species ferment individual amino acids, typically glutamate, glycine, alanine, cysteine, histidine, serine, or threonine. The biochemistry behind these fermentations is quite complex, but the metabolic strategy is not. In virtually all cases, the amino acids are metabolized to eventually yield a fatty acid-CoA derivative, typically acetyl (C_2), butyryl (C_4), or caproyl (C_6). From these, ATP is produced by substrate-level phosphorylation (Table 21.1). Other products of amino acid fermentation include NH_3 and CO_2.

Some clostridia ferment only an amino acid pair. In this situation, one amino acid functions as the electron *donor* and is oxidized, whereas the other amino acid is the electron *acceptor* and is reduced. This coupled amino acid fermentation is known as a **Stickland reaction**. For instance, *Clostridium sporogenes* catabolizes a mixture of glycine and alanine; in this reaction alanine is the electron donor and glycine is the electron acceptor (**Figure 21.7**). Various amino acids that can function as electron donors or acceptors in Stickland reactions are listed in Figure 21.7. The products of the Stickland reaction are NH_3, CO_2, and a carboxylic acid with one fewer carbons than the amino acid that was oxidized (Figure 21.7).

Many of the products of amino acid fermentation by clostridia are foul-smelling substances, and the odor that results from putrefaction is mainly a result of clostridial activity. In addition to fatty acids, other odoriferous compounds produced include hydrogen sulfide (H_2S), methylmercaptan (from sulfur amino acids), cadaverine (from lysine), putrescine (from ornithine), and ammonia. Purines and pyrimidines, released from the degradation of nucleic acids, lead to many of the same fermentation products and yield ATP from the hydrolysis of fatty acid-CoA derivatives (Table 21.1) produced in their respective fermentative pathway.

Clostridium kluyveri Fermentation

Another species of *Clostridium* also ferments a mixture of substrates in which one is the donor and one is the acceptor, as in the Stickland reaction. However this organism, *C. kluyveri*, does not ferment amino acids but instead, *ethanol* plus *acetate*: Ethanol is the electron donor, and acetate is the electron acceptor. The reaction is the caproate/butyrate fermentation shown in Table 21.2.

The ATP yield in the caproate/butyrate fermentation is low, 1 ATP/6 ethanol fermented. However, *C. kluyveri* has a selective advantage over all other organisms in its unique ability to oxidize a highly reduced fermentation product (ethanol) and couple it to the reduction of another common fermentation product (acetate), reducing it to even longer-chain fatty acids. The single ATP produced in this reaction comes from

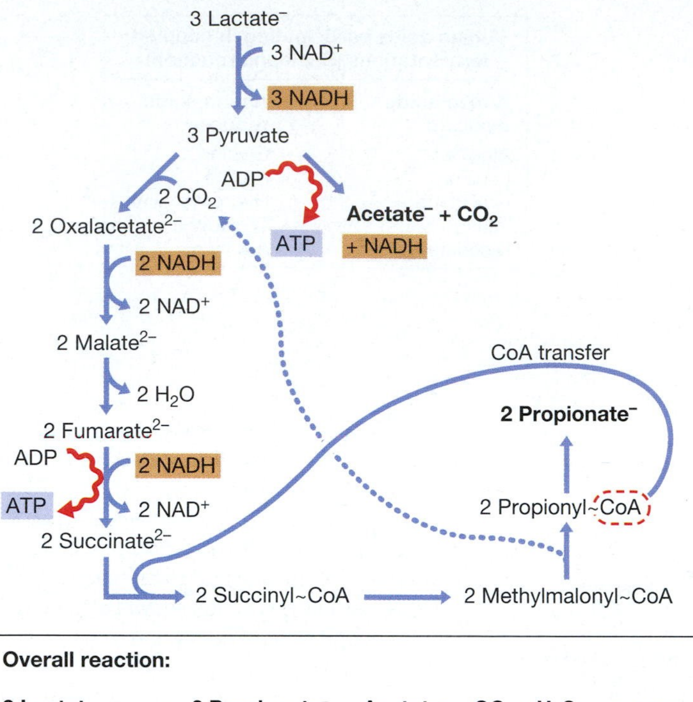

Figure 21.8 The propionic acid fermentation of *Propionibacterium*. Products are shown in bold. The four NADH made from the oxidation of three lactate are reoxidized in the reduction of oxalacetate and fumarate, and the CoA group from propionyl-CoA is exchanged with succinate during the formation of propionate.

substrate-level phosphorylation during conversion of a fatty acid-CoA formed in the pathway to the free fatty acid. The fermentation of *C. kluyveri* is an example of a **secondary fermentation**, which is essentially a fermentation of fermentation products. We see another example of this now.

Propionic Acid Fermentation

The propionic acid bacterium *Propionibacterium* and some related prokaryotes produce *propionic acid* as a major fermentation product, starting with either glucose or lactate as substrate. However, lactate, a fermentation product of the lactic acid bacteria, is probably the major substrate for propionic acid bacteria in nature, where these two groups live in close association. *Propionibacterium* is an important component in the ripening of Swiss (Emmentaler) cheese, to which the propionic and acetic acids produced give the unique bitter and nutty taste, and the CO_2 produced forms bubbles leaving the characteristic holes (eyes) in the cheese.

Figure 21.8 shows the reactions leading from lactate to propionate. When glucose is the starting substrate, it is first catabolized to pyruvate by the glycolytic pathway. Then pyruvate, produced either from glucose or from the oxidation of lactate, is carboxylated to form methylmalonyl-CoA, leading to the formation of oxalacetate and, eventually, propionyl-CoA (Figure 21.8). The latter reacts with succinate in a step catalyzed by the enzyme CoA transferase, producing succinyl-CoA and propionate. This results in a lost opportunity for ATP production but avoids the energetic costs of having to activate succinate with ATP to form succinyl-CoA. The succinyl-CoA is then isomerized to methylmalonyl-CoA and the cycle is complete; propionate is formed and CO_2 regenerated (Figure 21.8).

NADH is oxidized in the steps between oxalacetate and succinate. Notably, the reaction in which fumarate is reduced to succinate is linked to electron transport and the formation of a proton motive force that yields ATP by oxidative phosphorylation (Figure 21.8). The propionate pathway also converts some lactate to acetate plus CO_2, which allows for additional ATP to be made (Figure 21.8). Thus in metabolism of the propionic acid bacteria, both substrate-level and oxidative phosphorylation take place.

Propionate is also formed in the fermentation of succinate by the bacterium *Propionigenium* but by a completely different mechanism than that described here for *Propionibacterium*. *Propionigenium*, to be considered next, is phylogenetically and ecologically unrelated to *Propionibacterium*, but energetic aspects of its metabolism are of considerable interest from the standpoint of bioenergetics.

21.3 MiniReview

Clostridia ferment sugars, amino acids, and other organic compounds. *Propionibacterium* produces propionate and acetate in a secondary fermentation of lactate.

- Compare the mechanism for energy conservation in *C. acetobutylicum* and *Propionibacterium*.
- What are the substrates for the *C. kluyveri* fermentation?

21.4 Fermentations without Substrate-Level Phosphorylation

The fermentation of certain compounds yields insufficient energy to synthesize ATP by substrate-level phosphorylation (that is, less than −32 kJ, Table 21.1), yet energy conservation leading to ATP synthesis still occurs. In these cases, catabolism of the compound is linked to ion pumps that establish a proton motive force or sodium motive force across the cytoplasmic membrane. Examples of this include fermentation of the C_4 dicarboxylic acid and citric acid cycle intermediate succinate by *Propionigenium modestum* and the C_2 dicarboxylic acid oxalate by *Oxalobacter formigenes*.

Propionigenium modestum

Propionigenium modestum was first isolated in anoxic enrichment cultures lacking alternative electron acceptors and fed succinate as electron donor. *Propionigenium* inhabits marine and freshwater sediments and can also be isolated from the human oral cavity. The organism is a gram-negative short rod and, phylogenetically, is a species of actinobacteria (Section 16.4). During studies of the physiology of *P. modestum*,

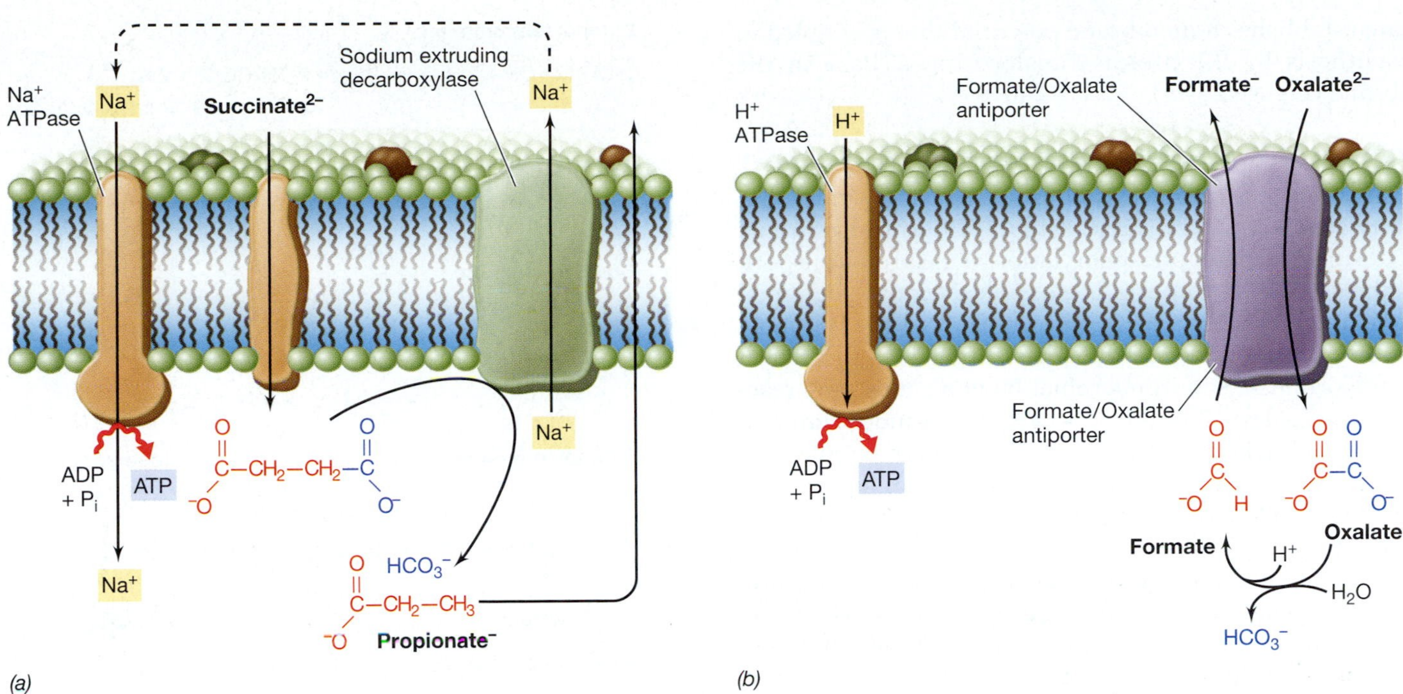

Figure 21.9 The unique fermentations of succinate and oxalate. *(a)* Succinate fermentation by *Propionigenium modestum*. Sodium export is linked to the energy released by succinate decarboxylation, and a sodium-translocating ATPase produces ATP. *(b)* Oxalate fermentation by *Oxalobacter formigenes*. Oxalate import and formate export by a formate-oxalate antiporter consume cytoplasmic protons. ATP synthesis is linked to a proton-driven ATPase. All substrates and products are shown in bold.

it was shown to catabolize succinate under strictly anoxic conditions:

$$\text{Succinate}^{2-} + H_2O \longrightarrow \text{propionate}^- + HCO_3^- \qquad \Delta G^{0\prime} = -20.5 \text{ kJ}$$

This reaction yields insufficient free energy to directly couple to ATP synthesis by substrate-level phosphorylation (Table 21.1), but nonetheless supports growth of the organism. Energy conservation in *Propionigenium* is linked to the decarboxylation of succinate by a membrane-bound decarboxylase, yielding propionate. This reaction releases sufficient free energy to drive the export of a sodium ion (Na^+) across the cytoplasmic membrane, establishing a sodium motive force. A sodium-translocating ATPase uses the sodium motive force to drive ATP synthesis (**Figure 21.9a**).

In a related reaction, *Malonomonas* (*Deltaproteobacteria*, Sections 15.17 and 15.18) decarboxyates the C_3 dicarboxylic acid malonate, forming acetate plus CO_2. As for *Propionigenium*, energy metabolism in *Malonomonas* is linked to a sodium pump and sodium-driven ATPase. However, the mechanism of malonate decarboxylation is more complex than that of *Propionigenium* and involves many additional proteins. Interestingly, however, the energy yield of malonate fermentation by *Malonomonas* is even lower than that of *P. modestum*, −17.4 kJ per malonate oxidized. *Sporomusa*, an endospore-forming bacterium (Section 16.2) and also an acetogen (Section 21.9), as well as a few other *Bacteria*, are also capable of fermenting malonate.

Oxalobacter formigenes

Oxalobacter formigenes is a bacterium present in the intestinal tract of animals, including humans. It catabolizes oxalate and produces formate. Oxalate degradation by *O. formigenes* is thought to be important in humans for preventing the accumulation of oxalate in the body, a substance which can trigger the synthesis of calcium oxalate kidney stones. Like *P. modestum*, *O. formigenes* is a gram-negative strict anaerobe that is a member of the *Betaproteobacteria*. *O. formigenes* carries out the following reaction:

$$\text{Oxalate}^{2-} + H_2O \longrightarrow \text{formate}^- + HCO_3^- \qquad \Delta G^{0\prime} = -26.7 \text{ kJ}$$

As in the catabolism of succinate by *P. modestum*, insufficient energy is available from this reaction to drive ATP synthesis by substrate-level phosphorylation (Table 21.1). However, the reaction supports growth of the organism because the decarboxylation of oxalate is exergonic and forms formate, which is excreted from the cell. The internal consumption of protons during the oxidation of oxalate and production of formate is, in effect, a proton pump. That is, a divalent molecule (oxalate) enters the cell while a univalent molecule (formate) is excreted. The continued exchange of oxalate for

UNIT 4

formate establishes a membrane potential that is coupled to ATP synthesis by the proton-translocating ATPase in the membrane (Figure 21.9*b*).

Energetics of *P. modestum* and *O. formigenes*

The unique aspect of the fermentations of *Propionigenium, Malonomonas,* and *Oxalobacter* is that ATP synthesis occurs without substrate-level phosphorylation *or* electron transport. Nevertheless, ATP synthesis is possible because the small amount of energy released can be coupled to the pumping of an ion across the membrane. These organisms thus teach us an important lesson in microbial bioenergetics: Any reaction that yields less than the 32 kJ required (under standard conditions) to make one ATP (Table 21.1) cannot be ruled out as a growth-supporting reaction for a bacterium. If the reaction can be coupled to an ion pump, ATP production remains a possibility.

A minimal requirement for an energy-conserving reaction is that it must yield sufficient free energy to pump a single ion across the membrane. This is estimated to be about 15 kJ. Reactions that yield fewer kJ than this in theory cannot fuel ion pumps and should therefore not be potential energy-conserving reactions. However, as we will see in the next section, there are examples known that push this theoretical limit even lower and whose energetics are still incompletely understood. These are the syntrophs, prokaryotes living on the energetic edge of life.

Ethanol fermentation:

$$2\ CH_3CH_2OH + 2\ H_2O \rightarrow 4\ H_2 + 2\ CH_3COO^- + 2\ H^+ \quad \Delta G^{0\prime} = +19.4\ \text{kJ/reaction}$$

Methanogenesis:

$$4\ H_2 + CO_2 \rightarrow CH_4 + 2\ H_2O \quad \Delta G^{0\prime} = -130.7\ \text{kJ/reaction}$$

Coupled reaction:

$$2\ CH_3CH_2OH + CO_2 \rightarrow CH_4 + 2\ CH_3COO^- + 2\ H^+ \quad \Delta G^{0\prime} = -111.3\ \text{kJ/reaction}$$

(a) Reactions

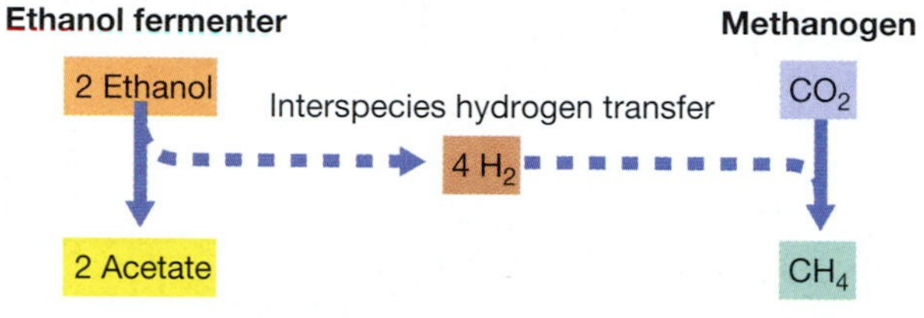

(b) Syntrophic transfer of H_2

Figure 21.10 Syntrophy: Interspecies H_2 transfer. Shown is the fermentation of ethanol to methane and acetate by syntrophic association of an ethanol-oxidizing bacterium and a H_2-consuming partner (methanogen). *(a)* Reactions involved. The two organisms thus share the energy released in the coupled reaction. *(b)* Nature of the syntrophic transfer of H_2.

21.4 MiniReview

The physiology of *Propionigenium, Oxalobacter,* and *Malonomonas* is linked to decarboxylation reactions that pump sodium ions or protons across the membrane. The reactions catalyzed by these organisms yield insufficient energy to make ATP by substrate-level phosphorylation.

- Why does *P. modestum* require sodium for growth?
- How can *Oxalobacter* be a medically beneficial organism for humans?

21.5 Syntrophy

There are many examples in microbiology of **syntrophy**, a situation in which two different organisms team up to degrade a substance—and conserve energy doing it—that neither can degrade individually. Most syntrophic reactions are secondary fermentations (Section 21.3), in which organisms ferment the fermentation products of other anaerobes. We will see in Section 24.2 how syntrophy is a key to the anoxic catabolism that leads to the production of methane. Here we consider the microbiology and energetic aspects of syntrophy.

Hydrogen Consumption in Syntrophic Reactions

The heart of most syntrophic reaction is H_2 production by one partner linked to H_2 consumption by the other. The H_2 consumer can be any of a number of physiologically distinct organisms: denitrifying bacteria, ferric iron-reducing bacteria, sulfate-reducing bacteria, acetogens, methanogens, or even anoxygenic phototrophic bacteria. These types of syntrophic reactions have also been called *interspecies H_2 transfer* to highlight the fact that H_2 metabolism is the key to the whole process.

Consider syntrophy involving ethanol fermentation to acetate with the eventual production of methane (**Figure 21.10**). As can be seen, the ethanol fermenter carries out a reaction that has an unfavorable (that is, positive) standard free-energy change ($\Delta G^{0\prime}$). However, the H_2 produced by the ethanol fermenter can be used as an electron donor for methanogenesis by a methanogen. When the two reactions are summed, the overall reaction is exergonic (Figure 21.10) and supplies the energy needed for growth of both partners in the syntrophic mixture.

Another example of syntrophy is the oxidation of butyrate to acetate plus H_2 by the fatty acid-oxidizing syntroph *Syntrophomonas* (**Figure 21.11**):

$$\text{Butyrate}^- + 2\ H_2O \longrightarrow 2\ \text{acetate}^- + H^+ + 2\ H_2 \qquad \Delta G^{0\prime} = +48.2\ \text{kJ}$$

The free energy change of this reaction is highly unfavorable, and in pure culture *Syntrophomonas* will not grow on butyrate. But if H_2 is consumed by a partner organism,

Syntrophomonas will grow on butyrate in coculture with the H_2 consumer. How can chemical reactions whose standard free energy changes are positive support growth of an organism? We examine this conundrum now.

Energetics of H_2 Transfer

In a syntrophic relationship, the removal of H_2 by a partner organism affects the energetics of the reaction. And, because H_2 can be consumed to such low levels, energetic calculations need to take these extremely low levels of H_2 into account. A review of the principles of free energy given in Appendix 1 indicates that the actual concentration of reactants and products in a reaction can have a major effect on energetics. This is not a significant issue in many reactions because the products are removed in roughly equal amounts. However, this is not the case with H_2. Because it is such a powerful electron donor for anaerobic respirations, H_2 is quickly consumed in anoxic habitats. This serves to maintain H_2 below 10^{-4} atm (∞ Table 24.2). At these vanishingly low levels of H_2, energetic calculations are dramatically affected.

For convenience, the $\Delta G^{0\prime}$ of a reaction is calculated on the basis of standard conditions—1-molar concentration of products and reactants. By contrast, the related term ΔG is used to calculate free energy changes on the basis of the actual concentrations of products and reactants present (Appendix 1 explains how to calculate ΔG). At very low levels of H_2, the energetics of the oxidation of ethanol or fatty acids to acetate plus H_2, a reaction that is endergonic under standard conditions, becomes exergonic. For example, if the concentration of H_2 is kept extremely low from consumption by the partner organism, ΔG for the oxidation of butyrate by *Syntrophomonas* yields about -18 kJ/mol (Figure 21.11*a*).

Energetics in Syntrophs

ATP synthesis by syntrophs is likely driven by both substrate-level and oxidative phosphorylations. Substrate-level phosphorylation can theoretically occur during the conversion of acetyl-CoA (generated by beta-oxidation of ethanol or the fatty acid) to acetate (Figure 21.11*a*), although the -18 kJ of energy released (ΔG) should be insufficient for this. However, the energy released is sufficient to produce a fraction of an ATP, so it is possible that two rounds of butyrate oxidation (Figure 21.11*a*) are required to yield one ATP by substrate-level phosphorylation.

Syntrophomonas has other metabolic capabilities as well, as the organism can carry out anaerobic respiration (Section 21.6) by the disproportionation of unsaturated fatty acids. (Disproportionation is a process in which some molecules of a substrate are oxidized while some are reduced). For example, crotonate, an intermediate in syntrophic butyrate metabolism (Figure 21.11*a*), supports growth of *Syntrophomonas* in pure culture. Under these conditions some of the crotonate is oxidized to acetate and some is reduced to butyrate (Figure 21.11*b*). Because crotonate reduction by *Syntrophomonas* is coupled to the formation of a proton motive force, as occurs in other anaerobic respirations that employ organic electron acceptors (such as fumarate reduction to succinate, Section 21.12), it is possible that somewhere during syntrophic metabolism (Figure 21.11*a*) a reaction is coupled to formation of a proton motive force as well.

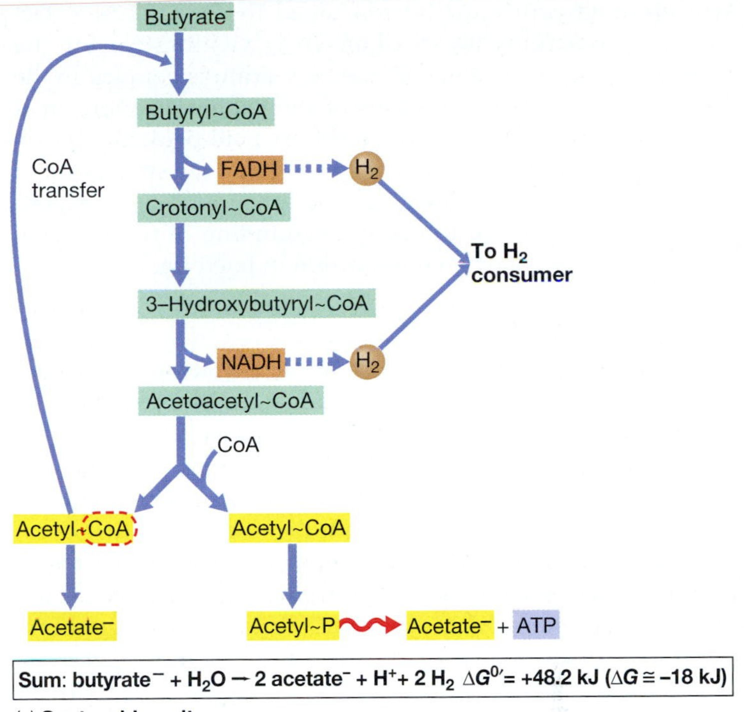

(*a*) **Syntrophic culture**

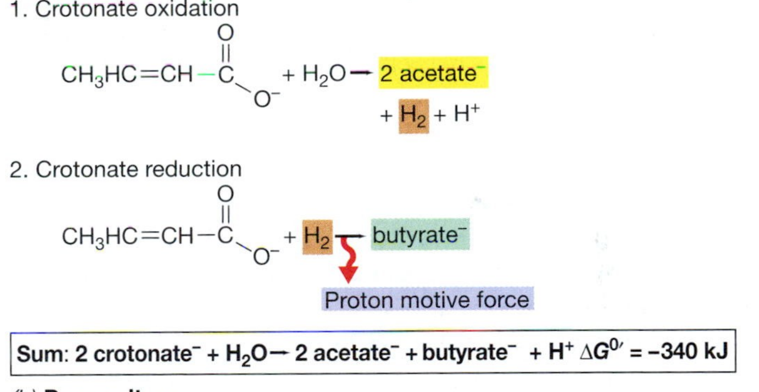

(*b*) **Pure culture**

Figure 21.11 Energetics of growth of *Syntrophomonas* in syntrophic culture and in pure culture. (*a*) In syntrophic culture, growth requires aH_2-consuming organism, such as a methanogen. H_2 production is driven by reverse electron flow because the E_0' of FAD/FADH and of NAD^+/NADH are more electropositive than that of $2\ H^+/H_2$. (*b*) In pure culture, energy conservation is linked to anaerobic respiration with crotonate reduction to butyrate.

Regardless of how ATP is made during syntrophic growth, additional energetic burdens exist for organisms in this growth mode. This is because H_2(E_0' -0.42 V) is produced from more electropositive electron donors such as FADH (E_0' -0.22 V) and NADH (E_0' -0.32 V), which are generated during fatty acid oxidation reactions (Figure 21.11*a*). Thus, some fraction of the ATP generated by *Syntrophomonas* during syntrophic growth must be consumed to drive reverse electron flow reactions (Section 20.4) yielding H_2 for the hydrogen consumer. When this energy drain is coupled to the inherently poor energetic yields of syntrophic reactions, it is clear that syntrophic alcohol- and fatty acid-oxidizing bacteria are living on the "edge of existence" from the standpoint of their overall bioenergetics. Even today these organisms pose a significant challenge to our understanding of the minimal requirements for energy conservation in bacteria.

Ecology of Syntrophs

Ecologically, syntrophic bacteria are key links in the anoxic portions of the carbon cycle. Syntrophs use the fermentation products of primary fermenters and release a key product for methanogens, acetogens, and other H_2 consumers. Without syntrophs, a bottleneck would develop in anoxic environments in which alternative electron acceptors other than CO_2 were limiting. By contrast, when conditions are oxic or alternative electron acceptors abundant, syntrophic relationships are unnecessary. For example, if O_2 or NO_3^- were available as electron acceptors, the energetics of the oxidation of a fatty acid or an alcohol is so favorable that cooperative arrangements with other organisms for the degradation of these substrates is not needed. Thus, syntrophy is characteristic of anoxic processes in which (1) the energy available is only very small, (2) one or more products is continually removed, and (3) the organisms are highly specialized for exploiting energetically marginal reactions.

21.5 MiniReview

In syntrophy two organisms work together to degrade some compound that neither can degrade alone. In this process H_2 produced by one organism is consumed by the partner. H_2 consumption can affect the energetics of the reaction carried out by the H_2 producer, allowing it to make ATP where it otherwise could not.

- Give an example of interspecies H_2 transfer. Why can it be said that both organisms benefit in this example?
- Predict how ATP is made during the syntrophic degradation of ethanol shown in Figure 21.10.

II ANAEROBIC RESPIRATION

In the next several sections we survey anaerobic respiration and related processes and see the many ways by which prokaryotes can conserve energy using electron acceptors *other* than O_2.

21.6 Anaerobic Respiration: General Principles

We examined the process of *aerobic* respiration in some detail in Chapter 5. As we noted there, molecular oxygen (O_2) functions as a terminal electron acceptor, accepting electrons from electron carriers by way of an electron transport chain. However, we also noted that other electron acceptors could be used instead of O_2, in which case the process is called **anaerobic respiration**. Here we consider some of these processes.

Bacteria that carry out anaerobic respiration employ electron transport systems containing cytochromes, quinones, iron-sulfur proteins, and other typical electron transport proteins. Their respiratory systems are thus similar to those of aerobes. In some organisms, such as the denitrifying bacteria, which are for the most part facultative aerobes, anaerobic respiration competes with aerobic respiration. In such cases, if O_2 is present, the bacteria respire aerobically, and genes encoding anaerobic processes are repressed. However, when O_2 is depleted from the environment, the bacteria respire anaerobically, and the alternate electron acceptor is reduced. Other organisms carrying out anaerobic respiration are obligate anaerobes and are unable to use O_2.

Alternative Electron Acceptors and the Redox Tower

The energy released from the oxidation of an electron donor using O_2 as electron acceptor is greater than if the same compound is oxidized with an alternate electron acceptor (Figure 5.10). These energy differences are apparent if the reduction potentials of each acceptor are examined (**Figure 21.12**). Because the O_2/H_2O couple is the most electropositive listed, more energy is available when O_2 is used than when another electron acceptor is used. Other electron acceptors that are near the O_2/H_2O couple are Fe^{3+}, NO_3^-, and NO_2^-. Examples of more electronegative acceptors are SO_4^{2-}, S^0, and CO_2. A summary of the most common types of anaerobic respiration is given in Figure 21.12.

Assimilative and Dissimilative Metabolism

Inorganic compounds such as NO_3^-, SO_4^{2-}, and CO_2 are reduced by many organisms as sources of cellular nitrogen, sulfur, and carbon, respectively. The end products of such reductions are primarily amino groups ($—NH_2$), sulfhydryl groups ($—SH$), and organic carbon compounds, respectively. When an inorganic compound such as NO_3^-, SO_4^{2-}, or CO_2 is reduced for use in biosynthesis, it is said to be *assimilated*, and the reduction process is called *assimilative* metabolism. Assimilative metabolism of NO_3^-, SO_4^{2-}, and CO_2 is conceptually and physiologically quite different from the reduction of these electron acceptors for energy metabolism in anaerobic respiration. To distinguish these two kinds of reductive processes, the use of these compounds as electron acceptors in energy metabolism is called *dissimilative* metabolism.

Assimilative and dissimilative metabolisms differ markedly. In assimilative metabolism, only enough of the compound (NO_3^-, SO_4^{2-}, or CO_2) is reduced to satisfy the needs for biosynthesis. The products are eventually converted to cell material in the form of macromolecules. In dissimilative metabolism, a large amount of the electron acceptor is reduced, and the reduced product is excreted into the environment.

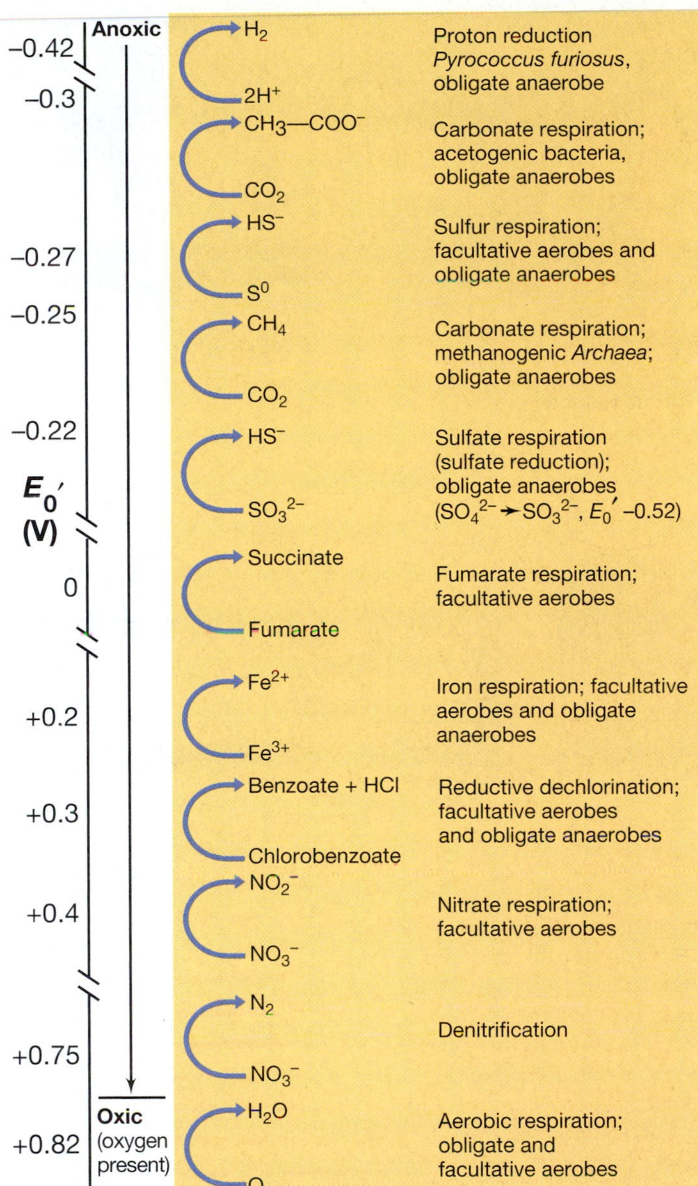

Figure 21.12 Major forms of anaerobic respiration. The redox couples are arranged in order from most electronegative E_0' (top) to most electropositive E_0' (bottom). See Figure 5.10 to compare how the energy yields of these anaerobic respirations vary.

Many organisms carry out assimilative metabolism of compounds such as NO_3^-, SO_4^{2-}, and CO_2, whereas only a restricted variety of primarily prokaryotic organisms carry out dissimilative metabolism. We focus on the latter here.

21.6 MiniReview

Although oxygen is the most widely used electron acceptor in energy-yielding metabolism, certain other compounds can be used as electron acceptors. Anaerobic respiration is less energy efficient than aerobic respiration but can proceed in environments where oxygen is absent.

Table 21.4 Oxidation states of key nitrogen compounds

Compound	*Oxidation state of N atom*
Organic N (—NH_2)	−3
Ammonia (NH_3)	−3
Nitrogen gas (N_2)	0
Nitrous oxide (N_2O)	+1 (average per N)
Nitrogen oxide (NO)	+2
Nitrite (NO_2^-)	+3
Nitrogen dioxide (NO_2)	+4
Nitrate (NO_3^-)	+5

- What is anaerobic respiration?
- With H_2 as electron donor, why is the reduction of NO_3^- a more favorable reaction than the reduction of S^0? (You may need to refer to Figure 5.10 for help here.)

21.7 Nitrate Reduction and Denitrification

Inorganic nitrogen compounds are some of the most common electron acceptors in anaerobic respiration. **Table 21.4** summarizes the various forms of inorganic nitrogen with their oxidation states. One of the most common alternative electron acceptors is nitrate, NO_3^-, which can be reduced to N_2O, NO, and N_2. Because these products of nitrate reduction are all gaseous, they can easily be lost from the environment, a process called **denitrification** (**Figure 21.13**).

Denitrification is the main means by which gaseous N_2 is formed biologically. As a source of nitrogen, N_2 is much less available to plants and microorganisms than is nitrate, so for

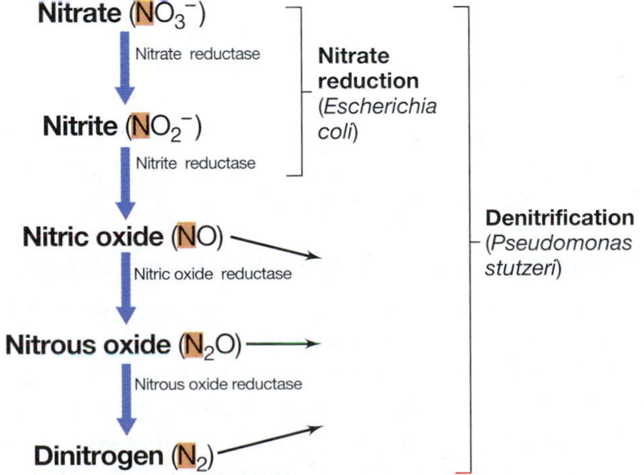

Figure 21.13 Steps in the dissimilative reduction of nitrate. Some organisms can carry out only the first step. All enzymes involved are derepressed by anoxic conditions. Also, some prokaryotes are known that can reduce NO_3^- to NH_4^+ in dissimilative metabolism.

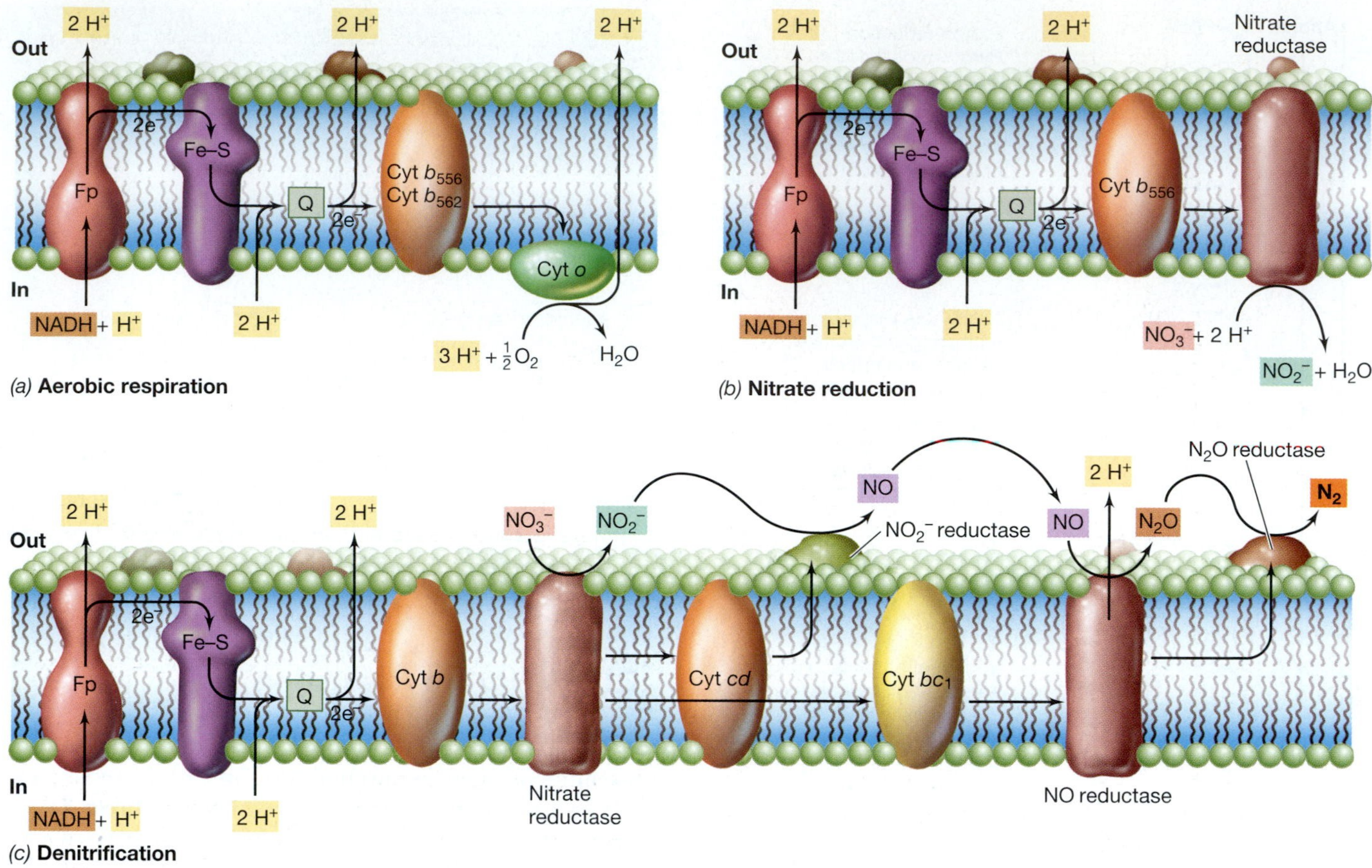

Figure 21.14 Respiration and anaerobic respiration. Electron transport processes in the membrane of *Escherichia coli* when *(a)* O_2 or *(b)* NO_3^- is used as an electron acceptor and NADH is the electron donor. Fp, flavoprotein; Q, ubiquinone. Under high-oxygen conditions, the sequence of carriers is Cyt $b_{562} \rightarrow$ Cyt $o \rightarrow O_2$. However, under low-oxygen conditions (not shown), the sequence is Cyt $b_{568} \rightarrow$ Cyt $d \rightarrow O_2$. Note how more protons are translocated per two electrons oxidized aerobically during electron transport reactions than anaerobically with nitrate as electron acceptor, because the aerobic terminal oxidase (Cyt *o*) can pump one proton. *(c)* Scheme for electron transport in membranes of *Pseudomonas stutzeri* during denitrification. Nitrate and nitric oxide (NO) reductases are integral membrane proteins, whereas nitrite (NO_2^-) and nitrous oxide (N_2O) reductases are periplasmic enzymes.

agricultural purposes, at least, denitrification is a detrimental process. For sewage treatment (∞ Section 36.2), however, denitrification is beneficial because it converts NO_3^- to N_2; this transformation decreases the load of fixed nitrogen in the sewage treatment effluent that can stimulate algal growth in receiving waters (∞ Section 23.6).

Biochemistry of Dissimilative Nitrate Reduction

The enzyme in the first step of dissimilative nitrate reduction, nitrate reductase, is a molybdenum-containing membrane-integrated enzyme whose synthesis is repressed by molecular oxygen. All subsequent enzymes of the pathway (**Figure 21.14**) are coordinately regulated and thus also repressed by O_2. But, in addition to anoxic conditions, nitrate must also be present before these enzymes are fully expressed.

The first product of nitrate reduction is nitrite (NO_2^-), and the enzyme nitrite reductase reduces it to nitric oxide (NO) (Figure 21.14*c*). Some organisms can reduce NO_2^- to ammonia (NH_3) in a dissimilative process, but the production of gaseous products—denitrification—is of greatest global significance. This is because it consumes a fixed form of nitrogen (NO_3^-) and produces gaseous nitrogen compounds, some of which are of environmental significance. For example, N_2O can be converted to NO by sunlight, and NO reacts with ozone (O_3) in the upper atmosphere to form nitrite (NO_2^-). Nitrite returns to Earth as acid rain (nitrous acid, HNO_2). The remaining biochemical steps in denitrification are shown in Figure 21.14*c*.

The biochemistry of dissimilative nitrate reduction has been studied in detail in several organisms, including *Escherichia coli*, in which NO_3^- is reduced only to NO_2^-, and *Paracoccus denitrificans* and *Pseudomonas stutzeri*, in which denitrification occurs. The *E. coli* nitrate reductase accepts electrons from a *b*-type cytochrome, and a comparison of the electron transport chains in aerobic versus nitrate-respiring cells of *E. coli* is shown in Figure 21.14*a,b*.

Because of the reduction potential of the NO_3^-/NO_2^- couple (+0.43 V), only two proton-translocating steps occur during nitrate reduction, whereas three protons are pumped in aerobic respiration ($\frac{1}{2}$ O_2/H_2O, +0.82 V). In *P. denitrificans* and *P. stutzeri*, nitrogen oxides are formed from nitrite by the enzymes nitrite reductase, nitric oxide reductase, and nitrous oxide reductase, as summarized in Figure 21.14*c*. During electron transport, a proton motive force is established, and ATPase functions to produce ATP in the usual fashion. Additional ATP is available when NO_3^- is reduced to N_2 because the NO reductase is linked to proton extrusion (Figure 21.14*c*). **www.microbiologyplace.com Online Tutorial 21.1: Electron Transport Processes under Aerobic and Anaerobic Conditions**

Other Properties of Denitrifying Prokaryotes

Most denitrifying prokaryotes are phylogenetically members of the *Proteobacteria* (Chapter 15) and, physiologically, are facultative aerobes. Aerobic respiration occurs when air is present, even if nitrate is also present in the medium. Many denitrifying bacteria also reduce other electron acceptors anaerobically, such as ferric iron (Fe^{3+}) and certain organic electron acceptors (Section 21.12). In addition, some denitrifying bacteria can grow by fermentation. Thus, the denitrifying bacteria are quite metabolically diverse in alternative energy-generating mechanisms. Interestingly, at least one eukaryote has been shown to be a denitrifyer. The protist *Globobulimina pseudospinescens*, a shelled amoeba (foraminifera, Section 18.11), can denitrify and likely employs this form of metabolism to survive in its habitat, anoxic marine sediments.

21.7 MiniReview

Nitrate is a commonly used electron acceptor in anaerobic respiration. Nitrate reduction is catalyzed by the enzyme nitrate reductase, reducing nitrate to nitrite. Many bacteria that use nitrate in anaerobic respiration eventually produce N_2, a process called *denitrification*.

- For *Escherichia coli*, why is more energy released in aerobic respiration than during NO_3^- reduction?
- Where is the dissimilative nitrate reductase found in the cell? What unusual metal does it contain?
- Why does *Pseudomonas stutzeri* derive more energy from NO_3^- respiration than does *Escherichia coli*?

21.8 Sulfate and Sulfur Reduction

Several inorganic sulfur compounds are important electron acceptors in anaerobic respiration. A summary of the oxidation states of the key sulfur compounds is given in **Table 21.5**. Sulfate, the most oxidized form of sulfur, is one of the major anions in seawater and is reduced by the sulfate-reducing bacteria, a group that is widely distributed in nature. The end product of sulfate reduction is hydrogen sulfide, H_2S, an important natural product that participates in many biogeochemical processes (Section 24.4). Species in the genus *Desulfovibrio* have been widely used for the study of sulfate reduction.

Table 21.5 Sulfur compounds and electron donors for sulfate reduction

Compound	*Oxidation state of S atom*
Oxidation states of key sulfur compounds	
Organic S (R—SH)	−2
Sulfide (H_2S)	−2
Elemental sulfur (S^0)	0
Thiosulfate ($S_2O_3^{2-}$)	+2 (average per S)
Sulfur dioxide (SO_2)	+4
Sulfite (SO_3^{2-})	+4
Sulfate (SO_4^{2-})	+6
Some electron donors used for sulfate reduction	
H_2	Acetate
Lactate	Propionate
Pyruvate	Butyrate
Ethanol and other alcohols	Long-chain fatty acids
Fumarate	Benzoate
Malate	Indole
Choline	Various hydrocarbons

UNIT 4

Assimilative and Dissimilative Sulfate Reduction

Again, as with nitrogen, it is necessary to distinguish between assimilative and dissimilative metabolism. Many organisms, including plants, algae, fungi, and most prokaryotes, use sulfate as a sulfur source for biosynthetic needs. The ability to use sulfate as an electron acceptor for energy-generating processes, however, involves the large-scale reduction of SO_4^{2-} and is restricted to the sulfate-reducing bacteria. In assimilative sulfate reduction, the H_2S formed is immediately converted into organic sulfur in the form of amino acids and other organic sulfur compounds, but in dissimilative sulfate reduction the H_2S is excreted.

Biochemistry and Energetics of Sulfate Reduction

As the reduction potentials in Table A1.2 and Figure 21.12 show, sulfate is a much less favorable electron acceptor than is O_2 or NO_3^-. However, sufficient free energy to make ATP is available from sulfate reduction when an electron donor that yields NADH or FADH is oxidized. Table 21.5 lists some of the electron donors used by sulfate-reducing bacteria. H_2, lactate, and pyruvate are widely used by sulfate-reducing bacteria, whereas the others have more restricted use. Many morphological and physiological types of sulfate-reducing bacteria are known, and with the exception of *Archaeoglobus* (Section 17.7), a species of *Archaea*, all known sulfate-reducers are *Bacteria* (Section 15.18)

The reduction of SO_4^{2-} to H_2S requires eight electrons (Table 21.5) and proceeds through a number of intermediate stages. Sulfate is chemically quite stable and cannot be reduced without first being activated. Sulfate is activated by ATP. The enzyme ATP sulfurylase catalyzes the attachment of the sulfate ion to a phosphate of ATP, leading to the formation of *adenosine phosphosulfate (APS)* as shown in **Figure 21.15**.

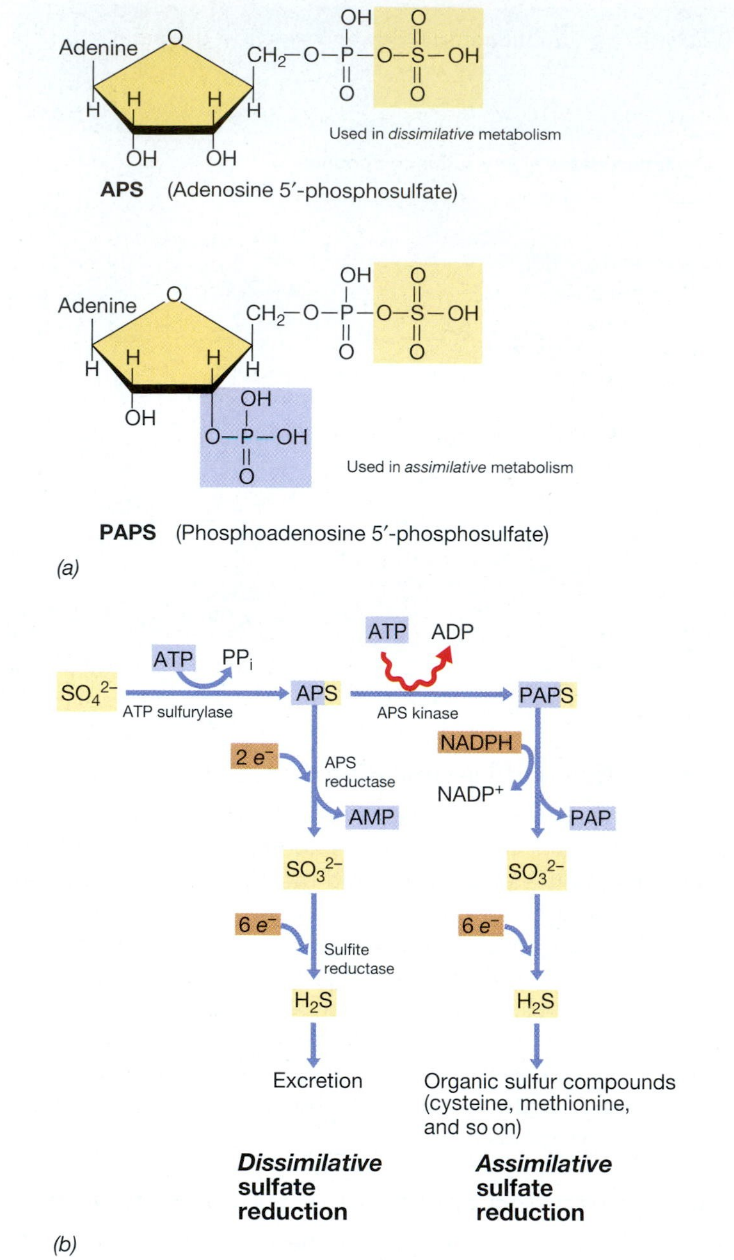

Figure 21.15 Biochemistry of sulfate reduction: Activated sulfate. *(a)* Two forms of active sulfate can be made, adenosine 5′-phosphosulfate (APS) and phosphoadenosine 5′-phosphosulfate (PAPS). Both are derivatives of adenosine diphosphate (ADP), with the second phosphate of ADP being replaced by sulfate. *(b)* Schemes of assimilative and dissimilative sulfate reduction.

In dissimilative sulfate reduction, the sulfate in APS is reduced directly to sulfite (SO_3^{2-}) by the enzyme APS reductase with the release of AMP. In assimilative reduction, another phosphate is added to APS to form *phosphoadenosine phosphosulfate (PAPS)* (Figure 21.15*a*), and only then is the sulfate reduced. However, in both cases the product of sulfate reduction is sulfite, SO_3^{2-}. Once SO_3^{2-} is formed, sulfide is formed by the enzyme sulfite reductase (Figure 21.15*b*).

During dissimilative sulfate reduction, electron transport reactions leading to proton motive force formation occur, and

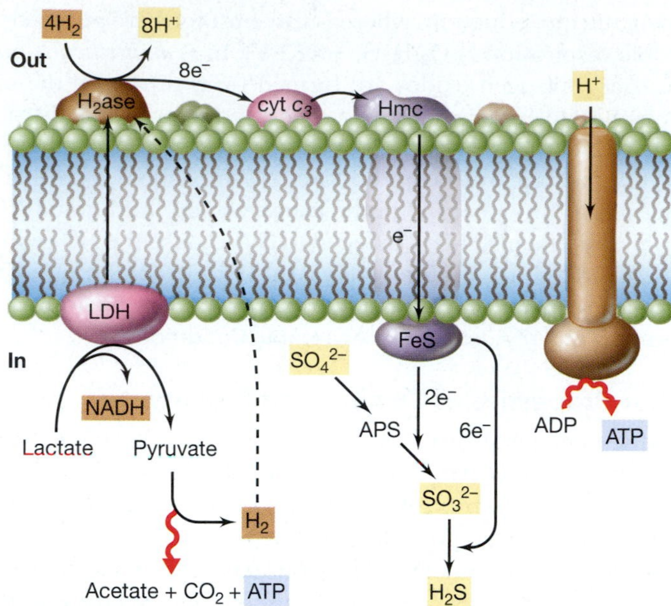

Figure 21.16 Electron transport and energy conservation in sulfate-reducing bacteria. In addition to external hydrogen (H_2), H_2 originating from the catabolism of organic compounds such as lactate and pyruvate can fuel hydrogenase. The enzymes hydrogenase (H_2ase), cytochrome (cyt) c_3, and a cytochrome complex (Hmc) are periplasmic proteins. A separate protein shuttles electrons across the cytoplasmic membrane from Hmc to a cytoplasmic iron-sulfur protein (FeS) that supplies electrons to APS reductase (forming SO_3^{2-}) and sulfite reductase (forming H_2S, Figure 21.15*b*). LDH, lactate dehydrogenase.

this drives ATP synthesis by ATPase. A major electron carrier involved is cytochrome c_3, a periplasmic low-potential cytochrome (**Figure 21.16**). Cytochrome c_3 accepts electrons from a periplasmically located hydrogenase and transfers these electrons to a membrane-associated protein complex. This complex, called *Hmc,* carries the electrons across the cytoplasmic membrane, thus making them available to APS reductase and sulfite reductase, which are cytoplasmic enzymes (Figure 21.16).

The enzyme hydrogenase plays a central role in sulfate reduction whether *Desulfovibrio* is growing on H_2, *per se,* or on an organic compound, like lactate. This is because lactate is converted through pyruvate to acetate (the latter is mainly excreted because *Desulfovibrio* is a nonacetate-oxidizing sulfate reducer; Section 15.18) with the production of H_2. The H_2 produced crosses the cytoplasmic membrane and is oxidized by the periplasmic hydrogenase to initiate a proton motive force (Figure 21.16). Growth yields of sulfate-reducing bacteria indicate that one ATP is produced for each SO_4^{2-} reduced to HS^-. With H_2 as electron donor, the reaction is

$$4\,H_2 + SO_4^{2-} + H^+ \longrightarrow HS^- + 4\,H_2O \qquad \Delta G^{0\prime} = -152 \text{ kJ}$$

When lactate or pyruvate is the electron donor, not only is ATP produced from the proton motive force, but additional

ATP is produced during the oxidation of pyruvate to acetate plus CO_2 via acetyl-CoA and acetylphosphate (Figure 21.3).

Acetate Use and Autotrophy

Many sulfate-reducing bacteria can oxidize acetate to CO_2 to obtain electrons for sulfate reduction (Section 15.18):

$$CH_3COO^- + SO_4^{2-} + 3\,H^+ \longrightarrow 2\,CO_2 + H_2S + 2\,H_2O$$

$$\Delta G^{0\prime} = -57.5 \text{ kJ}$$

The mechanism for acetate oxidation is the *acetyl-CoA pathway,* a series of reversible reactions used by many anaerobes for acetate synthesis or acetate oxidation. This pathway employs the key enzyme *carbon monoxide dehydrogenase* (Section 21.9). A few sulfate-reducing bacteria can also grow autotrophically with H_2. When growing under these conditions, the organisms use the acetyl-CoA pathway for incorporating CO_2 into cell material. The acetate-oxidizing sulfate-reducing bacterium *Desulfobacter* lacks acetyl-CoA pathway enzymes and oxidizes acetate through the citric acid cycle (Figure 5.22), but this seems to be the exception rather than the rule.

Sulfur Disproportionation

Certain sulfate-reducing bacteria can disproportionate sulfur compounds of intermediate oxidation state. We discussed the process of disproportionation in Section 21.5. For example, *Desulfovibrio sulfodismutans* can disproportionate thiosulfate as follows:

$$S_2O_3^{2-} + H_2O \longrightarrow SO_4^{2-} + H_2S$$

$$\Delta G^{0\prime} = -21.9 \text{ kJ/reaction}$$

Note that in this reaction one sulfur atom of $S_2O_3^{2-}$ becomes more oxidized (forming SO_4^{2-}), while the other becomes more reduced (forming H_2S). The oxidation of thiosulfate by *D. sulfodismutans* is coupled to proton motive force formation that is used by the organism to make ATP by ATPase. Other reduced sulfur compounds such as sulfite (SO_3^{2-}) and sulfur (S^0) can also be disproportionated. These forms of metabolism allow sulfate-reducing bacteria to recover energy from sulfur intermediates produced from the oxidation of H_2S by sulfur chemolithotrophs that coexist with them in nature (Sections 20.10 and 24.4) and also from intermediates generated in their own metabolism during sulfate reduction.

Phosphite Oxidation

At least one sulfate-reducing bacterium can couple phosphite (HPO_3^-) oxidation to sulfate reduction. The reaction is chemolithotrophic, and the products are phosphate and sulfide:

$$4\,HPO_3^- + SO_4^{2-} + H^+ \longrightarrow 4\,HPO_4^{2-} + HS^-$$

$$\Delta G^{0\prime} = -364 \text{ kJ}$$

This bacterium, *Desulfotignum phosphitoxidans,* requires only CO_2 for its carbon needs (it is thus an autotroph). It is a strict anaerobe, which it must be because phosphite spontaneously oxidizes in air. The natural sources of phosphite in nature are unknown. However, as *D. phosphitoxidans* can use phosphite as sole electron donor, it is likely produced in anoxic environments in nature, probably from the degradation of organophosphates. Along with sulfur disproportionation (also a chemolithotrophic process) and H_2 utilization, phosphite oxidation underscores the diversity of chemolithotrophic reactions carried out by sulfate-reducing bacteria.

Sulfur Reduction

Some organisms produce H_2S in anaerobic respiration but are unable to reduce sulfate; these are the elemental sulfur reducers. Sulfur-reducing bacteria carry out the reaction

$$S^0 + 2\,H \longrightarrow H_2S$$

The electrons for this process can come from H_2 or from various organic compounds. The first sulfur-reducing organism to be discovered was *Desulfuromonas acetoxidans* (Section 15.18). This organism oxidizes acetate, ethanol, and a few other compounds to CO_2, coupled with the reduction of S^0 to H_2S. The physiology of dissimilative sulfur-reducing bacteria is not as well understood as that of sulfate-reducing bacteria, but it is known that sulfur reducers lack the capacity to activate sulfate to APS, and presumably this prevents them from using sulfate as an electron acceptor. *Desulfuromonas* contains high levels of several types of cytochromes, including an analog of cytochrome c_3, a key electron carrier in sulfate-reducing bacteria. Because the oxidation of acetate to CO_2 releases less energy than that needed to make an ATP by substrate-level phosphorylation, it is clear that oxidative phosphorylation plays a major role in the energetics of these organisms. A variety of other bacteria can use sulfur as an electron acceptor, including some species of the genera *Wolinella* and *Campylobacter*.

21.8 MiniReview

The sulfate-reducing bacteria reduce sulfate to hydrogen sulfide. This process requires activation of sulfate by ATP to form the compound adenosine phosphosulfate (APS). Electron donors for sulfate reduction include H_2, various organic compounds, and even phosphite. Disproportionation of sulfur compounds is an additional energy-yielding strategy for certain members of this group.

- What are the names of the following sulfur compounds: S^0, SO_4^{2-}, SO_3^{2-}, $S_2O_3^{2-}$, H_2S?
- How is sulfate converted to sulfite during dissimilative sulfate reduction?
- Why is H_2 of importance to sulfate-reducing bacteria?
- Give an example of the disproportionation of sulfur compounds.

UNIT 4

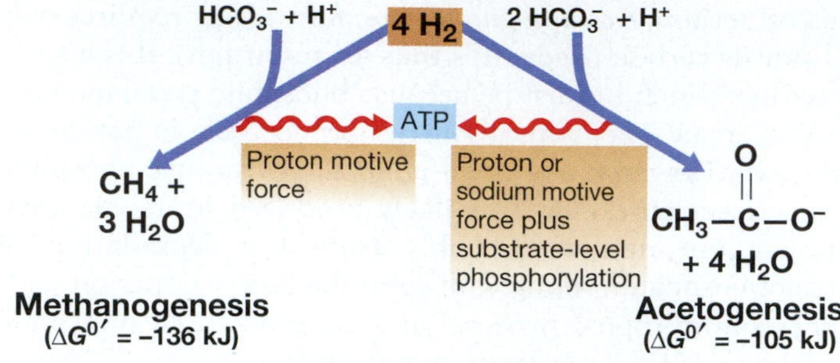

Figure 21.17 The contrasting processes of methanogenesis and acetogenesis. Note the difference in free energy released in the reactions.

21.9 Acetogenesis

Carbon dioxide, CO_2, is common in nature and typically abundant in anoxic habitats because it is a major product of the energy metabolisms of chemoorganotrophs. Two major groups of strictly anaerobic prokaryotes use CO_2 as an electron acceptor in energy metabolism; one of these groups is the *acetogens* (sometimes called *homoacetogens* to indicate that acetate is the sole product). We discuss the other group, the methanogens, in the next section. H_2 is a major electron donor for both of these organisms, and an overview of their energy metabolism, **acetogenesis** and **methanogenesis**, is shown in Figure 21.17. Both processes are linked to ion pumps, either of H^+ or Na^+, which fuel ATPases in the membrane. Acetogenesis also conserves energy in substrate-level phosphorylation reactions.

Organisms and Pathway

Acetogens carry out the reaction

$$4\ H_2 + H^+ + 2\ HCO_3^- \longrightarrow CH_3COO^- + 4\ H_2O \qquad \Delta G^{0\prime} = -105\ kJ$$

In addition to H_2, electron donors for acetogenesis include C_1 compounds, sugars, organic and amino acids, alcohols, and certain nitrogen bases, depending on the organism. Many acetogens can also reduce NO_3^- and $S_2O_3^{2-}$. However, CO_2 reduction is probably the major reaction of ecological significance.

A major unifying thread among acetogens is the pathway of CO_2 reduction. Acetogens reduce CO_2 to acetate by the **acetyl-CoA pathway**. Table 21.6 lists the major groups of organisms that produce acetate or oxidize acetate via the acetyl-CoA pathway. Acetogens such as *Acetobacterium woodii* and *Clostridium aceticum* can grow either chemoorganotrophically by fermentation of sugars (reaction 1) or chemolithotrophically and autotrophically through the reduction of CO_2 to acetate with H_2 (reaction 2) as electron donor. In either case, the sole product is acetate:

$$(1)\ C_6H_{12}O_6 \longrightarrow 3\ CH_3COO^- + 3\ H^+$$

$$(2)\ 2\ HCO_3^- + 4\ H_2 + H^+ \longrightarrow CH_3COO^- + 4\ H_2O$$

Table 21.6 Organisms employing the acetyl-CoA pathway of CO_2 fixation

I. Acetate synthesis, the result of energy metabolism
Acetoanaerobium noterae
Acetobacterium woodii
Acetobacterium wieringae
Acetogenium kivui
Acetitomaculum ruminis
Clostridium aceticum
Clostridium formicaceticum
Moorella thermoacetica
Desulfotomaculum orientis
Sporomusa paucivorans
Eubacterium limosum (also produces butyrate)
Treponema primitia (from termite hindguts)
II. Acetate synthesis in autotrophic metabolism
Autotrophic homoacetogenic bacteria
Autotrophic methanogens
Autotrophic sulfate-reducing bacteria
III. Acetate oxidation in energy metabolism
Reaction: Acetate + 2 $H_2O \rightarrow 2\ CO_2$ + 8 H
Group II sulfate reducers (other than *Desulfobacter*)
Reaction: Acetate $\rightarrow CO_2 + CH_4$
Acetotrophic methanogens (*Methanosarcina, Methanosaeta*)

Acetogens ferment glucose via the glycolytic pathway, converting glucose to two molecules of pyruvate and two molecules of NADH (the equivalent of 4 H). From this point, two molecules of acetate are produced:

$$(3)\ 2\ pyruvate^- \longrightarrow 2\ acetate^- + 2\ CO_2 + 4\ H$$

The third acetate of reaction (1) comes from reaction (2), using the two molecules of CO_2 generated in reaction (3), plus the four electrons generated from glycolysis and the four electrons generated from the oxidation of two pyruvates to two acetates [reaction (3)]. Starting from pyruvate, then, the overall production of acetate can be written as

$$2\ pyruvate^- + 4\ H \longrightarrow 3\ acetate^- + H^+$$

Most acetogenic bacteria that produce and excrete acetate in energy metabolism are gram-positive *Bacteria*, and many are species of *Clostridium* or *Acetobacterium* (Table 21.6). A few other gram-positive and many different gram-negative *Bacteria* and *Archaea* use the acetyl-CoA pathway for autotrophic purposes, reducing CO_2 to acetate as a source of cell carbon.

The acetyl-CoA pathway functions in autotrophic growth for certain sulfate-reducing bacteria and is also used by the methanogens, most of which grow autotrophically on H_2 + CO_2 (∞ Sections 17.4 and 21.10). By contrast, some bacteria employ the reactions of the acetyl-CoA pathway primarily in the reverse direction as a means of oxidizing acetate to CO_2. These include acetotrophic methanogens (Section 17.4) and sulfate-reducing bacteria (Sections 15.18 and 21.8).

Reactions of the Acetyl-CoA Pathway

Unlike other autotrophic pathways such as the Calvin cycle (Section 20.6) or the reverse citric acid or hydroxypropionate cycles (Section 20.7), the acetyl-CoA pathway of CO_2 fixation is not a cycle. Instead it catalyzes the reduction of CO_2 along two linear pathways; one molecule of CO_2 is reduced to the methyl group of acetate, and the other molecule of CO_2 is reduced to the carbonyl group. The two C_1 units are then assembled at the end to form acetyl-CoA (**Figure 21.18**).

A key enzyme of the acetyl-CoA pathway is *carbon monoxide (CO) dehydrogenase*. CO dehydrogenase is a complex enzyme that contains the metals Ni, Zn, and Fe as cofactors. CO dehydrogenase catalyzes the reaction

$$CO_2 + H_2 \longrightarrow CO + H_2O$$

and the CO produced ends up in the *carbonyl* position of acetate (Figure 21.18). The methyl group of acetate originates from the reduction of CO_2 by a series of reactions requiring the coenzyme *tetrahydrofolate* (Figure 21.18). The methyl group is then transferred from tetrahydrofolate to an enzyme that contains vitamin B_{12} as cofactor (Figure 21.18). In the final step of the pathway, CH_3 is combined with CO by CO dehydrogenase to form acetyl-CoA. Conversion of acetyl-CoA to acetate plus ATP completes the reaction series (Figure 21.18).

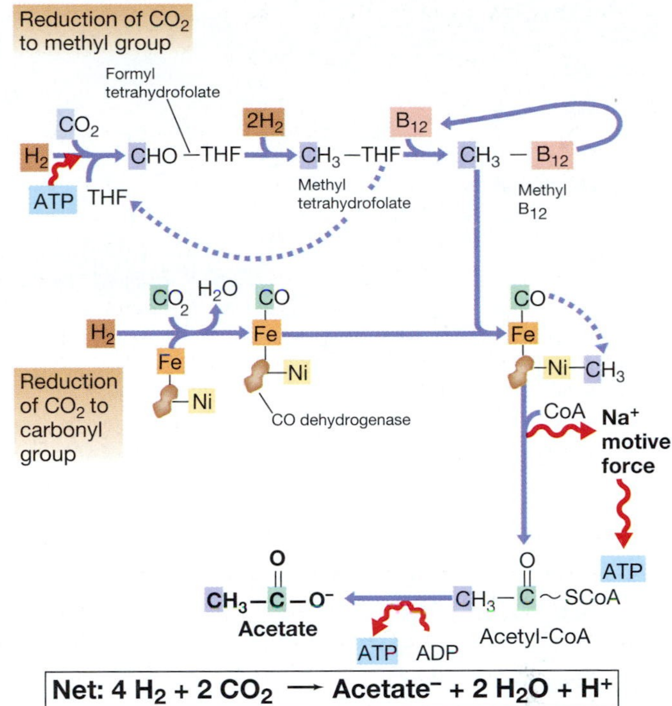

Figure 21.18 Reactions of the acetyl-CoA pathway. Carbon monoxide (CO) is bound to an Fe atom in CO dehydrogenase and the CH_3 group to a nickel atom in an organic nickel compound in CO dehydrogenase. Note how the formation of acetyl-CoA is coupled to the generation of a Na^+ motive force that drives ATP synthesis and that ATP is also synthesized in the conversion of acetyl-CoA to acetate. THF, tetrahydrofolate; B_{12}, vitamin B_{12} in an enzyme-bound intermediate.

Energy Conservation in Acetogenesis

Energy conservation in acetogenesis is the result of both substrate-level and oxidative phosphorylations (Figure 21.18). ATP is synthesized during the conversion of acetyl-CoA to acetate plus ATP (Section 21.3). There are also energy-conserving steps when a sodium motive force is established across the cytoplasmic membrane during acetogenesis. This energized state of the membrane allows for energy conservation from a Na^+ ATPase. We saw a similar situation in the succinate fermenter *Propionigenium*, where succinate decarboxylation was linked to Na^+ export (Section 21.4).

21.9 MiniReview

Acetogens are anaerobes that reduce CO_2 to acetate, usually with H_2 as electron donor. The mechanism of acetate formation is the acetyl-CoA pathway, a series of reactions widely distributed in obligate anaerobes as either a mechanism of autotrophy or for acetate catabolism.

- Draw the structure of acetate and identify the carbonyl group and the methyl group. What key enzyme of the acetyl-CoA pathway produces the carbonyl group of acetate?
- How do homoacetogens make ATP from the synthesis of acetate?
- If the catabolism of fructose by glycolysis yields only two molecules of acetate, how does *Clostridium aceticum* ferment fructose by this pathway and produce three molecules of acetate?

21.10 Methanogenesis

The biological production of methane—methanogenesis—is carried out by a group of strictly anaerobic *Archaea* called the **methanogens**. The reduction of CO_2 with H_2 is a major pathway of methanogenesis and so we focus on this process and compare it with other forms of anaerobic respiration. We considered the basic properties, phylogeny, and taxonomy of the methanogens in Section 17.4; here we focus on their biochemistry and bioenergetics. Methanogenesis is a complex series of biochemical reactions that employ novel coenzymes. Because of this, we begin our discussion with a consideration of the coenzymes and then move on to the actual production of methane.

C_1 Carriers in Methanogenesis

The key coenzymes in methanogenesis can be divided into two classes: (1) those that carry the C_1 unit from the initial

UNIT 4

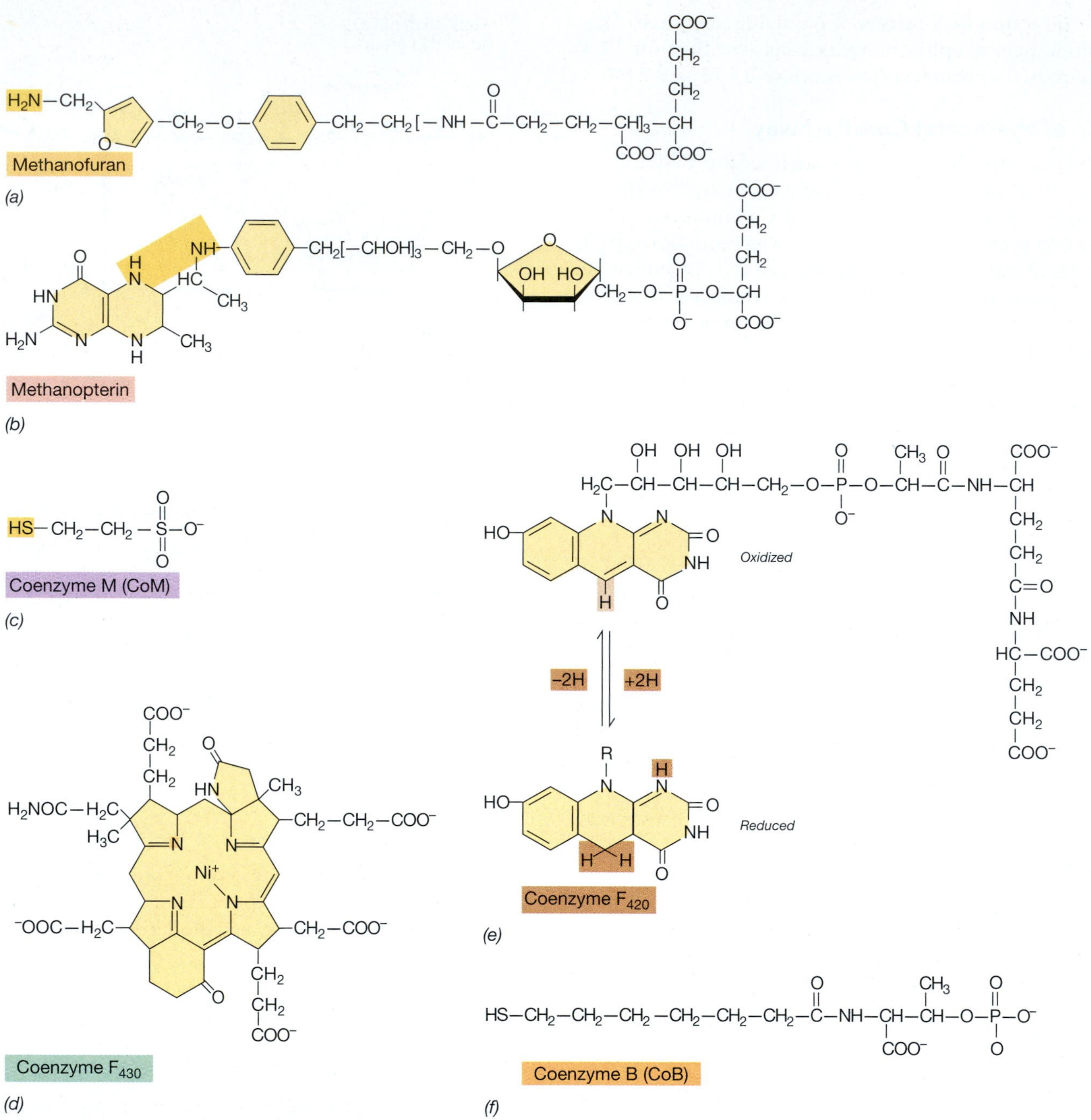

Figure 21.19 Coenzymes of methanogenesis. The atoms shaded in brown or yellow are the sites of oxidation–reduction reactions (F_{420}—brown) or the position to which the C_1 moiety is attached during the reduction of CO_2 to CH_4 (methanofuran, methanopterin, and coenzyme M—yellow). The colors used to highlight a particular coenzyme (CoB is orange, for example) are also in Figures 21.21–21.23 to follow the reactions in each figure.

substrate, CO_2, to the final product and (2) those that supply the electrons necessary for the reduction of CO_2 to CH_4 (**Figure 21.19**, and see Figure 21.21).

The coenzyme methanofuran is required for the first step of methanogenesis. Methanofuran contains the five-membered furan ring and an amino nitrogen atom that binds CO_2 (Figure 21.19*a*). Methanopterin (Figure 21.19*b*) is a methanogenic coenzyme that resembles the vitamin folic acid (Figure 27.16*c*) and plays a role analogous to tetrahydrofolate (a coenzyme that catalyzes C_1 transformations, Table 5.3), by carrying the C_1 unit in the intermediate steps of CO_2 reduction to CH_4. Coenzyme M (CoM) (Figure 21.19*c*) is a small molecule required for the terminal step of methanogenesis, the conversion of a methyl group (CH_3) to CH_4. Although not a C_1 carrier, the nickel-containing tetrapyrrole coenzyme F_{430} (Figure 21.19*d*) is also needed for the terminal step of methanogenesis as part of the methyl reductase enzyme complex (discussed later).

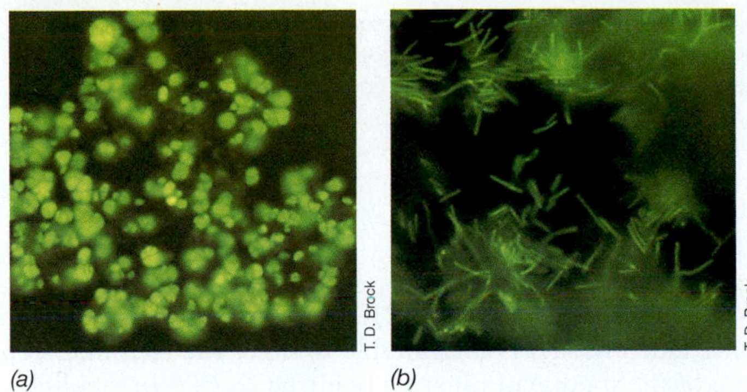

Figure 21.20 Fluorescence due to the methanogenic coenzyme F_{420}. *(a)* Autofluorescence in cells of the methanogen *Methanosarcina barkeri* due to the presence of the unique electron carrier F_{420}. A single cell is about 1.7 μm in diameter. The organisms were made visible with blue light in a fluorescence microscope. *(b)* F_{420} fluorescence in cells of the methanogen *Methanobacterium formicicum*. A single cell is about 0.6 μm in diameter.

Redox Coenzymes

The coenzymes F_{420} and 7-mercaptoheptanoylthreonine phosphate, also called coenzyme B (CoB), are electron donors in methanogenesis. Coenzyme F_{420} (Figure 21.19*e*) is a flavin derivative, structurally resembling the flavin coenzyme FMN (Figure 5.16). F_{420} plays a role in methanogenesis as the electron donor in several steps of CO_2 reduction (see Figure 21.21). The oxidized form of F_{420} absorbs light at 420 nm and fluoresces blue-green. Such fluorescence is useful for the microscopic identification of a methanogen (**Figure 21.20**). CoB is required for the terminal step of methanogenesis catalyzed by the *methyl reductase enzyme complex*. As shown in Figure 21.19*f*, the structure of CoB resembles the vitamin pantothenic acid (which is part of acetyl-CoA) (Figure 5.13).

Methanogenesis from CO_2 + H_2

Electrons for the reduction of CO_2 to CH_4 come mainly from H_2, but formate, carbon monoxide, and even certain organic compounds such as alcohols can also supply the electrons for CO_2 reduction in some methanogens. **Figure 21.21** shows the steps in CO_2 reduction by H_2:

1. CO_2 is activated by a methanofuran-containing enzyme and reduced to the formyl level. The immediate electron donor is ferredoxin, a redox protein with a very low reduction potential (Section 5.11).
2. The formyl group is transferred from methanofuran to an enzyme containing methanopterin (MP in Figure 21.21). It is subsequently dehydrated and reduced in two separate steps to the methylene and methyl levels. The immediate electron donor is reduced F_{420}.
3. The methyl group is transferred from methanopterin to an enzyme containing CoM.
4. Methyl-CoM is reduced to methane by methyl reductase; in this reaction, F_{430} and CoB are intimately involved. Coenzyme F_{430} removes the CH_3 group from CH_3-CoM, forming a Ni^{2+}–CH_3 complex. This complex is reduced by CoB, generating CH_4 and a disulfide complex of CoM and CoB (CoM-S—S-CoB).
5. Free CoM and CoB are regenerated by the reduction of CoM-S—S-CoB with H_2. As we will see, it is this reaction that is coupled to energy conservation in methanogenesis.

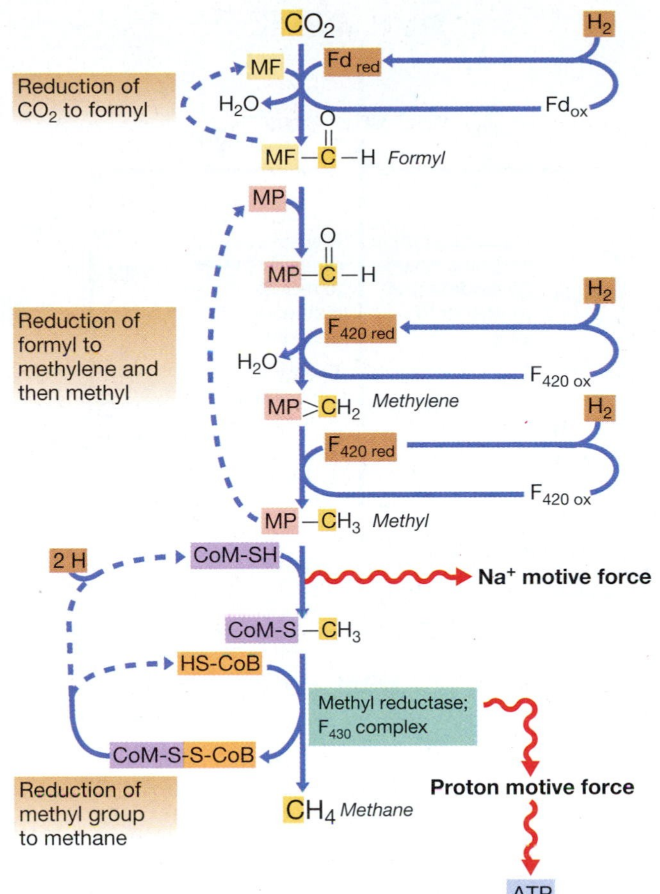

Figure 21.21 Methanogenesis from CO_2 plus H_2. The carbon atom reduced is shown in yellow, and the source of electrons is highlighted in brown. See Figure 21.19 for the structures of the coenzymes and see the text for discussion of the reversible Na^+ pump. MF, Methanofuran; MP, methanopterin; CoM, coenzyme M; F_{420red}, reduced coenzyme F_{420}; F_{430}, coenzyme F_{430}; Fd, ferredoxin; CoB, coenzyme B.

Methanogenesis from Methyl Compounds and Acetate

We learned in Section 17.4 that methane can be formed from methylated compounds as well as from H_2 + CO_2. Methyl compounds such as methanol are catabolized by donating methyl groups to a corrinoid protein to form CH_3-corrinoid (**Figure 21.22**). Corrinoids are the parent structures of compounds such as vitamin B_{12} and contain a porphyrin-like corrin ring with a central cobalt atom (Figure 25.12*a*). The CH_3-corrinoid complex then transfers the methyl group to CoM, yielding CH_3-CoM from which methane is formed in the same way as in the terminal step of CO_2 reduction just

UNIT 4

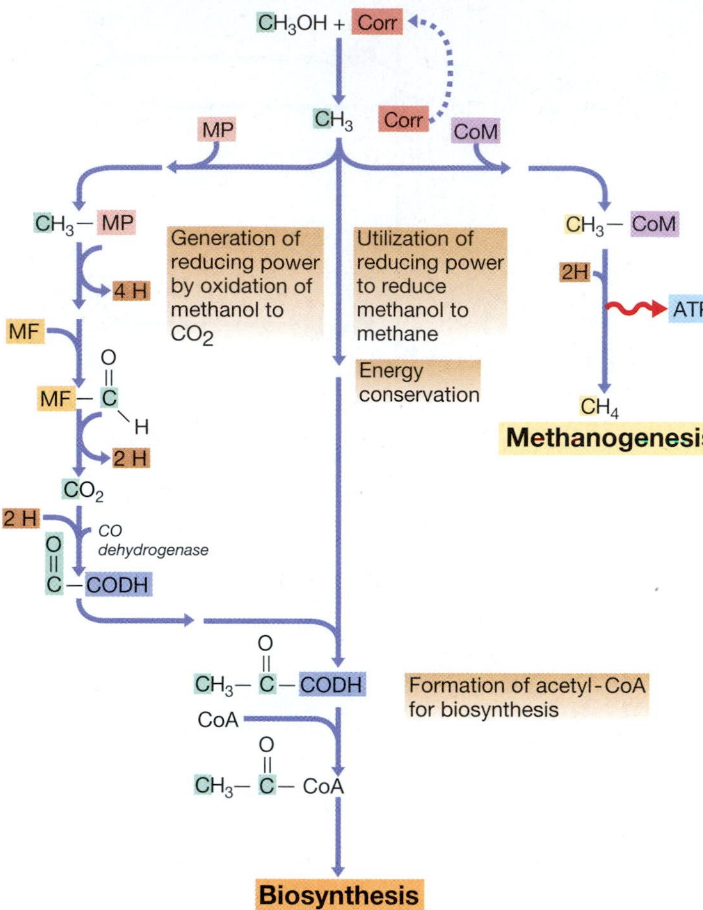

(a) **Methanol to CH_4**

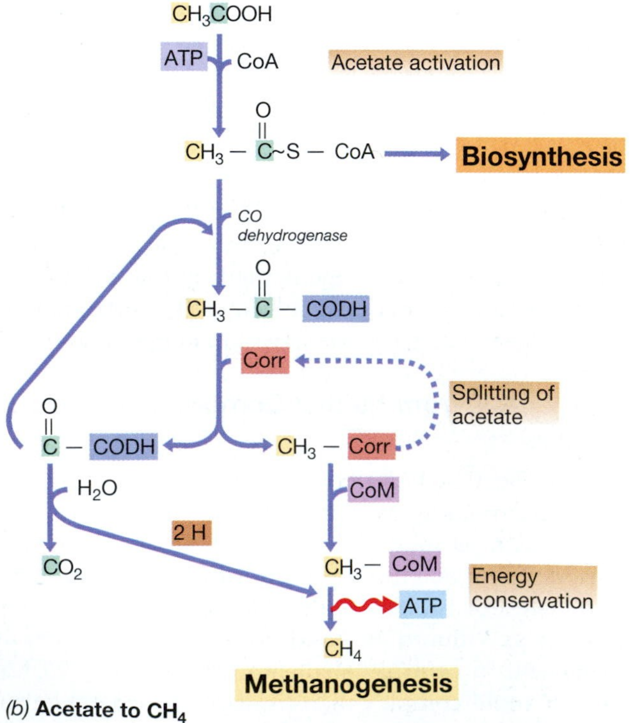

(b) **Acetate to CH_4**

Figure 21.22 Methanogenesis from methanol and acetate. Both reaction series contain parts of the acetyl-CoA pathway. For growth on methanol, most methanol carbon is converted to CH_4, and a smaller amount is converted to either CO_2 or, via formation of acetyl-CoA, is assimilated into cell material. Abbreviations and color-coding are as in Figures 21.19 and 21.21: Corr, corrinoid-containing protein; CODH, carbon monoxide dehydrogenase.

described (compare Figures 21.21 and 21.22*a*). If reducing power (such as H_2) is unavailable to drive the terminal step, some of the methanol must be oxidized to CO_2 to yield electrons for this purpose. This occurs by reversal of steps in methanogenesis (Figure 21.22*a*).

When acetate is the substrate for methanogenesis, it is first activated to acetyl-CoA, which interacts with the carbon monoxide dehydrogenase of the acetyl-CoA pathway (Section 21.9). The methyl group of acetate is then transferred to the corrinoid enzyme to yield CH_3-corrinoid, and from there it goes through the CoM-mediated terminal step of methanogenesis. Simultaneously, the CO group is oxidized to yield the final products of acetate catabolism, CH_4 plus CO_2 (Figure 21.22*b*).

Autotrophy

Autotrophy in methanogens occurs via the acetyl-CoA pathway discussed in Section 21.9. As we have seen, parts of this pathway are integrated into the catabolism of methanol and acetate (Figure 21.22). However, methanogens lack the tetrahydrofolate-driven series of reactions of the acetyl-CoA pathway that lead to the production of a methyl group (Figure 21.18). But this is not a problem because methanogens either derive methyl groups directly from their electron donors (Figure 21.22) or make methyl groups during methanogenesis from $H_2 + CO_2$ (Figure 21.21). Thus methyl groups are abundant, and the removal of some for biosynthesis places little strain on bioenergetics. The carbonyl group of the acetate produced during autotrophic growth of methanogens is derived from the enzyme CO dehydrogenase, and the terminal step in acetate synthesis is as described for acetogens (Section 21.9 and Figure 21.18).

Energy Conservation in Methanogenesis

Under standard conditions, the free energy change in the reduction of CO_2 to CH_4 with H_2 is −131 kJ/mol. This is sufficient for the synthesis of at least one ATP. As mentioned, energy conservation in methanogenesis is linked to the terminal step, the methyl reductase step (Figure 21.21). The interaction of CoB with CH_3—CoM in this step forms CH_4 and a heterodisulfide, CoM-S—S-CoB. This heterodisulfide is reduced by F_{420} to regenerate the active form of the coenzymes, CoM-SH and CoB-SH (Figure 21.21). This reduction, carried out by the enzyme heterodisulfide reductase, is exergonic and is coupled to the pumping of protons across the membrane, creating a proton motive force (**Figure 21.23**). Electron flow to the heterodisulfide reductase requires a unique membrane-integrated electron carrier, a compound called *methanophenazine* (Figure 21.23). In the electron transport process methanophenazine is alternately reduced by F_{420} and then oxidized by a *b*-type cytochrome; the latter is the electron donor to the heterodisulfide reductase (Figure 21.23).

From a conceptual standpoint, ATP synthesis in methanogenesis is simply another example of an anaerobic respiration; membrane-mediated events drive the formation of a proton motive force. However, a major difference is that the electron acceptor, CO_2 in methanogenesis from $H_2 + CO_2$, is reduced in a series of reactions distinct from electron transport events. That is, no membrane-bound "terminal oxidase" reduces the CO_2 such as when, for example, nitrate is reduced in denitrification (Section 21.7).

Methanogenesis from methyl compounds is also linked to the heterodisulfide reductase proton pump, but an additional factor is involved. As previously mentioned, in the absence of H_2, methanogenesis from compounds such as CH_3OH requires that some of the CH_3OH be oxidized to CO_2 to generate the electrons needed for methyl reduction to methane (Figure 21.22). At least one step in this reaction series requires an energy input and is driven by the energy from a sodium motive force that forms during the conversion of CH_3-MP to CH_3-CoM, which is an exergonic reaction (Figure 21.21).

In methanogens we thus see two mechanisms of energy conservation: (1) a proton motive force linked to the methylreductase reaction and used to drive ATP synthesis, and (2) a sodium motive force formed during methanogenesis and used to drive methyl group oxidation during growth on methylated compounds.

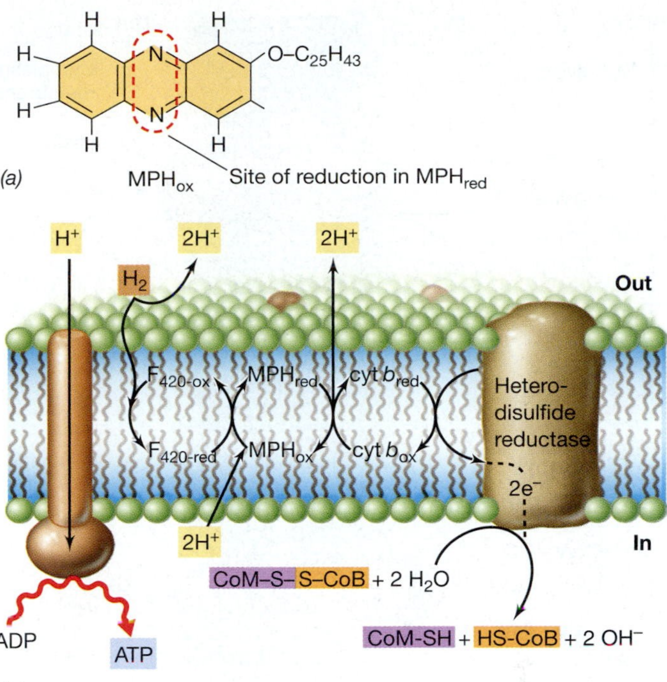

Figure 21.23 Energy conservation in methanogenesis. *(a)* Structure of methanophenazine (MPH in part b), an electron carrier in the electron transport chain leading to ATP synthesis; the central ring of the molecule can be alternately reduced and oxidized. *(b)* Steps in electron transport. Electrons originating from H_2 reduce F_{420} and then methanophenazine. The latter, through a cytochrome of the *b* type, reduces heterodisulfide reductase with the extrusion of protons to the outside of the membrane. In the final step, heterodisulfide reductase reduces Co-M-S—S-CoB to HS-CoM and HS-CoB. See Figure 21.19 for the structures of CoM and CoB.

21.10 MiniReview

Methanogenesis is the biological production of CH_4 from CO_2 plus H_2 or from methylated compounds. Unique coenzymes are required for methanogenesis, and the process is strictly anaerobic. Energy conservation in methanogenesis is linked to both proton and sodium motive forces.

- What coenzymes function as C_1 carriers in methanogenesis? As electron donors?
- Why are the steps in CH_3OH catabolism during methanogenesis different if H_2 is present from those when it is not present?
- How is a proton motive force produced in methanogenesis?

21.11 Proton Reduction

Perhaps the simplest of all anaerobic respirations is one carried out by the hyperthermophile *Pyrococcus furiosus*. *P. furiosus* is a member of the *Archaea* and grows optimally at 100°C (Section 17.11) on sugars and small peptides as electron donors. *P. furiosus* was originally thought to use the glycolytic pathway, because typical fermentation products such as acetate, CO_2, and H_2 were produced from glucose. However, analyses of sugar metabolism in this organism revealed an unusual and enigmatic situation.

During a key step of glycolysis, the oxidation of glyceraldehyde 3-phosphate, the synthesis of 1,3-bisphosphoglyceric acid, an intermediate with two energy-rich phosphate bonds each of which eventually yields ATP, is bypassed in *P. furiosus*, yielding 3-phosphoglyceric acid directly from glyceraldehyde 3-phosphate (**Figure 21.24**). This prevents *P. furiosus* from making ATP by substrate-level phosphorylation at the 1, 3-bisphosphoglyceric acid to 3-phosphoglyceric acid step, one of two sites of energy conservation in the glycolytic pathway (Figure 5.15). This yields *P. furiosus* a net of 0 ATP from glycolytic steps that normally yield 2 ATP. How can *P. furiosus* ferment glucose and ignore the most important energy-yielding steps?

Protons as Electron Acceptors

The riddle of energy conservation in *P. furiosus* was solved when it was discovered that the electron acceptor for the oxidation of 3-phosphoglyceric acid in glycolysis was not the usual acceptor, NAD^+, but instead, the protein ferredoxin (Figure 21.24). Ferredoxin has a much lower E_0' than that of $NAD^+/NADH$, about the same as that of the 2 H^+/H_2 couple, −0.42 V. Reduced ferredoxin is reoxidized by transferring electrons to protons to form H_2 in a reaction that is coupled to energy conservation from proton pumping (Figure 21.24).

UNIT 4

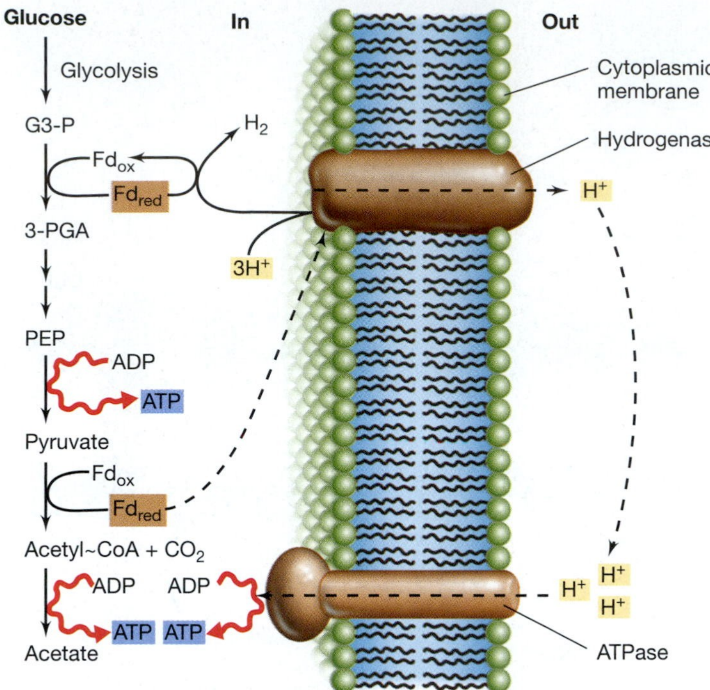

Figure 21.24 Modified glycolysis and proton reduction in anaerobic respiration in the hyperthermophile *Pyrococcus furiosus*. Hydrogen (H_2) production is linked to proton pumping by a hydrogenase that receives electrons from reduced ferredoxin (Fd_{red}). All intermediates from G3-P downward in the pathway are present in two copies. Compare this figure with classical glycolysis in Figure 5.15. G3-P, glyceraldehyde 3-phosphate; 3-PGA, 3-phosphoglycerate; PEP, phosphoenolpyruvate.

H_2 production by fermentative anaerobes is not unusual; H_2 is typically produced during the oxidation of pyruvate to acetate plus CO_2. This allows for ATP to be synthesized by substrate-level phosphorylation, and this also occurs in *P. furiosus* (Figure 21.24). But in addition, the H_2 released from ferredoxin is coupled to the pumping of protons across the membrane by a membrane-integrated hydrogenase. This establishes a proton motive force that drives ATP synthesis from ATPase (Figure 21.24).

Although proton reduction by *P. furiosus* does not involve an electron transport chain *per se,* it can still be considered a form of anaerobic respiration because protons are the net electron acceptor. It thus differs from the proton pumping associated with decarboxylation reactions, such as those of *Oxalobacter,* where the free energy released during decarboxylation is coupled directly to proton translocation (Figure 21.9*b*). In *P. furiosus,* a proton is pumped during hydrogenase activity, analogous to how terminal electron carriers such as cytochromes pump protons in aerobic or other respiratory processes (Figures 5.20 and 21.14).

Growth Yields and Evolution

Measurements of growth yields of *P. furiosus* on glucose indicate that, despite being unable to conserve energy from the main reactions in glycolysis, the organism actually synthesizes more ATP from glucose than many other glucose fermenters! Two ATP are produced by substrate-level phosphorylation during the conversion of two acetyl-CoA to acetate, and about one additional ATP is produced from H_2 production by hydrogenase (Figure 21.24).

Whether proton reduction by prokaryotes is more widespread than this special case is unknown. However, the ancient phylogeny of *Pyrococcus* (Figure 17.1), coupled to its hot, anoxic habitat, similar to that of early Earth (Section 14.1), suggests that proton reduction, a bioenergetic mechanism that requires only a single membrane protein other than ATPase, might have been a very early form of anaerobic respiration, perhaps even nature's first proton pump.

21.11 MiniReview

Pyrococcus furiosus ferments glucose in an unusual fashion, reducing protons in an anaerobic respiration linked to ATPase activity.

- From the standpoint of energy, what advantage does *P. furiosus* have over other organisms that ferment glucose and produce H_2?

21.12 Other Electron Acceptors

In addition to the electron acceptors for anaerobic respiration discussed thus far, ferric iron (Fe^{3+}), manganic ion (Mn^{4+}), chlorate (ClO_3^-), and various organic compounds are important electron acceptors for bacteria in nature (**Figure 21.25**). Diverse bacteria are able to reduce these acceptors, especially Fe^{3+}, and many are able to reduce other acceptors, such as NO_3^- and S^0 (Sections 21.7 and 21.8), as well.

Ferric Iron Reduction

Ferric iron is an electron acceptor for energy metabolism in certain chemoorganotrophic and chemolithotrophic prokaryotes. Because Fe^{3+} is abundant in nature, its reduction is a major form of anaerobic respiration. The reduction potential of the Fe^{3+}/Fe^{2+} couple is somewhat electropositive ($E_0' = +0.2$ V at pH 7), and thus, Fe^{3+} reduction can be coupled to the oxidation of several organic and inorganic electron donors. Various organic compounds, including aromatic compounds, can be oxidized anaerobically by ferric iron reducers; electrons travel through an electron transport chain that terminates in a ferric iron reductase system, reducing Fe^{3+} to Fe^{2+}. The electron flow establishes a proton motive force that drives ATP synthesis by ATPase.

Much research on the energetics of ferric iron reduction has been done with the gram-negative bacterium *Shewanella putrefaciens,* in which Fe^{3+}-dependent anaerobic growth occurs with various organic electron donors. Other important Fe^{3+} reducers include *Geobacter, Geospirillum,* and *Geovibrio,* and several hyperthermophilic *Archaea* (Chapter 17).

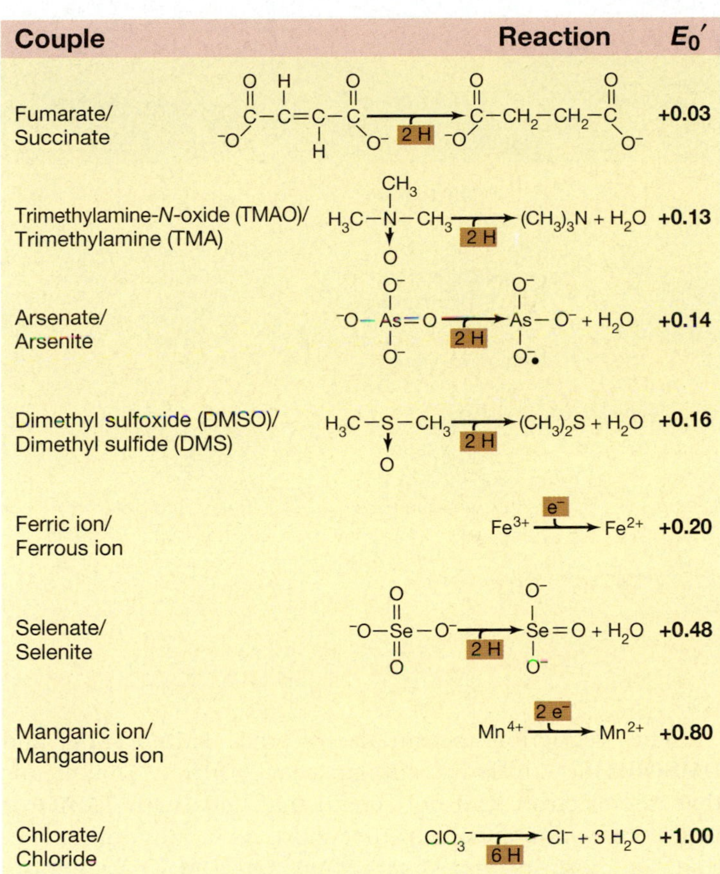

Couple
Reaction
E_0'
Fumarate/ Succinate
2 H
+0.03
Trimethylamine-N-oxide (TMAO)/ Trimethylamine (TMA)
$H_3C-N(CH_3)(\rightarrow O)-CH_3 \xrightarrow{2\,H} (CH_3)_3N + H_2O$
+0.13
Arsenate/ Arsenite
2 H
+0.14
Dimethyl sulfoxide (DMSO)/ Dimethyl sulfide (DMS)
$H_3C-S(\rightarrow O)-CH_3 \xrightarrow{2\,H} (CH_3)_2S + H_2O$
+0.16
Ferric ion/ Ferrous ion
$Fe^{3+} \xrightarrow{e^-} Fe^{2+}$
+0.20
Selenate/ Selenite
2 H
+0.48
Manganic ion/ Manganous ion
$Mn^{4+} \xrightarrow{2\,e^-} Mn^{2+}$
+0.80
Chlorate/ Chloride
$ClO_3^- \xrightarrow{6\,H} Cl^- + 3\,H_2O$
+1.00

Figure 21.25 Some alternative electron acceptors for anaerobic respirations. Note the reaction and E_0' of each redox pair.

Geobacter metallireducens has been a model for study of the physiology of Fe^{3+} reduction. *Geobacter* oxidizes acetate with Fe^{3+} as an acceptor as follows:

$$\text{Acetate}^- + 8\,Fe^{3+} + 4\,H_2O \longrightarrow 2\,HCO_3^- + 8\,Fe^{2+} + 9\,H^+$$

$$\Delta G^{0\prime} = +48.2\text{ kJ}$$

Geobacter can also use H_2 or organic electron donors, including the aromatic hydrocarbon toluene. This is of environmental significance because toluene from accidental spills or leakage from hydrocarbon storage tanks often contaminates iron-rich anoxic aquifers, and organisms such as *Geobacter* may be natural cleanup agents in such environments.

Reduction of Manganese (Mn^{4+}) and Other Inorganic Substances

Manganese has a number of oxidation states, of which Mn^{4+} and Mn^{2+} are the most biologically relevant. The dissimilative reduction of Mn^{4+} to Mn^{2+} is catalyzed by various bacteria, mostly chemoorganotrophs. *Shewanella putrefaciens* and a few other bacteria grow anaerobically on acetate or several other nonfermentable carbon sources with Mn^{4+} as electron acceptor. The reduction potential of the Mn^{4+}/Mn^{2+} couple is extremely high (Figure 21.25); thus, several compounds can donate electrons to Mn^{4+} reduction. This is also the case for chlorate (ClO_3^-) (Figure 21.25). Several chlorate-reducing and perchlorate (ClO_4^{2-})-reducing bacteria have been isolated, and most of them are facultative and thus also capable of aerobic growth.

Figure 21.26 Biomineralization during arsenate reduction by the sulfate-reducing bacterium *Desulfotomaculum auripigmentum*. Left, appearance of culture bottle after inoculation. Right, following growth for two weeks and biomineralization of arsenic trisulfide, As_2S_3. Center, synthetic sample of As_2S_3.

UNIT 4

Other inorganic substances can function as electron acceptors for anaerobic respiration. These include selenium and arsenic compounds (Figure 21.25). Although usually not abundant in natural systems, arsenic and selenium compounds are occasional pollutants and can support anoxic growth of various bacteria. The reduction of SeO_4^{2-} to SeO_3^{2-} and eventually to Se^0 (metallic selenium) is an important method of selenium removal from water and has been used as a means of cleaning—a process called *bioremediation* (∞ Section 24.9)—selenium-contaminated soils. By contrast, the reduction of arsenate to arsenite can actually create a toxicity problem. Some groundwaters flow through strata containing insoluble arsenate minerals. However, if the arsenate is reduced to arsenite by bacteria, the arsenite becomes more mobile and can contaminate groundwater. This has caused a serious problem of arsenic contamination of well water in some developing countries, such as Bangladesh, in recent years.

Other forms of arsenate reduction are beneficial. For example, the sulfate-reducing bacterium *Desulfotomaculum* can reduce arsenate (AsO_4^{3-}) to arsenite (AsO_3^{3-}), along with SO_4^{2-} (to HS^-). During this process a mineral complex of arsenic and sulfide (As_2S_3) precipitates spontaneously (**Figure 21.26**). The mineral is formed both intracellularly and extracellularly, and the process is an example of *biomineralization,* the formation of a mineral by bacterial activity. In this case As_2S_3 formation (Figure 21.26) also functions as a means of detoxifying what would otherwise be a toxic compound (arsenic), and such activities may have practical applications for the microbial cleanup of arsenic-containing toxic wastes and groundwater.

Table 21.7 Characteristics of major genera of bacteria capable of reductive dechlorination

	Genus			
Property	**Dehalobacter**	**Desulfomonile**	**Desulfitobacterium**	**Dehalococcoides**
Electron donors	H_2	H_2, formate, pyruvate, lactate, benzoate	H_2, formate pyruvate, lactate	H_2, lactate
Electron acceptors	Trichloroethylene, tetrachloroethylene	Metachlorobenzoates, tetrachloroethylene, SO_4^{2-}, SO_3^{2-}, $S_2O_3^{2-}$	Ortho-, meta- or para-chlorophenols, NO_3^-, fumarate, SO_3^{2-}, $S_2O_3^{2-}$, S^0	Trichloroethylene, tetrachloroethylene
Product of reduction of tetrachloroethylene	Dichloroethylene	Dichloroethylene	Trichloroethylene	Ethene
Other properties[a]	Contains cytochrome *b*	Contains cytochrome c_3; requires organic carbon source; can grow by fermentation of pyruvate	Can also grow by fermentation	Lacks peptidoglycan
Phylogeny[b]	Related to gram-positive *Bacteria*	Related to *Deltaproteobacteria*	Related to gram-positive *Bacteria*	Related to green non-sulfur *Bacteria* (*Chloroflexus* group)

[a]All organisms are obligate anaerobes.
[b]See Chapters 14–16.

Organic Electron Acceptors

Several organic compounds can be electron acceptors in anaerobic respirations. Of those listed in Figure 21.25, the compound that has been most extensively studied is *fumarate,* a citric acid cycle intermediate, which is reduced to succinate. The role of fumarate as an electron acceptor for anaerobic respiration derives from the fact that the fumarate–succinate couple has a reduction potential near 0 V (Figure 21.25), which allows coupling of fumarate reduction to NADH, FADH, or H_2 oxidation. Bacteria able to use fumarate as an electron acceptor include *Wolinella succinogenes* (which can grow on H_2 as electron donor using fumarate as electron acceptor), *Desulfovibrio gigas* (a sulfate-reducing bacterium that can also grow under non-sulfate-reducing conditions), some clostridia, *Escherichia coli,* and many other bacteria.

Trimethylamine oxide (TMAO) (Figure 21.25) is an important organic electron acceptor. Trimethylamine oxide is a product of marine fish, where it functions as a means of excreting excess nitrogen. Various bacteria can reduce TMAO to trimethylamine (TMA), which has a strong odor and flavor (the odor of spoiled seafood is due primarily to TMA produced by bacterial action). Certain facultatively aerobic bacteria are able to use TMAO as an alternate electron acceptor. In addition, several phototrophic purple nonsulfur bacteria are able to use TMAO as an electron acceptor for anaerobic metabolism in darkness.

A compound similar to TMAO is dimethyl sulfoxide (DMSO), which is reduced by bacteria to dimethyl sulfide (DMS). DMSO is a common natural product and is found in both marine and freshwater environments. DMS has a strong, pungent odor, and bacterial reduction of DMSO to DMS is signaled by this characteristic odor. Bacteria, including *Campylobacter, Escherichia,* and many phototrophic purple bacteria, are able to use DMSO as an electron acceptor in energy generation.

The reduction potentials of the TMAO/TMA and DMSO/DMS couples are similar, near +0.15 V. This means that any electron transport chain that ends with TMAO or DMSO reduction must be rather short. As in fumarate reduction, in most instances of TMAO and DMSO reduction, cytochromes of the *b* type (reduction potentials near 0 V) have been identified as terminal oxidases.

Halogenated Compounds as Electron Acceptors: Reductive Dechlorination

Several chlorinated compounds can function as electron acceptors for anaerobic respiration in the process called **reductive dechlorination** (also called *dehalorespiration*). For example, the sulfate-reducing bacterium *Desulfomonile* grows anaerobically on H_2 or organic compounds as electron donors with chlorobenzoate as an electron acceptor:

$$\underset{\text{3-Chlorobenzoate}}{C_7H_4O_2Cl^-} + 2\,H \longrightarrow \underset{\text{Benzoate}}{C_7H_5O_2^-} + HCl$$

Besides *Desulfomonile,* which is also a sulfate-reducing bacterium (Table 21.5), several other bacteria can reductively dechlorinate, and some of these are restricted to chlorinated compounds as electron acceptors (**Table 21.7**). Many of the chlorinated compounds reduced are toxic to fish and other animal life, whereas many of the products of reductive dechlorination are less toxic or even completely nontoxic. For example, the bacterium *Dehalococcoides* reduces tri- and tetrachloroethylene to the harmless gas ethene (Table 21.7).

The dehalorespiring organism *Dehalobacterium* converts the toxic compound dichloromethane (CH_2Cl_2) into the fatty acids acetate and formate. Thus, reductive dechlorination is not only a form of energy metabolism, but also an environmentally significant bioremedial process. Many reductive dechlorinators are also capable of reducing nitrate or various reduced sulfur compounds (Table 21.7), and thus the group consists of both specialist and opportunist species.

21.12 MiniReview

Besides inorganic nitrogen and sulfur compounds and CO_2, other substances, both organic and inorganic, can function as electron acceptors for anaerobic respiration. These include in particular Fe^{3+}, Mn^{4+}, fumarate, and certain chlorinated compounds.

- With H_2 as electron donor, why is reduction of Fe^{3+} a more favorable reaction than reduction of fumarate?
- Give an example of biomineralization.
- Where in nature might TMAO-degrading bacteria be abundant? Why?
- What is reductive dechlorination?

21.13 Anoxic Hydrocarbon Oxidation Linked to Anaerobic Respiration

Hydrocarbons are organic compounds containing only carbon and hydrogen and are highly insoluble in water. We will see later in this chapter that *aerobic* hydrocarbon oxidation is a common process in nature (Section 21.15). However, both aliphatic and aromatic hydrocarbons can be oxidized to CO_2 under anoxic conditions as well. Anoxic hydrocarbon oxidation is catalyzed by anaerobically respiring bacteria, including, in particular, denitrifying and sulfate-reducing.

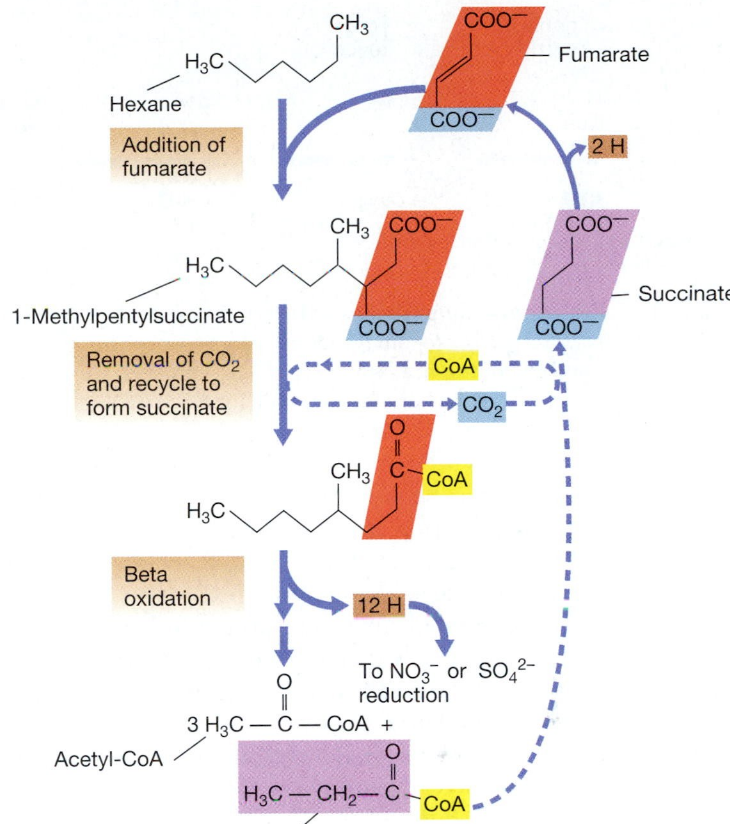

Figure 21.27 Anoxic catabolism of the aliphatic hydrocarbon hexane. The addition of fumarate provides the oxygen atoms necessary to form a fatty acid derivative that can be catabolized by beta-oxidation (see Figure 21.43) to yield acetyl-CoA. Electrons (H) generated from hexane catabolism are used to reduce sulfate or nitrate in anaerobic respirations.

Aliphatic Hydrocarbons

Aliphatic hydrocarbons are straight-chain saturated or unsaturated compounds, and many are substrates for denitrifying and sulfate-reducing bacteria. Saturated aliphatic hydrocarbons as long as C_{20} have been shown to support growth, although shorter chain hydrocarbons are more readily catabolized. The mechanism of anoxic hydrocarbon degradation has been well studied in hexane (C_6H_{14}) metabolism in denitrifying bacteria. The mechanism is probably the same for catabolism of hydrocarbons containing more than six carbon atoms by denitrifying bacteria and is also likely to be the mechanism for aliphatic hydrocarbon degradation in sulfate-reducing bacteria.

Hexane is a saturated aliphatic hydrocarbon. In anoxic hexane metabolism by *Azoarcus*, a species of *Proteobacteria*, hexane is attacked on carbon atom 2 by an enzyme that attaches a molecule of the citric acid cycle intermediate, fumarate, forming an intermediate called *1-methylpentylsuccinate* (**Figure 21.27**). This compound now contains oxygen atoms and can be further catabolized anaerobically (Figure 21.27). After the addition of coenzyme A, a series of reactions occurs that includes beta-oxidation (see Figure 21.43) and regeneration of fumarate. The electrons generated during beta-oxidation travel through an electron transport chain and generate a proton motive force. At the end of the chain, either nitrate (in denitrifying bacteria) or sulfate (in sulfate-reducing bacteria) is reduced (Sections 21.7 and 21.8, respectively).

Aromatic Hydrocarbons

Aromatic hydrocarbons can be degraded anaerobically by a restricted group of denitrifying, phototrophic, ferric iron-reducing, and sulfate-reducing bacteria. However, if the aromatic compound already has an oxygen atom on it, many other organisms can catabolize it anaerobically, including fermentative organisms (Table 21.3). When we examine the aerobic catabolism of aromatic compounds (Section 21.15), we will see that the biochemical mechanism employs ring oxidation (see Figure 21.31). However, under anoxic conditions, the catabolism of aromatic compounds proceeds by ring reduction rather than ring oxidation. Benzoate catabolism has been the focus of much of the work in this area, and the purple phototrophic bacterium *Rhodopseudomonas palustris*, an organism capable of catabolizing a wide variety of aromatic compounds,

UNIT 4

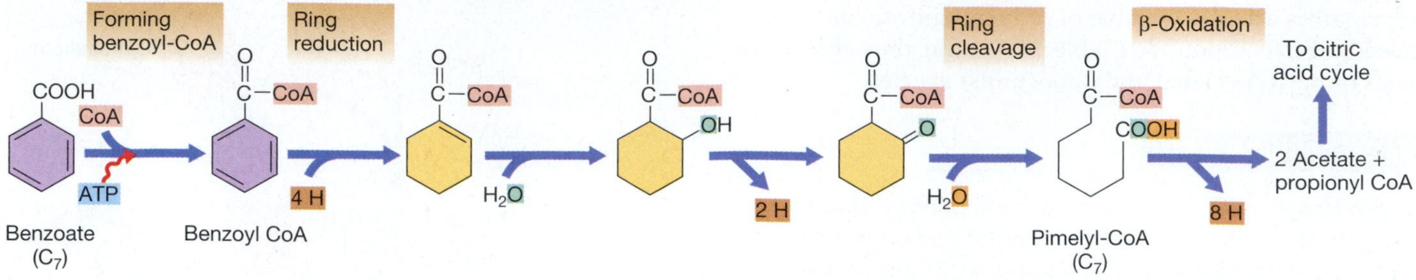

Figure 21.28 Anoxic degradation of benzoate by the benzoyl-CoA pathway. This pathway operates in the purple phototrophic bacterium *Rhodopseudomonas palustris* and many other facultative bacteria, both phototrophic and chemotrophic. Note that all intermediates of the pathway are bound to coenzyme A. The acetate produced is further catabolized in the citric acid cycle.

has been a model experimental organism (**Figure 21.28**). Ring reduction is followed by ring cleavage to yield a straight-chain fatty acid or dicarboxylic acid that can be further catabolized to intermediates of the citric acid cycle (Figure 21.28).

For the anoxic catabolism of the aromatic hydrocarbon toluene, oxygen needs to be added to the compound before ring reduction reactions can proceed. This occurs by the addition of fumarate, just as it does in aliphatic hydrocarbon catabolism (Figure 21.27). The reaction series eventually yields benzoyl-CoA, which is then further catabolized by ring reduction as shown in Figure 21.28. Benzene (C_6H_6) can also be catabolized by nitrate-reducing bacteria, likely by a mechanism similar to that of toluene. Multiring aromatic hydrocarbons such as naphthalene ($C_{10}H_8$) can be degraded by certain sulfate-reducing and denitrifying bacteria. Growth on these substrates is very slow, and oxygenation of the hydrocarbon occurs by the addition of a molecule of CO_2 to the ring to form a carboxylic acid derivative.

Anoxic Oxidation of Methane

Methane (CH_4) is the simplest hydrocarbon. In freshwater ecosystems, methane, produced in the anoxic sediments, is oxidized to CO_2 by methanotrophs when it reaches oxic zones. These methanotrophs require O_2 for the catabolism of methane because the first step in methane oxidation employs a monooxygenase enzyme (Section 21.16 and Figure 21.32). However, methane can also be oxidized under *anoxic* conditions in marine sediments by cell aggregates called *consortia* that contain both sulfate-reducing bacteria and *Archaea* phylogenetically related to methanogens (**Figure 21.29**). However, the archaeal component does not function in the consortium as a methano*gen,* but instead as a methano*troph,* oxidizing methane as an electron donor. Electrons from methane oxidation are presumably transferred in some way to the sulfate reducer, which uses them to reduce sulfate to H_2S (Figure 21.29).

Details of the mechanism of anoxic oxidation of methane (AOM) are unknown, but it appears that the methanotroph first activates methane in some way and then oxidizes it to CO_2 by reversing the steps of methanogenesis, a series of reactions that would be highly exergonic (Section 21.10). Electrons are generated during the oxidative steps, but in what form the electrons are released to the sulfate reducer is unknown. Electrons are not released as H_2. Instead, electrons from the oxidation of methane are shuttled from the methanotroph to the sulfate reducer as an organic compound such as acetate or formate or possibly some sort of diffusible electron carrier (Figure 21.29*b*). Alternatively, it is possible that the archaeal component of the consortium is both a methanotroph *and* a sulfate reducer and completes the entire process

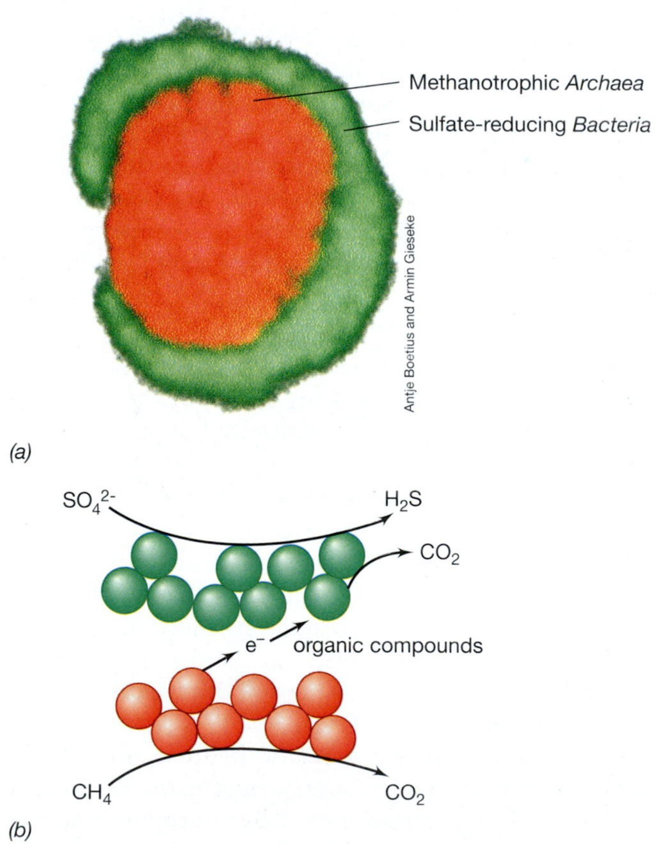

Figure 21.29 Anoxic methane oxidation. *(a)* Methane-oxidizing cell aggregates from marine sediments. The aggregates contain methanogenic *Archaea* (red) surrounded by sulfate-reducing bacteria (green). Each cell type has been stained by a different FISH probe. The aggregate is about 30 μm in diameter. *(b)* Possible mechanism for the cooperative degradation of methane. An organic compound or some other carrier of reducing power is likely the mechanism of electron transfer from methanotroph to sulfate reducer.

of AOM itself. In this case, the sulfate-reducing bacterium would exist in the consortium for reasons other than consuming electrons from AOM.

Regardless of mechanism, AOM yields only a very small amount of energy, and how this energy is split between the methanotroph and the sulfate reducer (assuming that the consortium is obligatory) is unknown. Biochemical studies of the consortia have shown that the methanotrophic component contains a modified form of the methanogenic coenzyme F_{430}. Recall that F_{430} plays a key role in the terminal step of methanogenesis (Figures 21.19 and 21.21). The modified F_{430} probably functions in the reverse role during AOM. Whether the methanotrophic component of the consortia is also able to make methane (that is, function as a methanogen) is unknown.

AOM is not limited to methanogen/sulfate-reducing bacteria consortia. Methane-oxidizing denitrifying consortia are present in anoxic environments where CH_4 and nitrate coexist in significant amounts. Conceivably, then, AOM could be coupled to other forms of anaerobic respiration as well, such as ferric iron reduction. The link to sulfate reduction was the first to be discovered and would naturally predominate in marine sediments because sulfate reduction is the dominant form of anaerobic respiration that occurs there (Section 24.4).

21.13 MiniReview

Hydrocarbons can be oxidized under anoxic conditions by bacteria. Whether the starting reactant is an aromatic or aliphatic hydrocarbon, oxygen must be added to the molecule. This occurs by the addition of fumarate. Aromatic compounds are catabolized by ring reduction and cleavage. Subsequently, both types of molecules are oxidized to intermediates that can be catabolized in the citric acid cycle. Methane can be oxidized under anoxic conditions by consortia containing sulfate-reducing bacteria and methanogen-like *Archaea*.

- Why is toluene a hydrocarbon whereas benzoate is not?
- How is hexane oxygenated during anoxic catabolism?
- What is AOM and which organisms participate in the process?

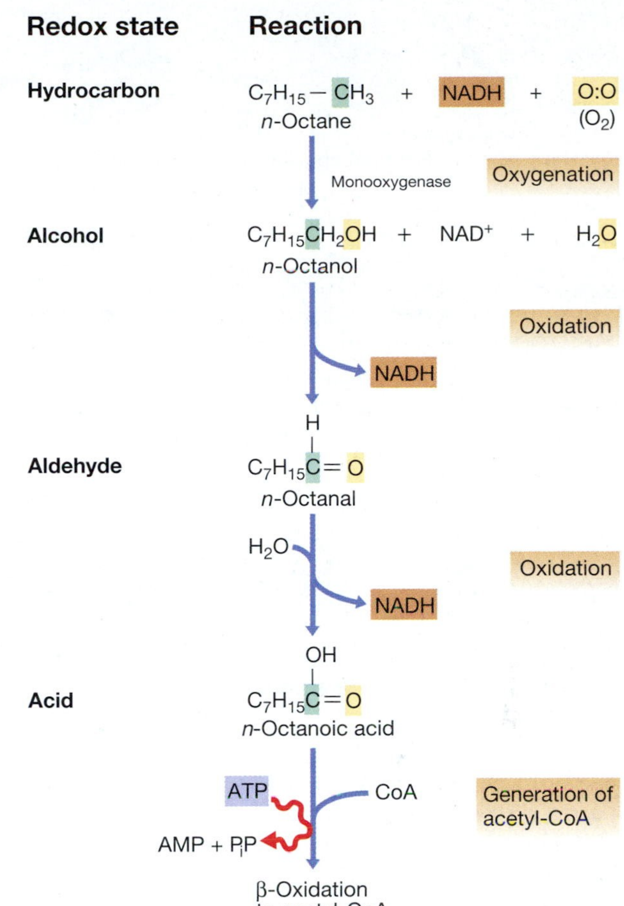

Figure 21.30 Monooxygenase activity. Steps in oxidation of an aliphatic hydrocarbon, the first of which is catalyzed by a monooxygenase. Some sulfate-reducing and denitrifying bacteria can degrade aliphatic hydrocarbons under anoxic conditions. For a description of beta-oxidation, see Figure 21.43.

III AEROBIC CHEMOORGANOTROPHIC PROCESSES

Many organic compounds are catabolized aerobically, and we survey some major aerobic processes here. We begin with a consideration of the O_2 requirements for some of these reactions.

21.14 Molecular Oxygen as a Reactant in Biochemical Processes

We previously discussed the role of molecular oxygen (O_2) as an *electron acceptor* in energy-generating reactions (Sections 5.11 and 5.12). Although this is by far the most important role of O_2 in cellular metabolism, O_2 plays an important role as a *direct reactant* in certain types of anabolic and catabolic processes.

Oxygenases

Oxygenases are enzymes that catalyze the incorporation of atoms of oxygen from O_2 into organic compounds. There are two classes of oxygenases: *dioxygenases*, which catalyze the incorporation of both atoms of O_2 into the molecule, and *monooxygenases*, which catalyze the incorporation of only one of the two oxygen atoms of O_2 to an organic compound; the second atom of O_2 is reduced to water. For most monooxygenases, the electron donor is NADH or NADPH (**Figure 21.30**). In the example of ammonia monooxygenase discussed previously (Section 20.12), the electron donor was cytochrome *c*.

Several types of reactions in living organisms require O_2 as a reactant. One of the best examples is O_2 in sterol biosynthesis. The formation of the fused sterol ring system (Figure 4.6*a*) requires O_2. Such a reaction obviously cannot take place under

UNIT 4

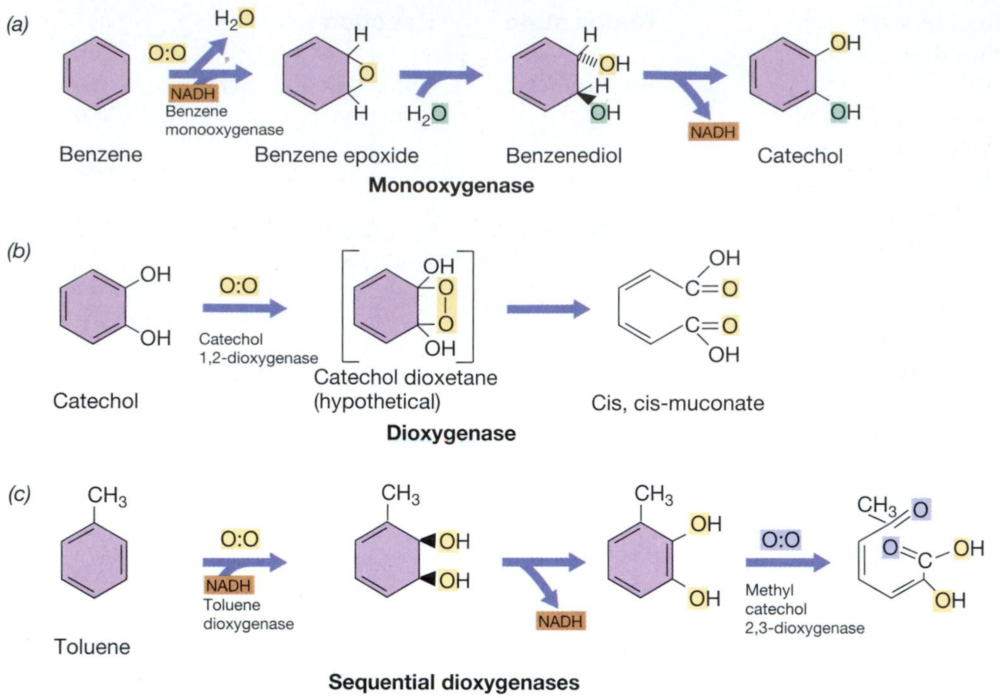

Figure 21.31 Roles of oxygenases in catabolism of aromatic compounds. Monooxygenases introduce one atom of oxygen from O_2 into a substrate, whereas diooxygenases introduce both atoms of oxygen. *(a)* Hydroxylation of benzene to catechol by a monooxygenase in which NADH is an electron donor. *(b)* Cleavage of catechol to *cis,cis*-muconate by an intradiol ring-cleavage dioxygenase. *(c)* The activities of a ring-hydroxylating dioxygenase and an extradiol ring-cleavage dioxygenase in the degradation of toluene. The oxygen atoms that each enzyme introduces are distinguished by different colors. Catechol and related compounds are common intermediates in aerobic aromatic catabolism.

anoxic conditions, so organisms that grow anaerobically must either grow without sterols or obtain the needed sterols preformed from their environment. The requirement of O_2 in biosynthesis is of evolutionary significance, as O_2 was originally absent from the atmosphere of Earth when life first evolved. Oxygen became available on Earth only after the proliferation of cyanobacteria, approximately 2.7 billion years before the present (Section 14.3).

21.14 MiniReview

In addition to its role as an electron acceptor, oxygen is also a chemical reactant in certain biochemical processes. Enzymes called oxygenases introduce O_2 into a biochemical compound.

- How do monooxygenases differ in function from dioxygenases?

21.15 Aerobic Hydrocarbon Oxidation

We saw in Section 21.13 how hydrocarbons could be degraded under anoxic conditions. Here we consider the aerobic oxidation of hydrocarbons. Low-molecular-weight hydrocarbons are gases, whereas those of higher molecular weight are liquids or solids. Hydrocarbon consumption can be a natural process or can be a directed process for the cleaning up of spilled hydrocarbons from human activities (oil bioremediation, Section 24.8). Either way, the aerobic catabolism of hydrocarbons can be a very rapid process owing to the metabolic advantage of having O_2 available as an electron acceptor.

Aliphatic Hydrocarbon Metabolism

A number of bacteria and several molds and yeasts can use hydrocarbons as electron donors to support growth under aerobic conditions. The initial oxidation step of saturated aliphatic hydrocarbons by these organisms requires molecular oxygen (O_2) as a reactant, and one of the atoms of the oxygen molecule is incorporated into the oxidized hydrocarbon, typically at a terminal carbon atom. This reaction is carried out by a monooxygenase (Section 21.14), and a typical reaction sequence is that shown in Figure 21.30. The end product of the reaction sequence is acetyl-CoA, and this is catabolized in the citric acid cycle along with the production of electrons for the electron transport chain. The sequence is repeated to progressively degrade hydrocarbon chains, and in most cases the hydrocarbon is oxidized completely to CO_2.

Aromatic Hydrocarbons

Many aromatic hydrocarbons can also be used as electron donors under aerobic conditions by microorganisms, of which bacteria of the genus *Pseudomonas* have been the best studied. The metabolism of these compounds, some of which can be quite large molecules, typically has as its initial stage the formation of catechol or a structurally related compound via attack by oxygenase enzymes, as shown in **Figure 21.31**.

Single-ring compounds such as catechol are called *starting substrates* because oxidative catabolism proceeds only after the large polyaromatic molecules have been converted to these more simple forms. Once a compound like catechol is formed it can be further degraded to compounds that can enter the citric acid cycle: succinate, acetyl-CoA, and pyruvate. Several steps in the aerobic catabolism of aromatic hydrocarbons require oxygenases. Figures 21.31*a–c* show four different oxygenase-catalyzed reactions, one using a monooxygenase, two using a ring-cleaving dioxygenase, and one using a ring-hydroxylating dioxygenase. As in aerobic aliphatic hydrocarbon catabolism, aromatic compounds, whether single or multiple ringed, are typically oxidized completely to CO_2.

21.15 MiniReview

Aerobic hydrocarbon metabolism is widespread in nature, and oxygenase enzymes are key to these catalyses. Unlike in anaerobic aromatic catabolism, aerobic aromatic degradation proceeds by pathways that employ ring oxidation.

- Draw the chemical structure for benzene. Do the same for benzoate. Are these compounds aliphatic or aromatic?
- What fundamental difference exists in the anaerobic degradation of an aromatic compound compared with its aerobic metabolism? Give an example of this.

21.16 Methylotrophy and Methanotrophy

Methane and many other C_1 compounds can be catabolized by **methylotrophs**. Methylotrophs use C_1 compounds or other organic compounds that lack C—C bonds as electron donors and also as carbon sources (Section 15.6). We focus here on the physiology of this process, using methane as an example.

Biochemistry of Methane Oxidation

The steps in methane oxidation to CO_2 can be summarized as

$$CH_4 \longrightarrow CH_3OH \longrightarrow CH_2O \longrightarrow HCOO^- \longrightarrow CO_2 \text{ (methane} \longrightarrow \text{methanol} \longrightarrow \text{formaldehyde} \longrightarrow \text{formate} \longrightarrow CO_2)$$

Methanotrophs are those methylotrophs that can use CH_4. Methanotrophs assimilate either all or one-half of their carbon (depending on the pathway used) at the oxidation state of formaldehyde. We will see later that this affords a major energy savings compared with the carbon assimilation of autotrophs, which also assimilate C_1 units, but from CO_2 rather than from organic compounds.

Reactions and Bioenergetics of Aerobic Methanotrophy

The initial step in the oxidation of methane under oxic conditions requires an enzyme called *methane monooxygenase* (MMO). As we discussed in Section 21.14, oxygenases catalyze the incorporation of oxygen from O_2 into carbon compounds (and into some nitrogen compounds, Section 20.12) and help metabolize hydrocarbons. Methanotrophy has been well studied in *Methylococcus capsulatus*. This organism contains two MMOs, one cytoplasmic and the other membrane integrated. The electron donor for the cytoplasmic MMO is NADH, and NADH is probably the electron donor for the membrane-integrated MMO as well (**Figure 21.32**).

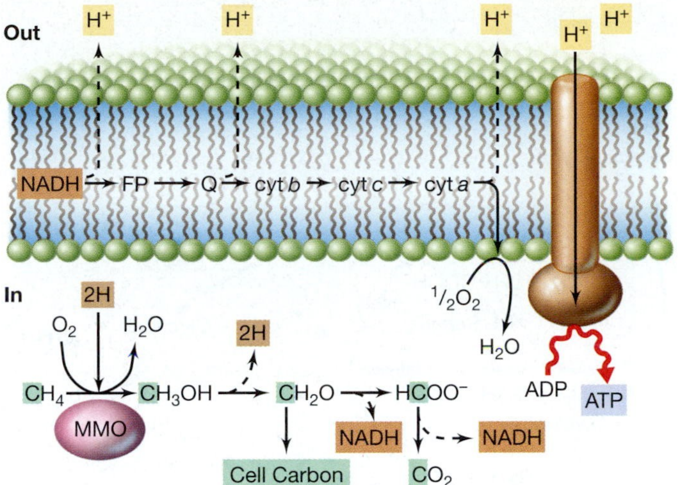

Figure 21.32 Oxidation of methane by methanotrophic bacteria. Methane (CH_4) is converted to methanol (CH_3OH) by the enzyme methane monooxygenase (MMO). A proton motive force is established from electron flow in the membrane, and this fuels ATPase. Note how carbon for biosynthesis comes from formaldehyde (CH_2O). Although not depicted as such, MMO is actually a membrane-associated enzyme and methanol dehydrogenase is periplasmic. FP, flavoprotein; Cyt, cytochrome; Q, quinone.

UNIT 4

In the MMO reaction, an atom of oxygen is introduced into CH_4, and CH_3OH and H_2O are the products. Reducing power for the first step comes from later oxidative steps in the pathway. These include CH_3OH, CH_2O, and $HCOO^-$ oxidation, each of which is a two-electron oxidation. CH_3OH is oxidized by a periplasmic dehydrogenase, yielding formaldehyde and NADH. Once CH_2O is formed it is oxidized to CO_2 by either of two different pathways. One pathway involves enzymes that employ the coenzyme tetrahydrofolate, a coenzyme widely involved in C_1 transformations (Section 5.1). The second and totally independent pathway employs the coenzyme methanopterin. Recall that methanopterin is a C_1 carrier during methanogenesis, the reduction of CO_2 to CH_4, and that methanogens are *Archaea* (Section 21.10). It is of interest that methanopterin is also employed by methanotrophs, which are phylogenetically *Bacteria*, to carry out the reverse reaction, the oxidation of formaldehyde to formate. This is likely an example of horizontal gene flow (Section 13.11), where methanogens and methanotrophs have shared genes that encode proteins common to their biochemistry. However, regardless of the CH_2O oxidation pathway employed, electrons from the oxidation of CH_2O to CO_2 enter the electron transport chain generating a proton motive force from which ATP is synthesized (Figure 21.32).

C_1 Assimilation into Cell Material

As was noted in Section 15.6, two physiological groups of methanotrophs are known, type I and type II, and each group employs its own specific pathway for C_1 incorporation into cell material. The **serine pathway**, utilized by type II methanotrophs, is outlined in **Figure 21.33**. In this pathway, a two-carbon unit, acetyl-CoA, is synthesized from one molecule of formaldehyde (produced from the oxidation of CH_3OH,

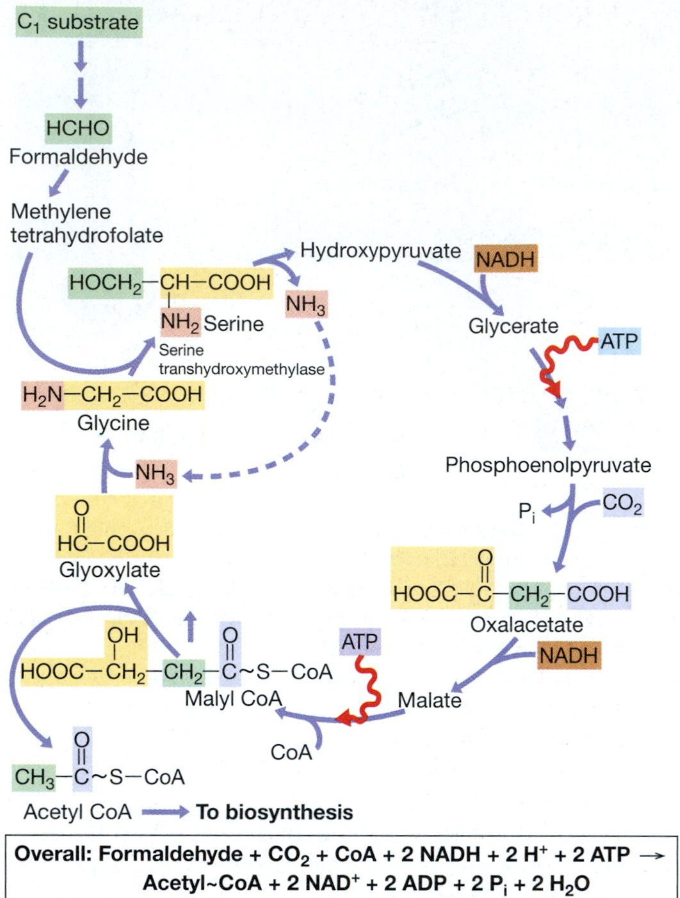

Figure 21.33 The serine pathway for the assimilation of C_1 units into cell material by methylotrophic bacteria. The product of the pathway, acetyl-CoA, is used as the starting point for making new cell material. The key enzyme of the pathway is serine transhydroxymethylase.

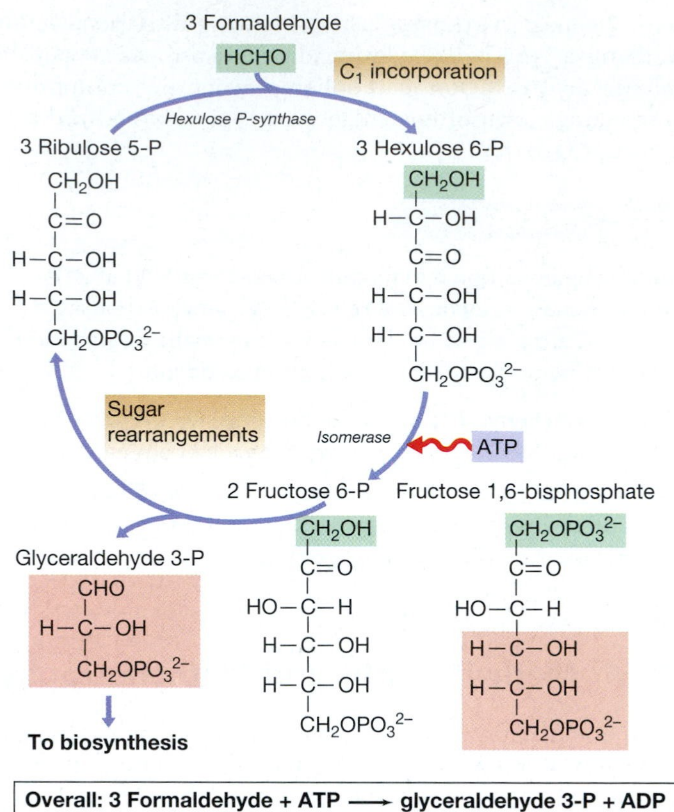

Figure 21.34 The ribulose monophosphate pathway for assimilation of C_1 units by methylotrophic bacteria. Three molecules of formaldehyde are needed to complete the cycle, with the net result being one molecule of glyceraldehyde 3-P. The key enzyme of this pathway is hexulose P-synthase. The sugar rearrangements require enzymes of the pentose phosphate pathway (Figure 21.39).

Figure 21.32) and one molecule of CO_2. The serine pathway requires reducing power and energy in the form of two molecules each of NADH and ATP, respectively, for each acetyl-CoA synthesized. The serine pathway employs a number of enzymes of the citric acid cycle and one enzyme, *serine transhydroxymethylase,* unique to the pathway (Figure 21.33).

The **ribulose monophosphate pathway**, used by type I methanotrophs, is outlined in **Figure 21.34**. This pathway is more efficient than the serine pathway because all of the carbon for cell material is derived from formaldehyde. And, because formaldehyde is at the same oxidation level as cell material, no reducing power is needed. The ribulose monophosphate pathway requires one molecule of ATP for each molecule of glyceraldehyde 3-phosphate synthesized (Figure 21.34). Consistent with the lower energy requirements of the ribulose monophosphate pathway, the cell yield (grams of cells produced per mole of CH_4 oxidized) of type I methanotrophs is higher than for type II methanotrophs.

The enzymes *hexulosephosphate synthase,* which condenses one molecule of formaldehyde with one molecule of ribulose 5-P, and *hexulose 6-P isomerase* (Figure 21.34) are unique to the ribulose monophosphate pathway. The remaining enzymes of this pathway are common enzymes in sugar transformations in many different organisms. It should also be noted that the substrate for the initial reaction in this pathway, ribulose 5-P, is very similar to the C_1 acceptor in the Calvin cycle, ribulose 1,5-bisphosphate (Section 20.6), an indication that these two cycles likely share evolutionary roots.

21.16 MiniReview

Methanotrophy is the use of CH_4 as a source of both carbon and electrons. The enzyme methane monooxygenase is a key enzyme in the catabolism of methane. In methanotrophs C_1 units are assimilated into cell material at the oxidation level of formaldehyde by either the ribulose monophosphate pathway or the serine pathway.

- What are the energy and reducing power requirements for the ribulose monophosphate pathway? For the serine pathway? Write a single sentence that explains why the energy requirements of the two pathways differ.
- Why does the oxidation of CH_4 to CH_3OH require reducing power?
- Which pathway, the Calvin cycle or the ribulose monophosphate pathway, requires the greater energy input? Why?

21.17 Hexose, Pentose, and Polysaccharide Metabolism

Sugars and polysaccharides are common substrates for chemoorganotrophs, and we consider their catabolism here.

Hexose and Polysaccharide Utilization

Sugars containing six carbon atoms, called *hexoses,* are the most important electron donors for many chemoorganotrophs and are also important structural components of microbial cell walls, capsules, slime layers, and storage products (⚭ Chapter 4). The most common sources of hexose in nature are listed in **Table 21.8**, from which it can be seen that most are polysaccharides, although a few are disaccharides. Cellulose and starch are two of the most abundant natural polysaccharides.

Although both starch and cellulose are composed entirely of glucose, the glucose units are bonded differently (Table 21.8; ⚭ Figure 3.6*b*), and this profoundly affects their properties. Cellulose is much more insoluble than starch and is usually less rapidly digested. Cellulose forms long fibrils, and organisms that digest cellulose are often found attached directly to these fibrils (**Figure 21.35**). In this way cellulase, the enzyme required to degrade cellulose, can contact its substrate and begin the digestive process. Many fungi are able to digest cellulose, and these are mainly responsible for the decomposition of plant materials on the forest floor. Among bacteria, however, cellulose digestion is restricted to relatively few groups, of which the gliding bacteria such as *Sporocytophaga* and *Cytophaga* (**Figures** 21.35 and **21.36**), clostridia, and actinomycetes are the most common.

Anoxic digestion of cellulose is carried out by a few *Clostridium* species, which are common in lake sediments, animal intestinal tracts, and systems for anoxic sewage digestion. Cellulose digestion is also a major process in the rumen

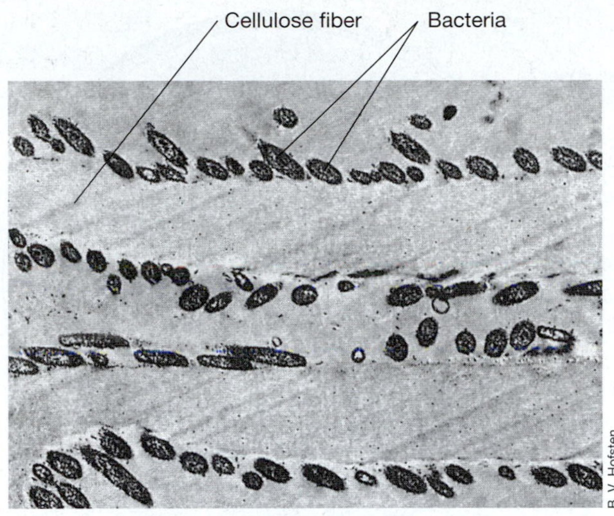

Figure 21.35 Cellulose digestion. Transmission electron micrograph showing attachment of the cellulose-digesting bacterium *Sporocytophaga myxococcoides* to cellulose fibers. Cells are about 0.5 μm in diameter (⚭ Figure 16.36*d*).

of ruminant animals where *Fibrobacter* and *Ruminococcus* species actively degrade cellulose (⚭ Section 24.10).

Starch is digestible by many fungi and bacteria; this is illustrated for a laboratory culture in **Figure 21.37**. Starch-digesting enzymes, called *amylases,* are of considerable practical utility in industrial situations where starch must be digested, such as the textile, laundry, paper, and food industries, and fungi and bacteria are the commercial sources of these enzymes (⚭ Section 25.9).

All polysaccharides used as growth substrates are first enzymatically hydrolyzed to monomeric or oligomeric units. In contrast, polysaccharides formed within cells as storage products (⚭ Section 4.10) are broken down not by *hydro*lysis, but

Table 21.8 Naturally occurring polysaccharides yielding hexose and pentose sugars[a]

Substance	*Composition*	*Sources*	*Catabolic enzymes*
Cellulose	Glucose polymer (β-1,4-)	Plants (leaves, stems)	Cellulases (β, 1-4-glucanases)
Starch	Glucose polymer (α-1,4-)	Plants (leaves, seeds)	Amylase
Glycogen	Glucose polymer (α-1,4- and α-1,6-)	Animals (muscle) and microorganisms (granules)	Amylase, phosphorylase
Laminarin	Glucose polymer (β-1,3-)	Marine algae (Phaeophyta)	β-1,3-Glucanase (laminarinase)
Paramylon	Glucose polymer (β-1,3-)	Algae (Euglenophyta and Xanthophyta)	β-1,3-Glucanase
Agar	Galactose and galacturonic acid polymer	Marine algae (Rhodophyta)	Agarase
Chitin	*N*-Acetylglucosamine polymer (β-1,4-)	Fungi (cell walls) Insects (exoskeletons)	Chitinase
Pectin	Galacturonic acid polymer (from galactose)	Plants (leaves, seeds)	Pectinase (polygalacturonase)
Dextran	Glucose polymer	Capsules or slime layers of bacteria	Dextranase
Xylan	Heteropolymer of xylose and other sugars (β-1,4- and α-1,2 or α-1,3 side groups)	Plants	Xylanases
Sucrose	Glucose-fructose disaccharide	Plants (fruits, vegetables)	Invertase
Lactose	Glucose-galactose disaccharide	Milk	β-Galactosidase

[a]Each of these is subject to degradation by microorganisms.

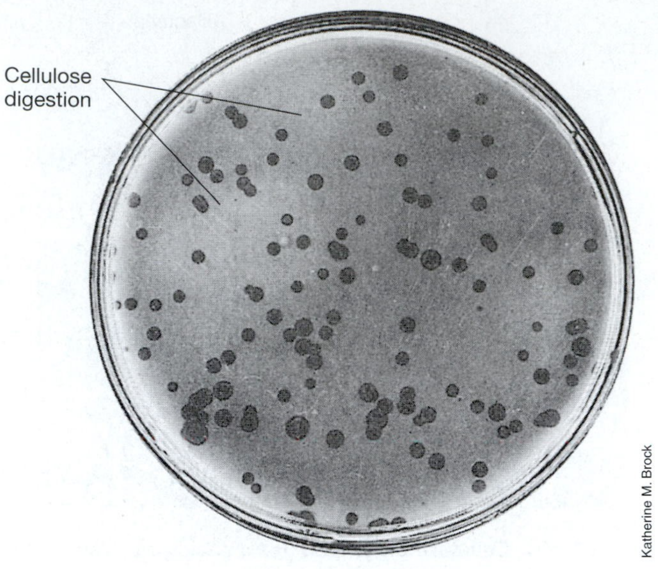

Figure 21.36 ***Cytophaga hutchinsonii* colonies on a cellulose-agar plate.** Areas where cellulose has been hydrolyzed are clear.

by *phosphorolysis*. This involves the addition of inorganic phosphate and results in the formation of hexose phosphate rather than free hexose. It may be summarized as follows for the degradation of starch, an α-1,4 polymer of glucose:

$$(C_6H_{12}O_6)_n + P_i \longrightarrow (C_6H_{12}O_6)_{n-1} + \text{glucose 1-phosphate}$$

Because glucose 1-phosphate can be easily converted to glucose 6-phosphate—a key intermediate in glycolysis—with no energy expenditures, phosphorolysis represents a net energy savings to the cell.

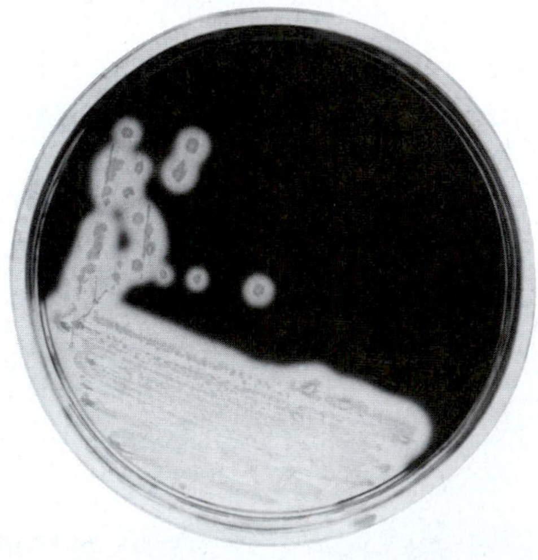

Figure 21.37 **Hydrolysis of starch by *Bacillus subtilis*.** After incubation, the starch agar plate was flooded with Lugol's iodine solution. Where starch has been hydrolyzed, the characteristic purple-black color of the starch-iodine complex is absent. Starch is hydrolyzed at some distance from the colonies because cells of *B. subtilis* produce the extracellular enzyme (exoenzyme) amylase, which diffuses into the surrounding medium.

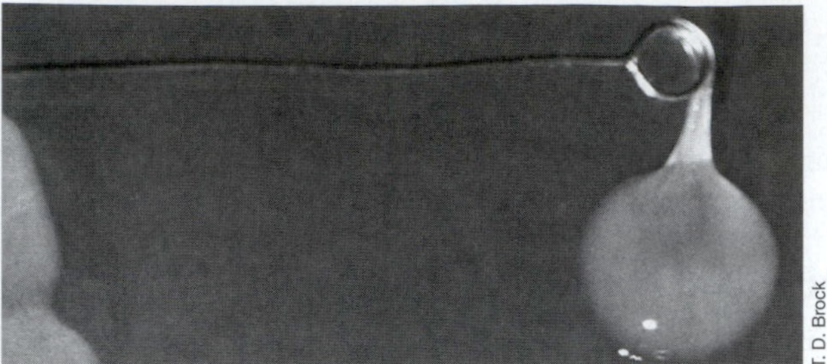

Figure 21.38 **Slime formation.** A slimy colony formed by the dextran-producing bacterium *Leuconostoc mesenteroides* growing on a sucrose-containing medium. When the same organism is grown on glucose, the colonies are small and not slimy because dextran (a branched polysaccharide of glucose) synthesis specifically requires sucrose (∞ Section 28.3).

Disaccharides

Many microorganisms can use disaccharides for growth (Table 21.8). *Lactose* utilization by microorganisms is of considerable economic importance because milk-souring organisms produce lactic acid from lactose. *Sucrose,* the common disaccharide of higher plants, is usually first hydrolyzed to its component monosaccharides (glucose and fructose) by the enzyme invertase, and the monomers are then metabolized by the glycolytic pathway. *Cellobiose* (β-1,4-diglucose), a major product of cellulose digestion by cellulase, is degraded by cellulolytic bacteria but can also be degraded by many bacteria that are unable to degrade the cellulose polymer itself.

The microbial polysaccharide *dextran* is synthesized by some bacteria using the enzyme dextransucrase and sucrose as starting material:

$$n \text{ sucrose} \longrightarrow \underset{\text{dextran}}{(\text{glucose})_n} + n \text{ fructose}$$

Dextran is formed in this way by the bacterium *Leuconostoc mesenteroides* and a few others, and the polymer formed accumulates around the cells as a massive slime layer or capsule (**Figure 21.38**). Because sucrose is required for dextran formation, no dextran is formed when the bacterium is cultured on a medium containing glucose or fructose (∞ Section 28.3). In nature, when cells die that contain dextran or other polysaccharide capsules, these materials once again become available for attack by fermentative or other chemoorganotrophic microorganisms.

The Pentose Phosphate Pathway

Pentose sugars are often available in nature. But if they are not available, they must be synthesized, because they form the backbone of the nucleic acids (∞ Section 3.5). Pentoses are made from hexose sugars, and the major pathway for this process is the **pentose phosphate pathway**.

Figure 21.39 summarizes the pentose phosphate pathway. Several important features should be noted. First, glucose can be converted into a pentose by loss of one carbon

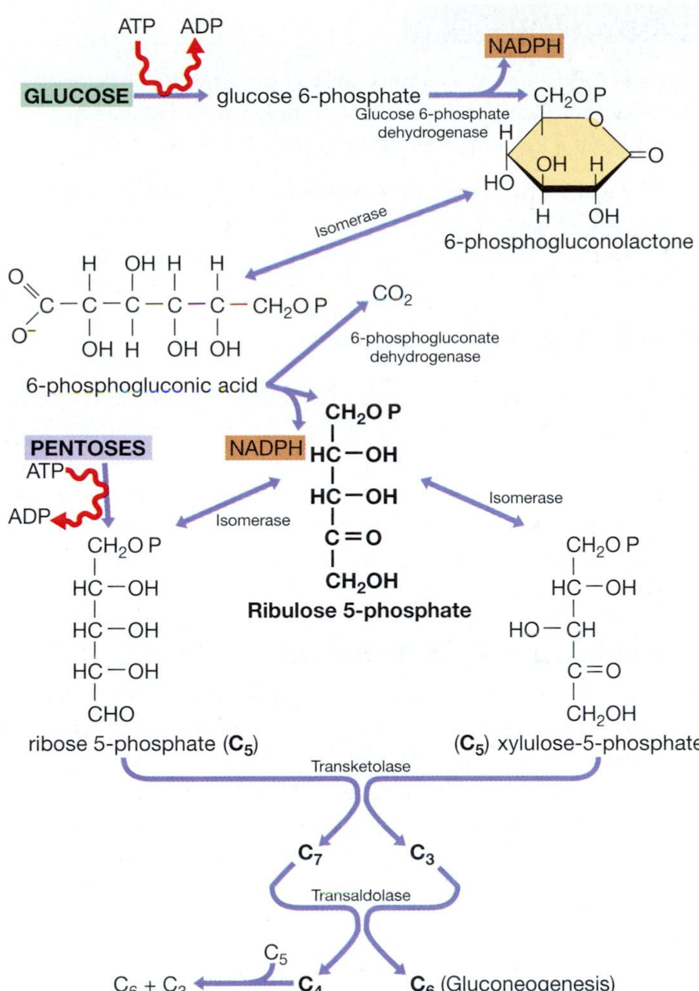

Figure 21.39 The pentose phosphate pathway. The pathway is used to: (1) form pentoses from hexoses; (2) form hexoses from pentoses (gluconeogenesis); (3) catabolize pentoses as electron donors; and (4) generate NADPH. Some key enzymes of the pathway are indicated.

atom by oxidation followed by decarboxylation. This generates NADPH and the key intermediate of the pathway, *ribulose 5-phosphate* (Figure 21.39). From the latter, ribose (and from it deoxyribose, Section 5.16) are formed to supply the cell with nucleic acid precursors. Pentose sugars as electron donors can also feed into the pentose phosphate pathway, typically becoming phosphorylated to form ribose phosphate or a related compound (Figure 21.39) before being further catabolized.

A second important feature of the pentose phosphate pathway is the generation of sugar diversity. A variety of sugar derivatives, including C_4, C_5, C_6, and C_7, are formed in reactions of the pathway (Figure 21.39). This allows for pentose sugars to eventually yield hexoses for either catabolic purposes or for biosynthesis (gluconeogenesis, Section 5.15).

A final important aspect of the pentose phosphate pathway is that it generates NADPH (Figure 21.39). The reduced form of this redox coenzyme is used by the cell for reductive biosyntheses. Although most cells have a mechanism for converting NADH into NADPH, the pentose phosphate pathway is the major means for direct synthesis of this important coenzyme.

21.17 MiniReview

Polysaccharides are abundant in nature and can be broken down into hexose or pentose monomers and used as sources of both carbon and reductant. Starch and cellulose are common polysaccharides. The pentose phosphate pathway is the major means for generating pentose sugars for biosynthesis.

- What is phosphorolysis?
- What disaccharides are very common in nature?
- What functions does the pentose phosphate pathway play in the cell?

21.18 Organic Acid Metabolism

Various organic acids can be metabolized as carbon sources and electron donors by microorganisms. The acids of the citric acid cycle, such as *citrate, malate, fumarate,* and *succinate,* are common natural products formed by plants and are also fermentation products of microorganisms. Because the citric acid cycle has major biosynthetic as well as energetic functions (Section 5.13), the complete cycle or major portions of it are nearly universal in microorganisms. Thus, it is not surprising that many microorganisms are able to use citric acid cycle intermediates as electron donors and carbon sources.

Glyoxylate Cycle

Unlike the utilization of organic acids containing four to six carbons, two- or three-carbon acids cannot be used as growth substrates by the citric acid cycle alone. The same is true for substrates such as hydrocarbons and lipids, degraded via beta-oxidation to acetyl-CoA (Section 21.19). The citric acid cycle can continue to operate only if the acceptor molecule, the four-carbon acid oxalacetate, is regenerated at each turn; any removal of carbon compounds for biosynthetic reactions would prevent completion of the cycle (Figure 5.22).

When acetate is used, the oxalacetate needed to continue the cycle is produced through the **glyoxylate cycle** (**Figure 21.40**), so named because glyoxylate is a key intermediate. This cycle is composed of most citric acid cycle reactions plus two additional enzymes: *isocitrate lyase,* which splits isocitrate into succinate and glyoxylate, and *malate synthase,* which converts glyoxylate and acetyl-CoA to malate (Figure 21.40).

Biosynthesis through the glyoxylate cycle occurs as follows. The splitting of isocitrate into succinate and glyoxylate allows the succinate molecule (or another citric acid cycle intermediate derived from it) to be removed for biosynthesis because glyoxylate (C_2) combines with acetyl-CoA (C_2) to yield malate (C_4). Malate can be converted to oxalacetate to maintain the citric acid cycle after the C_4 intermediate (succinate)

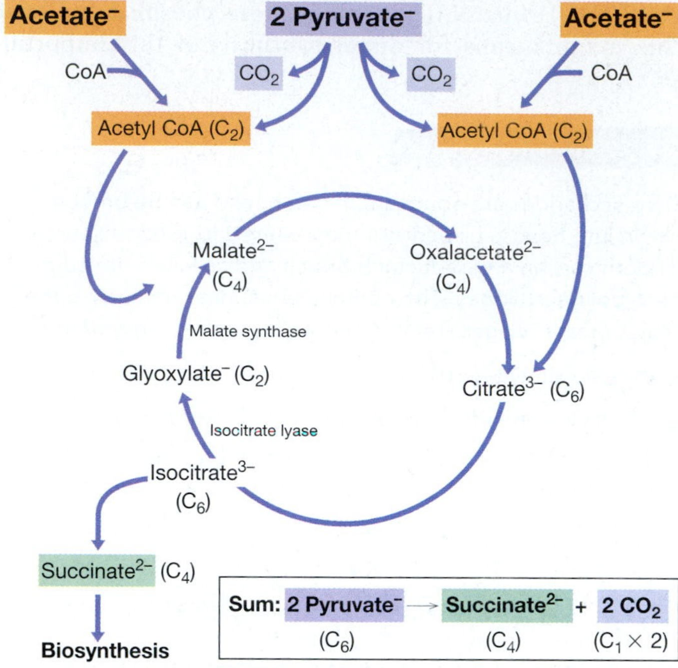

Figure 21.40 The glyoxylate cycle, leading to the synthesis of oxalacetate from acetate. Two unique enzymes, isocitrate lyase and malate synthase, operate along with most citric acid cycle enzymes. In addition to growth on pyruvate, the glyoxylate cycle also functions during growth on acetate.

has been drawn off. Succinate is used in the production of porphyrins (needed for cytochromes, chlorophyll, and other tetrapyrroles). Succinate can also be oxidized to oxalacetate as a carbon skeleton for C_4 amino acids, or it can be converted (via oxalacetate and phosphoenolpyruvate) to glucose.

Pyruvate and C_3 Utilization

Three-carbon compounds such as pyruvate or compounds that can be converted to pyruvate (for example, lactate or carbohydrates) also cannot be catabolized through the citric acid cycle alone. Because some of the citric acid cycle intermediates are used for biosynthesis, the oxalacetate needed to keep the cycle going is synthesized from pyruvate or phosphoenolpyruvate by the addition of a carbon atom from CO_2. In some organisms this step is catalyzed by the enzyme pyruvate carboxylase:

$$\text{Pyruvate} + \text{ATP} + CO_2 \longrightarrow \text{oxalacetate} + \text{ADP} + P_i$$

whereas in others it is catalyzed by phosphoenolpyruvate carboxylase:

$$\text{Phosphoenolpyruvate} + CO_2 \longrightarrow \text{oxalacetate} + P_i$$

These reactions replace oxalacetate that is lost when intermediates of the citric acid cycle are removed for use in biosynthesis, and the cycle can continue to function.

21.18 MiniReview

Organic acids are usually metabolized through the citric acid cycle or the glyoxylate cycle. Isocitrate lyase and malate synthase are the key enzymes of the glyoxylate cycle.

- Why is the glyoxylate cycle necessary for growth on acetate but not on succinate?

21.19 Lipid Metabolism

Lipids are abundant in nature. The cytoplasmic membranes of all cells contain lipids, and many organisms produce lipid storage materials and contain lipids in their cell walls. These substances are biodegradable and are excellent substrates for microbial energy-yielding metabolism. When cells die, their lipids are thus catabolized, with CO_2 being the final product.

Fat and Phospholipid Hydrolysis

Fats are esters of glycerol and fatty acids (Section 3.4) and are readily available from the release of lipids from dead organisms. Microorganisms use fats only after hydrolysis of the ester bond, and extracellular enzymes called *lipases* are responsible for the reaction (**Figures 21.41** and **21.42**). Lipases attack fatty acids of various chain lengths. Phospholipids are

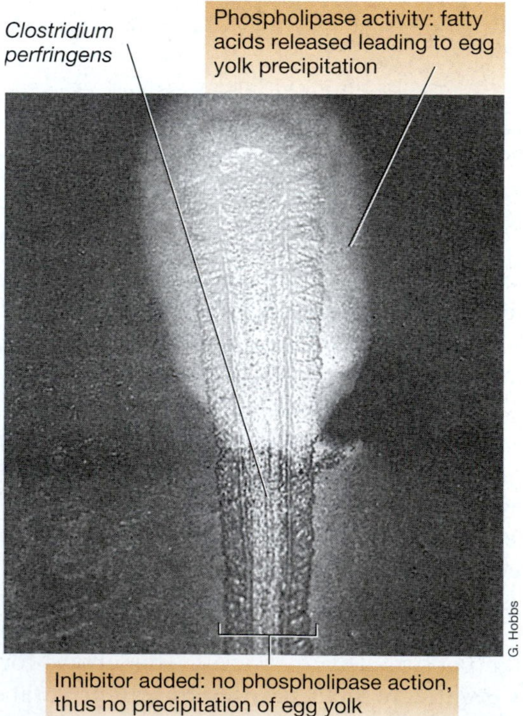

Figure 21.41 Phospholipase activity. Enzyme activity of phospholipase around a streak of *Clostridium perfringens* growing on an agar medium containing egg yolk. On half of the plate an inhibitor of phospholipase was added, preventing activity of the enzyme.

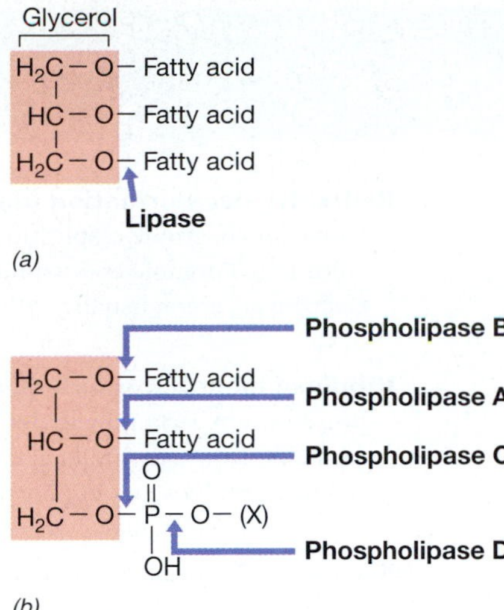

Figure 21.42 Lipases. *(a)* Activity of lipases on a fat. *(b)* Phospholipase activity on phospholipid. The cleavage sites of the four distinct phospholipases A, B, C, and D are shown. X refers to a number of small organic molecules that may be at this position in different phospholipids. Compare this diagram to the more complete figure of a phospholipid in Figure 3.7.

hydrolyzed by enzymes called *phospholipases,* each of which is given a different letter designation, depending on which ester bond they cleave in the lipid (Figure 21.42). Phospholipases A and B cleave *fatty acid* esters, whereas phospholipases C and D cleave *phosphate* esters and hence are different classes of enzymes. The result of lipase activity is the release of free fatty acids and glycerol; these substances can then be metabolized by chemoorganotrophic microorganisms.

Fatty Acid Oxidation

Fatty acids are oxidized by *beta-oxidation,* a series of reactions in which two carbons of the fatty acid are split off at a time (**Figure 21.43**). The fatty acid is first activated with coenzyme A; oxidation results in the release of acetyl-CoA and the formation of a new fatty acid two carbon atoms shorter (Figure 21.43). The process of beta-oxidation is then repeated, and another acetyl-CoA molecule is released.

There are two separate dehydrogenation reactions in beta-oxidation. In the first, electrons are transferred to flavin-adenine dinucleotide (FAD) forming FADH, whereas in the second they are transferred to NAD^+, forming NADH. Most fatty acids in a cell have an even number of carbon atoms, and complete oxidation yields acetyl-CoA. If odd-chain or branched-chain fatty acids are catabolized, propionyl-CoA or a branched chain fatty acid-CoA remains after beta-oxidation, and these are either further metabolized to acetyl-CoA by ancillary reactions or excreted from the cell. The acetyl-CoA formed is then oxidized by the citric acid cycle or is converted to hexose and other cell constituents via the glyoxylate cycle (Figure 21.40).

Because they are highly reduced, fatty acids are excellent electron donors. For example, the oxidation of the 16-carbon

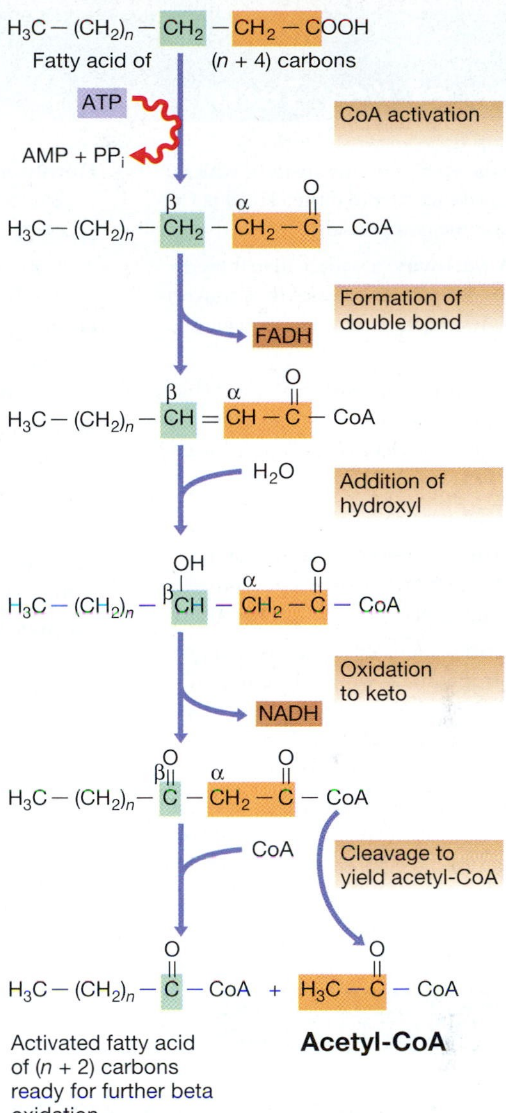

Figure 21.43 Beta-oxidation. Beta-oxidation of a fatty acid leading to the successive formation of acetyl-CoA.

saturated fatty acid palmitic acid results in the net synthesis of 129 ATP molecules. These include electron transport phosphorylation from electrons generated during the formation of acetyl-CoA from beta-oxidations and from oxidation of the acetyl-CoA units themselves through the citric acid cycle (Figure 5.22).

21.19 MiniReview

Fats are metabolized by hydrolysis by lipases or phospholipases to fatty acids plus glycerol. The fatty acids are oxidized by beta-oxidation reactions to acetyl-CoA, which is then oxidized to CO_2 by the citric acid cycle.

- What are the functions of phospholipases?
- What is meant by the term beta oxidation? How many electrons are released for every acetyl-coA produced?

Review of Key Terms

Acetogenesis energy metabolism in which acetate is produced from either H_2 plus CO_2 or from organic compounds

Acetyl-CoA pathway a widely distributed pathway in obligate anaerobes that converts $H_2 + CO_2$ into acetate or oxidizes acetate to CO_2

Anaerobic respiration respiration in which some substance, such as SO_4^{2-} or NO_3^-, is used as a terminal electron acceptor instead of O_2

Anoxic oxygen-free

Denitrification anaerobic respiration in which NO_3^- or NO_2^- is reduced to nitrogen gases, primarily N_2

Fermentation anaerobic catabolism of an organic compound in which the compound serves as both an electron donor and an electron acceptor and in which ATP is usually produced by substrate-level phosphorylation

Glyoxylate cycle a series of reactions including some citric acid cycle reactions that are used for aerobic growth on C_2 or C_3 organic acids

Homofermentative producing only lactic acid from the fermentation of glucose

Heterofermentative producing a mixture of products, typically lactate, ethanol, plus CO_2, from the fermentation of glucose

Hydrogenase an enzyme, widely distributed in anaerobic microorganisms, capable of oxidizing or evolving H_2

Methanogen an organism that produces methane (CH_4)

Methanogenesis the biological production of methane

Methanotroph an organism that can oxidize methane

Methylotroph an organism capable of growth on compounds containing no C—C bonds; some methylotrophs are methanotrophic

Oxygenase an enzyme that catalyzes the incorporation of oxygen from O_2 into organic or inorganic compounds

Pentose phosphate pathway a major metabolic pathway for the production and catabolism of pentoses (C_5 sugars)

Reductive dechlorination (dehalorespiration) an anaerobic respiration in which a chlorinated organic compound is used as an electron acceptor, usually with the release of Cl^-

Ribulose monophosphate pathway a reaction series in certain methylotrophs in which formaldehyde is assimilated into cell material using ribulose monophosphate as the C_1 acceptor molecule

Secondary fermentation a fermentation where the substrates are the fermentation products of some other organism

Serine pathway a reaction series in certain methylotrophs in which formaldehyde plus CO_2 is assimilated into cell material by way of the amino acid serine

Stickland reaction the fermentation of an amino acid pair

Syntrophy a process whereby two or more microorganisms cooperate to degrade a substance neither can degrade alone

Review Questions

1. Define the term substrate-level phosphorylation. How does it differ from oxidative phosphorylation? Assuming an organism is facultative, what basic nutritional conditions dictate whether the organism obtains energy from substrate-level rather than oxidative phosphorylation (Section 21.1)?
2. Although many different compounds are theoretically fermentable, in order to support a fermentative process, most organic compounds must eventually be converted to one of a relatively small group of molecules. What are these molecules and why must they be produced (Sections 21.1–21.3)?
3. Give an example of a fermentation that does not employ substrate-level phosphorylation (Section 21.4). How is energy conserved in this fermentation?
4. Why is syntrophy also called "interspecies H_2 transfer" (Section 21.5)?
5. Why is NO_3^- a better electron acceptor for anaerobic respiration than is SO_4^{2-} (Section 21.6)?
6. In *Escherichia coli,* synthesis of the enzyme nitrate reductase is repressed by oxygen. On the basis of bioenergetic arguments, why do you think this repression phenomenon might have evolved (Section 21.7)?
7. Why is hydrogenase a constitutive enzyme in *Desulfovibrio* (Section 21.8)?
8. Compare and contrast acetogens with methanogens in terms of (1) substrates and products of their energy metabolism, (2) ability to use organic compounds as electron donors in energy metabolism, and (3) phylogeny (Sections 21.9 and 21.10; you may also want to review material in Section 17.4 before preparing your answer).
9. Why can it be said that in glycolysis in *Pyrococcus furiosus,* both fermentation and anaerobic respiration are occurring at the same time (Section 21.11).
10. Compare and contrast ferric iron reduction with reductive dechlorination in terms of (1) product of the reduction and (2) environmental significance (Section 21.12).
11. How can denitrifying and sulfate-reducing bacteria degrade hydrocarbons without the participation of oxygenase enzymes (Sections 21.13–21.15)?
12. How do monooxygenases differ from dioxygenases in the reactions they catalyze (Sections 21.14)?
13. How does a methanotroph differ from a methanogen? How do type I and type II methanotrophs differ in their carbon assimilation patterns (Section 21.16)?

14. Compare and contrast the conversion of cellulose and intracellular starch to glucose units. What enzymes are involved and which process is the more energy efficient (Section 21.17)?

15. What is the major function of the glyoxalate cycle (Section 21.18)?

16. What is the product of the beta-oxidation of a fatty acid? How is this product oxidized to CO_2 (Section 21.19)?

Application Questions

1. When methane is made from CO_2 (plus H_2) or from methanol (in the absence of H_2), various steps in the pathway shown in Figures 21.21 and 21.22 are used. Compare and contrast methanogenesis from these two substrates and discuss why they must be metabolized in opposite directions.

2. Although dextran is a glucose polymer, glucose cannot be used to make dextran. Explain. How is dextran synthesis important in oral hygiene (Section 28.3)?

3. Why can't a fatty acid such as butyrate be fermented in pure culture, whereas its anaerobic catabolism under other conditions occurs readily in pure culture? What are these conditions and why do they now allow for butyrate catabolism? How then can butyrate be fermented in mixed culture?

22

Methods in Microbial Ecology

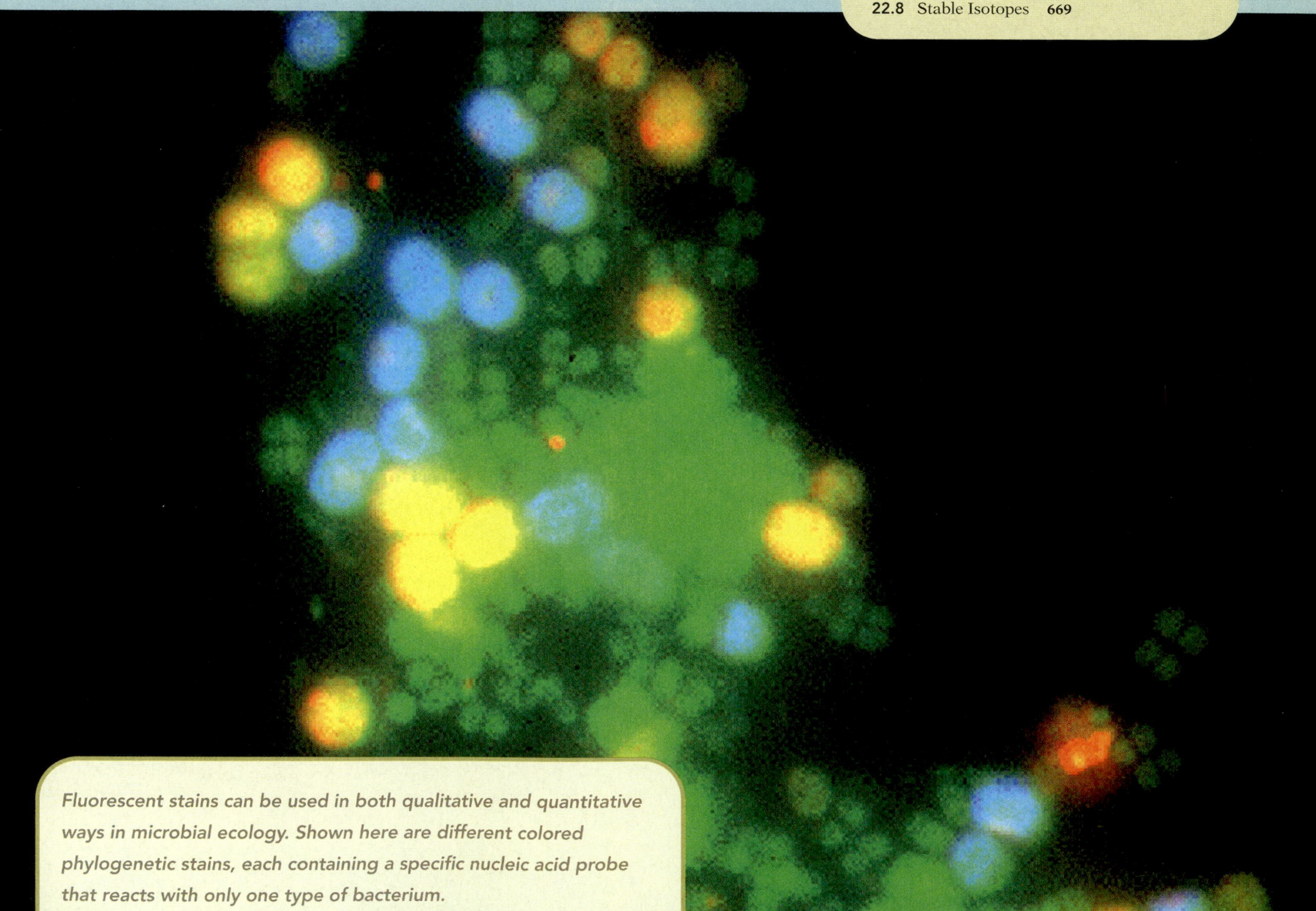

Fluorescent stains can be used in both qualitative and quantitative ways in microbial ecology. Shown here are different colored phylogenetic stains, each containing a specific nucleic acid probe that reacts with only one type of bacterium.

We learned in Chapter 1 that in nature, microorganisms exist in populations of interactive cells called *microbial communities*. The science of **microbial ecology** deals with how microbial communities interact with each other and their environment.

The major components of microbial ecology are *biodiversity* and *microbial activity*. To study biodiversity, microbial ecologists must be able to identify and quantify microorganisms directly in their habitats. Knowing this is often helpful for isolating organisms of interest, another important goal of microbial ecology. To study microbial activity, techniques must be available to measure the important metabolic processes that microorganisms carry out in their microbial habitats. This chapter discusses several important tools that microbial ecologists use to make these measurements.

In this chapter we consider the enrichment and isolation of microorganisms, general and specific cell-staining methods, gene isolation and characterization, and methods for measuring the activities of microorganisms *in situ* (in the environment). Chapter 23 outlines the basic principles of microbial ecology and examines the types of environments that microorganisms inhabit. Chapter 24 completes our coverage of microbial ecology with a consideration of nutrient cycles and the microbiology of specialized microbial habitats, for example, the rumen of ruminant animals and the root nodules of leguminous plants. With the tools to be described in this chapter, microbial ecologists can define the structure and function of virtually any microbial community.

I CULTURE-DEPENDENT ANALYSES OF MICROBIAL COMMUNITIES

Microbial ecology uses a variety of both culture methods and culture-independent methods. We begin by considering the possibilities for laboratory culture of organisms from nature. Both classical and rather high-tech methods exist for the isolation of microorganisms from natural samples, and we examine both here.

22.1 Enrichment and Isolation

One way to study a microbial ecosystem is to isolate microorganisms from it and study their properties in laboratory cultures. Then, from knowledge of the organisms present and their metabolic capacities, conclusions can be drawn concerning how the microbial ecosystem functions. Isolation is also important because it yields cultures for experimental laboratory studies and for applications in the fields of biotechnology and industrial and environmental microbiology.

Isolation requires separating different organisms out from the microbial community, a procedure called *enrichment*, and then obtaining the organisms in pure culture. In an **enrichment**

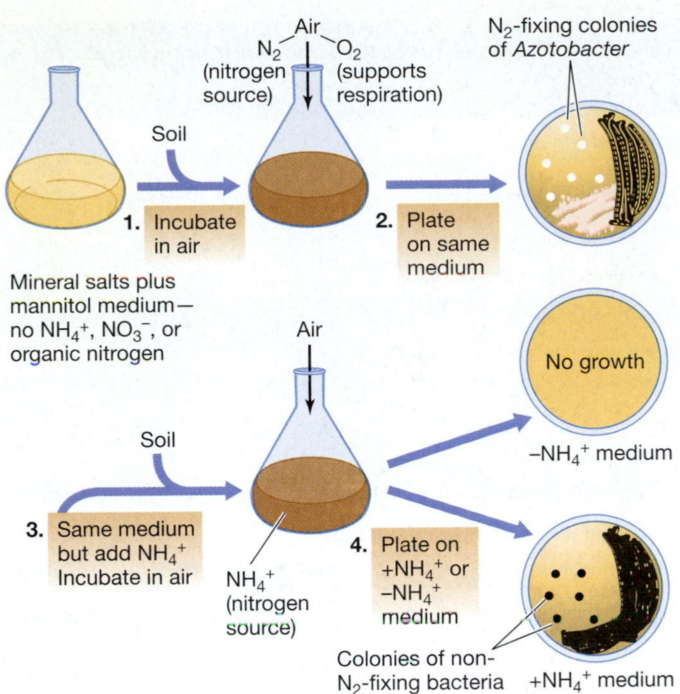

Figure 22.1 The isolation of *Azotobacter*. Selection for aerobic N_2-fixing bacteria usually results in the isolation of *Azotobacter* or its relatives. By contrast, enrichment with fixed forms of nitrogen such as NH_4^+ rarely results in N_2-fixing bacteria because the selective pressure for nitrogen fixation is lost. See Section 1.9 and Figure 1.17 for more on the historical importance of the bacterium *Azotobacter*.

culture, a medium and a set of incubation conditions are established that are selective for the desired organism and counterselective for undesired organisms. Effective enrichment cultures duplicate as closely as possible the resources and conditions of a particular ecological niche. Literally hundreds of different enrichment strategies have been devised, and **Tables 22.1** and **22.2** provide an overview of some successful ones.

Inocula

Successful enrichment cultures require that an appropriate inoculum containing the organism of interest be available. Thus, enrichment cultures begin by sampling the appropriate habitat (Tables 22.1 and 22.2). Enrichment cultures are established by placing the inoculum into selective media and incubating under specific conditions. In this way, many common prokaryotes can be isolated. For example, the great Dutch microbiologist Martinus Beijerinck, who conceptualized the enrichment culture technique (∞ Section 1.9), first isolated the N_2-fixing bacterium *Azotobacter* using a classic enrichment strategy (**Figure 22.1**). Because *Azotobacter* is a rapidly growing bacterium capable of N_2 fixation in air, enrichment using media devoid of ammonia or other forms of fixed nitrogen and incubated in air selects strongly for this organism. Non–nitrogen fixers or anaerobic nitrogen-fixing bacteria are counterselected in this instance (Figure 22.1).

Table 22.1 Some enrichment culture methods for phototrophic and chemolithotrophic bacteria

Light-phototrophic bacteria: main C source, CO_2			
Incubation in air		***Organisms enriched***	***Inoculum***
N_2 as nitrogen source		Cyanobacteria	Pond or lake water; sulfide-rich muds; stagnant water; raw sewage; moist, decomposing leaf litter; moist soil exposed to light
NO_3^- as nitrogen source, 55°C		Thermophilic cyanobacteria	Hot spring microbial mat
Anoxic incubation			
H_2 or organic acids; N_2 as sole nitrogen source		Purple nonsulfur bacteria, heliobacteria	Same as above plus hypolimnetic lake water; pasteurized soil (heliobacteria); microbial mats for thermophilic species
H_2S as electron donor		Purple and green sulfur bacteria	
Fe^{2+}, NO_2^- as electron donor		Purple bacteria	
Dark-chemolithotrophic bacteria: main C source, CO_2 (medium must lack organic C)			
Incubation in air: aerobic respiration			
Electron donor	***Electron acceptor***	***Organisms enriched***	***Inoculum***
NH_4^+	O_2	Ammonia-oxidizing bacteria (*Nitrosomonas*)	Soil, mud; sewage effluent
NO_2^-	O_2	Nitrite-oxidizing bacteria (*Nitrobacter, Nitrospira*)	
H_2	O_2	Hydrogen bacteria (various genera)	
H_2S, S^0, $S_2O_3^{2-}$	O_2	*Thiobacillus* spp.	
Fe^{2+}, low pH	O_2	*Acidithiobacillus ferrooxidans*	
Anoxic incubation			**Inoculum**
S^0, $S_2O_3^{2-}$	NO_3^-	*Thiobacillus denitrificans*	Mud, lake sediments, soil
H_2	NO_3^- + yeast extract	*Paracoccus denitrificans*	

Enrichment Culture Outcomes

To achieve success with enrichment cultures, attention to both the culture medium and the incubation conditions are important. That is, both *resources* (nutrients) and *conditions* (temperature, pH, osmotic considerations, and the like) must be optimized in an enrichment culture in order to have the best chance of obtaining the organism of interest (Table 23.1). Some enrichment cultures yield nothing. This may be because the organism capable of growing under the enrichment conditions specified is absent from the habitat (that is, not every organism is everywhere). Alternatively, a negative enrichment could occur because even though the organism of interest exists in the habitat sampled, resources and conditions for its laboratory culture were insufficient to enrich it. Thus enrichment cultures can prove a positive (that an organism with certain capacities exists in a particular environment) but never a negative (that such an organism does not). Moreover, the isolation of the desired organism from an enrichment culture says nothing about the ecological importance or abundance of the organism in its habitat; a positive enrichment proves only that the organism was present in the sample and, in theory, requires that only a single viable cell be present.

The Winogradsky Column

The **Winogradsky column** is an artificial microbial ecosystem that serves as a long-term source of bacteria for enrichment cultures. Winogradsky columns have been used routinely to isolate purple and green phototrophic bacteria, sulfate-reducing bacteria, and many other anaerobes (**Figure 22.2**). Named for the famous Russian microbiologist Sergei Winogradsky (Microbial Sidebar, "Winogradsky's Legacy," Chapter 20, and Section 1.9), the column was first used by him in the late nineteenth century to study soil microorganisms.

A Winogradsky column is prepared by filling a glass cylinder about one-half full with organically rich, preferably sulfide-containing, mud (Figure 22.2) into which carbon substrates have been mixed. The substrates will determine which organisms are enriched, and fermentative substrates (such as glucose) that can lead to excessive gas formation are avoided. The mud is supplemented with $CaCO_3$ and $CaSO_4$ as a buffer and as a source of sulfate, respectively. The mud is packed tightly in the container (taking care to avoid trapping air) and then covered with lake, pond, or ditch water (or seawater if a marine column is established). The top of the cylinder is covered to prevent evaporation (Figure 22.2), and the container placed near a window that receives diffuse sunlight to develop for a period of months.

In a typical Winogradsky column a diverse community of organisms develops. Algae and cyanobacteria appear quickly in the upper portions of the water column; by producing O_2 these organisms help to keep this zone of the column oxic. Decomposition processes in the mud lead to the production of organic acids, alcohols, and H_2, suitable substrates for

Table 22.2 Some enrichment culture methods for chemoorganotrophic and strictly anaerobic bacteria[a]

Dark-chemoorganotrophic bacteria

Incubation in air: aerobic respiration			
Electron donor and nitrogen source	***Electron acceptor***	***Typical organisms enriched***	***Inoculum***
Lactate + NH_4^+	O_2	*Pseudomonas fluorescens*	Soil, mud; lake sediments; decaying vegetation; pasteurize inoculum (80°C for 15 min) for all *Bacillus* enrichments
Benzoate + NH_4^+	O_2	*Pseudomonas fluorescens*	
Strach + NH_4^+	O_2	*Bacillus polymyxa*, other *Bacillus* spp.	
Ethanol (4%) + 1% yeast extract, pH 6.0	O_2	*Acetobacter, Gluconobacter*	
Urea (5%) + 1% yeast extract	O_2	*Sporosarcina ureae*	
Hydrocarbons (e.g., mineral oil, gasoline, toluene) + NH_4^+	O_2	*Mycobacterium, Nocardia, Pseudomonas* (Figure 24.19)	
Cellulose + NH_4^+	O_2	*Cytophaga, Sporocytophaga* (Figure 16.36)	
Mannitol or benzoate, N_2 as N source	O_2	*Azotobacter*	
Anoxic incubation: anaerobic respiration			
Electron donor	***Electron acceptor***	***Organisms enriched***	***Inoculum***
Organic acids	KNO_3	*Pseudomonas* (denitrifying species)	Soil, mud; lake sediments
Yeast extract	KNO_3	*Bacillus* (denitrifying species)	
Organic acids	Na_2SO_4	*Desulfovibrio, Desulfotomaculum*	
Acetate, propionate, butyrate	Na_2SO_4	Fatty-acid oxidizing sulfate reducers	As above; or sewage digestor sludge; rumen contents; marine sediments
Acetate, ethanol	S^0	*Desulfuromonas*	
Acetate	Fe^{3+}	*Geobacter, Geospirillum*	
Acetate	ClO_3^-	Various chlorate-reducing bacteria	
H_2	Na_2CO_3	Methanogens (chemolithotrophic species only), homoacetogens	Mud, sediments, sewage sludge
CH_3OH	Na_2CO_3	*Methanosarcina barkeri*	
CH_3NH_2 or CH_3OH	KNO_3	*Hyphomicrobium*	
Hydrocarbons	Na_2SO_4 or KNO_3	Anoxic hydrocarbon-degrading bacteria	Freshwater or marine sediments
Anoxic incubation: fermentation			
Electron donor and nitrogen source	***Electron acceptor***	***Organisms enriched***	***Inoculum***
Glutamate or histidine	No exogenous electron acceptors added	*Clostridium tetanomorphum* or other proteolytic *Clostridium* species	Mud, lake sediments; rotting plant or animal material; dairy products (lactic and propionic acid bacteria); rumen or intestinal contents (enteric bacteria); sewage sludge; soil; pasteurize inoculum for *Clostridium* enrichments
Strach + NH_4^+	None	*Clostridium* spp.	
Strach + N_2 as N source	None	*Clostridium pasteurianum*	
Lactate + yeast extract	None	*Veillonella* spp.	
Glucose or lactose + NH_4^+	None	*Escherichia, Enterobacter*, other fermentative organisms	
Glucose + yeast extract (pH 5)	None	Lactic acid bacteria (*Lactobacillus*)	
Lactate + yeast extract	None	Propionic acid bacteria	
Succinate + NaCl	None	*Propionigenium*	
Oxalate	None	*Oxalobacter*	

[a]All media must contain an assortment of mineral salts including N, P, S, Mg^{2+}, Mn^{2+}, Fe^{2+}, Ca^{2+}, and other trace elements (Sections 5.1–5.3). Certain organisms may have requirements for vitamins or other growth factors. This table is meant as an overview of enrichment methods and does not speak to the effect incubation temperature might have in isolating thermophilic (high temperature), hyperthermophilic (very high temperature), and psychrophilic (low temperature) species, or the effect that extremes of pH or salinity might have, assuming an appropriate inoculum was available.

UNIT 4

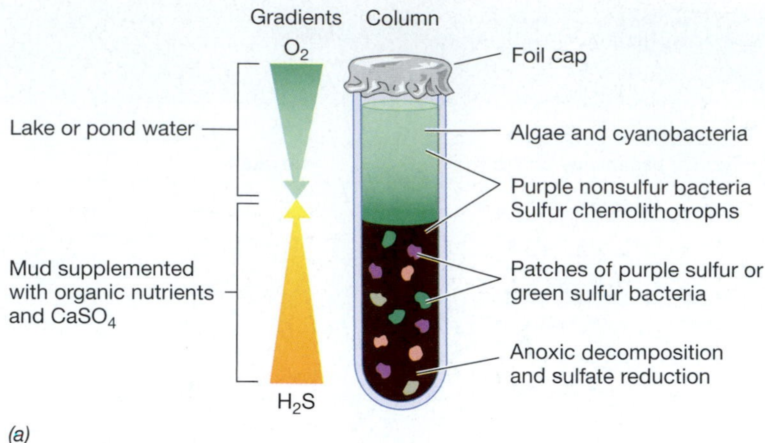

Figure 22.2 The Winogradsky column. *(a)* Schematic view of a typical column. The column is incubated in a location that receives subdued sunlight. Chemoorganotrophic bacteria grow throughout the column, aerobes and microaerophiles in the upper regions, anaerobes in the zones containing H_2S. Anoxic decomposition leading to sulfate reduction creates the gradient of H_2S. Green and purple sulfur bacteria stratify according to their tolerance for H_2S. *(b)* Photo of Winogradsky columns that have remained anoxic up to the top; each column had a bloom of a different phototrophic bacterium. Left to right: *Thiospirillum jenense, Chromatium okenii,* (both purple sulfur bacteria), and *Chlorobium limicola* (green sulfur bacteria).

sulfate-reducing bacteria (Section 15.18). Sulfide from the sulfate reducers triggers the development of purple and green sulfur bacteria (anoxygenic phototrophs, Sections 15.2 and 16.15) that use sulfide as a photosynthetic electron donor. These organisms typically grow in patches in the mud on the sides of the column but may bloom in the water itself if oxygenic phototrophs are scarce (Figure 22.2*b*). The pigmented material can be sampled with a pipette for microscopy, isolation, and characterization (Tables 22.1 and 22.2).

Winogradsky columns have been used to enrich a variety of prokaryotes, both aerobes and anaerobes. Besides supplying a ready source of inocula for enrichment cultures, columns can also be supplemented with an unusual compound to test the hypothesis that an organism or organisms exist in the inoculum that can degrade it. Once a crude enrichment has been established in the column, culture media can be inoculated and pure cultures pursued, as will be discussed in the next section. **www.microbiologyplace.com Online Tutorial 22.1: Enrichment Cultures**

Enrichment Bias

Although the enrichment culture technique is a powerful tool, in typical enrichments there exists a bias, and sometimes a very severe bias, in the outcome. This bias is usually most profound in liquid enrichment cultures where the most rapidly growing organisms for the chosen set of conditions dominate. However, we now know that the most rapidly growing organisms in laboratory cultures are, from the standpoint of abundance, often only minor components of the microbial ecosystem instead of the most ecologically relevant organisms carrying out the process of interest. This is because the levels of resources available in laboratory cultures are typically much higher than what is found in nature, and the conditions in the habitat, including both the types and proportions of different organisms present and the physical and chemical conditions, are nearly impossible to reproduce and maintain for long periods in laboratory cultures.

Enrichment bias can be demonstrated by comparing the results obtained in dilution cultures (see later) with classical liquid enrichment. Dilution of an inoculum followed by liquid enrichment or plating often yields different organisms than liquid enrichments established with the same but undiluted inocula. It is thought that dilution of the inoculum eliminates quantitatively insignificant but rapidly growing "weed" species, and this allows other, frequently more abundant, organisms in the community to develop. Dilution of the inoculum is thus a common practice in enrichment culture microbiology today.

22.1 MiniReview

The enrichment culture technique is a means of obtaining microorganisms from natural samples. Specific reactions can be studied using enrichment methods by choosing media and incubation conditions selective for organisms able to carry out the reaction.

- Describe the enrichment strategy behind Beijerinck's isolation of *Azotobacter*.
- Why is sulfate (SO_4^{2-}) added to a Winogradsky column?
- What is enrichment bias?

22.2 Isolation in Pure Culture

A pure culture—one containing a single kind of microorganism—can be obtained from enrichment cultures in many ways. Common isolation procedures include the streak plate, the agar shake, and liquid dilution. For organisms that form colonies on agar plates, the streak plate is quick, easy, and the method of choice (**Figure 22.3a**); if a well-isolated colony is picked and restreaked several successive times, a pure culture can usually be obtained. With proper incubation facilities (for example, anoxic jars for anaerobes, Section 6.17), it is possible to purify both aerobes and anaerobes on agar plates by the streak plate method.

Agar Dilution Tubes and Most-Probable Number (MPN) Technique

In the agar dilution tube method a mixed culture is diluted in tubes of molten agar, resulting in colonies embedded in the agar rather than growing on the surface of an agar plate (Figure 22.3*b*). Dilution tubes are useful for purifying anaerobic organisms such as phototrophic sulfur bacteria and sulfate-reducing bacteria from samples taken from Winogradsky columns. A culture is purified by successive dilutions of cell suspensions in tubes of molten agar medium (Figure 22.3*b*). By repeating this procedure using a colony from the highest dilution tube as inoculum for a new set of dilutions, pure cultures are eventually obtained. A related procedure called the *roll tube method* uses tubes containing a thin layer of agar on their inner surface. The agar can then be streaked for isolated colonies. Because the tubes can be flushed with an oxygen-free gas while streaking, the roll tube method is primarily used for the isolation of anaerobic prokaryotes.

Another purification procedure is to serially dilute an inoculum in a liquid medium until the final tube in the series shows no growth. Using ten-fold serial dilutions, for example, the last tube showing growth should have originated from ten or fewer cells. By repeating this process several times, pure cultures can be obtained. This method, known as the **most-probable number (MPN) technique** (**Figure 22.4**), has been used for estimating the numbers of microorganisms in foods, wastewater, and other samples in which cell numbers need to be assessed routinely. An MPN count can be done using highly selective media and incubation conditions to target one or a small group of organisms, such as a particular pathogen, or using complex media to get an estimate of viable and cultivable cell numbers (but see the caveat that applies to such estimates in Section 6.10). **www.microbiologyplace.com Online Tutorial 22.2: Serial Dilutions and a Most-Probable-Number Analysis**

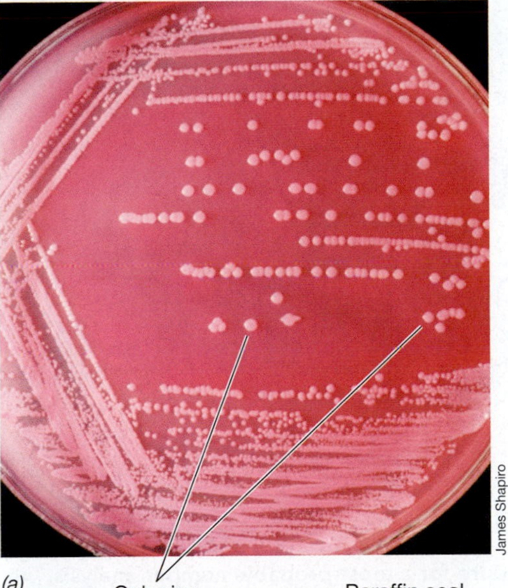

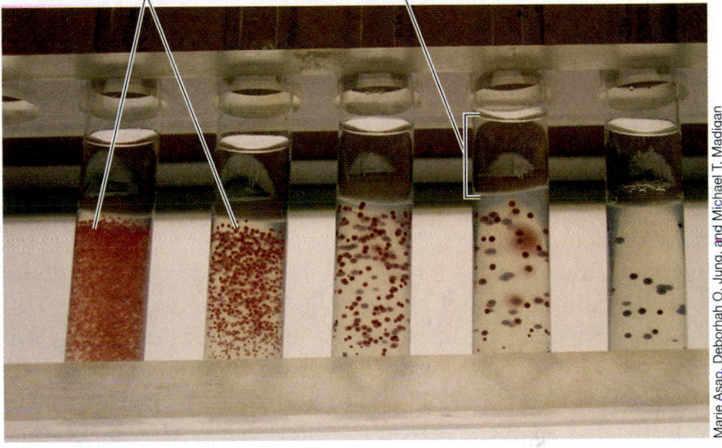

Figure 22.3 Pure culture methods. *(a)* Organisms that form distinct colonies on plates are usually easy to purify. *(b)* Colonies of phototrophic purple bacteria in agar dilution tubes. A dilution series was established from left to right, eventually yielding well-isolated colonies. The tubes are sealed with sterile mineral oil to maintain anaerobiosis.

Criteria for Purity

Regardless of the methods used to purify a culture, once a putative pure culture has been obtained, it is essential to verify its purity. This should be done through a combination of (1) microscopy, (2) observation of colony characteristics on plates or in shake tubes, and (3) tests of the culture for growth in other media. In the latter regard, it is important to test the culture for growth in media in which it is predicted that the desired organism will grow poorly or not at all but that contaminants will grow vigorously. In the final analysis, the microscopic observation of a single morphological type of cell that displays uniform staining characteristics (for example, in a Gram stain) coupled with uniform colony characteristics and the absence of contamination in growth tests with various culture media, is good evidence that a culture is *axenic* (pure).

UNIT 4

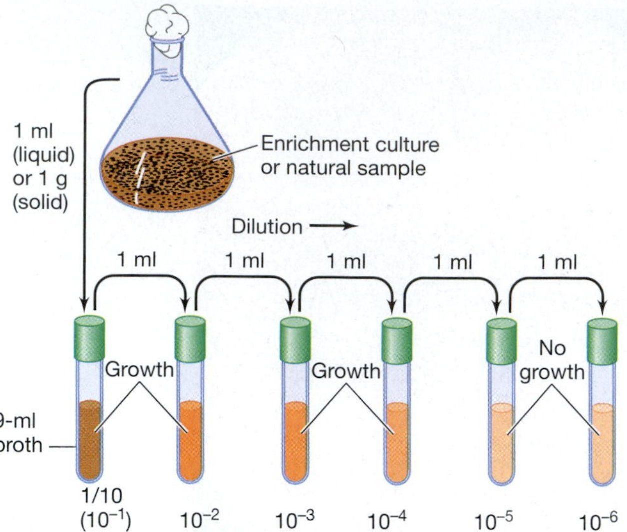

Figure 22.4 Procedure for a most-probable number analysis. An existing enrichment culture or a natural sample of water, soil, etc., is added to a selective culture medium, and serial dilutions are made. The last tube showing growth (in the example shown it is the 10^{-4} dilution) should have developed from 10 or fewer cells. Use of several replicate tubes at each dilution improves accuracy of the final MPN obtained.

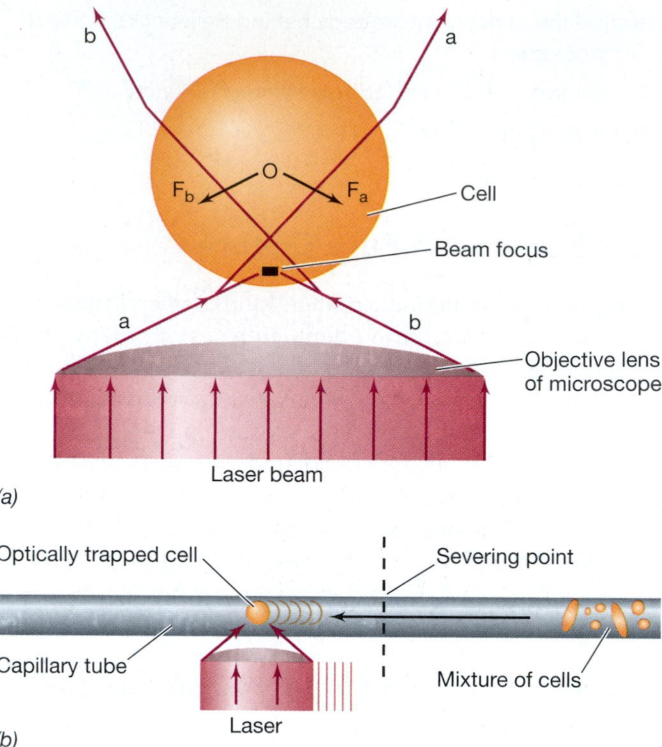

Figure 22.5 The laser tweezers for the isolation of single cells. *(a)* Principle of the laser tweezers. A strongly focused laser beam on a very small object such as a cell creates downward radiation forces (F_a, F_b) that allow the cell to be dragged in any direction. *(b)* Isolation. The laser beam can lock onto a single cell present in a mixture in a capillary tube and drag the optically trapped cell away from the other cells. Once the desired cell is far enough away from the other cells, the capillary is severed and the cell is flushed into a tube of sterile medium.

Probing for Pure Cultures—The Laser Tweezers

In addition to the more classical methods just described, technological advances in microscopy have yielded new tools for obtaining pure cultures, such as the **laser tweezers**.

Laser tweezers consist of an inverted light microscope equipped with a strongly focused infrared laser and a micromanipulation device (**Figure 22.5**). Trapping a single cell is possible because the laser beam creates a force that pushes down on a microbial cell (or other small object) and holds it in place. Then when the laser beam is moved, the trapped cell moves along with it. If a mixed sample is in a capillary tube, a single cell can be optically trapped and moved away from contaminating organisms. The cell can then be isolated by breaking the tube at a point between the cell and the contaminants and flushing the cell into a small tube of sterile medium (Figure 22.5).

Laser tweezers are especially useful for isolating slow-growing bacteria that may be overgrown in enrichment culture or for organisms present in such low numbers that they would be missed using dilution-based enrichment methods. Coupled to the use of dyes that can identify particular organisms in a microscope field (Section 22.4), laser tweezers can be used to select organisms of interest from a mixture for purification and further laboratory study.

22.2 MiniReview

Once a successful enrichment culture has been established, pure cultures can often be obtained by conventional microbiological procedures, including streak plates, agar shakes, and dilution methods. Laser tweezers allow one to isolate a cell from a microscope field and move it away from contaminants.

- How does the agar shake method differ from streaking to obtain isolated colonies?
- In an MPN (Figure 22.4), why is the actual number of cultivable cells per sample likely to be between the reciprocal of the highest dilution showing growth and the next highest dilution?
- How might you isolate a morphologically unique bacterium present in an enrichment culture in relatively low numbers?

II CULTURE-INDEPENDENT ANALYSES OF MICROBIAL COMMUNITIES

Microbial ecologists quantify cells in a microbial habitat to estimate relative abundances. Cell stains are necessary to obtain these types of data, and we detail these methods here. Organisms in natural environments can also be detected by assaying their genes. Genes encoding either ribosomal RNA (Sections 14.6–14.9) or enzymes that support a specific physiology are the usual targets in these studies. Environmental genomics (Section 22.6) is a method for assessing the entire gene complement of a habitat, revealing both the biodiversity and metabolic capabilities of the microbial community at the same time.

22.3 General Staining Methods

Several staining methods are suitable for quantifying microorganisms in natural samples. Although these methods do not reveal clues about the physiology or phylogeny of the cells, they are nonetheless reliable and widely used by microbial ecologists for measuring total cell numbers. One method to be described here also allows for an assessment of cell viability.

Fluorescent Staining Using DAPI or Acridine Orange

Fluorescent dyes can be used to stain microorganisms from virtually any microbial habitat. **DAPI** (4′, 6-diamido-2-phenylindole) is a popular stain for this purpose, as is the dye **acridine orange**. Cells stained with DAPI fluoresce bright blue, whereas cells stained with acridine orange stain orange or greenish-orange (**Figure 22.6**). Both stains bind to DNA and are strongly fluorescent when exposed to ultraviolet radiation (DAPI absorption maximum, 400 nm) or green light (acridine orange absorption maximum, 500 nm), making the microbial cells in the sample readily visible and easy to enumerate (Figure 22.6).

DAPI and acridine orange staining are widely used for the enumeration of microorganisms in environmental, food, and clinical samples. Depending on the sample, background staining is occasionally a problem with fluorescent stains, but because DAPI and acridine orange stain nucleic acids, they are for the most part unreactive with inert matter. Thus, for many samples, soil as well as aquatic, DAPI/acridine orange staining can give a reasonable estimate of the cell numbers present. For dilute aquatic samples, cells can be stained after filtering.

Staining by DAPI and acridine orange is nonspecific; *all* microorganisms in a sample will stain. Although this may at first seem to be a desirable property, it is not necessarily so. For example, DAPI and acridine orange fail to differentiate between living and dead cells or between different species of microorganisms, and thus they cannot be used to assess cell viability or to track specific organisms in an environment.

Viability Staining

Viability staining allows one to differentiate live from dead cells. Viability stains thus both enumerate and yield viability data at the same time. The basis of staining as either live or dead lies with whether or not the cytoplasmic membrane of a cell is intact. Two dyes, colored green and red, are added to a sample; the green fluorescing dye penetrates all cells, viable or not, whereas the red dye, which contains the chemical propidium iodide, penetrates only those cells whose cytoplasmic membrane is no longer intact and are therefore dead. Thus, when viewed microscopically, green cells are scored as alive and red cells as dead, yielding an instant assessment of both abundance and viability (**Figure 22.7**).

Marc Mussman and Michael Wagner

(a)

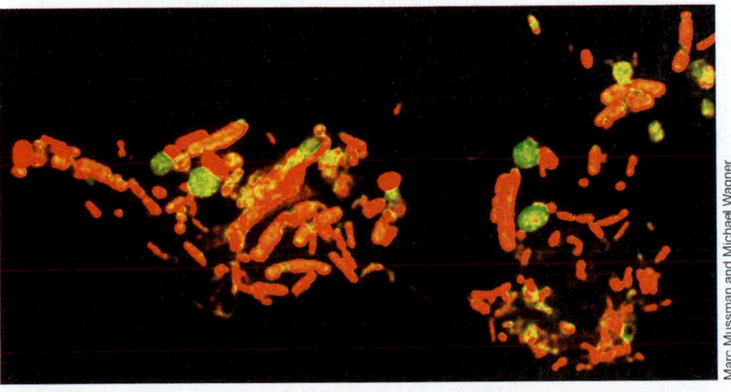

Marc Mussman and Michael Wagner

(b)

Figure 22.6 Nonspecific fluorescent stains. *(a)* DAPI, and *(b)* acridine orange. Both photomicrographs show microbial communities inhabiting activated sludge in a municipal wastewater treatment plant. With acridine orange, cells containing low RNA levels stain green.

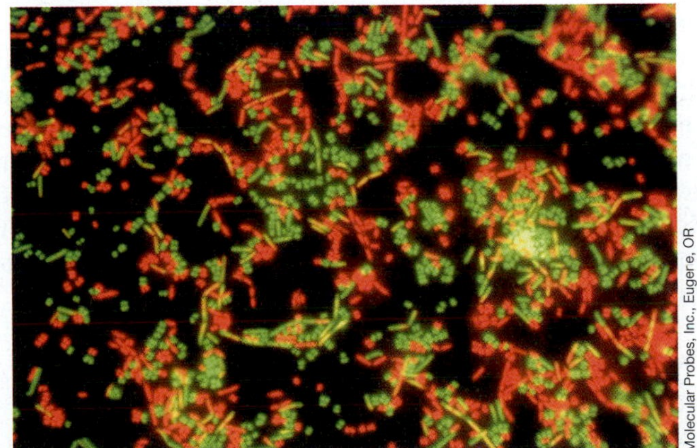

Molecular Probes, Inc., Eugene, OR

Figure 22.7 Viability staining. Live (green) and dead (red) cells of *Micrococcus luteus* (cocci) and *Bacillus cereus* (rods) stained by the LIVE/DEAD BacLight™ Bacterial Viability Stain.

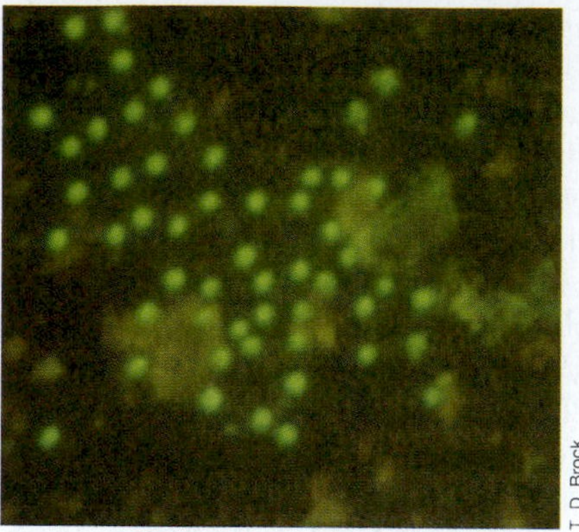

Figure 22.8 Fluorescent antibodies as a cell tag. Cells of the hyperthermophile *Sulfolobus acidocaldarius* (*Archaea*) attached to the surface of solfatara soil particles visualized by the fluorescent antibody technique. Cells are about 1 μm in diameter.

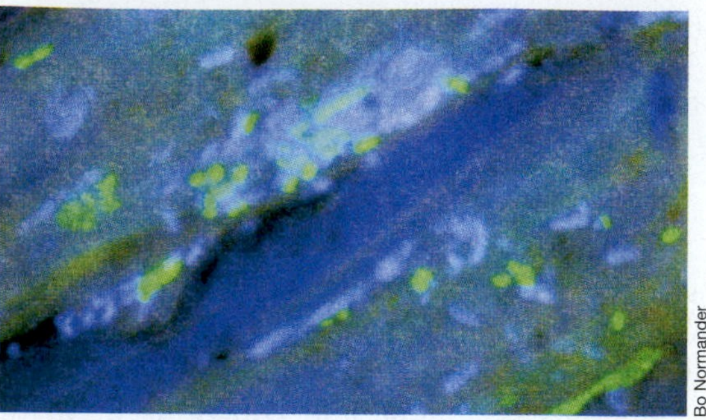

Figure 22.9 The green fluorescent protein. Cells of *Pseudomonas fluorescens* genetically engineered to express the green fluorescent protein and observed by confocal laser microscopy. The cells are attached to barley roots and fluoresce green. The blue-staining cells are part of the normal flora of barley roots and have been stained with DAPI (Figure 22.6*a*).

Although useful for research that uses laboratory cultures, the live/dead staining method is not suitable for use in the direct microscopic examination of samples from many natural habitats because of problems with nonspecific staining of background materials. However, procedures have been developed for use of the live/dead stain in viability analyses of aquatic environments; a water sample is filtered and the filters are stained with the live/dead stain and examined microscopically. Thus in aquatic microbiology, live/dead staining is often used to measure the viability of cell populations in the water column of lakes or the oceans or in streams, rivers, and other aquatic environments.

Fluorescent Antibodies

Staining techniques can be even more specific if fluorescent antibodies are used. Antibodies are specific biological reagents, and we discuss the use of fluorescent antibodies in Section 32.9. The great specificity of antibodies against cell surface constituents can be exploited as a means of identifying or tracking a specific organism in a habitat that contains a mixture of many organisms, such as soil (**Figure 22.8**), or in clinical samples. The method requires the preparation of specific antibodies against the organism of interest, often a time-consuming and laborious procedure, and in most cases requires that a pure culture of the desired organism be available. For pathogenic microorganisms, however, highly specific antibodies can be purchased commercially and are widely used in clinical microbiology laboratories for the microscopic diagnosis of infectious diseases.

Green Fluorescent Protein as a Cell Tag

Bacterial cells can be altered by genetic engineering to make them autofluorescent. As was discussed earlier (∞ Section 12.10), a gene encoding a protein called the **green fluorescent protein (GFP)** can be inserted into the genome of virtually any cultured bacterium. When the GFP gene is expressed, cells fluoresce green when observed with ultraviolet microscopy (**Figure 22.9**).

Although not useful for the study of natural populations of microorganisms (because these cells lack the GFP gene), GFP-tagged cells can be introduced into an environment, such as plant roots (Figure 22.9), and then tracked over time by microscopy. Using this method, microbial ecologists can study microbial competition between the native microflora and the GFP-tagged introduced strain and can assess the effect of perturbations of an environment on the survivability of the introduced strain. However, GFP requires oxygen to become fluorescent, and thus GFP is not suitable for use as a cell tag in anoxic habitats.

The gene for GFP has also been used extensively in laboratory cultures of various bacteria as a *reporter gene*. When fused with an operon under the control of a specific repressor protein, transcriptional control of the genes of interest can be studied using fluorescence as the assay of transcription. That is, when the genes containing the fused GFP gene are transcribed, the GFP gene is also transcribed, and cells fluoresce green (∞ Section 12.10 and Figure 12.14).

Limitations of Microscopy

The microscope is an essential tool for exploring microbial diversity and for enumerating and identifying microorganisms in natural samples. However, microscopy alone does not suffice for the study of microbial diversity for many reasons. Small cells can be a major problem. Prokaryotes vary greatly in size (∞ Section 4.2 and Table 4.1), and some cells are near the limits of resolution of the light microscope. When observing natural samples, such cells can easily be overlooked, especially if the sample contains high levels of particulate matter or high numbers of large cells. Also, it is often difficult to differentiate live cells from dead cells or cells from nonliving matter in natural samples. However, the biggest limitation with the microscopic methods we have discussed thus far is that none of them reveal the phylogenetic diversity of the microorganisms in the habitat under study.

We will see in the next section and get a preview here (**Figure 22.10**) that powerful staining methods exist that can

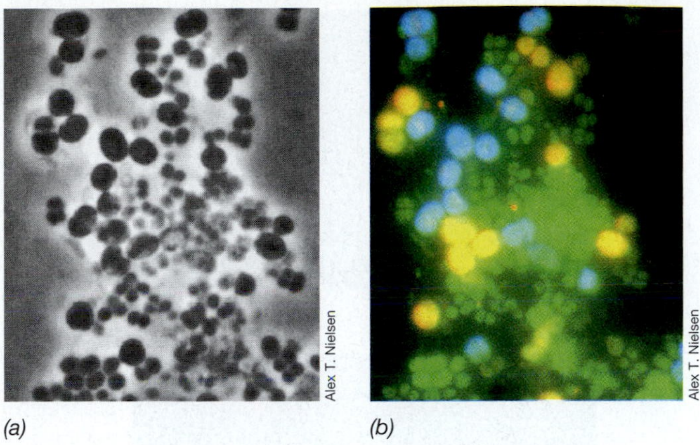

Figure 22.10 Morphology and genetic diversity. The photomicrographs shown here, *(a)* phase contrast and *(b)* phylogenetic FISH, are of the same field of cells. Although the large oval-shaped cells are of a rather unusual morphology and size for prokaryotic cells and look similar by phase-contrast microscopy, the phylogenetic stains reveal that they are genetically distinct (one stains yellow and one stains blue). The oval cells stained in blue or yellow are about 2.25 μm in diameter. The green cells in pairs or clusters are about 1 μm in diameter.

reveal the phylogeny of organisms observed in a natural sample. These methods have revolutionized microbial ecology. They have helped microbiologists overcome the major limitation of the light microscope in microbial ecology, that is, identifying what you are observing from a phylogenetic perspective. These methods have also taught microbial ecologists an important lesson: When observing unstained or nonspecifically stained natural populations of microorganisms under the microscope, one must remember that the sample almost certainly contains a genetically diverse community, even if many cells "look" the same (Figure 22.10).

22.3 MiniReview

DAPI and acridine orange are general stains for quantifying microorganisms in natural samples. Some stains can differentiate live versus dead cells, and fluorescent antibodies that are specific for one or a small group of related cells can be prepared. The green fluorescent protein makes cells autofluorescent and is a means for tracking cells introduced into the environment. In natural samples, morphologically identical cells may actually be genetically distinct.

- Why is it incorrect to say that the GFP is a "staining" method?
- How do stains like DAPI differ from fluorescent antibodies in terms of specificity?

22.4 FISH

Because of their great specificity, nucleic acid probes are powerful tools for identifying and quantifying microorganisms. Recall that a **nucleic acid probe** is a DNA or RNA oligonucleotide complementary to a sequence in a target gene or RNA; when the probe and the target come together, they hybridize (Section 12.2). As discussed in Section 14.9, nucleic acid probes can be made fluorescent by attaching fluorescent dyes to them. The fluorescent probes can be used to identify organisms that contain a nucleic acid sequence complementary to the probe. This technique is called **fluorescence *in situ* hybridization (FISH)**, and different applications are described here, including methods that target phylogeny or gene expression.

Phylogenetic Staining Using FISH

Phylogenetic FISH stains are fluorescing oligonucleotides complementary in base sequence to sequences in ribosomal RNA (16S or 23S rRNA in prokaryotes or 18S rRNA in eukaryotes) (Section 14.9). Phylogenetic stains penetrate the cell without lysing it and hybridize with RNA in the ribosomes. Because ribosomes are scattered throughout the cell in prokaryotes, the entire cell becomes fluorescent (Figure 14.17).

Phylogenetic stains can be designed to be very specific and react with only one species or a handful of related microbial species, or they can be made more general and react with, for example, all cells of a given phylogenetic domain. Using FISH, one can then identify or track an organism of interest or a domain of interest in a natural sample. For example, if one wished to determine the percentage of a given microbial population that were *Archaea*, an archaeal-specific phylogenetic stain could be used in combination with DAPI (Section 22.3) to assess *Archaea* and total numbers, respectively, and a percentage could then be derived by calculation.

FISH technology can also use multiple phylogenetic probes. Using a suite of probes, each designed to react with a particular organism or group and each containing its own fluorescent dye, FISH can characterize the phylogenetic breadth of a habitat in a single experiment (**Figure 22.11**). If FISH is combined with confocal microscopy (Section 2.3), it is possible to explore microbial populations with depth, as, for example, in a biofilm (Section 23.4 and Figure 23.5). Besides microbial ecology, FISH is an important tool in the food industry and in clinical diagnostics for the microscopic detection of specific pathogens.

ISRT-FISH and CARD-FISH

Besides characterizing the phylogenetic diversity of a habitat, FISH can be used to measure *gene expression* in organisms in a natural sample (**Figure 22.12**). Because the target in this case, mRNA, is much less abundant than is rRNA in the ribosomes of a cell, standard FISH techniques cannot be applied. Instead, either the target (mRNA) or the signal (fluorescence) must be amplified for FISH to be applicable for gene expression studies.

A method that amplifies the target is called *in situ reverse transcription (ISRT)-FISH*. This method uses a DNA probe to trap specific mRNA produced by cells in a natural sample. Following this, the enzyme reverse transcriptase is added to carry out reverse transcription; this produces a complementary DNA (cDNA) from the mRNA (Section 10.12). The cDNA is

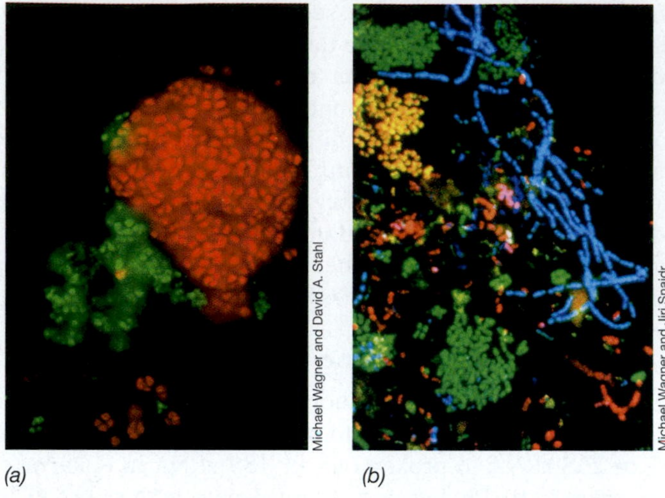

Figure 22.11 FISH analysis of sewage sludge. *(a)* Nitrifying bacteria. Red, ammonia-oxidizing bacteria; green, nitrite-oxidizing bacteria. *(b)* Confocal laser scanning micrograph of a sewage sludge sample. The sample was treated with three phylogenetic FISH probes, each containing a different fluorescent dye (green, red, or purple) that targeted a different group of *Proteobacteria*. Green-, red-, or purple-stained cells reacted with only a single probe; blue- and yellow-stained cells reacted with multiple probes to give the different colors.

then amplified by the polymerase chain reaction (PCR, Section 12.8 and see Section 22.5) and detected by hybridization to the fluorescent probe (Figure 22.12).

ISRT-FISH is used to explore factors that control gene expression in natural populations of microorganisms and to observe how experimental perturbations in an environment affect the transcription of specific genes. In combination with phylogenetic staining methods, ISRT-FISH can also be used to detect which groups of organisms in a habitat are metabolically active. This is based on the assumption that cells that are transcribing specific metabolic genes are using the enzymes that the mRNA encodes. In such a study, ISRT-FISH would be used to assess gene expression, whereas FISH would be used to reveal the phylogeny of the metabolically active cells.

A FISH method that enhances the signal instead of the target is called *catalyzed reporter deposition FISH (CARD-FISH)*. In CARD-FISH the specific nucleic acid probe contains a molecule of the enzyme peroxidase conjugated to it instead of a fluorescent dye. Once hybridization has occurred, the preparation is treated with a highly fluorescent dye called *tyramide*. The tyramide is converted by the activity of the peroxidase into a very reactive intermediate that binds to adjacent protein, and this amplifies the signal sufficiently to be detected by fluorescence microscopy. Each molecule of peroxidase activates many molecules of tyramide and thus even very low copy number targets can be visualized. Besides detecting mRNA, CARD-FISH is also useful in phylogenetic studies of prokaryotes that may be growing very slowly, for example organisms inhabiting the open oceans where cold temperatures and low nutrient concentrations combine to limit growth rates. Such cells have few ribosomes compared with more actively growing cells, and thus standard FISH often yields only a weak signal.

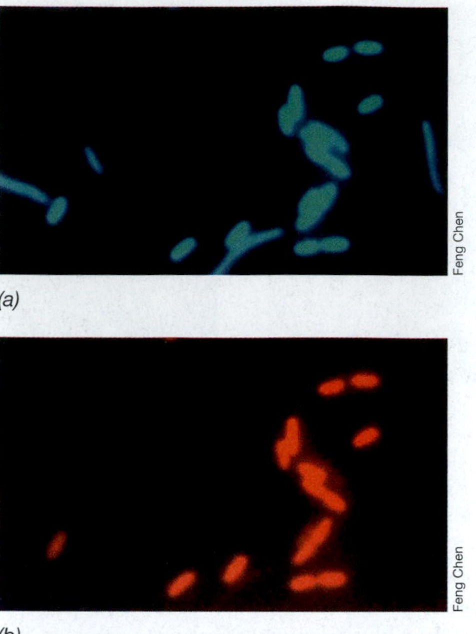

Figure 22.12 *In situ* reverse transcription (ISRT). The two fluorescent photomicrographs show a mixed culture of the bacteria *Microbulbifer hydrolyticus* (short, fat rods) and *Sagittula stellata* (long, thin rods). Reverse transcription of an mRNA produced only by *M. hydrolyticus* formed a cDNA that was subsequently amplified by PCR and hybridized to a fluorescently labeled probe. *(a)* All cells stained with DAPI, a nonspecific stain (Figure 22.6*a*). *(b)* Cells stained with the ISRT probe.

22.4 MiniReview

FISH methods have harnessed the power of nucleic acid probes and are thus highly specific in their staining properties. These include phylogenetic stains and reverse transcription fluorescent *in situ* hybridization (ISRT-FISH).

- What structure in the cell is the target for fluorescent probes in phylogenetic FISH?
- FISH and ISRT-FISH reveal different things about cells in nature. Explain.
- How might CARD-FISH be better able to detect starving microorganisms in an environment than could FISH?

22.5 Linking Specific Genes to Specific Organisms Using PCR

Many microbial biodiversity studies forgo the isolation of organisms or even quantifying or identifying them microscopically using the stains described in the previous sections. Instead, specific genes are used as the measure of biodiversity.

Specific genes are often linked to specific organisms. Therefore, detection of a specific *gene* in an environmental sample implies that the specific *organism* linked to this gene is present there. The major techniques employed in this type of microbial community analysis are the polymerase chain reaction (PCR), denaturing gradient gel electrophoresis (DGGE) or molecular cloning, DNA sequencing and analysis, and a variety of related methods. In addition, as we will see, entire genomes can also be used as a measure of the biodiversity of microbial communities.

PCR and Microbial Community Analysis

We discussed the principle of PCR in Chapter 12 (Section 12.8). Recall the major steps in PCR: (1) two nucleic acid primers are hybridized to a complementary sequence in a target gene, (2) DNA polymerase copies the target gene, and (3) multiple copies of the target gene are made by repeated melting of complementary strands, hybridization of primers, and new synthesis (Figure 12.11). From a single copy of a gene, several million copies can be made (**Figure 22.13**).

Which genes should be used as target genes? Because genes encoding small subunit ribosomal RNA are phylogenetically informative and techniques for their analysis are well developed (Sections 14.5–14.9), they are widely used in molecular biodiversity studies. Alternatively, metabolic genes that encode proteins unique to a specific organism or group of related organisms can be the target genes. In a typical community analysis experiment, total DNA is isolated from a microbial habitat (Figure 22.13). Commercially available kits that yield high-purity DNA from soil and other complex habitats are available for this purpose. The DNA obtained is a mixture of genomic DNA from all of the microorganisms that were in the sample from the habitat (Figure 22.13). However, from this mixture, PCR amplifies the targeted gene of interest and makes multiple copies of each gene variant, called a **phylotype**. Different forms of the same gene are to be expected if the target gene is widely distributed, for example, rRNA genes. Each species in the environment will have its own unique sequence for rRNA, and PCR will amplify each phylotype. These then have to be sorted out, and we consider this next.

Denaturing Gradient Gel Electrophoresis: Separating Very Similar Genes

The results of a community sampling experiment are bands on a gel characteristic of the target gene in size (**Figures** 22.13 and **22.14**). Only a single band appears; however, because the amplified target gene came from a mixture of different cells, the phylotypes need to be sorted out before they are sequenced. One way to do this is by molecular cloning (Figure 22.13, Section 12.3). Alternatively, one can use **denaturing gradient gel electrophoresis (DGGE)**, a method that separates genes of the same *size* that differ in their melting (denaturing) profile because of differences in their *base sequence* (Figures 22.13 and 22.14*b*).

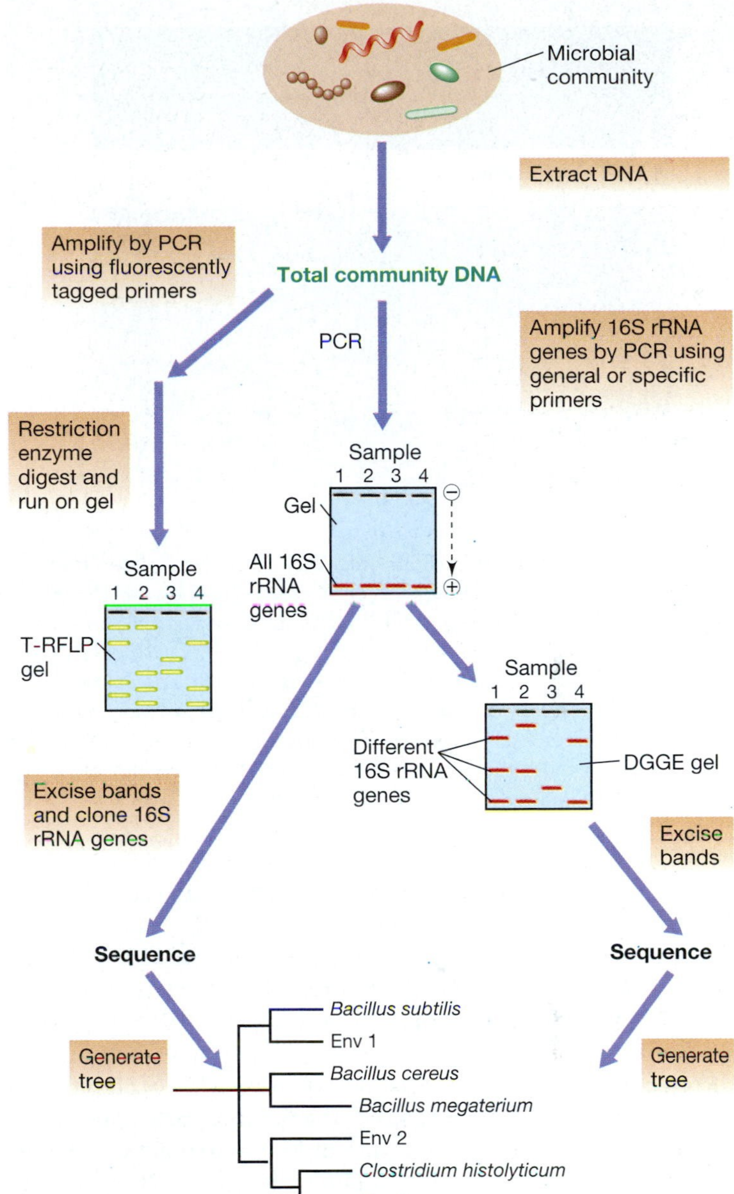

Figure 22.13 Steps in single gene biodiversity analysis of a microbial community. From total community DNA, 16S rRNA genes are amplified using, in the DGGE example, primers that target only gram-positive *Bacteria*. The PCR bands are excised and the different 16S rRNA genes separated by either cloning or DGGE. Following sequencing, a phylogenetic tree is generated. *Env* indicates an environmental sequence (phylotype). In T-RFLP analyses, the number of bands indicates the number of phylotypes. In DGGE or T-RFLP gels, bands that run to the same location usually indicate the same phylotype.

DGGE employs a gradient of a DNA denaturant, such as a mixture of urea and formamide. When a double-stranded DNA fragment moving through the gel reaches a region containing sufficient denaturant, the strands begin to "melt"; at this point, their migration stops (Figures 22.13 and 22.14*b*). Differences in base sequence cause differences in the melting properties of DNA. Thus, the different bands observed in a

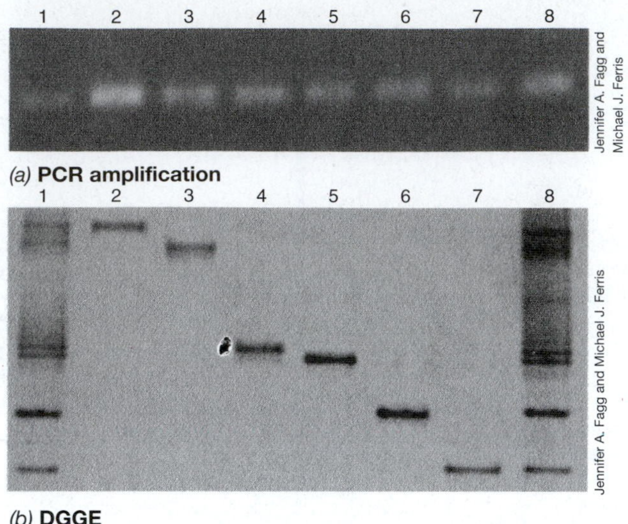

Figure 22.14 PCR and DGGE gels. Bulk DNA was isolated from a microbial community and amplified by PCR using primers for 16S rRNA genes of *Bacteria* (*a*; lanes 1 and 8). Six PCR products all yielded a single band on the gel but actually consisted of six distinct 16S rRNA gene sequences as determined by DGGE (*b*; lanes 1 and 8). Each of these six bands was then purified, reamplified by PCR (*a*; lanes 2–7), and then run on DGGE (*b*; lanes 2–7). Note how all bands run to the same location in the PCR gel because they are all of the same size, but to different locations on the DGGE gel because they have different sequences.

DGGE gel are phylotypes that can differ in base sequence by as little as a single base change.

Once DGGE has been performed, the individual bands are excised and sequenced (Figures 22.13 and 22.14). With 16S rRNA as the target gene, for example, the DGGE pattern immediately reveals the number of phylotypes (distinct 16S rRNA genes) present in a habitat (Figure 22.14*b*). Following sequencing of each DGGE band, the actual species present in the community can be determined by phylogenetic analyses (∞ Sections 14.6 and 14.7) (Figure 22.13*b*). If PCR primers specific for genes other than 16S rRNA are used, such as a metabolic gene, the phylotypes of this specific gene that exist in the sample can also be assessed. Thus, although the number of bands on a DGGE gel is an overview of the biodiversity in a habitat, sequence analysis is required to orient the phylotypes on a phylogenetic tree (Figures 22.13 and 22.14*b*).

T-RFLP

Another method for rapid microbial community analysis is *terminal restriction fragment length polymorphism* (*T-RFLP*). In this method a target gene (usually rRNA genes) is amplified by PCR from community DNA using a primer set in which one of the primers is end-labeled with a fluorescent dye. The PCR products are then treated with a restriction enzyme (∞ Section 12.1) that cuts the DNA at specific sequences. This generates a series of DNA fragments, the number of which depends on how many restriction cut sites are present in the DNA. The fluorescently labeled terminal fragments are then separated on a gel, and the pattern obtained is a function of the rRNA sequence variation in the community sampled.

DGGE and T-RFLP both measure single gene diversity, but in different ways. While the pattern of bands on a DGGE gel reflects the number of sequence variants in a single gene of the same size (Figure 22.13), the pattern of bands on a T-RFLP gel reflects the variability of a single gene as measured by differences in restriction enzyme cut sites, which are ultimately due to variations in DNA sequence.

Results of PCR Phylogenetic Analyses

Phylogenetic analyses of microbial communities have yielded surprising results. For example, using the gene encoding 16S rRNA as the target gene, analyses of natural microbial communities typically show that several phylogenetically distinct prokaryotes are present whose rRNA gene sequences differ from those of all known laboratory cultures (Figure 22.13). Moreover, there are additional methods that allow for a quantitative assessment of each phylotype, and with rather few exceptions, it turns out that the most abundant species in a natural community are ones that have so far defied laboratory culture! These results make it clear that our knowledge of microbial diversity from enrichment cultures is very incomplete and that enrichment bias (Section 22.1) is likely a serious problem in culture-dependent biodiversity studies. Microbial ecologists estimate that at present less than 0.5% of the microbial diversity revealed by molecular community analyses currently exists in the form of laboratory cultures.

Phylochips

We previously considered the use of DNA chips (microarrays) for assessing overall gene expression in microorganisms (∞ Section 13.6 and Figure 13.7). However, more targeted microarrays can be constructed for rapid analyses in biodiversity studies. These phylogenetic microarrays, called *phylochips*, have been developed for screening microbial communities for specific groups or phyla of prokaryotes (**Figure 22.15**).

Phylochips are constructed by affixing rRNA- or rRNA gene-targeted oligonucleotide probes to the chip surface in a known pattern. Each phylochip can be made as specific or broad as required for the study, and several thousand different probes can be added to a single phylochip. For example, if a goal was to assess the diversity of sulfate-reducing bacteria in a sulfidic environment, such as a microbial mat (Section 22.7 and see Figure 22.20) in a single experiment, a phylochip containing oligonucleotides complementary to specific sequences in the 16S rRNA genes of all known sulfate-reducing bacteria (over 100 species) would be arranged in a known pattern on the phylochip. Then, following the isolation of total community DNA from the mat sample and PCR amplification and fluorescence labeling of the 16S rRNA genes, the environmental DNA is hybridized with the probes on the phylochip and the species present determined by observing which probes hybridized sample DNA (Figure 22.15). Alternatively, rRNA is extracted directly from the microbial community, labeled

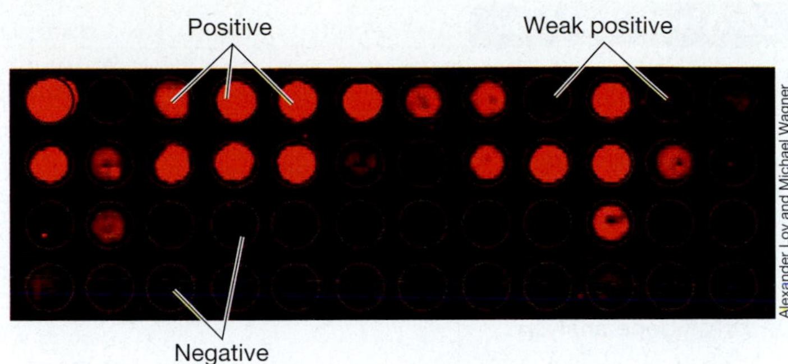

Figure 22.15 Phylochip analysis of sulfate-reducing bacteria diversity. Each spot on the microarray shown has an oligonucleotide complementary to a sequence in the 16S rRNA of a different species of sulfate-reducing bacteria. Following hybridization with 16S rRNA genes PCR-amplified from a microbial community and then fluorescently labeled, the presence or absence of each species is signaled by fluorescence (shown as positive or weak positive) or nonfluorescence (shown as negative), respectively.

with a fluorescent dye, and hybridized directly to the phylochip without an amplification step.

Phylochips circumvent many of the time-consuming steps, such as PCR, DGGE, cloning, and sequencing, that are part of the molecular microbial community analyses considered earlier (Figure 22.14). Moreover, instead of using 16S rRNA as the target, chips can be made to carry probes targeting genes that encode a key metabolic function, such as nitrogen fixation or ammonia oxidation, to address whether nitrogen-fixing or nitrifying bacteria, respectively, are present in the sample, and if so, which ones. Thus, the phylochip is another important tool for the culture-independent assessment of microbial biodiversity.

22.5 MiniReview

PCR can be used to amplify specific target genes such as small subunit rRNA genes or key metabolic genes. DGGE can identify different versions of these genes present in the different species inhabiting a natural sample. Phylochips combine microarray and phylogenetic technologies.

- What could you conclude from PCR/DGGE analysis of a sample that yielded one band by PCR and one band by DGGE? One band by PCR and four bands by DGGE?
- What surprising finding has come out of many molecular studies of natural habitats using 16S rRNA as the target gene?
- What is a phylochip and what can it tell you?

22.6 Environmental Genomics

A more encompassing approach to the molecular study of microbial communities is **environmental genomics**, also called **metagenomics**. In this approach, random shotgun sequencing (Section 12.6) of total DNA cloned from a microbial community, or community DNA previously cloned in large cloning vectors such as bacterial artificial chromosomes (BACs Section 12.15) before sequencing, is used to potentially reveal the entire gene complement of that community (**Figure 22.16**).

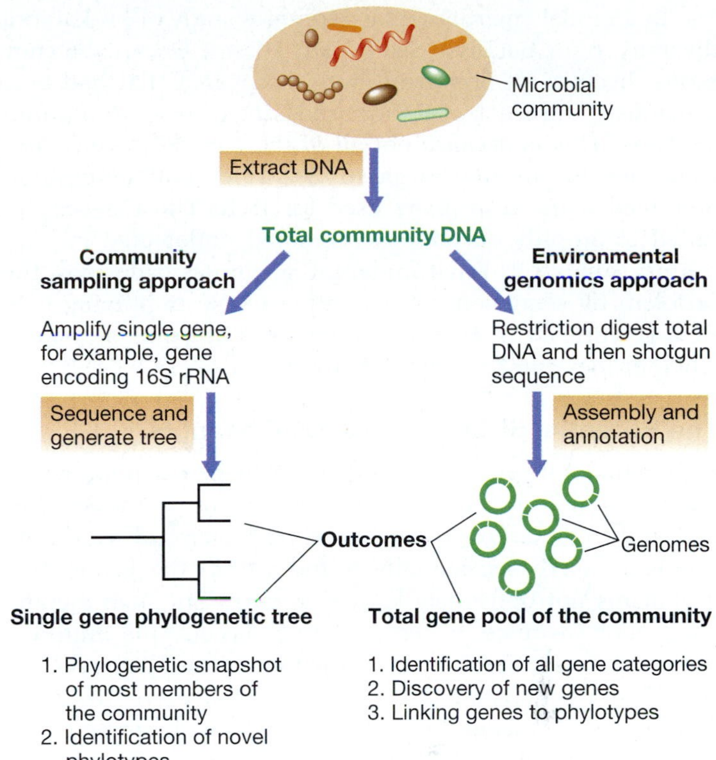

Figure 22.16 Single gene versus environmental genomics approaches to microbial community analysis. Note that in the environmental genomics approach, all community DNA can be sequenced. Genomes are assembled and annotated but complete "finished" genomes are not obtained. Thus, not every gene is accounted for and the genomes will contain some gaps. Total gene recovery is variable and depends, among other factors, on the complexity of the habitat; low diversity habitats typically show better overall recovery.

The goal of environmental genomics is not to generate complete and finished genome sequences, as has been done for many cultured microorganisms. Instead, the idea is to detect as many genes as possible that encode recognizable proteins and then, if possible, to determine the phylogenetic group(s) to which they belong. This can be accomplished by sequencing overlaps to the genes that include phylogenetic markers such as 16S rRNA genes. It is important to distinguish environmental genomics from molecular community analysis, discussed in the previous section. Microbial community analyses typically focus on the diversity of a *single* gene. By contrast, in environmental genomics, theoretically *all* genes in the microbial community can be sampled, and this can reveal features of a microbial community that may be overlooked in single gene approaches. In practice, gene recovery in environmental genomics is not 100%; nevertheless, the technology can yield an impressive picture of the gene pool of an environment.

UNIT 4

In a model environmental genomics study of prokaryotic diversity in the Sargasso Sea (North Atlantic Ocean), an enormous diversity of organisms was discovered that had been previously missed by rRNA-based single gene community analyses. This is because not all of the 16S rRNA genes that were present in the Sargasso Sea microbial community amplified with the primers used for PCR. Those genes that failed to amplify, of course, remained undetected in community analyses. Environmental genomics sidesteps this problem by sequencing DNA without first amplifying it by PCR (Figure 2.15). Genes and genomes are thus sequenced whether they can be amplified or not.

The Potential of Environmental Genomics

Environmental genomics can detect new genes in known organisms and known genes in new organisms. In the Sargasso Sea study for example, genes encoding proteins of known metabolisms were occasionally found within the genomes of organisms not previously known to carry out such metabolisms. For instance, genes encoding the enzyme ammonia monooxygenase, the key enzyme of ammonia-oxidizing *Bacteria* (Sections 15.3 and 20.12), were discovered within the genomes of species of *Archaea*. This indicated that ammonia-oxidizing *Archaea* were almost certain to exist. Not long after this discovery was made, microbiologists were successful in isolating nitrifying *Archaea* from the marine environment (Section 17.12).

In a second example from the Sargasso Sea study, genes encoding proteorhodopsin, the light-mediated proton pump known from certain *Proteobacteria* (Sections 17.3 and 23.9), were found within the genomes of several new phylogenetic lineages of *Bacteria*. Because cultures of all of these organisms have not yet been obtained, the presence of proteorhodopsin in these groups would not have been suspected otherwise. This discovery points to light as being important to the physiology and ecology of these organisms and suggests new strategies for their enrichment and isolation. These examples show how environmental genomics offers microbial ecologists a powerful method for exploring microbial and physiological diversity not attainable in other ways.

Environmental genomics is a powerful tool for assessing both the phylogenetic and metabolic diversity of microbial communities, major goals of microbial ecology. However, in addition to genome analyses, other, more targeted genomics approaches can be used to analyze microbial communities. These include in particular the use of microarrays for assessing gene expression (Section 13.6). Once genome sequences are available for some major organisms from a microbial community, one can design microarrays to measure the transcription of any gene of interest. This allows the experimenter to measure gene expression in known organisms as they interact with other organisms in the microbial community. In Section 22.5 we introduced the phylochip, another use of microarrays for assessing microbial diversity in natural environments (Figure 22.15).

22.6 MiniReview

Environmental genomics is based on cloning, sequencing, and analysis of the collective genomes of the organisms present in a microbial community.

- What is a metagenome?
- How do environmental genomic approaches differ from microbial community analyses, such as that based on 16S rRNA gene analysis?

III MEASURING MICROBIAL ACTIVITIES IN NATURE

So far in this chapter our discussion has focused on measuring microbial *diversity*. We now turn to how microbial ecologists measure microbial *activity;* that is, what microorganisms are actually *doing* in their environment. The techniques to be described include the use of radioisotopes, microelectrodes, and stable isotopes.

Activity measurements in a natural sample are collective estimates of the physiological reactions occurring in the entire microbial community, although one technique to be discussed later, FISH-MAR, allows for a more targeted assessment of physiological activity. Activity measurements reveal both the types and rates of major metabolic reactions in a habitat. In conjunction with biodiversity estimates, these help define the structure and function of the microbial ecosystem, the ultimate goal of microbial ecology. Activity measurements can also provide valuable information for the design of enrichment cultures.

22.7 Chemical Assays, Radioisotopic Methods, and Microelectrodes

In many studies direct chemical measurements of microbial reactions are sufficient for assessing microbial activity in an environment. For example, the fate of lactate oxidation by sulfate-reducing bacteria in a sediment sample can be tracked easily. If sulfate-reducing bacteria are present and active in a sediment sample, then lactate added to the sediment will be consumed and SO_4^{2-} will be reduced to H_2S (**Figure 22.17*a***). Lactate, sulfate, and sulfide can all be measured with fairly high sensitivity using simple chemical assays.

However, when very high sensitivity is required, turnover rates need to be determined, or the fate of portions of a molecule is to be followed, radioisotopes are very useful. For instance, if measuring photoautotrophy is the goal, the light-dependent uptake of $^{14}CO_2$ into microbial cells can be measured (Figure 22.17*b*). If sulfate reduction is of interest, the rate of conversion of $^{35}SO_4^{2-}$ to $H_2{}^{35}S$ can be assessed (Figure 22.17*c*). Heterotrophic activities can be measured by tracking the release of $^{14}CO_2$ from ^{14}C-labeled organic compounds (Figure 22.17*d*), and so on.

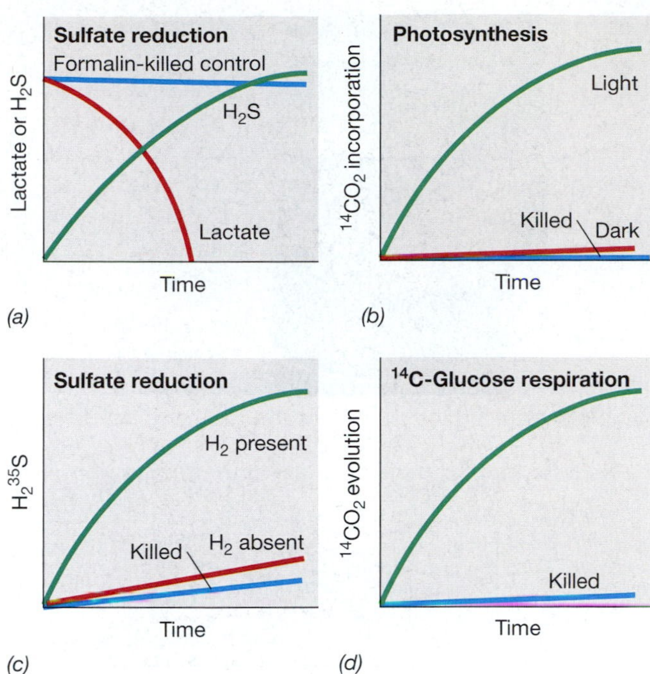

Figure 22.17 Microbial activity measurements. *(a)* Chemical assay for lactate and H_2S transformations during sulfate reduction. *(b–d)* Radioisotopic measurements. *(b)* Photosynthesis measured with $^{14}CO_2$. *(c)* Sulfate reduction measured with $^{35}SO_4^{2-}$. *(d)* Production of $^{14}CO_2$ from ^{14}C-glucose.

Isotope methods are widely used in microbial ecology. To be valid, however, these must employ proper controls, because some transformations might be due to abiotic processes. The *killed cell* control is the key control in such experiments. That is, it is absolutely essential to show that the transformation being measured stops when chemical agents or heat treatments that kill organisms are applied to the sample. Formalin at a final concentration of 4% is commonly used as a chemical sterilant in microbial ecology studies. This kills all cells, and transformations of radiolabeled materials in the presence of 4% formalin can be ascribed to abiotic processes (Figure 22.17).

Radioisotopes in Combination with FISH: FISH-MAR

Radioisotopes can be used as measures of microbial activity in a microscopic technique called **microautoradiography (MAR)**. In this method, cells from a microbial community are exposed to a radioisotope, such as an organic compound (for heterotrophs) or CO_2 (for autotrophs). Following exposure, cells are affixed to a slide and the slide is dipped in photographic emulsion. When left in darkness for a period, radioactive decay from the incorporated substrate exposes silver grains in the emulsion. These appear as black dots within and around the cells (**Figure 22.18a**).

MAR can be done simultaneously with FISH (Section 22.4) in *FISH-MAR*, a powerful technique that combines identification with activity measurements. FISH-MAR allows a microbiologist to identify which organisms in a natural sample are metabolizing a particular radiolabeled substance

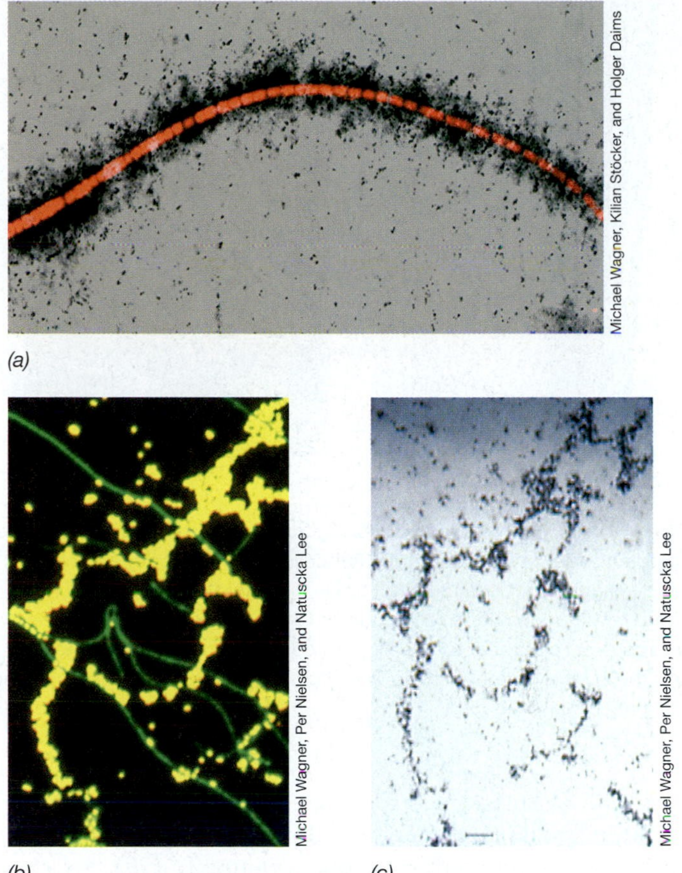

Figure 22.18 FISH-MAR. Fluorescence *in situ* hybridization (FISH) combined with microautoradiography (MAR). *(a)* An uncultured filamentous cell belonging to the *gammaproteobacteria* (as revealed by FISH) is shown to be an autotroph (as revealed by MAR-measured uptake of $^{14}CO_2$). Radioactive decay of the ^{14}C exposes the photographic emulsion, generating the black dots. *(b)* Uptake of ^{14}C-glucose by a mixed culture of *Escherichia coli* (yellow cells) and *Herpetosiphon aurantiacus* (filamentous, green cells). *(c)* MAR of the same field of cells shown in *(b)*. Incorporated radioactivity exposes the film and shows that glucose was assimilated by *E. coli* but not *H. aurantiacus*.

(by MAR) and also to identify these organisms (by FISH) (Figure 22.18*b*). FISH-MAR thus goes a step beyond phylogenetic identification; the technique reveals physiological information on the organisms as well. Such data are useful for not only understanding the activity of the microbial ecosystem but also for guiding enrichment cultures. For example, knowledge of the phylogeny and morphology of an organism metabolizing a particular substrate in a natural sample can be used to design an enrichment protocol to isolate the organism.

Microelectrodes

Small glass electrodes, called **microelectrodes**, can be used to study the activity of microorganisms in nature. Microelectrodes have been developed that measure pH, oxygen, N_2O, CO_2, H_2, or H_2S. As the name implies, microelectrodes are very small, their tips ranging in diameter from 2 to 100 μm

UNIT 4

Figure 22.19 Microelectrodes. *(a)* Schematic drawing of an oxygen microelectrode. The platinum rod functions as a cathode, and when voltage is applied, O_2 is reduced to H_2O, generating a current. The current resulting from the reduction of O_2 at the gold surface of the cathode is proportional to the oxygen concentration in the sample. Note the scale of the electrode. *(b)* Microelectrodes being used in a hot spring microbial mat.

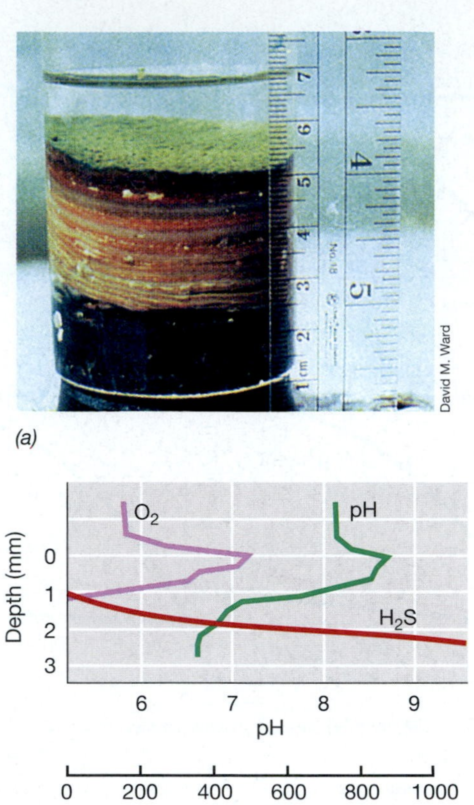

Figure 22.20 Microbial mats and the use of microelectrodes to study them. *(a)* Photograph of a core taken of a hot spring microbial mat sample similar to that used in the experiment summarized in part *(b)*. Upper layer (dark green) contains cyanobacteria, beneath which are several layers of anoxygenic phototrophic bacteria (orange), primarily *Chloroflexus* and *Roseiflexus*. The mat is about 2 cm thick. *(b)* Oxygen, sulfide, and pH profiles in a hot spring microbial mat. Note the millimeter scale on the ordinate.

(**Figure 22.19*a***). The electrodes are carefully inserted into the habitat in small increments to follow microbial activities with depth across a very narrow profile (Figure 22.19*b*).

Microelectrodes have been used extensively in the study of microbial activities, particularly photosynthesis, in microbial mats. Microbial mats are layered microbial communities typically containing cyanobacteria in the uppermost layers, anoxygenic phototrophic bacteria in middle layers (the depth at which the mat becomes light-limited), and chemoorganotrophic, especially sulfate-reducing, bacteria in the lower layers (**Figures** 22.19*b* and **22.20**). Microbial mats develop in many aquatic environments, including hot springs (Figure 22.20*a*) and other extreme environments, and in marine intertidal zones.

With the use of O_2 microelectrodes, oxygen concentrations in microbial mats (Figure 22.19*b*) or soil particles (Figure 23.3) can be very accurately measured over extremely fine intervals. A micromanipulator is used to immerse the electrodes gradually through the sample such that measurements can be taken every 50–100 μm (Figures 22.19*b* and 22.20*b*). Using a bank of microelectrodes, each sensitive to a different chemical, simultaneous measurements of several transformations in a habitat can be made. For example, O_2 and H_2S measurements are often taken simultaneously because these substances form in microbial mats as a result of photosynthesis and sulfate reduction, respectively (Figure 22.20*b*). Near the zone where O_2 and H_2S begin to mix, intense metabolic activity by phototrophic and chemolithotrophic sulfur bacteria may consume these substrates rapidly over very short vertical distances. Detecting the rate of these changes reveals the zone(s) of greatest microbial activity.

Using microelectrodes in microbial mats, one can perturb the mat by adding an organic compound or shading the mat surface and then follow the effect this has in real time on the production or consumption of substances of interest. From such studies, it is possible to explore the types of microbial activities that take place in the mat and make predictions about the habitat that can be addressed experimentally. In this way, it is often possible to model the metabolic interactions of a microbial environment and use this information to dissect microbial activities in other, more complex microbial habitats.

22.7 MiniReview

The activity of microorganisms in natural samples can be assessed very sensitively using radioisotopes or microelectrodes or both. The measurements obtained give the net activity of the microbial community.

- Why are radioisotopes so useful in measuring microbial activities?
- What does the acronym MAR stand for, and how are MAR reactions detected?
- Explain how a microelectrode works.

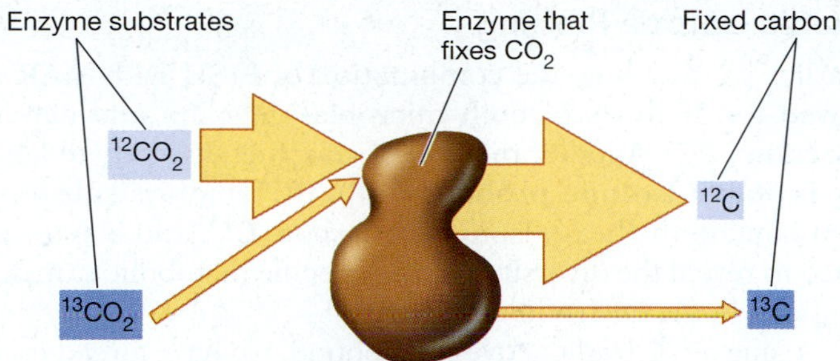

Figure 22.21 Mechanism of isotopic fractionation using carbon as an example. Enzymes that fix CO_2 preferentially fix the lighter isotope (^{12}C). This results in fixed carbon being enriched in ^{12}C and depleted in ^{13}C relative to the starting substrate. The extent of ^{13}C depletion is calculated as an isotopic fractionation. The size of the arrows indicates the relative abundance of each isotope of carbon.

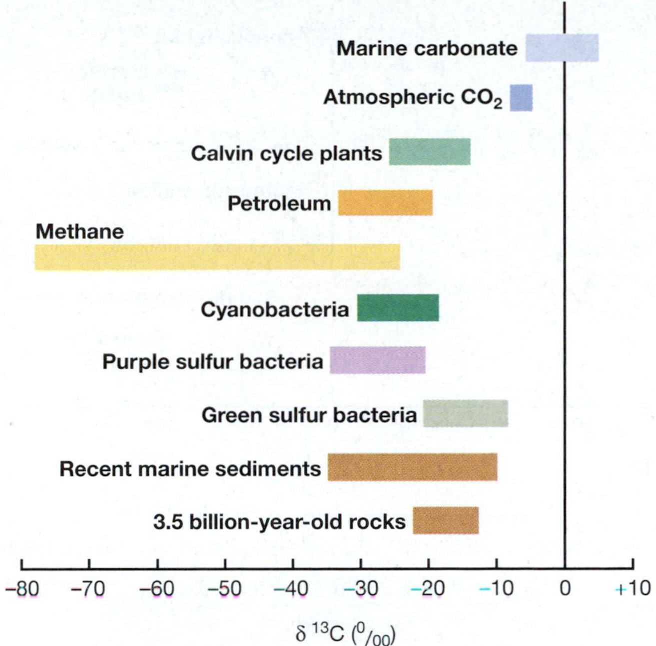

Figure 22.22 Isotopic geochemistry of ^{13}C and ^{12}C. The values are given in parts per thousand (%) and were calculated using the formula:

$$\frac{(^{13}C/^{12}C \text{ sample}) - (^{13}C/^{12}C \text{ standard})}{(^{13}C/^{12}C \text{ standard})} \times 1000$$

The standard is a belemnite sample from the PeeDee rock formation. Note that carbon fixed by autotrophic organisms is enriched in ^{12}C and depleted in ^{13}C. Methane shows extreme isotopic fractionation.

22.8 Stable Isotopes

For many of the chemical elements different isotopes exist, varying in their number of neutrons. Certain isotopes are unstable and break down as a result of radioactive decay. Others, called *stable isotopes*, are not radioactive but are metabolized differentially by microorganisms and can be used to study microbial transformations in nature. We describe here two methods in which stable isotopes can yield information on microbial activities, *stable isotope fractionation* and *stable isotope probing*.

Isotope Fractionation

The two elements that have proven most useful for stable isotope studies in microbial ecology are carbon and sulfur. Carbon exists in nature primarily as ^{12}C, but a small amount (about 5%) exists as ^{13}C. Likewise, sulfur exists primarily as ^{32}S, although some sulfur is found as ^{34}S. The natural abundance of these isotopes changes when C or S is metabolized by microorganisms because enzymes typically favor the *lighter* isotope. That is, relative to the lighter isotope, the heavier isotope is discriminated against when both are metabolized by an enzyme (**Figure 22.21**).

For example, when CO_2 is fed to a phototrophic organism, the cellular carbon becomes *enriched* in ^{12}C and *depleted* in ^{13}C, relative to an abiotic carbon standard. Likewise, the sulfur atom in H_2S produced from the bacterial reduction of SO_4^{2-} is "lighter" than H_2S that has formed geochemically. These processes, known as **isotopic fractionation** (Figure 22.21), are uniquely biological and can be used as a measure of whether a particular transformation has been catalyzed by microorganisms or not.

Use of Isotopic Fractionation in Microbial Ecology

The isotopic composition of a material reveals its past biological or geological origin. For example, plant material and petroleum (which is derived from plant material) have similar isotopic compositions (**Figure 22.22**). Carbon from both plants and petroleum is isotopically lighter than an abiotic standard because the carbon was fixed as CO_2 by a biochemical pathway that discriminated against $^{13}CO_2$ (Figures 22.21 and 22.22). Moreover, methane produced by methanogenic *Archaea* (Section 17.4) is isotopically extremely light, indicating that methanogens discriminate strongly against $^{13}CO_2$ when they reduce CO_2 to CH_4. By contrast, isotopically much heavier marine carbonates are clearly of geological origin (Figure 22.22).

Because of the differences in the proportion of ^{12}C and ^{13}C in carbon of biological versus geological origin, the $^{13}C/^{12}C$ ratio of rocks of different ages have been used as evidence for or against past biological activity in them. Interestingly, organic carbon in rocks as old as 3.5 billion years shows some sign of fractionation (Figure 22.22), supporting the idea that autotrophic life existed at this time. Indeed, we now believe that the first life on Earth appeared somewhat before this, about 3.8–3.9 billion years ago (Sections 1.4 and 14.2).

The activity of sulfate-reducing bacteria is easy to recognize from their fractionation of $^{34}S/^{32}S$ in sulfides (**Figure 22.23**). As compared with a sulfide standard, sedimentary sulfide is highly enriched in ^{32}S (Figure 22.23). Fractionation during sulfate reduction allows one to identify biologically produced sulfur and has been widely used to trace the activities of sulfur-cycling prokaryotes through geological time. Sulfur isotopic analyses have also been used as evidence for the lack of life on the Moon. For example, the data in Figure 22.23

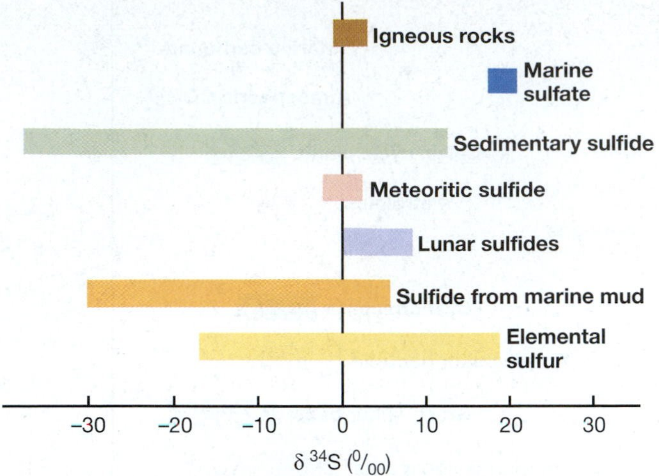

Figure 22.23 Isotopic geochemistry of ^{34}S and ^{32}S. The values are given in parts per thousand (‰) and were calculated using the formula:

$$\frac{(^{34}S/^{32}S \text{ sample}) - (^{34}S/^{32}S \text{ standard})}{(^{34}S/^{32}S \text{ standard})} \times 1000$$

The standard is an iron sulfide mineral from the Canyon Diablo meteorite. Note that sulfide and sulfur of biogenic origin are enriched in ^{32}S and depleted in ^{34}S.

show that the isotopic composition of sulfides in lunar rocks closely approximates that of the abiotic sulfide standard and not that of biogenic sulfide.

Oxygen isotopic analyses ($^{18}O/^{16}O$) have been used to trace Earth's transition from an anoxic to an oxic environment because Earth's molecular oxygen originated from oxygenic photosynthesis ($H_2O \rightarrow \frac{1}{2} O_2 + 2H$) by cyanobacteria (Sections 14.3 and 24.1). Using ratios of oxygen isotopes, it has been possible to show that Earth was oxygenated gradually over geological time and that current oxygen levels (~21%) have been present for only the past 750 million years or so (Figure 14.7).

Stable Isotope Probing

Earlier we saw how the combination of FISH with MAR allowed for analyses of both microbial diversity and activity (Section 22.7). Another method of coupling diversity to activity is **stable isotope probing** (SIP). SIP employs substrates that contain the heavy isotope of carbon, C^{13}, and is typically used to reveal the diversity behind specific metabolic transformations in the environment.

Using a ^{14}C (radioactive) compound, we have already seen how the rate of conversion of the compound to $^{14}CO_2$, for example, can be used as a measure of the activity of this process in an environment (Section 22.7 and Figure 22.17). However, the results of such an experiment tell you nothing about the organisms that carried out the process. SIP solves this problem by producing ^{13}C-labeled DNA from the organism(s) that carried out a specific metabolic transformation. The DNA can then be used to identify the organism(s) or for genomic analyses.

How is an SIP experiment done? Let's say the goal is to characterize organisms capable of catabolizing aromatic compounds in lake sediment. Using benzoate as a model aromatic compound, some ^{13}C-benzoate is added to a sediment sample, the sample incubated for an appropriate period, and then bulk DNA from the sample isolated by standard techniques (**Figure 22.24**). As we have seen before (Figure 22.13), such DNA would originate from all of the organisms in the microbial community. However, those organisms that incorporated and metabolized ^{13}C-benzoate would incorporate some of the ^{13}C into their DNA; this would make their DNA isotopically heavier than the DNA in cells that did not incorporate the benzoate. ^{13}C-DNA is heavier, albeit only slightly, than ^{12}C-DNA, but the difference is enough that ^{13}C-DNA can be separated from ^{12}C-DNA by a special type of centrifugation technique (Figure 22.24).

Once the ^{13}C-DNA is purified, it can then be analyzed for genes of interest. Returning to our benzoate example, if the goal was to characterize the phylogeny of the organism(s) catabolizing the benzoate, PCR amplification of 16S rRNA genes

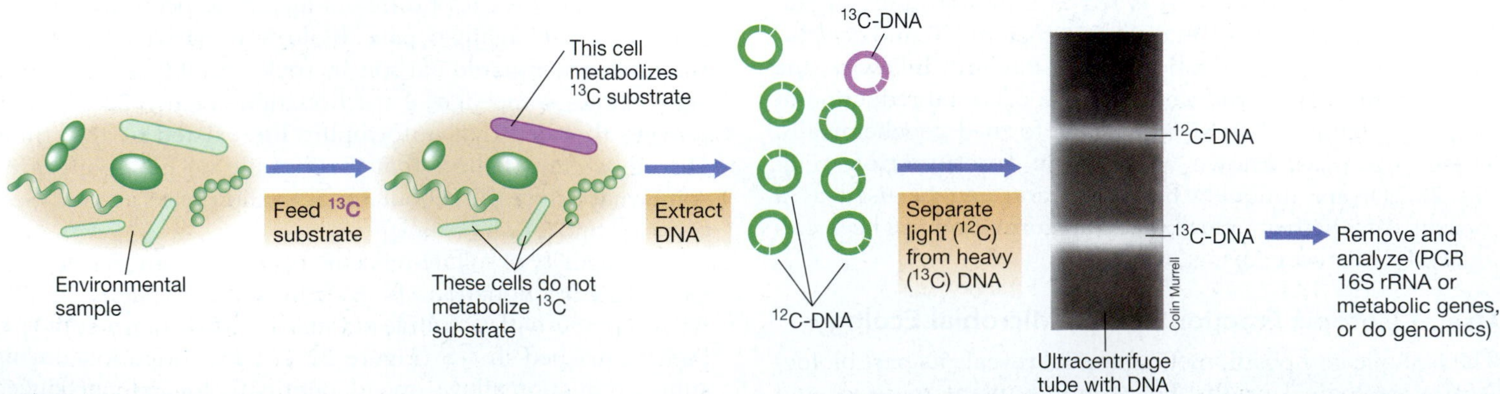

Figure 22.24 Stable isotope probing. The microbial community in an environmental sample is fed a specific ^{13}C-substrate. Organisms that can metabolize the substrate will produce ^{13}C-DNA as they grow and divide; ^{13}C-DNA can be separated from lighter (^{12}C) DNA by density gradient centrifugation (photo). The isolated DNA is then subject to specific gene or entire genomic analyses.

in the ^{13}C-DNA could be used to do so (Figures 22.13 and 22.14). However, in addition to phylogenetic analyses, many other genes could be targeted once the ^{13}C-DNA is obtained. For example, genes linked to specific metabolisms or genes other than rRNA genes that are linked to certain organisms or groups of organisms could be amplified and sequenced. Moreover, full-blown genomic analyses could also be undertaken (Figures 22.16 and 22.24).

SIP of *RNA* instead of *DNA* is also possible. It has been shown with bacterial cultures that ^{13}C from a labeled substrate actually ends up in RNA more quickly than in DNA. Thus, RNA SIP allows for shorter incubation times than DNA SIP. This helps reduce the problem of "cross-feeding," a situation in which the original ^{13}C substrate is catabolized by one organism, but then some of the label is excreted in a form that can be assimilated by other organisms. The DNA from these organisms then becomes labeled, making it appear that they catabolized the original test substrate. If RNA SIP is done, the ^{13}C-RNA is purified and then converted to DNA by reverse transcription using the retrovirus enzyme reverse transcriptase (RT-PCR, Section 12.8). Once a DNA copy is available, it is analyzed in the same way as for DNA obtained by standard SIP analyses.

22.8 MiniReview

Isotopic fractionation can reveal the biological origin of various substances. Fractionation is a result of the activity of enzymes that discriminate against the heavier form of an element when binding their substrates.

- How can the $^{12}C/^{13}C$ composition of a substance reveal its biological or geological origin?
- What is the simplest explanation for why lunar sulfides are isotopically heavy?
- How can stable isotope probing reveal the identity of an organism that carries out a particular process?

With an understanding of some of the methods that are used in microbial ecology, we now move on to Chapter 23 to explore the discipline of microbial ecology itself. There we will consider microbial communities and some of their major activities. In many cases we will use one or more of the tools described in this chapter to help illustrate important ecological principles or identify microbial species in diverse microbial communities. In Chapter 24 we will consider major nutrient cycles along with some special, and in some cases spectacular, microbial habitats.

Review of Key Terms

Acridine orange a nonspecific fluorescent dye used to stain microbial cells in a natural sample

Denaturing gradient gel electrophoresis (DGGE) an electrophoretic technique capable of separating nucleic acid fragments of the same size that differ in base sequence

DAPI a nonspecific fluorescent dye used to stain microbial cells in a natural sample to obtain total cell numbers

Enrichment bias a problem with enrichment cultures in which "weed" species tend to dominate in the enrichment, often to the exclusion of the most abundant or ecologically significant organisms in the inoculum

Enrichment culture a means of obtaining laboratory cultures of microorganisms from a natural sample by using highly selective culture methods

Environmental genomics (metagenomics) the use of genomic methods (sequencing and analyzing genomes) to characterize natural microbial communities

Fluorescence *in situ* hybridization (FISH) a method employing a fluorescent dye covalently bonded to a specific nucleic acid probe for identifying or tracking organisms in the environment

Green fluorescent protein (GFP) a fluorescing protein that renders a cell green and that can be used to track a genetically modified organism in nature

Isotopic fractionation discrimination by enzymes against the heavier isotope of the various isotopes of C or S, leading to enrichment of the lighter isotopes

Laser tweezers a device used to obtain pure cultures in which a single cell is optically trapped with a laser beam and moved away from surrounding cells into sterile growth medium

Microautoradiography (MAR) measurement of the uptake of radioactive substrates by visually observing the cells in an exposed photographic emulsion

Microbial ecology study of the interaction of microorganisms with each other and their environment

Microelectrode a small glass electrode for measuring pH or specific compounds such as O_2 or H_2S that can be immersed into a microbial habitat at microscale intervals

Most-probable number (MPN) technique serial dilution of a natural sample to determine the highest dilution yielding growth

Nucleic acid probe an oligonucleotide, usually 10–20 bases in length, complementary in base sequence to a nucleic acid sequence in a target gene or RNA

Phylotype in microbial community analysis, one unique sequence of a phylogenetic marker gene

Stable isotope probing a method for characterizing an organism that incorporates a particular substrate by feeding the substrate in ^{13}C form and then isolating ^{13}C-enriched DNA and analyzing the genes

Winogradsky column a glass column packed with mud and overlaid with water to mimic an aquatic environment, in which various bacteria develop over a period of months

Review Questions

1. What is the basis of the enrichment culture technique? Why is an enrichment medium usually suitable for the enrichment of only a certain group or groups of organisms (Section 22.1)?
2. What is the principle of the Winogradsky column and what types of organisms does it serve to enrich? How might a Winogradsky column be used to enrich organisms present in an extreme environment, like a hot spring microbial mat (Section 22.1)?
3. Describe the principle of MPN for enumerating bacteria from a natural sample (Section 22.2).
4. Why would the laser tweezers be a method superior to dilution and liquid enrichment for obtaining an organism present in a sample in low numbers (Section 22.2)?
5. Compare and contrast the use of fluorescent dyes and fluorescent antibodies for use in enumerating microbial cells in natural environments. What advantages and limitations do each of these methods have (Section 22.3)?
6. Can nucleic acid probes in microbial ecology be as sensitive as culturing methods? What advantages do nucleic acid methods have over culture methods? What disadvantages (Section 22.4)?
7. How does ISRT-FISH work? What can it tell us about a microbial community that FISH alone cannot (Section 22.4)?
8. What is the green fluorescent protein? In what ways does a green fluorescing cell differ from a cell fluorescing from, for example, staining with a phylogenetic stain (Section 22.4)?
9. How can a phylogenetic picture of a microbial community be obtained without culturing its inhabitants (Section 22.5)?
10. After PCR amplification of total community DNA using a specific primer set, why is it necessary to either clone or run DGGE on the products before sequencing them (Section 22.5)?
11. Give an example of how environmental genomics has discovered a known metabolism in a new organism (Section 22.6).
12. What are the major advantages of radioisotopic methods in the study of microbial ecology? What type of controls (discuss at least two) would you include in a radioisotopic experiment to show $^{14}CO_2$ incorporation by phototrophic bacteria or to show $^{35}SO_4^{2-}$ reduction by sulfate-reducing bacteria (Section 22.7)?
13. What can FISH-MAR tell you that FISH alone cannot (Section 22.7)?
14. Will autotrophic organisms contain more or less ^{12}C in their organic compounds than was present in the CO_2 that fed them (Section 22.8)?

Application Questions

1. Design an experiment for measuring the activity of sulfur-oxidizing bacteria in soil. If only certain species of the sulfur-oxidizers present were metabolically active, how could you tell this? How would you prove that your activity measurement was due to biological activity?
2. You wish to know whether *Archaea* exist in a lake water sample but are unsuccessful in culturing any. Using techniques described in this chapter, how could you determine whether *Archaea* existed in the sample, and if they did, what proportion of the cells in the lake water were *Archaea*?
3. Design an experiment to solve the following problem. Determine the rate of methanogenesis ($CO_2 + 4\ H_2 \rightarrow CH_4 + H_2O$) in anoxic lake sediments and whether or not it is H_2-limited. Also, determine the morphology of the dominant methanogen (recall that these are *Archaea*, ∞ Section 13.4). Finally, calculate what percentage the dominant methanogen is of the total archaeal and total prokaryotic populations in the sediments. Don't forget to specify necessary controls.
4. Design a SIP experiment that would allow you to determine which organism(s) in a lake water sample were capable of oxidizing methane (CH_4). Assume that four different species could do this. How would you combine SIP with other molecular analyses to identify these four species?

23

Microbial Ecosystems

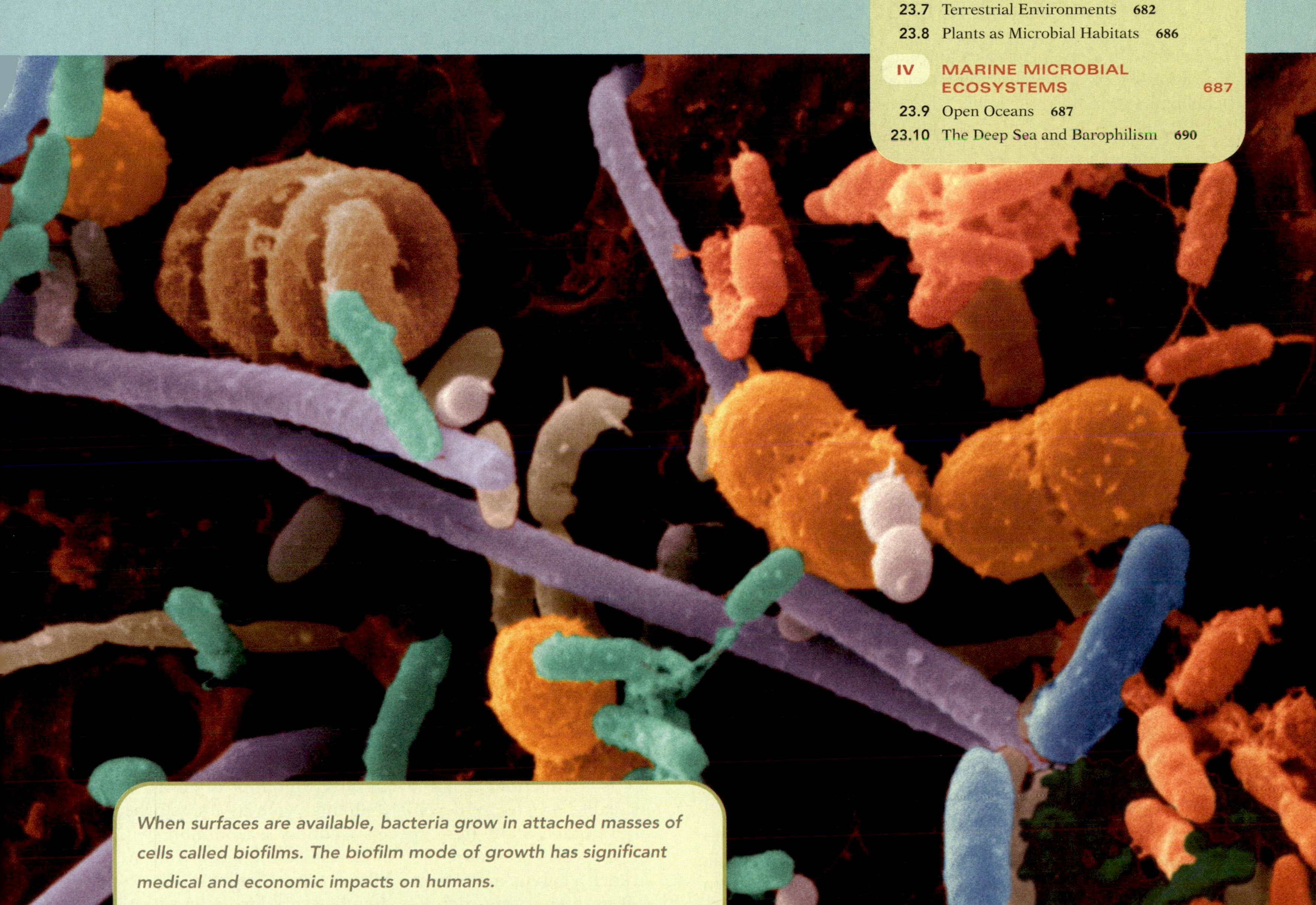

When surfaces are available, bacteria grow in attached masses of cells called biofilms. The biofilm mode of growth has significant medical and economic impacts on humans.

Microorganisms do not live alone in nature; they interact with other organisms and with their environment. In so doing, microorganisms carry out many essential activities that support all life on Earth. In this chapter we discuss some of the major principles of microbial ecology and then explore some typical microbial habitats—soil, freshwater, plants, and the oceans. The human body as a microbial habitat is explored in Chapter 28. In Chapter 24 we examine more specialized microbial habitats and activities that play essential roles in the biology of plants and animals.

I PRINCIPLES OF MICROBIAL ECOLOGY

We begin with some principles and terminology of microbial ecology and look at how organisms interact with each other in beneficial and in harmful ways.

23.1 Ecological Concepts

In Chapter 1 we learned that microorganisms form microbial communities in ecosystems. An **ecosystem** is the sum total of all of the organisms and abiotic factors in a particular environment. An ecosystem contains many different **habitats**, portions of the ecosystem best suited to one or a small number of microbial populations. Examples of ecosystems include soil, water, and the animal body, among others.

Despite their small size, microorganisms account for roughly half of all biomass on Earth. No natural habitat on Earth supports higher organisms but not microorganisms, whereas many microbial habitats are unsuitable for higher organisms. Microorganisms are ubiquitous on the surface of and even deep within the earth, and the breadth of microbial ecosystems far exceeds those that support higher life forms. Collectively, microorganisms show great metabolic diversity and are the primary catalysts of nutrient cycles in nature. The *types* of microbial activities possible in an ecosystem depend on the species composition, population sizes, and physiological state of the microorganisms in each habitat. The *rates* of microbial activities are controlled by the nutrients and growth conditions that prevail in their habitat. Under optimal conditions, microbial activities can profoundly impact an ecosystem, diminishing or enhancing the activities of other microorganisms and the macroorganisms that live there.

Symbioses: Parasitism

Many microorganisms establish relationships with other organisms. **Symbiosis** is defined as a relationship between two or more organisms that share a particular ecosystem. **Parasitism** is a form of symbiosis in which one member in the relationship is harmed in the process. A parasite is an organism that obtains its nutrients from another organism, the host. Pathogenic (disease-causing) microorganisms can live a parasitic lifestyle but are usually not considered parasites, *per se*, because most of them have other modes of growth and survival. By contrast, an organism such as *Giardia*, an intestinal protist of warm-blooded animals that causes violent intestinal symptoms (Section 36.6), is a true parasite because it can only reproduce in its host. Parasites rarely kill their hosts, although they may cause long-term harm to them.

Symbioses: Mutualism and Commensalism

Mutualism is a symbiotic relationship in which both species benefit. By contrast, **commensalism** is a symbiotic relationship in which one species benefits while the other is neither harmed nor helped. We will see many examples of mutualism in the next chapter in which we examine some classical microbial mutualisms, such as the root nodule–legume association or the rumen of ruminant animals.

Clear-cut cases of commensalism, on the other hand, are probably common but more difficult to identify. What may appear to be a commensal relationship may actually benefit both partners in some way. Bacteria that inhabit the animal gut are mostly commensals, growing on the nutrients released from food eaten by the animal. However, laboratory studies with mice reared aseptically from birth to be free of microorganisms have shown that although many gut microorganisms are of no benefit to the animal, others excrete essential nutrients, such as vitamins. Many of the microorganisms that inhabit the rhizosphere (root zones) of soil are probably commensals, benefiting from organic compounds excreted by the plant roots.

Species Diversity and Abundance in Microbial Habitats

The species that inhabit a given ecosystem are those best adapted to growth with the nutrients and conditions that prevail there. The diversity of microbial species in an ecosystem can be expressed in two ways. One is *species richness*, the total number of different species present (**Figure 23.1*a***). Microbial species richness can be defined in molecular terms by the diversity of phylotypes (Section 22.5) in the ecosystem. *Species abundance*, by contrast, is the proportion of each species in the ecosystem (Figure 23.1*b*). A major goal of microbial ecology is to understand species richness and abundance in an ecosystem, along with the major microbial activities that occur. Once these are known, ecologists can model the microbial ecosystem by perturbing it in some way and observing whether predicted changes match experimental results.

The microbial species richness and abundance of a habitat is a function of the kinds and amounts of nutrients available. In some microbial habitats, such as undisturbed organic-rich soils, high species richness is common, with most species present at only moderate abundance. Nutrients in such a habitat are of many different types, and this helps select for high species richness. In other habitats, such as some extreme environments, species richness is often very low and abundance of one or a few species very high. This is because the conditions in the environment exclude all but a handful of species, and key nutrients are present at such high levels that

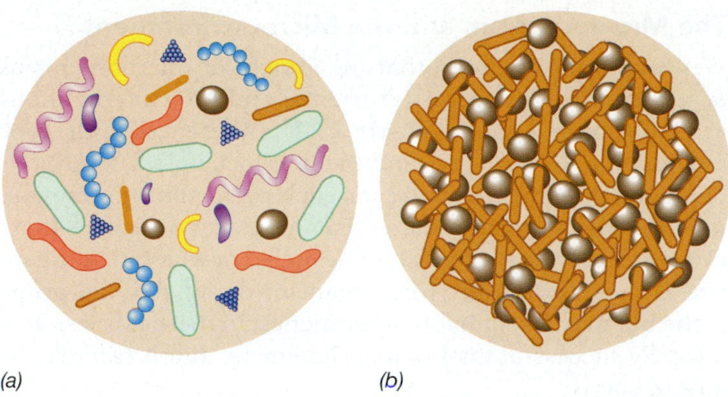

Figure 23.1 Microbial species diversity: Richness versus abundance. *(a)* This community shows high species richness and low to moderate abundance (the different cell morphologies are meant to indicate distinct species). *(b)* This community shows low richness but high abundance.

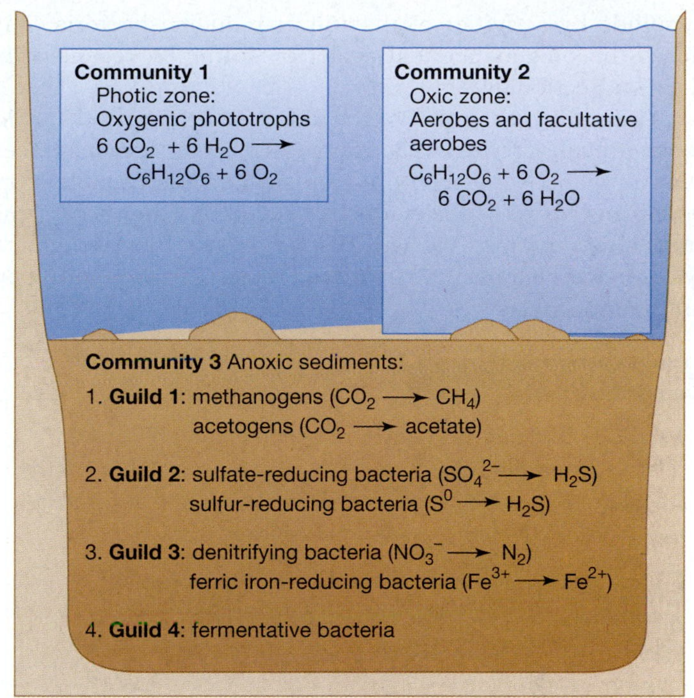

Figure 23.2 Populations, guilds, and communities. Microbial communities consist of populations of cells of different species. In a freshwater lake ecosystem, for example, there would likely be the communities here. For the sediment community, major guilds are indicated. The reduction of CO_2, SO_4^{2-}, S^0, NO_3^-, and Fe^{3+} are examples of anaerobic respirations.

the highly adapted species can grow to high cell densities. Bacteria that catalyze acid mine runoff from the oxidation of iron are a good example here. These organisms thrive in highly acidic, iron-rich but organic-poor waters, where low pH and lack of organic matter limit species richness. However, the elevated levels of Fe^{2+} present, which is oxidized to Fe^{3+} in energy-yielding reactions (∞ Section 20.11), fuel high species abundance. We examine the activities of iron-oxidizing organisms in acidic environments in Sections 24.5 and 24.6.

23.1 MiniReview

Ecosystems contain microbial communities whose components can interact in various ways. Symbiotic interactions may benefit one or both partners in the interaction. Species richness and abundance are aspects of species diversity in an ecosystem.

- How does a parasitic microorganism differ from a commensal?
- What is the difference between species richness and species abundance?

23.2 Microbial Ecosystems and Biogeochemical Cycling

In any habitable ecosystem, individual microbial cells grow to form populations. Metabolically related microbial populations are called **guilds**, and sets of guilds form microbial communities (**Figure 23.2**). Microbial communities interact with macroorganisms and abiotic factors in the ecosystem in a way that defines the workings of that ecosystem.

Energy enters ecosystems in the form of sunlight, organic carbon, and reduced inorganic substances. Light is used by phototrophs (∞ Sections 20.1–20.5) to make ATP and synthesize new organic matter (Figure 23.2). In addition to carbon, new organic matter contains nitrogen, sulfur, phosphorus, iron, and the other elements of life (∞ Section 5.1). This newly synthesized organic material along with organic matter that enters the ecosystem from the outside (called *allochthonous* organic matter) fuels the catabolic activities of heterotrophic organisms. Organic matter is then oxidized to CO_2 by respiration or fermented to various reduced substances. If they are present and metabolically active in the ecosystem, chemolithotrophs obtain their energy from inorganic electron donors, such as H_2, Fe^{2+}, S^0, or NH_3 (∞ Chapter 21) and contribute to the synthesis of new organic matter through their autotrophic activities (Figure 23.2).

Biogeochemical Cycles

Microorganisms play an essential role in the cycling of elements including in particular, carbon, sulfur, nitrogen, and iron elements. The study of these chemical transformations is called **biogeochemistry**. A biogeochemical cycle defines the transformations of a key element that is catalyzed by either biological or chemical agents. Biogeochemical cycles typically proceed by oxidation–reduction reactions (∞ Section 5.6) as the element moves through the ecosystem. Sulfur, for example, in the form of hydrogen sulfide (H_2S) is oxidized by a variety of microorganisms, both phototrophic and chemotrophic, to sulfur (S^0) and sulfate (SO_4^{2-}). Sulfate is a key nutrient for plants. However, sulfate can also be reduced to sulfide by activities of the sulfate-reducing bacteria. This closes the biogeochemical cycle for sulfur by regenerating H_2S (Section 24.4).

Many different microorganisms can be involved in biogeochemical cycling reactions, and in most instances, microorganisms are the only biological agents capable of regenerating forms of the elements needed by other organisms, particularly plants. We will discuss many biogeochemical cycles in the next chapter, including those of carbon, nitrogen, sulfur, and iron.

23.2 MiniReview

Microbial communities consist of guilds of metabolically related organisms. Microorganisms play major roles in energy transformations and biogeochemical processes that result in the recycling of elements essential to living systems.

- How does a microbial guild differ from a microbial community?
- What is a biogeochemical cycle?

II THE MICROBIAL HABITAT

As we have said, a habitat is the place within an ecosystem that is occupied by a particular microorganism or group of microorganisms. Conditions in microbial habitats are often very specific but are subject to rapid change due to inputs and outputs to the habitat and to microbial activity or physical disturbance.

23.3 Environments and Microenvironments

Microorganisms are found in every habitat that will sustain life, and the habitats themselves can be exceedingly diverse. Besides the common habitats of soil and water, microorganisms thrive in extreme environments and also reside on the surface of and within higher organisms. In some cases, microorganisms actually live intracellularly within higher organisms. Thus we can safely say that habitats are as diverse as the microorganisms that live in them.

The Microorganism and the Microenvironment

We learned in Chapter 6 that the growth of microorganisms depends on the resources (nutrients) and growth conditions available in their habitat (**Table 23.1**). Differences in the type and quantity of resources and the physiochemical conditions of a habitat define the **niche** for each particular microorganism. Ecological theory states that for every organism there exists at least one niche, the *prime niche,* in which that organism is most successful. The organism dominates the prime niche but may also inhabit other niches; in these it is less ecologically successful than in its prime niche, but it still may be able to survive.

Because microorganisms are so small, their niches will also be small. The term **microenvironment** is used to describe the niche where a microorganism actually lives and metabolizes. A microbial niche will be very small. For example, for a typical 3-μm rod-shaped bacterium, a distance of 3 mm is huge. It is equivalent to that which a human would experience over a distance of 2 km! Across this 3-mm distance, several microenvironments can exist, each of which offers opportunities for different physiological types of microorganisms to thrive. Consider, for example, the distribution of an important microbial nutrient such as oxygen (O_2) in a soil particle. Microelectrodes (Section 22.7) can be used to measure oxygen concentrations throughout small soil particles. As shown in data from an actual microelectrode experiment (**Figure 23.3**), soil particles are not homogeneous in terms of their oxygen content but instead contain many microenvironments. The outer zones of the soil particle may be fully **oxic** while the center, only a very short distance away (in human terms, but of course a great distance from a microbial standpoint) remains completely **anoxic** (O_2-free) (Figure 23.2). Thus, anaerobic organisms could thrive near the center of the particle, microaerophiles (aerobes that require very low oxygen levels) further out, and obligately aerobic organisms in the outermost fully oxic region of the particle. Facultatively aerobic bacteria could be distributed throughout the particle.

Physiochemical conditions in a microenvironment are subject to rapid change, both temporally and spatially. The oxygen concentrations shown in the soil particle in Figure 23.3 represent "instantaneous" measurements. Measurements taken following a period of intense microbial respiration or following a disturbance due to wind, rain, or disruption by soil animals, could differ dramatically from those shown.

Table 23.1 Major resources and conditions that govern microbial growth in nature

Resources	*Conditions*
Carbon (organic, CO_2)	Temperature: cold → warm → hot
Nitrogen (organic, inorganic)	Water potential: dry → moist → wet
Other macronutrients (S, P, K, Mg)	pH: 0 → 7 → 14
Micronutrients (Fe, Mn, Co, Cu, Zn, Mn, Ni)	O_2: oxic → microoxic → anoxic
O_2 and other electron acceptors (NO_3^-, SO_4^{2-}, Fe^{3+}, etc.)	Light: bright light → dim light → dark
Inorganic electron donors (H_2, H_2S, Fe^{2+}, NH_4^+, NO_2^-, etc.)	Osmotic conditions: freshwater → marine → hypersaline

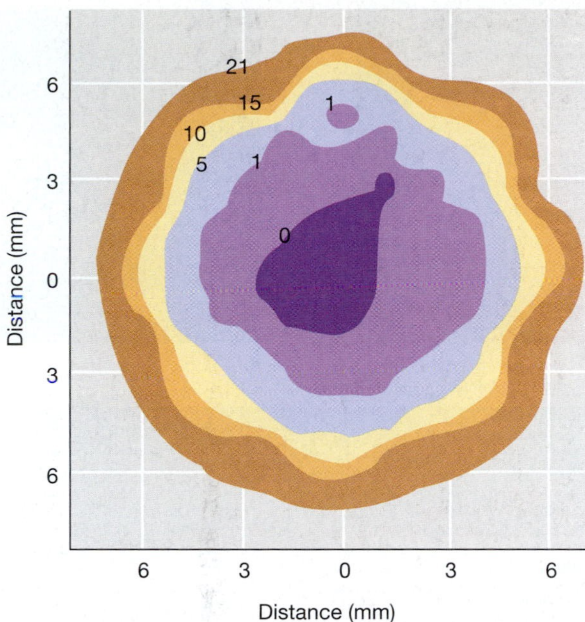

Figure 23.3 Oxygen microenvironments. Contour map of O_2 concentrations in a soil particle. The axes show the dimensions of the particle. The numbers on the contours are O_2 concentrations (in percentage, air is 21% O_2). In terms of oxygen relationships for microorganisms, each zone can be considered a different microenvironment.

During such disturbances, certain populations may temporarily dominate the activities in the soil particle and grow to high numbers while others remain dormant or nearly so. However, if the microenvironments shown in Figure 23.3 are eventually reestablished, the microbial activities characteristic of the original soil particle will eventually return as well.

Nutrient Levels and Growth Rates

Resources (Table 23.1) typically enter an ecosystem intermittently. A large pulse of nutrients—for example, an input of leaf litter or the carcass of a dead animal—may be followed by a period of nutrient deprivation. Because of this, microorganisms in nature often face a "feast-or-famine" existence. It is thus common for them to produce storage polymers as reserve materials when resources are abundant. Examples of storage materials are poly-β-hydroxyalkanoates, polysaccharides, polyphosphate, and so on (∞ Section 4.10).

Extended periods of exponential microbial growth in nature are probably rare. Growth typically occurs in spurts, linked closely to the availability and nature of resources. Because physiochemical conditions in nature are rarely optimal for microbial growth all at the same time, growth rates of microorganisms in nature are usually well below the maximum growth rates recorded in the laboratory. For instance, the generation time of *Escherichia coli* in the intestinal tract of a healthy adult eating at regular intervals is about 12 hours (two doublings per day), whereas in pure culture it grows much faster, with a minimum generation time of 20 minutes under the best conditions (∞ Figure 7.18). Estimates have shown that typical soil bacteria grow in nature at less than 1% of the maximal growth rate measured in the laboratory. On average, these slow growth rates reflect the fact that (1) resources or growth conditions (Table 23.1) are suboptimal; (2) the distribution of nutrients throughout the microbial habitat is not uniform; and (3) except for rare instances, microorganisms in nature grow in mixed populations rather than pure culture. An organism that grows rapidly in pure culture may grow much slower in a natural environment where it must compete with other organisms that are equally or better suited to the resources and growth conditions available.

Microbial Competition and Cooperation

Competition among microorganisms for resources in a habitat may be intense, with the outcome dependent on several factors, including rates of nutrient uptake, inherent metabolic rates and, ultimately, growth rates. A typical habitat contains a mixture of different species (Figures 23.1 and 23.2), with the density of each population dependent on how closely its niche resembles its prime niche.

Some microorganisms work together to carry out transformations that neither can accomplish alone. These types of microbial partnerships, called *syntrophy* (∞ Section 21.5), are symbioses that are crucial to anoxic carbon cycling, as will be described in Section 24.2. Metabolic cooperation can also be seen in the activities of organisms that carry out complementary metabolisms. For example, in Chapter 20 we discussed metabolic transformations that are carried out by two distinct groups of organisms, such as those of the nitrifying bacteria. Together, the nitrifying bacteria oxidize NH_3 to NO_3^-, although neither group (the ammonia-oxidizers nor the nitrite-oxidizers) is capable of doing this alone (∞ Section 20.12). Because the product of the ammonia-oxidizing bacteria (NO_2^-) is the substrate for the nitrite-oxidizing bacteria, such organisms typically live in nature in tight association within their habitats (∞ Figure 22.11*a*).

23.3 MiniReview

The microenvironment is the place within a niche in which microorganisms live and are active. Microorganisms in nature often live a feast-or-famine existence such that only the best-adapted species reach high populations in a given niche. Cooperation among microorganisms is also important in many microbial interrelationships.

- What aspects define the niche of a particular microorganism?
- Why can many different physiological groups of organisms live in a single habitat?

23.4 Biofilms: Microbial Growth on Surfaces

Surfaces are important microbial habitats for at least two reasons. For one, nutrients adsorb to surfaces and thus often contain more resources than are available to planktonic cells

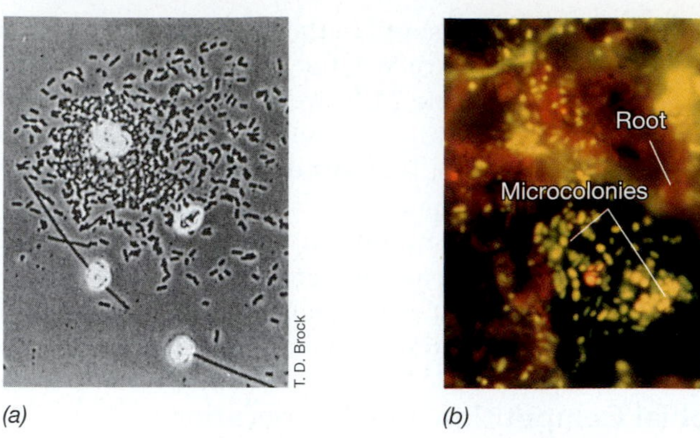

Figure 23.4 Microorganisms on surfaces. *(a)* Bacterial microcolonies developing on a microscope slide that was immersed in a river. The bright particles are mineral matter. The short, rod-shaped cells are about 3 μm long. *(b)* Fluorescence photomicrograph of a natural microbial community living on plant roots in soil. Note microcolony development. The preparation has been stained with acridine orange.

(cells that live a floating existence, Section 23.6). And second, surfaces are areas to which microbial cells can attach. Attachment is a means for cells to remain in a favorable habitat and not be washed away. As a consequence, microbial numbers and activities are typically quite high on surfaces.

Microscope slides can be used as experimental surfaces to which organisms can attach and grow. A slide can be immersed in a microbial habitat, left for a period of time, and then retrieved and examined microscopically (**Figure 23.4*a***). Microcolonies readily develop on such surfaces much as they do on natural surfaces in nature. In fact, periodic microscopic examination of immersed microscope slides has been used to measure growth rates of attached organisms in nature.

A surface may also be a nutrient, such as a particle of organic matter. On such surfaces the attached microorganisms catabolize nutrients directly from the surface of the particle. For example, plant roots become heavily colonized by soil bacteria living on organic exudates from the plant, and fluorescent stains can be used to detect and enumerate them (Figure 23.4*b*).

Biofilms

As bacterial cells grow on surfaces they tend to form **biofilms**—assemblages of bacterial cells attached to a surface and enclosed in an adhesive matrix excreted by the cells (**Figure 23.5**). The matrix is typically a mixture of polysaccharides, but can contain proteins and even nucleic acids. Biofilms trap nutrients for microbial growth and help prevent the detachment of cells on surfaces present in flowing systems (Figure 23.5*c*).

Biofilms typically contain several porous layers, and the cells in each layer can be examined by scanning laser confocal microscopy (Section 2.3) (Figure 23.5*b*). Biofilms may contain only one or two species or, more commonly, several species of bacteria. The biofilm that forms around a tooth surface, for example, contains several hundred different phylotypes, including species of both *Bacteria* and *Archaea*.

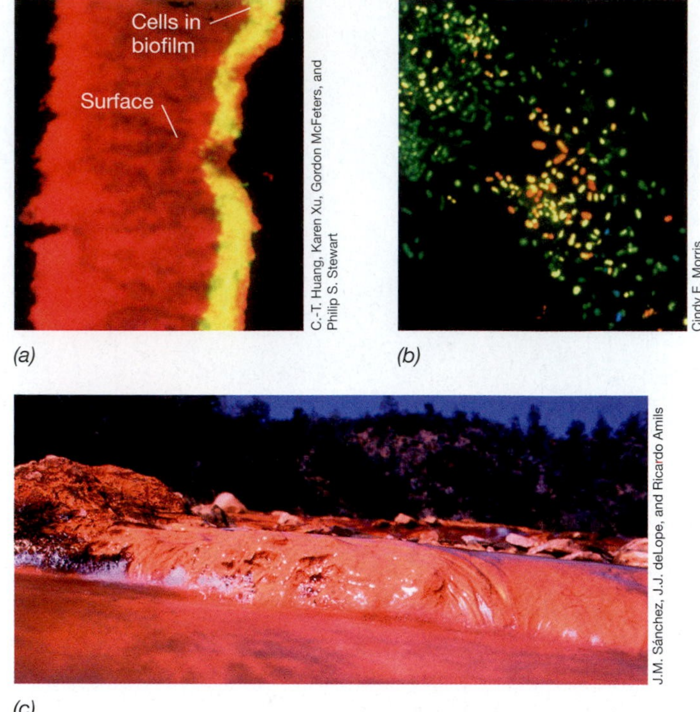

Figure 23.5 Examples of microbial biofilms. *(a)* A cross-sectional view of an experimental biofilm made up of cells of *Pseudomonas aeruginosa*. The yellow layer (about 15 μm in depth) contains cells and is stained by a reaction showing activity of the enzyme alkaline phosphatase. *(b)* Confocal laser scanning microscopy of a natural biofilm (top view) on a leaf surface. The color of the cells indicates their depth in the biofilm: red, cells on the surface; green, 9 μm depth; blue, 18 μm depth. *(c)* A biofilm of iron-oxidizing prokaryotes attached to rocks in the iron-rich Rio Tinto, Spain. As Fe^{2+}-rich water passes over and through the biofilm, iron-oxidizing microorganisms oxidize Fe^{2+} to Fe^{3+} to obtain energy.

Biofilms are thus functional microbial communities and not just cells trapped in a sticky matrix. We contrasted growth in the biofilm form with that of planktonic growth in Chapter 6 (Microbial Sidebar, "Microbial Growth in the Real World"). Wherever submerged surfaces are present in natural environments, biofilm growth is almost always more extensive and diverse than that of the liquid that surrounds the surface.

Biofilm Formation

How do biofilms form? Attachment of a cell to a surface is a signal for the expression of biofilm-specific genes. These genes encode proteins that synthesize intercellular signaling molecules and initiate matrix formation (**Figure 23.6*a***). Once committed to biofilm formation, a previously swimming cell loses its flagella and becomes nonmotile. Although the mechanism is yet to be discovered, in some way bacteria "sense" a suitable surface and coordinate events that lead to the biofilm growth mode. How surface sensing takes place is an area of active research, but the actual switch from planktonic to biofilm growth is triggered by the synthesis of cyclic dimeric guanosine monophosphate (c-di-GMP), a derivative of the

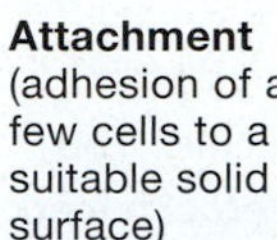

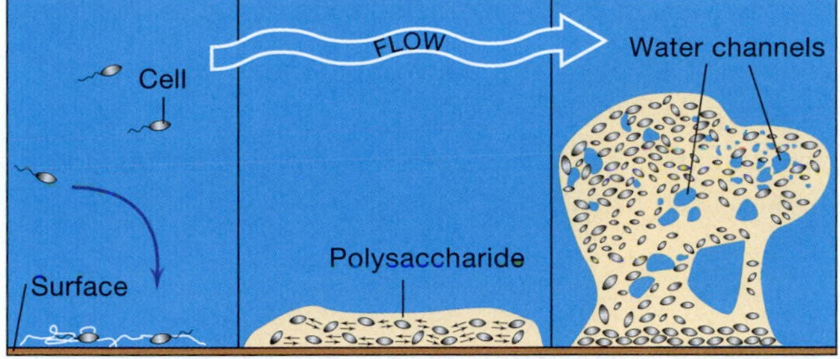

(a)

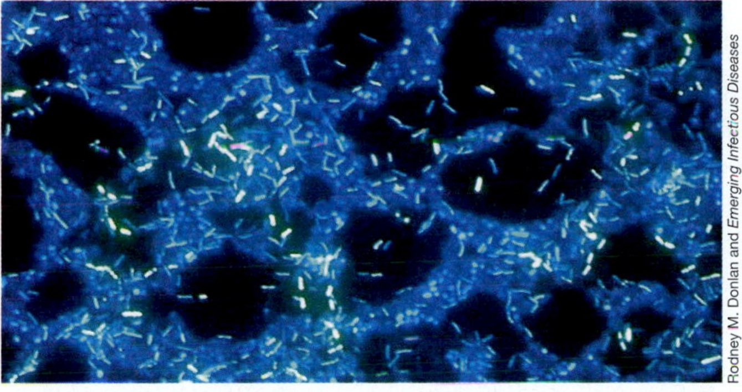

(b)

Figure 23.6 Biofilm formation. *(a)* Biofilms begin with the attachment of a few cells that then grow and communicate with other cells. The matrix is formed and becomes more extensive as the biofilm grows. *(b)* Photomicrograph of a DAPI-stained biofilm that developed on a stainless steel pipe. Note the water channels.

(a)

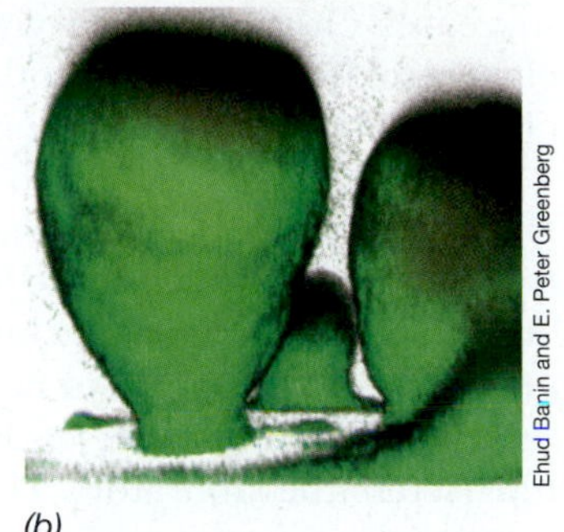

(b)

Figure 23.7 Biofilms of *Pseudomonas aeruginosa*. *(a)* Small "buttons" of *P. aeruginosa* cells attached to a surface in the early stages of biofilm formation. *(b)* A mature biofilm "mushroom." The largest mushroom is about 120 μm high. Compare with Figure 23.6*a*.

nucleotide guanosine triphosphate. C-di-GMP is made by a series of proteins associated with membrane-integrated sensory proteins that in some way detect an opportunity for surface-associated growth. It is thought that c-di-GMP functions by both triggering biofilm-specific gene expression and by activating enzymes in the cell that synthesize matrix material.

Intercellular communication is critical in the development and maintenance of a biofilm. In *Pseudomonas aeruginosa*, a notorious biofilm former (**Figure 23.7**), the major intercellular signaling molecules are compounds called *acylated homoserine lactones*. As these molecules accumulate, they signal adjacent *P. aeruginosa* cells (a mechanism called *quorum sensing*, ∞ Section 8.10) that the population of this species is enlarging, and then the biofilm develops (Figure 23.7*a*). Over time, large "mushrooms" of *P. aeruginosa* form that are over 100 μm high and contain billions of cells enmeshed in a sticky polysaccharide matrix (Figure 23.7*b*).

P. aeruginosa biofilms can occur in the disease cystic fibrosis. This genetic disorder is often associated with a tenacious *P. aeruginosa* biofilm that forms in the lungs, leading to symptoms of pneumonia. However, like most biofilms, the cystic fibrosis biofilm contains more than one bacterial species (∞ Microbial Sidebar, "Microbial Growth in the Real World," Chapter 6). So, in addition to *intra*species signaling, *inter*species signaling probably also occurs in biofilms to coordinate the events necessary for the formation and maintenance of the structure.

23.4 MiniReview

Surfaces are ideal habitats for bacterial growth. Biofilms are structured surface microbial communities embedded in an attaching matrix.

- Why are surfaces attractive microbial habitats?
- What observations indicate that biofilm formation is a cooperative effort among cells?

23.5 Biofilms: Advantages and Control

The biofilm lifestyle confers many advantages on microbial populations, and once formed, biofilms can be very difficult to control.

Why Do Bacteria Form Biofilms?

At least four reasons underlie the formation of biofilms. First, biofilms are a means of microbial self-defense. Biofilms resist physical forces that could sweep away unattached cells. Moreover, biofilms resist phagocytosis by cells of the immune system and the penetration of toxic molecules, such as antibiotics. All of these advantages improve the chances for survival of cells in the biofilm. Second, biofilm formation allows cells to remain in a favorable niche. Biofilms attached to nutrient-rich surfaces, such as animal tissues, or to surfaces in flowing systems, such as a rock in a stream (Figure 23.5*c*), fix bacterial cells in locations where nutrients are more abundant or are constantly replenished. Third, biofilms form because they allow bacterial cells to live in close association with each other. As we have already seen in the case of *Pseudomonas aeruginosa* (Figure 22.7) and the biofilm that forms in cystic fibrosis patients, this facilitates intercellular communication and increases chances for survival. Moreover, when cells are in close proximity to one another, opportunities for genetic exchange (Chapter 11) are more available.

Finally, biofilms seem to be the typical way bacterial cells grow in nature, where nutrient concentrations are often one to two orders of magnitude lower than they are in laboratory culture media. Indeed, the biofilm may be the default mode of growth for prokaryotes in natural environments, which differ dramatically in nutrient levels from the rich liquid culture media used in the laboratory. Thus, when a surface is available in a natural environment, planktonic growth may only be the norm for those bacteria adapted to life at extremely low nutrient concentrations (discussed in Sections 23.6 and 23.9)

Biofilm Control

Biofilms have significant implications in human medicine and commerce. In the body, bacterial cells within a biofilm are protected from attack by the immune system, and antibiotics and other antimicrobial agents often fail to pierce the biofilm. Biofilms have been implicated in medical and dental conditions, including periodontal disease, kidney stones, tuberculosis, Legionnaire's disease, and *Staphylococcus* infections. Medical implants are excellent surfaces for biofilm development. These include both short-term devices, such as a urinary catheter, as well as long-term implants, such as artificial joints. It is estimated that 10 million people a year in the United States experience biofilm infections from implants or intrusive medical procedures. Biofilms explain why routine oral hygiene is so important for maintaining dental health. Dental plaque is a typical biofilm and contains acid-producing bacteria responsible for dental caries (Section 28.3).

In industrial situations biofilms can slow the flow of water, oil, or other liquids through pipelines and can accelerate corrosion of the pipes themselves. Biofilms also initiate the degradation of submerged objects, such as structural components of offshore oil platforms, boats, and shoreline installations. The safety of drinking water may be compromised by biofilms that develop in water distribution pipes, many of which in the United States are nearly 100 years old. Although water pipe biofilms mostly contain harmless microorganisms, if pathogens successfully colonize the biofilm, standard chlorination practices may be insufficient to kill them. Periodic releases of cells can then lead to outbreaks of disease. There is some concern that *Vibrio cholerae*, the causative agent of cholera (Section 36.5), may be propagated in this manner.

Biofilm control is big business, and thus far, only a limited number of tools exist to fight biofilms. Collectively, industries commit huge financial resources to treating pipes and other surfaces to keep them free of biofilms. New antimicrobial agents that can penetrate biofilms, as well as drugs that prevent biofilm formation by interfering with intercellular communication, are being developed. A class of chemicals called *furanones*, for example, has shown promise as biofilm preventatives in tests on abiotic surfaces. Furanones are also stable and some are relatively nontoxic, so they may have applications as antibiofilm agents in human medicine as well.

23.5 MiniReview

Biofilms confer several protective advantages on cells. The excretory products of which biofilms are composed can lead to the destruction of inert as well as living surfaces, and currently, few highly effective antibiofilm agents are available.

- Why might a biofilm be a good habitat for bacterial cells living in a flowing system?
- Give an example of a medically relevant biofilm that most humans likely harbor.

II FRESHWATER, SOIL, AND PLANT MICROBIAL ECOSYSTEMS

Major microbial ecosystems include soils and freshwater, the latter consisting of lakes, ponds, and rivers and streams, and the surfaces of plants. These ecosystems can vary greatly in physical structure, nutrient composition, and temperature, and all of these factors can influence the diversity and abundance of the microorganisms present.

23.6 Freshwater Environments

Freshwater environments are highly variable in the resources and conditions (Table 23.1) available for microbial growth. Both oxygen-producing and oxygen-consuming organisms are present in aquatic environments, and the balance between photosynthesis and respiration controls the oxygen and carbon cycles in nature.

Primary Production

Oxygenic phototrophs suspended freely in the water are called *phytoplankton* and include algae and cyanobacteria. The term "plankton" means "floating" and these organisms can exist throughout the water column of lakes, sometimes accumulating in large numbers at a particular depth. Phototrophs attached to the bottom or sides of a lake or stream are *benthic* species.

Because oxygenic phototrophs obtain their energy from light and use water as an electron donor to reduce CO_2 to organic matter (Section 20.5), they are called **primary producers**. The activity of guilds of heterotrophic microorganisms in aquatic ecosystems (Figure 23.2) depends to a major extent on the rate of primary production. Oxygenic phototrophs produce new organic material as well as oxygen. If photosynthetic activity is very high, the excessive organic matter can lead to O_2 depletion and anoxic conditions. This in turn triggers anaerobic metabolisms, such as anaerobic respirations and fermentations. Like phytoplankton, anoxygenic photosynthetic prokaryotes can also fix CO_2 into organic material. But these organisms use reduced substances, such as H_2S or H_2, as photosynthetic electron donors (Section 20.4).

Nevertheless, organic matter produced by anoxygenic phototrophs can diffuse into oxic waters and increase respiration, accelerating the spread of anoxic conditions.

Oxygen Relationships in Lakes

Although oxygen is one of the most plentiful gases in the atmosphere (~21% of air), it has only limited solubility in water, and in a large body of water its exchange with the atmosphere is slow. Significant photosynthetic production of oxygen occurs only in the surface layers of a lake or ocean, where light is available (Figure 23.2). Organic matter that is not consumed in surface layers sinks to the depths and is decomposed by anaerobic chemoorganotrophs, including those that carry out fermentation and anaerobic respirations.

Once oxygen is consumed in freshwater lakes, the deep layers become anoxic. Strictly aerobic organisms such as higher plants and animals cannot exist in anoxic waters. Instead, the bottom layers contain anaerobic prokaryotes and a few anaerobic eukaryotes, in particular, certain protists. In addition, a transition occurs in anoxic waters from respiratory metabolisms to fermentative and methanogenic metabolisms, with important consequences for the carbon and other nutrient cycles (∞ Sections 24.1–24.5).

Whether a body of water actually becomes depleted of oxygen depends on several factors, including the amount of organic matter present and mixing of the water column. If organic matter is sparse, as it is in pristine lakes or in the open ocean, there may be insufficient substrate available for heterotrophs to consume all the oxygen. The microorganisms present in such environments are typically oligotrophs, organisms adapted to growth under very dilute conditions (Section 23.9). Where currents are strong or there is turbulence because of wind mixing, the water column may be well mixed, and consequently oxygen may be transferred to the deeper layers.

In many lakes in temperate climates the water column becomes stratified during the summer, with the warmer and less dense surface layers, called the **epilimnion**, separated from the colder and denser bottom layers (the **hypolimnion**) (**Figure 23.8**). The *thermocline* is the transition zone from epilimnion to hypolimnion. After stratification sets in during early summer, the bottom layers become anoxic (Figure 23.8). Such lakes often contain high levels of dissolved organic matter from inorganic nutrients that run off the land surrounding the lakes. The nutrients trigger phytoplankton blooms that increase the organic content of the lake.

In temperate lakes in the late fall and early winter, the surface waters become colder and thus more dense than the bottom layers. This causes the cold water to sink and the lake to "turn over," leading to reaeration of the bottom waters. This annual cycle allows the bottom waters to pass from oxic to anoxic and back to oxic. Microbial activity and community composition is altered with these changes in oxygen content, but other factors that accompany fall turnover, especially changes in temperature and nutrient levels, govern microbial diversity and activity as well.

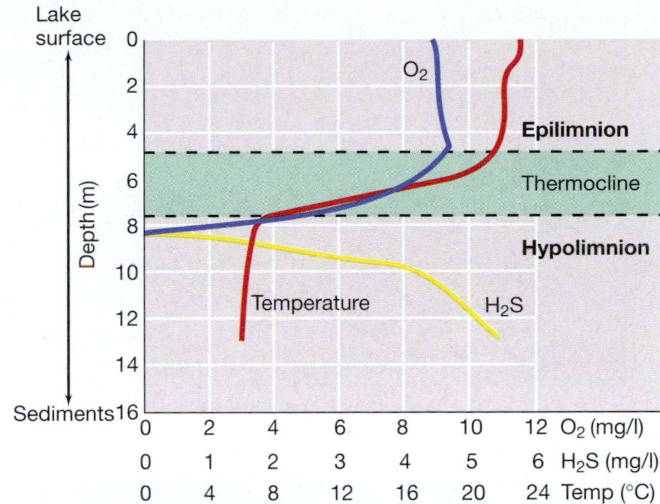

Figure 23.8 Development of anoxic conditions in a temperate lake due to summer stratification. The colder bottom waters are more dense and contain H_2S from bacterial sulfate reduction. The thermocline is the zone of rapid temperature change. As surface waters cool in the fall and early winter they reach the temperature and density of hypolimnetic waters and sink, displacing bottom waters and effecting lake turnover. Data from a small freshwater lake in northern Wisconsin (USA).

Rivers and Streams

Oxygen levels in rivers are of particular interest, especially rivers that receive inputs of organic matter from sewage and agricultural and industrial pollution. Even though a river may be well mixed because of rapid water flow and turbulence, large organic inputs can lead to a marked oxygen deficit from bacterial respiration (**Figure 23.9*a***). As the water moves away from a point source input, for example, an input of sewage, organic matter is gradually consumed and the oxygen content returns to previous levels.

Oxygen depletion in freshwater ecosystems is undesirable because many aquatic animals die under even very temporary anoxia. Furthermore, anoxia results in the production by anaerobic bacteria of odoriferous compounds (for example, amines, H_2S, mercaptans, fatty acids), some of which are also toxic to higher organisms. As in lakes, nutrient inputs to rivers and streams from sewage or other pollutants can also trigger mass blooms of cyanobacteria, algae and aquatic plants (Figure 23.9*b*). The increased load of organic matter that results from these blooms eventually leads to oxygen depletion when the biomass dies and is mineralized by heterotrophic organisms.

Biochemical Oxygen Demand

The microbial oxygen-consuming capacity of a body of water is called its **biochemical oxygen demand (BOD)**. The BOD of water is determined by taking a sample, aerating it well to saturate the water with dissolved O_2, placing it in a sealed bottle, incubating for a standard period of time in the dark (usually 5 days at 20°C), and determining the residual

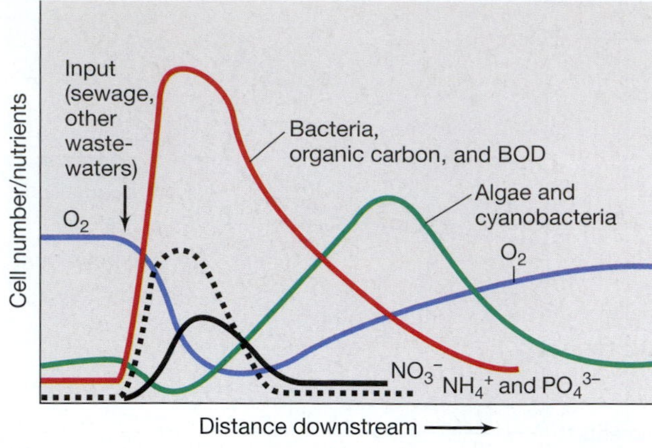

(a)

(b)

Figure 23.9 Effect of the input of organic-rich wastewaters into aquatic systems. *(a)* In a river, bacterial numbers increase and O_2 levels decrease with the spike of organic matter. The rise in numbers of algae and cyanobacteria is primarily a response to inorganic nutrients, especially PO_4^{3-}. *(b)* Photo of a eutrophic (nutrient-rich) lake, Lake Mendota, Madison, Wisconsin (USA), showing algae, cyanobacteria, and aquatic plants that bloom in response to nutrient pollution from agricultural runoff.

oxygen in the water at the end of incubation. A BOD determination gives a measure of the amount of organic material in the water that can be oxidized by the microorganisms present there. As a river recovers from an input of organic matter or from excessive primary production, the initially high BOD becomes lower and is accompanied by a corresponding increase in dissolved oxygen in the ecosystem (Figure 23.9*a*).

We thus see that in freshwater habitats the oxygen and carbon cycles are linked, with the levels of organic carbon and oxygen being inversely related. Although oxygenic photosynthesis produces O_2, the corresponding production of organic matter leads to O_2 deficiencies. Anoxic aquatic environments, which are typically rich in organic material, are the end result of respiratory processes that remove dissolved oxygen from the ecosystem, leaving the remaining organic material to be mineralized by organisms employing the anaerobic energy metabolisms we discussed in Chapter 21.

23.6 MiniReview

In aquatic ecosystems phototrophic microorganisms are the main primary producers. Most of the organic matter produced is consumed by bacteria, which can lead to depletion of oxygen in the environment. BOD is a measure of the oxygen-consuming properties of a water sample.

- What is a primary producer?
- In a freshwater lake, where is the epilimnion and where is the hypolimnion?
- Will addition of organic matter to a water sample increase or decrease its BOD?

23.7 Terrestrial Environments

The word *soil* refers to the loose outer material of Earth's surface, a layer distinct from bedrock that lies underneath (**Figure 23.10**). Soil develops over long periods of time through complex interactions among the parent material (rock, sand, glacial drift, and so on), the topography, climate, and living organisms. Soils can be divided into two broad groups—*mineral soils* and *organic soils*—depending on whether they are derived from the weathering of rock and other inorganic materials or from sedimentation in bogs and marshes, respectively. Our discussion concentrates on mineral soils, the predominant soil in most terrestrial environments.

Soil Composition

Soils are composed of at least four components. These include (1) inorganic mineral matter, typically 40% or so of the soil volume; (2) organic matter, usually about 5%; (3) air and water, roughly 50%; and (4) living organisms, both microorganisms and macroorganisms. In soil, particles of various sizes are present. Soil scientists classify soil particles on the basis of size: Those in the range of 0.1–2 mm in diameter are called *sand,* those between 0.002 and 0.1 mm *silt,* and those less than 0.002 mm in diameter *clay*. Different textural classes of soil are then given names such as "sandy clay" or "silty clay" based on the percentage of sand, silt, and clay they contain. A soil in which no one particle size dominates is called a *loam*.

Soil Formation

Soils form as a result of combined physical, chemical, and biological processes. An examination of almost any exposed rock reveals the presence of algae, lichens, or mosses. These organisms are dormant or nearly so on dry rock but grow when moisture becomes available. They are phototrophic and produce organic matter, which supports the growth of chemoorganotrophic bacteria and fungi. Populations of chemoorganotrophs increase as the degree of plant cover increases, and eventually, microbial communities develop. Carbon dioxide produced during respiration becomes dissolved in

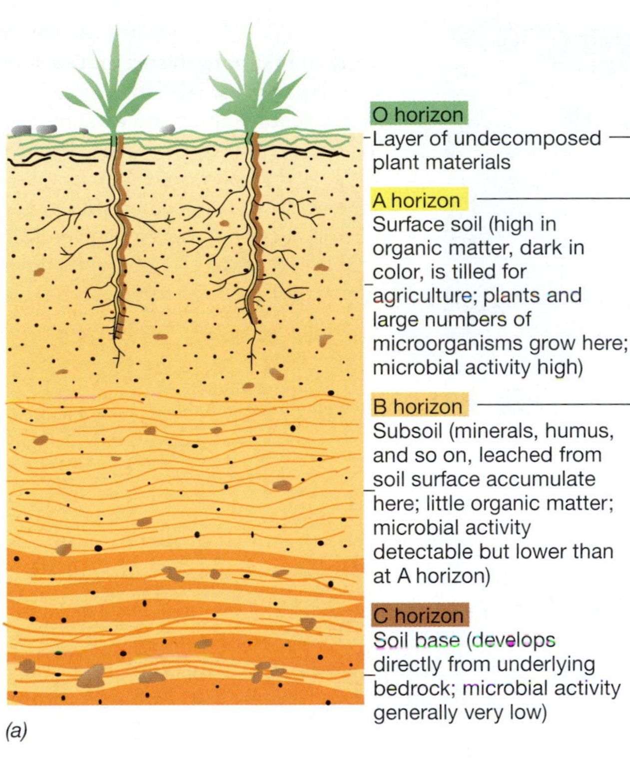
O horizon
Layer of undecomposed plant materials
A horizon
Surface soil (high in organic matter, dark in color, is tilled for agriculture; plants and large numbers of microorganisms grow here; microbial activity high)
B horizon
Subsoil (minerals, humus, and so on, leached from soil surface accumulate here; little organic matter; microbial activity detectable but lower than at A horizon)
C horizon
Soil base (develops directly from underlying bedrock; microbial activity generally very low)
(a)

Michael T. Madigan
(b)

Figure 23.10 Soil. *(a)* Profile of a mature soil. The soil horizons are zones as defined by soil scientists. *(b)* Photo of a soil profile showing O, A, and B horizons. This soil from Carbondale, Illinois (USA) is rich in clay and is very compact. Such soils are not as well drained as those containing sand as a major component.

water to form carbonic acid (H_2CO_3), which slowly dissolves the rock, especially rocks containing limestone ($CaCO_3$). In addition, many chemoorganotrophs excrete organic acids, which also promote the dissolution of rock into smaller particles.

Freezing, thawing, and other abiotic processes assist in soil formation by forming cracks in the rocks. With the combination of the particles generated and organic matter, a crude soil forms in these crevices and pioneering plants develop. The plant roots penetrate farther into crevices and increase the fragmentation of the rock, and their excretions promote the development of a microflora in the rhizosphere (the soil that surrounds plant roots, Figure 23.4*b*). When the plants die, their remains are added to the crude soil and become nutrients for more extensive microbial development. Minerals are rendered soluble, and as water percolates, it carries some of these substances deeper into the soil.

As weathering proceeds, the soil increases in depth, thus permitting the development of larger plants and small trees. Soil animals appear and play an important role in keeping the upper layers of the soil mixed and aerated. Eventually, the movement of materials downward results in the formation of soil layers, called a *soil profile* (Figure 23.10). The rate of development of a typical soil profile depends on climatic and other factors, but it can take hundreds to thousands of years.

Soil as a Microbial Habitat

The most extensive microbial growth takes place on the surfaces of soil particles, usually within the rhizosphere (Figure 23.4). As we have seen in Section 23.2, even a small soil aggregate can contain many different microenvironments (compare **Figures** 23.3 and **23.11**), and thus support the growth of several different types of microorganisms. To examine soil particles directly for microorganisms, fluorescence microscopes are used, the organisms in the soil being stained with a fluorescent dye (Figure 23.3*b*). To observe a specific microorganism in a soil particle, fluorescent antibody staining (Figure 22.8) or FISH (Section 22.4)

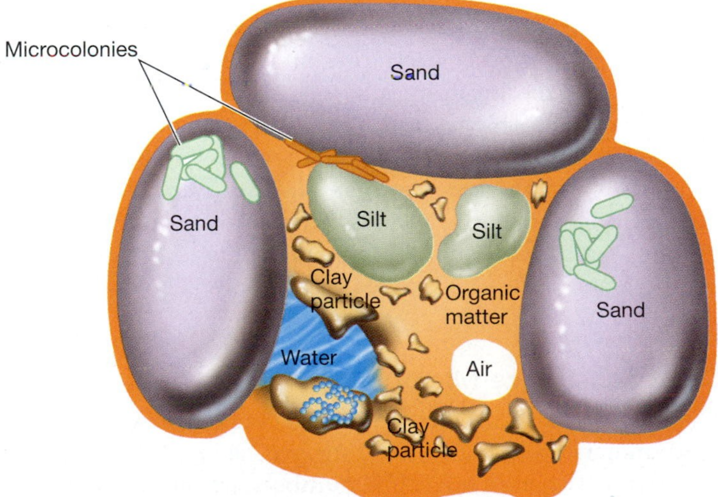
Microcolonies
Sand
Sand
Silt
Silt
Sand
Clay particle
Organic matter
Water
Air
Clay particle

Figure 23.11 A soil microbial habitat. Very few microorganisms are free in the soil solution; most of them reside in microcolonies attached to the soil particles. Note the relative size difference between sand, clay, and silt particles.

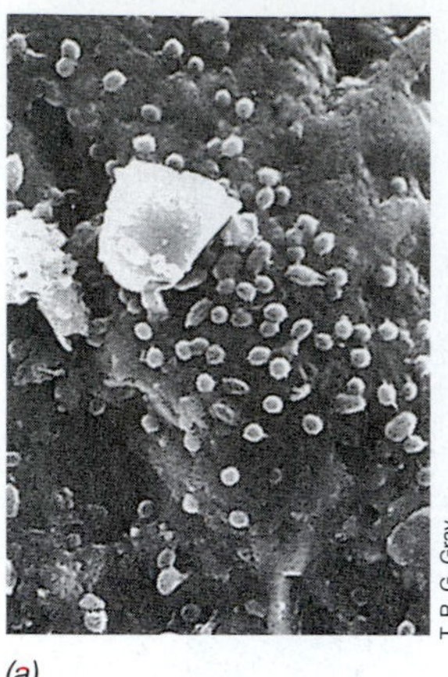

(a)

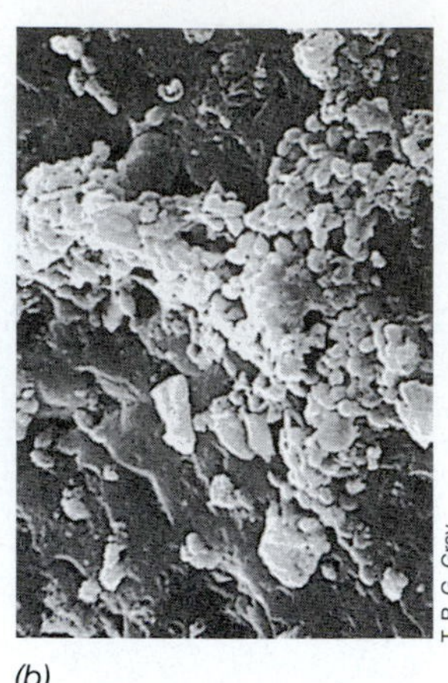

(b)

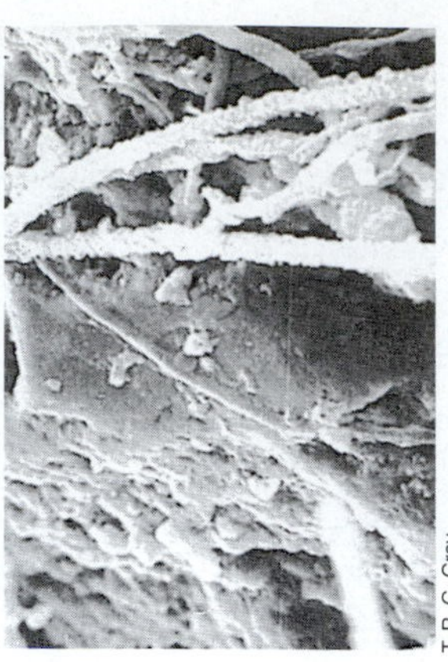

(c)

Figure 23.12 Scanning electron microscopy of microorganisms on the surface of soil particles. *(a)* A microcolony of coccobacilli. *(b)* Actinomycete spores. The cells in (a) and the spores in (b) are about 1–2 μm wide. *(c)* Fungal hyphae. The hyphae are about 4 μm wide and are coated with mineral matter. Microbial cell numbers in a given soil can vary widely and are a reflection of the levels of nutrients and water, and the degree of aeration.

can also be used. Microorganisms can also be observed on soil surfaces directly with the scanning electron microscope (**Figure 23.12**).

One of the major factors affecting microbial activity in soil is the availability of water, and we have previously discussed the importance of water to microbial growth (Section 6.16). Water is a highly variable component of soil, and soil water content depends on soil composition, rainfall, drainage, and plant cover. Water is held in the soil in two ways—by adsorption onto surfaces or as free water in thin sheets or films between soil particles (Figure 23.11).

The water present in soils has materials dissolved in it, and the mixture is called the *soil solution*. In well-drained soils, air penetrates readily, and the oxygen concentration of the soil solution can be high. In waterlogged soils, however, the only oxygen present is that dissolved in water, and this can be rapidly consumed by the resident microflora. Such soils then become anoxic, and, as described for freshwater ecosystems (Section 23.6), show profound changes in their biological properties.

The other major factor affecting microbial activity in soils is the extent of the resources present. The greatest microbial activity is in the organic-rich surface layers, especially in and around the rhizosphere (Figure 23.4*b* and 23.10). The numbers and activity of soil microorganisms depend to a great extent on the kinds and amounts of nutrients present. The limiting nutrient in soils is often not carbon but instead inorganic nutrients such as phosphorus and nitrogen, key components of several classes of macromolecules (Sections 3.1, 3.2, and 5.1).

Deep Subsurface Microbiology

The deep soil subsurface, which can extend for several hundred meters below the soil surface, is not a biological wasteland. Although microbial numbers are much lower than in the A horizon soil (Figure 23.10), a variety of microorganisms, primarily prokaryotes, inhabit deep subsurface soils. For example, in samples collected aseptically from bore holes drilled down to 300 m, diverse populations of both *Bacteria* and *Archaea* have been found, including anaerobes such as sulfate-reducing bacteria, methanogens, and acetogens (Sections 21.8–21.10), as well as various aerobes and facultative aerobes.

Microorganisms in the deep subsurface have access to nutrients because groundwater flows through their habitats, but activity measurements indicate that the metabolic rates of these bacteria are rather low in their buried habitats (Microbial Sidebar, "Microbial Life Deep Underground"). Compared to microorganisms in the upper layers of soil, the biogeochemical significance of deep subsurface microorganisms may thus be minimal. However, over very long periods and because the volume of deep subsurface soils is so vast, the collective metabolic activities of these buried microorganisms are likely responsible for mineralizing large amounts of organic matter and releasing metabolic products into groundwater.

Microbial bioremediation (cleaning up polluted soils and waters, Section 24.9) is a method for removing toxic substances such as aromatic and agricultural chemicals leached from soil into groundwaters and the deep subsurface. Two approaches have been considered. In one method, inorganic nutrients, especially nitrogen and phosphorus, are added to stimulate biodegradation of toxic organic chemicals by the resident microflora; in the other method, large numbers of specific bacterial cells are introduced into the deep subsurface to accelerate the biodegradation process.

23.7 MiniReview

Soil is a complex habitat with numerous microenvironments and niches. Microorganisms are present in the soil primarily attached to soil particles. The most important factor influencing microbial activity in surface soil is the availability of water, whereas in deep soil (the subsurface environment) nutrient availability plays a major role.

- Differentiate between mineral and organic soils.
- What factors govern the extent and type of microbial activity in soils?
- Which region of soil is the most microbially active?

Microbial Sidebar

Microbial Life Deep Underground

Microbiologists studying the deep terrestrial subsurface have found viable prokaryotic cells at depths of several *thousand* meters below the surface. How are these microorganisms making a living in these unusual habitats? Initial indications were that these buried microorganisms were chemoorganotrophs that were metabolizing the organic carbon deposited within the sediments. However, studies of deep basalt aquifers[1] have shown that sluggish chemoorganotrophs are not the only viable organisms deep underground. Chemolithotrophic prokaryotes are present there, too, and probably dominate microbial life underground.

Basalts are iron-rich volcanic rocks essentially devoid of organic matter. In certain basalts up to 1500 m deep from the Columbia River Basin (Washington, USA), large numbers of anaerobic, chemolithotrophic *Bacteria* and *Archaea* have been discovered (**Figure 1*a***), including sulfate-reducing bacteria, methanogens, and acetogens.[1] Carbon stable isotope analyses showed that the methanogens were responsible for the methane (CH_4) present in the rocks. Measurements showed a strong enrichment in the lighter isotope of carbon (^{12}C) of the methane, which is indicative of biological methanogenesis (∞ Section 22.8). If the methanogens deep underground are active, it is likely that the other physiological groups are active as well. This is because in an organic-poor environment the common metabolic thread uniting these organisms is molecular hydrogen (H_2).

H_2 is an excellent electron donor for the energy-yielding metabolisms of methanogens, sulfate-reducing bacteria, and acetogens (∞ Sections 21.8–21.10). H_2 is a common product of the fermentative catabolism of organic matter. But if basalts contain very little organic material, where does the H_2 to support metabolism of the H_2 consumers come from? H_2 in the Columbia River basalts apparently originates from the chemical interaction of water with iron minerals in the rocks (see proposed reaction in Figure 1*b*). Such reactions are known from inorganic chemistry, and in laboratory studies in which crushed Columbia River basalt was mixed with sterile water under anoxic conditions, H_2 evolved rapidly.[2] H_2 was also detected directly in groundwater percolating through the basalts. It therefore appears that H_2 formed within basalts is the electron donor for the anaerobic prokaryotes found there. If true, these organisms would be surviving strictly geochemically and totally divorced from any reliance on phototrophically generated organic materials. This is because both their electron acceptors (CO_2 in methanogens and acetogens and SO_4^{2-} in sulfate-reducing bacteria) and electron donor (H_2) are derived from inorganic materials.

As sometimes happens in science, further research has questioned the Columbia River basalt findings. Although there is not much organic matter in these basalts, there is some, and thus fermentation could be the source of some of the H_2 used by the subsurface chemolithotrophs,[2] as it is in anoxic surface communities.

How ecologically significant are H_2-based microbial ecosystems in the deep subsurface? This is an unanswered question. However, the fact that we now suspect that total prokaryotic numbers in Earth's deep subsurface are very large (∞ Section 1.3), larger in fact than the total number of prokaryotes present in all other environments, suggests that these buried microorganisms may be major catalysts for the cycling of carbon on Earth.

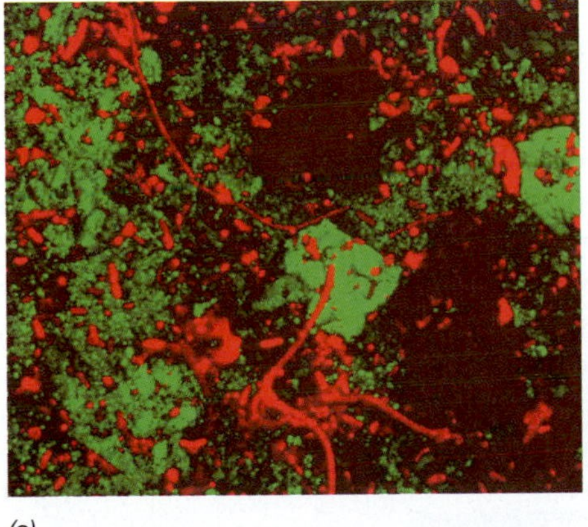
Todd O. Stevens

(a)

Energy-yielding process	Reaction
Methanogenesis:	$4\,H_2 + CO_2 \rightarrow CH_4 + 2\,H_2O$
Acetogenesis:	$4\,H_2 + 2\,HCO_3^- + H^+ \rightarrow CH_3COO^- + 4\,H_2O$
Sulfate reduction:	$4\,H_2 + SO_4^{2-} + H^+ \rightarrow HS^- + 4\,H_2O$
Inorganic H_2 production:	$FeO + H_2O \rightarrow H_2 + FeO_2$

(b)

Figure 1 Microbial life in the deep subsurface. (a) *Laser confocal photomicrograph of a microbial biofilm attached to the surface of basalt chips from a depth of 1500 meters. Green is reflected light from the basalt surface and red is from Nile red–stained bacterial cells. The cells in the biofilm were grown on H_2 from basalt as depicted in the bottom reaction of Fig. 1 b.* (b) *Key metabolic reactions of anaerobic prokaryotes growing in anoxic deep basalt aquifers.*

[1]Stevens, T.O., and J.P. McKinley. 1995. Lithoautotrophic microbial ecosystems in deep basalt aquifers. *Science* 270:450–454.
[2]Anderson, R.T., F.H. Chapelle, and D.R. Lovely. 1998. Evidence against hydrogen-based microbial ecosystems in basalt aquifers. *Science* 281:976–977.

23.8 Plants as Microbial Habitats

As microbial habitats, plants are vastly different from animals and also from any of the habitats we have considered thus far. Compared with warm-blooded animals, plants vary greatly in temperature, both diurnally and throughout the year. Compared with the complex circulatory system of an animal, the internal communication system of a plant is poorly developed, and so transfer of microorganisms within the plant is relatively inefficient. Compared with terrestrial and freshwater microbial ecosystems, the surfaces of plants are fully exposed to oxygen and may be rich in organic matter from exudates produced by the plant. And finally, because plants need exposure to sunlight, microbial communities on the surface of plants are exposed to potentially harmful ultraviolet radiation and are subject to far greater photooxidative effects than are microorganisms in most other communities.

Rhizosphere and Phyllosphere

The aboveground parts of the plant, especially the leaves and stems, are subjected to intermittent drying. For this reason, many plants contain waxy coatings on their leaves that function to both retain moisture and keep out microorganisms. The roots, by contrast, are a major area for microbial activity. Here moisture is less variable, nutrient concentrations and microbial activities are typically high, and the microbial communities are the most diverse and abundant.

The **rhizosphere** is the region immediately outside the root (Figure 23.10*b*); it is a zone where microbial activity is usually high. The *rhizoplane* is the actual root surface. Microbial numbers are almost always higher in the rhizosphere and rhizoplane than in regions of the soil devoid of roots, often many times higher. This is because roots excrete significant amounts of sugars, amino acids, hormones, and vitamins. Bacteria and fungi often form microcolonies on the root surface in order to have ready access to these nutrients (**Figure 23.13*b*,*c***).

The *phyllosphere* is the surface of the plant leaf. Under conditions of high humidity, as in rain forests in tropical and temperate zones, the microflora of leaves may be quite high and include fungi (Figure 23.13*a*) as well as bacteria. Many phyllosphere bacteria on plant leaves or stems fix nitrogen (Sections 20.14 and 24.15) and share the fixed nitrogen with the plant; the plant in turn provides carbohydrates and other nutrients to the bacteria.

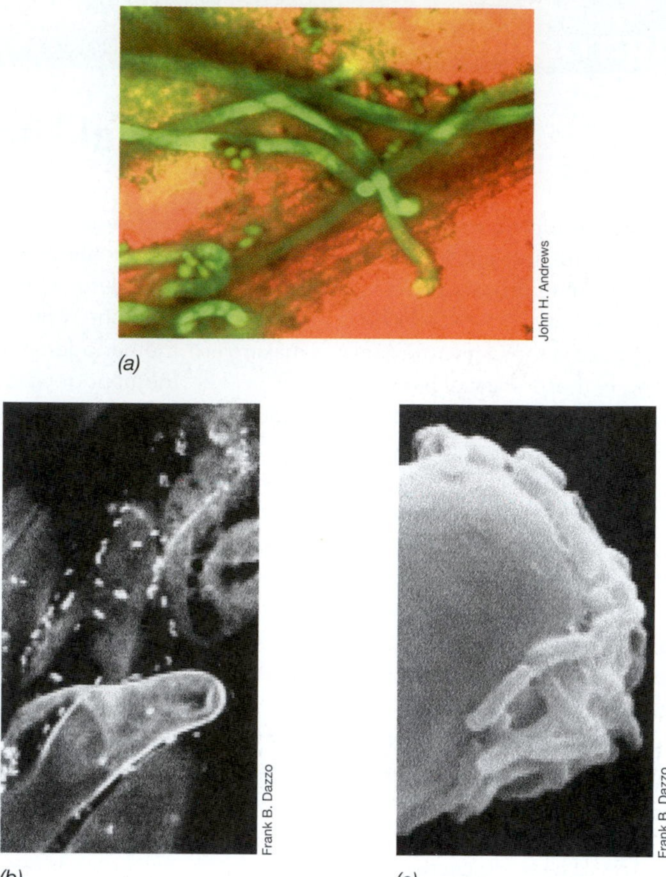

Figure 23.13 Examples of phyllosphere and rhizosphere microbial life. *(a)* Fluorescent micrograph of the fungus *Aureobasidium pullulans* on the surface of a leaf from an apple tree. Cells were stained by FISH, and both fungal filaments and spores stain green. A fungal filament is about 7 μm in diameter. *(b)* Laser-scanning confocal micrograph of bacterial cells on the rhizosphere/rhizoplane of clover. *(c)* Scanning electron micrograph of cells of *Rhizobium leguminosarum* biovar *trifolii* attached to the root hair of clover. The bacterial cells in *(b)* and *(c)* are about 1 μm in diameter.

Phytopathogens

A number of microorganisms, including bacteria, fungi, and viruses, as well as some microscopic worms (nematodes), cause diseases in plants. Plant pathogens are called **phytopathogens**. Some key phytopathogenic bacteria include certain species of *Pseudomonas, Xanthomonas, Xylella,* and *Erwinia*. Major fungal pathogens include *Phytophora* and *Fusarium,* and both RNA and DNA viruses are known to infect plants.

Phytopathogens can cause disease in plants in different ways. Some cause the breakdown of key plant components, such as photosynthetic pigments. Others cause direct physical damage to the plant, opening up the possibility for further infection by plant insect pests. Other phytopathogens cause growth-related problems manifested in reduced biomass, seeds, fruits, or other plant components. The healthy and unstressed plant is a natural physical barrier to microbial infection. However, plant stress due to drought, high temperatures, nutrient limitations, insect grazing, or other factors can increase plant susceptibility to microbial infection and disease.

23.8 MiniReview

Key microbial habitats on plants include the rhizoplane/rhizosphere and the phyllosphere.

- Why might soil near the roots of plants be a more attractive environment for bacteria than soil some distance away from any plant roots?
- What is a phytopathogen?

IV MARINE MICROBIAL ECOSYSTEMS

The oceans differ from freshwater environments in many ways, in particular, in salinity, average temperature, depth, and nutrient status. By using the molecular tools of microbial ecology, especially genetic stains and gene sequencing (Sections 22.4–22.6), much new information is emerging about marine microorganisms. We focus here on microorganisms in two habitats: (1) the open oceans and (2) the deep sea.

23.9 Open Oceans

Compared with many freshwater environments, nutrient levels in the open ocean (called the *pelagic zone*) are often very low. This is especially true of key inorganic nutrients for phototrophic organisms such as nitrogen, phosphorus, and iron. In addition, water temperatures in the oceans are cooler and more constant seasonally than those of most freshwater lakes. The activity of marine phototrophs is limited by these factors, and thus overall microbial cell numbers are typically lower in the oceans than in freshwaters. However, because the oceans are so large, the collective carbon dioxide sequestration and oxygen production from oxygenic photosynthesis that takes place in the oceans are major factors in Earth's carbon balance. Indeed, it is now thought that marine microbial communities heavily influence everything from food chains to global climate.

Salinity is more or less constant in the pelagic zone but is more variable in coastal areas. Nearshore waters typically contain higher microbial numbers than pelagic waters due to the influx of nutrients and other pollutants from runoff of adjoining land and from the activities of humans. These inputs, along with the upwelling of nutrient-rich deep nearshore waters support higher populations of phototrophic microorganisms (**Figure 23.14**); these in turn support higher densities of heterotrophic bacteria and aquatic animals, such as fish and shellfish. Marine bays and inlets that receive sewage or industrial runoff can have very high phytoplankton and bacterial populations. If the pollution is severe enough, shallow marine waters can become intermittently anoxic from the removal of O_2 by respiration and the production of H_2S by sulfate-reducing bacteria.

Primary Productivity: *Prochlorococcus*

Much of the primary productivity in the open oceans, even at significant depths, is due to photosynthesis by **prochlorophytes**, tiny prokaryotic phototrophs that contain chlorophylls *a* and *b*, or *a* and *d*, which are phylogenetically related to cyanobacteria (Section 16.8). The organism *Prochlorococcus* contains chlorophylls *a* and *b* and is a particularly important primary producer (**Figure 23.15**). Because *Prochlorococcus* lacks phycobilins, the accessory pigments of the cyanobacteria (Section 20.3), suspensions of *Prochlorococcus* cells are olive green (as are green algae) rather than the more blue-green color of cyanobacteria

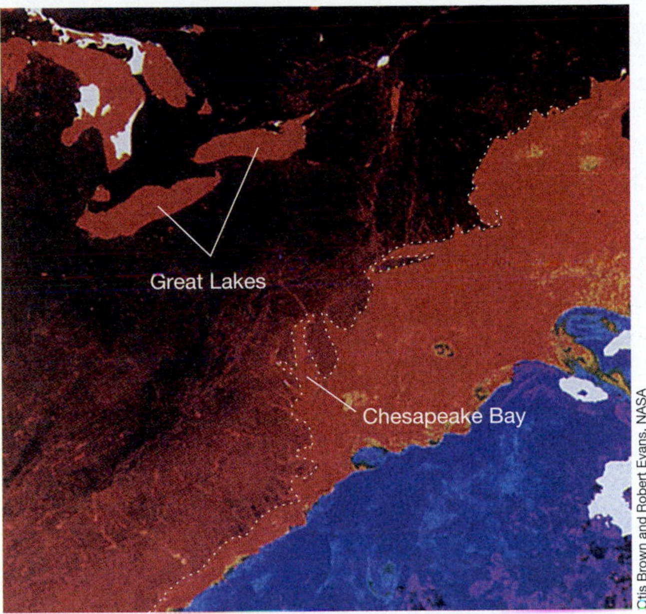

Figure 23.14 Distribution of chlorophyll in the western North Atlantic Ocean as recorded by satellite. The east coast of the United States from the Carolinas to northern Maine is shown in dotted outline. Areas rich in phytoplankton are shown in red (>1 mg chlorophyll/m^3); blue and purple areas have lower chlorophyll concentrations (<0.01 mg/m^3). Note the high primary productivity of coastal areas and the Great Lakes.

(Figure 23.15*b*). *Prochlorococcus* is the dominant oxygenic phototroph in tropical and subtropical oceans worldwide. Cells of *Prochlorococcus* reach densities of 10^5/ml, accounting for over 40% of the biomass of marine phototrophs and up to 50% of net primary production. The prochlorophyte *Acaryochloris* contains chlorophylls *a* and *d* and is present in both marine waters and inland hypersaline lakes (Figure 16.29*c*).

Interestingly, at least four strains (*ecotypes*, Section 14.13) of *Prochlorococcus* are known, and each inhabits its own range of depths in pelagic marine waters. The different *Prochlorococcus* ecotypes are physiologically and genetically distinct and are adapted to photosynthesize at different light intensities. *Prochlorococcus* is thus distributed in both surface waters and deeper waters to depths up to 200 m, near the bottom of the photic zone where light intensities are very low (Figure 23.19). The genome sequence of *Prochlorococcus* is known, and amazingly, the organism contains a rather small genome (about 1.7 Mbp), the smallest genome known for an oxygenic phototrophic organism.

Other Pelagic Phototrophs

In tropical and subtropical oceans, the planktonic filamentous marine cyanobacterium *Trichodesmium* (**Figure 23.16a**) is a widespread and abundant phototroph. Cells of *Trichodesmium* form tufts of filaments that constitute a significant fraction of the biomass suspended in these waters. *Trichodesmium* is a nitrogen-fixing cyanobacterium, and the production of fixed nitrogen by this organism is thought to be a major link in the

UNIT 4

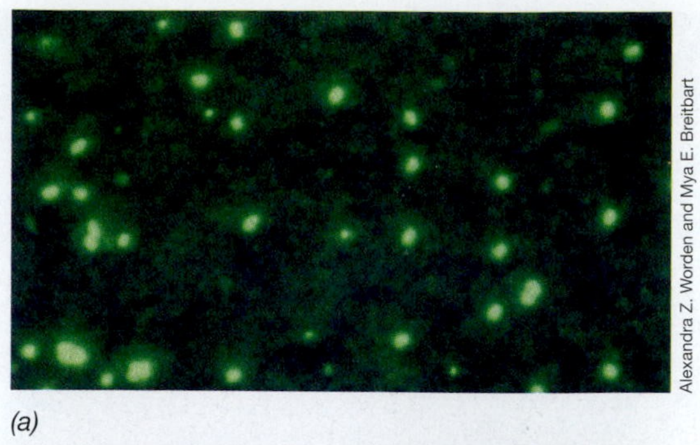

(a)

(b)

Figure 23.15 ***Prochlorococcus*, the most abundant oxygenic phototroph in the oceans.** *(a)* FISH-stained cells of *Prochlorococcus* in a marine water sample. *(b)* A cell suspension of *Prochlorococcus* showing the olive green color of the chlorophyll *a*- and *b*-containing cells.

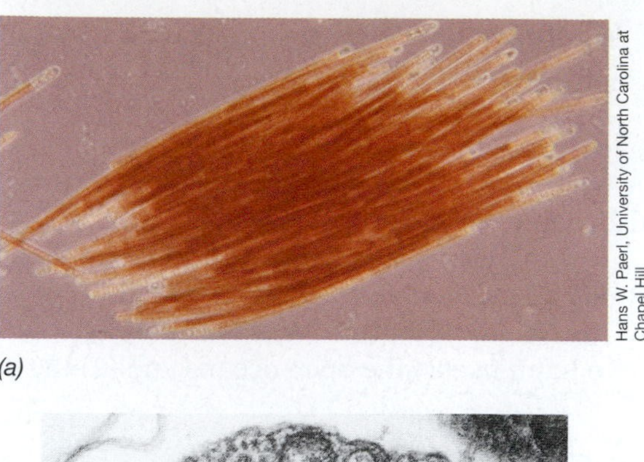

(a)

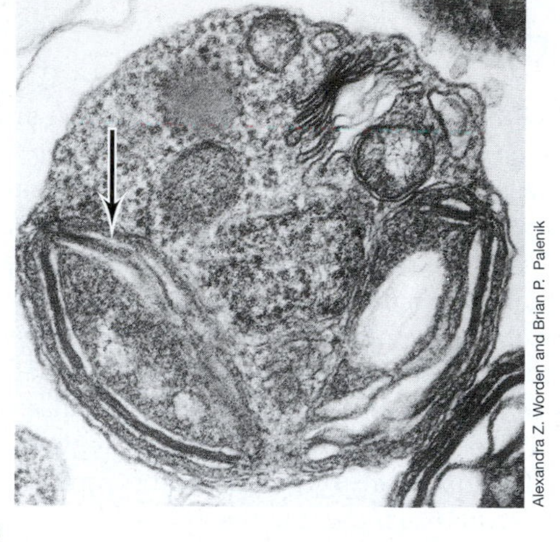

(b)

Figure 23.16 ***Trichodesmium* and *Ostreococcus*.** *(a)* The nitrogen-fixing cyanobacterium *Trichodesmium*. Cells form tufts of filaments that fix nitrogen in tropical marine waters worldwide. A cell is about 6 μm in diameter. *(b)* Transmission electron micrograph of a cell of *Ostreococcus*, a very small alga (eukaryote), found in substantial numbers in marine coastal waters. The arrow points to the chloroplast. An *Ostreococcus* cell is about 0.7 μm in diameter.

marine nitrogen cycle (Section 24.3). *Trichodesmium* contains phycobilins, absent from prochlorophytes, and thus differs in its absorption properties from these organisms (Section 20.3).

Very small phototrophic eukaryotes also inhabit coastal and pelagic marine waters, and some of these are among the smallest eukaryotic cells known. *Ostreococcus*, for example, is a very small member of the *Prasinophyceae*, a family of very ancient green algae (Section 18.21). Cells of *Ostreococcus* are cocci that measure only about 0.7 μm in diameter (Figure 23.16*b*); this is even smaller than a cell of *Escherichia coli*! Interestingly, although cells of *Ostreococcus* and *Prochlorococcus* are roughly of the same dimensions and they are both oxygenic phototrophs, the genome of *Ostreococcus* is 12.6 Mbp (distributed over 20 chromosomes), which is more than seven times the size of the *Prochlorococcus* genome. In many marine waters, about 10^4 small eukaryotic cells are present per milliliter, and it is likely that these are all phototrophic species, either *Ostreococcus* or relatives.

Nonphototrophic Microorganisms in the Pelagic Zone: *Pelagibacter*

Despite vanishingly low levels of organic carbon, significant populations, from 10^5–10^6/ml, of very small planktonic heterotrophic prokaryotes are present in pelagic marine waters. The most abundant of all of these is *Pelagibacter*, a species of *Alphaproteobacteria* (Chapter 15). Cells of *Pelagibacter* are rods that measure only 0.2–0.5 μm, near the limits of resolution of the light microscope (**Figure 23.17**). How do these tiny organisms make a living?

Pelagibacter is an oligotroph, as most open ocean prokaryotes are. An **oligotroph** is an organism that grows best at very low concentrations of nutrients. *Pelagibacter* is a chemoorganotroph and grows in laboratory culture only up to the densities found in nature. However, in addition to respiring organic materials, *Pelagibacter* has genes that encode a form of the visual pigment rhodopsin. The cells use this pigment to convert light energy into ATP. In Section 17.3 we discussed the

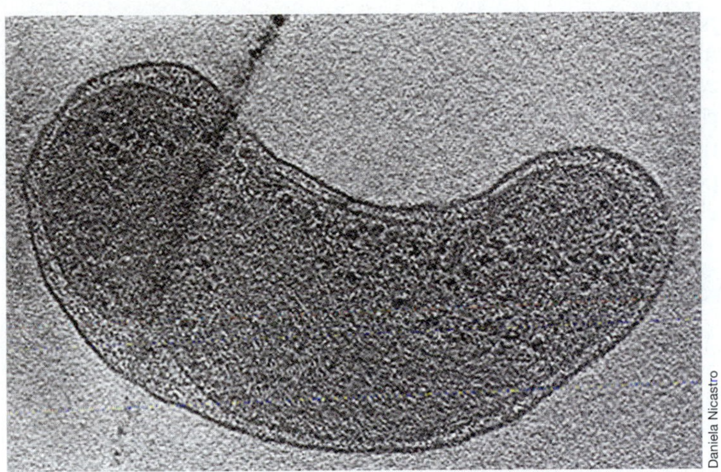

Figure 23.17 ***Pelagibacter ubique*****, the most abundant prokaryote in the ocean.** Electron micrograph taken by electron tomography, a technique for introducing a three-dimensional effect onto the image. A single cell of *P. ubique* is about 0.2 μm in diameter.

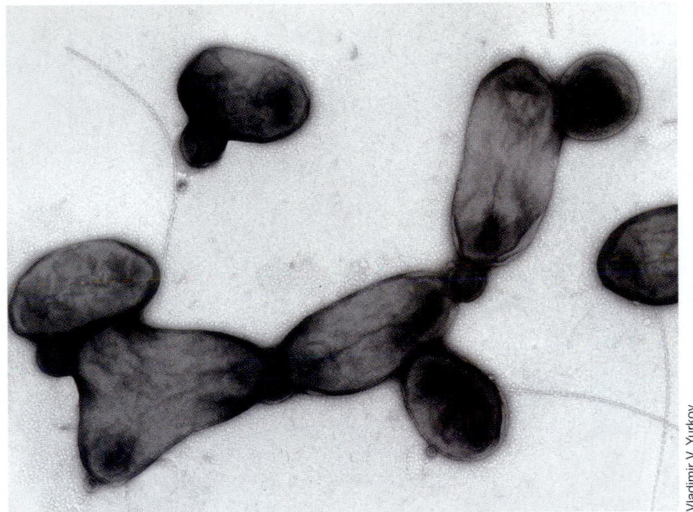

Figure 23.18 Aerobic anoxygenic phototrophic bacteria. Transmission electron micrograph of negatively stained cells of *Citromicrobium*. Cells of this marine aerobic anoxygenic phototroph produce photopigments only under oxic conditions and divide by both budding and binary fission yielding morphologically unusual and irregular cells.

now well-studied case of *bacteriorhodopsin*, present in the extreme halophile *Halobacterium (Archaea)*, and how this molecule functions in ATP synthesis as a simple light-driven proton pump. The form of rhodopsin in *Pelagibacter* and probably many other pelagic prokaryotes is very similar to bacteriorhodopsin and has been called **proteorhodopsin** ("proteo" referring to *Proteobacteria*). It is thought that proteorhodopsin supplements the energy metabolism of *Pelagibacter*, such that it does not have to rely on scarce organic carbon for both a carbon source and an energy source. However, despite the fact that *Pelagibacter* and other proteorhodopsin-containing prokaryotes use light energy to drive the synthesis of ATP, they are not considered "phototrophic" in the classical sense; they lack the chlorophyll pigments and photocomplexes that underlie the process of photosynthesis (Sections 20.1–20.4) and are thus still considered chemotrophic bacteria.

The genome of *Pelagibacter* is very small, only 1.3 Mbp. This is the smallest known genome for a free-living bacterium (Chapter 13). The genome encodes an unusually high number of ABC-type transport systems, transporters that have an extremely high affinity for their substrates (Section 4.5), and various other enzymes helpful for an organism living in a very dilute environment. Environmental genomic studies of pelagic waters (Sargasso Sea study, Section 22.6) have revealed a great abundance of *Pelagibacter* genes, confirming direct measurements using FISH that indicate that this organism is the dominate prokaryote in open ocean waters worldwide.

Aerobic Anoxygenic Phototrophs

Besides oxygenic phototrophs and organisms containing proteorhodopsin, other prokaryotes that use light energy are present in marine waters. These are the aerobic anoxygenic phototrophs. Like purple anoxygenic phototrophs (Section 15.2), these organisms contain bacteriochlorophyll *a*. However, unlike purple anoxygenic phototrophs that photosynthesize only under *anoxic* conditions, aerobic anoxygenic phototrophs catalyze photosynthetic light reactions only under *aerobic* conditions.

Aerobic anoxygenic phototrophs include organisms such as *Erythrobacter*, *Roseobacter*, and *Citromicrobium* (**Figure 23.18**), all genera of *Alphaproteobacteria*. Aerobic anoxygenic phototrophs synthesize ATP by photophosphorylation when oxygen is present, but they are unable to grow autotrophically and thus rely on organic carbon for their carbon source. So, as with *Pelagibacter* and its proteorhodopsin, aerobic anoxygenic phototrophs likely use the ATP produced in their photocomplexes to supplement their otherwise chemotrophic metabolism.

Surveys have shown that a great diversity of aerobic anoxygenic phototrophs exist in marine waters, especially nearshore waters. Oligotrophic and highly oxic freshwater lakes are also habitats for these interesting phototrophic bacteria.

Distribution of *Archaea* and *Bacteria* in Pelagic Waters

Numbers of prokaryotes in the open oceans decrease with depth. In surface waters cell numbers average about 10^6 cells/ml. Below 1000 m, however, total numbers fall to between 10^3 and 10^5/ml. The distribution of *Bacteria* and *Archaea* with depth has been tracked in pelagic waters by FISH (Section 22.4). In general, species of *Bacteria* predominate in upper waters (<1000 m), and numbers are about equal or show a slight predominance of *Archaea* in lower waters (**Figure 23.19**). The *Archaea* present in lower waters are almost exclusively species of *Crenarchaeota*, a phylum of *Archaea* that includes the hyperthermophiles (Sections 17.9–17.12). Extrapolating from the data of Figure 23.19, it has been estimated that 1.3×10^{28} and 3.1×10^{28} cells of

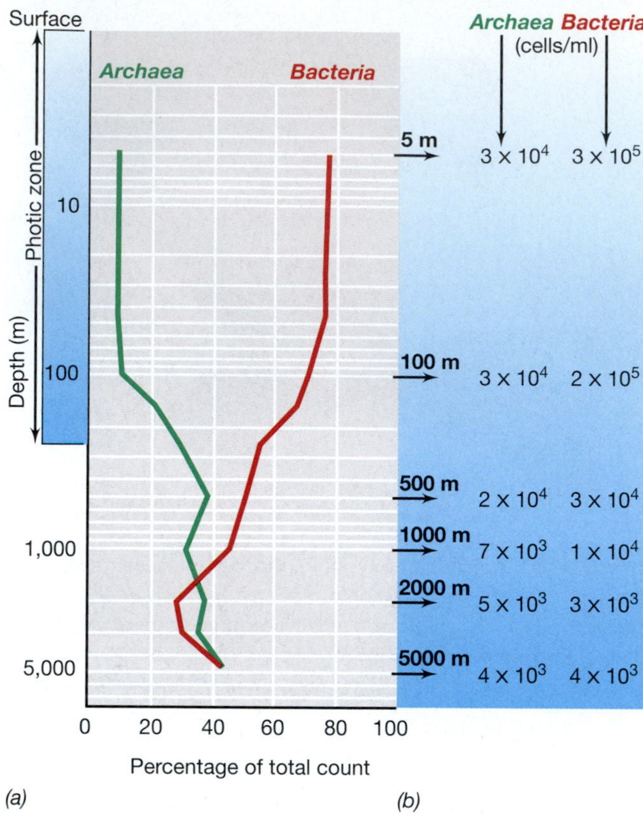

Figure 23.19 Percentage of total prokaryotes belonging to *Archaea* and *Bacteria* in North Pacific Ocean water. *(a)* Distribution of *Archaea* and *Bacteria* with depth. *(b)* Absolute numbers of *Archaea* and *Bacteria* with depth (per milliliter). Adapted from data published in *Nature* 409:507–510 (2000).

Archaea and *Bacteria*, respectively, are present in the world's oceans. This indicates that the oceans contain the largest microbial biomass on the surface of the Earth (∞ Section 1.3).

Marine Viruses

Viruses are the most abundant microorganisms in the oceans. In typical seawater they can number over 10^7 virion particles per milliliter (∞ Section 13.14 and Figure 13.15). Most of these are bacteriophages, which infect species of *Bacteria*, and archaeal viruses, which infect species of *Archaea*. The number of viral particles in seawater is about tenfold greater than average prokaryotic cell numbers in seawater, suggesting that they are actively infecting their hosts, replicating, and being released into seawater. In coastal waters where bacterial cell numbers are higher, viral numbers are also higher, as many as 10^8/ml.

Presumably marine viruses help to maintain host cell populations at the levels that are observed, but they could also have other important functions. These include genetic exchange between prokaryotic cells (∞ Section 11.9–11.15) and lysogeny, the state in which a virus genome integrates within the cellular genome; this often confers new genetic properties on the cell (∞ Section 10.10). The discovery that some of the viruses that infect *Prochlorococcus* (Figure 23.15), the most abundant oxygenic phototroph in the oceans (see earlier discussion), carry photosynthesis genes in their genomes is an example of how an important metabolic property could be transferred between ecotypes by viral shuttles. Although the genetic diversity of marine viruses is just now being recognized, it has been estimated that the genetic diversity of marine viral genomes could surpass even that of all prokaryotic cells! Thus, marine waters are clearly a reservoir of genetic diversity.

23.9 MiniReview

Marine waters are more nutrient deficient than many freshwaters, yet substantial numbers of prokaryotes inhabit the oceans. Many of these use light to drive ATP synthesis, either by photosynthesis or proton pumps driven by rhodopsins. Species of *Bacteria* tend to predominate in marine surface waters, whereas *Archaea* are more prevalent in deeper waters. Viruses are the most abundant of all microorganisms in the oceans.

- How does the organism *Prochlorococcus* contribute to both the carbon and oxygen cycles in the oceans?
- What is proteorhodopsin and why was it so named?
- Compared with *Escherichia coli*, list some of the unusual properties of the bacterium *Pelagibacter*.

23.10 The Deep Sea and Barophilism

Visible light penetrates no further than about 300 m in pelagic waters; this illuminated region is called the *photic zone* (Figure 23.19). Beneath the photic zone, down to a depth of about 1000 m, there is still considerable biological activity. However, water at depths greater than 1000 m is, by comparison, much less biologically active and is known as the **deep sea**. Greater than 75% of all ocean water is deep-sea water, lying primarily at depths between 1000 and 6000 m. The deepest waters in the oceans lie below 10,000 m. However, because holes this deep are very rare, they make up only a very small proportion of all pelagic waters.

Conditions in the Deep Sea

Organisms that inhabit the deep sea face three major environmental extremes: (1) low temperature, (2) high pressure, and (3) low nutrient levels. In addition, deep-sea waters are completely dark, such that photosynthesis is impossible. Thus microorganisms that inhabit the deep sea are chemotrophic and are able to grow under high pressure and oligotrophic conditions in the cold.

Below depths of about 100 m ocean water temperatures stay constant at 2–3°C. We discussed the responses of microorganisms to changes in temperature in Sections 6.12–6.14. As would be expected, bacteria isolated from marine waters below 100 m are *psychrophilic* (cold-loving) or *psychrotolerant*. Deep-sea microorganisms must also be able to withstand the enormous hydrostatic pressures associated with great depths. Pressure increases by 1 atm for every 10 m

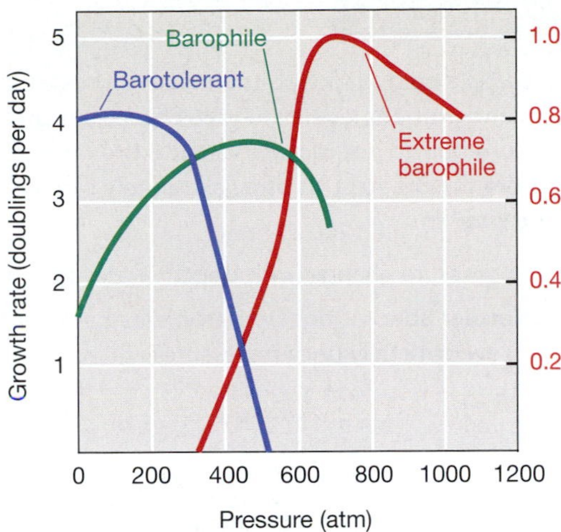

Figure 23.20 Growth of barotolerant, barophilic, and extremely barophilic bacteria. The extreme barophile (*Moritella*) was isolated from the Mariana Trench (Figure 23.21). Note the slower growth rate of the extreme barophile (right ordinate) as compared to the barotolerant and barophilic bacteria (left ordinate) and its inability to grow at low pressures.

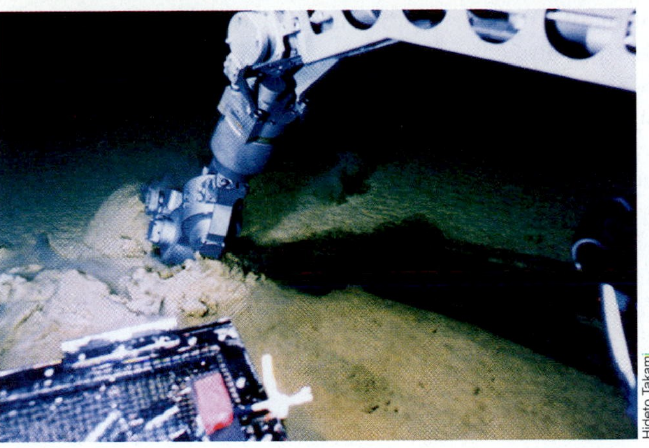

Figure 23.21 Sampling the deep sea. The unmanned submersible *Kaiko* collecting a sediment sample on the sea floor of the Mariana Trench (off the Philippines, Pacific Ocean) at a depth of 10,897 m. The tubes of sediment are used for enrichment and isolation of barophilic bacteria.

depth in a water column. Thus, an organism growing at a depth of 5000 m must be able to withstand pressures of 500 atm. We will see that microorganisms are remarkably tolerant of high hydrostatic pressures; many species can withstand pressures of 500 atm and in some cases far more than this.

Barotolerant and Barophilic Bacteria

Different physiological responses to pressure are observed in different deep-sea prokaryotes. Some organisms simply tolerate high pressure but don't grow best under pressure; these are called **barotolerant**. By contrast, others actually grow best under pressure; these are called **barophilic**. Organisms isolated from surface waters down to about 3000 m are typically barotolerant. In barotolerant organisms, higher metabolic rates are observed at 1 atm than at 300 atm, although growth rates at the two pressures may be similar (**Figure 23.20**). However, barotolerant isolates typically do not grow at pressures above 500 atm.

By contrast, cultures derived from samples taken at greater depths, 4000–6000 m, are typically barophilic, growing optimally at pressures of around 300–400 atm. However, although barophiles grow best under pressure, they can still grow at 1 atm (Figure 23.20). In even deeper waters (10,000 m), **extreme barophiles** are present. These organisms have absolute and major pressure requirements for growth. For example, the bacterium *Moritella*, isolated from the Mariana Trench (Pacific Ocean, >10,000 m depth), grows optimally at a pressure of 700–800 atm and nearly as well at 1035 atm, the pressure it experiences in its natural habitat (**Figures** 23.20 and **23.21**).

Unlike barotolerant or barophilic prokaryotes, *Moritella* does not grow at pressures of less than about 400 atm; it has an *obligate* requirement for pressure (Figure 23.20). Interestingly, however, *Moritella* can tolerate moderate periods of decompression. However, viability is lost when a culture of the organism is left for several hours in a decompressed state. *Moritella* is also temperature sensitive. Its optimal growth temperature is its environmental temperature (2°C), and temperatures above 10°C significantly affect viability.

Molecular Effects of High Pressure

Pressure affects cellular physiology and biochemistry in many ways. Pressure decreases the affinity of enzymes for their substrates. Thus, the enzymes of extreme barophiles must fold in such a way as to minimize pressure-related effects. Other pressure-sensitive targets include protein synthesis and transport. An organism grown under high pressure has a higher proportion of unsaturated fatty acids in its cytoplasmic membrane than one grown at 1 atm. Unsaturated fatty acids allow membranes to remain functional and keep from gelling at high pressures or at low temperatures (∞ Section 6.13). The rather slow growth rates of extreme barophiles such as *Moritella* compared with other marine bacteria (Figure 23.20) are likely due to the combined effects of pressure and low temperature; the latter slows down the reaction rates of enzymes, and this has a direct effect on cell growth (∞ Sections 6.12 and 6.13).

What are the secrets behind a barophilic lifestyle? In gram-negative barophiles capable of growth at both 1 atm and 500–600 atm (Figure 23.21), it has been shown that growth at high pressure is accompanied primarily by changes in the protein composition of the outer membrane (∞ Section 4.7). In particular, a specific outer membrane protein called *OmpH* (*o*uter *m*embrane *p*rotein *H*) is synthesized in cells grown under high pressure but not in cells grown at 1 atm. OmpH is a type of *porin*. Porins are proteins that form channels for the diffusion of organic molecules through the outer membrane and into the periplasm (∞ Section 4.7 and Figure 4.23).

UNIT 4

Presumably, the porin present in the outer membrane of cells grown at 1 atm cannot function properly at high pressure, and thus a new type of porin must be synthesized. Surprisingly, relatively few other proteins seem to be controlled by pressure, as most proteins present in cells grown at 1 atm are also present in cells grown under high pressure. This indicates that alterations in only a few key proteins may be sufficient to allow for growth under significant hydrostatic pressure. This suggests that the evolutionary changes necessary to convert surface marine bacteria into deep-sea-dwelling organisms are probably relatively few.

23.10 MiniReview

The deep sea is a cold, dark habitat where hydrostatic pressure is high and nutrient levels are low. Barophiles grow best under pressure but do not require it, but extreme barophiles, obtained from the greatest depths, absolutely require high pressure for growth.

- How does pressure change with depth in a water column?
- What molecular adaptations do barophiles have that allow them to grow optimally under pressure?

Review of Key Terms

Anoxic an oxygen-free environment

Barophilic growing best under a pressure greater than 1 atm

Barotolerant able to grow under elevated pressures but growing best at 1 atm

Biochemical oxygen demand (BOD) the microbial oxygen-consuming properties of a water sample

Biofilm colonies of microbial cells encased in a porous matrix and attached to a surface

Biogeochemistry study of biologically mediated chemical transformations

Commensalism a type of symbiosis in which only one of two organisms in a relationship benefits

Deep sea marine waters below a depth of 1000 m

Ecosystem a community of organisms and their natural environment

Epilimnion the warmer and less dense surface waters of a stratified lake

Extreme barophile a barophilic organism unable to grow at a pressure of 1 atm and typically requiring several hundred atmospheres of pressure for growth

Guild a population of metabolically related microorganisms

Habitat a portion of an ecosystem where a microbial community could reside

Hypolimnion the colder, more dense, and often anoxic bottom waters of a stratified lake

Microenvironment the immediate environmental surroundings of a microbial cell or group of cells

Mutualism a type of symbiosis in which both organisms in the relationship benefit

Niche in ecological theory, an organism's residence in a community, including both biotic and abiotic factors

Oligotroph an organism that grows only or grows best at very low levels of nutrients

Oxic an oxygen-containing environment, typically with a high E_0'

Parasitism a symbiotic relationship between two organisms in which the host organism is harmed in the process

Phytopathogen a microorganism that causes plant disease

Primary producer an organism that uses light to synthesize new organic material from CO_2

Prochlorophyte a prokaryotic phototroph that contains chlorophylls *a* and either *b* or *d*, and lacks phycobilins

Proteorhodopsin a light-sensitive protein present in some open ocean *Bacteria* that catalyzes ATP formation

Rhizosphere the region immediately adjacent to plant roots

Symbiosis a relationship between two or more organisms

Review Questions

1. Distinguish between a parasite, mutualist, and commensal (Section 23.1).
2. List some of the key resources and conditions that microorganisms need to thrive in their habitats (Section 23.2).
3. Explain why both obligately anaerobic and obligately aerobic bacteria can be isolated from the same soil sample (Section 23.3).
4. The surface of a rock in a flowing stream will often contain a biofilm. What advantages could be conferred on bacteria growing in a biofilm compared with growth within the flowing stream (Section 23.4)?
5. How can biofilms complicate treatment of infectious diseases (Section 23.5)?
6. How and in what way does an input of organic matter, such as sewage, affect the oxygen content of a river or stream (Section 23.6)?
7. In what soil horizon are microbial numbers and activities the highest, and why (Section 23.7)?
8. What part of a plant is typically the most heavily colonized by microorganisms (Section 23.8)?
9. List some major prokaryotes found in the open oceans and describe how they make ATP (Section 23.9).

10. Assuming proteorhodopsin is a functional analog of bacteriorhodopsin, describe how this molecule might benefit a cell living in a very organically deficient environment such as the open ocean (Section 23.9).

11. What is the difference between barotolerant and barophilic bacteria? Between these two groups and extreme barophiles? What properties do barotolerant, barophilic, and extremely barophilic bacteria have in common (Section 23.10)?

Application Questions

1. Imagine a sewage plant that is releasing sewage containing high levels of ammonia and phosphate, but very little organic carbon. Which types of organisms might be triggered by this sewage? How would the oxygen profile near and beyond the point source differ from that shown in Figure 23.9*a*?

2. Look at the data of Figure 23.19. Keeping in mind that the open ocean waters are highly oxic, predict the possible metabolic lifestyles of open ocean *Archaea* and *Bacteria*. Why might proteorhodopsin be present only in one group of organisms but not the other?

24

Nutrient Cycles, Bioremediation, and Symbioses

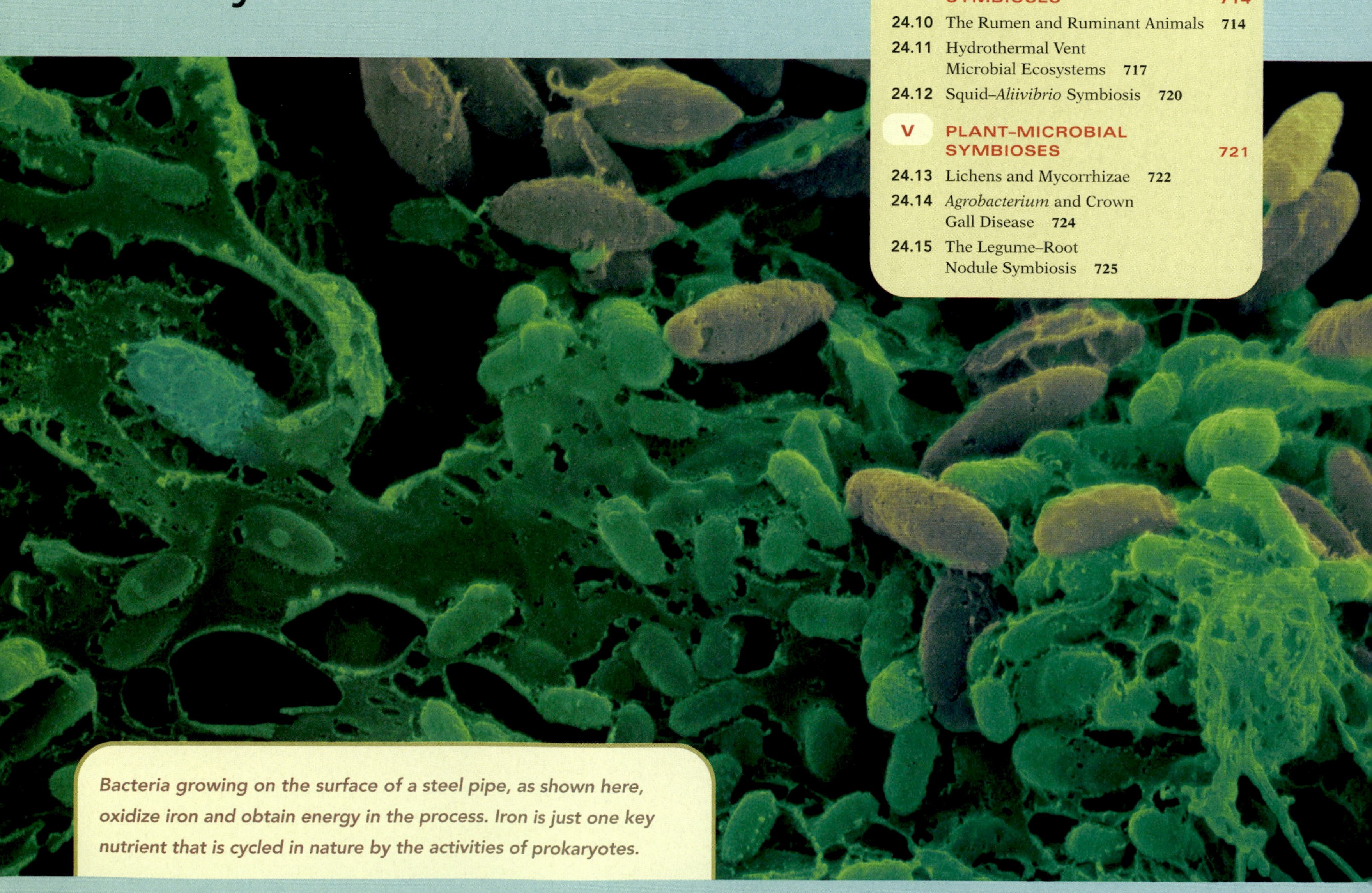

Bacteria growing on the surface of a steel pipe, as shown here, oxidize iron and obtain energy in the process. Iron is just one key nutrient that is cycled in nature by the activities of prokaryotes.

In the previous chapter we examined common microbial habitats to set the stage for the study of some major microbial activities in this chapter. This chapter has three main themes: nutrient cycles, bioremediation, and symbioses. Using the research tools described in Chapter 22 and keeping in mind the types and dynamics of microbial habitats we covered in Chapter 23, we are now ready to tackle microbial biogeochemistry.

I THE CARBON AND OXYGEN CYCLES

Global carbon cycling requires the activities of both microorganisms and macroorganisms. Major areas of interest in the carbon cycle are the magnitude of carbon reservoirs, the major sources and sinks for CO_2 and CH_4, and the rates of carbon cycling within and between compartments.

24.1 The Carbon Cycle

On a global basis, carbon is cycled through all of Earth's major carbon reservoirs: the atmosphere, the land, the oceans and other aquatic environments, sediments and rocks, and biomass (**Figure 24.1**). As we have already seen for freshwater environments, the carbon and oxygen cycles are intimately linked (Section 23.6), and this is true of soil and all other environments on Earth as well.

Carbon Reservoirs

The largest carbon reservoir is in the sediments and rocks of Earth's crust (**Table 24.1**), but the turnover time is so long that flux out of this compartment is insignificant on a human time scale. From the viewpoint of living organisms, a large amount of organic carbon is found in land plants. This is the carbon of forests, grasslands, and crop agriculture, and constitutes major sites of photosynthetic CO_2 fixation. However, more carbon is present in dead organic material, called **humus**, than in living organisms. Humus is a complex mixture of organic materials that is derived from dead soil microorganisms that have resisted decomposition along with resistant plant organic materials. Some humic substances are fairly stable, with a global turnover time of several decades, but certain other humic components decompose much more rapidly than this.

The most rapid means of global transfer of carbon is via the CO_2 of the atmosphere. Carbon dioxide is removed from the atmosphere primarily by photosynthesis of land plants and marine microorganisms and is returned to the atmosphere by respiration of animals and chemoorganotrophic microorganisms. The single most important contribution of CO_2 to the atmosphere is microbial decomposition of dead organic material, including humus. In recent eras, however, human activities have added enormous amounts of new CO_2 to the atmosphere from previously buried carbon by the burning of fossil fuels. For example, in the past 40 years alone, CO_2 levels in the atmosphere have risen nearly 15%. This has in large part triggered a period of steadily increasing global temperatures called *global warming*.

Importance of Photosynthesis in the Carbon Cycle

The only major ways in which new organic carbon is synthesized on Earth are via phototrophic and chemolithotrophic CO_2 fixation; most organic carbon comes from photosynthesis. Phototrophic organisms are therefore the foundation of the carbon cycle (Figure 24.1). However, phototrophic organisms are abundant in nature only in habitats where light is available. Thus, the deep sea and other permanently dark habitats are devoid of indigenous phototrophs.

Oxygenic phototrophic organisms can be divided into two groups: *plants* and *microorganisms*. Plants are the dominant phototrophic organisms of terrestrial environments, whereas phototrophic microorganisms dominate aquatic environments.

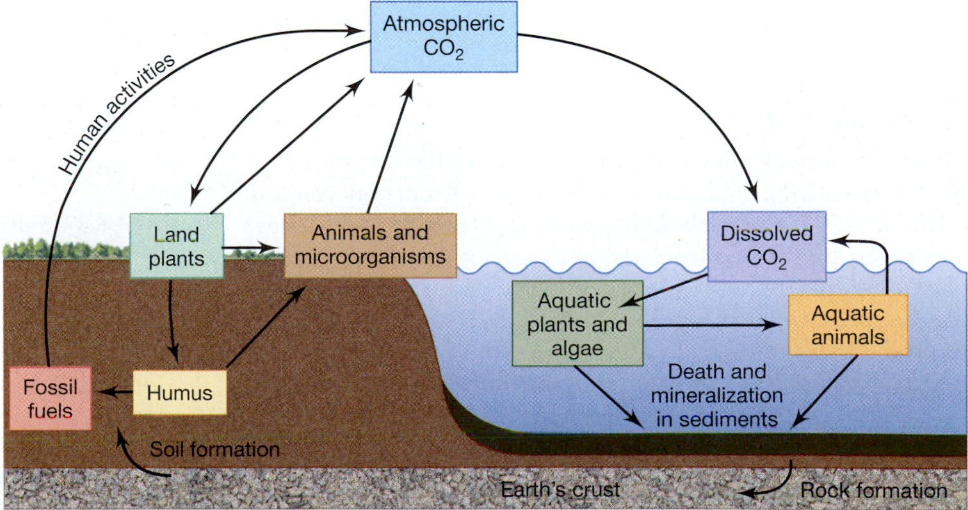

Figure 24.1 The carbon cycle. The carbon and oxygen cycles are closely connected, as oxygenic photosynthesis both removes CO_2 and produces O_2 and respiratory processes both produce CO_2 and remove O_2. By far the greatest reservoir of carbon on Earth is in rocks and sediments, and most of this is inorganic carbon (carbonates).

Table 24.1 Major carbon reservoirs on Earth

Reservoir	Carbon (gigatons)[a]	Percent of total carbon on Earth
Oceans	38×10^3 (>95% is inorganic C)	0.05
Rocks and sediments	75×10^6 (>80% is inorganic C)	>99.5[b]
Terrestrial biosphere	2×10^3	0.003
Aquatic biosphere	1–2	0.000002
Fossil fuels	4.2×10^3	0.006
Methane hydrates	10^4	0.014

[a]One gigaton is 10^9 tons. Data adapted from *Science* 290:291–295 (2000).
[b]Much of the organic carbon is in prokaryotic cells.

The redox cycle for carbon (**Figure 24.2**) begins with photosynthesis:

$$CO_2 + H_2O \xrightarrow[\text{light}]{} (CH_2O) + O_2$$

Here (CH_2O) represents organic matter at the oxidation state of cell material, such as polysaccharides (the main form in which photosynthesized organic matter is stored in the cell). Phototrophic organisms also carry out respiration, both in the light and the dark. The overall equation for respiration is the reverse of oxygenic photosynthesis:

$$(CH_2O) + O_2 \xrightarrow[\text{light or dark}]{} CO_2 + H_2O$$

where (CH_2O) again represents storage polysaccharides. If a phototrophic organism is to increase in cell number or mass, then the rate of photosynthesis must exceed the rate of respiration. If the organism grows, then some of the carbon fixed from CO_2 into polysaccharide has become the starting material for biosynthesis. In this way, autotrophic organisms build biomass from CO_2, and this biomass eventually supports the organic carbon needs of all heterotrophic organisms.

Decomposition

Photosynthetically fixed carbon is eventually degraded by microorganisms, and two major forms of carbon remain: *methane* (CH_4) and *carbon dioxide* (CO_2) (Figure 24.2). These two gases are formed from the activities of methanogens and chemoorganotrophs, respectively. In anoxic habitats CH_4 is produced from the reduction of CO_2 with H_2 and from certain organic compounds such as acetate. However, virtually any organic compound can eventually be converted to CH_4 from the combined activities of methanogens and other bacteria (syntrophs); H_2 generated from the fermentative degradation of organic compounds by syntrophs gets consumed by methanogens and converted to CH_4 as discussed in the next section. Methane produced in anoxic habitats is insoluble and

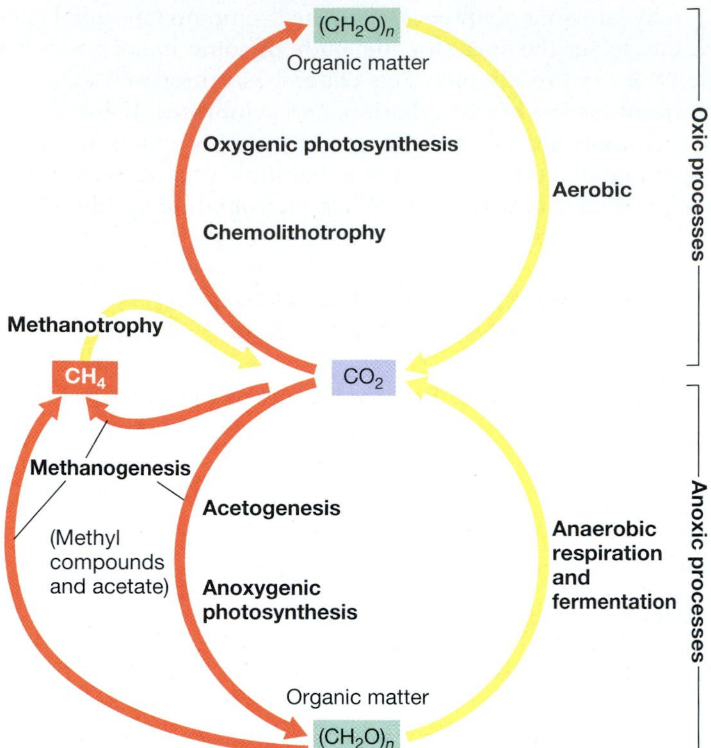

Figure 24.2 Redox cycle for carbon. The figure contrasts autotrophic ($CO_2 \rightarrow$ organic compounds) and heterotrophic (organic compounds $\rightarrow CO_2$) processes. Yellow arrows indicate oxidations; red arrows indicate reductions.

flows to oxic environments where it is oxidized to CO_2 by methanotrophs (Figure 24.2). Hence, all organic carbon eventually returns to CO_2, and the carbon cycle is complete.

The balance between the oxidative and reductive portions of the carbon cycle is critical: The metabolic products of some organisms are the substrates for others. Thus, the cycle needs to keep in balance if it is to continue as it has for billions of years. Any significant changes in levels of gaseous forms of carbon may have serious global consequences, as we are already experiencing in the form of global warming from the increasing CO_2 levels in the atmosphere from deforestation and the burning of fossil fuels. Total CO_2 released by microbial activities far exceeds that produced from decomposition by higher eukaryotes, and this is especially true of anoxic environments, which we consider next.

24.1 MiniReview

The oxygen and carbon cycles are interconnected through the complementary activities of autotrophic and heterotrophic organisms. Microbial decomposition is the single largest source of CO_2 released to the atmosphere.

- How is new organic matter made in nature?
- In what ways are oxygenic photosynthesis and respiration related?

24.2 Syntrophy and Methanogenesis

Biological methanogenesis is central to carbon cycling in anoxic habitats. Methanogenesis is carried out by a group of *Archaea*, the methanogens, which are strict anaerobes. We discussed the biochemistry of methanogenesis in Section 21.10 and methanogens themselves in Section 17.4. Most methanogens can use CO_2 as a terminal electron acceptor in anaerobic respiration, reducing it to CH_4 with H_2 as electron donor. Only a very few other substrates, acetate being chief among them, are directly converted to CH_4 by methanogens. To convert most organic compounds to CH_4, methanogens must therefore team up with partner organisms that can supply them with methanogenic precursors. This is the job of the syntrophs.

Anoxic Decomposition and Syntrophy

In Section 21.5 we discussed the energetics of syntrophy, a process in which two or more organisms cooperate in the anaerobic degradation of organic compounds. Here we consider the interactions of syntrophic bacteria with their partner organisms and their significance for the anoxic carbon cycle. Our focus is on anoxic freshwater sediments, where methanogenesis is typically a major process.

Polysaccharides, proteins, lipids, and nucleic acids from dead organisms find their way into anoxic habitats. Following hydrolysis, the monomers are excellent electron donors for energy metabolism. For the breakdown of a typical polysaccharide such as cellulose (**Figure 24.3** and **Table 24.2**), the process begins with *cellulolytic bacteria*; these organisms hydrolyze cellulose into cellobiose (glucose–glucose) and then into glucose. The glucose is fermented by *primary* fermenters to short-chain fatty acids (acetate, propionate, and butyrate) and to alcohols, H_2, and CO_2. H_2 is quickly removed by methanogens, along with the acetate, but the bulk of the organic carbon remains in the form of fatty acids and alcohols; these cannot be directly catabolized by methanogens. The catabolism of these compounds requires **syntrophy** and the important fermentative activities of syntrophic bacteria (Section 21.5) (Figure 24.3).

Role of the Syntrophs

Key bacteria in the conversion of organic materials to methane are the syntrophs. These organisms are *secondary* fermenters. Syntrophs ferment the products of primary fermenters, producing primarily H_2, CO_2, and acetate. For example, *Syntrophomonas wolfei* oxidizes C_4 to C_8 fatty acids yielding acetate, CO_2 (if the fatty acid contained an odd number of carbon atoms), and H_2 (Table 24.2). Other species of *Syntrophomonas* use fatty acids up to C_{18} in length, including some unsaturated fatty acids. *Syntrophobacter wolinii* specializes in propionate (C_3) fermentation, generating acetate, CO_2, and H_2, and *Syntrophus gentianae* degrades aromatic compounds such as benzoate to acetate, H_2, and CO_2 (Table 24.2). However, the syntrophs are unable to carry out these syntrophic reactions in pure culture; their growth requires a H_2-consuming partner organism. This requirement has to do with the energetics of syntrophic processes.

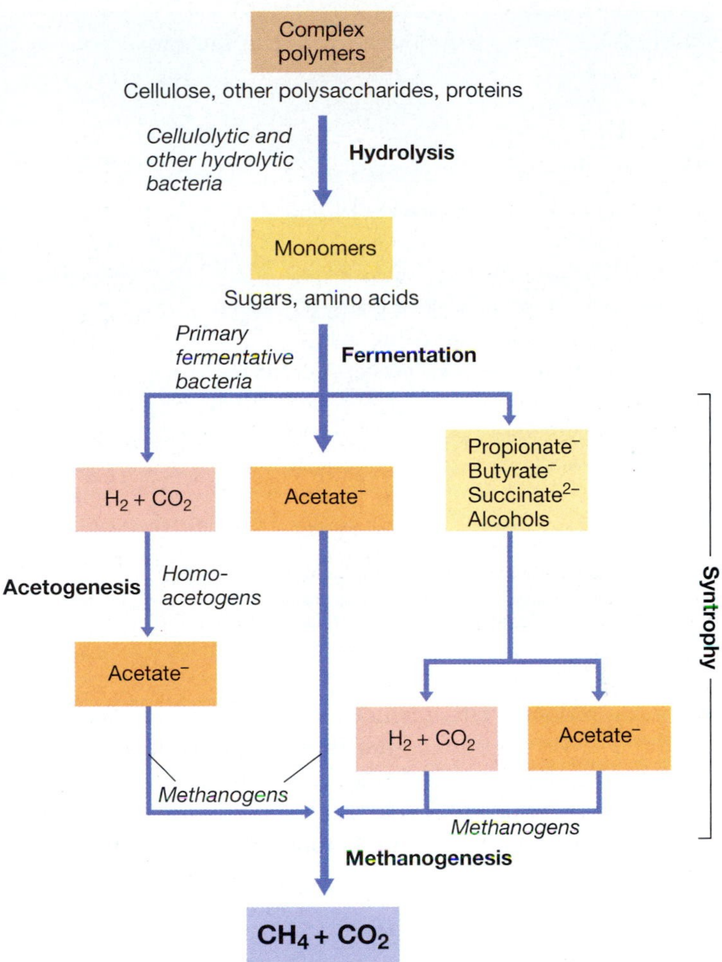

Figure 24.3 Anoxic decomposition. Shown is the overall process of anoxic decomposition, in which various groups of fermentative anaerobes cooperate in the conversion of complex organic materials to methane (CH_4) and CO_2. This picture holds for environments in which sulfate-reducing bacteria play only a minor role, for example, in freshwater lake sediments, sewage sludge bioreactors, or the rumen.

As was described in Section 21.5, H_2 consumption by a partner organism is absolutely essential for growth of the syntrophs. When the reactions in Table 24.2 are written with all reactants at standard conditions (solutes, 1 M; gases, 1 atm, 25°C), the reactions yield free-energy changes that are positive in arithmetic sign. That is, the $\Delta G^{0\prime}$ (Section 5.6) of these reactions is endergonic (Table 24.2). But H_2 consumption dramatically affects the energetics, making the reaction exergonic and allowing energy to be conserved. This can be seen in Table 24.2, where the ΔG values (free-energy change measured under actual conditions in the habitat) are exergonic if H_2 concentrations are kept very low through consumption by a partner organism.

The final products of the syntrophic partnership are CO_2 plus CH_4 (Figure 24.3), and virtually any organic compound that enters a methanogenic habitat will eventually be converted

Table 24.2 **Major reactions occurring in the anoxic conversion of organic compounds to methane**[a]

Reaction type	*Reaction*	*Free-energy change (kJ/reaction)* $\Delta G^{0\prime b}$	ΔG^{c}
Fermentation of glucose to acetate, H_2, and CO_2	Glucose + 4 $H_2O \rightarrow$ 2 Acetate$^-$ + 2 HCO_3^- + 4 H^+ + 4 H_2	−207	−319
Fermentation of glucose to butyrate, CO_2, and H_2	Glucose + 2 $H_2O \rightarrow$ Butyrate$^-$ + 2 HCO_3^- + 2 H_2 + 3 H^+	−135	−284
Fermentation of butyrate to acetate and H_2	Butyrate$^-$ + 2 $H_2O \rightarrow$ 2 Acetate$^-$ + H^+ + 2 H_2	+48.2	−17.6
Fermentation of propionate to acetate, CO_2, and H_2	Propionate$^-$ + 3 $H_2O \rightarrow$ Acetate$^-$ + HCO_3^- + H^+ + H_2	+76.2	−5.5
Fermentation of ethanol to acetate and H_2	2 Ethanol + 2 $H_2O \rightarrow$ 2 Acetate$^-$ + 4 H_2 + 2 H^+	+19.4	−37
Fermentation of benzoate to acetate, CO_2, and H_2	Benzoate$^-$ + 7 $H_2O \rightarrow$ 3 Acetate$^-$ + 3 H^+ + HCO_3 + 3 H_2	+70.1	−18
Methanogenesis from H_2 + CO_2	4 H_2 + HCO_3^- + $H^+ \rightarrow CH_4$ + 3 H_2O	−136	−3.2
Methanogenesis from acetate	Acetate$^-$ + $H_2O \rightarrow CH_4$ + HCO_3^-	−31	−24.7
Acetogenesis from H_2 + CO_2	4 H_2 + 2 HCO_3^- + $H^+ \rightarrow$ Acetate$^-$ 4 H_2O	−105	−7.1

[a]Data adapted from Zinder, S. 1984. Microbiology of anaerobic conversion of organic wastes to methane: Recent developments. *Am. Soc. Microbiol.* 50:294–298.

[b]Standard conditions: solutes, 1 M; gases, 1 atm, 25°C.

[c]Concentrations of reactants in typical anoxic freshwater ecosystem: fatty acids, 1 m*M*; HCO_3^-, 20 m*M*; glucose, 10 μ*M*; CH_4, 0.6 atm; H_2, 10^{-4} atm. For calculating ΔG from $\Delta G^{0\prime}$, refer to Appendix 1.

to these products. This includes even complex aromatic and aliphatic hydrocarbons. Additional organisms other than those shown in Figure 24.3 may be involved in such degradations, but eventually fatty acids and alcohols will be generated, and they will be converted to methanogenic substrates by the syntrophs.

Methanogenic Symbionts

In addition to the great diversity of free-living methanogens, endosymbiotic methanogens have also been found in certain protists. Several protists, including free-living aquatic amoebas and flagellates present in certain insect guts, have been shown to harbor methanogens. In termites, for example, methanogens are present primarily within cells of trichomonal protists inhabiting the termite hindgut (**Figure 24.4**). Methanogenic symbionts of protists are rod-shaped species of the genus *Methanobacterium* or *Methanobrevibacter* (Section 17.4), but their exact relationship to free-living methanogens is unclear. In the termite hindgut, endosymbiotic methanogens (along with acetogens, see later text) are thought to benefit their protist hosts by consuming H_2 generated from glucose fermentation by cellulolytic protists.

Global Methanogenesis

Although levels of methanogenesis are high only in anoxic environments such as swamps, lake sediments, and marshes, or in the rumen (Section 24.10), there is some production of methane in habitats that would otherwise be considered oxic, such as forest and grassland soils. In such habitats methane is produced in anoxic microenvironments, for example in the midst of soil crumbs. In addition, certain plants make methane as well, although the mechanism of methane formation and the global input from this source is uncertain.

Table 24.3 lists the methane outputs of different habitats. Note that biogenic production of methane far exceeds production from gas wells and other abiogenic sources. Eructation by ruminants and CH_4 released from termites, paddy fields, and natural wetlands are the largest sources of biogenic methane (Table 24.3).

Acetogenic Habitats

A competing H_2-consuming process to methanogenesis is *acetogenesis* (formation of acetate, Section 21.9). In some habitats, for example the rumen, acetogens appear to compete only poorly with methanogens, and thus methanogenesis is the dominant H_2-consuming process. But in other habitats, such as the termite hindgut, where methane is released by methanogens living inside protists (Figure 24.4), acetogenesis is quantitatively a much more important process.

Based on simple energetic considerations, methanogenesis from H_2 is more favorable than acetogenesis (−131 kJ versus −105 kJ, respectively), and thus methanogens should have a competitive advantage in all habitats in which the two processes compete. However, in termites they do not. There are at least three reasons why this may be so. First, in some way acetogens may be able to position themselves in the termite gut nearer to the source of H_2 than methanogens and thus consume the majority of H_2 produced from cellulose fermentation before it arrives to the endosymbiotic methanogens (Figure 24.4*c*). Second, unlike methanogens, acetogens can ferment glucose (obtained from cellulose). And third, termites eat wood, which contains a high lignin content, and this type of material may be degraded in such a way that favors the production of methoxylated aromatic compounds, which are also substrates for acetogens (Section 21.9). So, despite the fact that termites are methanogenic (Table 24.3), carbon and electron flow favor acetogenesis in this anoxic habitat. Other habitats in which acetogenesis is a major process include very cold environments, such as

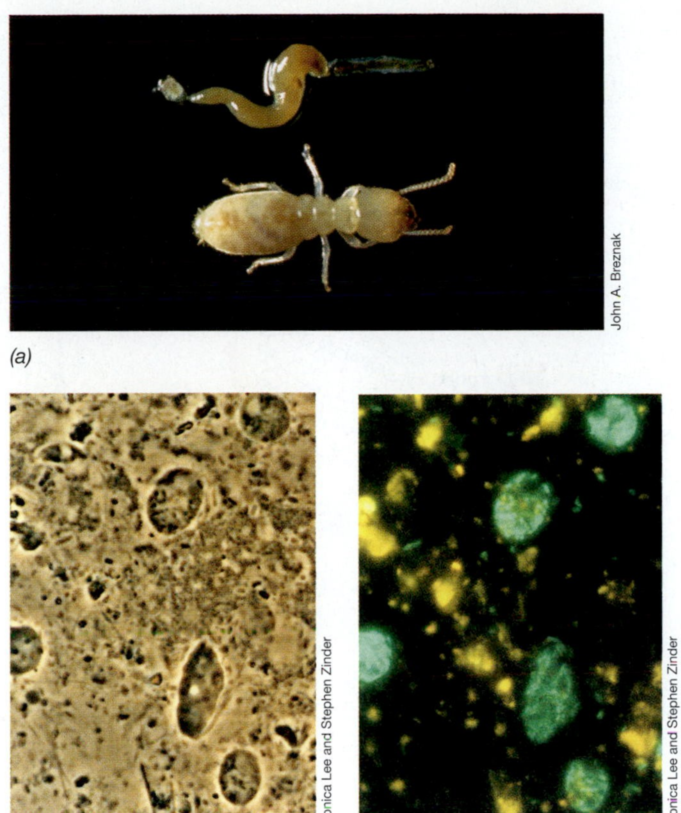

Figure 24.4 Termites and their carbon metabolism. *(a)* A common eastern (USA) subterranean termite worker larva shown beneath a hindgut extracted from a separate worker. The animal is about 0.5 cm long. Acetogenesis is the major form of anoxic carbon metabolism in these termites although methanogenesis also occurs. *(b, c)* Microorganisms from the hindgut of the termite, *Zootermopsis angusticolis*; a single microscope field was photographed by two different methods. *(b)* Phase contrast. *(c)* Epifluorescence, showing color typical of methanogens due to the high content of the fluorescent coenzyme F_{420} (compare with Figure 21.20). The methanogens are inside cells of the protist *Tricercomitis* sp. Plant particles fluoresce yellow. The average diameter of a protist cell is 15–20 μm.

permafrost soils, where the very cold temperatures seem to be better tolerated by acetogens than by methanogens.

Methanogenesis versus Sulfidogenesis

Methanogenesis and acetogenesis are extensive processes in anoxic freshwater sediments and terrestrial environments but not in marine sediments. This is because sulfate-reducing bacteria are abundant in marine sediments and outcompete methanogens and acetogens for H_2. The basis for this lies in the fact that the energetics of sulfate reduction with H_2 is better than the reduction of CO_2 with H_2 to either CH_4 or acetate; for biochemical reasons, this favors sulfate reduction. In freshwater, however, where sulfidogenesis is typically very low because sulfate is limiting, methanogenesis and acetogenesis dominate.

Table 24.3 Estimates of CH_4 released into the atmosphere[a]

Source	*CH_4 emission (10^{12} g/year)*	
Biogenic		
Ruminants	80–100	
Termites	25–150[b]	
Paddy fields	70–120	
Natural wetlands	120–200	
Landfills	5–70	
Oceans and lakes	1–20	
Tundra	1–5	
Abiogenic		
Coal mining	10–35	
Natural gas flaring and venting	10–30	
Industrial and pipeline losses	15–45	
Biomass burning	10–40	
Methane hydrates	2–4	
Volcanoes	0.5	
Automobiles	0.5	
Total	350–820	
Total biogenic	302–665	81–86% of total
Total abiogenic	48–155	13–19% of total

[a]Data adapted from estimates in Tyler, S. C. 1991. The global methane budget, pp. 7–58, in E. J. Rogers and W. B. Whitman (eds.), *Microbial Production and Consumption of Greenhouse Gases: Methane, Nitrogen Oxides, and Halomethanes*, American Society for Microbiology, Washington, DC.

[b]More recent estimates indicate that the lower value is probably the more accurate.

24.2 MiniReview

Under anoxic conditions, organic matter is degraded to CH_4 and CO_2. CH_4 is formed primarily from the reduction of CO_2 by H_2 and from acetate, both supplied by syntrophic bacteria; these organisms depend on H_2 consumption as the basis of their energetics. On a global basis, biogenic CH_4 is a much larger source than abiogenic CH_4.

- What kinds of organisms can grow in coculture with *Syntrophomonas*?
- What is the final product of acetogenesis? What anoxic habitat shows greater acetogenesis than methanogenesis?
- Why is methanogenesis from H_2 not a major process in marine sediments?

II NITROGEN, SULFUR, AND IRON CYCLES

In addition to carbon, many other key elements are metabolized by microorganisms. These include in particular *nitrogen, sulfur,* and *iron,* and we examine the cycling of these key nutrients now.

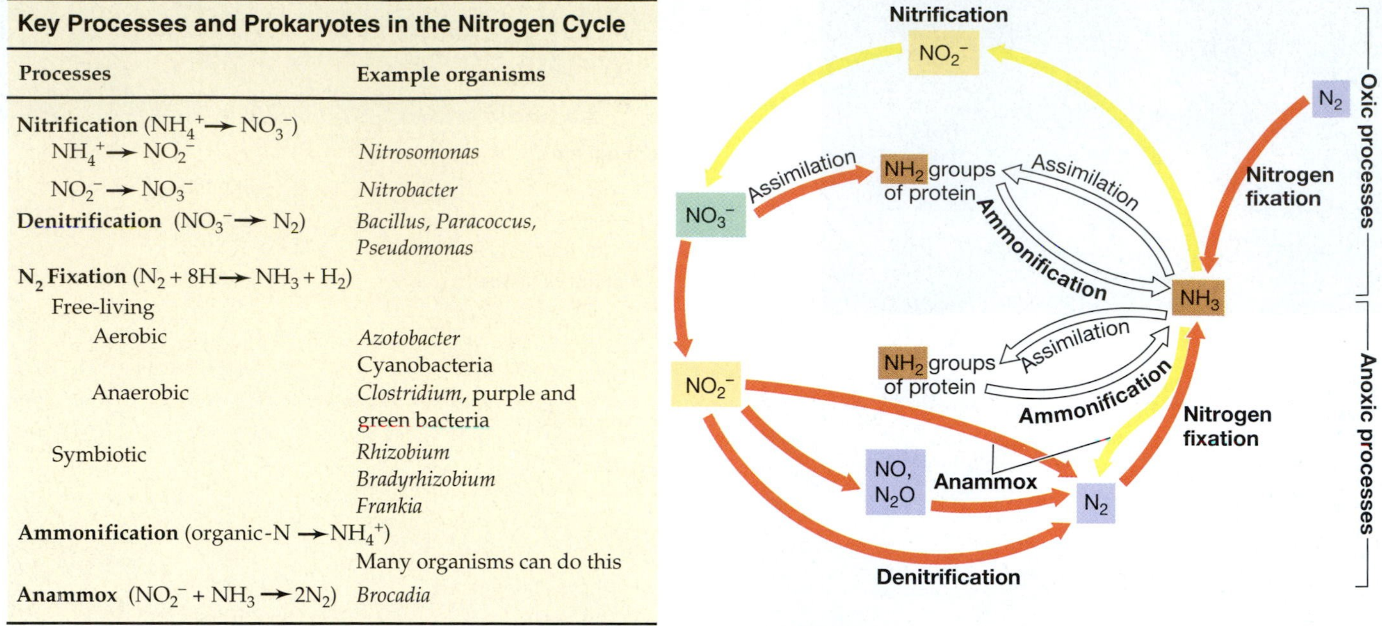

Key Processes and Prokaryotes in the Nitrogen Cycle

Processes	Example organisms
Nitrification ($NH_4^+ \rightarrow NO_3^-$)	
$NH_4^+ \rightarrow NO_2^-$	*Nitrosomonas*
$NO_2^- \rightarrow NO_3^-$	*Nitrobacter*
Denitrification ($NO_3^- \rightarrow N_2$)	*Bacillus, Paracoccus, Pseudomonas*
N_2 Fixation ($N_2 + 8H \rightarrow NH_3 + H_2$)	
Free-living	
Aerobic	*Azotobacter* Cyanobacteria
Anaerobic	*Clostridium*, purple and green bacteria
Symbiotic	*Rhizobium* *Bradyrhizobium* *Frankia*
Ammonification (organic-N $\rightarrow NH_4^+$)	Many organisms can do this
Anammox ($NO_2^- + NH_3 \rightarrow 2N_2$)	*Brocadia*

Figure 24.5 Redox cycle for nitrogen. Oxidation reactions are shown by yellow arrows and reductions by red arrows. Reactions without redox change are in white. The anammox reaction is $NH_3 + NO_2^- + H^+ \rightarrow N_2 + 2H_2O$.

24.3 The Nitrogen Cycle

The element nitrogen, N, a key constituent of cells, exists in a number of oxidation states. We have discussed four major microbial nitrogen transformations thus far: nitrification, denitrification, anammox, and nitrogen fixation (Sections 20.12–20.14 and 21.7). These and other key nitrogen transformations are summarized in the redox cycle shown in **Figure 24.5**.

Nitrogen Fixation

Nitrogen gas (N_2) is the most stable form of nitrogen and is a major reservoir for nitrogen on Earth. However, only a relatively small number of prokaryotes are able to use N_2 as a cellular nitrogen source by *nitrogen fixation* ($N_2 + 8\ H \rightarrow 2\ NH_3 + H_2$) (Section 20.14). The nitrogen recycled on Earth is mostly in fixed forms of nitrogen, such as ammonia (NH_3) and nitrate (NO_3^-). In many environments, however, the short supply of such compounds puts a premium on biological nitrogen fixation. We revisit this important process later in this chapter when we describe symbiotic N_2 fixation in Section 24.15.

Denitrification

We discussed the role of nitrate as an alternative electron acceptor in anaerobic respiration in Section 21.7. Under most conditions, the end product of nitrate reduction is N_2, NO, or N_2O. The reduction of nitrate to gaseous nitrogen compounds, called *denitrification* (Figure 24.5), is the main means by which gaseous N_2 is formed biologically. On the one hand, denitrification is a detrimental process. For example, if fields fertilized with potassium nitrate fertilizer become waterlogged following heavy rains, anoxic conditions can develop and denitrification can be extensive; this removes fixed nitrogen from the soil. On the other hand, denitrification can aid in wastewater treatment (Section 36.2). By removing nitrate, denitrification minimizes algal growth when the treated sewage is discharged into lakes and streams.

The production of N_2O and NO by denitrification can have other environmental consequences. N_2O can be photochemically oxidized to NO in the atmosphere. NO reacts with ozone (O_3) in the upper atmosphere to form nitrite (NO_2^-), and this returns to Earth as nitric acid (HNO_2). Thus, denitrification contributes to both ozone destruction and acid rain, leading to increased passage of ultraviolet radiation to the surface of Earth and acidic soils, respectively. Increases in soil acidity can change microbial community structure and function and, ultimately, soil fertility, impacting both plant diversity and agricultural yields of crop plants.

Ammonia Fluxes, Nitrification, and Anammox

Ammonia is released during the decomposition of organic nitrogen compounds such as amino acids and nucleotides, a process called *ammonification* (Figure 24.5). At neutral pH, ammonia exists as ammonium ion (NH_4^+). Much of the ammonium released by aerobic decomposition in soils is rapidly recycled and converted to amino acids in plants and microorganisms. However, because ammonia is volatile, some of it can be lost from alkaline soils by vaporization, and there are major losses of ammonia to the atmosphere in areas with dense animal populations (for example, cattle feedlots). On a global basis, however, ammonia constitutes only about 15% of

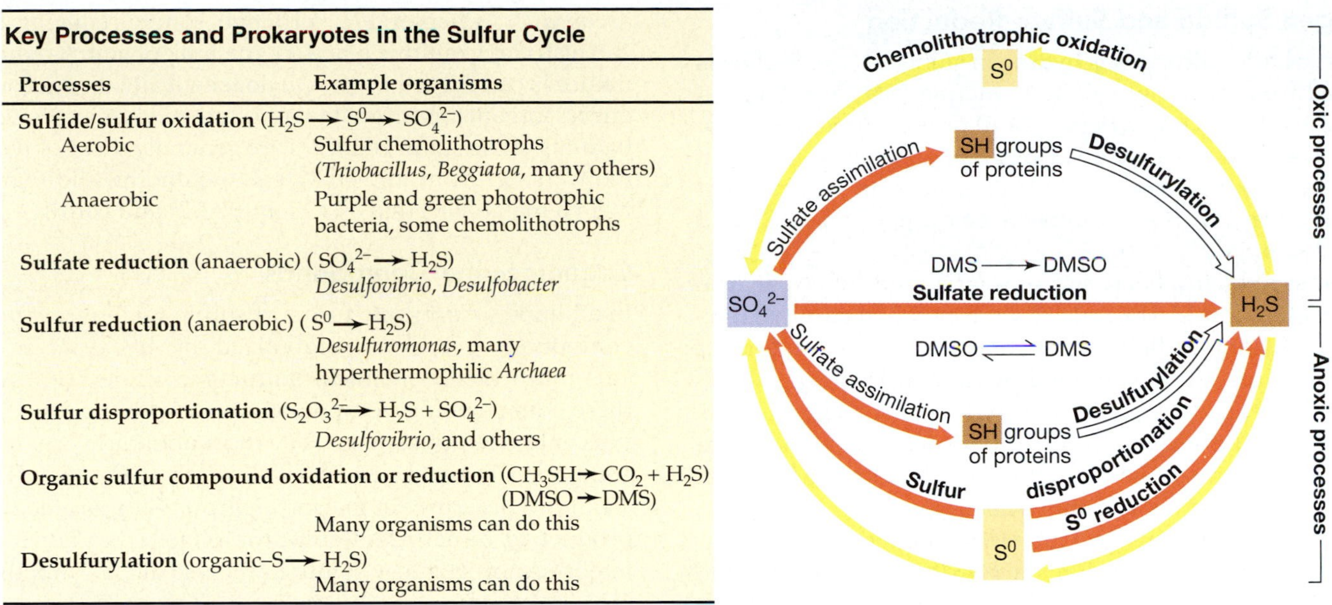

Key Processes and Prokaryotes in the Sulfur Cycle

Processes	Example organisms
Sulfide/sulfur oxidation ($H_2S \rightarrow S^0 \rightarrow SO_4^{2-}$)	
Aerobic	Sulfur chemolithotrophs (*Thiobacillus*, *Beggiatoa*, many others)
Anaerobic	Purple and green phototrophic bacteria, some chemolithotrophs
Sulfate reduction (anaerobic) ($SO_4^{2-} \rightarrow H_2S$)	*Desulfovibrio*, *Desulfobacter*
Sulfur reduction (anaerobic) ($S^0 \rightarrow H_2S$)	*Desulfuromonas*, many hyperthermophilic *Archaea*
Sulfur disproportionation ($S_2O_3^{2-} \rightarrow H_2S + SO_4^{2-}$)	*Desulfovibrio*, and others
Organic sulfur compound oxidation or reduction ($CH_3SH \rightarrow CO_2 + H_2S$) (DMSO → DMS)	Many organisms can do this
Desulfurylation (organic–S → H_2S)	Many organisms can do this

Figure 24.6 Redox cycle for sulfur. Oxidations are shown by yellow arrows and reductions by red arrows. Reactions without redox changes are in white. DMSO, dimethylsulfoxide; DMS, dimethylsulfide.

the nitrogen released to the atmosphere, the rest being primarily N_2 or N_2O from denitrification.

Nitrification, the oxidation of NH_3 to NO_3^-, occurs readily in well-drained soils at neutral pH through the activities of the nitrifying bacteria (Sections 15.3 and 20.12) (Figure 24.5). Whereas denitrification *consumes* nitrate, nitrification *produces* nitrate. If materials high in protein, such as manure or sewage, are added to soils, the rate of nitrification increases. Although nitrate is readily assimilated by plants, it is very water-soluble and is rapidly leached or denitrified from soils receiving high rainfall. Consequently, nitrification is not beneficial to plant agriculture. Ammonia, on the other hand, is positively charged and strongly adsorbed to negatively charged clay-rich soils.

Anhydrous ammonia is used extensively as a nitrogen fertilizer, and chemicals are commonly added to the fertilizer to inhibit nitrification. One common inhibitor is a substituted pyridine called *nitrapyrin* (2-chloro-6-trichloromethylpyridine, also known as N-SERVE). Nitrapyrin specifically inhibits the first step in nitrification, the oxidation of NH_3 to NO_2^- (Section 20.12). This effectively inhibits both steps in the nitrification process because the second step, $NO_2^- \rightarrow NO_3^-$, depends on the first. The addition of nitrification inhibitors to anhydrous ammonia has greatly increased the efficiency of crop fertilization and has helped prevent the pollution of waterways from nitrate leached from fertilized soils.

Ammonia can be catabolized anaerobically by *Brocadia* and related organisms in the process called *anammox*. In this reaction, ammonia is oxidized with nitrite (NO_2^-) as electron acceptor, forming N_2 as the final product (Figure 24.5), which is released to the atmosphere. Although a major process in sewage and sediments, anammox is not significant in well-drained soils. The microbiology and biochemistry of anammox was discussed in Section 20.13.

24.3 MiniReview

The principal form of nitrogen on Earth is nitrogen gas (N_2), which can be used as a nitrogen source only by nitrogen-fixing bacteria. Ammonia produced by nitrogen fixation or by ammonification can be assimilated into organic matter or oxidized to nitrate. Denitrification and anammox cause losses of nitrogen from the biosphere.

- What is nitrogen fixation and why is it important to the nitrogen cycle?
- How do the processes of nitrification and denitrification differ? How do nitrification and anammox differ?
- How does the compound nitrapyrin benefit both agriculture and the environment?

24.4 The Sulfur Cycle

Sulfur transformations by microorganisms are even more complex than those of nitrogen because of the large number of oxidation states of sulfur and the fact that several transformations of sulfur occur abiotically. Sulfate reduction and chemolithotrophic sulfur oxidation were covered in Sections 21.8 and 20.10, respectively. The redox cycle for microbial sulfur transformations is shown in **Figure 24.6**.

Although a number of oxidation states of sulfur are possible, only three are significant in nature, −2 (sulfhydryl, R—SH, and sulfide, HS^-), 0 (elemental sulfur, S^0), and +6 (sulfate, SO_4^{2-}). The bulk of the sulfur on Earth is in sediments and rocks in the form of sulfate minerals, primarily gypsum ($CaSO_4$) and sulfide minerals (pyrite, FeS_2), although the oceans constitute the most significant reservoir of sulfur (as sulfate) in the biosphere.

UNIT 4

Hydrogen Sulfide and Sulfate Reduction

A major volatile sulfur gas is hydrogen sulfide (H_2S). Sulfide is produced from bacterial sulfate reduction ($SO_4^{2-} + 8\ H^+ \rightarrow H_2S + 2\ H_2O + 2\ OH^-$) (Figure 24.6) or is emitted from geochemical sources in sulfide springs and volcanoes. Although H_2S is volatile, the form of sulfide present in an environment is pH dependent: H_2S predominates below pH 7 and HS^- and S^{2-} predominate above pH 7.

Sulfate-reducing bacteria are a large and highly diverse group (Section 15.18) and are widespread in nature. However, in many anoxic habitats, such as freshwaters and many soils, sulfate reduction is limited by the low levels of sulfate. Moreover, because organic electron donors (or H_2, which is a product of the fermentation of organic compounds) are needed to support sulfate reduction, it only occurs where significant amounts of organic material are present.

In marine sediments, the rate of sulfate reduction is typically carbon-limited and can be greatly increased by the addition of organic matter. This is important because the disposal of sewage, sewage sludge, and garbage in the oceans can lead to marked increases in organic matter in the sediments and trigger sulfate reduction. Because H_2S is toxic to many plants and animals, formation of HS^- by sulfate reduction is potentially detrimental (sulfide is toxic because it combines with the iron of cytochromes and blocks respiration). Sulfide is commonly detoxified in the environment by combination with iron, forming the insoluble minerals FeS and FeS_2 (pyrite). The black color of sulfidic sediments is due to these metal sulfide minerals (Figure 15.49*g*).

Sulfide and Elemental Sulfur Oxidation/Reduction

Under oxic conditions, sulfide rapidly oxidizes spontaneously at neutral pH. Sulfur-oxidizing chemolithotrophic bacteria, most of which are aerobes, can catalyze the oxidation of sulfide. However, because of the rapid chemical reaction, significant amounts of sulfide are oxidized by bacteria only in areas in which H_2S emerging from anoxic areas meets O_2 from oxic areas. In addition, if light is available, there can also be anoxic oxidation of sulfide, catalyzed by the phototrophic purple and green sulfur bacteria (Sections 15.2 and 16.15).

Elemental sulfur (S^0) is chemically stable but is readily oxidized by sulfur-oxidizing chemolithotrophic bacteria such as *Thiobacillus* and *Acidithiobacillus* (Sections 15.4 and 20.10). Elemental sulfur is insoluble, and thus the bacteria that oxidize it must attach to the sulfur crystals to obtain their substrate (Figure 20.26*b*). The oxidation of elemental sulfur forms sulfuric acid (H_2SO_4), and thus sulfur oxidation characteristically lowers the pH in the environment. Elemental sulfur is sometimes added to alkaline soils to lower the pH, reliance being placed on the ubiquitous thiobacilli to carry out the acidification process.

Elemental sulfur can be reduced as well as oxidized. Sulfur reduction to sulfide (a form of anaerobic respiration) is a major ecological process, especially among hyperthermophilic *Archaea* (Chapter 17). Although sulfate-reducing bacteria can also carry out this reaction, the bulk of sulfur reduction in nature is carried out by the phylogenetically distinct sulfur reducers, organisms incapable of sulfate reduction. However, the habitats of the sulfur reducers are generally those of the sulfate reducers, so from an ecological standpoint, the two groups form a metabolic guild (Figure 23.2) and coexist.

Organic Sulfur Compounds

In addition to *inorganic* forms of sulfur, various *organic* sulfur compounds are also metabolized by bacteria, and these enter into biogeochemical sulfur cycling as well. Many of these foul-smelling compounds are highly volatile and can thus enter the atmosphere. The most abundant organic sulfur compound in nature is *dimethyl sulfide* ($CH_3{-}S{-}CH_3$); it is produced primarily in marine environments as a degradation product of dimethylsulfoniopropionate (Figure 6.26), a major osmoregulatory solute in marine algae. Dimethylsulfoniopropionate can be used as a carbon source and electron donor by microorganisms and is catabolized to dimethyl sulfide and acrylate. The latter, a derivative of the fatty acid propionate, is used to support growth.

Dimethyl sulfide released to the atmosphere undergoes photochemical oxidation to methane sulfonate ($CH_3SO_3^-$), SO_2, and SO_4^{2-}. By contrast, dimethyl sulfide produced in anoxic habitats can be transformed microbially in at least three ways: (1) by methanogenesis (yielding CH_4 and H_2S), (2) as an electron donor for photosynthetic CO_2 fixation in phototrophic purple bacteria (yielding dimethyl sulfoxide, DMSO), and (3) as an electron donor in energy metabolism in certain chemoorganotrophs and chemolithotrophs (also yielding DMSO). DMSO can be an electron acceptor for anaerobic respiration (Section 21.12), producing dimethyl sulfide. Many other organic sulfur compounds affect the global sulfur cycle, including methanethiol (CH_3SH), dimethyl disulfide ($H_3C{-}S{-}S{-}CH_3$), and carbon disulfide (CS_2), but on a global basis, dimethyl sulfide is the most significant.

24.4 MiniReview

Bacteria play major roles in both the oxidative and reductive sides of the sulfur cycle. Sulfur- and sulfide-oxidizing bacteria produce sulfate, whereas sulfate-reducing bacteria consume sulfate, producing hydrogen sulfide. Because sulfide is toxic and reacts with various metals, sulfate reduction is an important biogeochemical process. Dimethyl sulfide is the major organic sulfur compound of ecological significance in nature.

- Is H_2S a substrate or a product of the sulfate-reducing bacteria? Of the chemolithotrophic sulfur bacteria?
- Why does the bacterial oxidation of sulfur result in a pH drop?
- What organic sulfur compound is most abundant in nature?

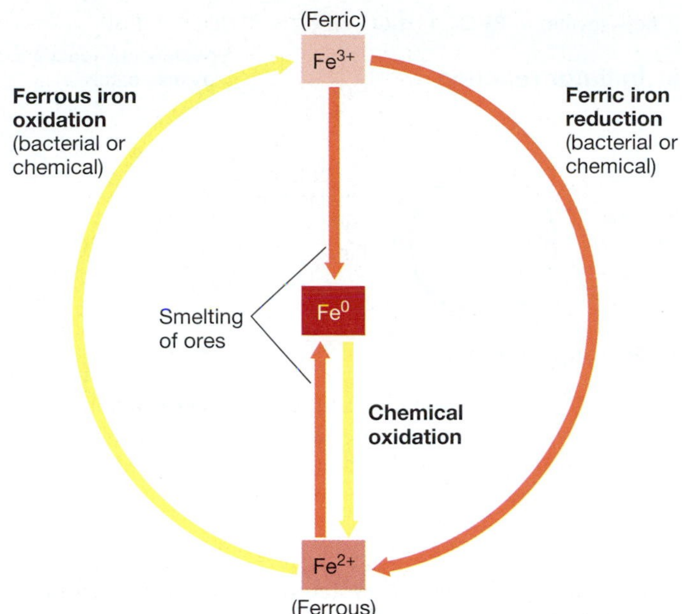

Figure 24.7 Redox cycle for iron. The major forms of iron in nature are Fe^{2+} and Fe^{3+}; Fe^0 is primarily a product of human activities in the smelting of iron ores. Oxidations are shown by yellow arrows and reductions by red arrows.

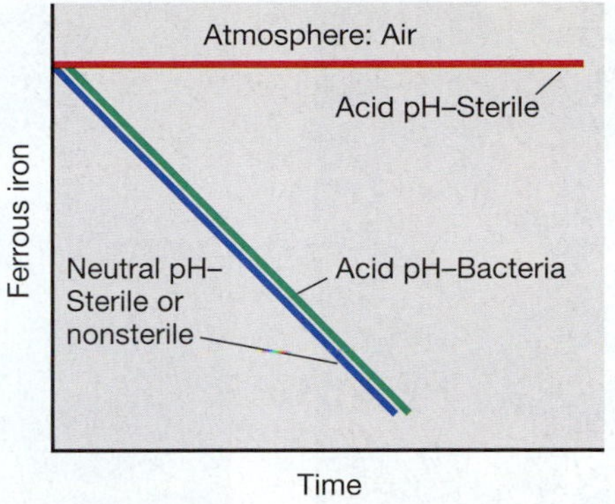

(a)

(b)

Figure 24.8 Oxidation of ferrous iron (Fe^{2+}). *(a)* Oxidation of ferrous iron as a function of pH and the presence of the bacterium *Acidithiobacillus ferrooxidans*. Note how Fe^{2+} is stable under acidic conditions in the absence of bacterial cells. *(b)* A microbial mat containing acidophilic green algae and various iron-oxidizing prokaryotes in the Rio Tinto, Spain. The river is highly acidic and contains high levels of dissolved metals, in particular Fe^{2+}. The red-brown precipitates contain $Fe(OH)_3$.

24.5 The Iron Cycle

Iron is one of the most abundant elements in Earth's crust. On the surface of Earth, iron exists naturally in two oxidation states, ferrous (Fe^{2+}) and ferric (Fe^{3+}). A third oxidation state, Fe^0, is a major product of human activities only, from the smelting of iron ores to form cast iron. In nature, iron cycles primarily between the ferrous and ferric forms. The redox reactions in the iron cycle include both oxidations and reductions. Fe^{3+} is reduced both chemically and as a form of anaerobic respiration, and Fe^{2+} is oxidized both chemically and as a form of chemolithotrophic metabolism (**Figure 24.7**).

Bacterial Iron Reduction

Some organisms can use ferric iron as an electron acceptor in anaerobic respiration (Section 21.12). Ferric iron reduction is common in waterlogged soils, bogs, and anoxic lake sediments. Movement of iron-rich groundwater from anoxic bogs or waterlogged soils can result in the transport of large amounts of ferrous iron. When this iron-laden water reaches oxic regions, the ferrous iron is oxidized chemically or by iron bacteria. Ferric compounds then precipitate, leading to the formation of brown iron deposits (Figure 20.28):

$$Fe^{2+} + \tfrac{1}{4}O_2 + 2\tfrac{1}{2}H_2O \longrightarrow Fe(OH)_3 + 2\,H^+$$

The ferric hydroxide precipitate can interact with other nonbiological substances, such as humics (Section 24.1), to reduce Fe^{3+} back to Fe^{2+} (Figure 24.7). Ferric iron can also form complexes with various organic constituents. In this way it becomes solubilized and once again available to ferric iron-reducing bacteria as an electron acceptor.

Ferrous Iron and Pyrite Oxidation at Acid pH

The only electron acceptor able to oxidize Fe^{2+} abiotically is O_2. In neutral pH habitats Fe^{2+} can be oxidized by iron bacteria such as *Gallionella* and *Leptothrix* (Section 15.15). However, this occurs primarily at interfaces between anoxic ferrous-rich groundwaters and air. The most extensive bacterial iron oxidation occurs at acidic pH, where Fe^{2+} is stable to spontaneous oxidation. In extremely acidic habitats, the acidophilic chemolithotroph *Acidithiobacillus ferrooxidans* and related acidophilic iron oxidizers oxidize Fe^{2+} to Fe^{3+} (**Figure 24.8**). Very little energy is generated in the oxidation of ferrous to ferric iron (Section 20.11), and so these bacteria must oxidize large amounts of iron in order to grow; consequently,

Figure 24.9 Pyrite and coal. *(a)* Pyrite in coal that can be oxidized by sulfur- and iron-oxidizing bacteria. Shown is a section through a piece of coal from the Black Mesa formation in northern Arizona (USA). The gold-colored spherical discs (about 1 mm in diameter) are particles of pyrite, FeS_2. *(b)* A coal seam in a surface coal-mining operation. Exposing the coal to oxygen and moisture stimulates the activities of iron-oxidizing bacteria growing on the pyrite in the coal.

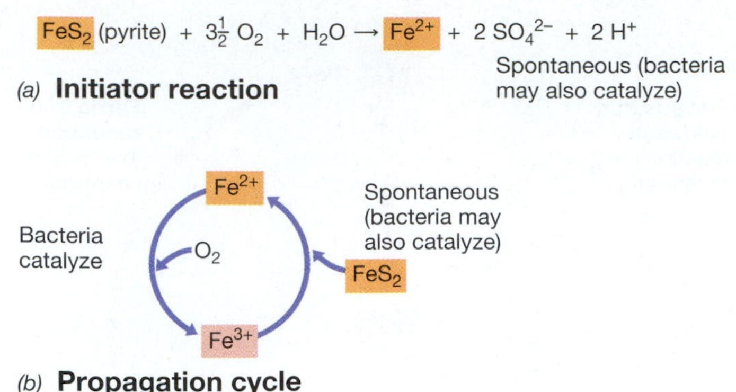

Figure 24.10 Role of iron-oxidizing bacteria in oxidation of the mineral pyrite. *(a)* The primarily nonbiological initiator reaction *(b)* The propagation cycle, which includes biotic and abiotic components.

even a small population can precipitate a large amount of iron. *A. ferrooxidans* is a strict acidophile and is very common in acid mine drainages and in acidic springs; it is responsible for most of the ferric iron precipitated under moderately acidic (pH 2–4) conditions.

A. ferrooxidans and *Leptospirillum ferrooxidans* live in environments in which sulfuric acid is the dominant acid and large amounts of sulfate are present. At 20–30°C and moderately acidic pH, *A. ferrooxidans* dominates; at 30–50°C and more acidic pH (1–2), *L. ferrooxidans* is the dominant organism. Under these conditions, ferric iron does not precipitate as the hydroxide but as a complex sulfate mineral called *jarosite* $[HFe_3(SO_4)_2(OH)_6]$. Jarosite is a yellowish or brownish precipitate and is one of the major pollutants in acid mine drainage, an unsightly yellow solution called "yellow boy" by U.S. coal miners (∞ Figure 20.28 and Figure 24.8). *A. ferrooxidans* and *L. ferrooxidans* are phylogenetically and morphologically distinct, but are also distinct in their metabolism. Whereas *L. ferrooxidans* can grow only on Fe^{2+}, *A. ferrooxidans* grows chemolithotrophically on either Fe^{2+} or S^0 as electron donors.

One of the most common forms of iron in nature is **pyrite** (FeS_2). Pyrite forms from the reaction of sulfur with ferrous sulfide (FeS) as an insoluble crystalline mineral; pyrite is very common in bituminous coals and in many ore bodies (**Figure 24.9**). The bacterial oxidation of pyrite is of great significance in the development of acidic conditions in coal-mining operations (Figure 24.9*b*). Additionally, oxidation of pyrite by bacteria is of considerable importance in the microbial leaching of ores described in the next section.

The oxidation of pyrite is a combination of chemically and bacterially catalyzed reactions. Two electron acceptors are involved in this process: molecular oxygen (O_2) and ferric ions (Fe^{3+}). When pyrite is first exposed, as in a mining operation (Figure 24.9*b*), a slow chemical reaction with O_2 occurs, as shown in **Figure 24.10a**. This reaction, called the *initiator reaction,* leads to the oxidation of sulfide to sulfate and the development of acidic conditions under which the ferrous iron released is stable in the presence of oxygen. *A. ferrooxidans* and *L. ferrooxidans* then catalyze the oxidation of ferrous iron to ferric iron. The Fe^{3+} formed under these acidic conditions, being soluble, reacts spontaneously with more pyrite and oxidizes it to ferrous ions plus sulfuric acid:

$$FeS_2 + 14\,Fe^{3+} + 8\,H_2O \longrightarrow 15\,Fe^{2+} + 2\,SO_4^{2-} + 16\,H^+$$

The ferrous ions formed are again oxidized to ferric ions by the bacteria, and these ferric ions again react with more pyrite. Thus, there is a progressive, rapidly increasing rate at which pyrite is oxidized, called the *propagation cycle* (Figure 24.10*b*). Under natural conditions some of the ferrous iron generated by the bacteria leaches away, being carried by groundwater into surrounding streams. However, because oxygen is present in the aerated drainage, bacterial oxidation of the ferrous iron takes place in these outflows and an insoluble ferric precipitate is formed.

Acid Mine Drainage

Bacterial oxidation of sulfide minerals is the major factor in **acid mine drainage**, an environmental problem in coal-mining regions (**Figure 24.11**). Mixing of acidic mine waters into natural waters in rivers and lakes seriously degrades water quality because both the acid and the dissolved metals are toxic to aquatic organisms. In addition, such polluted waters are unsuitable for human consumption and industrial use. The breakdown of pyrite ultimately leads to the formation of sulfuric acid and ferrous iron, and pH values can be lower than 1. The acid formed attacks other minerals associated with the coal and pyrite, causing breakdown of the rock fabric. A major rock-forming element, aluminum, is soluble only at low pH, and often high levels of Al^{3+}, which

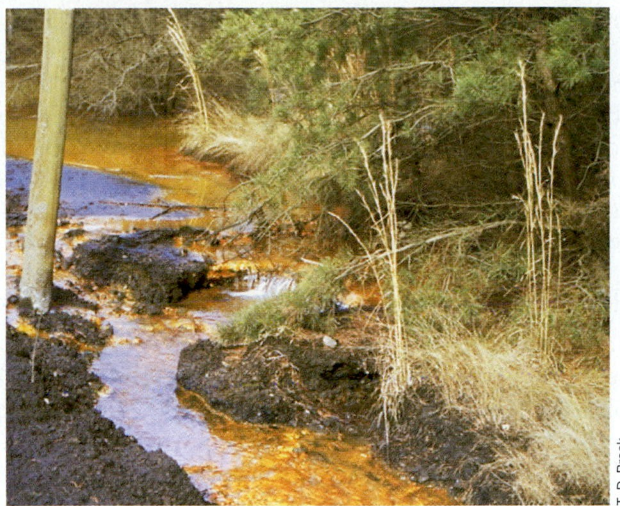

Figure 24.11 Acid mine drainage from a bituminous coal region. Note the yellowish-red color ("yellow boy") due to precipitated iron oxides (Figure 20.28).

can be highly toxic to aquatic organisms, are present in acid mine waters.

The oxygen requirement for the oxidation of ferrous to ferric iron explains how acid mine drainage develops. As long as the coal is unmined, pyrite cannot be oxidized because air, water, and the bacteria cannot reach it. However, when the coal seam is exposed (Figure 24.9*b*), it quickly becomes contaminated with *A. ferrooxidans,* and O_2 and water are introduced, making oxidation of pyrite possible. The acid formed can then leach into the surrounding streams (Figure 24.11).

Where acid mine drainage is extensive, a strongly acidophilic species of *Archaea, Ferroplasma,* is typically present. This aerobic iron-oxidizing prokaryote is capable of growth at pH 0 and at temperatures to 50°C. *Ferroplasma acidarmanus* forms slimy cell masses attached to pyrite surfaces in iron ore deposits. At Iron Mountain, California (**Figure 24.12**), a particularly well-studied acid mine drainage site, the thick biofilm of *F. acidarmanus* maintains the pH near 0. With iron concentrations of nearly 30 g/l at this site, the *F. acidarmanus* mat is constantly bathed in fresh substrate (Fe^{2+}) from which more acid is generated by the reactions described previously. *Ferroplasma* is a cell wall–less prokaryote and is phylogenetically related to *Thermoplasma* (Figure 24.12*b* and Section 17.5).

24.5 MiniReview

Iron exists naturally in two oxidation states, ferrous (Fe^{2+}) and ferric (Fe^{3+}). Bacteria reduce ferric iron in anoxic environments and oxidize ferrous iron aerobically at acidic pH. Ferrous iron oxidation is common in coal-mining regions where it causes a type of pollution called acid mine drainage.

- What oxidation state is iron in the mineral $Fe(OH)_3$? FeS? How is $Fe(OH)_3$ formed?
- Why does biological Fe^{2+} oxidation under oxic conditions occur mainly at acidic pH?

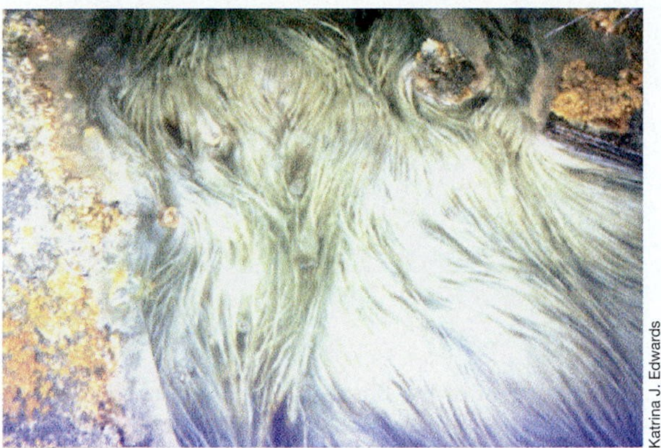

(a)

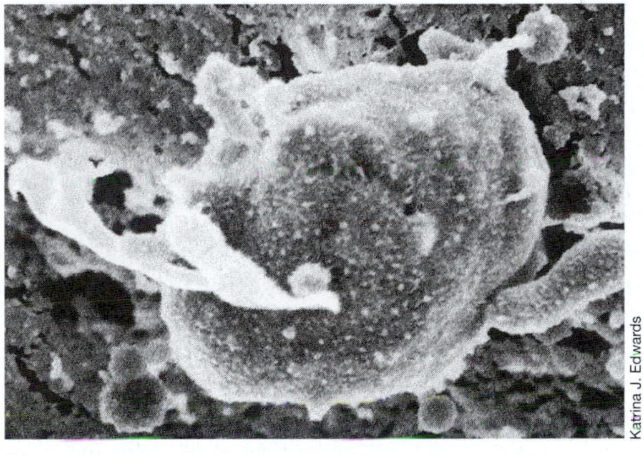

(b)

Figure 24.12 *Ferroplasma acidarmanus.* An extremely acidophilic iron-oxidizing archaeon responsible for severe acid mine drainage. *(a)* Streamers of *F. acidarmanus* cells growing in an acid (pH near 0) mine drainage stream, Iron Mountain, CA. *(b)* Scanning electron micrograph of a cell of *F. acidarmanus* among mineral matter.

UNIT 4

III MICROBIAL BIOREMEDIATION

The biogeochemical capacities of microorganisms seem almost limitless, and it has often been said that microorganisms are Earth's greatest chemists. As such, microorganisms have been used to extract valuable metals from low-grade ores (microbial leaching) and as agents for cleaning up the environment. The term **bioremediation** refers to the cleanup of oil, toxic chemicals, or other pollutants from the environment by microorganisms. Bioremediation is a cost-effective way of cleaning up pollutants, and in some cases, is the only practical way to get the job done.

24.6 Microbial Leaching of Ores

The acid production and mineral dissolution by acidophilic bacteria discussed in the previous section can be put to use in the mining of metal ores. Sulfide forms insoluble minerals with many metals, and many ores mined as sources of these metals are sulfides. If the concentration of metal in the ore is

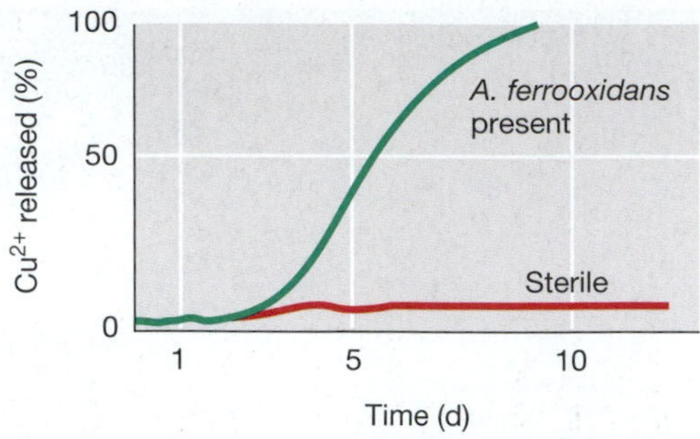

Figure 24.13 Effect of the bacterium *Acidithiobacillus ferrooxidans* on the leaching of copper from the mineral covellite (CuS). The leaching was done in a laboratory column and the acidic leach solution contained inorganic nutrients necessary for growth of the bacterium. The leaching activity was monitored by assaying for soluble copper in the leach solution at the bottom of the column. The leach solution was continuously recirculated, maintaining an essentially closed system.

low, it may not be economically feasible to concentrate the mineral by conventional means. Under these conditions, **microbial leaching** is practiced. Leaching is especially useful for copper ores because copper sulfate, formed during the oxidation of copper sulfide ores, is very water-soluble. Indeed, approximately one-fourth of all of the copper mined worldwide is obtained by leaching.

Acidithiobacillus ferrooxidans and other metal-oxidizing chemolithotrophic bacteria can catalyze the oxidation of sulfide minerals, thus aiding in solubilization of the metal. The relative oxidation rates of a copper mineral in the presence and absence of bacteria are illustrated in Figure 24.13. The susceptibility to oxidation varies among minerals, and those minerals that are most readily oxidized are most amenable to microbial leaching. Thus, iron and copper sulfide ores such as pyrrhotite (FeS) and covellite (CuS) are readily leached, whereas lead and molybdenum ores are much less so.

The Leaching Process

In microbial leaching, low-grade ore is dumped in a large pile (the leach dump) and a dilute sulfuric acid solution (pH 2) is percolated down through the pile (Figure 24.14). The liquid emerging from the bottom of the pile (Figure 24.14*b*) is rich in dissolved metals and is transported to a precipitation plant (Figure 24.14*c*) where the desired metal is precipitated and purified (Figure 24.14*d*). The liquid is then pumped back to the top of the pile and the cycle repeated. As needed, acid is added to maintain the low pH.

We illustrate copper ore leaching with the example of covellite (CuS), in which copper has a valence of +2. As shown in Figure 24.15, *A. ferrooxidans* oxidizes the sulfide in CuS to SO_4^{2-}, thus releasing copper as Cu^{2+}. However, this reaction can also occur spontaneously. Indeed, the key reaction in copper leaching is the chemical oxidation of CuS with ferric ions generated by the bacterial oxidation of ferrous ions (Figure 24.15). In any copper ore, pyrite (FeS_2) is also present,

(a) (b)

(c)

(d)

Figure 24.14 The leaching of low-grade copper ores using iron-oxidizing bacteria. *(a)* A typical leaching dump. The low-grade ore has been crushed and dumped in a large pile in such a way that the surface area exposed is as high as possible. Pipes distribute the acidic leach water over the surface of the pile. The acidic water slowly percolates through the pile and exits at the bottom. *(b)* Effluent from a copper leaching dump. The acidic water is very rich in Cu^{2+}. *(c)* Recovery of copper as metallic copper (Cu^0) by passage of the Cu^{2+}-rich water over metallic iron in a long flume. *(d)* A small pile of metallic copper removed from the flume, ready for further purification.

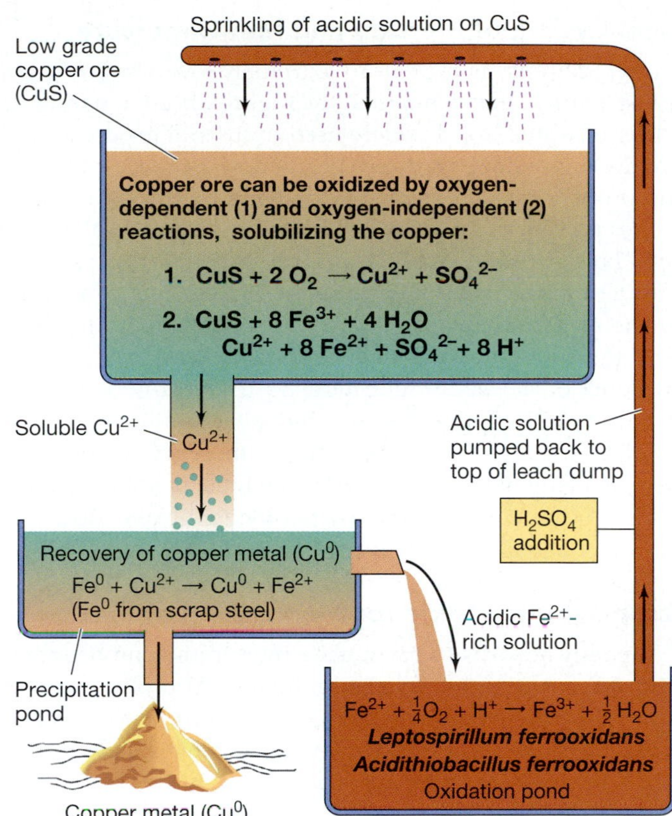

Figure 24.15 Arrangement of a leaching pile and reactions in the microbial leaching of copper sulfide minerals to yield metallic copper. Reaction 1 occurs both biologically and chemically. Reaction 2 is strictly chemical and is the most important reaction in copper-leaching processes. For Reaction 2 to proceed, it is essential that the Fe^{2+} produced from the oxidation of sulfide in CuS to sulfate be oxidized back to Fe^{3+} by iron chemolithotrophs.

and its oxidation by bacteria leads to the formation of Fe^{3+} (Figure 24.15). The spontaneous reaction of CuS with Fe^{3+} forms Cu^{2+} plus Fe^{2+}. Importantly, this reaction can take place deep in the leach dump where conditions are anoxic.

Metal Recovery

The precipitation plant is where the Cu^{2+} from the leaching solution is recovered (Figure 24.14*c, d*). Shredded scrap iron, Fe^0, is added to the precipitation pond to recover copper from the leach liquid by the reaction shown in the lower part of Figure 24.15, and this once again results in the formation of Fe^{2+}. This Fe^{2+}-rich liquid is transferred to a shallow oxidation pond where *A. ferrooxidans* grows and oxidizes the Fe^{2+} to Fe^{3+}. This now ferric-rich liquid is pumped to the top of the pile and the Fe^{3+} used to oxidize more CuS (Figure 24.15). The entire leaching operation is thus maintained by the oxidation of Fe^{2+} to Fe^{3+} by the iron-oxidizing bacteria.

High temperature can be a problem with leaching operations. *A. ferrooxidans* is a mesophile, but temperatures inside a leach dump often rise spontaneously as a result of heat generated by microbial activities. Thus, thermophilic iron-oxidizing chemolithotrophs such as thermophilic *Thiobacillus*

Figure 24.16 Gold bioleaching. Gold leaching tanks in Ghana (Africa). Within the tanks, a mixture of *Acidithiobacillus ferrooxidans, Acidithiobacillus thiooxidans,* and *Leptospirillum ferrooxidans* solubilizes the pyrite/arsenic mineral containing trapped gold and releases the gold.

species, *Leptospirillum ferrooxidans,* and *Sulfobacillus* or, at higher temperatures (60–80°C), the organism *Sulfolobus* (∞ Section 17.10), are also important in the leaching of ores.

Other Microbial Leaching Processes: Uranium and Gold

Microorganisms are also used in the leaching of uranium and gold ores. *A. ferrooxidans* can oxidize U^{4+} to U^{6+} with O_2 as an electron acceptor. However, uranium leaching probably depends more on the abiotic oxidation of U^{4+} by Fe^{3+}, with *A. ferrooxidans* contributing mainly through the reoxidation of Fe^{2+} to Fe^{3+} as in copper leaching (Figure 24.15). The reaction observed is

$$\underset{(U^{4+})}{UO_2} + \underset{(Fe^{3+})}{Fe_2(SO_4)_3} \longrightarrow \underset{(U^{6+})}{UO_2SO_4} + 2\ \underset{(Fe^{2+})}{FeSO_4}$$

Unlike UO_2, the oxidized uranium mineral is soluble and is recovered by other processes.

Gold is typically found in deposits associated with minerals containing arsenic and pyrite. *A. ferrooxidans* and its relatives can attack and solubilize the arsenopyrite minerals, releasing the trapped gold (Au):

$$2\ FeAsS[Au] + 7\ O_2 + 2\ H_2O + H_2SO_4 \longrightarrow Fe_2(SO_4)_3 + 2\ H_3AsO_4 + [Au]$$

The gold is then complexed with cyanide by traditional gold-mining methods. Unlike copper, where leaching occurs in a huge leach dump (Figure 24.14*a*), gold leaching takes place in relatively small and enclosed bioreactor tanks (**Figure 24.16**), where more than 95% of the trapped gold is released. Moreover, although arsenic and cyanide are toxic residues from the mining process, both are removed in the gold-leaching bioreactor. Arsenic is removed as a ferric precipitate, and cyanide (CN^-) is removed by its microbial oxidation to CO_2 plus urea in later stages of the gold recovery process. Small-scale microbial gold leaching has thus

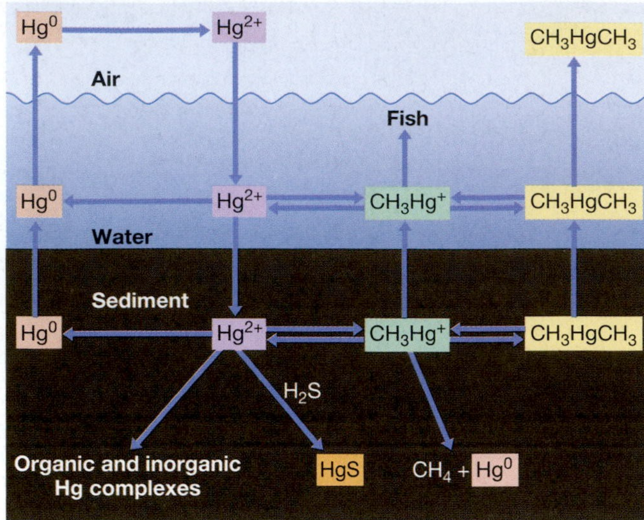

Figure 24.17 Biogeochemical cycling of mercury. The major reservoirs of mercury are in water and in sediments where it can be concentrated in animal tissues or precipitated out as HgS, respectively. The forms of mercury commonly found in aquatic environments are each shown in a different color.

become popular as an alternative to the more costly and environmentally devastating conventional gold-mining techniques. Pilot processes are also being developed for the bioleaching of zinc, lead, and nickel using bioreactor methods.

24.6 MiniReview

Bacterial solubilization of copper is a process called microbial leaching. Leaching is important in the recovery of copper, uranium, and gold from low-grade ores. Bacterial oxidation of ferrous iron is the key reaction in the microbial leaching process because the ferric iron itself can oxidize metals in the ores.

- How can CuS be oxidized under anoxic conditions?
- Why is it important to keep the sprinkling solution in the copper ore leaching process acidic?
- From the standpoint of metal oxidation, how does the copper leaching process differ from the gold leaching process?

24.7 Mercury and Heavy Metal Transformations

Metals are typically present in low concentrations in rocks, soils, waters, and the atmosphere. Some of these are trace metals needed by cells (for example, cobalt, copper, zinc, nickel, molybdenum). In high concentrations, however, these same metals can be toxic. Of these, several are sufficiently volatile to contaminate air as well. These include mercury, lead, arsenic, cadmium, and selenium. Because of environmental concern and significant microbial involvement, we focus our discussion on the biogeochemistry of the element mercury.

Global Cycling of Mercury and Methylmercury

Although mercury is present in extremely low concentrations in most natural environments, averaging about 1 nanogram per liter of water, it is a widely used industrial product and is an active ingredient of many pesticides. Because of its unusual propensity to concentrate in living tissues and its high toxicity, mercury is of considerable environmental importance. For example, the mining of mercury ores and the burning of fossil fuels release about 40,000 *tons* of mercury into the environment each year; even more mercury is released in abiotic geochemical processes. Mercury is also a by-product of the electronics industry (especially battery and wiring production), the chemical industry, and the burning of municipal wastes. The major form of mercury in the atmosphere is elemental mercury (Hg^0), which is volatile and is oxidized to mercuric ion (Hg^{2+}) photochemically. Most mercury enters aquatic environments as Hg^{2+} (**Figure 24.17**).

Microbial Redox Cycle for Mercury

Hg^{2+} readily adsorbs to particulate matter and can be metabolized from there by microorganisms. Microbial activity results in the methylation of mercury, yielding *methylmercury*, CH_3Hg^+ (Figure 24.17). Methylmercury is extremely toxic to animals because it can be absorbed through the skin and is a potent neurotoxin. But in addition, methylmercury is soluble and can be concentrated in the food chain, primarily in fish, or can be further methylated by microorganisms to yield the volatile compound dimethylmercury, $CH_3—Hg—CH_3$.

Both methylmercury and dimethylmercury accumulate in animals, especially in muscle tissues. Methylmercury is about 100 times more toxic than Hg^0 or Hg^{2+}, and its accumulation seems to be a particular problem in freshwater lakes and marine coastal waters where enhanced levels have been observed in recent years in fish caught for human consumption. Mercury can also cause liver and kidney damage in humans and other animals.

Several other mercury transformations occur on a global scale, including reactions carried out by sulfate-reducing bacteria ($H_2S + Hg^{2+} \rightarrow HgS$) and methanogens ($CH_3Hg^+ \rightarrow CH_4 + Hg^0$) (Figure 24.17). The solubility of HgS is very low, so in anoxic sulfate-reducing sediments, most mercury is present as HgS. But upon aeration, HgS is oxidized, primarily by thiobacilli, leading to the formation of Hg^{2+} and SO_4^{2-}, and eventually, methylmercury.

Mercuric Resistance

At sufficiently high concentrations, Hg^{2+} and CH_3Hg^+ can be toxic not only to higher organisms but also to microorganisms. However, several bacteria can carry out the biotransformation of toxic mercury to nontoxic forms. In mercury-resistant gram-negative bacteria the NADPH-linked enzyme mercuric reductase reduces Hg^{2+} to Hg^0. The Hg^0 produced in this reaction is volatile but is nontoxic to humans and microorganisms. Bacterial conversion of Hg^{2+} to Hg^0 then allows for more CH_3Hg^+ to be converted to Hg^{2+}.

In the gram-negative bacterium *Pseudomonas aeruginosa*, genes for mercury resistance reside on a plasmid. These genes, called *mer* genes, are arranged in an operon and are under control of the regulatory protein MerR (**Figure 24.18**). MerR functions as both a repressor and an activator of transcription. In the absence of Hg^{2+}, MerR functions as a repressor and binds to the operator region of the *mer* operon, thus preventing transcription of *merTPCAD*. However, when Hg^{2+} is present, it forms a complex with MerR, which then activates transcription of the *mer* operon.

The protein MerP is a periplasmic Hg^{2+}-binding protein. MerP binds Hg^{2+} and transfers it to a membrane protein MerT, which transports Hg^{2+} into the cell for reduction by mercuric reductase (Figure 24.18*b*). The final result is reduction of Hg^{2+} to Hg^0 (which, as discussed, is volatile and nontoxic) and release of the Hg^0 from the cell (Figure 24.18*b*).

Microbial Resistance to Other Heavy Metals

Plasmids isolated from diverse gram-positive and gram-negative *Bacteria* encode resistance to heavy metals. Certain antibiotic resistance plasmids also have genes for resistance to mercury and arsenic. The mechanism of resistance to any specific metal varies. For example, arsenate and cadmium resistances are due to the activity of enzymes that pump out any arsenate or cadmium that enters the cell, thus preventing the metals from accumulating and denaturing proteins. Other resistances involve redox changes in the metal. Bacteria naturally resistant to high levels of several different metals often contain multiple plasmids encoding the resistances. Such multiple-resistant organisms are common in wastewater effluents of the metal-processing industry or in mining operations, where heavy metals are typically leached out along with iron or copper (Section 24.6).

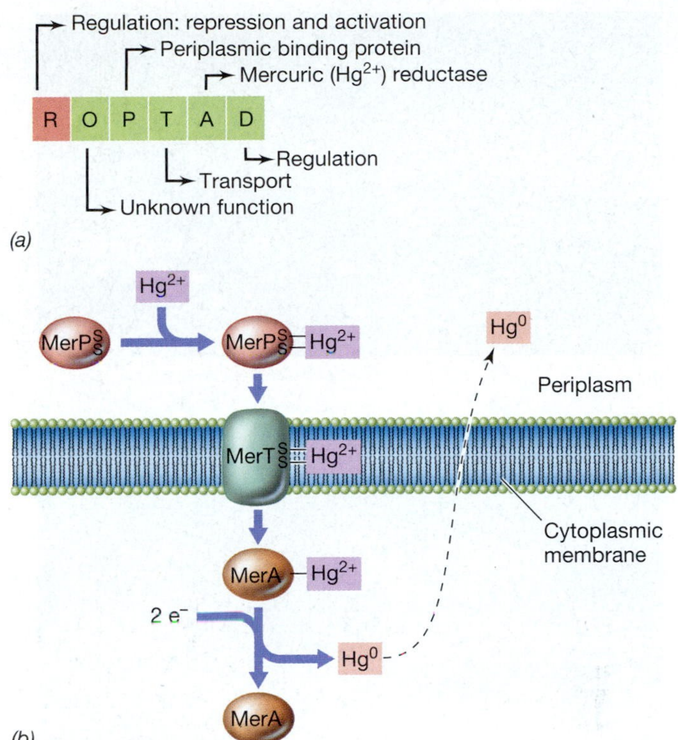

Figure 24.18 Mechanism of Hg^{2+} reduction to Hg^0 in *Pseudomonas aeruginosa*. *(a)* The *mer* operon. MerR can function as either a repressor (absence of Hg^{2+}) or transcriptional activator (presence of Hg^{2+}). *(b)* Transport and reduction of Hg^{2+}. Hg^{2+} is bound by cysteine residues in both proteins MerP and MerT.

24.7 MiniReview

A major toxic form of mercury is methylmercury, which can yield Hg^{2+}, which is reduced by bacteria to Hg^0. The ability of bacteria to resist the toxicity of heavy metals is often due to the presence of specific plasmids that encode enzymes capable of detoxifying or pumping out the metals.

- What forms of mercury are most toxic to organisms?
- How is mercury detoxified by bacteria?

24.8 Petroleum Biodegradation

Petroleum is a rich source of organic matter, and because of this, microorganisms readily attack hydrocarbons when petroleum is brought into contact with air and moisture. Under some circumstances, such as in bulk storage tanks, microbial growth is undesirable. However, in oil spills, bioremediation is desirable and can be promoted by the addition of inorganic nutrients. Indeed, the importance of prokaryotes to petroleum bioremediation has been demonstrated in several major crude oil spills in recent years (**Figure 24.19**).

The biochemistry of petroleum catabolism was covered in Sections 21.13 and 21.15. Both anoxic and oxic biodegradation is possible. Under oxic conditions we emphasized the important role that oxygenase enzymes play in introducing oxygen atoms into the hydrocarbon. Our discussion here will thus focus on aerobic processes, because it is only when oxygen is present that oxygenase enzymes are active and hydrocarbon oxidation a rapid process.

Hydrocarbon Decomposition

Diverse bacteria, fungi, and a few cyanobacteria and green algae can oxidize petroleum products aerobically. Small-scale oil pollution of aquatic and terrestrial ecosystems from human as well as natural activities is common. Oil-oxidizing microorganisms develop rapidly on oil films and slicks.

Oil-oxidizing activity is most extensive if temperature and inorganic nutrient (primarily nitrogen and phosphorous) concentrations are optimal. Because oil is insoluble in water and is less dense, it floats to the surface and forms slicks. There, hydrocarbon-degrading bacteria attach to the oil droplets (**Figure 24.20**) and eventually decompose the oil and disperse the slick. Certain oil-degrading specialist species, such as *Alcanivorax borkumensis*, a bacterium that grows only on hydrocarbons, fatty acids, or pyruvate, produce glycolipid surfactants that help break up the oil and solubilize it. Once solubilized, the oil can be taken up more readily and catabolized as an energy source.

(a)

(b)

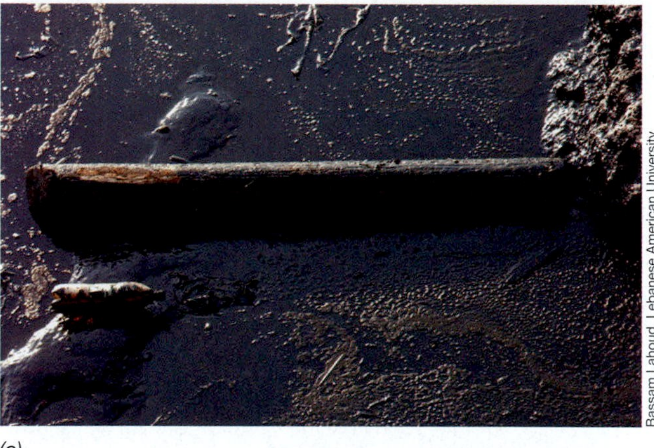

(c)

Figure 24.19 Environmental consequences of large oil spills and the effect of bioremediation. *(a)* A contaminated beach along the coast of Alaska containing oil from the *Exxon Valdez* spill of 1989. *(b)* The center rectangular plot (arrow) was treated with inorganic nutrients to stimulate bioremediation of spilled oil by microorganisms, whereas areas to the left and right were untreated. *(c)* Oil spilled into the Mediterranean Sea from the Jiyyeh (Lebanon) power plant that flowed to the port of Byblos during the 2006 war in Lebanon.

In large oil spills, volatile hydrocarbon fractions evaporate quickly, leaving medium- to longer-chain aliphatic and aromatic components for cleanup crews and microorganisms to tackle. Microorganisms consume oil by oxidizing it to CO_2. When bioremediation activities have been promoted by inorganic nutrient application, oil-oxidizing bacteria can grow quickly after an oil spill (Figure 24.20*b*). Under ideal bioremediation conditions, up to 80% of the nonvolatile components in oil can be oxidized within a year of a spill. However, certain oil fractions, such as those containing branched-chain and polycyclic hydrocarbons, remain in the environment much longer. Spilled oil that travels to marine sediments is more slowly degraded, and it can have a significant long-term impact on fisheries and related activities that depend on unpolluted waters for productive yields.

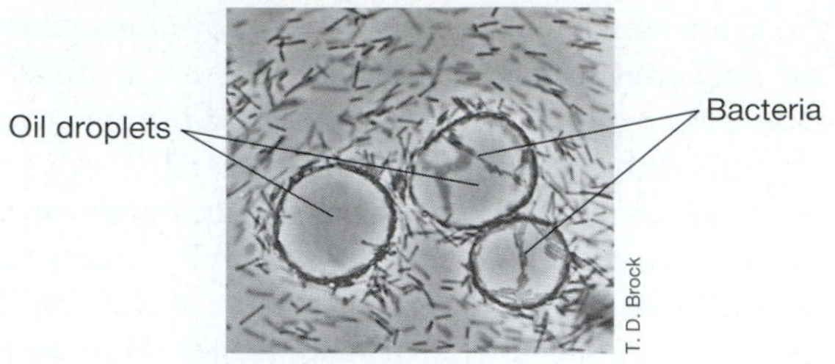

Figure 24.20 Hydrocarbon-oxidizing bacteria in association with oil droplets. The bacteria are concentrated in large numbers at the oil–water interface but are actually not within the droplet itself.

Interfaces where oil and water meet often form on a large scale. For example, it is virtually impossible to keep moisture from accumulating inside bulk-fuel storage tanks because water forms in a layer beneath the petroleum. Gasoline and crude oil storage tanks (**Figure 24.21**) are thus potential habitats for hydrocarbon-oxidizing microorganisms. If sufficient sulfate is present in the water, sulfate-reducing bacteria can grow in the tanks, consuming hydrocarbons under anoxic conditions (Section 21.13). The sulfide (H_2S) produced is highly corrosive and causes pitting and subsequent leakage of the tanks along with souring of the fuel.

Petroleum Production

In addition to microbial *degradation* of petroleum, some microorganisms can also *produce* petroleum, particularly certain green algae. For example, as the colonial alga *Botryococcus braunii* grows, it excretes long-chain (C_{30}–C_{36})

Figure 24.21 Oil biodegradation. Bulk fuel-storage tanks often support massive microbial growth at oil–water interfaces.

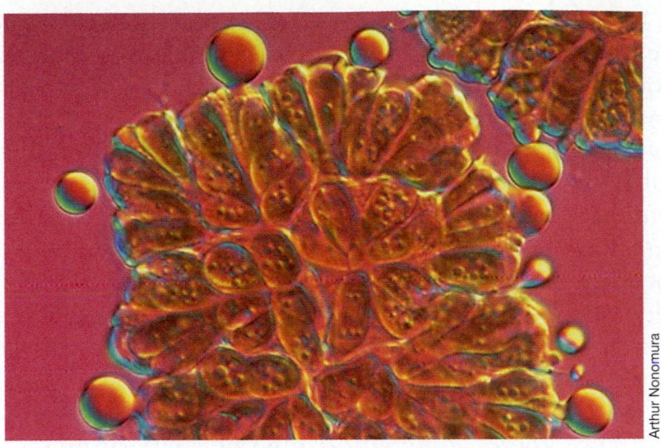

Figure 24.22 Phototrophic oil production. Photomicrograph by Nomarski interference contrast of cells of the green alga *Botryococcus braunii*. Note excreted oil droplets at the margin of the cells.

hydrocarbons that have the consistency of crude oil (**Figure 24.22**). In *B. braunii* about 30% of the cell dry weight is petroleum, and there has been some interest in using this and other oil-producing algae as renewable sources of petroleum. There is even evidence that oil in certain types of oil shale formations originated from green algae like *B. braunii* that grew in lakebeds in ancient times.

24.8 MiniReview

Hydrocarbons are subject to microbial oxidation. Hydrocarbon-oxidizing bacteria bioremediate spilled oil, and their activities are assisted by addition of inorganic nutrients. Some algae can produce hydrocarbons.

- What is bioremediation?
- Why would the addition of inorganic nutrients stimulate oil degradation whereas the addition of glucose would not?

24.9 Biodegradation of Xenobiotics

A **xenobiotic** compound is a synthetic chemical that is not naturally occurring. Xenobiotics include pesticides, polychlorinated biphenyls (PCBs), munitions, dyes, and chlorinated solvents, among other things. Some xenobiotics are structurally related to natural compounds and can sometimes be degraded slowly by enzymes that degrade the structurally related natural compounds. However, many xenobiotics differ chemically in such a major way from anything organisms have naturally experienced that they degrade extremely slowly, if at all. We focus here on pesticides as an example of the potential of microbial xenobiotics degradation.

Pesticide Catabolism

Some of the most widely distributed xenobiotics are pesticides, which are common components of toxic wastes. Over

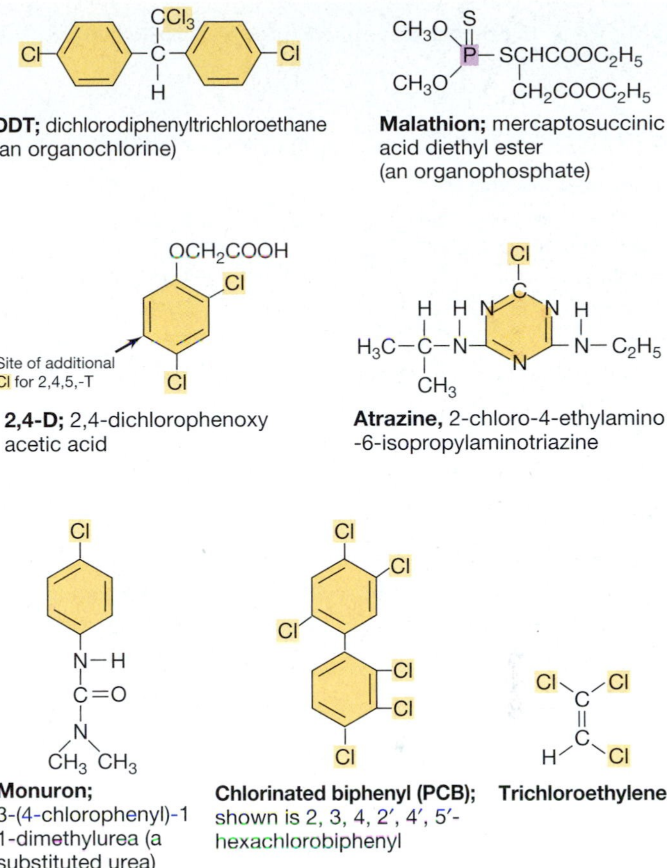

Figure 24.23 Examples of xenobiotic compounds. Although none of these compounds exist naturally, microorganisms exist that can break them down.

1000 pesticides have been marketed for chemical pest control purposes. Pesticides include *herbicides, insecticides,* and *fungicides*. Pesticides display a wide variety of chemistries, including chlorinated, aromatic, and nitrogen- and phosphorus-containing compounds (**Figure 24.23**). Some of these substances can be used as carbon sources and electron donors by microorganisms, whereas others cannot.

If a xenobiotic substance can be biodegraded, it will eventually disappear from a habitat. Such degradation in the soil is desirable because toxic accumulations of the compound are avoided. However, even closely related compounds may differ remarkably in their degradability, as shown in **Table 24.4**. Some compounds persist relatively unaltered for years in soils whereas others are significantly degraded in only weeks or months. For example, some of the chlorinated insecticides (Figure 24.23) are so recalcitrant that they can persist for several years in soil (Table 24.4).

The relative persistence rates of xenobiotics are only approximate because environmental factors—temperature, pH, aeration, and organic matter content of the soil—also influence decomposition. Moreover, disappearance of a pesticide from an ecosystem does not necessarily mean that it was degraded by microorganisms; pesticides can also be removed by volatilization, leaching, or spontaneous chemical breakdown.

Table 24.4 Persistence of herbicides and insecticides in soils

Substance	*Time for 75–100% disappearance*
Chlorinated insecticides	
DDT [1,1,1-trichloro-2,2-bis-(ρ-chlorophenyl)ethanne][a]	4 years
Aldrin	3 years
Chlordane	5 years
Heptachlor	2 years
Lindane (hexachlorocyclohexane)	3 years
Organophosphate insecticides	
Diazinon	12 weeks
Malathion[a]	1 week
Parathion	1 week
Herbicides	
2,4-D (2,4-dichlorophenoxyacetic acid)[a]	4 weeks
2,4,5-T (2,4,5-trichlorophenoxyacetic acid)	20 weeks
Dalapon	8 weeks
Atrazine[a]	40 weeks
Simazine	48 weeks
Propazine	1.5 years

[a]Structure shown in Figure 24.23.

Substrate availability also governs microbial attack of xenobiotic compounds. Many xenobiotics are quite hydrophobic and poorly soluble in water. Adsorption of these compounds to organic matter and clay in soils and sediments prevents access by organisms. Thus the addition of surfactants or emulsifiers often increases bioavailability, and ultimately biodegradation, of the xenobiotic compound.

Many bacteria and fungi metabolize pesticides and herbicides. Some pesticides can be both carbon and energy sources and are oxidized completely to CO_2; however, other compounds are attacked only slightly, if at all. Some compounds may be degraded either partially or totally provided that some other organic material is present as primary energy source. This is a phenomenon called **cometabolism**. However, if the breakdown is only partial, the microbial degradation product may be even more toxic than the original compound. Thus, from an environmental standpoint, cometabolism of xenobiotics is not always a good thing.

Reductive Dechlorination

We previously described a form of anaerobic respiration called **reductive dechlorination**. In this process, chlorinated organic compounds are used as terminal electron acceptors under anoxic conditions (∞ Section 21.12). A laboratory model for studying reductive dechlorination is the reduction of chlorobenzoate to benzoate by the bacterium *Desulfomonile*:

$$C_7H_4O_2Cl^- + 2\,H \longrightarrow C_7H_5O_2^- + HCl$$

However, from a bioremediation standpoint, other chlorinated compounds are ecologically much more important than chlorobenzoate. For example, reductive dechlorination of the compounds dichloro-, trichloro-, and tetrachloro- (perchloro-) ethylene, chloroform, dichloromethane, and polychlorinated bi-phenyls (Figure 24.23) is observed, as well as the dehalogenation of several brominated and fluorinated compounds. These toxic compounds, some of which have been linked to cancer (particularly trichloroethylene, Figure 24.23), are widely used as industrial solvents, degreasing agents, and insulators in electrical transformers. They enter anoxic environments through accidental spills or from slow leakage of storage containers or abandoned electrical transformers. Eventually these compounds migrate into groundwater, where they are the most frequently detected groundwater contaminants in the United States. There is currently great interest in stimulating the bioremediation activities of reductive dechlorinators as a strategy for the removal of these highly toxic compounds from anoxic environments.

Aerobic Dechlorination

Aerobic dechlorination of chlorinated xenobiotics is also possible. For example, the pseudomonad *Burkholderia* will grow on the pesticide 2,4,5-T (Figure 24.23), dechlorinating the molecule and releasing Cl^- (**Figure 24.24**). The aerobic biodegradation of chlorinated aromatic compounds occurs by way of oxygenase enzymes (∞ Section 21.14). In the aerobic catabolism of 2,4,5-T, following dechlorination, a dioxygenase enzyme breaks down the aromatic ring. This generates compounds that can be metabolized by the citric acid cycle (Figure 24.24*b*).

Although the aerobic breakdown of chlorinated xenobiotics is undoubtedly of ecological importance, reductive dechlorination is of particular importance because of the rapidity with which anoxic conditions develop in polluted microbial habitats and the biochemical constraints (oxygen requirement, Figure 24.24*b*) this puts on aerobic organisms that could otherwise degrade the compound.

Biodegradable Plastics

A major area of environmental concern besides the biodegradation of toxic wastes such as pesticides and other chlorinated hydrocarbons is the biodegradation of solid wastes—in particular, plastics. The plastics industry currently produces over 40 *billion* kilograms of plastic per year, almost half of which is discarded in landfills. Plastics are xenobiotic polymers of various types; polyethylene, polypropylene, and polystyrene are typical examples (**Figure 24.25**). Many of these synthetic polymers remain essentially unaltered for decades in landfills and refuse dumps. This problem has fueled the search for biodegradable alternatives—called **biopolymers**—as replacements for some of the synthetic polymers now in use.

Several biopolymers with plastic-like properties are known. Photobiodegradable plastics are polymers whose structure is altered by exposure to ultraviolet radiation (from

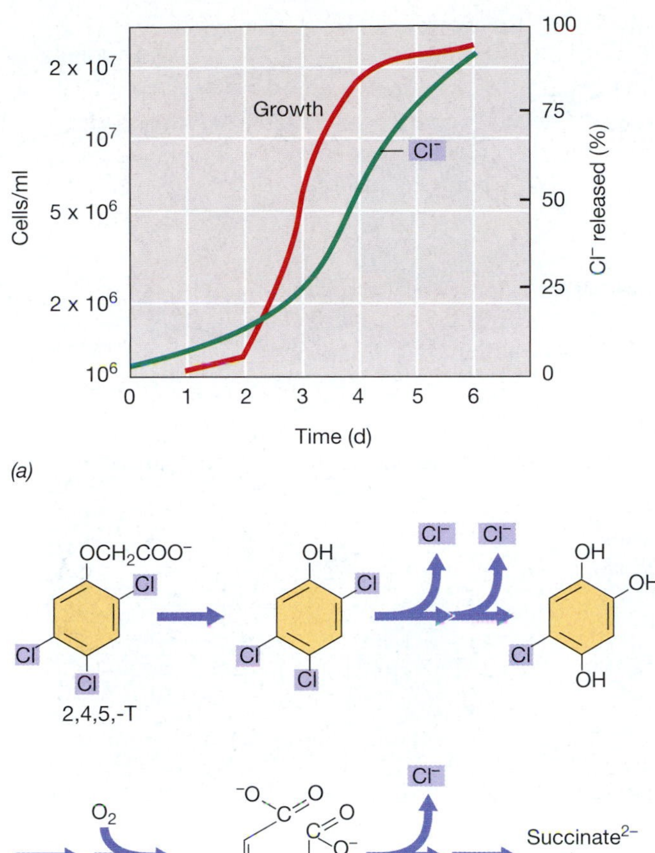

Figure 24.24 Biodegradation of the herbicide 2,4,5-T. *(a)* Growth of *Burkholderia cepacia* on 2,4,5-T as sole source of carbon and energy. The strain was enriched from nature using a chemostat to keep the concentration of herbicide low. Growth here is aerobic on 1.5 g/l of 2,4,5,-T. The release of chloride from the molecule is indicative of biodegradation. *(b)* Pathway of aerobic 2,4,5-T biodegradation.

sunlight), generating modified polymers more amenable to microbial attack. Starch-based plastics incorporate starch (a glucose polymer) as a link to short fragments of a second biodegradable polymer. This design accelerates biodegradation, because starch-digesting bacteria in soil attack the starch, releasing polymer fragments that are then degraded by other microorganisms. The glucose released from starch biodegradation can also be used as a carbon and energy source, thus supporting cometabolism of the polymer.

Microbial plastics are based on the carbon storage polymer poly-β-hydroxybutyrate and related compounds (∞ PHAs, Section 4.10). These storage polymers have many of the desirable properties of synthetic plastics and can be biosynthesized in various chemical forms. A *copolymer* containing approximately equal amounts of poly-β-hydroxybutryate and poly-β-hydroxyvalerate (**Figure 24.26a**) has had the greatest market success thus far. However, because they are more cost effective, petroleum- based plastics make up virtually the entire plastics market today.

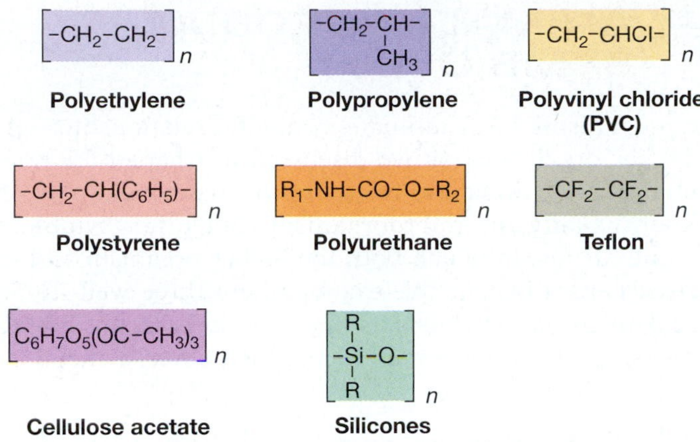

Figure 24.25 Chemistry of synthetic polymers. Shown is the monomeric structure of common synthetic polymers.

24.9 MiniReview

Many chemically synthesized compounds such as insecticides, herbicides, and plastics are called xenobiotics but can often be degraded by a bacterium. Both aerobic and anaerobic mechanisms are known.

- Which chemical class of pesticides are the most recalcitrant to microbial attack?
- What is reductive dechlorination and how does it differ from the reactions shown in Figure 24.24?
- What advantages do biopolymers have over synthetic polymers for the plastics industry?

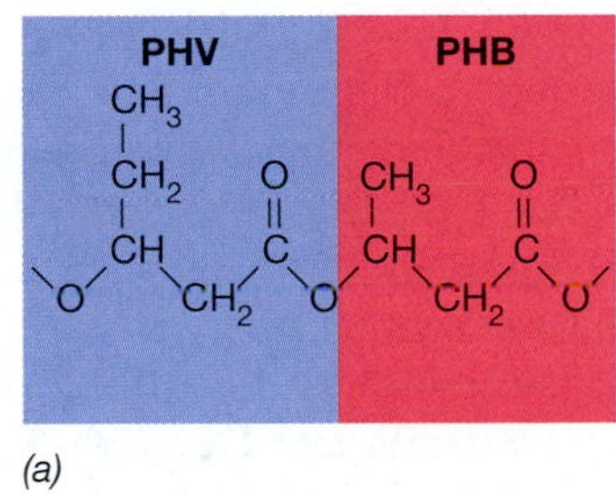

Figure 24.26 Bacterial plastics. *(a)* Copolymer of poly-β-hydroxybutyrate (PHB) and poly-β-hydroxyvalerate (PHV). *(b)* A brand of shampoo previously marketed in Germany and packaged in a bottle made of the PHB/PHV copolymer. Because this material is a natural product, the bottle readily degrades both aerobically and anaerobically.

IV ANIMAL–MICROBIAL SYMBIOSES

Microorganisms form intimate symbiotic relationships with all higher organisms. As we discussed in Chapter 23, some symbioses are commensal relationships that benefit only one partner, usually the microorganism. But other symbioses are mutualisms, in which both the higher organism and the microorganism benefit. We examine here three well-studied animal–microbial mutualisms, focusing in each on how the partners in the symbioses interact to each other's benefit.

24.10 The Rumen and Ruminant Animals

Ruminants are herbivorous mammals that possess a special digestive organ, the **rumen**, within which cellulose and other plant polysaccharides are digested with the help of microorganisms. Some of the most important domestic animals—cows, sheep, and goats—are ruminants. Microbial fermentation of sugars released from these polysaccharides produces fatty acids that feed the ruminants. Because the human food economy depends to a great extent on these animals, rumen microbiology is of considerable economic significance and importance.

Rumen Anatomy and Activity

The bulk of the organic matter in terrestrial plants is present in insoluble polysaccharides, of which cellulose is the most important. Mammals—and indeed almost all animals—lack the enzymes necessary to digest cellulose. But ruminants can metabolize cellulose by using microorganisms as digestive agents. Unique features of the rumen as a site of cellulose digestion are its relatively large size (100–150 liters in a cow, 6 liters in a sheep) and its position in the alimentary tract before the acidic stomach. The warm and constant temperature (39°C), constant pH (5.5–7, depending on when the animal last fed), and anoxic environment of the rumen are also important factors in overall rumen function.

Figure 24.27a shows the relationship of the rumen to other parts of the ruminant digestive system. The digestive processes and microbiology of the rumen have been well studied, in part because it is possible to create a sampling port, called a *fistula,* into the rumen of a cow (Figure 24.27*b*) or a sheep and remove samples for analysis. Food first enters the reticulum and is quickly pumped into the rumen. There it is mixed with saliva containing bicarbonate (HCO_3^-) and churned in a rotary motion. This peristaltic action grinds the cellulose into a fine suspension that assists in microbial attachment. The food mass is then transferred back into the reticulum, where it is formed into small clumps called *cuds;* these are regurgitated into the mouth and chewed again. The now finely divided solids, well mixed with saliva, are swallowed again, but this time the material passes mainly to the omasum and from there to the abomasum, an organ more like a true (acidic) stomach. In the abomasum, chemical digestive processes begin that continue in the small and large intestine.

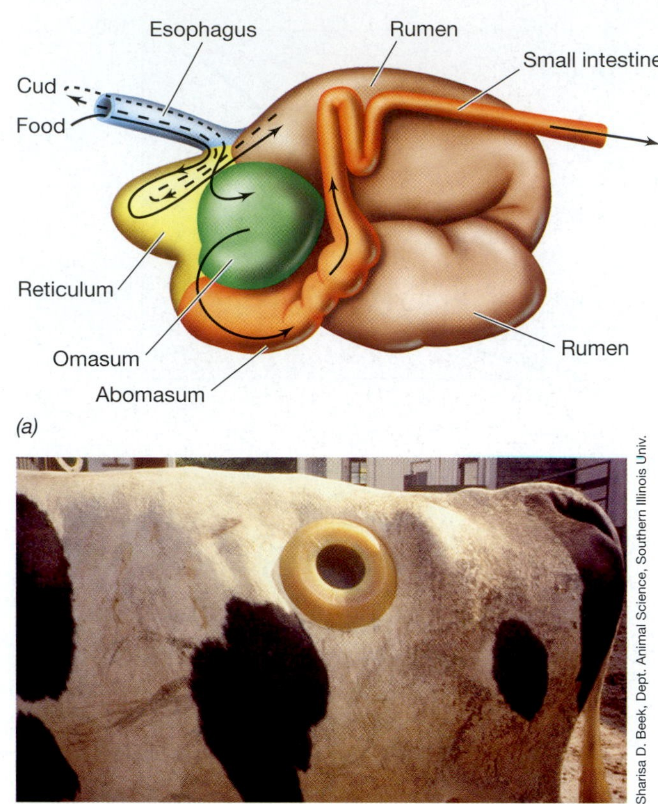

Figure 24.27 The rumen. *(a)* Schematic diagram of the rumen and gastrointestinal system of a cow. Food travels from the esophagus to the rumen and is then regurgitated and travels to the reticulum, omasum, abomasum, and intestines, in that order. The abomasum is an acidic vessel, analogous to the stomach of monogastric animals like pigs and humans. *(b)* Photo of a fistulated Holstein cow. The fistula, shown unplugged, is a sampling port that allows access to the rumen.

Microbial Fermentation in the Rumen

Food remains in the rumen about 9–12 hours. During this period cellulolytic microorganisms hydrolyze cellulose to free glucose. The glucose then undergoes bacterial fermentation with the production of **volatile fatty acids** (VFAs), primarily *acetic, propionic,* and *butyric,* and the gases CO_2 and CH_4 (**Figure 24.28**). These fatty acids pass through the rumen wall into the bloodstream and are oxidized by the animal as its main source of energy.

The rumen contains enormous numbers of prokaryotes (10^{10}–10^{11}/g of rumen contents). In addition to their digestive functions, rumen microorganisms synthesize amino acids and vitamins that are the main source of these essential nutrients for the animal. Most of the bacteria are adhered tightly to plant materials and feed particles. These materials proceed through the gastrointestinal tract of the animal where they undergo further digestive processes similar to those of non-ruminant animals. Microbial cells from the rumen are digested in the acidic abomasum and thus are a major source of protein and vitamins for the animal. Because this microbial protein is recovered and used by the animal, a ruminant is nutritionally superior to a nonruminant when subsisting on foods, such as grasses, that are deficient in protein.

Rumen Bacteria

Because the rumen is anoxic, anaerobic bacteria naturally dominate. Furthermore, because cellulose is converted to CO_2 and CH_4 in a multistep microbial food chain, various anaerobes participate in the process (**Table 24.5**). Several different rumen bacteria hydrolyze cellulose to sugars and ferment the sugars to volatile fatty acids. *Fibrobacter succinogenes* and *Ruminococcus albus* are the two most abundant cellulolytic rumen anaerobes. Although both organisms produce cellulases, *Fibrobacter,* a gram-negative bacterium, contains a periplasmic cellulase (Section 4.9). Because of this, cells of *Fibrobacter* must remain attached to the cellulose fibril while digesting it. *Ruminococcus,* by contrast, excretes cellulase into the rumen (thus cellulase is an *exoenzyme*), where it degrades cellulose outside the bacterial cell.

If a ruminant is gradually switched from cellulose to a diet high in starch (grain, for instance), the starch-digesting bacteria *Ruminobacter amylophilus* and *Succinomonas amylolytica* grow to high numbers in the rumen. On a low-starch diet these organisms are typically minor constituents. If an animal is fed legume hay, which is high in pectin, then the pectin-digesting bacterium *Lachnospira multiparus* (Table 24.5) becomes an abundant member of the rumen microbial community.

Some of the fermentation products of the saccharolytic rumen microflora are used as energy sources by secondary fermenters in the rumen. Thus, succinate is fermented to propionate plus CO_2 (Figure 24.28) by the bacterium *Schwartzia,* and lactate is fermented to acetic and other acids by *Selenomonas* and *Megasphaera* (Table 24.5). Hydrogen (H_2) produced in the rumen by fermentative processes never accumulates because it is quickly consumed by methanogens. Despite the high VFA content of the rumen, syntrophs (Section 24.2) do not play a role there because the animal itself is the sink for fatty acids (Figure 24.28). That is, with propionate and butyrate being consumed by the animal, syntrophic metabolisms are unnecessary in the rumen.

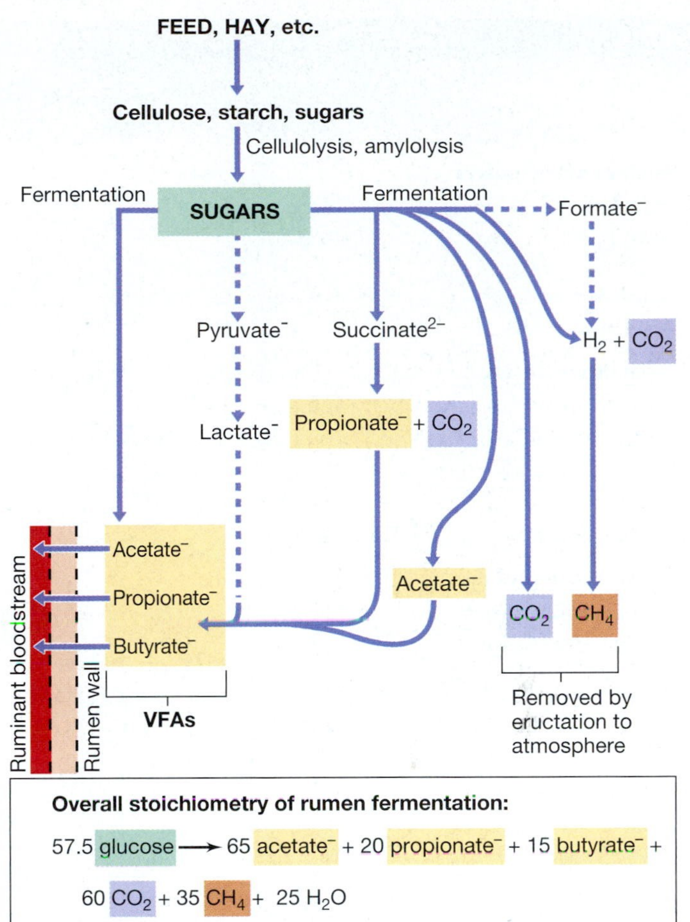

Figure 24.28 Biochemical reactions in the rumen. The major starting substrate, glucose, and end products are highlighted; dashed lines indicate minor pathways. Approximate steady-state rumen levels of volatile fatty acids (VFAs) are acetate, 60 mM; propionate, 20 mM; butyrate, 10 mM. VFAs are consumed by the ruminant and converted into animal proteins.

Dangerous Changes in the Rumen Microflora

Changes in the microbial composition of the rumen can cause illness or even death of the animal. For example, if a cow is changed abruptly from forage to a grain diet, the gram-positive bacterium *Streptococcus bovis* grows rapidly in the rumen. The normal level of *S. bovis,* about 10^7 cells/g, is insignificant in terms of total rumen bacterial numbers. But if large amounts of grain are fed abruptly, numbers of *S. bovis* can quickly rise to dominate the rumen microflora at over 10^{10} cells/g. This occurs because *S. bovis* grows rapidly on starch but is not cellulolytic; grain contains high levels of starch, whereas grasses contain mainly cellulose.

Because *S. bovis* is a lactic acid bacterium (Sections 16.1 and 21.2), when its populations are large its starch-fermenting metabolism elevates levels of lactic acid in the rumen. Lactic acid is a much stronger acid than the VFAs produced during normal rumen function. Lactate production thus acidifies the rumen below its lower functional limit of about pH 5.5, and this disrupts rumen activities. Rumen acidification, a condition called *acidosis,* causes inflammation of the rumen epithelium, and severe acidosis can cause hemorrhaging (bleeding) in the rumen followed by infections of other organs from rumen microflora that escape from the rumen.

Despite problems with *S. bovis,* ruminants such as cattle can be fed a diet exclusively of grain. However, to avoid acidosis, animals are switched from forage to grain gradually over a period of a few days. The slow introduction of starch selects for VFA-producing, starch-degrading bacteria (Table 24.5) instead of *S. bovis,* and thus normal rumen functions continue and the animal remains healthy. Animals suffering from acidosis can be treated by removing the source of starch and forcing strong buffering agents, such as calcium carbonate, into the rumen to help return rumen pH back to normal levels.

UNIT 4

Table 24.5 Characteristics of some rumen prokaryotes

Organism	*Gram stain*	*Phylogenetic domain*[a]	*Morphology*	*Motility*	*Fermentation products*
Cellulose decomposers					
Fibrobacter succinogenes[b]	Negative	B	Rod	–	Succinate, acetate, formate
Butyrivibrio fibrisolvens[c]	Negative	B	Curved rod	+	Acetate, formate, lactate, butyrate, H_2, CO_2
Ruminococcus albus[b]	Positive	B	Coccus	–	Acetate, formate, H_2, CO_2
Clostridium lochheadii	Positive	B	Rod (endospores)	+	Acetate, formate, butyrate, H_2, CO_2
Starch decomposers					
Prevotella ruminicola	Negative	B	Rod	–	Formate, acetate, succinate
Ruminobacter amylophilus	Negative	B	Rod	–	Formate, acetate, succinate
Selenomonas ruminantium	Negative	B	Curved rod	+	Acetate, propionate, lactate
Succinomonas amylolytica	Negative	B	Oval	+	Acetate, propionate, succinate
Streptococcus bovis	Positive	B	Coccus	–	Lactate
Lactate decomposers					
Selenomonas lactilytica	Negative	B	Curved rod	+	Acetate, succinate
Megasphaera elsdenii	Positive	B	Coccus	–	Acetate, propionate, butyrate, valerate, caproate, H_2, CO_2
Succinate decomposer					
Schwartzia succinovorans	Negative	B	Rod	+	Propionate, CO_2
Pectin decomposer					
Lachnospira multiparus	Positive	B	Curved rod	+	Acetate, formate, lactate, H_2, CO_2
Methanogens					
Methanobrevibacter ruminantium	Positive	A	Rod	–	CH_4 (from H_2 + CO_2 or formate)
Methanomicrobium mobile	Negative	A	Rod	+	CH_4 (from H_2 + CO_2 or formate)

[a]B, *Bacteria*; A, *Archaea*
[b]These species also degrade xylan, a major plant cell wall polysaccharide (Section 21.17).
[c]Also degrades starch

Rumen Protists and Fungi

The rumen has, in addition to prokaryotes, a characteristic fauna (about 10^6/ml), composed almost exclusively of ciliated protists (Section 18.9). Many of these protists are obligate anaerobes, a property that is rare among eukaryotes. Although these protists are not essential for the rumen fermentation, they contribute to the overall process. In fact, some protists are able to hydrolyze cellulose and starch and ferment glucose with the production of the same VFAs formed by the bacteria (Figure 24.28 and Table 24.5). Rumen protists also ingest rumen bacteria as food sources and are thought to play a role in controlling bacterial densities in the rumen.

Anaerobic fungi also inhabit the rumen and play a role in ruminal digestive processes. Rumen fungi are typically species that alternate between a flagellated and a thallus form, and studies with pure cultures of them show that they can ferment cellulose to VFAs. *Neocallimastix,* for example, is an obligately anaerobic fungus that ferments glucose to formate, acetate, lactate, ethanol, CO_2, and H_2. Although a eukaryote, this fungus lacks mitochondria and cytochromes and thus lives an obligately fermentative existence. However, *Neocallimastix* cells contain a redox organelle called the *hydrogenosome*; this mitochondrial analogue evolves H_2 and has thus far been found only in certain protists (Section 18.2). Rumen fungi play an important role in the degradation of polysaccharides other than cellulose as well, including a partial degradation of lignin (the strengthening agent in the cell walls of woody plants), hemicellulose (a derivative of cellulose that contains pentoses and other sugars), and pectins (a polysaccharide containing a mixture of C_6 and C_5 uronic acids).

Other Herbivorous Animals: Cecal Animals

The familiar ruminants are cows and sheep. Goats, camels, buffalo, deer, reindeer, caribou, and elk are also ruminants. However, although horses and rabbits are herbivorous mammals, they are not ruminants. Instead, these animals have only one stomach but use an organ called the *cecum,* a digestive organ located posterior to the small intestine and anterior to the large intestine, as their cellulolytic fermentation vessel. The cecum contains cellulolytic microorganisms, and cellulose is digested there. Nutritionally, ruminants are superior to cecal animals in that the cellulolytic microflora of the ruminant eventually passes through a true (acidic) stomach. As it does, it is killed and becomes a protein source for the animal. By contrast, in horses and rabbits the cellulolytic microflora is passed out of the animal in the feces because the cecum is located in the digestive tract posterior to the acidic stomach.

24.10 MiniReview

Ruminants have a digestive organ called the rumen that specializes in cellulose digestion. Bacteria, protists, and fungi in the rumen produce volatile fatty acids that are used by the ruminant. In addition to their role in the digestive process, rumen microorganisms synthesize vitamins and amino acids and are also a major source of protein for the ruminant.

- What physical and chemical conditions prevail in the rumen?
- What are VFAs and of what value are they to the ruminant?
- Why is the metabolism of *Streptococcus bovis* of special concern to ruminant nutrition?
- How do cecal animals differ from ruminants in their digestive tract anatomy?

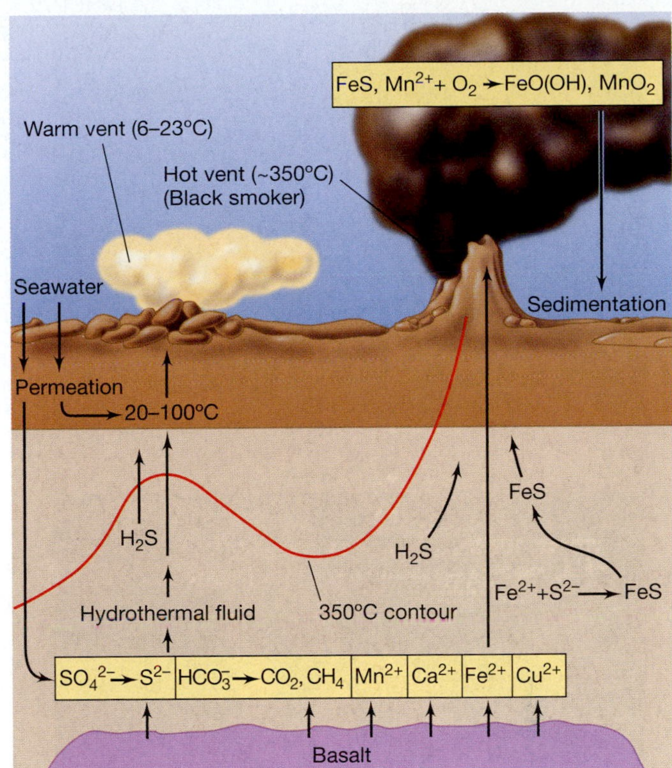

Figure 24.29 Hydrothermal vents. Schematic diagram showing geological formations and major chemical species at warm vents and black smokers. In warm vents, the hot hydrothermal fluid is cooled by cold 2–3°C seawater permeating the sediments. In black smokers, hot hydrothermal fluid near 350°C reaches the seafloor directly.

24.11 Hydrothermal Vent Microbial Ecosystems

Although we have previously painted a picture of the deep sea as a remote, low-temperature, high-pressure environment suitable only for slow-growing barotolerant and barophilic bacteria (∞ Section 23.10), there are some amazing exceptions. Thriving animal communities have been found clustered about thermal springs in deep-sea waters throughout the world. These springs are located about 2500 m from the ocean surface in regions of the seafloor where volcanic magma has caused the floor to drift apart. Seawater seeping into these cracked regions mixes with hot minerals and is emitted from the springs. These underwater hot springs are called **hydrothermal vents** (**Figure 24.29**).

Two major types of hydrothermal vents have been discovered. Warm vents emit hydrothermal fluid at temperatures of 6–23°C. Hot vents, called **black smokers** because the mineral-rich hot water forms a dark cloud of precipitated material upon mixing with cold seawater, emit hydrothermal fluid at 270–380°C. As we will now see, an amazing diversity of prokaryotes lives in and about these undersea hot springs.

Animals Living at Hydrothermal Vents

Diverse invertebrate communities develop near hydrothermal vents, including tube worms over 2 m in length and large clams and mussels (**Figure 24.30**). Photosynthesis cannot support these animal communities because they exist below the photic zone. However, hydrothermal fluids contain large amounts of reduced inorganic materials, including H_2S, Mn^{2+}, H_2, and CO; some vents contain high levels of NH_4^+ instead of H_2S. All of these are good electron donors for chemolithotrophic prokaryotes (∞ Chapter 20), and hydrothermal vent animals can exist in permanent darkness because they are nourished through a symbiotic association with these autotrophic bacteria. The bacteria actually live within the animal tissues and supply organic carbon to the animals in exchange for a safe residence and ready access to the electron donors needed for their energy metabolism.

Prokaryotes in Hydrothermal Vents

Large numbers of sulfur-oxidizing chemolithotrophs such as *Thiobacillus, Thiomicrospira, Thiothrix,* and *Beggiatoa* (∞ Section 15.4) live near sulfide-emitting hydrothermal vents. Some vents support nitrifying, hydrogen-oxidizing, iron- and manganese-oxidizing bacteria, or methylotrophic bacteria, the latter presumably growing on the methane and carbon monoxide (CO) emitted from the vents. **Table 24.6** summarizes the inorganic electron donors and electron acceptors that are thought to play a role in hydrothermal vent chemolithotrophic metabolisms.

Using the powerful tools of environmental genomics (∞ Sections 13.13 and 22.6), studies of prokaryotic diversity near hydrothermal vents have revealed an enormous number of different *Bacteria*. These communities are dominated by species of *Proteobacteria,* in particular, *Epsilonproteobacteria* (∞ Section 15.19). *Alpha-*, *Delta-*, and *Gammaproteobacteria*

UNIT 4

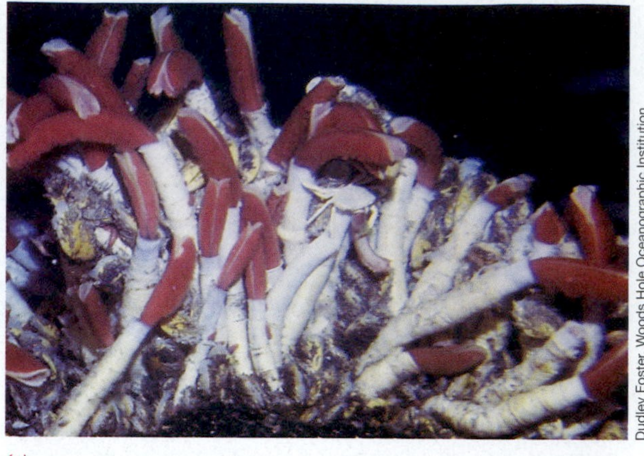

Dudley Foster, Woods Hole Oceanographic Institution

(a)

James Agvair, Woods Hole Oceanographic Institution

(b)

Carl Wirsen, Woods Hole Oceanographic Institution

(c)

Figure 24.30 Invertebrates living near deep-sea thermal vents. *(a)* Tube worms (family *Pogonophora*), showing the sheath (white) and plume (red) of the worm bodies. *(b)* Close-up photograph showing worm plume. *(c)* Mussel bed in vicinity of a warm vent. Note yellow deposition of elemental sulfur. See Table 24.6 for a list of chemolithotrophic prokaryotes found near hydrothermal vents.

are also abundant, whereas *Betaproteobacteria* are all but absent. Many *Epsilon*- and *Gammaproteobacteria* oxidize sulfide and sulfur as electron donors, with either oxygen or nitrate as elctron acceptors. By contrast, most *Deltaproteobacteria* specialize in anaerobic metabolisms using sulfur compounds as electron acceptors. Chapter 15 covers each of these groups in more detail.

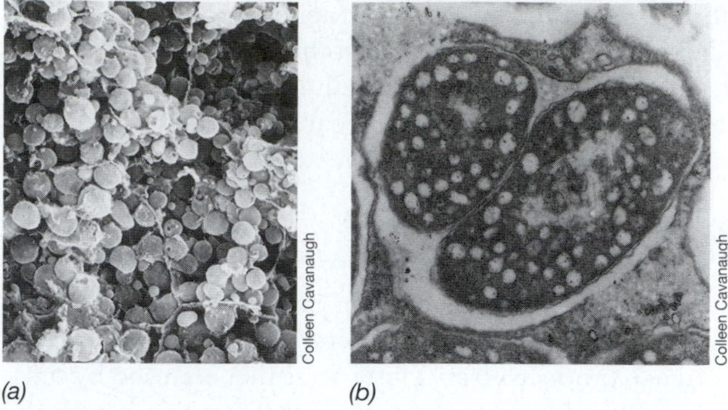

Colleen Cavanaugh

(a) (b)

Figure 24.31 Chemolithotrophic sulfur-oxidizing bacteria associated with the trophosome tissue of tube worms from hydrothermal vents. *(a)* Scanning electron microscopy of trophosome tissue showing spherical chemolithotrophic sulfur-oxidizing bacteria. Cells are 3–5 μm in diameter. *(b)* Transmission electron micrograph of bacteria in sectioned trophosome tissue. The cells are frequently enclosed in pairs by an outer membrane of unknown origin. Reprinted with permission from *Science* 213:340–342 (1981), © AAAS.

In contrast to *Bacteria,* the diversity of *Archaea* near hydrothermal vents is quite limited. Estimates of the number of unique phylotypes (a phylotype is a 16S rRNA sequence that differs from all other sequences by at least 3%; Sections 14.12 and 22.5) indicate that the diversity of *Bacteria* near hydrothermal vents is about ten times that of *Archaea*. Most of the *Archaea* detected near hydrothermal vents are either methanogens (Section 17.4) or representatives of the marine *Crenarchaeota* and *Euryarchaeota* groups (Section 17.12)

Nutrition of Animals Living at Hydrothermal Vents

As mentioned, some chemolithotrophs, rather than living free in the vicinity of vents, have evolved associations with hydrothermal vent animals. For example, the 2-m-long tube worms (Figure 24.30) lack a mouth, gut, or anus, but contain an organ consisting primarily of spongy tissue called the *trophosome*. This structure, which constitutes half the worm's weight, is filled with sulfur granules and large populations of spherical sulfur-oxidizing prokaryotes (**Figure 24.31**). Bacterial cells taken from trophosome tissue show activity of enzymes of the Calvin cycle, a major pathway for autotrophy (Section 20.6), but interestingly, contain enzymes of the reverse citric acid cycle, a second autotrophic pathway (Section 20.7), as well, and also show a suite of sulfur-oxidizing enzymes necessary to obtain energy from reduced sulfur compounds (Section 20.10). The tube worms are thus nourished by organic compounds produced from CO_2 and excreted by the sulfur chemolithotrophs.

Along with tube worms, giant mussels (Figure 24.30*c*) are also common inhabitants of warm vents, and sulfur-oxidizing bacterial symbionts have been found in the gill tissues of these animals. Phylogenetic analyses (Chapter 14) have shown that each vent animal harbors a single species of bacterial symbiont and that different bacterial symbionts inhabit different

Table 24.6 Chemolithotrophic prokaryotes present in the vicinity of deep-sea hydrothermal vents[a]

Chemolithotroph	Electron donor	Electron acceptor	Product from donor
Sulfur-oxidizing	HS^-, S^0, $S_2O_3^{2-}$	O_2, NO_3^-	S^0, SO_4^-
Nitrifying	NH_4^+, NO_2	O_2	NO_2^-, NO_3^-
Sulfate-reducing	H_2	S^0, SO_4^{2-}	H_2S
Methanogenic	H_2	CO_2	CH_4
Hydrogen-oxidizing	H_2	O_2, NO_3^-	H_2O
Iron and manganese-oxidizing	Fe^{2+}, Mn^{2+}	O_2	Fe^{3+}, Mn^{4+}
Methylotrophic[b]	CH_4, CO	O_2	CO_2

[a] Sections 15.3–15.5 and 20.8–20.13 for a discussion of chemolithotrophic metabolism.
[b] Sections 15.6 and 21.16 for a discussion of methylotrophy.

species of vent animal. Although fairly closely related to free-living sulfur chemolithotrophs, no animal symbionts have yet been obtained in laboratory culture.

The red plume of the tube worm (Figure 24.30*b*) is rich in blood vessels and is used to trap and transport inorganic substrates to the bacterial symbionts. The tube worms contain unusual hemoglobins that bind H_2S as well as O_2; these carry both substrates to the trophosome where they are released to the bacterial symbiont. The CO_2 content of tube worm blood is also high, about 25 mM, and presumably this is released in the trophosome as a carbon source for the symbionts. In addition, stable isotope analyses (Section 22.8) of elemental sulfur from the trophosome have shown that its $^{34}S/^{32}S$ composition is the same as that of the sulfide emitted from the vent. This ratio is distinct from that of seawater sulfate and is further proof that geothermal sulfide is actually entering the worm in large amounts.

Other marine animals have established bacterial symbioses for their nutrition, as well. For example, methanotrophic symbionts are present in giant clams that live near natural gas seeps at relatively shallow depths in the Gulf of Mexico (Figure 15.15). Although not autotrophs (CH_4 is an organic compound), the clam methanotrophs do provide nutrition to the animals; the symbionts use CH_4 as their electron donor and carbon source and excrete organic carbon to the clams.

Superheated Water: Black Smokers

Because of the huge hydrostatic pressure, water does not boil at the depths of hydrothermal vents until it reaches a temperature greater than 450°C. At some vent sites superheated hydrothermal fluid is emitted at temperatures as high as 350°C. These fluids contain metal sulfides, especially iron sulfides, and cool quickly as they mix with cold seawater. The precipitated metal sulfides form a tower called a "chimney" above the source (**Figure 24.32**), from which prokaryotes can be isolated.

Although it is clear that prokaryotes do not live in the superheated hydrothermal fluid itself, thermophilic and hyperthermophilic organisms live in the *gradients* that form as the hot water cools by mixing with cold seawater. For example, the walls of smoker chimneys are teeming with hyperthermophiles such as *Methanopyrus*, a species of *Archaea* that oxidizes H_2 and makes methane up to 110°C (Sections 17.6). Phylogenetic FISH staining (Section 14.9) has detected cells of both *Bacteria* and *Archaea* in smoker chimney walls (**Figure 24.33**). The most thermophilic of all known prokaryotes, species of *Pyrolobus* and *Pyrodictium* (Sections 17.11 and 17.13–17.15), were isolated from black smoker chimney walls.

When smokers plug up from mineral debris, hyperthermophiles presumably drift away to colonize active smokers and somehow become integrated into the growing chimney wall. Surprisingly, although requiring very high temperatures for growth, hyperthermophiles are remarkably tolerant of cold temperatures and oxygen. Thus, transport of cells from one vent site to another in cold oxic seawater apparently is not a problem.

UNIT 4

Figure 24.32 A hydrothermal vent black smoker emitting sulfide- and mineral-rich water at temperatures of 350°C. The walls of the black smoker chimneys display a steep temperature gradient and contain several types of prokaryotes.

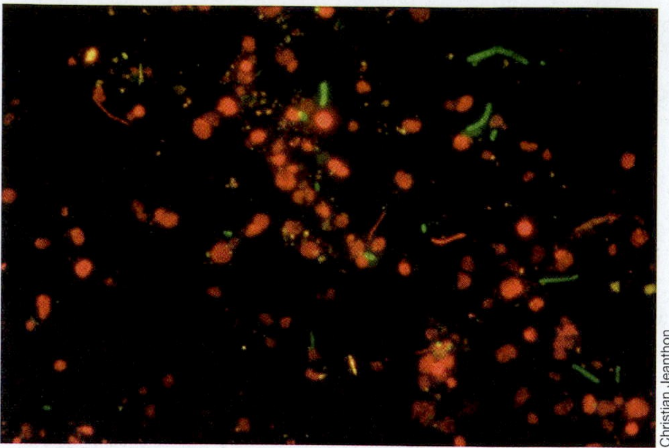

Figure 24.33 Phylogenetic FISH staining of black smoker chimney material. Snake Pit vent field, Mid-Atlantic Ridge (3500 m deep). A green fluorescing dye was conjugated to a probe that reacts with the 16S rRNA of all *Bacteria* and a red dye to a 16S rRNA probe for *Archaea*. The hydrothermal fluid going through the center of this chimney was 300°C.

24.11 MiniReview

Hydrothermal vents are deep-sea hot springs where volcanic activity generates fluids containing large amounts of inorganic electron donors that can be used by chemolithotrophic bacteria. Some of these organisms establish intimate mutualistic relationships with unusual marine invertebrates.

- How does a warm hydrothermal vent differ from a black smoker, both chemically and physically?
- How do giant tube worms receive their nutrition?
- Why is 350°C water emitting from a black smoker not boiling water?

24.12 Squid–*Aliivibrio* Symbiosis

A fascinating mutualistic symbiosis has evolved between the marine gram-negative bacterium *Aliivibrio fischeri* and the Hawaiian bobtail squid, *Euprymna scolopes,* a small marine invertebrate (an adult is about 4 cm long, **Figure 24.34*a***). The bobtail squid sequesters large populations of the bioluminescent *A. fischeri* (Section 15.12) in a light organ located on its ventral side. The bacteria emit light that resembles moonlight penetrating marine waters, and this is thought to camouflage the squid from predators that strike from beneath, thus conferring a survival advantage on the squid. Several other species of *Euprymna* inhabit marine waters near Japan, Australia, and in the Mediterranean, and these contain *Aliivibrio* symbionts as well.

The Squid–*Aliivibrio* System as a Model Symbiosis

Many features of the *E. scolopes–Aliivibrio* symbiosis have made it an important model for how animal–bacterial symbioses are established. These features include in particular the fact that the animals can be grown in the laboratory and that there is only a single bacterial species in the symbiosis rather than the huge number inhabiting the mammalian large intestine (Section 28.4) or the rumen (Section 24.10). In addition, the symbiosis is not an essential one; both the squid and its bacterial partner can be cultured apart from each other in the laboratory. This allows juvenile squid to be grown without bacterial symbionts and then experimentally colonized. Using colonization assays, experiments can be done to study specificity in the symbiosis, the number of bacterial cells needed to initiate an infection, the capacity of genetically defined mutants of *A. fischeri* to initiate infection of the squid, and many other aspects of the relationship. Moreover, because the genome of *A. fischeri* has been sequenced, this also allows the powerful techniques of microbial genomics to be employed in studies of the symbiosis.

(a)

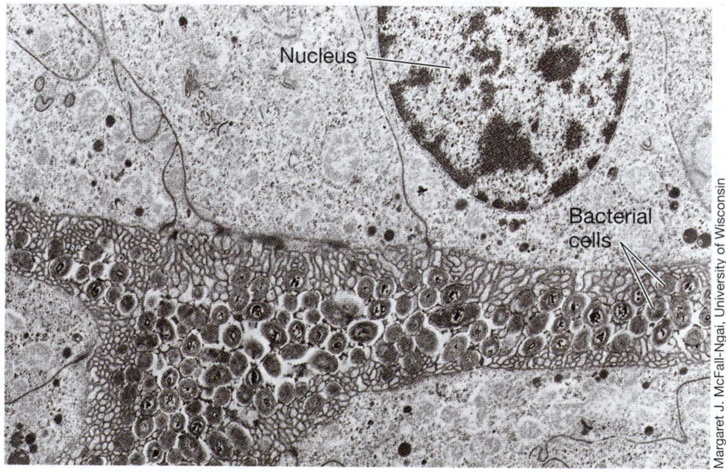

(b)

Figure 24.34 Squid–*Aliivibrio* symbiosis. *(a)* The Hawaiian bobtail squid, *Euprymna scolopes*. An animal is about 4 cm long. *(b)* Thin-sectioned transmission electron micrograph through the *E. scolopes* light organ showing the dense population of bioluminescent *Aliivibrio fischeri* cells.

Establishing the Squid–*Aliivibrio* Symbiosis

Juvenile squid hatched from eggs do not contain cells of *A. fischeri*. Thus, transmission of bacterial cells to juvenile squid is a horizontal (environmental) rather than a vertical (transovarian) event. Cells of *A. fischeri* from surrounding seawater begin to colonize tissues in the juveniles almost immediately after they emerge from the eggs by entering an animal through ciliated ducts that end in the immature light organ. Amazingly, the light organ becomes colonized specifically with *A. fischeri* and not with any of the many other species of gram-negative bacteria present in seawater. Even if large numbers of other species of bioluminescent bacteria are offered to juvenile squid along with low numbers of *A. fischeri*, only *A. fischeri* establishes residence in the light organ. This implies that the animal in some way recognizes *A. fischeri* cells to the exclusion of other species.

The squid–*Aliivibrio* symbiosis develops in several stages. Contact of the squid with any bacterial cells triggers recognition in a very general way. Upon contact with peptidoglycan (a component of the cell wall of *Bacteria*, Section 4.6), the young squid secretes mucus from its developing light organ. The mucus is the first layer of specificity in the symbiosis, as it makes gram-negative but not gram-positive bacteria aggregate. Within the aggregates of gram-negative cells that may contain only low numbers of *A. fischeri*, this bacterium somehow outcompetes the other gram-negative bacteria to eventually form a monoculture. All of these events occur within 2 h of a juvenile's hatching from an egg. The highly motile *A. fischeri* cells present in the aggregate then migrate into ducts that lead to the light organ tissues. Once there, they lose their flagella and become nonmotile and divide to form dense populations (Figure 24.34*b*). The light organ in a mature animal contains between 10^8 and 10^9 *A. fischeri* cells. Following colonization, the bacterial cells trigger a series of developmental events that lead to maturation of the host light organ.

Specific colonization of *A. fischeri* in the squid is also assisted by nitric oxide (NO). Nitric oxide is a well-known defense response of animal cells to attack by bacterial pathogens; the gas is a strong oxidant and causes oxidative damage to bacterial cells sufficient to kill them (Section 29.2). Nitric oxide produced by the squid is incorporated into the mucus aggregates and is present in the light organ itself. As *A. fischeri* colonizes the light organ, NO levels diminish rapidly. It appears that cells of *A. fischeri* can tolerate exposure to NO and consume it through the activity of NO-inactivating enzymes. The inability of other gram-negative bacteria in the mucus aggregates to detoxify NO would help explain the sudden enrichment of *A. fischeri* prior to actual colonization of the light organ. Then, continued production of NO in the light organ would help perpetuate *A. fischeri* and prevent colonization by other bacterial species.

Propagating the Symbiosis

The squid matures into an adult in about 2 months and then lives a strictly nocturnal existence in which it feeds on small crustaceans and other seafood. During the day, the animal buries itself and remains quiescent in the sand. Each morning the squid nearly empties its light organ of *A. fischeri* cells and proceeds to grow a new population of the bacterium before beginning its nighttime feeding routine. The bacterial cells grow rapidly in the light organ; by midafternoon, the structure contains the dense populations of *A. fischeri* cells required for the production of visible light. The actual emission of light requires a certain density of cells and is controlled by the regulatory mechanism called *quorum sensing* (Section 9.6) The diurnal expulsion of bacterial cells is thought to be a mechanism for seeding the environment with cells of the bacterial symbionts. This, of course, increases the chances for colonization of the next generation of juvenile squid.

A. fischeri grows much faster in the light organ than in the open ocean, presumably because it is supplied with nutrients by the squid. Thus *A. fischeri* benefits from the symbiosis by having an alternative habitat to seawater and one in which rapid growth to high populations is possible. Upon expulsion of *A. fischeri* cells from the light organ, numbers of the organism become enriched among the prokaryotic community. Isolation studies have shown that *A. fischeri* is not a particularly abundant marine bacterium. Thus, enrichment of *A. fischeri* through its symbiotic relationship with the squid probably helps this organism maintain larger populations in seawater than it would be capable of in the free-living state. Because the competitive success of a microbial species is to some degree a function of population size (Section 23.1), this boost in cell numbers may confer an important ecological advantage on *A. fischeri* in its marine habitat.

UNIT 4

24.12 MiniReview

The Hawaiian bobtail squid contains a light-emitting organ on its underside that is filled with cells of a single prokaryote, the bioluminescent marine bacterium *Aliivibrio fischeri*. This symbiotic relationship is highly specific and benefits both the animal and the bacterium.

- Of what value is the squid–*Aliivibrio* symbiosis to the squid? To the bacterium?
- What features of the squid–*Aliivibrio* symbiosis make it an ideal model for studying animal–bacterial symbioses?

V PLANT–MICROBIAL SYMBIOSES

We close this chapter with four examples of interactions among organisms: two will deal with plants and bacteria, one with fungi and plants, and the third is an exclusively microbial symbiosis. Many beneficial symbiotic associations exist; lichens, mycorrhizae, and the root nodules of leguminous plants are examples. However, there are also associations between plants and microorganisms that cause plant diseases. As an example of this, we consider crown gall, which is interesting because of the unique features of the disease and the microbial mode of transmission.

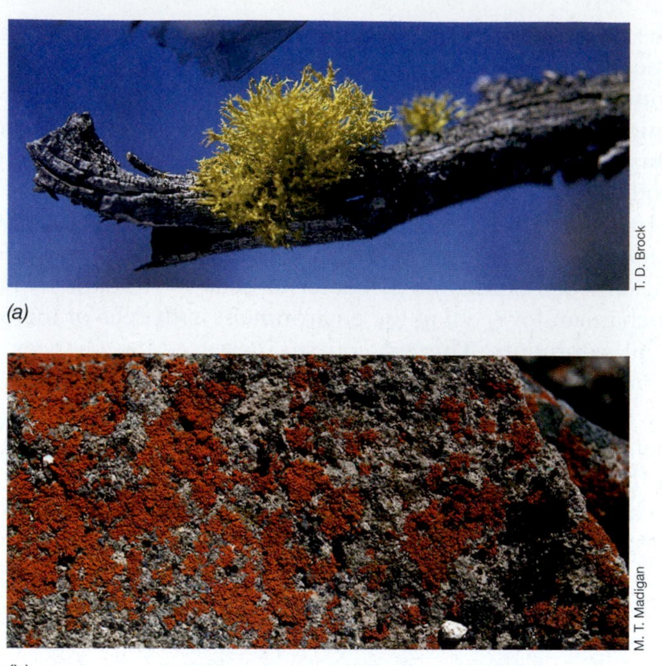

(a)

(b)

Figure 24.35 Lichens. *(a)* A lichen growing on a branch of a dead tree. *(b)* Lichens coating the surface of a large rock.

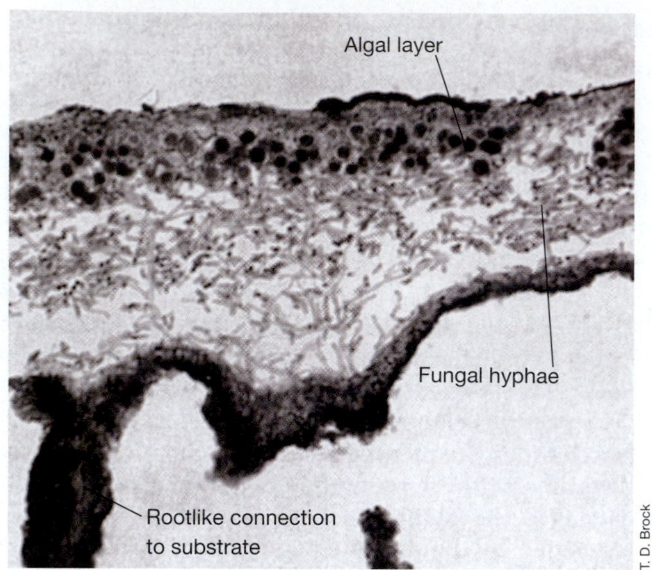

Figure 24.36 Lichen structure. Photomicrograph of a cross section through a lichen. The algal layer is positioned within the lichen structure so as to receive the most sunlight.

24.13 Lichens and Mycorrhizae

Lichens are leafy or encrusting microbial symbioses often found growing on bare rocks, tree trunks, house roofs, and the surfaces of bare soils (**Figure 24.35**). Lichens consist of a mutualistic relationship between two microorganisms, a fungus and an alga (or cyanobacterium), rather than between a microorganism and a macroorganism. The alga is the phototrophic partner and produces organic matter, which is then used for nutrition of the fungus. The fungus, unable to carry out photosynthesis, provides a firm anchor within which the phototrophic component can grow protected from erosion by rain or wind. Lichens are typically found on surfaces where other organisms do not grow, and their success in colonizing such environments is due to the mutualistic relationship.

Lichen Structure and Ecology

Lichens consist of a tight association of fungal cells within which cells of the phototroph are embedded (**Figure 24.36**). The morphology of a lichen is primarily determined by the fungus, and many fungi are able to form lichen associations. Diversity among the phototrophs is much lower, and many different kinds of lichens may have the same phototrophic component. Lichens that contain cyanobacteria frequently harbor N_2-fixing species, organisms such as *Anabaena* or *Nostoc* (∞ Section 16.7). The phototrophic cells are typically present in defined layers or clumps within the lichen structure (Figure 24.36).

The fungus clearly benefits from associating with the alga, but how does the phototroph benefit? Lichen acids, complex organic compounds excreted by the fungus, promote the dissolution and chelation of inorganic nutrients needed by the phototroph. Another role of the fungus is to protect the phototroph from drying; most of the habitats in which lichens live are dry (rock, bare soil, roof tops), and fungi are, in general, much better able to tolerate dry conditions than are algae. The fungus actually facilitates the uptake of water for the phototroph.

Lichens grow extremely slowly. For example, a lichen 2 cm in diameter growing on the surface of a rock may actually be several years old. Measurements of lichen growth vary from 1 mm or less per year to over 3 cm per year, depending on the organisms composing the symbiosis and the amount of rainfall and sunlight received.

Mycorrhizae

Mycorrhizae are mutualistic associations of plant roots and fungi. There are two classes of mycorrhizae. In *ectomycorrhizae,* fungal cells form an extensive sheath around the outside of the root with only a little penetration into the root tissue itself (**Figure 24.37**). In *endomycorrhizae,* the fungal mycelium becomes deeply embedded within the root tissue.

Ectomycorrhizae are found mainly in forest trees, especially conifers, beeches, and oaks, and are most highly developed in boreal and temperate forests. In such forests, almost every root of every tree is mycorrhizal. The root system of a mycorrhizal tree such as a pine (genus *Pinus*) is composed of both long and short roots. The short roots, which are characteristically dichotomously branched in *Pinus* (Figure 24.37*a*), show typical fungal colonization, and long roots are also frequently colonized. Endomycorrhizae are even more common than ectomycorrhizae. Arbuscular mycorrhizae, a type of endomycorrhizae, are found in the roots of over 80% of all terrestrial plant species so far examined.

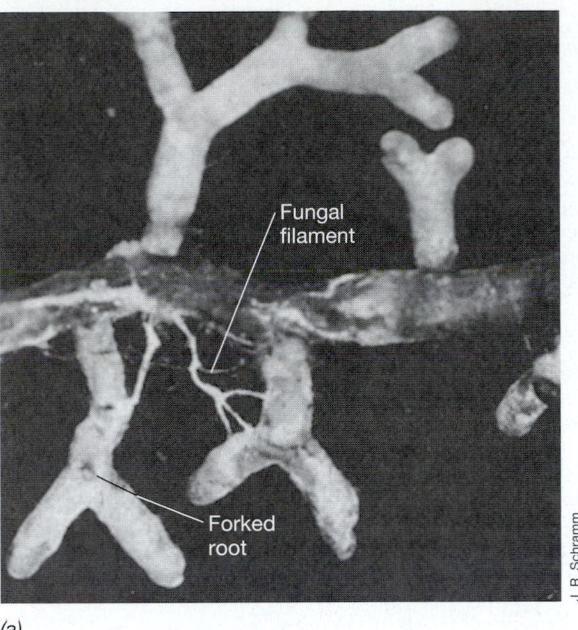

Figure 24.37 Mycorrhizae. *(a)* Typical ectomycorrhizal root of the pine, *Pinus rigida*, with filaments of the fungus *Thelophora terrestris*. *(b)* Seedling of *Pinus contorta* (lodgepole pine), showing extensive development of the absorptive mycelium of its fungal associate *Suillus bovinus*. This grows in a fanlike formation from the ectomycorrhizal roots to capture nutrients from the soil. The seedling is about 12 cm high.

Most mycorrhizal fungi do not catabolize cellulose and other leaf litter polymers. Instead, they catabolize simple carbohydrates and typically have one or more vitamin requirements. They obtain their carbon from root secretions and get inorganic minerals from the soil. Mycorrhizal fungi are rarely found in nature except in association with roots, and many are probably obligate symbionts. Mycorrhizal fungi produce plant growth substances that induce morphological alterations in the roots, stimulating formation of the mycorrhizal state. However, despite the close relationship between fungus and root, a single species of pine can form a mycorrhizal association with over 40 species of fungi.

The beneficial effect of the mycorrhizal fungus on the plant is best observed in poor soils, where trees that are mycorrhizal thrive but nonmycorrhizal ones do not. For example, if trees planted in prairie soils, which ordinarily lack a suitable fungal inoculum, are artificially inoculated at the time of planting, they grow much more rapidly than uninoculated trees (**Figure 24.38**). The mycorrhizal plant can absorb nutrients from its environment more efficiently and thus has a competitive advantage. This improved nutrient absorption is due to the greater surface area provided by the fungal mycelium. For example, in the pine seedling shown in Figure 24.37*b*, the ectomycorrhizal fungal mycelium makes up the overwhelming part of the absorptive capacity of the plant root system.

In addition to helping plants absorb nutrients, mycorrhizae also play a significant role in supporting plant diversity. Field experiments have clearly shown a positive correlation between the abundance and diversity of mycorrhizae in a soil and the extent of the plant diversity that develops in it. Thus, mycorrhizae are a true mutualistic symbiosis: The mycorrhizal plant is better able to function physiologically and compete successfully in a species-rich plant community, and the fungus benefits from a steady supply of organic nutrients.

Figure 24.38 Effect of mycorrhizal fungi on plant growth. Six-month-old seedlings of Monterey pine (*Pinus radiata*) growing in pots containing prairie soil: left, nonmycorrhizal; right, mycorrhizal.

Figure 24.39 Crown gall. Photograph of a crown gall tumor (arrow) on a tobacco plant caused by the crown gall bacterium *Agrobacterium tumefaciens.* The disease usually does not kill the plant but may weaken it and make it more susceptible to drought and other diseases.

24.13 MiniReview

Lichens are a symbiotic association between a fungus and an oxygenic phototroph. Mycorrhizae are formed from fungi that associate with plant roots and improve their ability to absorb nutrients. Mycorrhizae have a great beneficial effect on plant health and competitiveness.

- How do endomycorrhizae differ from ectomycorrhizae?
- Why are mycorrhizal associations with plants considered a type of symbiosis?

24.14 *Agrobacterium* and Crown Gall Disease

Some microorganisms develop parasitic symbioses with plants. The genus *Agrobacterium,* a close relative of the root nodule bacterium *Rhizobium* (Section 24.15), is such an organism, causing the formation of tumorous growths on diverse plants. The two species of *Agrobacterium* most widely studied are *Agrobacterium tumefaciens,* which causes *crown gall* disease, and *Agrobacterium rhizogenes,* which causes *hairy root* disease.

The Ti Plasmid

Although plants often form a benign accumulation of tissue called a *callus* when wounded, the growth in crown gall (**Figure 24.39**) is different in that it is uncontrolled growth,

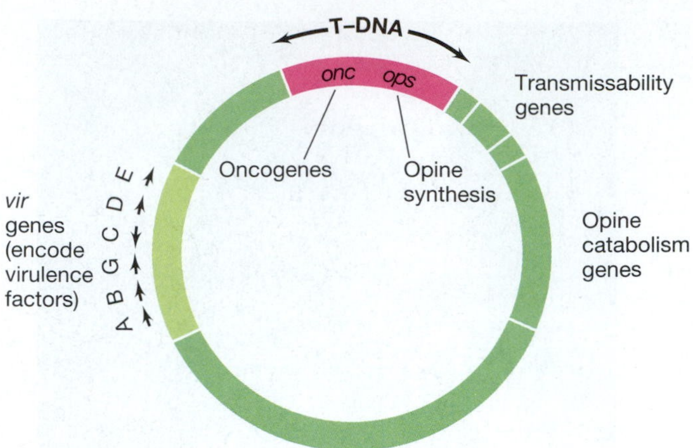

Figure 24.40 Structure of the Ti plasmid of *Agrobacterium tumefaciens.* T-DNA is the region transferred to the plant. Arrows indicate the direction of transcription of each gene. The entire Ti plasmid is about 200 kbp of DNA, and the T-DNA is about 20 kbp.

resembling a tumor in animals. *A. tumefaciens* cells induce tumor formation only when a large plasmid called the **Ti** (*t*umor *i*nduction) **plasmid** (**Figure 24.40**) is present in *Agrobacterium*. In *A. rhizogenes,* a similar plasmid called the *Ri plasmid* is necessary for induction of hairy root disease.

Following infection, a part of the Ti plasmid called the *transferred DNA* (T-DNA), is integrated into the plant's genome. T-DNA carries the genes for tumor formation and also for the production of a number of modified amino acids called *opines*. Octopine [N^2-(1,3-dicarboxyethyl)-L-arginine] and nopaline [N^2-(1,3-dicarboxypropyl)-L-arginine] are two common opines. Opines are produced by plant cells transformed by T-DNA and are a source of carbon and nitrogen, and sometimes phosphate, for the parasitic *Agrobacterium* cells. These nutrients are the beneficial side of the symbiosis for the bacterium.

Recognition and T-DNA Transfer

To initiate the tumorous state, cells of *Agrobacterium* must first attach to a wound site on the plant, but little is known about any specificity that may be required. However, following attachment, the synthesis of cellulose microfibrils by the bacteria helps anchor them to the wound site and forms bacterial aggregates on the plant cell surface. This sets the stage for plasmid transfer from bacterium to plant.

The general structure of the Ti plasmid is shown in Figure 24.40. Although a number of genes are needed for infectivity, only a small region of the Ti plasmid, the *T-DNA,* is actually transferred to the plant. The T-DNA contains genes that induce tumorigenesis. The *vir* genes on the Ti plasmid encode proteins that are essential for T-DNA transfer. Transcription of *vir* is induced by metabolites synthesized by wounded plant tissues. Examples of inducers include the phenolic compounds acetosyringone, *p*-hydroxybenzoic acid, and vanillin. The transmissibility genes on the Ti plasmid (Figure 24.40) allow the plasmid to be transferred by conjugation from one bacterial cell to another.

The *vir* genes are the key to T-DNA transfer. The *virA* gene encodes a protein kinase (VirA) that interacts with inducer molecules and then phosphorylates the product of the *virG* gene (**Figure 24.41**). VirG is activated by phosphorylation and functions to activate other *vir* genes. The product of the *virD* gene (VirD) has endonuclease activity and nicks DNA in the Ti plasmid in a region adjacent to the T-DNA. The product of the *virE* gene is a DNA-binding protein that binds the single strand of T-DNA generated from endonuclease activity and transports this small fragment of DNA into the plant cell. VirB, located in the bacterial cytoplasmic membrane, mediates transfer of the single strand of DNA between bacterium and plant. T-DNA transfer from bacterium to plant (Figure 24.41) thus resembles bacterial conjugation (Section 11.12).

The T-DNA then becomes inserted into the genome of the plant. Tumorigenesis (*onc*) genes on the Ti plasmid (Figure 24.40) encode enzymes for plant hormone production and at least one key enzyme of opine biosynthesis. Expression of these genes leads to tumor formation and opine production. The Ri plasmid responsible for hairy root disease also contains *onc* genes. However, in this case the genes confer increased auxin responsiveness to the plant, and this may promote overproduction of root tissue and the symptoms of the disease. The Ri plasmid also encodes several opine biosynthetic enzymes.

Genetic Engineering with the Ti Plasmid

From the standpoint of microbiology and plant pathology, in both crown gall and hairy root disease there are intimate interactions between the plant and the bacterium that lead to genetic exchange from bacterium to plant. That is, Ti is a natural plant transformation system. Thus, in recent years the interest in the Ti–crown gall system has shifted away from the disease itself to applications of this natural genetic exchange process in plant biotechnology.

Several modified Ti plasmids that lack disease genes but that can still transfer DNA to plants have been developed by genetic engineering. These have been used for the construction of genetically modified (transgenic) plants. Many transgenic plants have been constructed thus far, including crop plants carrying genes for resistance to herbicides, insect attack, and drought. We discuss the use of the Ti plasmid as a vector in plant biotechnology in Section 26.10.

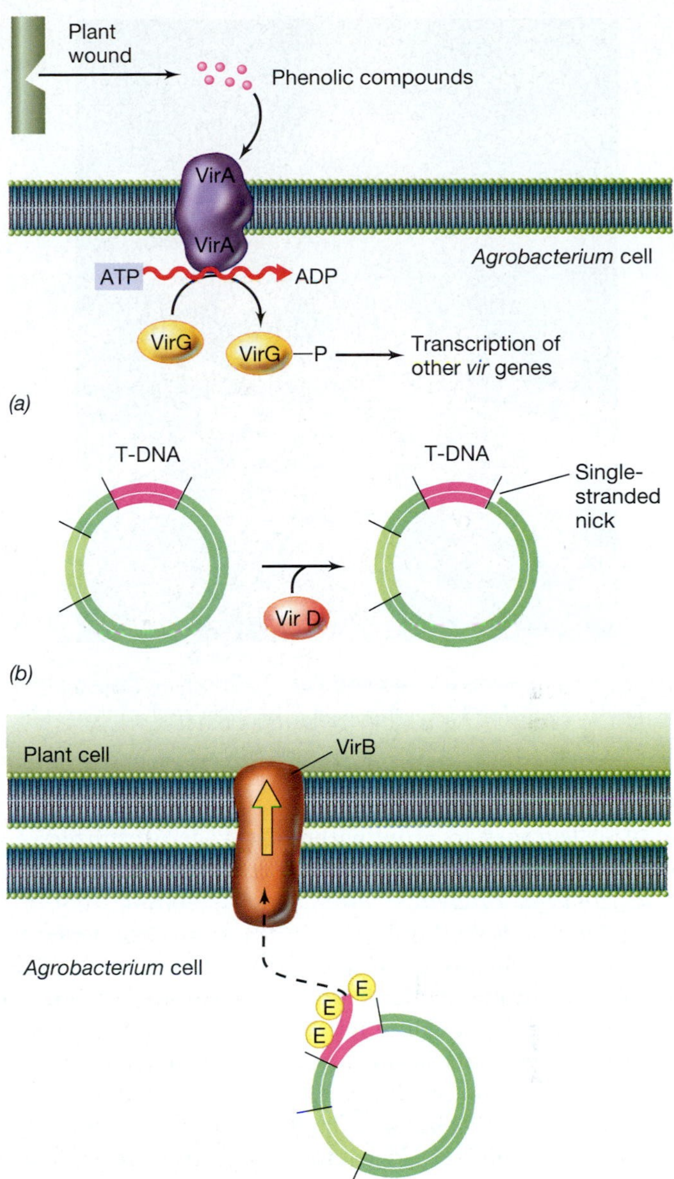

Figure 24.41 Mechanism of transfer of T-DNA to the plant cell by *Agrobacterium tumefaciens*. *(a)* VirA activates VirG by phosphorylation, and VirG activates transcription of other *vir* genes. *(b)* VirD is an endonuclease. *(c)* VirB functions as a conjugation bridge between the *Agrobacterium* cell and the plant cell, and VirE is a single-strand binding protein that assists in T-DNA transfer. Plant DNA polymerase produces the complementary strand to the transferred single strand of T-DNA.

24.14 MiniReview

The crown gall bacterium *Agrobacterium* enters into a unique relationship with plants. Part of a plasmid (the Ti plasmid) in the bacterium can be transferred into the genome of the plant, initiating crown gall disease. The Ti plasmid has also been used for the genetic engineering of crop plants.

- What are opines and why are they produced?
- How do the *vir* genes differ from *T-DNA* in the Ti plasmid?
- How has an understanding of crown gall disease benefited the area of plant molecular biology?

24.15 The Legume–Root Nodule Symbiosis

One of the most important plant bacterial mutualisms is that between leguminous plants and nitrogen-fixing bacteria. Legumes are plants that bear their seeds in pods and are the third largest family of flowering plants. This large group includes such agriculturally important plants as soybeans, clover, alfalfa, beans, and peas. These plants are key commodities for the food and agricultural industries, and the

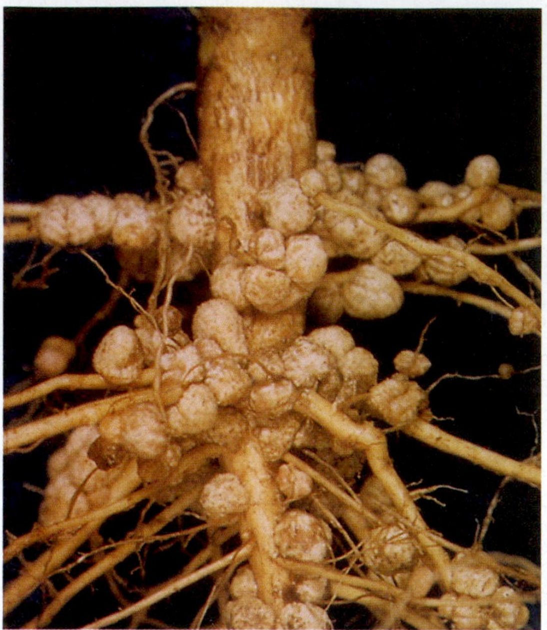

Figure 24.42 Soybean root nodules. The nodules develop from infection by *Bradyrhizobium japonicum*. The main stem of this soybean plant is about 0.5 cm in diameter.

ability of legumes to grow without nitrogen fertilizer saves farmers millions of dollars in fertilizer costs yearly.

Rhizobium, Bradyrhizobium, Sinorhizobium, Mesorhizobium, Azorhizobium, and *Photorhizobium* are genera of gram-negative *Alphaproteobacteria* that can grow in soil or can infect leguminous plants and establish a symbiotic relationship; these organisms are collectively called *rhizobia*. Infection of the roots of a legume by these bacteria leads to the formation of **root nodules** (Figure 24.42) that fix nitrogen (Section 20.14). Nitrogen fixation in root nodules is of enormous agricultural importance as it leads to significant increases in combined nitrogen in the soil. Because unfertilized bare soils are often nitrogen-deficient, nodulated legumes grow well in areas where other plants grow poorly (Figure 24.43).

Figure 24.43 Effect of nodulation on plant growth. A field of unnodulated (left) and nodulated (right) soybean plants growing in nitrogen-poor soil.

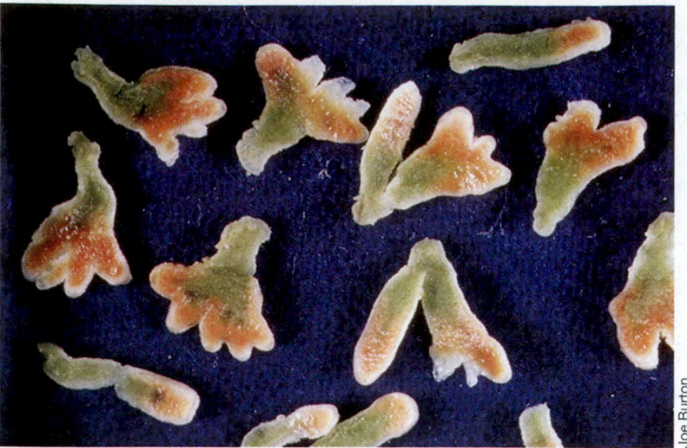

Figure 24.44 Root nodule structure. Sections of root nodules from the legume *Coronilla varia*, showing the reddish pigment leghemoglobin.

Leghemoglobin and Cross-Inoculation Groups

In the absence of the correct symbiotic bacterium, a legume cannot fix nitrogen. In pure culture, rhizobia are able to fix N_2 alone when grown under microaerophilic conditions. The rhizobia need some O_2 to generate energy for N_2 fixation, but their nitrogenases, like those of other nitrogen-fixing organisms, are inactivated by O_2. In the nodule, precise O_2 levels are controlled by the O_2-binding protein **leghemoglobin**. This iron-containing protein is present in healthy N_2-fixing nodules (Figure 24.44) and is induced through the interaction of the plant host and the bacterial symbiont. Leghemoglobin functions as an "oxygen buffer," cycling between the oxidized (Fe^{3+}) and reduced (Fe^{2+}) forms to keep unbound O_2 within the nodule low. The ratio of leghemoglobin-bound O_2 to free O_2 in the root nodule is on the order of 10,000:1.

There is a marked specificity between the species of legume and rhizobial species in establishing the symbiotic state. A single rhizobial species is able to infect certain species of legumes and not others. A group of related legumes that may be infected by a particular species of rhizobia is called a *cross-inoculation group* (Table 24.7). For example, there is a clover group, bean group, alfalfa group, and so on. If the correct strain is used, leghemoglobin-rich, nitrogen-fixing root nodules result (Figure 24.44).

Steps in Root Nodule Formation

How root nodules form is now fairly well understood (Figure 24.45). The steps are as follows:

1. Recognition of the correct partner by both plant and bacterium and attachment of the bacterium to the root hairs;
2. Excretion of nod factors by the bacterium;
3. Bacterial invasion of the root hair;
4. Travel to the main root via the infection thread;

Table 24.7 Major cross-inoculation groups of leguminous plants

Host plant	*Nodulated by*
Pea	*Rhizobium leguminosarum* biovar *viciae*[a]
Bean	*Rhizobium leguminosarum* biovar *phaseoli*[a]
Bean	*Rhizobium tropici*
Lotus	*Mesorhizobium loti*
Clover	*Rhizobium leguminosarum* biovar *trifolii*[a]
Alfalfa	*Sinorhizobium meliloti*
Soybean	*Bradyrhizobium japonicum*
Soybean	*Bradyrhizobium elkanii*
Soybean	*Sinorhizobium fredii*
Sesbania rostrata (a trophical legume)	*Azorhizobium caulinodans*

[a]Several varieties (biovars) of *Rhizobium leguminosarum* exist, each capable of nodulating a different legume.

5. Formation of modified bacterial cells, bacteroids, within the plant cells and development of the nitrogen-fixing state; and
6. Continued plant and bacterial division, forming the mature root nodule.

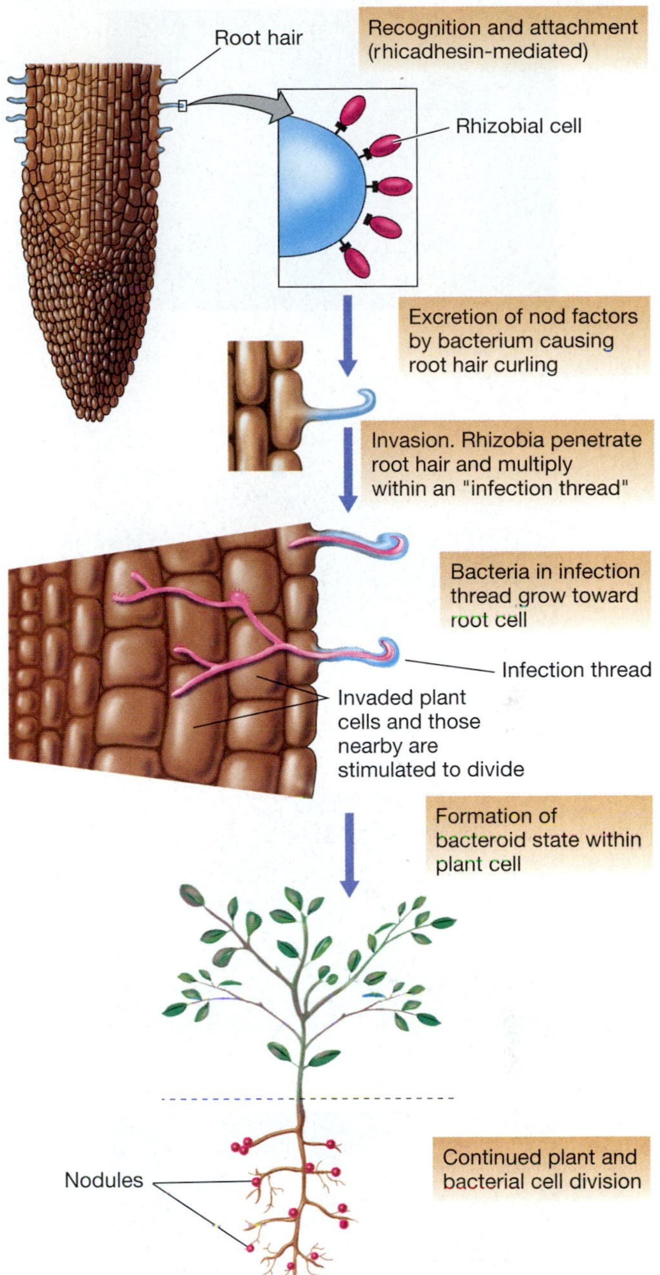
Root hair
Recognition and attachment (rhicadhesin-mediated)
Rhizobial cell
Excretion of nod factors by bacterium causing root hair curling
Invasion. Rhizobia penetrate root hair and multiply within an "infection thread"
Bacteria in infection thread grow toward root cell
Infection thread
Invaded plant cells and those nearby are stimulated to divide
Formation of bacteroid state within plant cell
Nodules
Continued plant and bacterial cell division

Figure 24.45 Steps in the formation of a root nodule in a legume infected by *Rhizobium*. Formation of the bacteroid state is a prerequisite for nitrogen fixation.

Attachment and Infection

The roots of leguminous plants secrete organic compounds that stimulate the growth of a diverse rhizosphere microflora. If rhizobia of the correct cross-inoculation group are in the soil, they will form large populations and eventually attach to root hairs extending from the roots of the plant (Figure 24.45). A specific adhesion protein called *rhicadhesin* is present on the cell surfaces of rhizobia. Other substances, such as carbohydrate-containing proteins called *lectins* and specific receptors in the plant cytoplasmic membrane, also play roles in plant–bacterium attachment.

If attachment occurs, rhizobial cells penetrate into the root hairs. Following the initial binding of bacterium to root hair, the root hair curls as a result of substances excreted by the bacterium. Following curling, the bacterium enters the root hair and induces formation by the plant of a cellulosic tube, called the **infection thread** (Figure 24.45 and see Figure 24.46*a*), which spreads down the root hair. Root cells adjacent to the root hairs subsequently become infected by rhizobia, and plant cell division occurs. Continued plant cell division leads to formation of a tumor-like nodule (**Figure 24.46**; Figures 24.42 and 24.44).

Bacteroids

The rhizobia multiply rapidly within the plant cells and become transformed into swollen, misshapen, and branched cells called **bacteroids**. Bacteroids become surrounded by portions of the plant cytoplasmic membrane to form a structure called the *symbiosome* (Figure 24.46*e*), and only after the symbiosome forms does nitrogen fixation begin. Nitrogen-fixing nodules can be detected by acetylene reduction to ethylene (∞ Figure 20.39).

When the plant dies, the nodule deteriorates, releasing bacteria into the soil. Bacteroids are incapable of division, but a small number of dormant rhizobial cells are always present in the nodule. These now proliferate, using some of the products of the deteriorating nodule as nutrients. The bacteria can then initiate the infection the next growing season or maintain a free-living existence in the soil.

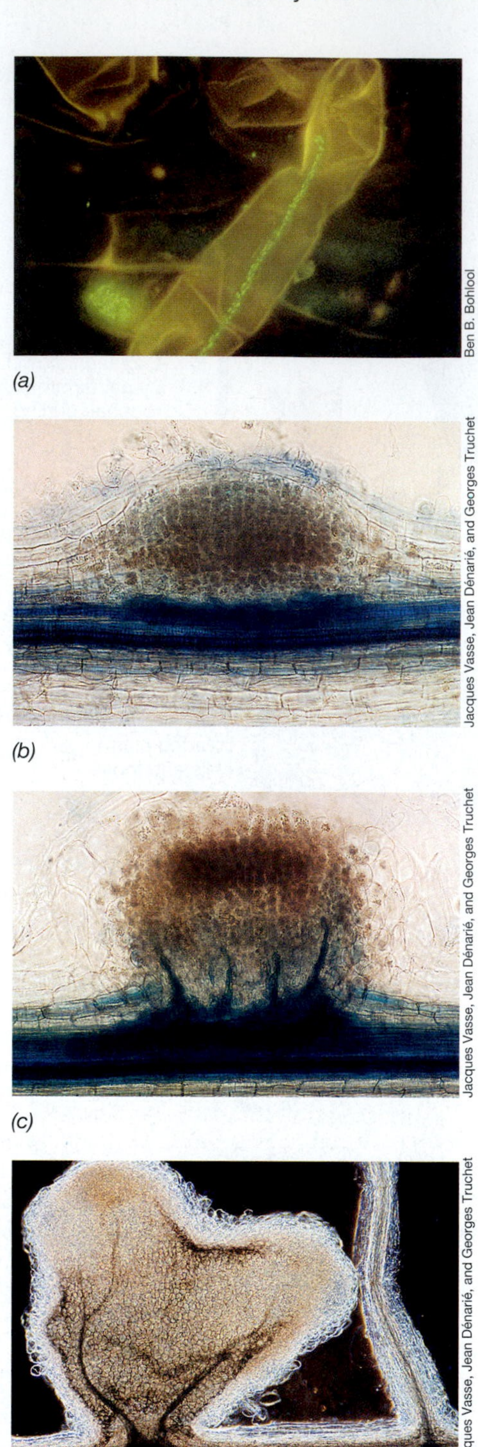

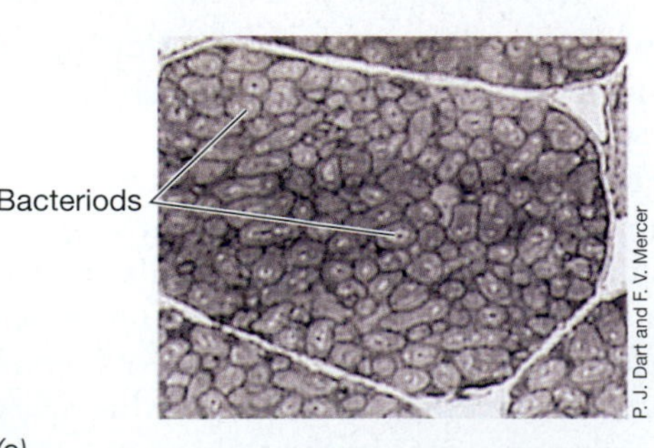

Figure 24.46 The infection thread and formation of root nodules. *(a)* An infection thread formed by cells of *Rhizobium leguminosarum* biovar *trifolii* on a root hair of white clover (*Trifolium repens*). The infection thread consists of a cellulosic tube through which bacteria move to root cells. *(b–d)* Nodules from alfalfa roots infected with cells of *Sinorhizobium meliloti* shown at different stages of development. Cells of both *R. leguminosarum* biovar *trifolii* and *S. meliloti* are about 2 μm long. The time course of nodulation events from infection to effective nodule is about 1 month in both soybean and alfalfa. *(e)* Cross-section through a root nodule of white clover showing the bacteroid form of *R. leguminosarum* biovar *trifolii* in the symbiosome. Bacteroids are about 2 μm long. Photos *(b–d)* reprinted with permission from *Nature* 351:670–673 (1991), © Macmillan Magazines Ltd.

Nodule Formation: Nod Genes, Nod Proteins, and Nod Factors

Bacterial genes that direct the steps in nodulation of a legume are called *nod genes*. In *Rhizobium leguminosarum* biovar *viciae*, ten nod genes have been identified. The *nodABC* genes encode proteins that produce oligosaccharides called **nod factors**; these induce root hair curling and trigger plant cell division, eventually leading to formation of the nodule.

Nod factors consist of a backbone of *N*-acetylglucosamine to which various substituents are linked (**Figure 24.47**). Host specificity is in part determined by the structure of the nod factor produced by a given species of *Rhizobium*. Besides the *nodABC* genes, which are universal and whose products synthesize the nod backbone, each cross-inoculation group contains nod genes that encode proteins that chemically modify the nod factor backbone to form the species-specific nod factor (Figure 24.47).

In *R. leguminosarum* biovar *viciae*, *nodD* encodes the regulatory protein NodD, and it controls transcription of other

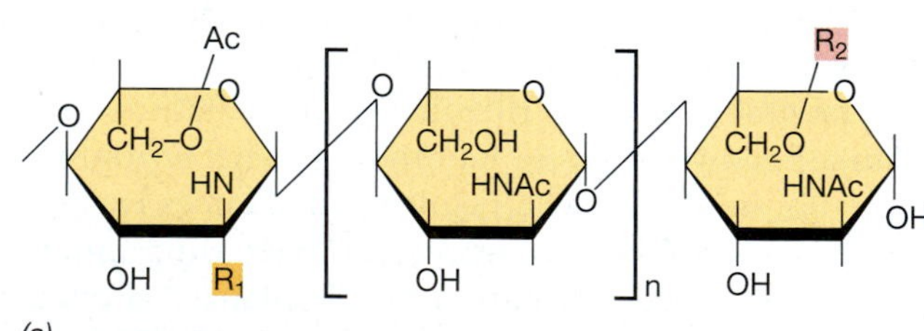

(a)

Species	R_1	R_2
Sinorhizobium meliloti	C16:2 or C16:3	SO_4^{2-}
Rhizobium leguminosarum biovar *viciae*	C18:1 or C18:4	H or Ac

(b)

Figure 24.47 Nod factors. *(a)* General structure of the nod factors produced by *Sinorhizobium meliloti* and *Rhizobium leguminosarum* biovar *viciae* and *(b)* a table of the structural differences (R_1, R_2) that define the precise nod factor of each species. The central hexose unit can repeat up to three times. C16:2, palmitic acid with two double bonds; C16:3, palmitic acid with three double bonds; C18:1, oleic acid with one double bond; C18:4, oleic acid with four double bonds; Ac, acetyl.

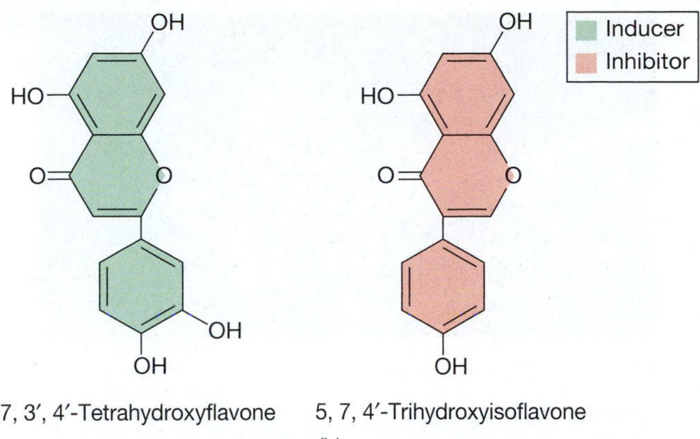

Figure 24.48 Plant flavonoids and nodulation. Structures of flavonoid molecules that are *(a)* an inducer of *nod* gene expression and *(b)* an inhibitor of *nod* gene expression in *Rhizobium leguminosarum* biovar *viciae,* the species that nodulates peas. Note the similarities in the structures of the two molecules. The common name of the structure shown in *(a)* is *luteolin,* and it is a flavone derivative. The structure in *(b)* is called *genistein,* and it is an isoflavone derivative.

nod genes. After interacting with inducer molecules, NodD promotes transcription and is thus a type of positive regulatory protein (Section 9.4). NodD inducers are plant flavonoids, organic molecules that are widely excreted by plants (**Figure 24.48**). Interestingly, some flavonoids that are structurally very closely related to *nodD* inducers in *R. leguminosarum* biovar *viciae,* inhibit induction of nod genes in other rhizobial species (Figure 24.48). This indicates that part of the specificity observed between plant and bacterium in the *Rhizobium*–legume symbiosis lies in the chemistry of the flavonoids excreted by each species of legume.

Biochemistry of Root Nodules

As discussed in Section 20.14, nitrogen fixation requires the enzyme nitrogenase. Nitrogenase from bacteroids shows the same biochemical properties as the enzyme from free-living N_2-fixing bacteria, including O_2 sensitivity and the ability to reduce acetylene as well as N_2. Bacteroids are dependent on the plant for the electron donor for N_2 fixation. The major organic compounds transported across the symbiosome membrane and into the bacteroid proper are citric acid cycle intermediates—in particular, the C_4 organic acids *succinate, malate,* and *fumarate* (**Figure 24.49**). These are used as electron donors for ATP production and, following conversion to pyruvate, as the ultimate source of electrons for the reduction of N_2 (Figure 20.36).

The product of N_2 fixation is ammonia, and the plant assimilates most of this ammonia by forming organic nitrogen compounds. The ammonia-assimilating enzyme glutamine synthetase is present in high levels in the plant cell cytoplasm and can convert glutamate and ammonia into glutamine (Figure 5.26*b*). This and a few other organic nitrogen compounds transport fixed nitrogen throughout the plant. **www.microbiologyplace.com Online Tutorial 24.1: Root Nodule Bacteria and Symbiosis with Legumes**

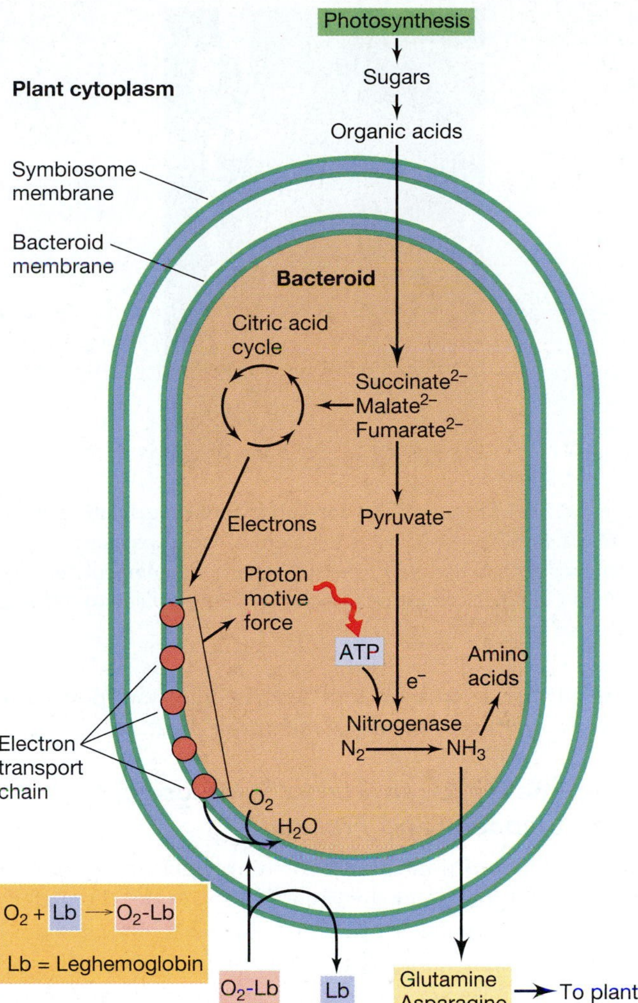

Figure 24.49 The root nodule bacteroid. Schematic diagram of major metabolic reactions and nutrient exchanges in the bacteroid. The symbiosome is a collection of bacteroids surrounded by a single membrane originating from the plant (Figure 24.46e).

Stem-Nodulating Rhizobia

Although most leguminous plants form nitrogen-fixing nodules on their *roots,* a few legume species bear nodules on their *stems*. Stem-nodulated leguminous plants are widespread in tropical regions where soils are often nitrogen-deficient because of leaching and intense biological activity. The best-studied system is the tropical aquatic legume *Sesbania,* which is nodulated by the bacterium *Azorhizobium caulinodans* (**Figure 24.50**). Stem nodules typically form in the submerged portion of the stems or just above the water level (Figure 24.50). The general sequence of events by which stem nodules form in *Sesbania* resembles that of root nodules: attachment, formation of an infection thread, and bacteroid formation.

Some stem-nodulating rhizobia produce bacteriochlorophyll *a* and thus have the potential to carry out anoxygenic photosynthesis (Section 20.4). Bacteriochlorophyll-containing rhizobia, grouped in the genus *Photorhizobium,* are widespread in nature, particularly in association with tropical

UNIT 4

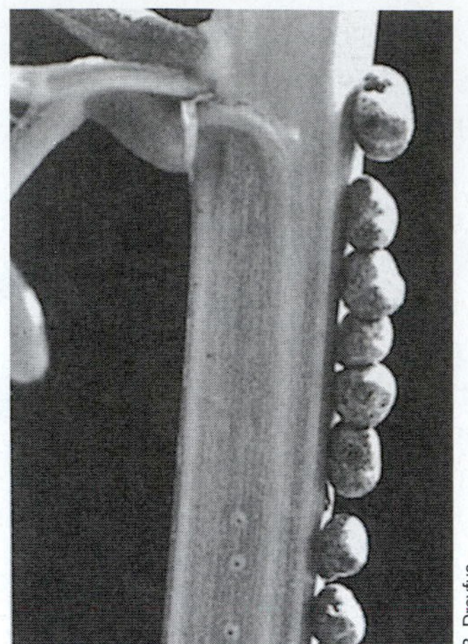

Figure 24.50 Stem nodules formed by stem-nodulating *Azorhizobium*. The photograph shows the stem of the tropical legume *Sesbania rostrata*. On the left side of the stem are uninoculated sites, on the right identical sites inoculated with stem-nodulating rhizobia.

legumes. In these species, light energy likely supports at least part of the energy needs for N_2 fixation.

Nonlegume Nitrogen-Fixing Symbioses: *Azolla–Anabaena* and *Frankia*

Various nonleguminous plants form nitrogen-fixing symbioses with bacteria other than rhizobia. For example, the water fern *Azolla* contains a species of heterocystous N_2-fixing cyanobacteria called *Anabaena azollae* within small pores of its fronds (**Figure 24.51**). *Azolla* has been used for centuries to enrich rice paddies with fixed nitrogen. Before planting rice, the farmer allows the surface of the rice paddy to become densely covered with *Azolla*. As the rice plants grow, they eventually crowd out the *Azolla*, leading to death of the fern and release of its nitrogen, which is assimilated by the rice plants. By repeating this process each growing season, the farmer can obtain high yields of rice without applying nitrogenous fertilizers.

Figure 24.51 *Azolla–Anabaena* symbiosis. *(a)* Intact association showing a single plant of *Azolla pinnata*. The diameter of the plant is approximately 1 cm. *(b)* Cyanobacterial symbiont *Anabaena azollae* as observed in crushed leaves of *A. pinnata*. Single cells of *A. azollae* are about 5 μm wide. Note the spherical-shaped heterocysts (lighter color, arrows), the site of nitrogen fixation in the cyanobacterium.

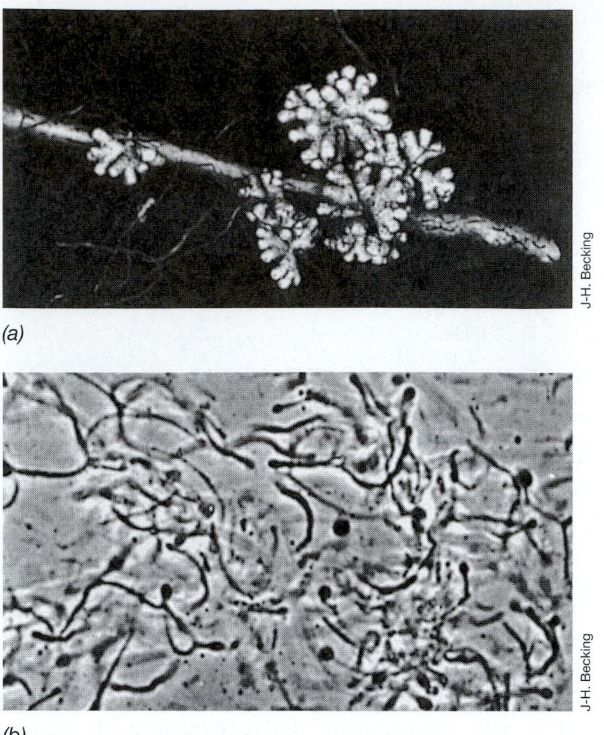

Figure 24.52 *Frankia* nodules and *Frankia* cells. *(a)* Root nodules of the common alder *Alnus glutinosa*. *(b) Frankia* culture purified from nodules of *Comptonia peregrina*. Note vesicles (black spherical structures) on the tips of hyphal filaments.

The alder tree (genus *Alnus*) has nitrogen-fixing root nodules (**Figure 24.52*a***) that harbor a filamentous, streptomycete-like, nitrogen-fixing organism called *Frankia*. Although when assayed in cell extracts the nitrogenase of *Frankia* is sensitive to molecular oxygen, like cells of *Azotobacter*, cells of *Frankia* fix N_2 at full oxygen tensions. This is because *Frankia* protects its nitrogenase from O_2 by localizing the enzyme in terminal swellings on the cells called *vesicles* (Figure 24.52*b*). The vesicles contain thick walls that retard O_2 diffusion, thus maintaining the O_2 tension within vesicles at levels compatible with nitrogenase activity. In this regard, *Frankia* vesicles resemble the heterocysts produced by some filamentous cyanobacteria as localized sites of N_2 fixation (Section 16.7).

Alder is a characteristic pioneer tree able to colonize nutrient-poor soils, probably because of its ability to enter into a symbiotic nitrogen-fixing relationship with *Frankia*. A number of other small or bushy, woody plants are nodulated by *Frankia*. However, unlike the *Rhizobium*–legume relationship, a single strain of *Frankia* can form nodules on several different species of plants, suggesting that the *Frankia*–root nodule symbiosis is less specific than that of leguminous plants.

24.15 MiniReview

One of the most agriculturally important plant–microbial symbioses is that between legumes and nitrogen-fixing bacteria. The bacteria induce the formation of root nodules within which nitrogen fixation occurs. The plant provides the energy source needed by the root nodule bacteria, and the bacteria provide fixed nitrogen for the plant. Other nitrogen-fixing symbioses include the water fern *Azolla* and the nodule-forming *Frankia*.

- What is leghemoglobin and what is its function?
- What is a bacteroid and what occurs within it?
- What are the major similarities and differences between *Rhizobium* and *Frankia*?

Review of Key Terms

Acid mine drainage acidic water containing H_2SO_4 derived from the microbial oxidation of iron sulfide minerals

Bacteroid morphologically misshapen cells of rhizobia, inside a leguminous plant root nodule; can fix N_2

Biopolymers polymeric materials consisting of biologically produced (and thus biodegradable) substances

Bioremediation the cleanup of oil, toxic chemicals, and other pollutants by microorganisms

Black smoker an extremely hot (250–350°C) deep-sea hot spring emitting both hot water and various minerals

Cometabolism metabolism of a compound in the presence of a second organic compound, which is used as the primary energy source

Humus dead organic matter

Hydrothermal vent a deep-sea warm or hot spring

Infection thread in the formation of root nodules, a cellulosic tube through which *Rhizobium* cells can travel to reach and infect root cells

Leghemoglobin an O_2-binding protein found in root nodules

Lichen a fungus and an alga (or cyanobacterium) living in symbiotic association

Microbial leaching the removal of valuable metals such as copper from sulfide ores by microbial activities

Mycorrhizae a symbiotic association between a fungus and the roots of a plant

Nod factors oligosaccharides produced by root nodule bacteria that help initiate the plant–bacterial symbiosis

Pyrite a common iron-containing ore, FeS_2

Reductive dechlorination removal of Cl as Cl^- from an organic compound by reducing the carbon atom from C—Cl to C—H

Root nodule a tumorlike growth on plant roots that contains symbiotic nitrogen-fixing bacteria

Rumen the first vessel in the multichambered stomach of ruminant animals in which cellulose digestion occurs

Syntrophy a process whereby two or more microorganisms cooperate to degrade a substance neither can degrade alone

Ti plasmid a conjugative plasmid present in the bacterium *Agrobacterium tumefaciens* that can transfer genes into plants

Volatile fatty acids (VFAs) the major fatty acids (acetate, propionate, and butyrate) produced during fermentation in the rumen

Xenobiotic a synthetic compound not naturally occurring in nature

UNIT 4

Review Questions

1. Why can it be said that the oxygen and carbon cycles are interconnected (Section 24.1)?
2. How can organisms such as *Syntrophobacter* and *Syntrophomonas* grow when their metabolism is based on thermodynamically unfavorable reactions? How does coculture of these syntrophs with certain other bacteria allow them to grow (Section 24.2)?
3. Compare and contrast the processes of nitrification and denitrification in terms of the organisms involved, the environmental conditions that favor each process, and the changes in nutrient availability that accompany each process (Section 24.3).
4. What organisms are involved in cycling sulfur compounds anoxically? If sulfur chemolithotrophs had never evolved, would there be a problem in the microbial cycling of sulfur compounds? What organic sulfur compounds are of interest in nature (Section 24.4)?
5. Why are most iron-oxidizing chemolithotrophs obligate aerobes, and why are most iron oxidizers acidophilic (Section 24.5)?
6. How is *Acidithiobacillus ferrooxidans* useful in the mining of copper ores? What crucial step in the indirect oxidation of copper ores is carried out by *A. ferrooxidans*? How is copper recovered from copper solutions produced by leaching (Section 24.6)?
7. How is mercury detoxified by the *mer* system (Section 24.7)?
8. What physical and chemical conditions are necessary for the rapid microbial degradation of oil in aquatic environments? Design an experiment that would allow you to test which conditions optimized the oil oxidation process (Section 24.8).
9. What are xenobiotic compounds and why might microorganisms have difficulty catabolizing them (Section 24.9)?

10. What is a rumen and how do the digestive processes operate in the ruminant digestive tract? What are the major benefits and the disadvantages of a rumen system? How does a cecal animal compare with a ruminant (Section 24.10)?

11. What evidence from hydrothermal vents exists to support the idea that prokaryotes are growing at extremely high temperatures (Section 24.11)?

12. How is the correct bacterial symbionts selected in the squid–*Aliivibrio* symbiosis (Section 24.12)?

13. How do mycorrhizae improve the growth of trees (Section 24.13)?

14. Compare and contrast the production of a plant tumor by *Agrobacterium tumefaciens* and a root nodule by a *Rhizobium* species. In what ways are these structures similar? In what ways are they different? Of what importance are plasmids to the development of both structures (Sections 24.14 and 24.15)?

15. Describe the steps in the development of root nodules on a leguminous plant. What is the nature of the recognition between plant and bacterium and how do nod factors help control this? How does this compare with recognition in the *Agrobacterium*–plant system (Section 24.15)?

Application Questions

1. Compare and contrast a lake ecosystem with a hydrothermal vent ecosystem. How does energy enter each ecosystem? What is the basis of primary production in each ecosystem? What nutritional classes of organisms exist in each ecosystem, and how do they feed themselves?

2. ^{14}C-Labeled cellulose is added to a vial containing a small amount of sewage sludge and sealed under anoxic conditions. A few hours later $^{14}CH_4$ appears in the vial. Discuss what has happened to yield such a result.

3. Imagine that you have discovered a new animal that consumes only grass in its diet. You suspect it to be a ruminant and have available a specimen for anatomical inspection. If this animal is a ruminant, describe the position and basic components of the digestive tract you would expect to find and any key microorganisms and substances you might look for.

4. Acid mine drainage is in part a chemical process and in part a biological process. Discuss the chemistry and microbiology that lead up to acid mine drainage and point out the key reaction(s) that are biological. How many ways can you think of to prevent acid mine drainage?

25

Industrial Microbiology

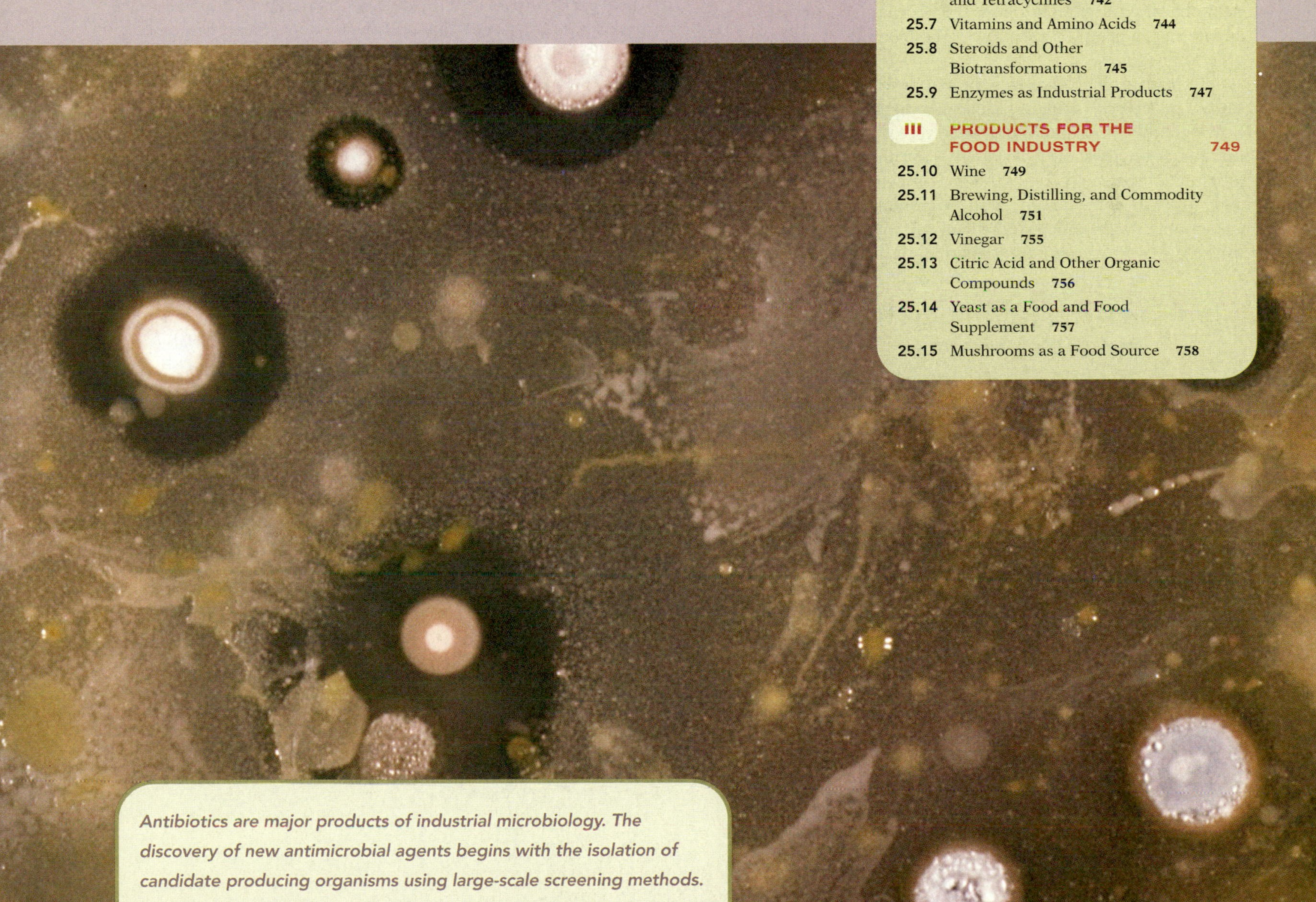

Antibiotics are major products of industrial microbiology. The discovery of new antimicrobial agents begins with the isolation of candidate producing organisms using large-scale screening methods.

I INDUSTRIAL MICROORGANISMS AND PRODUCT FORMATION

Industrial microbiology uses microorganisms, typically grown on a large scale, to produce commercial products or to carry out important chemical transformations. Industrial processes that employ microorganisms are enhancements of metabolic reactions that the microorganisms are already capable of carrying out, with the main goal being the *overproduction* of the product of interest. Industrial microbiology thus differs from microbial biotechnology, to be discussed in the next chapter. In biotechnology, microorganisms are *genetically engineered* to produce substances they would otherwise not be able to produce.

Industrial microbiology originated with alcoholic fermentation processes, such as those for making beer and wine. Subsequently, microbial processes were developed for the production of pharmaceuticals (such as antibiotics), food additives (such as amino acids and vinegar), enzymes, chemicals (such as alcohol and citric acid), and microbial cells themselves. In this chapter we see how industrial microbiologists have made microbial products economical on an industrial scale.

25.1 Industrial Microorganisms and Their Products

The major organisms used in industrial microbiology are fungi (yeasts and molds) (∞ Sections 18.13–18.19) and certain prokaryotes, in particular species of the genus *Streptomyces* (∞ Section 16.6). Industrial microorganisms can be thought of as metabolic specialists, capable of synthesizing one or more products in high yield. Industrial microbiologists often use classical genetic methods to select high-yielding microbial variants, with the goal being to increase the yield of the product to the point that an economically feasible process is possible. Thus the behavior of the actual production strain may be far removed from that of the original wild-type strain.

Properties of a Useful Industrial Microorganism

A microorganism used in an industrial process must have other features besides just being able to produce the substance of interest in high yield. First and foremost, the organism must be capable of growth and product formation in large-scale culture. Moreover, it should produce spores or some other reproductive cell form so that it can be easily inoculated into the large vessels used to grow the producing organism on an industrial scale. It must also grow rapidly and produce the desired product in a relatively short period of time.

An industrially useful organism must also be able to grow in a liquid culture medium obtainable in bulk quantities at a low price. Many industrial microbiological processes use waste carbon from other industries as major or supplemental ingredients for large-scale culture media. These include *corn steep liquor* (a product of the corn wet-milling industry that is rich in nitrogen and growth factors) and *whey* (a waste liquid of the dairy industry containing lactose and minerals).

An industrial microorganism should not be pathogenic, especially to humans or economically important animals or plants. Because of the high cell densities in industrial microbial processes and the virtual impossibility of avoiding contamination of the environment outside the growth vessel, a pathogen would present potentially disastrous problems.

Finally, an industrial microorganism should be amenable to genetic manipulation because increased yields are often obtained by means of mutation and classical genetic selection techniques. A genetically stable and easily manipulable microorganism is thus a clear advantage for an industrial process.

Examples of Industrial Products

Microbial products of industrial interest (**Figure 25.1**) include the microbial cells themselves—for example, yeast cultivated for food, baking, or brewing and substances produced by cells. Examples of the latter include enzymes such as glucose isomerase, important in the production of high-fructose syrups; pharmacologically active agents such as antibiotics, steroids, and alkaloids; specialty chemicals and food additives such as aspartame (a food and drink sweetener); and **commodity chemicals**—inexpensive chemicals produced in bulk—such as ethanol, citric acid, and many others.

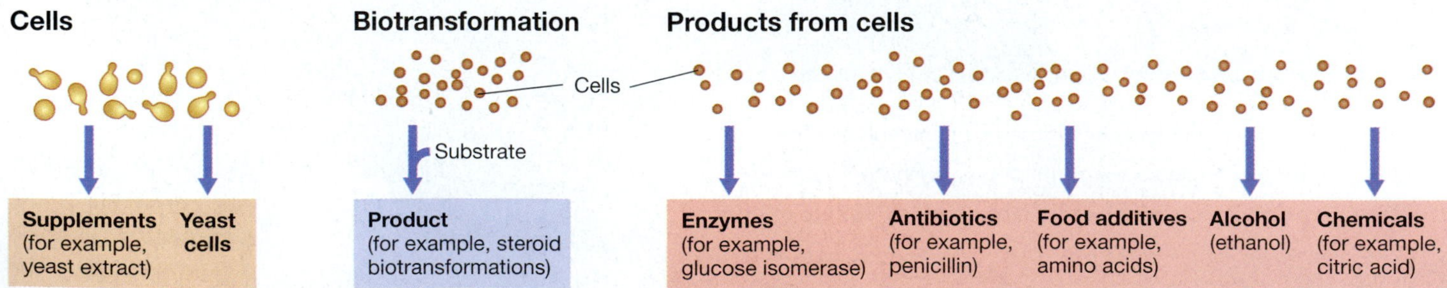

Figure 25.1 Products of industrial microbiology. The products may be the cells themselves or products made from cells. In biotransformation, cells convert a specific substance from one form to another.

25.1 MiniReview

An industrial microorganism must produce the product of interest in high yield, grow rapidly on inexpensive culture media available in bulk quantities, be amenable to genetic manipulation, and be nonpathogenic. Industrial products are many and include both cells and substances made by cells.

- Why should industrial microorganisms be genetically manipulable?
- List three important products of industrial microbiology.

25.2 Primary and Secondary Metabolites

In Section 6.7 we discussed microbial growth and described the various stages: *lag, exponential,* and *stationary*. Here we describe microbial growth and product formation in an industrial context; we will see that some products are made during exponential growth and others only after exponential growth has ceased.

There are two types of microbial metabolites of interest to industrial microbiology, primary and secondary. A **primary metabolite** forms during the exponential growth phase of the microorganism. By contrast, a **secondary metabolite** forms near the end of the growth phase, frequently at, near, or in the stationary phase of growth (**Figure 25.2**).

A typical primary metabolite is alcohol. Ethyl alcohol is a product of the fermentative metabolism of yeast and certain bacteria (Section 5.10) and is formed as part of energy metabolism. Because organisms can grow only if they produce energy, ethanol forms in parallel with growth (Figure 25.2*a*). In many industrial processes however, the desired product is not made during the exponential phase but only in the stationary phase. Secondary metabolites are some of the most complex and important metabolites of industrial interest (Figure 25.2*b*). Secondary metabolites typically share the following characteristics:

1. They are not essential for growth and reproduction.
2. Their formation is highly dependent on growth conditions, especially on the composition of the growth medium. Secondary metabolite formation is frequently repressed.
3. They are often produced as a group of closely related compounds. For instance, a single strain of a species of *Streptomyces* can produce over 30 related, yet chemically distinct antibiotics.
4. They are often overproduced, sometimes in huge amounts. By contrast, primary metabolites, linked as they are to energy and growth, usually cannot be significantly overproduced.
5. They are often produced by spore-forming microorganisms during the sporulation process itself. Virtually all antibiotics, for example, are produced by either fungi or spore-forming prokaryotes.

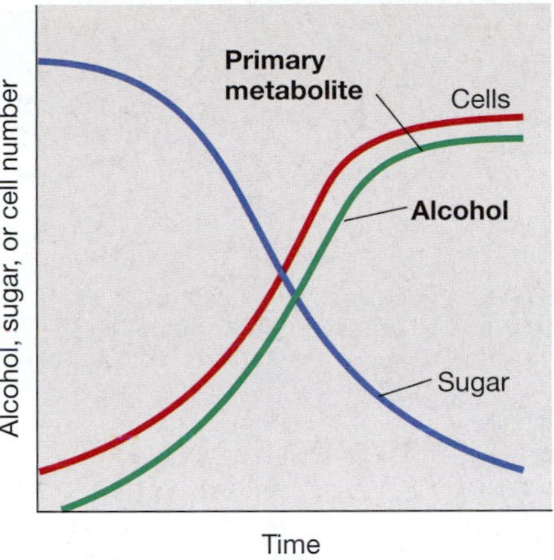

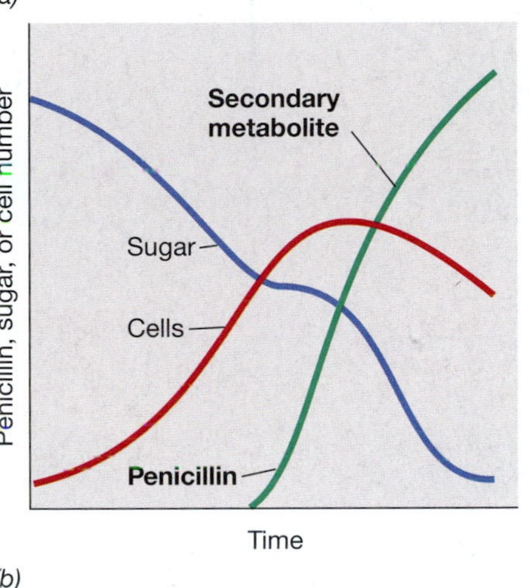

Figure 25.2 Contrast between production of primary and secondary metabolites. *(a)* Formation of alcohol by yeast—an example of a primary metabolite. *(b)* Penicillin production by the mold *Penicillium chrysogenum*—an example of a secondary metabolite. Note how penicillin is not made until after mid-exponential phase.

Primary and Secondary Metabolism Pathways

Most secondary metabolites are complex organic molecules that require a large number of specific enzymatic reactions for synthesis. For instance, at least 72 separate enzymatic steps are involved in the synthesis of the antibiotic tetracycline (Section 25.6), and there are over 25 reactions in the synthesis of erythromycin. None of these reactions occur during primary metabolism. However, the metabolic pathways for these secondary metabolites arise out of primary metabolism because their starting materials originate from the major biosynthetic pathways. **Figure 25.3** shows the interrelationship of the main primary metabolic pathway for aromatic amino acid synthesis (Section 5.16) with the secondary metabolic pathways for some antibiotics. As can be seen, certain antibiotics originate from primary metabolites in the amino acid biosynthetic pathway (Figure 25.3).

UNIT 5

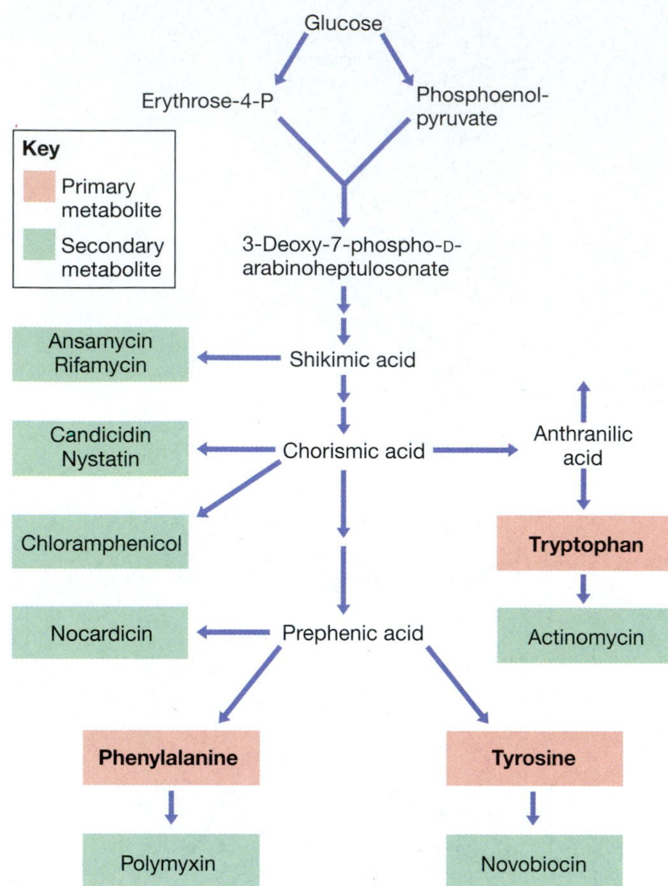

Figure 25.3 Aromatic antibiotics. Relationship of the primary metabolic pathway for the synthesis of aromatic amino acids (pink) and formation of secondary metabolite antibiotics containing aromatic rings (green). This is a composite scheme of processes in various microorganisms. No one organism produces all these secondary metabolites, and many individual steps exist between amino acid and antibiotic in all cases.

25.2 MiniReview

Primary and secondary metabolites are produced during the exponential phase of growth or near the onset of stationary phase, respectively. Many economically valuable microbial products are secondary metabolites.

- Is penicillin a primary or a secondary metabolite? Why?
- What type of metabolite, primary or secondary, can be more easily overproduced? Why?

25.3 Characteristics of Large-Scale Fermentations

The vessel in which an industrial microbiology process is carried out is called a **fermentor**. In industrial microbiology, the term **fermentation** refers to *any* large-scale microbial process, whether or not it is a fermentation in a biochemical sense. In fact, most industrial fermentations are *aerobic* processes, whereas glycolysis, a typical fermentation, is an *anaerobic* process (Section 5.10). Fermentors vary in size from the small 5- to 10-liter laboratory scale to the enormous 500,000-liter industrial scale (**Figure 25.4**). The size of the fermentor used depends on the process and how it is operated. A summary of fermentor sizes for some common microbial fermentations is given in **Table 25.1.**

Industrial fermentors can be divided into two major classes, those for anaerobic processes and those for aerobic processes. Anaerobic fermentors require little special equipment except for a means to remove heat generated during the fermentation. Aerobic fermentors, however, require much more elaborate equipment to ensure that mixing and adequate aeration are achieved. Because most industrial fermentations are aerobic, we focus on aerobic fermentors here.

Construction of an Aerobic Fermentor

Large-scale industrial fermentors are almost always constructed of stainless steel. Such a fermentor is essentially a large cylinder, closed at the top and bottom, into which various pipes and valves have been fitted (Figure 25.4*b*). Because sterilization of the culture medium and removal of heat are vital for successful operation, the fermentor is fitted with an external cooling jacket through which steam (for sterilization) or water (for cooling) can be run. For very large fermentors, sufficient heat cannot be transferred through the jacket and so internal coils must be provided through which either steam or cooling water can be piped (Figure 25.4).

A critical part of the fermentor is the aeration system. With large-scale equipment, transfer of oxygen throughout the growth medium is critical, and elaborate precautions must be taken to ensure proper aeration. Oxygen is poorly soluble in water, and in a fermentor with a high density of microbial cells, there is a tremendous oxygen demand by the culture. Because of this, two different devices are used to ensure adequate aeration: an aerator, called a *sparger*, and a stirring device, called an *impeller* (Figure 25.4*b*).

The sparger is typically just a series of holes or a nozzle through which filter-sterilized air (or oxygen-enriched air) can be passed into the fermentor under high pressure. The air enters the fermentor as a series of tiny bubbles from which the oxygen passes by diffusion into the liquid. In small fermentors use of a sparger alone may be sufficient to ensure adequate aeration.

Table 25.1 Fermentor sizes for various industrial processes

Size of fermentor (liters)	*Product*
1–20,000	Diagnostic enzymes, substances for molecular biology
40–80,000	Some enzymes, antibiotics
100–150,000	Penicillin, aminoglycoside antibiotics, proteases, amylases, steroid transformations, amino acids, wine, beer
200,000–500,000	Amino acids (glutamic acid), wine, beer

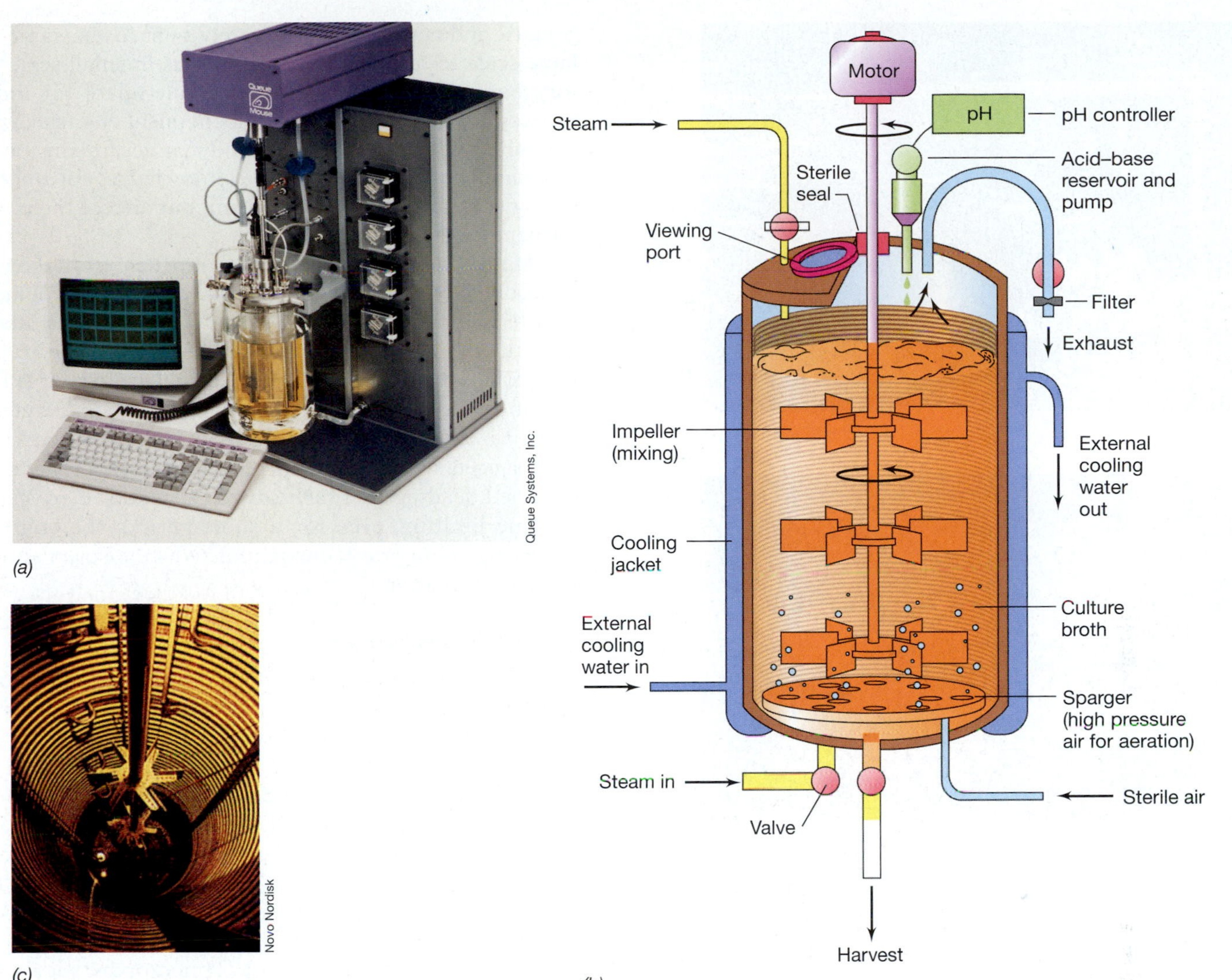

Figure 25.4 **Fermentors.** *(a)* A small research fermentor; the volume is 5 liters. *(b)* Diagram of a fermentor, illustrating construction and facilities for aeration and process control. *(c)* The inside of an industrial fermentor, showing the impeller and internal heating and cooling coils.

But in industrial-size fermentors, stirring of the fermentor with an impeller is absolutely essential (Figure 25.4*c*). Stirring accomplishes two things: It mixes the gas bubbles through the liquid and mixes the organism through the liquid, thus ensuring uniform access of microbial cells to the nutrients.

Fermentation Control and Monitoring

It is essential that industrial fermentors be closely monitored during a production run. It is necessary to not only measure growth and product formation but also to control the process by altering environmental parameters as needed. Environmental factors that are typically controlled include temperature, oxygen, pH, cell mass, levels of key nutrients, and product concentration.

During a production run it is also essential that data on the process be obtained in real time (see Figure 25.5*b*). For instance, it may be necessary to alter one or more environmental parameters as the fermentation progresses or to feed a nutrient at a rate that exactly balances growth (see, for example, Figure 25.10). Computers are used to process environmental data as the fermentation proceeds and are programmed to respond accordingly by sending signals for nutrient additions, increases in flow of cooling water, stirring rate adjustments, and the like, at just the right time to maintain high product yield.

Computers are also used to model fermentation processes. Mathematical models can be used to test the effect of various parameters on growth and product yield quickly and interactively. With the models, industrial microbiologists can modify the parameters to predict how each will affect the process. In this way, many variations in the fermentation can be studied inexpensively on screen, rather than at the pilot plant stage (Section 25.4), or worse yet, in the actual production phase, where costs become a major factor.

(a)

(b)

Figure 25.5 Industrial scale fermentations. *(a)* A large industrial fermentation plant. Only the tops of the fermentors, which can be several stories high, are visible. *(b)* Computer control room for a large fermentation plant.

25.3 MiniReview

Large-scale industrial fermentations present several engineering problems. Aerobic processes require mechanisms for stirring and aeration. The microbial process must be continuously monitored to ensure satisfactory yields of the desired product.

- What types of devices are used to ensure proper aeration in a large-scale fermentation?
- What parameters in an industrial fermentation need to be monitored and what adjustments need to be made by computerized control?

25.4 Scale-Up of Industrial Fermentations

A very important aspect of industrial microbiology is the transfer of a process from small-scale laboratory equipment to large-scale commercial equipment, a process called **scale-up**. An understanding of the problems in scale-up is important because industrial processes rarely behave the same way in large-scale fermentors (**Figure 25.5**) as in small-scale laboratory equipment (**Figure 25.6a**). Scale-up of an industrial process is a major task of the biochemical engineer, one who is familiar with gas transfer, fluid dynamics, mixing, and thermodynamics. Scale-up requires knowledge not only of the biology of the producing organism, but also of the physics of fermentor design and operation.

Many of the challenges in scale-up arise from the need for aeration and mixing. These essential components of an industrial fermentation are much easier to control in a laboratory flask than in a large industrial fermentor. Oxygen transfer in particular is much more difficult to achieve in a large fermentor, and because most industrial fermentations are aerobic, effective oxygen transfer is critical. With the rich culture media used in industrial processes, a high cell density is obtained, leading to a high oxygen demand. If oxygen levels become limiting, even for a short period, the culture may experience temporary anoxia and reduce or even shut down product formation.

The Scale-Up Process

An industrial process is transferred from the laboratory to a commercial fermentor in several stages. The process begins in the laboratory flask, a small-scale operation but typically the first indication that a process of commercial interest is possible. From here, the process is transferred to the laboratory fermentor, a small-scale vessel, generally made of glass and 1 to 10 liters in size; the first efforts at scale-up are made in the laboratory fermentor (Figures 25.4*a* and 25.6*a*). In the laboratory fermentor it is also possible to test variations in media, temperature, pH, and so on, quickly and inexpensively.

When tests in the laboratory fermentor are successful, the process moves into the **pilot plant** stage, usually carried out in fermentors of 300- to 3000-liter capacity. Here the conditions more closely approach those of the commercial fermentor; however, cost is not yet a major factor. Finally, the process moves to the commercial fermentor itself, 10,000–500,000 liters in volume (Table 25.1, and Figures 25.5*a* and 25.6*b*). In all stages of scale-up, aeration is a key variable that is closely monitored. As scale-up proceeds from flask to commercial fermentor, besides keeping temperature, pH, and nutrients at appropriate levels, oxygen dynamics are carefully measured to determine how volume increases affect oxygen demand in the fermentation.

25.4 MiniReview

Scale-up is the process of gradually converting a useful industrial fermentation from laboratory scale to production scale. Aeration is a particularly critical aspect to monitor during scale-up studies.

- What are the differences in size among a typical laboratory fermentor, a pilot plant fermentor, and a commercial fermentor?

(a)

(b)

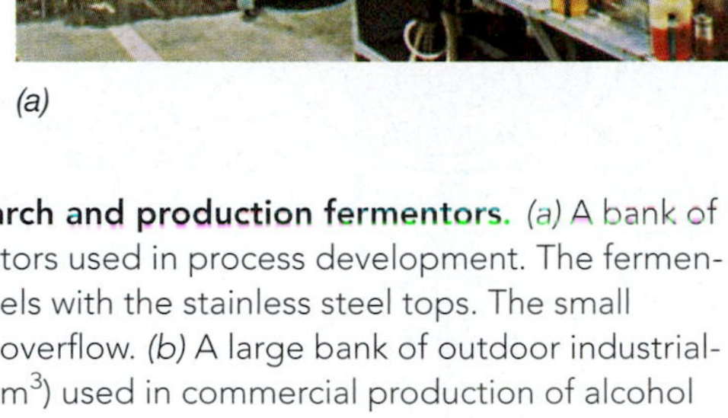

Figure 25.6 Research and production fermentors. *(a)* A bank of small research fermentors used in process development. The fermentors are the glass vessels with the stainless steel tops. The small plastic bottles collect overflow. *(b)* A large bank of outdoor industrial-scale fermentors (240 m^3) used in commercial production of alcohol in Japan.

II PRODUCTS FOR THE HEALTH INDUSTRY

We now consider some products of industrial microbiology, beginning with the antibiotics. Of the microbial products manufactured commercially, the most important for the health industry are the antibiotics. The development of antibiotics as agents for treatment of infectious disease has probably had more impact on the practice of medicine than any other single medical development. Antibiotic production is a huge industry worldwide and one in which many important aspects of large-scale microbial culture were perfected.

25.5 Antibiotics: Isolation and Characterization

Antibiotics are compounds produced by microorganisms that kill or inhibit the growth of other microorganisms. Antibiotics are typical secondary metabolites (Section 25.2). Most antibiotics in clinical use are produced by filamentous fungi or bacteria of the actinomycete group (∞ Section 16.6). **Table 25.2** lists the most important antibiotics produced by large-scale industrial fermentation today.

Search for New Antibiotics

Modern drug discovery relies heavily on computer modeling of drug–target interactions (∞ Section 27.13). However, in the past, and to a more limited extent today, new antibiotics are discovered by laboratory screening programs. In this approach, a large number of possible antibiotic-producing microorganisms are obtained from nature in pure culture (**Figure 25.7a**) and are then tested for antibiotic production by assaying for diffusible materials that are inhibitory to the growth of test

Table 25.2 Some antibiotics produced commercially[a]

Antibiotic	*Producing microorganism[b]*
Bacitracin	*Bacillus licheniformis* (EFB)
Cephalosporin	*Cephalosporium* spp. (F)
Chloramphenicol	Chemical synthesis (formerly produced microbially by *Streptomyces venezuelae*) (A)
Cycloheximide	*Streptomyces griseus* (A)
Cycloserine	*Streptomyces orchidaceus* (A)
Erythromycin	*Streptomyces erythreus* (A)
Griseofulvin	*Penicillium griseofulvin* (F)
Kanamycin	*Streptomyces kanamyceticus* (A)
Lincomycin	*Streptomyces lincolnensis* (A)
Neomycin	*Streptomyces fradiae* (A)
Nystatin	*Streptomyces noursei* (A)
Penicillin	*Penicillium chrysogenum* (F)
Polymyxin B	*Bacillus polymyxa* (EFB)
Streptomycin	*Streptomyces griseus* (A)
Tetracycline	*Streptomyces rimosus* (A)

[a]See Chapter 27 for structures and more information on these antibiotics.
[b]EFB, endospore-forming bacterium; F, fungus; A, actinomycete.

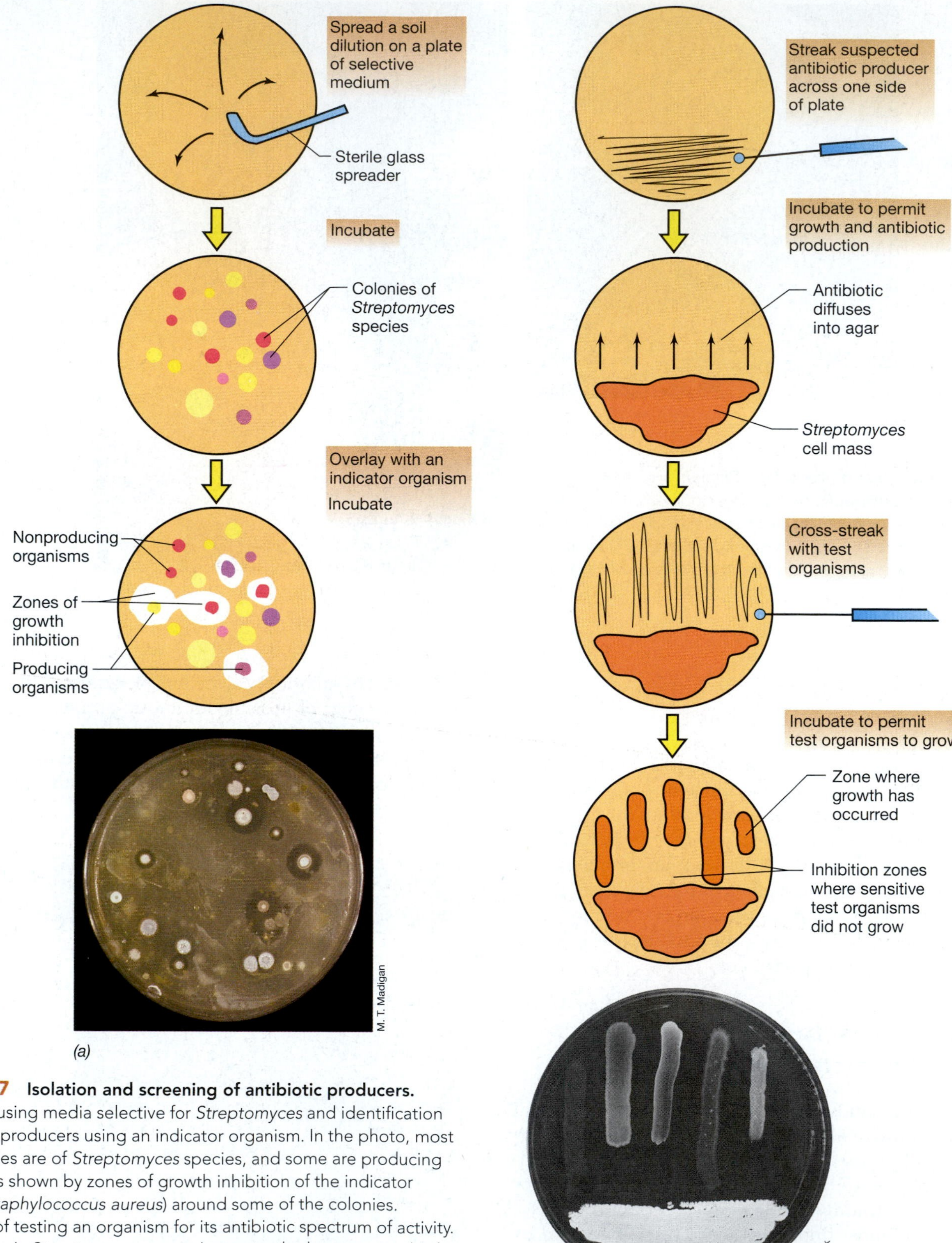

Figure 25.7 Isolation and screening of antibiotic producers. *(a)* Isolation using media selective for *Streptomyces* and identification of antibiotic producers using an indicator organism. In the photo, most of the colonies are of *Streptomyces* species, and some are producing antibiotics as shown by zones of growth inhibition of the indicator organism (*Staphylococcus aureus*) around some of the colonies. *(b)* Method of testing an organism for its antibiotic spectrum of activity. The producer (a *Streptomyces* species) was streaked across one-third of the plate and the plate was incubated. After good growth was obtained, the five species of test bacteria were streaked perpendicular to the *Streptomyces*, and the plate was further incubated. The failure of several species to grow near the mass growth of *Streptomyces* indicates that the *Streptomyces* produced an antibiotic active against these bacteria. Test species (left to right): *Escherichia coli, Bacillus subtilis, Staphylococcus aureus, Klebsiella pneumoniae, Mycobacterium smegmatis.*

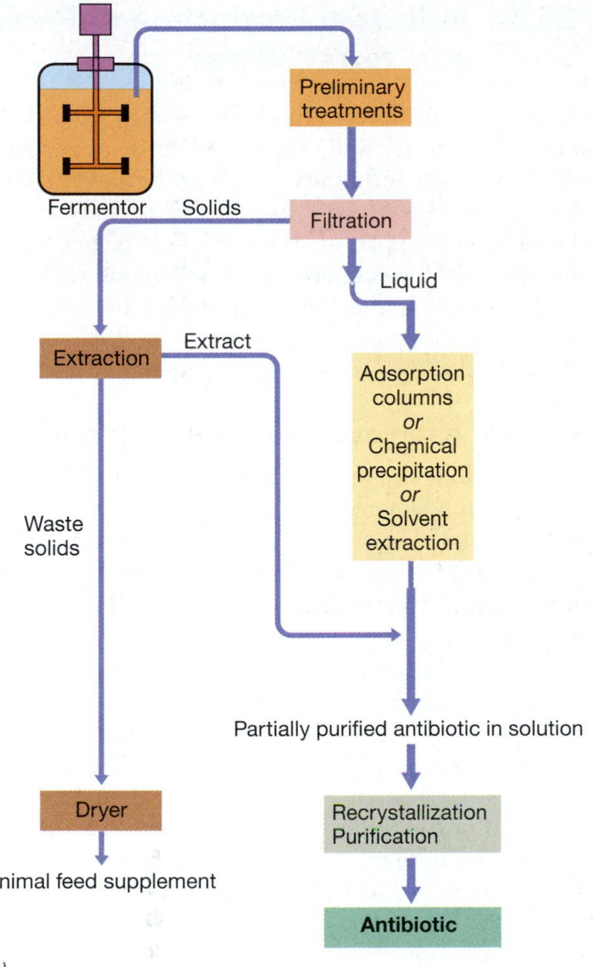

Figure 25.8 Purification of an antibiotic. *(a)* Overall process of extraction and purification. *(b)* Installation for the solvent extraction of an antibiotic from fermentation broth. Effective engineering is as important as good microbiology in the successful production of an antibiotic.

bacteria. Various test bacteria are used and are selected to be representative of, or related to, bacterial pathogens against which the antibiotics would actually be used.

The classical procedure for testing new microbial isolates for antibiotic production is the *cross-streak method* (Figure 25.7*b*). Those isolates that show evidence of antibiotic production are then studied further to determine if the compounds they produce are new. Most of the isolates obtained produce known antibiotics, so the industrial microbiologist must quickly identify such organisms and their products so that time and resources are not wasted in further study of them.

Once an organism producing a new antibiotic is discovered, the antibiotic is produced in sufficient amounts for structural analyses and then tested for toxicity and therapeutic activity in animals. Unfortunately, most new antibiotics fail these tests. However, a few prove to be medically useful and go on to be produced commercially. The time and costs in developing a new antibiotic, from discovery to clinical usage, averages 15 years and $1 billion ($US). This involves many levels of clinical trials, which alone can take several years to complete, analyze, and submit for U.S. Food and Drug Administration (FDA) approval. **www.microbiologyplace.com Online Tutorial 25.1: Isolation and Screening of Antibiotic Producers**

Purification

An antibiotic that is to be produced commercially must first be produced successfully in large-scale industrial fermentors (Section 25.4). The next challenge is to purify the product efficiently, and elaborate methods for extraction and purification of the antibiotic are often necessary (**Figure 25.8**). If the antibiotic is soluble in an organic solvent, it may be relatively simple to purify it by extraction into a small volume of the solvent. If the antibiotic is not solvent soluble, then it must be removed from the fermentation liquid by adsorption, ion exchange, or chemical precipitation (Figure 25.8). The goal is to eventually obtain a crystalline product of high purity. Depending on the process, further purification steps may be necessary to remove traces of microbial cells or cell products if they co-purify with the antibiotic.

Increasing the Yield

Rarely do antibiotic-producing strains just isolated from nature produce the desired antibiotic at sufficiently high concentration that commercial production can begin immediately. One of the major tasks of the industrial microbiologist is thus to isolate *high-yielding strains*. Strain selection involves mutagenesis of the initial culture, plating of mutant types, and testing of these mutants for antibiotic production. Genetic engineering can assist in this process. For example, gene amplification makes it possible to insert additional copies of genes of interest into a cell by means of a vector such as a plasmid (Chapter 26). Alterations in regulatory processes also may increase yields, but due to the complexity of antibiotic biosynthetic pathways, this can be a major challenge in itself.

Product yield is a critical issue with virtually all pharmaceuticals. Even after commercial production of an antibiotic or other product has begun, research often continues to identify or produce higher-yielding strains or to modify the process in some way so as to further increase yield. Although pharmaceuticals are produced on a small scale compared with commodity chemicals or agricultural products, there are nevertheless good economic reasons to seek the highest possible yields in the shortest period of time.

25.5 MiniReview

Industrial production of antibiotics begins with screening for antibiotic producers. Once new producers are identified, purification and chemical analyses of the antimicrobial agent are performed. If the new antibiotic is biologically active in experimental animals and later in human clinical trials, the industrial microbiologist may genetically modify the producing strain to increase yields to levels acceptable for commercial development.

- What is the natural habitat of most antibiotic-producing microorganisms?
- What is meant by the word "screening" in the context of finding new antibiotics?

25.6 Industrial Production of Penicillins and Tetracyclines

Once a new antibiotic has been characterized and proven medically effective and nontoxic in tests on experimental animals, it is ready for clinical trials on humans. If the new drug proves clinically effective and passes toxicity and other tests, it is given FDA approval and is ready to be produced commercially. For antibiotics such as penicillin and tetracycline, these hurdles were cleared long ago. Today, literally *tons* of these two antibiotics are produced for medical and veterinary use, and we thus focus on their production here.

β-Lactam Antibiotics: Penicillin and Its Relatives

The penicillins are a class of **β-lactam antibiotics** characterized by the β-lactam ring and are produced by fungi of the genera *Penicillium* and *Aspergillus* and by certain prokaryotes (**Figure 25.9**). Commercial penicillin is produced in the United States using high-yielding strains of the mold *Penicillium chrysogenum*.

Clinically useful penicillins are of several different types. The parent structure of all penicillins is 6-aminopenicillanic acid (6-APA), which consists of a thiazolidine ring with a condensed β-lactam ring (Figure 25.9). The 6-APA carries a variable side chain in position 6. If the penicillin fermentation is carried out without addition of side-chain precursors, the **natural penicillins**, a suite of related compounds, are produced (Figure 25.9). However, the fermentation can be targeted by adding to the culture broth a side-chain precursor so that only one type of penicillin is produced in greatest amount. The product formed under these conditions is called **biosynthetic penicillin** (Figure 25.9).

To produce the most useful penicillins, those with activity against gram-negative bacteria, a combined fermentation and chemical approach is used that leads to the production of **semisynthetic penicillins** (Figure 25.9). To produce semisynthetic penicillins, a natural penicillin is treated to yield 6-APA; the latter is then chemically modified by the addition of a side chain (Figure 25.9). Semisynthetic penicillins have many

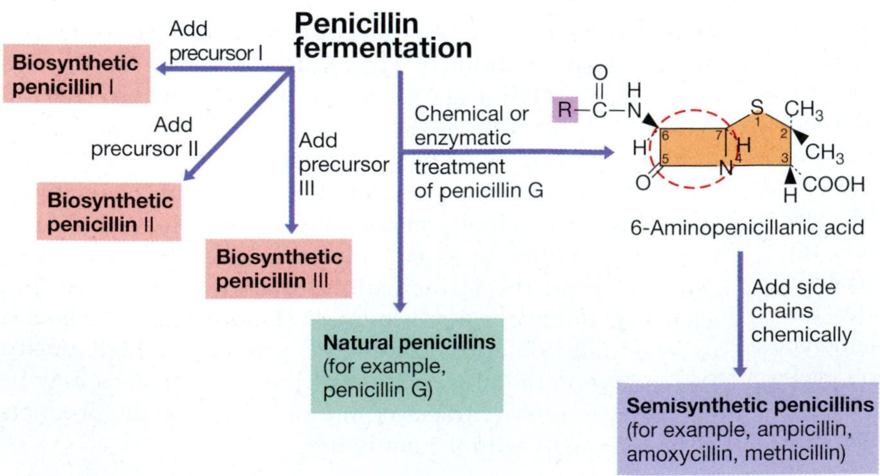

Figure 25.9 Industrial production of penicillins. The β-lactam ring is circled in red. The normal fermentation leads to the natural penicillins. If specific precursors are added during the fermentation, various biosynthetic penicillins are formed. Semisynthetic penicillins are produced by chemically adding a specific side chain to the 6-aminopenicillanic acid nucleus on the "R" group shown in purple. Semisynthetic penicillins are the most widely prescribed of all the penicillins today, primarily because of their broad spectrum of activity and ability to be taken orally.

significant clinical advantages. These include a typically broad spectrum of activity and the fact that most of them can be taken orally and thus do not require injection. The widely prescribed drug *ampicillin* is a good example here. For these reasons, semisynthetic penicillins make up the bulk of the penicillin market today.

Production of Penicillins

Penicillin G is produced in 40,000–200,000 liter fermentors. Penicillin production is a highly aerobic process, and efficient aeration is thus necessary. Penicillin is a typical secondary metabolite. During the growth phase, very little penicillin is produced, but once the carbon source has been nearly exhausted, the penicillin production phase begins. By supplying additional carbon and nitrogen at just the right times, the production phase can be extended for several days (**Figure 25.10**).

A major ingredient of penicillin production media is corn steep liquor. This substance supplies the fungus with nitrogen and growth factors. High levels of glucose repress penicillin production, but high levels of lactose do not, so lactose (from whey) is added to the corn steep liquor in large amounts as carbon source. As the lactose becomes limiting and cell densities in the fermentor become very high, "feedings" with glucose maximize penicillin yield (Figure 25.10). At the end of the production phase the cells are removed by filtration and the pH of the medium is lowered. The penicillin can then be extracted and concentrated into an organic solvent and, finally, crystallized.

Other β-Lactam Antibiotics

Other β-lactam antibiotics include the cephalosporins, drugs that also contain a β-lactam ring (Figure 25.9) but that form a dihydrothiazine ring system instead of the thiazolidine ring system of the penicillins. Cephalosporins were first discovered as products of the fungus *Cephalosporium acremonium*, but a number of other fungi as well as some prokaryotes also produce these antibiotics. In addition, a number of semisynthetic cephalosporins (analogous to the semisynthetic penicillins, Figure 25.9), are commercially produced. Cephalosporins are valued clinically not only because of their low toxicity but also because they are **broad-spectrum antibiotics**, useful against a wide variety of bacterial pathogens.

Production of Tetracyclines

The biosynthesis of **tetracyclines** has a large number of enzymatic steps. In the biosynthesis of chlortetracycline (**Figure 25.11**), for example, there are more than 72 intermediates.

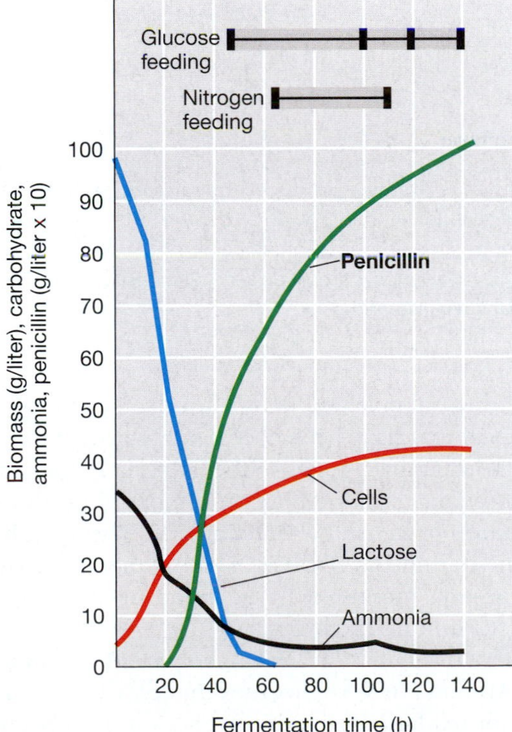

Figure 25.10 Kinetics of the penicillin fermentation with *Penicillium chrysogenum*. Note how penicillin is produced as cells are entering the stationary phase and most of the carbon and nitrogen is exhausted. Nutrient "feedings" keep penicillin production high.

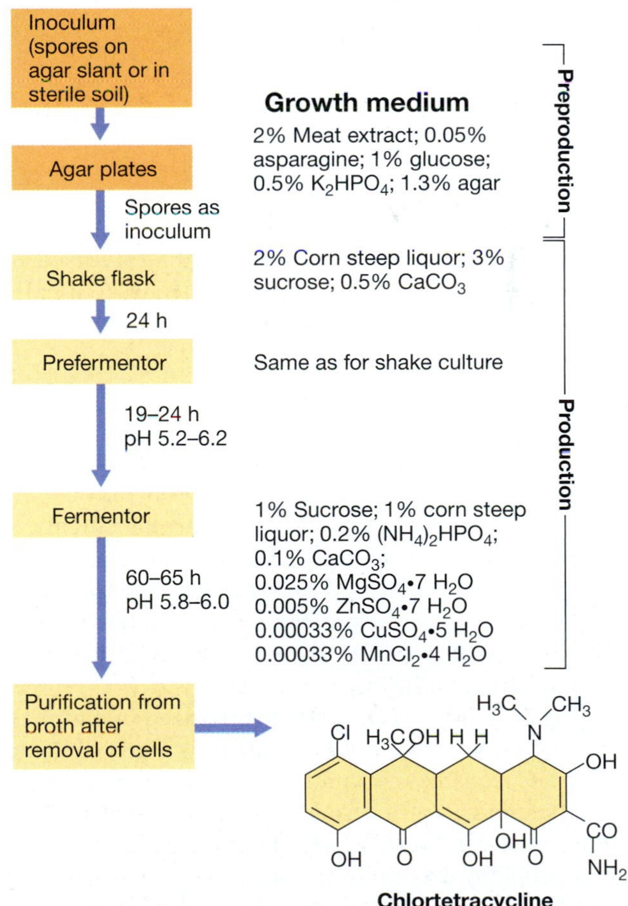

Figure 25.11 Production scheme for chlortetracycline using *Streptomyces aureofaciens*. The structure of chlortetracycline is shown on the bottom right. Glucose is used to grow the inoculum, but not for commercial production.

Genetic studies of *Streptomyces aureofaciens*, the producing organism in the chlortetracycline fermentation, have shown that a total of more than 300 genes are involved! With such a large number of genes, regulation of biosynthesis of this antibiotic is obviously quite complex. However, some key regulatory signals are known and are accounted for in the production scheme. For example, chlortetracycline synthesis is repressed by both glucose and phosphate. Phosphate repression is especially significant, and so the medium used in commercial production contains relatively low phosphate concentrations. Figure 25.11 shows a tetracycline production scheme. As in penicillin production, corn steep liquor is used, but sucrose rather than lactose is used as carbon source. Glucose is avoided because glucose strongly represses antibiotic production through catabolite repression (Section 9.9).

25.6 MiniReview

Major antibiotics of clinical significance include the β-lactam antibiotics penicillin and cephalosporin, and the tetracyclines. All of these are typical secondary metabolites, and their industrial production has been highly refined.

- What chemical structure is common to both penicillin and cephalosporin?
- In penicillin production, what is meant by the term semisynthetic? Biosynthetic?

25.7 Vitamins and Amino Acids

Vitamins and amino acids are nutrients that are used in the pharmaceutical, nutritional supplements (nutraceutical), and food industries. Of these, several are produced on an industrial scale by microorganisms.

Vitamins

Vitamins are used as supplements for human food and animal feeds, and production of vitamins is second only to that of antibiotics in terms of total sales of pharmaceuticals. Most vitamins are made commercially by chemical synthesis. However, a few are too complicated to be synthesized inexpensively but can be made in sufficient quantities relatively easily by microbial processes. Vitamin B_{12} and riboflavin are the most important of these vitamins.

Vitamin B_{12} (**Figure 25.12*a***) is synthesized in nature exclusively by microorganisms but is required as a growth factor by all animals. As a coenzyme, vitamin B_{12} plays an important role in animal biochemistry in certain types of methyl transfers and related processes. In humans, a major deficiency of vitamin B_{12} leads to a debilitating condition called *pernicious anemia*, characterized by low production of red blood cells and nervous system disorders. The requirements of animals for vitamin B_{12} are satisfied by food intake or by absorption of the vitamin produced in the animal gut by intestinal bacteria. Plants do not produce or use vitamin B_{12}, and thus it is important that strict vegetarians (vegans) supplement their diet with that vitamin.

(a) **B_{12}**

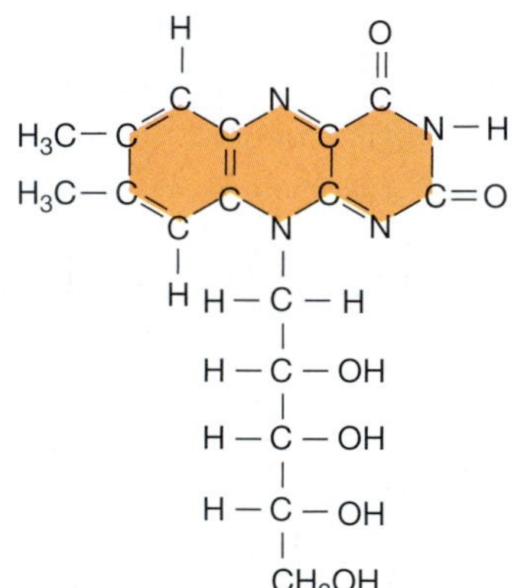

(b) **Riboflavin**

Figure 25.12 Vitamins produced by microorganisms on an industrial scale. *(a)* Vitamin B_{12}. Shown is the structure of cobalamin; note the central cobalt atom. The actual coenzyme form of vitamin B_{12} contains a deoxyadenosyl group attached to Co above the plane of the ring. *(b)* Riboflavin (vitamin B_2).

For industrial production of vitamin B_{12}, microbial strains are employed that have been specifically selected for their high yields of the vitamin. Species of the bacteria *Propionibacterium* and *Pseudomonas* are the main commercial producers, especially *Propionibacterium freudenreichii*. The metal cobalt is present in vitamin B_{12} (Figure 25.12*a*), and yields of the vitamin are greatly increased by addition of small amounts of cobalt to the culture medium.

Riboflavin (Figure 25.12*b*) is the parent compound of the flavins, FAD and FMN, coenzymes that play important roles in enzymes for oxidation–reduction reactions (Section 5.11). Riboflavin is synthesized by many microorganisms, including bacteria, yeasts, and fungi. The fungus *Ashbya gossypii* naturally produces large amounts of this vitamin (up to 7 g/l) and is therefore used for most of the microbial production processes. However, despite this good yield, there is significant

Table 25.3 Amino acids used in the food industry[a]

Amino acid[b]	*Annual production worldwide (metric tons)*	*Uses*	*Purpose*
L-Glutamate (monosodium glutamate, MSG)	370,000	Various foods	Flavor enhancer; meat tenderizer
L-Aspartate and L-alanine	5,000	Fruit juices	"Round off" taste
Glycine	6,000	Sweetened foods	Improves flavor; starting point for organic syntheses
L-Cysteine	700	Bread Fruit juices	Improves quality Antioxidant
L-Tryptophan + L-Histidine	400	Various foods, dried milk	Antioxidant, prevent rancidity; nutritive additives
Aspartame (made from L-phenylalanine + L-aspartic acid)	7,000	Soft drinks, chewing gum, many other "sugar-free" products	Low-calorie sweetener
L-Lysine	70,000	Bread (Japan), feed additives	Nutritive additive
DL-Methionine	70,000	Soy products, feed additives	Nutritive additive

[a]Data from Glazer, A. N., and H. Mikaido. 1995. *Microbial Biotechnology*, W. H. Freeman, New York.
[b]The structures of these amino acids are shown in Figure 3.12.

economic competition between the microbiological process and strictly chemical synthesis.

Amino Acids

Amino acids are extensively used in the food industry as feed additives, in the nutraceutical industry as nutritional supplements, and as starting materials in the chemical industry (**Table 25.3**). The most important commercial amino acid is glutamic acid, which is used as a flavor enhancer (monosodium glutamate, MSG). Two other important amino acids, aspartic acid and phenylalanine, make up the artificial sweetener **aspartame**, a nonnutritive sweetener of diet soft drinks and other foods sold as low-calorie or sugar-free products.

Lysine is an essential amino acid for humans and farm animals and is commercially produced by the bacterium *Brevibacterium flavum* for use as a food additive. Amino acids are used for the biosynthesis of proteins, and thus their production is strictly regulated, as was discussed in Chapter 9. However, for industrial production these regulatory mechanisms must be circumvented in order to obtain an *overproducing* strain capable of producing the amino acid economically.

The production of lysine in *B. flavum* is controlled at the level of the enzyme aspartokinase; excess lysine feedback inhibits the activity of this enzyme (**Figure 25.13**; the phenomenon of feedback inhibition was described in Section 5.18). However, overproduction of lysine can be obtained by isolating mutants of *B. flavum* in which aspartokinase is no longer subject to feedback inhibition. This is done by isolating mutants resistant to the lysine analog *S*-aminoethylcysteine (AEC); this compound binds to the allosteric site of aspartokinase and inhibits activity of the enzyme (Figure 25.13*b*). However, AEC-resistant mutants can be obtained easily and contain a modified form of aspartokinase whose allosteric site no longer recognizes AEC or lysine; feedback inhibition of this enzyme by lysine is thus greatly reduced. AEC-resistant mutants of *B. flavum* can produce over 60 g of lysine per liter in industrial fermentors, a concentration sufficiently high to make the process commercially viable.

25.7 MiniReview

Vitamins produced by industrial microbiology include vitamin B_{12} and riboflavin, and the major amino acids produced commercially are glutamate, aspartate, phenylalanine, and lysine. High yields of amino acids are obtained by modifying regulatory signals that control their synthesis such that overproduction occurs.

- Which amino acid is commercially produced in the greatest amounts?
- How can overcoming feedback inhibition improve the yield of an amino acid?

25.8 Steroids and Other Biotransformations

We discussed the role of sterols in eukaryotic membranes in Section 4.3. Steroids are derivatives of sterols and are important animal hormones that regulate various metabolic processes. Some steroids are also used as drugs in human medicine. Members of one group, the *corticosteroids*, reduce inflammation. Hence, these compounds are effective therapeutic agents in controlling the symptoms of topical inflammation, arthritis, and allergies. Members of another group, the estrogens and androgenic steroids, play a role in human fertility and can be used therapeutically in the control of fertility or as stimulants for building muscle mass.

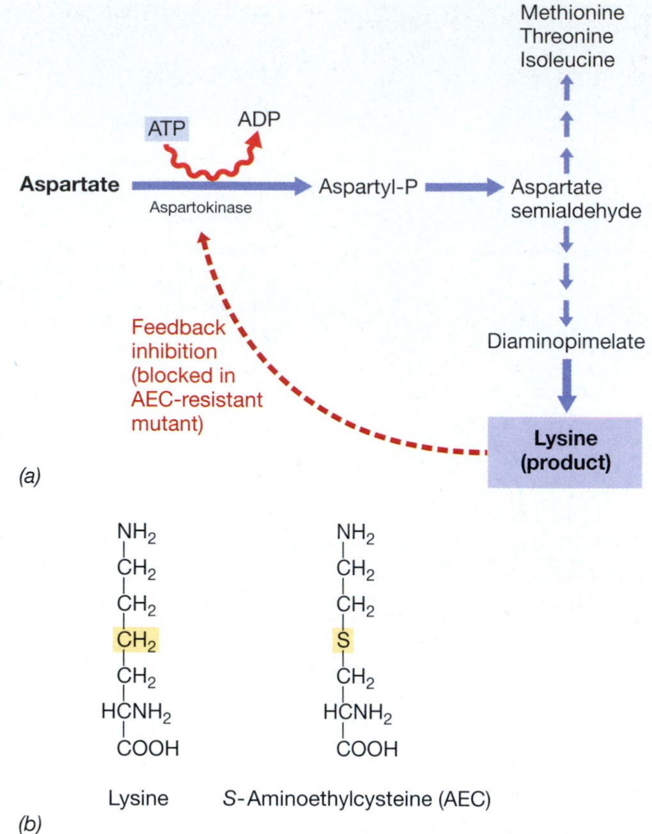

Figure 25.13 Industrial production of lysine using *Brevibacterium flavum*. *(a)* Biochemical pathway leading from aspartate to lysine; note that lysine can feedback-inhibit activity of the enzyme aspartokinase, leading to cessation of lysine production. *(b)* Structure of lysine and the lysine analog *S*-aminoethylcysteine (AEC). The molecules are identical except for the regions shown in color. AEC normally inhibits growth, but AEC-resistant mutants of *B. flavum* have an altered allosteric site on their aspartokinase and grow and overproduce lysine because feedback inhibition no longer occurs.

Steroids can be synthesized chemically, but this is often a complicated and expensive process. However, certain chemically difficult steps in steroid production can be catalyzed by microorganisms, and hence, commercial production of steroids typically has at least one microbial step. Such a targeted process is called a **biotransformation**. For a biotransformation, the organism is grown in large fermentors, and a steroid precursor is added at the appropriate phase of growth. The precursor is obtained from inexpensive starting materials, typically stigmasterol, a by-product of the soybean industry. Following an incubation period during which the sterol precursor is biotransformed, the fermentation broth is extracted and the sterol is purified.

Cortisone and Hydrocortisone

In the production of hydrocortisone and cortisone, corticosteroids used to reduce swelling and itching from minor skin irritations, the fungus *Rhizopus nigricans* carries out a key biotransformation, the stereospecific hydroxylation of a cortisone precursor (**Figure 25.14**). Most steroid biotransformations require hydroxylations of this type, and various fungi are used industrially to carry out the stereospecific hydroxylation of the steroid at a particular site on the steroid precursor.

Although most biotransformations are not beyond the capabilities of organic chemistry, microorganisms carry out the reactions more economically. Steroid production is currently a big business, as worldwide sales of the four major corticosteroids, hydrocortisone, cortisone, prednisone, and prednisolone, amount to over 800 tons/year.

25.8 MiniReview

Microbial biotransformation employs microorganisms to catalyze a specific step or steps in an otherwise strictly chemical synthesis.

- Give an example of a microbial biotransformation. Why is it necessary?
- Describe two ways in which a biotransformation differs from a typical fermentation, such as the production of antibiotics.

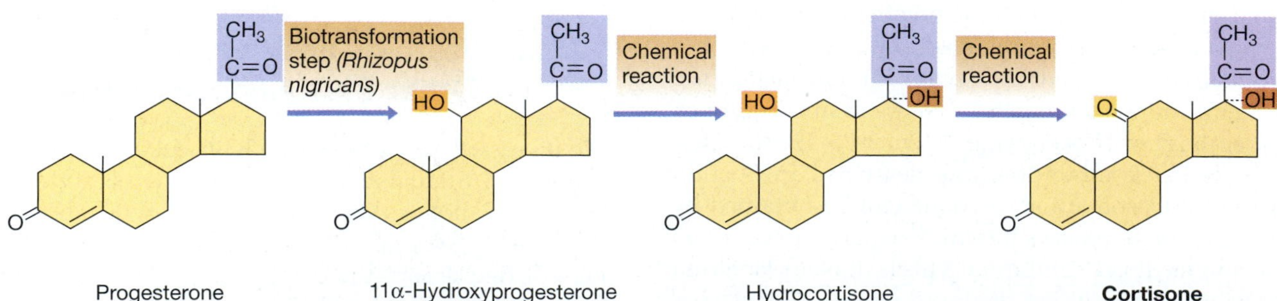

Figure 25.14 Cortisone production using a microorganism: an example of a biotransformation. The first reaction in cortisone production is a typical microbial biotransformation, the formation of 11 α-hydroxyprogesterone from progesterone. This specific oxidation, carried out by the fungus *Rhizopus nigricans*, bypasses a difficult chemical synthesis and is stereospecific. All the other steps from 11 α-hydroxyprogesterone to the steroid hormone cortisone are performed chemically.

Table 25.4 Microbial enzymes and their applications

Enzyme	Source	Application	Industry
Amylase (starch-digesting)	Fungi	Bread	Baking
	Bacteria	Starch coatings	Paper
	Fungi	Syrup and glucose manufacture	Food
	Bacteria	Cold-swelling laundry starch	Starch
	Fungi	Digestive aid	Pharmaceutical
	Bacteria	Removal of coatings (desizing)	Textile
	Bacteria	Removal of stains; detergents	Laundry
Protease (protein-digesting)	Fungi	Bread	Baking
	Bacteria	Spot removal	Dry cleaning
	Bacteria	Meat tenderizing	Meat
	Bacteria	Wound cleansing	Medicine
	Bacteria	Desizing	Textile
	Bacteria	Household detergent	Laundry
Invertase (sucrose-digesting)	Yeast	Soft-center candies	Candy
Glucose oxidase	Fungi	Glucose removal, oxygen removal	Food
		Test paper for diabetes	Pharmaceutical
Glucose isomerase	*Bacteria*	High-fructose corn syrup	Soft drink
Pectinase	Fungi	Pressing, clarification	Wine, fruit juice
Rennin	Fungi	Coagulation of milk	Cheese
Cellulase	*Bacteria*	Fabric softening, brightening; detergent	Laundry
Lipase	Fungi	Break down fat	Dairy, laundry
Lactase	Fungi	Breaks down lactose to glucose and galactose	Dairy, health foods
DNA polymerase	*Bacteria* *Archaea*	DNA replication in polymerase chain reaction (PCR) technique (Section 12.8)	Biological research; forensics

UNIT 5

25.9 Enzymes as Industrial Products

Every organism produces many enzymes, most of which are made in only small amounts and function within the cell. However, certain enzymes are produced in much larger amounts by some organisms, and instead of being held within the cell, are excreted into the medium. These extracellular enzymes, called **exoenzymes**, can digest insoluble polymers such as cellulose, protein, and starch. The products of digestion are then transported into the cell where they are used as nutrients for growth.

Enzymes are especially useful as industrial catalysts because of their high specificity for substrate. Moreover, in many cases (for example the sterols, just discussed), they catalyze reactions in a stereospecific manner, producing only one of two possible stereoisomers (Section 3.6). Various enzymes are used in the food and health industries, primarily as dietary supplements, and many others are produced for applications in the laundry and textile industries (**Table 25.4**).

Proteases, Amylases, and High-Fructose Syrup

Enzymes are produced industrially from both fungi and bacteria. The microbial enzymes produced in the largest amounts on an industrial basis are the bacterial **proteases**, used as additives in laundry detergents. Most laundry detergents contain enzymes, usually proteases, but also amylases, lipases, and reductases. Many of these enzymes are isolated from alkaliphilic bacteria, organisms that grow best at alkaline pH (Section 6.15). The main producers are species of *Bacillus*, such as *Bacillus licheniformis* (Table 25.4). These enzymes, which have pH optima between 9 and 10, remain active at the alkaline pH of laundry detergent solutions.

Other important enzymes manufactured commercially are amylases and glucoamylases, which are used in the production of glucose from starch. The glucose is then converted by a second enzyme, glucose isomerase, to fructose, which is much sweeter than glucose. The final product is *high-fructose syrup* produced from fructose-free but glucose-rich starting materials, such as corn, wheat, or potatoes. High-fructose syrups are widely used in the food industry to sweeten soft drinks, juices, and many other products. Worldwide production of high-fructose syrups is over 10 billion kilograms per year.

Extremozymes: Enzymes with Unusual Stability

In Chapters 2, 6, 16, and 17 we considered aspects of microbial growth at high temperature and saw that some prokaryotes, called *hyperthermophiles*, grow optimally at very high temperatures. Hyperthermophiles can grow at such high temperatures because they produce heat-stable macromolecules, including enzymes. The term **extremozyme** has been coined to describe enzymes that function at some environmental extreme, such as high or low temperature or pH (**Figure 25.15**). The organisms that produce extremozymes are called *extremophiles* (Table 2.1).

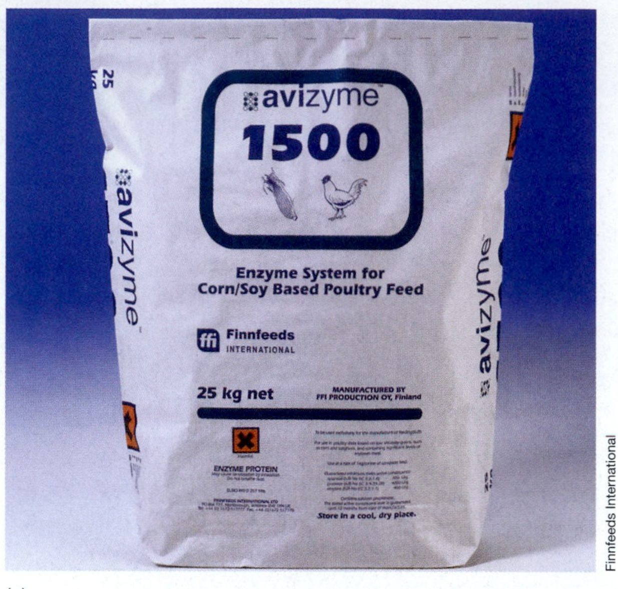

(a)

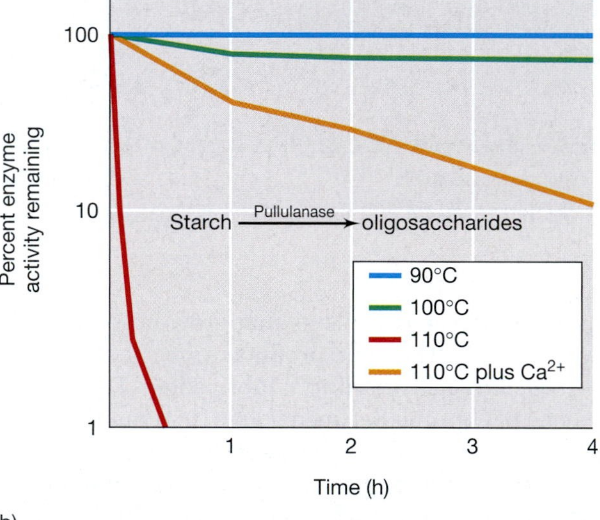

(b)

Figure 25.15 Examples of extremozymes, enzymes that function under environmentally extreme conditions. *(a)* Acid-tolerant enzymes. An enzyme mixture used as a feed supplement for poultry. The enzymes function in the bird's stomach to digest fibrous materials in the feed, thereby improving the nutritional value of the feed and promoting more rapid growth. *(b)* Thermostable enzymes. Thermostability of the enzyme pullulanase from *Pyrococcus woesei*, a hyperthermophile whose growth temperature optimum is 100°C. Calcium improves the heat stability of this enzyme; however, at 110°C, the enzyme denatures.

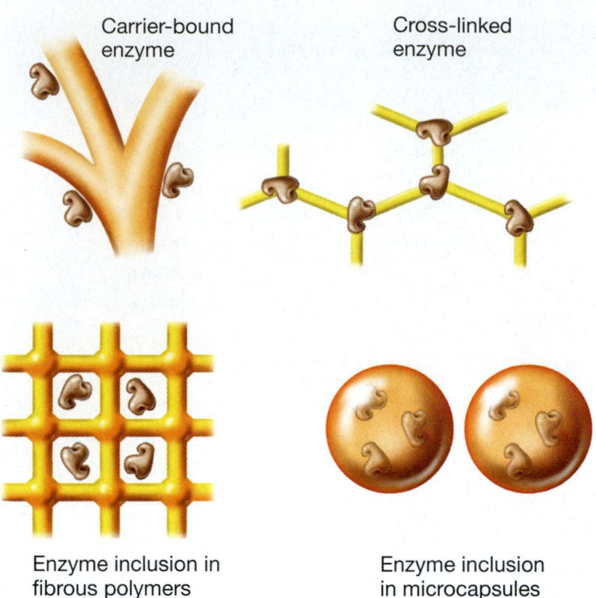

Figure 25.16 Procedures for the immobilization of enzymes. The procedure used varies with the enzyme, the product, and the production scale employed.

Many industrial catalyses operate best at high temperatures, and so extremozymes from hyperthermophiles are widely used in both industry and research. Besides the *Taq* and *Pfu* DNA polymerases used in the polymerase chain reaction (PCR) (∞ Section 12.8) thermostable proteases, amylases, cellulases, pullulanases (Figure 25.15*b*), and xylanases have been isolated and characterized from one or another species of hyperthermophile. However, it is not only thermal stable enzymes that have found a market. Cold-active enzymes (obtained from psychrophiles), enzymes that function at high salt (obtained from halophiles), and enzymes active at high or low pH (obtained from alkaliphiles and acidophiles, respectively) (Figure 25.15*a*) have been commercialized.

Immobilized Enzymes

For some industrial processes it is desirable to attach an enzyme to a solid surface to form an **immobilized enzyme**. Immobilization not only makes it easier to carry out the enzymatic reaction under large-scale continuous flow conditions, but also helps stabilize the enzyme to prevent denaturation. Different immobilized enzymes have been employed in industrial processes, and a good example is in the starch-processing industry, mentioned previously. Starch is converted to high-fructose corn syrup by sequential treatment with amylase and glucose isomerase, but typically conventional treatment converts only about 50% of the glucose to fructose. However, this yield can be increased significantly by removing the fructose on a continuous basis and recycling the remaining glucose over an immobilized enzyme column.

There are three ways in which enzymes can be immobilized (**Figure 25.16**):

1. ***Bonding of the enzyme to a carrier.*** The bonding can be through adsorption, ionic bonding, or covalent bonding. Carriers used include modified celluloses, activated carbon, clay minerals, aluminum oxide, and glass beads.
2. ***Cross-linking (polymerization) of enzyme molecules.*** Enzyme molecules are linked to each other by chemical reaction with a cross-linking agent such as dilute glutaraldehyde. The amino groups of the enzyme's amino acids react with the aldehyde.

3. ***Enzyme inclusion.*** Enzymes can be enclosed in microcapsules, gels, semipermeable polymer membranes, or fibrous polymers such as cellulose acetate.

Each of these methods for the use of immobilized enzymes has advantages and disadvantages, and the procedure used depends on the enzyme, the particular industrial application, and the scale of the operation.

25.9 MiniReview

Many microbial enzymes are used in the laundry industry to remove stains from clothing, and thermostable and alkali-stable enzymes have many advantages for this application. Enzymes from extremophiles are desirable for catalyses under extreme conditions. When an enzyme is used in a large-scale process, it is often useful to immobilize it by bonding it to an inert substrate.

- How are enzymes of use in the laundry industry?
- What key enzyme is needed to produce high-fructose syrups?
- What is an extremozyme?

III PRODUCTS FOR THE FOOD INDUSTRY

Although industrial microbiology as a discipline can trace its roots to the antibiotic era, antibiotics are only one class of products made on an industrial scale. Many food products are also made by industrial processes. We consider some of these now, beginning with the production of alcoholic beverages.

25.10 Wine

Most fruit juices and grains will undergo a natural fermentation by wild yeasts present on them. From these natural fermentations, particular strains of yeasts have been selected for use in the alcoholic beverage industry. We begin with wine production, a major worldwide industry that is growing rapidly with the influx of small specialty wineries, especially in the United States.

Wine Varieties

Most wine is made from grapes, and thus most wine is produced in parts of the world where quality grapes can be grown economically. These include many countries of the European Union, the United States (**Figure 25.17**), New Zealand and Australia, and South America.

There are a great variety of wines, and their quality and character vary considerably. *Dry* wines are wines in which the sugars of the juice are almost completely fermented, whereas in *sweet* wines some of the sugar is left or additional sugar is added after the fermentation. A *fortified* wine is one to which brandy or some other alcoholic spirit is added after the fermentation; sherry and port are the best-known fortified wines. A *sparkling* wine, such as champagne, is one in which considerable carbon dioxide (CO_2) is present, arising from a final fermentation by the yeast directly in the sealed bottle.

Wine Production

Wine production typically begins in the early fall with the harvesting of grapes. The grapes are crushed, and the juice, called *must*, is squeezed out. Depending on the grapes used and on how the must is prepared, either white or red wine may be produced (**Figure 25.18**). The yeasts that ferment wine are of two types: wild yeasts, which are present on the grapes as they are taken from the field and are transferred to the juice, and the cultivated wine yeast, *Saccharomyces ellipsoideus*, which is added to the juice to begin the fermentation. Wild yeasts are less alcohol-tolerant than commercial wine yeasts and can also produce undesirable compounds affecting quality of the final product. Thus, it is the practice in most wineries to kill the wild yeasts present in the must by adding "sulfites" (usually as sodium metabisulfite, $Na_2S_2O_5$) at a level of about 80 parts per million (mg/l). *S. ellipsoideus* is resistant to this concentration of sulfite and is added as a starter culture from a pure culture grown on sterilized grape juice.

The wine fermentation is carried out in fermentors of various sizes, from 200 to 200,000 liters, made of oak, cement, stone, or glass-lined metal (Figure 25.17*b*). The fermentor must be constructed so that the large amount of carbon dioxide produced during the fermentation can escape but air cannot enter, and this is accomplished by fitting the vessel with a special one-way valve.

Red and White Wines

A white wine is made either from white grapes or from the juice of red grapes from which the skins, containing the red coloring matter, have been removed. In the making of red wine, the skins, seeds, and pieces of stem, collectively called *pomace*, are left in during the fermentation. In addition to the color difference, red wine has a stronger flavor than white because of larger amounts of *tannins*, chemicals that are extracted into the juice from the grape skins during the fermentation.

In the production of a red wine, following about 5 days of fermentation, sufficient tannin and color have been extracted from the pomace that the wine can be drawn off for further fermentation in a new tank, usually for 1–2 weeks. The next step is called *racking;* the wine is separated from the sediment, which contains yeast cells and precipitate, and then stored at a lower temperature for aging, flavor development, and further clarification. The final clarification may be hastened by the addition of fining agents, materials such as casein, tannin, or bentonite clay that absorb particulates. Alternatively, the wine may be filtered through diatomaceous earth, asbestos, or membrane filters. The wine is then bottled and either stored for further aging, or sold.

Red wine is typically aged for months to several years (Figure 25.17*c*, *d*), but most white wine is sold without aging. During aging, complex chemical changes occur, including reduction of bitter components, resulting in improvement in flavor and odor, or *bouquet*. The final alcohol content of wine varies from 6% to 15% depending on the sugar content of the grapes, length of the fermentation, and strain of wine yeast used.

(a)

(b)

(c)

(d)

Figure 25.17 **Commercial wine making.** *(a)* Equipment for transporting grapes to the winery for crushing. *(b)* Large tanks where the main wine fermentation takes place. *(c)* Large barrel used for aging wine in a large winery. *(d)* Smaller barrels used in a small French winery. Wine may be aged in these wooden casks for years.

Malolactic Fermentation

Many high-quality dry red wines and a few white wines such as chardonnays are subjected to a secondary fermentation following the primary fermentation by yeast. This occurs before bottling and is called the *malolactic fermentation*. Full-bodied dry red wines are the typical candidates for malolactic fermentation.

In grape varieties used for the production of dry wines a considerable amount of malic acid can be present in the grapes. The malic acid content of the grape varies seasonally due to climatic and soil conditions. Malic acid is a sharp and rather bitter acid, but lactic acid is a softer, smoother acid. During the malolactic fermentation, malic acid is fermented to lactic acid and this makes the wine less acidic and fruity but more complex and palatable. Many other constituents are produced during the malolactic fermentation, including diacetyl (2, 3-butanedione), a major flavoring ingredient in butter; this imparts a soft, smooth character to the wine.

The malolactic fermentation is catalyzed by species of lactic acid bacteria (∞ Section 16.1), including *Lactobacillus*, *Pediococcus*, and *Oenococcus*. These organisms are extremely acid-tolerant and can carry out the fermentation even if the initial pH of the wine is as low as 3.3. Commercial wineries typically use starter cultures of selected malolactic fermentation organisms and then store the wine in barrels especially for this purpose. Inocula for future rounds of malolactic fermentation come from lactobacilli that stick to the insides of the barrel. The malolactic fermentation can take several weeks but is usually worth the wait, as the final product is often much smoother and far superior to the more sharp-tasting and bitter starting material.

25.10 MiniReview

Wine is a fermented beverage made from the fermentation of grape juice by yeast. Dozens of varieties of wine exist, each displaying regional and cultural characteristics. Complex chemical changes during the wine fermentation leave chemicals in addition to alcohol in the final product. Secondary fermentations and other aging methods are used to produce a smoother, more consistent final product.

- What production differences lead to a red wine versus a white wine?
- What occurs during the malolactic fermentation and why is it carried out?

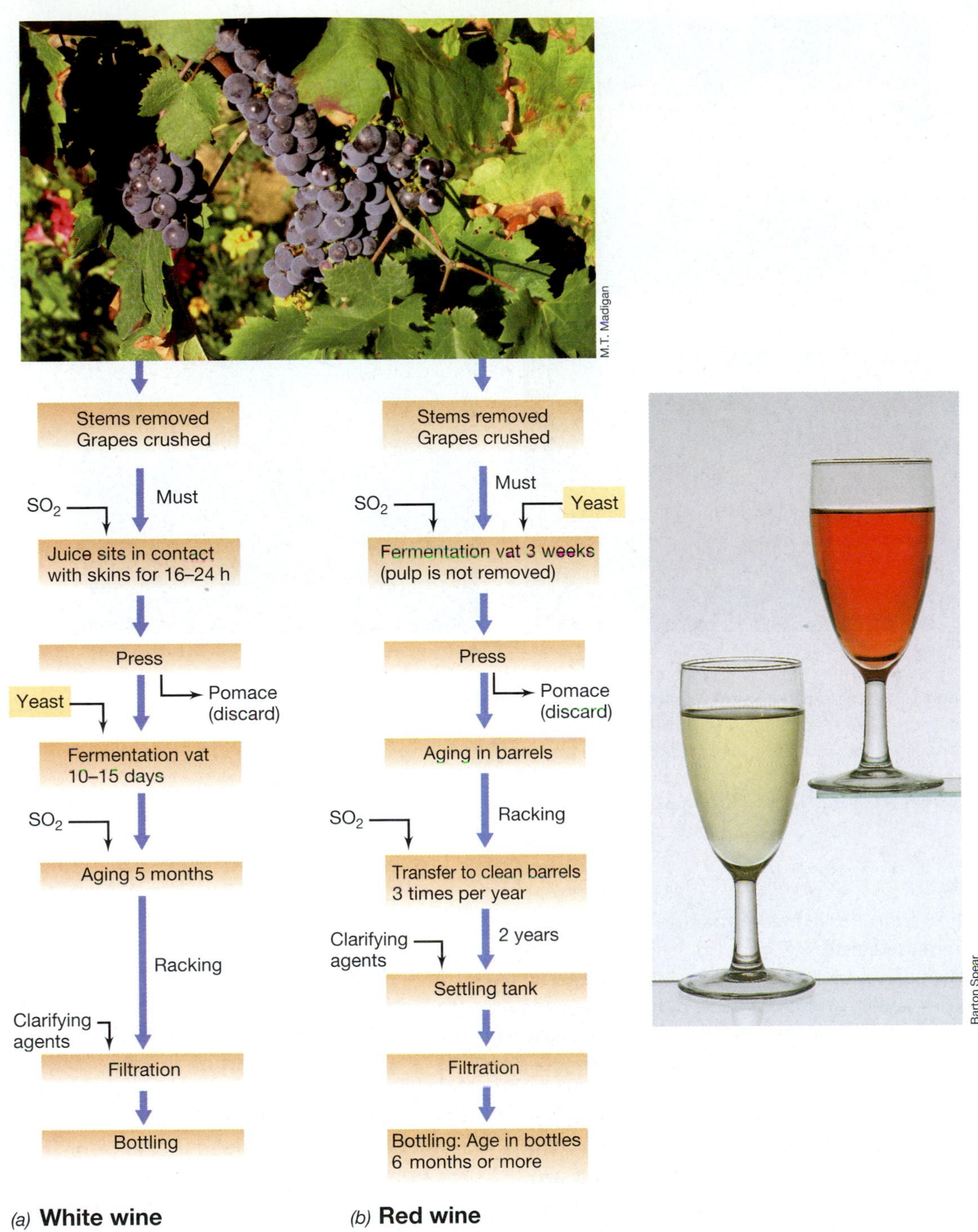

Figure 25.18 Wine production. *(a)* White wine. White wines vary from nearly colorless to straw-colored depending on the grapes used. *(b)* Red wine. Red wines vary in color from a faint red to a deep, rich burgundy. The photograph shows a glass of chenin blanc, a typical white wine (left), and a glass of a light red wine (rosé, right). Other typical varieties of white wine include Chablis, Rhine wine, sauterne, and chardonnay, and other typical red wines include burgundy, Chianti, claret, zinfandel, cabernet, and merlot.

25.11 Brewing, Distilling, and Commodity Alcohol

Beers and ales are popular alcoholic beverages produced worldwide from the fermentation of grains and other sources of starch rather than from the fermentation of grape and other fruit juices, as for wine. Although both the brewing and wine industries employ yeast to catalyze the fermentation itself, the amount of alcohol in brewed products is much lower than in wine and levels of CO_2 are typically much higher. Thus the two products—beer and wine—are quite different fermented beverages, and each has its own characteristic properties. And as for wines, regional and cultural differences can greatly influence the final brewed product.

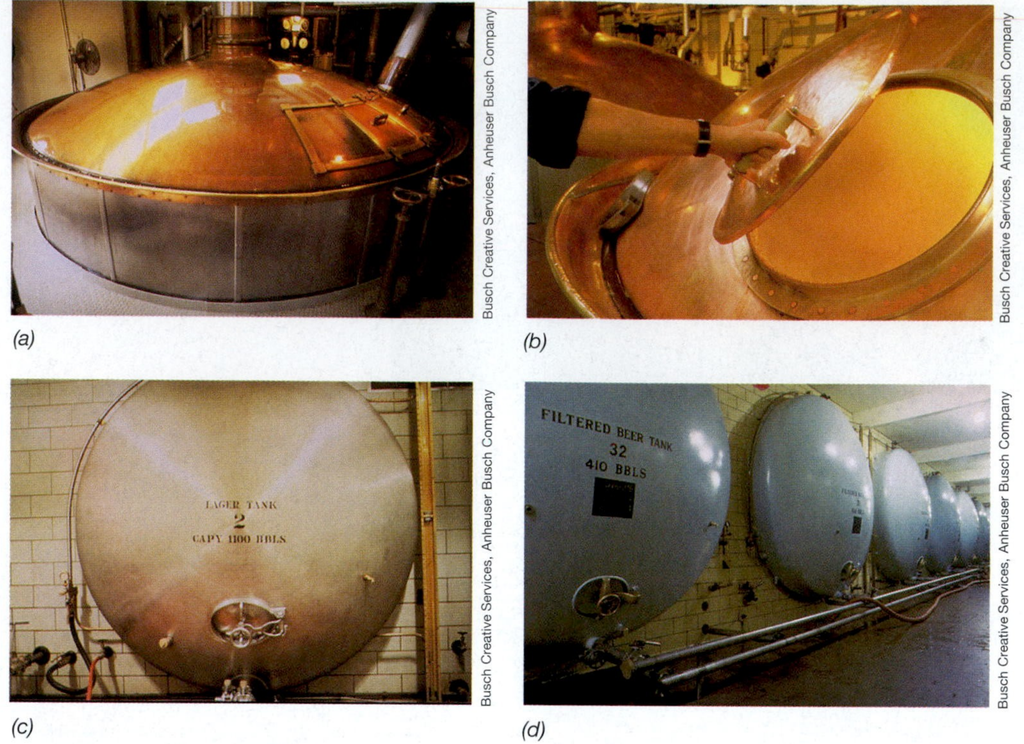

Figure 25.19 Brewing beer in a large commercial brewery. *(a, b)* The copper brew kettle is where the wort is mixed with hops and then boiled. From the brew kettle, the liquid passes to large fermentation tanks where yeast ferments glucose to ethanol plus CO_2. *(c)* If the beer is a lager, it is stored for several weeks at low temperature in tanks where particulate matter, including yeast cells, settle. *(d)* The beer is then filtered and placed in storage tanks from which it is packaged into kegs, bottles, or cans.

Making the Wort

Brewing is the term used to describe the manufacture of alcoholic beverages from malted grains (**Figure 25.19**). Typical malt beverages include beer, ale, porter, and stout. *Malt* is prepared from germinated barley seeds, and it contains natural enzymes that digest the starch of grains and convert it to sugar. Because brewing yeasts are unable to digest starch, the malting process is essential for the generation of fermentable substrates.

The fermentable liquid from which beer and ale are made is prepared by a process called *mashing*. The grain of the mash may consist only of malt, or other grains such as corn, rice, or wheat may be added. The mixture of ingredients in the mash is cooked and allowed to steep in a large mash tub at warm temperatures. During the heating period, enzymes from the malt cause digestion of the starches and liberate glucose, which will be fermented by the yeast. Proteins and amino acids are also liberated into the liquid, as are other nutrient ingredients necessary for the growth of yeast.

After cooking, the aqueous mixture, called *wort*, is separated by filtration. *Hops*, an herb derived from the female flowers of the hops plant, are added to the wort at this stage. Hops is a flavoring ingredient, but it also has antimicrobial properties, which help to prevent contamination in the subsequent fermentation. The wort is then boiled for several hours, usually in large copper kettles (Figure 25.19*a*, *b*), during which time desired ingredients are extracted from the hops, undesirable proteins present in the wort are coagulated and removed, and the wort is sterilized. The wort is filtered again, cooled, and transferred to the fermentation vessel.

The Fermentation Process

Brewery yeast strains are of two major types: *top* fermenting and *bottom* fermenting. The main distinction between the two is that top-fermenting yeasts remain uniformly distributed in the fermenting wort and are carried to the top by the CO_2 gas generated during the fermentation, whereas bottom-fermenting yeasts settle to the bottom. Top yeasts are used in the brewing of *ales*, and bottom yeasts are used to make *lager beers*. Bottom yeasts are given the species designation *Saccharomyces carlsbergensis*, and top yeasts are called *Saccharomyces cerevisiae*. Top yeasts usually ferment at higher temperatures (14–23°C) than bottom yeasts (6–12°C) and thus complete the fermentation in a shorter time (5–7 days for top fermentation versus 8–14 days for bottom fermentation).

After bottom yeasts complete lager beer fermentation, the beer is pumped off into large tanks where it is stored in the cold (about −1°C) for several weeks (Figure 25.19*c*). Following this, the beer is filtered and placed in storage tanks (Figure 25.19*d*) from which it is packaged. Top-fermented ale is stored for only short periods at a higher temperature (4–8°C), which assists in development of the characteristic ale flavor. Additional details on the brewing process are discussed in the Microbial Sidebar, "Home Brew."

Microbial Sidebar

Home Brew[a]

The amateur brewer can make many kinds of beer at home, from English bitters and India pale ale to German bock and Russian Imperial stout. The necessary equipment and supplies, including yeast, can be purchased from a local beer and winemakers shop. The Home Wine and Beer Trade Association, P.O. Box 1373, Valrico, FL 33595 (http://www.hwbta.org) can supply the address of a nearby shop.

The brewing process can be divided into three basic stages: (1) *making the wort*, (2) *carrying out the fermentation*, and (3) *bottling and aging*. The character of a brew depends on many factors: the proportion of malt, sugar, hops, and grain; the kind of yeast; the temperature and duration of the fermentation; and how the aging process is carried out. The fermentor itself consists of a 20-liter (5-gallon) narrow-necked glass jar that can be fitted with a tightly fitting closure. To brew a good quality beer, it is essential that *everything* be sterilized that comes into contact with the wort. This includes the fermentor, tubing, stirring spoon, and bottles. An easy procedure is to use a sterilizing rinse consisting of 50–60 ml of liquid bleach in 20 liters of water. Soak the items for 15 minutes and then rinse lightly with hot water or air dry.

1. **Making the wort** (**Figure 1**). In commercial brewing, the wort is made by extracting fermentable sugars and yeast nutrients from malt, sugar, and hops. Many home brewers make their own wort from malt, but a reasonably satisfactory beer can be made with hop-flavored malt extract purchased ready-made. Malt extracts come in a variety of flavors and colors, and the kind of beer depends on the type of malt extract used. A simple recipe for making the wort uses 2.25–2.75 kg (5–6 lb) of hop-flavored malt extract and 20 liters of water. The malt extract and 5–6 liters of water are brought to a boil for 15 minutes in an enamel or stainless steel container (aluminum heating kettles must be avoided because of the inhibiting action of metals leached from aluminum containers) (Figure 1). The hot wort is then poured into 14–15 liters of clean, cold water that has already been added to the fermentor. After the temperature has dropped below 30°C, the yeast is added to initiate the fermentation.
2. **Carrying out the fermentation** (**Figure 2**). Add two packs of fresh beer yeast to the cooled wort and cover the fermentor with a rubber stopper into which a plastic hose has been inserted. The hose is directed into a bucket containing water. During the initial 2–3 days of the fermentation, large amounts of CO_2 will be given off, which will exit through the hose. The water trap is to prevent wild yeasts or bacteria from the air from getting back into the fermentor. After about 3 days, the activity will diminish as the fermentable sugars are used up. At this time, the rubber stopper and hose are replaced with an inexpensive fermentation lock. The fermentation lock, which can be purchased at the home brew store, prevents contamination while permitting the small amount of gas still being produced to escape. Allow the beer to ferment for 7–10 days at 10–15°C or higher.
3. **Bottling and aging** (**Figure 3**). The fermentation should be allowed to proceed for the full 7–10 days, even if the vigorous fermentation action ceases earlier. By this time, most of the yeast should have settled to the bottom of the fermentor. Carefully siphon the beer off the yeast layer, allowing the beer but not the sediment to run into sanitized glass beer bottles. The bottles used should accept standard crown caps, and new, clean caps should be used. Before capping, add three-quarters of a teaspoon (but no more) of corn syrup to each 350- to 465-ml (12- to 16-ounce) bottle. Once the bottles are capped, turn each one upside down once to mix the syrup and then allow the beer to age upright at room temperature for at least 7–10 days. After this aging period, the beer may be stored at a cooler temperature.

All homemade beer will improve if it is allowed to age for several weeks. Aging tends to make beer smoother and remove bitter components. Using the same basic production equipment, several different types of beer can be made, each with its own distinctive taste and character (the reference listed in the footnote contains a number of beer recipes). Dark beers, which typically contain more alcohol than lighter beers, require more malt for their production and are usually brewed from a combination of different malts such as those obtained from darker varieties of

Bryon Burch

Figure 1

Bryon Burch

Figure 2

Bryon Burch

Figure 3

Home Brew *(continued)*

grain or those that have been roasted to caramelize the sugars and yield a darker color. A typical American-style light lager (**Figure 4**, left) contains about 3.5% alcohol (by volume), whereas a Munich-style dark (Figure 4, right) contains 4.25% alcohol, and bock beers contain about 5% alcohol.

The trend toward "individuality" in beer is evident not only by the growing number of home brewers, but also by the fact that major brewers in the United States are feeling more and more competition from new, usually very small, breweries called *microbreweries*. Although total production by a microbrewery may pale by comparison to that of a major brewer, the products themselves often have their own distinctive character and local appeal. Part of these differences has to do with the smaller scale on which the brewing takes place but also with the use of different sources of ingredients, water, yeast strains, and brewing procedures.

Barton Spear

Figure 4

[a]Source: Burch, B. 1992. *Brewing Quality Beers—The Home Brewer's Essential Guidebook,* 2nd edition. Joby Books, Fulton, CA

Distilled Alcoholic Beverages

Distilled alcoholic beverages are made by heating a previously fermented liquid to a temperature that volatilizes most of the alcohol. The alcohol is then condensed and collected, a process called *distilling*. A product much higher in alcohol content is obtained by this process than is possible by direct fermentation. Virtually any alcoholic liquid can be distilled, and each yields a characteristic distilled beverage. The distillation of malt brews yields *whiskey*, distilled wine yields *brandy*, distilled fermented molasses yields *rum*, distilled fermented grain or potatoes yields *vodka*, and distilled grain and juniper berries yields *gin* (**Figure 25.20**).

The distillate contains not only alcohol but also other volatile products arising either from the yeast fermentation or from the ingredients themselves. Some of these other products are desirable flavor ingredients, whereas others are undesirable. To eliminate the latter, the distilled product is almost always aged, usually in oak barrels. The aging process removes undesirable products and desirable new flavors and aromatic ingredients develop. The fresh distillate is typically colorless, whereas the aged product is often brown or yellow (Figure 25.20). The character of the final product is partly determined by the manner and length of aging, and aging times of 10 years or more are not uncommon for some distilled spirits.

Commodity Ethanol

Production of alcohol (ethanol) as a *commodity chemical* is a major industrial process, and today over 50 billion liters of alcohol are produced yearly worldwide from the fermentation of various feedstocks. In the United States most ethanol is obtained by yeast fermentation of glucose obtained from cornstarch. Ethanol is stripped from the fermentation broth by distillation (**Figure 25.21**). In other countries, for example in Brazil, a major ethanol producer, not only corn but also sugar cane, whey, sugar beets, and even wood chips and waste paper are used. For the latter the cellulose in wood is treated to release glucose, which is then fermented to alcohol. Various yeasts have been used in commodity ethanol production, including species of *Saccharomyces, Kluyveromyces*, and *Candida*, but most ethanol in the United States is produced by *Saccharomyces*.

Barton Spear

Figure 25.20 Distilled spirits. Gin and vodka (not shown) are colorless. Aging in wooden casks gives a distinctive amber or yellow color to distilled spirits. Left to right, dark rum, brandy, whiskey.

Figure 25.21 Ethanol production plant, Nebraska (USA). Here, glucose obtained from cornstarch is fermented by *Saccharomyces cerevisiae* to ethanol plus CO_2. The large tank in the left foreground is the ethanol storage tank, and the tanks and pipes in the background are for distilling ethanol from the fermentation broth.

Ethanol is used as an industrial solvent and increasingly as a gasoline supplement. In the United States *gasohol* is produced by adding ethanol to a final concentration of 10% in lead-free gasoline. The combustion of gasohol produces lower amounts of carbon monoxide and nitrogen oxides than pure gasoline. Hence gasohol is marketed as a cleaner-burning fuel and a gasoline extender and is widely available in the United States. The production of more ethanol-rich fuels such as E-85 (85% ethanol and 15% gasoline) is also growing in the United States, but this fuel can only be used in modified engines. However, E-85 fuel reduces emission of nitrogen oxides by nearly 90% and is thus a means for reducing important pollutants in the atmosphere and reducing dependence on conventional sources of oil.

Total ethanol production to meet fuel demands in the United States alone is scheduled to top 30 billion liters by the year 2012. The major downside to ethanol production is the fact that at present it takes about 25% more energy to produce a liter of ethanol than is present in the ethanol itself. However, because ethanol is a product of recently fixed carbon and not buried organic carbon (fossil fuels), the use of ethanol is considered a more sustainable and environmentally friendly way to supply the fuel needs of the immediate future.

25.11 MiniReview

Brewed products are made from malted grains or other starches and distilled spirits are made by the distillation of fermented products such as brews and wines. Commodity alcohol is used as a gasoline additive and industrial solvent.

- What are the major differences between a beer and an ale?
- How can yeast help to solve global energy problems?

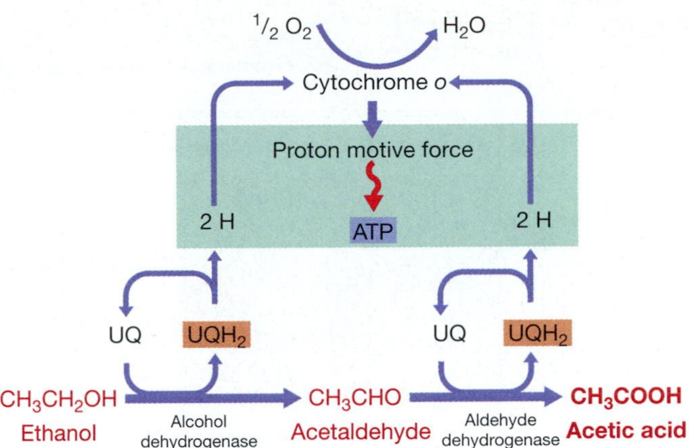

Figure 25.22 The chemistry of vinegar production. Oxidation of ethanol to acetic acid, the key process in the production of vinegar. UQ, ubiquinone.

25.12 Vinegar

Vinegar is the product resulting from the conversion of ethyl alcohol to the key ingredient, *acetic acid*, by the acetic acid bacteria. Key genera of acetic acid bacteria include *Acetobacter* and *Gluconobacter* (∞ Section 15.8). Vinegar can be produced from any substance that contains ethanol, although the usual starting material is wine, beer, fermented rice, or alcoholic apple juice (hard cider). Vinegar can also be produced from a mixture of pure alcohol in water, in which case it is called *distilled* vinegar, the term "distilled" referring to the alcohol from which the product is made rather than the vinegar itself. Vinegar is used as a flavoring agent in salads and other foods, and because of its acidity, it is also used in pickling. Meats and vegetables properly pickled in vinegar can be stored unrefrigerated for years.

The acetic acid bacteria are strictly aerobic bacteria that differ from most other aerobes in that some of them, such as species of *Gluconobacter*, do not oxidize their organic electron donors completely to CO_2 and water (**Figure 25.22**). Thus, when provided with ethyl alcohol as electron donor, they oxidize it only to acetic acid, which then accumulates in the medium. Acetic acid bacteria are quite acid-tolerant and are not killed by the low pH that they generate. There is a high oxygen demand during growth, and the main problem in the production of vinegar is to ensure sufficient aeration of the medium.

Vinegar Production

There are three different processes for the production of vinegar. The *open-vat (Orleans) method*, which is the original process, is still used in France where it was first developed. Wine is placed in shallow vats with considerable exposure to the air, and the acetic acid bacteria develop as a slimy layer on the top of the liquid. This process is not very efficient because the bacteria come in contact with both the air and the substrate only at the surface.

The second process is the *trickle (Quick vinegar) method*, in which contact between the bacteria, air, and substrate is

UNIT 5

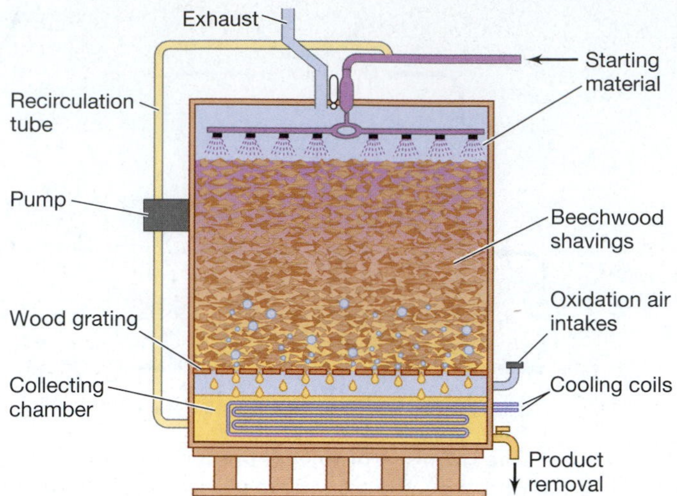

Figure 25.23 Diagram of a vinegar generator. The alcoholic juice is allowed to trickle through the wood shavings, and air is passed up through the shavings from the bottom. Acetic acid bacteria develop on the wood shavings and convert alcohol to acetic acid. The acetic acid solution accumulates in the collecting chamber and is recycled through the generator until the acetic acid content reaches at least 4%, the minimum for a product to be labeled as "vinegar."

increased by trickling the alcoholic liquid over beechwood twigs or wood shavings packed loosely in a vat or column while a stream of air enters at the bottom and passes upward. The bacteria grow on the surface of the wood shavings and thus are maximally exposed to both air and liquid. The vat is called a *vinegar generator* (**Figure 25.23**), and the whole process is operated in a continuous fashion. The life of the wood shavings in a vinegar generator is long, from 5 to 30 years, depending on the kind of alcoholic liquid used in the process.

Finally, the *bubble method* of vinegar production is basically a submerged fermentation such as was already described for antibiotic production. With proper aeration, the efficiency of the bubble method is high, and 90–98% of the alcohol is converted to acetic acid.

Although acetic acid can be easily made chemically from alcohol, the microbial product, *vinegar*, is a distinctive material, the flavor being affected by other substances present in the starting material and produced during the fermentation. For this reason the microbial process, especially the vinegar generator method, has not been supplanted by a chemical process.

25.12 MiniReview

The active ingredient in vinegar is acetic acid, which is produced by acetic acid bacteria oxidizing an alcohol-containing juice. Adequate aeration is the most important consideration in ensuring a successful vinegar process.

- Why is O_2 necessary in vinegar production?
- Why does vinegar produced by the trickle method have a more distinctive taste than vinegar produced by the bubble method?

25.13 Citric Acid and Other Organic Compounds

Some organic acids are produced by microorganisms in large amounts and can thus be commercial products. For example, *citric acid* is widely used in the food industry as a supplement in beverages, confectionaries, and other foods, and in the leavening of bread. Citrate is also used industrially to treat certain metals, as a detergent additive in place of phosphates, and in various pharmaceutical applications. Other acids commercially produced by microorganisms include itaconic acid, used in the manufacture of acrylic resins, and gluconic acid, used as calcium gluconate to treat calcium deficiencies in humans and industrially as a washing and water-softening agent. All of these acids are produced by species of fungi.

Important bacterially produced organic chemicals include sorbose, which is produced when *Acetobacter* oxidizes sorbitol and is used in the manufacture of *ascorbic acid* (vitamin C), and *lactic acid*, used in the food industry to acidify foods and beverages. Lactic acid is produced by various species of lactic acid bacteria (∞ Section 16.1).

We focus here on the citric acid fermentation. Historically, the citric acid fermentation is notable because it was the first aerobic industrial fermentation to be developed. As the citric acid fermentation was perfected, the lessons learned were applied to the penicillin and other antibiotic fermentations. Thus to some degree, we owe our current success with the large-scale production of antibiotics to the pioneering work done on the citric acid fermentation process. Overall, citric acid production is a big business, as over 550,000 tons are produced worldwide every year with a total value of over $1 billion.

Citric Acid Production: The Link to Iron

Citric acid is produced microbiologically by the mold *Aspergillus niger*. Although citric acid is normally considered in connection with the citric acid cycle, in certain organisms such as *A. niger* excretion of large amounts of citric acid can be obtained because the organism uses citrate as a chelator to scavenge iron for uptake into the cell. The citric acid fermentation is carried out aerobically in large fermentors, and a key requirement for high yields is that the medium be iron deficient (**Figure 25.24**). Therefore, the medium used for citric acid production is treated to remove most of the iron, and the fermentors themselves are constructed of stainless steel or are glass lined to prevent leaching of iron from the fermentor walls at the low pH values generated as citric acid accumulates.

Citric Acid Production: Growth Media, Conditions, and Purification

Growth media used for citric acid production contain various starting materials, including starch from potatoes, starch hydrolysates, glucose syrup from saccharified starch, sucrose, sugarcane syrup, sugarcane molasses, or sugar beet molasses. If starch is used, amylase (Table 25.4), formed by the producing fungus or added to the fermentation broth, hydrolyzes the starch to sugars. The sugars are catabolized through the

$$HO-C(CH_2-COOH)_2-COOH \rightleftharpoons HO-C(CH_2-COO^-)_2-COO^- \cdots Fe^{3+}$$

Citric acid ⇌ Citrate

(a)

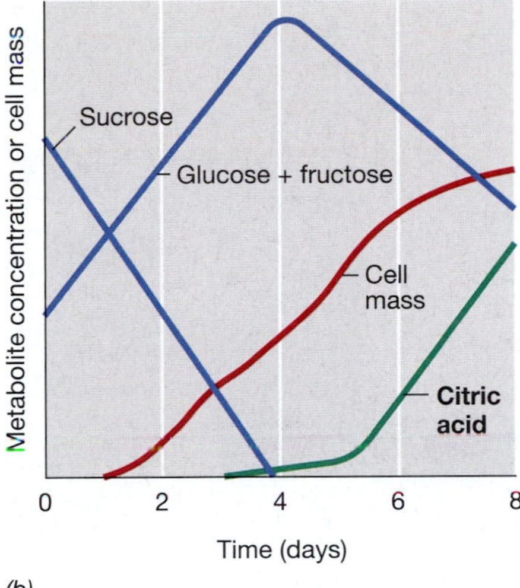

(b)

Figure 25.24 Citric acid fermentation. *(a)* Structure of citric acid. Note how the ionized form, citrate, contains three carboxylic acid groups, which can chelate ferric iron (Fe^{3+}). *(b)* Kinetics of the citric acid fermentation. Sucrose is enzymatically split by sucrase to yield glucose plus fructose.

glycolytic pathway and enter the citric acid cycle where citrate is produced (Figure 5.22).

Most citric acid today is produced by in large fermentors. However, because *A. niger* is a strict aerobe, it is crucial to this fermentation for the culture to be aerated without interruption. Citric acid is produced in this way as a typical secondary metabolite (Figure 25.2). During the growth phase, sucrose splits into glucose and fructose, and by the time stationary phase is reached, large amounts of these hexoses remain and are converted to citric acid to counter iron starvation (Figure 25.24).

Citric acid is purified following removal of fungal cells by adding lime (CaO); this precipitates citric acid as calcium citrate. The latter is concentrated by filtration and treated with sulfuric acid to form a solution of citric acid and calcium sulfate ($CaSO_4$, a solid). Following a second filtration to remove crystals of $CaSO_4$, crystals of citric acid form by evaporation.

25.13 MiniReview

A number of organic chemicals are produced commercially by microorganisms, of which the most important economically is citric acid, produced by *Aspergillus niger.*

- Why is citric acid considered a secondary metabolite?
- What is the relationship between iron and citric acid production by *A. niger*?

Table 25.5 Industrial uses of yeast and yeast products

Production of yeast cells
Baker's yeast for bread making
Dried food yeast, for food supplements
Dried feed yeast, for animal feeds
Yeast products
Yeast extract, for microbial culture media
B vitamins, vitamin D
Enzymes for food industry; invertase (sucrase), galactosidase
Biochemicals for research: ATP, NAD^+, RNA
Fermentaion products from yeast
Ethanol, for industrial alcohol and as a gasoline extender
Glycerol
Beverage alcohol
Beer, wine
Distilled beverages
Whiskey, brandy, vodka, rum

25.14 Yeast as a Food and Food Supplement

Yeasts are among the most extensively used microorganisms in industry. Besides their importance to the baking and beverage industries, yeasts are grown for the cells themselves or for compounds extracted from the cells (**Table 25.5**). However, unlike in the brewing and baking industries where *fermentation* is the key reaction of interest (Microbial Sidebar, "The Products of Yeast Fermentation and the Pasteur Effect," Chapter 5), the production of yeast cells requires *respiration* for maximum production of cell material. For food and food supplement yeast, derivatives of *Saccharomyces cerevisiae* are used almost exclusively, and we consider their industrial production here.

Yeast Cell Production

Yeast for baking or nutritional purposes is cultured in large aerated fermentors in a medium containing molasses as a major ingredient. Molasses contains large amounts of sugar as the source of carbon and energy and also contains minerals, vitamins, and amino acids used by the yeast. To make a complete medium for yeast growth, phosphoric acid (a phosphorus source) and ammonium sulfate (a source of nitrogen and sulfur) are added.

Fermentation vessels for yeast production range from 40,000 to 200,000 liters. Beginning with the pure stock culture, several intermediate stages are needed to scale up the inoculum to a size sufficient to inoculate the final stage (**Figure 25.25a**). Only a small amount of molasses is added to the commercial fermentor initially, and then as the yeast culture grows and consumes sugar, more molasses is added in controlled "feedings." This avoids sugar overload and switching of the metabolic process to the less energy-yielding fermentation.

UNIT 5

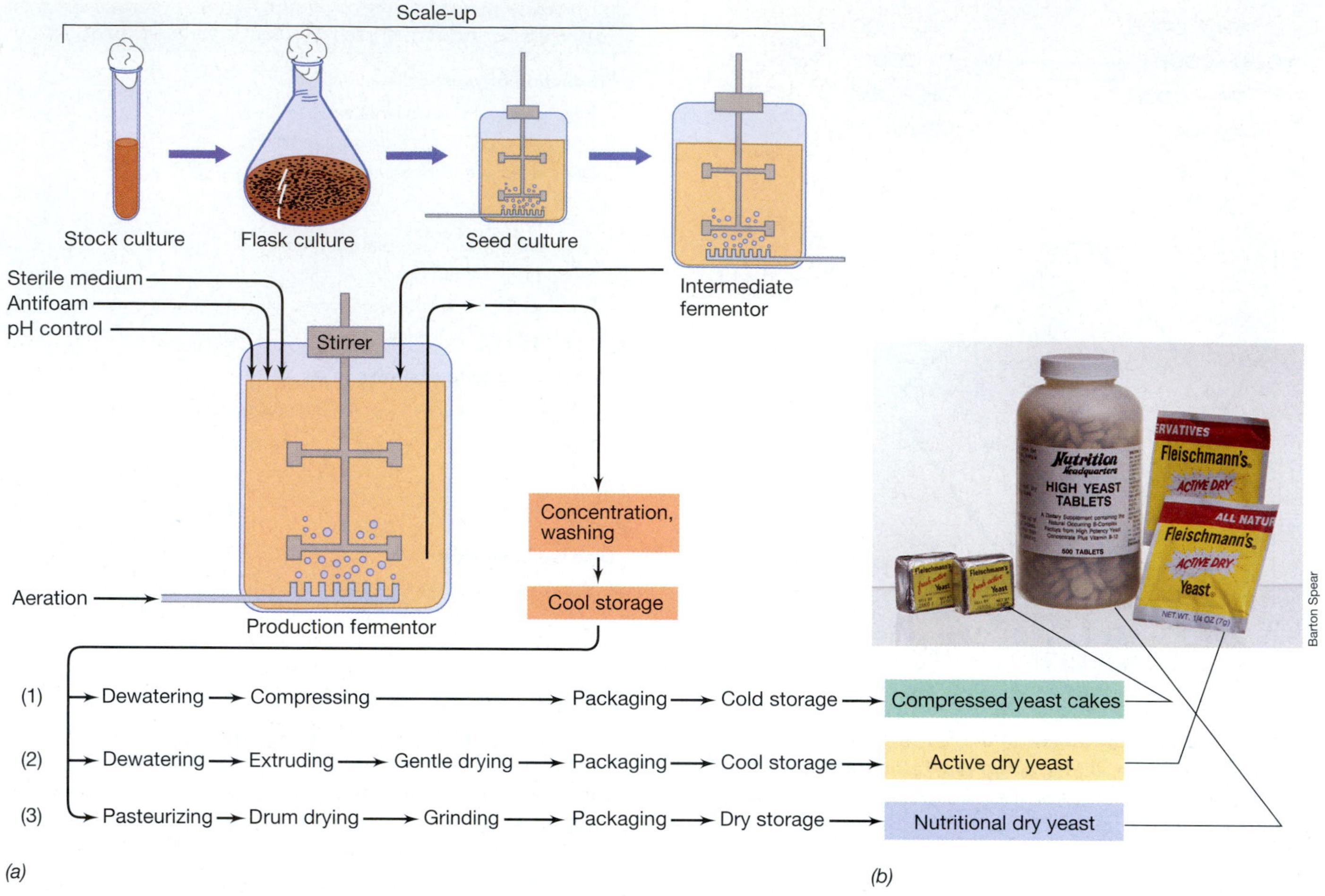

Figure 25.25 Industrial production of yeast cells. *(a)* Stages in production. Antifoaming agents are added to the fermentor to prevent aeration and stirring from creating excessive foam on the surface of the medium. *(b)* Common yeast products for consumers: yeast cakes, active dry yeast, nutritional yeast.

At the end of the growth period, the yeast cells are recovered from the broth by centrifugation. The cells are then washed with water until they are light in color. Baker's yeast is marketed in two ways, either as compressed cakes or as a dry powder. *Compressed yeast cakes* (Figure 25.25*b*) are made by mixing washed yeast cells with emulsifying agents that give them a suitable consistency and reasonable shelf life. The product is then formed into cubes or blocks of various sizes for domestic or commercial use and kept refrigerated.

Yeast marketed in the dry state for baking is called *active dry yeast* (Figure 25.25*b*). The washed yeast cells are mixed with additives and then dried under vacuum. The dried cells are then packed in airtight containers such as fiber drums, cartons, or bags, sometimes under a nitrogen atmosphere to promote long shelf life. In baking, active dry yeast does not exhibit as great a leavening action as compressed fresh yeast, but has a much longer shelf life.

Nutritional yeast, marketed as a food supplement (Figure 25.25*b*), contains heat-killed dried yeast cells. Yeast cells are rich in B vitamins and in protein, except for sulfur-containing amino acids. Yeast cells or yeast extracts are often added to wheat or corn flour to increase the nutritional value of these foods, and yeast cells are also sold in pelleted form as a health food.

25.14 MiniReview

Yeast cells are grown for use in the baking and food industries. Commercial yeast is produced in large-scale aerated fermentors using molasses as the main carbon and energy source.

- Why is it important when growing yeast for cells to maintain oxic conditions in the fermentor?
- What different forms of yeast are commercially available?

25.15 Mushrooms as a Food Source

Several kinds of fungi are sources of human food, of which the most important are the mushrooms. Mushrooms are a group of filamentous fungi that form large, edible fruiting bodies (**Figure 25.26**). The fruiting body is called the mushroom and is formed through the association of a large number of individual fungal hyphae to form a mycelium. We discussed the biology of mushrooms in Section 18.19, including the stage of mushroom development (Figure 18.28).

(a)

(b)

Figure 25.26 Commercial mushroom production. *(a)* A flush of the mushroom *Agaricus bisporus. (b)* The shiitake mushroom, *Lentinus edulus.*

Commercial Growth of Mushrooms

The mushroom commercially grown in most parts of the world is the basidiomycete *Agaricus bisporus*, and it is generally cultivated in "mushroom farms." The organism is grown in special beds, usually in buildings where temperature and humidity are carefully controlled and exposure to light severely limited (Figure 25.26*a*). Beds are prepared by mixing soil with a material very rich in organic matter, such as horse manure, and the beds are then inoculated with a pure culture of the mushroom fungus that has been grown in large bottles on an organic-rich medium.

In the mushroom bed, the mycelium grows and spreads through the substrate, and after several weeks it is ready for the next step, the induction of mushroom formation. This is accomplished by adding a layer of soil to the surface of the bed. The appearance of mushrooms on the surface of the bed is called a *flush* (Figure 25.26*a*), and for freshness the mushrooms must be collected immediately upon flushing. After collection they are packaged and kept cool until brought to market.

Another widely cultured mushroom is the shiitake, *Lentinus edulus*. The most widely cultivated mushroom in Asia, shiitake is now finding expanding demand in North America. Shiitake is a cellulose-digesting fungus that grows on hardwood trees and is cultivated on small logs (Figure 25.26*b*). The logs are soaked in water to hydrate them and then inoculated by inserting plugs of mushroom culture into small holes drilled in them. The fungus grows through the log, and after about a year forms a flush of fruiting bodies (Figure 25.26*b*). The shiitake mushroom is considered to have a superior taste to *Agaricus bisporus*, and because of this, it commands a substantially higher price.

UNIT 5

25.15 MiniReview

The most important food produced from a microorganism is the mushroom, which is produced not for its protein but for its flavor.

- Why are mushrooms considered microorganisms?
- What is a mushroom flush?

Review of Key Terms

Aspartame a nonnutritive sweetener composed of the amino acids aspartate and phenylalanine, the latter as a methyl ester

β-Lactam antibiotic a member of a group of antibiotics including penicillin that contain the four-membered heterocyclic β-lactam ring

Biosynthetic penicillin form of penicillin produced by supplying the organism with specific side-chain precursors

Biotransformation the use of microorganisms to carry out a chemical reaction that is more costly or not feasible nonbiologically

Brewing the manufacture of alcoholic beverages such as beer from the fermentation of malted grains

Broad-spectrum antibiotic an antimicrobial drug useful in treating a wide variety of bacterial diseases

Commodity chemical a chemical such as ethanol that has low monetary value and is thus sold primarily in bulk

Distilled alcoholic beverage a beverage containing alcohol concentrated by distillation

Exoenzyme an enzyme produced by a microorganism and then excreted into the environment

Extremozyme an enzyme able to function in the presence of one or more chemical or physical extremes, for example, high temperature or low pH

Fermentation in an industrial context, any large-scale microbial process, whether carried out aerobically or anaerobically

Fermentor the tank in which an industrial fermentation is carried out

Immobilized enzyme an enzyme attached to a solid support over which substrate is passed and converted to product

Industrial microbiology the large-scale use of microorganisms to produce products of commercial value

Natural penicillin the parent penicillin structure produced by cultures of *Penicillium* not supplemented with side-chain precursors

Pilot plant a relatively small-scale industrial fermentor; larger than a laboratory fermentor but not as large as an industrial production fermentor

Primary metabolite a metabolite excreted during the exponential growth phase

Protease an enzyme that can degrade proteins by hydrolysis

Scale-up conversion of an industrial process from a small laboratory setup to a large commercial fermentation

Secondary metabolite a metabolite excreted at the end of the exponential growth phase and into the stationary phase

Semisynthetic penicillin a penicillin produced using components derived from both microbial fermentation and chemical syntheses

Tetracycline a member of a class of antibiotics containing the four-membered naphthacene ring

Review Questions

1. In what ways do industrial microorganisms differ from conventional microorganisms? In what ways are they similar (Section 25.1)?
2. List three major types of industrial products that can be obtained with microorganisms and give two examples of each (Section 25.1).
3. Compare and contrast primary and secondary metabolites and give an example of each. List at least two molecular explanations for why some metabolites are secondary rather than primary (Section 25.2).
4. How does an industrial fermentor differ from a laboratory culture vessel? How does a fermen*tor* differ from a fermen*ter* (Section 25.3)?
5. Discuss the problems of scale-up from the viewpoints of aeration, sterilization, and process control. Why is sterility so important in an industrial fermentor (Section 25.4)?
6. List three examples of antibiotics that are important industrially. For each of these antibiotics, list the producing organisms, the general chemical structure, and the mode of action (Section 25.5).
7. Compare and contrast the production of natural, biosynthetic, and semisynthetic β-lactam antibiotics (Section 25.6).
8. Addition of what metal to the fermentation medium can markedly improve production of vitamin B_{12} (Section 25.7)?
9. What unusual characteristics must an organism have if it is to overproduce and excrete an amino acid such as lysine (Section 25.7)?
10. Define the term biotransformation and give an example. Explain why the chemical reactions involved in microbial biotransformations are preferably carried out microbially rather than chemically (Section 25.8).
11. List three different kinds of enzymes that are produced commercially. For each enzyme, list the organism used in commercial production, the activity of the enzyme, and how the enzyme is used commercially (Section 25.9).
12. What is high-fructose syrup, how is it produced, and what is it used for in the food industry (Section 25.9)?
13. What are extremozymes? What industrial uses do they have (Section 25.9)?
14. In what way is the manufacture of beer similar to the manufacture of wine? In what ways do these two processes differ? How does the production of distilled alcoholic beverages differ from that of beer and wine (Sections 25.10 and 25.11)?
15. What is the active ingredient in vinegar? From what substance is it made? Why is *Gluconobacter* suitable for the aerobic production of vinegar, whereas *Escherichia coli* is not (*Hint*: Think of metabolic differences between the two organisms here) (Section 25.12)?
16. Give two reasons why stainless steel fermentors are used in the industrial production of citric acid (Section 25.13).
17. Why are yeasts of such great industrial importance (Section 25.14)?
18. What part of the mushroom is actually consumed as food? What is contained within this structure (Sections 25.15)?

Application Questions

1. As a researcher in a pharmaceutical company you are assigned the task of finding and developing an antibiotic effective against a new bacterial pathogen. Outline a plan for this process, starting from isolation of the low-yield producing organism to high-yield industrial production of the new antibiotic.
2. A partially consumed bottle of an "organic" (containing no preservatives) red wine is recapped and stored under refrigeration for 2 months. On tasting the wine again, you notice a distinct bitter taste, making the wine undrinkable. Using information presented in this chapter and in Section 15.8, describe (a) what microbially mediated process occurred in the wine and (b) a very easy way in which this process could have been prevented.
3. You wish to produce high yields of the amino acid phenylalanine for use in production of the sweetener aspartame. The overproducing organism you wish to use is not subject to feedback inhibition by phenylalanine, but is subject to typical repression of phenylalanine biosynthesis enzymes by excess phenylalanine. Applying the principles of enzyme regulation studied in Chapter 9 and microbial genetics in Chapter 11, describe two classes of mutants you could isolate that would overcome this problem and detail the genetic lesions each would have.

26

Biotechnology

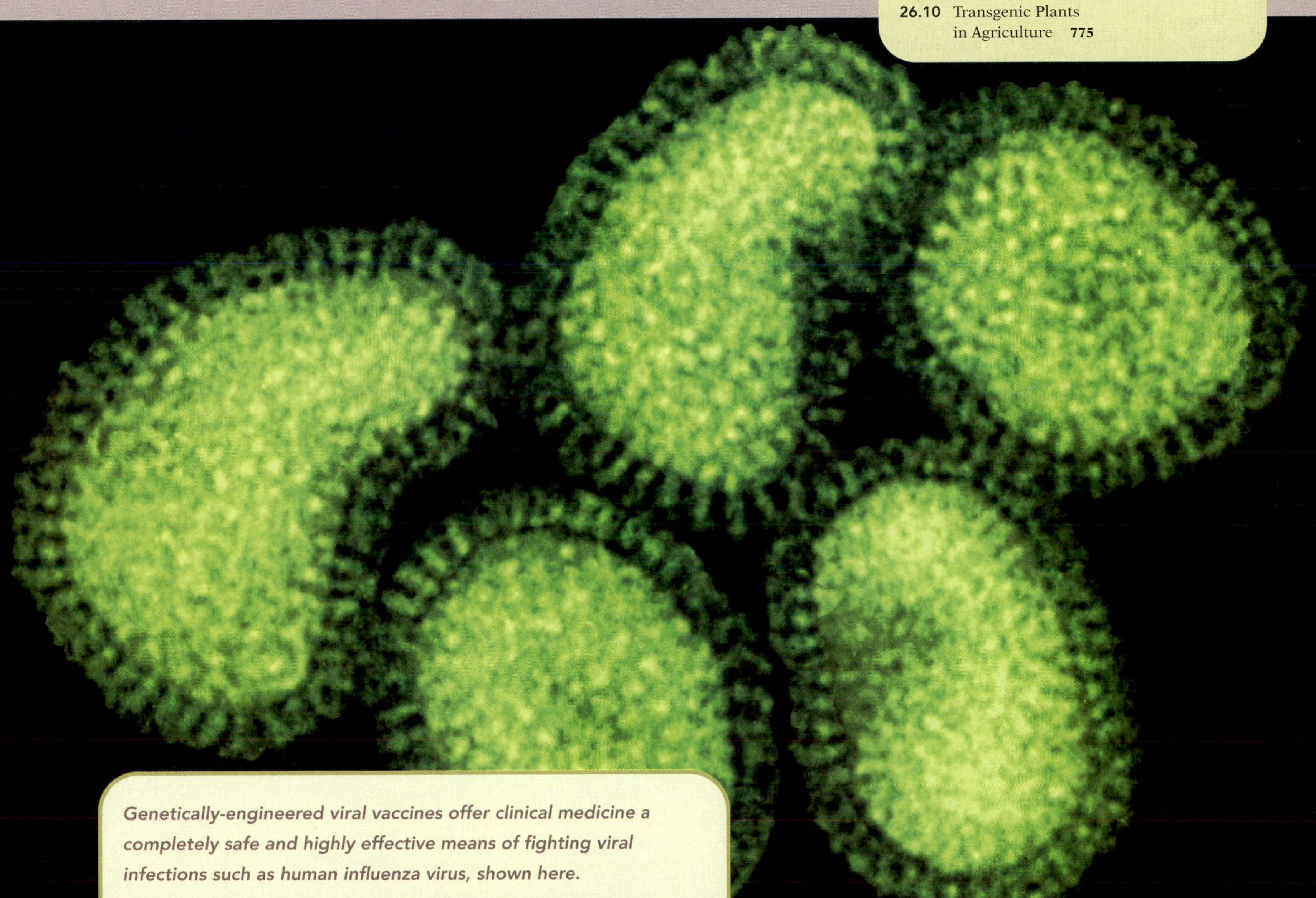

Genetically-engineered viral vaccines offer clinical medicine a completely safe and highly effective means of fighting viral infections such as human influenza virus, shown here.

Broadly defined, **biotechnology** is the use of living organisms for industrial or commercial applications. In this sense, biotechnology includes everything from the use of yeast in brewing to the use of leeches to heal wounds, techniques used by mankind since the dawn of civilization. At present, however, the word *biotechnology* implies that the organisms being used have been genetically altered by genetic engineering.

The basic techniques of **genetic engineering** were covered in Chapter 12. Here we are interested in its practical applications. The ability to manipulate DNA and control its expression is vital to biotechnology. In particular, the insertion and expression of foreign genes results in **genetically modified organisms (GMOs)**. In this chapter we look first at some valuable products that have emerged from biotechnology, such as human hormones, interferon, vaccines, and industrial enzymes. Then we consider the genetic alteration of higher organisms and their applications in agriculture and medicine.

I PRODUCTS FROM GENETIC ENGINEERING

We discuss here only a few of the many products that rely on genetic engineering. Relatively few products have been big commercial successes thus far, but many promising products are in development. As an aside, the needs of biotechnology have driven associated advances in molecular biology at a rapid pace such that many of the barriers to gene cloning of even 10 years ago are easily overcome today.

26.1 Overview of Biotechnology

Before the era of biotechnology, diabetics relied on insulin extracted from animals to control their blood sugar levels. In most cases this worked well, but a small percentage of those with diabetes showed immune reactions to the foreign (porcine or bovine) version of insulin. Genetic engineering allowed genuine human insulin to be produced by bacteria. Indeed, human insulin was the first commercialized product from genetic engineering.

Today several genetically engineered hormones and other human proteins are available for clinical use. These human proteins were originally produced by cloning human genes, inserting them into bacteria (typically *Escherichia coli*), and having the bacteria make the protein. However, several problems were encountered in this approach. An important problem arose because foreign proteins made in bacteria must be isolated by disrupting the bacterial cells and then purified. Traces of bacterial proteins that contaminate the desired protein(s) may elicit an immune response. Moreover, traces of lipopolysaccharide from the gram-negative bacterial cell outer membrane are toxic (endotoxin, ∞ Section 28.12).

Other major problems revolved around the challenge of expressing eukaryotic genes in bacteria. These include the fact that (1) the genes must be placed under control of a bacterial promoter (∞ Section 12.13); (2) the introns (∞ Section 8.8) must be removed; (3) codon bias (∞ Section 7.13) affects the efficiency of translation; and (4) many mammalian proteins are modified after translation, and bacteria lack the ability to perform most such modifications.

As a result, recent efforts in biotechnology have been to express mammalian proteins using genetically modified eukaryotic host cells. Both eukaryotic cells in culture and whole transgenic animals have been used. For example, transgenic goats have been used that secrete the protein of interest in their milk. Most recently, plants are being engineered to express mammalian proteins.

We present here only a few of the many applications of genetic engineering in biotechnology. These focus on applications of importance to medicine and agriculture. Many other applications are in development. These include the enhancement of industrial fermentations (∞ Chapter 25), the use of genetically engineered prokaryotes in bioremediation (∞ Section 24.9), and other areas of environmental biotechnology.

26.1 MiniReview

Genetic engineering allows for the cloning and expression of eukaryotic genes in prokaryotes. Hormones such as insulin and some other human proteins are now manufactured in this manner.

▮ What is the advantage of using genetic engineering to make insulin?

26.2 Expression of Mammalian Genes in Bacteria

The procedures for cloning and manipulating genes have been covered in Chapter 12 ("Genetic Engineering"). Here we are concerned with expressing cloned genes to manufacture a useful product. Expression vectors (∞ Section 12.13) are needed to express eukaryotic genes in bacteria. However, there are still obstacles to be faced, even if the mammalian gene has been cloned into an expression vector. One major issue is the presence of introns that disrupt the coding sequence of many eukaryotic genes, especially in higher organisms such as mammals (∞ Section 8.8). Introns must be spliced out for the genes to function; however, prokaryotic hosts lack the machinery to do so. To skirt this problem, introns are removed during the cloning process, before the genes are inserted into the host used for production. Typically, cloned mammalian genes no longer contain their original introns but consist of an uninterrupted coding sequence. The two major ways of achieving this are now described.

Cloning the Gene via mRNA

The standard way to obtain an intron-free eukaryotic gene is to clone it via its mRNA. Since introns are removed during the processing of mRNA, the mature mRNA carries an uninterrupted coding sequence. The isolated mRNA is used to make complementary DNA (cDNA) using the retroviral enzyme reverse transcriptase. Tissues expressing the gene of interest often contain large amounts of the desired mRNA, although other mRNAs are present as well. In certain situations, however, a single mRNA dominates in a tissue type, and extraction of bulk mRNA from that tissue provides a useful starting point for gene cloning.

In a typical mammalian cell, about 80–85% of the RNA is ribosomal RNA, 10–15% is transfer RNA, and only 1–5% is mRNA. However, eukaryotic mRNA is unique because of the poly(A) tails found at the 3′-end (Section 8.8), and this makes it easy to obtain, even though it is scarce. If a cell extract is passed over a chromatographic column containing strands of poly(T) linked to a cellulose support, most of the mRNA is separated from other RNAs by the specific pairing of A and T bases. The RNA is released from the column by a low-salt buffer, which gives a preparation greatly enriched in mRNA.

Once mRNA has been isolated, it is necessary to convert the genetic information into DNA. This is done by the enzyme reverse transcriptase. This enzyme is essential for retrovirus replication (Section 19.15) and copies information from RNA into DNA, a process called **reverse transcription** (**Figure 26.1**). Reverse transcriptase needs a primer to start DNA synthesis (the retrovirus itself uses a specific tRNA as a primer). When making DNA using mRNA as a template, a primer is used that is complementary to the poly(A) tail of the mRNA. This primer is hybridized with the mRNA, and reverse transcriptase is added.

Reverse transcriptase makes DNA that is complementary to the mRNA. As seen in Figure 26.1, the newly synthesized cDNA has a hairpin loop at its end. The loop forms because, after the enzyme completes copying the mRNA, it starts to copy the newly made DNA. This hairpin loop provides a convenient primer for synthesis of the second (complementary) strand of DNA by DNA polymerase I and is later removed by a single-strand-specific nuclease. The result is a linear, double-stranded DNA molecule, one strand of which is complementary to the original mRNA (Figure 26.1).

This double-stranded cDNA contains the coding sequences of interest but lacks introns. It can then be inserted into a plasmid or other vector for cloning. However, because the cDNA corresponds to the mRNA, it lacks a promoter and other upstream regulatory sequences that are not transcribed into RNA (Section 7.9). Special expression vectors with bacterial promoters and bacterial ribosome-binding sites are then used to obtain high levels of expression with genes cloned in this way (Section 12.13).

A cDNA library is a gene library (Section 12.3) consisting of cDNA versions of genes. It is made from the total mRNA extract of a eukaryotic cell. In practice, the library reflects only those genes expressed in the particular tissue under the chosen conditions.

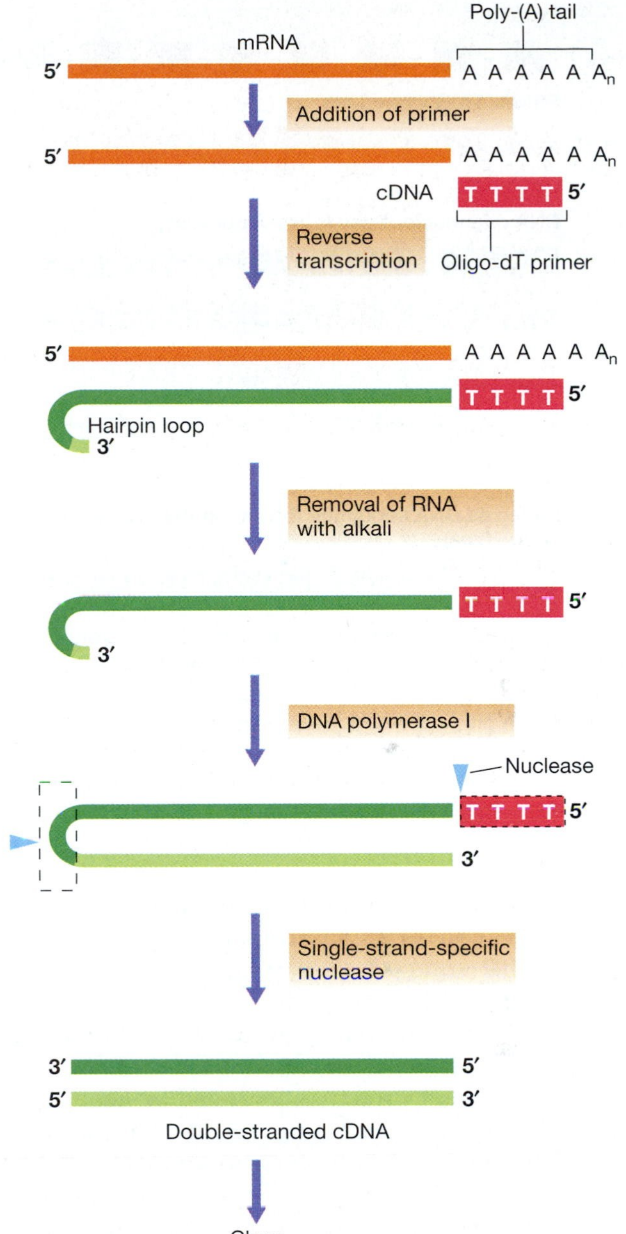

Figure 26.1 Complementary DNA (cDNA). Steps in the synthesis of cDNA from an isolated mRNA using the retroviral enzyme reverse transcriptase. The poly(A) tail is typical of eukaryotic mRNA.

Finding the Gene via the Protein

Knowing the sequence of a gene allows for the construction of a synthetic and complementary DNA molecule to use as a probe. This can then be used to find the gene in a gene library (Section 12.2). Knowledge of the amino acid sequence of a protein can also be used to construct a probe or even to synthesize a whole gene.

One can use the amino acid sequence of a protein to design and synthesize an oligonucleotide probe that encodes it. This process is illustrated in **Figure 26.2**. Unfortunately, degeneracy of the genetic code complicates this approach. Most amino acids are encoded by more than one codon (Table 7.4), and codon usage varies from organism to organism. Thus, the best region of a gene to synthesize as a probe is one that encodes part of the protein rich in amino acids specified by only a single codon (for example, methionine,

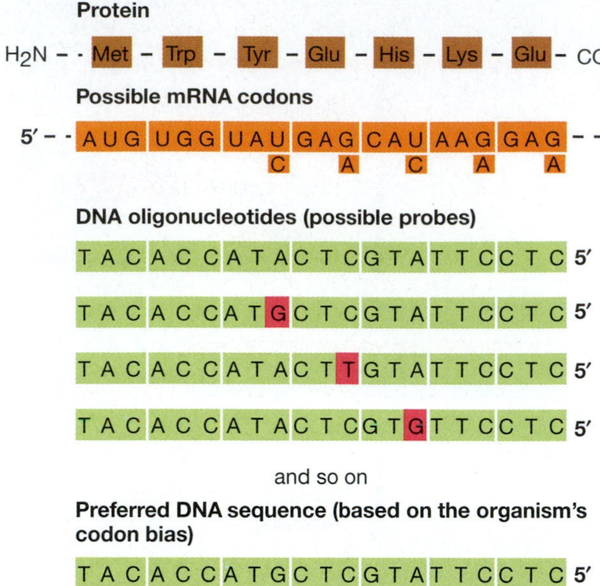

Figure 26.2 Deducing the best sequence of an oligonucleotide probe from the amino acid sequence of the protein. Because many amino acids are encoded by multiple codons, many nucleic acid probes are possible for a given polypeptide sequence. If codon usage by the target organism is known, a preferred sequence can be selected. Complete accuracy is not essential because a small amount of mismatch can be tolerated, especially with long probes.

AUG; tryptophan, UGG) or at most two codons (for example, phenylalanine, UUU, UUC; tyrosine, UAU, UAC; histidine, CAU, CAC). This strategy increases the chances that the probe will be nearly complementary to the mRNA or gene of interest. If the complete amino acid sequence of the protein is not known, partial sequence data may be used.

For certain small proteins there may be good reason to synthesize the entire gene. Many mammalian proteins (including high-value peptide hormones) are made by protease cleavage of larger precursors. Thus, to produce a short peptide hormone, such as insulin, it may be more efficient to construct an artificial gene that encodes just the final hormone rather than the larger precursor protein from which it is derived naturally. Chemical synthesis also allows synthesis of modified genes that may make useful new proteins. Techniques for the synthesis of DNA molecules are now well developed, and it is possible to synthesize genes encoding proteins over 200 amino acid residues in length (600 nucleotides). The synthetic approach was first used in a major way for production of the human hormone insulin in bacteria. It should also be noted that a constructed gene is free of introns, and thus its mRNA does not need processing. Also, promoters and other regulatory sequences can easily be built into the gene upstream of the coding sequences, and codon bias (Section 7.13) can be accounted for.

With these techniques many human and viral proteins have been expressed at high yield under the control of bacterial regulatory systems. These include insulin, somatostatin, virus capsid proteins, and interferon.

Protein Folding and Stability

The synthesis of a protein in a new host is sometimes accompanied by additional problems. For example, some proteins are susceptible to degradation by intracellular proteases and may be destroyed before they can be isolated. Moreover, some eukaryotic proteins are toxic to prokaryotic hosts, and therefore the host for the cloning vector may be killed before a sufficient amount of the product is synthesized. Further engineering of either the host or the vector may eliminate these problems.

Sometimes when foreign proteins are massively overproduced, they form inclusion bodies inside the host. Inclusion bodies consist of aggregated insoluble protein that is often misfolded or partly denatured, and they are often toxic to the host cell. Although inclusion bodies are relatively easy to purify because of their size, the protein they contain is often difficult to solubilize and may be inactive. One possible solution to this problem is to use a host that overproduces molecular chaperones that aid in folding (Section 7.17).

Fusion Proteins for Improved Purification

Protein purification can often be made much simpler if the target protein is made as a **fusion protein** along with a carrier protein encoded by the vector. To do this, the two genes are fused to yield a single coding sequence. A short segment that is recognized and cleaved by a commercially available protease is included between them. After transcription and translation, a single protein is made. This is purified by methods designed for the carrier protein. The fusion protein is then cleaved by the protease to release the target protein from the carrier protein. Fusion proteins simplify purification of the target protein because the carrier protein can be chosen to have ideal properties for purification.

Several fusion vectors are available to generate fusion proteins. **Figure 26.3** shows an example of a fusion vector that is also an expression vector. In this example, the carrier protein is the *Escherichia coli* maltose-binding protein, and the fusion protein is easily purified by methods based on its affinity for maltose. Once purified, the two portions of the fusion protein are separated by a specific protease (factor Xa, a protease whose natural role is to take part in blood clotting). In some cases the cloned target protein is released from the carrier protein by specific chemical treatment, rather than by protease cleavage. Fusion systems are also used for purposes other than achieving increased protein stability. One advantage of making a fusion protein is that the carrier protein can be chosen to contain the bacterial sequence encoding the *signal peptide*, a peptide rich in hydrophobic amino acids that enables transport of the protein across the cytoplasmic membrane (Section 7.17). This makes possible a bacterial expression system that not only makes mammalian proteins, but also secretes them. By employing the right strains and vectors, the desired protein can make up as much as 40% of the protein molecules in a cell.

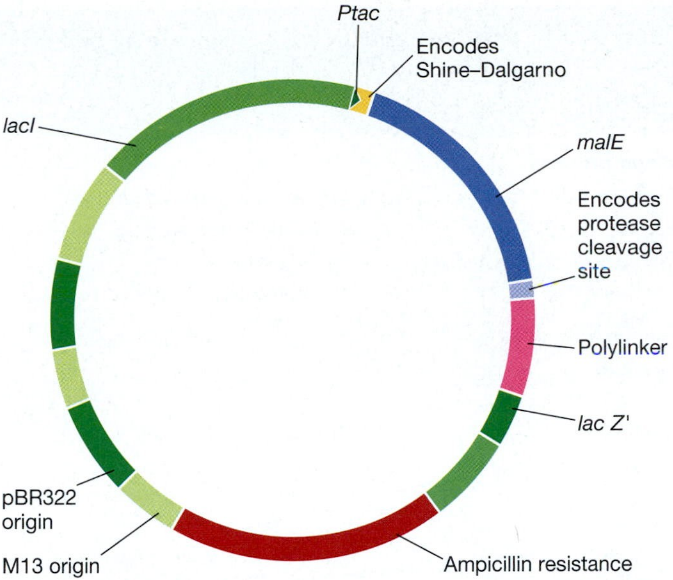

Figure 26.3 An expression vector for fusions. This vector was developed by New England Biolabs (Ipswich, MA). The gene to be cloned is inserted into the polylinker so it is in frame with the *malE* gene, which encodes maltose-binding protein. This insertion inactivates the gene for the alpha fragment of *lacZ*, which encodes β-galactosidase. The fused gene is under control of the hybrid *tac* promoter (*Ptac*). The plasmid also contains the *lacI* gene, which encodes the *lac* repressor. Therefore, an inducer must be added to the cells to turn on the *tac* promoter. The plasmid contains a gene conferring ampicillin resistance on its host. In addition to the plasmid origin of replication, there is a bacteriophage M13 origin. Therefore, this vector is a phagemid and can be propagated either as a plasmid or as a phage.

26.2 MiniReview

It is possible to achieve very high levels of expression of eukaryotic genes in prokaryotes. However, the expressed gene must be free of introns. This can be accomplished by using reverse transcriptase to synthesize cDNA from the mature mRNA encoding the protein of interest. It can also be accomplished by making an entirely synthetic gene using knowledge of the amino acid sequence of the protein of interest. Protein fusions are often used to stabilize or solubilize the cloned protein.

- What major advantage does cloning mammalian genes from mRNA or using synthetic genes have over PCR amplification and cloning of the native gene?
- How is a fusion protein made?

26.3 Production of Hormones

The most economically robust areas of biotechnology today are the production of human proteins and the use of genetically modified organisms in agriculture. Many mammalian proteins have high pharmaceutical value but are typically present in very low amounts in normal tissue, and it is therefore extremely costly to purify them. Even if the protein can be produced in cell culture, this is much more expensive and difficult than growing microbial cultures that produce it in high yield. Therefore, the biotechnology industry has genetically engineered microorganisms to produce many different mammalian proteins. Although insulin was the first human protein to be produced in this manner, the procedure had several unusual complications, because insulin consists of two short polypeptides held together by disulfide bonds. A more typical example is somatotropin, and we focus on this here.

Genetically Engineered Somatotropin

Growth hormone, or *somatotropin*, consists of a single polypeptide encoded by a single gene. Somatotropin from one mammalian species usually functions reasonably well in other species; indeed, transgenic animals have been made expressing foreign somatotropin genes, as discussed below.

Lack of somatotropin results in hereditary dwarfism. Because the human somatotropin gene was successfully cloned and expressed in bacteria, children showing stunted growth can be treated with recombinant human somatotropin (rHST). However, dwarfism may also be caused by lack of the somatotropin receptor. In this case administration of somatotropin has no effect. (People of the African pygmy tribes have normal levels of human somatotropin, but are rarely taller than 4 feet, 10 inches because they have defective growth hormone receptors.)

The somatotropin gene was cloned as cDNA from the mRNA as described in Section 26.2 (**Figure 26.4**). The cDNA was then expressed in a bacterial expression vector. The main problem with producing relatively short polypeptide hormones such as somatotropin is their susceptibility to protease digestion. This problem can be countered by using bacterial host strains defective for several proteases.

Recombinant bovine somatotropin (rBST) is used in the dairy industry (Figure 26.4). Injection of rBST into cows does not make them grow larger but instead stimulates milk production. The reason for this is that somatotropin has two binding sites. One binds to the somatotropin receptor and stimulates growth, the other to the prolactin receptor and promotes milk production. Excessive milk production by cows causes some health problems in the animals, including an increased frequency of infections of the udder and decreased reproductive capability.

When somatotropin is used to remedy human growth defects, it is desirable to avoid side effects from the hormone's prolactin activity (prolactin stimulates lactation). Site-directed mutagenesis (Section 12.9) of the somatotropin gene was used to genetically engineer somatotropin that no longer binds the prolactin receptor. To accomplish this, several amino acids needed for binding to the prolactin receptor were altered by mutation of the coding sequence. Thus it is possible not merely to make genuine human hormones, but also to alter their specificity and activity to make them better pharmaceuticals.

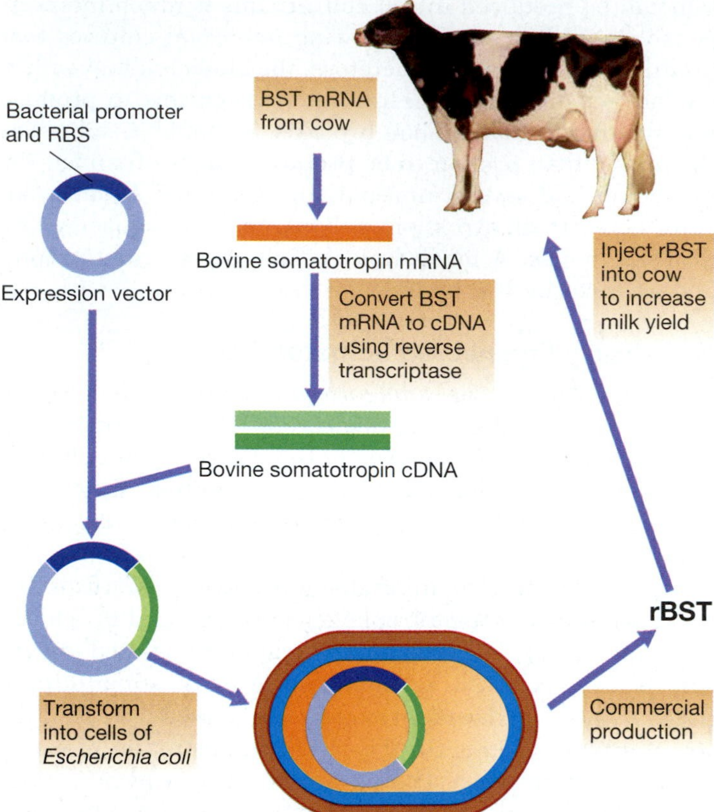

Figure 26.4 Cloning and expression of bovine somatotropin. The mRNA for somatotropin is obtained from an animal. The mRNA is converted to cDNA by reverse transcriptase. The cDNA version of the somatotropin gene is then cloned into a bacterial expression vector that has a bacterial promoter and ribosome-binding site (RBS). The construct is transformed into cells of *Escherichia coli*, and recombinant bovine somatotropin (rBST) is produced. Cows treated with rBST show increased milk production.

Table 26.1 A few therapeutic products made by genetic engineering

Product	*Function*
Blood proteins	
Erythropoietin	Treats certain types of anemia
Factors VII, VIII, IX	Promotes blood clotting
Tissue plasminogen activator	Dissolves blood clots
Urokinase	Promotes blood clotting
Human hormones	
Epidermal growth factor	Wound healing
Follicle-stimulating hormone	Treatment of reproductive disorders
Insulin	Treatment of diabetes
Nerve growth factor	Treatment of degenerative neurological disorders and stroke
Relaxin	Facilitates childbirth
Somatotropin (growth hormone)	Treatment of some growth abnormalities
Immune modulators	
α-Interferon	Antiviral, antitumor agent
β-Interferon	Treatment of multiple sclerosis
Colony-stimulating factor	Treatment of infections and cancer
Interleukin-2	Treatment of certain cancers
Lysozyme	Anti-inflammatory
Tumor necrosis factor	Antitumor agent, potential treatment of arthritis
Replacement enzymes	
β-glucocerebrosidase	Treatment of Gaucher disease, an inherited neurological disease
Therapeutic enzymes	
Human Dnase I	Treatment of cystic fibrosis
Alginate lyase	Treatment of cystic fibrosis

26.3 MiniReview

The first human protein made commercially using engineered bacteria was human insulin, but numerous other hormones and other human proteins are now being produced.

- Why is administration of a human hormone successful in some cases but not in others?
- What are the major problems when manufacturing proteins in bacteria?

26.4 Other Mammalian Proteins and Products

Many other mammalian proteins are produced by genetic engineering (**Table 26.1**). These include, in particular, an assortment of hormones and proteins for blood clotting and other blood processes. For example, tissue plasminogen activator (TPA) is a blood protein that scavenges and dissolves blood clots that can form in the final stages of the healing process. The clinical usefulness of TPA is primarily in heart patients or others suffering from poor circulation because of excessive clotting. TPA is administered following heart attacks and cardiac bypass, transplant, or other heart surgeries to prevent the development of clots that can be life threatening. Heart disease is a leading cause of death in many developed countries, especially in the United States, so microbially produced TPA is in high demand.

Blood clotting factors VII, VIII, and IX are important genetically engineered products. In contrast to TPA, these proteins are critically important for the *formation* of blood clots. Hemophiliacs suffer from a deficiency of one or more clotting factors and can therefore be treated with microbially produced clotting factors. Recombinant clotting factors take on added significance when one considers that hemophiliacs have in the past been treated with concentrated clotting factor extracts from pooled human blood, some of which was contaminated with viruses such as HIV and hepatitis C (Sections 34.12 and 34.15), putting hemophiliacs at high risk for contracting these diseases. Recombinant clotting factors have eliminated this health risk.

Certain proteins function as anticancer agents or immune modulators. Chief among these are the interferons, proteins made by animal cells in response to viral infection or immune activation in the case of gamma interferon (∞ Section 31.10). Although most interferon treatments have met with mixed success, it is still hoped that certain interferons may have specific clinical applications.

Enzymes

Some mammalian proteins made by genetic engineering are enzymes rather than hormones (Table 26.1). For instance, *human DNase I* is being produced and used to treat the buildup of DNA-containing mucus in patients with cystic fibrosis. The mucus forms because cystic fibrosis is accompanied by life-threatening lung infections by the bacterium *Pseudomonas aeruginosa* (∞ Microbial Sidebar, "Microbial Growth in the Real World: Biofilms." Chapter 6). The bacterial cells form biofilms (∞ Section 23.4) within the lungs that make drug treatment difficult. DNA is released when the bacterial cells lyse, and this fuels mucus formation. DNase digests the DNA and greatly decreases the viscosity of the mucus. There are more than 30,000 patients with cystic fibrosis in the United States alone. Treatment of cystic fibrosis with DNase was approved in 1994, and sales today of this life-saving enzyme exceed $100 million. A second enzyme, called *alginate lyase*, also produced by genetic engineering, shows promise in treating cystic fibrosis patients because it degrades the polysaccharide produced by *P. aeruginosa* cells. Like DNA from lysed cells, this polymer also contributes to lung mucus. Thus, its hydrolysis can relieve respiratory symptoms.

Not all the enzymes produced by genetic engineering have therapeutic uses. Many commercial enzymes (∞ Section 25.9) are produced in this way. Sometimes the benefits of genetic engineering can be quite unexpected. Rennin, which is used to make cheese, is an animal product and thus not consumed by strict vegetarians (vegans). However, "vegetarian cheese" containing recombinant rennin produced in a microorganism is being marketed and has found wide acceptance.

Further applications come from using site-directed mutagenesis (∞ Section 12.9) on existing cloned genes to generate new products with new properties. Certain molecules, such as many antibiotics, are synthesized in cells by biochemical pathways that use a series of enzymes (∞ Section 25.6). These enzymes can be modified by genetic engineering to produce modified forms of the antibiotics.

26.4 MiniReview

Many proteins are extremely expensive to obtain by traditional purification methods because they are found in human or animal tissues in only small amounts. Many of these can now be made in large amounts from the cloned gene in a suitable expression system.

- Contrast the activity of TPA and blood factors VII, VIII, and IX.
- Explain how a DNA-degrading enzyme can be useful in treating a bacterial infection, such as that which occurs with cystic fibrosis.

26.5 Genetically Engineered Vaccines

Vaccines are substances that elicit immunity to a particular disease when injected into an animal (∞ Section 30.5). Typically, vaccines are suspensions of killed or modified pathogenic microorganisms or viruses (or of specific components isolated from them). Often the component that elicits the immune response is a surface protein, for instance, a virus coat protein. Genetic engineering can be applied in many different ways to the production of vaccines. A list of vaccines, including some produced by genetic engineering, is shown in Table 30.2.

Recombinant Vaccines

Recombinant DNA techniques can be used to modify the pathogen itself. For instance, one can delete pathogen genes that encode virulence factors but leave those whose products elicit an immune response. This yields a recombinant, live, attenuated vaccine. Conversely, one can add genes from a pathogenic virus to another, relatively harmless virus, referred to as a carrier virus. Such vaccines are called **vector vaccines**. This approach induces immunity to the pathogenic viral disease. Indeed, one can even combine the two approaches. For example, a recombinant vaccine is used to protect poultry against both fowlpox (a disease that reduces weight gain and egg production) and Newcastle disease (a viral disease that is often fatal). The fowlpox virus (a typical pox virus; ∞ Section 19.13) was first modified to delete genes that cause disease, but not those that elicit immunity. Then immunity-inducing genes from the Newcastle virus were inserted. This resulted in a **polyvalent vaccine**, a single vaccine that immunizes against two different diseases.

Vaccinia virus is widely used to prepare live recombinant vaccines for human use (∞ Section 19.13). Vaccinia virus itself is generally not pathogenic for humans and has been used for over 100 years as a vaccine against the related smallpox virus. However, cloning genes into vaccinia virus requires a selective marker. This is provided by the gene for thymidine kinase. Vaccinia virus is unusual for a virus in that it carries its own thymidine kinase, an enzyme that can convert the base analog 5-bromodeoxyuridine to a nucleotide that is incorporated into DNA. However, this is a lethal reaction. Therefore, cells that express thymidine kinase (whether from the host cell genome or from a virus genome) are killed by 5-bromodeoxyuridine.

Genes to be put into vaccinia virus are first inserted into an *Escherichia coli* plasmid that contains a fragment of the vaccinia thymidine kinase (*tdk*) gene (**Figure 26.5**). The foreign DNA is inserted into the *tdk* gene, which is therefore disrupted. This recombinant plasmid is then transformed into animal cells whose own thymidine kinase gene is inactivated. The same cells are also infected with wild-type vaccinia virus (Figure 26.5). Homologous recombination occurs between the two versions of the *tdk* gene—one on the plasmid and the other on the virus. Some viruses therefore gain a disrupted *tdk* gene plus its foreign insert (Figure 26.5). Cells infected by wild-type virus, with active thymidine kinase, are killed by

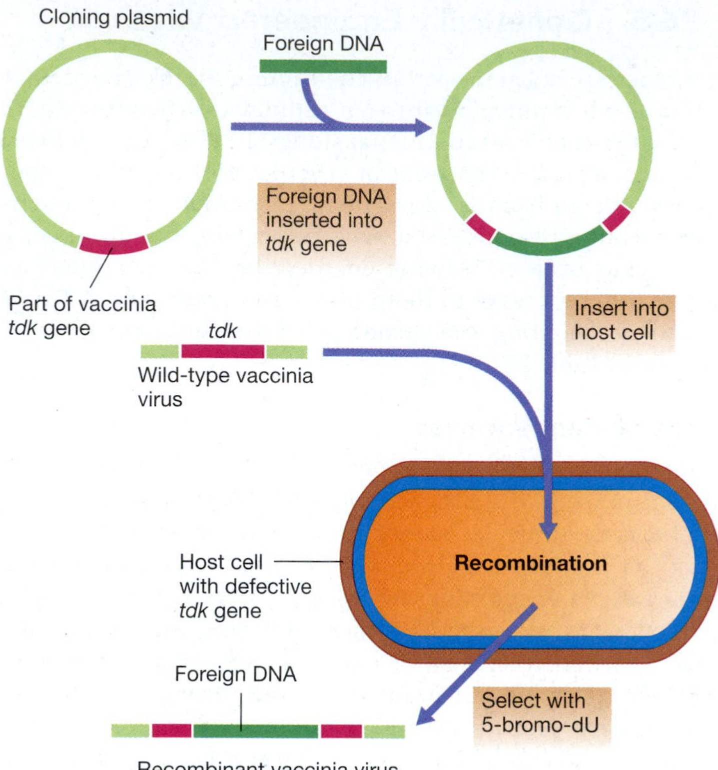

Figure 26.5 Production of recombinant vaccinia virus. Foreign DNA is inserted into a short segment of the thymidine kinase gene (*tdk*) from vaccinia virus carried on a plasmid. The plasmid with the insert and wild-type vaccinia virus are both put into the same host cell where they recombine. The cells are treated with 5-bromodeoxyuridine (5-bromo-dU), which kills cells with active thymidine kinase. Only recombinant vaccinia viruses whose *tdk* gene is inactivated by insertion of foreign DNA survive.

5-bromodeoxyuridine. Cells infected by recombinant vaccinia virus with a disrupted *tdk* gene grow long enough to yield a new generation of viral particles (Figure 26.5). The result is that viruses whose *tdk* gene contains a cloned insert of foreign DNA are selected.

Vaccinia virus does not actually need thymidine kinase to survive. Consequently, recombinant vaccinia viruses can still infect human cells and express any foreign genes they carry. Indeed, recombinant vaccinia viruses can be constructed to carry genes from multiple viruses (that is, they are polyvalent vaccines). Currently, several vaccinia vector vaccines have been developed and licensed for veterinary use, including one for rabies. Many other vaccinia vaccines are at the clinical trial stage. Vaccinia vaccines are relatively benign, yet highly immunogenic in humans, and their use will likely increase in the coming years.

Subunit Vaccines

Recombinant vaccines do not have to include the entire suite of proteins from the pathogenic organism. Subunit vaccines may contain only a specific protein or proteins from a pathogenic organism. For viruses this protein is often the coat protein(s), because these are typically highly immunogenic. The coat proteins are purified and used in high dosage to elicit a rapid and high level of immunity. Subunit vaccines are currently very popular because they can be used to produce large amounts of immunogenic proteins without the possibility that the purified products will contain the entire pathogenic organism, even in minute amounts.

The steps in preparing a viral subunit vaccine are as follows: fragmentation of viral DNA by restriction enzymes; cloning viral coat protein genes into a suitable vector; providing for proper promoters, reading frame, and ribosome-binding sites; and reinsertion and expression of the viral genes in a microorganism. Sometimes only certain portions of the protein are expressed rather than the entire protein, because immune cells and antibodies typically react with only small portions of the protein. (When using this approach against an RNA virus, the viral genome must be converted to a cDNA copy first.)

When *E. coli* is used as the cloning host, viral subunit vaccines are often poorly immunogenic and fail to protect in experimental tests of infection with the virus. The problem is that many viral coat proteins are posttranslationally modified, typically by the addition of sugar residues (glycosylation), by the animal host cells when the virus replicates. By contrast, the recombinant proteins produced by bacteria are unglycosylated, and glycosylation is necessary for the proteins to be immunologically active. To solve this problem, a eukaryotic host is used. For example, the first recombinant subunit vaccine approved for use in humans (against hepatitis B) was made in yeast. The gene encoding a surface protein from hepatitis B virus was cloned and expressed in yeast. The protein was produced and formed aggregates very similar to those found in patients infected with the virus. These aggregates were purified and used to effectively vaccinate humans against infection by hepatitis B virus.

Subunit vaccines against many viruses and other pathogens are currently being developed. Cultured insect or mammalian cells are often employed as hosts to prepare such recombinant vaccines. As we have said, to obtain the correct pattern of glycosylation or other modifications of the immunogenic protein, it is often important to use a eukaryotic host. However, vaccines with correct glycosylation can often be produced in eukaryotic hosts relatively unrelated to humans, such as plants or insect cells. Recently, both yeast and insect cells have themselves been genetically engineered by the insertion of human genes that catalyze glycosylation. The resulting host cells add human-type glycosylation patterns to the proteins they produce.

The Future of Recombinant Vaccines

Genetically engineered recombinant vaccines will likely become increasingly common for several reasons. They are safer than normal attenuated or killed vaccines because it is impossible to transmit the disease in the vaccine. They are also more reproducible because their genetic makeup can be carefully monitored.

In addition, recombinant vaccines can usually be prepared much faster than those produced by more traditional methods. For preparing vaccines for some diseases, such as influenza (∞ Section 34.9), time is of the essence. Recombinant vaccines using cloned influenza virus hemagglutinin genes can be made in just 2 or 3 months' time. This contrasts with the 6–9 months needed to make an attenuated intact virus influenza vaccine. Preparation time can be an important factor in responding to an epidemic caused by a new strain of virus, a common cause of outbreaks of influenza. Finally, recombinant vaccines are typically much less expensive than those produced by traditional methods (∞ Section 30.5).

DNA Vaccines

Although recombinant and traditional vaccines have been extremely successful in the fight against infectious diseases, in some cases vaccines are difficult to produce. However, a conceptually new approach to vaccine production is possible—**DNA vaccines**—also known as *genetic vaccines*.

DNA vaccines use the genome of the pathogen itself to immunize the individual. Defined fragments of the pathogen's genome or specific genes that encode immunogenic proteins are used. The key genes are cloned into a plasmid or viral vector and delivered by injection. When DNA is taken up by animal cells it may be degraded or it may be transcribed and translated. If the latter occurs and the protein produced is immunogenic, the animal will be effectively immunized against the pathogen. Thus, the immune response is made against the protein encoded by the vaccine DNA. The DNA itself is not immunogenic.

Some DNA vaccines have undergone clinical trials, for example, vaccines against HIV, hepatitis B, and several cancers. For unknown reasons, the DNA vaccines so far tested have not been potent enough to provide protective immunity to humans. It is hoped that future improvements will permit clinical use. Nonetheless, DNA vaccines have been licensed for use in animals (for example, a vaccine against West Nile virus for horses).

Unlike viral vaccines, DNA vaccines escape surveillance by the host immune system because nucleic acids themselves are poorly immunogenic. This prevents the animal from suffering autoimmune effects in which antibodies and immune cells attack host cells (∞ Section 30.7). DNA vaccines have the advantage that they are both safe and inexpensive. In addition, DNA is more stable than live vaccines, which avoids the need for refrigeration—an important practical point in using vaccines in developing countries.

26.5 MiniReview

Many recombinant vaccines have been produced or are under development. These include live recombinant, vector, subunit, and DNA vaccines.

- Explain why recombinant vaccines might be safer than some vaccines produced by traditional methods.
- What are the important differences between a recombinant, live, attenuated vaccine, a vector vaccine, a subunit vaccine, and a DNA vaccine?

26.6 Mining Genomes

Just as the total genetic content of an organism is its *genome*, so the collective genomes of an environment is known as its *metagenome*. Complex environments, such as fertile soil, contain vast numbers of uncultured bacteria and other microorganisms together with the viruses that prey on them (∞ Section 13.13). Taken together, these contain correspondingly vast numbers of novel genes. Indeed, most of the genetic information on Earth exists in microorganisms and their viruses that have not been cultured. We discussed the field of environmental genomics in Sections 13.13 and 22.6.

Environmental Gene Mining

Gene mining is the process of isolating potentially useful novel genes from the environment without culturing the organisms that carry them. Instead of culturing, DNA (or RNA) is isolated directly from environmental samples and cloned into suitable vectors to construct a metagenomic library (**Figure 26.6**). The nucleic acid includes genes that come from uncultured organisms as well as DNA from dead organisms that has been released into the environment but has not yet been degraded. If RNA is isolated, it must be converted to a DNA copy via reverse transcriptase (∞ Figure 26.1). However, isolating RNA has the disadvantage that it is more time

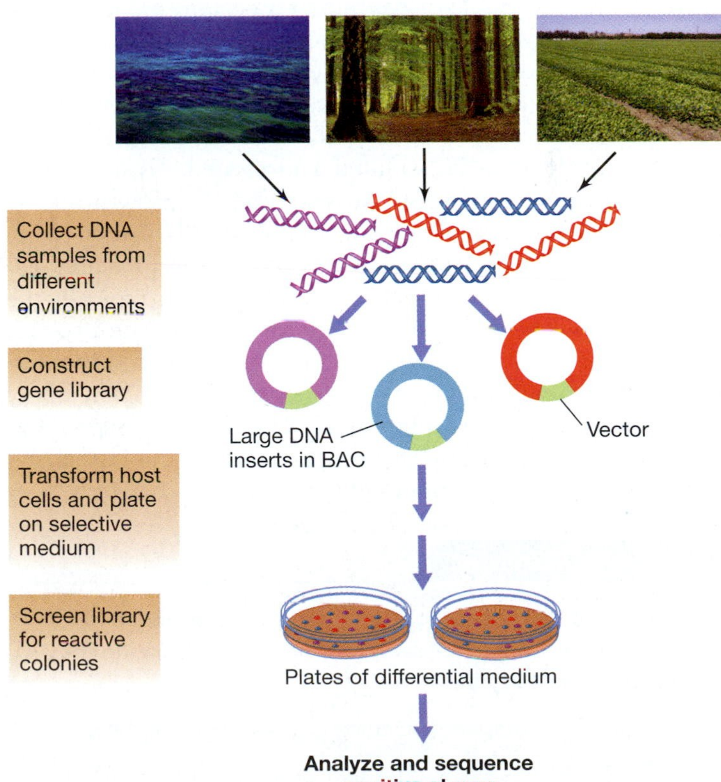

Figure 26.6 Metagenomic search for useful genes in the environment. DNA samples are obtained from different sites, such as agricultural soil, seawater, or forest soil. A clone library is constructed and screened for genes of interest. Possibly useful clones are analyzed further.

UNIT 5

consuming and limits the metagenomic library to those genes that are transcribed and therefore active in the environment sampled.

The metagenomic library is then screened by the same techniques as any other clone library. Metagenomics has identified novel environmental genes that encode enzymes that degrade pollutants and enzymes that make novel antibiotics. So far several lipases, chitinases, esterases, and other degradative enzymes with novel substrate ranges and other properties have been isolated by this approach. Such enzymes are used in industrial processes for various purposes (Section 25.9). Enzymes with improved resistance to industrial conditions, such as high temperature, high or low pH, and oxidizing conditions, are especially valuable and sought after.

Discovery of genes encoding entire metabolic pathways, such as for antibiotic synthesis, as opposed to single genes, requires vectors such as bacterial artificial chromosomes (BAC) that can carry large inserts of DNA (Section 12.15). BACs are especially useful for screening samples from rich environments, such as soil, where vast numbers of unknown genomes are present and hence large numbers of genes to screen.

Targeted Gene Mining

Metagenomics can be used to screen directly for enzymes with certain properties. Suppose one desired an enzyme or entire pathway capable of degrading a certain pollutant at a high temperature. The first step would be to find a hot environment polluted with the target compound. Assuming that microorganisms capable of degradation were present in the environment, a reasonable hypothesis, DNA from the environment would then be isolated and cloned. Host bacteria containing the clones would be screened for growth on the target compound. For convenience, this step is usually done in an *E. coli* host, on the assumption that thermostable enzymes will still show some activity at 40°C (this is typically the case). Once suspects have been identified, enzyme extracts can be tested *in vitro* at high temperatures. Recently, thermophilic cloning systems have been developed that allow direct selection at high temperature. These rely on expression vectors that can replicate in both *E. coli* and the hot spring thermophile, *Thermus thermophilus*.

26.6 MiniReview

In metagenomics, genes for useful products are cloned directly from environmental samples without first isolating the organisms that carry them.

- Explain why metagenomic cloning gives large numbers of novel genes.
- What are the advantages and disadvantages of isolating environmental RNA as opposed to DNA?

II TRANSGENIC ORGANISMS

A *transgene* is a gene from one organism that has been inserted into a different organism. Hence, a **transgenic organism** is one that contains a transgene. A related term is genetically modified organism (GMO). Strictly speaking, this refers to genetically engineered organisms whether or not they contain foreign DNA. However, in common usage, especially in agriculture, "GMO" is often used interchangeably with "transgenic organism."

On the one hand, the genetic engineering of higher organisms is not truly microbiology. On the other hand, much of the DNA manipulation is carried out using bacteria and their plasmids long before the engineered transgene is finally inserted into the plant or animal. Furthermore, vectors based on viruses are widely used in the genetic engineering of higher organisms. Therefore we emphasize the microbial systems that have contributed to the genetic manipulation of plants and animals. But before we consider transgenic animals and plants, we illustrate the overall approach by considering how bacteria can be modified on a large scale, not merely to yield one extra protein but to create whole new metabolic pathways.

26.7 Engineering Metabolic Pathways in Bacteria

Although proteins are large molecules, expressing a single protein in large amounts that is encoded by a single gene is relatively simple. By contrast, small metabolites are typically made in biochemical pathways employing several enzymes. In these cases, not only are multiple genes needed, but their expression must be regulated in a coordinated manner as well.

Pathway engineering is the process of assembling a new or improved biochemical pathway using genes from one or more organisms. Most efforts so far have modified and improved existing pathways rather than creating entirely new ones (but see the Microbial Sidebar, "Synthetic Biology and Bacterial Photography"). Because genetic engineering of bacteria is simpler than that of higher organisms, most pathway engineering has been done with bacteria. Engineered microorganisms are used to make products, including alcohol, solvents, food additives, dyes, and antibiotics. They may also be used to degrade agricultural waste, pollutants, herbicides, and other toxic or undesirable materials.

Pathway Engineering in the Production of Indigo

An example of pathway engineering is the production of indigo by *Escherichia coli* (**Figure 26.7**). Indigo is an important dye used for treating wool and cotton. Blue jeans, for example, are made of cotton dyed with indigo. In ancient times indigo and related dyes were extracted from sea snails. More recently, indigo was extracted from plants, but today it is synthesized chemically. However, the demand for indigo by the textile industry has spawned new approaches for its synthesis, including a biotechnological one.

Microbial Sidebar

Synthetic Biology and Bacterial Photography

The term "synthetic biology" refers to the use of genetic engineering to create novel biological systems out of available biological parts, often from several different organisms. An ultimate goal of synthetic biology is to synthesize a viable cell from component parts, a feat that will likely be possible in the near future. A major start in this direction occurred in 2007 when a team of synthetic biologists transferred the entire chromosome of one species of bacterium into another species of bacterium. The latter species then took on all of the properties of the species whose genome it contained.

An interesting example of synthetic biology on a smaller scale is the use of genetically modified *Escherichia coli* cells to produce photographs. The engineered bacteria are grown as a lawn on agar plates. When an image is projected onto the lawn, bacteria in the dark make a dark pigment whereas bacteria in the light do not. The result is a primitive black and white photograph of the projected image.

Construction of the photographic *E. coli* required the engineering and insertion of three genetic modules: (1) a light detector and signaling module; (2) a pathway to convert heme (already present in *E. coli*) into the photoreceptor pigment phycocyanobilin; and (3) an enzyme encoded by a gene whose transcription can be switched on and off to make the dark pigment (**Figure 1*a***). The light detector is a fusion protein. The outer half is the light-detecting part of the phytochrome protein from the cyanobacterium *Synechocystis*. This needs a special light-absorbing pigment, phycocyanobilin, which is not made by *E. coli*, hence the need to install the pathway to make phycocyanobilin.

The inner half of the light detector is the signal transmission domain of the EnvZ sensor protein from *E. coli*. EnvZ is part of a two-component regulatory system, its partner being OmpR (Section 9.5). Normally, EnvZ activates the DNA-binding protein, OmpR. Activated OmpR in turn activates target genes by binding to the promoter. In the present case, the hybrid protein was designed to activate OmpR in the dark but not in the light. This is because phosphorylation of OmpR is required for activation, and red light converts the sensor to a state in which phosphorylation is inhibited. Consequently the target gene is off in the light and on in the dark. When a mask is placed over the Petri plate containing a lawn of the engineered *E. coli* cells (Figure 1*b*), cells in the dark make a pigment that cells in the light do not, and in this way a "photograph" of the masked image develops (Figure 1*c*).

The pigment made by the *E. coli* cells is the result of the activity of an enzyme naturally found in this organism that functions in lactose metabolism, β-galactosidase. The target gene, *lacZ*, encodes this enzyme. In the dark, the *lacZ* gene is expressed and β-galactosidase is made. The enzyme cleaves a lactose analog called *S-gal* present in the growth medium to release galactose and a colored dye. In the light, the *lacZ* gene is not expressed, no β-galactosidase is made, and so no black dye is released. The difference in contrast between cells producing the dye and cells that are not generates the bacterial photograph (Figure 1*c*).

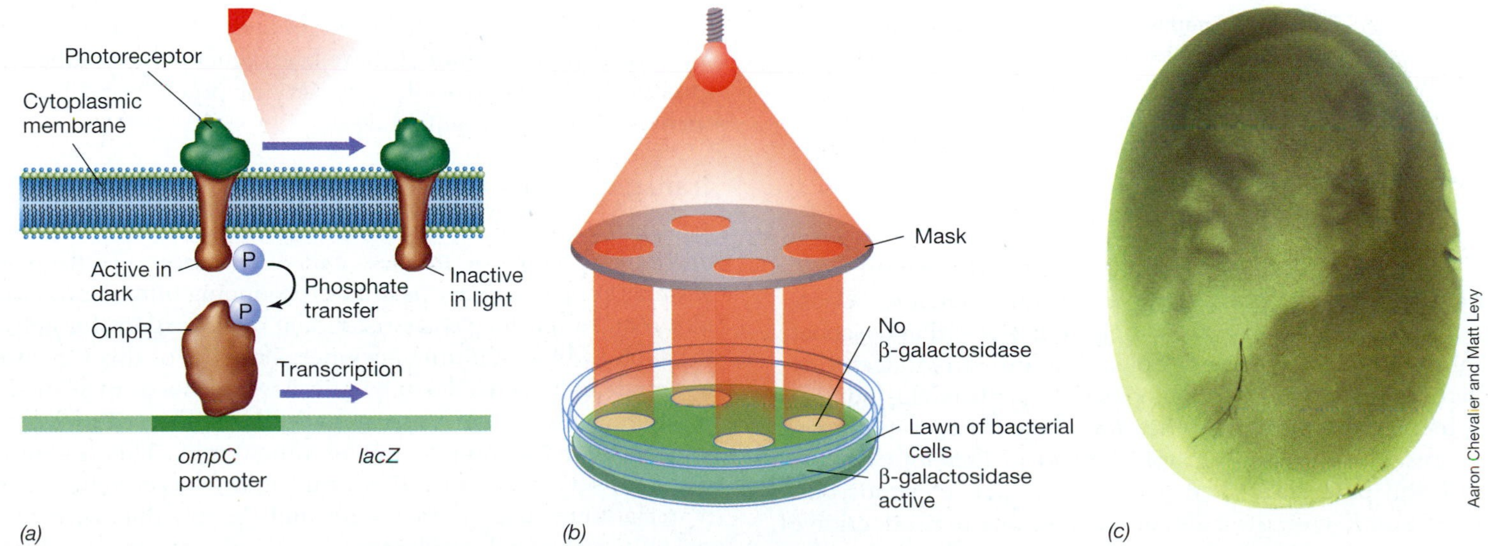

Figure 1 Bacterial Photography. (a) *Light-detecting* Escherichia coli *cells were genetically engineered using components from cyanobacteria and* E. coli *itself. Red light inhibits phosphate (P) transfer to the DNA-binding protein OmpR; phosphorylated OmpR is required to activate* lacZ *transcription (*lacZ *encodes β-galactosidase).* (b) *Set-up for making a bacterial photograph. The opaque portions of the mask correspond to zones where β-galactosidase is active and thus to the dark regions of the final image.* (c) *A bacterial photograph of a portrait of Charles Darwin.*

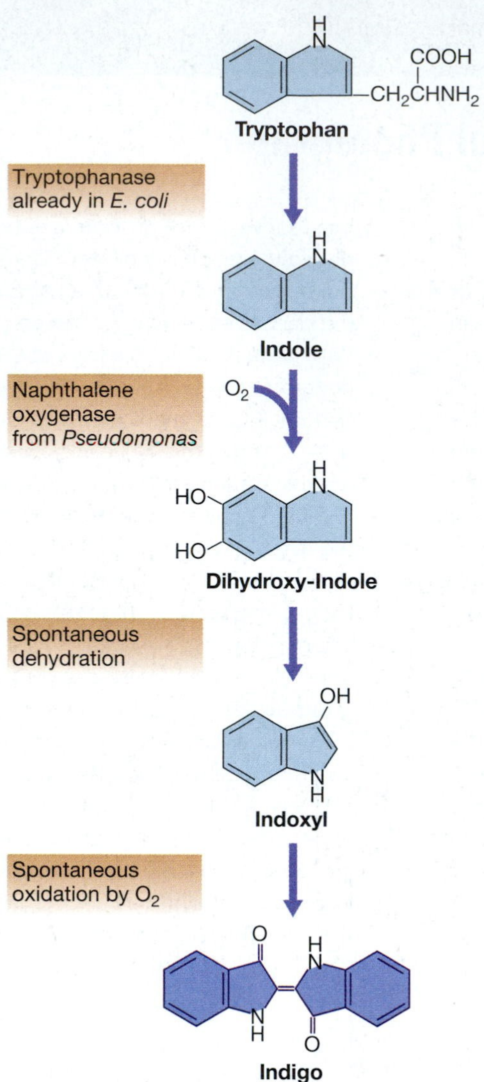

Figure 26.7 Engineered pathway for indigo production. *Escherichia coli* naturally expresses tryptophanase, which converts tryptophan into indole. Naphthalene oxygenase (originally from *Pseudomonas*) converts indole to dihydroxy-indole, which spontaneously dehydrates to indoxyl. Upon exposure to air, indoxyl forms indigo, which is blue.

Indigo is based on the indole ring system (Figure 26.7), which resembles naphthalene in overall structure. Consequently, oxygenase enzymes (Section 21.14) that oxygenate naphthalene also oxidize indole to its dihydroxy derivative, which then oxidizes spontaneously in air to yield indigo, a bright blue pigment. Enzymes for oxygenating naphthalene are present on several plasmids found in *Pseudomonas* and other soil bacteria. When genes from such plasmids were cloned into *E. coli*, the cells turned blue due to production of indigo; the blue cells had picked up the genes for the enzyme naphthalene oxygenase.

Although only the gene for naphthalene oxygenase was actually cloned during indigo pathway engineering, the indigo pathway consists of four steps, two enzymatic and two spontaneous (Figure 26.7). The enzyme that carries out the first step, the conversion of tryptophan to indole (tryptophanase) naturally occurs in *E. coli*. Therefore, for indigo production, tryptophan must be supplied. This can be accomplished by affixing cells of the recombinant *E. coli* to a solid support in a bioreactor and then trickling a tryptophan solution from waste protein or other sources over the suspended cells. Recirculating the material over the cells several times as is typically done in these types of immobilized cell industrial processes (Section 25.9) steadily increases indigo levels until the dye can be harvested.

26.7 MiniReview

In pathway engineering multiple genes that encode the enzymes for a metabolic pathway are assembled. These genes may come from one or more organisms but must be cloned and expressed in a synchronized manner.

- Explain why pathway engineering is more difficult than cloning and expressing a human hormone.
- How was *Escherichia coli* modified to produce indigo?

26.8 Genetic Engineering of Animals

By using microinjection to deliver cloned genes to fertilized eggs, followed by recombination of the foreign DNA into the genome, many foreign genes have been incorporated and expressed both in laboratory research animals and in commercially important animal species. The first transgenic animals developed were mice that were engineered as model systems for studying mammalian physiology. Genes for growth hormone from rats or humans were engineered for expression and were then inserted into mice where they were successfully expressed. The result, of course, was larger mice. More recently, farm animals have been genetically modified to improve yields or to avoid a variety of practical problems, such as pollution by animal waste.

Transgenic Animals in Pharming

Transgenic animals can be used to produce proteins of pharmaceutical value—a process called *pharming*. Transgenic animals are particularly useful for producing human proteins that require specific posttranslational modifications for activity, such as blood-clotting enzymes. Proteins of this type are not made in an active form by microorganisms or by plants.

Some proteins have been genetically engineered such that they are secreted in high yield in animal milk. This is convenient for several reasons. First, this allows larger volumes of material to be made more simply and cheaply than by bacterial culture. Second, a milk-processing industry already exists, so little new technology is needed to purify the protein. Third, milk is a natural product that most humans can tolerate, so purification away from possibly toxic bacterial proteins is unnecessary. Goats have proven useful for making several human proteins including tissue plasminogen activator, which is used to dissolve blood clots (Section 26.3).

Transgenic Animals in Medical Research

Transgenic animals have become increasingly important in basic biomedical research for studying gene regulation and developmental biology. For example, so-called "knockout" mice, which have both copies of a particular gene inactivated by genetic engineering, are used to analyze genes active in animal physiology. For instance, knockout mice that lack both copies of the gene for myostatin, a protein that slows muscle growth, develop massive muscles. In contrast, transgenic mice that overproduce myostatin show reduced muscle mass. Many other strains of knockout mice have been developed for use in medical research, and a 2007 Nobel Prize was awarded for the development of this important genetic tool.

Improving Livestock and Other Food Animals by Transgenics

Livestock may be engineered to improve their productivity, increase their nutritional value, and increase disease resistance. Occasionally, transgenic livestock are produced that do not necessarily have commercial value but demonstrate the feasibility of certain genetic techniques. For example, pigs that have been genetically engineered to express the reporter gene that encodes the green fluorescent protein (∞ Section 12.10) are, as expected, green (**Figure 26.8*a***).

One scheme to improve the nutrition of livestock is to insert entire metabolic pathways from bacteria into the animals. For example, genes that encode the enzymes of the metabolic pathway for making methionine, a required amino acid, could remove the need for this amino acid in the animals' diet. A notable technical success has been the insertion into pigs of a gene from *Escherichia coli* that helps degrade organic phosphate. The resulting Enviropig™ no longer needs phosphate supplements in its feed. However, most importantly, the manure from these animals is low in phosphate, and this prevents phosphate runoff from pig manure waste ponds into freshwater; such an influx of inorganic nutrients can trigger algal blooms and fish die-offs (∞ Section 23.6).

Pigs have also been genetically engineered to increase their levels of omega-3 fatty acids. These fatty acids reduce heart disease but are found in significant amounts only in cold-water fish, such as salmon, and a few other rare foods. To create transgenic pigs with an altered fatty acid profile, a gene from the roundworm *Caenorhabditis elegans*, called *fat1*, was inserted into the pigs. The enzyme encoded by *fat1* converts the less healthy but more common omega-6 fatty acids into omega-3 fatty acids. Such animals should be healthier for consumers, especially those who have dietary restrictions on fat or who are at high risk for heart disease. It will be several years before the omega-3-enriched pigs reach the consumer market, assuming they receive government approval.

Another interesting practical example of a transgenic animal is the "fast-growing salmon" (Figure 26.8*b*). These transgenic salmon are not actually larger than normal salmon but simply reach market size much faster. The gene for growth hormone in natural salmon is activated by light. Consequently, salmon grow rapidly only during the summer months. In the genetically engineered salmon, the promoter for the growth hormone gene was replaced with the promoter from another fish that grows at a more or less constant rate all year round. The result was salmon that make growth hormone constantly and thus grow faster.

Wu-Shinn Chih

(a)

Aqua Bounty Technologies

(b)

Figure 26.8 Transgenic animals. *(a)* A piglet (left) that has been genetically engineered to express the green fluorescent protein and thus fluoresces green under blue light. Control piglets are shown in the center and right. *(b)* Fast-growing salmon. The *AquAdvantage*™ Salmon (top) was engineered by Aqua Bounty Technologies (St. Johns, Newfoundland, Canada). Both the transgenic and the control fish are 18 months old and weigh 4.5 kg and 1.2 kg, respectively.

26.8 MiniReview

Genetic engineering can be used to develop transgenic animals capable of producing proteins of pharmaceutical value. It also creates animal models for human disease that may be useful in medical research. Most recently, attempts are being made to improve livestock for human consumption as opposed to merely using animals to manufacture individual proteins.

- What is pharming?
- Why are knockout mice useful in investigating human gene function?
- What environmental advantage does Enviropig™ have over normal pigs?

UNIT 5

26.9 Gene Therapy in Humans

Humans suffer from a variety of hereditary diseases. Because conventional genetic experiments cannot be done with humans as they can with other animals, our understanding of human genetics has lagged well behind that of many other organisms. However, the advent of genetic engineering has allowed us to fill in many of the gaps in our knowledge indirectly. Moreover, the human genome has been sequenced (∞ Section 13.1) and this database also greatly helps in identifying the causes of hereditary diseases.

Human Hereditary Diseases

A large number of human genetic diseases are known. A list can be found in the OMIM (Online Mendelian Inheritance in Man) database, available online at **http://www.ncbi.nlm.nih.gov/.** By using recombinant DNA technology, coupled with conventional genetic studies (following family inheritance, and so on), it is possible to localize particular genetic defects to specific regions of particular chromosomes. The region containing the presumed genetic defect can then be cloned and sequenced and the base sequence of the normal gene and the defective gene compared.

Even without knowledge of the enzyme defect, it is possible to obtain information about the genetic change. Many genes, including those for Huntington's disease, hemophilia types A and B, cystic fibrosis, Duchenne muscular dystrophy, multiple sclerosis, and breast cancer have been localized with these techniques and the mutations in the defective genes identified. With this in mind, how can genetic engineering be used to treat or cure these diseases?

Gene Therapy

The use of genetic engineering to treat human genetic diseases as well as attack cancer cells is known as **gene therapy**. In *replacement gene therapy*, a nonfunctional or dysfunctional gene in an individual is "replaced" by a functional gene. Strictly speaking, it is not the defective *gene* that is replaced but instead, its *function*. The therapeutic wild-type gene is inserted elsewhere in the genome and its gene product corrects the genetic disorder. Major obstacles to this approach exist in targeting the correct cells for gene therapy and in successfully inserting the required gene into cell lines that will perpetuate the genetic alteration.

The first genetic disease for which the use of gene therapy was approved was a form of severe combined immune deficiency (SCID). This disease is caused by the absence of adenosine deaminase (ADA), an enzyme of purine metabolism, in bone marrow cells and leads to a crippled immune system. The gene therapy approach used a retrovirus as a vector to carry a wild-type copy of the *ADA* gene. T cells (part of the immune system; ∞ Section 29.1) were removed from the patient and infected with the retrovirus carrying the *ADA* gene. The retrovirus also carries a marker gene, encoding resistance to the antibiotic neomycin, so that T cells carrying the inserted retrovirus could be selected for and identified. The corrected T cells were then placed back in the body. However, because T cells have a limited life span, the therapy must be repeated every few months. Consequently, in later treatment protocols for SCID the *ADA* gene was inserted into stem cells obtained from umbilical cord blood of newborn babies diagnosed with defects in the *ADA* gene. The engineered stem cells are then returned to the babies. Because stem cells continue to divide and provide a fresh supply of new T cells, this effects a long-term cure.

Several other gene therapy treatments, some using other virus vectors, are currently being tested with various levels of success. Since the first gene therapy experiment with ADA in 1990, there were no striking breakthroughs until 2000. Another form of SCID, caused by defects in a different gene, was successfully treated in several patients. It seems likely that this very rare form of the disease can now be successfully treated using gene therapy.

Technical Problems with Gene Therapy

Although gene therapy has tremendous practical potential, most applications remain a distant prospect. Some of the current difficulties are related to the vectors being used. Although using retroviral vectors gives stable integration of the transgene, the site of insertion is unpredictable and expression of the cloned gene is often transient. The vectors also have limited infectivity and are rapidly inactivated in the host. Many nonretroviral vectors, such as the adenovirus vector, have similar problems, and adverse reactions to the vector itself can also be a severe problem. However, some promising new vectors for gene therapy have emerged, including human artificial chromosomes (∞ Section 12.15) and highly modified retroviral vectors.

It is important to recall that in the gene therapy protocols being tested, the defective copy of the gene is not actually replaced, rather, its defective function is replaced. The retrovirus (containing the good copy of the gene) simply integrates somewhere into the human genome of the target cells. Actual gene replacements in germ line cells (cells that give rise to gametes) can be accomplished in experimental animals, although the techniques of isolating individual animals with these changes cannot readily be applied to humans. Moreover, attempts to change the germ cells of humans would also raise ethical questions that will likely keep these types of procedures, even if they have great medical promise, only a very long-range possibility.

26.9 MiniReview

A major hope of genetic engineering is in gene therapy, where functional copies of a gene can be inserted into an individual to treat genetic disease.

- Why is SCID such a serious disease and how can gene therapy help someone afflicted with SCID?
- What problems arise from using a retrovirus as a vector in gene therapy?
- A person treated successfully by gene therapy will still have a defective copy of the gene. Explain.

26.10 Transgenic Plants in Agriculture

Genetic improvement of plants by traditional methods of selection and breeding has a long history, but recombinant DNA technology has led to revolutionary changes. Genetic engineering can modify plant DNA and then use it to transform plant cells by either electroporation or particle gun methods (Section 12.11). Alternatively, one can use plasmids from the bacterium *Agrobacterium tumefaciens*, which naturally transfer DNA directly into the cells of certain types of plants (Section 24.14).

Unlike animals, there is no real separation of the germ line from the somatic cells in plants. Consequently plants can often be regenerated from just a single cell. Moreover, it is possible to culture plant cells *in vitro*. Therefore genetic engineering is normally done with plant cells growing in cell culture. After genetically altered clones have been selected, the cells are induced to grow back into whole plants by treatment with plant hormones.

Many successes in plant genetic engineering have already been achieved, and several transgenic plants are in agricultural production. The public knows these plants as *genetically modified (GM)* plants. In this section we discuss how foreign genes are inserted into plant genomes and how transgenic plants can be used.

The Ti Plasmid and Transgenic Plants

The gram-negative plant pathogen *A. tumefaciens* contains a large plasmid, called the **Ti plasmid**, that is responsible for its virulence. This plasmid contains genes that mobilize DNA for transfer to the plant, which as a result contracts crown gall disease (Section 24.14). The segment of the Ti plasmid DNA that is actually transferred to the plant is called **T-DNA**. The sequences at the ends of the T-DNA are essential for transfer, and the DNA to be transferred must be included between these ends.

One common Ti-vector system that has been used for the transfer of genes to plants is a two-plasmid system called a *binary vector*, which consists of a cloning vector plus a helper plasmid. The cloning vector contains the two ends of the T-DNA flanking a multiple cloning site plus an antibiotic resistance marker that can be used in plants. The plasmid also contains two origins of replication so that it can replicate in both *A. tumefaciens* and *Escherichia coli* (the latter is the host for cloning work) and another antibiotic resistance marker for selection in bacteria (**Figure 26.9**). The foreign DNA is inserted into the vector, which is then transformed into *E. coli* and then moved to *A. tumefaciens* by conjugation.

This cloning vector lacks the genes needed to transfer T-DNA to a plant. However, when placed in an *Agrobacterium* cell that contains a suitable helper plasmid, transfer of T-DNA to a plant can occur. The "disarmed" helper plasmid, called *D-Ti*, contains the virulence (*vir*) region of the Ti plasmid but lacks the T-DNA. Therefore it can direct the transfer of DNA into a plant, but no longer has genes that cause disease. Thus the helper plasmid supplies all the functions needed to transfer the T-DNA from the cloning vector. The cloned DNA and the kanamycin resistance marker of the vector are mobilized by D-Ti and transferred into a plant cell (Figure 26.9). Following integration into a plant chromosome, the foreign DNA can be expressed and confer new properties on the plant.

A number of transgenic plants have been produced using the Ti-plasmid of *A. tumefaciens*. The Ti system works well with broadleaf plants (dicots), including crops such as tomato, potato, tobacco, soybean, alfalfa, and cotton. It has also been used to produce transgenic trees, such as walnut and apple. The Ti system does not work with plants from the grass family (monocots, including the important crop plant, corn), but other methods of introducing DNA, such as the particle gun (Section 12.11), have been used successfully for them.

UNIT 5

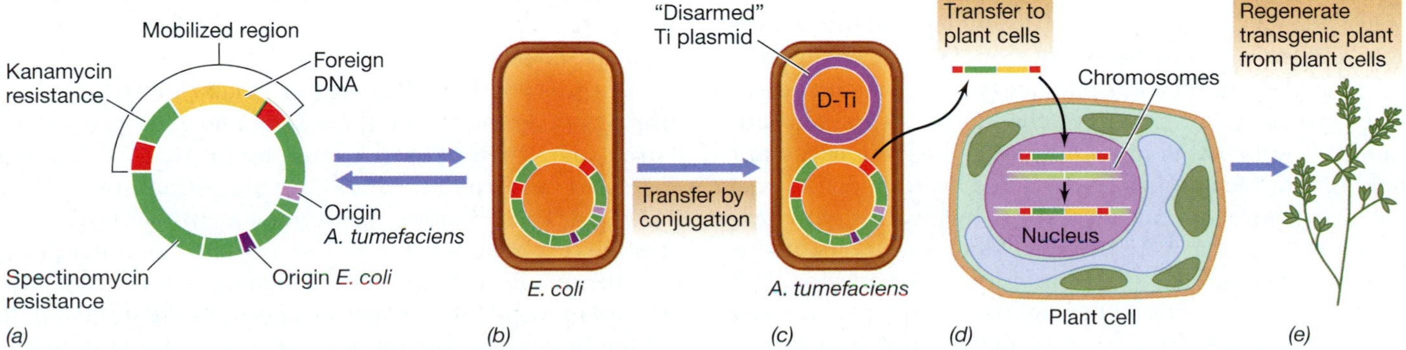

Figure 26.9 Production of transgenic plants using a binary vector system in *Agrobacterium tumefaciens*. *(a)* Generalized plant cloning vector containing ends of T-DNA (in red), foreign DNA, origins of replication, and resistance markers. *(b)* The vector is put into cells of *E. coli* for cloning and then transferred to *A. tumefaciens* by conjugation. *(c)* The resident Ti plasmid (D-Ti) used for transferring the vector to the plant has itself been genetically engineered to remove key pathogenesis genes. *(d)* However, D-Ti can mobilize the T-DNA region of the vector for transfer to plant cells grown in tissue culture. From the recombinant cell, whole plants can be regenerated. The biology of *A. tumefaciens* and the Ti plasmid is discussed in Section 24.14.

Stephen R. Padgette, Monsanto Company

Figure 26.10 Transgenic plants: herbicide resistance. The photograph shows a portion of a field of soybeans that has been treated with Roundup™, a glyphosate-based herbicide manufactured by Monsanto Company (St. Louis, MO). The plants on the right are normal soybeans; those on the left have been genetically engineered to express glyphosate resistance.

Herbicide and Insect Resistance

Major areas targeted for genetic improvement in plants include herbicide, insect, and microbial disease resistance, as well as improved product quality. The first GM crop to be grown commercially was tobacco grown in China in 1992 that was engineered for virus resistance. By the year 2005, the area planted with GM crops was estimated to have passed 1 billion acres (440 million hectares) worldwide. The main GM crops today are soybeans, corn, cotton, and canola. Almost all the GM soybeans and canola planted were herbicide resistant, whereas the corn and cotton were herbicide resistant or insect resistant, or both. In 2005 the United States grew over half the world's total of GM crops. Argentina, Canada, Brazil, and China were the other major producers, with the rest of the world accounting for less than 5% of the total.

Herbicide resistance is genetically engineered into a crop plant to protect it from herbicides applied to kill weeds. Many herbicides inhibit a key plant enzyme or protein necessary for growth. For example, the herbicide glyphosate (Roundup™) kills plants by inhibiting an enzyme necessary for making aromatic amino acids. Some bacteria contain an equivalent enzyme and are also killed by glyphosate. However, mutant bacteria were selected that were resistant to glyphosate and contained a resistant form of the enzyme. The gene encoding this resistant enzyme from *Agrobacterium* was cloned, modified for expression in plants, and transferred into important crop plants, such as soybeans. When sprayed with glyphosate, plants containing the bacterial gene are not killed (**Figure 26.10**). Thus glyphosate is used to kill weeds that compete for water and nutrients with the growing crop plants. Herbicide-resistant soybeans are now widely planted in the United States.

Insect Resistance: Bt Toxin

Insect resistance has also been genetically introduced into plants. One widely used approach is based on introducing the genes encoding the toxic protein of *Bacillus thuringiensis* into plants. *B. thuringiensis* produces a crystalline protein called *Bt-toxin* (∞ Section 16.2) that is toxic to moth and butterfly larvae. Many variants of Bt-toxin exist that are specific for different insects. Certain strains of *B. thuringiensis* produce additional proteins toxic to beetle and fly larvae and mosquitoes.

Several different approaches were used to enhance the efficacy of Bt-toxin for pest control in plants. One approach was to develop a single set of Bt-toxins that was effective against many different insects. This could be done because the protein is made up of separate structural regions (domains) that are responsible for specificity and toxicity. The toxic domain is highly conserved in all the various Bt-toxins. However, genetic engineers also made hybrid genes that encoded the toxic domain and one of several different specificity domains to yield a suite of toxins, each best suited for a particular plant or pest situation.

An effective approach to achieve Bt transgene expression and stability was to transfer the gene directly into the plant genome. For example, a natural Bt-toxin gene was cloned into a plasmid vector under control of a chloroplast rRNA promoter and then transferred into tobacco plant chloroplasts by microprojectile bombardment (∞ Section 12.11). With this methodology, transgenic plants that expressed this protein at levels that were extremely toxic to larvae from a number of insect species were obtained (**Figure 26.11**).

Although transgenic Bt-toxin looked at first to be a great agricultural success, some problems have arisen, in particular, the selection for insects resistant to Bt-toxin. Resistance to insecticides and herbicides is a common problem in agriculture, and the fact that a product has been produced by genetic engineering does not exempt it from this problem. In addition, Bt-toxin often kills nontarget insects, some of which may be helpful. Many approaches must be used for pest control in agriculture, and Bt-toxin is just one of many. Nevertheless, transgenic crops with Bt-toxin are still widely planted in the United States.

Bt-toxin is harmless to mammals, including humans, for several reasons. First, cooking and food processing destroy the toxin. Second, any toxin that is ingested is digested in the mammalian gastrointestinal tract. Third, Bt toxin works by binding to specific receptors in the insect intestine that are absent from the intestines of other groups of organisms. Binding promotes a change in conformation of the toxin, which then generates pores in the intestinal lining of the insect that disrupt the insect digestive system and kill the insect.

Other Uses of Plant Biotechnology

Not all genetic engineering is directed toward making plants disease resistant. Genetic engineering can also be used to develop GM plants that are more nutritious or that have more desirable consumer-oriented characteristics. For instance, the first GM food grown for sale in the U.S. market was a tomato in which spoilage was delayed. This increased the shelf life of the vegetable. In addition, transgenic plants can be genetically engineered to produce commercial or pharmaceutical products, as has been done with microorganisms and animals. For example, crop plants such as tobacco and tomatoes have been engineered to produce a number of products, such as the human protein *interferon*. Transgenic crop plants can also be used to produce human antibodies efficiently and inexpensively. These antibodies, called *plantibodies*, have potential as anticancer or antivirus drugs, and some are undergoing clinical trials. For example, transgenic tobacco plants have been used to make an antibody known as CaroRx that blocks bacteria that cause dental caries from attaching to teeth. CaroRx is made in high levels in tobacco leaves and is relatively easy to purify. So far clinical trials have shown it to be safe and effective. Plants are useful in producing these types of products because they typically modify proteins correctly and because crop plants can be efficiently grown and harvested in large amounts.

Crop plants are also being developed for the production of vaccines. For instance, a recombinant tobacco mosaic virus (∞ Section 19.6) has been engineered whose coat contains surface proteins of *Plasmodium vivax*, the organism that causes the disease malaria (∞ Section 35.5). The *P. vivax* proteins elicit an immune response in humans. Hence, this recombinant virus could be used to produce a malaria vaccine in large amounts and at low cost by simply harvesting infected tobacco or tomato plants and processing them for the immunogenic protein. Another interesting approach is to produce a vaccine in an edible plant product. Such *edible vaccines* now under development could immunize humans against diseases caused by enteric bacteria, including cholera and diarrhea (∞ Section 36.5).

A rather different kind of transgenic plant is the Amflora potato, developed by BASF, a German chemical company. The Amflora potato is not intended for eating. Unlike normal potatoes, which produce two types of starch, amylopectin and amylose, the Amflora potato makes only amylopectin, a raw material in the paper and adhesives industries. Using Amflora potatoes will avoid the expensive and energy-consuming purification of amylose from amylopectin. Approval of this transgenic crop is presently under consideration by the European Union.

Although public acceptance of GM crops remains high in the United States, there have been some concerns over the contamination of human food with GM corn, so far approved only for animal food. In some European Union countries there has been considerable public concern over GM organisms. Most concerns center around either the perception of adverse effects of foreign genes on humans or domesticated animals or the potential "escape" of transgenes from transgenic plants into native plants. At present, supporting evidence for either of these scenarios is not strong; however, there are indications that glyphosphate resistance genes are beginning to spread into the weed plant population. Thus, concerns remain about GM plants and have served to control the rate at which new transgenic plants enter the marketplace.

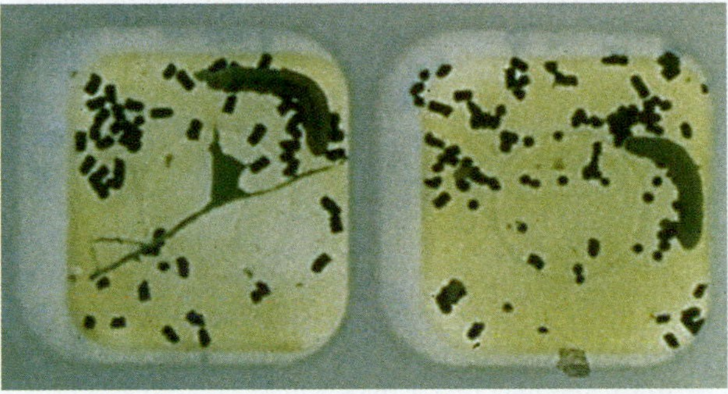

(a)

(b)

Kevin McBride, Calgene, Inc.

Figure 26.11 Transgenic plants: insect resistance. *(a)* The results of two different assays to determine the effect of beet armyworm larvae on tobacco leaves from normal plants. *(b)* The results of similar assays using tobacco leaves from transgenic plants that express Bt-toxin in their chloroplasts.

UNIT 5

26.10 MiniReview

Genetic engineering can make plants resistant to disease, improve product quality, and make crop plants a source of recombinant proteins and even vaccines. One commonly used cloning vector for plants is the Ti plasmid of the bacterium *Agrobacterium tumefaciens*. This plasmid can transfer DNA into plant cells. Commercial plants whose genomes have been modified using *in vitro* genetic techniques are called genetically modified (GM) organisms.

- What is a transgenic plant?
- Give an example of a genetically modified plant and describe how its genetic modification benefits plant agriculture.
- What advantages do plants have as vehicles for making antibodies?

Review of Key Terms

Biotechnology use of organisms, typically genetically altered, in industrial, medical, or agricultural applications

DNA vaccine a vaccine that uses the DNA of a pathogen to elicit an immune response

Fusion protein a protein that is the result of fusing two different proteins together by merging their coding sequences into a single gene

Genetic engineering the use of *in vitro* techniques in the isolation, manipulation, alteration, and expression of DNA (or RNA), and in the development of genetically modified organisms

Gene therapy treatment of a disease caused by a dysfunctional gene by introducing a functional copy of the gene

Genetically modified organism (GMO) an organism whose genome has been altered using genetic engineering; the abbreviation GM is also used in terms such as GM crops and GM foods

Pathway engineering the assembly of a new or improved biochemical pathway, using genes from one or more organisms

Polyvalent vaccine a vaccine that immunizes against more than one disease

Reverse transcription conversion of an RNA sequence into the corresponding DNA sequence

T-DNA the segment of the *Agrobacterium tumefaciens* Ti plasmid that is transferred into plant cells

Ti plasmid a plasmid in *Agrobacterium tumefaciens* capable of transferring genes from bacteria to plants

Transgenic organism a plant or animal with foreign DNA inserted into its genome

Vector vaccine a vaccine made by inserting genes from a pathogenic virus into a relatively harmless carrier virus

Review Questions

1. What is the significance of reverse transcriptase when cloning animal genes for expression in bacteria (Section 26.2)?
2. What classes of proteins are produced by biotechnology? How are the genes for such proteins obtained (Section 26.3)?
3. What is a subunit vaccine and why are subunit vaccines considered a safer way of conferring immunity to viral pathogens than attenuated virus vaccines (Section 26.5)?
4. How has metagenomics been used to find novel useful products (Section 26.6)?
5. What is pathway engineering? Why is it more difficult to produce an antibiotic than to produce a single enzyme via genetic engineering (Section 26.7)?
6. What is a knockout mouse? Why are knockout mice important for the study of human physiology and hereditary defects (Section 26.8)?
7. How has genetic engineering benefited the treatment of SCID and cystic fibrosis (Sections 26.4 and 26.9)?
8. How do transgenic plants differ from those modified by conventional breeding (Section 26.10)?
9. What is the Ti plasmid and how has it been of use in genetic engineering (Section 26.10)?
10. List several examples in which crop plants have been improved by genetic engineering. How have genetically engineered plants helped human medicine?

Application Questions

1. You have just discovered a protein in mice that may be an effective cure for cancer, but it is present only in tiny amounts. Describe the steps you would use to produce this protein in therapeutic amounts. Which host would you want to clone the gene into and why? Which host would you use to express the protein in and why?
2. Gene therapy is used to treat people who have a genetic disease and, if successful, will cure them. However, such people will still be able to pass on the genetic disease to their offspring. Explain. Why do you believe this might be an area of research that is not attracting as much attention as treatment of the individual?
3. Compare the advantages and disadvantages of using transgenic crops (such as Bt corn) as a source of human food. Consider various viewpoints, including that of the farmer, the environmentalist, and the consumer.

27

Microbial Growth Control

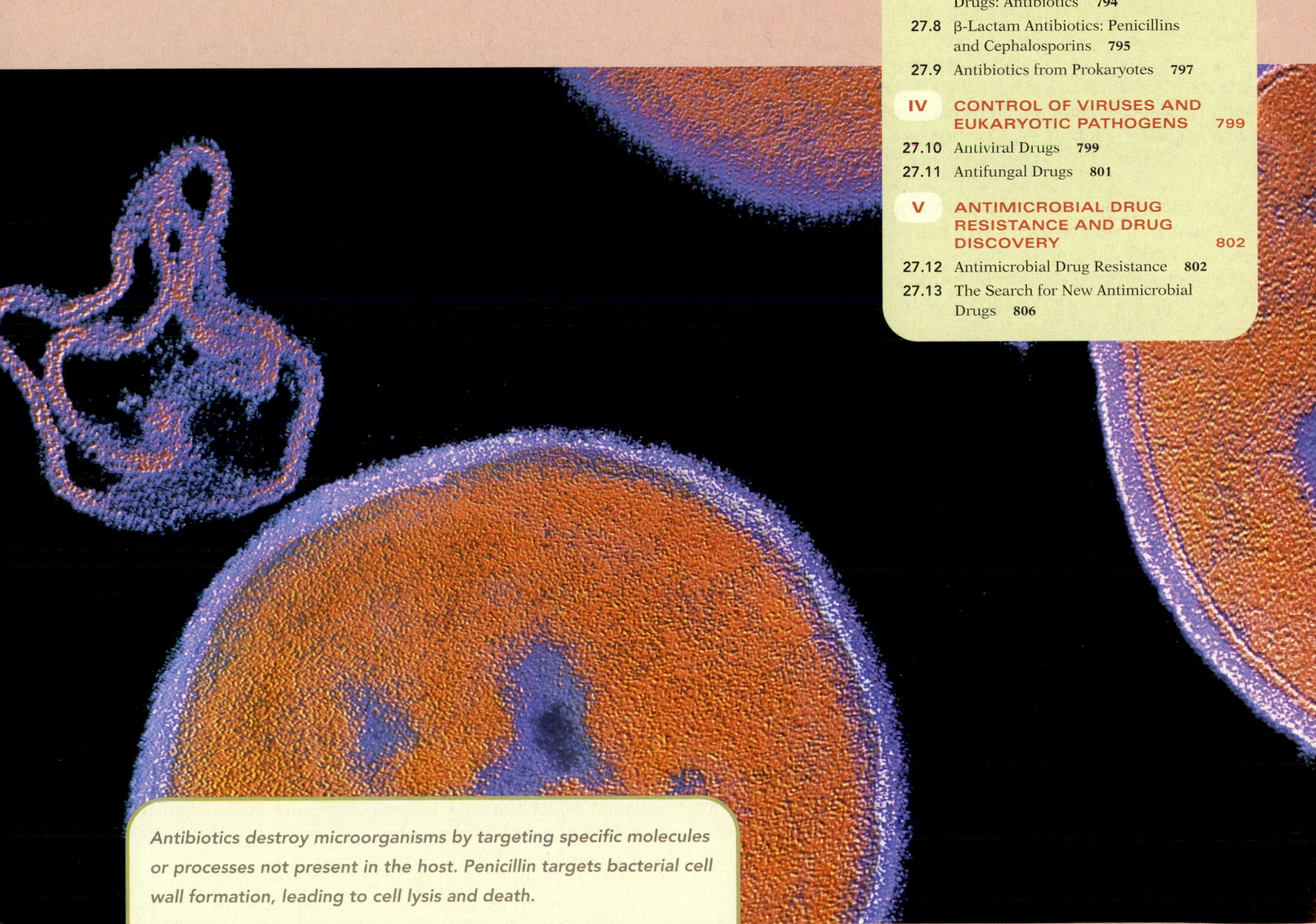

Antibiotics destroy microorganisms by targeting specific molecules or processes not present in the host. Penicillin targets bacterial cell wall formation, leading to cell lysis and death.

With this chapter we commence a new phase of this book and begin to study the relationships between microorganisms and humans. We start with the agents and methods used for control of microbial growth. A few agents eliminate microbial growth entirely by **sterilization**—the killing or removal of all viable organisms within a growth medium. Often, however, sterility is not attainable, but microorganisms can be effectively controlled by limiting their growth, the process of **inhibition**.

Methods for inhibiting rapid microbial growth include decontamination and disinfection. **Decontamination** is the treatment of an object or surface to make it safe to handle. For example, simply wiping a table after a meal removes contaminating microorganisms and their potential nutrients. **Disinfection**, in contrast, directly targets pathogens, although it may not eliminate all microorganisms. Specialized chemical or physical agents called *disinfectants* can kill microorganisms or inhibit microbial growth. Bleach (sodium hypochlorite) solution, for example, is a disinfectant used to clean and disinfect food preparation areas.

At times, it may be necessary to destroy all microorganisms. Sterilization, although difficult to achieve, completely eliminates all microorganisms, including viruses. Such measures are necessary, for instance, when making microbiological media or preparing surgical instruments. The goal for all decontamination and sterilization procedures is to either reduce or eliminate the microbial load.

Microbial control *in vivo* is much more difficult: Clinically useful bacteriocidal (bacteria killing) agents or bacteriostatic (bacteria inhibiting) agents prevent or reduce bacterial growth, while causing no harm to the host. This goal is achieved by a wide variety of selective natural and synthetic chemotherapeutic agents.

In this chapter we first examine methods of microbial control that are used *in vitro*. We then discuss antimicrobial drugs used in humans.

I PHYSICAL ANTIMICROBIAL CONTROL

Physical methods are often used to achieve microbial decontamination, disinfection, and sterilization. Heat, radiation, and filtration can destroy or remove microorganisms. These methods prevent microbial growth or decontaminate areas or materials harboring microorganisms. Here we discuss physical control mechanisms and present some practical examples.

27.1 Heat Sterilization

Perhaps the most widespread method used for controlling microbial growth is the use of heat, especially as a sterilization method. Factors that affect a microorganism's susceptibility to heat include the temperature and duration of the heat treatment, and whether the heat is moist or dry.

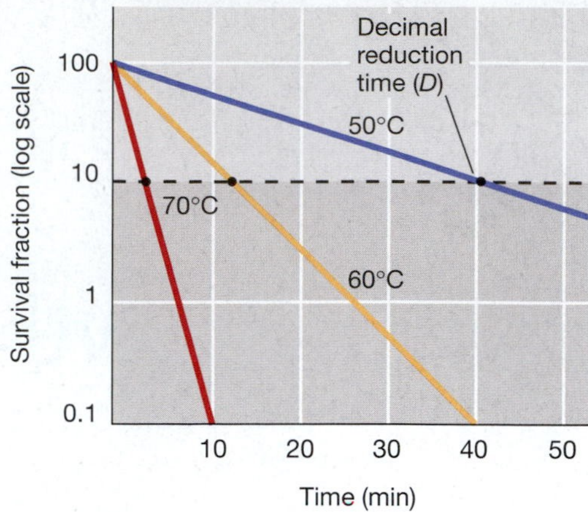

Figure 27.1 The effect of temperature over time on the viability of a mesophilic bacterium. The decimal reduction time, *D*, is the time at which only 10% of the original population of organisms remains viable at a given temperature. For 70°C, $D = 3$ min; for 60°C, $D = 12$ min; for 50°C, $D = 42$ min.

Measuring Heat Sterilization

All microorganisms have a maximum growth temperature beyond which viability decreases (∞ Section 6.12). Viability is lost because at very high temperatures most macromolecules lose structure and function, a process called *denaturation*. The effectiveness of heat as a sterilant is measured by the time required for a tenfold reduction in the viability of a microbial population at a given temperature. This is the *decimal reduction time* or *D*. Over the range of temperatures usually used in food preparation (cooking and canning), the relationship between *D* and temperature is exponential. Thus, when the logarithm of *D* is plotted against temperature, a straight line is obtained (**Figure 27.1**). The slope of the line indicates the sensitivity of the organism to heat under the conditions employed, and the graph can be used to calculate processing times to achieve sterilization, for instance in a canning operation (∞ Section 37.2). Death from heating is an exponential (first-order) function, proceeding more rapidly as the temperature rises, as shown in **Figure 27.2**. The time necessary to kill a defined fraction (for example, 90%) of viable cells is independent of the initial cell concentration. As a result, sterilization of a microbial population takes longer at lower temperatures than at higher temperatures. The time and temperature, therefore, must be adjusted to achieve sterilization for each specific set of conditions. The type of heat is also important: Moist heat has better penetrating power than dry heat and, at a given temperature, produces a faster reduction in the number of living organisms.

Determination of a decimal reduction time requires a large number of viable count measurements (∞ Section 6.10). An easier way to characterize the heat sensitivity of an organism is to measure the *thermal death time*, the time it takes to kill all cells at a given temperature. To determine the thermal death time, samples of a cell suspension are heated for different times, mixed with culture medium, and

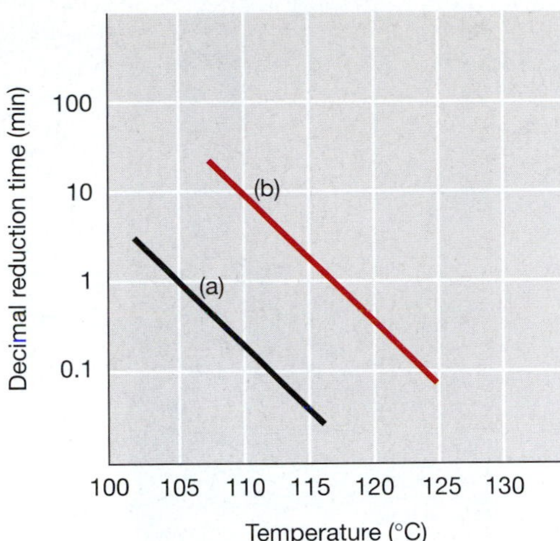

Figure 27.2 The relationship between temperature and the rate of killing in mesophiles and thermophiles. Data were obtained for decimal reduction times, *D*, at several different temperatures, as in Figure 27.1. For organism (*a*), a typical mesophile, exposure to 110°C for less than 20 sec resulted in a decimal reduction, while for organism (*b*), a thermophile, 10 min was required to achieve a decimal reduction.

incubated. If all the cells have been killed, no growth is observed in the incubated samples. The thermal death time depends on the size of the population tested; a longer time is required to kill all cells in a large population than in a small one. When the number of cells is standardized, it is possible to compare the heat sensitivities of different organisms by comparing their thermal death times at a given temperature.

Endospores and Heat Sterilization

Some bacteria produce highly resistant cells called *endospores* (Section 4.12). The heat resistance of vegetative cells and endospores from the same organism differs considerably. For instance, in the autoclave (see later) a temperature of 121°C is normally reached. Under these conditions, endospores may require 4–5 minutes for a decimal reduction, whereas vegetative cells may require only 0.1–0.5 min at 65°C. Because of this difference, effective heat sterilization procedures must be designed to destroy endospores.

Endospores can survive heat that would rapidly kill vegetative cells of the same species. A major factor in heat resistance is the amount and state of water within the endospore. During endospore formation, the protoplasm is reduced to a minimum volume as a result of the accumulation of Ca^{2+}-dipicolinic acid complexes and small acid-soluble spore proteins (SASP). This mixture forms a cytoplasmic gel, and a thick cortex then forms around the developing endospore. Contraction of the cortex results in a shrunken, dehydrated cell containing only 10–30% of the water of a vegetative cell (Section 4.12).

The water content of the endospore coupled with the concentration of SASPs determines its heat resistance. If endospores have a low concentration of SASPs and high water content, they have low heat resistance. Conversely, if they have a high concentration of SASPs and low water content, they have high heat resistance. Water moves freely in and out of endospores, so it is not the impermeability of the endospore coat that excludes water, but the gel-like material in the endospore protoplast.

The medium in which heating takes place also influences the killing of both vegetative cells and endospores. Microbial death is more rapid at acidic pH, and acid foods such as tomatoes, fruits, and pickles are much easier to sterilize than neutral pH foods such as corn and beans. High concentrations of sugars, proteins, and fats decrease heat penetration and usually increase the resistance of organisms to heat, whereas high salt concentrations may either increase or decrease heat resistance, depending on the organism. Dry cells and endospores are more heat-resistant than moist ones; consequently, heat sterilization of dry objects such as endospores always requires higher temperatures and longer times than sterilization of wet objects such as liquid bacterial cultures.

The Autoclave

The **autoclave** is a sealed heating device that allows the entrance of steam under pressure (**Figure 27.3**). Killing of heat-resistant endospores requires heating at temperatures above 100°C, the boiling point of water at normal atmospheric pressure. This is accomplished by applying steam under pressure (Figure 27.3*a*). The autoclave uses steam under 1.1 kilograms/square centimeter (kg/cm^2) [15 pounds/square inch (lb/in^2)] pressure, which yields a temperature of 121°C. At 121°C, the time to achieve sterilization is generally 10–15 minutes (Figure 27.3*b*).

If an object being sterilized is bulky, heat transfer to the interior is retarded, and the total heating time must be extended to ensure that the entire object is at 121°C for 10–15 minutes. Extended times are also required when large volumes of liquids are being autoclaved because large volumes take longer to reach sterilization temperatures. Note that it is not the *pressure* inside the autoclave that kills the microorganisms but the high *temperature* that can be achieved when steam is applied under pressure.

Pasteurization

Pasteurization uses precisely controlled heat to reduce the microbial load (microbial numbers) in milk and other heat-sensitive liquids. The process, named for Louis Pasteur (Section 1.7), was first used for controlling the spoilage of wine. Pasteurization does not kill all organisms and is therefore not synonymous with sterilization. Pasteurization of milk was originally developed to kill pathogenic bacteria, especially the organisms causing tuberculosis, brucellosis, Q fever, and typhoid fever. These pathogens are no longer common in even raw foods in developed countries, but use of pasteurization continues because it also controls commonly encountered pathogens today, including *Listeria monocytogenes*, *Campylobacter* species,

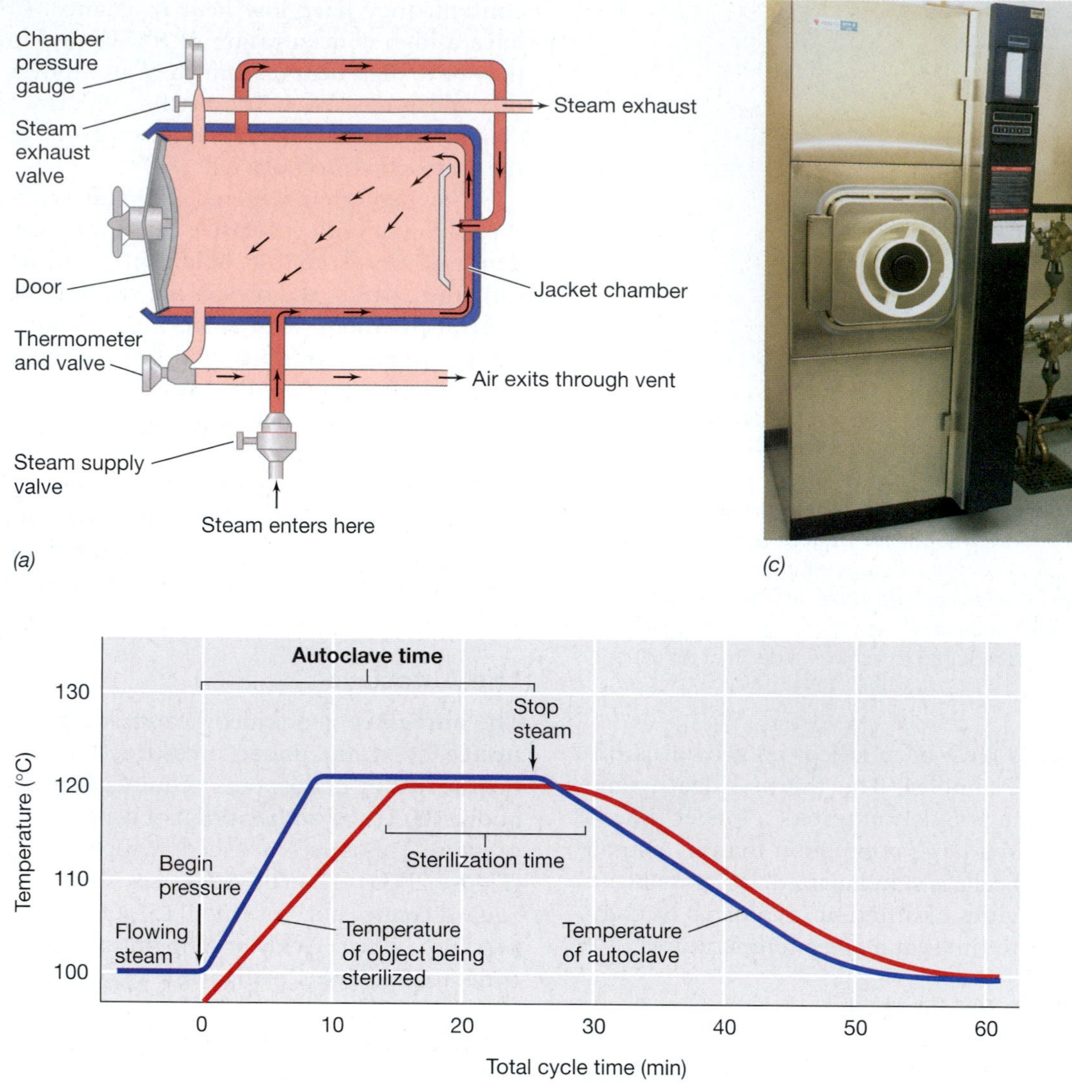

Figure 27.3 **The autoclave and moist heat sterilization.** *(a)* The flow of steam through an autoclave. *(b)* A typical autoclave cycle. The temporal heating profile of a fairly bulky object is shown. The temperature of the object rises and falls more slowly than the temperature of the autoclave. The temperature of the object must reach the target temperature and be held for 10–15 minutes to ensure sterility, regardless of the temperature and time recorded in the autoclave. *(c)* A modern research autoclave. Note the pressure-lock door and the automatic cycle controls on the right panel. The steam inlet and exhaust fittings are on the right side of the autoclave.

Salmonella, and *Escherichia coli* O157:H7; these pathogenic bacteria can be found in foods such as dairy products and juices (Sections 37.7–37.11). In addition, pasteurization retards the growth of spoilage organisms, dramatically increasing the shelf life of perishable liquids (Sections 37.1 and 37.2).

Pasteurization of milk is usually achieved by passing the milk through a heat exchanger. The milk is pumped through tubing that is in contact with a heat source. Careful control of the milk flow rate and the size and temperature of the heat source raises the temperature of the milk to 71°C for 15 seconds. The milk is then rapidly cooled. This process is aptly called flash pasteurization.

Milk can also be heated in large vats to 63–66°C for 30 minutes. However, this bulk pasteurization method is less satisfactory because the milk heats and cools slowly and must be held at high temperatures for longer times. Flash pasteurization, sometimes done at even higher temperatures and shorter times, alters the flavor less, kills heat-resistant organisms more effectively, and can be done on a continuous-flow basis, making it more adaptable to large dairy operations.

Figure 27.4 A biological safety cabinet. The cabinet shown has an ultraviolet (UV) radiation source (mercury vapor lamp), which is used for decontamination of the inside surfaces. The metal grid in the rear covers a HEPA (high-efficiency particulate air) filter. Air, drawn from outside the cabinet, is pumped through the HEPA filter. The filtered air, free from contaminants, including microorganisms, enters the cabinet. The air inside the hood is then drawn through the vents surrounding the front, and shunted back through the HEPA filter. Thus, the cabinet is designed to provide a contaminant-free workspace while protecting the investigator by preventing flow of air directly out of the cabinet.

J. Martinko

27.1 MiniReview

Sterilization is the killing of all organisms including viruses. Heat is the most widely used method of sterilization. The temperature must eliminate the most heat-resistant organisms, usually bacterial endospores. An autoclave permits applications of steam heat under pressure, achieving temperatures above the boiling point of water, and killing endospores. Pasteurization does not sterilize liquids, but reduces microbial load, kills most pathogens, and inhibits the growth of spoilage microorganisms.

- Why is heat an effective sterilizing agent?
- What steps are necessary to ensure the sterility of material that may be contaminated with bacterial endospores?
- Distinguish between the need to sterilize microbiological media and the need to pasteurize apple juice.

27.2 Radiation Sterilization

Heat is just one form of energy that can sterilize or reduce microbial load. Microwaves, ultraviolet (UV) radiation, X-rays, gamma rays (γ-rays), and electrons all can effectively reduce microbial growth if applied in the proper dose and time. However, each type of energy has a different mode of action. For example, the antimicrobial effects of microwaves are due, at least in part, to thermal effects. UV radiation between 220 and 300 nm in wavelength has sufficient energy to cause modifications or actual breaks in DNA, sometimes leading to disruption of DNA and death of the exposed organism (Section 11.7). This

Table 27.1 Radiation sensitivity of microorganisms and biological functions

Species or function	*Type of microorganism*	D10[a] (Gy)
Clostridium botulinum	Gram-positive anaerobic sporulating *Bacteria*	3300
Clostridium tetani	Gram-positive anaerobic sporulating *Bacteria*	2400
Bacillus subtilis	Gram-positive aerobic sporulating *Bacteria*	600
Escherichia coli O157:H7	Gram-negative *Bacteria*	300
Salmonella typhimurium	Gram-negative *Bacteria*	200
Lactobacillus brevis	Gram-positive *Bacteria*	1200
Deinococcus radiodurans	Gram-negative radiation-resistant *Bacteria*	2200
Aspergillus niger	Mold	500
Saccharomyces cerevisiae	Yeast	500
Foot-and-mouth	Virus	13,000
Coxsackie	Virus	4500
Enzyme inactivation		20,000–50,000
Insect deinfestation		1000–5000

[a]D10 is the amount of radiation necessary to reduce the initial population or activity level tenfold (one logarithm). Gy = grays. 1 gray = 100 rads. The lethal dose for humans is 10 Gy.

"near-visible" UV light is useful for disinfecting surfaces, air, and materials such as water that do not absorb the UV waves. For example, laboratory biological containment cabinets are equipped with a "germicidal" UV light to decontaminate the surface after use (**Figure 27.4**). UV radiation, however, cannot penetrate solid, opaque, or light-absorbing surfaces, limiting its use to disinfection of exposed surfaces.

UNIT 6

Ionizing Radiation

Ionizing radiation is electromagnetic radiation of sufficient energy to produce ions and other reactive molecular species from molecules with which the radiation particles collide. Ionizing radiation generates electrons, e^-, hydroxyl radicals, OH•, and hydride radicals, H• (Section 6.18). Each of these highly reactive molecules is capable of altering and disrupting macromolecules such as DNA and protein. The ionization and subsequent degradation of these biologically important molecules leads to the death of irradiated cells. Several radiation sources are useful for sterilization.

The unit of radiation is the *roentgen*, which is a measure of the energy output from a radiation source. The standard for biological applications such as sterilization is the absorbed radiation dose, measured in *rads* (100 erg/g) or *grays* (1 Gy = 100 rad). Some microorganisms are much more resistant to radiation than others. **Table 27.1** shows the dose of radiation necessary for a tenfold (one log) reduction in the numbers of selected microorganisms or biological functions. For example, the amount of energy necessary to achieve a tenfold

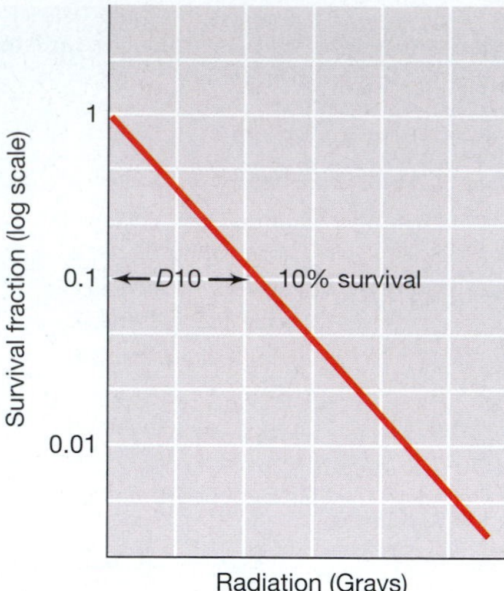

Figure 27.5 Relationship between the survival fraction and the radiation dose of a microorganism. The *D*10, or decimal reduction dose, can be interpolated from the data as shown.

reduction (*D*) of a radiation-sensitive bacterium such as *Escherichia coli* O157:H7 is 300 Gy. The *D* value is analogous to the decimal reduction time for heat sterilization: The relationship of the survival fraction plotted on a semilogarithmic scale versus the radiation dose in grays is essentially linear (**Figure 27.5** and compare to Figure 27.1).

In practice, this means that at a radiation dose of 300 Gy, 90% of *E. coli* O157:H7 in a given sample would be killed. A dose of 2 *D*, or 600 Gy, would kill 99% of the organism, and so on. A standard killing dose for radiation sterilization is 12 *D*. For destruction of radiation-resistant endospores of a bacterium such as *Clostridium botulinum*, for example, 39,600 Gy are required (Table 27.1). By contrast, the killing dose for *E. coli* O157:H7 is only 3600 Gy. In general, microorganisms are much more resistant to ionizing radiation than are multicellular organisms. For example, the lethal radiation dose for humans can be as low as 10 Gy if delivered over several minutes!

Radiation Practices

Common sources of ionizing radiation include cathode ray tubes that generate electron beams, X-ray machines, and radioactive nuclides ^{60}Co and ^{137}Cs, which are both relatively inexpensive by-products of nuclear fission. These sources produce either electrons, X-rays, or γ-rays, respectively, all of which have sufficient energy to efficiently inhibit microbial growth. In addition, X-rays and γ-rays penetrate solids and liquids, making them ideal for treatment of bulk items such as ground beef or cereal grains.

Radiation is currently used for sterilization and decontamination in the medical supplies and food industries. In the United States, the Food and Drug Administration has approved the use of radiation for sterilization of such diverse items as surgical supplies, disposable labware, drugs, and even tissue grafts (**Table 27.2**). However, because of the costs and hazards associated with radiation equipment, this type of sterilization is limited to large industrial applications or specialized facilities.

Table 27.2 Medical and laboratory products sterilized by radiation

Tissue grafts	*Drugs*	*Medical and laboratory supplies*
Cartilage	Chloramphenicol	Disposable labware
Tendon	Ampicillin	Culture media
Skin	Tetracycline	Syringes
Heart valve	Atropine	Surgical equipment
	Vaccines	Sutures
	Ointments	

Certain foods and food products are also routinely irradiated to ensure sterilization, pasteurization, or insect deinfestation. Radiation is approved by the World Health Organization and is used in the United States for decontamination of foods particularly susceptible to microbial contamination, especially fresh meat products such as hamburger and chicken and spices (Section 37.2). The use of radiation for these purposes is an established and accepted technology in many countries. However, the practice has not been readily accepted in some countries because of fears of possible radioactive contamination, alteration in nutritional value, production of toxic or carcinogenic products, and perceived "off" tastes in irradiated food.

27.2 MiniReview

Controlled doses of electromagnetic radiation effectively inhibit microbial growth. Ultraviolet radiation is used for decontaminating surfaces and materials that do not absorb light, such as air and water. Ionizing radiation, necessary to penetrate solid or light-absorbing materials, is used for sterilization and decontamination in the medical and food industries.

- Define the decimal reduction dose and the killing dose for radiation treatment of microorganisms.
- Why is ionizing radiation more effective than UV radiation for sterilization of food products?

27.3 Filter Sterilization

As we have seen, heat is an effective way to decontaminate most liquids and can even be used to treat gasses. Heat-sensitive liquids and gasses, however, must be treated by other methods. Filtration is a method that accomplishes decontamination and even sterilization without exposure to denaturing heat. The liquid or gas is passed through a filter, a device with pores too small for the passage of microorganisms, but large enough to allow the passage of the liquid or gas. The selection of filters for sterilization must account for the size range of the contaminants to be excluded. Some

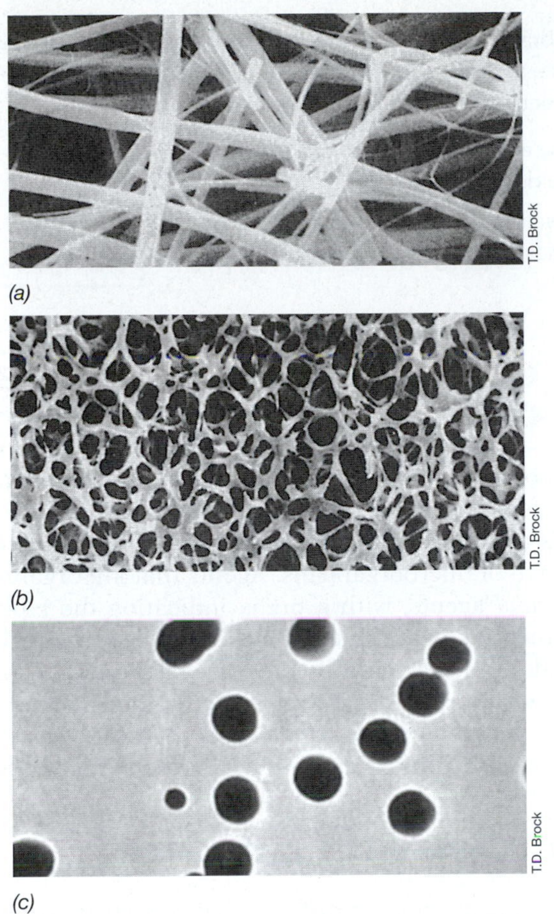

Figure 27.6 Microbiological filters. The structure of *(a)* a depth filter, *(b)* a conventional membrane filter, and *(c)* a Nucleopore filter.

microbial cells are greater than 10 μm in diameter, and the smallest bacteria are less than 0.3 μm in diameter. Historically, selective filtration methods were used to define and isolate viruses, most of which range from 25 nm to 200 nm (0.2 μm) in diameter. **Figure 27.6** illustrates major types of filters.

Depth Filters

A depth filter is a fibrous sheet or mat made from a random array of overlapping paper or borosilicate (glass) fibers (Figure 27.6*a*). The depth filter traps particles in the network of fibers throughout the depth of the structure. Because the filtration material is arranged randomly in a thick layer, depth filters resist clogging and are often used as prefilters to remove larger particles from liquid suspensions so that the final filter in the sterilization process is not clogged. Depth filters are also used for the filter sterilization of air in industrial processes. In the home, the filter used in forced air heating and cooling systems is a simple depth filter, designed to trap particulate matter such as dust, spores, and allergens.

Depth filters are important for biosafety applications. For example, manipulations of cell cultures, microbial cultures, and growth media require that contamination of both the operator and the experimental materials are minimized. These operations can be efficiently performed in a biological safety cabinet with airflow, both in and out of the cabinet, directed through a depth filter called a **HEPA filter** or *h*igh-*e*fficiency *p*articulate *a*ir filter (Figure 27.4). A typical HEPA filter is a single sheet of borosilicate (glass) fibers that has been treated with a water-repellant binder. The filter, pleated to increase the overall surface area, is mounted inside a rigid, supportive frame. HEPA filters come in various shapes and sizes, from several square centimeters for vacuum cleaners, to several square meters for biological containment hoods (Figure 27.4) and room air systems. Control of airborne particulate materials with HEPA filters allows the construction of "clean rooms" and isolation rooms for quarantine, as well as specialized biological safety laboratories (∞ Section 32.4). HEPA filters typically remove 0.3-μm test particles with an efficiency of at least 99.97%; they thus effectively remove both small and large particles, including most microorganisms, from the airstream.

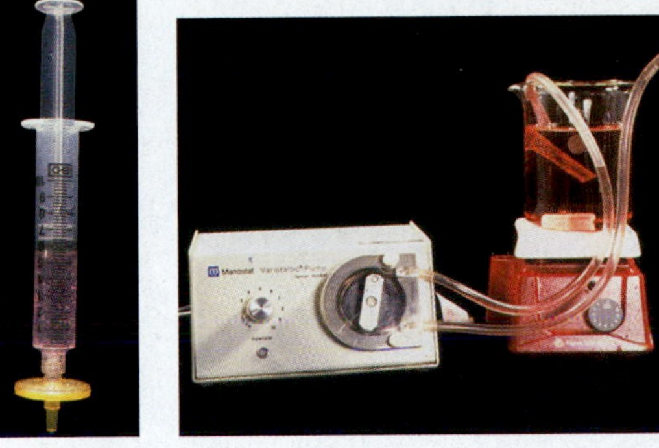

Figure 27.7 Membrane filters. Disposable, presterilized, and assembled membrane filter units. Left: a filter system designed for small volumes. Right: a filter system designed for larger volumes.

Membrane Filters

Membrane filters are the most common type of filters used for liquid sterilization in the microbiology laboratory (Figure 27.6*b*). Membrane filters are composed of high tensile strength polymers such as cellulose acetate, cellulose nitrate, or polysulfone, manufactured to contain a large number of tiny holes, or pores. By adjusting the polymerization conditions during manufacture, the size of the holes in the membrane (and thus the size of the molecules that can pass through) can be precisely controlled. The membrane filter differs from the depth filter, functioning more like a sieve and trapping particles on the filter surface. About 80–85% of the membrane surface area consists of open pores. The porosity provides for a relatively high fluid flow rate.

Membrane filters for the sterilization of a liquid are illustrated in **Figure 27.7**. Presterilized membrane filter assemblies for sterilization of small to medium volumes of liquids such as growth media are routinely used in research and clinical laboratories. Filtration is accomplished by using a syringe, pump, or vacuum to force the liquid through the filtration apparatus into a sterile collection vessel.

Another type of membrane filter in common use is the nucleation track (Nucleopore) filter. To make these filters, very

UNIT 6

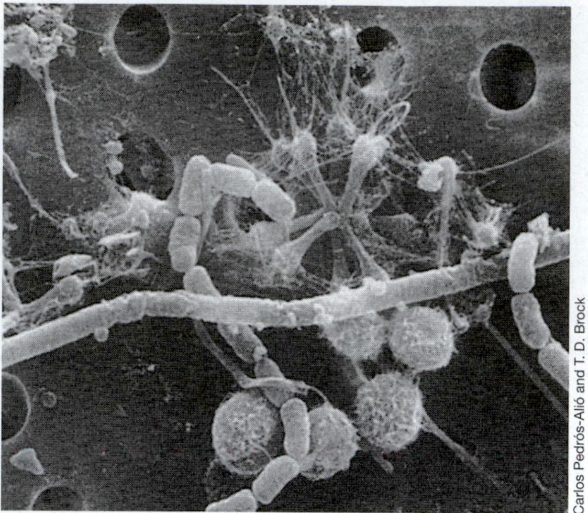

(a)

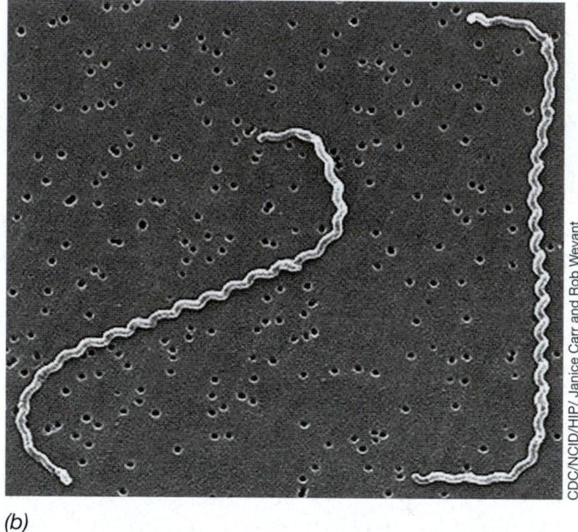

(b)

Figure 27.8 Scanning electron micrographs of bacteria trapped on Nucleopore membrane filters. *(a)* Aquatic bacteria and algae. The pore size is 5 μm. *(b) Leptospira interrogans*. The bacterium is about 0.1 μm in diameter and up to 20 μm in length. The pore size of the filter is 0.2 μm.

thin polycarbonate film (10 μm) is treated with nuclear radiation and then etched with a chemical. The radiation causes local damage to the film, and the etching chemical enlarges these damaged locations into holes. The size of the holes is precisely controlled by varying the strength of the etching solution and the etching time. A typical nucleation track filter therefore has very uniform holes (Figure 27.6*c*). Nucleopore filters are commonly used to isolate specimens for scanning electron microscopy. Microorganisms are removed from liquid and concentrated in a single plane on the filter, where they can be observed with the microscope (**Figure 27.8**).

27.3 MiniReview

Filters remove microorganisms from air or liquids. Depth filters, including HEPA filters, are used to remove microorganisms and other contaminants from liquids or air. Membrane filters are used for sterilization of heat-sensitive liquids, and nucleation filters are used to isolate specimens for electron microscopy.

- Why are filters used for sterilization of heat-sensitive liquids?
- Describe the use of depth filters for maintaining clean air in hospitals, laboratories, and the home.

II CHEMICAL ANTIMICROBIAL CONTROL

In the home, workplace, and laboratory, chemicals are routinely used to control microbial growth. An **antimicrobial agent** is a natural or synthetic chemical that kills or inhibits the growth of microorganisms. Agents that kill organisms are called *-cidal* agents, with a prefix indicating the type of microorganism killed. Thus, they are called **bacteriocidal**, **fungicidal**, and **viricidal** agents because they kill bacteria, fungi, and viruses, respectively. Agents that do not kill but only inhibit growth are called *-static* agents. These include **bacteriostatic**, **fungistatic**, and **viristatic** compounds.

27.4 Chemical Growth Control

Antimicrobial agents can differ in their selective toxicity. Nonselective agents have similar effects on all cells. Selective agents are more toxic for microorganisms than for animal tissues. Antimicrobial agents with selective toxicity are especially useful for treating infectious diseases because they kill selected microorganisms *in vivo* without harming the host. They are described later in this chapter. Here we discuss chemical agents that have relatively broad toxicity and are widely used for limiting microbial growth *in vitro*.

Effect of Antimicrobial Agents on Growth

Antimicrobial agents can be classified as bacteriostatic, bacteriocidal, and bacteriolytic by observing their effects on bacterial cultures (**Figure 27.9**). Bacteriostatic agents are frequently inhibitors of protein synthesis and act by binding to ribosomes. If the concentration of the agent is lowered, the agent is released from the ribosome and growth is resumed (Figure 27.9*a*). Many antibiotics work by this mechanism, and they will be discussed in Sections 27.6–27.9. Bacteriocidal agents bind tightly to their cellular targets and are not removed by dilution, killing the cell. The dead cells, however, are not destroyed, and total cell numbers remain constant (Figure 27.9*b*). Some *-cidal* agents are also *-lytic* agents, killing by cell lysis and release of cytoplasmic contents. Lysis decreases cell number and also culture turbidity (Figure 27.9*c*). Bacteriolytic agents include antibiotics that inhibit cell wall synthesis, such as penicillin, and chemicals such as detergents that rupture the cytoplasmic membrane.

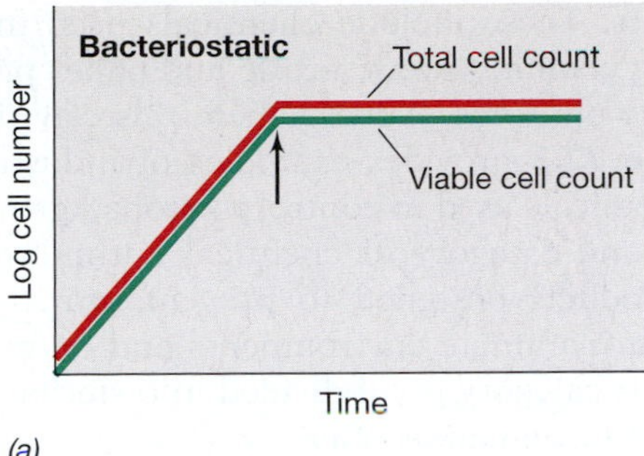

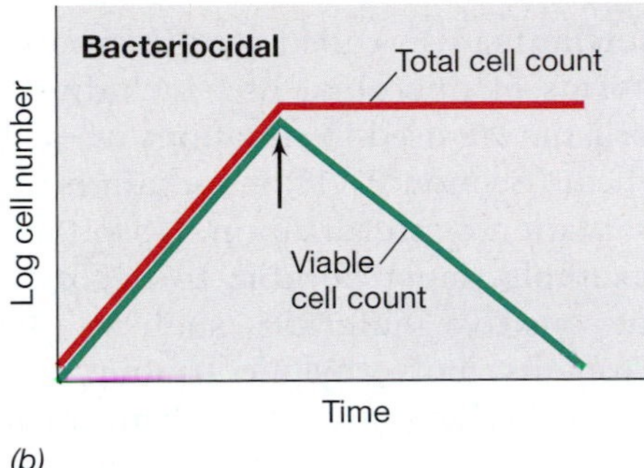

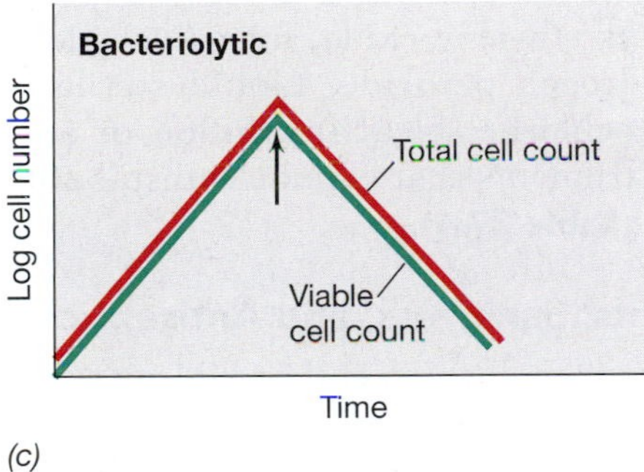

Figure 27.9 Bacteriostatic, bacteriocidal, and bacteriolytic antimicrobial agents. At the time indicated by the arrow, a growth-inhibitory concentration of each antimicrobial agent was added to an exponentially growing culture. Note the relationships between viable and total cell counts.

Measuring Antimicrobial Activity

Antimicrobial activity is measured by determining the smallest amount of agent needed to inhibit the growth of a test organism, a value called the **minimum inhibitory concentration (MIC)**. To determine the MIC for a given agent against a given organism, a series of culture tubes is prepared and inoculated. Each tube contains medium with a different concentration of the agent. After incubation, the tubes are checked for visible growth (turbidity). The MIC is the lowest concentration of agent that completely inhibits the growth of the test organism (**Figure 27.10**). This is called a *tube dilution technique*.

The MIC is not a constant for a given agent; it varies with the test organism, the inoculum size, the composition of the culture medium, the incubation time, and the conditions of incubation, such as temperature, pH, and aeration. When culture conditions are standardized, however, different antimicrobial agents can be compared to determine which is most effective against a given organism.

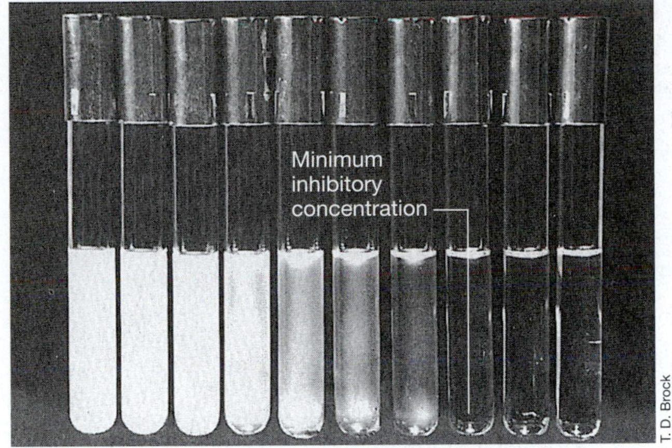

Figure 27.10 Antibiotic susceptibility assay by dilution methods. The assay defines the minimum inhibitory concentration (MIC). A series of increasing concentrations of antibiotic is prepared in the culture medium. Each tube is inoculated with a specific concentration of a test organism, followed by a defined incubation period. Growth, measured as turbidity, occurs in those tubes with antibiotic concentrations below the MIC.

Another common assay for antimicrobial activity is the *disk diffusion technique* (**Figure 27.11**). A Petri plate containing an agar medium is inoculated with a culture of the test organism. Known amounts of an antimicrobial agent are added to filter paper disks, which are then placed on the surface of the agar. During incubation, the agent diffuses from the disk into the agar, establishing a gradient; the further the chemical diffuses away from the filter paper, the lower is the concentration of the agent. At some distance from the disk, the effective MIC is reached. Beyond this point the microorganism grows, but closer to the disk, growth is absent. A *zone of inhibition* is created with a diameter proportional to the amount of antimicrobial agent added to the disk, the solubility of the agent, the diffusion coefficient, and the overall effectiveness of the agent. The disk diffusion technique and other growth-dependent methods are routinely used to test for pathogens' antibiotic susceptibility (Section 32.3).

27.4 MiniReview

Chemicals are often used to control microbial growth. Chemicals that kill organisms are called -cidal agents; those that inhibit growth are called -static agents; those that lyse organisms are called -lytic agents. Antimicrobial agents are tested for efficacy by determining their ability to inhibit growth in vitro.

- For antimicrobial agents, distinguish between the effects of -static, -cidal, and - lytic agents.
- Describe how the minimum inhibitory concentration of an antibacterial agent is determined.

UNIT 6

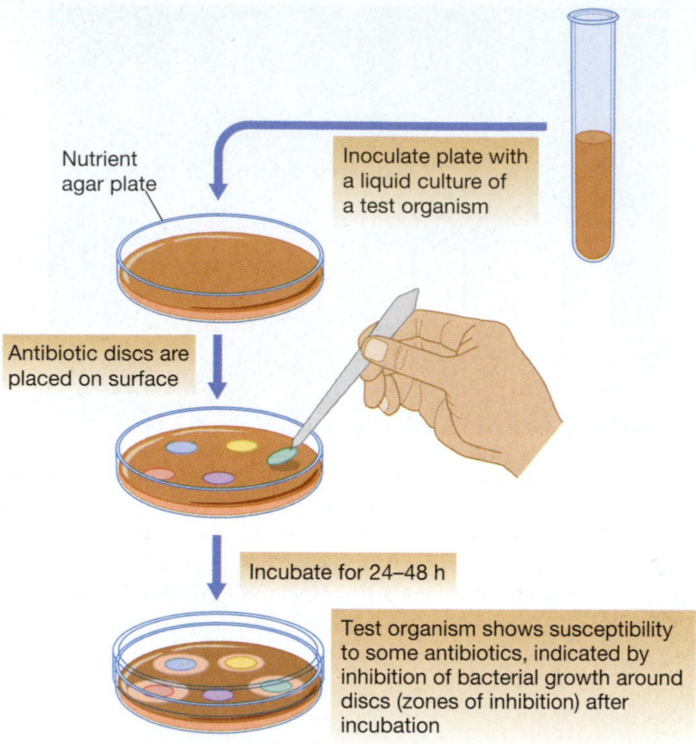
Nutrient agar plate
Inoculate plate with a liquid culture of a test organism
Antibiotic discs are placed on surface
Incubate for 24–48 h
Test organism shows susceptibility to some antibiotics, indicated by inhibition of bacterial growth around discs (zones of inhibition) after incubation

Figure 27.11 Measuring antimicrobial activity. Antibiotic diffuses from paper disks into the surrounding agar, inhibiting growth of susceptible microorganisms.

27.5 Chemical Antimicrobial Agents for External Use

Chemical antimicrobial agents are divided into two categories. The first category contains antimicrobial products used to control microorganisms in industrial and commercial environments. These include chemicals used in foods, air-conditioning cooling towers, textile and paper products, fuel tanks, and so on; some of these chemicals can affect human health. **Table 27.3** provides examples of industrial applications for chemicals used to control microbial growth.

The second category of chemical antimicrobial agents contains products designed to prevent growth of human pathogens in inanimate environments and on external body surfaces. This category is subdivided into sterilants, disinfectants, sanitizers, and antiseptics.

Sterilants

Chemical **sterilants**, also called **sterilizers** or **sporicides**, destroy all forms of microbial life, including endospores. Chemical sterilants are used in situations where it is impractical to use heat (Section 27.1) or radiation (Section 27.2) for decontamination or sterilization. Hospitals and laboratories, for example, must be able to decontaminate and sterilize heat-sensitive materials, such as thermometers, lensed instruments, polyethylene tubing, catheters, and reusable medical equipment such as respirometers. Usually, some form of cold sterilization is used for these purposes. Cold sterilization is performed in enclosed devices that resemble autoclaves, but which employ a gaseous chemical agent such as ethylene oxide, formaldehyde, peroxyacetic acid, or hydrogen peroxide. Liquid sterilants such as a sodium hypochlorite (bleach) solution or amylphenol are used for instruments that cannot withstand high temperatures or gas (**Table 27.4**).

Disinfectants, Sanitizers, and Antiseptics

Disinfectants are chemicals that kill microorganisms, but not necessarily endospores, and are used on inanimate objects. For example, disinfectants such as ethanol and

Table 27.3 Industrial uses of antimicrobial chemicals

Industry	*Chemicals*	*Use*
Paper	Organic mercurials, phenols[a], methylisothiazolinone	To prevent microbial growth during manufacture
Leather	Heavy metals, phenols[a]	Antimicrobial agents present in the final product inhibit growth
Plastic	Cationic detergents	To prevent growth of bacteria on aqueous dispersions of plastics
Textile	Heavy metals, phenols[a]	To prevent microbial deterioration of fabrics exposed in the environment, such as awnings and tents
Wood	Metal salts, phenols[a]	To prevent deterioration of wooden structures
Metal working	Cationic detergents	To prevent growth of bacteria in aqueous cutting emulsions
Petroleum	Mercurics, phenols[a], cationic detergents, methylisothiazolinone	To prevent growth of bacteria during recovery and storage of petroleum and petroleum products
Air conditioning	Chlorine, phenols[a], methylisothiazolinone	To prevent growth of bacteria (for example, *Legionella*) in cooling towers
Electrical power	Chlorine	To prevent growth of bacteria in condensers and cooling towers
Nuclear	Chlorine	To prevent growth of radiation-resistant bacteria in nuclear reactors

[a]Metallic (mercury, arsenic, and copper) compounds and phenolic compounds may produce environmentally hazardous waste products and create health hazards.

Table 27.4 Antiseptics, sterilants, disinfectants, and sanitizers

Agent	*Use*	*Mode of action*
Antiseptics		
Alcohol (60–85% ethanol or isopropanol in water)[a]	Topical antiseptic	Lipid solvent and protein denaturant
Phenol-containing compounds (hexachlorophene, triclosan, chloroxylenol, chlorhexidine)[b]	Soaps, lotions, cosmetics, body deodorants, topical disinfectants	Disrupts cell membrane
Cationic detergents, especially quaternary ammonium compounds (benzalkonium chloride)	Soaps, lotion, topical disinfectants	Interact with phospholipids of cytoplasmic membrane
Hydrogen peroxide[a] (3% solution)	Topical antiseptic	Oxidizing agent
Iodine-containing iodophor compounds in solution[a] (Betadine®)	Topical antiseptic	Iodinates tyrosine residues of proteins; oxidizing agent
Octenidine	Topical antiseptic	Disrupts cytoplasmic membrane
Sterilants, Disinfectants, and Sanitizers[c]		
Alcohol (60–85% ethanol or isopropanol in water)[a]	Disinfectant for medical instruments and laboratory surfaces	Lipid solvent and protein denaturant
Cationic detergents (quaternary ammonium compounds)	Disinfectant and sanitizer for medical instruments, food and dairy equipment	Interact with phospholipids
Chlorine gas	Disinfectant for purification of water supplies	Oxidizing agent
Chlorine compounds (chloramines, sodium hypochlorite, sodium chlorite, chlorine dioxide)	Disinfectant and sanitizer for dairy and food industry equipment, and water supplies	Oxidizing agent
Copper sulfate	Algicide disinfectant in swimming pools and water supplies	Protein precipitant
Ethylene oxide (gas)	Sterilant for temperature-sensitive materials such as plastics and lensed instruments	Alkylating agent
Formaldehyde	3%–8% solution used as surface disinfectant, 37% (formalin) or vapor used as sterilant	Alkylating agent
Glutaraldehyde	2% solution used as high-level disinfectant or sterilant	Alkylating agent
Hydrogen peroxide[a]	Vapor used as sterilant	Oxidizing agent
Iodine-containing iodophor compounds in solution[a] (Wescodyne®)	Disinfectant for medical instruments and laboratory surfaces	Iodinates tyrosine residues
Mercuric dichloride[b]	Disinfectant for laboratory surfaces	Combines with -SH groups
OPA (orthophalaldehyde)	High-level disinfectant for medical instruments	Alkylating agent
Ozone	Disinfectant for drinking water	Strong oxidizing agent
Peroxyacetic acid	Solution used as high-level disinfectant or sterilant	Strong oxidizing agent
Phenolic compounds[b]	Disinfectant for laboratory surfaces	Protein denaturant

[a]Alcohols, hydrogen peroxide, and iodine-containing iodophor compounds can act as antiseptics, disinfectants, sanitizers, or sterilants depending on concentration, length of exposure, and form of delivery.

[b]Use of heavy metal (mercury) compounds and phenolic compounds may produce environmentally hazardous waste products and may create health hazards.

[c]Many water-soluble antimicrobial compounds, with the exception of those containing heavy metals, can be used as sanitizers for food and dairy equipment and preparation areas, provided their use is followed by adequate draining before food contact.

cationic detergents are used to decontaminate floors, tables, bench tops, walls, and so on. These agents are important for infection control in, for example, hospitals and other medical settings. General disinfectants are used in households, swimming pools, and water purification systems (Table 27.4).

Sanitizers are agents that reduce, but may not eliminate, microbial numbers to levels considered to be safe. Food contact sanitizers are widely used in the food industry to treat surfaces such as mixing and cooking equipment, dishes, and utensils. Nonfood contact sanitizers are used to treat surfaces such as counters, floors, walls, carpets, and laundry (Table 27.4).

UNIT 6

Preventing Antimicrobial Drug Resistance

According to the Centers for Disease Control and Prevention (CDC) in Atlanta, Georgia (USA), widespread antimicrobial resistance is highly prevalent. For example, nearly 2 million patients in the United States develop hospital-acquired (nosocomial) infections each year. Nosocomial infections are difficult to treat because up to 70% of the infecting microorganisms are resistant to antimicrobial drugs. For *Staphylococcus aureus*, which causes over 10% of these infections in intensive care units, **Figure 1** shows a twofold increase in drug resistance to methicillin and oxacillin over a 13-year period. Since these *S. aureus* isolates are often resistant to other drugs as well, the appearance of methicillin-oxacillin-resistant *S. aureus* (MRSA) strains limits the choice of effective therapeutic drugs. Resistance in *Mycobacterium tuberculosis*, *Enterococcus faecium*, and *Candida albicans* are also of major concern.

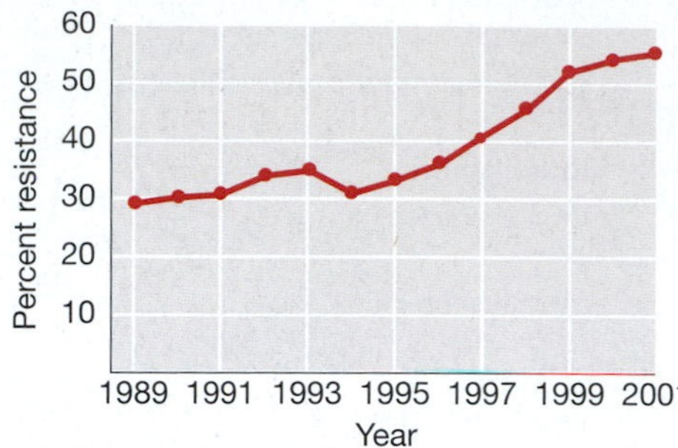

Figure 1 **Methicillin-resistant *Staphylococcus aureus* (MRSA).** *Nosocomial infections among intensive care patients, 1989–2001. Data from the CDC.*

The CDC has promoted a 12-step program to prevent resistance to antimicrobial agents, and the program is summarized below (see also **http://www.cdc.gov/drugresistance/healthcare/ha/12steps_HA.htm).** The program stresses the importance of preventing infection, rapidly and positively diagnosing and treating infections, using antimicrobial agents wisely, and preventing pathogen transmission.

1. **Immunize to prevent common diseases.** Keep immunizations up to date, especially for likely disease exposures. In addition to required vaccinations (Section 30.5), this should include yearly influenza vaccination for nearly everyone, meningitis vaccines, and pneumococcal immunizations for health-care providers or those exposed to large numbers of people, as in schools, colleges, and the military.
2. **Avoid unnecessary introduction of parenteral devices, such as catheters.** All present a risk of introducing infectious agents into the body. If such devices are necessary, remove them as soon as possible.
3. **Target the pathogen.** Attempt to culture the infectious agent while targeting antimicrobial drug treatment for the most likely pathogens. After positive culture results, adjust the therapy to target the known pathogen and its antibiotic susceptibility.
4. **Access the experts.** For serious infections, follow up with an infectious disease expert. Get a second opinion if conditions do not rapidly improve after treatment has begun.
5. **Practice antimicrobial control.** Be aware and current in knowledge of appropriate antimicrobial drugs and their use. Be sure the treatment offered is current and recommended for the pathogen.

Antiseptics and **germicides** are chemical agents that kill or inhibit growth of microorganisms and that are nontoxic enough to be applied to living tissues. Most of the compounds in this category are used for handwashing (Microbial Sidebar, "Preventing Antimicrobial Drug Resistance") or for treating surface wounds (Table 27.4). Under some circumstances, certain antiseptics are also effective disinfectants. Ethanol, for example, is categorized as an antiseptic, but can also be a disinfectant. This is largely dependent on the concentration of ethanol used and the exposure time, with disinfection generally requiring higher ethanol concentrations and exposure times of several minutes. The Food and Drug Administration in the United States regulates the formulation, manufacture, and use of antiseptics and germicides because these agents involve direct human exposure and contact.

Antimicrobial Efficacy

Several factors affect the efficacy of chemical antimicrobial agents. For example, many disinfectants are neutralized by organic material. These materials reduce effective disinfectant concentrations and microbial killing capacity. Furthermore, pathogens are often encased in particles or grow in large numbers as biofilms, covering surfaces of tissue or medical devices with several layers of microbial cells (Chapter 6, Microbial Sidebar, "Microbial Growth in the Real World: Biofilms"). Biofilms may slow or even completely prevent penetration of antimicrobial agents, reducing or negating their effectiveness.

Only sterilants are effective against bacterial endospores. Endospores are much more resistant to other agents than are

(continued)

6. **Use local data.** Obtain and understand the antibiotic susceptibility profile for the infectious agent from local healthcare sources.
7. **Treat infection, not contamination.** Antiseptic techniques must be followed to obtain appropriate samples from infected tissues. Contaminating organisms may be present on skin, catheters, or IV lines. Obtain cultures only from the site of infection.
8. **Treat infection, not colonization.** Treat the pathogen and not other colonizing microorganisms that are not causing disease. For example, cultures from normal skin and throat are often colonized with potential pathogens such as *Staphylococcus* species. These may have nothing to do with the current infection.
9. **Treat with the least exotic antimicrobial agent that will eliminate the pathogen.** Treatment with the latest broad-spectrum antibiotic, while efficacious, may not be warranted if other drugs are still effective. The more a drug is used, the greater the chance that resistant organisms will develop.

CDC/Kimberly Smith

Figure 2 **Handwashing.** *Handwashing is the easiest and one of the most important interventions to prevent pathogen spread in healthcare, home, and laboratory settings. This handwash station is in a clinical laboratory.*

For example, some *Enterococcus* isolates are already resistant to vancomycin, a relatively new broad-spectrum antibiotic, largely because the drug was over-prescribed to treat MRSA infections when it was first introduced.

10. **Monitor antimicrobial use.** Antimicrobial use should be discontinued as soon as the prescribed course of treatment is completed. If an infection cannot be diagnosed, treatment should be discontinued. For example, in the case of pharyngitis (sore throat), antibiotic treatment for *Streptococcus pyogenes* (strep throat) is often started before throat culture results are confirmed. If throat cultures are negative for *S. pyogenes*, treatment with antibiotics should be stopped. Antibiotics are ineffective for treatment of the viruses that are the most probable causes of pharyngitis.
11. **Isolate the pathogen.** Keep areas around infected persons free from contamination. Clean up and contain body fluids appropriately. Decontaminate linens, clothes, and other potential sources of contamination. In a hospital setting, infection control experts should be consulted.
12. **Break the chain of contagion.** Avoid contact with others, even at work or school, when sick. Maintain cleanliness, especially by handwashing (**Figure 2**), if you are sick, or are caring for sick persons.

vegetative cells because of their low water availability and reduced metabolism (Section 27.1). Some bacteria, such as *Mycobacterium tuberculosis*, the causal agent of tuberculosis, are resistant to the action of common disinfectants because of the waxy nature of their cell wall (Sections 16.5 and 34.5). Thus, the efficacy of antimicrobial treatment must be empirically determined under the actual conditions of use.

27.5 MiniReview

Sterilants, disinfectants, and sanitizers are used to decontaminate nonliving material. Antiseptics and germicides are used to reduce microbial growth on living tissues. Antimicrobial compounds have commercial, health care, and industrial applications.

- Distinguish between a sterilant, a disinfectant, a sanitizer, and an antiseptic.
- What disinfectants are routinely used for sterilization of water? Why are these disinfectants not harmful to humans?

III ANTIMICROBIAL AGENTS USED *IN VIVO*

Up to this point, we have considered the effects of physical and chemical agents used to inhibit microbial growth *outside* the human body. Most of the physical methods are too harsh, and most chemicals mentioned are too toxic to be used *inside*

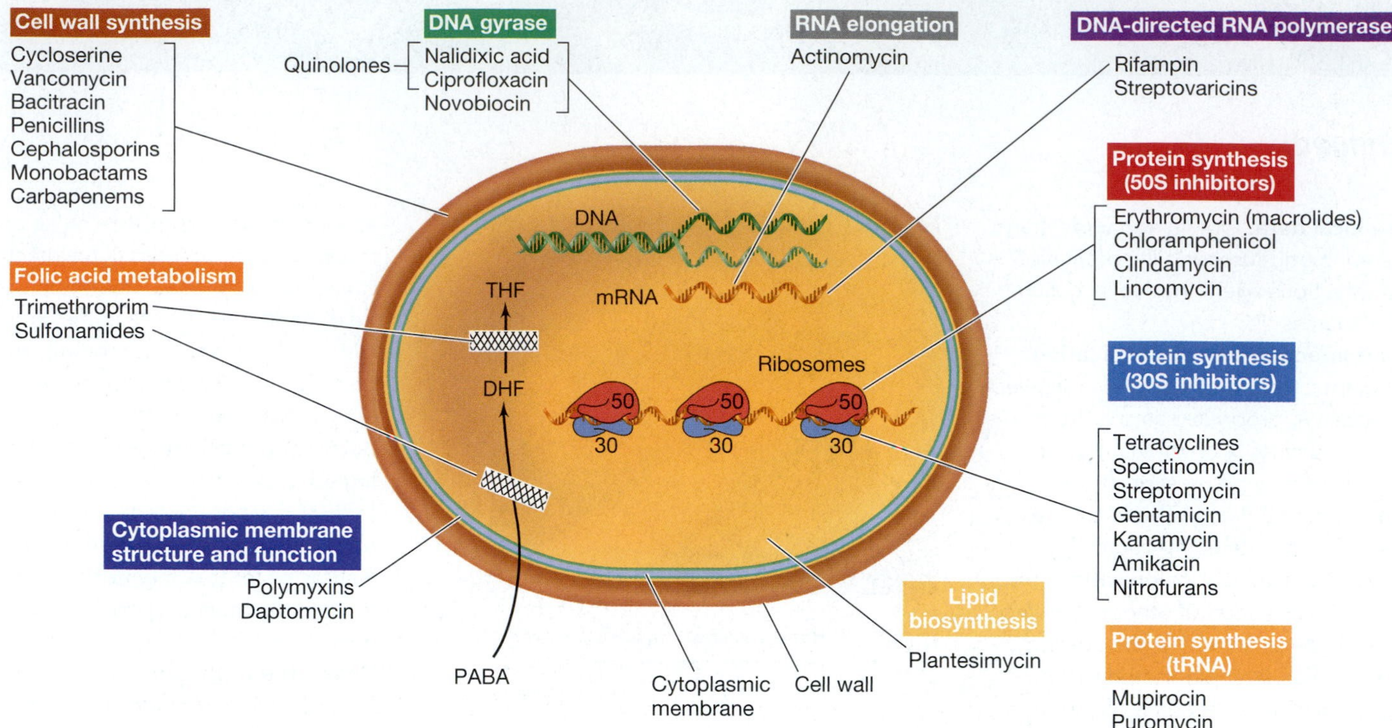

Figure 27.12 Mode of action of some major antimicrobial chemotherapeutic agents. Agents are classified according to their target structures in the bacterial cell. THF, tetrahydrofolate; DHF, dihydrofolate; mRNA, messenger RNA.

the body; even relatively mild antiseptics can be used only on the skin. For control of infectious disease, chemical compounds that can be used internally are required. Discovery and development of antimicrobial drugs has played a major role in clinical and veterinary medicine, as well as in agriculture.

Antimicrobial drugs are classified based on their molecular structure, mechanism of action (**Figure 27.12**), and spectrum of antimicrobial activity (**Figure 27.13**). Worldwide, more than 500 metric tons of various antimicrobial drugs are manufactured annually (**Figure 27.14**). Antimicrobial agents fall into two broad categories, *synthetic agents* and *antibiotics*. We first concentrate on synthetic antimicrobial compounds. We then discuss naturally produced antibiotics in the following sections.

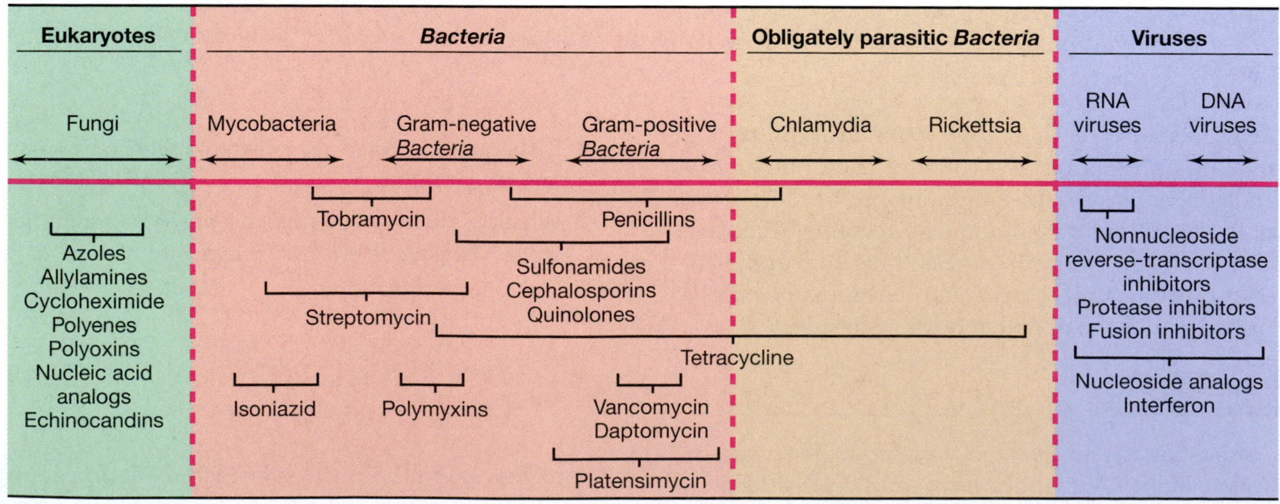

Figure 27.13 Antimicrobial spectrum of activity. Each antimicrobial chemotherapeutic agent affects a limited and well-defined group of microorganisms. A few agents are very specific and affect the growth of only a single genus. For example, isoniazid affects only organisms in the genus *Mycobacterium*.

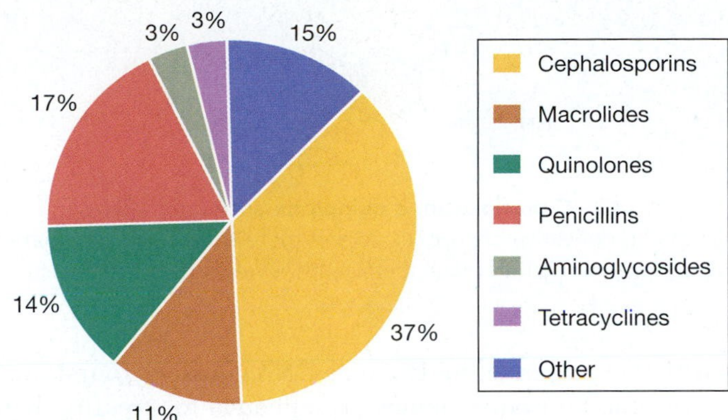

Figure 27.14 Annual worldwide production and use of antibiotics. Each year more than 500 metric tons of antimicrobial chemotherapeutic agents are manufactured.

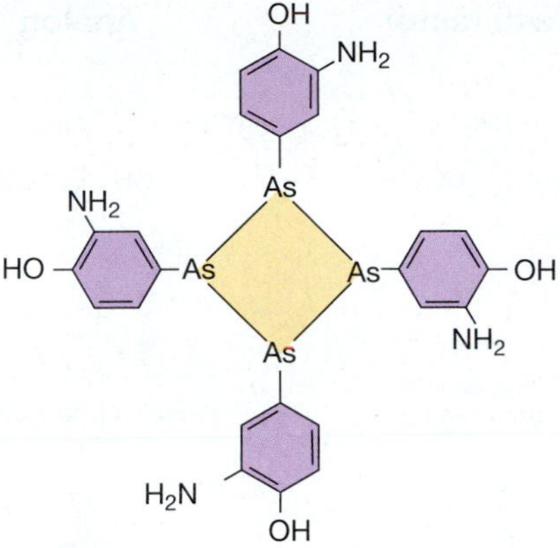

Figure 27.15 Salvarsan. This arsenic-containing compound was one of the first useful antimicrobial agents.

27.6 Synthetic Antimicrobial Drugs

Systematic work on antimicrobial drugs was first initiated by the German scientist Paul Ehrlich. In the early 1900s, Ehrlich developed the concept of **selective toxicity**, the ability to inhibit or kill pathogenic microorganisms without adversely affecting the host. In his search for a "magic bullet" that would kill only pathogens, Ehrlich tested large numbers of chemical dyes for selective toxicity and discovered the first effective antimicrobial drugs, of which Salvarsan, an arsenic-containing compound used for the cure of syphilis, was the most successful (**Figure 27.15**).

Growth Factor Analogs

One class of antimicrobial drugs functions by disrupting bacterial metabolism. We previously defined growth factors as specific chemical substances required in the medium because the organisms cannot synthesize them (∞ Section 5.1). A **growth factor analog** is a synthetic compound that is structurally similar to a growth factor, but subtle structural differences between the analog and the authentic growth factor prevent the analog from functioning in the cell. Analogs are known for many important biomolecules, including vitamins, amino acids, purines and pyrimidines, and other compounds. We begin by considering antibacterial growth factor analogs. Growth factor analogs effective for the treatment of viral and fungal infections will be discussed in Sections 27.10 and 27.11.

Sulfa Drugs

Discovered by Gerhard Domagk in the 1930s, the *sulfa drugs* were the first widely used growth factor analogs shown to specifically inhibit the growth of bacteria. The discovery of the first sulfa drug resulted from the large-scale screening of chemicals for activity against streptococcal infections in experimental animals.

Sulfanilamide, the simplest sulfa drug, is an analog of *p*-aminobenzoic acid, which is itself a part of the vitamin folic acid, a nucleic acid precursor (**Figure 27.16**). Sulfanilamide blocks the synthesis of folic acid, thereby inhibiting nucleic acid synthesis. Sulfanilamide is active only in bacteria because bacteria synthesize their own folic acid, whereas most animals obtain folic acid from their diet. Initially sulfa drugs were widely used for treatment of streptococcal infections (∞ Section 34.2). However, resistance to sulfonamides has been increasing because many formerly susceptible pathogens have developed an ability to take up folic acid from their environment. Antimicrobial therapy with sulfamethoxazole (a sulfa drug) plus trimethoprim, a related folic acid synthesis competitor, is still effective in many instances because the drug combination produces sequential blocking of the folic acid synthesis pathway. Resistance to this drug combination requires that two mutations in genes of the same pathway occur, a relatively rare event.

Isoniazid

Isoniazid (∞ Figure 34.11) is an important growth factor analog with a very narrow spectrum of activity (Figure 27.13). Effective only against *Mycobacterium*, isoniazid interferes with the synthesis of mycolic acid, a mycobacterial cell wall

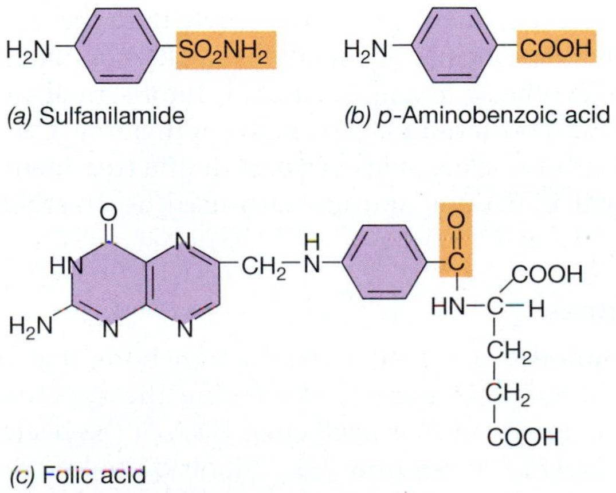

H_2N–(benzene ring)–SO_2NH_2

(a) Sulfanilamide

H_2N–(benzene ring)–$COOH$

(b) p-Aminobenzoic acid

(c) Folic acid

Figure 27.16 Sulfa drugs. *(a)* The simplest sulfa drug, sulfanilamide. *(b)* Sulfanilamide is an analog of *p*-aminobenzoic acid, a precursor of *(c)* folic acid, a growth factor.

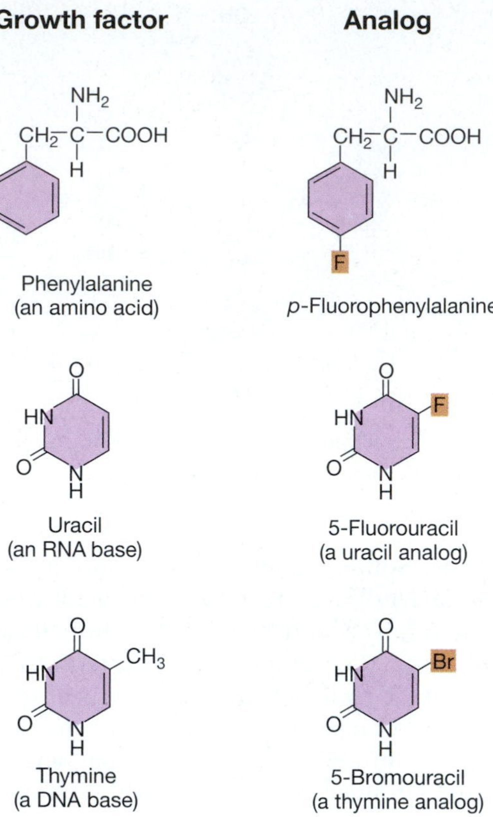

Figure 27.17 Growth factors and antimicrobial analogs. Structurally similar growth factors and their biologically active analogs are shown for comparison. Growth factors are discussed in Section 5.1.

component. A nicotinamide (vitamin) analog, isoniazid is the most effective single drug used for control and treatment of tuberculosis (Section 34.5).

Nucleic Acid Base Analogs

In the examples shown in **Figure 27.17**, analogs of nucleic acid bases have been formed by addition of a fluorine or bromine atom. Fluorine is a relatively small atom and does not alter the overall shape of the nucleic acid base, but changes the chemical properties such that the compound does not function in cell metabolism, thereby blocking nucleic acid synthesis. Examples include fluorouracil, an analog of uracil, and bromouracil, an analog of thymine. Growth factor analogs of nucleic acids are used in the treatment of viral and fungal infections and are also used as mutagens (Sections 27.10 and 27.11).

Quinolones

The **quinolones** are antibacterial compounds that interfere with bacterial DNA gyrase, preventing the supercoiling of DNA, a required step for packaging DNA in the bacterial cell (Figure 27.12) (Section 7.3). Fluoroquinolones such as ciprofloxacin (**Figure 27.18**) are routinely used to treat urinary tract infections in humans. Ciprofloxacin is also the drug of choice for treating anthrax because some strains of *Bacillus anthracis*, the causative agent of anthrax (Section 33.12), are resistant to penicillin. Because DNA gyrase is found in all *Bacteria*, the fluoroquinolones are effective for treating both gram-positive and gram-negative bacterial infections (Figure 27.13). Fluoroquinolones have also been widely used in the beef and poultry industries for prevention and treatment of respiratory diseases in animals.

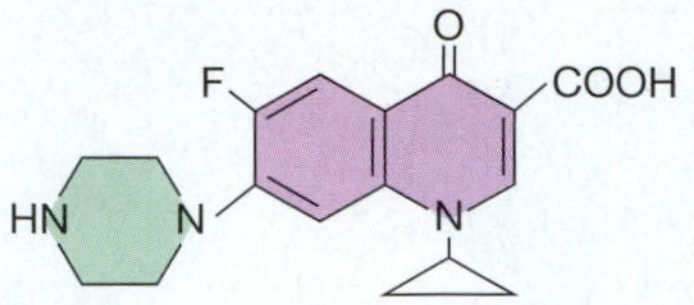

Figure 27.18 Ciprofloxacin, a quinolone antibiotic. Ciprofloxacin, a fluorinated derivative of nalidixic acid, is more soluble than the parent compound, allowing it to reach therapeutic levels in blood and tissues.

27.6 MiniReview

Synthetic antimicrobial agents are selective for *Bacteria*, viruses, and fungi. Growth factor analogs such as sulfa drugs, isoniazid, and nucleic acid analogs are synthetic metabolic inhibitors. Quinolones inhibit the action of DNA gyrase in *Bacteria*.

- Explain selective toxicity in terms of antibiotic therapy.
- Distinguish synthetic chemotherapeutic agents from antiseptics and disinfectants.
- Describe the action of any one of the growth factor analogs.

27.7 Naturally Occurring Antimicrobial Drugs: Antibiotics

Antibiotics are microbially produced antimicrobial agents. Antibiotics, produced by many different bacteria and fungi, apparently have the sole function of inhibiting or killing other microorganisms. Less than 1% of the thousands of known antibiotics are clinically useful, often because of toxicity or lack of uptake by host cells. However, the clinically useful antibiotics have had a dramatic impact on the treatment of infectious diseases. *Natural* antibiotics can often be artificially modified to enhance their efficacy. These are said to be *semisynthetic* antibiotics. The industrial-scale production of antibiotics was discussed in Chapter 25.

Antibiotics and Selective Antimicrobial Toxicity

The susceptibility of individual microorganisms to individual antimicrobial agents varies significantly (Figure 27.13). For example, gram-positive *Bacteria* and gram-negative *Bacteria* differ in their susceptibility to an individual antibiotic such as penicillin; gram-positive *Bacteria* are generally affected, whereas most gram-negative *Bacteria* are naturally resistant. Certain **broad-spectrum antibiotics** such as tetracycline, however, are effective against both groups. In general, a broad-spectrum antibiotic finds wider medical use than a narrow-spectrum antibiotic. An antibiotic with a limited spectrum of activity may, however, be quite valuable for the control of pathogens that fail to respond to other antibiotics. A good example is vancomycin, a narrow-spectrum glycopeptide antibiotic that is a highly effective bacteriocidal agent for gram-positive penicillin-resistant *Bacteria* from the genera *Staphylococcus, Bacillus,* and *Clostridium* (Figures 27.12 and 27.13).

Important targets of antibiotics in *Bacteria* are ribosomes, the cell wall, the cytoplasmic membrane, lipid biosynthesis enzymes, and DNA replication and transcription elements (Figure 27.12).

Antibiotics Affecting Protein Synthesis

Many antibiotics inhibit protein synthesis by interacting with the ribosome and disrupting translation (Figure 27.12). These interactions are quite specific and many involve binding to rRNA. Several of these antibiotics are medically useful, and several are also effective research tools because they block defined steps in protein synthesis (Section 7.15). For instance, streptomycin inhibits protein chain initiation, whereas puromycin, chloramphenicol, cycloheximide, and tetracycline inhibit protein chain elongation.

Even when two antibiotics inhibit the same step in protein synthesis, the mechanisms of inhibition can be quite different. For example, puromycin binds to the A site on the ribosome, and the growing polypeptide chain is transferred to puromycin instead of the amino acyl–tRNA complex. The puromycin–peptide complex is then released from the ribosome, prematurely halting elongation. By contrast, chloramphenicol inhibits elongation by blocking formation of the peptide bond (Section 7.15).

Many antibiotics specifically inhibit ribosomes of organisms from only one phylogenetic domain. For example, chloramphenicol and streptomycin specifically target the ribosomes of *Bacteria*, whereas cycloheximide only affects the cytoplasmic ribosomes of *Eukarya*. Since the major organelles (mitochondria and chloroplasts) in *Eukarya* also have ribosomes that are similar to those of *Bacteria* (that is, 70S ribosomes), antibiotics that inhibit protein synthesis in *Bacteria* also inhibit protein synthesis in these organelles. For example, tetracycline antibiotics inhibit 70S ribosomes, but are still medically useful because eukaryotic mitochondria are affected only at higher concentrations than are used for antimicrobial therapy.

Antibiotics Affecting Transcription

A number of antibiotics specifically inhibit transcription by inhibiting RNA synthesis (Figure 27.12). For example, rifampin and the streptovaricins inhibit RNA synthesis by binding to the β subunit of RNA polymerase. These antibiotics have specificity for *Bacteria*, chloroplasts, and mitochondria. Actinomycin inhibits RNA synthesis by combining with DNA and blocking RNA elongation. This agent binds most strongly to DNA at guanine–cytosine base pairs, fitting into the major groove in the double strand where RNA is synthesized.

Some of the most useful antibiotics are directed against unique structural features of *Bacteria*, such as their cell walls. We discuss these antibiotics and their targets in the next section.

27.7 MiniReview

Antibiotics are a chemically diverse group of antimicrobial compounds that are produced by microorganisms. Although many antibiotics are known and a few are clinically effective, most are not useful in humans or animals. Antibiotics function by inhibiting cellular processes and functions in the target microorganisms.

- Distinguish antibiotics from growth factor analogs.
- What is a broad-spectrum antibiotic?
- Identify the potential target sites for the antibiotics that inhibit protein synthesis and transcription.

27.8 β-Lactam Antibiotics: Penicillins and Cephalosporins

One of the most important groups of antibiotics, both historically and medically, is the β-lactam group. **β-lactam antibiotics** include the medically important penicillins, cephalosporins, and cephamycins. These antibiotics share a characteristic structural component, the *β-lactam ring* (**Figure 27.19**). Together, the penicillins and cephalosporins account for over one-half of all of the antibiotics produced and used worldwide (Figure 27.14).

Penicillins

In 1929, the British scientist Alexander Fleming characterized the first antibiotic, an antibacterial product of the fungus *Penicillium chrysogenum* called *penicillin* (Figure 27.19). Even with the widespread availability of sulfa drugs in the 1930s, most bacterial diseases were uncontrollable. However, in 1939, Howard Florey and his colleagues, motivated by the impending world war, developed a process for the large-scale production of **penicillin**. Penicillin G was the first clinically useful antibiotic. This new β-lactam antibiotic was dramatically effective in controlling staphylococcal and pneumococcal infections among military personnel and was more effective for treating streptococcal infections than sulfa drugs. By the end of World War II in 1945, penicillin became available for general use, and pharmaceutical companies began to look for other antibiotics, leading to drugs that revolutionized the treatment of infectious diseases.

Penicillin G is active primarily against gram-positive *Bacteria* because gram-negative *Bacteria* are impermeable to the antibiotic. Chemical modification of the penicillin G structure, however, significantly changes the properties of the

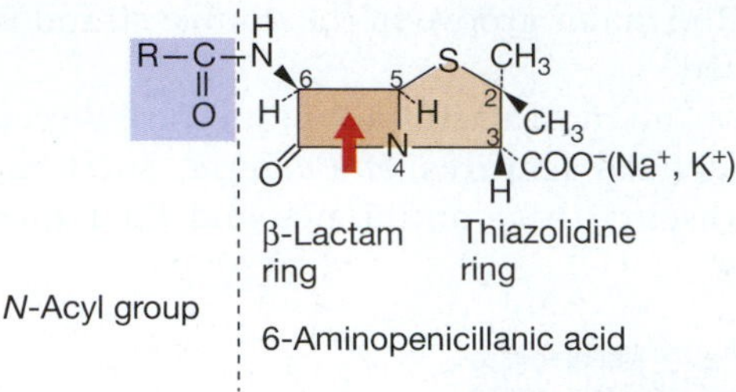

Designation	N-Acyl group
NATURAL PENICILLIN	
Benzylpenicillin (penicillin G) Gram-positive activity β-lactamase-sensitive	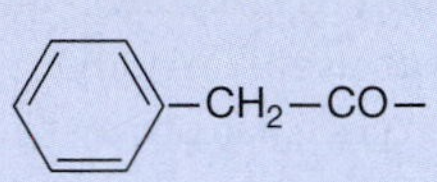 $-CH_2-CO-$
SEMISYNTHETIC PENICILLINS	
Methicillin acid-stable, β-lactamase-resistant	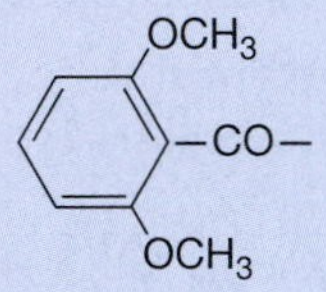 OCH_3 $-CO-$ OCH_3
Oxacillin acid-stable, β-lactamase-resistant	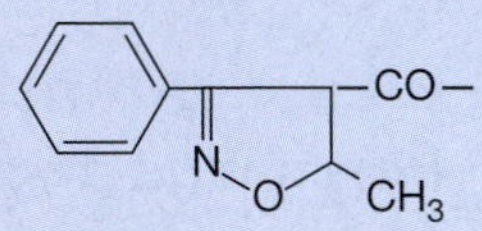 $-CO-$ N O CH_3
Ampicillin broadened spectrum of activity (especially against gram-negative *Bacteria*), acid-stable, β-lactamase-sensitive	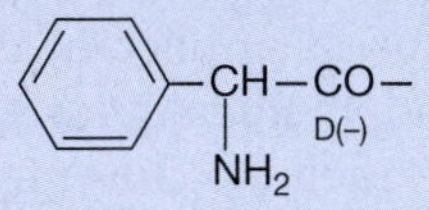 $-CH-CO-$ D(–) NH_2
Carbenicillin broadened spectrum of activity (especially against *Pseudomonas aeruginosa*), acid-stable but ineffective orally, β-lactamase-sensitive	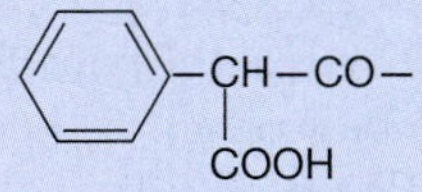 $-CH-CO-$ $COOH$

Figure 27.19 Penicillins. The red arrow (top panel) is the site of activity of most β-lactamase enzymes.

resulting antibiotic. Many chemically modified semisynthetic penicillins are quite effective against gram-negative *Bacteria*. Figure 27.19 shows the structures of some of the penicillins. For example, ampicillin and carbenicillin, semisynthetic penicillins, are effective against some gram-negative *Bacteria*. The structural differences in the *N*-acyl groups of these semisynthetic penicillins allow them to be transported inside the gram-negative outer membrane (∞ Section 4.7), where they inhibit cell wall synthesis. Penicillin G is also sensitive to β-lactamase, an enzyme produced by a number of penicillin-resistant *Bacteria* (Section 27.12). Oxacillin and methicillin are widely used β-lactamase-resistant semisynthetic penicillins.

Mechanism of Action

The β-lactam antibiotics are inhibitors of cell wall synthesis. An important feature of bacterial cell wall synthesis is *transpeptidation*, the reaction that results in the cross-linking of two glycan-linked peptide chains (∞ Section 6.4 and Figure 6.7). The transpeptidase enzymes bind to penicillin or other β-lactam antibiotics. Thus, these transpeptidases are called *penicillin-binding proteins* (PBPs). When PBPs bind penicillin, they cannot catalyze the transpeptidase reaction, but cell wall synthesis continues. As a result, the newly synthesized bacterial wall is no longer cross-linked and cannot maintain its strength. In addition, the antibiotic–PBP complex stimulates the release of autolysins, enzymes that digest the existing cell wall. The result is a weakened, self-degrading cell wall. Eventually the osmotic pressure differences between the inside and outside of the cell cause lysis. By contrast, vancomycin, also a cell wall synthesis inhibitor, does not bind to PBPs, but binds directly to the terminal D-alanyl-D-alanine peptide on the peptidoglycan precursors (∞ Figure 6.7); this effectively blocks transpeptidation.

Because the cell wall and its synthesis mechanisms are unique to *Bacteria*, the β-lactam antibiotics have very high selectivity and are not toxic to host cells. However, some individuals develop allergies to individual β-lactam compounds after repeated courses of antibiotic therapy.

Cephalosporins

The cephalosporins are another group of clinically important β-lactam antibiotics. Cephalosporins, produced by the fungus *Cephalosporium* sp. (∞ Section 25.6), differ structurally from the penicillins. They retain the β-lactam ring but have a six-member dihydrothiazine ring instead of the five-member thiazolidine ring. The cephalosporins have the same mode of action as the penicillins; they bind irreversibly to PBPs and prevent the cross-linking of peptidoglycan. Clinically important cephalosporins are semisynthetic antibiotics with a broader spectrum of antibiotic activity than the penicillins. In addition, cephalosporins are typically more resistant to the enzymes that destroy β-lactam rings, the β-lactamases. For example, ceftriaxone (**Figure 27.20**) is highly resistant to β-lactamases and has replaced penicillin for treatment of infections from *Neisseria gonorrhoeae* (gonorrhea) because many *N. gonorrhoeae* strains are now penicillin resistant (Section 27.12, ∞ Section 34.13).

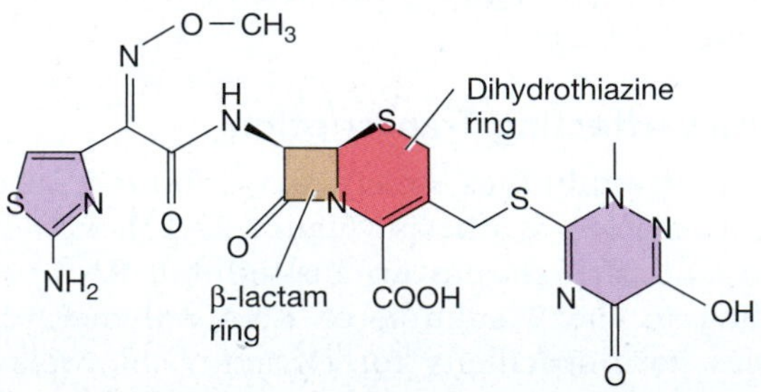

Figure 27.20 Ceftriaxone. Ceftriaxone is a β-lactam antibiotic that is resistant to most β-lactamases due to the adjacent six-member dihydrothiazine ring. Compare this structure to the five-member thiazolidine ring of the β-lactamase-sensitive penicillins (Figure 27.19).

27.8 MiniReview

The β-lactam compounds, including the penicillins and the cephalosporins, are the most important single class of clinical antibiotics. These antibiotics and their semisynthetic derivatives target cell wall synthesis in *Bacteria.* They have low host toxicity and collectively have a broad spectrum of activity.

- Draw the structure of the β-lactam ring and indicate the site of β-lactamase activity.
- How do the β-lactam antibiotics function?

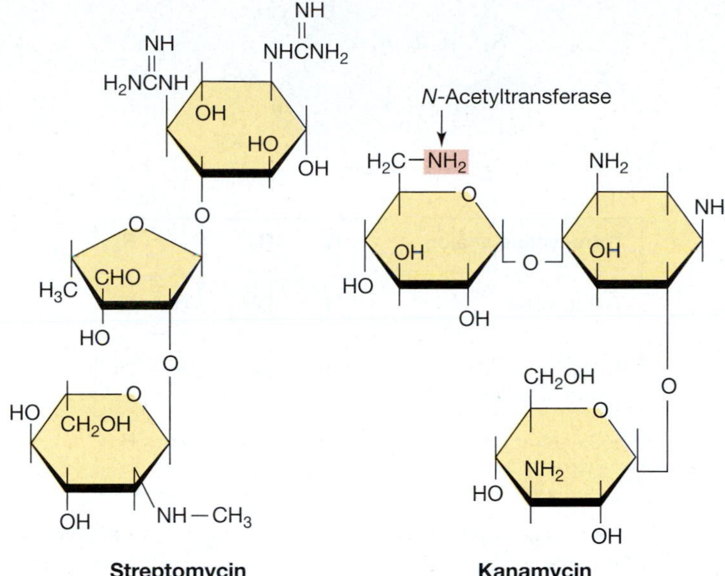

Figure 27.21 Aminoglycoside antibiotics: streptomycin and kanamycin. The amino sugars are in yellow. At the position indicated, kanamycin can be modified by a resistance plasmid that encodes *N*-acetyltransferase. Following acetylation, the antibiotic is inactive.

27.9 Antibiotics from Prokaryotes

Many antibiotics active against *Bacteria* are also produced by *Bacteria*. These include many antibiotics that have major clinical applications, and we discuss their general properties here.

Aminoglycosides

Antibiotics that contain amino sugars bonded by glycosidic linkage are called **aminoglycosides**. Clinically useful aminoglycosides include streptomycin (produced by *Streptomyces griseus*) and its relatives, kanamycin (**Figure 27.21**), neomycin, gentamicin, tobramycin, netilmicin, spectinomycin, and amikacin. The aminoglycosides target the 30S subunit of the ribosome, inhibiting protein synthesis (Figure 27.12), and are clinically useful against gram-negative *Bacteria* (Figure 27.13).

Streptomycin was the first effective antibiotic used for the treatment of tuberculosis. The aminoglycoside antibiotics, however, are not widely used today, and together the aminoglycosides account for only about 3% of the total of all antibiotics produced and used (Figure 27.14). Streptomycin, because of serious side effects such as neurotoxicity and nephrotoxicity (kidney toxicity), has been replaced by several synthetic antimicrobials for tuberculosis treatment. Bacterial resistance to aminoglycosides also develops readily. The use of aminoglycosides for treatment of gram-negative infections has decreased since the development of the semisynthetic penicillins (Section 27.8) and the tetracyclines (discussed later in this section). Aminoglycoside antibiotics are now considered reserve antibiotics used primarily when other antibiotics fail.

Macrolides

Macrolide antibiotics contain lactone rings bonded to sugars (**Figure 27.22**). Variations in both the lactone ring and the sugars result in a large number of macrolide antibiotics. The best-known macrolide is erythromycin (produced by *Streptomyces erythreus*). Other clinically useful macrolides include dirithromycin, clarithromycin, and azithromycin. The macrolides account for 11% of the total world production and use of antibiotics (Figure 27.14). Erythromycin is a broad-spectrum antibiotic that targets the 50S subunit of the bacterial ribosome, inhibiting protein synthesis (Figure 27.12 and Figure 27.13). Commonly used clinically in place of penicillin in patients allergic to penicillin or other β-lactam antibiotics, erythromycin is particularly useful for treating legionellosis (Section 36.7).

Tetracyclines

The **tetracyclines**, produced by several species of *Streptomyces* (Section 25.6), are an important group of antibiotics that find widespread medical use in humans. They were some of the first broad-spectrum antibiotics, inhibiting almost all gram-positive and gram-negative *Bacteria*. The basic structure of the tetracyclines consists of a naphthacene

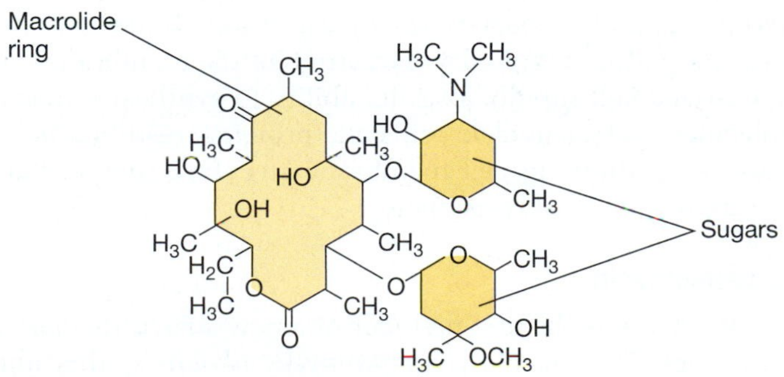

Figure 27.22 Erythromycin, a macrolide antibiotic. Erythromycin is a widely used broad-spectrum antibiotic.

UNIT 6

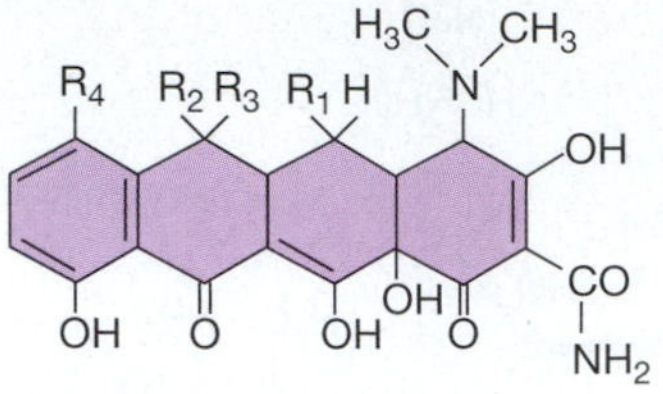

Tetracycline analog	R_1	R_2	R_3	R_4
Tetracycline	H	OH	CH_3	H
7-Chlortetracycline (aureomycin)	H	OH	CH_3	Cl
5-Oxytetracycline (terramycin)	OH	OH	CH_3	H

Figure 27.23 Tetracycline. Semisynthetic analogs of tetracycline are important antibiotics.

ring system (**Figure 27.23**). Substitutions to the basic naphthacene ring occur naturally and form new tetracycline analogs. Semisynthetic tetracyclines having substitutions in the naphthacene ring system have also been developed. Like erythromycin and the aminoglycoside antibiotics, tetracycline is a protein synthesis inhibitor, interfering with bacterial 30S ribosome subunit function (Figure 27.12).

The tetracyclines and the β-lactam antibiotics comprise the two most important groups of antibiotics in the medical field. The tetracyclines are also widely used in veterinary medicine, and in some countries are used as nutritional supplements for poultry and swine. Because extensive nonmedical uses of medically important antibiotics have contributed to widespread antibiotic resistance, this use is now discouraged.

Daptomycin

Daptomycin is another antibiotic produced by a member of the *Streptomyces* genus. This novel antibiotic is a cyclic lipopeptide (**Figure 27.24**) with a unique mode of action. Used mainly to treat infections by gram-positive *Bacteria* such as the pathogenic staphylococci and streptococci, daptomycin binds specifically to bacterial cytoplasmic membranes, forms a pore, and induces rapid depolarization of the membrane. The depolarized cell rapidly loses its ability to synthesize macromolecules such as nucleic acids and proteins, resulting in cell death. Alterations in cell membrane structure may account for rare instances of resistance.

Platensimycin

Platensimycin is the first member of a new structural class of antibiotics. Produced by *Streptomyces platensis*, this antibiotic (**Figure 27.25**) selectively inhibits a bacterial enzyme central to fatty acid biosynthesis, thus disrupting lipid biosynthesis. Platensimycin is effective against a broad range of

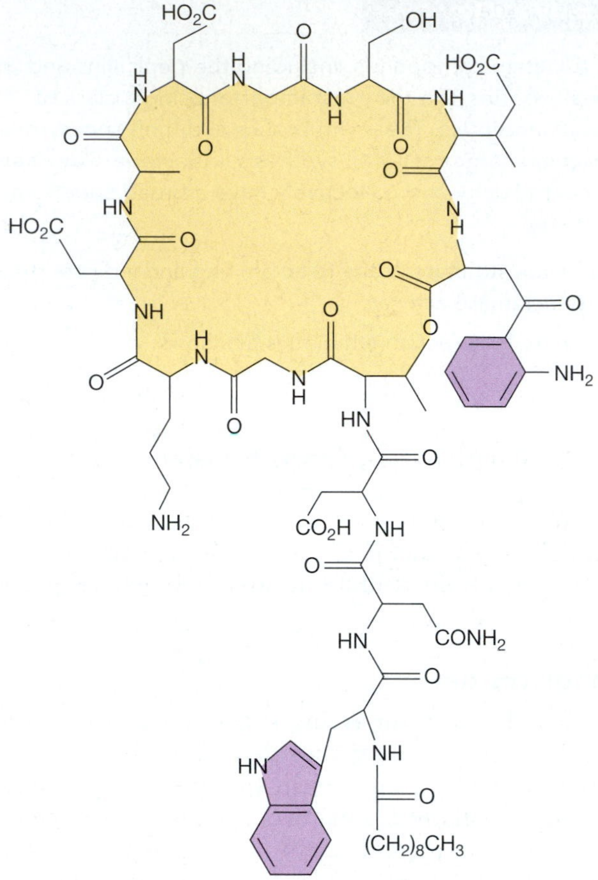

Figure 27.24 Daptomycin. Daptomycin is a cyclic lipopeptide that depolarizes cytoplasmic membranes in gram-positive *Bacteria*.

gram-positive *Bacteria*, including particularly difficult to treat infections caused by methicillin-resistant *Staphylococcus aureus* and vancomycin-resistant enterococci. Already shown to be effective in eradicating *S. aureus* infections in mice, this antibiotic shows no toxicity. Platensimycin has a unique mode of action, and there is no known potential for development of resistance by pathogens. We discuss the discovery of platensimycin in Section 27.13.

27.9 MiniReview

The aminoglycosides, macrolides, and tetracycline antibiotics are structurally complex molecules produced by *Bacteria* and are active against other *Bacteria*. These antibiotics selectively interfere with protein synthesis in *Bacteria*. Daptomycin and platensimycin are structurally novel antibiotics that target cytoplasmic membrane functions and lipid biosynthesis, respectively.

- What are the biological sources of aminoglycosides, tetracyclines, macrolides, daptomycin, and platensimycin?
- How does the activity of each class of antibiotics lead to death of the affected cell?

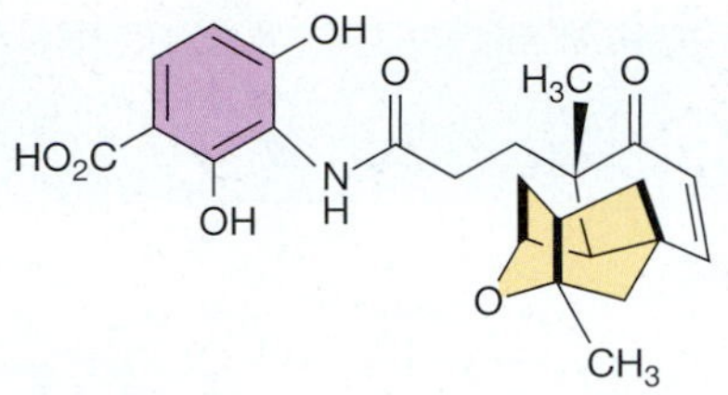

Figure 27.25 Platensimycin. Platensimycin selectively inhibits lipid biosynthesis in *Bacteria*.

IV CONTROL OF VIRUSES AND EUKARYOTIC PATHOGENS

Drugs that control growth of viruses and eukaryotic pathogens often affect eukaryotic host cells as well. As a result, selective toxicity for eukaryotic pathogens is very difficult to attain; only chemotherapeutic agents that preferentially affect pathogen-specific metabolic pathways or structural components are useful. There are a limited number of these drugs, and we discuss the important ones here.

27.10 Antiviral Drugs

Because viruses use their eukaryotic hosts to reproduce and perform metabolic functions, most antiviral drugs also target host structures, resulting in host toxicity. However, several agents are known that are more toxic for viruses than for the host, and a few agents specifically target viruses. Largely because of efforts to find effective measures to control infections with the human immunodeficiency virus (HIV), the cause of AIDS (Section 34.15), significant achievements have been made in the development and use of antiviral agents.

Antiviral Chemotherapeutic Agents

The most successful and commonly used agents for antiviral chemotherapy are the nucleoside analogs (**Table 27.5**). The first compound to gain universal acceptance in this category was zidovudine, or azidothymidine (AZT) (Figure 34.42). AZT inhibits retroviruses such as HIV (Section 10.12). Azidothymidine is chemically related to thymidine but is a dideoxy derivative, lacking the 3′-hydroxyl group. AZT inhibits multiplication of retroviruses by blocking reverse transcription and production of the virally encoded DNA intermediate. This inhibits multiplication of HIV. A number of other nucleoside analogs having similar mechanisms have been developed for the treatment of HIV and other viruses.

Nearly all nucleoside analogs, or **nucleoside reverse transcriptase inhibitors (NRTI)**, work by the same mechanism, inhibiting elongation of the viral nucleic acid chain by a nucleic acid polymerase. The nucleotide analog cidofovir works in the same way (Table 27.5). Because the normal cell function of nucleic acid replication is targeted, these drugs usually induce some host toxicity. Many NRTIs also lose their antiviral potency with time due to the emergence of drug-resistant viruses (Section 34.15).

Several other antiviral agents target the key enzyme of retroviruses, reverse transcriptase. Nevirapine, a **nonnucleoside reverse transcriptase inhibitor (NNRTI)**, binds directly to reverse transcriptase and inhibits reverse transcription. Phosphonoformic acid, an analog of inorganic pyrophosphate, inhibits normal internucleotide linkages, preventing synthesis of viral nucleic acids. As with the NRTIs, the NNRTIs generally induce some level of host toxicity because their action also affects normal host cell nucleic acid synthesis.

Protease inhibitors are another class of antiviral drugs that are effective for treatment of HIV (Table 27.5 and see Figure 27.30). These drugs prevent viral replication by binding the active site of HIV protease, inhibiting this enzyme from processing large viral proteins into their individual components, thus preventing virus maturation (Sections 19.15, 27.13, and 34.15).

A final category of anti-HIV drugs is represented by a single drug, enfuvirtide, a **fusion inhibitor** composed of a 36-amino acid synthetic peptide that binds to the gp41 membrane protein of HIV (Table 27.5 and Section 34.15). Binding of the gp41 protein by enfuvirtide stops the conformational changes necessary for the fusion of HIV and T lymphocyte membranes, thus preventing infection of cells by HIV.

Influenza Antiviral Agents

Two categories of drugs effectively limit influenza infection. The adamantanes amantadine and rimantadine are synthetic amines that interfere with an influenza A ion transport protein, inhibiting virus uncoating and subsequent replication. The neuraminidase inhibitors oseltamivir (Tamiflu®) and zanamivir (Relenza®) block the active site of neuraminidase in influenza A and B viruses, inhibiting virus release from infected cells. Zanamivir is used only for treatment of influenza, whereas oseltamivir is used for both treatment and prophylaxis. The adamantanes are less useful than the neuraminidase inhibitors because resistance to adamantanes develops rapidly in strains of influenza virus (Section 34.9).

Interferons

In studies of virus interference, a phenomenon in which infection with one virus interferes with subsequent infection by another virus, it was discovered that several small proteins are the cause of the interference; the proteins were named interferons. **Interferons** are small proteins that prevent viral multiplication by stimulating the production of antiviral proteins in uninfected cells. Interferons are formed in response to live virus, inactivated virus, and viral nucleic acids. Interferon is produced in large amounts by cells infected with viruses of low virulence, but little is produced against highly virulent viruses. Highly virulent viruses inhibit cell protein synthesis before interferon can be produced. Interferons are also induced by natural and synthetic double-stranded RNA (dsRNA) molecules. In nature, dsRNA exists only in virus-infected cells as the replicative form of RNA viruses such as rhinoviruses (cold viruses) (Section 34.8); the dsRNA from the infecting virus signals the animal cell to produce interferon.

Table 27.5 Antiviral chemotherapeutic compounds

Category/drug	*Mechanism of action*	*Virus affected*
Fusion inhibitor		
Enfuvirtide	Blocks HIV-T lymphocyte membrane fusion	HIV[a]
Interferons		
Interferon α	Induces proteins that inhibit viral replication	Broad spectrum (host specific)
Interferon β		
Interferon γ		
Neuraminidase inhibitors		
Oseltamivir (Tamiflu®)	Block active site of influenza neuraminidase	Influenza A and B
Zanamivir (Relenza®)		Influenza A and B
Nonnucleoside reverse transcriptase inhibitor (NNRTI)		
Nevirapine	Reverse transcriptase inhibitor	HIV
Nucleoside analogs		
Acyclovir	Viral polymerase inhibitors	Herpes viruses, *Varicella zoster*
Ganciclovir		Cytomegalovirus
Trifluridine		Herpesvirus
Valacyclovir		Herpesvirus
Vidarabine		Herpesvirus, vaccinia, hepatitis B virus
Abacavir (ABC)	Reverse transcriptase inhibitors	HIV
Didanosine (dideoxyinosine or ddI)		HIV
Emtricitabine (FTC)		HIV
Lamivudine (3TC)		HIV, hepatitis B virus
Stavudine (d4T)		HIV
Zalcitabine (ddC)		HIV
Zidovudine (AZT) (Figure 34.42)		HIV
Ribavirin	Blocks capping of viral RNA	Respiratory syncytial virus, influenza A and B, Lassa fever
Nucleotide analogs		
Cidofovir	Viral polymerase inhibitor	Cytomegalovirus, herpesviruses
Tenofovir (TDF)	Reverse transcriptase inhibitor	HIV
Protease inhibitors		
Amprenavir	Viral protease inhibitor	HIV
Indinavir (Figure 27.31)		HIV
Lopinavir		HIV
Nelfinavir		HIV
Saquinavir (Figure 27.31)		HIV
Pyrophosphate analog		
Phosphonoformic acid (Foscarnet)	Viral polymerase inhibitor	Herpesviruses, HIV, hepatitis B virus
RNA polymerase inhibitor		
Rifamycin	RNA polymerase inhibitor	Vaccinia, pox viruses
Synthetic amines		
Amantadine	Viral uncoating blocker	Influenza A
Rimantadine		Influenza A

[a]Human immunodeficiency virus

Interferons from virus-infected cells interact with receptors on uninfected cells, promoting the synthesis of antiviral proteins that function to prevent further virus infection. Interferons are cytokines and are produced in three molecular forms: *IFN-α* is produced by leukocytes, *IFN-β* is produced by fibroblasts, and *IFN-γ* is produced by immune lymphocytes (∞ Section 31.10). All three forms are effective viral inhibitors.

Interferon activity is *host*-specific rather than *virus*-specific. That is, interferon produced by a member of one species recognizes specific receptors only on cells from the same species. As a result, interferon produced by cells of an animal in response to, for example, a rhinovirus, could also inhibit multiplication of, for example, influenza viruses in cells within the same species, but has no effect on the multiplication of any virus in cells from other animal species.

Interferons have potential as possible antiviral and anticancer agents. Several approved recombinant interferons are available. However, the use of interferons as chemotherapeutic agents is not widespread because interferon must be delivered locally in high concentrations to stimulate the production of antiviral proteins in uninfected host cells. Thus, the clinical utility of these antiviral agents depends on our ability to deliver interferon to local areas in the host through injections or aerosols. Alternatively, appropriate interferon-stimulating signals such as stimulation with viral nucleotides, nonvirulent viruses, or even synthetic nucleotides, if given to host cells prior to viral infection, might stimulate natural production of interferon.

27.10 MiniReview

Effective antiviral agents selectively target virus-specific enzymes and processes. Clinically useful antiviral agents include nucleoside analogs and other drugs that inhibit nucleic acid polymerases and viral genome replication. Agents such as protease inhibitors interfere with viral maturation steps. Host cells also produce the antiviral interferon proteins that stop viral replication.

- Why are there relatively few effective antiviral chemotherapeutic agents? Why aren't such agents used to treat common viral illnesses such as colds?
- What steps in the viral maturation process are inhibited by nucleoside analogs? By protease inhibitors? By interferons?

27.11 Antifungal Drugs

Fungi, like viruses, pose special problems for the development of chemotherapy. Because fungi are *Eukarya*, much of their cellular machinery is the same as that of animals and humans. Thus, antifungal agents that act on metabolic pathways in fungi often affect corresponding pathways in host cells, making the drugs toxic. As a result, many antifungal drugs can be used only for topical (surface) applications. However, a few drugs are selectively toxic for fungi because they target unique fungal structures or metabolic processes. Fungal-specific drugs are becoming increasingly important as fungal infections in immunocompromised individuals become more prevalent (∞ Sections 34.15 and 35.8). We examine here the selective action and targets of several effective antifungal agents.

Ergosterol Inhibitors

Ergosterol in fungal cytoplasmic membranes replaces the cholesterol found in animal cytoplasmic membranes. Two types of antifungal compounds work by interacting with ergosterol or inhibiting its synthesis (**Table 27.6**). These include the polyenes, a group of antibiotics produced by species of

UNIT 6

Table 27.6 **Antifungal agents**

Category	*Target*	*Examples*	*Use*
Allylamines	Ergosterol synthesis	Terbenafine	Oral, topical
Aromatic antibiotic	Mitosis inhibitor	Griseofulvin	Oral
Azoles	Ergosterol synthesis	Clortrimazole	Topical
		Fluconazole	Oral
		Itraconazole	Oral
		Ketoconazole	Oral
		Miconazole	Topical
		Posoconazole	Experimental
		Ravuconazole	Experimental
		Voriconazole	Oral
Chitin synthesis inhibitor	Chitin synthesis	Nikkomycin Z	Experimental
Echinocandins	Cell wall synthesis	Caspofungin	Intravenous
Nucleic acid analogs	DNA synthesis	5-Fluorocytosine	Oral
Polyenes	Ergosterol synthesis	Amphotericin B	Oral, intravenous
		Nystatin	Oral, topical
Polyoxins	Chitin synthesis	Polyoxin A	Agricultural
		Polyoxin B	Agricultural

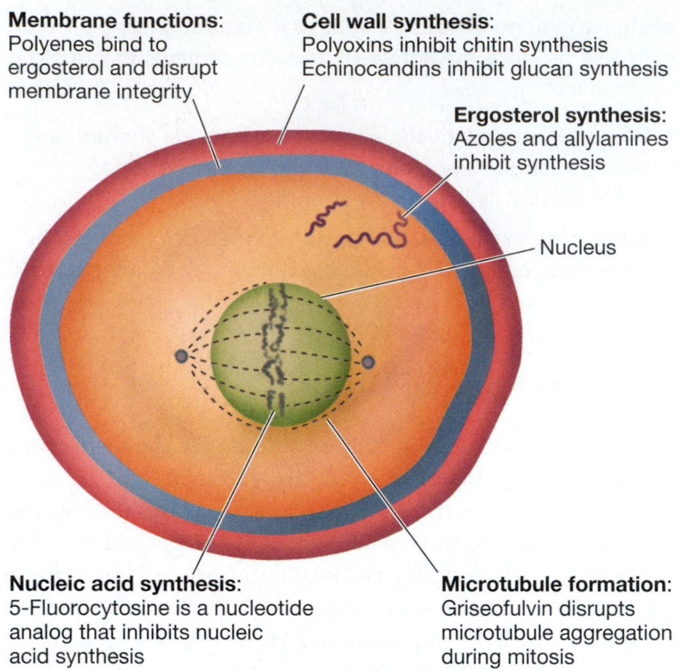

Membrane functions:
Polyenes bind to ergosterol and disrupt membrane integrity
Cell wall synthesis:
Polyoxins inhibit chitin synthesis
Echinocandins inhibit glucan synthesis
Ergosterol synthesis:
Azoles and allylamines inhibit synthesis
Nucleus
Nucleic acid synthesis:
5-Fluorocytosine is a nucleotide analog that inhibits nucleic acid synthesis
Microtubule formation:
Griseofulvin disrupts microtubule aggregation during mitosis

Figure 27.26 Action of some antifungal chemotherapeutic agents. Traditional antibacterial agents are generally ineffective because fungi are eukaryotic cells. The membrane and cell-wall targets shown here are unique structures not present in vertebrate host cells.

Streptomyces. Polyenes bind to ergosterol, disrupting membrane function, causing membrane permeability and cell death (**Figure 27.26**). A second major type of antifungal compound includes the azoles and allylamines, synthetic agents that selectively inhibit ergosterol biosynthesis and therefore have broad antifungal activity. Treatment with azoles results in abnormal fungal membranes, leading to membrane damage and alteration of critical membrane transport activities. Allylamines also inhibit ergosterol biosynthesis but are restricted to topical use because they are not readily taken up by animal cells and tissues.

Echinocandins

Echinocandins act by inhibiting 1,3 β-D glucan synthase, the enzyme that forms glucan polymers in the fungal cell wall (Figure 27.26 and Table 27.6). Because mammalian cells do not have 1,3 β-D glucan synthase (or cell walls), the action of these agents is specific, resulting in selective fungal cell death. These agents are used to treat infections with fungi such as *Candida* and some fungi that are resistant to other agents (∞ Sections 34.15 and 35.8.)

Other Antifungal Agents

Other antifungal drugs interfere with fungus-specific structures and functions (Table 27.6). For example, fungal cell walls contain chitin, a polymer of *N*-acetylglucosamine found only in fungi and insects. Several polyoxins inhibit cell wall synthesis by interfering with chitin biosynthesis. Polyoxins are widely used as agricultural fungicides, but are not used clinically. Other antifungal drugs inhibit folate biosynthesis, interfere with DNA topology during replication, or, in the case of drugs such as griseofulvin, disrupt microtubule aggregation during mitosis. Moreover, the nucleic acid analog 5-fluorocytosine is an effective nucleic acid synthesis inhibitor in fungi. Some very effective antifungal drugs also have other applications. For example, vincristine, vinblastin, and toxol are effective antifungal agents and also have anticancer properties.

Predictably, the use of antifungal drugs has resulted in the emergence of populations of resistant fungi and the emergence of "new" fungal pathogens. For example, *Candida* species, which are normally not pathogenic, are known to produce disease in immunocompromised individuals. Unfortunately, drug-resistant pathogenic *Candida* strains have developed in individuals who have been treated with antifungal drugs, and some are now resistant to all of the currently used antifungal agents (see Figure 27.29).

27.11 MiniReview

Antifungal agents fall into many chemical categories. Because fungi are *Eukarya*, selective toxicity is hard to achieve, but some effective chemotherapeutic agents are available. Treatment of fungal infections is an emerging human health issue.

- Why are there very few clinically effective antifungal agents?
- What factors are contributing to the rise in fungal infections?

V ANTIMICROBIAL DRUG RESISTANCE AND DRUG DISCOVERY

Antimicrobial drug resistance is a major problem when dealing with many pathogenic microorganisms, especially in healthcare settings. Here we explore some of the mechanisms for drug resistance in microorganisms and present strategies for developing new antimicrobial agents. Practical methods for preserving and enhancing the efficacy of currently used antimicrobial drugs are discussed in the Microbial Sidebar, "Preventing Antibiotic Resistance."

27.12 Antimicrobial Drug Resistance

Antimicrobial drug resistance is the acquired ability of a microorganism to resist the effects of a chemotherapeutic agent to which it is normally susceptible. No single antibiotic inhibits all microorganisms, and some form of antimicrobial drug resistance is an inherent property of virtually all microorganisms. As we have discussed, antibiotic producers are microorganisms, and in order to survive, the antibiotic-producing microorganism itself has had to develop resistance mechanisms to neutralize or destroy its own antibiotic. This reality is thought to be the origin of genes encoding antibiotic resistance. More widespread antimicrobial drug resistance can then occur rapidly from the spreading of these genes between and among microorganisms by horizontal gene transfer.

Resistance Mechanisms

For any of at least six different reasons, some microorganisms are naturally resistant to certain antibiotics. (1) The organism may lack the structure an antibiotic inhibits. For instance, some bacteria, such as the mycoplasmas, lack a bacterial cell wall and are therefore naturally resistant to penicillins. (2) The organism may be impermeable to the antibiotic. For example, most gram-negative *Bacteria* are impermeable to penicillin G and platensimycin. (3) The organism may be able to alter the antibiotic to an inactive form. Many staphylococci contain β-lactamases, an enzyme that cleaves the β-lactam ring of most penicillins (**Figure 27.27**). (4) The organism may modify the target of the antibiotic. (5) The organism may develop a resistant biochemical pathway. For example, many pathogens develop resistance to sulfa drugs that inhibit the production of folic acid in *Bacteria* (Section 27.6 and Figure 27.16). Resistant bacteria modify their metabolism to take up preformed folic acid from the environment, avoiding the need for the pathway blocked by the sulfa drugs. (6) The organism may be able to pump out an antibiotic entering the cell, a process called *efflux*. Some specific examples of bacterial resistance to antibiotics are shown in **Table 27.7**.

Antibiotic resistance can be genetically encoded by the microorganism on either the bacterial chromosome or on a plasmid called an *R* (for *resistance*) *plasmid* (∞ Sections 11.2 and 11.3) (Table 27.7). Because of widespread existing antibiotic resistance and continual emergence of new resistance, bacteria isolated from clinical specimens must be tested for antibiotic susceptibility using the MIC method or an agar diffusion method (Section 27.4). Details of the antibiotic susceptibility testing of clinical isolates are described in Section 32.3.

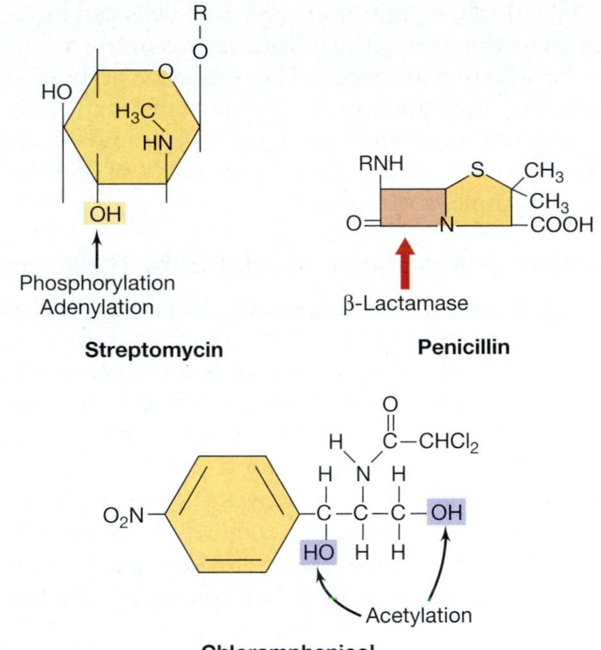

Figure 27.27 Sites at which antibiotics are attacked by enzymes encoded by R plasmid genes. Antibiotics may be selectively inactivated by chemical modification or cleavage. For the complete structure of streptomycin, see Figure 27.21, and for penicillin, Figure 27.19.

Table 27.7 Mechanisms of bacterial resistance to antibiotics

Resistance mechanism	*Antibiotic example*	*Genetic basis of resistance*	*Mechanism present in:*
Reduced permeability	Penicillins	Chromosomal	*Pseudomonas aeruginosa* Enteric *Bacteria*
Inactivation of antibiotic (for example, penicillinase; modifying enzymes such as methylases, acetylases, phosphorylases, and others)	Penicillins	Plasmid and chromosomal	*Staphylococcus aureus* Enteric *Bacteria* *Neisseria gonorrhoeae*
	Chloramphenicol	Plasmid and chromosomal	*Staphylococcus aureus* Enteric *Bacteria*
	Aminoglycosides	Plasmid	*Staphylococcus aureus*
Alteration of target (for example, RNA polymerase, rifamycin; ribosome, erythromycin, and streptomycin; DNA gyrase, quinolones)	Erythromycin Rifamycin Streptomycin Norfloxacin	Chromosomal	*Staphylococcus aureus* Enteric *Bacteria* Enteric *Bacteria* Enteric *Bacteria* *Staphylococcus aureus*
Development of resistant biochemical pathway	Sulfonamides	Chromosomal	Enteric *Bacteria* *Staphylococcus aureus*
Efflux (pumping out of cell)	Tetracyclines	Plasmid	Enteric *Bacteria*
	Chloramphenicol	Chromosomal	*Staphylococcus aureus* *Bacillus subtilis*
	Erythromycin	Chromosomal	*Staphylococcus* spp.

UNIT 6

In the laboratory, antibiotic-resistant cells can be isolated from cultures that were grown from strains uniformly susceptible to the selecting antibiotic. The resistance of these isolates is usually due to mutations in chromosomal genes. In most cases, antibiotic resistance mediated by chromosomal genes arises because of a modification of the *target* of antibiotic activity (for example, a ribosome).

Mechanism of Resistance Mediated by R Plasmids

Most drug-resistant bacteria isolated from patients contain drug-resistance genes located on R plasmids, rather than on the chromosome. Resistance is typically due to genes on the R plasmid that encode enzymes that modify and inactivate the drug (Figure 27.27) or genes that encode enzymes that prevent drug uptake or actively pump it out. For instance, the aminoglycoside antibiotics streptomycin, neomycin, kanamycin, and spectinomycin have similar chemical structures. Strains carrying R plasmids for these drugs make enzymes that phosphorylate, acetylate, or adenylylate the drug. The modified drug then lacks antibiotic activity.

For the penicillins, R plasmids encode the enzyme *penicillinase* (a β-lactamase that splits the β-lactam ring, inactivating the antibiotic). Chloramphenicol resistance is due to an R plasmid-encoded enzyme that acetylates the antibiotic. Many R plasmids contain several different resistance genes and can confer multiple antibiotic resistance on a cell previously sensitive to each individual antibiotic (Section 11.3).

Origin of Resistance Plasmids

Evidence indicates that the existence of R plasmids predated the antibiotic era. For example, a strain of *Escherichia coli* that was freeze-dried in 1946 contained a plasmid with genes conferring resistance to both tetracycline and streptomycin, even though neither of these antibiotics was used clinically until several years later. Also, R plasmid genes for resistance to semisynthetic penicillins were shown to exist before the semisynthetic penicillins had been synthesized.

Of perhaps even more ecological significance, R plasmids with antibiotic resistance genes have been detected in some nonpathogenic gram-negative soil bacteria. In the soil, R plasmids may confer selective advantages because major antibiotic-producing organisms (*Streptomyces* and *Penicillium*) are also soil organisms. Thus, it seems that R plasmids existed in the natural microbial population before antibiotics were discovered and used in medicine or agriculture.

Spread of Antimicrobial Drug Resistance

The widespread use of antibiotics in medicine, veterinary medicine, and agriculture provides highly selective conditions for the spread of R plasmids. The resistance genes on R plasmids confer an immediate selective advantage, and thus antibiotic resistance due to R plasmids is a predictable outcome of natural selection. The R plasmids and other sources of resistance genes pose significant limits on the long-term use of any single antibiotic as an effective chemotherapeutic agent.

Inappropriate use of antimicrobial drugs is the leading cause of rapid development of drug-specific resistance in disease-causing microorganisms. The discovery and clinical use of the many known antibiotics have been paralleled by the emergence of bacteria that resist them. **Figure 27.28** shows a correlation between the number of tons of antibiotics used and the numbers of bacteria resistant to each antibiotic.

There are numerous examples of the overuse of antibiotics resulting in development of resistance. Increasingly, the antimicrobial agent prescribed for treatment of a particular infection must be changed because of increased resistance of the microorganism causing the disease. A classic example is the development of resistance to penicillin in *Neisseria gonorrhoeae*, the bacterium that causes the sexually transmitted disease gonorrhea (Figure 27.28*b*). Prior to 1980, penicillin was the accepted treatment for gonorrhea, and penicillin had been in continuous use for this application since it became available in the 1940s. However, penicillin is no longer useful for treatment of gonorrhea because a large percentage of current clinical *N. gonorrhoeae* isolates now produce β-lactamase, conferring penicillin resistance. Virtually all of these resistant isolates have developed since 1980. The growing prevalence of fluoroquinolone resistance in men who have sex with men and in infection sources from Asia, Hawaii, and California prompted a change in first-line drug recommendations for treatment of gonorrhea from ciprofloxacin, a fluoroquinolone, to ceftriaxone (Figure 27.28*c*). Treatment guidelines are updated nearly every year to control continually emerging drug resistance in this organism (Section 34.13).

Antibiotics are still used in clinical practice far more often than necessary. Data indicate that antibiotic treatment is warranted in about 20% of individuals who are seen for infectious disease. Yet, antibiotics are prescribed up to 80% of the time. Furthermore, in up to 50% of cases, prescribed doses or duration of treatments are not correct. This is compounded by patient noncompliance: Many patients stop taking medications, particularly antibiotics, as soon as they feel better. For example, the emergence of isoniazid-resistant tuberculosis correlates with a patient's failure to take the oral medication daily for the full course of 6–9 months (Section 34.5). Exposure of virulent pathogens to sublethal doses of antibiotics for inadequate periods of time may select for drug-resistant strains.

Other recent studies, however, indicate that this trend is changing in the United States. Physicians prescribe about one-third fewer antibiotics for treatment of childhood infections than they did 10 years ago. This reduction has been done largely through efforts aimed at educating physicians, healthcare providers, and patients concerning the proper use of antibiotic therapy.

Indiscriminant, nonmedical use of antibiotics has also contributed to the emergence of resistant strains. Antibiotics are used in agriculture as supplements to animal feeds both as growth-promoting substances and as prophylactic additives to prevent the occurrence of disease in addition to their traditional use as a treatment for infections. Worldwide, about

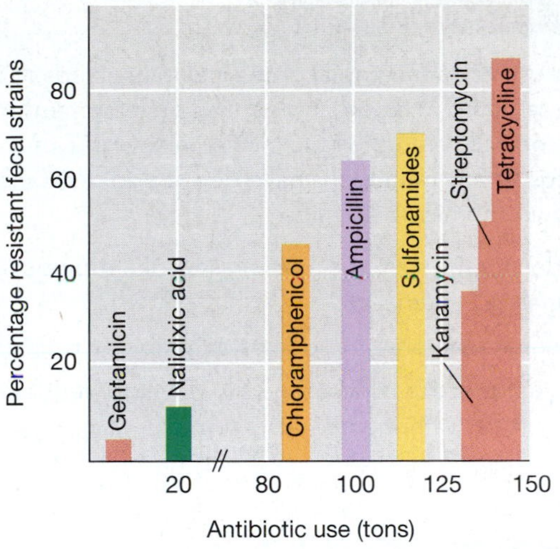

(a)

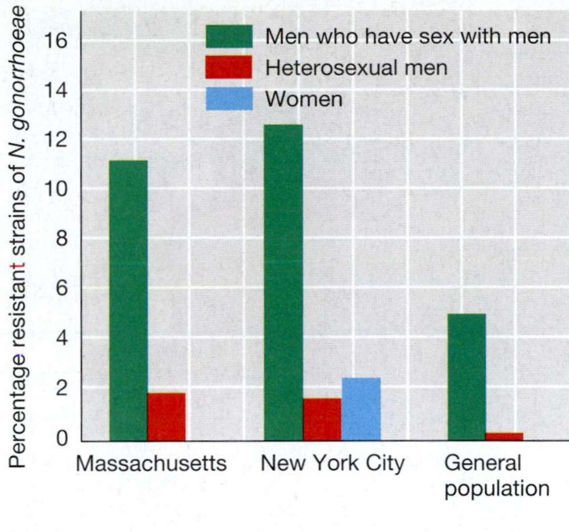

(c)

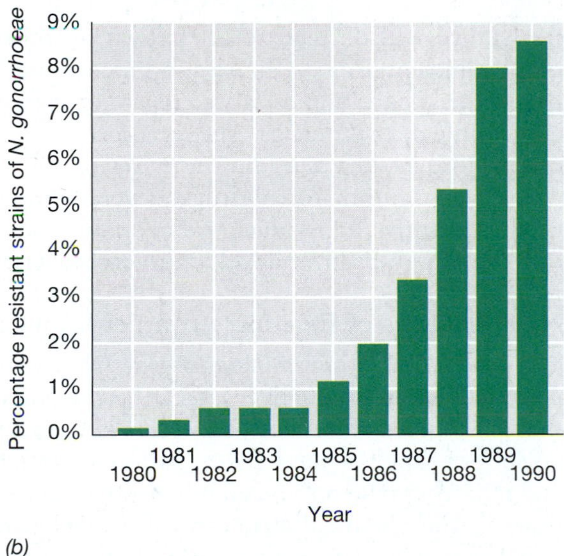

(b)

Figure 27.28 Patterns of drug resistance in pathogens. *(a)* The relationship between antibiotic use and the percentage of antibiotic-resistant bacteria isolated from diarrheal patients. Those agents that have been used in the largest amounts, as indicated by the amount produced commercially, are those for which drug-resistant strains are most frequent. *(b)* Percentage of reported cases of gonorrhea caused by drug-resistant strains. The actual number of reported drug-resistant cases in 1985 was 9,000. This number rose to 59,000 in 1990. Greater than 95% of reported drug-resistant cases were due to penicillinase-producing strains of *Neisseria gonorrhoeae*. Since 1990, penicillin has not been recommended for treatment of gonorrhea because of the emerging drug resistance. *(c)* The prevalence of fluoroquinolone-resistant *Neisseria gonorrhoeae* in certain populations in the United States in 2003. As of 2007, the fluoroquinolone ciprofloxacin is no longer recommended as a primary choice for treatment of *N. gonorrhoeae* infections.

(*Source*: Centers for Disease Control and Prevention, Atlanta, GA).

50% of all antibiotics made are used in animal agriculture applications. Antibiotics are also extensively used in aquaculture (fish farming) and even in fruit production! Antibiotics used far too frequently, over extended periods of time and in high doses in the food supply, are a proven source of food infection outbreaks due to the selection of antibiotic-resistant pathogens. For example, fluoroquinolones, a group of broad-spectrum antibiotics including clinically important therapeutic drugs such as ciprofloxacin, have been extensively used for over 20 years as growth-promoting and prophylactic agents in agriculture. As a result, fluoroquinolone-resistant *Campylobacter jejuni* have already emerged as foodborne pathogens (∞ Section 37.9), presumably because of the routine treatment of poultry flocks with fluoroquinolones to prevent respiratory diseases. Voluntary guidelines used by both poultry and drug producers are in place to monitor and reduce the use of fluoroquinolones. These measures may prevent development of resistance to new fluoroquinolones.

Antibiotic-Resistant Pathogens

Largely as a result of failures to properly use antibiotics and monitor resistance, almost all pathogenic microorganisms have developed resistance to some chemotherapeutic agents since widespread use of antimicrobial chemotherapy began in the 1950s (**Figure 27.29**). Penicillin and sulfa drugs, the first widely used chemotherapeutic agents, are not as widely used today because many pathogens have acquired some resistance. Even the organisms that are still uniformly sensitive to penicillin, such as *Streptococcus pyogenes* (the bacterium that causes strep throat, scarlet fever, and rheumatic fever), require larger doses of penicillin for successful treatment now as compared to a decade ago.

A few pathogens have developed resistance to all known antimicrobial agents (Figure 27.29). Among these are several isolates of methicillin-resistant *Staphylococcus aureus* (MRSA) (methicillin is a semisynthetic penicillin; Section 27.8). Although MRSA is usually associated with health-care

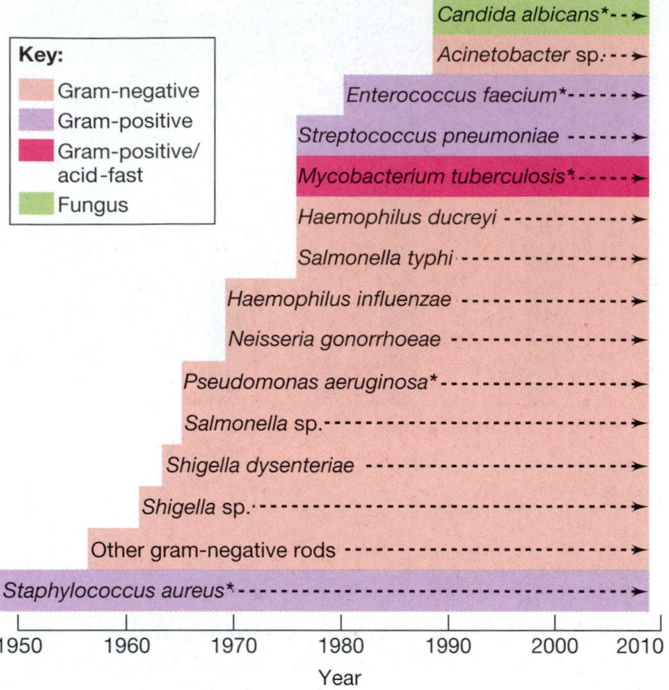

Figure 27.29 The appearance of antimicrobial drug resistance in some human pathogens. The asterisks indicate that some strains of these pathogens are now untreatable with known antimicrobial drugs.

settings, it also causes a significant number of community-associated infections (Section 33.7). An increasing number of independently derived MRSA strains have developed reduced susceptibility even to vancomycin and are termed "vancomycin intermediate *Staphylococcus aureus*" (VISA) strains (Section 34.10). Vancomycin-resistant *Enterococcus faecium* (VRE) and some isolates of *Mycobacterium tuberculosis* and *Candida albicans* have also developed resistance to all known antimicrobial drugs. Antibiotic resistance can be minimized if drugs are used only for treatment of susceptible diseases and are given in sufficiently high doses and for sufficient lengths of time that the microbial population is reduced before resistant mutants can develop. Combining two unrelated chemotherapeutic agents may also reduce resistance; it is less likely that a mutant strain resistant to one antibiotic will also be resistant to the second antibiotic. However, certain common R plasmids confer multiple drug resistance and make multiple antibiotic therapy less useful as a clinical treatment strategy.

A few reports suggest that if the use of a particular antibiotic is stopped, the resistance to that antibiotic will be reversed over the course of several years. On the other hand, antibiotic-resistant organisms may persist in the gut for some time. This information implies that the efficacy of some antibiotics may be reestablished by withdrawing the antibiotic from use, but only by following a carefully monitored plan of prudent use upon reintroduction. Finally, as we discuss below, new antimicrobial agents are actively being developed using various strategies for drug design and discovery.

27.12 MiniReview

The use of antimicrobial drugs inevitably leads to resistance in the targeted microorganisms. The development of resistance can be accelerated by the indiscriminate use of antimicrobial drugs. A few pathogens have developed resistance to all known antimicrobial drugs.

- Identify the six basic mechanisms of antibiotic resistance among bacteria.
- Identify the primary sources of antibiotic resistance genes.
- What practices encourage the development of antibiotic-resistant pathogens?

27.13 The Search for New Antimicrobial Drugs

Resistance will develop to all known antimicrobial drugs, given sufficient drug exposure and time. Conservative, appropriate use of antibiotics is necessary to prolong the useful clinical life of these drugs. The long-term solution to antimicrobial drug resistance, however, resides in our ability to develop new antimicrobial drugs. Several strategies are being used to identify and produce useful analogs of existing agents or to design or discover novel antimicrobial compounds.

New Analogs of Existing Antimicrobial Compounds

The production of new analogs of existing antimicrobial compounds is often productive, largely because new compounds that are structural mimics of older ones have a proven mechanism of action. In many cases, parameters such as solubility and affinity can be optimized by introducing minor modifications to the chemical structure of a drug without altering structures critical to drug action. The new compound may actually be more effective than the parent compound, and, because resistance is based on structural recognition, the new compound may not be recognized by resistance factors. For example, Figure 27.23 shows the structure of tetracycline and two bioactive derivatives. Using authentic tetracycline as the lead compound, systematic chemical substitutions at the four R group sites can generate an almost endless series of tetracycline analogs. Using this basic strategy, new tetracycline-related compounds (Section 27.9), new β-lactam antibiotics (Section 27.8), and new analogs of vancomycin (**Figure 27.30**), have been synthesized. Some of these derivatives are as much as 100 times more potent than the parent compound.

The application of automated chemistry methods to drug discovery has dramatically increased our ability to rapidly generate new antimicrobial compounds. These automated methods, called *combinatorial chemistry*, initiate systematic modifications of a known antimicrobial product to yield large numbers of new analogs. For instance, using automated combinatorial chemistry and starting with pure tetracycline (Figure 27.23), five different reagents might be used to introduce substitutions at the four different tetracycline R groups.

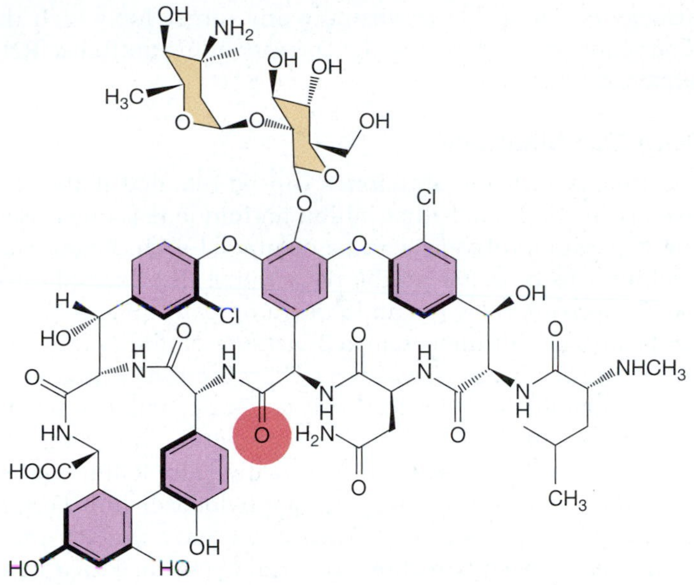

Vancomycin

Figure 27.30 Vancomycin. Intermediate drug resistance to the parent structure of vancomycin, shown above, has developed in recent years. However, modification at the position shown in red by substitution of a methylene ($=CH_2$) group for the carbonyl oxygen restores much of the lost activity.

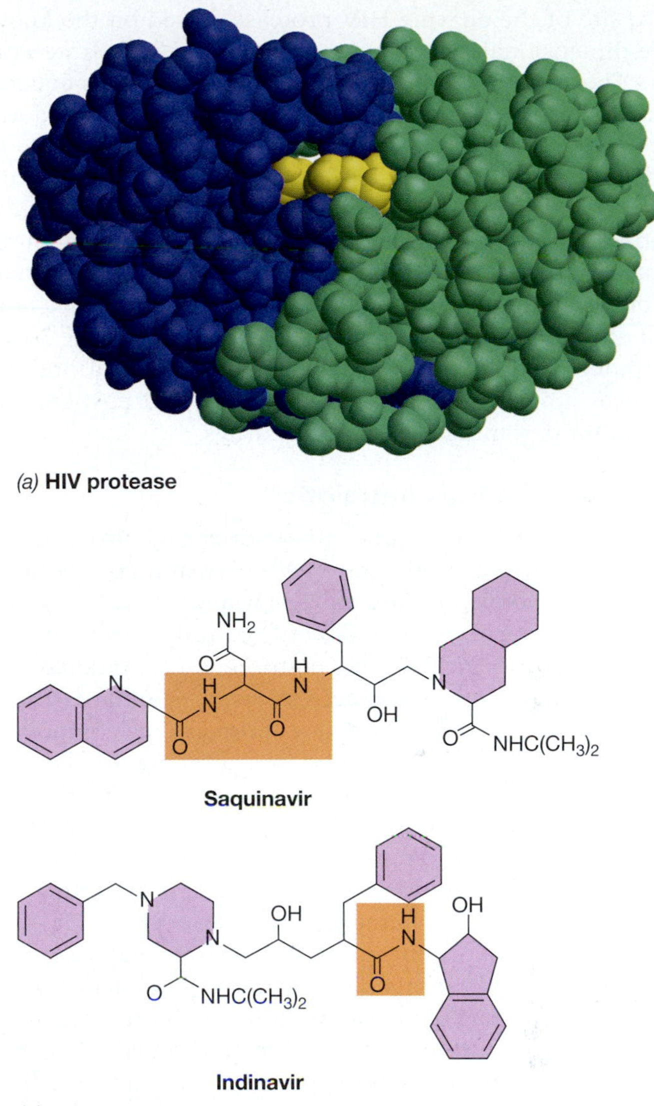

Figure 27.31 Computer-generated anti-HIV drugs. *(a)* The HIV protease homodimer. Individual polypeptide chains are shown in green and blue. A peptide (yellow) is bound in the active site. HIV protease cleaves an HIV precursor protein, a necessary step in virus maturation (Figure 19.26). Blocking of the protease site by the bound peptide inhibits precursor processing and HIV maturation. This structure is derived from information in the Protein Data Bank. *(b)* These anti-HIV drugs are peptide analogs called protease inhibitors that were designed by computer to block the active site of HIV protease. The areas highlighted in orange show the regions analogous to peptide bonds in proteins.

The substituted sites would yield 5 × 5 × 5 × 5 (five derivatives at each of four sites), or 625 different tetracycline derivatives from only six different reagents, all in a few hours time! These compounds would then be assayed for *in vitro* biological activity on different test organisms using automated testing methods for antibiotic susceptibility. The automated synthesis and screening processes dramatically shorten drug discovery time and increase the number of new candidate drugs by a factor of 10 or more each year.

According to pharmaceutical industry estimates, about 7 million candidate compounds must be screened to yield a single useful clinical drug. Drugs effective in the laboratory must then be tested for efficacy and toxicity in animals and finally in clinical trials in humans. Animal testing requires multiple trials over several years to ensure that the candidate drug is both effective and safe. Clinical trials in humans to check efficacy and safety take additional years for each drug. Each year, the pharmaceutical industry spends up to $4 billion on new antimicrobial drug development. The total time for discovery and development for each drug typically takes 10–25 years before it is approved for clinical use. The cost of discovery and development, from the laboratory through clinical trials, is estimated at over $500 million for each new drug approved for human use. This is a major reason why pharmaceutical drugs are so expensive.

Computer Drug Design

Novel antimicrobial compounds are much more difficult to identify than analogs of existing drugs because new antimicrobial compounds must work at unique sites in metabolism and biosynthesis or be structurally dissimilar to existing compounds to avoid inducing known resistance mechanisms.

Recent advances in computer technology and structural biology now make it possible to design a drug to interact with specific microbial structures. Thus, drug discovery can now begin at the computer, where new drugs can be rapidly synthesized and tested for binding and efficacy in the computer environment at relatively low cost.

One of the most dramatic successes in computer-directed drug design is the development of *saquinavir*, a protease inhibitor that is used to slow the growth of the human immunodeficiency virus (HIV) in infected individuals (**Figure 27.31**). Saquinavir was designed by computer to bind the active site of the enzyme HIV protease, based on the known three-dimensional structure of the protease–substrate complex. The HIV protease normally cleaves a virus-encoded precursor protein to produce the mature viral core and activate the reverse transcriptase enzyme necessary for replication (Section 19.15). Saquinavir is a peptide analog of the HIV precursor protein and displaces the protein, inhibiting virus maturation and slowing its growth in the human host. A number of other computer-designed protease inhibitors are in use as chemotherapeutic drugs for the treatment of AIDS (Table 27.5, Figure 27.31, and Figure 34.42). Computer design and testing based on structural and biochemical modeling is a practical, rapid, and cost-effective method for designing antimicrobial drugs.

Natural Products as Antibiotics

As the first antibiotics were discovered and brought into clinical use in the 1930s and 1940s, researchers developed standard methods to isolate more new antibiotics. Candidate drugs were routinely isolated from natural sources such as *Streptomyces* or *Penicillium* cultures and systematically screened for antimicrobial activity using standard MIC or agar diffusion methods to find new antimicrobial compounds. As time passed, the yield from these traditional methods decreased, supplanted by higher yields from the combinatorial chemistry and computer design methods discussed above. Most of the effective natural antibiotics produced at reasonable levels by antibiotic-producing microorganisms had already been isolated. Remaining effective antibiotics, presumably present in concentrations so low that they were ineffective against test organisms, could not be identified.

Platensimycin (Figure 27.25), however, is an exception to this rule. This antibiotic was recently discovered using a modification of direct natural product screening methods. Platensimycin represents a new class of antimicrobials that selectively inhibits bacterial lipid biosynthesis and is especially active against gram-positive pathogens, including MRSA, VISA, and VRE (Section 27.9). A key feature in the discovery of platensimycin was its selection using a novel method that may have broad applications for targeted drug discovery. To select an agent for a defined target, in this case, an enzyme in the lipid synthesis pathway of gram-positive bacteria, scientists introduced a defect in the β-ketoacyl-(acyl-carrier protein) synthase I/II (FabF/B) gene in *S. aureus* by using a strain expressing antisense FabF RNA (Section 9.14 for discussion of antisense RNA). The gene-specific antisense RNA decreased expression of FabF, reducing fatty acid synthesis and increasing the sensitivity of the crippled *S. aureus* strain to antibiotics that inhibit fatty acid synthesis. By screening 250,000 natural product extracts from 83,000 strains of potential antibiotic producers, the scientists were able to identify and isolate platensimycin from a soil microorganism, *Steptomyces platensis*. Although the screening of large numbers of strains is a huge task, the method identifies target-specific antibiotics present in low concentrations. This strategy is applicable to virtually any target for which the gene sequence (and, hence, the corresponding antisense RNA sequence) is known.

Drug Combinations

The efficacy of some antibiotics can be retained if they are given with compounds that inhibit antibiotic resistance. Several β-lactam antibiotics can be combined with β-lactamase inhibitors to preserve antibiotic activity. For example, the broad-spectrum β-lactam antibiotic ampicillin (Figure 27.20) can be mixed with sulbactam, a β-lactamase inhibitor. The inhibitor binds β-lactamase irreversibly, preventing degradation of the ampicillin and permitting it to disrupt cell wall formation in the affected cell. This combination preserves the effectiveness of the β-lactamase-sensitive ampicillin against β-lactamase producers such as staphylococci and certain gram-negative pathogens. Likewise, we have already mentioned the use of sulfamethoxazole-trimethoprim, a mixture of two folic acid synthesis inhibitors, to prevent the loss of efficacy through mutation and selection for resistance (Section 27.6).

Bacteriophages

Bacteriophages are viruses that infect bacteria (Sections 19.1–19.5). *Bacteriophage therapy* has been used on a limited basis for over 80 years to treat infections in animals and, in a few instances, in humans. Phages interact with individual bacterial cell surface components and show specificity for particular bacterial species. The attached phage enters the cell, replicates, and kills the bacterial host in the process. The efficacy and efficiency of these agents for human treatment is largely untested and somewhat controversial, although clinical trials of several products are ongoing. However, because bacteria can acquire resistance to a phage infection through mutations that alter receptors or reduce the susceptibility of the cell wall to phage enzymes, bacteriophage therapy will likely be susceptible to resistance, just as for most chemical antimicrobial agents.

27.13 MiniReview

New antimicrobial compounds are constantly being discovered and developed to deal with drug-resistant pathogens and to enhance our ability to treat infectious diseases. Analogs of existing drugs are often synthesized and used as next-generation antimicrobial compounds. Computer drug design is an important tool for drug discovery.

- Explain the advantages and disadvantages of developing new drugs based on existing drug analogs.
- How can computer drug design save time and money in the search for new drugs?
- Explain the use of antisense RNA for drug discovery.

Review of Key Terms

Aminoglycoside an antibiotic such as streptomycin, containing amino sugars linked by glycosidic bonds

Antibiotic chemical substance produced by a microorganism that kills or inhibits the growth of another microorganism

Antimicrobial drug resistance the acquired ability of a microorganism to grow in the presence of an antimicrobial drug to which the microorganism is usually susceptible

Antimicrobial agent a chemical compound that kills or inhibits the growth of microorganisms

Antiseptic (germicide) chemical agent that kills or inhibits growth of microorganisms and is sufficiently nontoxic to be applied to living tissues

Autoclave a sterilizer that destroys microorganisms with temperature and steam under pressure

Beta (β)-lactam antibiotic an antibiotic, including penicillin, that contains the four-membered heterocyclic β-lactam ring

Bacteriocidal agent an agent that kills bacteria

Bacteriostatic agent an agent that inhibits bacterial growth

Broad-spectrum antibiotic an antibiotic that acts on both gram-positive and gram-negative *Bacteria*

Decontamination treatment that renders an object or inanimate surface safe to handle

Disinfectant an antimicrobial agent used only on inanimate objects

Disinfection the elimination of microorganisms from inanimate objects or surfaces

Fungicidal agent an agent that kills fungi

Fungistatic agent an agent that inhibits fungal growth

Fusion inhibitor a peptide that blocks the fusion of viral and target cytoplasmic membranes

Germicide (antiseptic) chemical agent that kills or inhibits growth of microorganisms and is sufficiently nontoxic to be applied to living tissues

Growth factor analog a chemical agent that is related to and blocks the uptake of a growth factor

HEPA filter a *h*igh *e*fficiency *p*articulate *a*ir filter that removes particles, including microbes, from intake or exhaust air flow

Inhibition the reduction of microbial growth because of a decrease in the number of organisms present or alterations in the microbial environment

Interferon a cytokine protein produced by virus-infected cells that induces signal transduction in nearby cells, resulting in transcription of antiviral genes and expression of antiviral proteins

Minimum inhibitory concentration (MIC) the minimum concentration of a substance necessary to prevent microbial growth

Nonnucleoside reverse transcriptase inhibitor (NNRTI) nonnucleoside analog used to inhibit viral reverse transcriptase

Nucleoside reverse transcriptase inhibitor (NRTI) nucleoside analog used to inhibit viral reverse transcriptase

Pasteurization reduction of the microbial load in heat-sensitive liquids to kill disease-producing microorganisms and reduce the number of spoilage microorganisms

Penicillin a class of antibiotics that inhibit bacterial cell wall synthesis, characterized by a β-lactam ring

Protease inhibitor an inhibitor of a viral protease

Quinolone a synthetic antibacterial compound that interacts with DNA gyrase and prevents supercoiling of bacterial DNA

Sanitizer an agent that reduces microorganisms to a safe level, but may not eliminate them

Selective toxicity the ability of a compound to inhibit or kill pathogenic microorganisms without adversely affecting the host

Sterilant (sterilizer) (sporicide) a chemical agent that destroys all forms of microbial life

Sterilization the killing or removal of all living organisms and their viruses from a growth medium

Tetracycline an antibiotic characterized by the four-ring naphthacene structure

Viricidal agent an agent that stops viral replication and activity

Viristatic agent an agent that inhibits viral replication

UNIT 6

Review Questions

1. Why is the decimal reduction time (*D*) important in heat sterilization? How would the presence of bacterial endospores affect *D* (Section 27.1)?
2. Describe the effects of a lethal dose of ionizing radiation at the molecular level (Section 27.2).
3. What are the principal advantages of membrane filters? Of depth filters (Section 27.3)?
4. Why are Nucleopore filters used for isolating specimens for microscopy (Section 27.4)?
5. Describe the procedure for obtaining the minimum inhibitory concentration (MIC) for a chemical that is bacteriocidal for *Escherichia coli* (Section 27.5).
6. Contrast the action of disinfectants and antiseptics. Why can't disinfectants normally be used on living tissue (Section 27.6)?
7. Growth factor analogs are generally distinguished from antibiotics by a single important criterion. Explain (Section 27.7).
8. Describe the mode of action that characterizes a β-lactam antibiotic. Why are these antibiotics generally more effective against gram-positive bacteria than against gram-negative bacteria (Section 27.8)?
9. Distinguish between the modes of action of at least three of the protein synthesis-inhibiting antibiotics (Section 27.9).
10. Why do antiviral drugs generally exhibit host toxicity (Section 27.10)?

11. Identify the targets that allow selective toxicity of chemotherapeutic agents in fungi (Section 27.11).
12. Identify six mechanisms responsible for antibiotic resistance (Section 27.12).
13. Explain how application of antisense RNA methods can extend traditional natural product selection methods for antibiotic discovery (Section 27.13).

Application Questions

1. What are some potential drawbacks to the use of radiation in food preservation? Do you think these drawbacks could be manifested as health hazards? Why or why not? How would you distinguish between radiation-damaged and radiation-contaminated food?
2. Filtration is an acceptable means of pasteurization for some liquids. Design a filtration system for pasteurization of a heat-sensitive liquid. For a liquid of your choice, identify the advantages and disadvantages of a filtration system over a heat pasteurization system. Explain in terms of product quality, shelf life, and price.
3. Although growth factor analogs may inhibit microbial metabolism, only a few of these agents are useful in practice. Many growth factor analogs, including some in wide use, such as azidothymidine, exhibit significant host cell toxicity. Describe a growth factor analog that is effective and has low toxicity for host cells. Why is the toxicity low for the agent you chose? Also describe a growth factor analog that is effective against an infectious disease but exhibits toxicity for host cells. Why might a toxic agent such as AZT still be used in certain situations to treat infectious diseases? What precautions would you take to limit the toxic effects of such a drug while maximizing the therapeutic activity? Explain your answer.
4. Although many antibiotics demonstrate clear selective toxicity for *Bacteria*, many groups of *Bacteria* are innately resistant to their effects. Indicate why gram-negative *Bacteria* are resistant to the effects of most, but not all, antibiotics. Further explain why some antibiotics are useful against these organisms.
5. List the features of an ideal antiviral drug, especially with regard to selective toxicity. Do such drugs exist? What factors might limit the use of such a drug?
6. Like viruses, fungi present special chemotherapeutic problems. Explain the problems inherent in chemotherapy of both groups and explain whether or not you agree with the preceding statement. Give specific examples and suggest at least one group of chemotherapeutic agents that might target both types of infectious agents.
7. Explain the genetic basis of acquired resistance to β-lactam antibiotics in *Staphylococcus aureus*. Design experiments to reverse resistance to the β-lactam antibiotics. Do you think this can be done in the laboratory? Can your experiment be applied "in the field" to promote deselection of antibiotic-resistant organisms?
8. Design experiments to examine microorganisms for production of novel antibiotics. Which group or groups of microorganisms would you choose to screen for antibiotic production? Where could you find and isolate these organisms in a natural environment? What advantage, if any, would the production of an antibiotic provide for these organisms in nature? What *in vitro* methods would you use to test the efficacy of your potential new antibiotics? How might you increase the sensitivity of assays for natural products? Why are members of the genus *Streptomyces* still productive sources of novel antibiotics?

28

Microbial Interactions with Humans

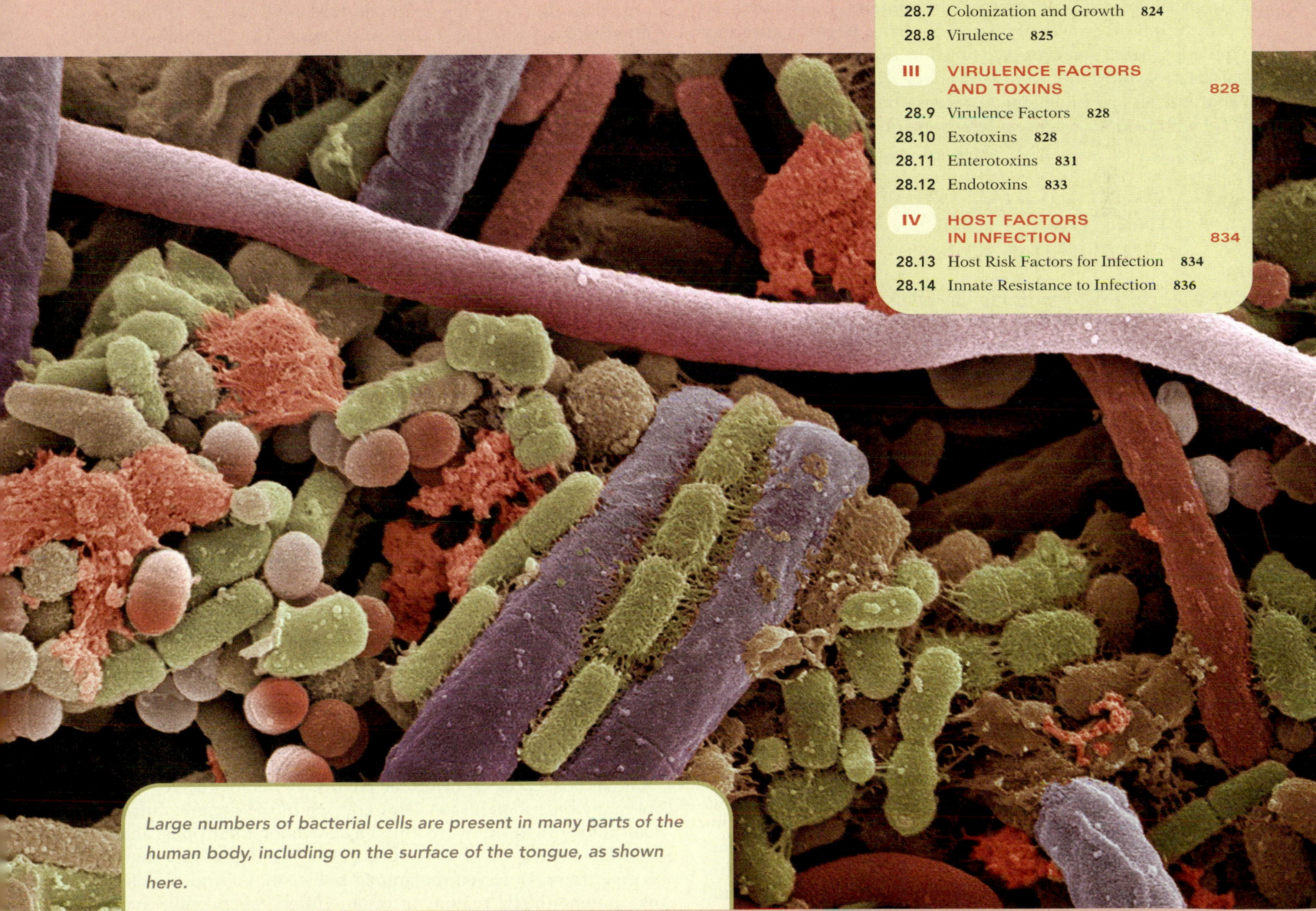

Large numbers of bacterial cells are present in many parts of the human body, including on the surface of the tongue, as shown here.

The human body has an extensive population of microorganisms, primarily bacteria, on the skin and the mucous membranes lining the mouth, gut, and excretory and reproductive systems. These microorganisms are often beneficial and sometimes necessary to maintain good health. However, other microorganisms called *pathogens* colonize, invade, and damage the human body through direct and indirect means. This is the process of infectious disease. Pathogens use several strategies to gain access to nutrients in a host. These strategies include the production of specialized attachment structures, unique growth factors, invasive enzymes, and potent biological toxins. These factors often lead to damage and occasionally death of the host.

In Chapter 27, we discussed countermeasures that destroy or inhibit growth of microorganisms. Here we introduce the nonspecific defenses developed by our bodies to suppress or destroy most microbial invaders. Nonspecific physical, anatomical, and biochemical defense mechanisms make microbial infectious disease a relatively infrequent event.

I BENEFICIAL MICROBIAL INTERACTIONS WITH HUMANS

After a brief overview of human–microbial interactions, we discuss microorganisms that inhabit the healthy human body and contribute to overall good health under normal circumstances.

28.1 Overview of Human–Microbial Interactions

Through normal everyday activities, the human body is exposed to countless microorganisms in the environment. In addition, hundreds of species and countless individual microbial cells, collectively referred to as the **normal microbial flora**, grow on or in the human body. Most, but not all, microorganisms are benign; a few contribute directly to our health, and even fewer pose direct threats to health.

Colonization by Microorganisms

Mammals *in utero* develop in a sterile environment and have no exposure to microorganisms. Starting with the birth process, **colonization**, growth of a microorganism after it has gained access to host tissues, begins as animals are exposed to microorganisms. The skin surfaces are readily colonized by many species. Likewise, the oral cavity and gastrointestinal tract acquire microorganisms through feeding and exposure to the mother's body, which, along with other environmental sources, initiates colonization of the skin, oral cavity, upper respiratory tract, and gastrointestinal tract. Different populations of microorganisms colonize individuals in different localities and at different times. For example, *Escherichia coli,* a normal inhabitant of the human and animal gut, colonizes the guts of infants in developing countries within several days after birth. Infants in developed countries, however, typically do not acquire *E. coli* for several months; the first microorganisms to colonize the gut of these infants would more typically be *Staphylococcus aureus* and other microorganisms associated with the skin. Genetic factors also play a role. Thus, the normal microbial flora is highly dependent on the conditions to which an individual is exposed. The normal flora is highly diverse in each individual and may differ significantly between individuals, even in a given population.

Pathogens

A **host** is an organism that harbors a **parasite**, another organism that lives on or in the host and causes damage. Microbial parasites are called **pathogens**. The outcome of a host–parasite relationship depends on **pathogenicity**, the ability of a parasite to inflict damage on the host. Pathogenicity differs considerably among potential pathogens, as does the resistance or susceptibility of the host to the pathogen. An **opportunistic pathogen** causes disease only in the absence of normal host resistance.

Pathogenicity varies markedly for individual pathogens. The quantitative measure of pathogenicity is called **virulence**. Virulence can be expressed quantitatively as the cell number that elicits disease in a host within a given time period. Neither the virulence of the pathogen nor the relative resistance of the host is a constant factor. The host–parasite interaction is a dynamic relationship between the two organisms, influenced by changing conditions in the pathogen, the host, and the environment.

Infection and Disease

Infection refers to any situation in which a microorganism is established and growing in a host, whether or not the host is harmed. **Disease** is damage or injury to the host that impairs host function. Infection is not synonymous with disease because growth of a microorganism on a host does not always cause host damage. Thus, species of the normal microbial flora have infected the host, but seldom cause disease. However, the normal flora sometimes cause disease if host resistance is compromised, as happens in diseases such as cancer and acquired immune deficiency syndrome (AIDS) (∞ Section 34.15).

Host–Parasite Interactions

Animal hosts provide favorable environments for the growth of many microorganisms. They are rich in the organic nutrients and growth factors required by chemoorganotrophs, and provide conditions of controlled pH, osmotic pressure, and temperature. However, the animal body is not a uniform environment. Each region or organ differs chemically and physically from others and thus provides a selective environment where the growth of certain microorganisms is favored. For example, the skin, respiratory tract, and gastrointestinal tract provide selective chemical and physical environments that support the growth of a highly diverse microflora. The

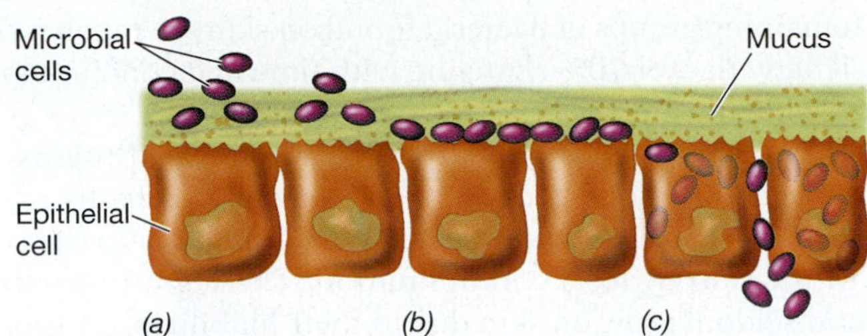

Figure 28.1 Bacterial interactions with mucous membranes. *(a)* Loose association. *(b)* Adhesion. *(c)* Invasion into submucosal epithelial cells.

relatively dry environment of the skin favors the growth of organisms that resist dehydration, such as the gram-positive bacterium *Staphylococcus aureus* (Section 34.10); the highly oxygenated environment of the lungs favors the growth of the obligately aerobic *Mycobacterium tuberculosis* (Section 34.5); and the anoxic environment of the large intestine supports growth of obligately anaerobic bacteria such as *Clostridium* and *Bacteroides* (Sections 16.2 and 16.12). Animals also possess defense mechanisms that collectively prevent or inhibit microbial invasion and growth. The microorganisms that successfully colonize the host have circumvented these defense mechanisms.

The Infection Process

Infections frequently begin at sites in the animal's **mucous membranes**. Mucous membranes consist of single or multiple layers of *epithelial cells*, tightly packed cells that interface with the external environment. They are found throughout the body, lining the urogenital, respiratory, and gastrointestinal tracts. Mucous membranes are frequently coated with a protective layer of viscous soluble glycoproteins called **mucus**. Microorganisms that contact host tissues at mucous membranes may associate loosely with the mucosal surface and are usually swept away by physical processes. Microorganisms may also adhere more strongly to the epithelial surface as a result of specific cell–cell recognition between pathogen and host. Tissue infection may follow, breaching the mucosal barrier and allowing the microorganism to invade deeper into submucosal tissues (**Figure 28.1**).

Microorganisms are almost always found on surfaces of the body exposed to the environment, such as the skin, and even on the mucosal surfaces of the oral cavity, respiratory tract, intestinal tract, and urogenital tract. They are not normally found on or in the internal organs or in the blood, lymph, or nervous systems of the body. The growth of microorganisms in these normally sterile environments indicates serious infectious disease.

Table 28.1 shows some of the major types of microorganisms normally found in association with body surfaces. Mucosal surfaces have a diverse microflora because they offer a sheltered, moist environment and a large overall surface area. For example, the specialized function of a mucosal organ such as the small intestine has a surface area of about 400 m^2 available for nutrient transport, and this entire surface is a potential site for microbial growth.

Table 28.1 Representative genera of microorganisms in the normal flora of humans

Anatomical site	*Genera or major groups*[a]
Skin	*Acinetobacter, Corynebacterium, Enterobacter, Klebsiella, Malassezia (f), Micrococcus, Pityrosporum (f), Propionibacterium, Proteus, Pseudomonas, Staphylococcus, Streptococcus*
Mouth	*Streptococcus, Lactobacillus, Fusobacterium, Veillonella, Corynebacterium, Neisseria, Actinomyces, Geotrichum (f), Candida (f), Capnocytophaga, Eikenella, Prevotella,* Spirochetes (several genera)
Respiratory tract	*Streptococcus, Staphylococcus, Corynebacterium, Neisseria, Haemophilus*
Gastrointestinal tract	*Lactobacillus, Streptococcus, Bacteroides, Bifidobacterium, Eubacterium, Peptococcus, Peptostreptococcus, Ruminococcus, Clostridium, Escherichia, Klebsiella, Proteus, Enterococcus, Staphylococcus Methanobrevibacter, Gram-positive bacteria, Proteobacteria, Actinobacteria, Fusobacteria*
Urogenital tract	*Escherichia, Klebsiella, Proteus, Neisseria, Lactobacillus, Corynebacterium, Staphylococcus, Candida (f), Prevotella, Clostridium, Peptostreptococcus, Ureaplasma, Mycoplasma, Mycobacterium, Streptococcus, Torulopsis (f)*

[a]This list is not meant to be exhaustive, and not all of these organisms are found in every individual. Some organisms are more prevalent at certain ages (adults vs. children). Distribution may also vary between sexes. Most of these organisms can be opportunistic pathogens under certain conditions. Several genera can be found in more than one body area. (f), fungi.

28.1 MiniReview

The animal body is a favorable environment for the growth of microorganisms, most of which do no harm. Microorganisms that cause harm are called pathogens. Pathogen growth initiated on host surfaces such as mucous membranes may result in infection and disease. The ability of a microorganism to cause or prevent disease is influenced by complex interactions between the microorganism and the host.

- Distinguish between infection and disease.
- Why might one area of the body be more suitable for microbial growth than another?

28.2 Normal Microbial Flora of the Skin

An average adult human has about 2 m^2 of skin surface that varies greatly in chemical composition and moisture content. **Figure 28.2** shows the anatomy of the skin and regions in which microorganisms may live. The skin surface (epidermis) is not a favorable place for abundant microbial growth, as it is subject to periodic drying.

UNIT 6

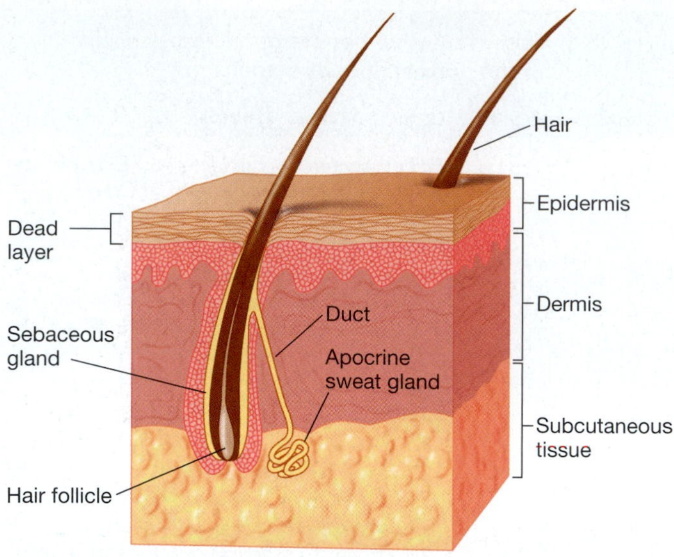

Figure 28.2 The human skin. Microorganisms are associated primarily with the sweat ducts and the hair follicles.

Most skin microorganisms are associated directly or indirectly with the apocrine (sweat) glands. These are secretory glands found mainly in the underarms, genital regions, the nipples, and the umbilicus. Inactive in childhood, they become fully functional at puberty. Microbial populations thrive on the surface of the skin in these warm, humid places in contrast to the poor growth observed on smooth, dry skin surfaces. Underarm odor develops as a result of bacterial activity in the apocrine secretions; aseptically collected apocrine secretions are odorless, but develop odor on inoculation with bacteria. Similarly, each hair follicle is associated with a sebaceous gland, a gland that secretes a lubricant fluid. Hair follicles provide an attractive habitat for microorganisms in the area just below the surface of the skin. The secretions of the skin glands are rich in microbial nutrients such as urea, amino acids, salts, lactic acid, and lipids. The pH of human secretions is almost always acidic, the usual range being between pH 4 and 6.

The Skin Microflora

The normal flora of the skin consist of either resident or transient populations of bacteria and fungi, mainly yeasts. Certain genera are highly conserved in healthy individuals, while others change over time. In all, over 180 species of *Bacteria* and several species of fungi are found on the skin when samples from many individuals are gathered. Samplings of individuals done several months apart indicate that certain members of the normal flora are extremely stable. The most common and stable resident microorganisms are gram-positive *Bacteria*, generally restricted to a few genera, including species of *Streptococcus* and *Staphylococcus*; various species of *Corynebacterium*; and *Propionibacterium*, including *Propionibacterium acnes*, which contributes to a skin condition called *acne* (Table 28.1). These four genera account for over one-half of all species found. The remaining groups of bacteria found on skin are much more transient, with over 70% changing with time and conditions on the individuals sampled.

Gram-negative *Bacteria* are occasional constituents of the normal skin flora because such intestinal organisms as *Escherichia coli* are continually being inoculated onto the surface of the skin by fecal contamination. These gram-negative *Bacteria* seldom grow on skin due to their inability to compete with gram-positive organisms that are better adapted to the dry conditions. The gram-negative rod *Acinetobacter*, however, is an exception and commonly colonizes skin.

Malassezia spp. are the most common fungi found on skin. At least five species of this yeast are typically found in healthy individuals. The lipophilic yeast *Pityrosporum ovalis* is occasionally found on the scalp. In the absence of host resistance, as in patients with AIDS or in the absence of normal bacterial flora, other yeasts such as *Candida* and other fungi sometimes grow and cause serious skin infections.

Although the resident microflora remains more or less constant, various environmental and host factors may influence its composition. (1) The *weather* may cause an increase in skin temperature and moisture, which increases the density of the skin microflora. (2) The *age* of the host has an effect; young children have a more varied microflora and carry more potentially pathogenic gram-negative *Bacteria* than do adults. (3) *Personal hygiene* influences the resident microflora; individuals with poor hygiene usually have higher microbial population densities on their skin. Organisms that cannot survive on the skin generally succumb from either the skin's low moisture content or low pH (due to organic acid content).

28.2 MiniReview

The skin is a generally dry, acidic environment that does not support the growth of most microorganisms. However, moist areas, especially around sweat glands, are colonized by gram-positive *Bacteria* and other skin normal flora. Environmental and host factors influence the quantity and makeup of the normal skin microflora.

- Compare the surface area of the skin to that of the mucosal tissues.
- Describe the properties of microorganisms that grow well on the skin.

28.3 Normal Microbial Flora of the Oral Cavity

The oral cavity is a complex, heterogeneous microbial habitat. Saliva contains microbial nutrients, but it is not an especially good growth medium because the nutrients are present in low concentration and saliva contains antibacterial substances. For example, saliva contains lysozyme, an enzyme that cleaves glycosidic linkages in peptidoglycan present in the bacterial cell wall, weakening the wall and causing cell lysis

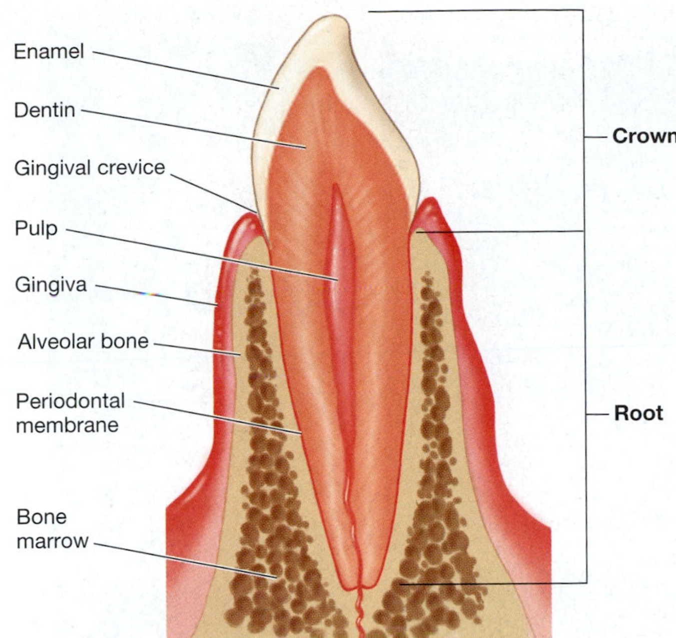

Figure 28.3 Section through a tooth. The diagram shows the tooth architecture and the surrounding tissues that anchor the tooth in the gum.

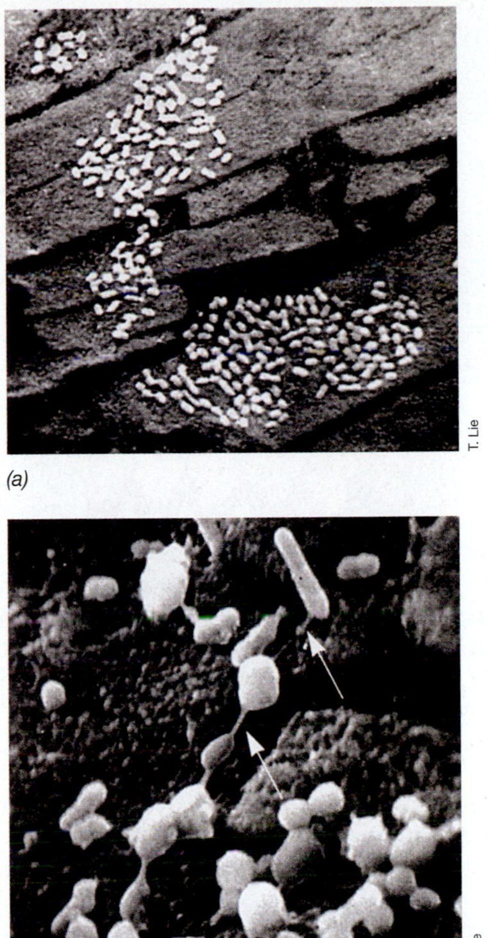

Figure 28.4 Microcolonies of bacteria. *(a)* The colonies are growing on a model tooth surface inserted into the mouth for 6 h. *(b)* Higher magnification of the preparation in *(a)*. Note the diverse morphology of the organisms present and the slime layer (arrows) holding the organisms together.

(∞ Section 4.6). Lactoperoxidase, an enzyme in both milk and saliva, kills bacteria by a reaction in which singlet oxygen is generated (∞ Section 6.18). Despite the activity of these antibacterial substances, food particles and cell debris provide high concentrations of nutrients near surfaces such as teeth and gums, creating favorable conditions for extensive local microbial growth, tissue damage, and disease.

The Teeth and Dental Plaque

The tooth consists of a mineral matrix of calcium phosphate crystals (enamel) surrounding living tooth tissue (dentin and pulp) (**Figure 28.3**). Bacteria found in the mouth during the first year of life (when teeth are absent) are predominantly aerotolerant anaerobes such as streptococci and lactobacilli. However, other bacteria, including some aerobes, occur in small numbers. When the teeth appear, the balance of the microflora shift toward anaerobes that are specifically adapted to growth on surfaces of the teeth and in the gingival crevices.

Bacterial colonization of tooth surfaces begins with the attachment of single bacterial cells. Even on a freshly cleaned tooth surface, acidic glycoproteins from the saliva form a thin organic film several micrometers thick. This film provides an attachment site for bacterial microcolonies (**Figure 28.4**). Streptococci (primarily *Streptococcus sanguis*, *S. sobrinus*, *S. mutans*, and *S. mitis*) can then colonize the glycoprotein film. Extensive growth of these organisms results in a thick bacterial layer called **dental plaque** (**Figures 28.5** and **28.6**).

If plaque continues to form, filamentous anaerobes such as *Fusobacterium* species begin to grow. The filamentous bacteria embed in the matrix formed by the streptococci and extend perpendicular to the tooth surface, making an ever-thicker bacterial layer. Associated with the filamentous bacteria are spirochetes such as *Borrelia* species (∞ Section 16.16), gram-positive rods, and gram-negative cocci. In heavy plaque, filamentous obligately anaerobic organisms such as *Actinomyces* may predominate. Thus, dental plaque can be considered a mixed-culture biofilm (∞ Sections 23.4 and 23.5), consisting of a relatively thick layer of bacteria from several different genera as well as accumulated bacterial products.

The anaerobic nature of the oral flora may seem surprising considering the intake of oxygen through the mouth. However, anoxia develops due to the metabolic activities of facultative bacteria growing on organic materials at the tooth surface. The plaque buildup produces a dense matrix that decreases oxygen diffusion to the tooth surface, forming an anoxic microenvironment. The microbial populations within

UNIT 6

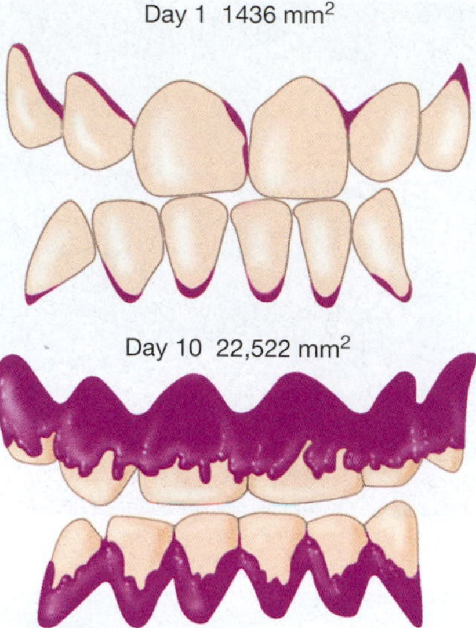

Figure 28.5 Distribution of dental plaque. Plaque is revealed by use of a disclosing agent on unbrushed teeth after 1 day (top) and 10 days (bottom). The stained areas indicate plaque. Plaque buildup starts near the gum line, beginning directly adjacent to the mucous membranes of the gingiva.

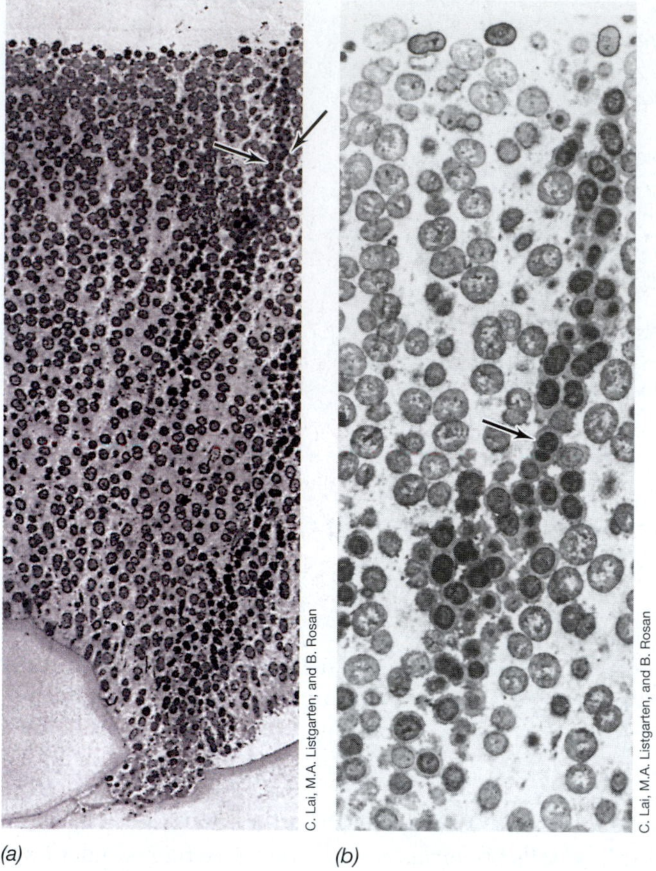

Figure 28.6 Electron micrographs of thin sections of dental plaque. The bottom of the photograph is the base of the plaque; the top is the portion exposed to the oral cavity. The total thickness of the plaque layer shown is about 50 μm. (*a*) Low-magnification electron micrograph. Organisms are predominantly streptococci. *Streptococcus sobrinus*, labeled by an antibody-microchemical technique, appears darker than the rest. *S. sobrinus* cells are seen as two distinct chains (arrows). (*b*) Higher-magnification electron micrograph showing the region with *S. sobrinus* cells (dark, arrow). Note the extensive slime layer surrounding the *S. sobrinus* cells. Individual cells are about 1 μm in diameter.

dental plaque exist in a microenvironment of their own making and maintain themselves in the face of wide variations in the conditions in the macroenvironment of the oral cavity.

Dental Caries

As dental plaque accumulates, the resident microflora produce locally high concentrations of organic acids that cause decalcification of the tooth enamel (Figure 28.3), resulting in **dental caries** (tooth decay). Thus, dental caries is an infectious disease. The smooth surfaces of the teeth are relatively easy to clean and resist decay. The tooth surfaces in and near the gingival crevice, however, can retain food particles and are the sites where dental caries typically begins.

Diets high in sucrose (table sugar) promote dental caries. Lactic acid bacteria ferment sugars to lactic acid. The lactic acid dissolves some of the calcium phosphate in localized areas, and proteolysis of the supporting matrix occurs through the action of bacterial proteolytic enzymes. Bacterial cells slowly penetrate further into the decomposing matrix. The structure of the calcified tooth tissue also plays a role in the extent of dental caries. For example, incorporation of fluoride into the calcium phosphate crystal tooth matrix increases resistance to acid decalcification. Consequently, fluorides in drinking water and dentifrices inhibit tooth decay.

Two bacteria implicated in dental caries are *Streptococcus sobrinus* and *Streptococcus mutans,* both lactic acid bacteria. *S. sobrinus* is probably the primary organism causing decay of smooth surfaces because of its specific affinity for salivary glycoproteins found on smooth tooth surfaces (Figure 28.6). *S. mutans,* found predominantly in crevices and small fissures, produces dextran, a strongly adhesive polysaccharide that it uses to attach to tooth surfaces (**Figure 28.7**). *S. mutans* produces dextran only when sucrose is present by activity of the enzyme dextransucrase:

$$n \text{ Sucrose} \xrightarrow{\text{Dextransucrase}} \text{Dextran } (n \text{ Glucose}) + n \text{ Fructose}$$

Susceptibility to tooth decay varies and is affected by genetic traits in the individual as well as by diet and other extraneous factors. For example, sucrose, highly cariogenic because it is a substrate for dextransucrase, is part of the diet of most individuals in developed countries. Studies of the distribution of the cariogenic oral streptococci show a direct correlation between the presence of *S. mutans* and *S. sobrinus* and the extent of dental caries. In the United States and Western Europe, 80–90% of all individuals are infected by *S. mutans,* and dental caries is nearly universal. By contrast, *S. mutans* is absent from the plaque of Tanzanian children,

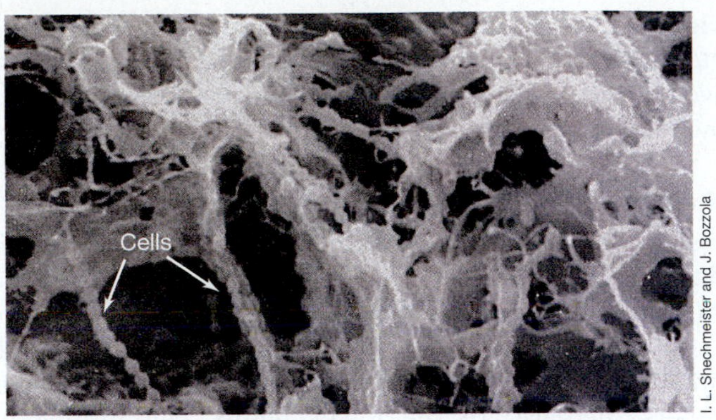

Figure 28.7 Scanning electron micrograph of the cariogenic bacterium *Streptococcus mutans*. The sticky dextran material holds the cells together as filaments. Individual cells are about 1 μm in diameter.

and dental caries does not occur, presumably because sucrose is almost completely absent from their diets.

Microorganisms in the mouth can also cause other infections. The areas along the periodontal membrane at or below the gingival crevice (periodontal pockets) (Figure 28.3) can be infected with microorganisms, causing inflammation of the gum tissues (gingivitis) leading to tissue and bone-destroying periodontal disease. Some of the genera involved include fusiform bacteria (long, thin, gram-negative rods with tapering ends) such as the facultative aerobe *Capnocytophaga*. The aerobe *Rothia*, and even strictly anaerobic methanogens, such as *Methanobrevibacter* (*Archaea*), may also be present.

28.3 MiniReview

Bacteria can produce adherent substances and growth on tooth surfaces, typically resulting in mixed-culture biofilms called plaque. Acid produced by microorganisms in plaque damages tooth surfaces, resulting in dental caries. Further infection can result in periodontal disease.

- How do anaerobic microorganisms become established in the mouth?
- Is dental caries an infectious disease? Give at least one reason for your answer.
- Identify the contribution of the lactic acid bacteria to tooth decay.

28.4 Normal Microbial Flora of the Gastrointestinal Tract

The human gastrointestinal tract consists of the stomach, small intestine, and large intestine (**Figure 28.8**). The gastrointestinal tract is responsible for digestion of food, absorption of nutrients, and the production of nutrients by the indigenous

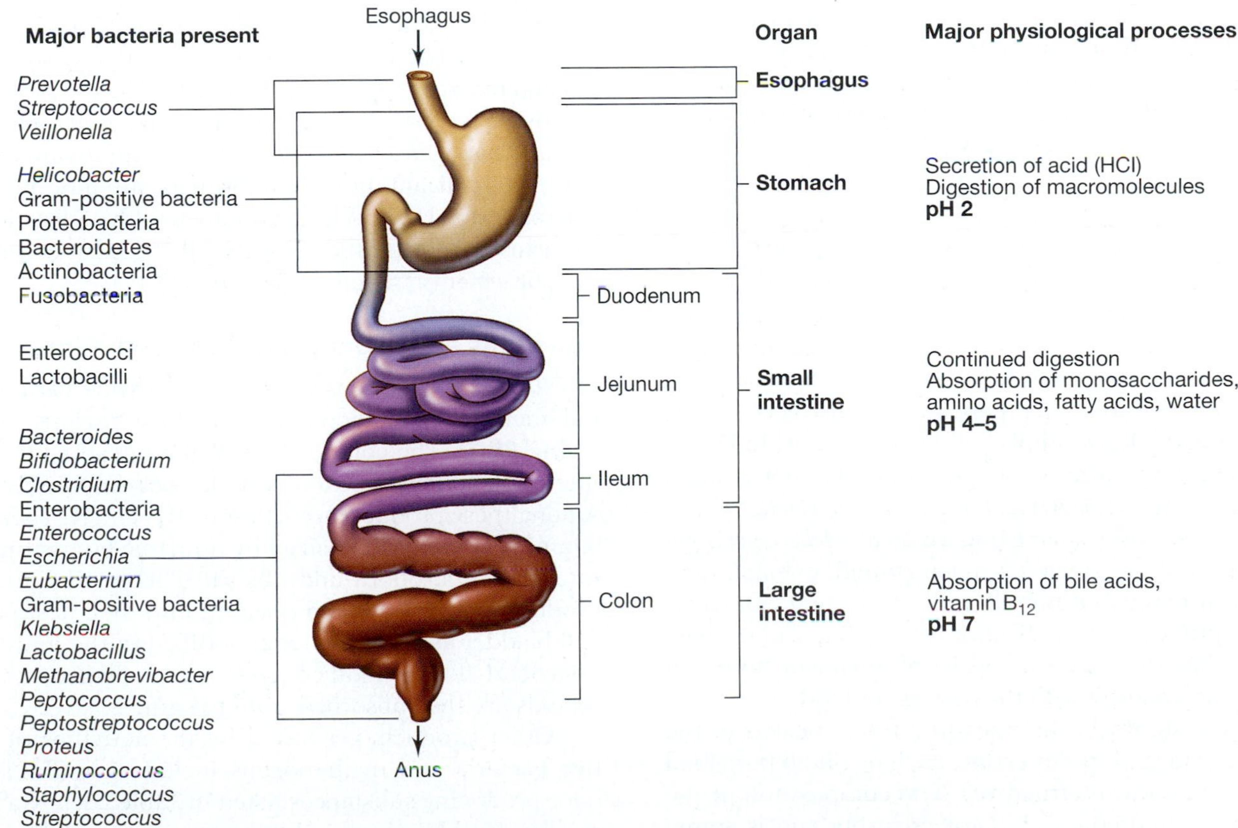

Figure 28.8 The human gastrointestinal tract. The distribution of representative nonpathogenic microorganisms often found in healthy adults.

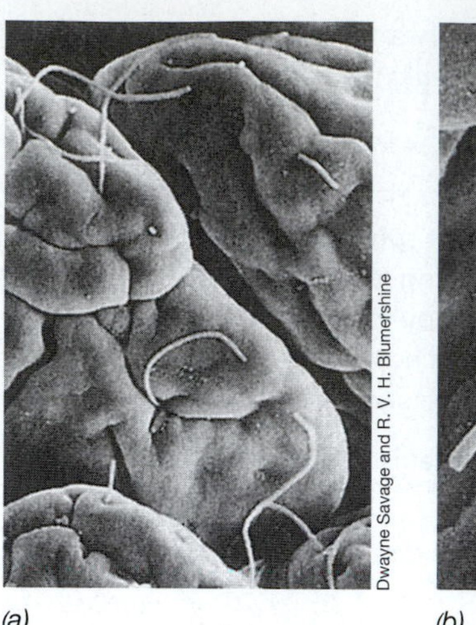

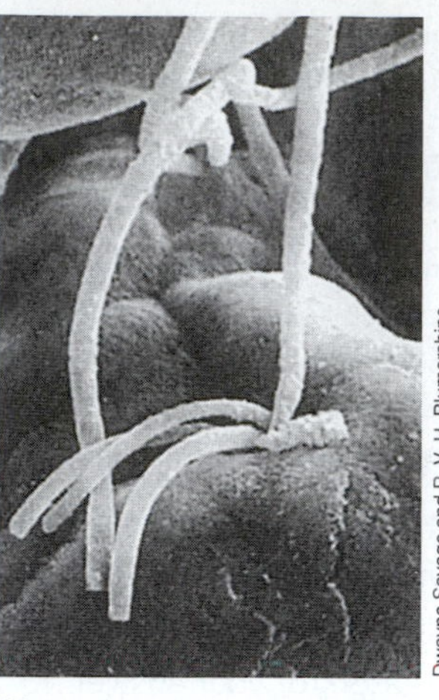

(a) (b)

Figure 28.9 Scanning electron micrographs of the microbial community on the surface of the epithelial cells in the mouse ileum. *(a)* An overview at low magnification. Note the long, filamentous fusiform bacteria lying on the surface. *(b)* Higher magnification, showing several filaments attached at a single depression. Note that the attachment is at the end of the filaments only. Individual cells are 10–15 μm long.

microbial flora. Starting with the stomach, the digestive tract is a column of nutrients mixed with microorganisms. The nutrients move one way through the column, encountering ever-changing populations of microorganisms. Here we examine the organisms as well as their functions and special properties.

Overall, about 10^{13} to 10^{14} microbial cells are present in the entire gastrointestinal tract. Our current view of the diversity and numbers of microorganisms that reside here has come from both standard culture methods and culture-independent molecular methods, such as microbial community analyses (Sections 14.9 and 22.3–22.6).

The Stomach

Because stomach fluids are highly acidic (about pH 2), the stomach is a chemical barrier to the entry of microorganisms into the gastrointestinal tract. However, microorganisms do populate this seemingly hostile environment. Studies using 16S rRNA sequences obtained from human stomach biopsies indicate that the stomach microbial population consists of several different phyla and a large number of bacterial taxa. Individuals clearly have very different populations, but all contain species of Gram-positive bacteria, *Proteobacteria, Bacteroidetes, Actinobacteria,* and *Fusobacteria* (Figure 28.8). *Helicobacter pylori,* the most common single organism found, colonizes the stomach wall in many, but not all, individuals and can cause ulcers in susceptible hosts (Section 34.11). Some of the bacteria that populate the stomach consist of organisms found in the oral cavity, introduced with the passage of food.

Distal to the stomach, the intestinal tract consists of the *small* intestine and the *large* intestine, each of which is divided into different anatomical structures. The composition of the intestinal flora in humans varies considerably and is somewhat dependent on diet. For example, persons who consume a considerable amount of meat show higher numbers of *Bacteroides* and lower numbers of coliforms and lactic acid bacteria than do individuals with a vegetarian diet. Representative microorganisms found in the gastrointestinal tract are shown in Figure 28.8.

The Small Intestine

The small intestine is separated into two parts, the *duodenum* and the *ileum,* with the *jejunum* connecting them. The duodenum, adjacent to the stomach, is fairly acidic and resembles the stomach in its microbial flora. From the duodenum to the ileum, the pH gradually becomes less acidic, and bacterial numbers increase. In the lower ileum, cell numbers of 10^5–10^7/gram of intestinal contents are common, even though the environment becomes progressively more anoxic. Fusiform anaerobic bacteria are typically present here, attached to the intestinal wall at one end (**Figure 28.9**).

The Large Intestine

The ileum empties into the *cecum,* the connecting portion of the large intestine. The *colon* makes up the rest of the large intestine. In the colon, bacteria are present in enormous numbers. The colon is a fermentation vessel, and many bacteria live here, using nutrients derived from the digestion of food (Table 28.1). Facultative aerobes such as *Escherichia coli* are present but in smaller numbers than other bacteria; total counts of facultative aerobes are less than 10^7/gram of intestinal contents. The facultative aerobes consume any remaining oxygen, making the large intestine strictly anoxic. This condition promotes growth of obligate anaerobes, including species of *Clostridium* and *Bacteroides*. The total number of obligate anaerobes in the colon is enormous. Bacterial counts of 10^{10} to 10^{11} cells/gram in distal gut and fecal contents are normal, with *Bacteroidetes* and gram-positive species accounting for greater than 99% of all bacteria. The methanogen *Methanobrevibacter smithii* can also be present in significant numbers. Protists are not found in the gastrointestinal tract of healthy humans although various protists can cause opportunistic infections if ingested in contaminated food or water (Sections 36.6 and 37.11).

Functions and Products of Intestinal Flora

Intestinal microorganisms carry out a wide variety of essential metabolic reactions that produce various compounds (**Table 28.2**). The composition of the intestinal flora and the diet influence the type and amount of compounds produced. Among these products are vitamins B_{12} and K. These essential vitamins are not synthesized by humans, but are made by the intestinal microflora and absorbed from the gut. Steroids, produced in the liver and released into the intestine from the gall bladder as bile acids, are modified in the intestine by the microbial flora; modified and now activated steroid compounds are then absorbed from the gut.

Other products generated by the activities of fermentative bacteria and methanogens include gas (flatus) and the odor-producing substances listed in Table 28.2. Normal adults expel several hundred milliliters of gas, of which about half is N_2 from swallowed air, from the intestines every day. Some foods metabolized by fermentative bacteria in the intestines

Table 28.2 Biochemical/metabolic contributions of intestinal microorganisms

Process	Product
Vitamin synthesis	Thiamine, riboflavin, pyridoxine, B_{12}, K
Gas production	CO_2, CH_4, H_2
Odor production	H_2S, NH_3, amines, indole, skatole, butyric acid
Organic acid production	Acetic, propionic, butyric acids
Glycosidase reactions	β-glucuronidase, β-galactosidase, β-glucosidase, α-glucosidase, α-galactosidase
Steroid metabolism (bile acids)	Esterified, dehydroxylated, oxidized, or reduced steroids

result in the production of hydrogen (H_2) and carbon dioxide (CO_2). Methanogens, found in the intestines of over one-third of normal adults, convert H_2 and CO_2 produced by fermentative bacteria to methane (CH_4). The methanogens in the rumen of cattle produce significant amounts of methane, up to a quarter of the total global production (Section 24.2).

During the passage of food through the gastrointestinal tract, water is absorbed from the digested material, which gradually becomes more concentrated and is converted to feces. Bacteria make up about one-third of the weight of fecal matter. Organisms living in the lumen of the large intestine are continuously displaced downward by the flow of material, and bacteria that are lost are continuously replaced by new growth. Thus, the large intestine resembles a chemostat (Section 6.8) in its action. The time needed for passage of material through the complete gastrointestinal tract is about 24 h in humans; the growth rate of bacteria in the lumen is one to two doublings per day. The total number of bacterial cells shed per day in human feces is on the order of 10^{13}.

Changing the Normal Flora

When an antibiotic is taken orally, it inhibits the growth of the normal flora as well as pathogens, leading to the loss of antibiotic-susceptible bacteria in the intestinal tract. This is often signaled by loose feces or diarrhea. In the absence of the full complement of normal flora, opportunistic microorganisms such as antibiotic-resistant *Staphylococcus, Proteus, Clostridium difficile,* or the yeast *Candida albicans* can become established. The retention of opportunistic pathogens can lead to a harmful alteration in digestive function or even to disease. For example, antibiotic treatment allows less susceptible microorganisms such as *C. difficile* to flourish, causing infection and colitis. After antibiotic therapy, however, the normal intestinal flora is typically reestablished quite quickly in adults. More rapid recolonization of the gut by desired species can be accomplished by administration of **probiotics**, live cultures of intestinal bacteria that, when administered to a host, may confer a health benefit. This is because a rapid recolonization of the gut may reestablish a competitive local flora and provide desirable microbial metabolic products (see the Microbial Sidebar, "Probiotics").

28.4 MiniReview

The stomach and the intestinal tract support a diverse population of microorganisms in a variety of nutritional and environmental conditions. The populations of microorganisms are influenced by the diet of the individual and by the unique physical conditions in each distinct anatomical area.

- Why might the small intestine be more suitable for growth of facultative aerobes than the large intestine?
- Identify several essential compounds made by indigenous intestinal microorganisms. What would happen if all microorganisms were completely eliminated from the body by the use of antibiotics?

28.5 Normal Microbial Flora of Other Body Regions

Each individual mucous membrane supports the growth of a specialized group of microorganisms. These organisms are part of the normal local environment and are characteristic of healthy tissue. In many cases, pathogenic microorganisms cannot colonize mucous membranes because of the competitive effects of the normal flora. Here we discuss two mucosal environments and their resident microorganisms.

Respiratory Tract

The anatomy of the respiratory tract is shown in **Figure 28.10**. In the **upper respiratory tract** (nasopharynx, oral cavity,

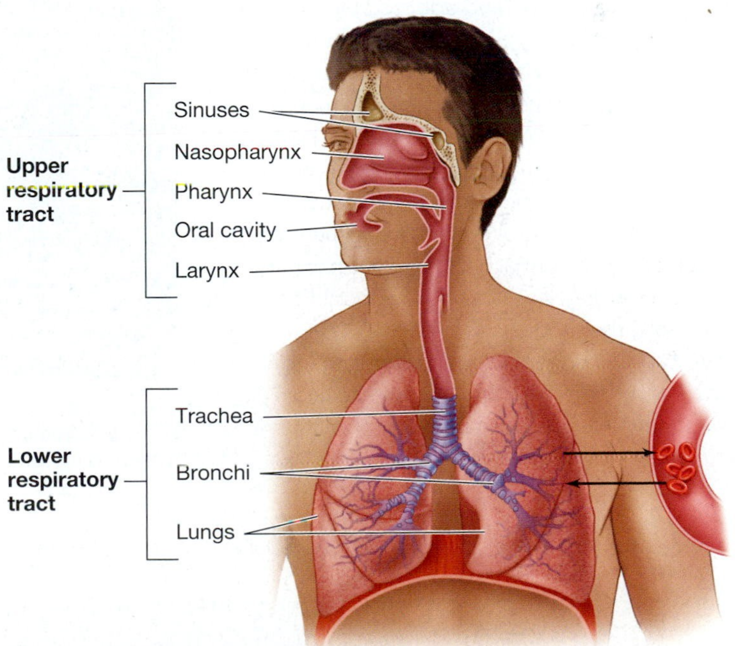

Figure 28.10 The respiratory tract. In healthy individuals the upper respiratory tract has a large variety and number of microorganisms. By contrast, the lower respiratory tract in a healthy person has few if any microorganisms.

UNIT 6

Microbial Sidebar

Probiotics[1]

The growth of microorganisms in and on the body is important for normal human development. These organisms, called *commensals* (Section 23.1), grow on and in the body and are likely essential to the well-being of all macroorganisms.

The microorganisms we acquire and retain constitute our normal flora and compete at various sites in the body with pathogens, preventing colonization by these organisms. Commensals that reside in the gut are active participants in the digestion of food and manufacture essential nutrients. In theory, the ingestion of selected microorganisms might be used to change or reestablish our gastrointestinal microbial flora to promote health, especially in individuals who experience major changes in their normal microflora due to disease, surgery, or other medical treatments, or whose normal flora changes for other reasons, such as poor diet. Intentionally ingested microorganisms used for this purpose are called *probiotics* (**Figure 1**).

As defined by the United Nations Food and Agricultural Organization and the World Health Organization and adopted by the American Academy of Microbiology, probiotics are suspensions of live microorganisms that, when administered in adequate amounts, confer a perceived health benefit on the host.

Are probiotics really useful? There is no reproducible scientific evidence that conclusively shows that the alteration of commensal populations in normal healthy adults has major, long-lasting, positive health effects. For example, the products shown in Figure 1 are directed toward replacing or reconstituting the intestinal flora of humans by ingesting what amounts to concentrated microbial cultures. As with the animal applications we discuss below, the products are directed at prevention or correction of digestive problems. While these products may confer short-term benefits, conclusive evidence for long-lasting establishment or reestablishment of an altered microbial flora is lacking.

Probiotics, however, are routinely used in production farm animals to prevent digestive problems. For example, strains of *Lactobacillus, Propionibacterium, Bacillus,* and *Saccharomyces* have been successfully used for this purpose. Conceivably, similar treatments in humans could have similar effects.

A number of human ailments have been shown to respond positively to probiotics administration, although the mechanisms are unknown. For example, the period of watery diarrhea from rotavirus infection in children can be shortened by administration of several different probiotics preparations. *Saccharomcyes* (yeast) may reduce the recurrence of diarrhea and shorten infections due to *Clostridium difficile*. Probiotic lactobacilli have also been used to treat urogenital infections in humans.

Evidence thus far suggests that the composition of the gut microflora can change rapidly when probiotics are administered. However, over the long term the gut microflora returns to its original state, indicating that the effects of probiotics are likely only short term. Thus, while probiotics may offer several benefits, especially for reestablishing the gut's normal flora following catastrophic changes, evidence for positive and lasting benefits is not well-established. In this regard there is considerable need for carefully designed and scientifically controlled studies to document the outcomes of probiotic treatment using standardized probiotic preparations containing known organisms administered in precise doses.

Figure 1 Some commercially available probiotics preparations. *The examples shown are marketed for use in humans.*

[1] *Source:* Walker, R., and M. Buckley. 2006. *Probiotic Microbes: The Scientific Basis.* American Academy of Microbiology.

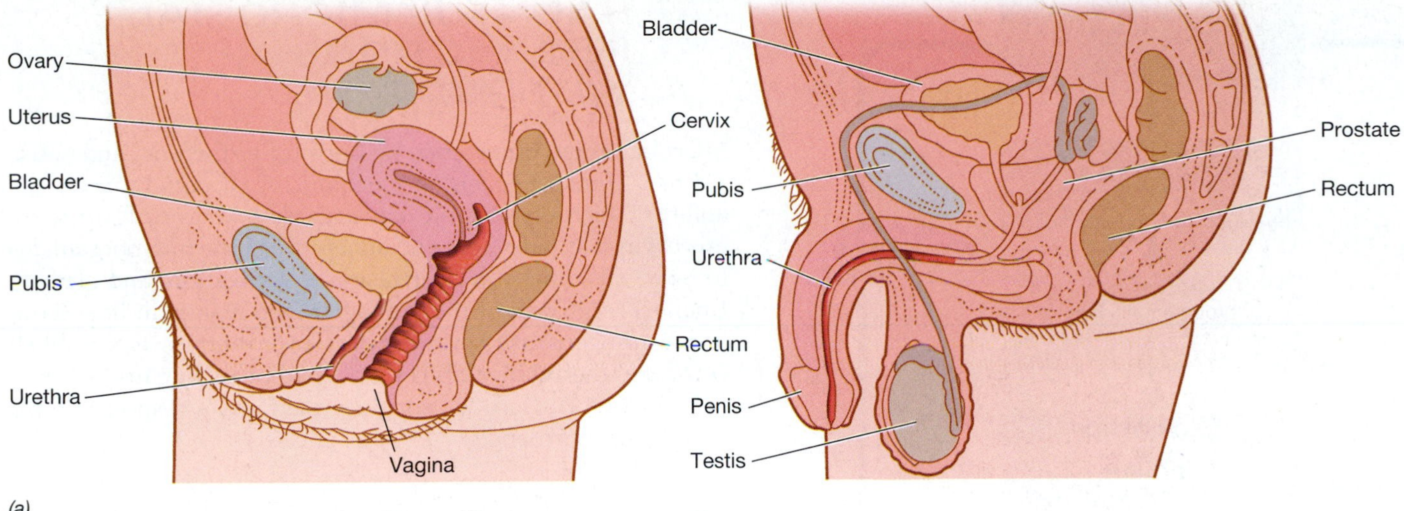

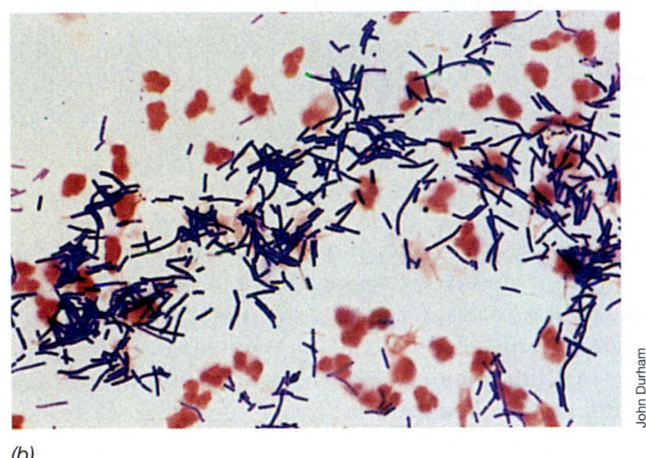

Figure 28.11 Microbial growth in the genitourinary tract. *(a)* The genitourinary tracts of the human female and male, showing regions (red) where microorganisms often grow. The upper regions of the genitourinary tracts of both males and females are sterile in healthy individuals. *(b)* Gram stain of *Lactobacillus acidophilus*, the predominant organism in the vagina of women between the onset of puberty and the end of menopause. Individual rod-shaped cells are 3–4 μm long.

larynx, and pharynx), microorganisms live in areas bathed with the secretions of the mucous membranes. Bacteria continually enter the upper respiratory tract from the air during breathing, but most are trapped in the nasal passages and expelled with the nasal secretions. A restricted group of microorganisms, however, colonizes respiratory mucosal surfaces in all individuals. The microorganisms most commonly found are staphylococci, streptococci, diphtheroid bacilli, and gram-negative cocci. Even potential pathogens such as *Staphylococcus aureus* and *Streptococcus pneumoniae* are often part of the normal flora of the nasopharynx of healthy individuals (Table 28.1). These individuals are *carriers* of the pathogens but do not normally develop disease, presumably because the other resident microorganisms compete successfully for resources and limit pathogen activities. The innate immune system and components of the adaptive immune system such as IgA (∞ Chapter 29) are particularly active at mucosal surfaces and may also inhibit the growth of pathogens.

The **lower respiratory tract** (trachea, bronchi, and lungs) has no resident microflora in healthy adults, despite the large number of organisms potentially able to reach this region during breathing. Dust particles, which are fairly large, settle out in the upper respiratory tract. As the air passes into the lower respiratory tract, the flow rate decreases markedly, and organisms settle onto the walls of the passages. The walls of the entire respiratory tract are lined with ciliated epithelium, and the cilia, beating upward, push bacteria and other particulate matter toward the upper respiratory tract where they are then expelled in the saliva and nasal secretions. Only particles smaller than about 10 μm in diameter reach the lungs. Nevertheless, some pathogens can reach these locations and cause disease, most notably pneumonias caused by certain bacteria or viruses (∞ Sections 34.2 and 34.7).

Urogenital Tract

In the male and female urogenital tracts (**Figure 28.11**), the bladder itself is typically sterile, but the epithelial cells lining the downstream urethra are colonized by facultatively aerobic gram-negative rods and cocci (Table 28.1). Potential pathogens such as *Escherichia coli* and *Proteus mirabilis,* normally present in small numbers in the body or local environment, can multiply in the urethra and become pathogenic under altered conditions such as changes in pH. Such organisms are a frequent cause of urinary tract infections, especially in women.

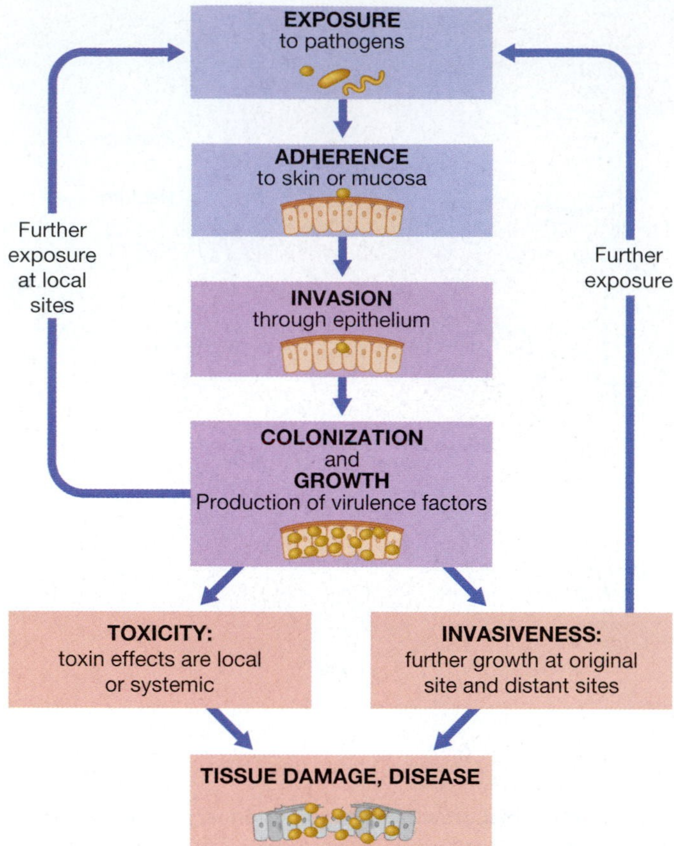

Figure 28.12 Microorganisms and mechanisms of pathogenesis. Following exposure to a pathogen, events in pathogenesis begin with specific adherence.

The vagina of the adult female is weakly acidic and contains significant amounts of glycogen. *Lactobacillus acidophilus,* a resident organism in the vagina, ferments the polysaccharide glycogen, producing lactic acid that maintains a local acidic environment (Figure 28.11*b*). Other organisms, such as yeasts (*Torulopsis* and *Candida* species), streptococci, and *E. coli,* may also be present. Before puberty, the female vagina is alkaline and does not produce glycogen, *L. acidophilus* is absent, and the flora consists predominantly of staphylococci, streptococci, diphtheroids, and *E. coli*. After menopause, glycogen production ceases, the pH rises, and the flora again resembles that found before puberty.

28.5 MiniReview

The presence of a population of normal nonpathogenic microorganisms in the respiratory and urogenital tracts is essential for normal organ function and often prevents the colonization of pathogens.

- Potential pathogens are often found in the normal flora of the upper respiratory tract. Why do they not cause disease in most cases?
- What is the importance of *Lactobacillus* found in the urogenital tract of normal adult women?

II HARMFUL MICROBIAL INTERACTIONS WITH HUMANS

Microbial interactions may be harmful to the host and cause disease. Here we examine mechanisms of *pathogenesis,* the ability of microorganisms to cause disease. Microbial pathogenesis begins with exposure and adherence of the microorganisms to host cells, followed by invasion, colonization, and growth. Unchecked growth of the pathogen then results in host damage. Pathogens use several different strategies to establish *virulence,* the relative ability of a pathogen to cause disease (**Figure 28.12**). Here we consider the factors responsible for establishing virulence.

28.6 Entry of the Pathogen into the Host

A pathogen must usually gain access to host tissues and multiply before damage can be done. In most cases, this requires that the organisms penetrate the skin, mucous membranes, or intestinal epithelium, surfaces that are normally microbial barriers.

Specific Adherence

Most microbial infections begin at breaks or wounds in the skin or on the mucous membranes of the respiratory, digestive, or genitourinary tract. Bacteria or viruses able to initiate infection often adhere specifically to epithelial cells (**Figure 28.13**) through macromolecular interactions on the surfaces of the pathogen and the host cell.

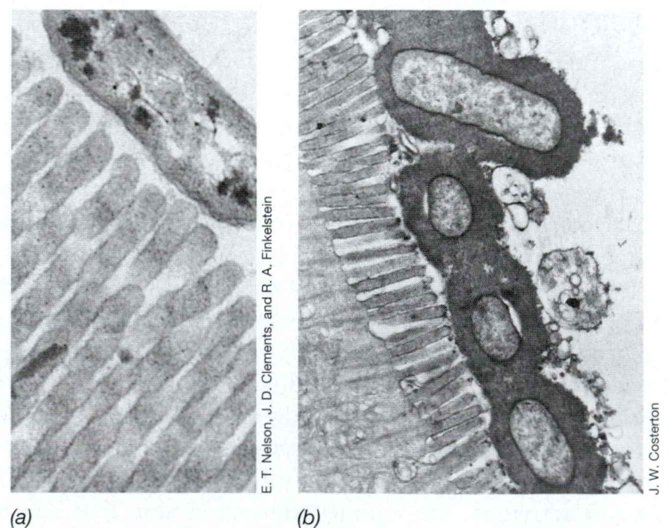

Figure 28.13 Adherence of pathogens to animal tissues. *(a)* Transmission electron micrograph of a thin section of *Vibrio cholerae* adhering to the brush border of rabbit villi in the intestine. Note the absence of a capsule. *(b)* Enteropathogenic *Escherichia coli* in a fatal model of infection in the newborn calf. The bacterial cells are attached to the brush border of calf intestinal villi through their distinct capsule. The rods are about 0.5 μm in diameter.

(a)

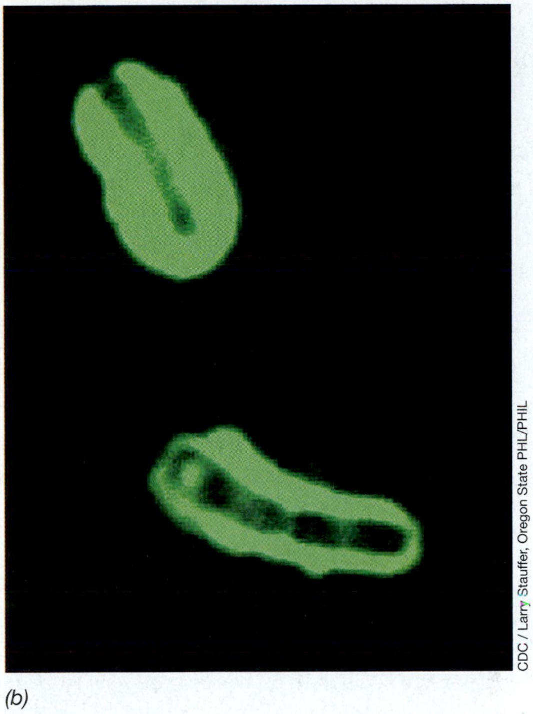

(b)

Figure 28.14 ***Bacillus anthracis* capsules.** *(a)* Capsules of *B. anthracis* on bicarbonate agar media. Encapsulated colonies are typically very large and mucoid in appearance. The individual encapsulated colonies are 0.5 cm in diameter. *(b)* Direct immunofluorescent stain of *B. anthracis* capsules. Antibodies coupled to fluorescein isothiocyanate (FITC) stain the capsule bright green, indicating that the capsule extends up to 1 μm from the cell, which is about 0.5 μm in diameter.

Most pathogens do not adhere to all epithelial cells equally, but selectively adhere to cells in a particular region of the body. For example, *Neisseria gonorrhoeae,* the pathogen that causes the sexually transmitted disease gonorrhea, adheres more strongly to urogenital epithelia. *N. gonorrhoeae* has a surface protein called Opa (*o*pacity *a*ssociated *p*rotein) that binds specifically to a host protein called CD66 found only on the surface of human epithelial cells. Thus *N. gonorrhoeae* interacts exclusively with host cells by binding a specific cell surface protein. The species of the host also influences specificity. In many cases, a bacterial strain that normally infects humans adheres more strongly to the appropriate human cells than to similar cells in another animal (for example, the rat) and vice versa.

Some macromolecules responsible for bacterial adherence are not covalently attached to the bacteria. These are usually polysaccharides, proteins, or protein–carbohydrate mixtures synthesized and secreted by the bacteria (∞ Section 4.9). A loose network of polymer fibers extending outward from a cell is called a **slime layer** (Figure 28.4*b*). A polymer coat consisting of a dense, well-defined layer surrounding the cell is called a **capsule** (**Figures** 28.13 and **28.14**). These structures may be important for adherence to other bacteria as well as to host tissues. In some cases, these structures can protect bacteria from host defense mechanisms such as phagocytosis (∞ Section 29.2).

Fimbriae and pili (∞ Section 4.9) are bacterial cell surface protein structures that may also function in the attachment process. For instance, the pili of *Neisseria gonorrhoeae* play a key role in attachment to urogenital epithelium, and fimbriated strains of *Escherichia coli* (**Figure 28.15**) are more frequent causes of urinary tract infections than strains lacking fimbriae. Among the best-characterized fimbriae are the type I fimbriae of enteric bacteria (*Escherichia, Klebsiella, Salmonella,* and *Shigella*). Type I fimbriae are uniformly distributed on the surface of cells. Pili are typically longer than fimbriae, with fewer pili found on the cell surface. Both pili

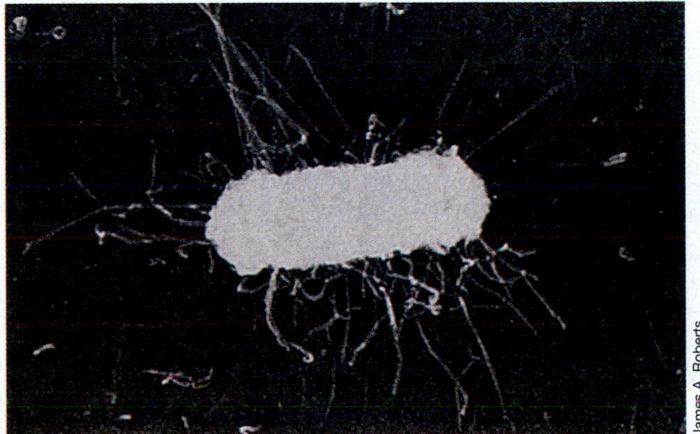

Figure 28.15 **Fimbriae.** Shadow-cast electron micrograph of the bacterium *Escherichia coli* showing type P fimbriae, which resemble type I fimbriae but are somewhat longer. The cell shown is about 0.5 μm wide.

UNIT 6

Table 28.3 **Major adherence factors used to facilitate attachment of microbial pathogens to host tissues[a]**

Factor	*Example*
Capsule/slime layer Figure 28.4, Figure 28.13, Figure 28.14	Pathogenic *Escherichia coli*—capsule promotes adherence to the brush border of intestinal villi
	Streptococcus mutans—dextran slime layer promotes binding to tooth surfaces
Adherence proteins	*Streptococcus pyogenes*—M protein on the cell binds to receptors on respiratory mucosa
	Neisseria gonorrhoeae—Opa protein on the cell binds to receptors on epithelium
Lipoteichoic acid (Figure 4.20)	*Streptococcus pyogenes*—facilitates binding to respiratory mucosal receptor (along with M protein)
Fimbriae (pili) (Figure 28.15)	*Neisseria gonorrhoeae*—pili facilitate binding to epithelium
	Salmonella species—type I fimbriae facilitate binding to epithelium of small intestine
	Pathogenic *Escherichia coli*—colonization factor antigens (CFAs) facilitate binding to epithelium of small intestine

[a]Most receptor sites on host tissues are glycoproteins or complex lipids such as gangliosides or globosides.

and fimbriae function by binding host cell surface glycoproteins, initiating attachment. Flagella can also increase adherence to host cells (see Figure 28.17).

Studies of diarrhea caused by enterotoxic strains of *E. coli* provide evidence for specific interactions between the mucosal epithelium and pathogens. Most strains of *E. coli* are normal nonpathogenic inhabitants of the cecum, the first part of the large intestine, and the colon (Figure 28.8). Several strains of *E. coli* are usually present in the body at the same time, and large numbers of these nonpathogens routinely pass through the body and are eliminated in feces. However, enterotoxic strains of *E. coli* contain genes encoding fimbrial CFA (colonization factor antigens); these are proteins that adhere specifically to cells in the small intestine. From here, they can colonize and produce enterotoxins that cause diarrhea as well as other illnesses (Section 28.11). Nonpathogenic strains of *E. coli* seldom express CFA proteins. Some major factors important in microbial adherence are shown in **Table 28.3**.

Invasion

A few microorganisms are pathogenic solely because of the toxins they produce. These organisms do not need to gain access to host tissues, and we will discuss them separately in Sections 28.10 and 28.11. However, most pathogens must penetrate the epithelium to initiate pathogenicity, a process called *invasion*. At the point of entry, usually at small breaks or lesions in the skin or in mucosal surfaces, growth is established. Growth may also begin on intact mucosal surfaces, especially if the normal flora is altered or eliminated, for example, by antibiotic therapy. Pathogens may then more readily colonize the tissue and begin the invasion process. Pathogen growth may also be established at sites distant from the original point of entry. Access to distant, usually interior, sites is through the blood or lymphatic circulatory system.

28.6 MiniReview

Pathogens gain access to host tissues by adherence at mucosal surfaces through interactions between pathogen and host macromolecules. Pathogen invasion starts at the site of adherence and may spread throughout the host via the circulatory or lymphatic systems.

- How do CFA molecules on *Escherichia coli* and Opa proteins on *Neisseria gonorrhoeae* influence adherence to mucosal tissues?
- How does adherence initiate invasion?

28.7 Colonization and Growth

If a pathogen gains access to tissues, it may multiply and colonize the tissue. Because the initial inoculum of a pathogen is usually too small to cause host damage, the pathogen must find appropriate nutrients and environmental conditions to grow in the host. Temperature, pH, and the presence or absence of oxygen affect pathogen growth, but the availability of microbial nutrients is most important.

Not all vitamins and growth factors are in adequate supply in all tissues at all times, even on a vertebrate host. Soluble nutrients such as sugars, amino acids, organic acids, and growth factors are limited, favoring organisms able to use host-specific nutrients. *Brucella abortus,* for example, grows very slowly in most tissues of infected cattle, but grows very rapidly in the placenta. The placenta is the only tissue that contains high concentrations of erythritol, a sugar that is readily metabolized by *B. abortus*. The erythritol enhances *B. abortus* growth, causing abortion in cattle (see Table 28.6).

Trace elements may also be in short supply and can influence establishment of the pathogen. For example, the concentration of iron greatly influences microbial growth (Section 5.1). Specific host proteins called *transferrin* and *lactoferrin* bind iron tightly and transfer it through the body. These proteins have such high affinity for iron that microbial iron deficiency may limit infection by many pathogens; an iron solution given to an infected animal greatly increases the virulence of some pathogens.

As we noted in Section 5.1, many bacteria produce iron-chelating compounds called *siderophores* that help them obtain iron, a growth-limiting micronutrient, from the environment. Siderophores from some pathogens are so efficient that they remove iron from animal iron-binding proteins. For example, *aerobactin,* a plasmid-encoded siderophore produced by certain strains of *Escherichia coli,* readily removes iron bound to transferrin. Likewise, *Neisseria* species produce a transferrin-specific receptor that binds to and removes iron from transferrin.

Localization in the Body

After initial entry, some pathogens remain localized, multiplying and producing a discrete focus of infection such as the boil that may arise from *Staphylococcus* skin infections (Section 34.10). Other pathogens may enter the lymphatic vessels and move to the lymph nodes, where they may be contained by the immune system. If a pathogen reaches the blood, either by direct infection or through the lymphatic vessels, it will be distributed to distant parts of the body, usually concentrating in the liver or spleen. Spread of the pathogen through the blood and lymph systems can result in a generalized (systemic) infection of the body, with the organism growing in a variety of tissues. If extensive bacterial growth in tissues occurs, some of the organisms are usually shed into the bloodstream in large numbers, a condition called **bacteremia**. Widespread infections of this type almost always start as a local infection in a specific organ such as the kidney, intestine, or lung.

28.7 MiniReview

A pathogen must gain access to nutrients and appropriate growth conditions before it can colonize and grow in substantial numbers in host tissue. Pathogens may grow locally at the site of invasion or may spread through the body.

- Why are colonization and growth necessary for the success of most pathogens?
- Identify host factors that limit or accelerate colonization and growth of a microorganism at selected local sites.

28.8 Virulence

Virulence is the relative ability of a pathogen to cause disease, and here we discuss some basic methods used to measure it. We then provide specific examples of particularly virulent organisms, highlighting properties that enhance their virulence.

Measuring Virulence

The virulence of a pathogen can be estimated from experimental studies of the LD_{50} (lethal dose$_{50}$), the dose of an agent that kills 50% of the animals in a test group. Highly virulent pathogens frequently show little difference in the number of cells required to kill 100% of the population as compared with the number required to kill 50% of the population. This is illustrated in **Figure 28.16** for experimental *Streptococcus* and *Salmonella* infections in mice. Only a few cells of virulent strains of *Streptococcus pneumoniae* are required to establish

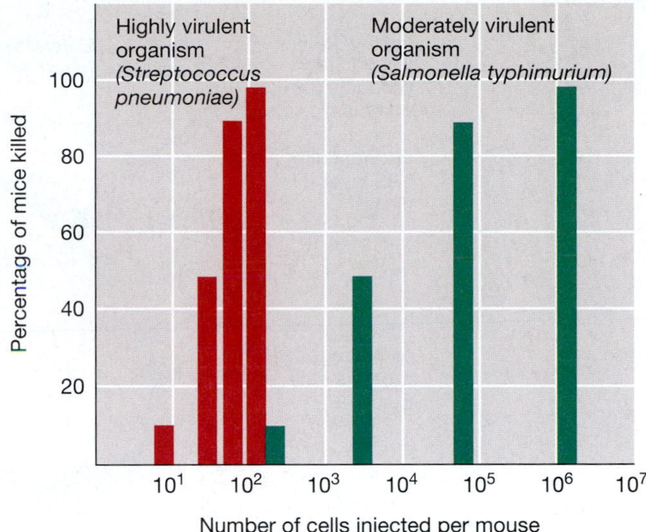

Figure 28.16 Microbial virulence. Differences in microbial virulence demonstrated by the number of cells of *Streptococcus pneumoniae* and *Salmonella typhimurium* required to kill mice.

a fatal infection and kill all animals in a test population of mice. In fact, the LD_{50} for *S. pneumoniae* is hard to determine in mice because so few organisms are needed to produce a lethal infection. By contrast, the LD_{50} for *Salmonella typhimurium,* a much less virulent pathogen, is much higher. The number of cells of *S. typhimurium* required to kill 100% of the population is more than 100 times greater than the number of cells needed to reach the LD_{50}.

When pathogens are kept in laboratory culture rather than isolated from diseased animals, their virulence is often decreased or even lost completely. Such organisms are said to be *attenuated*. **Attenuation** probably occurs because nonvirulent or weakly virulent mutants grow faster and, after successive transfers to fresh media, such mutants are selectively favored. Attenuation often occurs more readily when culture conditions are not optimal for the species. If an attenuated culture is reinoculated into an animal, the organism can regain its original virulence, but in many cases loss of virulence is permanent. Attenuated strains are often used for production of vaccines, especially viral vaccines (Section 30.5). Measles, mumps, and rubella vaccines, for example, consist of attenuated viruses.

Toxicity and Invasiveness

Virulence is due to the ability of a pathogen to cause host damage through toxicity and invasiveness. Each pathogen uses these properties to cause disease.

Toxicity is the ability of an organism to cause disease by means of a preformed toxin that inhibits host cell function or kills host cells. For example, the disease tetanus is caused by an exotoxin produced by *Clostridium tetani* (Section 28.10). *C. tetani* cells rarely leave the wound where they were first introduced, growing relatively slowly at the wound site. Yet

UNIT 6

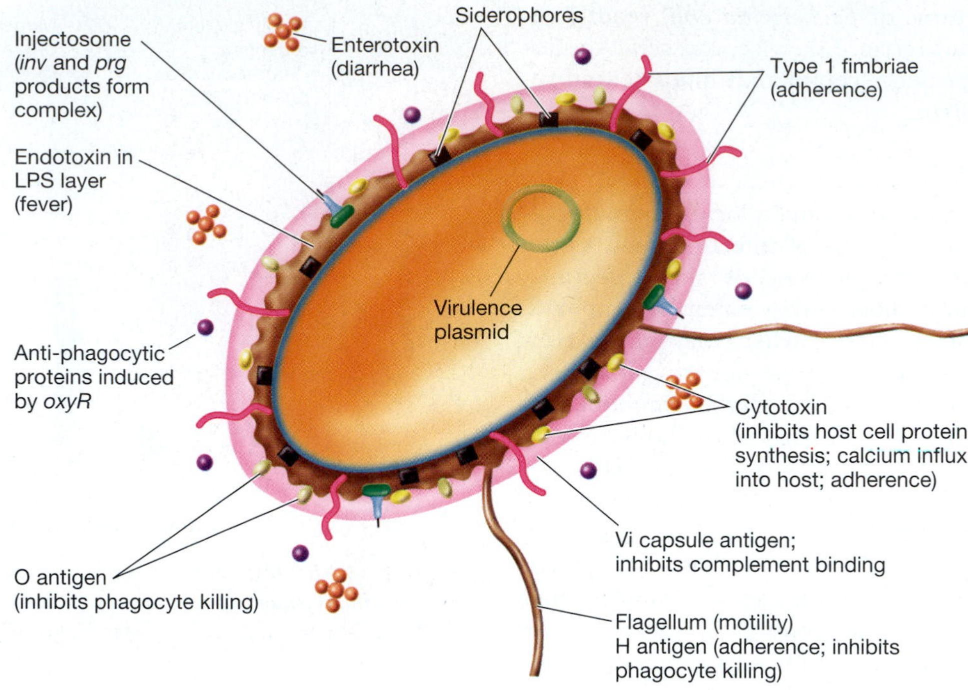

Figure 28.17 Virulence factors in *Salmonella* pathogenesis. Structural elements known to be important in pathogenesis are shown.

C. tetani is able to bring about disease because it produces tetanus toxin that moves to distant parts of the body, initiating irreversible muscle contraction and often death of the host.

Invasiveness is the ability of a pathogen to grow in host tissue in such large numbers that it inhibits host function. A microorganism may be able to produce disease through invasiveness even if it produces no toxin. For example, the major virulence factor for *Streptococcus pneumoniae* is the polysaccharide capsule that prevents the phagocytosis of pathogenic strains (Figure 34.3), defeating a major defense mechanism used by the host to prevent invasion. Encapsulated strains of *S. pneumoniae* grow in lung tissues in enormous numbers, where they initiate host responses that lead to pneumonia, interfere with lung function, and cause extensive host damage. Nonencapsulated strains are less invasive; they are quickly and efficiently ingested and destroyed by phagocytes, white blood cells that ingest and kill bacteria by a process called *phagocytosis* (Section 29.2).

Clostridium tetani and *Streptococcus pneumoniae* exemplify the extremes of toxicity and invasiveness, respectively. Most pathogens fall between these extremes and their virulence results from a combination of toxicity and invasiveness.

Virulence in *Salmonella*

Salmonella species employ a mixture of toxins, invasiveness, and other virulence factors to enhance pathogenicity (**Figure 28.17**). First, at least three toxins, *enterotoxin, endotoxin,* and *cytotoxin* contribute to their virulence (**Table 28.4**). We discuss enterotoxins and endotoxins in detail later in this chapter. *Salmonella* cytotoxin inhibits host cell protein synthesis and allows Ca^{2+} to escape from host cells, killing the cells.

Several virulence factors contribute to invasiveness in *Salmonella*. As we mentioned above, many pathogens including *Salmonella* produce iron-chelating siderophores, sequestering iron to aid in growth. The cell surface polysaccharide O antigen (Figure 4.22), the flagellar H antigen, and fimbriae enhance adherence. The capsular Vi polysaccharide inhibits complement binding and antibody-mediated killing (Section 29.11). The *inv* (invasion) genes of *Salmonella* encode at least ten different proteins that promote invasion. For example, *invH* encodes a surface adhesion protein. Other *inv* genes encode proteins important for trafficking of virulence proteins. The InvJ regulator protein controls assembly of structural proteins InvG, PrgH, PrgI, PrgJ, and PrgK that form an injection complex, called the *injectosome,* an organelle in the bacterial envelope that allows transfer of virulence proteins into host cells through a needlelike assembly.

Salmonella species readily establish infections through intracellular parasitism. *Salmonella* infections start with ingestion and passage of the bacterial cells through the stomach to the intestine. From here, *Salmonella* invades and replicates inside the cells that line the intestine as well as in phagocytes called *macrophages*. The genes that initiate the invasion process are contained on the chromosomal *Salmonella* pathogenicity island-1 (SPI-1). As with other pathogenicity islands, SPI-1 is a collection of virulence genes flanked by sequences suggesting a transposable genetic element. Another *Salmonella* pathogenicity island, SPI-2, contains genes that are responsible for systemic disease (Section 13.12).

Pathogenic *Salmonella* species also have other virulence factors. For example, the toxic oxygen products produced by host macrophages as an antibacterial defense can be neutralized by proteins encoded by the *Salmonella oxyR* gene. In addition, macrophage-produced antibacterial molecules

Table 28.4 Exotoxins and extracellular virulence factors produced by human pathogens

Organism	*Disease*	*Toxin or factor*[a]	*Action*
Bacillus anthracis	Anthrax	Lethal factor (LF) Edema factor (EF) Protective antigen (PA) (AB)	PA is the cell-binding B component, EF causes edema, LF causes cell death
Bacillus cereus	Food poisoning	Enterotoxin complex	Induces fluid loss from intestinal cells
Bordetella pertussis	Whooping cough	Pertussis toxin (AB)	Blocks G protein signal transduction, kills cells
Clostridium botulinum	Botulism	Neurotoxin (AB)	Flaccid paralysis (Figure 28.21)
Clostridium tetani	Tetanus	Neurotoxin (AB)	Spastic paralysis (Figure 28.22)
Clostridium perfringens	Gas gangrene, food poisoning	α-Toxin (CT)	Hemolysis (lecithinase, Figure 28.18*b*)
		β-Toxin (CT)	Hemolysis
		γ-Toxin (CT)	Hemolysis
		δ-Toxin (CT)	Hemolysis (cardiotoxin)
		κ-Toxin (E)	Collagenase
		λ-Toxin (E)	Protease
		Enterotoxin (CT)	Alters permeability of intestinal epithelium
Corynebacterium diphtheriae	Diphtheria	Diphtheria toxin (AB)	Inhibits protein synthesis in eukaryotes (Figure 28.20)
Escherichia coli (enterotoxigenic strains only)	Gastroenteritis	Enterotoxin (Shiga-like toxin) (AB)	Induces fluid loss from intestinal cells
Haemophilus ducreyi	Chancroid	Cytolethal distending toxin[b] (CDT) (AB)	Genotoxin (DNA lesions cause apoptosis in host cells)
Pseudomonas aeruginosa	*P. aeruginosa* infections	Exotoxin A (AB)	Inhibits protein synthesis
Salmonella spp.	Salmonellosis, typhoid fever, paratyphoid fever	Enterotoxin (AB)	Inhibits protein synthesis, lyses host cells
		Cytotoxin (CT)	Induces fluid loss from intestinal cells
Shigella dysenteriae	Bacterial dysentery	Shiga toxin (AB)	Inhibits protein synthesis
Staphylococcus aureus	Pyogenic (pus-forming) infections (boils, and so on), respiratory infections, food poisoning, toxic shock syndrome, scalded skin syndrome	α-Toxin (CT)	Hemolysis
		Toxic shock syndrome toxin (SA)	Systemic shock
		Exfoliating toxin A and B (SA)	Peeling of skin, shock
		Leukocidin (CT)	Destroys leukocytes
		β-Toxin (CT)	Hemolysis
		γ-Toxin (CT)	Kills cells
		δ-Toxin (CT)	Hemolysis, leukolysis
		Enterotoxin A, B, C, D, and E (SA)	Induce vomiting, diarrhea, shock
		Coagulase (E)	Induces fibrin clotting
Streptococcus pyogenes	Pyogenic infections, tonsillitis, scarlet fever	Streptolysin O (CT)	Hemolysin
		Streptolysin S (CT)	Hemolysin (Figure 28.18a)
		Erythrogenic toxin (SA)	Causes scarlet fever
		Streptokinase (E)	Dissolves fibrin clots
		Hyaluronidase (E)	Dissolves hyaluronic acid in connective tissue
Vibrio cholerae	Cholera	Enterotoxin (AB)	Induces fluid loss from intestinal cells (Figure 28.23)

[a]AB, A-B toxin; CT, cytolytic toxin; E, enzymatic virulence factor; SA, superantigen toxin.

[b]Cytolethal distending toxin is found in other gram-negative pathogens including *Actinobacillus actinomycetemocomitans*, *Campylobacter* sp., *Escherichia coli*, *Helicobacter* sp., *Salmonella typhi*, and *Shigella dysenteriae*.

called *defensins* can be neutralized by the products of the *Salmonella phoP* and *phoQ* genes. Thus, the oxy and pho gene products of *Salmonella* enhance pathogenicity by promoting intracellular invasion and neutralizing host defenses that normally inhibit intracellular bacterial growth.

Finally, several plasmidborne virulence factors can be spread between most *Salmonella* species as well as other enteric bacteria. For example, antibiotic resistance is encoded on a R plasmid (Section 11.3). Thus *Salmonella,* and most other pathogens, use a variety of factors and mechanisms to establish virulence and pathogenesis.

28.8 MiniReview

Virulence is determined by the invasiveness and toxicity of a pathogen. In most pathogens, a number of mechanisms and factors contribute to virulence. Attenuation is loss of virulence. *Salmonella* has many traits that enhance virulence.

- Distinguish between *toxicity* and *invasiveness.* Give examples of organisms that rely almost exclusively on toxicity or invasiveness to promote virulence.
- For *Salmonella*, identify structural genes involved in virulence and relate them to the toxic or invasive properties of the pathogen.

III VIRULENCE FACTORS AND TOXINS

Extracellular capsules and slime layers, cell wall and envelope material, and fimbriae and pili are all integral macromolecular components of microorganisms that may aid in pathogenesis and act as virulence factors. However, in addition to these, a number of microbial intracellular and extracellular components function solely as virulence factors. These specialized virulence factors are produced by different pathogens, but many share molecular characteristics and modes of action. Here we examine some key factors.

28.9 Virulence Factors

Many pathogens produce toxins, which we discuss in the following three sections. Many other virulence factors are enzymes that enhance pathogen colonization and growth. For example, streptococci, staphylococci, and certain clostridia produce hyaluronidase (Table 28.4), an enzyme that promotes spreading of organisms in tissues by breaking down the polysaccharide hyaluronic acid, a material that functions in animals as an intercellular cement. Hyaluronidase digests the intercellular matrix, enabling these organisms to spread from an initial infection site. Streptococci and staphylococci also produce proteases, nucleases, and lipases that degrade host proteins, nucleic acids, and lipids. Similarly, clostridia that cause gas gangrene produce collagenase, or κ-toxin (Table 28.4), which breaks down the tissue-supporting collagen network, enabling these organisms to spread through the body.

Fibrin, Clots, and Virulence

Fibrin is a protein that functions in the clotting of blood, and thus fibrin clots are often formed by the host at a site of microbial invasion. The clotting mechanism, triggered by tissue injury, isolates the pathogens, limiting infection to a local region. Some pathogens counter this process by producing fibrinolytic enzymes that dissolve the clots and make further invasion possible. One fibrinolytic substance produced by *Streptococcus pyogenes* is called *streptokinase* (Table 28.4).

By contrast, other pathogens produce enzymes that promote the formation of fibrin clots. These clots localize and protect the organism. The best-studied fibrin-clotting enzyme is *coagulase* (Table 28.4), produced by pathogenic *Staphylococcus aureus*. Coagulase causes fibrin to be deposited on *S. aureus* cells, protecting them from attack by host cells. The fibrin matrix produced as a result of coagulase activity probably accounts for the extremely localized nature of many staphylococcal infections, as in boils and pimples (Figure 34.24). Coagulase-positive *S. aureus* strains are typically more virulent than coagulase-negative strains.

28.9 MiniReview

Pathogens produce enzymes that enhance virulence by breaking down or altering host tissue to provide access and nutrients. Other pathogen-produced virulence factors protect the pathogen by interfering with normal host defense mechanisms. These factors enhance colonization and growth of the pathogen.

- What advantage does the pathogen gain by producing enzymes that affect structural components of host tissues?
- How does streptokinase enhance the virulence of *Streptococcus pyogenes*? How does coagulase enhance the virulence of *Staphylococcus aureus*?

28.10 Exotoxins

Exotoxins are toxic proteins released from the pathogen cell as it grows. These toxins travel from a site of infection and cause damage at distant sites. Table 28.4 provides a summary of the properties and actions of some of the known exotoxins as well as other extracellular virulence factors.

Exotoxins fall into one of three categories: the *cytolytic toxins,* the *AB toxins,* and the *superantigen toxins.* The cytolytic toxins work by degrading cytoplasmic membrane integrity, causing lysis. The AB toxins consist of two subunits, A and B. The B component generally binds to a host cell surface receptor, allowing the transfer of the A subunit across the targeted cell membrane, where it damages the cell. The

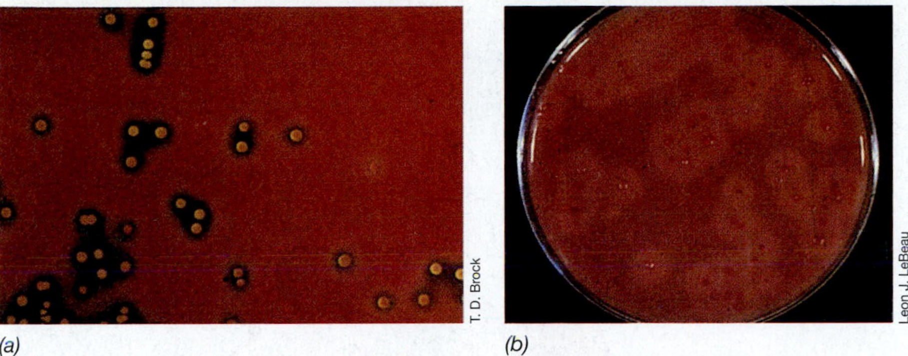

Figure 28.18 Hemolysis. *(a)* Zones of hemolysis around colonies of *Streptococcus pyogenes* growing on a blood agar plate. *(b)* Action of lecithinase, a phospholipase, around colonies of *Clostridium perfringens* growing on an agar medium containing egg yolk, a source of lecithin. Lecithinase dissolves the membranes of red blood cells, producing the cloudy zones of hemolysis around each colony.

superantigens work by stimulating large numbers of immune cells, resulting in extensive inflammation and tissue damage, as we will discuss later (Section 30.8).

Cytolytic Toxins

Various pathogens produce proteins that damage the host cytoplasmic membrane, causing cell lysis and death. Because the activity of these toxins is most easily observed with assays involving the lysis of red blood cells (erythrocytes), the toxins are called *hemolysins* (Table 28.4). However, they also lyse cells other than erythrocytes. The production of hemolysin is demonstrated in the laboratory by streaking the pathogen on a blood agar plate (a rich medium containing 5% whole blood). During growth of the colonies, hemolysin is released and lyses the surrounding red blood cells, releasing hemoglobin and creating a clearing, called *hemolysis* (**Figure 28.18**).

Some hemolysins attack the phospholipid of the host cytoplasmic membrane. Because the phospholipid lecithin (phosphatidylcholine) is often used as a substrate, these enzymes are called *lecithinases* or *phospholipases*. An example is the α-toxin of *Clostridium perfringens*, a lecithinase that dissolves membrane lipids, resulting in cell lysis (Table 28.4, Figure 28.18*b*; Figure 21.41). Because the cytoplasmic membranes of all organisms contain phospholipids, phospholipases sometimes destroy bacterial as well as animal cytoplasmic membranes.

Some hemolysins, however, are not phospholipases. Streptolysin O, a hemolysin produced by streptococci, affects the sterols of the host cytoplasmic membrane. *Leukocidins* (Table 28.4) lyse white blood cells and may decrease host resistance (Section 29.2).

Staphylococcal α-toxin (**Figure 28.19** and Table 28.4) kills nucleated cells and lyses erythrocytes. Toxin subunits first bind to the lipid bilayer. The subunits then oligomerize into nonlytic heptamers, now associated with the membrane. Following oligomerization, each heptamer undergoes conformational changes to produce a membrane-spanning pore, releasing the cell contents and allowing influx of extracellular material, disrupting cell function and causing cell death.

Protein Synthesis Inhibitor Toxins

Several pathogens produce AB exotoxins that inhibit protein synthesis. Diphtheria toxin, produced by *Corynebacterium diphtheriae,* is a good example of an AB toxin and is an important virulence factor (Section 34.3). Rats and mice are relatively resistant to diphtheria toxin, but human, rabbit, guinea pig, and bird cells are very susceptible, with only a single toxin molecule required to kill each cell. Diphtheria toxin is secreted by cells of *C. diphtheriae* as a single polypeptide. Fragment B specifically binds to a host cell receptor (**Figure 28.20**). After binding, proteolytic cleavage between fragment A and B allows entry of fragment A into the host cytoplasm. Here fragment A disrupts protein synthesis by blocking transfer of an amino acid from a tRNA to the growing polypeptide chain (Section 7.15). The toxin specifically inactivates elongation factor 2, a protein involved in growth of the polypeptide chain, by catalyzing the attachment of adenosine diphosphate (ADP) ribose from NAD^+. Following

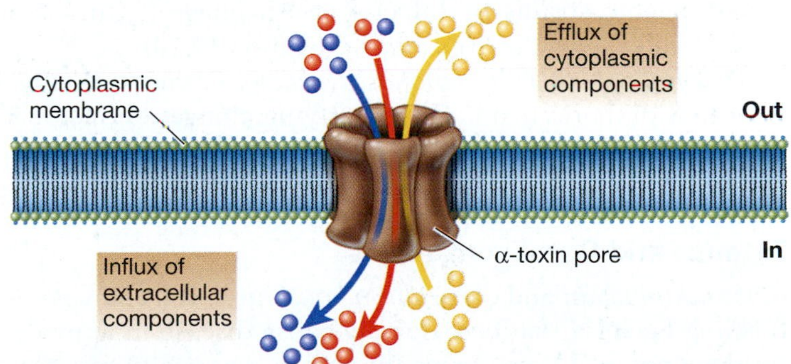

Figure 28.19 Staphylococcal α-toxin. Staphylococcal α-toxin is a pore-forming cytotoxin that is produced by growing *Staphylococcus* cells. Released as a monomer, seven identical protein subunits oligomerize in the cytoplasmic membrane of target cells. The oligomer forms a pore, releasing the contents of the cell and allowing the influx of extracellular material and the efflux of intracellular material. Eukaryotic cells swell and lyse. In erythrocytes, hemolysis occurs, visually indicating cell lysis.

UNIT 6

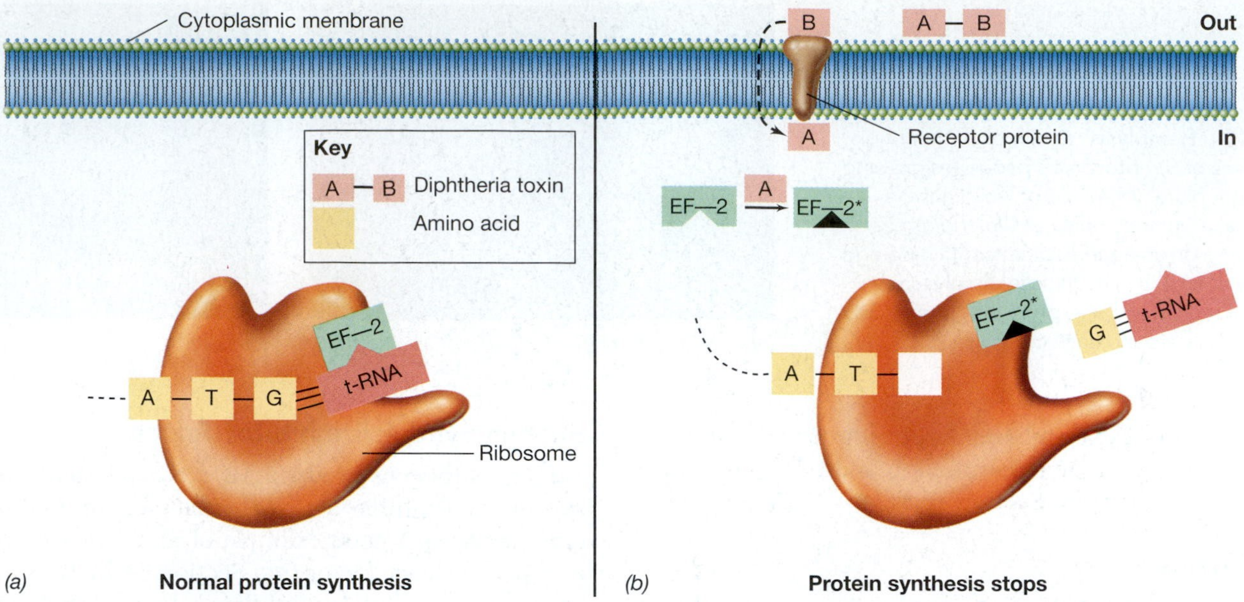

Figure 28.20 The action of diphtheria toxin from *Corynebacterium diphtheriae*. In a normal eukaryotic cell *(a)* elongation factor 2 (EF-2) binds to the ribosome, bringing an amino acid-charged t-RNA to the ribosome, causing protein elongation. In a cell affected by the diphtheria AB toxin *(b)*, the toxin binds to the cytoplasmic membrane receptor protein via the B portion. Cleavage between the A and B toxin components occurs, and the A peptide is internalized. The A peptide catalyzes the ADP-ribosylation of elongation factor 2 (EF—2*). This modified elongation factor no longer binds the ribosome and cannot aid transfer of amino acids to the growing polypeptide chain, resulting in shutdown of protein synthesis and death of the cell.

ADP-ribosylation, the activity of the modified elongation factor 2 decreases dramatically and protein synthesis stops.

Diphtheria toxin is formed only by strains of *C. diphtheriae* that are lysogenized by a bacteriophage called phage β; the *tox* gene in the phage genome encodes the toxin. Nontoxigenic, nonpathogenic strains of *C. diphtheriae* can be converted to pathogenic strains by infection with phage β (this is a process called *phage conversion*, Section 10.10).

Exotoxin A of *Pseudomonas aeruginosa* functions similarly to diphtheria toxin, also modifying elongation factor 2 by ADP-ribosylation (Table 28.4). www.microbiologyplace.com **Online Tutorial 28.1: Diphtheria and Cholera Toxin**

Tetanus and Botulinum Toxins

Clostridium tetani and *Clostridium botulinum*, soil endospore-forming bacteria that occasionally cause disease in animals, produce potent AB exotoxins that affect nervous tissue. Neither species is very invasive, and virtually all pathogenic effects are due to neurotoxicity. *C. botulinum* sometimes grows directly in the body, causing infant or wound botulism, and also grows and produces toxin in improperly preserved foods (Section 37.6). Death from botulism is usually from respiratory failure due to flaccid muscle paralysis. *C. tetani* grows in the body in deep wound punctures that become anoxic, and although *C. tetani* does not invade the body from the initial site of infection, the toxin can spread via the neural cells and cause spastic paralysis, the hallmark of tetanus, often leading to death (Section 35.9).

Botulinum toxin consists of seven related AB toxins that are the most potent biological toxins known. One milligram of botulinum toxin is enough to kill more than 1 million guinea pigs. Of the seven distinct botulinum toxins known, at least two are encoded on lysogenic bacteriophages specific for *C. botulinum*. The major toxin is a protein that forms complexes with nontoxic botulinum proteins to yield a bioactive protein complex. The complex then binds to presynaptic membranes on the termini of the stimulatory motor neurons at the neuromuscular junction, blocking the release of acetylcholine. Transmission of the nerve impulse to the muscle requires acetylcholine interaction with a muscle receptor and botulinum toxin prevents the poisoned muscle from receiving the excitatory signal (**Figure 28.21**). This prevents muscle contraction and leads to flaccid paralysis and death by suffocation, the outcome of botulism.

Tetanus toxin is also an AB protein neurotoxin. On contact with the central nervous system, this toxin is transported through the motor neurons to the spinal cord, where it binds specifically to ganglioside lipids at the termini of the inhibitory interneurons. The inhibitory interneurons normally work by releasing an inhibitory neurotransmitter, typically the amino acid glycine, which binds to receptors on the motor neurons. Normally, glycine from the inhibitory interneurons stops the release of acetylcholine by the motor neurons and inhibits muscle contraction, allowing relaxation of the muscle fibers. However, if tetanus toxin blocks glycine release, the motor neurons cannot be inhibited, resulting in tetanus,

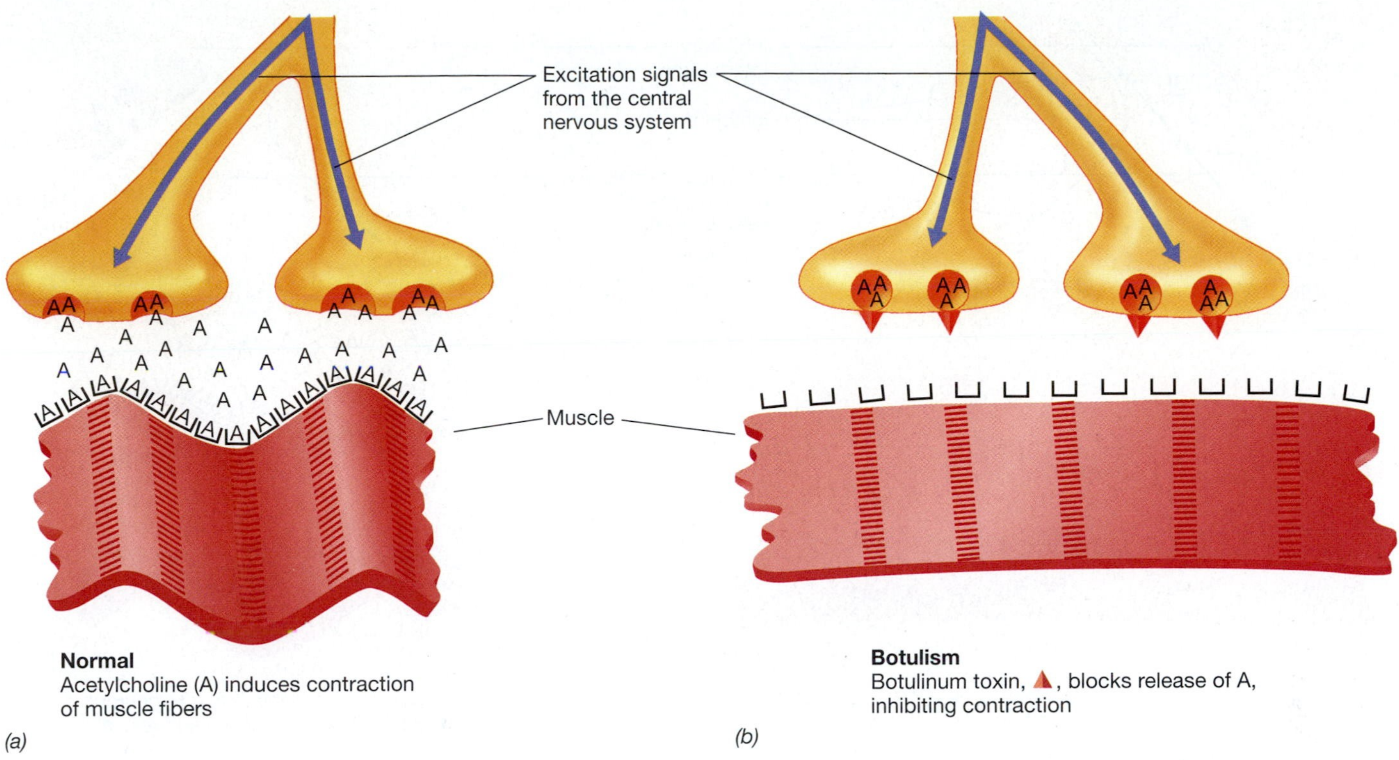
Excitation signals from the central nervous system
Muscle
Normal
Acetylcholine (A) induces contraction of muscle fibers
(a)
Botulism
Botulinum toxin, ▲, blocks release of A, inhibiting contraction
(b)

Figure 28.21 The action of botulinum toxin from *Clostridium botulinum*. *(a)* Upon stimulation of peripheral and cranial nerves, acetylcholine (A) is normally released from vesicles at the neural side of the motor end plate. Acetylcholine then binds to specific receptors on the muscle, inducing contraction. *(b)* Botulinum toxin acts at the motor end plate to prevent release of acetylcholine (A) from vesicles, resulting in a lack of stimulus to the muscle fibers, irreversible relaxation of the muscles, and flaccid paralysis.

continual release of acetylcholine, and uncontrolled contraction of the poisoned muscles (**Figure 28.22**). The outcome is a spastic, twitching paralysis, and affected muscles are constantly contracted. If the muscles of the mouth are involved, the prolonged contractions restrict the mouth's movement, resulting in the condition called *lockjaw* (trismus). If respiratory muscles are involved, prolonged contraction may result in death due to asphyxiation.

Tetanus toxin and botulinum toxin both block release of neurotransmitters involved in muscle control, but the symptoms are quite different and depend on the particular neurotransmitters involved.

28.10 MiniReview

Exotoxins contribute to the virulence of pathogens. Cytotoxins and AB toxins are potent exotoxins produced by microorganisms. Each exotoxin affects a specific host cell function. Bacterial exotoxins include some of the most potent biological toxins known.

- What key features are shared by all exotoxins? What features are unique to the AB exotoxins?
- Are bacterial growth and infection in the host necessary for the production of toxins? Explain and cite examples for your answer.

28.11 Enterotoxins

Enterotoxins are exotoxins whose activity affects the small intestine, generally causing massive secretion of fluid into the intestinal lumen resulting in vomiting and diarrhea. Generally acquired by ingestion of contaminated food or water, enterotoxins are produced by a variety of bacteria, including the food-poisoning organisms *Staphylococcus aureus, Clostridium perfringens,* and *Bacillus cereus,* and the intestinal pathogens *Vibrio cholerae, Escherichia coli,* and *Salmonella enteritidis*.

Cholera Toxin

The enterotoxin produced by *V. cholerae,* the organism that causes cholera (∞ Section 36.5), is the best understood enterotoxin. Cholera is characterized by massive fluid loss through the intestines, resulting in severe diarrhea characterized by life-threatening dehydration and electrolyte depletion. The disease starts by ingestion of *V. cholerae* in contaminated food or water. The organism travels to the intestine, where it colonizes and secretes the cholera toxin. Cholera toxin is an AB toxin consisting of an A subunit and five B subunits. In the gut, the B subunit binds specifically to GM1 ganglioside, a complex glycolipid found in the cytoplasmic membrane of

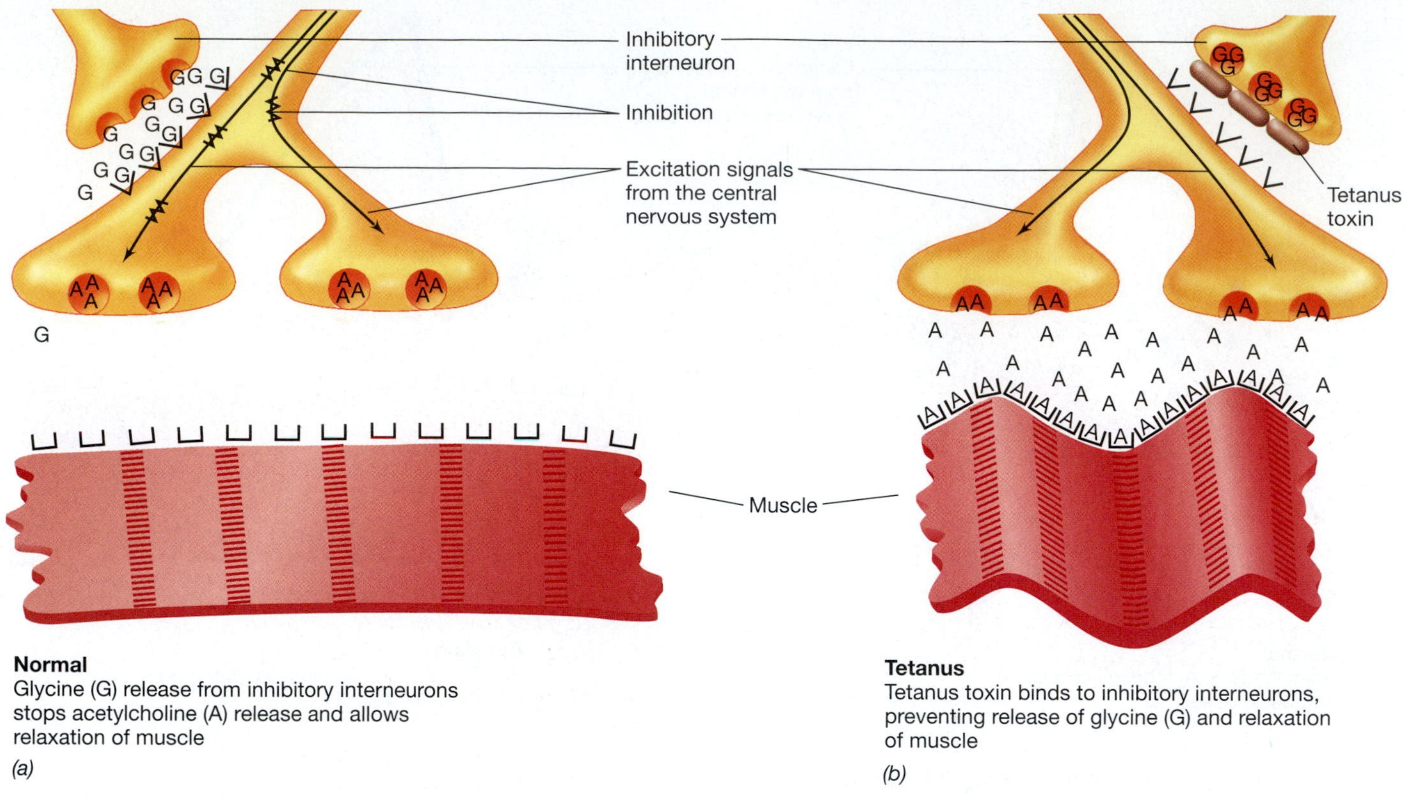
Inhibitory interneuron
Inhibition
Excitation signals from the central nervous system
Tetanus toxin
Muscle
Normal
Glycine (G) release from inhibitory interneurons stops acetylcholine (A) release and allows relaxation of muscle
(a)
Tetanus
Tetanus toxin binds to inhibitory interneurons, preventing release of glycine (G) and relaxation of muscle
(b)

Figure 28.22 The action of tetanus toxin from *Clostridium tetani*. *(a)* Muscle relaxation is normally induced by glycine (G) release from inhibitory interneurons. Glycine acts on the motor neurons to block excitation and release of acetylcholine (A) at the motor end plate. *(b)* Tetanus toxin binds to the interneuron to prevent release of glycine from vesicles, resulting in a lack of inhibitory signals to the motor neurons, constant release of acetylcholine to the muscle fibers, irreversible contraction of the muscles, and spastic paralysis.

intestinal epithelial cells (**Figure 28.23**). The B subunit targets the toxin specifically to the intestinal epithelium but has no role in alteration of membrane permeability; the toxic action is a function of the A chain, which crosses the cytoplasmic membrane and activates adenyl cyclase, the enzyme that converts ATP to cyclic adenosine monophosphate (cAMP).

As discussed in Section 9.9, cAMP is a mediator of many different regulatory systems in cells, including ion balance. The increased cAMP levels induced by the cholera enterotoxin induce secretion of chloride and bicarbonate ions from the mucosal cells into the intestinal lumen. This change in ion concentrations leads to the secretion of large amounts of water into the intestinal lumen (Figure 28.23). In acute cholera, the rate of water loss into the small intestine is greater than the possible reabsorption of water by the large intestine, resulting in massive net fluid loss. Cholera treatment is by oral fluid replacement with solutions containing electrolytes and other solutes to offset the dehydration-coupled ion imbalance.

Expression of cholera enterotoxin genes *ctxA* and *ctxB* is controlled by *toxR*. The *toxR* gene product is a transmembrane protein that controls cholera A and B chain production as well as other virulence factors, such as the outer membrane proteins and pili required for successful attachment and colonization of *V. cholerae* in the small intestine.

Other Enterotoxins

Enterotoxins produced by enterotoxigenic *Escherichia* and *Salmonella* are functionally, structurally, and evolutionarily related to cholera toxin. These toxins are also produced in the gut by colonizing bacteria. However, enterotoxins produced by some of the food-poisoning bacteria (*Staphylococcus aureus, Clostridium perfringens,* and *Bacillus cereus*) have quite different modes of action and are often ingested as preformed toxins produced from bacterial growth in contaminated food; growth of the pathogen in the host is unnecessary. *C. perfringens* enterotoxin is a cytotoxin and *S. aureus* enterotoxin is a superantigen (Table 28.4). Superantigens have a completely different mode of action, stimulating large numbers of immune lymphocytes and causing systemic as well as intestinal inflammatory responses (Section 30.8).

The toxin produced by *Shigella dysenteriae,* called *Shiga toxin,* and the Shiga-like toxin produced by enteropathogenic *E. coli* O157:H7 are protein synthesis-inhibiting AB toxins. The mode of action for this class of toxins is different, however, from that of the AB-type diphtheria toxin (Section 34.3), as they specifically kill small intestine cells, leading to bloody diarrhea and related intestinal symptoms.

1. Normal ion movement, Na^+ from lumen to blood, no net Cl^- movement

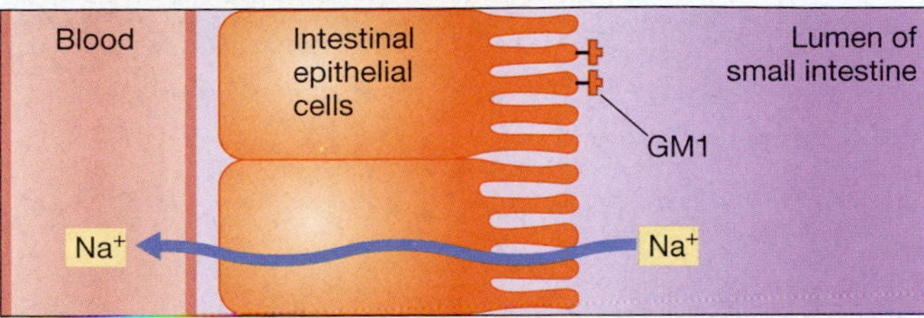

2. Colonization and toxin production

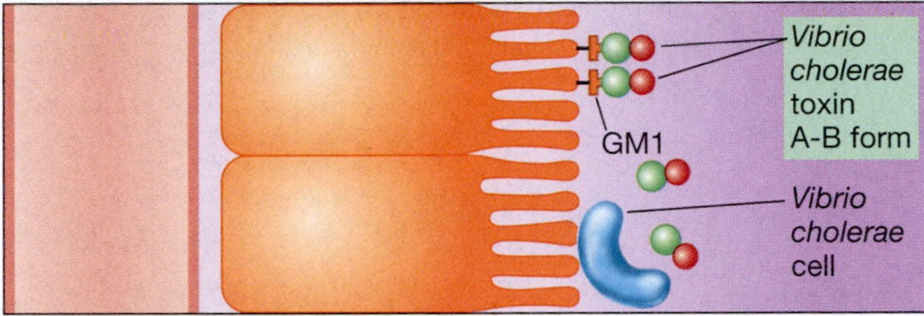

3. Activation of epithelial adenyl cyclase by cholera toxin

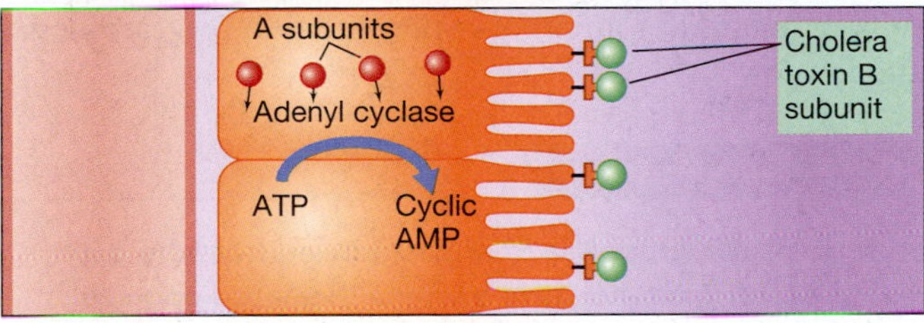

4. Na^+ movement blocked, net Cl^- movement to lumen

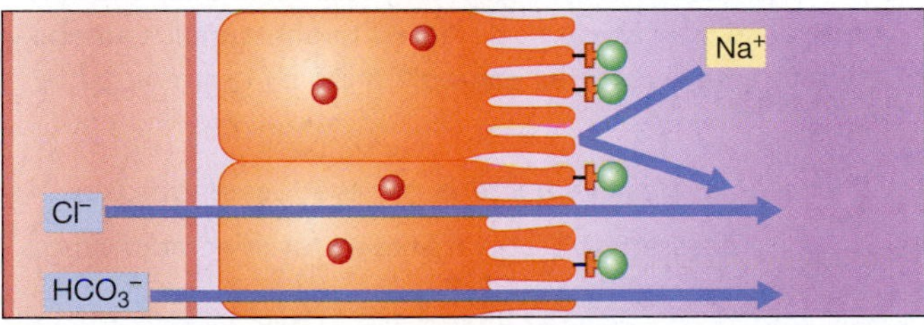

5. Massive water movement to the lumen

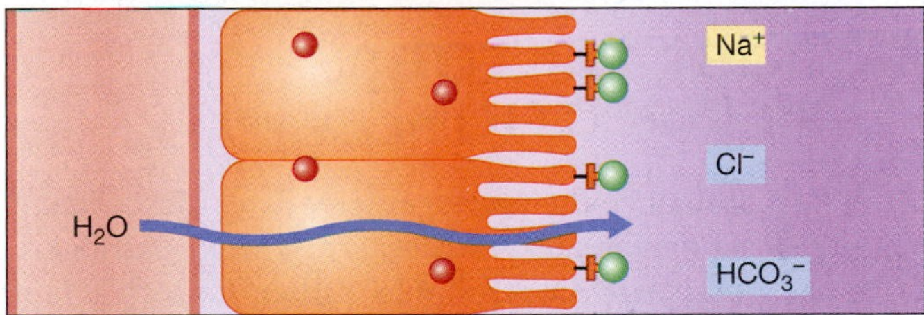

Figure 28.23 The action of cholera enterotoxin. Cholera toxin is a heat-stable AB enterotoxin that activates a second messenger pathway, disrupting normal ion flow in the intestine. (1) Ion movement in the normal intestine. (2) Adherence and colonization of the intestinal microvilli by *Vibrio cholerae*, followed by production and binding of the AB enterotoxin by interaction with the GM1 ganglioside on host cells. The B subunits (green) of AB toxin bind GM1, allowing the release of the A subunit (red) across the cell membrane. (3) The internalized A subunit activates adenyl cyclase, (4) blocking normal sodium (Na^+) influx, (5) causing massive loss of H_2O into the lumen, resulting in profuse diarrhea. Treatment for cholera is by ion replacement and rehydration. Antibiotic treatment may shorten the course of the disease by limiting *V. cholerae* growth, but does not affect the action of toxin that has already been produced.

28.11 MiniReview

Enterotoxins are exotoxins that affect the small intestine, causing changes in intestinal permeability that lead to diarrhea. Many enteric pathogens colonize the small intestine and produce AB enterotoxins. Food-poisoning bacteria often produce cytotoxins or superantigens.

- What key features are shared by all enterotoxins? By the AB enterotoxins?
- Describe the action of *Vibrio cholerae* toxin on the small intestine. Why does this lead to massive fluid loss?

28.12 Endotoxins

Many gram-negative *Bacteria* produce toxic lipopolysaccharides as part of the outer layer of their cell envelope (Section 4.7). These lipopolysaccharides are called **endotoxins**. In contrast to exotoxins, which are the secreted products of living cells, endotoxins are cell bound and released in large amounts only when the cells lyse. Endotoxins have been studied primarily in *Escherichia, Shigella,* and especially *Salmonella*. The properties of exotoxins and endotoxins are compared in **Table 28.5**.

Endotoxin Structure and Function

The structure of lipopolysaccharide (LPS) was diagrammed in Figures 4.22 and 4.23. LPS consists of three covalently linked subunits; the membrane-distal *O*-polysaccharide, lipid A, and a membrane-proximal core polysaccharide.

Endotoxins cause a variety of physiological effects. Fever is an almost universal result of endotoxin exposure because endotoxin stimulates host cells to release proteins called *endogenous pyrogens,* which affect the temperature-controlling center of the brain. In addition, endotoxins can cause diarrhea; rapid decrease in lymphocyte, leukocyte, and platelet numbers; release of cytokines; and generalized inflammation (Section 29.3). Large doses of endotoxin can cause death from hemorrhagic shock and tissue necrosis. The toxicity of endotoxins is, however, much lower than that of exotoxins. For instance, in mice the LD_{50} for endotoxin is 200–400 μg per animal, whereas the LD_{50} for botulinum toxin is about 25 picograms (pg), about 10 million times less!

Studies indicate that the lipid A portion of LPS is responsible for toxicity, and the polysaccharide fraction makes the complex water-soluble and immunogenic. Animal studies indicate that both the lipid and polysaccharide fractions are necessary for toxic effects to be observed.

Limulus Amebocyte Lysate Assay for Endotoxin

Because endotoxins are pyrogens, pharmaceuticals such as antibiotics and intravenous solutions must be free of endotoxin. An endotoxin assay of very high sensitivity has been developed using lysates of amebocytes from the horseshoe crab, *Limulus polyphemus*. Endotoxin specifically causes lysis

UNIT 6

Table 28.5 Properties of exotoxins and endotoxins

Property	*Exotoxins*	*Endotoxins*
Chemical properties	Proteins, excreted by certain gram-positive or gram-negative *Bacteria;* generally heat-labile	Lipopolysaccharide–lipoprotein complexes, released on cell lysis as part of the outer membrane of gram-negative *Bacteria;* extremely heat-stable
Mode of action; symptoms	Specific; usually binds to specific cell receptors or structures; either cytotoxin, enterotoxin, or neurotoxin with defined specific action on cells or tissues	General; fever, diarrhea, vomiting
Toxicity	Often highly toxic, sometimes fatal	Weakly toxic, rarely fatal
Immunogenicity response	Highly immunogenic; stimulate the production of neutralizing antibody (antitoxin)	Relatively poor immunogen; immune response not sufficient to neutralize toxin
Toxoid potential	Treatment of toxin with formaldehyde will destroy toxicity, but treated toxin (toxoid) remains immunogenic	None
Fever potential	Does not produce fever in host	Pyrogenic, often induces fever in host

of the amebocytes (**Figure 28.24**). In the standard *Limulus* amebocyte lysate (LAL) assay, *Limulus* amebocyte extracts are mixed with the solution to be tested. If endotoxin is present, the amebocyte extract forms a gel and precipitates, causing a change in turbidity. This reaction is measured quantitatively with a spectrophotometer and can detect as little as 10 pg/ml of LPS. The *Limulus* assay is used to detect endotoxin in clinical samples such as serum or cerebrospinal fluid. A positive test is presumptive evidence for infection by a gram-negative bacterium. Drinking water, water used for formulation of injectable drugs, and injectable aqueous solutions are routinely tested using the *Limulus* assay to identify and eliminate potential sources of contamination by gram-negative pathogens.

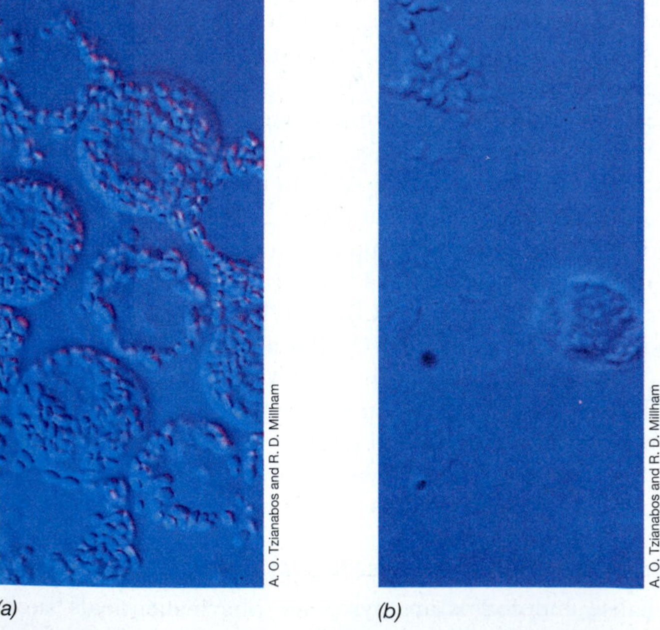

Figure 28.24 ***Limulus* amebocytes.** *(a)* Normal amebocytes from the horseshoe crab, *Limulus polyphemus. (b)* Amebocytes following exposure to bacterial lipopolysaccharide (LPS). LPS contained in test samples induces degranulation and lysis of the cells.

28.12 MiniReview

Endotoxins are lipopolysaccharides derived from the outer membrane of gram-negative bacteria. Released upon cell lysis, endotoxins cause fever and other systemic toxic effects in the host. Endotoxins are generally less toxic than exotoxins. The presence of endotoxin, detected by the *Limulus* amebocyte lysate assay, indicates contamination by gram-negative bacteria.

- Why do gram-positive bacteria not produce endotoxins?
- Why is it necessary to test water used for injectable drug preparations for endotoxin?

IV HOST FACTORS IN INFECTION

Host factors influence the pathogenicity of a microorganism. Certain risk factors related to diet, stress, and pathogen exposure are potentially controllable. Other host risk factors defined by, for example, age or genetics cannot be controlled. We conclude this chapter with a discussion of the passive physical and chemical barriers in human anatomy that limit infection and colonization. In the following chapter we deal with active host responses to pathogen contact.

28.13 Host Risk Factors for Infection

A number of factors contribute to the susceptibility of the host to infection and disease. Here we introduce some of the factors that provide passive resistance to infectious diseases. The failure of any of these mechanisms may result in invasion by pathogens and the onset of infectious disease.

Age as a Risk Factor

Age is an important factor for determining susceptibility to infectious disease. Infectious diseases are more common in the very young and in the very old. In the infant, for example, an intestinal microflora develops quickly, but the normal flora of an infant is not the same as that of an adult. Before the development of an adult flora, and especially in the days immediately following birth, pathogens have a greater opportunity to become established and produce disease. Thus, infants under the age of 1 year often acquire diarrhea caused by enteropathogenic strains of *Escherichia coli* or viruses such as Rotavirus (∞ Sections 37.8 and 37.11).

Infant botulism results from an intestinal infection with *Clostridium botulinum* (∞ Section 37.6). As the pathogen colonizes and grows, it secretes botulinum toxin, leading to flaccid paralysis (Section 28.10). Infant botulism, contracted after ingestion of *C. botulinum* from soil, air, or foods such as raw honey, is found almost exclusively in infants under 1 year of age, presumably because establishment of the normal intestinal flora in older children and adults prevents colonization by *C. botulinum*.

In individuals over 65 years of age, infectious diseases are much more common than in younger adults. For example, the elderly are much more susceptible to respiratory infections, particularly influenza (∞ Section 34.9), probably because of a declining ability to make an effective immune response to respiratory pathogens.

Anatomical changes associated with age may also encourage infection. Enlargement of the prostate gland, a common condition in men over the age of 50, frequently leads to a decreased urinary flow rate, allowing pathogens to colonize the male urinary tract more readily, leading to an increase in urinary tract infections (Figure 28.11).

Stress and Diet as Risk Factors

Stress can predispose a healthy individual to disease. In studies with rats and mice, physiological stressors such as fatigue, exertion, poor diet, dehydration, or drastic climate changes increase the incidence and severity of infectious diseases. For example, rats subjected to intense physical activity for long periods of time show a higher mortality rate from experimental *Salmonella* infections compared with rested control animals. Hormones that are produced under stress can inhibit normal immune responses and may play a role in stress-mediated disease. For example, cortisol, a hormone produced at high levels in the body in times of stress, is an anti-inflammatory agent that inhibits the activation of the immune response.

Diet plays a role in host susceptibility to infection. Inadequate diets low in protein and calories alter the normal flora, allowing opportunistic pathogens a better chance to multiply and increasing susceptibility of the host to known pathogens. For example, the number of *Vibrio cholerae* cells necessary to produce cholera in an exposed individual is drastically reduced if the individual is malnourished. The consumption of pathogen-contaminated food is an obvious way to acquire infections, and ingestion of pathogens with food can sometimes enhance the ability of the pathogen to cause disease. The number of organisms necessary to induce cholera, for example, is greatly reduced when the *V. cholerae* is ingested in food, presumably because the food neutralizes stomach acids that would normally destroy the pathogen on its way to colonizing the small intestine.

In some cases, absence of a particular dietary substance may prevent disease by depriving a pathogen of critical nutrients. The best example here is the effect sucrose has on the development of dental caries. As we saw in Section 28.3, dietary restriction of sucrose, along with good oral hygiene, can virtually eliminate tooth decay. Without dietary sucrose, the highly cariogenic *Streptococcus mutans* and *Streptococcus sobrinus* are unable to synthesize the dextran layer needed to keep the bacterial cells attached to the teeth.

The Compromised Host

A compromised host is one in whom one or more resistance mechanisms are inactive and in whom the probability of infection is therefore increased. Many hospital patients with noninfectious diseases (for example, cancer and heart disease) acquire microbial infections because they are compromised hosts (∞ Section 33.7). Such healthcare-associated infections are called **nosocomial infections** and affect up to 2 million individuals each year in the United States, causing up to 100,000 deaths. Healthcare procedures such as catheterization, hypodermic injection, spinal puncture, biopsy, and surgery may unintentionally introduce microorganisms into the patient. The stress of surgery may also lower patient resistance. Anti-inflammatory drugs given to reduce pain and swelling may also reduce host resistance. Organ transplant patients are treated with immunosuppressive drugs to suppress immune rejection of the transplant, but suppressed immunity also reduces the ability of the patient to resist infection.

Some factors can compromise host resistance even outside the hospital. Smoking, excess consumption of alcohol, intravenous drug use, lack of sleep, poor nutrition, and acute or chronic infection with another agent are conditions that can reduce host resistance. For example, infection with the human immunodeficiency virus (HIV) predisposes a patient to infections from microorganisms that are not pathogens in uninfected individuals. HIV causes AIDS by destroying one type of immune cell, the CD4 T lymphocytes, involved in the immune response. The reduction in CD4 T cells reduces immunity, and an opportunistic pathogen, a microorganism that does not cause disease in a normal host, can then cause serious disease or even death (∞ Sections 33.6 and 34.15).

Finally, certain genetic conditions compromise the host. For example, genetic diseases that eliminate important parts of the immune system predispose individuals to infections. Individuals with such conditions frequently die at an early age, not from the genetic condition itself, but from microbial infection.

28.13 MiniReview

Age, general health, prior or concurrent disease, genetic makeup, and lifestyle factors such as stress and diet contribute to susceptibility to infectious disease.

- Identify factors that control susceptibility to botulism in infants as compared to adults.
- Identify factors that influence susceptibility to infection and can be controlled by the host.

28.14 Innate Resistance to Infection

Hosts have innate resistance to most pathogens. Many pathogens are adapted to infect only certain hosts, and even closely related hosts with alterations in pathogen receptors or host metabolism can resist infection by individual pathogens. In addition, physical and chemical factors common to vertebrate hosts nonspecifically inhibit invasion by most pathogens (**Figure 28.25** and **Table 28.6**).

Natural Host Resistance

Under certain circumstances, closely related species, or even members of the same species, may have different susceptibilities to a particular pathogen. The ability of a particular pathogen to cause disease in an individual animal species is highly variable. In rabies, for instance, death usually occurs in all species of mammals once symptoms of the disease develop. Nevertheless, certain animal species are much more susceptible to rabies than others. Raccoons and skunks, for example, are extremely susceptible to rabies infection as compared with opossums, which rarely acquire the disease (Section 35.1). Anthrax infects a variety of animals, causing disease symptoms varying from mild pustules in cutaneous anthrax in humans to fatal blood poisoning in cattle. However, pulmonary, or airborne anthrax, such as that induced by weaponized strains used for bioterrorism (Sections 33.11 and 33.12), is almost universally fatal in humans. As another example of innate host resistance, diseases of warm-blooded animals are rarely transmitted to cold-blooded species, and vice versa. Presumably, the metabolic features of one group are not compatible with pathogens that infect the other.

Tissue Specificity

Most pathogens must first adhere and colonize at the site of exposure. Even if pathogens adhere to an exposure site, if the site is not compatible with their nutritional and metabolic needs, the organisms cannot colonize. Thus, if *Clostridium tetani* was ingested, tetanus would not result because the pathogen is either killed by the acidity of the stomach or cannot compete with the well-developed intestinal flora. If, on the other hand, *C. tetani* cells or endospores were introduced into a deep wound, the organism would grow and produce tetanus toxin in the anoxic zones created by local tissue death. Conversely, enteric bacteria such as *Salmonella* and *Shigella* do not cause wound infections but can successfully colonize the intestinal tract.

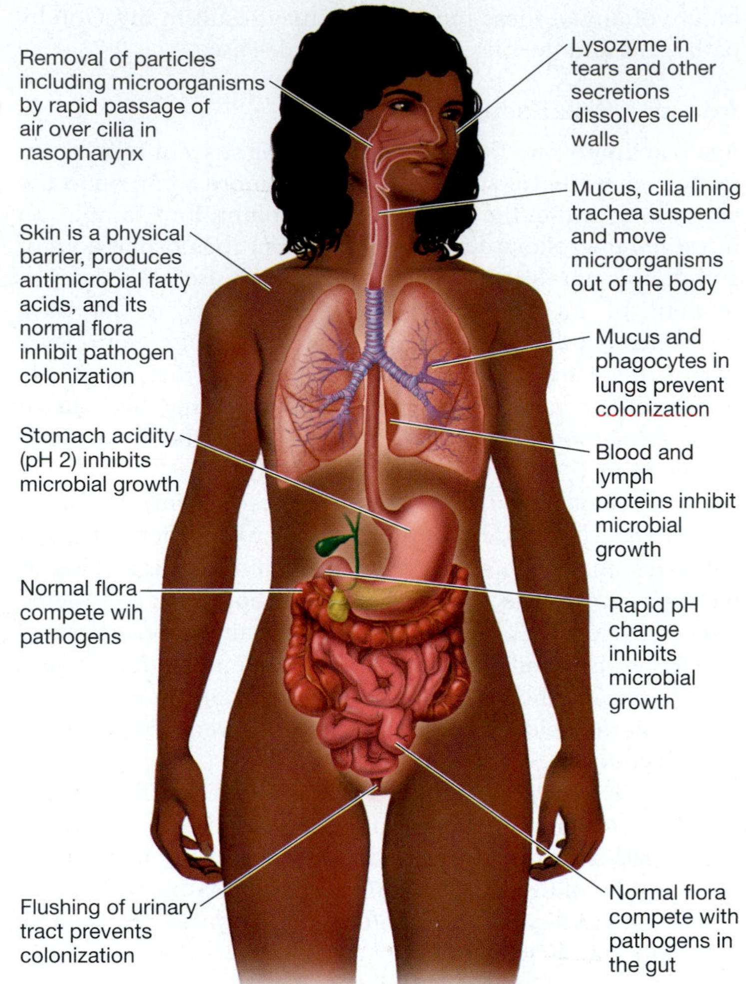
Removal of particles including microorganisms by rapid passage of air over cilia in nasopharynx
Lysozyme in tears and other secretions dissolves cell walls
Mucus, cilia lining trachea suspend and move microorganisms out of the body
Skin is a physical barrier, produces antimicrobial fatty acids, and its normal flora inhibit pathogen colonization
Mucus and phagocytes in lungs prevent colonization
Stomach acidity (pH 2) inhibits microbial growth
Blood and lymph proteins inhibit microbial growth
Normal flora compete wih pathogens
Rapid pH change inhibits microbial growth
Flushing of urinary tract prevents colonization
Normal flora compete with pathogens in the gut

Figure 28.25 Physical, chemical, and anatomical barriers to infection. These barriers provide natural resistance to colonization and infection by pathogens.

In some cases, pathogens interact exclusively with members of a few closely related host species because the hosts share tissue-specific receptors. HIV, for instance, infects only higher primates, including the great apes and humans. This is because the CXCR4 protein on human T cells and the CCR5 protein on human macrophages are also expressed in great apes. These proteins, the only cell surface receptors for HIV, bind the gp120 protein of HIV. Other animals lack these receptors, cannot bind HIV, and are thus protected from HIV infection (Section 34.15). Table 28.6 presents examples of tissue specificity.

Physical and Chemical Barriers

The structural integrity of tissue surfaces poses a barrier to penetration by microorganisms. In the skin and mucosal

Table 28.6 Tissue specificity in infectious disease

Disease	*Tissue infected*	*Organism*
Acquired immunodeficiency syndrome (AIDS)	T helper lymphocytes	Human immunodeficiency virus (HIV)
Botulism	Motor end plate	*Clostridium botulinum*
Cholera	Small intestine epithelium	*Vibrio cholerae*
Dental caries	Oral epithelium	*Streptococcus mutans, S. sobrinus, S. sanguis, S. mitis*
Diphtheria	Throat epithelium	*Corynebacterium diphtheriae*
Gonorrhea	Epithelium	*Neisseria gonorrhoeae*
Malaria	Blood (erythrocytes)	*Plasmodium* spp.
Pyelonephritis	Kidney medulla	*Proteus* spp.
Spontaneous abortion (cattle)	Placenta	*Brucella abortus*
Tetanus	Inhibitory interneuron	*Clostridium tetani*

tissues, potential pathogens must first adhere to tissue surfaces and then grow at these sites before traveling elsewhere in the body. Resistance to colonization and invasion is due to the production of host defense substances and to various anatomical mechanisms.

The skin is an effective barrier to the penetration of microorganisms. Sebaceous glands in the skin (Figure 28.2) secrete fatty acids and lactic acid, lowering the acidity of the skin to pH 5 and inhibiting colonization of many pathogenic bacteria (blood and internal organs are about pH 7.4). Microorganisms inhaled through the nose or mouth are removed by ciliated epithelial cells on the mucosal surfaces of the nasopharynx and trachea. Potential pathogens entering the stomach must survive its strong acidity (pH 2) and then successfully compete with the increasingly abundant resident microflora present in the small and large intestines. Finally, the lumen of the kidney, the surface of the eye, and some other tissues are constantly bathed with secretions containing lysozyme, an enzyme that can kill bacteria (Section 4.6).

28.14 MiniReview

Innate resistance factors, as well as physical, anatomical, and chemical barriers, prevent colonization of the host by most pathogens. Breakdown of these passive defenses may result in susceptibility to infection and disease.

- Identify physical and chemical barriers to pathogens. How might these barriers be compromised?
- How might preexisting infection compromise an otherwise healthy host?

UNIT 6

Review of Key Terms

Attenuation decrease or loss of virulence

Bacteremia the presence of microorganisms in the blood

Capsule dense, well-defined polysaccharide or protein layer closely surrounding a cell

Colonization growth of a microorganism after it has gained access to host tissues

Dental caries tooth decay resulting from bacterial infection

Dental plaque bacterial cells encased in a matrix of extracellular polymers and salivary products, found on the teeth

Disease injury to the host that impairs host function

Endotoxin the lipopolysaccharide portion of the cell envelope of certain gram-negative *Bacteria*, which is a toxin when solubilized

Enterotoxin protein released extracellularly by a microorganism as it grows that produces immediate damage to the small intestine of the host

Exotoxin protein released extracellularly by a microorganism as it grows that produces immediate host cell damage

Host an organism that harbors a parasite

Infection growth of organisms in the host

Invasiveness pathogenicity caused by the ability of a pathogen to enter the body and spread

Lower respiratory tract trachea, bronchi, and lungs

Mucous membrane layers of epithelial cells that interact with the external environment

Mucus soluble glycoproteins secreted by epithelial cells that coat the mucous membranes

Normal microbial flora microorganisms that are usually found associated with healthy body tissue

Nosocomial infection infection contracted in a healthcare-associated setting

Opportunistic pathogen an organism that causes disease in the absence of normal host resistance

Parasite an organism that grows in or on a host and causes disease

Pathogen an organism, usually a microorganism, that causes disease

Pathogenicity the ability of a pathogen to cause disease

Probiotic a live microorganism that, when administered to a host, may confer a health benefit

Slime layer a diffuse layer of polymer fibers, typically polysaccharides, that forms an outer surface layer on the cell

Toxicity pathogenicity caused by toxins produced by a pathogen

Upper respiratory tract the nasopharynx, oral cavity, and throat

Virulence the degree of pathogenicity displayed by a pathogen

Review Questions

1. Distinguish between a parasite and a pathogen. Distinguish between infection and disease. Identify organs in the human body that are normally colonized by microorganisms. Which organs are normally devoid of microorganisms? What do these organs have in common (Section 28.1)?
2. Distinguish between resident and transient microorganisms on the skin. How could you distinguish between resident and transient microorganisms experimentally (Section 28.2)?
3. Why are members of the genus *Streptococcus* instrumental in forming dental caries? Why are they more capable of causing caries than other organisms (Section 28.3)?
4. How do pH and oxygen affect the types of microorganisms that grow in each different region of the gastrointestinal tract (Section 28.4)?
5. Describe the relationship between *Lactobacillus acidophilus* and glycogen in the vaginal tract. What factors influence the differences between the normal vaginal flora of adult females as compared to that of prepubescent juvenile females (Section 28.5)?
6. Identify the role of the capsule and the fimbriae of bacteria in microbial adherence (Section 28.6).
7. Identify nutritional factors that may limit or accelerate growth of microorganisms in the body (Section 28.7).
8. Give an example of a microorganism that is pathogenic almost solely because of its toxin-producing ability. Define the toxin and its mode of action. Give an example of a microorganism that is pathogenic almost solely because of its invasive characteristics. What factors confer invasive qualities on this microorganism (Section 28.8)?
9. Identify the role of coagulase and streptokinase in virulence (Section 28.9).
10. Distinguish between AB toxins, cytotoxins, and superantigens. Give an example of each category of toxin. How does each toxin category promote disease (Section 28.10)?
11. Review the mode of action of cholera enterotoxin (Figure 28.23). What is the appropriate therapy for this disease, and why is antibiotic treatment generally not effective (Section 28.11)?
12. Describe the structure of a typical endotoxin. How does endotoxin induce fever? What microorganisms produce endotoxin (Section 28.12)?
13. Identify common factors that lead to host compromise. Indicate which factors are controllable by the host. Indicate which factors are not controllable by the host (Section 28.13).
14. In which body locations might pH values differ from standard body conditions? Which organisms might benefit or be inhibited by differences in body pH (Section 28.14)?

Application Questions

1. Mucous membranes are barriers against colonization and growth of microorganisms. However, mucous membranes, for example in the throat and the gut, are colonized with a variety of different microorganisms, some of which are potential pathogens. Explain how these potential pathogens are controlled under normal circumstances. Then describe at least one set of circumstances that might encourage pathogenicity.
2. Antibiotic therapy can significantly reduce the number of microorganisms residing in the gastrointestinal tract. What physiological symptoms might the reduction of normal flora produce in the host? Infection by opportunistic pathogens often follows long-term antimicrobial therapy. Many of these post-therapeutic infections are caused by the same microorganisms that produce opportunistic infections in individuals with AIDS. What pathogens might be involved? Why are individuals who have undergone antibiotic therapy particularly susceptible to these pathogens?
3. Design an experiment to increase the virulence and pathogenicity of *Streptococcus pneumoniae* (Hint: *S. pneumoniae* that is transferred for several passages in vitro loses its capsule and virulence for mice). Would an increase in virulence confer a selective advantage for the organism? Be sure to consider the natural habitat.
4. Coagulase is a virulence factor for *Staphylococcus aureus* that acts by causing clot formation at the site of *S. aureus* growth. Streptokinase is a virulence factor for *Streptococcus pyogenes* that acts by dissolving clots at the site of *S. pyogenes* growth. Reconcile these opposing strategies for enhancing pathogenicity.
5. Although mutants incapable of producing exotoxins are relatively easy to isolate, mutants incapable of producing endotoxins are much harder to isolate. From what you know of the structure and function of these types of toxins, explain the differences in mutant recovery.
6. Identify the potential for infectious disease problems in the case of burns to the body. What microorganisms are likely to be involved in burn infections? Why does the normal local microbial flora fail to protect burn victims from microbial infections?

29

Essentials of Immunology

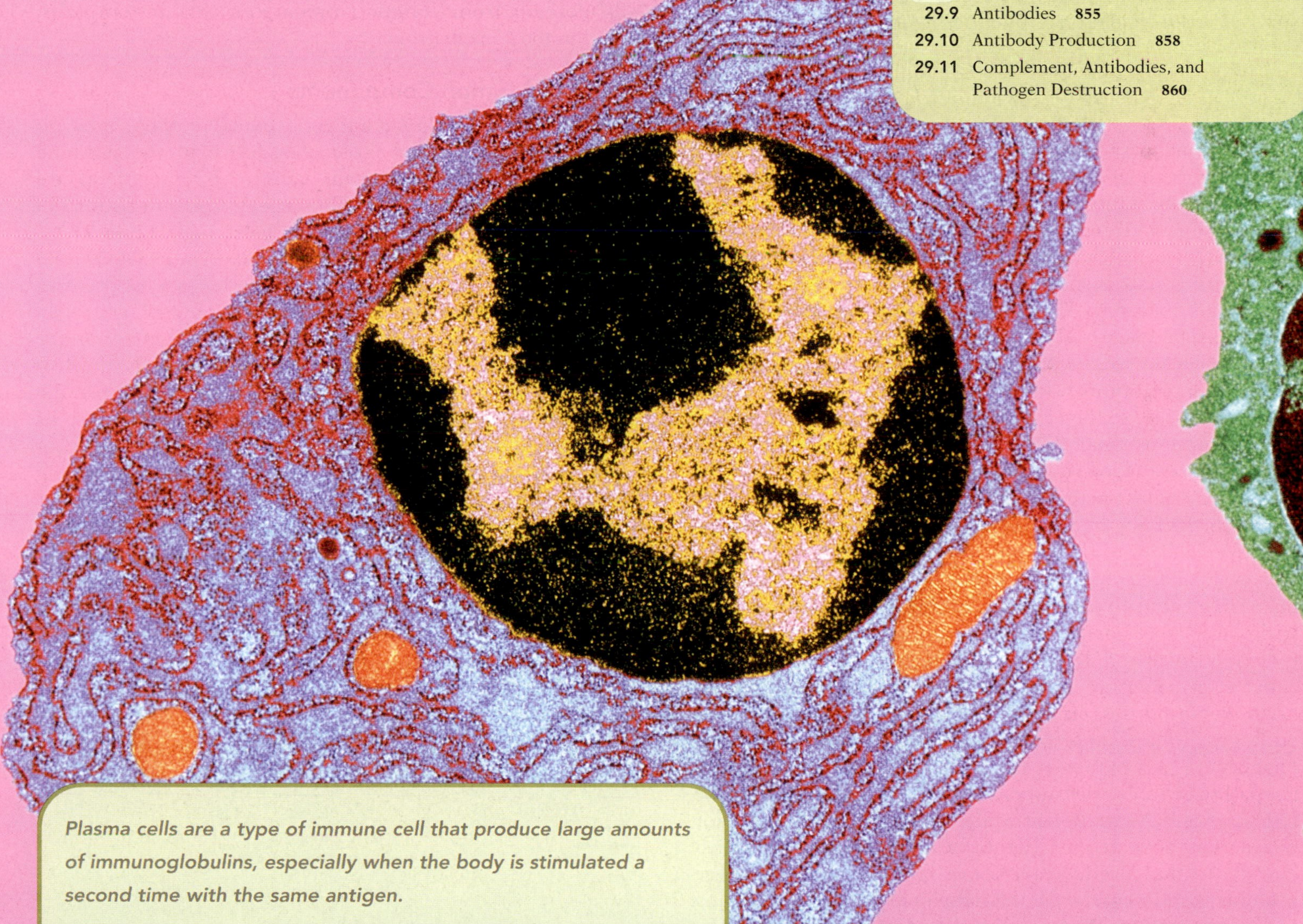

Plasma cells are a type of immune cell that produce large amounts of immunoglobulins, especially when the body is stimulated a second time with the same antigen.

We discussed passive physical and chemical protection against pathogen invasion, infection, and disease in Chapter 28. Now we shift our focus away from the microbiology of pathogens toward the active mechanisms used by multicellular organisms to resist pathogen infection and disease. This active ability to resist disease is called **immunity**.

Multicellular organisms use a variety of cells and their products to kill pathogens and neutralize their effects. The body has a built-in immune system that targets and destroys most common pathogens. A second part of the immune system adapts quickly and efficiently to target particularly dangerous individual pathogens such as new strains of bacteria or viruses. We will first see how our built-in immune system deals with most pathogens. We will then look at the more complex mechanisms the body uses to target individual pathogens. Together, these immune mechanisms have evolved to protect animals from dangerous nonself pathogens; our survival is dependent on a functioning immune system.

I OVERVIEW OF IMMUNITY

The immune response evolved to recognize and destroy dangerous pathogens. We start with **innate immunity**, the body's built-in ability to recognize and destroy pathogens or their products. Innate immunity is largely a function of phagocytes, cells that can engulf, kill, and digest most pathogens. Innate responses recognize a variety of common structural features found on foreign molecules and pathogens, and nearly all phagocytes have the ability to interact with most pathogens. Interactions with pathogens stimulate large numbers of phagocytes to activate a number of genes, leading to the transcription, translation, and expression of proteins that destroy the pathogen. Because many phagocytes are involved, innate immunity develops immediately upon contact with a pathogen.

Unfortunately, innate immune responses are not always effective, and dangerous infections sometimes still occur. However, phagocytes can activate another defense mechanism called **adaptive immunity**, or specific immunity, to deal with these infections. Adaptive immunity is the acquired ability to recognize and destroy a pathogen or its products and is activated by exposure of the immune system to the pathogen. Adaptive responses are directed at discrete molecules on pathogens. These molecules are called *antigens*. Certain phagocytes pass pathogen molecules to lymphocytes, where the pathogen molecules interact with specific receptors on the lymphocyte. The pathogen molecule– lymphocyte receptor interactions activate the lymphocytes to transcribe and translate genes that produce pathogen-specific proteins. These proteins then interact with the individual pathogen, marking it for destruction. A protective adaptive response usually takes several days to develop because only a few lymphocytes are available to interact with each pathogen; the strength of the adaptive response increases as the pathogen-reactive lymphocytes grow and their numbers multiply.

Here we introduce the cells involved in the innate and adaptive host responses to pathogens and other foreign substances. We begin with the cells and organs common to the entire immune system and then consider the cells and mechanisms involved in innate immunity. We finish with an overview of specific, or adaptive, immunity, the focus of the rest of the chapter.

29.1 Cells and Organs of the Immune System

Immunity results from the actions of cells that circulate through the blood and *lymph,* a fluid similar to blood that contains lymphocytes and proteins, but lacks red blood cells. Blood and lymph fluids interact directly or indirectly with every major organ system. All of the cells involved in immunity develop from common precursors called stem cells, found in the *bone marrow*.

Blood and Lymph Components

Blood consists of cellular and noncellular components, including many cells and molecules active in the immune response. The most numerous cells in human blood are erythrocytes (red blood cells), nonnucleated cells that function to carry oxygen from the lungs to the tissues (**Table 29.1**). However, about 0.1% of the cells in blood are nucleated white blood cells, or **leukocytes**. Leukocytes include phagocytic cells such as monocytes, as well as **lymphocytes**, cells specialized for antibody production and cell-mediated immunity.

As shown in **Figure 29.1**, pleuripotent stem cells, the progenitors of white blood cells, are produced and develop in the bone marrow. Stem cells differentiate to produce mature cells under the influence of soluble proteins called **cytokines**, proteins that influence many aspects of immunity, including growth of stem cells.

Whole blood is composed of **plasma**, a liquid containing proteins, other solutes, and suspended cells. Outside the body, whole blood or plasma quickly forms an insoluble clot, which is caused by the conversion of a soluble protein called *fibrinogen* to the insoluble *fibrin*. Plasma or whole blood remains liquid only when an anticoagulant is added. The addition of anticoagulants such as potassium citrate or

Table 29.1 Major cells found in normal human blood

Cell type	*Cells per milliliter*
Erythrocytes	$4.2–6.2 \times 10^9$
Leukocytes[a]	$4.5–11 \times 10^6$
Lymphocytes	$1.0–4.8 \times 10^6$
Myeloid cells	Up to 7.0×10^6

[a]Leukocytes include all nucleated blood cells and are subdivided into the lymphocytes and myeloid cells (monocytes and granulocytes).

Source: Henry, J. B. 1996. *Clinical Diagnosis and Management by Laboratory Methods*, 19th edition. W. B. Saunders, Philadelphia.

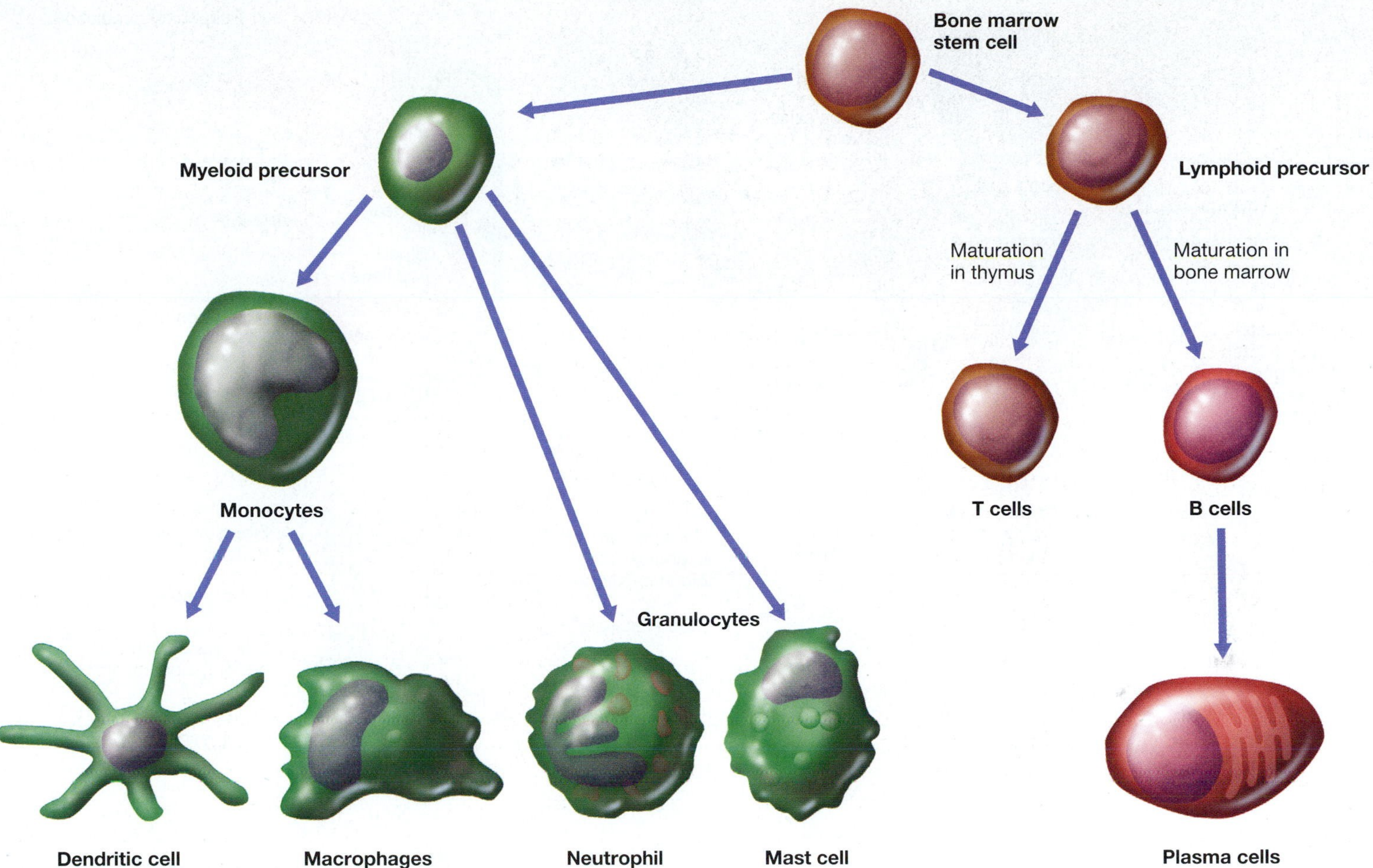

Figure 29.1 The origins of immune response cells. Immune response cells develop from pleuripotent stem cells in the bone marrow. These cells can develop into two immune cell precursors. The myeloid precursor may generate monocytes, which in turn develop into macrophages or dendritic cells, phagocyte cells involved in antigen uptake and display. The myeloid precursor can also develop into any one of a number of granulocytes, including neutrophils, mast cells, eosinophils, and basophils. Neutrophils are phagocytes. The others release their granule contents in response to pathogens, pathogen products, or damaged host cells. The lymphoid precursor generates T and B cells, the lymphocytes that participate directly in the adaptive immune response. Plasma cells derived from B cells produce antibodies.

heparin prevents the conversion of fibrinogen to fibrin, stopping the clotting process. When blood clots, the insoluble proteins trap the cells in a large, insoluble mass. The remaining fluid, called **serum**, contains no cells or clotting proteins. Serum does, however, contain a high concentration of other proteins, including soluble antibody proteins, and is widely used in immunological investigations. The use of serum antibodies to detect antigens is called *serology*.

Blood and Lymph Circulation

Blood is pumped by the heart through arteries and capillaries throughout the body and is returned through the veins (**Figure 29.2*a*,*b***). In the capillary beds, leukocytes and solutes pass to and from the blood into the lymphatic system, a separate circulatory system containing lymph (Figure 29.2*b*,*c*).

Lymph fluid drains from extravascular tissues into lymphatic capillaries, lymph ducts, and then into lymph nodes (Figure 29.2*d*) found throughout the lymph system. Lymph nodes contain high concentrations of lymphocytes and phagocytes, arranged in such a way that they encounter microorganisms and antigens as they enter the nodes via the lymph ducts. The mucosal-associated lymphoid tissue (MALT) is also a part of the lymphatic system and interacts with antigens and microorganisms that originate from the gut and bronchial mucosal tissues. The MALT has the same cellular components as the lymph nodes. The white pulp in the spleen also consists of organized concentrations of lymphocytes and phagocytes, arranged to filter the blood circulatory system. Collectively, the lymph nodes, MALT, and spleen are called *secondary lymphoid organs* and are the sites where antigens are concentrated and interact with antigen-presenting phagocytes and lymphocytes to generate an adaptive immune response (Figure 29.2*a*). Lymph fluid with antibodies and immune cells empties into the circulatory system via the thoracic lymph duct.

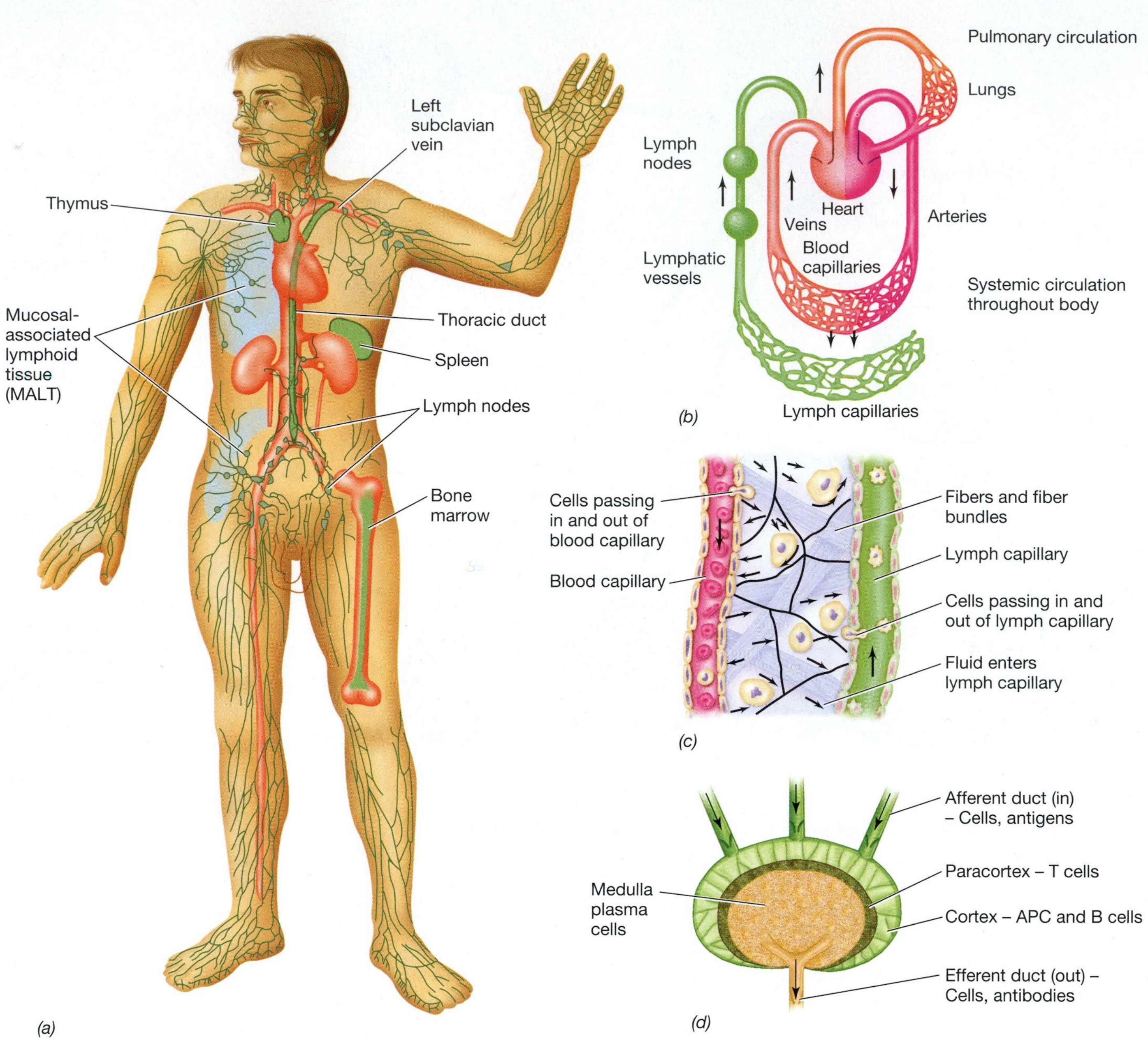

Figure 29.2 The blood and lymph systems. *(a)* The lymph system, showing the locations of major organs. The primary lymphoid organs are the bone marrow and thymus. The secondary lymphoid organs include the lymph nodes, spleen, and MALT. *(b)* Connections between the lymph and blood systems. Blood flows from the veins to the heart, to the lungs, and then through the arteries to the tissues. Lymph drains from the thoracic duct into the left subclavian vein of the blood circulatory system. *(c)* The exchange of cells between the blood and lymph systems is shown microscopically. Both blood and lymph capillaries are closed vessels, but cells pass from blood capillaries to lymph capillaries and back by a process known as extravasation. *(d)* A secondary lymphoid organ, the lymph node. The diagram identifies major anatomic areas and the immune cells present in each area. The MALT and spleen have analogous anatomic features.

Leukocytes

Leukocytes are nucleated white blood cells found in the blood and the lymph. Several distinct leukocytes (Table 29.1 and Figure 29.1) participate in innate or adaptive immune functions. Lymphocytes are specialized leukocytes involved exclusively in the adaptive immune response. Mature lymphocytes are concentrated in the lymph nodes and spleen where they interact with antigens. There are two types of lymphocytes, B cells (B lymphocytes) and T cells (T lymphocytes) (Figure 29.1). **B cells** originate and mature in the **bone marrow** and are the precursors of antibody-producing plasma cells. **T cells** begin their development in the bone marrow, but travel to the **thymus** to mature. The bone marrow and thymus in mammals are called **primary lymphoid organs** because they are the sites where stem cells and precursors develop into functional antigen-reactive lymphocytes (Figure 29.2*a*).

Myeloid cells are all derived from a shared myeloid precursor cell. The mature myeloid cells can be divided into two categories. The first category includes specialized phagocytic cells, the **antigen-presenting cells (APCs)**, that engulf, process, and present antigens to lymphocytes. These cells are the *monocytes, macrophages,* and *dendritic cells.* Monocytes are circulating precursors of macrophages, cells found in abundance in tissues, spleen, and lymph nodes (Figure 29.1). Dendritic cells are also phagocytes with antigen-presenting properties, and most have monocyte precursors.

A second category of cells derived from myeloid precursors is the granulocytes. Granulocytes contain cytoplasmic inclusions, or granules, that can be visualized using staining techniques. These granules contain toxins or enzymes that are released to kill target cells. The granulocytes include neutrophils, also called polymorphonuclear leukocytes, or PMNs. Neutrophils are phagocytic granulocytes. Other granulocytes, including mast cells, basophils, and eosinophils, also originate from the myeloid precursor. Release of the granules within these cells, a process called degranulation, can cause inflammation and allergy-like symptoms.

Leukocytes actively move throughout the body and pass from blood to interstitial spaces, then to lymphatic vessels, and back to the blood circulatory system, a process called extravasation (Figure 29.2*c*).

29.1 MiniReview

Cells involved in innate and adaptive immunity originate from bone marrow stem cells. The blood and lymph systems circulate cells and proteins that are important components of the immune response. Leukocytes participate in immune responses in all parts of the body.

- Describe the circulation of a leukocyte from the blood to the lymph and back to the blood.
- Trace the development of B cells, T cells, and macrophages from the common stem cell.

29.2 The Innate Immune Response

Pathogens sometimes breach the host physical and chemical barriers described in Chapter 28 (Section 28.14), leading to host infection. When infection starts, the immune system must be mobilized to protect the host from further damage, and the innate or nonspecific immune response is the first line of defense. Innate immunity begins when a phagocyte contacts a foreign substance such as a pathogen or a pathogen product. Several different phagocytes can engulf and destroy pathogens, often initiating complex host-mediated reactions collectively called *inflammation*. Here we look at the cells and molecules that recognize pathogens and initiate protective innate host responses.

Phagocytes

The first cell type active in the inflammatory response is usually a **phagocyte** (literally, a cell that eats). The primary function of a phagocyte is to engulf and destroy pathogens. In this process, some phagocytes also function as APCs, processing and displaying the peptide antigens that initiate the adaptive immune response.

Phagocytes include macrophages, monocytes, neutrophils, and immature dendritic cells. Found in tissues and fluids throughout the body, phagocytes are usually motile, moving by ameboid action. Most have inclusions called lysosomes, which contain bactericidal substances such as hydrogen peroxide, lysozyme, proteases, phosphatases, nucleases, and lipases. Phagocytes trap and engulf pathogens on surfaces such as blood vessel walls or fibrin clots. The membrane surrounding the pathogen pinches off and forms a phagosome. The phagosome, now containing the engulfed pathogen, then moves inside the cell and fuses with a lysosome to form a phagolysosome. The toxic substances and enzymes inside the phagolysosome usually kill and digest the engulfed microbial cell (**Figure 29.3**).

One group of phagocytes, the **neutrophils**, or **polymorphonuclear leukocytes (PMNs)**, are actively motile granulocytes containing large numbers of lysosomes (**Figure 29.4a**). Derived from the myeloid stem cell (Figure 29.1), neutrophils are found predominantly in the bloodstream and bone marrow, from where they migrate to sites of active infection in tissues. Neutrophils present in higher than normal numbers in the blood or at a site of inflammation indicate an active response to a current infection.

Monocytes are circulating precursors of macrophages, a major phagocytic cell type (Figures 29.4*a* and 29.3). Macrophages are large cells found in tissues such as lymph nodes and spleen, where they ingest and destroy most pathogens and foreign molecules as well as interact with lymphocytes to initiate adaptive immune responses. Macrophages also act as APCs, presenting peptide antigens to T cells, the first step in activating an adaptive immune response.

Dendritic cells (dendrocytes) (Figure 29.1) also have the dual function of phagocytosis and antigen presentation. Derived from the same monocyte progenitor as macrophages, immature dendritic cells are found throughout the body tissues where they function as very active phagocytes. When the dendritic cells ingest antigen, they migrate to the lymph nodes, where they present antigen to T cells. The specialized antigen-presenting properties of macrophages and dendritic cells are examined in Section 29.6.

Pathogen Recognition by Phagocytes

Phagocytes have a general pathogen-recognition system designed to trigger a timely and appropriate response, generally leading to destruction and containment of the pathogen. This system employs several evolutionarily conserved **pattern-recognition molecules (PRMs)**. PRMs are membrane-bound phagocyte proteins that recognize a

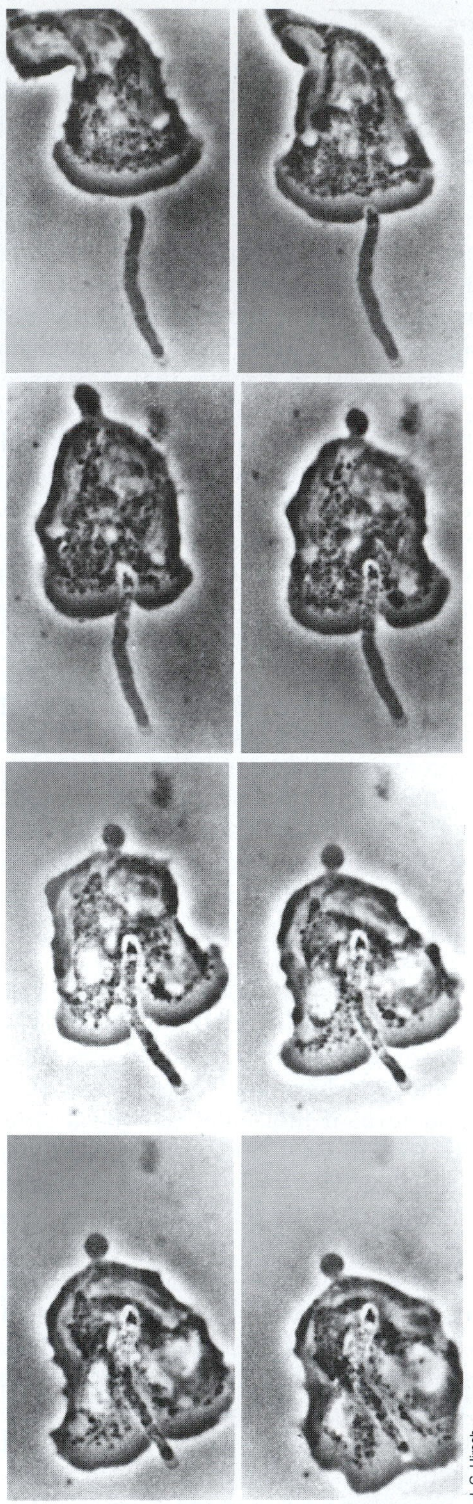

Figure 29.3 Phagocytosis. Time-lapse photomicrographs of the phagocytosis and digestion of a chain of *Bacillus megaterium* cells by a human macrophage, observed by phase-contrast microscopy. The bacterial chain is about 18–20 μm long. The macrophage is one of a group of cells that ingests and degrades pathogens and pathogen products.

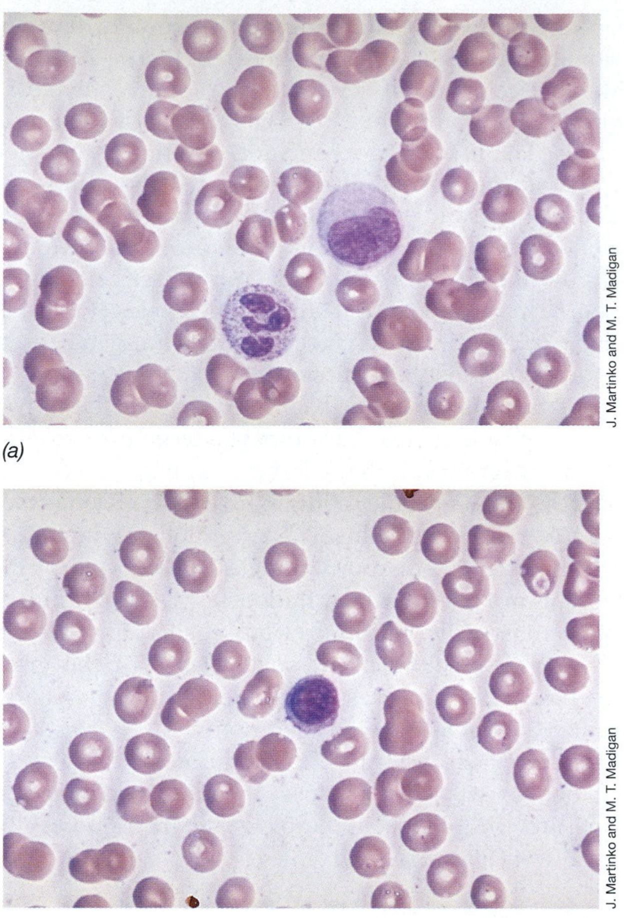

Figure 29.4 Major immune cell types. *(a)* The nucleated cell in the lower left center is a neutrophil (PMN), characterized by a segmented nucleus (violet stain) and granular cytoplasm. The nucleated cell to the right and slightly above the PMN is a monocyte. These phagocytes are 12–15 μm in diameter. The nonnucleated red blood cells are about 6 μm in diameter. *(b)* The nucleated cell is a circulating lymphocyte. The lymphocyte has almost no visible cytoplasm and is smaller than the phagocytes, about 10 μm in diameter.

pathogen-associated molecular pattern (PAMP), a unique structural component on a microbial cell or virus (area 1 of Figure 29.5). PRMs were first observed in phagocytes in *Drosophila*, the fruit fly, where they are called *Toll receptors*. Each **Toll-like receptor (TLR)** on a human phagocyte recognizes a specific PAMP. For example, TLR-4, a PRM on human phagocytes, responds to interactions with the lipopolysaccharide (LPS), a PAMP in the outer membrane of gram-negative bacteria (Section 4.7), inducing phagocyte activation and immunity to all gram-negative pathogens. Other TLRs recognize PAMPS such as the unmethylated CpG oligonucleotides and peptidoglycan molecules found only in bacteria. In addition to these phagocyte-associated PRMs, several soluble host molecules have similar functions, and we will discuss these unique, soluble PRMs in the context of their ability to activate complement, a system of interacting proteins that leads to

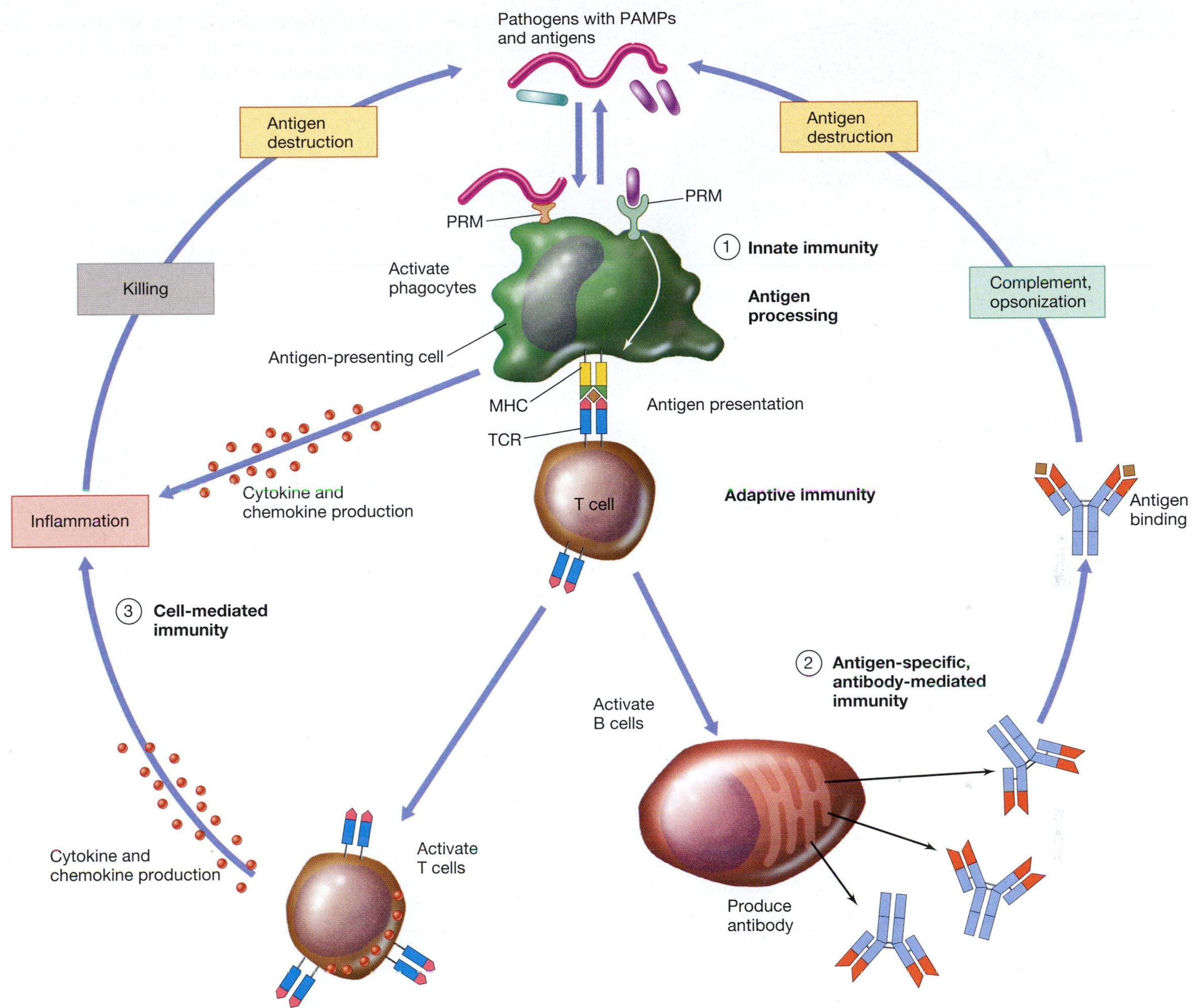

Figure 29.5 Overview of the immune response. Pathogens are targeted and destroyed by three immune mechanisms. ① Innate immunity results from interactions between pathogen-associated molecular patterns (PAMPs) found as cell surface components of pathogens and pattern recognition molecules (PRMs) found on phagocytes. Adaptive immunity, orchestrated by antigen-specific T cells, results in two distinct effector pathways: ② Antibody-mediated immunity results from soluble antigen-specific antibody proteins, products of antigen-stimulated B lymphocytes, and ③ Cell-mediated immunity is mediated by antigen-specific T cells.

enhanced phagocytosis or outright destruction of pathogens (Section 29.11). For a more extensive discussion of TLRs and other PRMs and PAMPs, see Section 31.1.

Interaction of a PAMP with the phagocyte PRM triggers a transmembrane signal that results in transcription of a number of cell proteins in the phagocyte. This transcriptional activation leads to production of toxic oxygen compounds that can cause pathogen death.

Oxygen-Dependent Activation of Genes in Phagocytes

The activation of certain genes in phagocytes enhances their phagocytic and pathogen-killing abilities. The activation of genes produces toxic oxygen-containing compounds, including hydrogen peroxide (H_2O_2), superoxide anions (O_2^-), hydroxyl radicals ($OH\cdot$), singlet oxygen ($^1O_2$), hypochlorous acid (HOCl), and nitric oxide (NO) (Section 6.18). The acidic conditions

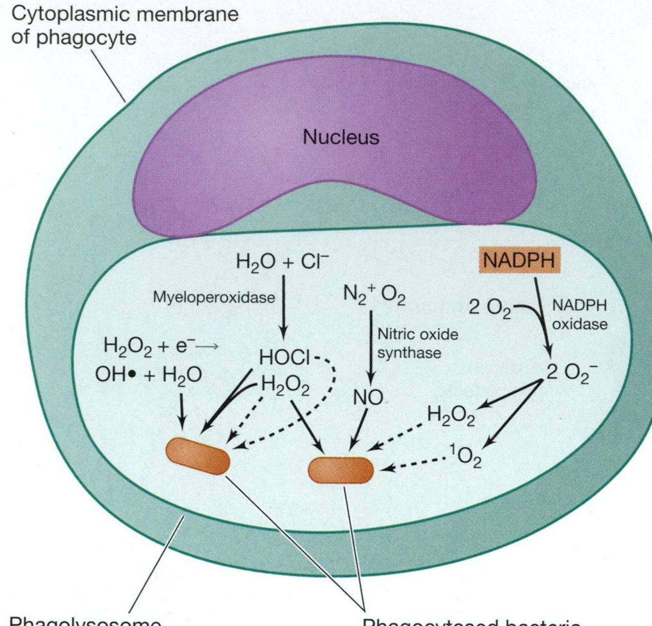

Figure 29.6 Action of phagocyte enzymes in generating toxic oxygen species. These include hydrogen peroxide (H_2O_2), the hydroxyl radical (OH•), hypochlorous acid (HOCl), the superoxide anion (O_2^-), singlet oxygen ($^1O_2$), and nitric oxide (NO). Formation of these toxic compounds requires a substantial increase in the uptake and utilization of molecular oxygen, O_2. This increase in oxygen uptake and consumption by activated phagocytes is known as the respiratory burst.

in the phagolysosome foster the generation and reactivity of these compounds. Phagocytic cells use toxic oxygen compounds to kill ingested bacterial cells by oxidizing key cellular constituents. The oxidations occur within the phagocytic cell itself, which is not damaged by the toxic oxygen products. Oxygen-mediated killing by phagocytes is summarized in **Figure 29.6**. Activated phagocytes take up and use larger than normal quantities of O_2 over a short time period to produce toxic oxygen compounds. This increased rate of O_2 uptake by activated phagocytes is called the *respiratory burst*.

Inhibiting Phagocytes

Some pathogens have developed mechanisms for neutralizing toxic phagocyte products, for killing the phagocytes, or for avoiding phagocytosis. For example, *Staphylococcus aureus* produces pigmented compounds called carotenoids that neutralize singlet oxygen and prevent killing (Section 34.10). Intracellular pathogens such as *Mycobacterium tuberculosis* (the cause of tuberculosis) grow and persist within phagocytic cells (Section 34.5). *M. tuberculosis* uses its cell wall glycolipids to absorb hydroxyl radicals and superoxide anions, the most lethal toxic oxygen species produced by phagocytes.

Some intracellular pathogens produce phagocyte-killing proteins called *leukocidins*. In such cases, the pathogen is ingested as usual, but the leukocidin kills the phagocyte and the pathogen is then released. The dead phagocytes make up much of the material of *pus*; organisms such as *Streptococcus pyogenes* and *S. aureus*, major leukocidin producers, are called *pyogenic* (pus-forming) pathogens. Localized infections by pyogenic bacteria often result in boils or abscesses.

Another important microbial defense against phagocytosis is the bacterial capsule (Section 4.9). Encapsulated bacteria are often highly resistant to phagocytosis, apparently because the capsule prevents adherence of the phagocyte to the bacterial cell. The clearest case of the importance of a capsule that prevents phagocytosis is that of *Streptococcus pneumoniae*. Fewer than ten cells of an encapsulated strain of *S. pneumoniae* can kill a mouse in a few days after injection (Figure 28.16). On the other hand, nonencapsulated strains are completely avirulent. Surface components other than capsules can also inhibit phagocytosis. For instance, pathogenic *S. pyogenes* produces M-protein, a substance that alters the surface of the bacterial cell and inhibits phagocytosis.

Antibodies or soluble PRMs that interact with capsules or other cell surface molecules often reverse the protective effect of bacterial defense mechanisms and enhance phagocytosis, a process known as *opsonization*. We will discuss opsonization in the context of complement activation in Section 29.11.

29.2 MiniReview

Phagocytes recognize pathogen-associated molecular patterns (PAMPs) via a family of membrane-bound pattern-recognition molecules (PRMs). Interaction of the PAMPs with PRMs activates phagocytes to produce oxygen-containing compounds and proteins that kill the pathogen or limit its effects. Many pathogens have developed mechanisms to inhibit phagocytes.

- Describe the cellular location and molecular specificity of PAMPs and PRMs.
- Identify toxic oxygen products and explain the respiratory burst that takes place in activated phagocytes.
- Identify at least one mechanism used by pathogens to inhibit phagocytosis.

29.3 Inflammation, Fever, and Septic Shock

Inflammation is a nonspecific reaction to noxious stimuli such as toxins and pathogens. Inflammation causes redness (erythema), swelling (edema), pain, and heat, usually localized at the site of infection. The molecular mediators of inflammation include a group of proteins called *cytokines* and *chemokines*. These proteins are produced by various immune cells, but especially by phagocytes and lymphocytes.

Inflammation is the usual outcome of either an innate or an adaptive immune response; both responses induce the same inflammatory mediators that recruit and activate the same effector cells and molecules. An effective inflammatory response isolates and limits tissue damage, destroying pathogen invaders as well as damaged cells at local sites. In some cases, however, inflammation can result in considerable damage to healthy host tissue.

Inflammatory Cells and Local Inflammation

The first inflammatory cells that arrive at the scene of an infection or tissue injury are neutrophils (**Figure 29.7a**). These phagocytes are attracted to the site of an active infection or tissue injury by soluble chemoattractants, such as the **interleukins**. For example, interleukin-8 (IL-8), a small protein in the chemokine family, is released from damaged host cells. Neutrophils interact with IL-8, activating genes that foster migration toward the cells secreting IL-8. The activated neutrophils migrate to the damaged cells and ingest them. They also secrete other chemokines such as MIP-α and MIP-β (macrophage inflammatory proteins). These chemokines recruit macrophages to the same site (Figure 29.7*b*), guiding the macrophages along the chemokine gradient toward the neutrophils at the site of infection or injury.

Interaction with chemokines also activates certain macrophage genes, leading to enhanced phagocytosis and the production and secretion of a number of cytokines (Figure 29.5). The major macrophage cytokines active in inflammation are IL-1, IL-6, and TNF-α (tumor necrosis factor-alpha). The chemokine and cytokine mediators released by injured cells and phagocytes contribute to inflammation. For example, the inflammatory cytokines IL-1, IL-6, and TNF-α increase vascular permeability, causing the swelling (edema), reddening (erythema), and local heating associated with inflammation. The edema stimulates local neurons, causing pain.

The usual outcome of the inflammatory response is a rapid localization and destruction of the pathogen by the recruited neutrophils and macrophages. As the pathogen is destroyed, the inflammatory cells are no longer stimulated, their numbers at the site are reduced, cytokine production decreases, chemoattraction stops, and inflammation subsides.

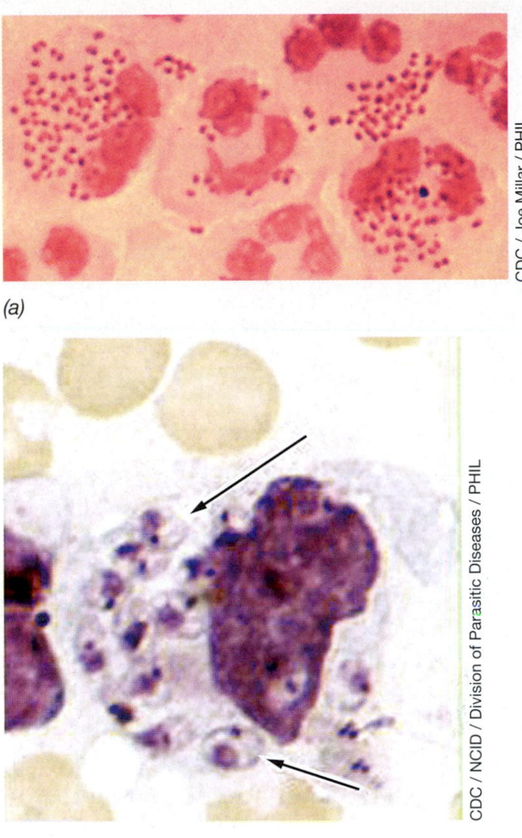

Figure 29.7 Phagocytes. *(a)* Neutrophils actively ingest and kill *Neisseria gonorrhoeae*. The neutrophils are about 12–15 μm in diameter. Note the multilobed nucleus in each cell. Not all neutrophils have ingested bacteria. *(b)* A skin macrophage that has taken up numerous *Leishmania* (arrows), a protozoan. Macrophages activated by T cells are capable of killing many intracellular parasites.

Systemic Inflammation and Septic Shock

In some cases, the inflammatory response fails to localize the pathogen and the reaction becomes widespread. Then inflammatory cells and mediators contribute to inflammation on a larger scale. An inflammatory response that spreads inflammatory cells and mediators through the entire circulatory and lymphatic systems can lead to septic shock, a life-threatening condition. The most common cause of septic shock is systemic infection by gram-negative enteric bacteria such as *Salmonella* or *Escherichia coli,* often caused by a ruptured or leaking bowel that releases the gram-negative organisms into the intraperitoneal cavity or the bloodstream. The primary infection is often cleared by the phagocytes or is treated successfully with antibiotics. However, the endotoxic outer membrane lipopolysaccharides (LPS) from these organisms interact with TLR-4, a PRM on the phagocytes, stimulating production of IL-1, IL-6, and IL-8, and TNF-α, which are released into the systemic circulation. The cytokines then induce systemic responses that mimic the localized inflammatory response. For example, IL-1, IL-6, and TNF-α are *endogenous pyrogens*, producing fever by stimulating release of prostaglandins in the brain. In the liver, the inflammatory cytokines induce production of a special class of proteins called *acute phase proteins*. These include C-reactive protein, complement proteins, serum amyloid protein, mannan-binding lectin, and fibrinogen. These proteins enhance opsonization and phagocytosis, usually accelerating pathogen destruction.

Systemic inflammatory reactions, however, may have very serious consequences. The systemic release of large quantities of endogenous pyrogens, instead of producing localized heating, induces uncontrollable high fever. In systemic inflammation, the same mechanism that causes local edema due to vasodilation and increased vascular permeability now causes massive efflux of fluids from the central vascular tissue, resulting in loss of systemic blood pressure and severe edema in the surrounding tissues. The resulting condition, termed septic shock, is characterized by the loss of blood volume as well as the high fever and causes death in up to 30% of affected individuals. Uncontrolled systemic inflammation can be more dangerous than the original infection.

29.3 MiniReview

Inflammation, characterized by pain, swelling (edema), redness (erythema), and heat, is a normal and generally desirable outcome due to activation of nonspecific immune response effectors. Uncontrolled systemic inflammation, called septic shock, can lead to serious illness and death.

- Identify the major symptoms of localized inflammation and of septic shock.
- Identify the molecular mediators of inflammation and define their individual roles.

29.4 The Adaptive Immune Response

The adaptive immune response is outlined in Figure 29.5. Phagocytes such as macrophages and dendritic cells, as well as B lymphocytes, take up and digest pathogens. These cells are called professional APCs; they process pathogen components into smaller pieces called *antigens*. The APCs then present peptide antigens to T lymphocytes. The T lymphocytes recognize the peptide antigen through antigen-specific, cell surface **T cell receptors (TCRs)** that interact specifically with a single peptide antigen. Some T cells, the T-cytotoxic (T_C) cells, directly attack and destroy antigen-bearing cells. Other antigen-activated T cells, the T-helper 1 (T_H1) cells, act indirectly by secreting cytokines that activate cells such as macrophages to destroy the antigen-bearing cells. This **cell-mediated immunity** leads to killing of pathogen-infected cells through recognition of pathogen antigens found on infected host cells.

Another subset of T cells, the T-helper 2 (T_H2) cells, interacts with antigen-specific B cells, stimulating them to make antibodies (immunoglobulins). Each B cell has a unique antigen receptor, an antibody, as a cell surface receptor, and the antibody is specific for a single antigen. Antigen-stimulated B cells produce soluble copies of the cell surface antibody; the soluble antibody proteins interact specifically with antigens in the body to neutralize antigens or target them for destruction. This **antibody-mediated immunity** is particularly effective against extracellular pathogens such as bacteria and soluble pathogen products such as toxins in the blood or lymph.

As we discussed, innate immunity is directed against common pathogen features. By contrast, adaptive immunity is directed to interactions with individual pathogen-specific macromolecules. Adaptive immunity is characterized by the properties of *specificity, memory,* and *tolerance*. None of these properties is found in the innate response.

Specificity

The **specificity** of the antigen–antibody or antigen–TCR interaction is dependent on the capacity of the lymphocyte cell receptor to interact with individual pathogen antigens. The innate host response challenges virtually any invading microorganism, even those pathogens the host has never before encountered. In the adaptive immune response, effective

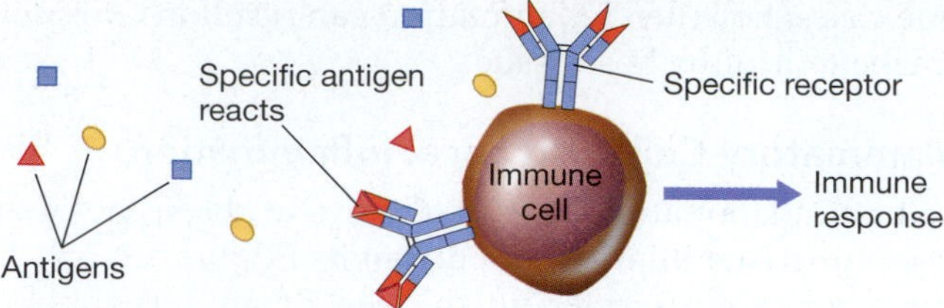

(a) **Specificity:** Immune cells recognize and react with individual molecules (antigens) via direct molecular interactions.

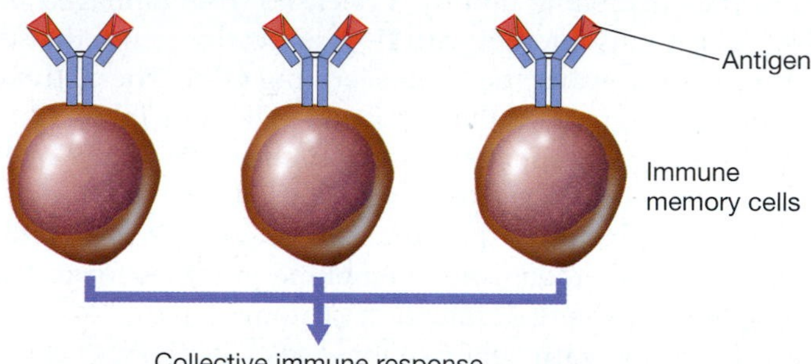

(b) **Memory:** The immune response to a specific antigen is *faster* and *stronger* upon subsequent exposure because the initial antigen exposure induced growth and division of antigen-reactive cells, resulting in multiple copies of antigen-reactive cells.

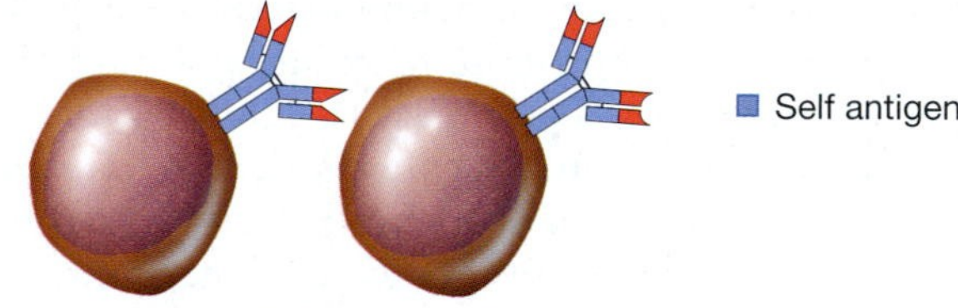

(c) **Tolerance:** Immune cells are not able to react with self antigen. Self-reactive cells are destroyed during development of the immune response.

Figure 29.8 The adaptive immune response. Key features of antibody-mediated and cell-mediated immunity are (*a*) specificity, (*b*) memory, and (*c*) tolerance.

immunity cannot be detected for several days after the first contact with the pathogen. However, once the adaptive immune response is triggered by antigen contact, it is exclusively and specifically directed to the eliciting pathogen through recognition of unique molecular antigen features (**Figure 29.8*a***).

Memory

The immune system must encounter antigen to stimulate production of detectable and effective antigen-specific antibodies or T cells, but a later exposure to the same antigen stimulates rapid production of large quantities of antigen-reactive T cells or antibodies. This capacity to respond more quickly and vigorously to subsequent exposures with the eliciting antigen is known as **immune memory** (Figure 29.8*b*). Immune memory allows the host to specifically resist previously encountered pathogens. We take advantage of immune memory by immunizing (inoculating, vaccinating) susceptible individuals with dead or weakened pathogens (or their products) to artificially stimulate and enhance immunity for a number of dangerous pathogens.

Tolerance

Tolerance is the acquired inability to make an adaptive immune response to an individual's own antigens. Because all macromolecules in the host are potential antigens, the immune system must avoid recognizing host macromolecules because they would be damaged if recognized by antibodies or T cells (Figure 29.8*c*). Thus, the adaptive immune response must develop the capacity to discriminate between *foreign* (nonself and dangerous) antigens and *host* (self and not dangerous) antigens.

29.4 MiniReview

Nonspecific phagocytes present antigen to specific T cells, triggering the production of T_H1, T_H2, and T_C cells as well as antibodies. T cells and antibodies react directly or indirectly to neutralize or destroy the antigen. The adaptive immune response is characterized by specificity for the antigen, the ability to respond more vigorously when reexposed to the same antigen (memory), and the ability to discriminate self-antigens from nonself antigens (tolerance).

- Identify the antigen-specific cells in the cell-mediated and antibody-mediated immune responses.
- What would be the outcome of a breakdown of immune specificity, memory, or tolerance with regard to a host response to a pathogen?

II ANTIGENS AND ANTIGEN PRESENTATION

The adaptive immune response recognizes a broad range of pathogen-derived macromolecules. The macromolecules are degraded and processed in host cells to produce molecules called antigens that are in turn presented to T lymphocytes. We first discuss antigens and then focus on the mechanisms of antigen processing and presentation to T cells.

29.5 Immunogens and Antigens

Antigens are substances that react with antibodies or TCRs. Most, but not all, antigens are **immunogens**, substances that induce an immune response. Here we examine the features of effective immunogens and then define the features of antigens that promote interactions with antibodies and TCRs.

Intrinsic Properties of Immunogens

Immunogens must meet several molecular criteria to effectively induce an adaptive immune response. Immunogens all share minimum intrinsic properties.

Molecular size is an important component of immunogenicity. For example, low-molecular-weight compounds called **haptens** cannot induce an immune response but can bind to antibodies. Because haptens are bound by antibodies, they are antigens even though they are not immunogenic. Haptens can include sugars, amino acids, and other low-molecular-weight organic compounds. When coupled to a larger protein carrier, haptens become effective immunogens. Most immunogens have a molecular weight of 10,000 or greater. Thus, sufficient molecular size is an indication of potential immunogenicity.

Complex, nonrepeating polymers such as proteins are usually effective immunogens. Complex carbohydrates can also be very effective immunogens. In contrast, nucleic acids, simple polysaccharides, and lipids, because they are composed of repeating monomers, tend to be poor immunogens. Thus, the *molecular complexity* of a substance is another predictor of immunogenicity.

Large, complex macromolecules in insoluble or aggregated form (for example, proteins precipitated by heating) are usually excellent immunogens. The insoluble material is readily taken up by phagocytes, leading to an immune response. By contrast, the soluble form of the same molecule is often a very poor immunogen because the soluble molecule is usually not ingested efficiently by phagocytes. Thus, *appropriate physical form* is another condition of immunogenicity.

Extrinsic Properties of Immunogens

Although many substances are intrinsically immunogenic, several *extrinsic* factors also influence immunogenicity. These include the *dose* of the immunogen, the *route* of administration, and the *foreign nature* of the immunogen with respect to the host.

The dose of an immunogen administered to a host can be important for an effective immune response, but a broad range of doses ordinarily provides satisfactory immunity. In general, doses of 10 μg to 1 g are effective in most mammals. Doses of immunogen higher than 1 g or lower than 10 μg may not stimulate an immune response; extremely high or low doses may actually suppress a specific immune response and cause tolerance.

The route of administration of an immunogen is also important. Immunizations given by parenteral (outside of the gastrointestinal tract) routes, usually by injection, are normally more effective than those given topically or orally. When given by oral or topical routes, antigens may be significantly degraded before contacting a phagocyte.

The final and most important extrinsic feature of an immunogen is that an effective immunogen must be foreign with respect to the host. The adaptive immune system recognizes and eliminates only foreign (nonself) antigens. Self-antigens are not recognized, and thus individuals are tolerant of their own self-molecules, even though the same antigens may have the capacity to be immunogens in other individuals of the same species.

Antigen Binding by Antibodies and T Cell Receptors

The antibody or TCR does not interact with the antigenic macromolecule as a whole, but only with a distinct portion of

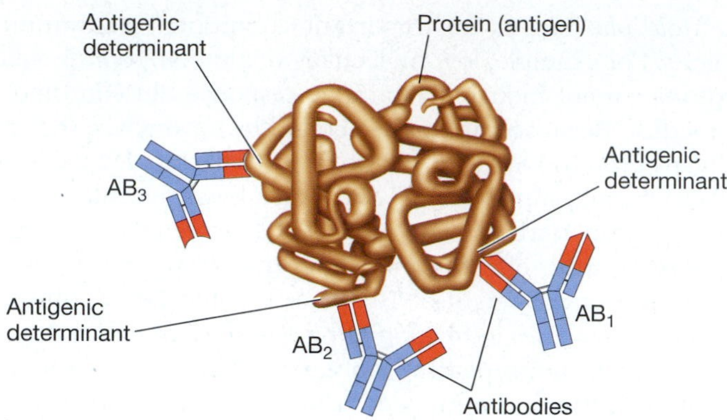

Figure 29.9 Antigens and antigenic determinants for antibodies. Antigens may contain several different antigenic determinants, each capable of reacting with a different specific antibody. The antigenic determinant recognized by AB_1 is a conformational determinant consisting of two different parts of the same polypeptide antigen. The polypeptide chain is folded to bring two distant parts of the antigen together to make a single determinant.

the molecule called an **antigenic determinant** or **epitope** (**Figure 29.9**). Antigenic determinants may include sugars, amino acids, and other organic molecules. Antibodies interact with accessible surface antigenic determinants. A sequence of four to six amino acids is the optimal size of an antigenic determinant on a protein. Thus, proteins, usually consisting of hundreds or even thousands of amino acids, are arrays of overlapping linear antigenic determinants. In many cases, antibodies recognize determinants composed of amino acids from two portions of the molecule that are distant in terms of their primary structure, but are brought together by protein folding due to the secondary, tertiary, or quaternary structures of a macromolecule. These conformational determinants add to the antigenic complexity of macromolecules. The surface of a bacterial cell or virus consists of a mosaic of proteins, polysaccharides, and other macromolecules, all with individual determinants. Although antibodies generally recognize determinants expressed on macromolecular surfaces, TCRs recognize determinants only after the immunogens have been partially degraded. Degraded (processed) immunogens are then presented to T cells on the surface of specialized APCs or target cells, as we will discuss in Section 29.6.

Antigen processing destroys the conformational structure of a macromolecule, generally breaking proteins into peptides of less than 20 amino acids in length. As a result, T cells recognize sequential linear determinants rather than the conformational determinants recognized by antibodies.

Antibodies or TCRs can distinguish between closely related determinants. For example, antibodies can distinguish between glucose and galactose sugars, which differ only in the orientation of a single hydroxyl group. However, specificity is not absolute, and an individual antibody or TCR may react to some extent with several different but structurally similar determinants. The antigen that induced the antibody or TCR is called the *homologous* antigen, and the noninducing antigens that react with the antibody are called *heterologous* antigens. The interaction between an antibody or TCR and a heterologous antigen is called a *cross-reaction*.

29.5 MiniReview

Immunogens are foreign macromolecules that induce an immune response. Molecular size, complexity, and physical form are intrinsic properties of immunogens. When foreign immunogens are introduced into a host in an appropriate dose and route, they initiate an immune response. Antigens are molecules recognized by antibodies or TCRs. Antibodies recognize linear and conformational determinants; TCRs recognize linear peptide determinants.

- Distinguish between immunogens and antigens.
- Identify the intrinsic and extrinsic features of an immunogen.
- Describe an antibody determinant and compare it to a TCR determinant.

29.6 Antigen Presentation to T Lymphocytes

T lymphocytes interact with antigens through receptors on their cell surface. In this section, we see that the TCR interacts with peptide antigen bound by a major histocompatibility complex (MHC) protein on an antigen-presenting phagocyte cell or on an infected target cell.

The T Cell Receptor (TCR)

The TCR is a membrane-spanning protein that extends from the T cell surface into the extracellular environment. Each T cell has thousands of copies of the same TCR on its surface. A functional TCR consists of two polypeptides, an α chain and a β chain. Each of these chains has a variable (V) domain and a constant (C) domain (**Figure 29.10**). The $V\alpha$ and $V\beta$ domains interact cooperatively to form the peptide antigen-binding site. As we will see in Section 29.7, the adaptive immune response can generate TCRs that will bind nearly every known peptide antigen. Other antigens, such as complex polysaccharides, are not recognized by TCRs, but these may be bound by the immunoglobulin receptors on B cells. TCRs recognize and bind a peptide antigen only when it is bound to a *self* protein, the major histocompatibility complex protein.

Major Histocompatibility Complex (MHC) Proteins

MHC proteins are encoded by a linked set of genes found in all vertebrates called the **major histocompatibility complex (MHC)**. The MHC proteins in humans, collectively called human leukocyte antigens or HLAs, were first identified as the major antigens targeted for immune-mediated organ transplant rejection. We now know, however, that MHC proteins function primarily as antigen-presenting molecules, binding pathogen-derived antigens and displaying these antigens for interaction with TCRs.

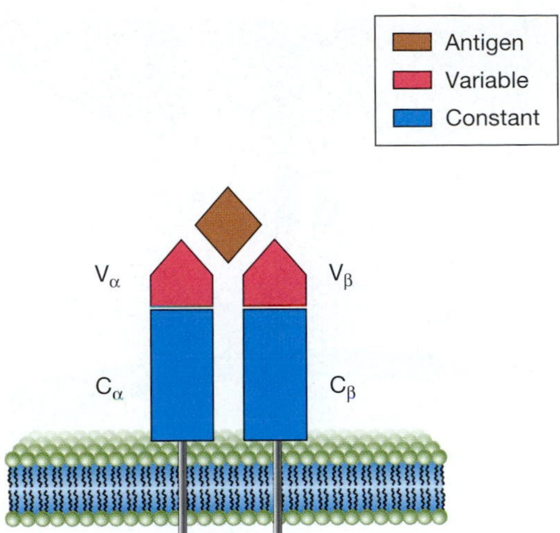

Figure 29.10 Structure of the T cell receptor (TCR). The V domains of the α chain and β chain combine to form the peptide antigen-binding site.

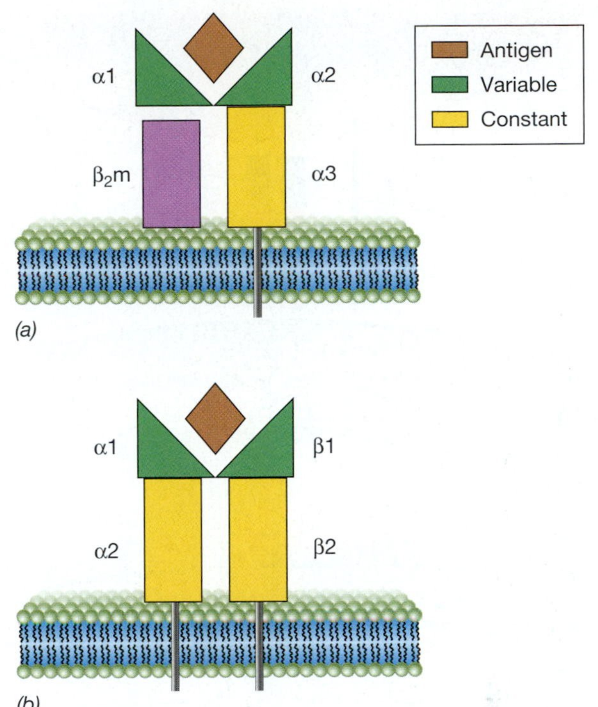

Figure 29.11 Structures of the MHC proteins. *(a)* Class I MHC protein. The α1 and α2 domains interact to form the peptide antigen-binding site. *(b)* Class II MHC protein. The α1 and β1 domains combine to form the peptide antigen-binding site.

MHC genes encode both class I and class II MHC proteins. **Class I MHC** proteins are found on the surfaces of all nucleated cells. **Class II MHC** proteins are found only on the surface of B lymphocytes, macrophages, and dendritic cells, all of which are dedicated APCs. This differential cellular distribution of class I and class II proteins is linked to their individual functions.

Class I MHC proteins consist of two polypeptides (**Figure 29.11a**), a membrane-embedded alpha chain encoded in the MHC gene region and a smaller protein called *beta-2 microglobulin* (*β2m*) encoded by a non-MHC gene. Class II MHC proteins consist of two noncovalently linked polypeptides called α and β. Like class I α chains, these polypeptides are embedded in the cytoplasmic membrane and project outward from the cell surface (Figure 29.11*b*).

In different members of the same species, MHC proteins are not structurally identical. Different individuals usually have subtle differences in the amino acid sequence of homologous MHC proteins. These genetically encoded MHC variants, of which there are over 200 in humans, are called *polymorphisms*. Polymorphisms in MHC proteins are the major antigenic barriers for tissue transplantation from one individual to another; tissue transplants not matched for MHC identity are recognized as nonself and are rejected. The detailed molecular structure and genetic organization of the MHC genes and proteins are presented in Chapter 31.

Antigen Presentation

The MHC proteins expressed on the cell surface reflect the composition of the proteins inside the cell. For example, a cell that contains no pathogens or foreign antigens will display MHC proteins complexed with self-peptides, those peptides derived from the normal catabolism of proteins during cell growth. On the other hand, cells that have ingested foreign proteins or pathogens and cells infected with viruses produce peptides that also interact with MHC proteins. In this case, the MHC proteins expressed on the cell surface are complexed with foreign peptides. These MHC proteins with embedded peptides function as molecular reference points that permit T cells to identify foreign antigens. T cells continually sample the molecular landscapes on the surface of other cells to identify cells carrying nonself antigens. The TCR on a given T cell binds only to MHC molecules having foreign antigens embedded in the MHC structure; a T cell cannot interact with a foreign antigen unless the antigen is presented in the context of an MHC protein. No T cells can react with the MHC–peptide complexes on uninfected cells because self-reactive T cells have been eliminated during the development of tolerance in the immune system.

How does this happen? Host cells can acquire nonself antigens through infection or phagocytosis. The host cells then degrade (process) the antigens to form small peptides. The processed antigen peptides are loaded into the MHC protein and the MHC–peptide complex is then inserted into the cytoplasmic membrane, to be recognized by T cells. Two distinct antigen-processing schemes are at work, one for MHC I antigen presentation and one for MHC II antigen presentation (**Figure 29.12**).

MHC I proteins present peptide antigens derived from pathogen proteins found in the cytoplasm of nonphagocytic cells (Figure 29.12*a*). The foreign pathogen proteins result from intracellular infections by viruses and other intracellular pathogens. For example, proteins derived from infecting viruses are taken up and digested in the cytoplasm in a structure called

UNIT 7

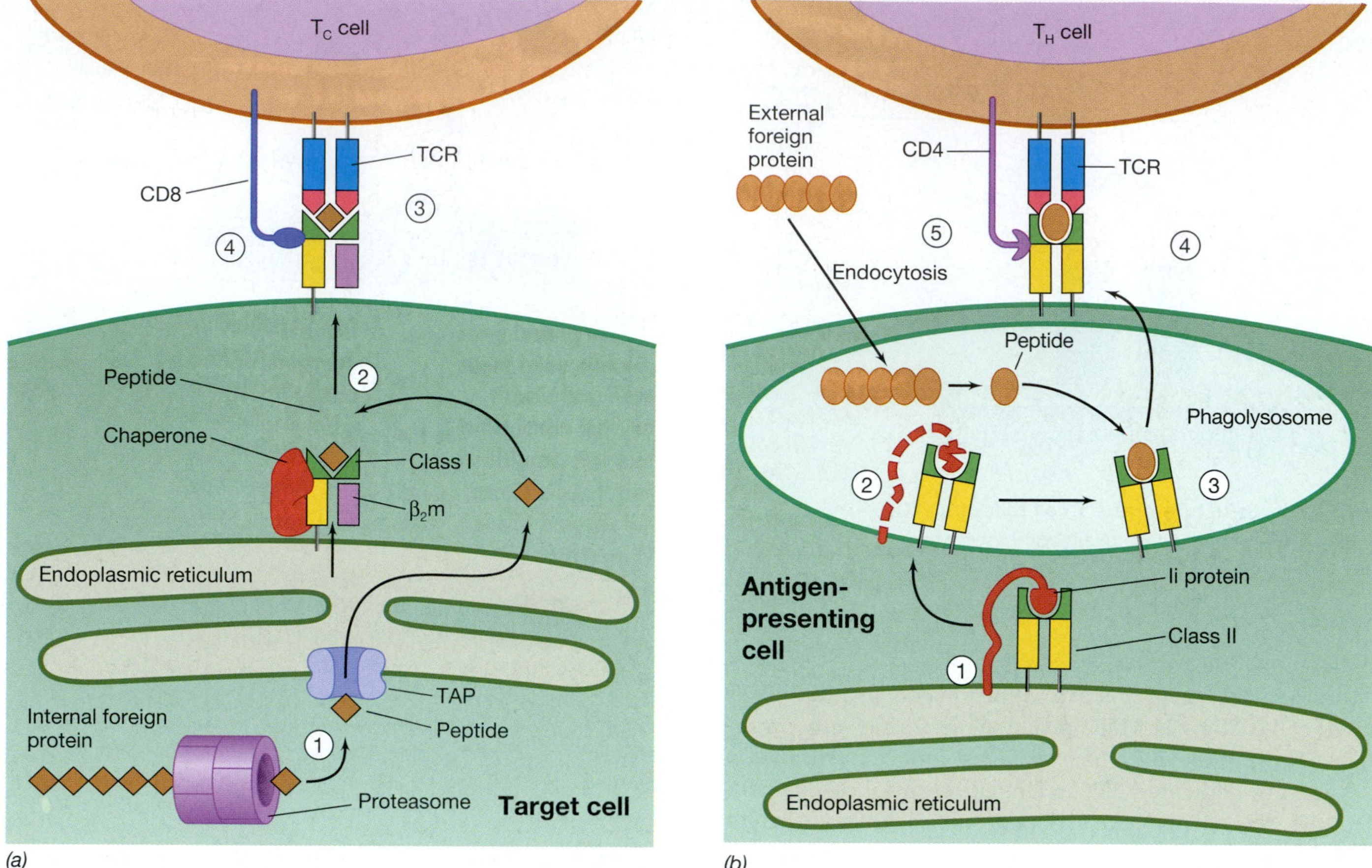

Figure 29.12 Antigen presentation by MHC I and MHC II proteins. *(a)* In the MHC I antigen presentation pathway, the membrane-bound MHC I proteins are made and assembled in the endoplasmic reticulum. Chaperone proteins stabilize MHC I until antigen is bound. ① Protein antigens manufactured within the cell, for instance from viruses, are degraded by the proteasome in the cytoplasm and transported across the endoplasmic reticulum membrane through a pore formed by the TAP proteins. ② The peptides then bind to MHC I, are transported to the cell surface, and ③ interact with T cell receptors (TCRs) on the surface of T$_c$ cells. ④ The CD8 coreceptor on the T$_C$ cell engages the class I MHC, resulting in a stronger complex. The T$_C$ cells then release cytokines and cytotoxins, killing the target cell. Any nucleated cell can act as a target cell for T cells recognizing peptide–MHC I complexes. *(b)* In the MHC II antigen presentation pathway, ① MHC II proteins are produced in the endoplasmic reticulum and are assembled with a blocking protein, Ii (invariant chain), preventing MHC II from complexing with peptides found in the endoplasmic reticulum. ② Lysosomes containing MHC II then fuse with phagosomes, forming phagolysosomes where the Ii and foreign proteins, imported from outside the cell by endocytosis, are digested. ③ The MHC II protein then binds to the digested foreign peptides, and the complex is transported to the cell surface, ④ where it interacts with TCRs and ⑤ the CD4 coreceptor on T$_H$ cells. The T$_H$ cells then release cytokines that act on other cells to promote an immune response. Only APCs can be targets for T cells recognizing the peptide–MHC II complex. The APCs are macrophages, dendritic cells, and B cells.

the *proteasome*. Peptides about ten amino acids long are transported into the endoplasmic reticulum (ER) through a pore formed by two proteins, called the *transporters associated with antigen processing (TAP)*. Once the peptides have entered the ER, they are bound by the MHC I protein, held in place near the TAP site by a group of proteins called *chaperones* until a peptide is bound. The MHC I–peptide complex is then released from the chaperones and moves to the cell surface where it integrates into the membrane and can be recognized by T cells. Thus, the MHC proteins act as a platform to which the foreign viral antigen is bound. Next, the TCR on the surface of a T cell interacts with both antigen (nonself) and MHC protein (self) on the surface of the target cell. This T cell–target cell interaction induces specialized T-cytotoxic (T$_C$) cells to produce cytotoxic proteins called perforins that kill the virus-infected target cell (Section 29.7).

The MHC class II proteins are in the second antigen presentation pathway (Figure 29.12*b*). MHC II proteins are expressed exclusively in the phagocytic APCs, where they function to present peptides from engulfed extracellular pathogens such as bacteria. MHC class II proteins are initially assembled in the ER, much like MHC I proteins. However, differing from the assembly pathway of MHC I proteins, a chaperone protein called Ii, or invariant chain, binds to the MHC II protein, blocking peptide loading inside the ER. These MHC II–Ii complexes are transported from the ER to lysosomes. After phagocytosis of a pathogen, the phagosome containing the foreign antigen fuses with the lysosome to form a phagolysosome, and the foreign antigens, as well as the Ii peptide, are digested by lysosomal enzymes. The foreign peptides, generally about 11–15 amino acids long (slightly larger than MHC I-binding peptides), are bound in the newly opened MHC II antigen-binding site. The complex is transported to the cytoplasmic membrane, where it is displayed on the cell surface to specialized **T-helper (T_H) cells**. The T_H cells, through the TCR, recognize the MHC II–peptide complex. This interaction activates the T_H cells to secrete cytokines, stimulating antibody production by B cells or causing inflammation.

CD4 and CD8 Coreceptors

In addition to the TCR, each T cell expresses a unique cell surface protein that functions as a coreceptor. T_H cells express a CD4 protein coreceptor, and T_C cells express a CD8 protein coreceptor (Figure 29.12). When the TCR binds to the peptide–MHC complex, the coreceptor on the T cell also binds to the MHC protein on the antigen-bearing cell, strengthening the molecular interactions between the cells and enhancing activation of the T cell. CD4 binds only to the class II protein, strengthening T_H cell interaction with APCs that express MHC II protein. Likewise, CD8 binds only to the MHC I protein, enhancing the binding of T_C cells to MHC I-bearing target cells. The CD4 and CD8 proteins are also used for *in vitro* tests as T cell markers to differentiate T_H ($CD4^+$) cells from T_C ($CD8^+$) cells (Section 29.9).

29.6 MiniReview

T cells interact with antigen-bearing cells including dedicated APCs and pathogen-infected cells. At the molecular level, TCRs bind peptide antigens presented by MHC proteins. These molecular interactions activate T cells to kill antigen-bearing cells or to produce cell-stimulating proteins known as cytokines.

- Identify the cells that preferentially display MHC I and MHC II proteins on their surface.
- Define the sequence of events for processing and presenting antigens from both intracellular and extracellular pathogens.

III T LYMPHOCYTES

Antigen presentation activates precursor T lymphocytes to differentiate into mature T cells responsible for antigen-specific cell-mediated killing, inflammatory responses, and "help" for antibody-producing B cells. In the absence of antigen-reactive T cells, there is little effective antigen-specific adaptive immunity and no immune memory.

29.7 T-Cytotoxic Cells and Natural Killer Cells

In the previous section we introduced two subsets of T lymphocytes, the T-cytotoxic cells and the T-helper cells. Here we examine the antigen-specific cell-killing function of the T-cytotoxic cells in detail. We also introduce the natural killer (NK) cell, a lymphocyte-like cell that uses another mechanism to target and kill cells infected with intracellular pathogens.

T-Cytotoxic (T_C) Cells

T_C cells, also known as cytotoxic T lymphocytes (CTLs), are $CD8^+$ T cells that directly kill cells that display foreign surface antigens. As we discussed in Section 29.6, T_C cells recognize foreign antigens embedded in MHC I proteins. A cell displaying a foreign antigen embedded in MHC I, such as a viral peptide displayed on a virus-infected cell, can be killed by T_C cells.

Contact between a T_C cell and the target cell is required for cell death. The contact is initiated by the TCR: the peptide–MHC I complex (Figure 29.12*a*). On contact with the target cell, granules in the T_C cell migrate to the contact site, where the contents of the granules are released (degranulation). The granules contain perforin, which enters the membrane of the target cell and forms a pore. In addition to the perforins, T_C granules contain granzymes, proteins that cause *apoptosis*, or programmed cell death. When granzymes enter the target cell through the pores created by perforins, the target cell undergoes apoptosis, characterized by death and degradation of the cell from within (**Figure 29.13**). The T_C cells, however, remain unaffected; their membranes are not damaged by perforin. The T_C cells kill only those cells displaying the foreign antigen because the granules are released only at the contact surface between the T_C and the antigen-bearing target cell. Cells lacking the antigen recognized by the T_C cells do not make contact and are not killed.

Natural Killer Cells

Natural killer cells (NK cells) are cytotoxic lymphocytes that are distinct from T cells and B cells. Nevertheless, NK cells resemble T_C cells in their ability to destroy cancer cells and cells infected with intracellular pathogens. NK cells also use perforin and granzymes to kill their targets. However, NK cells differ from T_C cells in that they kill targets in the *absence* of a specific protein. NK cells are capable of destroying cancer cells and virus-infected cells without prior exposure or contact with the foreign cells. In addition, NK numbers are not enhanced, nor do they exhibit memory after interaction with target cells.

Many tumor cells and virus-infected cells reduce or eliminate normal MHC I protein expression patterns to evade the antigen-specific immune response. The molecular target

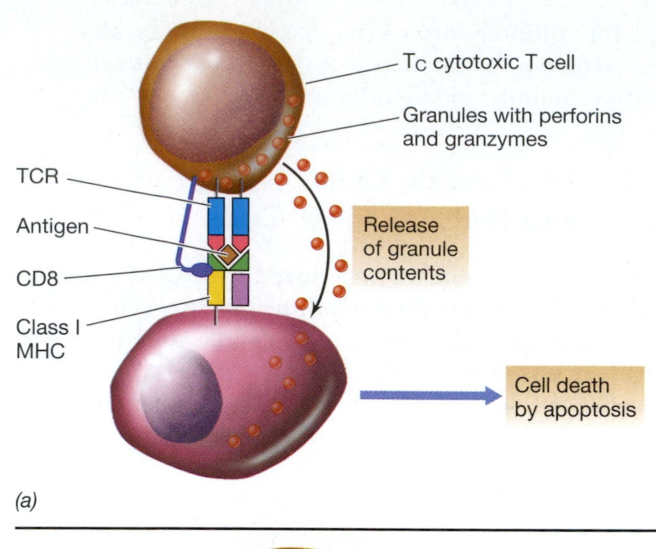

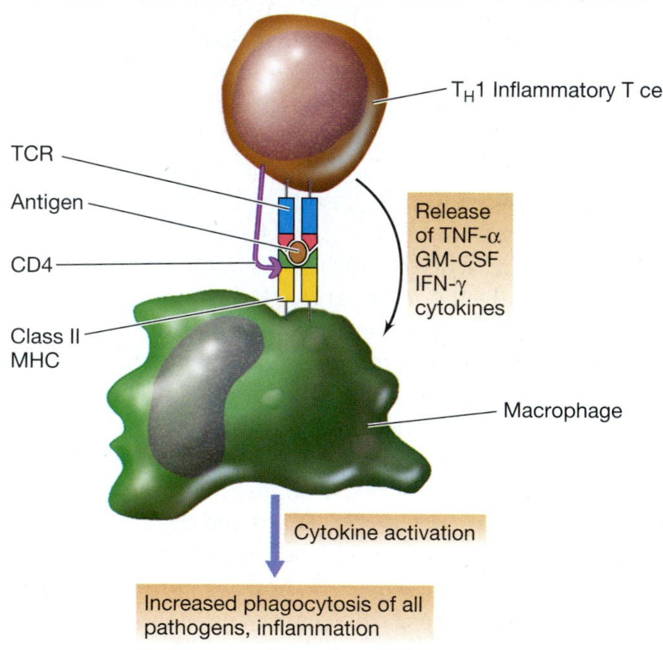

Figure 29.13 Effector T cells. *(a)* T cytotoxic cells, or T_C cells, are activated by antigens presented on any cell in the context of MHC I protein. The T_C cells respond by releasing granules that contain perforin and granzymes, cytotoxins that perforate the target cell and cause apoptosis, respectively. *(b)* T-inflammatory cells, or T_H1 cells, are activated by antigens presented on macrophages in the context of MHC II protein. Activated T_H1 cells produce cytokines that stimulate the macrophages to increase phagocyte activity and promote inflammation.

of NK cells, however, is the *absence* of appropriate MHC I proteins on target cells. As NK cells circulate and interact with the cells in the body, they recognize normal cells and their MHC I proteins through a set of special MHC I receptors. Binding of the NK receptors to MHC I deactivates an NK cell, turning off the perforin and granzyme killing mechanisms, a process called *licensing*. In the absence of licensing, the NK cell kills the unrecognized cell, thus destroying virus-infected or tumor cells that no longer express MHC proteins.

29.7 MiniReview

T-cytotoxic (T_C) cells recognize antigens on virus-infected host cells and tumor cells through antigen-specific TCRs. Antigen-specific recognition triggers killing via perforin and granzymes. Natural killer (NK) cells use the same effectors to kill virus-infected cells and tumors. NK cells, however, respond to the absence of normal MHC proteins on virus-infected cells and tumor cells. NK cells do not require antigen activation, nor do they exhibit memory.

- Identify and compare the targets and the recognition mechanisms used by T_C and NK cells.
- Describe the common effector system (the cell-killing system) used by T_C and NK cells.

29.8 T-Helper Cells: Activating the Immune Response

Here we focus on T_H cells. Interaction with MHC: peptide complexes activate $CD4^+$ T_H cells to produce cytokines. Cytokines, in turn, control the differentiation and activity of effector cells such as phagocytic macrophages and antibody-producing B cells. Undifferentiated T_H cells mature into two subsets, T_H1 cells and T_H2 cells. T_H1 cells play a role in macrophage activation and inflammation; T_H2 cells interact with B cells and stimulate antibody production.

T_H1 Cells and Macrophage Activation

Macrophages play a central role as APCs in both antibody-mediated and cell-mediated immunity. As was illustrated in Figure 29.12*b*, macrophages engulf antigens that are then processed and present antigen to T_H cells. In this process macrophages take up and kill foreign cells, an ability stimulated by T_H1 cells and the cytokines they produce. T_H1-activated macrophages kill intracellular bacteria that normally multiply in nonactivated macrophages or other cell types. Some bacteria survive and multiply within macrophages, although most bacteria taken into macrophages are killed and digested. Bacteria multiplying within macrophages include *Mycobacterium tuberculosis, Mycobacterium leprae,* and *Listeria monocytogenes,* the bacteria that cause tuberculosis, leprosy, and listeriosis, respectively. Animals given a moderate dose of *M. tuberculosis* are able to overcome the infection and develop resistance because of the T cell-mediated immune response. The active T cells are the T-inflammatory cells, the T_H1 subset. They activate macrophages and other nonspecific phagocytes by secreting cytokines, including IFN-γ (interferon gamma), GM-CSF (granulocyte-monocyte colony-stimulating factor), and TNF-α (Figure 29.13*b*). Surprisingly, such immunized animals also phagocytose and kill unrelated organisms such as *Listeria*. Macrophages in the immunized animal have thus been activated to kill any secondary invader as effectively as they resist and kill the original pathogen.

T_H1-activated macrophages not only kill pathogen-infected cells, but also help destroy tumor cells. For example,

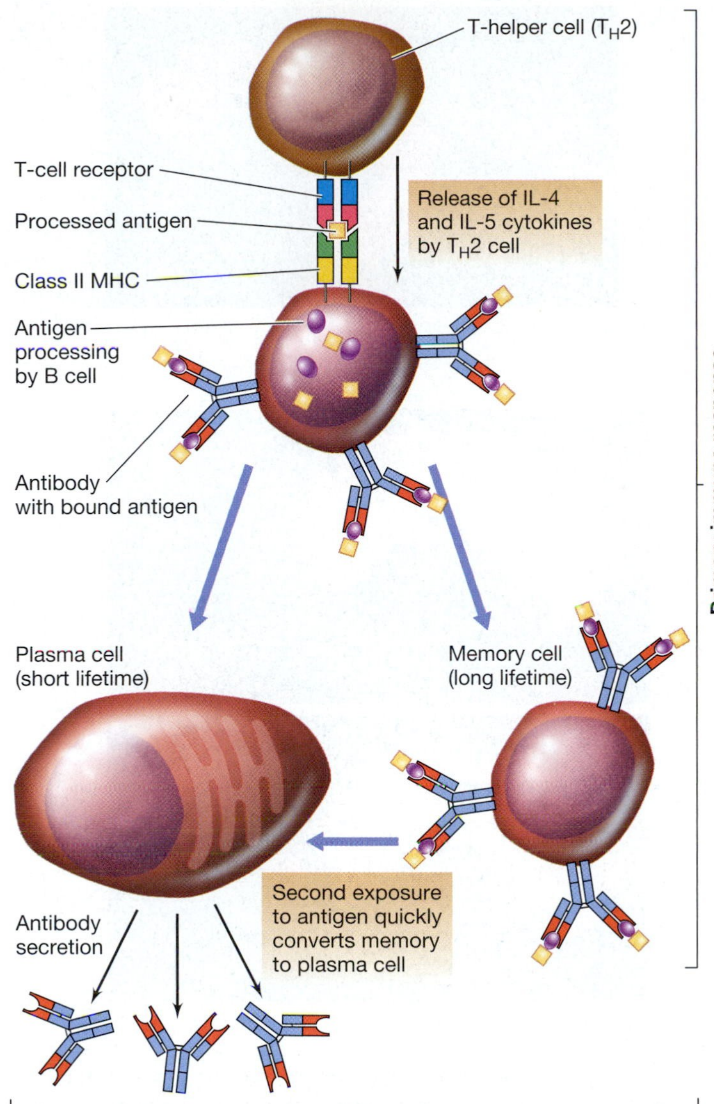

Figure 29.14 T cell–B cell interaction and antibody production. B cells initially function as antigen-presenting cells. They interact with antigen via the antigen-specific Ig receptor, promoting endocytosis of the antigen–antibody complex, and leading to antigen degradation and processing. After processing, antigen is presented to the T_H2 cell by the B cell's class II MHC molecule. The T_H2 cell is then activated to transcribe and translate genes for cytokines. The T cell cytokines then act on the same B cell to divide and form plasma cells (antibody producers) or memory cells. Plasma cells produce antibody. Memory cells quickly convert to plasma cells after a later antigen exposure.

tumor cells often produce tumor-specific antigens not found on normal cells. Tumor cells can be destroyed by macrophages activated by the T_H1 cells that react with the tumor-specific antigen. Transplantation rejection, a major problem encountered after organs or tissues are transplanted from one person to another, is also mediated by T_H1-activated macrophages. In this case, T_H1 cells recognize the nonself MHC proteins of the transplant, triggering macrophage activation and transplant destruction.

T_H2 Cells and B Cell Activation

T_H2 cells play a pivotal role in B cell activation and antibody production. As discussed in Section 29.3, B cells make antibodies. Mature B cells are coated with antibodies that act as antigen receptors. Antigen binds to the B cell antigen receptors, but the B cell does not immediately produce soluble antibodies. The bound antigen is first endocytosed and degraded in the B cell. Peptides from the degraded antigen are then presented on the B cell's MHC II protein (**Figures** 29.12*b* and **29.14**). In this way the B cell serves a dual role, first as an APC, and second as an antibody producer. In this role, the B cell takes up and processes antigen, and then presents the antigen to a T_H2 cell. The T_H2 cell responds by producing IL-4 (*inter*l*eukin-4*) and IL-5, providing the T cell help that activates the B cell. The activated B cell then produces and secretes antibodies, the second role of the B cell, as we will discuss in Section 29.10.

29.8 MiniReview

T_H1 and T_H2 cells are essential activators of cell-mediated and antibody-mediated immune responses. T_H1 inflammatory and T_H2 helper cells stimulate macrophage effector cells and B cells respectively through the action of cytokines.

- Describe the role of T_H1 cells in activation of macrophages.
- Describe the role of T_H2 cells in activation of B cells.

IV ANTIBODIES

Here we concentrate on the role of B cells and antibodies in immunity. Antibodies are proteins found in body fluids. They provide antigen-specific immunity that protects against extracellular pathogens and dangerous soluble proteins such as toxins. After considering the molecular structure of antibodies, we look at generation of diversity and production of immunoglobulins in B cells. We conclude by investigating the ability of antibodies to neutralize or destroy antigens within the animal body.

29.9 Antibodies

Antibodies, or **immunoglobulins (Ig)**, are protein molecules that interact specifically with antigenic determinants. They are found in the serum and other body fluids such as gastric secretions and milk. Serum containing antigen-specific antibody is called *antiserum*. Immunoglobulins (Igs) can be separated into five major classes on the basis of their physical, chemical, and immunological properties: *IgG, IgA, IgM, IgD,* and *IgE* (**Table 29.2**). In most individuals, about 80% of the serum immunoglobulins are the IgG proteins.

Immunoglobulin G Structure

IgG, the most common circulating antibody, has a molecular weight of about 150,000 and is composed of four polypeptide

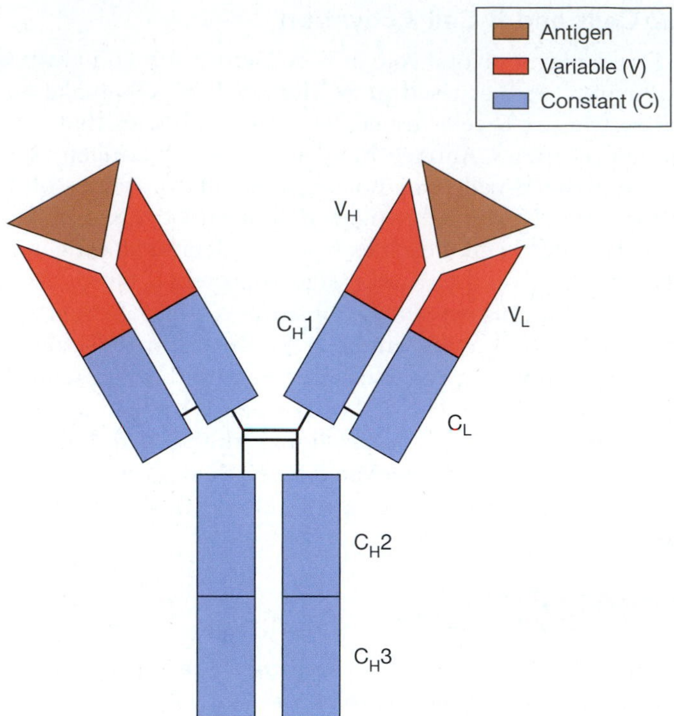

Figure 29.15 Immunoglobulin G structure. IgG consists of two heavy chains (50,000 molecular weight) and two light chains (25,000 molecular weight), with a total molecular weight of 150,000. One heavy and one light chain interact to form an antigen-binding unit. The variable domains of the heavy and light chain (V_H and V_L) bind antigen and show sequence differences in each different immunoglobulin. The constant domains (C_H1, C_H2, C_H3) are identical in all IgG proteins. The chains are covalently joined with disulfide bonds.

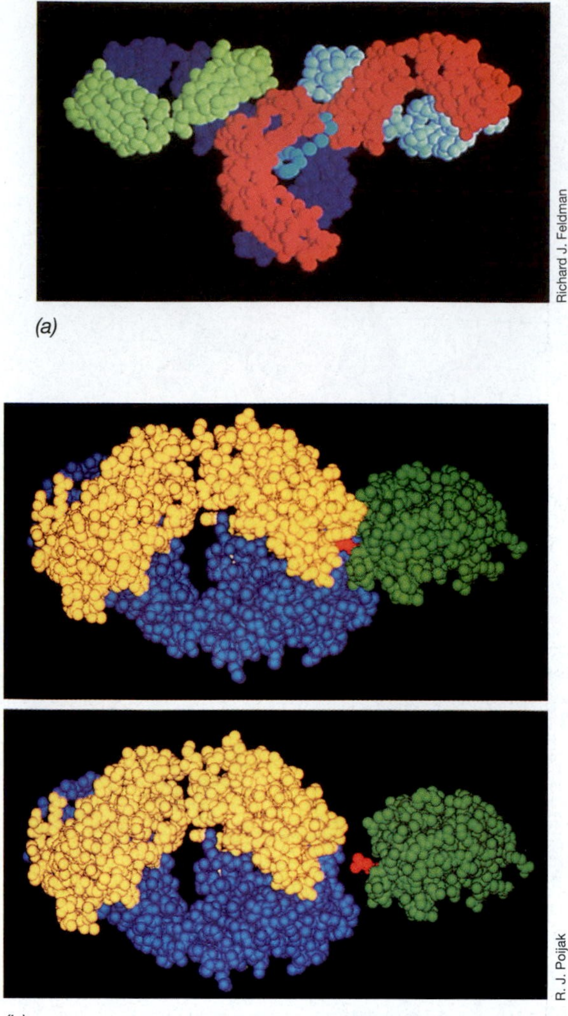

Figure 29.16 Immunoglobulin structure and the antigen-binding site. *(a)* Three-dimensional view of an IgG molecule. The heavy chains are shown in red and dark blue. The light chains are in green and light blue. *(b)* Space-filling structural model of the binding interactions between an antigen and an immunoglobulin. The antigen (lysozyme) is in green. The variable domain of the Ig heavy chain is shown in blue; the light-chain variable domain is shown in yellow. The amino acid shown in red is a glutamine in lysozyme. The glutamine fits into a pocket on the Ig molecule, but overall antigen–antibody interaction involves contacts between many other amino acids on the surfaces of both the Ig and the antigen. Reprinted with permission from *Science* 233:747 (1986) ©AAAS.

chains (**Figure 29.15**). Interchain disulfide bridges (S—S bonds) connect the individual chains. In each IgG protein, two identical light chains of 25,000 molecular weight are paired with two identical heavy chains of 50,000 molecular weight, giving a total molecular weight of 150,000. Each light chain has about 220 amino acids, and each heavy chain has about 440 amino acids. Each heavy chain interacts with a light chain to form a functional antigen-binding site. An IgG antibody, therefore, is *bivalent* because it contains two binding sites and can bind two identical determinants.

Heavy Chains and Light Chains

Each IgG heavy chain is composed of several distinct parts, called **domains** (Figure 29.15). A variable domain is connected to three constant domains, all of about 110 amino acids in length. The amino acid sequence in the variable domain differs in each different antibody. The variable domain binds antigen. The three constant domains of each heavy chain are identical in each different antibody protein of the same class.

Each IgG light chain consists of two equally sized parts, a variable and a constant domain. The variable domain of the light chain interacts with the variable domain of the heavy chain to bind antigen. The amino acid sequence in the constant domain does not differ among light chains of the same type.

The Antigen-Binding Site

The antigen-binding site of IgG and all other antibodies forms by cooperative interaction between the variable domains of both heavy and light chains (**Figure 29.16**). The variable domains of the two chains interact, forming a receptor that binds antigen strongly but noncovalently. The measurable strength of binding of antibody to antigen is called *binding affinity*. A high-affinity antibody binds tightly to antigen.

Table 29.2 Properties of human immunoglobulins

Class/ H chain[a]	Molecular weight/ formula[b]	Serum (mg/ml)	Antigen-binding sites	Properties	Distribution
IgG γ	150,000 2(H + L)	13.5	2	Major circulating antibody; four subclasses: IgG_1, IgG_2, IgG_3, IgG_4; IgG_1 and IgG_3 activate complement	Extracellular fluid; blood and lymph; crosses placenta
IgM μ	970,000 (pentamer) 5[2(H + L)] + J	1.5	10	First antibody to appear after immunization; strong complement activator	Blood and lymph; monomer is B cell surface receptor
	175,000 (monomer) 2(H + L)	0	2		
IgA α	150,000 2(H + L)	3.5	2	Important circulating antibody	Secretions (saliva, colostrum, cellular and blood fluids); monomer in blood and dimer in secretions
	385,000 (secreted dimer) 2[2(H + L)] + J + SC	0.05	4	Major secretory antibody	
IgD δ	180,000 2(H + L)	0.03	2	Minor circulating antibody	Blood and lymph; B lymphocyte surfaces
IgE ε	190,000 2(H + L)	0.00005	2	Involved in allergic reactions and parasite immunity	Blood and lymph; C_H4 binds to mast cells and eosinophils

[a]All immunoglobulins may have either λ or κ light chain types, but not both.
[b]Based on the number and arrangement of heavy (H) and light (L) chains in each functional molecule. J is joining protein present in serum IgM and secretory IgA. SC is the secretory component found in secreted IgA.

Each individual's immune system has the capacity to recognize, or bind, countless antigens, with each antigen being recognized by a unique antigen-binding site. To accommodate all possible antigens, each individual can produce billions of different antigen-binding sites in antibodies. How is this diversity in the antigen-binding site generated? As we discuss in the next section, new antibodies are constantly created through recombination and mutation events in approximately 300 genes that encode the light- and heavy-chain variable domains. The heavy-chain and light-chain genes together encode the unique antibody expressed on each B cell. Antigen interaction with the B cell antibody stimulates the B cell to produce and secrete soluble copies of the antibody.

Other Antibody Classes

Antibodies of the other classes differ from IgG. The heavy-chain constant domains of a given antibody molecule define its class as one of five, based on amino acid sequences. The five heavy chains are gamma (γ), alpha (α), delta (δ), mu (μ), or epsilon (ε). The constant domain sequences constitute three-fourths of the heavy chains of IgG, IgA, and IgD, respectively, and four-fifths of the heavy chains of IgM and IgE (**Figure 29.17**). Each antibody of the IgM class, for example, contains amino acids in its heavy-chain constant domains that constitute the *mu* sequence.

The structure of IgM is shown in **Figure 29.18**. IgM is usually found as an aggregate of five immunoglobulin molecules attached by at least one J (joining) chain. Each heavy chain of IgM contains a fourth constant domain (C_H4). IgM is the first class of Ig made in a typical immune response to a bacterial infection, but IgMs are generally of low affinity. Antigen-binding strength is enhanced to some degree, however, by the high *valence* of the pentameric IgM molecule; ten binding sites are available for interaction with antigen (Table 29.2

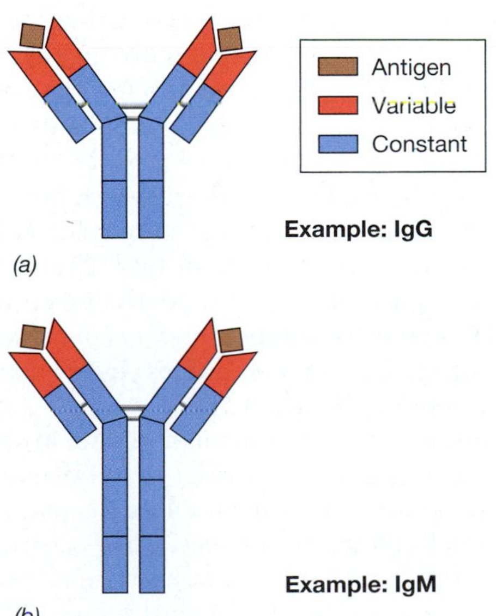

Figure 29.17 Immunoglobulin classes. All classes of Igs have V_H and V_L domains (red) that bind antigen (brown). *(a)* IgG, IgA, and IgD have three constant domains (blue). *(b)* IgM and IgE each have a fourth constant domain.

UNIT 7

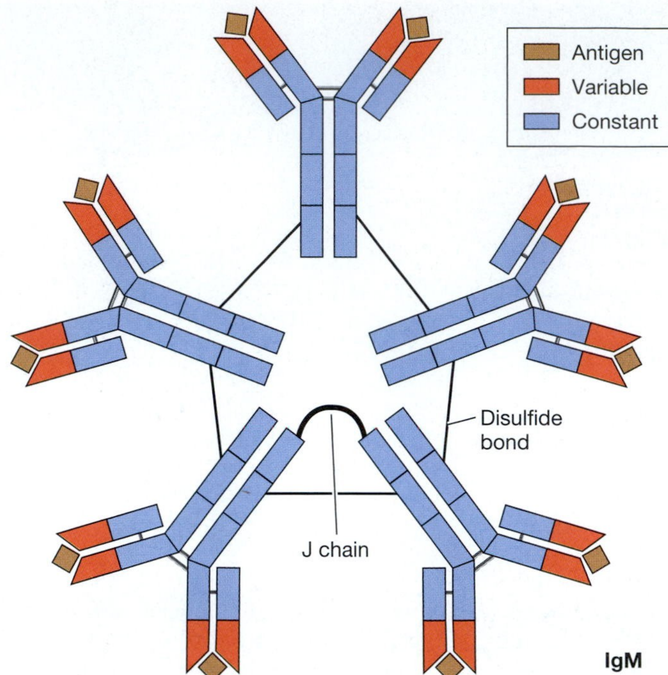

Figure 29.18 Immunoglobulin M. IgM is found in serum as a pentameric protein consisting of five IgM proteins covalently linked to one another via disulfide bonds and a J chain protein. Because it is a pentamer, IgM can bind up to 10 antigens, as shown.

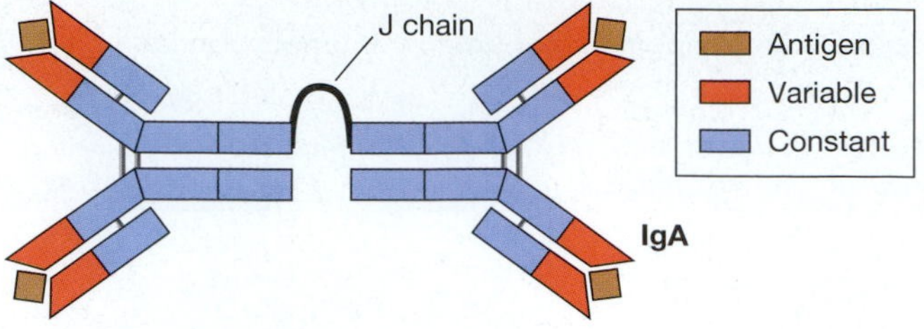

Figure 29.19 Immunoglobulin A. Secretory IgA (sIgA) is often found in body secretions as a dimer consisting of two IgA proteins covalently linked to one another via a joining (J) chain protein. A secretory component, not shown, aids in transport of IgA across mucosal membranes.

and Figure 29.18). The combined strength of binding by the multiple antigen-binding sites on IgM is called *avidity*. Thus, IgM has *low* affinity but *high* avidity for antigen. Up to 10% of serum antibodies are IgM. IgM monomers are also found on the surface of B cells, where they bind antigen.

Dimers of IgA are present in body fluids such as saliva, tears, breast milk colostrum, and mucosal secretions from the gastrointestinal, respiratory, and genitourinary tracts. These mucosal surfaces are associated with mucosal-associated lymphoid tissue (MALT) that produces IgA. The mucosal surfaces total about 400 m^2, and large amounts of secretory IgA are produced—about 10 g per day (**Figure 29.19**). By contrast, the serum IgG produced in an individual is about 5 g per day. Thus the total amount of *secretory* IgA produced by the body is higher than the amount of *serum* IgG. The secretory form consists of two IgA molecules covalently linked by a J chain peptide and a protein called the *secretory component* that aids in transport of IgA across membranes. IgA is also present in serum as a monomer (Table 29.2).

IgE is found in extremely small amounts in serum (about 1 of every 50,000 serum Ig molecules is an IgE). IgE functions as an antibody that binds to eosinophils, arming these granulocytes to target eukaryotic parasites like schistosomes and other worms. IgE is also important because it is the antibody involved in immediate-type hypersensitivities (allergies). The molecular weight of an IgE molecule is significantly higher than most other Igs (Table 29.2) because, like IgM, IgE has a fourth constant domain (Figure 29.17). This additional constant region functions to bind IgE to eosinophils and mast cell surfaces, a critical step for activating the protective and allergic reactions associated with these cell types (∞ Figure 30.5).

IgD, present in serum in low concentrations, has no known function. However, IgD, like IgM, is abundant on the surfaces of B cells, and IgD is especially abundant on memory B cells. Thus, IgD found on the surface of B cells may be important for the secondary antibody response.

29.9 MiniReview

Each Ig (antibody) protein consists of two heavy and two light chains. The antigen-binding site is formed by the interaction of variable regions of one heavy and one light chain. Each antibody class has different structural characteristics, expression patterns, and functional roles.

- Identify the antibody heavy- and light-chain domains that bind antigen.
- Differentiate among antibody classes using structural characteristics, expression patterns, and functional roles.

29.10 Antibody Production

In this section, we will examine a typical antibody response. At the cellular level, complex interactions between T and B cells produce effective antigen-specific antibody immunity. At the genetic level, B cells use unique genetic mechanisms to generate unique antigen-binding receptors. A predictable sequence of events leads to effective antibody production after antigen exposure.

T Cell–B Cell Interactions

Antibody production is a direct response to antigen exposure. Specific antibody responses involve interactions between T cells and B cells through their respective antigen-specific cell surface molecules, the TCR on the T cell, and the surface antibody on the B cell (Figure 29.14). A reactive B cell exposed to antigen for the first time binds the antigen through its antigen-specific surface receptor, membrane-bound antibody. The antigen–antibody complex is then internalized, and the antigen is processed to peptides for loading onto MHC II proteins.

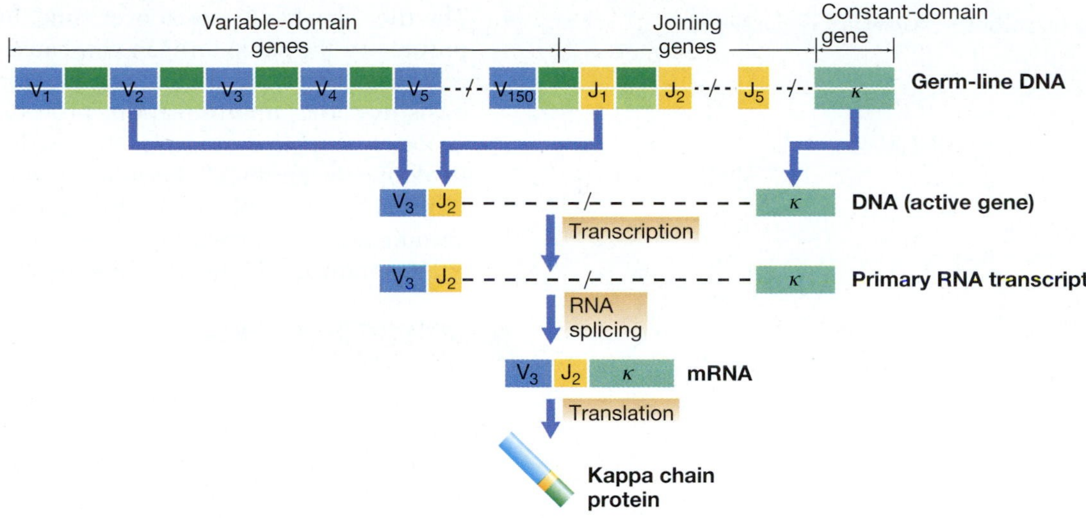

Figure 29.20 Immunoglobulin kappa chain gene rearrangement in human B cells. The gene segments are arranged in tandem in the kappa (κ) light-chain genes on chromosome 2. DNA rearrangements are completed in the maturing B cell. Any one of the 150 V (variable) sequences may combine with any one of the 5 J sequences. Thus, 750 (150 × 5) recombinations are possible, encoding 750 distinct kappa chains, but only one productive rearrangement occurs in each cell. An analogous process occurs in the heavy-chain genes in each B cell. The heavy-chain gene complex has even more gene segments, allowing greater numbers of recombinations and potential heavy chains, but again only one heavy-chain rearrangement is found in each mature B cell. The mature Ig is made by combining two light-chain proteins with two heavy-chain proteins.

Thus, the B cell first functions as an efficient APC using its surface antibody to capture an individual antigen. The MHC–antigen complex then moves to the cell surface, where the complex is displayed for interaction with a T_H2 cell having an antigen-specific TCR on its surface (Figure 29.12*b*). Formation of an MHC–antigen–TCR complex activates genes in the antigen-specific T_H2 cell, leading to cytokine production. The T_H2 cytokines stimulate the nearby antigen-specific B cell, activating it to clonally expand and differentiate into plasma cells that secrete antibodies.

Generation of Antigen Receptor Diversity

Each individual is capable of producing billions of different antibodies and TCRs, with each receptor aimed to specifically interact with one of the countless antigens in our environment. How does the immune system direct the production of all of these antigen-specific receptors? Immune receptor diversity is generated by a unique mechanism exclusive to only T and B cells. Antibody production starts with stepwise rearrangements of the Ig-encoding genes. During development of B cells in the bone marrow, both heavy and light chain genes rearrange. The genes are reassorted—individual gene pieces are mixed and matched in a variety of combinations—by gene splicing and rearrangements in the maturing cells, a process known as somatic recombination.

Figure 29.20 shows a typical rearrangement and expression pattern for the human light chain. The heavy-chain genes rearrange in an analogous, but more complex fashion. In each mature B cell, the final result is a single functional heavy-chain gene and a single functional light-chain gene. Each of these rearranged genes is transcribed, translated, and expressed to make an antibody protein on the surface of the B cell. Antigen exposure induces differentiation to plasma cells that produce soluble antibody copies. In addition, antigen exposure induces genetic hypermutation in B cells with productive antibody genes, further modifying and diversifying the expressed antibodies. The large number of possible gene rearrangements coupled with the somatic hypermutation events after antigen exposure ensure almost unlimited antibody diversity.

Similar rearrangements also occur during T cell development, resulting in the generation of considerable diversity in TCRs. However, T cells do not use hypermutation to expand diversity.

Antibody Production and Immune Memory

Starting with a B cell, antibody production begins with antigen exposure and culminates with the production and secretion of an antigen-specific antibody according to the following sequence:

1. Antigens are spread via the lymphatic and blood circulatory systems to nearby secondary lymphoid organs such as lymph nodes, spleen, or mucosal-associated lymphoid tissue (MALT) (Section 29.1 and Figure 29.2). The route of antigen exposure influences the class of the antibodies produced. Intravenously injected antigen travels via the blood to the spleen, where IgM, IgG, and serum IgA antibodies are formed. Antigen introduced subcutaneously, intradermally, topically, or intraperitoneally, is carried by the lymphatic system to the nearest lymph nodes, again

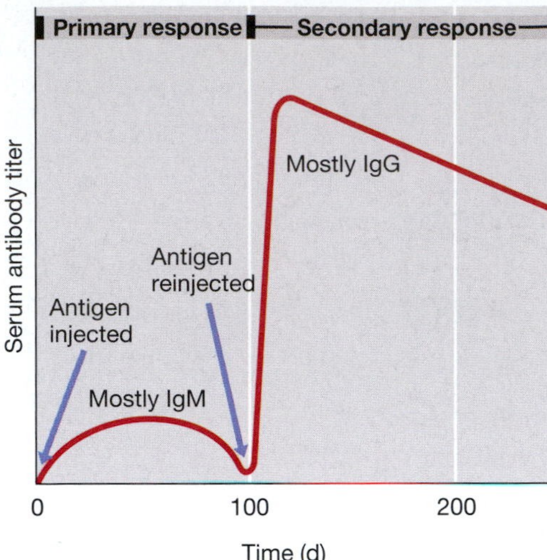

Figure 29.21 Primary and secondary antibody responses in serum. The antigen injected at day 0 and day 100 must be identical to induce a secondary response. The secondary response, also called a booster response, may be more than tenfold greater than the primary response. Note the class switch from IgM production in the primary response to IgG production in the secondary response.

stimulating production of IgM, IgG, and serum IgA. Antigen introduced to mucosal surfaces is delivered to the nearest MALT. For example, antigen delivered by mouth is delivered to the MALT in the intestinal tract, preferentially stimulating specific antibody production in the gut, with an antigen-specific secretory IgA response.

2. Following the initial antigen exposure, each antigen-stimulated B cell multiplies and differentiates to form antibody-secreting plasma cells and memory cells (Figure 29.14). **Plasma cells** are relatively short-lived (less than 1 week), but produce and secrete large amounts of mostly IgM antibody in the **primary antibody response** (**Figure 29.21**). There is a latent period before specific antibody appears in the blood, followed by a gradual increase in antibody titer (antibody quantity), and then a slow decrease in the primary antibody response.

3. The **memory B cells** generated by the initial exposure to antigen may live for years. If there is a later reexposure to the immunizing antigen, memory B cells need no T cell activation; they quickly transform to plasma cells and begin producing antibody. The second and each subsequent exposure to antigen causes the antibody titer to rise rapidly to a level often 10–100 times greater than the titer achieved following the first exposure. This rise in antibody titer is the **secondary antibody response**. The secondary response typifies immune memory, resulting in a more rapid, more abundant antibody response when compared with the primary response. The secondary response also induces a switch from mostly IgM to mostly IgG production, a phenomenon called *class switching* (Figure 29.21).

4. The titer slowly decreases over time, but subsequent exposures to the same antigen can cause another memory response. The rapid and strong memory response is the basis for the immunization procedure known as a "booster shot" (for example, the yearly rabies shot given to domestic animals). Periodic reimmunization maintains high levels of circulating antibody specific for a certain antigen, providing long-term active protection against individual infectious diseases.

29.10 MiniReview

Antibody production is initiated by antigen contact with an antigen-specific B cell. The antigen-reactive B cell processes the antigen and presents it to an antigen-specific T_H2 cell. The T_H2 cell becomes activated, producing cytokines that signal the antigen-specific B cell to clonally expand and differentiate to produce antibodies. Activated B cells live for years as memory cells and can rapidly expand and differentiate to produce large quantities (high titers) of antibodies after reexposure to antigen.

- How do B cells act as APCs?
- How do T_H2 cells activate antigen-specific B cells?
- Explain the rationale for periodic rabies reimmunizations in domestic animals.

29.11 Complement, Antibodies, and Pathogen Destruction

Complement is composed of a group of sequentially interacting proteins that play an important effector role in both innate and adaptive immunity. Complement activity can be initiated by interactions with antigen–antibody complexes but can also be initiated by innate immune mechanisms. Complement proteins, reacting with one another and with target cell components, cause lysis of pathogen cells or mark cells for recognition by phagocytes, accelerating their destruction.

Classical Complement Activation and Cell Damage

Complement is a group of proteins, many with enzymatic activity. These proteins react in a prescribed sequential order with antigen–antibody complexes on a target cell. The result can be cell membrane damage, leakage of contents, and lysis of the cell. The serum of all individuals contains complement, and most antigen-bound IgG or IgM antibodies can bind complement (Table 29.2).

The individual proteins of complement are designated C1, C2, C3, and so on. Classical activation of complement occurs when IgG or IgM antibodies bind antigens, especially on cell surfaces. The antibodies are said to *fix* (bind) the ever-present complement proteins. The complement proteins react in a defined sequence, with activation of one complement component leading to activation of the next, and so on (**Figure 29.22**). The key steps start with (1) binding of antibody to

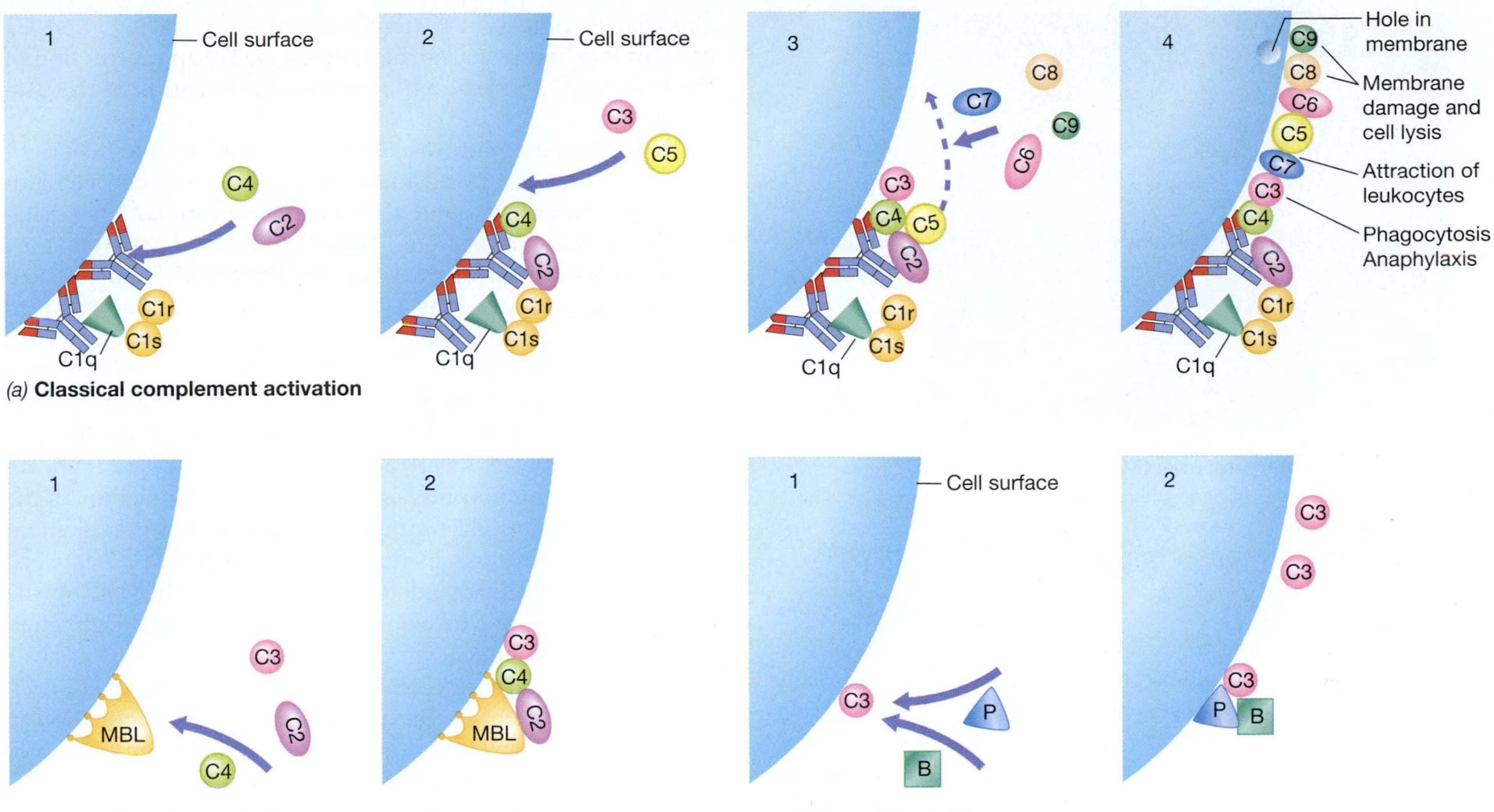

Figure 29.22 Complement activation. *(a)* The sequence, orientation, and activity of the components of the classical complement pathway as they interact to lyse a cell. 1. Binding of the antibody and the C1 protein complex (Clq, Clr, and Cls). 2. The C42 complex interacts with C3. 3. the C423 complex activates C5 that binds an adjacent membrane site. 4. Sequential binding of C6, C7, C8, and C9 to C5, producing a pore in the membrane. C5–C9 is the membrane attack complex (MAC). *(b)* The mannan-binding lectin (MBL) pathway. MBL binds to the mannose on the bacterial membrane, and anchors formation of C423. The membrane-bound C3 activates C5, as in 3 above, and initiates formation of the MAC (4 above). *(c)* The alternate pathway. C3 bound to the cell attracts protein B. The C3B complex is further stabilized on the membrane by factor P (properdin). C3B then acts on C3 in the blood, causing more C3 to bind to the membrane. Bound C3 then activates C5, as in 3 above, and initiates formation of the MAC (4 above). *(d)* A schematic view of the pore formed by complement components C5 through C9.

antigen (initiation); (2) binding of C1 components (C1q, Clr, and Cls) to the antibody–antigen complex, leading to C4–C2 binding at an adjacent membrane site and activation and binding of C3; (3) membrane-bound C3 catalyzes formation of a C5-C6-C7 complex at a second membrane site, and C8 and C9 are then deposited with the C5–C6–C7 complex, resulting in membrane damage and cell lysis (**Figure 29.23**). The C5-9 components, called the *membrane attack complex (MAC),* insert at the membrane site to form a pore.

By-products of complement activation include chemoattractants called anaphylatoxins. They cause inflammatory reactions at the site of complement deposition. For example, C3 is cleaved to C3a and C3b. C3b fixes to the target cell, as outlined above. The C3a cleavage product is released into the medium where it causes chemotactic attraction and activation of phagocytes, resulting in increased phagocytosis. Reactions involving the C5a cleavage product lead to T cell attraction and cytokine release.

When activated by specific antibody, complement lyses many gram-negative bacteria. Gram-positive bacteria, on the other hand, are not killed by complement and specific antibodies. Gram-positive bacteria can, however, be destroyed through a process known as opsonization.

Opsonization

A bacterial cell is more likely to be phagocytosed when antibody binds antigen on its surface. If complement binds to an antibody–antigen complex on the cell surface, the cell is even more likely to be ingested. This is because most phagocytes,

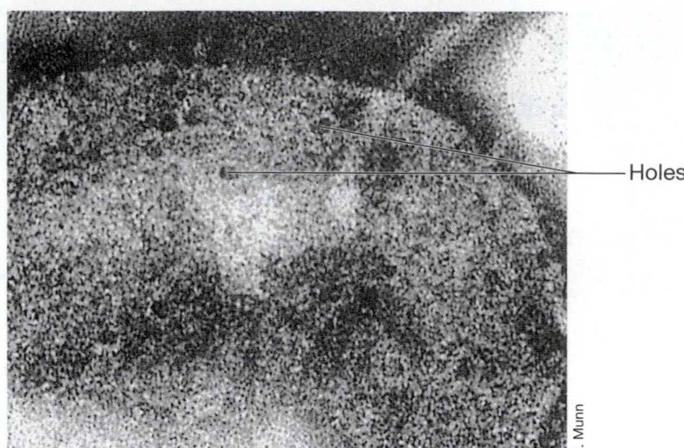

Figure 29.23 Complement activity on bacterial cells. Shown is an electron micrograph of *Salmonella paratyphi*. The negatively stained preparation shows holes created in the bacterial cell envelope as a result of a reaction involving cell envelope antigens, specific antibody, and complement.

including neutrophils, macrophages, and B cells, have antibody receptors (FcR) as well as C3 receptors (C3R). These receptors bind the antibody constant domain and C3 complement protein, respectively. Normal phagocytic processes are enhanced about tenfold by antibody binding and amplified another tenfold by C3 binding. This enhancement of phagocytosis by antibody or complement binding is called *opsonization*.

Antibodies bound to surface antigens on gram-positive *Bacteria* activate the classical complement pathway and promote opsonization, leading to enhanced phagocytosis and pathogen destruction.

Complement Activation by the Mannan-Binding Lectin and Alternate Pathways

Another method of activating the complement system is through the mannan-binding lectin (MBL) pathway. This pathway depends on the activity of a serum MBL protein that binds to unique mannose-containing polysaccharides found only on bacterial cell surfaces (Figure 29.22*b*). The MBL–polysaccharide complex resembles the C1 complexes of the classical complement system and fixes the C4, C2, and C3 components to the cell surface, again catalyzing formation of the C5-9 MAC and leading to lysis or opsonization of the bacterial cell.

The alternate pathway is a nonspecific complement activation mechanism using many of the classical complement pathway components and several serum proteins not associated with the classical complement pathway. Together they induce opsonization and activate the C5-9 MAC. The first step in alternate pathway activation is the binding of C3 produced by the classical or MBL pathway to the bacterial cell surface. C3 on the membrane can then bind the alternate pathway serum protein factor B. Factor P, or properdin, joins C3B to form the very stable C3BP complex, now fixed on the cell (Figure 29.22*c*). C3BP attracts more C3, which is then deposited on the membrane, initiating the same reaction as the membrane-bound C423 complex of the classical complement pathway. The result is formation of the C5-9 MAC and cell destruction. The C3 fixed to the surface of the cell by the alternate pathway can also enhance phagocytosis through opsonization via phagocyte C3 receptors.

Both the alternate pathway and the MBL pathway nonspecifically target bacterial invaders and lead to activation of the membrane attack complex and enhanced opsonization. Properdin, MBL, and complement proteins are part of the innate immune system. Neither the alternate pathway nor the MBL pathway require prior antigen exposure or the presence of antibodies for activation. Through the alternate and MBL pathways, C3 triggers formation of the C5-9 MAC or enhances opsonization via C3 receptors on phagocytes.

29.11 MiniReview

The complement system catalyzes bacterial cell destruction and opsonization. Complement is triggered by antibody interactions or by interactions with nonspecific activators such as mannan-binding lectin. Complement is a critical effector system in both innate and adaptive host defense.

- Which antibody classes fix complement?
- Identify the complement components that promote opsonization and those that induce cell lysis.

Review of Key Terms

Adaptive immunity the acquired ability to recognize and destroy an individual pathogen or its products that is dependent on previous pathogen exposure

Antibody a soluble protein, produced by differentiated B cells, that interacts with antigen; also called immunoglobulin

Antibody-mediated immunity (humoral immunity) immunity resulting from the action of antibodies

Antigen a molecule capable of interacting with specific components of the immune system

Antigenic determinant (epitope) the portion of an antigen that reacts with a specific antibody or T cell receptor

Antigen-presenting cell (APC) a macrophage, dendritic cell, or B cell that presents processed antigen peptides to a T cell

B cell a lymphocyte that has immunoglobulin surface receptors, differentiates to a plasma cell, produces immunoglobulin, and may present antigens to T cells

Bone marrow a primary lymphoid organ containing the pleuripotent precursor cells for all blood and immune cells, including B cells

Cell-mediated immunity immunity resulting from the action of antigen-specific T cells

Class I MHC protein antigen-presenting molecule found on all nucleated vertebrate cells

Class II MHC protein antigen-presenting molecule found primarily on macrophages, B cells, and dendritic cells

Complement a series of proteins that react in a sequential manner with antibody–antigen complexes, mannose-binding lectin, or alternate activation pathway proteins to amplify or potentiate target cell destruction

Cytokine a soluble immune response modulator produced by leukocytes

Domain a region of a protein having a defined structure and function

Hapten a low-molecular-weight molecule that combines with specific antibodies but that is incapable of eliciting an immune response by itself

Immune memory the capacity to respond more quickly and vigorously to second and subsequent exposures to an eliciting antigen

Immunity the ability of an organism to resist infection

Immunogen a molecule capable of eliciting an immune response

Immunoglobulin (Ig) a soluble protein produced by B cells and plasma cells that interacts with antigens; also called antibody

Inflammation a nonspecific reaction to noxious stimuli such as toxins and pathogens, characterized by redness (erythema), swelling (edema), pain, and heat, usually localized at the site of infection

Innate immunity the noninducible ability to recognize and destroy a pathogen or its products that is not dependent upon previous exposure to a pathogen or its products

Interleukin (IL) a soluble cytokine or chemokine secreted by leukocytes

Leukocyte (white blood cell) a nucleated cell found in the blood

Lymphocyte a leukocyte active in the adaptive immune response

Major histocompatibility complex (MHC) a genetic complex responsible for encoding several cell surface proteins important in antigen presentation

Memory B cells long-lived cells responsive to an individual antigen

Natural killer (NK) cell a specialized lymphocyte that recognizes and destroys foreign cells or infected host cells in a nonspecific manner

Neutrophil (polymorphonuclear leukocyte or PMN) a type of leukocyte exhibiting phagocytic properties, a granular cytoplasm (granulocyte), and a multilobed nucleus

Pathogen-associated molecular pattern (PAMP) a unique repeating structural component of a microbe or virus recognized by a pattern recognition molecule

Pattern recognition molecule (PRM) a membrane-bound protein that recognizes a pathogen-associated molecular pattern

Phagocyte a cell that recognizes, ingests, and degrades pathogens and pathogen products

Plasma the liquid portion of the blood with cells removed and clotting proteins inactivated

Plasma cell a differentiated B cell that produces large amounts of antibodies

Primary antibody response antibody made after initial exposure to antigen; mostly of the IgG class

Primary lymphoid organ an organ in which precursor lymphoid cells develop into mature lymphocytes

Secondary antibody response antibody made on subsequent exposure to antigen; mostly of the IgG class

Serum the liquid portion of the blood with clotting proteins and cells removed

Specificity the ability of the immune response to interact with individual antigens

T cell a lymphocyte responsible for antigen-specific cellular interactions in the adaptive immune response

T cell receptor (TCR) antigen-specific receptor protein on the surface of T cells

T helper (T_H) cells lymphocytes that interact with MHC-peptide complexes through their T cell receptor (TCR)

Thymus the primary lymphoid organ responsible for development of T cells

Tolerance inability to produce an immune response to specific antigens

Toll-like receptor (TLR) a pattern recognition molecule found on phagocytes that recognizes a pathogen-associated molecular pattern

UNIT 7

Review Questions

1. What is the origin of the phagocytic and antigen-specific cells involved in the immune response? Track the maturation of B cells and T cells (Section 29.1).
2. Identify some pathogen-associated molecular patterns (PAMPs) that are recognized by pattern recognition molecules (PRMs). What is the significance of the interactions between these molecules (Section 29.2)?
3. Explain how phagocytes engulf and kill microorganisms, with particular attention to oxygen-dependent mechanisms (Section 29.3).
4. Identify the most important features of the adaptive immune response (Section 29.4).
5. What molecules induce immune responses? What properties are necessary for a molecule to induce an immune response (Section 29.5)?
6. Describe the basic structure of class I and class II major histocompatibility complex (MHC) proteins. In what functional ways do they differ (Section 29.6)?
7. Differentiate between T_C cells and NK cells. What is the activation signal for each cell type (Section 29.7)?
8. How do T_H cells differ from T_C cells? Differentiate between the functional roles of T_H1 and T_H2 cells (Section 29.8).
9. Describe the structural and functional differences among the five major antibody classes (Section 29.9).

10. Identify the cell interactions involved in production of antibodies by B cells (Section 29.10).
11. Describe the complement cascade. Is the order of protein interactions important? Why or why not? Identify the components of the mannan-binding lectin pathway for complement activation. Identify the components of the alternate pathway for complement activation (Section 29.11).

Application Questions

1. Describe the potential problems that would arise if an individual had an acquired inability to phagocytose pathogens. Could the individual survive in a normal environment such as a college campus? What defects in the phagocyte might cause lack of phagocytosis? Explain.
2. Specificity and tolerance are necessary qualities for an adaptive immune response. However, memory seems to be less critical, at least at first glance. Define the role of immune memory and explain how the production and maintenance of memory cells might benefit the host in the long term.
3. Describe the potential problems that would arise if an individual had a hereditary deficiency that resulted in inability to present antigens to T_C cells. To T_H1 cells? To T_H2 cells? To all T cells? What molecules might be deficient in each individual? Could any of these individuals survive in a normal environment? Explain.
4. Antibodies of the IgA class are probably more prevalent than those of the IgG class. Explain why this is true and what advantage this provides for the host.
5. Complement is a critical humoral defense mechanism. Do you agree with this statement? Explain your answer. What might happen to individuals who lack complement component C3? C5? Properdin (alternate pathway)? Mannan-binding lectin (MBL)?

30

Immunity in Host Defense and Disease

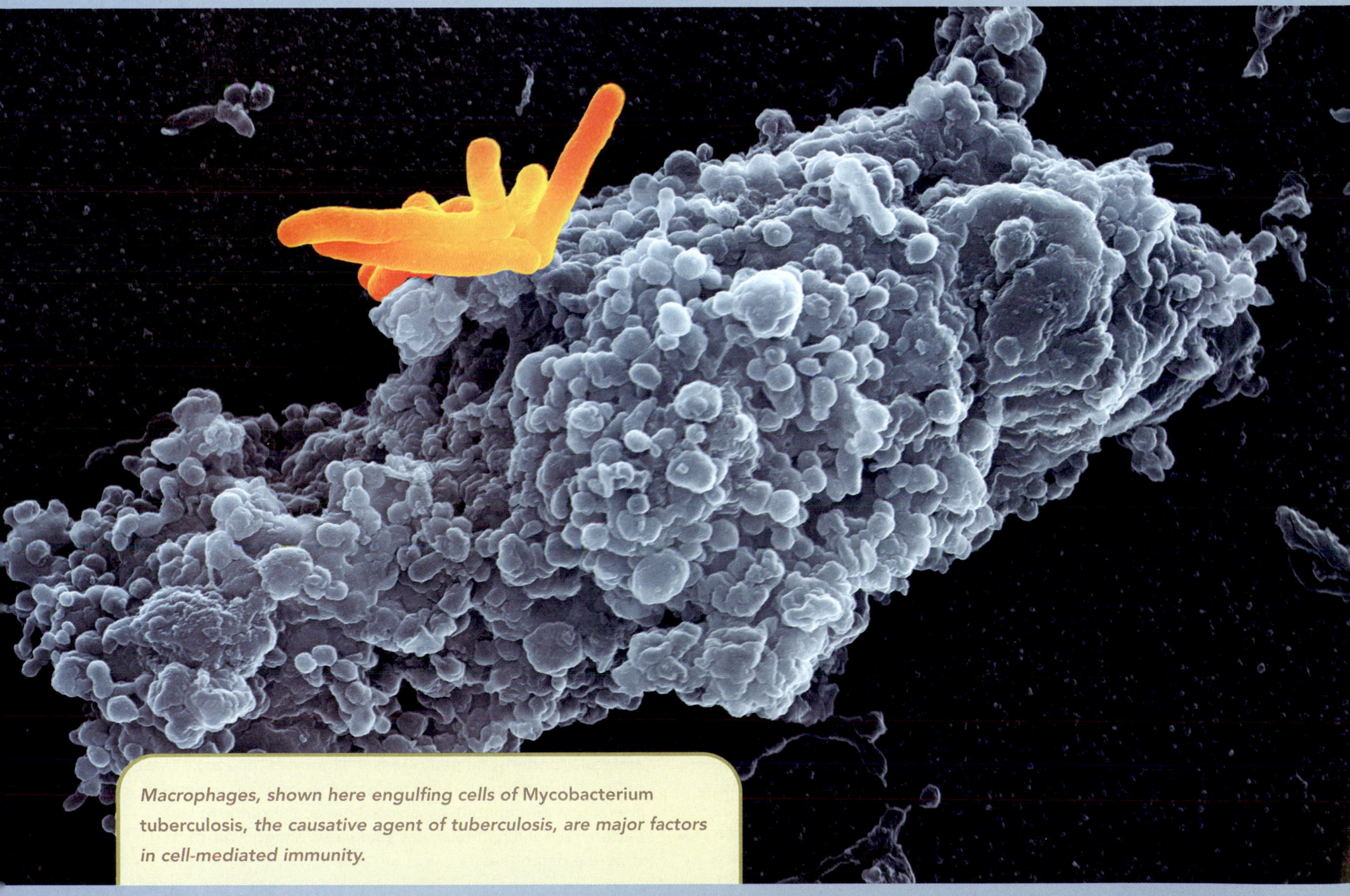

Macrophages, shown here engulfing cells of Mycobacterium tuberculosis, *the causative agent of tuberculosis, are major factors in cell-mediated immunity.*

Immunity is the ability to resist infection. We discussed the basic features of immunity in Chapter 29. Here we briefly review the principles of immunity and then discuss the protection afforded by immunity to pathogens. We then shift our focus to the manipulation of the immune response through immunization. We conclude by highlighting immune mechanisms that can lead to disease.

I IMMUNITY AND HOST DEFENSE

We start by examining the cells and molecules that protect against pathogens. We first consider the in-built, or innate, immune response and then expand our discussion to include the cells and molecules of the acquired, or adaptive, immune response.

30.1 Innate Immunity

Multicellular organisms must recognize and control infections by pathogens. To achieve this control, eukaryotes from plants to vertebrates have developed molecular recognition schemes that lead to rapid and effective activation of host defenses. These evolutionarily related and conserved mechanisms are collectively known as the innate immune system.

The Innate Immune Response to Pathogens

Innate immunity, or nonspecific immunity, is the noninducible ability to recognize and destroy an individual pathogen or its products. Most importantly, innate immunity does not require previous exposure to a pathogen or its products. The innate immune response is mediated by phagocytes.

The macromolecules in pathogens, especially those on the cell surface, display a variety of **pathogen-associated molecular patterns (PAMPs)**, structural components of a pathogen, generally consisting of repeating macromolecular subunits. An example of a PAMP is the repeating subunits in the lipopolysaccharide (LPS) common to all gram-negative bacterial outer membranes (∞ Section 4.7). **Leukocytes** (nucleated or white blood cells) known as phagocytes such as **macrophages** and **neutrophils** are generally the first line of defense against pathogens, especially those that the body has never before encountered (∞ Section 29.1). The phagocytes interact directly with PAMPs using specialized preformed receptors called **pattern recognition molecules (PRMs)** (**Figure 30.1**). A single PRM found on all phagocytes, for example, interacts with the LPS on all *Salmonella* spp., on all *Escherichia coli* strains, and on all *Shigella* spp. These interactions activate the phagocytes to ingest and destroy the targeted pathogens, a process called *phagocytosis* (∞ Section 29.2). A number of different PRMs are found on phagocytes, and each interacts with a PAMP shared by a number of pathogens. In addition to the LPS-reactive PRM discussed above, for example, another phagocyte PRM interacts with the peptidoglycan on gram-positive cells. The preformed PRMs are instantly available to confer immediate recognition and resistance to invasive pathogens.

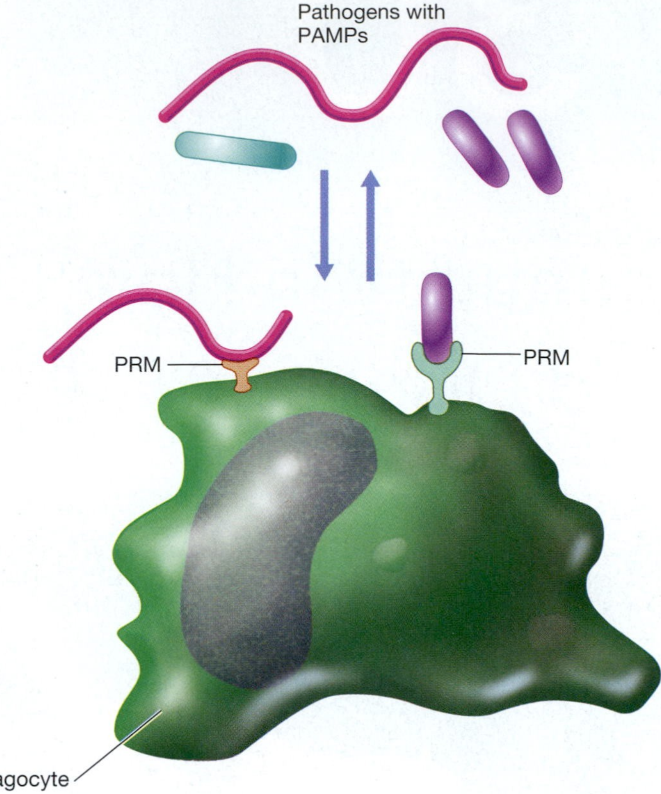

Figure 30.1 Innate immunity. Exposure to pathogens stimulates innate immunity mediated by phagocytes. Phagocytes interact with pathogens by recognizing pathogen-associated molecular patterns (PAMPs) with preformed pathogen recognition molecules (PRMs). The interactions activate the phagocyte to ingest and destroy the pathogen.

Innate immunity is an ancient response to infection. We know that the PRMs present in vertebrates, for example, have structural and evolutionary homologs in phylogenetic groups as distant as insects such as *Drosophila*. Functionally similar recognition systems involving phagocyte recognition and destruction of targeted pathogens are found in virtually all multicellular organisms.

30.1 MiniReview

Innate immunity is a natural protective response to infection characterized by recognition of common pathogen-associated molecular patterns on pathogens. Phagocytes recognize these patterns through preformed pathogen recognition molecules. Interaction of pathogen-associated molecular patterns with pattern recognition molecules stimulates phagocytes to destroy the pathogens.

- Identify a pathogen-associated molecular pattern shared by a group of microorganisms.
- Identify the groups of organisms that use the pattern recognition mechanisms to provide innate immunity to pathogens.

30.2 Adaptive Immunity and T Cells

In vertebrate animals, the phagocytes responsible for the innate response also initiate *adaptive* immunity. Adaptive or antigen-specific immunity involves the production of pathogen-specific receptors only after exposure to the pathogen or its products. T cells are the pathogen-specific cells that direct the adaptive immune response.

Adaptive Immunity

Adaptive immunity or antigen-specific immunity is the acquired ability to recognize and destroy an individual pathogen. A highly specific adaptive immune response develops only after exposure to the pathogen or its products. Adaptive immunity is directed toward an individual molecular component of the pathogen, called an **antigen**.

No effective adaptive immunity exists before exposure to antigen, but after the first antigen exposure, a *primary* immune response occurs, characterized by the stimulation of specialized antigen-reactive immune leukocytes called **lymphocytes**. These lymphocytes are divided into two populations called **T cells** and **B cells**. Each lymphocyte produces a unique protein that interacts with a single antigen and thus has **specificity** for that antigen. The unique proteins of T cells are the **T cell receptors (TCRs)** and those of B cells are **antibodies** or **immunoglobulins (Igs)**. After antigen contact, lymphocytes that react with antigen grow and multiply to produce **clones**, multiple copies of antigen-reactive lymphocytes, some of which may persist for years and confer long-term specific immunity.

As compared to innate immunity, the adaptive response is inducible only when triggered, for example, by a single polysaccharide component of a particular gram-negative LPS. Thus, an individual lymphocyte clone could interact with an LPS constituent on *Salmonella,* but will not interact with the LPS on *Escherichia coli* or *Shigella* spp. because these organisms have different terminal sugars on their polysaccharide. A second exposure to the same antigen activates the already expanded clones of antigen-reactive cells to more rapidly generate a very strong immune response that peaks within several days (∞ Section 29.10). The T cell and antibody products of this secondary immune response quickly target the pathogen for destruction. Thus, the adaptive immune response has **memory**, characterized by this rapid increase in immunity. Finally, the adaptive immune system exhibits **tolerance**, the acquired *inability* to make an immune response directed against self-antigens. Tolerance ensures that adaptive immunity is directed to outside agents that pose genuine threats to the host, and not to host proteins.

T Cells and Antigen Presentation

Adaptive immunity begins with the interactions of immune T lymphocytes (T cells) with antigens on infected cells. The infected cells that are first recognized by T cells may include the same phagocytes that were involved on the innate immune response (**Figure 30.2**). The T cell, with its TCR, can recognize antigen on host cell surfaces only when the antigens are

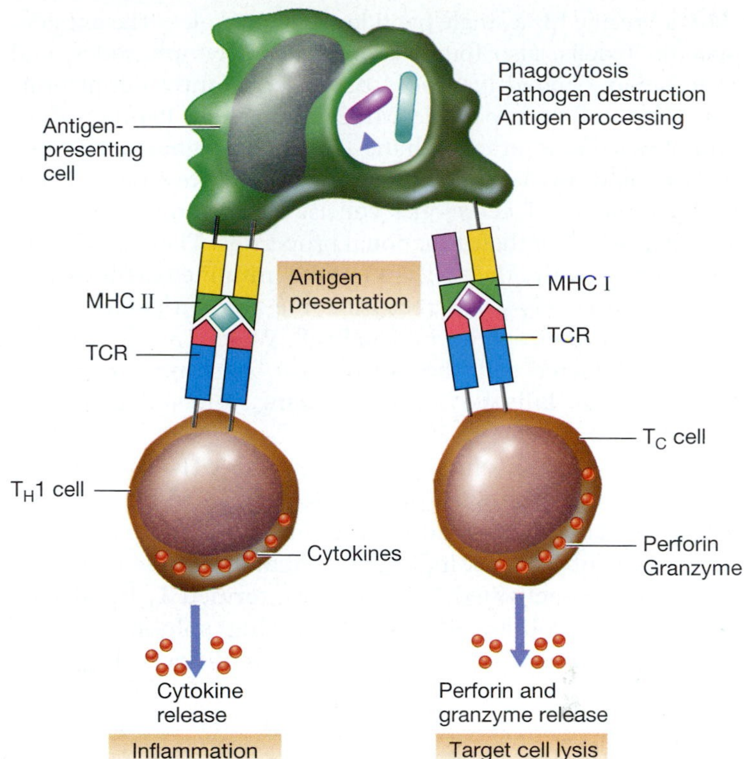

Figure 30.2 T cell immunity. Antigen-presenting cells such as the phagocytes in innate immunity ingest, degrade, and process antigens. They then present antigens to T cells that activate the adaptive immune response. Antigen-reactive T cells include T inflammatory cells (T_H1) and T-cytotoxic cells (T_C) cells.

presented on self proteins known as **major histocompatibility complex (MHC)** proteins. Inside the phagocyte, the MHC proteins interact with the small peptides derived from digested pathogens. The MHC-embedded peptides are then transported to the phagocyte surface, where the complex is displayed, a process called *antigen presentation* (∞ Section 29.6). For example, a phagocyte infected with influenza virus will display MHC proteins embedded with influenza peptides. These MHC-peptide complexes are targets for T cells.

All host cells display MHC I proteins, and several specialized phagocytes, called *antigen-presenting cells (APCs),* also display an additional antigen-presenting protein, MHC II. The APCs include macrophages, dendritic cells, and B cells. APCs are localized in spleen, lymph nodes, and mucosal-associated lymphoid tissue, the anatomical sites where the adaptive immune response begins (∞ Section 29.1). APCs are characterized by their ability to ingest bacteria, viruses, and other antigenic material by phagocytosis (macrophages and dendritic cells) or through internalization of molecular antigen bound to a surface receptor (B cells). After ingestion, the APCs degrade the antigens to small peptides and display the peptides with MHC on the cell surface.

T Lymphocyte Subsets

T cells interact with the peptide–MHC complex using the cell surface T cell receptor (TCR). Each T cell expresses a TCR that is specific for a single peptide–MHC complex. The antigen-specific T cells, also found in the spleen, lymph nodes, and mucosal-associated lymphoid tissue, constantly sample surrounding cells for peptide–MHC complexes. Peptide–MHC complexes that interact with the TCR send a signal to the T cell to grow and divide, producing antigen-reactive clones. These antigen-reactive T cell clones consist of three different T cell subsets, based on their functional properties. These T cell subsets interact with other cells to initiate immune reactions.

T-cytotoxic (T_C) cells recognize the antigen presented by an MHC I protein on an infected cell. When T_C cells interact with the infected cell, they secrete *perforins* and *granzymes,* proteins that kill the antigen-bearing infected cell (Figure 30.2 and ∞ Figure 29.12).

T-helper (T_H) cells interact with peptide–MHC II complexes on the surface of antigen-presenting cells (∞ Figure 29.12). This interaction causes differentiation of the T_H cells, resulting in two subsets that indirectly mediate immune reactions. These antigen-activated T cell subsets, termed T_H1 and T_H2, respond by proliferating and producing soluble proteins called **cytokines**. Cytokines then activate other cell types to initiate an immune response.

Differentiated antigen-specific T_H1 cells interact with peptide–MHC II complexes on the surface of macrophages (Figure 30.2). This interaction stimulates the T_H1 cell to produce cytokines that activate the macrophages, enhancing phagocytosis of the antigen-bearing cell and causing inflammatory reactions that limit the spread of infections (∞ Section 29.8). For example, *Mycobacterium tuberculosis* infects macrophages and other cells in the lung, causing tuberculosis. Activated macrophages kill *M. tuberculosis* inside the cell, limiting spread to other cells. An inflammatory reaction associated with *M. tuberculosis* is termed the *tuberculin reaction* and is used as a diagnostic test for *M. tuberculosis* exposure. This test uses *tuberculin,* an extract from *M. tuberculosis,* to attract immune T_H1 cells that produce cytokines, activating macrophages, and causing a localized red, hot, hardened, and swollen area that typifies **inflammation** and, usually, effective immunity (see Figure 30.6b; ∞ Section 34.5).

Differentiated T_H2 cells, the other T_H subset, also do not interact directly with the pathogen, but use cytokines to stimulate ("help") antigen-reactive B cells to produce antibodies, as we discuss below.

30.2 MiniReview

Adaptive immunity is triggered by the specific interactions of T cells with antigens presented on antigen-presenting cells. Peptide antigens are presented to T cells embedded in MHC proteins. T-cytotoxic, or T_C, cells kill antigen-bearing target cells directly. T helper cells act through cytokines to promote immune reactions. T_H1 cells initiate inflammation and immunity by activating macrophages.

- Explain the process of antigen presentation to T cells.
- Define the role of T_C and T_H1 cells in adaptive immunity.

30.3 Adaptive Immunity and Antibodies

Antibodies or immunoglobulins (Igs) are soluble proteins made by B cells in response to exposure to nonself antigens that are part of pathogens or are produced by them. Each antibody binds specifically to a single antigen. Antibody-mediated immunity controls the spread of infection through recognition of pathogens and their products in extracellular environments such as blood or mucus secretions.

Antibodies are produced by specialized lymphocytes called B cells. B cells have preformed antibodies on their surface, and each B cell displays multiple copies of a single antibody that is specific for a single antigen. To make more preformed antibodies, the B cells must first bind antigens through interactions with the surface antibodies. The surface antibody–antigen interaction induces the B cell to ingest the antigen-containing pathogen by phagocytosis. The B cell then kills and digests the pathogen, producing a battery of pathogen-derived peptide antigens. These antigens are then displayed, or presented, to the antigen-specific T_H2 cells (**Figure 30.3** and ∞ Section 29.6).

T_H2 cells do not interact directly with the pathogen, but stimulate ("help") other cells, in this case, the antigen-reactive B cells. T_H2 cells produce cytokines that stimulate B cells, which in turn respond by growing and dividing, establishing clones of the original antigen-reactive B cell. These activated B cells then differentiate into **plasma cells** that produce soluble antibodies (Figure 30.3 and ∞ Section 29.10). This initial antibody response, known as a **primary antibody response**, is generally detectable within five days, and antibodies reach peak quantities within several weeks. Subsequent exposure to the same antigen, for example by reinfection with the same pathogen, induces a memory or **secondary antibody response**, characterized by a faster development of higher quantities of antibodies (∞ Figure 29.21). The primary response is characterized by production of IgM antibodies, whereas the secondary response is characterized by production of IgG.

Several different classes of antibodies are distinguished from one another by their primary amino acid sequence. Each antibody class has a specific general function. IgM and IgG are found in blood. IgA is found in secretions from mucus membranes, such as in the lungs and gut, and IgE is involved in parasite immunity and allergies.

Antibodies released from the plasma cells interact with antigen on the pathogens. This interaction, however, does not directly kill the pathogen, but rather marks it for destruction by phagocytosis. Phagocytes have general antibody receptors called *Fc receptors* that bind to any antibody attached to an antigen. This interaction results in enhanced phagocytosis of the antibody-sensitized cells, a process known as *opsonization.*

Destruction can also be mediated by a group of proteins known collectively as *complement* (∞ Section 29.11). The complement proteins attach to IgM or IgG antibodies bound to the pathogen. The complement proteins, concentrated at the cell surface by the antibody, have two possible effects on

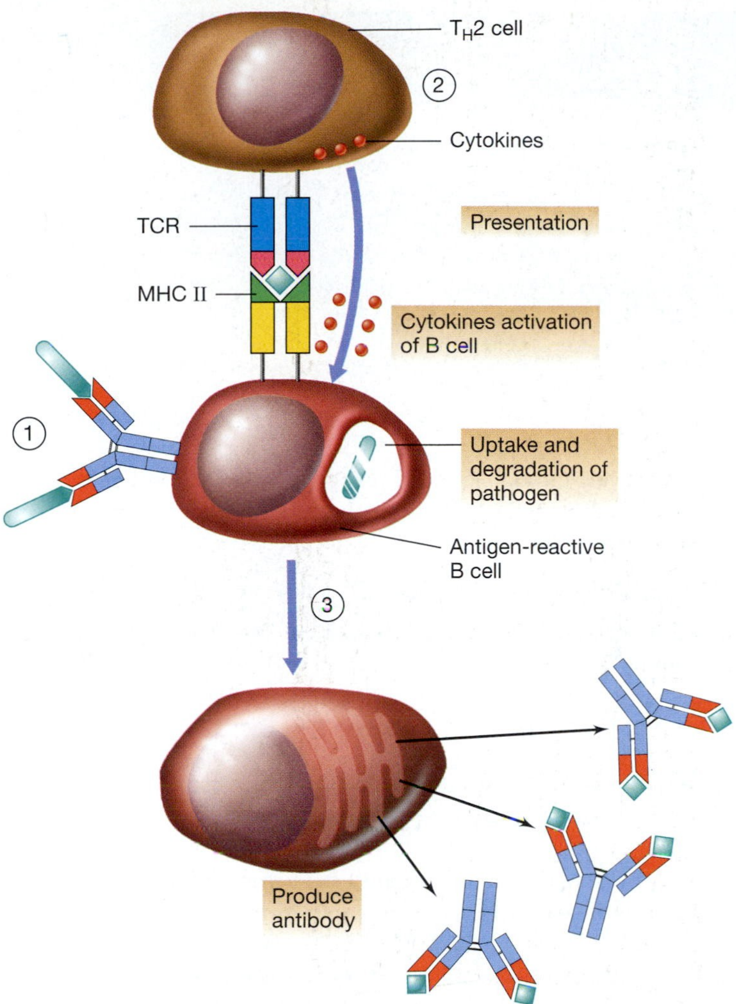

Figure 30.3 Antibody-mediated immunity. ① Antibody on B cells binds to a pathogen. The B cell ingests, degrades, and processes the pathogen. ② The B cell presents pathogen antigen to a T_H2 cell, activating it to produce cytokines that in turn influence the B cell to develop into a plasma cell ③. The plasma cell produces antibodies.

the pathogen. First, complement proteins can form a pore in the pathogen membrane, directly lysing all cells with attached antibody–complement complexes (Figure 29.22). This complement–antibody interaction affects only those cells with bound antibodies. For example, antibodies specific for cell surface proteins of *Salmonella* sp. interact only with *Salmonella*, and complement causes lysis of an antibody-sensitized *Salmonella* cell, but not of a nearby *Escherichia coli* cell that is not antibody-sensitized.

Many pathogens, such as the thick-walled gram-positive *Streptococcus* spp., are relatively resistant to complement-mediated lysis because the cell wall makes the membrane less accessible to complement proteins. However, antibodies to the external cell wall components can bind complement proteins, which are then bound by complement receptors found on the surface of phagocytes such as neutrophils and macrophages. This interaction results in enhanced opsonization and phagocytosis of the antibody-complement sensitized cells (Section 29.11).

The antibody response is highly specific for the eliciting antigen. Antibodies interact with antigens, triggering lysis and phagocytosis of pathogens through opsonization and complement binding.

30.3 MiniReview

T_H2 cells stimulate antigen-exposed B cells to differentiate to plasma cells. Plasma cells then produce antibodies. Antibodies are soluble, antigen-specific proteins that interact with antigens. Antibodies provide targets for interaction with proteins of the complement system, resulting in destruction of antigens through lysis or opsonization.

- Explain the process of antibody production starting with pathogen interaction with a B cell.
- Define the role of antibody and complement in pathogen destruction.

II IMMUNITY AND PREVENTION OF INFECTIOUS DISEASES

Immunity generated by natural exposure to pathogens is certainly an effective way to resist infections. Our knowledge of the immune response, however, makes it possible to initiate immunity by artificial means, thereby avoiding natural infections to induce immunity. We can now safely induce adaptive immune responses to various pathogens and their products. Artificial immunization remains our best weapon against many infectious diseases.

30.4 Natural Immunity

Both the innate and adaptive immune responses protect the host from infections by pathogens, and both innate and adaptive immunity are essential for survival. For example, individuals with genetic defects that prevent neutrophil or macrophage development produce nonfunctional phagocytes that cannot ingest and destroy pathogens. These phagocytes lack innate immune abilities and individuals with this defect cannot live without extraordinary intervention such as isolation from all environmental exposure. In a normal environment, they develop recurrent infections from bacteria, viruses, and fungi and die at an early age. Individuals who lack adaptive immunity have the same outcome, as we discuss next.

Adaptive Immunity and Prevention of Infectious Disease

Animals normally develop *natural active immunity* by acquiring a natural infection that initiates an adaptive immune response. Natural active immunity is the outcome of exposure to antigens through infection and usually results in protective immunity conferred by antibodies and T cells.

Table 30.1 Active and passive immunity

Active immunity	*Passive immunity*
Exposure to antigen; immunity achieved by injecting antigen or through infection	No exposure to antigen; immunity achieved by injecting antibodies or antigen-reactive T cells
Specific response made by individual achieving immunity	Specific immune response made in the secondary host which donates antibodies or T cells
Immune system activated to antigen; immune memory in effect	No immune system activation; no immune memory
Immune response can be maintained via stimulation of memory cells (i.e., booster immunization)	Immunity cannot be maintained and decays rapidly
Immune state develops over a period of weeks	Immunity develops immediately

The importance of active immunity in disease resistance is shown dramatically in individuals who are immunocompromised due to certain genetic disorders or infection. For example, *agammaglobulinemia* is a disease in which patients cannot produce antibodies because of genetic defects in their B cells. Such individuals suffer from recurrent, life-threatening bacterial infections, but develop normal immune responses to viruses. Thus, antibody-mediated immunity is essential for protection from extracellular pathogens, especially bacteria.

Individuals with *DiGeorge's syndrome,* a developmental defect that prevents maturation of the thymus and inhibits production of mature T cells, suffer from serious recurrent infections with viruses and other intracellular pathogens. Thus, the T cell immune-deficiency problems seen in DiGeorge's syndrome define the essential protective role for T cell immunity: protection from intracellular pathogens.

The general lack of an adaptive immune response is also observed in individuals with *acquired immunodeficiency syndrome (AIDS)*. In AIDS patients, infection with the human immunodeficiency virus (HIV), if not controlled, can cause nearly total depletion of T_H cells, resulting in a lack of effective antibody (B cell) and cellular (T cell) immunity. Such individuals suffer from recurrent, severe infections due to viral, bacterial, and fungal pathogens. Death from AIDS is characteristically due to secondary infections by one or more opportunistic pathogens (∞ Section 34.15).

In special cases, there is also *natural passive immunity*. For example, for several months after birth, newborns have maternal IgG antibodies, transferred across the placenta before birth, in their blood. IgA antibodies are also transferred in breast milk. The antibodies that are protective for the infant were made in the mother, thus their designation as passive antibodies. These preformed antibodies provide disease protection while the immune system of the newborn matures. Active and passive immunity are contrasted in **Table 30.1**.

30.4 MiniReview

Innate and adaptive immune responses are necessary for survival. A lack of innate immunity results in death due to recurrent, uncontrollable infections. Adaptive immunity develops naturally and actively through immune responses to infections, or naturally and passively through antibody transfer across the placenta or in breast milk.

- Describe the effect of a lack of innate immunity.
- Describe the effects of a lack of T cell immunity or B cell immunity.

30.5 Artificial Immunity and Immunization

The purposeful artificial induction of immunity to individual infectious diseases is a major weapon for the treatment and prevention of these diseases. There are two ways by which artificial immunity can be induced. An individual may be purposefully exposed to a controlled dose of harmless antigen to induce formation of antibodies, a type of immunity called *artificial active immunity* because the individual produces the antibodies. This process is commonly known as **vaccination**. Alternatively, an individual may receive injections of an *antiserum* (serum containing antibodies from the blood of an immune individual) or purified antibodies (immunoglobulin or gamma globulin) derived from an immune individual. This is called *artificial passive immunity* because the individual receiving the antibodies played no active part in antibody production.

In active immunity, introduction of antigen induces changes in the host: The immune system produces large quantities of antibodies and T cells in the primary response. A second ("booster") dose of the same antigen results in a faster response resulting in much higher levels of antibodies and T cells. This is known as a secondary immune response, or a memory response. Active immunity often remains throughout life as a result of immune memory (∞ Figure 29.21). A passively immunized individual never has more than the antibodies received in the initial injection, and these antibodies gradually disappear from the body. Moreover, a later exposure to the antigen does not elicit a secondary response.

Artificial passive immunity is usually therapeutic. Cells or antibodies from an immune individual are transferred to a nonimmune individual to prevent or cure active disease. For example, tetanus antiserum is administered to passively immunize an individual suspected of being exposed to *Clostridium tetani* due to an acute injury such as a car accident. Such an individual needs immediate immune protection, and cannot wait days to weeks for immunization to produce active immunity.

Artificial active immunity, as we discuss next, is often used as a prophylactic measure to protect a person against future attack by a pathogen. For example, immunization with tetanus toxoid actively immunizes individuals against future encounters with *Clostridium tetani* exotoxin, but is not an effective therapy for the trauma victim in the car accident above.

Immunization

The antigen or antigen mixture used to induce artificial active immunity is known as a **vaccine** or an immunogen, and the process of generating an artificial active immune response is **immunization**. Immunization with a vaccine designed to produce active immunity may introduce risks for infection and other adverse reactions. To reduce risks, pathogens or their products are often inactivated. For example, many immunogens consist of pathogenic bacteria killed by chemical agents such as phenol or formaldehyde, or physical agents such as heat. Formaldehyde is also used to inactivate viruses for vaccines, such as in the inactivated (Salk) polio vaccine. Likewise, the active form of many exotoxins cannot be used as an immunogen because of the toxic effects. Many exotoxins, however, can be modified chemically so they retain their antigenicity but are no longer toxic. Such a modified exotoxin is called a **toxoid**. Toxoids can be given safely in doses large enough to induce protective antitoxin immunity.

Immunization with live cells or virus is usually more effective than immunization with dead or inactivated material. It is often possible to isolate a mutant strain of a pathogen that has lost its virulence but still retains the immunizing antigens; strains of this type are called *attenuated strains* (Section 28.8). However, because attenuated strains of pathogens are still viable, some individuals, especially those who are immunocompromised, may acquire active disease caused by the live, attenuated immunizing pathogen. There have been serious cases of disease caused by vaccine-acquired infections in immunocompromised individuals, for example, from attenuated poliovirus vaccines and smallpox virus vaccines (Section 33.11).

A summary of vaccines available for use in humans is given in **Table 30.2**. Most effective viral vaccines are attenuated. Attenuated vaccines tend to provide long-lasting T cell-mediated immunity, as well as a vigorous antibody response and a strong secondary response upon reimmunization. However, the attenuated strains are difficult to select, standardize, and maintain. Live vaccines usually have a limited shelf life and require refrigeration for storage. Killed virus vaccines, on the other hand, tend to provide short-lived immune responses, without the development of a long-term memory response.

Bacterial vaccines are nearly always provided in an inactivated form. Inactivated bacterial vaccines induce long-term antibody-mediated protection without exposing recipients to the risk of infection.

Table 30.2 Vaccines for infectious diseases in humans

Disease	*Type of vaccine used*
Bacterial diseases	
Anthrax	Toxoid
Diphtheria	Toxoid
Tetanus	Toxoid
Pertussis	Killed bacteria (*Bordetella pertussis*) or acellular proteins
Typhoid fever	Killed bacteria (*Salmonella typhi*)
Paratyphoid fever	Killed bacteria (*Salmonella paratyphi*)
Cholera	Killed cells or cell extract (*Vibrio cholerae*)
Plague	Killed cells or cell extract (*Yersinia pestis*)
Tuberculosis	Attenuated strain of *Mycobacterium tuberculosis* (BCG)
Meningitis	Purified polysaccharide from *Neisseria meningitidis*
Bacterial pneumonia	Purified polysaccharide from *Streptococcus pneumoniae*
Typhus fever	Killed bacteria (*Rickettsia prowazekii*)
Haemophilus influenzae meningitis	Conjugated vaccine (polysaccharide of *Haemophilus influenzae* conjugated to protein)
Viral diseases	
Influenza	Inactivated virus
Hepatitis A	Recombinant DNA vaccine
Hepatitis B	Recombinant DNA vaccine or inactivated virus
Human papillomavirus (HPV)	Recombinant DNA vaccine
Measles	Attenuated virus
Mumps	Attenuated virus
Rubella	Attenuated virus
Polio	Attenuated virus (Sabin) or inactivated virus (Salk)
Rabies	Inactivated virus (human) or attenuated virus (dogs and other animals)
Rotavirus	Attenuated virus
Smallpox	Cross-reacting virus (vaccinia)
Varicella (chickenpox)	Attenuated virus
Yellow fever	Attenuated virus

Immunization Practices

Infants acquire natural passive immunity from maternal antibodies transferred across the placenta or in breast milk. As a result, infants are relatively immune to common infectious diseases during the first 6 months of life. However, infants should be immunized to prevent key infectious diseases as soon as possible so that their own active immunity can replace the maternal passive immunity. As discussed in Section 29.10, a single exposure to antigen does not lead to a high antibody *titer* or antibody quantity. After an initial immunization, a series of secondary or "booster" immunizations are given to produce a secondary response and a high antibody titer.

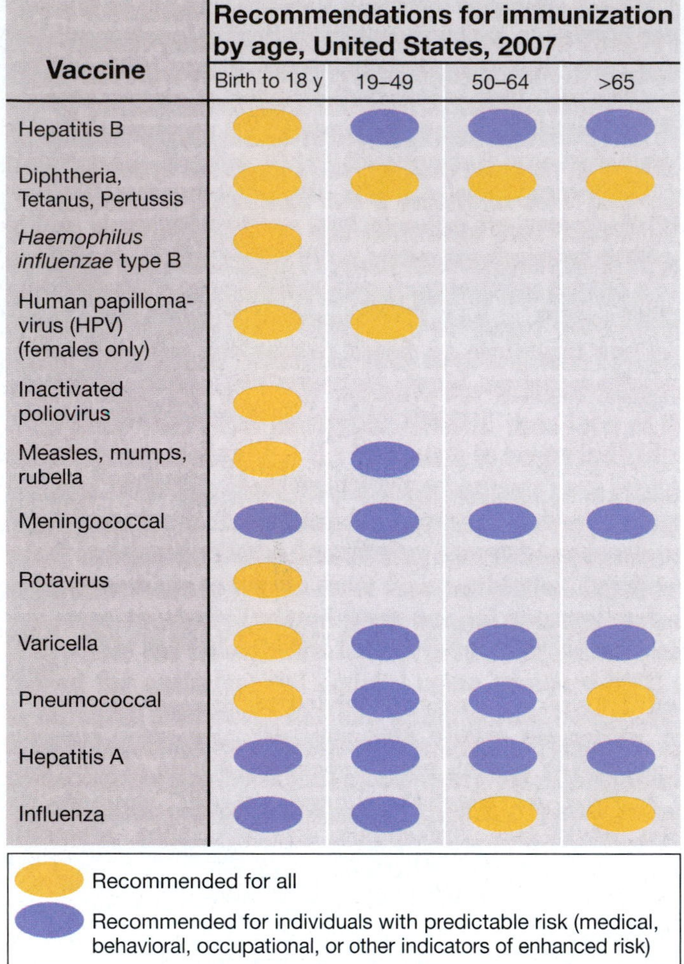

Vaccine
Recommendations for immunization by age, United States, 2007
Birth to 18 y
19–49
50–64
>65
Hepatitis B
Diphtheria, Tetanus, Pertussis
Haemophilus influenzae type B
Human papillomavirus (HPV) (females only)
Inactivated poliovirus
Measles, mumps, rubella
Meningococcal
Rotavirus
Varicella
Pneumococcal
Hepatitis A
Influenza
Recommended for all
Recommended for individuals with predictable risk (medical, behavioral, occupational, or other indicators of enhanced risk)

Figure 30.4 Recommended immunizations for children and adults in the United States. This course of immunizations is specified by the Centers for Disease Control and Prevention, Atlanta, GA, in 2007. For most available vaccines, a series of immunizations are required to initiate effective immunity. Periodic reimmunizations may be necessary to maintain long-term immunity. The CDC National Immunization Program website (http://www.cdc.gov/vaccines) has specific immunization recommendations for timing and dose of immunizations for all age groups. In addition, the website has special instructions for international travelers, women of child-bearing age, and persons with immunodeficiencies.

Current vaccine recommendations for children and adults in the United States are shown in **Figure 30.4**. For most vaccines, a series of immunizations are necessary to establish protective immunity and periodic reimmunization is often necessary to maintain immunity.

The importance of immunization in controlling infectious diseases is well established. For example, introduction of an effective vaccine into a population has reduced the incidence of formerly epidemic childhood diseases such as measles, mumps, and rubella (Figure 34.15) and has eliminated smallpox altogether (Section 33.11). The degree of immunity obtained by vaccination, however, varies greatly with the individual as well as with the quality and quantity of the vaccine. Lifelong immunity is rarely achieved by means of a single injection, or even a series of injections, and the immune cells and antibodies induced by immunization gradually disappear from the body. On the other hand, natural infections may stimulate memory or booster responses. In the complete absence of antigenic stimulation, the length of effective immunity varies considerably with different antigens. For example, protective immunity to tetanus from toxoid immunization may last many years. As a result, current recommendations call for reimmunization in adults only every 10 years to maintain protective immunity. Immunity induced by a particular influenza virus vaccine, however, disappears within a year or two without reimmunization through active infection or vaccination.

As we noted above, passive immunity is introduced by injecting preformed antibodies. The antibody-containing preparation is known as **serum** (also called an *antiserum* or an *antitoxin* if the antibodies are directed against a toxin). Antisera are obtained from immunized animals, such as horses, or from humans with high antibody titers. These individuals are said to be *hyperimmune*. The antiserum or antitoxin is standardized to contain a known antibody titer; a sufficient number of units of antiserum must be injected to neutralize any antigen that might be present in the body. The immunoglobulin fraction separated from serum pooled from a number of individuals is also used for passive immunization. Pooled sera contain a variety of antibodies induced by artificial or natural exposure to various antigens.

Immunizations, whether active or passive, benefit the individual. Immunization is also an important tool for public health disease control programs because infections spread poorly in populations with a large proportion of immune individuals. Smallpox, for example, has been eradicated, and polio, measles, whooping cough, diphtheria, and a number of other infectious diseases have been largely controlled through aggressive immunization programs (Section 33.8).

30.5 MiniReview

Immunity to infectious disease can be generated by passive or active, natural or artificial means. Immunization with antigen induces artificial active immunity and is widely used to prevent infectious diseases. Most agents used for immunization are either attenuated or inactivated pathogens or inactivated forms of microbial products such as toxins.

- Provide an example of natural passive immunity. How does natural passive immunity benefit the immunized individual?
- Provide an example of artificial active immunity. How does artificial active immunity benefit the immunized individual?

30.6 New Immunization Strategies

Most immunization preparations are produced from whole organisms or toxoids, as described in the previous section. However, there are several other methods for producing antigens suitable for immunization.

Synthetic and Genetically Engineered Immunizing Agents

The simplest alternate approach to vaccine development is the use of *synthetic peptides*. To make a vaccine, a peptide can be synthesized that corresponds to a known antigen on an infectious agent. For example, the structure of the protein antigen responsible for immunity to foot-and-mouth virus, an important animal pathogen (∞ Section 19.7), is known. A synthetic peptide of 20 amino acids constituting an important antigenic determinant of the protein has been made and attached to suitable carrier molecules. This synthetic vaccine evokes an excellent neutralizing antibody response to foot-and-mouth virus. However, as a general method, this approach has one major problem: The entire sequence, representing the complete antigenic profile of the protein, must be known to make an effective vaccine. The entire genomic sequences of a large number of pathogens are now known, however, providing the information necessary to identify the antigenic profile of each.

Molecular biology techniques can be used to make vaccines using information derived from genomics of pathogens. For example, genes that encode antigens from virtually any virus can be cloned into the vaccinia virus genome and expressed. Inoculation with the genetically engineered vaccinia virus can then be used to induce immunity to the product of the cloned gene. Such a preparation is called a *recombinant-vector vaccine*. This method depends on the identification and cloning of the gene that encodes the antigen and also on the ability of the vaccinia virus to express the cloned gene as an antigenic protein. An effective recombinant vaccinia–rabies vaccine has been developed for use in animals. Recombinant DNA methods to develop vaccines were discussed in Section 26.5.

Another immunization strategy involves the use of recombinant DNA proteins as immunogens. First, a pathogen gene must be cloned in a suitable microbial host that expresses the protein encoded by the cloned gene. The pathogen protein can then be harvested and used as a vaccine; such a vaccine is called a *recombinant-antigen vaccine*. For example, the current hepatitis B virus vaccine is a major hepatitis surface protein antigen (HbsAg) expressed by yeast cells. A vaccine that is effective against human papillomavirus (HPV) is also recombinant vaccine made in yeast cells (Microbial Sidebar, "The Promise of New Vaccines").

DNA Vaccines

A novel method for immunization is based on expression of cloned genes in host cells. Bacterial plasmids containing cloned DNA are injected intramuscularly into a host animal. After several weeks, the host responds with T_C cells, T_H1 cells, and antibodies directed to the protein encoded by the cloned DNA. Taken up by host cells, the DNA is transcribed and translated to produce immunogenic proteins, triggering a conventional immune response. These plasmids are called *DNA vaccines*.

DNA vaccine strategies may provide considerable advantages over conventional immunizations. For instance, because only a single foreign pathogen gene is normally cloned and injected, there is no chance of an infection as there might be with an attenuated vaccine. Second, genes for individual antigens such as a tumor-specific antigen, or even a single antigenic determinant, can be cloned, targeting the immune response to a particular cell component. The response can also be targeted directly to APCs by including a MHC class II promoter in the gene construct. The promoter assures selective expression in dendritic cells, B cells, and macrophages, the only cells capable of using the MHC II protein. The expressed and processed antigen can then be presented on both MHC I and MHC II proteins. Thus, a single bioengineered plasmid can encode an antigen and elicit a complete immune response, inducing immune T cells and antibodies.

In at least one case, an experimental DNA vaccine consisting of an engineered MHC–peptide complex protected mice from infection with a cancer-producing papillomavirus.

30.6 MiniReview

Alternate immunization strategies using bioengineered molecules eliminate exposure to microorganisms and, in some cases, even to protein antigen. Application of these strategies is providing safer vaccines targeted to individual pathogen antigens.

- Identify alternate immunization strategies already used for approved vaccines.
- What are the advantages of alternative immunization strategies as compared traditional immunization procedures?

III IMMUNE RESPONSE DISEASES

Immune reactions can cause host cell damage and disease. **Hypersensitivity** responses are inappropriate immune responses that result in host damage. Hypersensitivity diseases are categorized according to the antigens and effector mechanisms that produce disease (**Table 30.3**). Superantigens are proteins produced by certain bacteria and viruses that cause widespread stimulation of immune cells, resulting in host damage by activating massive inflammatory responses.

30.7 Allergy, Hypersensitivity, and Autoimmunity

Antibody-mediated **immediate hypersensitivity** is commonly called *allergy*. Cell-mediated reactions also cause disease in the form of **delayed hypersensitivity**. *Autoimmune diseases* are mediated by immune reactions directed against self antigens. These diseases are categorized as type I, II, III, or IV hypersensitivities based on symptoms and immune effectors (Table 30.3).

Table 30.3 Hypersensitivity

Classification	*Description*	*Immune mechanism*	*Time of latency*	*Examples*
Type I	Immediate	IgE sensitization of mast cells	Minutes	Reaction to bee venom (sting) Hay fever
Type II	Cytotoxic[a]	IgG interaction with cell surface antigen	Hours	Drug reactions (penicillin)
Type III	Immune complex	IgG interaction with soluble or circulating antigen	Hours	Systemic lupus erythematosus (SLE)
Type IV	Delayed type	T_H1 inflammatory cell activation of macrophages	Days (24–48 h)	Poison ivy Tuberculin test

[a]Autoimmune diseases may be caused by type II, type III, or type IV reactions.

Immediate Hypersensitivity

Immediate hypersensitivity, or type I hypersensitivity, is caused by release of vasoactive products from IgE antibody-coated mast cells (**Figure 30.5**). Immediate hypersensitivity reactions occur within minutes after exposure to antigen. Depending on the individual and the antigen, immediate hypersensitivity reactions can be very mild or can cause a life-threatening reaction called *anaphylaxis*. Antigens that cause type I hypersensitivities are called *allergens*.

About 20% of the population suffers from immediate hypersensitivity allergies to pollens, molds, animal dander, certain foods, insect venoms, and other agents (**Table 30.4**). Almost all allergens enter the body at the surface of mucous membranes such as the lungs or the gut where the allergens stimulate T_H2 cells to produce cytokines that preferentially induce B cells to make IgE antibodies. Rather than circulating like IgG or IgM, the allergen-specific IgE antibodies bind to IgE receptors on mast cells (Figure 30.3). Mast cells are nonmotile granulocytes (∞ Section 29.1) associated with the connective tissue adjacent to capillaries throughout the body. With any subsequent exposure to the immunizing allergen, the B cell-bound IgE molecules bind the antigen. Cross-linking of two or more IgEs by an antigen triggers the release of soluble allergic mediators from the mast cells, a process called *degranulation*. These

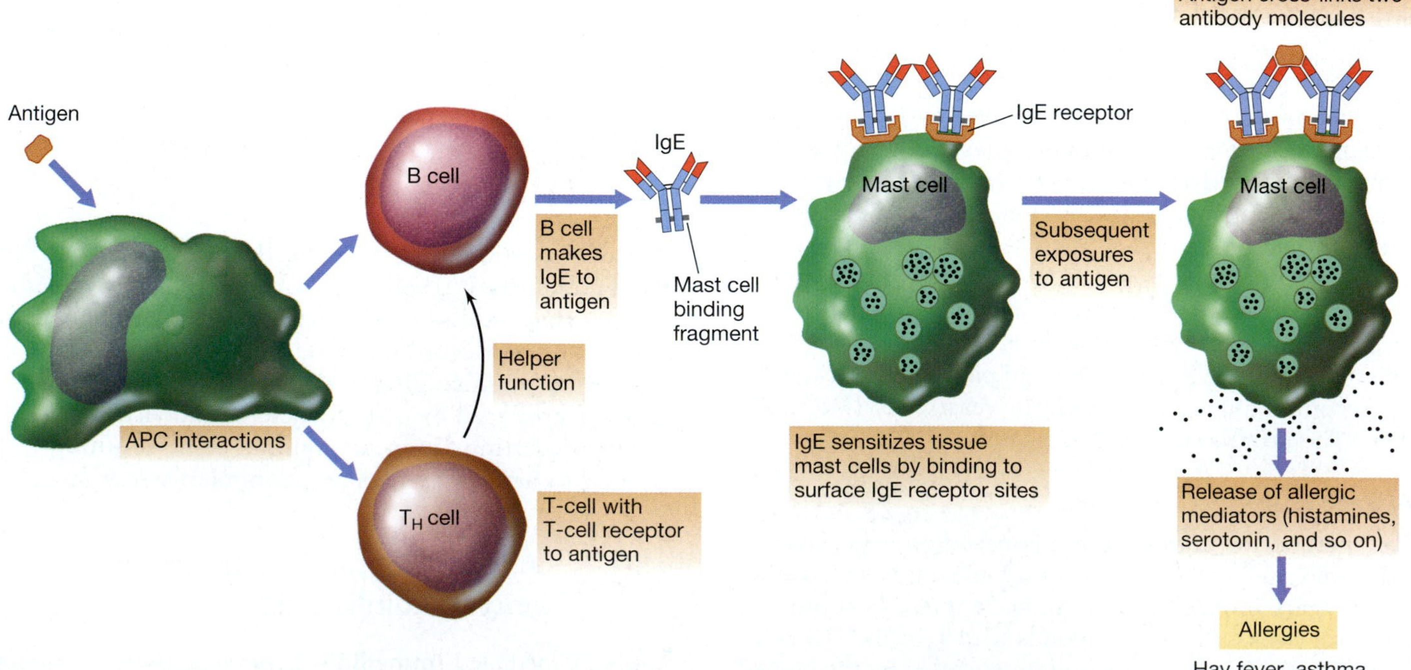

Figure 30.5 Immediate hypersensitivity. Certain antigens such as pollens stimulate IgE production. IgE binds to mast cells by means of a high-affinity surface receptor binding sensitizing the mast cell. Antigen contact cross-links surface IgE, causing release of soluble mediators such as histamine. These mediators produce symptoms ranging from mild allergies to life-threatening anaphylaxis.

Microbial Sidebar

The Promise of New Vaccines

A new vaccine has been licensed for protection against an infectious disease that is a major cause of cancer in women. This vaccine, called *Gardasil*, protects against infection by human papillomavirus (HPV). Gardasil has been licensed and recommended for women entering puberty and older. Various strains of HPV infect up to 75% of sexually active individuals and cause genital warts and vulvar, vaginal, and cervical cancers in infected women. The vaccine is targeted to HPV Types 6, 11, 16, and 18. Together, these strains account for 70% of cervical cancers and 90% of genital warts (**Figure 1**).

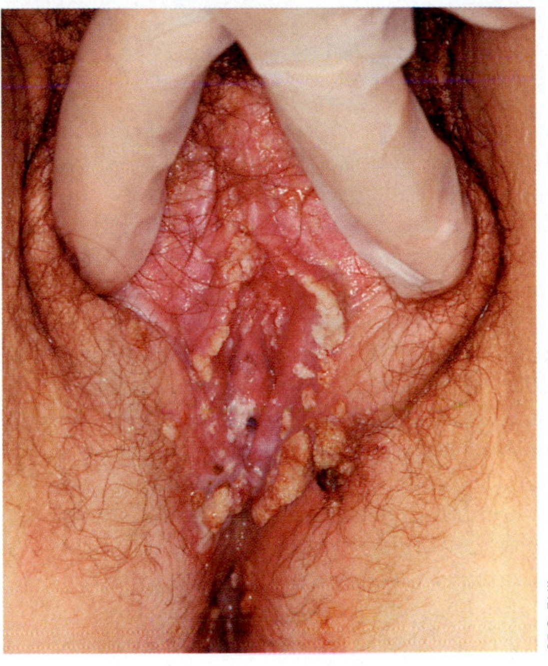

Figure 1 Perineal warts caused by infection with human papillomavirus

The quadrivalent HPV vaccine is a preparation of viruslike particles (VLPs) of the major L1 capsid proteins from these strains. The L1 proteins have been genetically engineered to be expressed by the yeast *Saccharomyces cerevisiae* (Section 26.5) and are released by disruption of the recombinant yeast cells as self-assembled VLPs. After purification, the VLPs are adsorbed onto a chemical adjuvant. The adjuvant immobilizes the VLPs, enhancing their ability to be taken up by phagocytes after injection.

Clinical trials have shown that the HPV vaccine is highly protective against viral infection, including genital warts (Figure 1) and all forms of cancers caused by the targeted HPVs. Because HPV is responsible for so much cervical cancer, risk in the immunized population will be significantly reduced. However, perhaps as important, the herd immunity (Section 33.5) resulting from immunization of a large proportion of the population will stop the spread of these viruses, providing protection even for individuals who are not immunized.

This vaccine is currently recommended for girls and women ranging in age from pre-puberty to the end of their reproductive years. While immunization of these susceptible individuals will certainly contribute to disease prevention, herd immunity can be raised to meaningful levels by immunizing all potential HPV sources such as the sexual partners of the women: The immunization of boys and men could provide the herd immunity necessary to eliminate transmission of this preventable STI.

Another newly released vaccine is protective for infection by Rotavirus, a common enterovirus that causes severe diarrhea, resulting in dehydration and even death in children worldwide. This intestinal disease can now be prevented by immunization with multiple doses of an oral, attenuated Rotavirus. An earlier Rotavirus vaccine was recalled due to postimmunization complications.

The HPV and Rotavirus vaccines are two examples of effective vaccine development and implementation, but a number of important infectious diseases still cannot be prevented by vaccination. This list includes tuberculosis, malaria, and HIV, arguably the three most important infectious diseases worldwide in terms of total disease and death. Over 1 million people die each year from each of these diseases. The current tuberculosis vaccine, BCG (Section 34.5), is considered inadequate and is not administered in many countries. No vaccine exists for the other diseases, although several vaccines are in development, and there are clinical trials for each of these. Finally, Gardasil is the only vaccine that is effective against any sexually transmitted infection in humans.

Even some effective vaccines have serious limitations. For example, influenza vaccines are only useful for one year because they are designed to target the strain-specific H and N antigens currently in circulation (Section 34.9). Development of a universal influenza vaccine that targets a common influenza virus antigen, M1, has been proposed, as in theory it should induce immunity to all influenza strains with a single vaccine. The immunogenicity and protection against influenza provided by M1 and other common antigens is, however, unproven.

Table 30.4 Immediate hypersensitivity allergens

Pollen and fungal spores (hay fever)
Insect venoms (bee sting)
Certain foods (strawberries, nuts, shellfish, etc.)
Animal dander
Mites in house dust

mediators cause allergic symptoms within minutes of antigen exposure. In general, these symptoms are relatively short-lived. After initial sensitization by an allergen, the allergic individual responds to each subsequent re-exposure to the antigen.

The primary chemical mediators released from mast cells are histamine and serotonin, modified amino acids that cause rapid dilation of blood vessels and contraction of smooth muscle, initiating the symptoms of systemic anaphylaxis. These symptoms include vasodilation (causing a sharp drop in blood pressure), severe respiratory distress caused by bronchial edema, flushed skin, mucus production, sneezing, and itchy, watery eyes. If severe cases of anaphylaxis are not treated immediately with adrenalin to counter smooth muscle contraction, increase blood pressure, and promote breathing, the person can die from *anaphylactic shock*. Fortunately, most allergic reactions are limited to mild local anaphylaxis with symptoms such as itchy, watery eyes. Less serious allergic symptoms are treated with drugs called *antihistamines* that neutralize the histamine mediators. Treatment for more serious symptoms may also include anti-inflammatory drugs such as steroids. Finally, immunization with escalating doses of the allergen may be done to shift antibody production from IgE to IgG. The IgG interacts with antigen and does not bind mast cells, preventing interactions with the relatively scarce IgE, stopping allergic symptoms, and inhibiting production of more IgE. This procedure is called *desensitization*.

Delayed-Type Hypersensitivity

Delayed-type hypersensitivity (DTH) or type IV hypersensitivity is cell-mediated hypersensitivity characterized by tissue damage due to inflammatory responses produced by T_H1 inflammatory cells (Table 30.3). Delayed-type hypersensitivity symptoms appear several hours after secondary exposure to the eliciting antigen, with a maximal response usually occurring in 24 to 48 hours. Typical antigens include certain microorganisms, a few self-antigens (**Table 30.5**), and several chemicals that covalently bind to the skin, creating new antigens. Hypersensitivity to these newly created antigens is known as *contact dermatitis* and results in, for example, skin reactions to poison ivy (**Figure 30.6*a***), jewelry, cosmetics, latex, and other chemicals. Within several hours after exposure to the agent, the skin feels itchy at the site of contact, and reddening and swelling appear, indicative of a general inflammatory response. Localized tissue destruction, often in the form of blistering, occurs.

An example of delayed-type hypersensitivity is the development of immunity to the causal agent of tuberculosis, *Mycobacterium tuberculosis* (Figure 30.6*b*). This protective cellular immune response was discovered by Robert Koch in his classic studies on tuberculosis (Section 1.8) and has been widely studied. Antigens derived from the bacterium, when injected subcutaneously into an animal previously infected with *M. tuberculosis*, elicit a characteristic skin reaction that develops fully only after a period of 24–48 hours. (By contrast, skin reactions due to IgE-mediated immediate hypersensitivity develop within minutes after antigen injection.) In the region of the introduced antigen, T_H1 cells become stimulated by the antigen and release cytokines that attract and activate large numbers of macrophages, which in turn produce a reaction at the site of injection. The reaction at the site of antigen injection is characterized by inflammation, including induration (hardening), edema (swelling), erythema

Table 30.5 Autoimmune diseases of humans

Disease	*Organ or area affected*	*Mechanism (hypersensitivity type)*[a]
Juvenile diabetes (insulin-dependent diabetes mellitus)	Pancreas	Cell-mediated immunity and autoantibodies against surface and cytoplasmic antigens of islets of Langerhans (II and IV)
Myasthenia gravis	Skeletal muscle	Autoantibodies against acetylcholine receptors on skeletal muscle (II)
Goodpasture's syndrome	Kidney	Autoantibodies against basement membrane of kidney glomeruli (II)
Rheumatoid arthritis	Cartilage	Autoantibodies against self IgG antibodies, which form complexes deposited in joint tissue, causing inflammation and cartilage destruction (III)
Hashimoto's disease (hypothyroidism)	Thyroid	Autoantibodies to thyroid surface antigens (II)
Male infertility (some cases)	Sperm cells	Autoantibodies agglutinate host sperm cells (II)
Pernicious anemia	Intrinsic factor	Autoantibodies prevent absorption of vitamin B_{12} (III)
Systemic lupus erythematosus	DNA, cardiolipin, nucleoprotein, blood clotting proteins	Autoantibody response to various cellular constituents results in immune complex formation (III)
Addison's disease	Adrenal glands	Autoantibodies to adrenal cell antigens (II)
Allergic encephalitis	Brain	Cell-mediated response against brain tissue (IV)
Multiple sclerosis	Brain	Cell-mediated and autoantibody response against central nervous system (II and IV)

[a]Refer to Table 30.3

(reddening), pain, and localized heating of the skin. The activated macrophages then ingest and destroy the invading antigen. This is a delayed-type hypersensitivity reaction and is the basis for the tuberculin test used to determine previous exposure to *M. tuberculosis* (Figure 30.6*b*).

Some microbial infections elicit delayed-type hypersensitivity reactions. In addition to tuberculosis, these include leprosy, brucellosis, psittacosis (all bacterial infections), mumps (viral), and coccidioidomycosis, histoplasmosis, and blastomycosis (fungal). In all of these cellular immune reactions, visible antigen-specific skin responses resembling the tuberculin reaction (Figure 30.4*b*) occur after injection of antigens derived from the pathogens, indicating previous exposure to the pathogen.

Autoimmune Diseases

T and B cells destined to react with self-antigens are normally eliminated during the process of lymphocyte maturation. However, in some individuals, T and B cells can be activated to produce immune reactions against self-proteins, leading to autoimmune diseases (Table 30.5). For example, T_H1-mediated delayed hypersensitivity can cause autoimmune responses directed against self-antigens, as is the case for allergic encephalitis. In Type I (juvenile) diabetes mellitus, T_H1 cells cause inflammatory reactions that destroy the beta cells of the pancreas. However, many autoimmune diseases are antibody-mediated, as we now discuss.

Some autoimmune diseases are caused by **autoantibodies**, antibodies that interact with self-antigens. In many cases, antibodies are directed to certain organ-specific antigens. For example, in *Hashimoto's disease*, autoantibodies are made against thyroglobulin, a product of the thyroid gland. The disease affects thyroid function and is classified as type II hypersensitivity: Antibodies interact with antigens found on the surface of host cells and initate destruction of the tissue (Table 30.5). In this case, antibodies to thyroglobulin fix complement and cause destruction of the thyroid gland cells. In juvenile diabetes, autoantibodies against the insulin-producing cells in the pancreas are observed, but tissue destruction occurs primarily through inflammatory reactions mediated by T_H1 cells.

Systemic lupus erythematosus (SLE) is an example of a disease caused by type III hypersensitivity: This disease and others like it are caused by autoantibodies directed against soluble, circulating self-antigens. In SLE, the antigens include nucleoproteins and DNA. Antibodies bind to soluble proteins, producing insoluble immune complexes. Disease results when circulating antigen–antibody complexes deposit in different body tissues such as the kidney, lungs, and spleen. Complement fixation and the resulting lytic and inflammatory responses cause local, often severe, cell damage at the site of the complex deposition. The complexes are deposited on organ membranes and bind complement, leading to destruction of the organ. Thus, type III hypersensitivity is an immune complex disorder (Table 30.5).

Organ-specific autoimmune diseases are sometimes more easily controlled clinically than diseases that affect multiple organs because the product of organ function, such as thyroxin in autoimmune hypothyroidism or insulin in juvenile diabetes, can often be supplied in pure form from another source. SLE, rheumatoid arthritis, and other diseases that affect multiple organs and sites can be controlled only by general immunosuppressive therapy, such as the use of steroid drugs. General immunosuppression, however, significantly increases chances of opportunistic infections.

Heredity influences the incidence, type, and severity of autoimmune diseases. Many autoimmune diseases correlate strongly with the presence of certain major histocompatibility

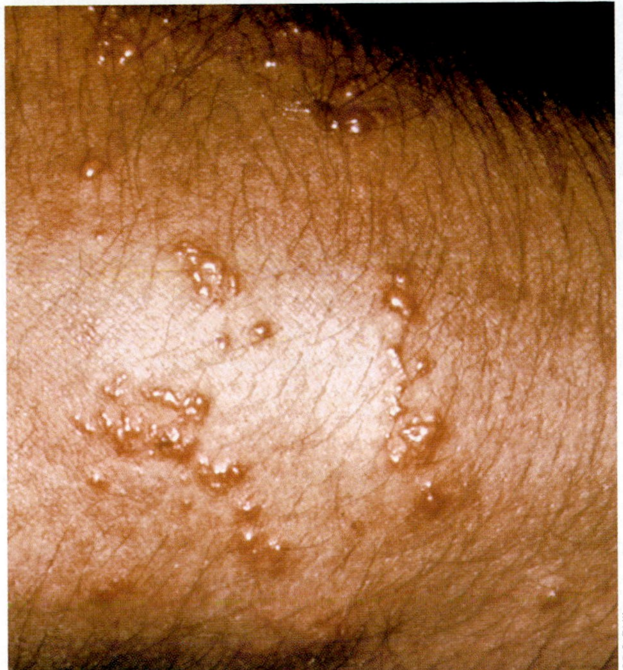

(a)

(b)

Figure 30.6 Cell-mediated immunity. *(a)* Poison ivy blisters on an arm. The raised rash appears maximally in sensitized individuals about 24–48 hours after exposure to plants of the genus *Rhus*. *(b)* A positive tuberculin test, typical for delayed hypersensitivity. The raised area of inflammation on the forearm is about 1.5 cm in diameter. Both reactions result from the action of macrophages activated by antigen-specific T_H1 cells.

complex (MHC) antigens (∞ Sections 29.6, 31.3, and 31.4). Studies of model autoimmune diseases in mice support such a genetic link, but the precise conditions necessary for developing autoimmunity may also depend on other factors, such as gender, age, and health status. Women, for example, are up to 20 times more likely to develop SLE than are men.

30.7 MiniReview

Hypersensitivity results when foreign antigens induce cellular or antibody immune responses, leading to host tissue damage. Autoimmunity results when the immune response is directed against self-antigens, resulting in host tissue damage.

- Discriminate between immediate hypersensitivity and delayed hypersensitivity with respect to antigens and immune effectors.
- Identify the two main categories of autoimmune disease with respect to antigens and immune effectors.

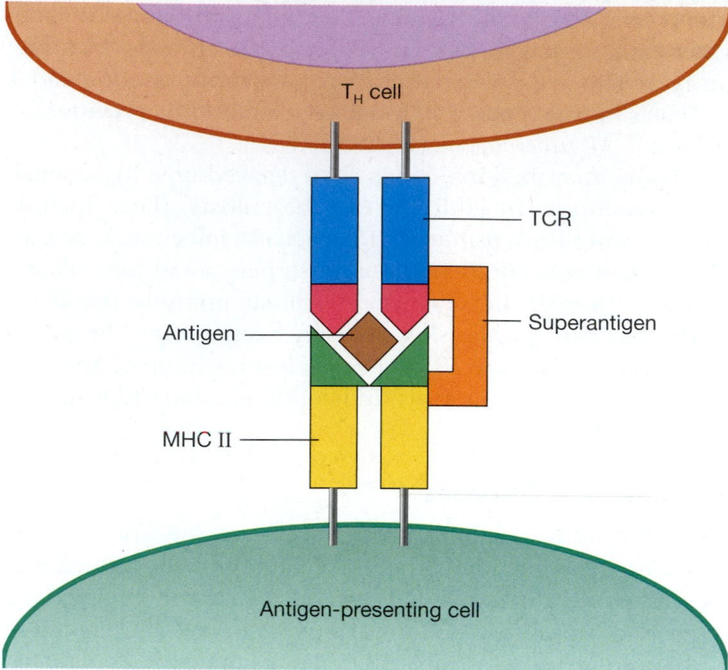

Figure 30.7 **Superantigens.** Superantigens act by binding to both the MHC protein and the TCR at positions outside the normal binding site. The superantigen binds to conserved regions of MHC and TCR proteins and can interact with large numbers of cells, causing massive T cell activation, cytokine release, and systemic inflammation.

30.8 Superantigens

We discussed the mechanisms of action for several different categories of bacterial toxins in Chapter 28. Most toxins interact directly with host cells to cause tissue damage. Endotoxins, for example, interact directly with many cell types, causing release of endogenous pyrogens and other soluble mediators and producing fever and general inflammation (∞ Section 29.3). Most exotoxins also interact directly with cells to cause cell damage (∞ Sections 28.10 and 28.11). However, certain exotoxins, the superantigens, act indirectly on host cells. These exotoxins do not act directly on host cell targets, but rather subvert the immune system to cause extensive host cell damage.

Superantigen Activation of T Cells

Superantigens are proteins capable of eliciting a very strong response because they activate more T cells than a normal immune response. Superantigens are produced by many viruses and bacteria and interact with TCRs. Streptococci and staphylococci, for example, produce several different and very potent superantigens (∞ Table 28.4). Superantigen interaction with the TCR, however, differs from conventional antigen-TCR binding.

Conventional foreign antigens bind to the TCR at a defined antigen-binding site. In a typical immune response, less than 0.01% of all available T cells interact with a conventional foreign antigen. However, superantigens bind to a site on the TCR that is outside the antigen-specific TCR binding site. A superantigen binds to all TCRs with a shared structure, and many different TCRs share the same structure outside the antigen-binding site. In some cases, superantigens can bind 5–25% of all T cells. The superantigens also bind to class II MHC molecules on APCs, again at a specific site outside the normal peptide-binding site (**Figure 30.7**). These cell surface interactions mimic conventional antigen presentation (∞ Section 29.6) and stimulate large numbers of T cells to grow and divide. These T cells, in turn, produce cytokines that stimulate other cells, such as macrophages and other phagocytes. Extensive cytokine production triggers a widespread cell-mediated response characterized by systemic inflammatory reactions, often resulting in fever, diarrhea, vomiting, mucus production, and systemic shock. In extreme cases, exposure to superantigens can be fatal. Superantigen shock is virtually indistinguishable from septic shock (∞ Section 29.3).

Notable superantigen diseases include *Staphylococcus aureus* food poisoning, characterized by fever, vomiting, and diarrhea. This common disease is caused by one of several superantigen staphylococcal enterotoxins. *S. aureus* also produces the superantigen responsible for *toxic shock syndrome*. *Streptococcus pyogenes* produces erythrogenic toxin, the superantigen responsible for scarlet fever (∞ Figure 34.6).

30.8 MiniReview

Superantigens bind and activate large numbers of T cells. Superantigen-activated T cells may produce systemic diseases characterized by systemic inflammatory reactions.

- Discriminate between normal and superantigen activation of T cells.
- Identify the binding site for superantigens on T cells and APCs.

Review of Key Terms

Adaptive immunity (antigen-specific immunity) the acquired ability to recognize and destroy a particular pathogen or its products, depending on previous exposure to the pathogen or its products

Antibody a soluble protein, produced by B cells, that interacts with antigen; also called immunoglobulin

Antigen a molecule capable of interacting with specific components of the immune system

Autoantibody an antibody that reacts to self antigens

B cell a lymphocyte that has immunoglobulin surface receptors, produces immunoglobulin, and may present antigens to T cells

Clone copy of an antigen-reactive lymphocyte, usually in large numbers

Cytokine a soluble protein produced by a leukocyte that modulates an immune response

Delayed hypersensitivity an inflammatory response mediated by T lymphocytes

Hypersensitivity an immune response leading to damage to host tissues

Immediate hypersensitivity an allergic response mediated by vasoactive products released from tissue mast cells

Immunization (vaccination) inoculation of a host with inactive or weakened pathogens or pathogen products to stimulate protective immunity

Immunoglobulin (Ig) a soluble protein produced by B cells and plasma cells that interacts with antigens; also called antibody

Inflammation a nonspecific reaction to noxious stimuli such as toxins and pathogens, characterized by redness (erythema), swelling (edema), pain, and heat (fever), usually localized at the site of infection

Innate immunity (nonspecific immunity) is the noninducible ability to recognize and destroy an individual pathogen or its products that does not rely on previous exposure to a pathogen or its products

Leukocyte a nucleated cell found in the blood (white blood cell)

Lymphocyte a subset of nucleated cells found in the blood that are involved in the immune response

Macrophage a large leukocyte found in tissues that has phagocytic and antigen-presenting capabilities

Major histocompatibility complex (MHC) a genetic region that encodes several proteins important for antigen processing and presentation. MHC I proteins are expressed on all cells. MHC II proteins are expressed only on antigen-presenting cells

Memory (immune memory) ability to rapidly produce large quantities of specific immune cells or antibodies after subsequent exposure to a previously encountered antigen

Neutrophil (polymorphonuclear leukocyte) (PMN) a type of leukocyte exhibiting phagocytic properties, a granular cytoplasm (granulocyte), and a multilobed nucleus

Pathogen-associated molecular pattern (PAMP) a repeating structural component of a microbe or virus recognized by a pattern recognition molecule (PRM)

Pattern recognition molecule (PRM) a membrane-bound protein that recognizes a pathogen-associated molecular pattern (PAMP)

Plasma cell differentiated B cell that produces antibodies

Primary antibody response antibodies made on first exposure to antigen; mostly of the IgM class

Secondary antibody response antibody made on subsequent exposure to antigen; mostly of the IgG class

Serum the liquid portion of the blood with clotting proteins and cells removed

Specificity the ability of the immune response to interact with individual antigens

Stem cell a cell that can develop into a number of cell types

Superantigen a pathogen product capable of eliciting an inappropriately strong immune response by stimulating greater than normal numbers of T cells

T cell a lymphocyte responsible for antigen-specific cellular interactions; T cells are divided into functional subsets including T_C cytotoxic T cells and T_H helper T cells. T_H cells are further subdivided into T_H1 inflammatory cells and T_H2 helper cells, which aid B cells in antibody formation

T cell receptor (TCR) antigen-specific receptor protein on the surface of T cells

Tolerance inability to produce an immune response to specific antigens

Toxoid an attenuated form of a toxin that retains immunogenicity while losing toxicity

Vaccination (immunization) inoculation of a host with inactive or weakened pathogens or pathogen products to stimulate protective immunity

Vaccine an inactivated or weakened pathogen, or an innocuous pathogen product used to stimulate protective immunity

UNIT 7

Review Questions

1. Identify the cells that express pattern recognition molecules (PRMs). How do PRMs associate with pathogen-associated molecular patterns (PAMPs) to promote innate immunity (Section 30.1)?
2. Identify the lymphocytes and the antigen-specific receptors involved in antibody-mediated adaptive immunity (Section 30.2).
3. Identify the lymphocytes and the antigen-specific receptors involved in cell-mediated adaptive immunity (Section 30.3).
4. List the diseases for which you have been immunized. List the diseases for which you may have acquired immunity naturally (Sections 30.4 and 30.5).
5. List the immunizations recommended for infants in the United States (Section 30.5).
6. Describe a biotechnology-based immunization strategy that has been adapted for an approved vaccine. Did this vaccine replace an existing vaccine? If so, what advantage does the

biotechnology vaccine have over the conventional vaccine (Section 30.6)?

7. Define the differences between immediate and delayed-type hypersensitivity in terms of immune effectors, target tissues, antigens, and clinical outcome (Section 30.7).

8. Describe the general mechanism used by superantigens to activate T cells. How does superantigen activation differ from T cell activation by conventional antigens (Section 30.8)?

Application Questions

1. Describe the relative importance of innate immunity compared to adaptive immunity. Is one more important than the other? Can we survive in a normal environment without immunity?

2. Many infectious diseases have no effective vaccines. Pick several of these diseases (for example, AIDS, malaria, the common cold) and explain why current vaccine strategies have not been effective. Prepare some alternate strategies for immunization against the diseases you have chosen.

3. Are superantigen reactions desirable for the host? Do they confer protection for the host or do they benefit the pathogen?

31

Molecular Immunology

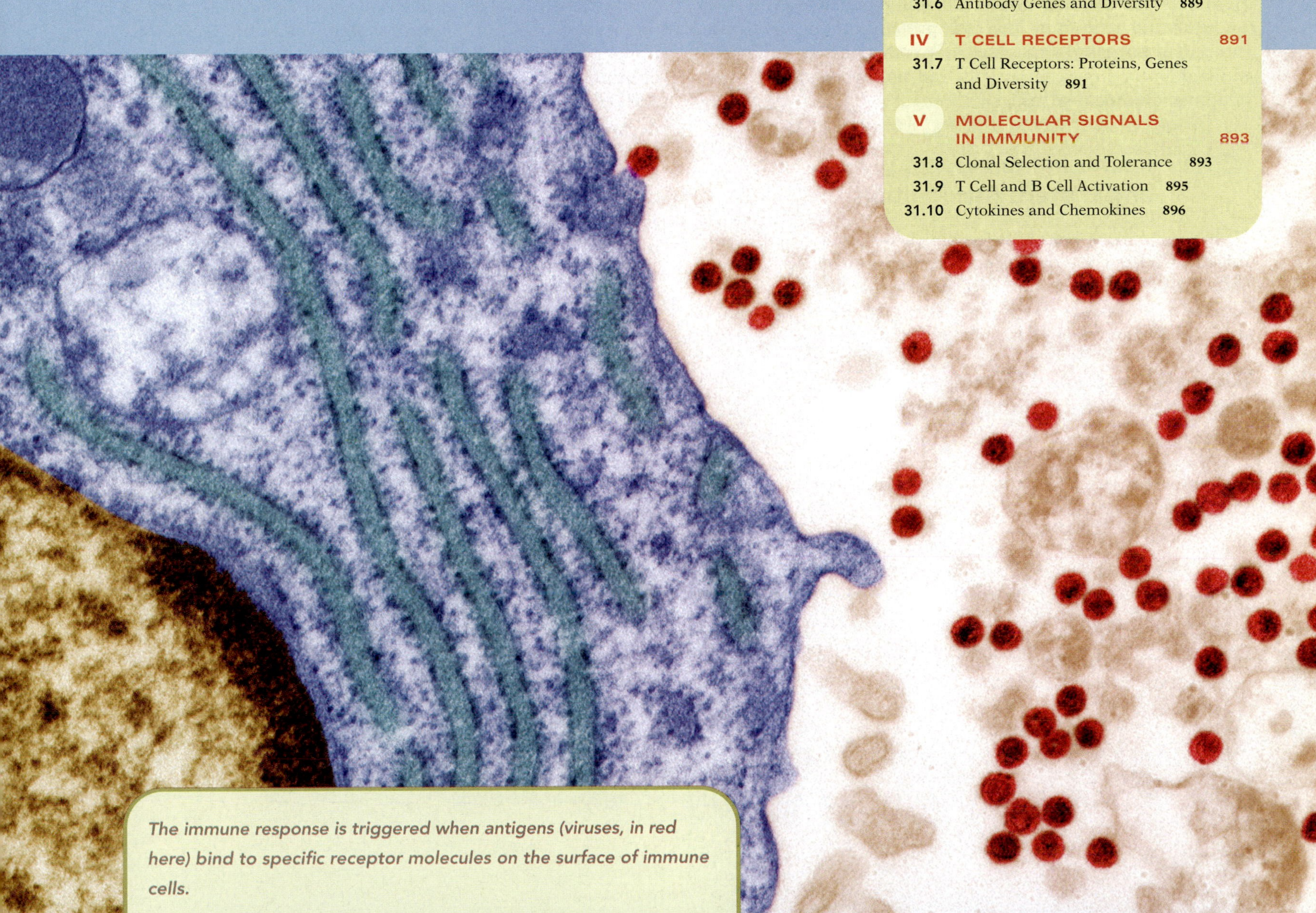

The immune response is triggered when antigens (viruses, in red here) bind to specific receptor molecules on the surface of immune cells.

We have seen that the immune response employs proteins to target pathogens. This chapter discusses these proteins and their complex molecular interactions. We will first examine proteins that target pathogens as part of the innate immune response. We then shift our attention to the antigen-binding proteins of the immunoglobulin superfamily. In the final section, we will investigate molecular interactions that control cell signaling and activate the immune response.

I RECEPTORS AND IMMUNITY

Here we first examine pattern recognition molecules (PRMs). These are innate immune response proteins that interact with molecular targets on pathogens. We then discuss the immunoglobulin supergene family and its involvement in the adaptive immune response.

31.1 Innate Immunity and Pattern Recognition

Multicellular organisms must recognize and control pathogen infection. A basic system for *innate* recognition of pathogens is widely distributed in nature. Organisms from primitive plants to vertebrate animals have molecular recognition mechanisms that rapidly and effectively activate host defenses.

Pathogen-Associated Molecular Patterns and Pattern Recognition Molecules

Pathogen-associated molecular patterns (PAMPs) are structural components common to a particular group of infectious agents. PAMPs are often macromolecules and include polysaccharides, proteins, nucleic acids, or even lipids. The lipopolysaccharide (LPS) of the gram-negative bacterial cell wall (Section 4.7) is an excellent example of a PAMP.

Pattern-recognition molecules (PRMs) are a group of soluble and membrane-bound host proteins that interact with PAMPs. Soluble PRMs such as the mannan-binding lectin and the complement proteins were previously discussed (Section 29.11) (**Table 31.1**). The PAMP recognized by mannan-binding lectin is the sugar mannose, found as a repeating subunit in bacterial polysaccharides (mannose on mammalian cells is inaccessible to mannan-binding lectin). Complement binds to a variety of pathogen cell surface components. C-reactive protein, another PRM, is an *acute phase protein* produced by the liver in response to inflammation. C-reactive protein interacts with the phosphorylcholine macromolecules of gram-positive bacterial cell walls. These PRMs all target pathogen surface PAMPs, leading to lysis of the targeted cell or opsonization.

Evolutionarily conserved, membrane-bound PRMs are found on the surfaces of macrophages, monocytes, dendritic cells, and neutrophils. These cells initiate phagocytosis (the engulfment and destruction of the pathogen). PRMs were first recognized in *Drosophila* (the fruit fly), where they were called Toll receptors (see the Microbial Sidebar, "*Drosophila* Toll Receptors—An Ancient Response to Infections"). Structural, functional, and evolutionary homologs of the Toll receptors, called **Toll-like receptors (TLRs)**, are widely expressed on mammalian phagocytes. At least nine TLRs in humans interact with a variety of cell surface and soluble PAMPs from viruses, bacteria, and fungi. Table 31.1 identifies some of the known TLRs, several other PRMs, and their associated PAMPs.

Table 31.1 Receptors and targets in the innate immune response

Pattern recognition molecules (PRM)	*Pathogen-associated molecular patterns (PAMP) and target organisms*	*Result of interaction*
Mannan-binding lectin[a] (soluble)	Mannose-containing cell surface microbial components, as in gram-negative bacteria	Complement activation
C-reactive protein (soluble)	Components of gram-positive cell walls	
TLR-1[b] (Toll-like receptor 1)	Lipoproteins in mycobacteria	Signal transduction, phagocyte activation, and inflammation[c]
TLR-2	Peptidoglycan on gram-positive bacteria; zymosan in fungi	
TLR-3	dsRNA in viruses	
TLR-4	LPS (lipopolysaccharide) in gram-negative bacteria	
TLR-5	Flagellin in bacteria	
TLR-6	Lipoproteins in mycobacteria; zymosan in fungi	
TLR-7	ssRNA in viruses	
TLR-8	ssRNA in viruses	
TLR-9	Unmethylated CpG oligonucleotides in bacteria	

[a]The soluble PRMs are produced by liver cells in response to inflammatory cytokines.
[b]The Toll-like receptors are membrane-integrated PRMs expressed in phagocytes. TLR-1, -2, -4, -5, and -6 are in the cytoplasmic membrane. TLR-3, -7, -8, and -9 are found in intracellular organelle membranes such as in lysosomes.
[c] Toll-like receptors are all involved in phagocyte activation via signal transduction.

Microbial Sidebar

Drosophila Toll Receptors—An Ancient Response to Infections

Multicellular organisms such as invertebrates and plants lack adaptive immunity but have a well-developed innate response to a wide variety of pathogens. Response to pathogens by the fruit fly, *Drosophila melanogaster* (**Figure 1**) have provided insight into innate immune mechanisms in many other groups of organisms. Several proteins required for fruit fly development are also important receptors for recognizing invading bacteria, functioning as pattern-recognition molecules (PRMs) that interact with pathogen-associated molecular patterns (PAMPs) on the macromolecules produced by the pathogen. The best example of a PRM is *Drosophila* Toll, a transmembrane protein involved in dorso-ventral axis formation as well as in the innate immune response of the fly.

Figure 1 ***Drosophila melanogaster*, the common fruit fly.** *In this insect the protein Toll, an analog of the Toll-like receptors of higher vertebrates, was first discovered.*

Toll immune signaling is initiated by the interaction of a pathogen or its components with the Toll protein displayed on the surface of phagocytes. *Drosophila* Toll, however, does not interact directly with the pathogen. Signal transduction events start with the binding of a PAMP such as the lipopolysaccharide (LPS) of gram-negative bacteria (Section 4.7) by one or more accessory proteins (Figure 31.1 shows the analogous TLR-4 system in humans). The LPS-accessory protein complex then binds to Toll. The membrane-integrated Toll protein initiates a signal transduction cascade, activating a nuclear transcription factor and inducing transcription of several genes that encode antimicrobial peptide. Toll-associated transcription factors induce expression of antimicrobial peptides, including drosomycin, an antifungal peptide, diptericin, active against gram-negative bacteria, and defensin, active against gram-positive bacteria. The peptides, produced in the liver-like fat body of *Drosophila*, are released into the fly's circulatory system where they interact with the target organism and cause cell lysis.

Structurally, the Toll proteins are related to lectins, a group of proteins found in virtually all multicellular organisms, including vertebrates and plants. Lectins interact specifically with certain oligosaccharide monomers. In humans, Toll-like receptors (TLRs) react with a wide variety of PAMPs. As with *Drosophila* Toll, human TLR-4 provides innate immunity against the gram-negative bacteria through indirect interactions with LPS, initiating a kinase signal cascade and activating the nuclear transcription factor NFκB. NFκB activates transcription of cytokines and other phagocyte proteins involved in the host responses (Figure 31.1).

Drosophila Toll is a functional, evolutionary, and structural ancestor to the Toll-like receptors in higher vertebrates, including humans. Toll and its homologs are evolutionarily ancient, highly conserved components of the innate immune system in animals and have even been found in plants.

Several TLRs interact with more than one PAMP. For example, TLR-4 is part of the innate immune response to gram-negative LPS and also responds to a host-response molecule called *heat shock protein*. Neither LPS nor heat shock protein interacts directly with the TLR-4, but rather interacts via a receptor protein that, in turn, interacts with TLR-4. In other cases, the TLR binds directly to the PAMP, as is the case for TLR-5 and its target, flagellin.

Interaction of a PAMP with the TLR triggers a transmembrane signal transduction event, initiating transcription of

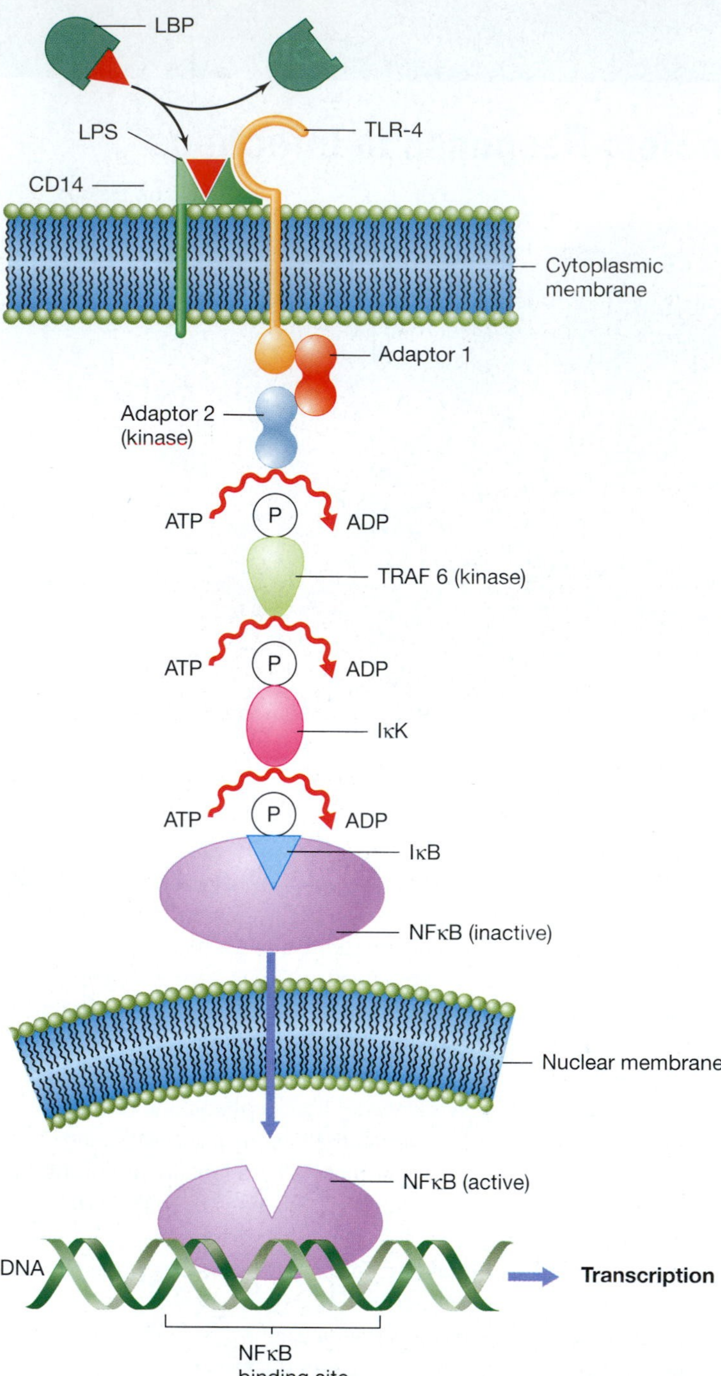

Figure 31.1 Signal transduction in innate immunity. Signal transduction is initiated when LPS, a PAMP, is bound by LBP (lipopolysaccharide-binding protein), which then transfers LPS to CD14 on the surface of a phagocyte. The LPS–CD14 complex then binds to the transmembrane TLR-4 receptor. The binding of TLR-4 initiates a series of reactions involving adaptor proteins and kinases, resulting in activation of NFκB (nuclear factor kappa B, a transcription factor). NFκB then diffuses across the nuclear membrane, binds to DNA, and initiates transcription of proteins essential for innate immunity.

DNA that leads to the synthesis of specific proteins and activation of the phagocyte. Activation can result in enhanced phagocytosis and killing of pathogens or contribute to inflammation (Sections 29.2 and 29.3).

Signal Transduction

Cytoplasmic membranes present a barrier to PAMPs. Eukaryotic cells must respond to the danger presented by these molecules without allowing the PAMPs to enter the cell. The cells transfer signals from the PAMPs across the membrane using the PRM membrane receptors to connect to common activation pathways called *signal transduction* pathways inside the cell. These pathways initiate transcription and translation of host-response proteins similarly to the membrane signal transduction mechanisms in prokaryotes (Section 9.5). For example, the pathway may be activated by the binding of LPS (a PAMP) to TLR-4 (a PRM) (**Figure 31.1**). TLR-4 then binds proteins in the cytosol, starting a cascade of reactions that activates transcription factors such as NFκB (nuclear factor kappa B), a protein that binds to specific regulatory sites on DNA, initiating transcription of downstream genes. Many of the NFκB-regulated genes encode innate host response proteins, such as the cytokines that initiate inflammation.

TLR-4 consists of three distinct protein domains, each with a separate function. The external domain of TLR-4 contains a binding site for LPS complexed with cell surface CD14 (Figure 31.1). A transmembrane domain in TLR-4 connects the external domain to a cytoplasmic domain. Binding of the CD14/LPS complex by the external domain of TLR-4 causes a change in the conformation of the third domain, a cytoplasmic domain, exposing a site that interacts with an adaptor protein. The adaptor protein, in turn, is altered and binds a second adaptor protein. The second adaptor protein initiates a *kinase* cascade, activating more proteins through ATP-mediated phosphorylation of TRAF6, another kinase that in turn activates IκK (inhibitor of kappa kinase). IκK then phosphorylates the IκB (inhibitor of kappa B) protein, causing it to dissociate from NFκB, which then diffuses across the nuclear membrane, binds to NFκB-binding motifs on DNA, and initiates transcription of downstream genes. As this example shows, signal transduction pathways induce transcriptional activation through ligand-receptor binding on the cell surface. The ligand-receptor interaction outside the cell induces the binding and concentration of the adapter proteins and kinase enzymes inside the cell. A single kinase enzyme can then catalyze phosphorylation of many targets, amplifying the effect of a single ligand-receptor interaction.

Signal transduction leading to activation of shared transcription factors and protein production is also the activation mechanism for phagocytes and lymphocytes in adaptive immunity. The binding of antigen to membrane Ig and TCRs on B and T cells initiates signal transduction cascades, again resulting in activation of NFκB and other transcription factors in the antigen-reactive lymphocytes.

31.1 MiniReview

Interactions between PAMPs and host PRMs are integral components of the innate immune response. PRMs interact with PAMPS shared by various pathogens, activating complement and phagocytes to target and destroy pathogens. These interactions initiate signal transduction cascades that activate effector cells.

- Identify the target of TLR-4 and the outcome for the host cell and the targeted pathogen.
- Define the general features of a signal transduction system starting with binding of a PAMP by a transmembrane PRM.

31.2 Adaptive Immunity and the Immunoglobulin Superfamily

The **immunoglobulin gene superfamily** consists of a large number of genes and their protein products that share structural, evolutionary, and functional features with immunoglobulin genes and proteins. The antigen-binding proteins in the adaptive immune response are part of this extended gene family.

As we discussed in Chapter 29, three different cell surface proteins interact directly with antigens during the adaptive immune response. These are the **immunoglobulins** (Igs or antibodies) produced by B cells that interact with antigens; the antigen-binding **T cell receptors** (TCRs) on the surface of T cells; and the proteins of the **major histocompatibility complex (MHC)** that process and present antigen. Each of these three antigen-binding proteins has a different location, structure, and function. MHC proteins, found on the surface of all cells, present antigens to TCRs found exclusively on T cells (Section 29.6). Igs, found on the surface of B lymphocytes and in serum and certain secretions, interact directly with extracellular antigens (Section 29.9).

Structure and Evolution of Antigen-Binding Proteins

Ig, TCR, and MHC proteins share structural features and have evolved by duplication and selection of primordial antigen receptors. Some important Ig superfamily proteins are shown in **Figure 31.2**. The proteins consist of a number of discrete **domains**, regions of the protein having a defined structure and function. Each protein has at least one domain with a highly conserved amino acid sequence called a *constant (C) domain*. The C domain typically has about 100 amino acids and an intrachain disulfide bond spanning 50–70 amino acids.

The *variable (V) domains* of TCR, Ig, and MHC proteins are about the same size as the constant domains, but V domain structures can be considerably different from one another and from the C domains. C domains provide structural integrity for the antigen-binding molecules, attach the V domains to the membrane, and give each protein its characteristic shape. C domains can also provide recognition sites for accessory molecules. For example, C domains of most IgG and all IgM proteins are targets for the C1q component of complement, a critical first step in initiating the complement activation sequence (Section 29.11). Likewise, MHC class I C domains bind to the accessory CD8 protein on T-cytotoxic (T_C) cells, and homologous MHC class II constant domains bind CD4 on T-helper (T_H) cells. MHC I-CD8 and MHC II-CD4 interactions are critical steps for T cell activation

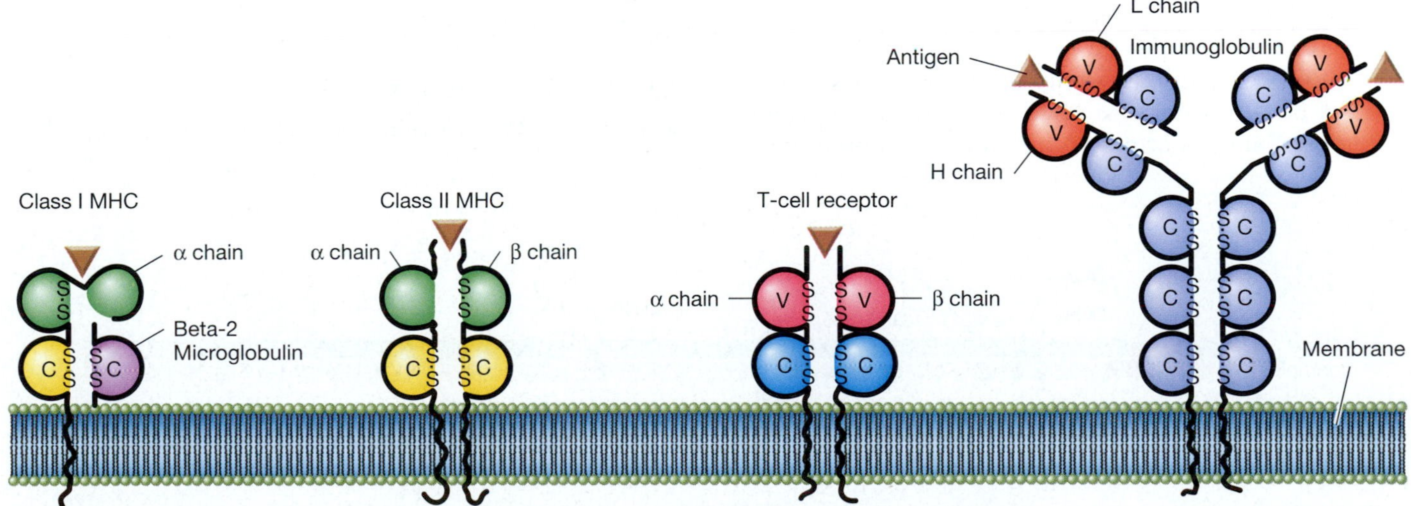

Figure 31.2 Immunoglobulin gene superfamily proteins. Constant domains (C) have homologous amino acid sequences and higher-order structures. The Ig-like C domains in each protein chain indicate evolutionary relationships that identify the proteins as members of the Ig gene superfamily. The variable domains (V) of Igs and TCRs are also Ig domains, but the peptide-binding domains of MHC class I and class II proteins are not identifiable Ig domains because their structures vary considerably from the basic features of the Ig domain.

(∞ Section 29.6). By contrast, V domains have evolved to interact with nonself antigens.

TCR, Ig, and MHC proteins each consist of two nonidentical polypeptides. The TCR consists of an α and a β chain. MHC proteins also consist of two different polypeptide chains, again designated α and β (∞ Section 29.6). Igs have a separate heavy and light chain (∞ Section 29.9). These heterodimers are expressed on a cell surface and bind antigens. However, the specific function of each of these molecules is quite different. Igs can be anchored on B cell surfaces where they bind to pathogens and their products such as toxins. Igs are also produced in large quantities as soluble serum proteins. TCRs, found exclusively on T lymphocytes, interact with antigenic peptides derived from processed pathogen proteins. These peptides are presented by the MHC proteins on target cells or APCs. The antigen-reactive T cells then kill the antigen-bearing cell or produce proteins that activate the immune response (∞ Sections 29.7 and 29.8).

31.2 MiniReview

The Ig gene superfamily encodes proteins that are evolutionarily, structurally, and functionally related to immunoglobulins. The antigen-binding Igs, TCRs, and MHC proteins are members of this family.

- Describe the structural features of the Ig constant domain.
- How do V domain structures differ from C domain structures?
- Identify the antigens bound by Igs, TCRs, and MHC proteins.

II THE MAJOR HISTOCOMPATIBILITY COMPLEX (MHC)

The major histocompatibility complex (MHC) is a group of genes found in all vertebrates. MHC proteins play a critical role in the presentation of processed antigens to other components of the immune system. The MHC spans about 4 Mbp on human chromosome 6 and is called the **human leukocyte antigen (HLA)** complex (**Figure 31.3**).

31.3 MHC Protein Structure

MHC proteins are the major antigen barriers for tissue transplantation, hence their name. Most individuals have slightly different MHC proteins, and tissue transplanted from one individual to another is usually rejected by an immune response triggered by MHC differences. This property of the MHC proteins hints at their natural function. MHC proteins are always expressed on cell surfaces in a complex with an embedded peptide. In normal cells, the embedded peptide is derived from breakdown products of cell metabolism. Thus, the MHC proteins hold embedded self-peptides. In cells infected with a virus, however, some of the embedded expressed peptides are derived from the virus. These viral peptides, when complexed with the MHC protein, look much like the slightly altered MHC proteins on a transplant. As a result, the virus–MHC complexes are recognized as nonself and are targeted for destruction by T cells. The MHC proteins are designed to present peptides to T cells for screening and potential targeting (∞ Section 29.6).

MHC proteins are divided into two structural classes. *Class I MHC proteins* are found on the surfaces of all nucleated cells. As a rule, the class I proteins present peptide antigens to T-cytotoxic cells (T_C). If class I-embedded peptides are recognized by T_C cells, the antigen-containing cell is targeted and directly destroyed (∞ Section 29.7). *Class II MHC proteins* are found only on the surface of B lymphocytes, macrophages, and dendritic cells, the professional APCs (∞ Section 29.6). Through the class II proteins, the APCs are involved in presentation of antigens to the T_H (T-helper) cells, stimulating either inflammatory reactions or antibody responses (∞ Section 29.8).

Class I MHC Proteins

A **class I MHC protein** consists of two polypeptides (**Figure 31.4**). The membrane-integrated alpha (α) chain is encoded

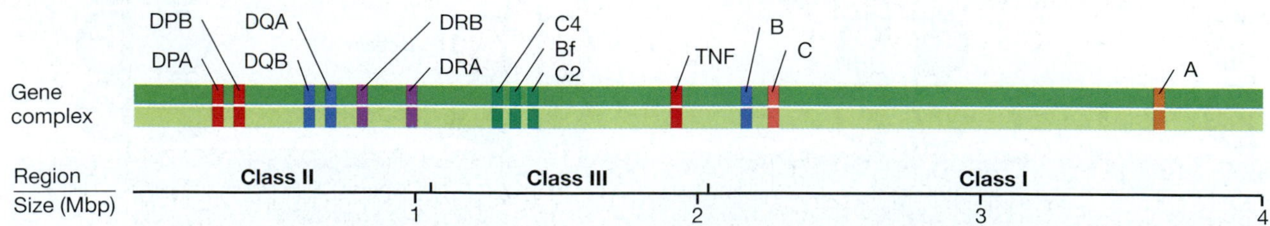

Figure 31.3 The human leukocyte antigen (HLA) gene map. The HLA complex, located on chromosome 6, is more than 4 million bases in length. Class II genes *DPA* and *DPB* encode class II proteins DPα and DPβ; *DQA* and *DQB* encode DQα and DQβ; *DRA* and *DRB* encode DRα and DRβ. Class III genes encode several proteins associated with immune recognition, as well as other proteins. *C4* and *C2* genes encode complement proteins C4 and C2 (∞ Section 29.11). The *TNF* gene encodes a cytokine, tumor necrosis factor. The class I MHC proteins HLA-B, HLA-C, and HLA-A are encoded by genes *B*, *C*, and *A*.

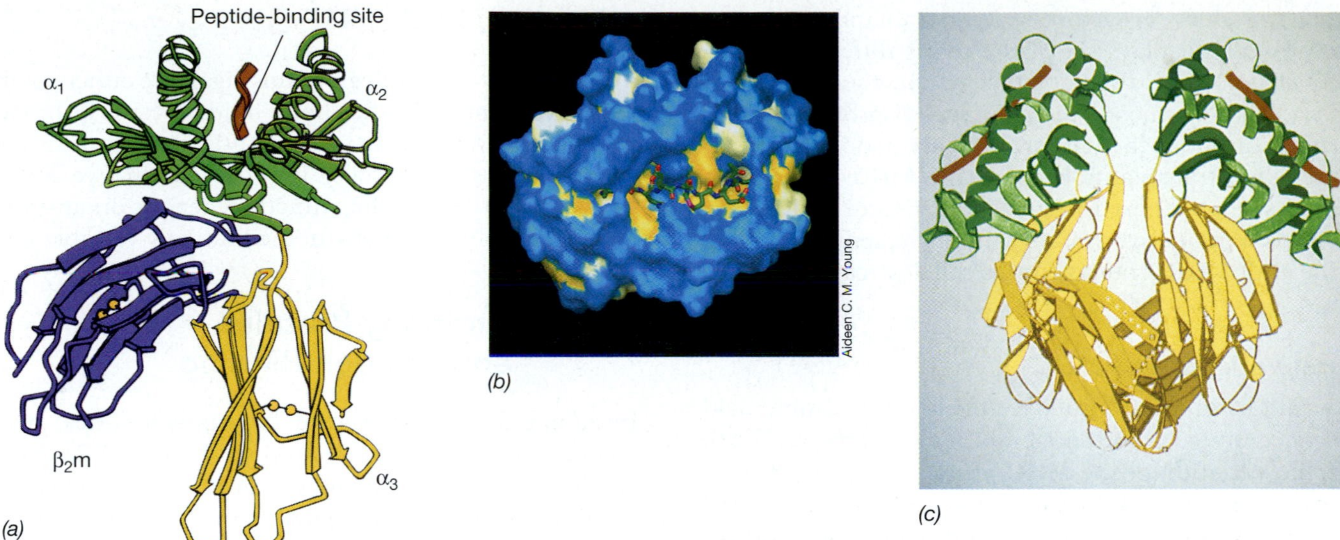

Figure 31.4 MHC protein structure. *(a)* The MHC class I protein. Beta-2 microglobulin (β2m) binds noncovalently to the α chain. *(b)* An MHC I protein with a bound peptide, as seen from above. A nine-amino acid peptide is shown as a carbon backbone structure, embedded in a space-filling model of a mouse MHC I protein. *(c)* A class II protein dimer. The peptides and their position in the binding sites of the associated MHC II proteins are in brown. A helical shape indicates alpha-helix protein structure and a flat shape indicates a beta-sheet (Figure 3.16).

in the MHC gene region on chromosome 6. The other class I polypeptide is the noncovalently associated beta-2 microglobulin (β_2m). The three-dimensional structure of class I MHC protein reveals a distinctive shape that suggests how this protein interacts with the antigen peptide and the TCR simultaneously. The class I α chain folds to form a groove, closed on both ends, between two α helices that straddle a β-sheet. In the endoplasmic reticulum, the MHC I groove is loaded with peptides that fit the groove, about eight to ten amino acids in length, derived from degraded endogenous proteins. For example, viral proteins produced inside the cell are degraded into peptides, loaded into class I MHC proteins, and transited to the cell surface to be recognized by TCRs on T_C cells (Sections 29.6 and 31.7).

Class II MHC Proteins

A **class II MHC protein** consists of two noncovalently linked, membrane-integrated polypeptides, α and β, found only on APCs. One α and one β polypeptide, expressed together, form a functional heterodimer (Figure 31.4). Class II proteins may be arranged in pairs or trimers that enhance their stability. The α1 and β1 domains of the class II protein interact to form a peptide-binding site similar to the class I peptide-binding site. However, the ends of the groove are open, permitting the class II protein to bind and display peptides that may be significantly longer than 10 amino acids. Class II binding peptides can be from 10 to 20 or more amino acids in length and are generally proteolytic fragments derived from exogenous pathogens processed by the APCs (Section 29.6). The APCs use the class II peptide complex to interact with TCRs on T_H cells, leading to T_H activation (Section 29.8).

31.3 MiniReview

Class I MHC proteins are expressed on all cells and function to present endogenous antigenic peptides to TCRs on T_C cells. Class II MHC proteins are expressed only on APCs. They function to present exogenously derived peptide antigens to TCRs on T_H cells.

- Compare the class I and class II MHC protein structures. How do they differ? How are they similar?
- Compare the peptide-binding sites of class I and class II MHC proteins. How do they differ? How are they similar?

31.4 MHC Polymorphism and Antigen Binding

There are at least three human MHC class I genes, *HLA-A*, *HLA-B*, and *HLA-C*, and all are highly polymorphic. **Polymorphism** is the occurrence in a population of multiple alleles (alternate forms of a gene) at a specific locus (the location of the gene on the chromosome) in frequencies that cannot be explained by recent random mutations. For example, there are 313, 559, and 150 different alleles at the HLA-A, HLA-B, and HLA-C loci, respectively, in the human population. Each person, however, has only two of these alleles at each locus; one allele is of paternal origin and one is of maternal origin. The two allelic variant proteins are expressed codominantly (equally).

Likewise, highly polymorphic alleles encode class II proteins at the *HLA-DR*, *HLA-DP*, and *HLA-DQ* loci. Again, the

UNIT 7

class II gene products are expressed codominantly. Thus, an individual usually displays six of the many different genetically and structurally distinct alleles that encode class I proteins and six that encode distinct class II proteins. These polymorphic variations in MHC proteins are major barriers to successful tissue transplants because the MHC proteins on the donor tissue (graft) are recognized as foreign antigens by the recipient's immune system. An immune response directed against the graft MHC proteins causes cell death and rejection of the graft.

Structural Variations of MHC

Allelic variations in MHC proteins translate into amino acid changes concentrated in the antigen-binding groove, and each polymorphic variation of the MHC protein binds a different set of peptide antigens. The peptides bound by a single MHC protein share a common structural pattern, or peptide **motif**, and each different MHC protein binds a different motif. For example, all of the eight-amino acid peptides bound by a certain class I protein may have a phenylalanine at position 5 and a leucine at position 8. Thus, all peptides with the sequence X-X-X-X-phenylalanine-X-X-leucine (where X is any amino acid) would be bound and presented by that MHC protein. A MHC class I protein encoded by a different MHC allele binds a different motif, with nine amino acids, and invariant acids tyrosine at position 2 and isoleucine at position 9 (X-tyrosine-X-X-X-X-X-X-isoleucine).

The invariant amino acids in each motif are *anchor residues*: They bind directly and specifically within an individual MHC peptide-binding groove. Thus, an individual MHC protein can bind and present many different peptides as long as the peptides contain the same anchor residues. Each MHC protein binds a different motif with different anchor residues. In this manner, the six possible MHC I proteins in an individual bind six different motifs and present a large number of different peptide antigens. MHC II proteins bind peptides in an analogous manner. As a result, within the human species, at least a few peptide antigens from each pathogen will display a motif that will be bound and presented by the MHC proteins. On the other hand, Igs and TCRs also bind antigens, but each Ig or TCR interacts very specifically with only a *single* antigen. These proteins employ a unique genetic mechanism to generate virtually unlimited diversity (Section 31.6).

31.4 MiniReview

MHC genes encode proteins used to present peptide antigens to T cells. Class I and class II MHC genes are highly polymorphic. MHC class I and class II alleles encode proteins that bind and present peptides with conserved structural motifs.

- Define polymorphism in MHC genes.
- How do individual MHC proteins present many different peptides to T cells?

III ANTIBODIES

Antibodies, or immunoglobulins (Igs), are either soluble proteins in serum and other body fluids, where Ig functions to neutralize and opsonize foreign antigens, or cell surface antigen receptors on B cells. In this section, we look at the structure, antigen-binding function, genetic organization, and the generation of diversity in the infinitely variable Igs.

31.5 Antibody Proteins and Antigen Binding

Immunoglobulins consist of four polypeptides, two heavy chains (H) and two light chains (L) (Figure 31.2). A complete antigen-binding unit includes a heavy chain–light chain heterodimer. The heavy and light chains are further divided into C (constant) and V (variable) domains, with the C domains responsible for common functions such as complement binding, and the V domains of one H and one L chains interact to form an antigen-binding site (**Figure 31.5**). Here we examine the structural features of the V domains and the antigen-binding site.

Variable Domains

Amino acid sequences are considerably different in the *variable* domains (V domains) of different Igs (Figure 31.5). Amino acid variability is especially apparent in several **complementarity-determining regions (CDR)**. The three CDRs in each of the V domains provide most of the molecular contacts with antigen. CDR1 and CDR2 differ somewhat between different immunoglobulins, but the CDR3s differ very dramatically from one another. The CDR3 of the heavy chain has a particularly complex structure, encoded within three distinct gene segments (see below). The CDR3 consists of the carboxy-terminal portion of the V domain, followed by a short "diversity" (D) segment of about 3 amino acids, and a longer "joining" (J) segment about 13–15 amino acids long. The light chain CDR3 is similar, but lacks the D segment. All of the heavy and light chain CDRs cooperate in antigen binding.

Antigen Binding

The Ig three-dimensional structure was shown in Figure 29.16. The principle of all antibody reactions lies in the specific combination of determinants on the antigen with the variable domains of the associated heavy and light chains. The antigen-binding site of an antibody molecule measures about 2 × 3 nm, large enough to accommodate a small portion of the antigen, called an **epitope**, about 10 to 15 amino acids long. Antigen binding is ultimately a function of the Ig folding pattern of the heavy and light polypeptide chains. The Ig folds of the V region bring all six CDRs (CDR1, 2, and 3 from both heavy and light chains) together at the end of the Ig protein. The result is a unique and specific antigen-binding site (∞ Figure 29.16 and Figure 31.5). In the next section, we

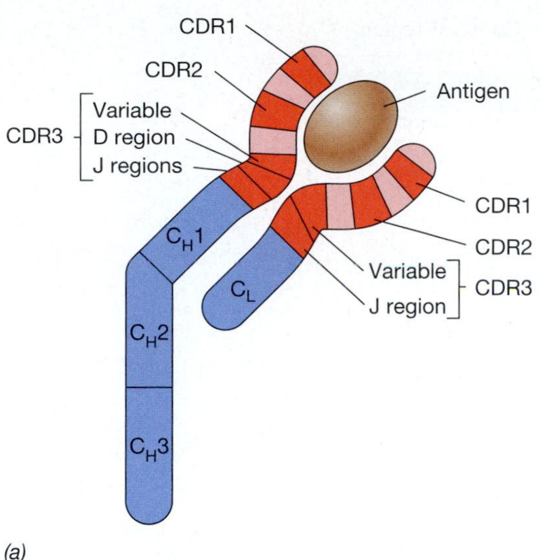

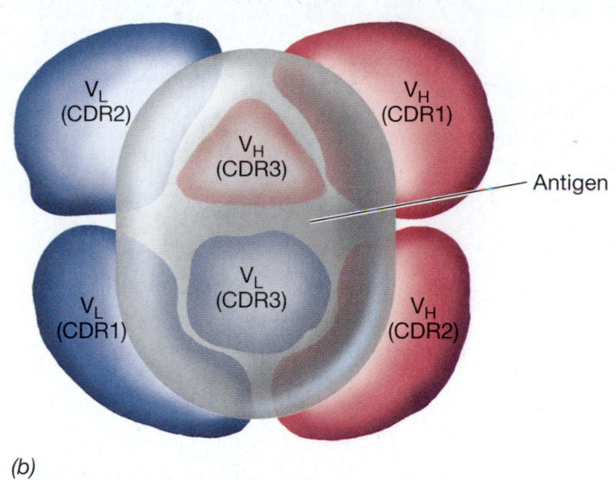

Figure 31.5 Antigen binding by immunoglobulin light and heavy chains. *(a)* One-half of an Ig is shown schematically, with a bound antigen. The variable (V) domains on the heavy (H) and light (L) chains are shown in red, with the antigen-binding CDR1, CDR2, and CDR3. C_H1, C_H2, and C_H3 are constant domains in the heavy chain, and C_L is the constant domain in the light chain. *(b)* The complementarity-determining regions (CDRs) from both heavy and light chains in *(a)* are conformed to make a single antigen-binding site, or pocket, on the Ig. The site is shown from above. Red binding areas are from the heavy chain, and blue binding areas are from the light chain. The highly variable CDR3s from both heavy and light chains cooperate at the center of the site. An antigen is shown in gray, overlaying the site and contacting all CDRs. The actual shape of the site may be a shallow groove or a deep pocket, depending on the antibody–antigen pair involved.

examine the genetic mechanisms that generate the tremendous diversity found in the Ig proteins.

31.5 MiniReview

The antigen-binding site of an Ig is composed of the V (variable) domains of one heavy chain and one light chain. Each heavy and light chain contains three complementarity-determining regions, or CDRs, that are folded together to form the antigen-binding site.

- Draw a complete Ig molecule and identify antigen-binding sites on the antibody.
- Describe antigen binding to the CDR 1, 2, and 3 regions of the heavy chain and light chain variable domains.

31.6 Antibody Genes and Diversity

For most proteins, one gene encodes one protein. However, this is not the case with the heavy and light chains of immunoglobulins. Because collectively, antibodies must recognize and bind a wide variety of molecular structures, the immune system must be able to generate almost unlimited antibody variation. Interestingly, this is done using a very limited number of genes. Somatic recombination, random heavy and light chain reassortment, coding for joint diversity, and hypermutation all contribute to the almost limitless diversity generated from a relatively small, fixed number of Ig genes.

Immunoglobulin Genes

The gene encoding each immunoglobulin H or L chain is constructed from several immunoglobulin gene segments, and each segment encodes a portion of the final gene. In each B cell, these gene segments undergo a series of somatic rearrangements (recombination followed by deletion of intervening sequences), forming a large number of antibody genes by recombining segments in all possible combinations. Molecular studies have verified this "genes in pieces" hypothesis by demonstrating that the V, D, and J gene segments that encode heavy chain variable domains, as well as the genes encoding C domains, are separated from one another in the genome. The gene segments, however, are brought together (somatically recombined) to form a single mature Ig heavy chain gene in each mature B cell (**Figure 31.6**). A single V gene encodes CDR1 and CDR2, whereas CDR3 is encoded by a mosaic of the 3′ end of the V gene, followed by the D and J genes.

In each B cell, only one protein-producing rearrangement occurs in the heavy and light chain genes. Called *allelic exclusion,* this mechanism ensures that each B cell produces only one Ig. Finally, the class-defining constant domains of Igs are encoded by separate C genes. Thus, four different gene segments, V, D, J, and C, recombine to form one functional heavy chain gene. Similarly, light chains are encoded by recombination products of light chain V and J genes.

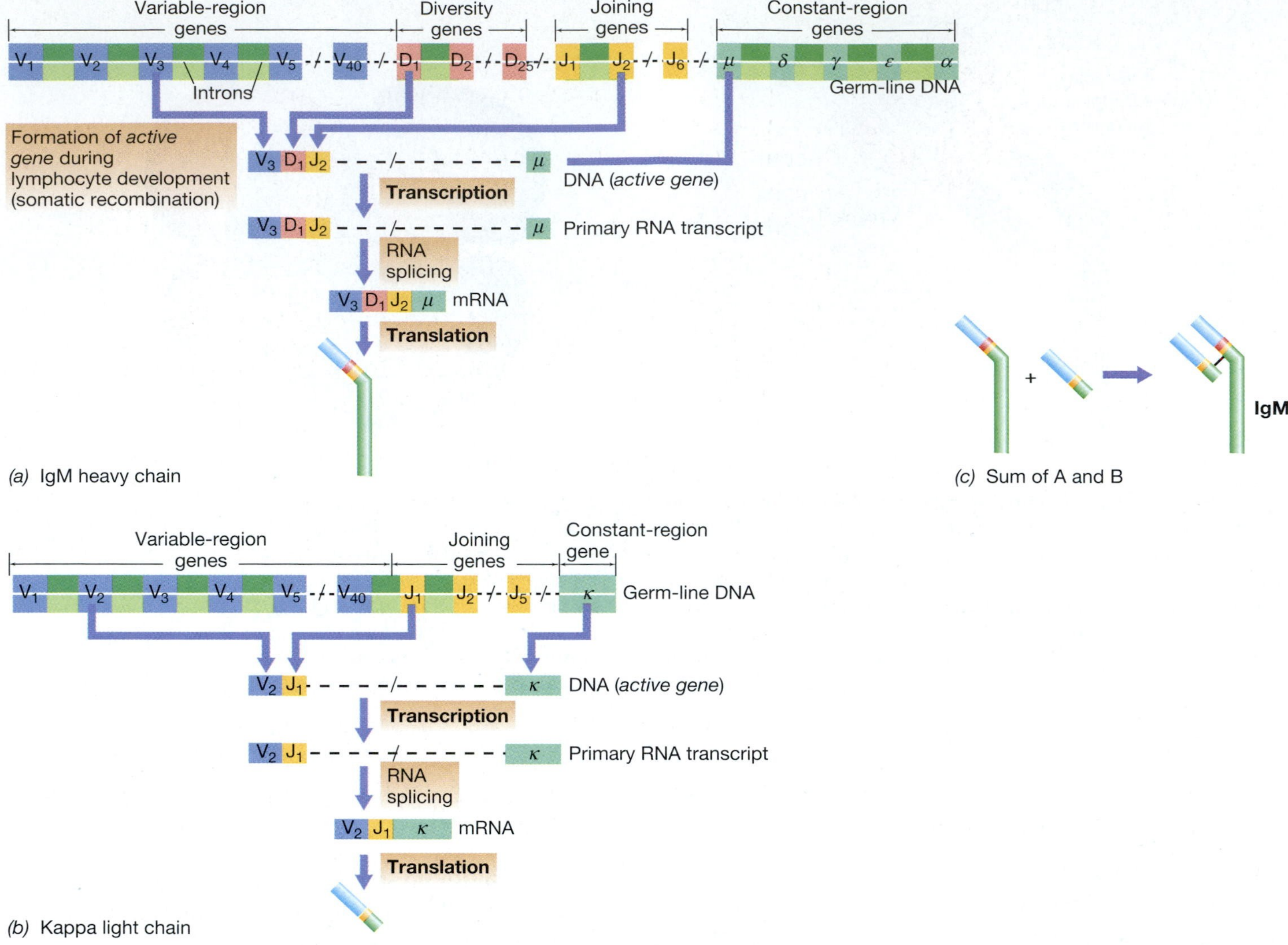

Figure 31.6 Immunoglobulin gene rearrangement in human B cells. Ig genes are arranged in tandem on three different chromosomes. *(a)* The heavy chain (H) gene complex on chromosome 14. The filled boxes represent Ig coding genes. The broken lines indicate intervening sequences and are not shown to scale. *(b)* The kappa (κ) light chain complex on chromosome 2. The lambda (λ) light chain genes are in a similar complex on chromosome 22. *(c)* Assembly of one-half of an antibody molecule.

The gene segments required for all Igs exist in all cells but undergo recombination only in developing B lymphocytes. As shown in Figure 31.6, each B cell contains multiple kappa (κ) and lambda (λ) light chain V and J genes arranged in tandem. Each B cell also contains tandem V genes, D genes, and J genes for the heavy chains. In addition, the heavy chain constant domain (C_H) genes and the light chain constant domain genes (C_L) are present. The V, D, J, and C genes are separated by noncoding sequences (introns) typical of gene arrangements in eukaryotes. Genetic recombination occurs in each B cell during the development of B lymphocytes. Randomly selected V, D, and J segments are recombined to form a functional heavy chain gene. On another chromosome, V and J are also randomly recombined to form a complete light chain gene. The active gene, still containing intervening sequence between the VDJ or VJ gene segments and the C gene segments, is transcribed, and the resulting primary RNA transcript is spliced to yield the final mRNA. The mRNA is then translated to make the heavy and light chains of the Ig molecule.

Reassortment and VDJ Joining

Up to this point, all Ig diversity has been generated from recombination of existing genes. The final light chain and heavy chain genes expressed by a given B cell result from chance reassortment of these rearranged heavy chains and light chains (Figure 31.6). In humans, for example, based on the numbers of genes at the kappa (κ) light chain loci, there are 40 V × 5 J possible rearrangements or 200 possible κ light chains.

For the alternative lambda (λ) light chain, there are 30 V × 4 J, or 120 possible chain combinations. There are approximately 40 V × 25 D × 6 J or 6,000 possible heavy chains. Assuming that each heavy chain and light chain has an equal chance to be expressed in each cell, there are 6,000 × 200 or 1,200,000 possible immunoglobulins with κ light chains and 6,000 × 120 or 720,000 possible immunoglobulins with λ chains. Thus, at least 1,920,000 possible antibodies can be expressed!

Additional diversity is generated by the DNA-joining mechanism. Joining of the V-D or D-J segments in the heavy chain or the V-J gene segments in the light chain is imprecise and frequently varies the sequence at these coding joints by a few nucleotides. Even more diversity is generated by additions of nucleotides at V-D and D-J coding joints on the heavy chain genes, and at V-J coding joints in light chain genes. Either random (N) or template-specific (P) nucleotides may be added. This N and P diversity at V domain coding joints changes or adds amino acids in the CDR3 of both heavy and light chains.

Hypermutation

Finally, antibody diversity is expanded even more in B cells by **somatic hypermutation**, the mutation of Ig genes at much higher rates than the mutation rates observed in other genes. Somatic hypermutation of Ig genes is typically evident after a second exposure to an immunizing antigen. As we saw, a second exposure to antigen results in a change in the predominant antibody class produced, with a switch from IgM to IgG production (∞ Section 29.10). Somatic hypermutation occurs only in the V regions of rearranged heavy and light chain genes. A generally random process, B cells bearing mutated receptors compete for available antigen. This process selects the better-binding B cell receptors having higher antigen-binding strength (affinity) than the original B cell receptor. This *affinity maturation* process is one of the factors responsible for a dramatically stronger secondary immune response (∞ Figure 29.21). The somatic mechanisms responsible for affinity maturation add virtually unlimited possibilities to the generation of Ig diversity, making the potential antibody repertoire almost limitless.

31.6 MiniReview

Immunoglobulin diversity is generated by several mechanisms. Somatic recombination of gene segments allows shuffling of the various Ig gene segments. Random reassortment of the heavy and light chain genes, imprecise joining of VDJ and VJ gene segments, and hypermutation mechanisms contribute to nearly unlimited immunoglobulin diversity.

- Describe the recombination events that produce a mature heavy chain gene.
- Describe other somatic events that further enhance antibody diversity.

IV T CELL RECEPTORS

TCRs are cell surface antigen receptors on T cells that recognize peptide antigens embedded in MHC proteins. In this section, we look at the structure, antigen-binding function, and genetic organization of the TCRs.

31.7 T Cell Receptors: Proteins, Genes and Diversity

TCR proteins are integrated into the cell surface membrane of T cells. TCRs consist of two polypeptides, the alpha (α) chain and the beta (β) chain. The α/β TCR specifically binds foreign peptides that are embedded in MHC molecules on the surface of APCs or target cells (∞ Section 29.6). Thus, the TCR binds both a self-MHC protein and a foreign peptide. The TCRs accomplish this dual binding function through a binding site composed of the V domains of the α chain and β chain. The α-chain and β-chain V domains of TCRs contain CDR1, CDR2, and CDR3 segments that interact directly with the MHC–peptide antigen complex.

TCR Proteins

The three-dimensional structure of the TCR–MHC peptide complex is shown in **Figure 31.7**. Both TCR and MHC proteins bind directly to peptide antigen. The MHC protein binds one face of the peptide, the MHC motif, whereas the TCR binds the other peptide face, the T cell epitope. The CDR regions of the TCR bind directly to the MHC–peptide complex, and each CDR has a specific binding function. The CDR3 regions of the TCR α chain and β chain bind with the epitope; the CDR1 and CDR2 regions of the TCR α and β chains bind mainly to the MHC proteins.

TCR Genes and Diversity

The TCR α and β chains are encoded by distinct gene segments for constant and variable domains. Much like Ig genes, TCR V region genes are arranged as a series of tandem segments. The α chain has about 80 variable (V) genes and 61 joining (J) segments, whereas the β chain has 50 variable (V) genes, 2 diversity (D) genes, and 13 joining (J) segments (**Figure 31.8**). The β chain V, D, and J genes and the α chain V and J genes undergo recombination to form functional V-region genes. As in Igs, additional diversity results from N and P diversity at V-D and D-J coding joints in the β chain and at the V-J coding joint in the α chain. Finally, the D region of the β chain can be transcribed in all three reading frames, leading to production of three separate transcripts from each D-region gene and creating greater diversity than would be expected from the D gene segments alone. As we discussed for reassortment of Ig H and L chains, individual α and β chains are produced by each T cell at random and joined to form a complete α:β heterodimer. Somatic hypermutation mechanisms do not generate TCR diversity. Potential TCR diversity,

UNIT 7

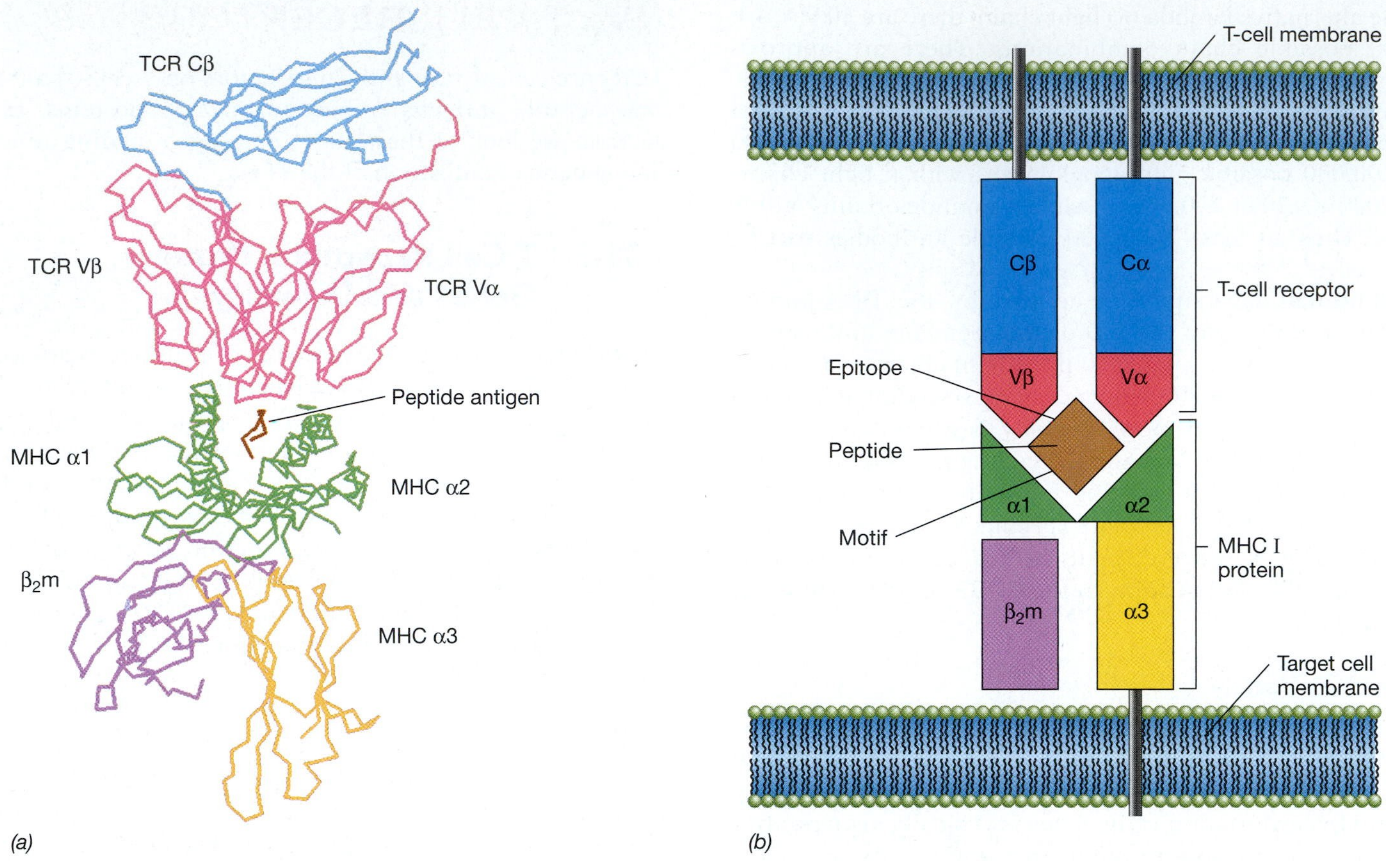

Figure 31.7 The TCR–peptide–MHC I protein complex. *(a)* A three-dimensional structure showing the orientation of TCR, peptide (brown), and MHC. This structure was derived from data deposited in the Protein Data Bank. *(b)* A diagram of the MHC–peptide–TCR structure. Note that the peptide is bound by both MHC and TCR proteins and has a distinct surface structure that interacts with each.

however, is still extensive, and on the order of 10^{15} different TCRs can be generated.

31.7 MiniReview

T cell receptors bind to peptide antigens presented by MHC proteins. The CDR3 regions of both the α chain and the β chain bind to the epitope, whereas the CDR1 and CDR2 regions bind to the MHC protein. The V domain of the β chain of the TCR is encoded by V, D, and J gene segments. The V domain of the α chain of the TCR is encoded by V and J gene segments. TCR diversity is generated by a variety of genetic mechanisms and results in practically unlimited TCR antigen-binding diversity.

- Distinguish among the functions of the TCR CDR1, CDR2, and CDR3 segments.
- Which diversity-generating mechanisms are unique to TCRs? Which mechanisms are unique to Igs?

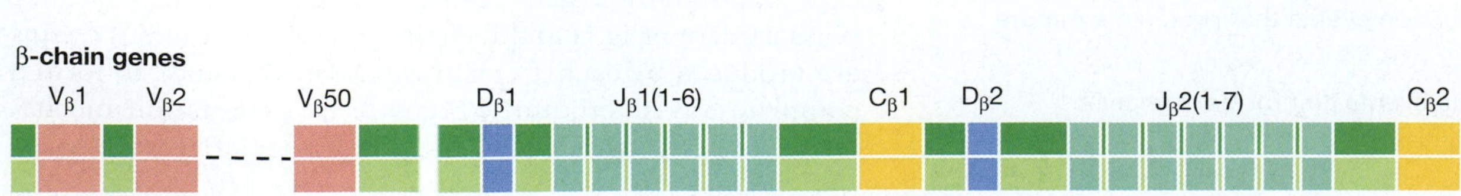

Figure 31.8 Organization of the human TCR α- and β-chain genes. The α-chain genes are located on chromosome 14, and the β-chain genes are on chromosome 6.

V MOLECULAR SIGNALS IN IMMUNITY

Here we first explore clonal selection, the mechanism by which antigen-reactive cells respond to foreign antigens while ignoring self-antigens. Next we examine the pairs of molecular signals that are responsible for activating the selected antigen-receptor bearing cells. Finally, we introduce cytokines and chemokines, soluble proteins produced by activated cells to recruit and activate other cells in the immune response.

31.8 Clonal Selection and Tolerance

T cells must be able to discriminate between the dangerous nonself antigens and the harmless self-antigens that compose our body tissues. Thus, T cells must acquire **tolerance**, or specific unresponsiveness to self-antigens. To acquire tolerance, immune lymphocytes are maintained that interact only with the nonself antigens derived from dangerous pathogens.

Clonal Selection

The **clonal selection** theory states that each antigen-reactive B cell or T cell has a cell surface receptor for a single antigen epitope. When stimulated by interaction with that antigen, each cell can replicate, and antigen-stimulated B and T cells grow and differentiate. As a result, antigen-stimulated cells divide, producing a pool of cells that express the same antigen-specific receptors. A *clone* comprises the identical progeny of the initial antigen-reactive cell (**Figure 31.9**). Cells that have not interacted with antigen do not multiply.

To respond to the seemingly infinite variety of antigens, a nearly infinite number of antigen-reactive cells are needed in the body. As we saw previously (Sections 31.6 and 31.7), the immune system generates a large number of antigen-specific B and T cell receptors. Inevitably, some of these receptors will have the potential to initiate immune reactions against self-antigens in the host. As a result, the immune system must eliminate or suppress these self-reactive clones while at the same time selecting clones that may be useful against nonself antigens.

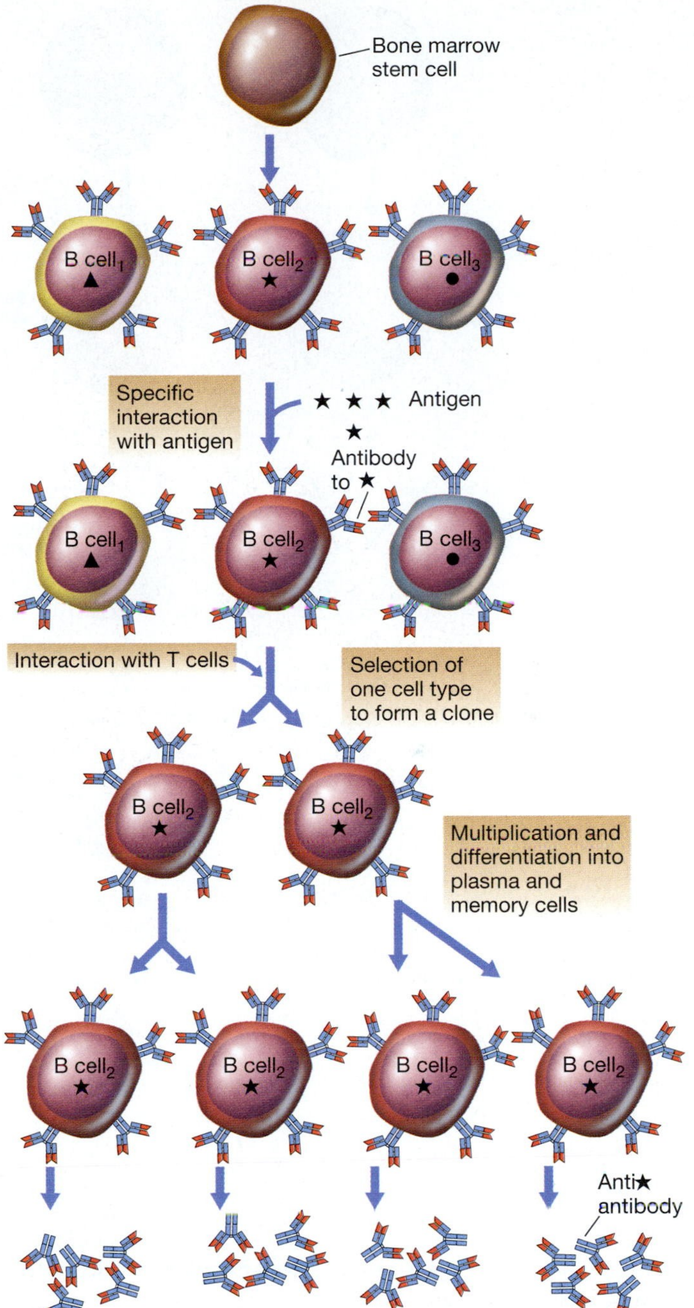

Figure 31.9 Clonal selection. Individual B cells, specific for a single antigen, proliferate and expand to form a clone after interaction with the specific antigen. The antigen drives selection and then proliferation of the individual antigen-specific B cell. Clonal copies of the original antigen-reactive cell have the same antigen-specific surface antibody. Continued exposure to antigen results in continued expansion of the clone.

T Cell Selection and Tolerance

T cells undergo immune selection for potential antigen-reactive cells and selection against those cells that react strongly with self-antigens. Selection against self-reactive clones results in the development of tolerance. The failure to develop tolerance may result in dangerous reactions to self-antigens, a condition called *autoimmunity* (∞ Section 30.7).

Lymphocytes that are destined to become T cells leave the bone marrow and enter the thymus, a primary lymphoid organ, via the bloodstream (**Figure 31.10**). During the process of T cell maturation in the thymus, immature T cells undergo a two-step selection to (1) select potential antigen-reactive cells (positive selection) and (2) eliminate cells that react with self-antigens (negative selection). **Positive selection** requires the interaction of new T cells in the thymus with the thymic self-antigens. Using their TCRs, some T cells bind to MHC–peptide complexes on the thymic tissue. The T cells that do not bind MHC–peptide complexes undergo a process called *apoptosis*, or programmed cell death. By contrast, those T cells that bind thymic MHC proteins receive survival signals and continue to divide and grow. Positive selection therefore

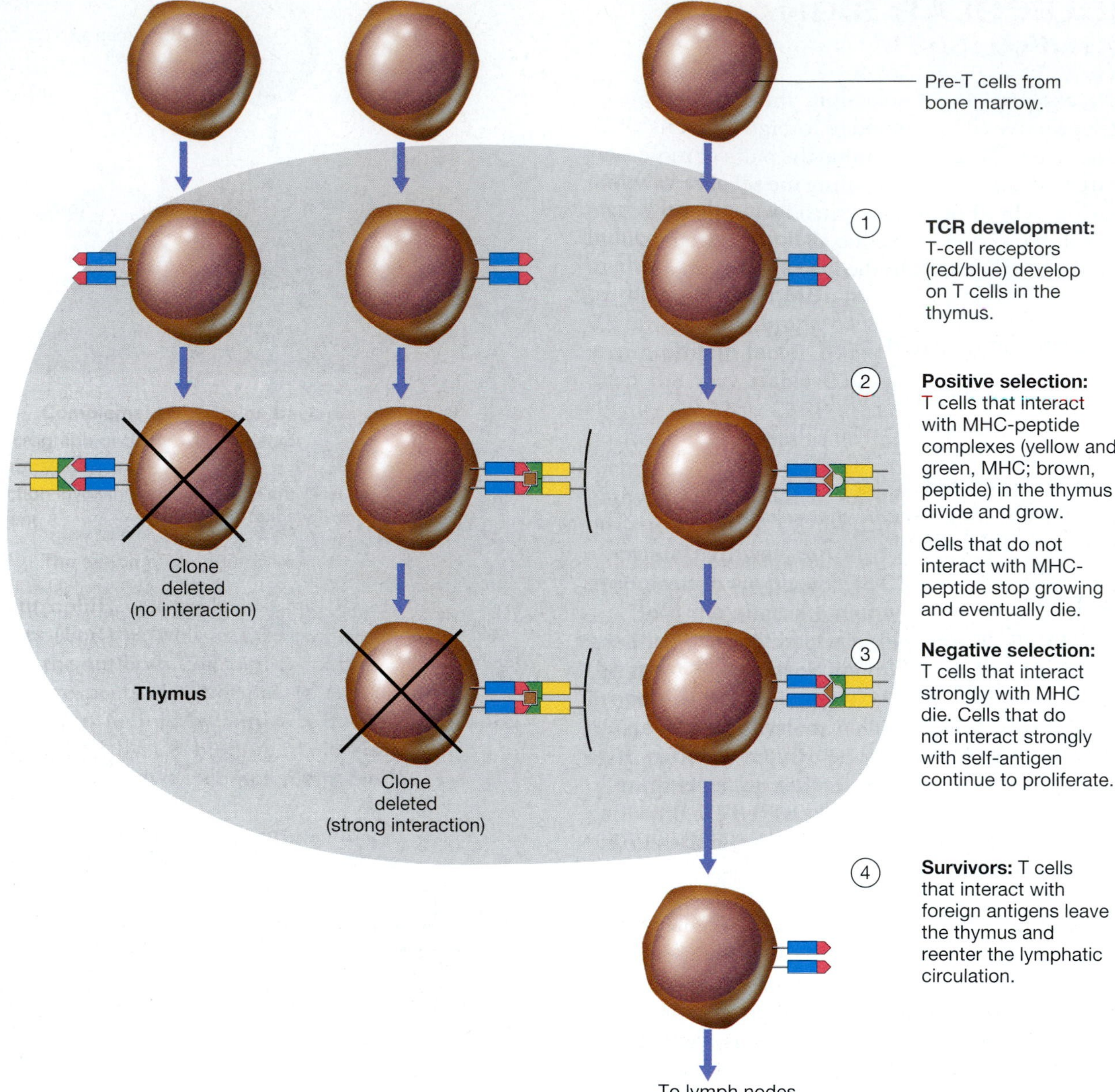
Pre-T cells from bone marrow.
1
TCR development: T-cell receptors (red/blue) develop on T cells in the thymus.
2
Positive selection: T cells that interact with MHC-peptide complexes (yellow and green, MHC; brown, peptide) in the thymus divide and grow.
Cells that do not interact with MHC-peptide stop growing and eventually die.
3
Negative selection: T cells that interact strongly with MHC die. Cells that do not interact strongly with self-antigen continue to proliferate.
4
Survivors: T cells that interact with foreign antigens leave the thymus and reenter the lymphatic circulation.
Clone deleted (no interaction)
Thymus
Clone deleted (strong interaction)
To lymph nodes

Figure 31.10 T cell selection and clonal deletion. T cells undergo selection for recognition of dangerous nonself antigens in the thymus.

retains T cells that recognize MHC–peptide and deletes T cells that do not recognize self-MHC proteins and would be unable to present peptides to T cells.

In the second stage of T cell maturation, **negative selection**, the positively selected T cells continue to interact with thymic MHC-peptide (⚭ Section 29.6). The peptide antigens in the thymus are of self origin. T cells that react with thymic self-antigens are potentially dangerous if they react strongly with these antigens (autoimmunity). These strongly self-reactive T cells bind tightly to thymus cells and cannot divide and eventually die. This two-stage thymic selection process for selecting self-tolerant, antigen-reactive T cells results in **clonal deletion**. Precursors of T cell clones that are either useless (do not bind) or harmful (bind too tightly) die; more than 99% of all T cells that enter the thymus do not survive the selection process.

The remaining selected T cells are destined to interact very strongly with nonself antigens. They are not destroyed in the thymus because their weak binding interactions with thymic self-antigens signal them to grow. The selected and growing T cells leave the thymus and migrate to the spleen, mucosal-associated lymphoid tissue, and lymph nodes where they can contact foreign antigens presented by B lymphocytes and other APCs.

B Cell Tolerance

The acquisition of immune tolerance in B cells is also necessary because antibodies produced by self-reactive B cells (autoantibodies) may cause autoimmunity and damage to host tissue. B cells also undergo a process of clonal deletion. Many self-reactive B cells are eliminated during development in the bone marrow, the primary lymphoid organ responsible for B cell development in mammals (Section 29.1).

In addition to clonal deletion, **clonal anergy** (unresponsiveness) also plays a role in final selection of the B cell repertoire. Some immature B cells are reactive to self-antigens but do not become activated, even when exposed to high concentrations of self-antigens. This is because B cell activation requires a second signal from T cells, as we shall now see. If an antigen-reactive T cell is eliminated during T cell selection, no second signal is generated and the B cell remains unresponsive.

31.8 MiniReview

The thymus is a primary lymphoid organ which provides an environment for the maturation of antigen-reactive T cells. Immature T cells that do not interact with MHC protein (positive selection) or react strongly with self-antigens (negative selection) are eliminated by clonal deletion in the thymus. T cells that survive positive and negative selection leave the thymus and can participate in an effective immune response. B cell reactivity to self-antigens is controlled through clonal deletion, selection, and anergy.

- Distinguish between positive and negative T cell selection.
- Identify the role of the thymic cells in T cell selection.
- For B cells, distinguish between clonal deletion and clonal anergy.

31.9 T Cell and B Cell Activation

T and B cells require two molecular signals for activation. Interaction with antigen through Igs or TCRs is the first signal. A second signal is necessary before T cells or B cells can respond to antigens.

T Cell Activation

As we have discussed, T cells that react with self-antigens are deleted in the thymus. However, many self-antigens are not expressed in the thymus. As a result, many T cell clones responsive to nonthymus antigens avoid clonal deletion in the thymus. These self-reactive T cells become anergic but may persist as unresponsive T cells. The key to maintaining clonal anergy in these potentially dangerous self-reactive T cells is the two-signal mechanism used to activate T cells after they leave the thymus.

When positively and negatively selected T cells leave the thymus, they migrate to the secondary lymphoid organs (lymph nodes, spleen, and mucosal-associated lymphoid tissue (Section 29.1). These T cells have not yet been exposed to antigen and are therefore naive or uncommitted T cells. Uncommitted T cells must be activated by an APC to become competent effector cells (Section 29.6).

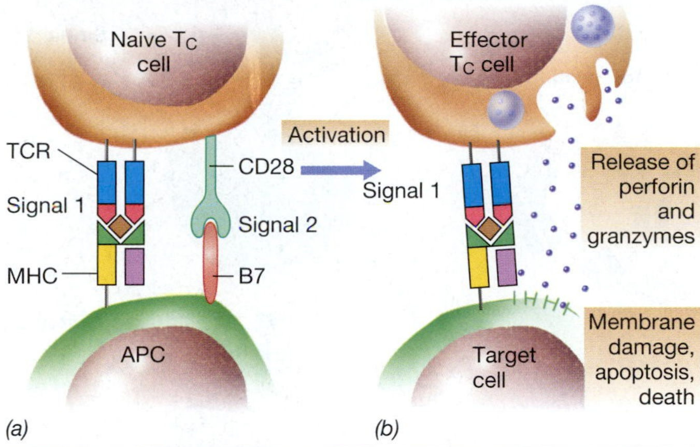

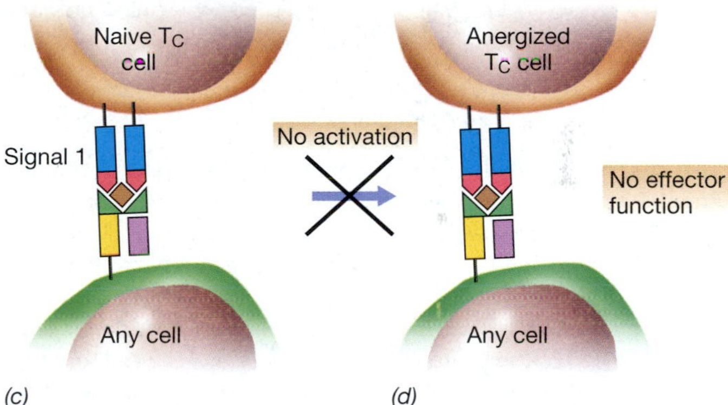

Figure 31.11 T cell activation: signal 1 and signal 2. *(a)* A naive T_C cell interacts via TCR with the peptide–MHC complex on an APC. This is signal 1. The T_C cell also has a CD28 protein that interacts with a B7 protein on the APC. This is signal 2. The simultaneous interactions of the T_C cell and APC via signal 1 and signal 2 activate the naive T cell. *(b)* The activated T_C cell is then capable of killing any target cell as long as signal 1 interactions take place. *(c)* A naive T_C cell interacts via the TCR with the peptide–MHC complex on any cell. Although the conditions for signal 1 (interactions via TCR with the peptide–MHC complex) are met, signal 2 cannot be generated because only APCs display the B7 protein. *(d)* In the absence of signal 2, the T_C cell becomes permanently unresponsive, or anergized.

The first step in activation of uncommitted T cells is binding of the peptide antigen–MHC protein complex on the APC by the TCR (**Figure 31.11**). This is signal 1 and is absolutely required for activation. Without signal 1, a T_C cell cannot be activated. The next step requires the interaction of two more proteins, one found on the APC, called B7, and one found only on T cells, called CD28. The binding of B7 to CD28 is signal 2 and activates the T_C cell, making it an effector cell. In the absence of signal 2, the T cell is not activated (Figure 31.11). A T_C cell that is activated will kill any target cell that displays antigen, even those cells that do not display CD28. After a T cell is activated, only signal 1 (peptide–MHC) is necessary to induce effector activity. An analgous situation occurs with T_H cells.

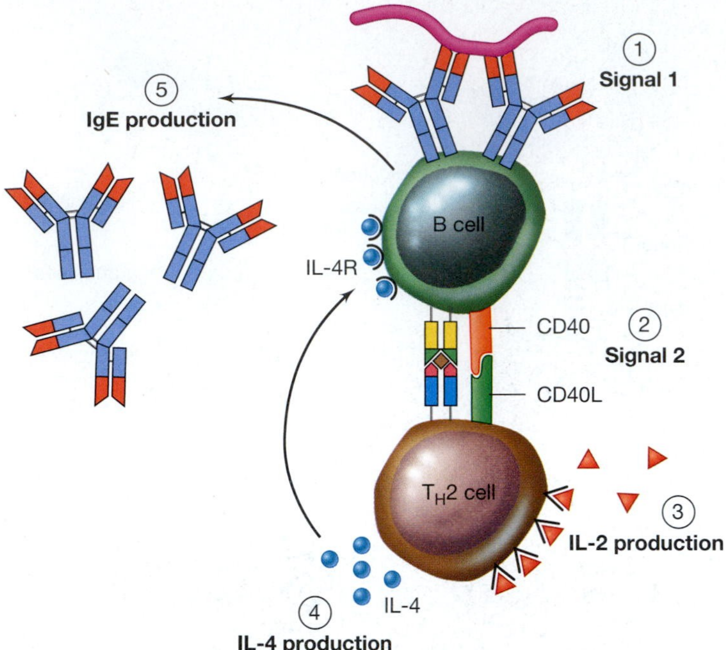

Figure 31.12 B cell activation: signal 1, signal 2, and cytokines. (1) Antigen binds and cross-links the Ig receptors on a naive B cell. This is signal 1, which stimulates the B cell to produce CD40 and express it on the cell surface. The B cell then processes the antigen and presents it to a T_H2 cell via MHC II. (2) The T_H2 cell interacts with the peptide–MHC II complex with its TCR. CD40L on the T_H2 cell then interacts with the B cell CD40. (3) These interactions stimulate the T_H2 cell to produce IL-2, which stimulates the same T_H2 cell (autocrine function). (4) The stimulated T_H2 cell can make a battery of cytokines, one of which is IL-4. IL-4 is signal 2 for the B cell. (5) In this case, the cytokine-stimulated B cells then produce IgE. The T_H2 cytokines also stimulate activation, differentiation, and expansion of both T and B lymphocytes.

T Cell Anergy

The requirement for a second activation signal has major implications for establishing and maintaining clonal anergy. For example, an uncommitted T_C cell that interacts with a self-antigen on a non-APC will receive only signal 1 because non-APCs do not display the B7 protein necessary to complete signal 2. In the absence of signal 2, this T_C is permanently anergized and can never be activated (Figure 31.11). Thus, the B7-CD28 second signal is absolutely required for activation. Absence of signal 2 in the presence of signal 1 induces permanent anergy. Uncommitted T_H1 and T_H2 lymphocytes are activated in the same way, also using the B7-CD28 coreceptor second signal.

B Cell Activation

The B cell also requires two independent signals for activation and antibody production. As compared to the activation signals for T cells, however, different signals activate B cells. As we have seen, B cells are responsible for antigen uptake, processing, and presentation (Section 29.6) as well as the production of specific antibodies. Signal 1 for the B cell is antigen binding and cross-linking of surface immunoglobulin (**Figure 31.12**). This antigen–Ig interaction generates a transmembrane signal transduction event, stimulating the B cell to express CD40 on its surface. Meanwhile, the B cell ingests the antigen bound on the Igs, processes the ingested antigen to peptides, and presents peptide antigen embedded in MHC II to neighboring T_H2 cells (Figure 29.6). Because of their proximity in the lymph nodes, a T_H2 cell with reactivity toward the presented antigen can interact with the antigen-presenting B cell. Interaction via the peptide–MHC II complex results in the expression of CD40L (CD40 ligand) by the T_H2 cells, which in turn binds to the B cell CD40. The CD40-CD40L interaction initiates a signal transduction event in the T_H2 cell, leading to transcription of a number of T cell proteins, including several soluble cytokines, as we discuss next.

These cytokines are signal 2 for the B cells. They interact with receptors on B cells and stimulate antibody production. After a B cell is activated, it no longer needs T cell interactions or cytokines to make antibody. These previously activated cells are called *memory B cells* and play a major role in the immune response during subsequent exposure to an antigen following a primary immune response (Section 29.10).

31.9 MiniReview

Many self-reactive T cells are deleted during development and maturation in the thymus. Uncommitted T cells are activated in the secondary lymphoid organs by first binding peptide–MHC with their TCRs (signal 1), followed by binding of the B7 APC protein to the CD28 T cell protein (signal 2). B cell activation is initiated by antigen interaction with surface immunoglobulin (signal 1), followed by interaction between the B cell CD40 protein with CD40L on the T cell to generate cytokine production (signal 2).

- Define signal 1 and signal 2 for an uncommitted T cell.
- Define activation signals for an uncommitted B cell.

31.10 Cytokines and Chemokines

Intercellular communication in the immune system is accomplished in many cases through a heterogeneous family of soluble proteins known as **cytokines** that are produced by leukocytes and other cells. Cytokines regulate cellular functions in immune cells and activate various cell types. The cytokines produced by lymphocytes are called lymphokines or interleukins (ILs).

Cytokines secreted from one cell bind specific receptors on other cells. Some cytokines bind to receptors on the cell that produced them. Thus, these cytokines have autocrine (self-stimulatory) abilities. Other cytokines bind to receptors on other cells. Cytokine-receptor binding generally activates a signal transduction pathway (Section 31.1), relaying information across the cytoplasmic membrane to control activities such as transcription and protein synthesis. These signals can ultimately result in cell growth, differentiation, and clonal proliferation.

Table 31.2 Properties of some major cytokines and chemokines

Cytokine	*Producers*	*Major targets*	*Effect*
IL-1[a]	Monocytes	T_H	Activation
IL-2	Activated T cells	T cells	Growth, differentiation
IL-3	T_H1	Hematopoietic stem cells	Growth factor
IL-4	T_H2	B cells	IgG and IgE synthesis
IL-5	T_H2	B cells	IgA synthesis
IL-10	T_H2	T_H1	Inhibits T_H1
IL-12	Macrophages, dendritic cells	T_H1, NK cells	Differentiation, activation
IFN-α[b]	Leukocytes	Normal cells	Antiviral
IFN-γ	T_H1	Macrophages	Activation
GM-CSF[c]	T_H1	Myeloid stem cells	Differentiation to granulocytes, monocytes
TGF-β[d]	T_H1 and T_H2	Macrophages	Inhibits activation
TNF-α[e]	T_H1, macrophages, NK cells	Macrophages	Activation
TNF-β	T_H1	Macrophages	Activation
Chemokines			
CXCL8 (IL-8)	Macrophages, fibroblasts, keratinocytes	Neutrophils, T cells	Attractant and activator
CCL2 (MCP-1[f])	Macrophages, fibroblasts, keratinocytes	Macrophages, T cells	Attractant and activator

[a]IL, interleukin; [b]IFN, interferon; [c]GM-CSF, granulocyte, monocyte-colony stimulating factor; [d]TGF, T cell growth factor; [e]TNF, tumor necrosis factor, [f]MCP, macrophage chemoattractant protein.

Table 31.2 lists some important cytokines, the cells that produce them, their most common target cells, and their most important biological effects. There are over 50 cytokines known, most of which are produced by either T cells or monocytes and macrophages. We now examine the activity of two cytokines involved in the induction of an antigen-specific antibody-mediated immune response.

IL-2, IL-4, and Antibody Production

B cells are responsible for antigen uptake, processing, and presentation as well as the production of specific antibodies. As we discussed in the previous section, B cells require two independent signals for activation and antibody production. B cells are activated by antigen surface immunoglobulin (signal 1) followed by interaction between the B cell CD40 with CD40L on the T cell (Figure 31.12). The activated T_H2 cell responds by producing IL-2, which is secreted and bound by the IL-2R on the surface of the T_H2 cells. Thus, IL-2 can activate the same cell that secreted it. Under the influence of IL-2, the cell divides, making clonal copies. In the process, the T_H2 cell also makes other cytokines such as IL-4. IL-4 then binds to the IL-4R on the original antigen-presenting B cell.

The IL-4–IL-4R interaction stimulates the B cell to differentiate into a plasma cell, which ultimately produces antibodies (∞ Section 29.10). The IL-4 generated by the responding T cell is the second signal (signal 2) necessary for initiation of antibody production. In addition, the IL-4 interaction may signal an immunoglobulin class switch. For example, IL-4–IL-4R interaction can switch antibody production by an affected B cell from IgM to IgE. Thus, IL-2 and IL-4 cytokines are soluble mediators and activators for T lymphocytes and B cells that interact to produce the antibody-mediated immune response. IL-4 not only controls activation of the B cell, but also controls the quality of the antibody response, inducing a class switch from IgM to IgE production.

Other Cytokines

Table 31.2 shows the activity of several other cytokines. Like IL-4, many cytokines affect cells involved in specific immunity. However, several cytokines do not affect T or B lymphocytes, but act on other cells; the cytokine-activated cells in turn serve as important modulators of innate host responses. For example, interferons (IFN-α and IFN-γ) are produced by leukocytes and inhibit viral replication in virtually any cell in the body. Tumor necrosis factors TNF-α and TNF-β can kill certain tumors if the TNF-producing cells have access to the tumor. TNF-α is also a critical activator of inflammation (∞ Section 29.3). Interferons and TNFs, produced by T cells as well as phagocytes, have no target cell specificity, but amplify the effects of immune cells.

Chemokines

Chemokines are a group of small proteins that function as chemoattractants for phagocytes and lymphocytes. They are produced by lymphocytes and many other cells in response to bacterial products, viruses, and other agents that cause damage to host cells. Chemokines attract phagocytes and T cells to the site of injury, stimulating an inflammatory response as well as potentiating a specific immune response.

About 40 chemokines are known. Perhaps the best-studied chemokines are CXCL8 (IL-8) and CCL2 (macrophage chemoattractant protein-1, or MCP-1) (Table 31.2). CXCL8 is produced by monocytes, macrophages, fibroblasts (connective tissue cells), keratinocytes (skin cells), endothelial cells, and other cells. CXC8 is secreted by the affected cells and binds to receptors on T cells and neutrophils, where it acts as a chemoattractant. This results in a neutrophil-mediated inflammatory response followed by a specific immune response by the attracted T cells. As is the case for the cytokine receptors, engaged chemokine receptors on the target cells act through signal transduction pathways (Section 31.1) to induce activation of the neutrophils or T cells.

MCP-1 is also produced by a variety of cells and attracts basophils, eosinophils, monocytes, dendritic cells, natural killer cells, and T cells, stimulating production of inflammatory mediators and potentially organizing an antigen-specific immune response. Thus, chemokines are potent initiators of nonspecific inflammatory reactions that can lead to recruitment of T cells and antigen-specific immune reactions.

31.10 MiniReview

Cytokines, produced by leukocytes and other cells, are soluble mediators that regulate interactions between cells. Several cytokines, such as IL-2 and IL-4, affect lymphocytes and are critical components in the generation of specific immune responses. Other cytokines, such as IFN-γ and TNF-α, affect a wide variety of cell types. Chemokines produced by various cells are released in response to injury and are strong attractants for nonspecific inflammatory cells and T cells.

- How do cytokines and chemokines differ with respect to their cell sources? Their cell targets?
- What events stimulate cytokine production? What events stimulate chemokine production?

Review of Key Terms

Antibody a soluble protein, produced by B cells, that interacts with antigen; also called immunoglobulin

Chemokine a small soluble protein that modulates inflammatory reactions and immunity in target cells

Class I MHC protein antigen-presenting molecule found on all nucleated vertebrate cells

Class II MHC protein antigen-presenting molecule found on macrophages, B cells, and dendritic cells (antigen-presenting cells)

Clonal anergy the inability to produce an immune response to specific antigens due to neutralization of effector cells

Clonal deletion for T cell selection in the thymus, the killing of useless or self-reactive clones

Clonal selection the production by a B or T cell of copies of itself after antigen interaction

Complementarity-determining region (CDR) a varying amino acid sequence within the variable domains of immunoglobulins or T cell receptors where most molecular contacts with antigen are made

Cytokine a small, soluble protein produced by a leukocyte that modulates inflammatory reactions and immunity in target cells

Domain a region of a protein having a defined structure and function

Epitope the portion of an antigen that is recognized by an immunoglobulin or a T cell receptor

Human leukocyte antigen (HLA) human leukocyte antigen encoded by major histocompatibility complex genes in humans

Immunoglobulin (Ig) a soluble protein, produced by B cells, that interacts with antigen; also called antibody

Immunoglobulin gene superfamily a family of genes that are evolutionarily, structurally, and functionally related to immunoglobulins

Major histocompatibility complex (MHC) a genetic region that encodes several proteins important for antigen presentation and other host defense functions

Motif in antigen presentation, a conserved amino acid sequence found in all peptides that bind to a given MHC protein

Negative selection in T cell selection, deletion of T cells that interact with self-antigens in the thymus (see clonal deletion)

Pathogen-associated molecular pattern (PAMP) a structural component of a pathogen or pathogen product that is recognized by a pattern recognition molecule (PRM)

Pattern recognition molecule (PRM) a protein that recognizes a pathogen-associated molecular pattern (PAMP), such as a component of a microbial cell surface structure

Polymorphism the occurrence in a population of multiple alleles at a locus in frequencies that cannot be explained by recent random mutations

Positive selection in T cell selection, the growth and development of T cells that interact with self MHC protein in the thymus

Somatic hypermutation the mutation of immunoglobulin genes at rates higher than those observed in other genes

T cell receptor (TCR) antigen-specific receptor protein on the surface of T cells

Tolerance inability to produce an immune response to a specific antigen

Toll-like receptor (TLR) a pattern recognition molecule of phagocytes, structurally and functionally related to Toll receptors in *Drosophila*

Review Questions

1. Identify at least one soluble pattern recognition molecule (PRM), its interacting pathogen-associated molecular pattern (PAMP), and the resulting host response. Identify at least one membrane-bound pattern recognition molecule (PRM), its interacting pathogen-associated molecular pattern (PAMP), and the resulting host response (Section 31.1).
2. Define the criteria used to assign a gene and its encoded protein to the Ig gene superfamily (Section 31.2).
3. Identify the major structural features of class I and class II MHC proteins (Section 31.3).
4. Polymorphism implies that each different MHC protein binds a different peptide motif. For the MHC class I polymorphisms, how many different MHC proteins are expressed in an individual? By the entire human population (Section 31.4)?
5. Which Ig chains are used to construct a complete antigen-binding site? Which domains? Which CDRs (Section 31.5)?
6. Calculate the total number of V_H and V_L domains that can be constructed from the available Ig genes. How many complete Ig proteins can be produced from the reassortment of all possible heavy chains and light chains (Section 31.6)?
7. Describe the interaction of the TCR with peptide antigen and MHC protein. Be sure to identify the roles of the CDRs in the TCR (Section 31.7).
8. In TCRs, diversity can be generated by recombination and reassortment events such as in Igs. As is the case in Igs, additional diversity is generated with somatic events such as N-region nucleotide additions and reading of the D (diversity) segment in all three reading frames. Explain these diversity-generating mechanisms (Section 31.7).
9. Explain positive and negative selection of T cells (Section 31.8).
10. What molecular interactions are necessary for activation of uncommitted T cells? For activation of uncommitted B cells? (Section 31.9)?
11. What are the chief effects of cytokines and chemokines? What are the chief differences between cytokines and chemokines (Section 31.10)?

Application Questions

1. Identify the consequences of a genetic mutation that eliminates a PRM by predicting the outcome for the host. Do this for at least one soluble PRM and one membrane-bound PRM.
2. Construct a table that lists the common features of proteins encoded by members of the Ig gene superfamily. For Igs, TCRs, and MHC proteins, identify the structural components that fit these common features.
3. Polymorphism implies that each different MHC protein binds a different peptide motif. However, for the **MHC** class I proteins, only 6 peptide motifs can be recognized in an individual, whereas over 350 motifs can be recognized by the entire human population. What advantage does this have for the population? For the individual?
4. While genetic recombination events are important for generating significant diversity in the antigen-binding site of Igs, postrecombination somatic events may be even more important in achieving overall Ig diversity. Do you agree or disagree with this statement? Explain.
5. What would happen to the T cell repertoire in the absence of positive selection? In the absence of negative selection?
6. What would be the result of activation of all peripheral T cells that contact antigen? How does the second signal scheme prevent this from happening?

32

Diagnostic Microbiology and Immunology

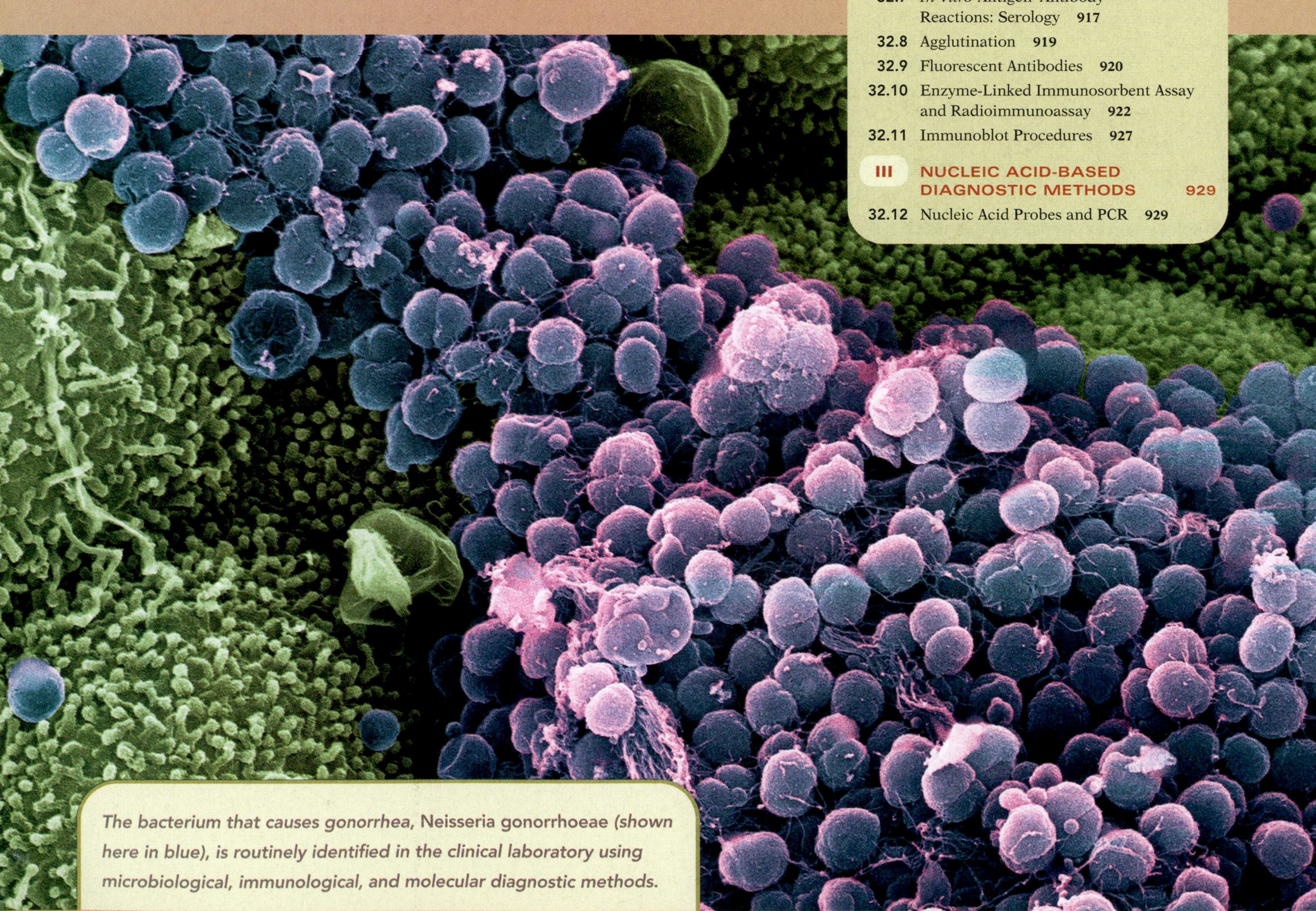

The bacterium that causes gonorrhea, Neisseria gonorrhoeae (shown here in blue), is routinely identified in the clinical laboratory using microbiological, immunological, and molecular diagnostic methods.

A major task of the clinical microbiologist is to identify the microorganisms that cause infectious disease. Clinical laboratories must be able to grow, isolate, and identify most routinely encountered pathogenic bacteria within 48 hours of sampling. Immunologic and molecular methods are also used to identify many pathogens. These methods are particularly important for the rapid identification of bacteria that are difficult to isolate or grow and for identification of viruses and protozoa.

I GROWTH-DEPENDENT DIAGNOSTIC METHODS

The isolation and growth of pathogens from host tissue is an important step in defining the cause of many infectious diseases. After a pathogen is positively identified and tested for antimicrobial drug susceptibility, a specific treatment plan can be formulated.

32.1 Isolation of Pathogens from Clinical Specimens

If a healthcare provider suspects a disease is caused by an infectious agent, samples of tissues or fluids are collected for microbiological, immunological, and molecular biological analyses (**Figure 32.1**). The samples may include blood, urine, feces, sputum, cerebrospinal fluid, or pus from a wound. Swabs may be used to obtain samples from suspected infected areas such as skin, nares, or throat (**Figure 32.2**). The swab is then used to inoculate the surface of an agar plate or a tube of liquid culture medium. In some cases, small pieces of tissue may be obtained for culture. **Table 32.1** summarizes recommendations for initial culture of organisms isolated from typical clinical specimens.

If clinically relevant organisms are to be isolated and identified, the specimen must be obtained and handled properly to ensure that the pathogen survives. First, the specimen must be obtained from the actual site of the infection. The sample must also be taken aseptically to avoid contamination with irrelevant microorganisms. Next, the sample size must be large enough to ensure a sufficient inoculum. Finally, the metabolic requirements for organism survival must be maintained during sampling, storage, and transport. For example, samples obtained from anoxic sites must be obtained, stored, and transported under anoxic conditions to ensure the survival of potential anaerobic pathogens. Once samples have been obtained, they must be analyzed as soon as possible.

Growth Media and Culture

Enrichment culture, the use of selected culture media and incubation conditions to isolate microorganisms from samples (∞ Section 22.1), is an important tool in the clinical laboratory. Most microorganisms of clinical importance can be grown, isolated, and identified using specialized growth

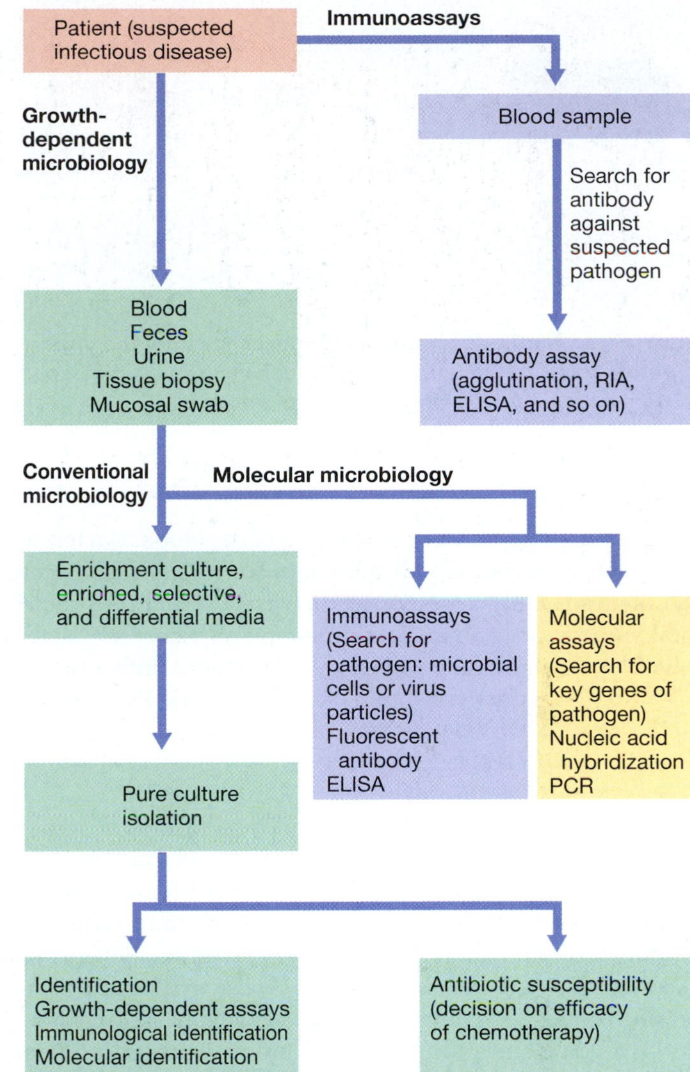

Figure 32.1 Laboratory identification of clinical pathogens. Diagnostic methods used for identification of infectious pathogens include growth-dependent microbiology assays, immunoassays, and molecular biology assays. Immunoassays can be used to measure patient immune responses, indicating pathogen exposure, or can be used to directly identify the pathogen in host tissue or culture.

media. Clinical samples are first grown on **general purpose media**, media such as blood agar that support the growth of most aerobic and facultatively aerobic organisms (∞ Figure 28.18). Organisms isolated from such media are often subcultured on more specialized media (Table 32.1). **Enriched media** containing specific growth factors enhance the growth of certain fastidious pathogens, such as *Neisseria gonorrhoeae*, the organism that causes gonorrhea. **Selective media** allow for some organisms to grow while inhibiting the growth of others due to inhibitory agents. Finally, **differential media** are specialized media that allow identification of organisms based on their growth and appearance on the medium, often based on color differences (see Figure 32.7).

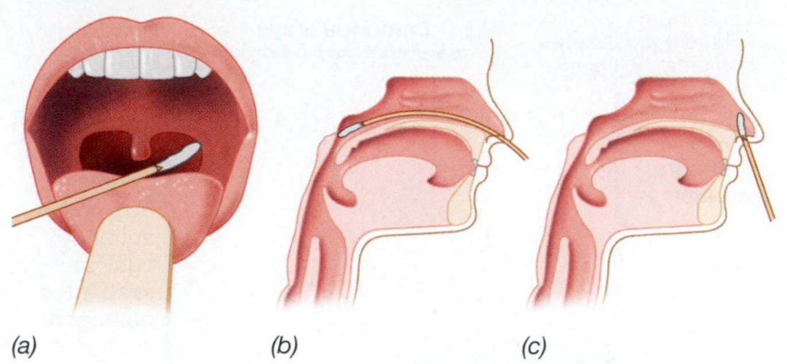

Figure 32.2 Methods for obtaining specimens from the upper respiratory tract. *(a)* Throat swab. *(b)* Nasopharyngeal swab passed through the nose. *(c)* Swabbing the inside of the nose.

Blood Cultures

Bacteremia is the presence of bacteria in the blood. Bacteremia is extremely uncommon in healthy individuals, normally occurring only transiently in response to invasive procedures such as dental surgery or trauma. The prolonged presence of bacteria in the blood is generally indicative of systemic infection. The most common pathogens found in blood include *Pseudomonas aeruginosa,* enteric bacteria, especially *Escherichia coli* and *Klebsiella pneumoniae,* and the gram-positive cocci *Staphylococcus aureus* and *Streptococcus pyogenes*.

Septicemia results from a blood infection by a virulent organism that enters the blood from a focus of infection, multiplies, and travels to various body tissues to initiate new infections. Septicemia can cause severe systemic symptoms, including fever and chills, followed by prostration. Severe cases of septicemia may result in *septic shock,* a life-threatening systemic condition characterized by severe reduction in blood pressure and multiple organ failures, including heart, kidneys, and lungs (Section 29.3). Blood cultures provide the only immediate way of isolating and identifying the causal agent. Bacteria from blood cultures are commonly detected by indicators of microbial growth in automated systems, microscopic examination, and subculture.

The standard blood culture procedure is to draw 20 ml of blood aseptically from a vein and inject it into two blood culture bottles containing an anticoagulant and a general purpose culture medium. Some blood culture systems employ a chemical that lyses red and white blood cells, releasing intracellular pathogens. One bottle is incubated in air and one is incubated under anoxic conditions (35°C). Both bottles are examined several times each hour for up to 5 days. Automated blood culture systems detect growth by monitoring carbon dioxide production and turbidity as often as every 10 minutes. Clinically significant bacteria are generally recovered within 2 days, but detectable growth of more fastidious organisms may take 3 to 5 days.

Up to 2–3% of blood cultures are contaminated by microorganisms introduced from the skin during blood sampling. Typically, these organisms include *Staphylococcus epidermidis,* coryneform bacteria, or propionibacteria. However, these organisms can also infect the heart (subacute bacterial endocarditis) or colonize intravascular devices such as artificial heart valves. Thus, the results of the blood culture must be reconciled with the clinical problem for an accurate diagnosis.

Urine Cultures

Urinary tract infections are very common, especially in women. Interpretation of microbiological findings from urine cultures can be confusing because the disease-causing agents are often members of the normal flora (for example, *E. coli*). In most cases, urinary tracts become infected by organisms

Table 32.1 Recommended enriched and selective media for primary isolation of pathogens

	Media[a]				
Specimen	***Blood agar***	***Enteric agar***	***CA***	***MTM***	***ANA***
Fluids from chest, abdomen, pericardium, joint	+	+	+	–	+
Feces: rectal or enteric transport swabs[b]	+	+	+	–	–
Surgical tissue biopsies	+	+	–	–	+
Throat, sputum, tonsil, nasopharynx, lung, lymph nodes	+	+	+	–	–
Urethra, vagina, cervix	+	+	+	+	–
Urine	+	+	–	–	–
Blood[c]	+	+	+	–	+
Wounds, abscesses, exudates	+	+	+	–	+

[a]Blood agar, 5% whole sheep blood added to trypticase soy agar; enteric agar, either eosin-methylene blue (EMB) agar or MacConkey agar; CA, chocolate (heated blood) agar; MTM, modified Thayer-Martin agar; ANA, anaerobic agar, thioglycollate-containing blood agar or supplemented thioglycollate agar incubated anaerobically.

[b]Special enteric pathogen media, SMAC (MacConkey agar with sorbitol), is also used to culture fecal and enteric samples. SMAC is a selective and differential medium used for the isolation and identification of sorbitol-negative enteric pathogens such as enteropathogenic *Escherichia coli.*

[c]Blood is cultured initially in broth. Depending on the Gram stain characteristics of isolates, subculturing is done on MacConkey agar (gram-negative) or chocolate agar (gram-positive).

Source: Adapted from Murray, P. R., E. J. Baron, J. H. Jorgenson, M. A. Pfaller, and R. H. Yolken. 2003. *Manual of Clinical Microbiology,* 8th edition. American Society for Microbiology, Washington, DC.

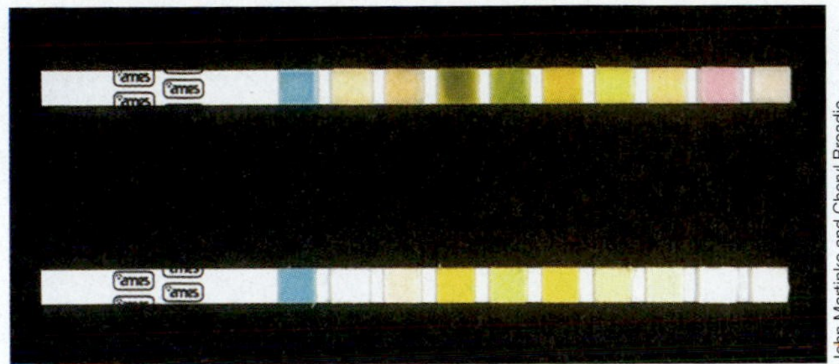

Figure 32.3 Urinalysis dipstick test. A control strip is shown underneath the test strip. From left to right, the strip measures abnormal levels of glucose, bilirubin, ketones, specific gravity, blood, pH, protein, urobilinogen, nitrite, and leukocytes (esterase) in a urine sample. Abnormal readings for esterase (trace positive, far right) and nitrite (strong positive, second from right) indicate bacteriuria. Subsequent culture of this sample indicated the presence of *Escherichia coli.*

ascending the urethra from the outside into the bladder. Urinary tract infections are also the most common form of healthcare-associated infections (∞ Section 33.7).

Significant urinary infection typically results in bacterial counts of 10^5 or more organisms per milliliter of a clean-voided midstream urine specimen. In the absence of infection, contamination of the urine from the external genitalia (almost unavoidable to some extent) results in less than 10^3 organisms per milliliter. The most common urinary tract pathogens are enteric bacteria, with *E. coli* accounting for about 90% of the cases. Other urinary tract pathogens include *Klebsiella, Enterobacter, Proteus, Pseudomonas, Staphylococcus saprophyticus,* and *Enterococcus faecalis*. *N. gonorrhoeae,* the causal agent of gonorrhea, does not grow in the urine itself, but on the urethral epithelium, and is diagnosed by methods discussed later.

Direct microscopic examination of urine may be used to indicate *bacteriuria,* the presence of abnormal numbers of bacteria in the urine. However, because nearly all urine contains some level of bacterial growth, bacteriuria is usually monitored with the use of commercially available dipstick tests. For example, one dipstick test monitors the reduction of nitrate by detecting the reduction product, nitrite; this form of anaerobic respiration is common among enteric bacteria (∞ Section 21.7). A positive test indicates high cell numbers and is indicated by a color change on the dipstick (**Figure 32.3**). Because significant nitrite production is produced in urine only when large numbers of organisms ($>10^5$ per milliliter) are present, the method is a very rapid check for urinary tract infections. Other dipstick tests for urinary tract infections, often used in conjunction with nitrate reduction, detect esterase (produced by leukocytes) and peroxidase (produced by a variety of bacteria). Positive dipstick tests indicate infection and are followed up with urine culture.

A Gram stain (∞ Figure 2.4) may also be done directly on urine samples exhibiting bacteriuria to identify the morphology of potential urinary tract pathogens. These include gram-negative rods including the enteric bacteria, gram-negative cocci such as *Neisseria,* and gram-positive cocci such as *Enterococcus*. The Gram stain and other direct staining methods are also useful for direct detection of bacteria in other body fluids such as sputum and wound exudates.

Figure 32.4 An eosin-methylene blue (EMB) agar plate. The plate shows a lactose fermenter, *Escherichia coli* (left), and a nonlactose fermenter, *Pseudomonas aeruginosa* (right). The green metallic sheen of the *E. coli* colonies is definitive for the lactose fermenters.

To culture potential urinary tract pathogens, two media types are normally used: (1) blood agar as a general purpose medium and (2) MacConkey or eosin-methylene blue agar (EMB), selective media for enteric bacteria (**Figure 32.4**). These selective and differential enteric media permit the initial differentiation of lactose fermenters from lactose nonfermenters and inhibit the growth of gram-positive organisms such as *Staphylococcus* species, common skin contaminants. Clinical microbiologists can often make a tentative identification of an isolate by observing the color and morphology of colonies of the suspected pathogen grown on various media as described in **Table 32.2**. This presumptive identification is followed by more detailed tests to make a positive identification. Urine cultures can be done quantitatively by counting colonies on blood agar or a selective agar medium, using a calibrated amount of urine, usually 1 μl, as the inoculum for a plate.

If no bacterial growth is obtained despite persistent urinary tract infection symptoms, a clinician may request direct cultures for fastidious organisms such as *N. gonorrhoeae, Chlamydia trachomatis, Branhamella* species, mycoplasma, or several anaerobic organisms.

UNIT 8

Fecal Cultures

Proper collection and preservation of feces is important for the isolation of intestinal pathogens. During storage, fecal

Table 32.2 Colony characteristics of frequently isolated gram-negative rods cultured on various clinically useful media

	Agar media[a]				
Organism	*EMB*	*MC*	*SS*	*BS*	*HE*
Escherichia coli	Dark center with greenish metallic sheen	Red or pink	Red to pink	Mostly inhibited	Yellow-pink
Enterobacter	Similar to *E. coli*, but colonies are larger	Red or pink	White or beige	Mucoid colonies with silver sheen	Yellow-pink
Klebsiella	Large, mucoid, brownish	Pink	Red to pink	Mostly inhibited	Yellow-pink
Proteus	Translucent, colorless	Transparent, colorless	Black center, clear periphery	Green	Clear
Pseudomonas	Translucent, colorless to gold	Transparent, colorless	Mostly inhibited	No growth	Clear
Salmonella	Translucent, colorless to gold	Translucent, colorless	Opaque	Black to dark green	Green or transparent with black centers
Shigella	Translucent, colorless to gold	Transparent, colorless	Opaque	Brown or inhibited	Green or transparent

[a]BS, Bismuth sulfite agar; EMB, eosin-methylene blue agar; MC, MacConkey agar; SS, *Salmonella-Shigella* agar; HE, Hektoen enteric agar.

Source: Adapted from Murray, P. R., E. J. Baron, J. H. Jorgenson, M. A. Pfaller, and R. H. Yolken. 2003. *Manual of Clinical Microbiology*, 8th edition. American Society for Microbiology, Washington, DC.

acidity increases, so extended delay between sampling and processing must be avoided. This is especially critical for the isolation of acid-sensitive *Shigella* and *Salmonella*.

Freshly collected fecal samples are placed in a vial containing phosphate buffer for transport to the lab. Bloody or pus-containing stools as well as stools from patients with suspected foodborne or waterborne infections are inoculated into a variety of selective media for isolation of individual bacteria. Intestinal eukaryotic pathogens are identified by observing cysts microscopically in the stool sample or through antigen-detection assays rather than by culture methods. Many laboratories also use a variety of selective and differential media and incubation conditions to identify *E. coli* O157:H7 and *Campylobacter,* two important intestinal pathogens typically acquired from contaminated food or water (Sections 37.8 and 37.9).

Wounds and Abscesses

Infections associated with traumatic injuries such as animal or human bites, burns, cuts, or the penetration of foreign objects must be carefully sampled to recover the relevant pathogen, and the results must be interpreted carefully. Wound infections and abscesses are frequently contaminated with normal flora, and swab samples from such lesions are frequently misleading. For abscesses and other purulent lesions, the best sampling method is to aspirate pus with a sterile syringe and needle following disinfection of the skin surface. Internal purulent lesions are sampled by biopsy or from tissues removed in surgery.

Several pathogens can be associated with wound infections. Because some of these are anaerobes, proper evaluation requires that samples be obtained, transported, and cultured under anoxic as well as oxic conditions. For example, potential pathogens commonly associated with purulent discharges from wound infections are *S. aureus,* enteric bacteria, *Pseudomonas aeruginosa,* and anaerobes such as *Bacteroides* and *Clostridium* species. The major isolation media are blood agar, several selective media for enteric bacteria (Tables 32.1 and 32.2), and blood agar containing additional supplements and reducing agents for obligate anaerobes. Gram stains from such specimens are examined directly by microscopy.

Genital Specimens and the Laboratory Diagnosis of Gonorrhea

In males, a purulent urethral discharge is the classic symptom of the sexually transmitted disease gonorrhea (Section 34.13). If no discharge is present, a sample can be obtained using a sterile narrow-diameter cotton swab that is inserted into the anterior urethra, left in place a few seconds to absorb any exudate, and then removed for culture and identification of *N. gonorrhoeae,* the causative agent. Alternatively, a sample of the first early morning urine of an infected individual usually contains viable cells of *N. gonorrhoeae*. In females, samples are usually obtained by swab from the cervix and the urethra.

Clinical microbiology procedures are central to the diagnosis of gonorrhea. *N. gonorrhoeae* (referred to clinically as gonococcus) colonizes mucosal surfaces of the urethra, uterine cervix, anal canal, throat, and conjunctiva. The organism is sensitive to drying and therefore is transmitted almost exclusively by direct person-to-person contact, usually by sexual intercourse. Public health measures to control gonorrhea include identification of asymptomatic carriers, and this requires microbiological analysis.

N. gonorrhoeae is usually found as gram-negative diplococci, but can be pleomorphic (Figure 15.21*a*). No similar microorganisms are observed among the normal flora of the urogenital tract. Thus, a vaginal or cervical smear

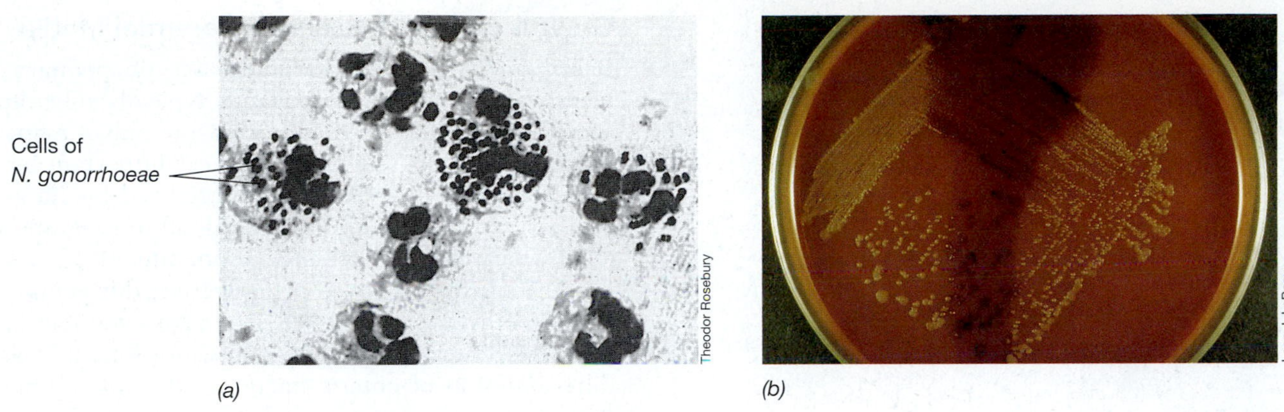

Figure 32.5 Identification of *Neisseria gonorrhoeae*. *(a)* Photomicrograph of *Neisseria gonorrhoeae* within human polymorphonuclear leukocytes from a urethral exudate. Note the paired diplococci (leader). *(b) N. gonorrhoeae* growing on Thayer-Martin agar. The plate has been stained in the middle with a reagent that turns colonies blue if cells contain cytochrome *c* (the oxidase test). *N. gonorrhoeae* colonies in contact with the reagent are blue, indicating that they are oxidase-positive.

showing gram-negative diplococci is a presumptive indicator of gonorrhea. In acute gonorrhea, microscopic examination of purulent discharges usually reveals phagocytosed gram-negative diplococci in neutrophils (**Figure 32.5*a***).

Most laboratory testing of urogenital samples for *N. gonorrhoeae* (and the often-associated *Chlamydia trachomatis*, Section 34.14) is done using a nucleic acid probe or DNA amplification via polymerase chain reaction (Section 32.12). However, specimens obtained from nonurogenital sources, such as the eyes and rectum, should also be cultured.

A nonselective enriched medium for the isolation of *N. gonorrhoeae* contains heat-lysed blood and is called *chocolate agar* because of the deep brown appearance. The heated blood interacts with the media components, absorbing compounds that are normally toxic for *N. gonorrhoeae*. One of several selective media used for primary isolation is modified Thayer-Martin (MTM) agar (Figure 32.5). This medium incorporates the antibiotics vancomycin, nystatin, trimethoprim, and colistin to suppress the growth of normal flora. These antibiotics have no effect on either *N. gonorrhoeae* or *N. meningitidis*, the cause of bacterial meningitis (Section 34.6).

Inoculated plates are incubated in a humid environment in an atmosphere containing 3–7% CO_2, required for growth of gonococci. The plates are examined after 32 and 48 hours and tested for their oxidase reaction because *Neisseria* species are oxidase-positive (Figure 32.5*b*). Oxidase-positive gram-negative diplococci growing on chocolate agar or selective media are presumed to be gonococci if the inoculum was derived from genitourinary sources. Definitive identification of *N. gonorrhoeae* requires determination of carbohydrate utilization patterns and immunological or nucleic acid probe tests.

Culture of Anaerobic Microorganisms

Obligate anaerobic bacteria are common causes of infection, and their identification requires special isolation and culture methods. In general, media for anaerobes do not differ greatly from those used for aerobes, except that they are (1) usually richer in organic constituents, (2) contain reducing agents (usually cysteine or thioglycollate) to remove oxygen, and (3) contain a redox indicator to indicate that conditions are anoxic. Specimen collection, handling, and processing are designed to exclude oxygen contamination because oxygen is toxic to obligate anaerobic organisms.

Several habitats in the body such as portions of the oral cavity and the lower intestinal tract are generally anoxic and support the growth of an anaerobic normal flora. Other parts of the body, however, can also become anoxic as a result of tissue injury or trauma, reducing blood supply and oxygen perfusion to the injured site. These anoxic sites can then be colonized by obligate anaerobes. In general, pathogenic anaerobic bacteria are part of the normal flora and are opportunistic pathogens. Two important exceptions are the pathogenic anaerobes *Clostridium tetani* (the cause of tetanus) and *Clostridium perfringens* (the cause of gas gangrene and one type of food poisoning), both endospore-forming bacteria that are predominantly soil organisms.

Isolation, growth, and identification of anaerobic pathogens are complicated by specimen contamination as well as the additional challenge of maintaining an anoxic growth environment during collection, transport, and culture. Samples collected by syringe aspiration or biopsy must be immediately placed in a tube containing oxygen-free gas, usually with a dilute salt solution containing a reducing agent such as thioglycollate and the redox indicator resazurin. Resazurin is colorless when reduced and becomes pink when oxidized, indicating oxygen contamination of the specimen. If an anaerobic transport tube is not available, the syringe itself can be used to transport the specimen; the needle is discarded and the syringe is plugged with a rubber stopper.

For anoxic incubation, agar plates are placed in a sealed jar, which is made anoxic either by replacing the atmosphere in the jar with an oxygen-free gas mixture (usually a mixture

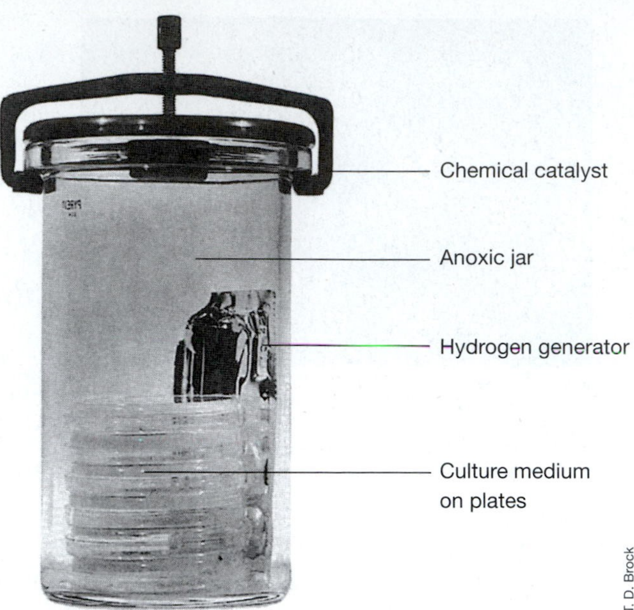

Figure 32.6 Sealed jar for incubating cultures under anoxic conditions. The catalyst and hydrogen generator packet produce and maintain a reducing (anoxic) environment.

of N_2 and CO_2) or by removing O_2 from the enclosed vessel by some chemical means. For example, as shown in **Figure 32.6**, hydrogen (H_2) is generated chemically in the jar. In the presence of a palladium catalyst, the H_2 combines with the free oxygen in the vessel, forming water and removing the contaminating oxygen. Alternate means for providing anoxic conditions include the use of culture media containing reducing agents or the use of anoxic "glove boxes" filled with an oxygen-free gas such as nitrogen or hydrogen (Figure 6.28*b*).

32.1 MiniReview

Appropriate sampling and culture techniques are necessary to isolate bacteria to identify potential pathogens. The selection of appropriate sampling and culture conditions requires knowledge of bacterial ecology, physiology, and nutrition.

- Why do urine cultures almost always test positive for bacterial growth?
- Describe the methods used to maintain optimum conditions for the isolation of anaerobic pathogens.

32.2 Growth-Dependent Identification Methods

If the inoculation of a general purpose medium results in bacterial growth, the clinical microbiologist must identify the organism or organisms present. Many clinical isolates can be identified using a series of growth-dependent assays. We consider some of these methods here.

Growth on Selective and Differential Media

Based on its growth characteristics on primary isolation media, a presumptive pathogen is typically subcultured onto specialized media designed to measure one of many different biochemical reactions. Some of these important biochemical tests are listed in **Table 32.3**. Specialized media are usually available as kits containing several, all in separate wells, and all of which can be inoculated at one time (**Figure 32.7**).

The media employed are selective, differential, or both. Eosin-methylene blue (EMB) agar, for example, is a widely used selective and differential medium for the isolation and differentiation of enteric bacteria. Methylene blue is a dye that inhibits the growth of gram-positive bacteria, and thus only gram-negative organisms can grow. EMB agar has an initial pH of 7.2 and contains lactose and sucrose, but not glucose, as energy sources. Acidification changes eosin from colorless to red or black. Strong lactose-fermenting bacteria such as *Escherichia coli* acidify the medium and the colonies appear black with a greenish sheen. Butanediol-producing enteric bacteria such as *Klebsiella* or *Enterobacter* produce less acid, and colonies on EMB are pink to red. Colonies of lactose nonfermenters, such as *Salmonella, Shigella,* and *Pseudomonas,* are translucent or pink (Figure 32.4). Thus, EMB preferentially selects for the growth of gram-negative bacteria and at the same time differentiates among common representatives.

Differential media incorporate biochemical tests to measure the presence or absence of enzymes involved in catabolism of a specific substrate or substrates. For example, fermentation of sugars is measured by incorporating pH indicator dyes that change color on acidification (Figure 32.7*a*). Production of hydrogen or carbon dioxide during sugar fermentation is assayed by observing gas production either in gas collection vials or in agar (Figure 32.7*a,b*). Hydrogen sulfide (H_2S) production is assayed by growth in a medium containing ferric iron. If sulfide is produced, ferric iron reacts with H_2S to form FeS, visible as a black precipitate (Figure 32.7*b*). Utilization of citric acid, a tricarboxylic acid, causes the pH to rise, and a dye in this test medium changes color as conditions become alkaline (Figure 32.7*c*). Hundreds of differential tests have been developed for clinical use, but only about 20 are used routinely (Figure 32.7*d*).

The biochemical reaction patterns for pathogens are stored in a computer databank. As the results of differential tests on an unknown pathogen are entered, the computer matches the characteristics of the unknown to metabolic patterns of known pathogens, allowing identification. As few as three or four key tests are sufficient to make an unambiguous identification of many pathogens. However, in some cases, more sophisticated identification procedures are required.

Identification and Diagnosis

Growth-dependent rapid identification systems are often used to identify enteric bacteria because these organisms are common causes of urinary tract and intestinal infections (Figure 32.7*d,e*). These systems consist of media that are selective and

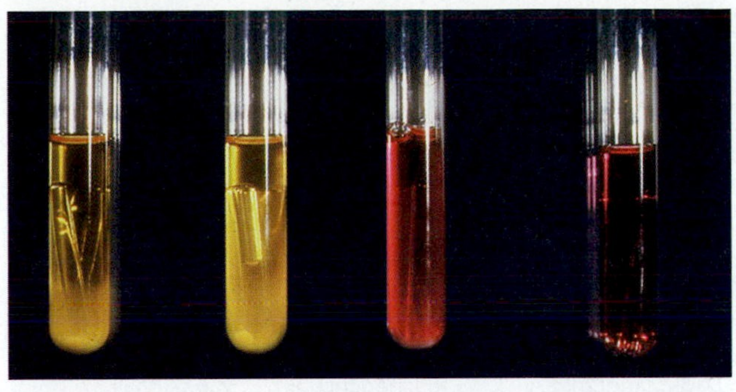

(a)

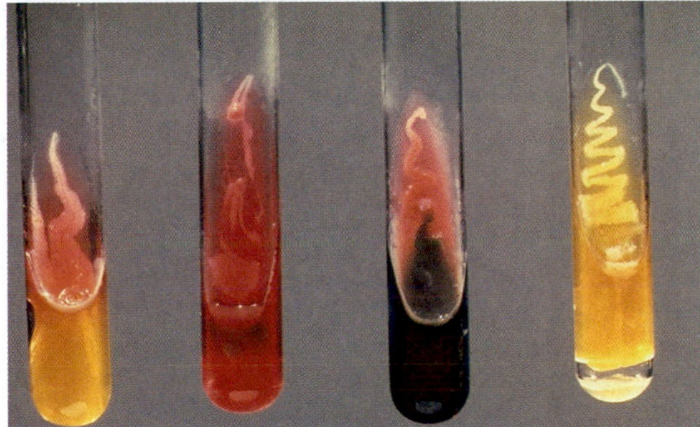

(b)

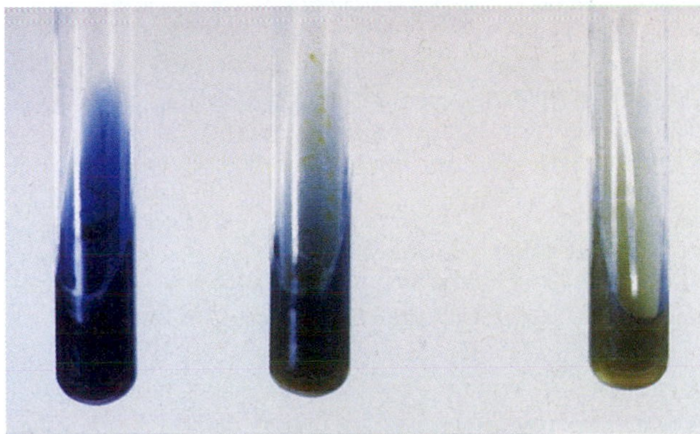

(c)

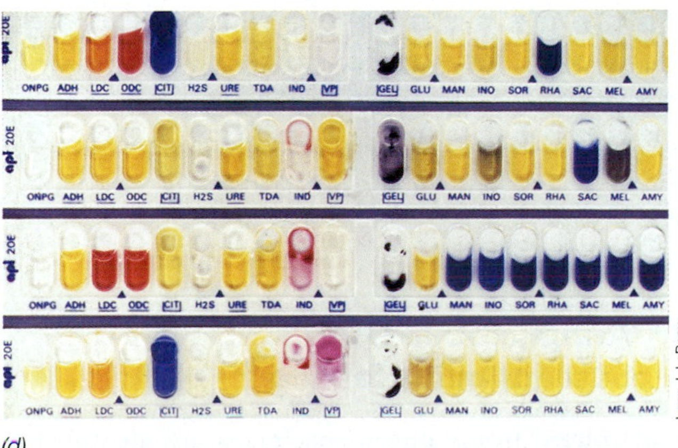

Leon J. LeBeau

(d)

Leon J. LeBeau

(e)

Figure 32.7 Growth-dependent diagnostic methods used for the identification of clinical isolates by color changes in various diagnostic media. *(a)* Use of a differential medium to assess sugar fermentation. Acid production is indicated by color change of the pH-indicating dye added to the liquid medium. If gas production occurs, a bubble appears in the inverted vial in each tube. From left to right: acid, acid and gas, negative, uninoculated. *(b)* A conventional diagnostic test for enteric bacteria in triple sugar iron (TSI) agar. The medium is inoculated both on the surface of the slant and by stabbing into the solid agar butt. The medium contains a small amount of glucose and a large amount of lactose and sucrose. Organisms able to ferment only the glucose cause acid formation only in the butt, whereas lactose or sucrose-fermenting organisms also cause acid formation throughout the slant. Gas formation is indicated by the breaking up of the agar in the butt. Hydrogen sulfide formation (either from protein degradation or from reduction of thiosulfate in the medium) is indicated by a blackening due to reaction of H_2S with ferrous iron in the medium. From left to right: fermentation of glucose only; no reaction; hydrogen sulfide formation; fermentation of glucose and another sugar. *(c)* Measurement of citrate utilization by *Salmonella* on Simmons citrate agar. The change in pH causes a change in the color of the indicator dye. From left to right: positive, negative, uninoculated. *(d)* Media kits used for the rapid identification of clinical isolates. The principle is the same as in *(a)*, but the whole arrangement has been miniaturized so that a number of tests can be run at the same time. Four separate strips, each with a separate culture, are shown. *(e)* Another arrangement of a miniaturized test kit. This one defines sugar utilization in nonfermentative organisms.

Table 32.3 Important clinical diagnostic tests for bacteria

Test	*Principle*	*Procedure*	*Most common use*
Carbohydrate fermentation	Acid and/or gas produced during fermentative growth with sugars or sugar alcohols	Broth medium with carbohydrate and phenol red as pH indicator; inverted tube for gas	Enteric bacteria differentiation
Catalase	Enzyme decomposes hydrogen peroxide, H_2O_2	Add a drop of H_2O_2 to dense culture and look for bubbles (O_2)	*Bacillus* (+) from *Clostridium* (−); *Streptococcus* (−) from *Micrococcus-Staphylococcus* (+)
Citrate utilization	Utilization of citrate as sole carbon source, results in alkalinization of medium	Citrate medium with bromthymol blue as pH indicator. Look for intense blue color (alkaline pH)	*Klebsiella-Enterobacter* (+) from *Escherichia* (−), *Edwardsiella* (−) from *Salmonella* (+)
Coagulase	Enzyme causes clotting of blood plasma	Mix dense liquid suspension of bacteria with plasma, incubate, and look for fibrin clot	*Staphylococcus aureus* (+) from *S. epidermidis* (−)
Decarboxylases (lysine, ornithine, arginine)	Decarboxylation of amino acid releases CO_2 and amine	Medium enriched with amino acids. Bromcresol purple pH indicator becomes purple (alkaline pH) if there is enzyme action	Aid in determining bacterial group among the enteric bacteria
β-Galactosidase (ONPG) test	Orthonitrophynyl-β-galactoside (ONPG) is an artificial substrate for the enzyme. When hydrolyzed, nitrophenol (yellow) is formed.	Incubate heavy suspension of lysed culture with ONPG. Look for yellow color	*Citrobacter* (+) from *Salmonella* (−). Identifying some *Shigella* and *Pseudomonas* species
Gelatin liquefaction	Many proteases hydrolyze gelatin and destroy the gel	Incubate in broth with 12% gelatin. Cool to check for gel formation. If gelatin is hydrolyzed, tube remains liquid on cooling	Aid in identification of *Serratia, Pseudomonas, Flavobacterium, Clostridium*
Hydrogen sulfide (H_2S) production	H_2S produced by breakdown of sulfur amino acids or reduction of thiosulfate	H_2S detected in iron-rich medium from formation of black ferrous sulfide (many variants: Kliger's iron agar and triple sugar iron agar also detect carbohydrate fermentation)	In enteric bacteria, to aid in identifying, *Salmonella, Edwardsiella,* and *Proteus*
Indole test	Tryptophan from proteins converted to indole	Detect indole in culture medium with dimethyl-aminobenzaldehyde (red color) or in colony smeared on paper containing dimethylaminocin-namaldehyde (spot test; blue color)	Distinguish *Escherichia* (+) from most *Klebsiella* (−) and *Enterobacter* (−); *Edwardsiella* (+) from *Salmonella* (−) *Proteus vulgaris* (+) from *Proteus mirabilis* (−)
Methyl red test	Mixed-acid fermenters produce sufficient acid to lower pH below 4.3	Glucose-broth medium. Add methyl red indicator to a sample after incubation	Differentiate *Escherichia* (+, culture red) from *Enterobacter* and *Klebsiella* (usually −, culture yellow)

differential for groups of important pathogens or even for single bacterial species. For example, kits containing multiple media have been developed for identification of *Staphylococcus aureus, Streptococcus pyogenes, Neisseria gonorrhoeae, Haemophilus influenzae,* and *Mycobacterium tuberculosis*. Other kits are available for identification of the pathogenic fungi (eukaryotes) *Candida albicans* and *Cryptococcus neoformans* (Section 35.8).

The clinical microbiologist decides which diagnostic tests to use based on the origin of the clinical specimen, the basic characteristics (morphology and Gram stain) of a pure culture of the specimen grown on general purpose media, and previous experience with similar cases.

32.2 MiniReview

Growth-dependent pathogen identification methods depend on observing metabolic reactions in specialized selective and differential media. Growth-dependent tests provide results needed for accurate pathogen identification.

- Distinguish between selective and differential media. Give an example of a medium used for each purpose.
- Suggest appropriate general purpose, selective, and differential media for isolation of pathogens from (1) a urine culture and (2) a blood culture.

32.3 Antimicrobial Drug Susceptibility Testing

Pathogens isolated from clinical specimens are identified to confirm medical diagnoses and to guide antimicrobial therapy. For many pathogens, appropriate and effective antimicrobial treatment is based on current experience and practices. For a select group of pathogens, however, decisions about appropri-

Table 32.3 *(continued)*

Test	Principle	Procedure	Most common use
Nitrate reduction	Nitrate (NO_3^-) as alternate electron acceptor, reduced to NO_2^- or N_2	Broth with nitrate. After incubation, detect nitrate with α-naphthylamine-sulfanilic acid (red color). If negative, confirm that NO_3^- is still present by adding zinc dust to reduce NO_3^- to NO_2^-. If no color after zinc, then $NO_3^- \rightarrow N_2$	Aid in identification of enteric bacteria (usually +)
Oxidase test	Cytochrome c oxidizes artificial electron acceptor: tetramethyl *p*-phenylenediamine (Kovac's reagent), or dimethyl *p*-phenylenediamine (Gordon and McLeod's reagent).	Colonies are smeared on paper impregnated with reagent. Oxidase-positive colonies produce dark purple-black color in 10–15 s with Kovac's reagent and blue color in 10–30 m with Gordon and McLeod's reagent	Differentiate *Neisseria* and *Moraxella* (+) from *Acinetobacter* (−), pseudomonads (+) and *Vibrionaceae* (+) from Enterobacteriaceae (−). To aid in identification of *Aeromonas* (+)
Oxidation-fermentation (O/F) test	Some organisms produce acid only when growing aerobically	Acid production in top part of sugar-containing culture tube; soft agar used to restrict mixing during incubation	Differentiate *Micrococcus* (acid produced aerobically only) from *Staphylococcus* (acid produced anaerobically). To characterize *Pseudomonas* (aerobic acid production) from enteric bacteria (acid produced anaerobically)
Phenylalanine deaminase test	Deamination produces phenylpyruvic acid, which is detected in a colorimetric test	Medium enriched in phynylalanine. After growth, add ferric chloride reagent and look for green color	Characterize the genera *Proteus* and the *Providencia*
Starch hydrolysis	Iodine-iodide gives blue color with starch	Grow organism on plate containing starch. Flood plate with Gram's iodine and look for clear zones around colonies	Identify typical starch hydrolyzers such as *Bacillus* spp.
Urease test	Urea, $H_2N—CO—NH_2$, split to $2\ NH_3 + CO_2$	Medium with 2% urea and phenol red indicator. Ammonia release raises pH, intense pink-red color	Distinguish *Klebsiella* (+) from *Escherichia* (−), and *Proteus* (+) from *Providencia* (−). To identify *Helicobacter pylori* (+)
Voges-Proskauer test	Acetoin produced from sugar fermentation	Chemical test for acetoin using α-naphthol	Separate *Klebsiella* and *Enterobacter* (+) from *Escherichia* (−). To characterize members of the genus *Bacillus*

ate antimicrobial therapy must be made on a case-by-case basis. Such pathogens include those for which antimicrobial drug resistance is common (for example, gram-negative enteric bacteria), those that cause life-threatening disease (for example, *Neisseria meningitidis* and bacterial meningitis), and those that require bacteriocidal rather than bacteriostatic drugs to prevent disease progression and tissue damage. This latter category might include organisms that cause bacterial endocarditis, where total and rapid killing of the pathogen is critical for patient survival.

We discussed some principles for the measurement of antimicrobial activity in Chapter 27. The susceptibility of a culture can be most easily determined by an agar diffusion method or by using a tube dilution technique to determine the *minimum inhibitory concentration (MIC)* of an agent that is necessary to inhibit growth (Section 27.4). U.S. Food and Drug Administration regulations control the automated instruments used for susceptibility testing in the United States. Procedures and standards, including experimental end points for each organism and antibiotic, are constantly updated by the Clinical and Laboratory Standards Institute, a nonprofit organization that develops and establishes voluntary consensus standards for antibiotic testing as well as other healthcare technologies (http://www.clsi.org).

The standard procedure for assessing antimicrobial activity is the *disk diffusion test* (**Figure 32.8*a–e***). Agar media are inoculated by evenly spreading a defined density of a suspension of a pure culture on the agar surface. Filter paper disks containing a defined quantity (micrograms per disk) of an antimicrobial agent are then placed on the inoculated agar. After a specified period of incubation, the diameter of the inhibition zone around each disk is measured. **Table 32.4** presents zone sizes for several antibiotics. Inhibition zone diameters are then interpreted into susceptibility categories based on zone size. Standards for the efficacy of different antimicrobial agents against different bacterial pathogens are provided by the Food and Drug Administration or the Clinical and Laboratory Standards Institute.

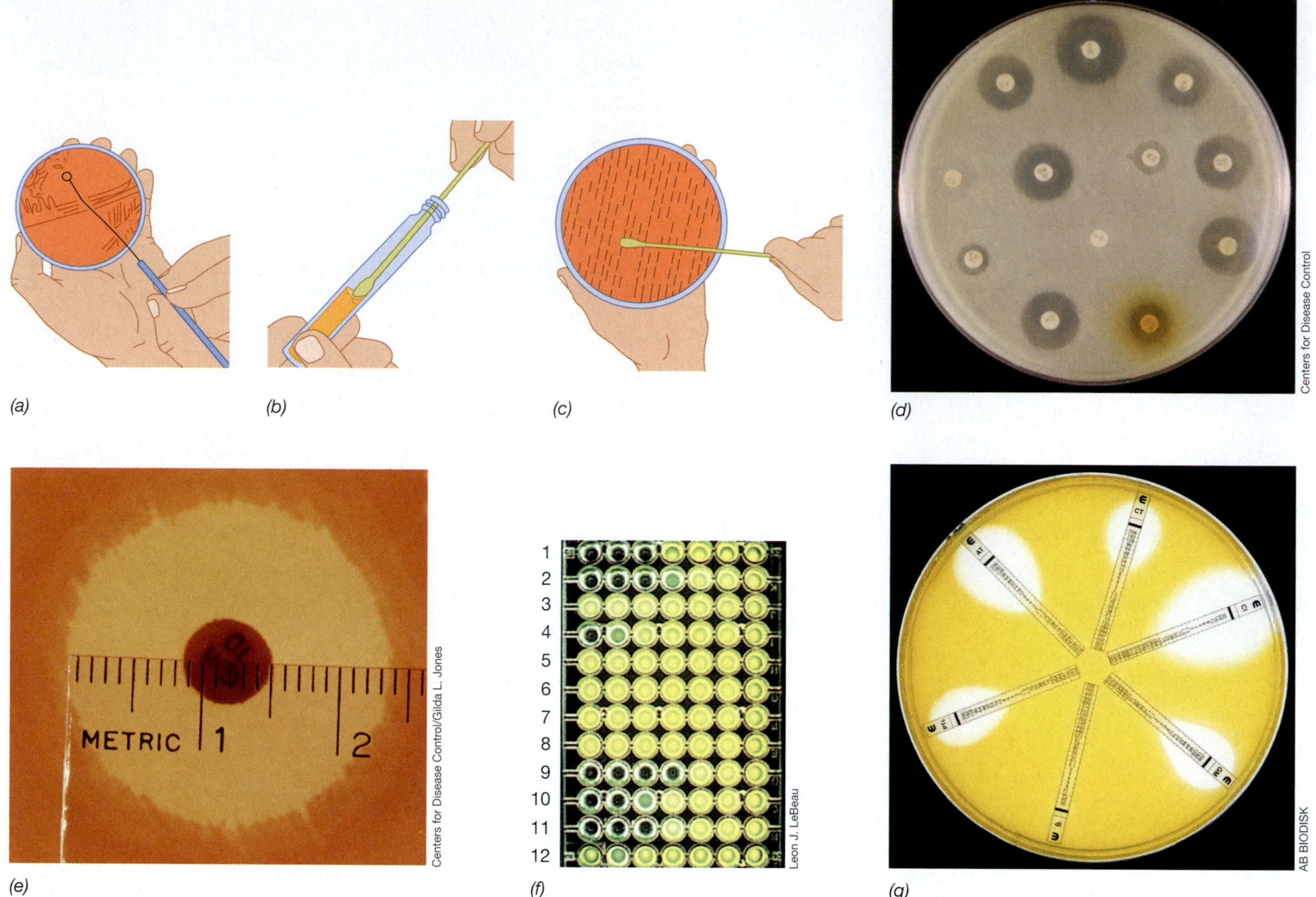

Figure 32.8 Antibiotic susceptibility testing. Methods for determining the susceptibility of an organism to antibiotics. The disk diffusion test: *(a)* isolated pure colonies are homogenized in a tube with an appropriate liquid medium to achieve a specified density as judged by comparison to a turbidity standard; *(b)* a sterile cotton swab is dipped into the bacterial suspension and excess fluid removed by pressing the swab against the side of the tube; *(c)* the swab is streaked evenly over the surface of an appropriate agar medium; *(d)* disks containing known amounts of different antibiotics are placed on the agar surface after inoculation with the bacterial culture. *(e)* after incubation, inhibition zones are observed and measured. From these data, the susceptibility category of the organism is determined by reference to an interpretive chart of zone sizes (Table 32.4). *(f)* Antibiotic susceptibility as determined by the broth dilution method. The organism is *Pseudomonas aeruginosa*. Each row has a different antibiotic. The microtiter plate enables automation of these tests. The end point is the first well with the lowest concentration of antibiotic that shows no visible bacterial growth. The highest concentration of antibiotic is in the well at the left; serial twofold dilutions are made in the wells to the right. For example, in rows 1 and 2, the end point is the third well. In row 3, the antibiotic is ineffective at the concentrations tested, since there is bacterial growth in all the wells. In row 4, the end point is in the first well. *(g)* Antibiotic susceptibility determined by the Etest (AB BIODISK, Solna, Sweden) for different antibiotics (from 8 o'clock, PTc-piperacillin/tazobactam; AT-aztreonam; CT-cefotaxime; CI-ciprofloxacin; GM-gentamicin; IP-imipenem). Each strip is calibrated in terms of the minimum inhibitory concentration (MIC) in μg/ml starting with the lowest concentration from the center of the plate. The lowest concentration of antibiotic that inhibits bacterial growth is the MIC value for that particular agent (Section 27.4). For example, the MIC for cefotaxime (CT) is 16 μg/ml. This organism is resistant to imipenem (IP); MIC $>$ 32 μg/ml.

The MIC procedure for antibiotic susceptibility testing employs an antibiotic dilution assay, either in culture tubes (Figure 27.10) or in the wells of a microtiter plate (Figure 32.8*f*). Wells containing serial dilutions of antibiotics are inoculated with a standard amount of a test organism. Growth in the presence of each antibiotic is then observed by measuring turbidity. Antibiotic susceptibility is usually expressed as the highest dilution (lowest concentration) of

Table 32.4 Standards for antimicrobial disk diffusion susceptibility tests[a]

		Inhibition zone diameter (mm)		
Antibiotic	**Amount on disk**	**Resistant**	**Intermediate**	**Susceptible**
Ampicillin[b]	10μg	13 or less	14–16	17 or more
Ampicillin[c]	10μg	28 or less	—	29 or more
Ceftriaxone	30μg	13 or less	14–20	21 or more
Chloramphenicol	30μg	12 or less	13–17	18 or more
Clindamycin	2μg	14 or less	15–20	21 or more
Erythromycin	15μg	13 or less	14–22	23 or more
Gentamicin	10μg	12 or less	13–14	15 or more
Methicillin[b]	5μg	9 or less	10–13	14 or more
Nitrofurantoin	300μg	14 or less	15–16	17 or more
Penicillin G[d]	10 units	28 or less	—	29 or more
Penicillin G[e]	10 units	14 or less	—	15 or more
Streptomycin	10μg	6 or less	7–9	10 or more
Sulfonamide	10μg	12 or less	13–16	17 or more
Tetracycline	30μg	14 or less	15–18	19 or more
Trimethoprim-sulfamethoxazole	1.25/23.75μg	10 or less	11–15	16 or more
Tobramycin	10μg	12 or less	13–14	15 or more
Vancomycin[f]	10μg	14 or less	15–16	17 or more
Vancomycin[g]	10μg	14 or less (test for MIC)	—	15 or more

[a]Standards are defined and updated by Clinical and Laboratory Standards Institute (CLSI), an international nonprofit organization that develops voluntary consensus standards for antibiotic testing and other healthcare technologies (http://www.clsi.org)

[b]For *Enterobacteriaceae*.

[c]For staphylococci and highly penicillin-sensitive organisms.

[d]For staphylococci.

[e]For organisms such as enterococci that may cause some systemic infections treatable with high doses of penicillin G.

[f]For enterococci.

[g]For staphylococci.

antibiotic that completely inhibits growth. This defines the value of the MIC.

Etest (AB BIODISK, Solna, Sweden) is a nondiffusion-based technique that employs a preformed and predefined gradient of an antimicrobial agent immobilized on a plastic strip. The concentration gradient covers a MIC range across 15 twofold dilutions. When applied to the surface of an inoculated agar plate, the gradient transfers from the strip to the agar and remains stable for a period that covers the wide range of critical times associated with the growth characteristics of different microorganisms. After overnight incubation or longer, an elliptical zone of inhibition centered along the axis of the strip develops. The MIC value (in micrograms per millileter) can be read at the point where the ellipse edge intersects the precalibrated Etest strip, providing a precise MIC (Figure 32.8g). This value can then be interpreted using current Food and Drug Administration or Clinical and Laboratory Standards Institute standards.

Many of the pathogens for which susceptibility testing is necessary are *healthcare-associated pathogens* or *nosocomial pathogens*, those acquired in hospitals and other healthcare settings. Hospital infection-control microbiologists generate and examine susceptibility data to generate periodic reports called **antibiograms**. These reports define the susceptibility of clinically isolated organisms to the antibiotics in current use. Antibiograms are used to monitor control of known pathogens, to track the emergence of new pathogens, and to identify the emergence of antibiotic resistance, all at the local level.

32.3 MiniReview

Antimicrobial drugs are widely used for the treatment of infectious diseases. Dangerous pathogens are tested for antimicrobial susceptibility to ensure appropriate therapy.

- Describe the disk diffusion test for antimicrobial susceptibility. For an individual organism and antimicrobial agent, what do the results indicate?
- What is the value of antimicrobial drug susceptibility testing for the microbiologist, the physician, and the patient, both for community-based and healthcare-associated infections?

32.4 Safety in the Microbiology Laboratory

Clinical microbiology laboratories present significant biological hazards for workers and those that come into contact with them. To prevent accidental laboratory infections, standard laboratory practices for handling clinical samples have been established. In the United States, every clinical and research institution that deals with human or primate tissue is required by law to have an occupational exposure control plan for handling bloodborne pathogens. This law was specifically designed to protect workers from infection by hepatitis B virus (HBV, the cause of infectious hepatitis, ∞ Section 34.12) and human immunodeficiency virus (HIV, the cause of AIDS, ∞ Section 34.15), but effectively limits infection by all pathogens because of the implementation of stringent infection controls in the laboratory.

The two most common causes of laboratory accidents are ignorance and carelessness. Training and enforcement of established safety procedures, however, can prevent most accidents. Unfortunately, most laboratory-acquired infections do not result from identifiable exposures or accidents, but rather from routine handling of patient specimens. Infectious aerosols generated during processing of specimens are the most common causes of laboratory infections. Clinical laboratories follow the safety rules outlined here (required by law in the United States) to minimize the exposure of healthcare workers to infectious agents and thereby reduce the numbers of non-accident-associated laboratory infections.

1. Laboratories handling hazardous materials must restrict access to laboratory workers and essential support personnel. These individuals must have knowledge of the biological risks in the laboratory and act accordingly.
2. Effective decontamination procedures must be in place. Infectious materials or wastes, including specimens, syringes and needles, inoculated media, bacterial cultures, tissue cultures, experimental animals, glassware, instruments, and surfaces must be must be fully decontaminated, without compromise. A 5.25% (full strength) chlorine bleach solution or other approved disinfectant should be used to decontaminate spilled infectious material. All potentially infectious waste must be burned in a certified incinerator or handled by a licensed waste handler.
3. Personnel working with hazardous infectious agents or vaccines (for example, rabies, polio, or diphtheria-pertussis-tetanus vaccines) must be properly vaccinated against the agent. Persons working with human or primate tissue must be vaccinated against HBV.
4. All clinical specimens should be considered infectious and handled appropriately. This is especially important for preventing laboratory-acquired hepatitis because of the relative frequency with which hepatitis viruses are present in blood specimens.
5. All pipetting must be done with mechanical pipetting devices (not by mouth).
6. Animals should be handled only by trained laboratory personnel. Anesthetics and tranquilizers should be used to avoid injury to both personnel and animals.
7. Laboratory personnel must wear laboratory coats or gowns, sealed shoes, rubber gloves, masks, eye protection, respiratory devices when needed, and other barrier devices appropriate for the level of exposure and the severity of the potential infection. Such barrier devices must also be properly decontaminated and stored after use. Laboratory personnel must also practice good personal hygiene with respect to hand washing. Eating and drinking, smoking, applying cosmetics or lip balm, or wearing contact lenses is never permitted in the clinical laboratory.
8. Because of the special risks associated with AIDS, all clinical (human) specimens should be treated as if they contain HIV. Protective gloves should be worn when handling specimens of any kind. Masks or full-face shields must be worn any time there is a possibility of generating an aerosol during specimen preparation. Needles must not be resheathed, bent, or broken; they should be placed in a labeled container designated expressly for this purpose that can be sealed and decontaminated before disposal.

These safety rules should be in effect in all laboratories that handle potential infectious agents. Specialized clinical laboratories may have additional rules on top of these to ensure a safe work environment, as discussed below. In the final analysis, however, it is the responsibility of laboratory personnel to assure safety in the workplace. Any clinical laboratory has potential biohazards and is even more dangerous for untrained personnel or those unwilling to take the necessary precautions to prevent laboratory-acquired infection.

Biological Containment and Laboratory Biosafety Levels

The level of containment used to prevent accidental infections or accidental environmental contamination (escape) in clinical, research, and teaching laboratories must be adjusted to counter the biohazard potential of the organisms handled in the laboratory. Laboratories are classified according to their containment potential, or *biosafety level (BL)*, and are designated as *BL1*, *BL2*, *BL3*, or *BL4*. Personnel in laboratories working at all biosafety levels must follow good laboratory practices that ensure basic cleanliness and limit contamination, and laboratory surfaces must be decontaminated after each work shift or whenever spills occur. Personnel cannot consume food or drink in the laboratory and must wash their hands when leaving the laboratory. Access to the laboratory is to be restricted to laboratory personnel at all times.

BL1 laboratories are the lowest level of containment. Work can be done on the open bench with organisms that present a low risk of infection; these organisms are not pathogens in normal individuals and include organisms such as *Bacillus subtilis*. An example of a BL1 facility is a microbiology student teaching laboratory that does not use pathogens.

BL2 laboratories are designed to contain organisms that present a moderate risk of infection due to accidental ingestion, percutaneous injection, or exposure to mucous membranes via aerosols. Work with pathogens such as *Escherichia coli* or *Streptococcus pyogenes* is done in a BL2 laboratory or sometimes at higher containment levels. Normal procedures may be performed on bench tops, but barrier protection devices such as face and eye protection, gloves, and lab coats or gowns are used when appropriate. Most microbiology research, clinical, and instructional laboratories maintain BL2 containment standards.

BL3 laboratories are designed to contain pathogens that have a very high potential for causing infections, especially from aerosols. For example, if laboratory personnel handle extremely infectious airborne pathogens such as *Mycobacterium tuberculosis*, the causative agent of tuberculosis, the laboratory should be fitted with special features such as negatively pressurized rooms and air filters to prevent accidental release of the pathogen from the laboratory. Biological safety cabinets (∞ Figure 27.4) are required for manipulations in a BL3 laboratory; work must not be done on the open bench. In some special cases, organisms that can normally be handled at BL2 must be handled at BL3. For example, *Staphylococcus aureus* can be handled on culture plates at BL2. However, when large quantities are grown, and especially when such quantities are centrifuged, work must be done in a BL3 facility to contain potential infectious aerosols. Specialized research and teaching facilities often have a BL3 laboratory.

BL4 laboratories are designed for maximum containment of life-threatening pathogens that have a high probability of transmission by aerosols and for which there is no effective immunization, treatment, or cure. Physical containment in BL4 facilities requires total isolation of the organism, such as manipulation through gloves in a sealed biological safety cabinet or by personnel wearing full-body, positive-pressure suits with air supplies (**Figure 32.9**). Some examples of pathogens that must be manipulated in a BL4 facility include hemorrhagic fever viruses (Lassa, Marburg, Ebola, ∞ Section 33.10), variola virus (smallpox, ∞ Section 33.11), and drug-resistant *Mycobacterium tuberculosis* (∞ Section 34.5). BL4 laboratories are specifically designed and equipped to handle the most dangerous pathogens. They are usually associated with government facilities such as the Centers for Disease Control and Prevention (Atlanta, Georgia, USA) or university laboratories that specialize in infectious disease.

32.4 MiniReview

Clinical laboratory safety requires training, planning, and care to prevent the infection of laboratory workers with pathogens. Specific precautions and appropriate decontamination procedures must be in place for materials such as live cultures, inoculated culture media, used hypodermic needles, and patient specimens.

- What are the major precautions necessary to prevent spread of a bloodborne pathogen to laboratory personnel?
- What are the major causes of laboratory infections?
- Identify the basic features of biological hazard containment in laboratories, from BL1 to BL4.

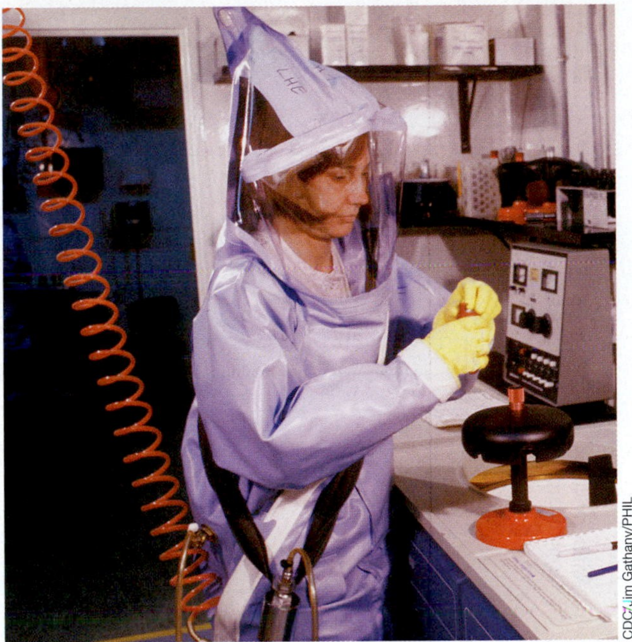

Figure 32.9 A worker in a BL4 (biological safety level-4) laboratory. BL4 is the highest level of biological control, affording maximum worker protection and pathogen containment. The worker has a whole-body sealed suit with an outside air supply and ventilation system. Air locks control all access to the laboratory. All material leaving the laboratory is autoclaved or chemically decontaminated.

II IMMUNOLOGY AND CLINICAL DIAGNOSTIC METHODS

Immunoassays are used in clinical, reference, and research laboratories to detect specific pathogens or pathogen products. When culture methods for pathogens are not routinely available or are prohibitively difficult to perform, as is the case with most viral infections, immunoassays often provide effective and relatively simple means to identify individual pathogens or exposure to pathogens.

32.5 Immunoassays for Infectious Disease

The immune response was discussed in Chapters 29–31. The major aspects of immunity were summarized in Figures 30.1–30.3. Many immunoassays utilize antibodies specific for pathogens or their products for *in vitro* tests designed to detect individual infectious agents. In some cases, as we now discuss, patient immune responses are monitored to obtain evidence for infection by a pathogen.

Antibody Titers, Skin Tests, and the Diagnosis of Infectious Disease

Isolation of a pathogen is not always possible or practical to confirm diagnosis of an infectious disease. An alternative approach that provides strong circumstantial evidence for

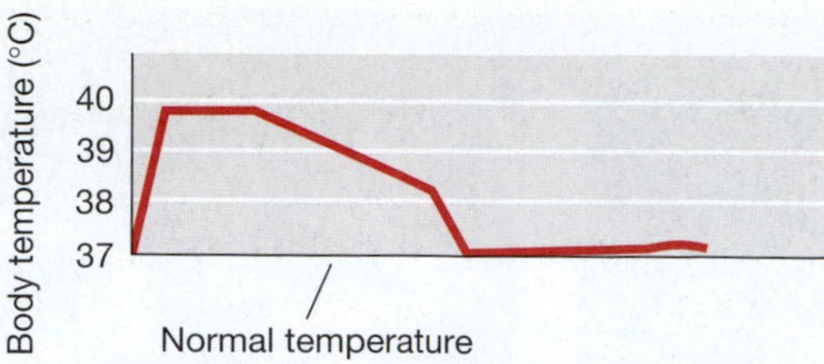

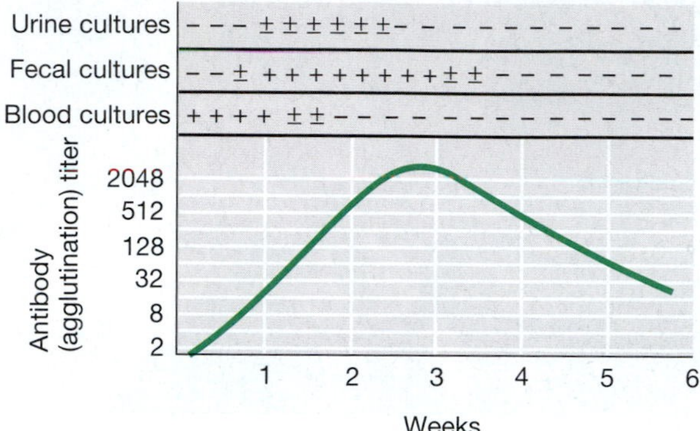

Figure 32.10 The course of infection in a typical untreated typhoid fever patient. Measurement of body temperature provides a measure of the course of clinical symptoms. The antibody titer was measured by determining the highest serum dilution (twofold series) causing agglutination of a test strain of *Salmonella typhi*. Titer is shown as the *reciprocal* of the highest dilution showing an agglutination reaction. Presence of viable bacteria in blood, feces, and urine was determined from periodic cultures. Note that the pathogen clears from the blood as the antibody titer rises, and clearance from feces and urine requires a longer time. Body temperature gradually drops to normal as the antibody titer rises. The data given do not represent a single patient but are a composite of the pattern seen in large numbers of patients.

infection by a particular pathogen is to measure antibody **titer** (quantity) against an antigen or antigens produced by the suspected pathogen. If an individual is infected with a suspected pathogen, the immune response—in this case, the antibody titer—to that pathogen should be elevated. Antibody titer can be measured by methods discussed in Sections 32.7–32.11. Tests such as agglutination, enzyme-linked immunosorbent assay, and radioimmunoassay are commonly used. Serial dilutions of patient serum are prepared and assayed by the chosen method. The titer is defined as the highest dilution (lowest concentration) of serum at which the antigen–antibody reaction is observed (**Figure 32.10**). These methods are called *serological tests* because they make use of patient serum antibodies.

A single measure of antibody titer indicates previous infection or exposure to a pathogen. For pathogens rarely found in a population, the presence of antibody may be sufficient to indicate ongoing, active infection. This is the case, for example, for HIV–AIDS (Section 34.15). Methods for determining HIV antibody levels are discussed in Sections 32.10 and 32.11.

In most cases, however, the mere presence of antibody does not indicate active infection. Antibody titers typically remain detectable for long periods after a previous infection has been resolved. To link an acute illness to a particular pathogen, it is essential to show a *rise* in antibody titer in serum samples taken from a patient during the acute disease and later during the convalescent phase of the disease. Frequently, the antibody titer is low during the acute stage of the infection and rises during convalescence (Figure 32.10). A rise in antibody titer is the best indication that the illness is due to the suspected pathogen. Unfortunately, not all infections result in formation of systemic immunity. In some cases, the presence of antibody in the serum may be due to a recent immunization. In fact, measurement of the rise in antibody titer following immunization is one of the best ways of determining that the immunization was effective.

Skin testing is another method for determining exposure to a pathogen. The most commonly used skin test is the *tuberculin test*, which consists of an intradermal injection of a soluble extract from cells of *Mycobacterium tuberculosis*. A positive inflammatory reaction at the site of injection within 48 hours indicates current infection or previous exposure to *M. tuberculosis*. This test identifies responses caused by pathogen-specific T_H1 cells (Figure 30.6*b*). Skin tests are routinely used for diagnosing tuberculosis, Hansen's disease (leprosy), some fungal diseases, and other infectious disease in which the antibody response is weak or nonexistent. Common immunodiagnostic tests for pathogens are shown in **Table 32.5**.

If a pathogen is extremely localized, there may be little induction of an immunological response and no rise in antibody titer or skin test reactivity, even if the pathogen is proliferating profusely at the site of infection. A good example is gonorrhea, caused by infection of mucosal surfaces with *Neisseria gonorrhoeae*. As we will discuss in Section 34.13, gonorrhea does not elicit a systemic or protective immune response, there is no antibody titer or skin test reactivity, and reinfection of individuals is common.

32.5 MiniReview

An immune response is often a natural outcome of infection. Specific immune responses, particularly antibody titers and skin tests, can be monitored to provide evidence for past infections, current infections, and convalescence.

- Explain the reasons for changes in antibody titer for a single infectious agent, from the acute phase through the convalescent phase of the infection.
- Describe the method, time frame, and rationale for the TB skin test. What component of the immune response does this test detect?

32.6 Polyclonal and Monoclonal Antibodies

The immune response to a pathogen typically results in the production of immunoglobulins (Igs) directed at numerous antigenic determinants present on the pathogen (Section

Table 32.5 Immunological procedures for identification of infectious agents

Pathogen/disease	*Antigen*	*Procedure*[a]
HIV (AIDS)	Human immunodeficiency virus (HIV)	ELISA Immunoblot
Borrelia burgdorferi (Lyme disease)	Flagellin Surface proteins	ELISA Immunoblot
Brucella (brucellosis)	Cell wall antigen	Agglutination
Candida albicans (yeast infections)	Soluble extract of fungal proteins	Skin test
Corynebacterium diphtheriae (diphtheria)	Toxin	Skin test (Schick test)
Influenza virus (influenza)	Influenza virus suspensions Nasopharynx cells containing influenza virus	Complement-based assay Immunofluorescence
Mycobacterium leprae (leprosy, or Hansen's disease)	Lepromin (soluble extract of bacterial proteins)	Skin test
Mycobacterium tuberculosis (tuberculosis)	Tuberculin (purified protein derivative, PPD)	Skin test
Neisseria meningitidis (meningitis)	Capsular polysaccharide	Passive hemagglutination (*N. meningitidis* polysaccharide adsorbed to red blood cells)
Pneumocystis carinii (lung infection)	*P. carinii* cells	Immunofluorescence
Rickettsial diseases (Q fever, typhus, Rocky Mountain spotted fever)	Killed rickettsial cells	Complement-based assay or cell agglutination tests ELISA
Salmonella (gastroenteritis)	O and H antigen	Agglutination (Widal test) ELISA
Streptococcus (group A) (strep throat, scarlet fever)	Streptolysin O (extoxin), DNase (extracellular protein)	Neutralization of hemolysis Neutralization of enzyme
Treponema pallidum (syphilis)	Cardiolipin-lecithin-cholesterol	Flocculation [Venereal Disease Research Laboratory (VDRL) test]
Vibrio cholerae (cholera)	O antigen	Agglutination Bacteriocidal test (in presence of complement) ELISA

[a]Immunofluorescence tests use preformed antibody to detect the presence of the indicated pathogen in a patient specimen. Skin tests for *C. albicans, M. tuberculosis,* and *M. leprae* indicate T_H1-mediated delayed-type hypersensitivity. The *C. diphtheriae* Schick test detects serum antibodies with a toxin-neutralization skin test. All other tests measure serum antibody levels.

29.10). Only a few of the many Igs are directed toward each antigen determinant. The resulting antiserum is a complex mixture of different antibodies called **polyclonal antibodies** derived from many individual B cells. The serum is called a *polyclonal antiserum*. Polyclonal antisera provide adequate immune protection to the host, but they are not precisely reproducible because they are the sum of the antibody response by an individual to a complex antigen for only a single occurence.

Hybridomas and Monoclonal Antibodies

Each Ig is produced by a single B lymphocyte (∞ Section 29.10). As a result, an *in vitro* B cell clone can produce limitless supplies of a single monospecific antibody; this is called a **monoclonal antibody**. Antibody-producing B cells, however, normally die after several weeks in cell culture (*in vitro*). To produce long-lived B cell clones, antibody-producing B cells are fused with B cell tumors called *myelomas*. Myelomas are capable of dividing indefinitely and are therefore immortal cell lines. The immortal cell lines that result from the B cell–myeloma fusion are hybrid cell lines called *hybridomas*. The hybridoma cell lines share the properties of both fusion partners. They grow indefinitely *in vitro* and produce antibodies (**Figure 32.11**).

To produce a monoclonal antibody, a mouse is immunized with the antigen of interest. During the next several weeks, antigen-specific B cells proliferate and begin producing antibodies in the mouse. Spleen or lymph node tissue, rich in B cells, is then removed from the mouse, and the B cells are fused with myeloma cells (Figure 32.11). Many cells fuse in culture and begin to grow, but only a small number are antibody-producing hybridomas. Hybridomas are selected from other cells by addition of *h*ypoxanthine, *a*minopterin, and *t*hymidine (HAT) to the *in vitro* cell culture medium. The HAT medium stops the growth of unfused myeloma cells because the myeloma cells, though able to grow indefinitely in cell culture, are unable to use the metabolites hypoxanthine and thymidine to bypass a metabolic block caused by aminopterin, a cell poison. By contrast, fused hybridoma cells can use hypoxanthine and thymidine to bypass the aminopterin block and grow normally in HAT medium; they receive the genes for use of hypoxanthine and thymidine from the B cell fusion partner. Unfused B cells therefore die in a few days because they cannot grow in culture. Following fusion, the antibody-producing hybridoma clones must be identified.

An enzyme-linked immunosorbent assay (ELISA) (Section 32.10) can be used to identify hybridomas that produce monoclonal antibodies. From a typical fusion, several distinct clones are isolated, each making a monoclonal antibody. Once the clones of interest are identified, they can be grown in the

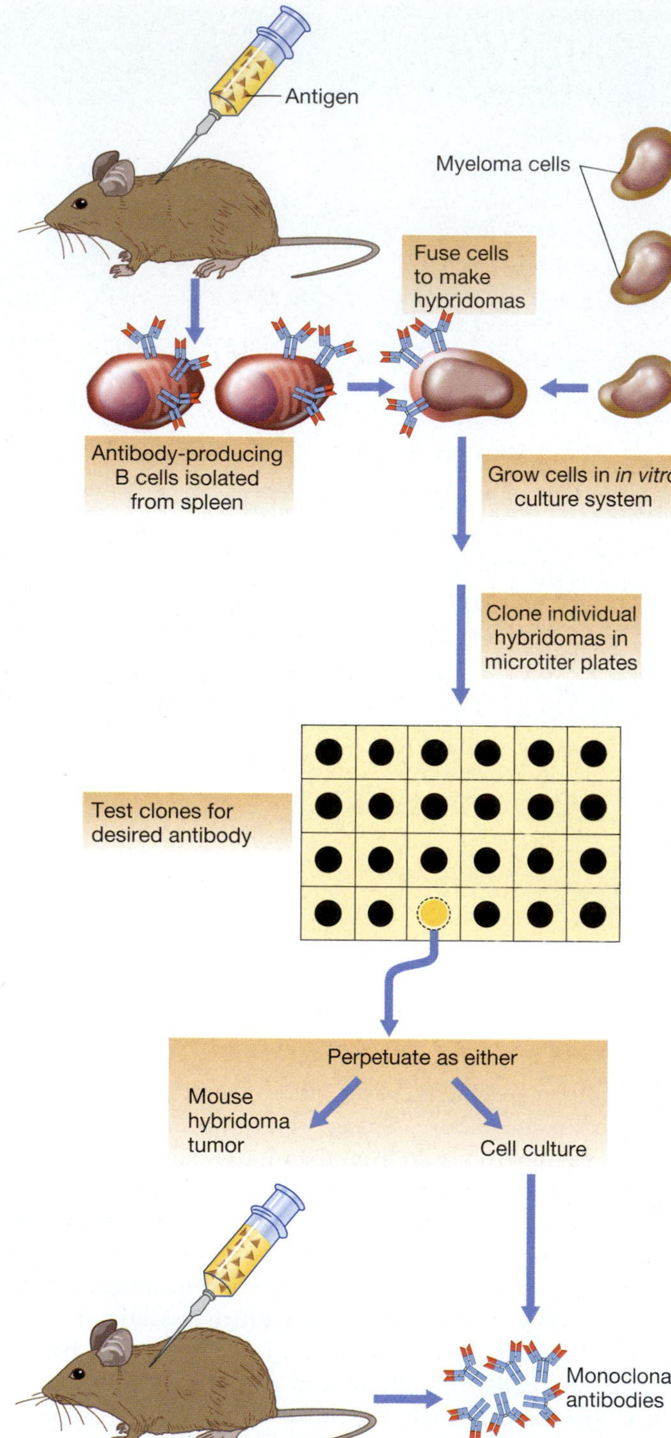

Figure 32.11 The hybridoma technique and production of monoclonal antibodies. The hybridoma can be indefinitely cultured or passed through animals as a tumor. The hybridoma cells can also be stored as frozen tumor cells and reconstituted when needed in tissue culture or in a suitable animal host.

mouse as an antibody-producing tumor, or they can be grown in cell culture. Antibody can be harvested from the tumor or from the culture supernatant of the cell culture. Hybridomas can grow indefinitely or can be stored as frozen cells and later reconstituted to provide the desired monoclonal antibodies.

Table 32.6 Characteristics of monoclonal and polyclonal antibody production

Polyclonal	*Monoclonal*
Contains many antibodies recognizing many determinants on an antigen	Contains a single antibody recognizing only a single determinant
Various classes of antibodies are present (IgG, IgM, and so on)	Single class of antibody produced
Can make a specific antibody using only a highly purified antigen	Can make a specific antibody using an impure antigen
Reproducibility and standardization difficult	Highly reproducible

Monoclonal antibodies have replaced polyclonal antibodies for many immunodiagnostic applications because they are more specific bioreagents. **Table 32.6** compares the properties of polyclonal antibodies and monoclonal antibodies.

Diagnostic and Therapeutic Uses

Monoclonal antibodies are widely used for clinical diagnostic tests, immunological typing of bacteria, and identification of cells containing foreign surface antigens (for example, a virus-infected cell). Monoclonal antibodies have also been used in genetic engineering for identifying and measuring levels of gene products not detectable by other methods and also for increasing the specificity of existing clinical tests, including blood and tissue typing.

Because of their specificity, monoclonal antibodies are also used to detect and treat human cancers. Malignant cells contain surface antigens not expressed by normal cells. These tumor antigens are unique, tumor-specific cell proteins. Monoclonal antibodies prepared against the tumor antigens specifically target the malignant cells and have been used as vehicles to deliver toxins directly to them. Tumor-specific monoclonal antibodies covalently linked to toxins are now undergoing clinical trials. The specificity of monoclonal antibody treatments may greatly improve cancer therapy by offering an alternative to chemical and radiation treatments that damage normal host cells as well as cancer cells. **www.microbiologyplace.com Online Tutorial 32.1: Producing Monoclonal Antibodies**

32.6 MiniReview

Polyclonal and monoclonal antibodies are used for research and diagnostic applications. Hybridoma technology reproducibly provides specific antibodies for a wide range of clinical, diagnostic, and research purposes.

- How can a polyclonal antibody preparation recognize a variety of antigenic determinants?
- What advantages do monoclonal antibodies have as compared with polyclonal antibodies? What are the advantages of polyclonal antibodies?

Table 32.7 Types of antigen-antibody reactions

Location of antigen	*Accessory factors required*	*Reaction observed*
Soluble	None	Precipitation
On cell or inert particle	None	Agglutination
Flagellum	None	Immobilization or agglutination
On bacterial cell	Complement	Lysis
On bacterial cell	Complement	Killing
On erythrocyte	Complement	Hemolysis
Toxin	None	Neutralization
Virus	None	Neutralization
On bacterial cell	Phagocyte, complement	Phagocytosis and opsonization

32.7 *In Vitro* Antigen–Antibody Reactions: Serology

The study of antigen–antibody reactions *in vitro* is called **serology**. Serological reactions are used for many diagnostic immunology tests. Antigen–antibody reactions rely on the specific interaction of antigenic determinants with the *variable* region of the antibody molecule (⚭ Section 29.9). Various serological tests are used to identify antigens, depending on the properties of the antigen and on the conditions chosen for reaction (**Table 32.7**).

Specificity and Sensitivity

The usefulness of a serological test for diagnostic purposes depends on the test's specificity and sensitivity. **Specificity** is the ability of an antibody preparation to recognize a single antigen. Optimal specificity implies that the antibody is specific for a single antigen, will not cross-react with any other antigen, and therefore will not provide false positive results. Specificity must be defined in terms of reactions with positive- and negative-control antigens. Specificity for each test must be determined experimentally and verified every time the test is used.

Sensitivity defines the lowest amount of an antigen that can be detected. The highest level of sensitivity requires that the antibody in a test be capable of identifying a single antigen molecule. High sensitivity prevents false-negative reactions. The sensitivity of some common tests in terms of the amount of antibody necessary to detect antigen is shown in **Table 32.8**. The amount of antigen detected by each test system is proportional to the amount of antibody used. For example, immune precipitation reactions require a large amount of antibody and generally detect 0.1–1.0 mg quantities of antigen. Thus, precipitation tests are the least sensitive serological tests. By contrast, ELISA tests (Section 32.10) require 100,000 times less antibody and can detect 1 million times less antigen (0.1–1.0 ng quantities) than precipitation tests. Thus, ELISA tests are among the most sensitive serological tests.

Neutralization

Neutralization is the interaction of antibody with antigen to block or distort the antigen sufficiently to reduce or eliminate its biological activity. Neutralization reactions can occur *in vitro* or *in vivo*.

For example, neutralization of a microbial toxin by specific antibody occurs when the toxin and specific antibody combine in such a way that the active portion of the toxin is blocked (**Figure 32.12**). Neutralization reactions can block the effects of many bacterial exotoxins, including many of those listed in Table 28.4. An antiserum containing an antibody that neutralizes a toxin is called an *antitoxin*. Antitoxin therapy is used to treat botulism, tetanus, and diphtheria, all diseases that result from bacterial exotoxins.

Neutralization reactions may also occur when viruses are bound by specific antibodies. For example, antibodies directed against the hemagglutinin and neuraminidase proteins of influenza viruses prevent the adsorption of the viruses to specific receptors on host cells (⚭ Section 34.9). Neutralization reactions can be used for *in vitro* testing using patient serum, but are not used in routine diagnostic laboratories because neutralization tests require biologically active test systems.

Precipitation

Precipitation results from the interaction of a soluble antibody with a soluble antigen to form an insoluble complex.

Table 32.8 Sensitivity of immunodiagnostic assays

Assay	*Sensitivity (μg antibody/ml)*[a]
Precipitin reaction	
In fluids	24–160
In gels (double immunodiffusion)	24–160
Agglutination reactions	
Direct	0.4
Passive	0.08
Radioimmunoassay (RIA)	0.0008–0.008
Enzyme-linked immunosorbent assay (ELISA)	0.0008–0.008
Immunofluorescence	8.0

[a]The smallest amount of antibody necessary to give a positive reaction in the presence of antigen.

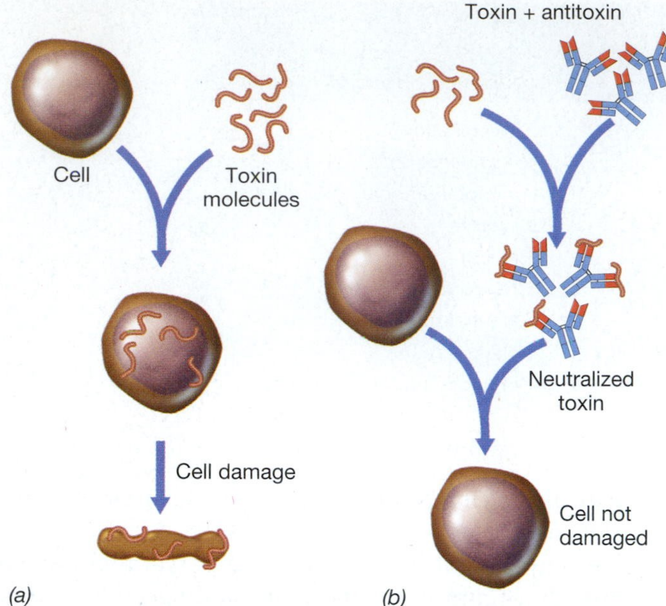

Figure 32.12 Neutralization of an exotoxin by an antitoxin antibody. *(a)* Untreated toxin results in cell destruction. *(b)* Antitoxin antibody neutralizes toxin and prevents cell destruction.

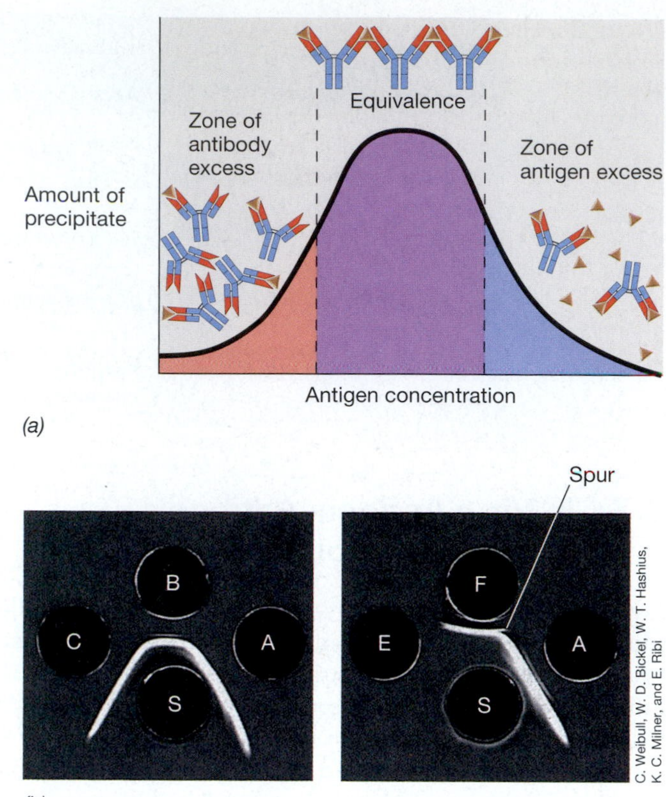

Figure 32.13 Precipitation reactions between soluble antigen and antibody. The graph *(a)* shows the extent of precipitation as a function of antigen and antibody concentration. *(b)* Precipitation in agar gel, a process called immunodiffusion. Wells labeled S contain antibodies to cells of *Proteus mirabilis.* Wells labeled A, B, and C contain soluble extracts of *Proteus mirabilis.* A line of identity is observed in the wells on the left. On the right, antigen E does not react, and antigen A shows partial identity with antigen F (see leader to spur).

Antibody molecules generally have two antigen-binding sites (that is, they are bivalent). Therefore, each antibody can bind two separate antigen molecules. If the antigen also has more than one available antibody-binding determinant, a precipitate may develop from aggregates of antibody and antigen molecules (**Figure 32.13**). Because they are easily observed *in vitro,* precipitation reactions are very informative serological tests, especially for the quantitative measurement of antibody concentrations. Precipitation occurs maximally, however, only when there are optimal proportions of the two reacting substances. The presence of either antigen or antibody in excess quantities results in the formation of soluble immune complexes.

Precipitation reactions carried out in agar gels, called *immunodiffusion tests,* are used to study the specificity of antigen–antibody reactions. Both antigen and antibody diffuse outward from separate wells cut in an agar gel, and precipitation bands form in the region where antibody interacts with antigen in optimal proportions (Figure 32.13*b*). The precipitation bands formed are characteristic for the reacting substances; two antigens reacting with an antiserum can be tested for molecular relationships by observing the bands formed when the two antigens are placed in adjacent wells equidistant from the antiserum well. For example, if two antigens in adjacent wells are identical, they will form a single, fused, precipitin band. This is called a *line of identity*. If, on the other hand, adjacent wells contain one antigen in common, but one well contains a second reacting antigen, a line of *partial identity* will form (Figure 32.13*b*). The extension of the precipitin line (representing a reaction between the antiserum and the second antigen) is called a *spur*. Immunodiffusion is used to assess the relatedness of proteins obtained from different sources.

Unfortunately, the readily visible precipitation reactions are not very sensitive. Microgram quantities of specific antibody are necessary to visualize a precipitate (Table 32.8), whereas the most useful diagnostic tests require sensitivity at the nanogram level. Consequently, assays based on precipitation reactions are normally used only in research and reference laboratories.

32.7 MiniReview

Antigen–antibody binding is the basis for a number of serological tests. Specificity and sensitivity define the accuracy of individual assays. Neutralization and precipitation reactions produce visible results involving antigen–antibody interactions.

- In serological reactions, high specificity prevents false-positive reactions. High sensitivity prevents false-negative reactions. Explain.
- Explain the principles of a neutralization reaction.
- What are the minimum antigen and antibody requirements for a precipitation reaction?

32.8 Agglutination

Agglutination is the visible clumping of a particulate antigen when mixed with antibodies specific for the particulate antigens. Agglutination tests are about 100 times more sensitive than precipitation tests (Table 32.8) and are widely used in clinical and diagnostic laboratories; they are simple to perform, highly specific, inexpensive, rapid, and reasonably sensitive. Standardized agglutination tests are used for the identification of blood group (red blood cell) antigens as well as many pathogens and pathogen products.

Direct Agglutination

Direct agglutination results when soluble antibody causes clumping due to interaction with an antigen that is an integral part of the surface of a cell or other insoluble particle. Direct agglutination procedures are used for the classification of antigens found on the surface of red blood cells (erythrocytes). Agglutination of red blood cells is called *hemagglutination* and is the basis for human blood typing.

Red blood cells exhibit a variety of cell surface antigens, and individuals vary considerably with respect to the antigens present on their red blood cells. The major human red blood cell surface antigens are called *A, B,* and *D*. D is also called *Rh*. A and B antigens and antibodies are the basis for the ABO blood-typing assay. Red blood cells carrying the antigen visibly clump when mixed with specific antisera (**Figure 32.14**). The antisera are obtained from human donors who have been immunized to A or B antigens by natural or artificial means.

For the A, B, and O blood types, individuals express codominant A and B alleles as one of the following antigen phenotypes: A, B, AB (one allele expressing the A antigen and one expressing the B antigen), or O (the absence of both A or B alleles). In addition, individuals make antibodies to most nonself blood group antigens. Type A individuals make antibodies to group B antigens, while type B individuals make antibodies to group A antigens. Type AB individuals have neither A nor B antibodies, but type O individuals have antibodies to both A and B antigens (Figure 32.14). These antibodies against A and B antigens are natural antibodies; they appear to be produced by most individuals in response to ubiquitous related antigen sources such as enteric bacteria and are not related to exposure to red blood cells from other individuals.

Blood typing using the A, B, and D antisera is done before blood transfusion to prevent red blood cell destruction that would occur if antibodies in the recipient's blood reacted with the red blood cells in the transfused blood, or vice versa. Antibody-coated red blood cells would likely undergo hemolysis (lysis of red blood cells) through the activity of complement (Section 29.11), resulting in severe anemia.

Passive Agglutination

Passive agglutination is the agglutination of soluble antigens or antibodies that have been adsorbed or chemically coupled to cells or insoluble particles such as latex beads or charcoal particles. The insolubilized antigen or antibody can then be detected by agglutination reactions. The cell or particle serves as an inert carrier. Passive agglutination reactions can be up to five times more sensitive than direct agglutination tests (Table 32.8), significantly increasing sensitivity.

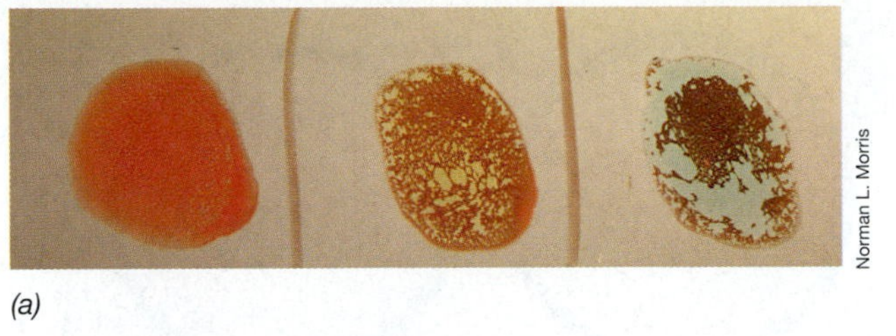

(a)

Blood type	Percentage of U.S. population	Serum: Anti A	Serum: Anti B
Type O	46	No aggl.	No aggl.
Type A	39	Aggl.	No aggl.
Type B	11	No aggl.	Aggl.
Type AB	4	Aggl.	Aggl.

(b)

Figure 32.14 Direct agglutination of human red blood cells for ABO blood typing. *(a)* The reaction on the left shows no agglutination. The reaction in the center shows the diffuse agglutination pattern that indicates a positive reaction for the B blood group. The reaction on the right shows the strong agglutination pattern with large, clumped agglutinates typical for the A blood group. *(b)* Table of expected blood grouping results for the U.S. population.

The agglutination of antigen-coated or antibody-coated latex beads by complementary antibody or antigen from a patient is a typical method of rapid diagnosis. Small (0.8 μm) latex beads coated with a specific antigen are mixed with patient serum on a microscope slide and incubated for a short period. If patient antibody binds the antigen on the bead surface, the milky white latex suspension will become visibly clumped, indicating a positive agglutination reaction. Latex agglutination is also used to detect bacterial surface antigens by mixing a small amount of a bacterial colony with antibody-coated latex beads. For example, a commercially available suspension of latex beads coated with antibodies to protein A and clumping factor, two proteins found exclusively on the surface of *Staphylococcus aureus* cells, is specific for identification of clinical isolates of *S. aureus*. Unlike traditional growth-dependent tests for *S. aureus*, the latex bead assay takes only 30 seconds (**Figure 32.15**) and can be used directly on a clinical sample, such as of a purulent infection suspected to be caused by *S. aureus*. Latex bead agglutination assays have also been developed to identify other common pathogens, such as *Streptococcus pyogenes, Neisseria gonorrhoeae, N. meningitidis, Haemophilus influenzae, Escherichia coli* O157: H7, and the fungi *Cryptococcus neoformans* and *Candida albicans*.

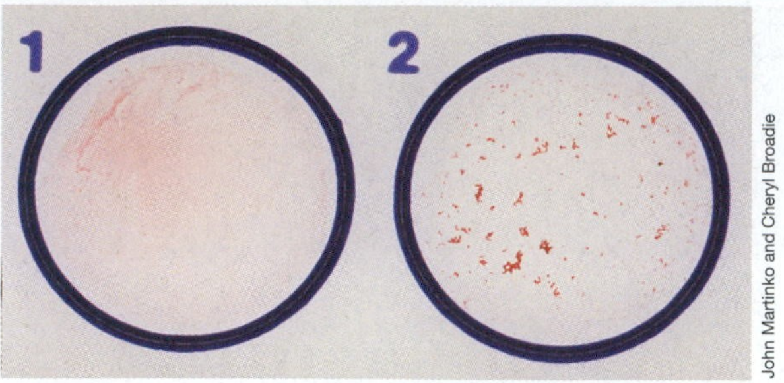

Figure 32.15 Latex bead agglutination test for *Staphylococcus aureus*. Panel 1 shows a negative control. Note the uniform pink color of the suspended latex beads coated with antibodies to protein A and clumping factor, two antigens found exclusively on the surface of *S. aureus* cells. Panel 2 shows the same suspension after a loopful of material from a bacterial colony was mixed into the suspension. The bright red clumps indicate a positive agglutination reaction took place and indicates that the colony is *S. aureus*.

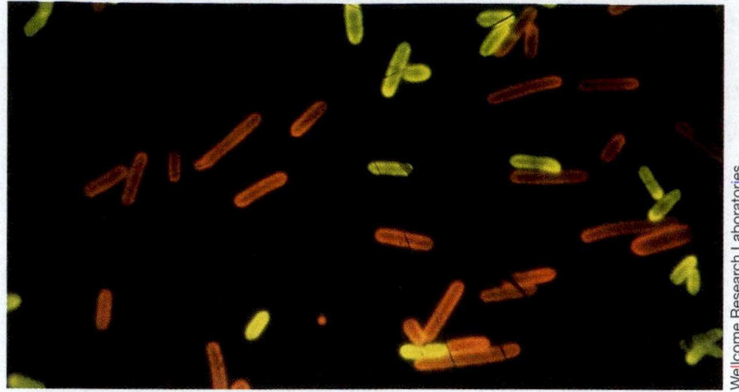

Figure 32.16 Fluorescent antibody reactions. Cells of *Clostridium septicum* were stained with antibody conjugated with fluorescein isothiocyanate, which fluoresces yellow-green. Cells of *C. chauvoei* were stained with antibody conjugated with rhodamine B, which fluoresces red-orange.

Latex agglutination tests are also used to detect serum antibodies directed against the body's own Ig, DNA, and other macromolecules. These self-reacting antibodies are associated with several autoimmune diseases and their detection is an important diagnostic finding when coupled with other clinical information (∞ Section 31.7).

Passive agglutination assays require no expensive equipment or particular expertise and can be highly specific and very sensitive. In addition, the cost-effective nature of the assays makes them suitable for large-scale screening programs. These tests are therefore widely used in clinical and research applications.

32.8 MiniReview

Direct agglutination tests are used for determination of blood types. Passive agglutination tests are available for identification of a variety of pathogens and pathogen-related products. Agglutination tests are rapid, relatively sensitive, highly specific, simple to perform, and inexpensive.

- Distinguish between direct and passive agglutination. Which tests are more sensitive?
- What advantages do agglutination tests have over other immunoassays? What disadvantages?

32.9 Fluorescent Antibodies

Antibodies can be chemically modified with fluorescent dyes. These modified antibodies can then be used to detect antigens on intact cells. **Fluorescent antibodies** are widely used for diagnostic and research applications.

Fluorescent Methods

Antibodies can be covalently modified by fluorescent dyes such as rhodamine B, which fluoresces red, or fluorescein isothiocyanate, which fluoresces yellow-green. The attached dyes do not alter the specificity of the antibody but make it possible to detect the antibody by use of a fluorescence microscope once it has bound to cell or tissue surface antigens (**Figure 32.16**). Cell-bound fluorescent antibodies emit a bright fluorescent color when excited with light of particular wavelengths. The emitted light is red-orange or yellow-green, depending on the dye used. Fluorescent antibodies are used in diagnostic microbiology because they permit the identification of a microorganism directly in a patient specimen (*in situ*), bypassing the need for the isolation and culturing of the organism (see below). The fluorescent antibody technique is also very useful in microbial ecology as a method for directly viewing and identifying microbial cells without prior isolation and culture (∞ Figure 22.8).

Distinct direct and indirect fluorescent antibody-staining methods are used. In the *direct method,* the antibody targeted against the surface antigen is itself covalently linked to the fluorescent dye. In the *indirect method,* the presence of a nonfluorescent antibody on the surface of a cell is detected by the use of a fluorescent antibody directed against the nonfluorescent antibody (**Figure 32.17**).

Applications

In a typical test using fluorescent antibodies, a specimen containing a suspected pathogen is allowed to react with a specific fluorescent antibody and observed with a fluorescent microscope. If the pathogen contains surface antigens reactive with the antibody, the pathogen cells fluoresce (**Figure 32.18**).

Fluorescent antibodies can be applied directly to infected host tissues, permitting diagnosis long before primary isolation techniques yield a suspected pathogen. For example, for diagnosing legionellosis, a form of infectious pneumonia, a positive identification can be made by staining biopsied lung tissue directly with fluorescent antibodies specific for cell wall antigens of *Legionella pneumophila* (∞ Figure 32.18*a*), the causative agent of the disease. Likewise, a direct fluorescent

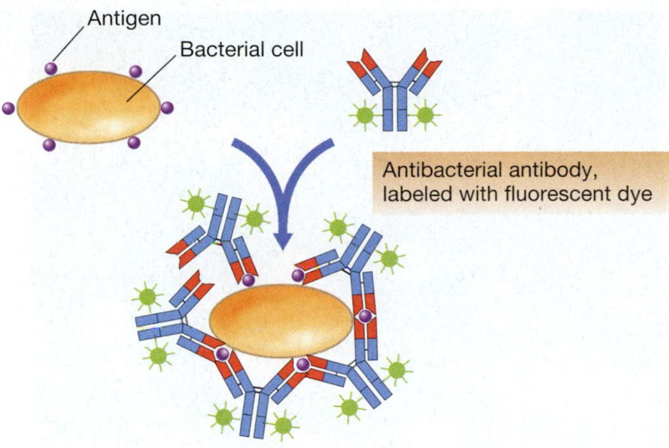

(a) **Direct immunofluorescence**

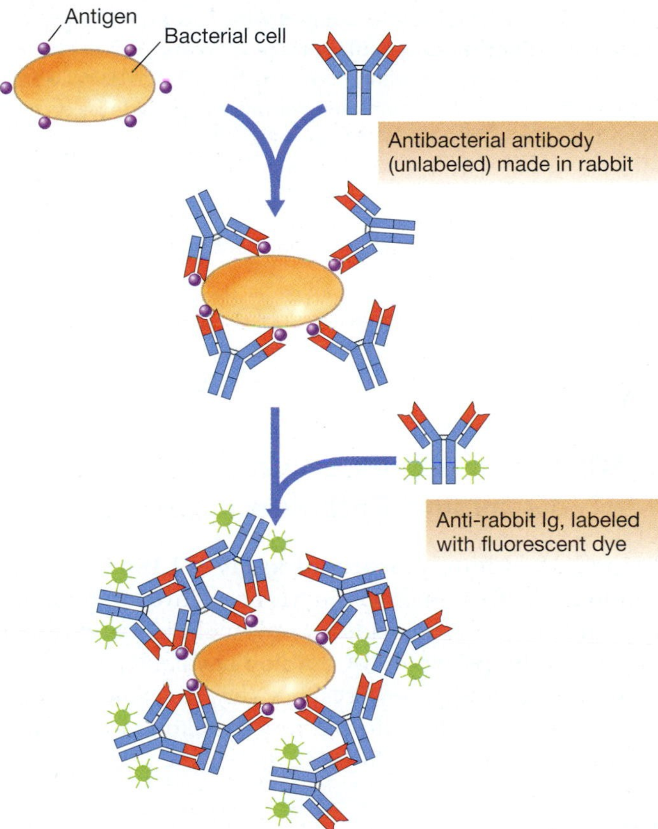

(b) **Indirect immunofluorescence**

Figure 32.17 Fluorescent antibody methods for detection of microbial surface antigens.

antibody test can be used against the capsule of *Bacillus anthracis* to confirm a diagnosis for anthrax (∞ Figure 28.14*b*). Direct fluorescent antibody tests are also used to help diagnose viral infections (Figure 32.18*b*). The common respiratory pathogens influenza A and B, parainfluenza, respiratory syncytial virus (RSV), and adenovirus can be identified from respiratory tract specimens by direct fluorescent antibody methods. Fluorescent antibody methods can also be used to identify viruses grown in tissue or organ culture.

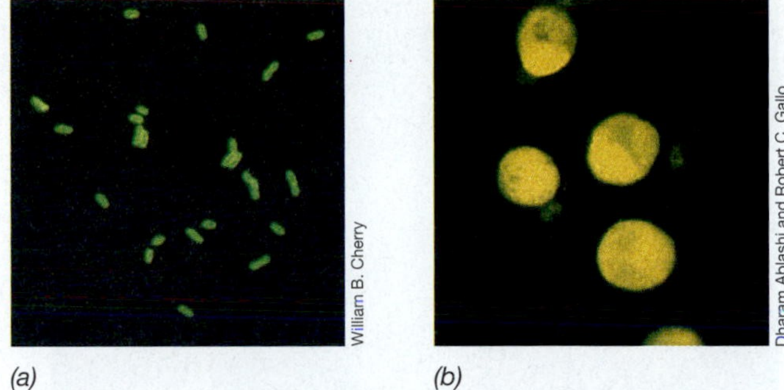

Figure 32.18 Fluorescent antibodies in clinical microbiology. *(a)* Immunofluorescent stained cells of *Legionella pneumophila*, the cause of legionellosis. The specimen was taken from biopsied lung tissue. The individual organisms are 2–5 μm in length. *(b)* Detection of virus-infected cells by immunofluorescence. Human B lymphotrophic virus (HBLV)-infected spleen cells were incubated with serum containing antibodies to HBLV. Cells were then treated with fluorescein isothiocyanate-conjugated anti-human IgG antibodies. HBLV-infected cells fluoresce bright yellow. Unstained cells in the background did not react with the serum. Individual cells are about 10 μm in diameter.

Fluorescent antibody assays are also used to aid in the diagnosis of noninfectious diseases. Fluorescent antibodies that interact with a particular antigen can be used to identify cell types expressing that antigen. For example, fluorescent antibodies directed against tumor-specific antigens found on malignant cells may be used to identify malignant cells and monitor the course of the disease as well as therapy (**Figure 32.19**).

Fluorescent antibodies can also be used to separate mixtures of cells into relatively pure populations or to define the numbers of individual cell types in complex mixtures such as blood. For example, fluorescent-labeled monoclonal antibodies directed against the CD4 and CD8 surface antigens of T lymphocytes are routinely used to identify and enumerate these cells in the blood (**Figure 32.20**). This assay is extremely important for patients with HIV-AIDS. The CD4 T cell number and CD4/CD8 ratio change during the progression of AIDS and is a diagnostic indicator of disease progression. Thus, by defining the CD4 numbers, the clinician can follow the progress of the disease and monitor therapy (∞ Section 34.15).

Cells labeled with fluorescent antibodies can be visualized, counted, and separated with an instrument called a fluorescence cytometer, or *fluorescence activated cell sorter (FACS)*. The FACS uses a laser beam to activate fluorescent antibody bound to cells, placing a charge on the labeled cells. An electric field is then applied to the cell mixture. Fluorescing and nonfluorescing cells are then deflected to opposite poles of the electric field, where each cell population is counted and deposited in a tube. The use of several antibodies, each labeled with a different fluorescent dye, can be used to simultaneously identify several cell markers. A typical application compares CD3 and CD4 surface proteins on T cells in healthy people and those with AIDS in **Figure 32.21**.

FACS analysis is also useful for research applications. For example, immunologists routinely use FACS methods to

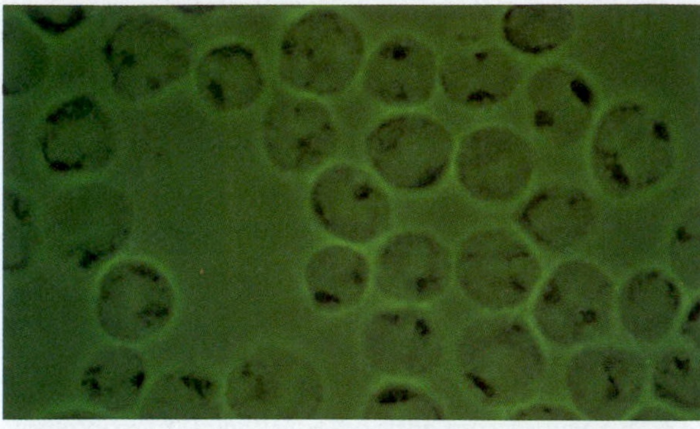

(a)

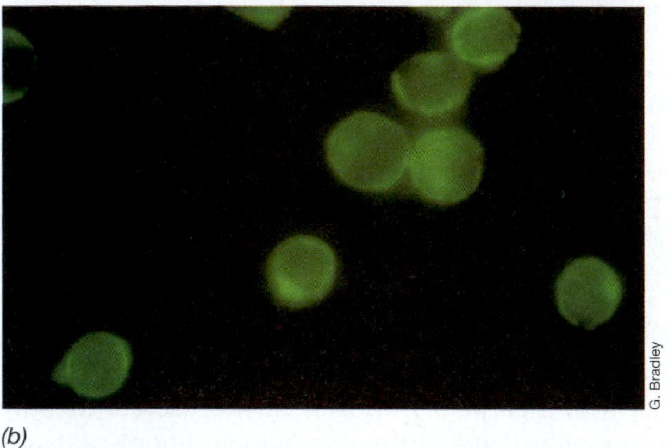

(b)

Figure 32.19 Fluorescent antibodies in noninfectious disease diagnostics. Human leukemic cells exhibit variable resistance to chemotherapeutic anticancer drugs. *(a)* Here leukemia cells appear indistinguishable. *(b)* When the cells in *(a)* are treated with a fluorescent monoclonal antibody that binds specifically to a protein found only on the surface of drug-resistant cells, the latter fluoresce, whereas drug-sensitive cells do not. Individual cells are about 10–12 μm in diameter.

separate complex mixtures of immune cells. They can then study the properties of the highly enriched cell populations.

Under appropriate conditions, fluorescent antibodies yield rapid, highly specific information about a variety of clinical conditions. However, antibodies to surface antigens may cross-react between and among some bacterial species, some of which may be members of the normal flora. This is a major problem among enteric bacteria, for example, where cell wall lipopolysaccharide antigens are very similar. The clinical microbiologist must therefore perform controls using nonspecific sera and confirm positive immunofluorescent findings with other immunological or microbiological tests.

32.9 MiniReview

Fluorescent antibodies are used for quick, accurate identification of pathogens and other antigenic substances in tissue samples, blood, and other complex mixtures. Fluorescent antibody-based methods are used for identification, quantitative enumeration, and sorting of a variety of prokaryotic and eukaryotic cell types.

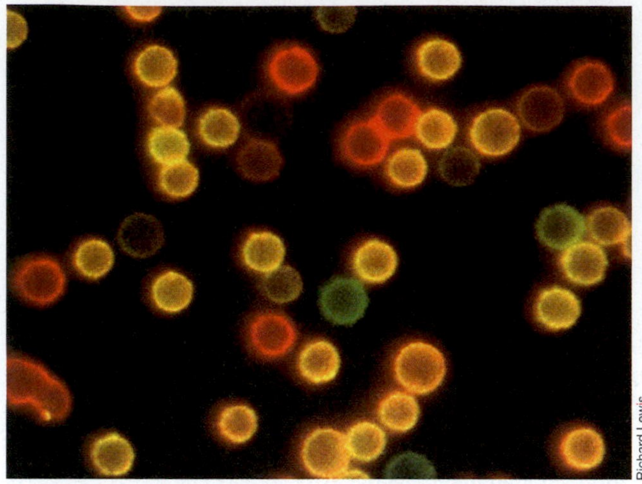

Figure 32.20 T lymphocytes stained with fluorescent-tagged monoclonal antibodies to specific surface markers. Yellow-green cells are T_C (CD8) cells; red-orange cells are T_H (CD4) cells. Individual cells are about 10–12 μm in diameter. Reprinted with permission from *Science* 239: Cover (Feb. 12, 1988), © AAAS.

- Explain and compare direct and indirect fluorescent antibody assays.
- How are fluorescent antibodies used to identify specific cells in complex mixtures such as blood?

32.10 Enzyme-Linked Immunosorbent Assay and Radioimmunoassay

Enzyme-linked immunosorbent assay (ELISA) and **radioimmunoassay (RIA)** methods are very sensitive immunological assays and are therefore widely used in clinical and research applications. ELISA and RIA employ covalently bonded enzymes and radioisotopes, respectively, to label antibody molecules, allowing detection of very small quantities of antigen–antibody complexes (Table 32.8).

ELISA

In ELISA, an enzyme is covalently attached to an antibody molecule, creating an immunological tool with high specificity and high sensitivity. The enzyme's catalytic properties and the antibody's specificity are unaltered. Typical bound enzymes include peroxidase, alkaline phosphatase, and β-galactosidase, all of which catalyze reactions that develop colored products that can be detected in very low amounts.

Two different ELISA methodologies are used, one for detecting antigen (direct ELISA) and the other for detecting antibodies (indirect ELISA). For detecting antigens such as virus particles from a blood or fecal sample, the *direct ELISA* method is used. The specimen is added to the wells of a microtiter plate previously coated with antibodies specific for the antigen to be detected. If present in the sample, the virus particle will be bound by the antibodies. After washing away

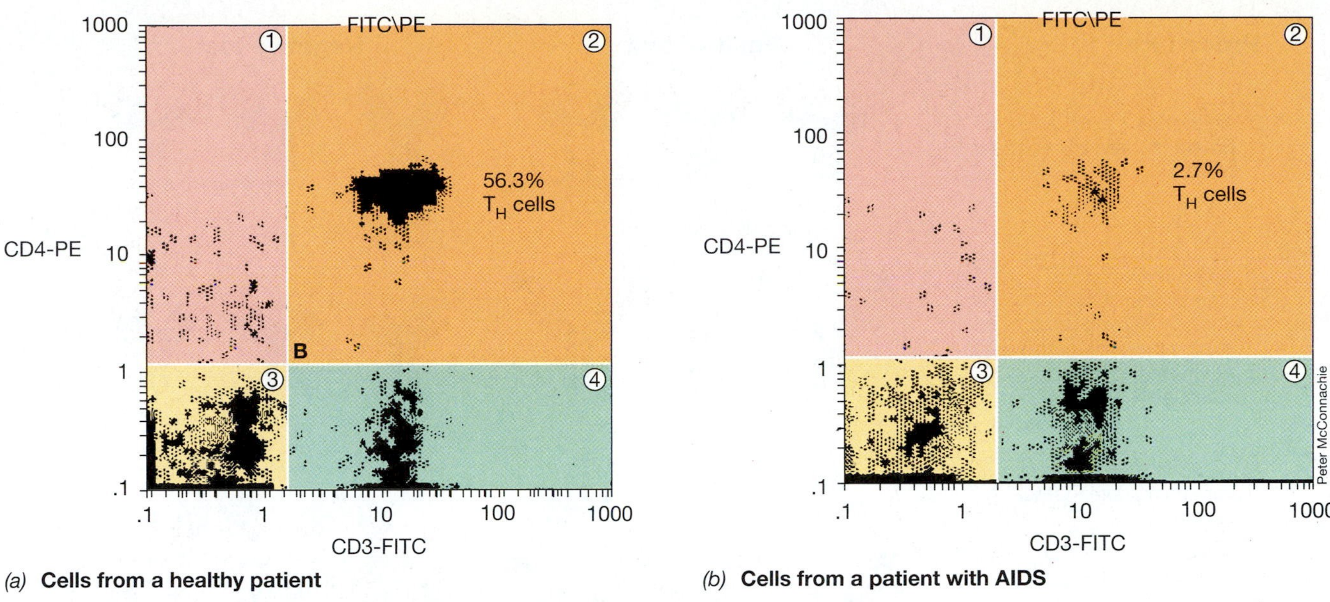

Figure 32.21 CD3 and CD4 cell enumeration. Peripheral blood cells from a healthy human *(a)* and from a human with acquired immunodeficiency syndrome (AIDS) *(b)* were assayed using a flow cytometer. Each dot represents a single cell. The cells were simultaneously labeled with monoclonal antibody to CD4 conjugated to phycoerythrin (PE) and with monoclonal antibody to CD3 conjugated to fluorescein isothiocyanate (FITC). CD3 is found on all T cells. CD4 is found only on T-helper (T_H) cells. Quadrant 3 shows cells that were stained with neither antibody. Quadrant 1 shows cells stained with only anti-CD4. Quadrant 4 shows cells stained with only anti-CD3. Quadrant 2 shows cells stained with both anti-CD3 and anti-CD4. For the healthy patient in *(a)*, 56.3% of the T cells were T_H cells, as shown by the dense staining pattern in quadrant 2. For the patient with AIDS *(b)*, only 2.7% of the total T cells were T_H cells, as indicated by the very light staining pattern in quadrant 2. Original data from Peter McConnachie, used with permission.

unbound material, a second antibody containing a conjugated enzyme is added. The second antibody is also specific for the antigen, and it binds to other exposed antigenic determinants. Following a wash, the enzyme activity of the bound material in each microtiter well is determined by adding the substrate for the enzyme. The enzyme catalyzes the conversion of the substrate to a colored product, which is detected with a spectrophotometer. The color produced is proportional to the amount of antigen present (**Figure 32.22**).

To detect antibodies in human serum, an indirect ELISA is employed. An indirect ELISA is widely used to detect antibodies to human immunodeficiency virus (HIV) in human body fluids. This test illustrates the principle features of all indirect ELISA tests and is discussed below (**Figure 32.23**).

Modified rapid ELISA procedures use reagents adsorbed to a fixed support material such as paper strips, nitrocellulose or plastic membranes, or plastic "dipsticks." These tests cause a color change on the strip or stick in a very short time. These rapid "point of care" tests are diagnostic aids for infectious diseases such as HIV–AIDS (⚭ Section 34.15) or "strep throat" (pharyngeal infection with *Streptococcus pyogenes*, ⚭ Section 34.2).

In addition, applications for rapid tests include pregnancy testing and drug testing (**Figure 32.24**). In most of these tests, a body fluid, generally urine or blood, is applied to the reagent-support matrix. After reacting with the body fluids, the support is washed and developed with a second reagent that identifies antigen (direct test) or antibody (indirect test such as that for HIV) bound to the matrix. These tests are especially valuable where a small number of samples are to be analyzed at the site where the urine or blood is collected (for instance, away from a clinical laboratory) and can be performed by individuals lacking certified clinical laboratory skills. Results can be reported at the site, avoiding the need for delays in patient care or for follow-up visits to obtain test results. The drawback to these tests, however, is that they tend to be less specific than more elaborate tests. As a result, these point-of-care tests often need to be confirmed by standard laboratory tests.

The HIV ELISA

The virus that causes HIV–AIDS, the human immunodeficiency virus (HIV) (⚭ Section 34.15), is transmitted by body fluids including blood. Sensitive, specific, rapid, and cost-effective screening tools are needed to routinely test blood samples from individuals exposed to HIV and to ensure that HIV is not being inadvertently transmitted during blood transfusions or through the transfer of blood products.

Initial infection with HIV leads to the production of antibodies to several HIV antigens, in particular, those of the HIV envelope. These antibodies can be detected by the HIV ELISA test, an indirect ELISA designed to measure antibodies to HIV present in serum (Figure 32.23).

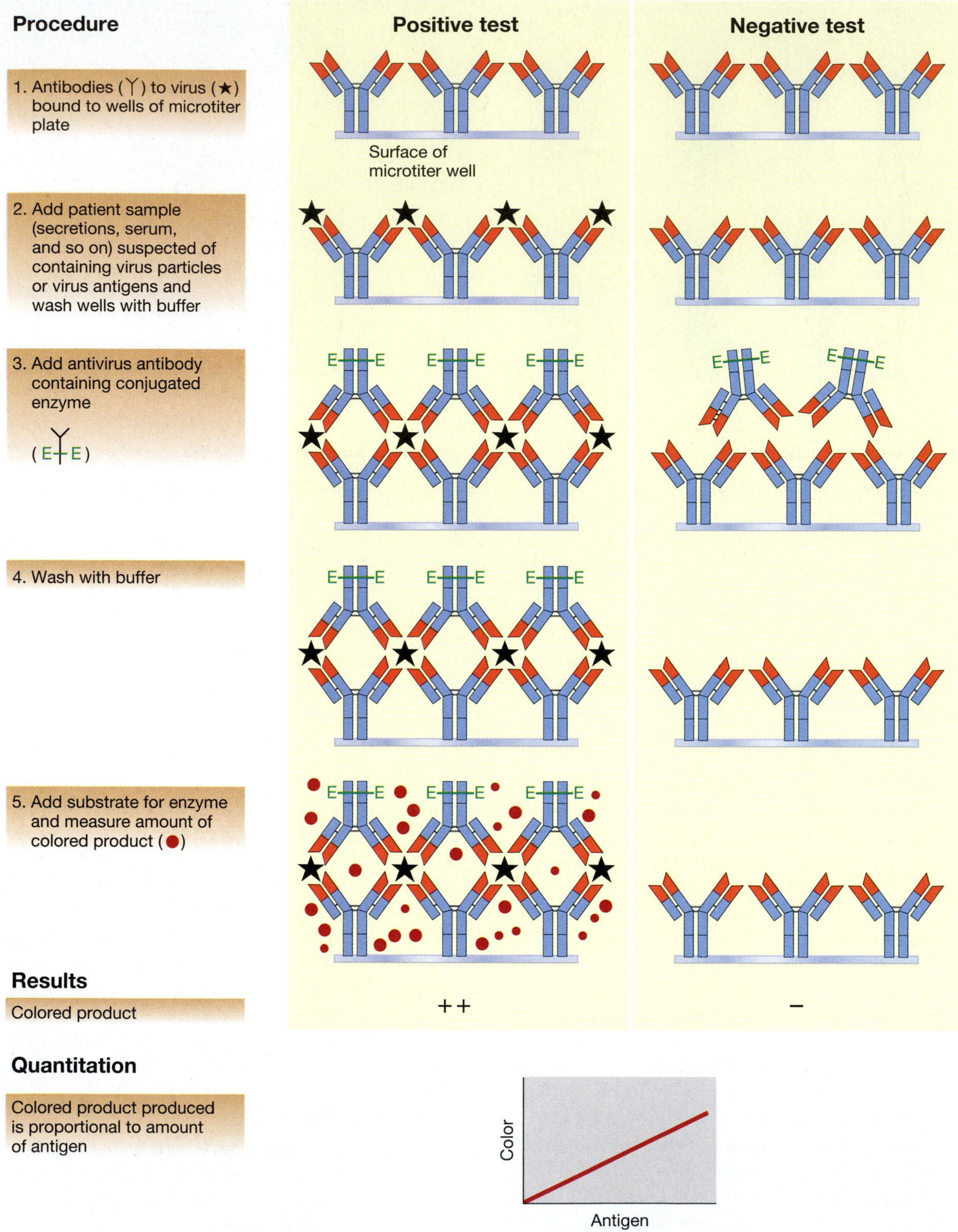
Procedure
Positive test
Negative test
1. Antibodies (Y) to virus (★) bound to wells of microtiter plate
Surface of microtiter well
2. Add patient sample (secretions, serum, and so on) suspected of containing virus particles or virus antigens and wash wells with buffer
3. Add antivirus antibody containing conjugated enzyme
(E-Y-E)
4. Wash with buffer
5. Add substrate for enzyme and measure amount of colored product (●)
Results
Colored product
++
–
Quantitation
Colored product produced is proportional to amount of antigen
Color
Antigen

Figure 32.22 The direct ELISA test. A direct ELISA test can be used to detect antigenic pathogen components or antigenic metabolites in blood, urine, and other body fluids.

To carry out an HIV ELISA test, microtiter plates are first coated with a preparation of disrupted HIV particles; about 200 ng of disrupted HIV is placed in each well. A diluted patient serum sample is then added, and the mixture is incubated to allow HIV-specific antibodies to bind to HIV antigens. To detect the presence of antigen–antibody complexes, a second antibody is then added. This second antibody is an enzyme-conjugated anti-human IgG preparation. The anti-human IgG antibodies bind to HIV-specific IgG antibodies now bound to the HIV antigen preparation. Next, the

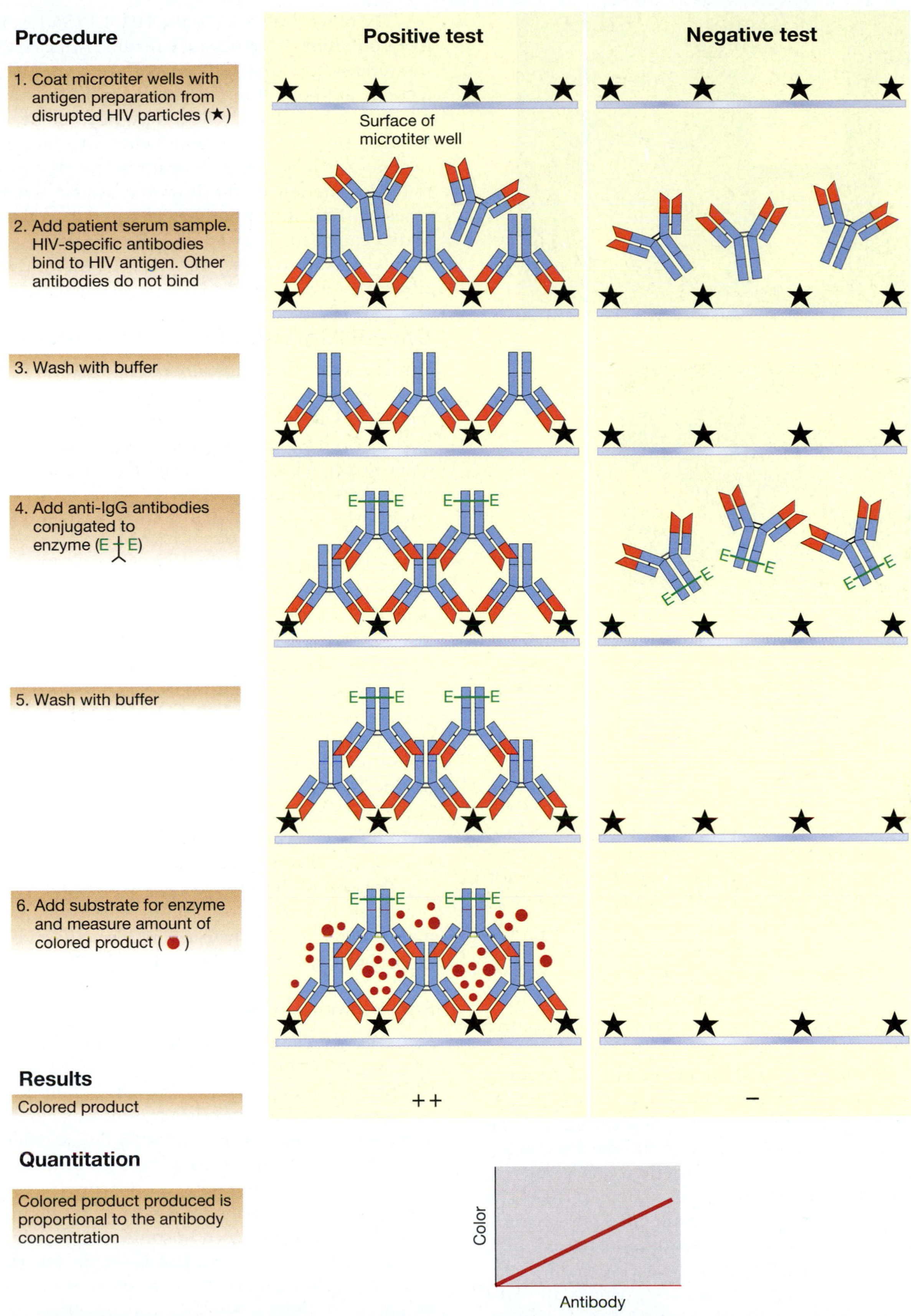
Procedure
Positive test
Negative test
1. Coat microtiter wells with antigen preparation from disrupted HIV particles (★)
Surface of microtiter well
2. Add patient serum sample. HIV-specific antibodies bind to HIV antigen. Other antibodies do not bind
3. Wash with buffer
4. Add anti-IgG antibodies conjugated to enzyme (E┼E)
5. Wash with buffer
6. Add substrate for enzyme and measure amount of colored product (●)
Results
Colored product
++
−
Quantitation
Colored product produced is proportional to the antibody concentration
Color
Antibody

Figure 32.23 Indirect ELISA test. An indirect ELISA test is used in many immunoassays including for detecting antibodies to human immunodeficiency virus (HIV), the causal agent of acquired immunodeficiency syndrome (AIDS).

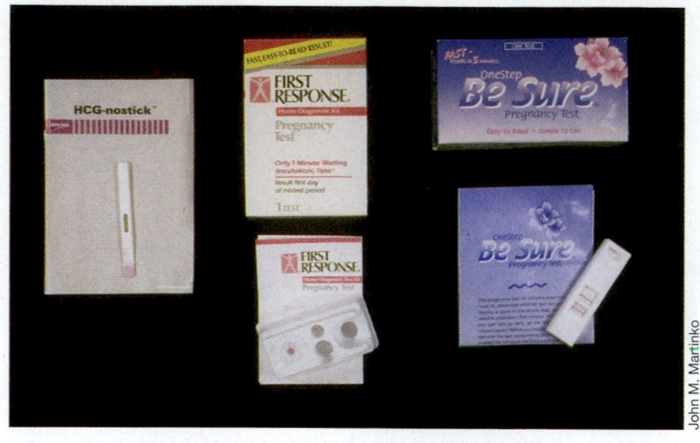

(a)

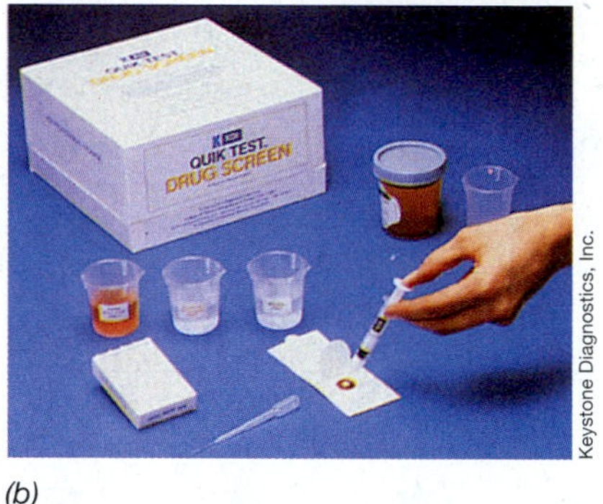

(b)

Figure 32.24 Rapid ELISA-based assay kits. Pregnancy test kits *(a)* and the drug-testing kit *(b)* are available for point-of-care testing for these applications. Other kits are used for diagnosis of infectious diseases including streptococcal sore throat and HIV-AIDS.

substrate for the conjugated enzyme is added and the enzyme activity is assayed. The color obtained in the enzyme assay is proportional to the amount of anti-human IgG antibody bound. The binding of the second antibody is an indication that antibodies from the patient's serum recognized the HIV antigens and that the patient has antibodies to HIV, therefore has been infected with HIV. Control sera (known to be HIV-positive or HIV-negative) are assayed in parallel with patient samples to establish specificity (positive control) and measure the extent of background absorbance in the assay (negative control).

The HIV ELISA test is a rapid, highly sensitive, extremely specific method for detecting exposure to HIV. Since ELISAs in general are highly adaptable to mass screening and automation, the HIV ELISA test is used as a standard screening method for blood. However, this test method can give erroneous results under certain circumstances. For example, the test occasionally gives false-positive results. Because a number of factors can contribute to these results, none of which are related to exposure to HIV, all positive HIV ELISA tests must be confirmed by an independent test, usually the HIV Western blot (immunoblot) test (Section 32.11). A positive HIV Western blot test after a positive HIV ELISA test is considered proof of HIV infection.

The other drawback to the HIV ELISA test is the possibility of obtaining false-negative results. After HIV exposure, the immune system may take 6 weeks to a year to produce a detectable antibody titer. Therefore, individuals who have been recently infected with HIV may not yet be producing detectable amounts of antibody when they are tested. Another reason for a false-negative result in the HIV ELISA test is the total destruction of the immune system seen in advanced cases of AIDS; if no immune cells are left in the body, no antibodies can be made and the ELISA test is not useful. However, at this stage of disease, an AIDS diagnosis can be based on clinical information.

Other ELISA Tests of Clinical Importance

Hundreds of other clinically useful ELISAs have been developed. Some of these are direct ELISAs for detecting antigens, including bacterial toxins such as cholera toxin, enteropathogenic *Escherichia coli* toxin, and *Staphylococcus aureus* enterotoxin. Viruses currently detected using direct ELISA techniques include rotavirus, hepatitis viruses, rubella virus, bunyavirus, measles virus, mumps virus, and parainfluenza virus.

Indirect ELISAs have been developed for detecting antibodies to a variety of clinically important bacteria. ELISAs for detecting serum antibodies to *Salmonella* (gastrointestinal diseases), *Yersinia* (plague), *Brucella* (brucellosis), a variety of rickettsias (Rocky Mountain spotted fever, typhus, Q fever), *Vibrio cholerae* (cholera), *Mycobacterium tuberculosis* (tuberculosis), *Mycobacterium leprae* (leprosy), *Legionella pneumophila* (legionellosis), *Borrelia burgdorferi* (Lyme disease), and *Treponema pallidum* (syphilis) have been developed, among others. ELISAs have also been developed for detecting antibodies to *Candida* (yeast) and a variety of eukaryotic pathogens, including those causing amebiasis, Chagas' disease, schistosomiasis, toxoplasmosis, and malaria.

The speed, low cost, lack of hazardous waste, long shelf life, high specificity, and high sensitivity of ELISA tests make them particularly useful immunodiagnostic tools. www.microbiologyplace.com **Online Tutorial 32.2: The ELISA Test**

Radioimmunoassay

Radioimmunoassay (RIA) employs radioisotopes as antibody or antigen conjugates instead of the the enzymes used in ELISA. The isotope iodine-125 (^{125}I) is commonly used as the conjugate because antibodies or antigens can be readily iodinated without disrupting their immune specificity. RIA is used clinically to measure rare serum proteins such as human growth hormone, glucagon, vasopressin, testosterone, and insulin present in humans in extremely small amounts (**Figure 32.25**). RIAs are also used in some tests for illegal drugs.

In most cases, a direct RIA is employed. The direct assay is a two-step procedure. First, a series of microtiter wells are prepared containing known concentrations of pure antigen (such as a hormone). Radiolabeled antigen-specific antibodies are added to the wells, and the radioactivity bound in each of these standard wells is measured. These data establish a stan-

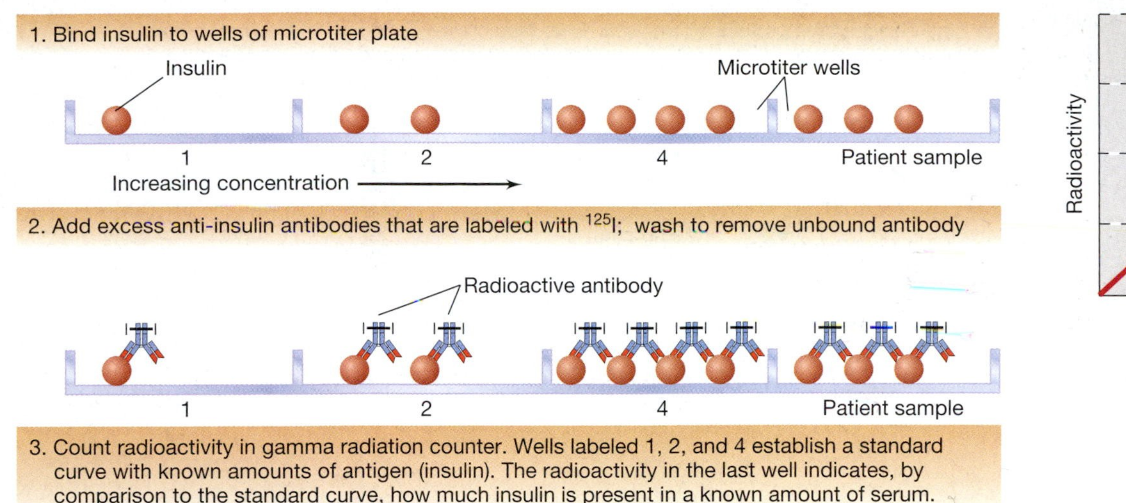

1. Bind insulin to wells of microtiter plate

Microtiter wells

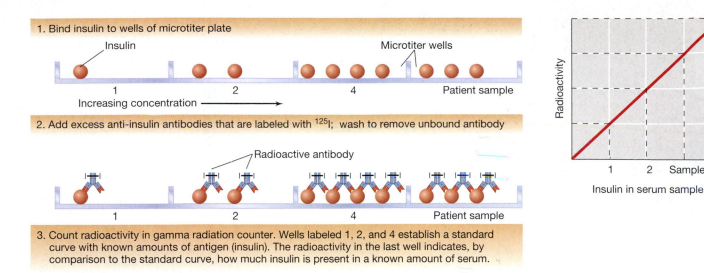

4
Insulin in serum sample

Figure 32.25 **Radioimmunoassay (RIA).** Using RIA to detect insulin levels in human serum. Following establishment of a standard curve, the insulin concentration in a serum sample can be estimated.

dard curve for antigen concentrations. Next, a sample from a patient is bound to another well, and radioactive antibodies are added and measured, as before. The amount of radioactivity bound by the patient sample is then compared to a standard plot generated from binding data obtained using the pure antigen, and the concentration of antigen in the patient sample is interpolated from the standard plot (Figure 32.25).

RIA has the same sensitivity range as ELISA (Table 32.8) and can also be performed very rapidly. However, the instruments used to detect radioactivity are quite specialized and expensive. RIA generates radioactive waste, and the radioactive decay time (half-life) of the radioisotopes used for detection may limit the useful life of a test kit. As a result, RIA is often used only when ELISA is not sufficiently specific or sensitive. For example, RIA is often more useful than ELISA for detecting levels of certain proteins and drug metabolites in serum because serum components inhibit some ELISA enzyme–substrate or antigen–antibody interactions. Thus, for certain applications, each test system has clear advantages.

32.10 MiniReview

ELISA and RIA are the most sensitive immunologic assays. Both techniques involve linking a detection system, either an enzyme or a radioactive molecule, to an antibody or antigen, significantly enhancing sensitivity. ELISA and RIA are used for clinical and research work; tests have been designed to detect either antibody (indirect tests) or antigen (direct tests) in many applications.

- Why are ELISA and RIA techniques more sensitive than immunoassays such as precipitation and agglutination?
- Compare ELISA and RIA with respect to their relative uses, advantages, and disadvantages. Are they more specific than other immunoassays? Explain.

32.11 Immunoblot Procedures

Antibodies can also be used in the **immunoblot** method to identify individual specific proteins associated with specific pathogens, even in complex mixtures such as cell lysates or blood. The immunoblot method employs three techniques: (1) the separation of proteins on polyacrylamide gels, (2) the transfer (blotting) of proteins from gels to a nitrocellulose or nylon membrane, and (3) identification of the proteins by specific antibodies. Protein blotting and the subsequent identification of the proteins by specific antibodies is also called the *Western blot* technique.

In the first step of an immunoblot, described in **Figure 32.26**, a protein mixture is subjected to electrophoresis on a polyacrylamide gel. This separates the proteins into several distinct bands, each of which represents a single protein of specific molecular weight. An electrophoretic process is then used to elute the proteins from the gel and transfer them to a membrane. Antibodies specific for the pathogen components are then added to the membrane blot. Following an incubation period to allow the antibodies to bind, a radioactive marker that binds antigen–antibody complexes is added. A common radioactive marker is *Staphylococcus* protein A labeled with ^{125}I; protein A has a strong affinity for antibody and binds firmly. Once the radioactive marker has bound, its location on the blot can be detected by exposing the membrane to X-ray film; the gamma rays emitted by the ^{125}I expose the film only at the bands that have formed the labeled antigen–antibody complexes (Figure 32.26).

4

Increasing concentration
2. Add excess anti-insulin antibodies that are labeled with ¹²⁵I; wash to remove unbound antibody
Radioactive antibody

4

3. Count radioactivity in gamma radiation counter. Wells labeled 1, 2, and 4 establish a standard curve with known amounts of antigen (insulin). The radioactivity in the last well indicates, by comparison to the standard curve, how much insulin is present in a known amount of serum.

Immunoblots are sometimes done using ELISA technology for detection of bound antigen–antibody complexes rather than radioisotopes. Following treatment of the blotted proteins with specific antibody, the membrane is washed and then treated with a second antibody, which binds to the first. Covalently attached to this second antibody is an enzyme. The original antigen–antibody complexes are visualized when the

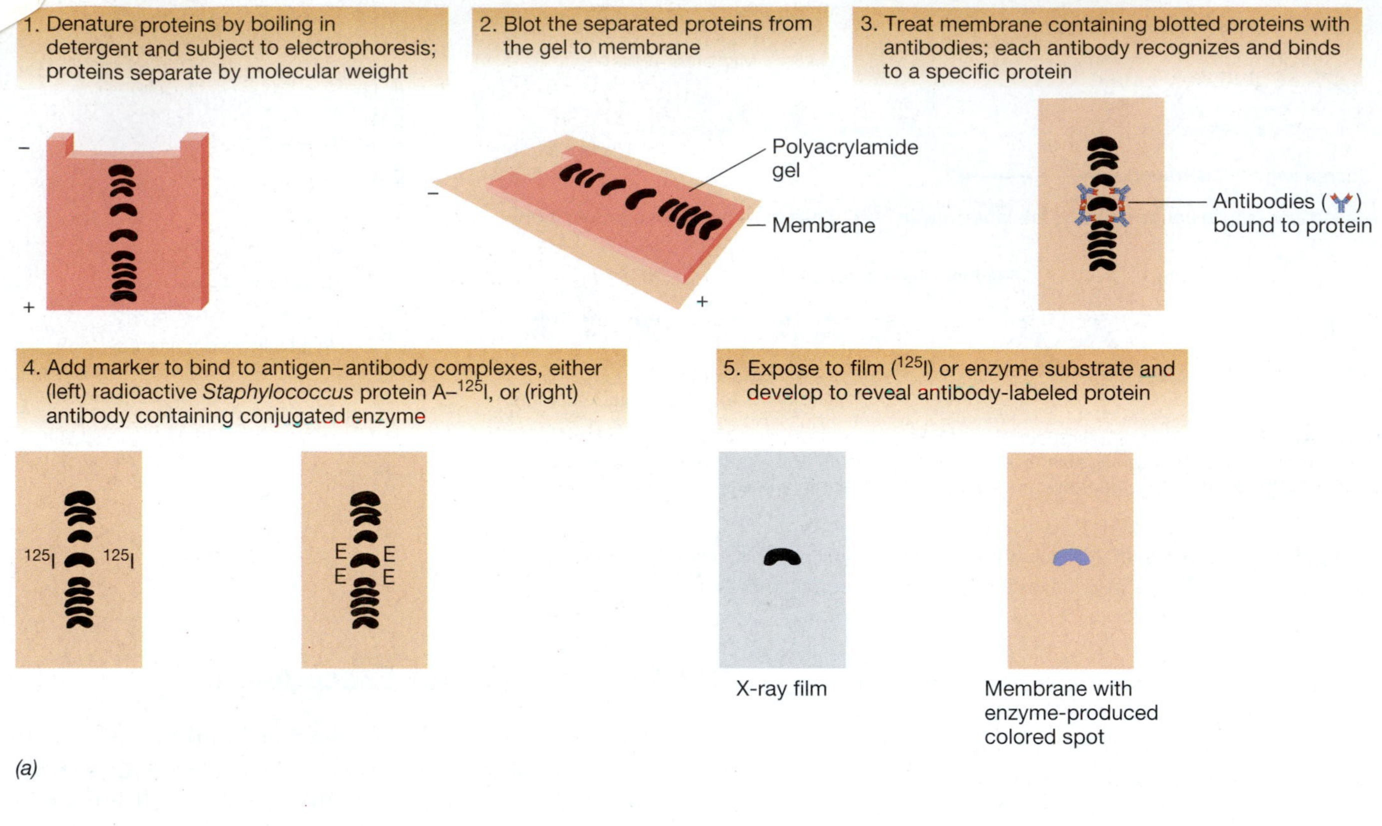
1. Denature proteins by boiling in detergent and subject to electrophoresis; proteins separate by molecular weight
2. Blot the separated proteins from the gel to membrane
3. Treat membrane containing blotted proteins with antibodies; each antibody recognizes and binds to a specific protein
Polyacrylamide gel
Membrane
Antibodies ( ) bound to protein
4. Add marker to bind to antigen–antibody complexes, either (left) radioactive *Staphylococcus* protein A–125I, or (right) antibody containing conjugated enzyme
5. Expose to film (125I) or enzyme substrate and develop to reveal antibody-labeled protein
125I
E
X-ray film
Membrane with enzyme-produced colored spot
(a)

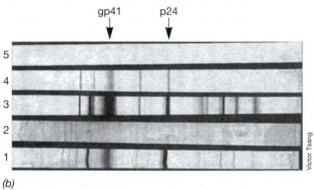

gp41 p24

5
4
3
2
1

Victor Tsang

(b)

Figure 32.26 The Western blot (immunoblot) and its use in the diagnosis of human immunodeficiency virus (HIV) infection. *(a)* Protocol for an immunoblot. *(b)* Developed HIV immunoblot. The molecules p24 (capsid protein) and gp41 (envelope glycoprotein) are diagnostic for HIV. Lane 1, positive control serum (from known AIDS patients); lane 2, negative control serum (from healthy volunteer); lane 3, strong positive from patient sample; lane 4, weak positive from patient sample; lane 5, reagent blank to check for background binding.

enzyme is exposed to substrate: The product of the enzyme reaction leaves a colored product on the membrane at any spot where the enzyme-labeled secondary antibodies are bound to the antigen-reactive antibodies. By comparing the location of the colored bands on the blot with the position of protein bands from control samples, a protein associated with a given pathogen can be positively identified.

The immunoblot procedure can be used to detect either antigen (direct evidence for pathogen presence by detecting pathogen antigens in patient samples) or antibody (indirect evidence for pathogen exposure by detecting antibodies to pathogen).

The HIV Immunoblot

Immunoblots are not used as HIV exposure screening tools because HIV immunoblots are generally less sensitive, but more laborious, time-consuming, and costly than the HIV ELISA. Immunoblots are, however, widely used for confirmation of HIV exposure. This is because the HIV ELISA, while very sensitive, occasionally yields false-positive results; the highly specific immunoblot is used to confirm positive ELISA results.

Like the HIV ELISA, the HIV immunoblot is designed to detect the presence of antibodies to HIV in a serum sample. To perform the immunoblot, a purified preparation of HIV is treated with the detergent sodium dodecyl sulfate to solubilize HIV proteins and inactivate the virus. The HIV proteins are then resolved by polyacrylamide gel electrophoresis and blotted from the gel onto membranes (Figure 32.26). This technique resolves at least seven major HIV proteins, and two of them, designated p24 and gp41, are used as specific diagnostic proteins for identification of HIV exposure. Protein p24 is an HIV capsid protein, and glycoprotein gp41 is an HIV envelope protein. Membrane strips with the blotted proteins are available commercially for use by clinical laboratories.

The membrane strips are incubated with the test serum sample. If the sample is HIV-positive, patient antibodies against HIV proteins will be present and will bind to the HIV proteins on the membrane (Figure 32.26). To detect whether antibodies from the serum sample have bound to HIV antigens, a detecting antibody, anti-human IgG conjugated to the enzyme peroxidase, is added to the strips. If the detecting antibody binds, the activity of the conjugated enzyme, after addition of substrate, will form a brown band on the strip at the site of antibody binding. The patient is HIV-positive if the position of the bands resolved by exposure to the patient serum and a positive control serum are identical; a control negative serum is also analyzed in parallel and must show no bands (Figure 32.26).

Although the intensity of the bands obtained in the HIV immunoblot varies somewhat from sample to sample (Figure 32.26*b*), the interpretation of an immunoblot is generally unequivocal, and thus the test is valuable for confirming positive ELISA results for HIV and for eliminating false positives. The immunoblot technique is also used to confirm the specificity of antibody tests for Lyme disease.

32.11 MiniReview

Immunoblot procedures detect antibodies to specific antigens or the antigens themselves. The antigens are separated by electrophoresis, transferred (blotted) to a membrane, and exposed to antibody. Immune complexes are visualized with enzyme-labeled or radioactive secondary antibodies. Immunoblots are extremely specific, but procedures are technically demanding, expensive, and time-consuming.

- What advantage does the immunoblot have over immunoassays such as ELISA and RIA?
- Why is the immunoblot not used for general screening for HIV exposure?

III NUCLEIC ACID-BASED DIAGNOSTIC METHODS

Extremely sensitive methods based on nucleic acid analyses are widely used in clinical microbiology to detect pathogens. These methods do not depend on pathogen isolation or growth or on the detection of an immune response to the pathogen.

32.12 Nucleic Acid Probes and PCR

Molecular methods use genotypic rather than phenotypic characteristics to identify specific pathogens. The success of genetic or DNA-based diagnostic procedures is based on several principles: (1) Nucleic acids can be readily isolated from infected tissues; (2) nucleic acids can be readily visualized and measured in very low amounts; (3) the nucleic acid sequence of a given pathogen's genome is unique, and so nucleic acid analysis can be used for unequivocal identification; and (4) nucleic acid sequences can be amplified to increase the amount of material available for analysis.

Nucleic Acid Probes in Diagnostics

One of the most powerful analytical tools available to clinical microbiologists involves nucleic acid hybridization (Section 12.2). Instead of detecting a whole organism or its products, hybridization detects the presence of specific DNA sequences associated with a specific organism. To identify a microorganism through DNA analysis, the clinical microbiologist must have a unique **nucleic acid probe** for that microorganism. Nucleic acid probes typically consist of a single strand of DNA with a sequence unique to the gene of interest. An oligonucleotide probe may be less than 100 bases or up to several kilobases in length. If a microorganism from a clinical specimen contains DNA or RNA sequences complementary to the probe, the probe will hybridize (following appropriate sample preparation to yield single-stranded DNA from the microorganism), forming a double-stranded molecule (**Figure 32.27**). To detect a reaction, the probe is labeled with a reporter molecule, a radioisotope, an enzyme, or a fluorescent compound that can be detected following hybridization. Depending on the reporter (radioisotopes are the most sensitive), as little as 0.25 μg of DNA per sample can be used for identification.

Nucleic acid probes offer several advantages over immunological assays. Nucleic acids are much more stable than proteins at high temperatures and at high pH and are more resistant to organic solvents and other chemicals. Because of the relative chemical stability of the target nucleic acids, nucleic acid probe technology can even be used to positively identify organisms that are no longer viable. Additionally, some nucleic acid probes may be more specific than antibodies and can detect single-nucleotide differences between DNA sequences.

Clinical Laboratory Probes

In most clinical probe assays, colonies from plates or samples of infected tissue are treated with strong alkali, usually NaOH, to lyse the cells and partially denature the pathogen DNA, forming single-stranded DNA molecules (Figure 32.27*a*). This mixture is then fixed to a matrix (filter or dipstick) or left in solution, and the labeled probe is added. Hybridization is then carried out by incubating at a temperature necessary to form a stable duplex between target DNA and probe DNA. The actual temperature used in each probe assay is governed by the length and nucleic acid composition of the probe and target DNA.

Following a wash to remove unhybridized probe DNA, the extent of hybridization is measured using the reporter molecule attached to the probe. This would require the measurement of radioactivity, enzyme activity, or fluorescence, depending on how the probe was labeled. Nucleic acid probes have been marketed for the identification of several major microbial pathogens and are in widespread use for the detection of *Neisseria gonorrhoeae* and *Chlamydia trachomatis* (Table 32.9; Sections 34.13 and 34.14).

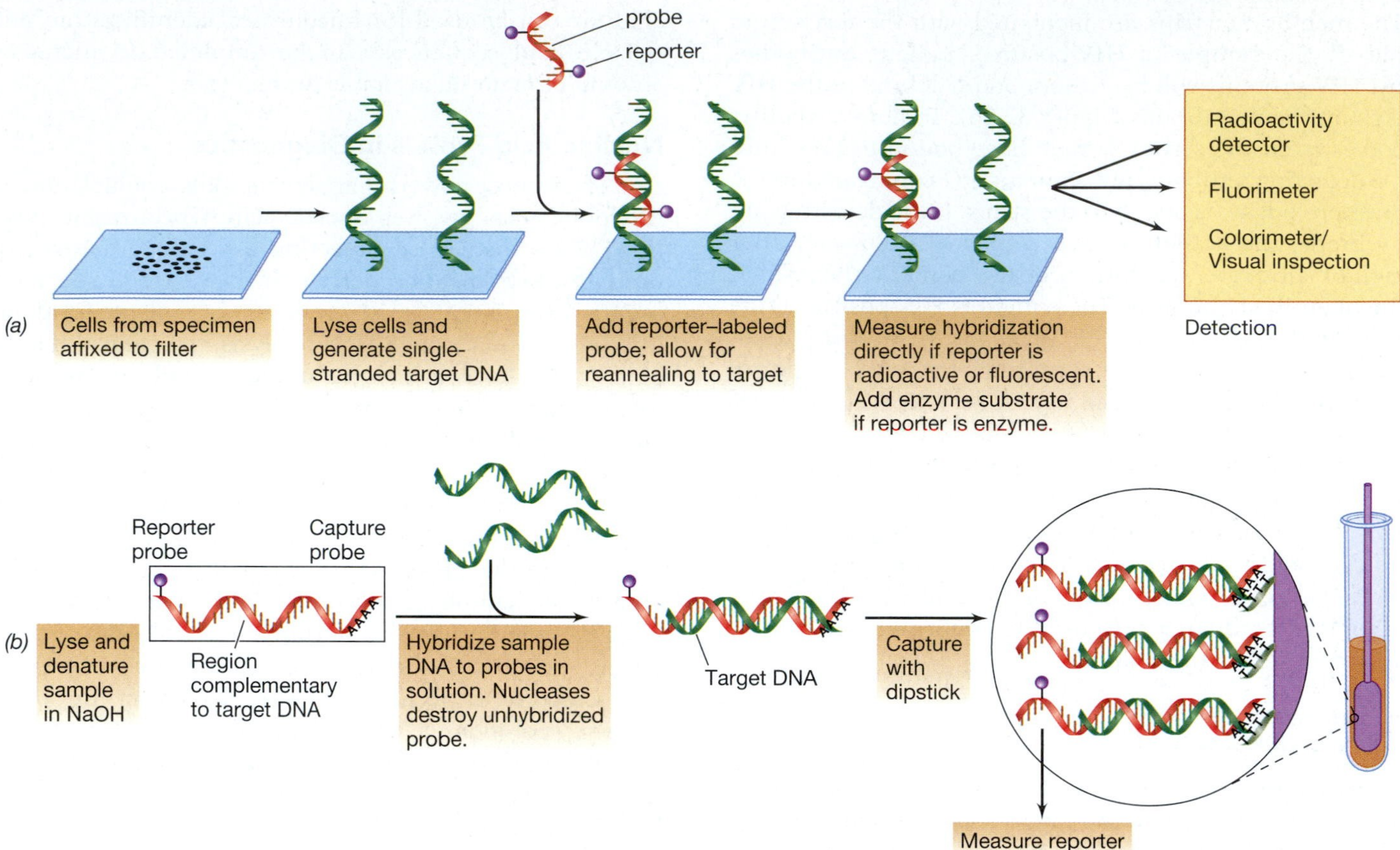

Figure 32.27 **Nucleic acid probe methodology in clinical diagnostics.** *(a)* Membrane filter assay. The detecting system (reporter) can be a radioisotope, a fluorescent dye, or an enzyme. *(b)* Dipstick assay. In the dipstick assay a dual reporter or capture probe is used. The capture probe contains a poly(dA) tail that hybridizes to a poly(dT) oligonucleotide affixed to the dipstick, binding the oligonucleotide–target-reporter complex. The complex can be detected as in *(a)* above.

In addition to their use in clinical diagnostics, molecular probes are widely used in food industries and food regulatory agencies. Probe detection systems can be used to monitor foods for contamination by pathogens such as *Salmonella* and *Staphylococcus*. In probe assays of food, an enrichment period is usually employed to allow low numbers of cells in the food to multiply to a detectable number. Probes designed for use in the food industry employ dipsticks precoated with pathogen-specific probe DNA to hybridize with pathogen DNA from the sample. Two-component probes that function as both a *reporter* probe and a *capture* probe are often used in these applications (Figure 32.27*b*). Following hybridization of the reporter to DNA from the target organism, the dipstick, which contains a sequence complementary to the capture probe (usually poly(dT) to capture poly(dA) on the probe), is inserted into the hybridization solution, where it binds the hybridized DNA. The detection system is then activated to visualize and quantify the hybridized DNA on the dipstick (Figure 32. 27*a*).

Advances in bacterial phylogenetics based on 16S ribosomal RNA (rRNA) sequences (∞ Section 14.9) have allowed for the construction of species-specific and even strain-specific nucleic acid probes. Typing based on differences in rRNA sequences is called *ribotyping* (∞ Section 14.9). Ribotyping reveals the unique DNA restriction patterns of the rRNA genes when DNA from a particular organism is digested by restriction endonucleases. The digested DNA is separated on an agarose gel, and a labeled rRNA probe is used to visualize the unique restriction patterns of the genes encoding rRNA. Within a species, and especially within a strain, the restriction pattern is highly conserved and is a molecular fingerprint for the organism (∞ Figure 14.18). Because all organisms have ribosomal RNA genes, ribotyping has been applied in a wide variety of clinical as well as phylogenetic studies.

Nucleic acid probes detecting less than 1 μg of nucleic acid per sample can identify DNA extracted from about 10^6 bacterial cells or about 10^8 virus particles. Although molecular probes are not as sensitive as direct culture (where as few as 1–10 cells per sample can be detected), probe methods are useful in situations where culture of organisms is difficult or even impossible. However, some applications of DNA technology rival the sensitivity of the culture method.

The Polymerase Chain Reaction in Diagnostics

In Section 12.8 we discussed how the polymerase chain reaction (PCR) formed copies of target genes, and PCR-based methods are now widely used in diagnostic microbiology to identify specific pathogens. Extremely short oligonucleotides

Table 32.9 Pathogens identified with nucleic acid and PCR methods

Pathogen	*Diseases*
Bacteria	
Campylobacter spp.	Food infections
Chlamydia trachomatis	Venereal syndromes; trachoma
Enterococcus spp.	Healthcare-associated infections
Escherichia coli (enteropathogenic strains)	Gastrointestinal disease
Haemophilus influenzae	Infectious meningitis
Legionella pneumophila	Pneumonia
Listeria monocytogenes	Listeriosis
Mycobacterium avium	Tuberculosis
Mycobacterium tuberculosis	Tuberculosis
Mycoplasma hominis	Urinary tract infection; pelvic inflammatory disease
Mycoplasma pneumoniae	Pneumonia
Neisseria gonorrhoeae	Gonorrhea
Neisseria meningitidis	Meningitis
Rickettsia spp.	Typhus, hemorrhagic fever, etc.
Salmonella spp.	Gastrointestinal disease
Shigella spp.	Gastrointestinal disease
Staphylococcus aureus	Purulent discharges (boils, blisters, pus-forming skin infections)
Streptococcus pyogenes	Scarlet fever; rheumatic fever; strep throat
Streptococcus pneumoniae	Pneumonia
Treponema pallidum	Syphilis
Fungi	
Blastomyces dermatitidis	Blastomycosis
Candida spp.	Candidiasis, thrush
Coccidioides immitis	Coccidioidomycosis
Histoplasma capsulatum	Histoplasmosis
Viruses	
Cytomegalovirus	Congenital viral infections
Epstein-Barr virus	Burkitt's lymphoma; mononucleosis
Hepatitis viruses A, B, C, D, E	Hepatitis
Herpes virus (types I and II)	Cold sores; genital herpes
Human immunodeficiency virus (HIV)	Acquired immunodeficiency syndrome (AIDS)
Human papilloma virus	Genital warts; cervical cancer
Influenza	Respiratory disease
Polyoma virus	Neurological disease
Rotavirus	Gastrointestinal disease
Protists	
Leishmania donovani	Leishmaniasis
Plasmodium spp.	Malaria
Pneumocystis carinii	Pneumonia
Trichomonas vaginalis	Trichomoniasis
Trypanosoma spp.	Trypanosomiasis

(typically 15–32 nucleotides in length) are usually sufficient as primers for PCR amplification of a specific gene or genes characteristic of a specific pathogen. For example, primers for a pathogen-specific gene might be used to examine DNA derived from suspected infected tissue, even in the absence of an observable, culturable pathogen. The presence of the appropriate amplified gene segment (**Figure 32.28**) confirms the presence of the pathogen. PCR methods are particularly useful for identifying viral and intracellular infections, where culturing the responsible agents may be very difficult or time-consuming.

Several pathogens for which either hybridization or PCR diagnostic methods are used for their identification are listed in **Table 32.9**. Various PCR applications are used in clinical laboratories. Currently, many PCR tests employ real-time PCR. As we discussed in Section 12.8 real-time PCR employs

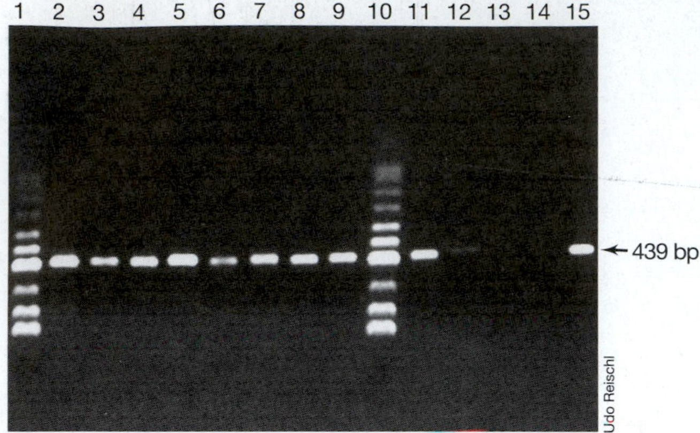

Figure 32.28 Polymerase chain reaction (PCR) analysis of patient sputum for *Mycobacterium tuberculosis* in the diagnosis of tuberculosis. Sputum samples from patients were used as a source of DNA. Amplification was initiated with a primer pair, which produced the indicated 439-base pair product when a pure culture of *M. tuberculosis* was used as the DNA source (lane 15). Lanes 2–9, 11, and 12 are from sputums positive for *M. tuberculosis* (lane 12 is a weak positive). Lanes 13 and 14 are from *M. tuberculosis*-negative sputum samples. Lanes 1 and 10 are molecular weight reference markers.

fluorescent-labeled PCR primers that yield an almost immediate result and avoids the need for post-PCR procedures such as nucleic acid purification and electrophoresis.

Another application of PCR technology, RT-PCR (reverse-transcriptase PCR) (∞ Section 12.8), is used, for example, to monitor the progression of HIV-AIDS and is discussed in Section 34.15. RT-PCR uses pathogen-specific RNA to produce cDNA directly from patient samples, and is used for detection of RNA retroviruses such as HIV (∞ Section 10.12).

32.12 MiniReview

Nucleic acid hybridization is used for identification of microorganisms. A nucleic acid sequence specific for the microorganism of interest must be available to design a probe. Perhaps the most widespread use of nucleic acid-based technology is the application of gene amplification (PCR) methods. Various DNA-based methodologies are currently used in clinical, food, and research laboratories.

- What advantage does nucleic acid hybridization have over standard culture methods for identification of microorganisms? What disadvantages?
- How can information about a microorganism be obtained with a nucleic acid probe or PCR assay in the absence of standard growth-dependent assays?

Review of Key Terms

Agglutination reaction between antibody and particle-bound antigen, resulting in visible clumping of the particles

Antibiogram a report indicating the susceptibility of clinically isolated microorganisms to the antibiotics in current use

Bacteremia the presence of bacteria in the blood

Differential media growth media that allow identification of microorganisms based on phenotypic properties

Enriched media media that allow metabolically fastidious microorganisms to grow because of the addition of specific growth factors

Enrichment culture the use of selected culture media and incubation conditions to isolate microorganisms from natural samples

Enzyme-linked immunosorbent assay (ELISA) immunoassay that uses antibodies to detect antigens or antibodies in body fluids

Fluorescent antibody covalent modification of an antibody molecule with a fluorescent dye that makes the antibody visible under fluorescent light

General purpose media growth media that support the growth of most aerobic and facultatively aerobic organisms

Immunoblot (Western blot) electrophoresis of proteins followed by transfer to a membrane and detection by addition of specific antibodies

Monoclonal antibody antibody made by a single B cell hybridoma clone

Neutralization interaction of antibody with antigen that reduces or blocks the biological activity of the antigen

Nucleic acid probe an oligonucleotide of unique sequence used as a hybridization probe for identifying specific genes

Polyclonal antibodies antibodies made by many different B cell clones

Precipitation reaction between antibody and a soluble antigen resulting in a visible, insoluble complex

Radioimmunoassay (RIA) assay that employs radioactive antibody or antigen to detect antigen or antibody binding

Selective media media that enhance the growth of certain organisms while retarding the growth of others due to an added media component

Sensitivity the lowest amount of antigen that can be detected

Septicemia blood infection

Serology the study of antigen–antibody reactions *in vitro*

Specificity the ability of an antibody to recognize a single antigen

Titer in an immunological context, the quantity of antibody present in a solution

Review Questions

1. Describe the standard procedure for obtaining and culturing a throat culture and a blood sample. What special precautions must be taken while obtaining the blood culture (Section 32.1)?
2. Why are bacteria nearly always cultured from a urine specimen? Why is the number of bacterial cells in urine of significance? What organism is responsible for most urinary tract infections? Why (Section 32.1)?
3. Why is it important to process clinical specimens as rapidly as possible? What special procedures and precautions are necessary for the isolation and culture of anaerobes (Section 32.1)?
4. Differentiate between selective and differential media. Is eosin-methylene blue agar a selective medium or a differential medium? How and why is it used in a clinical laboratory (Section 32.2)?
5. Describe the disk diffusion test for antibiotic susceptibility. Why should potential pathogens from patient isolates be tested by this method (Section 32.3)?
6. How are most laboratory-associated infections contracted? What action can be taken to prevent laboratory infections (Section 32.4)?
7. Why does the antibody titer rise after infection? Is a high antibody titer indicative of an ongoing infection? Explain. Why is it necessary to obtain an acute and a convalescent blood sample to monitor infections (Section 32.5)?
8. What advantages do monoclonal antibodies have over polyclonal antibody preparations, especially with regard to standardization of antibody preparations (Section 32.6)?
9. Describe a neutralization reaction with reference to microbial toxins and antisera (Section 32.7).
10. Agglutination tests are widely used for clinical diagnostic purposes. Why is this the case (Section 32.8)?
11. How are fluorescent antibodies used for the diagnosis of viral diseases? What advantages do fluorescent antibodies have over unlabeled antibodies (Section 32.9)?
12. Radioimmunoassay (RIA) and enzyme-linked immunosorbent assay (ELISA) tests are extremely sensitive, as compared with agglutination. Why is this the case (Section 32.10)?
13. Why is the immunoblot (Western blot) procedure used to confirm screening tests that are positive for human immunodeficiency virus (HIV) (Section 32.11)?
14. What information is essential for the design of a pathogen-specific nucleotide probe? Where can one obtain such information? Is this information available for all pathogens (Section 32.12)?

Application Questions

1. A blood culture is positive for *Staphylococcus epidermidis*. Explain the finding. Is it likely that the patient has *S. epidermidis* bacteremia? Prepare a list of possibilities and questions for a discussion with the physician in charge. What additional information will be needed to confirm or rule out a bacteremia?
2. With respect to both short-term and the long-term consequences, why is it a common medical practice to treat an infectious disease with antibiotics before isolating the suspected pathogen? After a pathogen has been isolated and identified, what further steps should be taken to confirm appropriate antibiotic susceptibility? Why are these measures rarely employed away from a hospital environment?
3. Explain the rationale for collecting serum specimens from patients during an acute infectious disease and about 2 weeks later (Section 32.5). What information would you expect to obtain from the serum of a recovering patient?
4. What are the advantages of rapid identification systems such as agglutination tests and immune-based detection systems such as ELISA tests as compared with growth-dependent clinical diagnostic procedures? What are the potential disadvantages of the rapid nonculture tests?
5. What are the major advantages of using DNA probes in diagnostic microbiology? What information is needed to design sequence-specific polymerase chain reaction (PCR) assay probes for a microorganism? Where can you find this information?
6. Define the procedures you would use to isolate and identify a new pathogen. Be sure to include growth-dependent assays, immunoassays, and molecular assays. Where would you report your findings? Which of your assays could be adapted to be used as a routine, high-throughput test for rapid clinical diagnosis?

33

Epidemiology

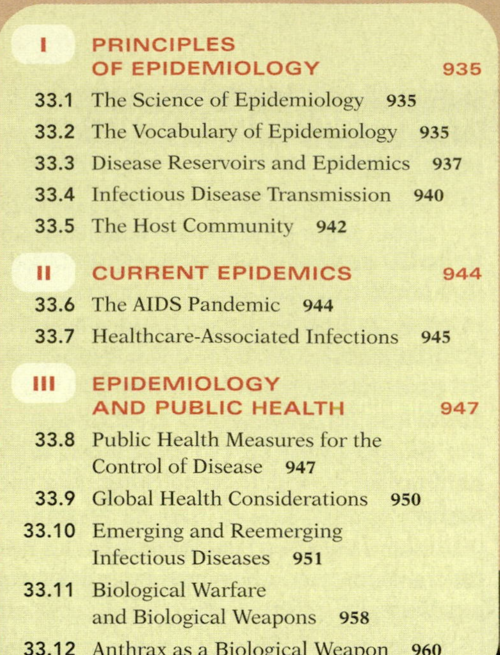

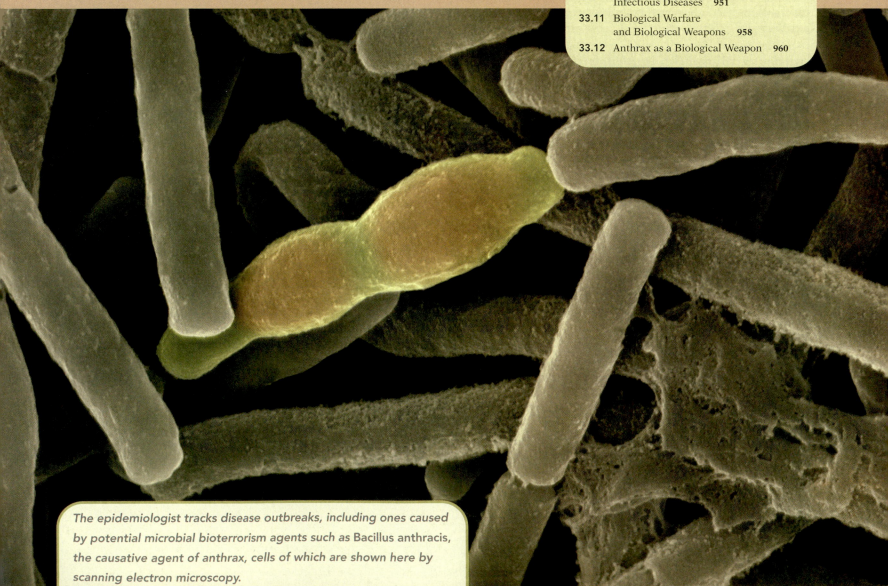

The epidemiologist tracks disease outbreaks, including ones caused by potential microbial bioterrorism agents such as Bacillus anthracis, *the causative agent of anthrax, cells of which are shown here by scanning electron microscopy.*

I PRINCIPLES OF EPIDEMIOLOGY

Individuals may acquire pathogens through infected *vectors* (living carriers) and *vehicles* (non-living carriers) and spread the pathogens to other members of a population. Here we consider how pathogens spread to individuals through populations. **Epidemiology** is the study of the occurrence, distribution, and determinants of health and disease in a population. It thus deals with public health. In the next four chapters we will survey the diseases themselves. Here we examine the principles of epidemiology and their application to the control of infectious diseases.

Many infectious diseases are adequately controlled in developed countries. In Chapter 1 we compared the current causes of death in the United States with those at the beginning of the twentieth century (∞ Figure 1.8). In most developed countries, infectious diseases cause far fewer deaths than noninfectious diseases. Worldwide, however, infectious diseases remain serious public health problems, accounting for nearly 30% of the 56 million annual deaths. Even in developed countries, new infectious diseases such as West Nile fever are emerging, and previously controlled diseases such as tuberculosis are reemerging. In the United States, deaths due to infectious diseases are increasing (**Figure 33.1**). Effective control of infectious diseases remains a worldwide challenge that requires scientific, medical, economic, sociological, political, and educational solutions.

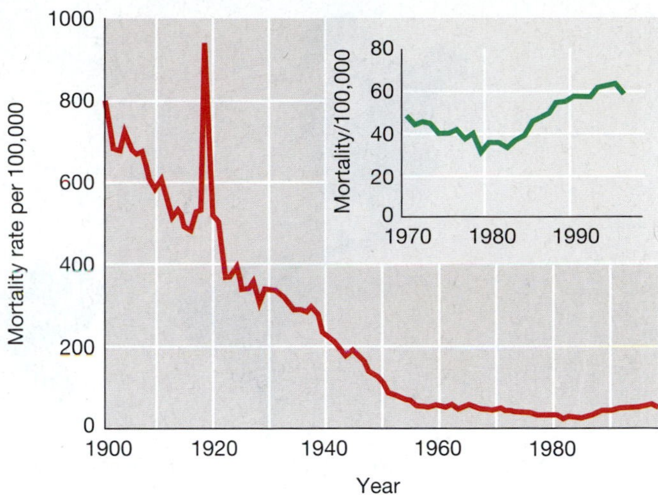

Figure 33.1 Deaths due to infectious disease in the United States. Although infectious disease death rates steadily declined throughout most of the twentieth century (except for the large numbers of deaths in 1918–1919 due to the influenza pandemic), the death rate has increased significantly since 1980. Adapted from Hughes, J. M. 2001. Emerging Infectious Diseases: A CDC Perspective. *Emerg. Infect. Dis.* 17: 494–496.

33.1 The Science of Epidemiology

To cause disease, a pathogen must grow and reproduce in the host. For this reason, epidemiologists track the natural history of pathogens. In many cases, an individual pathogen cannot grow outside the host; if the host dies, the pathogen also dies. Pathogens that kill the host before they move to a new host will become extinct. Most host-dependent pathogens must therefore adapt to coexist with the host.

A well-adapted pathogen lives in balance with its host, taking what it needs for existence and causing only a minimum of harm. Such pathogens may cause **chronic infections** (long-term infections) in the host. When equilibrium between host and pathogen exists, both host and pathogen survive. On the other hand, the host can be damaged when resistance of the host is low because of factors such as poor diet, age, and other stressors (∞ Section 28.13). In addition, new natural pathogens sometimes emerge for which the individual host, and sometimes the entire species, has not developed resistance. Such emerging pathogens often cause **acute infections**, characterized by rapid and dramatic onset. In these cases, pathogens can be selective forces in the evolution of the host, just as hosts, as they develop resistance, can be selective forces in the evolution of pathogens.

In cases where the pathogen is not dependent on the host for survival, the pathogen can cause devastating acute disease. Organisms in the genus *Clostridium*, for example, ubiquitous inhabitants of the soil, are occasional accidental human pathogens, causing life-threatening diseases such as tetanus, botulism, gangrene, and certain gastrointestinal diseases.

The epidemiologist traces the spread of a disease to identify its origin and mode of transmission. Epidemiologic data are obtained by collecting disease information in a population. With the goal of defining common factors for an illness, data are gathered from disease-reporting surveillance networks, clinical records, and patient interviews. This is in contrast to individual patient treatment and diagnosis in the clinic or laboratory. Knowledge of both the population dynamics and clinical problems associated with a given disease are important if public health measures to control diseases are to be effective.

33.1 MiniReview

Epidemiology is the study of the occurrence, distribution, and determinants of health and disease in a population. To understand infectious disease, effects on both populations and individuals must be studied. The interactions of pathogens with hosts are dynamic, affecting the long-term evolution and survival of all species involved.

- How does an epidemiologist differ from a microbiologist?
- Why do epidemiologists acquire population-based data for infectious diseases?

33.2 The Vocabulary of Epidemiology

A number of terms have specific meaning to the epidemiologist. A disease is an **epidemic** when it occurs in an unusually high number of individuals in a population at the same time;

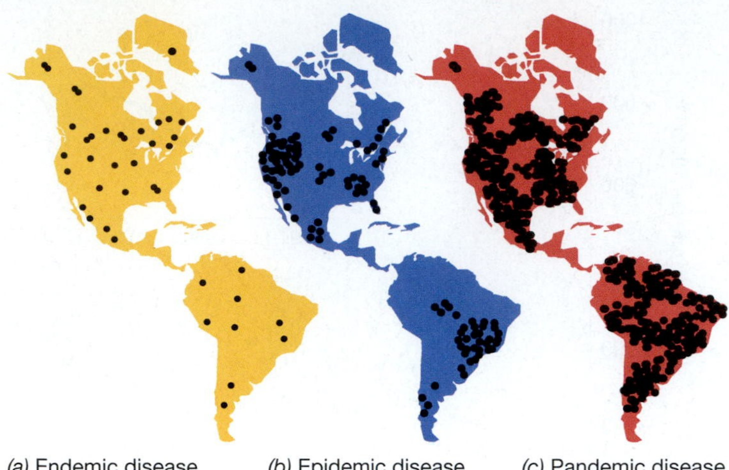

(a) Endemic disease (b) Epidemic disease (c) Pandemic disease

Figure 33.2 Endemic, epidemic, and pandemic disease. Each dot represents several cases of a particular disease. *(a)* Endemic diseases are always present in a population in a given geographical area. *(b)* Epidemic diseases show high incidence in a wider area, usually developing from an endemic focus. *(c)* Pandemic diseases are distributed worldwide. Diseases such as influenza are endemic in certain areas and develop into annual epidemics under appropriate circumstances, such as crowding. Epidemics may develop into pandemics.

a **pandemic** is a widespread, usually worldwide, epidemic (**Figure 33.2**). By contrast, an **endemic disease** is one that is constantly present, usually at low incidence, in a population. An endemic disease implies that the pathogen may not be highly virulent, or the majority of individuals in the selected population may be immune, resulting in low disease incidence. However, as long as an endemic situation exists, the infected individuals are reservoirs of infection, providing a source of viable infectious agents from which other individuals may be infected.

The **incidence** of a particular disease is the number of new cases of an individual disease in a population in a given time period. For example, in 2005 there were 41,953 new cases of AIDS in the United States, for an incidence of 14 new cases per 100,000 people per year. The **prevalence** of a given disease is the total number of new and existing disease cases reported in a population in a given time period. For example, there were 425,910 persons living with AIDS at the end of 2005 within the United States. Expressed another way, the prevalence of AIDS in this population was 176.2 per 100,000 in 2005. Thus, *incidence* provides a record of new cases of a disease, whereas *prevalence* indicates the total disease burden in a population.

Sporadic cases of disease may occur when individual cases are recorded in geographically separated areas, implying that the incidents are not related. A disease **outbreak**, on the other hand, occurs when a number of cases are observed, usually in a relatively short period of time, in an area previously experiencing only sporadic cases of the disease. Finally, diseased individuals who show no symptoms or only mild symptoms have what are called *subclinical infections*. Subclinically infected individuals are frequently **carriers** of a particular disease, because even though they themselves show few (or perhaps no) symptoms, they may be actively carrying and shedding the pathogen.

Mortality and Morbidity

The incidence and prevalence of disease, as determined from statistical analyses of illness and death records, is an indicator of the public health of a selected group such as the total global population or the population of a localized region, such as a city, state, or country. Public health conditions and concerns vary with location and time, and the assessment of public health at a given moment provides only a snapshot of a dynamic situation. Public health policies are designed to reduce incidence and prevalence of disease and can be adequately assessed only by examining public health statistics over longer time periods.

Mortality is the incidence of *death* in the population. Infectious diseases were the major causes of death in 1900 in developed countries, but they are now much less significant. Noninfectious "lifestyle" diseases such as heart disease and cancer are now much more prevalent and cause higher mortality than do infectious diseases (∞ Figure 1.8). However, the current situation could change rapidly if a breakdown in public health measures were to occur. In developing countries, infectious diseases are still major causes of mortality (**Table 33.1** and Section 33.10).

Morbidity refers to the incidence of *disease* in populations and includes both fatal and nonfatal diseases. Morbidity statistics define the public health of a population more precisely than mortality statistics because many diseases have relatively low mortality. The major causes of illness are quite different from the major causes of death. High morbidity diseases include acute respiratory diseases such as the common cold and acute digestive disorders. Both can be due to infectious agents but seldom directly cause death in the population of developed countries.

Disease Progression

In terms of clinical symptoms, the course of a typical acute infectious disease can be divided into stages:

1. *Infection:* The organism invades, colonizes, and grows in the host.
2. *Incubation period:* A period of time elapses between infection and the appearance of disease symptoms. Some diseases, like influenza, have very short incubation periods, measured in days; others, like AIDS, have longer ones, sometimes extending for years. The incubation period for a given disease is determined by inoculum size, virulence, and life cycle of the pathogen, resistance of the host, and distance of the site of entrance from the focus of infection. At the end of incubation, the first symptoms, such as headache and a feeling of illness, appear.
3. *Acute period:* The disease is at its height, with overt symptoms such as fever and chills.

Table 33.1 Worldwide deaths due to infectious diseases, 2002

Disease	*Deaths*	*Causative agent(s)*
Acute respiratory infections[a, b]	3,963,000	Bacteria, viruses, fungi
Acquired immunodeficiency syndrome (AIDS)	2,777,000	Virus
Diarrheal diseases	1,798,000	Bacteria, viruses
Tuberculosis[a]	1,566,000	Bacterium
Malaria	1,272,000	Protist
Measles[a]	611,000	Virus
Pertussis (whooping cough)[a]	294,000	Bacterium
Tetanus[a]	214,000	Bacterium
Meningitis, bacterial[a]	173,000	Bacterium
Hepatitis (all types)[a, c]	157,000	Viruses
Syphilis	153,000	Bacterium
Leishmaniasis	51,000	Protist
Trypanosomiasis (sleeping sickness)	48,000	Protist
Chlamydia	16,000	Bacterium
Schistosomiasis	15,000	Helminth
Chagas disease	14,000	Helminth
Japanese encephalitis	14,000	Virus
Dengue	13,000	Virus
Intestinal nematode infections	12,000	Helminth
Other communicable diseases	1,700,000	Various agents

Globally, there were about 57 million deaths from all causes in 2002. About 14.9 million deaths were from communicable infectious diseases, nearly all in developing countries. Data show the 20 leading causes of death due to infectious diseases. The world population in 2002 was estimated at 6.2 billion.

Data are from the World Health Organization (WHO), Geneva, Switzerland.

[a]Diseases for which effective vaccines are available.

[b]For some acute respiratory agents such as influenza and *Streptococcus pneumoniae* there are effective vaccines; for others, such as colds, there are no vaccines.

[c]Vaccines are available for hepatitis A virus and hepatitis B virus. There are no vaccines for other hepatitis agents.

4. *Decline period:* Disease symptoms are subsiding, any fever subsides, usually following a period of intense sweating, and a feeling of well-being develops. The decline period may be rapid (within 1 day), in which case it is said to occur by *crisis,* or it may be slower, extending over several days, in which case it is said to be by *lysis*.
5. *Convalescent period:* The patient regains strength and returns to normal.

During the later stages of the infection cycle, the immune mechanisms of the host become increasingly important, and in most cases complete recovery from the disease requires (and results in) active immunity.

33.2 MiniReview

An endemic disease is constantly present at low incidence in a specific population. An epidemic disease occurs in unusually high incidence in a specific population. Incidence is a record of new cases of a disease, whereas prevalence is a record of total cases of a disease in a population. Infectious diseases cause morbidity (illness) and may cause mortality (death). An infectious disease follows a predictable clinical pattern in the host.

- Distinguish between an endemic disease, an epidemic disease, and a pandemic disease.
- Distinguish between morbidity and mortality. Is host mortality advantageous for the pathogen?

33.3 Disease Reservoirs and Epidemics

Reservoirs are sites in which infectious agents remain viable and from which infection of individuals may occur. Reservoirs may be either animate or inanimate. **Table 33.2** lists some human infectious diseases with epidemic potential and their reservoirs. Some pathogens are primarily saprophytic (living on dead matter) and only incidentally infect humans and cause disease. For example, *Clostridium tetani* (the causal agent of tetanus) normally inhabits the soil. Infection of animals by this organism is an accidental event. That is, infection of a host is not essential for its continued existence, and in the absence of susceptible hosts, *C. tetani* would still survive in nature.

For many other pathogens, however, living organisms are the only reservoirs. In these cases, the reservoir host is essential for the life cycle of the infectious agent. A number of

Table 33.2 **Epidemic diseases: Agents, sources, reservoirs, and control**

Disease	*Causative agent*[a]	*Infection sources*	*Reservoirs*	*Control measures*
Common-source epidemics[b]				
Anthrax	*Bacillus anthracis* (B)	Milk or meat from infected animals	Cattle, swine, goats, sheep, horses	Destruction of infected animals
Bacillary dysentery	*Shigella dysenteriae* (B)	Fecal contamination of food and water	Humans	Detection and control of carriers; oversight of food handlers; decontamination of water supplies
Botulism	*Clostridium botulinum* (B)	Soil-contaminated food	Soil	Proper preservation of food
Brucellosis	*Brucella melitensis* (B)	Milk or meat from infected animals	Cattle, swine, goats, sheep, horses	Pasteurization of milk; control of infection in animals
Cholera	*Vibrio cholerae* (B)	Fecal contamination of food and water	Humans	Decontamination of public water sources; immunization
E. coli O157:H7 food infection	*Escherichia coli* O157:H7 (B)	Fecal contamination of food and water	Humans, cattle	Decontamination of public water sources; oversight of food handlers; pasteurization of beverages
Giardiasis	*Giardia* spp. (P)	Fecal contamination of water	Wild mammals	Decontamination of public water sources
Hepatitis	Hepatitis A, B, C, D, E (V)	Infected humans	Humans	Decontamination of contaminated fluids and fomites; immunization if available (A and B)
Legionnaire's disease	*Legionella pneumophila* (B)	Contaminated water	High-moisture environments	Decontamination of air conditioning cooling towers, etc.
Paratyphoid	*Salmonella paratyphi* (B)	Fecal contamination of food and water	Humans	Decontamination of public water sources; oversight of food handlers; immunization
Typhoid fever	*Salmonella typhi* (B)	Fecal contamination of food and water	Humans	Decontamination of public water sources; oversight of food handlers; pasteurization of milk; immunization
Host-to-host epidemics				
Respiratory diseases				
Diphtheria	*Corynebacterium diphtheriae* (B)	Human cases and carriers; infected food and fomites	Humans	Immunization; quarantine of infected individuals
Hantavirus pulmonary syndrome	Hantavirus (V)	Inhalation of contaminated fecal material; contact	Rodents	Control of rodent population and exposure
Hemorrhagic fever	Ebola virus (V)	Infected body fluids	Unknown	Quarantine of active cases
Meningococcal meningitis	*Neisseria meningitidis* (B)	Human cases and carriers	Humans	Exposure treated with sulfadiazine for susceptible strains; immunization
Pneumococcal pneumonia	*Streptococcus pneumoniae* (B)	Human carriers	Humans	Antibiotic treatment; isolation of cases for period of communicability
Tuberculosis	*Mycobacterium tuberculosis* (B)	Sputum from human cases; infected milk	Humans, cattle	Treatment with isoniazid; pasteurization of milk
Whooping cough	*Bordetella pertussis* (B)	Human cases	Humans	Immunization; case isolation
German measles	Rubella virus (V)	Human cases	Humans	Immunization; avoid contact between infected individuals and pregnant women

pathogens live only in humans, and maintenance of the pathogen requires person-to-person transmission. This is common for viral and bacterial respiratory pathogens and sexually transmitted pathogens. The staphylococci and streptococci are examples of human-restricted pathogens, as are the agents that cause diphtheria, gonorrhea, and mumps. As we shall see, pathogens that live their entire life cycle dependent on a single host species, especially humans, can be eradicated and many are controlled.

Zoonosis

A number of infectious diseases are caused by pathogens that propagate in both humans and animals. A disease that primarily infects animals but is occasionally transmitted to humans is called a **zoonosis**. Because public health measures for animal populations are much less developed than for humans, the infection rate for veterinary diseases may be higher when animal-to-animal transmission is the rule. Occasionally, transmission is from animal to human; person-to-person transfer of these pathogens is rare, but does occur. Factors leading to the emergence of zoonotic disease include the existence of the infectious agent, the proper environment for propagation and transfer of the agent, and the presence of the new susceptible host species. When there is animal-to-human transmission, a new infectious disease may suddenly emerge in the exposed human population. For examples, see the

Table 33.2 (continued)

Disease	Causative agent[a]	Infection sources	Reservoirs	Control measures
Respiratory diseases				
Influenza	Influenza virus (V)	Human cases	Humans, animals	Immunization
Measles	Measles virus (V)	Human cases	Humans	Immunization
Sexually transmitted diseases[c]				
Acquired immunodeficiency syndrome (AIDS)	Human immunodeficiency virus (HIV)	Infected body fluids, especially blood and semen	Humans	Treatment with metabolic inhibitors (not curative)
Chlamydia	*Chlamydia trachomatis* (B)	Urethral, vaginal, and anal secretions	Humans	Testing for organism during routine pelvic examinations; chemotherapy of carriers and potential contacts; case tracing and treatment
Genital warts, cervical cancer	Human papilloma-virus (HPV)	Urethral and vaginal secretions	Humans	Immunization
Gonorrhea	*Neisseria gonorrhoeae* (B)	Urethral and vaginal secretions	Humans	Chemotherapy of carriers and potential contacts; case tracing and treatment
Syphilis	*Treponema pallidum* (B)	Infected exudate or blood	Humans	Identification by serological tests; antibiotic treatment of seropositive individuals
Trichomoniasis	*Trichomonas vaginalis* (P)	Urethral, vaginal, and prostate secretions	Humans	Chemotherapy of infected individuals and contacts
Vectorborne diseases				
Epidemic typhus	*Rickettsia prowazekii* (B)	Bite from infected louse	Humans, lice	Control louse population
Lyme disease	*Borrelia burgdorferi* (B)	Bite from infected tick	Rodents, deer, ticks	Avoid tick exposure; treat infected individuals with antibiotics
Malaria	*Plasmodium* spp. (P)	Bite from *Anopheles* mosquito	Humans, mosquito	Control mosquito population; treat infected humans with antimalarial drugs
Plague	*Yersinia pestis* (B)	Bite from flea	Wild rodents	Control rodent populations; immunization
Rocky Mountain spotted fever	*Rickettsia rickettsii* (B)	Bite from infected tick	Ticks, rabbits, mice	Avoid tick exposure; treat infected individuals with antibiotics
Direct-contact diseases				
Psittacosis	*Chlamydia psittaci* (B)	Contact with birds or bird excrement	Wild and domestic birds	Avoid contact with birds; treat infected individuals with antibiotics
Rabies	Rabies virus (V)	Bite by carnivores, contact with infected neural tissue	Wild and domestic carnivores	Avoid animal bites; immunization of animal handlers and exposed individuals
Tularemia	*Francisella tularensis* (B)	Contact with rabbits	Rabbits	Avoid contact with rabbits; treat infected individuals with antibiotics

[a]B, Bacteria; V, virus; P, protist.
[b]Some common-source diseases can also be spread from host to host.
[c]Sexually transmitted diseases can also be controlled by effective use of condoms and by sexual abstinence.

Microbial Sidebar, "SARS as a Model of Epidemiological Success" for a discussion of severe acute respiratory syndrome (SARS) and other zoonotic epidemics, and see also the discussion of hantaviruses in Section 35.2.

In many cases, control of a zoonotic disease in the human population does not eliminate the disease as a potential public health problem. Eradication of the human form of a zoonotic disease can generally be achieved only through elimination of the disease in the animal reservoir. This is because maintenance of the pathogen in nature depends on animal-to-animal transfer, and humans are incidental, nonessential hosts. For example, plague is primarily a disease of rodents. Effective control of plague is achieved by control of the infected rodent population and the insect (flea) vector. These methods are more effective in preventing plague transmission than interventions such as vaccines in the incidental human host (∞ Section 35.7). Zoonotic bovine tuberculosis is indistinguishable from human tuberculosis. Often spread from infected cattle to humans, control was achieved primarily by identifying and destroying infected animals. Pasteurization of milk was also of considerable importance because milk was the main vehicle of bovine tuberculosis transmission to humans (∞ Section 34.5).

Certain infectious diseases, particularly those caused by organisms such as protists, have more complex life cycles, involving an obligate transfer from a nonhuman host to humans, followed by transfer back to the nonhuman host (for example, malaria, ∞ Section 35.5). In such cases, the disease

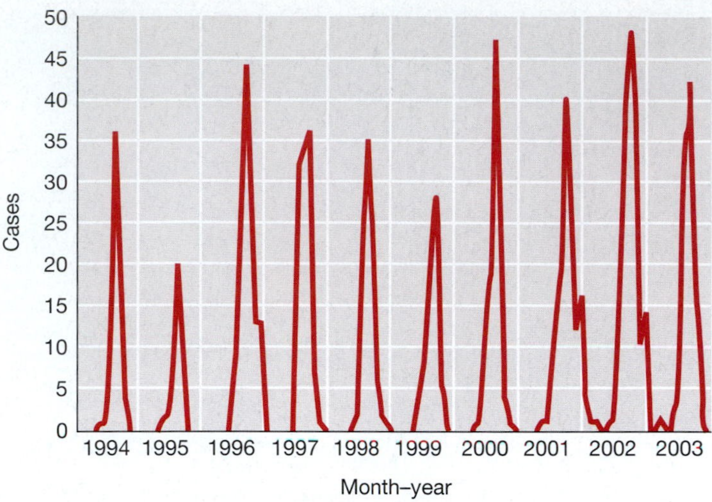

Figure 33.3 The incidence of California encephalitis in the United States by year and month of onset. Note the sharp rise in late summer, followed by a complete decline in winter. The disease cycle follows the yearly cycle of the mosquito vector prevalence. In 2003 there were 108 cases in 12 states. Data are from the Centers for Disease Control and Prevention, Atlanta, GA, USA.

may potentially be controlled in either humans or the alternate animal host.

Carriers

A *carrier* is a pathogen-infected individual showing no signs of clinical disease. Carriers are potential sources of infection for others. Carriers may be individuals in the incubation period of the disease, in which case the carrier state precedes the development of actual symptoms. Respiratory infections such as colds and influenza, for example, are often spread via carriers because they are unaware of their infection and so are not taking any precautions against infecting others. For such acute carriers, the carrier state lasts for only a short time. On the other hand, chronic carriers may spread disease for extended periods of time. Chronic carriers usually appear perfectly healthy. They may be individuals who have recovered from a clinical disease, but still harbor viable pathogens, or they may be individuals with inapparent infections.

Carriers can be identified in populations using diagnostic techniques such as culture or immunoassay surveys. For example, skin testing with *Mycobacterium tuberculosis* antigens tests for delayed hypersensitivity. This reaction, easily detected in the skin test, reveals exposure and previous or current infection with *M. tuberculosis* and is widely used to identify previous infection and carriers of tuberculosis (Section 34.5). Other diseases in which carriers are important for the spread of infection include hepatitis, typhoid fever, and AIDS. Culture or immunoassay surveys of food handlers and health-care workers are sometimes used to identify individuals who are carriers and pose a risk as common sources of infection.

A classic example of a chronic carrier was the woman known as Typhoid Mary, a cook in New York City in the early part of the twentieth century. Typhoid Mary (her real name was Mary Mallon) was employed as a cook during a typhoid fever epidemic in 1906. Investigations revealed that Mary was associated with a number of the typhoid outbreaks. She was the likely source of infection because her feces contained large numbers of the typhoid bacterium, *Salmonella typhi*. She remained a carrier throughout her life, probably because her gallbladder was infected and continuously secreted organisms into her intestine. She refused to have her gallbladder removed and was imprisoned. Released on the pledge that she would not cook or handle food for others, Mary disappeared, changed her name, and continued to cook in restaurants and public institutions, leaving behind epidemic outbreaks of typhoid fever. After several years, she was again arrested and imprisoned and remained in custody until her death in 1938.

33.3 MiniReview

Many pathogens exist only in humans and are maintained only by transmission from person to person. Some human pathogens, however, live mostly in soil, water, or animals. An understanding of disease reservoirs, carriers, and pathogen life cycles is critical for controlling disease.

- What is a disease reservoir?
- Distinguish between acute and chronic carriers. Provide an example of each.

33.4 Infectious Disease Transmission

Epidemiologists follow the transmission of a disease by correlating geographic, climatic, social, and demographic data with disease incidence. These correlations are used to identify possible modes of transmission. A disease limited to a restricted geographic location, for example, may suggest a particular vector; malaria, a disease of tropical regions, is transmitted only by mosquito species restricted to tropical regions. A marked seasonality or periodicity of a disease is often indicative of certain modes of transmission. Such is the case for influenza, where disease incidence increases dramatically when children enter school and come in close contact, increasing opportunities for person-to-person viral transmission.

Finally, pathogen survival depends on efficient host-to-host transmission. Pathogens often have modes of transmission that are related to the preferred habitat of the pathogen in the body. Respiratory pathogens are typically airborne, for example, whereas intestinal pathogens are spread through contaminated food or water. In some cases, environmental factors such as weather patterns may influence the survival of the pathogen. For example, California encephalitis, caused by single-stranded RNA bunyaviruses, occurs primarily during the summer and fall months and disappears every winter in a predictable cyclical pattern (Figure 33.3). The virus is transmitted from mosquito hosts that die during the winter months, causing the disease to disappear until the insect host reappears and retransmits the virus in the summer months. Virtually all mosquito-transmitted encephalitis viruses follow the same seasonal pattern.

Pathogens can be classified by their mechanism of transmission, but all mechanisms have these stages in common: (1) escape from the host, (2) travel, and (3) entry into a new host. Pathogen transmission can be by direct or indirect mechanisms.

Direct Host-to-Host Transmission

Host-to-host transmission often occurs when an infected host transmits a disease directly to a susceptible host without the assistance of an intermediate host or inanimate object. Upper respiratory infections such as the common cold and influenza are most often transmitted directly by droplets resulting from sneezing or coughing. Many of these droplets, however, do not remain airborne for long. Transmission, therefore, requires close, although not necessarily intimate, person-to-person contact.

Some pathogens are extremely sensitive to environmental factors such as drying and heat and are unable to survive for significant periods of time away from the host. These pathogens, transmitted only by intimate person-to-person contact such as exchange of body fluids in sexual intercourse, include those responsible for sexually transmitted diseases such as syphilis (*Treponema pallidum*) and gonorrhea (*Neisseria gonorrhoeae*).

Direct contact also transmits skin pathogens such as staphylococci (boils and pimples) and fungi (ringworm). These pathogens, however, are relatively resistant to environmental conditions such as drying, and they often spread by indirect means as well.

Indirect Host-to-Host Transmission

Indirect transmission of an infectious agent can be facilitated by either living or inanimate agents. Living agents transmitting pathogens are called **vectors**. Commonly, arthropod insects (mites, ticks, or fleas) or vertebrates (dogs, cats, or rodents) act as vectors. Arthropod vectors may not be hosts for the pathogen, but may carry the agent from one host to another. Many arthropods obtain their nourishment by biting and sucking blood, and if the pathogen is present in the blood, the arthropod vector may ingest the pathogen and transmit it when biting another individual. In some cases viral pathogens replicate in the arthropod vector, which is then considered an alternate host. Such is the case for West Nile virus (∞ Section 35.6). Such replication leads to an increase in pathogen numbers, increasing the probability that a subsequent bite will lead to infection.

Inanimate agents such as bedding, toys, books, and surgical instruments can also transmit disease. These inanimate objects are collectively called **fomites**. Food and water are potential disease **vehicles**. Fomites can also be disease vehicles, but major epidemics originating from a single-vehicle source are typically traced to food or water because these are consumed in large amounts by many individuals in a population.

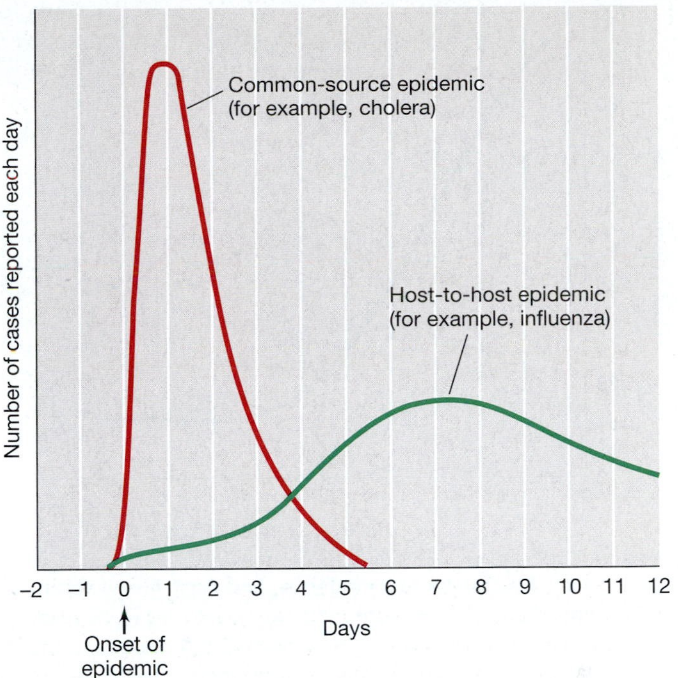

Figure 33.4 Origins of epidemics. The shape of the epidemic curve identifies the likely origin. In a common-source epidemic, such as from contaminated food or water, the curve is characterized by a sharp rise to a peak, with a rapid decline, which is less abrupt than the rise. Cases continue to be reported for a period approximately equal to the duration of one incubation period of the disease. In a host-to-host epidemic, the curve is characterized by a relatively slow, progressive rise, and cases continue to be reported over a period equivalent to several incubation periods of the disease.

Epidemics

Major epidemics are usually classified as *common-source* or *host-to-host* epidemics. These two types of epidemics are contrasted in **Figure 33.4**. Table 33.2 summarizes the key epidemiological features of major epidemic diseases.

A **common-source epidemic** arises as the result of infection (or intoxication) of a large number of people from a contaminated common source such as food or water. Such epidemics are often caused by a breakdown in the sanitation of a central food or water distribution system. Foodborne and waterborne common-source epidemics are primarily intestinal diseases; the pathogen leaves the body in fecal material, contaminates food or water supplies due to improper sanitary procedures, and then enters the intestinal tract of the recipient during ingestion. Waterborne and foodborne diseases are generally controlled by public health measures, which we discuss further in Chapters 36 and 37. A classic example of a common-source epidemic is that of cholera. In 1855 the British physician John Snow showed that cholera spreads through drinking water. His classic studies of water distribution systems in London clearly demonstrated that cholera is spread by fecal contamination of a water supply. In the case of cholera, the infectious agent, the bacterium *Vibrio cholerae*, was transmitted through consumption of the contaminated common-source vehicle, water (∞ Figure 36.11).

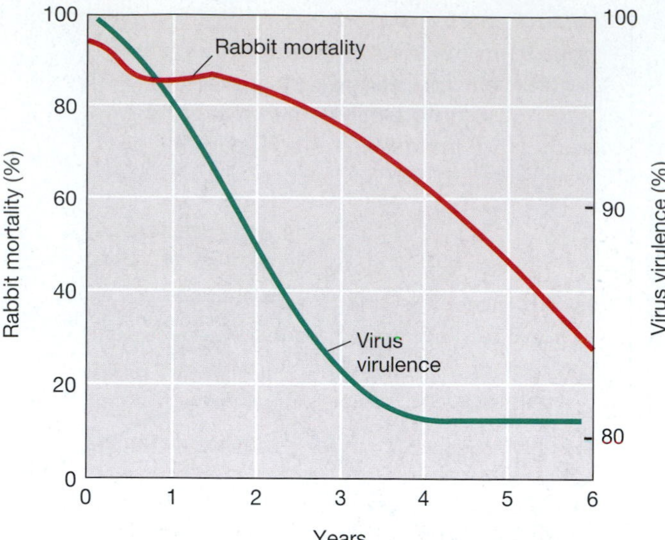

Figure 33.5 Myxoma virus, virulence, and Australian rabbit susceptibility. Data were collected after myxoma virus was introduced into Australia in 1950. Virus virulence is given as the average mortality in standard laboratory rabbits for virus recovered from the field each year. Rabbit susceptibility was determined by removing young feral rabbits from dens and infecting them with a virus strain of moderately high virulence. The test strains killed 90–95% of normal laboratory rabbits.

The disease incidence for a common-source outbreak is characterized by a rapid rise to a peak because a large number of individuals become ill within a relatively brief period of time (Table 33.4). Assuming that the pathogen-contaminated common source is discovered and sanitized, the incidence of a common-source illness also declines rapidly, although the decline is less rapid than the rise. Cases continue to be reported for a period of time approximately equal to the duration of one incubation-period of the disease.

In a **host-to-host epidemic**, the disease incidence shows a relatively slow, progressive rise (Table 33.4) and a gradual decline. Cases continue to be reported over a period of time equivalent to several incubation periods of the disease. A host-to-host epidemic can be initiated by the introduction of a single infected individual into a susceptible population, with this individual infecting one or more people. The pathogen then replicates in susceptible individuals, reaches a communicable stage, and is transferred to other susceptible individuals, where it again replicates and becomes communicable. Influenza and chickenpox are examples of diseases that are typically spread in host-to-host epidemics. Chapter 34 discusses these and a number of other diseases propagated by host-to-host transmission.

33.4 MiniReview

Infectious diseases can be transmitted directly from one host to another, or indirectly by living vectors or inanimate objects (fomites) and common vehicles such as food and water. Epidemics may be of common-source or host-to-host origin.

- Distinguish between direct and indirect transmission of disease. Cite at least one example of each.
- Distinguish between a common-source epidemic and a host-to-host epidemic. Cite at least one example of each.

33.5 The Host Community

The colonization of a susceptible, unimmunized host by a pathogen may first lead to explosive infections, transmission to uninfected hosts, and an epidemic. As the host population develops resistance, however, the spread of the pathogen is checked, and eventually a balance is reached in which host and pathogen are in equilibrium. In an extreme case, failure to reach equilibrium could result in death and eventual extinction of the host species. If the pathogen has no other host, then the extinction of the host also results in extinction of the pathogen. Thus, the evolutionary success of a pathogen may depend on its ability to establish a balanced equilibrium with the host, rather than its ability to destroy the host. In most cases, the evolution of the host and the pathogen affect one another; that is, the host and parasite *coevolve*.

Coevolution of a Host and a Parasite

A classic example of host and pathogen coevolution occurred when a virus was intentionally introduced for purposes of controlling feral rabbits in Australia. Rabbits introduced into Australia from Europe in 1859 spread until they were overrunning large parts of the continent and causing massive crop and vegetation damage.

Myxoma virus was introduced into Australia in 1950 to control the rabbit population. The virus is extremely virulent and usually causes a fatal infection. It spreads rapidly through mosquitoes and other biting insects. Within several months, the virus spread over a large area, rising to a peak in the summer when the mosquito vectors were present and declining in the winter. During the first year of the epidemic, over 95% of the infected rabbits died. However, when virus isolated from infected rabbits was characterized for virulence in newborn feral and laboratory rabbits, the viral isolates from the field had decreased virulence and the resistance of the feral rabbits had increased dramatically. Within 6 years, rabbit mortality dropped to about 84% (**Figure 33.5**). In time, all of the surviving feral rabbits acquired the resistance factors. By the 1980s the rabbit population in Australia was nearing the premyxomatosis levels, with widespread environmental destruction and pressure on native plants and animals.

In 1995, Australian authorities began controlled releases of another highly virulent rabbit pathogen, the rabbit hemorrhagic disease virus (RHDV), a single-stranded, positive-sense RNA virus (Section 19.7). Because RHDV is spread by direct contact and kills animals within days of initial infection, authorities believed the infections would kill all rabbits in a local population, preventing the emergence of resistance. Thus, they reasoned, the virus–host relationship could be maintained in favor of the pathogen more reliably than the arthropod-borne myxomatosis virus. Initial reports indicated that RHDV was very effective at reducing local rabbit populations. However, natural infection of some rabbits by an indigenous hemorrhagic fever virus conferred immune cross-resistance to the introduced RHDV. This unpredictable immune response limited the effectiveness of the control program in certain areas of Australia. Again, the host developed resistance to the control agent, moving the host–pathogen balance toward equilibrium.

Although coevolution of host and pathogen may be common in diseases that rely on host-to-host transmission, for pathogens that do not rely on host-to-host transmission, as we mentioned for *Clostridium,* there is no selection for decreased virulence to support mutual coexistence. Vector-borne pathogens normally transmitted by the bite of arthropods or ticks are also under no evolutionary pressure to spare the human host. As long as the vector can obtain its blood meal before the host dies, the pathogen can maintain a high level of virulence, decimating the human host in the process of infection. For example, the malaria parasites *Plasmodium* spp. show antigenic variations in their coat proteins that aid in avoiding the immune response of the host. This genetic ability to avoid the host responses increases pathogen virulence without regard to the susceptibility of the host. However, as we shall see for the disease malaria, the host may develop disease-specific resistance under the constant evolutionary pressure exerted by a highly virulent pathogen (Section 35.5).

Other evidence for the phenomenon of continually increasing pathogen virulence comes from studies of supervirulent diarrheal diseases in newborns. In hospital situations, *Escherichia coli* can cause severe diarrheal illness and even death, and virulence seems to increase with each passage of the pathogen through a hospital patient. The *E. coli* organisms replicate in one host and are then transferred to another patient through carriers such as healthcare providers or on fomites such as soiled bedding and furniture. Even if the host dies or cannot contact others to transfer the disease, the virulent *E. coli* strain infects others through transmission by means other than the person-to-person route. Extraordinary efforts such as completely washing the nursery and furniture with disinfectant and transferring staff are sometimes necessary to interrupt the cycle of these highly virulent infections.

Herd Immunity

Herd immunity is the resistance of a group to infection due to immunity of a high proportion of the members of the group. Assessment of herd immunity is important for understanding the development of epidemics. In general, if a high proportion of individuals in a group are immune to an infectious agent, then the whole population will be protected. A higher proportion of individuals must be immune to prevent an epidemic by a highly virulent agent or one with a long period of infectivity and a lower proportion for a less virulent agent or one with a brief period of infectivity. In the absence of immunity, even poorly infective agents can be transmitted person-to-person if susceptible hosts have repeated or constant contact with an infected individual. Such appears to be the case for the transmission of H5N1 avian influenza between and among humans (Section 34.9).

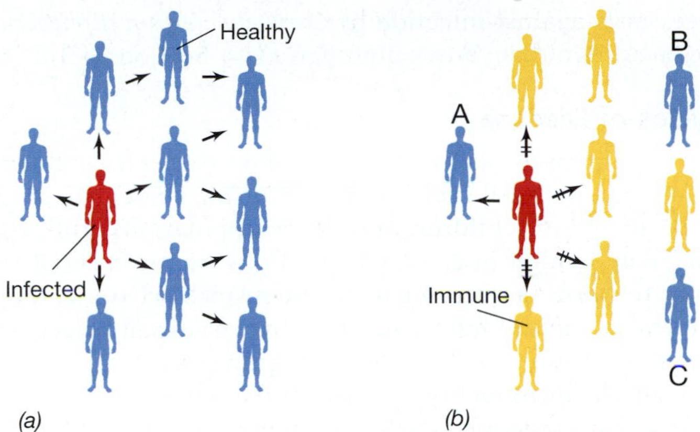

Figure 33.6 Herd immunity and transmission of infection. Immunity in some individuals protects nonimmune individuals from infection. *(a)* In an unprotected population, an infected individual can successfully infect (arrows) all of the healthy individuals. Newly infected individuals will in turn transfer the pathogen directly to other healthy individuals. *(b)* For a moderately transmissible pathogen such as *Corynebacterium diphtheriae* (diphtheria) in a population of moderate density, the infected individual cannot transfer the organism to all susceptible individuals because resistant individuals, immune by virtue of previous exposure or immunization, break the cycle of pathogen transmission. Even if healthy individual A becomes infected, other healthy individuals, B and C, are protected. For a moderately transmissible pathogen such as *C. diphtheriae*, 70% immunity confers resistance to the entire population. For highly transmissible pathogens such as chickenpox (varicella), higher levels of herd immunity, on the order of 90%, are needed to stop transmission.

The proportion of the population that must be immune to prevent infection in the rest of the population can be estimated from data derived from immunization programs. For example, for poliovirus immunization in the United States, studies of polio incidence in large populations indicate that if a population is 70% immunized, polio will be essentially absent from the population. The immunized individuals protect the rest of the population because they cannot acquire and pass on the pathogen, thus breaking the cycle of infection (**Figure 33.6**). For highly infectious diseases such as influenza and measles, up to 90–95% of the population must be immune to confer herd immunity.

A value of about 70% of the population immunized has also been estimated to confer herd immunity for diphtheria, but studies of several small diphtheria outbreaks indicate that in densely populated areas a much higher proportion of susceptible individuals must be immunized to prevent an epidemic. With diphtheria, an additional complication arises because immunized persons can still harbor the pathogen and can thus be chronic carriers. This is because immunization protects against the effects of the diphtheria toxin, but not necessarily against infection by *Corynebacterium diphtheriae*, the bacterium that causes diphtheria (Section 34.3).

Cycles of Disease

Certain diseases occur in cycles. For example, influenza occurs in an annual cyclic pattern, causing epidemics propagated in school children and other populations. Influenza infectivity is high in crowded situations such as schools because the virus is transmitted by the respiratory route. Major epidemic strains of influenza virus change virtually every year, and as a result most children are highly susceptible to infection. On the introduction of virus into a school, an explosive, propagated epidemic results. Virtually every individual becomes infected and then becomes immune. As the immune population increases, the epidemic subsides.

33.5 MiniReview

For most epidemic diseases, hosts and pathogens coevolve to reach a steady state that favors the continued survival of both host and pathogen. When a large proportion of a population is immune to a given disease, disease spread is inhibited. Disease cycles occur when a large, recurring, nonimmune population is exposed to a pathogen.

- Explain coevolution of host and pathogen. Cite a specific example.
- How does herd immunity prevent a nonimmune individual from acquiring a disease? Give an example.

II CURRENT EPIDEMICS

Here we examine data collected by national and worldwide disease-surveillance programs that provide a picture of emerging disease patterns for AIDS and healthcare-associated infections.

33.6 The AIDS Pandemic

Acquired immunodeficiency syndrome (AIDS) is a viral disease that attacks the immune system (Section 34.15). The first reported cases of AIDS were diagnosed in the United States in 1981. Through 2005, in the United States 956,666 cases have been reported, with 550,394 deaths. A total of 44,198 new AIDS cases were reported in United States in 2005, and 38,000 or more new cases have been reported every year since 1989 (**Figure 33.7**).

Worldwide, from 1981 through 2003, at least 70 million individuals have been infected with human immunodeficiency virus (HIV), the virus that causes AIDS. A total of more than 25 million people have already died from AIDS, and 40 million are currently living with the disease. Globally, another 5 million individuals are infected each year. North America has about 1 million HIV-infected individuals, and over 476,000 individuals are now living with HIV/AIDS in the United States. Sub-Saharan Africa has 26.6 million infected people (**Table 33.3**). In the African countries of Botswana and Swaziland, about 40% of the adult population is infected with HIV. AIDS caused about 3 million deaths in 2003, with 2.3 million of those deaths in sub-Saharan Africa.

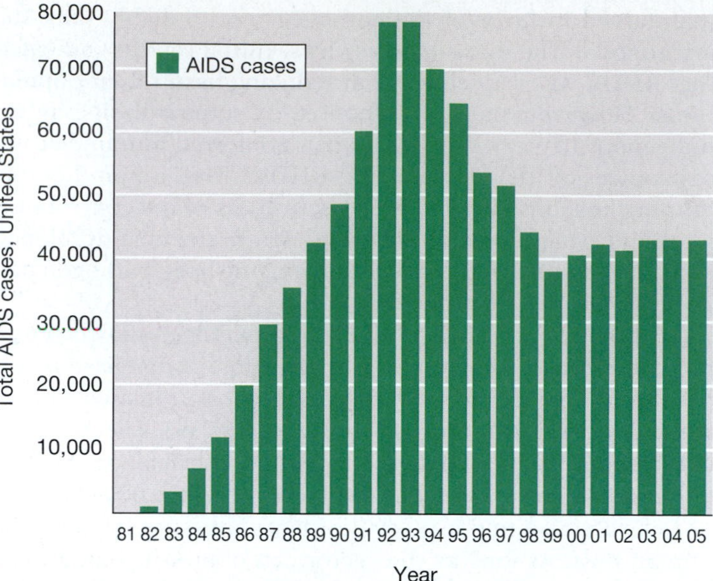

Figure 33.7 Annual diagnosed cases of acquired immunodeficiency syndrome (AIDS) since 1981 in the United States. Cumulatively, there have been 956,666 cases of AIDS through 2005. Data are from the HIV/AIDS Surveillance Report, Centers for Disease Control and Prevention, Division of HIV/AIDS Prevention-Surveillance and Epidemiology.

Tracking the Epidemic

Initial case studies in the United States suggested an unusually high AIDS prevalence among homosexual men and intravenous drug abusers. This indicated a transmissible agent, presumably transferred during sexual activity or by blood-contaminated needles. Individuals receiving blood or

Table 33.3 HIV/AIDS infections, worldwide, 2003

Location	*HIV/AIDS Infections*
North America	1 million
Caribbean	460,000
Latin America	1.6 million
Western Europe	600,000
Eastern Europe and Central Asia	1.5 million
North Africa and Middle East	600,000
Sub-Saharan Africa	26.6 million
East Asia and Pacific	1 million
South and Southeast Asia	6.4 million
Australia and New Zealand	15,000

The total number of individuals infected with HIV/AIDS is estimated to be 40 million. Data are from the World Health Organization and the Joint United Nations Programme on HIV/AIDS.

blood products were also at high risk: Hemophiliacs who required infusions of blood products and a small number of individuals who received blood transfusions or tissue transplants before 1982 (when blood-screening procedures were implemented) acquired AIDS. Today, fewer than 1% of the total current AIDS cases can be attributed to these modes of transmission.

Soon after the discovery of HIV, laboratory ELISA and Western blot tests (∞ Sections 32.10 and 32.11) were developed to detect antibodies to the virus in serum. Extensive surveys of HIV incidence and prevalence defined the spread of HIV and ensured that new cases would not be transmitted by blood transfusions. The pattern illustrated in **Figure 33.8** is typical of an agent transmissible by sexual activity or by blood. The identification of well-defined high-risk groups implied that HIV was not transmitted from person to person by casual contact, such as the respiratory route, or by contaminated food or water. Instead, body fluids, primarily blood and semen, were identified as the vehicles for transmission of HIV.

Figure 33.8 shows that the number of AIDS cases is disproportionately high in homosexual men in the United States, but the patterns in women and in certain racial and ethnic groups indicate that homosexuality is not a prerequisite for acquiring AIDS. Among women, for example, heterosexual women are the largest risk group, whereas in African-American and Hispanic men, intravenous drug use is linked to HIV infection nearly as often as homosexual activity. In fact, if we consider all risk groups, heterosexual activity is the fastest growing risk factor for new AIDS cases among adults.

This cohort of individuals who are at high risk for acquiring AIDS indicates that virtually all who acquire HIV today share two specific behavior patterns. First, they engage in activities (sex or drug use) that involve transfer of body fluids, usually semen or blood. Second, they exchange body fluids with multiple partners, either through sexual activity or through needle-sharing drug activity (or both). Thus, with each encounter they increase the probability of exchanging body fluids with an HIV-infected individual and therefore their chance of acquiring HIV infection.

The incidence of AIDS in hemophiliacs and blood transfusion recipients has been virtually eliminated. This is due to rigorous screening of the blood supply and also because many blood clotting factors needed by hemophiliacs can withstand a heat treatment sufficient to inactivate HIV or are available as genetically engineered products. In 2005, there were 111 cases of pediatric AIDS in the United States. HIV can be transmitted to the fetus by infected mothers and probably also in mothers' milk. Infants born to HIV-infected mothers have maternally derived antibodies to HIV in their blood. However, a positive diagnosis of HIV infection in infants must wait a year or more after birth because about 70% of infants showing maternal HIV antibodies at birth are later found not to be infected with HIV.

Epidemiological studies of AIDS in Africa indicate that transmission of HIV by heterosexual transmission of HIV is the norm. In some regions, fewer men than women are infected with HIV. The identification of high-risk groups such as prostitutes has led to the development of health education campaigns. These campaigns inform the public of HIV transmission methods and define high-risk behaviors. Because no cure or effective immunization for AIDS is available, public health education remains the most effective approach to the control of AIDS and spread of HIV infection. We discuss the pathology and therapy of AIDS in Section 34.15.

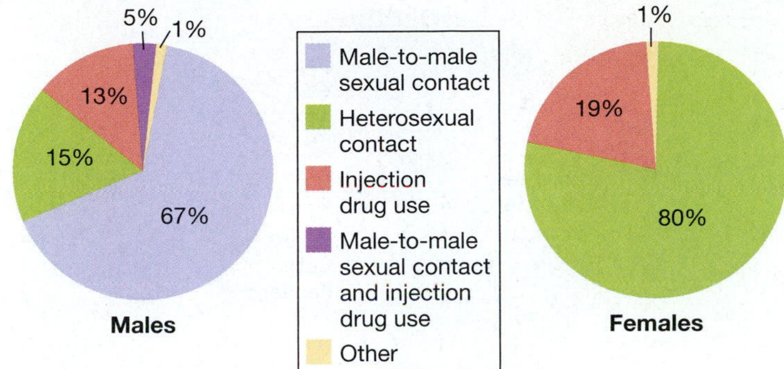

Figure 33.8 Distribution of AIDS cases by risk group and sex in adolescents and adults in the United States, 2005. Data were collected from 28,037 males and 9893 females diagnosed with HIV/AIDS in 2005.
Source: Centers for Disease Control and Prevention.

33.6 MiniReview

HIV/AIDS will continue to be a major worldwide public health problem. There is no effective cure or immunization to prevent AIDS. HIV/AIDS transmission control depends on public health surveillance and education.

- Describe the major risk factors for acquiring HIV infection. Tailor your answer to your country of origin.
- Estimate the total number of individuals in the United States who now have AIDS and predict how many will be living with AIDS in the next 2 years.

33.7 Healthcare-Associated Infections

A healthcare-associated infection (HAI) is a local or systemic condition resulting from an infectious agent or its products that occurs during admission to a healthcare facility and that was not present on admission. HAIs cause significant morbidity and mortality. About 5% of patients admitted to healthcare facilities acquire HAIs, also called **nosocomial infections** (*nosocomium* is the Latin word for "hospital"). In all, there are about 1.7 million nosocomial infections each year in the United States, leading directly or indirectly to about 99,000 deaths.

Some nosocomial infections are acquired from patients with communicable diseases, but others are caused by pathogens that are selected and maintained within the hospital environment. Cross-infection from patient to patient or

UNIT 8

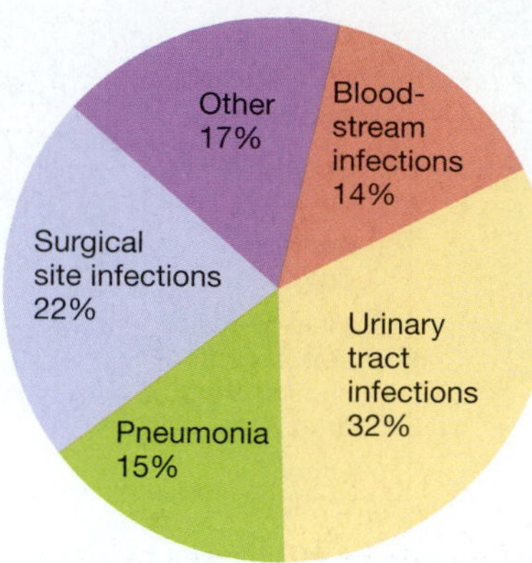

Figure 33.9 Healthcare-associated infections, by site. About 1.7 million healthcare-associated infections occur annually in the United States.

Source: Klevens et al., Estimated Health Care-Associated Infections and Deaths in U.S. Hospitals, 2002. *Public Health Reports* 122: 160–166, 2007.

from healthcare personnel to patient presents a constant hazard. Nosocomial pathogens are often found as normal flora in either patients or healthcare staff.

The Hospital Environment

Infectious diseases are spread easily and rapidly in hospitals for several reasons. (1) Many patients have low resistance to infectious disease because of their illness (they are compromised hosts; Section 28.13). For example, intensive care units provide care for the most acute illnesses and account for 24.5% of total HAIs. (2) Healthcare facilities treat infectious disease patients, and these patients may be pathogen reservoirs. (3) Multiple patients in rooms and wards increase the chance of cross-infection. (4) Healthcare personnel move from patient to patient, increasing the probability of transfer of pathogens. (5) Healthcare procedures such as hypodermic injection, spinal puncture, and removal of tissue samples (biopsy) or fluids (drawing blood), breach the skin barrier and may introduce pathogens into the patient. (6) In maternity wards of hospitals, newborn infants are unusually susceptible to certain infections because they lack well-developed defense mechanisms. (7) Surgical procedures expose internal organs to sources of contamination, and the stress of surgery often diminishes the resistance of the patient to infection. (8) Certain therapeutic drugs, such as steroids used for controlling inflammation, increase the susceptibility to infection. (9) Use of antibiotics to control infections selects for antibiotic-resistant organisms (Section 27.12).

Infection Sites

The most common sites of HAIs are shown in **Figure 33.9**. Of the 99,000 estimated deaths caused by HAIs in 2002, 36,000 were from pneumonia, 31,000 from bloodstream infections, 13,000 from urinary tract infections, 8,000 from surgical site infections, and 11,000 for all other sites. This distribution and the numbers of people infected are similar to annual patterns.

Table 33.4 Number of intensive care unit nosocomial infections in the United States, by site and organism

	Bloodstream	Pneumonia	Urinary Tract
Pathogen	**Number**	**Number**	**Number**
Enterobacter spp.	1,083	4,444	1,560
Escherichia coli	514	1,725	5,393
Klebsiella pneumoniae	735	2,865	1,891
Haemophilus influenzae		1,738	
Pseudomonas aeruginosa	841	6,752	3,365
Staphylococcus aureus	2,758	7,205	497
Staphylococcus spp.	8,181		838
Enterococcus spp.	2,967	682	4,226
Candida albicans	1,090	1,862	4,856
Other pathogens	3,774	12,537	8,075
Total number[a]	**21,943**	**39,810**	**30,701**
Total %	**23.7**	**43.1**	**33.2**

[a]The total number of nosocomial infections in intensive care units during a recent 8-year time period was 92,454.

Source: National Nosocomial Infections Surveillance System Report, Centers for Disease Control and Prevention, Atlanta, Georgia, USA.

Healthcare-associated Pathogens

Healthcare-associated pathogens preferentially infect several sites in the body, notably the urinary tract, blood, and the respiratory tract. A relatively small number of pathogens cause the majority of nosocomial infections at these sites (**Table 33.4**).

One of the most important and widespread hospital pathogens is *Staphylococcus aureus*. It is the most common cause of pneumonia and the third most common cause of blood infections. *S. aureus* is also particularly problematic in nurseries. Many hospital strains of *S. aureus* are unusually virulent and are resistant to common antibiotics, making treatment very difficult. *S. aureus* and other staphylococci together constitute the largest cause of nosocomial blood infections and are also very prevalent in wound infections. Most staphylococci are found in the upper respiratory tract or on the skin, where they are a part of the normal flora in many individuals, including hospital patients and personnel.

Escherichia coli is the most common cause of urinary tract infections in hospitals, but *Enterococcus* species, *Pseudomonas aeruginosa, Candida albicans,* and *Klebsiella pneumoniae* infections are also very common. *Enterococcus, E. coli,* and *K. pneumoniae* are normally found only in the human body. But *Candida* and *Pseudomonas* are opportunistic pathogens; they are commonly found in the environment but cause disease only in individuals with compromised defenses. Isolates of *P. aeruginosa* from healthcare-associated infections are often resistant to many different antibiotics, complicating treatment. *E. coli*, *Staphylococcus*, and *Enterococcus* also have potential for multiple drug resistance.

33.7 MiniReview

Patients in healthcare facilities are unusually susceptible to infectious disease and are exposed to various infectious agents. Treatment of nosocomial infections is complicated by reduced host resistance; this leads to opportunistic infections, as well as the presence of antibiotic-resistant pathogens.

- Why are patients in healthcare facilities more susceptible than normal individuals to pathogens?
- What is the source of HAI pathogens?

III EPIDEMIOLOGY AND PUBLIC HEALTH

Here we identify some of the methods used to identify, track, contain, and eradicate infectious diseases within populations. We also identify some important current and future threats from infectious diseases.

33.8 Public Health Measures for the Control of Disease

Public health refers to the health of the general population and to the activities of public health authorities in the control of disease. The incidence and prevalence of many infectious diseases has dropped dramatically over the past century, especially in developed countries, because of universal improvements in basic living conditions. Better nutrition, access to clean potable water, improved public sewage treatment, less crowded living conditions, and lighter workloads have contributed immeasurably to disease control, primarily by reducing risk factors. Several diseases, including smallpox, typhoid fever, diphtheria, brucellosis, and poliomyelitis, have been controlled by active, disease-specific public health measures such as quarantine and vaccination.

Controls Directed against the Reservoir

When the disease reservoir is primarily in domestic animals, the infection of humans can be prevented if the disease is eliminated from the infected animal population. Immunization or destruction of infected animals may eliminate the disease in animals and, consequently, in humans. These procedures have nearly eliminated brucellosis and bovine tuberculosis in humans. These procedures have also been used to control bovine spongiform encephalitis (mad cow disease) in cattle in the United Kingdom, Canada, and the United States. In the process, the health of the domestic animal population is also improved.

When the disease reservoir is a wild animal, eradication is much more difficult. Rabies, for example, is a disease that occurs in both wild and domestic animals but is transmitted to domestic animals primarily by wild animals. Thus, control of rabies in domestic animals and in humans can be achieved by immunization of domestic animals. However, because the majority of rabies cases occur in wild rather than domestic animals, at least in the United States (∞ Section 35.1), eradication of rabies would require the immunization or destruction of all wild animal reservoirs, including such diverse species as raccoons, bats, skunks, and foxes. Although oral rabies immunization is practical and recommended for rabies control in restricted wild animal populations, its efficacy is untested in large, diverse animal populations such as the wild animal reservoir in the United States.

If insects such as the mosquito vectors that transmit malaria and West Nile fever are the disease reservoir, effective control of the disease can be accomplished by eliminating the reservoir with insecticides or other agents. The use of toxic or carcinogenic chemicals, however, must be balanced with environmental concerns. In some cases the elimination of one public health problem only creates another. For example, the insecticide dichlorodiphenyltrichloroethane (DDT) is very effective against mosquitoes and is credited with eradicating yellow fever and malaria in North America. However, its use is currently banned in the United States because of environmental concerns. DDT is still used in many developing countries to control mosquito-borne diseases, but its use is declining worldwide.

When humans are the disease reservoir (for example, AIDS), control and eradication can be difficult, especially if there are asymptomatic carriers. On the other hand, certain diseases that are limited to humans have no asymptomatic phase. If these can be prevented through immunization or treatment with antimicrobial drugs, the disease can be eradicated if those who have contracted the disease and all possible contacts are strictly quarantined, immunized, and treated. Such a strategy was successfully employed by the World Health Organization to eradicate smallpox and is currently being used to eradicate polio (discussed below).

Controls Directed against Transmission of the Pathogen

The transmission of pathogens in food or water can be eliminated by preventing contamination of these sources. Water purification methods have dramatically reduced the incidence of typhoid fever. Food protection laws have greatly decreased the probability of transmission of pathogens to humans. For example, the pasteurization of milk has largely controlled bovine tuberculosis in humans.

Transmission of respiratory pathogens is difficult to prevent. Attempts at chemical disinfection of air have been unsuccessful. Air filtration is a viable method but is limited to small, enclosed areas. In Japan, many individuals wear facemasks when they have upper respiratory infections to prevent transmission to others, but such methods, although effective, are voluntary and are difficult to institute as public health measures.

Immunization

Smallpox, diphtheria, tetanus, pertussis (whooping cough), measles, mumps, rubella, and poliomyelitis have been controlled primarily by means of immunization. Effective vaccines are available for a number of other infectious diseases (Table 30.2). As we discussed in Section 33.5, 100% immunization is not necessary for disease control in a population, although the percentage needed to ensure disease control varies with the infectivity and virulence of the pathogen and with the living conditions of the population (for example, crowding).

Measles epidemics offer an example of the effects of herd immunity. The occasional resurgence of the highly contagious measles virus emphasizes the importance of maintaining appropriate immunization levels for a given pathogen. Until 1963, the year an effective measles vaccine was licensed, nearly every child in the United States acquired measles through natural infections, resulting in over 400,000 annual cases. After introduction of the vaccine, the number of annual measles infections dropped precipitously (Figure 34.15). Case numbers reached a low of 1,497 by 1983. However, by 1990, the percentage of children immunized against measles fell to 70%, and the number of new cases rose to 27,786. Within 3 years, a concerted effort to increase measles immunization levels to above 90% virtually eliminated indigenous measles transmission in the United States, and a total of only 312 measles cases were reported in 1993. Currently, about 100 cases of measles are reported each year in the United States, over half due to infections imported by visitors from other countries.

In the United States, virtually all children are now adequately immunized, but up to 80% of adults lack effective immunity to important infectious diseases because immunity from childhood vaccinations declines with time. When childhood diseases occur in adults, they can have devastating effects. For example, if a woman contracts rubella (a vaccine-preventable viral disease) during pregnancy, the fetus may develop serious developmental and neurological disorders. Measles, mumps, and chickenpox are also more serious diseases in adults than in children.

All adults are advised to review their immunization status and check their medical records (if available) to ascertain dates of immunizations. Tetanus immunizations, for example, must be renewed at least every 10 years to provide effective immunity. Surveys of adult populations have shown that more than 10% of adults under the age of 40 and over 50% of those over 60 are not adequately immunized. Measles immunity in adults should also be reviewed. People born before 1957 probably had measles as children and are immune. Those born after 1956 may have been immunized, but the effectiveness of early vaccines was variable and effective immunity may not be present, especially if the immunization was given before 1 year of age. Reimmunization for polio is not recommended for adults unless they are traveling to countries in western Africa and Asia, where polio may still be endemic or where recent outbreaks of polio have occurred.

General recommendations for immunization were discussed in Section 30.5 and those for specific infections will be discussed in Chapters 34 through 37.

Table 33.5 Reportable infectious agents and diseases in the United States

Diseases caused by *Bacteria*
- Anthrax
- Botulism
- Brucellosis
- Chancroid
- *Chlamydia trachomatis*
- Cholera
- Diphtheria
- Ehrlichiosis
- Enterohemorrhagic *Escherichia coli*
- *Escherichia coli* O157:H7
- Gonorrhea
- *Haemophilus influenzae*, invasive disease
- Hansen's disease (leprosy)
- Hemolytic uremic syndrome
- Legionellosis
- Listeriosis
- Lyme disease
- Meningococcal disease
- Pertussis
- Plague
- Psittacosis
- Q fever
- Rocky Mountain spotted fever
- Salmonellosis
- Shigellosis
- Streptococcal diseases, invasive, Group A
- Streptococcal toxic shock syndrome
- *Streptococcus pneumoniae*, drug-resistant and invasive disease
- Syphilis
- Tetanus
- Toxic shock syndrome
- Tuberculosis
- Tularemia
- Typhoid fever
- Vancomycin Intermediate *Staphylococcus aureus* (VISA)
- Vancomycin Resistant *Staphylococcus aureus* (VRSA)

Diseases caused by fungi (molds, yeast)
- Coccidiomycosis
- Cryptosporidiosis

Diseases caused by viruses
- Acquired immunodeficiency syndrome (AIDS) and pediatric HIV infection
- Encephalitis/meningitis (mosquito-borne)
  - California serogroup
  - Eastern equine
  - Powassan
  - St. Louis
  - Western equine
  - West Nile
- Hantavirus pulmonary syndrome
- Hepatitis A, B, C
- HIV infection
  - Adult
  - Pediatric (<13 yrs)
- Measles
- Mumps
- Poliomyelitis, paralytic
- Rabies, animal, human
- Rubella, acute and congenital syndrome
- Severe acute respiratory syndrome (SARS)
- Smallpox
- Varicella
- Yellow fever

Diseases caused by protists
- Cyclosporiasis
- Malaria
- Giardiasis

Disease caused by a helminth
- Trichinosis

Quarantine

Quarantine restricts the movement of a person with active infection to prevent spread of the pathogen to other people. The length of quarantine for a given disease is the longest period of communicability for that disease. To be effective, quarantine measures must prevent the infected individual from contacting unexposed individuals. Quarantine is not as severe a measure as strict isolation, which is used in hospitals for unusually infectious and dangerous diseases.

By international agreement, six diseases require quarantine: smallpox, cholera, plague, yellow fever, typhoid fever,

Table 33.6 National Center for Infectious Diseases (NCID) surveillance systems for infectious disease notification and tracking in the United States

Surveillance System (acronym)	*Disease Surveillance Responsibility*
121 Cities Mortality Reporting System	Influenza, pneumonia, all deaths
Active Bacterial Core Surveillance	Invasive bacterial diseases
BaCon Study	Bacterial contamination associated with blood transfusion
Border Infectious Disease Surveillance Project (BIDS)	Infectious disease along the U.S.-Mexican border
Dialysis Survey Network (DSN)	Vascular access infections and bacterial resistance in hemodialysis patients
Electronic Foodborne Outbreak Investigation and Reporting System (EFORS)	Foodborne outbreaks
EMERGEncy ID NET	Emerging infectious diseases
Foodborne Diseases Active Surveillance Network (FOODNET)	Foodborne disease
Global Emerging Infections Sentinel Network (GeoSentinel)	Global emerging diseases
Gonococcal Isolate Surveillance Project (GISP)	Antimicrobial resistance in *Neisseria gonorrhoeae*
Health Alert Network (HAN)	Health threat notification network, especially for bioterrorism
Integrated Disease Surveillance and Response (IDSR)	World Health Organization (WHO/AFRO) initiative for infectious diseases in Africa
Intensive Care Antimicrobial Resistance Epidemiology (ICARE)	Antimicrobial resistance and antimicrobial use in healthcare settings
International Network for the Study and Prevention of Emerging Antimicrobial Resistance (INSPEAR)	Global emergence of drug-resistant organisms
Laboratory Response Network (LRN)	Bioterrorism, chemical terrorism, and public health emergencies
Measles Laboratory Network	Measles in the Americas and the Caribbean
National Antimicrobial Resistance Monitoring System: Enteric Bacteria (NARMS)	Antimicrobial resistance in human nontyphoid *Salmonella*, *Escherichia coli* O157:H7, and *Campylobacter* isolates from agricultural and food sources
National Malaria Surveillance	Malaria in the United States
National Molecular Subtyping Network for Foodborne Disease Surveillance (PulseNet)	Molecular fingerprinting of foodborne bacteria
National Nosocomial Infections Surveillance System (NNIS)	Healthcare-associated infections
National Notifiable Diseases Surveillance System (NNDSS)	Reportable infectious diseases (see Table 33.4)
National Respiratory and Enteric Virus Surveillance System (NREVSS)	Respiratory syncytial virus (RSV), human parainfluenza viruses, respiratory and enteric adenoviruses, and rotavirus
National Surveillance System for Health Care Workers (NaSH)	Healthcare worker occupational infections
National Tuberculosis Genotyping and Surveillance Network	Tuberculosis genotyping repository
National West Nile Virus Surveillance System	West Nile virus
Public Health Laboratory Information System (PHLIS)	Notifiable diseases
Select Agent Program (SAP)	Regulate potential bioterrorism agents
Surveillance for Emerging Antimicrobial Resistance Connected to Healthcare (SEARCH)	Emerging antimicrobial resistance in healthcare settings
Unexplained Deaths and Critical Illnesses Surveillance System	Emerging infectious diseases worldwide
United States Influenza Sentinel Physicians Surveillance Network	260 clinical sites that report incidence and prevalence of influenza infections
Viral Hepatitis Surveillance Program (VHSP)	Viral hepatitis
Waterborne-Disease Outbreak Surveillance System	Waterborne diseases

and relapsing fever. Each is a very serious, particularly communicable disease. Spread of certain other highly contagious diseases such as Ebola hemorrhagic fever and meningitis may also be controlled by quarantine as outbreaks occur.

Surveillance

Surveillance is the observation, recognition, and reporting of diseases as they occur. **Table 33.5** lists the diseases currently under surveillance in the United States. Several of the epidemic diseases listed in Table 33.2 and Table 33.8 are not on the surveillance list. However, many other diseases are surveyed through regional laboratories that identify index cases—those cases of disease that exhibit new syndromes or characteristics or are linked to new pathogens, indicating high potential for new epidemics.

The **Centers for Disease Control and Prevention (CDC)** in the United States, through the National Center for Infectious Diseases (NCID), operates a number of surveillance programs, as shown in **Table 33.6**. Many diseases are reportable to more than one surveillance program. Although redundant reporting may at first seem unnecessary, a disease may fall into several categories that affect health-care plans and policies. For example, reporting of vancomycin-resistant staphylococci to the National Nosocomial Infections

Surveillance System (NNIS) and to CDC as a notifiable disease (Table 33.5) provides a national database. Using this information, a hospital infection team can formulate and implement plans for isolation, diagnosis, and drug-susceptibility testing of staphylococcal infections to identify antibiotic resistant strains, to stop their spread, and to begin appropriate treatment.

Pathogen Eradication

A concerted disease eradication program was responsible for the eradication of naturally occurring smallpox. Smallpox was a disease with a reservoir consisting solely of the individuals with acute smallpox infections, and transmission was exclusively person to person. Infected individuals transmitted the disease through direct contact with previously unexposed individuals. Although smallpox, a viral disease, cannot be treated once acquired, immunization practices were very effective; vaccination with the related vaccinia virus conferred virtually complete immunity. The World Health Organization (WHO) implemented the smallpox eradication plan in 1967. Because of the success of vaccination programs worldwide, endemic smallpox was then confined to Africa, the Middle East, and the Indian subcontinent. WHO workers then vaccinated everyone in remaining endemic areas. Each subsequent outbreak or suspected outbreak was targeted by WHO teams who traveled to the outbreak site, quarantined individuals with active disease, and vaccinated all contacts. To break the chain of possible infection, they then immunized everyone who had contact with the contacts. This aggressive policy eliminated the active natural disease within a decade, and in 1980, WHO proclaimed the eradication of smallpox.

Polio, another viral disease that is largely preventable with an effective vaccine, is also targeted for eradication (endemic polio has been eradicated from the Western Hemisphere). Using much the same strategy to target polio as was used for smallpox, WHO undertook a massive immunization program in the 1990s, concentrating efforts in remaining endemic areas. By 2006, known endemic polio was restricted to Nigeria, India, Pakistan, and Afghanistan. Of 1,500 polio cases reported in 2006, 1,393 were in these countries. The remaining cases were spread among 11 countries in Africa, the Middle East, the Indian subcontinent, and Indonesia. Individual outbreaks are treated with massive regional immunization.

Hansen's disease (leprosy), another disease restricted to humans, is also targeted for eradication. Active cases of Hansen's disease can now be effectively treated with a multidrug therapy that cures the patient and also prevents spread of *Mycobacterium leprae,* the causal agent (∞ Section 34.5).

Other communicable diseases are candidates for eradication. These include Chagas' disease (treat active cases and destroy the insect vector of the *Trypanosoma cruzi* parasite in the American tropics) and dracunculiasis (treat drinking water to prevent transmission of *Dracunculus medinensis,* the Guinea helminth parasite in Africa, Saudi Arabia, Pakistan, and other places in Asia). Eradication of syphilis may be possible because the disease is found only in humans and is treatable. Rabies might be eradicated with oral baits that provide immunization of wild carnivores that constitute the reservoir.

33.8 MiniReview

Food and water purity regulations, vector control, immunization, quarantine, disease surveillance, and pathogen eradication are public health measures that reduce the incidence of communicable diseases.

- Compare public measures for controlling infectious disease caused by insect reservoirs and by human carriers.
- Identify public health methods used to halt the spread of an epidemic disease.
- Outline the steps taken to eradicate smallpox and polio.

33.9 Global Health Considerations

The World Health Organization has divided the world into six geographic regions for the purpose of collecting and reporting health information such as causes of morbidity and mortality. These geographic regions are Africa, the Americas (North America, the Caribbean, Central America, and South America), the eastern Mediterranean, Europe, Southeast Asia, and the Western Pacific. Here we compare mortality data from a relatively developed region, the Americas, to that from a developing region, Africa.

Infectious Disease in the Americas and Africa: A Comparison

The current worldwide population is now about 6,225,000,000. In 2002, about 57,029,000 individuals died, for an overall mortality rate of 9.2 deaths per 1,000 inhabitants per year. About 14.9 million, or 26%, of these deaths were attributable to infectious diseases. About 853 million people live in the Americas. Each year there are about 6 million deaths, or about 7 deaths per 1,000 inhabitants per year. In Africa, there are about 672 million people and about 10.7 million annual deaths, or about 15.9 deaths per 1,000 inhabitants per year. Although these statistics alone are cause for concern, examination of the causes of mortality in these regions is even more disturbing.

Figure 33.10 indicates that most African deaths are due to infectious diseases, whereas in the Americas, cancer and cardiovascular diseases are the leading causes of mortality. In Africa, there are about 6.7 million annual deaths due to infectious diseases, over ten times as many as in the Americas. The African death toll due to infectious diseases is 12% of the total deaths in the world and 45% of all worldwide deaths due to infectious diseases. In developed countries, the dramatic reduction in death rates from infection observed over the last century (Figure 1.8) is undoubtedly due to a number of advances in public health. Lack of resources in developing countries limits access to adequate sanitation, safe food and water, immunizations, health care, and medicines.

Travel to Endemic Areas

The high incidence of disease in many parts of the world is a concern for people traveling to such areas. However, travelers can be immunized against many of the diseases that are endemic in foreign countries. Recommendations for immunization for those traveling abroad are shown in **Table 33.7**. By international agreement, immunization certificates for yellow fever are required for travel to or from areas with endemic yellow fever. These areas include much of equatorial South America and Africa. Most other nonstandard immunizations are recommended only for people who are expected to be at high risk. In many parts of the world, travelers may be exposed to diseases for which there are no effective immunizations (for example, AIDS, Ebola hemorrhagic fever, dengue fever, amebiasis, encephalitis, malaria, and typhus). Travelers should take precautions such as avoiding unprotected sex, avoiding insect and animal bites, drinking only water that has been properly treated to kill all microorganisms, eating properly stored and prepared food, and undergoing antibiotic and chemotherapeutic programs for prophylaxis or for suspected exposure.

33.9 MiniReview

Infectious diseases account for 26% of all mortality worldwide. Most infectious diseases occur in developing countries. Infection control can be accomplished by application of public health measures.

- Contrast mortality due to infectious diseases in Africa and the Americas.
- List a series of infectious diseases for which you have not been immunized and with which you could come into contact next year.

33.10 Emerging and Reemerging Infectious Diseases

Infectious diseases are global, dynamic health problems. Here we examine some recent patterns of infectious disease, some reasons for the changing patterns, and the methods used by epidemiologists to identify and deal with new threats to public health.

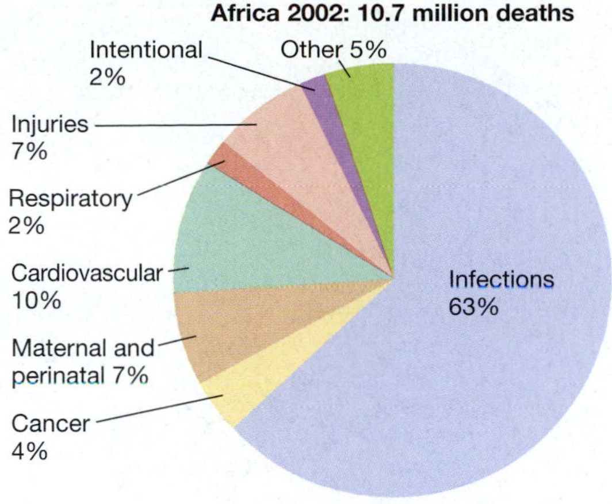

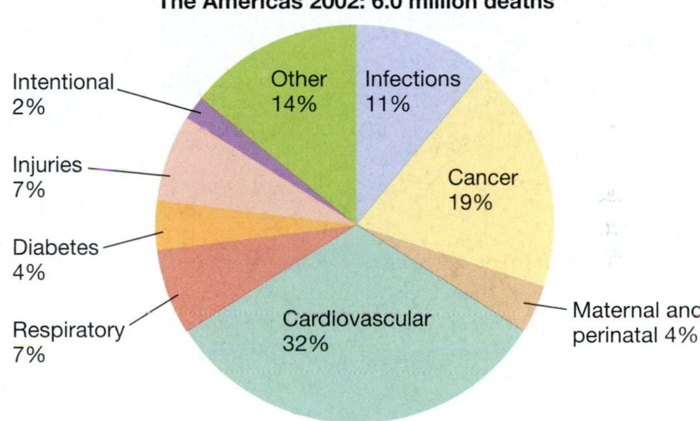

Figure 33.10 Causes of death in Africa and the Americas, 2002, by percentage of cause. There were 10.7 million deaths in Africa, 6.7 million due to infectious diseases. There were 6 million deaths in the Americas, 623,000 due to infectious diseases. Intentional deaths include murder, suicide, and war.

Emerging and Reemerging Diseases

The worldwide distribution of diseases can change dramatically and rapidly. Alterations in the pathogen, the environment, or the host population contribute to the spread of new diseases, with potential for high morbidity and mortality. Diseases that

Table 33.7 Immunizations required or recommended for international travel[a]

Disease	*Destination*	*Recommendation[b]*
Yellow fever	Tropical and subtropical countries, especially in sub-Saharan Africa and South America	*Immunization required* for entry and exit from endemic regions
Rabies	Rural, mountainous, and upland areas	*Immunization recommended* if direct contact with wild carnivores is anticipated
Typhoid fever	Many African, Asian, Central, and South American countries	*Immunization recommended* in areas endemic for typhoid fever

[a]*National Center for Infectious Diseases Travelers' Health*, U.S. Department of Health and Human Services, http://www.cdc.gov/travel/
[b]Vaccinations are generally recommended for diphtheria, pertussis, hepatitis A, hepatitis B, tetanus, polio, measles, mumps, rubella, and influenza as appropriate for the age of the traveler as well as the destination. Many U.S. citizens are immunized against these diseases through normal immunization practices. Requirements for specific vaccinations for each country are found at the website. Recommendations are also made for other appropriate infectious disease prevention measures, such as prophylactic drug therapy for malaria and plague prevention when visiting endemic areas. Yellow fever immunizations are required for travel to or from endemic areas.

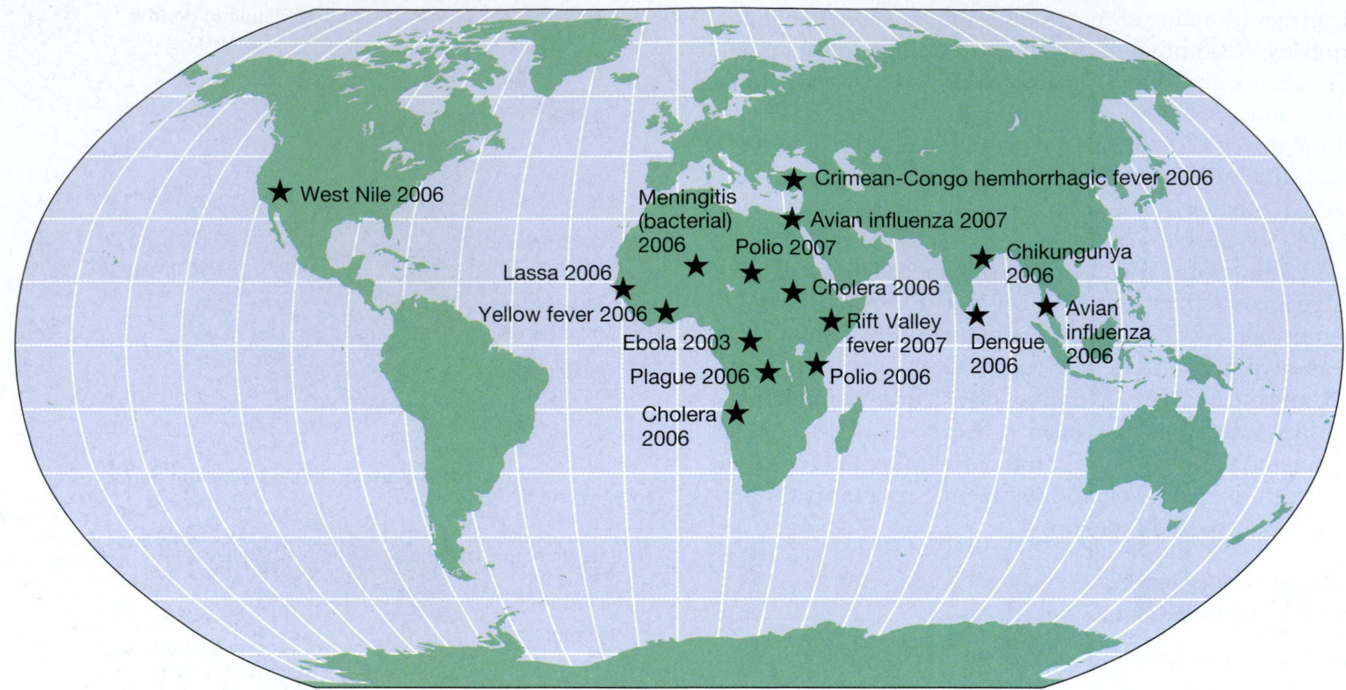

Figure 33.11 **Recent outbreaks of emerging and reemerging infectious diseases.** Emerging and reemerging diseases are first recognized as local epidemics. Recent human outbreaks of significant but rare diseases are recorded here. All of these diseases are capable of producing widespread epidemics and pandemics. Established pandemic diseases such HIV/AIDS and predictable annual epidemic diseases such as human influenza are not shown.

suddenly become prevalent are **emerging** diseases. Emerging infections are not limited to "new" diseases but also include **reemerging** diseases that were previously under control; reemerging diseases are especially a problem when antibiotics become less effective and public health systems fail. Recent dramatic examples of global emerging and reemerging disease are shown in **Figure 33.11**. Diseases with potential for emergence or reemergence are described in **Table 33.8**. In addition, the epidemic diseases listed in Table 33.2 have the potential to emerge or reemerge as widespread epidemics and pandemics.

Emerging epidemic diseases are not a new phenomenon. Some of the diseases that rapidly and sometimes catastrophically emerged in the past were syphilis (caused by *Treponema pallidum*) and plague (caused by *Yersinia pestis*). In the Middle Ages, up to one-third of all humans were killed by the plague epidemics that swept Europe, Asia, and Africa. Influenza caused a devastating worldwide pandemic in 1918–1919, claiming up to 230 million lives. In the 1980s, legionellosis (caused by *Legionella pneumophila*), acquired immunodeficiency syndrome (AIDS), and Lyme disease emerged as major new diseases. Current emerging pathogens in the United States include West Nile virus. Health officials worldwide are concerned about the potential for rapid emergence of pandemic influenza developing from avian influenza.

Emergence Factors

Some factors responsible for emergence of new pathogens are (1) human demographics and behavior; (2) technology and industry; (3) economic development and land use; (4) international travel and commerce; (5) microbial adaptation and change; (6) breakdown of public health measures; and (7) abnormal natural occurrences that upset the usual host–pathogen balance.

The demographics of human populations have changed dramatically in the last two centuries. In 1800, less than 2% of the world's population lived in urban areas. By contrast, today nearly one-half of the world's population lives in cities. The numbers, sizes, and population density in modern urban centers make disease transmission much easier. For example, dengue fever (Table 33.8) is now recognized as a serious hemorrhagic disease in tropical cities, largely due to the spread of dengue virus in the mosquito *Aedes aegypti*. The disease now spreads as an epidemic in tropical urban areas. Prior to 1950, dengue fever was rare, presumably because the virus was not easily spread among a more dispersed, smaller population.

Human behavior, especially in large population centers, also contributes to disease spread. For example, sexually promiscuous practices in population centers have been a major contributing factor to the spread of hepatitis and AIDS.

Technological advances and industrial development have a generally positive impact on living standards worldwide, but in some cases these advances have contributed to the spread of diseases. For example, although tremendous technological advances have been made in healthcare during the twentieth century, there has been a dramatic increase in nosocomial infections (Section 33.7). Antibiotic resistance in microorganisms

Table 33.8 Emerging and reemerging epidemic infectious diseases

Agent	*Disease and symptoms*	*Mode of transmission*	*Cause(s) of emergence*
Bacteria, Rickettsias, and Chlamydias			
Bacillus anthracis	Anthrax: respiratory distress, hemorrhage	Inhalation or contact with endospores	Bioterrorism
Borrelia burgdorferi	Lyme disease: rash, fever, neurological and cardiac abnormalities, arthritis	Bite of infective *Ixodes* tick	Increase in deer and human populations in wooded areas
Campylobacter jejuni	Campylobacter enteritis: abdominal pain, diarrhea, fever	Ingestion of contaminated food, water, or milk; fecal-oral spread from infected person or animal	Increased recognition; consumption of undercooked poultry
Chlamydia trachomatis	Trachoma, genital infections, conjunctivitis, infant pneumonia	Sexual intercourse	Increased sexual activity; changes in sanitation
Escherichia coli O157:H7	Hemorrhagic colitis; thrombocytopenia; haemolytic uremic syndrome	Ingestion of contaminated food, especially undercooked beef and raw milk	Development of a new pathogen
Haemophilus influenzae biogroup *aegyptus*	Brazilian purpuric fever; purulent conjunctivitis, fever, vomiting	Discharges of infected persons; flies are suspected vectors	Possible increase in virulence due to mutation
Helicobacter pylori	Gastritis, peptic ulcers, possibly stomach cancer	Contaminated food or water, especially unpasteurized milk; contact with infected pets	Increased recognition
Legionella pneumophila	Legionnaires' disease: malaise, myalgia, fever, headache, respiratory illness	Air-cooling systems, water supplies	Recognition in an epidemic situation
Mycobacterium tuberculosis	Tuberculosis: cough, weight loss, lung lesions; infection can spread to other organ systems	Sputum droplets (exhaled through a cough or sneeze) of a person with active disease	Immunosuppression, immunodeficiency
Neisseria meningitidis	Bacterial meningitis	Person-to-person contact	Urbanization, breakdown or lack of local public health surveillance
Staphylococcus aureus	Abscesses, pneumonia, endocarditis, toxic shock	Contact with the organism in a purulent lesion or on the hands	Recognition in an epidemic situation; possibly mutation
Streptococcus pyogenes	Scarlet fever, rheumatic fever, toxic shock	Direct contact with infected persons or carriers; ingestion of contaminated foods	Change in virulence of the bacteria; possibly mutation
Vibrio cholerae	Cholera: severe diarrhea, rapid dehydration	Water contaminated with the feces of infected persons; food exposed to contaminated water	Poor sanitation and hygiene; possibly introduced via bilge water from cargo ships
Viruses			
Chikungunya virus (CHIKV)	Debilitating fever, nausea, muscle pain, chronic fatigue	Bite of an infected mosquito (*Aedes* spp. in Africa and Asia)	Poor mosquito control; outdoor exposure; rapid spread to nonimmune populations
Dengue	Hemorrhagic fever	Bite of an infected mosquito (primarily *Aedes aegypti*)	Poor mosquito control; increased urbanization in tropics; increased air travel
Filoviruses (Marburg, Ebola)	Fulminant, high mortality, hemorrhagic fever	Direct contact with infected blood, organs, secretions, and semen	Unknown; in Europe and the United States, virus-infected monkeys shipped from developing countries via air
Hendravirus	Respiratory and neurological disease in horses and humans	Contact with infected bats, horses	Human intrusion into natural environment
Hantaviruses	Abdominal pain, vomiting, hemorrhagic fever	Inhalation of aerosolized rodent urine and feces	Human intrusion into virus or rodent ecological niche
Hepatitis B	Nausea, vomiting, jaundice; chronic infection leads to hepatocellular carcinoma and cirrhosis	Contact with saliva, semen, blood, or vaginal fluids of an infected person; mode of transmission to children not known	Probably increased sexual activity and intravenous drug abuse; transfusion (before 1978)
Hepatitis C	Nausea, vomiting, jaundice; chronic infection leads to hepatocellular carcinoma and cirrhosis	Exposure (percutaneous) to contaminated blood or plasma; sexual transmission	Recognition through molecular virology applications; blood transfusion practices, especially in Japan
Hepatitis E	Fever, abdominal pain, jaundice	Contaminated water	Newly recognized

Table 33.8 Emerging and reemerging epidemic infectious diseases *(continued)*

Agent	*Disease and symptoms*	*Mode of transmission*	*Cause(s) of emergence*
Viruses			
Human immunodeficiency viruses: HIV-1 and HIV-2	HIV disease, including AIDS: severe immune system dysfunction, opportunistic infections	Sexual contact with or exposure to blood or tissues of an infected person; vertical transmission	Urbanization; changes in lifestyle or mores; increased intravenous drug use; international travel; medical technology (transfusions and transplants)
Human papillomavirus	Skin and mucous membrane lesions (genital warts); strongly linked to cancer of the cervix and penis	Direct contact (sexual contact or contact with contaminated surfaces)	Increased surveillance and reporting
Human T cell lymphotrophic viruses (HTLV-I and HTLV-II)	Leukemias and lymphomas	Vertical transmission through blood or breast milk; exposure to contaminated blood products; sexual transmission	Increased intravenous drug abuse; medical technology (transfusion and transplantation)
Influenza	Fever, headache, cough, pneumonia	Airborne; especially in crowded, enclosed spaces	Animal-human virus reassortment; antigenic shift
Lassa	Fever, headache, sore throat, nausea	Contact with urine or feces of infected rodents	Urbanization and conditions favoring infestation by rodents
Measles	Fever, conjunctivitis, cough, red blotchy rash	Airborne; direct contact with respiratory secretions of infected persons	Deterioration of public health infrastructure supporting immunization
Monkeypox	Rash, lymphadenopathy, pulmonary distress	Direct contact with infected primates and other hosts	Travel to endemic areas, consumption and handling of infected primates and other hosts
Nipah virus	Hemorrhagic fever	Close contact with bats and pigs in Malaysia	Exposure to infected animals
Norwalk and Norwalk-like agents	Gastroenteritis; epidemic diarrhea	Most likely fecal-oral; vehicles may include drinking and swimming water, and uncooked foods	Increased recognition
Rabies	Acute viral encephalomyelitis	Bite of a rabid animal; contact with infected neural tissue	Introduction of infected host reservoir to new areas
Rift Valley	Febrile illness	Bite of an infective mosquito	Importation of infected mosquitoes and/or animals; development (dams, irrigation)
Rotavirus	Enteritis: diarrhea, vomiting, dehydration, and low-grade fever	Primarily fecal-oral; fecal-respiratory transmission can also occur	Increased recognition
Venezuelan equine encephalitis	Encephalitis	Bite of an infective mosquito	Movement of mosquitoes and hosts (horses)
West Nile virus	Meningitis, encephalitis	*Culex pipiens* mosquito and avian hosts	Agricultural development, increase in mosquito breeding areas, rapid spread to nonimmune populations
Yellow fever	Fever, headache, muscle pain, nausea, vomiting	Bite of an infective mosquito (*Aedes aegypti*)	Lack of effective mosquito control and widespread vaccination; urbanization in tropics; increased air travel
Protists and Fungi			
Candida	Candidiasis: fungal infections of the gastrointestinal tract, vagina, and oral cavity	Endogenous flora; contact with secretions or excretions from infected persons	Immunosuppression; medical devices (catheters); antibiotic use
Cryptococcus	Meningitis; sometimes infections of the lungs, kidneys, prostate, liver	Inhalation	Immunosuppression
Cryptosporidium	Cryptosporidiosis: infection of epithelial cells in the gastrointestinal and respiratory tracts	Fecal-oral, person to person, waterborne	Development near watershed areas; immunosuppression
Giardia lamblia	Giardiasis; infection of the upper small intestine, diarrhea, bloating	Ingestion of fecally contaminated food or water	Inadequate control in some water supply systems; immunosuppression; international travel
Microsporidia	Gastrointestinal illness, diarrhea; wasting in immunosuppressed persons	Unknown; probably ingestion of fecally contaminated food or water	Immunosuppression; recognition

Table 33.8 (continued)

Agent	Disease and symptoms	Mode of transmission	Cause(s) of emergence
Protists and Fungi			
Plasmodium	Malaria	Bite of an infective *Anopheles* mosquito	Urbanization; changing protist biology; environmental changes; drug resistance; air travel
Pneumocystis carinii	Acute pneumonia	Unknown; possibly reactivation of latent infection	Immunosuppression
Toxoplasma gondii	Toxoplasmosis; fever, lymphadenopathy, lymphocytosis	Exposure to feces of cats carrying the protists; sometimes foodborne	Immunosuppression; increase in cats as pets
Other Agents			
Bovine prions	Bovine spongiform encephalitis (BSE, animal) and variant Creutzfeld-Jacob disease (vCJD, human)	Foodborne	Consumption of contaminated beef

is another negative outcome of modern healthcare practices. For example, vancomycin-resistant enterococci and staphylococci and multiple drug-resistant *Streptococcus pneumoniae* are important emerging pathogens in developed countries.

Transportation, bulk processing, and central distribution methods have become increasingly important for quality assurance and economy in the food industry. However, these same factors can increase the potential for common-source epidemics when sanitation measures fail. For example, a single meat-processing plant spread *Escherichia coli* O157:H7 (Table 33.8) to at least 500 individuals in four states in the United States. The contaminated food source, ground beef, was recalled and the epidemic was eventually stopped, but not before several people died. There was a similar incident with spinach contaminated by *Escherichia coli* O157:H7 in runoff from a dairy farm in 2006. The *E. coli*-contaminated spinach was distributed nationally by a single packing plant and caused illness, kidney failure, and a few deaths; this prompted a United States Food and Drug Administration recommendation that fresh spinach not be consumed for a time (∞ Section 37.8 and Microbial Sidebar, "Spinach and *Escherichia coli* O157:H7," in Chapter 37).

Economic development and changes in land use can also promote disease spread. For example, Rift Valley fever, a mosquito-borne viral infection, has been on the increase since the completion of the Aswan High Dam in Egypt in 1970. The dam flooded 2 million acres, and the enlarged shoreline increased breeding grounds for mosquitoes at the edge of the new reservoir. The first major epidemic of Rift Valley fever developed in Egypt in 1977, when an estimated 200,000 people became ill and 598 died. There have been several epidemic outbreaks in the area since then, and the disease has become endemic near the reservoir.

Lyme disease, the most common vectorborne disease in the United States, is on the rise largely due to changes in land use patterns (∞ Section 35.4). Reforestation and the resulting increase in populations of deer and mice (the natural reservoirs for the disease-producing *Borrelia burgdorferi*) have resulted in greater numbers of infected ticks, the arthropod vector. In addition, larger numbers of homes and recreational areas in and near forests increase contact between the infected ticks and humans, consequently increasing disease incidence.

International travel and commerce also affect the spread of pathogens. For example, filoviruses (*Filoviridae*), a group of RNA viruses, cause fevers culminating in hemorrhagic disease in infected hosts. These untreatable viral diseases typically have a mortality rate above 20%. Most outbreaks have been restricted to equatorial central Africa, where the natural primate hosts and other vectors live. Travel of potential hosts to or from endemic areas is usually implicated in disease transmission. For example, one of the filoviruses was imported into Marburg, Germany, with a shipment of African green monkeys used for laboratory work. The virus quickly spread from the primate host to some of the human handlers. Twenty-five people were initially infected, and six more developed disease as a result of contact with the human cases. Seven people died in this outbreak of what became known as the *Marburg virus*. Another shipment of laboratory monkeys brought a different filovirus to Reston, Virginia, in the United States. Fortunately, the virus was not pathogenic for humans, but due to its respiratory transmission mode, the Reston virus infected and killed most of the monkeys at the Reston facility within days. These two filoviruses are closely related to the Ebola virus (Table 33.8).

Sporadic Ebola outbreaks in central Africa, often characterized by mortality rates greater than 50%, highlight a group of viral hemorrhagic fever pathogens for which there is no immunity or therapy. These pathogens could potentially be spread via air travel throughout the world in a matter of days. A highly contagious respiratory agent such as the Reston virus that also has the high mortality potential of the Ebola virus could devastate population centers worldwide in a matter of weeks.

Table 33.9 Virulence factors encoded by bacteriophages, plasmids, and transposons

Genetic element	*Organism*	*Virulence factors*
Bacteriophage	*Streptococcus pyogenes*	Erythrogenic toxin
	Escherichia coli	Shiga-like toxin
	Staphylococcus aureus	Enterotoxins A, D, E, staphylokinase, toxic shock syndrome toxin-1 (TSST-1)
	Clostridium botulinum	Neurotoxins C, D, E
	Corynebacterium diphtheriae	Diphtheria toxin
Plasmid	*Escherichia coli*	Enterotoxins, pili colonization factor, hemolysin, urease, serum resistance factor, adherence factors, cell invasion factors
	Bacillus anthracis	Edema factor, lethal factor, protective antigen, poly-D-glutamic acid capsule
	Yersinia pestis	Coagulase, fibrinolysin, murine toxin
Transposon	*Escherichia coli*	Heat-stable enterotoxins, aerobactin siderophores, hemolysin and pili operons
	Shigella dysenteriae	Shiga toxin
	Vibrio cholerae	Cholera toxin

Pathogen adaptation and change can contribute to pathogen emergence. For example, nearly all RNA viruses, including influenza, HIV, and the hemorrhagic fever viruses, undergo rapid, unpredictable genetic mutations. Because RNA viruses lack correction mechanisms for errors made during RNA replication, they incorporate genome mutations at an extremely high rate compared with most DNA viruses. The RNA viruses can present major epidemiological problems because of their easily changeable genomes.

Bacterial genetic mechanisms are capable of enhancing virulence and promoting emergence of new epidemics. Virulence-enhancing factors are often carried on mobile genetic elements such as bacteriophages, plasmids, and transposons. **Table 33.9** lists some virulence factors carried on these mobile genetic elements that contribute to pathogen emergence.

Drug resistance is another factor in the reemergence of some bacterial and viral pathogens. Although several drugs are effective against certain viral diseases, resistance to these drugs is very common, especially among the RNA viruses. For example, many strains of HIV develop resistance to azidothymidine (AZT) unless it is used in combination with other drugs (∞ Section 34.15).

A breakdown of public health measures is sometimes responsible for the emergence or reemergence of diseases. For instance, cholera (caused by *Vibrio cholerae*) can be adequately controlled, even in endemic areas, by providing proper sewage disposal and water treatment. However, in 1991 an outbreak of cholera due to contaminated municipal water supplies in Peru was one of the first indications that the current cholera pandemic had reached the Americas (∞ Section 36.5). In 1993, the municipal water supply of Milwaukee, Wisconsin, was contaminated with the chlorine-resistant protist *Cryptosporidium,* resulting in over 400,000 cases of intestinal disease, 4,000 of which required hospitalization. Enhanced filtration systems rid the water supply of the pathogen (∞ Section 36.6).

Inadequate public vaccination programs can lead to the resurgence of previously controlled diseases. For example, recent outbreaks of diphtheria in the former Soviet Union resulted from inadequate immunization of susceptible children due to the breakdown in public health infrastructures. Pertussis, another vaccine-preventable childhood respiratory disease, has increased recently in Eastern Europe and in the United States due to inadequate immunization among adults.

Finally, abnormal natural occurrences sometimes upset the usual host–pathogen balance. For example, hantavirus is a well-known human pathogen that occurs naturally in rodent populations, including some laboratory animals (∞ Section 35.2). A number of fatal cases of hantavirus infection and disease were reported in 1993 in the American Southwest and were linked to exposure to wild animal droppings. The likelihood of exposure to mice and droppings was increased due to a larger than normal wild mouse population resulting from near-record rainfall, a long growing season, and a mild winter. The pathogen density was increased by favorable environmental conditions. These factors enhanced exposure probability for susceptible human hosts.

Addressing Emerging Diseases

Many of the emerging diseases we consider here are absent from the official notifiable disease list for the United States (Table 33.5). How then do public health officials define emerging diseases and prevent major epidemics? The keys for addressing emerging diseases are recognition of the disease and intervention to prevent pathogen transmission.

The first step in disease recognition is surveillance. Epidemic diseases that exhibit particular clinical syndromes warrant intensive public health surveillance. These syndromes are (1) acute respiratory diseases, (2) encephalitis and aseptic meningitis, (3) hemorrhagic fever, (4) acute diarrhea, (5) clusterings of high fever cases, (6) unusual clusterings of any disease or deaths, and (7) resistance to common drugs or treatment. Thus, new diseases are primarily recognized because of their epidemic incidence, clusterings, and syndromes. As the prevalence and pathology of an emerging disease are recognized, the disease is added to the notifiable disease list.

Table 33.10 Bioterrorism agents and diseases

Agent (disease)
Bacteria and rickettsias
Bacillus anthracis (anthrax)
Brucella sp. (brucellosis)
Burkholderia mallei (glanders)
Burkholderia pseudomallei (melioidosis)
Chlamydia psittaci (psittacosis)
Vibrio cholerae (cholera)
Clostridium botulinum toxin (botulism[a])
Clostridium perfringens (Epsilon toxin[a])
Coxiella burnetii (Q fever)
Escherichia coli O157:H7 (gastrointestinal disease)
Francisella tularensis (tularemia)
Yersinia pestis (plague)
Staphylococcus aureus enterotoxin B[a]
Salmonella Typhi (typhoid fever)
Salmonella sp. (salmonellosis)
Shigella (shigellosis)
Rickettsia prowazekii (typhus)
Viral agents
Variola major (smallpox)
Alphaviruses (viral encephalitis)
Venezuelan equine encephalitis virus
Eastern equine encephalitis virus
Western equine encephalitis virus
Nipah virus
Viral hemorrhagic fevers viruses filoviruses; Ebola, Marburg arenaviruses; Lassa, Machupo hantaviruses
Protists
Cryptosporidium parvum (waterborne gastroenteritis)
Plants
Ricinus communis (ricin toxin from castor bean[a])

[a]Preformed toxin; all other agents require infection.

Source: Information is from the Centers for Disease Control and Prevention, Atlanta, GA, USA.

For example, AIDS was recognized as a disease in 1981 and became a reportable disease in 1984. Likewise, outbreaks of gastrointestinal disease due to enteropathogenic *Escherichia coli* O157:H7 have increased in recent years, and the strain became a reportable disease in 1995 (Table 33.5).

Intervention to prevent spread of emerging infections must be a public health response employing various methods. Disease-specific intervention is the key to controlling individual outbreaks. Methods such as quarantine, immunization, and drug treatment must be applied to contain and isolate outbreaks of specific diseases. Finally, for vectorborne and zoonotic diseases, the nonhuman host or vector must be identified to allow intervention in the life cycle of the pathogen and interrupt transfer to humans. International public health surveillance and intervention programs were instrumental in controlling the emergence of severe acute respiratory syndrome (SARS), a disease that emerged rapidly, explosively, and unpredictably from a zoonotic source (see the Microbial Sidebar, "SARS as a Model of Epidemiological Success").

33.10 MiniReview

Changes in host, vector, or pathogen conditions, whether natural or artificial, can result in conditions that encourage the explosive emergence or reemergence of infectious diseases. Global surveillance and intervention programs must be in place to prevent new epidemics and pandemics.

- What factors are important in the emergence or reemergence of potential pathogens?
- Indicate general and specific methods that would be useful for dealing with emerging infectious diseases.

33.11 Biological Warfare and Biological Weapons

Biological warfare is the use of biological agents to incapacitate or kill a military or civilian population in an act of war or terrorism. Biological weapons have been used against targets in the United States, and biological weapon-making facilities are suspected to be in the hands of several governments as well as extremist groups.

Characteristics of Biological Weapons

Biological weapons are organisms or toxins that are (1) easy to produce and deliver, (2) safe for use by the offensive soldiers, and (3) able to incapacitate or kill individuals under attack in a reproducible and consistent manner. Many organisms or biological toxins fit these rather general criteria, and we discuss several of these below.

Although biological weapons are potentially useful in the hands of conventional military forces, the greatest likelihood of biological weapons use is probably by terrorist groups. This is in part due to the availibility and low cost of producing and propagating many of the organisms useful for biological warfare. Biological weapons are accessible to nearly every government and well-financed private organization.

Candidate Biological Weapons

Virtually all pathogenic bacteria or viruses are potentially useful for biological warfare, and several of the most likely candidate organisms are relatively simple to grow and disseminate. Commonly considered biological weapons agents are listed in **Table 33.10**. The most commonly mentioned candidate as a biological weapon is *Bacillus anthracis,* the causal agent of anthrax. We discuss anthrax in the next section.

Other important candidates as bacterial biological weapons include *Yersinia pestis,* the organism responsible for plague, *Brucella abortus* (fever and bacteremia; brucellosis),

Microbial Sidebar

SARS as a Model of Epidemiological Success

Handling of the Severe Acute Respiratory Disease (SARS) epidemic early in this decade is an excellent example of epidemiological success. Like many other rapidly emerging diseases, SARS was viral and zoonotic in origin. Such characteristics have the potential to trigger explosive disease in humans when the infectious agents cross host species barriers. In many cases, the original viruses have been traced back to an animal host, but in others, the original host is unknown or is so ubiquitous that adequate vector control is nearly impossible, and thus disease persists. For example, West Nile virus is transmitted through mosquitoes that feed on infected birds. Although public health officials knew from the outset that West Nile disease would be seasonal and related to mosquitoes and infected birds, they could not prevent its spread. Thus human West Nile cases spread quickly across the United States over a 5-year period, starting in Florida in 2001, and are still with us today.

In contrast to West Nile disease, a different scenario surrounds the SARS epidemic. The SARS epidemic originated in late 2002 in Guandong Province, China. By the following February, the virus had spread to 32 countries. Global travel provided the major vehicle for SARS dissemination. The etiology of SARS was quickly traced to a coronavirus derived from an animal source. The coronavirus entered the human food chain through exotic food animals such as civet cats. The SARS coronavirus (SARS-CoV), shown in **Figure 1**, originated in bats. Civet cats consumed fruit contaminated by the bats and acquired the virus in this way. SARS-CoV likely evolved over an extended period of time in bats and developed, quite by accident, the ability to infect civet cats and then humans.

Much like common cold viruses, SARS-CoV is a relatively hardy, easily spread RNA virus that is difficult to contain. Once in humans, SARS-CoV is very contagious because it can be spread in several ways, including person to person by sneezing and coughing and by contact with contaminated fomites or feces. Ordinarily, a new coldlike virus would be of little concern, but SARS-CoV causes infections with significant morbidity and mortality. There have been about 8,500 known SARS-CoV infections and over 800 deaths, for an overall mortality rate of nearly 10%. In persons over 65 years of age, the mortality rate approached 50%, attesting to SARS-CoV virulence as a human pathogen. About 20% of all SARS cases were in healthcare workers, demonstrating the high infectivity of the virus. Standard containment and infection control methods practiced by healthcare personnel were not effective in controlling spread of the disease. When this was realized, SARS patients were confined for the course of the disease in strict isolation in negative-pressure rooms. To prevent infection, healthcare workers wore respirators when working with SARS patients or when handling fomites (bed linens, eating utensils, and so on) contaminated with SARS-CoV.

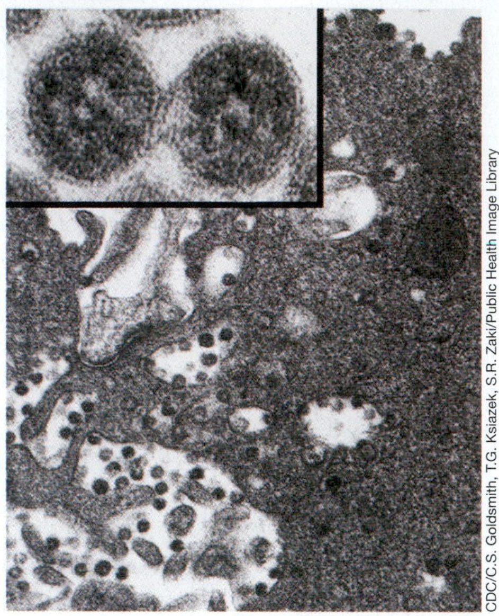

CDC/C.S. Goldsmith, T.G. Ksiazek, S.R. Zaki/Public Health Image Library

Figure 1 Severe acute respiratory virus syndrome corona virus (SARS-CoV). *The upper left panel shows isolated SARS-CoV virions. An individual virion is 133 nm in diameter. The large panel shows coronaviruses within the cytoplasmic membrane-bound vacuoles and in the rough endoplasmic reticulum of host cells. The virus replicates in the cytoplasm and exits the cell through the cytoplasmic vacuoles.*

The recognition and containment of the clinical disease was the start of an international response involving clinicians, scientists, and public officials. Almost immediately, travel to and from the endemic area was restricted, limiting further outbreaks. SARS-CoV isolation was achieved rapidly, and this information was used to develop the PCR tests used to track the disease. As laboratory work progressed, epidemiologists traced the virus back to the civet food source in China and stopped further transmission to humans by restricting the sale of civets and other foods from wild sources. These actions collectively stopped the outbreak.

SARS is an example of a serious infection that emerged very rapidly from a unique source. However, rapid isolation and characterization of the SARS pathogen, nearly instant development of worldwide notification procedures and diagnostic tests, and a concerted effort to understand the biology and genetics of this novel pathogen quickly controlled the disease; there has not been another case of SARS since early 2004. The rapid emergence of SARS, and the equally rapid and successful international effort to identify and control the outbreak, provide a model for the control of emerging epidemics.

As international travel and trade expand, the chances for propagation and rapid dissemination of new exotic diseases will continue to increase. We should therefore anticipate the emergence of other serious infectious zoonotic diseases, including pandemic influenza. We hope that the lessons learned from the SARS epidemic will pay dividends when other emerging diseases appear.

Francisella tularensis ("rabbit fever"), and *Salmonella* (foodborne and waterborne illnesses). Viral pathogens with biological weapons potential include smallpox, hemorrhagic fever viruses, and encephalitis viruses. These agents cause diseases associated with significant morbidity, and some have very high mortality rates.

Bacterial toxins such as botulinum toxin from *Clostridium botulinum* are also potential biological weapons. Large amounts of the preformed toxin delivered to a population through a common vehicle such as drinking water could have devastating consequences: The lethal dose of botulinum toxin for a human is 2 μg or less.

Smallpox

Smallpox virus has intimidating potential as a biological warfare agent because it can be easily spread by contact or aerosol spray and it has a mortality rate of 30% or more. However, its potential for use as a biological weapon is considered low, partly because the only known stocks of smallpox virus are in guarded repositories in the United States and Russia, but a finite possibility remains for terrorist groups or military forces to gain access to smallpox virus. Because of this, the United States government has made provisions to immunize frontline healthcare and public safety personnel for smallpox. Although an extremely effective smallpox vaccine exists using the closely related vaccinia virus as the immunogen, this vaccine has not been in general use for almost 30 years because wild smallpox was eradicated worldwide by 1977. In addition, the vaccine can have serious side effects. As a result, over 90% of the current worldwide population is now inadequately vaccinated and susceptible to the disease. Preparations for a potential smallpox attack in the United States have included recommendations for immunization of select individuals: persons having close contact with smallpox patients; workers evaluating, caring for, or transporting smallpox patients; laboratory personnel handling clinical specimens from smallpox patients; and other persons such as housekeeping personnel who might contact infectious materials from smallpox patients.

Although vaccinia immunization is very effective, it carries significant risk. Normal vaccine reactions include formation of a pustule, with a scab that falls off in 2 to 3 weeks, leaving a small scar. Many people have mild adverse reactions such as fevers and rashes. Vaccination is not currently recommended for persons with eczema or other chronic or acute skin conditions, heart disease, pregnant women, and those with reduced immune competence, such as individuals using anti-inflammatory steroid medications and those with HIV/AIDS.

About 1 in 1,000 vaccinated individuals develop serious complications from the vaccine. These include myocarditis and erythema multiforme, a toxic or allergic response to the vaccine. Generalized vaccinia (systemic vaccinia infection) occasionally occurs in individuals with skin conditions such as eczema. Life-threatening progressive vaccinia sometimes occurs in vaccinated individuals, usually in those who are immunosuppressed due to therapy or disease. On average, one to two people per million who receive the vaccine will die from a vaccinia virus complication.

Delivery of Biological Weapons

Most organisms suitable for biological weapons use can be spread as an aerosol, providing simple, rapid, widespread dissemination leading to infection. Examples of several aerosol exposures are instructive.

In 1962, one of the last outbreaks of smallpox in a developed country occurred in Germany. A German worker developed smallpox after returning from Pakistan, a country with endemic smallpox. The individual was immediately hospitalized and quarantined, but the patient had a cough, and the aerosolized virus caused illness in 19 vaccinated individuals; at least one individual died from the resulting infection.

There have been planned bioterrorist attacks in the United States and other countries even before the anthrax attacks of 2001 (Section 33.12). In 1984 in The Dalles, Oregon (United States), cultists inoculated a salad bar with a *Salmonella typhimurium* culture in aerosol form at ten local restaurants, causing 751 cases of foodborne salmonellosis in a region that usually has less than 10 cases per year. In 1995, a radical political group released Saran nerve gas into a Tokyo subway, killing several people and injuring scores of others. Although this was a chemical weapon, this group also possessed anthrax cultures, bacteriological media, drone airplanes, and spray tanks.

Delivery of preformed bacterial toxins such as botulinum toxin or staphylococcal enterotoxin to large populations is somewhat impractical because most potent exotoxins are proteins that would lose effectiveness as they are diluted or are destroyed in common sources such as drinking water. However, delivery of toxins could be aimed at selected individuals and small groups, or delivered randomly to instigate panic.

Prevention and Response to Biological Weapons

Proactive measures against the deployment of biological weapons have already begun with periodic planned efforts to update the international agreements of the 1972 Biological and Toxic Weapons Convention. The fifth and most recent update was in 2002. At the practical level, governments are now supporting the large-scale production and distribution of vaccines along with the development of strategic and tactical plans to prevent and contain biological weapons.

The United States government, through the Centers for Disease Control and Prevention, has devised and enhanced the Select Agent Program surveillance systems to monitor possession and use of potential bioterrorism agents. The CDC Laboratory Response Network and the Health Alert Network have been upgraded to enhance their diagnostic capabilities and increase the reporting abilities of local and regional healthcare centers to rapidly identify bioterrorism events as well as emerging diseases.

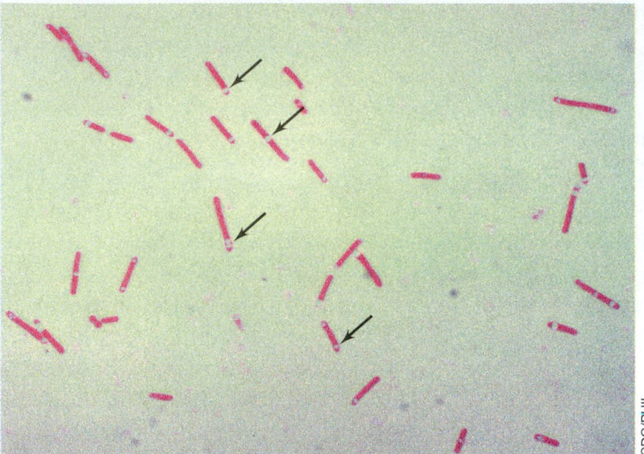

(a)

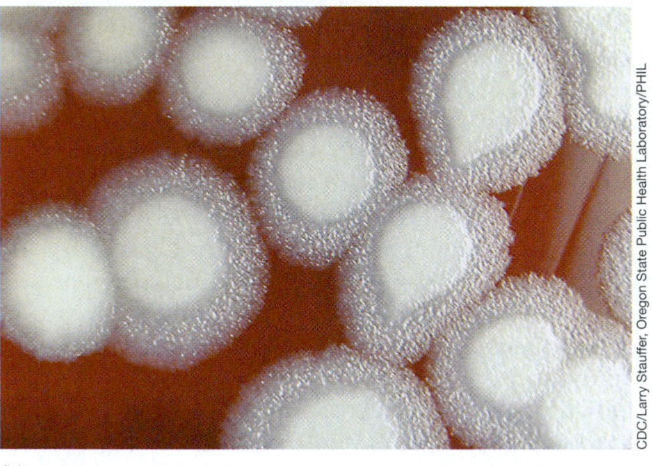

(b)

Figure 33.12 ***Bacillus anthracis.*** *(a) B. anthracis* is a gram-positive endospore-forming rod approximately 1 μm in diameter and 3–4 μm in length. Note the developing endospores (arrows). *(b) B. anthracis* colonies on blood agar. The nonhemolytic colonies take on a characteristic "ground glass" appearance.

33.11 MiniReview

Bioterrorism is a threat in a world of rapid international travel and easily accessible technical information. Biological agents can be used as weapons by military forces or by terrorist groups. Aerosols or common sources such as food and water are the most likely modes of inoculation. Prevention and containment measures rely on a well-prepared public health infrastructure.

- What characteristics make a pathogen or its products particularly useful as a biological weapon?
- Identify two infectious agents that could be effective biological weapons. How could the agents be disseminated?

33.12 Anthrax as a Biological Weapon

Bacillus anthracis is a preferred agent for biowarfare and bioterrorism. Here we discuss its unique properties, the diseases it causes, and methods for prevention, diagnosis, and treatment.

Biology and Growth

Bacillus anthracis is a ubiquitous saprophytic soil inhabitant. It grows as an aerobic gram-positive rod, 1 μm in diameter and 3–4 μm in length. As with other species of the genus *Bacillus*, *B. anthracis* produces endospores resistant to heat and drying (**Figure 33.12*a***). Endospore formation enhances the ability to disseminate *B. anthracis* in aerosols. Viable endospores are sometimes recovered from contaminated animal products such as hides and fur. Growth on blood agar results in large colonies with a characteristic "ground glass" appearance (Table 33.12*b*). Strains having a poly-D-glutamic acid capsule are resistant to phagocytosis.

Infection and Pathogenesis

Bacillus anthracis endospores are the standard means of acquiring anthrax. The disease usually affects domestic animals, especially ungulates—cows, sheep, and goats. The number of infections in animals, although considerable, is not known. The animals acquire the disease from plants or soil in pastures. In humans and animals, there are three forms of the disease. *Cutaneous anthrax* is contracted when abraded skin is contaminated by *B. anthracis* endospores (**Figure 33.13*a***). *Gastrointestinal anthrax* is contracted from consumption of endospore-contaminated plants or anthrax-infected carcasses. Cutaneous anthrax cases are rare in the United States. Human gastrointestinal anthrax is rarely seen. *Pulmonary anthrax* is contracted when the endospores are inhaled. Inhalation of the endospores or the live bacteria results in pulmonary infections characterized by pulmonary and cerebral hemorrhage (Figure 33.13*b*). Untreated pulmonary anthrax infections have a mortality rate of nearly 100%. Pulmonary anthrax cases, even in agricultural workers, are extremely rare. The last naturally acquired pulmonary anthrax case in the United States occurred in 1976. However, several cases of pulmonary anthrax were identified in 2001 due to bioterrorism events.

Pathogenesis results from inhalation of 8000–50,000 endospores of an encapsulated toxigenic strain. Pathogenic *B. anthracis* produces three proteins—*protective antigen* (PA), *lethal factor* (LF), and *edema factor* (EF). PA and LF form *lethal toxin*. PA and EF form *edema toxin*. PA is the cell-binding B component of these AB-type toxins (∞ Table 28.4). EF cause edema, and LF causes cell death. Growth of *B. anthracis* in the lymph nodes and lymphatic tissues draining the lungs leads to edema and cell death, culminating in tissue destruction, shock, and death.

Clinical symptoms can start with sore throat, fever, and muscle aches. After several days, symptoms include difficulty in breathing, followed by systemic shock. Fatality rates can approach 90% even when exposure is recognized and treatment is started, and can be nearly 100% in cases for which treatment is not started until after the onset of symptoms.

Weaponized Anthrax

The term *weaponized* is applied to strains and preparations of *B. anthracis*, usually in endospore form, that exhibit properties that enhance dissemination and use as biological weapons. Such strains and preparations were developed in several countries in the post-World War II era, but overt development of new biological weapons was halted by international treaty in 1972. The physical characteristics of the weaponized anthrax preparations typically include a small particle size, usually interspersed with a very fine particulate agent such as talc. This small-particle, powdery form ensures that the endospores will spread easily by air currents. Thus, opening an envelope containing endospores or releasing the powder–endospore mixture into a ventilation system or other air current has the potential to contaminate surrounding areas and personnel.

A weaponized form of anthrax was used in a series of bioterrorism attacks in the United States in 2001. These incidents were carried out by mailing envelopes or packages containing weaponized anthrax endospores. The attacks were apparently directed at the news media (Florida) and the government (Washington, DC area). A third focus of attack, the Pennsylvania-New Jersey-New York area, had no defined single target, but disrupted mail service in the Northeast; some anthrax-contaminated mail facilities were still not in use 2 years later. In all, there were 22 anthrax infections. Eleven were cutaneous anthrax. Of 11 cases of inhalation anthrax, 5 cases resulted in death. The bioterrorists were never identified.

The incidents in the United States were not the first or the most serious anthrax biological weapons infections. In a previous incident, *B. anthracis* spores were inadvertently released into the atmosphere from a biological weapons facility in Sverdlovsk, Russia, in 1979. Less than 1 g of endospores was released, and everyone in the area surrounding the facility was immunized and given prophylactic antibiotic therapy as soon as the first anthrax case was diagnosed. However, 77 individuals outside the facility contracted pulmonary anthrax and 66 died.

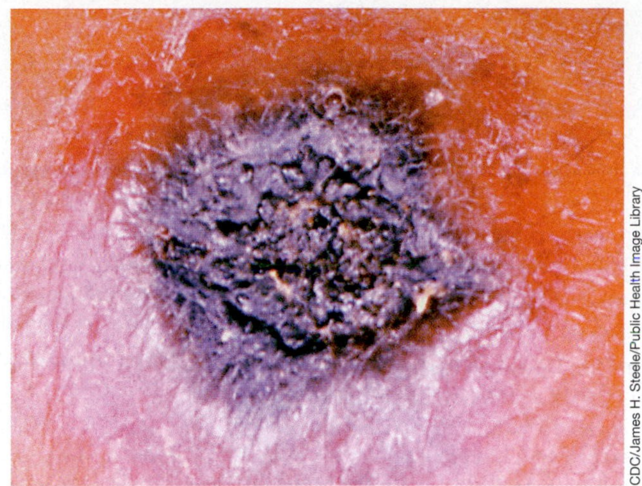

(a)

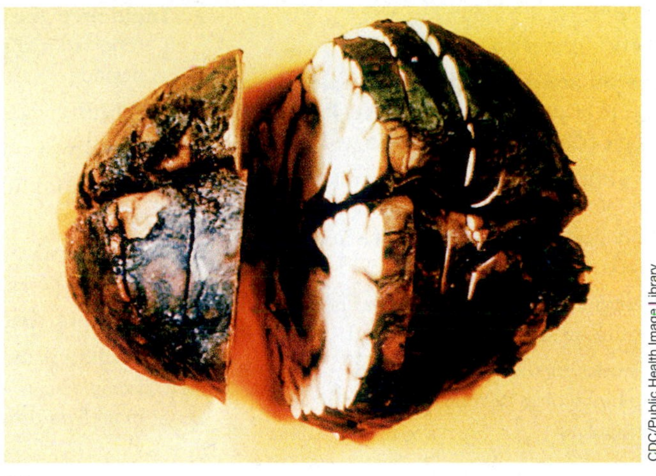

(b)

Figure 33.13 Anthrax. *(a)* Cutaneous anthrax. The blackened lesion on the forearm of a patient, about 2 cm in diameter, results from tissue necrosis. Cutaneous anthrax, even when untreated, usually is a localized, nonlethal infection. *(b)* Inhalation anthrax. The fixed and sectioned human brain shows hemorrhagic meningitis (dark coloration) due to a fatal case of inhalation anthrax.

Vaccination, Prophylaxis, Treatment, and Diagnosis

Vaccination for anthrax has thus far been restricted to individuals who are considered at risk. This includes agricultural animal workers and military personnel. The current vaccine, called *anthrax vaccine adsorbed* (AVA), is prepared from a cell-free *B. anthracis* culture filtrate.

Treatment of *B. anthracis* infection, which seems to have a minimum incubation time of about 8 days, is usually done with antibiotics. Ciprofloxacin, a broad-spectrum quinolone antibiotic, is used against strains that are penicillin-resistant, including many laboratory and biological weapons strains. Ciprofloxacin is also used as a prophylactic measure to treat potentially exposed individuals.

Rapid diagnostic tests are available to detect microbial endospores. However, positive identification of *B. anthracis* relies on culture techniques and direct observation of either infected tissues or cultured organisms. The characteristic ground-glass appearance on blood agar, coupled with the isolation of gram-positive endospore-forming rods growing in extended chains, is presumptive evidence for *B. anthracis* (Table 33.12).

33.12 MiniReview

Bacillus anthracis has emerged as an important pathogen because of its use as a biological weapon. Highly infective weaponized endospore preparations have been used as bioterror agents. Inhalation anthrax has a fatality rate of over 90% in untreated individuals. Effective treatment relies on timely observation and diagnosis of symptoms. Treatment for inhalation anthrax does not guarantee survival.

- What factors contribute to the preferred use of *B. anthracis* as a biological weapon?
- Indicate the steps you would use to identify the use of *B. anthracis* in a bioterror attack. Indicate treatment steps for potential victims.

Review of Key Terms

Acute infection short-term infection usually characterized by dramatic onset

Biological warfare the use of biological agents to incapacitate or kill humans

Carrier subclinically infected individual who may spread a disease

Centers for Disease Control and Prevention (CDC) an agency of the United States Public Health Service that tracks disease trends, provides disease information to the public and to healthcare professionals, and forms public policy regarding disease prevention and intervention

Chronic long-term infection

Common-source epidemic an epidemic resulting from infection of a large number of people from a single contaminated source

Emerging infection infectious disease whose incidence has increased recently or whose incidence threatens to increase in the near future

Endemic disease a disease that is constantly present, usually in low numbers

Epidemic the occurrence of a disease in unusually high numbers in a localized population

Epidemiology the study of the occurrence, distribution, and determinants of health and disease in a population

Fomite an inanimate object that, when contaminated with a viable pathogen, can transfer the pathogen to a host

Herd immunity resistance of a population to a pathogen as a result of the immunity of a large portion of the population

Host-to-host epidemic an epidemic resulting from person-to-person contact, characterized by a gradual rise and fall in number of cases

Incidence the number of new disease cases reported in a population in a given time period

Morbidity incidence of illness in a population

Mortality incidence of death in a population

Nosocomial infection healthcare-associated infection

Outbreak the occurrence of a large number of cases of a disease in a short period of time

Pandemic a worldwide epidemic

Prevalence the total number of new and existing disease cases reported in a population in a given time period

Public health the health of the population as a whole

Quarantine the practice of restricting the movement of individuals with highly contagious serious infections to prevent spread of the disease

Reemerging infection infectious disease, thought to be under control, that produces a new epidemic

Reservoir a source of viable infectious agents from which individuals may be infected

Surveillance observation, recognition, and reporting of diseases as they occur

Vector a living agent that transfers a pathogen (note alternative usage in Chapter 26)

Vehicle nonliving source of pathogens that infect large numbers of individuals; common vehicles are food and water

Zoonosis a disease that occurs primarily in animals but can be transmitted to humans

Review Questions

1. List the five most common causes of mortality due to infectious diseases throughout the world. Are any of these diseases preventable by immunization (Section 33.1)?
2. Distinguish between mortality and morbidity, prevalence and incidence, and epidemic and pandemic, as these terms relate to infectious disease (Section 33.2).
3. Explain the difference between a chronic carrier and an acute carrier of an infectious disease (Section 33.3).
4. Give examples of host-to-host transmission of disease via direct contact. Also give examples of indirect host-to-host transmission of disease via vector agents and fomites (Section 33.4).
5. How can immunity to a pathogen by a large proportion of the population protect the nonimmune members of the population from acquiring a disease? Will this herd immunity work for diseases that have a common source, such as water? Why or why not (Section 33.5)?
6. Identify the major risk factors for acquiring human immunodeficiency virus (HIV) infection in the United States. Does this pattern hold for all geographic regions (Section 33.6)?
7. Healthcare environments are conducive to the spread of infectious diseases. Review the reasons for the enhanced spread of infection in healthcare facilities. What are the sources of most healthcare-associated infections (Section 33.7)?
8. Describe the major medical and public health measures developed in the twentieth century that were instrumental for controlling the spread of infectious diseases in developed countries (Section 33.8).
9. Compare the role of infectious diseases on mortality in developed and developing countries (Section 33.9).
10. Review the major reasons for the emergence of new infectious diseases. What methods are available for identifying and controlling the emergence of new infectious diseases (Section 33.10)?
11. Describe the general properties of an effective biological warfare agent. How does smallpox meet these criteria? Identify other organisms that meet the basic requirements for a bioweapon (Section 33.11).
12. Describe the use of *Bacillus anthracis* as a biological weapon. Devise a plan to protect yourself against a *B. anthracis* attack (Section 33.12).

Application Questions

1. Smallpox, a disease that was limited to humans, was eradicated. Plague, a disease with a zoonotic reservoir in rodents (Table 33.2) can never be eradicated. Explain this statement and why you agree or disagree with the possibility of eradicating plague on a global scale. Devise a plan to eradicate plague in a limited environment such as a town or city. Be sure to use methods that involve the reservoir, the pathogen, and the host.
2. Acquired immunodeficiency syndrome (AIDS) is a disease that can be eliminated because it is propagated by person-to-person contact and there are no known animal reservoirs. Do you agree or disagree with this statement? Explain your answer. Design a program for eliminating AIDS in a developed country and in a developing country. How would these programs differ? What factors would work against the success of your program, both in terms of human behavior and in terms of the AIDS disease itself? Why are the numbers of HIV-infected and AIDS patients continuing to grow, especially in developing countries? HIV/AIDS incidence (new cases) in developed countries has been virtually unchanged in this century (Figure 33.7). The numbers of individuals living with AIDS, however, is increasing. Explain this contradiction.
3. Travel to developing countries involves some exposure to infectious diseases. What general precautions should you take before, during, and after visits to developing countries? Where can you obtain information concerning infectious diseases in a specific foreign country? When you return from a foreign country, are you a disease risk to your family or your associates? Explain.
4. Identify a specific pathogen that would be a suitable agent for effective biological warfare. Describe the properties of the pathogen in the context of its use as a biological weapon. Describe equipment and other resources necessary for growing large amounts of the pathogen. Identify a suitable delivery method. Since you will propagate and deliver the pathogen, describe the precautions you will take to protect yourself. Now reverse your role. As a public health official at your university, describe how you would recognize and diagnose the disease caused by the agent. Indicate the measures you would take to treat the illnesses caused by the agent. How could you best limit the damage? Would quarantine and isolation methods be useful? What about immunization and antibiotics?

34

Person-to-Person Microbial Diseases

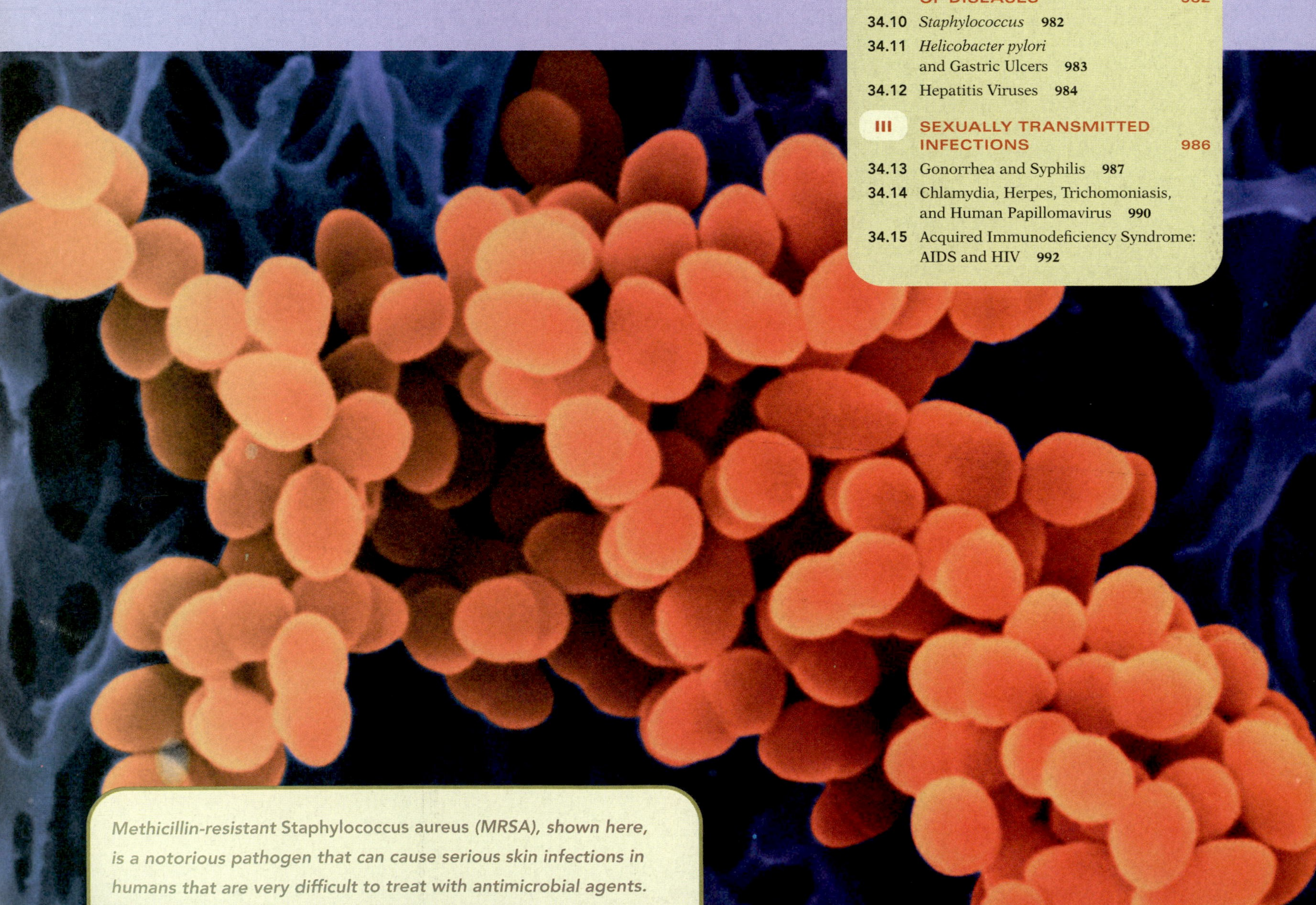

Methicillin-resistant Staphylococcus aureus *(MRSA), shown here, is a notorious pathogen that can cause serious skin infections in humans that are very difficult to treat with antimicrobial agents.*

Perhaps more than a million microbial species exist in nature, but only a few hundred species cause disease. Most microorganisms grow and metabolize independent of interactions with other organisms, and many microorganisms are closely associated with plants or animals, including humans, in beneficial relationships. However, pathogenic species have profoundly negative effects on host organisms. In the next four chapters, representative human pathogens and their biology are examined, as well as the pathology, diagnosis, treatment, and prevention of the diseases they cause. Our coverage is organized based on the pathogen's mode of transmission, which presents infectious disease in the context of the ecology of the pathogen. In this chapter we consider diseases transmitted from person to person. In Chapters 35 through 37, diseases whose modes of transmission require animal or arthropod vectors, or common sources such as soil, water, and food, are examined.

Connections between and among seemingly unrelated microorganisms will be made in our disease coverage by examining pathogens according to their modes of transmission and the diseases they cause. For example, influenza virus and streptococci cause diseases with overlapping symptoms, although the causal agents, one viral and one bacterial, are very different. Here these pathogens are discussed together because they are spread from person to person via a respiratory route. Using this approach, we will establish the connections between biologically diverse, but ecologically and pathogenically related, disease agents.

I AIRBORNE TRANSMISSION OF DISEASES

Aerosols, such as those generated by a human sneeze (**Figure 34.1**), are important vehicles for person-to-person transmission of many infectious diseases. Most respiratory diseases are spread almost exclusively in this fashion. For example, *Mycobacterium tuberculosis*, the bacterium that causes the disease tuberculosis, has spread in this way to infect at least one-third of the world's population. In addition, influenza and cold viruses are passed so easily by the respiratory route that virtually everyone is infected, in the case of colds sometimes several times a year.

34.1 Airborne Pathogens

Microorganisms found in air are derived from soil, water, plants, animals, people, and other sources. In outdoor air, soil organisms predominate. Indoors, the concentration of microorganisms is considerably higher than outdoors, especially for organisms that originate in the human respiratory tract.

Most microorganisms survive poorly in air. As a result, pathogens are effectively transmitted among humans only over short distances. Certain pathogens, however, survive under dry conditions and can remain alive in dust for long periods of time. Gram-positive bacteria (*Staphylococcus*,

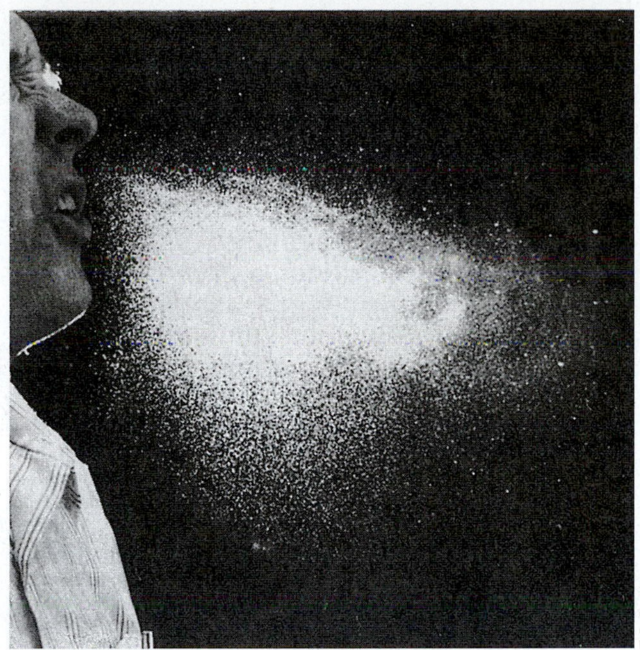
T. D. Brock

Figure 34.1 **High speed photograph of an unstifled sneeze.**

Streptococcus) are in general more resistant to drying than gram-negative bacteria because of their thick, rigid cell wall. Likewise, the waxy layer of *Mycobacterium* cell walls (Section 34.5) resists drying and promotes survival. The endospores of endospore-forming bacteria are extremely resistant to drying but are not generally passed from human to human in the endospore form.

Large numbers of moisture droplets are expelled during sneezing (Figure 34.1), and a sizeable number are expelled during coughing or simply talking. Each infectious droplet is about 10 μm in diameter and may contain one or two microbial cells or virions. The initial speed of the droplet movement is about 100 m/sec (more than 325 km/h) in a sneeze and ranges from 16 to 48 m/sec during coughing or shouting. The number of bacteria in a single sneeze varies from 10,000 to 100,000. Because of their small size, the moisture droplets evaporate quickly in the air, leaving behind a nucleus of organic matter and mucus to which bacterial cells are attached.

Respiratory Infections

Humans breathe about 500 million liters of air in a lifetime, much of it containing microorganism-laden dust. The speed at which air moves through the respiratory tract varies, and in the lower respiratory tract the rate is quite slow. As air slows down, particles in it stop moving and settle. Large particles settle first and the smaller ones later; only particles smaller than 3 μm travel as far as the bronchioles in the lower respiratory tract (**Figure 34.2**). Of course, most pathogens are much smaller than this, and different organisms characteristically colonize the respiratory tract at different levels. The upper and lower respiratory tracts offer decidedly different environments, favoring different microorganisms.

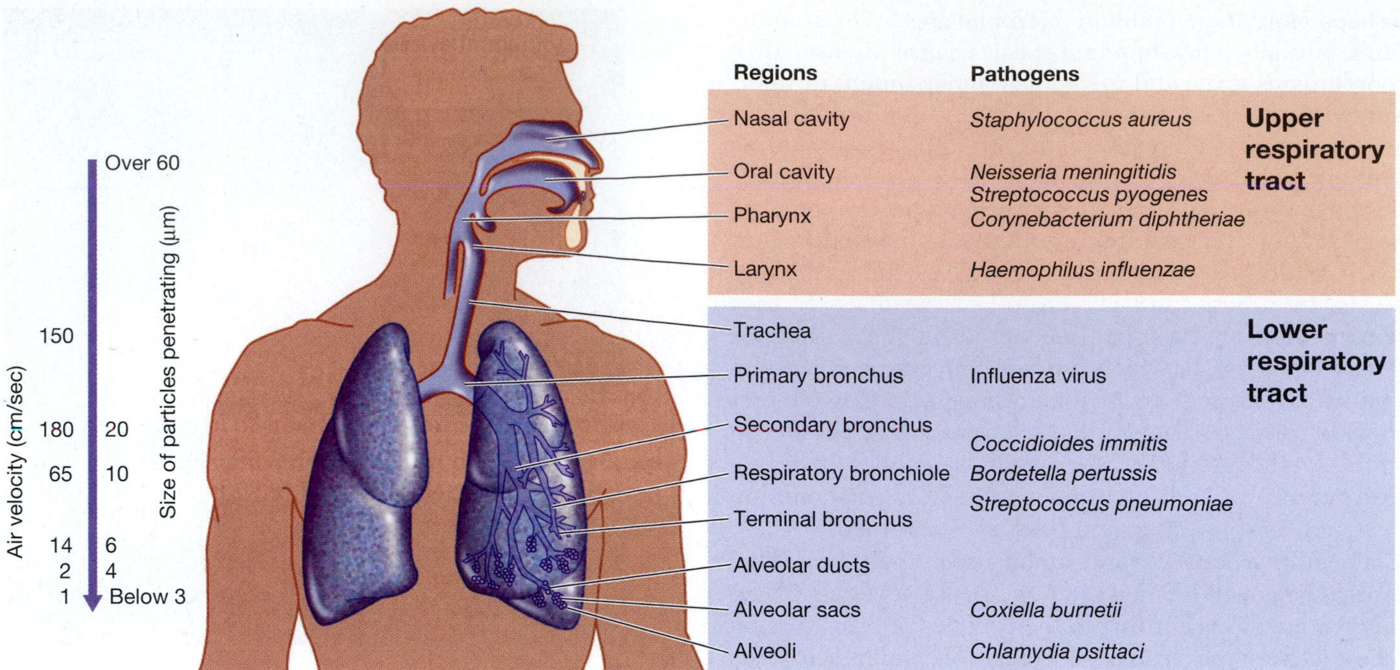

Figure 34.2 The respiratory system of humans. The microorganisms listed generally initiate infections at the indicated locations.

Bacterial and Viral Pathogens

Most human respiratory pathogens are transmitted from person to person because humans are the only reservoir for these pathogens; pathogen survival thus depends on person-to-person transmission. A few respiratory pathogens such as *Legionella pneumophila* (Legionnaire's disease) are transmitted primarily from water or soil and thus do not require person-to-person propagation; we discuss these in Chapter 36. Bacterial and viral respiratory infections, serious in themselves, often initiate secondary problems that can be life threatening. Thus, accurate and rapid diagnosis and treatment of respiratory infections, where possible, is necessary to limit host damage. Many bacterial and viral pathogens can be controlled by immunization. Most respiratory bacterial pathogens respond readily to antibiotic therapy, but many viral pathogens are resistant to treatment.

34.1 MiniReview

Bacterial and viral respiratory pathogens are transmitted in air. Most respiratory pathogens are transferred from person to person via respiratory aerosols generated by coughing, sneezing, talking, or breathing.

- Identify the physical features of gram-positive bacteria that allow them to survive for long periods in air and dust.
- Identify pathogens more commonly found in the upper respiratory tract. Identify pathogens more commonly found in the lower respiratory tract.

34.2 Streptococcal Diseases

The bacteria *Streptococcus pyogenes* and *Streptococcus pneumoniae* are important human respiratory pathogens; both organisms are transmitted by the respiratory route. *S. pneumoniae* is found in the respiratory flora of up to 40% of healthy individuals, and endogenous strains can cause severe respiratory disease in compromised individuals.

Streptococci are nonsporulating, homofermentative, aerotolerant, anaerobic gram-positive cocci (Section 16.1). Cells of *S. pyogenes* typically grow in elongated chains (Figure 16.3*b*). Pathogenic strains of *S. pneumoniae* typically grow in pairs or short chains, and virulent strains produce an extensive polysaccharide capsule (**Figure 34.3**).

Streptococcus pyogenes: Epidemiology and Pathogenesis

Streptococcus pyogenes, also called *group A Streptococcus* (GAS), is frequently isolated from the upper respiratory tract of healthy adults. Although numbers of endogenous *S. pyogenes* are usually low, if host defenses are weakened or a new, highly virulent strain is introduced, acute suppurative (pus-forming) infections are possible. *S. pyogenes* is the cause of streptococcal pharyngitis, better known as "strep throat" (Figure 34.2). Most isolates from clinical cases of streptococcal pharyngitis produce a toxin that lyses red blood cells in culture media, a condition called *β-hemolysis* (Figure 28.18). Streptococcal pharyngitis is characterized by a severe sore throat, enlarged tonsils with exudate, tender cervical lymph nodes, a mild fever, and general malaise. *S. pyogenes*

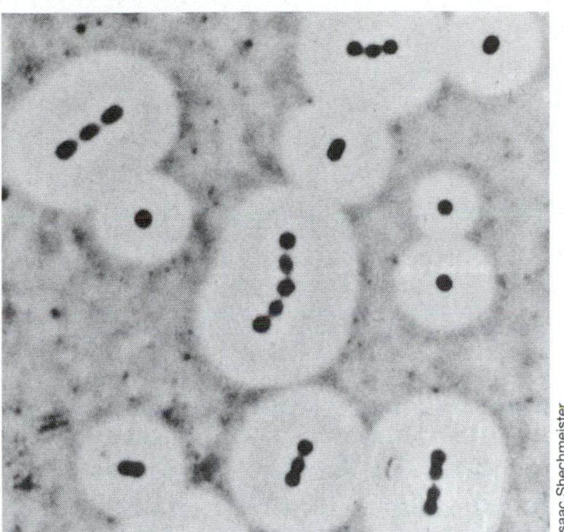

Isaac Shechmeister

Figure 34.3 The gram-positive pathogen *Streptococcus pneumoniae*. India ink negative stain. An extensive capsule surrounds the cells. The cells are about 0.5 μm in diameter.

can also cause related infections of the inner ear (otitis media), the mammary glands (mastitis), infections of the superficial layers of the skin (pyoderma or impetigo) (impetigo can also be caused by *Staphylococcus aureus*) (**Figure 34.4**), and erysipelas, an acute streptococcal skin infection (**Figure 34.5**).

About half of the clinical cases of severe sore throat are due to *Streptococcus pyogenes,* with most others due to viral infections. An accurate, rapid determination of the cause of the sore throat is important. If the sore throat is due to *S. pyogenes,* rapid, complete treatment of streptococcal sore throat is important because untreated streptococcal infections can lead to serious diseases such as scarlet fever, rheumatic fever, acute glomerulonephritis, and streptococcal toxic shock syndrome. On the other hand, if the sore throat is due to a virus, treatment with antibacterial drugs (antibiotics) will be useless, and may promote antibiotic resistance (∞ Section 27.12).

Certain GAS strains carry a lysogenic bacteriophage that encodes streptococcal pyrogenic exotoxin A (SpeA), SpeB, SpeC, and SpeF. These exotoxins are responsible for most of the symptoms of streptococcal toxic shock syndrome and **scarlet fever** (**Figure 34.6**). Spe's are superantigens that recruit massive numbers of T cells to the infected tissues (∞ Section 30.8). Toxic shock results when the activated T cells secrete cytokines, which activate large numbers of cells, causing systemic inflammation and tissue destruction.

Occasionally GAS causes fulminant (sudden and severe) invasive systemic infection such as cellulitis, a skin infection in subcutaneous layers, and necrotizing fasciitis, a rapid and progressive disease resulting in extensive destruction of subcutaneous tissue, muscle, and fat. Necrotizing fasciitis is responsible for the dramatic reports of "flesh-eating bacteria." In these cases, SpeA, SpeB, SpeC, and SpeF, as well as the bacterial cell surface M protein, function as superantigens. These diseases cause inflammation and extensive tissue destruction resulting in death in about 15% of the estimated 11,000 cases per year. In all of these cases, timely and adequate treatment of the GAS infection stops production of the superantigen and its effects.

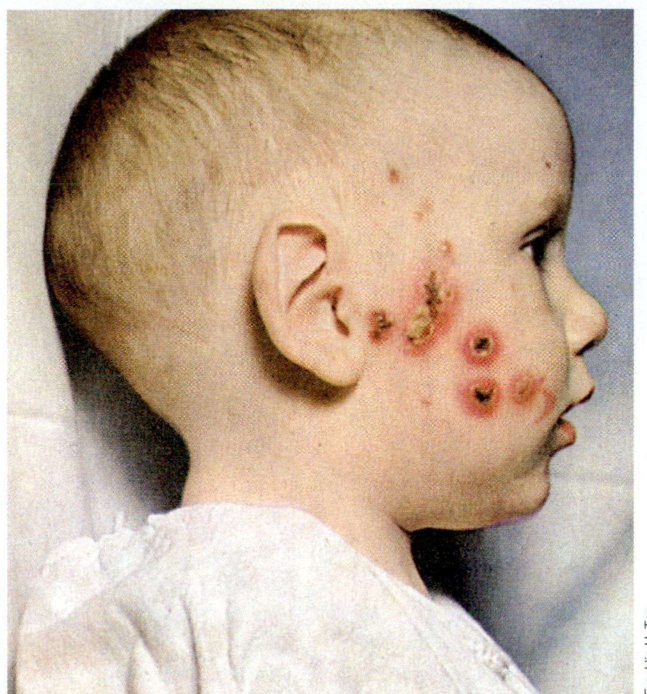

Franklin H. Top

Figure 34.4 Typical lesions of impetigo. Impetigo is commonly caused by *Streptococcus pyogenes* or *Staphylococcus aureus*.

Other Streptococcal Syndromes

Untreated or insufficiently treated *S. pyogenes* infections may lead to other diseases, even in the absence of active infection.

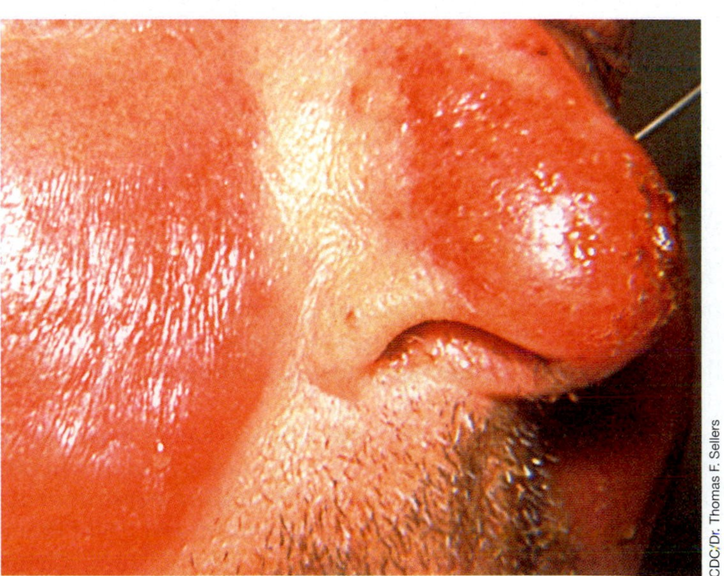

CDC/Dr. Thomas F. Sellers

Figure 34.5 Erysipelas. Erysipelas is a *Streptococcus pyogenes* infection of the skin, shown here on the nose and cheeks, characterized by redness and distinct margins of infection.

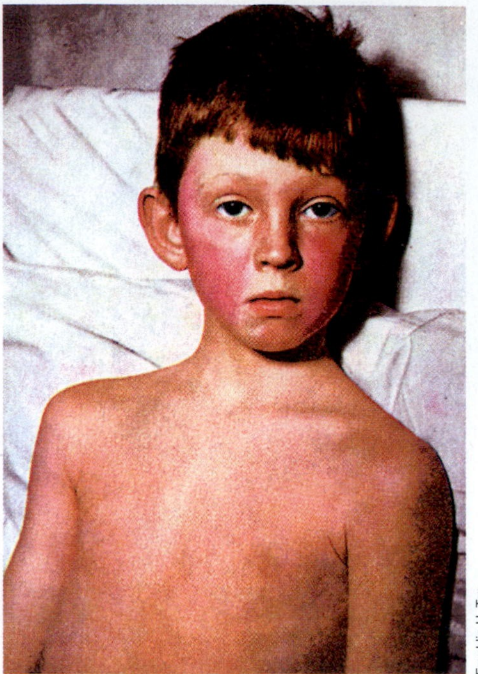

Figure 34.6 Scarlet fever. The typical rash of scarlet fever results from the action of the erythrogenic toxin produced by *Streptococcus pyogenes*.

These severe nonsuppurative (non-pus-forming) poststreptococcal diseases usually occur about 1 to 4 weeks after the onset of a streptococcal infection. The immune response to the invading pathogen produces antibodies that cross-react with host tissue antigens on the heart, joints, and kidneys, resulting in damage to these tissues. The most serious of these diseases is **rheumatic fever** caused by rheumatogenic strains of *S. pyogenes*. These strains contain cell surface antigens that are similar to heart valve and joint antigens. Rheumatic fever is an autoimmune disease, with antibodies directed against streptococcal antigens also reacting with heart valve and joint antigens (∞ Section 30.7). Damage to host tissues may be permanent, and it is often exacerbated by later streptococcal infections that lead to recurring bouts of rheumatic fever.

Another nonsuppurative disease is acute poststreptococcal glomerulonephritis, a painful kidney disease. This immune complex disease develops following infection with *S. pyogenes* due to the formation of streptococcal antigen–antibody complexes in the blood. The immune complexes lodge in the glomeruli (filtration membranes of the kidney), causing inflammation of the kidney (nephritis) accompanied by severe pain. Within several days, the complexes are usually dissolved and the patient returns to normal. Unfortunately, even timely antibacterial treatment may not prevent glomerulonephritis. Only a few strains of *S. pyogenes*—so-called nephritogenic strains—produce this painful disease, but up to 15% of infections with nephritogenic strains cause glomerulonephritis (∞ Section 30.7).

Because infection induces strain-specific immunity, reinfection by a particular *S. pyogenes* strain is rare. However, there are over 80 different strains defined by distinct cell surface M proteins. Thus, an individual can be infected multiple times by different *S. pyogenes* strains. There are no available vaccines to prevent *S. pyogenes* infections.

Diagnosis of *Streptococcus pyogenes*

Because serious host damage can follow a streptococcal sore throat, several rapid antigen detection (RAD) systems have been developed for identification of *S. pyogenes*. Surface antigens are first extracted by enzymatic or chemical means directly from a swab of the patient's throat. The antigens are then detected using antibodies specific for surface proteins of *S. pyogenes* with immunological methods such as latex bead agglutination, fluorescent antibody staining, and enzyme-linked immunosorbent assay (ELISA), methods described in Chapter 32. Using these methods, clinical specimens are processed and analyzed quickly, sometimes in just a few minutes. These rapid diagnostic procedures allow the physician to initiate antibiotic therapy immediately to prevent serious GAS infections and complications such as rheumatic fever.

A more accurate confirmation of GAS infection is a positive culture from the throat grown on sheep blood agar (∞ Figure 28.18). Although the RAD tests are nearly as specific as throat cultures, they can be up to 40% less sensitive, leading to false-negative reports. Throat cultures take up to two days to process, hence the popularity of the RAD tests. Serology tests are the most sensitive tests available for identifying recent streptococcal infections. Patients are examined for the presence or increase of antibodies (rise in titer) to streptococcal antigens. The detection of new antibodies or an increase in the quantity of existing antibodies confirms a recent streptococcal infection (∞ Section 32.5).

Streptococcus pneumoniae

The other major pathogenic streptococcal species, *Streptococcus pneumoniae,* causes invasive lung infections that often develop as secondary infections to other respiratory disorders. Strains of *S. pneumoniae* that are encapsulated are particularly pathogenic because they are potentially very invasive. Cells invade alveolar tissues (lower respiratory tract) of the lung, where the capsule enables the cells to resist phagocytosis and elicit a strong host inflammatory response. Reduced lung function, called *pneumonia,* can result from accumulation of recruited phagocytic cells and fluid. The *S. pneumoniae* cells can then spread from the focus of infection as a bacteremia, sometimes resulting in bone infections, inner ear infections, and endocarditis. Untreated invasive pneumococcal disease has a mortality rate of about 30%. Even with aggressive antimicrobial treatment, individuals hospitalized with pneumococcal pneumonia have up to 10% mortality.

Laboratory diagnosis of *S. pneumoniae* is based on the culture of gram-positive diplococci from either patient sputum or blood. There are over 90 different serotypes (antigenic capsule variants), and, as for *S. pyogenes,* infection induces immunity to only the infecting serotype of *S. pneumoniae*.

Prevention and Treatment

An effective vaccine is available for prevention of infection by at least two-thirds of the 90 known strains of *S. pneumoniae,* including all common pathogenic strains. The vaccine consists of a mixture of the capsular polysaccharides from the most prevalent pathogenic strains. The vaccine is recommended for the elderly, healthcare providers, individuals with compromised immunity, and others at high risk for respiratory infections (Section 30.5).

Penicillin and its many derivatives are the agents of choice for treating GAS infections. Erythromycin and other antibacterial drugs are used in individuals who have acquired penicillin allergies. Most strains of *S. pneumoniae* respond quickly to penicillin therapy. However, penicillin-resistant strains are common among those causing healthcare-associated infections. Thus, individual isolates must be tested for penicillin susceptibility. Erythromycin is the drug of choice for penicillin-resistant organisms, but fluoroquinolones, cephalosporin, ceftriaxone, cefotaxime, or vancomycin may also be used. However, some strains have acquired resistance to each of these drugs, and some strains have acquired multiple drug resistance, underscoring the need to test each isolate individually. Invasive disease such as pneumonia caused by drug-resistant *S. pneumoniae* is now a reportable disease in the United States, and over 2,500 cases are reported annually. Almost half of these are now in children less than 5 years of age.

34.2 MiniReview

Diseases caused by streptococci include streptococcal pharyngitis and pneumococcal pneumonia. Occasionally, *Streptococcus pyogenes* infections develop from pharyngitis into serious conditions such as scarlet fever and rheumatic fever. Pneumonia caused by *Streptococcus pneumoniae* is a serious disease with high mortality. Definitive diagnosis for both pathogens is by culture. Infections with both pathogens are treatable with antimicrobial drugs, but drug-resistant strains of *S. pneumoniae* are common.

- How does *S. pyogenes* infection cause rheumatic fever?
- What is the primary virulence factor for *S. pneumoniae*?

34.3 *Corynebacterium* and Diphtheria

Corynebacterium diphtheriae causes *diphtheria,* a severe respiratory disease that typically infects children. Diphtheria is preventable and treatable. *C. diphtheriae* is a gram-positive, nonmotile, aerobic bacterium that forms irregular rods that may appear as club-shaped cells during growth (**Figure 34.7a**) (Section 16.4).

Epidemiology and Pathology

Corynebacterium diphtheriae enters the body via the respiratory route, with cells lodging in the throat and tonsils. The organism is usually spread from healthy carriers or infected individuals to susceptible individuals by airborne droplets. Previous infection or immunization provides resistance to the effects of the potent diphtheria exotoxin. Although the mechanism of adherence of *C. diphtheriae* cells to upper respiratory tract tissues is not clear, the organism produces a neuraminidase capable of splitting *N*-acetylneuraminic acid (a component of glycoproteins found on animal cell surfaces), and this may enhance the invasion process. The inflammatory response of throat tissues to *C. diphtheriae* infection results in formation of a characteristic lesion called a *pseudomembrane* (Figure 34.7*b*), which consists of damaged host cells and cells of *C. diphtheriae*. Pathogenic strains of *C. diphtheriae* lysogenized by bacteriophage β produce a powerful exotoxin, diphtheria toxin. Diphtheria toxin inhibits eukaryotic protein synthesis and thus kills cells (Figure 28.20).

Tissue death due to absorption of the toxin causes the appearance of the pseudomembrane in the patient's throat. The pseudomembrane may block the passage of air, and death

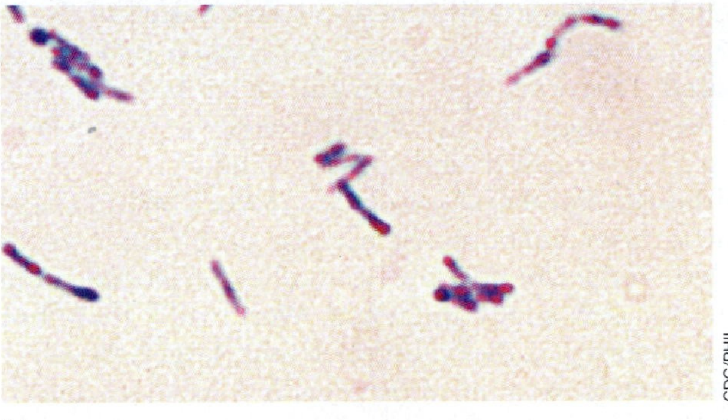

(a)

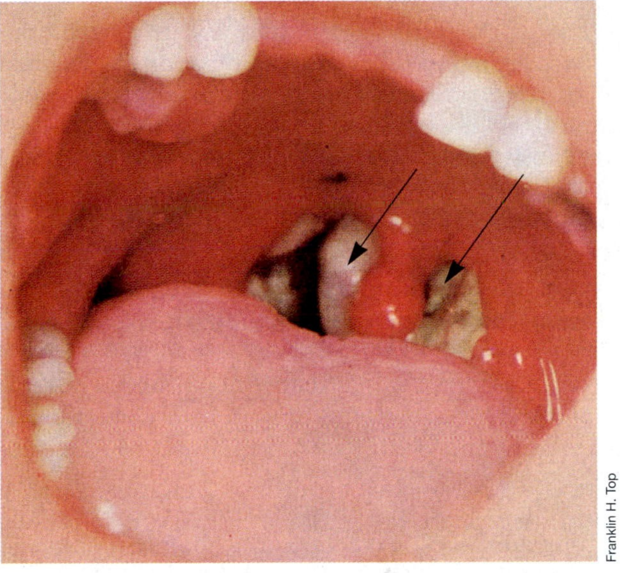

(b)

Figure 34.7 Diphtheria. *(a)* Cells of *Corynebacterium diphtheriae* showing typical club-shaped appearance. The gram-positive cells are 0.5–1.0 μm in diameter and may be several micrometers in length. *(b)* Pseudomembrane (arrows) in an active case of diphtheria caused by the bacterium *C. diphtheriae*.

from diphtheria is usually due to a combination of the effects of partial suffocation and tissue destruction by exotoxin. In untreated infections, the toxin can cause systemic damage to the heart (about 25% of diphtheria patients develop myocarditis), kidneys, liver, and adrenal glands.

Although diphtheria was once a major childhood disease, it is now rarely encountered because an effective vaccine is available. In the United States and other developed countries, the disease is virtually unknown. Worldwide, there are still 5,000 fatal cases of diphtheria per year, largely because of a lack of effective immunization programs in Africa and Southeast Asia.

Diagnosis, Prevention, and Treatment

Corynebacterium diphtheriae isolated from the throat is diagnostic for diphtheria. Nasal or throat swabs are used to inoculate blood agar, tellurite medium, or the selective Loeffler's medium that inhibits the growth of most other respiratory pathogens.

Prevention of diphtheria is accomplished with a highly effective vaccine. The vaccine is prepared by treating diphtheria exotoxin with formalin to yield an immunogenic toxoid preparation. Diphtheria toxoid is part of the DTaP (diphtheria toxoid, tetanus toxoid, and acellular pertussis) vaccine. Alternative vaccines with a reduced dose of diphtheria toxoid have been approved for adolescents and adults because they induce adequate immunity with fewer side effects in those populations (∞ Section 30.5).

A patient diagnosed with diphtheria is treated with antibiotics. Penicillin, erythromycin, and gentamicin are generally effective for stopping *C. diphtheriae* growth and further toxin production, but do not alter the effects of preformed toxin. Diphtheria antitoxin (an antiserum produced in horses) contains neutralizing antibodies, but is available only for serious acute cases of diphtheria. Early administration of both antibiotics and antitoxin is necessary for effective treatment of the acute disease.

34.3 MiniReview

Diphtheria is an acute respiratory disease caused by the gram-positive bacterium *Corynebacterium diphtheriae*. Early childhood immunization is effective for preventing this very serious respiratory disease.

- Is the pathogenesis of diphtheria due to infection?
- How can the spread of diphtheria be prevented?

34.4 *Bordetella* and Pertussis

Pertussis is a serious respiratory disease caused by infection with *Bordetella pertussis,* a small, gram-negative, aerobic coccobacillus that is a member of the Betaproteobacteria (∞ Table 15.1). *B. pertussis* was identified as the cause of pertussis in 1906 by the French microbiologist Jean Baptiste Bordet and the Belgian microbiologist Octave Gengou.

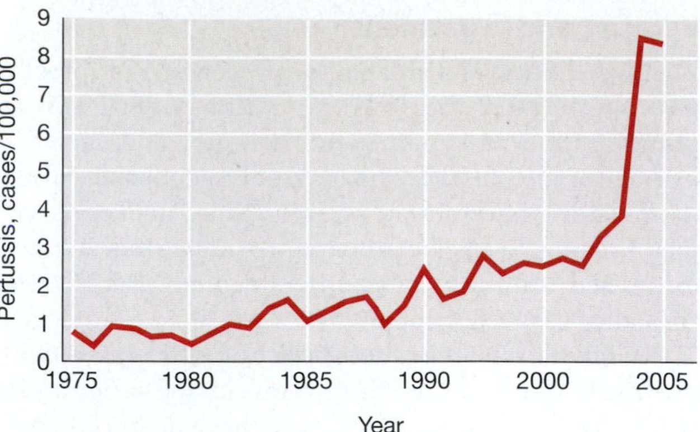

Figure 34.8 Pertussis in the United States. The incidence of pertussis, per 100,000 population, caused by respiratory infection with *Bordetella pertussis*, is increasing rapidly. There were 25,616 cases of pertussis in 2005, mostly in infants and school-age children, triple the number of 2001. Data are from the Centers for Disease Control and Prevention, Atlanta, GA, USA.

Epidemiology and Pathology

Pertussis, also known as **whooping cough**, is an acute, highly infectious respiratory disease now observed frequently in school-age children under 19 years of age. Infants less than 6 months of age, who are too young to be effectively vaccinated, have the highest incidence of disease and also have the most severe disease. *B. pertussis* attaches to cells of the upper respiratory tract by producing a specific adherence factor called *filamentous hemagglutinin antigen,* which recognizes a complementary molecule on the surface of host cells. Once attached, *B. pertussis* grows and produces *pertussis exotoxin.* This potent toxin induces synthesis of cyclic adenosine monophosphate (cyclic AMP), which is at least partially responsible for the events that lead to host tissue damage.

B. pertussis also produces an endotoxin, which may induce some of the symptoms of whooping cough. Clinically, whooping cough is characterized by a recurrent, violent cough that can last up to 6 weeks. The spasmodic coughing gives the disease its name, for a whooping sound results from the patient inhaling deep breaths to obtain sufficient air. Worldwide, there are up to 50 million cases and 300,000 pertussis deaths each year, most in developing countries.

Recent Trends with Pertussis

In the United States there has been a consistent upward trend of *B. pertussis* infections and disease since the 1980s, reversing a trend that started with the introduction of an effective pertussis vaccine. In 1976, the year of lowest prevalence and incidence, there were only 1,010 reported cases of pertussis. By contrast, in 2005, there were 25,616 cases (**Figure 34.8**). In the United States pertussis causes about 14 deaths per year. About 60% of recent cases were in adolescents and adults who lacked appropriate immunity. About 13% of cases were in children less than 6 months of age who had not yet received all of the recommended doses of pertussis vaccine. The threat of this very communicable disease remains high, as illustrated by recent sporadic outbreaks in cities in the United States and epidemic outbreaks in Afghanistan. Up to 32% of coughs lasting 1 to 2 weeks or longer may be caused by *B. pertussis*. Worldwide, research indicates that immunization programs should still be targeted to children, but immunization of adolescents and adults should also be a priority because vaccinated individuals lose effective immunity within 10 years. Current vaccination in a large percentage of all members of the population is necessary to build herd immunity (∞ Section 33.5) and prevent serious infections by this endemic respiratory pathogen.

Diagnosis, Prevention, and Treatment

Diagnosis of whooping cough can be made by fluorescent antibody staining of a nasopharyngeal swab specimen or by actual culture of the organism. For best recovery of *B. pertussis,* a nasopharyngeal aspirate is inoculated directly onto a blood–glycerol–potato extract agar plate (although not selective, this rich medium supports good recovery of *B. pertussis*). The β-hemolytic colonies containing small gram-negative coccobacilli are tested for *B. pertussis* by a latex bead agglutination test or are stained with a fluorescent antibody specific for *B. pertussis* for positive identification. A polymerase chain reaction (PCR) test is considered the most sensitive and preferred diagnostic test. Improved diagnostic and reporting techniques may be one reason for the recent observed increase in pertussis cases in the United States, but the disease may still be underreported, especially in adolescents and adults.

A vaccine consisting of proteins derived from *B. pertussis* is part of the routinely administered DTaP vaccine. This vaccine is normally given to children at appropriate intervals beginning soon after birth (∞ Section 30.5). The acellular pertussis vaccine has fewer side effects than the older pertussis vaccines and has caused no deaths. It is also recommended for adolescents and certain populations of adults (healthcare and child-care workers) as well as young children.

Cultures of *B. pertussis* are killed by ampicillin, tetracycline, and erythromycin, although antibiotics alone do not seem to be sufficient to kill the pathogen *in vivo:* A patient with whooping cough remains infectious for up to 2 weeks following commencement of antibiotic therapy, indicating that the immune response may be more important than antibiotics for eliminating *B. pertussis* from the body.

34.4 MiniReview

In the United States, there has been a recent increase in the number of annual cases of whooping cough. Inadequately immunized children, adolescents, and adults are at high risk for acquiring and spreading pertussis.

- What measures can be taken to decrease the current incidence of pertussis in a population?
- Indicate potential problems with the use of pertussis vaccines.

34.5 *Mycobacterium*, Tuberculosis, and Hansen's Disease

Tuberculosis (TB) is caused by the gram-positive, acid-fast bacillus *Mycobacterium tuberculosis* (∞ Section 16.5). The famous German microbiologist Robert Koch isolated and described the causative agent in 1882 (∞ Section 1.8). A related *Mycobacterium* species, *Mycobacterium leprae,* causes Hansen's disease (leprosy). All mycobacteria share acid-fast properties due to the waxy mycolic acid constituent of their cell wall. Mycolic acid allows these organisms to retain carbolfuchsin, a red dye, even after washing in 3% hydrochloric acid in alcohol (**Figure 34.9**; ∞ Section 16.5).

Epidemiology

Mycobacterium tuberculosis is easily transmitted by the respiratory route; even normal conversation can spread the organism from person to person. At one time, tuberculosis was the most important infectious disease of humans and accounted for one-seventh of all deaths worldwide. Presently, over 14,000 new cases of tuberculosis and over 700 deaths occur each year in the United States. Worldwide, TB still accounts for almost 1.6 million deaths per year, about 11% of all deaths due to infectious disease (∞ Table 33.1). Up to one-third of the world's population has been infected with *M. tuberculosis*. Many new TB cases in the United States occur in acquired immunodeficiency syndrome (AIDS) patients.

Pathology

The interaction of the human host and the bacterium *M. tuberculosis* is determined both by the virulence of the strain

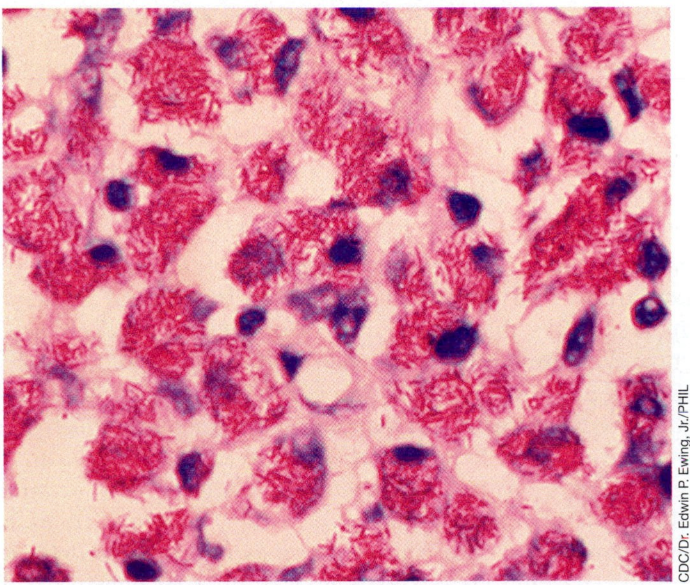
CDC/Dr. Edwin P. Ewing, Jr./PHIL

Figure 34.9 ***Mycobacterium avium* in an acid-fast stained lymph node biopsy from a patient with AIDS.** Multiple bacilli, stained red with carbolfuchsin, are evident inside each cell. The individual rods are about 0.4 μm in diameter and up to 4 μm in length.

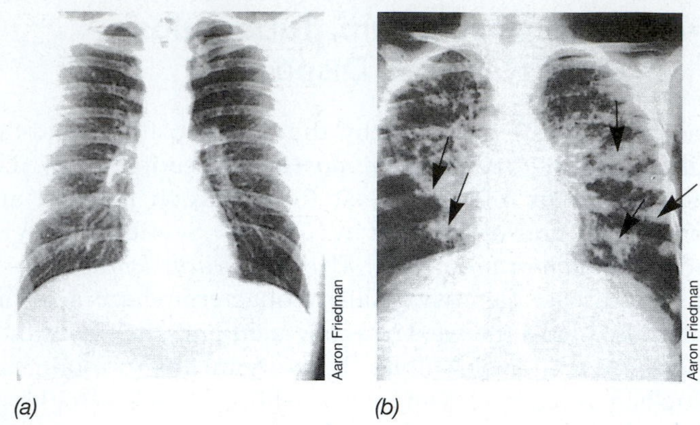

Figure 34.10 Tuberculosis X-ray. *(a)* Normal chest X-ray. The faint white lines are arteries and other blood vessels. *(b)* Chest X-ray of an advanced case of pulmonary tuberculosis; white patches (arrows) indicate areas of disease. These patches, or tubercles as they are called, may contain viable cells of *Mycobacterium tuberculosis*. Lung tissue and function is permanently destroyed by these lesions.

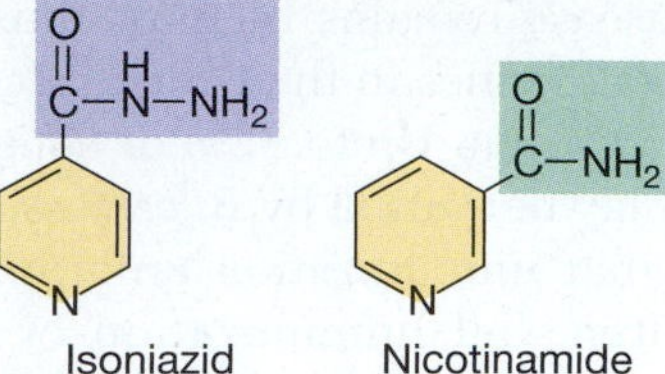

Figure 34.11 Structure of isoniazid (isonicotinic acid hydrazide). Isoniazid is an effective chemotherapeutic agent for tuberculosis. Note the structural similarity to nicotinamide.

and the resistance of the host. Cell-mediated immunity plays a critical role in the prevention of active disease after infection. Tuberculosis can be classified as a *primary* infection (initial infection) or *postprimary* infection (reinfection). Primary infection typically results from inhalation of droplets containing viable *M. tuberculosis* bacteria from an individual with an active pulmonary infection. The inhaled bacteria settle in the lungs and grow. The host responds with an immune response to *M. tuberculosis,* resulting in a delayed-type hypersensitivity reaction (∞ Section 30.7) and the formation of aggregates of activated macrophages, called *tubercles* (∞ Figure 1.16). Mycobacteria often survive and grow within the macrophages, even with an ongoing immune response. In individuals with low resistance, the bacteria are not controlled and the pulmonary infection becomes acute, leading to the extensive destruction of lung tissue, the spread of the bacteria to other parts of the body, and death. In these cases, *M. tuberculosis* survives both the low pH and the effects of the oxidative antibacterial products found in the lysosomes of phagocytes such as macrophages.

In most cases of TB, however, an obvious acute infection does not occur. The infection remains localized, is usually inapparent, and appears to end. But this initial infection hypersensitizes the individual to the bacteria or their products and consequently alters the response of the individual to subsequent *M. tuberculosis* exposures. A diagnostic skin test, called the **tuberculin test**, can be used to measure this hypersensitivity. In a hypersensitive individual, tuberculin, a protein extract from *M. tuberculosis,* elicits a local immune inflammatory reaction within 1–3 days at the site of an intradermal injection. The reaction is characterized by induration (hardening) and edema (swelling) (∞ Figure 30.6*b*). An individual exhibiting this reaction is said to be *tuberculin-positive,* and many healthy adults show positive reactions as a result of previous inapparent infections. A positive tuberculin test does not indicate active disease but only that the individual has been exposed to the organism in the past and has generated a cell-mediated immune response against *M. tuberculosis*.

For most individuals, this cell-mediated immunity is protective and lifelong. However, some tuberculin-positive patients develop postprimary tuberculosis through reinfection from outside sources or as a result of reactivation of bacteria that have remained dormant in lung macrophages, often for years. Because of the latent nature of *M. tuberculosis* infection, individuals who have a positive tuberculin test are generally treated with antimicrobial agents for long periods of time. Factors such as age, malnutrition, overcrowding, stress, and hormonal changes may reduce effective immunity in untreated individuals and allow reactivation of dormant infections.

Secondary mycobacterial infections often progress to chronic infections that result in destruction of lung tissue, followed by partial healing and calcification at the infection site. Chronic postprimary tuberculosis often results in a gradual spread of tubercular lesions in the lungs. Bacteria are found in the sputum in individuals with active disease, and areas of destroyed tissue can be seen in X-rays (**Figure 34.10**).

Prevention and Treatment

Individuals who have active cases of TB may spread the disease simply by coughing or speaking. Because TB is highly contagious, the United States Occupational Safety and Health Administration has stringent requirements for the protection of healthcare workers who are responsible for tuberculosis patient care. For example, patients with infectious tuberculosis must be hospitalized in negative-pressure rooms. In addition, healthcare workers who have patient contact must be provided with personally fitted face masks having high-efficiency particulate air (HEPA) filters to prevent the passage of *M. tuberculosis* cells in sputum or on dust particles.

Antimicrobial therapy of tuberculosis has been a major factor in control of the disease. Streptomycin was the first effective antibiotic, but the real revolution in TB treatment came with the discovery of isonicotinic acid hydrazide, called *isoniazid* (INH) (**Figure 34.11**). This drug, specific for mycobacteria, is effective, inexpensive, relatively nontoxic, and readily absorbed when given orally. Although the mode of action of isoniazid is not completely understood, it affects the synthesis of mycolic acid by *Mycobacterium*. Mycolic acid is a lipid that complexes with peptidoglycan in the mycobacterial cell wall. Isoniazid probably functions as a growth factor analog of the structurally related molecule, nicotinamide. As such, isoniazid would be incorporated in place of nicotinamide and inactivate enzymes required for mycolic acid synthesis.

Treatment of mycobacteria with very small amounts of isoniazid (as little as 5 picomoles [pmol] per 10^9 cells) results in complete inhibition of mycolic acid synthesis, and continued incubation results in loss of outer areas of the cell wall, a loss of cellular integrity, and death. Following treatment with isoniazid, mycobacteria lose their acid-fast properties, in keeping with the role of mycolic acid in this staining property.

Treatment is typically achieved with daily doses of isoniazid and rifampin for 2 months, followed by biweekly doses for a total of 9 months. This treatment eradicates the tubercle bacilli and prevents emergence of antibiotic-resistant organisms. Failure to complete the entire prescribed treatment may allow the infection to be reactivated, and reactivated organisms are often resistant to the original treatment drugs. Incomplete treatment encourages antibiotic resistance because a high rate of spontaneous mutations in surviving *M. tuberculosis* promotes rapid acquisition of resistance to single antibiotics. To ensure treatment and thus discourage development of antibiotic-resistant organisms, direct observation of treatment may be necessary for noncompliant individuals. In populations such as hospitals and nursing homes, where resistant strains are most likely to be present, patients are routinely treated with up to 4 antimycobacterial drugs for 2 months, followed by rifampin–isoniazid treatment for a total of 6 months. Multiple drug therapy reduces the possibility that strains will emerge having resistance to more than one drug.

Resistance of *M. tuberculosis* to isoniazid and other drugs, however, is increasing, especially in AIDS patients. A number of strains that are resistant to both isoniazid and rifampin have already emerged. Treatment of these strains, called *multi-drug-resistant tuberculosis strains* (MDR TB), requires the use of second-line tuberculosis drugs that are generally more toxic, less effective, and more costly than rifampin and isoniazid. A World Health Organization (WHO) survey of MDR TB strains indicate that up to 20% are extensively drug-resistant (XDR TB) strains. XDR TB strains have resistance to virtually all TB drugs, including the second-line drugs. Preventing emergence of these strains requires better diagnostic and drug susceptibility tests in addition to new anti-TB treatment drugs and regimens.

In many countries, immunization with an attenuated strain of *Mycobacterium bovis*, the *bacillus Calmette-Guerin (BCG)* strain, is routine for prevention of tuberculosis. However, in the United States and other countries where the prevalence of tuberculosis is low, immunization with BCG is discouraged. The live BCG vaccine induces a delayed-type hypersensitivity response, and all individuals who receive it develop a positive tuberculin test. This compromises the tuberculin test as a diagnostic and epidemiologic indicator for the spread of *M. tuberculosis* infection.

Mycobacterium leprae and Hansen's Disease (Leprosy)

Mycobacterium leprae, discovered by the Norwegian scientist G. A. Hansen in 1873, is the causative agent of Hansen's disease, also known as *leprosy*. *M. leprae* is the only *Mycobacterium* species that has not been grown on artificial media. The armadillo is the only experimental animal that has been successfully used to grow *M. leprae* and achieve symptoms similar to those in the human disease.

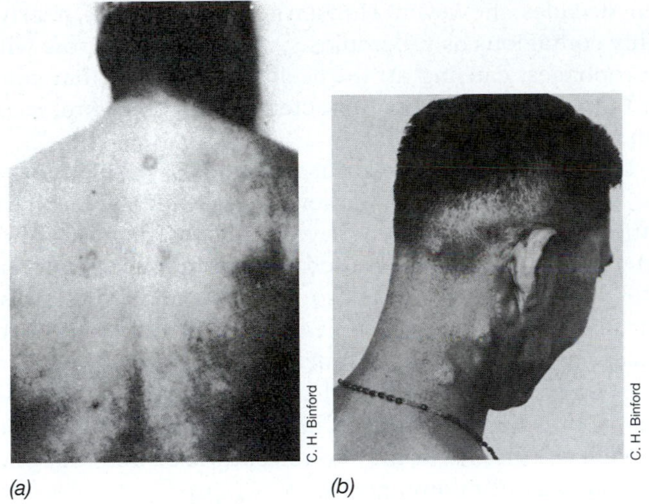

Figure 34.12 Lepromatous leprosy lesions on the skin. Lepromatous leprosy is caused by infection with *Mycobacterium leprae*. The lesions can contain up to 10^9 bacterial cells per gram of tissue, indicating an active uncontrolled infection with a poor prognosis.

The most serious form of Hansen's disease is characterized by folded, bulblike lesions on the body, especially on the face and extremities (**Figure 34.12**). These lesions are due to the growth of *M. leprae* cells in the skin. The lesions contain up to 10^9 bacterial cells per gram of tissue. Like other mycobacteria, *M. leprae* from the lesions stain deep red with carbol-fuchsin in the acid-fast staining procedure, providing a rapid, definitive demonstration of active infection. This *lepromatous* form of Hansen's disease has a very poor prognosis. In severe cases the disfiguring lesions lead to destruction of peripheral nerves and loss of motor function.

Many Hansen's disease patients exhibit less-pronounced lesions from which no bacterial cells can be recovered. These individuals have the *tuberculoid* form of the disease. Tuberculoid Hansen's disease is characterized by a vigorous delayed-type hypersensitivity response (∞ Section 30.7) and a good prognosis for spontaneous recovery. Hansen's disease of either form, and the continuum of intermediate forms, is treated using a multiple drug therapy (MDT) protocol, which includes some combination of dapsone (4,4′-sulfonylbisbenzeneamine), rifampin, and clofazimine. As in tuberculosis, drug-resistant strains have appeared, especially after treatment with single drugs or inadequate treatment. Extended drug therapy of up to 1 year with a MDT protocol is required for eradication of the organism.

The pathogenicity of *M. leprae* is due to a combination of delayed hypersensitivity and the invasiveness of the organism. Transmission is by both direct contact and respiratory routes, and incubation times vary from several weeks to years, or even decades; however, Hansen's disease is not nearly as highly contagious as tuberculosis. *M. leprae* cells grow within macrophages, causing an intracellular infection that can result in the large numbers of bacteria within the skin, leading to the characteristic lesions.

In many areas of the world, the incidence of Hansen's disease is very low. Worldwide, there are over 750,000 new cases of the disease reported each year. About 100 cases are reported annually in the United States, mostly in southern states, among immigrants from the Caribbean islands or Central America. Ninety percent of worldwide cases are in Madagascar, Mozambique, Tanzania, and Nepal. Up to 2 million people are permanently disabled as a result of Hansen's disease, but, because of the chronic nature and long latent period of the disease, it may go unrecognized and unreported in as many as 12 million people.

Other Pathogenic *Mycobacterium* Species

A common pathogen of dairy cattle, *Mycobacterium bovis,* is pathogenic for humans as well as other animals. *M. bovis* enters humans via the intestinal tract, typically from the ingestion of unpasteurized milk. After a localized intestinal infection, the organism eventually spreads to the respiratory tract and initiates the classic symptoms of tuberculosis. *M. bovis* is a different organism from *M. tuberculosis,* although the genomes of the two organisms are highly similar. Although the genome of *M. bovis* has several gene deletions compared with that of *M. tuberculosis,* there is no observed difference in their infectivity and pathogenesis in humans. Pasteurization of milk and elimination of diseased cattle have eradicated bovine-to-human transmission of tuberculosis in developed countries.

A number of other *Mycobacterium* species are also occasional human pathogens. For example, *M. kansasii, M. scrofulaceum, M. chelonae,* and a few other mycobacterial species can cause disease. Respiratory disease due to the *M. avium* complex of organisms (including *M. avium* and *M. intracellulare*) is particularly dangerous in AIDS patients or other immune-compromised individuals; these opportunistic pathogens rarely infect healthy individuals (Figure 34.9).

34.5 MiniReview

Tuberculosis is one of the most prevalent and dangerous diseases in the world. Its incidence is increasing in developed countries in part because of the emergence of drug-resistant strains of *Mycobacterium tuberculosis*. The pathology of tuberculosis and Hansen's disease is influenced by the cellular immune response.

- Why is *Mycobacterium tuberculosis* a widespread respiratory pathogen?
- Describe factors that contribute to drug resistance in mycobacterial infections.
- Identify other species of the genus *Mycobacterium* that can cause human disease.

34.6 *Neisseria meningitidis,* Meningitis, and Meningococcemia

Meningitis is an inflammation of the meninges, the membranes that line the central nervous system, especially the spinal cord and brain. Meningitis can be caused by viral, bacterial, fungal, or protist infections. Here we will deal with infectious bacterial meningitis caused by *Neisseria meningitidis* and a related infection, **meningococcemia**.

Neisseria meningitidis, often called *meningococcus,* is a gram-negative, nonsporulating, obligately aerobic, oxidase-positive, encapsulated diplococcus (∞ Section 15.10, **Figure 34.13**) about 0.6–1.0 μm in diameter. At least 13 pathogenic strains of *N. meningitidis* are recognized, based on antigenic differences in their capsular polysaccharides.

Epidemiology and Pathology

Meningococcal meningitis often occurs in epidemics, usually in closed populations such as military installations and college campuses. It typically strikes older school-age children and young adults. Up to 30% of individuals carry *N. meningitidis* in the nasopharynx with no apparent harmful effects. In epidemic situations, the prevalence of carriers may rise to 80%. The trigger for conversion from the asymptomatic carrier state to pathogenic acute infection is unknown.

In an acute meningococcus infection, the bacterium is transmitted to the host, usually via the airborne route, and attaches to the cells of the nasopharynx. Once there the organism gains access to the bloodstream, causing bacteremia and upper respiratory tract symptoms. The bacteremia sometimes leads to fulminant meningococcemia, characterized by septicemia, intravascular coagulation, shock, and death in over 10% of cases. Meningitis is another possible serious outcome of infection. Meningitis is characterized by sudden onset of headache, vomiting, and stiff neck, and can progress to coma

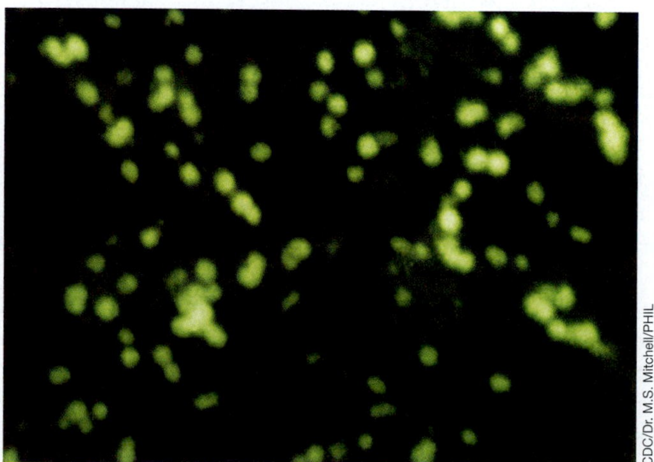

Figure 34.13 Fluorescent antibody stain of *Neisseria meningitidis*. The organism causes meningitis and meningococcemia. This specimen is from the cerebrospinal fluid of an infected patient. The individual cocci are about 0.6–1.0 μm in diameter.

and death in a matter of hours. Up to 3% of acute meningococcal meningitis victims die.

In the United States, there were 1,245 cases of serious meningococcal disease in 2005, the lowest number of cases since 1977. The pattern of decreased incidence indicates the success of widespread vaccination in susceptible populations. However, the mortality rate in recent years was over 10%.

Diagnosis, Prevention, and Treatment

Specimens isolated from nasopharyngeal swabs, blood, or cerebrospinal fluid are inoculated onto modified Thayer-Martin medium (Figure 32.5), a selective medium that suppresses the growth of most normal flora but allows the growth of pathogenic neisseria, *N. meningitidis* and *Neisseria gonorrhoeae*. Colonies showing a gram-negative diplococcus morphology and a positive oxidase test are presumptively identified as *Neisseria* (Table 32.3). Due to the rapid onset of life-threatening symptoms, preliminary diagnosis is often based on clinical symptoms and treatment is started before culture tests confirm infection with *N. meningitidis*.

Penicillin G is the drug of choice for the treatment of *N. meningitidis* infections. However, resistant strains have been reported. Chloramphenicol is the accepted alternative agent for treatment of infections in penicillin-sensitive individuals. A number of broad-spectrum cephalosporins are also effective.

Naturally occurring strain-specific antibodies acquired by subclinical infections are effective for preventing infections in most adults. Vaccines consisting of purified polysaccharides or polysaccharides conjugated to proteins from the most prevalent pathogenic strains are available and are used to immunize susceptible individuals. The vaccines are used to prevent infection in certain susceptible populations such as military recruits and students living in dormitories. In addition, rifampin is often used as a chemoprophylactic antimicrobial drug to prevent disease in close contacts of infected individuals.

Other Causes of Meningitis

A number of other organisms can also cause meningitis. Acute meningitis is usually caused by one of the pyogenic bacteria such as *Staphylococcus, Streptococcus,* or *Haemophilus influenzae. H. influenzae* primarily infects young children. An effective vaccine for preventing *H. influenzae* meningitis is available and is required in the United States for school-age children (Section 30.5).

Several viruses also cause meningitis. Among these are herpes simplex virus, lymphocytic choriomeningitis virus, mumps virus, and a variety of enteroviruses. In general, viral meningitis is less severe than bacterial meningitis.

34.6 MiniReview

Neisseria meningitidis is a common cause of meningococcemia and meningitis in young adults and occasionally occurs in epidemics in closed populations such as schools and military installations. Bacterial meningitis and meningococcemia are serious diseases with very high mortality rates. Treatment and prevention strategies are in place to deal with epidemic outbreaks. Effective vaccines are available for the most prevalent pathogenic strains.

- Identify the symptoms and causes of meningitis.
- Describe the infection by *Neisseria meningitidis* and the resulting development of meningococcemia.

34.7 Viruses and Respiratory Infections

As we discussed in Section 27.10, viruses are less easily controlled by chemotherapeutic means than bacteria because the propagation of viruses depends on host cell functions. Many chemotherapeutic agents that specifically attack viruses cause side effects in host cells. The most prevalent human infectious

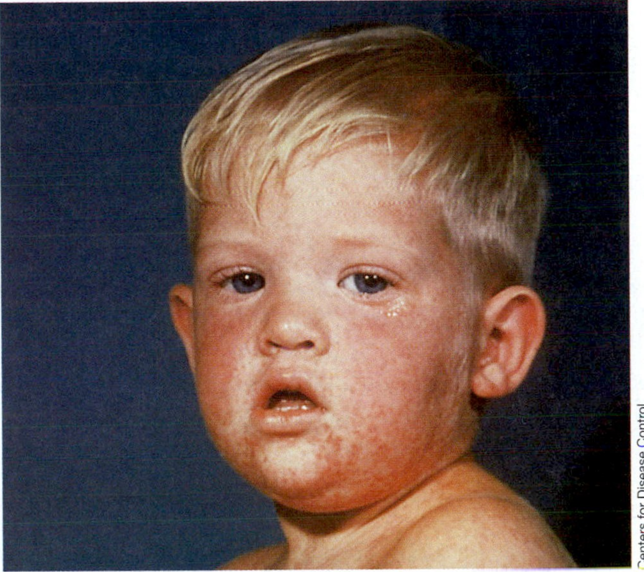

(a)

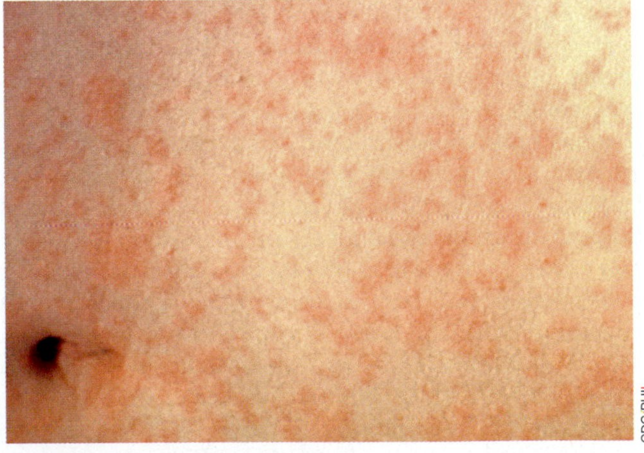

(b)

Figure 34.14 Measles in children. *(a)* The light pink rash starts on the head and neck, and *(b)* spreads to the chest, trunk, and limbs. Discrete papules coalesce into blotches as the rash progresses for several days.

UNIT 9

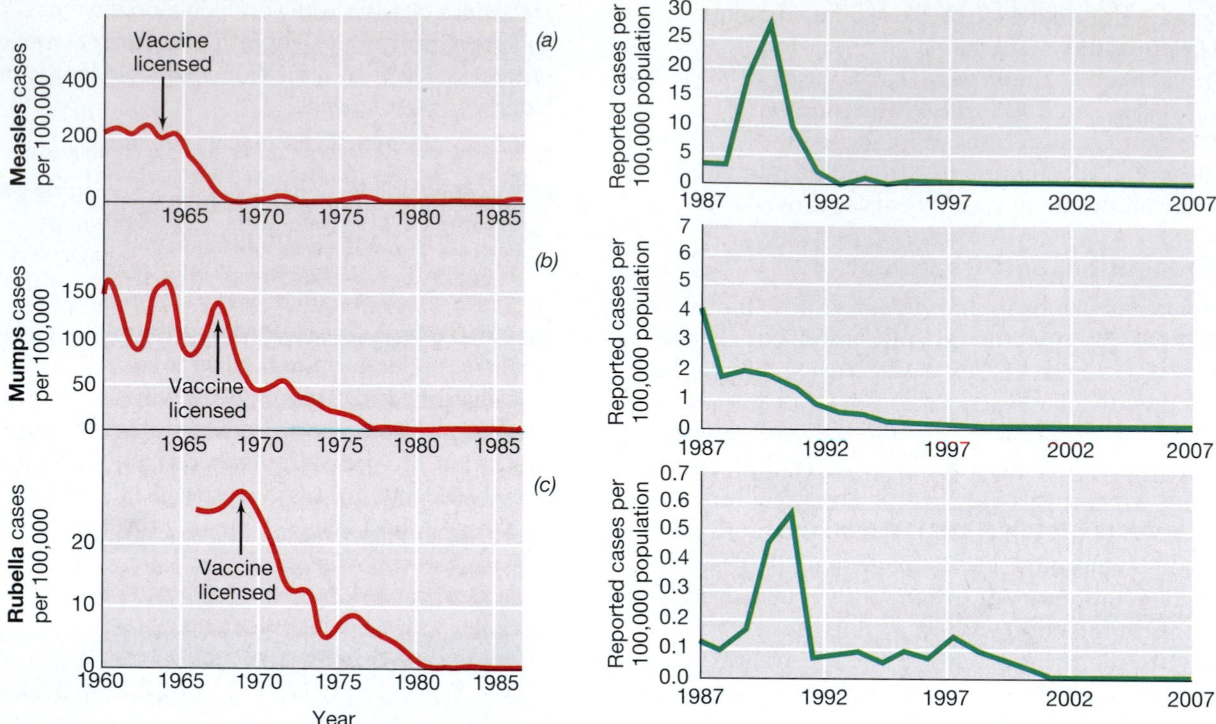

Figure 34.15 **Viral diseases and vaccines.** Major childhood viral diseases are now controlled by the MMR (measles, mumps, rubella) vaccine in the United States. Data are from the Centers for Disease Control and Prevention, Atlanta, GA, USA.

diseases are caused by viruses. Most viral diseases are acute, self-limiting infections, but some can be problematic in healthy adults. A few serious viral diseases such as smallpox and rabies have been effectively controlled by immunization. We begin here by describing measles, mumps, rubella, and chickenpox, all viral diseases transmitted in infectious droplets by an airborne route.

Measles

Measles (*rubeola* or *7-day measles*) usually affects susceptible children as an acute, highly infectious, often epidemic disease. The measles virus is a paramyxovirus, a negative-strand RNA virus (Section 19.8) that enters the nose and throat by airborne transmission, quickly leading to systemic viremia. Symptoms of infection start with nasal discharge, redness of the eyes, cough, and fever. As the disease progresses, fever and cough appear and rapidly intensify, and a rash appears (**Figure 34.14**); symptoms generally persist for 7–10 days. Circulating antibodies to measles virus are measurable about 5 days after initiation of infection, and both serum antibodies and cytotoxic T lymphocytes combine to eliminate the virus from the system. Possible postinfection complications include inner ear infection, pneumonia, and, in rare cases, measles encephalomyelitis. Encephalomyelitis can cause neurological disorders and a form of epilepsy and has a mortality rate of nearly 20%.

Although once a common childhood illness, measles generally occurs now only in rather isolated outbreaks because of widespread immunization programs begun in the mid-1960s (**Figure 34.15a**). Worldwide, however, there are over 600,000 annual deaths, mostly in children. Because the disease is highly infectious, all public school systems in the United States require proof of immunization before a child can enroll. Active immunization is done with an attenuated virus preparation as part of the MMR (measles, mumps, rubella) vaccine (Figure 30.4). A childhood case of measles generally confers lifelong immunity to reinfection.

Mumps

Mumps, like measles, is caused by a paramyxovirus and is also highly infectious. Mumps is spread by airborne droplets, and the disease is characterized by inflammation of the salivary glands, leading to swelling of the jaws and neck (**Figure 34.16**). The virus spreads through the bloodstream and may infect other organs, including the brain, testes, and pancreas. Severe complications may include encephalitis and, very rarely, sterility. The host immune response produces antibodies to mumps virus surface proteins, and this generally leads to a quick recovery. An attenuated vaccine is highly effective for preventing mumps (Figure 30.4). Hence, the prevalence of mumps in developed countries is very low, with disease usually restricted to individuals who did not receive the MMR vaccine (Figure 34.15*b*).

Rubella

Rubella (*German measles* or *3-day measles*) is caused by a single-stranded, positive-sense RNA virus of the togavirus group (Section 10.11). Disease symptoms resemble measles but are generally milder. Rubella is less contagious than true measles, and thus a large proportion of the population has never been infected. However, during the first 3 months of

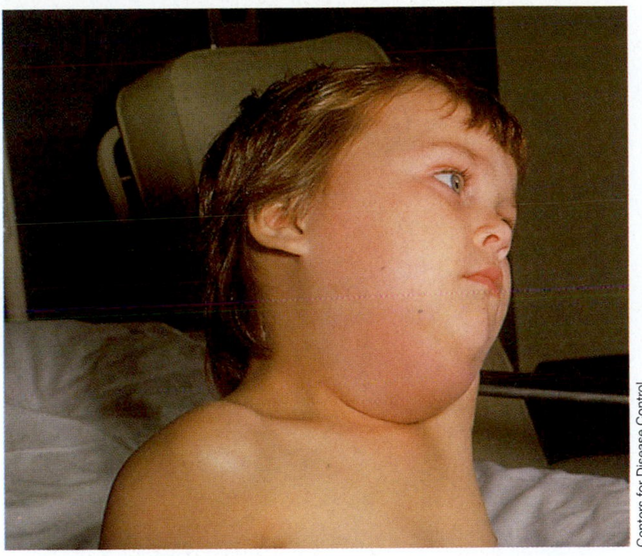

Figure 34.16 Mumps. Glandular swelling characterizes infection with the mumps virus.

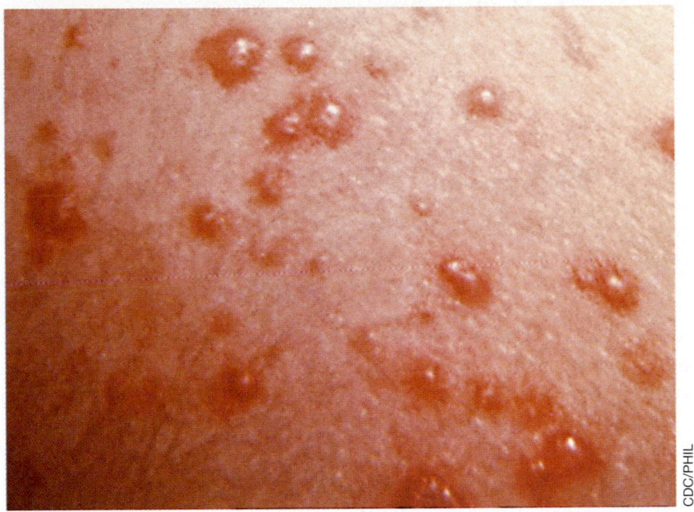

Figure 34.17 Chickenpox. Mild papular rash associated with the infection by varicella-zoster virus (VZV), the herpesvirus that causes chickenpox.

pregnancy, rubella virus can infect the fetus by placental transmission and cause serious fetal abnormalities. Rubella can cause stillbirth, deafness, heart and eye defects, and brain damage in live births. Thus, women should not be immunized with the rubella vaccine or contract rubella during pregnancy. For this reason, routine childhood immunization against rubella should be practiced. An attenuated virus is administered as part of the MMR vaccine (∞ Figure 30.4). The low number of cases for the last 5 years, coupled with the high degree of protection from the vaccine and the relatively low infectivity of the virus, suggest that rubella is no longer endemic in the United States (Figure 34.15*c*).

Chickenpox and Shingles

Chickenpox (*varicella*) is a common childhood disease caused by the varicella-zoster virus (VZV), a DNA herpesvirus (∞ Section 19.12). VZV is highly contagious and is transmitted by infectious droplets, especially when susceptible individuals are in close contact. In schoolchildren, for example, close confinement during the winter months leads to the spread of chickenpox through airborne droplets from infected classmates and through contact with contaminated fomites. The virus enters the respiratory tract, multiplies, and is quickly disseminated via the bloodstream, resulting in a systemic papular rash that quickly heals, rarely leaving disfiguring marks (**Figure 34.17**). An attenuated virus vaccine is now used in the United States. The reported annual incidence of chickenpox, now about 32,000 cases per year, is about one-fifth of the number of cases reported prior to 1995, the year the vaccine was licensed for use. The disease has always been underreported.

The VZV virus establishes a lifelong latent infection in nerve cells. The virus occasionally migrates from this reservoir to the skin surface, causing a painful skin eruption referred to as *shingles* (*zoster*). Shingles most commonly strikes immunosuppressed individuals or the elderly. The prophylactic use of human hyperimmune globulin prepared against the virus is useful for preventing the onset of symptoms of shingles. Such therapy is advised only for patients where secondary infections occasionally associated with shingles, such as pneumonia or encephalitis, may be life-threatening. To prevent shingles, a vaccine is recommended for individuals over 60 years of age. The vaccine stimulates antibody and cytotoxic T cell immunity to VZV, keeping VZV from migrating out of nerve ganglia to skin cells.

34.7 MiniReview

Viral respiratory diseases are highly infectious and may cause serious health problems. However, the common childhood viral diseases measles, mumps, rubella, and chickenpox are all controllable with appropriate immunizations.

- Describe the potential serious outcomes of infection by each of these viruses.
- Identify the effects of immunization on the incidence of measles, mumps, rubella, and chickenpox.

34.8 Colds

Colds are the most common of infectious diseases. As shown in **Figure 34.18**, people acquire about ten colds for every other infectious disease, except influenza. Colds are viral infections that are transmitted via droplets spread from person to person in coughs, sneezes, and respiratory secretions. Colds are usually of short duration, and the symptoms are milder than other respiratory diseases such as influenza. **Table 34.1** compares the symptoms of colds and influenza.

The Common Cold

Each person averages over three colds per year throughout his or her lifetime (Figure 34.18). Cold symptoms include rhinitis

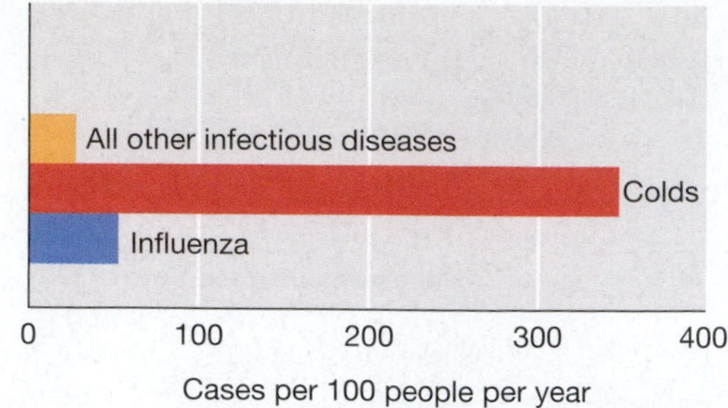

Figure 34.18 Colds and influenza. These viral diseases are the leading causes of acute infectious disease in the United States. This pattern is typical for recent years. Colds and influenza cause much higher morbidity as compared to all other infectious diseases.

(inflammation of the nasal region, especially the mucous membranes), nasal obstruction, watery nasal discharges, and a general feeling of malaise, usually without fever. *Rhinoviruses*, positive-sense, single-stranded RNA viruses of the picornavirus group (**Figure 34.19*a*** and Section 19.8), are the most common causes of colds. At least 115 different rhinoviruses have been identified. About 25% of colds are due to infections with other viruses. *Coronaviruses* (Figure 34.19*b*) cause 15% of all colds in adults. Adenoviruses, coxsackie viruses, respiratory syncytial viruses (RSV), and orthomyxoviruses are collectively responsible for about 10% of colds.

Each of these viruses may also cause more serious disease. For example, one adenovirus strain produces a severe and sometimes lethal respiratory infection. Colds generally induce a specific, local, neutralizing IgA antibody response. However, the number of potential infectious agents makes immunity due to previous exposure very unlikely. The sheer numbers of viruses that might cause a cold also preclude the development of useful vaccines.

Aerosol transmission of the virus is probably the major means of spreading colds, although experiments with volunteers suggest that direct contact and fomite contact are also methods of transmission. Most antiviral drugs are ineffective against the common cold, but a pyrazidine derivative (**Figure 34.20*a***) has shown promise for preventing colds after virus exposure. In addition, new experimental antiviral drugs are being designed based on information derived from three-dimensional structures. For example, the antirhinovirus drug WIN 52084 (Figure 34.20*b*) binds to the virus, changing its three-dimensional surface configuration and disrupting rhinovirus binding to the host cell receptor, ICAM-1 (intercellular adhesion molecule-1), thus preventing infection. Interferon-α, a cytokine, is also effective in preventing the onset of colds. Thus, there are several experimental possibilities for cold prevention and treatment, although none are

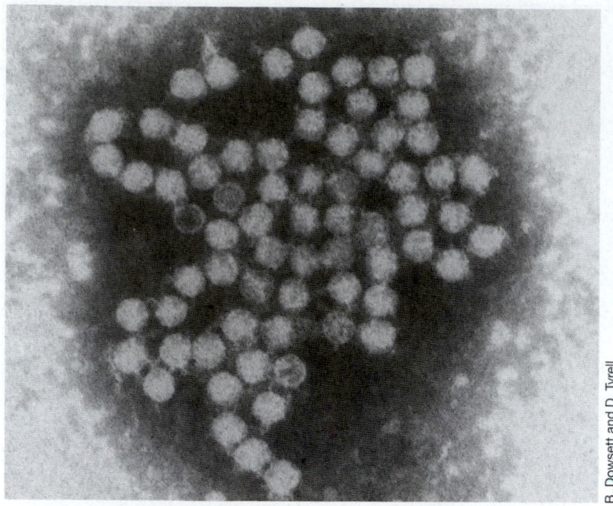

(a)

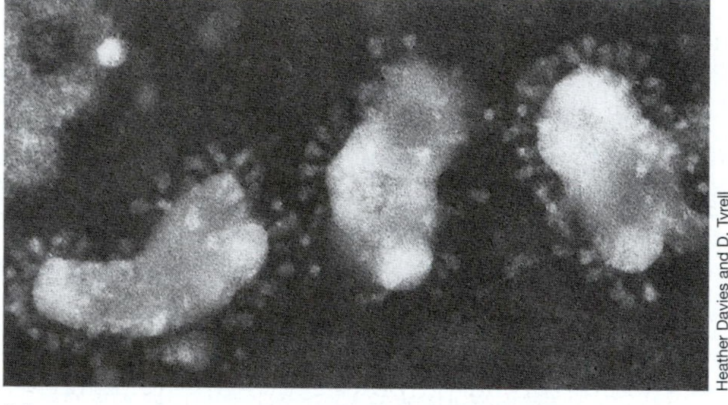

(b)

Figure 34.19 Common cold viruses. Transmission electron micrographs. *(a)* Human rhinovirus. Each rhinovirus virion is about 30 nm in diameter. *(b)* Human coronavirus. Each coronavirus virion is about 60 nm in diameter.

Table 34.1 Colds and influenza

Symptoms	*Cold*	*Influenza*
Fever	Rare	Common (39–40°C) sudden onset
Headache	Rare	Common
General malaise	Slight	Common; often quite severe; can last several weeks
Nasal discharge	Common and abundant	Less common; usually not abundant
Sore throat	Common	Less common
Vomiting and/or diarrhea	Rare	Common in children

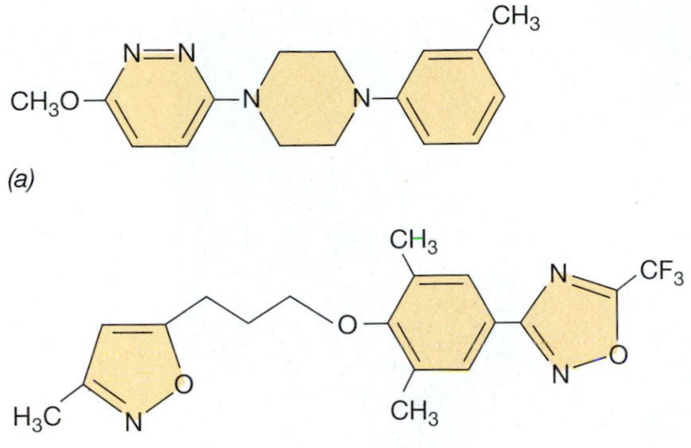

(a)

(b)

Figure 34.20 Experimental antirhinovirus drugs. *(a)* The structure of 3-methoxy-6-[4-(3-methylphenyl)]-1-piperazinyl. *(b)* The structure of WIN 52084, a receptor-blocking drug.

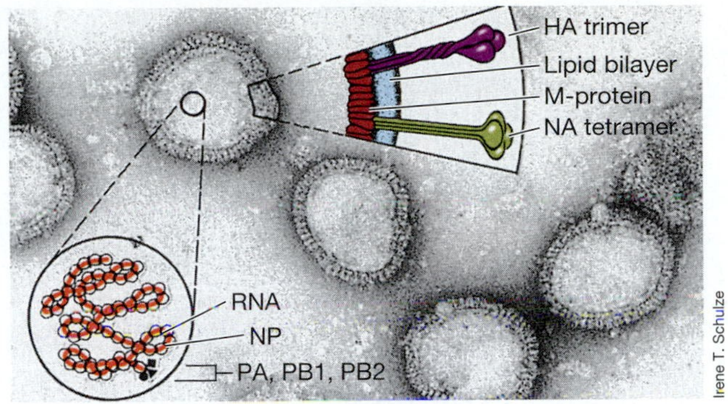

Figure 34.21 Electron micrograph of influenza virus. The photo shows the location of the major viral coat proteins and the nucleic acid. Each virion is about 100 nm in diameter. HA, hemagglutinin (three copies make up the HA coat spike); NA, neuraminidase (four copies make up the NA coat spike); M, coat protein; NP, nucleoprotein; PA, PB1, PB2, other internal proteins, some of which may have enzymatic functions.

widely accepted as effective and safe. Because colds are generally brief and self-limiting, treatment is aimed at controlling symptoms, especially nasal discharges, with antihistamine and decongestant drugs.

34.8 MiniReview

Colds are the most common infectious viral diseases. Usually caused by a rhinovirus, colds are generally mild and self-limiting diseases. Each infection induces specific, protective immunity, but the large number of viral cold pathogens precludes complete protective immunity or vaccines.

- Define the cause and symptoms of common colds.
- Discuss the possibilities for effective treatment and prevention of colds.

34.9 Influenza

Influenza is caused by an RNA virus of the orthomyxovirus group (∞ Section 19.8). Influenza virus is a single-stranded, negative-sense, helical RNA genome surrounded by an envelope made up of protein, a lipid bilayer, and external glycoproteins (**Figure 34.21**). There are three different types of influenza viruses: influenza A, influenza B, and influenza C. Here we consider only influenza A because it is the most important human pathogen.

The influenza A virus genome is single-stranded RNA that is arranged in a highly unusual manner. As discussed in Section 10.8, the influenza virus genome is *segmented*, with genes found on each of eight distinct fragments of single-stranded RNA (∞ Figure 19.16*b*). This arrangement allows the reassortment of gene fragments between different strains of influenza virus if more than one strain of influenza infects a single cell at one time. During virus maturation in the host cell, the viral RNA segments are packaged randomly. Thus, in a host cell infected with two strains of influenza virus, the new virions contain a random assortment of RNA segments from each infecting virus. Uniquely reassorted complete viruses are new viral strains. Unique reassortments result in **antigenic shift**, a major change in a protein coat antigen resulting from the total replacement of an RNA segment. Antigenic shift can immediately and substantially alter one of the two glycoproteins important in the attachment and eventual release of virus from host cells, hemagglutinin (HA or H) and neuraminidase (NA or N) (Figure 34.21). The H and N glycoprotein antigens can also acquire minor antigenic changes due to point mutations in the coding sequences that alter one or more amino acids. These limited mutations create slightly altered antigens, a phenomenon called **antigenic drift**.

Influenza Epidemiology

Human influenza virus is transmitted from person to person through the air, primarily in droplets expelled during coughing and sneezing. The virus infects the mucous membranes of the upper respiratory tract and occasionally invades the lungs. Symptoms include a low-grade fever lasting for 3–7 days, chills, fatigue, headache, and general aching (Table 34.1). Recovery is usually spontaneous and rapid. Most of the serious consequences of influenza infection occur from bacterial secondary infections in persons whose resistance has been lowered by the influenza infection. Especially in infants and elderly people, influenza is often followed by bacterial pneumonia; death, if it occurs, is usually due to the bacterial infection. Annually, influenza causes 3–5 million cases of severe illness and is implicated in 250,000–500,000 deaths worldwide.

Most infected individuals develop protective immunity to the infecting virus, making it impossible for a strain of similar antigenic type to cause widespread infection until the virus encounters another susceptible population. Immunity is largely dependent on the production of secretory IgA antibodies

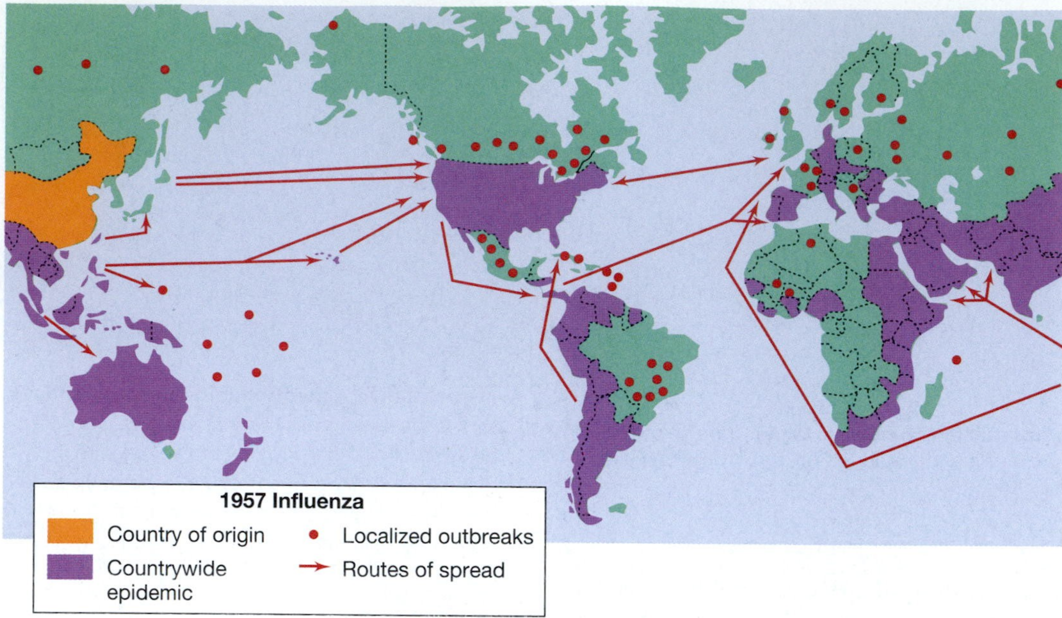

Figure 34.22 An influenza pandemic. The spread of the Asian influenza pandemic of 1957 is shown. The original epidemic focus of this pandemic was probably in China. Evidence suggests that agricultural practices involving poultry and swine, and human interactions with these animals, allowed the reassortment of influenza viral genomes from the three host species, producing a new strain for which there was no immune memory in humans.

directed to antigenic determinants of the hemagglutinin (H) and neuraminidase (N) glycoproteins.

Influenza exists in human populations as an endemic viral disease, and there are widespread outbreaks every year from late autumn through the winter. Antigenic drift results in a reduction of immunity in the population and is responsible for the recurrence of epidemics, severe localized influenza outbreaks, occurring in a 2- to 3-year cycle. Pandemics, worldwide epidemics, are much less frequent, being from 10 to 40 years apart, and are the result of antigenic shift.

The 1957 outbreak of the so-called Asian flu developed into a pandemic (Figure 34.22). The pandemic strain was a virulent mutant virus, differing antigenically from all previous strains. Immunity to this strain was not present, and the virus spread rapidly throughout the world. It first appeared in the interior of China in February 1957 and by April had spread to Hong Kong. From Hong Kong, the virus was carried to San Diego, California, by sailors on naval ships. In May, an outbreak occurred in Newport, Rhode Island, on a naval vessel. From that time, outbreaks continuously occurred in various parts of the United States. The peak incidence occurred in October, when 22 million new cases developed.

In the pandemic of 1918, influenza A "Spanish Flu" killed 230 million people worldwide including several million in the United States (Figure 33.1). Although there have been many pandemics, none has been as catastrophic as the 1918 flu. The virulence of the 1918 influenza is not fully understood, but appears to be due to the host response to the novel pathogen. This pathogen apparently stimulated production and release of large amounts of inflammatory cytokines, resulting in catastrophic systemic inflammation and disease in susceptible individuals.

The pandemics of 1918 and 1957 resulted from the reassortment in swine of avian influenza viruses and human influenza viruses. Swine cells have receptors for both avian and human orthomyxoviruses and can bind and propagate both avian and human influenza strains. If swine are infected with both human and avian strains at the same time, the two unrelated viruses can reassort, resulting in antigenically unique viruses (antigenic shift) that can infect all humans because of the total lack of immunity. Such reassortment from animal strains and infection into humans occurs periodically but unpredictably, continually raising the possibility of a rapidly emerging, highly virulent influenza strain for which there is no preexisting immunity in the human population.

A new avian influenza strain, the influenza A H5N1 strain, also called *avian influenza,* appeared in Hong Kong in 1997, apparently jumping directly from the avian host to humans (Figure 34.23). H5N1 has now been reported in birds throughout Asia, Europe, the Middle East, and North Africa. This strain reemerged in China and Vietnam in 2003 and has now caused infections and deaths in several countries in Southeast Asia, Africa, and the Middle East. H5N1 is still spread directly from avian hosts, usually domestic chickens or ducks, to humans. At this time, avian influenza can be spread human to human only after prolonged close contact. There have been 313 human cases of avian flu worldwide, with 191 deaths through June 2007. There is some indication that H5N1 has infected swine, setting the

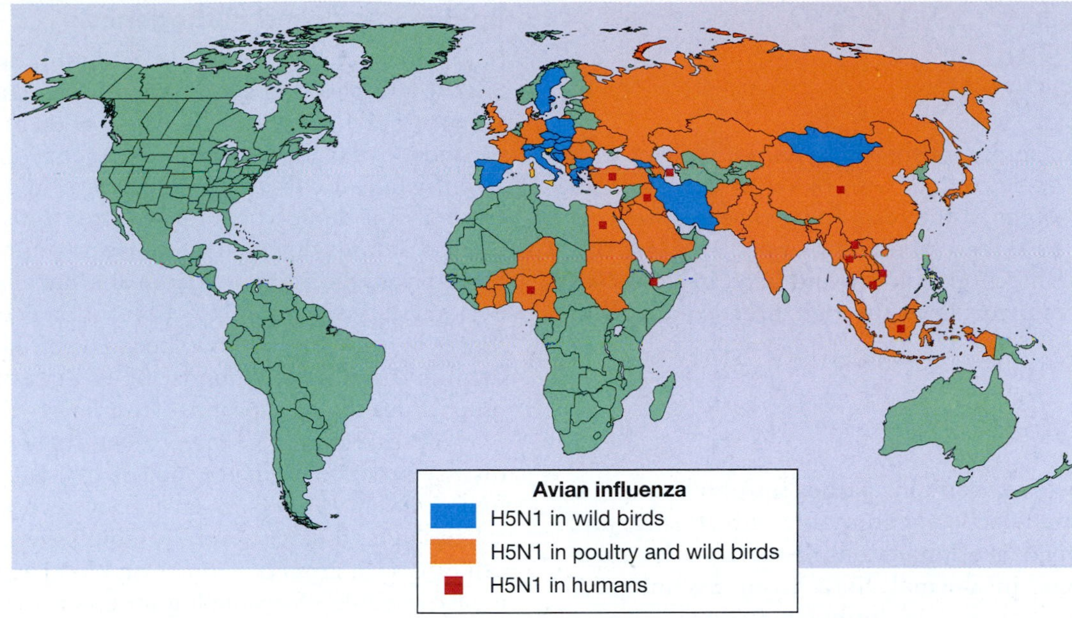

Figure 34.23 Avian influenza. Cumulative occurrences of confirmed avian influenza worldwide in both domestic and wild birds by 2007. Locations of human outbreaks, in red, are also shown.
(*Source*: World Health Organization.)

stage for reassortment with human influenza strains that also infect swine. This reassortment could create a new and highly infective virus for which there is no immunity in humans, starting a new influenza pandemic. Plans are being developed to provide appropriate vaccines and support for a potential pandemic initiated by this and other emergent influenza strains **(http://www.cdc.gov/flu/pandemic/cdcplan.htm)**.

Influenza Prevention and Treatment

Influenza epidemics can be controlled by immunization. However, the choice of appropriate vaccines is complicated by the large number of existing strains and the ability of existing strains to undergo antigenic drift and antigenic shift. When new strains evolve, vaccines are not immediately available, but through careful worldwide surveillance, samples of the major emerging strains of influenza virus are usually obtained before there are epidemic outbreaks. In the United States, candidate strains chosen at the end of each influenza season are grown in embryonated eggs and inactivated. The inactivated viral strains are mixed to prepare a vaccine used for immunization prior to the next influenza season.

Influenza immunization is recommended for those individuals most likely to acquire the disease and develop serious secondary illnesses. Influenza immunization is recommended for everyone over 50 years of age, for those suffering from chronic debilitating diseases (for example, AIDS patients, chronic respiratory disease patients, and so on), and for health-care workers. Effective artificial immunity from the inactivated influenza vaccine lasts only a few years and is strain-specific. Therefore, immunization preparations are updated annually.

Influenza A may also be controlled by use of the chemicals amantadine and rimantadine, synthetic amines that inhibit viral replication. The H5N1 avian influenza is not, however, susceptible to these drugs. The neuraminidase inhibitors oseltamivir and zanamivir (Table 27.5) block release of newly replicated virions of influenza A and B and H5N1 avian virus. These drugs are used to treat ongoing influenza and shorten the course and severity of infection. They are most effective when given very early in the course of the infection. Amantadine, rimantadine, and oseltamivir also prevent the onset and spread of influenza.

Treatment of influenza symptoms with aspirin, especially in children, is not recommended. Aspirin treatment of influenza has been linked to development of *Reye's syndrome,* a rare but occasionally fatal complication involving the central nervous system.

34.9 MiniReview

Influenza outbreaks occur annually due to the plasticity of the influenza genome. Influenza epidemics and pandemics occur periodically. Surveillance and immunization are used to prevent influenza.

- Distinguish between antigenic drift and antigenic shift in influenza.
- Discuss the possibilities for effective immunization programs for influenza and compare them to the possibilities for immunization for colds.

II DIRECT CONTACT TRANSMISSION OF DISEASES

Some pathogens are spread primarily by direct contact with an infected person or by contact with blood or excreta from an infected person. Many of the respiratory diseases we have discussed can also be spread by direct contact. Here we discuss staphylococcal infections, ulcers, and hepatitis, diseases spread primarily person to person through direct contact with infected individuals.

34.10 *Staphylococcus*

The genus *Staphylococcus* contains pathogens of humans and other animals. Staphylococci commonly infect skin and wounds. Most staphylococcal infections result from the transfer of staphylococci in normal flora from an infected, asymptomatic individual to a susceptible individual.

Staphylococci are nonsporulating gram-positive cocci about 0.8–1.0 μm in diameter. They divide in several planes to form irregular clumps (⚭ Figure 16.1). They are resistant to drying and are readily dispersed in dust particles through the air and on surfaces. In humans, two species are important: *Staphylococcus epidermidis,* a nonpigmented species usually found on the skin or mucous membranes, and *Staphylococcus aureus,* a yellow-pigmented species. Both species are potential pathogens, but *S. aureus* is more commonly associated with human disease. Both species are frequently present in the normal microbial flora of the upper respiratory tract and the skin (Figure 34.2).

Epidemiology and Pathogenesis

Staphylococci cause diseases including acne, boils (**Figure 34.24**), pimples, impetigo, pneumonia, osteomyelitis, carditis, meningitis, and arthritis. Many of these diseases cause the production of pus, so they are said to be *pyogenic* (pus-forming). Healthy individuals are often carriers, and resident staphylococci in the upper respiratory tract or skin seldom cause disease. Infants during the first week of life often become colonized from the mother or from another close human contact. Serious staphylococcal infections often occur when the resistance of the host is low because of hormonal changes, debilitating illness, wounds, or treatment with steroids or other drugs that compromise immunity.

Those strains of *S. aureus* most frequently causing human disease produce virulence factors (⚭ Section 28.9). At least four different *hemolysins* have been recognized, and a single strain often produces several. Hemolysins cause the red blood cell lysis seen around colonies on blood agar plates. *S. aureus* is also capable of producing an enterotoxin associated with foodborne illness (Section 37.5). All staphylococci also produce catalase, an enzyme that converts H_2O_2 to H_2O and O_2. Catalase is not considered a virulence factor, but the catalase test distinguishes staphylococci from streptococci, which do not produce catalase.

Another substance produced by *S. aureus* is *coagulase,* an enzyme that causes fibrin to coagulate and form a clot (⚭ Section 28.9). The production of coagulase is generally associated with pathogenicity. Clotting induced by coagulase results in the accumulation of fibrin around the bacterial cells, making it difficult for host defense agents to come into contact with the bacteria and preventing phagocytosis (Figure 34.24). Most *S. aureus* strains also produce *leukocidin,* a

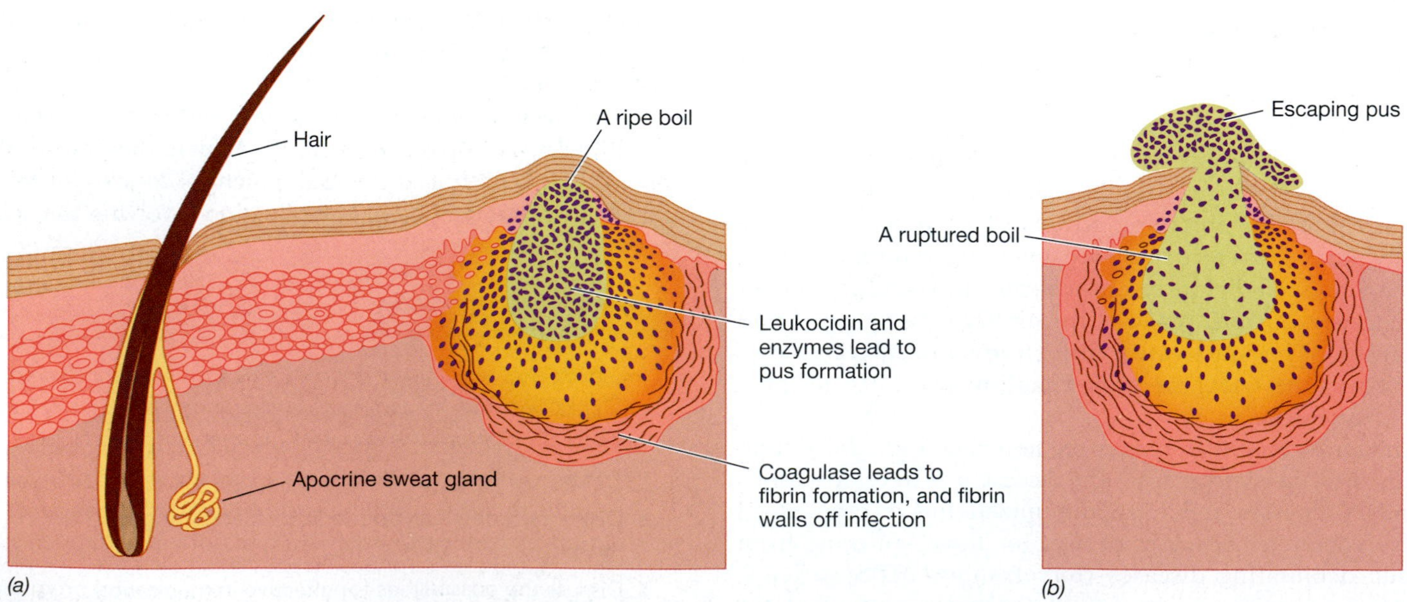

Figure 34.24 The structure of a boil. *(a)* Staphylococci initiate a localized infection of the skin and become walled off by coagulated blood and fibrin through the action of coagulase. *(b)* The rupture of the boil releases pus and bacteria.

protein that destroys leukocytes. Production of leukocidin in skin lesions such as boils and pimples results in considerable host cell destruction and is one of the factors responsible for pus. Some strains of *S. aureus* also produce other extracellular virulence factors, including proteolytic enzymes, hyaluronidase, fibrinolysin, lipase, ribonuclease, and deoxyribonuclease (Table 28.4).

Certain strains of *S. aureus* have been implicated as the agents responsible for **toxic shock syndrome (TSS)**, a serious outcome of staphylococcal infection characterized by high fever, rash, vomiting, diarrhea, and occasionally death. TSS was first recognized in menstruating women and was associated with use of highly absorbent tampons. In menstruating females, blood and mucus in the vagina can become colonized by *S. aureus* from the skin, and the presence of a tampon concentrates this material, creating ideal microbial growth conditions. Largely through education and alterations in materials used in tampons, toxic shock syndrome due to tampon use is now relatively rare, with fewer than 100 cases per year being reported. Over 70% of TSS cases, however, result in death. Toxic shock syndrome is now seen in both men and women and is usually initiated by staphylococcal infections following surgery.

The symptoms of TSS result indirectly from an exotoxin called *toxic shock syndrome toxin-1* (TSST-1). TSST-1 is a superantigen (Section 30.8) released by growing staphylococci, and causes a massive T cell reaction, culminating in the inflammatory response characteristic of superantigen reactions. A TSS-like disease may also be caused by superantigens from other pathogenic bacteria, including *Streptococcus pyogenes* (Section 34.2).

Staphylococcal *enterotoxin A*, another superantigen, causes a form of food poisoning. After ingestion of toxin-contaminated food, the toxin stimulates T cells localized along the intestine, resulting in a massive T cell response and release of inflammatory mediators. The final outcome is the severe but short-lived diarrhea and vomiting associated with staphylococcal food poisoning (Section 37.5).

Treatment and Prevention

Extensive use of antibiotics has resulted in the selection of resistant strains of *S. aureus* and *S. epidermidis*. Healthcare-associated (nosocomial) infections from antibiotic-resistant staphylococci often occur in patients whose resistance is lowered due to other diseases, surgical procedures, or drug therapy (Section 33.7). Patients often acquire staphylococci from healthcare personnel who are asymptomatic carriers of drug-resistant strains. As a result, appropriate antimicrobial drug therapy for *S. aureus* infections is a major problem in healthcare environments. For example, over 100,000 cases of infection with methicillin-resistant *S. aureus* (MRSA) are reported each year in healthcare facilities, but some studies estimate that the total number of MRSA infections may exceed 1 million in these facilities each year. More serious invasive MRSA infections have been a problem in healthcare settings since the 1960s. About 100,000 invasive MRSA infections occur annually. Many of these serious infections are still associated with healthcare settings, but up to 15% are acquired in the community by individuals who have no association with healthcare as either patients or contacts. Although some community-acquired staphylococcal infections are still treatable with penicillin, MRSA and other antibiotic-resistant *S. aureus* strains are becoming more common. Disease-producing isolates of *S. aureus*, regardless of their origin, must therefore be checked individually for antibiotic susceptibility.

Prevention of staphylococcal infections is problematic because most individuals are asymptomatic carriers, and diseases such as acne and impetigo can be transmitted by simple contact with contaminated fingers. In hospital environments such as surgical wards and nurseries, carriers of known pathogenic strains must be either excluded or treated with topical or systemic antimicrobial drugs to eliminate the pathogens.

34.10 MiniReview

Although staphylococci are usually harmless inhabitants of the upper respiratory tract and skin, several serious diseases can result from pyogenic infection or from the actions of staphylococcal superantigen exotoxins.

- What is the normal habitat of *Staphylococcus aureus*? How does *S. aureus* spread from person to person?
- Identify the mechanisms of staphylococcal food poisoning and toxic shock syndrome.

34.11 *Helicobacter pylori* and Gastric Ulcers

Helicobacter pylori is a gram-negative, highly motile, spiral-shaped bacterium (**Figure 34.25**) that is related to *Campylobacter* (Section 37.9). The organism is 2.5–3.5 μm long and 0.5–1.0 μm in diameter and has one to six polar flagella at one end. *H. pylori*, first identified in human intestinal biopsies in 1983, is a pathogen associated with gastritis, ulcers, and gastric cancers. This organism colonizes the non-acid-secreting mucosa of the stomach and the upper intestinal tract, including the duodenum (Figure 28.8). Genetic studies of different strains of *H. pylori* have shown that this organism has been intimately associated with humans at least since humans first migrated from Africa.

Epidemiology

Up to 80% of gastric ulcer patients have concomitant *H. pylori* infections, and up to 50% of asymptomatic adults in developing countries are chronically infected. Person-to-person contact and ingestion of contaminated food or water are the probable transmission methods for *H. pylori*. Although there is no known nonhuman reservoir of *H. pylori*, the organism has occasionally been recovered from cats kept as household pets, indicating that it can be spread to or from animals in

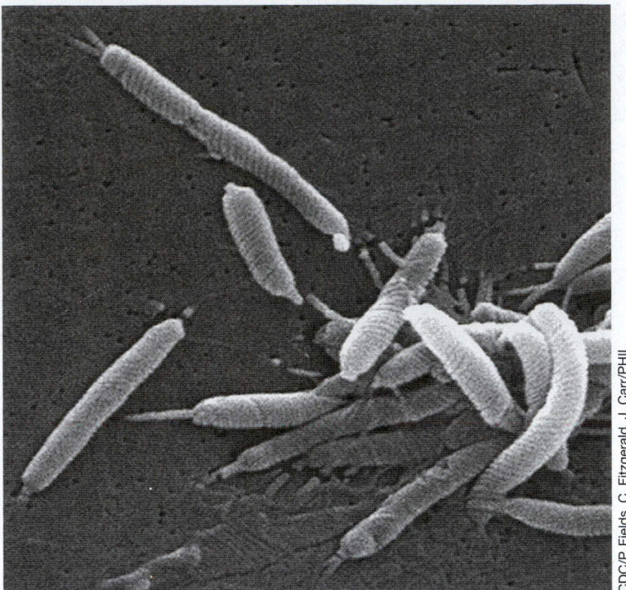

Figure 34.25 ***Helicobacter.*** Scanning electron micrograph. Cells range in size from 1.5 to 10 μm in length and 0.3–1 μm in diameter. The organism in the lower left is 3.2 μm in length. Note the sheathed flagella.

close contact with humans. Infection occurs in high incidence in certain families, and the overall prevalence in the population increases with age. These factors suggest a host-to-host type of transmission. However, infections with *H. pylori* sometimes also occur in epidemic clusters, suggesting that there may be a common source such as food or water.

Pathology, Diagnosis, and Treatment

Current evidence indicates that *H. pylori* is a major preventable and treatable cause of many gastric ulcers. The bacterium is slightly invasive and colonizes the surfaces of the gastric mucosa, where it is protected from the effects of stomach acids by the gastric mucus layer. After mucosal colonization, a combination of pathogen products and host responses cause inflammation, tissue destruction, and ulceration. Pathogen products such as vacA (a cytotoxin), urease, and lipopolysaccharide may contribute to localized tissue destruction and ulceration. Antibodies to *H. pylori* are usually present in infected individuals but are not protective and do not prevent colonization. Individuals who acquire *H. pylori* tend to have chronic infections unless they are treated with antibiotics. Chronic gastritis due to untreated *H. pylori* infection may lead to the development of gastric cancers.

Clinical signs of *H. pylori* infection include belching and stomach (epigastric) pain. Definitive diagnosis requires the recovery and culture or observation of *H. pylori* from a gastric ulcer biopsy. Serum antibodies indicate *H. pylori* infection, but because infections seem to be chronic and antibodies may persist for months after a given infection, *H. pylori* antibodies are not reliable indicators of acute, active disease. A simple *in vivo* diagnostic test, the urease test, is available for *H. pylori*. In this test, the patient ingests ^{13}C or ^{14}C-labeled urea (H_2N–CO–NH_2). If urease is present, the urea will be hydrolyzed into CO_2 and 2 NH_2. The presence of labeled CO_2 in the patient's exhaled breath indicates the presence of urease, produced almost exclusively by *H. pylori*. Recovery of *H. pylori* organisms or antigens from the stool is also indicative of infection.

Evidence for a causal association between *H. pylori* and gastric ulcers comes from antibiotic treatments for the disease. Most patients relapse within 1 year after long-term treatment of ulcers with antacid preparations. However, by treating ulcers as an infectious disease, permanent cures are often obtained. *H. pylori* infection is usually treated with a combination of drugs including the antibacterial compound metranidazole, an antibiotic such as tetracycline or amoxicillin, and a bismuth-containing antacid preparation. The combination treatment, administered for 14 days, abolishes the *H. pylori* infection and provides a long-term cure. For their contributions to unraveling the connection between *H. pylori* and peptic and duodenal ulcers, the Australian scientists Robin Warren and Barry Marshall were awarded the 2005 Nobel Prize in Physiology or Medicine.

34.11 MiniReview

Helicobacter pylori infection appears to be the most common cause of gastric ulcers. Gastric ulcers are now treated as an infectious disease, promoting a permanent cure.

- Describe the infection by *H. pylori* and the resulting development of an ulcer.
- Describe evidence indicating that *H. pylori* infections are spread from person to person and also from common sources.

34.12 Hepatitis Viruses

Hepatitis is a liver inflammation commonly caused by an infectious agent. Hepatitis sometimes results in acute illness followed by destruction of functional liver anatomy and cells, a condition known as **cirrhosis**. Hepatitis due to an infection can cause chronic or acute disease, and some forms lead to liver cancer. Although many viruses and a few bacteria can cause hepatitis, a restricted group of viruses is often associated with liver disease. Hepatitis viruses are diverse, and none are genetically related, but all infect cells in the liver. **Table 34.2** characterizes the five known hepatitis viruses.

Epidemiology

Hepatitis A virus (HAV) is transmitted from person to person or by ingestion of fecally contaminated food or water. Often called *infectious hepatitis,* the virus usually causes mild, even subclinical infections, but rare cases of severe liver disease occur. The most significant food vehicles for hepatitis A are shellfish, usually oysters and clams harvested from water polluted by human fecal material. In recent years, HAV has also

Table 34.2 Hepatitis viruses

Disease	*Virus and genome*	*Vaccine*	*Clinical illness*	*Transmission route*
Hepatitis A	*Hepatovirus* (HAV) ssRNA	Yes	Acute	Enteric
Hepatitis B	*Orthohepadnavirus* (HBV) dsDNA	Yes	Acute, chronic, oncogenic	Parenteral, sexual
Hepatitis C	*Hepacivirus* (HCV) ssRNA	No	Chronic, oncogenic	Parenteral
Hepatitis D	*Deltavirus* (HDV) ssRNA	No	Fulminant, only with HBV	Parenteral
Hepatitis E	Calciviridae family (HEV) ssRNA	No	Fulminant disease in pregnant women	Enteric
Hepatitis G	Flaviviridae family (HGV) ssRNA	No	Asymptomatic	Parenteral

been transmitted in fresh produce. In 2003 a significant outbreak in the eastern United States was traced to eating raw or undercooked green onions. The general trend for numbers of HAV infections has moved downward (**Figure 34.26**), partly due to the availability of an effective vaccine. HAV causes more cases of viral hepatitis than any other virus, and over 30% of individuals in the United States have antibodies to HAV, indicating a past infection.

Infection due to *hepatitis B virus* (HBV) is often called *serum hepatitis*. HBV is a hepadnavirus, a partially double-stranded DNA virus (Section 19.15). The mature virus particle containing the viral genome is called a *Dane particle* (**Figure 34.27**). HBV causes acute, often severe disease that can lead to liver failure and death. Chronic HBV infection can lead to cirrhosis and liver cancer. HBV is usually transmitted by a parenteral (outside the gut) route, such as blood transfusion or through shared hypodermic needles contaminated with infected blood. HBV may also be transmitted through exchanges of body fluids, as in sexual intercourse. The number of new HBV infections is decreasing, again due to an effective vaccine. However, over 150,000 people worldwide and nearly 5,000 people in the United States die each year due to complications such as cancers generated by chronic HBV infection.

Hepatitis D virus (HDV) is a defective virus that lacks genes for its own protein coat (Section 19.15). HDV is also transmitted by parenteral routes, but because it is a defective virus, it cannot replicate and express a complete virus unless the cell is also infected with HBV: the HDV genome replicates independently but uses the protein coat of HBV for expression. Thus, HDV infections are always coinfections with HBV.

Hepatitis C virus (HCV) is also transmitted parenterally. HCV generally produces a mild or even asymptomatic disease at first, but up to 85% of individuals develop chronic hepatitis, with up to 20% leading to chronic liver disease and cirrhosis. Chronic infection leads to hepatocarcinoma (liver cancer) in 3–5% of infected individuals. The latency period for development of cancer can be several decades after the primary infection. The reported new cases of HCV in the United States (Figure 34.26) are only a fraction of the approximately 25,000 new infections annually. Large numbers of HCV-related deaths occur annually due to chronic HCV infections that develop into liver cancer. HCV-induced liver disease is the most common liver disease currently seen in clinical settings in the United States and accounts for up to 10,000 of the 25,000 annual deaths due to liver cancer, other chronic liver disease, and cirrhosis.

Hepatitis E virus (HEV) transmits hepatitis via an enteric route. HEV causes an acute, self-limiting hepatitis that varies in

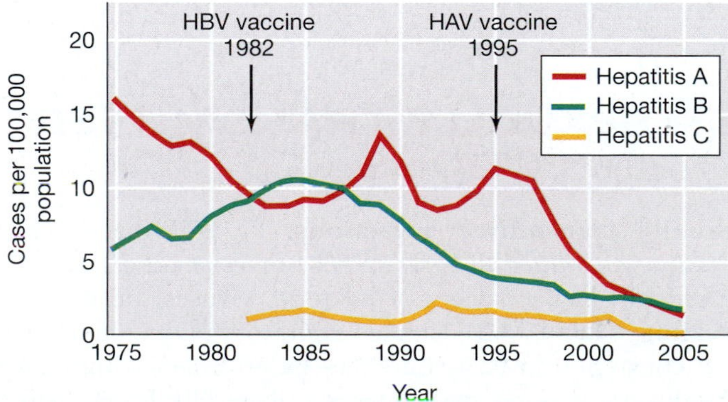

Figure 34.26 Hepatitis in the United States. The incidence of hepatitis is shown by viral agent. In 2005 there were 4,488 reported cases of hepatitis A, 5,119 reported cases of hepatitis B, and 652 reported cases of hepatitis C. Data obtained from the Centers for Disease Control and Prevention, Atlanta, GA, USA.

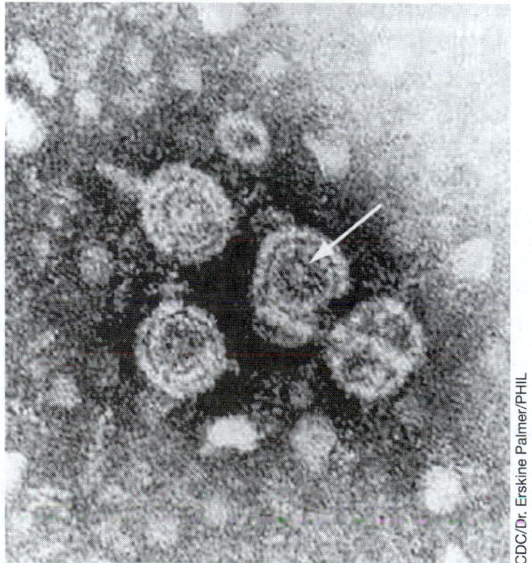

Figure 34.27 Hepatitis B virus (HBV). The arrow indicates a complete HBV particle, which is about 42 nm in diameter and is called a Dane particle.

severity from case to case but is often the cause of rapid fulminant disease in pregnant women. HEV is endemic in Mexico as well as in tropical and subtropical regions of Africa and Asia.

Hepatitis G virus (HGV) is commonly found in the blood of patients with other forms of acute hepatitis, but HGV alone seems to cause very mild disease or is completely asymptomatic. Screening for HGV shows that up to 8.1% of blood donors may be positive for HGV, but because HGV is not associated with demonstrable clinical disease, the significance of these findings is not clear.

Pathology and Diagnosis

Hepatitis is an acute disease of the liver. Symptoms include fever; jaundice (production and release of excess bilirubin by the liver due to destruction of liver cells, resulting in yellowing of the skin); icterus (yellowing of the whites of the eyes); hepatomegaly (liver enlargement); and cirrhosis (breakdown of the normal liver tissue architecture, including fibrosis). Mild hepatitis is characterized by relatively minor elevation of liver enzymes such as alanine aminotransferase (ALT). Fulminant disease is characterized by a rapid onset of severe symptoms such as jaundice and cirrhosis and is often a life-threatening condition. All hepatitis viruses cause similar acute clinical diseases and cannot be readily distinguished based on the clinical findings alone. Chronic hepatitis infections, usually caused by HBV or HCV, are often asymptomatic or produce very mild symptoms but can cause serious liver disease, even in the absence of hepatocarcinoma.

Diagnosis of hepatitis is based primarily on clinical findings and laboratory tests that determine liver function problems. Cirrhosis is diagnosed by visual examination of biopsied liver tissue. Virus-specific assays are also used to confirm diagnosis, identify the infectious agent, and determine a course of treatment. Direct culture of hepatitis viruses is usually not used for identification purposes, and HCV and HGV have not been successfully cultured.

Many of the molecular diagnostic tools discussed in Chapter 32 are used to diagnose hepatitis. Some of the most common methods for determining hepatitis identity are enzyme-linked immunosorbent assay (ELISA) tests. Most hepatitis ELISAs identify viral proteins in blood specimens. However, indirect ELISA tests are available that indicate the presence of IgM or IgG antibodies to HBV. IgM is associated with the primary immune response to HBV, and IgG is associated with the secondary response to HBV. Therefore, identification of the antibody class can determine whether the HBV infection is a new infection (IgM) or a chronic or latent infection (IgG). Other immune-based tests used for the detection of hepatitis viruses include immunoblots, immunoelectron microscopy, and immunofluorescence. Polymerase chain reaction tests and dot-blot DNA hybridization tests are also used for the detection of the viral genome in blood or in liver tissue obtained by biopsy.

Prevention and Treatment

Infection with HAV or HBV can be prevented with effective vaccines. HBV vaccination is recommended and in most cases is required for school-age children in the United States. No effective vaccines are available for the other hepatitis viruses.

Universal precautions are standards developed for personnel handling infectious waste and body fluids. Mandated by law for patient care and clinical laboratory facilities, they are designed to prevent infection by all parenterally transmitted hepatitis viruses (HBV, HCV, HDV, and HGV) as well as HIV (human immunodeficiency virus). The precautions prescribe a high level of vigilance and aseptic handling and containment procedures to deal with patients, body fluids, and infected waste materials. In addition to person-to-person transmission, HAV can be spread through contamination of common sources such as food and water. Epidemic outbreaks of hepatitis A can be prevented by maintaining pathogen-free food and water supplies.

Pooled human immune gamma globulin can be used to prevent HAV infection if given soon after exposure. For post-exposure prevention of HBV infection, specific hepatitis B immune globulin, coupled with administration of the HBV vaccine, has been effective.

Most treatment of hepatitis is supportive, providing rest and time to allow liver damage to resolve and be repaired. In some cases, antiviral drugs are effective for treatment. Interferon α is effective against HCV when combined with the drug ribavirin in some patients. HBV can be treated with the antiviral drugs foscarnet, ribavirin, lamivudine, and ganciclovir.

34.12 MiniReview

Viral hepatitis can result in cirrhosis, an acute liver disease. HBV and HCV can cause chronic infections leading to liver cancer. Vaccines are available for HAV and HBV. The incidence and prevalence of hepatitis has decreased significantly in the last 20 years in the United States, but viral hepatitis is still a major public health problem because of the high infectivity of the viruses and the lack of effective treatment options.

- Describe the mode of transmission for hepatitis A virus, hepatitis B virus, and hepatitis C virus.
- Describe potential prevention and treatment methods for hepatitis A virus and hepatitis B virus.

III SEXUALLY TRANSMITTED INFECTIONS

Sexually transmitted infections, or **STIs**, also called *sexually transmitted diseases (STDs)* or *venereal diseases*, are caused by a wide variety of bacteria, viruses, protists, and even fungi (**Table 34.3**). Unlike respiratory pathogens that are shed constantly in large numbers by an infected individual, sexually transmitted pathogens are generally found only in body fluids from the genitourinary tract that are exchanged during sexual activity. This is because sexually transmitted pathogens are typically very sensitive to drying and other environmental stresses such as heat and light. Their habitat, the

Table 34.3 Sexually transmitted diseases and treatment guidelines

Disease	*Causative organism(s)*[a]	*Recommended treatment*[b]
Gonorrhea	*Neisseria gonorrhoeae* (B)	Cefixime or ceftriaxone, *and* azithromycin or doxycycline
Syphilis	*Treponema pallidum* (B)	Benzathine penicillin G
Chlamydia trachomatis infections	*Chlamydia trachomatis* (B)	Doxycycline or azithromycin
Nongonococcal urethritis	*C. trachomatis* (B) or *Ureaplasma urealyticum* (B) or *Mycoplasma genitalium* (B) or *Trichomonas vaginalis* (P)	Azithromycin
Lymphogranuloma venereum	*C. trachomatis* (B)	Doxycycline
Chancroid	*Haemophilus ducreyi* (B)	Azithromycin
Genital herpes	Herpes simplex type 2 (V)	No known cure; symptoms can be controlled with acyclovir, valcyclovir, and other antiviral drugs
Genital warts	Human papillomavirus (HPV) (certain strains)	No known cure; symptomatic warts can be removed surgically, chemically, or by cryotherapy
Trichomoniasis	*Trichomonas vaginalis* (P)	Metronidazole
Acquired immunodeficiency syndrome (AIDS)	Human immunodeficiency virus (HIV)	No known cure; nucleotide base analogs, protease inhibitors, fusion inhibitors, and nonnucleoside reverse transcriptase inhibitors may slow disease progression
Pelvic inflammatory disease	*N. gonorrhoeae* (B) or *C. trachomatis* (B)	Cefotetan
Vulvovaginal candidiasis	*Candida albicans* (F)	Butoconazole

[a]B, bacterium; V, virus; P, protist; F, fungus.

[b]Recommendations of the U.S. Department of Health and Human Service, Public Health Service. For many treatment plans, there are a number of alternatives.

human genitourinary tract, is a protected, moist environment. Thus, these pathogens preferentially and sometimes exclusively colonize the genitourinary tract.

Diagnosis and treatment of STIs is very challenging for both social and biological reasons. First, up to one-third of all STIs are in teenagers with multiple sex partners, making it difficult to identify the infection source and stop its spread. Second, many STIs have minor symptoms; infected individuals often do not seek treatment. Third, social stigmas attached to STIs prevent many individuals from seeking prompt treatment. Prompt effective treatment of STIs, however, is important for a number of reasons. First, most STIs are curable, and virtually all are controllable with appropriate medical intervention. Second, delay or lack of treatment can lead to long-term problems such as infertility, cancer, heart disease, degenerative nerve disease, birth defects, stillbirth, or destruction of the immune system.

Because transmission of STIs is limited to intimate physical contact, generally during sexual intercourse, their spread can be controlled by sexual abstinence (no exchange of body fluids) or by the use of barriers such as condoms that stop the exchange of body fluids during sexual activity.

STIs are very common and continue to pose social as well as medical problems. Here we discuss some prevalent STIs.

34.13 Gonorrhea and Syphilis

Gonorrhea and *syphilis* are preventable, treatable bacterial STIs. Because of differences in their symptoms, the overall pattern of disease differs between the two. Gonorrhea is very prevalent, and often asymptomatic, especially in women. The disease is often unrecognized and remains untreated. Syphilis, on the other hand, now has a low prevalence (**Figure 34.28**). This is partly because syphilis exhibits very obvious symptoms in its primary stage and infected individuals usually seek immediate treatment.

Gonorrhea

Neisseria gonorrhoeae, often called the *gonococcus,* causes gonorrhea. *N. gonorrhoeae* is a gram-negative, nonsporulating, obligately aerobic, oxidase-positive diplococcus related biochemically and phylogenetically to *Neisseria meningitidis* (∞ Section 15.10 and Section 34.6). *N. gonorrhoeae* is very sensitive to drying and normally does not survive away from the mucus membranes of the genitourinary tract (**Figure 34.29**). Gonococci are killed rapidly by drying, sunlight, and ultraviolet light. Because of its extreme sensitivity to environmental conditions, *N. gonorrhoeae* can be transmitted only by intimate person-to-person contact. The pathogen enters the body by way of the mucus membranes of the genitourinary tract.

The symptoms of gonorrhea are quite different in the male and female. In females, gonorrhea is characterized by a mild vaginitis that is difficult to distinguish from vaginal infections caused by other organisms, and thus, the infection may easily go unnoticed. Complications from untreated gonorrhea in females can lead, however, to a condition known as pelvic inflammatory disease (PID). PID is a chronic inflammatory

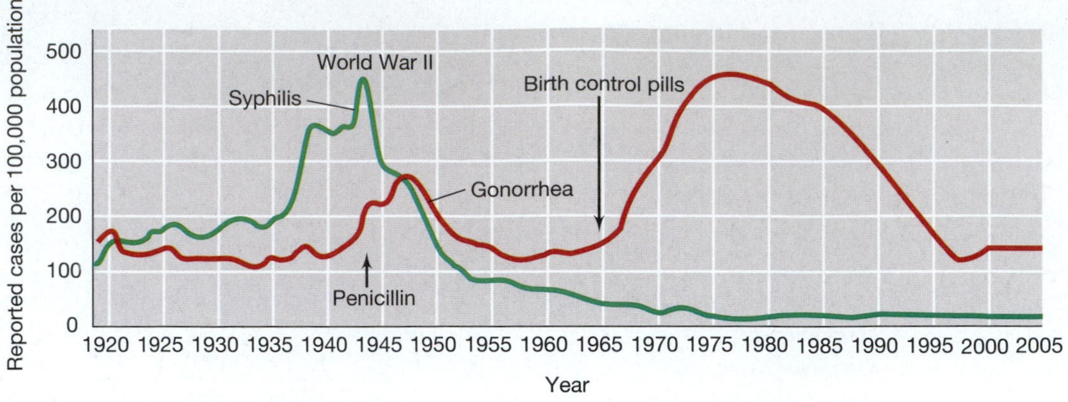

Figure 34.28 Reported cases of gonorrhea and syphilis in the United States. Note the downward trend in disease incidence after the introduction of antibiotics and the upward trend in the incidence of gonorrhea after the introduction of birth control pills. In 2005 there were 339,593 cases of gonorrhea and only 8,724 cases of primary and secondary syphilis in the United States.

disease that can lead to long-term complications such as sterility. In the male, the organism causes a painful infection of the urethral canal. Complications from untreated gonorrhea affecting both males and females include damage to heart valves and joint tissues due to immune complex deposition.

In addition to gonorrhea, *N. gonorrhoeae* also causes eye infections in newborns. Infants born of infected mothers may acquire eye infections during birth. Therefore, prophylactic treatment of the eyes of all newborns with an ointment containing erythromycin is generally mandatory to prevent gonococcal infection in infants. We discussed the clinical microbiology and diagnosis of gonorrhea in Section 32.1.

Treatment of gonorrhea with penicillin has been successful in the past. However, strains of *N. gonorrhoeae* resistant to penicillin arose in the 1980s and are now widespread. The quinolones ciprofloxacin, oflaxacin, or levofloxacin were also used, but by 2006, about 14% of *N. gonorrhoeae* strains isolated nationwide had developed resistance. Strains resistant to penicillin and quinolones respond to alternative antibiotic therapy with a single dose of the β-lactam antibiotics cefixime or ceftriaxone. An antichlamydial agent, normally azithromycin or doxycycline, is often given at the same time because nearly 50% of gonorrhea patients are also infected with the harder to diagnose sexually transmitted pathogen, *Chlamydia trachomatis* (Table 34.3; Section 34.14).

The incidence of gonococcus infection remains relatively high for the following reasons: (1) Acquired immunity does not exist; hence repeated reinfection is possible. Although antibodies are produced, they are either not protective, or they are strain-specific and provide no cross-immunization. Within a single *N. gonorrhoeae* strain, there may be antigenic switches that change opacity protein antigens (Opa) and surface pilin antigens. These changes create new serotypes and thwart effective immunity. (2) The use of oral contraceptives alters the local mucosal environment in favor of the pathogen. Oral contraceptives induce the body to mimic pregnancy, which results, among other things, in a lack of glycogen production in the vagina and a rise in the vaginal pH. Lactic acid bacteria normally found in the adult vagina (∞ Figure 28.11) fail to develop under such circumstances, facilitating colonization by *N. gonorrhoeae* transmitted from an infected partner. (3) Symptoms in the female are so mild that the disease may be unrecognized, and an infected female with multiple partners can infect many males. The disease can be controlled if the sexual contacts of infected persons are quickly identified and treated. But it is often difficult to obtain this information and even more difficult to arrange treatment, as social stigmas associated with sexually transmitted infections are often obstacles to obtaining medical care.

Figure 34.29 The causative agent of gonorrhea, *Neisseria gonorrhoeae*. The scanning electron micrograph of the microvilli of human fallopian tube mucosa shows how cells of *N. gonorrhoeae* attach to the surface of epithelial cells. Note the distinct diplococcus morphology. Cells of *N. gonorrhoeae* are about 0.8 μm in diameter.

Syphilis

Syphilis is caused by a spirochete, *Treponema pallidum. T. pallidum* is about 10–15 μm in length and extremely thin, about 0.15 μm in diameter (**Figure 34.30**). The spirochete is extremely sensitive to environmental stress; therefore, syphilis is normally transmitted from person to person by intimate sexual contact. The biology of the spirochetes and the genus *Treponema* is discussed in Section 16.16.

Syphilis is often transmitted at the same time as gonorrhea. However, syphilis is potentially more serious than gonorrhea. For example, worldwide, syphilis kills over 150,000 people per year, whereas gonorrhea directly kills only about 1,000 people per year. Largely because of differences in the symptoms and pathobiology of the two diseases, however, the incidence of syphilis in the United States is much lower than the incidence of gonorrhea (Figure 34.28).

The syphilis spirochete does not pass through unbroken skin, and initial infection most probably takes place through tiny breaks in the epidermal layer. In the male, initial infection is usually on the penis; in the female it is most often in the vagina, cervix, or perineal region. In about 10% of cases, infection is extragenital, usually in the oral region. During pregnancy, the organism can be transmitted from an infected woman to the fetus; the disease acquired by the infant is called **congenital syphilis**.

Syphilis is an extremely complex disease and, in an individual patient, may progress into any of three stages, but the disease always begins with a localized infection called *primary syphilis*. In primary syphilis, *T. pallidum* multiplies at the initial site of entry, and a characteristic primary lesion called a *chancre* forms within 2 weeks to 2 months (**Figure 34.31**). Dark-field microscopy of the syphilitic chancre exudate reveals the actively motile spirochetes (Figure 34.30*a*). In most cases the chancre heals spontaneously and *T. pallidum* disappears from the site. Some cells, however, spread from the initial site to various parts of the body, such as the mucous membranes, the eyes, joints, bones, or central nervous system, where extensive multiplication occurs. A hypersensitivity reaction to the treponemes often takes place, revealed by the development of a generalized skin rash; this rash is the key symptom of *secondary syphilis*. At first, the secondary rash papules may contain *T. pallidum*, making them highly infectious. Eventually the spirochetes are cleared from the secondary lesions and infectivity is reduced.

The subsequent course of the disease in the absence of treatment is highly variable. About one-fourth of infected

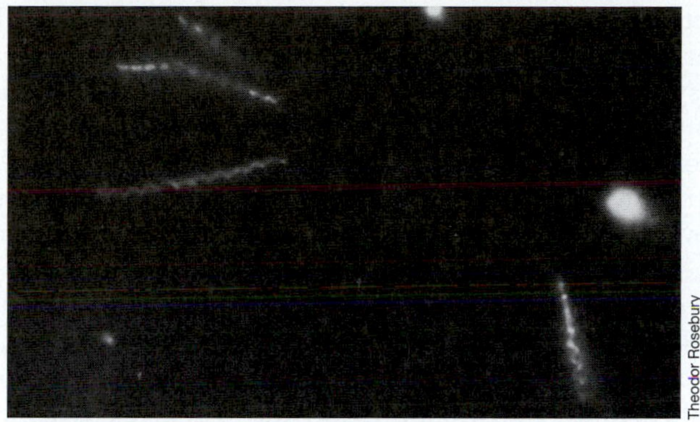

Theodor Rosebury

(a)

Centers for Disease Control

(b)

Figure 34.30 The syphilis spirochete, *Treponema pallidum*. *(a)* Dark-field microscopy of an exudate. *Treponema pallidum* cells measure 0.15 μm wide and 10–15 μm long. *(b)* Shadow-cast electron micrograph of a cell of *T. pallidum*. The endoflagella are typical of spirochetes.

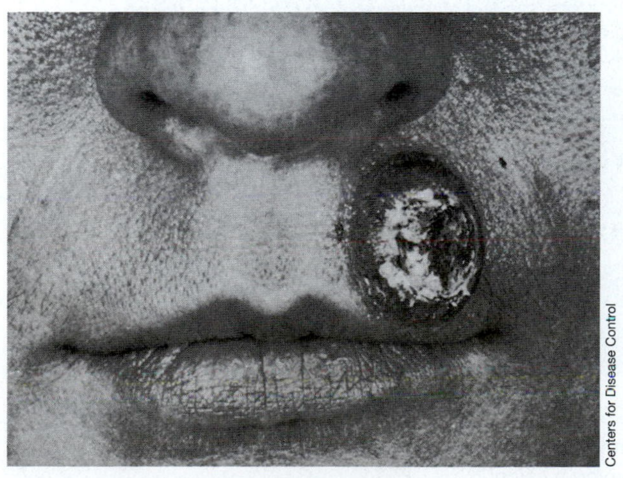

Centers for Disease Control

(a)

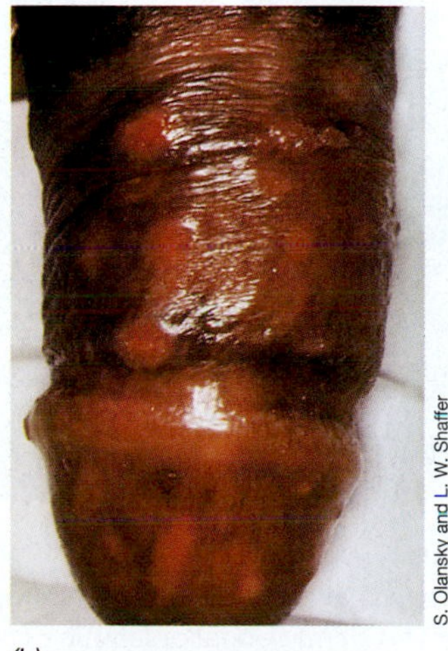

S. Olansky and L. W. Shaffer

(b)

Figure 34.31 Primary syphilis lesions. *(a)* Chancre on lip. *(b)* Several chancres on penis. The chancre is the characteristic lesion of primary syphilis at the site of infection by *Treponema pallidum*. Patients who acquire such lesions generally seek medical intervention, and the obvious chancre hastens diagnosis and treatment.

individuals undergo a spontaneous cure as demonstrated by a decrease in antibody titer. Another one-fourth exhibit no further symptoms, although static or elevated antibody titers indicate a persistent, active infection. About half of untreated patients develop *tertiary syphilis,* with symptoms ranging from relatively mild infections of the skin and bone to serious and even fatal infections of the cardiovascular system or central nervous system. Involvement of the nervous system can cause generalized paralysis or other severe neurological damage. Relatively low numbers of *T. pallidum* are present in individuals with tertiary syphilis, and most of the symptoms probably result from inflammation due to delayed hypersensitivity reactions to the spirochetes.

The clinical immunology and microbiology as well as the laboratory diagnosis of syphilis were discussed in Section 32.1. The single most important physical sign of a primary syphilis infection, the chancre, is diagnostic for the disease. Infected individuals generally seek treatment for syphilis because of the highly visible chancre.

Penicillin is highly effective in syphilis therapy, and the primary and secondary stages of the disease can usually be controlled by a single injection of benzathine penicillin G. In tertiary syphilis, penicillin treatment must be extended for longer periods of time. The incidence of primary and secondary syphilis in the United States has decreased significantly over the last two decades and is near the lowest levels since record keeping began.

34.13 MiniReview

Gonorrhea and syphilis, caused by *Neisseria gonorrhoeae* and *Treponema pallidum*, respectively, are STIs with potential serious consequences if infections are not treated. Although the incidence of these diseases has generally declined in recent years, there are still over 330,000 new cases of gonorrhea and nearly 8,000 new cases of syphilis annually in the United States.

- Explain at least one potential reason for the high incidence of gonorrhea as compared with syphilis.
- Describe the progression of untreated gonorrhea and untreated syphilis. Do treatments produce a cure for each disease?

34.14 Chlamydia, Herpes, Trichomoniasis, and Human Papillomavirus

Chlamydia, herpes, trichomoniasis, and human papillomavirus infections are important STIs. These diseases are very prevalent in the population and are much more difficult to diagnose and treat than are syphilis and gonorrhea.

Chlamydia

A number of sexually transmitted diseases can be ascribed to infection by the obligate intracellular bacterium *Chlamydia trachomatis* (**Figure 34.32** and Section 16.9). The total incidence of sexually transmitted *C. trachomatis* infections probably greatly outnumbers the incidence of gonorrhea. Over 900,000 cases are reported, and there may be over 4 million new *C. trachomatis* sexually transmitted infections every year. This organism is the cause of the most prevalent STI and reportable communicable disease in the United States. *C. trachomatis* also causes a serious eye infection called *trachoma,* but the strains of *C. trachomatis* responsible for STIs are distinct

(a)

(b)

Figure 34.32 Cells of *Chlamydia trachomatis* (arrows) attached to human fallopian tube tissues. *(a)* Cells attached to the microvilli of a fallopian tube. *(b)* A damaged fallopian tube containing a cell of *C. trachomatis* (arrow) in the lesion.

from those causing trachoma. Chlamydial infections may also be transmitted congenitally to the newborn in the birth canal, causing newborn conjunctivitis and pneumonia.

Nongonococcal urethritis (NGU) due to *C. trachomatis* is one of the most frequently observed sexually transmitted diseases in males and females, but the infections are often inapparent. In a small percentage of cases, chlamydial NGU leads to serious acute complications, including testicular swelling and prostate inflammation in men, and cervicitis, pelvic inflammatory disease, and fallopian tube damage in women; cells of *C. trachomatis* attach to microvilli of fallopian tube cells, enter, multiply, and eventually lyse the cells (Figure 34.32). Untreated NGU can cause infertility.

Chlamydia NGU is relatively difficult to diagnose by traditional isolation and identification methods, but the organism can be cultured. To speed diagnoses, however, tests have been developed for identifying *C. trachomatis* from a vaginal or pelvic swab or from discharges. These include nucleic acid probe tests, nucleic acid amplification tests, fluorescent antibody tests, and enzyme-linked immunosorbent assay (ELISA) tests for detecting *C. trachomatis* genes and antigens. If a chlamydial infection is suspected even in the absence of a positive diagnostic test, treatment is initiated with azithromycin or doxycycline (Table 34.3).

Chlamydial NGU is frequently observed as a secondary infection following gonorrhea. If both *Neisseria gonorrhoeae* and *C. trachomatis* are transmitted to a new host in a single event, treatment of gonorrhea with cefixime or ceftriaxone is usually successful but does not eliminate the chlamydia. Although cured of gonorrhea, such patients are still infected with chlamydia and eventually experience an apparent recurrence of gonorrhea that is instead a case of chlamydial NGU. Thus, patients treated for gonorrhea are also given azithromycin or doxycycline to treat the potential coinfection with the usually undiagnosed *C. trachomatis*.

Lymphogranuloma venereum is a sexually transmitted disease caused by distinct strains of *C. trachomatis* (LGV 1, 2, and 3). The disease occurs most frequently in males and is characterized by infection and swelling of the lymph nodes in and about the groin. From the infected lymph nodes, chlamydial cells may travel to the rectum and cause a painful inflammation of rectal tissues called proctitis. Lymphogranuloma venereum has the potential to cause regional lymph node damage and the complications of proctitis. It is the only chlamydial infection that invades beyond the epithelial cell layer.

Herpes

Herpesviruses are a large group of complex double-stranded DNA viruses (∞ Section 19.12), many of which are human pathogens. A subgroup of herpesviruses, the herpes simplex viruses, is responsible for cold sores and genital infections.

Herpes simplex virus type 1 (*HSV-1*) infects the epithelial cells around the mouth and lips, causing cold sores (fever blisters) (**Figure 34.33a**). HSV-1 may, however, occasionally infect other body sites including the anogenital regions. HSV-1 is spread via direct contact or through saliva. The incubation period of HSV-1 infections is short (3–5 days), and the lesions heal without treatment in 2–3 weeks. The virus is most likely spread primarily by contact with infectious lesions. Latent herpes infections are common, with the virus persisting in low numbers in nerve tissue. Recurrent acute herpes infections are due to a periodic triggering of virus activity by unknown or indeterminant causes such as coinfections and stress. Oral herpes caused by HSV-1 is quite common and apparently has no harmful effects on the host beyond the discomfort of the oral blisters.

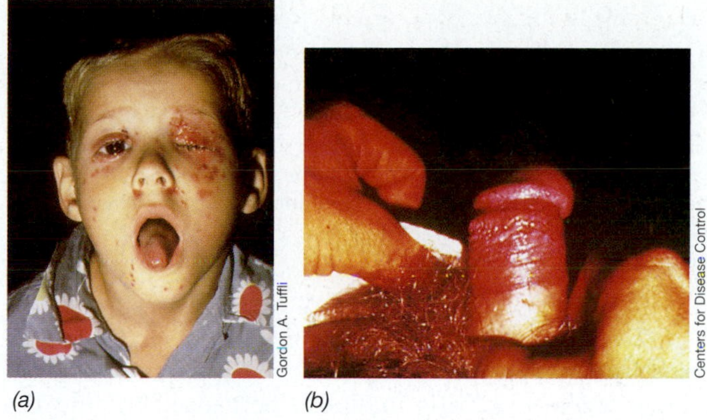

Figure 34.33 Herpesvirus. *(a)* A severe case of herpes blisters on the face due to infection with herpes simplex virus type 1. *(b)* Herpes simplex virus type 2 infection on the penis.

Herpes simplex virus 2 (*HSV-2*) infections are associated primarily with the anogenital region, where the virus causes painful blisters on the penis of males (Figure 34.33*b*) or on the cervix, vulva, or vagina of females. HSV-2 infections are generally transmitted by direct sexual contact, and the disease is most easily transmitted when active blisters are present but may also be transmitted during asymptomatic periods, even when the infection is presumably latent. HSV-2 occasionally infects other sites such as the mucous membranes of the mouth. HSV-2 can also be transmitted to a newborn at birth by contact with herpetic lesions in the birth canal. The disease in the newborn varies from latent infections with no apparent damage to systemic disease resulting in brain damage or death. To avoid herpes infections in newborns, delivery by caesarean section is advised for pregnant women with genital herpes infections. The long-term effects of genital herpes infections are not yet fully understood. However, studies have indicated a significant correlation between genital herpes infections and cervical cancer in females.

Genital herpes infections are presently incurable, although a limited number of drugs have been successful in controlling the infectious blister stages. The guanine analog acyclovir (**Figure 34.34**), given orally and also applied topically, is particularly effective in limiting the shed of active virus from blisters and promoting the healing of blistering lesions. Acyclovir, valcyclovir, and vidarabine are nucleoside analogs that interfere with herpesvirus DNA polymerase, inhibiting viral DNA replication.

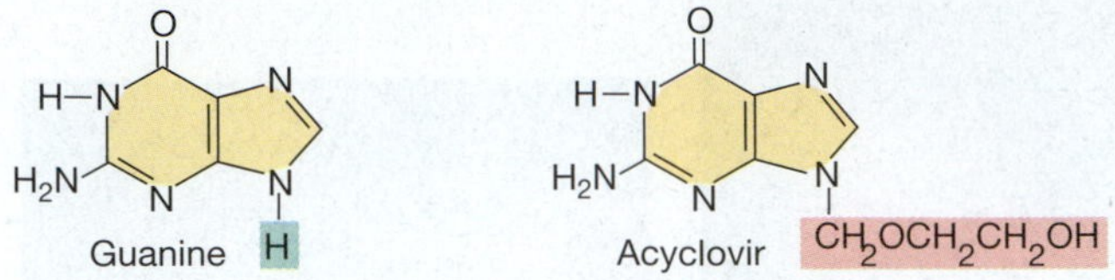

Figure 34.34 Guanine and the guanine analog acyclovir. Acyclovir has been used therapeutically to control genital herpes (HSV-2) blisters.

Trichomoniasis

Nongonococcal urethritis may also be caused by infections with the protist *Trichomonas vaginalis* (**Figure 34.35**). *T. vaginalis* does not produce the resting cells or cysts important to the life cycle of many protists. As a result, transmission is usually from person to person, generally by sexual intercourse. However, cells of *T. vaginalis* can survive for 1–2 hours on moist surfaces, 30–40 minutes in water, and up to 24 hours in urine or semen. Thus, *T. vaginalis* is sometimes transmitted by contaminated toilet seats, sauna benches, and paper towels. *T. vaginalis* infects the vagina in women, the prostate and seminal vesicles of men, and the urethra of both males and females.

Many cases of trichomoniasis are totally asymptomatic in males. In women trichomoniasis is characterized by a vaginal discharge, vaginitis, and painful urination. The infection is more common in females; surveys indicate that 25–50% of sexually active women are infected; only about 5% of men are infected because of the killing action of prostatic fluids. The male partner of an infected female should be examined for *T. vaginalis* and treated if necessary because sexually active asymptomatic males can transmit the infection to several females. Trichomoniasis is diagnosed by observation of the motile protists in a wet mount of fluid discharged from the patient. The antiprotozoal drug *metronidazole* is particularly effective in treating trichomoniasis (Table 34.3).

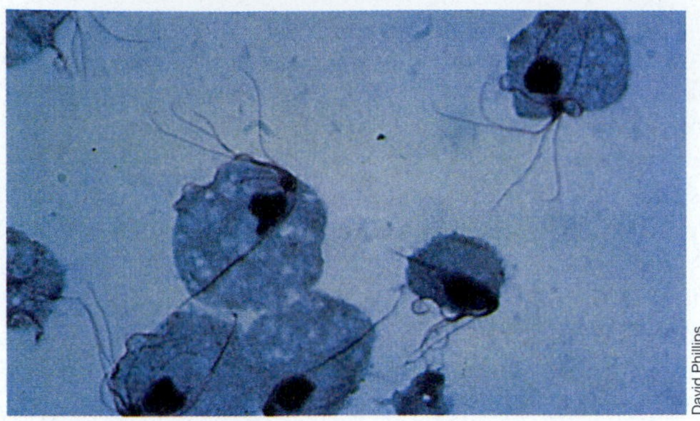

Figure 34.35 ***Trichomonas vaginalis.*** This flagellated protist causes trichomonas, a common sexually transmitted infection.

Human Papillomavirus

Human papillomaviruses (HPV) comprise a family of double-stranded DNA viruses. Of more than 100 different strains, about 30 are transmitted sexually, and several of these are known to cause genital warts and cervical cancer (Chapter 30, Microbial Sidebar, "The Promise of New Vaccines"). About 20 million people in the United States are infected, and up to 80% of women over age 50 have had at least one HPV infection. Over 6 million people acquire new HPV infections annually. Almost 10,000 women develop cervical cancer, with about 3,700 deaths each year.

Most HPV infections are asymptomatic, with some progressing to cause genital warts. Others cause cervical neoplasia (abnormalities in cells of the cervix), and a few progress to cervical cancers. Most HPV infections resolve spontaneously but, as with most viral infections, there is no adequate treatment or cure for active infections. Because of their potential as oncogenic viruses, a HPV vaccine has been developed. The vaccine is designed to provide immunity to the most oncogenic (cancer-causing) viral strains. The HPV vaccine is currently recommended for females 11–26 years of age. The vaccine may also be recommended in the future for males because they can develop anal and penile cancers from HPV infections and because they are HPV carriers.

34.14 MiniReview

Chlamydia, the most prevalent STI, is caused by infection with the bacterium *Chlamydia trachomatis*. Untreated chlamydial nongonococcal urethritis causes serious complications in males and females. Herpes lesions can also be transmitted sexually and are caused by herpes simplex virus type 1 and herpes simplex virus type 2. HSV-2 is generally associated with sexual transmission and infection of the anogenital regions. There is no cure for herpes infections. *Trichomonas vaginalis* is a protist responsible for trichomoniasis, another STI. Human papillomaviruses cause widespread STIs that may lead to cancer. There is an effective HPV vaccine. In general, these STIs are widespread and are more difficult to diagnose and treat than gonorrhea or syphilis.

- Describe pertinent clinical features and treatment protocols for chlamydia, herpes, trichomoniasis, and human papillomavirus.
- Why are these diseases more difficult to diagnose than gonorrhea or syphilis?

34.15 Acquired Immunodeficiency Syndrome: AIDS and HIV

Acquired immunodeficiency syndrome (AIDS) was recognized as a distinct disease in 1981. More than 950,000 cases of AIDS have been reported since then in the United States alone, and more than 530,000 people have died from AIDS (Section 33.6). At least 476,000 and perhaps 1.4 million people are living with the infection by human immunodeficiency virus (HIV), the causative agent of AIDS. Worldwide, more than 70 million people have been infected with HIV and at least 5 million are infected each year. About 3 million

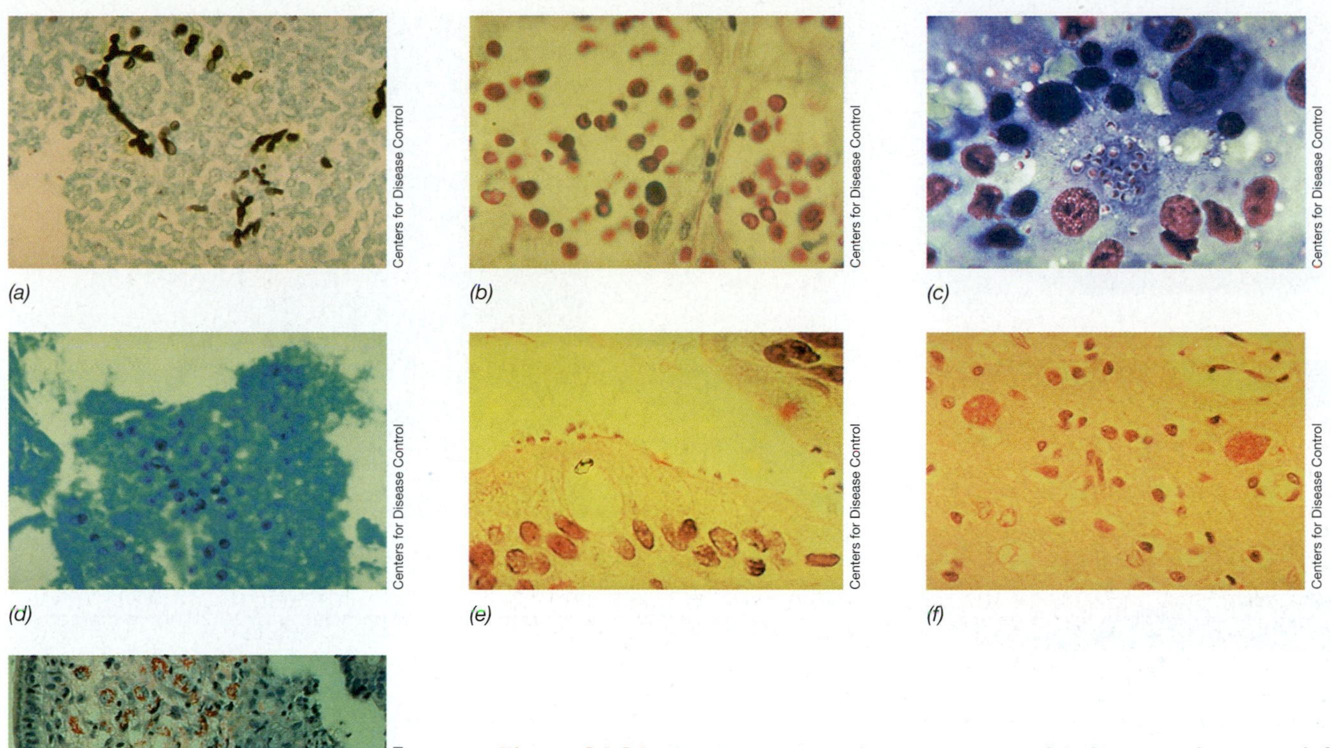

Figure 34.36 Opportunistic pathogens associated with acquired immunodeficiency syndrome (AIDS). *(a) Candida albicans*, from heart tissue of patient with systemic *Candida* infection. *(b) Cryptococcus neoformans*, from liver tissue of a patient with cryptococcosis. *(c) Histoplasma capsulatum*, from liver tissue of patient with histoplasmosis. *(d) Pneumocystis carinii*, from patient with pulmonary pneumocytosis. *(e) Cryptosporidium* sp. from small intestine of a patient with cryptosporidiosis. *(f) Toxoplasma gondii*, from brain tissue of patient with toxoplasmosis. *(g) Mycobacterium* spp. infection of the small bowel (acid-fast stain).

people die each year, and over 25 million people have already died from AIDS.

HIV

HIV is divided into two types, *HIV-1* and *HIV-2*. HIV-1 is genetically similar, but distinct from HIV-2. HIV-2, discovered in West Africa in 1985, is less virulent than HIV-1 and causes a milder AIDS-like disease. Currently, more than 99% of global AIDS cases are due to HIV-1, and thus we focus on HIV-1 here.

HIV-1 is a retrovirus (∞ Section 19.15). The genome contains 9,749 nucleotides in each of its two identical single-stranded RNA molecules. Using the viral RNA as a template, reverse transcriptase in the intact virion catalyzes formation of a complementary single-stranded DNA molecule. The enzyme then converts the complementary DNA (cDNA) into double-stranded DNA, which integrates into the host cell genome (∞ Figure 10.24).

The number of HIV-infected individuals will continue to rise unless curative treatment measures or prevention methods are discovered. In the United States, the numbers of newly diagnosed HIV infections are increasing, and declines in HIV morbidity and mortality attributable to combination antiretroviral therapy have ended, reversing the trends of decreasing AIDS prevalence and severity seen in the 1990s. We have already considered the epidemiology of AIDS (∞ Section 33.6) and some diagnostic methods for identifying and tracking HIV infection (∞ Sections 32.10–32.11). Here we focus on the pathogenesis of AIDS, the usual course of HIV infection, and the effects of HIV on the immune system.

A Definition of AIDS

AIDS was first suspected of being a disease affecting the immune system because a large number of opportunistic infections were observed in certain populations (∞ Section 33.6). **Opportunistic infections** are infections typically observed only in individuals with a dysfunctional immune system. The following definition of AIDS was adopted in 1993 by the Centers for Disease Control and Prevention (Atlanta, Georgia, USA) and is used to define AIDS in the United States.

The case definition for acquired immunodeficiency syndrome (AIDS) includes those who (a) test positive for human immunodeficiency virus (HIV) and (b) meet one of the two following criteria.

1. A CD4 T cell number of less than 200/mm^3 of whole blood (the normal count is 600–1,000/mm^3) or a CD4 T cell/total lymphocytes percentage of less than 14%.
2. A CD4 T cell number of more than 200/mm^3; *and* any of the following conditions (**Figure 34.36**): fungal diseases including candidiasis coccidiomycosis, cryptococcosis, histoplasmosis, isosporiasis, *Pneumocystis carinii* pneumonia, cryptosporidiosis, or toxoplasmosis

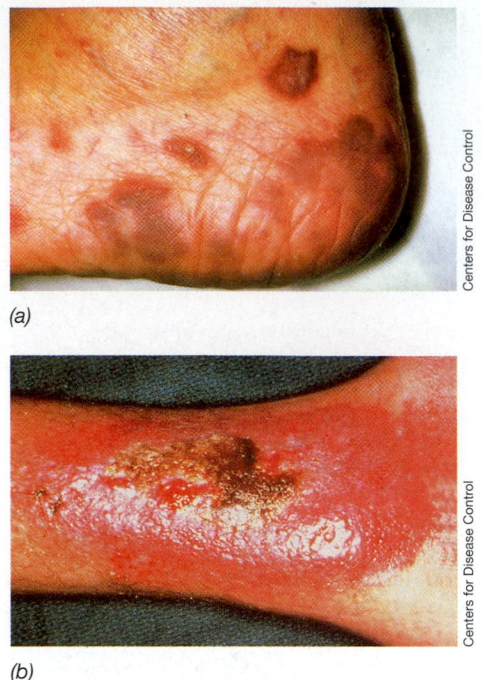

Figure 34.37 Kaposi's sarcoma. Lesions are shown as they appear on *(a)* the heel and lateral foot, and *(b)* the distal leg and ankle.

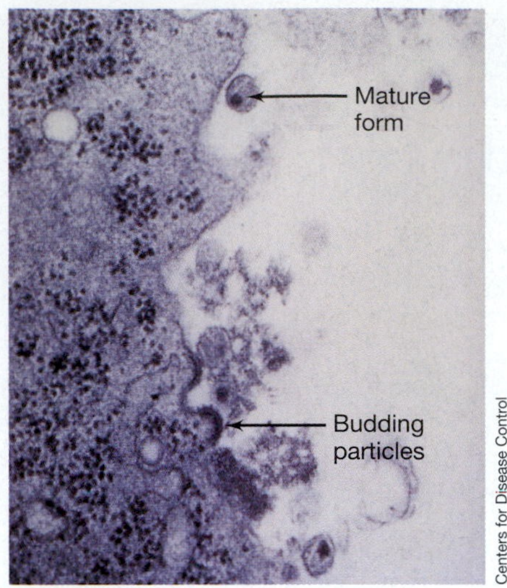

Figure 34.38 Human lymphocyte releasing HIV. Transmission electron micrograph of a thin section. Cells were from a hemophiliac patient who developed AIDS. HIV particles are 90–120 nm in diameter.

of the brain; bacterial diseases including pulmonary tuberculosis or other *Mycobacterium* spp. infections, or recurrent *Salmonella* septicemia; viral diseases including cytomegalovirus infection, HIV-related encephalopathy, HIV wasting syndrome, chronic ulcers, or bronchitis due to herpes simplex; or progressive multifocal leukoencephalopathy, malignant diseases such as invasive cervical cancer, Kaposi's sarcoma, Burkitt's lymphoma, primary lymphoma of the brain, or immunoblastic lymphoma; or recurrent pneumonia due to any agent.

The most common opportunistic disease in AIDS patients is pneumonia caused by the protist *P. carinii*. A frequent nonmicrobial disease in AIDS patients is Kaposi's sarcoma, an atypical cancer observed in high frequency in HIV-infected homosexual males. Kaposi's sarcoma is a cancer of the cells lining the blood vessels and is characterized by purple patches on the surface of the skin, especially in the extremities (**Figure 34.37**). Kaposi's sarcoma is caused by coinfection of HIV and human herpesvirus 8 (HHV-8) and is 20,000-fold more prevalent in AIDS patients than in the general population.

In summary, an individual has AIDS if he or she tests positive for HIV or HIV antibodies and has a drastically reduced T-helper lymphocyte count, or has at least one of a number of opportunistic infections or atypical cancers.

HIV Pathogenesis

HIV infects cells that contain the CD4 cell surface protein. The two cell types most commonly infected are macrophages and a class of lymphocytes called T-helper (T_H) cells, both of which are important components of the immune system. Infected macrophages and T cells produce and release large numbers of HIV particles, which in turn infect other cells that display CD4 (**Figure 34.38**).

In addition to CD4, HIV must interact with coreceptors on target cells. HIV infection normally occurs first in macrophages, a type of antigen-presenting cell (APC) that has a very low level of CD4 on its surface (**Figure 34.39**). At the cell surface, the macrophage CD4 molecule binds to the gp120 protein of HIV. The viral gp120 protein then interacts with another macrophage protein, the membrane-spanning chemokine receptor CCR5. CCR5 is a coreceptor for HIV and, together with CD4, forms the docking site where the HIV envelope fuses with the host cytoplasmic membrane; this allows the insertion of the viral nucleocapsid into the cell. The CCR5 coreceptor is required for HIV binding to macrophages. Individuals who express a variant CCR5 protein do not bind HIV and do not acquire HIV infections. After HIV has infected the macrophage APCs, a different form of gp120 is made, which in turn binds to a different coreceptor, the CXCR4 chemokine receptor on T cells.

HIV then enters and destroys the CD4 T-helper lymphocytes, the T_H1 and T_H2 cells that provide cell-mediated inflammatory responses and B cell help (∞ Section 29.8). Thus, HIV infection starts in macrophages and progresses to a T cell infection. The result of HIV infection is the systematic destruction of macrophages and T cells, leading to a catastrophic breakdown of immunity; opportunistic infections can then develop unchecked.

In persons with AIDS, CD4 lymphocytes are greatly reduced in number. However, HIV infection does not immediately kill the host cell. HIV can exist as a provirus and not an infectious virion; under these conditions, the reverse-transcribed HIV genome is integrated (as DNA) into host chromosomal DNA. The cell may show no outward sign of infection, and HIV DNA can remain latent for long periods, replicating as the host DNA replicates. Eventually, virus synthesis occurs and new HIV particles are produced and

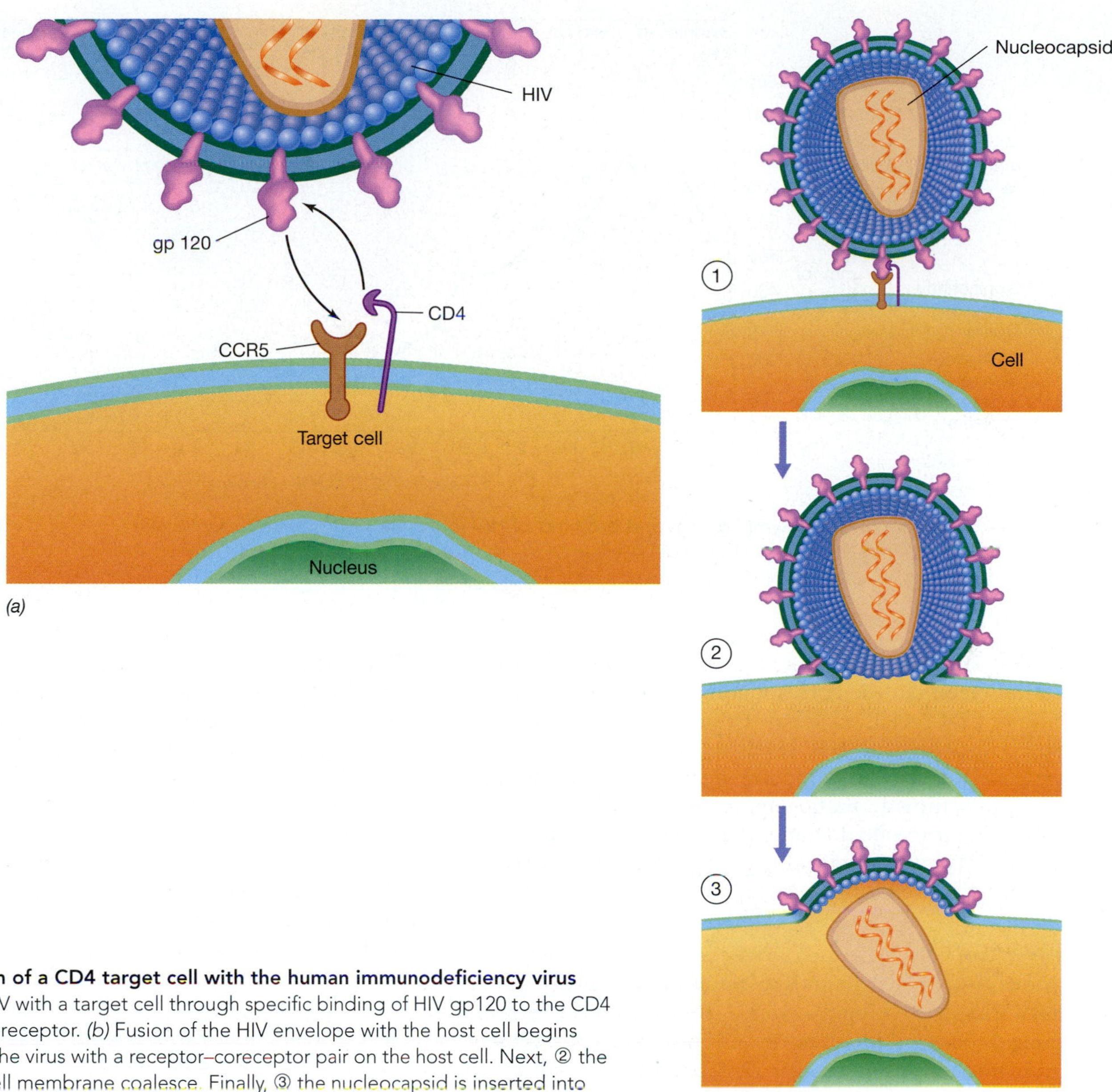

Figure 34.39 Infection of a CD4 target cell with the human immunodeficiency virus (HIV). *(a)* Interaction of HIV with a target cell through specific binding of HIV gp120 to the CD4 receptor and the CCR5 coreceptor. *(b)* Fusion of the HIV envelope with the host cell begins with ① the interaction of the virus with a receptor–coreceptor pair on the host cell. Next, ② the viral envelope and host cell membrane coalesce. Finally, ③ the nucleocapsid is inserted into the host cell, beginning the active viral infection.

released from the cell. T cells producing HIV no longer divide and eventually die.

Destruction of CD4 cells is accelerated following the processing of HIV antigens by infected T cells. Such cells insert molecules of gp120 from HIV particles into their cell surfaces. The embedded gp120 protein on the infected cells then binds CD4 on uninfected T cells. The infected and uninfected cells can then fuse to produce multinucleate giant cells called *syncytia*. One HIV-infected T cell may eventually bind and fuse with up to 50 uninfected T cells. Shortly after syncytia form, the fused cells lose immune function and die.

The HIV infection results in a progressive decline in CD4 cell number. In a healthy human, CD4 cells constitute about 70% of the total T cell pool; in AIDS patients, the number of CD4 cells steadily decreases, and by the time opportunistic infections begin to appear, CD4 cells may be almost absent (∞ Figure 32.21 and **Figure 34.40**).

Outcomes of HIV Infection

The reduction of CD4 T cells has serious consequences for AIDS patients. As the number of CD4 cells declines, cytokine production falls, leading to the functional reduction of the immune response; the antibody-mediated and cell-mediated immune responses in AIDS patients are gradually destroyed. Systemic infections by opportunistic fungi and mycobacteria point to a loss of T_H1 cell-mediated immunity. Opportunistic viral and bacterial infections associated with AIDS indicate a decline in antibody production due to the loss of the T_H2 cells necessary to stimulate antibody production by B cells.

The progression of untreated HIV infection to AIDS follows a predictable pattern. During the clinical latency period, a very active infection is in progress. First, there is an intense immune response to HIV: About 1 billion virions are destroyed each day and HIV numbers drop. However, this

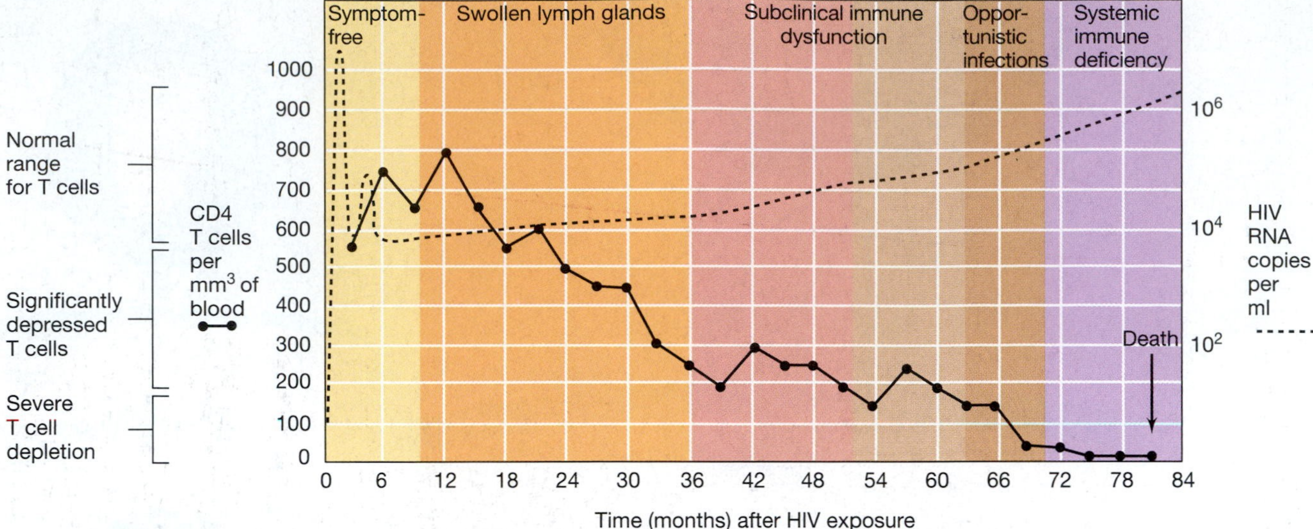

Figure 34.40 Decline of CD4 T lymphocytes and progress of HIV infection. During the typical progression of untreated AIDS, there is a gradual loss in the number and functional ability of the CD4 T cells, while the viral load, measured as HIV-specific RNA copies per milliliter of blood, gradually increases after an initial decline.

means that HIV is replicating at a very high rate, and this replication results in the corresponding destruction of about 100 million CD4 T cells each day. Eventually, the immune response is overwhelmed, HIV levels increase, and the T cells are completely destroyed, crippling the immune response and allowing opportunistic pathogens to initiate infections. The example in Figure 34.40 documents T cell destruction and the increase in HIV over a typical time course for an untreated HIV infection culminating in AIDS.

Diagnosing HIV

HIV infection can be diagnosed by immunological means. The HIV-ELISA (Figure 32.23) is used for large-scale screening of donated blood to prevent transfusion-associated HIV transmission. About 0.25% (2–3 per 1,000) of all blood donated by volunteer donors in the United States tests HIV-positive in the ELISA assay. A positive HIV-ELISA must be confirmed by a second procedure called an immunoblot (Western blot), a technique that combines the analytical tools of protein purification and immunology (Figure 32.26).

Point-of-care rapid tests are also available to identify individuals infected with HIV. These tests are used for preliminary screening of blood or other body fluids for the presence of antibodies to HIV. One test uses a single drop of patient blood and a single reagent. The reagent is a bioengineered antibody in which one binding site is directed to a red blood cell antigen, while the other site is directed to the gp41 HIV surface antigen. In a positive test, the bifunctional antibody cross-links the red blood cells to the HIV, resulting in a visible agglutination. In another test, saliva is used as a source of secretory antibody to HIV. The saliva is expelled onto a cartridge containing immobilized HIV antigens. A second antibody, reactive with the bound antibody and conjugated to an enzyme, is then added. After addition of the enzyme substrate, a positive reaction shows a colored product. The rapid tests are designed to provide maximum convenience, speed (minutes instead of the hours or days required for ELISA or immunoblot), extended shelf life, portability, and ease of use and interpretation. In general, however, the rapid tests are not as sensitive or specific as the standard HIV-ELISA and HIV-immunoblot tests (Sections 32.10 and 32.11).

These tests, no matter how sensitive or specific, fail to detect HIV-positive individuals who have just recently acquired the virus and have not yet made a detectable antibody response. This period may be more than 6 weeks after exposure to HIV. Despite this drawback, these tests ensure the general safety of the blood supply, and the risk of contracting HIV through contaminated blood or blood products is now very low. Sexual contact with multiple partners and group intravenous drug use are the major risk factors for acquiring HIV infection (Figure 33.8).

Several laboratory tests detect HIV RNA directly and quantitatively from blood samples. These tests use a virus-specific reverse transcriptase–polymerase chain reaction (RT-PCR). The RT-PCR estimates the number of viruses present in the blood, or the **viral load**. The RT-PCR test indicates the magnitude of HIV replication and correlates with the rate of CD4 T cell destruction, a direct indicator of the magnitude of destruction of the immune system. The RT-PCR test is not routinely used to screen for HIV because it is costly and technically demanding. After initial diagnosis of infection, however, the RT-PCR test is used to monitor progression of AIDS and the effectiveness of chemotherapy (**Figure 34.41**).

The prognosis for an untreated HIV-infected individual is poor. Opportunistic pathogens or malignancies eventually kill most AIDS patients. Long-term studies of AIDS patients indicate that the average person infected with HIV progresses through several stages of decreasing immune function, with CD4 cells dropping from a normal range of 600–1,000/mm^3 of blood to near zero over a period of 5–7 years (Figure 34.40).

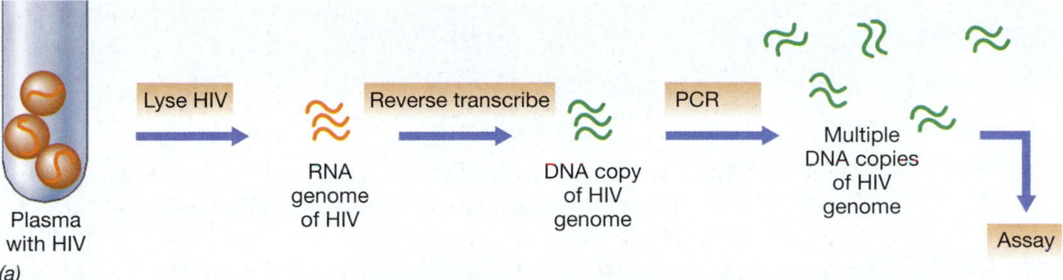

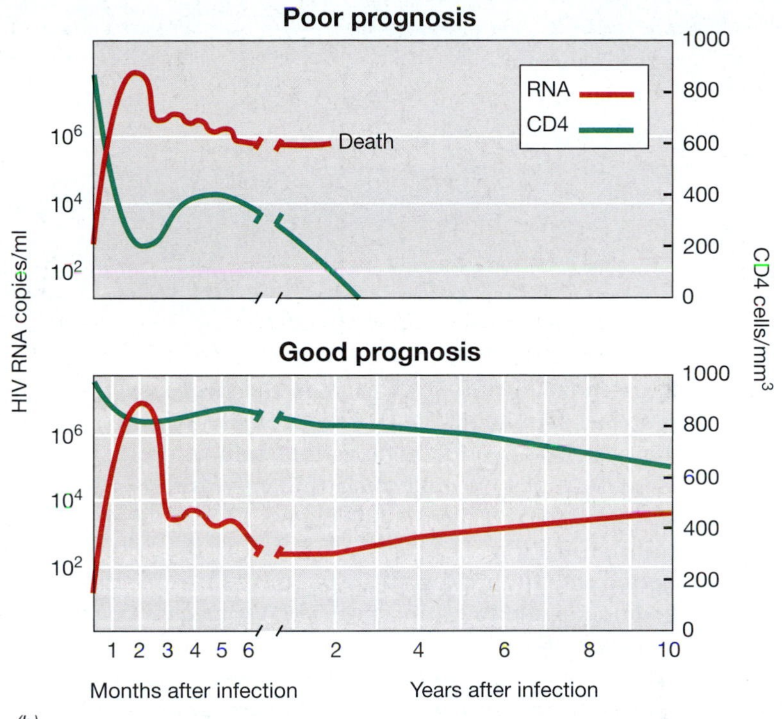

Figure 34.41 Monitoring of HIV load. *(a)* Detection of HIV by the RT-PCR test (reverse transcription-polymerase chain reaction). The HIV copies obtained are compared quantitatively with DNA copies from a control template that is amplified in the same PCR amplification. HIV load is expressed as the number of HIV copies per milliliter of patient plasma. *(b)* Time course for HIV infection as monitored by HIV RT-PCR. Progression of infection is estimated based on viral load at successive times after infection. CD4 T cell counts are measured in cells per cubic millimeter. In the upper panel, a viral load greater than 10^4 copies per milliliter correlates with below normal CD4 cell numbers (normal = 600–1,500/mm^3), indicating a poor prognosis and early death of the patient. In the lower panel, a viral load less than 10^4 copies per milliliter correlates with normal CD4 cell numbers, indicating a good prognosis and extended survival of the patient. Data are adapted from the Centers for Disease Control and Prevention, Atlanta, GA, USA.

Although the rate of decline in immune function varies from one HIV-infected individual to another, it is rare for an HIV-positive individual to live for more than 10 years without some form of antiretroviral drug therapy.

Treatment

Several drugs have been identified that delay the symptoms of AIDS and significantly prolong the life of those infected with HIV (**Table 34.4**). Therapy is aimed at reducing the viral load of HIV-infected individuals to below detectable levels. The strategy used to accomplish this aim is called *highly active antiretroviral therapy* (HAART) and is carried out by giving at least three antiretroviral drugs at once to inhibit the development of drug-resistant HIV. Multi-drug therapy, however, is not a cure for HIV infection. In individuals who have no detectable viral load after drug treatment, a significant viral load returns if therapy is interrupted or discontinued, or if multiple drug resistance develops.

Effective anti-HIV drugs fall into four categories. The first of these is a group of nucleoside analogs that functions as reverse transcriptase inhibitors. Reverse transcriptase is the enzyme that converts the single-stranded RNA genetic information into complementary DNA. The oldest effective anti-HIV drug, *azidothymidine* (AZT), is an effective inhibitor of HIV replication because it closely resembles the nucleoside thymidine but lacks the correct attachment site for the next base in a replicating nucleotide chain, resulting in termination of the growing DNA chain. Thus, AZT and the other nucleoside analogs are **nucleoside reverse transcriptase inhibitors (NRTIs)**, and they stop HIV replication (**Figure 34.42*a***, Table 34.4). When these drugs are given to HIV-infected individuals, the result is a rapid decrease in viral load. If only a single NRTI is administered, within several weeks drug-resistant strains of HIV arise as a result of mutation and selection in almost all patients. Although this process seems very rapid, HIV replicates very quickly (see above), and as little as four single-nucleotide mutations result in resistance to a given nucleoside analog.

The second category of anti-HIV drugs is the **non-nucleoside reverse transcriptase inhibitors (NNRTIs)** (Figure 34.42*b*; Table 34.4). NNRTIs directly inhibit the activity of reverse transcriptase by interacting with the protein and altering the conformation of the catalytic site. Unfortunately, a single mutation in the reverse transcriptase gene is often sufficient to reduce the effectiveness of these drugs.

Table 34.4 Chemotherapeutic agents for HIV/AIDS treatment

Drug name	*Drug class and mechanism of activity*
Azidothymidine (AZT, ZDV, or Zidovudine) Dideoxycytidine (ddC or zalcitabine) Dideoxyinosine (ddI or didanosine) Stavudine (d4T) Lamivudine (3TC)	*Nucleoside reverse transcriptase inhibitors (NRTIs)*; Nucleoside analogs that inhibit reverse transcriptase; nucleotide chain synthesis terminator; increases survival time and reduces incidence of opportunistic infection in AIDS patients; toxic to bone marrow cells; may be used in combination with other drugs in multiple drug treatment (HAART) protocols.
Efavirenz Nevirapine Delavirdine	*Nonnucleoside reverse transcriptase inhibitors (NNRTIs)*; bind directly to reverse transcriptase and disrupt the catalytic site; do not compete with nucleosides; may be used in combination with other drugs in multiple-drug treatment (HAART) protocols.
Indinavir Nelfinavir Saquinavir Ritinovir	*Protease inhibitors*; computer-designed peptide analogs designed to bind to the active site of the HIV protease, inhibiting processing of viral polypeptides and virus maturation; may be used in combination with other drugs in highly active antiretroviral therapy (HAART) protocols.
Enfuvirtide	*Fusion inhibitor*; synthetic polypeptide that binds to the gp41 protein and inhibits the fusion of HIV membranes with host cytoplasmic membranes.
Elvitegravir Raltegravir	*Integrase inhibitors*; experimental drugs that inhibit integration of HIV DNA into host DNA

Another category of anti-HIV drugs is the **protease inhibitor** (Figure 34.42*c*; Table 34.4). The protease inhibitors are computer-designed peptide analogs that inhibit processing of viral polypeptides by binding to the active site of the processing enzyme, HIV protease (∞ Figure 27.31); this inhibits virus maturation. However, as with the other enzyme-targeted chemotherapy strategy (reverse transcriptase; see above), a single mutation in the HIV protease gene can cause drug resistance.

A final category of approved anti-HIV drugs is the drug *enfuvirtide,* a **fusion inhibitor** composed of a 36–amino acid synthetic peptide that functions by binding to the gp41 membrane protein of HIV (Table 34.4). Binding of the protein stops the conformational changes necessary for the fusion of the viral envelope and CD4 cytoplasmic membranes. Enfuvirtide is prescribed to HIV-positive individuals who have developed antiretroviral drug resistance or increased viral load after conventional therapy.

Integrase inhibitors are a new category of anti-HIV drugs. Not yet approved for general use, these drugs interfere with the integration of viral dsDNA into host cell DNA, thus interrupting the HIV replication cycle. Two integrase inhibitors, *eltegravir* and *raltegravir,* are undergoing clinical trials to treat patients who have developed resistance to all other classes of HIV drugs (Table 34.4).

Because of drug resistance, a typical recommended HAART protocol for treatment of an individual with established HIV infection includes at least one protease inhibitor or one NNRTI plus a combination of two NRTIs (Table 34.4). Multiple-drug therapy reduces the possibility that a drug-resistant virus could emerge because the virus would need to develop resistance to three dru gs simultaneously. This combination therapy is then monitored by RT-PCR to track changes in viral load. An effective HAART protocol reduces viral load to nondetectable levels (less than 500 copies of HIV per milliliter of blood) within several days (Figure 34.41). The therapy

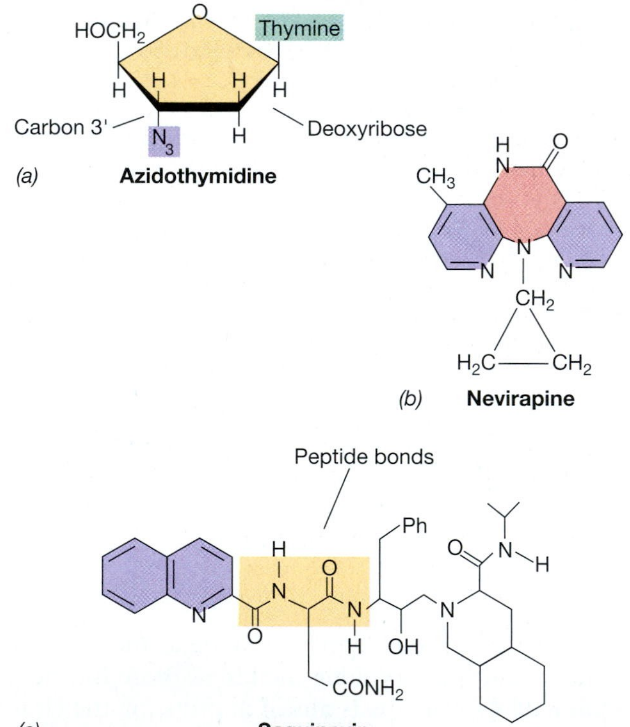

Figure 34.42 HIV/AIDS chemotherapeutic drugs. *(a)* Azidothymidine (AZT), a nucleoside reverse transcriptase inhibitor. This nucleoside analog is missing the OH group on the 3′ carbon, causing nucleotide chain elongation to terminate when the analog is incorporated, inhibiting virus replication. *(b)* Nevirapine, a non-nucleoside reverse transcriptase inhibitor, binds directly to the catalytic site of HIV reverse transcriptase, also inhibiting elongation of the nucleotide chain. *(c)* Saquinavir, a protease inhibitor, was designed by computer modeling to fit the active site of the HIV protease. Saquinavir is a peptide analog: The tan highlighted area shows the region analogous to peptide bonds. Blocking the activity of HIV protease prevents the processing of HIV proteins and maturation of the virus.

is continued and monitored for viral load indefinitely. If the viral load again reaches detectable limits, the drug cocktail is changed because an increase in viral load indicates the emergence of drug-resistant HIV.

In addition to drug resistance, some of the antiviral drugs are toxic to the host. In many cases, nucleoside analogs are not well tolerated by patients, presumably because they interfere with host functions such as cell division (Table 34.4). In general, the NNRTIs and the protease inhibitors are better tolerated because they only interfere with virus-specific functions. However, drug resistance and host toxicity are major problems in all forms of HIV therapy. Thus, new chemotherapeutic agents and drug protocols are constantly being developed and tailored to the needs of individual patients.

AIDS Immunization

The genetic variability of HIV has thus far hampered the development of an AIDS vaccine. One vaccine strategy is to induce antibodies to the envelope protein, gp120. These antibodies would then block CD4–gp120 interactions and inhibit infection. A gp120 subunit vaccine is now in phase III human clinical trials, the step before general public use. The vaccine consists of two gp120 variants known to be predominant in the local population in Thailand, where it is being tested. Unfortunately, this approach probably will not be successful for a universal vaccine because the gene encoding gp120 mutates frequently, forming antigenic variants of the protein that would not be recognized by antibodies made to the vaccine form of gp120.

Clinical immunization trials are proceeding with subunit vaccines (∞ Sections 26.5 and 30.6). In HIV subunit vaccines, genes for several HIV envelope proteins have been engineered into vaccinia virus or adenovirus particles. Using these harmless viruses as expression vectors and vehicles for delivery of HIV antigens, several vaccines elicit a strong antibody and cellular immune response to HIV.

Other potential immunization candidates include killed intact HIV. These inactivated vaccines are restricted to use in HIV-infected individuals because inactivation procedures may not deactivate 100% of the HIV; it would be unethical to expose uninfected individuals to even a small risk of HIV infection. Some laboratories are also exploring the possibilities of producing live attenuated virus for use as an immunizing agent. This strategy is supported by the finding that individuals infected with HIV-2, a related virus that causes a milder form of AIDS with a very long latent period, prevents infection with HIV-1, the strain responsible for AIDS. However, there are potential risks with this strategy. For example, integrated virus could cause cancer, mutations might reactivate virulence in attenuated vaccines, and so on.

Several HIV immunization protocols are undergoing clinical trials, but no immunization has yet been found to be protective and safe. Additionally, HIV immunization would most probably not be useful for treating most patients that already have HIV infection and AIDS because these individuals lack significant immune function and could not respond to a vaccine. Thus, despite considerable knowledge of the mechanisms of HIV infection and a clear understanding of the AIDS disease process, there is no medical intervention strategy for the prevention or cure of HIV/AIDS.

AIDS Prevention

Public education and avoidance of high-risk behavior remain the major tools used to prevent HIV/AIDS. HIV spread is linked to promiscuous sexual activities and other activities that involve exchange of body fluids, which include not only male homosexuality but also female prostitution and intravenous drug use. In the United States the fastest growing method of transmission is between heterosexual partners. Prevention requires avoidance of the high-risk behaviors associated with exchange of body fluids. The United States Surgeon General has issued a report that makes specific recommendations that individuals can follow if they wish to reduce the likelihood of HIV infection. Among the recommendations are the following:

1. Avoid mouth contact with penis, vagina, or rectum.
2. Avoid all sexual activities that could cause cuts or tears in the linings of the rectum, vagina, or penis.
3. Avoid sexual activities with individuals from high-risk groups. These include prostitutes (both male and female); those who have multiple partners, particularly homosexual men and bisexual individuals; and intravenous drug users.
4. If a person has had sex with a member of one of the high-risk groups, a blood test should be done to determine if infection with HIV has occurred. Keep in mind that the blood test should be repeated at intervals for a year or more because of the lag time in the immune response. If the test is positive, the sexual partners of the HIV-positive individual must be protected by use of a condom during sexual intercourse. The use of condoms is also recommended for all extramarital sexual activity.

For more information about prevention of STIs and HIV/AIDS, contact the American Social Health Association STI Resources Hotline at 800-227-8922. The ASHA website is http://www.ashastd.org/contactus.cfm. **www.microbiology-place.com Online Tutorial 34.1 HIV Replication**

34.15 MiniReview

AIDS is one of the most prevalent infectious diseases in the human population. Human immunodeficiency virus destroys the immune system, and opportunistic pathogens then kill the host. There is no effective cure for HIV infection or AIDS. However, several antiviral drugs slow or stop the progress of AIDS. There is no effective vaccine for HIV. Prevention for the spread of HIV infection requires education and avoidance of high-risk behavior.

- Review the definition of AIDS. What symptoms of AIDS are shared by all AIDS patients?
- What is the current treatment for HIV infection? Is it effective?

Review of Key Terms

Antigenic drift minor change in influenza virus antigens due to gene mutation

Antigenic shift major change in influenza virus antigen due to gene reassortment

Cirrhosis breakdown of normal liver architecture resulting in fibrosis

Congenital syphilis syphilis contracted by an infant from its mother during birth

Fusion inhibitor a synthetic polypeptide that binds to viral glycoproteins, inhibiting fusion of viral and host cell membranes

Hepatitis a liver inflammation commonly caused by an infectious agent

Human papillomavirus (HPV) a sexually transmitted virus that causes genital warts, cervical neoplasia, and cancer

Meningitis inflammation of the meninges (brain tissue), sometimes caused by *Neisseria meningitidis* and characterized by sudden onset of headache, vomiting, and stiff neck, often progressing to coma within hours

Meningococcemia rapidly progressing severe disease caused by *Neisseria meningitidis* and characterized by septicemia, intravascular coagulation, and shock

Non-nucleoside reverse transcriptase inhibitor (NNRTI) a non-nucleoside compound that inhibits the action of viral reverse transcriptase by binding directly to the catalytic site

Nucleoside reverse transcriptase inhibitor (NRTI) a nucleoside analog compound that inhibits the action of viral reverse transcriptase by competing with nucleosides

Opportunistic infection an infection usually observed only in an individual with a dysfunctional immune system

Pertussis (whooping cough) a disease caused by an upper respiratory tract infection with *Bordetella pertussis*, characterized by a deep persistent cough

Protease inhibitor (PI) a compound that inhibits the action of viral protease by binding directly to the catalytic site, preventing viral protein processing

Rheumatic fever an inflammatory autoimmune disease triggered by an immune response to infection by *Streptococcus pyogenes*

Scarlet fever characteristic reddish rash resulting from an exotoxin produced by *Streptococcus pyogenes*

Sexually transmitted infection (STI) an infection that is usually transmitted by sexual contact

Toxic shock syndrome (TSS) acute systemic shock resulting from a host response to an exotoxin produced by *Staphylococcus aureus*

Tuberculin test a skin test for previous infection with *Mycobacterium tuberculosis*

Viral load a quantitative assessment of the amount of virus in a host organism, usually in the blood

Review Questions

1. Why do gram-positive bacteria cause respiratory diseases more frequently than gram-negative bacteria (Section 34.1)?
2. What are the typical symptoms of a streptococcal respiratory infection? Why should streptococcal infections be treated promptly (Section 34.2)?
3. Describe the causal agent and the symptoms of diphtheria. Link the disease symptoms to toxicity and explain your reasoning. How is diphtheria prevented and treated (Section 34.3)?
4. Describe the causal agent and the symptoms of whooping cough. Describe the changes in vaccine technology that have led to safer pertussis vaccines. Why is the incidence of this disease on the rise (Section 34.4)?
5. Describe the process of infection by *Mycobacterium tuberculosis*. Does infection always lead to active tuberculosis? Why or why not? Why are individuals in the United States not vaccinated with the BCG vaccine (Section 34.5)?
6. Describe the symptoms of meningococcemia and meningitis. How are these diseases treated? What is the prognosis for each (Section 34.6)?
7. Compare and contrast measles, mumps, and rubella. Include a description of the pathogen, major symptoms encountered, and any potential consequences of these infections. Why is it important that women be vaccinated against rubella before puberty (Section 34.7)?
8. Why are colds such common respiratory diseases? Why aren't vaccines used to prevent colds (Section 34.8)?
9. Why is influenza such a common respiratory disease? How are influenza vaccines chosen and made (Section 34.9)?
10. Distinguish between pathogenic staphylococci and those that are part of the normal flora (Section 34.10).
11. Describe the evidence linking *Helicobacter pylori* to gastric ulcers. How would you treat an ulcer patient (Section 34.11)?
12. Describe the major pathogenic hepatitis viruses. How are they related to one another? How is each spread (Section 34.12)?
13. Why did the incidence of gonorrhea rise dramatically in the mid-1960s, while the incidence of syphilis actually decreased at the same time (Section 34.13)?
14. For the sexually transmitted diseases chlamydia, herpes, trichomoniasis, and human papillomavirus, describe the incidence of each. In each case, is treatment possible, and, if so, is it an effective cure? Why or why not (Section 34.14)?
15. Describe how human immunodeficiency virus (HIV) effectively shuts down both humoral immunity and cell-mediated immunity. Why are vaccines to HIV so difficult to develop (Section 34.15)?

Application Questions

1. How can an epidemic of whooping cough be controlled? How can it be prevented? Since the incidence of this disease is no longer decreasing, apply your prevention methods to the current "mini-epidemic" (Section 34.4). Compare the incidence of pertussis to that of diphtheria. Since childhood immunizations generally include both vaccines, why is the incidence of pertussis rising whereas that of diphtheria has remained very low? *Hint*: There may be several reasons, including available vaccines and vaccine practices, herd immunity, and increased susceptibility in various populations.
2. Why does active tuberculosis often lead to a permanent reduction in lung capacity, whereas most other respiratory diseases cause only temporary respiratory problems? Worldwide, the prevalence of tuberculosis infection is very high, but active disease is much lower. Please explain.
3. Your college roommate goes home for the weekend, becomes extremely ill, and is diagnosed with bacterial meningitis at a local hospital. Because he was away, university officials are not aware of his illness. What should you do to protect yourself against meningitis? Should you notify university health officials?
4. Measles, mumps, and rubella were once very common childhood diseases. However, outbreaks of these diseases are now regarded as serious incidents requiring immediate attention from public health officials. Explain this shift in attitudes in the context of disease prevalence, availability of vaccines, and the potential health consequences of each disease in a college population.
5. Discuss the molecular biology of antigenic shift in influenza viruses and comment on the immunologic consequences for the host. Why does antigenic shift prevent the production of a single universally effective vaccine for influenza control? Next, compare antigenic shift to antigenic drift. Which mechanism is more important for the evolution of the influenza virus? Which causes the greatest antigenic change? Which creates the biggest problems for vaccine developers? Why?
6. Arrange the hepatitis viruses in order of disease severity, both in the short term and in the long term.
7. As the director of your dormitory's public health advisory group, you are charged to present information on chlamydia, herpes, trichomoniasis, and human papillomavirus, all STIs. Besides this textbook, where can you get reliable information about STIs? Present information on prevention, symptoms, and treatment for each STI. Will your program for each disease overlap? For each of the diseases, discuss the practical, social, legal, and public health issues that must be considered to identify and notify the sexual partners of infected individuals.

35

Vectorborne and Soilborne Microbial Diseases

The malarial parasite Plasmodium *is a typical vectorborne pathogen transmitted by mosquitoes. Following infection of a human, plasmodial cells grow in red blood cells, as shown here, causing anemia and the other symptoms associated with the disease malaria.*

Animals are vectors that transmit serious, sometimes fatal diseases such as rabies and hantavirus syndrome. Insects are efficient vectors for transmission of human diseases such as plague and malaria, diseases that have killed millions of people and have altered the course of human history and even human evolution. Lyme disease (tick-borne) and West Nile fever (mosquito-borne) are important vectorborne diseases in the United States at present. Fungal diseases and tetanus present unique problems because pathogens that live in soil cannot be effectively eliminated or contained.

Our focus in this chapter is diseases transmitted by these routes: animals, insects, and soil. The host for animal-transmitted pathogens is a nonhuman vertebrate. Infected animal populations can transmit infections to humans. Vectorborne pathogens are spread to new hosts via the bite of an arthropod vector that fed on an infected host. Humans are often accidental hosts in the life cycle of vectorborne pathogens, but they may also be a disease reservoir, as is the case for malaria. Soilborne pathogens include various fungi and species of the genera *Clostridium* and *Bacillus*.

I ANIMAL-TRANSMITTED DISEASES

A **zoonosis** is an animal disease transmissible to humans, generally by direct contact, aerosols, or bites. Immunization and veterinary care control many infectious diseases in domesticated animals, preventing much zoonotic disease transfer to humans. However, feral (wild) animals are not immunized, nor do they receive veterinary care. Diseases in animals may be **enzootic**, present endemically in certain populations, or **epizootic**, with incidence reaching epidemic proportions. Epizootic diseases often occur on a periodic, sometimes cyclic basis. Because of the unusually high prevalence of diseased animals in epizootic situations, the potential for transferring disease from infected animals to humans increases. We focus our discussion on two important zoonotic diseases transmitted to humans by contact with infected vertebrates—rabies and hantavirus syndromes.

35.1 Rabies

Rabies occurs primarily as an epizootic disease in animals but is spread as a zoonotic disease to humans under certain conditions. The major enzootic reservoirs of rabies in the United States are wild animals, primarily raccoons, skunks, coyotes, foxes, and bats. A small number of rabies cases are also seen in domestic animals (**Figure 35.1**).

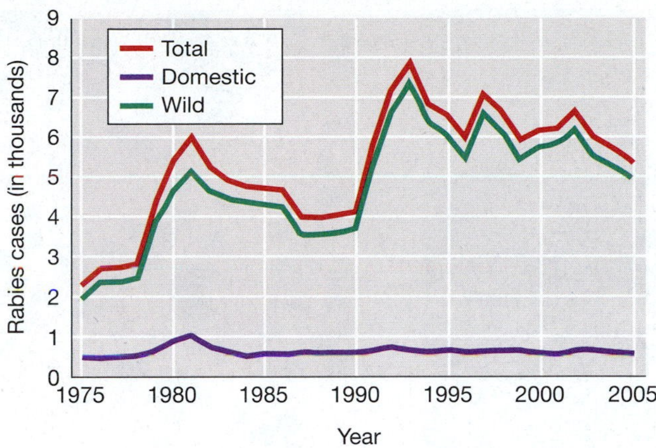

Figure 35.1 Rabies cases in wild and domestic animals in the United States. Rabies is an enzootic disease in wild animal populations, especially in raccoons in the eastern United States. The peaks in the numbers of infected wild animals are epizootic events. An epizootic event kills a large portion of the host population, and this smaller population supports less rabies transmission. The disease numbers fall, and the disease again becomes enzootic. As the population recovers from the epizootic, rabies case numbers rise, endemic rabies becomes an epizootic disease, a large number of hosts die, and the cycle repeats. Over 500 cases of rabies are reported annually in domestic animals, nearly all acquired from contact with wild animals. Data are from the Centers for Disease Control and Prevention, Atlanta, GA, USA.

Epidemiology and Pathology

Rabies is a major preventable infectious disease in humans worldwide. Nevertheless, over 50,000 people die every year from rabies, primarily in developing countries where it is endemic in domestic animals such as dogs. Annually, about 1 million people worldwide receive rabies postexposure prophylactic treatment after animal bites. In the United States, over 20,000 individuals receive postexposure prophylaxis annually.

Rabies is caused by a rhabdovirus, a negative-strand RNA virus (Section 19.8) that infects cells in the central nervous system of most warm-blooded animals, almost invariably leading to death if not treated. The virus (**Figure 35.2*a***), present in the saliva of rabid animals, enters the body through a wound from a bite or through contamination of mucous membranes by infected saliva. Rabies virus multiplies at the site of inoculation and then travels to the central nervous system. The incubation period for the onset of symptoms is highly variable, depending on the animal, the size, location, and depth of the wound, and the number of viral particles transmitted in the bite. In dogs, the incubation period averages 10–14 days. In humans, 9 months or more may pass before rabies symptoms become apparent.

The virus proliferates in the brain (especially in the thalamus and hypothalamus), leading to fever, excitation, dilation of the pupils, excessive salivation, and anxiety. A fear of swallowing (hydrophobia) develops from uncontrollable spasms of the throat muscles. Death eventually results from respiratory paralysis. In humans, an *untreated* rabies infection that becomes symptomatic is almost always fatal.

Diagnosis and Treatment of Rabies

Rabies is diagnosed in the laboratory by examining tissue samples. Fluorescent antibodies or other monoclonal antibody

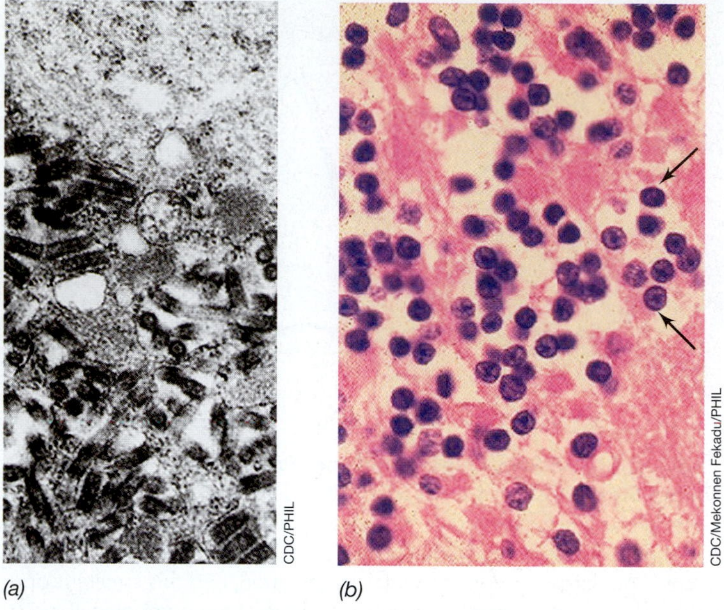

Figure 35.2 Rabies virus. *(a)* The bullet-shaped rabies viruses shown in this transmission electron micrograph of a tissue section from an infected animal are about 75 × 180 nm. *(b)* Pathology of rabies. Tissue section from the brain of a human rabies victim. The section is stained with hematoxylin and eosin. Rabies virus causes characteristic cytoplasmic inclusions called Negri bodies, which contain rabies virus antigens. They are seen here as dark-stained, sharply differentiated, roughly spherical masses of about 2–10 μm in diameter (arrows).

tests that recognize rabies virus in brain or corneal tissue are used for confirming a clinical diagnosis of rabies, either in a potentially rabid animal or in postmortem examination of a human (Figure 35.2*a*). Characteristic virus inclusion bodies in the cytoplasm of nerve cells, called *Negri bodies* (Figure 35.2*b*), obtained by biopsy or postmortem sampling, further confirm rabies virus infection. Reverse transcriptase–polymerase chain reaction (RT-PCR) testing and sequencing can also be used to identify rabies virus in clinical specimens.

Prevention of rabies in humans must start immediately after contact with potentially rabid animals because rabies is generally lethal otherwise. Guidelines for treating possible human exposure to rabies are shown in **Table 35.1**. A wild or stray animal suspected of being rabid should be captured, sacrificed, and immediately examined for evidence of rabies. If a domestic animal, generally a dog, cat, or ferret, bites a human, especially if the bite is unprovoked, the animal is typically held in quarantine for 10 days to check for clinical signs of rabies. If the animal is wild, exhibits rabies symptoms, or a determination cannot be made after 10 days, the patient is passively immunized with rabies immune globulin (purified anti-rabies virus antibodies obtained from a hyperimmune individual) injected at both the site of the bite and intramuscularly. The patient is also immunized with a rabies virus vaccine. Because of the very slow progression of rabies in humans, this combination of passive and active immune therapy is nearly 100% effective, stopping the onset of the active disease. The famous French microbiologist Louis Pasteur developed the first rabies vaccine over 120 years ago (Section 1.7).

Rabies spread is prevented largely through immunization. Inactivated rabies vaccines are used in the United States for both human and domestic animal immunizations, and inactivated and attenuated virus preparations are also used worldwide. Prophylactic rabies immunization is recommended for individuals at high risk, such as veterinarians, animal control personnel, animal researchers, and individuals who work in rabies research or rabies vaccine production laboratories.

Rabies Prevention

The rabies treatment strategy has been extremely successful, and fewer than three cases of human rabies are reported in the United States each year, nearly always the result of bites by wild animals. Because domestic animals often have exposure to wild animals, all dogs and cats should be vaccinated against rabies beginning at 3 months of age, and booster inoculations should be given yearly or every 3 years. Other

Table 35.1 Guidelines for treating possible human exposure to rabies virus

Unprovoked bite by a domestic animal	
Animal suspected of rabies	*Animal not suspected of rabies*
1. Sacrifice animal and test for rabies.	1. Hold for 10 days. If no symptoms, do not treat human.
2. Begin treatment of human immediately.[a]	2. If symptoms develop, treat human immediately.[a]
Bite by wild carnivore (for example, skunk, bat, fox, raccoon, coyote)	
Regard animal as rabid	
1. Sacrifice animal and test for rabies.	
2. Begin treatment of human immediately.[a]	
Bite by wild rodent, squirrel, livestock, rabbit	
Consult local or state public health officials about possible recent cases of rabies transmitted by these animals (these animals rarely transmit rabies). If no reports, do not treat human.	

[a]All bites should be thoroughly cleansed with virucidal soap and water. Treatment for previously unvaccinated individuals is generally a combination of rabies immunoglobulin and rabies vaccine. Previously vaccinated individuals are not given rabies immunoglobulin but are given another course of rabies vaccine.

domestic animals, including large farm animals, are often immunized with rabies vaccines.

However, the key to effective rabies prevention and possible eradication, at least in the United States, lies in control of the disease in the large rabies virus reservoir in wild animals (Figure 35.1). If all or even most members of the disease reservoir are immune, the disease can be stopped and possibly eradicated. Currently, subunit vaccines (Section 26.5) consisting of rabies virus genes that encode rabies coat proteins expressed in vaccinia virus or canary poxvirus are available. Because these are oral vaccines, they have been included in food "baits" and have been used to immunize local populations of susceptible wild animals, reducing the incidence and spread of rabies in limited geographic areas. This strategy has the potential to control rabies in the wild animal reservoir and eventually eradicate the disease.

35.1 MiniReview

Rabies occurs primarily in wild animals and is an important enzootic and epizootic disease that can cause serious zoonotic infections in humans, especially in developing countries. In the United States rabies is transmitted from the wild animal reservoir to domestic animals or, very rarely, to humans. Vaccination of domestic and wild animals is important for the control of rabies.

- What is the procedure for treating a human bitten by an animal if the animal cannot be found?
- What major advantage does an oral vaccine have over a parenteral (injected) vaccine for rabies control in wild animals?

35.2 Hantavirus Syndromes

Hantaviruses cause several severe diseases including **hantavirus pulmonary syndrome (HPS)**, an acute respiratory and cardiac disease, and **hemorrhagic fever with renal syndrome (HFRS)**, an acute disease characterized by shock and kidney failure. Both diseases are caused by hantavirus transmission from infected rodents.

Hantavirus Biology

Hantavirus is named for Hantaan, Korea, the site of a hemorrhagic fever outbreak where the virus was first recognized as a human pathogen. A significant hantavirus outbreak in the United States occurred near the Four Corners region of Arizona, Colorado, New Mexico, and Utah, in 1993. The outbreak resulted from rapid growth of the deer mouse (*Peromyscus maniculatis*) population in the area in the spring of 1993. The outbreak caused 32 deaths in 53 infected adults, defining the potential danger of outbreaks due to diseases that are directly transmitted from animal reservoirs, sometimes under new or unusual circumstances. In total, there have been 453 cases of HPS with 160 deaths (35%) from 1993 to 2006 in the United States, most in western states.

The genus *Hantavirus* is a member of the *Bunyaviridae*, a family of enveloped segmented, negative-strand RNA viruses (Section 19.8) (**Figure 35.3**). The family includes viruses that cause either HPS or HFSR. Hantaviruses are related to hemorrhagic fever viruses such as Lassa fever virus and Ebola virus (Section 33.10), and all are occasionally transmitted to humans from animal reservoirs. Because of the life-threatening implications of hantavirus infections and the lack of effective cures or immunizations, hantaviruses are handled with biosafety level-4 precautions (BL4; Section 32.4). Human cases of hantavirus and other BL4 viral pathogens are investigated by the Special Pathogens Branch of the CDC (Microbial Sidebar, "Special Pathogens and Viral Hemorrhagic Fevers").

Hantaviruses persistently infect a number of rodents, including mice and rats of several species, lemmings, and voles, and are occasionally found in other animals. HFRS strains are

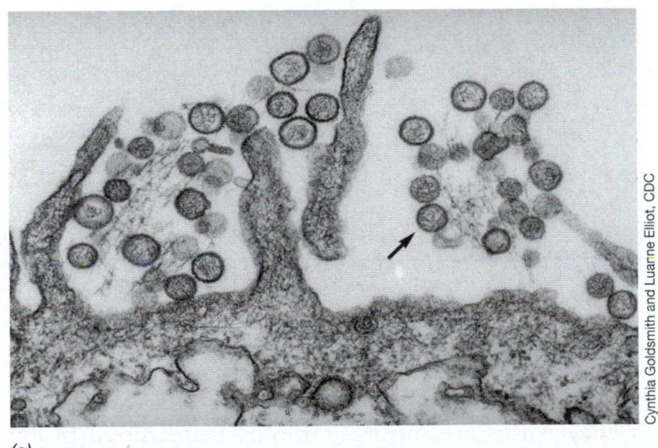

(a)

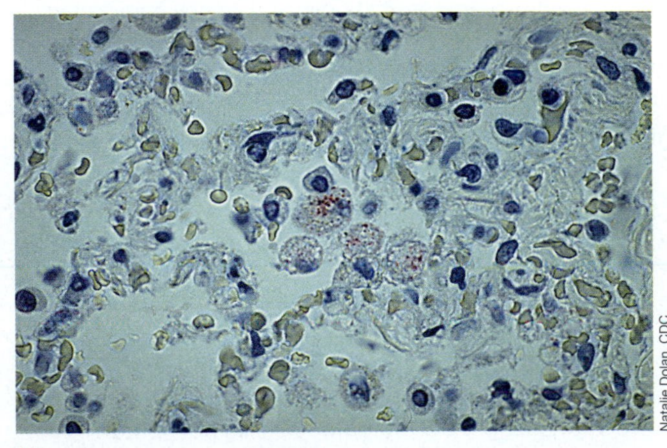

(b)

Figure 35.3 Hantavirus. *(a)* An electron micrograph of the Sin Nombre hantavirus. The arrow indicates one of several virions. The virus is approximately 0.1 μm in diameter. *(b)* Immunostaining of Andes hantavirus antigens in alveolar macrophages. Each granular dark blue-stained area indicates cellular infection of an individual macrophage (approximately 15 μm in diameter).

UNIT 9

Microbial Sidebar

Special Pathogens and Viral Hemorrhagic Fevers

The Special Pathogens Branch of the Centers for Disease Control and Prevention (CDC) in Atlanta, Georgia (USA), specializes in the handling of a subgroup of dangerous pathogens, the hemorrhagic fever viruses. These agents cause viral hemorrhagic fevers (VHF) and include the hantaviruses (discussed in this chapter) and the filoviruses, such as Ebola (**Figure 1**), discussed in Section 33.10 in the context of emerging infectious diseases. These viruses warrant an entire program because they are some of the most lethal infectious agents known.

Hemorrhagic fever viruses are handled under biosafety level-4 (BL4) standards. BL4 is the highest level of biological containment available and is used only for work with agents that pose a high risk of life-threatening disease (Figure 32.9). The VHF agents are RNA viruses transmitted by the aerosol route from animal or arthropod hosts. They are not normally human pathogens and do not depend on humans for their transmission. Outbreaks of VHFs occur irregularly, and infection from human to human by the aerosol route is inefficient. Most human-to-human transmission is the result of prolonged contact with an infected individual or with his or her blood or waste.

VHFs are characterized by severe symptoms that affect multiple organ systems. The virus damages the overall vascular system, and body functions such as oxygen and waste transport and temperature regulation go out of control. Hemorrhaging (bleeding), for which the viruses are named, is rarely the cause of death in VHFs. Instead, extensive damage to organ systems is typically the cause of death.

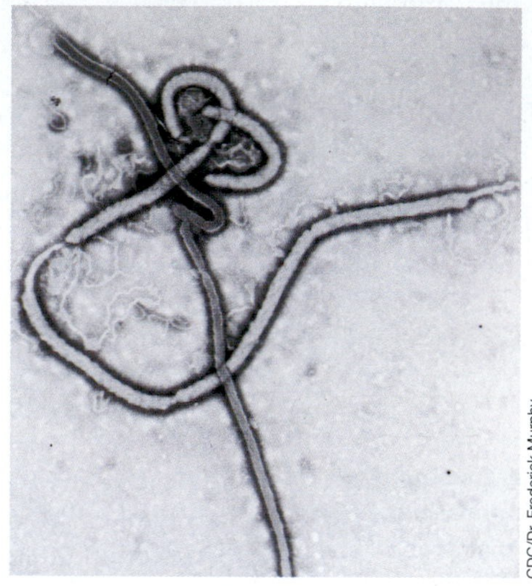

Figure 1 Ebola virus. *Transmission electron micrograph of a negatively stained preparation of Ebola virus.*

In the United States, the only endemic VHFs are caused by hantaviruses. With hantaviruses we understand the vectors and hosts and how to prevent human infections; rodent infestations and human contact with rodent wastes must be limited because hantavirus can survive for long periods in dried feces, saliva, or urine. However, with many hemorrhagic fever viruses, the vectors and mechanisms of transmission are unknown. For example, Ebola virus (Figure 1) is endemic in central Africa and spreads among humans, but the virus originates in a still unidentified animal population. Bats have been implicated as possible hosts, but conclusive proof is lacking.

Hantavirus diseases occur in two forms. Hantavirus hemolytic uremic syndrome causes about 30 deaths per year (15% mortality) in the United States. A total of 453 cases of hantavirus pulmonary syndrome and 160 deaths (35%) have been reported from 1993 to 2006. By contrast, there have been seven major Ebola outbreaks in Africa since 1976, the latest in 2007. In humans, Ebola hemorrhagic fever is devastating. In outbreaks involving more than 15 cases, mortality ranged from 29% to 88%. In total, 1,709 people have acquired Ebola and 1,146 have died (67% mortality). Obviously, if Ebola were to infect individuals in a densely populated area, the results could be devastating.

The Special Pathogens Branch of the CDC works with hemorrhagic fever viruses and outbreaks in this country and worldwide. They are charged with managing infected patients, developing diagnostic tools to identify the viruses, and gathering scientific and clinical information about the viruses, their diseases and distribution. Their goal is to predict outbreaks, quickly identify them when they occur, predict viral behavior during the outbreak, and implement adequate measures to stop the outbreak. They are the first and only line of defense for identifying new and established VHF pathogens when outbreaks occur. For more information about this branch of the CDC and the viruses and diseases they study, consult their website at **http://www.cdc.gov/ncidod/dvrd/spb/index.htm**.

more commonly implicated in outbreaks in the Eastern Hemisphere and Europe. Up to 200,000 cases per year are recognized, chiefly in China, Korea, and Russia. The HPS strains are more prevalent in the Western Hemisphere. Continued investigation of hantaviruses will likely identify a number of other pathogenic strains.

Epidemiology and Pathology

Hantaviruses are most commonly transmitted by inhalation of virus-contaminated rodent excreta. Humans seem to be an accidental host and are infected only when they come into contact with rodents or their waste. For example, the Four Corners outbreak started when a mild winter was followed by abundant spring rains, producing increased vegetation. The vegetation provided abundant food and triggered a large rodent population in 1993. Humans were therefore more likely to be exposed to the mice or their droppings and acquire HPS. The virus is most commonly spread via aerosols in the form of dust generated from mouse droppings or dried urine. However, there are rare reports of person-to-person transmission, as well as a few incidents where HPS or HFRS was spread by a rodent bite.

HPS is characterized by a sudden onset of fever, myalgia (muscle pain), thrombocytopenia (reduction in the number of blood platelets), leukocytosis (an increase in the number of circulating leukocytes), and pulmonary capillary leakage. Death occurs within several days in about 35% of cases, usually due to shock and cardiac complications precipitated by pulmonary edema (leakage of fluid into the lungs, causing suffocation and heart failure). These symptoms are typical of the Sin Nombre hantavirus, which caused the Four Corners outbreak, but other symptoms may be evident, depending on the strain of virus causing the disease. For example, the Bayou strain common in rodents in the southeastern United States also causes kidney failure. The HFRS strains common in Eurasia differ considerably with regard to the severity of human disease they cause, with some strains causing as little as 1% mortality.

Diagnosis, Treatment, and Prevention

If hantavirus from candidate infections can be grown in tissue culture, the strain can be identified by serological techniques including a virus plaque-reduction neutralization assay. In this assay, patient serum is tested for antibodies that inhibit the formation of viral plaques in tissue culture. More commonly, ELISAs (enzyme-linked immunosorbent assays; ∞ Section 32.10) are performed on patient blood to identify antibodies, indicating exposure and an immune response. The presence of the viral genome, indicating infection, can be detected with RT-PCR (∞ Sections 12.8 and 32.12) from patient tissue or blood specimens.

There is no virus-specific treatment or vaccine for hantaviruses. Infection can, however, be prevented by avoiding contact with rodents and rodent habitat, because the mode of transmission is usually through exposure to rodent excreta. Destruction of mouse habitat, restricting food supplies (for example, keeping food in sealed containers), and aggressive rodent extermination measures are the accepted means of control. The long-term prognosis for disease eradication is poor because surveys have shown that a considerable percentage of rodents in a given geographical area are infected with the local hantavirus strain. For example, retrospective serological testing of deer mice in the Four Corners area in 1993 indicated that 30% of the local mouse population carried the Sin Nombre hantavirus.

35.2 MiniReview

Hantaviruses are present worldwide in rodent populations and cause serious zoonotic diseases such as hantavirus pulmonary syndrome (HPS) and hemorrhagic fever with renal syndrome (HFRS) in humans. In the United States, hantavirus infections cause significant mortality.

- Why are hantaviruses considered a major public health problem in the United States?
- Describe the spread of hantaviruses to humans. What are some effective measures for preventing infection by hantaviruses?

II ARTHROPOD-TRANSMITTED DISEASES

Pathogens can be spread to hosts from the bite of a pathogen-infected arthropod vector. In many cases, such as in the rickettsial illnesses, humans are accidental hosts for the pathogen. However, infected humans are required hosts in the life cycle of other pathogens, as is the case for malaria.

35.3 Rickettsial Diseases

The **rickettsias** are small bacteria that have a strictly intracellular existence in vertebrates, usually mammals, and are also associated at some point in their life cycle with blood-sucking arthropods such as fleas, lice, or ticks. We discussed the biology of rickettsias in Section 15.13. Rickettsias cause diseases in humans and animals, of which the most important are typhus fever, Rocky Mountain spotted fever, and ehrlichiosis. Rickettsias take their name from Howard Ricketts, a scientist at the University of Chicago who first discovered them and who died from infection with the rickettsia that causes typhus fever, *Rickettsia prowazekii*. Rickettsias have not been cultured in artificial media but can be cultured in laboratory animals, lice, mammalian tissue culture cells, and the yolk sac of chick embryos. In animals, growth takes place primarily in phagocytes.

Comparisons of the sequence of the 1.1-Mbp genome of *R. prowazekii* indicates that these intracellular parasites are closely related to human mitochondria. Like the mitochondria, the rickettsial genome contains only a minimal set of genes necessary for intracellular dependency. For example, the rickettsias lack most of the genes necessary for independent energy metabolism and structural biosynthesis. On the other hand, the rickettsial genome contains virulence genes closely related to the *virB* operon of the plant pathogen *Agrobacterium tumefaciens* (∞ Section 24.14). This operon encodes components of virulence factors for DNA transfer and protein export; in rickettsias, these factors are probably required so that the pathogen can use host systems to its advantage.

Rickettsias are divided into three groups, based loosely on the clinical diseases they cause. The groups are (1) the *typhus group*, typified by *R. prowazekii*; (2) the *spotted fever group*, typified by *Rickettsia rickettsii*; and (3) the *ehrlichiosis group*, characterized by *Ehrlichia chaffeensis*. Here, we examine these three pathogens as examples of the three groups.

Figure 35.4 The human louse, *Pediculus humanus*. The female louse, about 3 mm long, can carry *Rickettsia prowazekii*, the agent that causes typhus. In addition, the body louse can carry *Borrelia recurrentis*, the agent of relapsing fever, and *Bartonella quintana*, the agent of trench fever.

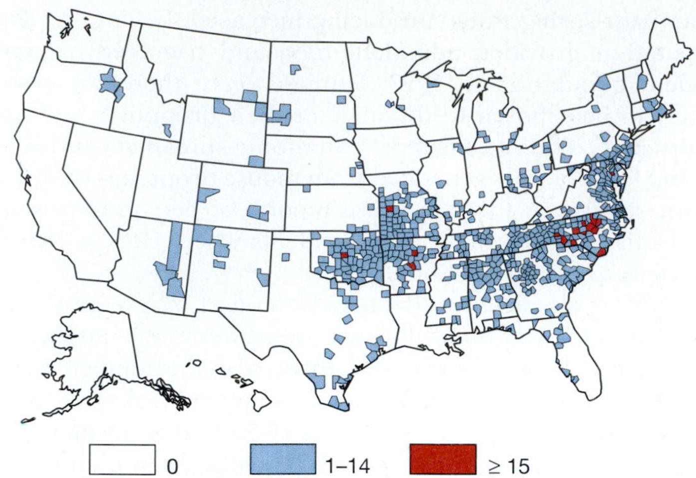

Figure 35.5 Rocky Mountain Spotted Fever in the United States. The 1,936 cases reported in 2005 are shown by county of origin. Cases were concentrated in the eastern and mid-south states west to Oklahoma.

The Typhus Group: *Rickettsia prowazekii*

Typhus is caused by *R. prowazekii*. Epidemic typhus is transmitted from human to human by the common body or head louse (**Figure 35.4**). Humans are the only known mammalian host for typhus. During World War I, an epidemic of typhus spread throughout Eastern Europe and caused almost 3 million deaths. Typhus has historically been a problem among military troops during wartime. Because of the unsanitary, cramped conditions characteristic of wartime military infantry operations, lice are spread easily among soldiers and typhus is spread in epidemic proportions. Up until World War II, typhus caused more military deaths than combat.

Cells of *R. prowazekii* are introduced through the skin when a puncture caused by a louse bite becomes contaminated with louse feces, the major source of rickettsial cells. During an incubation period of 1–3 weeks, the organism multiplies inside cells lining the small blood vessels. Symptoms of typhus (fever, headache, and general body weakness) then begin to appear. Five to nine days later, a characteristic rash is observed in the armpits and generally spreads over the body except for the face, palms of the hands, and soles of the feet. Complications from untreated typhus include damage to the central nervous system, lungs, kidneys, and heart. Epidemic typhus has a mortality rate of 6–30%. Tetracycline and chloramphenicol are most commonly used to control *R. prowazekii*.

Rickettsia typhi, the organism that causes murine typhus, is another important pathogen in the typhus group.

The Spotted Fever Group: *Rickettsia rickettsii*

Rocky Mountain spotted fever was first recognized in the western United States in about 1900 but is more prevalent today in the mid-South region (**Figure 35.5**). Rocky Mountain spotted fever is caused by *R. rickettsii* and is transmitted to humans by various ticks, most commonly the dog and wood ticks. The incidence of Rocky Mountain spotted fever is relatively low, but over 1,000 people acquire the disease every year in the United States. Humans acquire the pathogen from tick fecal matter, which enters the body during a bite, or by rubbing infectious material into the skin by scratching.

Cells of *R. rickettsii,* unlike other rickettsias, grow within the nucleus of the host cell as well as in host cell cytoplasm (**Figures 35.6*a*** and 35.6*b*). Following an incubation period of 3–12 days, characteristic symptoms, including fever and a severe headache, occur. Within 3–5 days, a rash breaks out on the whole body (Figure 35.6*c*). Gastrointestinal problems such as diarrhea and vomiting are usually observed as well, and the clinical symptoms of Rocky Mountain spotted fever may persist for over 2 weeks if the disease is untreated. Tetracycline or chloramphenicol generally promotes a prompt recovery from Rocky Mountain spotted fever if administered early in the course of the infection, and treated patients have less than 1% mortality. There is about 30% mortality in untreated cases.

Ehrlichiosis and Tickborne Anaplasmosis

The *Ehrlichia* and related genera (Section 15.13) are responsible for two emerging tickborne diseases in the United States, *human monocytic ehrlichiosis (HME)* and *human granulocytic anaplasmosis (HGA)* (formerly called human granulocytic ehrlichiosis). The rickettsias that cause HME are *Ehrlichia chaffeensis* and *Neorickettsia sennetsu*. The rickettsias that cause HGA are *Ehrlichia ewingii* and *Anaplasma phagocytophilum*.

The onset of these clinically indistinguishable rickettsial diseases is characterized by flulike symptoms that can include fever, headache, malaise, and frequently leukopenia (decreased number of leukocytes) or thrombocytopenia. Laboratory findings frequently document changes in liver function, characterized by an increase in the enzyme hepatic transaminase. In addition, peripheral blood leukocytes have visible inclusions of cells, a diagnostic indicator for the diseases

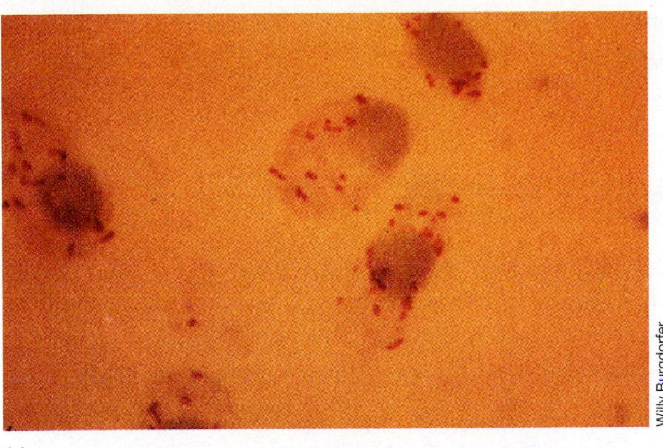

(a)

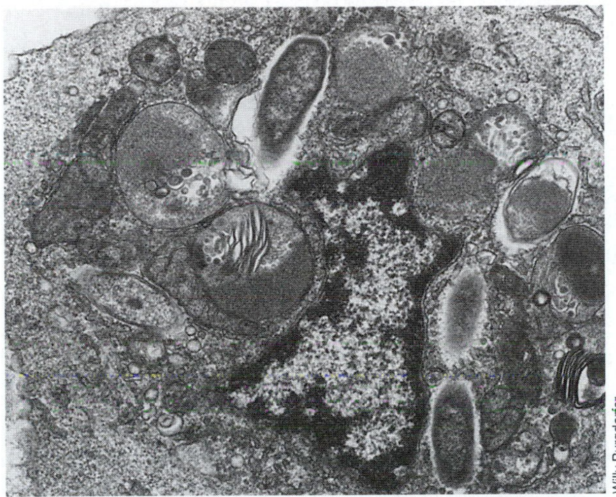

(b)

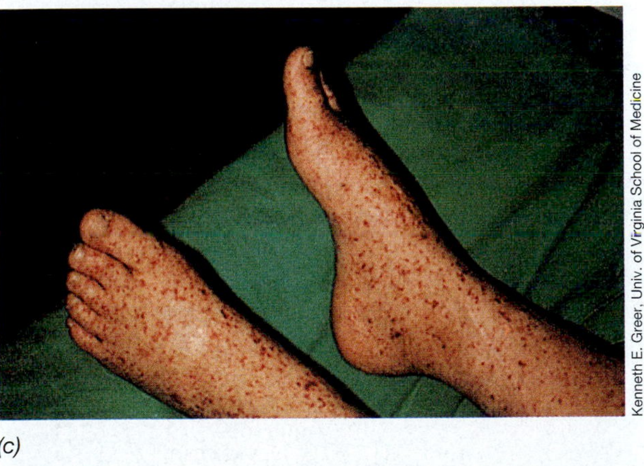

(c)

Figure 35.6 ***Rickettsia rickettsii*** **and Rocky Mountain Spotted Fever.** *(a)* Cells of *R. rickettsii*, growing in the cytoplasm and nucleus of tick hemocytes. Individual cells are about 0.4 μm in diameter. *(b)* Transmission electron micrograph of *R. rickettsii* in a granular hemocyte of an infected wood tick (*Dermacentor andersoni*). *(c)* Rocky Mountain spotted fever rash on the feet. The whole body rash is indicative of Rocky Mountain spotted fever and helps distinguish this disease from typhus, in which the rash does not cover the whole body. Interestingly, the vast majority of cases of Rocky Mountain spotted fever in the United States are not found in western states but instead in central and eastern states (Figure 35.5).

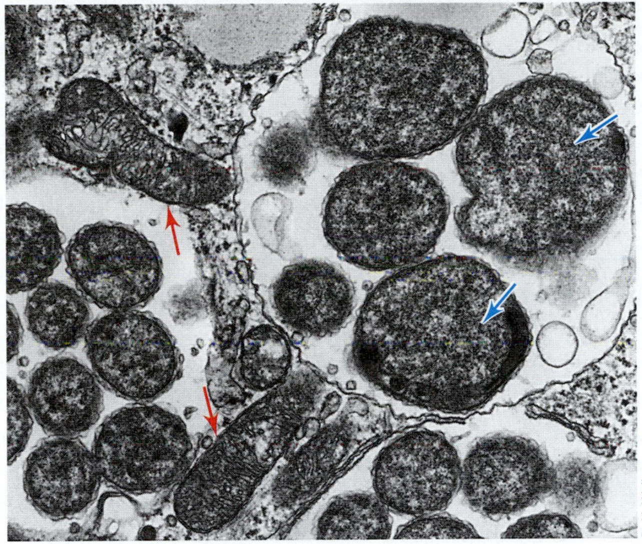

Figure 35.7 ***Ehrlichia chaffeensis*, the causative agent of human monocytic ehrlichiosis (HME).** The electron micrograph shows inclusions in a human monocyte that contains large numbers of *E. chaffeensis* cells. The blue arrows indicate two of the many bacteria in each inclusion. The *E. chaffeensis* cells are about 300–900 nm in diameter. Mitochondria are shown with red arrows.

(**Figure 35.7**). The symptoms, except for the inclusions, are similar to other rickettsial infections, and the diseases can range from subclinical to fatal in outcome. Long-term complications for progressive untreated cases may include respiratory and renal insufficiency and serious neurological involvement.

Laboratory diagnosis of these rickettsial diseases is based on an indirect fluorescent antibody assay of patient serum and also on PCR tests of whole blood or serum to detect the presence of rickettsial DNA. Rickettsias can be observed in inclusions in granulocytes from the blood of patients with ehrlichiosis.

HGA and HME are spread by the bites of infected ticks. The mammalian reservoirs include deer and possibly rodents, in addition to the human hosts. Retrospective serological analyses in areas with relatively high incidence of tickborne disease indicate that HGA may be a more prevalent disease than Rocky Mountain spotted fever. Many HGA infections are not properly identified because of the variable nature of the symptoms. However, since 1999 HGA and HME have been reportable diseases in the United States. HGA causes about 60% of total ehrlichiosis cases, and HME is responsible for 40%. Together, over 1,000 cases are reported each year, but this number is undoubtedly lower than the actual number of cases that occur. HGA and HME will be reported more frequently as physicians become more familiar with these emerging tickborne diseases.

As with other tickborne illnesses, humans are exposed to ehrlichiosis during outdoor activities in tick-infested areas. Golfers, hikers, and others who are recreationally or occupationally exposed to tick habitat are most prone to infection. Prevention of ehrlichiosis involves reducing exposure to ticks and tick bites by avoiding tick habitat, wearing tick-proof clothing, and applying appropriate insect repellents such as those containing diethyl-*m*-toluamide (DEET). At the community level, tick densities can be successfully reduced through areawide application of acaricides (chemicals specifically toxic for ticks and related arthropods) and removal of tick habitat, such as leaves and brush. Doxycycline, a semisynthetic tetracycline, is the antibiotic of choice for the treatment of HGA and HME.

Other Rickettsial Diseases

Q fever is a pneumonia-like infection caused by an obligate intracellular parasite, *Coxiella burnetii,* a bacterium related to the rickettsias (Section 15.13). Although not transmitted to humans directly by an insect bite, the agent of Q fever is transmitted to animals such as sheep, cattle, and goats by insect bites. Various arthropod species are reservoirs and vectors for infection. Domestic animals generally have inapparent infections, but may shed large quantities of *C. burnetii* cells in their urine, feces, milk, and other body fluids. Infected animals or contaminated animal products such as meat and milk are potential sources for human infection. The resulting influenza-like illness may progress to include prolonged fever, headache, chills, chest pains, pneumonia, and endocarditis.

Laboratory diagnosis of infection with *C. burnetti* can be made by immunologic tests designed to measure host antibodies to the pathogen. A complement fixation test and an immunofluorescence antibody test are widely used. *C. burnetii* infections respond to tetracycline, and therapy is usually begun quickly in any suspected human case of Q fever in order to prevent endocarditis and related heart damage. Finally, Q fever is a potential biological warfare agent (Section 33.11).

Scrub typhus, or *tsutsugamushi disease,* is restricted to Asia, the Indian subcontinent, and Australia and is caused by *Orientia tsutsugamushi*. Although the disease is similar to typhus, *O. tsutsugamushi* is transmitted by mites to rodent hosts. Humans are only occasional accidental hosts.

Diagnosis and Control

In the past, rickettsial infections have been difficult to diagnose because the characteristic rash associated with many rickettsial diseases may be mistaken for measles, scarlet fever, or adverse drug reactions. Clinical confirmation of rickettsial diseases has now been greatly aided by the use of specific immunological and molecular biology reagents. These include antibody-based tests that detect rickettsial surface antigens by latex bead agglutination assays, immunofluorescent antibody assays, ELISA, and by PCR-based nucleic acid assays.

Control of most rickettsial diseases requires control of the vectors: lice, fleas, and ticks. For humans traveling in wooded or grassy areas, the use of insect repellants containing DEET usually prevents tick attachment. Firmly attached ticks should be removed gently with forceps, care being taken to remove all the mouthparts of the insect. A solvent such as ethanol applied to a tick with a saturated swab usually expedites removal. Although a vaccine is available for the prevention of typhus, the few cases reported do not warrant its general administration in the United States. No vaccines are currently available for the prevention of the more prevalent tickborne infections, Rocky Mountain spotted fever, HGA, or HME.

35.3 MiniReview

Rickettsias are obligate intracellular parasitic bacteria transmitted to hosts by arthropod vectors. Most rickettsial infections can be controlled by antibiotic therapy, but prompt recognition and diagnosis of these diseases remains difficult.

- What are the arthropod vectors and animal hosts for typhus, Rocky Mountain spotted fever, ehrlichiosis, and anaplasmosis?
- What precautions can be taken to prevent rickettsial infections?

35.4 Lyme Disease

Lyme disease is an emerging tickborne disease that affects humans and other animals. Lyme disease was named for Old Lyme, Connecticut, where cases were first recognized, and is currently the most prevalent tickborne disease in the United States. Lyme disease is caused by a spirochete, *Borrelia burgdorferi* (**Figure 35.8**; Section 16.16), which is spread primarily by the deer tick, *Ixodes scapularis* (**Figure 35.9**). The ticks that carry *B. burgdorferi* feed on the blood of birds, domesticated animals, various wild animals, and humans.

Epidemiology

Deer and the white-footed field mouse are prime mammalian reservoirs of *B. burgdorferi* in the northeastern United States.

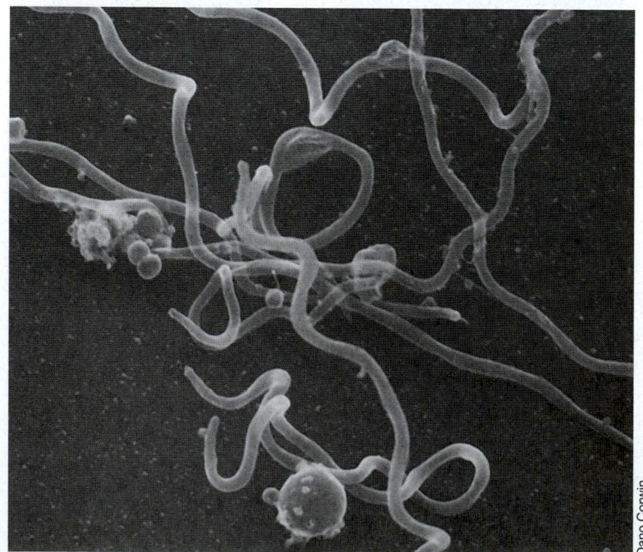

Figure 35.8 Scanning electron micrograph of the Lyme spirochete, *Borrelia burgdorferi*. The diameter of a single cell is approximately 0.4 μm.

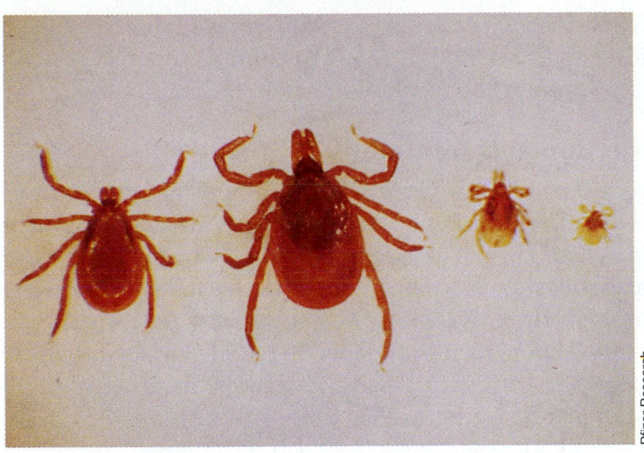

Figure 35.9 Deer ticks (*Ixodes scapularis*), the major vectors of Lyme disease. Left to right, male and female adult ticks, nymph, and larva forms. The length of an adult female is about 3 mm. All forms feed on humans and are capable of transmitting *Borrelia burgdorferi.*

In other parts of the country, different species of rodents and ticks transmit the Lyme spirochete. For example, in the western United States, *Ixodes pacificus* and the wood rat are the common vector and host.

Lyme disease has also been identified in Europe and Asia. In Europe, the tick vector is *Ixodes ricinus*, which may also harbor *Borrelia garinii*, another organism that causes a Lyme disease-like illness. In Asian countries, *Borrelia afzelii* is transmitted by *Ixodes persulcatus*. In all cases, different local rodent host reservoirs have been identified. Thus, Lyme disease seems to have a broad geographic distribution and is transmitted to humans by closely related *Borrelia* pathogens and tick vectors that have rodent and mammalian hosts and reservoirs.

Ixodes spp. are smaller than many other ticks, making them easy to overlook (Figure 35.9). Unlike the vectors of other tickborne diseases, a very high percentage of the deer ticks (up to 50% in certain regions of the northeast) carry *B. burgdorferi*. Extended contact with the infected tick vectors increases the probability of disease transmission.

In the United States most cases of Lyme disease have been reported from the Northeast and upper Midwest, but cases have been observed in nearly every state. The number of Lyme disease cases is rising and up to 24,000 are reported each year (**Figure 35.10**).

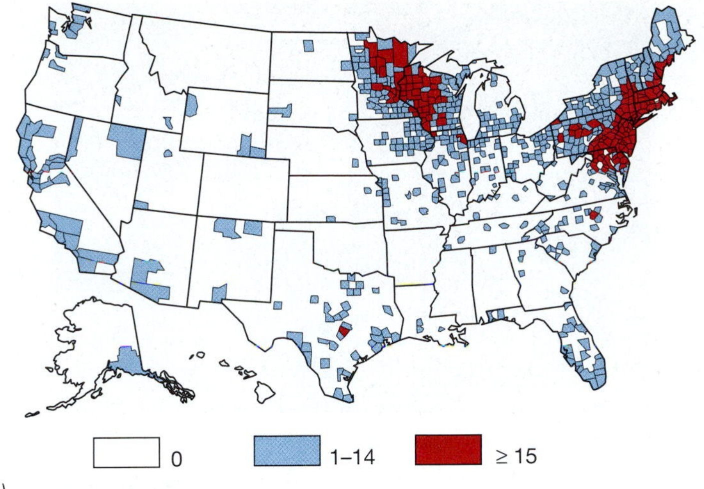

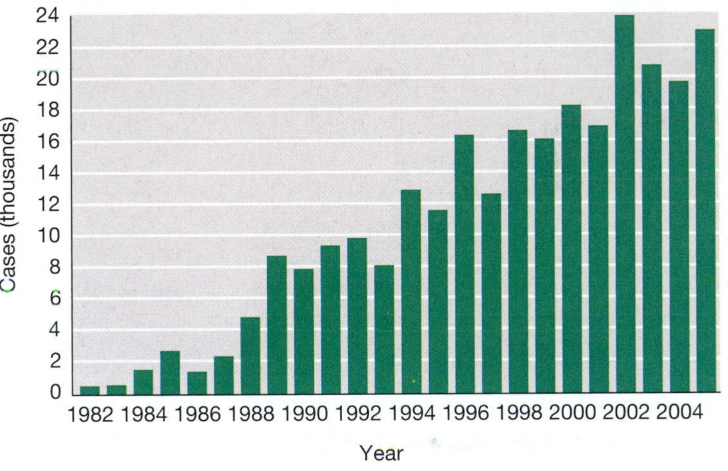

Figure 35.10 Lyme disease in the United States. *(a)* Lyme disease in the United States in 2005. There were 23,305 cases concentrated in the Northeast and upper Midwest. *(b)* Number of reported cases of Lyme disease by year in the United States. Lyme disease is reported through the National Notifiable Diseases Surveillance System of the Centers for Disease Control and Prevention.

Pathology of Lyme Disease

Cells of *B. burgdorferi* are transmitted to humans while the tick is obtaining a blood meal (**Figure 35.11a**). A systemic infection develops, leading to the acute symptoms of Lyme disease, which include headache, backache, chills, and fatigue. In about 75% of cases, a large rash known as *erythema migrans* is observed at the site of the tick bite (Figure 35.11*b*). During this acute stage, Lyme disease is treatable with tetracycline or penicillin. Untreated Lyme disease may progress to a chronic stage weeks to months after the initial tick bite. Chronic Lyme disease is characterized by arthritis in 40–60% of patients. There is neurological involvement such as palsy, weakness in the limbs, and facial ticks in 15–20% and heart damage in about 8% of patients. In untreated cases, cells of *B. burgdorferi* infecting the central nervous system may lie dormant for long periods before causing additional chronic symptoms, including visual disturbances, facial paralysis, and seizures.

No toxins or other virulence factors have yet been identified in Lyme disease pathogenesis. In many respects, the latent symptoms of Lyme disease, especially the neurological involvement, resemble the symptoms of chronic syphilis, caused by a different spirochete, *Treponema pallidum* (∞ Section 34.13). Unlike syphilis, however, Lyme disease is not spread by human contact. Small numbers of *B. burgdorferi* cells are, however, shed in the urine of infected individuals, and Lyme disease can occasionally spread from domestic animal populations, particularly cattle, through infected urine.

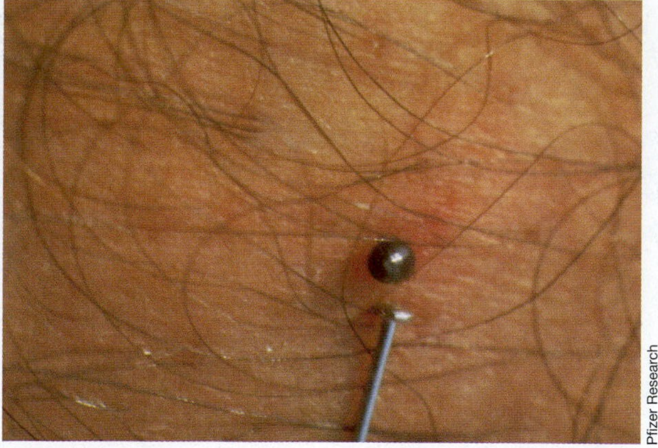

(a)

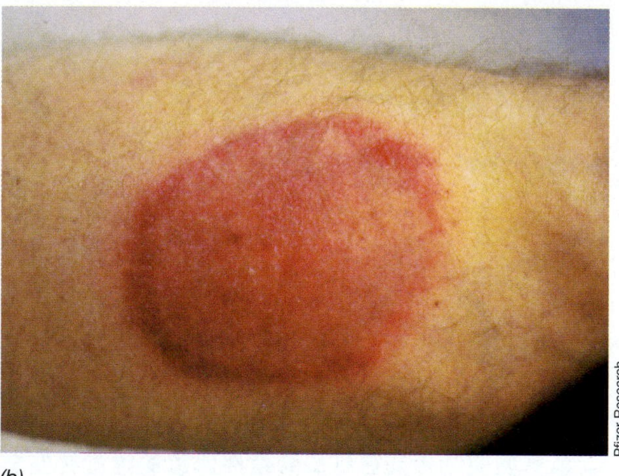

(b)

Figure 35.11 Lyme disease infection. *(a)* Deer tick obtaining a blood meal from a human. *(b)* Characteristic circular rash associated with Lyme disease. The rash, known as *erythema migrans* (EM), typically starts at the site of the bite and grows in a circular fashion over a period of several days. This typical EM example is about 5 cm in diameter.

Diagnosis

Antibodies appear in response to *B. burgdorferi* 4–6 weeks after infection and can be detected by an indirect ELISA or a fluorescent antibody assay. However, the most definitive serological test for Lyme disease is the Lyme Western blot (∞ Section 32.11). Because antibodies to the Lyme spirochete antigens persist for years after infection, the presence of antibodies does not necessarily indicate recent infection and may not confer immunity to further infection.

A PCR assay has also been developed for the detection of *B. burgdorferi* in body fluids and tissues. Although rapid and sensitive, the PCR assay cannot differentiate between live *B. burgdorferi* in active disease and dead *B. burgdorferi* found in treated or inactive disease. *B. burgdorferi* can also be cultured from nearly 80% of the original erythema migrans lesions (Figure 35.11*b*), but culture is usually not done because of the very slow growth of *B. burgdorferi,* even on highly specialized media.

In the end, Lyme disease is usually diagnosed clinically. If a patient has Lyme disease symptoms and other findings such as facial ticks or arthritis, has had recent tick exposure, and exhibits erythema migrans, a presumptive diagnosis of Lyme disease is made and antibiotic treatment is initiated.

Prevention and Treatment

Prevention of Lyme disease requires proper precautions to prevent tick attachment. In tick-infested areas such as woods, tall grass, and brush, it is advisable to wear protective clothing such as shoes, long pants, and a long-sleeved shirt with a snug collar and cuffs. Tucking the pants into tight-fitting socks worn with boots forms an effective barrier to tick attachment. After spending time in a tick-infested environment, individuals should check themselves carefully for ticks and gently remove any attached ticks (including the head). Insect repellants containing DEET are very effective. An effective human Lyme disease vaccine was withdrawn from the market in 2002 due to poor sales. Lyme disease vaccines are available, however, for immunization of susceptible domestic animals.

Treatment of early acute Lyme disease can be with doxycycline, amoxicillin (a β-lactam antibiotic), or cefuroxime axetil, normally for 14 days. For patients having neurological or cardiac symptoms due to *B. burgdorferi* infection, parenteral ceftriaxone is indicated. This β-lactam antibiotic crosses the blood–brain barrier, affecting spirochetes in the central nervous system. There is no indication that antibiotic treatment beyond recommended doses and times is useful to treat delayed sequelae, such as Lyme arthritis.

35.4 MiniReview

Lyme disease is the most prevalent arthropod-borne disease in the United States today. It is transmitted from several mammalian host vectors to humans by ticks. Prevention and treatment of Lyme disease are straightforward, but accurate and timely diagnosis of infection is essential.

- What are the primary symptoms of Lyme disease?
- Outline methods for prevention of *Borrelia burgdorferi* infection.
- What antibiotics can be used to treat Lyme disease?

35.5 Malaria

Malaria is a disease caused by *Plasmodium* spp., a group of protists that are members of the alveolate group (∞ Section 18.9). *Plasmodium* spp. cause malaria-like diseases in warm-blooded hosts; the complex protist life cycle includes an arthropod mosquito vector (**Figure 35.12**). The malaria protists are important human pathogens. Malaria has played an important role in the development and spread of human culture and has even affected human evolution. Malaria is still a significant human disease even though several effective treatments are available. Estimates set the incidence of malaria at up to 350 million people worldwide, and each year over 1 million of these will die, making malaria one of

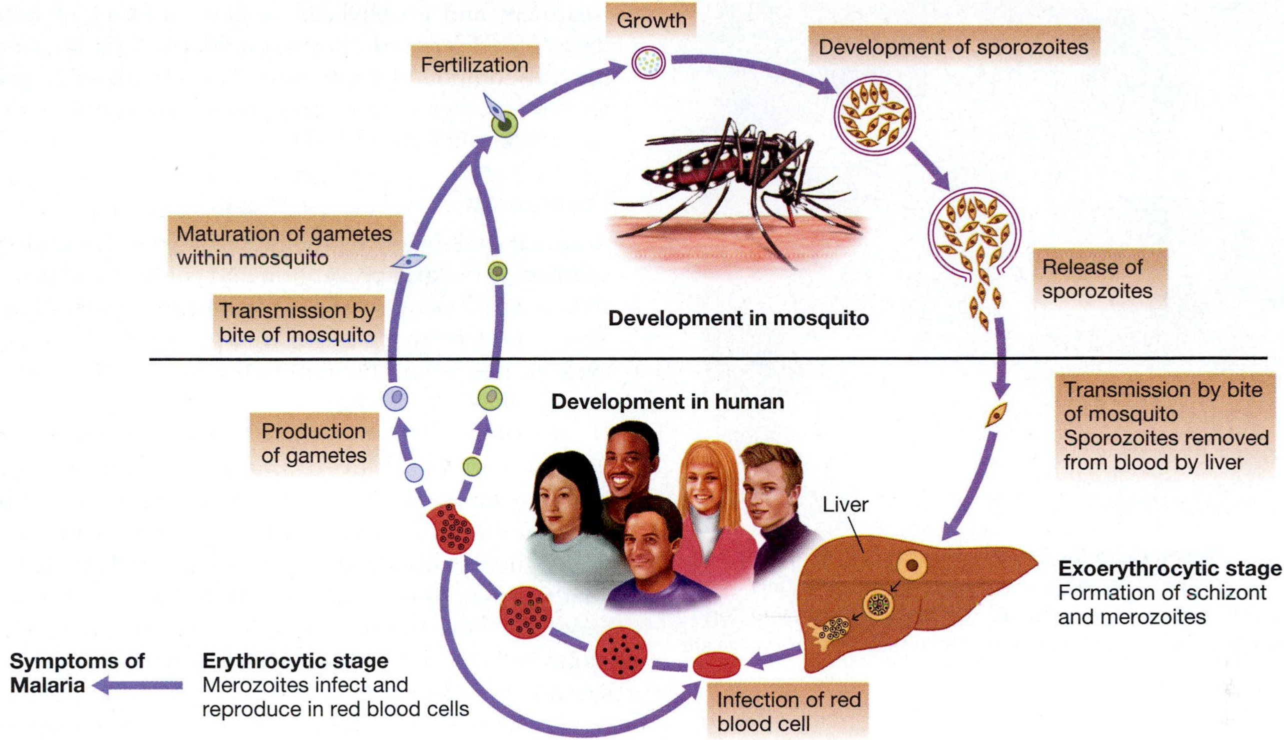

Figure 35.12 **Life cycle *Plasmodium vivax.*** The protist genus *Plasmodium* comprises the malarial pathogens, all of which have a life cycle dependent on growth in both a warm-blooded host and the mosquito vector. Transmission of the protist to and from the warm-blooded host is done by the bite of a mosquito.

the most common causes of death due to infectious disease worldwide (Table 33.1).

Four species, *Plasmodium vivax, P. falciparum, P. ovale,* and *P. malariae,* infect humans and cause malaria. The most widespread disease is caused by *P. vivax,* the most serious disease is caused by *P. falciparum*. Humans are the only reservoirs for these four species. The protists carry out part of their life cycle in the human reservoir and part in the female *Anopheles* mosquito, the only vector that transmits malaria. The vector spreads the protist from person to person.

Epidemiology

Anopheles mosquitoes and *Plasmodium* spp. live predominantly in the tropics and subtropics. In general, malaria is not a disease of temperate or colder regions. For example, malaria did not exist in the northern regions of North America prior to settlement by Europeans, but was a major problem in the southern United States, where appropriate mosquito habitat existed. The disease is associated with wet low-lying areas where mosquitoes breed. The term *malaria* is derived from Italian meaning "bad air."

The life cycle of the malaria protist is complex (Figure 35.12). First, the human host is infected by plasmodial *sporozoites,* small, elongated cells produced in the mosquito that localize in the salivary gland of the insect. The mosquito injects saliva (containing an anticoagulant) along with the sporozoites into the human host when obtaining a blood meal. The sporozoites travel through the bloodstream to the liver, where they remain quiescent or replicate and become enlarged in a stage called the *schizont*. The schizonts then segment into a number of small cells called *merozoites,* which then leave the liver, again entering the blood circulation. Some of the merozoites then infect red blood cells (erythrocytes).

The plasmodial life cycle in erythrocytes proceeds with division, growth, and release of merozoites; this results in destruction of the host red blood cells. *P. vivax* growth in red cells usually repeats at synchronized intervals of 48 hours. During this 48-hour period, the host experiences the defining clinical symptoms of malaria, characterized by chills followed by fever of up to 40°C (104°F). The chill–fever pattern coincides with the release of *P. vivax* merozoites from the erythrocytes during the synchronized asexual reproduction cycle. Vomiting and severe headache may accompany the chill–fever cycles, and over the longer term, characteristic symptomatic malaria generally alternates with asymptomatic periods. Because of the destruction of red blood cells, malaria generally causes anemia and some enlargement of the spleen (splenomegaly).

Not all merozoites liberated from red blood cells are able to infect other erythrocytes. The cells that cannot infect erythrocytes are called *gametocytes* and infect only mosquitoes. The gametocytes are ingested when another *Anopheles* mosquito takes a blood meal from an infected person; they mature within the mosquito into *gametes*. Two gametes fuse,

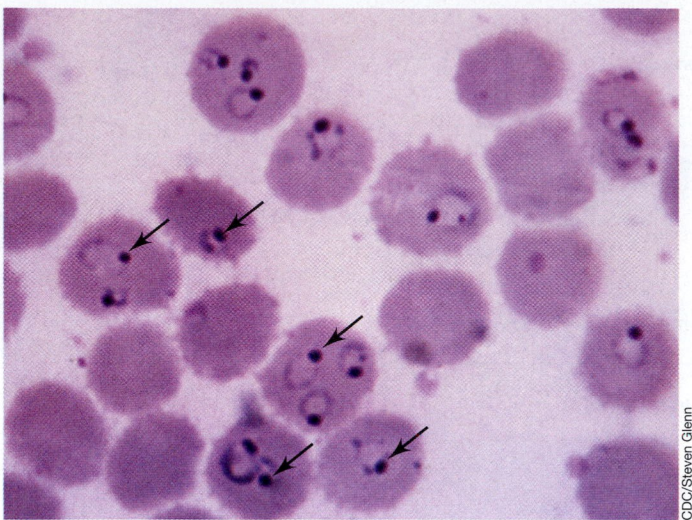

Figure 35.13 ***Plasmodium falciparum.*** This parasitic protist is one of several in the *Plasmodium* genus that causes malaria. Organisms (arrows) are shown growing inside human red blood cells. Uninfected red blood cells are about 6 μm in diameter. Infected red blood cells are slightly enlarged.

and a zygote forms. The zygote then migrates by amoeboid motility to the outer wall of the insect's intestine where it enlarges and forms a number of sporozoites. These are released and reach the salivary gland of the mosquito from where they can be inoculated into another human, and the cycle begins again.

Diagnosis and Treatment

Conclusive diagnosis of malaria in humans requires the identification of *Plasmodium*-infected erythrocytes in blood smears (**Figure 35.13**). Fluorescent nucleic acid stains, nucleic acid probes, PCR assays, and antigen-detection methods (rapid diagnostic tests, RDTs) may all be used to verify *Plasmodium* infections or to differentiate between infections with various *Plasmodium* species.

Prophylaxis for travel to endemic areas and treatment of malaria are usually accomplished with chloroquine. Chloroquine is the drug of choice for treating merozoites within red cells, but does not kill sporozoites, merozoites, and gametes outside the cells. The closely related drug primaquine, however, eliminates sporozoites, merozoites, and gametes outside the cells. Treatment with both chloroquine and primaquine produces a cure. Even in individuals who have undergone drug treatment, however, malaria may recur years after the primary infection. Apparently, small numbers of sporozoites survive in the liver and can reinitiate malaria months or years later by releasing merozoites.

In many parts of the world *Plasmodium* strains have developed resistance to chloroquine or primaquine or both, and some strains have developed resistance to other drugs as well. In areas with known drug-resistant strains, mefloquin or doxycycline is prescribed for prophylaxis; malarone, a combination of two drugs, atovaquone and proguanil, is recommended for both treatment and prophylaxis. A new category of anti-malarial drugs are comprised of synthetic derivatives of artemisinin, a natural compound containing reactive peroxide groups that form free radicals. These compounds are active *in vivo* and are now undergoing clinical trials.

Prevention

Anti-malarial drug treatment is an expensive and short-term solution to malaria prevention and control, and drug-resistant strains of *Plasmodium* spp. complicate matters even further. The most effective control measure is to interrupt the life cycle of the protist by eliminating one of the obligate hosts, the *Anopheles* mosquito.

Two approaches to mosquito control are possible: (1) *elimination of habitat* by drainage of swamps and similar breeding areas, or (2) *elimination of the mosquito* by insecticides, followed by treatment of patients with anti-malarial drugs, thereby breaking the *Plasmodium* life cycle. During the 1930s, about 33,000 miles of ditches were constructed in 16 southern states in the United States, removing 544,000 acres of mosquito breeding area. Millions of gallons of oil were also spread on swamps to reduce the oxygen supply to mosquito larvae. With the discovery of the insecticide dichlorodiphenyltrichloroethane (DDT), chemical control of both larvae and adult mosquitoes was possible. During World War II, the Public Health Service organized an Office of Mosquito Control in War Areas, and because many U.S. military bases were in the southern states, this organization carried out an extensive eradication program in the United States as well as overseas. In 1946 there were 48,610 cases of malaria in the United States when Congress established a 5-year malaria eradication program using drug prophylaxis and treatment for individuals and DDT treatment of mosquito infestations. By 1953 there were only 1,310 malaria cases. In 1935 there were about 4,000 deaths from malaria; in 1952 there were only 25 deaths.

Although the overall public health threat from malaria in the United States is now minimal, very low numbers of endemic malaria cases have resurfaced in recent years as far north as New York City. Malaria incidence also increases due to cases imported by soldiers or immigrants from malaria-endemic areas. On average, there are about 1,500 cases of malaria and five deaths in the United States each year. Most are imported.

In other parts of the world, eradication has been much slower, but the same control measures are used. Reduction of mosquito habitat, control of mosquitoes by insecticides, and treatment of infected individuals with drugs both for cure and prophylaxis are still the major strategies for controlling malaria. Several malaria vaccines are in development, including synthetic peptide vaccines, recombinant particle vaccines, and DNA vaccines (Section 26.5).

Malaria and Human Evolution

Malaria has been endemic in Africa for thousands of years. In West Africans, resistance to malaria caused by *P. falciparum* is associated with an altered red blood cell protein, hemoglobin S, whose amino acid sequence differs from that of normal hemoglobin A at only a single amino acid site in the protein. In the beta chain of hemoglobin S, the neutral amino acid valine is substituted for the glutamic acid of hemoglobin A. As a result, hemoglobin S binds oxygen less efficiently than hemoglobin A. Under conditions of low oxygen concentration, hemoglobin S forms long, thin aggregates that cause the red cell to change from a biconcave round cell to an elongated C-shaped cell, called a *sickle cell*. Individuals who are homozygous for the sickle cell trait are particularly susceptible to changes in oxygen concentrations and suffer from the severely debilitating disease, **sickle cell anemia**.

Individuals who are heterozygous for hemoglobin S have what is called *sickle cell trait,* but also have resistance to malaria as compared to normal individuals. In heterozygotes hemoglobin S can still produce sickled cells, but not as readily as in homozygotes. However, the growth of *P. falciparum* inside the red blood cell causes the heterozygous cells to sickle more easily than in uninfected heterozygous cells. The aggregated hemoglobin S in sickled cells apparently disrupts the red blood cell cytoplasmic membrane, allowing potassium to diffuse from the cell. *P. falciparum* cannot grow in the low-potassium environment of the disrupted cell. Thus, persons with the sickle cell trait can live a more or less normal life and are resistant to malaria.

In certain Mediterranean regions where malaria is endemic, resistance to *P. falciparum* is associated with a deficiency in the red blood cells of the enzyme glucose-6-phosphate dehydrogenase (G6PD), an enzyme that acts as an intracellular antioxidant (reducing) compound. The faulty G6PD leads to higher levels of intracellular oxidants such as the H_2O_2 produced inside the red blood cell by the growing *P. falciparum*. The increased levels of oxidants, normally removed by the activity of functional G6PD, limit *Plasmodium* growth by damaging the cytoplasmic membrane.

In many Mediterranean populations, a diverse group of genetic abnormalities affects hemoglobin production and efficiency. These are known collectively as the **thalassemias**. The thalassemias are also statistically and geographically associated with increased resistance to malaria, and, like the G6PD deficiency, are associated with an increase in oxidants in red blood cells.

Hemoglobin S, G6PD deficiency, and thalassemias result from genetic mutations. These mutations cause red blood cell and oxygen-processing deficiencies that are deleterious in normal human populations. In individuals and populations exposed to *Plasmodium* infections and malaria, however, these mutations are positively selected; although the mutations cause red blood cell abnormalities and oxygen-processing deficiencies, they confer resistance to malaria and enhance the survival of those individuals carrying the mutation.

Another case in which *Plasmodium* spp. influence evolution involves the major histocompatibility complex (MHC) and the immune system. As discussed in Chapter 29, the MHC class I and class II proteins present antigens to T cells for initiation of an immune response. In malaria-prone equatorial West Africa, individuals are very likely to have one particular MHC class I gene and one particular set of class II genes. These selected MHC genes, common in the West African population, are virtually unknown in other human populations. Individuals who express these genes have as much resistance to severe malaria as those with the hemoglobin S trait. The MHC proteins encoded by these selected genes are exceptionally good antigen-presenting molecules for certain malarial antigens and initiate a strong protective immune response to *Plasmodium* spp. infection. As is the case with the hemoglobin variants, *Plasmodium* is a selective agent for MHC genes that enhance host survival. Individuals with selected MHC genes that confer malaria resistance have a measurable survival advantage and are more likely to live and pass the resistance-conferring genes on to their descendents.

Thus, *Plasmodium* infection causing malaria has been a selective agent in human evolution. Other pathogens, such as *Mycobacterium tuberculosis* (tuberculosis, Section 34.5) and *Yersinia pestis* (plague, Section 35.7), may also have promoted selective changes in humans, but in no case is the evidence as clear as it is for malaria.

35.5 MiniReview

Infections with *Plasmodium* spp. cause malaria, a widespread, mosquito-borne disease common in tropical and subtropical regions of the world. Malaria is a major cause of morbidity and mortality in developing countries and is a selection factor for resistance genes in humans. The disease is preventable with a combination of public health and chemotherapy measures.

- What are the natural reservoirs and vectors for *Plasmodium* species? How can malaria be prevented or eradicated?
- Review genetic mechanisms responsible for malaria resistance. Why are anti-malarial genes not found in all humans?

35.6 West Nile Virus

West Nile virus (WNV) causes **West Nile fever**, a rapidly emerging human viral disease transmitted through the bite of a mosquito (**Figure 35.14**). The virus can invade the nervous system of its warm-blooded host. WNV is a member of the flavivirus group and has a symmetrical, enveloped icosahedral capsid (Figure 35.14*b*) containing a positive-sense, single-stranded RNA genome of about 11,000 nucleotides (Section 19.8).

Epidemiology

WNV infection in humans was first identified in Uganda (Africa) in 1937. By the 1950s, the virus had spread to Egypt and Israel. In the 1990s there were WNV outbreaks in horses, birds, and humans in African and European countries. In 1999, the

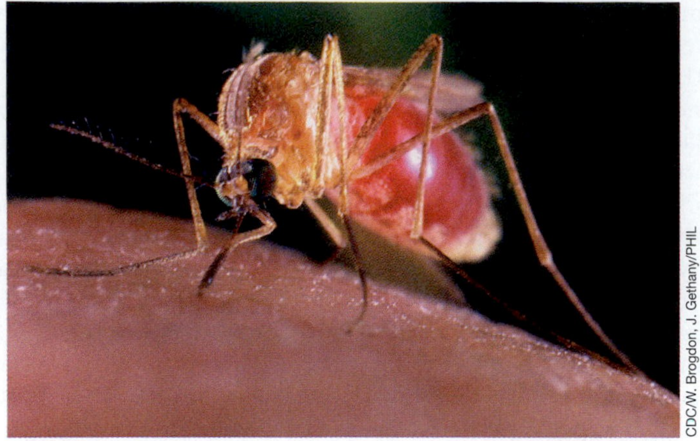

(a)

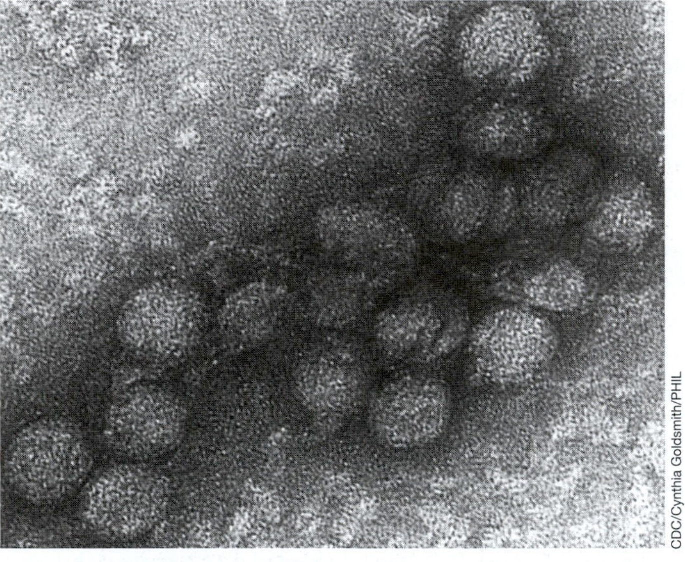

(b)

Figure 35.14 **West Nile virus.** *(a)* The mosquito *Culex quinquefasciatus,* shown here engorged with human blood, is a West Nile virus vector. *(b)* An electron micrograph of the West Nile virus. The icosohedral virion is about 40–60 nm in diameter.

first cases were reported in the United States in the northeast, around New York. From 1999 through 2001, there were 149 confirmed cases of human WNV disease, including 18 deaths. Moving with the seasonal appearance and disappearance of the mosquito vectors, by 2002 this emerging disease had shifted from the East Coast to the Midwest, with a peak reported number of cases of 884 in Illinois and nationwide case totals of 4,156. The disease continued to move across North America. In 2006 there were 9,186 confirmed cases, now centered in upper midwest and western states. The highest incidence of WNV disease now occurs in the mountain states of the West, with Idaho having 996 cases (**Figure 35.15*a***).

WNV Transmission and Pathology

WNV normally causes active disease in birds and is transferred to susceptible hosts by the bite of an infected mosquito. A number of mosquito species are known vectors, and at least 130 species of birds are known host reservoirs. The infected birds develop a viremia lasting 1–4 days, and survivors develop lifelong immunity. Mosquitoes feeding on viremic birds are infected and can then infect susceptible birds, renewing the cycle. The incidence of disease in the avian population in a given area decreases as susceptible avian hosts die or recover and develop immunity. However, the mosquito vectors transmit the WNV to new susceptible hosts in new areas, moving the epidemic in a wavelike fashion across the continent, as discussed above. The highest incidence of neuroinvasive human disease has also been centered in the western United States (Figure 35.15*b*). For all practical purposes, WNV infection is now endemic in the contiguous United States and in North American bird populations.

Humans and other animals are terminal hosts because they do not develop the viremia necessary to infect mosquitoes. The human mortality rate for diagnosed infections is very significant at about 4% (4,269 cases with 177 deaths in 2007), and horses have mortality rates of up to 40%. Most human infections are asymptomatic or very mild. After an incubation period of 3–14 days, about 20% of infected individuals develop West Nile fever, a mild illness lasting 3–6 days. Fever may be accompanied by headache, nausea, myalgia, rash, lymphadenopathy (swelling of lymph nodes), and malaise. Less than 1% of infected individuals develop serious neurological diseases such as West Nile encephalitis or meningitis. Again, however, diagnosed cases have a much higher rate of neurological complications, with 1,459 cases of neuroinvasive disease out of the 4,269 reported cases in 2006 (Figure 35.15*b*). Adults over age 50 appear to be more susceptible to neurological complications than are others. Diagnosis of WNV disease includes assessment of clinical symptoms followed by confirmation with a positive ELISA test for WNV antibodies in serum.

Prevention and Control of WNV

Like St. Louis encephalitis virus and other mosquito-borne viruses that cause encephalitis, transmission of WNV is seasonal in the United States and is dependent on exposure to the mosquito population. The primary means of control for WNV spread is by limiting exposure to the disease vector. Individuals should avoid mosquito habitat, remain indoors between dusk and dawn (the prime hours for mosquito activity), wear appropriate mosquito-resistant clothing, and apply insect repellents containing DEET, as for Lyme disease. At the community level, WNV spread can be controlled by destruction of mosquito habitat and application of appropriate insecticides. There is no effective human vaccine, although several candidates are in development. Veterinary vaccines are available and widely used in horses, but their efficacy is uncertain. Treatment, as for most viral illnesses, is rest, fluids, and symptomatic relief of fever and pain. There are no antiviral drugs known to be effective *in vivo* against WNV.

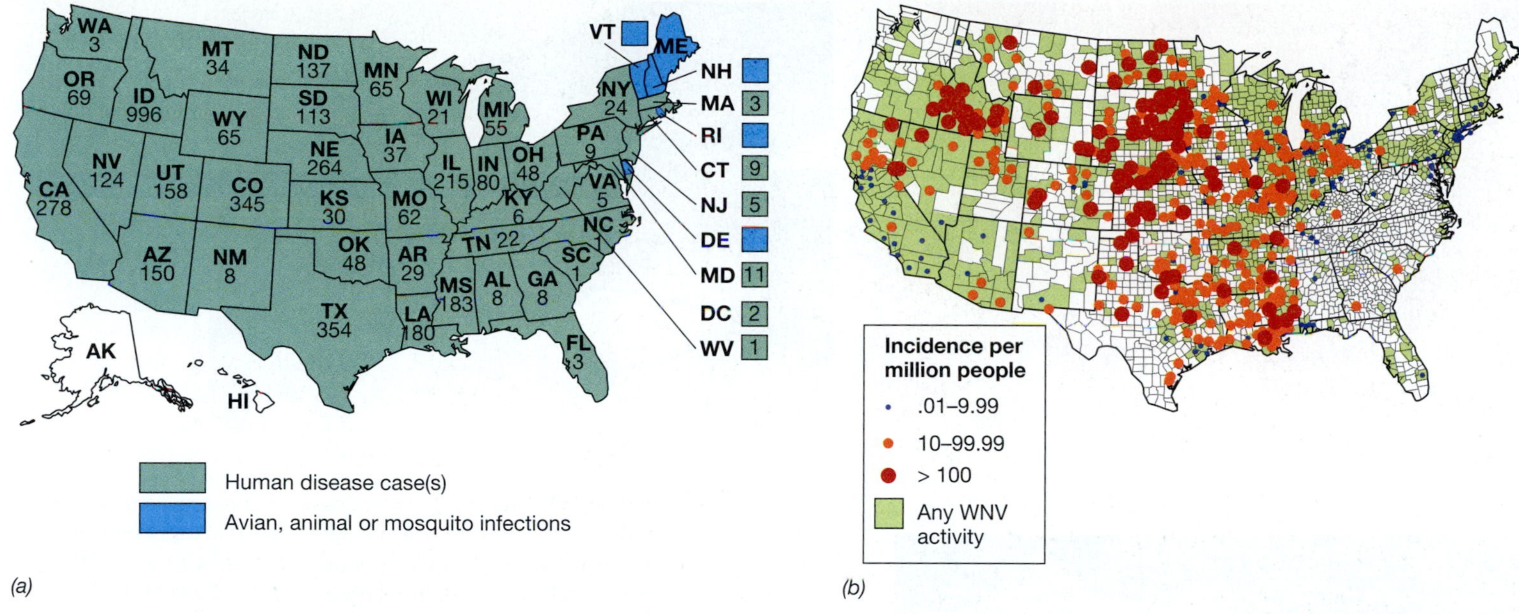

Figure 35.15 West Nile virus in the United States in 2006. *(a)* The virus caused 4,269 cases of human disease and 177 deaths. From 1999 through 2006, the disease in the United States shifted from the East Coast, through the Midwest, and to the upper Midwest and West. *(b)* There were 1,459 cases of neuroinvasive WNV disease in 2006. The incidence of neuroinvasive disease, shown per million inhabitants, was highest in the upper Midwest and Idaho. Data are from the Centers for Disease Control and Prevention, Atlanta, GA, USA.

35.6 MiniReview

West Nile fever is a mosquito-borne viral disease. In the natural cycle of the disease, birds are infected with West Nile virus by the bite of infected mosquitoes. Humans and other vertebrates are occasional terminal hosts. Most human infections are asymptomatic and undiagnosed, but complications in diagnosed infections cause about 3.5% mortality due to encephalitis and meningitis.

- Identify the vector and reservoir for West Nile virus.
- Trace the progress of West Nile virus in the United States since 1999.

35.7 Plague

Pandemic **plague** has caused more human deaths than any other infectious disease except for malaria and tuberculosis. Plague killed as much as one-third of Europe's population in individual pandemics in the Middle Ages.

Plague is caused by *Yersinia pestis,* a gram-negative facultatively aerobic rod-shaped bacterium (**Figure 35.16**) and a member of the enteric bacteria group (Section 15.11). Plague is a disease of domestic and wild rodents; rats are the primary disease reservoir. Humans are accidental hosts and are not critical for the maintenance of the disease. Fleas are intermediate hosts and vectors, spreading plague between the mammalian hosts (**Figure 35.17**). Most infected rats die soon after symptoms appear, but the small proportion of survivors develops a chronic infection, providing a persistent reservoir of virulent *Y. pestis*.

Epidemiology

Most cases of human plague in the United States are in the southwestern states, where the disease, called *sylvatic plague,* is endemic among wild rodents. Plague is transmitted by *Xenopsylla cheopis,* the rat flea, which ingests *Y. pestis* cells by sucking blood from an infected animal. Cells multiply in the flea's intestine and are transmitted to a healthy animal in the next bite.

As the disease spreads, rat mortality becomes so great that infected fleas seek new hosts, including humans. Once in humans, cells of *Y. pestis* typically travel to the lymph nodes, where they cause swelling. The regional and pronounced swollen lymph nodes are called *buboes* and for this reason, the disease is often called *bubonic plague* (Figure 35.16*b*). The buboes become filled with *Y. pestis,* and encapsulated *Y. pestis* prevents phagocytosis and destruction by cells of the immune system. Secondary buboes form in peripheral lymph nodes, and cells eventually enter the bloodstream, causing septicemia. Multiple local hemorrhages produce dark splotches on the skin, giving plague its historical name, the "Black Death" (Figure 35.16*c*). If not treated prior to the septicemic stage, the symptoms of plague (lymph node swelling and pain, prostration, shock, and delirium) usually progress and cause death within 3–5 days.

Pathology of Plague

The pathogenesis of plague is not clearly understood, but cells of *Y. pestis* produce virulence factors that contribute to the

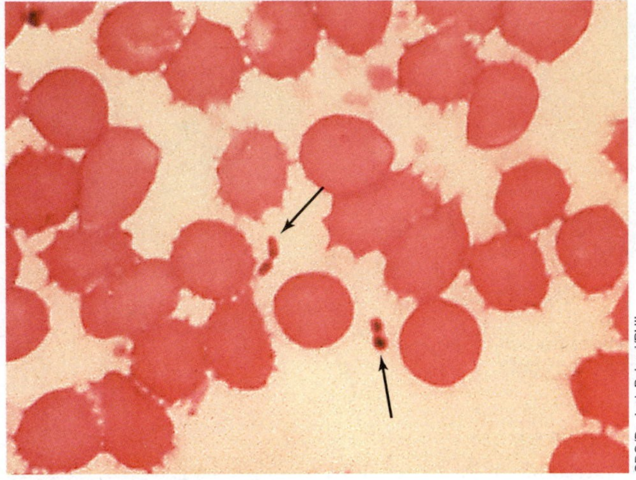

(a)

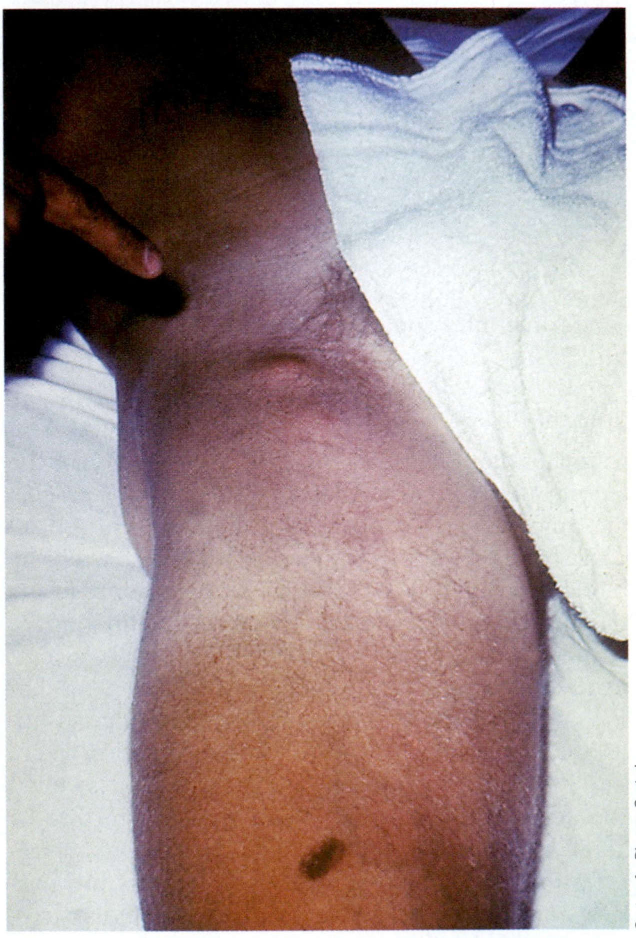

(b)

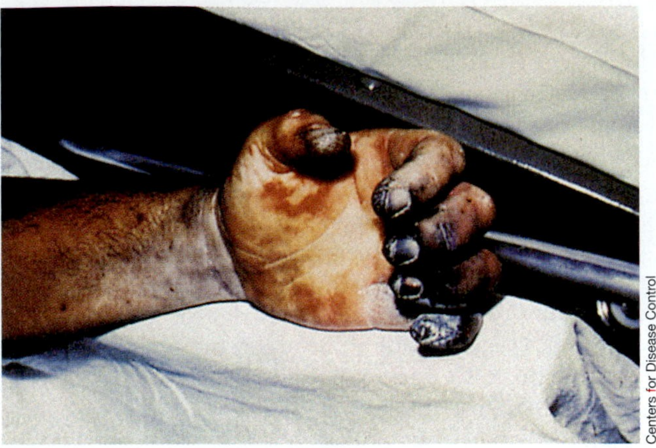

(c)

Figure 35.16 Plague in humans. *(a) Yersinia pestis*, the causative agent of plague is a gram-negative rod, about 2 μm in length and up to 1 μm in diameter. The organisms in this blood smear (arrows) show the characteristic bipolar staining pattern. *(b)* A bubo formed in the groin. *(c)* Gangrene and sloughing of skin in the hand of a plague victim.

disease process. The V and W antigens of *Y. pestis* cell walls are protein–lipoprotein complexes that inhibit phagocytosis. An exotoxin called *murine toxin,* because of its extreme toxicity for mice, is produced by virulent strains of *Y. pestis*. Murine toxin is a respiratory inhibitor that blocks mitochondrial electron transport at coenzyme Q. It produces systemic shock, liver damage, and respiratory distress in mice. Although murine toxin is highly active in only certain animal species, it may be involved in human plague because these symptoms are also seen in affected humans. *Y. pestis* also produces a highly immunogenic endotoxin that may also play a role in the disease process.

Pneumonic plague occurs when cells of *Y. pestis* are either inhaled directly or via the blood or lymphatic circulation. Symptoms are usually absent until the last day or two of the disease when large amounts of bloody sputum are produced. Untreated individuals rarely survive more than 2 days. Pneumonic plague is highly contagious and can spread rapidly via the person-to-person respiratory route if infected individuals are not immediately quarantined. *Septicemic plague* is the rapid spread of *Y. pestis* throughout the body via the bloodstream without the formation of buboes and usually causes death before a diagnosis can be made.

Treatment and Control

Bubonic plague can be successfully treated if rapidly diagnosed. *Y. pestis* infection is treated with streptomycin or gentamycin, given parenterally. Alternatively, doxycycline, ciprofloxacin, or chloramphenicol may be given intravenously. If treatment is started promptly, mortality from bubonic plague can be reduced to 1–5% of those infected. Pneumonic and septicemic plague can also be treated, but these forms progress so rapidly that antibiotic therapy, even if begun when symptoms first appear, is usually too late: up to 90% of untreated pneumonic plague victims die. There have been only 22 cases of plague in the United States since 2000. Mortality is about 10%. Worldwide, there are usually fewer than 1,500 confirmed cases and 300 deaths per year. *Y. pestis* is an organism that could be used for a bioterrorism

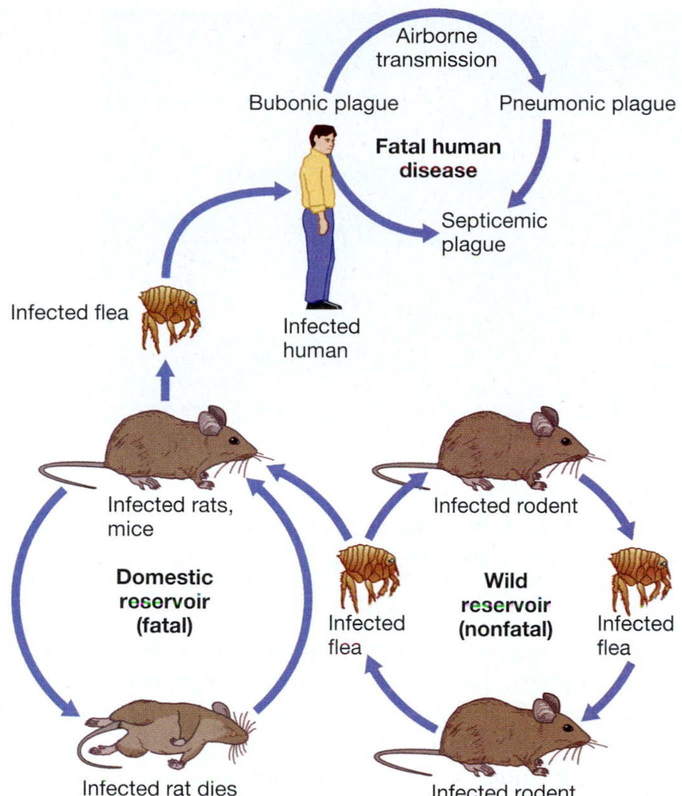

Figure 35.17 The epidemiology of plague due to *Yersinia pestis.* Plague in most wild rodents is generally a mild, self-limiting infection. Plague in rats and humans is frequently fatal. Infected fleas desert the dead host and look for another host, such as a human, an accidental host.

attack (Section 33.11), and oral doxycycline and ciprofloxacin are recommended as prophylactic antibiotics in this setting.

Plague control is accomplished through surveillance and control of animal reservoirs (rats), vectors (rat fleas), and human contacts. Plague-infected animal populations must be destroyed when identified, since a flea can still obtain a blood meal from a dead rat. Undoubtedly, improved public health practices and the control of rodent populations have limited human exposure to plague.

35.7 MiniReview

Plague is largely confined to individuals who come into contact with rodent populations and their parasitic fleas, the endemic reservoirs for *Yersinia pestis.* A disseminated systemic infection or a pneumonic infection leads to rapid death, but the bubonic form is treatable with antibiotics if quickly diagnosed.

- Distinguish among sylvatic, bubonic, septicemic, and pneumonic plague.
- What are the insect vector, the natural host reservoir, and the treatment for plague?

III SOILBORNE DISEASES

Several pathogenic microorganisms live in soil. Fungi are ubiquitous soil microorganisms, and a few are pathogens. Some bacteria are also important soilborne pathogens. In contrast to many person-to-person or vectorborne pathogens, soilborne pathogens are accidental agents of infection, with no life cycle dependency on the accidental host. Soil is an unlimited reservoir of these pathogens, and thus these pathogens cannot be eliminated.

35.8 Fungal Diseases

Fungi in some form grow in nearly every ecological niche, but are most commonly found in nature as free-living saprophytes. Some fungi cause accidental, often opportunistic and sometimes serious infections and disease. Individuals who have impaired immunity due to drug treatment or diseases such as AIDS are more susceptible to opportunistic fungal pathogens. Increases in fungal infections are due mainly to these factors.

The fungi include the eukaryotic organisms commonly known as yeasts, which normally grow as single cells (**Figure 35.18*a***), and molds (mycelial forms), which grow in branching filaments (hyphae) with or without septa (cross walls) (Figure 35.18*b*). The taxonomy and biological diversity of these organisms were discussed in Sections 18.13–18.19. Fortunately, most fungi are harmless to humans. Only about 50 species cause human disease, and in healthy individuals the overall incidence of serious fungal infections is rather low, although certain superficial fungal infections are quite common.

Epidemiology

Fungi cause disease through three major mechanisms. First, some fungi trigger immune responses that result in allergic (hypersensitivity) reactions following exposure to specific fungal antigens. Reexposure to the same fungi, whether growing on the host or in the environment, may cause allergic symptoms. For example, *Aspergillus* spp. (Figure 18.27), a common saprophyte often found in nature as a leaf mold, produces potent allergens, often causing asthma and other hypersensitivity reactions. *Aspergillus* also has other mechanisms for producing disease.

A second fungal disease-producing mechanism involves the production and activity of *mycotoxins*, a large, diverse group of fungal exotoxins. The best-known examples of mycotoxins are the *aflatoxins* (**Figure 35.19**) produced by *Aspergillus flavus*, a species that commonly grows on improperly stored food, such as grain. Aflatoxins are highly toxic and carcinogenic, inducing tumors at high frequency in some animals, especially in birds that feed on contaminated grain. The direct role of aflatoxins in human disease is not well defined.

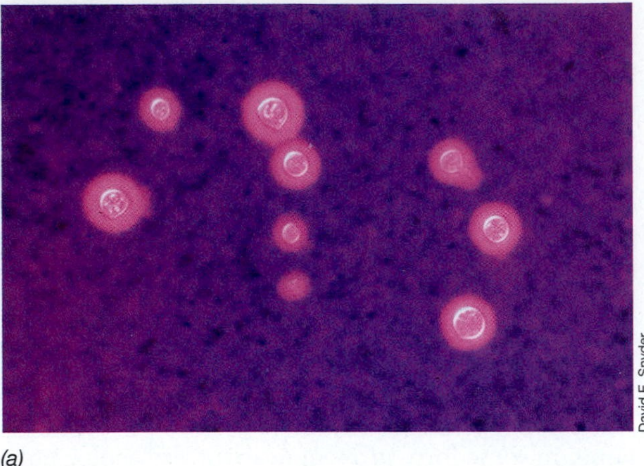

David E. Snyder

(a)

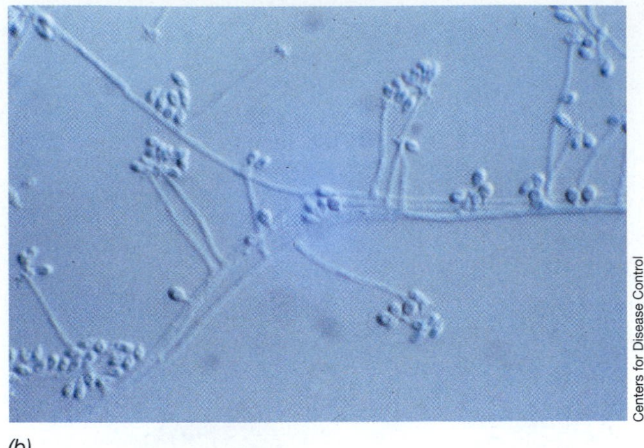

Centers for Disease Control

(b)

Figure 35.18 Typical forms of pathogenic fungi. *(a)* Yeast form of *Cryptococcus neoformans*, stained with India ink to show the capsule. The cells are from 4 to 20 μm in diameter. *(b) Sporothrix schenckii*, showing the branching, or hyphae, characteristic of the mold form of fungi. The round conidia are about 2 μm in diameter.

Mycoses

The third fungal disease-producing mechanism is through infections called **mycoses**. The growth of a fungus on or in the body is called a **mycosis** (plural, mycoses). Mycoses are fungal infections that range in severity from relatively innocuous, superficial lesions to serious, life-threatening diseases. Mycoses fall into three categories. The first of these are the *superficial mycoses*. In these diseases, fungi colonize the skin, hair, or nails, and infect only the surface layers (**Figure 35.20*a***). **Table 35.2** lists some of the fungi that cause superficial mycoses. In general, these diseases are benign and self-limiting. Some, such as *Trichophyton* infections of the feet (athlete's foot), are quite common. Spread is by personal contact with an infected person, by contact with contaminated surfaces such as bathtubs, shower stalls, or floors, or by contact with contaminated shared articles such as towels or bed linens. Treatment for severe cases is with topical application of miconazole nitrate or griseofulvin. Griseofulvin can also be administered orally. After entering the bloodstream, it passes to the skin where it can inhibit fungal growth.

The *subcutaneous mycoses* are a second category of fungal infections. They involve deeper layers of skin (Figure 35.20*b*) and a different group of organisms (Table 35.2). One disease in this category is sporotrichosis, an occupational hazard of agricultural workers, miners, and others who come into contact with the soil. The causative organism, *Sporothrix schenckii,* is a ubiquitous saprophyte on wood and in soil. The lesions are usually initiated by infection at a small wound or abrasion site, and *S. schenckii* can readily be isolated from the lesion and cultured *in vitro*. Treatment is with oral potassium iodide or oral ketoconazole.

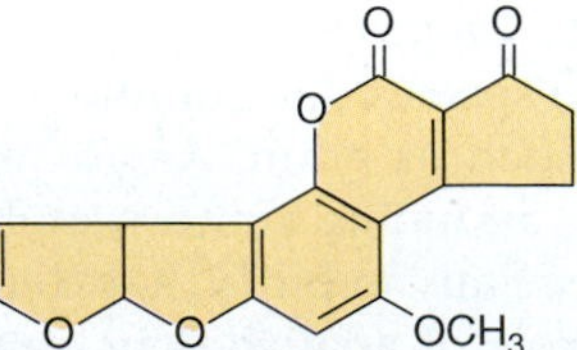

Figure 35.19 Structure of aflatoxin B1. This toxin is one of a group of related compounds produced by *Aspergillus flavus*.

The *systemic mycoses* are the third and most serious category of fungal infections. They involve fungal growth in internal organs of the body and are subclassified as primary or secondary infections. A *primary* infection is one resulting directly from the fungal pathogen in an otherwise normal, healthy individual. A *secondary* infection is one in a host that harbors a predisposing condition, such as antibiotic therapy or immunosuppression.

In the United States, the most widespread primary fungal infections are histoplasmosis, caused by *Histoplasma capsulatum,* and coccidioidomycosis (San Joaquin Valley fever), caused by *Coccidioides immitis*. Both of these organisms normally live in soil and both cause respiratory disease. The host becomes infected by inhaling airborne spores that germinate and grow in the lungs. Histoplasmosis is primarily a disease of rural areas in the midwestern United States, especially in the Ohio and Mississippi River valleys. Most cases are mild and are often mistaken for more common respiratory infections. San Joaquin Valley fever is generally restricted to the desert regions of the southwestern United States. The fungus lives in desert soils, and the spores are disseminated on dry, windblown particles that are inhaled. In some areas in the southwestern United States, as many as 80% of the inhabitants may be infected, although most individuals suffer no apparent ill effects.

A number of systemic fungal infections, including histoplasmosis and coccidioidomycosis, are especially serious and common in individuals whose immune systems have been

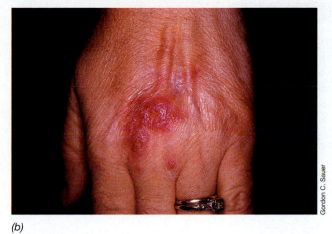

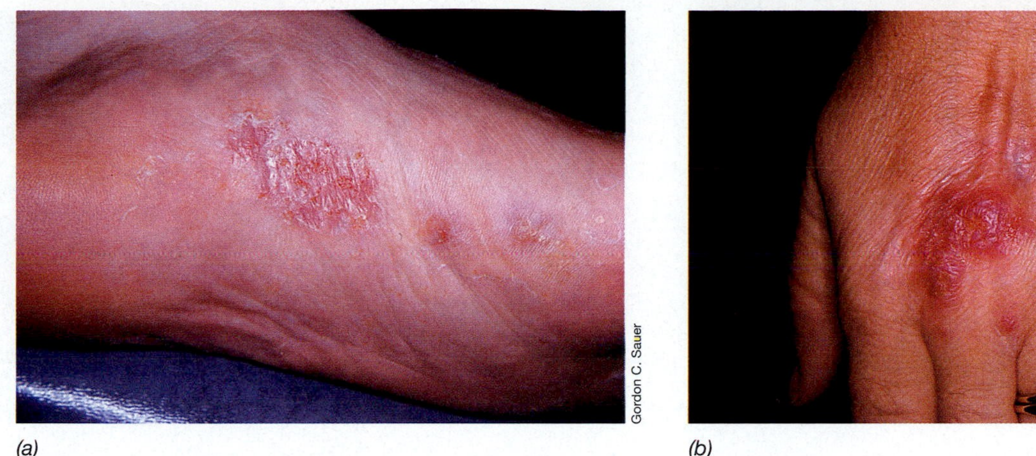

Figure 35.20 Fungal infections. *(a)* Superficial mycosis of the foot (athlete's foot) due to infection with *Trichophyton rubrum*. *(b)* Sporotrichosis, a subcutaneous infection due to *Sporothrix schenckii*.

impaired, for example, by acquired immunodeficiency syndrome (AIDS) or by immunosuppressive drugs. These fungi are opportunistic pathogens; they cause serious infections only in individuals who have impaired defense mechanisms. These are secondary fungal diseases because normal individuals either do not get the disease or generally have a less severe form. Examples of other fungi involved as secondary opportunistic pathogens are given in Table 35.2.

Treatment and Control

Effective chemotherapy against systemic fungal infections is difficult because most antibiotics that inhibit fungi (which are eukaryotes) also affect their hosts (Section 27.11). For example, one of the most effective antifungal agents, amphotericin B, is widely used to treat systemic fungal infections of humans but may cause serious side effects such as kidney toxicity.

Control of infections by elimination of fungal pathogens from the environment is impractical. As with many common-source pathogens, control of fungal growth is very difficult because there is a limitless reservoir. Exposure to fungi cannot be eliminated, but risks can be reduced by decontamination and air filtration in restricted local environments.

Table 35.2 Some pathogenic fungi and the diseases they cause

Disease	*Causal organism*	*Site*
Superficial mycoses (dermatomycoses)		
Ringworm	*Microsporum*	Scalp of children
Favus	*Trichophyton*	Scalp
Athlete's foot	*Epidermophyton, Trichophyton*	Between toes, skin
Jock itch	*Trichophyton, Epidermophyton*	Genital region
Subcutaneous mycoses		
Sporotrichosis	*Sporothrix schenckii*	Arms, hands
Chromoblastomycosis	Several fungal genera	Legs, feet
Systemic mycoses		
Aspergillosis	*Aspergillus* spp.[a]	Lungs
Blastomycosis	*Blastomyces dermatitidis*	Lungs, skin
Candidiasis	*Candida albicans*[b]	Oral cavity, intestinal tract
Coccidioidomycosis	*Coccidioides immitis*[b]	Lungs
Cryptococcosis	*Cryptococcus neoformans*[b]	Lungs, meninges
Histoplasmosis	*Histoplasma capsulatum*[b]	Lungs
Pneumocystis pneumonia	*Pneumocystis carinii*[b]	Lungs

[a]*Aspergillus* can also cause allergies, toxemia, and limited infections.
[b]An opportunistic pathogen frequently implicated in the pathogenesis of AIDS.

35.8 MiniReview

Certain soilborne fungi produce disease in humans. Superficial, subcutaneous, and systemic mycoses are difficult to control because of a lack of specific antifungal drugs and the ubiquitous nature of the pathogens. Fungal infections may cause serious systemic disease, often in individuals with impaired immunity, such as in AIDS patients.

- Describe superficial, subcutaneous, and systemic mycoses.
- Distinguish between a primary and a secondary fungal disease.

35.9 Tetanus

Tetanus is a serious, often life-threatening disease. Although tetanus is preventable through immunization, about 500 individuals have acquired tetanus in the United States within the last decade and about 75 of these have died. Worldwide, tetanus causes over 200,000 deaths per year, even though it is a preventable infectious disease.

Biology and Epidemiology

Tetanus is caused by an exotoxin produced by *Clostridium tetani*, an obligately anaerobic, endospore-forming rod (Section 16.2). The natural reservoir of *C. tetani* is soil, where it is a ubiquitous resident, although it is occasionally found in the gut of mammals, as are other *Clostridium* species.

Cells of *C. tetani* normally gain access to the body through a soil-contaminated wound, typically a deep puncture. In the wound, anoxic conditions allow germination of endospores, growth of the organism, and production of a potent exotoxin, the *tetanus toxin*. The organism is noninvasive; its sole method of causing disease is through the action of tetanus toxin on host cells. The incubation time is variable and may take from 4 days to several weeks, depending on the number of endospores inoculated at the time of injury. Tetanus is not transmitted from person to person.

Pathogenesis

We have already examined the activity of tetanus toxin at the cellular and molecular level (Section 28.10). The toxin directly affects the release of inhibitory signaling molecules in the nervous system. These inhibitory signals control the "relaxation" phase of muscle contraction. The overall result is rigid paralysis of the voluntary muscles, often called *lockjaw* because it is observed first in the muscles of the jaw and face (**Figure 35.21**). Death is usually due to respiratory failure, and mortality is relatively high (15% over the last decade in the United States).

Diagnosis, Control, Prevention, and Treatment

Diagnosis of tetanus is based on exposure, clinical symptoms, and, rarely, identification of the toxin in the blood or tissues

Figure 35.21 A soldier dying from tetanus. Note the rigid paralysis. This painting by Charles Bell is in the Royal College of Surgeons, Edinburgh, Scotland.

of the patient. The organism may also be cultured from the wound, but success is highly variable.

The natural reservoir of *C. tetani* is the soil. Because *C. tetani* is an accidental pathogen in humans and is not dependent on humans or other animals for its propagation, there is no possibility for eradication. Therefore, control measures must focus on prevention.

Tetanus is a preventable disease. The existing toxoid vaccine is completely effective for disease prevention. Virtually all tetanus cases occur in individuals who were inadequately immunized. Individuals from 25–59 years of age are the fastest growing age group for contracting tetanus, presumably because public health immunization programs target infants, school-age individuals, and seniors 60 years of age and older.

Appropriate treatment of serious cuts, lacerations, and punctures includes administration of a "booster" tetanus toxoid immunization. If the wound is severe and is contaminated by soil, treatment should also include administration of an antitoxin preparation, especially if the patient's immunization status is unknown or is out of date. The tetanus antitoxin is typically pooled human anti-tetanus immunoglobulin (approved for human use worldwide) or a preparation of antibodies to tetanus made in horses (approved for use in many developing countries). Both of these preparations work by binding and neutralizing the tetanus exotoxin. The antitoxin is generally given intramuscularly, but intrathecal injection (injection into the sheath surrounding the spinal cord) is superior because the antitoxin can then get to the affected nerve root much more efficiently (Section 28.10). These measures prevent active tetanus from occurring.

Acute symptomatic tetanus is treated with antibiotics, usually penicillin, to stop growth and toxin production by *C. tetani* and antitoxin to prevent binding of newly released toxin to cells. Supportive therapy such as sedation, administration of muscle relaxants, and mechanical respiration may be necessary to control the effects of paralysis. Treatment cannot provide a reversal of symptoms, because toxin that is already bound to tissues cannot be neutralized. Even with antitoxin, antibiotics, and supportive therapy, tetanus patients have significant morbidity and mortality.

Other Endospore-Forming Pathogens in Soil

Several other species of *Clostridium* and *Bacillus,* all endospore-forming organisms, are pathogens and all are normally found in soil, making their eradication impossible. All cause disease because of their production of potent exotoxins. *C. tetani* is found almost exclusively in soil, but *C. botulinum, C. difficile,* and *C. perfringens* are occasionally found in the gut of humans and other animals as part of the normal microbial flora. *C. perfringens* and *C. botulinum* are important potential pathogens, but unlike *C. tetani,* cause diseases transmitted by the foodborne route (Section 37.6) rather than directly from soil. *C. difficile* is a commensal of the human colon and occasionally causes diarrhea. *Bacillus anthracis,* an important veterinary pathogen that has also been used in biowarfare, causes anthrax, and is also typically found in soils (Section 33.12).

35.9 MiniReview

Clostridium tetani is a ubiquitous soilborne microorganism that can cause tetanus, a disease characterized by toxin production and rigid paralysis. Tetanus has significant morbidity and mortality. Tetanus is preventable with appropriate immunization. Treatment for acute tetanus includes antibiotics, active and passive immunization, and supportive therapy.

- Describe infection by *C. tetani* and the elaboration of tetanus toxin.
- Describe the steps necessary to prevent tetanus in an individual who has sustained a puncture wound.

Review of Key Terms

Enzootic an endemic disease present in an animal population

Epizootic an epidemic disease present in an animal population

Hantavirus pulmonary syndrome (HPS) an acute viral disease characterized by respiratory pneumonia, transmitted by rodent hantavirus

Hemorrhagic fever with renal syndrome (HFRS) an acute viral disease characterized by shock and kidney failure, obtained by transmission of hantavirus from rodents

Lyme disease a tick-transmitted disease caused by the spirochete *Borrelia burgdorferi*

Malaria an insect-transmitted disease characterized by recurrent episodes of fever and anemia caused by the protist *Plasmodium* spp., usually transmitted between mammals through the bite of the *Anopheles* mosquito

Mycosis (plural, **mycoses**) infection caused by a fungus

Plague an endemic disease in rodents caused by *Yersinia pestis* that can be transferred to humans through the bite of a flea

Rabies a usually fatal neurological disease caused by the rabies virus usually transmitted by the bite or saliva of an infected animal

Rickettsias obligate intracellular bacteria of the genus *Rickettsia* responsible for diseases including typhus, Rocky Mountain spotted fever, and ehrlichiosis

Rocky Mountain spotted fever a tick-transmitted disease caused by *Rickettsia rickettsii,* characterized by fever, headache, rash, and gastrointestinal symptoms

Sickle cell anemia a genetic trait that confers resistance to malaria but causes a reduction in the efficiency of red blood cells by reducing the oxygen-binding affinity of hemoglobin

Thalassemia a genetic trait that confers resistance to malaria but causes a reduction in the efficiency of red blood cells by altering a red blood cell enzyme

Tetanus a disease characterized by rigid paralysis of the voluntary muscles, caused by an exotoxin produced by *Clostridium tetani*

Typhus a louse-transmitted disease caused by *Rickettsia prowazekii,* characterized by fever, headache, weakness, rash, and damage to the central nervous system and internal organs

West Nile fever a neurological disease caused by West Nile virus, a virus transmitted by mosquitoes from birds to humans

Zoonosis an animal disease transmitted to humans

Review Questions

1. Identify the animals most likely to carry rabies in the United States. Which immunization programs are in place for the treatment of rabies? Which immunization programs are in place for the prevention of rabies (Section 35.1)?
2. Describe the conditions that may cause emergence of hantavirus pulmonary syndrome (HPS). How can HPS be prevented (Section 35.2)?
3. Identify the three major categories of organisms that cause rickettsial diseases. For typhus, Rocky Mountain spotted fever, and ehrlichiosis, identify the most common reservoir and vector (Section 35.3).
4. Identify the most common reservoir and vector for Lyme disease in the United States. How can the spread of Lyme disease be controlled? How can Lyme disease be treated (Section 35.4)?

5. Malaria symptoms include fever followed by chills. These symptoms are related to activities of the pathogen. Describe the growth stages of *Plasmodium* spp. in the human host and relate them to the fever–chill pattern. Why might a person of western European origin be more susceptible to malaria than a person of African or Mediterranean origin (Section 35.5)?
6. Explain the life cycle and the role of the mosquito in spreading West Nile virus. What animals are the primary hosts? Are humans productive alternative hosts? Explain (Section 35.6).
7. For a potentially serious disease like bubonic plague, why aren't vaccines provided for the general population? Identify public health measures used to control this disease (Section 35.7).
8. Identify the natural source of most fungal pathogens. How can fungal exposure be controlled? What particular problems, especially in terms of therapy, do fungi pose for the clinician (Section 35.8)?
9. Describe the invasiveness and toxicity of *Clostridium tetani*. Discuss the major mechanism of pathogenesis for tetanus and define measures for prevention and treatment (Section 35.9).

Application Questions

1. Describe the sequence of events you would take if a child received a bite (provoked or unprovoked) from a stray dog with no record of rabies immunization. Present one scenario where you were able to capture and detain the dog and another for a dog that escaped. How would these procedures differ from a situation in which the child was bitten by a dog that had documented, up-to-date rabies immunizations?
2. Oral histories from Native Americans indicate the presence of hantavirus pulmonary syndrome prior to the "discovery" and definition of HPS in 1993. Explain these findings in terms of emerging zoonotic diseases and human land use practices.
3. Discuss at least three common properties of the disease agents and review the disease process for Rocky Mountain spotted fever, typhus, and ehrlichiosis. Why is ehrlichiosis emerging as an important rickettsial disease? Compare its emergence to that of Lyme disease.
4. Malaria eradication has been a goal of public health programs for at least 100 years. What factors preclude our ability to eradicate malaria? If an effective vaccine was developed, could malaria be eradicated? Compare this possibility to the possibility of eradicating plague.
5. Devise a plan to prevent the spread of West Nile virus to humans in your community. Identify the costs involved in such a plan, both at the individual level and at the community level. Find out if a mosquito abatement program is active in your community. What methods, if any, are used in your area for the reduction of mosquito populations?

36

Wastewater Treatment, Water Purification, and Waterborne Microbial Diseases

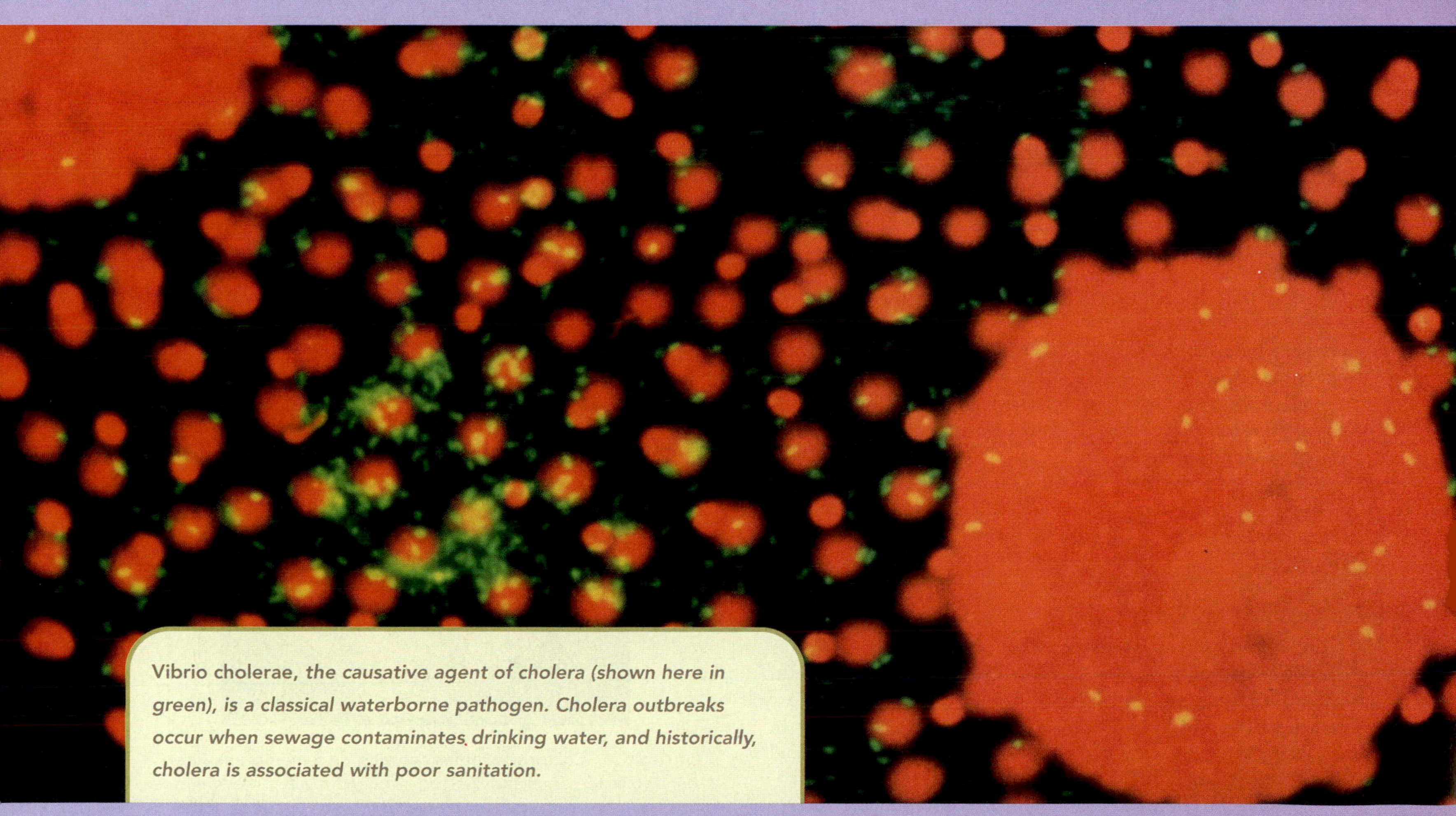

Vibrio cholerae, the causative agent of cholera (shown here in green), is a classical waterborne pathogen. Cholera outbreaks occur when sewage contaminates drinking water, and historically, cholera is associated with poor sanitation.

Water free of biological and chemical contaminants is essential for public health, and procedures to monitor and assess water quality are thus necessary. Water treatment, including disinfection of drinking water and remediation of wastewater, is necessary to ensure high-quality water. Water quality, however, is sometimes compromised, even in large-scale public wastewater and drinking water systems. Lapses in water quality can promote dramatic and even life-threatening spread of infectious disease. This chapter examines standard methods of water monitoring, treatment, and remediation and some common waterborne diseases.

I WASTEWATER MICROBIOLOGY AND WATER PURIFICATION

Water is the most important potential common source of infectious diseases and can also be a source for chemically induced intoxications. This is because a single water source often serves large numbers of people, as, for example, in large cities. Everyone in these circumstances must use the available water, and contaminated water has the potential to spread disease to all exposed individuals. Water purity is therefore the most important single factor for ensuring public health. The methods commonly used to assess water quality depend on standard microbiological and chemical techniques. In addition to physical and chemical purification procedures, water treatment and purification schemes use microorganisms to identify, remove, and degrade pollutants.

36.1 Public Health and Water Quality

Even water that looks perfectly transparent and clean may be contaminated with pathogenic microorganisms and may pose a serious health hazard. It is impractical to screen water for every pathogenic organism that may be present, and a few non-pathogenic microorganisms are generally tolerable, and even unavoidable, in a water supply. However, water supplies can be sampled for the presence of specific *indicator microorganisms*, the presence of which signals potential contamination with pathogens.

Coliforms and Water Quality

A widely used indicator for microbial water contamination is the **coliform** group of microorganisms. Coliforms are useful indicators of water contamination because many of them inhabit the intestinal tract of humans and other animals in large numbers. Thus, their presence in water indicates fecal contamination. Coliforms are defined as facultatively aerobic, gram-negative, non-spore-forming, rod-shaped bacteria that ferment lactose with gas formation within 48 hours at 36°C. This operational definition of the coliform group includes taxonomically unrelated microorganisms. Many coliforms, however, are members of the enteric bacterial group (∞ Section 15.11). For example, the coliform group includes the usually harmless *Enterobacter, Escherichia coli,* a common intestinal organism and occasional pathogen, and *Klebsiella pneumoniae,* a less common pathogenic intestinal inhabitant.

In general, the presence of total coliforms, or especially *E. coli,* in a water sample indicates fecal contamination and makes the water unsafe for human consumption. When excreted into water, the coliforms eventually die, but they do not die as quickly as some pathogens. The coliforms and the pathogens behave similarly during water purification.

Testing for Coliforms and *Escherichia coli*

Several procedures are used to test for coliforms and *E. coli* in water samples. All tests assay the growth of organisms recovered from water samples. Common methods of enumerating the samples include the *most-probable-number* (*MPN*) procedure and the *membrane filter* (*MF*) procedure. The MPN procedure employs liquid culture medium in test tubes to which samples of drinking water are added. Growth in the culture vessels indicates microbial contamination of the water supply. For the MF procedure, at least 100 ml of the water sample is passed through a sterile membrane filter, trapping any bacteria on the filter surface. The filter is placed on a plate of eosin-methylene blue (EMB) culture medium, which is selective for coliform organisms (**Figure 36.1**; ∞ Figure 32.4). Following incubation, coliform colonies are counted, and from this value the number of coliforms in the original water sample can be calculated.

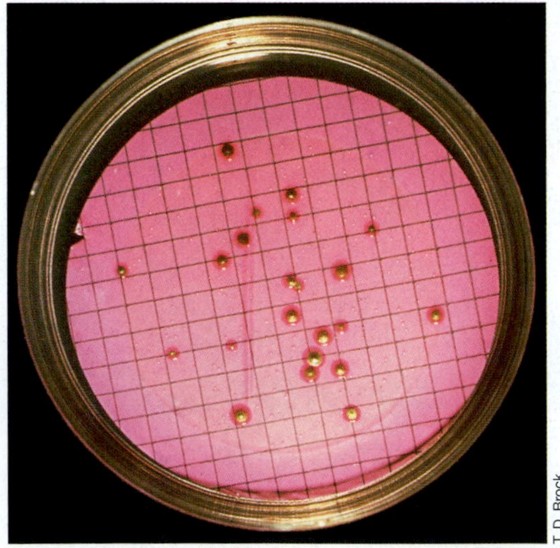

T. D. Brock

Figure 36.1 Coliform colonies growing on a membrane filter. A drinking water sample was passed through the filter. The filter was then placed on eosin-methylene blue (EMB) medium that is both selective and differential for lactose-fermenting bacteria (coliforms). The dark, shiny appearance of the colonies is characteristic of coliforms. Each colony developed from one viable coliform cell present in the original sample.

Current selective media methods can detect not only total coliforms, but can also specifically identify *E. coli* at the same time. Designated as defined substrate tests, they are generally faster and usually more accurate than the older EMB agar tests. Defined substrate tests are based on the ability of coliforms and *E. coli* to metabolize certain substrates. For example, all coliforms, including *E. coli,* metabolize 4-methylumbelliferyl-β-D-galactopyranoside (MUGal) using the enzyme β-galactosidase. If coliforms are present in a sample, MUGal is metabolized to produce a fluorescent product visible under ultraviolet (UV) light (**Figure 36.2**). To distinguish total coliforms from *E. coli,* another enzyme–substrate reaction is used at the same time. Strains of *E. coli* but not other coliforms produce the enzyme β-glucuronidase, which metabolizes indoxyl β-D-glucuronide (IBDG) to a blue compound. The blue compound colors only growing *E. coli* colonies. As a result, *E. coli* colonies fluoresce and are also dark blue, and this differentiates them from colonies of other coliforms (Figure 36.2). The test uses a membrane filter method and media containing both MUGal and IBDG (called *MI media*). The filter is overlaid on MI agar, incubated, and examined within 24 hours for blue colonies (*E. coli*) and fluorescent colonies (total coliforms).

In properly regulated drinking water supply systems, coliform and *E. coli* tests should be negative. A positive test indicates that a breakdown has occurred in the purification or distribution system. Drinking water standards in the United States are specified under the Safe Drinking Water Act, which provides a framework for the development of drinking water standards. For the MF technique, 100-ml samples are filtered. To be considered safe, the number of coliform bacteria in drinking water samples cannot exceed any of the following levels: (1) 1 per 100 ml as the arithmetic mean of all samples examined per month; (2) 4 per 100 ml in more than one sample when fewer than 20 are examined per month; or (3) 4 per 100 ml in more than 5% of the samples when 20 or more samples are examined per month. Water utilities report coliform test results to the United States Environmental Protection Agency, and if they do not meet the prescribed standards, the utilities must notify the public and take steps to correct the problem. Many smaller communities and even large cities sometimes fail to meet these standards.

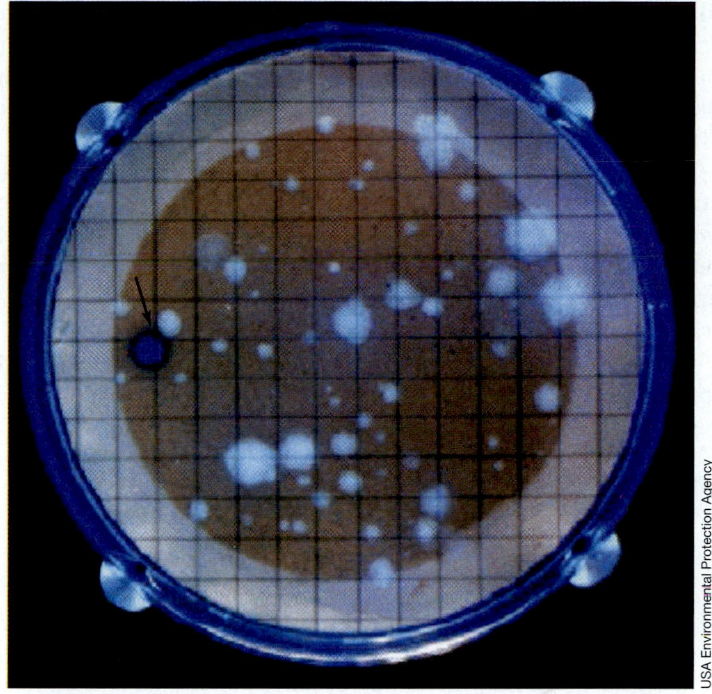

Figure 36.2 Total coliforms and *Escherichia coli.* A filter exposed to a drinking water sample was incubated at 35°C for 24 hours on MI media and examined under UV light. The single *E. coli* colony appears dark blue (arrow). The other colonies are coliforms that fluoresce white to light blue.

Public Health and Drinking Water Purification

The incidence of waterborne disease in developed countries is now so low that it is difficult to directly measure the effect of current treatment practices. Most intestinal infections in developed countries are no longer transmitted by water but by food, another major common source of infection. Effective water treatment practices, however, were not in place until the twentieth century.

Coliform-counting culture methods were developed and adapted in about 1905. At the time, water purification was limited to *filtration* to reduce turbidity. Although filtration significantly decreased the microbial load of water, many microorganisms still passed through the filters. In about 1910, **chlorine** came into use as a disinfectant for large water supplies. Chlorine gas was an effective and inexpensive general disinfectant for drinking water, and its use quickly reduced the incidence of waterborne disease. **Figure 36.3**

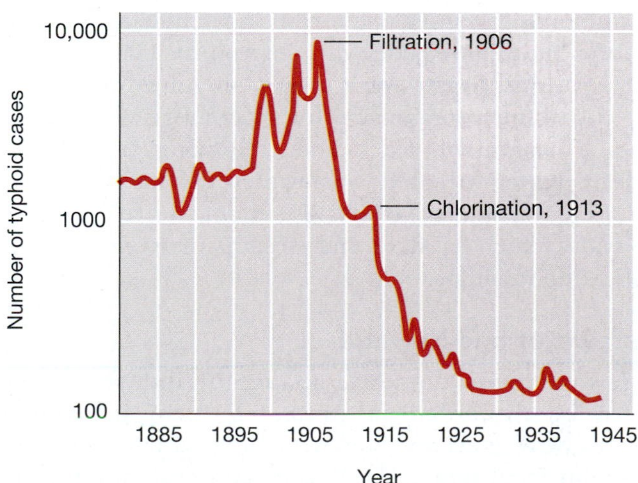

Figure 36.3 The effect of water purification on the incidence of waterborne disease. The graph shows the incidence of typhoid fever in Philadelphia (Pennsylvania, USA). Note the dramatic reduction in the incidence of typhoid fever after the introduction of both filtration and chlorination.

illustrates the dramatic drop in incidence of typhoid fever (caused by *Salmonella typhi*) in a major American city after purification procedures using filtration and chlorination were introduced. Similar results were obtained in other cities.

Major improvements in public health in the United States, starting near the beginning of the twentieth century, were largely due to the establishment of the new water treatment procedures used in large-scale, publicly operated wastewater and drinking water treatment plants. The effectiveness of filtration and chlorination was monitored by the coliform test. Public works engineering and microbiology were the most important contributors to the dramatic advances in public health in developed countries in the twentieth century.

36.1 MiniReview

Drinking water quality is determined by counting coliform bacteria using standardized techniques. This method is a reliable indicator of fecal contamination in water supplies. Filtration and chlorination of water significantly decreases microbial numbers. Application of water monitoring and purification methods to drinking water is the most important public health measure ever devised.

- Why do the bacterial colonies recovered from drinking water that grow on eosin-methylene blue media indicate fecal contamination of the water supply?
- What general procedures are used to reduce microbial numbers in water supplies?

36.2 Wastewater and Sewage Treatment

Wastewater is domestic sewage or liquid industrial waste that cannot be discarded in untreated form into lakes or streams due to public health, economic, environmental, and aesthetic considerations. Wastewater treatment relies on industrial-scale use of microorganisms for bioconversion. Wastewater enters a treatment plant and, following treatment, the **effluent water**—treated wastewater discharged from the wastewater treatment facility—is suitable for release into surface waters such as lakes and streams or to drinking water purification facilities.

Wastewater and Sewage

Wastewater from domestic sewage or industrial sources cannot be discarded in untreated form into lakes or streams. **Sewage** is liquid effluent contaminated with human or animal fecal materials. Wastewater commonly contains potentially harmful inorganic and organic compounds as well as pathogenic microorganisms. Wastewater treatment can use physical, chemical, and biological (microbiological) processes to remove or neutralize contaminants.

On average, each person in the United States uses 100–200 gallons of water every day for washing, cooking, drinking, and sanitation. Wastewater collected from these activities must be treated to remove contaminants before it can be released into surface waters. About 16,000 publicly owned treatment works (POTW) operate in the United States. Most POTWs are fairly small, treating 1 million gallons (3.8 million liters) or less of wastewater per day. Collectively, however, these plants treat about 32 billion gallons of wastewater daily. Wastewater plants are usually constructed to handle both domestic and industrial wastes. Domestic wastewater is made up of sewage, "gray water" (the water resulting from washing, bathing, and cooking), and wastewater from food processing.

Industrial wastewater includes liquid discharged from the petrochemical, pesticide, food and dairy, plastics, pharmaceutical, and metallurgical industries. Industrial wastewater may contain toxic substances; some plants are required by the United States Environmental Protection Agency (EPA) to pretreat toxic or heavily contaminated discharges before they enter POTWs. Pretreatment may involve mechanical processes in which large debris is removed. However, some wastewaters are pretreated biologically or chemically to remove highly toxic substances such as cyanide; heavy metals such as arsenic, lead, and mercury; or organic materials such as acrylamide, atrazine (a herbicide), and benzene. These substances are converted to less toxic forms by treatment with chemicals or microorganisms capable of neutralizing, oxidizing, precipitating, or volatilizing these wastes. The pretreated wastewater can then be released to the POTW.

Wastewater Treatment and Biochemical Oxygen Demand

The goal of a wastewater treatment facility is to reduce organic and inorganic materials in wastewater to a level that no longer supports microbial growth and to eliminate other potentially toxic materials. The efficiency of treatment is expressed in terms of a reduction in the **biochemical oxygen demand (BOD)**, the relative amount of dissolved oxygen consumed by microorganisms to completely oxidize all organic and inorganic matter in a water sample (∞ Section 23.6). High levels of organic and inorganic materials in the wastewater result in a high BOD.

Typical values for domestic wastewater, including sewage, are approximately 200 BOD units. For industrial wastewater, for example, from sources such as dairy plants, the values can be as high as 1500 BOD units. An efficient wastewater treatment facility reduces the high BOD levels to less than 5 BOD units in the final treated water. Wastewater facilities are designed to treat both low-BOD sewage and high-BOD industrial wastes.

Treatment is a multistep operation employing a number of independent physical and biological processes (**Figure 36.4**).

Primary, secondary, and sometimes *tertiary* treatments are employed to reduce biological and chemical contamination in the wastewater, and each level of treatment employs more complex technologies.

Primary Wastewater Treatment

Primary wastewater treatment uses only physical separation methods to separate solid and particulate organic and inorganic materials from wastewater. Wastewater entering the treatment plant is passed through a series of grates and screens that remove large objects. The effluent is left for several hours to allow solids to settle to the bottom of the separation reservoir (**Figure 36.5**).

Municipalities that provide only primary treatment discharge extremely polluted water with high BOD into adjacent waterways because high levels of soluble and suspended organic matter and other nutrients remain in water following primary treatment. These nutrients can trigger undesirable microbial growth (∞ Figure 23.9), further reducing water quality. Therefore, most treatment plants employ secondary and even tertiary treatments to reduce the organic content of the wastewater before release to natural waterways. Secondary treatment processes use both aerobic and anaerobic microbial digestion to further reduce organic nutrients in wastewater.

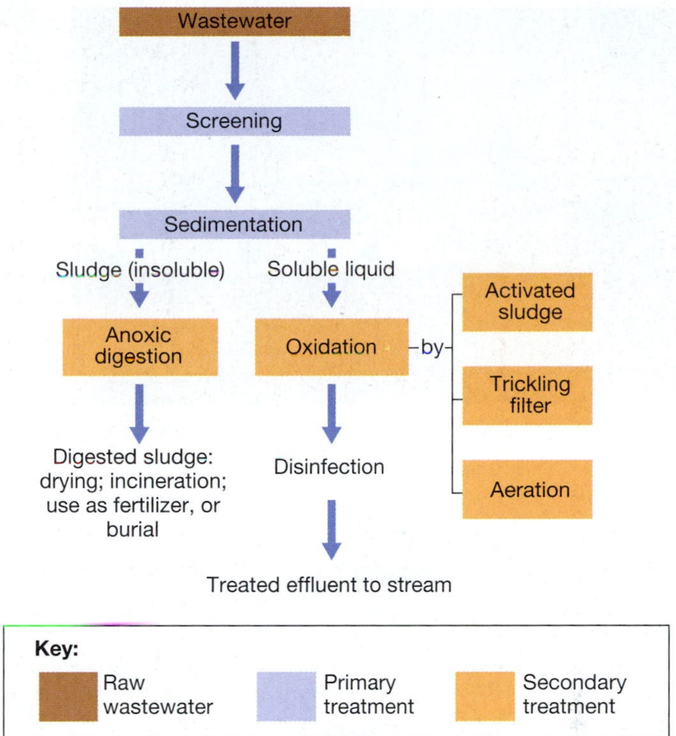

Figure 36.4 Wastewater treatment processes. Effective water treatment plants use the primary and secondary treatment methods shown here. Tertiary treatment to achieve less than 5 BOD units in effluent water is seldom done.

Anoxic Secondary Wastewater Treatment

Anoxic secondary wastewater treatment involves a series of digestive and fermentative reactions carried out by various prokaryotes under anoxic conditions. Anoxic treatment is typically used to treat wastewater containing large quantities of insoluble organic matter (and therefore having a very high BOD) such as fiber and cellulose waste from food and dairy plants. The anoxic degradation process itself is carried out in large enclosed tanks called *sludge digesters* or *bioreactors* (**Figure 36.6**). The process requires the collective activities of many different types of prokaryotes. The major reactions are summarized in Figure 36.6*c*.

First, anaerobes use polysaccharidases, proteases, and lipases to digest macromolecular and suspended waste into soluble components. These soluble components are then fermented to yield a mixture of fatty acids, H_2, and CO_2; the fatty acids are further fermented by syntrophic bacteria (∞ Sections 21.5 and 24.2) to produce acetate, CO_2, and H_2. These products are then used as substrates by methanogenic *Archaea* (∞ Section 17.4), fermenting acetate to produce methane and carbon dioxide, the major products of anoxic sewage treatment (Figure 36.6*c*). The methane is burned off or used as fuel to heat and power the wastewater treatment plant.

Aerobic Secondary Wastewater Treatment

Aerobic secondary wastewater treatment uses digestive reactions carried out by microorganisms under aerobic conditions to treat wastewater containing low levels of organic materials. In general, wastewaters that originate from residential sources can be treated efficiently using only aerobic treatment. Several kinds of aerobic decomposition processes are used for wastewater treatment, but the *trickling filter* and *activated sludge* methods are the most common. A trickling

Figure 36.5 Primary treatment of wastewater. Wastewater is pumped into the reservoir (left) where solids settle. As the water level rises, the water spills through the grates to successively lower levels. Water at the lowest level, now virtually free of solids, enters the spillway (arrow) and is pumped to a secondary treatment facility.

(a)

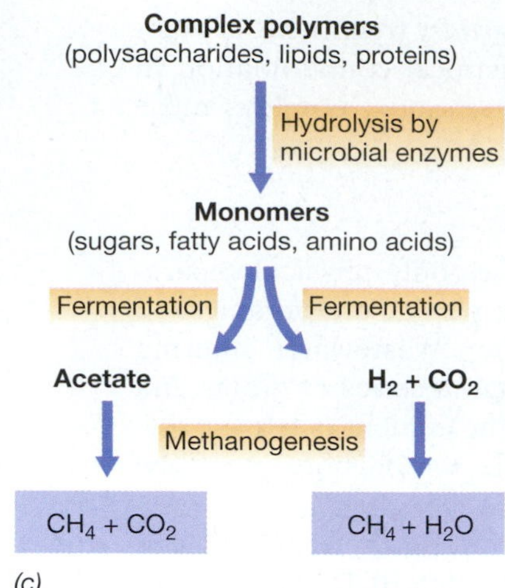

(c)

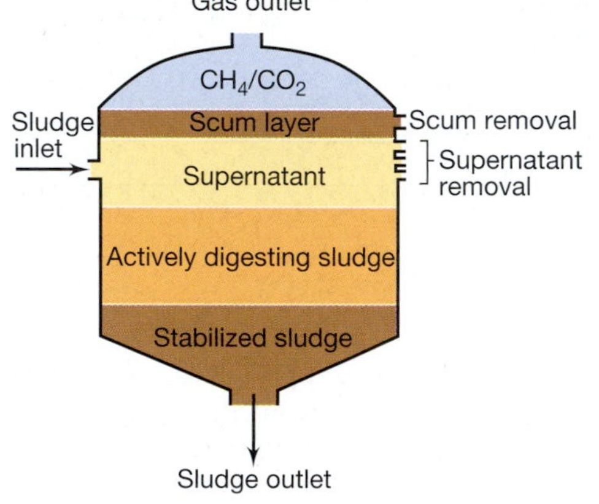

(b)

Figure 36.6 Anoxic secondary wastewater treatment. *(a)* Anoxic sludge digester. Only the top of the tank is shown; the remainder is underground. *(b)* Inner workings of a sludge digester. *(c)* Major microbial processes in anoxic sludge digestion. Methane (CH_4) and carbon dioxide (CO_2) are the major products of anaerobic biodegradation.

filter (**Figure 36.7*a***) is a bed of crushed rocks, about 2 m thick. Wastewater is sprayed on top of the rocks and slowly passes through the bed. The organic material in the wastewater adsorbs to the rocks, and microorganisms grow on their large exposed surface. The complete mineralization of organic matter to carbon dioxide, ammonia, nitrate, sulfate, and phosphate takes place in the extensive microbial biofilm that develops on the rocks.

The most common aerobic treatment system is the activated sludge process. Here, the wastewater to be treated is mixed and aerated in large tanks (Figure 36.7*b*). Slime-forming bacteria, including *Zoogloea ramigera,* among others, grow and form flocs (aggregated masses) (**Figure 36.8**). The biology of *Zoogloea* is discussed in Section 15.7. Protists, small animals, filamentous bacteria, and fungi attach to the flocs. Oxidation then proceeds as with the trickling filter. The aerated effluent containing the flocs is pumped into a holding tank or clarifier where the flocs settle. Some of the floc material (called activated sludge) is then returned to the aerator as inoculum for new wastewater, and the rest is pumped to the anoxic sludge digester (Figure 36.6) or is removed, dried, and burned or used for fertilizer.

Wastewater normally stays in an activated sludge tank for 5–10 hours, a time too short for complete oxidation of all organic matter. However, during this time much of the soluble organic matter is adsorbed to the floc and incorporated by the microbial cells. The BOD of the liquid effluent is considerably reduced (up to 95%) when compared to the incoming wastewater; most of the material with high BOD is now in the settled flocs. The flocs can then be transferred to the anoxic sludge digestor for conversion to CO_2 and CH_4.

Most treatment plants chlorinate the effluent after secondary treatment to further reduce the possibility of biological contamination. The treated effluent can then be discharged into streams or lakes. In the eastern United States, many wastewater treatment facilities use UV radiation to disinfect effluent water. Ozone (O_3), a strong oxidizing agent that is an effective bactericide and viricide, is also used for wastewater disinfection at over 40 plants in the United States.

Tertiary Wastewater Treatment

Tertiary wastewater treatment is any physicochemical or biological process employing bioreactors, precipitation, filtration, or chlorination procedures similar to those employed for drinking water purification (Section 36.3). Tertiary treatment sharply reduces the levels of inorganic nutrients, especially phosphate, nitrite, and nitrate, from the final effluent.

Wastewater receiving tertiary treatment essentially contains no nutrients and cannot support extensive microbial growth. Tertiary treatment is the most complete method of treating sewage but has not been widely adopted due to the costs associated with such complete nutrient removal.

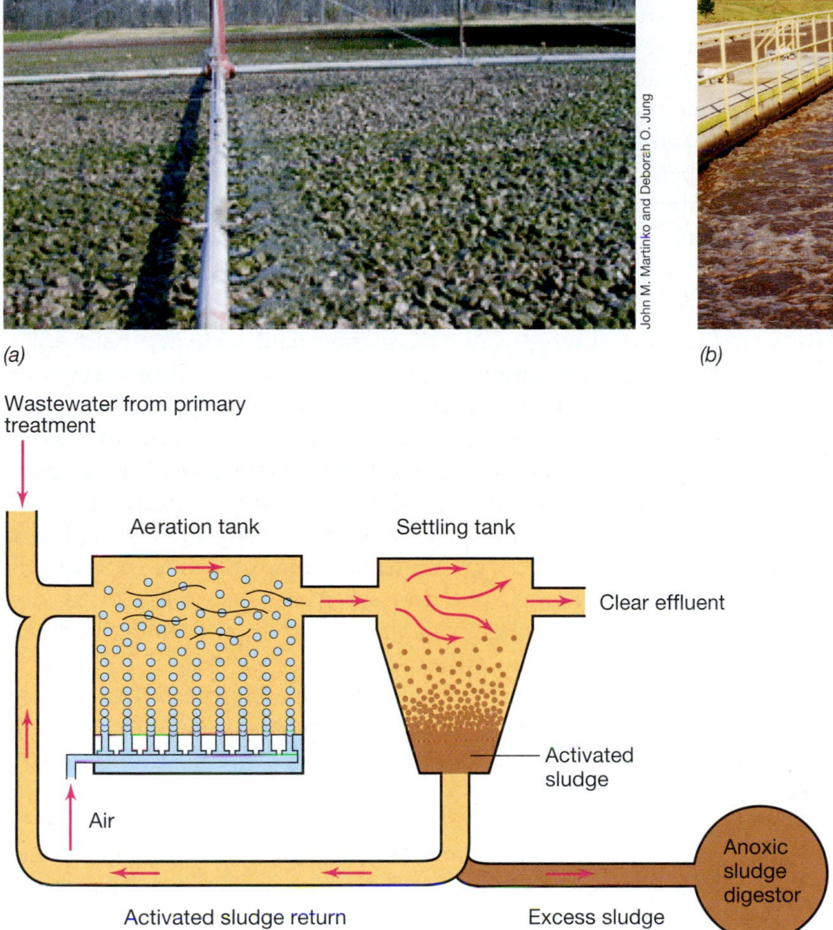

Figure 36.7 Aerobic secondary wastewater treatment processes. A treatment facility for a small city, Carbondale, Illinois (USA). *(a)* Trickling filter. The booms rotate, distributing wastewater slowly and evenly on the rock bed. The rocks are 10–15 cm in diameter, and the bed is 2 m deep. *(b–c)* The activated sludge process. *(b)* Aeration tank of an activated sludge installation in a metropolitan wastewater treatment plant. The tank is 30 m long, 10 m wide, and 5 m deep. *(c)* Wastewater flow through an activated sludge installation. Recirculation of activated sludge to the aeration tank introduces microorganisms responsible for oxic digestion of the organic components of the wastewater. The anoxic sludge digester is detailed in Figure 36.6.

36.2 MiniReview

Sewage and industrial wastewater treatment reduces the BOD (biochemical oxygen demand) to acceptable levels. Primary, secondary, and tertiary wastewater treatment uses physical, biological, and physicochemical processes. After secondary or tertiary treatment, effluent water may be suitable for release directly to a water purification plant.

- What is biochemical oxygen demand (BOD)? Why is BOD reduction necessary in wastewater treatment?
- Identify primary, secondary (anoxic and oxic), and tertiary wastewater treatment methods.
- Other than treated water, what are the final products of wastewater treatment? How might these end products be used?

36.3 Drinking Water Purification

Wastewater treated by secondary methods can usually be discharged into rivers and streams. However, such water is not **potable** (safe for human consumption). The production of potable water requires further treatment to remove potential pathogens, eliminate taste and odor, reduce nuisance chemicals such as iron and manganese, and decrease **turbidity**, which is a measure of suspended solids. **Suspended solids** are small particles of solid pollutants that resist separation by ordinary physical means.

Physical and Chemical Purification

A typical city drinking water treatment installation is shown in **Figure 36.9*a***. Figure 36.9*b* shows the process that purifies **raw water** (also called **untreated water**) that flows through the treatment plant. Raw water is first pumped from the source, in this case a river, to a sedimentation basin where anionic **polymers**, alum (aluminum sulfate), and chlorine are added. **Sediment** including soil, sand, mineral particles, and other large particles, settles out. The sediment-free water is then pumped to a **clarifier** or **coagulation basin**, which is a large holding tank where **coagulation** takes place. The alum and anionic polymers form large particles from the much smaller suspended solids. After mixing, the particles continue to interact, forming large, aggregated masses, a process called **flocculation**. The large, aggregated particles (floc) settle out by gravity, trapping microorganisms and absorbing suspended organic matter and sediment.

After coagulation and flocculation, the clarified water undergoes **filtration** through a series of filters designed to remove organic or inorganic solutes, as well as any suspended particles and microorganisms. The filters typically consist of thick layers of sand, activated charcoal, and ion exchange filtration media. When combined with previous purification steps, the filtered water is free of all particulate matter, most organic and inorganic chemicals, and nearly all microorganisms.

Richard Unz

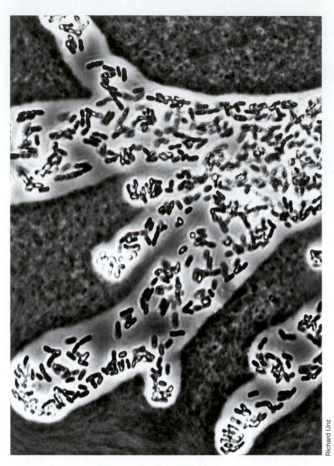

Figure 36.8 A wastewater floc formed by the bacterium *Zoogloea ramigera*. Floc formed in the activated sludge process consists of a large number of small, rod-shaped cells of *Z. ramigera* surrounded by a polysaccharide slime layer, arranged in characteristic fingerlike projections in this negative stain with India ink.

Disinfection

Clarified, filtered water must then be disinfected before it is released to the supply system as pure, potable **finished water**. Chlorination is the most common method of disinfection. In sufficient doses, chlorine kills most microorganisms within 30 minutes. A few pathogenic protists such as *Cryptosporidium*, however, are not easily killed by chlorine treatment (Section 36.6). In addition to killing microorganisms, chlorine oxidizes and effectively neutralizes many organic compounds, improving water taste and smell because most taste- and odor-producing chemicals are organic compounds. Chlorine is added to water either from a concentrated solution of sodium hypochlorite or calcium hypochlorite, or as chlorine gas from pressurized tanks. Chlorine gas is commonly used in large water treatment plants because it is most amenable to automatic control.

Chlorine is consumed when it reacts with organic materials. Therefore, sufficient quantities of chlorine must be added to water containing organic materials so that a small amount,

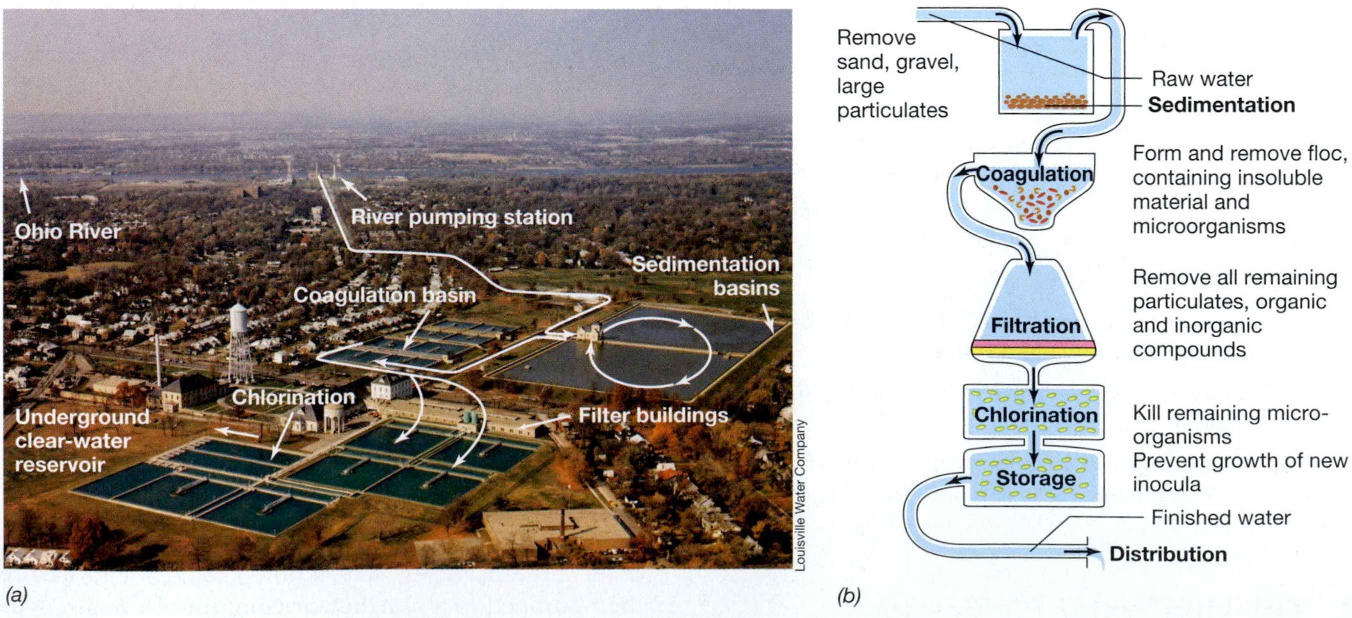

Figure 36.9 Water purification plant. *(a)* Aerial view of a water treatment plant in Louisville, Kentucky (USA). The arrows indicate direction of flow of water through the plant. *(b)* Schematic overview of a typical community water purification system.

called the *chlorine residual,* remains. The chlorine residual reacts to kill any remaining microorganisms. The water plant operator performs chlorine analyses on the treated water to determine the level of chlorine to be added. A chlorine residual level of 0.2–0.6 μg/ml is suitable for most water supplies. After chlorine treatment, the now potable water is pumped to storage tanks from which it flows by gravity or pumps through a **distribution system** of storage tanks and supply lines to the consumer. Residual chlorine levels ensure that the finished water will reach the consumer without becoming contaminated (assuming that there is no catastrophic failure, such as a broken pipe, in the distribution system). Chlorine gas, even when dissolved in water, is extremely volatile and dissipates within hours from treated water. To maintain residual chlorine levels throughout the distribution system, most municipal water treatment plants also introduce ammonia gas with the chlorine to form the stable, nonvolatile chlorine-containing compound **chloramine**, $HOCl + NH_3 \rightarrow NH_2Cl + H_2O$.

UV radiation is also used as an effective means of disinfection. As we discussed in Section 27.2, UV radiation is used to treat secondarily treated effluent from water treatment plants. In Europe, UV irradiation is commonly used for drinking water applications and is being considered for use in the United States. For disinfection purpose, UV light is generated from mercury vapor lamps. Their major energy output is at 253.7 nm, a wavelength that is bacteriocidal and viricidal. However, standard application of UV radiation does not kill cysts and oocysts of protists such as *Giardia* and *Cryptosporidium,* important eukaryotic pathogens in water (Section 36.6).

There are several advantages when using UV radiation over chemical disinfection procedures like chlorination. First, UV irradiation is a physical process that introduces no chemicals to the water. Second, short contact times allow it to be used in existing flow systems, keeping capital costs very low. Third, numerous studies indicate that no disinfection byproducts are formed. Especially in smaller systems where finished water is not pumped long distances or held for long periods (reducing the need for residual chlorine), UV disinfection may be preferable to chlorination.

36.3 MiniReview

Drinking water plants employ industrial-scale physical and chemical systems that remove or neutralize biological, inorganic, and organic contaminants from natural, community, and industrial sources. Water purification plants employ clarification, filtration, and disinfection processes to produce potable water. Finished potable water is free of chemical and biological contamination.

- Trace the treatment of water through a drinking water treatment plant, from the inlet to the final distribution point (faucet).
- What specific purposes do sedimentation, coagulation, filtration, and disinfection accomplish in the drinking water treatment process?

II WATERBORNE MICROBIAL DISEASES

Common-source infectious diseases are caused by microbial contamination of materials shared by a large number of individuals. The most important common source of infectious disease is contaminated water. This is because the failure of a single step in the drinking water purification process may result in the exposure of thousands or even millions of individuals to an infectious agent.

Common-source waterborne diseases are a very significant source of morbidity and mortality, especially in developing countries. Even in developed countries, breakdowns in water treatment plants or the lack of access to clean water in times of emergency can contribute to the propagation of waterborne diseases.

Bacteria, viruses, and protists cause waterborne infectious diseases. Waterborne diseases begin as infections. Water may cause infection even if only a small number of microorganisms are present. The exact numbers of pathogens necessary to cause disease are functions of the virulence of the pathogen and the general ability of the host to resist infection.

36.4 Sources of Waterborne Infection

Human pathogens can be transmitted through improperly treated water used for drinking and cooking. Another common source of disease transmission is through pathogen-contaminated water used for swimming and bathing.

Potable Water

Because everyone consumes water through drinking and cooking, water is a common source of pathogen dissemination and has a very high potential for the catastrophic spread of epidemic disease. As we have already discussed, water supplies in developed countries usually meet rigid quality standards, limiting the spread of waterborne diseases. There are, however, occasional waterborne disease outbreaks in developing countries due to lapses in water quality. Isolated outbreaks affecting low numbers of individuals also occur from consumption of contaminated water from nonregulated sources (such as private wells) or from consumption of untreated water from streams or lakes. These sources may be contaminated by fecal material from humans or animals.

Microorganisms transmitted in drinking water generally grow in the intestines and leave the body in feces; water may then be polluted by feces. If a new host consumes the water, the pathogen may colonize the host's intestine and cause disease. From 1971 to 2002 in the United States, 764 drinking water-associated disease outbreaks occurred—an average of 24 per year (**Figure 36.10*a***). Various bacterial and protist pathogens are occasionally transmitted in drinking water (**Table 36.1**). We discuss giardiasis and cryptosporidiosis in Section 36.6 and *Legionella* in Section 36.7, but *Escherichia coli* strain O157:H7 and the gastrointestinal virus infections are discussed in Chapter 37, where we introduce these

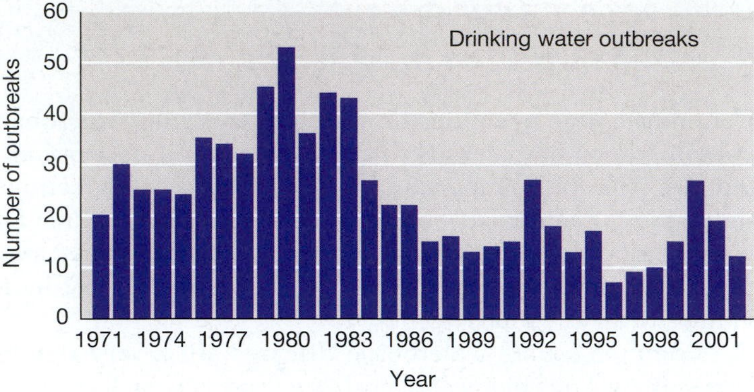

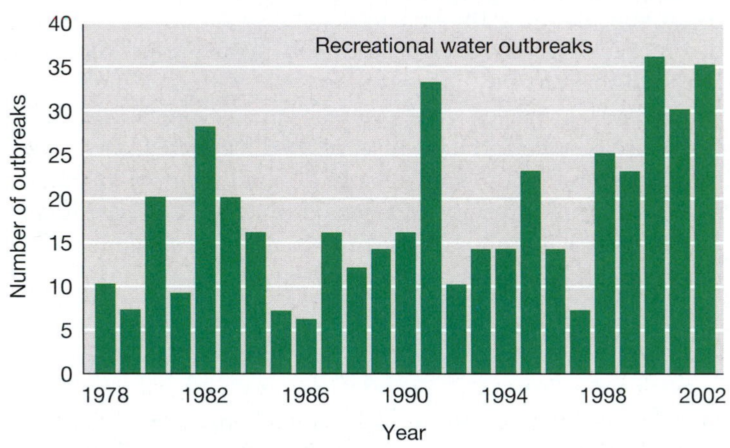

Figure 36.10 Waterborne disease outbreaks. *(a)* Drinking water. Data reported from 1971–2002 by the CDC. Of 764 outbreaks, about 12% were due to chemical contamination of drinking water supplies. The other 88% were due to biological agents including bacteria, viruses, and protists. *(b)* Recreational water. Data reported from 1978–2002 by the CDC. Of 446 total outbreaks, almost all outbreaks were due to biological agents such as bacteria, viruses, and protists.

pathogens as agents of foodborne infections, another common-source mode of transmission.

Recreational Water

Recreational waters include freshwater recreational areas such as ponds, streams, and lakes, as well as public swimming and wading pools. Recreational waters can also be sources of waterborne disease, and historically cause disease outbreaks at levels roughly comparable to those caused by drinking water (Figure 36.10*b*).

The operation of public swimming and wading pools is regulated by state and local health departments. The United States Environmental Protection Agency establishes limits for bacteria in recreational freshwaters (monthly geometric mean for all samples of $< 33/100$ ml for enterococci or $< 126/100$ ml for *Escherichia coli*) and marine waters ($< 35/100$ ml for enterococci). Local and state governments have the authority to set standards above or below these guidelines, and many states use a single-sample maximum as well as the geometric mean for setting standards and defining levels of contamination that constitute violations. For example, the state of Indiana (USA) standard is 125 *E. coli*/100 ml as a geometric mean, with a single-sample maximum of 235/100 ml. Thus, waters that exceed 235 *E. coli*, even if their geometric mean count was not greater than 125, would be in violation of Indiana's water standards. Private swimming pools, spas, and hot tubs are unregulated and are occasional sources of outbreaks of waterborne diseases.

Over a 24-year period (1978–2002), there have been 446 waterborne disease outbreaks from recreational waters in the United States, an average of almost 19 outbreaks per year (Figure 36.10*b*). **Table 36.2** categorizes recreational water outbreaks according to the diseases produced in recent years.

Table 36.1 Infectious disease outbreaks associated with drinking water in the United States[a]

Disease	*Agent*	*Outbreaks*	*Cases*
Meningoencephalitis	*Naegleria fowleri*	1	2
Giardiasis	*Giardia*	3	18
Cryptosporidiosis	*Cryptosporidium*	1	10
Legionellosis	*Legionella*	6	80
Acute gastrointestinal illness	*Escherichia coli* O157:H7	1	2
	Campylobacter jejuni	1	13
	Yersinia enterocolitica and *C. jejuni*	1	12
	Norovirus	5	727
	Unknown	7	119

[a]Compiled from data provided by the Centers for Disease Control and Prevention for 2001–2002. There were 26 outbreaks and 981 cases of infectious disease due to drinking water contamination by infectious agents. Regulated community-owned water systems were implicated in 10 outbreaks including all 6 outbreaks involving *Legionella*. Seven outbreaks were due to noncommunity water systems such as those in some schools, churches, and lodges. Individual water supply systems such as wells accounted for the remaining 11 outbreaks.

Table 36.2 Infectious disease outbreaks associated with recreational water in the United States[a]

Disease	*Outbreaks*	*Cases*
Gastroenteritis[b]	30	1,137
Dermatitis[c]	21	436
Meningoencephalitis[d]	8	8
Other[e]	2	72

[a]Compiled from data provided by the Centers for Disease Control and Prevention for 2001–2002.

[b]Most cases of gastroenteritis were due to *Cryptosporidium, Escherichia coli* O157:H7, or a Norovirus.

[c]Most cases of dermatitis were caused by *Pseudomonas aeruginosa*.

[d]Meningoencephalitis was caused by the amoeba *Naegleria fowleri*. All cases were fatal.

[e]Other diseases include Pontiac fever due to infection by *Legionella* (1 outbreak, 68 cases) and acute respiratory infections of unknown cause (1 outbreak, 4 cases).

Waterborne Infections in Developing Countries

Worldwide, waterborne infections are a much larger problem than in the United States and other developed countries. Developing countries often have inadequate water and sewage treatment facilities, and access to safe, potable water is limited. As a result, diseases such as cholera (Section 36.5), typhoid fever, and amoebiasis (Section 36.8) are important public health problems worldwide.

36.4 MiniReview

Contaminated drinking water and recreational water are sources of waterborne pathogens. In the United States, the number of disease outbreaks due to either of these sources is relatively small in relation to the large number of exposures to water. Worldwide, lack of adequate water treatment facilities and access to clean water contribute significantly to the spread of infectious diseases.

- Identify the bacteria, protists, and viruses commonly responsible for disease outbreaks due to drinking water contamination.
- Identify the bacteria, protists, and viruses commonly responsible for disease outbreaks due to recreational water contamination.

36.5 Cholera

Cholera is a severe diarrheal disease that is now largely restricted to the developing parts of the world. Cholera is an example of a major waterborne disease that can be controlled by application of appropriate water treatment measures.

Biology and Epidemiology

Cholera is caused by *Vibrio cholerae,* a gram-negative, curved rod-shaped *Proteobacterium* (Section 15.12) typically transmitted through ingestion of contaminated water. As with many waterborne diseases, cholera is also associated with food consumption. For example, in the Americas, consumption of raw shellfish and raw vegetables has been associated with cholera. Presumably, vegetables washed in contaminated water and shellfish beds contaminated by untreated sewage transmitted the disease.

Since 1817, cholera has swept the world in seven major pandemics. Two distinct strains of *V. cholerae* are recognized, known as the *classic* and the *El Tor* biotypes. The *V. cholerae* O1 El Tor biotype started the seventh pandemic in Indonesia in 1961, and its spread continues to the present. This pandemic has caused over 5 million cases of cholera and more than 250,000 deaths. As is typical for infectious diseases, the highest prevalence of cholera is in developing countries, especially in Africa. In 1992, a genetic variant known as *V. cholerae* O139 Bengal arose in Bangladesh and caused an extensive epidemic. *V. cholerae* O139 has continued to spread since 1992, causing several major epidemics, and may be the agent of an eighth pandemic.

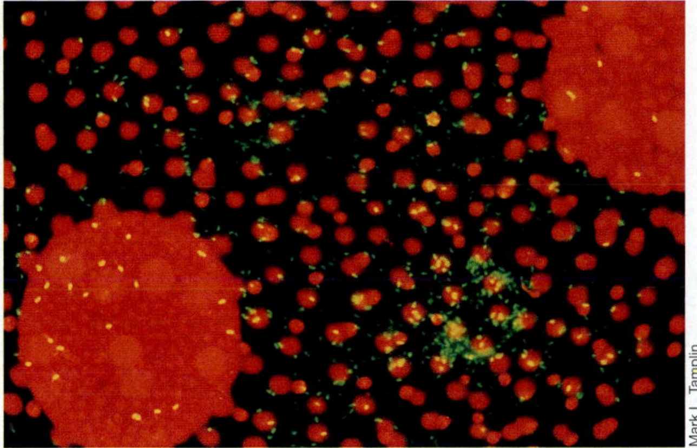

Figure 36.11 Cells of *Vibrio cholerae* attached to the surface of *Volvox*, a freshwater alga. The sample was from a cholera-endemic area in Bangladesh. The *V. cholerae* cells are stained green by a monoclonal antibody to bacterial cell surface proteins. The red color is due to the fluorescence of chlorophyll *a* in the algae.

Cholera is endemic in Africa, Southeast Asia, the Indian subcontinent, and Central and South America. Epidemic cholera occurs frequently in areas where sewage treatment is either inadequate or absent. Worldwide, there were nearly 132,000 reported cases and 2,272 deaths reported in 2005, with over 125,000 cases originating in Africa. According to the World Health Organization, only 5–10% of total cases are reported, so the total incidence of cholera is probably over 1 million cases per year. Even in developed countries, the disease is a threat. A handful of cases are reported each year in the United States, rarely caused by drinking water. Many recent cases (all 5 in 2004 and 8 of 12 in 2005) are imported, often in food. For the few possibly endemic cases, raw shellfish seems to be the most common vehicle, presumably because *V. cholerae* appears to be free-living in coastal waters in endemic areas, adhering to marine microflora (**Figure 36.11**).

Pathogenesis

After a person ingests a substantial number of *V. cholerae* cells, the cells take up residence in the small intestine. Studies in human volunteers have shown that stomach acidity is responsible for the large inoculum needed to initiate cholera. The ingestion of 10^8–10^9 cholera vibrios is generally required to cause disease, but human volunteers given bicarbonate to neutralize gastric acidity developed cholera when given as few as 10^4 cells. Even lower cell numbers can initiate infection if *V. cholerae* is ingested with food, presumably because the food protects the vibrios from stomach acidity.

V. cholerae attaches to epithelial cells in the small intestine where it grows and releases enterotoxin (Figure 28.23). Cholera enterotoxin causes severe diarrhea that can result in dehydration and death unless the patient is given fluid and electrolyte therapy. The enterotoxin causes fluid losses of up to 20 liters (20 kg or 44 lb) per day. The mortality rate from *untreated* cholera is typically 25–50% and can be much higher under conditions of severe crowding and malnutrition.

Figure 36.12 **A fecal sample from a cholera patient.** The "rice-water" stool is nearly liquid. The solid material that has settled in a bottom layer is mucus. The stool from cholera patients is essentially isotonic with blood, containing high amounts of Na^+, K^+, and HCO_3^- (bicarbonate) ions, as well as large numbers of *Vibrio cholerae* cells.

Diagnosis and Prevention of Cholera

Cholera is diagnosed by the presence of the gram-negative comma-shaped *V. cholerae* bacilli in the "rice water" stools (nearly liquid feces) of patients with severe diarrhea (**Figure 36.12**).

Immunization is not normally recommended for cholera prevention. Natural infection and the current vaccines provide variable, short-term immunity. No vaccine protects against the *V. cholerae* O139 serotype. Public health measures such as adequate sewage treatment and a reliable source of safe drinking water are the most important measures for preventing cholera. *V. cholerae* is eliminated from wastewater during proper sewage treatment and drinking water purification procedures. For individuals traveling in cholera-endemic areas, careful attention to personal hygiene and avoidance of untreated water or ice, raw foods, fish, and shellfish offer protection against contracting cholera.

Treatment of Cholera

Cholera treatment is simple, effective, and inexpensive. Intravenous or oral liquid and electrolyte replacement therapy (20 g glucose; 4.2 g NaCl; 4.0 g $NaHCO_3$; 1.8 g KCl, dissolved in 1 liter of water) is the most effective means of cholera treatment. Oral treatment is preferred because no special equipment or sterile precautions are necessary. Effective electrolyte and fluid replacement reduces the mortality rate to about 1%. Streptomycin or tetracycline may shorten the course of infection and the shedding of viable cells, but antibiotics are of little benefit without simultaneous fluid and electrolyte replacement.

36.5 MiniReview

Vibrio cholerae causes cholera, an acute diarrheal disease that causes severe dehydration. Cholera occurs in pandemics. The current focus of the more than 1 million annual cases of cholera is in Africa. In endemic areas, appropriate precautions to avoid contaminated water and food are reasonable preventative measures. Treatment with oral rehydration and electrolytes is the most efficient and effective way to treat the disease, reducing overall mortality to about 1%.

- Define the most likely ways for acquiring cholera.
- Identify specific, effective methods for preventing and treating cholera.

36.6 Giardiasis and Cryptosporidiosis

Giardiasis and cryptosporidiosis are diseases caused by the protists *Giardia intestinalis* and *Cryptosporidium parvum*, respectively. These organisms continue to be problematic even in well-regulated water supplies because they are found in nearly all surface waters and are highly resistant to chlorine.

Giardiasis

Giardia intestinalis is a flagellated protist (Section 18.7) that is usually transmitted to humans in fecally contaminated water, although foodborne and sexual transmission of giardiasis has also been documented. Giardiasis is an acute gastroenteritis caused by this organism. The protist cells, called *trophozoites* (**Figure 36.13*a***), produce a resting stage called a **cyst** (Figure 36.13*b*). The cyst has a thick protective wall that allows the pathogen to resist drying and chemical disinfection. After a person ingests the cysts in contaminated water, the cysts germinate, attach to the intestinal wall, and cause the symptoms of giardiasis; an explosive, foul-smelling, watery diarrhea, intestinal cramps, flatulence, nausea, weight loss, and malaise. Symptoms may be acute or chronic. The foul-smelling diarrhea and the absence of blood or mucus in the stool distinguish giardiasis from bacterial or viral diarrheas. Many infected individuals exhibit no symptoms but act as carriers, indicating that *G. intestinalis* can also establish itself in a stable, symptom-free relationship with its host.

G. intestinalis was the infectious agent in only 1 of the 26 recent drinking water infectious disease outbreaks in the

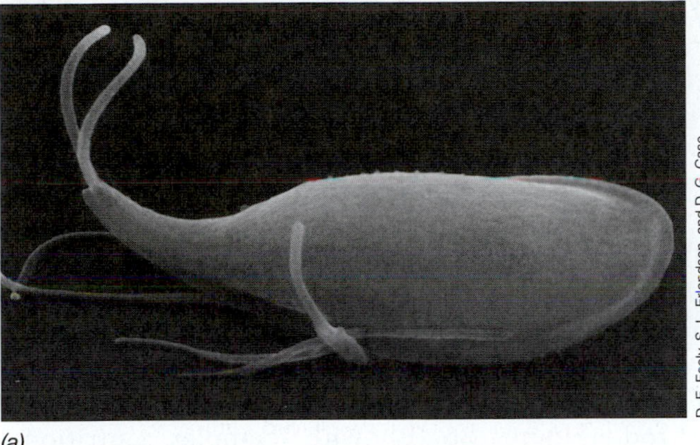

(a)

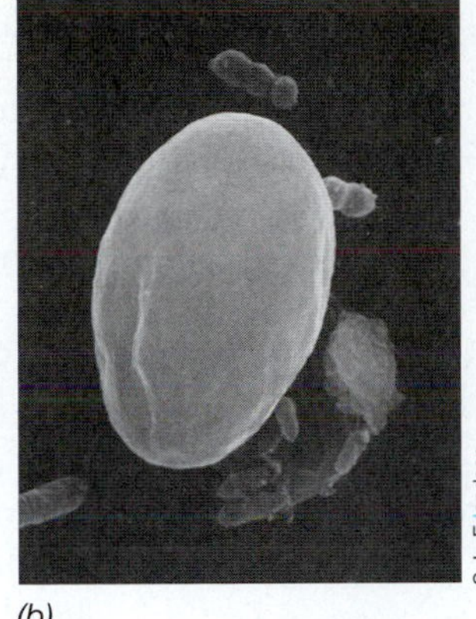

(b)

Figure 36.13 The parasite *Giardia*. Scanning electron micrographs of *(a)* a motile trophozoite. The trophozoite is about 15 μm in length. *(b)* A giardial cyst. The cyst is about 11 μm in length.

United States (Table 36.1). The thick-walled cysts are resistant to chlorine and UV radiation, and most past outbreaks have been associated with water systems that used only chlorination as a means of water purification. Water subjected to proper clarification and filtration followed by chlorination or other disinfection (Section 36.3) is generally free of *Giardia* cysts. Giardiasis can be contracted from ingestion of water from infected swimming pools or lakes. *Giardia* cysts have been found in 97% of surface water sources (lakes, ponds, and streams) in the United States. Isolated cases of giardiasis have been associated with untreated drinking water in wilderness areas. Beavers and muskrats are frequent carriers of *Giardia* and may transmit cells or cysts to water supplies, making the water a possible source of human infection. As a safety precaution, water consumed from rivers and streams, for example, during a camping or hiking trip, should be filtered and treated with iodine or chlorine, or filtered and boiled. Boiling is the preferred method to assure that water is free of pathogens.

Laboratory diagnostic methods include the demonstration of *Giardia* cysts in the stool or the demonstration of *Giardia* antigens in the stool using a direct ELISA (enzyme-linked immunosorbent assay). The drugs quinacrine, furazolidone, and metronidazole are useful in treating acute giardiasis.

Cryptosporidiosis

The protist *Cryptosporidium parvum* lives as a parasite in warm-blooded animals. The parasites are small 2–5 μm round cells that invade and grow intracellularly in mucosal epithelial cells of the stomach and intestine (**Figure 36.14*a***). The protist produces thick-walled, chlorine-resistant, infective cells called *oocysts,* which are shed into water in high numbers in the feces of infected warm-blooded animals (Figure 36.14*b*). The infection is passed on when other animals consume the fecally contaminated water. *Cryptosporidium* cysts are highly resistant to chlorine (up to 14 times more resistant than chlorine-resistant *Giardia*) and UV radiation disinfection. Therefore sedimentation and filtration methods must be used to remove *Cryptosporidium* from water supplies. From 2001 to 2003, *Cryptosporidium* was responsible for 11 of the 30 recreational waterborne disease outbreaks.

C. parvum was responsible for the largest single common-source outbreak of a waterborne disease ever recorded in the United States. In the spring of 1993 in Milwaukee, Wisconsin, over 403,000 people in the population of 1.6 million developed a diarrheal illness that was traced to the municipal water supply. Spring rains and runoff from surrounding farmland had drained into Lake Michigan and overburdened the water purification system, leading to contamination by *C. parvum*. The protist is a significant intestinal parasite in dairy cattle, the likely source of the outbreak.

Cryptosporidiosis is usually a self-limiting mild diarrhea that subsides in 2 weeks or less in normal individuals. However, individuals with impaired immunity, such as that caused

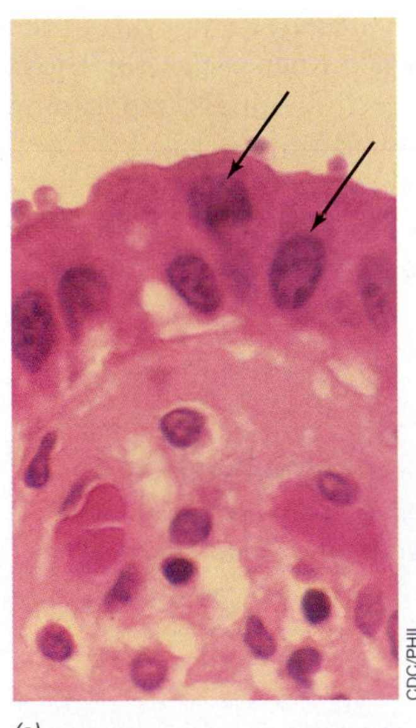

(a)

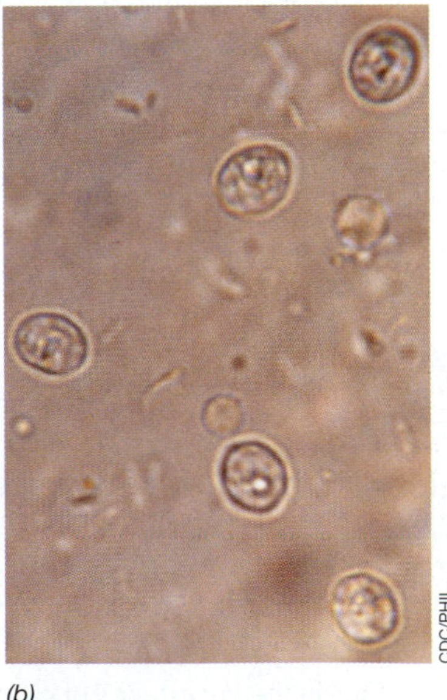

(b)

Figure 36.14 *Cryptosporidium*. *(a)* The arrows point to two of the many intracellular trophozoites imbedded in human gastrointestinal epithelium. The trophozoites are 2–5 μm in diameter. *(b)* The thick-walled oocysts are 3–5 μm in diameter in this fecal sample.

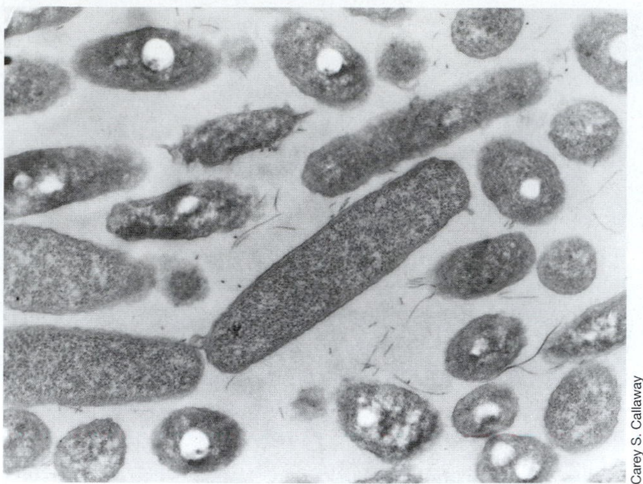

Figure 36.15 ***Legionella pneumophila.*** Transmission electron micrograph of a thin section of *L. pneumophila* cells. Cells are 0.3–0.6 μm in diameter and up to 2 μm in length. The clear areas in the cells are polymers of β-hydroxybutyrate, a carbon storage polymer.

by AIDS, or the very young or old can develop serious complications. In the Milwaukee outbreak, about 4,400 people required hospital care, and several died of complications from the disease, including severe dehydration.

The Milwaukee outbreak highlights the vulnerability of water purification systems, the need for constant water monitoring and surveillance, and the consequences of the failure of a large water supply system. In addition to the toll of human morbidity and mortality, the epidemic cost an estimated $96 million in medical costs and lost productivity.

Laboratory diagnostic methods for cryptosporidiosis include the demonstration of *Cryptosporidium* oocysts in the stool (Figure 36.14*b*). Treatment is unnecessary for those with uncompromised immunity. For individuals undergoing immunosuppressive therapy (for example, prednisone), discontinuation of immunosuppressive drugs is recommended. Immunocompromised individuals should be given supportive therapy such as intravenous fluids and electrolytes.

36.6 MiniReview

Giardiasis and cryptosporidiosis are spread by the chlorine-resistant cysts of *Giardia* and *Cryptosporidium* in drinking water and recreational water contaminated by the feces of infected humans or animals. Infection with either protist can cause acute gastrointestinal illness and may lead to more serious disease in compromised individuals.

- Explain the importance of cysts in the survival, infectivity, and chlorine resistance of both *Giardia* and *Cryptosporidium*.
- Why are protists often associated with waterborne diseases, even in developed countries? Outline steps to reduce their impact.

36.7 Legionellosis (Legionnaires' Disease)

Legionella pneumophila, the bacterium that causes legionellosis, is an important waterborne pathogen normally transmitted in aerosols rather than through drinking or recreational waters.

Biology and Epidemiology

Legionella pneumophila was first discovered as the pathogen that caused an outbreak of pneumonia during an American Legion convention in Philadelphia (USA) in the summer of 1976. *L. pneumophila* is a thin, gram-negative obligately aerobic rod (**Figure 36.15**) with complex nutritional requirements, including an unusually high iron requirement. *L. pneumophila* can be detected by immunofluorescence techniques. The organism can be isolated from terrestrial and aquatic habitats as well as from legionellosis patients.

L. pneumophila is present in small numbers in lakes, streams, and soil. It is relatively resistant to heating and chlorination, so it can spread through water distribution systems. It is commonly found in large numbers in cooling towers and evaporative condensers of large air conditioning systems. The pathogen grows in the water and is disseminated in humidified aerosols. Human infection is by way of airborne droplets, but the infection is not spread from person to person. Further evidence for this is the fact that multiple outbreaks of legionellosis tend to peak in mid- to late summer months when air conditioners are extensively used.

L. pneumophila has also been found in hot water tanks and whirlpool spas where it can grow to high numbers in warm (36–45°C), stagnant water. Epidemiological studies indicate that *L. pneumophila* infections occur at all times of the year, primarily as a result of aerosols generated by heating/cooling systems and common practices such as showering or bathing. Overall, the incidence of reported cases of legionellosis had been about 4–6 cases per million in the United States, but in the last several years the incidence has risen to nearly 8 cases per million. This may be a result of increase in infections or an increase in recognition and reporting. Formerly, up to 90% of actual cases were probably not diagnosed or properly reported (**Figure 36.16**). Prevention of legionellosis can be accomplished by improving the maintenance and design of water-dependent cooling and heating systems and water delivery systems. The pathogen can be eliminated from water supplies by hyperchlorination or by heating to warmer than 63°C.

Pathogenesis

In the body, *L. pneumophila* invades and grows in alveolar macrophages and monocytes as an intracellular parasite. Infections are often asymptomatic or produce a mild cough, sore throat, mild headache, and fever. These mild, self-limiting cases, called *Pontiac fever,* are generally not treated and resolve in 2–5 days. Elderly individuals whose resistance has been previously compromised, however, often acquire more serious infections resulting in pneumonia. Certain serotypes of *L. pneumophila* (more than 10 are known) are strongly

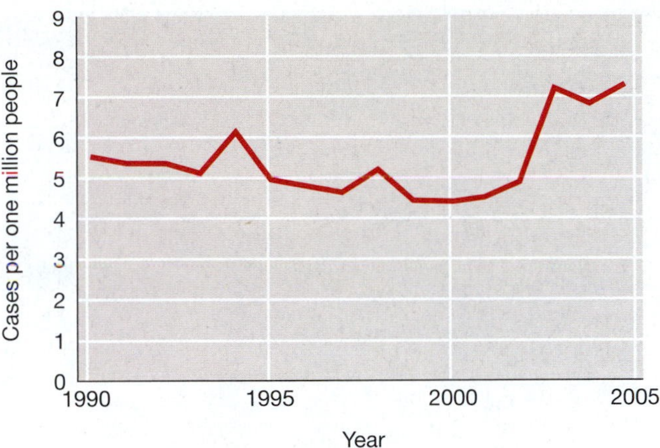

Figure 36.16 Incidence of Legionnaire's Disease in the United States. In 2005, there were 2,301 reported cases, the most reported in a single year. Data were obtained from the Centers for Disease Control and Prevention, Atlanta, GA, USA.

associated with the pneumonic form of the infection. Prior to the onset of pneumonia, intestinal disorders are common, followed by high fever, chills, and muscle aches. These symptoms precede the dry cough and chest and abdominal pains typical of legionellosis. Death occurs in up to 10% of cases and is usually due to respiratory failure.

Diagnosis and Treatment

Clinical detection of *L. pneumophila* is usually done by culture from bronchial washings, pleural fluid, or other body fluids. Serological (antibody) tests are used as retrospective evidence for *Legionella* infection. As an aid in diagnosis, *L. pneumophila* antigens can sometimes be detected in patient urine. *Legionella pneumophila* can be treated with the antibiotics rifampin and erythromycin. Intravenous administration of erythromycin is the treatment of choice.

36.7 MiniReview

Legionella pneumophila is a respiratory pathogen that causes legionellosis and Pontiac fever. *L. pneumophila* grows to high numbers in warm water and is spread via aerosols. The prevalence of legionellosis is increasing and infections are underreported.

- Indicate the source of *Legionella pneumophila*.
- Identify specific measures for control of *Legionella pneumophila*.

36.8 Typhoid Fever and Other Waterborne Diseases

Various bacteria, viruses, and protists can transmit common-source waterborne diseases. These diseases are a significant source of morbidity, especially in developing countries.

Typhoid Fever

On a global scale, probably the most important pathogenic bacteria transmitted by the water route are *Salmonella typhi*, the organism causing typhoid fever, and *Vibrio cholerae*, the organism causing cholera, just discussed (Section 36.7).

Although *S. typhi* may also be transmitted by contaminated food (∞ Section 37.7) and by direct contact from infected individuals, the most common and serious means of transmission worldwide is through water. Fortunately, typhoid fever has been virtually eliminated in developed countries, primarily due to effective water treatment procedures. In the United States, there are fewer than 400 cases in most years, but, as was described in Section 36.1, typhoid fever was a major public health threat before drinking water was routinely filtered and chlorinated (Figure 36.2). However, breakdown of water treatment methods, contamination of water during floods, earthquakes, and other disasters, or cross-contamination of water supply pipes from leaking sewer lines can propagate epidemics of typhoid fever, even in developed countries.

Viruses

Viruses can also be transmitted in water and cause human disease. Quite commonly, enteroviruses such as poliovirus, norovirus, and hepatitis A virus are shed into the water in fecal material. The most serious of these is poliovirus, but wild poliovirus has been eliminated from the Western Hemisphere and is endemic only in Nigeria, Afghanistan, Pakistan, and India. Although viruses can survive in water for relatively long periods, they are inactivated by disinfection with agents such as chlorine. The maintenance of 0.6 parts per million (ppm) of chlorine in water (Section 36.3) ensures the neutralization of viruses in the water supply.

Amoebiasis

Certain amoebas inhabit the tissues of humans and other vertebrates, usually in the oral cavity or intestinal tract, and some of these are pathogenic. We discussed the general properties of amoeboid protists in Section 18.12.

Worldwide, *Entamoeba histolytica* is a common pathogenic protist transmitted to humans, primarily by contaminated water and occasionally through contaminated food (**Figure 36.17**). *E. histolytica* is an anaerobic amoeba; the trophozoites lack mitochondria. Like *Giardia*, the trophozoites of *E. histolytica* produce cysts. Cysts ingested by humans germinate in the intestine, where amoebic cells grow both on and in intestinal mucosal cells. Many infections are asymptomatic, but continued growth may lead to invasion and ulceration of the intestinal mucosa, causing diarrhea and severe intestinal cramps. With further growth the amoebae can invade the intestinal wall, a condition called *dysentery*, characterized by intestinal inflammation, fever, and the passage of intestinal exudates, including blood and mucus.

If not treated, invasive trophozoites of *E. histolytica* can invade the liver and occasionally the lung and brain. Growth

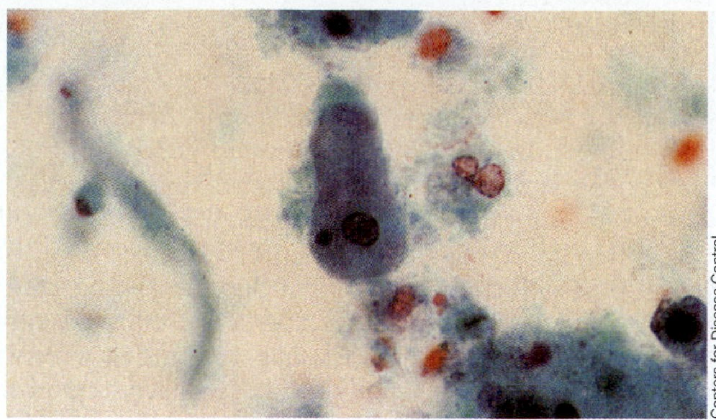

Figure 36.17 The trophozoite of *Entamoeba histolytica*, the amoeba that causes amoebiasis. Note the discrete, darkly stained nucleus. The small red structures are red blood cells. The trophozoites range from 12–60 μm in length.

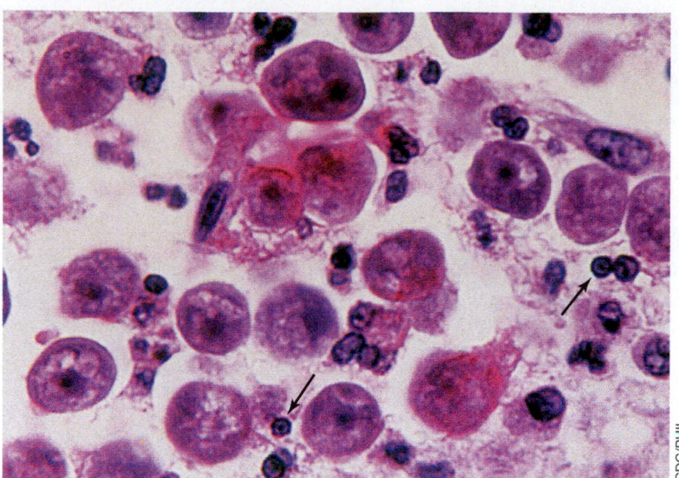

Figure 36.18 Trophozoites of *Naegleria fowleri* in brain tissue. This amoeba causes meningoencephalitis. Oval to round and amoeboid (irregularly shaped) trophozoites (arrows) are present as dark-stained structures with densely stained nuclei. There is extensive destruction of the surrounding brain tissue. Individual trophozoites are 10–35 μm long.

in these tissues can cause severe abscesses and death. Worldwide, up to 100,000 individuals die each year from invasive amoebic dysentery. The disease is extremely common in tropical and subtropical countries worldwide, with at least 50 million people developing symptomatic diarrhea annually and up to tenfold more having asymptomatic disease. In the United States, there are several hundred cases per year, with most occurring near international borders in the Southwest.

E. histolytica amoebiasis can be treated with the drugs dehydroemetine for invasive disease and diloxanide furoate for certain asymptomatic cases, as in immunocompromised individuals, but amoebicidal drugs are not universally effective. Spontaneous cures do occur, suggesting that the host immune system plays a role in ending the infection. However, protective immunity is not an outcome of primary infection, and reinfection is common. The disease is kept at very low incidence in regions that practice adequate sewage treatment. Amoebic infestation due to exposure to improperly treated sewage and the use of untreated surface waters for drinking purposes are the usual causes of amoebiasis.

Demonstration of *E. histolytica* cysts in the stool, trophozoites in tissue, or the positive results for antibodies to *E. histolytica* in the blood from an ELISA (enzyme-linked immunosorbent assay) are used for the laboratory diagnosis of amoebiasis.

Naegleria fowleri can also cause amoebiasis, but in a very different form. *N. fowleri* is a free-living amoeba found in soil and in water runoff. *N. fowleri* infections usually result from swimming or bathing in warm, soil-contaminated water sources such as hot springs or lakes and streams in the summer. This free-living amoeba enters the body through the nose and burrows directly into the brain. Here, the organism propagates, causing extensive hemorrhage and brain damage (**Figure 36.18**). This condition is called **meningoencephalitis**. Death usually results within a week. In the past 18 years, there have been 30 outbreaks of meningoencephalitis from recreational waters in the United States. From 1999 to 2003, there were 12 outbreaks, each a single individual who was infected by swimming or wading in a lake, pond, or stream in summer. All cases resulted in death.

Prevention can be accomplished by avoiding swimming in shallow, warm, fresh water in summer, such as farm ponds and shallow lakes and rivers. Swimmers are advised to avoid stirring up bottom sediments, the natural habitat of the pathogen.

Diagnosis of *N. fowleri* infection requires observation of the amoebae in the cerebrospinal fluid. If a definitive diagnosis can be done quickly, the drug amphotericin B is used to treat infections.

36.8 MiniReview

Typhoid fever, viral infections, and amoebiasis are important waterborne diseases. Waterborne typhoid fever and viral illnesses, common diseases in developing countries, can be controlled by effective water treatment. Amoebic dysentery caused by *Entamoeba histolytica* affects millions of people worldwide. Meningoencephalitis is a rare but usually fatal condition caused by *Naegleria fowleri* amoebiasis.

- Explain the impact of effective water hygiene on the spread of human diseases such as typhoid fever and polio spread by fecal contamination of water supplies.
- Describe public health measures that could be used to eliminate or reduce the number of cases of amoebiasis due to *Entamoeba histolytica* or meningoencephalitis due to *Naegleria fowleri*.

Review of Key Terms

Aerobic secondary wastewater treatment digestive reactions carried out by microorganisms under aerobic conditions to treat wastewater containing low levels of organic materials

Anoxic secondary wastewater treatment digestive and fermentative reactions carried out by microorganisms under anoxic conditions to treat wastewater containing high levels of insoluble organic materials

Biochemical oxygen demand (BOD) the relative amount of dissolved oxygen consumed by microorganisms for complete oxidation of organic and inorganic material in a water sample

Chloramine a disinfectant chemical manufactured on site by combining chlorine and ammonia at precise ratios

Chlorine a chemical used in its gaseous state to disinfect water; a residual level is maintained throughout the distribution system

Clarifier (coagulation basin) a reservoir in which the suspended solids of raw water are coagulated and removed

Coagulation the formation of large insoluble particles from much smaller, colloidal particles by the addition of aluminum sulfate and anionic polymers

Coliforms facultatively aerobic, gram-negative, non-spore-forming, lactose-fermenting bacteria

Cyst an infectious form of a protist that is encased in a thick-walled, chemically and physically resistant coating

Distribution system water pipes, storage reservoirs, tanks, and other equipment used to deliver drinking water to consumers or store it before delivery

Effluent water treated wastewater discharged from a wastewater treatment facility

Filtration the removal of suspended particles from water by passing it through one or more permeable membranes or media (e.g., sand, anthracite, or diatomaceous earth)

Finished water water delivered to the distribution system after treatment

Flocculation the water-treatment process after coagulation that uses gentle stirring to cause suspended particles to form larger, aggregated masses (floc)

Meningoencephalitis invasion, inflammation, and destruction of brain tissue by the amoeba *Naegleria fowleri* or another pathogen

Polymer in water purification, a chemical in liquid form used as a coagulant in the clarification process to flocculate a suspension

Potable drinkable; safe for human consumption

Primary wastewater treatment physical separation of wastewater contaminants, usually by separation and settling

Raw water surface water or groundwater that has not been treated in any way (also called untreated water)

Sediment soil, sand, minerals, and other large particles found in raw water

Sewage liquid effluents contaminated with human or animal fecal material

Suspended solid small particle of solid pollutant that resists separation by ordinary physical means

Tertiary wastewater treatment physicochemical processing of wastewater to reduce levels of inorganic nutrients

Turbidity a measurement of suspended solids in water

Untreated water surface water or groundwater that has not been treated in any way (also called raw water)

Wastewater liquid derived from domestic sewage or industrial sources which cannot be discarded in untreated form into lakes or streams

Review Questions

1. Define the term *coliform* and explain the coliform test. Why is the coliform test used to assess the purity of drinking water (Section 36.1)?
2. Trace wastewater treatment in a typical plant, from incoming water to release. What is the overall reduction in the BOD for typical household wastewater? What is the overall reduction in the BOD for typical industrial wastewater (Section 36.2)?
3. Identify (stepwise) the process of purifying drinking water. What important contaminants are targeted by each step in the process (Section 36.3)?
4. Why are common sources of infection, such as contaminated water sources, a significant threat to public health (Section 36.4)?
5. Why are antibiotics ineffective for the treatment of cholera? What methods are useful for treating cholera victims (Section 36.5)?
6. Giardiasis and cyptosporidiosis remain significant public health problems even in areas with stringent water-quality standards. Explain (Section 36.6).
7. Describe the main features of legionellosis. What distinguishes this disease from other waterborne diseases (Section 36.7)?
8. Indicate the methods that are used to control salmonellosis, norovirus infection, and amoebiasis in water systems in developed countries (Section 36.8).

Application Questions

1. Why is reduction in the BOD of wastewater a primary goal of wastewater treatment? What are the consequences of releasing wastewater with a high BOD into local water sources such as lakes or streams?
2. In the United States, the federal government has defined, by law, a strict set of drinking water standards. Federal recreational water standards, however, are recommendations, and local governments can set more or less stringent standards. Explain why recreational water standards are flexible and devise recreational water standards for the area in which you live.
3. Worldwide, we are in the midst of the seventh cholera pandemic, and the eighth pandemic may be starting. Using sources such as the World Health Organization and the Centers for Disease Control and Prevention, define the status of the current pandemic with regard to its geographic distribution, endemic areas, and most recent outbreaks. Comment on methods that could be used to decrease the spread of cholera and control the annual epidemic outbreaks that occur in endemic areas. Can cholera be eradicated?
4. As a visitor to a country in which cholera is an endemic disease, what specific steps would you take to reduce your risk of cholera exposure? Will these precautions also prevent you from contracting other waterborne diseases? Which ones? Identify waterborne diseases for which your precautions may not prevent infection.
5. Why are surface waters contaminated with the cysts of various protists? What steps might public health officials take to remedy this problem?
6. Discuss the wastewater and drinking water treatment schemes that must be in place to control such diseases as typhoid fever. Is it possible to eliminate *Salmonella typhi,* as has been effectively done for poliovirus? Would it be possible to eliminate *Naegleria fowleri* or *Entamoeba histolytica*? Explain.
7. Noroviruses are frequently causes of acute gastrointestinal disease outbreaks. How can control of these viruses be accomplished?

37

Food Preservation and Foodborne Microbial Diseases

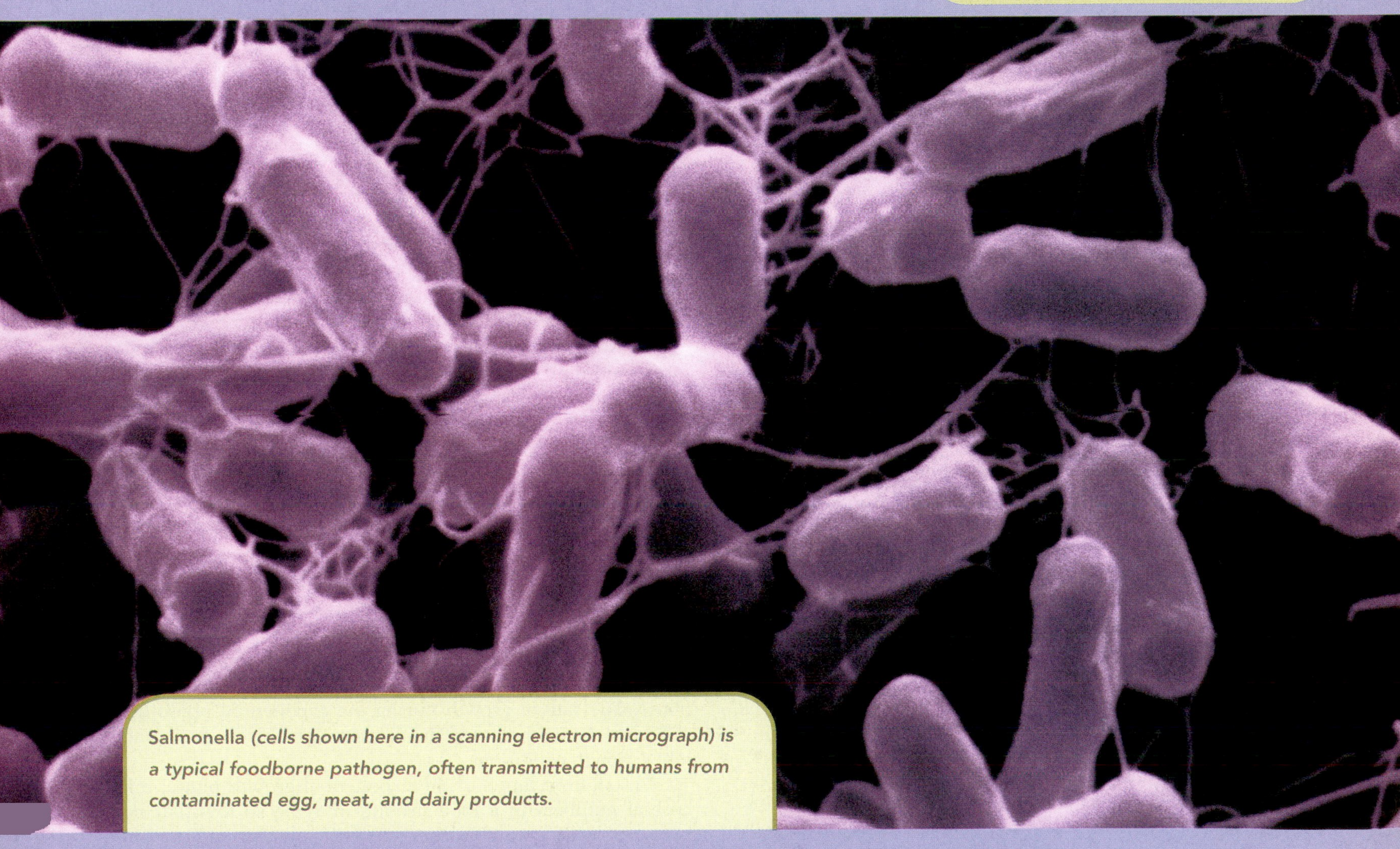

Salmonella (cells shown here in a scanning electron micrograph) is a typical foodborne pathogen, often transmitted to humans from contaminated egg, meat, and dairy products.

Humans are constantly exposed to microorganisms in air, water, and food. Fresh foods, prepared foods, and even preserved foods are often contaminated with spoilage microorganisms as well as a few pathogenic microorganisms. On the other hand, microbial activity is important for the production of some foods. For example, dairy products such as cheese, buttermilk, sour cream, and yogurt are all produced by microbial fermentation. Sauerkraut is a fermented vegetable food. Certain sausages, pates, and liver spreads are produced by microbial fermentation. Cider vinegar is produced by the activities of the acetic acid bacteria, and alcoholic beverages are produced by fermentation. We discussed the use of microorganisms to produce various fermented foods on an industrial scale in Chapter 25.

Here we examine food preservation methods that use microbial processes to limit unwanted microbial growth and food spoilage. Next, we discuss microbial products or microorganisms that cause food poisoning and food infection.

I FOOD PRESERVATION AND MICROBIAL GROWTH

Microorganisms are important spoilage agents in foods, causing food shortages and economic loss. Various methods, some involving desirable microbial growth, are used for controlling spoilage organisms.

37.1 Microbial Growth and Food Spoilage

Microorganisms, including a few human pathogens, colonize and grow on common foods. Most foods provide a suitable medium for the growth of various microorganisms, and microbial growth usually reduces food quality and availability.

Food Spoilage

Food spoilage is any change in the appearance, smell, or taste of a food product that makes it unacceptable to the consumer. Spoiled food may still be safe to eat, but is generally regarded as unpalatable and will not be purchased or readily consumed. Food spoilage causes losses to producers, distributors, and consumers in the form of reduced quality and quantity and higher prices.

Foods consist of organic materials and can be looked at as nutrients for the growth of chemoorganotrophic bacteria. The physical and chemical characteristics of the food determine its degree of susceptibility to microbial activity. With respect to spoilage, foods are classified into three major categories: (1) **perishable food**, including many fresh food items; (2) **semiperishable food**, such as potatoes and nuts; and (3) **stable** or **nonperishable food**, such as flour and sugar (**Table 37.1**).

Table 37.1 Food classification by storage potential

Food classification	*Examples*
Perishable	Meats, fish, poultry, eggs, milk, most fruits and vegetables
Semiperishable	Potatoes, some apples, and nuts
Nonperishable	Sugar, flour, rice, and dry beans

These food categories differ greatly with regard to their *moisture content*, which is related to water activity, a_w (Section 6.16). Water activity is a measure of the availability of water for use in metabolic processes. Nonperishable foods have low water activity and can generally be stored for considerable lengths of time without spoilage. Perishable and semiperishable foods, by contrast, typically have higher water activities. Thus, these foods must be stored under conditions that inhibit microbial growth.

Fresh foods are spoiled by a variety of both bacteria and fungi. The chemical properties of foods vary widely, and each type of fresh food is typically colonized and spoiled by a relatively small group of microorganisms. The spoilage organisms are those that gain access to the food and can use the available nutrients (**Table 37.2**).

For example, enteric bacteria such as *Salmonella*, *Shigella*, and *Escherichia*, all potential pathogens sometimes found in the gut of animals, often contaminate and spoil meat. At slaughter, intestinal contents containing live bacteria can leak, contaminating meat products. These organisms can also occasionally contaminate produce through fecal contamination of water supplies. Likewise, lactic acid bacteria, the most common microorganisms in dairy products, are the major spoilers of milk and milk products. *Pseudomonas* species are found in both soil and animals and cause the spoilage of fresh foods of all types.

Growth of Microorganisms in Foods

Microbial growth in foods follows the normal pattern for bacterial growth (Section 6.7). The lag phase may be of variable duration in a food, depending on the properties of both the contaminating microorganism and the food. The time required for the population density to reach a significant level in a given food product depends on both the size of the initial inoculum and the rate of growth during the exponential phase. The rate of growth during the exponential phase depends on the temperature, the nutrient value of the food, and other growth conditions.

Throughout much of the exponential growth phase, microbial numbers in a food product may be so low that no measurable effect can be observed, with only the last few population doublings leading to observable spoilage. Thus, for much of the period of microbial growth in a food there is no visible or easily detectable change in food quality; spoilage is usually observed only when the microbial population density is high.

Table 37.2 Microbial spoilage of fresh food[a]

Food product	Type of microorganism	Common spoilage organisms, by genus
Fruits and vegetables	Bacteria	*Erwinia*, ***Pseudomonas*, *Corynebacterium*** (mainly vegetable pathogens; rarely spoil fruit)
	Fungi	*Aspergillus, Botrytis, Geotrichum, Rhizopus, Penicillium, Cladosporium, Alternaria, Phytophthora*, various yeasts
Fresh meat, poultry, and seafood	Bacteria	*Acinetobacter, Aeromonas*, ***Pseudomonas***, *Micrococcus, Achromobacter, Flavobacterium*, ***Proteus, Salmonella, Escherichia, Campylobacter, Listeria***
	Fungi	*Cladosporium, Mucor, Rhizopus, Penicillium, Geotrichum*, ***Sporotrichium, Candida***, *Torula, Rhodotorula*
Milk	Bacteria	***Streptococcus***, *Leuconostoc, Lactococcus, Lactobacillus*, ***Pseudomonas, Proteus***
High-sugar foods	Bacteria	***Clostridium, Bacillus***, *Flavobacterium*
	Fungi	*Saccharomyces, Torula, Penicillium*

[a]The organisms listed are the most commonly observed spoilage agents of fresh, perishable foods. Genera in boldface include species that are possible human pathogens.

37.1 MiniReview

Most foods will spoil due to the growth of contaminating microorganisms. The potential for microbial food spoilage depends on the nutrient value and water content of the food. Microbial spoilage limits the shelf life of perishable and semi-perishable foods. Various microorganisms induce spoilage, and some food spoilage microorganisms are also pathogens.

- List the major categories of food with respect to water availability. Suggest storage conditions that inhibit spoilage microorganisms for each food type.
- Identify at least three bacterial genera that cause both food spoilage and human disease.

37.2 Food Preservation

Food storage and preservation methods slow the growth of microorganisms that produce spoilage and foodborne disease.

Cold

One of the most crucial factors affecting microbial growth is temperature (Section 6.12). In general, a lower storage temperature results in less microbial growth and slower spoilage. However, a number of psychrotolerant (cold-tolerant) microorganisms can grow, albeit slowly, at refrigerator temperatures (3–5°C). Therefore, storage of perishable food products for long periods of time (more than several days) is possible only at temperatures below freezing. However, because freezing and subsequent thawing alters the physical structure, freezing is not an acceptable preservation method for many fresh foods, but is widely used for the preservation of meats and many fruits and vegetables. Freezers providing a temperature of −20°C are most commonly used. At −20°C, storage for weeks or months is possible, but microorganisms can still grow in pockets of liquid water trapped within the frozen mass. For long-term storage, temperatures of −80°C (the temperature of solid CO_2, "dry ice") are necessary. Maintenance of such low temperatures is expensive and consequently is not used for routine food storage.

Pickling and Acidity

Another factor affecting microbial growth in food is pH. Foods vary somewhat in pH, but most are neutral or acidic. Microorganisms differ in their ability to grow under acidic conditions, but conditions of pH 5 or less inhibit the growth of most spoilage organisms. Therefore, weak acids are often used in food preservation in a process called **pickling**. Vinegar, a dilute acetic acid fermentation product of the acetic acid bacteria, is usually added in the pickling process. Pickling methods usually mix the vinegar with large amounts of salt or sugar to decrease water availability (see below) and further inhibit microbial growth. Common pickled foods include cucumbers (sweet, sour, and dill pickles), peppers, meats, fish, and fruits.

Drying and Dehydration

As previously mentioned, **water activity**, or a_w, is a measure of the availability of water for use by microorganisms in metabolic processes. The a_w of pure water is 1.00; the water molecules are loosely ordered and rearrange freely. When solute is added, the a_w is decreased. As water molecules reorder around the solute, the free rearrangement of the solute-bound water molecules becomes energetically unfavorable. The microbial cells must then compete with solute for the reduced amount of free water. In general, bacteria are poor competitors for the remaining free water, but fungi are good competitors. In practice this means that high concentrations of solutes such as sugars or salts, which greatly reduce a_w, typically inhibit bacterial growth. For example, most bacteria are inhibited by a concentration of 7.5% NaCl (a_w of 0.957), with the exception of some of the gram-positive cocci, such as *Staphylococcus* species. On the other hand, molds compete well for free water under conditions of low a_w and often grow well in high-sugar foods such as syrups.

(a)

(b)

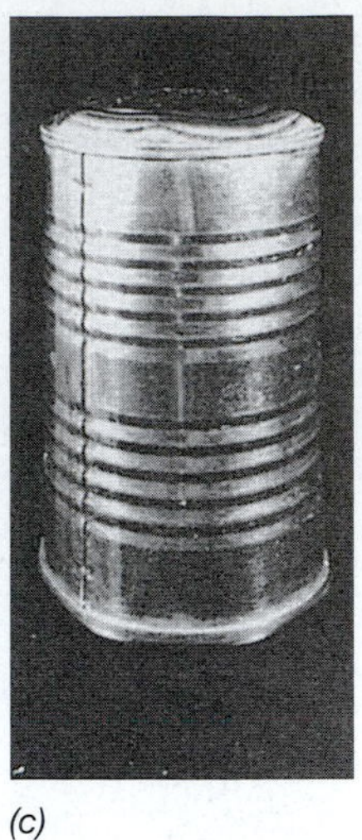
(c)

(d)

T. D. Brock

Figure 37.1 Changes in sealed tin cans as a result of microbial spoilage. *(a)* Normal can; the top of the can is pulled slightly in due to the negative pressure (vacuum) inside. *(b)* Swelling resulting from minimal gas production. The top of the can bulges slightly. *(c)* Severe swelling due to extensive gas production. *(d)* The can shown in *(c)* was dropped, and the gas pressure resulted in a violent explosion, tearing the lid apart.

A number of commercially important foods are preserved by the addition of salt or sugar. Sugar is used mainly in fruits (jams, jellies, and preserves). Salted products are primarily meats and fish. Sausage and ham are preserved using various curing salts, including NaCl. Some meats also undergo a smoking process. The preserved meat products vary widely in a_w, depending on how much salt is added and how much the meat has been dried. Some cured meat products such as country ham or jerky can be kept at room temperature for extended periods of time. Others with higher a_w require refrigeration for long-term storage

Microbial growth in foods can also be controlled by lowering the water content by drying. Drying is used to preserve highly perishable foods such as meat, fish, milk, vegetables, fruit, and eggs. The least damaging physical method used to dry foods is the process of **lyophilization (freeze-drying)**, in which foods are frozen and water is then removed under vacuum. This method is very expensive, however, and is used mainly for specialized applications such as preparation of military rations that may need to be stored for long periods under adverse conditions.

Spray drying is the process of spraying, or atomizing, liquids such as milk in a heated atmosphere. The atomization produces small droplets, increasing the surface area-to-volume ratio of the liquid, promoting rapid drying without destroying the food. This technology is widely used in the production of powdered milk, certain concentrated liquid dairy products such as evaporated milk, and concentrated food ingredients such as liquid flavorings.

Heating

Heat is used to reduce the bacterial load of a food product or to actually sterilize it and is especially useful for the preservation of liquids and wet foods. *Pasteurization,* a process in which liquids are heated to a specified temperature for a precise time, was described in Section 27.1. Pasteurization does not sterilize liquids, but reduces the bacterial load of both spoilage organisms and pathogens and significantly extends the shelf life of the liquid. **Canning** is a process in which food is sealed in a container such as a can or glass jar and then heated. In theory, canning should sterilize the food product, although this is not true 100% of the time (see next paragraph). However, when properly sealed and heated, the average can of food should remain stable and unspoiled indefinitely at any temperature.

The temperature–time relationships for canning depend on the type of food, its pH, the size of the container, and the consistency or density of the food. Because heat must completely penetrate the food within the can, effective heating times must be longer for large cans or very dense foods. Acid foods can often be canned effectively by heating just to boiling, 100°C, whereas nonacid foods must be heated to autoclave temperatures (121°C). For some foods in large containers, times of 20–50 minutes must be used. Heating times long enough to guarantee absolute sterility of every can would make most foods unpalatable and could also degrade nutritional value. Even properly canned foods, therefore, may not be sterile. The process used for commercial canning, called *retort canning,* employs equipment similar to an autoclave to apply steam under pressure (∞ Section 27.1).

If live microorganisms remain in a can, growth of organisms can produce extensive amounts of gas and build pressure, resulting in bulges or, in severe cases, explosion (**Figure 37.1**). The environment inside a can is anoxic, and some of the anaerobic bacteria that grow in canned foods are toxin producers of the genus *Clostridium* (Section 37.6). Food from a bulging can, therefore, should never be eaten. On the other hand, the lack of obvious gas production is not an absolute guarantee that canned food is safe to consume.

Aseptic Food Processing

Several foods in the United States, and many more in Asia and Europe, are now prepared and packaged under aseptic conditions. Foods processed and packaged aseptically should be sterile and can be stored at room temperature for months or

Table 37.3 Chemical food preservatives

Chemical	*Foods*
Sodium or calcium propionate	Bread
Sodium benzoate	Carbonated beverages, fruit, fruit juices, pickles, margarine, preserves
Sorbic acid	Citrus products, cheese, pickles, salads
Sulfur dioxide, sulfites, bisulfites	Dried fruits and vegetables, wine
Formaldehyde (from food-smoking process)	Meat, fish
Ethylene and propylene oxides	Spices, dried fruits, nuts
Sodium nitrite	Smoked ham, bacon

longer without spoilage. Aseptic processing involves flash heating or cooking foods to sterility and then packaging them in aseptic containers, usually cardboard cartons lined with foil and plastic. The process may require "clean room" conditions similar to those in a hospital operating room. For example, incoming room air needs to be filtered to limit contamination from spores and bacteria in the atmosphere. Special equipment is required to flash-heat and deliver the product aseptically into sterile packaging materials.

In the United States, liquids such as fruit juices (juice boxes) and milk substitutes are often processed in this way. In many European countries, milk products are flash-heated to 133°C and packaged aseptically. Perishable food products prepared aseptically can be stored at room temperature for at least 6 months. This developing technology increases product shelf life almost indefinitely and eliminates the need for refrigeration for many products. However, the equipment and processing plants necessary to achieve and maintain aseptic food processing are very expensive.

Chemical Preservation

Over 3,000 different compounds are used as food additives. These chemical additives are classified by the U.S. Food and Drug Administration as "generally recognized as safe" (GRAS) and find wide application in the food industry for enhancing or preserving texture, color, freshness, or flavor. A small number of these compounds are used to control microbial growth in food (**Table 37.3**). Many of these microbial growth inhibitors, such as sodium propionate and sodium benzoate, have been used for many years with no evidence of human toxicity. Others, such as nitrites (a carcinogen precursor), and ethylene oxide and propylene oxide (mutagens), are more controversial food additives because studies suggest that these compounds may adversely affect human health. The use of spoilage-retarding additives, however, significantly extends the useful shelf life of finished foods. Chemical food additives contribute significantly to an increase in quantity and in the perceived quality of available food items.

Because of time and cost required for testing of any chemical proposed as a food preservative or additive, it is unlikely that many new chemicals will be added to the list of safe and approved chemical food preservatives listed in Table 37.3.

Table 37.4 Irradiated foods by category, dose, and purpose

Food category	*Dose[a] (kGy)*	*Purpose*
Fresh meat: ground beef	4.50	Reduce bacterial pathogens
Herbs, spices, enzymes, and flavorings	30.00	Sterilize
Pork	1.00	Reduce *Trichinella spiralis* protist
Meats used in NASA[b] space flight program	44.0	Sterilize
Poultry	4.50	Reduce bacterial pathogens
Wheat flour	0.50	Inhibit mold
White flour	0.15	

[a]The highest value for recommended doses is given. One kGy (kiloGray) is one thousand grays. One gray, an SI unit, is one joule of radiation absorbed by one kilogram of matter and is also equivalent to 100 rad.

[b]National Aeronautics and Space Administration, USA.

Irradiation

Irradiation of food with ionizing irradiation is an effective method for reducing contamination by bacteria, fungi, and even insects (∞ Section 27.2). **Table 37.4** lists foods for which radiation treatment has been approved. Foods such as spices are routinely irradiated. In the United States, fresh meat products such as hamburger and poultry can now be irradiated to limit contamination by *Escherichia coli* O157:H7 (hamburger) and other pathogens such as *Campylobacter jejuni* (poultry).

For food irradiation, gamma rays generated from ^{60}Co or ^{137}Cs sources, or from high-energy electrons produced by linear accelerators, are used as radiation sources. Alternatively, beta rays can be generated from an electron gun, analogous to but significantly more powerful than the electron beam generated by the cathode ray gun used in an older television. In addition, X-rays can be generated with electron beams focused on metal foil. X-rays have much greater penetrating power than beta rays and are therefore useful for treating large-volume food preparations. Beta ray and X-ray sources can be switched on and off at will and do not require a radioactive source.

Irradiated food products receive a controlled radiation dose. This dose varies considerably for each food category and purpose. For example, a dose of 44 kiloGrays (kGys) is used to sterilize meat products used on United States NASA space flights and is nearly ten times higher than the dose of 4.5 kGys used for control of pathogens in hamburger (Table 37.4). In the United States, a consumer product information label must be affixed to foods that are irradiated (**Figure 37.2**).

Figure 37.2 The radura, the international symbol for radiation. Packaging of foods treated with radiation must be labeled with the radura as well as the statement "treated by irradiation" or "treated with radiation."

37.2 MiniReview

Microbial growth in foods must be limited to reduce spoilage and prevent disease. Foods vary considerably in their sensitivity to microbial growth, depending on nutrient content, water availability, and pH. The growth of microorganisms in perishable foods can be controlled by refrigeration, freezing, canning, pickling, dehydration, chemical preservation, or irradiation.

- Identify food spoilage microorganisms that are also pathogens.
- Identify physical and chemical methods used for food preservation. How does each method limit growth of microorganisms?

37.3 Fermented Foods

Many common foods and beverages are preserved, produced, or enhanced through the direct actions of microorganisms. Microbial processes can produce significant alterations in raw products, and the product is called a *fermented food*. **Fermentation** is the anaerobic catabolism of organic compounds, generally carbohydrates, in the absence of an external electron acceptor (∞ Section 5.10). Important bacteria in the fermented foods industry include the lactic acid bacteria, the acetic acid bacteria, and the propionic acid bacteria (**Table 37.5**). These bacteria do not grow below about pH 3.5, so food fermentation is a self-limiting process.

One of the most common fermented foods is yeast bread, in which the fermentation of simple sugars and grain carbohydrates by the yeast *Saccharomyces cerevisiae* (Table 37.5) produces carbon dioxide, raising the bread and producing the holes in the finished loaf (**Figure 37.3**). Fermented beverage products such as wine, beer, and whiskey were discussed in Sections 25.10 and 25.11 as industrial applications of microbial processes.

Table 37.5 Fermented foods and primary fermentation microorganisms[a]

Food category	*Primary fermenting microorganism*
Dairy foods	
Cheeses	*Lactococcus*
	Lactobacillus
	Streptococcus thermophilus
Fermented milk products	
Buttermilk	*Lactococcus*
Sour cream	*Lactococcus*
Yogurt	*Lactobacillus*
	Streptococcus thermophilus
Alcoholic beverages	*Zymomonas*
	Saccharomyces[b]
Yeast breads	*Saccharomyces cerevisiae*[c]
Meat products	
Dry sausages (pepperoni, salami)	*Pediococcus*
and semidry sausages	*Lactobacillus*
(summer sausage, bologna)	*Micrococcus*
	Staphylococcus
Vegetables	
Cabbage (sauerkraut)	*Leuconostoc*
	Lactobacillus
Cucumbers (pickles)	Lactic acid bacteria
Soy sauce	*Aspergillus*
	Tetragenococcus halophilus
	Yeasts

[a]Unless otherwise noted, these are all members of the lactic acid bacteria. *Zymomonas* is related to *Pseudomonas*.
[b]Yeast. A variety of *Saccharomyces* species are used in alcohol fermentations.
[c]Baker's yeast.

Dairy Products

Fermented dairy products were originally developed as methods to preserve milk, an economically important fresh food that normally undergoes rapid spoilage (Section 37.2). Dairy products include cheese and other fermented milk products such as yogurt, buttermilk, and sour cream (Figure 37.3). Milk contains the disaccharide lactose. Lactose can be hydrolyzed by the enzyme lactase into glucose and galactose. These monosaccharides are fermented to the final product of lactic acid (∞ Figure 5.15) by the lactic acid bacteria (Table 37.5). This fermentation reaction produces a significant decrease in pH from the neutral or slightly basic pH of raw milk to a pH of less than 5.3 in cheeses and less than 4.6 in other fermented milk products.

Starter cultures of lactic acid bacteria are introduced into raw milk, and fermentation proceeds for a time period depending on the desired product (Table 37.5). For some cheeses, a second inoculum may be introduced to produce a second

Figure 37.3 Fermented foods. Bread, sausage meats, cheeses, other dairy products, and vegetables are all food products that are produced or enhanced by fermentation reactions catalyzed by microorganisms.

fermentation. For example, following lactic acid fermentation, Swiss-style (Emmentaler) cheeses are reinoculated with *Propionibacterium*. The secondary fermentation catabolizes lactic acid to propionic acid, acetic acid, and CO_2. The carbon dioxide produces the large holes (eyes) that characterize Swiss-type cheeses. Secondary fermentations with *Lactobacillus* and the mold *Penicillium roqueforti* produce the blue veins and distinctive taste and aroma of blue cheese. Each cheese type is produced under carefully controlled conditions. Time of fermentation, temperature, the extent of aging, and the types of fermenting microorganisms must be rigidly controlled to ensure a distinctive and reproducible product.

Meat Products

Meat products fall into several categories. *Sausages* are generally made from pork, beef, or poultry meats. The most common are the dry sausages, such as salami and pepperoni, and the semidry sausages such as bolognas and summer sausages (Figure 37.3).

Sausages are made using a uniformly blended mixture of meat, salt, and seasonings. A starter culture of lactic acid bacteria is added, and fermentation proceeds to reduce the pH of the mixture to below 5. After fermentation, sausages are often smoked and dried to a moisture content of about 30%. After final processing, dry sausages can be held at room temperature for extended periods of time. The semidry sausages have a final moisture content of about 50% and are less resistant to spoilage, so they are generally refrigerated.

Fish, often mixed with rice, shrimp, and spices, are also fermented to make fish pastes and fish-flavored products.

Vegetables and Vegetable Products

The variety of specialty fermented vegetable foods is practically endless. The most economically important fermented vegetable foods are sauerkraut (fermented cabbage) and some types of pickles (fermented cucumbers). Olives, onions, tomatoes, peppers, and many fruits are also fermented.

Vegetables are often fermented in salt brine to enhance preservation and flavor. The salt also helps prevent contamination with unwanted organisms, the desired fermentative organisms being salt-tolerant. Fermentation may also improve digestibility by breaking down plant tissues. For example, fermented legume products (peas, beans, lentils) have a marked reduction in the flatulence-producing oligosaccharides that characterize fresh legumes.

Soy sauce is a complex fermentation product made by fermentation of soybeans and wheat. Cooked soybeans are mixed with an enzyme preparation derived from a culture of *Aspergillus* grown on rice or other cereals (Table 37.5). This starter culture is spread on a wheat–soybean mixture and is then grown for 2–3 days. This preparation is known as koji and is mixed with brine (17–19% NaCl) and the fermentation allowed to proceed for 2–4 months or more. Various microorganisms, including *Lactobacillus* and *Pediococcus* and several yeasts and molds, produce fermentation products that contribute to the desirable characteristics of the final product. After fermentation, the liquid sauce is decanted, filtered, pasteurized, and bottled as soy sauce.

37.3 MiniReview

Microbial fermentation is used for preparing, preserving, and enhancing foods including breads, dairy products, meats, and vegetables.

- Identify important dairy, meat, and vegetable products that are produced or enhanced by microbial fermentation processes.
- Identify the microbial group or groups that are most important for food fermentations.

II MICROBIAL SAMPLING AND FOOD POISONING

Inadequate preservation and decontamination of food may allow the growth of pathogens, resulting in foodborne diseases with significant morbidity and mortality. Like waterborne diseases, foodborne illnesses are common-source diseases. A single contaminated food source from a food-processing plant or a restaurant may affect a large number of people. Each year in the United States, nearly 25,000 estimated foodborne disease outbreaks cause an estimated 13 million foodborne illnesses. Most outbreaks are due to improper food handling and preparation at the consumer level and affect small numbers of individuals, usually in the home. Occasionally, there are outbreaks that affect large numbers of individuals caused by breakdowns in safe food handling and preparation at food-processing and distribution plants. Most foodborne outbreaks and diseases remain unreported because the connection between food and illness is not made.

Table 37.6 Annual foodborne disease estimates for the United States[a]

Organism	*Disease*[b]	*Number per year*	*Foods*
Bacteria			
Bacillus cereus	FP and FI	27,000	Rice and starchy foods, high-sugar foods, meats, gravies, pudding, dry milk
Campylobacter jejuni	FI	1,963,000	Poultry, dairy
Clostridium perfringens	FP and FI	248,000	Cooked and reheated meats and meat products
Escherichia coli O157:H7	FI	63,000	Meat, especially ground meat
Other enteropathogenic *Escherichia coli*	FI	110,000	Meat, especially ground meat
Listeria monocytogenes	FI	2,500	Meat and dairy
Salmonella spp.	FI	1,340,000	Poultry, meat, dairy, eggs
Staphylococcus aureus	FP	185,000	Meat, desserts
Streptococcus spp.	FI	50,000	Dairy, meat
Yersinia enterocolitica	FI	87,000	Pork, milk
All other bacteria	FP and FI	102,000	
Total bacteria		***4,177,500***	
Protists			
Cryptosporidium parvum	FI	30,000	Raw and undercooked meat
Cyclospora cayetanensis	FI	16,000	Fresh produce
Giardia lamblia	FI	200,000	Contaminated or infected meat
Toxoplasma gondii	FI	113,000	Raw and undercooked meat
Total Protists		***359,000***	
Viruses			
Noroviruses	FI	9,200,000	Shellfish, many other foods
All other viruses	FI	82,000	
Total viruses		***9,282,000***	
Total annual foodborne diseases		***13,818,500***	

[a]Estimates are based on data provided by the centers for Disease Control and Prevention, Atlanta, GA, USA, and are typical of the recent years.
[b]FP, food poisoning: FI, food infection.

37.4 Foodborne Diseases and Microbial Sampling

Table 37.6 summarizes the most prevalent foodborne diseases in the United States and the microorganisms that cause them. These common diseases can be separated into two categories, *food poisoning* and *food infection*. Specialized microbial sampling techniques are necessary to isolate the pathogens responsible for foodborne diseases.

Foodborne Diseases

Food poisoning, also called **food intoxication**, is disease that results from ingestion of foods containing preformed microbial toxins. The microorganisms that produced the toxins do not have to grow in the host and are often not alive at the time the contaminated food is consumed. The illness results from the ingestion and action of a bioactive toxin. We previously discussed some of these toxins, notably the exotoxin of *Clostridium botulinum* (Figure 28.21) and the superantigen toxins of *Staphylococcus* and *Streptococcus* (Section 30.8). By contrast to food poisoning, food infection is a microbial infection resulting from the ingestion of pathogen-contaminated food followed by growth of the pathogen in the host. We discuss major foodborne infections in Sections 37.7–37.11.

Microbial Sampling for Foodborne Disease

In addition to nonpathogenic microorganisms, pathogenic microorganisms may be present in fresh foods. Rapid diagnostic methods that do not require pathogen growth or culture have been developed to detect important food pathogens such as *Escherichia coli* O157:H7, *Salmonella*, *Staphylococcus*, and *Clostridium botulinum*. Molecular and immunology-based tests are used to identify both toxin and pathogen contamination of foods and other products such as drugs and cosmetics. We discussed in Section 32.12 the use of nucleic acid probes and the polymerase chain reaction for the detection of specific pathogens, including foodborne pathogens.

The presence of a foodborne pathogen or toxin, however, may not be sufficient to link particular foods or pathogens to a specific foodborne disease outbreak. The causal organism should be isolated and identified to complete investigations of foodborne illness outbreaks. Isolation and growth of pathogens from nonliquid foods usually requires preliminary treatment to suspend microorganisms embedded or entrapped within the food. A standard method uses a specialized sterile blender called a *stomacher*. The stomacher blends and homogenizes using paddles to crush and extrude the food sample. The food should then be examined as soon after sampling as possible; if examination cannot begin within 1 hour of sampling, the food should be refrigerated. Frozen food should be thawed in the original container in a refrigerator and examined or cultured as soon as thawing is complete. In addition to recovering agents from food, foodborne pathogens must be recovered from the patients to establish a cause-and-effect relationship between the pathogen and the illness. In many cases, fecal samples can be cultured to recover organisms for testing.

Food or patient samples can be inoculated onto enriched media, followed by transfer to differential or selective media for isolation and identification, as was described for the isolation of human pathogens (Section 32.2). Final identification of foodborne pathogens by growth characteristics as well as the use of molecular and genetic methods such as PCR, nucleic acid probes, nucleic acid sequencing, and ribotyping may be used to link specific organisms to foodborne diseases and individual illnesses.

In the United States foodborne outbreaks are reportable to the Centers for Disease Control and Prevention. Identification of particular organisms responsible for foodborne disease outbreaks is particularly important, and a nationwide reporting system called *PulseNet* is used to report and compare pulse-field gel electrophoresis (PFGE) DNA fingerprints from foodborne disease organisms. Tracking the characteristics of foodborne illnesses and identifying the causative agents often allow epidemiologists to pinpoint the common source of contaminated food.

37.4 MiniReview

Foodborne diseases include food poisoning and food infection. Food poisoning results from the action of microbial toxins, and food infections are due to the growth of microorganisms in the body. Specialized techniques are used to sample microorganisms in food.

- Distinguish between food infection and food poisoning.
- Describe microbial sampling procedures for solid foods such as meat.

37.5 Staphylococcal Food Poisoning

Food poisoning is often caused by toxins produced by the bacterium *Staphylococcus aureus*. *Staphylococcus* spp. are small, gram-positive cocci (**Figure 37.4**) (Section 16.1).

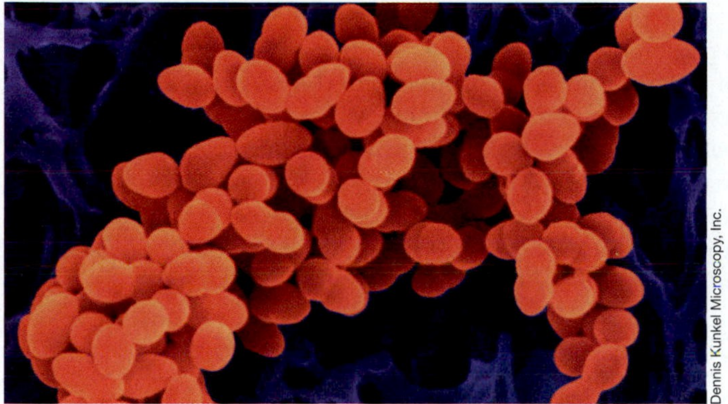

Figure 37.4 ***Staphylococcus aureus.*** In this colorized scanning electron micrograph, the individual gram-positive cocci are about 0.8 μm in diameter. Staphylococci divide in multiple planes, producing the appearance that give the genus its name (from the Greek *staphyle*, bunch of grapes).

As we discussed in Section 34.10, staphylococci are normal members of the local flora of the skin and upper respiratory tract of nearly all humans and are often opportunistic pathogens. *S. aureus* is frequently associated with food poisoning because it can grow in many common foods, and some strains produce several heat-stable enterotoxins. If the enterotoxin is consumed in food, gastroenteritis characterized by nausea, vomiting, and diarrhea, occurs within 1–6 hours.

Epidemiology

Each year an estimated 185,000 cases of staphylococcal food poisoning occur in the United States (Table 37.6). The foods most commonly involved are custard- and cream-filled baked goods, poultry, meat and meat products, gravies, egg and meat salads, puddings, and creamy salad dressings. If such foods are refrigerated immediately after preparation, they usually remain safe because *S. aureus* grows poorly at low temperatures. However, foods of this type are often kept at room temperature in kitchens or outdoors at picnics. The food, if inoculated with *S. aureus* from an infected food handler, supports rapid bacterial growth and enterotoxin production. Even if the toxin-containing foods are reheated before eating, the heat-stable toxin may remain active. Live *S. aureus* need not be present in foods causing illness: The illness is solely due to the preformed toxin.

Staphylococcal Enterotoxins

S. aureus produce at least seven different but related enterotoxins. Most strains of *S. aureus* produce only one or two of these toxins, and some strains are nonproducers. However, any one of these toxins can cause staphylococcal food poisoning. These enterotoxins are further classified as *superantigens*. Superantigens stimulate large numbers of T cells, which in turn release intercellular mediators called *cytokines*. In the intestine, superantigens activate a general inflammatory response that causes gastroenteritis and significant fluid loss (diarrhea and vomiting) (Section 30.8).

UNIT 9

The most common *S. aureus* enterotoxin is enterotoxin A, a protein encoded by *entA*, a chromosomal gene. Comparison of *entA* with other *S. aureus* enterotoxin genes shows that the toxins are genetically related. Although the *entA* gene is on the bacterial chromosome, *S. aureus* enterotoxins type B and C may be encoded on plasmids, transposons, or lysogenic bacteriophages. These movable genetic elements can propagate toxin production to nontoxigenic strains of *Staphylocccus* by lateral gene flow mechanisms (Section 13.11)

Diagnosis, Treatment, and Prevention

Assays based on the detection of either enterotoxin (ELISA detection of enterotoxin) or *S. aureus* exonuclease (an enzyme that degrades DNA) are used to detect enterotoxins and *S. aureus* metabolites in food, respectively. These qualitative tests simply confirm the presence or absence of enterotoxin or the past presence of *S. aureus* above the detection limits of the assay. To obtain quantitative data and determine the extent of bacterial contamination, bacterial plate counts are required. For staphylococcal counts, a high-salt medium (either sodium chloride or lithium chloride at a final concentration of 7.5%) is used. Compared to most bacteria present in foods, staphylococci thrive in habitats with a high salt content and low water activity.

The symptoms of *S. aureus* food poisoning can be quite severe but are typically self-limiting, usually resolving within 48 hours as the toxin is shed from the body. Severe cases may require treatment for dehydration. Treatment with antibiotics is not useful because staphylococcal food poisoning is caused by a preformed toxin, not an active bacterial infection. Staphylococcal food poisoning can be prevented by proper sanitation and hygiene in food production, food preparation, and food storage. As a rule, foods susceptible to colonization by *S. aureus* and kept for several hours at temperatures above 4°C should be discarded rather than eaten.

37.5 MiniReview

Staphylococcal food poisoning results from the ingestion of preformed enterotoxin, a superantigen produced by *Staphylococcus aureus* when growing in foods. In some cases, *S. aureus* cannot be cultured from toxin-containing food.

- Identify the symptoms and mechanism of staphylococcal food poisoning.
- Will antibiotic treatment affect the outcome or the severity of staphylococcal food poisoning? Explain.

37.6 Clostridial Food Poisoning

Clostridium perfringens and *Clostridium botulinum* cause serious food poisoning. Members of the genus *Clostridium* are anaerobic endospore forming rods (Section 16.2). Canning and cooking procedures kill living organisms but do not necessarily kill endospores. Under appropriate anaerobic conditions, the endospores can then germinate and toxin is produced.

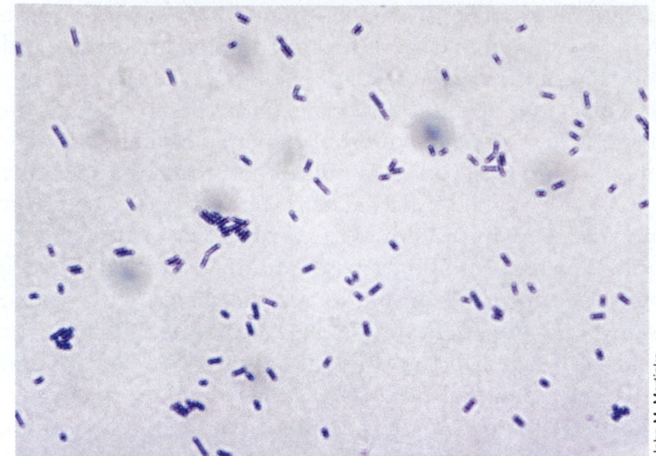

Figure 37.5 ***Clostridium perfringens.*** The Gram stain shows individual gram-positive rods about 1 μm in diameter.

Clostridium perfringens Food Poisoning

C. perfringens is an anaerobic, gram-positive endospore-forming rod commonly found in soil (**Figure 37.5**). It also lives in small numbers in the intestinal tract of many animals and humans and is therefore found in sewage. *C. perfringens* is the most prevalent reported cause of food poisoning in the United States, with an estimated 248,000 annual cases (Table 37.6).

Perfringens food poisoning requires the ingestion of a large dose of *C. perfringens* (>10^8 cells) in contaminated cooked or uncooked foods, especially high-protein foods such as meat, poultry, and fish. Large numbers of *C. perfringens* can grow in meat dishes cooked in bulk. In such food preparations, heat penetration is often insufficient, and surviving *C. perfringens* endospores germinate under anoxic conditions, such as in a sealed container. The *C. perfringens* grows quickly in the meat, especially if the food is left to cool at 20–40°C for short time periods. However, the toxin is not yet present.

After consumption of the contaminated food, the living *C. perfringens* begin to sporulate in the intestine, which coincides with production of the perfringens enterotoxin (Table 28.4). When ingested, perfringens enterotoxin alters the permeability of the intestinal epithelium, leading to nausea, diarrhea, and intestinal cramps, usually with no fever. The onset of perfringens food poisoning begins about 7–15 hours after consumption of the contaminated food but usually resolves within 24 hours, and fatalities are rare.

Diagnosis, Treatment, and Prevention

Diagnosis of perfringens food poisoning is made by isolation of *C. perfringens* from the feces or, more reliably, by a direct enzyme-linked immunosorbent assay (ELISA) to detect *C. perfringens* enterotoxin in feces. Because *C. perfringens* food poisoning is self-limiting, antibiotic treatment is not indicated. Supportive therapy can be used in serious cases. Prevention of perfringens food poisoning requires measures to prevent contamination of raw and cooked foods and control of cooking and canning procedures to ensure proper heat treatment of all foods. Cooked foods should be refrigerated as soon as possible to rapidly lower temperatures and inhibit *C. perfringens* growth.

Botulism

Botulism is a severe, often fatal, food poisoning that occurs following the consumption of food containing the exotoxin produced by *C. botulinum*. This bacterium normally inhabits soil or water, but its endospores may contaminate raw foods before harvest or slaughter. If the foods are properly processed so that the *C. botulinum* endospores are removed or killed, no problem arises; but if viable endospores are present, they may germinate and produce toxin. Even a small amount of the resultant neurotoxin can be dangerous.

We discussed the nature and activity of botulinum toxin in Section 28.10 (Figure 28.21). Botulinum toxin is a neurotoxin that causes flaccid paralysis, usually affecting the autonomic nerves that control body functions such as respiration and heartbeat. At least seven distinct botulinum toxins are known. However, because the toxins are destroyed by heat (80°C for 10 minutes), thoroughly cooked food, even if contaminated with toxin, can be totally harmless.

Most cases of foodborne botulism are caused by eating foods that are not cooked after processing (**Figure 37.6**). For example, nonacid, home-canned vegetables (e.g., home-canned corn and beans) are often used without cooking when making cold salads. Smoked and fresh fish, vacuum-packed in plastic, are also often eaten without cooking. Under such conditions, viable *C. botulinum* endospores may germinate, and the vegetative cells may produce sufficient toxin to cause severe food poisoning. An average of 24 cases of foodborne botulism have occurred per year in the United States from 2000–2005.

The majority of botulism cases occur following infection with *C. botulinum*. For example, infant botulism occurs when neonates ingest endospores of *C. botulinum* (Figure 37.6). In some cases, raw honey is the vehicle, but more often the source cannot be identified. If the infant's normal flora is not developed or if the infant is undergoing antibiotic therapy, endospores can germinate in the infant's intestine, triggering *C. botulinum* growth and toxin production. Most cases of infant botulism occur between the first week of life and 2 months of age, rarely occurring in children older than 6 months when the normal intestinal flora is more developed. Over 60% of all botulism cases in the United States are in infants. An average of 85 cases of infant botulism have occurred per year in the United States from 2000 to 2005. Wound botulism can also occur from infection, presumably from endospores introduced by introduction of contaminating material by a parenteral route. Wound botulism is most commonly associated with illicit injectable drug use and in the United States has averaged 26 cases per year from 2000 to 2005.

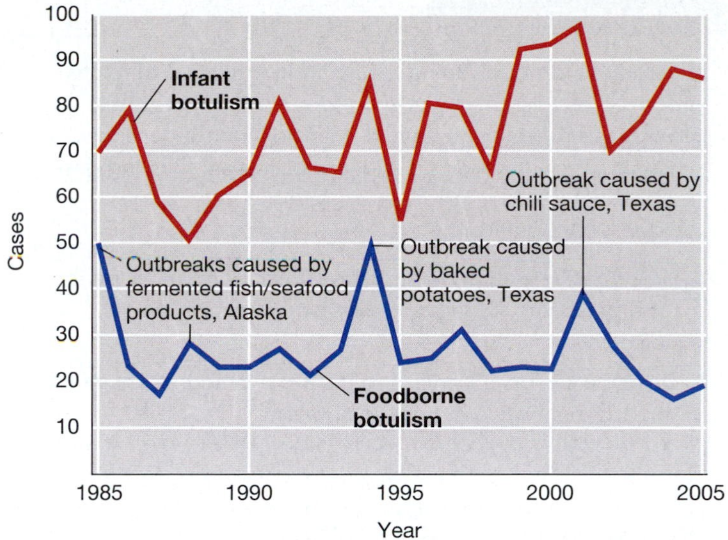

Figure 37.6 Botulism in the United States. Both foodborne and infant botulism is shown. In years with high numbers of cases, major outbreaks that account for the increase are indicated. Data are from the Centers for Disease Control and Prevention, Atlanta, GA, USA.

All forms of botulism are quite rare, with at most six cases occurring per 10 million individuals per year in the United States in recent years. Botulism, however, is a very serious disease because of the high mortality associated with the disease. Over the last decade there have been fewer than 155 cases each year, but about 25% of all cases were fatal. Death occurs from respiratory paralysis or cardiac arrest due to the paralyzing action of the botulinum neurotoxin.

Diagnosis, Treatment, and Prevention

Diagnosis of botulism is by demonstrating botulinum toxin in patient serum or by finding toxin or live *C. botulinum* in suspected food products. Laboratory findings are coupled with clinical observations, including neurological signs of localized paralysis (impaired vision and speech) beginning 18–24 hours after ingestion of contaminated food. Treatment involves administration of botulinum antitoxin if the diagnosis is early, and mechanical ventilation for flaccid respiratory paralysis. In infant botulism, *C. botulinum* and toxin are often found in bowel contents. Infant botulism is usually self-limiting, and most infants recover with only supportive therapy, such as assisted ventilation. Antitoxin administration is not recommended. Respiratory failure causes occasional deaths.

Prevention of botulism requires maintaining careful controls over canning and preservation methods. Susceptible foods should be heated to destroy endospores; boiling for 20 minutes destroys the toxin. Home-prepared foods are the most common source of foodborne botulism outbreaks. In addition, feeding honey to children under 2 years of age is not recommended because honey is an occasional source of *C. botulinum* endospores.

37.6 MiniReview

Clostridium food poisoning results from ingestion of toxins produced by microbial growth in foods or from microbial growth and toxin production in the body. Perfringens food poisoning is quite common and is usually a self-limiting gastrointestinal disease. Botulism is a rare but very serious disease, with significant mortality.

- Describe the events that lead to *Clostridium perfringens* food poisoning. What is the likely outcome of the poisoning?
- Describe the development of botulism in adults and infants. What is the likely outcome of botulism?

III FOOD INFECTION

Food infection results from ingestion of food containing sufficient numbers of viable pathogens to cause infection and disease in the host. Food infection is very common (Table 37.6), and we begin with a common bacterial cause, *Salmonella*. Many food infection agents can also cause the waterborne diseases.

37.7 Salmonellosis

Salmonellosis is a gastrointestinal disease typically caused by foodborne *Salmonella* infection. Symptoms begin after the pathogen colonizes the intestinal epithelium. *Salmonella* are gram-negative facultatively aerobic motile rods related to *Escherichia coli* and other enteric bacteria (Section 15.11). *Salmonella* normally inhabits the animal intestine and is thus found in sewage.

The accepted species name for the pathogenic members of the genus is *Salmonella enterica*. Based on nucleic acid analyses, there are seven evolutionary groups or subspecies of *Salmonella enterica*. Most human pathogens fall into group I, designated as a single subspecies, *S. enterica* subspecies *enterica*. Finally, each subspecies may be divided into serotypes (serological variations defined by antibodies to cell surface antigens). Thus, the organism formally named *Salmonella enterica* subspecies *enterica* serotype Typhi is usually called *Salmonella* Typhi. *Salmonella* Typhi causes the serious human disease typhoid fever but is very rare in the United States. Most of the 500 or so foodborne cases per year are imported from other countries. A number of other *S. enterica* serotypes also cause foodborne gastroenteritis. In all, over 1,400 serotypes cause disease in humans. *Salmonella* Typhimurium and *Salmonella* Enteriditis serotypes most commonly cause foodborne salmonellosis in humans.

Epidemiology and Pathogenesis

The incidence of salmonellosis has been steady over the last decade, with about 40,000–45,000 documented cases each year (**Figure 37.7**). However, less than 4% of salmonellosis cases are probably reported, and the incidence of salmonellosis may be over 1 million every year (Table 37.6).

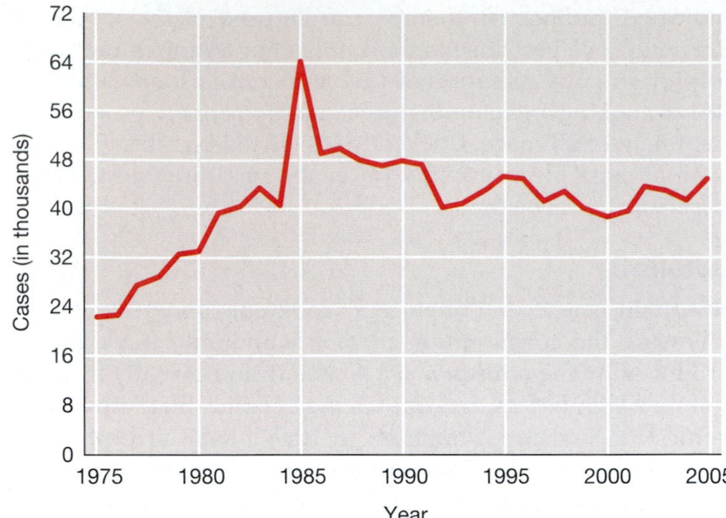

Figure 37.7 Salmonellosis in the United States, 1975–2005. Most cases of salmonellosis are foodborne. The total number of reported cases in 2005 was 45,322. Only about 4% of all cases of salmonellosis are identified and reported. Thus, over 1 million cases of salmonellosis actually occur each year. The high incidence in 1985 was caused by contamination of pasteurized milk that was mixed with raw (unprocessed) milk in a major dairy plant in Illinois. Data are from the Centers for Disease Control and Prevention, Atlanta, GA, USA.

The ultimate sources of the foodborne salmonellas are the intestinal tracts of humans and warm-blooded animals, and several mechanisms may introduce these organisms into the food supply. The organism may reach food by fecal contamination from food handlers. Food production animals such as chickens, pigs, and cattle may also harbor *Salmonella* serotypes that are pathogenic to humans and may pass the bacteria to finished fresh foods such as eggs, meat, and dairy products. *Salmonella* food infections are often traced to products such as custards, cream cakes, meringues, pies, and eggnog made with uncooked eggs. Other foods commonly implicated in salmonellosis outbreaks are meats and meat products such as meat pies, cured but uncooked sausages and meats, poultry, milk, and milk products.

The most common salmonellosis is enterocolitis. Ingestion of food containing viable *Salmonella* results in colonization of the small and large intestine. Onset of the disease occurs 8–48 hours after ingestion. Symptoms include the sudden onset of headache, chills, vomiting, and diarrhea, followed by a fever that lasts a few days. The disease normally resolves without intervention in 2–5 days. After recovery, however, patients may shed *Salmonella* in feces for several weeks. Some patients recover and remain asymptomatic, but shed organisms for months or even years; they are chronic carriers (Section 33.3). A few serotypes of *Salmonella* may also cause septicemia (a blood infection) and enteric or typhoid fever, a disease characterized by systemic infection and high fever lasting several weeks. Mortality can approach 15% in untreated typhoid fever.

The pathogenesis of *Salmonella* infections starts with uptake of the organisms from the gut. *Salmonella* ingested in food or water invades phagocytes and grows as an intracellular pathogen, spreading to adjacent cells as host cells die. As we showed in Figure 28.17, after invasion, pathogenic *Salmonella* uses a combination of endotoxin, enterotoxin, and cytotoxin to damage and kill host cells, leading to the classic symptoms of salmonellosis.

Diagnosis, Treatment, and Prevention

Diagnosis of foodborne salmonellosis is made by observation of clinical symptoms, history of recent food consumption, and by culture of the organism from feces. Selective media (Section 32.2) and tests for the presence of *Salmonella* are commonly used on animal food products, such as raw meat, poultry, eggs, and powdered milk, because *Salmonella* from food production animals is the usual source of food contamination.

For enterocolitis, treatment is usually unnecessary, and antibiotic treatment does not shorten the course of the disease or eliminate the carrier state. Antibiotic treatment, however, significantly reduces the length and severity of septicemia and typhoid fever. Mortality due to typhoid fever can be reduced to less than 1% with appropriate antibiotic therapy. Multi-drug-resistant *Salmonella* are a significant clinical problem.

Properly cooked foods heated to at least to 70°C for 10 minutes are generally safe if consumed immediately, held at 50°C, or stored immediately at 4°C. Any foods that become contaminated by an infected food handler can support the growth of *Salmonella* if the foods are held for long periods of time, especially without heating or refrigeration. *Salmonella* infections are more common in summer than in winter, probably because warm environmental conditions generally favor the growth of microorganisms in foods.

Although local laws and enforcement vary, because of the lengthy carrier state, infected individuals are often banned from work as food handlers until their feces are negative for *Salmonella* in three successive cultures.

37.7 MiniReview

More than 1 million cases of salmonellosis occur every year in the United States. Infection results from ingestion of *Salmonella* introduced into food from food production animals or food handlers.

- Describe salmonellosis food infection. How does it differ from food poisoning?
- How might *Salmonella* contamination of food production animals be contained?

37.8 Pathogenic *Escherichia coli*

Most strains of *Escherichia coli* are not pathogenic and are common members of the enteric microflora in the human colon. A few strains, however, are potential foodborne pathogens. All pathogenic strains are intestinal pathogens and several are characterized by their ability to produce potent enterotoxins. There are about 200 known pathogenic *E. coli* strains. Several cause life-threatening diarrheal disease and urinary tract infections. The pathogenic strains are divided into categories based on the type of toxin they produce and the specific diseases they cause.

Enterohemorrhagic *Escherichia coli* (EHEC)

Enterohemorrhagic *E. coli* (EHEC) produce *verotoxin,* an enterotoxin similar to one produced by *Shigella dysenteriae,* the Shiga toxin (Table 28.4). After a person ingests food or water containing one well-known EHEC strain, *E. coli* O157:H7, the organism grows in the small intestine and produces verotoxin. Verotoxin causes both hemorrhagic (bloody) diarrhea and kidney failure. *E coli* O157:H7 causes at least 60,000 infections and 50 deaths from foodborne disease in the United States each year (Table 37.6). This pathogen is the leading cause of hemolytic uremic syndrome and kidney failure, with 221 cases reported in 2005, about half in children under 5 years of age. The most common cause of this infection is the consumption of contaminated uncooked or undercooked meat, particularly mass-processed ground meat.

In several major outbreaks in the United States caused by *E. coli* O157:H7, infected ground beef from regional distribution centers was the source of the contaminated meat. Infected meat products caused disease in several states. Another outbreak was caused by processed and cured, but uncooked beef in ready-to-eat sausages. The source of contamination was the beef, and the *E. coli* O157:H7 probably originated from slaughtered beef carcasses. In 2001, there were 16 documented food infection outbreaks in the United States due to *E. coli* O157:H7. Five of these were conclusively linked to contaminated beef.

In 2003, the Food Safety and Inspection Service of the U.S. Department of Agriculture reported that there were 20 positive results of 6584 samples (0.03%) of ground beef analyzed for *E. coli* O157:H7. *E. coli* O157:H7 has also been implicated in food infection outbreaks from dairy products, fresh fruit, and raw vegetables. Contamination of the fresh foods by fecal material, typically from cattle carrying the *E. coli* O157:H7 strain, has been implicated in several of these cases (Microbial Sidebar, "Spinach and *Escherichia coli* O157:H7").

Because *E. coli* O157:H7 grows in the intestines and is found in fecal material, it is also a potential source of waterborne gastrointestinal disease. There have been several cases of serious *E. coli* O157:H7 infections from public swimming areas contaminated with feces (Table 36.2). Several outbreaks have also been reported in day-care facilities, where the presumed route of exposure is by oral–fecal contamination.

Other Pathogenic *Escherichia coli*

Children in developing countries often contract diarrheal disease caused by *E. coli*. *E. coli* also can be the cause of "traveler's diarrhea," an extremely common enteric infection causing watery diarrhea in travelers to developing countries. The primary causal

Microbial Sidebar

Spinach and *Escherichia coli* O157:H7

In 2006 an outbreak of *Escherichia coli* O157:H7-associated illness occurred in the United States that was linked to the consumption of ready-to-eat fresh bagged spinach. The outbreak was quickly traced to a vegetable-processing facility in California. First linked to the spinach product in September, the outbreak caused at least 199 infections. Of these, 102 individuals were hospitalized and 31 developed hemolytic uremic syndrome. At least three deaths were attributed to the outbreak.

The remarkably short duration and rapid end to this epidemic—the first case was confirmed in late August and the last reported in early October—is a testament to efficiency and cooperation among public health facilities across the country. We discussed surveillance networks for infectious disease information in Chapter 33. Here we see these networks—PulseNet and FoodNet (CDC)—in action. In this case, the spinach was distributed nationwide, and most cases were not in the West. The two states affected most were Wisconsin, with 49 cases, and Ohio, with 25 cases. California, the state where the contamination actually occurred, had only 2 cases attributed to this outbreak.

Public health officials were able to identify the strain of *E. coli* O157:H7 (**Figure 1**) and determine its origin. They conclusively linked the outbreak to the bagged spinach, traced it back to the processing plant, and eventually traced it to an agricultural field in the vicinity of the processing plant. DNA from the organisms isolated from regional outbreaks was typed using a molecular method called *pulsed-field gel electrophoresis*. The patterns obtained were then compared, and it could be shown that the same organism was responsible for the disease cases observed and was present in the suspected lots of bagged spinach sold throughout the country.

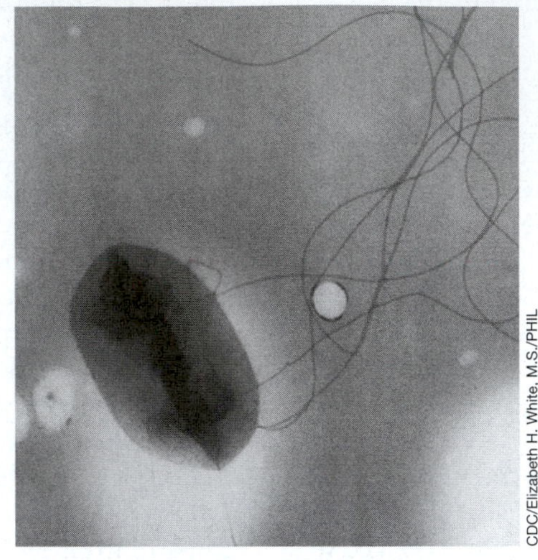

Figure 1 *Negative stained transmission electron micrograph of a cell of* Escherichia coli *O157:H7. The cell is about 1.1 μm in diameter.*

The precise source of the outbreak, although it has been traced to a field near the processing plant, remains unknown. Feral pigs and domestic cattle are present in the vicinity of the identified field, and contaminated wells or surface waters used for irrigation may have introduced the pathogen into the fields and eventually into the spinach. However, whatever the ultimate source, it was almost certainly animal in origin, as *E. coli* is an enteric organism found naturally only in the intestine of warm-blooded animals.

The encouraging part of this story is that the spinach epidemic, although serious and even deadly for some, was discovered, contained, and stopped very quickly. However, this incident also shows how centralized food-processing facilities can quickly spread disease to large and very distant populations. Food hygiene standards in these large processing and distribution facilities must be maintained at the highest possible level and microbiological analyses of food products must be carefully and consistently monitored to prevent future outbreaks of foodborne illnesses.

agents are the enterotoxigenic *E. coli* (ETEC). The ETEC strains usually produce one of two heat-labile diarrhea-producing enterotoxins. In studies done with U.S. citizens traveling in Mexico, the infection rate with ETEC is often greater than 50%. The prime vehicles are foods such as fresh vegetables (for example, lettuce in salads) and water. The very high infection rate in travelers is due to contamination of local public water supplies. The local population is usually resistant to the infecting strains, presumably because they have acquired resistance to the endemic ETEC strains. Secretory IgA antibodies in the bowel prevent colonization of the pathogen in local residents, but the organism readily infects the nonimmune travelers and causes disease.

Enteropathogenic *E. coli* (EPEC) strains cause diarrheal diseases in infants and small children but do not cause invasive disease or produce toxins. Enteroinvasive *E. coli* (EIEC) strains cause invasive disease in the colon, producing watery, sometimes bloody diarrhea. The EIEC strains are taken up by phagocytes, but escape lysis in the phagolysosomes, grow in the cytoplasm, and move into other cells. This invasive disease causes diarrhea and is common in developing countries.

Diagnosis and Treatment

E. coli O157:H7 illness is a reportable infectious disease in the United States. The general pattern established for diagnosis, treatment, and prevention of infection by *E. coli* O157:H7 reflects current procedures used for all of the pathogenic *E. coli* strains. Laboratory diagnosis requires culture from the feces and identification of the O (lipopolysaccharide) and H (flagellar) antigens and toxins by serology. Identification of strains is also done using DNA analyses such as restriction fragment-length polymorphism and pulse-field gel electrophoresis. *E. coli* O157:H7 outbreaks are reported through the PulseNet division of the CDC.

Treatment of *E. coli* O157:H7 and other EHEC includes supportive care and monitoring of renal function, blood hemoglobin, and platelets. Antibiotics may be harmful because they may cause the release of large amounts of verotoxin from dying *E. coli* cells. For other pathogenic *E. coli* infections, treatment usually involves supportive therapy and, for severe cases and invasive disease, antimicrobial drugs to shorten and eliminate infection.

Prevention of *Escherichia coli* Disease

The most effective way to prevent infection with foodborne EHEC O157:H7 is to make sure that meat is cooked thoroughly, which means that it should appear gray or brown and juices should be clear. As we discussed above (Section 37.2), the United States has approved the irradiation of ground meat as an acceptable means of eliminating or reducing food infection bacteria, largely because *E. coli* O157:H7 has been implicated in several foodborne epidemics. To process foods such as hamburger, large-scale production plants may mix and grind meat from hundreds or even thousands of animals together; the grinding process could distribute the pathogens from a single infected animal throughout all the meat. Penetrating radiation is considered the only effective means to ensure decontamination.

In general, proper food handling, water purification, and appropriate hygiene prevent the spread of pathogenic *E. coli*. Traveler's diarrhea can be prevented by avoiding consumption of local water and uncooked foods.

37.8 MiniReview

Enteropathogenic *Escherichia coli* cause many food infections. Contamination of foods from fecal material spreads the pathogenic strains. Good hygiene practices and specific measures such as irradiation of ground beef can curb spread of these pathogens.

- Describe the pathology of *Escherichia coli* food infections due to EHEC, ETEC, EPEC, and EIEC strains.
- Why is *E. coli* O157:H7 considered a dangerous and reportable pathogen?

37.9 *Campylobacter*

Campylobacter spp. are the most common cause of bacterial foodborne infections in the United States. Cells of *Campylobacter* species are gram-negative, motile, curved rods to spiral-shaped bacteria that grow at reduced oxygen tension as microaerophiles (Sections 6.17 and 15.14). Several pathogenic species, *Campylobacter jejuni* (**Figure 37.8**), *C. coli*, and *C. fetus*, are recognized. *C. jejuni* and *C. coli* account for almost 2 million annual cases of bacterial diarrhea (Table 37.6). *C. fetus* is a major cause of sterility and spontaneous abortion in cattle and sheep.

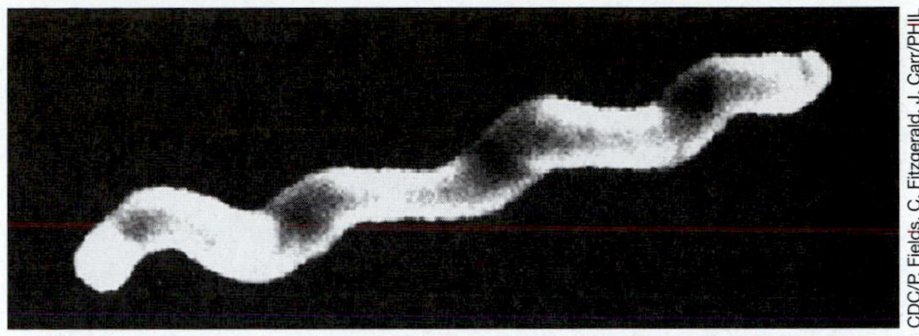

Figure 37.8 ***Campylobacter jejuni.*** The gram-negative curved rod shown in this scanning electron micrograph is about 1 μm in diameter.

Epidemiology and Pathology

Campylobacter is transmitted to humans via contaminated food, most frequently in poultry, pork, raw shellfish, or in surface waters. *C. jejuni* is a normal resident in the intestinal tract of poultry; virtually all chickens and turkeys normally have this organism. According to the U.S. Department of Agriculture, up to 90% of turkey and chicken carcasses and over 30% of hog carcasses may be contaminated with *Campylobacter*. Beef, on the other hand, is rarely a vehicle for this pathogen. *Campylobacter* species also infect domestic animals such as dogs, causing a milder form of diarrhea than that observed in humans. *Campylobacter* infections in infants are frequently traced to infected domestic animals, especially dogs.

After a person ingests cells of *Campylobacter*, the organism multiplies in the small intestine, invades the epithelium, and causes inflammation. Because *C. jejuni* is sensitive to gastric acid, cell numbers as high as 10^4 may be required to initiate infection. However, ingestion of the pathogen directly in food, or ingestion by individuals taking medication to reduce stomach acid production, may reduce this number to less than 500 bacteria. *Campylobacter* infection causes a high fever (usually greater than 104°F or 40°C), headache, malaise, nausea, abdominal cramps, and profuse diarrhea with watery, frequently bloody, stools. The disease subsides in about 7–10 days. Spontaneous recovery from *Campylobacter* infections is often complete, but relapses occur in up to 25% of cases.

Diagnosis, Treatment, and Prevention

Diagnosis of *Campylobacter* food infection requires isolation of the organism from stool samples and identification by growth-dependent tests or immunological assays. Serious *C. jejuni* infections are often seen in infants. In these cases, diagnosis is important; selective media and specific immunological methods have been developed for positive identification of this organism. Erythromycin treatment and quinolone treatment may be useful early in severe diarrheal disease. Adequate personal hygiene, proper washing of uncooked poultry (and any kitchenware coming in contact with uncooked poultry), and thorough cooking of meat eliminate *Campylobacter* contamination.

37.9 MiniReview

Campylobacter infection is the most prevalent foodborne bacterial infection in the United States. Though usually self-limiting, this disease affects nearly 2 million people per year.

- Describe the pathology of *Campylobacter* food infection. What is the likely outcome?
- How might *Campylobacter* contamination of food production animals be controlled?

37.10 Listeriosis

Listeria monocytogenes causes **listeriosis**, a gastrointestinal food infection that may lead to bacteremia and meningitis. *L. monocytogenes* is a short, gram-positive, non-spore-forming coccobacillus that is acid-, salt- and cold-tolerant and facultatively aerobic (Section 16.1) (**Figure 37.9**).

Epidemiology and Pathology

L. monocytogenes is found widely in soil and water; virtually no food source is safe from possible *L. monocytogenes* contamination. Food can become contaminated at any stage during food growth or processing. Food preservation by refrigeration, which ordinarily slows microbial growth, is ineffective in limiting growth of this psychrotolerant organism. Ready-to-eat meats, fresh soft cheeses, unpasteurized dairy products, and inadequately pasteurized milk are the major food vehicles for this pathogen, even when foods are properly stored at refrigerator temperature (4°C).

L. monocytogenes is an intracellular pathogen. It enters the body through the gastrointestinal tract with ingestion of contaminated food. Uptake of the pathogen by phagocytes results in growth and proliferation of the bacterium, lysis of the phagocyte, and spread to surrounding cells such as fibroblasts. Immunity to *L. monocytogenes* is mainly T_H1 cell-mediated. Particularly susceptible populations include the elderly, pregnant women, neonates, and immunosuppressed individuals (for example, transplant patients undergoing steroid therapy and AIDS patients).

Although exposure to *L. monocytogenes* is undoubtedly very common, there are only about 2,500 estimated cases of clinical listeriosis each year, and fewer than 1,000 are reported. Nearly all diagnosed cases require hospitalization. Acute listeriosis is rare and is characterized by septicemia, often leading to meningitis, with a mortality rate of 20% or higher. About 32 listeriosis deaths are reported annually.

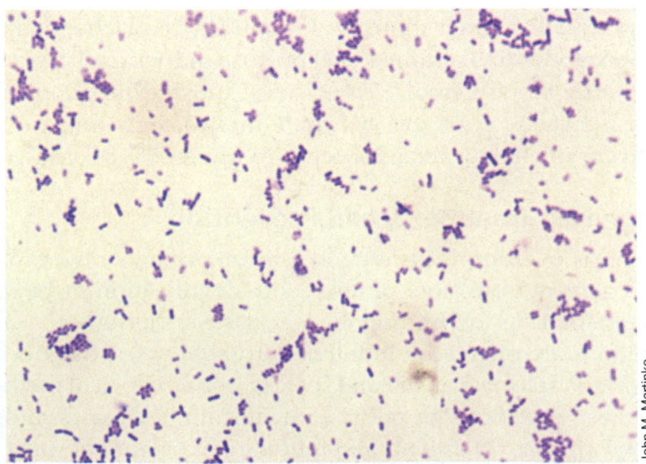

Figure 37.9 ***Listeria monocytogenes.*** This Gram stain shows gram-positive coccobacilli, about 0.5 μm in diameter.

Diagnosis, Treatment, and Prevention

Listeriosis is diagnosed by culturing *L. monocytogenes* from the blood or spinal fluid. *L. monocytogenes* can be identified in food by direct culture or by molecular methods such as ribotyping and the polymerase chain reaction (PCR). All clinical isolates are analyzed by pulsed-field gel electrophoresis to determine molecular subtypes. The subtype patterns are reported to PulseNet at CDC. Intravenous antibiotic treatment with penicillin, ampicillin, or trimethoprim plus sulfamethoxazole is recommended for invasive disease.

Prevention measures include recalling contaminated food and taking steps to limit *L. monocytogenes* contamination at the food-processing site. Because *L. monocytogenes* is susceptible to heat and radiation, raw food and food-handling equipment can be readily decontaminated. However, without pasteurizing the finished food product, the risk of contamination cannot be eliminated because of the widespread distribution of the pathogen.

Individuals who are immunocompromised should avoid unpasteurized dairy products and ready-to-eat processed meats. Spontaneous abortion is a frequent outcome of listeriosis. Therefore, to protect the fetus, pregnant women should also avoid foods that may transmit *L. monocytogenes*.

37.10 MiniReview

Listeria monocytogenes is an environmentally ubiquitous microorganism. In healthy individuals, *Listeria* seldom causes infection. However, in immunocompromised individuals, *Listeria* can cause serious disease and even death.

- What is the likely outcome of *Listeria* exposure in normal individuals?
- What populations are most susceptible to *Listeria* infection? Why?

37.11 Other Foodborne Infectious Diseases

Other microorganisms and infectious agents (for example, prions) contribute to foodborne diseases, and we consider a few of them here.

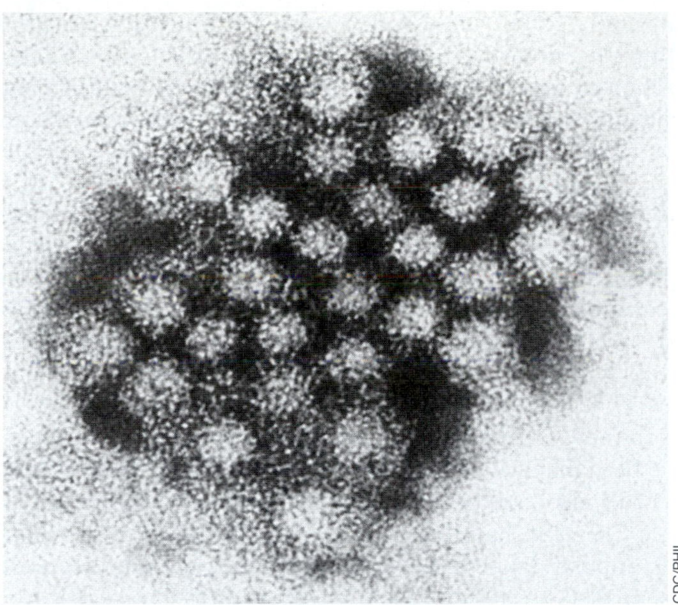

Figure 37.10 Human norovirus. The virus was isolated from a patient with diarrhea. Individual norovirus particles have an indistinct rough outer edge and are about 27 nm in diameter.

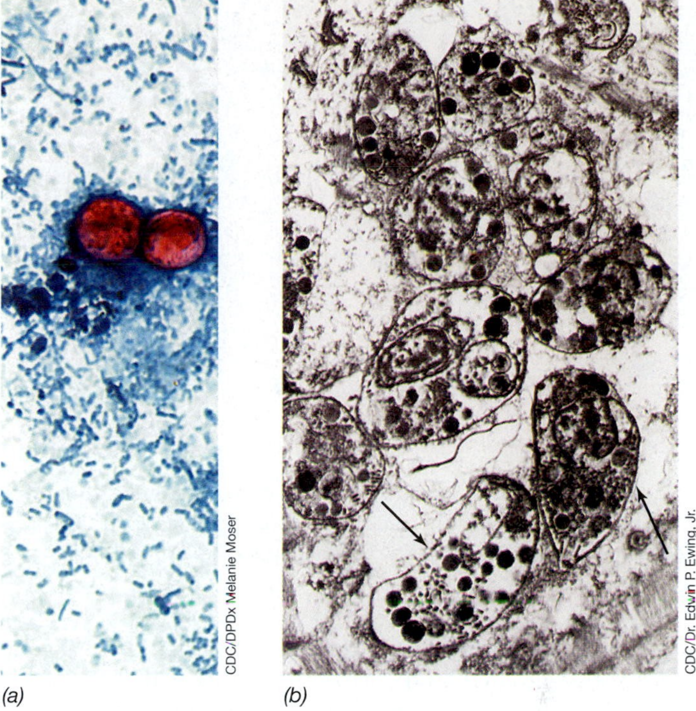

Figure 37.11 Protists transmitted in food. *(a) Cyclospora cayetanensis* oocysts in a stool sample from an affected patient. The oocysts, stained red with safranin, are about 8–10 μm in diameter. *(b)* Tachyzoites of *Toxoplasma gondii*, an intracellular parasite. In this transmission electron micrograph, the tachyzoites are in a cystlike structure in a cardiac myocyte. Tachyzoites are generally elongated to crescent in form, about 4–7 μm long by 2–4 μm wide.

Bacteria

Table 37.6 lists several bacteria that cause human foodborne disease that we have not covered in this chapter. *Yersinia enterocolitica* is commonly found in the intestines of domestic animals and causes foodborne infections due to contaminated meat and dairy products. The most serious consequence of *Y. enterocolitica* infection is enteric fever, a severe life-threatening infection. *Bacillus cereus* produces two enterotoxins that cause diarrhea and vomiting. The organism grows in foods that are cooked and left to cool slowly at room temperature, such as rice, pasta, meats, or sauces. Endospores of this gram-positive rod germinate and toxin is produced. Reheating may kill the *B. cereus,* but the toxin may remain active. *B. cereus* may also cause a food infection similar to that caused by *Clostridium perfringens*. *Shigella* spp. cause nearly 100,000 cases of severe foodborne invasive gastroenteritis called *shigellosis* each year. Several members of the *Vibrio* genus cause food poisoning in persons who consume contaminated shellfish.

Viruses

The largest number of annual foodborne infections are thought to be caused by viruses. In general, viral foodborne illness consists of gastroenteritis characterized by diarrhea, often accompanied by nausea and vomiting. Recovery is spontaneous and rapid, usually within 24–48 hours ("24-hour bug"). Noroviruses (**Figure 37.10**) are responsible for most of these mild foodborne infections in the United States (Table 37.6), accounting for over 9 million of the estimated 13 million annual cases of foodborne disease. Rotavirus, astrovirus, and hepatitis A collectively cause 100,000 cases of foodborne disease each year. These viruses inhabit the gut and are often transmitted to food or water with fecal matter. As with many foodborne infections, proper food handling, handwashing, and a source of clean water to prepare fresh foods are essential to prevent infection.

Protists

Important foodborne protist diseases are listed in Table 37.6. Protists including *Giardia lamblia, Cryptosporidium parvum* (Figures 36.13 and 36.14) and *Cyclospora cayetanensis* (**Figure 37.11*a***) can be spread in foods contaminated by fecal matter in untreated water used to wash, irrigate, or spray crops. Fresh foods such as fruits are often implicated as the source of these protists. We discussed giardiasis and cryptosporidiosis as waterborne diseases (Section 36.6). Cyclosporiasis is an acute gastroenteritis and is an important emerging disease. In the United States, most cases are acquired by eating fresh produce imported from other countries.

Toxoplasma gondii (Figure 37.11*b*) is a protist spread through cat feces, but is also found in raw or undercooked meat. In most individuals, toxoplasmosis is a mild, self-limiting gastroenteritis. However, prenatal infection can lead to serious acute toxoplasmosis resulting in tissue involvement, cyst formation, and complications such as myocarditis, blindness, and stillbirth in the fetus. Immunocompromised individuals such

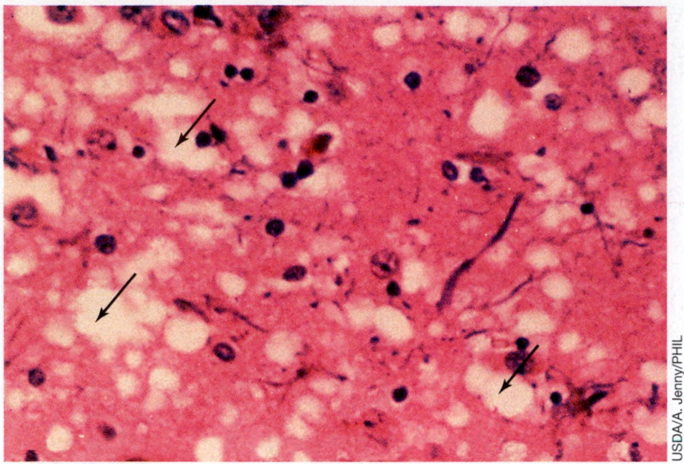

Figure 37.12 A brain section from a cow with bovine spongiform encephalopathy (BSE). The vacuoles appearing as holes (arrows) give the brains of infected animals a distinct spongelike appearance.

as AIDS patients may develop acute toxoplasmosis. *T. gondii* grows intracellularly and forms structures called *tachyzoites* that eventually lyse the cell and infect nearby cells, resulting in tissue destruction. Tachyzoites can cross the placenta and thereby infect the fetus. Toxoplasma infections in compromised hosts can be treated with the antiprotist drug, pyrimethamine.

Prions, BSE, and nvCJD Disease

Prions are proteins, presumably of host origin, that adopt novel conformations, inhibiting normal protein function and causing degeneration of neural tissue (Section 10.15). Human prion diseases are characterized by a number of neurological symptoms including progressive depression, loss of motor coordination, and dementia.

A foodborne prion disease in humans known as "new variant Creutzfeldt-Jakob Disease" (nvCJD) has been linked to consumption of meat products from cattle afflicted with *bovine spongiform encephalopathy (BSE),* a prion disease commonly called "mad cow disease." The nvCJD is a slow-acting degenerative nervous system disorder with a latent period that extends for years after exposure to the BSE prion. Nearly 200 people in Great Britain and other European countries have acquired nvCJD. However, nvCJD linked to domestic meat consumption has not been observed in the United States. BSE prions consumed in the meat products from affected cattle trigger human protein analogs to assume an altered conformation, resulting in protein dysfunction and disease (Figure 10.28). The terminal stages of both BSE and nvCJD are characterized by large vacuoles in brain tissue, giving the brain a "spongy" appearance, from which BSE derives its name (**Figure 37.12**).

In the United Kingdom and Europe, about 180,000 cattle have been diagnosed with BSE. Brains of slaughtered animals are routinely tested for BSE in the United States, and several cattle with BSE have been found in Canadian and U.S. herds. In Europe and North America, all cattle known or suspected to have BSE have been destroyed. Bans on feeding cattle meat and bone meal appear to have stopped the development of new cases of BSE in Europe and have kept the incidence of this disease very low in North America. The infecting prions were probably transferred to food production animals through meat and bone meal feed derived from infected cattle or other animals not approved for human consumption.

Diagnosis of BSE is done by testing using a prion-susceptible mouse strain or by immunohistochemical or micrographic analysis of biopsied neural tissue (Figure 37.12).

37.11 MiniReview

Over 200 different infectious agents cause foodborne disease. Viruses cause the most foodborne illnesses. Bacteria, protists, and prions, however, also cause significant foodborne illness.

- Identify the viruses, bacteria, and protists most likely to cause foodborne illnesses.
- How might prion contamination of food production animals be prevented in the United States?

Review of Key Terms

Botulism food poisoning due to ingestion of food containing botulinum toxin produced by *Clostridium botulinum*

Canning the process of sealing food in a closed container and heating to destroy living organisms and endospores

Fermentation the anaerobic catabolism of organic compounds, generally carbohydrates, in the absence of an external electron acceptor

Food infection disease caused by active infection resulting from ingestion of pathogen-contaminated food

Food poisoning (food intoxication) disease caused by the ingestion of food that contains preformed microbial toxins

Food spoilage a change in the appearance, smell, or taste of a food that makes it unacceptable to the consumer

Irradiation the exposure of food to ionizing radiation for the purpose of inhibiting growth of microorganisms and insect pests or to retard ripening

Listeriosis gastrointestinal food infection caused by *Listeria monocytogenes* that may lead to bacteremia and meningitis

Lyophilization (freeze-drying) the removal of all water from frozen food under vacuum

Nonperishable (stable) food food of low water activity that has an extended shelf life and is resistant to spoilage by microorganisms

Perishable food fresh food generally of high water activity that has a very short shelf life due to potential for spoilage by growth of microorganisms

Pickling the process of acidifying food to prevent microbial growth and spoilage

Salmonellosis enterocolitis caused by any of over 2000 variants of *Salmonella* spp.

Semiperishable food food of intermediate water activity that has a limited shelf life due to potential for spoilage by growth of microorganisms

Water activity (a_w) the availability of water for use in metabolic processes

Review Questions

1. Identify and define the three major food categories with respect to perishability (Section 37.1).
2. Identify the major methods used to preserve food. Provide an example of a food preserved by each method (Section 37.2).
3. Identify the major categories of fermented foods (Section 37.3).
4. Distinguish between foodborne infection and foodborne poisoning (Section 37.4).
5. Outline the pathogenesis of staphylococcal food poisoning. Suggest methods for prevention of this disease (Section 37.5).
6. Identify the two major types of clostridial food poisoning. Which is most prevalent? Which is most dangerous? Why (Section 37.6)?
7. What are the possible sources of *Salmonella* spp. that cause food infections (Section 37.7)?
8. What measures can control the inoculation and growth of *Escherichia coli* O157:H7 in ground meat in food-processing and preparation settings (Section 37.8)?
9. *Campylobacter* causes more foodborne infections than any other bacterium. Identify at least one reason why this is true (Section 37.9).
10. Identify the food sources of *Listeria monocytogenes* infections. Identify the individuals who are at high risk for listeriosis (Section 37.10).
11. Why are viral agents so commonly associated with foodborne disease (Section 37.11)?

Application Questions

1. Identify optimum storage conditions for perishable, semiperishable, and nonperishable food products. Consider economic factors such as the cost of preservatives, storage space, and the intrinsic value of the food item.
2. For a food of your choice, devise a way to preserve the food by lowering the water activity without drying.
3. Perfringens food poisoning involves ingestion of *Clostridium perfringens* followed by growth and sporulation in the intestine of the host. Sporulation triggers toxin production. Is this disease truly a food poisoning, or might it be classified as a food infection? Explain.
4. Improperly handled potato salads are often the source of staphylococcal food poisoning or salmonellosis. Explain the means by which a potato salad could become inoculated with either *Staphylococcus aureus* or *Salmonella* spp.
5. *Clostridium botulinum* requires an anoxic environment for production of botulinum toxin. Identify methods of food preservation that create the anoxic environment necessary for growth of *C. botulinum*. Conversely, identify methods of food preservation that create an aerobic environment and prevent the growth of *C. botulinum*. What other factors influence the growth of *C. botulinum*?
6. Indicate the precautions necessary to prevent infection with pathogenic *Escherichia coli*. Concentrate on *E. coli* O157:H7 and safe food handling, cooking, and consumption practices.
7. Devise a plan to eliminate *Campylobacter* from a poultry flock or from the finished poultry product. Explain the benefits of *Campylobacter*-free poultry and explain the problems that your plan might encounter.
8. Listeriosis normally occurs only when there is a breakdown in T_H1 cell-mediated immunity. Indicate why this is so. Devise a vaccine to protect against listeriosis. Would your vaccine be an inactivated bacterial strain or product, or would it be an attenuated organism? Explain. Would your vaccine be of use in the listeriosis-prone population?
9. Indicate reasons for the high incidence of viral foodborne disease, especially with noroviruses. Devise a plan to eliminate noroviruses from the food supply.
10. Indicate the problems inherent in tracking a latent infectious agent such as the BSE prion. Can prion diseases be eliminated? If so, how?

Appendix 1 Energy Calculations in Microbial Bioenergetics

The information in Appendix 1 is intended to help calculate changes in free energy accompanying chemical reactions carried out by microorganisms. It begins with definitions of the terms required to make such calculations and proceeds to show how knowledge of redox state, atomic and charge balance, and other factors are necessary to calculate free-energy problems successfully.

I. DEFINITIONS

1. ΔG^0 = standard free-energy change of the reaction at 1 atm pressure and 1 *M* concentrations; ΔG = free-energy change under the conditions specified; $\Delta G^{0\prime}$ = free-energy change under standard conditions at pH 7.
2. Calculation of ΔG^0 for a chemical reaction from the free energy of formation, $G_f{}^0$, of products and reactants:

$$\Delta G^0 = \sum \Delta G_f{}^0 \text{ (products)} - \sum \Delta G_f{}^0 \text{ (reactants)}$$

   That is, sum the $\Delta G_f{}^0$ of products, sum the $\Delta G_f{}^0$ of reactants, and subtract the latter from the former.
3. For energy-yielding reactions involving H^+, converting from standard conditions (pH 0) to cellular conditions (pH 7):

$$\Delta G^{0\prime} = \Delta G^0 + m\Delta G_f{}^0(H^+)$$

   where *m* is the net number of protons in the reaction (*m* is negative when more protons are consumed than formed) and $\Delta G_f{}^0$ (H^+) is the free energy of formation of a proton at pH 7 (−39.83 kJ) at 25°C.
4. Effect of concentrations on ΔG: With soluble substrates, the concentration ratios of products formed to exogenous substrates used are generally equal to or greater than 10^{-2} at the beginning of growth and equal to or less than 10^{-2} at the end of growth. From the relation between ΔG and the equilibrium constant (see item 8), it can be calculated that ΔG for the free-energy yield in practical situations differs from the free-energy yield under standard conditions by at most 11.7 kJ, a rather small amount, and so for a first approximation, standard free-energy yields can be used in most situations. However, with H_2 as a product, H_2-consuming bacteria present may keep the concentration of H_2 so low that the free-energy yield is significantly affected. Thus, in the fermentation of ethanol to acetate and H_2 by syntrophic bacteria ($C_2H_5OH + H_2O \longrightarrow C_2H_3O_2{}^- + 2\ H_2 + H^+$), the $\Delta G^{0\prime}$ at 1 atm H_2 is +9.68 kJ, but at 10^{-4} atm H_2 it is −36.03 kJ. With H_2-consuming bacteria present, therefore, the ethanol fermentation becomes exergonic. (See also item 9.)
5. Reduction potentials: by convention, electrode equations are written in the direction, oxidant + $ne^- \longrightarrow$ reductant (that is, as reductions), where *n* is the number of electrons transferred. The standard potential (E_0) of the hydrogen electrode, $2\ H^+ + 2\ e^- \longrightarrow H_2$, is set by definition at 0.0 V at 1 atm pressure of H_2 gas and 1.0 *M* H^+ at 25°C. $E_0{}'$ is the standard reduction potential at pH 7. See also Table A1.2.
6. Relation of free energy to reduction potential:

$$\Delta G^{0\prime} = -nF\Delta E_0{}'$$

   where *n* is the number of electrons transferred, *F* is the Faraday constant (96.48 kJ/V), and $\Delta E_0{}'$ is the $E_0{}'$ of the electron-*accepting* couple minus the $E_0{}'$ of the electron-*donating* couple.
7. Equilibrium constant, *K*. For the generalized reaction $aA + bB \rightleftharpoons cC + dD$,

$$K = \frac{[C]^c[D]^d}{[A]^a[B]^b}$$

   where A, B, C, and D represent reactants and products; *a*, *b*, *c*, and *d* represent number of molecules of each; and brackets indicate concentrations. This is true only when the chemical system is in equilibrium.
8. Relation of equilibrium constant, *K*, to free-energy change. At constant temperature, pressure, and pH,

$$\Delta G = \Delta G^{0\prime} + RT \ln K$$

   where *R* is a constant (8.29 J/mol/°K) and *T* is the absolute temperature (in °K).
9. Two substances can react in a redox reaction even if the standard potentials are unfavorable, provided that the concentrations are appropriate.

Assume that normally the reduced form of A would donate electrons to the oxidized form of B. However, if the concentration of the reduced form of A was low and the concentration of the reduced form of B was high, it would be possible for the reduced form of B to donate electrons to the oxidized form of A. Thus, the reaction would proceed in the direction opposite that predicted from standard potentials. A practical example of this is the utilization of H^+ as an electron acceptor to produce H_2. Normally, H_2 production in fermentative bacteria is not extensive because H^+ is a poor electron acceptor; the $E_0{}'$ of the $2\ H^+/H_2$ pair is −0.41 V. However, if the concentration of H_2 is kept low by continually removing it (a process done by methanogenic *Archaea*, which use $H_2 + CO_2$ to produce methane, CH_4, or by many other anaerobes capable of consuming H_2 anaerobically), the potential will be more positive and then H^+ will be a suitable electron acceptor.

II. OXIDATION STATE OR NUMBER

1. The oxidation state of an element in an elementary substance (for example, H_2, O_2) is zero.
2. The oxidation state of the ion of an element is equal to its charge (for example, $Na^+ = +1$, $Fe^{3+} = +3$, $O^{2-} = -2$).
3. The sum of oxidation numbers of all atoms in a neutral molecule is zero. Thus, H_2O is neutral because it has two H at +1 each and one O at −2.
4. In an ion, the sum of oxidation numbers of all atoms is equal to the charge on that ion. Thus, in the OH^- ion, O(−2) + H(+1) = −1.
5. In compounds, the oxidation state of O is virtually always −2 and that of H is +1.
6. In simple carbon compounds, the oxidation state of C can be calculated by adding up the H and O atoms present and using

the oxidation states of these elements as given in item 5, because in a neutral compound the sum of all oxidation numbers must be zero. Thus, the oxidation state of carbon in methane, CH_4, is -4 (4 H at $+1$ each $= +4$); in carbon dioxide, CO_2, the oxidation state of carbon is $+4$ (2 O at -2 each $= -4$).

7. In organic compounds with more than one C atom, it may not be possible to assign a specific oxidation number to each C atom, but it is still useful to calculate the oxidation state of the compound as a whole. The same conventions are used. Thus, the oxidation state of carbon in glucose, $C_6H_{12}O_6$, is zero (12 H at $+1 = 12$; 6 O at $-2 = -12$) and the oxidation state of carbon in ethanol, C_2H_6O, is -2 each (6 H at $+1 = +6$; one O at -2).
8. In all oxidation-reduction reactions there is a balance between the oxidized and reduced products. To calculate an oxidation-reduction balance, the number of molecules of each product is multiplied by its oxidation state. For instance, in calculating the oxidation-reduction balance for the alcoholic fermentation, there are two molecules of ethanol at $-4 = -8$ and two molecules of CO_2 at $+4 = +8$ so the net balance is zero. When constructing model reactions, it is useful to first calculate redox balances to be certain that the reaction is possible.

III. CALCULATING FREE-ENERGY YIELDS FOR HYPOTHETICAL REACTIONS

Energy yields can be calculated either from free energies of formation of the reactants and products or from differences in reduction potentials of electron-donating and electron-accepting partial reactions.

Calculations from Free Energy

Free energies of formation are given in **Table A1.1**. The procedure to use for calculating energy yields of reactions follows.

1. *Balancing reactions.* In all cases, it is essential to ascertain that the coupled oxidation-reduction reaction is balanced. Balancing involves three things: (a) the *total number of each kind of atom* must be identical on both sides of the equation; (b) there must be an *ionic balance* so that when positive and negative ions are added up on the right side of the equation, the total ionic charge (whether positive, negative, or neutral) exactly balances the ionic charge on the left side of the equation; and (c) there must be an *oxidation-reduction balance* so that all the electrons removed from one substance are transferred to another substance. In general, when constructing balanced reactions, one proceeds in the reverse of the three steps just listed. Usually, if steps (c) and (b) have been properly handled, step (a) becomes correct automatically.
2. *Examples:* (a) What is the balanced reaction for the oxidation of H_2S to SO_4^{2-} with O_2? First, decide how many electrons are involved in the oxidation of H_2S to SO_4^{2-}. This can be most easily calculated from the oxidation states of the compounds, using the rules given previously. Because H has an oxidation state of $+1$, the oxidation state of S in H_2S is -2. Because O has an oxidation state of -2, the oxidation state of S in SO_4^{2-} is $+6$ (because it is an ion, using the rules given in items 4 and 5 of the previous section). Thus, the oxidation of H_2S to SO_4^{2-} involves an *eight-electron transfer* (from -2 to $+6$). Because each O atom can accept two electrons (the oxidation state of O in O_2 is zero, but in H_2O is -2), this means that two molecules of molecular oxygen, O_2, are required to provide sufficient electron-accepting capacity. Thus, at this point, we know that the reaction requires 1 H_2S and 2 O_2 on the left side of the equation, and 1 SO_4^{2-} on the right side. To achieve an ionic balance, we must have two positive charges on the right side of the equation to balance the two negative charges of SO_4^{2-}. Thus, 2 H^+ must be added to the right side of the equation, making the overall reaction.

$$H_2S + 2\,O_2 \longrightarrow SO_4^{2-} + 2\,H^+$$

By inspection, it can be seen that this equation is also balanced in terms of the total number of atoms of each kind on each side of the equation.

(b) What is the balanced reaction for the oxidation of H_2S to SO_4^{2-} with Fe^{3+}as electron acceptor? We have just ascertained that the oxidation of H_2S to SO_4^{2-} is an eight-electron transfer. Because the reduction of Fe^{3+} to Fe^{2+} is only a one-electron transfer, 8 Fe^{3+} will be required. At this point, the reaction looks like

$$H_2S + 8\,Fe^{3+} \longrightarrow 8\,Fe^{2+} + SO_4^{2-} \text{ (not balanced)}$$

We note that the ionic balance is incorrect. We have 24 positive charges on the left and 14 positive charges on the right (16+ from Fe, 2− from sulfate). To equalize the charges, we add 10 H^+on the right. Now our equation looks like:

$$H_2S + 8\,Fe^{3+} \longrightarrow 8\,Fe^{2+} + 10\,H^+ + SO_4^{2-} \text{ (not balanced)}$$

To provide the necessary hydrogen for the H^+ and oxygen for the sulfate, we add 4 H_2O to the left and find that the equation is now balanced:

$$H_2S + 4\,H_2O + 8\,Fe^{3+} \longrightarrow 8\,Fe^{2+} + 10\,H^+ + SO_4^{2-} \text{ (balanced)}$$

In general, in microbiological reactions, ionic balance can be achieved by adding H^+ or OH^- to the left or right side of the equation, and because all reactions take place in an aqueous medium, H_2O molecules can be added where needed. Whether H^+ or OH^- is added generally depends on whether the reaction is taking place under acidic or alkaline conditions.
3. *Calculation of energy yield for balanced equations from free energies of formation.* Once an equation has been balanced, the free-energy yield can be calculated by inserting the values for the free energy of formation of each reactant and product from Table A1.1 and using the formula in item 2 of the first section of this appendix.

For instance, for the equation

$$H_2S + 2\,O_2 \longrightarrow SO_4^{2-} + 2\,H^+$$

$$G_f{}^0 \text{ values} \longrightarrow (-27.87) + (0)(-744.6) + 2\,(-39.83) \text{ (assuming pH 7)}$$

$$\Delta G^{0\prime} = -796.39 \text{ kJ}$$

The $G_f{}^0$ values for the products (right side of equation) are summed and subtracted from the $G_f{}^0$ values for the reactants (left side of equation), taking care to ensure that the arithmetic signs are correct. From the data in Table A1.1, a wide variety of free-energy yields for reactions of microbiological interest can be calculated.

Calculation of Free-Energy Yield from Reduction Potential

Reduction potentials of some important redox pairs are given in **Table A1.2**. The amount of energy that can be released from two half reactions can be calculated from the *differences* in reduction potentials of the two reactions and from the number of electrons transferred. The farther apart the two half reactions are, and the greater the number of electrons transferred, the more energy released.

The conversion of potential difference to free energy is given by the formula $\Delta G^{0\prime} = -nF\Delta E_0{}'$, where n is the number of electrons, F is the Faraday constant (96.48 kJ/V), and $\Delta E_0{}'$ is the difference in potentials. Thus, the 2 H^+/H_2 couple has a potential of −0.41 V and the $\frac{1}{2}O_2/H_2O$ pair has a potential of +0.82 V, and so the potential difference is 1.23 V, which (because two electrons are involved) is equivalent to a free-energy yield (ΔG^0) of −237.34 kJ. On the other hand, the potential difference between the $2H^+/H_2$ and the NO_3^-/NO_2^- reactions is less, 0.84 V, which is equivalent to a free-energy yield of −162.08 kJ.

Because many biochemical reactions are two-electron transfers, it is often useful to give energy yields for two-electron reactions, even if more electrons are involved. Thus, the SO_4^{2-}/H_2 redox pair involves eight electrons, and complete reduction of SO_4^{2-} with H_2 requires 4 H_2 (equivalent to eight electrons). From the reduction potential difference between 2 H^+/H_2 and SO_4^{2-}/H_2S (0.19 V), a free-energy yield of −146.64 kJ is calculated, or −36.66 kJ per two electrons. By convention, reduction potentials are given for conditions in which equal concentrations of oxidized and reduced forms are present. In actual practice, the concentrations of these two forms may be quite different. As discussed earlier in this appendix (Section I, item 9), it is possible to couple half reactions even if the potential difference is unfavorable, providing the concentrations of the reacting species are appropriate.

Table A1.1 Free energies of formation (G_f^0) for some substances (kJ/mol)[a]

Carbon compound	*Carbon compound*	*Metal*	*Nonmetal*	*Nitrogen compound*
CO, −137.34	Glutamine, −529.7	Cu^+, +50.28	H_2, 0	N_2, 0
CO_2, −394.4	Glyceraldehyde, −437.65	Cu^{2+}, +64.94	H^+, 0 at pH 0;	NO, +86.57
CH_4, −50.75	Glycerate, −658.1	CuS, −49.02	−39.83 at pH 7	NO_2, +51.95
H_2CO_3, −623.16	Glycerol, −488.52	Fe^{2+}, −78.87	(−5.69 per pH unit)	NO_2^-, −37.2
HCO_3^-, −586.85	Glycine, −314.96	Fe^{3+}, −4.6	O_2, 0	NO_3^-, −111.34
CO_3^{2-}, −527.90	Glycolate, −530.95	$FeCO_3$, −673.23	OH^-, −157.3 at pH 14;	NH_3, −26.57
Acetaldehyde, −139.9	Glyoxalate, −468.6	FeS_2, −150.84	−198.76 at pH 7;	NH_4^+, −79.37
Acetate, −369.41	Guanine, +46.99	$FeSO_4$, −829.62	−237.57 at pH 0	N_2O, +104.18
Acetone, −161.17	α-Ketoglutarate, −797.55	PbS, −92.59	H_2O, −237.17	
Alanine, −371.54	Lactate, −517.81	Mn^{2+}, −227.93	H_2O_2, −134.1	
Arginine, −240.2	Lactose, −1515.24	Mn^{3+}, −82.12	PO_4^{3-}, −1026.55	
Aspartate, −700.4	Malate, −845.08	MnO_4^-, −506.57	Se^0, 0	
Benzene, +124.5	Mannitol, −942.61	MnO_2 −456.71	H_2Se, −77.09	
Benzoic acid, −245.6	Methanol, −175.39	$MnSO_4$, −955.32	SeO_4^{2-}, −439.95	
n-Butanol, −171.84	Methionine, −502.92	HgS, −49.02	S^0, 0	
Butyrate, −352.63	Methylamine, −40.0	MoS_2, −225.42	SO_3^{2-}, −486.6	
Caproate, −335.96	Oxalate, −674.04	ZnS, −198.60	SO_4^{2-}, −744.6	
Citrate, −1168.34	Palmitic acid, −305		$S_2O_3^{2-}$, −513.4	
o-Cresol, −37.1	Phenol, −47.6		H_2S, −27.87	
Crotonate, −277.4	*n*-Propanol, −175.81		HS^-, +12.05	
Cysteine, −339.8	Propionate, −361.08		S^{2-}, +85.8	
Dimethylamine, −3.3	Pyruvate, −474.63			
Ethanol, −181.75	Ribose, −757.3			
Formaldehyde, −130.54	Succinate, −690.23			
Formate, −351.04	Sucrose, −370.90			
Fructose, −951.38	Toluene, +114.22			
Fumarate, −604.21	Trimethylamine, −37.2			
Gluconate, −1128.3	Tryptophan, −112.6			
Glucose, −917.22	Urea, −203.76			
Glutamate, −699.6	Valerate, −344.34			

[a]Values for free energy of formation of various compounds can be found in Dean, J. A. 1973. *Lange's Handbook of Chemistry*, 11th edition. McGraw-Hill, New York; Garrels, R. M., and C. L. Christ. 1965. *Solutions, Minerals, and Equilibria*. Harper and Row, New York; Burton, K. 1957. In Krebs, H. A., and H. L. Komberg. Energy transformation in living matter, *Ergebnisse der Physiologie* (appendix): Springer-Verlag, Berlin; and Thauer, R. K., K. Jungermann, and H. Decker. 1977. Energy conservation in anaerobic chemotrophic bacteria. *Bacteriol Rev* 41:100–180.

Table A1.2 Microbiologically important reduction potentials[a]

Redox pair	E_0' *(V)*
SO_4^{2-}/HSO_3^-	−0.52
CO_2/formate	−0.43
$2H^+/H_2$	−0.41
$S_2O_3^{2-}/HS^- + HSO_3^-$	−0.40
Ferredoxin ox/red	−0.39
Flavodoxin ox/red[b]	−0.37
NAD^+/NADH	−0.32
Cytochrome c_3 ox/red	−0.29
CO_2/acetate$^-$	−0.29
S^0/HS^-	−0.27
CO_2/CH_4	−0.24
FAD/FADH	−0.22
SO_4^{2-}/HS^-	−0.217
Acetaldehyde/ethanol	−0.197
Pyruvate$^-$/lactate$^-$	−0.19
FMN/FMNH	−0.19
Dihydroxyacetone phosphate/glycerolphosphate	−0.19
$HSO_3^-/S_3O_6^{2-}$	−0.17
Flavodoxin ox/red[b]	−0.12
HSO_3^-/HS^-	−0.116
Menaquinone ox/red	−0.075
APS/AMP + HSO_3^-	−0.060
Rubredoxin ox/red	−0.057
Acrylyl-CoA/propionyl-CoA	−0.015
Glycine/acetate$^-$ + NH_4^+	−0.010
$S_4O_6^{2-}/S_2O_3^{2-}$	+0.024
Fumarate^{2-}/succinate^{2-}	+0.033
Cytochrome *b* ox/red	+0.035
Ubiquinone ox/red	+0.113
AsO_4^{3-}/AsO_3^{3-}	+0.139
Dimethyl sulfoxide (DMSO)/dimethylsulfide (DMS)	+0.16
$Fe(OH)_3 + HCO_3^-/FeCO_3$ (Fe^{3+}/Fe^{2+}, pH 7)	+0.20
$S_3O_6^{2-}/S_2O_3^{2-} + HSO_3^-$	+0.225
Cytochrome c_1 ox/red	+0.23
NO_2^-/NO	+0.36
Cytochrome a_3 ox/red	+0.385
Chlorobenzoate$^-$ /benzoate$^-$ + HCl	+0.297
NO_3^-/NO_2^-	+0.43
SeO_4^{2-}/SeO_3^{2-}	+0.475
Fe^{3+}/Fe^{2+} (pH 2)	+0.77
Mn^{4+}/Mn^{2+}	+0.798
O_2/H_2O	+0.82
ClO_3^-/Cl^-	+1.03
NO/N_2O	+1.18
N_2O/N_2	+1.36

[a]Data from Thauer, R. K., K. Jungermann, and K. Decker, 1977. Energy conservation in anaerobic chemotrophic bacteria. *Bacteriol. Rev.* 41:100–180.

[b]Separate potentials are given for each electron transfer in this potentially two-electron transfer.

Appendix 2 Bergey's Manual of Systematic Bacteriology, Second Edition

LIST OF GENERA AND HIGHER ORDER TAXA[a]

Domain *Archaea*
Phylum AI. *Crenarchaeota*
Class I. *Thermoprotei*
Order I. *Thermoproteales*
Family I. *Thermoproteaceae*
Genus I. *Thermoproteus*
Genus II. *Caldivirga*
Genus III. *Pyrobaculum*
Genus IV. *Thermocladium*
Genus V. *Vulcanisaeta*
Family II. *Thermofilaceae*
Genus I. *Thermofilum*
Order II. *Caldisphaerales*
Family I. *Caldisphaeraceae*
Genus I. *Caldisphaera*
Order III. *Desulfurococcales*
Family I. *Desulfurococcaceae*
Genus I. *Desulfurococcus*
Genus II. *Acidilobus*
Genus III. *Aeropyrum*
Genus IV. *Ignicoccus*
Genus V. *Staphylothermus*
Genus VI. *Stetteria*
Genus VII. *Sulfophobococcus*
Genus VIII. *Thermodiscus*
Genus IX. *Thermosphaera*
Family II. *Pyrodictiaceae*
Genus I. *Pyrodictium*
Genus II. *Hyperthermus*
Genus III. *Pyrolobus*
Order IV. *Sulfolobales*
Family I. *Sulfolobaceae*
Genus I. *Sulfolobus*
Genus II. *Acidianus*
Genus III. *Metallosphaera*
Genus IV. *Stygiolobus*
Genus V. *Sulfurisphaera*
Genus VI. *Sulfurococcus*
Phylum AII. *Euryarchaeota*
Class I. *Methanobacteria*
Order I. *Methanobacteriales*
Family I. *Methanobacteriaceae*
Genus I. *Methanobacterium*
Genus II. *Methanobrevibacter*
Genus III. *Methanosphaera*
Genus IV. *Methanothermobacter*
Family II. *Methanothermaceae*
Genus I. *Methanothermus*
Class II. *Methanococci*
Order I. *Methanococcales*
Family I. *Methanococcaceae*
Genus I. *Methanococcus*
Genus II. *Methanothermococcus*
Family II. *Methanocaldococcaceae*
Genus I. *Methanocaldococcus*
Genus II. *Methanotorris*
Class III. *Methanomicrobia*
Order I. *Methanomicrobiales*
Family I. *Methanomicrobiaceae*
Genus I. *Methanomicrobium*
Genus II. *Methanoculleus*
Genus III. *Methanofollis*
Genus IV. *Methanogenium*
Genus V. *Methanolacinia*
Genus VI. *Methanoplanus*
Family II. *Methanocorpusculaceae*
Genus I. *Methanocorpusculum*
Family III. *Methanospirillaceae*
Genus I. *Methanospirillum*
Genera incertae sedis[b]
Genus I. *Methanocalculus*
Order II. *Methanosarcinales*
Family I. *Methanosarcinaceae*
Genus I. *Methanosarcina*
Genus II. *Methanococcoides*
Genus III. *Methanohalobium*
Genus IV. *Methanohalophilus*
Genus V. *Methanolobus*
Genus VI. *Methanomethylovorans*
Genus VII. *Methanimicrococcus*
Genus VIII. *Methanosalsum*
Family II. *Methanosaetaceae*
Genus I. *Methanosaeta*
Class IV. *Halobacteria*
Order I. *Halobacteriales*
Family I. *Halobacteriaceae*
Genus I. *Halobacterium*
Genus II. *Haloarcula*
Genus III. *Halobaculum*
Genus IV. *Halobiforma*
Genus V. *Halococcus*
Genus VI. *Haloferax*
Genus VII. *Halogeometricum*
Genus VIII. *Halomicrobium*
Genus IX. *Halorhabdus*
Genus X. *Halorubrum*
Genus XI. *Halosimplex*
Genus XII. *Haloterrigena*
Genus XIII. *Natrialba*
Genus XIV. *Natrinema*
Genus XV. *Natronobacterium*
Genus XVI. *Natronococcus*
Genus XVII. *Natronomonas*
Genus XVIII. *Natronorubrum*
Class V. *Thermoplasmata*
Order I. *Thermoplasmatales*
Family I. *Thermoplasmataceae*
Genus I. *Thermoplasma*
Family II. *Picrophilaceae*
Genus I. *Picrophilus*
Family III. *Ferroplasmaceae*
Genus I. *Ferroplasma*
Class VI. *Thermococci*
Order I. *Thermococcales*
Family I. *Thermococcaceae*
Genus I. *Thermococcus*
Genus II. *Palaeococcus*
Genus III. *Pyrococcus*
Class VII. *Archaeoglobi*
Order I. *Archaeoglobales*
Family I. *Archaeoglobaceae*
Genus I. *Archaeoglobus*
Genus II. *Ferroglobus*
Genus III. *Geoglobus*
Class VIII. *Methanopyri*
Order I. *Methanopyrales*
Family I. *Methanopyraceae*
Genus I. *Methanopyrus*
Domain *Bacteria*
Phylum BI. *Aquificae*
Class I. *Aquificae*
Order I. *Aquificales*
Family I. *Aquificaceae*
Genus I. *Aquifex*
Genus II. *Calderobacterium*
Genus III. *Hydrogenobaculum*
Genus IV. *Hydrogenobacter*
Genus V. *Hydrogenothermus*
Genus VI. *Persephonella*
Genus VII. *Sulfurihydrogenibium*
Genus VIII. *Thermocrinis*
Genera incertae sedis[b]
Genus I. *Balnearium*
Genus II. *Desulfurobacterium*
Genus III. *Thermovibrio*
Phylum BII. *Thermotogae*
Class I. *Thermotogae*
Order I. *Thermotogales*
Family I. *Thermotogaceae*
Genus I. *Thermotoga*
Genus II. *Fervidobacterium*
Genus III. *Geotoga*
Genus IV. *Marinitoga*
Genus V. *Petrotoga*
Genus VI. *Thermosipho*
Phylum BIII. *Thermodesulfobacteria*
Class I. *Thermodesulfobacteria*
Order I. *Thermodesulfobacteriales*
Family I. *Thermodesulfobacteriaceae*
Genus I. *Thermodesulfobacterium*
Genus II. *Thermodesulfatator*
Phylum BIV. *Deinococcus-Thermus*
Class I. *Deinococci*
Order I. *Deinococcales*
Family I. *Deinococcaceae*
Genus I. *Deinococcus*
Order II. *Thermales*
Family I. *Thermaceae*
Genus I. *Thermus*
Genus II. *Marinithermus*
Genus III. *Meiothermus*
Genus IV. *Oceanithermus*
Genus V. *Vulcanithermus*
Phylum BV. *Chrysiogenetes*
Class I. *Chrysiogenetes*
Order I. *Chrysiogenales*
Family I. *Chrysiogenaceae*
Genus I. *Chrysiogenes*
Phylum BVI. *Chloroflexi*
Class I. *Chloroflexi*
Order I. *Chloroflexales*
Family I. *Chloroflexaceae*
Genus I. *Chloroflexus*
Genus II. *Chloronema*
Genus III. *Heliothrix*
Genus IV. *Roseiflexus*

[a]The list of genera and higher order taxa shown here are the organisms recognized as of 2005. Genera or higher-order taxa in quotation marks are recognized taxa whose names have not yet been validated. Because bacterial taxonomy is a work in progress, updates to the list shown here occur as new genera and species are described and as new data support new taxonomic arrangements. For further discussion on bacterial taxonomy, see Sections 14.10–14.14. For a current list of validly published genus and species names of prokaryotes, refer to http://www.bacterio.cict.fr/

[b]Taxa of uncertain affiliation. *Incertae sedis,* Latin for uncertain position.

Family II. *Oscillochloridaceae*
Genus I. *Oscillochloris*
Order II. *Herpetosiphonales*
Family I. *Herpetosiphonaceae*
Genus I. *Herpetosiphon*
Class II. *Anaerolieae*
Order I. *Anaerolinaeles*
Family I. *Anaerolinaceae*
Genus I. *Anaerolinea*
Genus II. *Caldilinea*
Phylum BVII. *Thermomicrobia*
Class I. *Thermomicrobia*
Order I. *Thermomicrobiales*
Family I. *Thermomicrobiaceae*
Genus I. *Thermomicrobium*
Phylum BVIII. *Nitrospirae*
Class I. *Nitrospira*
Order I. *Nitrospirales*
Family I. *Nitrospiraceae*
Genus I. *Nitrospira*
Genus II. *Leptospirillum*
Genus III. *Magnetobacterium*
Genus IV. *Thermodesulfovibrio*
Phylum BIX. *Deferribacteres*
Class I. *Deferribacteres*
Order I. *Deferribacterales*
Family I. *Deferribacteraceae*
Genus I. *Deferribacter*
Genus II. *Denitrovibrio*
Genus III. *Flexistipes*
Genus IV. *Geovibrio*
Genera incertae sedis[b]
Genus I. *Synergistes*
Genus II. *Caldithrix*
Phylum BX. *Cyanobacteria*
Class I. *Cyanobacteria*
Subsection I. *Subsection 1*
Family I. Family 1.1
Form genus I. *Chamaesiphon*[c]
Form genus II. *Chroococcus*
Form genus III. *Cyanobacterium*
Form genus IV. *Cyanobium*
Form genus V. *Cyanothece*
Form genus VI. *Dactylococcopsis*
Form genus VII. *Gloeobacter*
Form genus VIII. *Gloeocapsa*
Form genus IX. *Gloeothece*
Form genus X. *Microcystis*
Form genus XI. *Prochlorococcus*
Form genus XII. *Prochloron*
Form genus XIII. *Synechococcus*
Form genus XIV. *Synechocystis*
Subsection II. *Subsection 2*
Family I. Family 2.1
Form genus I. *Cyanocystis*
Form genus II. *Dermocarpella*
Form genus III. *Stanieria*
Form genus IV. *Xenococcus*
Family II. Family 2.2
Form genus I. *Chroococcidiopsis*
Form genus II. *Myxosarcina*
Form genus III. *Pleurocapsa*
Subsection III. *Subsection 3*
Family I. Family 3.1
Form genus I. *Arthrospira*
Form genus II. *Borzia*
Form genus III. *Crinalium*
Form Genus IV. *Geitlerinema*
Genus V. *Halospirulina*
Form genus VI. *Leptolyngbya*
Form genus VII. *Limnothrix*
Form genus VIII. *Lyngbya*
Form genus IX. *Microcoleus*
Form genus X. *Oscillatoria*
Form genus XI. *Planktothrix*
Form genus XII. *Prochlorothrix*
Form genus XIII. *Pseudanabaena*
Form genus XIV. *Spirulina*
Form genus XV. *Starria*
Form Genus XVI. *Symploca*
Genus XVII. *Trichodesmium*
Form genus XVIII. *Tychonema*
Subsection IV. *Subsection 4*
Family I. Family 4.1
Form genus I. *Anabaena*
Form genus II. *Anabaenopsis*
Form genus III. *Aphanizomenon*
Form genus IV. *Cyanospira*
Form genus V. *Cylindrospermopsis*
Form genus VI. *Cylindrospermum*
Form genus VII. *Nodularia*
Form genus VIII. *Nostoc*
Form genus IX. *Scytonema*
Family II. Family 4.2
Form genus I. *Calothrix*
Form genus II. *Rivularia*
Form genus III. *Tolypothrix*
Subsection V. *Subsection 5*
Family I. Subsection 5.1
Form genus I. *Chlorogloeopsis*
Form genus II. *Fischerella*
Form genus III. *Geitleria*
Form genus IV. *Iyengariella*
Form genus V. *Nostochopsis*
Form genus VI. *Stigonema*
Phylum BXI. *Chlorobi*
Class I. *Chlorobia*
Order I. *Chlorobiales*
Family I. *Chlorobiaceae*
Genus I. *Chlorobium*
Genus II. *Ancalochloris*
Genus III. *Chlorobaculum*
Genus IV. *Chloroherpeton*
Genus V. *Pelodictyon*
Genus VI. *Prosthecochloris*
Phylum BXII. *Proteobacteria*
Class I. *Alphaproteobacteria*
Order I. *Rhodospirillales*
Family I. *Rhodospirillaceae*
Genus I. *Rhodospirillum*
Genus II. *Azospirillum*
Genus III. *Inquilinus*
Genus IV. *Magnetospirillum*
Genus V. *Phaeospirillum*
Genus VI. *Rhodocista*
Genus VII. *Rhodospira*
Genus VIII. *Rhodovibrio*
Genus IX. *Roseospira*
Genus X. *Skermanella*
Genus XI. *Thalassospira*
Genus XII. *Tistrella*
Family II. *Acetobacteraceae*
Genus I. *Acetobacter*
Genus II. *Acidiphilium*
Genus III. *Acidisphaera*
Genus IV. *Acidocella*
Genus V. *Acidomonas*
Genus VI. *Asaia*
Genus VII. *Craurococcus*
Genus VIII. *Gluconacetobacter*
Genus IX. *Gluconobacter*
Genus X. *Kozakia*
Genus XI. *Muricoccus*
Genus XII. *Paracraurococcus*
Genus XIII. *Rhodopila*
Genus XIV. *Roseococcus*
Genus XV. *Rubritepids*
Genus XVI. *Stella*
Genus XVII. *Teichococcus*
Genus XVIII. *Zavarzinia*
Order II. *Rickettsiales*
Family I. *Rickettsiaceae*
Genus I. *Rickettsia*
Genus II. *Orientia*
Family II. *Anaplasmataceae*
Genus I. *Anaplasma*
Genus II. *Aegyptianella*
Genus III. *Cowdria*
Genus IV. *Ehrlichia*
Genus V. *Neorickettsia*
Genus VI. *Wolbachia*
Genus VII. *Xenohaliotis*
Family III. *Holosporaceae*
Genus I. *Holospora*
Genera incertae sedis[b]
Genus I. *Caedibacter*
Genus II. *Lyticum*
Genus III. *Odyssella*
Genus IV. *Pseudocaedibacter*
Genus V. *Symbiotes*
Genus VI. *Tectibacter*
Order III. *Rhodobacterales*
Family I. *Rhodobacteraceae*
Genus I. *Rhodobacter*
Genus II. *Ahrensia*
Genus III. *Albidovulum*
Genus IV. *Amaricoccus*
Genus V. *Antarctobacter*
Genus VI. *Gemmobacter*
Genus VII. *Hirschia*
Genus VIII. *Hyphomonas*
Genus IX. *Jannaschia*
Genus X. *Ketogulonicigenium*
Genus XI. *Leisingera*
Genus XII. *Maricaulis*
Genus XIII. *Methylarcula*
Genus XIV. *Oceanicaulis*
Genus XV. *Octadecabacter*
Genus XVI. *Pannonibacter*
Genus XVII. *Paracoccus*
Genus XVIII. *Pseudorhodobacter*
Genus XIX. *Rhodobaca*
Genus XX. *Rhodothalassium*
Genus XXI. *Rhodovulum*
Genus XXII. *Roseibium*
Genus XXIII. *Roseinatronobacter*
Genus XXIV. *Roseivivax*
Genus XXV. *Roseobacter*
Genus XXVI. *Roseovarius*
Genus XXVII. *Rubrimonas*
Genus XXVIII. *Ruegeria*
Genus XXIX. *Sagittula*
Genus XXX. *Silicibacter*
Genus XXXI. *Staleya*
Genus XXXII. *Stappia*
Genus XXXIII. *Sulfitobacter*
Order IV. *Sphingomonadales*
Family I. *Sphingomonadaceae*
Genus I. *Sphingomonas*
Genus II. *Blastomonas*
Genus III. *Erythrobacter*
Genus IV. *Erythromicrobium*

[c]The taxonomic position of the cyanobacteria in *Bergey's Manual* is left open. The term "form genus" refers to a group of cyanobacteria with very characteristic morphology found worldwide. However, not all isolates of such a type may actually fit into the same genus. In some cases, pure cultures of form genera are not available.

Genus V. *Erythromonas*
Genus VI. *Novosphingobium*
Genus VII. *Porphyrobacter*
Genus VIII. *Rhizomonas*
Genus IX. *Sandaracinobacter*
Genus X. *Sphingobium*
Genus XI. *Sphingopyxis*
Genus XII. *Zymomonas*
Order V. *Caulobacterales*
Family I. *Caulobacteraceae*
Genus I. *Caulobacter*
Genus II. *Asticcacaulis*
Genus III. *Brevundimonas*
Genus IV. *Phenylobacterium*
Order VI. *Rhizobiales*
Family I. *Rhizobiaceae*
Genus I. *Rhizobium*
Genus II. *Agrobacterium*
Genus III. *Allorhizobium*
Genus IV. *Carbophilus*
Genus V. *Chelatobacter*
Genus VI. *Ensifer*
Genus VII. *Sinorhizobium*
Family II. *Aurantimonadaceae*
Genus I. *Aurantimonas*
Genus II. *Fulvimarina*
Family III. *Bartonellaceae*
Genus I. *Bartonella*
Family IV. *Brucellaceae*
Genus I. *Brucella*
Genus II. *Mycoplana*
Genus III. *Ochrobactrum*
Family V. *Phyllobacteriaceae*
Genus I. *Phyllobacterium*
Genus II. *Aminobacter*
Genus III. *Aquamicrobium*
Genus IV. *Fluvibacter*
Genus V. *Candidatus Liberibacter*[d]
Genus VI. *Mesorhizobium*
Genus VII. *Nitratireductor*
Genus VIII. *Pseudaminobacter*
Family VI. *Methylocystaceae*
Genus I. *Methylocystis*
Genus II. *Albibacter*
Genus III. *Methylopila*
Genus IV. *Methylosinus*
Genus V. *Terasakiella*
Family VII. *Beijerinckiaceae*
Genus I. *Beijerinckia*
Genus II. *Chelatococcus*
Genus III. *Methylocapsa*
Genus IV. *Methylocella*
Family VIII. *Bradyrhizobiaceae*
Genus I. *Bradyrhizobium*
Genus II. *Afipia*
Genus III. *Agromonas*
Genus IV. *Blastobacter*
Genus V. *Bosea*
Genus VI. *Nitrobacter*
Genus VII. *Oligotropha*
Genus VIII. *Rhodoblastus*
Genus IX. *Rhodopseudomonas*
Family IX. *Hyphomicrobiaceae*
Genus I. *Hyphomicrobium*
Genus II. *Ancalomicrobium*
Genus III. *Ancylobacter*
Genus IV. *Angulomicrobium*
Genus V. *Aquabacter*
Genus VI. *Azorhizobium*
Genus VII. *Blastochloris*
Genus VIII. *Devosia*
Genus IX. *Dichotomicrobium*
Genus X. *Filomicrobium*
Genus XI. *Gemmiger*
Genus XII. *Labrys*
Genus XIII. *Methylorhabdus*
Genus XIV. *Pedomicrobium*
Genus XV. *Prosthecomicrobium*
Genus XVI. *Rhodomicrobium*
Genus XVII. *Rhodoplanes*
Genus XVIII. *Seliberia*
Genus XIX. *Starkeya*
Genus XX. *Xanthobacter*
Family X. *Methylobacteriaceae*
Genus I. *Methylobacterium*
Genus II. *Microvirga*
Genus III. *Protomonas*
Genus IV. *Roseomonas*
Family XI. *Rhodobiaceae*
Genus I. *Rhodobium*
Genus II. *Roseospirillum*
Order VII. *Parvularculales*
Family I. *Parvularculaceae*
Genus I. *Parvularcula*

Class II. *Betaproteobacteria*

Order I. *Burkholderiales*
Family I. *Burkholderiaceae*
Genus I. *Burkholderia*
Genus II. *Cupriavidus*
Genus III. *Lautropia*
Genus IV. *Limnobacter*
Genus V. *Pandoraea*
Genus VI. *Paucimonas*
Genus VII. *Polynucleobacter*
Genus VIII. *Ralstonia*
Genus IX. *Thermothrix*
Genus X. *Wautersia*
Family II. *Oxalobacteraceae*
Genus I. *Oxalobacter*
Genus II. *Duganella*
Genus III. *Herbaspirillum*
Genus IV. *Janthinobacterium*
Genus V. *Massilia*
Genus VI. *Oxalicibacterium*
Genus VII. *Telluria*
Family III. *Alcaligenaceae*
Genus I. *Alcaligenes*
Genus II. *Achromobacter*
Genus III. *Bordetella*
Genus IV. *Brackiella*
Genus V. *Derxia*
Genus VI. *Kerstersia*
Genus VII. *Oligella*
Genus VIII. *Pelistega*
Genus IX. *Pigmentiphaga*
Genus X. *Sutterella*
Genus XI. *Taylorella*
Family IV. *Comamonadaceae*
Genus I. *Comamonas*
Genus II. *Acidovorax*
Genus III. *Alicycliphilus*
Genus IV. *Brachymonas*
Genus V. *Caldimonas*
Genus VI. *Delftia*
Genus VII. *Diaphorobacter*
Genus VIII. *Hydrogenophaga*
Genus IX. *Hylemonella*
Genus X. *Lampropedia*
Genus XI. *Macromonas*
Genus XII. *Ottowia*
Genus XIII. *Polaromonas*
Genus XIV. *Ramlibacter*
Genus XV. *Rhodoferax*
Genus XVI. *Variovorax*
Genus XVII. *Xenophilus*
Genera incertae sedis[b]
Genus I. *Aquabacterium*
Genus II. *Ideonella*
Genus III. *Leptothrix*
Genus IV. *Roseateles*
Genus V. *Rubrivivax*
Genus VI. *Schlegelella*
Genus VII. *Sphaerotilus*
Genus VIII. *Tepidimonas*
Genus IX. *Thiomonas*
Genus X. *Xylophilus*
Order II. *Hydrogenophilales*
Family I. *"Hydrogenophilaceae"*
Genus I. *Hydrogenophilus*
Genus II. *Thiobacillus*
Order III. *Methylophilales*
Family I. *Methylophilaceae*
Genus I. *Methylophilus*
Genus II. *Methylobacillus*
Genus III. *Methylovorus*
Order IV. *Neisseriales*
Family I. *Neisseriaceae*
Genus I. *Neisseria*
Genus II. *Alysiella*
Genus III. *Aquaspirillum*
Genus IV. *Chromobacterium*
Genus V. *Eikenella*
Genus VI. *Formivibrio*
Genus VII. *Iodobacter*
Genus VIII. *Kingella*
Genus IX. *Laribacter*
Genus X. *Microvirgula*
Genus XI. *Morococcus*
Genus XII. *Prolinoborus*
Genus XIII. *Simonsiella*
Genus XIV. *Vitreoscilla*
Genus XV. *Vogesella*
Order V. *Nitrosomonadales*
Family I. *Nitrosomonadaceae*
Genus I. *Nitrosomonas*
Genus II. *Nitrosolobus*
Genus III. *Nitrosospira*
Family II. *Spirillaceae*
Genus I. *Spirillum*
Family III. *Gallionellaceae*
Genus I. *Gallionella*
Order VI. *Rhodocyclales*
Family I. *Rhodocyclaceae*
Genus I. *Rhodocyclus*
Genus II. *Azoarcus*
Genus III. *Azonexus*
Genus IV. *Azospira*
Genus V. *Azovibrio*
Genus VI. *Dechloromonas*
Genus VII. *Dechlorosoma*
Genus VIII. *Ferribacterium*
Genus IX. *Propionibacter*
Genus X. *Propionivibrio*
Genus XI. *Quadricoccus*
Genus XII. *Sterolibacterium*
Genus XIII. *Thauera*
Genus XIV. *Zoogloea*
Order VII. *Procabacteriales*
Family I. *Procabacteriaceae*
Genus I. *Procabacter*

Class III. *Gammaproteobacteria*

Order I. *Chromatiales*
Family I. *Chromatiaceae*
Genus I. *Chromatium*
Genus II. *Allochromatium*

[d]In bacterial taxonomy, candidatus status is for organisms known to exist by 16S rRNA gene sequencing and other key properties, but which are not yet in pure culture.

Genus III. *Amoebobacter*
Genus IV. *Halochromatium*
Genus V. *Isochromatium*
Genus VI. *Lamprobacter*
Genus VII. *Lamprocystis*
Genus VIII. *Marichromatium*
Genus IX. *Nitrosococcus*
Genus X. *Pfennigia*
Genus XI. *Rhabdochromatium*
Genus XII. *Rheinheimera*
Genus XIII. *Thermochromatium*
Genus XIV. *Thioalkalicoccus*
Genus XV. *Thiobaca*
Genus XVI. *Thiocapsa*
Genus XVII. *Thiococcus*
Genus XVIII. *Thiocystis*
Genus XIX. *Thiodictyon*
Genus XX. *Thioflavicoccus*
Genus XXI. *Thiohalocapsa*
Genus XXII. *Thiolamprovum*
Genus XXIII. *Thiopedia*
Genus XXIV. *Thiorhodococcus*
Genus XXV. *Thiorhodovibrio*
Genus XXVI. *Thiospirillum*
Family II. *Ectothiorhodospiraceae*
Genus I. *Ectothiorhodospira*
Genus II. *Alcalilimnicola*
Genus III. *Alkalispirillum*
Genus IV. *Arhodomonas*
Genus V. *Halorhodospira*
Genus VI. *Nitrococcus*
Genus VII. *Thioalkalispira*
Genus VIII. *Thialkalivibrio*
Genus IX. *Thiorhodospira*
Family III. *Halothiobacillaceae*
Genus I. *Halothiobacillus*
Order II. *Acidithiobacillales*
Family I. *Acidithiobacillaceae*
Genus I. *Acidithiobacillus*
Family II. *Thermithiobacillaceae*
Genus I. *Thermithiobacillus*
Order III. *Xanthomonadales*
Family I. *Xanthomonadaceae*
Genus I. *Xanthomonas*
Genus II. *Frateuria*
Genus III. *Fulvimonas*
Genus IV. *Luteimonas*
Genus V. *Lysobacter*
Genus VI. *Nevskia*
Genus VII. *Pseudoxanthomonas*
Genus VIII. *Rhodanobacter*
Genus IX. *Schineria*
Genus X. *Stenotrophomonas*
Genus XI. *Thermomonas*
Genus XII. *Xylella*
Order IV. *Cardiobacteriales*
Family I. *Cardiobacteriaceae*
Genus I. *Cardiobacterium*
Genus II. *Dichelobacter*
Genus III. *Suttonella*
Order V. *Thiotrichales*
Family I. *Thiotrichaceae*
Genus I. *Thiothrix*
Genus II. *Achromatium*
Genus III. *Beggiatoa*
Genus IV. *Leucothrix*
Genus V. *Thiobacterium*
Genus VI. *Thiomargarita*
Genus VII. *Thioploca*
Genus VIII. *Thiospira*
Family II. *Francisellaceae*
Genus I. *Francisella*
Family III. *Piscirickettsiaceae*
Genus I. *Piscirickettsia*
Genus II. *Cycloclasticus*
Genus III. *Hydrogenovibrio*
Genus IV. *Methylophaga*
Genus V. *Thioalkalimicrobium*
Genus VI. *Thiomicrospira*
Order VI. *Legionellales*
Family I. *Legionellaceae*
Genus I. *Legionella*
Family II. *Coxiellaceae*
Genus I. *Coxiella*
Genus II. *Aquicella*
Genus III. *Rickettsiella*
Order VII. *Methylococcales*
Family I. *Methylococcaceae*
Genus I. *Methylococcus*
Genus II. *Methylobacter*
Genus III. *Methylocaldum*
Genus IV. *Methylomicrobium*
Genus V. *Methylomonas*
Genus VI. *Methylosarcina*
Genus VII. *Methylosphaera*
Order VIII. *Oceanospirillales*
Family I. *Oceanospirillaceae*
Genus I. *Oceanospirillum*
Genus II. *Balneatrix*
Genus III. *Marinomonas*
Genus IV. *Marinospirillum*
Genus V. *Neptunomonas*
Genus VI. *Oceanobacter*
Genus VII. *Oleispira*
Genus VIII. *Pseudospirillum*
Genus IX. *Thalassolituus*
Family II. *Alcanivoraceae*
Genus I. *Alcanivorax*
Genus II. *Fundibacter*
Family III. *Hahellaceae*
Genus I. *Hahella*
Genus II. *Zooshikella*
Family IV. *Halomonadaceae*
Genus I. *Halomonas*
Genus II. *Carnimonas*
Genus III. *Chromohalobacter*
Genus IV. *Cobetia*
Genus V. *Deleya*
Genus VI. *Zymobacter*
Family V. *Oleiphilaceae*
Genus I. *Oleiphilus*
Family VI. *Saccharospirillaceae*
Genus I. *Saccharospirillum*
Order IX. *Pseudomonadales*
Family I. *Pseudomonadaceae*
Genus I. *Pseudomonas*
Genus II. *Azomonas*
Genus III. *Azotobacter*
Genus IV. *Cellvibrio*
Genus V. *Chryseomonas*
Genus VI. *Flavimonas*
Genus VII. *Mesophilobacter*
Genus VIII. *Rhizobacter*
Genus IX. *Rugamonas*
Genus X. *Serpens*
Family II. *Moraxellaceae*
Genus I. *Moraxella*
Genus II. *Acinetobacter*
Genus III. *Psychrobacter*
Family III. Incertae sedis[b]
Genus I. *Enhydrobacter*
Order X. *Alteromonadales*
Family I. *Alteromonadaceae*
Genus I. *Alteromonas*
Genus II. *Aestuariibacter*
Genus III. *Alishewanella*
Genus IV. *Colwellia*
Genus V. *Ferrimonas*
Genus VI. *Glaciecola*
Genus VII. *Idiomarina*
Genus VIII. *Marinobacter*
Genus IX. *Marinobacterium*
Genus X. *Microbulbifer*
Genus XI. *Moritella*
Genus XII. *Pseudoalteromonas*
Genus XIII. *Psychromonas*
Genus XIV. *Shewanella*
Genus XV. *Thalassomonas*
Family II. Incertae sedis[b]
Genus I. *Teredinibacter*
Order XI. *Vibrionales*
Family I. *Vibrionaceae*
Genus I. *Vibrio*
Genus II. *Allomonas*
Genus III. *Catenococcus*
Genus IV. *Enterovibrio*
Genus V. *Grimontia*
Genus VI. *Listonella*
Genus VII. *Photobacterium*
Genus VIII. *Salinivibrio*
Order XII. *Aeromonadales*
Family I. *Aeromonadaceae*
Genus I. *Aeromonas*
Genus II. *Oceanimonas*
Genus III. *Oceanisphaera*
Genus IV. *Tolumonas*
Family II. Incertae sedis: *Succinivibrionaceae*[b]
Genus I. *Succinivibrio*
Genus II. *Anaerobiospirillum*
Genus III. *Ruminobacter*
Genus IV. *Succinimonas*
Order XIII. *Enterobacteriales*
Family I. *Enterobacteriaceae*
Genus I. *Escherichia*
Genus II. *Alterococcus*
Genus III. *Arsenophonus*
Genus IV. *Brenneria*
Genus V. *Buchnera*
Genus VI. *Budvicia*
Genus VII. *Buttiauxella*
Genus VIII. *Calymmatobacterium*
Genus IX. *Cedecea*
Genus X. *Citrobacter*
Genus XI. *Edwardsiella*
Genus XII. *Enterobacter*
Genus XIII. *Erwinia*
Genus XIV. *Ewingella*
Genus XV. *Hafnia*
Genus XVI. *Klebsiella*
Genus XVII. *Kluyvera*
Genus XVIII. *Leclercia*
Genus XIX. *Leminorella*
Genus XX. *Moellerella*
Genus XXI. *Morganella*
Genus XXII. *Obesumbacterium*
Genus XXIII. *Pantoea*
Genus XXIV. *Pectobacterium*
Genus XXV. *Phlomobacter*
Genus XXVI. *Photorhabdus*
Genus XXVII. *Plesiomonas*
Genus XXVIII. *Pragia*
Genus XXIX. *Proteus*
Genus XXX. *Providencia*
Genus XXXI. *Rahnella*
Genus XXXII. *Raoultella*
Genus XXXIII. *Saccharobacter*
Genus XXXIV. *Salmonella*
Genus XXXV. *Samsonia*
Genus XXXVI. *Serratia*
Genus XXXVII. *Shigella*
Genus XXXVIII. *Sodalis*
Genus XXXIX. *Tatumella*
Genus XL. *Trabulsiella*
Genus XLI. *Wigglesworthia*

Genus XLII. *Xenorhabdus*
Genus XLIII. *Yersinia*
Genus XLIV. *Yokenella*
Order XIV. *Pasteurellales*
Family I. *Pasteurellaceae*
Genus I. *Pasteurella*
Genus II. *Actinobacillus*
Genus III. *Gallibacterium*
Genus IV. *Haemophilus*
Genus V. *Lonepinella*
Genus VI. *Mannheimia*
Genus VII. *Phocoenobacter*
Class IV. *Deltaproteobacteria*
Order I. *Desulfurellales*
Family I. *Desulfurellaceae*
Genus I. *Desulfurella*
Genus II. *Hippea*
Order II. *Desulfovibrionales*
Family I. *Desulfovibrionaceae*
Genus I. *Desulfovibrio*
Genus II. *Bilophila*
Genus III. *Lawsonia*
Family II. *Desulfomicrobiaceae*
Genus I. *Desulfomicrobium*
Family III. *Desulfohalobiaceae*
Genus I. *Desulfohalobium*
Genus II. *Desulfomonas*
Genus III. *Desulfonatronovibrio*
Genus IV. *Desulfothermus*
Family IV. *Desulfonatronumaceae*
Genus I. *Desulfonatronum*
Order III. *Desulfobacterales*
Family I. *Desulfobacteraceae*
Genus I. *Desulfobacter*
Genus II. *Desulfatibacillum*
Genus III. *Desulfobacterium*
Genus IV. *Desulfobacula*
Genus V. *Desulfobotulus*
Genus VI. *Desulfocella*
Genus VII. *Desulfococcus*
Genus VIII. *Desulfofaba*
Genus IX. *Desulfofrigus*
Genus X. *Desulfomusa*
Genus XI. *Desulfonema*
Genus XII. *Desulforegula*
Genus XIII. *Desulfosarcina*
Genus XIV. *Desulfospira*
Genus XV. *Desulfotignum*
Family II. *Desulfobulbaceae*
Genus I. *Desulfobulbus*
Genus II. *Desulfocapsa*
Genus III. *Desulfofustis*
Genus IV. *Desulforhopalus*
Genus V. *Desulfotalea*
Family III. *Nitrospinaceae*
Genus I. *Nitrospina*
Order IV. *Desulfarcales*
Family I. *Desulfarculaceae*
Genus I. *Desulfarculus*
Order V. *Desulfuromonales*
Family I. *Desulfuromonaceae*
Genus I. *Desulfuromonas*
Genus II. *Desulfuromusa*
Genus III. *Malonomonas*
Genus IV. *Pelobacter*
Family II. *Geobacteraceae*
Genus I. *Geobacter*
Genus II. *Trichlorobacter*
Order VI. *Syntrophobacterales*
Family I. *Syntrophobacteraceae*
Genus I. *Syntrophobacter*
Genus II. *Desulfacinum*
Genus III. *Desulforhabdus*
Genus IV. *Desulfovirga*
Genus V. *Thermodesulforhabdus*
Family II. *Syntrophaceae*
Genus I. *Syntrophus*
Genus II. *Desulfobacca*
Genus III. *Desulfomonile*
Genus IV. *Smithella*
Order VII. *Bdellovibrionales*
Family I. *Bdellovibrionaceae*
Genus I. *Bdellovibrio*
Genus II. *Bacteriovorax*
Genus III. *Micavibrio*
Genus IV. *Vampirovibrio*
Order VIII. *Myxococcales*
Suborder I. *Cystobacterineae*
Family I. *Cystobacteraceae*
Genus I. *Cystobacter*
Genus II. *Anaeromyxobacter*
Genus III. *Archangium*
Genus IV. *Hyalangium*
Genus V. *Melittangium*
Genus VI. *Stigmatella*
Family II. *Myxococcaceae*
Genus I. *Myxococcus*
Genus II. *Corallococcus*
Genus III. *Pyxicoccus*
Suborder II. Sorangineae
Family I. *Polyangiaceae*
Genus I. *Polyangium*
Genus II. *Byssophaga*
Genus III. *Chondromyces*
Genus IV. *Haploangium*
Genus V. *Jahnia*
Genus VI. *Sorangium*
Suborder III. *Nannocystineae*
Family I. *Nannocystaceae*
Genus I. *Nannocystis*
Genus II. *Plesiocystis*
Family II. *Haliangiaceae*
Genus I. *Haliangium*
Family III. *Kocueriaceae*
Genus I. *Kocueria*
Class V. *Epsilonproteobacteria*
Order I. *Campylobacterales*
Family I. *Campylobacteraceae*
Genus I. *Campylobacter*
Genus II. *Arcobacter*
Genus III. *Dehalospirillum*
Genus IV. *Sulfurospirillum*
Family II. *Helicobacteraceae*
Genus I. *Helicobacter*
Genus II. *Sulfurimonas*
Genus III. *Thiovulum*
Genus IV. *Wolinella*
Family III. *Nautiliaceae*
Genus I. *Nautilia*
Genus II. *Caminibacter*
Family IV. *Hydrogenimonaceae*
Genus I. *Hydrogenimonas*
Phylum BXIII. *Firmicutes*
Class I. *Clostridia*
Order I. *Clostridiales*
Family I. *Clostridiaceae*
Genus I. *Clostridium*
Genus II. *Acetivibrio*
Genus III. *Acidaminobacter*
Genus IV. *Alkaliphilus*
Genus V. *Anaerobacter*
Genus VI. *Anaerotruncus*
Genus VII. *Bryantella*
Genus VIII. *Caminicella*
Genus IX. *Caloramator*
Genus X. *Caloranaerobacter*
Genus XI. *Coprobacillus*
Genus XII. *Dorea*
Genus XIII. *Faecalibacterium*
Genus XIV. *Hespellia*
Genus XV. *Natronincola*
Genus XVI. *Oxobacter*
Genus XVII. *Parasporobacterium*
Genus XVIII. *Sarcina*
Genus XIX. *Soehngenia*
Genus XX. *Tepidibacter*
Genus XXI. *Thermobrachium*
Genus XXII. *Thermohalobacter*
Genus XXIII. *Tindallia*
Family II. *Lachnospiraceae*
Genus I. *Lachnospira*
Genus II. *Acetitomaculum*
Genus III. *Anaerofilum*
Genus IV. *Anaerostipes*
Genus V. *Butyrivibrio*
Genus VI. *Catenibacterium*
Genus VII. *Catonella*
Genus VIII. *Coprococcus*
Genus IX. *Johnsonella*
Genus X. *Lachnobacterium*
Genus XI. *Pseudobutyrivibrio*
Genus XII. *Roseburia*
Genus XIII. *Ruminococcus*
Genus XIV. *Shuttleworthia*
Genus XV. *Sporobacterium*
Family III. *Peptostreptococcaceae*
Genus I. *Peptostreptococcus*
Genus II. *Anaerococcus*
Genus III. *Filifactor*
Genus IV. *Finegoldia*
Genus V. *Fusibacter*
Genus VI. *Gallicola*
Genus VII. *Helcococcus*
Genus VIII. *Micromonas*
Genus IX. *Peptoniphilus*
Genus X. *Sedimentibacter*
Genus XI. *Sporanaerobacter*
Genus XII. *Tissierella*
Family IV. *Eubacteriaceae*
Genus I. *Eubacterium*
Genus II. *Acetobacterium*
Genus III. *Anaerovorax*
Genus IV. *Mogibacterium*
Genus V. *Pseudoramibacter*
Family V. *Peptococcaceae*
Genus I. *Peptococcus*
Genus II. *Carboxydothermus*
Genus III. *Dehalobacter*
Genus IV. *Desulfotobacterium*
Genus V. *Desulfonispora*
Genus VI. *Desulfosporosinus*
Genus VII. *Desulfotomaculum*
Genus VIII. *Pelotomaculum*
Genus IX. *Syntrophobotulus*
Genus X. *Thermoterrabacterium*
Family VI. *Heliobacteriaceae*
Genus I. *Heliobacterium*
Genus II. *Heliobacillus*
Genus III. *Heliophilum*
Genus IV. *Heliorestis*
Family VII. *Acidaminococcaceae*
Genus I. *Acidaminococcus*
Genus II. *Acetonema*
Genus III. *Allisonella*
Genus IV. *Anaeroarcus*
Genus V. *Anaeroglobus*
Genus VI. *Anaeromusa*
Genus VII. *Anaerosinus*
Genus VIII. *Anaerovibrio*
Genus IX. *Centipeda*
Genus X. *Dendrosporobacter*
Genus XI. *Dialister*
Genus XII. *Megasphaera*
Genus XIII. *Mitsuokella*
Genus XIV. *Papillibacter*

Genus XV. *Pectinatus*
Genus XVI. *Phascolarctobacterium*
Genus XVII. *Propionispira*
Genus XVIII. *Propionispora*
Genus XIX. *Quinella*
Genus XX. *Schwartzia*
Genus XXI. *Selenomonas*
Genus XXII. *Sporomusa*
Genus XXIII. *Succiniclasticum*
Genus XXIV. *Succinispira*
Genus XXV. *Veillonella*
Genus XXVI. *Zymophilus*
Family VIII. *Syntrophomonadaceae*
Genus I. *Syntrophomonas*
Genus II. *Acetogenium*
Genus III. *Aminobacterium*
Genus IV. *Aminomonas*
Genus V. *Anaerobaculum*
Genus VI. *Anaerobranca*
Genus VII. *Caldicellulosiruptor*
Genus VIII. *Carboxydocella*
Genus IX. *Dethiosulfovibrio*
Genus X. *Pelospora*
Genus XI. *Syntrophospora*
Genus XII. *Syntrophothermus*
Genus XIII. *Thermaerobacter*
Genus XIV. *Thermanaerovibrio*
Genus XV. *Thermohydrogenium*
Genus XVI. *Thermosyntropha*
Order II. *Thermoanaerobacteriales*
Family I. *Thermoanaerobacteriaceae*
Genus I. *Themoanaerobacterium*
Genus II. *Ammonifex*
Genus III. *Caldanaerobacter*
Genus IV. *Carboxydibrachium*
Genus V. *Coprothermobacter*
Genus VI. *Gelria*
Genus VII. *Moorella*
Genus VIII. *Sporotomaculum*
Genus IX. *Thermacetogenium*
Genus X. *Thermanaeromonas*
Genus XI. *Thermoanaerobacter*
Genus XII. *Thermoanaerobium*
Genus XIII. *Thermovenabulum*
Family II. *Thermodesulfobiaceae*
Genus I. *Thermodesulfobium*
Order III. *Halanaerobiales*
Family I. *Halanaerobiaceae*
Genus I. *Halanaerobium*
Genus II. *Halocella*
Genus III. *Halothermothrix*
Family II. *Halobacteroidaceae*
Genus I. *Halobacteroides*
Genus II. *Acetohalobium*
Genus III. *Halanaerobacter*
Genus IV. *Halonatronum*
Genus V. *Natroniella*
Genus VI. *Orenia*
Genus VII. *Selenihalanaerobacter*
Genus VIII. *Sporohalobacter*
Class II. *Mollicutes*
Order I. *Mycoplasmatales*
Family I. *Mycoplasmataceae*
Genus I. *Mycoplasma*
Genus II. *Eperythrozoon*
Genus III. *Haemobartonella*
Genus IV. *Ureaplasma*
Order II. *Entomoplasmatales*
Family I. *Entomoplasmataceae*
Genus I. *Entomoplasma*
Genus II. *Mesoplasma*
Family II. *Spiroplasmataceae*
Genus I. *Spiroplasma*
Order III. *Acholeplasmatales*
Family I. *Acholeplasmataceae*
Genus I. *Acholeplasma*
Genus II. *Phytoplasma*
Order IV. *Anaeroplasmatales*
Family I. *Anaeroplasmataceae*
Genus I. *Anaeroplasma*
Genus II. *Asteroleplasma*
Order V. Incertae sedis[b]
Family I. *Erysipelotrichaceae*
Genus I. *Erysipelothrix*
Genus II. *Bulleidia*
Genus III. *Holdemania*
Genus IV. *Solobacterium*
Class III. *Bacilli*
Order I. *Bacillales*
Family I. *Bacillaceae*
Genus I. *Bacillus*
Genus II. *Amphibacillus*
Genus III. *Anoxybacillus*
Genus IV. *Exiguobacterium*
Genus V. *Filobacillus*
Genus VI. *Geobacillus*
Genus VII. *Gracilibacillus*
Genus VIII. *Halobacillus*
Genus IX. *Jeotgalibacillus*
Genus X. *Lentibacillus*
Genus XI. *Marinibacillus*
Genus XII. *Oceanobacillus*
Genus XIII. *Paraliobacillus*
Genus XIV. *Saccharococcus*
Genus XV. *Salibacillus*
Genus XVI. *Ureibacillus*
Genus XVII. *Virgibacillus*
Family II. *Alicyclobacillaceae*
Genus I. *Alicyclobacillus*
Genus II. *Pasteuria*
Genus III. *Sulfobacillus*
Family III. *Caryophanaceae*
Genus I. *Caryophanon*
Family IV. *Listeriaceae*
Genus I. *Listeria*
Genus II. *Brochothrix*
Family V. *Paenibacillaceae*
Genus I. *Paenibacillus*
Genus II. *Ammoniphilus*
Genus III. *Aneurinibacillus*
Genus IV. *Brevibacillus*
Genus V. *Oxalophagus*
Genus VI. *Thermicanus*
Genus VII. *Thermobacillus*
Family VI. *Planococcaceae*
Genus I. *Planococcus*
Genus II. *Filibacter*
Genus III. *Kurthia*
Genus IV. *Planomicrobium*
Genus V. *Sporosarcina*
Family VII. *Sporolactobacillaceae*
Genus I. *Sporolactobacillus*
Genus II. *Marinococcus*
Family VIII. *Staphylococcaceae*
Genus I. *Staphylococcus*
Genus II. *Gemella*
Genus III. *Jeotgalicoccus*
Genus IV. *Macrococcus*
Genus V. *Salinicoccus*
Family IX. *Thermoactinomycetaceae*
Genus I. *Thermoactinomyces*
Family X. *Turicibacteraceae*
Genus I. *Turicibacter*
Order II. *Lactobacillales*
Family I. *Lactobacillaceae*
Genus I. *Lactobacillus*
Genus II. *Paralactobacillus*
Genus III. *Pediococcus*
Family II. *Aerococcaceae*
Genus I. *Aerococcus*
Genus II. *Abiotrophia*
Genus III. *Dolosicoccus*
Genus IV. *Eremococcus*
Genus V. *Facklamia*
Genus VI. *Globicatella*
Genus VII. *Ignavigranum*
Family III. *Carnobacteriaceae*
Genus I. *Carnobacterium*
Genus II. *Agitococcus*
Genus III. *Alkalibacterium*
Genus IV. *Allofustis*
Genus V. *Alloiococcus*
Genus VI. *Desemzia*
Genus VII. *Dolosigranulum*
Genus VIII. *Granulicatella*
Genus IX. *Isobaculum*
Genus X. *Lactosphaera*
Genus XI. *Marinilactibacillus*
Genus XII. *Trichococcus*
Family IV. *Enterococcaceae*
Genus I. *Enterococcus*
Genus II. *Atopobacter*
Genus III. *Melissococcus*
Genus IV. *Tetragenococcus*
Genus V. *Vagococcus*
Family V. *Leuconostocaceae*
Genus I. *Leuconostoc*
Genus II. *Oenococcus*
Genus III. *Weissella*
Family VI. *Streptococcaceae*
Genus I. *Streptococcus*
Genus II. *Lactococcus*
Family VII. Incertae sedis[b]
Genus I. *Acetoanaerobium*
Genus II. *Oscillospira*
Genus III. *Syntrophococcus*
Phylum BXIV. *Actinobacteria*
Class I. *Actinobacteria*
Subclass I. *Acidimicrobidae*
Order I. *Acidimicrobiales*
Suborder IV. *Acidimicrobineae*
Family I. *Acidimicrobiaceae*
Genus I. *Acidimicrobium*
Subclass II. *Rubrobacteridae*
Order I. *Rubrobacterales*
Suborder V. *Rubrobacterineae*
Family I. *Rubrobacteraceae*
Genus I. *Rubrobacter*
Genus II. *Conexibacter*
Genus III. *Solirubrobacter*
Genus IV. *Thermoleophilum*
Subclass III. *Coriobacteridae*
Order I. *Coriobacteriales*
Suborder VI. *Coriobacterineae*
Family I. *Coriobacteriaceae*
Genus I. *Coriobacterium*
Genus II. *Atopobium*
Genus III. *Collinsella*
Genus IV. *Cryptobacterium*
Genus V. *Denitrobacterium*
Genus VI. *Eggerthella*
Genus VII. *Olsenella*
Genus VIII. *Slackia*
Subclass IV. *Sphaerobacteridae*
Order I. *Sphaerobacterales*
Suborder VII. *Sphaerobacterineae*
Family I. *Sphaerobacteraceae*
Genus I. *Sphaerobacter*
Subclass V. *Actinobacteridae*
Order I. *Actinomycetales*
Suborder VIII. *Actinomycineae*
Family I. *Actinomycetaceae*
Genus I. *Actinomyces*
Genus II. *Actinobaculum*

Genus III. *Arcanobacterium*
Genus IV. *Mobiluncus*
Genus V. *Varibaculum*
Suborder IX. *Micrococcineae*
Family I. *Micrococcaceae*
Genus I. *Micrococcus*
Genus II. *Arthrobacter*
Genus III. *Citricoccus*
Genus IV. *Kocuria*
Genus V. *Nesterenkonia*
Genus VI. *Renibacterium*
Genus VII. *Rothia*
Genus VIII. *Stomatococcus*
Genus IX. *Yania*
Family II. *Bogoriellaceae*
Genus I. *Bogoriella*
Family III. *Rarobacteraceae*
Genus I. *Rarobacter*
Family IV. *Sanguibacteraceae*
Genus I. *Sanguibacter*
Family V. *Brevibacteriaceae*
Genus I. *Brevibacterium*
Family VI. *Cellulomonadaceae*
Genus I. *Cellulomonas*
Genus II.Oerskovia
Genus III. *Tropheryma*
Family VII. *Dermabacteraceae*
Genus I. *Dermabacter*
Genus II. *Brachybacterium*
Family VIII. *Dermatophilaceae*
Genus I. *Dermatophilus*
Genus II. *Kineosphaera*
Family IX. *Dermacoccaceae*
Genus I. *Dermacoccus*
Genus II. *Demetria*
Genus III. *Kytococcus*
Family X. *Intrasporangiaceae*
Genus I. *Intrasporangium*
Genus II. *Arsenicicoccus*
Genus III. *Janibacter*
Genus IV. *Knoellia*
Genus V. *Ornithinicoccus*
Genus VI. *Ornithinimicrobium*
Genus VII. *Nostocoidia*
Genus VIII. *Terrabacter*
Genus IX. *Terracoccus*
Genus X. *Tetrasphaera*
Family XI. *Jonesiaceae*
Genus I. *Jonesia*
Family XII. *Microbacteriaceae*
Genus I. *Microbacterium*
Genus II. *Agreia*
Genus III. *Agrococcus*
Genus IV. *Agromyces*
Genus V. *Aureobacterium*
Genus VI. *Clavibacter*
Genus VII. *Cryobacterium*
Genus VIII. *Curtobacterium*
Genus IX. *Frigoribacterium*
Genus X. *Leifsonia*
Genus XI. *Leucobacter*
Genus XII. *Mycetocola*
Genus XIII. *Okibacterium*
Genus XIV. *Plantibacter*
Genus XV. *Rathayibacter*
Genus XVI. *Rhodoglobus*
Genus XVII. *Salinibacterium*
Genus XVIII. *Subtercola*
Family XIII. *Beutenbergiaceae*
Genus I. *Beutenbergia*
Genus II. *Georgenia*
Genus III. *Salana*
Family XIV. *Promicromonosporaceae*
Genus I. *Promicromonospora*
Genus II. *Cellulosimicrobium*
Genus III. *Xylanibacterium*
Genus IV. *Xylanimonas*
Suborder X. *Corynebacterineae*
Family I. *Corynebacteriaceae*
Genus I. *Corynebacterium*
Family II. *Dietziaceae*
Genus I. *Dietzia*
Family III. *Gordoniaceae*
Genus I. *Gordonia*
Genus II. *Skermania*
Family IV. *Mycobacteriaceae*
Genus I. *Mycobacterium*
Family V. *Nocardiaceae*
Genus I. *Nocardia*
Genus II. *Rhodococcus*
Family VI. *Tsukamurellaceae*
Genus I. *Tsukamurella*
Family VII. *Williamsiaceae*
Genus I. *Williamsia*
Suborder XI. *Micromonosporineae*
Family I. *Micromonosporaceae*
Genus I. *Micromonospora*
Genus II. *Actinoplanes*
Genus III. *Asanoa*
Genus IV. *Catellatospora*
Genus V. *Catenuloplanes*
Genus VI. *Couchioplanes*
Genus VII. *Dactylosporangium*
Genus VIII. *Pilimelia*
Genus IX. *Spirilliplanes*
Genus X. *Verrucosispora*
Genus XI. *Virgisporangium*
Suborder XII. *Propionibacterineae*
Family I. *Propionibacteriaceae*
Genus I. *Propionibacterium*
Genus II. *Luteococcus*
Genus III. *Microlunatus*
Genus IV. *Propioniferax*
Genus V. *Propionimicrobium*
Genus VI. *Tessaracoccus*
Family II. *Nocardioidaceae*
Genus I. *Nocardioides*
Genus II. *Aeromicrobium*
Genus III. *Actinopolymorpha*
Genus IV. *Friedmanniella*
Genus V. *Hongia*
Genus VI. *Kribbella*
Genus VII. *Micropruina*
Genus VIII. *Marmoricola*
Genus IX. *Propionicimonas*
Suborder XIII. *Pseudonocardineae*
Family I. *Pseudonocardiaceae*
Genus I. *Pseudonocardia*
Genus II. *Actinoalloteichus*
Genus III. *Actinopolyspora*
Genus IV. *Amycolatopsis*
Genus V. *Crossiella*
Genus VI. *Kibdelosporangium*
Genus VII. *Kutzneria*
Genus VIII. *Prauserella*
Genus IX. *Saccharomonospora*
Genus X. *Saccharopolyspora*
Genus XI. *Streptoalloteichus*
Genus XII. *Thermobispora*
Genus XIII. *Thermocrispum*
Family II. *Actinosynnemataceae*
Genus I. *Actinosynnema*
Genus II. *Actinokineospora*
Genus III. *Lechevalieria*
Genus IV. *Lentzea*
Genus V. *Saccharothrix*
Suborder XIV. *Streptomycineae*
Family I. *Streptomycetaceae*
Genus I. *Streptomyces*
Genus II. *Kitasatospora*
Genus III. *Streptoverticillium*
Suborder XV. *Streptosporangineae*
Family I. *Streptosporangiaceae*
Genus I. *Streptosporangium*
Genus II. *Acrocarpospora*
Genus III. *Herbidospora*
Genus IV. *Microbispora*
Genus V. *Microtetraspora*
Genus VI. *Nonomuraea*
Genus VII. *Planobispora*
Genus VIII. *Planomonospora*
Genus IX. *Planopolyspora*
Genus X. *Planotetraspora*
Family II. *Nocardiopsaceae*
Genus I. *Nocardiopsis*
Genus II. *Streptomonospora*
Genus III. *Thermobifida*
Family III. *Thermomonosporaceae*
Genus I. *Thermomonospora*
Genus II. *Actinomadura*
Genus III. *Spirillospora*
Suborder XVI. *Frankineae*
Family I. *Frankiaceae*
Genus I. *Frankia*
Family II. *Geodermatophilaceae*
Genus I. *Geodermatophilus*
Genus II. *Blastococcus*
Genus III. *Modestobacter*
Family III. *Microsphaeraceae*
Genus I. *Microsphaera*
Family IV. *Sporichthyaceae*
Genus I. *Sporichthya*
Family V. *Acidothermaceae*
Genus I. *Acidothermus*
Family VI. *Kineosporiaceae*
Genus I. *Kineosporia*
Genus II. *Cryptosporangium*
Genus III. *Kineococcus*
Suborder XVII. *Glycomycineae*
Family I. *Glycomycetaceae*
Genus I. *Glycomyces*
Order II. *Bifidobacteriales*
Family I. *Bifidobacteriaceae*
Genus I. *Bifidobacterium*
Genus II. *Aeriscardovia*
Genus III. *Falcivibrio*
Genus IV. *Gardnerella*
Genus V. *Parascardovia*
Genus VI. *Scardovia*
Family II. Incertae sedis[b]
Genus I. *Actinobispora*
Genus II. *Actinocorallia*
Genus III. *Excellospora*
Genus IV. *Pelczaria*
Genus V. *Turicella*

Phylum BXV. *Planctomycetes*
Class I. *Planctomycetacia*
Order I. *Planctomycetales*
Family I. *Planctomycetaceae*
Genus I. *Planctomyces*
Genus II. *Gemmata*
Genus III. *Isosphaera*
Genus IV. *Pirellula*

Phylum BXVI. *Chlamydiae*
Class I. *Chlamydiae*
Order I. *Chlamydiales*
Family I. *Chlamydiaceae*
Genus I. *Chlamydia*
Genus II. *Chlamydophila*
Family II. *Parachlamydiaceae*
Genus I. *Parachlamydia*
Genus II. *Neochlamydia*
Family III. *Simkaniaceae*
Genus I. *Simkania*
Genus II. *Rhabdochlamydia*
Family IV. *Waddliaceae*
Genus I. *Waddlia*

Phylum BXVII. *Spirochaetes*
- **Class I. *Spirochaetes***
  - Order I. *Spirochaetales*
    - Family I. *Spirochaetaceae*
      - Genus I. *Spirochaeta*
      - Genus II. *Borrelia*
      - Genus III. *Brevinema*
      - Genus IV. *Clevelandina*
      - Genus V. *Cristispira*
      - Genus VI. *Diplocalyx*
      - Genus VII. *Hollandina*
      - Genus VIII. *Pillotina*
      - Genus IX. *Treponema*
    - Family II. *Serpulinaceae*
      - Genus I. *Serpulina*
      - Genus II. *Brachyspira*
    - Family III. *Leptospiraceae*
      - Genus I. *Leptospira*
      - Genus II. *Leptonema*

Phylum BXVIII. *Fibrobacteres*
- **Class I. *Fibrobacteres***
  - Order I. *Fibrobacterales*
    - Family I. *Fibrobacteraceae*
      - Genus I. *Fibrobacter*

Phylum BXIX. *Acidobacteria*
- **Class I. *Acidobacteria***
  - Order I. *Acidobacteriales*
    - Family I. *Acidobacteriaceae*
      - Genus I. *Acidobacterium*
      - Genus II. *Geothrix*
      - Genus III. *Holophaga*

Phylum BXX. *Bacteroidetes*
- **Class I. *Bacteroidetes***
  - Order I. Bacteroidales
    - Family I. *Bacteroidaceae*
      - Genus I. *Bacteroides*
      - Genus II. *Acetofilamentum*
      - Genus III. *Acetomicrobium*
      - Genus IV. *Acetothermus*
      - Genus V. *Anaerophaga*
      - Genus VI. *Anaerorhabdus*
      - Genus VII. *Megamonas*
    - Family II. *Rikenellaceae*
      - Genus I. *Rikenella*
      - Genus II. *Alistipes*
      - Genus III. *Marinilabilia*
    - Family III. *Porphyromonadaceae*
      - Genus I. *Porphyromonas*
      - Genus II. *Dysgonomonas*
      - Genus III. *Tannerella*
    - Family IV. *Prevotellaceae*
      - Genus I. *Prevotella*
- **Class II. *Flavobacteria***
  - Order I. *Flavobacteriales*
    - Family I. *Flavobacteriaceae*
      - Genus I. *Flavobacterium*
      - Genus II. *Aequorivita*
      - Genus III. *Arenibacter*
      - Genus IV. *Bergeyella*
      - Genus V. *Capnocytophaga*
      - Genus VI. *Cellulophaga*
      - Genus VII. *Chryseobacterium*
      - Genus VIII. *Coenonia*
      - Genus IX. *Croceibacter*
      - Genus X. *Empedobacter*
      - Genus XI. *Gelidibacter*
      - Genus XII. *Gillisia*
      - Genus XIII. *Mesonia*
      - Genus XIV. *Muricauda*
      - Genus XV. *Myroides*
      - Genus XVI. *Ornithobacterium*
      - Genus XVII. *Polaribacter*

Glossary

Only the major terms and concepts are included. If a term is not here, consult the index.

ABC (ATP-binding cassette) transporter A membrane transport system consisting of three proteins, one of which hydrolyzes ATP, one of which binds the substrate, and one of which functions as the transport channel through the membrane.

Abscess A localized infection characterized by production of pus.

Acetogenesis Energy metabolism in which acetate is produced from either H_2 plus CO_2 or from organic compounds.

Acetogen A bacterium that carries out acetogenesis.

Acetotrophic Acetate consuming.

Acetyl-CoA pathway A pathway of autotrophic CO_2 fixation widespread in obligate anaerobes including methanogens, acetogens, and sulfate-reducing bacteria.

Acetylene reduction assay Method of measuring activity of nitrogenase by substituting acetylene for the natural substrate of the enzyme, N_2. Acetylene is reduced to ethylene or ethane, depending on the nitrogenase system involved.

Acid-fastness A property of *Mycobacterium* species; cells stained with basic fuchsin dye resist decolorization with acidic alcohol.

Acid mine drainage Acidic water containing H_2SO_4 derived from the microbial oxidation of iron sulfide minerals.

Acidophile An organism that grows best at acidic pH values.

Acridine orange A nonspecific fluorescent dye used to stain microbial cells in a natural sample.

Activation energy Energy needed to make substrate molecules more reactive; enzymes function by lowering activation energy.

Activator protein A regulatory protein that binds to specific sites on DNA and stimulates transcription; involved in positive control.

Active immunity An immune state achieved by self-production of antibodies. Compare with *passive immunity*.

Active site The portion of an enzyme that is directly involved in binding substrate(s).

Active transport The energy-dependent process of transporting substances into or out of the cell in which the transported substances are chemically unchanged.

Acute infection Short-term infection usually characterized by dramatic onset.

Adaptive immunity (antigen-specific immunity) The acquired ability to recognize and destroy a particular pathogen or its products, depending on previous exposure to the pathogen or its products.

Adenosine triphosphate (ATP) a nucleotide that is the primary form in which chemical energy is conserved and utilized in cells.

Adherence This property allows cells to stick to host surfaces.

Aerobic secondary wastewater treatment Digestive reactions carried out by microorganisms under aerobic conditions to treat wastewater containing low levels of organic materials.

Aerobe An organism that grows in the presence of O_2; may be facultative, obligate, or microaerophilic.

Aerosol Suspension of particles in airborne water droplets.

Aerotolerant anaerobe An anaerobic microorganism whose growth is not inhibited by O_2.

Agglutination Reaction between antibody and particle-bound antigen resulting in visible clumping of the particles.

Agretope The portion of a processed antigen that is recognized by MHC protein.

Algae Phototrophic eukaryotic microorganisms.

Alkaliphile An organism that grows best at high pH.

Allele A sequence variant of a given gene.

Allergy A harmful immune reaction, usually caused by a foreign antigen in food, pollen, or chemicals, which results in immediate-type or delayed-type hypersensitivity.

Allosteric enzyme An enzyme that contains two combining sites, the active site (where the substrate binds) and the allosteric site (where an effector molecule binds).

Amoeboid movement A type of motility in which cytoplasmic streaming moves the organism forward.

Amino acid One of the 22 monomers that make up proteins; chemically, a two-carbon carboxylic acid containing an amino group and a characteristic substituent on the alpha carbon.

Aminoacyl-tRNA synthetase An enzyme that catalyzes the attachment of the correct amino acid to the correct tRNA.

Aminoglycoside An antibiotic such as streptomycin, containing amino sugars linked by glycosidic bonds.

Anabolic reactions (anabolism) The biochemical processes involved in the synthesis of cell constituents from simpler molecules, usually requiring energy.

Anaerobe An organism that grows in the absence of O_2; some may even be killed by O_2 (obligate or strict anaerobes).

Anaerobic respiration Use of an electron acceptor other than O_2 in an electron transport-based oxidation leading to a proton motive force.

Anammox The term that describes anoxic ammonia oxidation.

Anaphylatoxins The C3a and C5a fractions of complement that act to mimic some of the reactions of anaphylaxis.

Anaphylaxis (anaphylactic shock) A violent allergic reaction caused by an antigen–antibody reaction.

Anergy The inability to produce an immune response to specific antigens due to neutralization of effector cells.

Anoxic Oxygen-free. Usually used in reference to a microbial habitat.

Anoxic secondary wastewater treatment Digestive and fermentative reactions carried out by microorganisms under anoxic conditions to treat wastewater containing high levels of insoluble organic materials.

Anoxygenic photosynthesis Use of light energy to synthesize ATP by cyclic photophosphorylation without O_2 production.

Antenna pigments Light-harvesting chlorophylls or bacteriochlorophylls in photocomplexes that funnel energy to the reaction center.

Antibiogram A report indicating the sensitivity of clinically isolated microorganisms to the antibiotics in current use.

Antibiotic A chemical substance produced by a microorganism that kills or inhibits the growth of another microorganism.

Antibiotic resistance The acquired ability of a microorganism to grow in the presence of an antibiotic to which the microorganism is usually sensitive.

Antibody A protein, produced by differential B lymphocytes, present in serum or other body fluid that combines specifically with antigen. An immunoglobulin.

Antibody-mediated immunity Immunity resulting from direct interaction with antibodies; also called *Humoral immunity*.

Anticodon A sequence of three bases in transfer RNA that base-pairs with a codon during protein synthesis.

Antigen A molecule capable of interacting with specific components of the immune system.

Antigen-presenting cell (APC) A macrophage, dendritic cell, or B cell that presents processed antigen peptides to a T cell.

Antigenic determinant (epitope) The portion of an antigen that interacts with an immunoglobulin or T cell receptor. Also called an *epitope*.

Antigenic drift In influenza virus, minor changes in viral proteins (antigens) due to gene mutation.

Antigenic shift In influenza virus, major changes in viral proteins (antigens) due to gene reassortment.

Antimicrobial Harmful to microorganisms by either killing or inhibiting growth.

Antimicrobial agent A chemical that kills or inhibits the growth of microorganisms.

Antimicrobial drug resistance The acquired ability of a microorganism to grow in the presence of an antimicrobial drug to which the microorganism is usually susceptible.

Antiparallel In reference to double-stranded nucleic acids, the two strands run in opposite directions; one strand runs $5' \rightarrow 3'$, the complementary strand $3' \rightarrow 5'$.

Antiseptic (germicide) A chemical agent that kills or inhibits growth of microorganisms and is sufficiently nontoxic to be applied to living tissues.

Antiserum A serum containing antibodies.

Antitoxin An antibody that specifically interacts with and neutralizes a toxin.

Apoptosis Programmed cell death.

Archaea A phylogenetic domain of prokaryotes consisting of the methanogens, most extreme halophiles and hyperthermophiles, and extreme acidophiles such as *Thermoplasma*.

Artificial chromosome A single copy vector that can carry extremely long inserts of DNA and is widely used for cloning segments of large genomes.

Aseptic technique Manipulation of sterile instruments or culture media in such a way as to maintain sterility.

Aspartame A nonnutritive sweetener composed of the amino acids aspartate and phenylalanine, the latter as a methyl ester.

ATP Adenosine triphosphate, the principal energy carrier of the cell.

ATPase (ATP synthase) A multiprotein enzyme complex embedded in the cytoplasmic membrane that catalyzes the synthesis of ATP coupled to dissipation of the proton motive force.

Attenuation In a pathogen, a decrease or loss of virulence. Also, a mechanism for controlling gene expression. Typically, transcription is terminated after initiation but before a full-length mRNA is produced.

Autoantibody An antibody that reacts to self antigens.

Autoclave A sterilizer that destroys microorganisms with temperature and steam under pressure.

Autoimmunity Immune reactions of a host against its own self antigens.

Autoinducer Small signal molecule that takes part in quorum sensing.

Autoinduction A gene regulatory mechanism involving small, diffusible signal molecules.

Autolysis The lysis of a cell brought about by the activity of the cell itself.

Autoradiography Detection of radioactivity in a sample, for example, a cell or gel, by placing it in contact with a photographic film.

Autotroph An organism able to grow on CO_2 as sole source of carbon.

Auxotroph An organism that has developed a nutritional requirement through mutation. Contrast with a *Prototroph*.

B cell A lymphocyte that has immunoglobulin surface receptors, differentiates to a plasma cell, produces immunoglobulin, and may present antigens to T cells.

Bacteremia The transient appearance of bacteria in the blood.

Bacteria All prokaryotes that are not members of the domain *Archaea*.

Bacterial artificial chromosome (BAC) Circular artificial chromosome with bacterial origin of replication.

Bacteriochlorophyll Pigment of phototropic organisms consisting of light-sensitive magnesium tetrapyrroles.

Bacteriocidal agent An agent that kills bacteria.

Bacteriocins Agents produced by certain bacteria that inhibit or kill closely related species.

Bacteroid A swollen, deformed *Rhizobium* cell found in the root nodule; capable of nitrogen fixation.

Bacteriophage A virus that infects prokaryotic cells.

Bacteriorhodopsin A protein containing retinal that is found in the membranes of certain extremely halophilic *Archaea* and that is involved in light-mediated ATP synthesis.

Bacteriostatic agent An agent that inhibits bacterial growth.

Barophile An organism that lives optimally at high hydrostatic pressure.

Barotolerant An organism able to tolerate high hydrostatic pressure, although growing better at 1 atm.

Basal In a phylogenetic sense, an organism or group of organisms evolving early and giving rise to other groups.

Basal body The "motor" portion of the bacterial flagellum, embedded in the cell membrane and wall.

Base composition In reference to nucleic acids, the proportion of the total bases consisting of guanine plus cytosine or thymine plus adenine base pairs. Usually expressed as a guanine plus cytosine (GC) value, for example, 60% GC.

Batch culture A closed-system microbial culture of fixed volume.

β-lactam antibiotic An antibiotic such as penicillin that contains the four-membered heterocyclic β-lactam ring.

Binary fission Cell division whereby a cell grows by intercalary growth to twice its minimum size and then divides to form two cells.

Binomial system The system devised by Linnaeus for naming organisms by which an organism is given a genus name and a species epithet.

Biocatalysis The use of microorganisms to synthesize a product or carry out a specific chemical transformation.

Biochemical oxygen demand (BOD) The amount of dissolved oxygen consumed by microorganisms for complete oxidation of organic and inorganic material in a water sample.

Biofilm Microbial colonies encased in an adhesive, usually polysaccharide material and attached to a surface.

Biogeochemistry Study of microbially mediated chemical transformations of geochemical interest, for example, nitrogen or sulfur cycling.

Bioinformatics The use of computer programs to analyze, store, and access DNA and protein sequences.

Biological warfare The use of biological agents to kill or incapacitate a population.

Bioluminescence The enzymatic production of visible light by living organisms.

Biopolymers Polymeric materials consisting of biologically produced (and thus biodegradable) substances.

Bioremediation Use of microorganisms to remove or detoxify toxic or unwanted chemicals in an environment.

Biosynthesis The production of needed cellular constituents from other (usually simpler) molecules.

Biosynthetic penicillin Production of a particular form of penicillin by supplying the producing organism with specific side-chain precursors.

Biotechnology Use of organisms, typically genetically altered, in industrial, medical, or agricultural applications.

Biotransformation In industrial microbiology, use of microorganisms to convert a substance to a chemically modified form.

Black smoker A deep-sea hydrothermal vent emitting super-heated 250–400°C water and minerals.

Bone marrow A primary lymphoid organ containing the pleuripotent precursor cells for all blood and immune cells, including B cells.

Botulism Food poisoning due to ingestion of food containing botulinum toxin produced by *Clostridium botulinum*.

Brewing The manufacture of alcoholic beverages such as beer and ales from the fermentation of malted grains.

Broad-spectrum antibiotic An antibiotic that acts on both gram-positive and gram-negative *Bacteria*.

Calvin cycle The series of biosynthetic reactions by which most phototsynthetic organisms convert CO_2 to organic compounds.

Canning The process of sealing food in a closed container and heating to destroy living organisms.

Capsid The protein shell that surrounds the genome of a virus.

Capsomere Subunit of the virus capsid.

Capsule A dense, well-defined polysaccharide or protein layer closely surrounding a cell.

Carboxysome Polyhedral cellular inclusions of crystalline ribulose bisphosphate carboxylase (RubisCO), the key enzyme of the Calvin cycle.

Carcinogen A substance that causes the initiation of tumor formation. Frequently a mutagen.

Cardinal temperatures The minimum, maximum, and optimum growth temperatures for a given organism.

Carotenoid A hydrophobic accessory pigment present along with chlorophyll in photosynthetic membranes.

Carrier Subclinically infected individual who may spread a disease.

Cassette mutagenesis Creating mutations by the insertion of a DNA cassette.

Catabolic reactions (catabolism) The biochemical processes involved in the breakdown of organic or inorganic compounds, usually leading to the production of energy.

Catabolite repression The suppression of alternative catabolic pathways by a preferred source of carbon and energy.

Catalysis Increase in rate of a chemical reaction.

Catalyst A substance that promotes a chemical reaction without itself being changed in the end.

CD4 cells T-helper cells. They are targets for HIV infection.

Cell The fundamental unit of living matter.

Cell-mediated immunity An immune response generated by interactions with antigen-specific T cells. Compare with *Humoral immunity*.

Cell wall A rigid layer present outside the cytoplasmic membrane that confers structural strength on the cell and protection from osmotic lysis.

Centers for Disease Control and Prevention (CDC) An agency of the United States Public Health Service that tracks disease trends, provides disease information to the public and to healthcare professionals, and forms public policy regarding disease prevention and intervention.

Chaperonin (molecular chaperone) A protein that helps other proteins fold or refold from a partly denatured state.

Chemiosmosis The use of ion gradients, especially proton gradients, across membranes to generate ATP.

Chemokine A small soluble protein produced by a variety of cells that modulates inflammatory reactions and immunity in target cells.

Chemolithotroph An organism that obtains its energy from the oxidation of inorganic compounds.

Chemoorganotroph An organism that obtains its energy from the oxidation of organic compounds.

Chemostat A continuous culture device controlled by the concentration of limiting nutrient and dilution rate.

Chemotaxis Movement toward or away from a chemical.

Chemotherapeutic agent An antimicrobial agent that can be used internally.

Chemotherapy Treatment of infectious disease with chemicals or antibiotics.

Chitin A polymer of *N*-acetylglucosamine commonly found in the cell walls of fungi.

Chloramine A water purification chemical made by combining chlorine and ammonia at precise ratios.

Chlorination A highly effective disinfectant procedure for drinking water using chlorine gas or other chlorine-containing compounds as disinfectant.

Chlorine A chemical used in its gaseous state to disinfect water. A residual level is maintained throughout the distribution system.

Chlorophyll Pigment of phototrophic organisms consisting of light-sensitive magnesium tetrapyrroles.

Chloroplast The chlorophyll-containing organelle of phototrophic eukaryotes.

Chlorosome Cigar-shaped structure enclosed by a nonunit membrane and containing the light-harvesting bacteriochlorophyll (c, c_s, d, or e) in green sulfur bacteria and in *Chloroflexus*.

Chromogenic Producing color; for example, a chromogenic colony is a pigmented colony.

Chromosomal island Region of bacterial chromosome of foreign origin that contains clustered genes for some extra property such as virulence or symbiosis.

Chromosome A genetic element carrying genes essential to cellular function. Prokaryotes typically have a single chromosome consisting of a circular DNA molecule. Eukaryotes typically have several chromosomes, each containing a linear DNA molecule.

Chronic infection Long-term infection.

Cidal Lethal or killing.

Ciliate A protist characterized in part by rapid motility driven by numerous short appendages called cilia.

Cilium Short, filamentous structure that beats with many others to make a cell move.

Cirrhosis Breakdown of the normal liver architecture resulting in fibrosis.

Cistron A gene as defined by the *cis-trans* test; a segment of DNA (or RNA) that encodes a single polypeptide chain.

Citric acid cycle A cyclical series of reactions resulting in the conversion of acetate to CO_2 and NADH. Also called the *tricarboxylic acid cycle* or *Kreb's cycle*.

Clarifier (coagulation basin) A reservoir in which the suspended solids of raw water are coagulated and removed.

Class I MHC protein Antigen-presenting molecule found on all nucleated vertebrate cells.

Class II MHC protein Antigen-presenting molecule found on macrophages, B lymphocytes, and dendritic cells in vertebrates.

Clonal anergy The inability to produce an immune response to specific antigens due to neutralization of effector cells.

Clonal deletion For T cell selection in the thymus, the killing of useless or self-reactive clones.

Clonal selection A theory that each B or T lymphocyte, when stimulated by antigen, divides to form a clone of itself.

Clone In immunology, a copy of an antigen-reactive lymphocyte, usually in large numbers. Also, a number of copies of a DNA fragment obtained by allowing an inserted DNA fragment to be replicated by a phage or plasmid.

Cloning vectors Genetic elements into which genes can be recombined and replicated.

Coagulation The formation of large insoluble particles from much smaller, colloidal particles by the addition of aluminum sulfate and anionic polymers.

Coccoid Sphere-shaped.

Coccus A spherical bacterium.

Codon A sequence of three bases in messenger RNA that encodes a specific amino acid.

Codon bias Nonrandom usage of multiple codons encoding the same amino acid.

GLOSSARY

Codon usage The relative proportions of different codons encoding the same amino acid; it varies in different organisms.
Coenocytic The presence of multiple nuclei in fungal hyphae without septa.
Coenzyme A low-molecular-weight molecule that participates in an enzymatic reaction by accepting and donating electrons or functional groups. Examples: NAD^+, FAD.
Coliforms Gram-negative, nonsporing, facultatively aerobic rod that ferments lactose with gas formation within 48 hours at 35°C.
Colonial The growth form of certain protists and green algae in which several cells live together and cooperate for feeding, motility, or reproduction; an early form of multicellularity.
Colonization Multiplication of a microorganism after it has attached to host tissues or other surfaces.
Colony A macroscopically visible population of cells growing on solid medium, arising from a single cell.
Cometabolism The metabolic transformation of a substance while a second substance serves as primary energy or carbon source.
Commensalism A type of symbiosis in which only one of two organisms in a relationship benefits.
Commodity chemicals Chemicals such as ethanol that have low monetary value and thus are sold primarily in bulk.
Common-source epidemic An epidemic resulting from infection of a large number of people from a single contaminated source.
Compatible solutes Organic compounds (or potassium ions) that serve as cytoplasmic solutes to balance water relations for cells growing in environments of high salt or sugar.
Competence Ability to take up DNA and become genetically transformed.
Complement A series of proteins that react in a sequential manner with antibody–antigen complexes, mannose-binding lectin, or alternate activation pathway proteins to amplify or potentiate target cell destruction.
Complement fixation The consumption of complement by an antibody–antigen reaction.
Complementarity-determining region (CDR) A varying amino acid sequence within the variable domains of immunoglobulins or T cell receptors where most molecular contacts with antigen are made.
Complementary Nucleic acid sequences that can base-pair with each other.
Complex medium Culture media whose precise chemical composition is unknown. Also called *undefined media*.
Concatemer A DNA molecule consisting of two or more separate molecules linked end to end to form a long, linear structure.
Congenital syphilis Syphilis contracted by an infant from its mother during birth.
Conidia Asexual spores of fungi.
Conjugation Transfer of genes from one prokaryotic cell to another by a mechanism involving cell-to-cell contact.
Consensus sequence A nucleic acid sequence in which the base present in a given position is that base most commonly found when many experimentally determined sequences are compared.
Consortium A two-membered (or more) bacterial culture (or natural assemblage) in which each organism benefits from the others.
Contagious Transmissible.
Cortex The region inside the spore coat of an endospore, around the core.
Covalent bond A nonionic chemical bond formed by a sharing of electrons between two atoms.
Crenarchaeota A phylum of *Archaea* that contains both hyperthermophilic and cold-dwelling organisms.
Crista (plural, cristae) Inner membrane in a mitochondrion; site of respiration.
Culture A particular strain or kind of organism growing in a laboratory medium.
Culture medium An aqueous solution of various nutrients suitable for the growth of microorganisms.
Cutaneous Relating to the skin.
Cyanobacteria Prokaryotic oxygenic phototrophs containing chlorophyll *a* and phycobilins.
Cyclic AMP A regulatory nucleotide that participates in catabolite repression.
Cyst A resting stage formed by some bacteria and protists in which the whole cell is surrounded by a thick-walled chemically and physically resistant coating; not the same as a spore or endospore.
Cytochrome Iron-containing porphyrin complexed with proteins, which functions as an electron carrier in the electron transport system.
Cytokine A small, soluble protein produced by a leukocyte that modulates inflammatory reactions and immunity in target cells.
Cytoplasm The fluid portion of a cell, bounded by the cell membrane.
Cytoplasmic membrane A semipermeable barrier that separates the cell interior (cytoplasm) from the environment.
Cytoskeleton Cellular scaffolding typical of eukaryotic cells in which microtubules, microfilaments, and intermediate filaments define the cell's shape.

DAPI A nonspecific fluorescent dye used to stain microbial cells in a natural sample to obtain total cell numbers.
Decontamination Treatment that renders an object or inanimate surface safe to handle.
Deep sea Marine waters below a depth of 1,000 m.
Defective virus A virus that relies on another virus, the helper virus, to provide some of its components.
Defined medium Culture media whose exact chemical composition is known. Compare with *Complex medium*.
Degeneracy In relation to the genetic code, the fact that more than one codon can code for the same amino acid.
Delayed hypersensitivity An inflammatory response mediated by T lymphocytes.
Deletion Removal of a portion of a gene.
Denaturation Irreversible destruction of a macromolecule, as for example, the destruction of a protein by heat.
Denaturing gradient gel electrophoresis (DGGE) An electrophoretic technique that allows resolving nucleic acid fragments of the same size but that differ in sequence.
Dendritic cell A type of leukocyte having phagocytic and antigen-presenting properties, found in lymph nodes and spleen.
Denitrification Microbial conversion of nitrate into nitrogen gases under anoxic conditions.
Dental caries Tooth decay resulting from bacterial infection.
Dental plaque Bacterial cells encased in a matrix of extracellular polymers and salivary products, found on the teeth.
Deoxyribonucleic acid (DNA) A polymer of nucleotides connected via a phosphate–deoxyribose sugar backbone; the genetic material of cells and some viruses.
Derived In a phylogenetic sense, an organism or group of organisms evolving later, from a more basal group.
Desiccation Drying.
Dideoxynucleotide A nucleotide lacking the 3′-hydroxyl group on the deoxyribose sugar. Used in the Sanger method of DNA sequencing.
Differential media A growth medium that allows identification of microorganisms based on phenotypic properties.
Differentiation The modification of a cell in terms of structure and/or function occurring during the course of development.
Dipicolinic acid A substance unique to endospores that confers heat resistance on these structures.
Diploid In eukaryotes, an organism or cell with two chromosome complements, one derived from each haploid gamete.

Disease Injury to the host that impairs host function.
Disinfectant An antimicrobial agent used only on inanimate objects.
Disinfection The elimination of microorganisms from inanimate objects or surfaces.
Disproportionation The splitting of a chemical compound into two new compounds, one more oxidized and one more reduced than the original compound.
Distilled beverage A beverage containing alcohol concentrated by distillation.
Distribution system Water pipes, storage reservoirs, tanks, and other means used to deliver drinking water to consumers or store it before delivery.
Divisome A complex of proteins that directs cell division processes in prokaryotes.
DNA Deoxyribonucleic acid, the genetic material of cells and some viruses.
DNA cassette An artificially designed segment of DNA that usually carries a gene for resistance to an antibiotic or some other convenient marker and is flanked by convenient restriction sites.
DNA–DNA hybridization The experimental determination of genomic similarity by measuring the extent of hybridization of DNA from the genome of one organism with that of another.
DNA fingerprinting Use of DNA technology to determine the origin of DNA in a sample of tissue.
DNA gyrase An enzyme found in most prokaryotes that introduces negative supercoils in DNA.
DNA library See *gene library*.
DNA polymerase An enzyme that synthesizes a new strand of DNA in the $5' \rightarrow 3'$ direction using an antiparallel DNA strand as a template.
DNA vaccine A vaccine that uses the DNA of a pathogen to elicit an immune response.
Domain The highest level of biological classification. The three domains of biological organisms are the *Bacteria*, the *Archaea*, and the *Eukarya*. Also used to describe a region of a protein having a defined structure and function.
Doubling time The time needed for a population to double. See also *Generation time*.
Downstream position Refers to nucleic acid sequences on the 3′ side of a given site on the DNA or RNA molecule. Compare with *Upstream position*.

Early protein A protein synthesized soon after virus infection and before replication of the virus genome.
Ecology Study of the interrelationships between organisms and their environments.
Ecosystem A community of organisms and their natural environment.
Ecotype A population of genetically identical cells sharing a particular resource within an ecological niche.
Effluent water Treated wastewater discharged from a wastewater treatment facility.
Ehrlichiosis One of a group of emerging tick-transmitted diseases caused by rickettsias of the *Ehrlichia* genus.
Electron acceptor A substance that accepts electrons during an oxidation–reduction reaction.
Electron donor A compound that donates electrons in an oxidation–reduction reaction.
Electron transport phosphorylation Synthesis of ATP involving a membrane-associated electron transport chain and the creation of a proton motive force. Also called *Oxidative phosphorylation*.
Electrophoresis Separation of charged molecules in an electric field.
Electroporation The use of an electric pulse to enable cells to take up DNA.
ELISA Enzyme-linked immunosorbent assay. An immunoassay that uses specific antibodies to detect antigens or antibodies in body fluids. The antibody-containing complexes are visualized through enzymes coupled to the antibody. Addition of substrate to the enzyme–antibody–antigen complex results in a colored product.
Emerging infection Infectious disease whose incidence has increased recently or whose incidence threatens to increase in the near future.
Enantiomer One form of a molecule that is the mirror image of another form of the same molecule.
Endemic disease A disease that is constantly present in low numbers in a population, usually in low numbers. Compare with *Epidemic*.
Endergonic reaction A chemical reaction requiring an input of energy to proceed.
Endocytosis A process in which a particle such as a virus is taken intact into an animal cell. Phagocytosis and pinocytosis are two kinds of endocytosis.
Endoplasmic reticulum An extensive array of internal membranes in eukaryotes.
Endospore A differentiated cell formed within the cells of certain gram-positive bacteria that is extremely resistant to heat as well as to other harmful agents.
Endosymbiosis The engulfment of one cell type by another cell type and the subsequent and stable association of the two cells.
Endotoxin The lipopolysaccharide portion of the cell envelope of certain gram-negative *Bacteria*, which is a toxin when solubilized. Compare with *Exotoxin*.
Enriched media Media that allow metabolically fastidious organisms to grow because of the addition of specific growth factors.
Enrichment bias The skewing of enrichment culture results toward "weed" species.
Enrichment culture Use of selective culture media and incubation conditions to isolate specific microorganisms from natural samples.
Enteric bacteria A large group of gram-negative, rod-shaped *Bacteria* characterized by a facultatively aerobic metabolism and commonly found in the intestines of animals.
Enterotoxin A protein released extracellularly by a microorganism as it grows that produces immediate damage to the small intestine of the host.
Entropy A measure of the degree of disorder in a system; entropy always increases in a closed system.
Enveloped In reference to a virus, having a lipoprotein membrane surrounding the virion.
Environmental genomics (Metagenomics) Genomic analysis of pooled DNA from an environmental sample without first isolating or identifying the individual organisms.
Enzootic An endemic disease present in an animal population.
Enzyme A catalyst, usually composed of protein, that promotes specific reactions or groups of reactions.
Enzyme-linked immunosorbent assay (ELISA) Immunoassay that uses antibodies to detect antigens or antibodies in body fluids.
Epidemic A disease occurring in an unusually high number of individuals in a population at the same time. Compare with *Endemic*.
Epidemiology The study of the occurrence, distribution, and determinants of health and disease in a population.
Epilimnion The warmer and less dense surface waters of a stratified lake.
Epitope The portion of an antigen that is recognized by an immunoglobulin or a T cell receptor.
Epizootic An epidemic disease present in an animal population.
***Escherichia coli* O157:H7** An enterotoxigenic strain of *E. coli* spread by fecal contamination of animal or human origin to food and water.
Eukarya The phylogenetic domain containing all eukaryotic organisms.
Eukaryote A cell or organism having a unit membrane-enclosed nucleus and usually other organelles; a member of the *Eukarya*.

GLOSSARY

Euryarchaeota A phylum of *Archaea* that contains primarily methanogens, extreme halophiles, *Thermoplasma*, and some marine hyperthermophiles.

Evolution Descent with modification; DNA sequence variation and the inheritance of that variation.

Evolutionary distance In phylogenetic trees, the sum of the physical distance on a tree separating organisms; this distance is inversely proportional to evolutionary relatedness.

Exergonic reaction A chemical reaction that proceeds with the liberation of energy.

Exoenzyme An enzyme produced by a microorganism and then excreted into the environment.

Exon The coding sequence in a split gene. Contrast with *Introns*, the intervening noncoding regions.

Exotoxin A protein released extracellularly by a microorganism as it grows that produces immediate host cell damage. Compare with *Endotoxin*.

Exponential growth Growth of a microorganism where the cell number doubles within a fixed time period.

Exponential phase A period during the growth cycle of a population in which growth increases at an exponential rate.

Expression The ability of a gene to function within a cell in such a way that the gene product is formed.

Expression vector A cloning vector that contains the necessary regulatory sequences allowing transcription and translation of a cloned gene or genes.

Extein The portion of a protein that remains and has biological activity after the splicing out of any inteins.

Extracellular matrix (ECM) Proteins and polysaccharides that surround an animal cell and in which the cell is embedded.

Extreme barophile A barophilic organism unable to grow at a pressure of 1 atm and typically requiring several hundred atmospheres of pressure for growth.

Extreme halophile An organism whose growth is dependent on large amounts (generally $>10\%$) of NaCl.

Extremophile An organism that grows optimally under one or more chemical or physical extremes, such as high or low temperature or pH.

Facultative A qualifying adjective indicating that an organism is able to grow in either the presence or absence of an environmental factor (for example, "facultative aerobe").

FAME Fatty acid methyl ester.

Fatty acid An organic acid containing a carboxylic acid group and a hydrocarbon chain of various lengths; major components of lipids.

Feedback inhibition A decrease in the activity of the first enzyme of a biochemical pathway caused by buildup of the final product of the pathway.

Fermentation Catabolic reactions producing ATP in which organic compounds serve as both primary electron donor and ultimate electron acceptor, and ATP is produced by substrate-level phosphorylation.

Fermentation (industrial) A large-scale microbial process.

Fermenter An organism that carries out the process of fermentation.

Fermentor A growth vessel, usually quite large, used to culture microorganisms for the production of some commercially valuable product.

Ferredoxin An electron carrier of very negative reduction potential; small protein containing iron–sulfur clusters.

Fever An abnormal increase in body temperature.

Filamentous In the form of very long rods, many times longer than wide.

Filtration The removal of suspended particles from water by passing it through one or more permeable membranes or media (e.g., sand, anthracite, or diatomaceous earth).

Fimbria (plural fimbriae) Short, filamentous structures on a bacterial cell; although flagella-like in structure, generally present in many copies and not involved in motility. Plays a role in adherence to surfaces and in the formation of pellicles. See also *Pilus*.

Finished water Water delivered to the distribution system after treatment.

FISH Fluorescent *in-situ* hybridization; a process in which a cell is made fluorescent by labeling it with a specific nucleic acid probe that contains an attached fluorescent dye.

Fitness The capacity of an organism to survive and reproduce as compared to competing organisms.

Flagellum (plural flagella) A thin, filamentous organ of motility in prokaryotes that functions by rotating. In motile eukaryotes, the flagellum, if present, moves by a whiplike motion.

Flavoprotein A protein containing a derivative of riboflavin, which functions as electron carrier in the electron transport system.

Flocculation The water treatment process after coagulation that uses gentle stirring to cause suspended particles to form larger, aggregated masses (floc).

Fluorescent *in-situ* hybridization (FISH) A method employing a fluorescent dye covalently bonded to a specific nucleic acid probe for identifying or tracking organisms in the environment.

Fluorescent Having the ability to emit light of a certain wavelength when activated by light of another wavelength.

Fluorescent antibody Covalent modification of an antibody molecule with a fluorescent dye; the dye makes the antibody visible under fluorescent light.

Fomite Inanimate object that, when contaminated with a viable pathogen, can transfer the pathogen to a host population.

Food infection Disease caused by active infection resulting from ingestion of pathogen-contaminated food.

Food poisoning (food intoxication) Disease caused by the ingestion of food that contains preformed microbial toxins.

Food spoilage Any change in a food product that makes it unacceptable to the consumer.

Frameshift A type of mutation. Because the genetic code is read three bases at a time, if reading begins at either the second or third base of a codon, a faulty product usually results.

Free energy (G) Energy available to do work; $G^{0\prime}$ is free energy under standard conditions.

Fruiting body A macroscopic reproductive structure produced by some fungi (for example, mushrooms) and some *Bacteria* (for example, myxobacteria), each distinct in size, shape, and coloration.

FtsZ A protein that forms a ring along the mid-cell division plane to initiate cell division.

Fungi Nonphototrophic eukaryotic microorganisms that contain rigid cell walls.

Fungicidal agent An agent that kills fungi.

Fungistatic agent An agent that inhibits fungal growth.

Fusion inhibitor A synthetic polypeptide that binds to viral glycoproteins, inhibiting fusion of viral and host cell membranes.

Fusion protein A protein that is the result of fusing two different proteins together by merging their coding sequences into a single gene.

GC ratio In DNA (or RNA) from any organism, the percentage of the total nucleic acid that consists of guanine plus cytosine bases (expressed as mol% GC).

Gametes In eukaryotes, the haploid germ cells that result from meiosis.

Gas vesicle A gas-filled structure made of protein that, when present in the cytoplasm in large numbers, confers buoyancy on a cell.

Gel An inert polymer, usually made of agarose or polyacrylamide, used for separating macromolecules such as nucleic acids and proteins by electrophoresis.

Gel electrophoresis Technique for separation of nucleic acid molecules by passing an electric current through a gel made of agarose or polyacrylamide.

Gene A unit of heredity; a segment of DNA (or RNA in some viruses) specifying a particular protein or polypeptide chain, a tRNA or an rRNA.

Gene cloning See *Molecular cloning*.

Gene disruption (also called cassette mutagenesis or gene knockout) Inactivation of a gene by inserting into it a DNA fragment containing a selectable marker. The inserted fragment is called a DNA cassette.

Gene expression Transcription of a gene followed by translation of the resulting mRNA into protein(s).

Gene family Genes that are related in sequence to each other as the result of a common evolutionary origin.

Gene fusion A structure created by joining together segments of two separate genes, in particular when the regulatory region of one gene is joined to the coding region of a reporter gene.

Gene library A collection of cloned DNA fragments that contains all the genetic information for a particular organism.

Gene therapy Treatment of a disease caused by a dysfunctional gene by introduction of a normally functioning copy of the gene.

General purpose medium A growth medium that supports the growth of most aerobic and facultatively aerobic organisms.

Generation time The time required for a cell population to double. See also *Doubling time*.

Genetically modified organism (GMO) An organism whose genome has been altered using genetic engineering. The abbreviation is also used in constructions such as *GM crops* and *GM foods*.

Genetic code Correspondence between nucleic acid sequence and amino acid sequence of proteins.

Genetic element A structure that carries genetic information, such as a chromosome, a plasmid, or a virus genome.

Genetic engineering The use of *in vitro* techniques in the isolation, manipulation, alteration, and expression of DNA (or RNA) and in the development of genetically modified organisms.

Genetic map The arrangement of genes on a chromosome.

Genetics Heredity and variation of organisms.

Genome The total complement of genetic information of a cell or a virus.

Genomics The discipline involving mapping, sequencing, and analyzing genomes.

Genotype The complete genetic makeup of an organism; the complete description of a cell's genetic information. Compare with *Phenotype*.

Genus A taxonomic group of related species.

Germicide (antiseptic) A chemical agent that kills or inhibits growth of microorganisms and is sufficiently nontoxic to be applied to living tissues.

Glycocalyx A term that describes polysaccharide components outside of the bacterial cell wall; usually a loose network of polymer fibers extending outward from the cell.

Glycolysis Reactions of the Embden–Meyerhof pathway in which glucose is converted to pyruvate.

Glycosidic bond A type of covalent bond that links sugar units together in a polysaccharide.

Glyoxylate cycle A series of reactions including some citric acid cycle reactions that are used for aerobic growth on C_2 or C_3 organic acids.

Gonococcus *Neisseria gonorrhoeae*, the gram-negative diplococcus that causes the disease gonorrhea.

Gram-negative cell A prokaryotic cell whose cell wall contains relatively little peptidoglycan but has an outer membrane composed of lipopolysaccharide, lipoprotein, and other complex macromolecules.

Gram-positive cell Major phylogenetic lineage of prokaryotic cells that contains mainly peptidoglycans in their cell walls; stain purple in the Gram stain.

Gram stain A differential staining technique in which cells stain either pink (gram-negative) or purple (gram-positive), depending upon their structural and phylogenetic makeup.

Green fluorescent protein (GFP) A protein that fluoresces green and is widely used in genetic analysis.

Green sulfur bacteria Anoxygenic phototrophs containing chlorosomes and bacteriochlorophyll *c*, *d*, or *e* as light-harvesting chlorophyll.

Group translocation An energy-dependent transport system in which the substance transported is chemically modified during the transport process.

Growth In microbiology, an increase in cell number.

Growth factor analog A chemical agent that is related to and blocks the uptake or utilization of a growth factor.

Growth rate The rate at which growth occurs, usually expressed as the generation time.

Guild A group of metabolically related organisms.

HAART (highly active antiretroviral therapy) Treatment of HIV infection with two or more antiretroviral drugs at once to inhibit the development of drug resistance.

Habitat The location in nature where an organism resides.

Halophile An organism requiring salt (NaCl) for growth.

Halorhodopsin A light-driven chloride pump that accumulates Cl^- within the cytoplasm.

Halotolerant Capable of growing in the presence of NaCl, but not requiring it.

Hantavirus pulmonary syndrome (HPS) An emerging acute viral disease characterized by respiratory pneumonia, obtained by transmission of hantavirus from rodents.

Haploid An organism or cell containing only one set of chromosomes.

Hapten A low-molecular-weight substance not inducing antibody formation itself but still able to combine with a specific antibody.

Heat shock proteins Proteins induced by high temperature (or certain other stresses) that protect against high temperature, especially by refolding partially denatured proteins or by degrading them.

Heat shock response Response to high temperature that includes the synthesis of heat shock proteins together with other changes in gene expression.

Heliobacteria Anoxygenic phototrophs containing bacteriochlorophyll *g*.

Helix A spiral structure in a macromolecule that contains a repeating pattern.

Helper virus A virus that provides some necessary components for a defective virus.

Hemagglutination Agglutination of red blood cells.

Hemolysins Bacterial toxins capable of lysing red blood cells.

Hemolysis Lysis of red blood cells.

Hemorrhagic fever with renal syndrome (HFRS) An acute viral disease characterized by shock and kidney failure, obtained by transmission of hantavirus from rodents.

HEPA filter A high efficiency particulate air filter used in laboratories and industry to remove particles, including microorganisms, from intake or exhaust air flow.

Hepadnavirus A virus whose DNA genome replicates by way of an RNA intermediate.

Hepatitis Liver inflammation commonly caused by an infectious agent.

Herd immunity Resistance of a group to a pathogen as a result of the immunity of a large proportion of the group to that pathogen.

Herpes simplex The virus that causes both genital herpes and cold sores.

Heterocyst A differentiated cyanobacterial cell that carries out nitrogen fixation.

Heteroduplex A DNA double helix composed of single strands from two different DNA molecules.

Heterofermentative In reference to lactic acid bacteria, capable of making more than one fermentation product.

Heterotroph An organism that requires organic carbon as its carbon source; also a chemoorganotroph.

Hfr cell A cell with the F plasmid integrated into the chromosome.

Histones Basic proteins that protect and compact the DNA in eukaryotes and some *Archaea*.

Human leukocyte antigen (HLA) Antigen encoded by major histocompatibility complex genes in humans.

Homoacetogens *Bacteria* that produce acetate as the sole product of sugar fermentation or from $H_2 + CO_2$. Also called *Acetogens*.

Homofermentative In reference to lactic acid bacteria, producing only lactic acid as a fermentation product.

Homologous Related in sequence to an extent that implies common genetic ancestry; includes both orthologs and paralogs.

Homologous antigen An antigen that reacts with the antibody it has induced.

Horizontal gene transfer The transfer of genetic information between organisms as opposed to its vertical inheritance from parental organism(s).

Host (or host cell) An organism or cell type capable of supporting the growth of a virus or other parasite.

Host-to-host epidemic An epidemic resulting from host-to-host contact, characterized by a gradual rise and fall in disease incidence.

Human artificial chromosome (HAC) Artificial chromosome with human centromere sequence array.

Human granulocytic anaplasmosis (HGA) Rickettsiosis caused by *Ehrlichia ewingii* or *Anaplasma phagocytophilum*.

Human leukocyte antigen (HLA) A white blood cell (leukocyte) antigen encoded by major histocompatibility complex genes in humans.

Human monocytic ehrlichiosis (HME) Rickettsiosis caused by *Ehrlichia chaffeensis* or *Neorickettsia sennetsu*.

Human papillomavirus (HPV) A sexually transmitted virus that causes genital warts, cervical neoplasia, and cancer.

Humoral immunity An immune response involving antibodies.

Humus Dead organic matter.

Hybridization Base pairing of single strands of DNA or RNA from two different (but related) sources to give a hybrid double helix.

Hybridoma The fusion of an immortal (tumor) cell with a lymphocyte to produce an immortal lymphocyte.

Hydrogen bond A weak chemical bond between a hydrogen atom and a second, more electronegative element, usually an oxygen or nitrogen atom.

Hydrogenase An enzyme, widely distributed in anaerobic microorganisms, capable of taking up or evolving H_2.

Hydrogenosome An organelle of endosymbiotic origin in the cytoplasm of certain anaerobic eukaryotes that function to oxidize pyruvate to $H_2 + CO_2 +$ acetate.

Hydrolysis Breakdown of a polymer into smaller units, usually monomers, by addition of water; digestion.

Hydrophobic interactions Attractive forces between molecules due to the close positioning of nonhydrophilic portions of the two molecules.

Hydrothermal vents Warm or hot water-emitting springs associated with crustal spreading centers on the sea floor.

Hydroxypropionate pathway An autotrophic pathway found in *Chloroflexus* and a few *Archaea*.

Hypersensitivity An immune reaction causing damage to the host, caused either by antigen-antibody reactions or cellular immune processes. See *Allergy*.

Hyperthermophile A prokaryote having a growth temperature optimum of 80°C or higher.

Hypervariable region Variation in amino acid sequence within the variable domains of Igs or TCRs that provide most of the molecular contacts with antigen (also known as *complementarity-determining regions*).

Hypolimnion The colder, more dense, and often anoxic bottom waters of a stratified lake.

Icosahedron A geometrical shape occurring in many virus particles, with 20 triangular faces and 12 corners.

Immediate hypersensitivity An allergic response mediated by vasoactive products released from tissue mast cells.

Immobilized enzyme An enzyme attached to a solid support over which substrate is passed and converted to product.

Immune Able to resist infectious disease.

Immune memory The capacity to respond more quickly and vigorously to second and subsequent exposures to an eliciting antigen.

Immunity The ability of an organism to resist infection.

Immunization (vaccination) Inoculation of a host with inactive or weakened pathogens or pathogen products to stimulate protective immunity.

Immunoblot (Western blot) Electrophoresis of proteins followed by transfer to a membrane and detection by addition of specific antibodies. Compare with *Southern blot* and *Northern blot*.

Immunodeficiency Having a dysfunctional or completely nonfunctional immune system.

Immunogen A molecule capable of eliciting an immune response.

Immunoglobulin (Ig) A soluble protein produced by B cells and plasma cells that interacts with antigens; also called *Antibody*.

Immunoglobulin gene superfamily A family of genes that are evolutionarily, structurally, and functionally related to immunoglobulins.

Immunologic memory The ability to rapidly produce large quantities of specific immune cells or antibodies following reexposure to a previously encountered antigen.

In silico The use of computers to perform sophisticated analyses.

In vitro In glass, away from the living organism.

In vivo In the body, in a living organism.

Incidence The number of new disease cases reported in a population in a given time period.

Induced enzyme An enzyme subject to induction.

Induced mutation A mutation caused by external agents such as mutagenic chemicals or radiation.

Induction Production of an enzyme in response to a signal (often the presence of the substrate for the enzyme).

Infection Growth of an organism within a host.

Infection thread In the formation of root nodules, a cellulosic tube through which *Rhizobium* cells travel to reach and infect root cells.

Inflammation A nonspecific reaction to noxious stimuli such as toxins and pathogens, characterized by redness (erythema), swelling (edema), pain, and heat, usually localized at the site of infection.

Informational macromolecule Any large polymeric molecule that carries genetic information, including DNA, RNA, and protein.

Inhibition In reference to growth, the reduction of microbial growth because of a decrease in the number of organisms present or alterations in the microbial environment.

Innate immunity (nonspecific immunity) The noninducible ability to recognize and destroy an individual pathogen or its products that does not rely on previous exposure to a pathogen or its products.

Inoculum Cell material used to initiate a microbial culture.

Insertion A genetic phenomenon in which a piece of DNA is inserted into the middle of a gene.

Insertion sequence (IS) The simplest type of transposable element, which carries only genes involved in transposition.

Integrase The enzyme that inserts cassettes into an integrin.

Integrating vector A cloning vector that becomes integrated into a host chromosome.

Integration The process by which a DNA molecule becomes incorporated into another genome.

Integron A genetic element that accumulates and expresses genes carried on mobile cassettes.

Intein An intervening sequence in a protein; a segment of a protein that can splice itself out.

Intercalary growth In cell division, enlargement of a cell at several growing points.

Interferon Cytokine proteins produced by virus-infected cells that induce signal transduction in nearby cells, resulting in transcription of antiviral genes and expression of antiviral proteins.

Interleukin (IL) Soluble cytokine or chemokine mediator secreted by leukocytes.

Intermediate filament A filamentous polymer of fibrous keratin proteins, supercoiled into thicker fibers, that functions in maintaining cell shape and the positioning of certain organelles in the eukaryotic cell.

Interspecies hydrogen transfer The process by which organic matter is degraded by the interaction of several groups of microorganisms in which H_2 production and H_2 consumption are closely coupled.

Introns The intervening noncoding sequences in a split gene. Contrasted with *Exons,* the coding sequences.

Invasiveness Pathogenicity caused by the ability of a pathogen to enter the body and spread.

Ionophore A compound that can cause the leakage of ions across membranes.

Irradiation In food microbiology, the exposure of food to ionizing radiation to inhibit microorganisms and insect pests or to retard growth or ripening.

Isomers Two molecules with the same molecular formula but that differ structurally.

Isotopes Different forms of the same element containing the same number of protons and electrons but differing in the number of neutrons.

Isotopic fractionation Discrimination by enzymes against the heavier isotope of the various isotopes of C or S, leading to enrichment of the lighter isotopes.

Jaundice Production and release of excess bilirubin in the liver due to destruction of liver cells, resulting in yellowing of the skin and whites of the eye.

Joule (J) A unit of energy equal to 10^7 ergs, 1,000 Joules equal 1 kilojoule (kJ).

Kilobase (kb) A 1,000-base fragment of nucleic acid. A *kilobase pair* (kbp) is a fragment containing 1,000 base pairs.

Kinase An enzyme that adds a phosphoryl group, usually from ATP, to a compound.

Koch's postulates A set of criteria for proving that a given microorganism causes a given disease.

Lag phase The period after inoculation of a culture before growth begins.

Lagging strand The new strand of DNA that is synthesized in short pieces during DNA replication and then joined together later.

Laser tweezers A device used to obtain pure cultures in which a single cell is optically trapped with a laser and moved away from contaminating organisms into sterile growth medium.

Late protein A protein synthesized later in virus infection after replication of the virus genome.

Latent virus A virus present in a cell, yet not causing any detectable effect.

Lateral gene transfer Transfer of genes from a cell to another cell that is not its offspring. Also called *Horizontal gene transfer*.

Leaching Removal of valuable metals from ores by microbial action.

Leading strand The new strand of DNA that is synthesized continuously during DNA replication.

Leghemoglobin An O_2-binding protein found in root nodules.

Leukocidin A substance able to destroy phagocytes.

Leukocyte A nucleated cell found in the blood. A white blood cell.

Lichen A fungus and an alga (or a cyanobacterium) living in symbiotic association.

Lipid Water-insoluble organic molecules important in structure of the cytoplasmic membrane and (in some organisms) the cell wall. See also *Phospholipid*.

Lipopolysaccharide (LPS) Complex lipid structure containing unusual sugars and fatty acids found in most gram-negative *Bacteria* and constituting the chemical structure of the outer membrane.

Listeriosis Gastrointestinal food infection caused by *Listeria monocytogenes* that may lead to bacteremia and meningitis.

Lophotrichous Having a tuft of polar flagella.

Lower respiratory tract Trachea, bronchi, and lungs.

Luminescence Production of light.

Lyme disease An emerging tick-transmitted disease caused by the spirochete *Borrelia burgdorferi*.

Lymph A fluid similar to blood that lacks red blood cells and travels through a separate circulatory system (the lymphatic system) containing lymph nodes.

Lymphocyte A subset of leukocytes found in the blood that are involved in the adaptive immune response.

Lyophilization (freeze-drying) The process of removing all water from frozen food under vacuum.

Lysin An antibody that induces lysis.

Lysis Loss of cellular integrity with release of cytoplasmic contents.

Lysogen A prokaryote containing a prophage. See also *Temperate virus*.

Lysogeny A state following virus infection in which the viral genome is replicated as a provirus along with the genome of the host.

Lysogenic pathway A series of steps after virus infection, that leads to a state (lysogeny) in which the viral genome is replicated as a provirus along with that of the host.

Lysosome An organelle containing digestive enzymes for hydrolyses of proteins, fats, and polysaccharides.

Lytic pathway A series of steps after virus infection that lead to virus replication and the destruction (lysis) of the host cell.

Macromolecule A large molecule (polymer) formed by the connection of a number of small molecules (monomers); proteins, nucleic acids, lipids, and polysaccharides in a cell.

Macrophage A large leukocyte found in tissues that has phagocytic and antigen-presenting capabilities.

Magnetosome Small particle of Fe_3O_4 present in cells that exhibit magnetotaxis (magnetic bacteria).

Magnetotaxis Directed movement of bacterial cells by a magnetic field.

Major histocompatibility complex (MHC) A genetic region with genes that encode several proteins important for antigen presentation and other host defense functions.

Malaria An insect-transmitted disease characterized by recurrent episodes of fever and anemia caused by the protist *Plasmodium* spp., usually transmitted between mammals through the bite of the *Anopheles* mosquito.

Malignant In reference to a tumor, an infiltrating metastasizing growth no longer under normal growth control.

Mast cells Tissue cells adjoining blood vessels throughout the body that contain granules with inflammatory mediators.

Medium (plural media) In microbiology, the nutrient solution(s) used to grow microorganisms.

Megabase (Mb) One million nucleotide bases (or base pairs, abbreviated Mbp).

Meiosis Specialized form of nuclear division that halves the diploid number of chromosomes to the haploid number, for gametes of eukaryotic cells.

Membrane Any thin sheet or layer. See especially *Cytoplasmic membrane*.

Memory (immune memory) Ability to rapidly produce large quantities of specific immune cells or antibodies after subsequent exposure to a previously encountered antigen.

Memory B cell Long-lived B cell responsive to an individual antigen.

Meningitis Inflammation of the meninges (brain tissue), sometimes caused by *Neisseria meningitidis* and characterized by sudden onset of headache, vomiting, and stiff neck, often progressing to coma within hours.

Meningococcemia Fulminant disease caused by *Neisseria meningitidis* and characterized by septicemia, intravascular coagulation, and shock.

Meningoencephalitis Invasion, inflammation, and destruction of brain tissue by the amoeba *Naegleria fowlerii* or a variety of other pathogens.

Mesophile Organism living in the temperature range near that of warm-blooded animals and usually showing a growth temperature optimum between 25 and 40°C.

Messenger RNA (mRNA) An RNA molecule that contains the genetic information to encode one or more polypeptides.

Metabolism All biochemical reactions in a cell, both anabolic and catabolic.

Metabolome The total complement of small molecules and metabolic intermediates of a cell or organism.

Metagenome The total genetic complement of all the cells present in a particular environment.

Metazoa Multicellular animals.

Methanogen A methane-producing prokaryote; member of the *Archaea*.

Methanogenesis The biological production of methane (CH_4).

Methanotroph An organism capable of oxidizing methane.

Methylotroph An organism capable of oxidizing organic compounds that do not contain carbon–carbon bonds; if able to oxidize CH_4, also a methanotroph.

Microaerophile An organism requiring O_2 but at a level lower than that in air.

Microarray Small solid-state supports to which genes or portions of genes are affixed and arrayed spatially in a known pattern (also called gene chips).

Microautoradiography (MAR) Measurement of the uptake of radioactive substrates by visually observing the cells in an exposed photograph emulsion.

Microbial ecology The study of microorganisms in their natural environments.

Microbial leaching The removal of valuable metals such as copper from sulfide ores by microbial activities.

Microelectrode A small glass electrode for measuring pH or specific compounds such as O_2 or H_2S that can be immersed into a microbial habitat at microscale intervals.

Microenvironment The immediate physical and chemical surroundings of a microorganism.

Microfilament A filamentous polymer of the protein actin that helps maintain the shape of a eukaryotic cell.

Micrometer One-millionth of a meter, or 10^{-6} m (abbreviated μm), the unit used for measuring microorganisms.

Microorganism A microscopic organism consisting of a single cell or cell cluster, also including the viruses, which are not cellular.

Microtubule A filamentous polymer of the proteins α-tubulin and β-tubulin that functions in eukaryotic cell shape and motility.

Minimum inhibitory concentration (MIC) The minimum concentration of a substance necessary to prevent microbial growth.

Minus (negative)-strand nucleic acid An RNA or DNA strand that has the opposite sense of (would be complementary to) the mRNA of a virus.

Missense mutation A mutation in which a single codon is altered so that one amino acid in a protein is replaced with a different amino acid.

Mitochondrion Eukaryotic organelle responsible for the processes of respiration and electron transport phosphorylation.

Mitosis Normal form of nuclear division in eukaryotic cells in which chromosomes are replicated and partitioned into two daughter nuclei.

Mixotroph An organism that uses organic compounds as carbon sources but uses inorganic compounds as electron donors for energy metabolism.

Modification enzyme An enzyme that chemically modifies bases within a restriction enzyme recognition site and thus prevents the site from being cut.

Molds Filamentous fungi.

Molecular chaperone A protein that helps other proteins fold or refold properly.

Molecular clock A gene, such as for ribosomal RNA, whose DNA sequence can be used as a comparative temporal measure of evolutionary divergence.

Molecular cloning Isolation and incorporation of a fragment of DNA into a vector where it can be replicated.

Molecule Two or more atoms chemically bonded to one another.

Monoclonal antibody An antibody produced from a single clone of B cells. This antibody has uniform structure and specificity.

Monocytes Circulating white blood cells that contain many lysosomes and can differentiate into macrophages.

Monomer A building block of a polymer.

Monophyletic In phylogeny, a group descended from one ancestor.

Monotrichous Having a single polar flagellum.

Morbidity Incidence of illness in a population.

Morphology The shape of an organism.

Mortality Incidence of death in a population.

Most-probable number (MPN) Serial dilution of a natural sample to determine the highest dilution yielding growth.

Motif A conserved amino acid sequence found in all peptide antigens that bind to a given MHC protein.

Motility The property of movement of a cell under its own power.

Mucous membrane Layers of epithelial cells that interact with the external environment.

Mucus Soluble glycoproteins secreted by epithelial cells that coat the mucous membrane.

Multilocus sequence typing (MLST) A taxonomic tool for classifying organisms on the basis of gene sequence variations in several housekeeping genes.

Mushroom The aboveground fruiting body, or basidiocarp, of basidiomycete fungi.

Mutagen An agent that induces mutation, such as radiation or certain chemicals.

Mutant An organism whose genome carries a mutation.

Mutation An inheritable change in the base sequence of the genome of an organism.

Mutator strain A mutant strain in which the rate of mutation is increased.

Mutualism A type of symbiosis in which both organisms in the relationship benefit.

Mycorrhiza A symbiotic association between a fungus and the roots of a plant.

Mycosis Infection caused by fungi.
Myeloma A malignant tumor of a plasma cell (antibody-producing cell).

Natural killer (NK) cell A specialized lymphocyte that recognizes and destroys foreign cells or infected host cells in a nonspecific manner.
Natural penicillin The parent penicillin structure, produced by cultures of *Penicillium* not supplemented with side-chain precursors.
Negative control A mechanism for regulating gene expression in which a repressor protein prevents transcription of genes.
Negative selection In T cell selection, T cells that interact with self-antigens in the thymus are deleted. See *Clonal deletion*.
Negative strand A nucleic acid strand that has the opposite sense to (is complementary to) the mRNA.
Negative-strand virus A virus with a single-stranded genome that has the opposite sense to (is complementary to) the viral mRNA.
Neutralization Interaction of antibody with antigen that reduces or blocks the biological activity of the antigen.
Neutrophil (polymorphonuclear leukocyte) (PMN) A type of leukocyte exhibiting phagocytic properties, a granular cytoplasm (granulocyte), and a multilobed nucleus.
Neutrophile An organism that grows best around pH 7.
Niche In ecological theory, an organism's residence in a community, including both biotic and abiotic factors.
Nitrification The microbial oxidation of ammonia to nitrate (NH_3 to NO_3^-).
Nitrifying bacteria (Nitrifyers) Chemolithotrophic *Bacteria* and *Archaea* that catalyze nitrification.
Nitrogen fixation Reduction of nitrogen gas to ammonia ($N_2 + 8\ H \rightarrow 2\ NH_3 + H_2$) by the enzyme nitrogenase.
Nod factors Oligosaccharides produced by root nodule bacteria that help initiate the plant–bacterial symbiosis.
Nodule A tumorlike structure produced by the roots of symbiotic nitrogen-fixing plants. Contains the nitrogen-fixing microbial component of the symbiosis.
Noncoding RNA An RNA molecule that is not translated into protein.
Nonnucleoside reverse transcriptase inhibitor (NNRTI) A nonnucleoside compound that inhibits the action of retroviral reverse transcriptase by binding directly to the catalytic site.
Nonperishable (stable) foods Foods of low water activity that have an extended shelf life and are resistant to spoilage by microorganisms.
Nonpolar Possessing hydrophobic (water-repelling) characteristics and not easily dissolved in water.
Nonsense codon Another name for a stop codon.
Nonsense mutation A mutation in which the codon for an amino acid is changed to a stop codon.
Normal microbial flora Microorganisms that are usually found associated with healthy body tissue.
Northern blot A hybridization procedure where RNA is in the gel and DNA or RNA is the probe. Compare with *Southern blot* and *Western blot*.
Nosocomial infection Infection contracted in a hospital or healthcare setting.
Nucleic acid A polymer of nucleotides. See *Deoxyribonucleic acid* and *Ribonucleic acid*.
Nucleic acid probe A strand of nucleic acid that can be labeled and used to hybridize to a complementary molecule from a mixture of other nucleic acids. In clinical microbiology or microbial ecology, a short oligonucleotide of unique sequence used as a hybridization probe for identifying specific genes.
Nucleocapsid The complete complex of nucleic acid and protein packaged in a virus particle.
Nucleoid The aggregated mass of DNA that makes up the chromosome of prokaryotic cells.
Nucleoside A nucleotide minus phosphate.
Nucleoside reverse transcriptase inhibitor (NRTI) A nucleoside analog compound that inhibits the action of viral reverse transcriptase by competing with nucleosides.
Nucleosome Spherical complex of eukaryotic DNA plus histones.
Nucleotide A monomeric unit of nucleic acid, consisting of a sugar, a phosphate, and a nitrogenous base.
Nucleus A membrane-enclosed structure in eukaryotes containing the genetic material (DNA) organized in chromosomes.
Nutrient A substance taken by a cell from its environment and used in catabolic or anabolic reactions.

Obligate A qualifying adjective referring to an environmental condition always required for growth (for example, "obligate anaerobe").
Oligonucleotide A short nucleic acid molecule, either obtained from an organism or synthesized chemically.
Oligotrophic A habitat in which nutrients are in low supply.
Oncogene A gene whose expression causes formation of a tumor.
Open reading frame (ORF) A sequence of DNA or RNA that could be translated to give a polypeptide.
Operator A specific region of the DNA at the initial end of a gene, where the repressor protein binds and blocks mRNA synthesis.
Operon One or more genes transcribed into a single RNA and under the control of a single regulatory site.
Opportunistic infection An infection usually observed only in an individual with a dysfunctional immune system.
Opportunistic pathogen An organism that causes disease in the absence of normal host resistance.
Opsonization Promotion of phagocytosis by a specific antibody in combination with complement.
Organelle A unit membrane-enclosed structure such as the mitochondrion found in eukaryotic cells.
Ortholog Gene found in one organism that is similar to that in another organism but differs because of speciation. See also *Paralog*.
Osmophile An organism that grows best in the presence of high levels of solute, typically a sugar
Osmosis Diffusion of water through a membrane from a region of low solute concentration to one of higher concentration.
Outbreak The occurrence of a large number of cases of a disease in a short period of time.
Outer membrane A phospholipid- and polysaccharide-containing unit membrane that lies external to the peptidoglycan layer in cells of gram-negative *Bacteria*.
Overlapping genes Two or more genes in which part or all of one gene is embedded in the other.
Oxic Containing oxygen; aerobic. Usually used in reference to a microbial habitat.
Oxidation A process by which a compound gives up electrons (or H atoms) and becomes oxidized.
Oxidation-reduction (redox) reaction A pair of reactions in which one compound becomes oxidized while another becomes reduced and takes up the electrons released in the oxidation reaction.
Oxidative (electron transport) phosphorylation The nonphototrophic production of ATP at the expense of a proton motive force formed by electron transport.
Oxygenase An enzyme that catalyzes the incorporation of oxygen from O_2 into organic or inorganic compounds.
Oxygenic photosynthesis Use of light energy to synthesize ATP and NADPH by noncyclic photophosphorylation with the production of O_2 from water.

Palindrome A nucleotide sequence on a DNA molecule in which the same sequence is found on each strand but in the opposite direction.
Pandemic A worldwide epidemic.

Paralog A gene within an organism whose similarity to one or more other genes in the same organism is the result of gene duplication (compare with *Ortholog*).

Parasite An organism able to live in or on a host and cause disease.

Parasitism A symbiotic relationship between two organisms in which the host organism is harmed in the process.

Passive immunity Immunity resulting from transfer of antibodies or immune cells from an immune to a nonimmune individual.

Pasteurization Reduction of the microbial load in heat-sensitive liquids to kill disease-producing microorganisms and reduce the number of spoilage microorganisms.

Pathogen A disease-causing microorganism.

Pathogen-associated molecular pattern (PAMP) A unique repeating structural component of a microbe or virus recognized by a pattern recognition molecule.

Pathogenicity The ability of a pathogen to cause disease.

Pathogenicity island Region of bacterial chromosome of foreign origin that contains clustered genes for virulence.

Pattern recognition molecule (PRM) A membrane-bound protein that recognizes a pathogen-associated molecular pattern, such as a unique component of a microbial cell surface structure.

Pathway engineering The assembly of a new or improved biochemical pathway, using genes from one or more organisms.

Penicillin A class of antibiotics that inhibit bacterial cell wall synthesis; characterized by a β-lactam ring.

Pentose phosphate pathway A major metabolic pathway for the production and catabolism of pentoses (C_5 sugars).

Peptide bond A type of covalent bond joining amino acids in a polypeptide.

Peptidoglycan The rigid layer of the cell walls of *Bacteria,* a thin sheet composed of *N*-acetylglucosamine, *N*-acetylmuramic acid, and a few amino acids.

Perishable food Fresh food generally of high water activity that has a very short shelf life due to potential for spoilage by growth of microorganisms.

Periplasm The area between the cytoplasmic membrane and the outer membrane in gram-negative *Bacteria*.

Peritrichous flagellation In flagellar arrangements, having flagella attached to many places on the cell surface.

Peroxisome Organelles that function to rid the cell of toxic substances such as peroxides, alcohols, and fatty acids.

Pertussis (whooping cough) A disease caused by an upper respiratory tract infection with *Bordetella pertussis,* characterized by a deep persistent cough.

pH The negative logarithm of the hydrogen ion (H^+) concentration of a solution.

Phage See *Bacteriophage*.

Phagemid A cloning vector that can replicate either as a plasmid or as a bacteriophage.

Phagocyte One of a group of cells that recognizes, ingests, and degrades pathogens and pathogen products.

Phagocytosis A mechanism for ingesting particulate food in which a portion of the cytoplasmic membrane surrounds the particle and brings it into the cell.

Phenotype The observable characteristics of an organism, such as color, motility, or morphology. Compare with *Genotype*.

Phosphodiester bond A type of covalent bond linking nucleotides together in a polynucleotide.

Phospholipid Lipids containing a substituted phosphate group and two fatty acid chains on a glycerol backbone.

Photoautotroph An organism able to use light as its sole source of energy and CO_2 as its sole carbon source.

Photoheterotroph An organism using light as a source of energy and organic compounds as carbon source.

Photophosphorylation Synthesis of energy-rich phosphate bonds in ATP using light energy.

Photosynthesis The use of light energy to drive the incorporation of CO_2 into cell material. See also *Anoxygenic photosynthesis* and *Oxygenic photosynthesis*.

Phototaxis Movement of a cell toward light.

Phototroph An organism that obtains energy from light.

Phycobilin The light-capturing open chain tetrapyrrole component of phycobiliproteins.

Phycobiliprotein The accessory pigment complex in cyanobacteria that contains phycocyanin or phycoerythrin coupled to proteins.

Phycobilisome Aggregates of phycobiliproteins.

Phylogenetic probe An oligonucleotide, sometimes made fluorescent by attachment of a dye, complementary in sequence to some ribosomal RNA signature sequence.

Phylogeny The evolutionary (natural) history of organisms.

Phylotype In microbial community analysis, one unique sequence of a phylogenetic marker gene.

Phylum A major lineage of cells in one of the three domains of life.

Phytanyl A branched-chain hydrocarbon containing 20 carbon atoms, commonly found in the lipids of *Archaea*.

Phytopathogen A microorganism that causes plant disease.

Pickling The process of acidifying food, typically with acetic acid, to prevent microbial growth and spoilage.

Pilus (plural, pili) A fimbria-like structure that is present on fertile cells, both Hfr and F^+, and is involved in DNA transfer during conjugation. Sometimes called a *sex pilus*. See also *Fimbria*.

Pinocytosis In eukaryotes, phagocytosis of soluble molecules.

Plague An endemic disease in rodents caused by *Yersinia pestis* that is occasionally transferred to humans through the bite of a flea.

Plaque A zone of lysis or cell inhibition caused by virus infection on a lawn of cells.

Plasma The liquid portion of the blood with cells removed and clotting proteins deactivated.

Plasma cell A large, differentiated, short-lived B lymphocyte specializing in abundant (but short-term) antibody production.

Plasmid An extrachromosomal genetic element that is not essential for growth and has no extracellular form.

Plate count A viable counting method where the number of colonies on a plate is used as a measure of cell member.

Platelet A noncellular disc-shaped structure containing protoplasm found in large numbers in blood and functioning in the blood-clotting process.

Plus-strand nucleic acid An RNA or DNA strand that has the same sense as the mRNA of a virus.

Point mutation A mutation that involves a single base pair.

Polar Possessing hydrophilic characteristics and generally water-soluble.

Polar flagellation In flagellar arrangements, having flagella attached at one end or both ends of the cell.

Poly-β-hydroxybutyrate (PHB) A common storage material of prokaryotic cells consisting of a polymer of β-hydroxybutyrate (PHB) or other β-alkanoic acids (PHA).

Polyclonal antibodies A mixture of antibodies made by many different B cell clones.

Polyclonal antiserum A mixture of antibodies to a variety of antigens or to a variety of determinants on a single antigen.

Polymer A large molecule formed by polymerization of monomeric units. In water purification, a chemical in liquid form used as a coagulant to produce flocculation in the clarification process.

Polymerase chain reaction (PCR) Artificial amplification of a DNA sequence by repeated cycles of strand separation and replication.

Polymorphism The occurrence of multiple alleles at a locus in frequencies that cannot be explained by the occurrence of recent random mutations.

Polymorphonuclear leukocyte (PMN) Motile white blood cells containing many lysosomes and specializing in phagocytosis. Characterized by a distinct segmented nucleus. Also, a neutrophil.

Polynucleotide A polymer of nucleotides bonded to one another by phosphodiester bonds.
Polypeptide Several amino acids linked together by peptide bonds.
Polyprotein A large protein expressed from a single gene and subsequently cleaved to form several individual proteins.
Polysaccharide A long chain of monosaccharides (sugars) linked by glycosidic bonds.
Polyvalent vaccine A vaccine that immunizes against more than one disease.
Porins Protein channels in the outer membrane of gram-negative *Bacteria* through which small to medium-sized molecules can flow.
Porters Membrane proteins that function to transport substances into and out of the cell.
Positive control A mechanism for regulating gene expression in which an activator protein functions to promote transcription of genes.
Positive selection In T cell selection, T cells that interact with self MHC protein in the thymus are stimulated to grow and develop.
Positive strand A nucleic acid strand that has the same sense as the mRNA.
Positive-strand virus A virus with a single-stranded genome that has the same complementarity as the viral mRNA.
Potable In water purification, drinkable; safe for human consumption.
Precipitation A reaction between antibody and soluble antigen resulting in visible antibody–antigen complexes.
Prevalence The total number of new and existing disease cases reported in a population in a given time period.
Pribnow box The consensus sequence TATAAT located approximately 10 base pairs upstream from the transcriptional start site. A binding site for RNA polymerase.
Primary antibody response Antibodies made on first exposure to antigen; mostly of the class IgM.
Primary endosymbiosis Acquisition of the α-proteobacterial ancestor of the mitochondrion or of the cyanobacterial ancestor of the chloroplast by another kind of cell.
Primary lymphoid organ An organ in which precursor lymphoid cells develop into mature lymphocytes.
Primary metabolite A metabolite excreted during the exponential growth phase.
Primary producer An organism that synthesizes new organic material from CO_2. Also an *Autotroph*.
Primary structure In an informational macromolecule, such as a polypeptide or a nucleic acid, the precise sequence of monomeric units.
Primary transcript An unprocessed RNA molecule that is the direct product of transcription.
Primary wastewater treatment Physical separation of wastewater contaminants, usually by separation and settling.
Primer A short length of DNA or RNA used to initiate synthesis of a new DNA strand.
Prion An infectious protein whose extracellular form contains no nucleic acid.
Probe See *Nucleic acid probe*.
Probiotic A live microorganism that, when administered to a host, confers a health benefit.
Prochlorophyte A prokaryotic oxygenic phototroph that contains chlorophylls *a* and *b* but lacks phycobilins.
Prokaryote A cell or organism lacking a nucleus and other membrane-enclosed organelles and usually having its DNA in a single circular molecule. Members of the *Bacteria* and the *Archaea*.
Promoter The site on DNA where the RNA polymerase binds and begins transcription.
Prophage The state of the genome of a temperate virus when it is replicating in synchrony with that of the host, typically integrated into the host genome. See *provirus*.
Prophylactic Treatment, usually immunological or chemotherapeutic, designed to protect an individual from a future attack by a pathogen.
Prostheca A cytoplasmic extrusion from a cell such as a bud, hypha, or stalk.
Prosthetic group The tightly bound, nonprotein portion of an enzyme; not the same as a *Coenzyme*.
Protease inhibitor A compound that inhibits the action of viral protease by binding directly to the catalytic site, preventing viral protein processing.
Protein A polymeric molecule consisting of one or more polypeptides.
Protein splicing Removal of interviewing sequences from a protein.
Proteobacteria A large phylum of *Bacteria* that includes many of the common gram-negative bacteria, including *Escherichia coli*.
Proteome The total set of proteins encoded by a genome or the total protein complement of an organism.
Proteomics The large scale or genomewide study of the structure, function, and regulation of the proteins of an organism.
Proteorhodopsin A light-sensitive retinal-containing protein found in some marine *Bacteria* that catalyzes ATP formation.
Protist A unicellular eukaryotic microorganism; may be flagellated or aflagellate, phototrophic or nonphototrophic, and most lack cell walls; also referred to as protozoa.
Proton motive force An energized state of a membrane created by a protein gradient and usually formed through action of an electron transport chain. See also *Chemiosmosis*.
Protoplasm The complete cellular contents, cytoplasmic membrane, cytoplasm, and nucleus/nucleoid of a cell.
Protoplast A cell from which the wall has been removed.
Prototroph The parent from which an auxotrophic mutant has been derived. Contrast with *Auxotroph*.
Protozoa Unicellular eukaryotic microorganisms that lack cell walls.
Provirus (prophage) The genome of a temperate virus when it is replicating in step with, and often integrated into, the host chromosome.
Psychrophile An organism able to grow at low temperatures and showing a growth temperature optimum of <15°C.
Psychrotolerant Able to grow at low temperature but having a growth temperature optimum of >20°C.
Public health The health of the population as a whole.
Pure culture A culture containing a single kind of microorganism.
Purine One of the nitrogen bases of nucleic acids that contain two fused rings; adenine and guanine.
Purple nonsulfur bacteria A group of phototrophic bacteria containing bacteriochlorophyll *a* or *b* that grows best as photoheterotrophs and has a relatively low tolerance for H_2S.
Purple sulfur bacteria A group of phototrophic bacteria containing bacteriochlorophylls *a* or *b* and characterized by the ability to oxidize H_2S and store elemental sulfur inside the cells (or in the genera *Ectothiorhodospira* and *Halorhodospira*, outside the cell).
Pyogenic Pus-forming; causing abscesses.
Pyrimidine One of the nitrogen bases of nucleic acids that contain a single ring; cytosine, thymine, and uracil.
Pyrite A common iron ore, FeS_2.
Pyrogenic Fever-inducing.

Quarantine The practice of restricting the movement of individuals with highly contagious serious infections to prevent spread of the disease.
Quaternary structure In proteins, the number and arrangement of individual polypeptides in the final protein molecule.
Quinolones Synthetic antibacterial compounds that interact with DNA gyrase and prevent supercoiling of bacterial DNA.
Quorum sensing A regulatory system that monitors the population size and controls gene expression based on cell density.

Rabies A usually fatal neurological disease caused by the rabies virus that is usually transmitted by the bite or saliva of an infected carnivore.

Raw water Surface water or groundwater that has not been treated in any way (also called untreated water).

Radioimmunoassay (RIA) An immunological assay employing radioactive antibody or antigen for the detection of antigen or antibody binding.

Radioisotope An isotope of an element that undergoes spontaneous decay with the release of radioactive particles.

Reaction center A photosynthetic complex containing chlorophyll (or bacteriochlorophyll) and other components, within which occurs the initial electron transfer reactions of photophosphorylation.

Reading-frame shift See *Frameshift*.

Recalcitrant Resistant to microbial attack.

Recombinant DNA A DNA molecule containing DNA originating from two or more sources.

Recombination The process by which DNA molecules from two separate sources exchange sections or are brought together into a single DNA molecule.

Redox See *Oxidation–reduction reaction*.

Reduction A process by which a compound accepts electrons to become reduced.

Reduction potential (E_0') The inherent tendency, measured in volts, of the oxidized compound of a redox pair to become reduced.

Reductive dechlorination Removal of Cl as Cl^- from an organic compound by reducing the carbon atom from C—Cl to C—H.

Reemerging infection Infectious disease, thought to be under control, that produces a new epidemic.

Regulation Processes that control the rates of synthesis of proteins, such as induction and repression.

Regulatory nucleotide A nucleotide that functions as a signal rather than being incorporated into RNA or DNA.

Regulon A set of operons that are all controlled by the same regulatory protein (repressor or activator).

Replacement vector A cloning vector, such as a bacteriophage, in which some of the DNA of the vector can be replaced with foreign DNA.

Replication Synthesis of DNA using DNA as a template.

Replicative form A double-stranded DNA molecule that is an intermediate in the replication of single-stranded DNA viruses.

Replication fork The site on the chromosome where DNA replication occurs and where the enzymes replicating the DNA are bound to untwisted, single-stranded DNA.

Reporter gene A gene incorporated into a vector because the product it encodes is easy to detect.

Repression Preventing the synthesis of an enzyme in response to a signal.

Repressor protein A regulatory protein that binds to specific sites on DNA and blocks transcription; involved in negative control.

Reservoir A source of viable infectious agents from which individuals may be infected.

Resolution In microbiology, the ability to distinguish two objects as distinct and separate under the microscope.

Respiration Catabolic reactions producing ATP in which either organic or inorganic compounds are primary electron donors and organic or inorganic compounds are ultimate electron acceptors.

Response regulator protein One of the members of a two-component system; a regulatory protein that is phosphorylated by a sensor protein (see *Sensor kinase protein*).

Restriction enzymes (restriction endonucleases) Enzymes that recognize and cleave specific DNA sequences, generating either blunt or single-stranded (sticky) ends.

Restriction map A map showing the location of restriction enzyme cut sites on a segment of DNA.

Retrovirus A virus whose RNA genome has a DNA intermediate as part of its replication cycle.

Reverse citric acid cycle A mechanism for autotrophy in green sulfur bacteria and several nonphototrophic prokaryotes.

Reverse electron transport The energy-dependent movement of electrons against the thermodynamic gradient to form a strong electron donor from a weaker electron donor.

Reverse DNA gyrase A topoisomerase present in all hyperthermophilic prokaryotes that introduces positive supercoils in DNA.

Reverse transcription The process of copying information found in RNA into DNA.

Reversion Alteration in DNA that reverses the effects of a prior mutation.

Rheumatic fever An inflammatory autoimmune disease triggered by an immune response to infection by *Streptococcus pyogenes*.

Rhizosphere The region immediately adjacent to plant roots.

Ribonucleic acid (RNA) A polymer of nucleotides connected via a phosphate–ribose backbone; involved in protein synthesis or as genetic material of some viruses.

Ribosomal Database Project (RDP) A large database of small subunit ribosomal RNA gene sequences that can be retrieved electronically and used in comparative ribosomal RNA gene sequence studies.

Ribosomal RNA (rRNA) Type of RNA found in the ribosome; some rRNAs participate actively in the process of protein synthesis.

Ribosome Structure composed of RNAs and proteins upon which new proteins are made.

Riboswitch An RNA domain, usually in an mRNA molecule, that can bind a specific small molecule and alter its secondary structure; this in turn controls translation of the mRNA.

Ribotyping A means of identifying microorganisms from analysis of DNA fragments generated from restriction enzyme digestion of genes encoding their 16S rRNA.

Ribozyme An RNA molecule that can catalyze a chemical reaction.

Ribulose monophosphate pathway A reaction series in certain methylotrophs in which formaldehyde is assimilated into cell material using ribulose monophosphate as the C_1 acceptor molecule.

Rickettsia Obligate intracellular bacteria that cause disease, including typhus, Rocky Mountain spotted fever, and ehrlichiosis.

RNA Ribonucleic acid, functions in protein synthesis as messenger RNA, transfer RNA, and ribosomal RNA.

RNA editing Changing the coding sequence of an RNA molecule by altering, adding, or removing bases.

RNA interference (RNAi) A response that is triggered by the presence of double-stranded RNA and results in the degradation of ssRNA homologous to the inducing dsRNA.

RNA life A hypothetical ancient life form lacking DNA and protein, in which RNA had both a genetic coding and a catalytic function.

RNA polymerase An enzyme that synthesizes RNA in the $5' \rightarrow 3'$ direction using an antiparallel $3' \rightarrow 5'$ DNA strand as a template.

RNA processing The conversion of a precursor RNA to its mature form.

RNA replicase An enzyme that can produce RNA from an RNA template.

Rocky Mountain spotted fever A tick-transmitted disease caused by *Rickettsia rickettsii*, causing fever, headache, rash, and gastrointestinal symptoms.

Rolling circle replication A mechanism, used by some plasmids and viruses of replicating circular DNA, which starts by nicking and unrolling one strand and using the other, still circular strand as a template for DNA synthesis.

Root nodule A tumorlike growth on certain plant roots that contains symbiotic nitrogen-fixing bacteria.

RubisCO The acronym for ribulose bisphosphate carboxylase, a key enzyme of the Calvin cycle.

Rumen The forestomach of ruminant animals in which cellulose digestion occurs.

S-layer A paracrystalline outer wall layer composed of protein or glycoprotein and found in many prokaryotes.

Salmonellosis Enterocolitis caused by any of more than 2,000 variants of *Salmonella* species.

Sanitizers Agents that reduce, but may not eliminate, microbial numbers to a safe level.

Scale-up Conversion of an industrial process from a small laboratory setup to a large commercial fermentation.

Scarlet fever Disease characterized by high fever and a reddish skin rash resulting from an exotoxin produced by cells of *Streptococcus pyogenes*.

Screening A procedure that permits the identification of organisms by phenotype or genotype, but does not inhibit or enhance the growth of particular phenotypes or genotypes.

Secondary antibody response Antibody made on second (subsequent) exposure to antigen; mostly of the class IgG.

Secondary endosymbiosis Acquisition by a mitochondrion-containing eukaryotic cell of the chloroplasts of a red or green algal cell.

Secondary fermentation A fermentation where the substrates are the fermentation products of some other organism.

Secondary metabolite A product excreted by a microorganism near the end of the exponential growth phase during the stationary phase.

Secondary structure The initial pattern of folding of a polypeptide or a polynucleotide, usually the result of hydrogen bonding.

Secondary treatment In sewage treatment, either the aerobic or anaerobic decomposition of sewage following the removal of nondegradable objects by primary treatment.

Secretion vector A DNA vector in which the protein product is both expressed in and secreted (excreted) from the cell.

Sediment In water purification, soil, sand, minerals, and other large particles found in raw water. In large bodies of water (lakes, the oceans), the materials (mud, rock, and the like) that form the bottom surface of the water body.

Selection Placing organisms under conditions where the growth of those with a particular phenotype or genotype will be favored or inhibited.

Selective medium A growth medium that enhances the growth of certain organisms while inhibiting the growth of others due to an added media component.

Selective toxicity The ability of a compound to inhibit or kill pathogenic microorganisms without adversely affecting the host.

Self-splicing intron An intron that possesses ribozyme activity and splices itself out.

Semiconservative replication DNA synthesis yielding new double helices, each consisting of one parental and one progeny strand.

Semiperishable food Food of intermediate water activity that has a limited shelf life due to potential for spoilage by growth of microorganisms.

Semisynthetic penicillin A natural penicillin that has been chemically altered.

Sensitivity In immunodiagnostics, the lowest amount of antigen that can be detected in an immunological assay.

Sensor kinase protein One of the members of a two-component system; a kinase found in the cell membrane that phosphorylates itself in response to an external signal and then passes the phosphoryl group to a response regulator protein (see *Response regulator protein*).

Septicemia Infection of the bloodstream by microorganisms.

Sequencing In reference to nucleic acids, deducing the order of nucleotides in a DNA or RNA molecule.

Serine pathway A reaction series in certain methylotrophs in which formaldehyde is assimilated into cell material by way of the amino acid serine.

Serology The study of antigen–antibody reactions *in vitro*.

Serum Fluid portion of blood remaining after the blood cells and materials responsible for clotting are removed.

Sewage Liquid effluents contaminated with human or animal fecal material.

Sexually transmitted infection (STI) An infection that is usually transmitted by sexual contact.

Shine–Dalgarno sequence A short stretch of nucleotides on a prokaryotic mRNA molecule upstream of the translational start site that binds to ribosomal RNA and thereby brings the ribosome to the initiation codon on the mRNA.

Short interfering RNA (siRNA) Short double-stranded RNA molecules that trigger RNA interference.

Shotgun cloning Making a gene library by random cloning of DNA fragments.

Shotgun sequencing Sequencing of DNA from previously cloned small fragments of a genome in a random fashion followed by computational methods to reconstruct the entire genome sequence.

Shuttle vector A cloning vector that can replicate in two different organisms; used for moving DNA between unrelated organisms.

Sickle cell anemia A genetic trait that confers resistance to malaria but causes a reduction in the number and efficiency of red blood cells by reducing the oxygen-binding affinity of hemoglobin.

Siderophore An iron chelator that can bind iron present at very low concentrations.

Signal sequence A special N-terminal sequence of approximately 20 amino acids that signals that a protein should be exported across the cytoplasmic membrane.

Signature sequence Short oligonucleotides of defined sequence SSU ribosomal RNA characteristic of specific organisms or a group of phylogenetically related organisms; useful for constructing probes.

Silent mutation A change in DNA sequence that has no effect on the phenotype.

Simple transport system A transporter that consists of only a membrane-spanning protein and typically driven by energy from the proton motive force.

Single-cell protein Protein derived from microbial cells for use as food or a food supplement.

Site-directed mutagenesis A technique whereby a gene with a specific mutation can be constructed *in vitro*.

16S rRNA A large polynucleotide (~1,500 bases) that functions as a part of the small subunit of the ribosome of prokaryotes (*Bacteria* and *Archaea*) and from whose sequence evolutionary relationships can be obtained; eukaryotic counterpart, 18S rRNA.

Slime layer A diffuse layer of polymer fibers, typically polysaccharides, that forms an outer surface layer on the cell.

Slime molds Nonphototrophic eukaryotic microorganisms lacking cell walls, which aggregate to form fruiting structures (cellular slime molds) or simply masses of protoplasm (acellular slime molds).

Small subunit (SSU) RNA Ribosomal RNA from the 30S ribosomal subunit of *Bacteria* and *Archaea* or the 40S ribosomal subunit of eukaryotes, that is, 16S or 18S ribosomal RNA, respectively.

Solfatara A hot, sulfur-rich, generally acidic environment commonly inhabited by hyperthermophilic *Archaea*.

Somatic hypermutation The mutation of immunoglobulin genes at rates higher than those observed in other genes.

Southern blot A hybridization procedure where DNA is in the gel and RNA or DNA is the probe. Compare with *Northern blot* and *Western blot*.

Species Defined in microbiology as a collection of strains that all share the same major properties and differ in one or more significant properties from other collections of strains; defined phylogenetically as monophyletic, exclusive groups based on DNA sequence.

Specificity The ability of the immune response to interact with individual antigens.

Spheroplast A spherical, osmotically sensitive cell derived from a bacterium by loss of some but not all of the rigid wall layer. If all the rigid wall layer has been completely lost, the structure is called a *protoplast*.

Spirilla Spiral-shaped cells (singular, spirillum).
Spirochete A slender, tightly coiled gram-negative bacterium characterized by possession of endoflagella used for motility.
Spliceosome A complex of ribonucleoproteins that catalyze the removal of introns from RNA primary transcripts.
Splicing The RNA-processing step by which introns are removed and exons joined.
Spontaneous generation The hypothesis that living organisms can originate from nonliving matter.
Spontaneous mutation A mutation that occurs "naturally" without the help of mutagenic chemicals or radiation.
Spore A general term for resistant resting structures formed by many prokaryotes and fungi.
Sporozoa Nonmotile parasitic protozoa.
Stalk An elongate structure, either cellular or excreted, that anchors a cell to a surface.
Stable isotope probing (SIP) Characterizing an organism that incorporates a particular substrate by feeding the substrate in ^{13}C form and then isolating ^{13}C-enriched DNA and analyzing the genes.
Start codon A special codon, usually AUG, that signals the start of a protein.
Static Inhibitory.
Stationary phase The period during the growth cycle of a microbial population in which growth ceases.
Stem cell A cell that can develop into a number of final cell types.
Stereoisomers Mirror image forms of two molecules having the same molecular and structural formulas.
Sterilant (sterilizer) (sporicide) A chemical agent that destroys all foms of microbial life.
Sterile Free of all living organisms and viruses.
Sterilization The killing or removal of all living organisms and their viruses from a growth medium.
Sterols Hydrophobic multiringed structures that strengthen the cytoplasmic membrane of eukaryotic cells and a few prokaryotes.
Stickland reaction Fermentation of an amino acid pair in which one amino acid serves as an electron donor and a second serves as an electron acceptor.
Stop codon A codon that signals the end of a protein.
Strain A population of cells of a single species all descended from a single cell; a clone.
Stringent response A global regulatory control that is activated by amino acid starvation or energy deficiency.
Stroma The inner membrane surrounding the lumen of the chloroplast.
Stromatolite A laminated microbial mat, typically built from layers of filamentous and other microorganisms that can become fossilized.
Substrate The molecule that undergoes a specific reaction with an enzyme.
Substrate-level phosphorylation Synthesis of high-energy phosphate bonds through reaction of inorganic phosphate with an activated organic substrate.
Sulfate-reducing and sulfur-reducing bacteria Two groups of anaerobic *Bacteria* that respire anaerobically with SO_4^{2-} and S^0, respectively, as electron acceptors, producing H_2S.
Superantigen A pathogen product capable of eliciting an inappropriately strong immune response by stimulating greater than normal numbers of T cells.
Supercoil Highly twisted form of circular DNA.
Superoxide anion (O_2^-) A derivative of O_2 capable of oxidative destruction of cell components.
Suppressor A mutation that restores a wild-type phenotype without altering the original mutation, usually arising by mutation in another gene.
Surveillance Observation, recognition, and reporting of diseases as they occur.
Suspended solid A small particle of solid pollutant that resists separation by ordinary physical means.
Symbiosis A relationship between two organisms.
Synthetic DNA A DNA molecule that has been made by a chemical process in a laboratory.
Syntrophy A nutritional situation in which two or more organisms combine their metabolic capabilities to catabolize a substance not capable of being catabolized by either one alone.
Systematics The study of the diversity of organisms and their relationships; includes taxonomy and phylogeny.
Systemic Not localized in the body; an infection disseminated widely through the body.

T cell A lymphocyte responsible for antigen-specific cellular interactions. T cells are divided into functional subsets including T_C (cytotoxic) T cells and T_H (helper) T cells. T_H cells are further subdivided into T_H1 (inflammatory) and T_H2 (helper cells) that aid B cells in antibody formation.
T cell receptor The antigen-specific receptor protein on the surface of T lymphocytes.
T-DNA The segment of the *Agrobacterium* Ti plasmid that is transferred to plant cells.
Taxis Movement toward or away from a stimulus.
Taxonomy The science of identification, classification, and nomenclature.
Teichoic acid A phosphorylated polyalcohol found in the cell wall of some gram-positive *Bacteria*.
Telomerase An enzyme complex that replicates DNA at the end of eukaryotic chromosomes.
Temperate virus A virus whose genome is able to replicate along with that of its host without causing cell death in a state called lysogeny.
Termination Stopping the elongation of an RNA molecule at a specific site.
Tertiary structure The final folded structure of a polypeptide that has previously attained secondary structure.
Tertiary wastewater treatment Physicochemical processing of wastewater to reduce levels of inorganic nutrients.
Tetanus A disease involving rigid paralysis of the voluntary muscles caused by an exotoxin produced by *Clostridium tetani*.
Tetracycline A member of a class of antibiotics characterized the four-membered naphthacene ring.
Thalassemia A genetic trait that confers resistance to malaria, but causes a reduction in the efficiency of red blood cells by altering a red blood cell enzyme.
T-helper (T_H) cells Lymphocytes that interact with MHC-peptide complexes through their T cell receptor (TCR).
Thermocline Zone of water in a stratified lake in which temperature and oxygen concentration drop precipitously with depth. If stratification is based on chemical differences, called the *chemocline*.
Thermophile An organism with a growth temperature optimum between 45 and 80°C.
Thermosome A heat-shock (chaperonin) protein complex that functions to refold partially heat-denatured proteins in hyperthermophiles.
Thylakoid Layer of membranes containing the photosynthetic pigments in chloroplasts and in cyanobacteria.
Thymus the primary lymphoid organ responsible for development of T cells.
Ti plasmid A conjugative plasmid present in the bacterium *Agrobacterium tumefaciens* that can transfer genes into plants.
Titer In immunology, the quantity of antibody present in a solution.
Tolerance Inability to produce an immune response to a specific antigen.
Toll-like receptor (TLR) One of a family of pattern recognition molecules (PRMs) found on phagocytes, structurally and functionally

related to Toll receptors in *Drosophila,* that recognize a pathogen-associated molecular pattern (PAMP).

Toxic shock syndrome (TSS) Acute systemic shock resulting from host response to an exotoxin produced by *Staphylococcus aureus*.

Toxicity Pathogenicity caused by toxins produced by a pathogen.

Toxigenicity The degree to which an organism is able to elicit toxic symptoms.

Toxin A microbial substance able to induce host damage.

Toxoid A toxin modified so that it is no longer toxic but is still able to induce antibody formation.

Transcription Synthesis of an RNA molecule complementary to one of the two strands of a double-stranded DNA molecule.

Transcriptome The complement of mRNAs produced in an organism under a specific set of conditions.

Transduction Transfer of host genes from one cell to another by a virus.

Transfection The transformation of a prokaryotic cell by DNA or RNA from a virus. Used also to describe the process of genetic transformation in eukaryotic cells.

Transfer RNA (tRNA) A small RNA molecule used in translation that possesses an anticodon at one end and has the corresponding amino acid attached to its other end.

Transformation Transfer of genetic information via free DNA. Also, a process, sometimes initiated by infection with certain viruses, whereby a normal animal cell becomes a cancer cell.

Transgenic organism A plant or animal with foreign DNA inserted into its genome.

Transition A mutation in which a pyrimidine base is replaced by another pyrimidine or a purine is replaced by another purine.

Translation The synthesis of protein using the genetic information in a messenger RNA as a template.

Transmissible spongiform encephalopathy (TSE) Degenerative disease of the brain caused by prion infection.

Transpeptidation The formation of peptide bonds between the short peptides present in the cell wall polymer, peptidoglycan.

Transposable element A genetic element with the ability to move (transpose) from one site to another on host DNA molecules.

Transposase An enzyme that catalyzes the insertion of DNA segments into other DNA molecules.

Transposon A type of transposable element that, in addition to genes involved in transposition, carries other genes; often genes conferring selectable phenotypes such as antibiotic resistance.

Transposon mutagenesis Insertion of a transposon into a gene; this inactivates the host gene, leading to a mutant phenotype, and also confers the phenotype associated with the transposon gene.

Transversion A mutation in which a pyrimidine base is replaced by a purine or vice versa.

Tuberculin test A skin test for previous infection with *Mycobacterium tuberculosis.*

Turbidity A measurement of suspended solids in water.

Two-component regulatory system A regulatory system containing a sensor protein and a response regulator protein (see *Sensor kinase protein* and *Response regulator protein*).

Typhus A louse-transmitted disease caused by *Rickettsia prowazekii,* causing fever, headache, weakness, rash, and damage to the central nervous system and internal organs.

Universal phylogenetic tree A tree that shows the evolutionary position of representatives of all domains of living organisms.

Untreated water Surface water or groundwater that has not been treated in any way (also called raw water).

Upper respiratory tract The nasopharynx, oral cavity, and throat.

Upstream position Refers to nucleic acid sequences on the 5′ side of a given site on a DNA or RNA molecule. Compare with *Downstream position*.

Vaccination (immunization) Inoculation of a host with inactive or weakened pathogens or pathogen products to stimulate protective immunity.

Vaccine An inactivated or weakened pathogen, or an innocuous pathogen product used to stimulate protective immunity.

Vacuole A small space in a cell that contains fluid and is surrounded by a membrane. In contrast to a vesicle, a vacuole is not rigid.

Vector An agent, usually an insect or other animal, able to carry pathogens from one host to another.

Vector (as in cloning vector) A self-replicating DNA molecule that is used to carry cloned genes or other DNA segments for genetic engineering.

Vector vaccine A vaccine made by inserting genes from a pathogenic virus into a relatively harmless carrier virus.

Vehicle Nonliving source of pathogens that infect large numbers of individuals; common vehicles are food and water.

Viable Alive; able to reproduce.

Viable count Measurement of the concentration of live cells in a microbial population.

Viral load The number of viral genome copies in the tissue of an infected host, providing a quantitative assessment of the amount of virus in the host.

Viricidal agent An agent that stops viral replication and activity.

Virion A virus particle; the virus nucleic acid surrounded by a protein coat and in some cases other material.

Viristatic agent An agent that inhibits viral replication.

Viroid Small, circular, single-stranded RNA that causes certain plant diseases.

Virulence Degree of pathogenicity produced by a pathogen.

Virulent virus A virus that lyses or kills the host cell after infection; a nontemperate virus.

Virus A genetic element containing either RNA or DNA and that replicates in cells; has an extracellular form.

Volatile fatty acids (VFAs) The major fatty acids (acetate, propionate, and butyrate) produced during fermentation in the rumen.

Wastewater Liquid derived from domestic sewage or industrial sources, which cannot be discarded in untreated form into lakes or streams.

Water activity (a_w) An expression of the relative availability of water in a substance. Pure water has an a_w of 1.000.

Western blot See *Immunoblot*.

West Nile fever A neurological disease caused by West Nile virus, a virus transmitted by mosquitoes from birds to humans.

Wild type A strain of microorganism isolated from nature. The usual or native form of a gene or organism.

Winogradsky column A glass column packed with mud and overlaid with water to mimic an aquatic environment in which various bacteria develop over a period of months.

Wobble In reference to protein synthesis, less rigid form of base pairing allowed only in codon–anticodon pairing.

Xenobiotic A completely synthetic chemical compound not naturally occurring on Earth.

Xerophile An organism adapted to growth at very low water potentials.

Yeast artificial chromosome (YAC) A genetically-engineered chromosome with yeast origin of replication and CEN sequence.

Yeast The single-celled growth form of various fungi.

Zoonosis A disease, primarily of animals, that is occasionally transmitted to humans.

Zygote In eukaryotes, the single diploid cell resulting from the union of two haploid gametes.

Photo Credits

The following photos appear in the text without on-page photo credits:

Figure 1-6 NASA
Figure 1-20 Yuri Arcurs/Shutterstock
Figure 26-6a Klagyi/Shutterstock
Figure 26-6b Puchan/Shutterstock
Figure 26-6c Karen Lau/Shutterstock
Chapter 1 Opener D. E. Caldwell
Chapter 2 Opener Gernot Arp, Christian Boeker, and Carl Zeis, Jena
Chapter 3 Opener Brent Selinger, Pearson Benjamin Cummings
Chapter 4 Opener Nicholas Blackburn
Chapter 5 Opener James A. Shapiro, University of Chicago
Chapter 6 Opener John Gosink and James T. Staley
Chapter 7 Opener G. Murti/Photo Researchers, Inc.
Chapter 8 Opener SciMAT/Photo Researchers, Inc.
Chapter 9 Opener Timothy C. Johnston
Chapter 10 Opener Jack Parker
Chapter 11 Opener Peter T. Borgia
Chapter 12 Opener Jason A. Kahana and Pamela A. Silver
Chapter 13 Opener CAMR/A. Barry Dowsett / Photo Researchers, Inc.
Chapter 14 Opener B. Boonyaratanakornkit & D.S. Clark, G. Vrdoljak
Chapter 15 Opener M Huttël
Chapter 16 Opener Richard W. Castenholz, University of Oregon
Chapter 17 Opener Francisco Rodriguez-Valera
Chapter 18 Opener F. Forsdyke
Chapter 19 Opener CDC / PHIL
Chapter 20 Opener Norbert Pfennig
Chapter 21 Opener Dr. Dennis Kunkel/Visuals Unlimited
Chapter 22 Opener Alex T. Nielsen
Chapter 23 Opener SciMAT/Photo Researchers, Inc.
Chapter 24 Opener Dennis Kunkel/Dennis Kunkel Microscopy
Chapter 25 Opener M. T. Madigan
Chapter 26 Opener Dr. Linda Stannard, UCT/Photo Researchers, Inc.
Chapter 27 Opener CNRI/Photo Researchers, Inc.
Chapter 28 Opener Steve Gschmeissner/Photo Researchers, Inc.
Chapter 29 Opener Dr. Gopal Murti/PhototakeUSA.com
Chapter 30 Opener SPL/Photo Researchers, Inc.
Chapter 31 Opener Steve Gschmeissner/Photo Researchers, Inc.
Chapter 32 Opener SPL/Photo Researchers, Inc.
Chapter 33 Opener Dennis Kunkel/Dennis Kunkel Microscopy
Chapter 34 Opener Dennis Kunkel/Dennis Kunkel Microscopy
Chapter 35 Opener LSHTM/Photo Researchers, Inc.
Chapter 36 Opener Mark L. Tamplin
Chapter 37 Opener SPL/Photo Researchers, Inc.

Index

INDEX

INDEX

PHYLOGENY OF THE LIVING WORLD
ARCHAEA

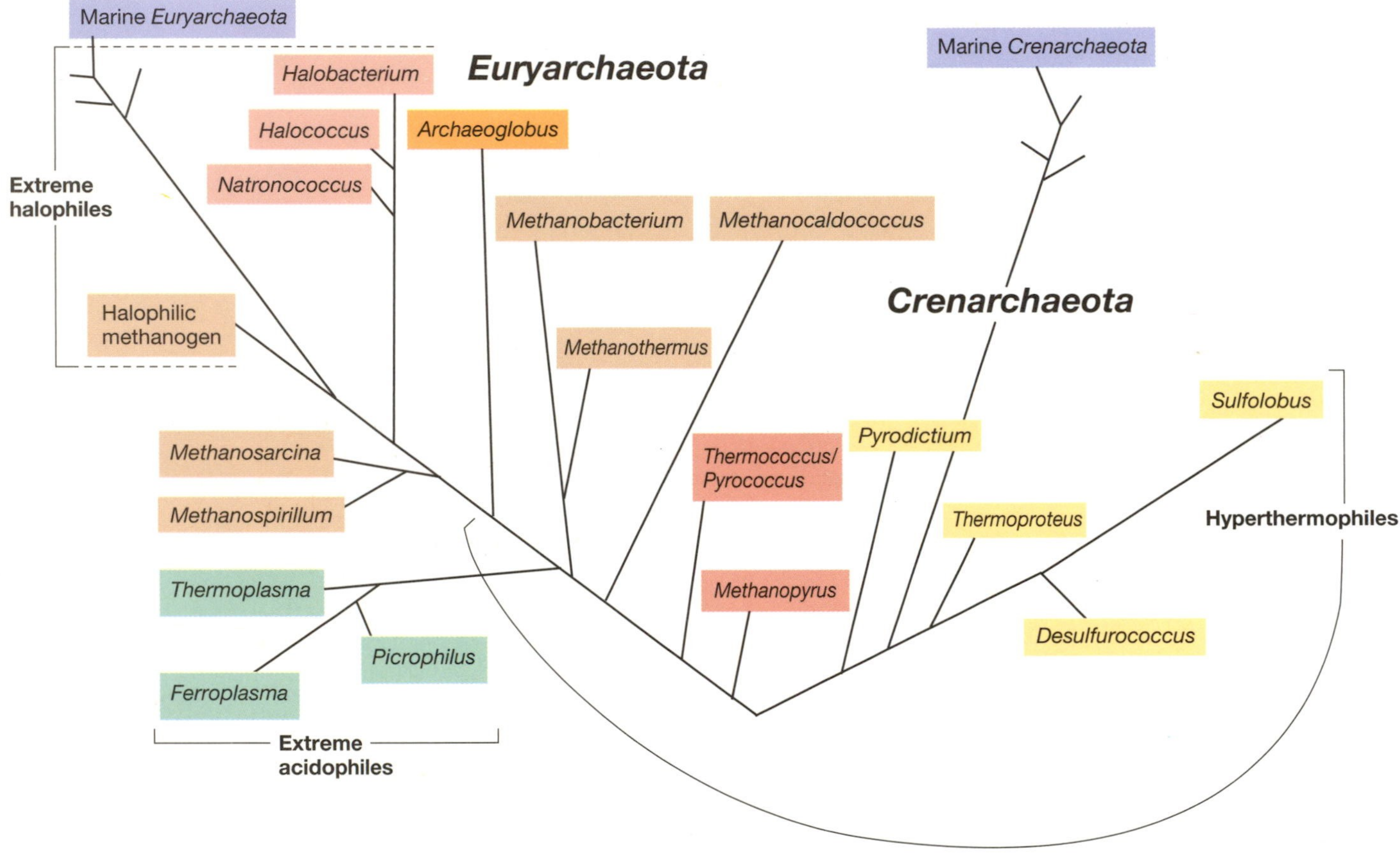

Phylogenetic Tree of *Archaea* This tree is derived from 16S ribosomal RNA sequences. Two major phyla of *Archaea* can be defined: the *Crenarchaeota*, which consists of both hyperthermophiles and cold-dwelling species, and the *Euryarchaeota*, which includes the methanogens, extreme halophiles, acidophiles, alkaliphiles, and cold-dwelling marine species. See Sections 14.5–14.9 for further information on ribosomal RNA-based phylogenies. *Data for the tree obtained from the Ribosomal Database project* **http://rdp.cme.msu.edu**